Circuit Analysis with PSpice

A Simplified Approach

Circuit Analysis with PSpice

A Simplified Approach

Nassir H. Sabah
American University of Beirut, Lebanon

CRC Press
Taylor & Francis Group
Boca Raton London New York

CRC Press is an imprint of the
Taylor & Francis Group, an **informa** business

CRC Press
Taylor & Francis Group
6000 Broken Sound Parkway NW, Suite 300
Boca Raton, FL 33487-2742

First issued in paperback 2021

CRC Press is an imprint of Taylor & Francis Group, an Informa business

ISBN 13: 978-0-367-78216-0 (pbk)
ISBN 13: 978-1-4987-9604-0 (hbk)

Library of Congress Cataloging-in-Publication Data

Names: Sabah, Nassir H., author.
Title: Circuit analysis with PSpice : a simplified approach / Nassir H. Sabah.
Description: Boca Raton : Taylor & Francis, CRC Press, 2017. | Includes
bibliographical references and index.
Identifiers: LCCN 2016033747 | ISBN 9781498796040 (hardback : alk. paper) |
ISBN 9781315402222 (e-book)
Subjects: LCSH: Electric circuit analysis--Data processing. | PSpice.
Classification: LCC TK454 .S229 2017 | DDC 621.38150285/53--dc23
LC record available at https://lccn.loc.gov/2016033747

Visit the Taylor & Francis Web site at
http://www.taylorandfrancis.com

and the CRC Press Web site at
http://www.crcpress.com

To those who shaped my learning over the years:

my parents, my teachers, and my students.

The elegant beauty of technical explanations is in their simplicity, clarity, and insight they provide.

Brief Contents

Contents

Part II: Topics in Circuit Analysis

Preface

This book is more than a textbook on electric circuits. It is a veritable learning reference that presents electric circuit analysis in a simplified manner, without sacrificing rigor and thoroughness. The book is a sequel to the author's *Electric Circuits and Signals*, CRC Press, 2008. The electric signal material has been omitted and circuit analysis is treated in a more simplified and expanded form. The book differs from other textbooks on electric circuits in its pedagogy and organization, as expounded later, particularly in the following respects:

1. Strong emphasis on (a) simple, clear, careful, and comprehensive explanations of the basic concepts in circuit analysis (simplicity is not to be construed as superficiality; what is meant is simple and clear, but in-depth, explanations); (b) a sound understanding of fundamentals, enhanced by physical and insightful interpretations of circuit behavior; and (c) extensive use of PSpice® (OrCAD, PSpice, SPECTRA for OrCAD, and Cadence are registered trademarks of Cadence Design Systems, Inc., San Jose, California), as detailed later in a section on PSpice simulations.
2. Effective problem solving based on (a) a systematic, logical, and imaginative approach, having the acronym ISDEPIC, formulated by the author and refined over the past several years through interaction with students, and (b) presenting a variety of topics and examples that foster problem-solving skills by encouraging the student to view a problem in different ways, particularly fresh and original ways, founded on a sound understanding of fundamentals. The author firmly believes that a course on electric circuits provides an excellent opportunity to nurture problem-solving skills, as a central objective of quality engineering education. That is why some topics, such as exploitation of symmetry in electric circuits, are included, although they are of limited practical importance.
3. Substantive application of the substitution theorem and of duality to facilitate circuit analysis and enhance the understanding of circuit behavior.
4. Some original contributions to circuit analysis by the author, such as (a) using the substitution theorem to replace dependent sources by independent sources when applying superposition, which greatly simplifies analysis of circuits that include dependent sources; (b) circuit equivalence, as a unifying concept that encompasses a variety of topics, ranging from simple series–parallel combinations of resistances to source transformation and Thevenin's theorem; and (c) the concept of effective magnetic flux, which allows dealing with leakage flux simply and conveniently, rather than skirt this seemingly awkward issue.

P.1 Pedagogy

The underlying theme throughout the book is presenting circuit analysis logically, coherently, and justifiably, yet simply and clearly, and not as a set of procedures that are to be followed without really understanding the "why?" in terms of critical thinking, logical reasoning, and sound understanding of fundamentals.

The following features exemplify this approach to circuit analysis:

1. It is emphasized from the very beginning that circuits obey two universal conservation laws: conservation of energy and conservation of charge, which imply, respectively, conservation of power and conservation of current. Kirchhoff's laws are simply an expression of these conservation laws and not some sacrosanct laws that are peculiar to electric circuits. They are convenient to apply in lieu of the more fundamental conservation laws because they are linear in voltage and current.
2. The rationale behind the node-voltage and mesh-current methods is explained as having Kirchhoff's voltage law automatically satisfied by the assignment of node voltages and having Kirchhoff's current law automatically satisfied by the assignment of mesh currents.
3. Circuit simplification techniques and effective problem-solving methodologies are strongly emphasized to help the student analyze electric circuits intelligently, understand their behavior, and gain insight into this behavior. These topics are thoroughly discussed *before* the node-voltage and mesh-current methods—because in the

author's experience, once students learn these routine, general methods, they tend to preferentially apply them to all circuits, even simple ones. This deprives students of the opportunity to understand circuit behavior and to foster their problem-solving skills. Another reason for deemphasizing the node-voltage and mesh-current methods is that these methods were originally developed to facilitate analysis of more complicated circuits. But the responses of such circuits are more conveniently derived nowadays by PSpice simulation.

4. In conformity with conventional practice, practically all the main circuit concepts and procedures are presented for the dc state to begin with. Some textbooks then discuss the transient behavior of *RC*, *RL*, and *RLC* circuits before the sinusoidal steady state. In this book, the sinusoidal steady state is discussed immediately following the dc state. The reason for this is that phasor analysis is presented as a means of allowing direct application of all the concepts and techniques developed for the dc state to the sinusoidal steady state. It is only logical, therefore, to consider the sinusoidal steady state immediately following the dc state.
5. Magnetic coupling is discussed in a comprehensive, realistic, and not oversimplified manner, using the concept of effective flux. Magnetic flux linkage is properly made use of as a basic quantity. It is emphasized that ideal transformers, irrespective of the number of windings and how they are interconnected, obey two fundamental, general principles: (a) the same volts/turn are induced in every winding, and (b) zero, net mmf acts on the core.
6. Duality is emphasized as a means of unifying in many respects the analysis of (a) series and parallel circuits of all types and (b) capacitive and inductive circuits.
7. Simplified and generalized methods are presented for deriving the responses of first-order and second-order circuits in the time domain.
8. The role of the transient response is clearly explained as a means of providing a smooth transition from the initial value of a given response to its steady-state, final value.
9. The basic, noninverting, and inverting op amp configurations are discussed in terms of the very fundamental concept of feedback. It is explained very simply and clearly how negative feedback, but not positive feedback, allows stable operation at any point in the linear region of the input–output characteristic of the op amp. It is stressed that this requires some circuit connection between the op amp output and the inverting input, a feature that is present in all non-switching-type op amp circuits.
10. The four basic types of frequency responses (low-pass, high-pass, bandpass, and bandstop) are all derived from a series *RLC* circuit to highlight the interrelations between these responses. It is emphasized that second-order, passive *RC* circuits cannot have a Q larger than 0.5, corresponding to critical damping.
11. The rationale for Butterworth and active filters is clearly explained.
12. Complex power and maximum power transfer under general conditions are included in Part II, after considering power due to periodic functions. The conservation of complex power is simply and clearly explained.
13. The impulse and step responses of *RC*, *RL*, and *RLC* circuits are discussed systematically and logically, with physical interpretations.
14. The concepts of equivalent capacitance and equivalent inductance are applied in a simple and imaginative manner to derive the responses of capacitive and inductive circuits to sudden changes, with or without initial energy storage.
15. Convolution is treated as an operation in the time domain that is important in its own right and that follows directly from the impulse response. The physical interpretation and significance of convolution are emphasized, particularly the special cases of convolution of staircase functions and convolution with the impulse and step functions.
16. Responses to periodic inputs, the Laplace transform, the Fourier transform, and two-port circuits are covered rather comprehensively.
17. Numerous references are made, whenever appropriate, to MATLAB® commands as a very useful aid to circuit analysis.

P.2 Organization

The book is divided into two parts. Part I covers what is conventionally considered as basic electric circuit analysis and constitutes a first course on electric circuits. Part II consists of a number of additional topics that can be selectively added in a second course. Operational amplifiers are not included in Part I, because they are not considered part of basic electric circuits. They are included as the first chapter of Part II in connection with active filters, where they belong. They could be added to a first course on electric circuits, if desired. Some sections and examples in both Parts I and II are marked with a star to indicate that they may be skipped in a more limited coverage of the material.

More than 430 exercises are included at the ends of most sections of chapters, or within sections. These exercises are of two types: (1) Primal exercises that are simple, straightforward applications of the main concepts discussed and are intended to allow students to practice direct applications of concepts and help them gain some self-confidence in doing so and (2) exercises that are not labeled "Primal" and that serve to extend some aspects of the topics discussed, or to verify some simple assertions made in the text, and not discussed in detail for the sake of brevity or avoidance of tedious repetition.

More than 175 solved examples are included throughout the book to illustrate the topic being discussed. In almost all examples, a PSpice simulation is added after the solution, followed by problem-solving tips, whenever appropriate, to emphasize some useful problem-solving techniques.

A "Learning Checklist" is added at the end of the main body of every chapter so as to serve both as a summary and as a check on the understanding of the main concepts and ideas presented in the chapter. The Learning Checklist is followed by a list of all the problem-solving tips in the solved examples of the chapter.

More than 1500 problems are included at the ends of chapters for students to test their understanding of the material and apply the problem-solving skills they have acquired. Some of these are of the "short-solution" type that test for the understanding of a specific concept, without involving much calculation. Other problems are of the "long-solution" type that require the logical formulation of a number of sequential calculation steps in order to obtain the required results. In general, the exercises and problems are ordered in increasing level of "challenge." Design-type problems are included as a group at the ends of some chapters, wherever appropriate. Another group of problems, labeled "Probing Further," are added at the ends of some chapters in order to examine some more advanced or specific topics. Answers are given following all exercises and problems that are not intended to verify or prove something.

P.3 PSpice Simulations

More than 100 PSpice simulations are included in the book, as listed after the Preface. The simulations are used to verify the results of analytical solutions and to graphically illustrate these results, wherever applicable. The simulation procedure is described in every case. The circuit, as entered, is shown, the entries in the simulation profile are indicated, and the graphical or analytical results are presented. An appendix on PSpice simulation is included, which is more than adequate for the simulations covered in a course on electric circuits. The appendix includes much useful information on PSpice simulations that is not found in any single reference on PSpice simulations that the author is aware of.

The PSpice program used is OrCAD 16.6 Lite version. PSpice Lite can be downloaded by students from the Cadence web page, free of charge. The simulation files of the PSpice simulations listed after the Preface can be downloaded from the book's web page that can be accessed at: https://www.crcpress.com/ to enable students to actually perform the simulations. Additional files will be made available at this website in the future for the PSpice simulation of problems at the ends of chapters.

P.4 Solutions Manual and Classroom Presentations

A solutions manual for all exercises and problems, as well as Class Presentations, are available to qualifying instructors adopting this book, and may be requested through the CRC Press website. The Class Presentations consist of a Microsoft Word® file for every chapter that presents, in the form of colored, bulleted text and figures, the main ideas and concepts discussed in the given chapter, together with the solved examples. The files are intended for projection in the classroom by instructors for use as a basis for explaining the material. The advantages of using Word files are the following: (1) the files can be easily modified by instructors as they deem appropriate for their own purposes and (2) top and bottom margins can be hidden, which allows seamless scrolling, up and down, through the whole file.

Acknowledgments

The author is indebted as usual to his students for their valuable interactions and for the many unanticipated or challenging questions they asked. The author is also indebted to his colleagues, who taught the electric circuit courses with him, for their esteemed comments and suggestions. The author gratefully acknowledges CRC Press for their permission to use material from his book *Electric Circuits and Signals*, CRC Press, 2008. The author also expresses his sincere appreciation of the efforts of CRC Press and their associates in producing and promoting this book, particularly Nora Konopka, Publisher, Engineering and Environmental Sciences, for her invaluable and steadfast support and understanding. The valuable and professional contributions of Richard Tressider, Project Editor, and Vinithan Sedumadhavan, of SPi Global, are gratefully acknowledged. Special thanks to John Gandour for his artistic cover design.

Author

Nassir Sabah is a professor of electrical and computer engineering at the American University of Beirut, Lebanon. He received his BSc (Hons. Class I) and his MSc in electrical engineering from the University of Birmingham, UK, and his PhD in biophysical sciences from the State University of New York (SUNY/Buffalo). He served as chairman of the Electrical Engineering Department, director of the Institute of Computer Studies, and dean of the Faculty of Engineering and Architecture, at the American University of Beirut. In these capacities, he was responsible for the development of programs, curricula, and courses in electrical, biomedical, communications, and computer engineering. Professor Sabah has extensive professional experience in the fields of electrical engineering, electronics, and computer systems, with more than 35 years teaching experience in electric circuits, electronics, neuroengineering, and biomedical engineering. He has more than 100 technical publications, mainly in neurophysiology, biophysics, and biomedical instrumentation. He has served on numerous committees and panels in Lebanon and the Middle East. Professor Sabah is a fellow of the Institution of Engineering and Technology, UK, and a member of the American Society of Engineering Education.

List of PSpice Simulations

Convention for Voltage and Current Symbols

The following convention for current and voltage symbols is adhered to in this book as much as possible:

- Capital letter with capital subscript denotes dc, or average, quantity. Example: V_O.
- Capital letter with lowercase subscript denotes rms value of an alternating quantity, its Fourier transform, or its Laplace transform. In some cases, the capital subscript is used, as when referring to a circuit element to avoid confusion with nodes or terminals. Examples: I_o, $V_i(\omega)$, $I_C(s)$, $V_{Th}(s)$.
- Capital letter with m subscript denotes the peak value of a sinusoidal quantity. Example: $I_m \sin \omega t$.
- Lowercase letter with capital subscript denotes a total instantaneous quantity. Example: v_{SRC}.
- Lowercase letter with lowercase subscript denotes a small signal of zero average value. Example: i_y.
- Boldface, not italicized, symbol of voltage, current, or power denotes a phasor. Example: $\mathbf{V}_\mathrm{b}$.
- Double subscript in a voltage symbol denotes a voltage drop from the node or terminal designated by the first subscript to the node or terminal designated by the second subscript. Example: V_{ab}. Nodes or terminals are denoted by lowercase subscripts or numbers.
- Double subscript in a current symbol denotes a current flowing from the node or terminal designated by the first subscript to the node or terminal designated by the second subscript. Example: i_{ab}.
- Non-italic subscripts are used for denoting phases in three-phase systems, for "rms," "max," and "min" in subscripts.

Part I

Basic Concepts in Circuit Analysis

1

Preliminaries to Circuit Analysis

Objective and Overview

This chapter introduces some basic notions on electric circuits before embarking on circuit analysis in the following chapters.

The chapter begins by explaining what electric circuits are, what they are used for, and what conservation laws they obey. The primary circuit variables of current and voltage are defined with reference to a useful and easy-to-follow, hydraulic analogy. The significance of direction of current and polarity of voltage is emphasized because of the key roles these play in circuit analysis. The relation of current and voltage to power and energy is derived, and active and passive circuit elements are characterized by the way they handle energy. The three passive circuit parameters of resistance, capacitance, and inductance are justified as accounting for three basic attributes of the electromagnetic field, namely, energy dissipation and energy storage in the electric and magnetic fields. The chapter concludes with an examination of the idealizations and approximations made in the circuits approach.

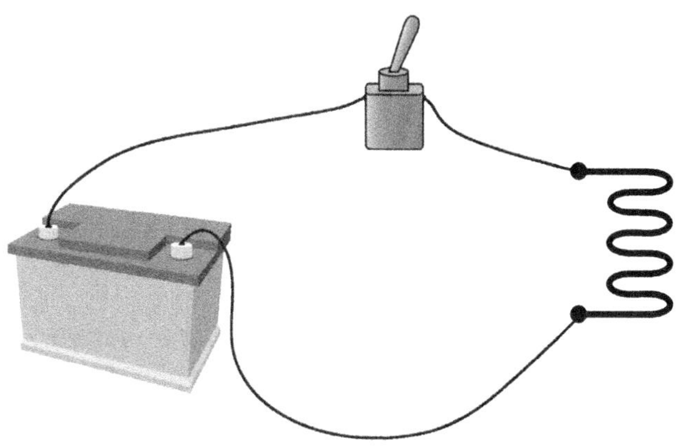

FIGURE 1.1
An electric circuit.

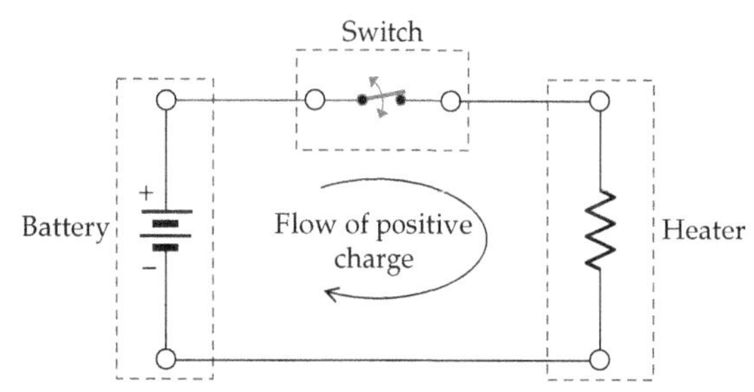

FIGURE 1.2
Circuit diagram for Figure 1.1.

1.1 What Are Electric Circuits and What Are They Used For?

Definition: *An electric circuit is an interconnection of components that affect electric charges in some characteristic manner.*

An example is a battery connected to a heater through a switch, as illustrated diagrammatically in Figure 1.1. Figure 1.2 is the corresponding circuit diagram in terms of symbols for the three components. When the switch is in the closed position, as shown, it allows electric charges to flow through the heater. In doing so, the charges impart some of their energy to the heater, thereby generating heat and raising the temperature of the heater metal. The battery restores energy to the electric charges, thereby allowing them to flow continuously through the circuit. Opening the switch interrupts the flow of charges and turns off the heater. Electrical installations in buildings provide many other examples of electric circuits, including lighting, air conditioning, alarm, and remote control systems. Electronic circuits, consisting of electrical and electronic components, are at the heart of electronic equipment of all kinds.

Electric circuits are used in two ways:

1. To perform some useful task, as in the case of the heater of Figure 1.1 or in the case of electrical installations in buildings or in the case of electronic equipment
2. To model or emulate the behavior of some component or system, as explained in Section 1.8. The modeling is not restricted to electric or electronic components or systems but can be applied to mechanical, thermal, and fluidic systems.

1.2 What Laws Govern the Behavior of Electric Circuits?

Concept: *The behavior of electric circuits is governed by two fundamental conservation laws: conservation of energy and conservation of charge.*

Energy is conserved in the sense that it can neither be created out of nothing nor be destroyed into nothing. It can only be converted from one form to another. A solar cell converts light energy to electric energy. An electric motor converts electric energy to mechanical energy. Strictly speaking, the universal conservation law is for mass + energy, but since conservation of mass does not play a role in the behavior of electric circuits, it is energy alone that is conserved.

Similarly, electric charges can be neither created nor destroyed. Materials or objects in their natural state are electrically neutral, that is, they contain equal quantities of positive and negative charges. These can be separated through expenditure of energy. In a battery, for example, energy-consuming reactions detach electrons from their parent atoms and raise their energies so that they flow through an external circuit connected to the battery. *Because they are conserved, electric charges always flow in closed paths.* If they did not flow in a closed path, then charges will start at a location where they are being created and end up in a location where they are destroyed, in violation of conservation of charge.

In principle, it is possible to analyze the behavior of electric circuits in terms of energy and charge. However, this is seldom done in practice. It is much more convenient, as explained in Section 2.7, to analyze electric circuits using two common **circuit variables**, namely, *electric current* and *voltage*.

1.3 What Is Electric Current?

To explain the meaning of electric current, a useful hydraulic analogy can be invoked. Consider water flowing down from a reservoir aboveground through some form of a water-driven turbine connected to a mechanical load (Figure 1.3). A motor-driven pump recirculates the water from the turbine outlet back to the reservoir. The system of Figure 1.3 can be described as a "hydraulic circuit" and is analogous to the electric circuit of Figures 1.1 and 1.2. The pump and reservoir are analogous to the combination of battery and switch. The pump raises the potential energy of water and can be used to turn the flow on and off. The reservoir stores water at a higher potential energy with respect to ground level. As a power-consuming load, the turbine and its load are analogous to the heater. In flowing from the reservoir through the turbine, the potential energy of water is converted to kinetic energy, which in turn is converted by the turbine to mechanical energy. The pump, driven from a source of energy, such as the electricity supply or an internal combustion engine, utilizes this energy to raise the potential energy of the water back to the level of the reservoir.

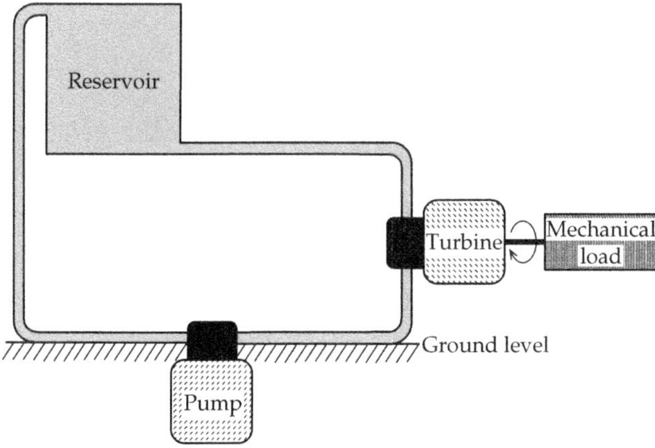

FIGURE 1.3
Hydraulic analogy.

A close analogy exists between the flow of water in Figure 1.3 and the flow of electric charge in Figure 1.2. More specifically, the *volume of water* that flows past a designated location in Figure 1.3, such as the outlet of the reservoir, over a specified interval is analogous to the *quantity of charge* that flows past a designated location in Figure 1.2, such as a terminal of the battery, over the same interval. The rate of flow of water in the hydraulic case is analogous to the rate of flow of charge in the electric circuit. The rate of flow of electric charge is the value of the electric current, or simply the **current**. In general, current is defined as follows.

Definition: *The current at any given point in an electric circuit and at a specified instant of time is the rate of flow of electric charge past the given point at that instant.*

To express this relation quantitatively, the units of charge and current must be specified. The unit of charge in the standard SI (*Système International*, in French, Appendix A) units is the **coulomb**, denoted by the symbol C, and the standard unit of current is the **ampere**, denoted by the symbol A, where a current of one ampere is a rate of flow of one coulomb per second.

If the flow of water in Figure 1.3 is steady, that is, it is not changing with time, the rate of flow is constant. Under these conditions, the volume of water that crosses any given location in the hydraulic circuit increases linearly with time:

$$\text{Volume of flow} = (\text{Constant rate of flow}) \times \text{Time} \quad (1.1)$$

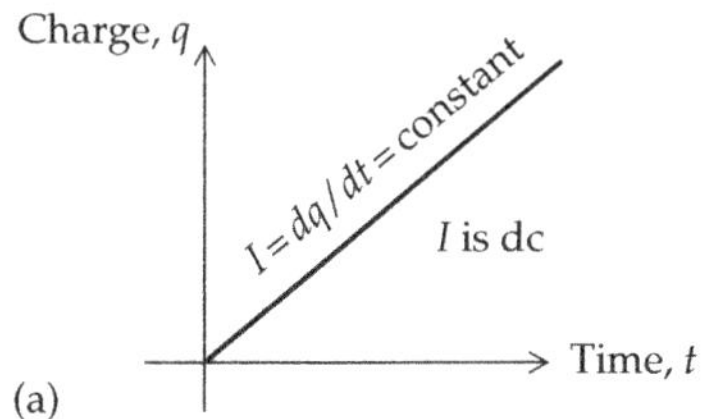

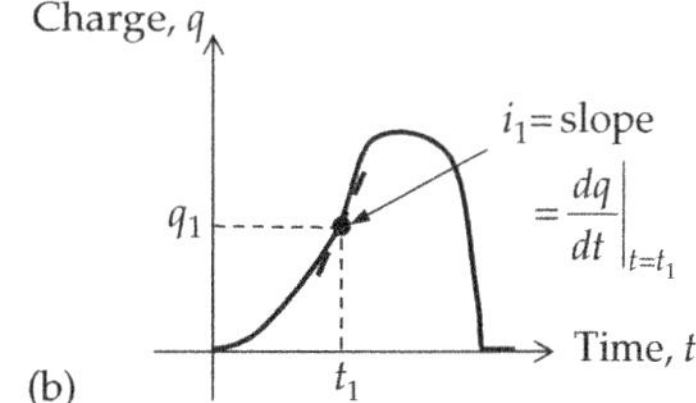

FIGURE 1.4
(a) Constant rate of flow of charge and (b) varying rate of flow of charge.

where the volume of flow is, say, in liters, the rate of flow is in liters/second, and time is in seconds.

Similarly, if the rate of flow of charge in the electric circuit of Figure 1.2 is steady, the current is constant, and the quantity of charge that crosses any given point in the circuit increases linearly with time, as illustrated in Figure 1.4a. A current that is constant with respect to time is a **direct current**, or **dc** current.

In general, the rate of flow of charge may vary with time, in which case an **instantaneous current** is defined at any particular instant of time as the slope of the charge vs. time graph at that instant. That is,

$$i = \frac{dq}{dt} \tag{1.2}$$

In Figure 1.4b, for example, where q is shown to vary arbitrarily with time, the current i_1 at the instant of time t_1 is the slope, dq/dt, of the q vs. t graph at $t = t_1$. In Figure 1.4a, the slope is constant and is equal to the dc current I.

By convention, dc currents are denoted by italic capital letters and instantaneous currents by italic small letters, with capital subscripts in both cases, as may be required (Convention for Voltage and Current Symbols, p. xxxi). Thus, I_O and I_{SRC} are dc currents, whereas i_O and i_{SRC} are instantaneous currents.

1.4 What Is the Direction of Current?

Convention: *It is assumed in circuit analysis that the direction of current is the same as that of the flow of positive electric charges. This assigned positive direction is indicated by an arrow associated with the current symbol.*

The reason for this convention is purely historical. It was postulated in the eighteenth century at a time when the nature of current carriers was not known. By current carriers is meant the charges whose rate of flow equals the current. It is now known that in most metals, current carriers are primarily negative charges, in the form of conduction electrons, that is, electrons that have detached from their parent atoms and are free to move under the influence of an applied electric field. In semiconductors and some metals, current carriers can be what are effectively positive charges, or *holes*, as they are called. In a conducting liquid, or electrolyte, current carriers are positively charged ions and negatively charged ions. In a gas, current carriers are positively charged ions, negatively charged ions, or electrons. Nevertheless, the convention in circuit analysis is that the direction of current is that of the flow of *assumed positive electric charges, irrespective of the sign of the charges that actually carry the current*. This is convenient and does not cause any confusion if applied consistently. If negatively charged current carriers flow in a given direction, then we can simply consider the current to be due to an equal flow of positive charges in the direction opposite to that of the flow of the negatively charged current carriers. This is explained more fully in Example 1.1.

Unless explicitly stated otherwise, it will henceforth be assumed that current carriers are positive charges and that the direction of current is that of the flow of positive electric charges, as indicted in Figure 1.2. It should be emphasized that current always has a direction, just as hydraulic flow has a direction. *It is meaningless to specify a current without indicating its direction.*

Example 1.1: Steady Flow of Electric Charges

(a) Consider positive electric charges flowing continuously in the positive x-direction in a conducting medium of cross-sectional area A, as illustrated in Figure 1.5. If the

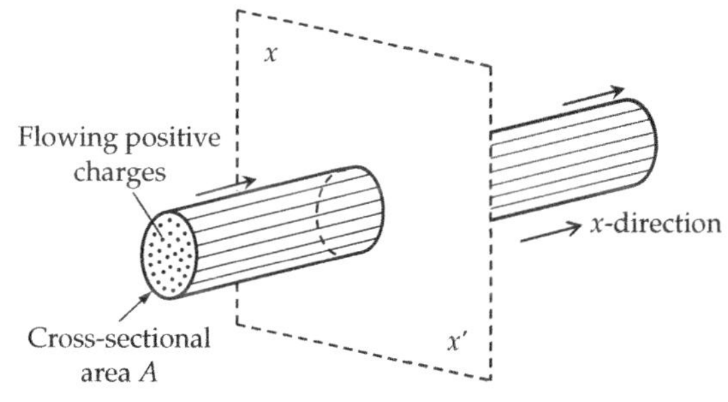

FIGURE 1.5
Figure for Example 1.1.

rate of flow is constant at 0.5 C/s, what is the current in amperes and in milliamperes (mA), both in magnitude and direction? (b) If the positive electric charges flow at a constant rate of 0.5 C/s in the negative x-direction, what is the current in the positive x-direction? (c) If negative charges flow in the positive x-direction at a constant rate of 0.5 C/s, what is the current in the positive x-direction?

Solution:

(a) According to the discussion of Section 1.3, q is the quantity of charge that flows past a specified location in the pathway of flow, such as the plane xx' in Figure 1.5 and in Figure 1.6a. If the rate of flow is constant at 0.5 C/s, then q increases linearly with time, that is, $q = 0.5t$, in accordance with Equation 1.1 and as illustrated in Figure 1.4a. According to Equation 1.2, the current is constant and is equivalent to a dc current of $I_{px+} = dq/dt = 0.5$ A. Its direction is that of the flow of positive charge, that is, in the positive x-direction, as indicated by the current arrow in Figure 1.6a. To convert this current to mA, it is multiplied by the number of mA in 1 A, which is 10^3. Thus, $I_{px+} = (0.5 \text{ A}) \times \text{mA/A}$. The 'A' unit cancels out, giving $I_{px+} = 0.5 \times 10^3 \equiv 500$ mA.

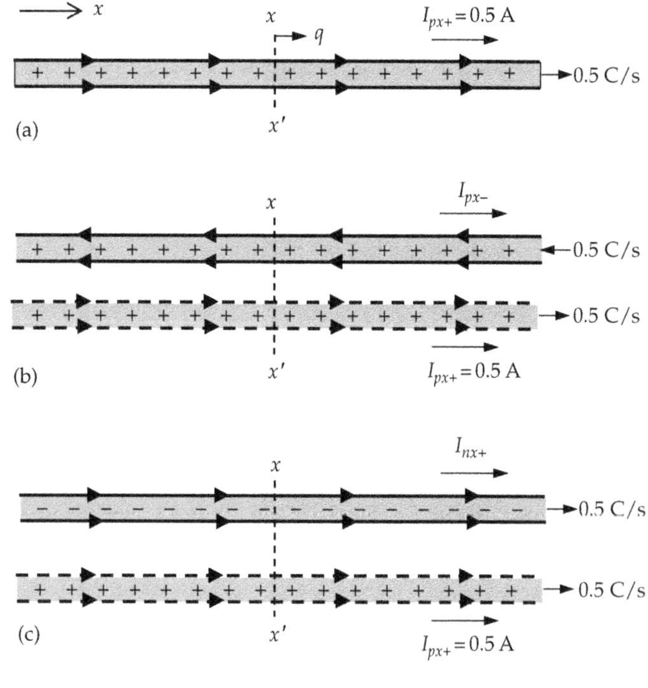

FIGURE 1.6
(a) Positive charges flowing in the positive x-direction, (b) upper trace, positive charges flowing in the negative x-direction; lower trace as in (a), and (c) upper trace, negative charges flowing in the positive x-direction; lower trace as in (a).

(b) Let the required current in the positive x-direction, due to positive charges flowing in the negative x-direction, be denoted as I_{px-} (Figure 1.6b). Suppose we add to this flow another flow of positive charges in the positive x-direction at a constant rate of 0.5 C/s, equivalent to the current I_{px+} (Figure 1.6b). As a result, there is no net flow of charge in either direction past the reference location xx' in Figure 1.6b. This means that q is zero and the total current in the positive x-direction is zero. That is, $I_{px+} + I_{px-} = 0$ so that $I_{px-} = -I_{px+}$. In other words, the current in the positive x-direction due to positive charges moving in the negative x-direction at a constant rate of 0.5 C/s is −0.5 A.

(c) Let the required current in the positive x-direction due to the flow of negative charges in this direction be denoted as I_{nx+} (Figure 1.6c). Suppose that we add to this flow an equal flow of positive charges also in the positive x-direction at the same rate of 0.5 C/s, equivalent to the current I_{px+} (Figure 1.6c). It can now be argued that the equal quantities of positive and negative charges flowing in the same direction at equal rates will completely neutralize one another. This means that there will be no net flow of charge in either direction past the reference location xx' in Figure 1.6c. The total current is therefore zero. That is, $I_{px+} + I_{nx+} = 0$ so that $I_{nx+} = -I_{px+}$. In other words, the current due to the flow of negative charges in the positive x-direction at a constant rate of 0.5 C/s is −0.5 A.

The three currents are indicated in Figure 1.6d. The following should be noted:

1. In terms of assignment in the positive x-direction, I_{px+}, I_{px-}, and I_{nx+} *are all in the same direction, as symbols*. But in terms of *numerical values*, I_{px+} has a positive value, whereas I_{px-} and I_{nx+} have negative values. This means that the conventional current, due to the flow of positive charges, is in the positive x-direction in the case of I_{px+} and in the negative x-direction in the case of I_{px-} and I_{nx+}.
2. Both I_{px-} and I_{nx+} have been *arbitrarily assigned a positive direction* in the positive x-direction, as stipulated in this example. Had they been assigned a positive direction in the negative x-direction, the current values of I_{px-} and I_{nx+} would be +0.5 A instead of −0.5 A.

Alternatively, it could be argued that q in case (a) is due to the movement of positive charge in the positive x-direction, which makes q positive. By the same token, q in both cases (b) and (c) is negative. According to Equation 1.2, the value of the resulting current is negative so that the currents I_{px-} and I_{nx+} have negative values.

Problem-Solving Tip

- Always check the units on both sides of an equation, and always specify the units of the results of calculations.

Primal Exercise 1.1

What is the current in the positive x-direction in the preceding example if negative charges move in the negative x-direction at a constant rate of 0.5 C/s?

Ans. 0.5 A.

Since i is the slope of the q vs. t graph, in accordance with Equation 1.2, it follows from this equation that

$$q = \int i dt \tag{1.3}$$

In other words, q is the area under the i vs. t graph. In Figure 1.7a, for example, q increases linearly from 0 at $t = 0$ to a peak value of 6 μC at $t = 1$ ms and then decreases linearly back to zero at $t = 1.5$ ms. The current, being the slope of the q vs. t graph, is constant at a positive value of 6 μC/1 ms, or 6 mA, during the interval from 0 to 1 ms (Figure 1.7b). The current then reverses direction and becomes −6 μC/0.5 ms = −12 mA during the interval from 1 to 1.5 ms. The current returns to zero at $t = 1.5$ ms. The area under the i vs. t graph increases linearly from zero at $t = 0$ and reaches a peak value of 6 mA × 1 ms = 6 μC at $t = 1$ ms. The area is negative during the interval from 1 to 1.5 ms and subtracts from the positive area. At $t = 1.5$ ms, the positive and negative areas are equal in magnitude so that the net area is zero, corresponding to a q of zero at $t > 1.5$ ms. The negative current is in a direction opposite to that of the positive current so that at $t = 1.5$ ms as much charge has flowed in one direction as in the opposite direction, and the net flow of charge is zero.

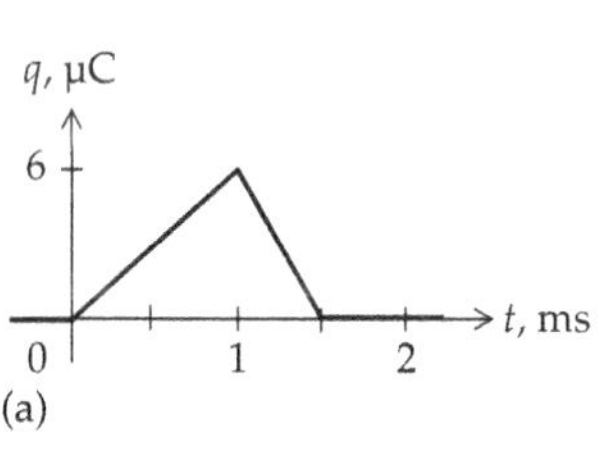

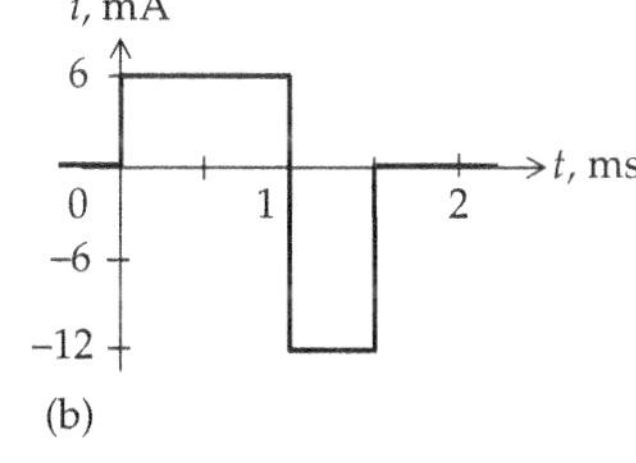

FIGURE 1.7
Relation between current and charge. (a) Variation of charge with time and (b) corresponding variation of current with time.

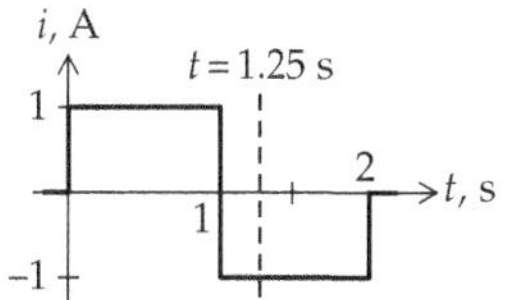

FIGURE 1.8
Figure for Primal Exercise 1.3.

Primal Exercise 1.2

Rework the example of Figure 1.7, assuming that the charge increases linearly from zero to 15 mC in 0.5 ms and then decreases linearly to zero at $t = 2$ ms.

Ans. $i = 30$ A, $0 < t < 0.5$ ms, and $i = -10$ A, $0.5 < t < 2$ ms.

Primal Exercise 1.3

The current i through a device varies with time as shown in Figure 1.8. Determine the charge that passes through the device between $t = 0$ and $t = 1.25$ s in the direction of i.

Ans. 0.75 C.

*Example 1.2: Time-Varying Flow of Electric Charges

Suppose that the flow of charge is given by $q = (1 - \cos t)$ C, $0 \leq t \leq 2\pi$ s, as illustrated in Figure 1.9a. It is required to follow q and i over the interval from $t = 0$ to $t = 2\pi$ s.

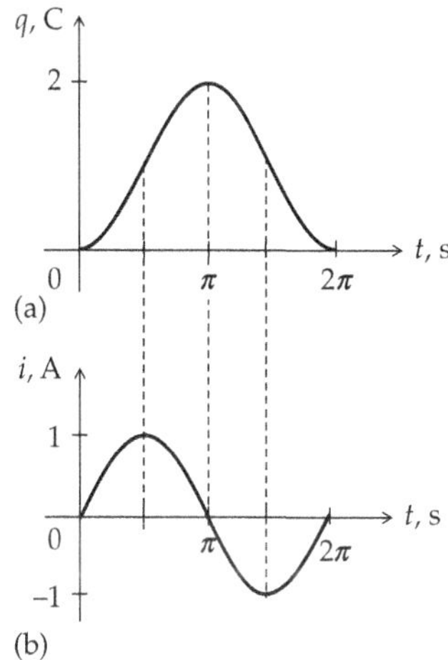

FIGURE 1.9
(a) Variation of charge with time and (b) corresponding variation of current with time.

* Sections and Examples whose titles are marked with this symbol may be skipped in a more limited coverage of the material.

Analysis: Recall that q is the quantity of charge that flows past an arbitrary location xx' in the positive x-direction between a time $t = 0$ and a time $t > 0$.

1. $0 \leq t \leq \pi/2$ s: Charge starts to flow at $t = 0$ in the positive x-direction and q increases with time (Figures 1.9a and 1.10a). The rate of flow, equal to the slope dq/dt of the q vs. t graph, is the current i, where $i = dq/dt = \sin t$ increases from 0 at $t = 0$ to a maximum positive value of 1 A at $t = \pi/2$ s (Figure 1.9b). The direction of i is that of the flow, that is, in the positive x-direction.
2. $\pi/2 \leq t < \pi$ s: Charge continues to flow in the positive x-direction, and q continues to increase, but at a decreasing rate of flow, so dq/dt decreases in magnitude but remains positive. Current continues to be in the positive x-direction but its magnitude decreases from its maximum positive value of +1 A toward zero (Figures 1.9b and 1.10a).
3. $t = \pi$ s: The flow stops momentarily, $dq/dt = i = 0$ (Figures 1.9b and 1.10b), and q reaches its maximum value of 2 C. The maximum quantity of charge has passed xx' in the positive x-direction.
4. $\pi < t \leq 3\pi/2$ s: q decreases, which means that charge now flows in the negative x-direction so as to reduce the net quantity of charge that has flowed in the positive x-direction past xx' (Figure 1.10c). The magnitude of the rate of flow in the negative x-direction, $|dq/dt|$, increases with time, reaching a maximum at $t = 3\pi/2$ s. Since dq/dt is negative, the current is in the negative x-direction and its magnitude increases to a maximum of 1 A at $t = 3\pi/2$ s (Figure 1.9b).
5. $3\pi/2 \leq t < 2\pi$ s: Charge continues to flow in the negative x-direction but at a decreasing magnitude of the rate of flow. The current continues to be in the negative x-direction but its magnitude decreases from its maximum positive value of +1 A toward zero (Figures 1.9b and 1.10c).
6. $t = 2\pi$ s: q is back to zero, which means that as much charge has flowed in the negative x-direction past xx' as has flowed in the positive x-direction. Since $dq/dt = 0$ at this instant, $i = 0$.

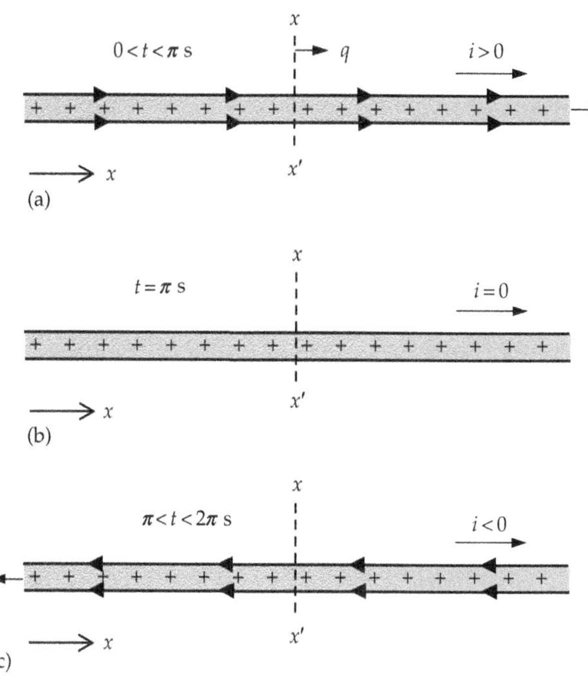

FIGURE 1.10
(a) Positive charges flowing in the positive x-direction, $0 < t < \pi$ s, (b) charge flow stops momentarily at $t = \pi$ s, and (c) positive charges flowing in the negative x-direction, $\pi < t < 2\pi$ s.

q starts at zero at $t = 0$ and ends at zero at $t = 2\pi$ s, which means that a quantity of charge moves in the positive x-direction past xx' during the time between $t = 0$ and $t = \pi$ s, and this same quantity moves back in the negative x-direction between $t = \pi$ and $t = 2\pi$ s. Because there is no *net flow* of charge in the negative x-direction at any time, q does not go negative. There is current during charge movement, the current i having a positive value when charge moves in the positive x-direction and a negative value when charge moves in the positive x-direction.

Note that the arrow associated with i in Figure 1.10 is in the positive x-direction, which means that the assigned positive direction of i is in the positive x-direction. However, i has a positive value when positive charges flow in the positive x-direction and a negative value when positive charges flow in the negative x-direction.

The area under the i vs. t graph increases from zero at $t = 0$ and reaches a maximum at $t = \pi$, in accordance with the q graph. As t increases beyond π, a negative area is added to the positive area of the positive half-cycle of i, so the net positive area decreases and becomes zero at $t = 2\pi$ s, when q is zero again.

*Example 1.3: Expression for Current through a Conducting Medium

Consider a conducting medium of cross-sectional area A m^2 through which positively charged particles are flowing, as illustrated in Figure 1.5 and reproduced in Figure 1.11. The concentration of particles is n particles/m^3, each particle having a charge e C and moving at a constant speed of u m/s. It is required to (a) derive an expression for the current I flowing through the conductor and (b) determine I if $n = 10^5$ particles/cm^3, $A = 1$ cm^2, $e = \pm 1$ μC, and $u = \pm 10$ cm/s, considering all possible sign combinations.

Solution:

(a) If the particles are assumed to be moving in the positive x-direction, then $u = dx/dt$ in the positive

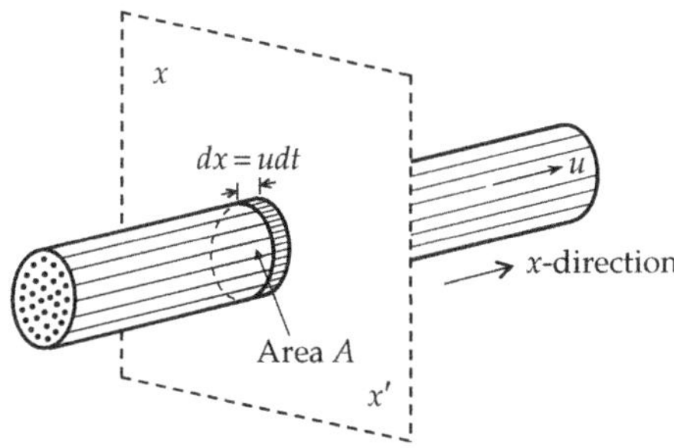

FIGURE 1.11
Figure for Example 1.3.

x-direction. Consider a disk of thickness $dx = udt$ m that lies immediately in front of the plane xx' (Figure 1.11). The volume of the disk is $Audt$ m^3, and the charge enclosed by the disk is

$$dq = e\left(\frac{\text{C}}{\text{Particle}}\right)\times n\left(\frac{\text{Particles}}{\text{m}^3}\right)\times Audt\,(\text{m}^3)$$
$$= enAudt\ C \tag{1.4}$$

Note that "particle" and "m^3" cancel out in the middle part of Equation 1.4, leaving the unit of charge C on the right-hand-side (RHS). In a time dt, the charges move a distance $dx = udt$ so that all the charge dq enclosed by the disk will move past the plane xx'. The charge that moves past the plane xx' per second is dq/dt. Dividing both sides of Equation 1.4 by dt, it follows from Equation 1.4 that the current I is

$$I = \frac{dq}{dt} = Aneu \tag{1.5}$$

(b) Substituting positive numerical values, with cm units converted to m units, and microcoulomb converted to coulomb, $I = (10^{-4}\ \text{m}^2)(10^5 \times 10^6\ \text{particles/m}^3)(10^{-6}\ \text{C/particle})(10 \times 10^{-2}\ \text{m/s}) = 1$ A. The positive numerical value of I is in accordance with the convention that if positive electric charges flow in the positive x-direction, the conventional direction of current is also in the positive x-direction. If positive electric charges flow at a constant rate in the negative x-direction and $u = -10$ cm/s, $I = 1 \times 10^5 \times 10^{-6} \times (-10) = -1$ A. The negative value of I signifies that the current is in the negative x-direction, as expected. If the particles are negatively charged, that is, $e = -1$ μC, and $u = 10$ cm/s, $I = 1 \times 10^5 \times (-10^{-6}) \times (10) = -1$ A. This is in accordance with the conclusion in Example 1.1, part (c), that if negative electric charges flow in the positive x-direction, the current is in the negative x-direction. Finally, if both e and u are negative, then the current is in the positive x-direction.

★ Exercise 1.4

The concentration of conduction electrons in a 1 mm diameter copper wire is 8.4×10^{28} electrons/m^3, the charge of an electron being -1.6×10^{-19} C. Determine the current if the net velocity of electrons in the positive x-direction is 0.02 cm/s. What is the direction of current?

Ans. −2.11 A, the negative sign indicating that the current is in the negative x-direction.

1.5 What Is Voltage?

Consider a volume Λ of water at a height h aboveground, where h is measured between the ground and the center of mass of Λ (Figure 1.12a). The potential energy of this volume of water with respect to the assumed ground level is

$$\text{PE} = (\Lambda\rho)hg \tag{1.6}$$

where
- ρ is the density of water so that $\Lambda\rho$ is the mass of the volume Λ of water
- g is the acceleration due to gravity (9.81 m/s^2 at sea level)

The potential energy per unit volume is then

$$\left(\frac{\text{PE}}{\Lambda}\right) = \rho hg \tag{1.7}$$

which is in fact the pressure, above atmospheric, at the assumed ground level due to a column of water of height h. The larger the PE/Λ, the greater is the amount of work that the volume of water Λ can do in falling

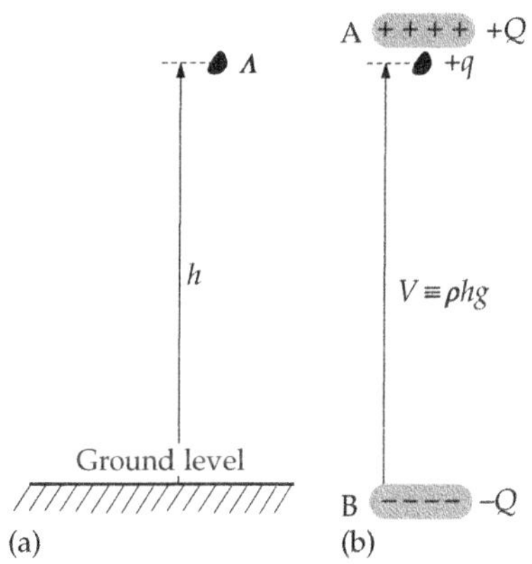

FIGURE 1.12
Analogy for voltage. (a) Volume of water Λ at a height h above ground and (b) a charge $+q$ close to a positively charged body 'A'.

through the height h. The work could be used to perform some useful task, such as driving a turbine, as in Figure 1.3.

In a similar manner, a quantity of charge $+q$, analogous to the volume Λ, located close to a positively charged body 'A' of charge $+Q$ and some distance away from a negatively charged body 'B', of charge $-Q$, possesses electric potential energy (Figure 1.12b). The reason for this electric potential energy is that work must be done in bringing $+q$ close to the positively charged body 'A', against the repulsion by this body and the attraction by the negatively charged body 'B'. This work is stored as potential energy in the system. If unconstrained, the charge $+q$ will move away from 'A' toward 'B', because of the repulsion by 'A' and the attraction by 'B'. Similarly, the volume of water Λ, if unconstrained, will fall toward ground level under the influence of gravity. In both cases, unconstrained Λ and $+q$ move from a region of higher potential energy to a region of lower potential energy and can do useful work in the process through conversion of the loss of potential energy to another form of energy. Dividing the electric potential energy of a given charge by the charge gives a quantity denoted as **voltage** and is analogous to the pressure ρhg in the case of water.

Definition: *The voltage between two points is the change in electric potential energy of a charged particle as it moves from one of these points to the other, divided by the charge of the particle.*

The unit of voltage in SI units is the **volt**, denoted by the symbol V, and is such that if a particle having a charge of +1 C moves down a voltage of 1 V, it loses 1 J of electric potential energy.

Equation 1.6 can be more generally expressed as a product:

$$\text{Gravitational potential energy difference between two points} = \text{Volume} \times \text{Pressure difference between the two points}$$

By analogy,

$$\text{Electric potential energy difference between two points} = \text{Charge} \times \text{Voltage difference between the two points}$$

It follows that if a particle of charge +1 C moves through a voltage difference v V, the change in electric potential energy (PE) of the particle is directly proportional to the voltage difference, that is, change in PE = (1 C) × (v V) = v J; if a particle of charge δq C moves through a voltage difference v V, the change in potential energy of the particle is also directly proportional to the charge of the particle, that is, $\delta w = v\delta q$, or $v = \delta w/\delta q$. In the limit, as δq approaches zero,

$$v = \frac{dw}{dq} \tag{1.8}$$

If w varies linearly with q, such as $w = kq$, where k is a constant, then $dw/dq = w/q = k$. If, however, w varies nonlinearly with q, as for q vs. t in Figure 1.4b, then Equation 1.8 defines v at a particular value of q in this more general case.

If a particle of charge $+q$ C moves to a point that is v V higher in voltage, it gains qv J in potential energy, and the voltage is $(+qv \text{ J})/(+q) = v$ V. On the other hand, if a particle of charge $-q$ C moves to a point v V higher in voltage, it *loses* qv J in potential energy. The change in potential energy is Δw = (final PE − initial PE) = $-qv$ J, and the voltage is $(-qv \text{ J})/(-q) = v$ V, independent of the sign of the charge.

A voltage that is constant with respect to time is a **dc** voltage; dc voltages are denoted by italic capital letters and instantaneous voltages by italic small letters, with capital subscripts in both cases, as may be required. Thus, V_O and V_{SRC} are dc voltages, whereas v_O and v_{SRC} are instantaneous voltages.

Potential energy is expressed with respect to some chosen reference that is assigned an arbitrary value, usually zero. For example, gravitational potential energy may be expressed relative to a reference value of zero at sea level. Similarly, voltage is always expressed with respect to some reference, usually that of the earth, which is assigned a voltage of zero. In electric circuits, what is of interest is the voltage difference between two points in the circuit. A battery voltage of 3 V, for example, means that the voltage difference between the positive and negative terminals of the battery is 3 V. *It is meaningless in an electric circuit to speak of the voltage at a certain point without specifying the point with respect to which this voltage is assigned.* When the circuit is connected to ground at some point, this is indicated by the ground symbol shown in Figure 1.13a, as in PSpice. If the circuit is not grounded, the reference point for voltages in the circuit is commonly indicated by the triangle symbol shown in Figure 1.13b. Table 1.1 lists the electrical quantities discussed so far and their counterparts in the hydraulic analogy.

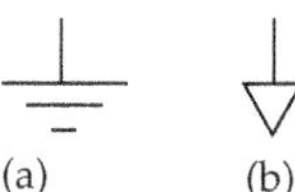

FIGURE 1.13
(a) Ground symbol and (b) symbol for voltage reference other than ground.

TABLE 1.1
Hydraulic Analogy

Electrical Quantity	Hydraulic Quantity
Quantity of charge	Volume of flow
Current	Rate of flow
Voltage	Pressure

1.6 What Is Voltage Polarity?

Just as current has direction, voltage, or more correctly the voltage difference between two points, has a polarity that indicates which of the two points is at a more positive voltage with respect to the other point. In the case of a battery, for example, the positive (+) and negative (–) terminals are marked on the battery in some appropriate manner. Similarly in an electric circuit, a plus sign next to a point and a minus sign next to another point, as in Figure 1.14, indicate that point 'a', having the + sign, is assigned a more positive voltage than point 'b' having the – sign. *Again, it is meaningless to specify a voltage between two points in a circuit, such as v_X in Figure 1.14 without indicating the polarity or direction of the voltage.* v_X in Figure 1.14 is *assigned* the positive polarity indicated. According to this assignment, then in going from 'a' to 'b', a *voltage drop* v_X is encountered, whereas in going from 'b' to 'a', a *voltage rise* v_X is encountered. If the numerical value of v_X is a positive number, point 'a' is in fact positive with respect to point 'b'. If the numerical value of v_X is a negative number, point 'b' is in fact positive with respect to point 'a'.

Primal Exercise 1.5

Assume that $v_X = -5$ V in Figure 1.14. (a) In which direction is the actual voltage drop, and (b) in which direction is the actual voltage rise?

Ans. (a) Actual voltage drop from 'b' to 'a'; (b) actual voltage rise from 'a' to 'b'.

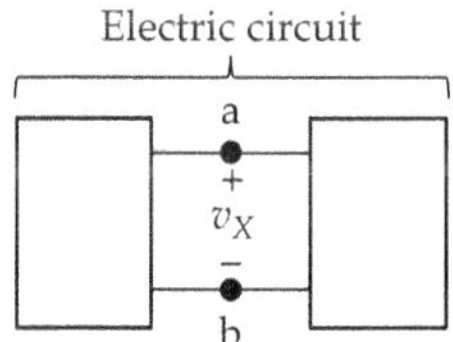

FIGURE 1.14
Voltage polarity.

Primal Exercise 1.6

(a) If a charge of +0.1 C moves to a point that is 3 V higher in voltage, by how much does its electric potential energy increase or decrease? (b) If a charge of –0.2 C moves to a point that is 5 V higher in voltage, by how much does its electric potential energy increase or decrease? (c) Are the movements in (a) and (b) those of an unconstrained particle, or do they require expenditure of energy *on* the particle?

Ans. (a) Electric potential energy of the particle increases by 0.3 J; (b) electric potential energy of the particle decreases by 1 J; (c) energy is expended on the particle in (a), whereas the motion in (b) is that of an unconstrained particle.

1.7 How Are Energy and Power Related to Voltage and Current?

It is seen from the preceding discussion that current carriers can possess two types of energy: potential energy that depends on the voltage and kinetic energy that depends on the current, that is, the rate of flow of the carriers, and hence their velocity. More will be said about this in future chapters.

Power p is defined as the rate at which energy w is generated or expended:

$$p = \frac{dw}{dt} \quad \text{or} \quad w = \int p dt \tag{1.9}$$

Substituting from Equations 1.8 and 1.2,

$$p = \frac{dw}{dt} = \frac{dw}{dq} \times \frac{dq}{dt} = vi \tag{1.10}$$

with v in volts and i in amperes, p in Equation 1.10 is in joules per second, or **watts**, denoted by the symbol W. The physical interpretation of Equation 1.10 is that a current of i A represents a flow of charge of i C/s. As each 1 C moves through a voltage difference v V, its electric potential energy changes by v J/C. The change in electric potential energy of the total charge in J/s is $(v \text{ J/C}) \times i(\text{C/s}) = vi$ J/s, which is the power in watts.

What does conservation of energy in a circuit mean? It means that at any instant of time, the total energy delivered by sources of energy in the circuit, such as batteries, must be equal to the total energy absorbed in energy-absorbing elements in the circuit, such as heating elements or lamps:

$$w_{delivered} = w_{absorbed} \quad \text{at every instant } t \tag{1.11}$$

Otherwise, if the energy delivered at any instant of time exceeds, say, that absorbed, then at least in principle, the system can be stopped at that instant, resulting in an excess of energy delivered over that absorbed. This means that energy can be extracted from the system at no energy cost, which violates conservation of energy.

Conservation of energy implies conservation of power. For if the power delivered or absorbed during a time interval Δt is p, the corresponding energy is $p\Delta t$. Substituting in Equation 1.11,

$$\left(p_{delivered}\right)\Delta t = \left(p_{absorbed}\right)\Delta t \tag{1.12}$$

Canceling out Δt,

$$p_{delivered} = p_{absorbed} \tag{1.13}$$

Power is conserved in the circuit as a whole *at every instant of time.*

Example 1.4: Power Delivered or Absorbed

Assume that the lamp in Figure 1.15 is rated at 0.75 W, 3 V, that is, it absorbs 0.75 W when a voltage of 3 V is impressed across its terminals. This is the total power that is converted primarily to light but also to some heat that is inevitably generated in the process. It is required to determine (a) the current through the lamp and its direction based on power absorption, (b) the power delivered by the battery, and (c) the direction of current through the battery based on power delivery.

Solution:

(a) Since terminals 'a' and 'b' of the battery are directly connected to terminals a′ and b′ of the lamp, respectively, by a wiring connection, the voltage across the lamp is also 3 V, the same as the battery voltage. Because the lamp absorbs 0.75 W, it follows from Equation 1.10 that $I = (0.75\text{ W})/(3\text{ V}) = 0.25$ A. The direction of I is that of the voltage drop from a′ to b′ because power absorption by the lamp implies that the assumed positive charges that carry I lose energy as they flow through the lamp. They must therefore flow through the lamp from terminal a′ that is at a higher voltage to terminal b′ that is at a lower voltage.

(b) By conservation of power in the circuit as a whole, the battery must deliver 0.75 W, neglecting the insignificant power dissipated in the connecting wires between 'a' and a′ and between 'b' and b′. Power dissipation in the connecting wires is always neglected in electric circuits, unless explicitly stated otherwise.

(c) As the assumed positive charges that carry I flow through the battery, they must regain the energy that was lost in flowing through the lamp. Hence, they must flow in the battery from the negatively charged terminal toward the positively charged terminal, that is, in the direction of a voltage rise through the battery.

Problem-Solving Tip

- Always mark on the circuit diagram the directions of all currents and the polarities of all voltages.

The argument made in the preceding example concerning power absorption and delivery underlies an important, general concept, usually referred to as the **passive sign convention**:

Concept: *If power is absorbed in a given circuit element, or component, current flows through that element in the direction of a voltage drop across the terminals of the given element, whereas if power is delivered by a given circuit element, current flows through that element in the direction of a voltage rise across the terminals of the given element.*

The physical justification is simply that positive charges flowing through a circuit element, or component, in the direction of a voltage drop lose electric potential energy as they flow through the element, as illustrated in Figure 1.16a. This loss of energy implies absorption of power in the circuit element. Conversely, positive charges flowing through a circuit element in the direction of a voltage rise (Figure 1.16b) gain electric potential energy as they flow through the element. This gain of energy implies delivery of power by the circuit element to the rest of the circuit.

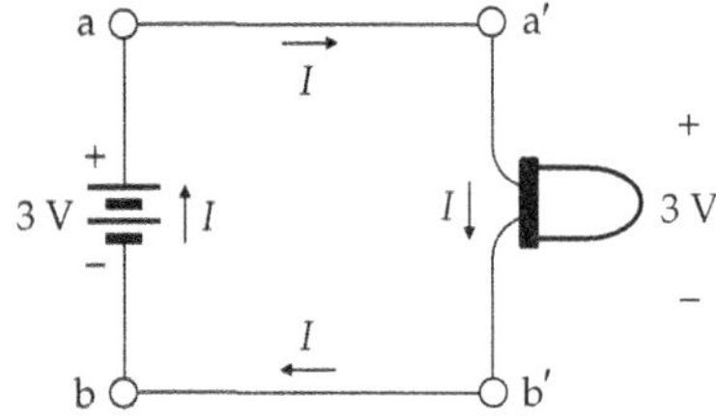

FIGURE 1.15
Figure for Example 1.4.

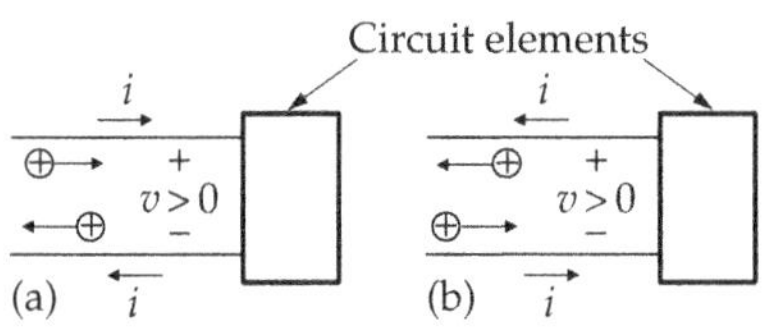

FIGURE 1.16
Positive charges flowing through a circuit element in the direction of a voltage drop (a) and in the direction of a voltage rise (b).

1.7.1 Positive and Negative Values of Circuit Variables

It will be seen in future chapters that before a circuit is analyzed, it is necessary to assign voltages and currents in various parts of the circuit, even though the actual directions of these voltages and currents are not yet known. What is done, therefore, is that the directions of these voltages and currents are *assigned arbitrarily*, using appropriate symbols, and subject to constraints imposed by circuit laws and the *v–i* relations of the circuit elements, as will be explained later. After the circuit is analyzed, if the numerical value of a given voltage or current is positive, then the actual direction of that voltage or current is the same as that assumed. On the other hand, if the numerical value of a given voltage or current is negative, the actual direction of that voltage or current is opposite to that assumed. Hence, negative values of voltages and currents are perfectly natural in electric circuits. This notion is not at all unusual. Suppose, for example, that the voltage of a battery needs to be measured but the battery terminals are unmarked. In this case, the positive lead of a voltmeter is connected *arbitrarily* to one of the battery terminals and the negative lead of the voltmeter to the other terminal of the battery. If the voltmeter reads a positive voltage, the positive terminal of the battery is that connected to the positive lead of the voltmeter. If the voltmeter reads a negative voltage, the positive terminal of the battery is that connected to the negative lead of the voltmeter.

Similarly, a negative value of power absorbed by a given element means that this element actually delivers power, and a negative value of delivered power by a given element means that this element actually absorbs power.

To summarize, consider Figure 1.17a, which shows the assigned positive direction of current I_A through an element 'A', and the assigned positive polarity of the voltage V_A across 'A'. I_A is arbitrarily assigned in the direction of a voltage drop V_A. The implication is that in the expression for power, $P_A = V_AI_A$, positive values of V_A and I_A make P_A positive. Since I_A is in the direction of a voltage drop V_A, then in accordance with the passive sign convention and its physical interpretation, a positive P_A is power absorbed by 'A'. Thus, if $I_A = 2$ A and $V_A = 9$ V (Figure 1.17b), then $P_A = +18$ W of power absorbed by 'A'. On the other hand, if either $V_A = -9$ V or $I_A = -2$ A (Figure 1.17c and d), then the power absorbed by 'A' is $P_A = -18$ W, which means that 'A' actually delivers 18 W, since the power absorbed is negative.

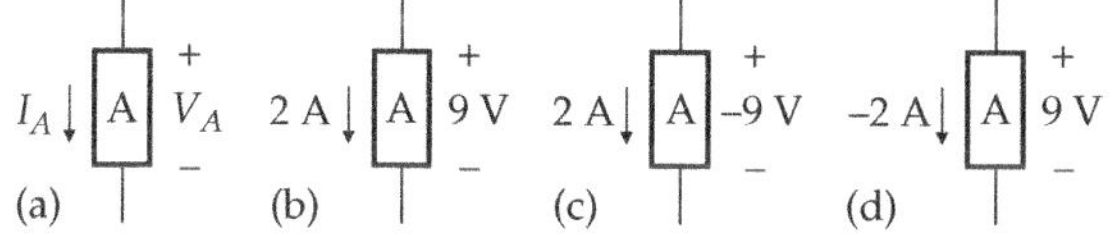

FIGURE 1.17
Power absorbed related to current and voltage. (a) Assigned positive direction of current through a circuit element 'A' is that of a voltage drop across 'A'. According to numerical values, the element actually absorbs power in (b) and delivers power in (c) and (d).

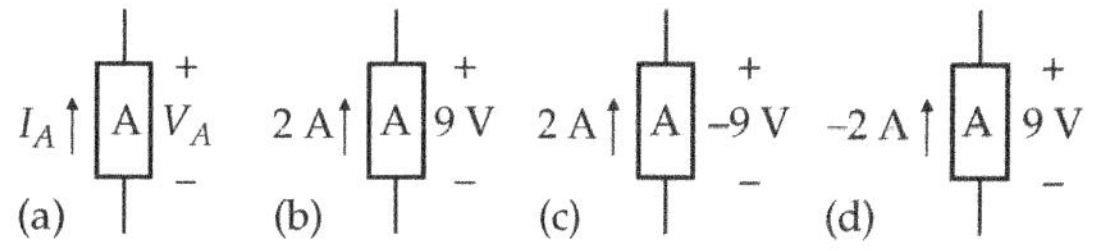

FIGURE 1.18
Power delivered related to current and voltage. (a) Assigned positive direction of current through a circuit element 'A' is that of a voltage rise across 'A'. According to numerical values, the element actually delivers power in (b) and absorbs power in (c) and (d).

On the other hand, suppose that the assigned positive direction of I_A is reversed to that of a voltage rise across 'A' (Figure 1.18a). Then a positive value of $P_A = V_AI_A$ is power delivered, in accordance with the passive sign convention and its physical interpretation. Thus, if $I_A = 2$ A and $V_A = 9$ V (Figure 1.18b), then $P_A = +18$ W is power delivered by 'A'. On the other hand, if either $V_A = -9$ V or $I_A = -2$ A (Figure 1.18c and d), then the power delivered by 'A' is $P_A = -18$ W, which means that 'A' actually absorbs 18 W, since the power delivered is negative.

Primal Exercise 1.7

Consider Example 1.4 and Figure 1.15. (a) What is the current that flows from terminal 'a' to terminal 'b' through the battery or from terminal b' to terminal a' through the lamp? (b) What is the voltage drop from terminal 'b' to terminal 'a' or the voltage rise from terminal 'a' to terminal 'b'? (c) How much is the power delivered by the lamp and that absorbed by the battery?

Ans. (a) −0.25 A; (b) −3 V; (c) −0.75 W.

Primal Exercise 1.8

The instantaneous power p absorbed by a device is shown in Figure 1.19. Determine the average power absorbed by the device over the interval 0–5 s. (*Note:* The average of a waveform over a specified interval is the net area under the graph of the function over the given interval divided by the interval.)

Ans. 0.5 W.

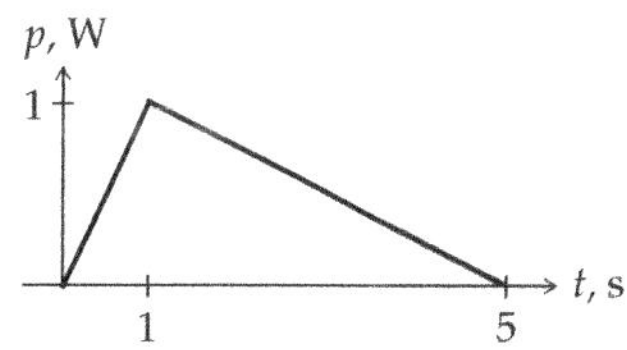

FIGURE 1.19
Figure for Primal Exercise 1.8.

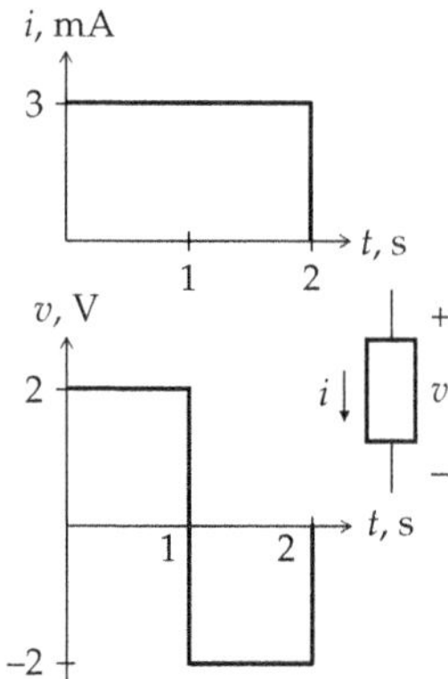

FIGURE 1.20
Figure for Primal Exercise 1.9.

Primal Exercise 1.9

The voltage across a device and the current through it in the direction of a voltage drop is as shown in Figure 1.20. Determine (a) the power absorbed or delivered by the device during the intervals $0 < t < 1$ s and $1 < t < 2$ s and (b) the energy absorbed or delivered by the device during the intervals $0 < t < 1$ s, $1 < t < 2$ s, and $0 < t < 2$ s.

Ans. (a) 6 mW absorbed, 6 mW delivered; (b) 6 mJ absorbed, 6 mJ delivered, 0.

1.8 What Are Ideal Circuit Elements and How Do They Handle Energy?

In circuit analysis, a fundamental conceptual step of great importance is taken, namely, that of representing physical or "practical" electric components, such as the battery and heater in Figure 1.2, in terms of a limited set of idealized, or "abstract", **circuit elements**. These ideal circuit elements can be divided into two general categories, **active** and **passive**, depending on how they handle energy.

Definition: *Active circuit elements are capable of delivering energy to an electric circuit through conversion of energy, ultimately from some nonelectrical source of energy. Passive circuit elements are incapable of doing so.*

The battery in Figure 1.2 can be represented by an *active* circuit element because it delivers energy to the heater through conversion of the energy of chemical reactions to electric energy. A solar cell or an electromechanical generator can be represented by an *active* circuit element because they convert solar energy and mechanical energy, respectively, to electric energy. All these active circuit elements can be represented using **ideal sources**, as discussed in detail in Chapter 2.

On the other hand, passive circuit elements are incapable of delivering energy in an electric circuit through conversion from another source of energy. Instead, they can absorb energy for one of the following purposes:

1. Energy consumption or conversion of energy in the electric circuit to nonelectric energy, which could be, for example, heat, light, sound, or mechanical energy. Energy-consuming devices can often be conveniently represented in electric circuits by **ideal resistors**, which dissipate electric energy as heat, as in the case of the heater in Figure 1.2. The property that allows a resistor to dissipate energy is the **resistance**, denoted by the symbol *R*.
2. Temporary storage in the form of electric energy or, more specifically, potential energy of current carriers. The circuit element that stores this energy is an **ideal capacitor**, and the property associated with this electric energy storage is the **capacitance**, denoted by the symbol *C* (note that italic *C* is capacitance, whereas nonitalic C is coulomb).
3. Temporary storage in the form of magnetic energy or, more specifically, kinetic energy of current carriers. The circuit element that stores this energy is an **ideal inductor**, and the property associated with this magnetic energy storage is the **inductance**, denoted by the symbol *L*.

Resistance, capacitance, and inductance are referred to as **circuit parameters**. Figure 1.21 illustrates the classification of the ideal circuit elements, and the circuit symbols for an ideal resistor, capacitor, and inductor. The designation "circuit element" will henceforth exclusively refer to an ideal circuit element.

The three passive circuit elements are "ideal" in the following respects:

1. Each element performs its designated function only. Thus, ideal resistors only dissipate energy.

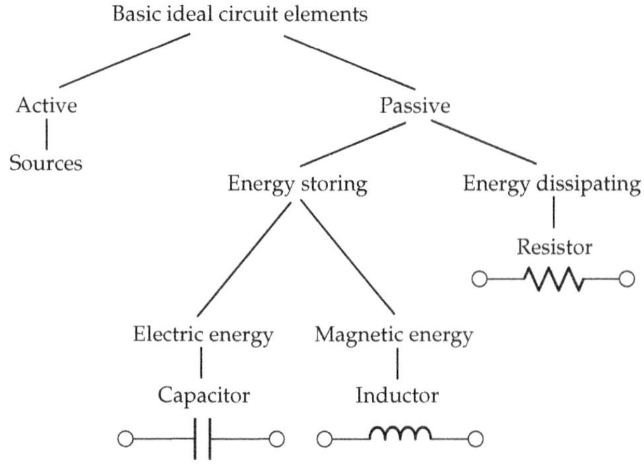

FIGURE 1.21
Basic circuit elements.

They neither store electric energy nor store magnetic energy. Ideal capacitors only store electric energy. They neither dissipate energy nor store magnetic energy. Ideal inductors only store magnetic energy. They neither dissipate energy nor store electric energy. Practical circuit elements are not ideal in this sense. For example, a practical inductor made from a coil of wire dissipates energy because of the finite resistance of the wire. Moreover and because the function performed by each of the ideal passive circuit elements is of a fundamentally different nature from that performed by the other two elements, none of the three ideal circuit elements can be represented in terms of the other two. In this sense, the ideal circuit elements are also described as *basic*.

2. The circuit parameters R, L, and C in a given circuit are constant, independent of any current or voltage with which they are associated, and are also independent of time. It will be shown later that the constancy of R, L, and C implies linearity of the governing relations for these elements. Because of this, basic electric circuits comprising ideal circuit elements are referred to as **linear time-invariant** (LTI) circuits.

It is quite legitimate to question the usefulness of working with such ideal elements. The justification is that consideration of idealized models is a central first step in the engineering approach, which has proved so successful in advancing technology. These idealized models, or "equivalent circuits", as they are often referred to in electrical engineering, employ ideal elements to reproduce, as a first approximation, the essential behavior of the system under consideration, without extraneous details that would otherwise hamper unnecessarily the analysis of system behavior. This is the second application of electric circuits mentioned in Section 1.1. For example, if we are only interested in the current and power in the circuit of Figure 1.15, we need not consider the chemical reactions in the battery, nor the electric field inside the battery and how it varies from one terminal to the other, nor the internal resistance or the opposition to the flow of current carriers through the battery, etc. As a first approximation, we can simply represent the battery as an ideal voltage source of the same voltage as that measured between the terminals of the battery. Similarly, if the lamp is a light-emitting diode (LED), we need not consider how light is produced in the semiconductor, nor the intensity of the light and its spectral distribution, nor the heat generated in the process, etc. As a power-absorbing device, the LED may be replaced by an ideal resistor that absorbs the same power.

Such idealized models may be used to investigate and improve the behavior of the system under consideration. If a model does not reproduce the actual behavior of the system accurately enough under certain conditions, it is refined in order to do so. For example, the resistance of a practical inductor can be accounted for, at low frequencies, by adding an ideal resistor; a practical transformer or an operational amplifier can be represented by equivalent circuits composed of ideal circuit elements, and the behavior of circuits that include these elements can be analyzed using their equivalent circuits, as will be discussed in future chapters. These equivalent circuits can be modified to account for various imperfections in transformers or operational amplifiers or to emulate the behavior of these components at high frequencies, for example.

It is seen that the study of circuits composed of ideal circuit elements is a crucial first step in understanding the behavior of various electrical and electronic systems. This will be our main concern in this book. Such understanding is essential for designing systems that perform desired tasks and for improving the performance of these systems.

*1.9 Why Resistance, Capacitance, and Inductance?

This section explains the significance of the three circuit parameters in terms of fundamental attributes of the electromagnetic field.

Fundamentally, an electric field is associated with electric charges at rest or in motion, and a magnetic field is associated with moving electric charges, that is, electric current. A rigorous analysis of systems involving electric charges and currents should therefore be carried out in terms of electromagnetic fields. This, however, is a daunting task even for relatively simple cases. Circuit analysis provides a much easier but nevertheless *approximate* solution that is adequate for a great number of cases encountered in practice. The nature of these approximations is examined in the following section.

If circuit analysis is to solve some electromagnetic problems, even in an approximate manner, it must account for some basic attributes of electromagnetic fields. Once launched by some source of electromagnetic energy, the electromagnetic field has three basic attributes, namely, (1) energy dissipation due to energy losses in the medium through which the field propagates; (2) electric energy stored in the electric component of the electromagnetic field, which represents the energy expended in establishing the electric field; and (3) magnetic energy stored in the magnetic component of the electromagnetic field, which represents the energy expended in establishing the magnetic field. In an electric circuit, these three attributes can be accounted for, respectively, by the circuit parameters of resistance, capacitance, and inductance. This implies that an electrical or electronic component or system can be represented, or modeled, by means of an appropriate electric circuit that includes these elements.

⋆1.10 What Are the Approximations Implicit in Basic Electric Circuits?

This section explains the nature of two main approximations implicit in basic electric circuits, when modeling physical systems, namely, the lumped-parameter representation and the neglect of wave propagation in a circuit.

In order to appreciate the nature of the approximations implicit in basic electric circuits, consider a cylindrical rod of resistive material, one end of which is connected to a battery of voltage V_B and the other end connected to ground (Figure 1.22a). A current I flows under the influence of the battery voltage. The hydraulic analogy is a narrow pipe through which water is driven. One end of the pipe is connected to a pump, the other end of the pipe being open to the atmosphere, above an open reservoir, say. The pipe would present considerable resistance to water flow because of friction, mainly between water and the inner walls of the pipe. Hence, the pump applies some hydraulic pressure at one end of the pipe in order to overcome this friction and drive water through the pipe. The pressure applied by the pump is analogous to the battery voltage, and the friction along the pipe is analogous to the resistance along the rod, this resistance being uniformly distributed along the length of the rod. The energy applied by the pump to overcome friction is analogous to the energy delivered by the battery to overcome the resistance of the rod. The power input to the rod is $p = V_BI$ and is dissipated as heat in the rod.

Just as the pressure drops along the pipe, from that established by the pump at one end to zero at the open end, the voltage along the rod, with respect to ground, drops from V_B at one end of the rod to zero at the end connected to ground. Associated with this voltage is an electric field between the rod and ground that progressively decreases with the voltage along the rod, as illustrated in Figure 1.22a. The energy stored in the electric field between the rod and ground can be represented in circuit terms by a capacitance that is distributed along the rod. There will also be some energy in the electric field due to the voltage drop along the rod itself, but this can be generally neglected compared to the energy stored in the electric field between the rod and ground when the voltage drop along the rod is not too large.

In addition, there will be a magnetic field along the length of the rod due to the current I, as illustrated in Figure 1.22b. The energy stored in the magnetic field can be represented in circuit terms by an inductance that is uniformly distributed along the rod. The question arises as to how can the distributed resistance, capacitance, and inductance be represented along the rod and taken into consideration when analyzing the flow of current through the rod.

Mathematically, an infinitesimal length dx of the rod can be considered to have infinitesimal resistance, inductance, and capacitance (Figure 1.23a). A differential equation is then written for current or voltage and the system analyzed accordingly. However, this is rather complicated, because the resulting differential equation is a partial differential equation involving both time and distance. Hence, this approach is not used in basic circuit analysis but is used in more advanced treatments, as in the case of transmission lines and waveguides. Instead, the rod can be divided into a number of segments, say, three, and each segment

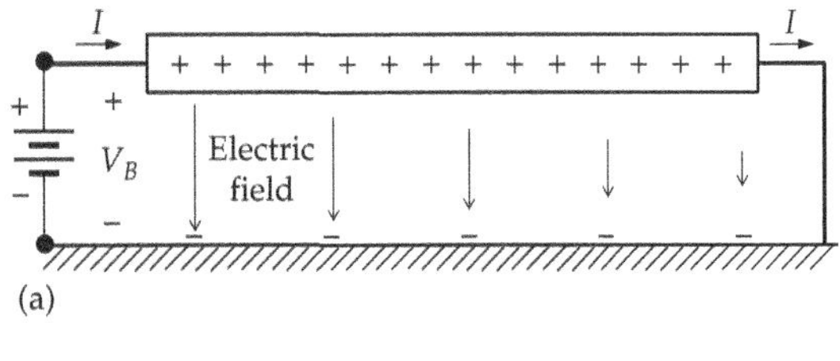

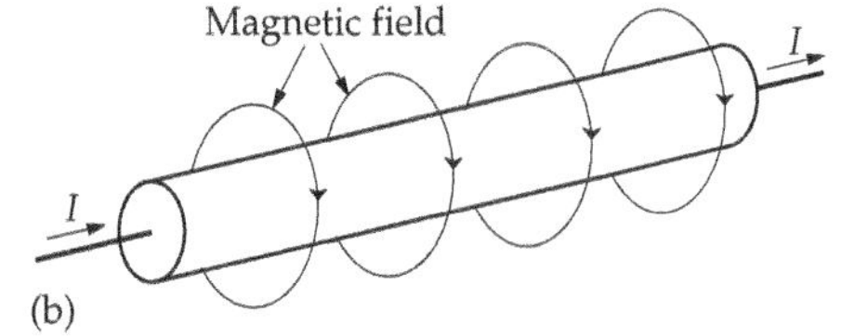

FIGURE 1.22
Current flow through conducting rod. (a) Current flow through electric rod grounded at one end, showing the electric field between the rod and ground and (b) magnetic field due to current in the rod.

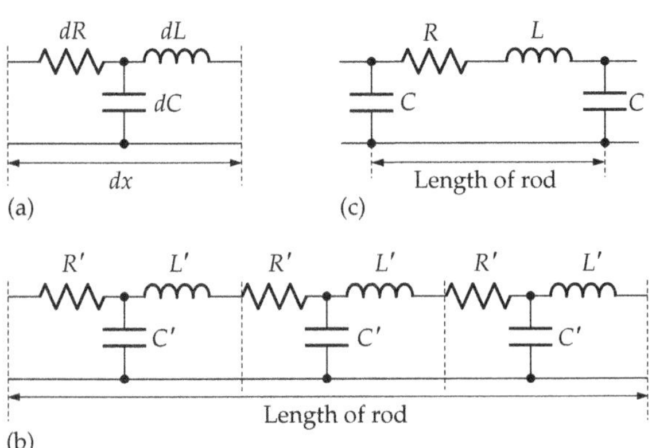

FIGURE 1.23
Approximations to distributed circuit parameters. (a) Infinitesimal segment of rod, (b) rod divided into three sections, and (c) approximate circuit representation of rod.

is represented by a *lumped* resistance, capacitance, and inductance representing, respectively, the total power dissipation, the total energy stored in the electric field, and the total energy stored in the magnetic field, in each segment (Figure 1.23b). Because of this lumping, representations such as that of Figure 1.23b are referred to as **lumped-parameter** representations. This is in contrast to **distributed-parameter** representations that take into account the distributed nature of resistance, capacitance, and inductance. Evidently, the larger the number of segments in Figure 1.23b, the better is the approximation of the lumped-parameter representation to the distributed-parameter representation, but the more complicated is the analysis. As a first approximation, the resistive rod can be represented by a single resistance and inductance, with capacitance at each end (Figure 1.23c). Moreover, if the electric and magnetic stored energies are small compared to the energy dissipation, the resistive rod can be represented by a single, ideal resistor.

Another aspect of the circuit approximation that is implicit in Figure 1.22a is that the electromagnetic field takes a finite time to propagate from one end of the rod to the other. That is, if the battery is suddenly connected to one end of the rod in Figure 1.22a, it takes a finite time before the effects of the electric and magnetic fields are felt at the other end of the rod (see Problem P1.2). The *propagation delay* due to the wave nature of the electromagnetic field is neglected in basic circuit analysis, which is tantamount to assuming that when an excitation is applied at one end of a circuit, its influence is immediately felt at the other end. This is justified as long as the propagation delay is small compared to the period of the voltage or current signal, where the period is the reciprocal of the signal frequency. This is tantamount to assuming that the physical dimensions of the circuit under consideration are small compared to the wavelength of the electromagnetic wave, where the wavelength is the ratio of the speed of propagation of the electromagnetic wave to the signal frequency. If the frequency is, say, 1 GHz (10^9 Hz), which is of the order of frequencies used in many communication systems, and assuming that the speed of propagation is nearly half that of the speed of light in vacuum (3×10^8 m/s), depending on the electric and magnetic properties of the medium, the wavelength is 15 cm. If the longest dimension of the circuit exceeds, say, one-tenth of this, or 1.5 cm, then the lumped-circuit representation can introduce a significant error. On the other hand, the wavelength at a frequency of 50 Hz is 6000 km, assuming the electromagnetic field propagates at the speed of light along overhead power lines in air. The lumped-circuit representation would generally be acceptable in this case for overhead power lines of a few hundred kilometers in length.

Learning Checklist: What Should Be Learned from This Chapter

- An electric circuit is an interconnection of components that affect electric charges in some characteristic manner.
- Electric circuits are used in two ways: (1) to perform some useful task, as in the case of the heater of Figure 1.1 or in the case of electrical installations in buildings, or (2) to model or emulate the behavior of some component or system.
- The behavior of electric circuits is governed by two fundamental conservation laws: conservation of energy and conservation of charge.
- The current at any given point in an electric circuit and at a specified instant of time is the rate of flow of electric charge past the given point at that instant.
- According to the hydraulic analogy,

  Volume of water ↔ Quantity of charge

  Rate of flow of water ↔ Current
- In general, instantaneous current is defined as $i = dq/dt$, where q is in coulombs, t is in seconds, and i is in amperes. The instantaneous current at a particular instant of time is the slope of the graph of charge vs. time at that instant. Conversely, charge is the area under the graph of current vs. time.
- A current that is constant with respect to time is as a dc current.
- It is assumed in circuit analysis that the direction of current is that of the flow of positive electric charges, irrespective of the sign of the charges that actually carry the current.
- Current always has a direction, indicated by an arrow that points in the assigned positive direction of current. That is, positive charges flowing in the assigned positive direction represent a current having a positive numerical value.
- A negative value of current arises either from positive charges moving in the direction opposite to that of the assigned positive direction of current or from negative charges moving in the same direction as the assigned positive direction of current.
- The voltage between two points is the change in electric potential energy of a charged particle as it moves from one of these points to the other, divided by the charge of the particle. The voltage is independent of the sign of the charge.
- According to the hydraulic analogy,

  Pressure ↔ Voltage

- In general, voltage is defined as $v = dw/dq$ and is a voltage difference between two points. A voltage that is constant with respect to time is a dc voltage.
- Voltage always has a polarity, or direction, indicated by a plus sign next to the point that is assigned a higher, or more positive, voltage relative to the point that is assigned a lower, or less positive or more negative, voltage, indicated by a negative sign next to this point.
- Current carriers can possess two types of energy: potential energy that depends on the voltage and kinetic energy that depends on the current.
- Power p is defined as the rate at which energy is generated or expended. In general, $p = dw/dt = vi$.
- Conservation of energy implies conservation of power in the circuit as a whole, which means that the total power delivered in the circuit is equal to the total power absorbed in the circuit at every instant of time.
- According to the passive sign convention, if power is absorbed in a given circuit element, current flows through that element in the direction of a voltage drop across the terminals of the given element, whereas if power is delivered by a given circuit element, current flows through that element in the direction of a voltage rise across the terminals of the given element.
- Negative values of current, voltage, and power are quite natural in circuit analysis and indicate that the direction of current, polarity of voltage, and direction of power flow are, respectively, opposite to that assumed to be positive.
- Active circuit elements are capable of supplying energy to an electric circuit through conversion of energy, ultimately from some nonelectrical source of energy. Passive circuit elements are incapable of doing so.
- In electric circuits, active circuit elements are represented by ideal sources, whereas energy-consuming devices may be represented by ideal resistors. Ideal resistors, capacitors, and inductors are passive circuit elements.
- The properties of resistance, inductance, and capacitance are referred to as circuit parameters.
- Idealized models employ ideal circuit elements to reproduce, as a first step, the essential behavior of the system under consideration, without extraneous details. They can be refined to reproduce more accurately the behavior of the system under some specified conditions.
- The properties of resistance, capacitance, and inductance represent, respectively, the three basic attributes of the electromagnetic field, namely, energy dissipation, energy stored in the electric field, and energy stored in the magnetic field. Each of these ideal, passive circuit elements represents exclusively one of the fundamental attributes of the electromagnetic field.
- Ideal circuit elements are basic in the sense that none of these elements can be represented in terms of the other ideal circuit elements. Moreover, the values of resistance, capacitance, and inductance are independent of voltage or current and are constant with respect to time. Hence, circuits in which all resistances, capacitances, and inductances have constant values are referred to as LTI circuits.
- Basic circuits involve two main approximations: (1) the distributed nature of resistance, capacitance, and inductance is ignored, and (2) propagation delay of the electromagnetic field is ignored, which is justifiable as long as the dimensions of the circuit are small compared to the wavelength of the electromagnetic wave.

Problem-Solving Tips

1. Always check the units on both sides of an equation, and always specify the units of the results of calculations.
2. Always mark on the circuit diagram the directions of currents and the polarities of voltages.

Problems

Current

P1.1 A belt that is 75 cm wide is moving at a speed of 10 m/s and has a surface charge of 4 μC/m^2. Determine the current carried by the belt.

Ans. 30 μA.

P1.2 A lamp assembly is connected to a battery and a switch by a wire 2 m long and 1 mm^2 cross section. If a current of 1 A flows in the wire, determine the time it takes a conduction electron to travel along the 2 m length of wire from the switch to the lamp assembly, assuming the concentration of conduction electrons in the wire is $8.4 \times 10^{28}/\text{m}^3$ and the charge on an electron is -1.6×10^{-19} C (refer to Equation 1.5). Note that the delay in the turning on of the lamp assembly after the switch is closed does not depend on the time it takes a conduction electron to travel from the switch to the lamp assembly, since these electrons are present all along the wire at any given time. The delay depends on how fast the electric field propagates, after the switch is closed, so as to act on the conduction electrons at the end of the wire in contact with the lamp assembly. The speed of propagation of the electric field along the wire is of the order of the speed of light.

Ans. 7 h and 28 min.

P1.3 A steady beam of alpha particles (nuclei of helium atoms) having a uniform cross section carries a current of 0.5 μA, each alpha particle having a positive charge that is twice the magnitude of the charge on an electron (i.e., $+3.2 \times 10^{-19}$ C). (a) If the beam is directed perpendicular to a flat surface, determine the number of alpha particles that strike the surface every second. (b) If the velocity of the alpha particles is 1.5×10^7 m/s and the cross section of the beam is 2 cm², determine the concentration of particles in the beam (refer to Equation 1.5).

Ans. (a) 1.563×10^{12} particles; (b) 5.21×10^8 particles/m³.

P1.4 A fuse is rated at 1 A. (a) What is the minimum steady rate of flow of conduction electrons per second that will blow the fuse? Assume the charge per electron is -1.6×10^{-19} C. (b) Is the direction of flow relevant, considering that the heating effect of current is proportional to the square of the current value? (c) If the concentration of conduction electrons in the metal of the fuse is 10^{28} electrons/m³ and the cross-sectional area of the fuse wire is 0.0025 mm², what is the average velocity of conduction electrons that will give the rate of flow in (a)?

Ans. (a) 6.25×10^{18} electrons/s; (b) no; (c) 0.25 m/s.

P1.5 In a one-dimensional flow of current through a semiconductor, positively charged holes move in the positive x-direction at a steady rate of 5×10^{18} holes/min, and electrons move in the negative x-direction at a steady rate of 2.5×10^{18} electrons/min. Determine the total current in mA in (a) the positive x-direction and (b) the negative x-direction. (c) What would be the current if the holes and electrons move in the same direction? Note that a hole carries a single electronic unit of positive charge (1.6×10^{-19} C).

Ans. (a) 20 mA; (b) −20 mA; (c) 20/3 mA in the direction of movement.

P1.6 In electroplating an object with silver, the object is made the cathode and a silver plate the anode. Silver ions in solution are attracted to the cathode, gain electrons from the cathode, and are deposited on the object as silver atoms (Figure P1.6). (a) How many silver atoms are deposited by a current of 10 A flowing for 1 h? (A silver ion carries a single electronic unit of positive charge of 1.6×10^{-19} C.) (b) If a gram molecular weight of silver is 0.1079 kg and contains 6.025×10^{23} atoms (Avogadro's number), what is the mass of silver deposited?

Ans. (a) 22.5×10^{22} silver atoms; (b) 40.3 g.

P1.7 The charge q along a wire varies with time as shown in Figure P1.7, where $q = 2\sin(\pi t/2)$ C, $0 \le t \le 1$ s. Determine the variation of the current i with time.

Ans. $i(t) = \pi\cos(\pi t/2)$ A, $0 \le t \le 1$ s; $i(t) = 0$, $1 \le t \le 2$ s; $i(t) = -2$ A, $2 \le t \le 4$ s; $i(t) = 1$ A, $4 \le t \le 6$ s.

P1.8 Consider that Figure P1.8 represents the variation with time of the current i in A through a wire. Determine the variation of q as a function of t during each interval and verify the values at the end of each interval by calculating the area involved.

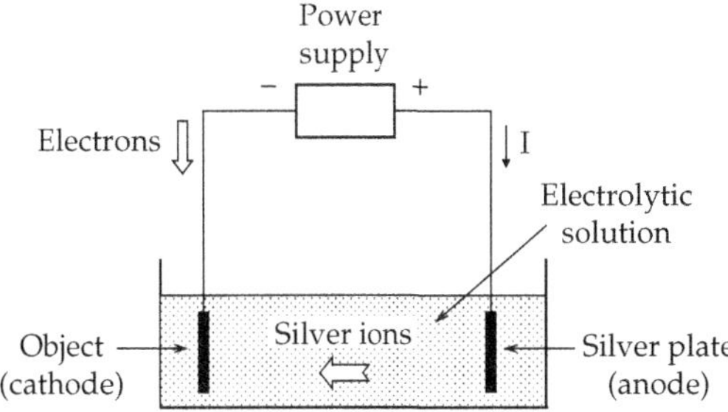

FIGURE P1.6

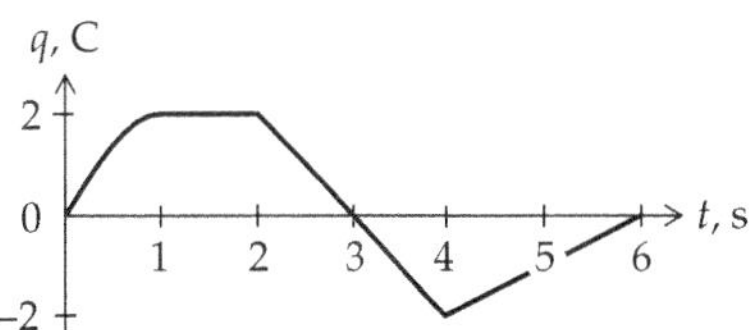

FIGURE P1.7

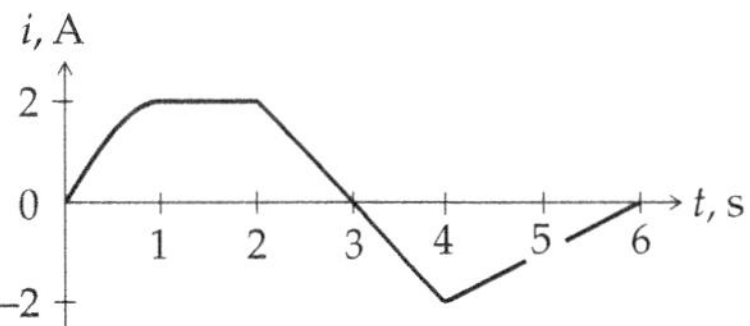

FIGURE P1.8

Ans. $q(t) = \frac{4}{\pi}\left(1 - \cos\left(\frac{\pi t}{2}\right)\right)$C, $0 \le t \le 1$ s; $q(t) = 2(t-1) + \frac{4}{\pi}$C, $1 \le t \le 2$ s; $q(t) = -t^2 + 6t - 6 + \frac{4}{\pi} =$ C, $2 \le t \le 4$ s; $q(t) = \frac{t^2}{2} - 6t + 18 + \frac{4}{\pi}$C, $4 \le t \le 6$ s; $q(t) = \frac{4}{\pi}$C, $t \ge 6$ s.

Voltage

P1.9 Given an electric field ξ V/m that is constant in the x-direction and a charge $+q$ C located at the origin and free to move in the x-direction (Figure P1.9), (a) what is the magnitude and direction of the force F acting on q? (b) If q moves under the influence of F a distance d m, how much work is done by F? (c) Assuming the voltage at the origin to be zero, what is the voltage V_d at $x = d$ m, bearing in mind that $\xi = -dv/dx$? (d) How is the loss in electric potential energy related to the work done by F? (e) Assuming the charge has a mass m kg and zero velocity at the origin, show that the kinetic energy of the charge at $x = d$ is equal to the loss in electric potential energy.

Ans. (a) $F = q\xi$ N in the positive x-direction; (b) Fd J; (c) $V_d = -\xi d$ V; (d) they are both equal to $q\xi d$ J.

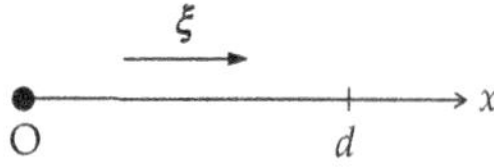

FIGURE P1.9

P1.10 An electron is released at the origin of the x-axis with a velocity of 1.33×10^6 m/s in the positive x-direction. If a uniform electric field of 200 V/m is directed along the x-axis, as in Figure P1.9, determine the value of x at which the electron comes to rest. Assume an electron has a charge of -1.6×10^{-19} C and a mass of 9.1×10^{-31} kg.

Ans. 2.52 cm.

P1.11 Electrons are emitted from a heated metal plate 'A' at a constant rate of 6.25×10^{14} electrons/s, with zero kinetic energy. They are accelerated toward a parallel metal plate 'B' that is separated from 'A' by 5 mm. Plates 'A' and 'B' are connected to an external power supply that maintains 'B' at a constant voltage of +10 V with respect to 'A'. (a) How much potential energy does an electron gain or lose in going from 'A' to 'B'? (b) What happens to this potential energy? (c) What is the velocity of the electron when it arrives at 'B'? Assume an electron has a charge of -1.6×10^{-19} C and a mass of 9.1×10^{-31} kg.

Ans. (a) Loses potential energy of 1.6×10^{-18} J; (b) it is converted to kinetic energy; (c) 1.88×10^6 m/s.

P1.12 Consider Problem P1.11. (a) What is the total kinetic energy of the electrons that arrive at 'B' during 1 s? (b) The accelerated electrons are "collected" at 'B', where they flow through plate 'B' to the positive plate of the battery. What is the magnitude and direction of the resulting current through the power supply? (c) What happens to the kinetic energy of the electrons once they are collected at 'B'? (d) How much power is expended by the power supply in order to keep the voltage between 'A' and 'B' at 10 V? (e) How is this power related to the kinetic energy given up by the electrons?

Ans. (a) 1 mJ; (b) 100 μA, in the direction of a voltage rise through the power supply; (c) converted to heat; (d) 1 mW; (e) the power is equal to the rate at which kinetic energy is given up in J/s.

P1.13 Consider the system of Figure P1.13, where the voltage V_{AB} V between the two metal plates is maintained constant by the battery. Let there be a mechanism for moving positive charge from the lower plate to the upper plate. (a) How much work, W, is done in moving an amount of charge $+q$ C? (b) What is the graph of W vs. q?

Ans. (a) $W = qV_{AB}$ J; (b) a straight line of slope V_{AB} passing through the origin.

P1.14 Suppose that the battery in Figure P1.13 is removed (Figure P1.14), but that positive charge is still moved from the lower plate to the upper plate. It is required to determine the work done in moving a quantity of charge q_F that establishes a voltage $V_{AB} = Kq_F$, where K is a constant. Assume that when the charge on the upper plate is $+q$ C, with an equal and opposite charge $-q$ C on the lower plate, the voltage between the plates

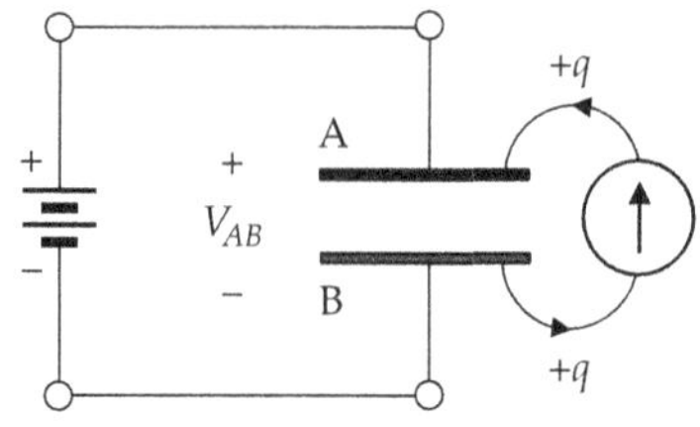

FIGURE P1.13

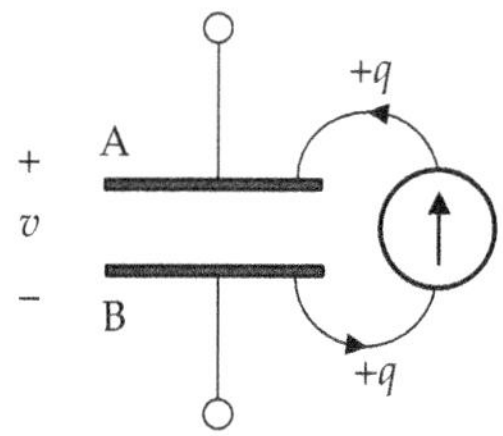

FIGURE P1.14

is $v = Kq$ V. Let an infinitesimal charge dq be moved, where dq is small enough to keep v approximately constant while dq is moved. (a) What is the work dW that is done in moving the charge dq? (b) Substitute Kq for v and determine W by integrating between 0 and q_F. (c) Express W as a function of V_{AB}. (d) What is the graph of W vs. q in this case?

Ans. (a) $dW = v\,dq$ J; (b) $W = \frac{1}{2}Kq_F^2$ J; (c) $W = \frac{1}{2}V_{AB}^2/K$ J; (d) parabola centered at the origin.

P1.15 If q in Problem P1.14 varies sinusoidally with time as $q(t) = q_m \sin\omega t$ C, determine the time variations of i, v, and w.

Ans. $i(t) = \omega q_m \cos\omega t$ A; $v(t) = Kq_m \sin\omega t$; $w(t) = \frac{K}{4}q_m^2(1 - \cos 2\omega t)$ J.

Power and Energy

P1.16 How much energy does a 100 W lamp consume in 1 h?

Ans. 360 kJ.

P1.17 The voltage v across an element 'A' is a rectangular waveform of 5 V amplitude and 4 s duration (Figure P1.17). The current i through 'A' in the direction of the voltage drop v is a biphasic pulse that has an amplitude of +2 A, $0 < t < 2$ s and an amplitude of −2 A, $2 < t < 4$ s. Determine the power delivered or absorbed by 'A' during each 2 s interval. What is the total charge that has passed through 'A' at $t = 4$ s? How is this related to the net power delivered or absorbed?

Ans. $0 < t < 2$ s, 'A' absorbs 10 W; $2 < t < 4$ s, 'A' delivers 10 W; 0 charge, and hence zero net power delivered or absorbed.

P1.18 The voltage drop across a certain device and the current through it in the direction of the voltage drop (Figure P1.18) are given by

$$v(t) = \sin\pi t/2 \text{ V},$$
$$i(t) = \cos\pi t/2 \text{ A}$$

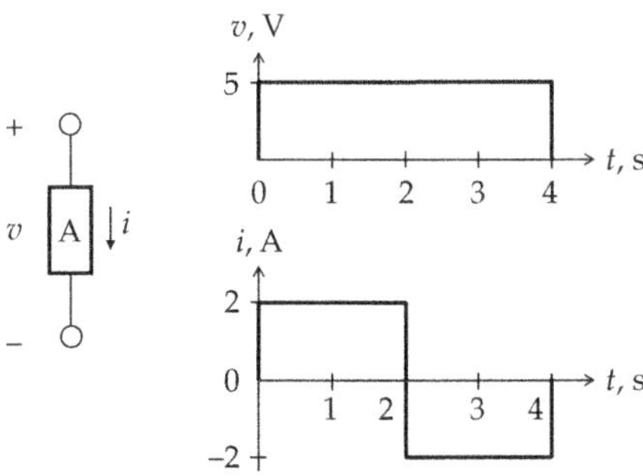

FIGURE P1.17

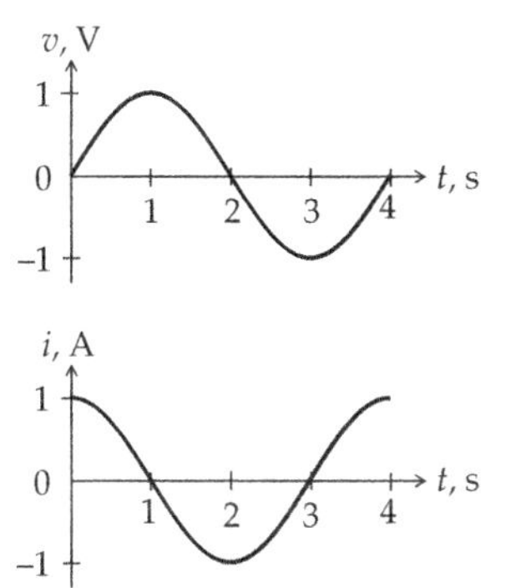

FIGURE P1.18

(a) Determine whether the device absorbs or delivers power during each of the quarter cycles of one period. (b) Derive the power at any instant as a function of time. How do you interpret positive values and negative values of power? What is the maximum magnitude of instantaneous power? What is the average power over a 4 s period?

Ans. (a) Power is absorbed during the intervals $0 \leq t \leq 1$ s and $2 \leq t \leq 3$ s, and power is delivered during the intervals $1 \leq t \leq 2$ s and $3 \leq t \leq 4$ s. (b) $p(t) = \sin\pi t/2 \cos\pi t/2$ W $= 0.5\sin\pi t$ W; $p > 0$ is power absorbed and $p < 0$ is power delivered; maximum is 0.5 W and average is zero.

P1.19 An A-size, 1.5 V battery is rated at 3 ampere-hours (Ah). During continuous use, with the battery supplying a current of 100 mA, the battery voltage stays substantially constant at 1.50 V for the first 20 h. During the next 10 h, the voltage drops linearly to 1.25 V, while the current drops linearly to 80 mA (Figure P1.19). At this point, the battery is no longer considered useful. (a) What was the useful Ah capacity of the battery, and (b) how much energy was delivered by the battery during the 30 h?

Ans. (a) 2.9 Ah; (b) 15.26 kJ.

P1.20 The voltage drop across a certain device and the current through it in the direction of the voltage drop are shown in Figure P1.20. Determine (a) the charge q through the device at the end of each 1 s interval from $t = 0$ to $t = 6$ s, (b) the instantaneous power $p(t)$ during the aforementioned intervals, and (c) the total energy consumed by the device.

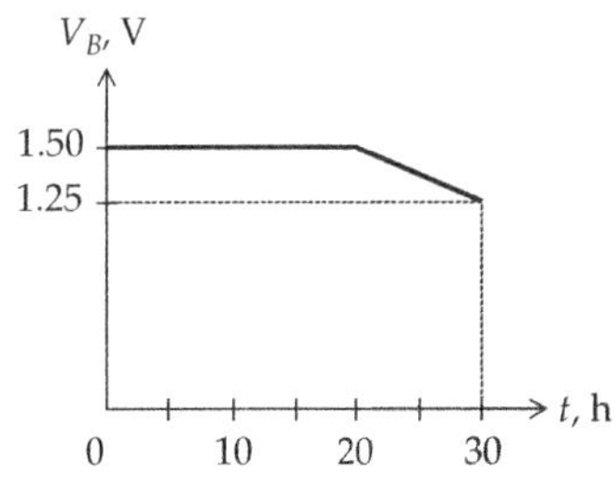

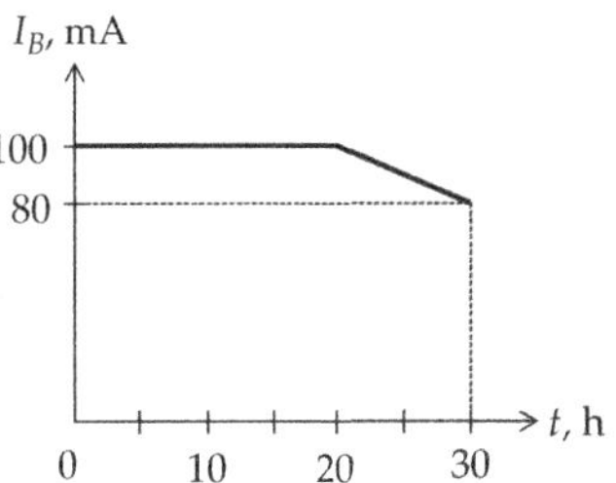

FIGURE P1.19

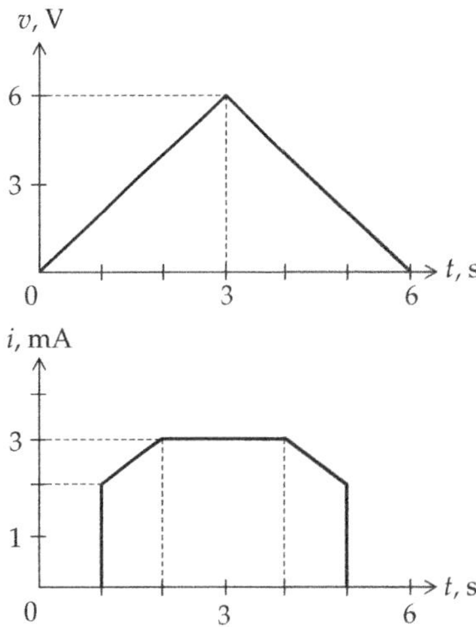

FIGURE P1.20

Ans. (a) 0 at 1 s, 2.5 mC at $t = 2$ s, 5.5 mC at $t = 3$ s, 8.5 mC at $t = 4$ s, 11 mC at $t = 5$ s and at $t = 6$ s; (b) $0 \leq t \leq 1$ s: $p(t) = 0$; $1 \leq t \leq 2$ s: $p(t) = 2t(t+1) = 2t^2 + 2t$ mW; $2 \leq t \leq 3$ s: $p(t) = 2t \times 3 = 6t$ mW; $3 \leq t \leq 4$ s: $p(t) = (-2t+12) \times 3 = -6t + 36$ mW; $4 \leq t \leq 5$ s: $p(t) = (-2t+12)(-t+7) = 2t^2 - 26t + 84$ mW; $5 \leq t \leq 6$: $p(t) = 0$; (c) 45.3 mJ.

P1.21 The voltage drop across a certain device and the current through it in the direction of the voltage drop are given by

$$v(t) = 2t + 1\text{ V}, \quad i(t) = 4 - 2t\text{ mA}, \quad 0 < t < 4\text{ s}$$

v and i are zero elsewhere.

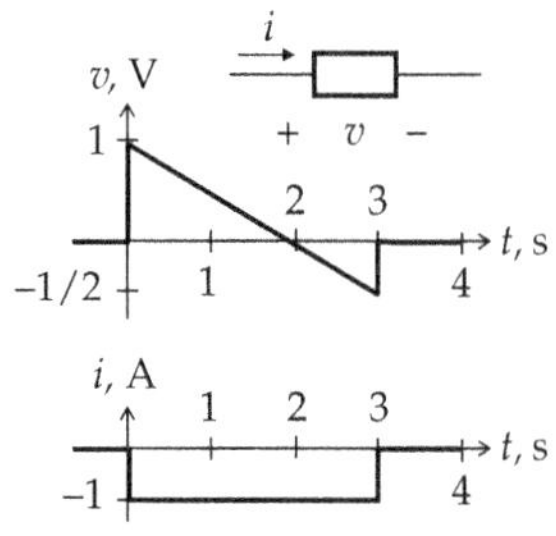

FIGURE P1.23

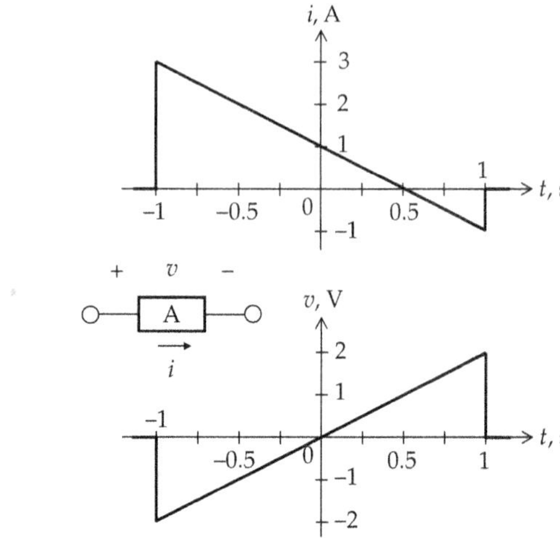

FIGURE P1.24

Sketch the variation of $v(t)$ and $i(t)$ with time. (a) At what instant of time is maximum instantaneous power absorbed by the device? (b) At what instant is it zero? (c) What is the energy delivered to the element at $t = 2$ s and at $t = 4$ s? (d) Over what time interval does the device absorb power and over what time interval does it deliver power?

Ans. (a) 0.75 s; (b) 2 s; (c) 9.3 mJ, −21.3 mJ; (d) power is absorbed by device for $t \leq 2$ s and is delivered by device for $2 \leq t \leq 4$ s.

P1.22 The voltage drop v V across a certain device and the current i A through it in the direction of the voltage drop are related by

$$i = 8 - 2v^2, \quad 0 \leq v \leq 2\text{ V}$$
$$i = 0, \quad v \leq 0 \quad \text{and} \quad v \geq 2\text{ V}$$

(a) Determine the power absorbed by the load when $v = 1$ V and when $v = 2$ V; (b) at what value of v is the instantaneous power a maximum? (c) If $v(t) = 2e^{-t}$ V, $t \geq 0$ s, what is the total charge that passes through the device from $t = 0$ to $t = 2$ s?

Ans. (a) 6 W, 0; (b) $2\sqrt{3}/3$ V; (c) 12.07 C.

P1.23 The voltage $v(t)$ across a device and the current $i(t)$ through the device are as shown in Figure P1.23. Determine the largest value of the magnitude of the energy absorbed or delivered by the device during the interval $0 < t < 3$ s.

Ans. 1 J delivered.

P1.24 The current $i(t)$ through a circuit element 'A' and the voltage $v(t)$ across it are as shown in Figure P1.24. (a) Determine the total charge passing through 'A'; (b) derive $p(t)$, sketch it, and indicate in which intervals is power absorbed or delivered; (c) determine the total energy absorbed or delivered by 'A' for $-1\text{ s} \leq t \leq 1$ s.

Ans. (a) 2 C; (b) $p(t) = 2t - 4t^2$, power is absorbed for $0 \leq t \leq 0.5$ s and is delivered for $-1\text{ s} \leq t \leq 0$ and $0.5\text{ s} \leq t \leq 1$ s; (c) 8/3 J delivered.

2

Fundamentals of Resistive Circuits

Objective and Overview

This chapter introduces (1) the two ideal circuit elements of dc circuits, namely, resistors and sources; (2) the two basic circuit laws, namely, Kirchhoff's current law (KCL) and Kirchhoff's voltage law (KVL); and (3) the two basic connections between circuit elements, namely, series and parallel connections.

The nature of electrical resistance is first explained using a simplified model, following which, the very basic Ohm's law defining an ideal resistor is presented. Ideal, independent and dependent, voltage and current sources are then discussed, with emphasis on their defining and essential properties.

Kirchhoff's laws are introduced as laws derived from conservation of charge and conservation of energy, but which provide a much simpler means of analyzing circuit behavior. Series and parallel connections of circuit elements are then discussed and linked to Kirchhoff's laws.

PSpice simulations are introduced in this chapter and are included in all numerical examples, whenever appropriate, to illustrate and verify the results of analytical solutions. This chapter concludes with a very helpful problem-solving approach and illustrating it with examples.

*2.1 Nature of Resistance

Concept: *Resistance is fundamentally due to impediments to the movement of current carriers in a conductor under the influence of an applied electric field.*

A sample of a metallic conductor typically consists of a large number of crystals in which the rest positions of the metal atoms at 0 K are arranged in a regular manner that is characteristic of the type of crystal. At temperatures above 0 K, (1) the crystal atoms vibrate, in randomly oriented directions, about their rest positions, with an amplitude of vibration that increases with temperature, and (2) some electrons, referred to as **conduction electrons** have sufficient energy to detach from their parent atoms and move freely in the crystal, in randomly oriented directions, at thermal velocities of the order of 10^7 cm/s at room temperature. Because of this randomness, there is no net current in any particular direction over a long enough interval of time, in the absence of an applied voltage.

When a voltage is applied to a conductor, an electric field is established in the conductor, which exerts a force on the conduction electrons. Quantum mechanics provides a rigorous basis for describing the movement of electrons in a crystal in terms of the wave nature of electrons and the scattering of these waves by the vibrating crystal atoms.

For our purposes, we will use a simpler, much less rigorous description based on considering conduction electrons as particles that are accelerated by the electric field due to the applied voltage. In moving under the influence of the electric field, conduction electrons "collide" with the vibrating crystal atoms. This is illustrated diagrammatically in Figure 2.1 for an electric field ξ applied in the negative x-direction at zero time. A conduction electron experiences a force $-q\xi$, where q is the *magnitude* of the electronic charge, and the negative sign accounts for the negative charge of electrons. Being in the negative x-direction, ξ has a negative value, which means that the force $-q\xi$ on an electron is in the positive x-direction. The conduction electron will therefore move in the positive x-direction with an acceleration $-q\xi/m^*$, where m^* is an effective electron mass that takes into account the forces exerted by the crystal on the conduction electron. In the absence of collisions, the velocity of the conduction electron increases linearly with time, theoretically without limit, as indicated by the dashed line through the origin. The current due to the total movement of conduction electrons will be in the negative x-direction, in terms of the conventional, assumed flow of positive charges, and will also theoretically increase without limit. No resistance to movement will be experienced by conduction electrons under these conditions. However,

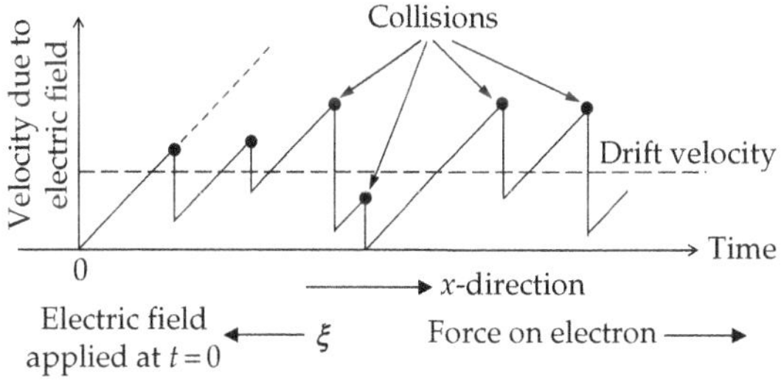

FIGURE 2.1
One-dimensional collision model.

as a conduction electron moves in the crystal, it collides with a crystal atom. Consequently, the conduction electron loses kinetic energy and hence velocity. After the collision, the conduction electron picks up velocity due to acceleration by the electric field until it collides again with a crystal atom, and so on (Figure 2.1). The collisions occur at random intervals and their net effect is that they prevent the velocities of conduction electrons from increasing without limit. Instead, there will be an average velocity of conduction electrons in the direction opposite to that of the electric field, referred to as a **drift velocity**. This drift velocity results in a corresponding **drift current** in the direction of the electric field, that is, in the negative x-direction in Figure 2.1. For currents normally encountered in practice, the drift velocity is quite small, less than a fraction of a centimeter per second, superposed on the much larger, random thermal velocity.

It is seen that this "collision model" provides a simple explanation for electrical resistance in a conductor as being due to the limiting of the increase in the velocities of conduction electrons, and hence the current, because of collisions between vibrating crystal atoms and conduction electrons moving under the influence of an applied voltage.

In addition, the collision model can explain two phenomena associated with current flow in a conductor: (1) the increase of resistance with temperature and (2) the heating effect of current. A rise in temperature increases the amplitude of vibration of crystal atoms, which increases the probability of collisions. The average interval between successive collisions is reduced, and the increase in velocity between collisions is thereby reduced, which decreases the average increase in the velocities of conduction electrons. This is reflected as a reduction in current, for a given applied voltage, and hence an increase in resistance. Moreover, when conduction electrons collide with crystal atoms, they lose kinetic energy while crystal atoms gain kinetic energy. The amplitude of vibration of crystal atoms therefore increases, which is reflected as a rise in temperature of the conductor. The heating of the material that is associated with current flow is referred to as **Joule heating**.

2.2 Ideal Resistor

Definition: *An ideal resistor is a purely dissipative circuit element that obeys Ohm's law.*

As mentioned in Section 1.8, an ideal resistor only dissipates energy. It does not store electric or magnetic energy. According to **Ohm's law**, the voltage drop across an ideal resistor is directly proportional to the current through the resistor. Thus,

$$v = Ri \tag{2.1}$$

where R is the resistance of the given resistor. When v is in volts and i is in amperes, R is in ohms and is denoted by Ω, the Greek capital omega. For an ideal resistor, R is a constant, independent of voltage, current, time, and temperature.

According to Equation 2.1, the plot of v against i is a straight line of slope R passing through the origin, irrespective of the magnitude or direction of current (Figure 2.2a). The graphical symbol for a resistor is illustrated in Figure 2.2b together with the direction of i that is associated with the polarity of v in Equation 2.1. It is important to note that *the current through an ideal resistor is always in the direction of the voltage drop across the resistor*, so as to give a positive value of R in Equation 2.1. Thus, with v and i in Equation 2.1 assigned the positive directions indicated in Figure 2.2b, a positive value of i is in the direction of a positive value of voltage drop v, which gives a positive value of R in Equation 2.1. This is also in accordance with the passive sign convention (Section 1.7); when the assumed positive charges flow down a voltage drop, they lose electric potential energy, which is converted to heat dissipation in the resistor.

Equation 2.1 may be written as

$$i = \frac{1}{R}v = Gv \tag{2.2}$$

where $G = 1/R$ is the reciprocal of resistance, or **conductance**. The reciprocal of one ohm is one siemens, denoted by the symbol S.

The power dissipated in a resistor can be derived from Equation 1.10, by substituting for v from Equation 2.1, or for i from Equation 2.2, in the expression $p = vi$. This gives

$$p = vi = Ri^2 = \frac{v^2}{R} = Gv^2 \tag{2.3}$$

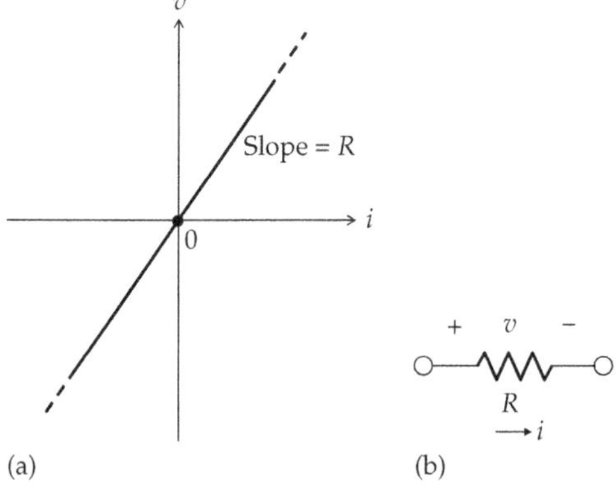

FIGURE 2.2
(a) v–i relation for an ideal resistor and (b) assigned positive directions of v and i for an ideal resistor.

When v is in volts, i is in amperes, R is in ohms, and G is in siemens, p is in watts.

R for a metal increases somewhat with temperature, the temperature coefficient of resistance for metallic conductors being approximately 0.004/°C, which means that the resistance increases by about 0.4% per °C rise in temperature.

It should be noted that whereas an increase in resistance with temperature is true of all conductors, the number of current carriers in metals does not increase significantly with temperature. On the other hand, the number of current carriers in semiconductors, such as silicon, increases very rapidly with temperature. As a result, the current, at a given voltage, markedly increases with temperature in semiconductors, which is reflected as an apparent *decrease* in resistance with temperature.

Example 2.1: Cold and Hot Resistance of Heating Element

Consider a 120 W, 12 V, heating element. It is required to determine (a) the resistance and conductance of the element at the rated voltage of 12 V, (b) the current under these conditions, and (c) the resistance of the element at room temperature (20°C), assuming that the working temperature of the element is 1200°C and its temperature coefficient of resistance is 0.0045/°C.

Solution:

(a) Applying Equation 2.3, $120 = (12)^2/R$, which gives $R = 1.2\ \Omega$ and $G = 1/1.2 = 0.83$ S.

(b) $I = 120\ \text{W}/12\ \text{V} = 10$ A. As a check, $RI^2 = 1.2 \times (10)^2 = 120$ W.

(c) The variation of resistance with temperature can be expressed as $R_2 = R_1[1 + \alpha_m (T_2 - T_1)]$, where α_m is the temperature coefficient of resistance and R_1 and R_2 are the resistances at temperatures T_1 and T_2, respectively, with $T_2 > T_1$. Substituting $R_2 = 1.2\ \Omega$, $T_1 = 20$°C, $T_2 = 1200$°C, and α_m 0.0045/°C gives $R_1 = \dfrac{1.2}{1 + 0.0045(1200 - 20)} = 0.19\ \Omega$ at 20°C, which is about one-sixth of the resistance at 1200°C.

Primal Exercise 2.1

A heater element draws 3 A at 24 V. Determine the resistance of the heater element, its conductance, and the power dissipated.

Ans. 8 Ω, 0.125 S, 72 W.

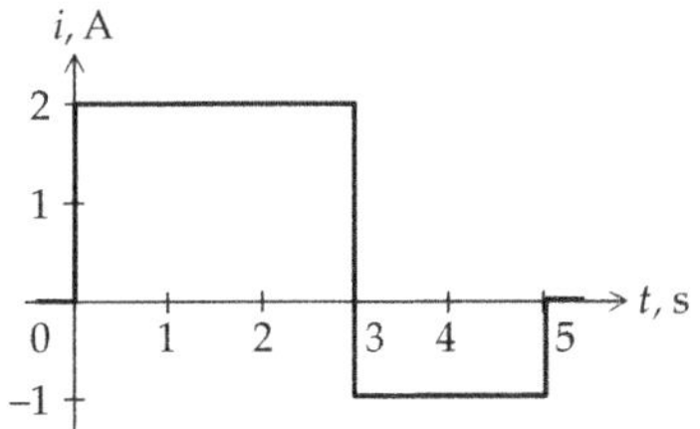

FIGURE 2.3
Figure for Exercise 2.3.

Primal Exercise 2.2

A 5 Ω resistor dissipates 180 W. Determine the current through the resistor and the voltage across it. Check that using the power relations of Equation 2.3 in terms of V, R, and G give 180 W.

Ans. 6 A, 30 V.

Primal Exercise 2.3

A current having the waveform shown in Figure 2.3 is applied to a 5 Ω resistor. Determine (a) the total energy dissipated in the resistor and (b) the average power dissipated in the resistor.

Ans. (a) 70 J; (b) 14 W.

2.3 Short Circuit and Open Circuit

Definition: *A short circuit is a connection of zero resistance, or infinite conductance. An open circuit has infinite resistance, or zero conductance.*

When $R = 0$, Ohm's law gives $v = Ri = 0$ for all finite i. In other words, the voltage across a short circuit is zero for all valid, that is finite, values of current through the short circuit. Hence, a plot of v against i for a short circuit is a horizontal line that coincides with the i-axis (Figure 2.4a). In fact, all wiring connections between circuit elements in a circuit diagram are implicitly

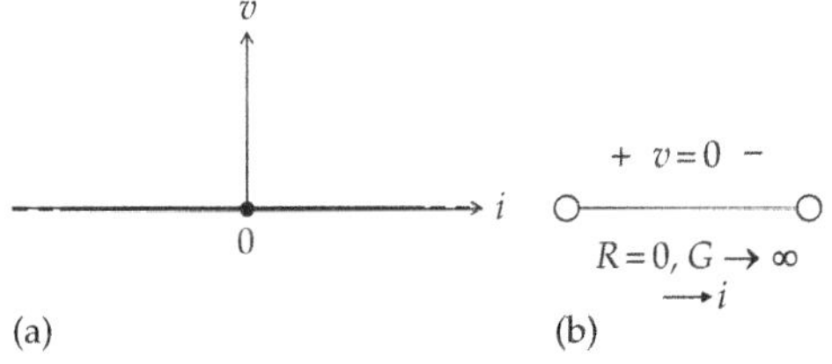

FIGURE 2.4
(a) v–i relation for a short circuit and (b) symbol for a specified short circuit.

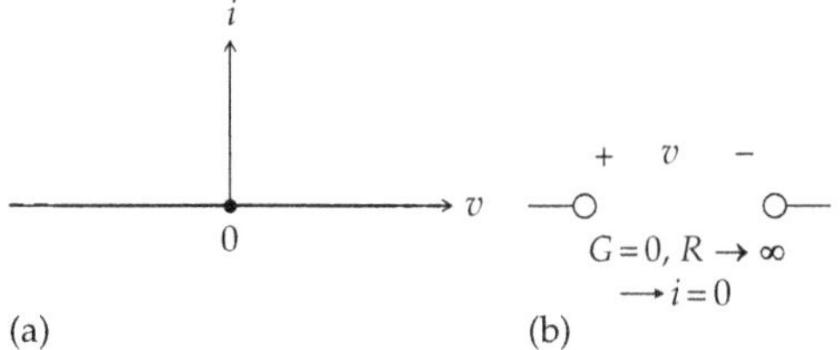

FIGURE 2.5
(a) v–i relation for an open circuit and (b) symbol for a specified open circuit.

assumed to be short circuits. When a short circuit is to be explicitly indicated, it will be represented as a wiring connection between two terminals (Figure 2.4b). When $R = 0, G = 1/R \to \infty$.

When $G = 0$, Ohm's law gives $i = Gv = 0$ for all finite v. In other words, the current through an open circuit is zero for all valid, that is finite, values of voltage across the open circuit. Hence, a plot of i against v for an open circuit is a horizontal line that coincides with the v-axis (Figure 2.5a). When an open circuit is to be explicitly indicated, it will be represented as an absence of any connection between two terminals (Figure 2.5b). When $G = 0, R = 1/G \to \infty$.

Short circuits and open circuits are useful idealizations representing limiting values of resistance ($R \to 0$ and $R \to \infty$) that can only be approximated in practice.

Primal Exercise 2.4

What is the power dissipated in (a) a short circuit and (b) an open circuit? Justify your answer using Equation 2.3.

Ans. (a) 0; (b) 0.

Primal Exercise 2.5

Determine R_L that gives the largest voltage across terminals 'ab' in Figure 2.6.

Ans. $R_L \to \infty$, that is, terminals 'ab' open circuited.

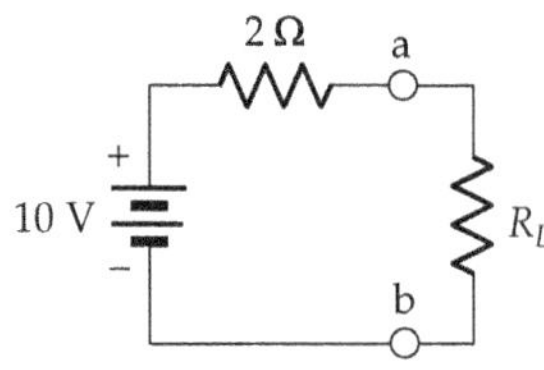

FIGURE 2.6
Figure for Exercise 2.5.

2.4 Ideal, Independent Voltage Source

Definition: *An ideal, independent voltage source maintains a specified voltage v_{SRC} between its terminals, irrespective of the current through the source, where v_{SRC} is independent of any voltage or current in the circuit.*

A battery denoted by the symbol used in Figures 1.2 and 1.15 is an example of a dc, ideal, independent voltage source. The "ideal" attribute in the definition refers to v_{SRC} being maintained irrespective of the current through the source, and the "independent" attribute refers to v_{SRC} being independent of any voltage or current in the circuit.

The fact that an ideal voltage source maintains a specified voltage v_{SRC} across its terminals irrespective of the current through the source means that the plot of v as a function of i for the source is a horizontal line displaced by v_{SRC} from the i-axis (Figure 2.7a). Ideal, independent voltage sources are conventionally represented by a circle symbol, as in Figure 2.7b, including dc, ideal, independent sources. But we will retain the battery symbol for the latter, as in PSpice, for clarity when using dc voltage sources. As in the case of the battery, the plus and minus signs indicate the assigned positive direction of source voltage in Figure 2.7b.

Two important considerations should be emphasized in connection with ideal, independent voltage sources. The first is that, according to Figure 2.7a, either v_{SRC} or i can be positive, negative, or zero. In Figure 2.8a, the 12 V battery is connected to a variable resistor R_{var}, that is, a resistor whose resistance can be varied, as symbolized by the arrow. As R_{var} is varied, the current I through the battery varies in accordance with Ohm's law, but the ideal battery maintains a voltage of V_{SRC} = 12 V between its terminals for all values of I, such as I = 2 A,

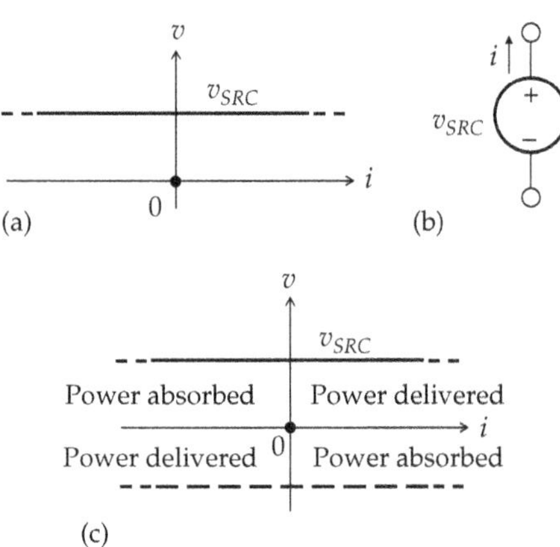

FIGURE 2.7
(a) v–i relation for an ideal, independent voltage source, (b) source symbol, and (c) power delivered or absorbed by the source based on the relative signs of v_{SRC} and i.

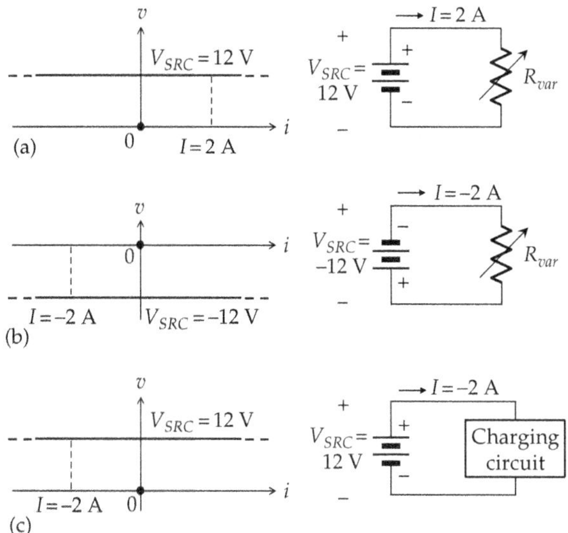

FIGURE 2.8
(a) Power delivered by an ideal, independent voltage source, (b) power delivered by the source with reversed polarity of the source, and (c) power absorbed by the source.

as indicated in Figure 2.8a. The power delivered by the battery is $V_{SRC} \times I = 12 \times 2 = 24$ W. With v and i both positive, power is delivered by the battery, as indicated in the first quadrant in Figure 2.7c. V_{SRC} is maintained at 12 V when $I = 0$, that is, when the resistor is disconnected from the battery. In Figure 2.8b, the polarity of the battery is reversed, without reversing the assigned positive directions of V_{SRC} and I. The numerical values of V_{SRC} and I are both negative, in accordance with Ohm's law, so that current actually flows into the battery at the negative terminal. On the v–i graph, both v and i assume negative values, but as R_{var} is varied, the battery voltage is maintained at 12 V magnitude. Moreover, with both v and i reversed, the product vi does not change sign; the battery still delivers power, as indicated in the third quadrant in Figure 2.7c.

In Figure 2.8c, the battery, assumed to be rechargeable, is connected to a charging circuit. The current through the battery is now reversed, so I assumes a negative value, considered in Figure 2.8c to be −2 A, but V_{SRC} is still maintained at 12 V.

It should be clarified that in Figure 2.8a through c, the assigned positive directions of I and V_{SRC} are the same, with I being in the direction of a voltage rise through the battery. This means that, *in terms of the symbols* V_{SRC} *and* I, the product $P = V_{SRC}I$ is the power delivered by the battery in the three cases, in accordance with the passive sign convention (Section 1.7). When numerical values are substituted, $V_{SRC} = 12$ V and $I = 2$ A in Figure 2.8a. The product is +24 W, so that this power is actually delivered by the battery. In Figure 2.8b, $V_{SRC} = -12$ V and $I = -2$ A. The product is again +24 W, which signifies that the battery actually delivers this power. In Figure 2.8c, $V_{SRC} = 12$ V and $I = -2$ A. P is now equal to −24 W, which means that the battery actually absorbs 24 W.

It should be noted, in accordance with Figure 2.8, that whereas the voltage of an ideal, independent voltage source is that specified for the source and is independent of the rest of the circuit, *the magnitude and sign of the source current depend on both the source voltage and the circuit to which the source is connected.*

The second important consideration is that ideal, independent voltage sources are in general time varying. This is illustrated for a sinusoidally varying voltage in Figure 2.9a and for a pulse voltage in Figure 2.9b. Note that although the variation of v_{SRC} vs. t can be quite arbitrary, depending on the type of source under consideration, the variation of v vs. i for all ideal voltage sources is a horizontal line, as in Figure 2.7a. For example, if $v_{SRC} = 10\sin t$ V (Figure 2.9a), where t is in seconds, then the v_{SRC} line in Figure 2.7a will move up or down in accordance with the value of v_{SRC} at any particular instant of time. This is illustrated in Figure 2.10,

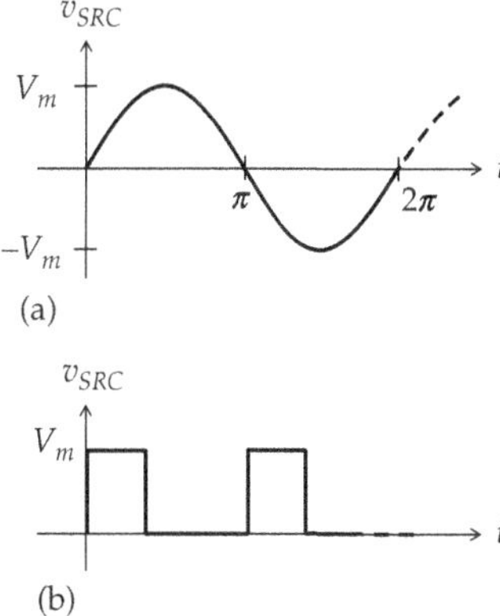

FIGURE 2.9
Time-varying voltage sources. (a) Sinusoidally-varying source voltage and (b) pulse source voltage.

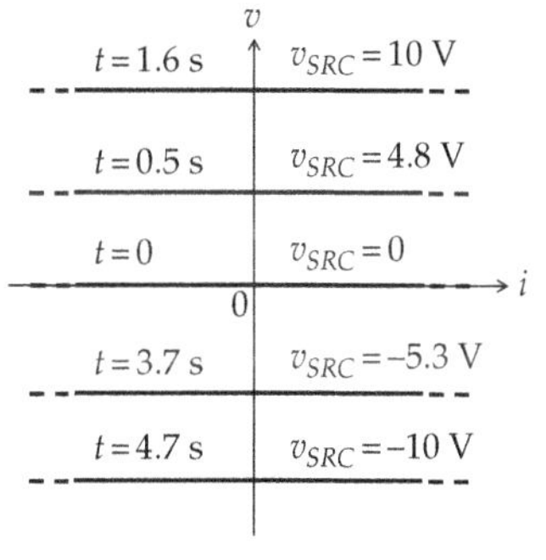

FIGURE 2.10
Sinusoidal time variation of v–i characteristic of ideal voltage source.

where some values of 10sin$t$ are shown for corresponding values of t. Thus, at $t = 0$, $v_{SRC} = 0$, and the v_{SRC} line coincides with the horizontal axis. As t increases during the positive half-cycle, the v_{SRC} line moves upward, is at 4.8 V at $t = 0.5$, and at the peak value of 10 V at $t = \pi/2 \cong 1.6$ s. It then moves downward, is at −5.3 V at $t = 3.7$ s, and at −10 V at $t = 3\pi/2 \cong 4.7$ s, and so on. *At any instant of time, v_{SRC} is independent of the source current*. When the value of v_{SRC} is positive, then according to the assigned positive direction of v_{SRC} in Figure 2.7b, the terminal of the voltage source adjacent to the plus sign is at a positive voltage with respect to the terminal adjacent to the minus sign; the converse is true when the value of v_{SRC} is negative.

It is important to note that if in Figure 2.7a, $v_{SRC} = 0$, then the v_{SRC} line coincides with the horizontal axis, so that Figure 2.7a becomes identical to Figure 2.4a for a short circuit. It follows that *when $v_{SRC} = 0$, an ideal voltage source is equivalent to a short circuit*. In other words, an ideal voltage source is set to zero by replacing it with a short circuit.

An ideal, independent voltage source is said to provide an **electromotive force (emf)** that provides a kind of "driving force" for the current by doing work on the assumed positive charges as they flow in the direction of a voltage rise through the source. The emf is equal to the source voltage. Thus, a 6 volt battery is said to provide an emf of 6 V.

Primal Exercise 2.6

Given an ideal, independent voltage source of 12 V having a source current of 2 A, determine the power and whether it is delivered or absorbed by the source when the source current is in the direction of (a) a voltage rise across the source or (b) a voltage drop across the source.

Ans. (a) 24 W delivered; (b) 24 W absorbed.

Primal Exercise 2.7

A 24 V battery, when being recharged, draws a current of 5 A from the charging circuit. Determine the power that is (a) absorbed by the battery, (b) delivered by the charging circuit, (c) delivered by the battery, and (d) absorbed by the charging circuit.

Ans. (a) and (b) 120 W; (c) and (d) −120 W.

2.5 Ideal, Independent Current Source

Definition: *An ideal, independent current source maintains a specified current i_{SRC} through the source irrespective of the voltage across the source, where i_{SRC} is independent of any voltage or current in the circuit.*

The "ideal" attribute refers to i_{SRC} being maintained irrespective of the voltage across the source, and the "independent" attribute refers to i_{SRC} being independent of any voltage or current in the circuit.

Although not as common as voltage sources, independent current sources can be derived from independent voltage sources using electronic circuits such as operational amplifiers (Chapter 13). A current source can be approximated by a voltage source in series with a large resistance (Section 3.6).

The fact that an ideal current source maintains the specified current i_{SRC} through the source irrespective of the voltage v across the source means that the plot of i as a function of v for the source is a horizontal line displaced by i_{SRC} from the v-axis (Figure 2.11a). The conventional symbol for an ideal, independent current source is a circle with an arrow that points in the assigned positive direction of source current i_{SRC} (Figure 2.11b); that is, if the value of i_{SRC} is a positive quantity, i_{SRC} is in the direction of the arrow. The symbol of Figure 2.11c will be used in this book for a dc, ideal, independent current source, for clarity when using such sources. The polarity of the battery in the source symbol is that of a voltage

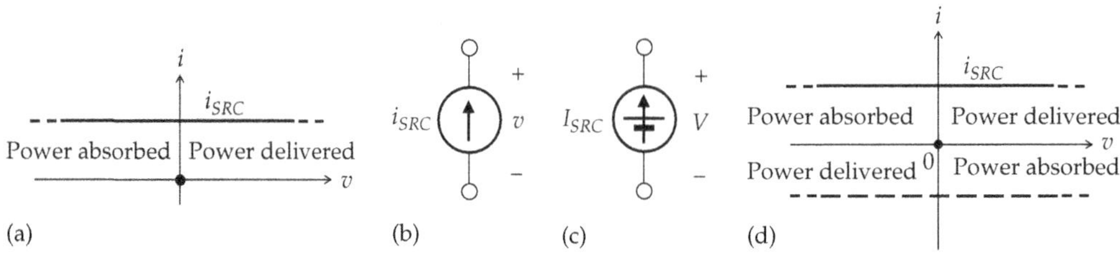

FIGURE 2.11
Ideal, independent current source. (a) i–v relation, (b) the symbol for an ideal, independent current source, (c) symbol for an ideal, independent, dc current source, and (d) power delivered or absorbed by an ideal, independent current source based on the relative signs of i_{SRC} and v.

rise in the direction of the arrow, which is appropriate for power delivery.

The following features of an ideal, independent current source are exactly analogous to those of an ideal, independent voltage source:

1. i_{SRC}, and the voltage v across the source, can be positive, negative, or zero.
2. For a given i_{SRC}, the value and sign of v depend on both i_{SRC} and the circuit to which the source is connected.
3. An ideal current source can deliver or absorb power, depending on the relative directions of i_{SRC} and v (Figure 2.11d).
4. i_{SRC} is in general a function of time but is constant with respect to the voltage across the source. That is, at any instant of time, i_{SRC} is independent of the voltage across the source, even though i_{SRC} may be time varying.

In a manner analogous to the case of a voltage source, if Figure 2.11a is compared with Figure 2.5a, it is seen that *when* $i_{SRC} = 0$, *an ideal current source is equivalent to an open circuit.* In other words, an ideal current source is set to zero by replacing it with an open circuit.

Fundamentally, a voltage source excites a circuit by raising the electric potential energy, that is, the voltage, of current carriers. On the other hand, a current source excites a circuit by imparting kinetic energy, that is, velocity, to current carriers, since current is the rate of flow of electric charges, which depends on velocity. Short-circuiting a voltage source results in infinite current, since the source voltage appears across the zero resistance of the short circuit. The resulting current is, from Ohm's law, $V_{SRC}/0 \rightarrow \infty$. Hence, short-circuiting an ideal voltage source is invalid. The hydraulic analogy is moving water in a reservoir at a height h above ground level instantly down to ground level, which would involve infinite rate of flow, or velocity.

Similarly, open-circuiting a current source results in infinite voltage, since the source current flows through an open circuit of zero conductance. The resulting voltage is, from Ohm's law, $I_{SRC}/0 \rightarrow \infty$. Hence, open-circuiting an ideal current source is invalid. The hydraulic analogy is instantly blocking the flow of water that is issuing from a pump that forces water to flow at a certain velocity, corresponding to I_{SRC}. This would involve an infinite force, corresponding to an infinite voltage.

Primal Exercise 2.8

A 0.1 A current source is connected to a 9 V battery as shown in Figure 2.12. Determine (a) the current through the battery, both in magnitude and direction; (b) the voltage across the current source, both in magnitude and polarity; (c) the power delivered or absorbed by the current source; and (d) the power delivered or absorbed by the battery. Repeat (a) through (d) with the polarity of the current source reversed.

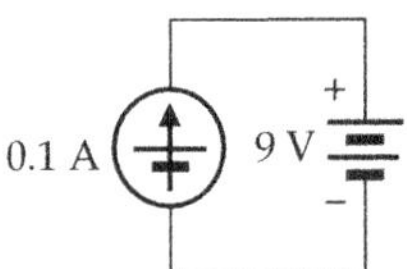

FIGURE 2.12
Figure for Exercise 2.8.

Ans. (a) 0.1 A directed from the + terminal to the − terminal of the battery; (b) 9 V of the same polarity as the battery; (c) current source delivers 0.9 W; (d) battery absorbs 0.9 W; (a′) 0.1 A directed from the '−' terminal to the '+' terminal of the battery; (b′) 9 V of the same polarity as that of the battery; (c′) current source absorbs 0.9 W; (d′) battery delivers 0.9 W.

Primal Exercise 2.9

Given an ideal, independent current source having $i_{SRC} = 4\sin t$ A, t is in s, the voltage across the source being $V_S = 10$ V dc. The directions of i_{SRC} and V_S are as indicated in Figure 2.11b. Determine the value of (a) the maximum instantaneous power delivered by the source; (b) the maximum instantaneous power absorbed by the source, specifying the times at which they occur; and (c) the average power delivered or absorbed by the source over a period.

Ans. (a) 40 W delivered at the positive peaks of the sinusoid, $t = \pi/2$ s plus an integer multiple of 2π s; (b) 40 W absorbed at the negative peaks of the sinusoid, $t = 3\pi/2$ s plus an integer multiple of 2π s; (c) 0.

2.6 Ideal, Dependent Sources

Concept: *Dependent sources behave exactly like independent sources in all respects except that the value of a dependent source is specified as a linear function of a voltage or a current in the circuit, other than that of the source itself.*

Dependent sources are invariably encountered in the equivalent circuits of electronic devices, such as operational amplifiers (Chapter 13) and transistors.

Ideal, dependent sources are represented by a diamond symbol.

2.6.1 Ideal, Dependent Voltage Sources

Depending on whether the source value is specified in terms of a voltage or a current, a dependent voltage source can be (1) a **voltage-controlled voltage source** (VCVS) or (2) a **current-controlled voltage source** (CCVS). Figure 2.13 illustrates a circuit having two dependent voltage sources. The value of the VCVS is specified as twice the voltage V_X across the 6 Ω resistor, and the value of the CCVS is specified as three times I_Y, the current through the 1.5 Ω resistor. In both cases, the plus and minus signs indicate the assigned positive direction of the source voltage, as in the case of independent voltage sources.

It should be noted that being ideal sources, *the specified source voltage of an ideal, dependent voltage source is independent of the current through the source itself*, as in Figure 2.7a. That is why in the case of the CCVS in Figure 2.13, for example, the source voltage depends on the current elsewhere in the circuit, such as I_Y. It should not depend on the current through the source itself, because this would violate the definition of an ideal voltage source as having the source voltage independent of the current through the source. If the source voltage of a CCVS depends on the current through the source, it can be shown (Section 4.5) that such a "source" is not really a source and can be replaced by a resistor. Similarly, if V_X in Figure 2.13 is the voltage across the VCVS of source voltage $2V_X$, then $V_X = 2V_X$, which makes $V_X = 0$, so that the source is equivalent to a short circuit. On the other hand, if the source value is V_X, and there is no other V_X in the circuit except across the source, then the source is in fact an independent source of value V_X.

Primal Exercise 2.10

Given $I_Y = 2.4$ A and $V_X = 1.8$ V in Figure 2.13, determine (a) the magnitude and direction of the current through the 6 Ω resistor; (b) the magnitude and polarity of the voltage across the 1.5 Ω resistor; (c) the power delivered or absorbed by the battery and by the VCVS; (d) the total power dissipated in the resistors; (e) the power delivered or absorbed by the CCVS, considering that power is conserved in the circuit as whole; and (f) the magnitude and direction of current through the CCVS. (g) Is charge conserved at the upper and lower junctions of three elements in the circuit? Justify your answer.

Ans. (a) 0.3 A in the direction of the voltage drop V_X; (b) 3.6 V drop in the direction of I_Y; (c) 2.7 W delivered by the battery, 8.64 W absorbed by VCVS; (d) 9.18 W; (e) 15.12 W delivered by the CCVS; (f) 2.1 A in the direction of a voltage rise through the CCVS. (g) Yes; the number of coulombs entering the junction per second is equal to the number of coulombs leaving the junction per second.

2.6.2 Ideal, Dependent Current Sources

Depending on whether the source value is specified in terms of a voltage or a current, a dependent current source can be (1) a **voltage-controlled current source** (VCCS) or (2) a **current-controlled current source** (CCCS). Figure 2.14 illustrates a circuit having two dependent current sources. The value of the VCCS is specified as twice V_X across the 1 Ω resistor, and the value of the CCCS is specified as three times I_Y, the current through the 2 Ω resistor. In both cases, the arrows indicate the assigned positive direction of the source, as in the case of independent current sources.

It should be noted that being ideal sources, *the specified source current of an ideal, dependent current source is independent of the voltage across the source itself*, as in Figure 2.11a. The specified source current should not depend on the voltage across the source itself, as this would violate the definition of an ideal current source as having the source current independent of the voltage across the source. If the source current of a VCCS depends on the voltage across the source, it can be shown (Section 4.5) that such a "source" is not really a source and can be replaced by a resistor.

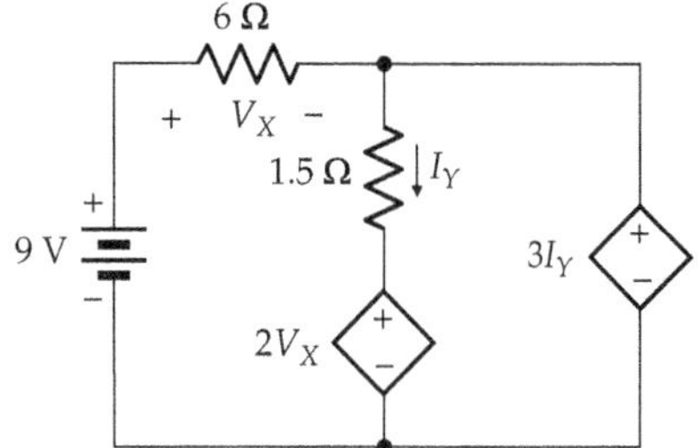

FIGURE 2.13
Ideal, dependent voltage sources.

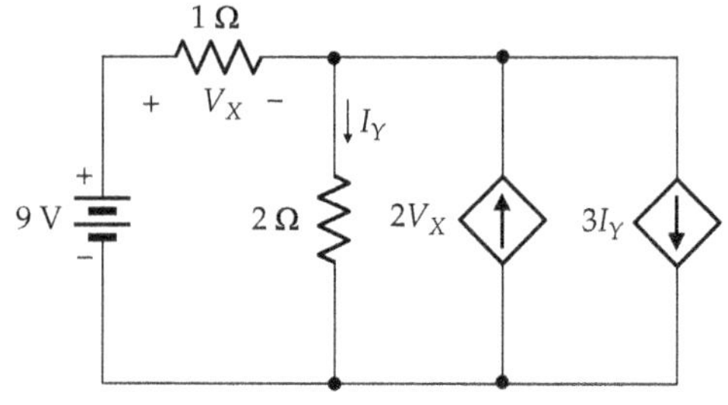

FIGURE 2.14
Ideal, dependent current sources.

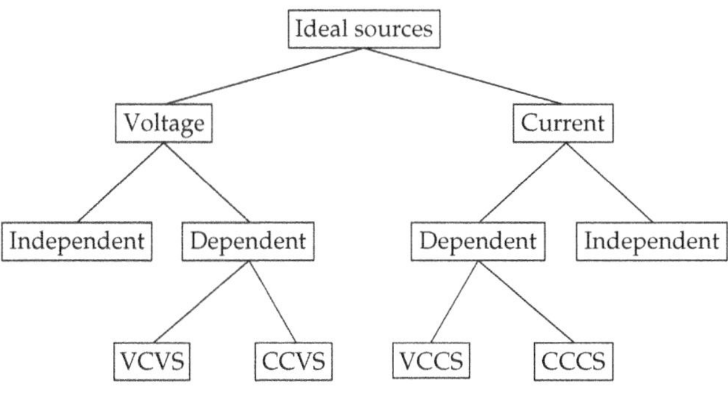

FIGURE 2.15
Classification of ideal sources.

Fundamentally, independent sources excite a circuit by delivering energy through conversion from some nonelectrical source of energy, such as solar energy or chemical energy of batteries, or mechanical energy from prime movers. On the other hand, dependent sources do not convert energy from nonelectrical sources. They can deliver or absorb energy, just like independent sources, but they are incapable of exciting a circuit on their own, without the presence of independent sources somewhere in the circuit (Section 4.1). They affect the voltages and currents in a circuit, thereby altering the power distribution in the circuit, by effectively modifying the values of some resistances in the circuit (Section 5.1). But the ultimate source of energy is the independent sources in a circuit, which convert energy from some nonelectrical source to potential energy or kinetic energy of current carriers.

Figure 2.15 summarizes the classification of ideal sources.

Primal Exercise 2.11

Given $I_Y = 2.7$ A and $V_X = 3.6$ V in Figure 2.14, determine (a) the magnitude and direction of the current through the 1 Ω resistor, (b) the magnitude and polarity of the voltage across the 2 Ω resistor, (c) the power delivered or absorbed by each source, and (d) the power dissipated in each resistor. (e) Is energy conserved in the circuit as whole? Is charge conserved at the upper and lower junctions of the four elements connected to these junctions?

Ans. (a) 3.6 A in the direction of the voltage drop V_X; (b) 5.4 V drop in the direction of I_Y; (c) battery delivers 32.4 W, VCCS delivers 38.88 W, CCCS absorbs 43.74 W; (d) 12.96 W in the 1 Ω resistor and 14.58 W in the 2 Ω resistor; (e) yes, yes.

Exercise 2.12

Argue as was done in connection with Figure 2.13 that I_Y cannot be the current through the CCCS itself in Figure 2.14.

Ans. If I_Y is the current through the CCCS, then $I_Y = 3I_Y$, so that $I_Y = 0$, and the CCCS is equivalent to an open circuit.

2.7 Nomenclature and Analysis of Resistive Circuits

Before considering the analysis of resistive circuits, some nomenclature should be explained.

A **node** is the electrical junction, or connection point, between a number of circuit elements. Because the term is quite general, it is sometimes necessary to distinguish between two types of nodes: an **inessential node** between just two circuit elements and an **essential node** between three or more circuit elements. In Figure 2.16, for example, 'a', 'b', and 'c' are nodes, but 'a' is an inessential node, whereas 'b' and 'c' are essential nodes. Note that 'b' and b′ are one and the same node, the connection bb′ being a short circuit. Similarly, 'c', c′, and c′′ are one and the same node, the connection c′cc′′ being a short circuit. Short-circuit wiring connections are used in electric circuit diagrams for clarity.

A **path** is a set of one or more adjoining circuit elements that may be traversed in succession without passing through the same node more than once. A **branch** is a path that connects two nodes, whereas an **essential branch** is the set of adjoining circuit elements traversed in going from one essential node to an adjacent essential node, without passing through another essential node. In Figure 2.16, all the individual circuit elements are branches. Each of the current source and the 3 Ω resistor is an essential branch between essential nodes 'b' and 'c'. The combination of the battery and the 6 Ω resistor is also an essential branch between nodes 'b' and 'c'.

A closed path in a circuit is a **loop**. A loop may enclose other loops, but if it doesn't, it is referred to as a **mesh**. In Figure 2.16, the battery, the 6 Ω resistor, and the 3 Ω resistor constitute a mesh, as do the 3 Ω resistor and the current source. The path consisting of the battery, the 6 Ω resistor, and the current source is a loop that encloses the two meshes.

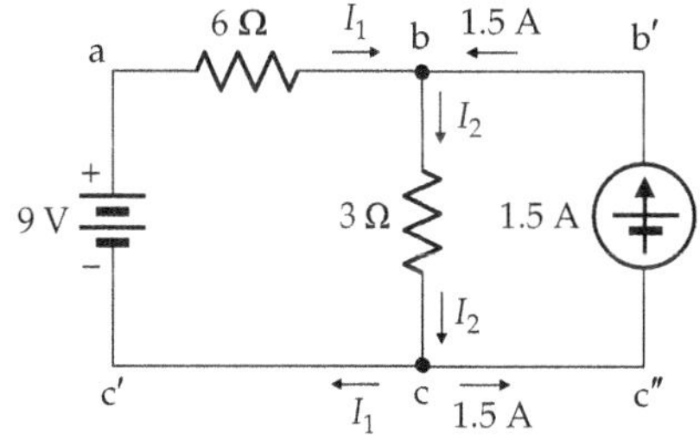

FIGURE 2.16
Nomenclature of electric circuits.

Analysis of resistive circuits means, in general, determining the voltages and currents in a circuit, given the values of the sources and resistances in the circuit or given some specified conditions in the circuit. In Figure 2.16, the values of the resistances, the voltage source, and the current source are specified. It is required to determine the values of the remaining currents and voltages in the circuit. Once I_1 and I_2 are known, the voltages readily follow from Ohm's law. Thus, the voltage drop across the 6 Ω resistor is $6I_1$ in the direction of I_1, where I_1 is also the current through the battery. The voltage drop across the 3 Ω resistor is $3I_2$ in the direction of I_2 and is also the voltage across the current source.

In principle, I_1 and I_2 can be determined from two simultaneous equations in these variables. It can be readily shown (Problem P2.61) that conservation of charge at the essential nodes provides one equation, whereas conservation of power in the circuit provides another equation. However, the power equation is quadratic in I_1 and I_2. This is generally the case, because the power dissipated in a resistor is proportional to the square of the current through the resistor, or the square of the voltage across the resistor. Consequently, using power as a primary circuit variable results in nonlinear equations that are rather awkward to work with, particularly in more complicated circuits. It is advantageous, therefore, to be able to write circuit equations that are linear in current and voltage. This is provided by Kirchhoff's current and voltage laws, which are discussed in the following section.

Primal Exercise 2.13

Given the circuit of Figure 2.17, specify (a) the number of nodes, (b) the number of essential nodes, (c) the number of branches, (d) the number of essential branches, (e) the number of meshes, and (f) the number of loops, other than meshes.

Ans. (a) 7; (b) 4; (c) 9; (d) 6; (e) 3; (f) 4.

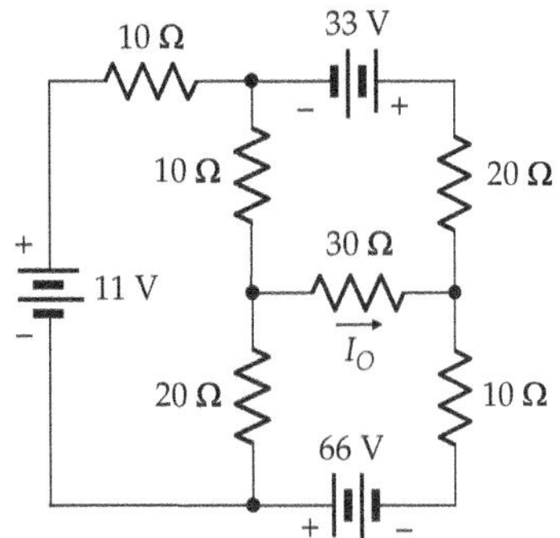

FIGURE 2.17
Figure for Primal Exercise 2.13.

2.8 Kirchhoff's Laws

2.8.1 Kirchhoff's Current Law

Statement: *At any instant of time, the sum of currents entering a node is equal to the sum of currents leaving the node.*

KCL is illustrated at a node 'n' in Figure 2.18a, where the current values indicated may be obtained through analysis, simulation, or measurement at node 'n'. These values are in accordance with KCL, since the sum of currents entering the node is 1 + 4 = 5 A and the sum of currents leaving the node is 2 + 3 = 5 A.

KCL is simply an expression of conservation of current, at any instant of time and at every node in a given circuit. It is a direct consequence of conservation of charge. The charge entering a node during an interval Δt is

$$\text{Charge entering node} = \left(\sum_{\text{Entering}} i\right)\Delta t \tag{2.4}$$

where the summation is over all the currents entering the node. The charge leaving the node during the interval Δt is

$$\text{Charge leaving node} = \left(\sum_{\text{Leaving}} i\right)\Delta t \tag{2.5}$$

where the summation is over all the currents leaving the node. From conservation of charge, the charge entering the node during the interval Δt equals the charge leaving the node during the same time interval. Equating the right-hand sides of Equations 2.4 and 2.5 and cancelling Δt gives the statement of KCL:

$$\sum_{\text{Entering}} i = \sum_{\text{Leaving}} i \tag{2.6}$$

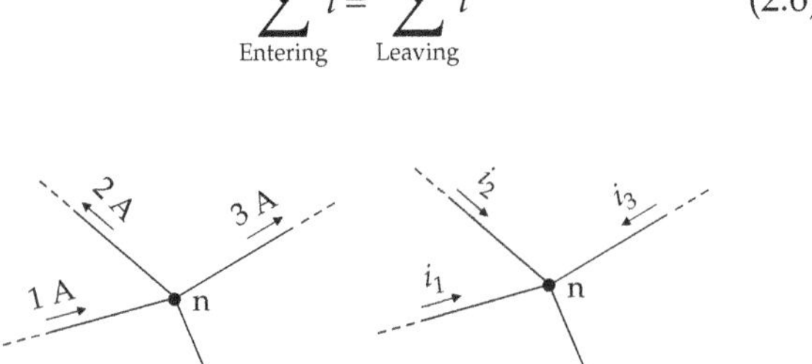

FIGURE 2.18
Kirchhoff's current law. Currents entering node 'n' expressed as numerical values (a), or as symbols (b).

It will be noted that just as conservation of energy implies conservation of its time derivative, which is power (Section 1.7), so does conservation of charge imply conservation of its time derivative, which is current.

KCL applies not only to known current values at a node, as in Figure 2.18a, but also to unknown currents at a node having arbitrarily assigned directions. Suppose, for example, that unknown currents are assigned as all entering node 'n' (Figure 2.18b). Then, according to KCL,

$$i_1 + i_2 + i_3 + i_4 = 0 \tag{2.7}$$

But when the numerical values of the currents are determined, these values will satisfy KCL. It may be found, for example, that $i_1 = 1$ A, $i_2 = -2$ A, $i_3 = -3$ A, and $i_3 = 4$ A. Substituting in Equation 2.7,

$$1 - 2 - 3 + 4 = 0 \quad \text{or} \quad 1 + 4 = 2 + 3 \tag{2.8}$$

as before.

Because it is an expression of conservation of charge, KCL applies to whole circuit elements and to combinations of circuit elements. It is always assumed that the current that enters a circuit element is equal to the current that leaves it. This applies to sources, resistors, inductors, and whole capacitors. It would not apply to only one of the plates of a capacitor, because charge can accumulate on one plate, while the other plate acquires an opposite, induced charge. So there will be a conduction current, due to motion of current carriers, entering one plate but no conduction current leaving this plate. But KCL applies to the whole capacitor, including both plates, as explained in more detail in Section 7.1.

KCL can also be usefully applied to whole circuits or to any combination of circuit elements that are part of a circuit, as will be illustrated in future examples. For now, consider the two interconnected circuits 'A' and 'B' in Figure 2.19a. If one of the circuits is enclosed by a surface 'S', then KCL requires that the total current entering 'S' must be equal to the total current leaving 'S'. If 'A' has two connections, the current entering 'A' through one connection must be equal in magnitude but opposite in direction to the current leaving 'A' through the other connection (Figure 2.19a). If either of these currents is zero, the current in the other connection must be zero. If there is only one connection between the two circuits, the current in the connection must be zero (Figure 2.19b).

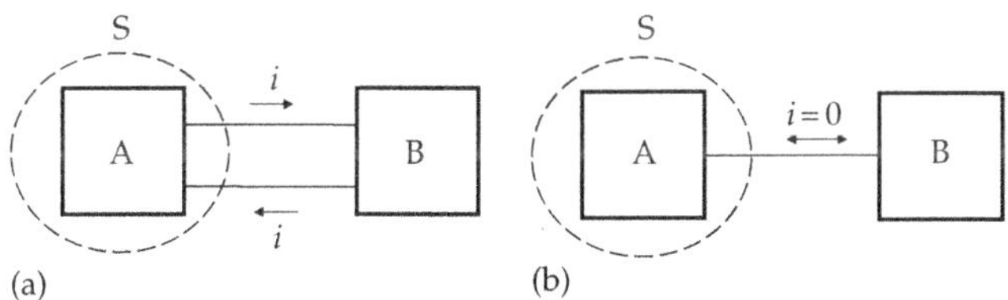

FIGURE 2.19
Kirchhoff's current law applied to a closed surface S.

Primal Exercise 2.14

If in Figure 2.18b $i_1 = 1.5$ A, $i_2 = -2$ A, and $i_3 = 1.25$ A, determine i_4 that satisfies KCL.

Ans. −0.75 A.

2.8.2 Kirchhoff's Voltage Law

Before illustrating KVL, it should be emphasized that *voltages along a path add algebraically*. This is consistent with the definition of voltage as electric potential energy per unit charge, and the fact that changes in potential energy can be added algebraically. Consider, for example, the path from ground to node 'c' in Figure 2.20. Suppose that a small positive charge $+\delta q$ C is moved along the path, where δq is small enough so as not to significantly disturb the circuit. If the charge is moved from ground, where the voltage and potential energy are assumed to be zero, to node 'a' through the 6 V source, the electric potential energy of δq at 'a' is $6\delta q$ J. If moved to node 'b' through a voltage rise of 4 V across the resistor, the potential energy of δq increases by $4\delta q$ J to become $10\delta q$ J. The voltage at node 'b' is $10\delta q/\delta q = 10$ V, with respect to ground, consistent with the voltage at node 'b' being the sum of 6 V and 4 V. If moved to node 'c' through a voltage drop of 2 V across the resistor, the potential energy of δq decreases by $2\delta q$ J to become $8\delta q$ J. The voltage at node 'c' is $8\delta q/\delta q = 8$ V, with respect to ground, consistent with the voltage at node 'c' being the algebraic sum (6 + 4 − 2) V of the voltages across the individual branches along the path.

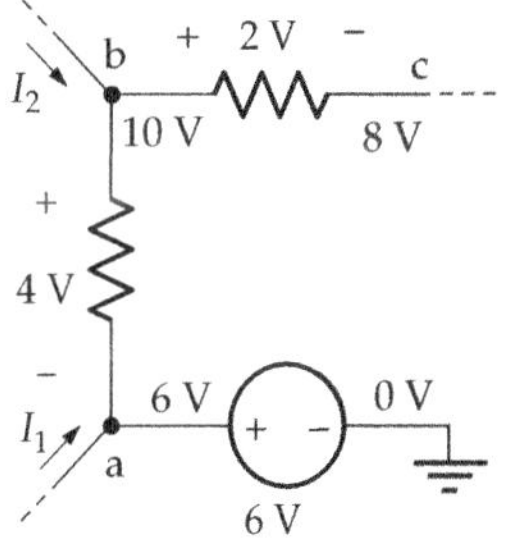

FIGURE 2.20
Algebraic addition of voltages along a path.

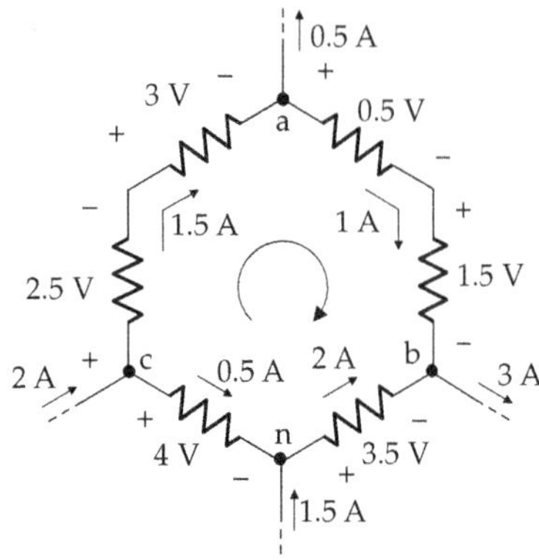

FIGURE 2.21
Kirchhoff's voltage law.

KVL can be stated as follows:

Statement: *At any instant of time, the sum of voltage rises around any loop is equal to the sum of voltage drops around the loop.*

KVL can be illustrated by a mesh in a circuit, as in Figure 2.21. The currents in the resistors are indicated, and the resistance values are assumed to give the voltage differences shown. Which of these voltage differences are voltage rises and which are voltage drops depends on the sense in which the mesh is traversed, clockwise or anticlockwise, but this does not affect KVL. If the mesh is traversed clockwise, starting at node 'b', there are two voltage rises, 3.5 and 4 V, and four voltage drops, 2.5, 3, 0.5, and 1.5 V. To satisfy KVL, the sum of the voltage rises around the mesh must be equal to the sum of the voltage drops. Thus, 3.5 + 4 = 7.5 V = 2.5 + 3 + 0.5 + 1.5.

The interpretation of KVL follows from conservation of energy. Suppose a small charge $+\delta q$ is taken around the mesh, where δq is small enough so as not to significantly affect the currents and voltages in the mesh. The work done in taking δq around the mesh is δq multiplied by the sum of all the voltage rises around the mesh, whereas the work done by δq is δq multiplied by the sum of all the voltage drops around the mesh. By conservation of energy, these must be equal; otherwise, energy either just vanishes in the mesh or can be continuously extracted from the mesh at no energy cost. For example, if the sum of all the voltage rises around the mesh in Figure 2.21 is 7.5 V, but the sum of all the voltage drops around the mesh is, say 9 V, a net amount of energy equal to $1.5\delta q$ can be extracted from the circuit in each traversal of the mesh, which violates conservation of energy. It follows that

$$\delta q \sum \text{Voltage rises} = \delta q \sum \text{Voltage drops} \tag{2.9}$$

Cancelling δq from both sides of Equation 2.9 gives the statement of KVL.

It must not be assumed that KVL alone is an expression of conservation of energy, because the currents that produce the voltage rises and voltage drops across the circuit elements around the mesh must satisfy KCL. Hence, *KCL and KVL together are an expression of conservation of energy in a circuit.*

Although KVL was illustrated in the preceding discussion using a mesh, the same argument applies to any loop in the circuit.

To minimize the possibility of error in writing KVL around a mesh or a loop, the following procedure is helpful and will be illustrated with reference to Figure 2.21:

1. *Choose a node as the starting point and as the endpoint in traversing the loop, such as node 'n' at the bottom of the loop.*
2. *Decide on the sense in which the loop is to be traversed, say clockwise.*
3. *Traverse the loop in the chosen direction proceeding through all the circuit elements in succession.*
4. *As each circuit element is crossed, record the voltage across the circuit element, assigning it a positive sign if it is a voltage rise and a negative sign if it is a voltage drop.*
5. *Set the algebraic sum of the recorded voltages equal to zero.*

Thus, in Figure 2.21, starting at node 'n' and going clockwise, the first voltage encountered is a 4 V rise, so it is recorded as +4. The next voltage encountered is a 2.5 voltage drop, so it is recorded as −2.5. Proceeding in this manner around the loop and setting the algebraic sum equal to zero,

$$+4 - 2.5 - 3 - 0.5 - 1.5 + 3.5 = 0 \tag{2.10}$$

It is seen that this is equivalent to writing 4 + 3.5 = 2.5 + 3 + 0.5 + 1.5, as before, and is in accordance with the statement of KVL.

KVL can also be applied to an open path in a circuit to determine the voltage between the two ends of the path. In Figure 2.22, for example, current and resistance

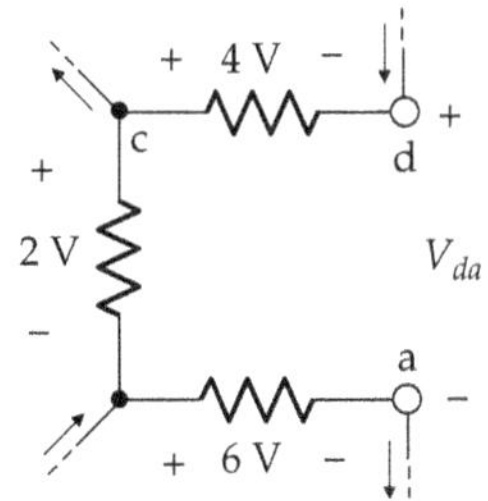

FIGURE 2.22
Kirchhoff's voltage law applied to an open path.

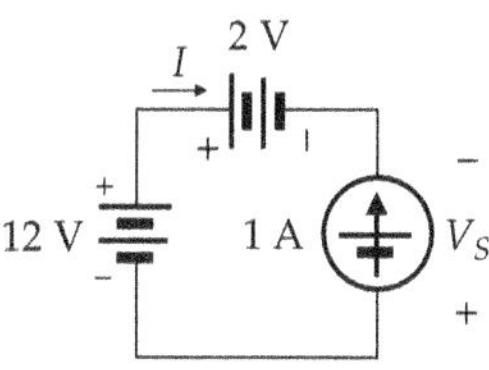

FIGURE 2.23
Figure for Primal Exercise 2.17.

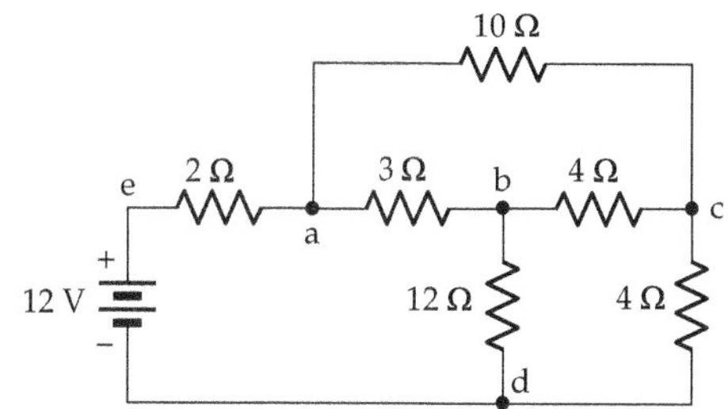

FIGURE 2.24
Figure for Example 2.2.

values are such that the voltages across the resistors are as indicated. Starting at node 'a' and going clockwise, KVL gives

$$+6 + 2 - 4 - V_{da} = 0 \tag{2.11}$$

so that $V_{da} = 4$ V. It should be noted that by convention, a voltage with a double subscript, such as V_{da}, is the *voltage drop* from the node denoted by the first subscript to the node denoted by the second subscript. Evidently, V_{da} is equally a *voltage rise* from the node denoted by the second subscript 'a' to the node denoted by the first subscript 'd'. In going from node 'd' to node 'a', the voltage drop is V_{da} and is entered with a negative sign in Equation 2.11.

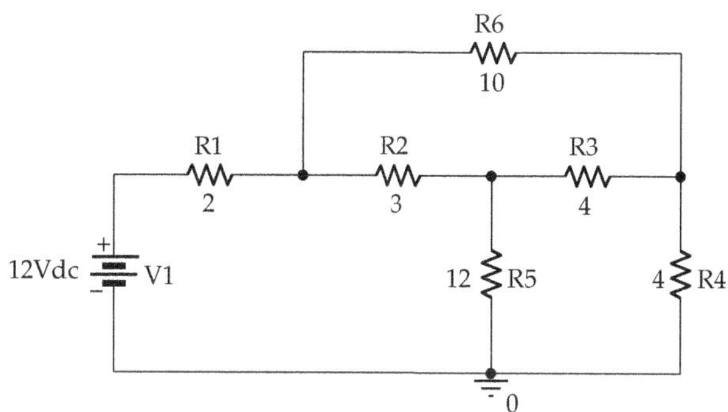

FIGURE 2.25
Figure for Example 2.2.

Primal Exercise 2.15

Verify KCL at the six nodes in Figure 2.21.

Primal Exercise 2.16

Suppose that in Figure 2.21, the currents are doubled, which doubles the voltages across the resistors. Verify KVL when going around the circuit, (a) clockwise and (b) counterclockwise.

Primal Exercise 2.17

Determine in Figure 2.23 (a) I, (b) V_S, and (c) the power delivered or absorbed by each source.

Ans. (a) −1 A; (b) −10 V; (c) 12 V source absorbs 12 W, 2 V source delivers 2 W, current source delivers 10 W.

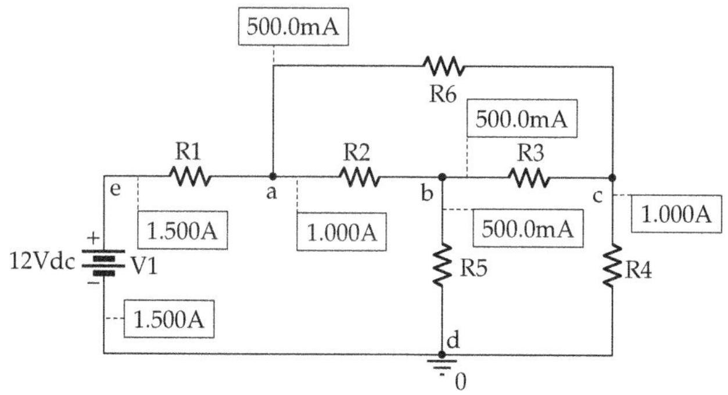

FIGURE 2.26
Figure for Example 2.2.

Example 2.2: Verification of KCL and KVL

It is required to simulate the circuit of Figure 2.24 and to verify KCL and KVL.

Simulation: Appendix C explains the basics of using the educational version of PSpice in an introductory course on electric circuits. The circuit is entered in the Schematic page of PSpice as illustrated in Figure 2.25. The battery is entered as VDC from the SOURCE library, and the resistors are entered as R from the ANALOG library. In the Simulation Settings, 'Bias Point' is selected under 'Analysis type' and then 'General Settings' under 'Options'. After the simulation is run, pressing the 'I' button displays the currents, as in Figure 2.26, in which the nodes have been labeled for clarity. The I blocks can be dragged to more convenient locations, which also displays dotted-line connections between each

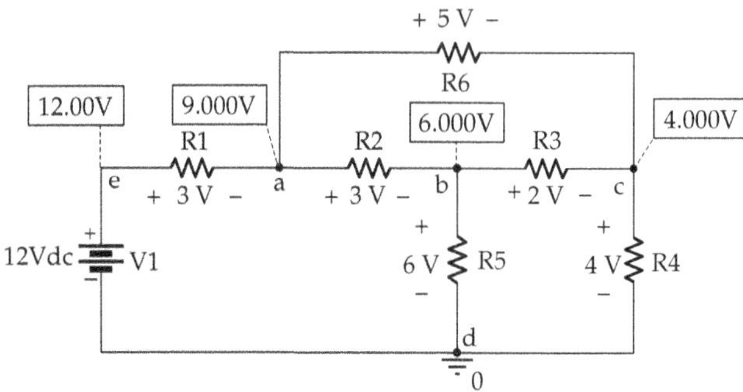

FIGURE 2.27
Figure for Example 2.2.

circuit element and the I block that indicates the current through that element. The location of a dotted line at one terminal of a circuit element signifies that current enters the circuit element at that terminal. Thus, the 1.5 A current flows through the battery in the direction 'de' and through R1 in the direction 'ea'. It is seen that 1.5 A enter node 'a' and leave this node through R2 and R6, in accordance with KCL. Similarly, KCL is satisfied at every other node in the circuit. Pressing the 'V' button displays the node voltages *with respect to a ground voltage of zero*, as indicated in Figure 2.27. The voltage drop across each resistor, which is equal to the difference between the node voltages at the ends of the resistor, has been added for clarity. The circuit has three meshes, and KVL is satisfied around each mesh. Thus, if mesh d–e–a–b–d is traversed clockwise, the algebraic sum of the voltage rises and the voltage drops is 12 – 3 – 3 – 6 = 0. The same is true of the other meshes b–a–c–b and d–b–c–d. KVL is also satisfied around any loop in the circuit, such as d–e–a–c–b–d. The algebraic sum of the voltage rises and voltage drops around this loop is 12 – 3 – 5 + 2 – 6 = 0.

Pressing the 'W' button displays the *power absorbed* in each circuit element as in Figure 2.28, where 'W' is the symbol for power in PSpice. A positive value of power

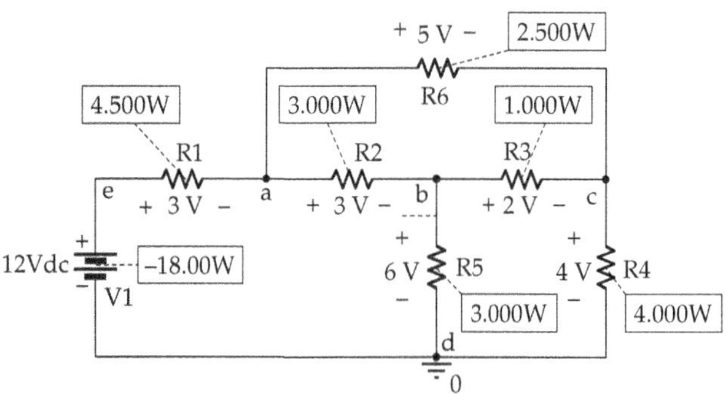

FIGURE 2.28
Figure for Example 2.2.

is power absorbed, which means that power delivered by a source is negative, as indicated by the –18 W for the battery. It is seen that the total power absorbed by the resistors is +18 W, in accordance with conservation of power.

Problem-Solving Tips

- The solution to any circuit problem can be checked by making sure that Ohm's law is satisfied for every resistor, KCL is satisfied at every node, and KVL is satisfied around every mesh.
- If a circuit has N essential nodes, then after writing KCL for $(N - 1)$ essential nodes, KCL at the remaining node should automatically be satisfied if KCL was written correctly at the other nodes. This is a useful check on KCL.

The second problem-solving tip of the preceding example can be illustrated for the circuit of Figure 2.24 by labeling the branch currents, as in Figure 2.29. There are four essential nodes: 'a', 'b', 'c', and 'd'. KCL for the first three nodes gives the following: node 'a': $I_1 = I_2 + I_3$; node 'b': $I_3 = I_4 + I_5$; and node 'c': $I_2 + I_4 = I_6$.

When these three equations are added, I_2, I_3, and I_4 cancel out, leaving $I_1 = I_5 + I_6$, which is KCL for node 'd'. Thus, $N = 4$, and $N - 1 = 3$, so that only three independent KCL equations can be written. KCL for the remaining node is not an independent equation but can be used as a check on the KCL equations for the other nodes. The number of independent KCL and KVL equations that can be written for any circuit is derived in Problem P2.62.

Primal Exercise 2.18

Refer to the circuit of Figure 2.17. Specify (a) the number of independent KCL equations and (b) the number of independent KVL equations that can be written for the circuit. Verify the relation of Problem P2.62.

Ans. (a) 3; (b) 3; 6 = 3 + (4 – 1).

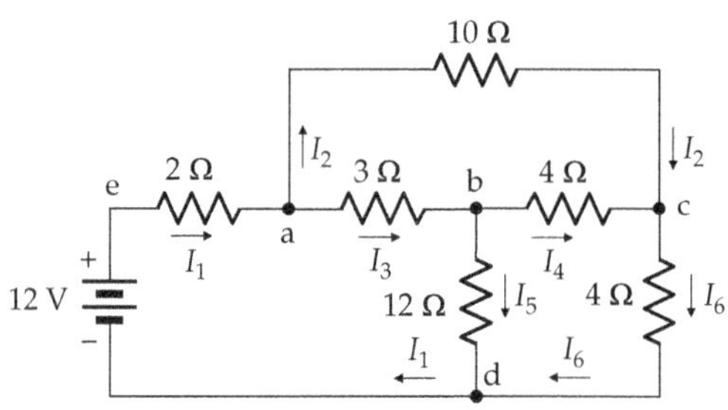

FIGURE 2.29
Independence of KCL Equations.

Example 2.3: Application of Ohm's Law, KCL, and KVL

It is required to determine V_O in Figure 2.30.

Solution:

V_O can be obtained by a step-by-step application of Ohm's and Kirchhoff's laws without introducing any additional, unknown circuit variables.

Step 1: Because the output terminals between which V_O is specified are open-circuited, no current flows in or out at these terminals (Figure 2.31).

Step 2: From KCL at node 'a', the current flowing through the 10 Ω resistor is 2 A.

Step 3: From Ohm's law, the voltage drop from node 'a' to node 'c' is 2 × 10 = 20 V.

Step 4: From KCL at node 'c', the current through the 20 Ω resistor is 0. This can also be deduced by enclosing the 2 A source and 10 Ω resistor by a surface, as in Figure 2.19a, and noting that since no current enters this surface at node 'a', then no current leaves this surface at node 'c'.

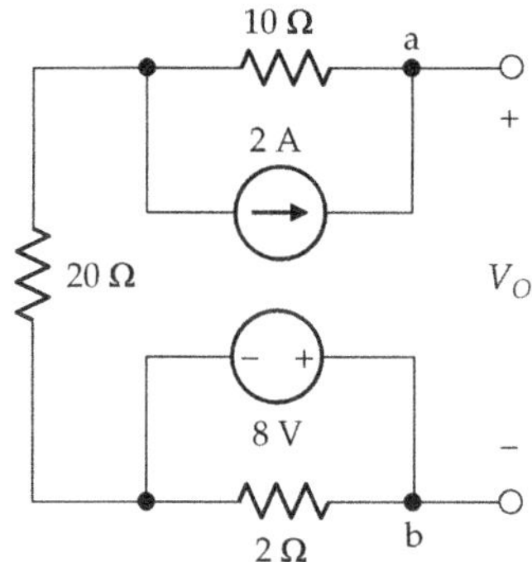

FIGURE 2.30
Figure for Example 2.3.

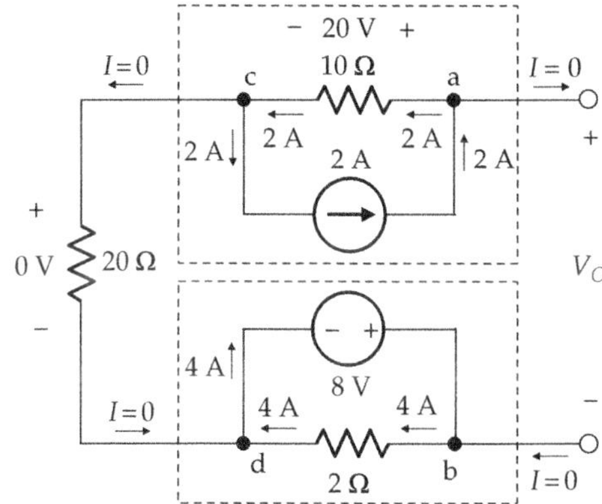

FIGURE 2.31
Figure for Example 2.3.

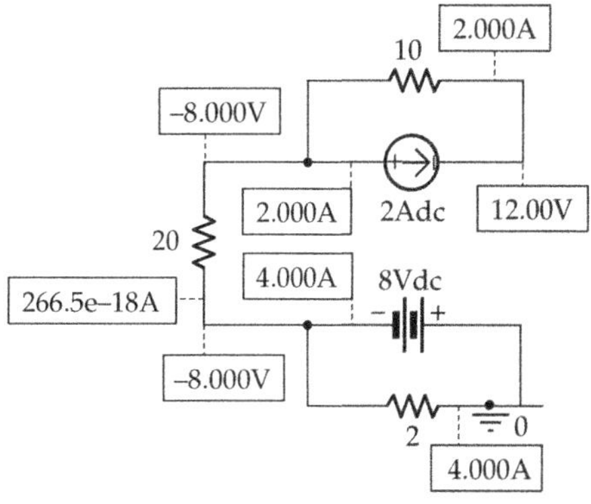

FIGURE 2.32
Figure for Example 2.3.

Step 5: From Ohm's law, the voltage across the 20 Ω resistor is zero.

Step 6: Applying KVL in going through nodes 'b', 'd', 'c', 'a', and back to 'd': $-8 + 0 + 20 - V_O = 0$, which gives $V_O = 12$ V.

We could continue and apply Ohm's law to the 2 Ω resistor to determine that 4 A flow through this resistor in the direction of the voltage drop of 8 V. From KCL at node 'd' or node 'b', the current in the 8 V source is 4 A in the direction of a voltage rise through the source.

Simulation: The circuit is entered as in Figure 2.32. After selecting 'Bias Point' under 'Analysis type' in the Simulation Settings and running the simulation, pressing the I and V buttons displays the currents and voltages indicated in Figure 2.32. The zero current in the 20 Ω resistor is indicated as 266.5E–18 A, which denotes 266.5×10^{-18} A.

Problem-Solving Tip

- Always mark on the circuit diagram the directions of currents of interest and the polarities of voltages of interest, bearing in mind that the current through an ideal resistor is in the direction of the voltage drop across the resistor.

2.9 Series and Parallel Connections

2.9.1 Series Connection

The four elements 'A', 'B', 'C', and 'D' in Figure 2.33a are connected in series. Geometrically, the most salient feature of the series connection is that *the elements are connected in succession, end to end, without branching at any of the nodes between the elements*, as diagrammatically illustrated by the long, thick arrow in Figure 2.33a. Electrically, the nodes between the elements are inessential nodes 'a', 'b', and 'c'. Conservation of charge requires that the current entering

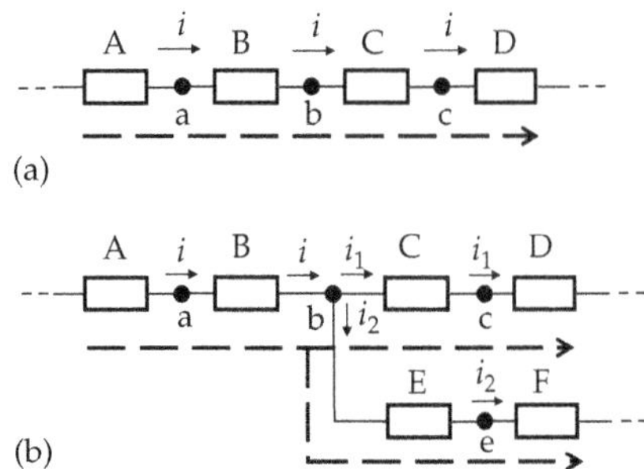

FIGURE 2.33
(a) Four elements connected in series and (b) series connection between elements 'B' and 'C' broken by the branch connecting to element 'E'.

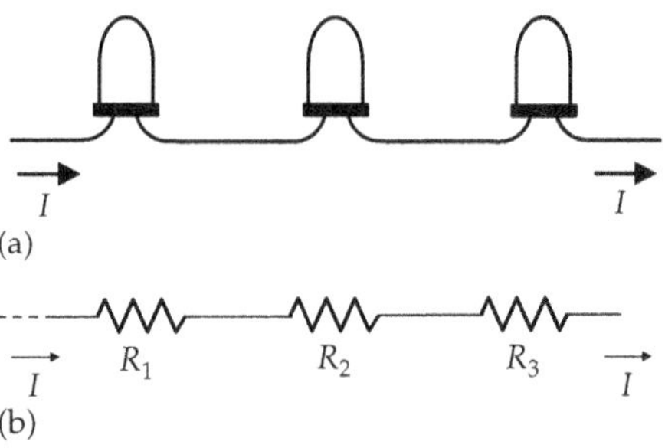

FIGURE 2.35
(a) Series connection of three lamps and (b) the lamps in (a) represented by resistors.

an inessential node is the same as the current leaving it. It follows that *the same current flows through all the series-connected elements* as illustrated in Figure 2.33a.

A simple test for elements in series is to check that the elements can be traversed in succession along an unbranched path, as is the case with elements 'A', 'B', 'C', and 'D' in Figure 2.33a. In contrast, elements 'A', 'B', 'C', and 'D' in Figure 2.33b are no longer in series, because the path through these elements divides at node 'b' into two branches as shown. Node 'b' is now an essential node at which the current i divides into i_1 and i_2. However, elements 'A' and 'B' in Figure 2.33b are still in series, as are elements 'C' and 'D', as well as 'E' and 'F'.

Note that the series connection has the following features:

1. Because the current through series-connected elements is the same, *KCL is automatically satisfied in a series connection.*
2. *Voltages add algebraically along the path through series-connected elements.* In Figure 2.34a, for example, the two resistors and the 6 V battery are connected in series, the current through the elements being assumed to be 1 A. With the voltage of node 'a' considered 0, the voltage of node 'c' is $5 + 3 = 8$ V, and the voltage of node 'd' is $8 - 6 = 2$ V. KVL can be written as $5 + 3 - 6 - V_{da} = 0$, as in Figure 2.22, which gives $V_{da} = 2$ V.
3. If one of the series-connected elements is removed from the circuit, the current in the series connection is zero. This is because when an element is removed from the series connection, the element is replaced by an open circuit, as in Figure 2.34b.

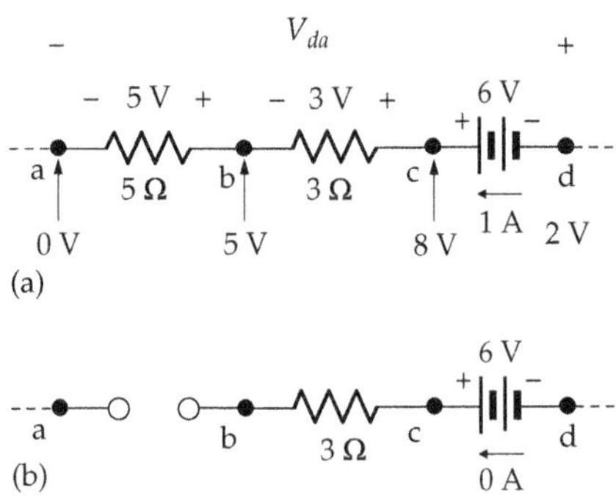

FIGURE 2.34
(a) Algebraic addition of voltages along a path and (b) removal of one of the series-connected elements.

Figure 2.35a shows three lamps connected in series, as is often done in decorative lighting. The lamps are equivalent to three resistors in series (Figure 2.35b). If a lamp is removed, the remaining lamps are turned off because the current is interrupted, that is, reduced to zero. A nonelectrical example of a series connection is a number of railroad cars coupled together, end to end, to a locomotive. In this case, the analog of the same current in the series connection is the same velocity at which the whole train moves.

2.9.2 Parallel Connection

The three elements 'A', 'B', and 'C' in Figure 2.36a are connected in parallel. Geometrically, the most salient feature of the parallel connection is that *one end of each element, marked with 'x' in Figure 2.36a, is connected to a common node, 'a', whereas the other end of each of these elements, marked with 'z' in Figure 2.36a, is connected to another common node, 'b'.* This implies that *the closed path through any two paralleled elements traverses these two elements only, and no other elements,*

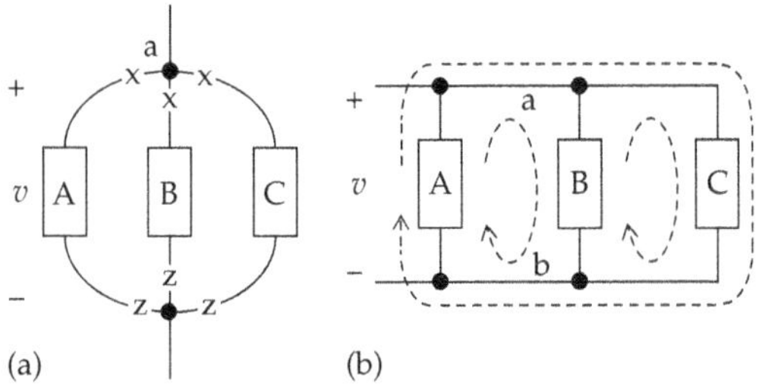

FIGURE 2.36
(a) Parallel connection of three lamps and (b) conventional representation of the parallel-connection in (a).

as diagrammatically illustrated by the dashed curves in Figure 2.36b, which is Figure 2.36a redrawn in the conventional manner, for clarity. Electrically, nodes 'a' and 'b' are essential nodes, and *the same voltage v between these nodes appears across each of the parallel-connected elements.*

Because students often have some difficulty in identifying circuit elements that are in parallel, we will elaborate a little on the parallel connection. It is seen from Figure 2.36 that either one of the following two tests can be applied to identify circuit elements in parallel:

1. Mark with 'x' the ends of all elements connected together at the same node and then mark with 'z' the other end of each of the elements; if the z-marked ends are all connected to the same node, the marked elements are in parallel.
2. Any two elements are in parallel if the closed path through them does not traverse any other element.

For example, suppose we wish to identify which elements in Figure 2.37a are in parallel. We may begin by marking with 'x' the ends of elements 'A', 'B', 'C', and 'D' connected together at node 'c'. We then mark with 'z' the other end of each of these elements and check which of these z-marked ends are connected to the same node. We note that the z-marked ends of elements 'A' and 'B' are connected to node 'a', so these elements are in parallel. Similarly, the z-marked ends of elements 'C' and 'D' are connected to node 'b', so these elements are in parallel. However, elements 'A' and 'B' are not in parallel with elements 'C' and 'D', because the z-marked ends of 'A' and 'B' are not connected to the same node as the z-marked ends of 'C' and 'D', due to the presence of element 'E'. In the absence of this element (Figure 2.37b), the four elements are in parallel.

As a further check, we note that elements 'A' and 'B' in Figure 2.37a form a mesh that does not include any

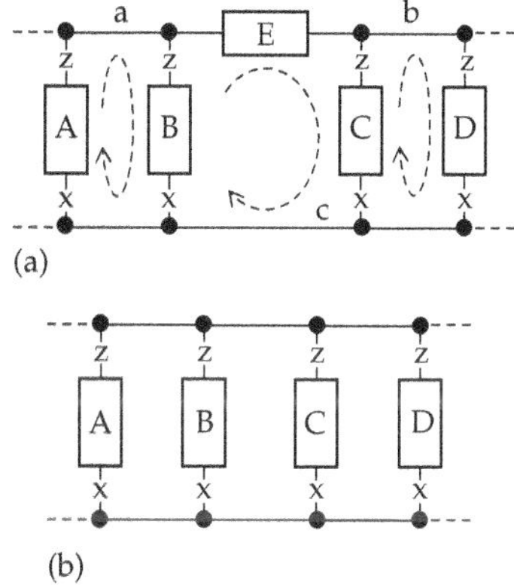

FIGURE 2.37
(a) Nonparallel connection of elements 'B' and 'C' because of the presence of element 'E' and (b) removal of element 'E' makes the four elements in parallel.

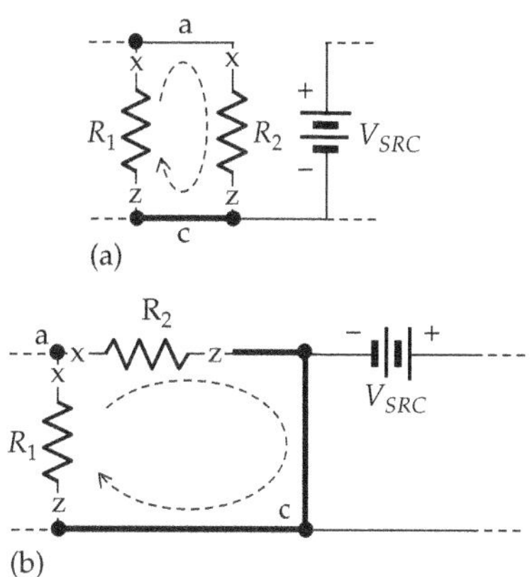

FIGURE 2.38
(a) R_1 and R_2 connected in parallel and (b) R_1 and R_2 remain in parallel when the circuit in (a) is redrawn.

other element, so these elements are in parallel, as are elements 'C' and 'D'. However, elements 'B' and 'C' are not in parallel, since the mesh that includes these two elements also includes element 'E'. Similarly, elements 'A' and 'C', 'A' and 'D', and 'B' and 'D' in Figure 2.37a are not in parallel. In Figure 2.37b, any mesh or loop formed by any two of the four elements does not include any other element, so these four elements are in parallel.

As another example, consider Figure 2.38a. Resistors R_1 and R_2 are evidently in parallel, in accordance with the aforementioned tests. The circuit of Figure 2.38a may be redrawn as in Figure 2.38b. The circuit is the same, for if the vertical thick line in Figure 2.38b is collapsed, the circuit reduces to that of Figure 2.38a. R_1 and R_2 are still in parallel. If their ends that are connected at node 'a' are marked with an 'x', and their other ends are marked with a 'z', the z-marked ends are connected at the same node 'c', although this connection is a 'short' connection in Figure 2.38a and is a 'long' connection in Figure 2.38b. Both connections are indicated by a thick line in the figures. They are wiring connections of zero resistance, as is the wiring connection between R_1 and R_2 at node 'a' in Figure 2.38a. Moreover, the mesh formed by R_1 and R_2 does not include any other element.

The following should be noted about the parallel connection:

1. In Figure 2.39, two resistors and a 1 A source are connected in parallel, the voltage across the parallel combination being assumed to be 6 V. Ohm's law gives a current of $6/2 = 3$ A in the 2 Ω resistor, and a current of $6/3 = 2$ A in the 3 Ω resistor, both in the direction of the voltage drop of 6 V. It is seen that *KVL is automatically satisfied in the mesh or loop formed by any two paralleled elements.* Thus, starting at node 'a', for example, and moving through any of the elements to node 'b', involves a voltage drop of 6 V. Moving back to node 'a' through one

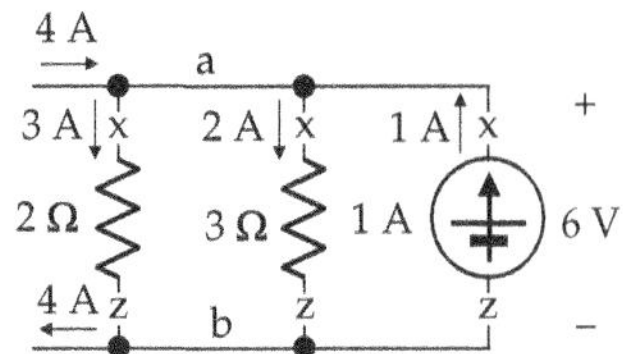

FIGURE 2.39
KVL around parallel-connected elements.

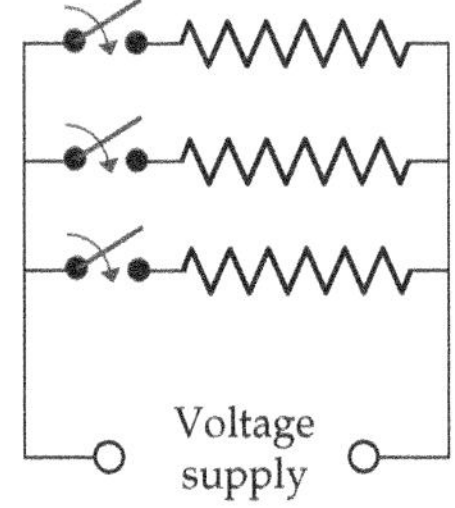

FIGURE 2.40
Parallel connection of heating elements.

of the other two elements involves an equal voltage rise of 6 V. KVL is therefore satisfied.

2. *The currents through the paralleled elements add algebraically at either node between which the elements are paralleled.* In Figure 2.39, the current entering node 'a' from the rest of the circuit is 4 A and is equal to the algebraic sum of the branch currents, 3 + 2 − 1 = 4 A, in accordance with KCL, as is the 4 A current leaving node 'b' to the rest of the circuit.
3. In order to have zero current between the end nodes 'a' and 'b' in Figure 2.39, *all* the paralleled elements must be removed from the circuit.

The heating elements of a multielement electric heater are typically connected in parallel across the mains voltage supply. Figure 2.40 diagrammatically illustrates three resistive heating elements connected to the voltage supply through switches that allow energizing a single element, or two elements, or three elements, at a time.

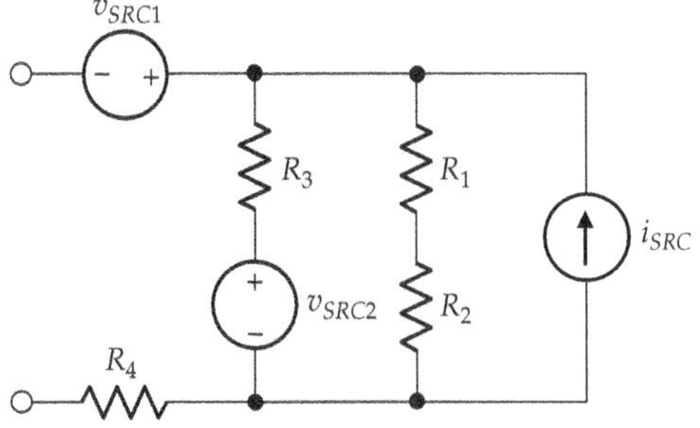

FIGURE 2.41
Series and parallel connections of circuit elements.

The series and parallel connections allow building circuits of any desired complexity. In Figure 2.41, for example, R_1 and R_2 are connected in series, as are R_3 and v_{SRC2}; i_{SRC} is connected in parallel with these series combinations. The resulting series–parallel combination is in turn connected in series with R_4 and v_{SRC1}.

Primal Exercise 2.19

Figure 2.42 illustrates a rear window heater used for defogging/defrosting in some cars. It consists of nine resistance wires connected as shown between the battery terminals. Draw a circuit diagram of the heater in terms of resistors labeled with the same numbers as the resistance wires, and describe the connection.

Ans. The circuit diagram is as shown in Figure 2.43. The resistance wires are connected as three sets in series, each set consisting of three equal resistances in parallel.

Primal Exercise 2.20

Determine V_X in Figure 2.44.

Ans. 1 V.

Primal Exercise 2.21

Element 'A' in Figure 2.45 absorbs 4 W. Determine the power delivered or absorbed by the 2 A source.

Ans. 8 W absorbed.

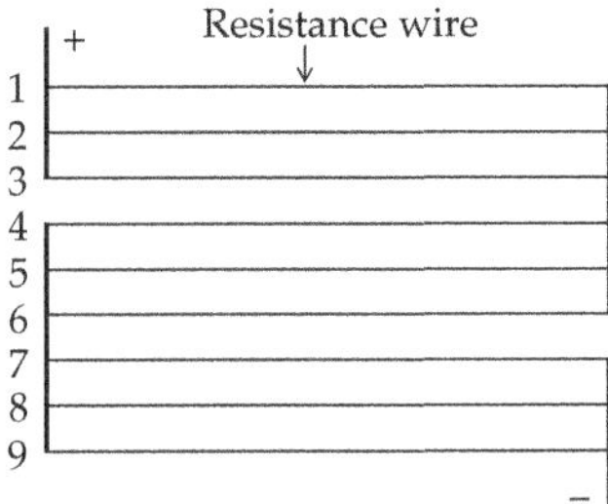

FIGURE 2.42
Figure for Primal Exercise 2.19.

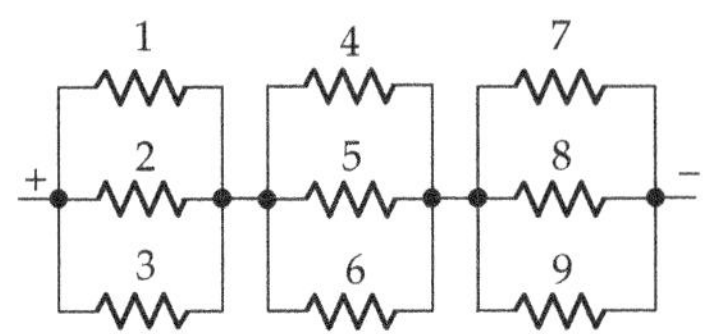

FIGURE 2.43
Figure for Primal Exercise 2.19.

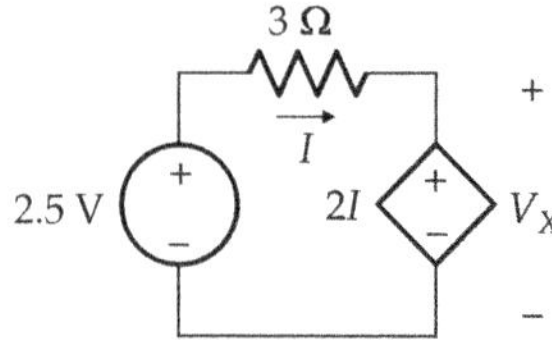

FIGURE 2.44
Figure for Primal Exercise 2.20.

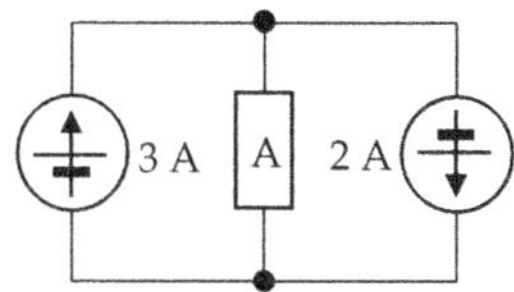

FIGURE 2.45
Figure for Primal Exercise 2.21.

2.10 Problem-Solving Approach

The essence of engineering is to produce *effective* solutions to practical problems through appropriate design of equipment, structures, or systems. The most effective engineering solution to a given problem is the one that is reached in minimum time, at minimum cost and effort, and which achieves the desired objectives in the simplest possible manner. Simplicity is crucial in engineering design, because the simplest solution is almost invariably the most reliable, durable, and economical. Imagination and creativity can make a decisive difference in effective problem solving.

It is important therefore to foster the mindset and approach that is conducive to effective problem solving. Problems on electric circuits provide an opportunity toward this goal that should be exploited. This theme is emphasized throughout the book.

The steps listed here are intended as a guide to an effective problem-solving approach based at this stage on Ohm's law, KCL, and KVL. Depending on the particular problem under consideration, not all of the steps may be applicable in every case. Some of these steps will be updated in future chapters in the light of circuit analysis concepts and methodologies introduced in each of these chapters. By the end of Chapter 6, all the main circuit analysis methodologies would have been presented.

Step 1—Initialize: Generally, this involves the following:

(a) Mark on the circuit diagram all the given values of circuit parameters, currents, and voltages, as well as the unknowns to be determined, keeping in mind that every current must have a direction and every voltage must have a polarity.

(b) Label the nodes, such as 'a', 'b', 'c', etc., as this is usually helpful.

(c) If the solution requires that a given value of current or voltage be satisfied, *assume this value from the very beginning*, as this can considerably facilitate the solution.

Step 2—Simplify: Try and reduce the circuit to a simpler form, if possible. This may be done, for example, by redrawing the circuit or by replacing series and parallel combinations of circuit elements by an equivalent circuit element, as discussed in Chapter 3.

Step 3—Deduce: Determine any values of current or voltage that *follow immediately from direct application of Ohm's law, KCL, or KVL, without introducing any additional unknowns*. The given unknowns, such as required outputs, as well as controlling voltages and currents of dependent sources, may be used in expressing Ohm's law, KCL, or KVL in this step.

The results of applying Ohm's law, KCL, and KVL should be entered on the circuit diagram itself, including directions of currents and polarities of voltages. This way, the correctness of these circuit laws can be readily checked visually. Bear in mind that the current through an ideal resistor that obeys Ohm's law is always in the direction of the voltage drop across the resistor.

KCL, KVL in Step 3 are often sufficient to solve the problem, as in Examples 2.3, 2.4, and 2.5. Moreover, it may be possible to repeat Steps 2 and 3 alternately in some cases. If Step 3 does not provide the solution, proceed to Step 4.

Step 4—Explore: Consider the nodes and meshes in the circuit to see if KCL or KVL can be expressed using *a single, unknown current or voltage, and if this unknown can then be directly determined from KCL or KVL*. If so, this step provides the solution, because once an unknown current or voltage is determined, other required values can be derived using KCL, KVL, or Ohm's law. If Step 4 does not provide the solution, proceed to Step 5.

Step 5—Plan: Think carefully about the problem in the light of circuit fundamentals and circuit analysis techniques. Imagination, creativity, and experience in problem solving can play a decisive role in this step. Try to think of alternative solutions and select what seems to be the simplest and most direct solution.

Step 6—Implement: Carry out your planned solution, bearing in mind the following considerations:

(a) Keep the number of unknown variables to a minimum, as this minimizes the likelihood of careless mistakes. Make use of any given variables such as an unknown variable to be determined or the controlling currents or voltages of dependent sources.

(b) Mentally ascertain the correctness of equations or relations as you write them or copy them, in order to minimize the likelihood of careless mistakes. If in doubt about the correctness of an equation or relation, check the units on both sides of the equation or relation or the units of numerators and denominators of expressions.

(c) Keep track of units, and label all calculated values with the appropriate units.

Step 7—Check your calculations and results.

(a) Check that your results make sense, in terms of magnitude and sign.

(b) Check that Ohm's law is satisfied for every resistor, that KCL is satisfied at every node, and that KVL is satisfied around every mesh.

(c) A good way to check your results is to seek an alternative solution to the problem and see if it gives the same results.

(d) Whenever feasible, check your results with PSpice simulation. This is a valuable habit to acquire. For this reason, PSpice simulation is strongly emphasized throughout the book.

The preceding steps will henceforth be referred to by the acronym ISDEPIC.

Example 2.4: Illustration of ISDEPIC Approach

The circuit of Figure 2.16 is analyzed in accordance with ISDEPIC.

Solution:

1. *Initialize*: The circuit is reproduced in Figure 2.46a, indicating values of all the circuit elements.
2. *Simplify*: The circuit is in a simple enough form.
3. *Deduce*: It is seen that 1.5 A enter the upper node and leave the lower node, as in Figure 2.46b. No more deductions can be made from immediate application of Ohm's law, KCL, or KVL.
4. *Explore*: KCL at either node can be expressed by introducing a single unknown variable at either node, which can then be determined from KVL around the mesh on the LHS. The variable could be the current in the 3 Ω resistor or in the 6 Ω resistor. If an unknown current I_X is assigned entering the upper node through the 6 Ω resistor (Figure 2.46c), then from KCL, the current leaving the node through the 3 Ω resistor is $(I_X + 1.5)$ A. Ohm's law and KVL around the mesh on the left allow writing an equation in I_X that can be used for determining I_X. Thus, moving clockwise around the mesh, starting at the negative terminal of the battery, KVL gives $+9 - 6I_X - 3(I_X + 1.5) = 0$, so that $I_X = 0.5$ A. Once I_X is known, all the other unknown currents and voltages can be determined.

   Alternatively, an unknown voltage V_X may be assigned to the voltage between the nodes (Figure 2.46d). From Ohm's law, the current leaving the node through the 3 Ω resistor is $V_X/3$. From KVL, the voltage drop across the 6 Ω resistor in going from the upper node to the positive terminal of the battery is $(V_X - 9)$ V, and the current leaving the node through the 6 Ω resistor is $(V_X - 9)/6$ A. From KCL, $1.5 = V_X/3 + (V_X - 9)/6$, which gives $V_X = 6$ V.

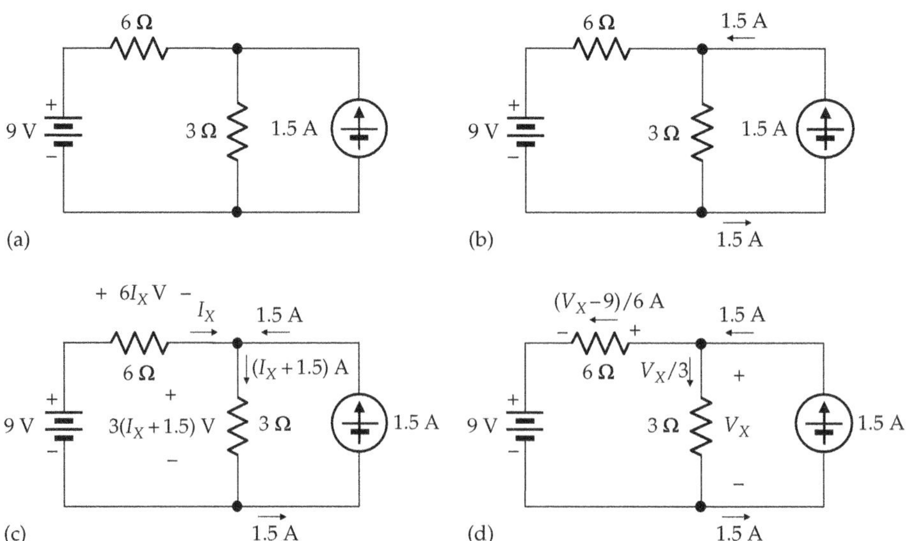

FIGURE 2.46
Figure for Example 2.4.

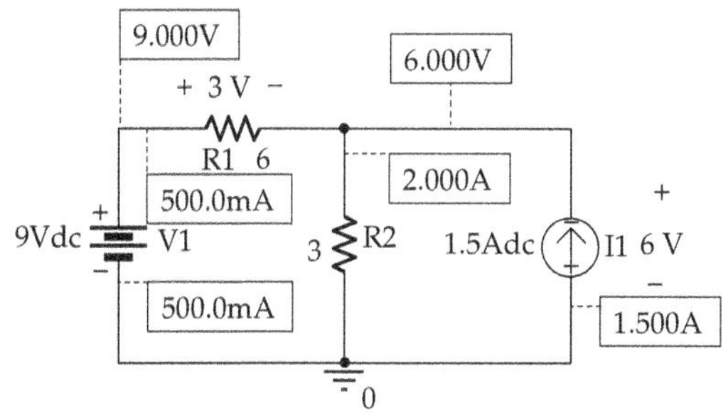

FIGURE 2.47
Figure for Example 2.4.

Note how much simpler is the solution compared to that based on conservation of power (Problem P2.61).

Simulation: The circuit is entered as in Figure 2.47. 'Bias Point' is selected under 'Analysis type' in the Simulation Settings. After the simulation is run, pressing the I and V buttons displays the currents and voltages indicated in Figure 2.47. It is seen that Ohm's law, KCL, and KVL are satisfied.

Problem-Solving Tip

- Circuits having only two essential nodes can generally be analyzed by applying KCL at either node.

Primal Exercise 2.22

Show that power is conserved in the circuit of Figure 2.46 based on the values of currents and voltages, and verify by PSpice simulation.

Ans. Power delivered is 4.5 W by voltage source and 9 W by current source; power dissipated is 12 W in the 3 Ω resistor and 1.5 W in the 6 Ω resistor.

Example 2.5: Illustration of ISDEPIC Approach

It is required to determine I_X in Figure 2.48a.

Solution:

The circuit has two essential nodes, as in Example 2.4, but with the addition of a VCVS controlled by the voltage V_A. It will be analyzed by applying ISDEPIC.

1. *Initialize*: The circuit is already marked with given values and the required I_X. The nodes are labeled 'a' and 'b'.
2. *Simplify*: The circuit is in a simple enough form.
3. *Deduce*: 2 A and a current $V_A/4$ A leave node 'a' (Figure 2.48b). The voltage drop across the 10 Ω resistor is $10I_X$. KCL at node 'a' is $I_X = 2 + V_A/4$. A second equation involving I_X and V_A can be derived from KVL around the mesh on the RHS.

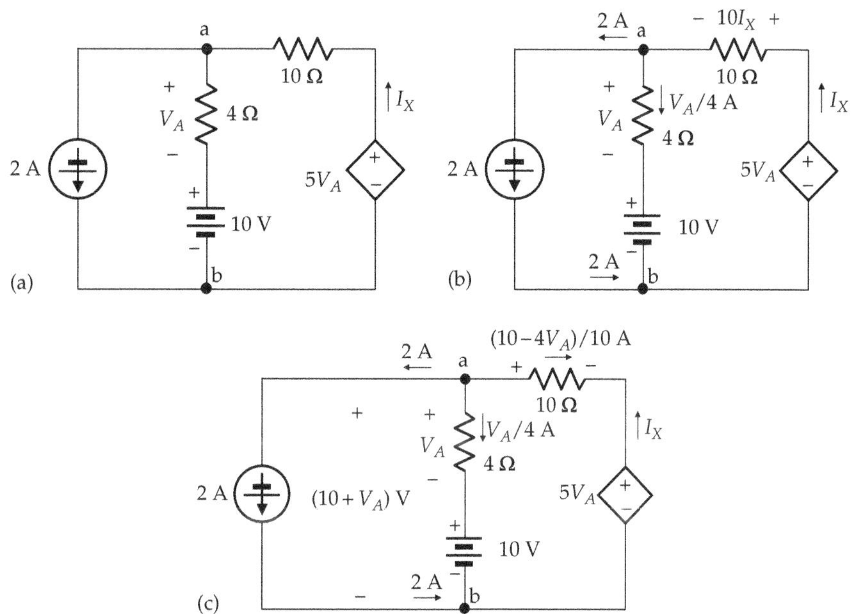

FIGURE 2.48
Figure for Example 2.5.

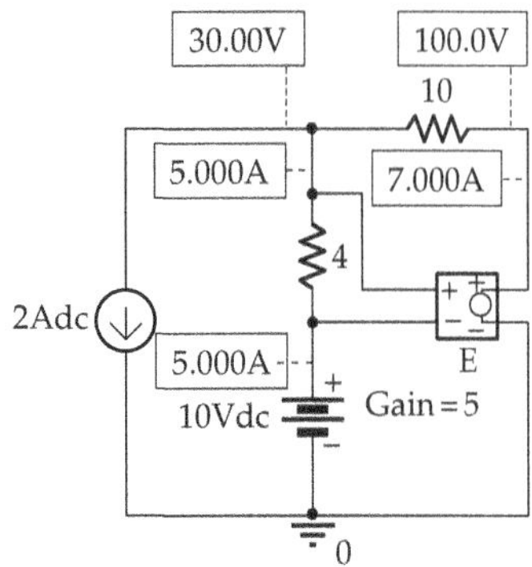

FIGURE 2.49
Figure for Example 2.5.

Starting at node 'b' and going clockwise, KVL gives $+10 + V_A + 10I_X - 5V_A = 0$, or $4V_A - 10I_X = 10$. Solving these equations for V_A and I_X gives $I_X = 7$ A and $V_A = 20$ V.

Since the circuit is a two-essential node circuit, it can alternatively be analyzed by writing a single KCL equation in one unknown. KVL can be used to express I_X directly in terms of V_A. Thus, $V_{ab} = (10 + V_A)$ (Figure 2.48c). The voltage drop from node 'a' to the positive terminal of the VCVS is $V_{ab} - 5V_A = 10 + V_A - 5V_A = 10 - 4V_A$, and the current leaving node 'a' through the 10 Ω resistor is $(10 - 4V_A)/10 = 1 - 0.4V_A$. From KCL at node 'a', $2 + 0.25V_A + 1 - 0.4V_A = 0$, which gives $V_A = 20$ V. Hence, $I_X = -(1 - 0.4V_A) = 7$ A.

Simulation: The circuit is entered as in Figure 2.49. The VCVS is entered from the ANALOG library as part number E having four terminals: two for the voltage source and two for the controlling voltage. The multiplier for this voltage is entered by double clicking on the default 'Gain =1' and changing the value from 1 to 5 in the 'Display Properties' window. 'Bias Point' is selected under 'Analysis type' in the Simulation Settings. After the simulation is run, pressing the I and V buttons displays the currents and voltages indicated in Figure 2.49. It is seen that $I_X = 7$ A, and the voltage across the VCVS is 100 V.

Problem-Solving Tip

- Controlling currents or voltages of dependent sources are often convenient to use as unknown variables in analyzing circuits.

Primal Exercise 2.23

Show that power is conserved in the circuit of Figure 2.48 based on the values of currents and voltages, and verify by PSpice simulation.

Ans. Dependent source delivers 700 W, voltage source absorbs 50 W, and current source absorbs 60 W. Power dissipated is 100 W in the 4 Ω resistor and 490 W in the 6 Ω resistor.

Example 2.6: Illustration of ISDEPIC Approach

It is required to determine R in Figure 2.50a so that 5 A flow in the short circuit between nodes 'b' and 'd' in the direction indicated.

Solution:

1. *Initialize*: To determine R one should not seek to derive a relation between R and the current in the short circuit, and then set this current to 5 A to find R. This would be a waste of time and effort. Instead, the required value of current is *assumed to begin with*, and the circuit is analyzed accordingly. This is an example of the initialization step 1(c) mentioned under the general problem-solving approach.
2. *Simplify*: The circuit configuration of Figure 2.50a is referred to as a lattice configuration. Because it

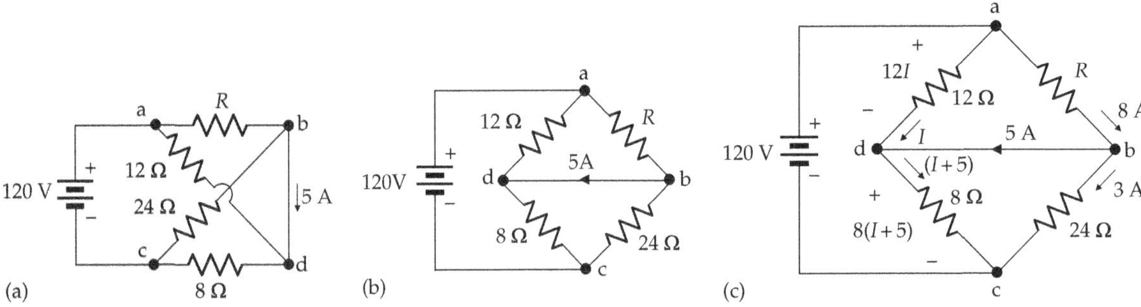

FIGURE 2.50
Figure for Example 2.6.

contains a crossover connection, it is not easy to visualize circuit behavior. The crossover connection is encountered in the zigzag path from node 'a' through nodes 'b', 'c', 'd', and back to 'a'. The crossover connection is removed by relocating the four nodes so as to have a straight path around the loop 'abcda' without a zigzag. The four nodes can be placed in a clockwise sense, as in Figure 2.50b, and the elements between the nodes connected accordingly. Note that nodes 'b' and 'd' are one and the same node, which means that the 12 Ω resistor is in parallel R and the 8 and 24 Ω resistors are in parallel. Moreover, the two parallel combinations are in series across the voltage source. Nevertheless, 'b' and 'd' have been separated to show the 5 A current.

3. *Deduce*: No deductions can be made from immediate application of Ohm's law, KCL, or KVL.
4. *Explore*: KCL at nodes 'a' or 'c' is not helpful because none of the currents at these nodes is known. KVL around the meshes is also not helpful for the same reason. KCL at node 'b' is not helpful either because assigning a single current at this node does not allow using KVL to determine this current, since R is unknown. However, if a current I is assigned entering at node 'd' from the 12 Ω resistor (Figure 2.50c), the current leaving node 'd' through the 8 Ω resistor is $(I + 5)$ A. I can then be determined from Ohm's law and KVL around the mesh on the LHS. Thus, the voltage drop across the 12 Ω resistor is $12I$ V, and the voltage drop across the 8 Ω resistor is $8(I + 5)$ V. Going around the mesh on the LHS in the clockwise sense, starting at node 'c', KVL gives $120 - 12I - 8(I + 5) = 0$, so that $I = 4$ A. To determine R, we note that the voltage across the 8 and 24 Ω resistors is the same, since these resistors are in parallel, as noted earlier. The voltage across the 8 Ω resistor is $8(4 + 5) = 72$ V. The current through the 24 Ω resistor is therefore $72/24 = 3$ A. From KCL at node 'b', the current in R is $(5 + 3) = 8$ A. Since R is in parallel with the 12 Ω resistor, the voltage across it is $12I = 48$ V. Hence, $R = 48/8 = 6$ Ω.

Simulation: The circuit is entered as in Figure 2.51. R is entered as 6 Ω, and the simulation is used to verify that a current of 5 A flows from node 'b' to node 'd'. To make PSpice display this current, a 1 μΩ resistor is inserted in place of the short circuit. The value of this resistance is too small to significantly affect the results. 'Bias Point' is selected under 'Analysis type' in the Simulation Settings. After the simulation is run, pressing the I and V buttons displays the currents and voltages indicated in Figure 2.51.

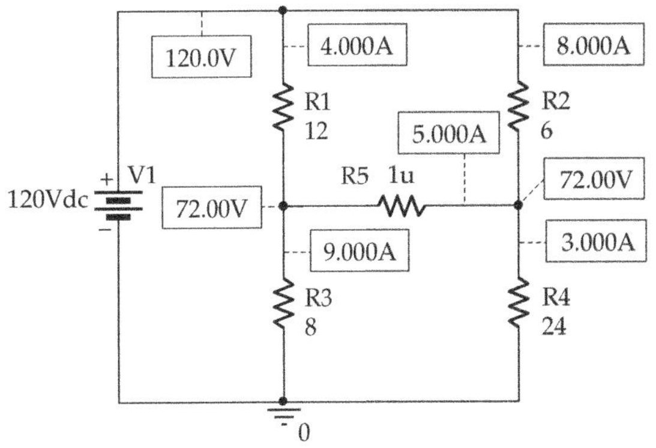

FIGURE 2.51
Figure for Example 2.6.

Problem-Solving Tip

- A circuit with rather awkward-looking connections can be redrawn, after labeling of nodes, for easier visualization of the connections.

Learning Checklist: What Should Be Learned from This Chapter

- Electrical resistance is fundamentally due to impediments to the movement of current carriers in a conductor in the presence of an applied electric field. According to the "collision" model, resistance arises because of repeated collisions between the vibrating crystal atoms and conduction electrons moving under the influence of the applied electric field. The collision model can also account for (1) the increase of resistance with temperature and (2) the heating effect of electric current.
- An ideal resistor is a purely dissipative circuit element that obeys Ohm's law: $v = Ri$, where R, the resistance, is a constant that is independent of current, voltage, time, or temperature. When v is in volts and i is in amperes, R is in ohms.
- Ohm's law can be expressed as $i = Gv$, where $G = 1/R$ is the conductance. When R is in ohms, G is in siemens.
- The current through an ideal resistor is always in the direction of the voltage drop across the resistor, so as to give a positive value of R in the expression for Ohm's law.
- The power dissipated in a resistor is $p = vi = Ri^2 = v^2/R = Gv^2$.

- A short circuit is a connection of zero resistance, or infinite conductance. An open circuit has infinite resistance, or zero conductance.
- An ideal voltage source, whether an independent or a dependent, ideal voltage source, maintains a specified source voltage between its terminals, irrespective of the current through the source. The specified source voltage could be positive, negative, or zero.
  1. Whereas the source voltage is solely that specified for the source, the current through the source depends on both the source voltage and the rest of the circuit to which the voltage source is connected. The source current can be positive, negative, or zero.
  2. The source voltage is in general a function of time, but does not vary with the current through the source.
  3. The ideal voltage source can deliver or absorb power, depending on the relative directions of source voltage and source current.
  4. When the source voltage is zero, the ideal voltage source is equivalent to a short circuit.
- In an ideal, independent voltage source, the source voltage is specified independently of any voltage or current in the circuit.
- An ideal, dependent voltage source behaves exactly like an ideal, independent voltage source, except that the specified source voltage depends on a voltage or a current other than that of the source itself. There are thus two types of dependent voltage sources: a VCVS and a CCVS.
- An ideal current source, whether an independent or a dependent, ideal current source, maintains a specified current through the source, irrespective of the voltage across the source. The specified source current could be positive, negative, or zero.
  1. Whereas the source current is solely that specified for the source, the source voltage depends on the both the source current and the rest of the circuit to which the current source is connected. The source voltage can be positive, negative, or zero.
  2. The source current is in general a function of time, but does not vary with the voltage across the source.
  3. The ideal current source can deliver or absorb power, depending on the relative directions of source voltage and source current.
  4. When the source current is zero, the ideal current source is equivalent to an open circuit.
- In an ideal, independent current source, the source voltage is specified independently of any voltage or current in the circuit.
- An ideal, dependent current source behaves exactly like an independent current source, except that the specified source current depends on a voltage or a current other than that of the source itself. There are thus two types of dependent current sources: a VCCS and a CCCS.
- A node is the connection point between a number of circuit elements. An inessential node is a node between just two circuit elements, whereas an essential node is a node between three or more circuit elements.
- A path is a set of one or more adjoining circuit elements that may be traversed in succession without passing through the same node more than once. A branch is a path that connects two nodes, whereas an essential branch is the set of adjoining circuit elements traversed in going from one essential node to an adjacent essential node, without passing through another essential node.
- A loop is a closed path in a circuit. A mesh is a loop that does not enclose any other loop.
- Although electric circuits can be analyzed based on conservation of current and conservation of power, this is awkward in practice because of the quadratic dependence of power on current or voltage. Kirchhoff's laws (KCL and KVL) are much more convenient to apply because they are linear in current and voltage.
- According to KCL, the sum of currents entering a node at any instant of time is equal to the sum of currents leaving the node at that instant.
  1. KCL is a direct expression of conservation of current.
  2. KCL applies not only to known current values at a node but also to unknown currents at a node having arbitrarily assigned directions.
  3. KCL can also be applied to whole circuits or to any combination of circuit elements that are part of a circuit.
- Voltages along a path add algebraically. This is consistent with the definition of voltage as electric potential energy per unit charge, and the fact that changes in potential energy can be added algebraically.
- According to KVL, the sum of voltage rises around any loop at any instant of time is equal to the sum of voltage drops around the loop at that instant.

1. To minimize the possibility of error in writing KVL around a mesh or a loop, a systematic procedure can be followed in writing KVL.
2. KVL can also be applied to an open path in a circuit to determine the voltage between the two ends of the path.

- KVL and KCL together are an expression of conservation of energy.
- The following are the features of a series connection of circuit elements:
  1. The elements are connected in succession, end to end, without branching at any of the nodes between the elements.
  2. The same current flows through all the elements, so KCL is automatically satisfied.
  3. Voltages add algebraically along the path through series-connected elements.
  4. If one of the series-connected elements is removed from the circuit, the current in the series connection is zero.
- The following are the features of a parallel connection of circuit elements:
  1. One end of each element is connected to a common node, whereas the other end of each of these elements is connected to another common node.
  2. Any two paralleled elements form a mesh or a loop that does not include any additional elements.
  3. The same voltage appears across all the parallel-connected elements, so KVL is automatically satisfied.
  4. Currents add algebraically at the nodes between which the elements are paralleled.
- The series and parallel connections can be used to build resistive circuits of any desired complexity.
- A problem-solving approach, ISDEPIC, can be applied as a series of steps that can be very helpful in analyzing a given circuit and arriving at the solution systematically and efficiently.

Problem-Solving Tips

1. The solution to any circuit problem can be checked by making sure that Ohm's law is satisfied for every resistor, KCL is satisfied at every node, and KVL is satisfied around every mesh.
2. If a circuit has N essential nodes, then after writing KCL for $(N - 1)$ essential nodes, KCL at the remaining essential node should be automatically satisfied if KCL was written correctly at the other nodes.
3. Always mark on the circuit diagram the directions of currents of interest and the polarities of voltages of interest, bearing in mind that the current through an ideal resistor is in the direction of the voltage drop across the resistor.
4. Circuits having only two essential nodes can generally be analyzed by applying KCL at either node.
5. Controlling currents or voltages of dependent sources are often convenient to use as unknown variables in analyzing circuits.
6. A circuit with rather awkward-looking connections can be redrawn, after labeling of nodes, for easier visualization of the connections.

Problems

Apply ISDEPIC and verify solutions by PSpice simulation whenever feasible.

Resistors

P2.1 A 1.5 MΩ resistor is rated at 1/2 W. Determine the maximum voltage that can be applied to the resistor without exceeding its power rating.

Ans. 866.0 V.

P2.2 Four 60 W and 120 V lamps are to be connected in parallel to a 240 V supply, using a resistor R to drop 120 V, so that the voltage across each lamp is 120 V, as illustrated in Figure P2.2. Determine (a) the current of each lamp, (b) the current through R, (c) the value of R, and (d) the power rating of the resistor R.

Ans. (a) 0.5 A; (b) 2 A; (c) 60 Ω; (d) 240 W.

P2.3 The voltage across a resistor is $60\sin 100\pi t$ V when the current through it is $4\sin 100\pi t$ A, in the direction of a voltage drop. Determine (a) the resistance value, (b) the instantaneous power $p(t)$ dissipated in the resistor, and (c) the average power dissipated in the resistor, which is the time integral of $p(t)$ over a period, divided by the period. (d) Is the average power in the resistor equal to the product of the average voltage across the resistor and the average current through it? Explain.

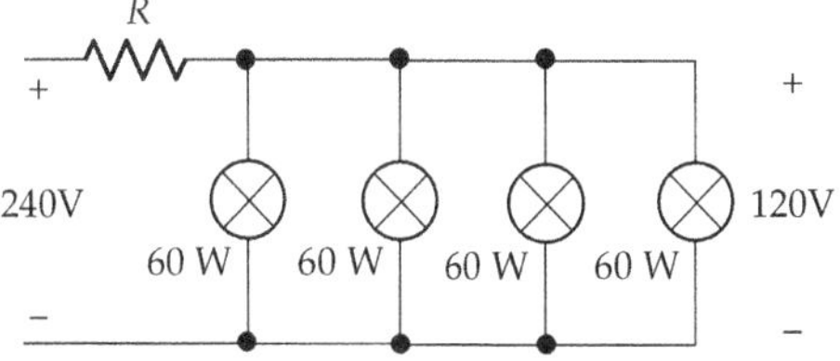

FIGURE P2.2

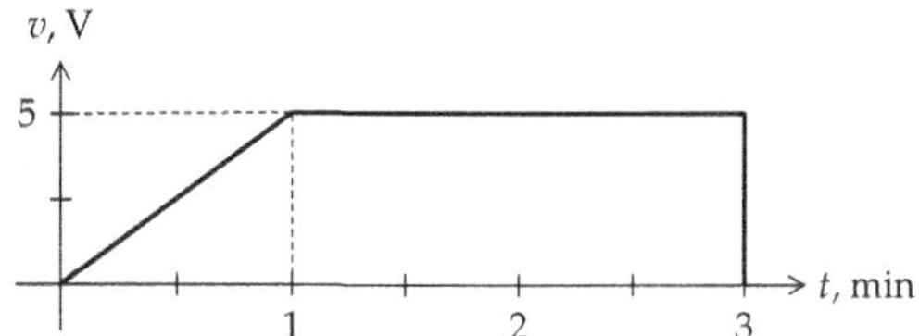

FIGURE P2.4

Ans. (a) 15 Ω; (b) 120(1 − cos200πt) W; (c)120 W; (d) No. The average of the voltage and the current is zero, because they are alternately positive and negative in successive half-cycles. But negative voltage multiplied by negative current gives a positive value of power, so that the power does not average to zero.

P2.4 The voltage shown in Figure P2.4 is applied across a 5 Ω resistor. Determine (a) the resistor current; (b) $p(t)$, $0 \leq t \leq 1$ min; and (c) the energy dissipated in the resistor at $t = 3$ min.

Ans. (a) $t/60$ A, $0 \leq t \leq 60$ s; 1 A, $60 \leq t < 180$ s; 0, $t > 80$ s; (b) $t^2/720$ W, t is in s; (c) 700 J.

P2.5 A voltage $v(t) = 10\cos 100\pi t$ V is applied across a 10 Ω resistor. (a) Sketch $p(t)$. (b) Determine the average power dissipated in the resistor and the energy dissipated during half a cycle of $v(t)$.

Ans. (a) $p(t) = 5(1 + \cos 200\pi t)$ W; (b) 5 W, 0.05 J.

P2.6 The triangular voltage waveform of Figure P2.6 is applied to a 100 Ω resistor. Determine (a) the resistor current, (b) $p(t)$, and (c) the average power dissipation.

Ans. (a) $i(t) = 0.1t$ A, $0 \leq t \leq 1$min, $i(t) = -0.1t + 0.2$A, $1 \leq t \leq 3$min; $i(t) = 0.1t - 0.4$A, $3 \leq t \leq 4$min; (b) $p(t) = \frac{v^2}{R} = t^2$W, $0 \leq t \leq 1$min, $p(t) = \frac{(-10t + 20)^2}{100}$W, $1 \leq t \leq 3$min, $p(t) = \frac{(10t - 40)^2}{100}$W, $3 \leq t \leq 4$min; (c) $\frac{1}{3}$W.

P2.7 The resistance of a copper power line is 60 Ω at 20°C, when not carrying any current. Its resistance, when carrying its rated current, is 70 Ω. Determine the temperature of the conductor under these conditions, assuming that the temperature coefficient of copper is 0.0039/°C.

Ans. 62.7°C.

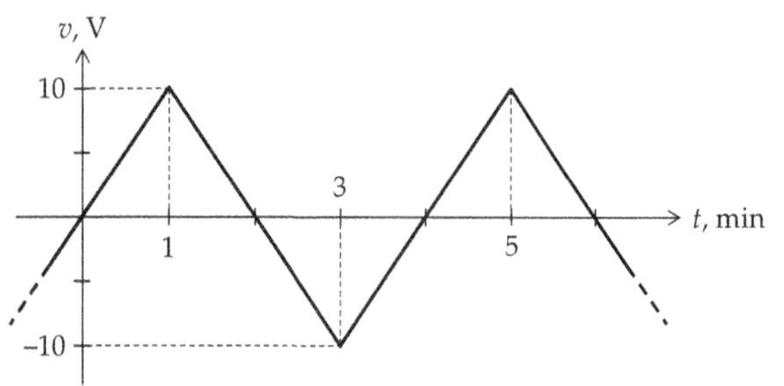

FIGURE P2.6

P2.8 A *pn* junction diode has an exponential *i–v* relation of the form: $i = 10^{-9}\left(e^{20v} - 1\right)$ A, where v is in volts. Determine the diode current for (a) $V = 0.7$ V and (b) $V = -0.7$ V. Note that the *i–v* relation is highly asymmetric.

Ans. (a) 1.20 mA; (b) −1 nA.

Sources and Kirchhoff's Laws

P2.9 Determine the average power delivered or absorbed by the current source in Figure P2.9, assuming $i_{SRC} = 2 + 2\cos 100\pi t$ A.

Ans. 2 W delivered.

P2.10 Determine the voltage across each current source and the current through each voltage source in Figure P2.10.

Ans. 40 V across 5 A source, 15 V across 10 A source, 10 A through 25 V source, 5 A through 40 V source.

P2.11 Determine V_X in Figure P2.11 and the power absorbed or delivered by each source.

Ans. V_X = 40 V; 50 V source delivers 250 W; dependent source absorbs 50 W; 5 A source absorbs 200 W.

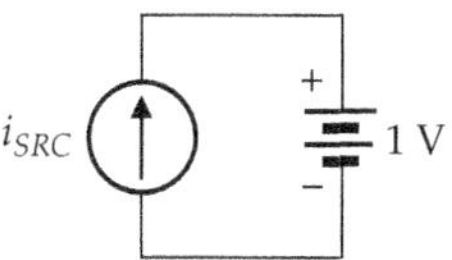

FIGURE P2.9

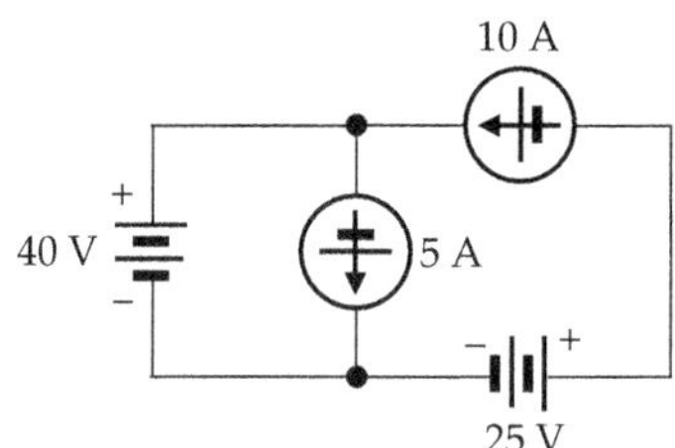

FIGURE P2.10

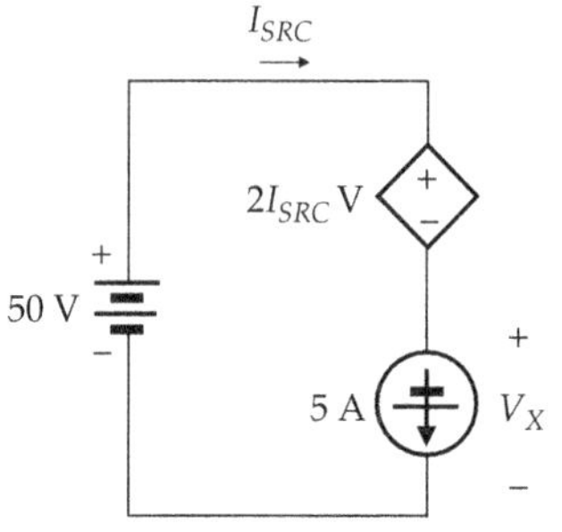

FIGURE P2.11

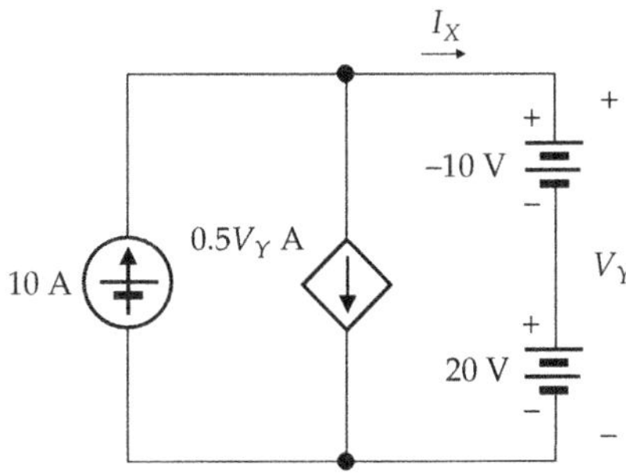

FIGURE P2.12

P2.12 Determine I_X and V_Y in Figure P2.12 and the power absorbed or delivered by each source.

Ans. $I_X = 5$ A; $V_Y = 10$ V; 10 A source delivers 100 W; dependent source absorbs 50 W; −10 V source delivers 50 W; 20 V source absorbs 100 W.

P2.13 Determine in Figure P2.13 the voltage across each current source, the current through each voltage source, and the power delivered or absorbed by each source.

Ans. 5 A though 5 V source and dependent source, 10 A through 10 V source; 45 V across 5 A source, 60 V across 10 A source; 600 W delivered by 10 A source, 25 W absorbed by 5 V source, 225 W absorbed by 5 A source, 100 W absorbed by 10 V source, 250 W absorbed by dependent source.

P2.14 Determine in Figure P2.14 the voltage across each current source, the current through each voltage source, and the power delivered or absorbed by each source.

Ans. $I_X = 16$ A, 6 A through 20 V source, 12 V across VCCS, and 8 V across CCVS; 20 V source delivers 120 W; 10 A source delivers 200 W; VCCS absorbs 192 W; and CCVS absorbs 128 W.

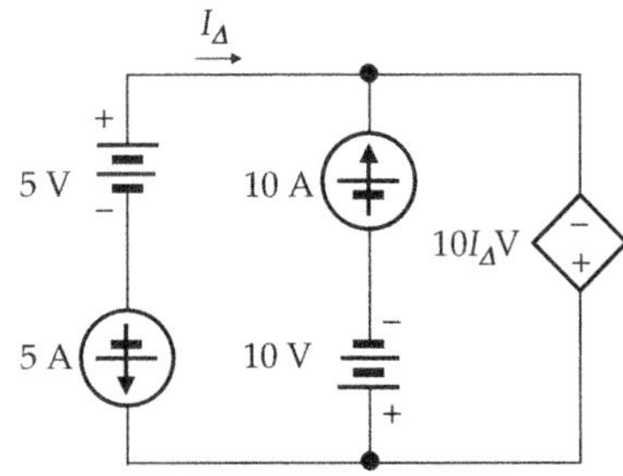

FIGURE P2.13

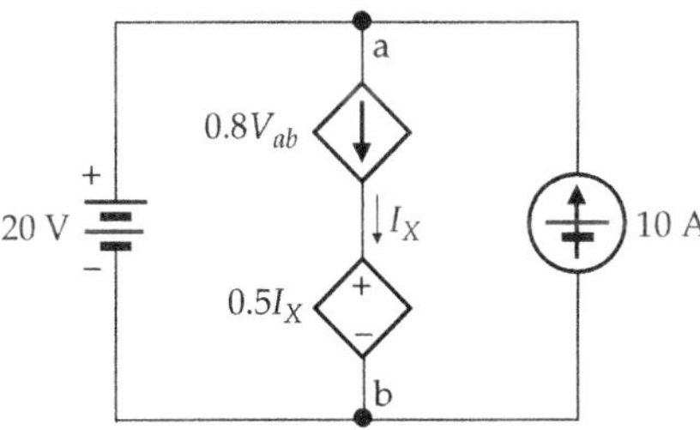

FIGURE P2.14

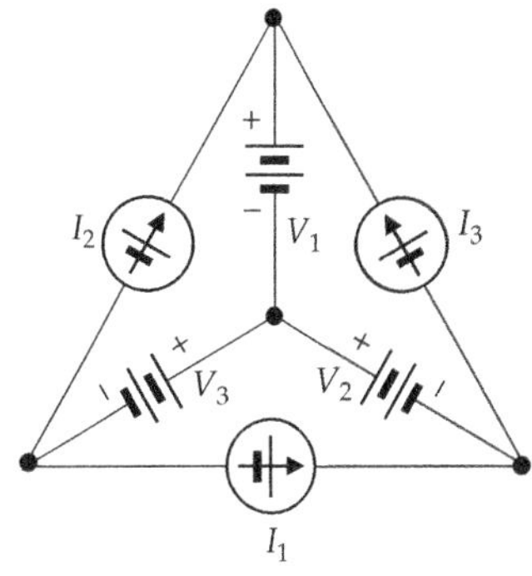

FIGURE P2.15

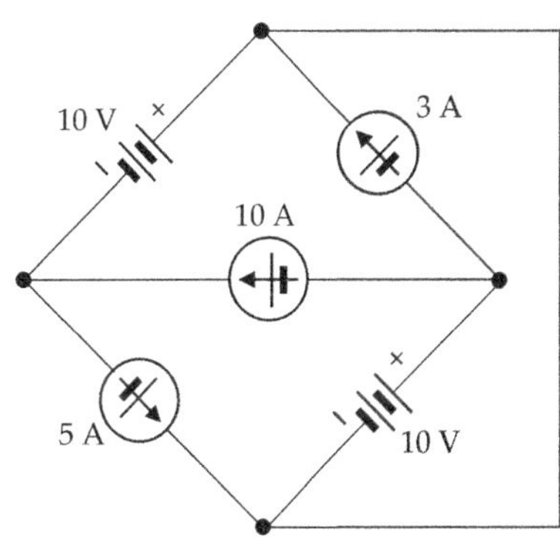

FIGURE P2.16

P2.15 Determine the total power delivered or absorbed by each source in Figure P2.15, assuming the voltage sources are 1 V each, $I_1 = 2$ A, $I_2 = 1$ A, and $I_3 = 1$ A.

Ans. V_1 absorbs 2 W; V_2 delivers 1 W; V_3 absorbs 3 W; I_1 neither absorbs nor delivers power; I_2 delivers 2 W; I_3 delivers 2 W.

P2.16 Determine in Figure P2.16 the voltage across each current source, the current through each voltage source, and the power delivered or absorbed by each source.

Ans. 5 A through upper 10 V source, 13 A through lower 10 V source, 10 V across 3 A source, 20 V across 10 A source, and 10 V across 5 A source. Upper 10 V source delivers 50 W; lower 10 V source delivers 130 W; 5 A source delivers 50 W; 10 A source absorbs 200 W; 3 A source absorbs 30 W.

Resistive Circuits and Kirchhoff's Laws

Avoid introducing additional unknowns whenever possible.

P2.17 Determine V_O in Figure P2.17.

Ans. 0.5 V.

P2.18 Determine I_{SRC} that will make $I_S = 0$ in Figure P2.18.

Ans. 0.5 mA.

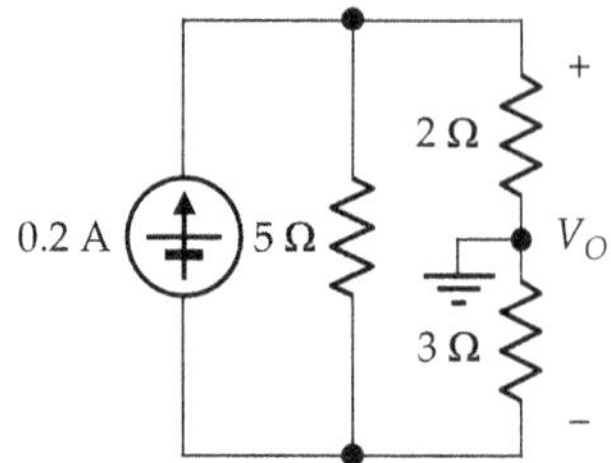

FIGURE P2.17

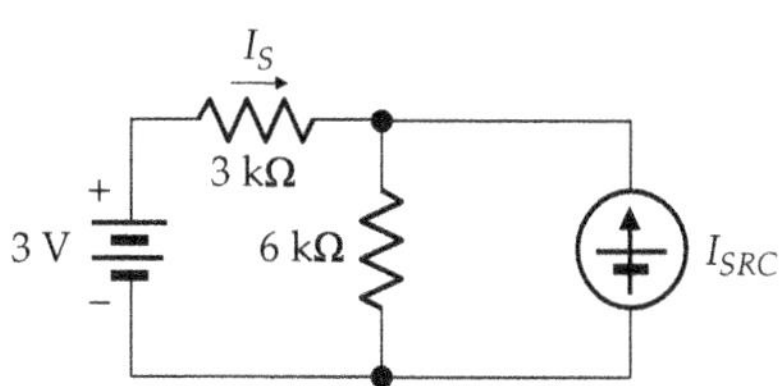

FIGURE P2.18

P2.19 Determine the power delivered or absorbed by each source in Figure P2.19.

Ans. Current source delivers 100 W; voltage source absorbs 60 W.

P2.20 Determine the power delivered or absorbed by each source in Figure P2.20.

Ans. Current source absorbs 120 W; voltage source delivers 300 W.

P2.21 Determine I_X in Figure P2.21.

Ans. 60 A.

P2.22 Determine I_X in Figure P2.22.

Ans. 4.5 A.

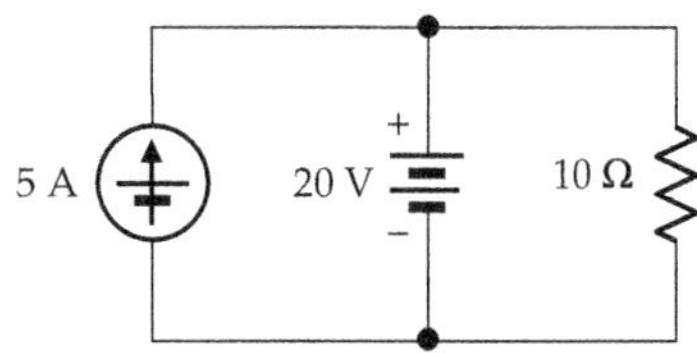

FIGURE P2.19

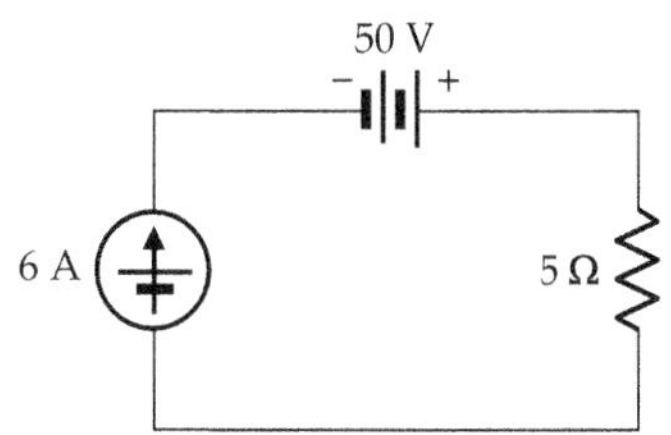

FIGURE P2.20

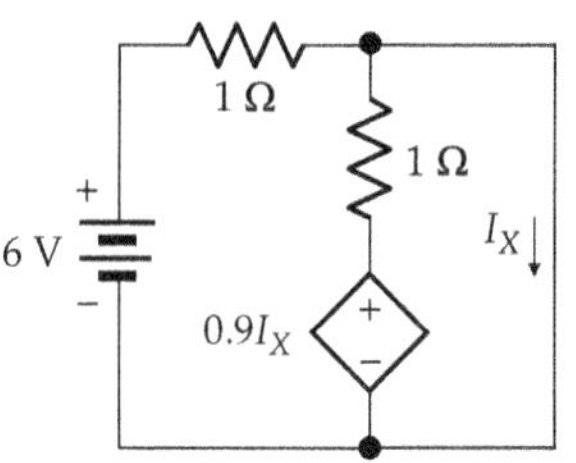

FIGURE P2.21

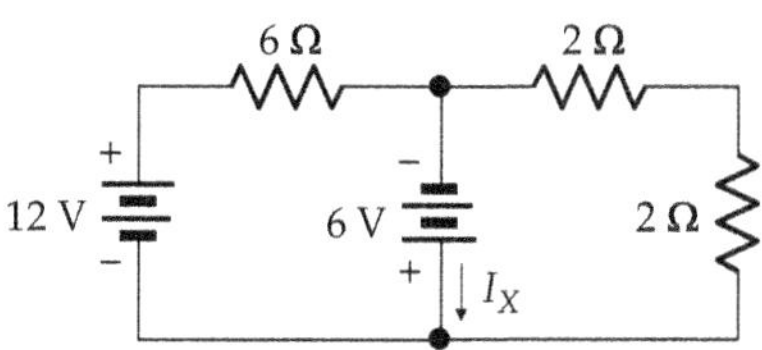

FIGURE P2.22

P2.23 Determine V_X in Figure P2.23, where R_1 and R_2 need not be specified.

Ans. 20 V.

P2.24 Determine R in Figure P2.24.

Ans. 6.25 Ω.

P2.25 Determine the power delivered or absorbed by the current source in Figure P2.25, given that the voltage source does not absorb or deliver power.

Ans. 9 W delivered.

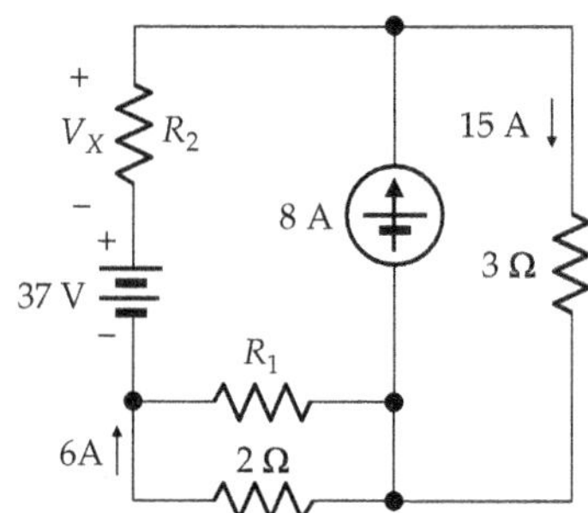

FIGURE P2.23

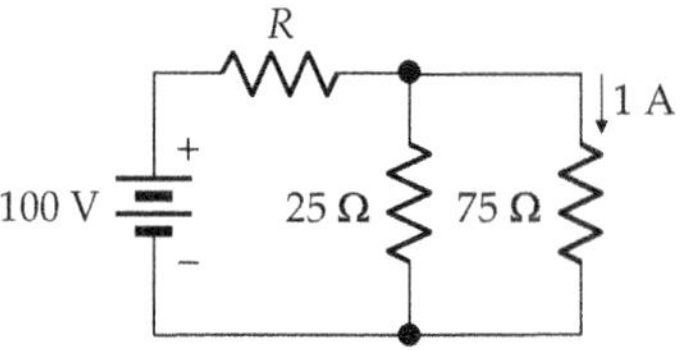

FIGURE P2.24

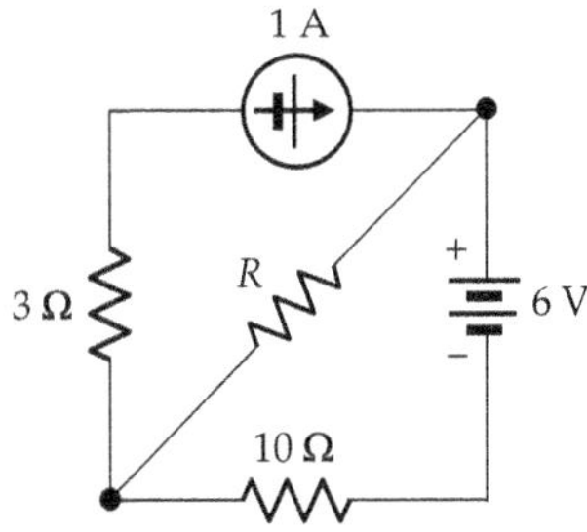

FIGURE P2.25

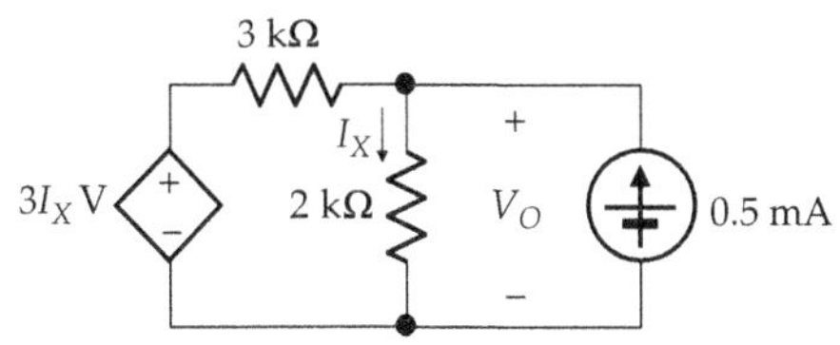

P2.26 Determine the power delivered or absorbed by the voltage source in Figure P2.25, given that the current source does not absorb or deliver power.

Ans. 5.4 W delivered.

P2.27 Determine the power delivered or absorbed by each source and the power absorbed by each resistor, in Figure P2.27.

Ans. Current source delivers 2800 W; voltage source absorbs 1500 W. Power absorbed by 20 Ω resistor is 500 W and that absorbed by 2 Ω resistor is 800 W.

P2.28 Determine the power delivered or absorbed by each source and the power absorbed by each resistor, in Figure P2.28.

Ans. Current source delivers 8000 W; voltage source absorbs 3600 W. Power absorbed by 0.2 Ω resistor is 2000 W and that absorbed by 1.5 Ω resistor is 2400 W.

P2.29 Determine V_O in Figure P2.29.

Ans. 1.5 V.

P2.30 Determine the power delivered or absorbed by each source and the power absorbed by each resistor, in Figure P2.30.

Ans. Independent source delivers 100 W; dependent source absorbs 240 W. Power absorbed by 2.5 Ω resistor is 360 W and that absorbed by 15 Ω resistor is 400 W.

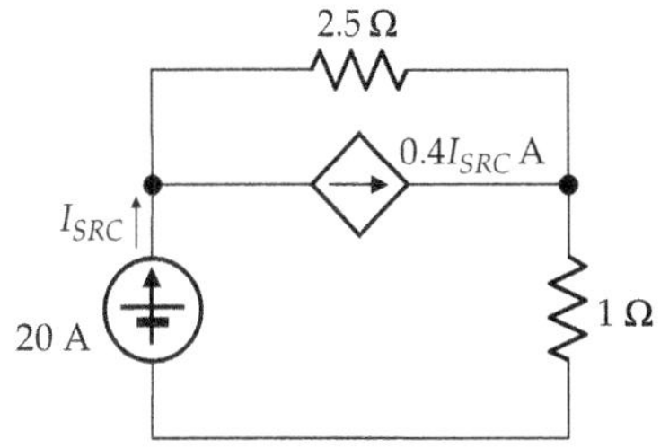

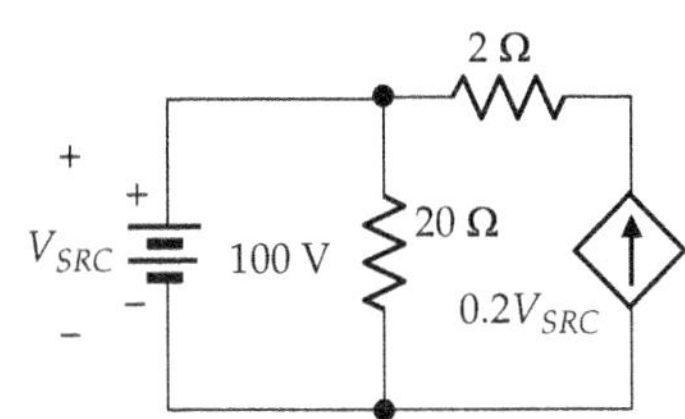

FIGURE P2.27

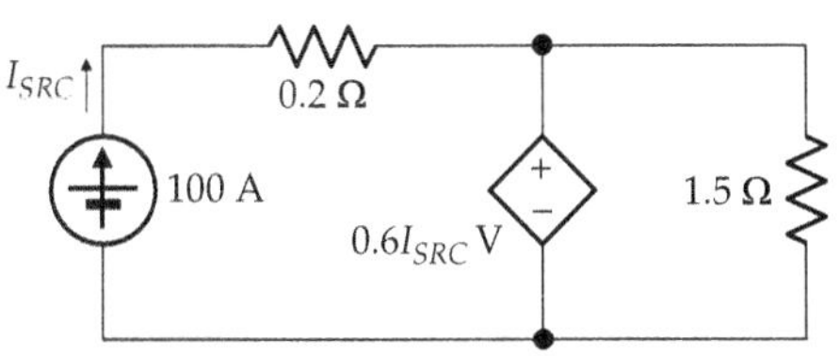

FIGURE P2.28

FIGURE P2.29

FIGURE P2.30

P2.31 Determine the power delivered or absorbed by each source and the power absorbed by each resistor, in Figure P2.31.

Ans. Independent source delivers 600 W; dependent source absorbs 120 W. Power absorbed by the 7.5 Ω resistor is 120 W, and the power absorbed by the 10 Ω resistor is 360 W.

P2.32 Determine the power delivered or absorbed by the independent source in Figure P2.32.

Ans. 0.5 W delivered.

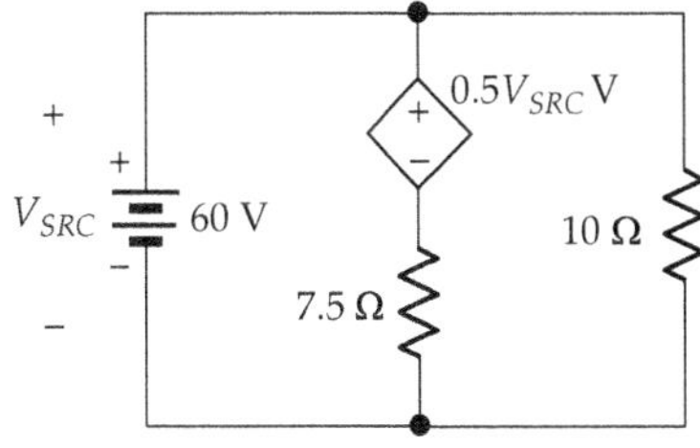

FIGURE P2.31

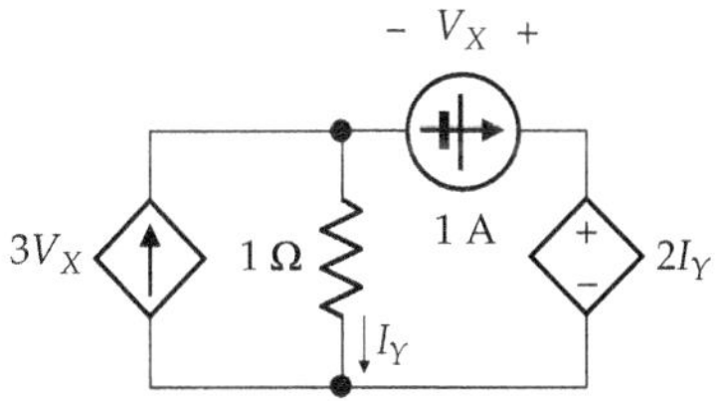

FIGURE P2.32

P2.33 Determine the power delivered or absorbed by the dependent source in Figure P2.33.

Ans. 4 W delivered.

P2.34 Determine V_X and I_S in Figure P2.34.

Ans. 2 V, 24 mA.

P2.35 Determine the power delivered or absorbed by the dependent source in Figure P2.35.

Ans. The source neither delivers nor absorbs power.

P2.36 Determine I in Figure P2.36.

Ans. −6/7 A.

P2.37 Determine I_X in Figure P2.37.

Ans. 2 A.

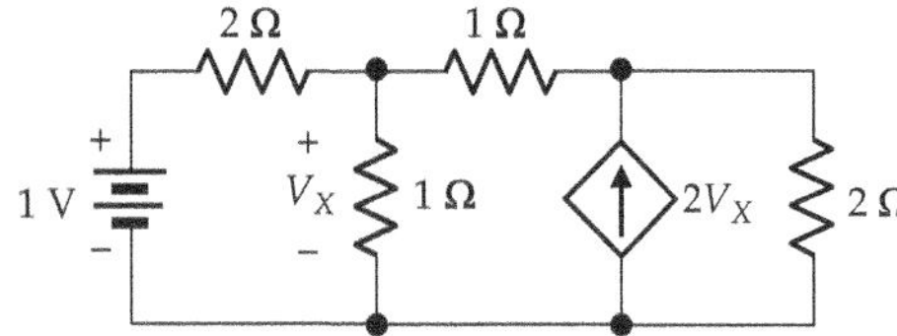

FIGURE P2.33

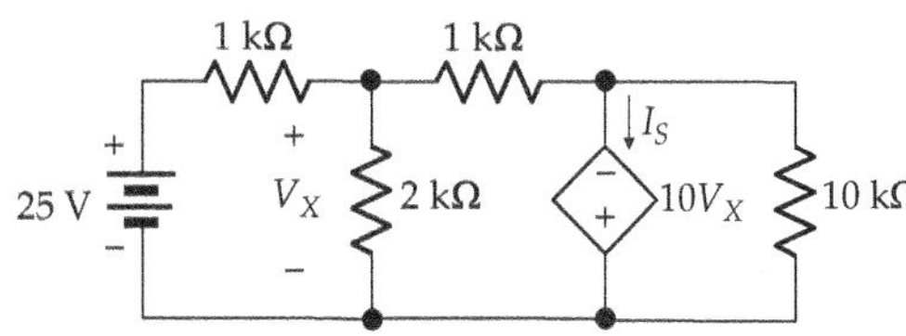

FIGURE P2.34

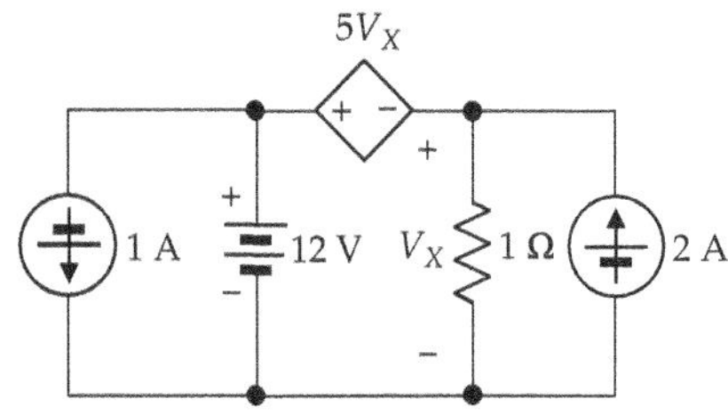

FIGURE P2.35

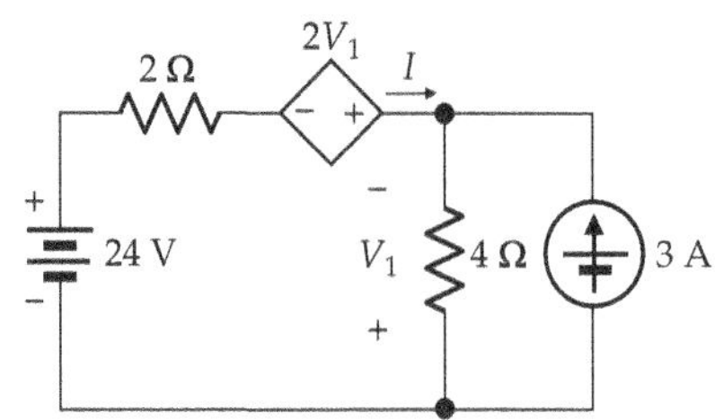

FIGURE P2.36

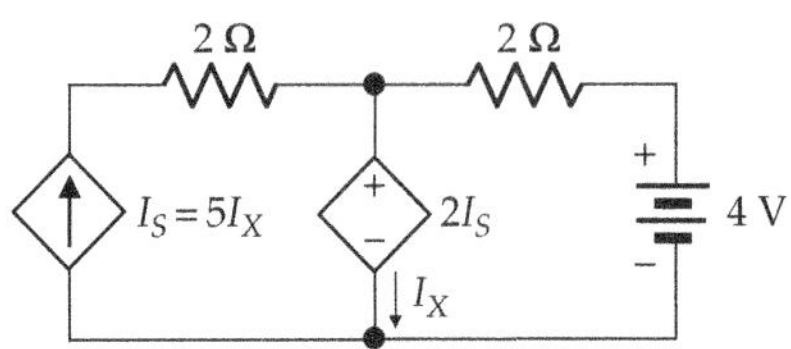

FIGURE P2.37

P2.38 Determine I_X in Figure P2.38.

Ans. 7 A.

P2.39 Determine K in Figure P2.39 so that 50 W is dissipated in the 2 Ω resistor.

Ans. 2.

P2.40 Determine V_{ab} in Figure P2.40.

Ans. 1 V.

P2.41 Determine the power delivered or absorbed by the dependent source in Figure P2.41.

Ans. 30 W absorbed.

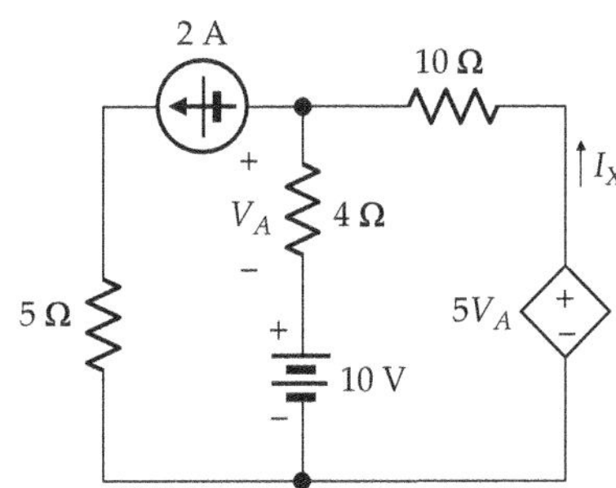

FIGURE P2.38

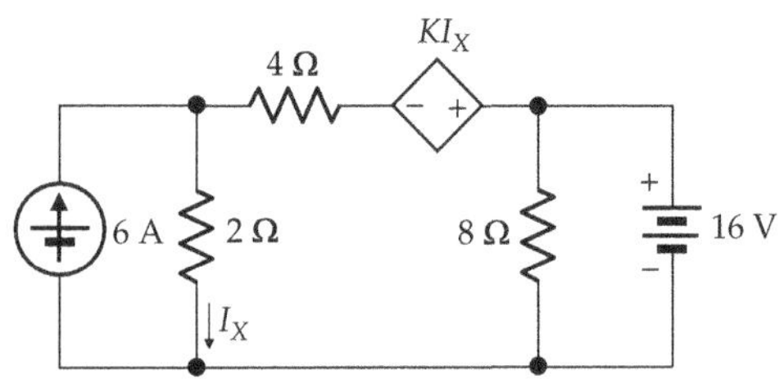

FIGURE P2.39

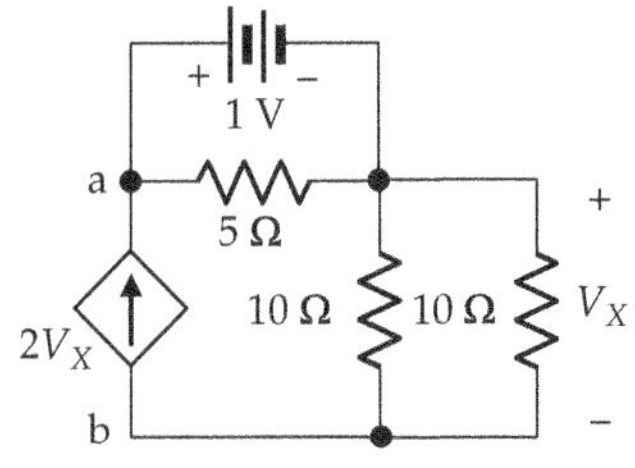

FIGURE P2.40

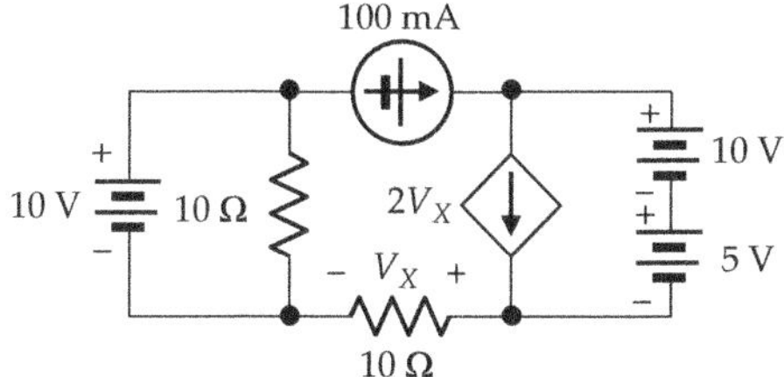

FIGURE P2.41

P2.42 Determine the power dissipated in the 4 Ω resistor in Figure P2.42.

Ans. 100 W.

P2.43 Determine the power delivered or absorbed by the dependent voltage source in Figure P2.43.

Ans. Absorbs 60 W.

P2.44 Determine I_X and V_Y in Figure P2.44.

Ans. 1/3 A, 50/3 V.

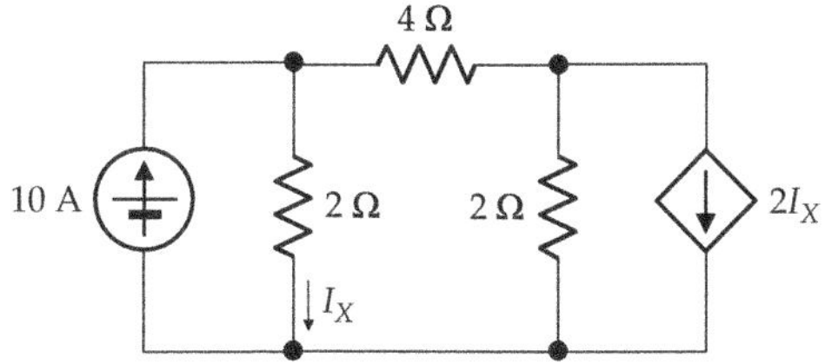

FIGURE P2.42

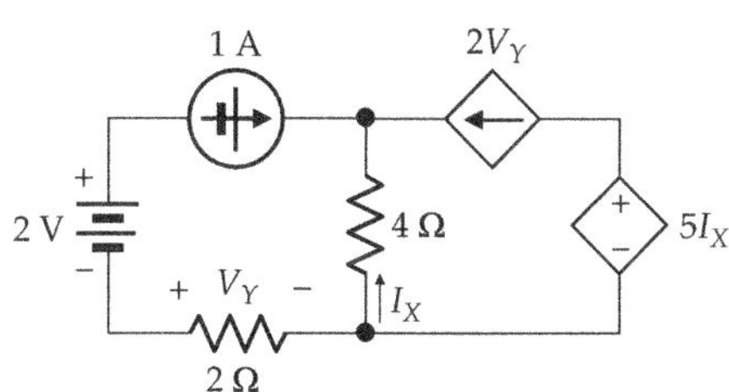

FIGURE P2.43

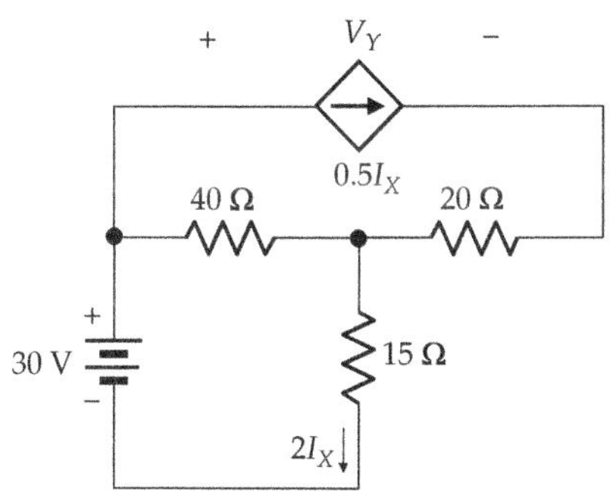

FIGURE P2.44

P2.45 Determine the power delivered or absorbed by the dependent source in Figure P2.45.

Ans. 0.

P2.46 Determine I_X in Figure P2.46.

Ans. 2 A.

P2.47 Determine the power delivered or absorbed by the dependent source in Figure P2.47.

Ans. 14 W delivered.

P2.48 Determine the power delivered or absorbed by each source in Figure P2.48.

Ans. Voltage source delivers 1000 W; current source neither absorbs nor delivers power.

P2.49 Determine V_{bc} in Figure P2.49, assuming all resistances are 1 kΩ.

Ans. 4 V.

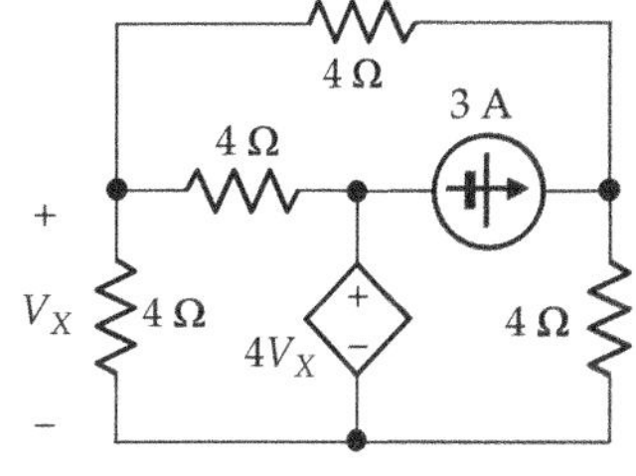

FIGURE P2.45

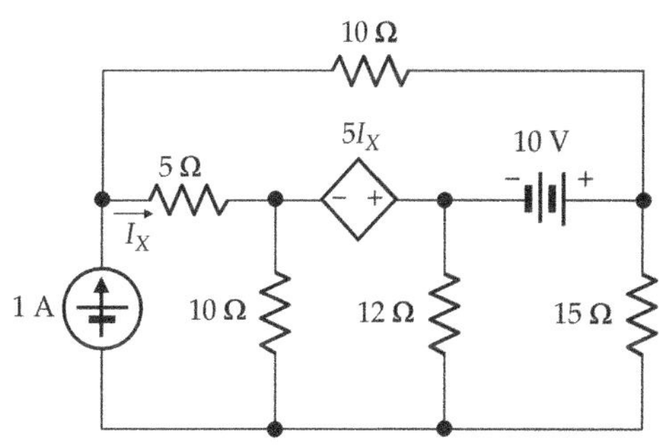

FIGURE P2.46

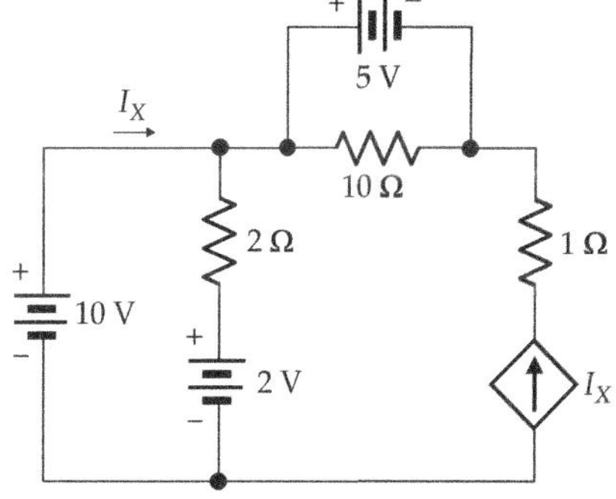

FIGURE P2.47

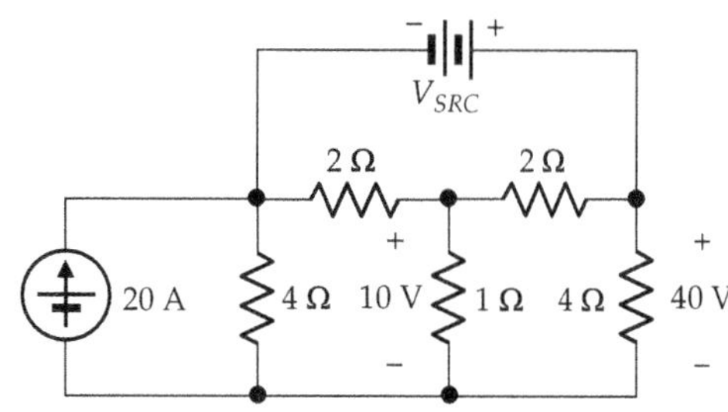

FIGURE P2.48

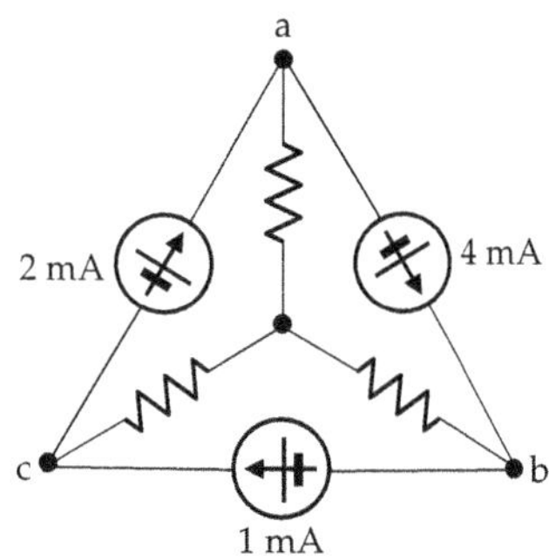

FIGURE P2.49

P2.50 Determine R, V_X, and V_Y in Figure P2.50, given that the current in the top connection is zero.

Ans. 3 Ω, $V_X = 10$ V, $V_Y = -5$ V.

P2.51 Determine I_O in Figure P2.51.

Ans. 1 A.

P2.52 Determine I_X in Figure P2.52.

Ans. 0.75 A.

P2.53 Determine V_X in Figure P2.53.

Ans. 75 V.

P2.54 Determine the ratio ρ/α in Figure P2.54 in terms of R so that $I_1 = I_2$.

Ans. R.

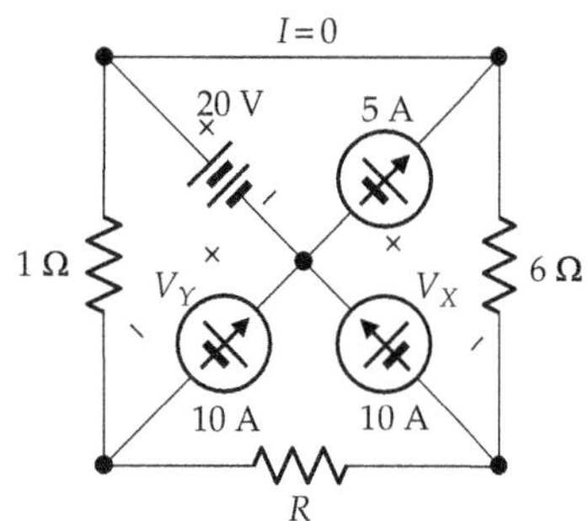

FIGURE P2.50

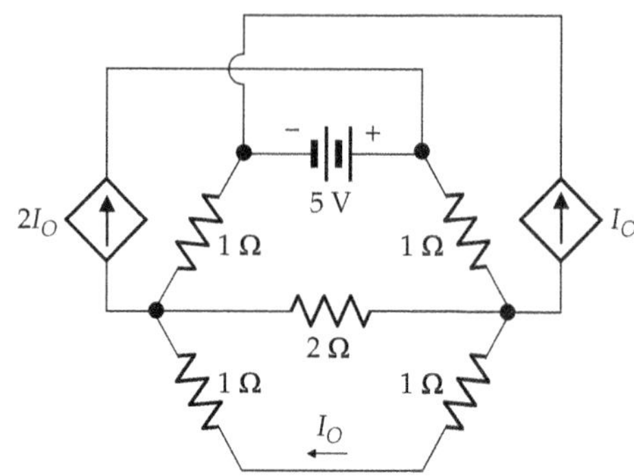

FIGURE P2.51

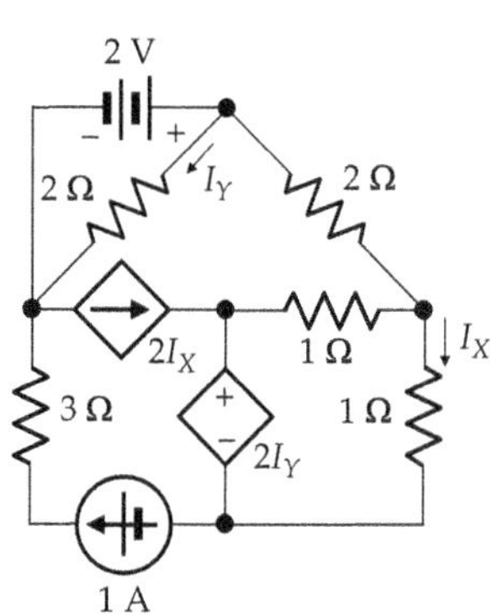

FIGURE P2.52

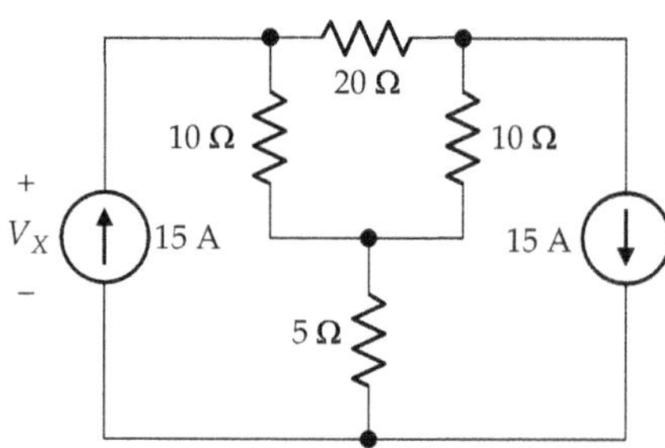

FIGURE P2.53

P2.55 Determine V_X in Figure P2.55.

Ans. −15 V.

P2.56 Determine V_X and I_Y in Figure P2.56.

Ans. $V_X = 0$ V, $I_Y = 3$ A.

P2.57 Determine R in Figure P2.57 so that the two sources deliver the same power.

Ans. 20 Ω.

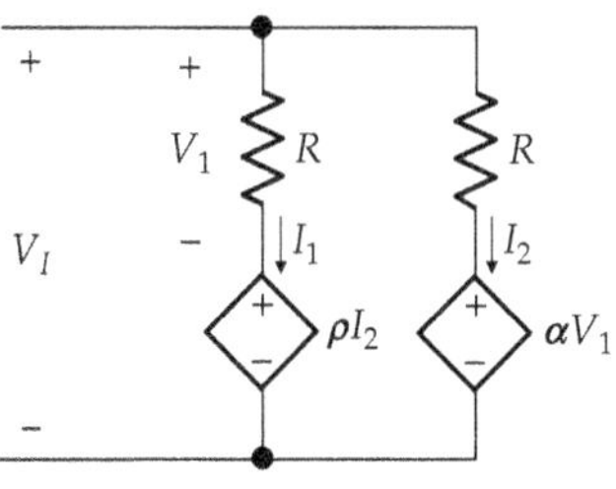

FIGURE P2.54

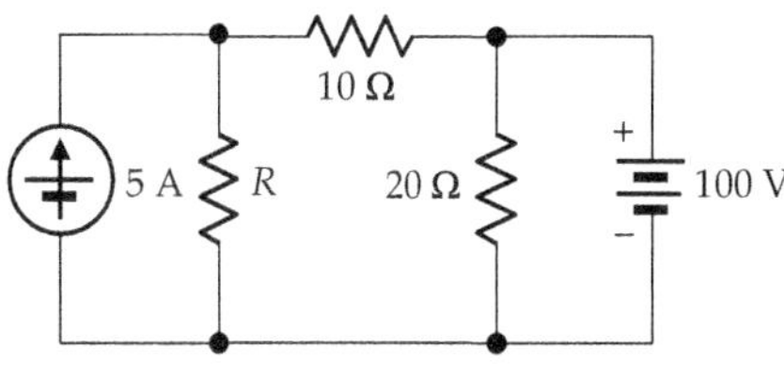

FIGURE P2.57

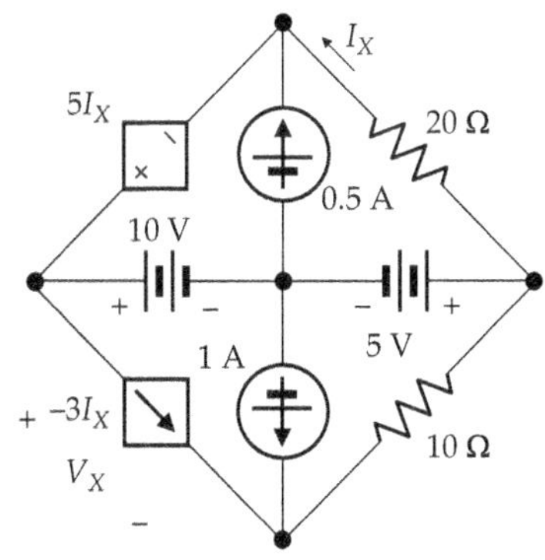

FIGURE P2.55

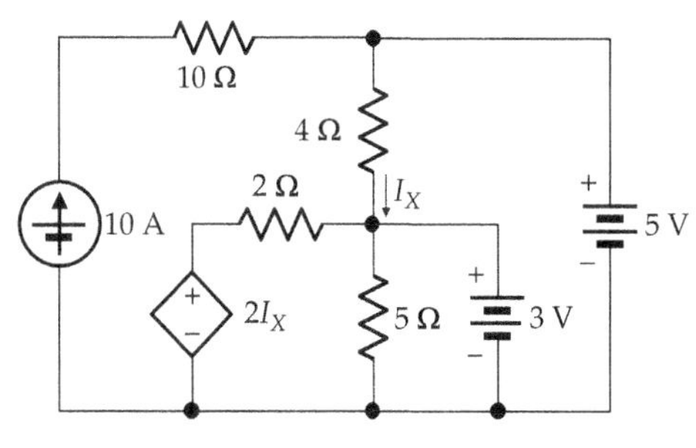

FIGURE P2.58

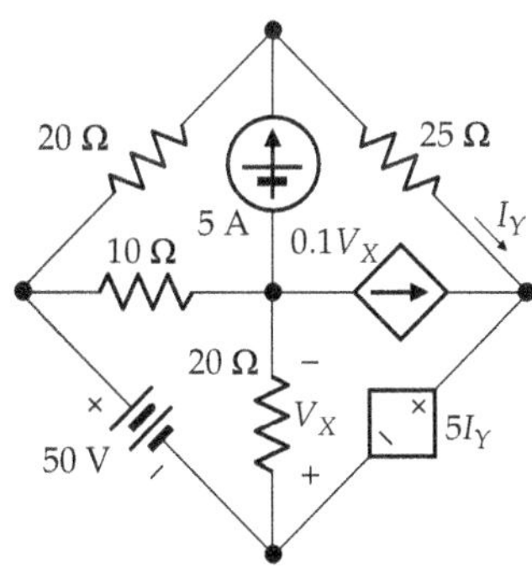

FIGURE P2.56

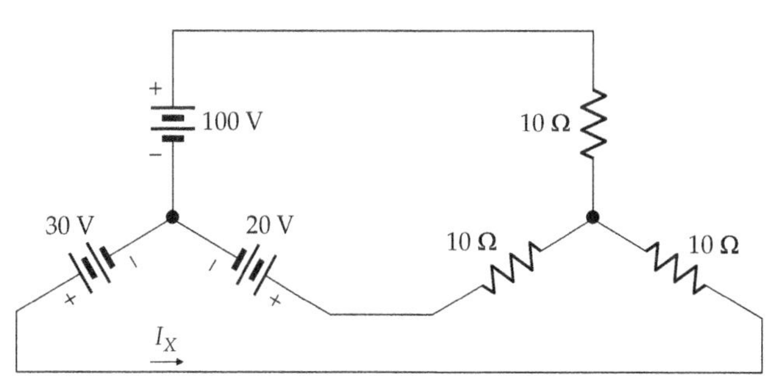

FIGURE P2.59

P2.58 Determine the power delivered or absorbed by the 3 V source in Figure P2.58.

Ans. 3.3 W delivered.

P2.59 Determine I_X in Figure P2.59.

Ans. −2 A.

P2.60 (a) Determine $i_R(t)$, $v_R(t)$, and $v(t)$ in Figure P2.60. (b) Determine $p_2(t)$, the power delivered or absorbed by $i_{SRC2}(t)$, and specify the time intervals over which this source delivers or absorbs power.

Ans. Power is absorbed $-0.7 \le t \le 0.7$ s and is delivered power for $-1 \le t \le -0.7$ s and $1 \ge t \ge 0.7$ s.

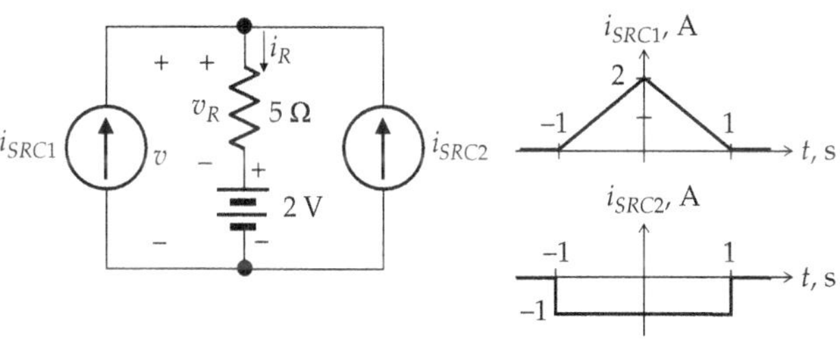

FIGURE P2.60

Probing Further

P2.61 Referring to Figure 2.16, (a) argue that conservation of charge at either of the essential nodes yields the equation: $I_1 + 1.5 = I_2$; (b) deduce from conservation of power and Ohm's law that $9I_1 + 4.5I_2 = 6I_1^2 + 3I_2^2$; and (c) solve these equations to obtain $I_1 = 0.5$ A and $I_2 = 2$ A. Compare with Example 2.4.

P2.62 Consider the circuit of Figure P2.62 having four essential nodes ($N = 4$), three meshes ($M = 3$), and six essential branches ($B = 6$). Remove three branches so that the four essential nodes remain connected by three branches in an open path, without any loops. Clearly, $B_1 = N - 1$, where B_1 is the number of remaining branches, because N will always exceed B_1 by 1 under these conditions. Now add the remaining $(B - B_1)$ branches one at a time, noting that each added branch forms a new loop. Deduce that $B = M + N - 1$. Try this on any other circuit. Note that the result is perfectly general and gives the number of independent KCL equations $(N - 1)$ and the number of independent KVL equations (M) that can be written for the circuit.

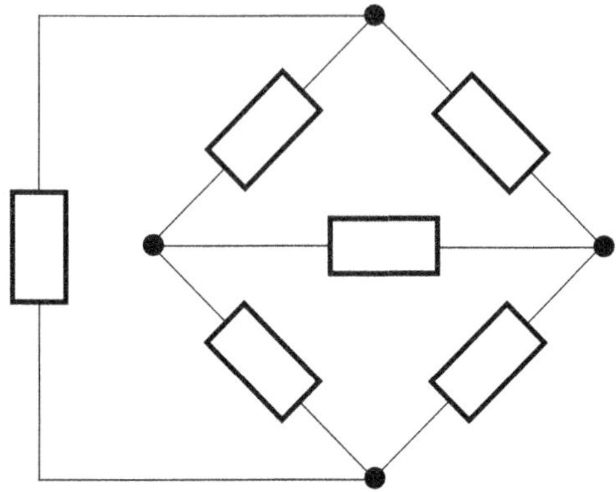

FIGURE P2.62

3

Circuit Equivalence

Objective and Overview

Circuit equivalence is ubiquitous in circuit analysis because of its usefulness, although it is often not explicitly referred to as such. The chapter begins by defining circuit equivalence and examining its general implications. The simple case of series and parallel connection of resistors is considered to begin with and applied to deriving the relation between resistivity and resistance and to the star–delta transformation. The series and parallel connections of ideal sources are contrasted with those of resistors. This leads to a discussion of linear-output sources that have a resistor as an integral part of the source. Circuit equivalence is then applied to deriving the very useful transformation between linear-output, voltage, and current sources.

In addition to highlighting the concept of circuit equivalence, this chapter presents some important deductions concerning circuit behavior.

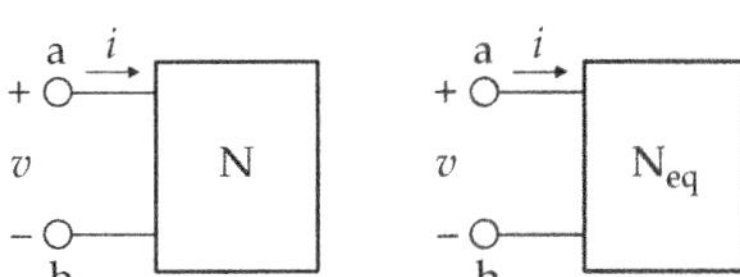

FIGURE 3.1
Equivalent circuits.

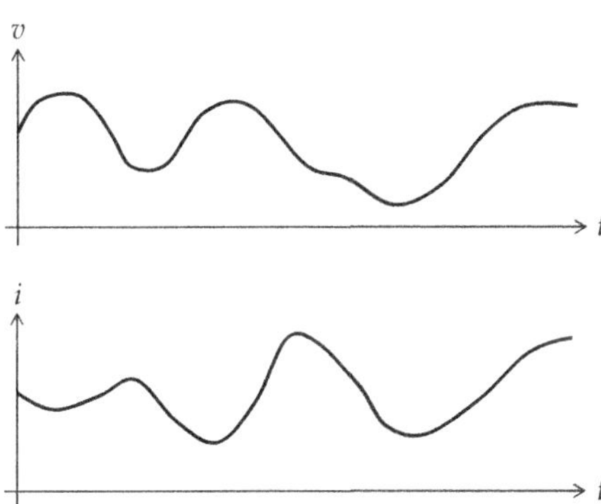

FIGURE 3.2
Time-varying inputs.

3.1 Circuit Equivalence and Its Implications

Definition: *Two circuits are equivalent at a given pair of terminals if the circuits have the same voltage–current relation at these terminals.*

Circuit 'N_{eq}' in Figure 3.1 is equivalent to circuit 'N' at terminals 'ab' if for any arbitrary v applied between terminals 'ab' to the two circuits, the resulting current i at these terminals is the same in both circuits. In other words, the v–i relation at terminals 'ab' is the same for the two circuits. This implies that 'N_{eq}' can be substituted for 'N', terminal for terminal, without affecting v and i at the terminals. Herein lies the usefulness of circuit equivalence, for 'N_{eq}' can be simple enough so that, when substituted for 'N' in a circuit that contains 'N', the analysis of the circuit is considerably facilitated, as will be demonstrated on many occasions.

Circuit equivalence has the following implications:

1. If v is a given function of time (Figure 3.2), then i is in general a different function of time but is the same in both circuits. This follows from having the same v–i relation, for if i is the same in both circuits for a given v, then as v varies with time, i will vary with time in the same way in both circuits.
2. The same power is delivered or absorbed at the terminals of equivalent circuits. This follows from the fact that if v and i are the same at the given terminals, then their product, the instantaneous power, is the same.
3. Just as the product of v and i is the same at the given terminals, their ratio must also be the same. As will be clarified later, the ratio v/i can be interpreted as an **input resistance** R_{in} looking into the given terminals if two conditions are satisfied: (1) the ratio is independent of time, as when v and i are dc quantities, for example, and (2) neither N or N_{eq} contain independent sources or these sources are set to zero.

3.2 Series and Parallel Connection of Resistors

3.2.1 Series Connection of Resistors

Figure 3.3a illustrates a series connection of three resistors to which a test source voltage v_T is applied, resulting in the flow of a test current i_T in the circuit. It is of interest

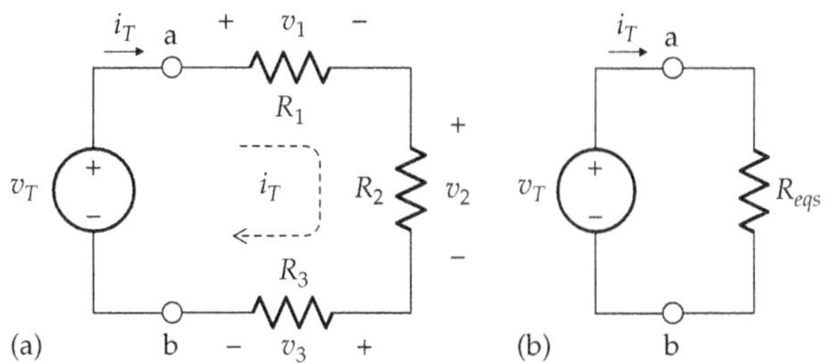

FIGURE 3.3
Series-connected resistors (a) and equivalent series resistor (b).

to determine (1) the equivalent series resistance R_{eqs} between terminals 'ab' and (2) the voltages v_1, v_2, and v_3 across the individual resistors in terms of v_T.

To analyze the circuit, we note that KCL is automatically satisfied by the series connection (Section 2.9). It remains to satisfy KVL and Ohm's law. Starting from node 'b' and going clockwise, KVL gives

$$v_T - v_1 - v_2 - v_3 = 0$$

or

$$v_T = v_1 + v_2 + v_3 \tag{3.1}$$

Substituting from Ohm's law for the voltage across each resistor,

$$v_T = R_1 i_T + R_2 i_T + R_3 i_T$$

or

$$v_T = (R_1 + R_2 + R_3) i_T \tag{3.2}$$

From the definition of circuit equivalence, the equivalent series resistance R_{eqs} is such that if the same voltage v_T is applied at the terminals of the equivalent resistor as at terminals 'ab' of the series combination, the same current i_T flows through the resistor (Figure 3.3b). That is,

$$v_T = R_{eqs} i_T \tag{3.3}$$

Comparing Equations 3.2 and 3.3, it is seen that

$$R_{eqs} = R_1 + R_2 + R_3 \tag{3.4}$$

If each of the resistances in Equation 3.4 is replaced by its reciprocal conductance,

$$\frac{1}{G_{eqs}} = \frac{1}{G_1} + \frac{1}{G_2} + \frac{1}{G_3} \tag{3.5}$$

Although derived for the case of three resistors, Equations 3.4 and 3.5 can be readily generalized to any number of resistors in series. In particular, the following applies:

1. If n identical resistors, each of resistance R, are connected in series, $R_{eqs} = nR$.
2. Since the resistances have positive values, it is evident from Equation 3.4 that R_{eqs} *is larger than the largest of the series-connected resistances.*

Summary: *In a series connection of resistors, (1) the equivalent series resistance is the sum of the individual resistances, (2) the equivalent series resistance is larger than the largest individual resistance, and (3) the reciprocal of the equivalent series conductance is the sum of the reciprocals of the individual conductances.*

Applying Ohm's law to any of the individual resistors, say, R_1,

$$v_1 = R_1 i_T \tag{3.6}$$

Dividing Equation 3.6 by Equation 3.2,

$$\frac{v_1}{v_T} = \frac{R_1}{R_1 + R_2 + R_3} \tag{3.7}$$

Similarly,

$$\frac{v_2}{v_T} = \frac{R_2}{R_1 + R_2 + R_3} \tag{3.8}$$

and

$$\frac{v_3}{v_T} = \frac{R_3}{R_1 + R_2 + R_3} \tag{3.9}$$

In words, the ratio of the voltage across any of the individual resistors to the total voltage across the series combination is the same as the ratio of the resistance in question to the total resistance of the series combination.

If any two of Equations 3.7 through 3.9 are divided by one another, v_T and $(R_1 + R_2 + R_3)$ cancel out so that

$$\frac{v_1}{v_2} = \frac{R_1}{R_2}, \quad \frac{v_1}{v_3} = \frac{R_1}{R_3}, \quad \frac{v_2}{v_3} = \frac{R_2}{R_3} \tag{3.10}$$

In words, the ratio of the voltages across any two resistors is the same as the ratio of the resistances. This also follows directly from Ohm's law, in that if the same current flows through any two resistors, then the ratio of the voltages across each resistor must be the same as the ratio of the resistances.

Summary: *In a series connection of resistors to which a voltage v is applied and the voltage across any of the individual* *resistors R_j is denoted by v_j, the ratio of v_j to v is the same as the ratio of R_j to the total series resistance.*

This is a useful result known as **voltage division**. It also follows from circuit equivalence that the same power $p = vi$ is dissipated in R_{eqs} as in the series combination.

Example 3.1: Series-Connected Resistors

Consider three resistors of 1, 2, and 3 Ω connected in series. It is required to (a) determine R_{eqs}, (b) derive the conductance of each resistor and G_{eqs}, (c) verify that the reciprocal of G_{eqs} is R_{eqs}, (d) determine i_T assuming $v_T = 3$ V, (e) verify that the power dissipated in R_{eqs} is the sum of the powers dissipated in each of the three resistors, and (f) determine v_1, v_2, and v_3 using both forms of Ohm's law (Equations 2.1 and 2.2) and Equations 3.7 through 3.9. Verify that the voltages are in the ratio of the resistances.

Solution:

(a) Let $R_1 = 1\ \Omega$, $R_2 = 2\ \Omega$, and $R_3 = 3\ \Omega$. It follows from Equation 3.4 that $R_{eqs} = 1 + 2 + 3 = 6\ \Omega$.

(b) $G_1 = 1/1 = 1$ S, $G_2 = 1/2 = 0.5$ S, and $G_3 = 1/3$ S, $\frac{1}{1}+\frac{1}{1/2}+\frac{1}{1/3}=\frac{1}{G_{eqs}}$; $6=\frac{1}{G_{eqs}}$; $G_{eqs} = 1/6$ S.

(c) $1/G_{eqs} = 6\ \Omega = R_{eqs}$. For a given series connection of resistors, the total resistance should be the same whether expressed in terms of resistance or conductance.

(d) Applying Ohm's law to R_{eqs}, $3 = 6i_T$, so $i_T = 0.5$ A.

(e) The power dissipated in R_{eqs} is $\frac{(3)^2}{6} = 1.5$ W $= 6 \times (0.5)^2$. The powers dissipated in R_1, R_2, and R_3 are, respectively, $1 \times (0.5)^2 = 0.25$ W, $2 \times (0.5)^2 = 0.5$ W, and $3 \times (0.5)^2 = 0.75$ W. The total power is 1.5 W, which is the same as the power dissipated in R_{eqs}.

(f) $$v_1 = 1\Omega \times 0.5\,\text{A} = \frac{0.5\,\text{A}}{1\text{S}} = 0.5\,\text{V} = \frac{1}{1+2+3}\times 3\,\text{V};$$

$$v_2 = 2\Omega \times 0.5\,\text{A} = \frac{0.5\,\text{A}}{(1/2)\text{S}} = 1\,\text{V} = \frac{2}{1+2+3}\times 3\,\text{V};$$

$$v_3 = 3\Omega \times 0.5\,\text{A} = \frac{0.5\,\text{A}}{(1/3)\text{S}} = 1.5\,\text{V} = \frac{3}{1+2+3}\times 3\,\text{V};$$

$$v_1 + v_2 + v_3 = 3\,\text{V} = v_T.$$

It is seen that $v_2 = 2v_1$, $v_3 = 2v_1 = 1.5v_2$, in the same ratio as the resistances.

Simulation: The circuit is entered as in Figure 3.4, assuming that $v_T = 3$ V. After selecting 'Bias Point' under 'Analysis type', in the Simulation Settings and running the simulation, pressing the I, V, and W buttons displays the currents, voltages, and powers dissipated, as indicated in Figure 3.4.

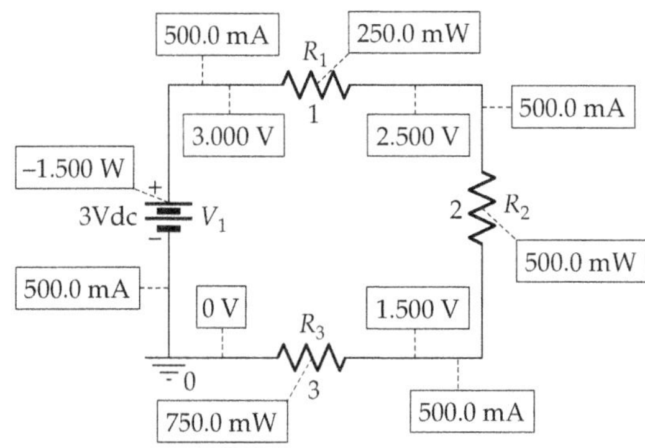

FIGURE 3.4
Figure for Example 3.1.

Primal Exercise 3.1

Determine R_{eqs}, G_1, G_2, G_3, and G_{eqs} for a series connection of $R_1 = 5\ \Omega$, $R_2 = 20\ \Omega$, and $R_3 = 25\ \Omega$.

Ans. $R_{eqs} = 50\ \Omega$, $G_1 = 0.2$ S, $G_2 = 0.05$ S, $G_3 = 0.04$ S, $G_{eqs} = 0.02$ S.

Primal Exercise 3.2

If a fourth resistor of 4 Ω is added to the resistors of Example 3.1, determine (a) R_{eqs}, (b) G_4 and G_{eqs}, (c) i_T assuming $v_T = 6$ V, and (d) the voltages across the individual resistors.

Ans. (a) 10 Ω; (b) $G_4 = 0.25$ S, 0.1 S; (c) 0.6 A; (d) $v_1 = 0.6$ V, $v_2 = 1.2$ V, $v_3 = 1.8$ V, $v_4 = 2.4$ V.

3.2.2 Parallel Connection of Resistors

Figure 3.5a illustrates a parallel connection of three resistors to which a test source current i_T is applied, resulting in a test voltage v_T across the parallel combination. It is of interest to determine (1) the equivalent parallel

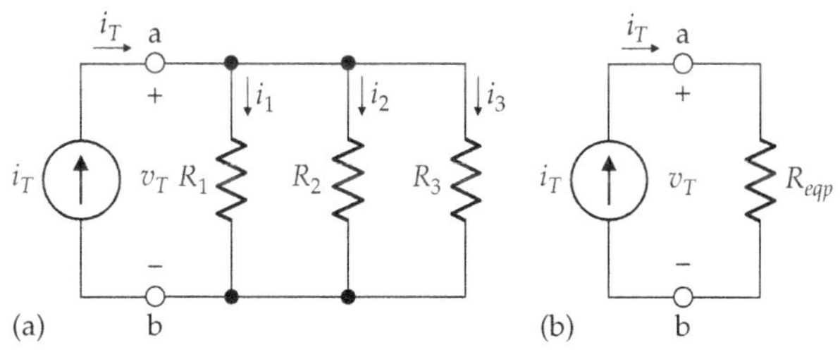

FIGURE 3.5
Parallel-connected resistors (a) and equivalent parallel resistor (b).

resistance R_{eqp} and (2) the currents i_1, i_2, and i_3 through the individual resistors in terms of i_T.

To analyze the circuit, we note that KVL is automatically satisfied by the parallel connection (Section 2.9). It remains to satisfy KCL and Ohm's law. KCL at either essential node gives

$$i_T = i_1 + i_2 + i_3 \quad (3.11)$$

Substituting from Ohm's law for the current through each resistor,

$$i_T = G_1 v_T + G_2 v_T + G_3 v_T$$

or

$$i_T = (G_1 + G_2 + G_3) v_T \quad (3.12)$$

where$G_1 = 1/R_1$, $G_2 = 1/R_2$, and $G_3 = 1/R_3$.

From the definition of circuit equivalence, the equivalent parallel resistance R_{eqp} is such that if the same current i_T flows at the terminals of the equivalent resistor as at terminals 'ab' of the parallel combination, the same voltage v_T appears across the resistor (Figure 3.3b). That is,

$$i_T = G_{eqp} v_T \quad (3.13)$$

where $G_{eqp} = 1/R_{eqp}$ in Figure 3.5b. Comparing Equations 3.12 and 3.13, it follows that

$$G_{eqp} = G_1 + G_2 + G_3 \quad (3.14)$$

If each of the conductances in Equation 3.14 is replaced by its reciprocal resistance,

$$\frac{1}{R_{eqp}} = \frac{1}{R_1} + \frac{1}{R_2} + \frac{1}{R_3} \quad (3.15)$$

Although derived for the case of three resistors, Equations 3.14 and 3.15 can be readily generalized to any number of resistors in parallel. In particular, the following applies:

1. If n identical resistors, each of resistance R, are connected in parallel, $R_{eqp} = R/n$.
2. Since the resistances have positive values, it is seen from Equation 3.15 that $1/R_{eqp}$ is larger than the largest reciprocal of the individual resistances, which is the reciprocal of the smallest resistance, say, R_1. If $1/R_{eqp} > 1/R_1$, then $R_{eqp} < R_1$, that is, R_{eqp} *is smaller than the smallest paralleled resistance.*

Summary: *In a parallel connection of resistors, (1) the equivalent parallel conductance is the sum of the individual conductances, (2) the reciprocal of the equivalent parallel resistance is the sum of the reciprocals of the individual resistances, and (3) the equivalent parallel resistance is smaller than the smallest individual resistance.*

Applying Ohm's law to any of the individual resistors, say, R_1,

$$i_1 = G_1 v_T \quad (3.16)$$

Dividing Equation 3.16 by Equation 3.11,

$$\frac{i_1}{i_T} = \frac{G_1}{G_1 + G_2 + G_3} \quad (3.17)$$

Similarly,

$$\frac{i_2}{i_T} = \frac{G_2}{G_1 + G_2 + G_3} \quad (3.18)$$

and

$$\frac{i_3}{i_T} = \frac{G_3}{G_1 + G_2 + G_3} \quad (3.19)$$

In words, the ratio of the current through any of the individual resistors to the total current through the parallel combination is the same as the ratio of the conductance in question to the total conductance of the parallel combination.

If any two of Equations 3.17 and 3.19 are divided by one another, i_T and $(G_1 + G_2 + G_3)$ cancel out so that

$$\frac{i_1}{i_2} = \frac{G_1}{G_2}, \quad \frac{i_1}{i_3} = \frac{G_1}{G_3}, \quad \frac{i_2}{i_3} = \frac{G_2}{G_3} \quad (3.20)$$

In words, the ratio of the currents through any two resistors is the same as the ratio of the conductances. This also follows directly from Ohm's law, $i = Gv$, in that if the same voltage appears across any two resistors, then the ratio of the currents through each resistor must be the same ratio as the conductances.

Summary: *In a parallel connection of resistors in which the total current is i and the current through any of the individual resistors R_j is denoted by i_j, the ratio of i_j to i is the same as the ratio of $G_j = 1/R_j$ to the total parallel conductance.*

This is a useful result known as **current division**. It also follows from circuit equivalence that the same power $p = vi$ is dissipated in R_{eqp} as in the parallel combination.

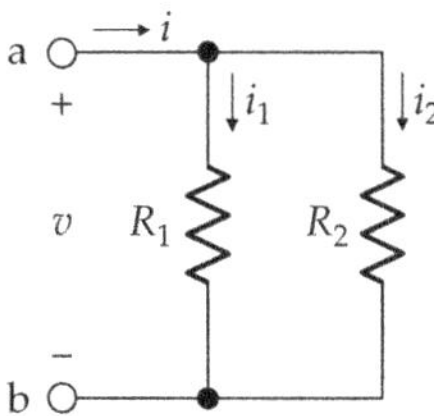

FIGURE 3.6
Two paralleled resistors.

The case of two paralleled resistors (Figure 3.6) is of special interest, as it often occurs in practice. Equation 3.15 reduces to

$$\frac{1}{R_{eqp}} = \frac{1}{R_1} + \frac{1}{R_2} \tag{3.21}$$

or

$$R_{eqp} = R_1 \| R_2 = \frac{R_1 R_2}{R_1 + R_2} \tag{3.22}$$

where the parallel lines || denote a parallel connection. In words, the equivalent resistance of two paralleled resistors is the product of the resistances divided by their sum. When R_{eqp} of more than two resistors is to be calculated manually, it is usually more convenient to apply Equation 3.22 to two resistances at a time or to use Equation 3.14. On the other hand, for calculations using computer programs, it is often more convenient to use a generalization of Equation 3.15 (Problem P3.65).

From Ohm's law applied in Figure 3.6,

$$v = R_1 i_1 = R_2 i_2 = R_{eqp} i = \frac{R_1 R_2}{R_1 + R_2} i \tag{3.23}$$

It follows that

$$\frac{i_1}{i} = \frac{R_2}{R_1 + R_2}, \quad \frac{i_2}{i} = \frac{R_1}{R_1 + R_2}, \quad \frac{i_1}{i_2} = \frac{R_2}{R_1} \tag{3.24}$$

These relations also follow from Equations 3.17 through 3.19 by setting $G_3 = 0$ and replacing each of G_1 and G_2 by its reciprocal resistance R_1 and R_2, respectively. Note that the ratio of either branch current to the total current is the same as the ratio of the resistance in the *other* branch to the sum of the two resistances. Another way of remembering this when the numerical values of R_1 and R_2 are given is to bear in mind that the *smaller current flows in the branch having the larger resistance* in order to equalize the voltage across the two resistors.

Example 3.2: Parallel-Connected Resistors

Consider three resistors of 2, 3, and 6 Ω connected in parallel. It is required to (a) determine R_{eqp} by applying (i) Equation 3.22 first to the 3 and 6 Ω resistors and then to the 2 Ω resistor and (ii) Equation 3.15 in the form $R_{eqp} = R_1R_2R_3/(R_1R_2 + R_2R_3 + R_3R_1)$, (b) derive the conductance of each resistor and G_{eqp}, (c) verify that the reciprocal of G_{eqp} is R_{eqp}, (d) determine v_T assuming i_T = 3 A, (e) verify that the power dissipated in R_{eqp} is the sum of the powers dissipated in each of the three resistors, and (f) determine i_1, i_2, and i_3 using Ohm's law and verify Equations 3.17 through 3.20.

Solution:

(a) Let $R_1 = 2$ Ω, $R_2 = 3$ Ω, and $R_3 = 6$ Ω; (i) the parallel resistance of 3 and 6 Ω is, from Equation 3.22, $\frac{3\times 6}{3+6} = 2$ Ω. This in parallel with 2 Ω gives $R_{eqp} = 1$ Ω;
(ii) $R_{eqp} = \frac{2\times 3\times 6}{2\times 3 + 3\times 6 + 6\times 2} = 1$ Ω.

(b) The conductances are $G_1 = 1/2$ S, $G_2 = 1/3$ S, and $G_3 = 1/6$ S, $G_{eqp} = 1$ S.

(c) $1/G_{eqp} = 1\ \Omega = R_{eqp}$. For a given parallel connection of resistors, the parallel resistance should be the same whether expressed in terms of resistance or conductance.

(d) Applying Ohm's law to R_{eqp}, $v_T = 1 \times 3 = 3$ V.

(e) The power dissipated in R_{eqp} is $\frac{(3)^2}{1} = 9\text{ W} = 1 \times (3)^2$.
The powers dissipated in R_1, R_2, and R_3 are, respectively, $\frac{(3)^2}{2} = 4.5$ W, $\frac{(3)^2}{3} = 3$ W, and $\frac{(3)^2}{6} = 1.5$ W.
The total power is 9 W, which is the same as the power dissipated in R_{eqp}.

(f) $i_1 = \frac{3\text{ V}}{2\,\Omega} = 1.5\text{ A}, \quad i_2 = \frac{3\text{ V}}{3\,\Omega} = 1\text{ A}, \quad i_3 = \frac{3\text{ V}}{6\,\Omega} = 0.5\text{ A}$
$i_1 + i_2 + i_3 = 3\text{ A} = i_T$. It is seen that the current ratios are the same as those of the conductances:
$\frac{i_1}{i} = \frac{1.5}{3} = 0.5 = \frac{1/2\text{S}}{1\text{S}}; \frac{i_2}{i} = \frac{1}{3} = \frac{1/3\text{S}}{1\text{S}}; \frac{i_3}{i} = \frac{0.5}{3} = \frac{1}{6} = \frac{1/6\text{S}}{1\text{S}};$
$\frac{i_1}{i_2} = \frac{1.5}{1} = \frac{1/2\text{S}}{1/3\text{S}}; \frac{i_1}{i_3} = \frac{1.5}{0.5} = \frac{1/2\text{S}}{1/6\text{S}}; \frac{i_2}{i_3} = \frac{1}{0.5} = \frac{1/3\text{S}}{1/6\text{S}}.$

Simulation: The circuit is entered as in Figure 3.7, assuming i_T = 3 A. After selecting 'Bias Point' in the Simulation Settings and running the simulation, pressing the I, V, and W buttons displays the currents, voltages, and powers dissipated, as indicated in Figure 3.7.

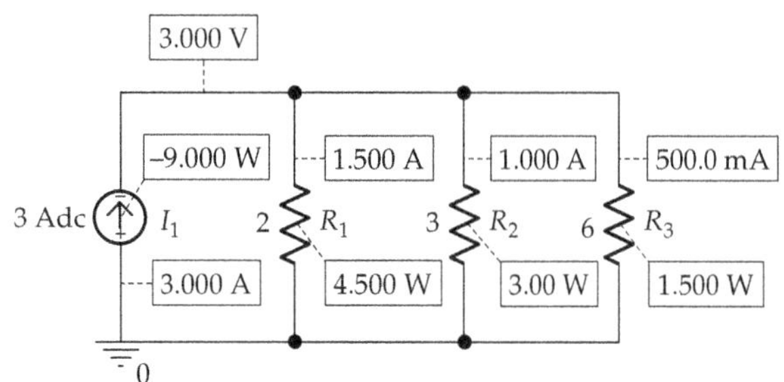

FIGURE 3.7
Figure for Example 3.2.

Problem-Solving Tip

- The parallel resistance of more than two resistors can be conveniently determined from paralleled resistances, two at a time.

Primal Exercise 3.3

Determine R_{eqp}, G_1, G_2, G_3, and G_{eqp} for a parallel connection of $R_1 = 2\ \Omega$, $R_2 = 12\ \Omega$, and $R_3 = 24\ \Omega$.

Ans. $R_{eqp} = 1.6\ \Omega$, $G_1 = 0.5$ S, $G_2 = 0.0833$ S, $G_3 = 0.0417$ S, $G_{eqp} = 0.625$ S.

Primal Exercise 3.4

Consider the defogger/defroster of Primal Exercise 2.19, whose resistive elements are connected as in Figure 3.8. If all the resistances are 1.5 Ω, determine the total equivalent resistance R_{eq}, the current drawn from the 12 V car battery, and the power dissipated.

Ans. 1.5 Ω, 8 A, 96 W.

Primal Exercise 3.5

Determine the effective resistance for the following combinations: (a) a short circuit in series with a resistor *R*, (b) an open circuit in parallel with *R*, (c) a short circuit in series with an open circuit, and (d) a short circuit in parallel with an open circuit.

Ans. (a) *R*; (b) *R*; (c) infinite; (d) zero.

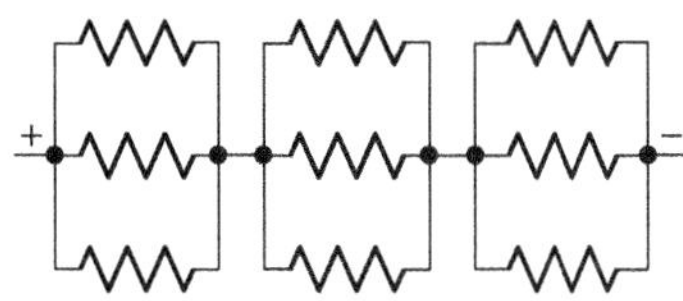

FIGURE 3.8
Figure for Primal Exercise 3.4.

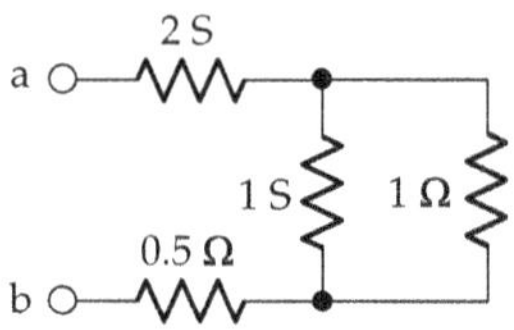

FIGURE 3.9
Figure for Primal Exercise 3.6.

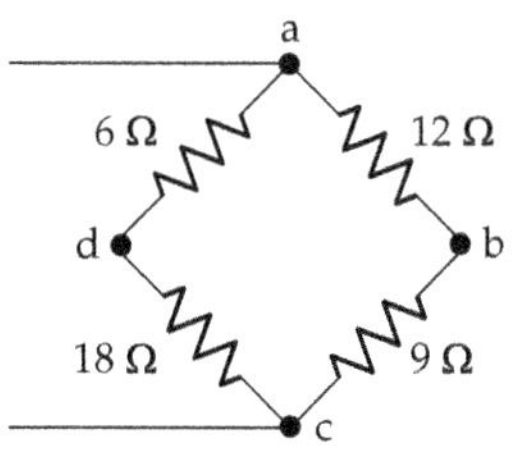

FIGURE 3.10 Figure for Primal Exercise 3.7.

Primal Exercise 3.6

Determine R_{eq} between terminals 'ab' in Figure 3.9.

Ans. 1.5 Ω.

Primal Exercise 3.7

Determine the equivalent resistance between nodes 'a' and 'c' in Figure 3.10 assuming (a) an open circuit between nodes 'b' and 'd', or (b) a short circuit between these nodes.

Ans. (a) 11.2 Ω; (b) 10 Ω.

Primal Exercise 3.8

Repeat Example 3.2 with a fourth resistor of 1 Ω added in parallel with the three resistors.

Ans. (a) 0.5 Ω, $G_4 = 1$ S, $G_{eqp} = 2$ S; (d) 1.5 V; (e) $i_1 = 0.75$ A, $i_2 = 0.5$ A, $i_3 = 0.25$ A, $i_4 = 1.5$ A.

Example 3.3: Voltage and Current Division

Given a 10 V battery applied to a 2 Ω resistor in series with a 3 Ω resistor (Figure 3.11a), it is required to determine (a) by voltage division the voltage across the 3 Ω resistor in Figure 3.11a and when a 6 Ω resistor is connected in parallel with the 3 Ω resistor (Figure 3.11b), and (b) by current division the currents in the 3 and 6 Ω resistors in Figure 3.11b.

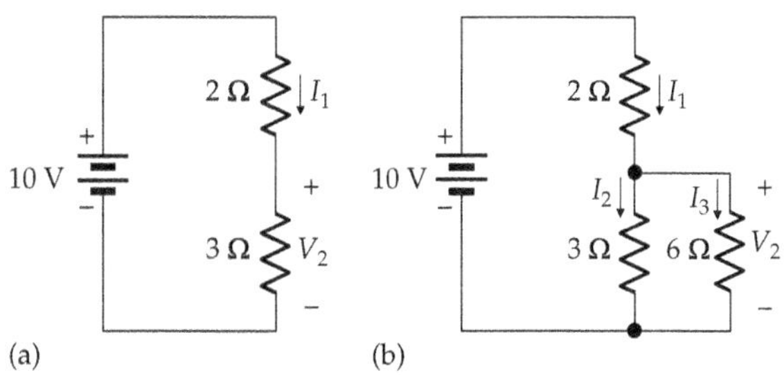

FIGURE 3.11
(a) Voltage divider and (b) voltage divider supplying a resistive load.

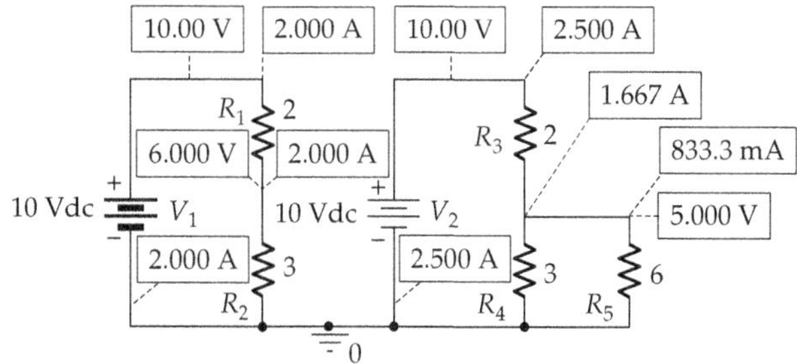

FIGURE 3.12
Figure for Example 3.3.

Solution:

(a) The same current I_1 flows in the battery and the two resistors in Figure 3.11a. Since only two resistors are connected in series, omitting R_3 in Equation 3.7 gives $\frac{V_2}{V_T} = \frac{R_2}{R_1 + R_2}$. Setting V_T = 10 V, R_1 = 2 Ω, and R_2 = 3 Ω gives $V_2 = 10 \times \frac{3}{2+3} = 6$ V. If a 6 Ω resistor is connected with parallel with the 3 Ω resistor (Figure 3.11b), the current I_1 divides into two components, I_2 and I_3. If the 3 Ω and the 6 Ω resistors are replaced by their equivalent parallel resistance, $R_{eqp} = \frac{3 \times 6}{3+6} = 2$ Ω, I_1 can then be considered to flow in R_{eqp}. Hence, from voltage division, $\frac{V_2}{V_T} = \frac{R_{eqp}}{R_1 + R_{eqp}}$, which gives $V_2 = 10 \times \frac{2}{2+2} = 5$ V. It is important to note that the voltage division Equations 3.7 through 3.9 *are applicable only when the same current flows in series-connected resistors*. With R_2 replaced by R_{eqp}, the same current I_1 flows in R_1 and R_{eqp}, so that the voltage division relations can be applied. Voltage division *should not* be applied in Figure 3.11b between the 2 Ω resistor and either the 3 Ω resistor alone or the 6 Ω resistor alone.

(b) I_1 in Figure 3.11b is 10/(2 + 2) = 2.5 A. From current division, I_2 = 2.5 × 6/(6 + 3) = 5/3 = 1.667 A, and I_3 = 2.5 × 3/(6 + 3) = 5/6 = 0.883 A. From Ohm's law, $v_2 = 3I_2 = 6I_3 = R_{eqp}I_1 = 5$ V.

Simulation: The two circuits are entered as in Figure 3.12. After selecting 'Bias Point' under 'Analysis type' in the Simulation Settings and running the simulation, pressing the *I* and *V* buttons displays the currents and voltages, respectively, as in Figure 3.12.

Problem-Solving Tips

- Never apply voltage division except to resistors in series, that is, resistors that carry the same current.
- In current division the largest current flows in the branch having the smallest resistance, and conversely.
- Never apply current division except to resistors in parallel.

Note that just as voltage division should be applied only to resistors in series, *current division should only be applied to resistors in parallel*. In Figure 3.13, for example, current division cannot be applied to I_1, I_2, and I_3, as was done in Figure 3.11b, because the 3 and 6 Ω resistors are not in parallel.

Primal Exercise 3.9

If V_{ac} = 42 V in Figure 3.10, determine V_{bd} using voltage division.

Ans. −13.5 V.

Primal Exercise 3.10

Determine: (a) R_{eqp} of the 10 and 5 Ω resistors in Figure 3.14; (b) V_O using voltage division; (c) I_S; (d) I_1 and I_2 using current division.

Ans. (a) 10/3 Ω; (b) 10 V; (c) 3 A; (d) I_1 = 1 A, I_2 = 2 A.

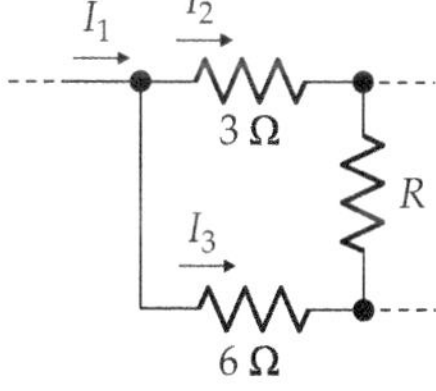

FIGURE 3.13
Current division should not be applied to resistors that are not in parallel.

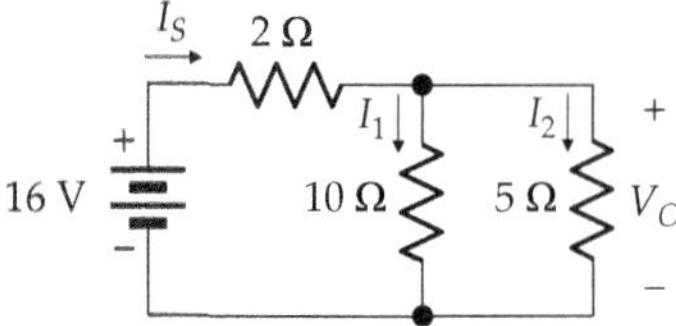

FIGURE 3.14
Figure for Primal Exercise 3.10.

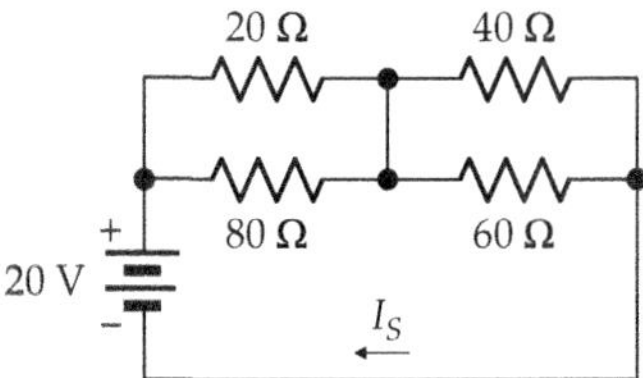

FIGURE 3.15
Figure for Primal Exercise 3.11.

Primal Exercise 3.11

Determine the current I_S in Figure 3.15 and the power dissipated in the 20 Ω resistor.

Ans. 0.5 A; 3.2 W.

★3.3 Resistivity

Definition: *The resistivity, or specific resistance, of a given material is the resistance between two opposite faces of a cube of the material multiplied by the length of the side of the cube. The reciprocal of resistivity is conductivity.*

The resistivity ρ of a given material is an intrinsic property of the material that is indicative of its electrical resistive property. It is independent of the size or shape of the conducting material, just as density is an intrinsic property indicative of the "heaviness" or mass property of the material irrespective of its size and shape. It follows from the definition of ρ that ρ is expressed as Ω-unit length, with reference to a cube of the material having a side of unit length. In the case of copper at room temperature, for example, $\rho \cong 1.7 \times 10^{-6}$ Ωcm if the cube is of 1 cm side, or 1.7×10^{-6} Ωcm × (mm/cm) = 1.7×10^{-5} Ω mm if the cube is of 1 mm side. Conductivity is expressed as siemens/unit length.

A useful relation involving resistivity is the following expression for the resistance R between two opposite ends of a block of length L units and having a uniform cross-sectional area of A square units:

$$R = \rho \frac{L}{A} = \frac{L}{\sigma A} \qquad (3.25)$$

where $\sigma = 1/\rho$ is the conductivity. The expression highlights the fact that resistance is directly proportional to the length of the current path through the conductor and inversely proportional to the cross-sectional area through which the current flows. Increasing the length of path means more collisions between conduction electrons and crystal atoms, which increases the resistance. A larger cross-sectional area provides more paths for the current to follow, which increases the current for a given applied voltage and hence decreases the resistance. ρ and σ in Equation 3.25 are independent of the length of path or cross-sectional area.

To derive Equation 3.25 we note that, according to the discussion of Section 2.1, current flows in a medium under the influence of the electric field ξ and in the direction of this field. The total current depends on the extent of the medium; hence, a more appropriate measure of the effect of ξ on current flow is the current per unit area of flow, denoted as the current density J. Consider a small cube of the material oriented so that one of its sides is parallel to ξ (Figure 3.16a). From the relation $\xi = -dv/dx$, the voltage drop from the rear face of the cube to its front face is $-dv = \xi\Delta x$. The current through the cube, in the direction of ξ, is the current density multiplied by the area of the face, that is, $J(\Delta x)^2$. If the resistance in ohms between the rear and front faces of the cube is denoted by ρ', Ohm's law applied to the cube is $\xi\Delta x = \rho' J(\Delta x)^2$, which gives

$$\rho' = \frac{\xi\Delta x}{J(\Delta x)^2} = \frac{\xi}{J\Delta x} \qquad (3.26)$$

The resistance ρ' depends on Δx, which makes it unsuitable as an intrinsic measure of resistance. However, both ξ and J are "local" quantities that do not depend on the size or shape of the medium, which makes the ratio ξ/J a suitable measure of the intrinsic resistance of the medium. The ratio ξ/J is denoted as the resistivity ρ of

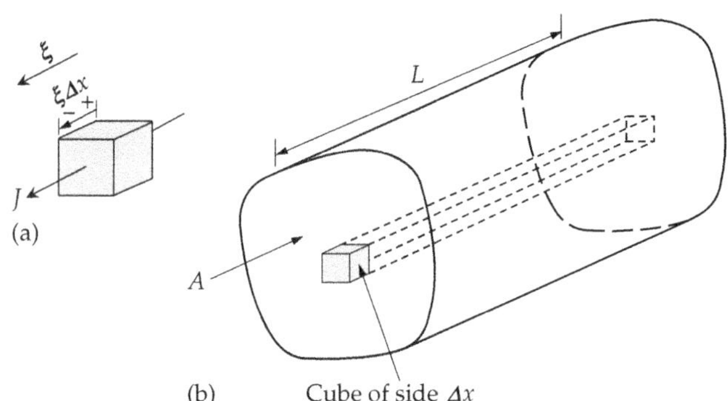

FIGURE 3.16
Derivation of expression for resistivity. (a) Cube of side Δx having a side parallel to the electric field and (b) conducting block of length L and uniform cross-sectional area A.

the medium and is equal to $\rho'\Delta x$, as in the definition of resistivity. Thus,

$$\rho = \frac{\xi}{J} = \rho'\Delta x \tag{3.27}$$

Note that the relation $\xi = \rho J$, or $J = \sigma\xi$, is referred to as the microscopic form of Ohm's law, or Ohm's law at a point in a conductor.

Consider next a block of a conductor of length L units and a uniform cross section of area A square units (Figure 3.16b). Let the block be divided into small cubes of side Δx oriented parallel to the longitudinal axis of the block. The resistance of each cube between opposite faces is $\rho' = \rho/\Delta x$. A strip of these cubes that extends along the length of the block will have $L/\Delta x$ of these cubes side by side. The resistances of these cubes add in series, which makes the resistance between the two ends of the strip $(L/\Delta x) \times (\rho/\Delta x) = \rho L/(\Delta x)^2$. The block will have $A/(\Delta x)^2$ strips side by side across the block, with the resistances of the strips combining in parallel. The resistance R between the two ends of the block will therefore be

$$R = \frac{\left(\dfrac{\rho L}{(\Delta x)^2}\right)}{\left(\dfrac{A}{(\Delta x)^2}\right)} = \frac{\rho L}{A} \tag{3.28}$$

Exercise 3.12

A 1 cm cube of material has a resistance of 2.5 kΩ between opposite faces. Determine the resistance of a rectangular block of this material that is 50 cm long and of 10 cm² cross-sectional area.

Ans. 12.5 kΩ.

3.4 Star–Delta Transformation

An application of series–parallel equivalence is the star–delta, or Y–Δ, transformation. Consider three resistors R_a, R_b, and R_c connected in delta (Figure 3.17a)

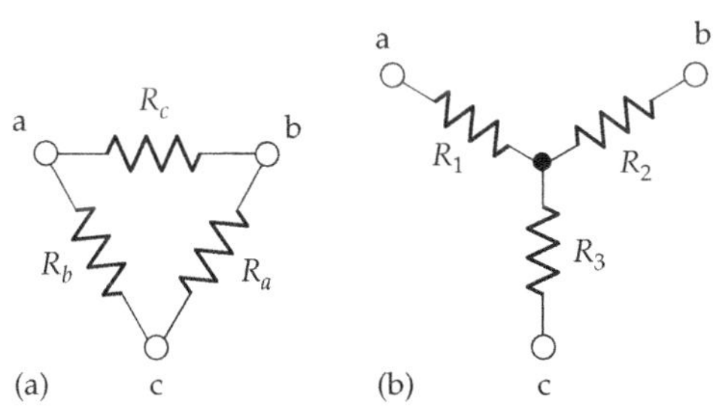

FIGURE 3.17 Resistors connected in Δ (a) and in Y (b).

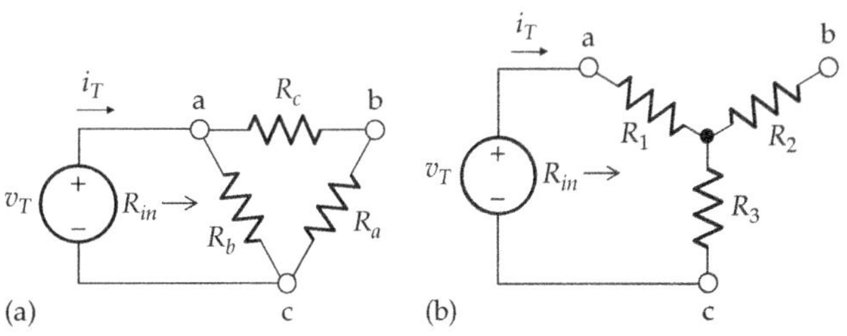

FIGURE 3.18
Equivalence between resistors connected in Δ and in Y. Test source applied between two terminals of Δ-connected resistors (a), and between the corresponding terminals of Y-connected resistors (b).

and another three resistors R_1, R_2, and R_3 connected in Y (Figure 3.17b). It is desired to derive the relations between these two sets of resistances so that the two circuits are equivalent between any two of the three terminals.

According to the definition of circuit equivalence, if a test source v_T is applied between, say, terminals 'a' and 'c' of the Δ-circuit and this same source is applied between terminals 'a' and 'c' of the Y-circuit, the same v_T–i_T relation is obtained in both circuits, as illustrated in Figure 3.18.

According to the third implication of circuit equivalence discussed in Section 3.1, the ratio of v to i is the same at the corresponding terminals of equivalent circuits. Moreover, since the Δ- and Y-circuits consist entirely of resistors, without any independent sources, the ratio of v_T to i_T is evidently some resistance, referred to as the input resistance R_{in} between terminals 'a' and 'c', or the *resistance seen by the source* between these terminals. It follows that the Δ- and Y-circuits in Figure 3.17 are equivalent if they have the same input resistance between terminals 'ab', 'bc', and 'ca'. However, an additional consideration in this case is the state of the third terminal when determining the input resistance between a given pair of terminals. Equivalence requires that whatever is done with the third terminal, it should be the same in the two circuits. A convenient condition to impose on the third terminal is to simply leave it open-circuited in both cases, although this is by no means the only possibility (Problem P3.66).

In Figure 3.18b, the resistance R_{ac} between terminals 'ac', with terminal 'b' open, is that of R_1 and R_3 in series, that is, $(R_1 + R_3)$. In Figure 3.18a, the resistance between terminals 'ab', with terminal 'b' open, is that of R_b in parallel with the combination of R_a and R_c in series. That is, $R_{ac} = R_b(R_a + R_c)/(R_a + R_b + R_c)$. Equating the resistances in both cases,

$$R_1 + R_3 = \frac{R_b(R_a + R_c)}{R_a + R_b + R_c} \tag{3.29}$$

Similarly, equating the resistances between terminals 'bc', with terminal 'a' open,

$$R_2 + R_3 = \frac{R_a(R_b + R_c)}{R_a + R_b + R_c} \tag{3.30}$$

Equating the resistances between terminals 'ab', with terminal 'c' open,

$$R_1 + R_2 = \frac{R_c(R_a + R_b)}{R_a + R_b + R_c} \tag{3.31}$$

Equations 3.29 through 3.31 are three independent equations. If R_a, R_b, and R_c of the delta circuit are given, these equations can be solved for R_1, R_2, and R_3 of the equivalent star circuit to give

$$R_1 = \frac{R_b R_c}{R_a + R_b + R_c} \tag{3.32}$$

$$R_2 = \frac{R_a R_c}{R_a + R_b + R_c} \tag{3.33}$$

$$R_3 = \frac{R_a R_b}{R_a + R_b + R_c} \tag{3.34}$$

Conversely, if the resistances R_1, R_2, and R_3 of the star circuit are given, Equations 3.29 through 3.31 can be solved for R_a, R_b, and R_c of the equivalent delta circuit to give

$$R_a = \frac{R_1R_2 + R_2R_3 + R_3R_1}{R_1} \tag{3.35}$$

$$R_b = \frac{R_1R_2 + R_2R_3 + R_3R_1}{R_2} \tag{3.36}$$

$$R_c = \frac{R_1R_2 + R_2R_3 + R_3R_1}{R_3} \tag{3.37}$$

To help apply the preceding relations in a systematic way, the two equivalent circuits are superimposed, as in Figure 3.19. Each Y-resistance is the product of the Δ-resistances on either side of it divided by the sum of the three Δ-resistances. Conversely, each Δ-resistance is the sum of the products of the Y-resistances, two at a time, divided by the Y-resistance that is at a right angle to the given Δ-resistance.

When the three resistances in either configuration are equal, the preceding relations reduce to

$$R_\Delta = 3R_Y \quad \text{and} \quad R_Y = R_\Delta/3 \tag{3.38}$$

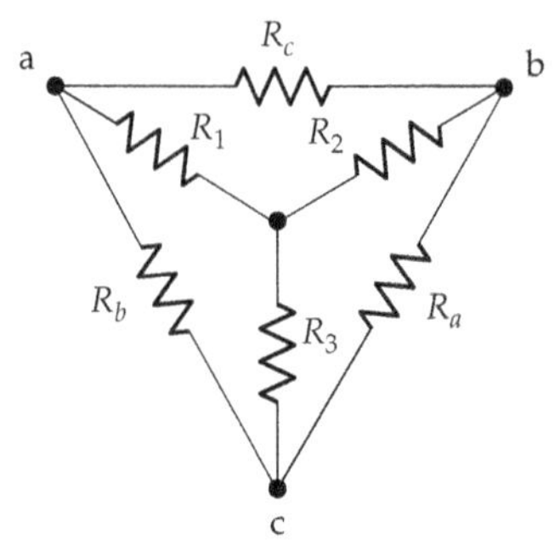

FIGURE 3.19
Δ–Y transformation.

An interpretation of these relations is that the Δ-circuit is more of a parallel circuit, whereas the Y-circuit is more of a series circuit. Since resistances in series give a larger equivalent resistance, whereas resistances in parallel give a smaller equivalent resistance, then if the two circuits are to be equivalent, the Δ-circuit should have the larger resistances.

Primal Exercise 3.13

Assume that in Figure 3.19 the resistances connected in Y are 1 Ω each and the resistances connected in Δ are 3 Ω each. Determine the resistance between nodes 'a' and 'b'.

Ans. 1 Ω.

Primal Exercise 3.14

Three 6 Ω resistors are connected in Δ. Determine (a) the equivalent Y-circuit and (b) the power dissipated in each circuit if a 12 V source is connected between any two corresponding terminals in each circuit, with the third terminal left open.

Ans. (a) $R_Y = 2$ Ω; (b) 36 W; from circuit equivalence, the product vi is the same.

Example 3.4: Delta–Star Transformation

It is required to determine R_{eq} between terminals 'bd' in the circuit of Figure 3.20a.

Solution:

Either of the two sets of delta-connected resistors could be transformed to its equivalent star circuit. The delta circuit between terminals 'acd' is transformed to its equivalent star circuit using Equations 3.32 through 3.34, where the sum of the resistances in delta is 10 + 25 + 15 = 50 Ω. It follows that $R_1 = \frac{25 \times 10}{50} = 5$ Ω, $R_2 = \frac{25 \times 15}{50} = 7.5$ Ω, and $R_3 = \frac{10 \times 15}{50} = 3$ Ω. The delta circuit is replaced by

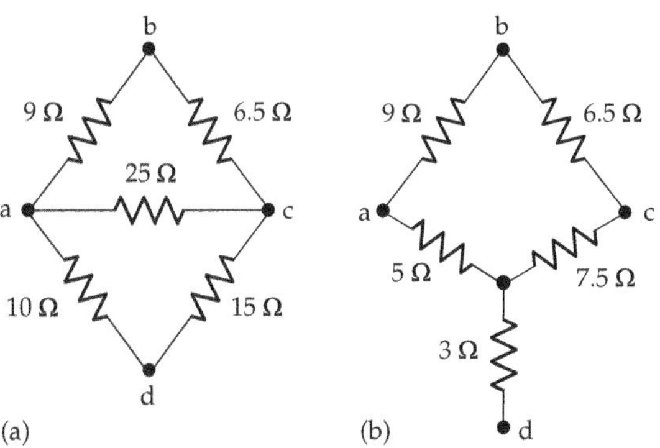

FIGURE 3.20
Figure for Example 3.4.

the equivalent star circuit, terminal for terminal, resulting in the circuit of Figure 3.20b. The 5 Ω resistance is added to the 9 Ω to give 14 Ω, and the 7.5 Ω resistance is added to the 6.5 Ω to give also 14 Ω. The two 14 Ω resistors in parallel give 7 Ω, which is added to the 3 Ω to give 10 Ω between terminals 'bd'.

It should be noted that although nodes 'b' and 'd' in Figure 3.20 are not shown connected to anything, it must be borne in mind that R_{eq} is formally determined, by measurement or PSpice simulation, by connecting a test source, such as v_T, and determining the resulting current i_T. R_{eq} is then the ratio v_T/i_T. *It must not be assumed that nodes 'b' and 'd' are open-circuited when determining* R_{eq} and therefore wrongly conclude that the 9 and 6.5 Ω resistors are in series in Figure 3.20a and that the 10 and 15 Ω resistors are also in series.

Simulation: The circuit is entered as in Figure 3.21. Note that the required equivalent resistance can be conveniently determined by applying a source current of 1 A between the two terminals in question. The voltage between these terminals is then numerically equal to the required resistance. After selecting 'Bias Point' under 'Analysis type' in the Simulation Settings and running the simulation, pressing the V button displays the voltages shown. V_{bd} = 10 V, so R_{eq} = (10 V)/(1 A) = 10 Ω.

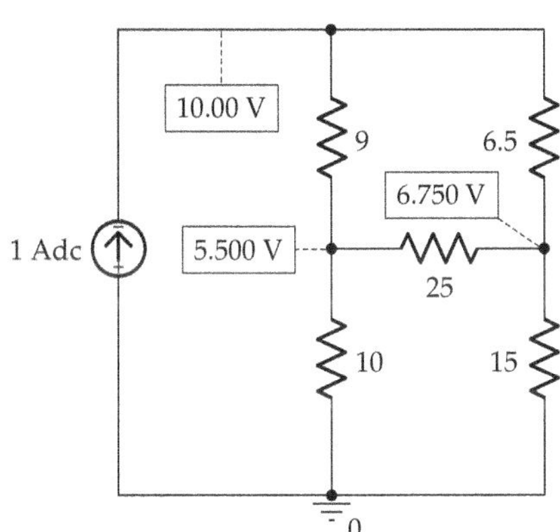

FIGURE 3.21
Figure for Example 3.4.

Exercise 3.15

Simulate the circuit of Figure 3.20b as in Example 3.4 and verify the voltages at terminals 'a', 'b', and 'c'.

Primal Exercise 3.16

Determine R_{eq} between terminals 'bd' in the circuit of Figure 3.20a by considering that the 9, 10, and 25 Ω resistors are connected in star and transforming them to the equivalent Δ.

Ans. 10 Ω.

3.5 Series and Parallel Connections of Ideal Sources

Having derived in Section 3.2 the equivalent series and parallel resistances of resistors connected in series or in parallel, we will examine in this section how ideal sources, the other basic circuit element introduced in Chapter 2, combine in series or in parallel.

3.5.1 Ideal Voltage Sources

Figure 3.22a illustrates a series connection of two ideal independent sources, whereas Figure 3.22b illustrates a series connection of an ideal independent source v_{SRC}

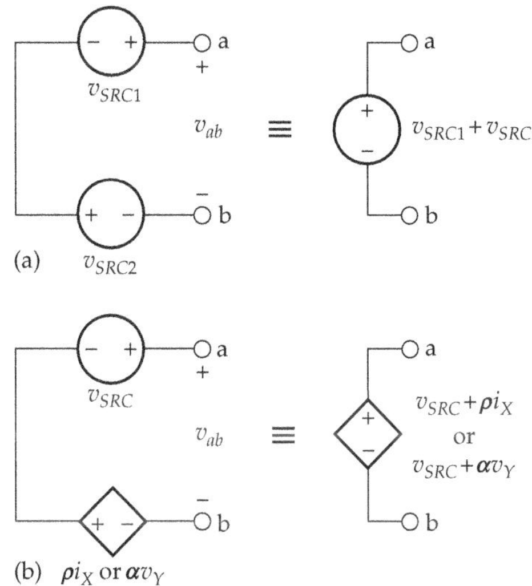

FIGURE 3.22
(a) Series connection of two ideal, independent voltage sources and (b) series connection of an ideal, independent voltage source and an ideal dependent voltage source.

with either a CCVS or a VCVS. In going from terminal 'b' to terminal 'a', the voltages along the series path add algebraically (Section 2.8), so that, for the polarities shown,

$$v_{ab} = v_{SRC1} + v_{SRC2}$$

or

$$v_{ab} = v_{SRC} + \rho i_X$$

or

$$v_{ab} = v_{SRC} + \alpha v_Y \tag{3.39}$$

where i_X and v_Y are, respectively, current and voltage somewhere in the circuit. Evidently, these relations can be generalized to any number of series-connected, independent or dependent ideal voltage sources of any source voltage and any polarity, in any combination. Moreover, because the sources are ideal, the terminal voltage v_{ab} is independent of the current through the series combination.

A similar generalization cannot be made in the case of paralleled ideal voltage sources. The following concept applies in this case:

Concept: *Ideal voltage sources should not be paralleled, unless they have the same source voltage and the same polarity. Otherwise, conservation of energy is violated.*

To justify this, consider two ideal voltage sources of unequal source voltages connected in parallel (Figure 3.23a). Such a connection evidently violates KVL because $v_{SRC1} - v_{SRC2} \neq 0$. More fundamentally, such a connection violates conservation of energy. Thus, if a charge q is taken clockwise around the circuit, the work done in taking q up a voltage rise v_{SRC1} is qv_{SRC1}, and the work done by the charge in moving down a voltage drop v_{SRC2} is qv_{SRC2}. If $qv_{SRC2} > qv_{SRC1}$, for example, then it would be possible, at least in principle, to extract energy continuously from the circuit, at no energy cost, simply by moving q around the circuit, in violation of conservation of energy. Such a connection is invalid in electric circuits and is also not allowed in PSpice. Another manifestation of this invalidity is that a finite voltage difference, $v_{SRC2} - v_{SRC1}$, in this case, divided by zero resistance in the circuit, results in an infinite current.

The parallel connection of any number of ideal, independent voltage sources of the same voltage and polarity is equivalent to that of a single ideal, independent voltage source of the same voltage and polarity (Figure 3.23b).

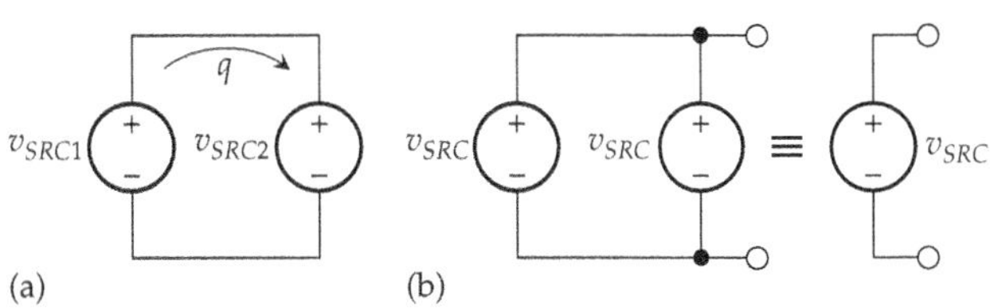

FIGURE 3.23
Parallel connections of ideal voltage sources. (a) Invalid connection if $v_{SRC1} \neq v_{SRC2}$ and (b) valid connection if the source voltages are equal.

3.5.2 Ideal Current Sources

Concept: *Ideal current sources should not be connected in series, unless they have the same source current and in the same direction. Otherwise, conservation of charge is violated.*

Consider two ideal current sources of unequal source currents connected in series (Figure 3.24a). Such a connection evidently violates KCL because $i_{SRC1} - i_{SRC2} \neq 0$. More fundamentally, such a connection violates conservation of charge. Thus, i_{SRC1} C/s enter node 'a' between the two sources and i_{SRC2} C/s leave the node. If, for the sake of argument, $i_{SRC1} > i_{SRC2}$, then $i_{SRC1} - i_{SRC2}$ C/s will simply vanish at node 'a' in violation of conservation of charge. Such a connection is invalid in electric circuits and is also not allowed in PSpice but for a different reason, namely, the occurrence of a 'floating' node (Appendix C).

Any number of identical, ideal, independent current sources connected in series in the same direction is equivalent to a single ideal, independent current source of the same current and in the same direction, since the terminal current is the same in both cases (Figure 3.24b).

When ideal current sources are connected in parallel, the source currents add algebraically. This applies to any number of parallel-connected, independent or dependent, ideal current sources of any source current and polarity, in any combination. Figure 3.25a illustrates the case of two independent sources, whereas Figure 3.25b illustrates the case of an independent source and a VCCS or a CCCS connected in parallel.

Primal Exercise 3.17

Determine the current I_S through the voltage sources in Figure 3.26 and the voltage V_S across the current sources.

Ans. 5 A, −5 V.

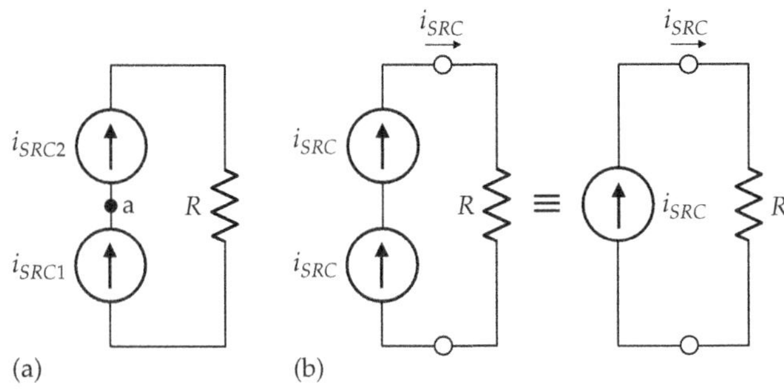

FIGURE 3.24
Series connections of ideal current sources. (a) Invalid connection if $i_{SRC1} \neq i_{SRC2}$ and (b) valid connection if the source currents are equal.

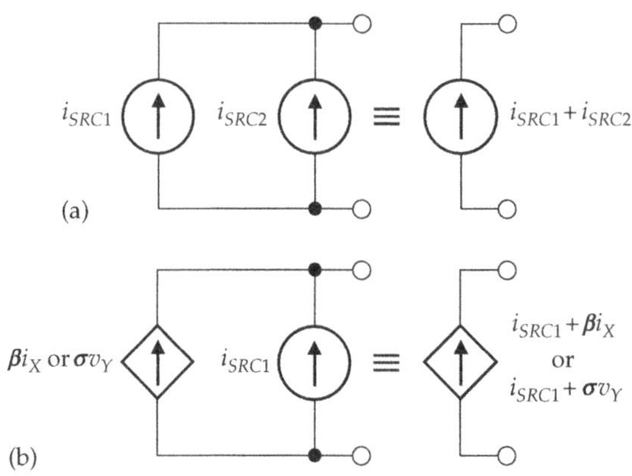

FIGURE 3.25
(a) Parallel connection of two ideal, independent current sources and (b) parallel connection of an ideal, independent current source and an ideal dependent current source.

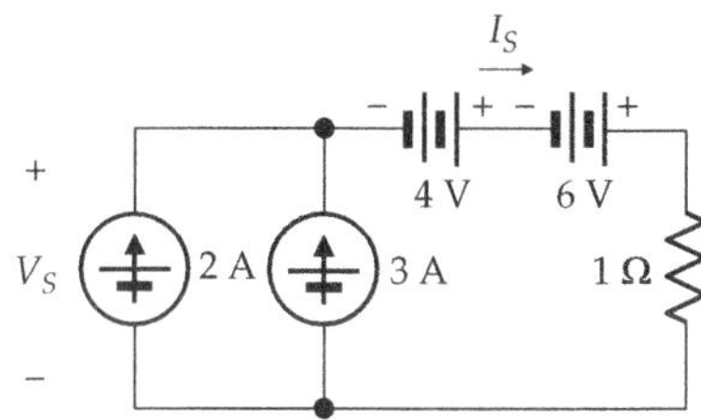

FIGURE 3.26
Figure for Primal Exercise 3.17.

3.6 Linear-Output Sources

Circuit equivalence was applied in the preceding sections to combinations of ideal resistors on their own and to combinations of ideal sources on their own. Circuit equivalence will be applied in this section to linear-output sources, consisting of an ideal voltage source in series with a resistor or an ideal current source in parallel with a resistor. The practical importance of linear-output sources is that they can be used to approximate the behavior of some nonideal sources.

3.6.1 Linear-Output Voltage Source

Voltage sources encountered in practice are nonideal in the sense that the voltage across the source terminals decreases as the current delivered by the source increases. If this decrease in voltage is directly proportional to the current delivered by the source or can be approximated as such, then the nonideal source can be represented, between the source terminals, by an ideal voltage source of voltage v_{SRC} in series with an ideal resistor R_{src}, as illustrated in Figure 3.27a. R_{src} is an integral part of the source, as symbolized by the dashed rectangle, and the actual terminals of the nonideal source are now 'ab'. This representation is noteworthy in that a combination of two ideal circuit elements is used to simulate the behavior of a nonideal component.

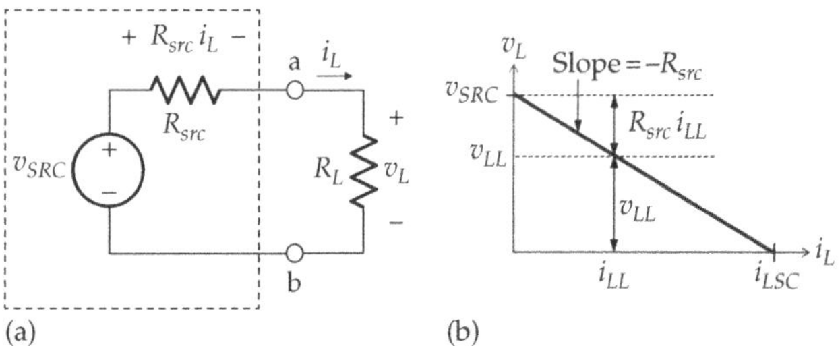

FIGURE 3.27
(a) Linear-output voltage source connected to a load and (b) v–i plot of source output.

Since the circuit in Figure 3.27a is a series circuit, KCL is automatically satisfied. From Ohm's law, the voltage drop across R_{src} is $R_{src}i_L$, and KVL gives

$$v_{SRC} - R_{src}i_L - v_L = 0$$

or

$$v_L = v_{SRC} - R_{src}i_L \tag{3.40}$$

v_{SRC} is now termed the **open-circuit voltage** of the source, because it is the voltage that appears between the source terminals 'ab' when these terminals are open-circuited, which makes $i_L = 0$ and $v_L = v_{SRC}$; R_{src} is the **source resistance** looking into terminals 'ab' with $v_{SRC} = 0$, that is, with the ideal source element set to zero by replacing it with a short circuit. R_{src} is also referred to as the **internal resistance** of the source.

As i_L increases from zero, the voltage drop across R_{src} increases in direct proportion to i_L and subtracts from the open-circuit voltage v_{SRC}; v_L at the source terminals decreases linearly with i_L, the slope being $-R_{src}$, in accordance with Equation 3.40 and illustrated in Figure 3.27b. For any set of values v_{LL} and i_{LL}, the vertical segments with arrows at both ends (Figure 3.27b) represent v_L and $R_{src}i_{LL}$. These two segments add up to v_{SRC}, in accordance with Equation 3.40. If the output terminals are short-circuited, $v_L = 0$. The current intercept in Figure 3.27b is $i_{LSC} = v_{SRC}/R_{src}$ and is termed the **short-circuit current** at these terminals.

Because of the linear v_L–i_L relation at the source terminals 'ab', we will refer to this type of nonideal voltage source as a **linear-output voltage source**.

Equation 3.40 relates the terminal variables v_L and i_L to the source quantities v_{SRC} and R_{src} only, independently of the load. This v_L–i_L relation is therefore referred to as the **source characteristic**, with R_{src} being the source resistance. If $R_{src} = 0$, the source characteristic is a horizontal line having a voltage intercept v_{SRC}. The terminal voltage is v_{SRC}, independently of the source current, as for an ideal voltage source. This underlies the following concept that we will often use:

Concept: *An ideal, independent or dependent, voltage source has zero source resistance.*

Note that this concept is also in accordance with the previous deduction that an ideal voltage source behaves as a short circuit if $v_{SRC} = 0$ (Figure 2.4), since a short circuit has zero resistance. We are now asserting that the source resistance is still zero when $v_{SRC} \neq 0$.

It also follows that an ideal voltage source must not be short-circuited, since $R_{src} = 0$ makes the resulting current infinite. In a nonideal source, the source resistance limits the short-circuit current.

Practical voltage sources are commonly paralleled. This is because any practical voltage source is designed to deliver a maximum current or power. When sources are paralleled, the total current delivered is the sum of the currents delivered by the individual sources, and more power can therefore be delivered than by any of the individual sources. Generally, paralleled voltage sources have nominally the same open-circuit voltage and source resistance, so they share the load equally.

Primal Exercise 3.18

A discharged 12 V car battery having an open-circuit voltage of 11.7 V is energized by paralleling it with a fully charged battery of open-circuit voltage 12.6 V. Determine the current that initially flows upon paralleling the two batteries, assuming that each battery has an internal, or source, resistance of 45 mΩ and that the connecting cables are of negligible resistance.

Ans. 10 A.

Primal Exercise 3.19

The voltage of a car battery drops from 12 to 8 V when supplying 80 A to the engine starter. Determine the internal, or source, resistance of the battery. What is the resistance of the starter and the connecting cables?

Ans. Battery internal resistance is 50 mΩ; 0.1 Ω.

3.6.2 Linear-Output Current Source

Current sources encountered in practice are nonideal in the sense that the voltage across the source terminals decreases as the current delivered by the source increases. If this decrease in voltage is directly proportional to the current delivered by the source or can be approximated as such, then the nonideal source can be represented, between the source terminals, by an ideal current source of current i_{SRC} in parallel with an ideal resistor R_{src} (Figure 3.28a).

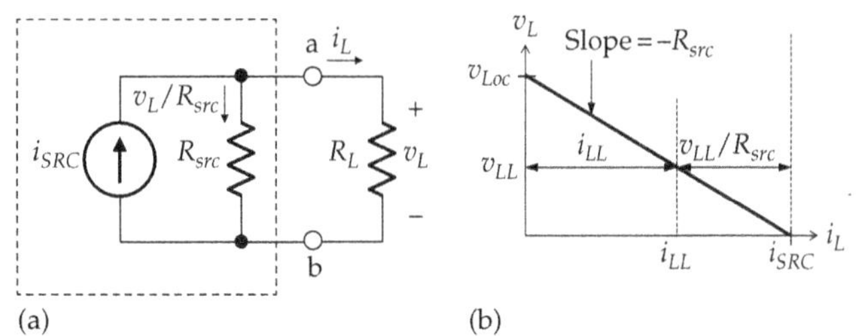

FIGURE 3.28
(a) Linear-output current source connected to a load and (b) v–i plot of source output.

Since the circuit in Figure 3.28a is a parallel circuit, KVL is automatically satisfied. From Ohm's law, the current through R_{src} is v_L/R_{src}, so that KCL at node 'a' gives $i_{SRC} = i_L + v_L/R_{src}$. Multiplying both sides of the equation by R_{src} and rearranging,

$$v_L = R_{src}i_{SRC} - R_{src}i_L \tag{3.41}$$

Equation 3.41 is plotted in Figure 3.28b. When $v_L = 0$, $i_L = i_{SRC}$ and is the current intercept. When $i_L = 0$, the voltage intercept is $v_{LOC} = R_{src}i_{SRC}$ and is the open-circuit voltage at terminals 'ab'. The slope of the line is $-R_{src}$, where R_{src} is the source resistance looking into terminals 'ab', with $i_{SRC} = 0$, that is, with the ideal current source set to zero by replacing it with an open circuit.

The interpretation of Equation 3.41 is that when $i_L = 0$, i_{SRC} flows through R_{src}, producing a voltage drop $R_{src}i_{SRC}$ at the open-circuited terminals 'ab'. As i_L increases from zero, it subtracts from i_{SRC}, so less current flows in R_{src} and v_L is reduced. When terminals 'ab' are short-circuited, $v_L = 0$, no current flows in R_{src}, so that i_{SRC} flows through the short circuit. For any set of values v_{LL} and i_{LL}, the horizontal segments with arrows at both ends in Figure 3.28b represent i_{LL} and v_{LL}/R_{src}.

Because of the linear v_L–i_L relation at the source terminals 'ab', we will refer to this type of nonideal current source as a **linear-output current source**. Equation 3.41 relates the terminal variables v_L and i_L to the source parameters i_{SRC} and R_{src} only, independently of the load. As in the case of the linear-output voltage source, it is referred to as the source characteristic, with R_{src} being the source resistance. If i_{SRC} remains constant in Figure 3.28b as R_{src} increases, V_{Loc} increases. As $R_{src} \to \infty$, the source characteristic becomes a vertical line through i_{SRC}. This underlies the following concept that we will often use:

Concept: *An ideal, independent or dependent, current source has zero source conductance or infinite source resistance.*

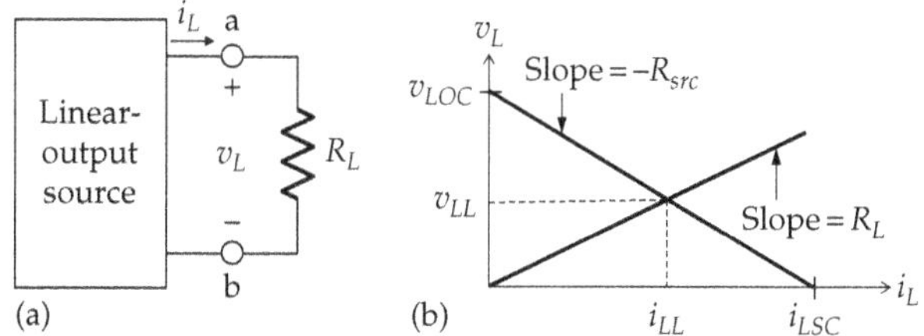

FIGURE 3.29 Graphical analysis of load connected to a linear-output source. (a) Linear-output source connected to a load and (b) graphical construction for determining output voltage and current.

Note that this concept is also in accordance with the previous deduction that an ideal current source behaves as an open circuit if $i_{SRC} = 0$ (Figure 2.5), since an open circuit has infinite resistance. We are again asserting that R_{src} remains infinite when $i_{SRC} \neq 0$.

It also follows that an ideal current source must not be open-circuited, as the resulting voltage across the source is infinite. In a nonideal source, the source resistance limits the open-circuit voltage.

Before ending this discussion, it may be noted that when a linear-output source is connected to a load resistance R_L, a simple graphical analysis can be made. Figure 3.29a illustrates a linear-output source connected to a resistor R_L. Figure 3.29b shows the source characteristic for the linear-output source in terms of v_L and i_L, representing the equation

$$v_L = v_{LOC} - R_{src}i_L \tag{3.42}$$

where

v_{LOC} is the open-circuit voltage at terminals 'ab' when $i_L = 0$

R_{src} is the source resistance looking into these terminals with the independent source set to zero

$I_{LSC} = v_{LOC}/R_{src}$ is the short-circuit current when $v_L = 0$

Equation 3.42 is independent of R_L, depending only on the source. However, v_L and i_L are related by Ohms' applied to R_L:

$$v_L = R_L i_L \tag{3.43}$$

Equation 3.43 is represented in Figure 3.29b as a line of slope R_L passing through the origin and referred to as the **load line**. Equations 3.42 and 3.43 are two equations in the two unknowns v_L and i_L. The intersection of the two lines representing these equations gives the particular values of v_{LL} and i_{LL} that are the solution to the two equations, because these values satisfy both equations.

3.6.3 Transformation of Linear-Output Sources

An important and useful procedure is the transformation between a linear-output voltage source and a linear-output current source. This transformation follows from a comparison of Equations 3.40 and 3.41. These equations are identical if (1) $v_{SRC} = R_{src}i_{SRC}$ and (2) R_{src} is the same in both cases. Under these conditions, the v_L–i_L relation is the same for both sources at terminals 'ab', which means that the two sources are equivalent at these terminals. That is, one type of linear-output source can be substituted for the other, terminal for terminal, without affecting the rest of the circuit.

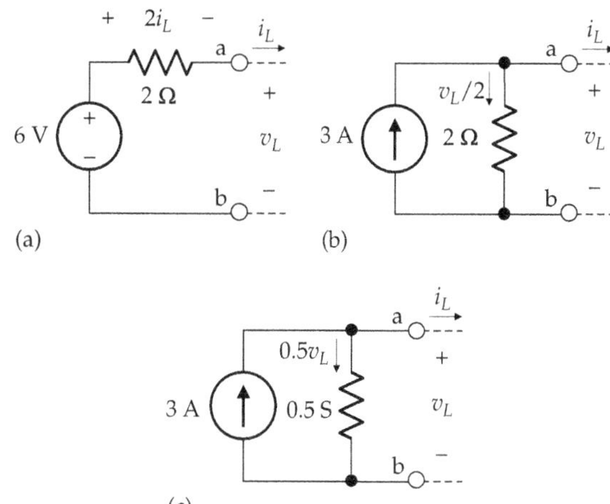

FIGURE 3.30
Transformation of linear-output sources. (a) Linear-output voltage source, (b) equivalent linear-output current source with source resistance, and (c) source resistance in (b) replaced by source conductance.

The transformation can be illustrated by a voltage source of open-circuit voltage $v_{SRC} = 6$ V and source resistance $R_{src} = 2\ \Omega$ connected to some load (Figure 3.30a). Equation 3.40 becomes

$$v_L = 6 - 2i_L \tag{3.44}$$

Consider next a current source having $i_{SRC} = v_{SRC}/R_{src} = 6/2 = 3$ A, in accordance with the relation $v_{SRC} = R_{src}i_{SRC}$ (Figure 3.30b). The source resistance $R_{src} = 2\ \Omega$ is in parallel with i_{SRC}, as in Figure 3.28a. Equation 3.41 gives $v_L = 6 - 2i_L$, which is the same as Equation 3.44. This means that the two sources are equivalent at terminals 'ab'. Note that the $2\ \Omega$ source resistance could alternatively be represented as a conductance of 0.5 S in parallel with i_{SRC} (Figure 3.30c).

The procedure for transforming a linear-output voltage source to an equivalent linear-output current source, or conversely, is illustrated in Figure 3.31 and can be summarized as follows:

1. *To transform a linear-output voltage source of open-circuit voltage v_{SRC} to a linear-output current source, the short-circuit current of the current source is given by $i_{SRC} = v_{SRC}/R_{src}$.*
2. *To transform a linear-output current source of short-circuit current i_{SRC} to a linear-output voltage source, the open-circuit voltage of the linear-output voltage source is given by $v_{SRC} = R_{src}i_{SRC}$.*

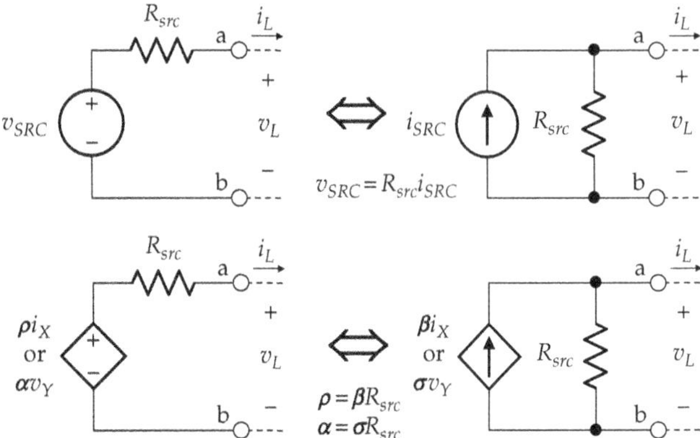

FIGURE 3.31
Transformation of independent and dependent linear-output sources.

3. *In both cases, the source resistance R_{src} in series with the ideal voltage source element is the same R_{src} in parallel with the ideal current source element.*

The transformation applies to dependent sources as well as independent sources, as emphasized in Figure 3.31. A CCVS of voltage ρi_X is transformed to a CCCS of current $(\rho/R_{src})i_X$, and conversely. Similarly, a VCVS of voltage αv_Y is transformed to a VCCS of current $(\alpha/R_{src})v_Y$, and conversely.

The following should be noted concerning source transformation:

1. Because of circuit equivalence, source transformation must preserve polarities of voltages and directions of currents at the terminals of the two sources. To ascertain this in a given circuit, the two sources are temporarily disconnected from the rest of the circuit, so that $i_L = 0$ in Figure 3.30, for example. The assigned positive direction of the ideal voltage source element in Figure 3.30a makes terminal 'a' positive with respect to terminal 'b'. Similarly, in Figure 3.30b, the assigned positive direction of the ideal current source element is such that i_{SRC} flowing in R_{src} makes terminal 'a' positive with respect to terminal 'b'.
2. The equivalence between the two sources applies at the specified terminals *only*. It does not apply inside the sources. For example, the power delivered by the ideal source elements is quite different in the two cases, as illustrated in Example 3.5.
3. An ideal voltage source *cannot* be transformed to an ideal current source, and conversely. Thus, $R_{src} = 0$ for an ideal voltage source, which makes $i_{SRC} = v_{SRC}/R_{src} \to \infty$. Similarly, $R_{src} \to \infty$ for an ideal current source, which makes $v_{SRC} = R_{src}i_{SRC} \to \infty$. Infinite values of v_{SRC} and i_{SRC} imply that the corresponding sources are not valid. This is in accordance with ideal voltage and current sources being basic circuit elements, where a basic circuit element cannot be represented in terms of other basic circuit elements, as mentioned in Section 1.8. Nonideal sources are not basic circuit elements. They can therefore be represented in terms of basic circuit elements, and linear-output voltage sources can be transformed to linear-output current sources, and conversely.

•The following deductions follow from the discussion of linear-output sources:

1. The ideal termination for a nonideal voltage source is an open circuit, in the sense that under these conditions, $v_L = v_{LOC} = v_{SRC}$ (Equation 3.40). In other words, the terminal voltage has its largest value, which is the open-circuit voltage of the source. However, the power delivered by the source is zero.
2. The ideal termination for a nonideal current source is a short circuit, in the sense that under these conditions, $i_L = I_{LSC} = i_{SRC}$ (Equation 3.41). In other words, the terminal current has its largest value, which is the short-circuit current of the source. However, the power delivered by the source is zero.
3. If $v_L = R_L i_L$ is substituted in Equation 3.40, i_L is given by

$$i_L = \frac{v_{SRC}}{R_{src} + R_L} \tag{3.45}$$

If $R_L \ll R_{src}$, then $i_L \cong v_{SRC}/R_{src}$, independently of R_L. But this same situation applies when a load R_L is connected to an ideal current source of current $i_{SRC} = v_{SRC}/R_{src}$ (Figure 3.32a), in that the ideal current source delivers i_{SRC} to R_L independently of R_L. In other words, a voltage source having $R_{src} \gg R_L$ approximates an ideal current source as far as the load is concerned (Figure 3.32). It should be noted that having $R_{src} \gg R_L$ requires that for a given $i_L = i_{SRC}$, v_{SRC} has a large value nearly equal to $R_{src}i_L$.

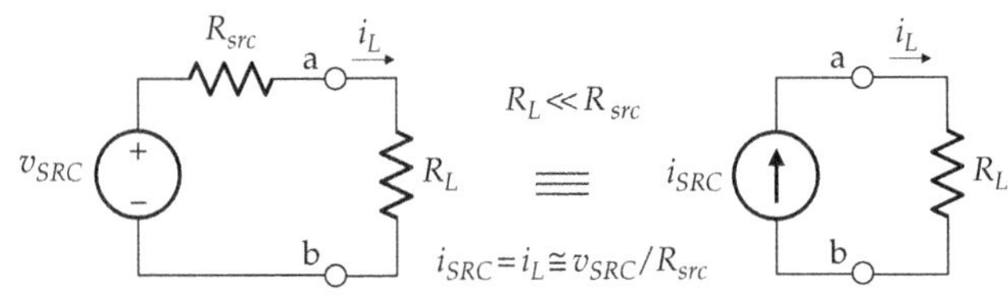

FIGURE 3.32
Approximation of an ideal current source.

Example 3.5: Transformation of Linear-Output Sources

Consider a linear-output voltage source of 12 V open-circuit voltage and 1.5 Ω source resistance supplying a 4.5 Ω load (Figure 3.33a). It is required to transform this source to an equivalent current source, verify that the load voltage and current are the same, and show that equivalence applies at the source terminals only.

Solution:

The load current in Figure 3.33a is $i_L = 12/(1.5+4.5) = 2$ A. The load voltage is $v_L = 4.5 \times 2 = 9$ V. This also follows from voltage division, since $v_L = \dfrac{4.5}{1.5+4.5} \times 12 = 9$ V. The linear-output current source has $i_{SRC} = 12/1.5 = 8$ A in parallel with 1.5 Ω (Figure 3.33b). From current division, $i_L = \dfrac{1.5}{1.5+4.5} \times 8 = 2$ A, which makes $v_L = 9$ V in Figure 3.33b. Note that $v_L = 8 \times (1.5||4.5) = 8 \times 1.125 = 9$ V.

Equivalence does not apply inside the sources. Thus, the voltage across the 1.5 Ω resistor in Figure 3.33a is 3 V, and the power dissipated in it is 6 W. The current in the 1.5 Ω resistor in Figure 3.33b is 6 A, and the power dissipated in it is 54 W. The power delivered by the ideal voltage source in Figure 3.33a is 24 W, whereas the power delivered by ideal current source in Figure 3.33b is 72 W. However, the power delivered by each source minus the power dissipated in the source resistance is the power delivered to the load, which must be the same in both cases, because of equivalence. Thus, $24-6 = 72-54 = 18$ W $= 9 \times 2$.

Simulation: The circuit is entered as in Figure 3.34. After selecting 'Bias Point' under 'Analysis type' in the Simulation Settings and running the simulation, pressing the I, V, and W buttons displays the currents, voltages, and powers, respectively (Figure 3.34).

Problem-Solving Tip

- In source transformation, equivalence does not apply inside the sources. This implies that the power delivered or absorbed by the ideal source, or the power dissipated in the source resistance, is not preserved under source transformation.

Primal Exercise 3.20

Transform the linear-output current source in Figure 3.35 to a linear-output voltage source between terminals 'ab'.

Ans. A voltage source of 5 V in series with 10 Ω.

Primal Exercise 3.21

Consider a current source of 6 A short-circuit current and 2 Ω source resistance. (a) Transform this source to a voltage source and back, (b) verify that the two equivalent

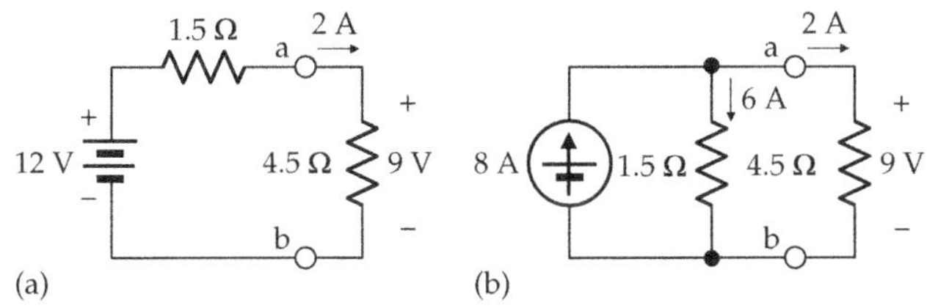

FIGURE 3.33
Figure for Example 3.5.

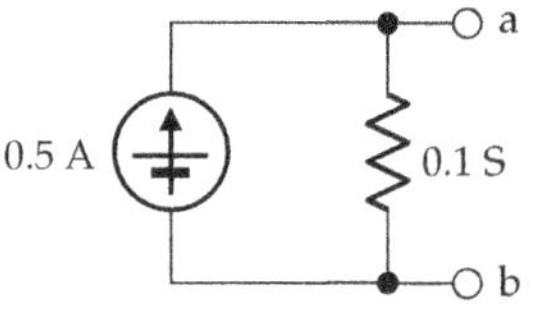

FIGURE 3.35
Figure for Primal Exercise 3.20.

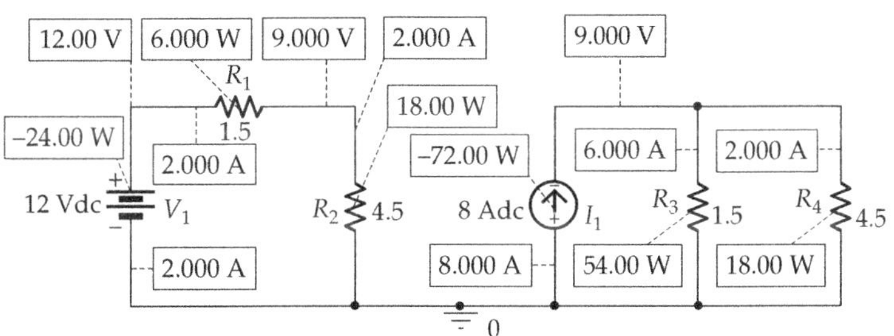

FIGURE 3.34
Figure for Example 3.5.

sources (i) have the same voltage across a 4 Ω load and will deliver the same power to the load, and (ii) have the same open-circuit voltage and short-circuit current at the source terminals.

Ans. (a) $V_{SRC} = 12$ V, $R_{src} = 2$ Ω; (b) $V_L = 8$ V, $P_L = 16$ W, 12 V, 6 A.

Example 3.6: Circuit Analysis Using Source Transformation

It is required to determine V_O in Figure 3.36.

Solution:

1. *Initialize*: All given values and the required V_O are entered. The nodes are labeled.
2. *Simplify*: The ideal current source and the 2 kΩ resistor in parallel with it can be transformed to the equivalent voltage source, whose source resistance can be combined with the 1 kΩ resistance. To ascertain the polarity of the transformed voltage source, the ideal current source and its 2 kΩ resistor are isolated from the rest of the circuit, as in Figure 3.37a. The 6 mA source current flowing in the 2 kΩ resistor makes node 'a' positive with respect to node 'b'. Hence, the polarity of the ideal voltage source in the transformed linear-output source should be as in Figure 3.37b. The justification for isolating the sources in order to determine polarity is that the transformed sources are equivalent under all terminal conditions, including isolation from the rest of the circuit. When the 2 kΩ resistance of the equivalent linear-output voltage source is combined with the 1 kΩ resistance, the circuit becomes as in Figure 3.38.
3. *Deduce*: The current in the 2 kΩ resistor in the middle is $V_O/2$ mA.
4. *Explore*: The circuit is now a two-essential-node circuit that can be analyzed in terms of KCL. The total current leaving node 'b' can be expressed in terms of V_O as

$$\frac{V_O}{2} + \frac{V_O - 12}{2} + \frac{V_O - 12}{3} = 0$$

which gives $V_O = 7.5$ V.

5. *Check*: An alternative solution is to transform the two linear-output voltage sources in Figure 3.38 to their equivalent linear-output current sources, as in Figure 3.39a. The circuit is now a parallel circuit having V_O across the paralleled elements. The two ideal current sources in parallel and in the same direction are combined into a single 10 mA source. R_{eqp} for the three resistors is derived by first combining the two 2 kΩ resistances in parallel into a 1 kΩ resistance that is paralleled with the 3 kΩ resistance to give $R_{eqp} = (3 \times 1)/(3 + 1) = 0.75$ kΩ, as in Figure 3.39b. It follows from this figure that $V_O = 7.5$ V.

Simulation: The circuit is entered as in Figure 3.40. After selecting 'Bias Point' under 'Analysis type' in the Simulation Settings and running the simulation, pressing I and V buttons displays the currents and voltages indicated in Figure 3.40.

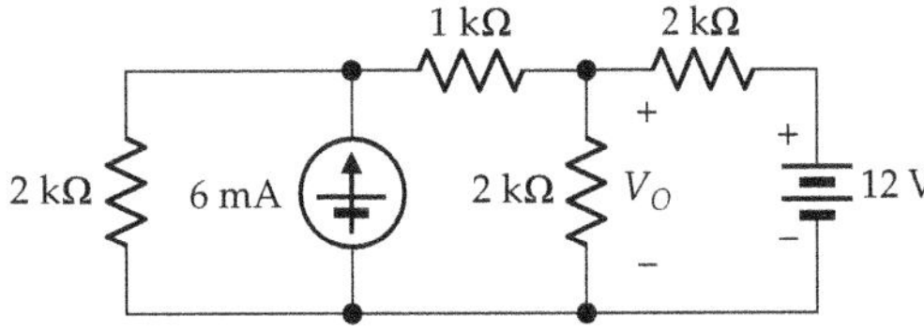

FIGURE 3.36
Figure for Example 3.6.

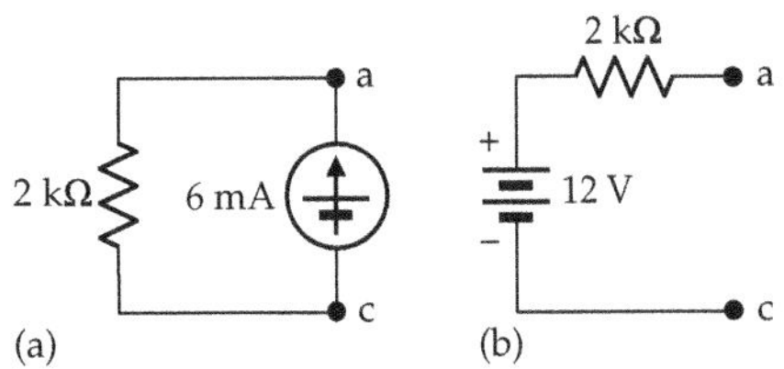

FIGURE 3.37
Figure for Example 3.6.

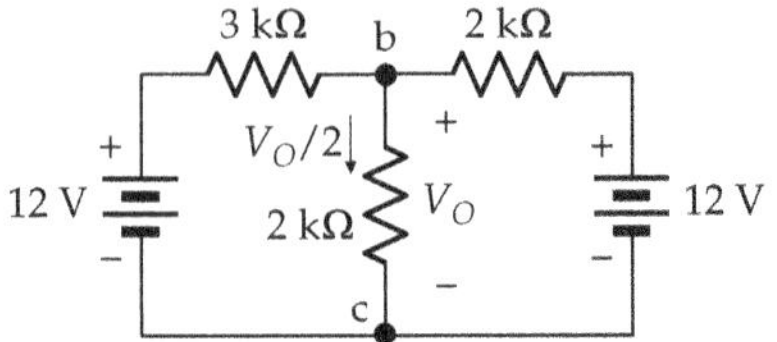

FIGURE 3.38
Figure for Example 3.6.

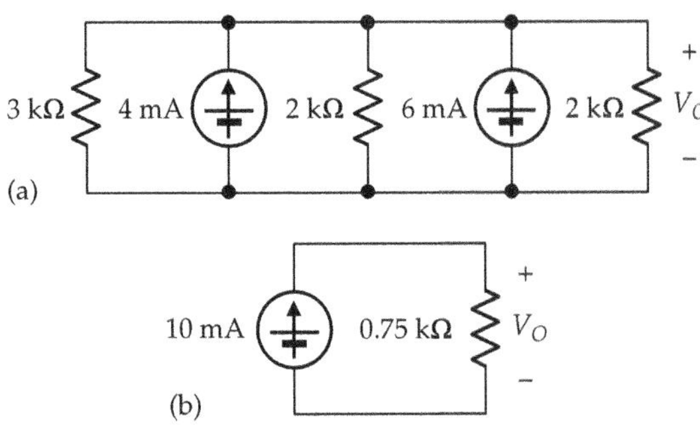

FIGURE 3.39
Figure for Example 3.6.

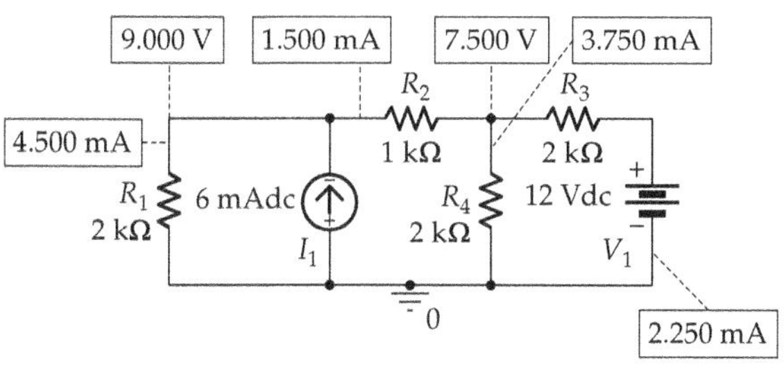

FIGURE 3.40
Figure for Example 3.6.

Primal Exercise 3.22

Determine I_X in Figure 3.41 using source transformation.

Ans. 0.8 A.

Primal Exercise 3.23

Determine the power dissipated in the 5 Ω resistor in Figure 3.42 by transforming the current sources to voltage sources.

Ans. 0.2 W.

Primal Exercise 3.24

Determine the current I in Figure 3.43 by transforming both linear-output current sources to their equivalent linear-output voltage sources.

Ans. 5 A.

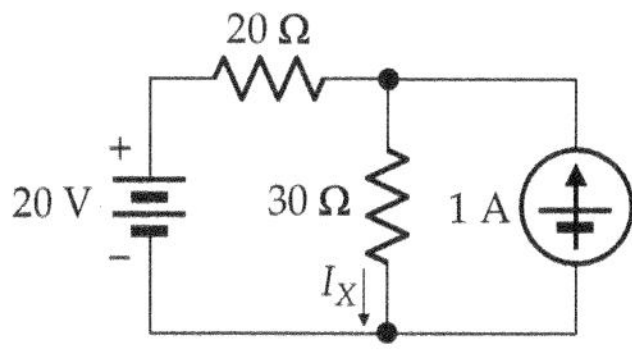

FIGURE 3.41
Figure for Primal Exercise 3.22.

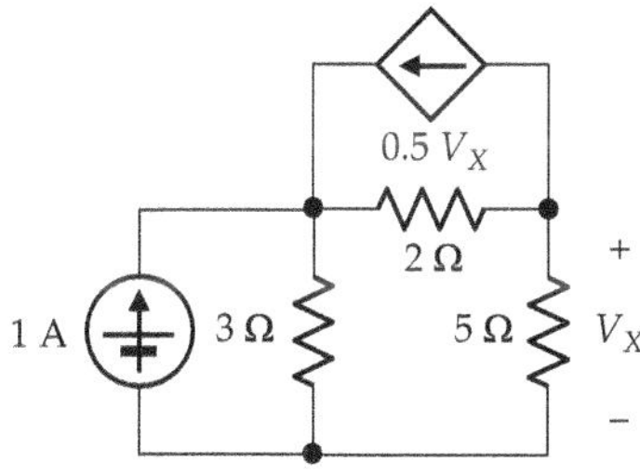

FIGURE 3.42
Figure for Primal Exercise 3.23.

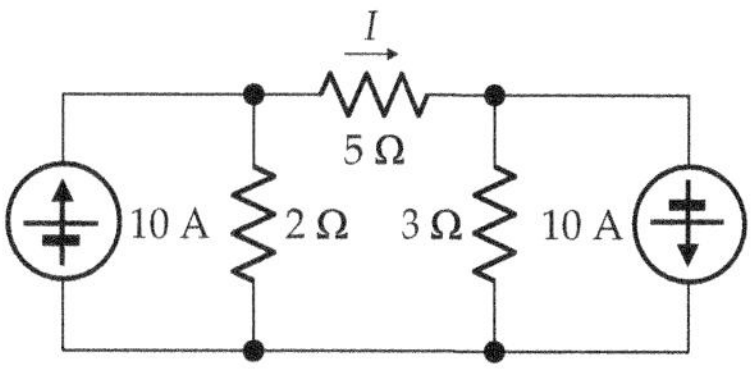

FIGURE 3.43
Figure for Primal Exercise 3.24.

3.7 Problem-Solving Approach Updated

The main procedural steps of the ISDEPIC approach are summarized and updated below in the light of the material covered in this chapter:

Step 1—Initialize:

(a) Mark on the circuit diagram all given values of circuit parameters, currents, and voltages, as well as the unknowns to be determined.
(b) Label the nodes, as this may be generally helpful.
(c) If the solution requires that a given value of current or voltage be satisfied, assume this value from the very beginning.

Step 2—Simplify: Consider as may be appropriate

(a) Redrawing the circuit
(b) Replacing series and parallel combinations of circuit elements by an equivalent circuit element
(c) Applying star–delta transformation
(d) Applying source transformation

Step 3—Deduce: Determine any values of current or voltage that follow immediately from direct application of Ohm's law, KCL, or KVL, without introducing any additional unknowns. If Step 3 does not provide the solution, proceed to Step 4.

Step 4—Explore: Consider the nodes and meshes in the circuit to see if KCL or KVL can be expressed using a single unknown current or voltage *and* if this unknown can then be directly determined from KCL or KVL. If Step 4 does not provide the solution, proceed to Step 5.

Step 5—Plan: Think carefully and creatively about the problem in the light of circuit fundamentals and circuit analysis techniques. Try to think of alternative solutions and select what seems to be the simplest and most direct solution.

Step 6—Implement: Carry out your planned solution.

Step 7: Check your calculations and results.

(a) Check that your results make sense, in terms of magnitude and sign.
(b) Check that Ohm's law is satisfied across every resistor, that KCL is satisfied at every node, and that KVL is satisfied around every mesh.
(c) Seek an alternative solution to see if it gives the same results.
(d) Whenever feasible, check the results with PSpice simulation.

Learning Checklist: What Should Be Learned from This Chapter

- Two circuits are equivalent at a given pair of terminals if the circuits have the same voltage–current relation at these terminals.
  1. If v and i are the voltage and current at the corresponding terminals of two equivalent circuits, then the time course of $v(t)$ and $i(t)$ is the same, the quotient v/i is the same, and the instantaneous power or the product vi is the same for both circuits.
- In a series connection of resistors:
  1. The equivalent series resistance R_{eqs} is the sum of the individual resistances.
  2. The reciprocal of the equivalent series conductance is the sum of the reciprocals of the individual conductances.
  3. R_{eqs} is larger than the largest of the series-connected resistors.
- In a series connection of resistors to which a voltage v is applied and the voltage across any of the individual resistors R_j is denoted by v_j, the ratio of v_j to v is the same as the ratio of R_j to the total series resistance.
- In a parallel connection of resistors:
  1. The equivalent parallel conductance is the sum of the individual conductances.
  2. The reciprocal of the equivalent parallel resistance R_{eqp} is the sum of the reciprocals of the individual resistances.
  3. R_{eqp} is smaller than the smallest paralleled resistance.
- In a parallel connection of resistors in which the total current is i and the current through any of the individual resistors R_j is denoted by i_j, the ratio of i_j to i is the same as the ratio of $G_j = 1/R_j$ to the total parallel conductance.
- The resistivity, or specific resistance, of a given material is the resistance between two opposite faces of a cube of the material multiplied by the length of the side of the cube. The resistivity of a given material is an intrinsic property of the material that is indicative of its electrical resistive property. It is independent of the size or shape of the conducting material.
- The reciprocal of resistivity is conductivity.
- Resistance is directly proportional to the length of the current path through a conductor and inversely proportional to the cross-sectional area through which the current flows.
- Δ-connected resistors can be substituted for the equivalent Y-connected resistors, and conversely, terminal for terminal, without affecting the voltages and currents in the rest of the circuit.
- Ideal, independent and dependent, voltage sources can be connected in series without restrictions on the source voltages and polarities. The source voltages add algebraically along the path.
- Ideal voltage sources should not be paralleled, unless they have the same source voltage and the same polarity. Otherwise, conservation of energy is violated.
- Ideal current sources should not be connected in series, unless they have the same source current and in the same direction. Otherwise, conservation of charge is violated.
- Ideal, independent and dependent, current sources can be connected in parallel without restrictions on the source currents and directions. The source currents add algebraically at the nodes between which the sources are paralleled.
- If the voltage at the source terminals of a nonideal voltage source decreases linearly with the current delivered by the source, the nonideal voltage source can be represented by an ideal voltage source in series with an ideal resistor. This series combination is referred to as a linear-output voltage source.
  1. The linear-output voltage source is characterized by (a) its open-circuit voltage, which is the voltage of the ideal source element, and (b) its source resistance, or internal resistance, measured between the source terminals with the ideal voltage source set to zero.
- If the voltage at the source terminals of a nonideal current source decreases linearly with the current supplied by the source, the nonideal current source can be represented by an ideal current source in parallel with an ideal resistor. This parallel combination is referred to as a linear-output current source.

1. The linear-output current source is characterized by (a) its short-circuit current, which is the current of the ideal source element, and (b) its source resistance, or internal resistance, measured between the source terminals with the ideal current source set to zero.

- A linear-output voltage source of open-circuit voltage v_{SRC} and source resistance R_{src} can be transformed to a linear-output current source of short-circuit current v_{SRC}/R_{src} in parallel with R_{src}. Conversely, a linear-output current source of short-circuit current i_{SRC} and source resistance R_{src} can be transformed to an linear-output voltage source of open-circuit voltage $R_{src}i_{SRC}$ and source resistance R_{src}.
  1. The transformation should preserve polarities of voltages and directions of currents at the terminals of the two sources.
  2. The transformation applies at the specified terminals only and does not apply inside the sources.
- An ideal, independent or dependent, voltage source has zero source resistance, whereas an ideal, independent or dependent, current source has zero source conductance or infinite source resistance.
- The ideal termination for a nonideal voltage source is an open circuit, whereas the ideal termination for a nonideal current source is a short circuit. However, no power is delivered by the sources under these conditions.
- A voltage source having $R_{src} \gg R_L$ approximates an ideal current source as far as the load R_L is concerned, since the current through R_L is almost independent of R_L.

Problem-Solving Tips

1. The parallel resistance of more than two resistors is conveniently determined from paralleled resistances, two at a time.
2. Never apply voltage division except to resistors in series, that is, resistors that carry the same current.
3. In current division the largest current flows in the branch having the smallest resistance, and conversely.
4. Never apply current division except to resistors in parallel.
5. In source transformation, equivalence does not apply inside the sources. This implies that the power delivered or absorbed by the ideal source, or the power dissipated in the source resistance, is not preserved under source transformation.

Problems

Apply ISDEPIC and verify solutions by PSpice simulation whenever feasible.

Equivalent Resistance

P3.1 Determine R_{eq} in Figure P3.1.
Ans. 30 Ω.

P3.2 Determine G_{eq} in Figure P3.2.
Ans. 6 mS.

P3.3 Determine R_{eq} in Figure P3.3.
Ans. 0.45 kΩ.

P3.4 Determine R_{eq} in Figure P3.4.
Ans. 8 Ω.

P3.5 Determine R_{eq} in Figure P3.5.
Ans. 6 Ω.

P3.6 Determine R_{eq} in Figure P3.6.
Ans. 5 Ω.

P3.7 Determine R_{eq} in Figure P3.7.
Ans. 4.5 Ω.

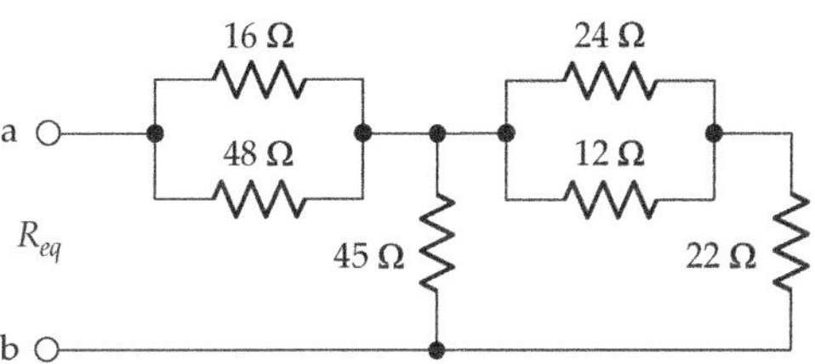

FIGURE P3.1

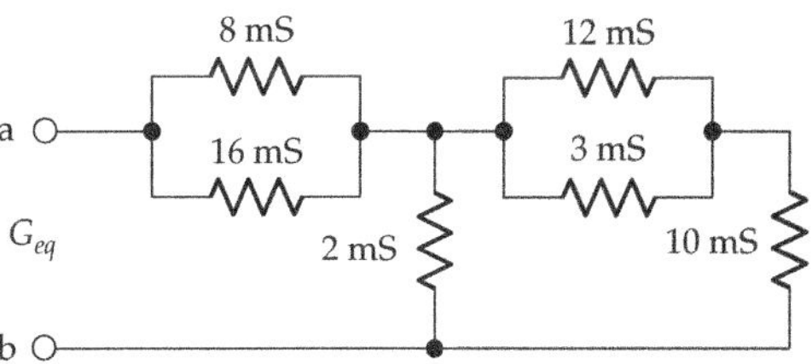

FIGURE P3.2

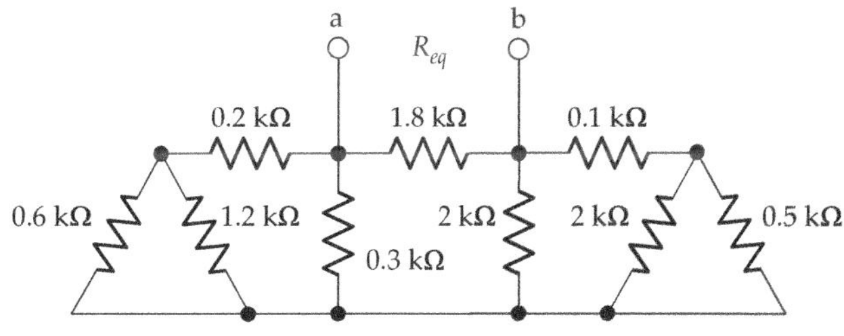

FIGURE P3.3

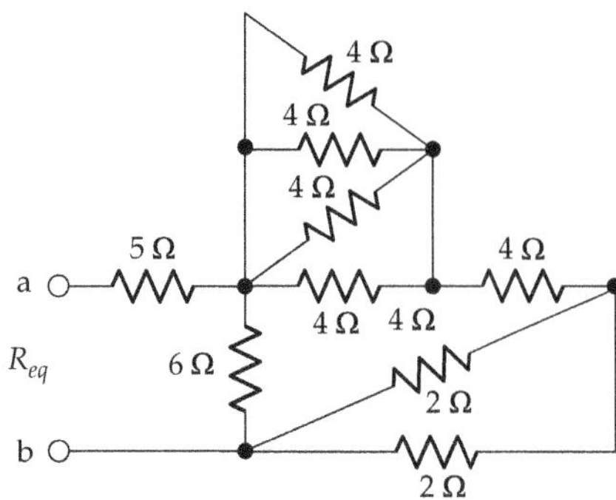

FIGURE P3.4

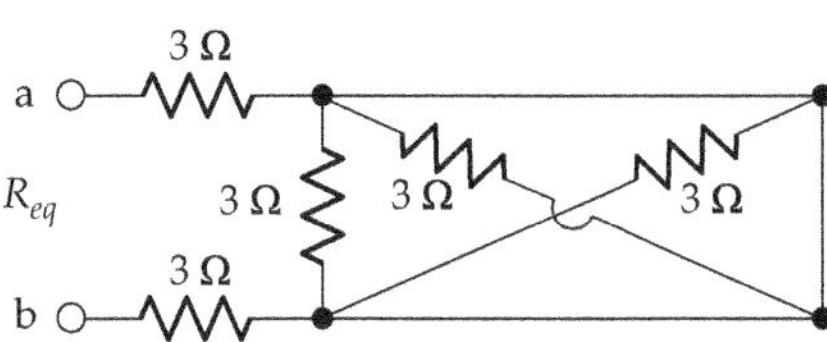

FIGURE P3.5

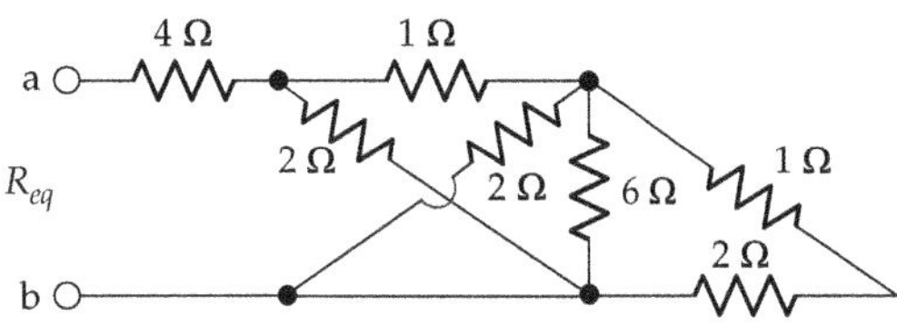

FIGURE P3.6

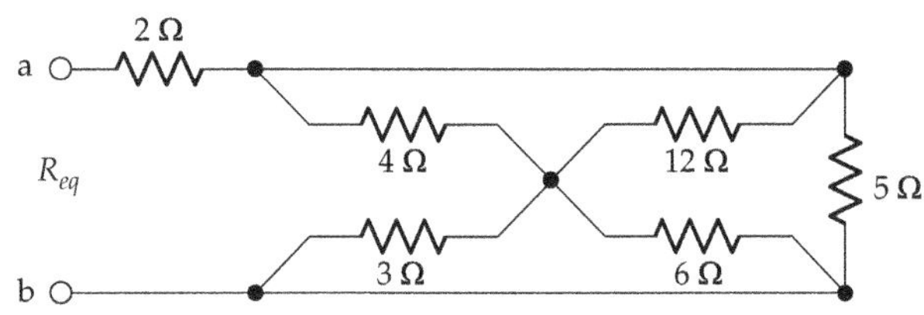

FIGURE P3.7

P3.8 Determine the resistance R_{be} between terminals 'b' and 'e' in Figure P3.8.

Ans. 15 Ω.

P3.9 Determine R_{eq} in Figure P3.9 when terminals 'cd' are (a) open-circuited and (b) short-circuited.

Ans. (a) 80/9 Ω; (b) 76/9 Ω.

P3.10 Determine G_{eq} between terminals 'ab' in Figure P3.9 if each resistance is replaced by a conductance having the same numerical value in S and with terminals 'cd' (a) open-circuited and (b) short-circuited.

Ans. (a) 171/20 S; (b) 9 S.

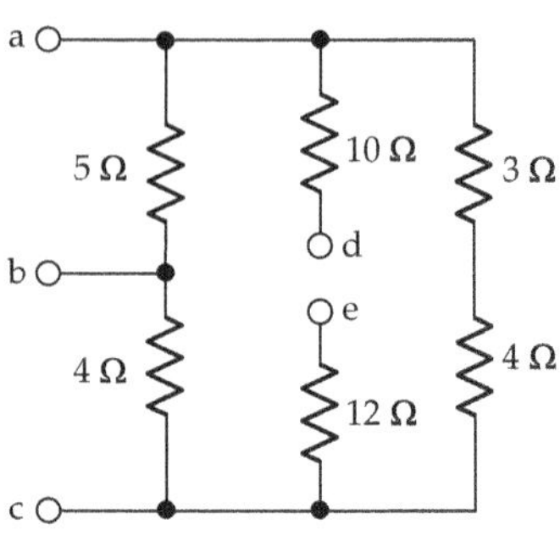

FIGURE P3.8

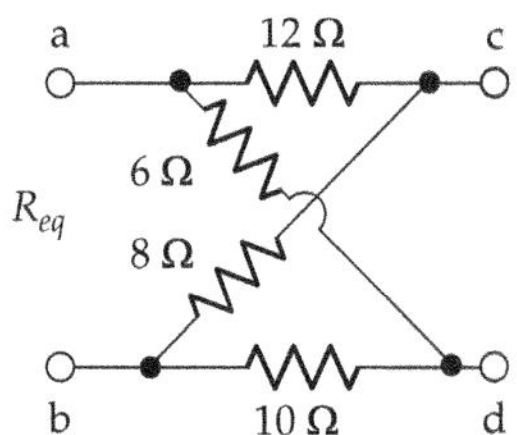

FIGURE P3.9

P3.11 Determine (a) R_{eq} between terminals 'ab' in Figure P3.11, assuming all resistances are 1 Ω, and (b) G_{eq} between terminals 'ab', assuming all conductances are 0.5 S.

Ans. (a) 0.25 Ω; (b) 2 S.

P3.12 Determine R_{eq} between terminals 'ab' in Figure P3.12.

Ans. 100/3 Ω.

P3.13 Determine G_{eq} between terminals 'ab' in Figure P3.13, where G is a conductance.

Ans. $7G/6$.

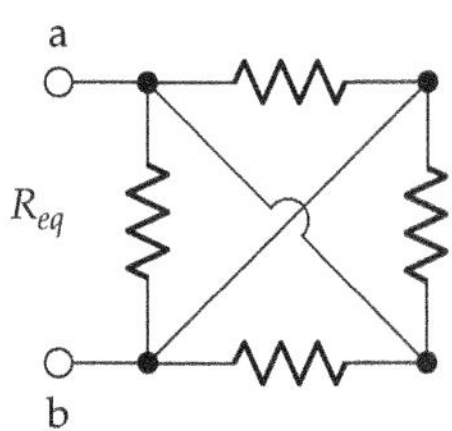

FIGURE P3.11

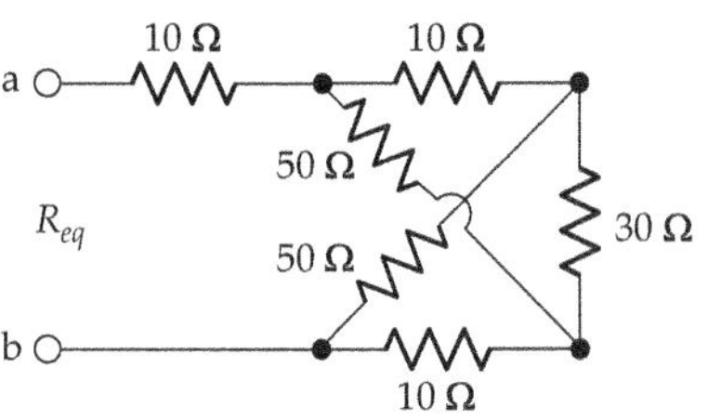

FIGURE P3.12

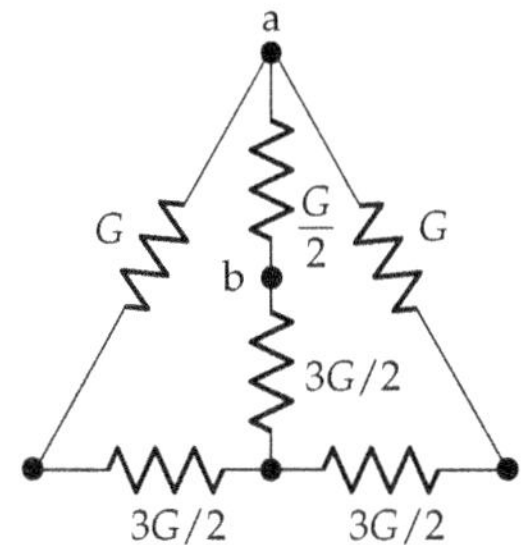

FIGURE P3.13

P3.14 Determine R_{eq} in Figure P3.14.

Ans. 10 Ω.

P3.15 Determine R_{eq} between terminals 'ab' in Figure P3.15, assuming all resistances are 1 Ω.

Ans. 1.2 Ω.

P3.16 Determine R_{eq} in Figure P3.16.

Ans. 5 Ω.

P3.17 Determine R_{eq} in Figure P3.17.

Ans. 3 Ω.

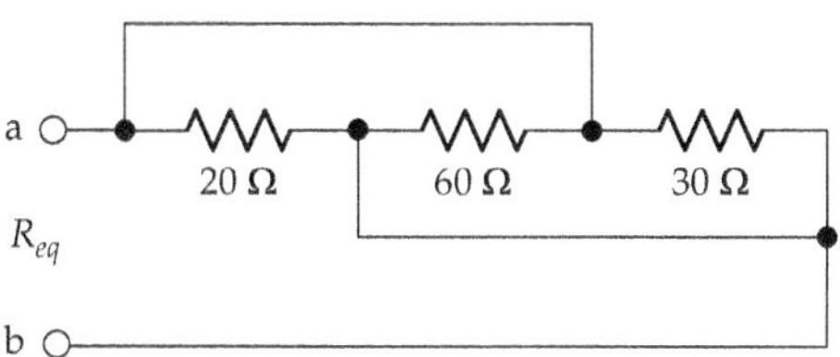

FIGURE P3.14

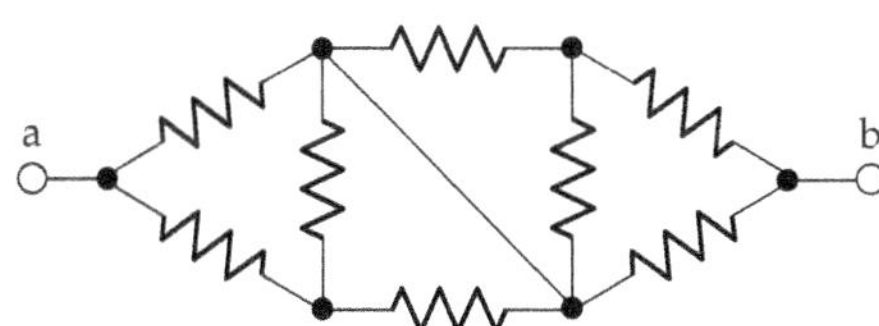

FIGURE P3.15

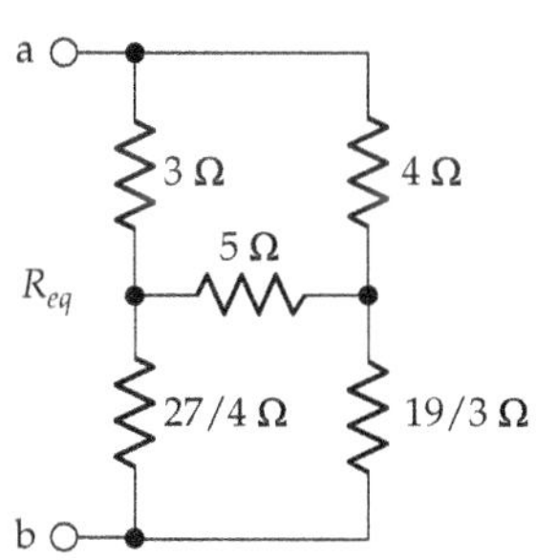

FIGURE P3.16

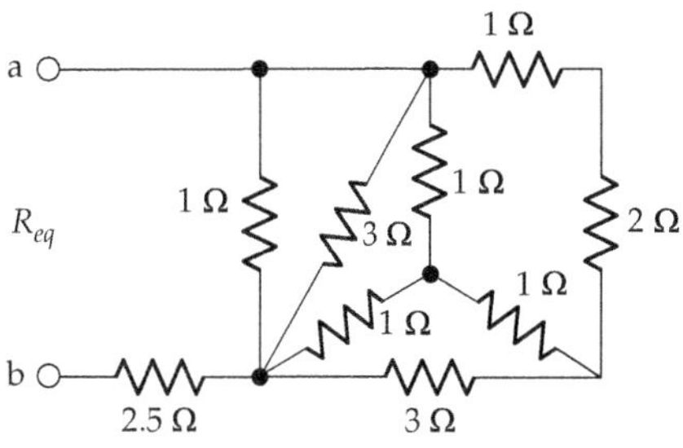

FIGURE P3.17

P3.18 Determine each of the resistors of the equivalent delta between terminals 'a', 'b', and 'c' in Figure P3.18, the value expressed in S.

Ans. 0.5 S.

P3.19 Determine V_O in Figure P3.19 given that the six, unmarked Y-connected resistors are 2 Ω each.

Ans. 2 V.

P3.20 Determine R_{in} in Figure P3.20.

Ans. 4.75 kΩ.

P3.21 Determine R_{in} in Figure P3.21.

Ans. Infinite.

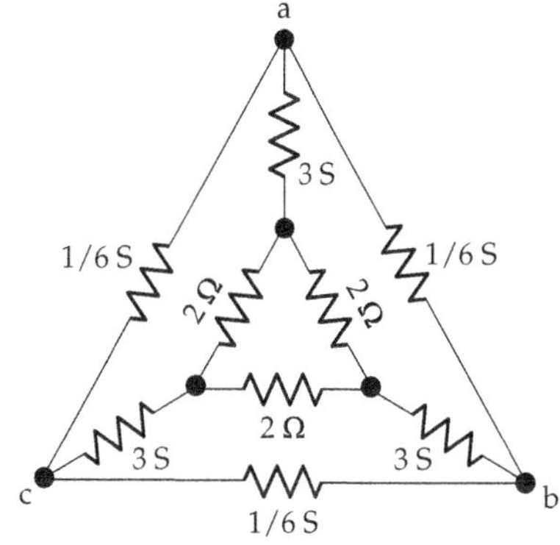

FIGURE P3.18

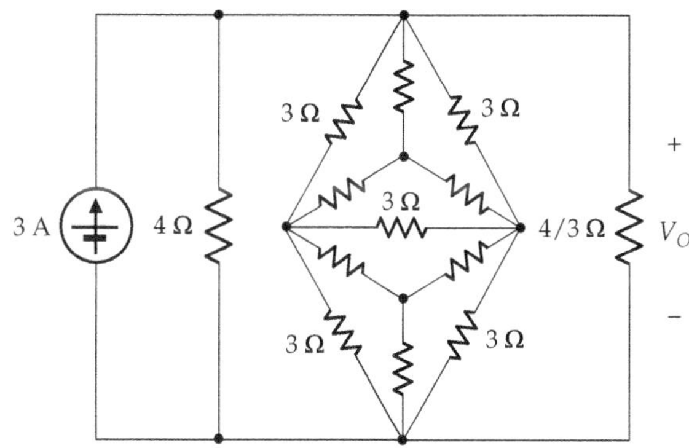

FIGURE P3.19

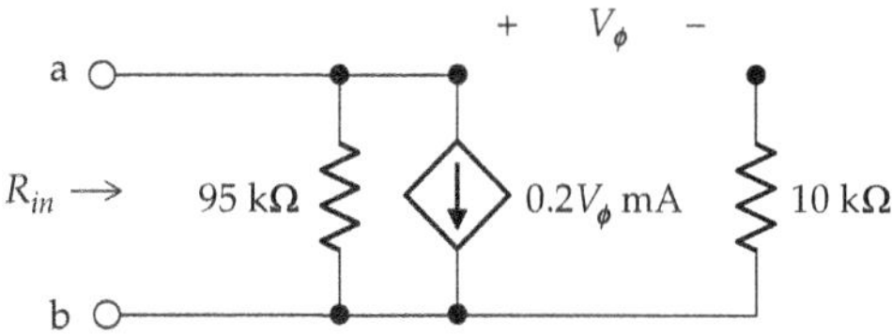

FIGURE P3.20

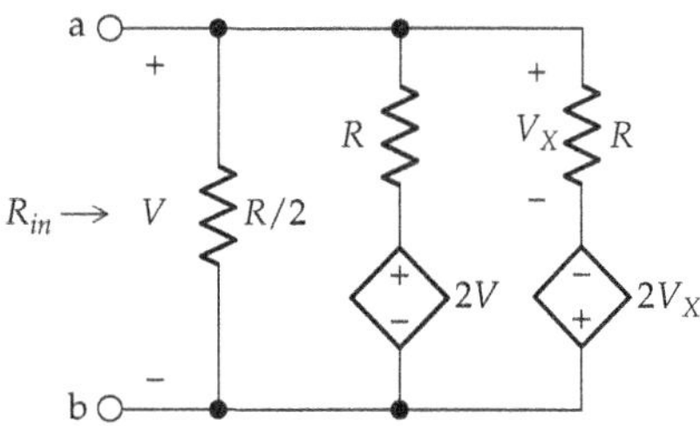

FIGURE P3.21

P3.22 n resistors are connected in series, the ith resistance being $2^{-(i-1)}$ where $i = 1, 2, \ldots, n$. Determine the total series resistance as $n \to \infty$.

Ans. 2 Ω.

Voltage Division, Current Division, and Source Transformation

P3.23 Determine V_O in Figure P3.23 by applying (a) voltage division and (b) source transformation and current division.

Ans. 24 V.

P3.24 Determine V_O in Figure P3.24.

Ans. 2 V.

P3.25 Determine V_O in Figure P3.25 using voltage division.

Ans. 7 V.

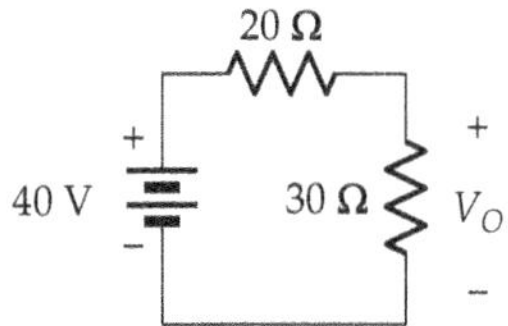

FIGURE P3.23

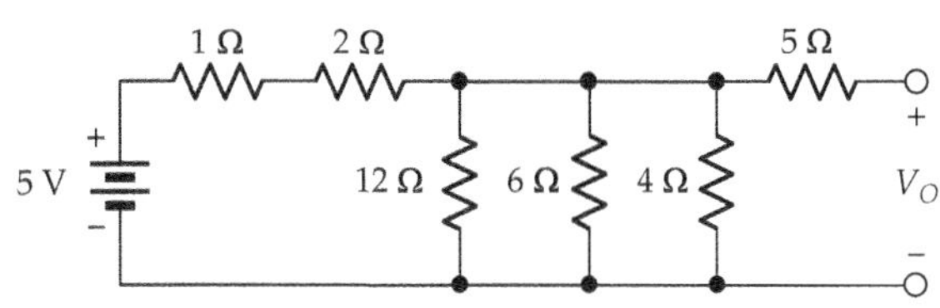

FIGURE P3.24

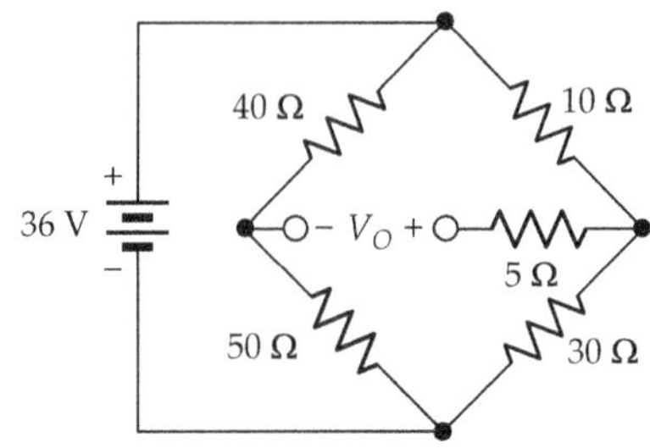

FIGURE P3.25

P3.26 Determine I_{SRC} in Figure P3.26.

Ans. −4 mA.

P3.27 Determine I_{SC} in Figure P3.27.

Ans. 0.2 A.

P3.28 Determine V_O by successive voltage division at nodes 'a', 'b', and then 'c' in Figure P3.28.

Ans. 16 V.

P3.29 Determine I_O by successive current division at nodes 'c', 'b', and then 'a' in Figure P3.29.

Ans. 25/3 mA.

P3.30 Determine V_X and I_Y in Figure P3.30.

Ans. 8 V, 0.375 A.

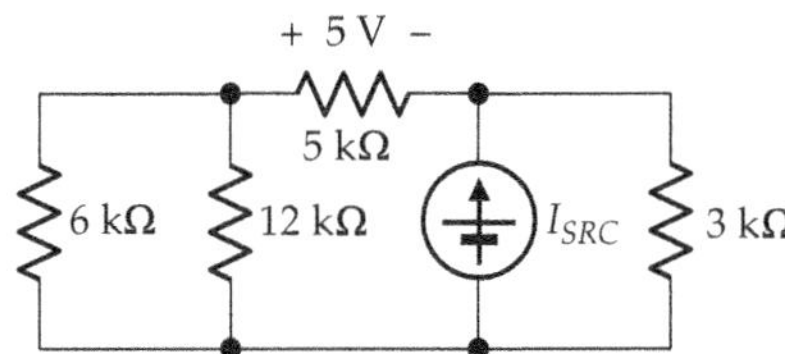

FIGURE P3.26

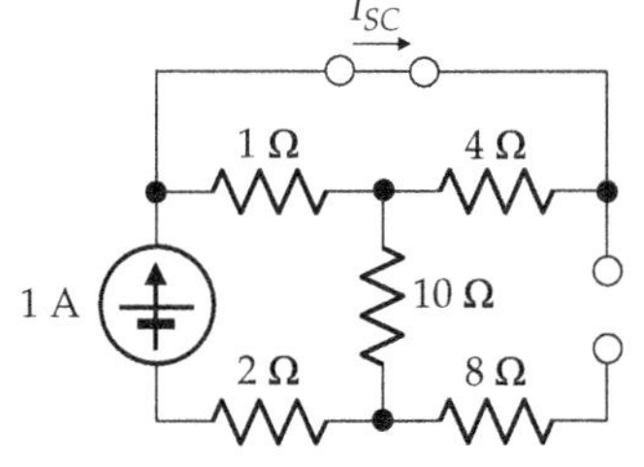

FIGURE P3.27

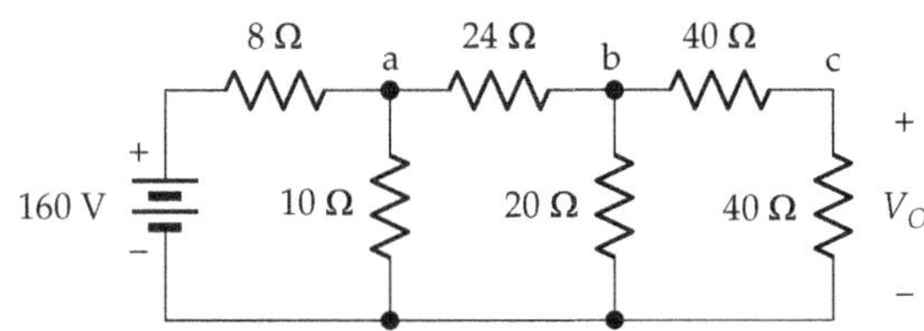

FIGURE P3.28

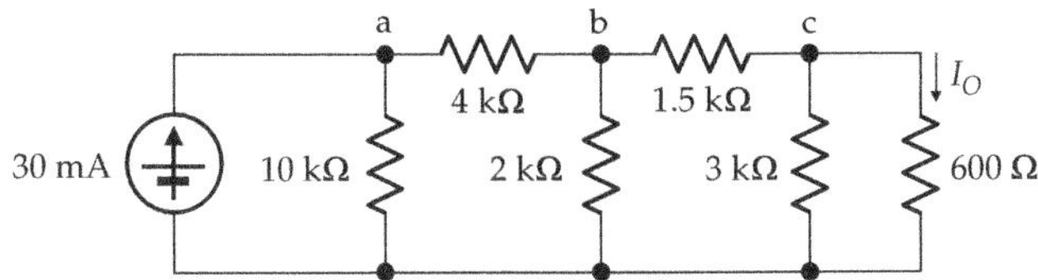

FIGURE P3.29

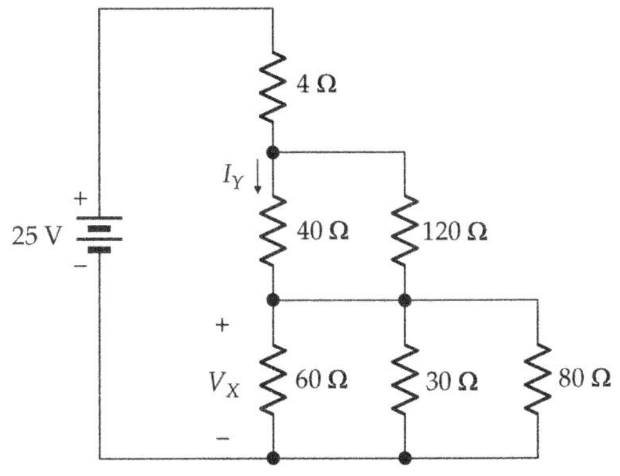

FIGURE P3.30

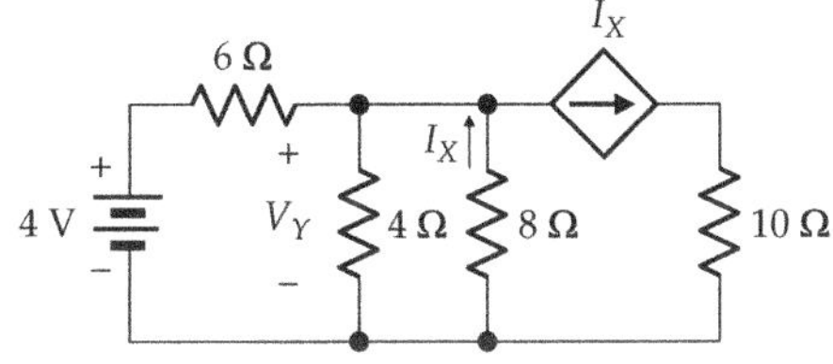

FIGURE P3.31

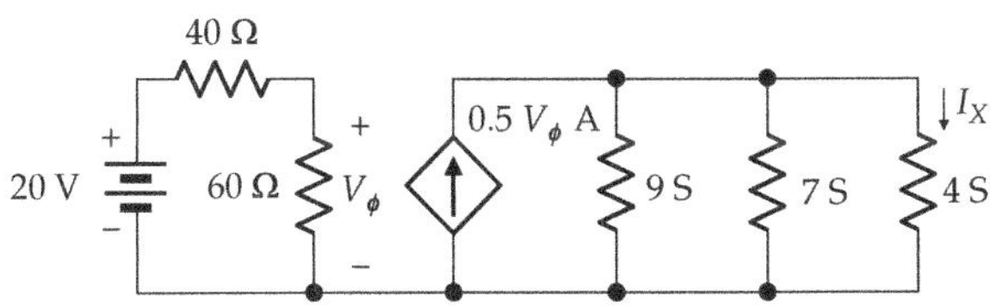

FIGURE P3.32

P3.31 Determine V_Y in Figure P3.31.

Ans. 1.6 V.

P3.32 Determine I_X in Figure P3.32.

Ans. 1.2 A.

P3.33 Determine the ratio of the power delivered by the ideal voltage source in Figure P3.33 to that delivered by the ideal current source of the equivalent linear-output source.

Ans. R_{src}/R_L.

P3.34 A nonideal voltage source has an open-circuit voltage V_{SRC}. When connected to a load resistor that draws a current of 1 A, the power dissipated in the load is four times the power dissipated in the source resistance. Determine the short-circuit current of the equivalent nonideal current source.

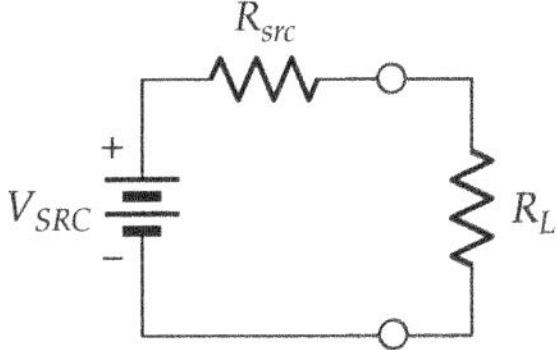

FIGURE P3.33

Ans. 5 A.

P3.35 Determine I_X in Figure P3.35 by transforming the dependent voltage source to its equivalent current source.

Ans. 1.5 V.

P3.36 Determine I_X and V_O in Figure P3.36 by transforming (a) the current source to a voltage source and (b) the voltage source to a current source. Note that I_X and V_O can be identified with respect to terminals 'ab'.

Ans. 1 A, 3 V.

P3.37 Determine R_x in Figure P3.37 so that $I_X = 0$.

Ans. 5 Ω.

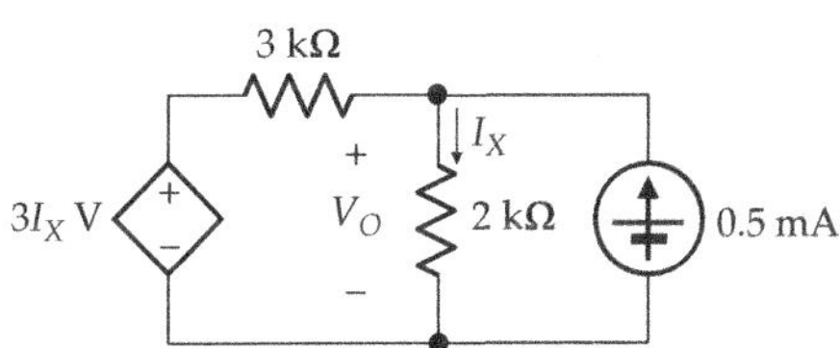

FIGURE P3.35

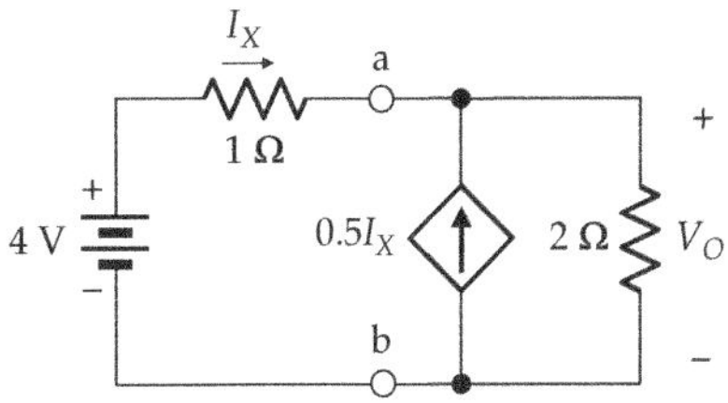

FIGURE P3.36

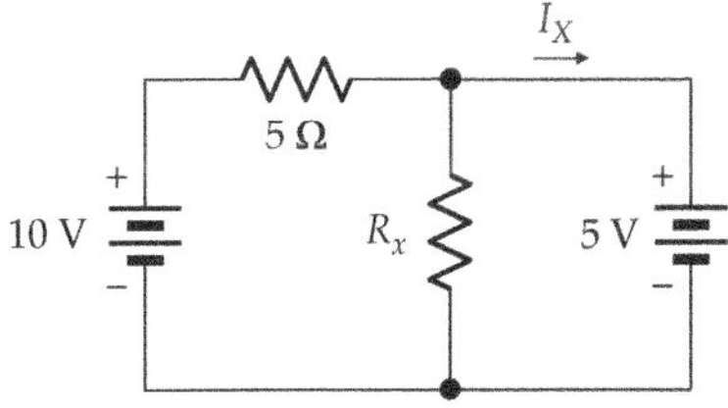

FIGURE P3.37

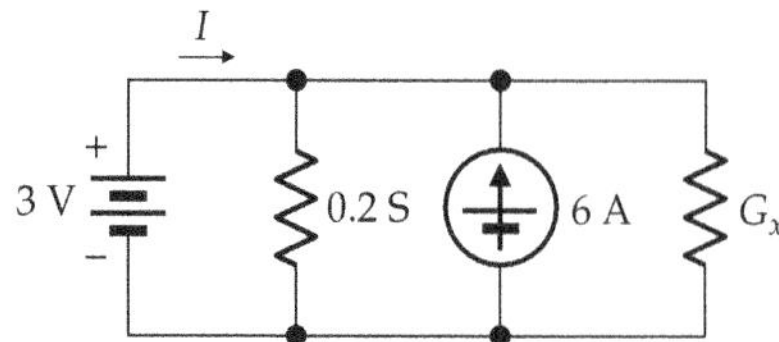

FIGURE P3.38

P3.38 Determine G_x in Figure P3.38 so that $I = 0$.
Ans. 1.8 S.

P3.39 Determine V_O in Figure P3.39.
Ans. 36 V.

P3.40 Determine V_O in Figure P3.40.
Ans. 6 V.

P3.41 Determine V_O in Figure P3.41.
Ans. 40 V.

P3.42 Determine I_O in Figure P3.42.
Ans. −1 A.

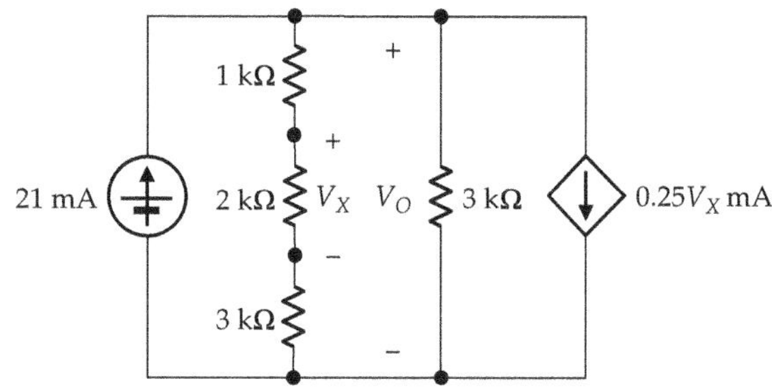

FIGURE P3.39

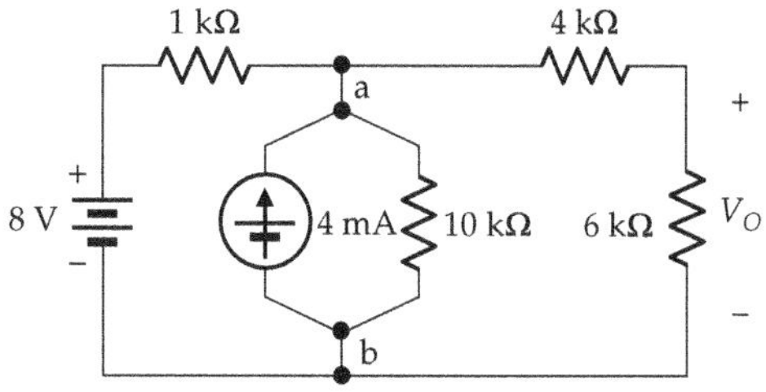

FIGURE P3.40

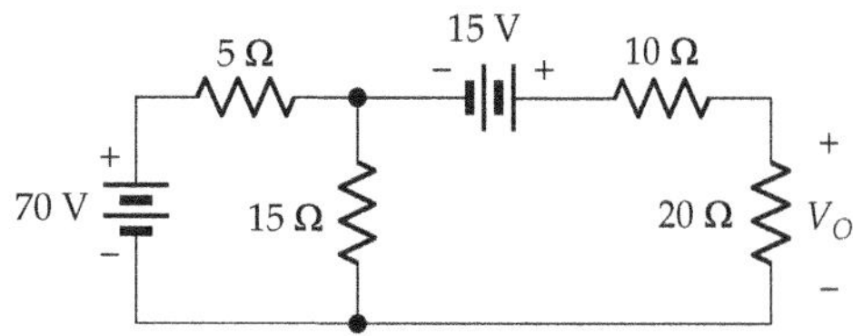

FIGURE P3.41

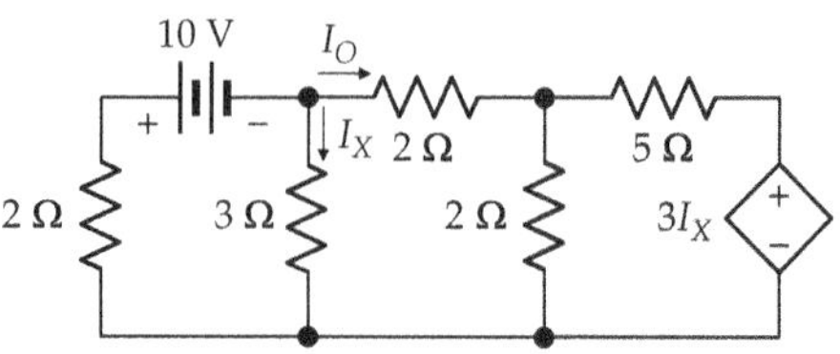

FIGURE P3.42

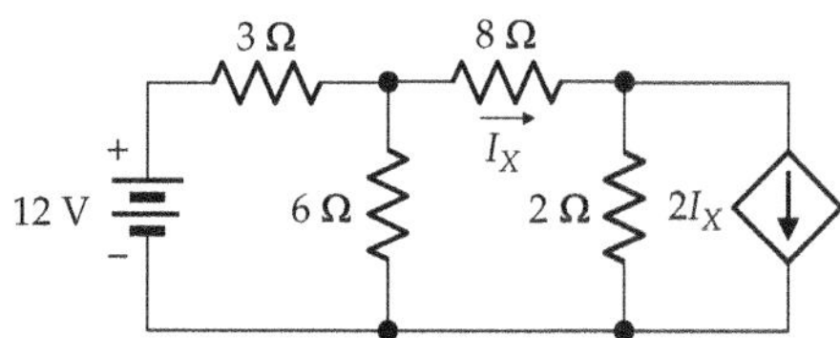

FIGURE P3.43

P3.43 Determine I_X in Figure P3.43 using source transformation and KVL.
Ans. 1 A.

P3.44 Determine V_O in Figure P3.44 using source transformation.
Ans. 16 V.

P3.45 Determine V_O and V_S in Figure P3.45.
Ans. $V_O = 30$ V, $V_S = 225$ V.

P3.46 Determine V_O in Figure P3.46.
Ans. 2/3 V.

P3.47 Determine R and I_X in Figure P3.47.
Ans. 3.2 Ω, 2 A.

P3.48 Determine V_S in Figure P3.48.
Ans. 35 V.

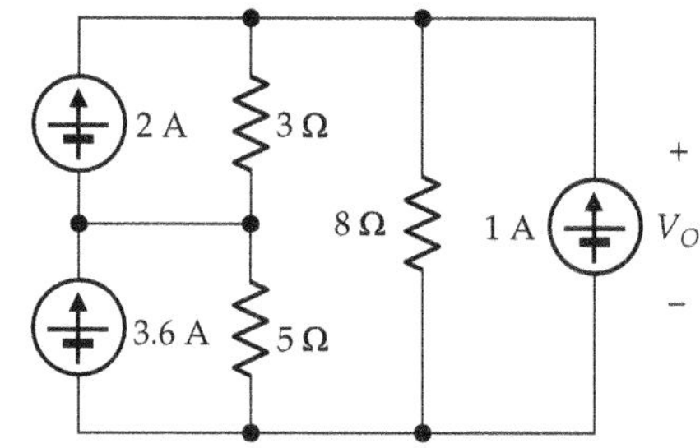

FIGURE P3.44

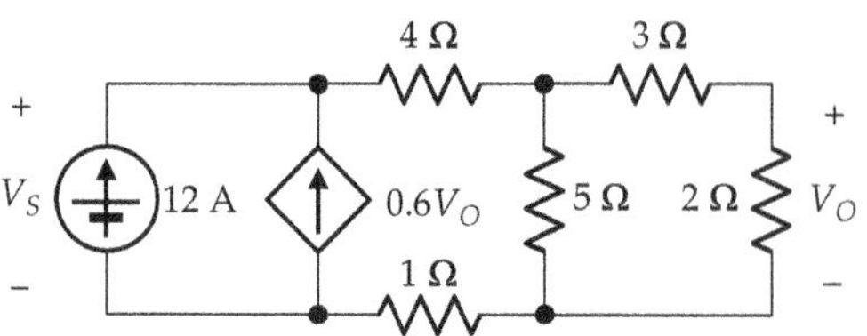

FIGURE P3.45

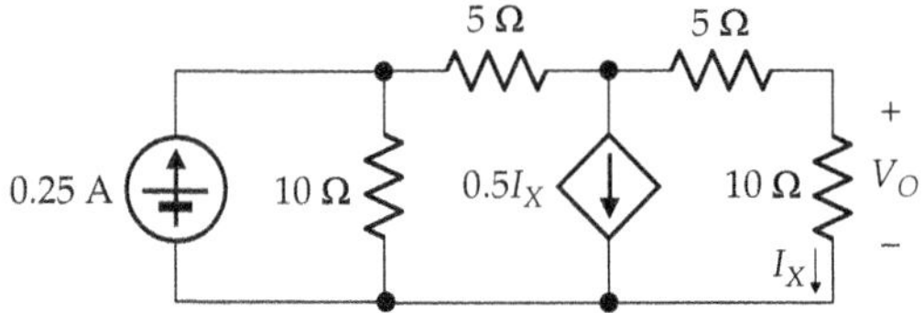

FIGURE P3.46

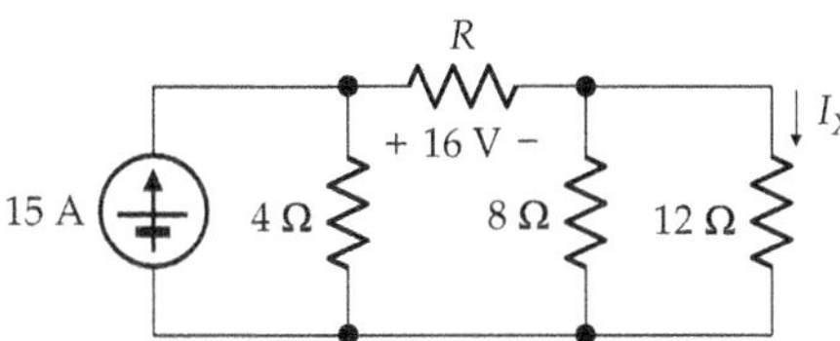

FIGURE P3.47

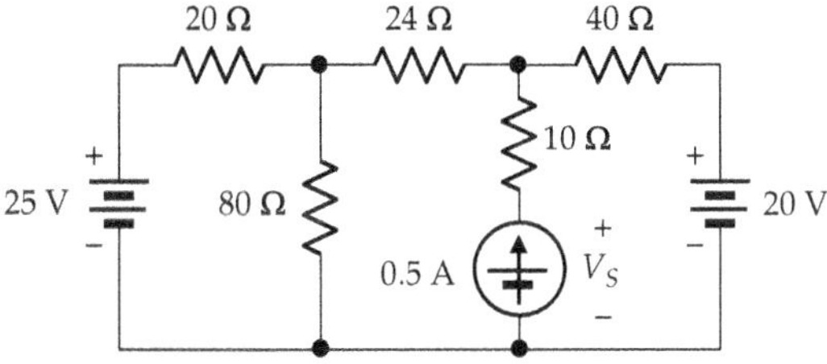

FIGURE P3.48

General Resistive Circuits

P3.49 R_X in Figure P3.49 can be varied between 0 and ∞. Determine (a) the largest value of I_S and (b) the smallest value of I_S, as R_X is varied over its full range.

Ans. 125 mA when $R_X = 0$, 80 mA when $R_X \to \infty$.

P3.50 Determine V_{SRC} and R in Figure P3.50 given that (a) when terminals 'a' and 'b' are open-circuited, the power delivered by V_{SRC} is 12 W and (b) when terminals 'a' and 'b' are short-circuited, the short-circuit current from 'a' to 'b' is 0.5 A.

Ans. 9 V, 4.5 Ω.

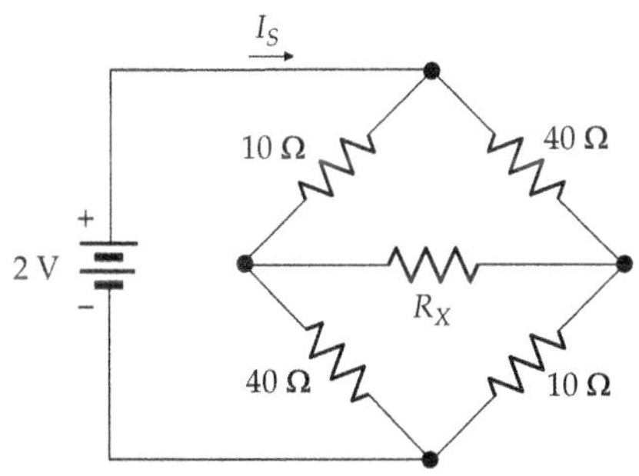

FIGURE P3.49

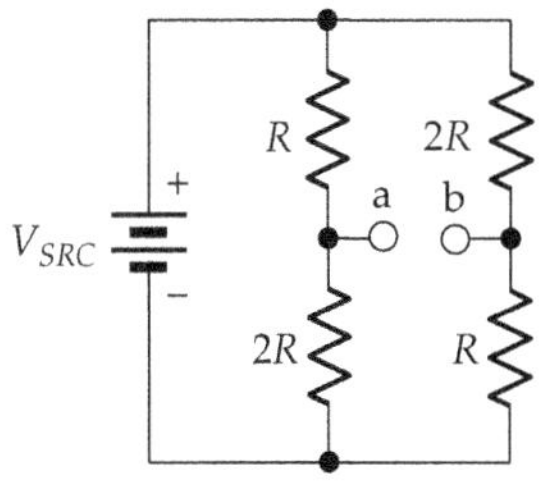

FIGURE P3.50

P3.51 Determine in Figure P3.51 (a) R_{eq} between terminals 'ab', (b) the total power absorbed by the resistors, (c) the power delivered or absorbed by each source, and (d) the battery voltage that will make the current source neither absorb or deliver power.

Ans. (a) 7 Ω; (b) 28 W; (c) voltage source absorbs 10 W, current source delivers 38 W; (d) 14 V, of polarity opposite to that shown.

P3.52 Determine I_{SRC} in Figure P3.52 so that no current flows in R_L.

Ans. 3 mA.

P3.53 Determine I_X in Figure P3.53.

Ans. 1.5 A.

P3.54 Determine V_X in Figure P3.54.

Ans. 1.5 V.

P3.55 Determine the power delivered or absorbed by the 4 V source in Figure P3.55.

Ans. 12 W delivered.

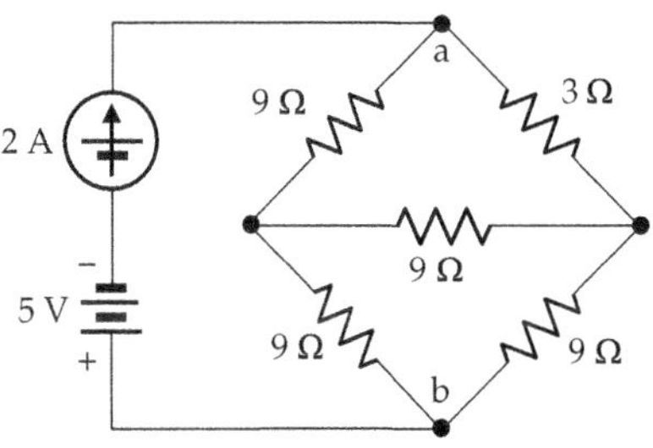

FIGURE P3.51

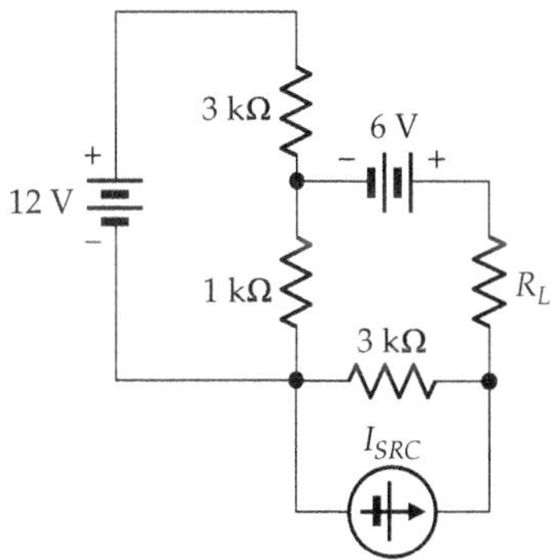

FIGURE P3.52

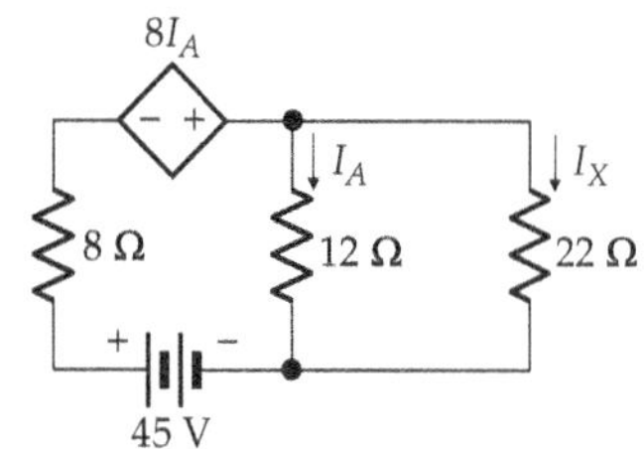

FIGURE P3.53

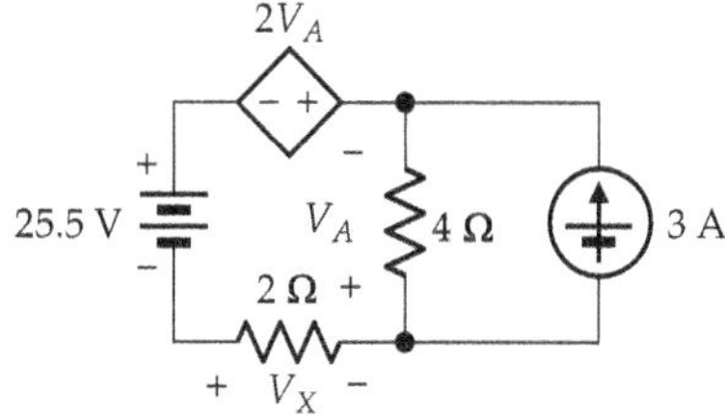

FIGURE P3.54

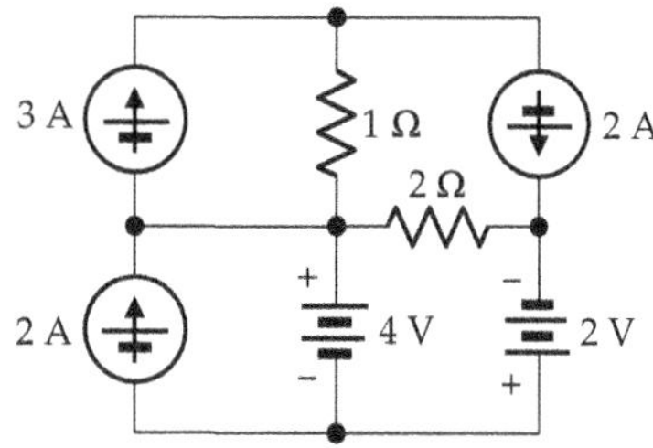

FIGURE P3.55

P3.56 Determine I_X in Figure P3.56.

Ans. −0.5 A.

P3.57 Determine R in Figure P3.57 such that no power is delivered or absorbed by the 2 V source.

Ans. 16 Ω.

P3.58 Determine V_X in Figure P3.58.

Ans. 50 V.

P3.59 Determine I_S in Figure P3.59.

Ans. 2 A.

P3.60 Determine I_S in Figure P3.60.

Ans. 5 A.

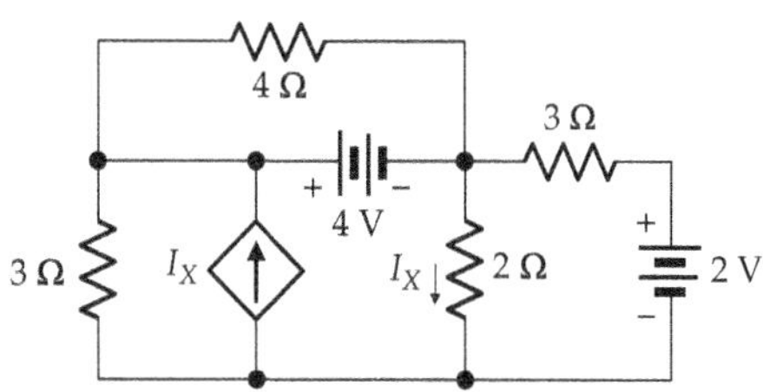

FIGURE P3.56

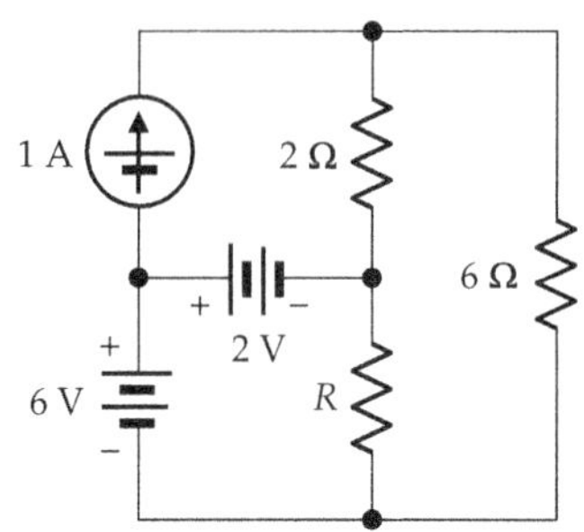

FIGURE P3.57

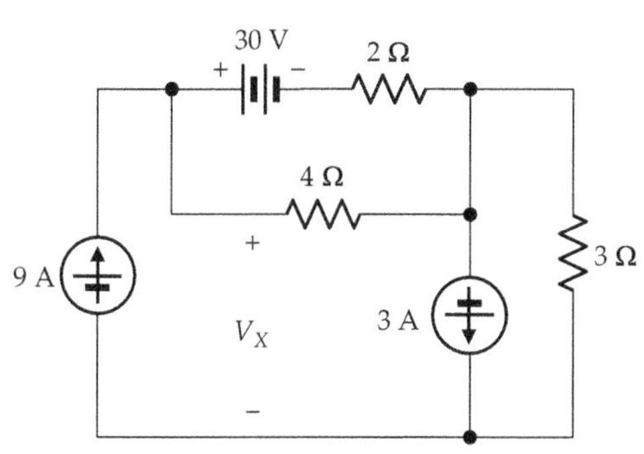

FIGURE P3.58

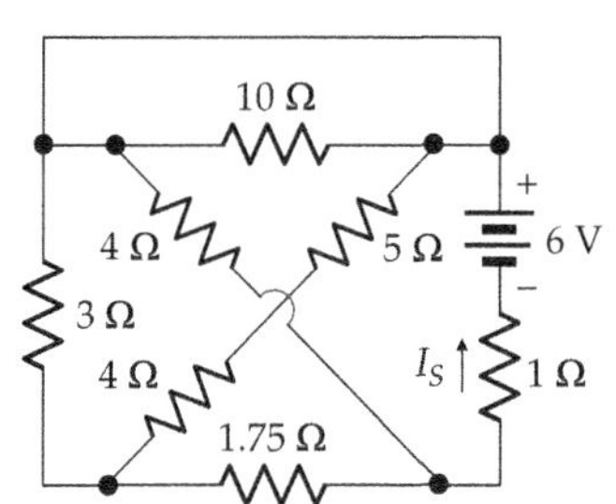

FIGURE P3.59

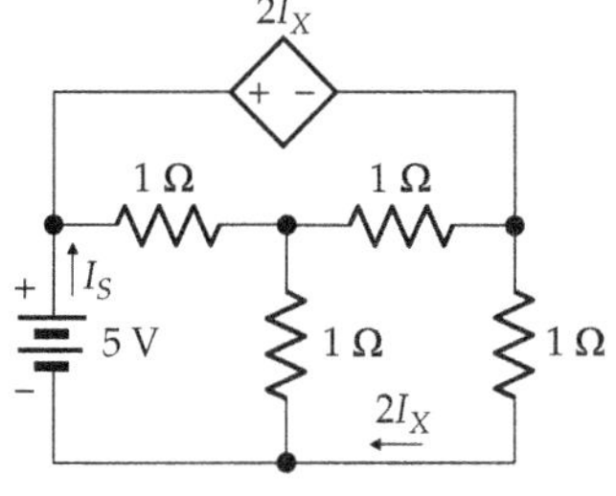

FIGURE P3.60

P3.61 Determine V_O in Figure P3.61.

Ans. 12 V.

P3.62 Determine V_Y in Figure P3.62.

Ans. 140 V.

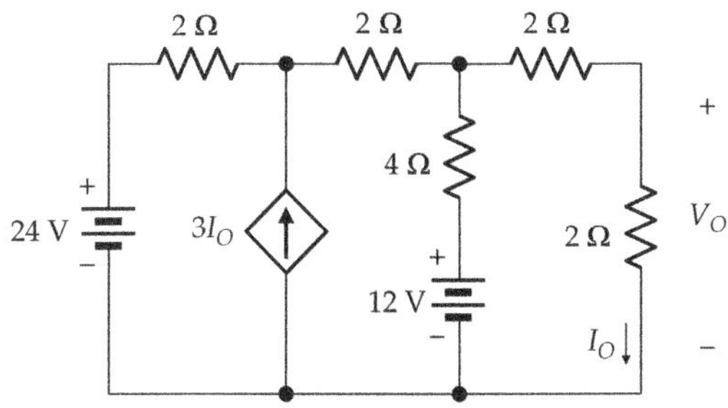

FIGURE P3.61

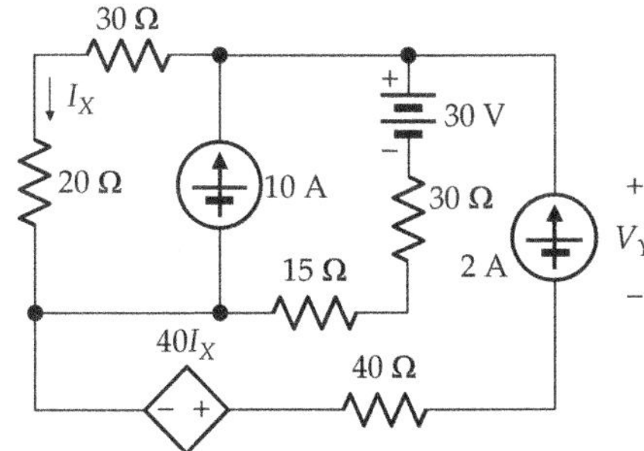

FIGURE P3.62

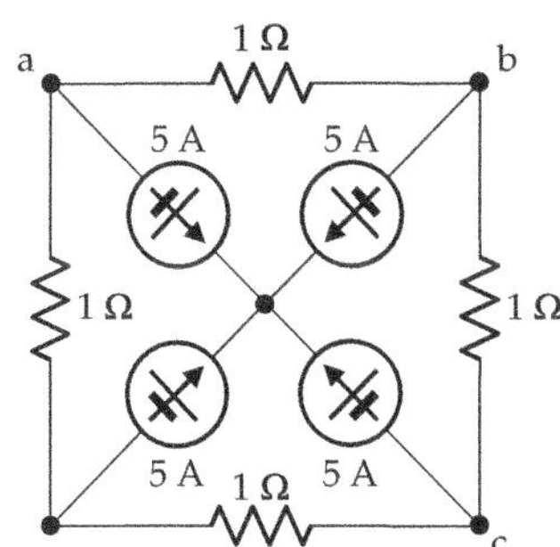

FIGURE P3.63

P3.63 (a) Argue that the circuit in Figure P3.63 is not valid, (b) reverse the directions of the upper two 5 A sources, and determine the power dissipated in the resistor between nodes (i) 'a' and 'b' and (ii) 'b' and 'c'.

Ans. (b) (i) 0, (ii) 25 W.

P3.64 (a) Argue that the circuit in Figure P3.64 is not valid, (b) replace the lower battery with a resistance R that dissipates 2 W, determine R, (c) determine the power delivered or absorbed by each battery, (d) show that no net power is absorbed or delivered by the three current sources, and (e) note that the voltage across each current source is indeterminate. Explain why.

Ans. (b) 2 Ω; (c) voltage source in branch 'ab' delivers 2 W, voltage source in branch 'ac' neither absorbs nor delivers energy; (d) power delivered by voltage sources equals that dissipated in R; (e) an ideal current source can have any voltage across it, and KVL only defines differences between voltages across individual current sources.

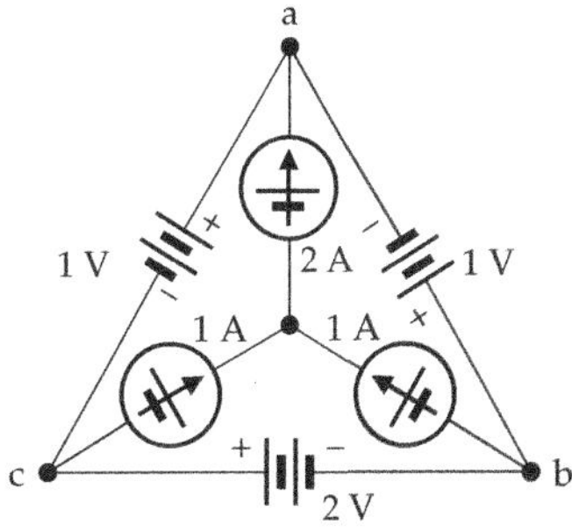

FIGURE P3.64

Probing Further

P3.65 Consider Equation 3.15 for the case of paralleled resistors. Show that for n resistors in parallel

$$\frac{1}{R_{eqp}} = \frac{1}{R_1} + \frac{1}{R_2} + \frac{1}{R_3} + \cdots + \frac{1}{R_{n-1}} + \frac{1}{R_n}$$

Deduce that

$$R_{eqp} = \frac{\prod_n R_i}{\sum_n \prod_{n-1} R_j}$$

where the numerator is the product of all the resistances and the denominator is the sum of n terms, each term consisting of the product of the different combinations of resistances $(n - 1)$ at a time.

P3.66 Suppose that equivalence between the Δ- and Y-circuits of Figure 3.17 is to be derived based on the resistance seen between any two terminals with the third terminal short-circuited to the terminal just preceding it in the sequence 'abcabc', that is, by deriving R_{ab} with terminal 'c' shorted to 'b', R_{bc} with terminal 'a' shorted to 'c', and R_{ca} with terminal 'b' shorted to 'a'. Show that the required relation between the resistors is the same as that of Equations 3.32 through 3.37.

4

Circuit Theorems

Objective and Overview

The chapter presents some theorems that apply to electric circuits and that center primarily around Thevenin's theorem.

Thevenin's theorem takes circuit equivalence to its extreme, by representing any LTI circuit, between any two given terminals, by a linear-output voltage source in the case of resistive circuits. Thevenin's theorem is arguably the most important theorem in circuit analysis, both from theoretical and practical viewpoints. It is therefore discussed at length with several examples that highlight some of its aspects.

The discussion of Thevenin's equivalent circuit (TEC) is naturally followed by a discussion of its current-source counterpart, namely, Norton's equivalent circuit (NEC). NEC has the added significance that some circuits may have an NEC, but not a TEC, just as the converse is also true.

The chapter ends with the substitution theorem, which is a useful theorem that simplifies the analysis of some types of circuits and can be readily proved using Thevenin's theorem. A particular form of the substitution theorem, the source absorption theorem, is presented as a useful tool for replacing dependent sources by resistors in some cases, which again simplifies circuit analysis in these cases.

Discussion in this chapter and in Chapters 5 and 6 is mostly restricted to the dc state.

4.1 Excitation by Dependent Sources

Before presenting Thevenin's theorem, the following concept is discussed:

Concept: *Dependent sources alone do not excite a circuit.*

To illustrate this concept, consider the circuit of Figure 4.1. From KCL at the upper essential node,

$$i_Y + 2i_Y = \frac{v_X}{6}$$

or

$$v_X = 18i_Y \tag{4.1}$$

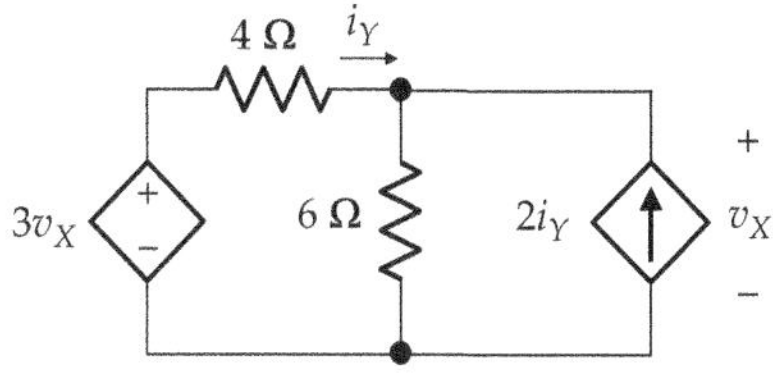

FIGURE 4.1
Excitation by dependent sources.

From KVL around the mesh on the left,

$$3v_X - 4i_Y - v_X = 0$$

or

$$v_X = 2i_Y \tag{4.2}$$

Substituting for i_Y from Equation 4.2 in Equation 4.1,

$$v_X = 9v_X$$

or

$$8v_X = 0 \tag{4.3}$$

It follows that $v_X = 0$, which also makes $i_Y = 0$.

Although demonstrated for a particular circuit, it is true in general that if dependent sources are the only sources in a circuit, the circuit is relaxed, that is, all voltages and currents in the circuit are zero.

Although they alone do not excite a circuit, dependent sources do, of course, affect the voltages and currents in the circuit. But the ultimate source of energy is the independent sources in the circuit. In the absence of these independent sources, there is no excitation in the circuit.

4.2 Thevenin's Theorem

In the context of resistive circuits, Thevenin's theorem can be stated as follows:

Statement: *A circuit consisting of ideal resistors and sources is equivalent, at a specified pair of terminals, to a linear-output voltage source.*

The equivalent circuit, consisting of an ideal voltage source in series with a resistor, is known as TEC. The open-circuit voltage, or the voltage of the ideal voltage source element, is referred to as the **Thevenin voltage**, V_{Th}, and the source resistance as the **Thevenin resistance**, R_{Th}. The open-circuit voltage and source resistance were defined for a linear-output voltage source in Section 3.6.

Thevenin's theorem takes circuit equivalence to the extreme, in that it reduces any LTI circuit at a given pair of terminals to the simplest possible equivalent, namely, an ideal voltage source in series with an ideal resistor.

Thevenin's theorem can be illustrated by the simple voltage divider circuit of Figure 4.2a supplying a load R_L. The circuit is a two-essential-node circuit that can be analyzed using KCL. The current leaving node 'a' is

$$I_L + \frac{V_L}{R_2} + \frac{V_L - V_{SRC}}{R_1} = 0 \quad (4.4)$$

Equation 4.4 can be rearranged as

$$V_L = \frac{R_2}{R_1 + R_2} V_{SRC} - \frac{R_1 R_2}{R_1 + R_2} I_L \quad (4.5)$$

Now let us replace the voltage divider by an ideal voltage source V_{Th} in series with an ideal resistor R_{Th} (Figure 4.2b). KVL gives

$$V_L = V_{Th} - R_{Th} I_L \quad (4.6)$$

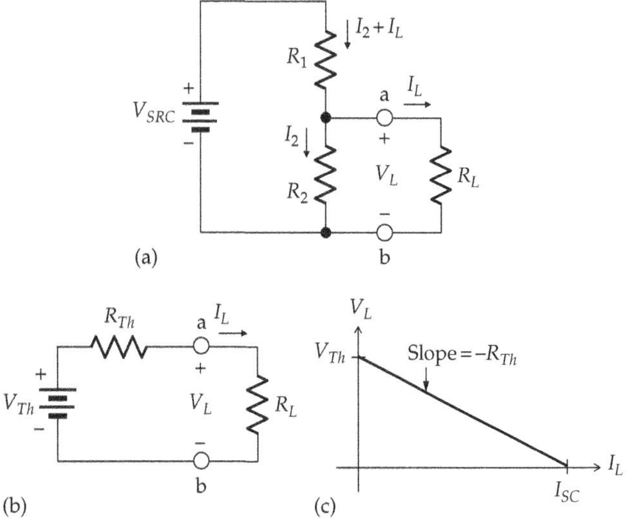

FIGURE 4.2
TEC of voltage divider. (a) Voltage divider supplying a load, (b) voltage divider represented by its TEC, and (c) the plot of the v–i relation at the load.

Equation 4.6 is of the same form as Equation 4.5, in accordance with Thevenin's theorem. Moreover, the two equations become identical if V_{Th} and R_{Th} are given by

$$V_{Th} = \frac{R_2}{R_1 + R_2} V_{SRC} \quad \text{and} \quad R_{Th} = \frac{R_1 R_2}{R_1 + R_2} \quad (4.7)$$

Under these conditions, V_{Th} in series with R_{Th} is equivalent to the voltage divider circuit at terminals 'ab', since they have the same V_L–I_L relation (Figure 4.2c). R_{Th} is the source resistance, and V_{Th} is the open-circuit voltage at terminals 'ab', when $I_L = 0$.

Although shown to apply for the voltage divider circuit, Thevenin's theorem in fact applies to a circuit of any complexity consisting of ideal sources and resistors. Evidently, replacing a complex circuit by its TEC between a given pair of terminals greatly simplifies the analysis of the overall circuit, as will be demonstrated on many occasions.

4.2.1 Derivation of TEC

In the preceding discussion, V_{Th} and R_{Th} were determined by deriving the V_L–I_L relation at the given pair of terminals, which is unnecessarily complicated. A simpler procedure is suggested by the nature of TEC itself and the V_L–I_L relation. Since the plot of V_L vs. I_L is a straight line (Figure 4.2c), this line is uniquely determined by specifying its slope, whose magnitude is R_{Th}, and a point through which the line passes, such as the intercept on the voltage axis. This intercept is V_{Th} and equals V_L, when $I_L = 0$. It can, therefore, be determined directly from the circuit as the open-circuit voltage at terminals 'ab'. Thus, if R_L is removed from the circuit (Figure 4.3a), it follows from voltage division that

$$V_{Th} = \frac{R_2}{R_1 + R_2} V_{SRC} \quad (4.8)$$

which is the same as Equation 4.7.

R_{Th} can be determined in one of two ways: either directly or as the ratio of the voltage intercept V_{Th} in

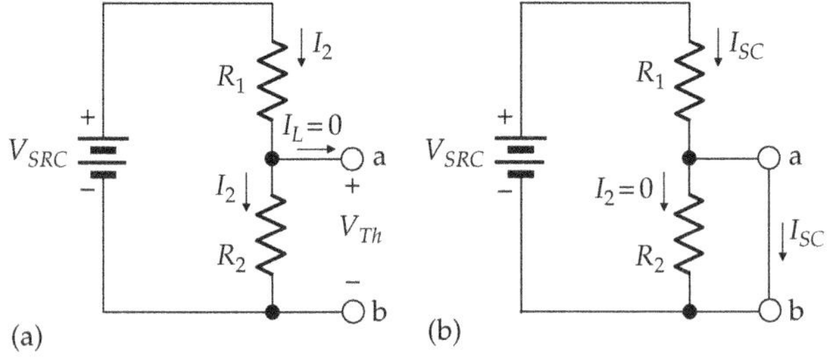

FIGURE 4.3
Derivation of TEC. Voltage divider with output terminals open circuited (a) and short circuited (b).

Figure 4.2c to the current intercept, which is the short-circuit current I_{SC}. When terminals 'ab' are short-circuited (Figure 4.3b), it follows from Ohm's law that

$$I_{SC} = \frac{V_{SRC}}{R_1} \tag{4.9}$$

The ratio V_{Th}/I_{SC} is, from Equations 4.8 and 4.9, $R_1R_2/(R_1 + R_2)$, the same as the expression for R_{Th} in Equation 4.7.

The direct method of determining R_{Th} is suggested by Figure 4.2b. If V_{Th} is set to zero, that is, replaced by a short-circuit (Section 2.4) and a test voltage source V_T is applied between terminals 'ab', then $R_{Th} = V_T/I_T$ (Figure 4.4a). That is, R_{Th} is R_{eq} seen by the test source between terminals 'ab'. Alternatively, a test current source I_T may be applied and R_{Th} is again given by V_T/I_T (Figure 4.4b). But what does setting V_{Th} to zero imply in the original circuit? It implies *setting all independent sources to zero*, for this removes excitation from the circuit and reduces all currents and voltages in the circuit to zero, including V_{Th}. As mentioned in Section 4.1, dependent sources alone do not excite the circuit. Hence, V_{Th} becomes zero when all independent sources are set to zero while leaving dependent sources unchanged. Note that although dependent sources alone do not excite the circuit, they do affect the relations between voltages and currents in the circuit in the presence of independent sources by effectively altering the values of resistances, including R_{Th}. Setting dependent sources to zero will therefore alter R_{Th}, whereas setting independent sources to zero removes excitation from the circuit, without altering R_{Th}. Recall that an ideal voltage source is set to zero by replacing it with a short circuit (Section 2.4) and that an ideal current source is set to zero by replacing it with an open circuit (Section 2.5).

It should be noted that applying a test source to determine R_{Th}, as in Figure 4.4a and b, is a formal and general method of determining R_{eq} between any two terminals of a circuit, with independent sources set to zero. That was, in fact, the method used for determining R_{eqs} and R_{eqp} in Chapter 3 (Figures 3.3 and 3.5). Using a test source is the *only* generally applicable method for

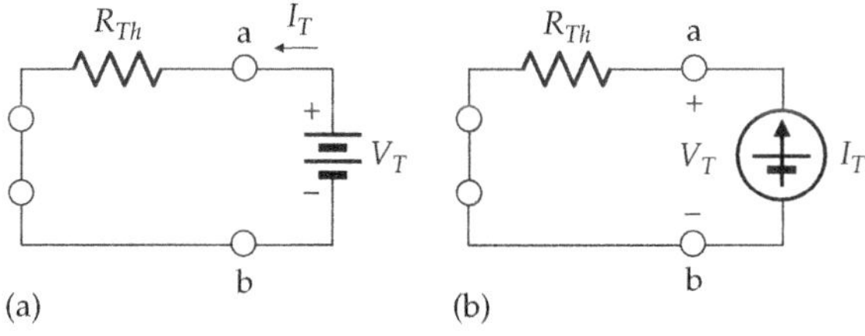

FIGURE 4.4
Alternative derivation of TEC. Determination of R_{Th} by applying a test voltage source (a), or a test current source (b), with $V_{Th} = 0$.

determining R_{eq} in the presence of dependent sources. However, *in the absence of dependent sources*, a test source need not be explicitly applied. Equivalently, R_{eq}, and hence R_{Th}, can be determined more directly using series/parallel combinations of resistors, star-delta transformation, etc.

The procedure for deriving TEC can be summarized as follows:

1. *Determine V_{Th} as the open-circuit voltage at the specified terminals.*
2. *Determine the short-circuit current I_{SC} at the specified terminals, which gives R_{Th} as V_{Th}/I_{SC}.*
3. *Set all independent sources in the given circuit to zero, leaving dependent sources unchanged. R_{Th} is the resistance R_{eq} looking into the specified terminals. Formally, this resistance is obtained by applying a test voltage source or a test current source and determining R_{eq} as the ratio of the voltage at the source terminals to the source current. In the absence of dependent sources, this effectively reduces to determining R_{eq} directly from series/parallel combinations of resistors, and using star-delta transformations, if necessary.*

The following should be noted concerning this procedure:

1. Since $V_{Th} = R_{Th}I_{SC}$, only two of the three quantities in this relation need be determined through the aforementioned three steps. However, it is useful for checking purposes to determine all three of these quantities independently.
2. Moreover, some of the aforementioned three steps many be easier to implement than others. Thus, setting independent sources to zero can make the circuit particularly simple.
3. In some cases, $V_{Th} = 0$, which means that $I_{SC} = 0$. It follows that V_{Th}/I_{SC} is 0/0, which is indeterminate. In this case, R_{Th} can only be determined by Step 3 of the aforementioned procedure. This is illustrated by Example 4.3.
4. A potential ambiguity in deriving TEC at a pair of terminals is whether or not to include in TEC a branch, such as a resistor R, that is connected between the given pair of terminals. The ambiguity is resolved in this book by the way the resistor is drawn with respect to the given terminals or by the way TEC is required. In Figure 4.5a, for example, R is drawn beyond the terminals at which TEC is required, as was done in Figure 4.2a. The implication is that R should *not* be included in TEC. Even without drawing the terminals in this manner, requiring

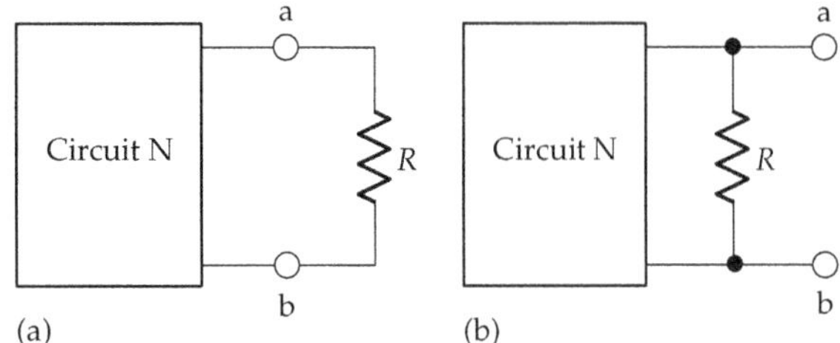

FIGURE 4.5
In determining TEC between terminals 'ab', R is not included in (a) but is included in (b).

"TEC seen by R," or "TEC looking into terminals 'ab'" means that R should not be included in TEC. On the other hand, if the given terminals are beyond R, as in Figure 4.5b, the implication is that R should be included in TEC. Requiring "TEC between terminals 'ab'" is unambiguous in this case.

5. Even in the case of Figure 4.5b, it may be advantageous to remove R and determine an intermediate TEC from the simpler circuit that results when R is removed. After obtaining this TEC, R is connected to the terminals of this TEC and the final TEC derived. This procedure is illustrated by Example 4.4.
6. If $R_L = R_{Th}$ (Figure 4.2b), the ideal voltage source V_{Th} transfers maximum power to R_L. This is proved in Section 17.4 and extended to more general cases.

⋆4.2.2 Derivation of TEC with PSpice

Although V_{Th}, I_{SC}, and R_{Th} can be derived from two separate simulations, it is possible, and more convenient, to derive TEC from a single simulation. The basis for this procedure is instructive and can be explained with reference to Figure 4.6a. A test current source I_T is applied at terminals 'ab', between which TEC is to be derived. This TEC, consisting of V_{Th} and R_{Th}, is shown between these terminals, which signifies that the original circuit is left as is, that is, with the *independent sources retained*, so that $V_{Th} \neq 0$. KVL gives

$$V_T = V_{Th} + R_{Th} I_T \tag{4.10}$$

If I_T is varied between 0 and 1 A, and V_T is plotted against I_T, a straight line graph is obtained having a voltage intercept V_{Th} at $I_T = 0$. Let V_{T_1} denote V_T at $I_T = 1$ A. From Equation 4.10, the difference $(V_{T_1} - V_{Th})$ is numerically equal to R_{Th} when $I_T = 1$ A (Figure 4.6b). I_T is conveniently varied in PSpice over a desired range of values, using the "DC Sweep" feature, as explained in Example 4.1.

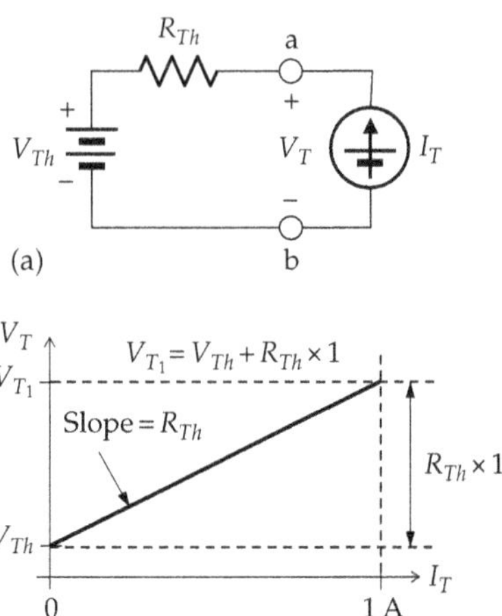

FIGURE 4.6
Derivation of TEC with PSpice. (a) Test current source applied to TEC and (b) graphical construction for determining TEC.

Example 4.1: Application of TEC

It is required to determine I_L in Figure 4.7 by deriving TEC looking into terminals 'ab'.

Solution:

When terminals 'ab' are open-circuited by removing the 50 Ω resistor, the circuit becomes as in Figure 4.8a.

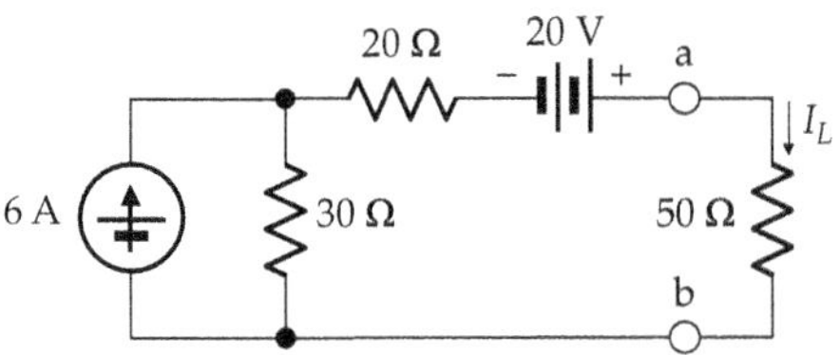

FIGURE 4.7
Figure for Example 4.1.

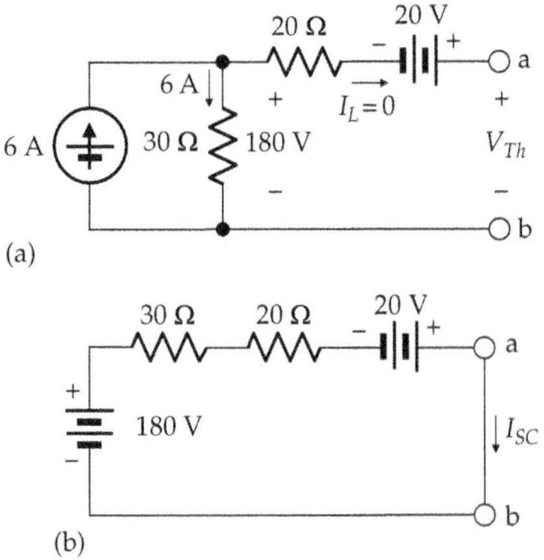

FIGURE 4.8
Figure for Example 4.1.

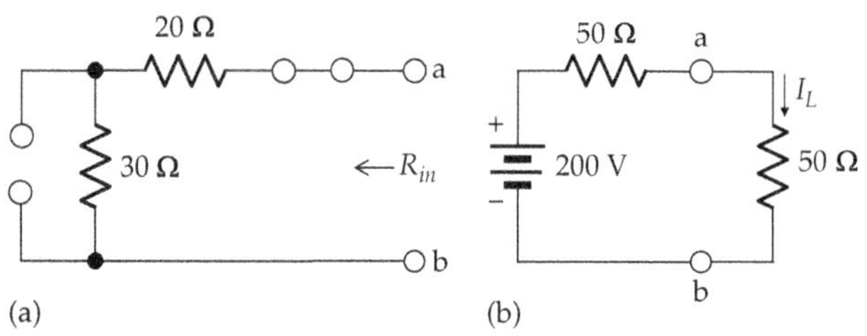

FIGURE 4.9
Figure for Example 4.1.

With no current in the 20 Ω resistor, the 6 A source current flows through the 30 Ω resistor, producing a voltage of 180 V across this resistor. From KVL, $V_{Th} = 20 + 180 = 200$ V.

When terminals 'ab' are short-circuited, the short-circuit current I_{SC} can be determined by transforming the ideal 6 A source in parallel with 30 Ω to an ideal voltage source of 30 × 6 = 180 V in series with 30 Ω, as in Figure 4.8b. I_{SC} now flows through the 30 and 20 Ω resistors. Applying KVL, starting from node 'b' and going clockwise, $180 - 30I_{SC} - 20I_{SC} + 20 = 0$. This gives $I_{SC} = 200/50 = 4$ A. It follows that $R_{Th} = V_{Th}/I_{SC} = 200/4 = 50$ Ω.

Since the circuit does not have dependent sources, it is not necessary to apply explicitly a test source and determine the ratio of the voltage of the test source to the current through the source. R_{Th} can be determined in this case as R_{eq} between terminals 'ab', or R_{in}, the input resistance looking into terminals 'ab', with independent sources set to zero. The 20 V source is replaced by a short circuit and the 6 A source is replaced by an open circuit, as in Figure 4.9a. The resistance looking into terminals 'ab' is seen to be 30 + 20 = 50 Ω, as determined previously. TEC between terminals 'ab' is therefore a 200 V source in series with 50 Ω (Figure 4.9b). When the 50 Ω resistor is connected between terminals 'ab' I_L is given by $I_L = 200/(50 + 50) = 2$ A.

Simulation: The circuit is entered as in Figure 4.10. An IDC I2 is connected between terminals 'ab' of the circuit. Its default value of 0A need not be changed. A voltage marker is placed at terminal 'a' of the circuit.

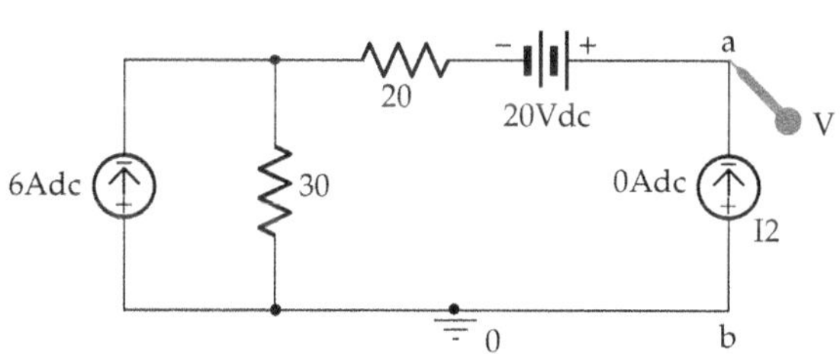

FIGURE 4.10
Figure for Example 4.1.

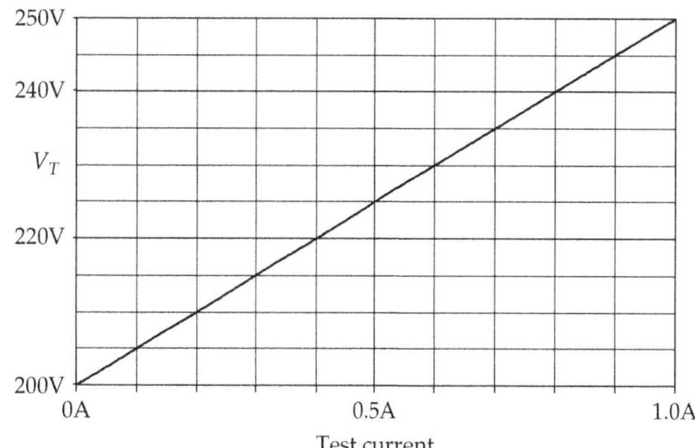

FIGURE 4.11
Figure for Example 4.1.

In the Simulation Settings, 'Analysis type' is 'DC Sweep', 'Primary Sweep' is selected under 'Options', 'Current source' is selected as 'Sweep variable', and I2 is entered in the 'Name' field. 'Sweep type' is 'Linear', 'Start value' is 0, 'End value' is 1, and 'Increment' is 1m, which is small enough to give a large number of points (1000) and hence a smooth line. When the simulation is run, Figure 4.11 is displayed. Cursor 1 is positioned at 1 A on the horizontal axis, and cursor 2 is positioned at the origin. In the cursor window, V_{Th} is read as Y2 = 200.000, and $R_{Th} \times 1$ is read as Y1 − Y2 = 50.000.

Exercise 4.1

Verify that applying a test voltage source or a test current source in Figure 4.9a gives the same R_{Th}.

Example 4.2: Derivation of TEC

It is required to derive TEC seen by the 26 Ω load in Figure 4.12.

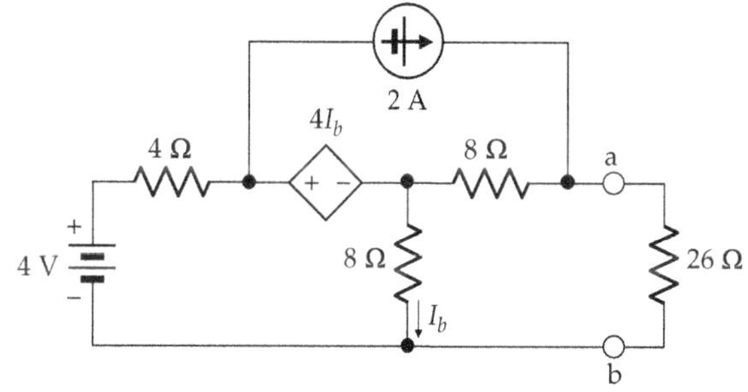

FIGURE 4.12
Figure for Example 4.2

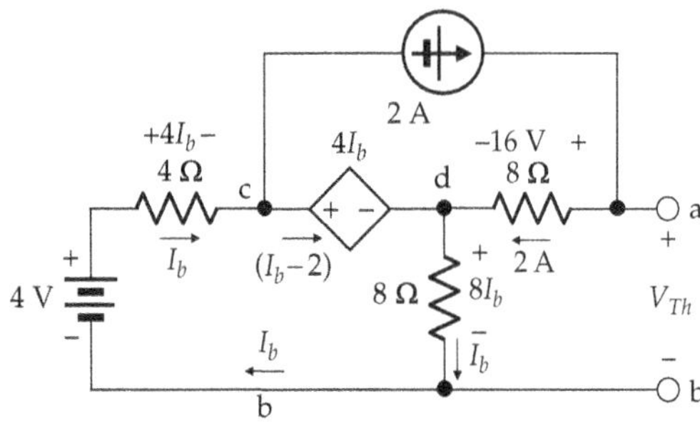

FIGURE 4.13
Figure for Example 4.2.

Solution:

Terminals 'ab' are open-circuited by removing the 26 Ω load, so that V_{Th} is now the voltage between these terminals (Figure 4.13). The 2 A current flows in the 8 Ω resistor between nodes 'a' and 'd', and I_b flows through the 4 Ω resistor and the 4 V source. The current leaving node 'c', through the CCVS is, from KCL, (I_b-2). Node 'd' can then be used to check that KCL is satisfied in the circuit. The current entering node 'd' is $(I_b - 2 + 2) = I_b$, the same as the current leaving the node. As emphasized previously, it is good practice in problem-solving not to use additional variables and to mark the currents and voltages on the circuit diagram. I_b is determined from KVL around mesh 'bcdb'. By going clockwise around this mesh, starting from node 'b', KVL gives $4 - 4I_b - 4I_b - 8I_b = 0$, so that $I_b = 4/16 = 0.25$ A, and $V_{Th} = 8 \times 2 + 8 \times 0.25 = 18$ V. Note that there is no point in taking KVL around the mesh that includes the 2 A source because the voltage across this source is an additional unknown.

When terminals 'ab' are short-circuited, the circuit becomes as in Figure 4.14. Because $V_{da} = V_{db}$, and the resistances in the branches 'da' and 'db' are equal, it follows that the current I_{da} is also I_b. From KCL at node 'a', the current in the short circuit is $I_{SC} = (2 + I_b)$. From KCL at node 'b', the current in the branch 'bc' is $(2 + 2I_b)$, and from KCL at node 'c', the current in the CCVS is $2I_b$. Again, KCL at node 'd' can serve as a check on KCL in the circuit.

By going clockwise around the mesh 'bcdb', KVL gives $4 - 4(2 + 2I_b) - 4I_b - 8I_b = 0$. Hence, $I_b = -4/20I_b = -0.2$ A, so that $I_{SC} = (2 + I_b) = 1.8$ A. It follows that $R_{Th} = 18/1.8 = 10\ \Omega$. It should be noted that *the assigned positive direction of I_{SC} should be consistent with that of V_{Th}, in accordance with Ohm's law. Otherwise, the sign of R_{Th} will be incorrect.* The positive direction of I_{SC} is that of the voltage drop V_{Th} at the open-circuited terminals. If a resistor is connected across the terminals, the positive direction of current through the resistor is that of a voltage drop between the terminals. Reducing the resistance to zero will not change the positive direction of current. It follows that the positive direction of I_{SC} is that of a voltage drop V_{Th}.

To determine R_{Th} by applying a test source, the independent sources are set to zero, so that the 4 V source is replaced by a short circuit and the 2 A source by an open circuit, as shown in Figure 4.15. With a 1 A test source applied, KCL is satisfied at node 'd' by having a current $(1 - I_b)$ leaving this node through the CCVS and the 4 Ω resistor. Node 'b' can be used to check KCL. By going clockwise around the mesh 'bcdb', KVL gives $4(1 - I_b) - 4I_b - 8I_b = 0$. Hence, $I_b = 4/16 = 0.25$ A, and $V_T = 8 \times 1 + 8 \times 0.25 = 10$ V. It follows that $R_{Th} = (10\ \text{V})/(1\ \text{A}) = 10\ \Omega$, as before.

Simulation: The circuit is entered as in Figure 4.16. Proceeding as in Example 4.1, Figure 4.17 is displayed when the simulation is run. It is seen that $V_{Th} = 18$ V and $R_{Th} = 10\ \Omega$.

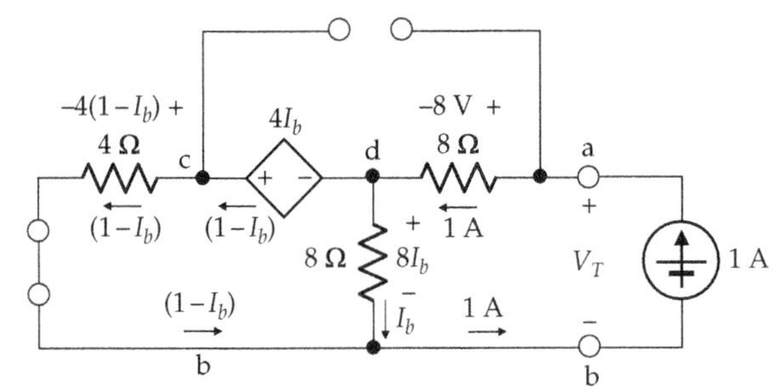

FIGURE 4.15
Figure for Example 4.2.

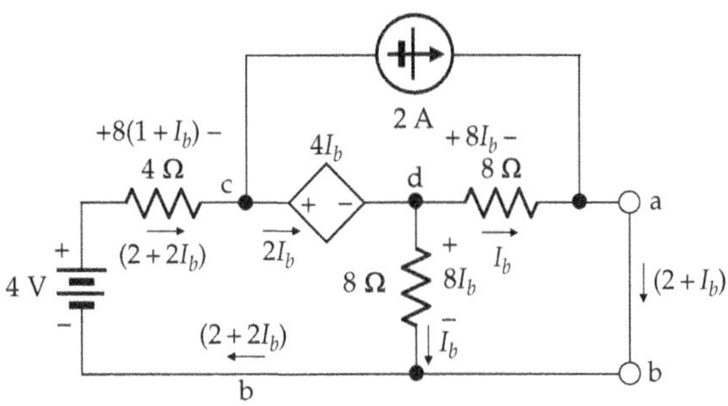

FIGURE 4.14
Figure for Example 4.2.

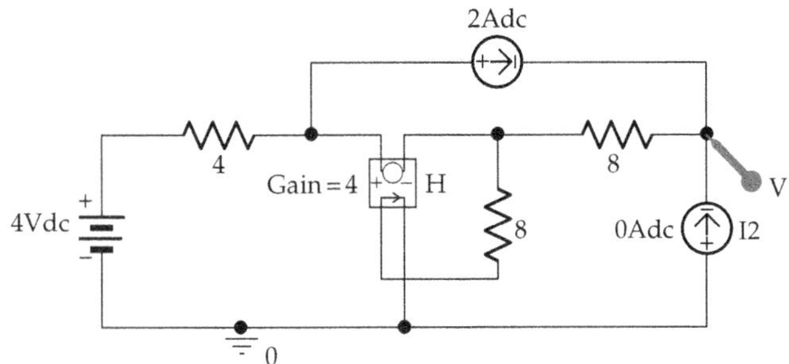

FIGURE 4.16
Figure for Example 4.2.

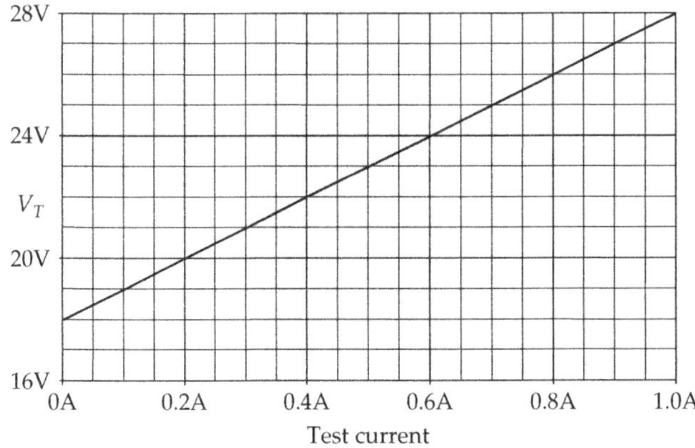

FIGURE 4.17
Figure for Example 4.2.

Problem-Solving Tip

- In deriving TEC, the assigned positive direction of I_{SC} should be in the direction of the voltage drop V_{Th} so as to obtain the correct sign of $R_{Th} = V_{Th}/I_{SC}$, in accordance with Ohm's law.

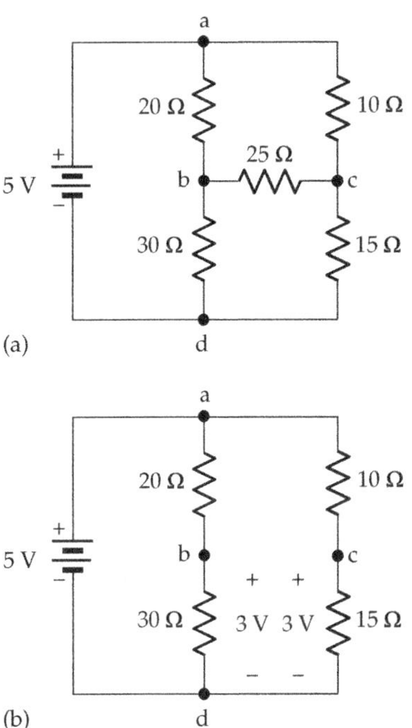

FIGURE 4.18
Figure for Example 4.3.

Exercise 4.2

Determine R_{Th} in Figure 4.15 by applying a 1 V test source rather than a 1 A source.

Example 4.3: Derivation of TEC for a Bridge Circuit

It is required to derive TEC seen by the 25 Ω resistor in Figure 4.18a.

Solution:

As mentioned previously, the circuit configuration is a bridge circuit, since the 25 Ω resistor between nodes 'b' and 'c' is a "crossover" element, like a bridge. To derive TEC seen by the 25 Ω resistor, the resistor is removed and V_{bc} determined (Figure 4.18b). The voltages V_{bd} and V_{cd} in Figure 4.18b can be determined from voltage division, since the 10 Ω resistor is in series with the 15 Ω resistor, and the 20 Ω resistor is in series with the 30 Ω resistor. Hence, $V_{bd} = 5 \times 30/(20 + 30) = 3$ V, and $V_{cd} = 5 \times 15/(10 + 15) = 3$ V.

When the two middle nodes of the bridge circuit are at the same voltage, that is, $V_{bd} = V_{cd}$, the bridge is said to be "balanced," as is discussed more fully in Appendix 5A.

With the bridge balanced, $V_{Th} = V_{bc} = V_{bd} - V_{cd} = 0$. Moreover, when nodes 'b' and 'c' are at the same voltage, then the current through any resistor connected between these nodes is zero, because there is no voltage to drive such a current. The current remains zero as the resistance is reduced to zero, that is, when nodes 'b' and 'c' are short-circuited. It follows that $I_{SC} = 0$. However, having V_{Th} and I_{SC} equal to zero *does not* mean that $R_{Th} = 0$, because $R_{Th} = V_{Th}/I_{SC}$ is indeterminate and could be finite.

To determine R_{Th}, therefore, the resistance looking into terminals 'bc' should be derived, with the independent voltage source set to zero, that is, replaced by a short circuit (Figure 4.19a). To make it easier to visualize the connections, it is helpful to redraw the circuit as in Figure 4.19b after relocating the short circuit between the two resistive branches. Clearly, the resistance between nodes 'c' and 'b' is $(10||15 + 20||30) = 18$ Ω. With $V_{Th} = 0$, TEC reduces to an 18 Ω resistor.

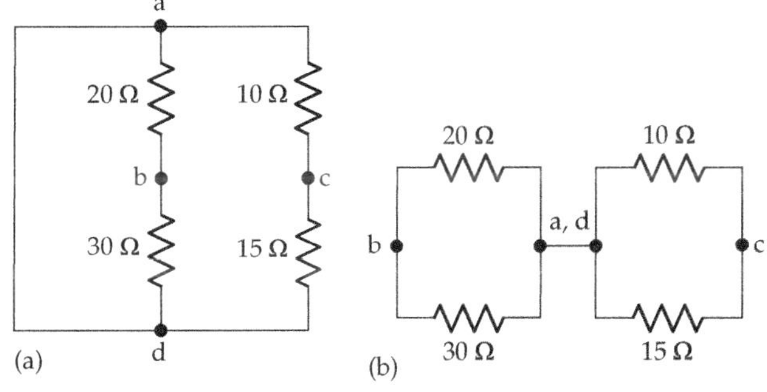

FIGURE 4.19
Figure for Example 4.3.

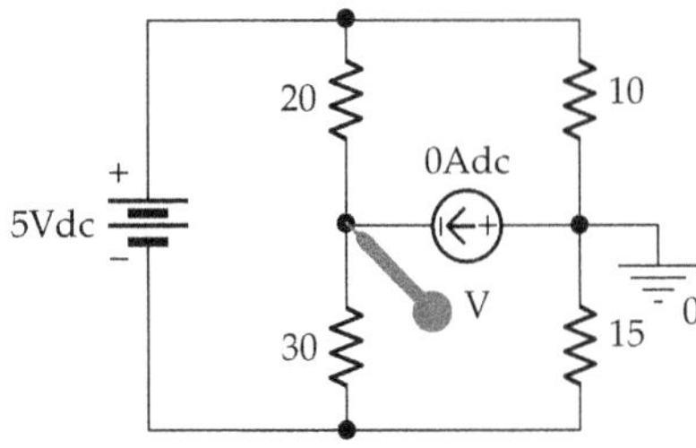

FIGURE 4.20
Figure for Example 4.3.

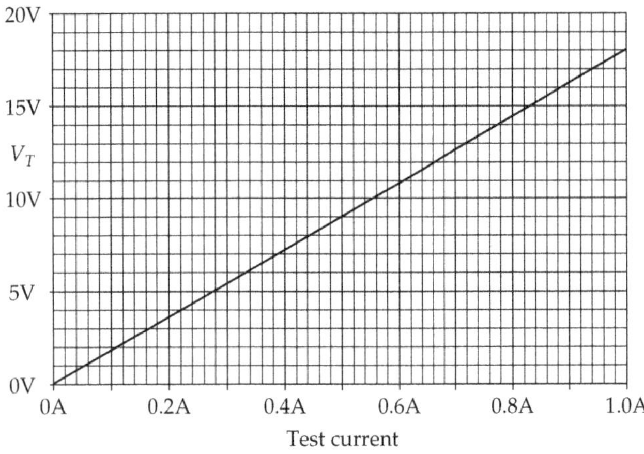

FIGURE 4.21
Figure for Example 4.3.

Simulation: The circuit is entered as in Figure 4.20. Proceeding as in Example 4.1, Figure 4.21 is displayed when the simulation is run. It is seen that $V_{Th} = 0$ and $R_{Th} = 18\ \Omega$.

Example 4.4: Application of TEC

It is required to determine I_X in Figure 4.22 using TEC.

Solution:

TEC will be derived in two steps. The first step is to determine TEC as seen by the 3 A source in parallel with the 6 Ω resistor, as illustrated in Figure 4.23a. The motivation for this step is that the parallel combination of the 3 A source and the 6 Ω resistor is connected between the terminals where TEC is required. Under these conditions, this parallel combination can be temporarily removed, resulting in a considerably simpler circuit for which an intermediate TEC can be derived more easily. The 3 A source is then added to this intermediate TEC and a new TEC derived, from which I_X is determined by adding the 6 Ω resistor.

It is seen from voltage division in Figure 4.23a that $V_{ac} = 9 \times 6/9 = 6$ V, and $V_{bc} = 9 \times 3/9 = 3$ V. It follows that

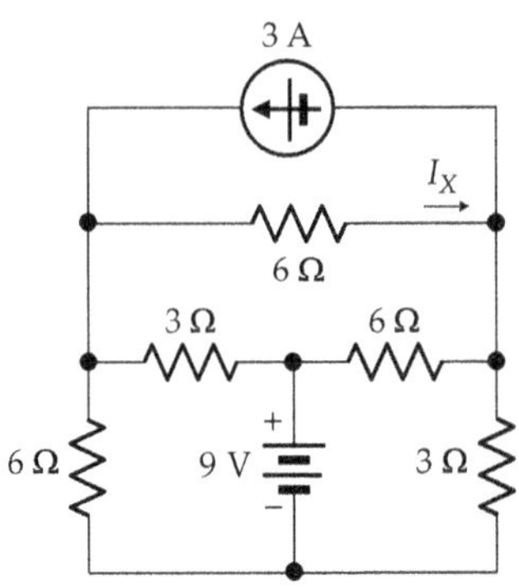

FIGURE 4.22
Figure for Example 4.4.

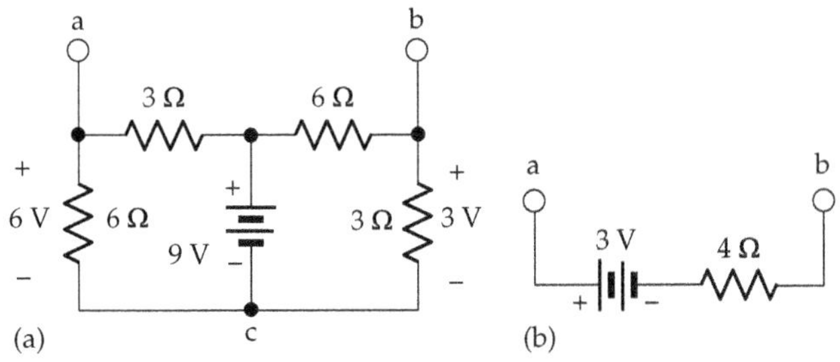

FIGURE 4.23
Figure for Example 4.4.

$V_{Th1} = V_{ac} - V_{bc} = 6 - 3 = 3$ V. R_{Th1} is most easily found by determining the resistance between terminals 'ab' with the 9 V source set to zero, that is, replaced by a short circuit. This makes $R_{Th1} = R_{ab} = (6||3) + (6||3) = 2 + 2 = 4\ \Omega$. The intermediate TEC will therefore consist of $V_{Th1} = 3$ V in series with $R_{Th1} = 4\ \Omega$ (Figure 4.23b).

The next step is to connect the 3 A current source and derive a second TEC as seen by the 6 Ω resistor (Figure 4.24a). The 3 A current now flows through the 3 V source and the 4 Ω resistor, so that the open-circuit voltage $V_{ab} = V_{Th2} = 3 + 12 = 15$ V. When the 3 A source is replaced by an open circuit and the 3 V source by a short circuit, the resistance seen between terminals 'ab' is $R_{Th2} = R_{ab} = 4\ \Omega$ as before (Figure 4.24b). When the 6 Ω resistor is connected to terminals 'ab', I_X that flows is $15/(6 + 4) = 1.5$ A (Figure 4.24c).

Simulation: Although the current I_X can be derived directly by simulating the circuit of Figure 4.22 without invoking TEC, it is instructive to derive by simulation TEC as seen by the 6 Ω resistor. The circuit is entered as in Figure 4.25. Proceeding as in Example 4.1, Figure 4.26 is displayed when the simulation is run. It is seen that $V_{Th} = 15$ V and $R_{Th} = 4\ \Omega$, as determined previously.

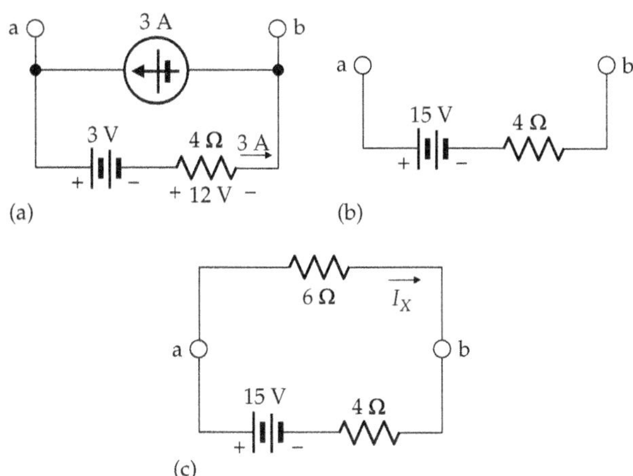

FIGURE 4.24
Figure for Example 4.4.

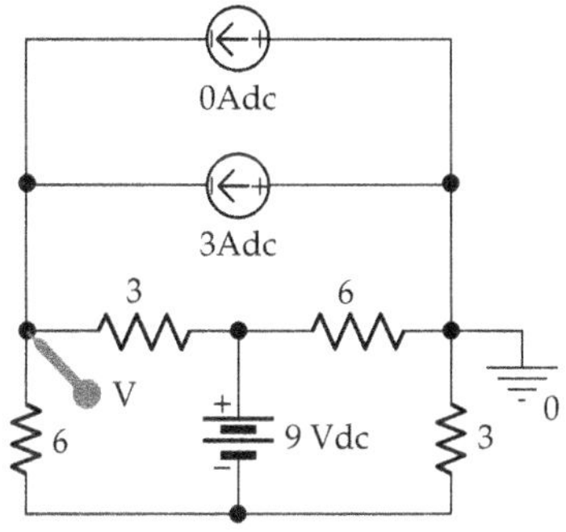

FIGURE 4.25
Figure for Example 4.4.

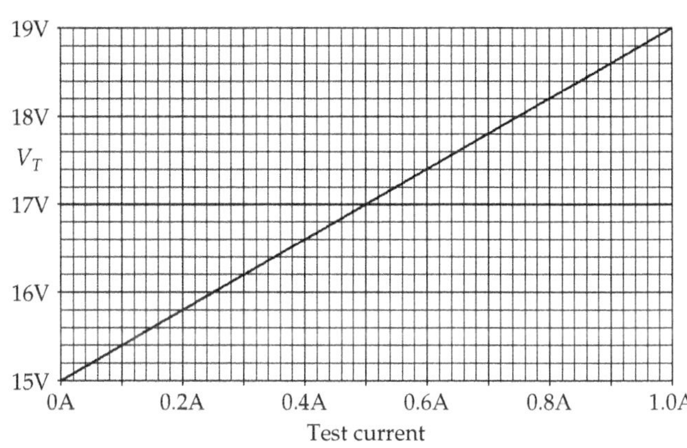

FIGURE 4.26
Figure for Example 4.4.

Problem-Solving Tips

- When a voltage or a current in a circuit is required, it can often be conveniently determined by deriving TEC at the terminals associated with this voltage or current.
- In deriving TEC, it is often advantageous to temporarily remove elements that appear in parallel or in series with the terminals between which TEC is required, derive an intermediate TEC, and then restore the removed elements to this intermediate TEC in order to derive the final TEC.

Exercise 4.3

Determine R_{Th} in Figure 4.23a by deriving I_{SC}.

Primal Exercise 4.4

Derive TEC between nodes 'a' and 'b' in Figure 4.27: (a) without including the 60 Ω resistance between these nodes and (b) including this resistance.

Ans. (a) 12 V in series with 30 Ω; (b) 8 V in series with 20 Ω.

Primal Exercise 4.5

Derive TEC between terminals 'a' and 'b' in Figure 4.28.

Ans. 6 V, 2 Ω.

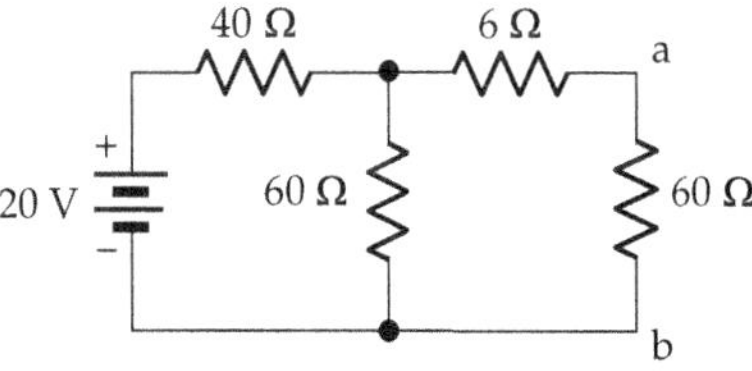

FIGURE 4.27
Figure for Primal Exercise 4.4.

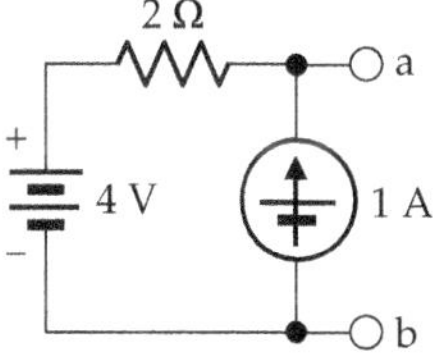

FIGURE 4.28
Figure for Primal Exercise 4.5.

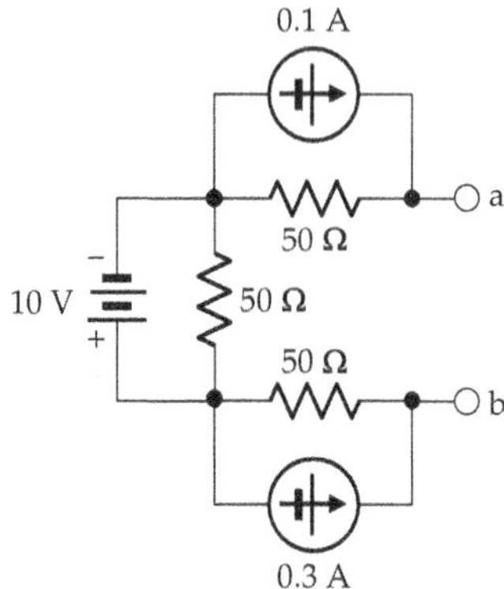

FIGURE 4.29
Figure for Primal Exercise 4.6.

Primal Exercise 4.6

(a) Derive TEC looking into terminals 'ab' in Figure 4.29; (b) determine I_{SC} between terminals 'ab' by transforming the current sources to their equivalent voltage sources.

Ans. (a) $V_{Th} = V_{ab} = -20$ V, $R_{Th} = 100\ \Omega$; (b) $I_{SC} = I_{ab} = -0.2$ A.

4.3 Norton's Theorem

In the context of resistive circuits, Norton's theorem can be stated as follows:

Statement: *A circuit consisting of ideal resistors and sources is equivalent, at a specified pair of terminals, to a linear-output current source.*

It is seen that NEC is in fact the linear-output current source equivalent of TEC. The two equivalent circuits are related by source transformation, as illustrated in Figure 4.30. The ideal current source I_N is referred to as **Norton's current** and is the short-circuit current of TEC, that is, V_{Th}/R_{Th}, in accordance with source transformation. The source resistance that is in parallel with I_N is **Norton's resistance** R_N and is the same as R_{Th} in TEC.

It is sometimes more convenient to derive NEC rather than TEC, as in Example 4.5. Moreover, some circuits may have an NEC but not a TEC, or conversely, as in the case of ideal voltage sources and ideal current sources. Thus, an ideal voltage source can be considered to be its own TEC, with $R_{Th} = 0$. $I_N = V_{Th}/0 \rightarrow \infty$, which means that NEC does not exist. Similarly, an ideal current source can be regarded as its own NEC, with R_N infinite, so that $V_{Th} \rightarrow \infty$, which means that TEC does not exist. This is in accordance with the fact that an ideal voltage source cannot be transformed to an ideal current source, and conversely, as explained in Section 3.6. Circuits that have TEC but not NEC generally reduce to an ideal voltage source between the terminals involved, whereas circuits that have NEC but not TEC generally reduce to an ideal current source between these terminals. Examples of these are given in the problems at the end of the chapter.

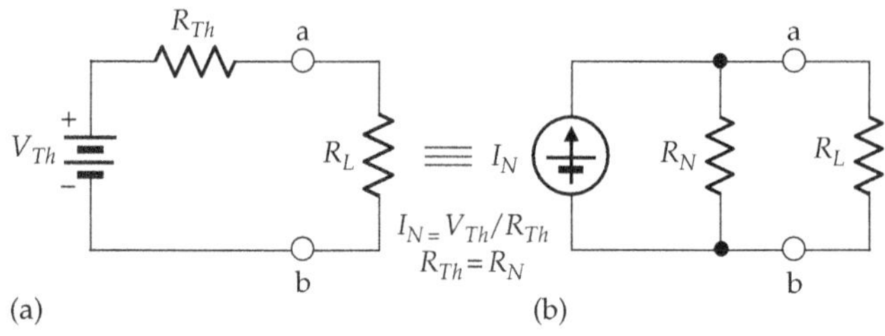

FIGURE 4.30
Derivation of NEC. (a) TEC looking into terminals 'ab' and (b) its equivalent transformed source (NEC) between terminals 'ab'.

The procedure for deriving NEC is essentially the same as that for TEC. When independent sources are set to zero in Figure 4.30, the ideal voltage source in TEC is replaced by a short circuit and the ideal current source in NEC is replaced by an open circuit. The resistance looking into terminals 'ab' is $R_{Th} = R_N$ in both cases. In the case of TEC, V_{Th} is generally determined directly, and the short-circuit current, I_N, is determined as an alternative method for finding R_{Th}. In the case of NEC, I_N is generally determined directly, and the open-circuit voltage, V_{Th}, is determined as an alternative method for finding R_N.

*4.3.1 Derivation of NEC with PSpice

NEC can be derived in a single simulation, analogous to that described for TEC, and explained in Figure 4.31a.

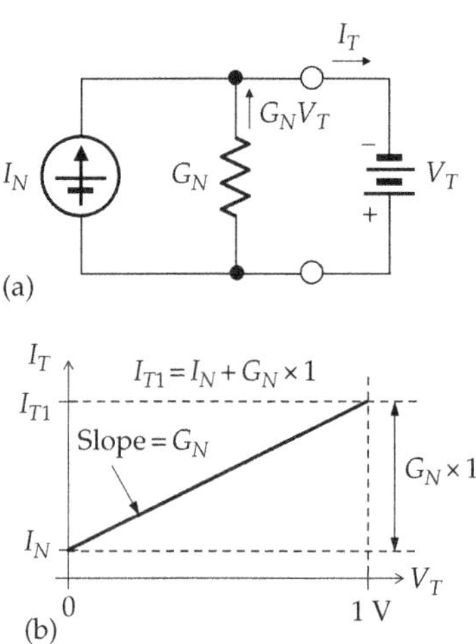

FIGURE 4.31
Derivation of NEC with PSpice. (a) Test current voltage applied to NEC and (b) graphical construction for determining NEC.

A test source V_T is connected between these terminals with the circuit left as is, that is, with the independent sources retained, so that $I_N \neq 0$. KCL gives

$$I_T = I_N + G_N V_T \tag{4.11}$$

If V_T is varied between 0 and 1 V, and I_T is plotted against V_T, a straight line graph is obtained having a current intercept I_N at $V_T = 0$. The difference between I_{T1}, which is I_T at $V_T = 1$ V, and I_N is numerically equal to G_N (Figure 4.31b). V_T is conveniently varied in PSpice over a desired range of values, using the "DC Sweep" feature.

Example 4.5: Application of NEC

It is required to determine I_L in Figure 4.32 using NEC.

Solution:

When terminals 'ab' are short-circuited, $V_X = 0$ and the VCVS becomes a short circuit (Figure 4.33a). To clarify the evaluation of I_{SC}, the circuit can be redrawn as in Figure 4.33b. It is seen that the 15 and 10 Ω resistors are in parallel, so that current division can be applied to the

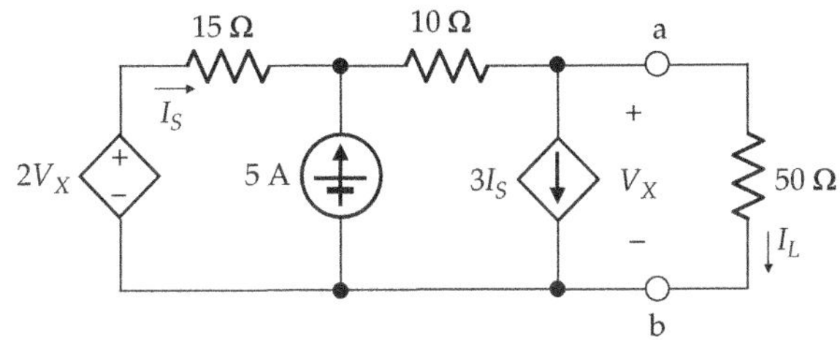

FIGURE 4.32
Figure for Example 4.5.

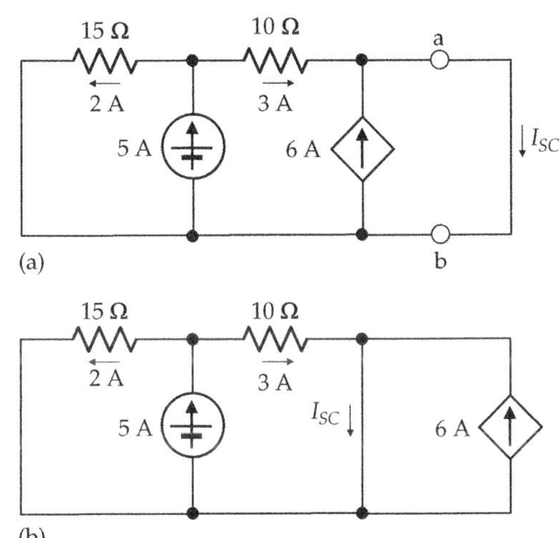

FIGURE 4.33
Figure for Example 4.5.

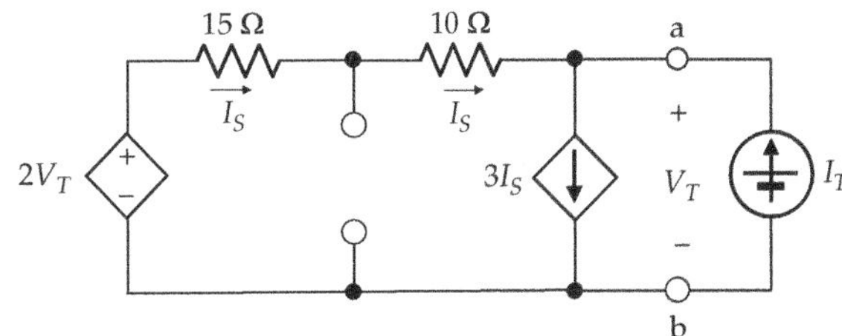

FIGURE 4.34
Figure for Example 4.5.

5 A source. Accordingly, the currents in these two resistors are 2 and 3 A as shown. $I_S = -2$ A and the CCCS becomes 6 A directed upward. It follows from KCL that $I_{SC} = I_N = 3 + 6 = 9$ A. Note that the 6 A source current only adds to the current in the short circuit between nodes 'a' and 'b' in Figure 4.33a and does not affect current division.

To determine R_N, a test source I_T is applied, with the 5 A current source replaced by an open circuit (Figure 4.34). KVL around the outer loop gives

$$2V_T - 25I_S - V_T = 0$$

or

$$V_T = 25I_S \tag{4.12}$$

From KCL at node 'a',

$$I_S + I_T = 3I_S$$

or

$$I_T = 2I_S \tag{4.13}$$

Dividing Equation 4.13 by Equation 4.12,

$$R_N = \frac{V_T}{I_T} = 12.5\ \Omega \tag{4.14}$$

NEC therefore consists of a 9 A source in parallel with a 12.5 Ω resistor (Figure 4.35a). Note that the direction of the 9 A source is such that the short-circuit current is directed from 'a' to 'b', as in Figure 4.33a. When the

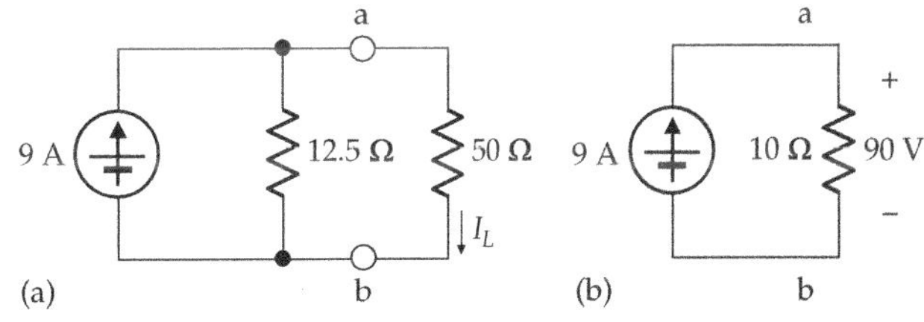

FIGURE 4.35
Figure for Example 4.5.

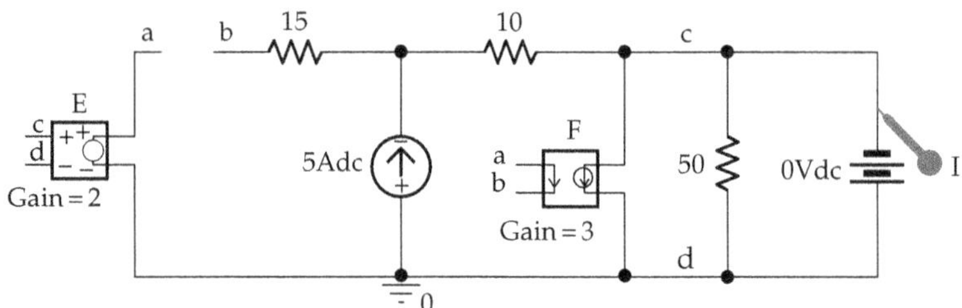

FIGURE 4.36
Figure for Example 4.5.

50 Ω resistor is connected between terminals 'ab', it follows from current division that

$$I_L = \frac{12.5}{12.5+50} \times 9 = 1.8 \text{ A} \tag{4.15}$$

The voltage V_{ab} is 90 V, as in Figure 4.35b where 12.5||50 = 10 Ω. It may be noted that determining the open-circuit voltage in Figure 4.32 in order to work with TEC is slightly more complicated than determining I_N as in Figure 4.33a (Exercise 4.7).

Simulation: The circuit is entered as in Figure 4.36, in accordance with the method explained in connection with Figure 4.31. The 50 Ω resistor is included in order to facilitate finding I_L, as explained later, but could be left out for the purpose of determining NEC between terminals 'ab'. Note the alternative way of connecting dependent sources in Figure 4.35 in order to avoid making the somewhat awkward connections to the control terminals of dependent sources. This is to label the appropriate nodes using the net alias feature of PSpice, as described in Appendix C. PSpice considers nodes having the same label to be connected together, as shown in Figure 4.36. A DC sweep is performed as described in Example 4.1 but sweeping a voltage source instead of a current source. The DC sweep gives the plot of Figure 4.37, from which,

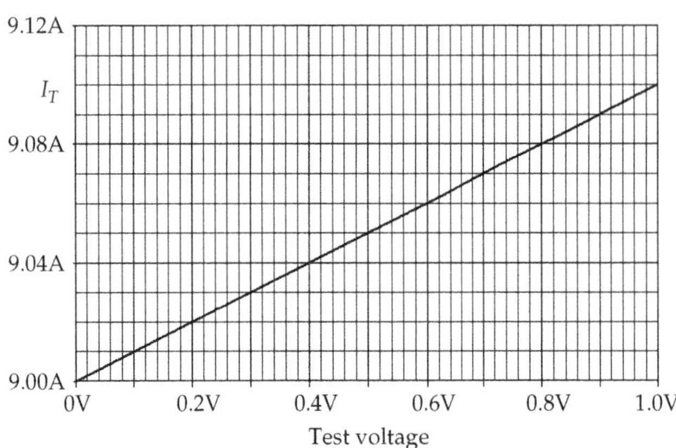

FIGURE 4.37
Figure for Example 4.5.

I_N = 9 A and R_N = 1/0.1 = 10 Ω, this being the parallel resistance of 12.5 and 50 Ω (Figure 4.35b). I_L is determined from V_{ab} = 9 × 10 = 90 V in Figure 4.35a and b. It follows from Figure 4.35a that I_L = 90/50 = 1.8 A.

Exercise 4.7

Determine V_{Th} directly from the circuit of Figure 4.32.

Primal Exercise 4.8

Determine NEC between nodes 'a' and 'b' in Figure 4.38: (a) without including the 10 Ω resistance between these nodes and (b) including this resistance.

Ans. (a) 10 A in parallel with 15 Ω; (b) 10 A in parallel with 6 Ω.

Primal Exercise 4.9

Determine V_{Th}, I_N, and G_N looking into terminals 'ab' in Figure 4.39.

Ans. 6 V, 6 mA, 1 mS.

Primal Exercise 4.10

Derive NEC between terminals 'ab' in Figure 4.40. Note how much easier it is to derive I_N compared to V_{Th}.

Ans. 6 mA, in parallel with 1 kΩ.

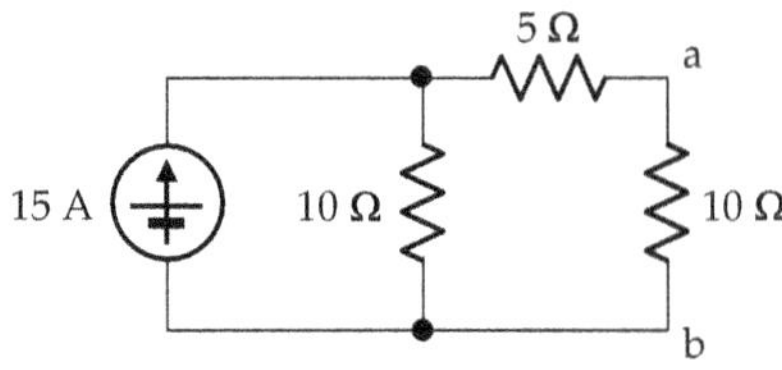

FIGURE 4.38
Figure for Primal Exercise 4.8.

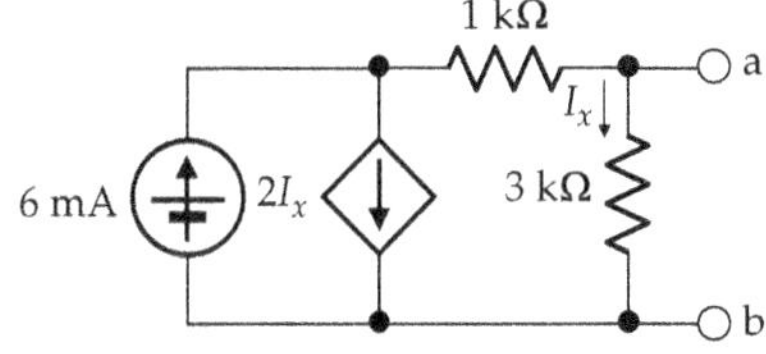

FIGURE 4.39
Figure for Primal Exercise 4.9.

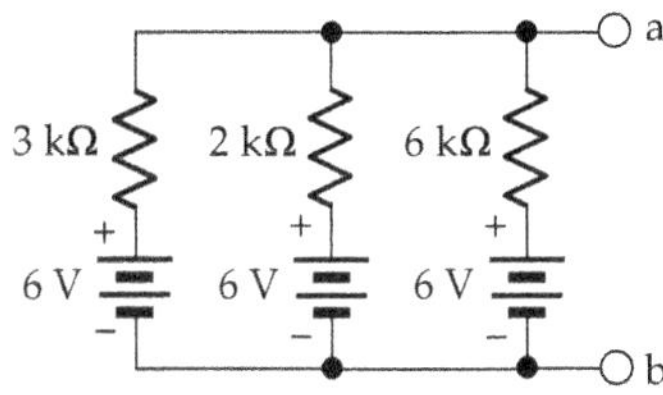

FIGURE 4.40
Figure for Primal Exercise 4.10.

4.4 Substitution Theorem

Consider a circuit 'N' connected at terminals 'ab' to a circuit 'N_A' having a designated voltage V_A across 'ab' (Figure 4.41a), where V_A could be of known or unknown value. Let I_X be the current flowing from 'N' to 'N_A'. According to the substitution theorem, 'N_A' can be replaced by an independent voltage source V_A (Figure 4.41b), without affecting I_X. This can be readily justified if 'N' is represented between terminals 'ab' by its TEC, as in Figure 4.41c and d. It is evident from these figures that KVL is the same in both cases, namely,

$$V_{Th} - R_{Th}I_X = V_A$$

which gives

$$I_X = \frac{V_{Th} - V_A}{R_{Th}} \quad (4.16)$$

In other words, replacing 'N_A' by an independent voltage source V_A does not affect 'N', since I_X remains the same.

Similarly, suppose that 'N_A' has a designated current I_A at the common terminals 'ab' (Figure 4.42a), where I_A could be of known or unknown value. Let V_X be the voltage across terminals 'ab'. According to the substitution theorem, an independent current source I_A can be substituted for 'N_A' (Figure 4.42b), without affecting V_X. Again, this can be justified if 'N' is represented between terminals 'ab' by its TEC, as in Figure 4.42c and d. It is evident from these figures that KVL is the same in both cases and gives

$$V_X = V_{Th} - R_{Th}I_A \quad (4.17)$$

In other words, replacing 'N_A' by an independent current source I_A does not affect 'N', since V_X remains the same. The substitution theorem can be stated as follows:

Statement: *A circuit having a designated voltage V across it can be replaced by an ideal, independent voltage source V,*

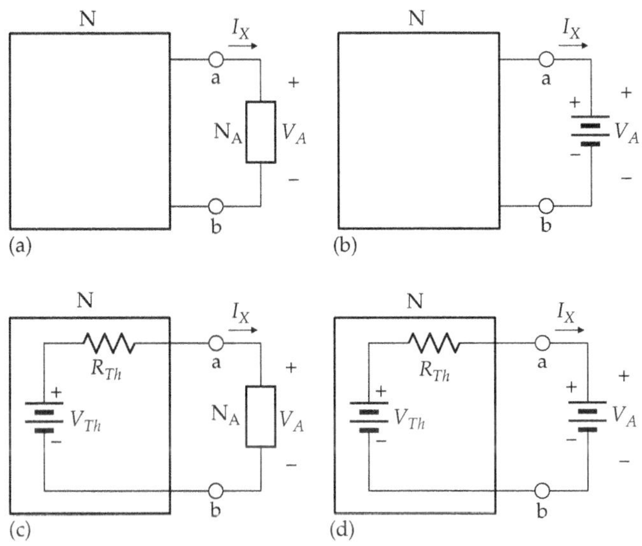

FIGURE 4.41
Substitution theorem in terms of a voltage source. Circuit 'N_A' in (a) having a designated voltage V_A is replaced in (b) by an ideal voltage source of source voltage V_A. Circuit 'N' in (a) and (b) is replaced by its TEC in (c) and (d), respectively.

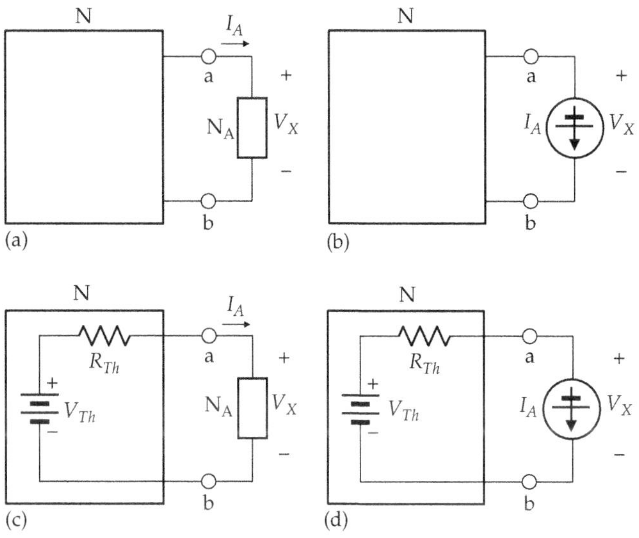

FIGURE 4.42
Substitution theorem in terms of a current source. Circuit 'N_A' in (a) having a designated current I_A is replaced in (b) by an ideal current source of source current I_A. Circuit 'N' in (a) and (b) is replaced by its TEC in (c) and (d), respectively.

without affecting the rest of the circuit. Similarly, a circuit having a designated current I through it can be replaced by an ideal, independent current source I, without affecting the rest of the circuit. The designated V or I could be a numerical value, or V and I could be symbols for unknown values.

There is no restriction on the nature of the circuit 'N$_A$' that is being replaced by an independent source. It could be a single resistor, a dependent source, or any valid combination of independent sources, dependent sources, and resistors. In fact, according to the substitution theorem, a designated voltage between any two nodes in a circuit can be replaced by an ideal voltage source of the same voltage as the designated voltage. Similarly, a current in any branch in a circuit can be replaced by an ideal current source of the same current as the designated current. The substitution theorem is illustrated by Example 4.6; it is particularly useful in connection with superposition, discussed in the following chapter.

Exercise 4.11

Justify the substitution theorem by replacing circuit N by its NEC: (a) in Figure 4.41 and (b) in Figure 4.42.

Example 4.6: Application of Substitution Theorem

Given a known bridge circuit connected to a circuit 'N$_A$' of unknown component values, as illustrated in Figure 4.43. The bridge circuit is inaccessible for measurements, but the voltage across 'N$_A$' can be measured by means of a voltage-measuring device (a voltmeter) and is found to be 15 V, of the polarity indicated. It is required to determine I_S, the current drain on the 6 V battery.

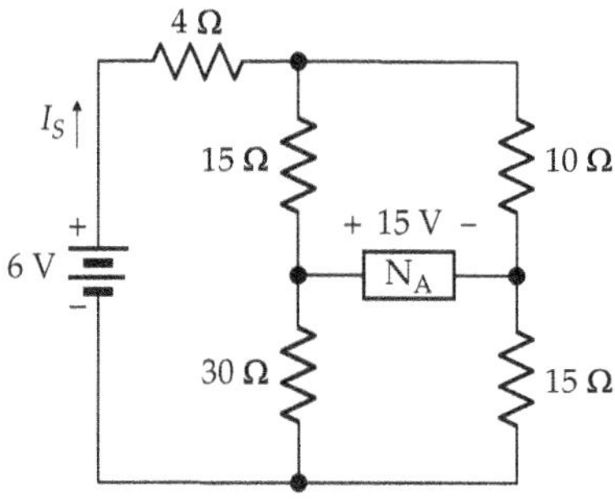

FIGURE 4.43
Figure for Example 4.6.

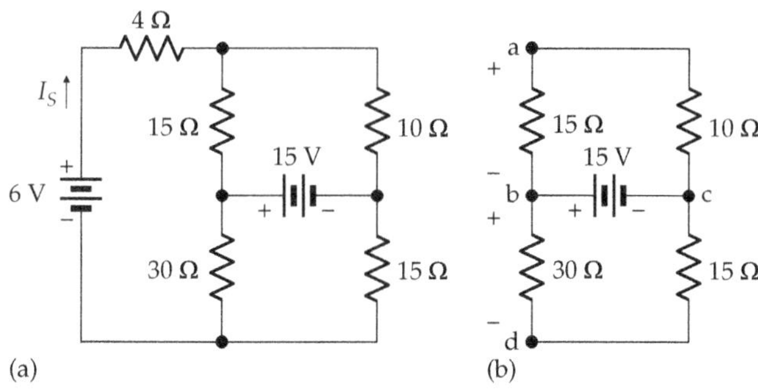

FIGURE 4.44
Figure for Example 4.6.

Solution:

This may look like an impossible problem, but I_S can be readily determined by means of the substitution theorem. According to this theorem, 'N$_A$' can be replaced by a 15 V independent source, without disturbing the circuit (Figure 4.44a). I_S can be conveniently determined by deriving TEC between terminals 'ad' (Figure 4.44b). $V_{Th} = V_{ad} = V_{ab} + V_{bd}$. From voltage division, $V_{ab} = \dfrac{-15}{15+10}15 = -9$ V, and $V_{bd} = \dfrac{30}{30+15}15 = 10$ V. It follows that $V_{Th} = V_{ad} = -9 + 10 = 1$ V.

R_{Th} is determined as R_{eq} between terminals 'a' and 'd' with the 15 V source set to zero (Figure 4.45a), which makes nodes 'b' and 'c' one and the same. The resistance between terminals 'a' and 'b' is $15 \| 10 = \dfrac{15\times 10}{15+10} = 6\ \Omega$. The resistance between terminals 'c' and 'd' is $30 \| 15 = \dfrac{30\times 15}{30+15} = 10\ \Omega$. Hence, $R_{Th} = 10 + 6 = 16\ \Omega$. Replacing the load circuit between terminals 'a' and 'd' by its TEC, the circuit becomes as shown in Figure 4.45b. It follows from KVL that $(6 - 1) = (4 + 16)I_S$, which gives $I_S = 5/20 = 0.25$ A. Note that deriving TEC for the bridge circuit in Figure 4.44b illustrates a useful general application of TEC, namely, simplifying a

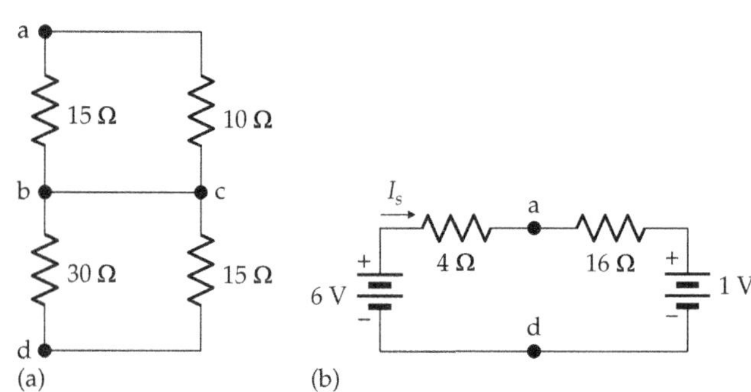

FIGURE 4.45
Figure for Example 4.6.

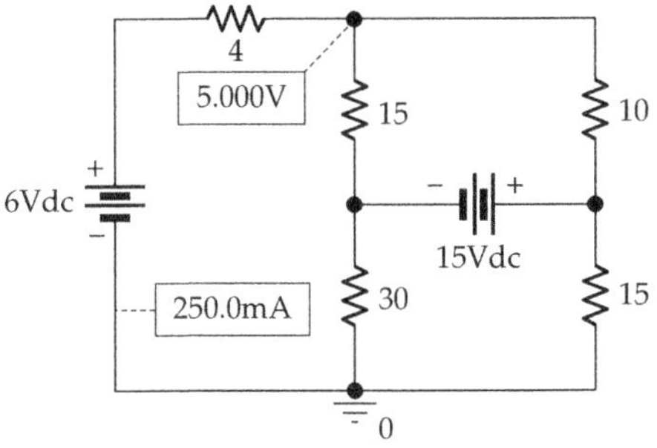

FIGURE 4.46
Figure for Example 4.6.

circuit as part of the solution to a given problem (see Problem P4.18).

Simulation: The circuit is entered as in Figure 4.46. After selecting 'Bias Point' under 'Analysis type' in the simulation profile and running the simulation, pressing the I and V buttons displays the currents and voltages indicated in Figure 4.46. It is seen that $V_{ad} = 5$ V and $I_S = 0.25$ A.

Problem-Solving Tip

- Use TEC to simplify a circuit as part of the solution to a given problem.

Exercise 4.12

Determine R_{Th} in Example 4.6 by applying (a) a 1 A test source and (b) a 1 V test source.

Primal Exercise 4.13

Consider the circuit of Figure 4.47. Determine (a) the independent voltage source that can replace the 5 Ω resistor without affecting the current I in the circuit and (b) the independent current source that can replace this resistor without affecting V_{ab}.

Ans. (a) 5 V, with node 'a' positive with respect to 'b'; (b) 1 A directed from node 'a' to 'b'.

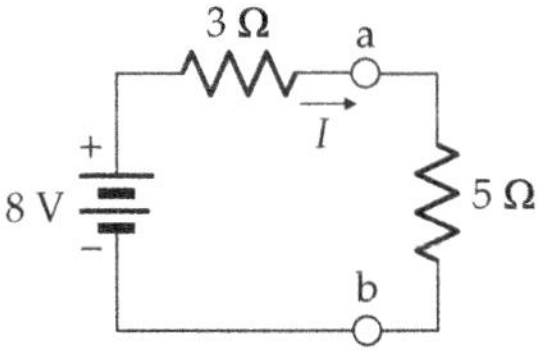

FIGURE 4.47
Figure for Primary Exercise 4.13.

4.5 Source Absorption Theorem

The source absorption theorem is a special case of the substitution theorem that can be usefully applied in some cases involving dependent sources, particularly in transistor circuits. In the definition of dependent sources (Section 2.6), it was stated that the controlling variable is a current or voltage *elsewhere* in the circuit, which excludes the controlling variable being that of the source itself, or a quantity proportional to it. In these cases, the dependent source can be conveniently replaced by a resistor.

Concept: *If a direct proportionality exists between the voltage across a dependent source and the source current, the dependent source can be replaced by a resistor having a resistance equal to the ratio of the voltage across the source to the source current.*

To justify this, consider the dependent voltage source of Figure 4.48a, where the source voltage is proportional to the current through the source, $V = \rho I$. If the dependent source is replaced by a resistor having $R = \rho I/I = \rho$, then for the same I through the two circuit elements, the voltage V across them is the same. The dependent voltage source having $V = \rho I$ is therefore equivalent to a resistor R and can be replaced by this resistor between the same terminals.

The dependent current source of Figure 4.48b has $I = \sigma V$, where V is the voltage across the source. The source can be replaced by a resistor having $R = V/\sigma V = 1/\sigma$. The two circuit elements are equivalent since, for the same voltage across them, the current through them is the same.

Note that in Figure 4.48a and b, a positive value of R corresponds to having the current in the dependent source in the direction of a voltage drop across the

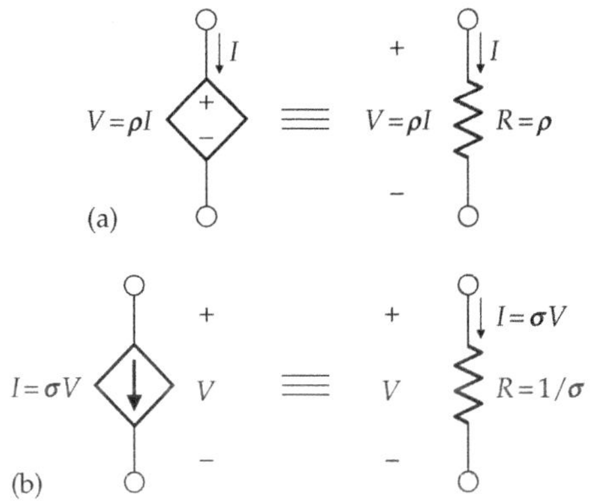

FIGURE 4.48
Source absorption theorem. (a) Dependent voltage source equivalent to a resistor and (b) dependent current source equivalent to a resistor.

source. If the current through the source is in the direction of a voltage rise across the source, the resistance is negative. Just as a positive resistance dissipates power, a negative resistance does the opposite; it delivers power. In this sense, it acts as source, but it differs from an ideal source in that it has a finite, nonzero resistance. Recall that an ideal voltage source has zero resistance, whereas an ideal current source has infinite resistance.

Example 4.7: Output Resistance of Transistor Circuit

A case that is often encountered in transistor circuits is that of Figure 4.49a, where R_{eq} between terminals 'ab' is required.

Solution:

The current source $g_m v_x$ is transformed to a voltage source $g_m v_x r_o$ in series with r_o (Figure 4.49b). The current i through R_x is v_x/R_x, which makes the source voltage $g_m v_x r_o$ proportional to the current v_x/R_x through the source. The dependent source can therefore be replaced by a resistance whose value is the source voltage divided by the source current. This resistance is $(g_m v_x r_o)/(v_x/R_x) = g_m r_o R_x$; R_{eq} between terminals 'ab' is then the sum of the three resistances in Figure 4.49c:

$$R_{eq} = r_o + g_m r_o R_x + R_x \tag{4.18}$$

Simulation: The circuit is entered as in Figure 4.50 using $R_x = 1\ \text{k}\Omega$, $r_o = 200\ \text{k}\Omega$, and $g_m = 4\ \text{mA/V}$. A 1 μA dc current source is applied so that the voltage across the source in volts is numerically equal to the resistance in megohms seen by the source. After selecting 'Bias Point' under 'Analysis type' in the simulation profile and running the simulation, pressing the V button displays the voltages indicated in Figure 4.50. It is seen that $R_{eq} = 1.001\ \text{M}\Omega$, in accordance with Equation 4.18.

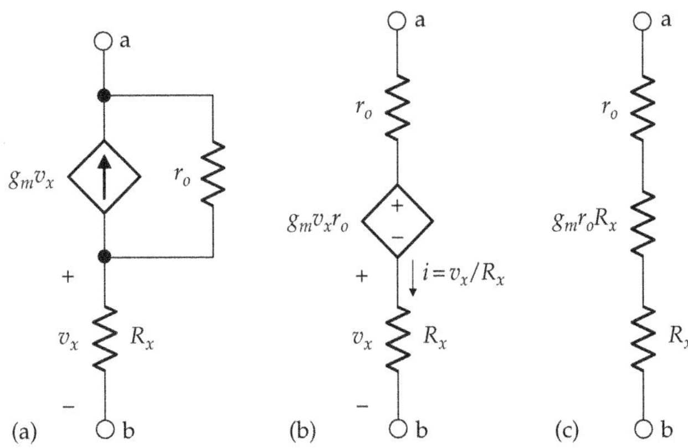

FIGURE 4.49
Figure for Example 4.7.

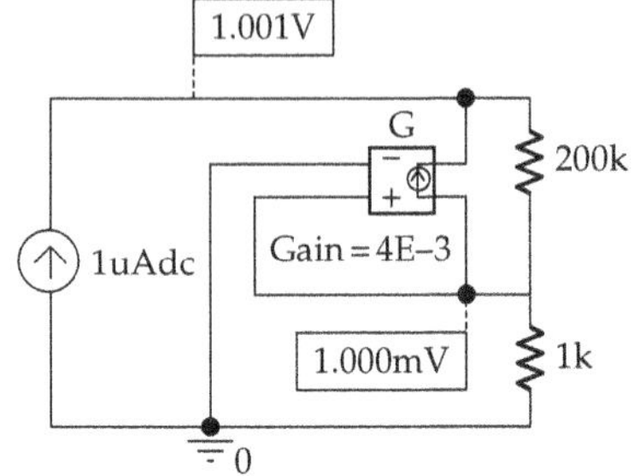

FIGURE 4.50
Figure for Example 4.7.

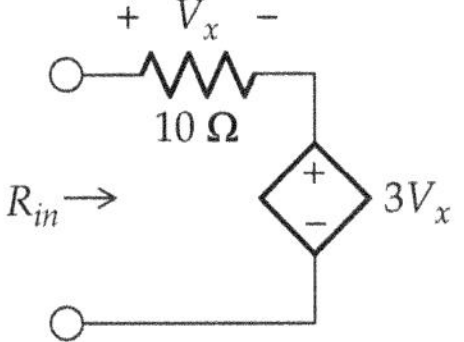

FIGURE 4.51
Figure for Primal Exercise 4.15.

Exercise 4.14

Derive R_{eq} in Example 4.7 by applying (a) a 1 A test source, and (b) a 1 V test source.

Primal Exercise 4.15

Determine R_{in} in Figure 4.51 using the source absorption theorem based on the current that flows through the source.

Ans. 40 Ω.

4.6 Problem-Solving Approach Updated

The main procedural steps of the ISDEPIC approach are summarized and updated as follows in the light of the material covered in this chapter:

Step 1—Initialize:

(a) Mark on the circuit diagram all the given values of circuit parameters, currents and voltages, as well as the unknowns to be determined.
(b) Label the nodes, as this may be generally helpful.
(c) If the solution requires that a given value of current or voltage be satisfied, assume this value from the very beginning.

Step 2—Simplify: Consider, as may be appropriate:

(a) Redrawing the circuit.
(b) Replacing series and parallel combinations of circuit elements by an equivalent circuit element.
(c) Applying star-delta transformation.
(d) Applying source transformation.
(e) Using TEC or NEC to simplify part of the circuit. In applying TEC or NEC, remove temporarily any elements in series or in parallel with the given terminals and derive an intermediate TEC or NEC.
(f) Applying the source absorption theorem.

Step 3—Deduce: Determine any values of current or voltage that follow immediately from direct application of Ohm's law, KCL, or KVL, without introducing any additional unknowns. If Step 3 does not provide a solution, proceed to Step 4.

Step 4—Explore: Examine each of the following alternatives, as may be applicable:

(a) Consider the nodes and meshes in the circuit to see if KCL or KVL can be expressed using a single unknown current or voltage *and* if this unknown can then be directly determined from KCL or KVL.
(b) Use TEC or NEC to determine a voltage or a current through a given circuit element.
(c) Apply the substitution theorem.

If Step 4 does not provide the solution, proceed to Step 5.

Step 5—Plan: Think carefully and creatively about the problem in the light of circuit fundamentals and circuit analysis techniques. Consider alternative solutions and select what seems to be the simplest and most direct solution.

Step 6—Implement: Carry out your planned solution.

Step 7: Check your calculations and results.

(a) Check that your results make sense, in terms of magnitude and sign.
(b) Check that Ohm's law is satisfied across every resistor, that KCL is satisfied at every node, and that KVL is satisfied around every mesh.
(c) Seek an alternative solution to see if it gives the same result.
(d) Whenever feasible check the results with PSpice simulation.

Learning Checklist: What Should Be Learned from This Chapter

- Dependent sources alone do not excite a circuit. They affect currents and voltages in the circuit by effectively modifying the values of some resistances in the circuit.
- Thevenin's Theorem: A circuit consisting of ideal resistors and sources is equivalent, at a specified pair of terminals, to a linear-output voltage source. The voltage of the ideal voltage source element is referred to as the Thevenin voltage, V_{Th}, and the source resistance as the Thevenin resistance, R_{Th}.
- V_{Th} is determined as the open-circuit voltage at the specified terminals. R_{Th} can be determined as V_{Th}/I_{SC}, where I_{SC} is the short-circuit current between the specified terminals.
- The procedure for deriving TEC can be summarized as follows:
  1. Determine V_{Th} as the open-circuit voltage at the specified terminals.
  2. Determine the short-circuit current I_{SC} at the specified terminals, which gives R_{Th} as V_{Th}/I_{SC}.
  3. Set all independent sources in the given circuit to zero, leaving dependent sources unchanged. R_{Th} is the resistance R_{eq} looking into the specified terminals. Formally, this resistance is obtained by applying a test voltage source or a test current source and determining R_{eq} as the ratio of the voltage at the source terminals to the source current. In the absence of dependent sources, this effectively reduces to determining R_{eq} directly from series/parallel combinations of resistors, and using star-delta transformations, if necessary.
- Norton's Theorem: A circuit consisting of ideal resistors and sources is equivalent, at a specified pair of terminals, to a linear-output current source.
- NEC follows from TEC through source transformation.
- According to the substitution theorem, a circuit having a designated voltage V across it can be replaced by an ideal, independent voltage source V, without affecting the rest of the circuit. Similarly, a circuit having a designated current I through it can be replaced by an ideal, independent current source I, without affecting the rest of the circuit. The designated V or I could be a numerical value, or V and I could be symbols for unknown values.

- There is no restriction on the nature of the circuit that is being replaced by an independent voltage or current source in accordance with the substitution theorem. It could be a single resistor, a dependent source, or any valid combination of independent sources, dependent sources, and resistors.
- If a direct proportionality exists between the voltage across a dependent source and the source current, the dependent source can be replaced by a resistor having a resistance equal to the ratio of the voltage across the source to the source current. The resistance value is positive when the source current is in the direction of a voltage drop across the source.

Problem-Solving Tips

1. In deriving TEC, the assigned positive direction of I_{SC} should be in the direction of the voltage drop V_{Th} so as to obtain the correct sign of $R_{Th} = V_{Th}/I_{SC}$, in accordance with Ohm's law.
2. When a voltage or a current in a circuit is required, it can often be conveniently determined by deriving TEC at the terminals associated with this voltage or current.
3. Use TEC to simplify a circuit as part of the solution to a given problem.
4. In deriving TEC it is often advantageous to temporarily remove elements that appear in parallel or in series with the terminals between which TEC is required, derive an intermediate TEC, then restore the removed elements to this intermediate TEC in order to derive the final TEC.

Problems

Apply ISDEPIC and verify solutions by PSpice simulation whenever feasible.

TEC and NEC

P4.1 Derive TEC looking into terminals 'ab' in Figure P4.1.

Ans. $V_{Th} = V_{ab} = 6$ V, $R_{Th} = 20\ \Omega$.

P4.2 Determine $V_{Th} = V_{ab}$, I_{SC}, and R_{Th} independently between terminals 'ab' in Figure P4.2.

Ans. 48 V, 2.75 A, $R_{Th} = 192/11\ \Omega$.

P4.3 Use TEC to determine R_L in Figure P4.3 so that $V_O = V_{SRC}/6$.

Ans. 4/3 Ω.

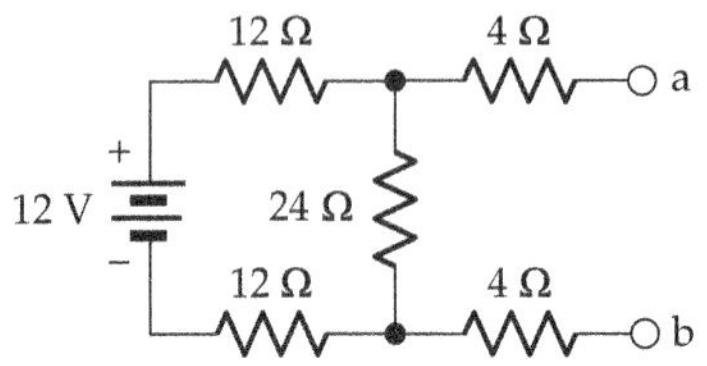

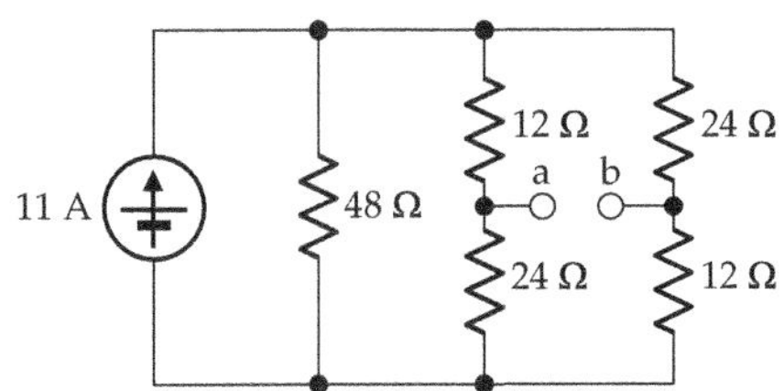

FIGURE P4.1

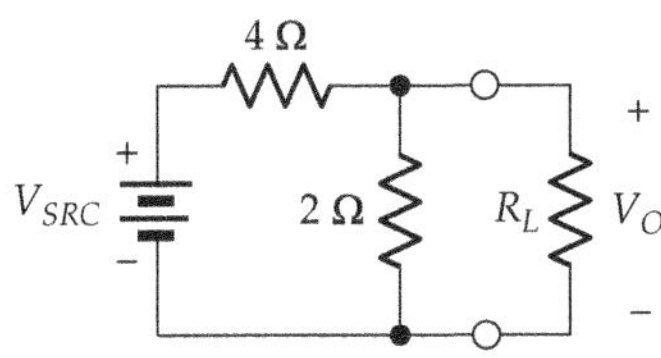

FIGURE P4.2

FIGURE P4.3

P4.4 Derive TEC looking into terminals 'ab' in Figure P4.4.

Ans. $V_{Th} = V_{ab} = 12$ V, $R_{Th} = 6\ \Omega$.

P4.5 Derive TEC and NEC looking into terminals 'ab' in Figure P4.5.

Ans. TEC is an ideal 5 V source, $V_{ab} = 5$ V; NEC does not exist.

P4.6 Determine I_X in Figure P4.6 in two ways: (a) by deriving TEC for each half-circuit and combining the two TECs; (b) By deriving a single TEC between the two terminals through which I_X flows.

Ans. 0.75 A.

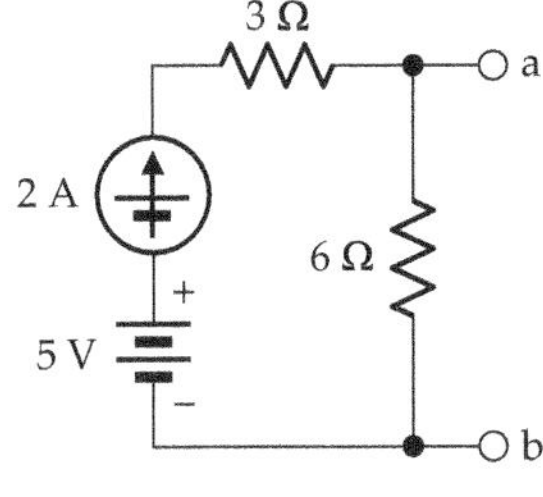

FIGURE P4.4

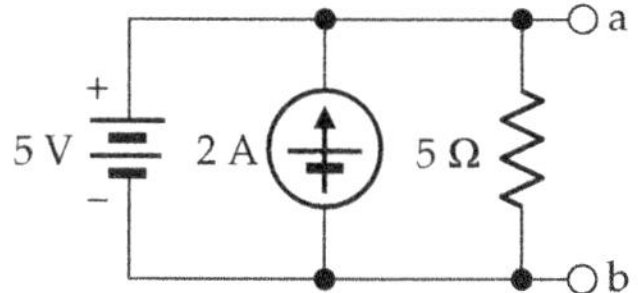

FIGURE P4.5

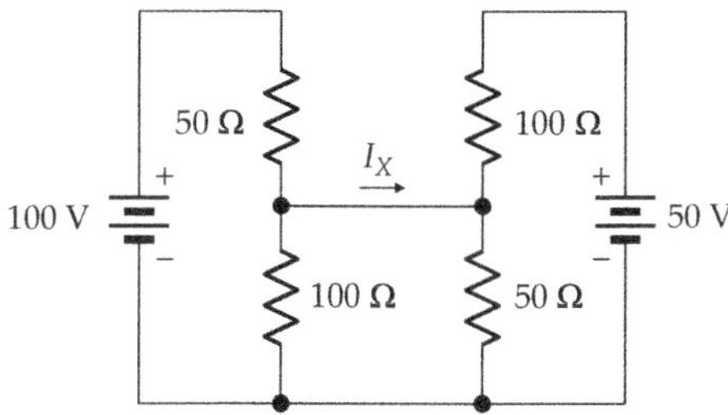

FIGURE P4.6

P4.7 (a) Determine V_{SRC} in Figure P4.7 by deriving TEC between terminals 'bc'. (b) Determine I_{SRC}, V_X, and V_Y.

Ans. (a) 120 V; (b) 12 A, $V_X = 48$ V, $V_Y = 72$ V.

P4.8 Derive TEC looking into terminals 'ab' in Figure P4.8.

Ans. $V_{Th} = V_{ab} = 10$ V, $R_{Th} = 10\ \Omega$.

P4.9 Derive NEC looking into terminals 'ab' in Figure P4.9.

Ans. $I_N = I_{ab} = 1$ A, $R_N = 20/3\ \Omega$.

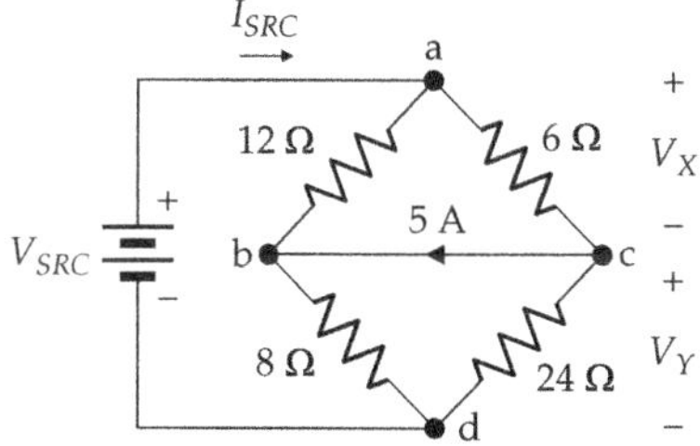

FIGURE P4.7

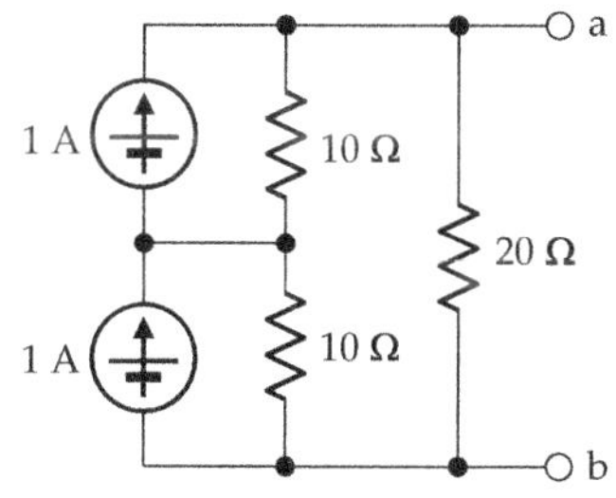

FIGURE P4.8

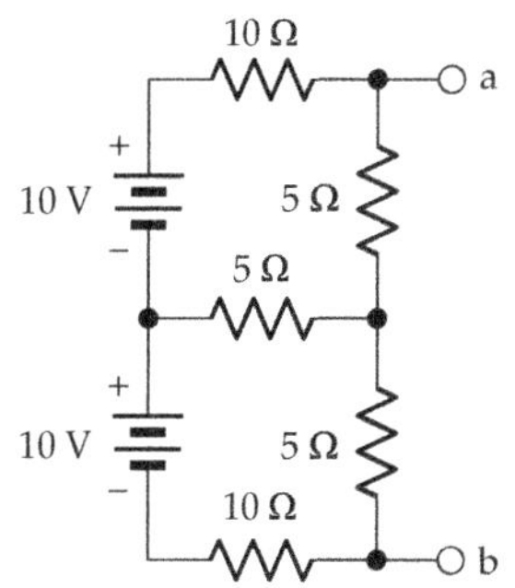

FIGURE P4.9

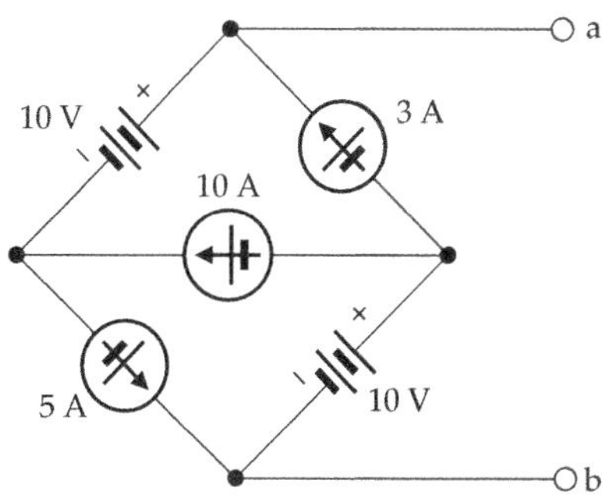

FIGURE P4.10

P4.10 Derive TEC and NEC looking into terminals 'ab' in Figure P4.10.

Ans. NEC is an ideal 8 A source directed from node 'a' to node 'b'. TEC does not exist.

P4.11 Derive TEC looking into terminals 'ab' in Figure P4.11.

Ans. $V_{Th} = V_{ab} = 0$, $R_{Th} = 25\ \Omega$.

P4.12 Derive TEC looking into terminals 'ab' in Figure P4.12.

Ans. $V_{Th} = V_{ab} = 16$ V, $R_{Th} = 8\ \Omega$.

P4.13 Derive NEC looking into terminals 'ab' in Figure P4.13.

Ans. $I_N = I_{ab} = 4.4$ A, $G_N = 0.04$ S.

P4.14 Derive TEC as seen by the 50 Ω resistor in Figure P4.14.

Ans. $V_{Th} = V_{ab} = 20$ V, $R_{Th} = 10\ \Omega$.

P4.15 Determine V_O in Figure P4.15 using TEC.

Ans. 20 V.

P4.16 Determine I_O in Figure P4.16 using NEC.

Ans. 20 A.

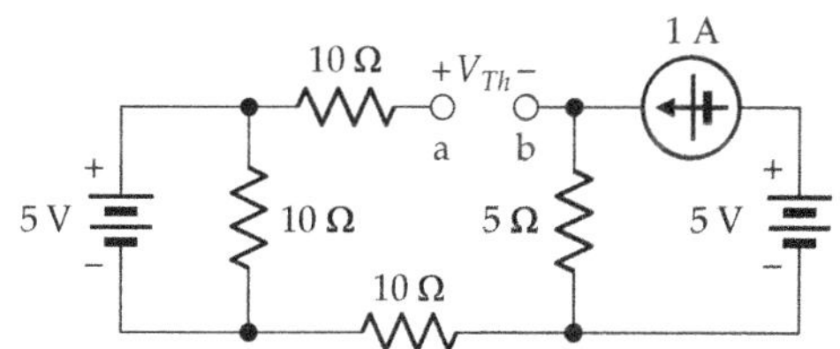

FIGURE P4.11

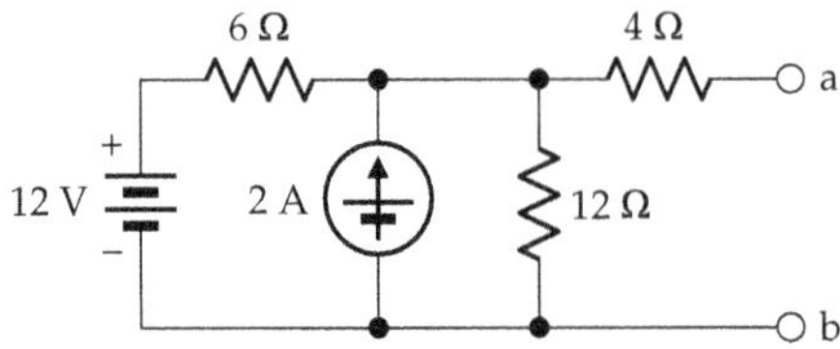

FIGURE P4.12

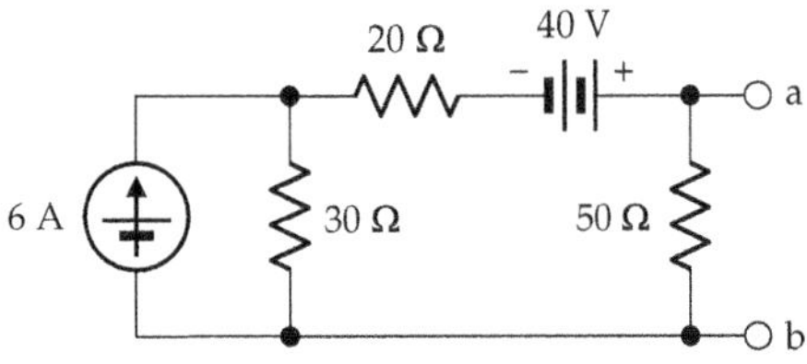

FIGURE P4.13

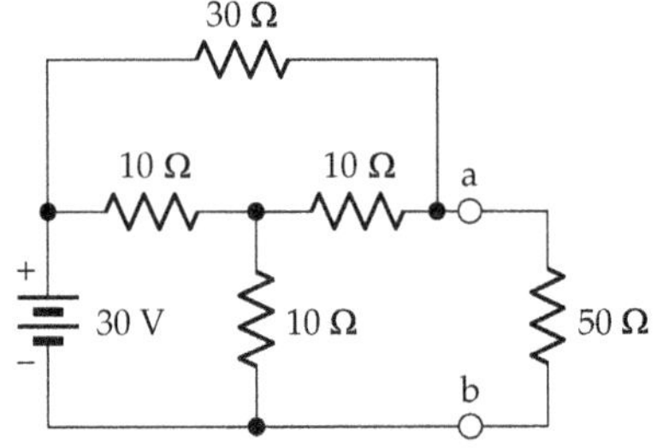

FIGURE P4.14

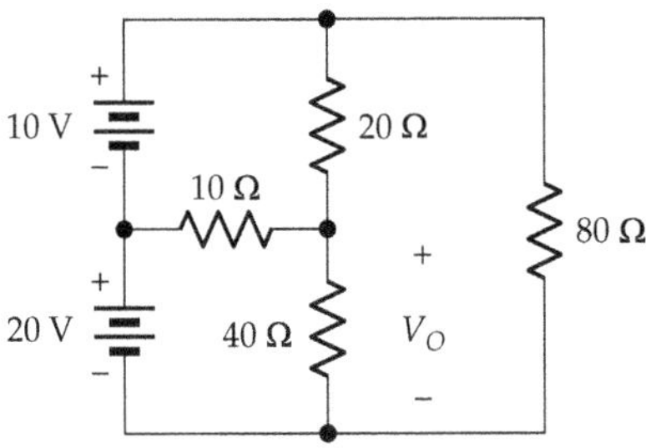

FIGURE P4.15

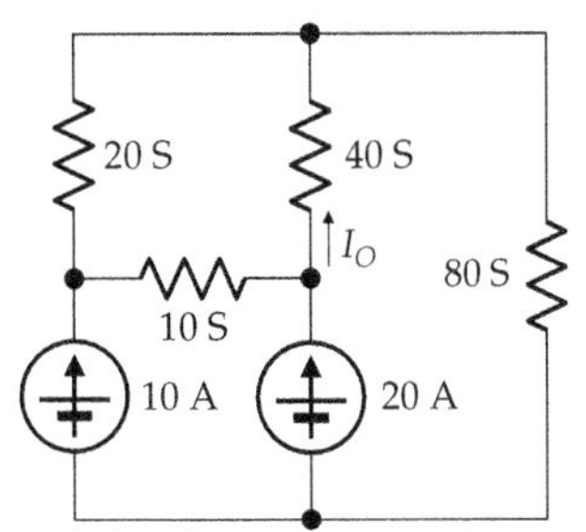

FIGURE P4.16

P4.17 Derive TEC looking into terminals 'ab' in Figure P4.17.
Ans. $V_{Th} = V_{ab} = 7$ V, $R_{Th} = 1.5$ Ω.

P4.18 Derive TEC looking into terminals 'ab' in Figure P4.18.
Ans. $V_{Th} = 0$, $R_{Th} = 1$ Ω.

P4.19 Derive NEC looking into terminals 'ab' in Figure P4.19.
Ans. $I_N = 0$, $R_N = 20/3$ Ω.

P4.20 Derive TEC and NEC looking into terminals 'ab' in Figure P4.20.
Ans. $V_{Th} = 0 = I_N$, $R_{Th} = R_N = 25$ Ω.

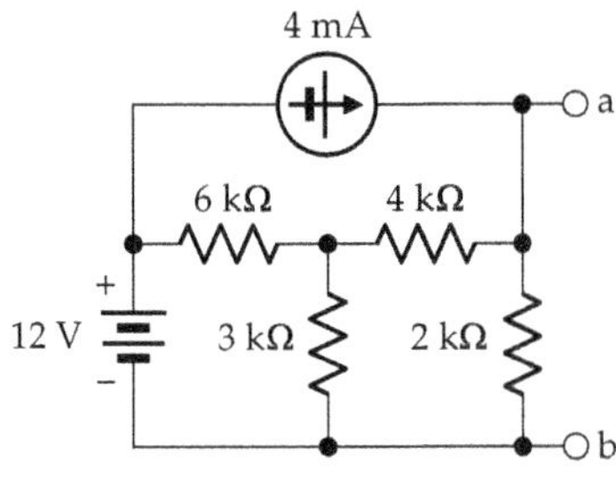

FIGURE P4.17

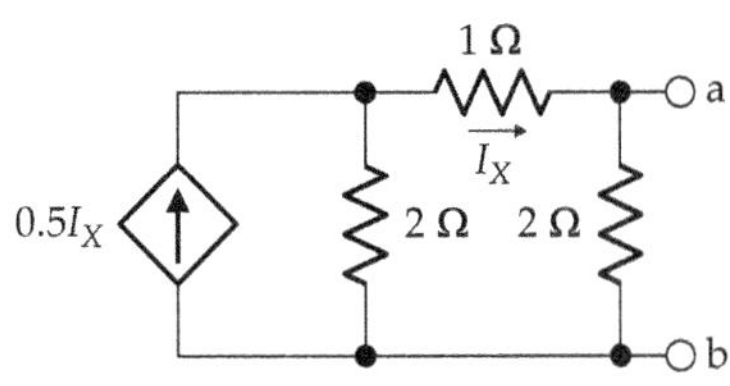

FIGURE P4.18

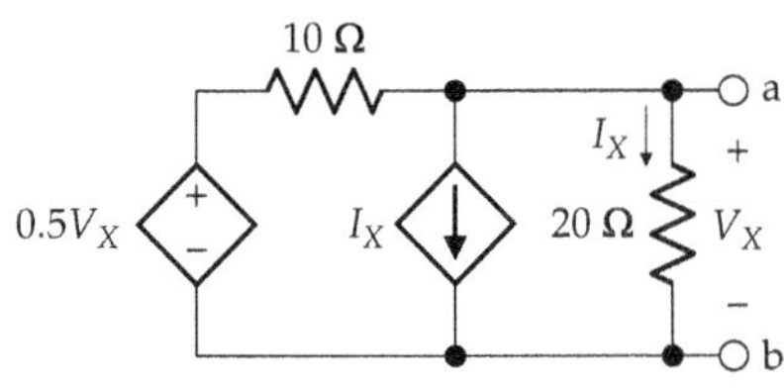

FIGURE P4.19

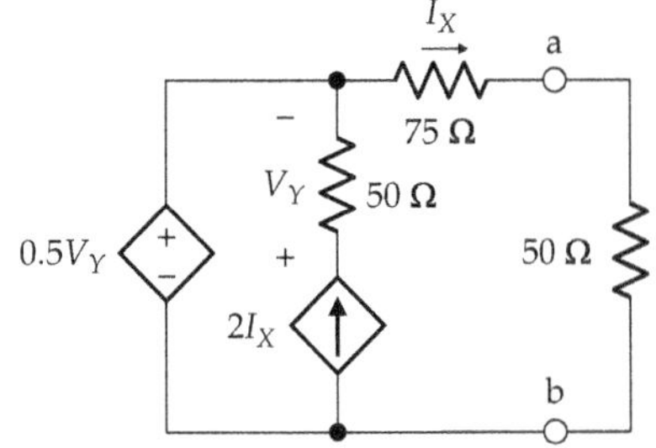

FIGURE P4.20

P4.21 Connect a resistor R_L between terminals 'ab' in Figure P4.21 and show that the voltage V_{ab} is independent of R_L. Deduce that TEC looking into terminals 'ab' is an ideal voltage source. Verify this deduction by determining V_{Th} and R_{Th} looking into terminals 'ab'.

Ans. $V_{Th} = V_{ab} = 24$ V, $R_{Th} = 0$.

P4.22 Derive TEC looking into terminals 'ab' in Figure P4.22.

Ans. $V_{Th} = V_{ab} = 4$ V, $R_{Th} = 4\ \Omega$.

P4.23 Derive TEC looking into terminals 'ab' in Figure P4.23.

Ans. $V_{Th} = V_{ab} = -100/3$ V, $R_{Th} = 1000/3\ \Omega$.

P4.24 Derive TEC and NEC looking into terminals 'ab' in Figure P4.24, assuming (a) $\alpha = 1$, and (b) $\alpha = 2$.

Ans. (a) $V_{Th} = 0 = I_N$, $R_{Th} = 1\ \Omega = R_N$; (b) $V_{Th} = V_{ab} = -5$ V, $R_{Th} = 0$, NEC does not exist.

P4.25 Derive TEC looking into terminals 'ab' in Figure P4.25.

Ans. $V_{Th} = V_{ab} = 10$ V, $R_{Th} = 10\ \Omega$.

P4.26 Derive TEC looking into terminals 'ab' in Figure P4.26.

Ans. $V_{Th} = V_{ab} = 3$ V, $R_{Th} = 75\ \Omega$.

P4.27 Derive NEC looking into terminals 'ab' in Figure P4.27.

Ans. $I_N = I_{ab} = 0.3$ A, $G_N = 0.025$ S.

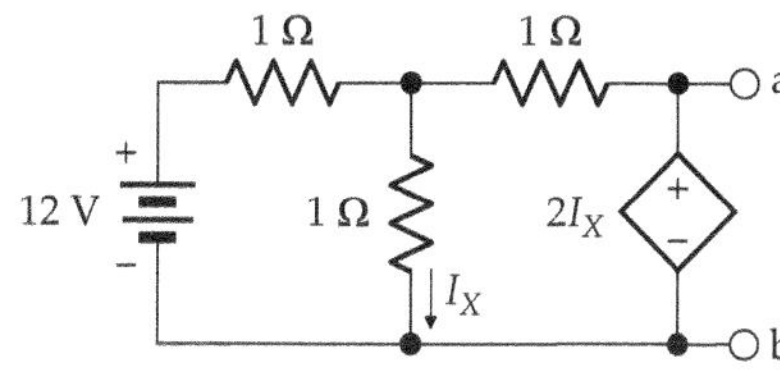

FIGURE P4.21

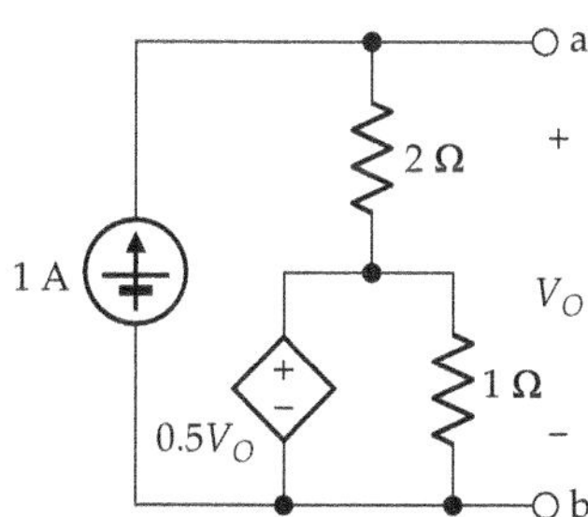

FIGURE P4.22

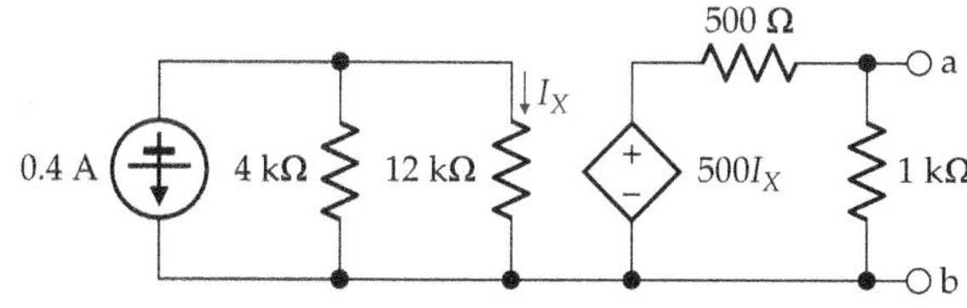

FIGURE P4.23

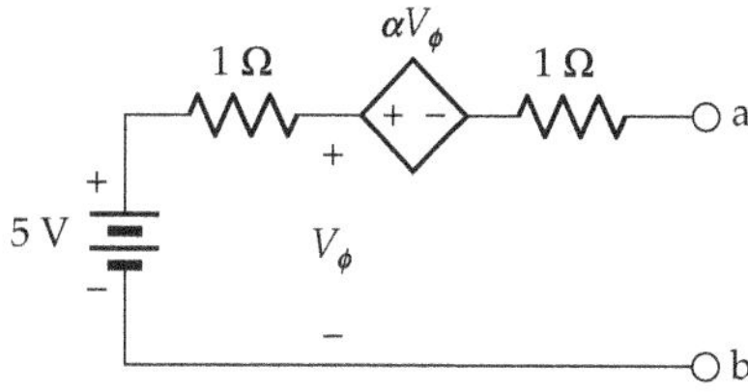

FIGURE P4.24

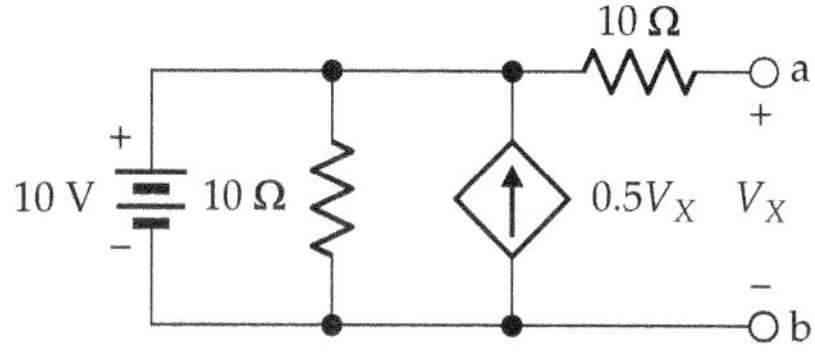

FIGURE P4.25

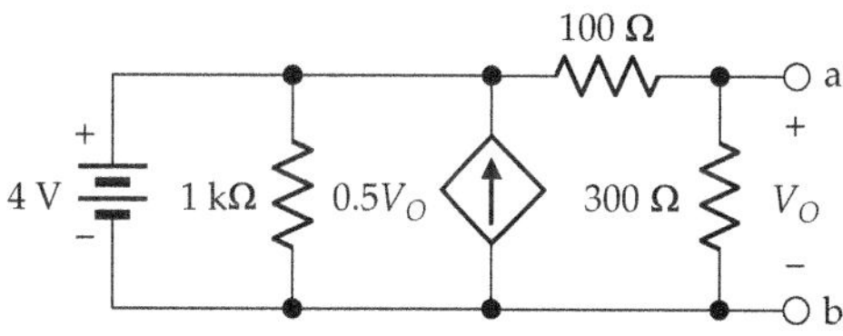

FIGURE P4.26

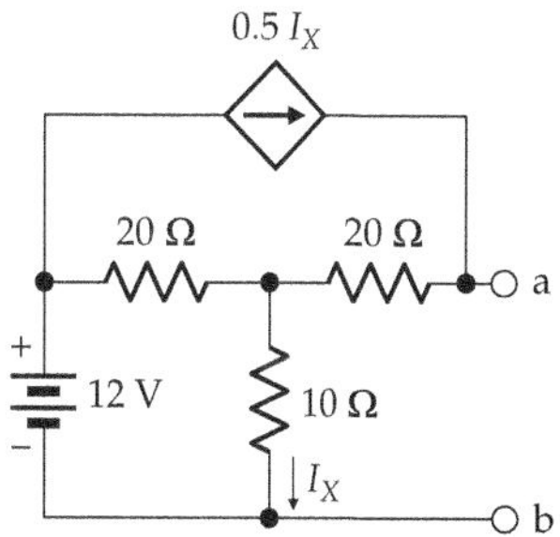

FIGURE P4.27

P4.28 Derive TEC looking into terminals 'ab' in Figure P4.28.

Ans. $V_{Th} = V_{ab} = 40$ V, $R_{Th} = 0$.

P4.29 Determine R so that Norton's current between nodes 'ab' in Figure P4.29 is zero.

Ans. 1 Ω.

P4.30 Determine in Figure P4.30 (a) TEC between node 'c' and the reference node, that is, including the 4 Ω resistor, and (b) V_O using TEC as seen by the 4 Ω resistor and taking I_X into account.

Ans. (a) TEC is a source of –10/3 V in series with a resistor of –4/3 Ω; (b) TEC is a source of 10 V in series with –16 Ω, which gives $V_0 = -10/3$ V.

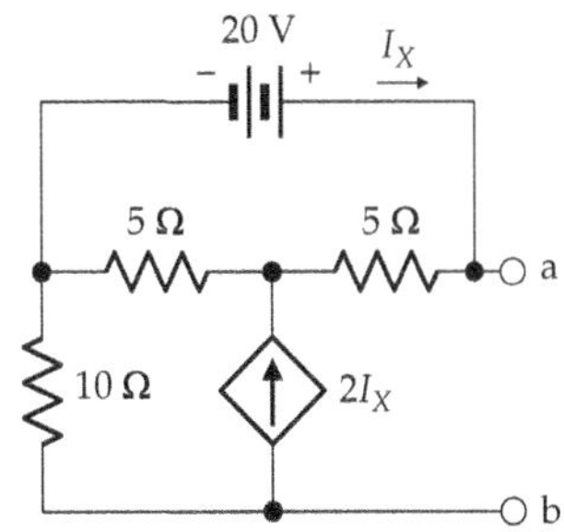

FIGURE P4.28

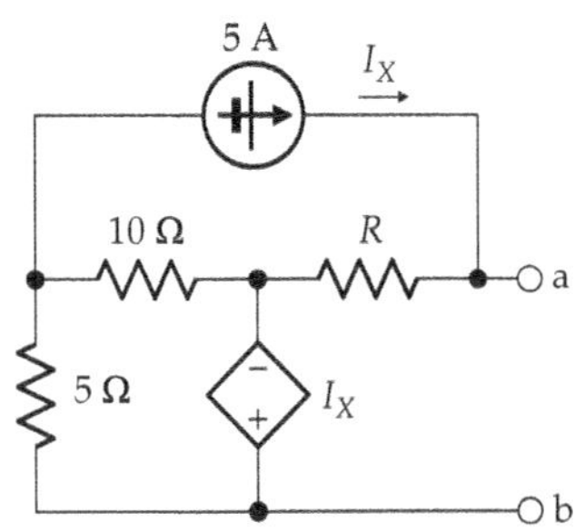

FIGURE P4.29

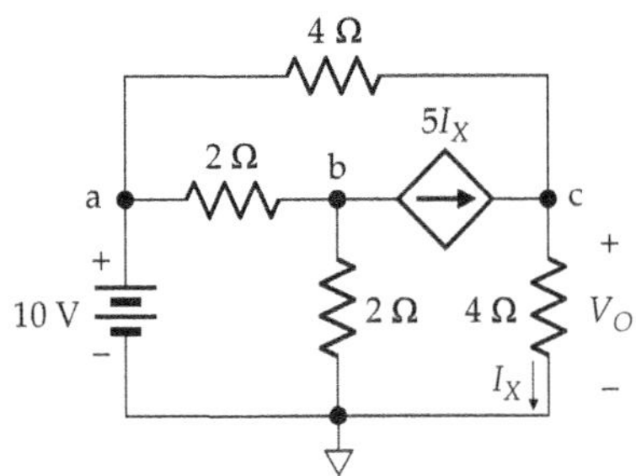

FIGURE P4.30

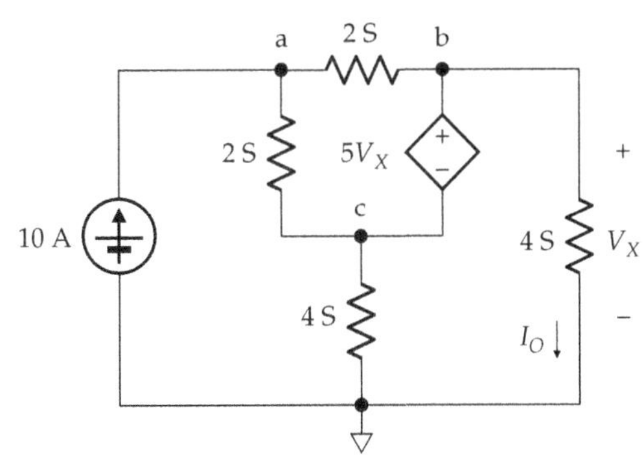

FIGURE P4.31

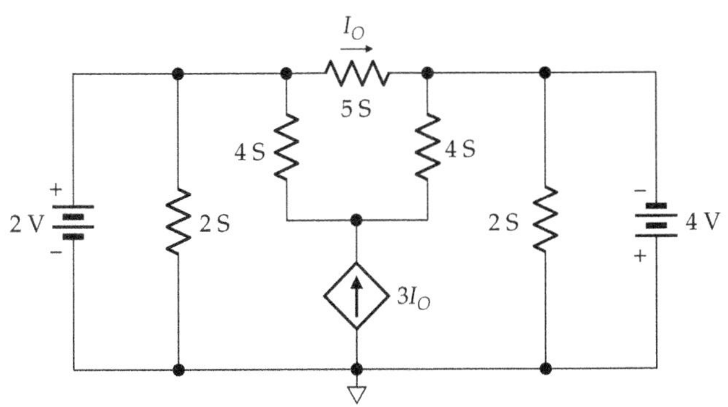

FIGURE P4.32

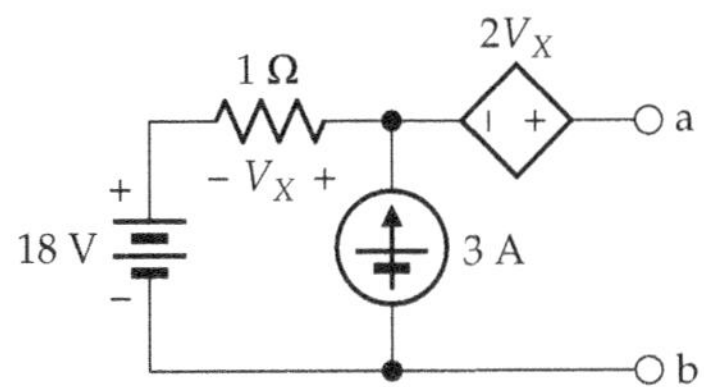

FIGURE P4.33

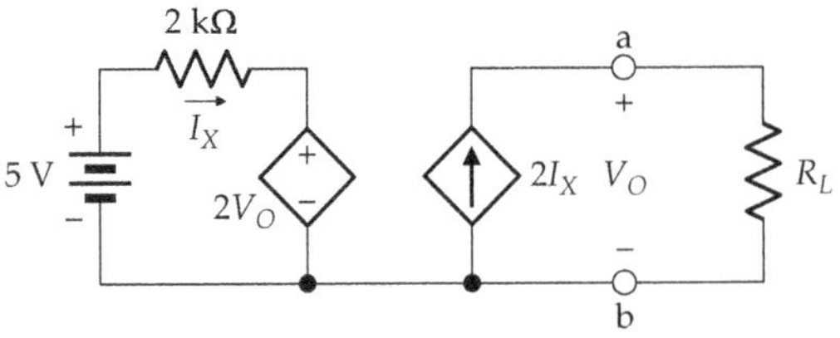

FIGURE P4.34

P4.31 Determine I_O in Figure P4.31 using NEC.

Ans. –10/3 A.

P4.32 Determine I_O in Figure P4.32 using TEC. Note that the circuit does not possess an NEC.

Ans. 30 A.

P4.33 Determine TEC looking into terminals 'ab' in Figure P4.33.

Ans. $V_{Th} = V_{ab} = 27$ V, $R_{Th} = 3\ \Omega$.

P4.34 Derive TEC as seen by R_L in Figure P4.34.

Ans. $V_{Th} = V_{ab} = 2.5$ V, $R_{Th} = 0.5$ kΩ.

P4.35 Derive NEC looking into terminals 'ab' in Figure P4.35.

Ans. $I_N = I_{ab} = 0.5$ A, $R_N = 10\ \Omega$.

P4.36 Derive TEC between terminals 'ab' in Figure P4.36.

Ans. $V_{Th} = V_{ab} = 5$ V, $R_{Th} = 5\ \Omega$.

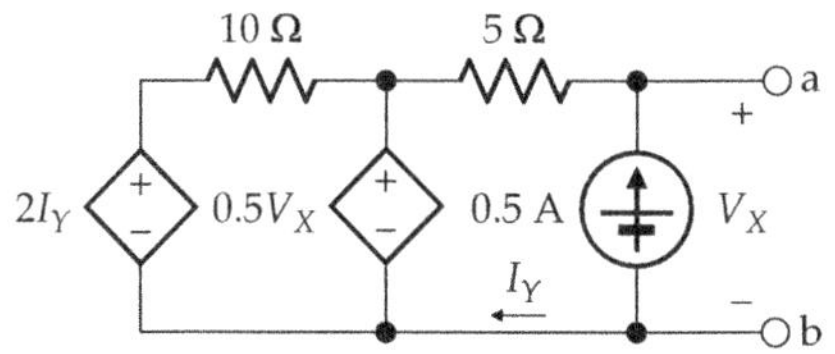

FIGURE P4.35

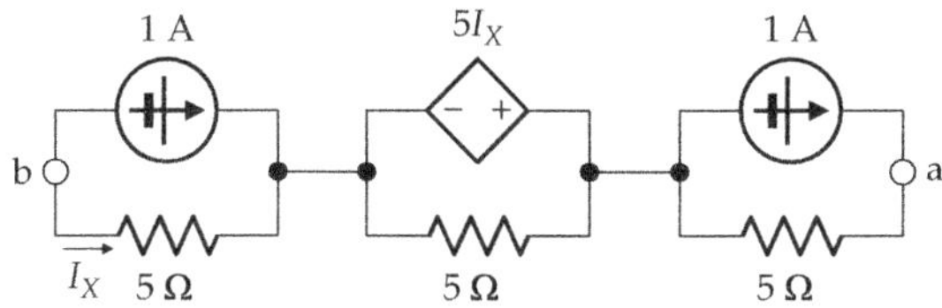

FIGURE P4.36

P4.37 Derive TEC looking into terminals 'ab' in Figure P4.37.
Ans. $V_{Th} = V_{ab} = 12$ V, $R_{Th} = 80\ \Omega$.

P4.38 Derive TEC between nodes 'ab' in Figure P4.38.
Ans. $V_{Th} = V_{ab} = 20$ V, $R_{Th} = 8\ \Omega$.

P4.39 Determine V_O in Figure P4.39 using TEC.
Ans. 15.51 V.

P4.40 Derive TEC looking into terminals 'ab' in Figure P4.40.
Ans. 0 V, $R_{Th} = 18/7\ \Omega$.

P4.41 Determine V_O in Figure P4.41 using NEC. Note that the circuit does not possess a TEC.
Ans. 30 V.

P4.42 Determine I_O in Figure P4.42 using NEC.
Ans. 15.51 A.

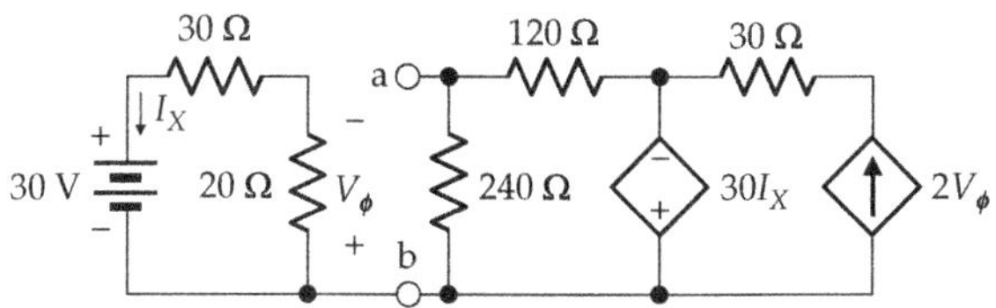

FIGURE P4.37

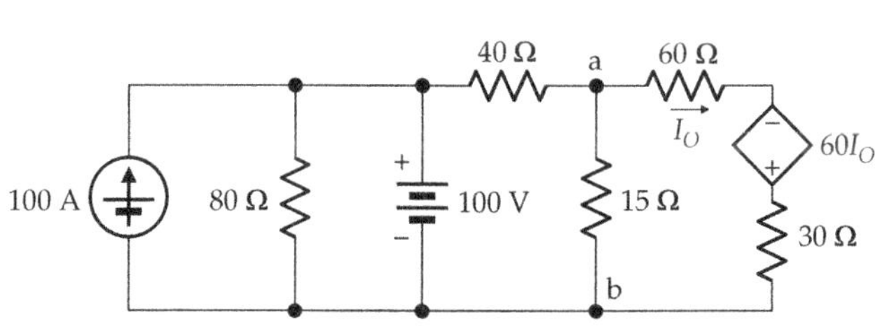

FIGURE P4.38

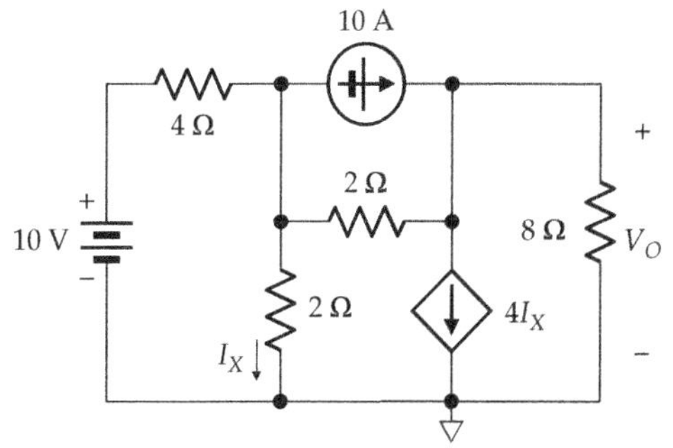

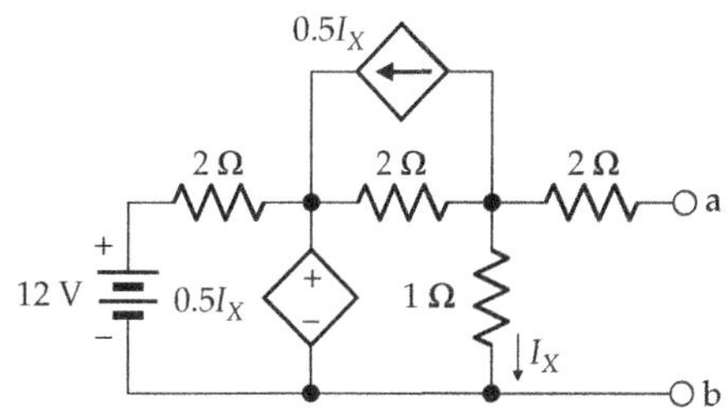

FIGURE P4.39

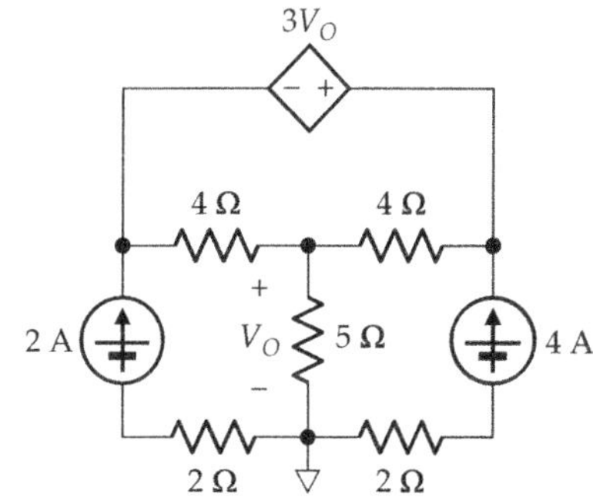

FIGURE P4.40

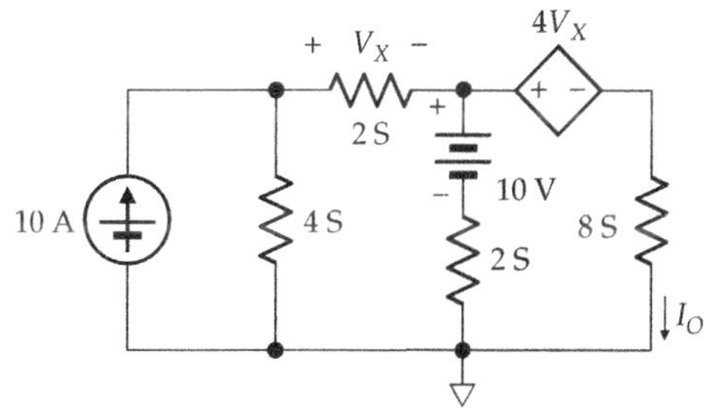

FIGURE P4.41

FIGURE P4.42

P4.43 Determine TEC looking into terminals 'ab' in Figure P4.43.

Ans. $V_{Th} = V_{ab} = 4$ V, $R_{Th} = 1/3\ \Omega$.

P4.44 Determine I_O in Figure P4.44 using TEC.

Ans. −0.65 A.

P4.45 Derive TEC looking into terminals 'ab' in Figure P4.45. Verify by deriving an intermediate TEC with the 4 Ω resistor and the 1 V–3 Ω branch removed.

Ans. $V_{Th} = V_{ab} = 4$ V, $R_{Th} = 2\ \Omega$.

P4.46 Derive TEC between terminals 'ab' in Figure P4.46.

Ans. $V_{Th} = V_{ab} = 80$ V, $R_{Th} = 10\ \Omega$.

P4.47 Derive TEC looking into terminals 'ab' in Figure P4.47.

Ans. $V_{Th} = V_{ab} = -1/3$ V, $R_{Th} = 8/9$ kΩ.

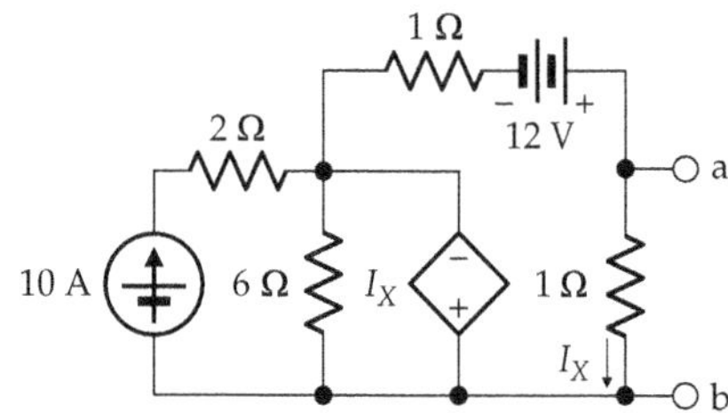

FIGURE P4.43

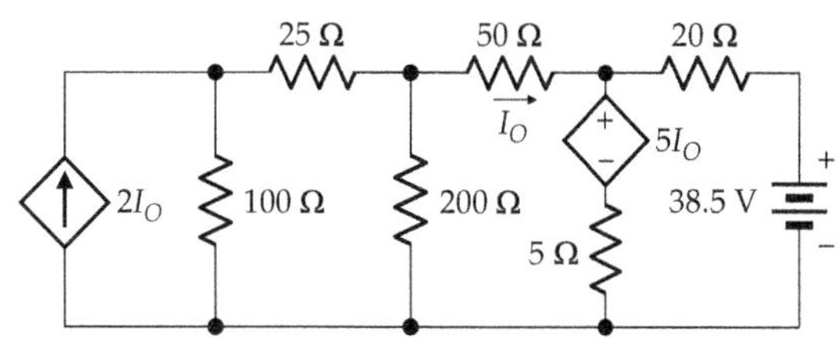

FIGURE P4.44

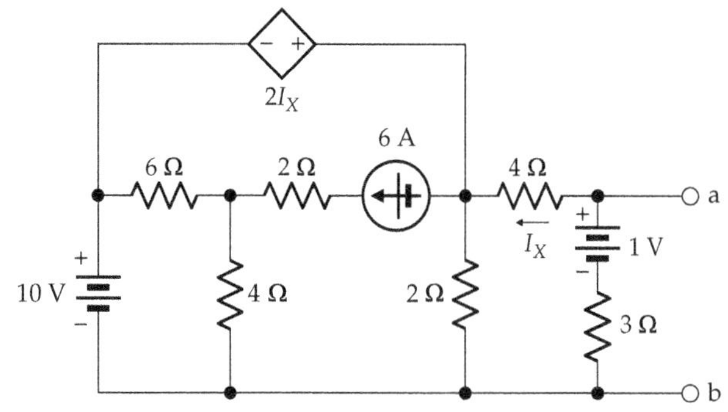

FIGURE P4.45

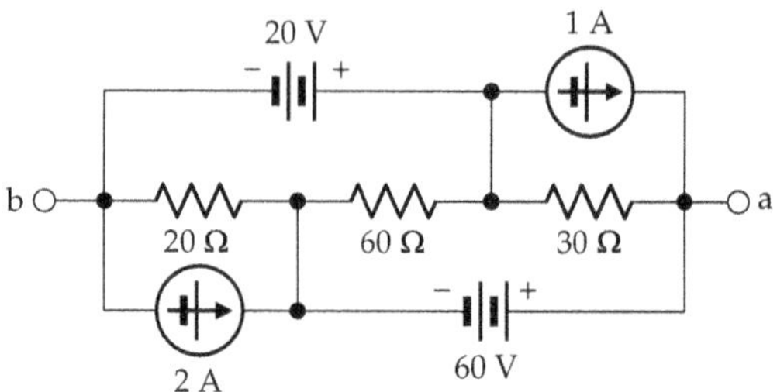

FIGURE P4.46

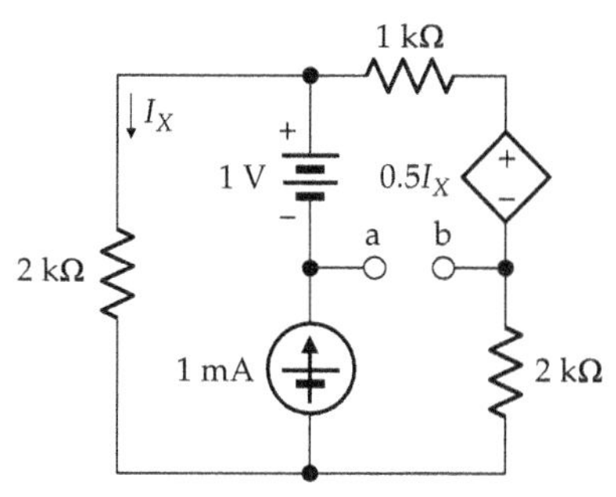

FIGURE P4.47

P4.48 Derive TEC looking into terminals 'ab' in Figure P4.48, (a) keeping the 4 Ω resistor in place, (b) temporarily removing this resistor. Note that although these TECs are different, they give the same V_{ab} and the same I_X.

Ans. (a) $V_{Th} = V_{ab} = 20$ V, $R_{Th} = 4\ \Omega$; (b) $V_{Th} = 12.5$ V, $R_{Th} = -1.5\ \Omega$.

P4.49 Derive TEC between terminals 'ab' in Figure P4.49.

Ans. $V_{Th} = V_{ab} = 6$ V, $R_{Th} = 10\ \Omega$.

P4.50 Derive TEC looking into terminals 'ab' in Figure P4.50.

Ans. $V_{Th} = V_{ab} = 1$ V, $R_{Th} = 4\ \Omega$.

P4.51 Derive NEC between terminals 'ab' in Figure P4.51, assuming all resistances are 2 Ω.

Ans. $I_N = I_{ab} = -41/33$ A, $R_N = 66/23\ \Omega$.

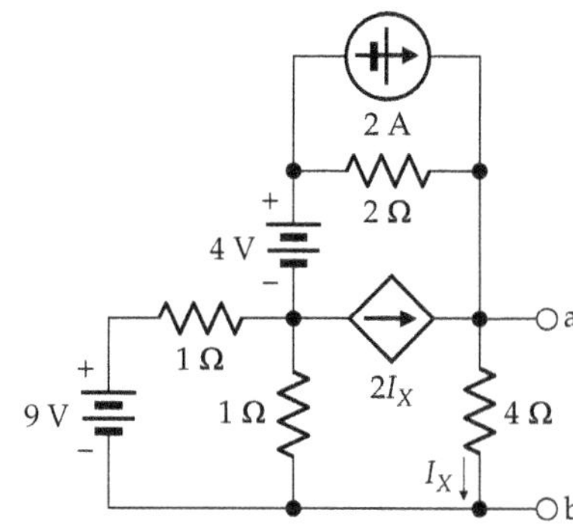

FIGURE P4.48

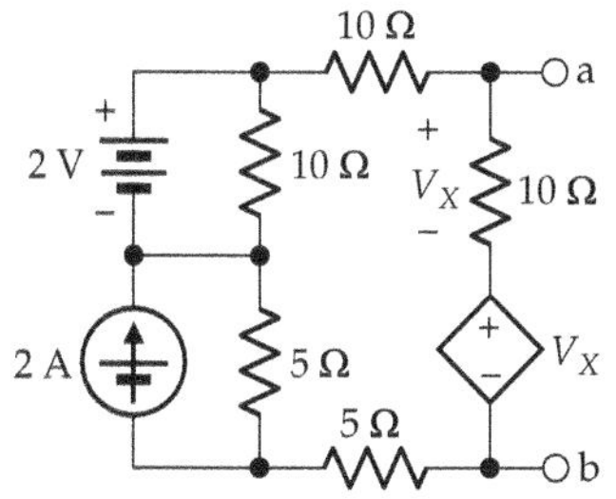

FIGURE P4.49

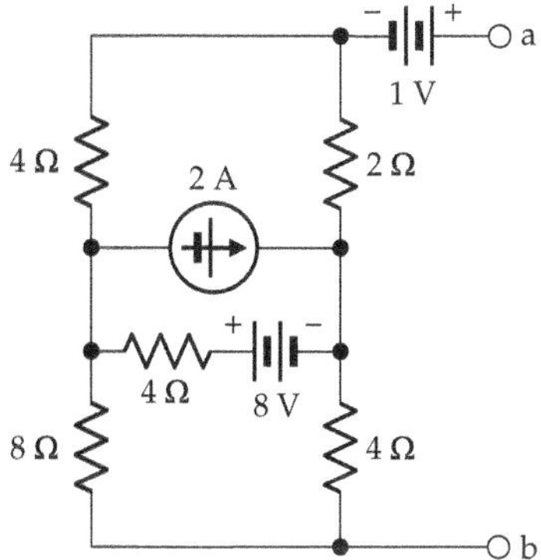

FIGURE P4.50

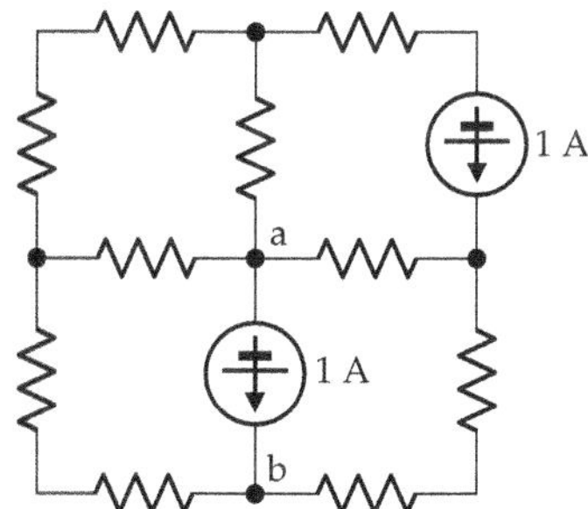

FIGURE P4.51

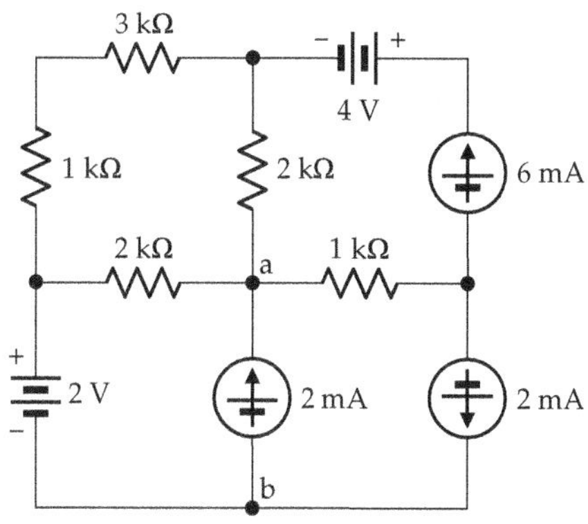

FIGURE P4.52

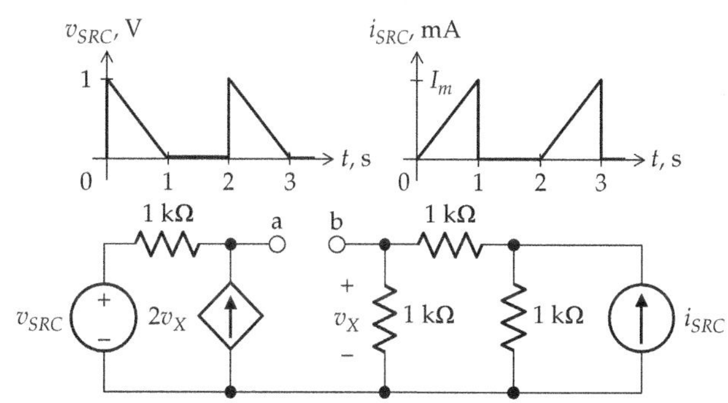

FIGURE P4.53

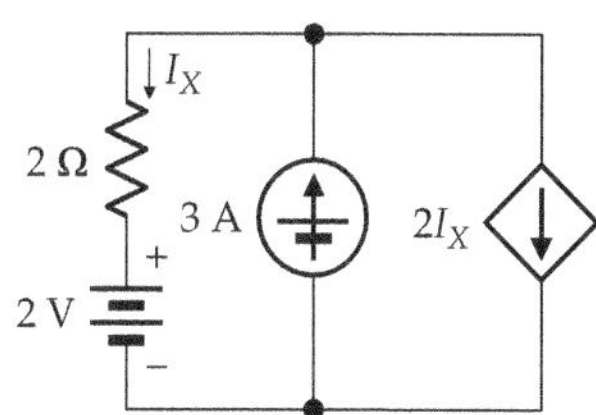

FIGURE P4.54

P4.52 Derive TEC between terminals 'ab' in Figure P4.52.

Ans. $V_{Th} = V_{ab} = -1$ V, $R_{Th} = 1.5$ kΩ.

P4.53 Use TEC to determine I_m in Figure P4.53 so that V_{ab} is a square waveform.

Ans. 3 mA.

Substitution and Source Absorption Theorems

P4.54 Determine, according to the substitution theorem, (a) the independent voltage source, (b) the independent current source, and (c) the resistance that can replace the dependent current source in Figure P4.54 without affecting the rest of the circuit.

Ans. (a) 4 V; (b) 2 A; (c) 2 Ω.

P4.55 Determine V_X in Figure P4.55 by using the substitution theorem, where 'N_A' is an unspecified circuit that passes a current of 0.5 A.

Ans. 15 V.

P4.56 Determine V_O in Figure P4.56 by using the substitution theorem and by deriving NEC between nodes 'ab', where 'N_A' is an unspecified circuit having a voltage of 12.5 V across it.

Ans. –10/3 V.

P4.57 Determine I_O in Figure P4.57 by using the substitution theorem and by deriving TEC between nodes 'ab', where 'N_A' is an unspecified circuit that passes a current of 12.5 A.

Ans. –10/3 A.

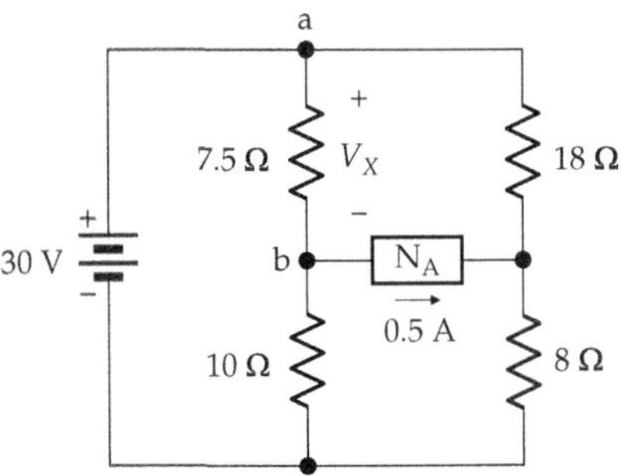

FIGURE P4.55

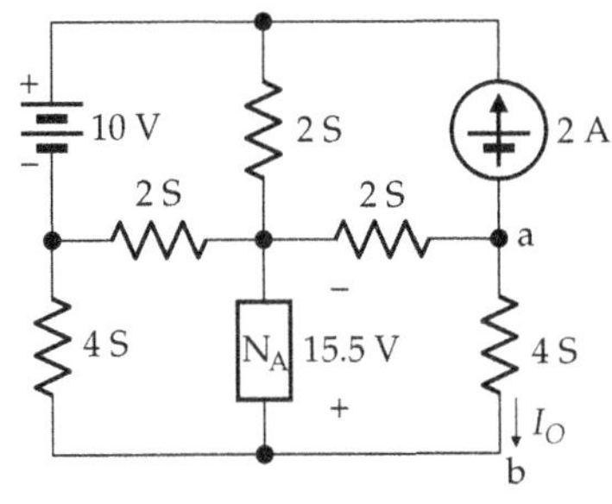

FIGURE P4.58

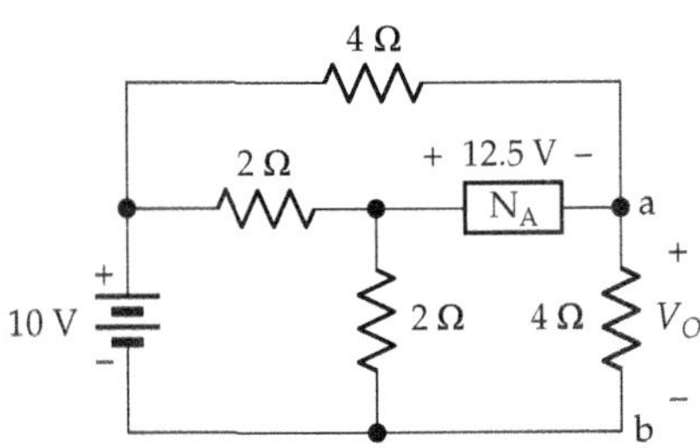

FIGURE P4.56

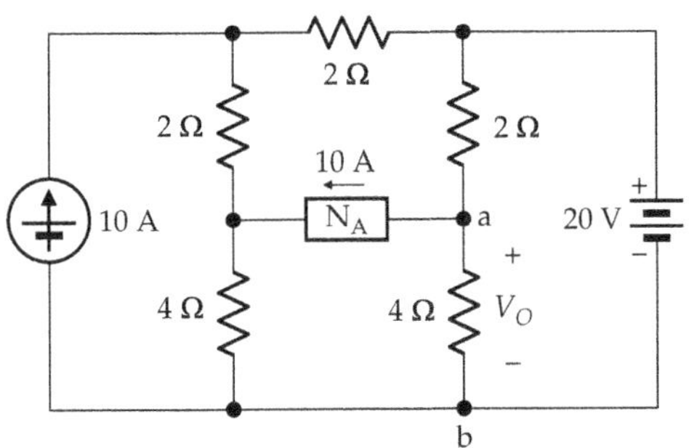

FIGURE P4.59

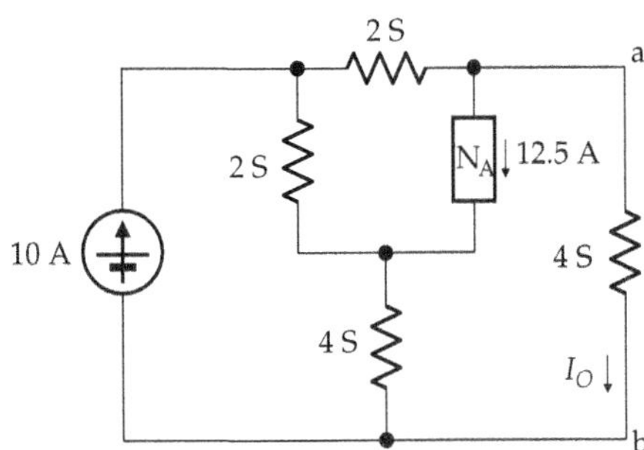

FIGURE P4.57

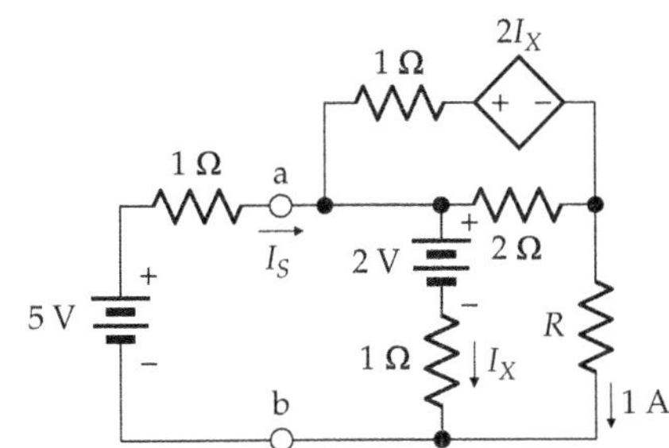

FIGURE P4.60

P4.58 Determine I_O in Figure P4.58 by using the substitution theorem and by deriving NEC between nodes 'ab', where 'N_A' is an unspecified circuit having a voltage of 15.5 V across it.

Ans. −22 A.

P4.59 Determine V_O in Figure P4.59 by using the substitution theorem and by deriving TEC between nodes 'ab', where 'N_A' is an unspecified circuit passing a current of 10 A.

Ans. 0.

P4.60 Determine I_S in Figure P4.60 by deriving TEC looking into terminals 'ab', given that the current in the resistor R is 1 A.

Ans. 2 A.

P4.61 Redo Example 4.6 assuming a current of 1.4 A in 'N_A' directed from left to right.

Ans. 105/281 = 0.37 A.

P4.62 Determine R_{in} in Figure P4.62 by applying the source absorption theorem.

Ans. 40 Ω.

P4.63 Determine R_{in} in Figure P4.63 by applying the source absorption theorem, where I_X is in amperes.

Ans. 100 Ω.

P4.64 Determine R_{in} in Figure P4.64 by applying the source absorption theorem.

Ans. 1.25 Ω.

P4.65 Determine G_{in} looking into terminals 'ab' in Figure P4.65.

Ans. 6 S.

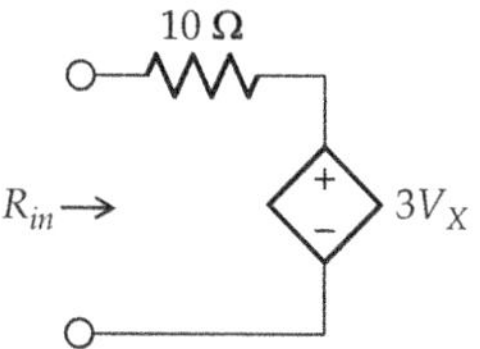

FIGURE P4.62

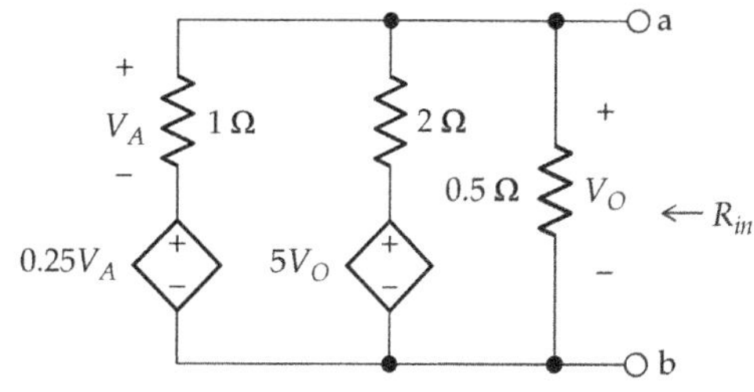

FIGURE P4.64

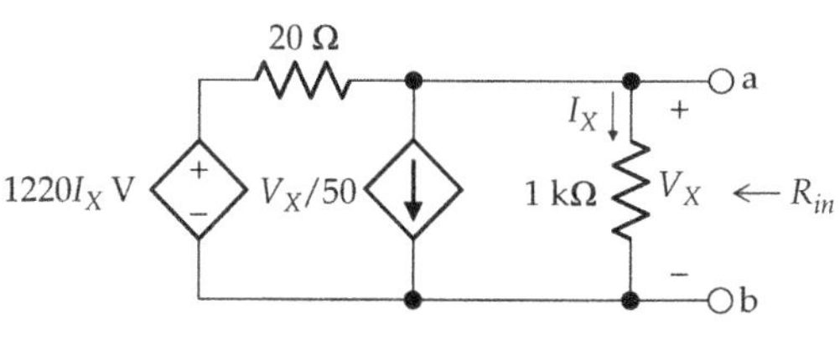

FIGURE P4.63

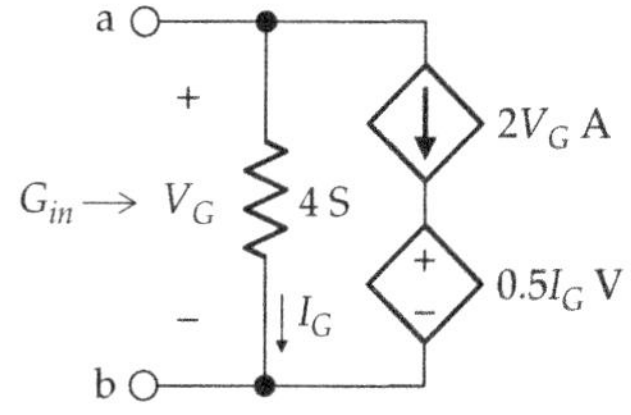

FIGURE P4.65

5

Circuit Simplification

Objective and Overview

This chapter is concerned with various procedures and techniques that simplify circuit analysis, either by reducing a given circuit to a simpler form or by following certain methodologies that facilitate obtaining the desired circuit response. We have already encountered circuit simplification in previous chapters, as in the equivalent series or parallel connections of resistors, star-delta transformation, and Thevenin's equivalent circuit (TEC). However, the main objective in these cases was pursuing circuit equivalence rather than circuit simplification per se. By the end of this chapter, all the main circuit simplification techniques would have been presented.

The chapter begins with the fundamental concept of superposition, its implications, and its application in circuit analysis, including the use of the substitution theorem in conjunction with superposition. This is followed by the method of output scaling, according to which a convenient output is arbitrarily assumed, and voltages and currents are determined by working backward toward an applied source, then scaling all voltages and currents in accordance with the value of this source. Output scaling is followed by the technique of removal of redundant elements, which are elements that either do not carry current or do not affect the responses of interest.

Two other simplification techniques, discussed next, are partitioning of circuits by sources and source rearrangement. These techniques can simplify the analysis of some types of circuits and provide useful insight into their behavior. The chapter ends by considering circuits that possess symmetry of a form that can be exploited to greatly simplify the analysis.

5.1 Superposition

Definition: *If an input x_1 to a given system produces an output y_1 and an input x_2 produces an output y_2, then the system obeys superposition if an input $(x_1 + x_2)$, that is, the sum of the two inputs, produces an output $(y_1 + y_2)$, that is, the sum of the outputs due to each input acting alone. The same applies for more than two inputs.*

Superposition is a defining property of linear systems, as it is an essential attribute of linearity. Consider, for example, an ideal resistor that obeys Ohm's law, $v = Ri$. With R being constant, this is a linear relation in v and i. A current i_1 produces a voltage $v_1 = Ri_1$, and a current i_2 produces a voltage $v_2 = Ri_2$. The voltage produced by a current $(i_1 + i_2)$ is

$$v = R(i_1 + i_2) = Ri_1 + Ri_2 = v_1 + v_2 \tag{5.1}$$

It is seen that superposition is obeyed in this case because of linearity of Ohm's law. On the other hand, if $v = ki^2$, where k is a constant, the v–i relation is not linear. A current i_1 produces a voltage $v_1 = ki_1^2$, and a current i_2 produces a voltage $v_2 = ki_2^2$. The voltage produced by a current $(i_1 + i_2)$ is

$$v = k(i_1 + i_2)^2 = ki_1^2 + 2ki_1i_2 + ki_2^2 \neq ki_1^2 + ki_2^2 \tag{5.2}$$

Superposition is not obeyed in this case, because the nonlinearity introduces an additional nonlinear product term $2ki_1i_2$.

To see how superposition can be applied in circuit analysis, consider the simple two-essential-node circuit of Figure 5.1, in which it is required to determine V_X. It is assumed that V_{SRC} and I_{SRC} are given, but their symbols will be retained to begin with. KCL at node 'a' gives

$$I_{SRC} = \frac{V_X}{10} + \frac{V_X - V_{SRC}}{5} \tag{5.3}$$

or

$$V_X = \frac{2}{3}V_{SRC} + \frac{10}{3}I_{SRC} \tag{5.4}$$

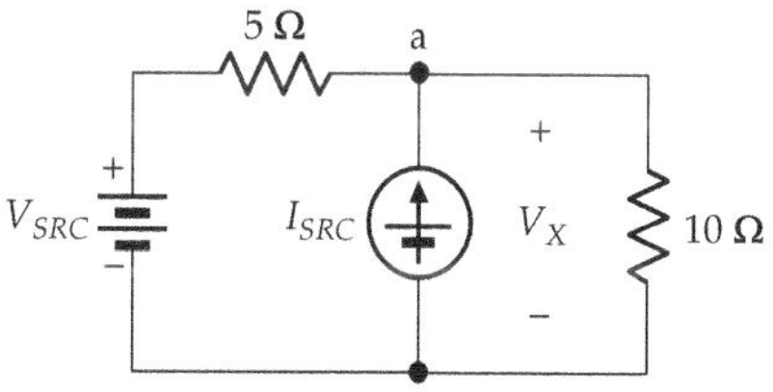

FIGURE 5.1
Circuit for superposition.

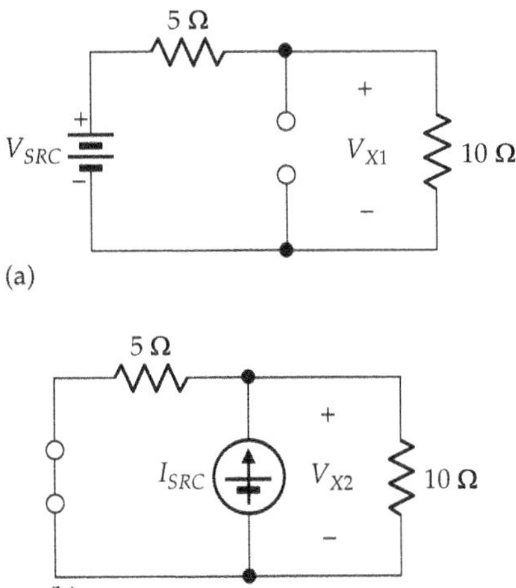

FIGURE 5.2
Application of one source at a time. Components of the desired response V_X, due to the voltage source acting alone (a) and due to the current source acting alone (b).

It is seen that V_X is the sum of two components V_{X1} and V_{X2}, where $V_{X1} = (2/3)V_{SRC}$ is due to V_{SRC} acting alone, with $I_{SRC} = 0$, and $V_{X2} = (10/3)I_{SRC}$ is due to I_{SRC} acting alone, with $V_{SRC} = 0$. This suggests that V_{X1} can be obtained from the circuit by setting $I_{SRC} = 0$, that is, replacing the current source with an open circuit, as in Figure 5.2a. From voltage division, V_{X1} is indeed $(2/3)V_{SRC}$. Similarly, if V_{SRC} is set to zero, by replacing the voltage source with a short circuit, as in Figure 5.2b, V_{X2} is indeed $(50/15)I_{SRC} = (10/3)I_{SRC}$. It follows that V_X in Figure 5.1 can be obtained by applying each independent source alone, with the other independent source set to zero.

The preceding argument illustrates the essence of using superposition in circuit analysis and can be generalized to a circuit containing any number of independent sources. The desired variable is the sum of components, each of which is obtained by applying one independent source at a time, with the other independent sources set to zero. Evidently, the advantage of this procedure is that the circuit resulting from applying one source at a time is simpler to analyze than the original circuit.

Superposition can be defined in slightly more general terms than stated earlier. A system is said to obey superposition if an input $(ax_1 + bx_2)$ produces an output $(ay_1 + by_2)$, where a and b are constants, y_1 is the response to x_1 acting alone, and y_2 is the response to x_2 acting alone. LTI circuits satisfy both definitions. For in an LTI circuit, multiplying an input acting alone by a constant multiplies the output due to this input by the same constant, as can be seen from Equation 5.4. Hence, ax_1 and bx_2 can be considered as two inputs x_1' and x_2' that produce outputs $y_1' = ay_1$ and $y_2' = by_2$, respectively, so that the sum of the two inputs produces an output that is the sum of the two outputs.

Example 5.1: Governing Relation for Zener Diode Circuit

A voltage regulator diode, or Zener diode, is commonly used to supply a variable load current at a nominally constant voltage from an unregulated dc voltage supply that may vary between specified limits. The diode can be represented over its normal operating range by a linear-output voltage source consisting of a battery V_{Z0} in series with a resistance r_Z. The diode is shown connected in parallel with the load R_L in Figure 5.3a and is supplied from the unregulated dc supply V_I through a series resistor, R_S. It is required to derive an expression for the load voltage V_L as a function of V_I, V_{Z0}, and I_L, in terms of R_S and r_Z.

Solution:

The required relation can be readily derived using superposition and the substitution theorem. The circuit is shown in Figure 5.3b, where, according to the substitution theorem, V_I, as a designated voltage between two nodes, is represented by a battery and R_L is replaced by an independent current source I_L. Superposition can then be applied to V_L with each of the independent sources alone, while the other sources set to zero.

Applying V_I alone (Figure 5.4a) gives, by voltage division,

$$V_{L1} = \frac{r_Z}{R_S + r_Z} V_I \tag{5.5}$$

With V_{Z0} applied alone (Figure 5.4b), V_{L2} is the voltage across R_S. By voltage division,

$$V_{L2} = \frac{R_S}{R_S + r_Z} V_{Z0} \tag{5.6}$$

With I_L applied alone (Figure 5.4c), V_{L3} is the voltage across the parallel combination of R_s and r_Z, with a negative sign, because I_L produces a voltage drop from the lower node to the upper node. Thus,

$$V_{L3} = -\frac{R_S r_Z}{R_S + r_Z} I_L \tag{5.7}$$

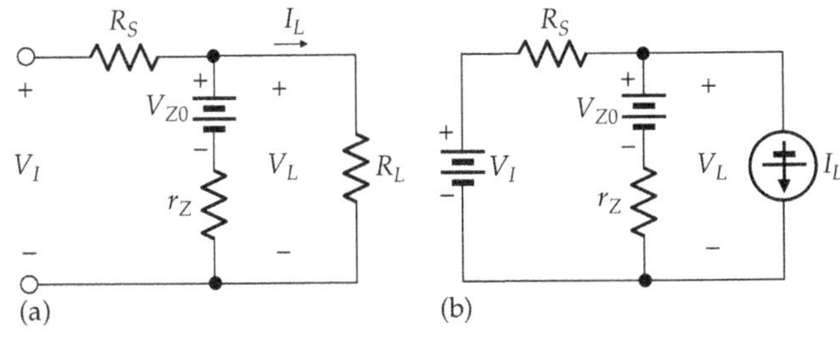

FIGURE 5.3
Figure for Example 5.1.

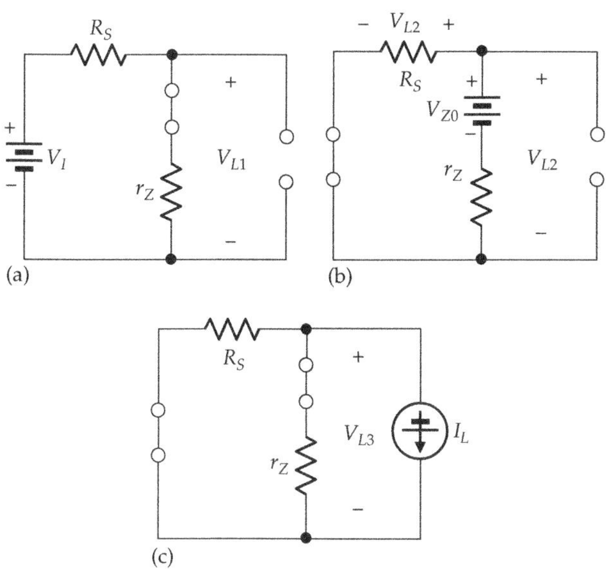

FIGURE 5.4
Figure for Example 5.1.

V_L is then the sum of the three components:

$$V_L = \frac{r_Z}{R_S + r_Z} V_I + \frac{R_S}{R_S + r_Z} V_{Z0} - \frac{R_S r_Z}{R_S + r_Z} I_L \quad (5.8)$$

It is seen that if $r_Z \ll R_S$, as is the case in practice when the diode is operating in its normal regulating range, the first and third terms on the RHS of Equation 5.8 can be neglected compared to the middle term, which gives $V_L \cong V_{Z0}$, independently of V_I and I_L.

Problem-Solving Tip

- The substitution theorem can be used to replace a branch current or voltage by an independent source, which allows application of superposition.

Exercise 5.1

Derive Equation 5.8 from KVL and KCL applied to the circuit of Figure 5.3a. Note the relative ease of applying superposition and the substitution theorem.

Primal Exercise 5.2

Determine in Figure 5.5 (a) the component of V_X due to the battery acting alone, (b) the component of V_X due to I_{SRC} = 2 A acting alone, (c) the components in (a) and (b) if the battery voltage and I_{SRC} are doubled, and (d) I_{SRC} that makes $V_X = 0$, (i) using superposition, (ii) considering V_X to be zero and determining I_{SRC} using KCL and KVL.

Ans. (a) 6 V; (b) –4 V; (c) 12 V, –8 V; (d) 3 A.

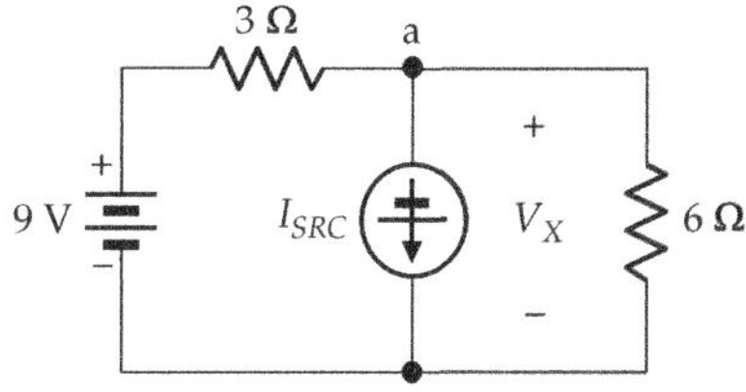

FIGURE 5.5
Figure for Primal Exercise 5.2.

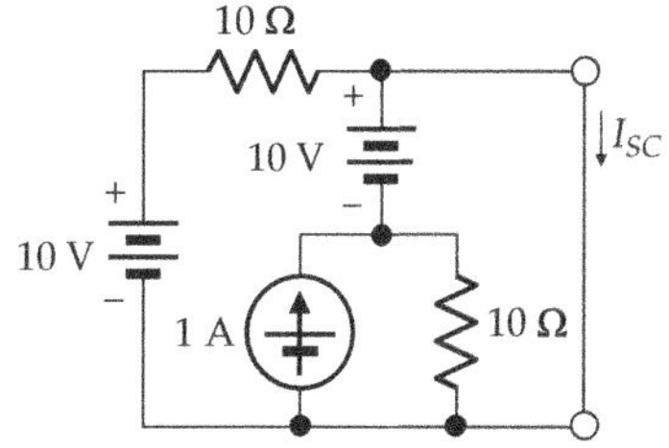

FIGURE 5.6
Figure for Primal Exercise 5.3.

Primal Exercise 5.3

Determine I_{SC} in Figure 5.6 by applying (a) the current source alone, with the voltage sources set to zero and (b) the two voltage sources together, with the current source set to zero. Note that superposition can be applied with some sources applied together if it is convenient to do so.

Ans. 3 A.

5.1.1 Dependent Sources

How is superposition applied in the presence of dependent sources? To answer this question, consider the circuit of Figure 5.1 with a dependent source added as in Figure 5.7. From KCL at node 'a',

$$I_S + I_{SRC} = \frac{V_X - \rho I_S}{10} \quad (5.9)$$

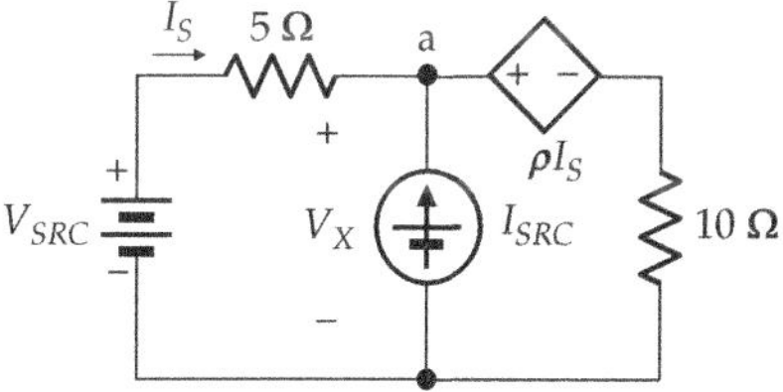

FIGURE 5.7
Superposition with dependent source.

where $I_S = (V_{SRC} - V_X)/5$. Substituting for I_S in Equation 5.9 and rearranging, we get

$$V_X = \frac{(10+\rho)}{(15+\rho)} V_{SRC} + \frac{50}{(15+\rho)} I_{SRC} \qquad (5.10)$$

Equation 5.10 is of the same form as Equation 5.4, but with the resistive coefficients of V_{SRC} and I_{SRC} modified by ρ. The dependent source does not contribute a component to V_X, as do V_{SRC} and I_{SRC}. This is consistent with the conclusion in Section 4.1, that dependent sources alone do not excite a circuit. Note that the modification of resistances by dependent sources is in accordance with the special case exemplified by the source absorption theorem (Section 4.5), whereby the dependent source is itself replaced by a resistance.

The fact that Equation 5.10 is of the same form as Equation 5.4 indicates that superposition can be applied in the same manner, by applying V_{SRC} and I_{SRC} one at a time, while keeping the dependent source in place. In Figure 5.8a, with I_{SRC} replaced by an open circuit, KVL gives $V_{SRC} - 5I_{S1} - \rho I_{S1} - 10I_{S1} = 0$ or $I_{S1} = V_{SRC}/(15 + \rho)$. Moreover, $V_{X1} = V_{SRC} - 5I_{S1}$. Substituting for I_{S1} gives $V_{X1} = V_{SRC}(10 + \rho)/(15 + \rho)$, as in Equation 5.10. If $V_{SRC} = 12$ V, $I_{SRC} = 2$ A, and $\rho = 5$, then $V_{X1} = 9$ V.

In Figure 5.8b, with V_{SRC} replaced by a short circuit, KVL around the outer loop gives $5I_{S2} + \rho I_{S2} + 10(I_{S2} + I_{SRC}) = 0$. From Ohm's law, $V_{X2} = -5I_{S2}$. Substituting for I_{S2} gives $V_{X2} = 50I_{SRC}/(15 + \rho)$, as in Equation 5.6. Substituting numerical values, $V_{X2} = 5$ V, so that $V_X = V_{X1} + V_{X2} = 14$ V.

Comparing Figures 5.2 and 5.8, it is seen that the presence of the dependent source increases the complexity of

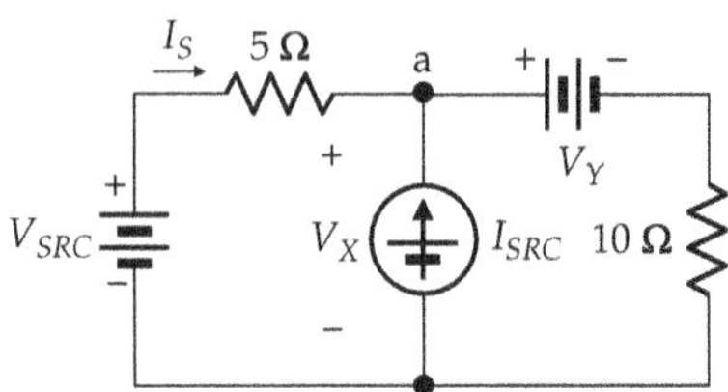

FIGURE 5.9
Dependent replaced by independent source.

the circuits involving one independent source at a time. These circuits become much simpler if the dependent source can also be treated like an independent source. Can this be done? The answer is yes, using the substitution theorem. For according to this theorem, a dependent voltage source of designated voltage ρI_S can be replaced by an independent voltage source of the same voltage. This is done in Figure 5.9, where the dependent source has been replaced by an independent source V_Y that is assumed to have the same numerical value as ρI_S. Clearly, if I_S is known, replacing the dependent source ρI_S by an independent source V_Y of the same source voltage does not change the values of the voltages and currents in the circuit. But since I_S is not known, V_Y can be considered at this stage to be just a symbol, like V_{SRC} and I_{SRC}. We will now show that the same V_X is obtained as before.

Although superposition can be applied to the required variable V_X, it is generally simpler and more systematic, when replacing a dependent source by an independent source, to apply superposition to the controlling variable of the dependent source, which is I_S in this case. The procedure is illustrated in Figure 5.10. If V_{SRC} is applied alone, with I_{SRC} and

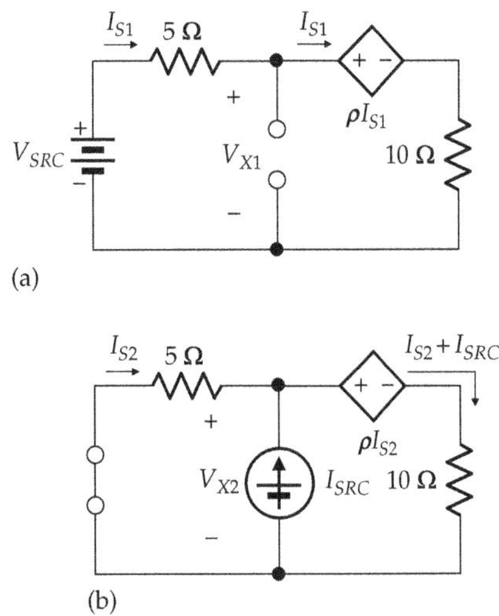

FIGURE 5.8
Superposition with dependent source retained. Components of the desired response V_X, due to the voltage source acting alone (a) and due to the current source acting alone (b), with the dependent source left unchanged.

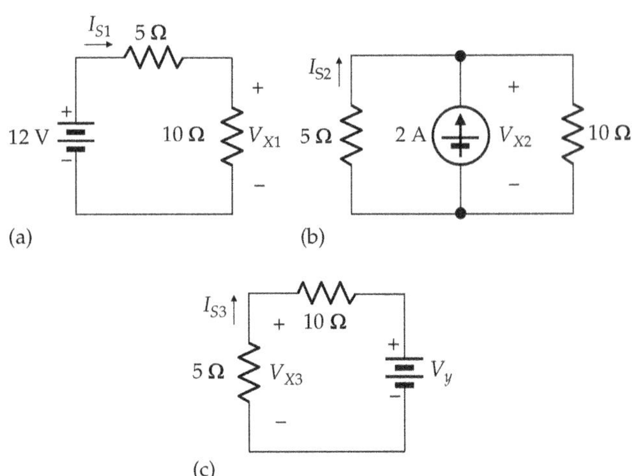

FIGURE 5.10
Application of one independent source at a time. Components of the desired response I_S, due to the voltage source acting alone (a), due to the current source acting alone (b), and due to the dependent source replaced by an independent source (c).

V_Y set to zero, the circuit becomes as in Figure 5.10a, from which it follows that $I_{S1} = 12/15 = 4/5$ A. If I_{SRC} is applied alone, with V_{SRC} and V_Y set to zero, the circuit becomes as in Figure 5.10b, from which it follows that $I_{S2} = -2 \times (10/15) = -4/3$ A. Finally, V_{SRC} and I_{SRC} are set to zero with V_Y applied, as in Figure 5.10c. It is seen from this figure that $I_{S3} = -V_Y/15$. It follows from superposition that

$$I_S = \frac{4}{5} - \frac{4}{3} - \frac{V_Y}{15} \tag{5.11}$$

Substituting $V_Y = 5I_S$ and solving for I_S give $I_S = -2/5$ A. In the original circuit, $V_X = 12 - 5I_S$. Substituting for I_S gives $V_X = 14$ V, as before. Evidently, replacing the dependent source by an independent source simplifies the application of superposition, as it leads to simpler circuits for determining the individual components of the variable in question.

Although the preceding discussion is based on a specific circuit, the conclusions apply to LTI circuits in general.

Primal Exercise 5.4

Determine V_X in Figure 5.7 with the polarity of the dependent source reversed.

Ans. 16 V.

5.1.2 Procedure for Applying Superposition

The procedure for applying superposition in the absence of dependent sources can be summarized as follows:

1. *Select the desired voltage or current response as the circuit variable to which superposition will be applied.*
2. *A component of the desired response is obtained with each independent source acting alone, while the remaining independent sources are set to zero.*
3. *The desired response is the algebraic sum of the individual components.*

In the presence of a single dependent source, the procedure for applying superposition can be summarized as follows:

1. *Replace the dependent source with an independent source of unknown value.*
2. *Select the controlling variable of the dependent source as the circuit variable to which superposition will be applied.*
3. *A component of the controlling variable is obtained with each independent source acting alone, while the remaining independent sources are set to zero.*
4. *Apply the superposition equation and substitute for the unknown value of the independent source its value in terms of the controlling variable in the original circuit.*
5. *The controlling variable is determined from the superposition equation.*
6. *Once the controlling variable is determined, the desired circuit response can be found using KCL, KVL, and Ohm's law.*

When two dependent variables are present, the preceding procedure can be applied to the controlling variable of each of the dependent sources. The result is two superposition equations in the two controlling variables as unknowns. Once these variables are determined from the solution of these equations, the desired circuit responses can be found using KCL, KVL, and Ohm's law.

When three or more dependent sources are present, the superposition method does not have a decisive advantage over alternative methods discussed in preceding chapters and in Chapter 6.

Example 5.2: Superposition with Dependent Current Source

It is required to determine V_O in Figure 5.11a using superposition.

Solution:

In accordance with the aforementioned procedure, the dependent source is replaced by an independent current source I_Y of unknown value (Figure 5.11b) and superposition applied to I_O in the middle branch. When each independent source is acting alone, with the other sources set to zero, the resulting circuits are as shown in Figure 5.12. With the 40 V source applied alone (Figure 5.12a), the current through the source is $40/(20 + 10\|10) = 40/25 = 1.6$ A. By current division, $I_{O1} = 1.6/2 = 0.8$ A. With the 20 V source applied alone (Figure 5.12b), the current through the source is $20/(10 + 20\|10) = 60/50 = 1.2$ A. By current division, $I_{O2} = 1.2 \times 20/30 = 0.8$ A. With I_Y applied alone, the circuit can be redrawn as in Figure 5.12c. By current division, using ratios of conductances,

$$I_{O3} = \frac{0.1}{0.1+0.1+0.05} I_Y = 0.4 I_Y \tag{5.12}$$

The superposition equation is $I_O = 0.8 + 0.8 + 0.4I_Y$. Substituting $I_Y = 0.5I_O$ gives $I_O = 2$ A. From the given circuit, the voltage across the middle 10 Ω resistor is 20 V.

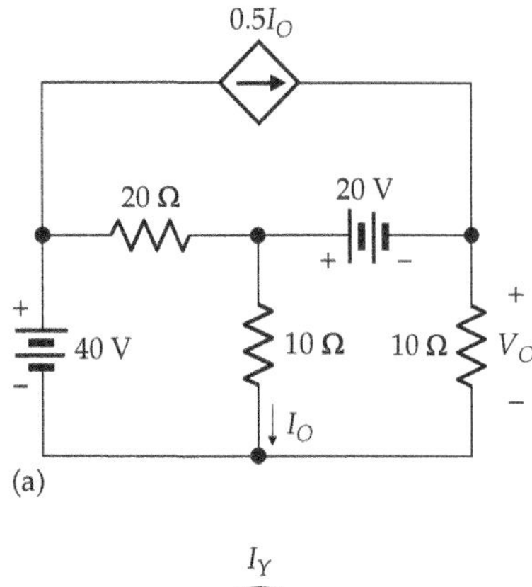

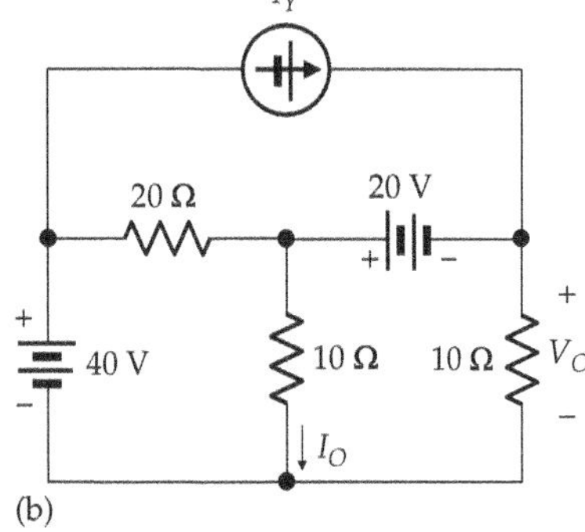

FIGURE 5.11
Figure for Example 5.2.

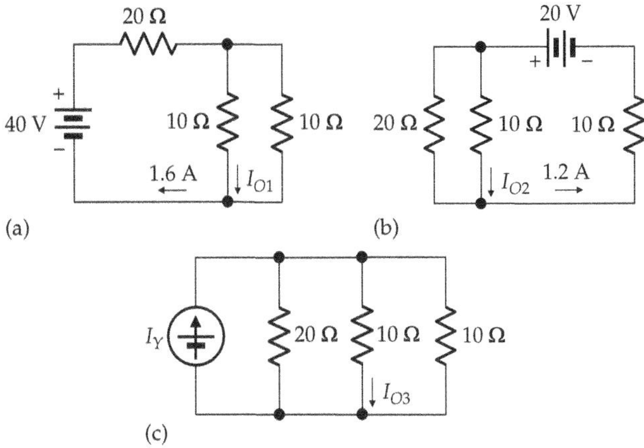

FIGURE 5.12
Figure for Example 5.2.

KVL around the mesh on the RHS gives $20 - 20 - V_O = 0$, or $V_O = 0$.

Simulation: The circuit is entered as in Figure 5.13. After selecting 'Bias Point' under 'Analysis type' in the Simulation Settings and running the simulation, pressing the I and V buttons displays the currents and voltages, respectively, indicated in Figure 5.13. V_O is $10 \times 20 \times 10^{-12} \cong 0$.

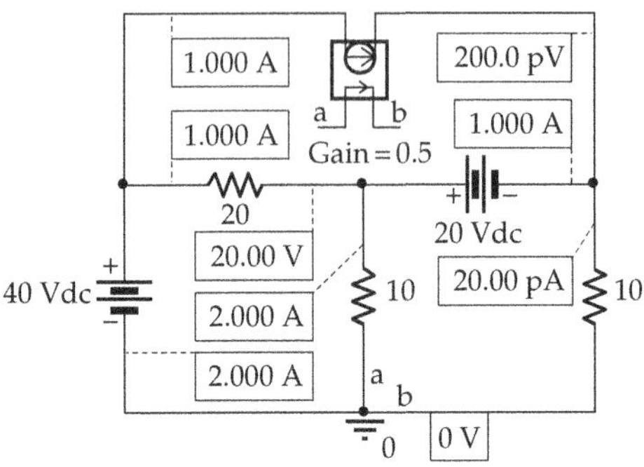

FIGURE 5.13
Figure for Example 5.2.

Exercise 5.5

Apply superposition in Figure 5.11 keeping the dependent source in place. Note how much simpler is applying superposition with the dependent source replaced by an independent source.

Example 5.3: Superposition with Dependent Voltage Source

Given that $I_X = 6$ A in Figure 5.14a, it is required to determine ρ and I_Y.

Solution:

From KCL at node 'd' in Figure 5.14b, $I_Y = 1$ A. From KVL around the loop 'bcdab', $6\rho - 2 \times 1 - 6 \times 3 + 15 = 0$, which gives $\rho = 5/6$.

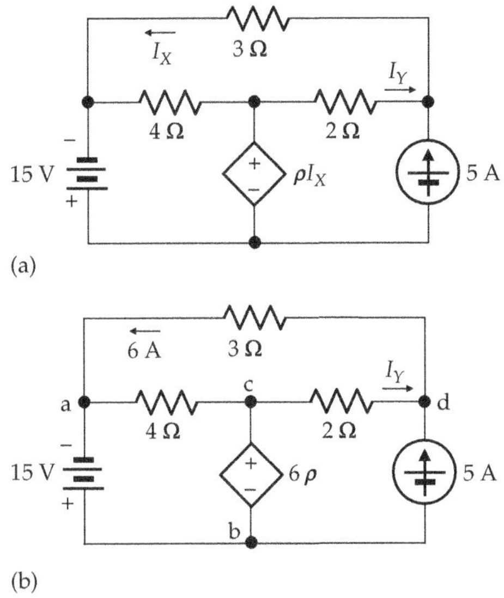

FIGURE 5.14
Figure for Example 5.3.

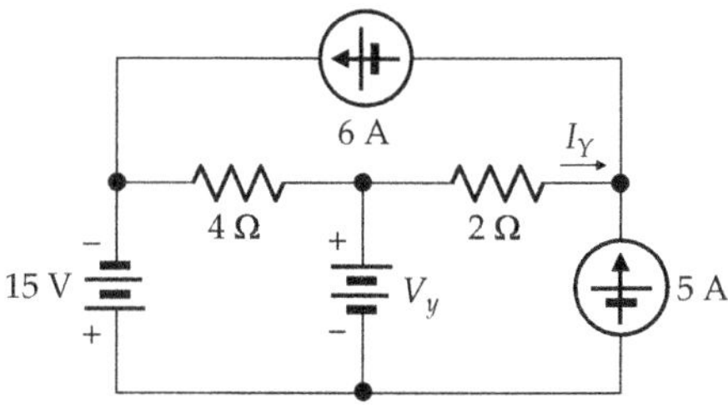

FIGURE 5.15
Figure for Example 5.3.

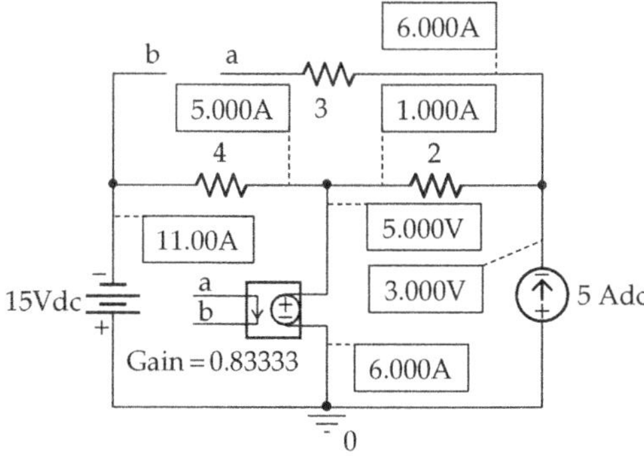

FIGURE 5.16
Figure for Example 5.3.

I_Y will next be derived as an illustration of superposition and the substitution theorem. The dependent source is replaced by an independent voltage source of unknown value V_Y, and the 3 Ω resistor is replaced by a 6 A source in accordance with the substitution theorem (Figure 5.15). Superposition can then be applied to I_Y with each of the independent sources acting alone.

When either voltage source is applied alone, the two current sources are replaced by open circuits, which makes both components of I_Y, due to the voltage sources, equal to zero. When the 6 A source is applied alone, with the 5 A source replaced by an open circuit, I_{Y1} = 6 A. Similarly, when the 5 A source is applied alone, with the 6 A source replaced by an open circuit, I_{Y2} = −5 A. It follows that I_Y = 1 A, as before.

Simulation: $\rho = 5/6$ is used in the simulation to verify that the current in the 6 Ω resistor is 6 A. The circuit is entered as in Figure 5.16. Note that 5/6 is entered as a decimal number since PSpice does not accept fractions in data entries. After selecting 'Bias Point' under 'Analysis type' in the Simulation Settings and running the simulation, pressing the I and V buttons displays the currents and voltages, respectively, indicated in Figure 5.16.

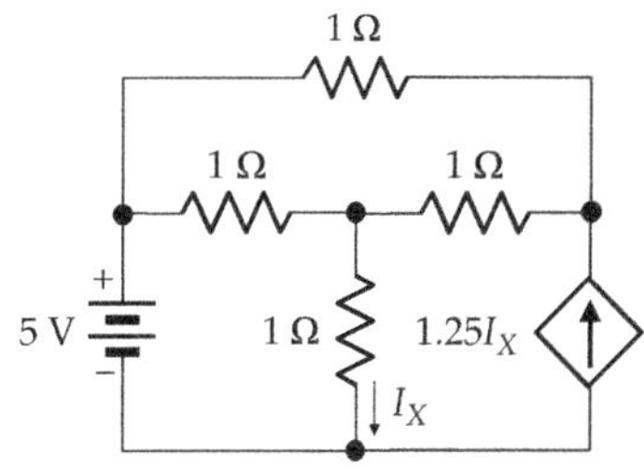

FIGURE 5.17
Figure for Primal Exercise 5.6.

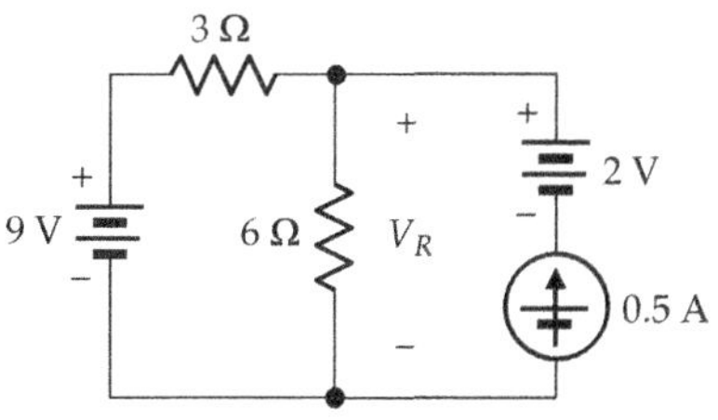

FIGURE 5.18
Figure for Primal Exercise 5.7.

Primal Exercise 5.6

Determine I_X in Figure 5.17 using superposition.

Ans. 4 A.

Primal Exercise 5.7

Determine V_R in Figure 5.18 using superposition.

Ans. 7 V.

5.1.3 Power with Superposition

Concept: *Because power is a nonlinear function of voltage or current, superposition cannot be applied directly to power.*

Consider, for example, the circuit of Figure 5.19, in which the power in the 60 Ω resistor is required as a function of V_{SRC1} and V_{SRC2}. V_X, the voltage across the 60 Ω,

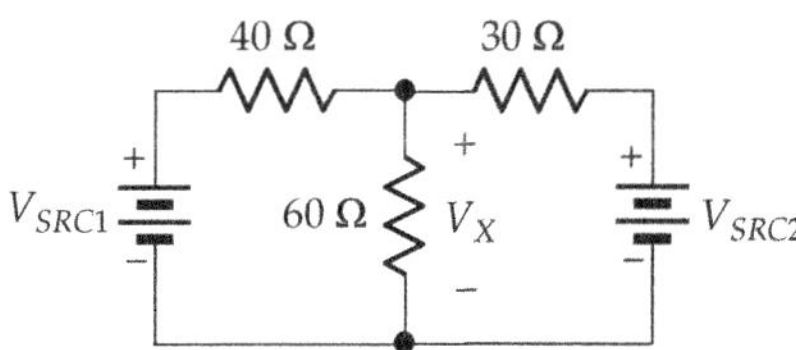

FIGURE 5.19
Power with superposition.

can be readily derived by superposition. Thus, if V_{SRC1} is applied alone, with V_{SRC2} replaced by a short circuit, it follows from voltage division that $V_{X1} = (30\|60)V_{SRC1}/(40 + 30\|60) = 20V_{SRC1}/(40 + 20) = V_{SRC1}/3$. Similarly, if V_{SRC2} is applied alone, with V_{SRC1} replaced by a short circuit, $V_{X2} = (40\|60)V_{SRC2}/(30 + 20\|60) = 24V_{SRC2}/(30 + 24) = V_{SRC2}/2.25$. Hence,

$$V_X = \frac{V_{SRC1}}{3} + \frac{V_{SRC2}}{2.25} \tag{5.13}$$

The power P_{X1} dissipated in the 60 Ω resistor due to V_{SRC1} acting alone is $(1/60)(V_{SRC1}/3)^2$, whereas the power P_{X2} dissipated in the 60 Ω resistor due to V_{SRC2} acting alone is $(1/60)(V_{SRC2}/3)^2$. The sum of these powers is

$$P_{X1} + P_{X2} = \frac{1}{60}\left(\frac{V_{SRC1}}{3}\right)^2 + \frac{1}{60}\left(\frac{V_{SRC2}}{2.25}\right)^2 \tag{5.14}$$

However, the true power dissipated in the 60 Ω resistor is

$$P_X = \frac{V_X^2}{60} = \frac{1}{60}\left(\frac{V_{SRC1}}{3} + \frac{V_{SRC2}}{2.25}\right)^2 \tag{5.15}$$

It is seen that $P_X \neq P_{X1} + P_{X2}$, because the square of the sum of $V_{SRC1}/3$ and $V_{SRC2}/2.25$ in Equation 5.15 is not equal to the sum of the squares of these terms in Equation 5.14. Hence, to obtain the correct value of power dissipated in a resistor using superposition, *the voltage across the resistor, or the current through the resistor, is first obtained by superposition, and the power can then be determined using the total value of the voltage or current.*

Exercise 5.8

Tabulate the powers dissipated in each of the resistors in Figure 5.19 due to each source applied alone, with the other source set to zero, and when both sources are applied together, assuming V_{SRC1} = 12 V and V_{SRC2} = 9 V. Compare the total power dissipated to the source power delivered in each case.

Primal Exercise 5.9

Determine the power dissipated in the 4 Ω resistor in Figure 5.20 when (a) either source is applied alone or (b) both sources are applied together.

Ans. (a) 16 W due to either source applied alone; (b) 64 W.

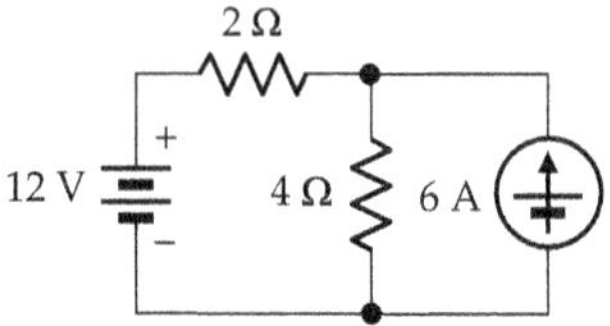

FIGURE 5.20
Figure for Primal Exercise 5.9.

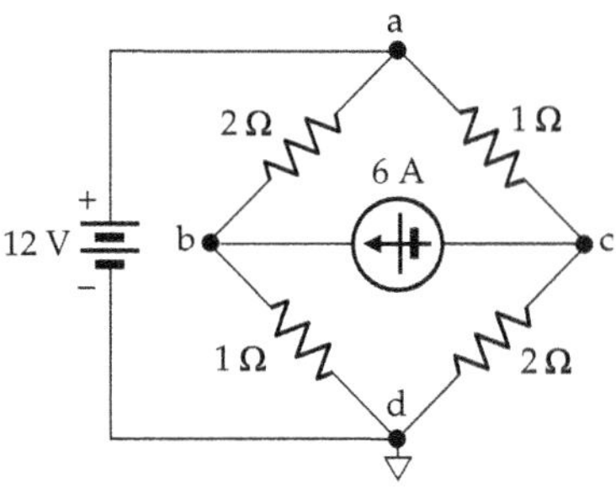

FIGURE 5.21
Figure for Primal Exercise 5.10.

Primal Exercise 5.10

Determine the power dissipated in either 2 Ω resistor in Figure 5.21 when (a) either source is applied alone or (b) both sources are applied together.

Ans. (a) 32 W due to voltage source, 8 W due to current source; (b) 8 W.

5.2 Output Scaling

Consider the circuit of Figure 5.7. If only V_{SRC} is present in the circuit and I_{SRC} is replaced by an open circuit, it follows from Equation 5.10 that

$$V_X = \frac{(10+\rho)}{(15+\rho)}V_{SRC} \tag{5.16}$$

On the other hand, if only I_{SRC} is present in the circuit and V_{SRC} is replaced by a short circuit, it follows from Equation 5.10 that

$$V_X = \frac{50}{(15+\rho)}I_{SRC} \tag{5.17}$$

In general, if a circuit is excited by a *single independent source*, any voltage or current response in the circuit can be expressed as

$$\text{Response} = f(R) \times \text{Excitation} \tag{5.18}$$

where $f(R)$ depends on the resistances and values of dependent sources in the circuit and is a constant for a given LTI circuit. The excitation could be due to an independent voltage source or an independent current source. If the excitation is multiplied by a constant factor K, the

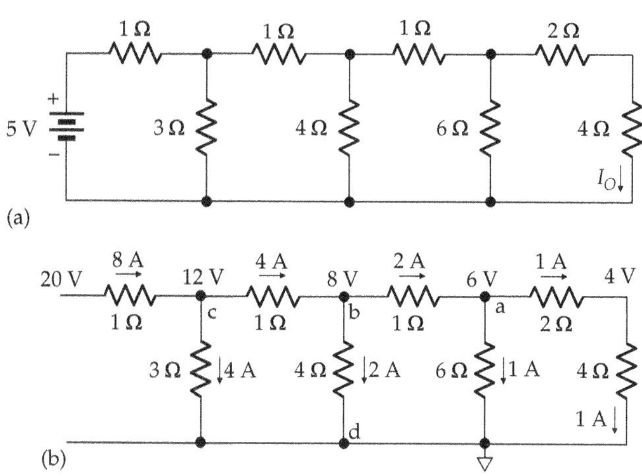

FIGURE 5.22
Output scaling. (a) Circuit in which I_O is to be determined and (b) $I_O = 1$ A is assumed and, working backward, all currents and voltages are evaluated, leading to the source terminals.

response is multiplied by the same factor. This can be made use of in some types of circuits by *assuming* a convenient output, working backward from this output to the source of excitation, and then scaling the output in accordance with the value of the given source.

As an example, consider the circuit of Figure 5.22a, where it is required to determine I_O. It is rather awkward to do this by any of the conventional methods. But it may be assumed quite arbitrarily that $I_O = 1$ A and then work backward toward the voltage at the source terminals (Figure 5.22b). The voltage of node 'a' with respect to the common node 'd' is 6 V. The current from node 'a' to node 'd' through the 6 Ω resistor is 1 A. From KCL at node 'a', the current flowing toward node 'a' through the 1 Ω resistor is 2 A. The voltage of node 'b' with respect to node 'd' is 8 V. Proceeding in this manner gives a voltage of 20 V at the source terminals. I_O per unit of applied excitation is 1/20 A/V, which means that when the excitation is 5 V, then $I_O = 5/20 = 0.25$ A in the original circuit.

The method of output scaling can be applied to circuits that are more general than the ladder circuit of Figure 5.22a, and which may include dependent sources. But the circuit must have a single independent source of excitation, and must be such that one can assume a certain output and then work backward toward the input by systematically determining all the currents and voltages along the way, without having to invoke additional variables. This is illustrated by Example 5.4.

Primal Exercise 5.11

Determine I_O in Figure 5.22a if the excitation is a 4 A source instead of a 5 V source.

Ans. $I_O = 0.5$ A.

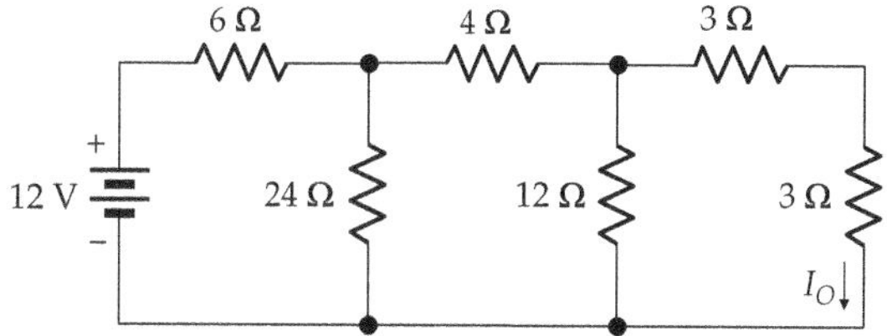

FIGURE 5.23
Figure for Primal Exercise 5.12.

Primal Exercise 5.12

Determine I_O in Figure 5.23.

Ans. $I_O = 0.5$ A.

Example 5.4: Output Scaling

It is required to determine I_O in Figure 5.24a using output scaling.

Solution:

We will assume a convenient value for I_O, such as 1 A, and work backward toward the source terminals using KCL, KVL, and Ohm's law.

Let the nodes be labeled as in Figure 5.24b with $I_O = 1$ A and node 'd' taken as reference. $V_c = 5$ V. The CCCS is 4 A, so that $I_{cb} = 3$ A, where I_{cb} is the current in the 3 Ω resistor that flows from node 'c' toward node 'b'.

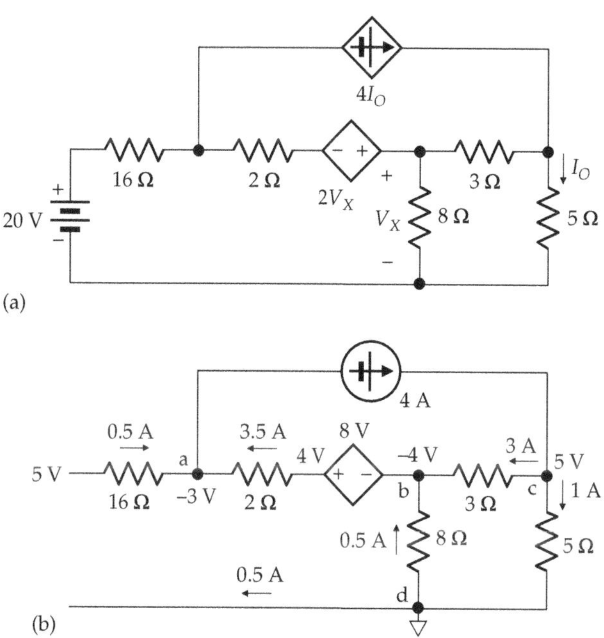

FIGURE 5.24
Figure for Example 5.4.

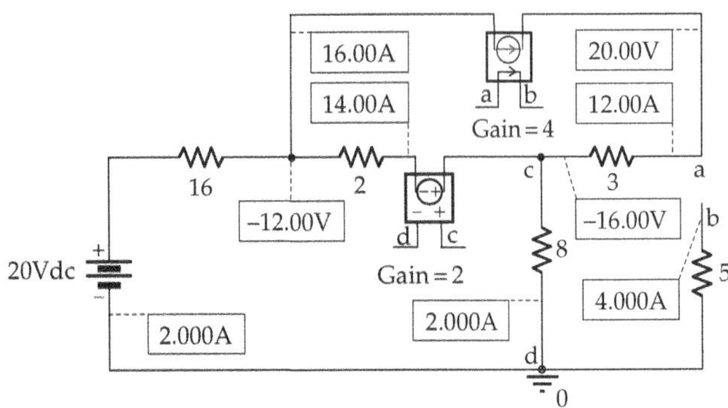

FIGURE 5.25
Figure for Example 5.4.

The voltage drop across the 3 Ω resistor is 9 V, so that $V_b = V_X = 5 - 9 = -4$ V, $I_{db} = 0.5$ A. The voltage of the VCVS is −8 V, so it can be taken as +8 V of reversed polarity. From KCL at node 'b', $I_{ba} = 3.5$ A. This makes $V_a = -4 + 8 - 2 \times 3.5 = -3$ V. KCL at node 'a' gives a current of 0.5 A flowing in the 16 Ω resistor toward node 'a'. The voltage at the input terminal is $-3 + 16 \times 0.5 = 5$ V. Since I_O per unit source voltage is 1/5 A/V, it follows that $I_O = 4$ A for a source voltage of 20 V, and all currents and voltages in Figure 5.24b are multiplied by 4.

Simulation: The circuit is entered as in Figure 5.25. After selecting 'Bias Point' under 'Analysis type' in the Simulation Settings and running the simulation, pressing the I and V buttons displays the currents and voltages, respectively, indicated in Figure 5.25.

5.3 Redundant Resistors

5.3.1 Redundant Resistors Connected to Sources

Concept: *A resistor in series with an ideal current source is redundant as far as the rest of the circuit is concerned, but affects the voltage across the source. Similarly, a resistor in parallel with an ideal voltage source is redundant as far as the rest of the circuit is concerned, but affects the current through the source.*

Consider a 2 A current source connected in series with a 3 Ω resistor to a circuit 'N' represented by its TEC between terminals 'ab' (Figure 5.26a). The current source forces 2 A into terminal 'a' and out of terminal 'b', producing a voltage $V_{ab} = 4 \times 2 + 8 = 16$ V. Removing the 3 Ω resistor does not affect the current supplied to circuit 'N', and hence does not affect V_{ab} (Figure 5.26b). The 3 Ω resistor is therefore redundant as far the rest of the circuit is concerned, which in this case is circuit 'N'. So what is the effect of the 3 Ω resistor? Its effect is on the voltage across the current source, which is 16 V in the absence of the 3 Ω resistor, and 22 V in its presence,

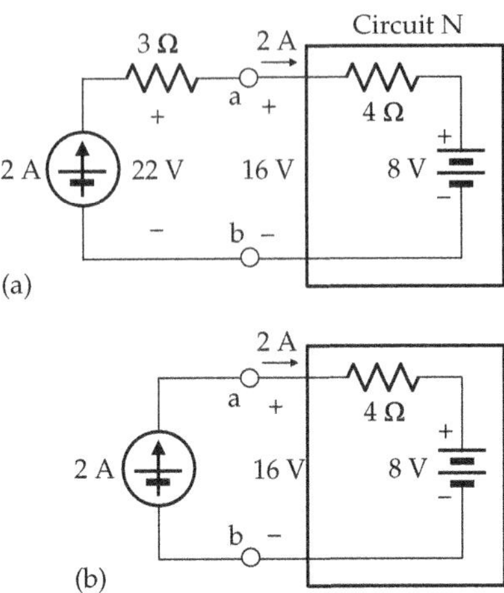

FIGURE 5.26
(a) Redundant resistor in series with current source and (b) resistor removed.

due to the 6 V drop across the resistor. Being an ideal current source, the source current is not affected by the voltage across it.

Figure 5.27a shows an ideal voltage source of 12 V connected in parallel with a 3 Ω resistor across terminals 'ab' of the same circuit 'N' that is now represented by its NEC between these terminals. The source impresses 12 V across the terminals, so that the current in R_N is 3 A and the input current at terminal 'a' is 1 A. Removing the 3 Ω resistor (Figure 5.27b) does not affect the voltage across terminals 'ab' and hence the input current to circuit 'N'. The 3 Ω resistor is therefore redundant as far as circuit 'N' is concerned. The effect of the 3 Ω resistor is on the current in the ideal voltage source. This current is 1 A without the 3 Ω resistor, and 5 A in its presence, because of the 4 A that it draws from the source. Being an ideal voltage source, the source voltage is not affected by the current through the source.

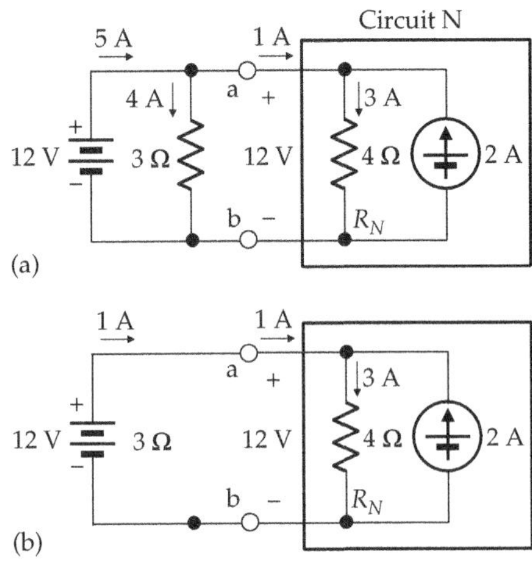

FIGURE 5.27
(a) Redundant resistor in parallel with voltage source and (b) resistor removed.

Example 5.5: Redundant Resistors Connected to Sources

It is required to determine TEC at terminals 'ab' in Figure 5.28a.

Solution:

The 8 Ω resistor in parallel with the 5 V source and the 3 Ω resistor in series with the VCCS are both redundant as far as deriving TEC at terminals 'ab' is concerned and can be removed from the circuit.

Another simplification that can be made is to temporarily remove the 6 and 2.6 Ω resistors, derive an intermediate TEC that does not include these resistors, and then add the resistors to this intermediate TEC and determine the required TEC, as was done in Example 4.4. Accordingly, TEC is first determined at terminals 'cd' in Figure 5.28b, without including the 6 and 2.6 Ω resistors. V_{Th1} across these terminals is also the controlling voltage for the VCCS. From KVL, $V_{Th1} = 2 \times 0.25V_{Th1} + 5$, which gives $V_{Th1} = 10$ V. If terminals 'ab' are short-circuited (Figure 5.28c), V_{Th1} becomes zero and the VCCS is replaced by an open circuit. Hence, $I_{SC} = 5/2 = 2.5$ A, and $R_{Th1} = 10/2.5 = 4$ Ω.

The second step is to connect the 6 and 2.6 Ω resistors to V_{Th1} and R_{Th1} (Figure 5.29a) and derive TEC between terminals 'ab'. With these terminals open-circuited, V_{Th} is the same as the voltage across the 6 Ω resistor, which by voltage division is $10 \times 6/(4 + 6) = 6$ V. With the 10 V source set to zero, the resistance looking into terminals 'ab' is $2.6 + 4\|6 = 2.6 + (4 \times 6)/(4 + 6) = 5$ Ω. TEC is as shown in Figure 5.29b.

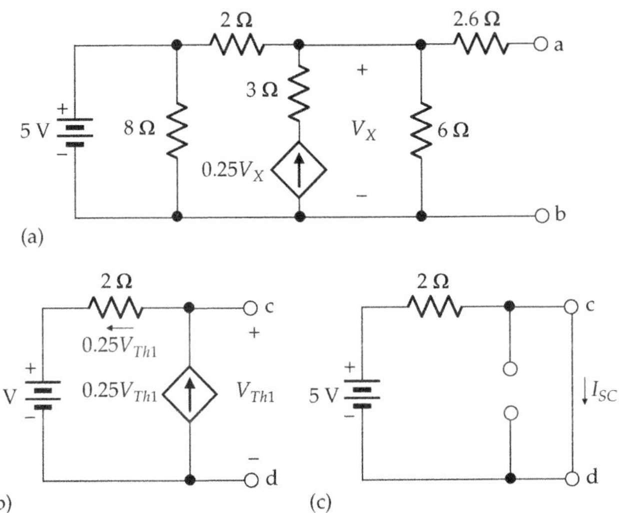

FIGURE 5.28
Figure for Example 5.5.

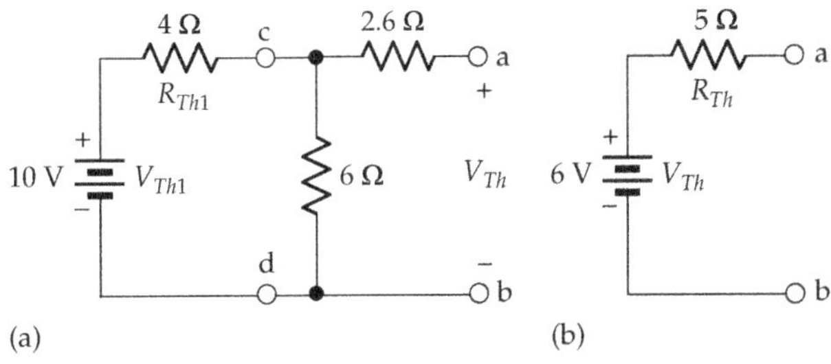

FIGURE 5.29
Figure for Example 5.5.

Simulation: The circuit is entered as in Figure 5.30, with the test current source applied at terminals 'ab' in the given circuit so as to determine TEC. Note that in order to apply the controlling voltage V_X in the required polarity, the positive terminal of the controlling voltage input should be connected to the upper line and the negative terminal to ground. But this involves awkward-looking cross connections, which can be avoided by reversing the connections of the controlling voltage, as shown, and changing the sign of the gain of the dependent source to –0.25. DC Sweep is used as explained in Example 4.1, the resulting graph being shown in Figure 5.31. It is seen that $V_{Th} = 6$ V and $R_{Th} = 5$ Ω.

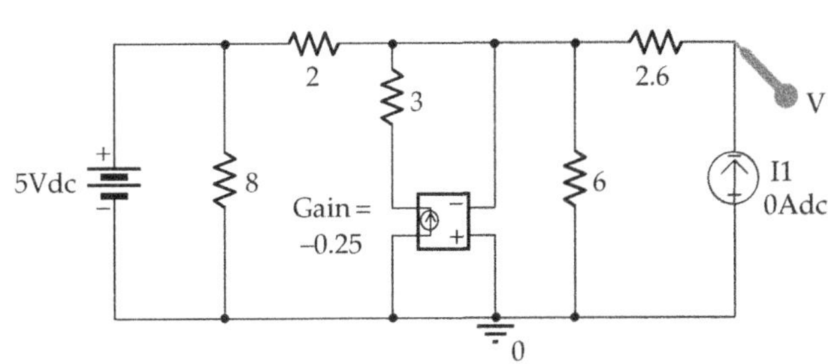

FIGURE 5.30
Figure for Example 5.5.

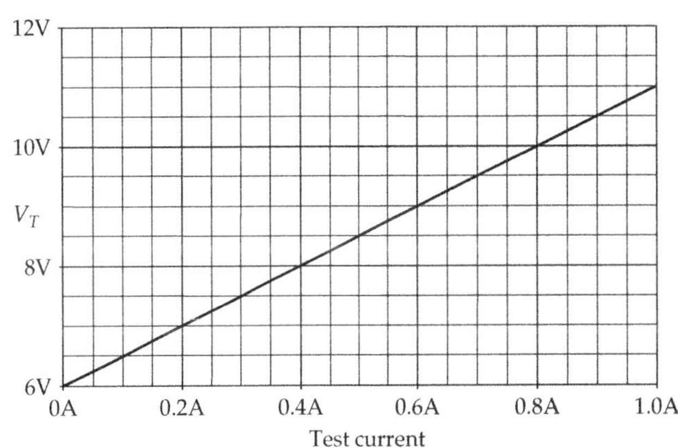

FIGURE 5.31
Figure for Example 5.5.

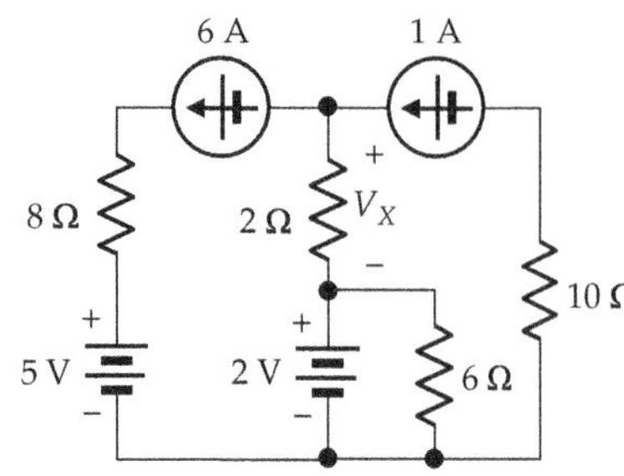

FIGURE 5.32
Figure for Primary Exercise 5.14.

Primal Exercise 5.13

Using the value derived for V_{Th} in Example 5.5, determine the voltage across the dependent current source and the current through the voltage source.

Ans. 10.5 V, 0.125 A.

Primal Exercise 5.14

Determine V_X in Figure 5.32.

Ans. −10 V.

5.3.2 Resistors Not Carrying Current

In Figure 5.33, the resistances are such that nodes 'b' and 'c' are at the same voltage with respect to node 'd', as can be readily verified. Thus, from voltage division,

$$V_{bd} = \frac{6}{24} \times 12 = 3 \text{ V} \quad \text{and} \quad V_{cd} = \frac{4}{16} \times 12 = 3 \text{ V}$$

The currents and voltages in the circuit are indicated in Figure 5.34. It is seen that none of these depends on any resistance R that may be connected between nodes 'b' and 'd'. This is because R does not carry any current,

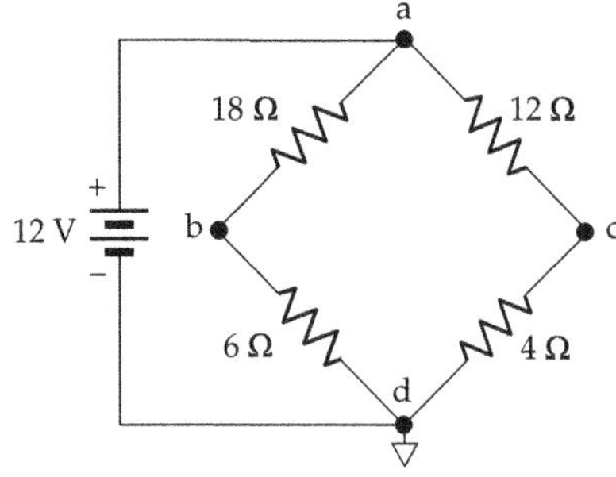

FIGURE 5.33
Balanced bridge.

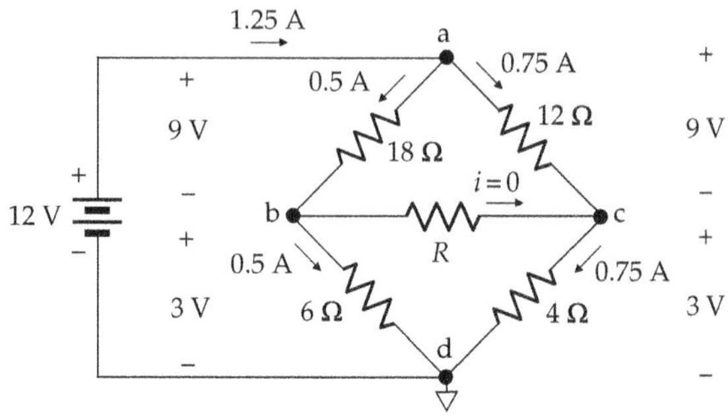

FIGURE 5.34
Redundant resistor not carrying current.

since its terminals are at the same voltage, so there is no voltage drop across it. It follows that R can have any value, without affecting the rest of the circuit. This includes extreme values of R, that is, an open circuit, as in Figure 5.33, or a short circuit, when nodes 'b' and 'c' are directly connected together.

The conditions depicted in Figures 5.33 and 5.34 lead to the following conclusions:

1. Resistor R in Figure 5.34, not carrying any current, is redundant. It can have any value, can be removed from the circuit, or can be replaced by a short circuit, without affecting the rest of the circuit.
2. Nodes 'b' and 'c' in Figure 5.33, being at the same voltage, can be connected together, without affecting the rest of the circuit, and no current flows in the connection.

It may be noted that the condition in Figure 5.33, when nodes 'b' and 'c' are at the same voltage is referred to as **bridge balance**, and the bridge is said to be balanced under these conditions. The resistive bridge circuit is of considerable practical importance and is known as a **Wheatstone bridge**. It is discussed in Appendix 5A.

The preceding conclusions can be generalized to the following concepts:

Concepts:

1. *A reduntant resistor can be removed from the circuit, without disturbing the rest of the circuit. When removing a redundant resistor the following applies:*
   (i) *A series resistor is replaced by a short circuit, as in the case of a resistor in series with an ideal current source.*
   (ii) *A parallel resistor is replaced by an open circuit, as in the case of a resistor in parallel with an ideal voltage source.*

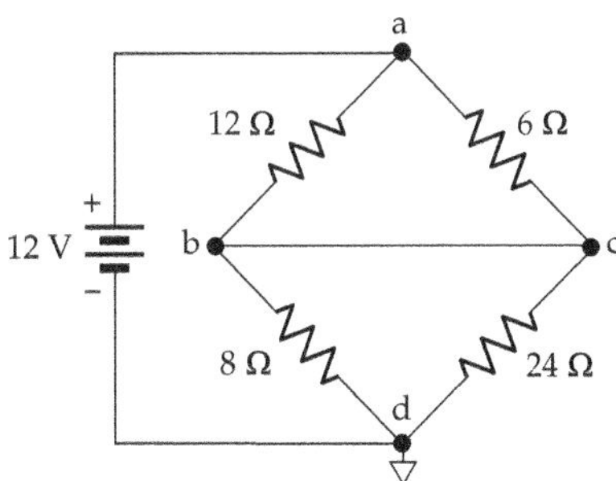

FIGURE 5.35
Bridge with short-circuited output.

(iii) *In cases where the redundant resistor does not carry current, as in the case of bridge balance, the resistor may be replaced by a short circuit or an open circuit, as can be readily checked.*

2. *Nodes that are at the same voltage, without being short-circuited together, can be left open-circuited or can be connected by a short circuit or by resistors of arbitrary values, without any current flowing in the connections and without affecting the rest of the circuit.*
3. *If several circuit elements are connected to nodes at the same voltage, the circuit elements can be reconnected between the nodes in any arbitrary manner without affecting the rest of the circuit, as long as the reconnection does not change the node voltages (see Problem P5.58).*

It should be emphasized that when two nodes are referred to as being at the same voltage, as in item '2' of the preceding concepts, these nodes are not connected together by a short circuit, which evidently makes them at the same voltage. If the short circuit is replaced by an open circuit, the two nodes will not in general be at the same voltage. A resistor connected between the nodes will therefore carry current and is *not* redundant. Consider, for example, the bridge circuit of Figure 5.35, where the output nodes 'b' and 'c' are short-circuited and are therefore at the same voltage with respect to node 'd'. However, if this short circuit is replaced by an open circuit, $V_{cd} = 12 \times 24/30 = 9.6$ V, and $V_{bd} = 12 \times 8/20 = 4.8$ V, so that a voltage difference exists between these nodes. If a resistor is connected between nodes 'b' and 'c', the resistor carries current and cannot be considered redundant. Note that current flows in the short circuit in Figure 5.35.

Primal Exercise 5.15

Determine the resistance seen by the 12 V source in Figure 5.33 and verify that it remains the same when nodes 'b' and 'c' are short-circuited.

Ans. 9.6 Ω.

5.4 Partitioning of Circuits by Ideal Sources

Concept: *If an ideal voltage source is connected in parallel with two circuits or an ideal current source is connected in series with two circuits, the two circuits can be separated, one from the other, as long as the two circuits are independent of one another, that is, they are not interconnected in any other way.*

Consider Figure 5.36a in which an ideal 12 V source is connected in parallel between two circuits 'N_1' and 'N_2', each represented by its TEC between terminals 'a_1' and 'b_1' of 'N_1' and 'a_2' and 'b_2' of 'N_2'. According to the foregoing concept, the ideal voltage source effectively separates the two circuits, one from the other, so that changes in one circuit do not affect the other circuit.

To verify this, note that KVL around the mesh consisting of 'N_1' and the 12 V source gives $16 = 4I_1 + 12$, where I_1 is the current flowing out of the terminal 'a_1'. This makes $I_1 = 1$ A, as shown. KVL around the mesh consisting of 'N_2' and the 12 V source gives $12 = 2I_2 + 8$, where I_2 is the current entering terminal 'a_2'. This makes $I_2 = 2$ A, as shown. Suppose, for the sake of argument, that 'N_1' is altered so that its V_{Th} is 18 V and its R_{Th} is 5 Ω (Figure 5.36b). KVL around the mesh on

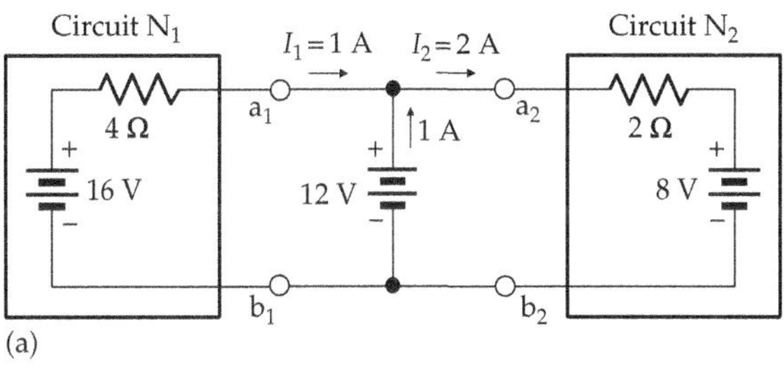

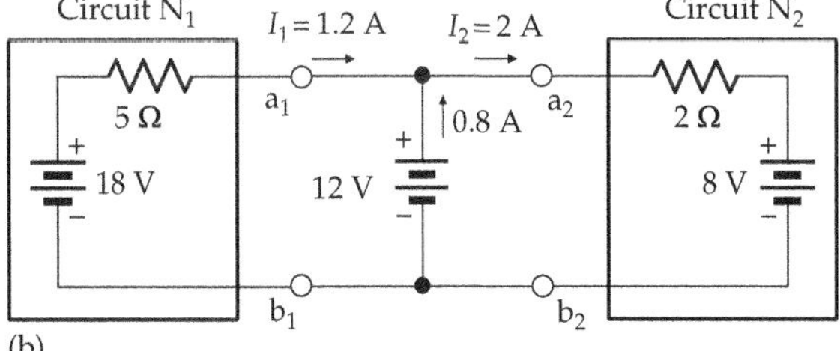

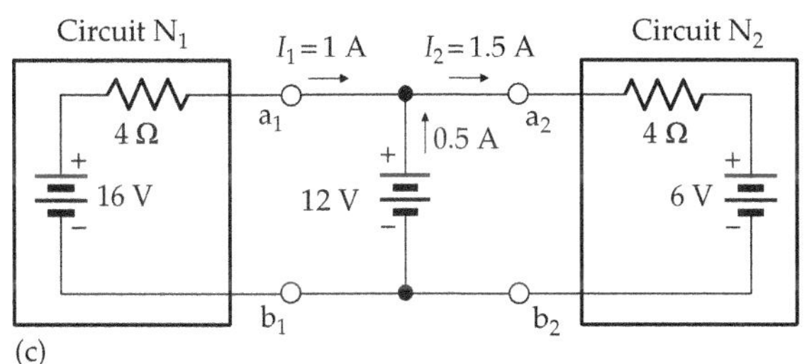

FIGURE 5.36
Circuit partitioning by voltage source. (a) Ideal voltage source connected between circuits 'N_1' and 'N_2', (b) change in 'N_1' does not affect 'N_2', and (c) change in 'N_2' does not affect 'N_1'.

the left gives $18 = 5I_1 + 12$, which makes $I_1 = 1.2$ A. KVL around the mesh on the right is unchanged, so that I_2 remains the same. With the voltage between terminals 'a$_2$' and 'b$_2$' remaining at 12 V, conditions in 'N$_2$' remain the same, despite changes in 'N$_1$'. The current in the 12 V source is reduced from 1 to 0.8 A, but because the source is assumed to be ideal, its voltage does not change, even if the current through it is reversed. The same argument applies if 'N$_2$' is changed, for example, so that its V_{Th} is 6 V and its R_{Th} is 4 Ω (Figure 5.36c). KVL around the mesh on the right gives $12 = 4I_2 + 6$, which makes $I_2 = 1.5$ A. KVL around the mesh on the left is unchanged, so that I_1 remains the same. With the voltage between terminals 'a$_1$' and 'b$_1$' remaining at 12 V, conditions in 'N$_1$' remain the same, despite changes in 'N$_2$'. The current in the 12 V source is reduced from 1 to 0.5 A, but because the source is assumed to be ideal, its voltage does not change. 'N$_1$' and 'N$_2$' are therefore effectively separated by the voltage source, so that each circuit can be considered independently of the other.

As a straightforward application of partitioning of a circuit by a voltage source, suppose that V_O is required in the circuit of Figure 5.37a. At first glance, this looks like a complex circuit, but a closer inspection reveals that the 8 V is connected in parallel between the left part of the circuit and the right part. The circuit is therefore partitioned into two separate circuits. The circuit on the right is shown in Figure 5.37b, after removal of the 10 Ω resistor connected in parallel with the 8 V source, because this resistor is redundant as far as V_O is concerned. From voltage division, $V_O = \dfrac{6}{6+2} \times 8 = 6$ V.

An analogous argument can be made for an ideal current source connected in series with two circuits.

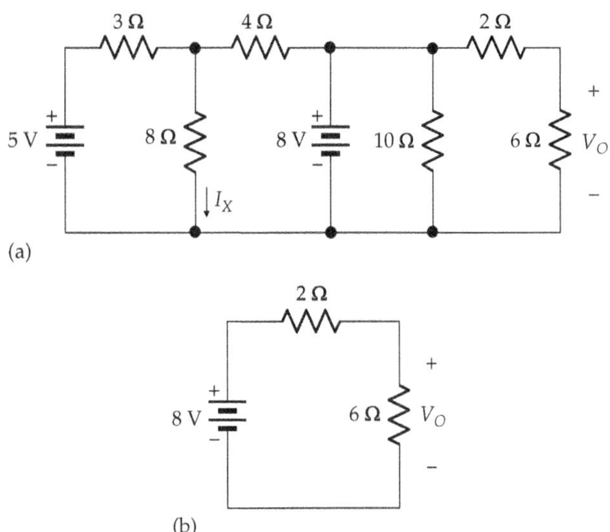

FIGURE 5.37
(a) Circuit partitioned by ideal voltage source and (b) circuit after simplification.

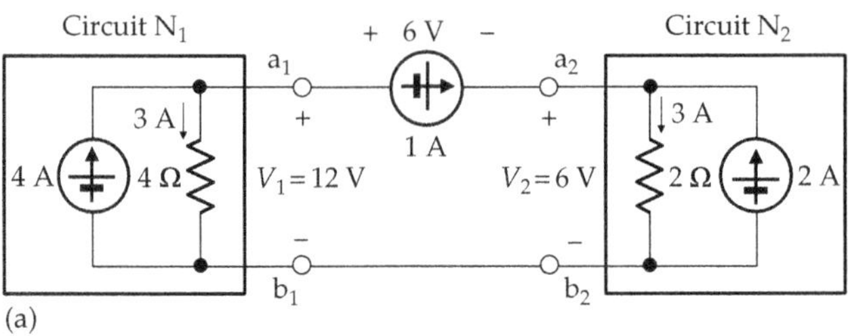

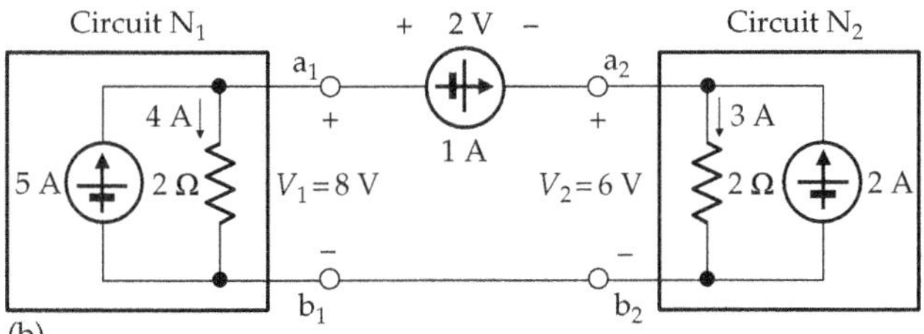

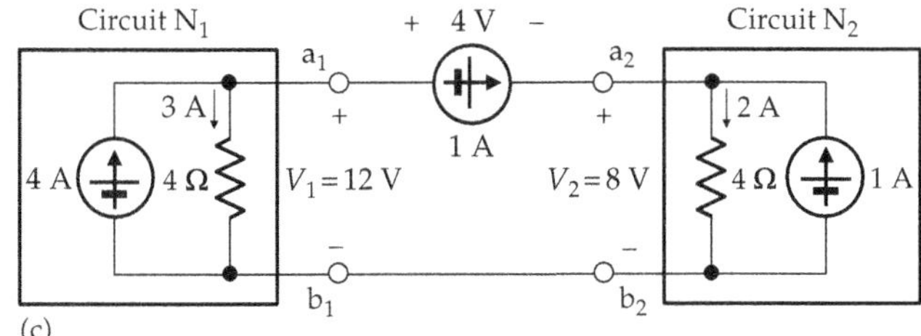

FIGURE 5.38
Circuit partitioning by current source. (a) Ideal current source connected between circuits 'N$_1$' and 'N$_2$', (b) change in 'N$_1$' does not affect 'N$_2$', and (c) change in 'N$_2$' does not affect 'N$_1$'.

Consider Figure 5.38a in which an ideal 1 A source is connected in series with two circuits 'N$_1$' and 'N$_2$', each represented by its NEC between terminals 'a$_1$' and 'b$_1$' of 'N$_1$' and 'a$_2$' and 'b$_2$' of 'N$_2$'. According to the foregoing concept, the ideal current source effectively separates the two circuits, one from the other, so that changes in one circuit do not affect the other circuit.

To verify this, note that from KCL at node 'a$_1$', the current flowing downward in the 4 Ω resistor of 'N$_1$' is 3 A, and by Ohm's law the voltage across the terminals of 'N$_1$' is $V_1 = 12$ V. Similarly, the current flowing downward in the 2 Ω resistor of 'N$_2$' is 3 A, and by Ohm's law the voltage across the terminals of 'N$_2$' is $V_2 = 6$ V. From KVL, the voltage across the 1 A source is $V_1 - V_2 = 6$ V, of the polarity shown.

Suppose, for example, that 'N$_1$' is altered so that its I_N is 5 A and its R_N is 2 Ω (Figure 5.38b). From KCL at node 'a$_1$', a current of 4 A flows downward through the 2 Ω resistor, so that $V_1 = 8$ V. The source current entering node 'a$_2$' is 1 A, the current flowing in the 2 Ω resistor of 'N$_2$' is 3 A, and $V_2 = 6$ V. With the terminal current and voltage at 'a$_2$b$_2$' remaining the same, the conditions in 'N$_2$' remain the same, despite changes in 'N$_1$'. The voltage across the 1 A source decreases from 6 to 4 V, but because the source is assumed to be ideal, its current does not change.

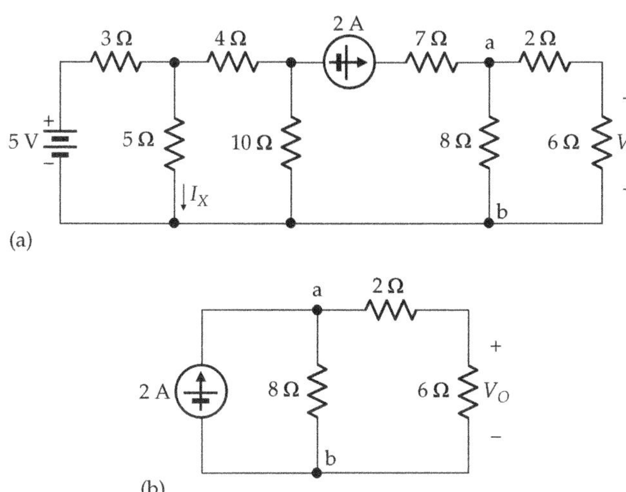

FIGURE 5.39
(a) Circuit partitioned by ideal current source and (b) circuit after simplification.

A similar argument applies if 'N$_2$' is altered, as in Figure 5.38c. The terminal voltage and current of 'N$_1$' remain the same, but the voltage V_{a1a2} across the current source changes. 'N$_1$' and 'N$_2$' are therefore effectively separated by the current source, so that each circuit can be considered independently of the other.

As an example of the partitioning of a circuit by a current source, consider the circuit of Figure 5.39a, where it is required to determine V_O. The 2 A series-connected source partitions the circuit into two separate circuits. The circuit on the right is shown in Figure 5.39b, after removal of the 7 Ω resistor connected in series with the 2 A source, because this resistor is redundant as far as V_O is concerned. The current source is reconnected across the 8 Ω resistor, which does not alter KCL at nodes 'a' and 'b'. From current division, the current in the 6 Ω resistor is $2 \times 8/(8 + 8) = 1$ A, which gives $V_O = 6$ V.

Before ending this section, it should be emphasized that the two separate circuits that resulted from partitioning the circuits of Figures 5.36a and 5.38a are independent, that is, they are not interconnected in any other way. Otherwise, each circuit could not be considered separately. The two circuits could be interconnected, for example, by one or more elements that bridge them or by having a dependent source in one circuit controlled by a voltage or a current in the other circuit.

Primal Exercise 5.16

(a) Replace the single 12 V source in Figure 5.36a by two 12 V sources, one connected to circuit 'N$_1$' between terminals 'a$_1$' and 'b$_1$', the other to circuit 'N$_2$' between terminals 'a$_2$' and 'b$_2$', the 'a' terminals being positive with respect to the 'b' terminals. Does this affect circuits 'N$_1$' and 'N$_2$'? Compare the net power delivered by the two sources with that delivered by the single source in Figure 5.36a. (b) Replace the 1 A current source in Figure 5.38a by two 1 A current sources, one directed from terminal 'a$_1$' to terminal 'b$_1$', the other directed from terminal 'b$_2$' to terminal 'a$_2$'. Does this affect circuits 'N$_1$' and 'N$_2$'? Compare the net power absorbed by the two sources with that absorbed by the single source in Figure 5.38a.

Ans. (a) No; 12 W are delivered in both cases; (b) no, 6 W absorbed in both cases.

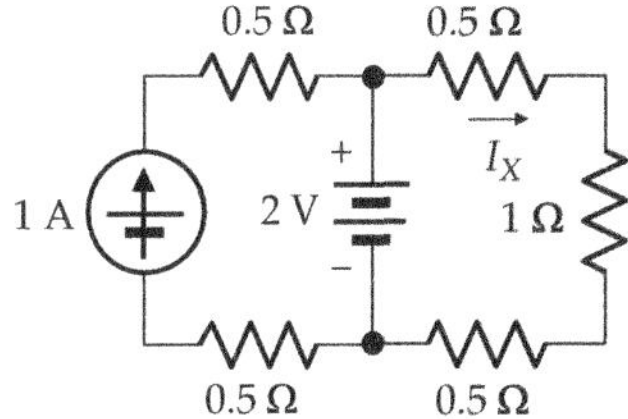

FIGURE 5.40
Figure for Primal Exercise 5.19.

Primal Exercise 5.17

Determine I_X in the part of the circuit on the left in Figure 5.37a. Note that the separated circuit is a two-essential-node circuit.

Ans. 11/17 A.

Primal Exercise 5.18

Determine I_X in the part of the circuit on the left in Figure 5.39a. Note that after source transformation of the 2 A source, the separated circuit is a two-essential-node circuit.

Ans. 100/127 A.

Primal Exercise 5.19

Determine I_X in Figure 5.40.

Ans. 1 A.

5.5 Source Rearrangement

Circuits can sometimes be simplified by rearranging sources. The basic idea will be illustrated by a simple example, before a more realistic case is considered in Example 5.6.

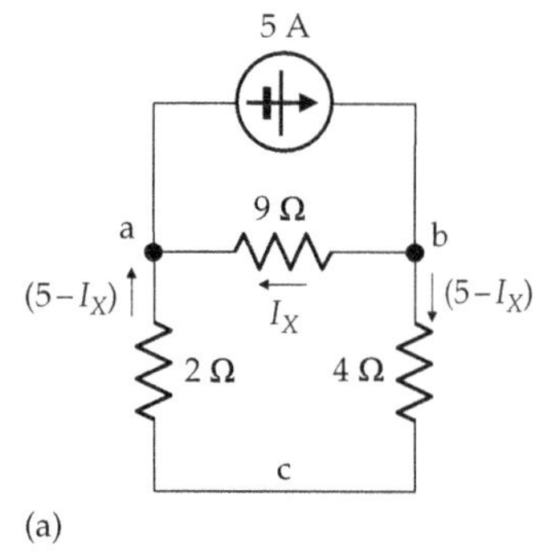

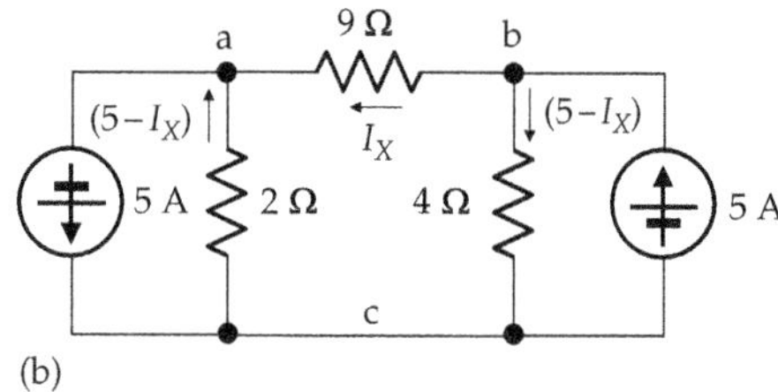

FIGURE 5.41
Source rearrangement. Current source in (a) replaced by two current sources in (b).

Suppose that I_X in Figure 5.41a is required. I_X in this simple case can be obtained from current division as $5 \times 6/15 = 2$ A. A more general procedure is to replace the 5 A source by two sources, as in Figure 5.41b. KCL at nodes 'a', 'b', and 'c' remains the same. Thus, at nodes 'a' and 'b': $5 - I_X + I_X = 5$, and at node 'c', the two 5 A sources cancel out. From KVL around the mesh of three resistors, $2(5 - I_X) - 9I_X + 4(5 - I_X) = 0$, which gives $I_X = 2$ A.

Example 5.6: Source Rearrangement

It is required to determine I_X in Figure 5.42 using source rearrangement.

Solution:

The 10 A source is split into two sources as in Figure 5.43. In Figure 5.42, a current of 10 A enters node 'b' from the source and a current of 10 A leaves node 'a' from the source. The same conditions are preserved in Figure 5.43. A source current of 10 A both enters and leaves node 'c' in Figure 5.43, so that the net source current at this node is zero, as in Figure 5.42.

The 10 V source can also be split into two 10 V sources in parallel, as in Figure 5.44a. The upper terminals of the two 10 V sources are evidently at the same voltage with respect to the lower node, so that no current flows in this connection (Section 5.3). The connection can therefore be removed, as in Figure 5.44b, without disturbing the rest of the circuit, which is now split into two subcircuits. I_X can be readily determined by deriving TEC seen by the 6 Ω resistor (Figure 5.45).

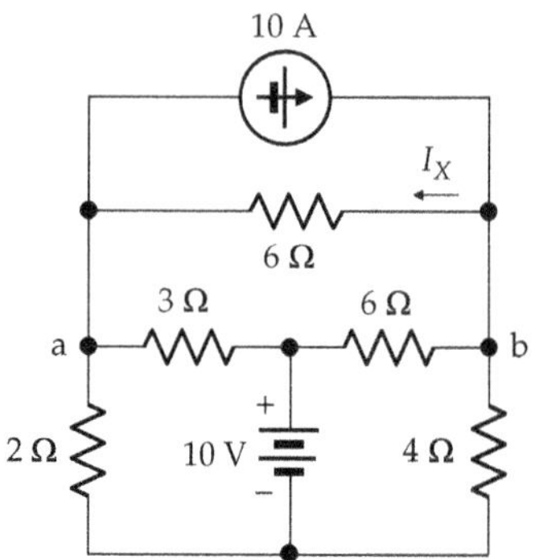

FIGURE 5.42
Figure for Example 5.6.

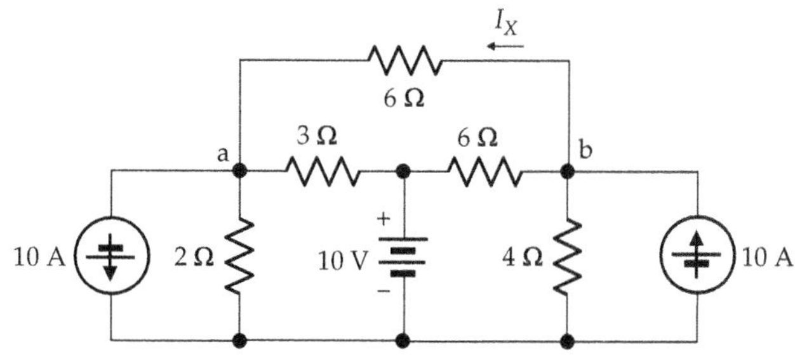

FIGURE 5.43
Figure for Example 5.6.

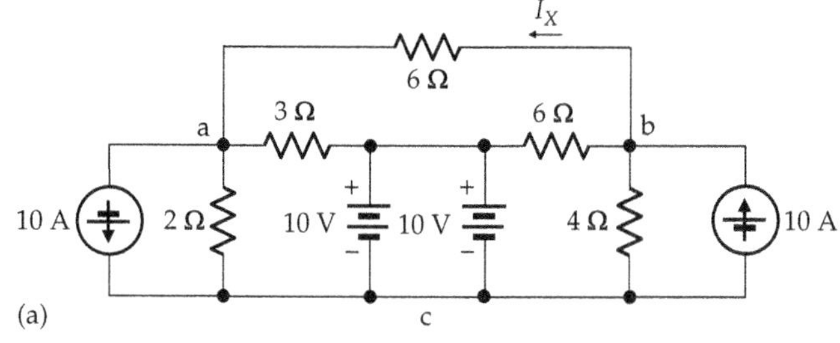

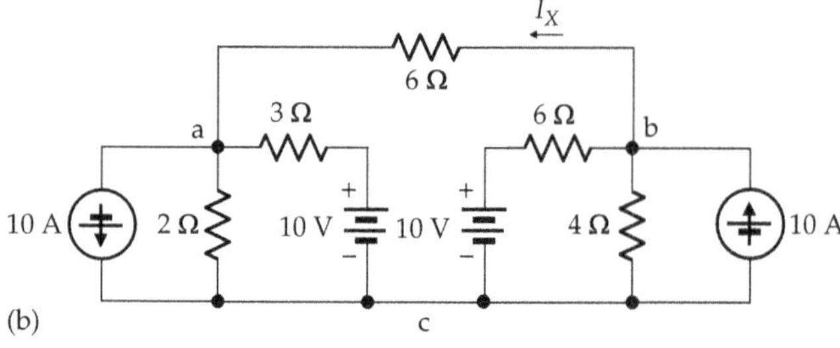

FIGURE 5.44
Figure for Example 5.6.

Superposition can be applied with the voltage sources acting alone and with the current sources acting alone. When the two 10 A sources are set to zero, $V_{ac} = 10 \times 2/5 = 4$ V, and $V_{bc} = 10 \times 4/10 = 4$ V. Hence, nodes 'a' and 'b' are at the same voltage, so that this component of V_{ba} is zero. If the 10 V sources are set to zero,

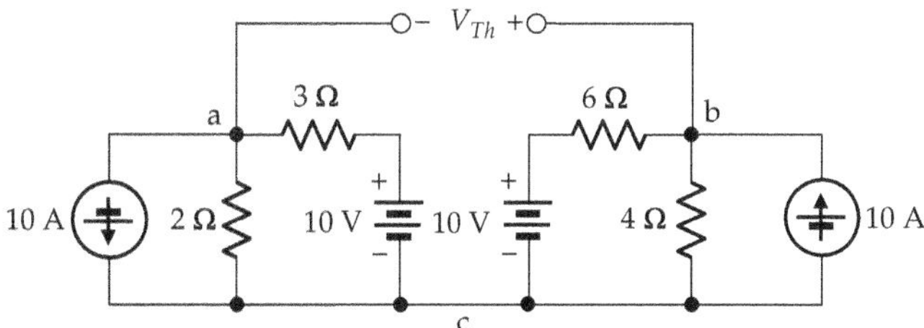

FIGURE 5.45
Figure for Example 5.6.

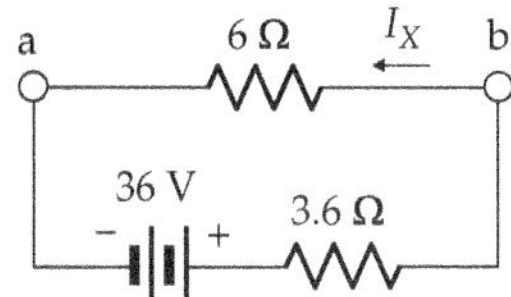

FIGURE 5.46
Figure for Example 5.6.

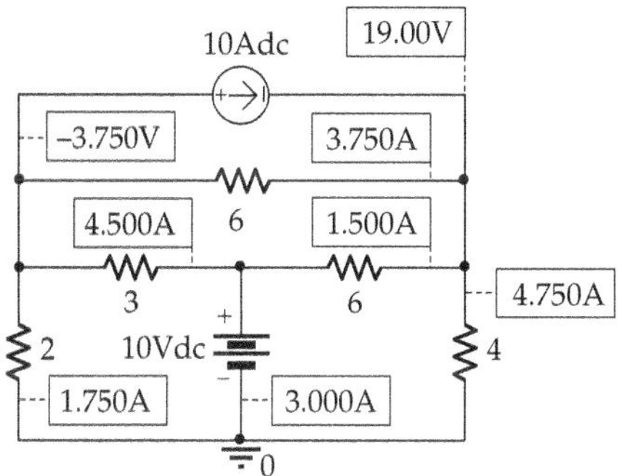

FIGURE 5.47
Figure for Example 5.6.

$V_{ac} = -10 \times 6/5 = -12$ V, and $V_{bc} = 10 \times 24/10 = 24$ V. It follows that $V_{ba} = V_{Th} = 36$ V. With all sources set to zero, $R_{Th} = (4||6) + (2||3) = 3.6$ Ω. TEC is therefore a source of 36 V in series with 3.6 Ω. When the 6 Ω resistor is reconnected (Figure 5.46), it is seen that $I_X = 36/9.6 = 3.75$ A.

Simulation: The circuit is entered as in Figure 5.47. After selecting 'Bias Point' under 'Analysis type' in the Simulation Settings and running the simulation, pressing the I and V buttons displays the currents and voltages, respectively, indicated in Figure 5.47.

⋆5.6 Exploitation of Symmetry

Symmetry in a circuit can generally be exploited to greatly simplify the analysis required for obtaining a given response. Consider, for example, Figure 5.48, in which it is desired to determine *I*. The circuit is symmetrical about a vertical line *XX′* through its midline.

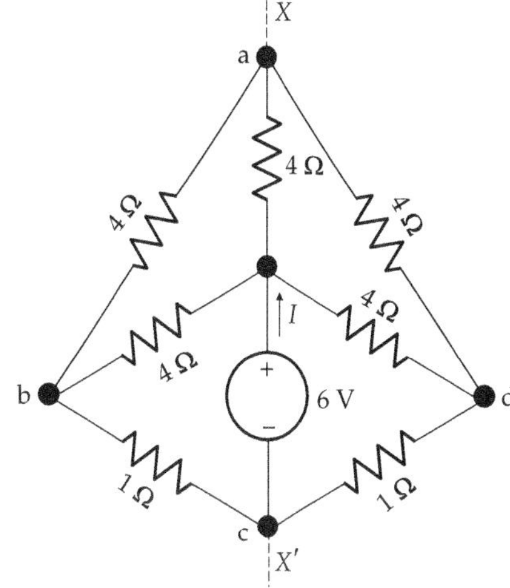

FIGURE 5.48
Symmetrical circuit.

This feature can be made use of in one of two ways to determine *I*.

The first method is based on the observation that because of symmetry, nodes 'b' and 'd' are at the same voltage with respect to node 'c'. They can therefore be connected together without disturbing the circuit, as explained in Section 5.3. In effect, this is equivalent to folding the circuit about the midline *XX′*. The 4 Ω resistor along *XX′* is not affected, but each of the other resistors is paralleled with a resistor of equal value, which halves the resistance in each case, as illustrated in Figure 5.49a. The three upper resistances are equivalent to 2 Ω in parallel with a combination of 2 Ω in series with 4 Ω, which is 1.5 Ω. This resistance is in series with 0.5 Ω across the voltage source, so that $I = 6/2 = 3$ A.

In the second method, the circuit is split along the midline into two symmetrical halves. The 4 Ω resistor, being common to the two halves, is replaced by two 8 Ω resistors,

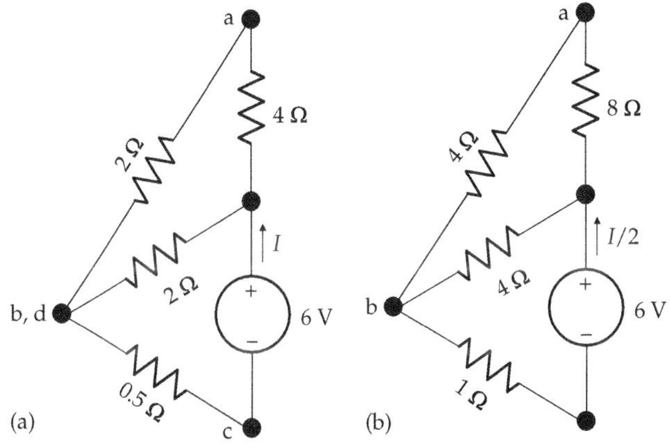

FIGURE 5.49
(a) Folding the circuit around the midvertical axis and (b) splitting the circuit at the midvertical axis.

which when paralleled give 4 Ω. The 6 V source is rearranged into two parallel sources of 6 V each, the current through each source being $I/2$. The left half of the circuit is illustrated in Figure 5.42b. As in Figure 5.49a, the resistance across the voltage source is then $1 + 4\|(4 + 8) = 1 + 3 = 4\ \Omega$, and $I/2 = 6/4 = 1.5$ A, or $I = 3$ A, as before.

Primal Exercise 5.20

Suppose that the 6 V source in Figure 5.48 is replaced by a 3 A current source. Determine the voltage across the source, using the two methods described in connection with Figure 5.49.

Ans. 6 V.

Example 5.7: Exploitation of Symmetry

It is required to determine the equivalent resistance between nodes 'ab' in Figure 5.50.

Solution:

To determine the resistance between nodes 'a' and 'b', a test source is applied between these nodes and the resulting test current or voltage determined, as in Figure 5.51. It may be suspected that the voltages at nodes 'c', 'd', and 'e' are halfway between the voltage of nodes 'a' and 'b', so that nodes 'c', 'd', and 'e' are at the same voltage. But how can this be verified in an easy manner? The answer is to apply a very useful problem-solving technique, namely, *make a reasonable assumption, and then check if this assumption leads to logically consistent results; if it does, then the assumption is justified.* In this case, let us *assume* that nodes 'c', 'd', and 'e' in Figure 5.51 are at the same voltage, which means that the 0.5 Ω resistors between these nodes are redundant, since they do not carry current, and can be removed without affecting the rest of the circuit. When these resistors are removed, the circuit becomes as in Figure 5.51. It follows from voltage division between nodes 'a' and 'b' that nodes 'c', 'd', and 'e' are at the same voltage with respect to node 'b', which means that any resistance, not only 0.5 Ω, can be connected

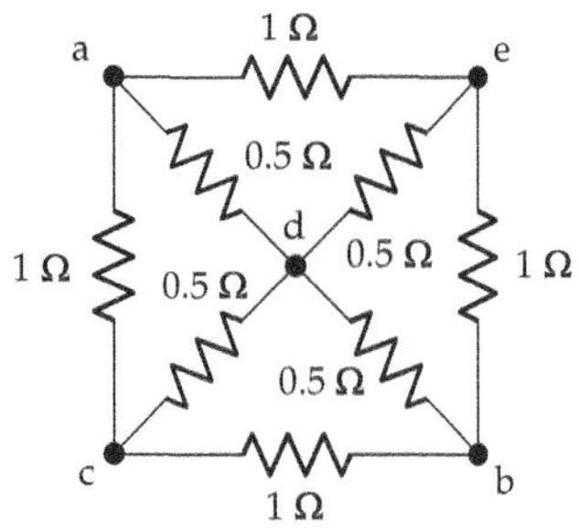

FIGURE 5.50
Figure for Example 5.7.

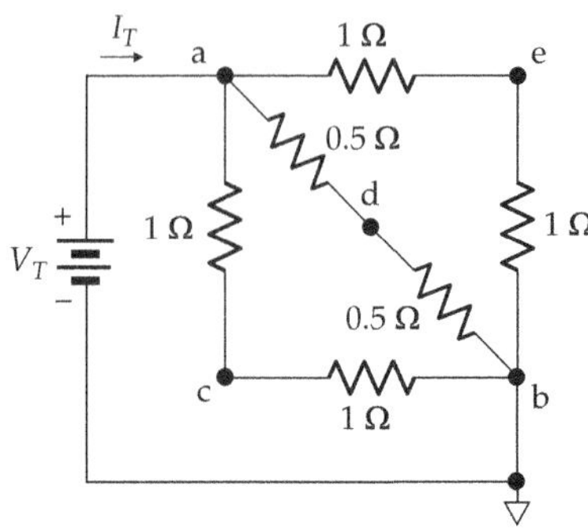

FIGURE 5.51
Figure for Example 5.7.

between these nodes without affecting the rest of the circuit. Moreover, the 0.5 Ω resistors could equally well be replaced by open circuits or short circuits. Hence, the initial assumption that nodes 'c', 'd', and 'e' are at the same voltage leads to a logically consistent conclusion, in that these nodes are indeed at the same voltage, so that the initial assumption is justified. Had the removal of the 0.5 Ω resistors led to unequal voltages of nodes 'c', 'd', and 'e', then this result is inconsistent with the assumption of equality of voltages at the nodes, which would invalidate the assumption.

It is seen from Figure 5.51 that V_T/I_T is three resistances in parallel: 2 Ω in parallel with 2 Ω gives 1 Ω. This, in parallel with 1 Ω, is 0.5 Ω.

Simulation: The circuit is entered as in Figure 5.52. A 1 A current source is used to indicate the resistance between nodes 'a' and 'b' as being numerically equal to the voltage of node 'a' with respect to ground at node 'b'. After selecting 'Bias Point' under 'Analysis type' in the Simulation Settings and running the simulation, pressing the I and V buttons displays the currents and voltages, respectively, indicated in Figure 5.52.

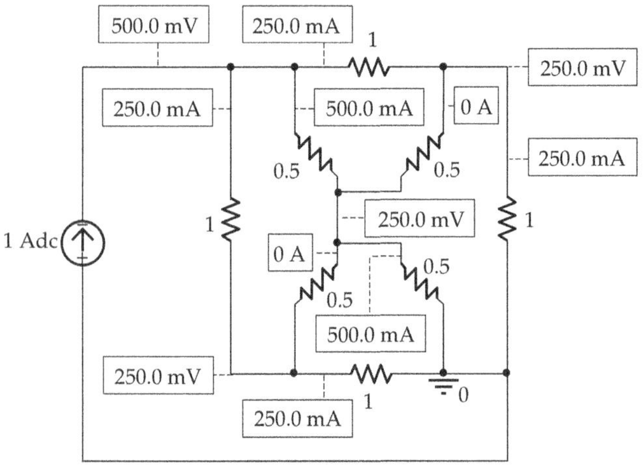

FIGURE 5.52
Figure for Example 5.7.

Problem-Solving Tips

- Always bear in mind that determining an equivalent resistance between two nodes is based on applying a test source and determining the resulting current or voltage. In the absence of dependent sources, the equivalent resistance can be found directly, as from series–parallel combinations of resistances.
- A useful problem-solving technique is to make a reasonable assumption and then check if this assumption leads to logically consistent results; if it does, then the assumption is justified.

Primal Exercise 5.21

Verify that short-circuiting nodes 'c', 'd', and 'e' in Figure 5.51 give $R_{ab} = 0.5\ \Omega$.

It should be noted that exploitation of symmetry in the preceding examples basically depends on having nodes of the same voltage about an axis of symmetry rather than having equal resistances about this axis. In Figure 5.53, for example, the resistance values have been changed compared to Figure 5.50. Applying the same argument as in Example 5.7 leads to the conclusion that if a test source is applied between nodes 'a' and 'b', the 0.5 Ω and 6 Ω resistors do not carry current and can be removed from the circuit. The resistance between nodes 'a' and 'b' is then $3\|6\|12 = 12/7\ \Omega$.

5.7 Problem-Solving Approach Updated

The main procedural steps of the ISDEPIC approach are summarized and updated as follows in the light of the material covered in this chapter:

Step 1—Initialize:

(a) Mark on the circuit diagram all given values of circuit parameters, currents, and voltages, as well as the unknowns to be determined.

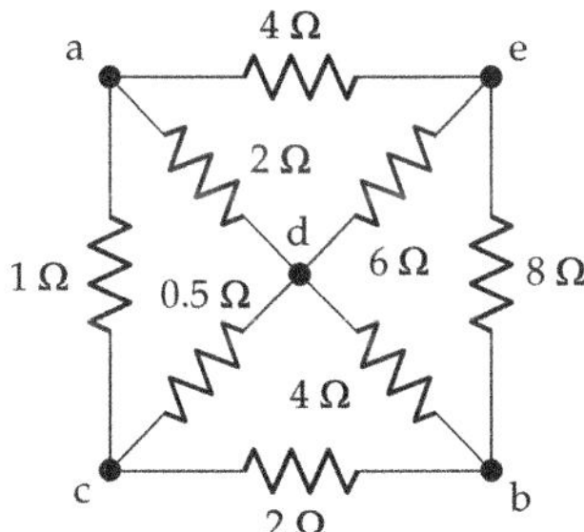

FIGURE 5.53
Equality of node voltages about an axis of symmetry.

(b) Label the nodes, as this may be generally helpful.
(c) If the solution requires that a given value of current or voltage be satisfied, assume this value from the very beginning

Step 2—Simplify: Consider, as may be appropriate,

(a) Redrawing the circuit
(b) Replacing series and parallel combinations of circuit elements by an equivalent circuit element
(c) Applying star-delta transformation
(d) Applying source transformation
(e) Using TEC or NEC to simplify part of the circuit
(f) Applying the source absorption theorem
(g) Removing resistors that are redundant as far as the variables of interest are concerned
(h) Partitioning the circuit in the presence of shunt-connected ideal voltage sources or series-connected ideal current sources
(i) Exploiting symmetry

Step 3—Deduce: Determine any values of current or voltage that follow immediately from direct application of Ohm's law, KCL, or KVL, without introducing any additional unknowns. If Step 3 does not provide a solution, proceed to Step 4.

Step 4—Explore: Examine each of the following alternatives, as may be applicable:

(a) Consider the nodes and meshes in the circuit to see if KCL or KVL can be expressed using a single unknown current or voltage, *and* if this unknown can then be directly determined from KCL or KVL.
(b) Use TEC or NEC to determine a voltage or a current through a given circuit element.
(c) Apply the substitution theorem.
(d) Apply superposition if more than one source is present in the circuit.
(e) Use output scaling.
(f) Rearrange sources.

If Step 4 does not provide the solution, proceed to Step 5.

Step 5—Plan: Think carefully and creatively about the problem in the light of circuit fundamentals and circuit analysis techniques. Consider alternative solutions and select what seems to be the simplest and most direct solution.

Step 6—Implement: Carry out your planned solution.

Step 7—Check your calculations and results.

(a) Check that your results make sense, in terms of magnitude and sign.
(b) Check that Ohm's law is satisfied across every resistor, that KCL is satisfied at every node, and that KVL is satisfied around every mesh.
(c) Seek an alternative solution to see if it gives the same result.
(d) Whenever feasible, check simulate with PSpice.

Learning Checklist: What Should Be Learned from This Chapter

- If an input x_1 to a given system produces an output y_1 and an input x_2 produces an output y_2, then the system obeys superposition if an input $(x_1 + x_2)$ that is the sum of the two inputs produces an output $(y_1 + y_2)$ that is the sum of the outputs due to each input acting alone. The same applies for more than two inputs.
- The procedure for applying superposition in the absence of dependent sources can be summarized as follows:
    1. Select the desired voltage or current response as the circuit variable to which superposition will be applied.
    2. A component of the desired response is obtained with each independent source acting alone, while the remaining independent sources are set to zero.
    3. The desired response is the algebraic sum of the individual components.
- In the presence of a single dependent source, the dependent source can be left in place and superposition applied to the desired response by applying the independent sources one at a time, or the following procedure adopted:
    1. Replace the dependent source with an independent source of unknown value.
    2. Select the controlling variable of the dependent source as the circuit variable to which superposition will be applied.
    3. A component of the controlling variable is obtained with each independent source acting alone, while the remaining independent sources are set to zero.
    4. Apply the superposition equation and substitute for the unknown value of the independent source its value in terms of the controlling variable in the original circuit.
    5. The controlling variable is determined from the superposition equation.
    6. Once the controlling variable is determined, the desired circuit response can be found using KCL and KVL.
- Because power is a nonlinear function of voltage or current, superposition cannot be applied directly to power. The voltage across the resistor, or the current through the resistor, is first obtained by superposition, and the power can then be determined using the total value of the voltage or current.
- In some circuits, such as those excited by a single independent source, a convenient output can be assumed, and the voltages and currents determined by working backward from the output to the source, using KCL, KVL, and Ohm's law. The output is then scaled in accordance with the value of the source.
    1. Scaling can be applied to more general circuits as long as it is possible to assume a certain output and then work backward toward the input by systematically determining all the currents and voltages along the way, without having to invoke additional variables.
- A resistor in series with an ideal current source is redundant as far as the rest of the circuit is concerned, but affects the voltage across the source. Similarly, a resistor in parallel with an ideal voltage source is redundant as far as the rest of the circuit is concerned, but affects the current through the source.
- A redundant resistor can be removed from the circuit, without disturbing the rest of the circuit. When removing a redundant resistor, a redundant series resistor is replaced by a short circuit, whereas a redundant parallel resistor is replaced by an open circuit. A redundant resistor that does not carry current may be replaced by a short circuit or an open circuit.
- Nodes that are at the same voltage, without being short-circuited together, can be left open-circuited or can be connected by a short circuit or by resistors of arbitrary values, without any current flowing in the connection and without affecting the rest of the circuit.
- If an ideal voltage source is connected in parallel with two circuits or an ideal current source is connected in series with two circuits, the two circuits can be separated, one from the other, as long as the two circuits are independent of one another, that is, they are not interconnected in any other way.

- A circuit can be simplified by replacing a current source by two current sources, without changing KCL at the nodes involved. A circuit can be simplified by replacing a voltage source by two voltage source, without changing KVL around the meshes involved.
- Symmetry in a circuit can generally be exploited to greatly simplify the analysis required for obtaining a given response.

Problem-Solving Tips

1. The substitution theorem can be used to replace a branch current or voltage by an independent source, which allows application of superposition.
2. Superposition can be advantageously applied in conjunction with the substitution theorem by replacing dependent sources, or branch currents or voltages, by independent sources.
3. A circuit should always be carefully inspected to see if it can be simplified through removal of redundant resistors, partitioning of circuits because of parallel location of an ideal voltage source, or series location of an ideal current source, or through source rearrangement, or exploitation of symmetry.
4. Always bear in mind that determining an equivalent resistance between two nodes is based on applying a test source and determining the resulting current or voltage. In the absence of dependent sources, the equivalent resistance can be found directly from series–parallel combinations of resistances, star-delta transformation, etc..
5. A useful problem-solving technique is to make a reasonable assumption, and then check if this assumption leads to logically consistent results; if it does, then the assumption is justified.

Appendix 5A: Wheatstone Bridge

Bridge circuits are commonly used for accurate measurements of circuit parameters. The basic principle is illustrated by the **Wheatstone bridge** for dc measurement of resistance (Figure 5.54). When the bridge is *balanced*, $V_{bc} = V_{dc}$, so that $V_O = 0$, and the detector 'D' does not draw any current, From voltage division, $V_{bc} = \frac{R_3}{R_2 + R_3} V_I$ and $V_{dc} = \frac{R_4}{R_1 + R_4} V_I$. Equating V_{bc} and V_{dc} gives

$$\frac{R_3}{R_2 + R_3} = \frac{R_4}{R_1 + R_4} \tag{5.19}$$

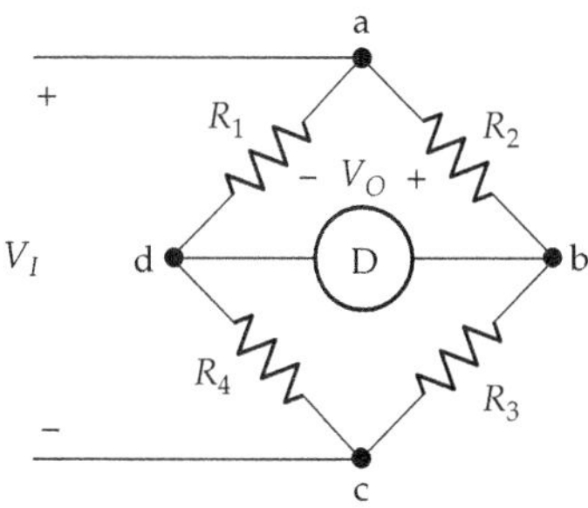

FIGURE 5.54
Wheatstone bridge.

Cross multiplying and simplifying,

$$R_1R_3 = R_2R_4 \quad \text{or} \quad \frac{R_2}{R_3} = \frac{R_1}{R_4} \quad \text{or} \quad \frac{R_1}{R_2} = \frac{R_4}{R_3} \tag{5.20}$$

The balance condition can be stated as (1) the product of resistances in diametrically opposite arms of the bridge is the same or (2) the ratio of resistances in adjacent, opposite arms of the bridge is the same.

For the accurate measurement of an unknown resistance say R_3, R_2 is an accurately known fixed, reference resistor, and R_1 and R_4 are high-quality resistors of the same type whose resistances can be varied in small enough steps. R_1 and R_4 are varied until bridge balance is achieved. R_3 is then determined from Equation 5.20. Two factors underlie the accuracy of this measurement:

1. A zero indication can be much more accurately determined than a nonzero value of current or voltage, because the scale around zero can be expanded to almost any desired degree of accuracy. A sensitive, current-measuring device, known as a **galvanometer** is normally used to indicate bridge balance. A method of measurement based on an indication of zero is referred to as a **null method**.
2. The ratio of resistances of two resistors of the same type, such as R_1 and R_4, is much more accurate than the absolute value of either one, because systematic, that is, nonrandom, errors in R_1 and R_4 tend to cancel out.

Problems

Apply ISDEPIC and verify solutions by PSpice simulation whenever feasible.

Superposition

P5.1 Determine V_X in Figure P5.1.

Ans. 80/3 V.

P5.2 Determine V_O, I_{SRC1}, and I_{SRC2} in Figure P5.2.

Ans. $I_{SRC1} = -1.3$ A, and $I_{SRC2} = 2.2$ A, $V_O = 18$ V.

P5.3 Determine I_X in Figure P5.3.

Ans. 50 mA.

P5.4 Determine the power delivered or absorbed by the 15 V source in Figure P5.4.

Ans. 17.5 W delivered.

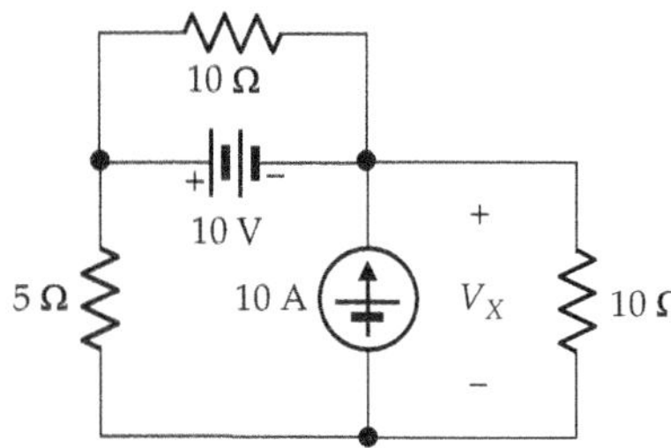

FIGURE P5.1

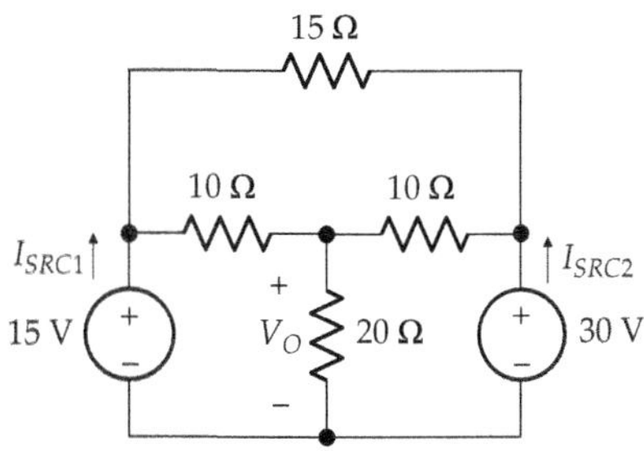

FIGURE P5.2

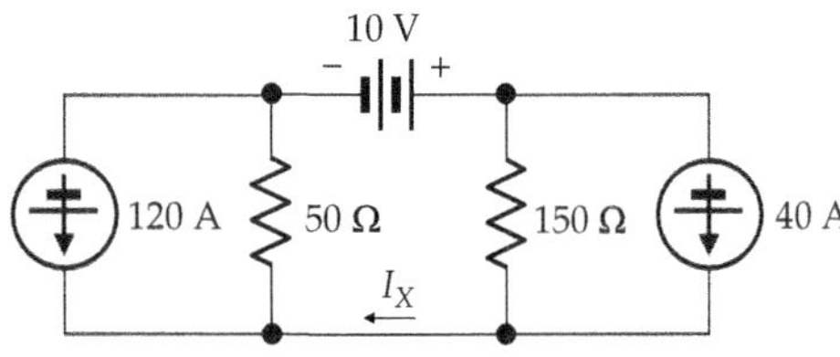

FIGURE P5.3

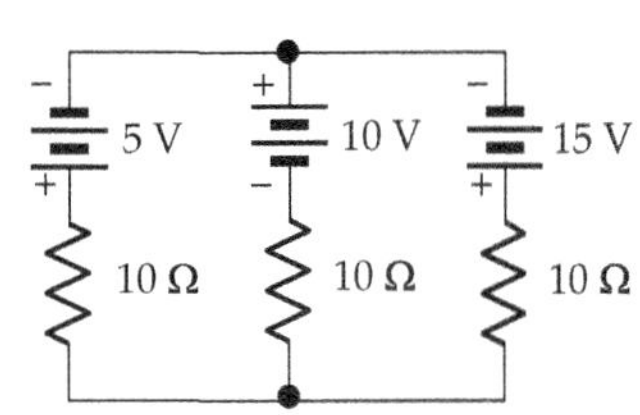

FIGURE P5.4

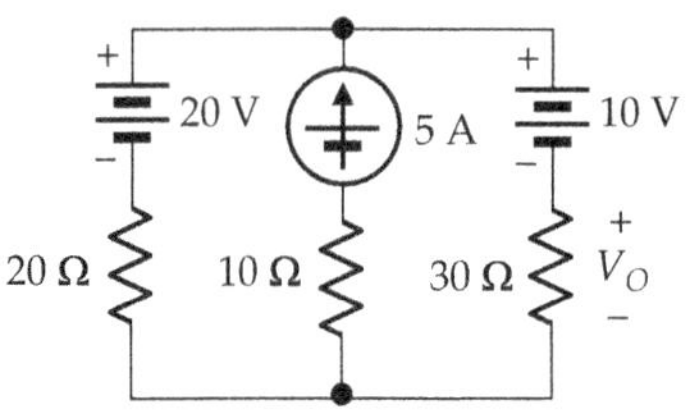

FIGURE P5.5

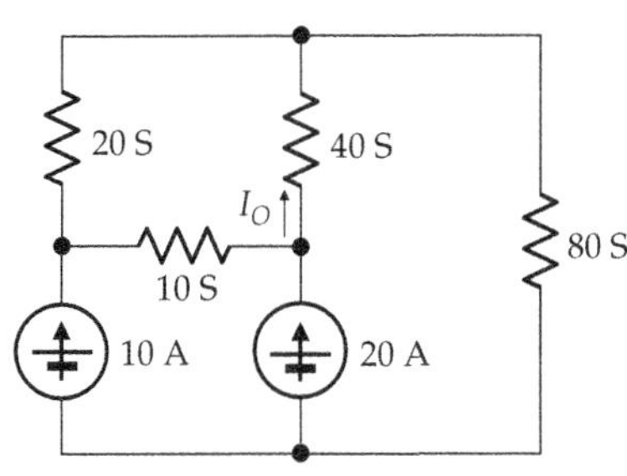

FIGURE P5.6

P5.5 Determine V_O in Figure P5.5.

Ans. 66 V.

P5.6 Determine I_O in Figure P5.6.

Ans. 20 A.

P5.7 Determine V_{SRC1} and V_{SRC2} in Figure P5.7.

Ans. $V_{SRC1} = -1.3$ V, $V_{SRC2} = 2.2$ V.

P5.8 Determine V_S and I_S in Figure P5.8.

Ans. 29 V, 1.6 A.

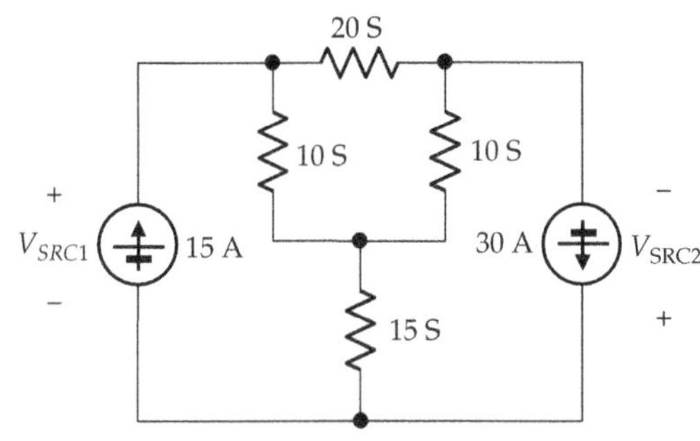

FIGURE P5.7

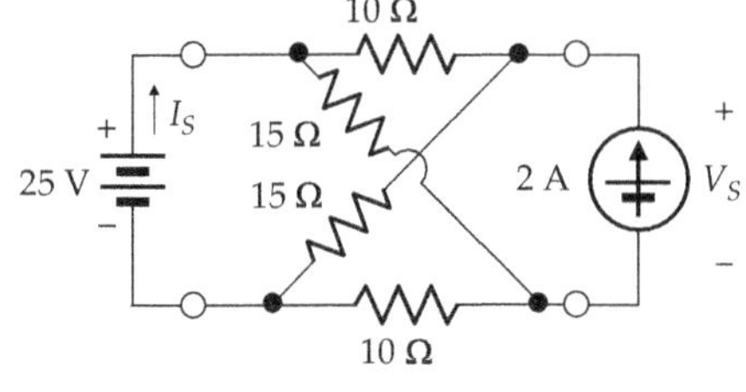

FIGURE P5.8

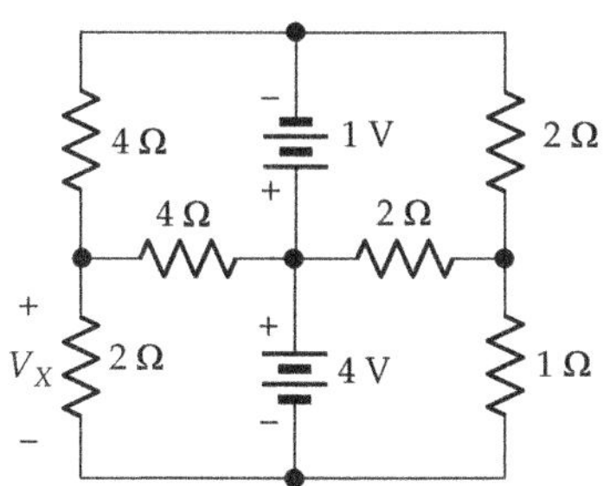

FIGURE P5.9

P5.9 Determine V_X in Figure P5.9.

Ans. 1.75 V.

P5.10 Determine I_X in Figure P5.10.

Ans. 0.5 A.

P5.11 Determine the power delivered or absorbed by the current source in Figure P5.11.

Ans. 180 W absorbed.

P5.12 A resistive circuit is excited by two identical, independent sources. When either source acts alone, with the other source set to zero, the power dissipated in a given resistor in the circuit is 1 W. Determine the power dissipated in this same resistor when both sources act together.

Ans. 0 if the sources act in opposition and 4 W if the sources augment one another.

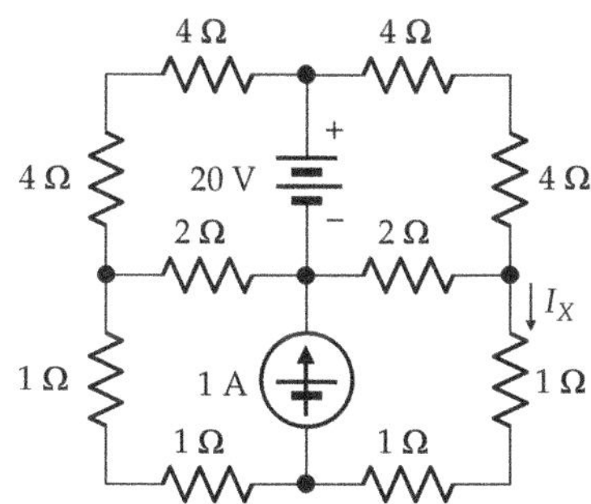

FIGURE P5.10

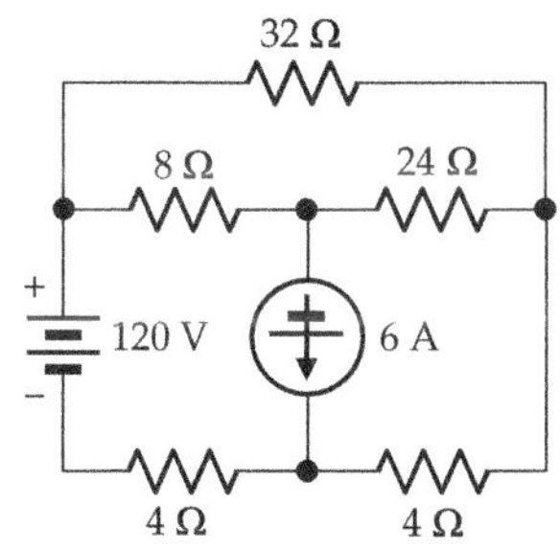

FIGURE P5.11

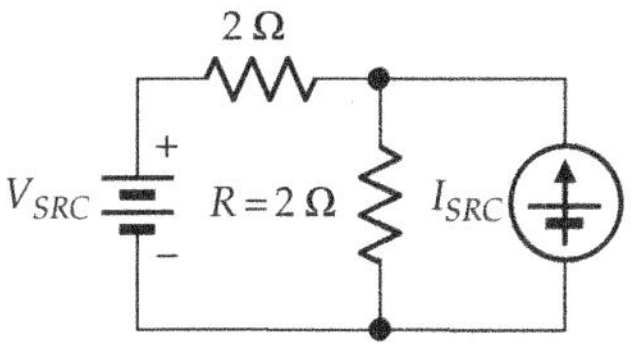

FIGURE P5.13

P5.13 In Figure P5.13, the power dissipated in R is 8 W if I_{SRC} is applied alone, and is 0.5 W if V_{SRC} is applied alone, with the other source set to zero. Determine the power dissipated in R if both sources are applied together, with polarities that will give (a) the largest current in R and (b) the smallest current in R.

Ans. (a) 12.5 W; (b) 4.5 W.

P5.14 The resistance values in Figure P5.14 are not specified. It is given that (a) when $V_{SRC} = 5$ V and $I_{SRC} = 1$ A, $V_{O1} = 2$ V and (b) when $V_{SRC} = 5$ V and $I_{SRC} = 0$, $V_{O2} = 1$ V. Determine V_O when $V_{SRC} = 0$ and $I_{SRC} = 2$ A.

Ans. 2 V.

P5.15 Determine I_X in Figure P5.15. Note that the 5 A source partitions the circuit into two subcircuits, the one having I_X being independent of the subcircuit consisting of the dependent source and the 10 Ω resistor.

Ans. 1.5 A.

P5.16 Determine I_X in Figure P5.16.

Ans. 0.

P5.17 Determine I_O in Figure P5.17.
Ans. −1 A.

P5.18 Determine I_O in Figure P5.18.

Ans. 0.1 A.

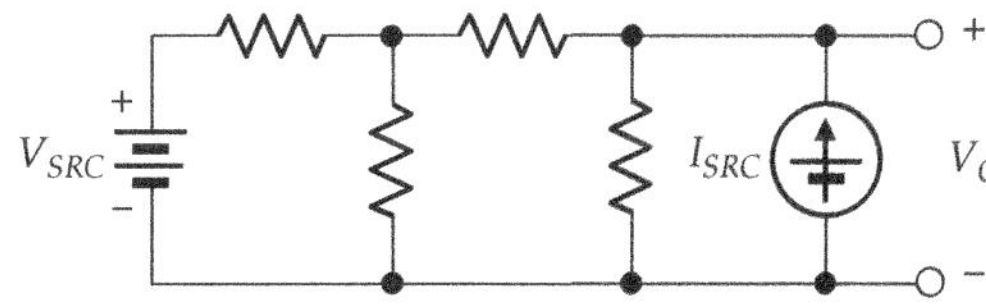

FIGURE P5.14

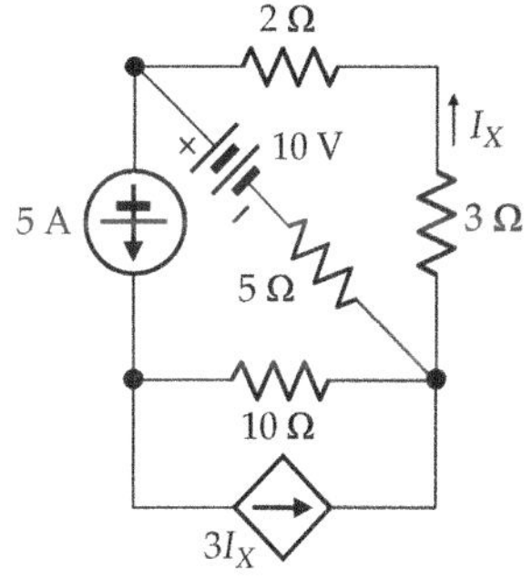

FIGURE P5.15

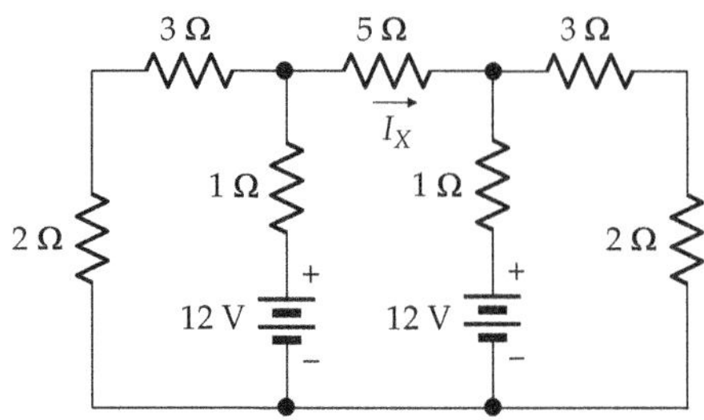

FIGURE P5.16

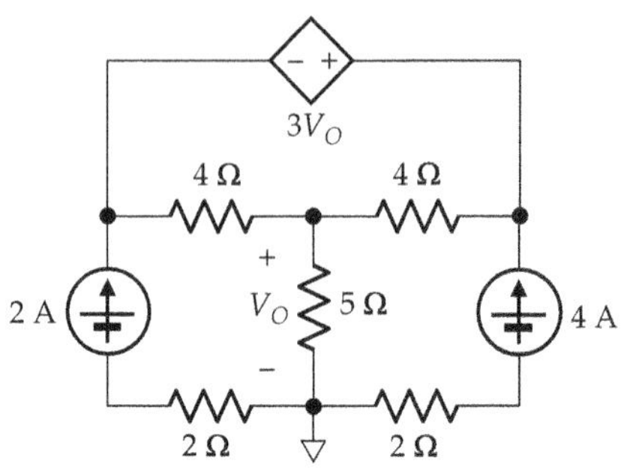

FIGURE P5.20

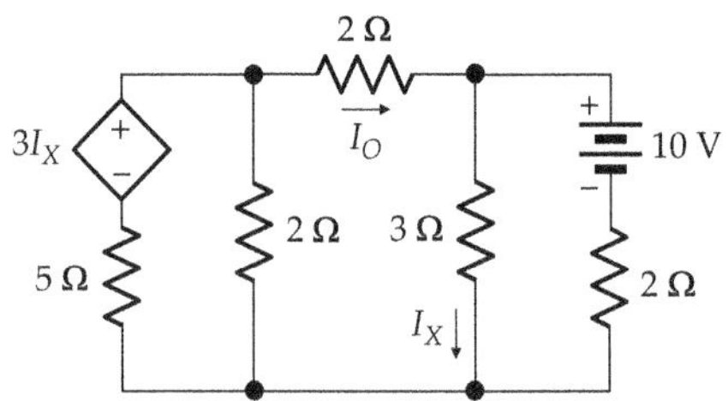

FIGURE P5.17

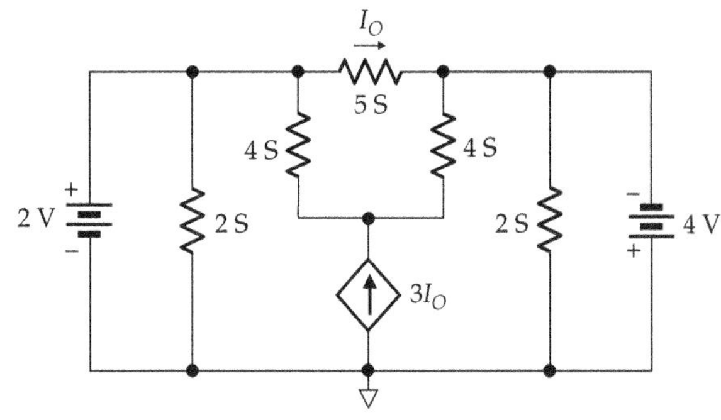

FIGURE P5.21

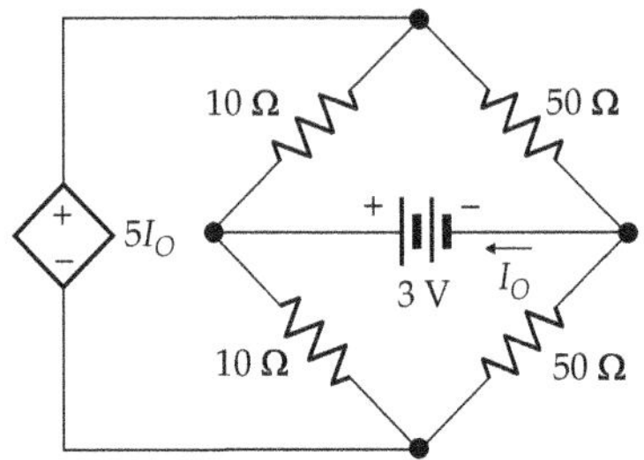

FIGURE P5.18

P5.19 Determine TEC looking into terminals 'ab' in Figure P5.19, given $R = 1\ \Omega$.

Ans. $V_{Th} = V_{ab} = 1$ V, $R_{Th} = 0.5\ \Omega$.

P5.20 Determine the power dissipated in the 5 Ω resistor due to each source in Figure P5.20 and the total power dissipated in this resistor. Note that it should be evident from superposition that the dependent source does not contribute to current in this resistor.

Ans. 20 W due to the 2 A source, 80 W due to 4 A source, 180 W.

P5.21 Determine the power dissipated in the 5 S resistor due to each source in Figure P5.21 and the total power dissipated in this resistor. Note that it should be evident from superposition that the dependent source does not contribute to current in this resistor.

Ans. 20 W due to 2 V source, 80 W due to 4 V source, 180 W.

P5.22 Determine I_X in Figure P5.22.

Ans. 2.5 A.

P5.23 Determine V_O in Figure P5.23.

Ans. −10/3 V.

P5.24 Determine I_O in Figure P5.24.

Ans. −10/3 A.

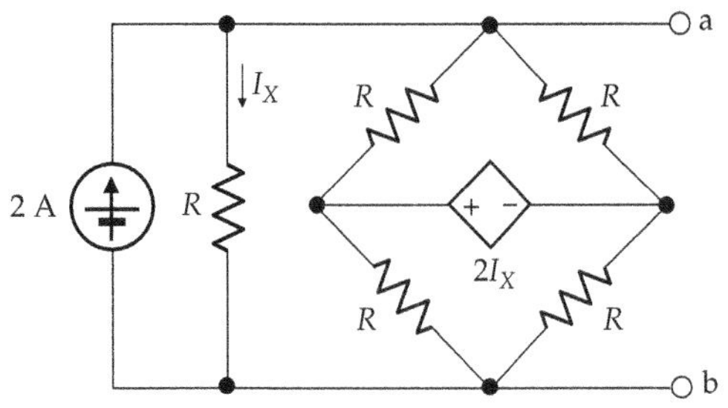

FIGURE P5.19

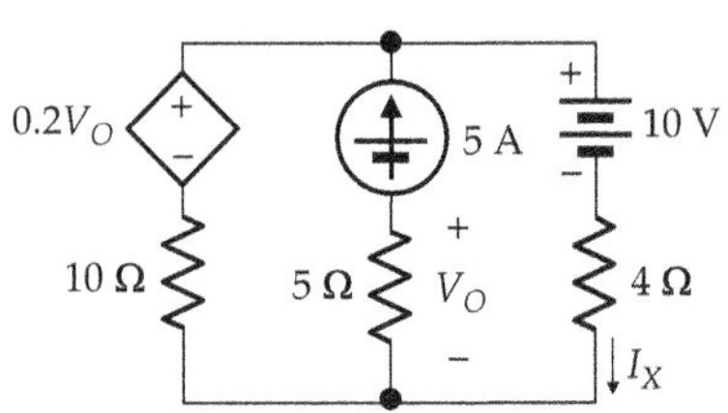

FIGURE P5.22

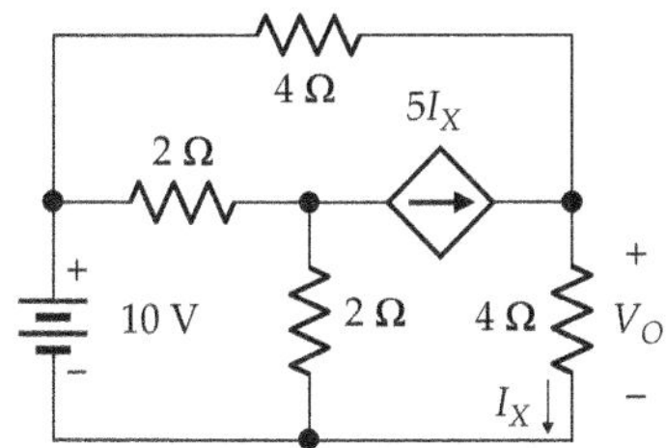

FIGURE P5.23

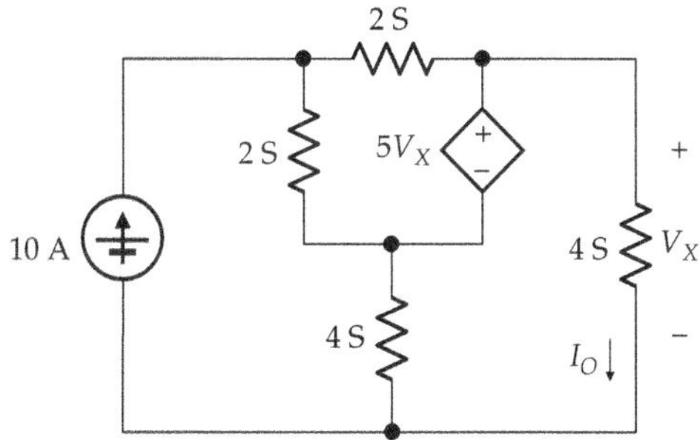

FIGURE P5.24

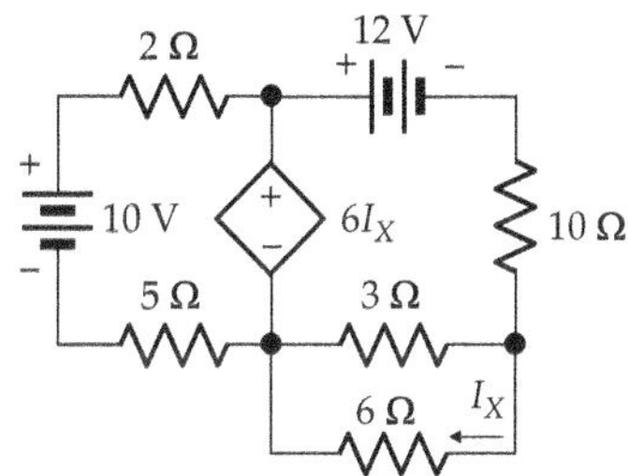

FIGURE P5.25

P5.25 Determine I_X in Figure P5.25.

Ans. −0.4 A.

P5.26 Determine V_O in Figure P5.26.

Ans. −11/9 = −1.22 V.

P5.27 Determine V_O in Figure P5.27.

Ans. $\frac{760}{49} = 15.51$ V.

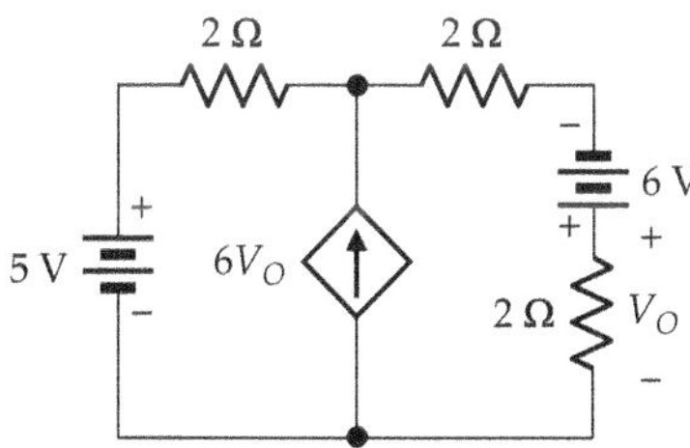

FIGURE P5.26

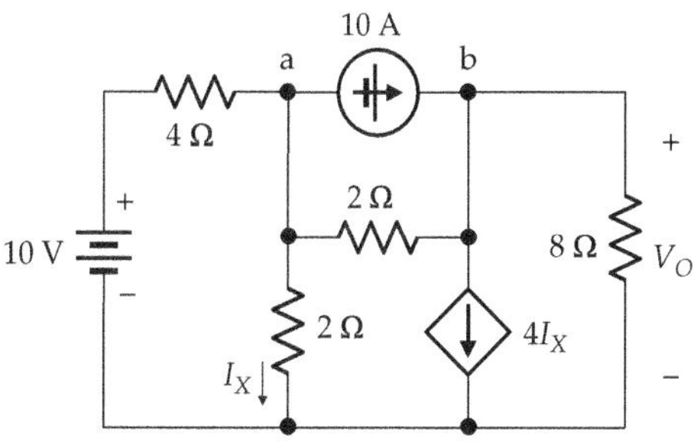

FIGURE P5.27

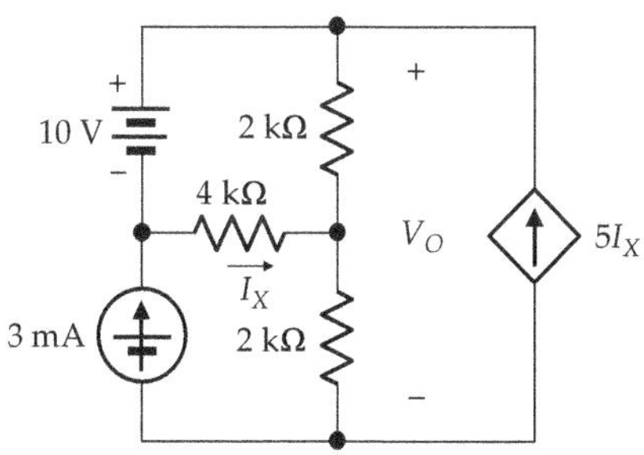

FIGURE P5.28

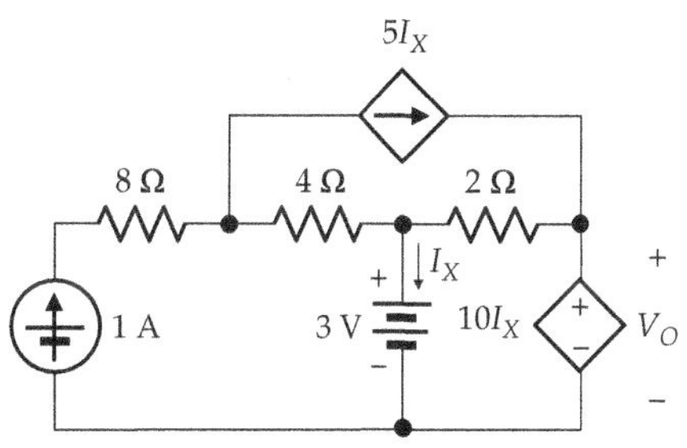

FIGURE P5.29

P5.28 Determine V_O in Figure P5.28.

Ans. 30 V.

P5.29 Determine in Figure P5.29 (a) V_O and (b) the power delivered or absorbed by the 1 A source.

Ans. (a) −5 V; (b) 25 W.

Miscellaneous Circuit Simplification

P5.30 Determine I_O in Figure P5.30 assuming all resistances are 1 Ω.

Ans. 5/41 A.

P5.31 Determine I_O in Figure P5.31 assuming all resistances are 1 Ω.

Ans. 10/13 A.

P5.32 Determine V_{SRC} in Figure P5.32 assuming all resistances are 1 Ω. Note that this is a special ladder known as the *R*-2*R* ladder, in which the resistance between each of

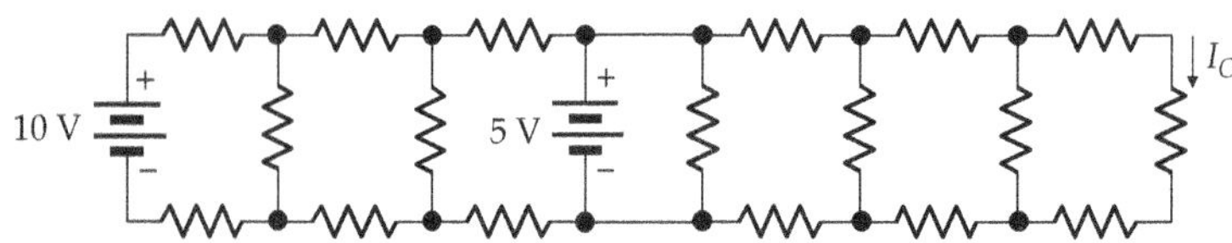

FIGURE P5.30

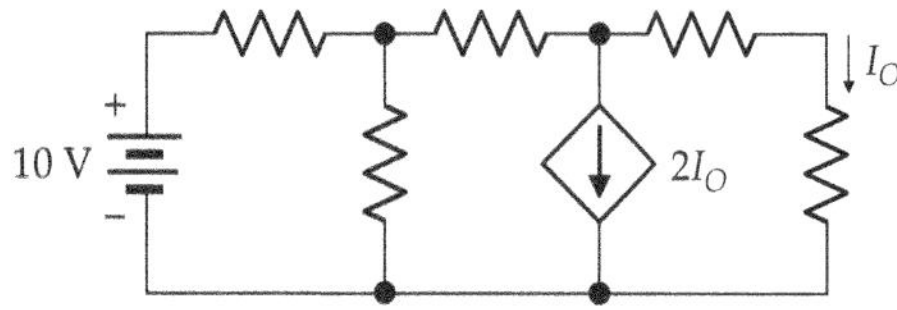

FIGURE P5.31

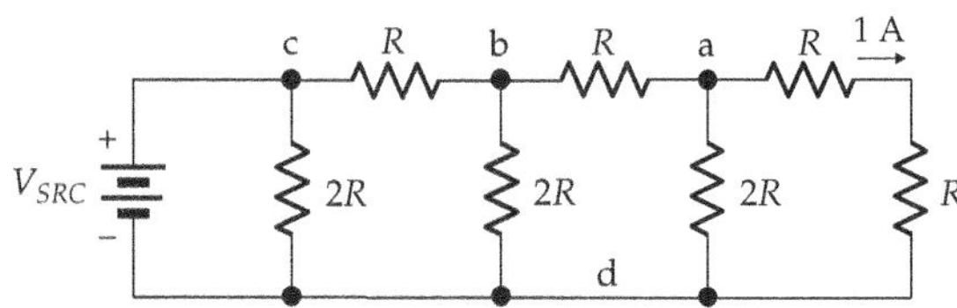

FIGURE P5.32

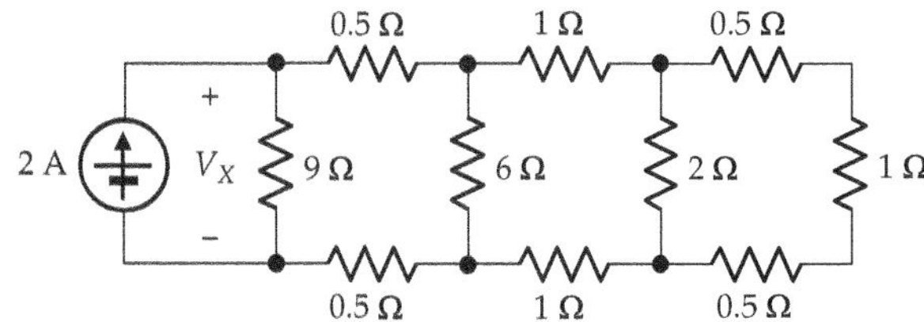

FIGURE P5.33

the upper essential nodes and the lower common node is 2*R* and the resistance looking to the right of each of the upper essential nodes is also 2*R*. Consequently, the voltages at the upper nodes are successively multiplied by 2 in going from right to left.

Ans. 8 V.

P5.33 Determine V_X in Figure P5.33.

Ans. 4.5 V.

P5.34 Determine V_O in Figure P5.34 using scaling.

Ans. 4 V.

P5.35 Determine the power delivered or absorbed by the current source in Figure P5.35.

Ans. 2.75 W delivered.

P5.36 Determine I_O in Figure P5.36.

Ans. 2/3 A.

P5.37 Determine I_{SRC1}, I_{SRC2}, and V_{SRC1} in Figure P5.37.

Ans. I_{SRC1} = 31 A, I_{SRC2} = 0.5 A, V_{SRC1} = 5 V.

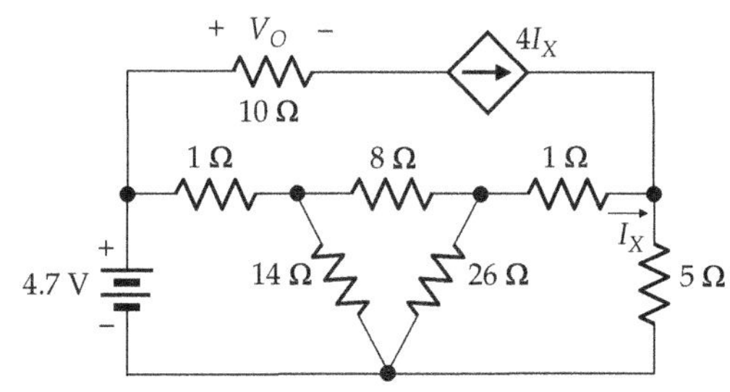

FIGURE P5.34

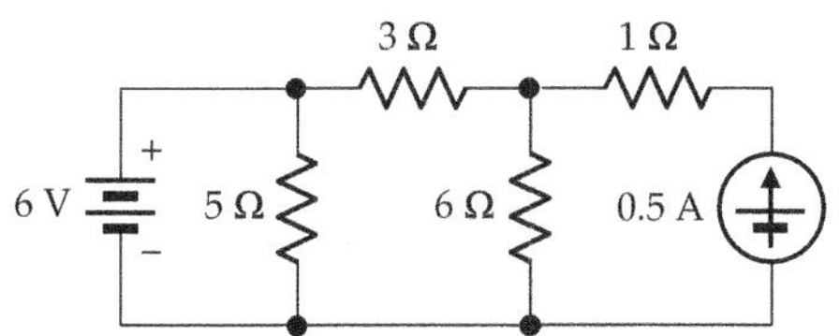

FIGURE P5.35

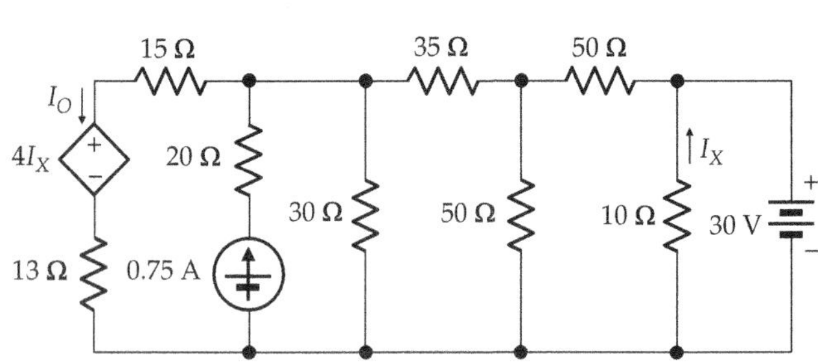

FIGURE P5.36

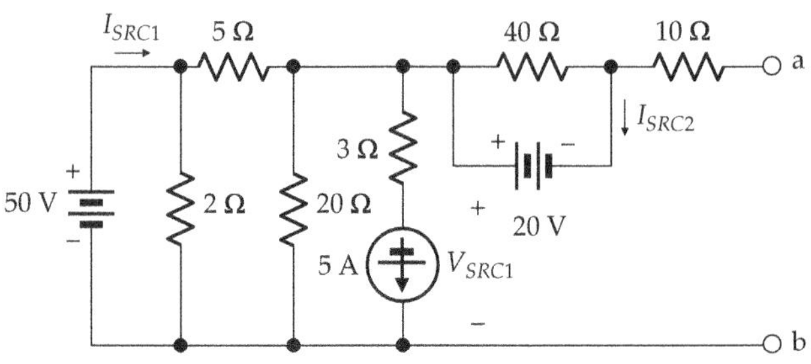

FIGURE P5.37

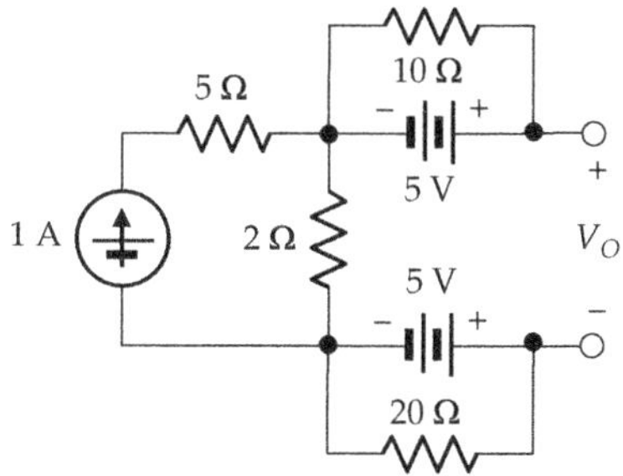

FIGURE P5.38

P5.38 Determine V_O in Figure P5.38.

Ans. 2 V.

P5.39 Determine V_A in Figure P5.39 so that $V_X = 0$.

Ans. 6 V.

P5.40 Determine I_O in Figure P5.40.

Ans. 2/3 A.

P5.41 Determine V_O in Figure P5.41.

Ans. 12 V.

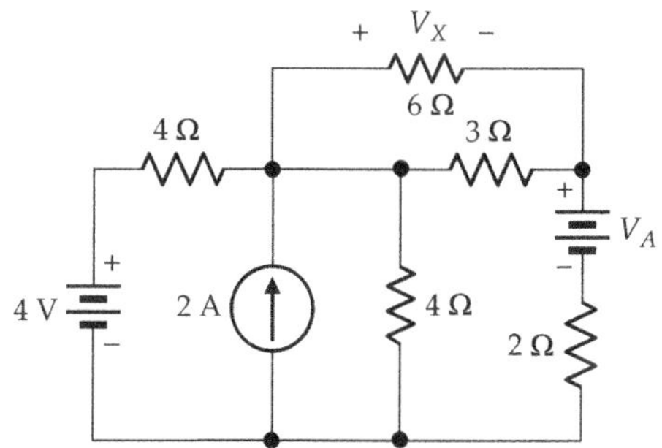

FIGURE P5.39

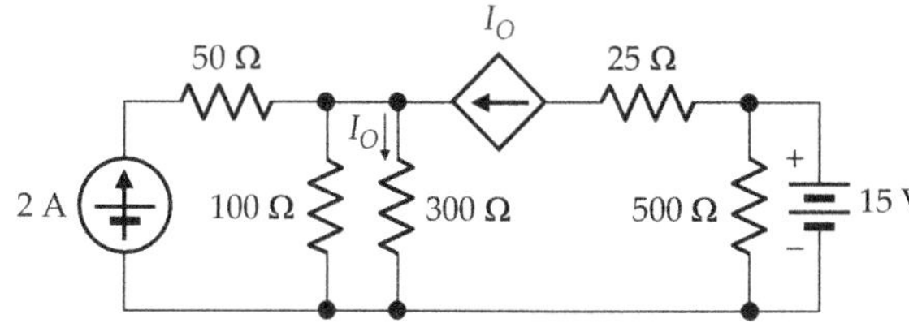

FIGURE P5.40

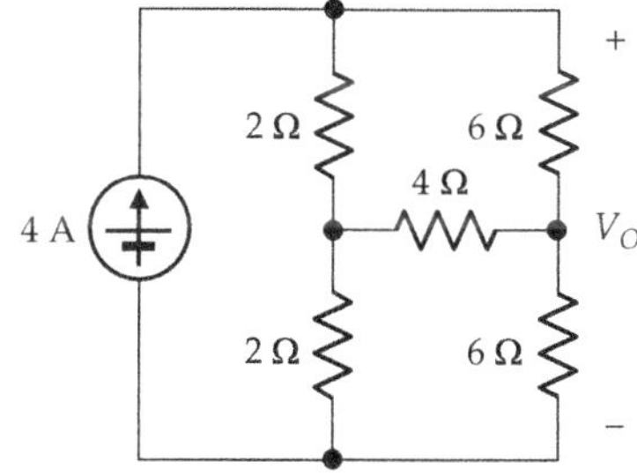

FIGURE P5.41

P5.42 Determine R_{eq} between terminals 'ab' in Figure P5.42, assuming all resistances are 1 Ω.

Ans. 3 Ω.

P5.43 Determine R_{eq} between terminals 'a' and 'b' in Figure P5.43, assuming all resistances are 1 Ω.

Ans. 0.8 Ω.

P5.44 Determine V_O in Figure P5.44 by rearranging the two current sources.

Ans. 5 V.

P5.45 Determine V_X in Figure P5.45.

Ans. 0.5 V.

P5.46 Determine I_{SRC} in Figure P5.46 assuming all resistances are 1 Ω.

Ans. 12 A.

P5.47 Determine I_{SRC} in Figure P5.47 assuming all resistances are 1 Ω.

Ans. 14 A.

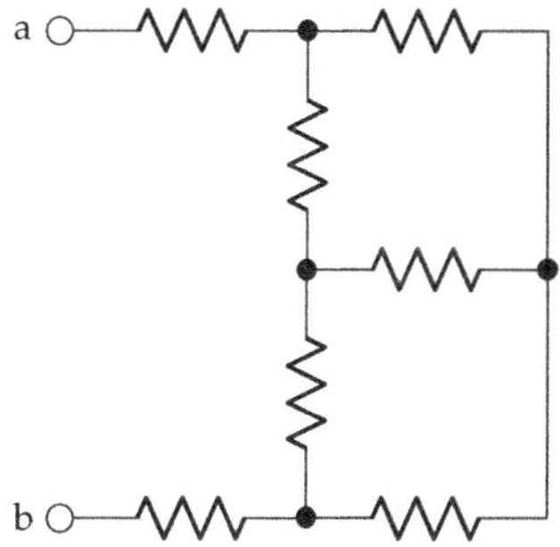

FIGURE P5.42

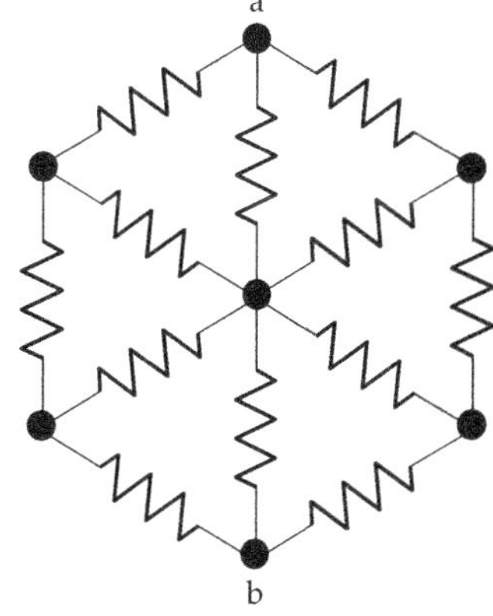

FIGURE P5.43

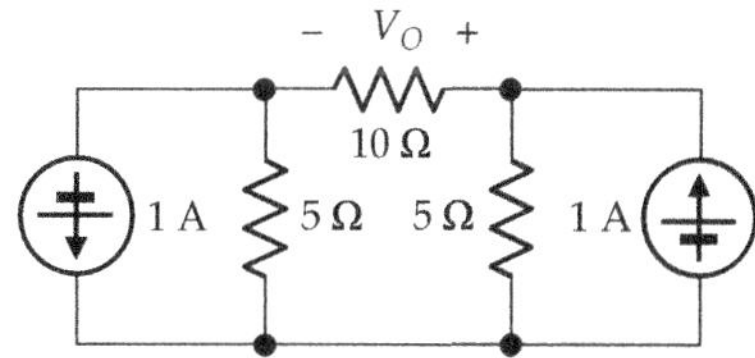

FIGURE P5.44

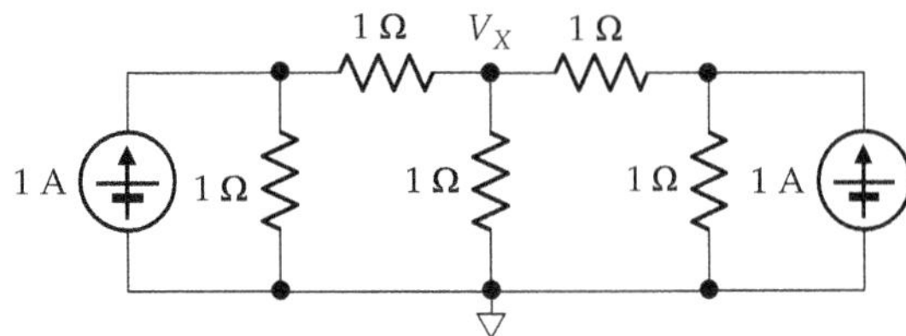

FIGURE P5.45

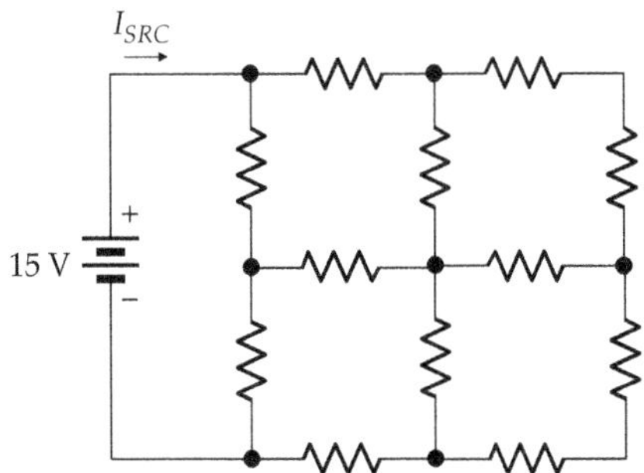

FIGURE P5.46

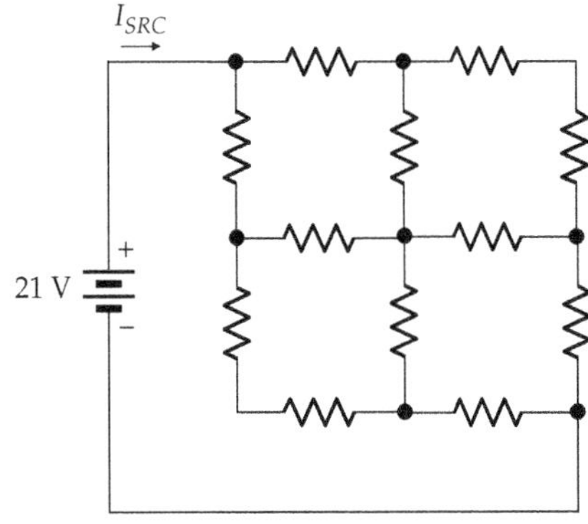

FIGURE P5.47

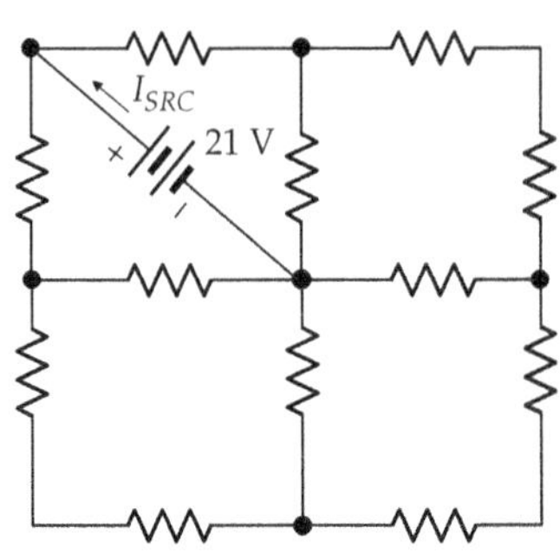

FIGURE P5.48

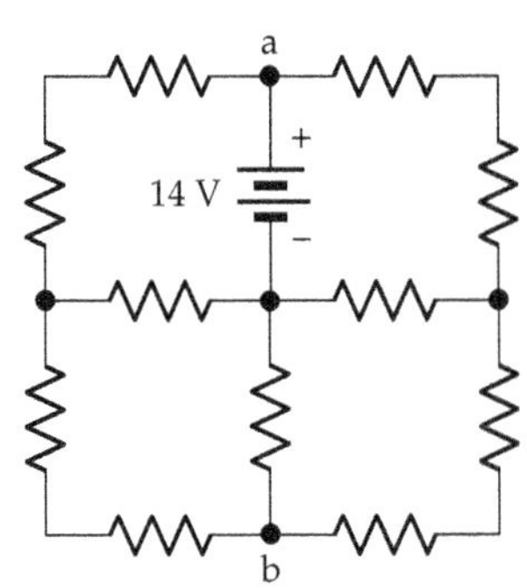

FIGURE P5.49

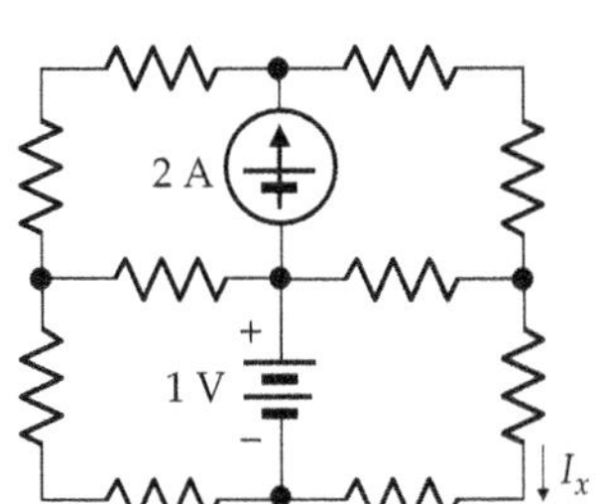

FIGURE P5.50

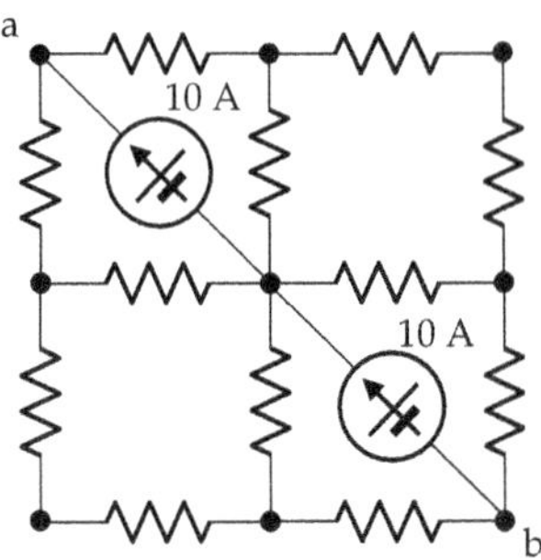

FIGURE P5.51

P5.48 Determine I_{SRC} in Figure P5.48 assuming all resistances are 2 Ω. (*Hint*: Identify redundant resistors based on symmetry.)

Ans. 12 A.

P5.49 Determine V_{ab} in Figure P5.49, assuming all resistances are 1 Ω.

Ans. 12 V.

P5.50 Determine I_X in Figure P5.50, assuming all resistances are 1 Ω.

Ans. 2/3 A.

P5.51 Determine V_{ab} in Figure P5.51, assuming all resistances are 1 Ω.

Ans. 15 V.

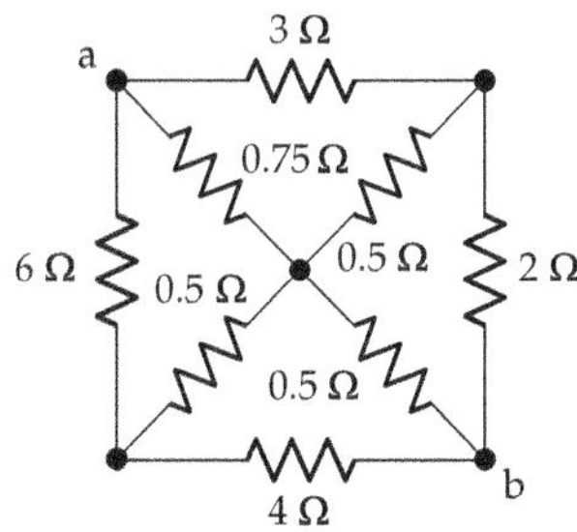

FIGURE P5.52

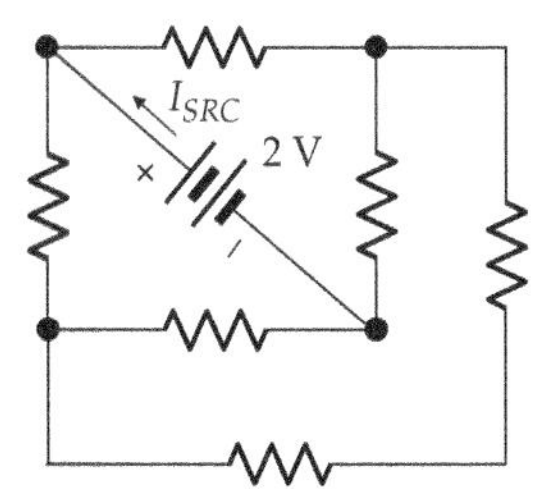

FIGURE P5.53

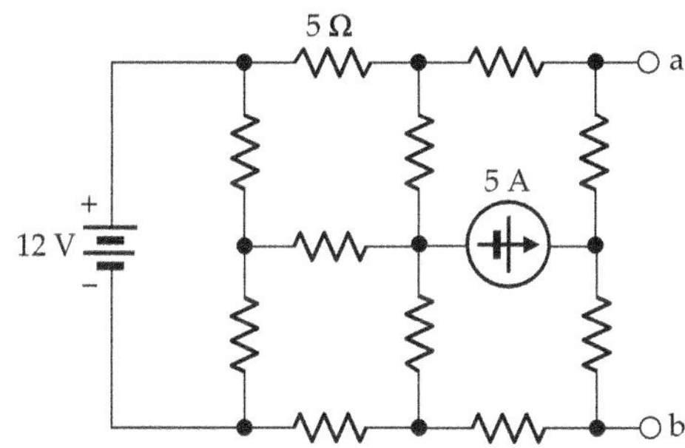

FIGURE P5.54

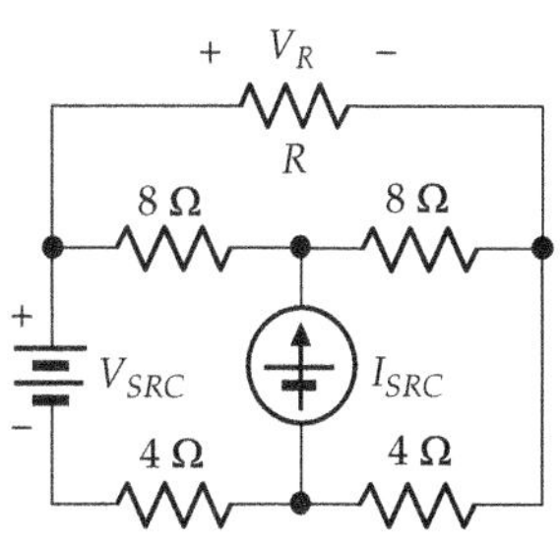

FIGURE P5.55

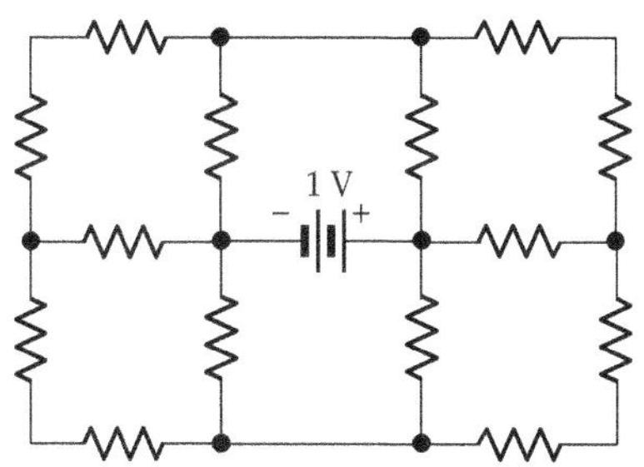

FIGURE P5.56

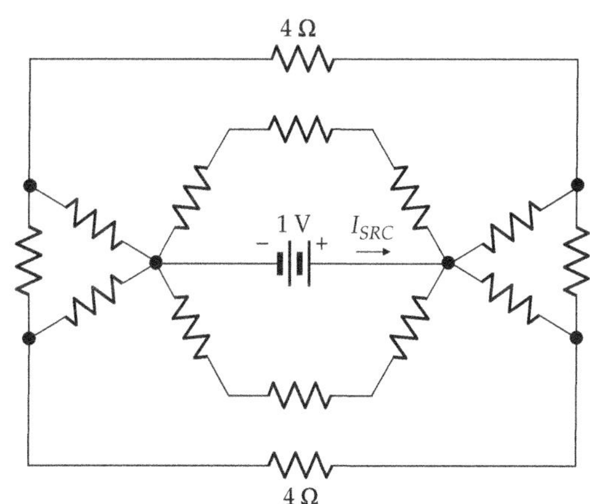

FIGURE P5.57

P5.52 Determine the resistance between terminals 'a' and 'b' in Figure P5.52.

Ans. 10/11 Ω.

P5.53 Determine I_{SRC} in Figure P5.53, assuming all resistances are 2 Ω.

Ans. 1 A.

P5.54 Determine TEC between terminals 'ab' in Figure P5.54, assuming all resistances are 5 Ω.

Ans. $V_{Th} = V_{ab} = 2.4$ V, $R_{Th} = 6$ Ω.

P5.55 R in Figure P5.55 is unspecified, but $V_R = 6$ V when $V_{SRC} = 48$ V and $I_{SRC} = 6$ A. Determine V_R when $V_{SRC} = 32$ V and $I_{SRC} = 8$ A.

Ans. 4 V.

P5.56 Determine the power delivered by the voltage source in Figure P5.56, assuming all resistances are 1 Ω.

Ans. 1.25 W.

P5.57 Determine I_{SRC} in Figure P5.57, assuming all resistances are 1 Ω, except for the two 4 Ω resistances indicated.

Ans. 1 A.

P5.58 Determine R_{ab} in Figure P5.58.

Ans. 8/3 Ω.

P5.59 Determine R_{ab} in Figure P5.59, assuming that all resistances are in ohms.

Ans. 18/11 Ω.

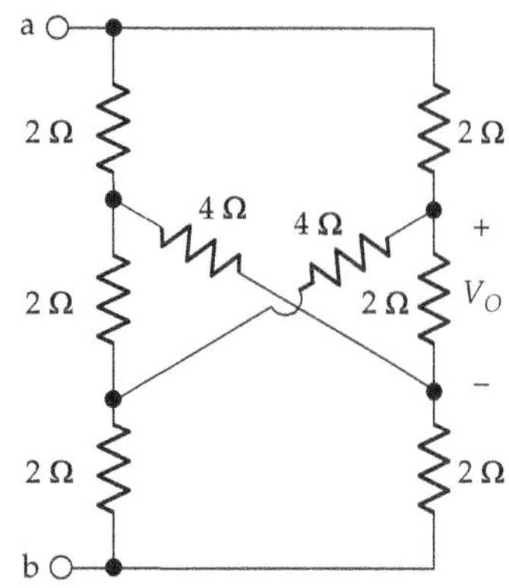

FIGURE P5.58

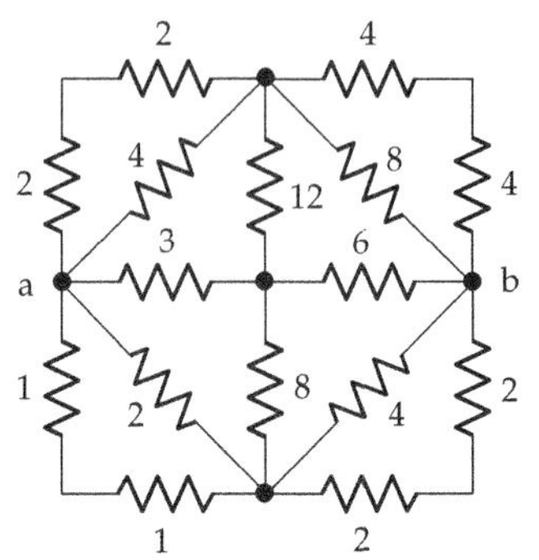

FIGURE P5.59

6

Circuit Equations

Objective and Overview

This chapter presents the systematic methods for analyzing circuits, namely, the node-voltage and mesh-current methods and discusses some of their variants.

Kirchhoff's laws and Ohm's law provide the number of equations necessary to analyze an electric circuit, but the number of equations can become unwieldy except in simple circuit configurations. Although the theorems, procedures, and techniques discussed in Chapters 3 through 5 can be invaluable in simplifying the analysis in many cases, there remains cases where these methods are not practicable. Systematic methods based on Kirchhoff's laws have therefore been developed to facilitate analysis of more complex circuits and reduce the likelihood of error in writing the equations that govern the behavior of electric circuits. The commonly used node-voltage and mesh-current methods are presented in this chapter, including some special considerations and generalizations.

6.1 Node-Voltage Method

Concept: *In the node-voltage method, voltages of essential nodes are assigned with respect to one of the essential nodes taken as a reference. This automatically satisfies Kirchhoff's voltage law (KVL) in every mesh in the circuit. Equations based on Kirchhoff's current law (KCL) are then written directly in terms of Ohm's law for each essential node other than the reference node.*

The node-voltage method is illustrated by the circuit in Figure 6.1a, which is excited by a current source and in which resistance values are expressed as conductances. The first step is to select one of the essential nodes as the reference node. This can be done quite arbitrarily, but it is usually convenient to select as reference the node that has the largest number of connections, which is usually a grounded node. Alternatively, the reference node may be selected as a node with respect to which a voltage is required so that only one unknown needs to be determined. If the node voltages are required with respect to a node other than the selected reference node, these can be derived very simply, as will be shown later. Node 'd' is chosen as the reference in Figure 6.1a, as indicated by the unfilled arrow.

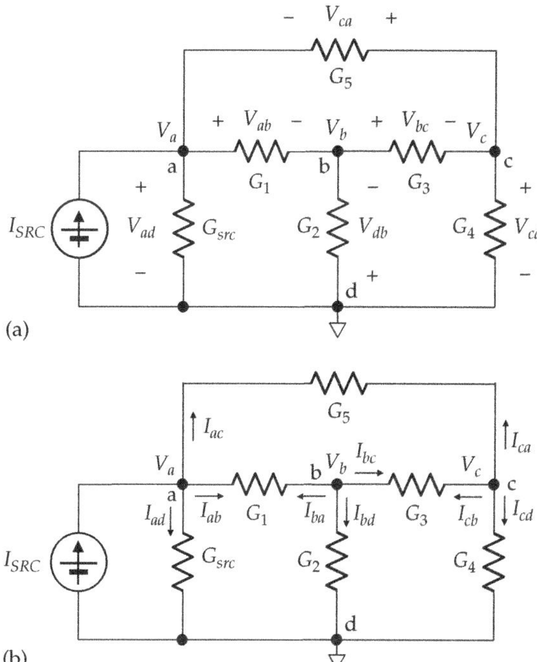

FIGURE 6.1
(a) Circuit to be analyzed by the node-voltage method and (b) branch currents between nodes are explicitly shown.

The reference node is assigned a voltage of zero, and the voltages of all the other essential nodes are expressed with reference to this node. To verify that KVL is satisfied around every mesh, consider the upper mesh as an example. The voltage drops around this mesh are denoted as V_{ab}, V_{bc}, and V_{ca} in Figure 6.1a. KVL around this mesh is

$$V_{ab} + V_{bc} + V_{ca} = 0 \tag{6.1}$$

The voltage drops V_{ab}, V_{bc}, and V_{ca} can be expressed in terms of the assigned node voltages as $V_{ab} = V_a - V_b$, $V_{bc} = V_b - V_c$, and $V_{ca} = V_c - V_a$. The LHS of Equation 6.1 becomes

$$(V_a - V_b) + (V_b - V_c) + (V_c - V_a) = 0$$

The node voltages cancel out in pairs, so they sum to zero, as required by KVL. The same is true of the other meshes in the circuit (Exercise 6.1).

With KVL satisfied, KCL is written for each of the essential nodes in terms of Ohm's law and any source currents entering or leaving the node. For node 'a', the total current leaving the node through the conductances connected to the node is $I_{ad} + I_{ac} + I_{ab}$ (Figure 6.1b), where $I_{ad} = G_{src}V_a$, $I_{ac} = G_5(V_a - V_c)$, and $I_{ab} = G_1(V_a - V_b)$. The source current entering node 'a' is I_{SRC}. Equating the current leaving the node through the conductances to the current entering the node from the source and collecting terms in the node voltages gives KCL for node 'a' as

$$(G_{src} + G_1 + G_5)V_a - G_1V_b - G_5V_c = I_{SRC} \quad (6.2)$$

The total current leaving node 'b' through the conductances connected to the node is $I_{ba} + I_{bc} + I_{bd}$, where $I_{ba} = G_1(V_b - V_a)$, $I_{bc} = G_3(V_b - V_c)$, and $I_{bd} = G_2V_b$. There is no source current entering node 'b'. Collecting terms in the node voltages gives KCL for node 'b' as

$$-G_1V_a + (G_1 + G_2 + G_3)V_b - G_3V_c = 0 \quad (6.3)$$

The total current leaving node 'c' through the conductances connected to the node is $I_{cb} + I_{ca} + I_{cd}$, where $I_{cb} = G_3(V_c - V_b)$, $I_{ca} = G_5(V_c - V_a)$, and $I_{cd} = G_4V_c$. There is no source current entering node 'c'. Collecting terms in the node voltages gives KCL for node 'c' as

$$-G_5V_a - G_3V_b + (G_3 + G_4 + G_5)V_c = 0 \quad (6.4)$$

Comparing Equations 6.2 through 6.4 reveals a pattern that allows writing the node-voltage equations by inspection:

1. In the equation for a given node, the coefficient multiplying the voltage of this node is the sum of all the conductances connected directly to that node. Thus, in the equation for node 'a' (Equation 6.2), V_a is multiplied by $(G_{src} + G_1 + G_5)$, the sum of the three conductances connected directly to node 'a'. Similarly, in the equation for node 'b' (Equation 6.3), V_b is multiplied by $(G_1 + G_2 + G_3)$, and in the equation for node 'c' (Equation 6.4), V_c is multiplied by $(G_3 + G_4 + G_5)$.These coefficients are known as the **self-conductances** of the nodes.
2. In the equation for a given node, the coefficient multiplying the voltage of each of the other nodes is the conductance that directly connects this node to the given node, with a minus sign. The minus sign arises from subtracting the voltages of the other nodes from the voltage of the given node. Thus, the current leaving node 'a' through G_1 is $(V_a - V_b)G_1 = G_1V_a - G_1V_b$, where G_1 directly connects node 'a' to node 'b'. In the equation for node 'a' (Equation 6.2), V_b is therefore multiplied by $-G_1$. The current leaving node 'a' through G_5 is $(V_a - V_c)G_5 = G_5V_a - G_5V_c$, where G_5 connects node 'a' to node 'c'. In the equation for node 'a' (Equation 6.2) V_c is therefore multiplied by $-G_5$. The same is true of the other node equations. These coefficients are known as the **mutual conductances** between the nodes. If there is no conductance that directly connects a certain node with the node in question, the corresponding mutual conductance is zero.

If the coefficients of the node voltages in Equations 6.2 through 6.4 are arranged in an array, as in Figure 6.2, the array has the following characteristic features that provide a useful check on the correctness of the node-voltage equations:

1. The self-conductances are the diagonal entries in the array from top left to bottom right.
2. The array is symmetrical with respect to this diagonal, as illustrated by the coefficients pointed to by the double-sided arrows in Figure 6.2. This symmetry is because the conductance is independent of the direction of current. For example, in the expression $I_{ab} = G_1(V_a - V_b)$, G_1 is the same as in the expression $I_{ba} = G_1(V_b - V_a)$, although $I_{ab} = -I_{ba}$. It will be shown in the next section that this symmetry is destroyed when the dependency relations of dependent sources are taken into account.
3. All the mutual conductances have a negative sign, as explained previously.
4. In any row or column, the mutual conductances are part of the self-conductance in that row or column. Thus, in the first row or first column in the array of Example 6.2, G_1 and G_5 are included in the self-conductance term $(G_{src} + G_1 + G_5)$. The remaining conductance is that between the given node and the reference node.

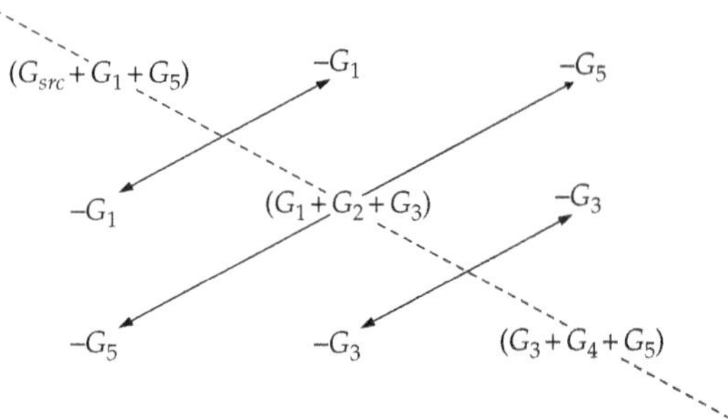

FIGURE 6.2
Symmetry of coefficients in node-voltage equations.

The procedure for writing the node-voltage equation for a given essential node 'n' can be summarized as follows:

1. *The voltage of node 'n' is multiplied by the sum of all the conductances connected directly to this node, which means that node 'n' constitutes one terminal for each of these conductances, whereas other nodes constitute the other terminals.*
2. *The voltage of every other node is multiplied by the conductance connected directly between this node and node 'n', with a negative sign. A conductance is directly connected between two nodes, if each node is a terminal for the conductance. If there is no such conductance, the coefficient is zero.*
3. *The LHS of the node-voltage equation for node 'n' is the sum of the terms from the preceding steps, ordered as the unknown node voltages. This sum is the total current leaving node 'n' through the conductances connected to this node.*
4. *The RHS of the equation is equal to the algebraic sum of source currents entering node 'n'. Thus, a source current entering node 'n' will have a positive sign, whereas a source current leaving node 'n' will have a negative sign.*

It must be remembered that in the node-voltage equations, KCL at any essential node, other than the reference node, is satisfied by having the LHS and the RHS in the following form:

Sum of currents leaving a node through conductances connecting this node to other nodes = Algebraic sum of currents entering the node from sources connected to the node

Note that a conductance connected to a node and is in series with an ideal, dependent or independent current source is redundant as far as the rest of the circuit is concerned. It does not affect the current entering or leaving the node, so it does not appear in the node-voltage equations (see Problems 5.28 and 5.29).

Example 6.1: Node-Voltage Method

It is required to analyze the circuit of Figure 6.3 using the node-voltage method.

Solution:

Taking node 'd' as a reference node and following the aforementioned standard procedure, the node-voltage equations are

$$\text{Node 'a'}: \quad (0.5+1/3+0.1)V_a-(1/3)V_b-0.1V_c=6 \quad (6.5)$$

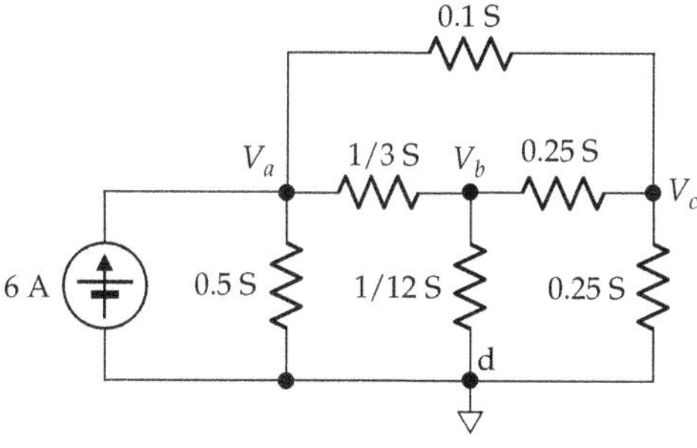

FIGURE 6.3
Figure for Example 6.1.

$$\text{Node 'b'}: \quad -(1/3)V_a+(1/3+0.25+1/12)V_b-0.25V_c=0 \quad (6.6)$$

$$\text{Node 'c'}: \quad -0.1V_a-0.25V_b+(0.25+0.25+0.1)V_c=0 \quad (6.7)$$

Equations 6.5 through 6.7 can be simplified to

$$(14/15)V_a-(1/3)V_b-0.1V_c=6 \quad (6.8)$$

$$-(1/3)V_a+(2/3)V_b-0.25V_c=0 \quad (6.9)$$

$$-0.1V_a-0.25V_b+(3/5)V_c=0 \quad (6.10)$$

Equations 6.8 through 6.10 can be solved by any of the usual methods for solving linear simultaneous equations or by using appropriate calculators to give: V_a = 9 V, V_b = 6 V, and V_c = 4 V. The determinant method for solving simultaneous equations is explained in Appendix E.

Linear simultaneous equations can be conveniently solved using MATLAB. To do so, the conductance coefficients on the LHS of the node-voltage equations are entered as a square matrix, and the source currents on the RHS of the node-voltage equations are entered as a column matrix. In the example under consideration, the matrix of coefficients is entered in MATLAB as follows:

$$C=[14/15,-1/3,-0.1;-1/3,2/3,-0.25;-0.1,-0.25,3/5].$$

In MATLAB, a matrix is entered between square brackets. Elements in a row are separated by commas, whereas rows are separated by semicolons. The matrix of source currents is entered as

$$S=[6;0;0].$$

The command C\S is equivalent to [inv(C)]*S and gives a column matrix of the node voltages in the order V_a, V_b, and V_c. MATLAB returns the solution to the equations as

9.000
6.000
4.000

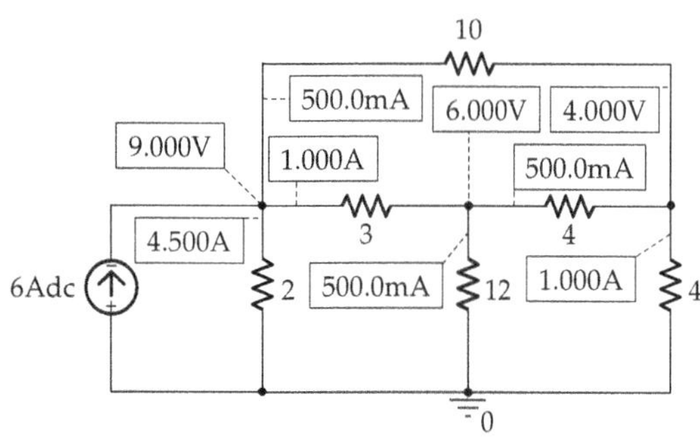

FIGURE 6.4
Figure for Example 6.1.

Simulation: The circuit is entered as in Figure 6.4. After selecting 'Bias Point' under 'Analysis type' in the Simulation Settings and running the simulation, pressing the I and V buttons displays the currents and voltages indicated in Figure 6.4.

Example 6.2: Node-Voltage Method with More than One Current Source

It is required to analyze the circuit of Figure 6.5 using the node-voltage method. The circuit is of the same form as that in Figure 6.3, but with different circuit parameters and with the uppermost conductance replaced by a current source.

Solution:

Recall that in writing the node-voltage equation for a given node as a KCL equation, the LHS of the equation represents current leaving the node through conductances, and the RHS represents source current entering the node. Hence, in writing the node-voltage equation for node 'a', a source current leaving the node must be entered on the RHS with a negative sign. Alternatively, it may be considered that the net source current entering node 'a' is (12 – 6) A. The equation for node 'a' is

$$\text{Node 'a'}: \ (0.5+0.25)V_a - 0.25V_b - 0\times V_c = 12-6 \quad (6.11)$$

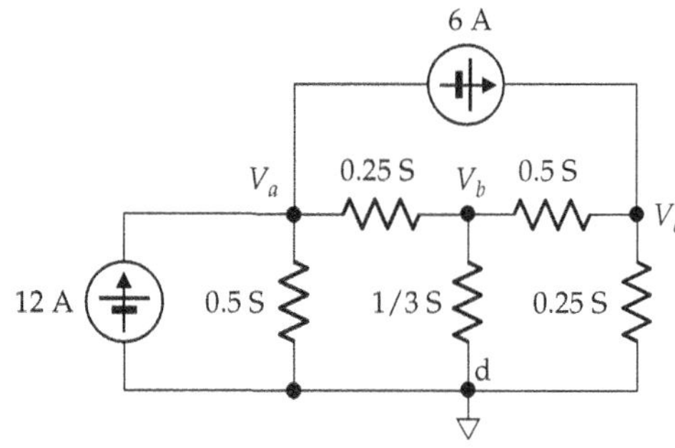

FIGURE 6.5
Figure for Example 6.2.

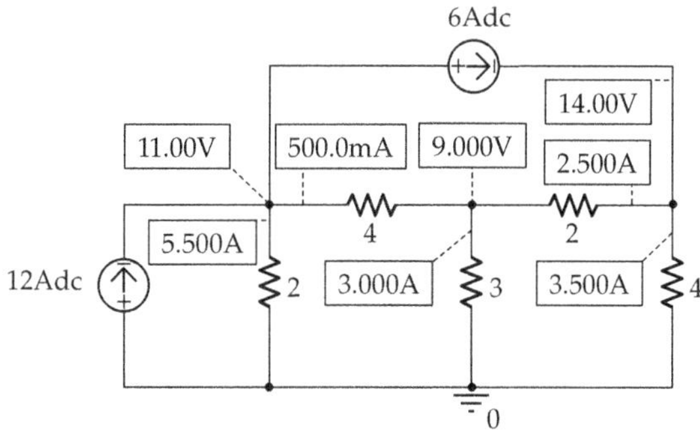

FIGURE 6.6
Figure for Example 6.2.

Note that since there is no conductance that directly connects node 'a' to node 'c', the mutual conductance between these two nodes is zero, which means that V_c no longer appears in the node-voltage equation for node 'a' nor does V_a appear in the node-voltage equation for node 'c'. The remaining node-voltage equations are

$$\text{Node 'b'}: \ -0.25V_a + (1/3+0.25+0.5)V_b - 0.5V_c = 0 \quad (6.12)$$

$$\text{Node 'c'}: \ 0\times V_a - 0.5V_b + (0.25+0.5)V_c = 6 \quad (6.13)$$

The solution to these equations gives V_a = 11 V, V_b = 9 V, and V_c = 14 V.

Simulation: The circuit is entered as in Figure 6.6. After selecting 'Bias Point' under 'Analysis type' in the Simulation Settings and running the simulation, pressing the I and V buttons displays the currents and voltages indicated in Figure 6.6.

Exercise 6.1

Verify that KVL is satisfied around the other meshes in Figure 6.1a.

Primal Exercise 6.2

Verify that KCL and Ohm's law are satisfied in Figure 6.6.

Primal Exercise 6.3

Reverse the direction of the 6 A source in the circuit of Figure 6.5, derive the node-voltage equations, and determine V_a, V_b, and V_c. Simulate the circuit and verify the values of the node voltages, KCL, and Ohm's law.

Ans. V_a = 25 V, V_b = 3 V, and V_c = –6 V.

6.1.1 Change of Reference Node

Consider Figure 6.1a. Any branch voltage is the difference between two node voltages. For example, $V_{ab} = V_a - V_b$. If the same quantity is added to both V_a and V_b, it will cancel out from the RHS, leaving V_{ab} the same. In Example 6.1, node 'd' is taken as a reference and the node voltages are $V_a = 9$ V, $V_b = 6$ V, and $V_c = 4$ V, with $V_d = 0$ because it is the reference node. Suppose that after finding the node voltages with respect to node 'd' as reference, we wish to determine the node voltages with respect to another node, say, node 'b', as reference. This means that V_b must be zero. To make $V_b = 0$ without changing the branch voltages, we simply subtract 6 V from all the node voltages, which gives $V_a = 3$ V, $V_b = 0$ V, $V_c = -2$ V, and $V_d = -6$ V. The branch voltages remain the same. Thus, with node 'd' as reference, $V_{ab} = 9 - 6 = 3$ V, and with node 'b' as reference, $V_{ab} = 3 - 0 = 3$ V. If the branch voltages remain the same, then the branch currents and KCL will also remain the same.

Concept: *In any circuit, the branch voltages and currents are independent of the choice of reference node.*

Primal Exercise 6.4

Redo Example 6.1 with node 'b' as reference and verify the new values of node voltages.

6.1.2 Nontransformable Voltage Source

When the node-voltage method is to be used in a circuit that has an ideal voltage source in series with a resistance, the combination is conveniently transformed to a current source in parallel with the same resistance. But when the ideal voltage source does not have a resistance in series with it, it cannot be transformed to a current source and must be left unaltered. The circuit of Figure 6.7, for example, is the same as that of Figure 6.5, but with the 6 A source replaced by a 3 V source that cannot be transformed to a current source. In applying the node-voltage method, an unknown current I_X is assigned an arbitrary direction through the voltage source and the standard procedure followed, treating I_X like a source current, in accordance with the substitution theorem. The node-voltage equations become

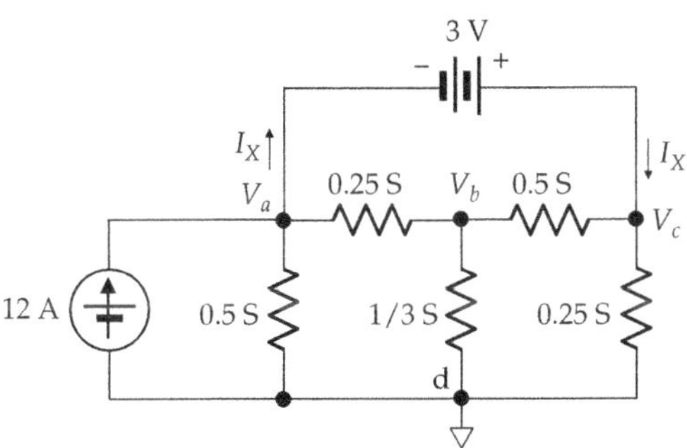

FIGURE 6.7
Nontransformable voltage source.

$$\text{Node 'a'}: \ (0.5+0.25)V_a - 0.25V_b - 0 \times V_c = 12 - I_X \quad (6.14)$$

$$\text{Node 'b'}: \ -0.25V_a + (1/3+0.25+0.5)V_b - 0.5V_c = 0 \quad (6.15)$$

$$\text{Node 'c'}: \ 0 \times V_a - 0.5V_b + (0.25+0.5)V_c = I_X \quad (6.16)$$

I_X can be eliminated by adding together Equations 6.14 and 6.16 for the two nodes between, which the voltage source is connected. The resulting equation is

$$0.75V_a - 0.75V_b + 0.75V_c = 12 \quad (6.17)$$

Equation 6.17 is sometimes referred to as the equation of a "supernode" that results from combining nodes 'a' and 'c'. The node-voltage equation for the supernode can be written following the usual procedure, without having to introduce an unknown source current. However, introducing such a current is more fundamental, transparent, and less likely to cause an error.

Adding two node-voltage equations to eliminate the unknown source current reduces the number of independent voltage equations derived by one. But an additional equation in the node voltages is provided by the relation between the node voltages and the source voltage. In Figure 6.7,

$$V_c - V_a = 3 \quad (6.18)$$

Equations 6.15, 6.17, and 6.18 are the three independent equations that can be solved to give $V_a = 11$ V, $V_b = 9$ V, and $V_c = 14$ V. These values are the same as in Example 6.2, because in this example $V_c - V_a = 3$, as for the voltage source between nodes 'a' and 'c' in Figure 6.7.

Exercise 6.5

Simulate the circuit of Figure 6.7 and verify that KCL, KVL, and Ohm's law are satisfied.

6.1.3 Dependent Sources in Node-Voltage Method

Dependent current sources are treated in exactly the same manner as independent current sources. Consider, for example, the circuit of Figure 6.8, which is the same as that of Figure 6.5 but with the 6 A independent

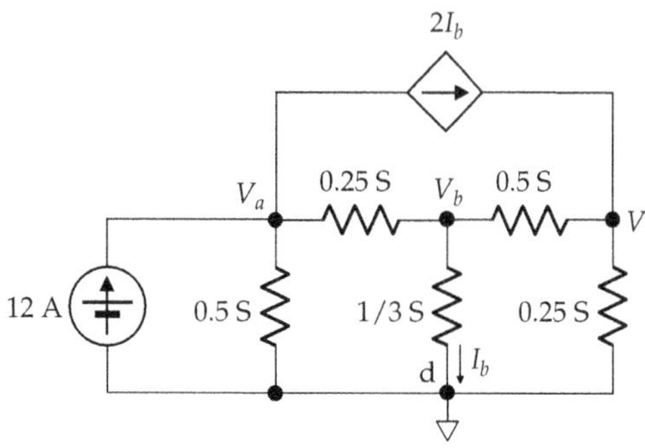

FIGURE 6.8
Dependent source in node-voltage equations.

source replaced by a dependent current source. The node-voltage equations are written in the usual manner as follows:

$$\text{Node 'a'}: \ (0.5+0.25)V_a - 0.25V_b - 0 \times V_c = 12 - 2I_b \tag{6.19}$$

$$\text{Node 'b'}: \ -0.25V_a + (1/3+0.25+0.5)V_b - 0.5V_c = 0 \tag{6.20}$$

$$\text{Node 'c'}: \ 0 \times V_a - 0.5V_b + (0.25+0.5)V_c = 2I_b \tag{6.21}$$

Note that in these equations, the net current entering node 'a' is $12 - 2I_b$ and the current entering node 'c' is $2I_b$. Leaving the 0 coefficient in the equations maintains the symmetry in the array of coefficients.

In order to solve the node-voltage equations, the controlling variable, I_b in this case, should be expressed in terms of the node voltages. In the circuit of Figure 6.8, $I_b = V_b/3$ A. Substituting and moving the term in V_b to the LHS, Equations 6.19 through 6.21 become

$$\text{Node 'a'}: \ (0.5+0.25)V_a - (0.25-2/3)V_b - 0 \times V_c = 12 \tag{6.22}$$

$$\text{Node 'b'}: \ -0.25V_a + (1/3+0.25+0.5)V_b - 0.5V_c = 0 \tag{6.23}$$

$$\text{Node 'c'}: \ 0 \times V_a - (0.5+2/3)V_b + (0.25+0.5)V_c = 0 \tag{6.24}$$

Solving these equations gives $V_a = 11$ V, $V_b = 9$ V, and $V_c = 14$ V, the same as in Example 6.2, because in this example $I_b = 3$ A, and $2I_b = 6$ A, the same as the independent source current. Note that the array of coefficients is symmetrical with respect to the diagonal in Equations 6.19 through 6.21, when $2I_b$ is on the RHS, but the symmetry is destroyed in Equations 6.22 and 6.23 when $2I_b$ is substituted for in terms of V_b and moved to the LHS.

In the case of dependent voltage sources, if the source is in series with a resistance, the dependent voltage source is transformed to a dependent current source in parallel with this resistance, and the standard procedure followed. If the dependent voltage source cannot be transformed to a current source, an unknown current is assigned to the voltage source, and the procedure explained in connection with Figure 6.7 is followed. Several examples of this type are included in problems at the end of this chapter.

Based on the preceding discussion, we can note that the effect of the dependent source in Figure 6.8 is to modify some of the conductance coefficients in the node-voltage equations, leaving only the values of independent source on the RHS of the equations. This is in accordance with the argument in Section 4.1, that dependent sources alone do not excite a circuit and with the argument in connection with Equation 5.10, that dependent sources modify the values of resistance coefficients in circuit equations.

Primal Exercise 6.6

Simulate the circuit of Figure 6.8 and verify that KCL, KVL, and Ohm's law are satisfied.

6.2 Mesh-Current Method

Concept: *In the mesh-current method, the unknown mesh currents are assigned in such a manner that KCL is automatically satisfied at every essential node. Equations based on KVL are then written for each mesh directly in terms of Ohm's law.*

The mesh-current method will be illustrated by the circuit of Figure 6.9. The first step is to assign a current to each mesh. Conventionally, mesh currents are assigned in a clockwise direction, as illustrated by I_1, I_2, and I_3 in the figure. Since the same mesh current enters and leaves

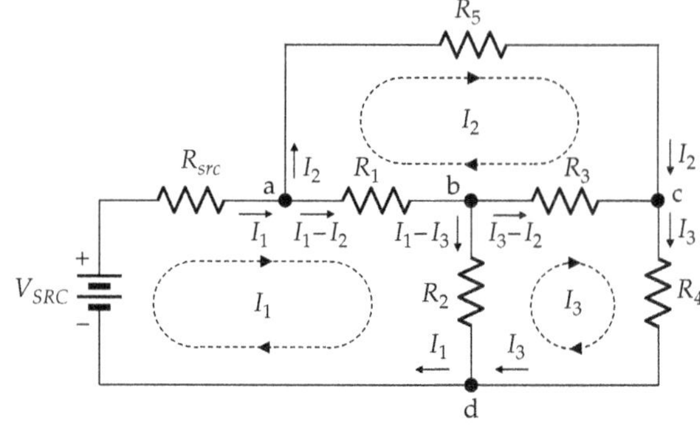

FIGURE 6.9
KCL satisfied by assignment of mesh currents.

any given node, KCL is automatically satisfied at each of the essential nodes 'a', 'b', 'c', and 'd'. To verify this, consider node 'a'. The mesh current I_1 enters node 'a' through R_{src} and leaves this node through R_1. The mesh current I_2 enters node 'a' though R_1 and leaves this node through R_5. The current leaving node 'a' through R_1 is therefore $(I_1 - I_2)$. Equating the current I_1 entering node 'a' through R_{src} to the total current leaving node 'a' through R_1 and R_5 gives $I_1 = I_1 - I_2 + I_2 = I_1$, which satisfies KCL. The same is true at every other node (Exercise 6.7).

The next step is to write KVL around each mesh. Figure 6.10 indicates the voltage drop across each resistor due to the mesh currents flowing through the resistor. Considering R_1, for example, the net current through R_1 is $(I_1 - I_2)$ in the direction of I_1, as indicated in Figure 6.9. The voltage drop across R_1 due to this current is $R_1(I_1 - I_2)$ in the direction of I_1. This voltage drop is $(R_1I_1 - R_1I_2)$, where the first term R_1I_1 is a voltage drop in R_1, in the direction of I_1, due to this current alone, and R_1I_2 is a voltage drop in the direction of I_2 but is a *voltage rise* in the direction of I_1 due to I_2 flowing in R_1. The sign of R_1I_2 is therefore negative in the expression for the net voltage drop $R_1(I_1 - I_2)$ in the direction of I_1. Similarly, the net voltage drop across R_2 is $R_2(I_1 - I_3)$ in the direction of I_1. The total voltage drop in the direction of I_1 due to the resistances in the mesh is therefore $R_{src}I_1 + R_1(I_1 - I_2) + R_2(I_1 - I_3)$. According to KVL, the total voltage drop due to the resistances in the mesh must be equal to the voltage rise due to any sources in the mesh. This voltage rise in mesh 1 is V_{SRC} in the direction of I_1. Equating the voltage drop to the voltage rise and collecting terms in the mesh currents give

$$\left(R_{src} + R_1 + R_2\right)I_1 - R_1I_2 - R_2I_3 = V_{SRC} \tag{6.25}$$

The term $(R_{src} + R_1 + R_2)I_1$ is the voltage drop in mesh 1 due to I_1 alone. As explained previously, the negative sign of the terms R_1I_2 and R_2I_3 is due to the fact that these terms represent voltage rises in mesh 1 in the direction of I_1 and would therefore have a negative sign as voltage drops in mesh 1 in the direction of I_1.

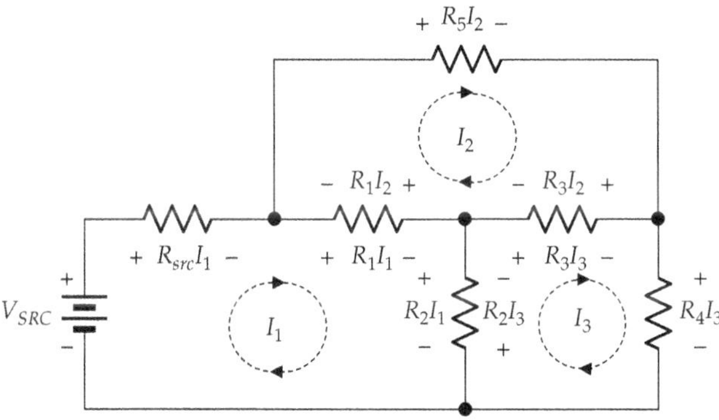

FIGURE 6.10
Mesh-current method.

In mesh 2, the net current in R_1 in the direction of I_2 is $(I_2 - I_1)$ and the voltage drop in R_1 in the direction of I_2 is $R_1(I_2 - I_1)$. Similarly, the net current in R_3 in the direction of I_2 is $(I_2 - I_3)$ and the voltage drop in R_3 in the direction of I_2 is $R_3(I_2 - I_3)$. The total voltage drop in the direction of I_2 due to the resistances in mesh 2 is $R_1(I_2 - I_1) + R_5I_2 + R_3(I_2 - I_3)$. As there are no sources in mesh 2, this total voltage drop must be equal to zero. Collecting terms in the mesh currents gives the mesh-current equation for mesh 2 as

$$-R_1I_1 + \left(R_1 + R_3 + R_5\right)I_2 - R_3I_3 = 0 \tag{6.26}$$

In mesh 3, the net current in R_2 in the direction of I_3 is $(I_3 - I_1)$ and the voltage drop in R_2 in the direction of I_3 is $R_2(I_3 - I_1)$. Similarly, the net current in R_3 in the direction of I_3 is $(I_3 - I_2)$ and the voltage drop in R_3 in the direction of I_3 is $R_3(I_3 - I_2)$. The total voltage drop in the direction of I_3 due to the resistances in mesh 3 is $R_2(I_3 - I_1) + R_4I_3 + R_3(I_3 - I_2)$. As there are no sources in mesh 3, this total voltage drop must be equal to zero. Collecting terms in the mesh currents gives the mesh-current equation for mesh 3 as

$$-R_2I_1 - R_3I_2 + \left(R_2 + R_3 + R_4\right)I_3 = 0 \tag{6.27}$$

Comparing Equations 6.25 through 6.27 reveals a pattern that allows writing the mesh-current equations by inspection:

1. In the equation for a given mesh, the coefficient multiplying the mesh current is the sum of all the resistances in the mesh. Thus, in the equation for mesh 1 (Equation 6.25), I_1 is multiplied by $(R_{src} + R_1 + R_2)$. Similarly, in the equation for mesh 2 (Equation 6.26), I_2 is multiplied by $(R_1 + R_3 + R_5)$, and in the equation for mesh 3 (Equation 6.27), I_3 is multiplied by $(R_2 + R_3 + R_4)$.These coefficients are known as the **self-resistances** of the meshes.
2. In the equation for a given mesh, the coefficient multiplying the current of each of the other meshes is the resistance that is common to the two meshes, with a minus sign. Thus, in the equation for mesh 1 (Equation 6.25), I_2 is multiplied by $-R_1$, where R_1 is the resistance that is common between meshes 1 and 2, and the minus sign is because R_1I_2 is a voltage rise in mesh 1 (Figure 6.10). I_3 is multiplied by $-R_2$, where R_2 is the resistance that is common between meshes 1 and 3. The same is true of the other mesh equations. These coefficients are known as the **mutual resistances** between the meshes. If there is no resistance that is common between a certain mesh with the mesh in question, the corresponding mutual resistance is zero.

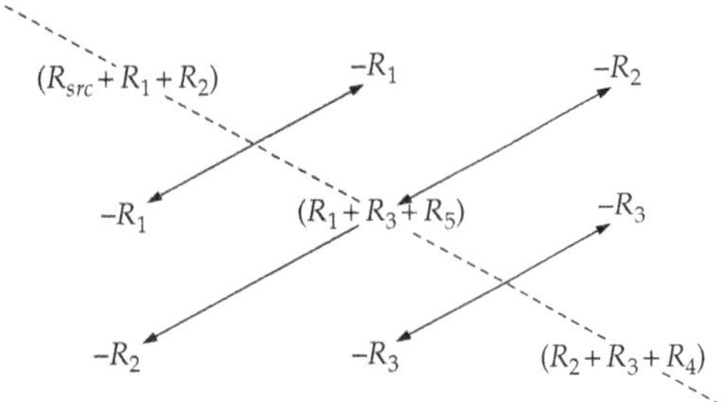

FIGURE 6.11
Symmetry of coefficients in mesh-current equations.

If the coefficients of the mesh currents in Equations 6.25 through 6.27 are arranged in an array, as illustrated in Figure 6.11, the array has the following features that are a useful check on the correctness of the mesh-current equations:

1. The self-resistances are the diagonal entries in the array from top left to bottom right.
2. The array is symmetrical with respect to this diagonal, as illustrated by the coefficients pointed to by the double-sided arrows in Figure 6.11. This symmetry is because the same resistance that is common between two meshes appears in the equation of each of the two meshes as a multiplier of the current in the other mesh, the resistance being independent of the direction of current. It will be shown in the next section that this symmetry is destroyed when the dependency relations of dependent sources are taken into account.
3. In any row or column, the mutual resistances are part of the self-resistance in that row or column. The remaining resistance is that exclusive to the mesh.

The procedure for writing the mesh-current equation for a given mesh 'n' can be summarized as follows:

1. *The current of mesh 'n' is multiplied by the sum of all the resistances in the mesh.*
2. *The current of every other mesh is multiplied by the resistance that is common between this mesh and mesh 'n', with a negative sign. If there is no such resistance, the coefficient is zero.*
3. *The LHS of the mesh-current equation for mesh 'n' is the sum of the terms from the preceding steps, ordered as the unknown mesh currents. This sum is the total voltage drop, in the direction of the mesh current, due to all the resistances in the mesh.*
4. *The RHS of the equation is equal to the algebraic sum of source voltages in mesh 'n'. A source voltage that is a voltage rise in the direction of the mesh current has a positive sign, whereas a source voltage that is a voltage drop in the direction of the mesh current has a negative sign.*

It must be remembered that in the mesh-current equations, KVL around any mesh is satisfied by having the LHS and the RHS in the following form:

Sum of voltage drops in the direction of the mesh current due to the resistances in the mesh = Algebraic sum of voltage rises in the direction of the mesh current due to voltage sources in the mesh.

Example 6.3: Mesh-Current Method

It is required to analyze the circuit of Figure 6.9 using the mesh-current method.

Solution:

The circuit is redrawn in Figure 6.12 with values of the circuit parameters indicated. Following the standard procedure, the mesh-current equations are written as follows:

$$\text{Mesh 1:}\quad (2+3+12)I_1 - 3I_2 - 12I_3 = 12 \qquad (6.28)$$

$$\text{Mesh 2:}\quad -3I_1 + (3+4+10)I_2 - 4I_3 = 0 \qquad (6.29)$$

$$\text{Mesh 3:}\quad -12I_1 - 4I_2 + (4+4+12)I_3 = 0 \qquad (6.30)$$

These equations reduce to

$$17I_1 - 3I_2 - 12I_3 = 12 \qquad (6.31)$$

$$-3I_1 + 17I_2 - 4I_3 = 0 \qquad (6.32)$$

$$-12I_1 - 4I_2 + 20I_3 = 0 \qquad (6.33)$$

The solution to these equations gives $I_1 = 1.5$ A, $I_2 = 0.5$ A, and $I_3 = 1$ A.

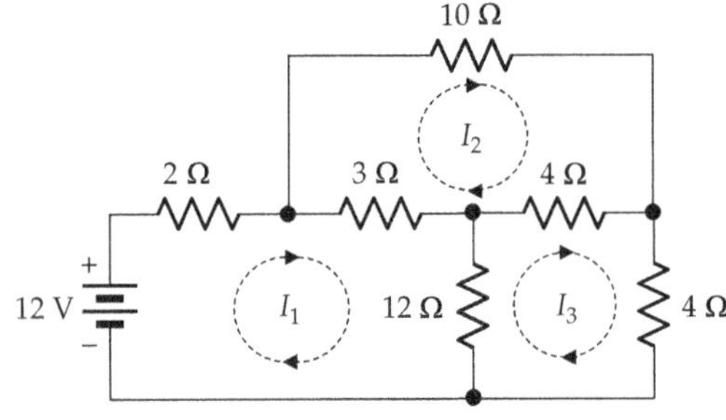

FIGURE 6.12
Figure for Example 6.3.

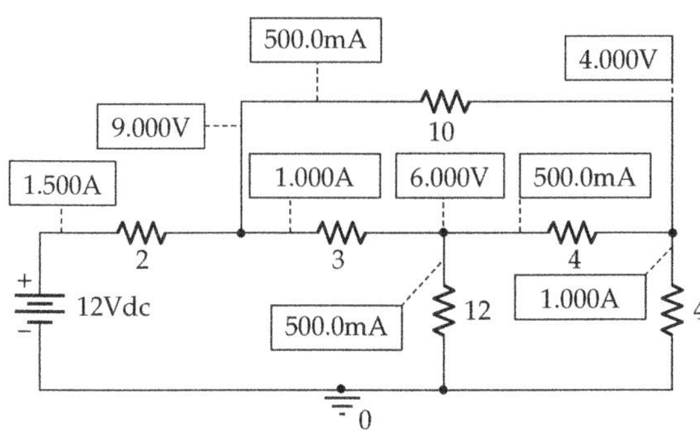

FIGURE 6.13
Figure for Example 6.3.

Simulation: The circuit is entered as in Figure 6.13. After selecting 'Bias Point' under 'Analysis type' in the Simulation Settings and running the simulation, pressing the I and V buttons displays the currents and voltages indicated in Figure 6.13.

Example 6.4: Mesh-Current Method with More than One Voltage Source

It is required to analyze the circuit of Figure 6.14 using the mesh-current method. The circuit is of the same form as that in Figure 6.12, but with different circuit parameters and with the 3 Ω resistance replaced by an 8 V source.

Solution:

Following the standard procedure and starting with mesh 1, the self-resistance of this mesh is (2 + 8) Ω; the mutual resistance with mesh 2 is zero, since there is no common resistance with this mesh; and the mutual resistance with mesh 3 is 8 Ω. The LHS of the mesh-current equation, which accounts for the total voltage drop in the direction of I_1 due to the resistances in mesh 1 is $(2+8)I_1 + 0 \times I_2 - 8I_3$. The source voltage of the 12 V source is a voltage rise in the direction of I_1, and the source voltage of the 8 V source is a voltage drop in the direction of I_1. The net voltage rise in the direction of I_1 due to the

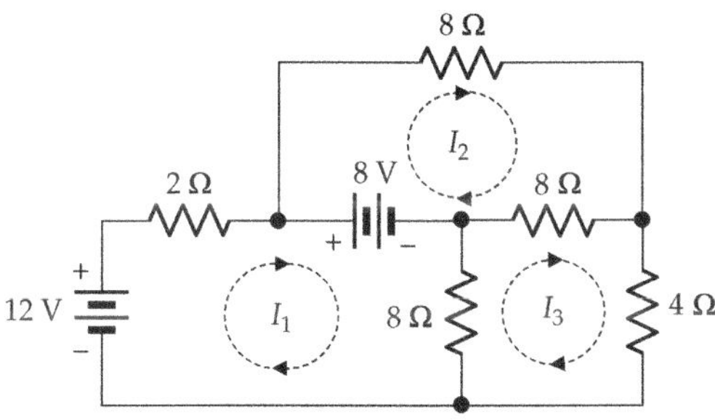

FIGURE 6.14
Figure for Example 6.4.

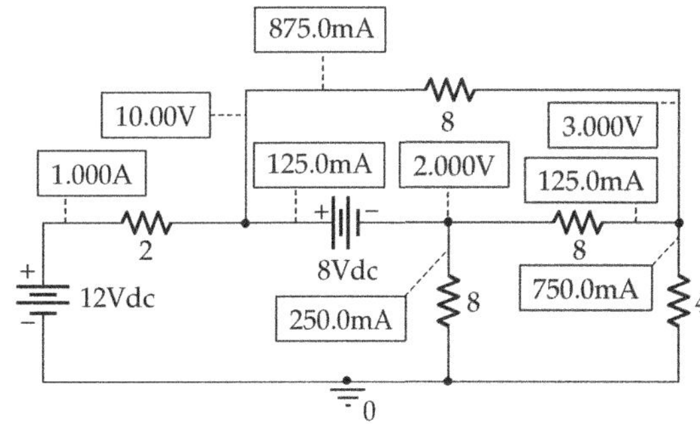

FIGURE 6.15
Figure for Example 6.4.

voltage sources in mesh 1 is (12 – 8) V. The mesh-current equation for mesh 1 is therefore

$$\text{Mesh 1:}\quad (2+8)I_1 - 0\times I_2 - 8I_3 = 12-8 \qquad (6.34)$$

The mesh-current equations for the other two meshes, in accordance with the standard procedure are

$$\text{Mesh 2:}\quad 0\times I_1 + (8+8)I_2 - 8I_3 = 8 \qquad (6.35)$$

$$\text{Mesh 3:}\quad -8I_1 - 8I_2 + 20I_3 = 0 \qquad (6.36)$$

These equations reduce to

$$10I_1 - 0\times I_2 - 8I_3 = 4 \qquad (6.37)$$

$$0\times I_1 + 16I_2 - 8I_3 = 8 \qquad (6.38)$$

$$-8I_1 - 8I_2 + 20I_3 = 0 \qquad (6.39)$$

The solution to these equations gives I_1 = 1 A, I_2 = 0.875 A, and I_3 = 0.75 A. Note the symmetry of the coefficients with respect to the diagonal.

Simulation: The circuit is entered as in Figure 6.15. After selecting 'Bias Point' under 'Analysis Type' in the Simulation Settings and running the simulation, pressing the I and V buttons displays the currents and voltages indicated in Figure 6.15. I_1 is the same as the current in the 12 V source, I_2 is the same as the current in the upper 8 Ω resistance, and I_3 is the same as the current in the 4 Ω resistance.

Exercise 6.7

Verify that KCL is satisfied at nodes 'b', 'c', and 'd' in Figure 6.9.

Primal Exercise 6.8

Verify that KVL and Ohm's law are satisfied in Figure 6.15.

Primal Exercise 6.9

Reverse the polarity of the 8 V source in the circuit of Figure 6.14, derive the mesh-current equations, and determine I_1, I_2, and I_3. Simulate the circuit and verify the values of the mesh currents, KVL, and Ohm's law.

Ans. $I_1 = 3$ A, $I_2 = 0.125$ A, and $I_3 = 1.25$ A.

*6.2.1 Generalization of Mesh-Current Method

It is sometimes convenient not to take all mesh currents in the same sense or to consider as a variable a loop current rather than a mesh current. Compared to Figure 6.12, for example, I_1 in Figure 6.16 is the current in the outer loop, I_2 is the current in the same mesh 2, and I_3 is the current in the same mesh 3 but in a counterclockwise sense. How can the equations relating I_1, I_2, and I_3 be written in the same form as the previously described mesh-current equations? Doing so enhances the understanding of how KVL is applied in a more general context.

Considering loop 1, the self-resistance of the loop is $(2 + 10 + 4)$ Ω. The voltage drop in the direction of I_1 due to the self-resistance of the loop is $(2 + 10 + 4)I_1$. The mutual resistance between loop 1 and mesh 2 is the 10 Ω resistance. But I_1 and I_2 flow in the same direction in this resistance so that the voltage across this resistance due to I_2 is a voltage drop $10I_2$ in loop 1 in the direction of I_1, just like the voltage drop $10I_1$ in this resistance due to I_1. The mutual resistance between loop 1 and mesh 3 is the 4 Ω resistance. But with I_1 and I_3 flowing in opposite directions in this resistance, I_3 produces a voltage rise $4I_3$ in loop 1, which *subtracts* from the total voltage drop in the direction of I_1 in loop 1. The net voltage drop in the direction of I_1 in loop 1 is therefore $(2 + 10 + 4)I_1 + 10I_2 - 4I_3 = 16I_1 + 10I_2 - 4I_3$. The voltage rise due to source voltages in loop 1 is 12 V. Hence, the loop-current equation for this loop is

$$16I_1 + 10I_2 - 4I_3 = 12 \tag{6.40}$$

Considering mesh 2, the voltage drop in the direction of I_2 due to the self-resistance of the mesh is $(10 + 4 + 3)I_2$. I_1 flowing in the 10 Ω resistance and I_3 flowing in the 4 Ω resistance common with mesh 3 both add to the voltage drop in mesh 2. The total voltage drop in the direction of I_2 in mesh 2 is therefore $(10 + 4 + 3)I_2 + 10I_1 + 4I_3 = 17I_2 + 10I_1 + 4I_3$. With no sources in mesh 2, the mesh-current equation for mesh 2 is

$$10I_1 + 17I_2 + 4I_3 = 0 \tag{6.41}$$

Considering mesh 3, the voltage drop in the direction of I_3 due to the self-resistance of the mesh is $(12 + 4 + 4)I_3$. I_1 flowing in the common 4 Ω resistance on the right produces a voltage rise in mesh 3, whereas I_2 flowing in the common 4 Ω resistance produces a voltage drop in mesh 3. The net voltage drop in the direction of I_3 in mesh 3 is therefore $(12 + 4 + 4)I_3 - 4I_1 + 4I_2 = 20I_2 - 4I_1 + 4I_2$. With no sources in mesh 3, the mesh-current equation for mesh 3 is

$$-4I_1 + 4I_2 + 20I_3 = 0 \tag{6.42}$$

Equations 6.40 through 6.42 are the three independent equations that can be solved to give $I_1 = 1.5$ A, $I_2 = -1$ A, and $I_3 = 0.5$ A. These values are in agreement with those derived for the same circuit in Example 6.3. The current in the 12 V source is 1.5 A, the current in the 10 Ω resistance is $I_1 + I_2 = 0.5$ A, and the current in the 4 Ω resistance on the side is $I_1 - I_3 = 1$ A.

When using loop currents, with or without mesh currents, the following should be noted:

1. The only modification from the standard procedure is that if the loop or mesh currents flow in the same direction in a mutual resistance, this resistance is written with a positive sign in the mesh-current equations.
2. The number of independent equations is the same as the number of meshes in the circuit.
3. The array of coefficients is symmetrical with respect to the diagonal, in the absence of dependent sources, as in Equations 6.40 through 6.42. The symmetry also applies in the presence of dependent sources, before the dependency relations are taken into account, as discussed in connection with Equations 6.22 and 6.23.

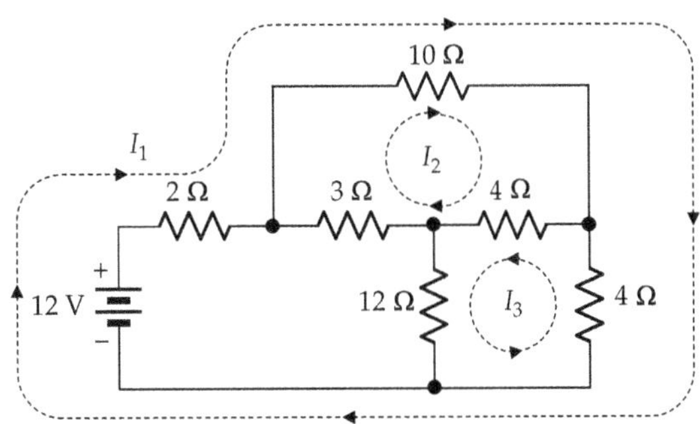

FIGURE 6.16
Generalization of mesh-current method.

Primal Exercise 6.10

Redo Example 6.4 using the same loop and mesh currents as in Figure 6.16.

6.2.2 Nontransformable Current Source

When the mesh-current method is to be used in a circuit that has an ideal current source in parallel with a resistance, the combination is conveniently transformed to a voltage source in series with the same resistance. But when the ideal current source does not have a resistance in parallel with it, it cannot be transformed to a voltage source and must be left unaltered. The circuit of Figure 6.17, for example, is the same as that of Figure 6.14, except that the 8 V source has been replaced by a current source that cannot be transformed to a voltage source. An unknown voltage V_X of arbitrary polarity is assumed across the current source and is treated like a source voltage, in accordance with the substitution theorem. The mesh-current equations become

$$\text{Mesh 1:} \quad (2+8)I_1 - 0 \times I_2 - 8I_3 = 12 - V_X \qquad (6.43)$$

$$\text{Mesh 2:} \quad 0 \times I_1 + (8+8)I_2 - 8I_3 = V_X \qquad (6.44)$$

$$\text{Mesh 3:} \quad -8I_1 - 8I_2 + (8+8+4)I_3 = 0 \qquad (6.45)$$

V_X can be eliminated by adding together Equations 6.43 and 6.44 for the two meshes between which the current source is connected, which gives

$$10I_1 + 16I_2 - 16I_3 = 12 \qquad (6.46)$$

Equation 6.46 is sometimes referred to as the equation of a "supermesh" that results from combining meshes 1 and 2. The mesh-current equation for the supermesh can be written following the usual procedure, without having to introduce an unknown source voltage. However, introducing such a voltage is more fundamental, transparent, and less likely to cause an error.

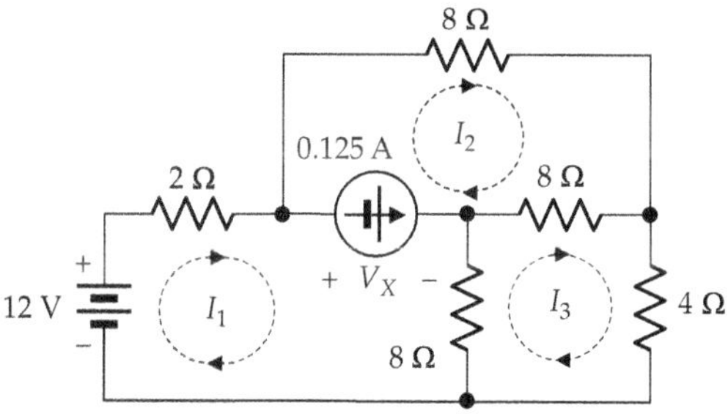

FIGURE 6.17
Nontransformable current source.

Adding two mesh-current equations reduces the number of independent mesh-current equations derived by one. But an additional equation is provided by the relation between the mesh currents and the source current. From Figure 6.17,

$$I_1 - I_2 = 0.125 \qquad (6.47)$$

Equations 6.45 through 6.47 are the three independent equations that can be solved to give $I_1 = 1$ A, $I_2 = 0.875$ A, and $I_3 = 0.75$ A. These values are the same as in Example 6.4, because $I_1 - I_2 = 0.125$ A, as for the current source between meshes 1 and 2.

Primal Exercise 6.11

Simulate the circuit of Figure 6.17 and verify that KCL, KVL, and Ohm's law are satisfied.

6.3 Dependent Sources in Mesh-Current Method

Dependent voltage sources are treated in exactly the same manner as independent voltage sources. Consider, for example, the circuit of Figure 6.18, which is the same as that of Figure 6.13 but with the 8 V independent source replaced by a dependent voltage source. The mesh-current equations are written in the usual way as follows:

$$\text{Mesh 1:} \quad (2+8)I_1 - 0 \times I_2 - 8I_3 = 12 - 32I_b \qquad (6.48)$$

$$\text{Mesh 2:} \quad 0 \times I_1 + (8+8)I_2 - 8I_3 = 32I_b \qquad (6.49)$$

$$\text{Mesh 3:} \quad -8I_1 - 8I_2 + (8+8+4)I_3 = 0 \qquad (6.50)$$

Note that in these equations, the net voltage rise in mesh 1 is $12 - 32I_b$ and the voltage rise in mesh 2 is $32I_b$. Leaving the 0 coefficient in the equations maintains the symmetry in the array of coefficients.

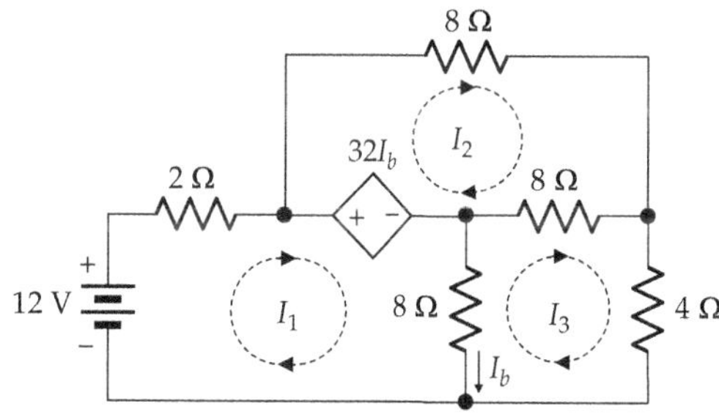

FIGURE 6.18
Dependent source in node-voltage equations.

In order to solve the mesh-current equations, the controlling variable, I_b in this case, should be expressed in terms of the mesh currents. In the circuit of Figure 6.18, $I_b = I_1 - I_3$. Substituting and moving the term in I_b to the LHS, Equations 6.48 through 6.50 become

$$\text{Mesh 1:} \quad (10+32)I_1 - 0\times I_2 - 40I_3 = 12 \qquad (6.51)$$

$$\text{Mesh 2:} \quad -32I_1 + (8+8)I_2 + 24I_3 = 0 \qquad (6.52)$$

$$\text{Mesh 3:} \quad -8I_1 - 8I_2 + (8+8+4)V_c = 0 \qquad (6.53)$$

Solving these equations gives $I_1 = 1$ A, $I_2 = 0.875$ A, and $I_3 = 0.75$ A, the same as in Example 6.4, because in this example $I_b = 0.25$ A and $32I_b = 8$ V, the same as the independent source voltage. Note that the array of coefficients is symmetrical with respect to the diagonal in Equations 6.48 through 6.50, when $32I_b$ is on the RHS, but the symmetry is destroyed in Equations 6.51 through 6.53 when $32I_b$ is substituted for in terms of I_1 and I_3 and moved to the LHS. Again, the effect of the dependent source is to modify some of the resistance coefficients in the mesh-current equations.

In the case of dependent current sources, if the source is in parallel with a resistance, the dependent current source is transformed to a dependent voltage source, and the standard procedure followed. If the dependent current source cannot be transformed to a current source, an unknown voltage is assigned across the current source, and the procedure explained in connection with Figure 6.17 is followed. Several examples of this type are included in problems at the end of this chapter.

Primal Exercise 6.12

Simulate the circuit of Figure 6.18 and verify that KCL, KVL, and Ohm's law are satisfied.

6.4 Problem-Solving Approach Updated

The main procedural steps of the ISDEPIC approach are summarized and updated in the following in light of the material covered in this chapter:

Step 1—Initialize:

(a) Mark on the circuit diagram all given values of circuit parameters, currents and voltages, as well as the unknowns to be determined.
(b) Label the nodes, as this may be generally helpful.
(c) If the solution requires that a given value of current or voltage be satisfied, assume this value from the very beginning.

Step 2—Simplify: Consider, as may be appropriate, the following:

(a) Redrawing the circuit.
(b) Replacing series and parallel combinations of circuit elements by an equivalent circuit element,
(c) Applying star-delta transformation.
(d) Removing resistors that are redundant as far as the variables of interest are concerned.
(e) Using shunt-connected ideal voltage sources or series-connected ideal current sources to partition the circuit.
(f) Exploiting symmetry.

Step 3—Deduce: Determine any values of current or voltage that follow immediately from direct application of Ohm's law, KCL, or KVL, without introducing any additional unknowns.

Step 4—Explore: Examine each of the following alternatives, as may be applicable:

(a) Consider the nodes and meshes in the circuit to see if KCL or KVL can be satisfied by assignment of a single unknown current or voltage, *and* if this unknown can then be directly determined from KCL or KVL.
(b) Apply source transformation.
(c) Use TEC or NEC to (i) determine a voltage or a current through a given circuit element, or (ii) simplify part of the circuit.
(d) Apply the substitution or source absorption theorems.
(e) Apply superposition if more than one source is present in the circuit.
(f) Use output scaling.
(g) Rearrange sources.

Step 5—Plan: Think carefully and creatively about the problem in the light of circuit fundamentals and circuit analysis techniques. Consider alternative solutions and select what seems to be the simplest and most direct solution, including the node-voltage or mesh-current method.

Step 6—Implement: Carry out your planned solution.

Step 7—Check: your calculations and results.

(a) Check that your results make sense, in terms of magnitude and sign.
(b) Check that Ohm's law is satisfied across every resistor, that KCL is satisfied at every node, and that KVL is satisfied around every mesh.

(c) Seek an alternative solution to see if it gives the same result.

(d) Whenever feasible, check simulation with PSpice.

Learning Checklist: What Should Be Learned from This Chapter

- In the node-voltage method, voltages of essential nodes are assigned with respect to one of the essential nodes taken as a reference. This automatically satisfies KVL in every mesh in the circuit. Equations based on KCL are then written directly in terms of Ohm's law for each essential node other than the reference node.
- The procedure for writing the node-voltage equation for a given essential node 'n' can be summarized as follows:
  1. The voltage of node 'n' is multiplied by the sum of all the conductances connected directly to this node, which means that node 'n' constitutes one terminal for each of these conductances.
  2. The voltage of every other node is multiplied by the conductance connected directly between this node and node 'n', with a negative sign. A conductance is directly connected between two nodes, if each node is a terminal for the conductance. If there is no such conductance, the coefficient is zero.
  3. The LHS of the node-voltage equation for node 'n' is the sum of the terms from the preceding steps, ordered as the unknown node voltages. This sum is the total current leaving node 'n' through the conductances connected to this node.
  4. The RHS of the equation is equal to the algebraic sum of source currents entering node 'n'. Thus, a source current entering node 'n' will have a positive sign, whereas a source current leaving node 'n' will have a negative sign.
- In any circuit, the branch voltages and currents are independent of the choice of reference node.
- In the mesh-current method, the unknown mesh currents are assigned in such a manner that KCL is automatically satisfied at every essential node. Equations based on KVL are then written for each mesh directly in terms of Ohm's law.
- The procedure for writing the mesh-current equation for a given mesh 'n' can be summarized as follows:
  1. The current of mesh 'n' is multiplied by the sum of all the resistances in the mesh.
  2. The current of every other mesh is multiplied by the resistance that is common between this mesh and mesh 'n', with a negative sign. If there is no such resistance, the coefficient is zero.
  3. The LHS of the mesh-current equation for mesh 'n' is the sum of the terms from the preceding steps, ordered as the unknown mesh currents. This sum is the total voltage drop, in the direction of the mesh current, due to all the resistances in the mesh.
  4. The RHS of the equation is equal to the algebraic sum of source voltages in mesh 'n'. A source voltage that is a voltage rise in the direction of the mesh current has a positive sign, whereas a source voltage that is a voltage drop in the direction of the mesh current has a negative sign.
- In writing the node-voltage and mesh-current equations, dependent sources are treated in exactly the same way as independent sources.

Problem-Solving Tips

1. A useful check on the node-voltage and mesh-current equations is that the array of coefficients on the left-hand side of the equations should be symmetrical about the diagonal, from top left to bottom right. For the purpose of this check, zero coefficients must be included and dependent sources should appear as sources on the right-hand side of the equations.
2. An additional check is that in any row or column, the mutual conductances (resistances) are part of the self-conductances (resistances) in that row or column.

Problems

Verify solutions by PSpice simulation.

Node-Voltage Method

Use the node-voltage method in Problems P6.1 through P6.30.

P6.1 The node-voltage equation for node 'b' in Figure P6.1 can be expressed as $AV_a + BV_b + CV_c = 2$, where A, B, and C are constants that depend on the conductances only. Determine C.

Ans. −0.25 S.

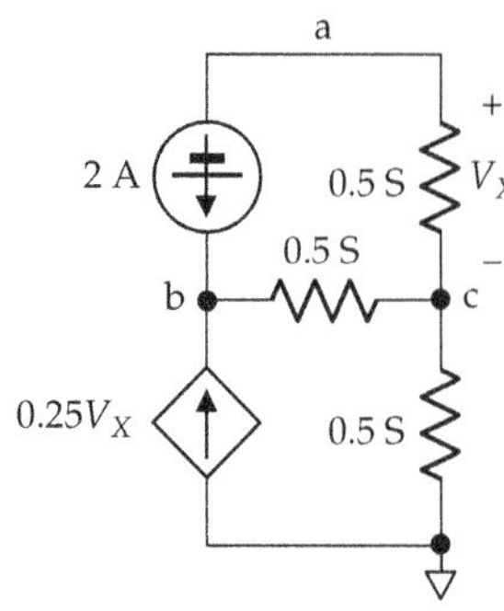

FIGURE P6.1

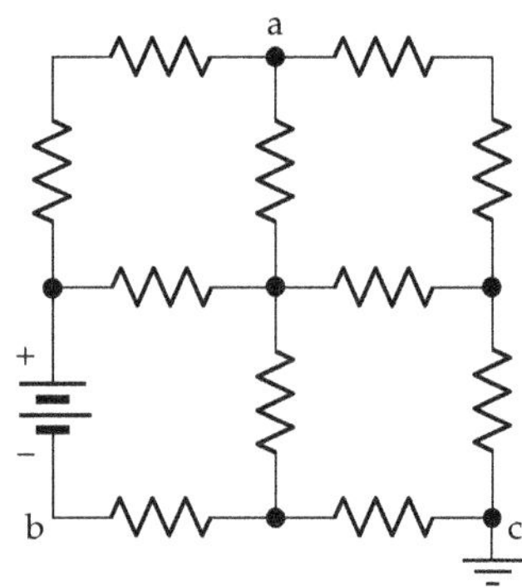

FIGURE P6.2

P6.2 Given that $V_a = 25$ V and $V_b = 12$ V in Figure P6.2, with node 'c' grounded, determine V_a if node 'b' is grounded instead of node 'c'.

Ans. 13 V.

P6.3 (a) Determine V_a in Figure P6.3 by transforming the voltage sources to current sources and writing the node-voltage equation for node 'a'. (b) Write the same node-voltage equation based on KCL, without transforming the sources.

Ans. 10 V.

P6.4 Determine the node voltages V_a and V_b in Figure P6.4.

Ans. $V_a = 0$, $V_b = -5/3$ V.

P6.5 (a) Determine V_O in Figure P6.5 and the voltages of the middle nodes, taking the lower node as reference. (b) Repeat (a) taking node 'n' as reference.

Ans. (a) V_O = 12 V, voltage of both nodes is 6 V; (b) V_O = 12 V, voltage of middle nodes is zero.

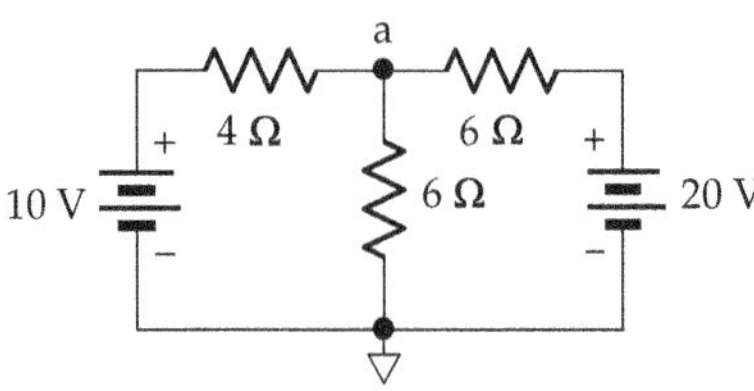

FIGURE P6.3

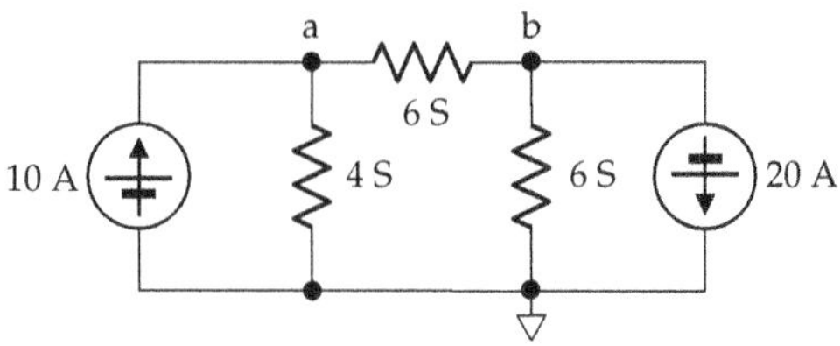

FIGURE P6.4

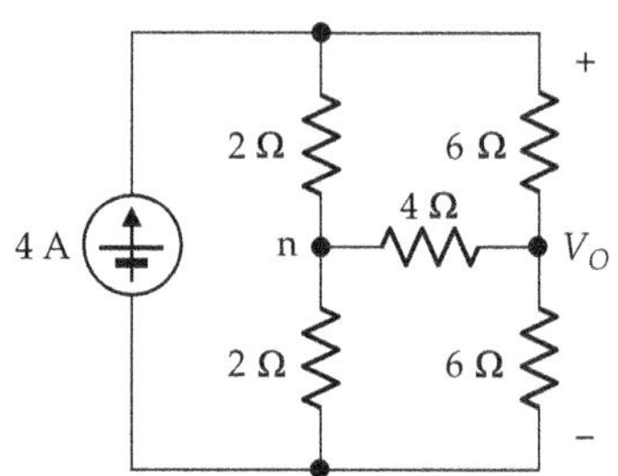

FIGURE P6.5

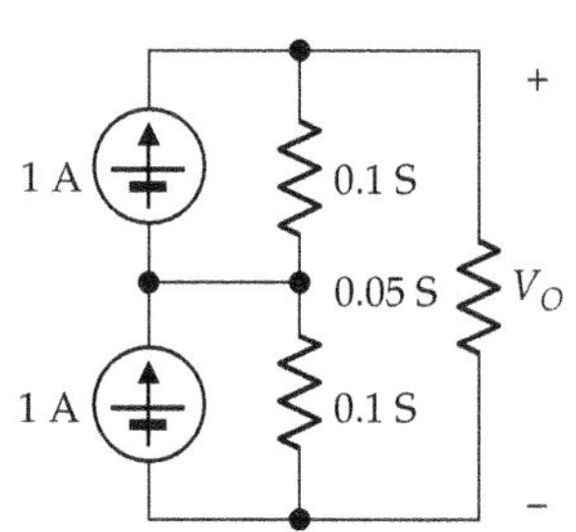

FIGURE P6.6

P6.6 (a) Determine V_O in Figure P6.6 and the voltage of the middle node, taking the lower node as reference. (b) Determine V_O taking the middle node as reference.

Ans. (a) V_O = 10 V, voltage of middle node = 5 V; (b) V_O = 10 V.

P6.7 (a) Determine I_O in Figure P6.7, taking the bottom node as reference. (b) Repeat (a), taking node 'n' as reference.

Ans. (a) and (b) I_O = 20 A.

P6.8 Determine V_L and I_A in Figure P6.8.

Ans. 12.1 V, 0.35 A.

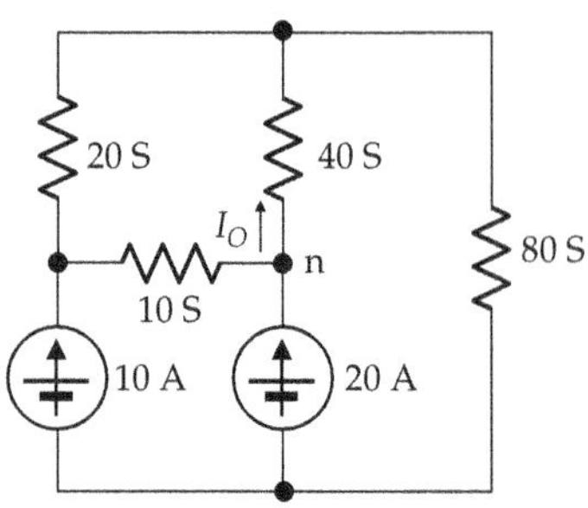

FIGURE P6.7

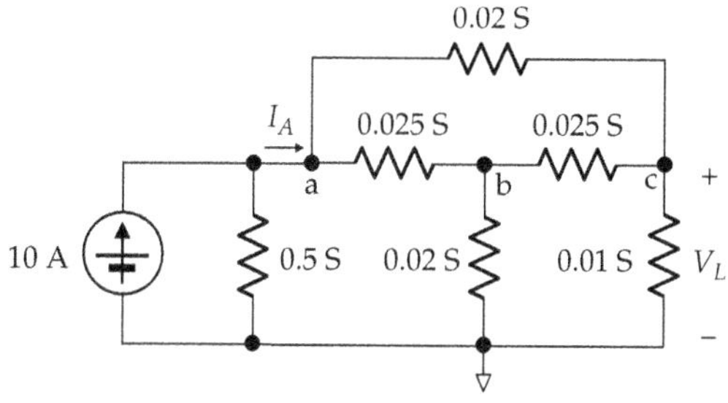

FIGURE P6.8

P6.9 Determine the power delivered or absorbed by the current sources in Figure P6.9.

Ans. 15 A source absorbs 19.5 W, 30 A source delivers 66 W.

P6.10 Determine the node voltages in Figure P6.10.

Ans. $V_a = 36.43$ V, $V_b = 23.57$ V, $V_c = 37.86$ V

P6.11 Determine V_O in Figure P6.11.

Ans. 1.818 V.

P6.12 Determine V_O in Figure P6.12.

Ans. 20 V.

P6.13 Determine the branch voltages in Figure P6.13.

Ans. $V_L = 200/11$ V, 350/11 V rise across the 5 A source, and 150/11 V rise across the dependent source.

P6.14 Determine V_O in Figure P6.14.

Ans. –10/3 V.

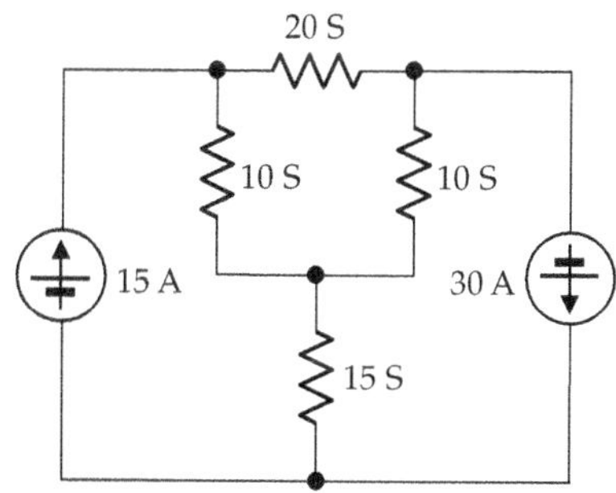

FIGURE P6.9

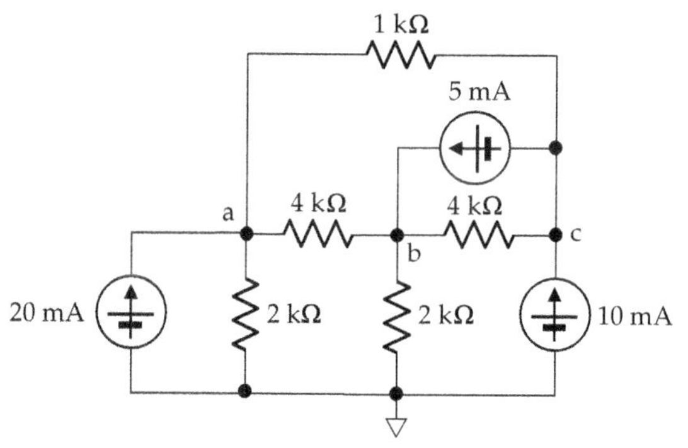

FIGURE P6.10

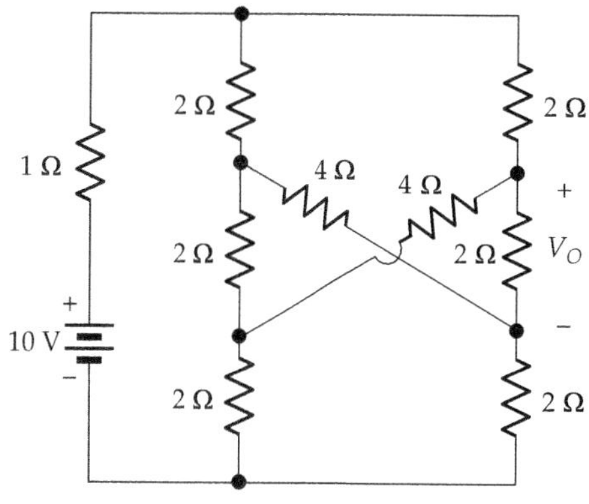

FIGURE P6.11

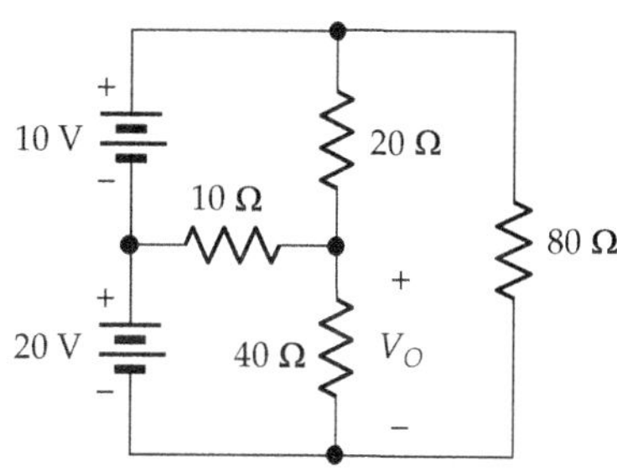

FIGURE P6.12

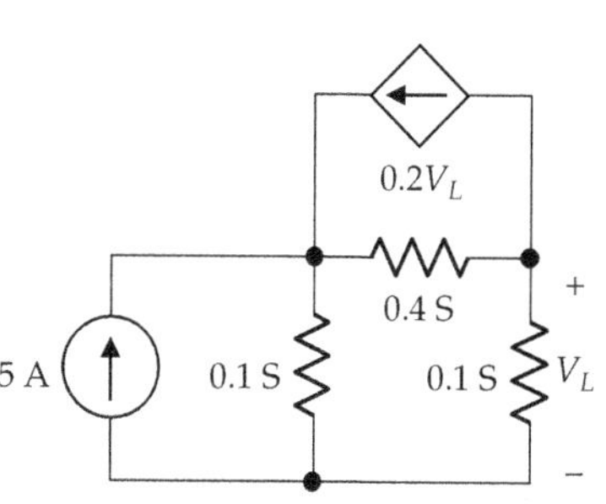

FIGURE P6.13

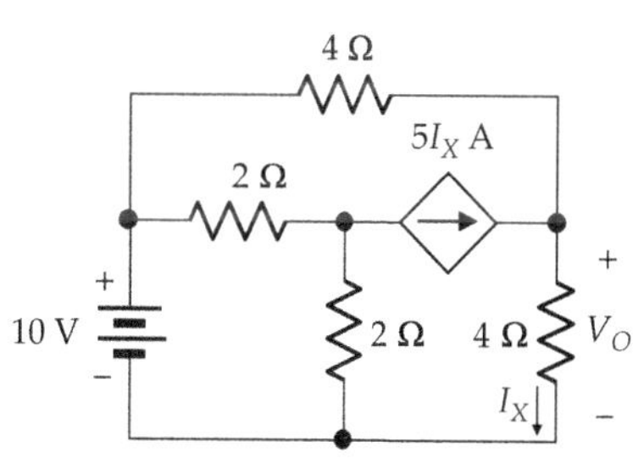

FIGURE P6.14

P6.15 Determine in Figure P6.15, (a) V_{ab} and (b) I_X, assuming all resistances are 1 Ω.

Ans. (a) −0.5 V; (b) −2 A.

P6.16 Determine V_O in Figure P6.16.

Ans. 30 V.

P6.17 Determine V_O in Figure P6.17.

Ans. −2 V.

P6.18 Determine V_O in Figure P6.18.

Ans. 0.

P6.19 Determine I_{SRC1} and I_{SRC2} in Figure P6.19.

Ans. I_{SRC1} = 1 A, I_{SRC2} = 95 A.

P6.20 Determine I_X in Figure P6.20.

Ans. 0.5 A.

P6.21 Determine V_O in Figure P6.21.

Ans. 760/49 = 15.51 V.

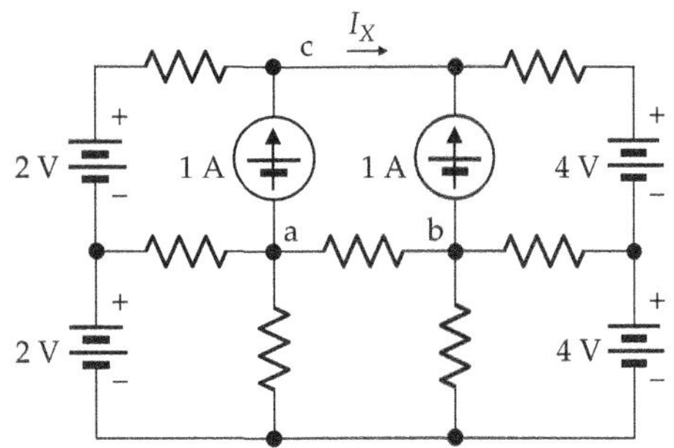

FIGURE P6.15

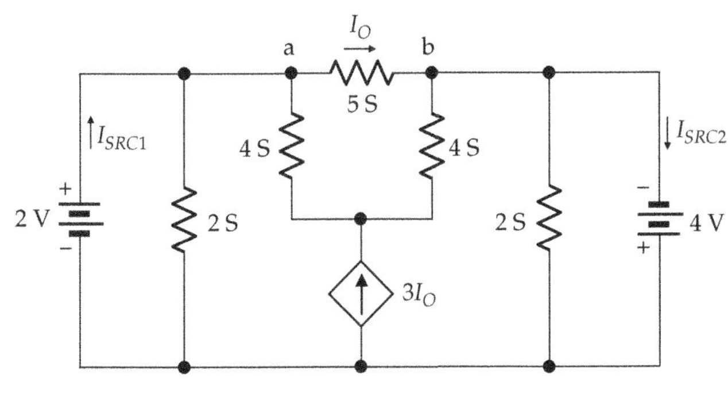

FIGURE P6.18

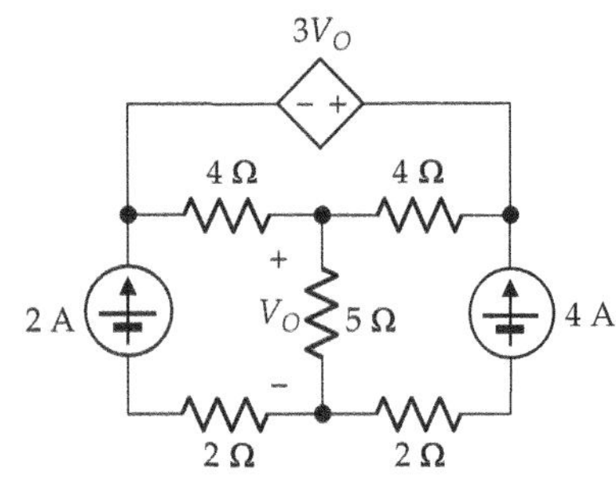

FIGURE P6.16

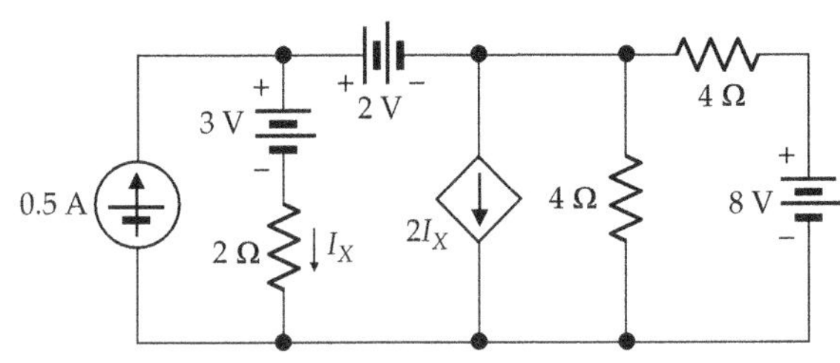

FIGURE P6.19

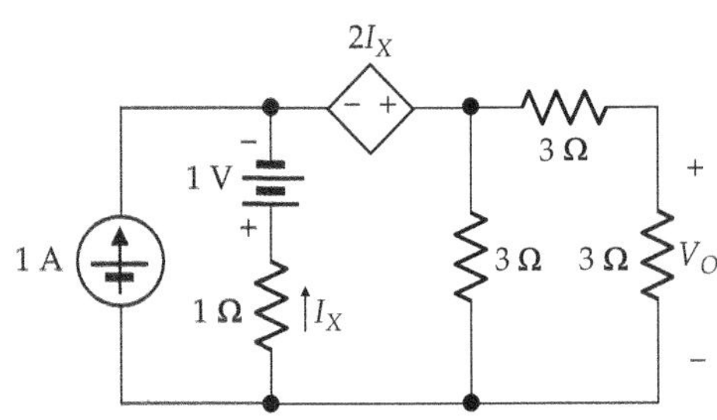

FIGURE P6.17

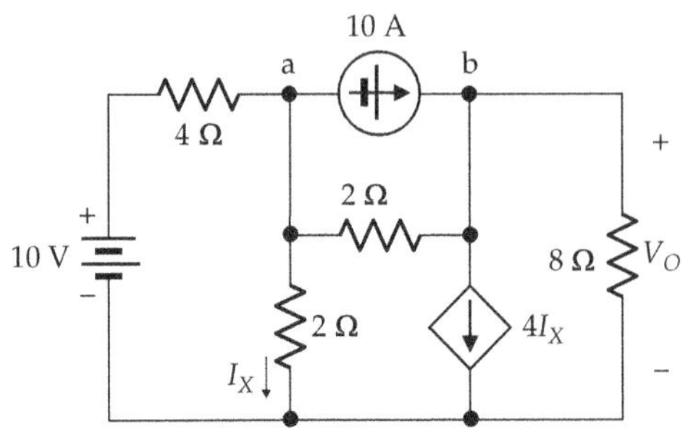

FIGURE P6.21

P6.22 Determine I_O in Figure P6.22.

Ans. 15.51 A.

P6.23 Determine I_O in Figure P6.23.

Ans. −10/3 A.

P6.24 Determine V_O in Figure P6.24.

Ans. 4 V.

P6.25 Determine I_O in Figure P6.25.

Ans. −22 A.

P6.26 Determine V_O in Figure P6.26 taking node 'a' as reference.

Ans. 1400/133 = 10.53 V.

P6.27 Determine the node voltages in Figure P6.27.

Ans. $V_a = 12$ V, $V_b = 34/3$ V; $V_c = 8$ V; $V_d = 20/3$ V.

P6.28 Determine V_O in Figure P6.28 assuming that all resistances are 2 Ω.

Ans. 4.90 V.

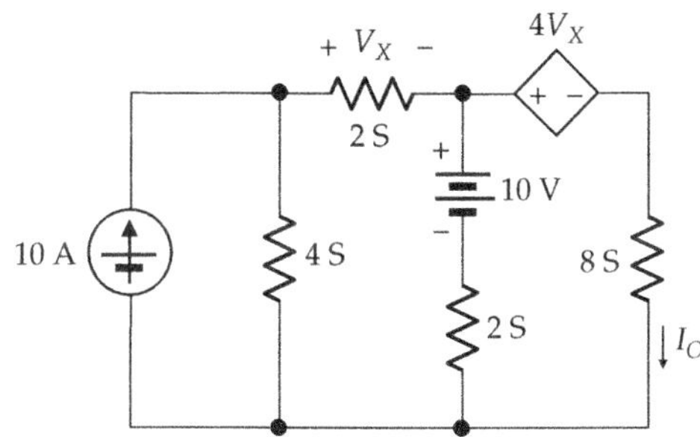

FIGURE P6.22

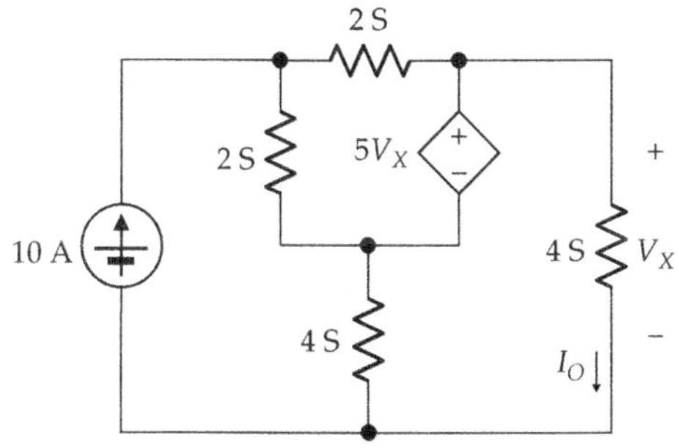

FIGURE P6.23

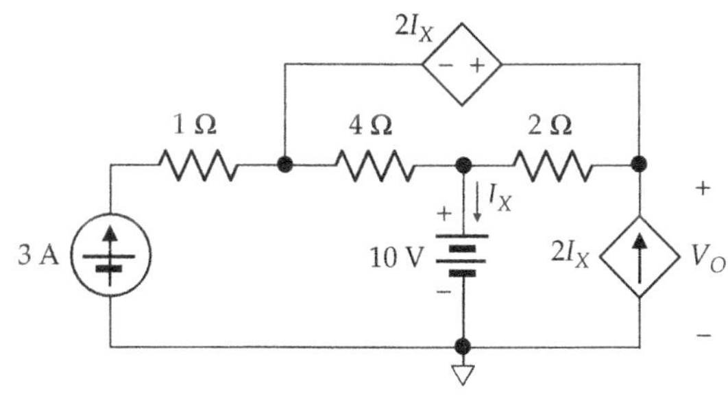

FIGURE P6.24

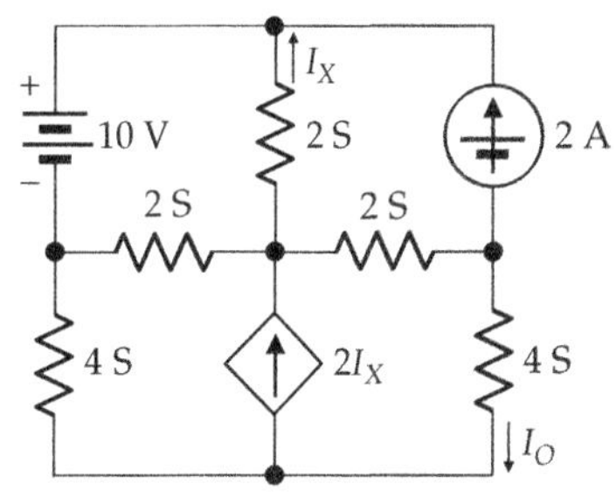

FIGURE P6.25

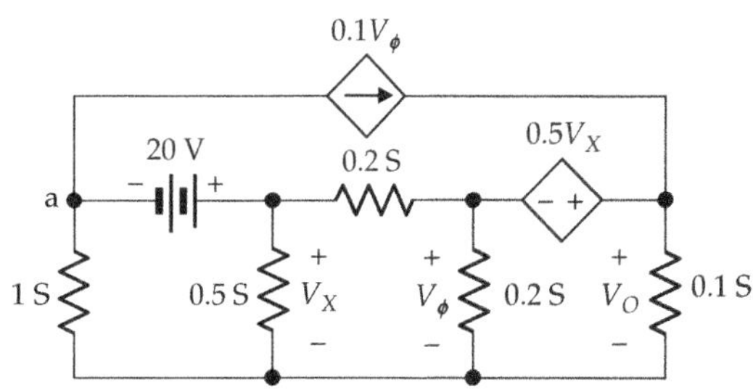

FIGURE P6.26

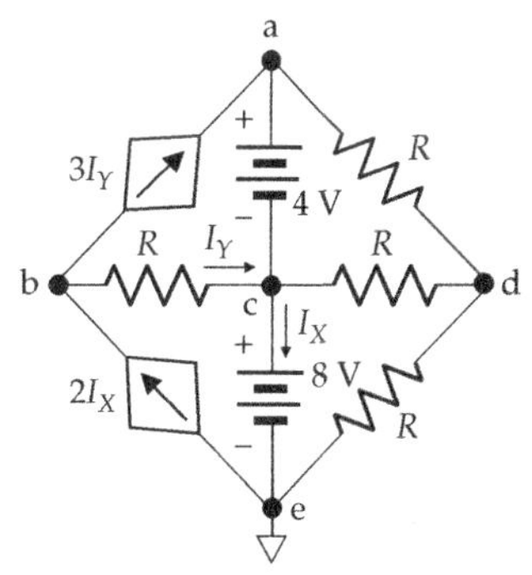

FIGURE P6.27

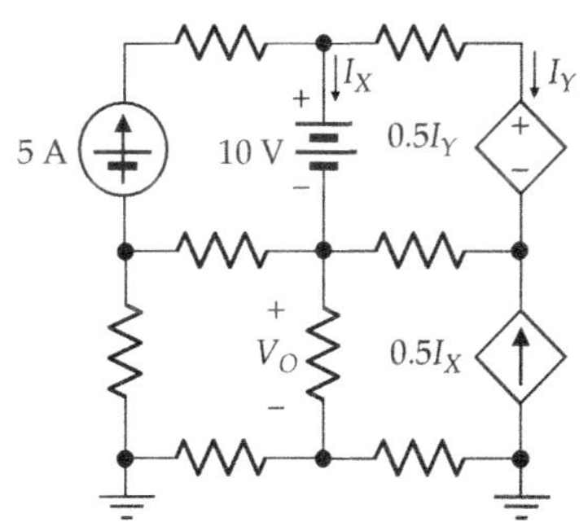

FIGURE P6.28

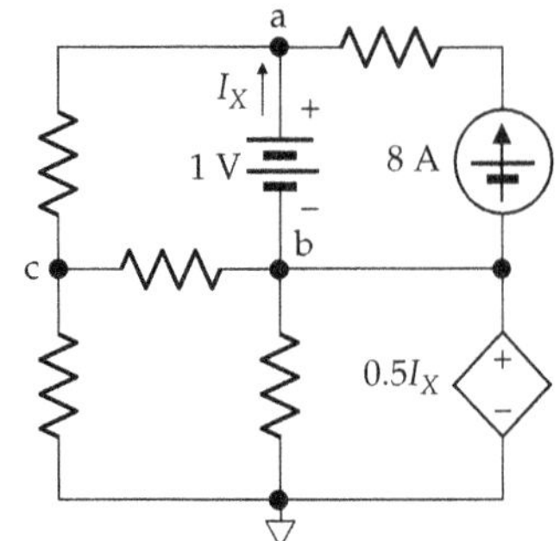

FIGURE P6.29

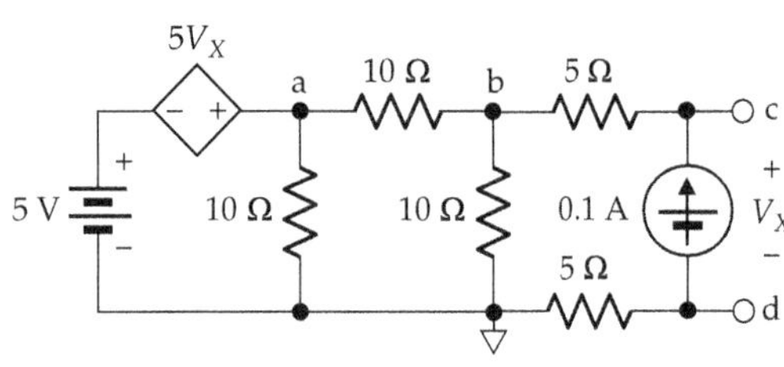

FIGURE P6.30

P6.29 Determine V_a, V_b, and V_c in Figure P6.29 assuming all resistances are 0.5 Ω.

Ans. $V_a = -4$ V, $V_b = -5$ V, $V_c = -3$ V.

P6.30 Determine V_{Th} and R_{Th} looking into terminals 'cd' in Figure P6.30.

Ans. $V_{Th} = V_{cd} = -8/3$ V, $R_{Th} = -10$ Ω.

Mesh-Current Method

Use the mesh-current method in Problems P6.31 through P6.55.

P6.31 Determine I_{SRC1} and I_{SRC2} in Figure P6.31.

Ans. $I_{SRC1} = 0$, $I_{SRC2} = 5/3$ A.

P6.32 Determine I_{SRC1} and I_{SRC2} in Figure P6.32.

Ans. $I_{SRC1} = -1.3$ A, $I_{SRC2} = 2.2$ A.

P6.33 Determine I_O in Figure P6.33.

Ans. 0.

P6.34 Determine V_O in Figure P6.34.

Ans. 12.5 V.

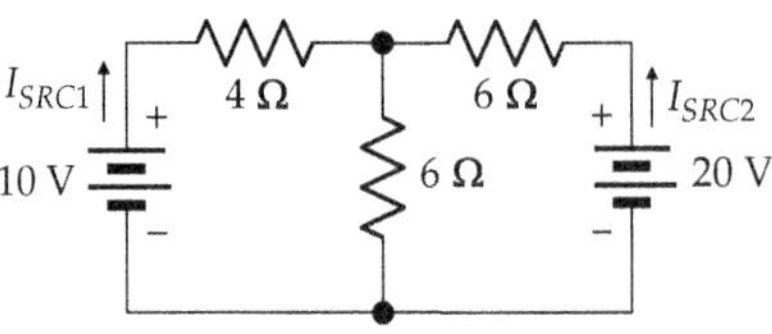

FIGURE P6.31

FIGURE P6.32

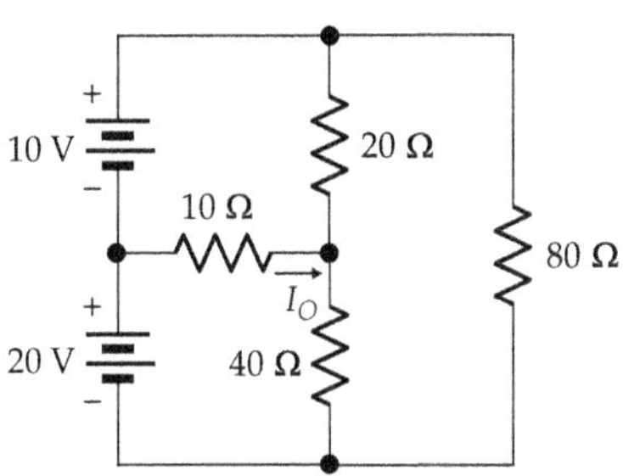

FIGURE P6.33

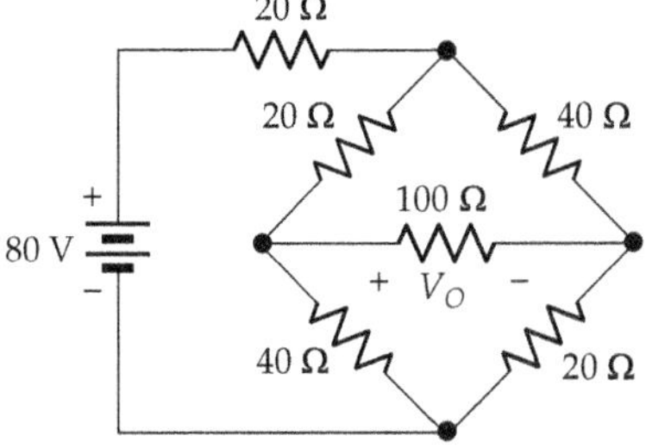

FIGURE P6.34

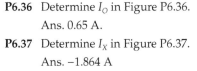

P6.35 Determine V_O in Figure P6.35.

Ans. 12.1 V.

P6.36 Determine I_O in Figure P6.36.

Ans. 0.65 A.

P6.37 Determine I_X in Figure P6.37.

Ans. −1.864 A

P6.38 Determine V_O in Figure P6.38.

Ans. 40 V.

P6.39 Determine I_X in Figure P6.39.

Ans. −1.43 mA.

P6.40 Determine V_X and V_Y in Figure P6.40, assuming all resistances are 2 Ω.

Ans. $V_X = 1.5$ V, $V_Y = -9.5$ V.

P6.41 Determine I_A in Figure P6.41.

Ans. 1.38 A.

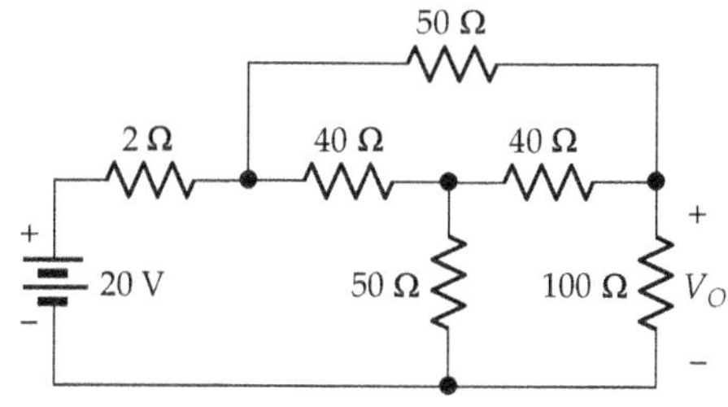

FIGURE P6.35

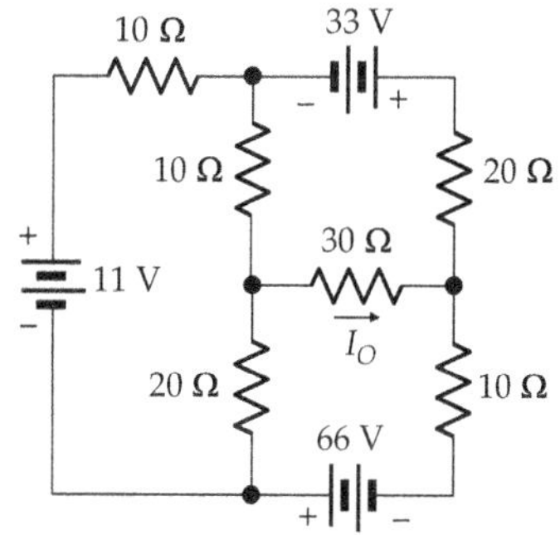

FIGURE P6.36

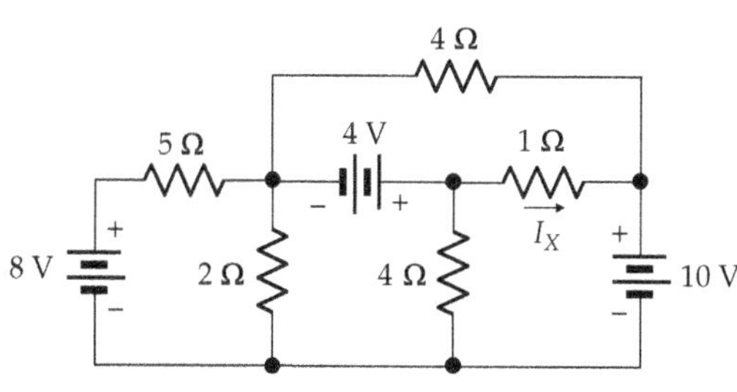

FIGURE P6.37

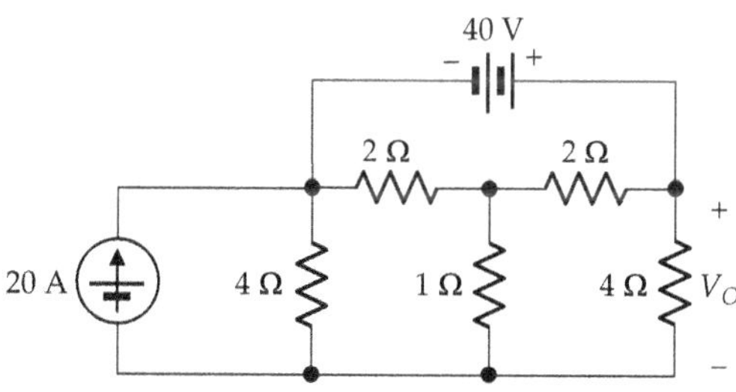

FIGURE P6.38

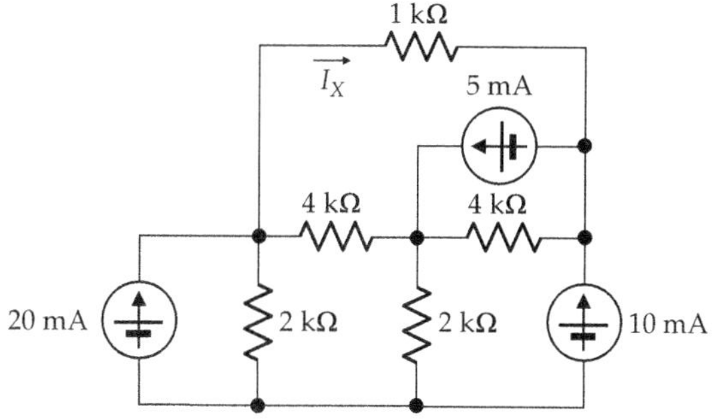

FIGURE P6.39

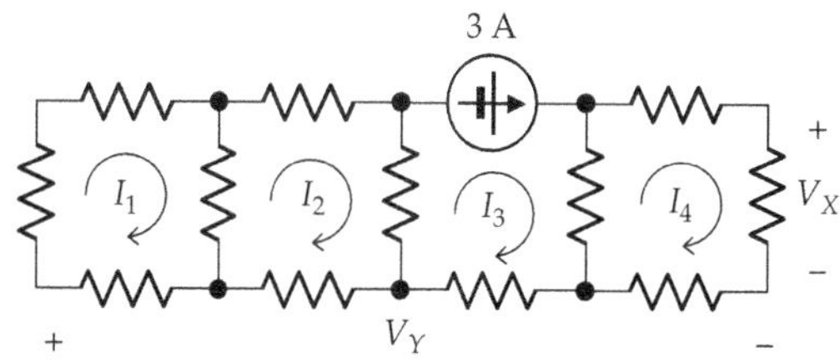

FIGURE P6.40

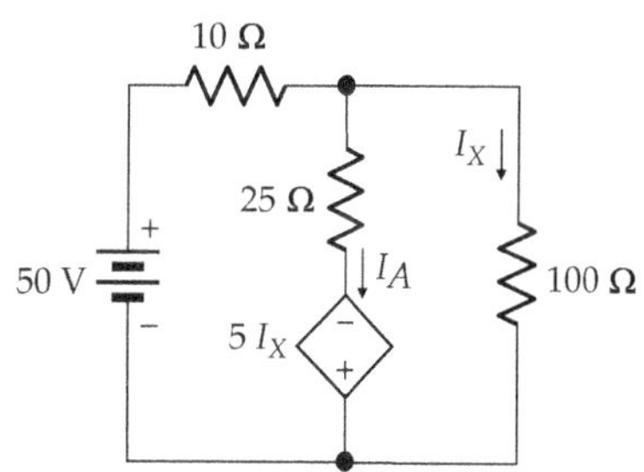

FIGURE P6.41

P6.42 Determine V_O in Figure P6.42.

Ans. 20 V.

P6.43 Determine I_O in Figure P6.43.

Ans. 15.51 A.

P6.44 Determine I_O in Figure P6.44.

Ans. –10/3 A.

P6.45 Determine the power delivered or absorbed by each current source in Figure P6.45.

Ans. 2 A source delivers 2 W and 4 A source delivers 380 W.

P6.46 Determine I_Y and V_S in Figure P6.46, assuming all resistances are 1 kΩ.

Ans. 22/25 mA, −62/25 V.

P6.47 Determine I_O in Figure P6.47.

Ans. –22 A.

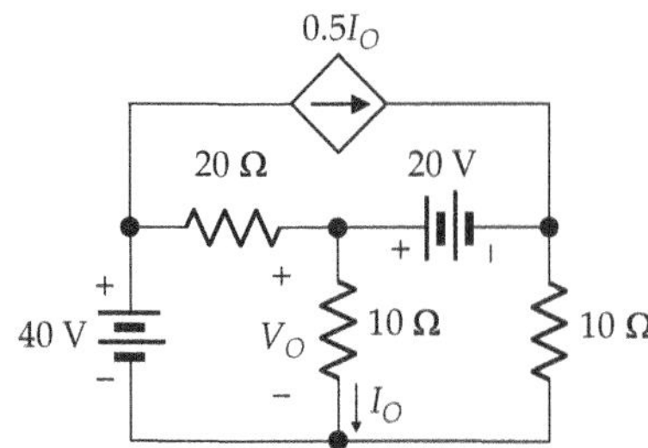

FIGURE P6.42

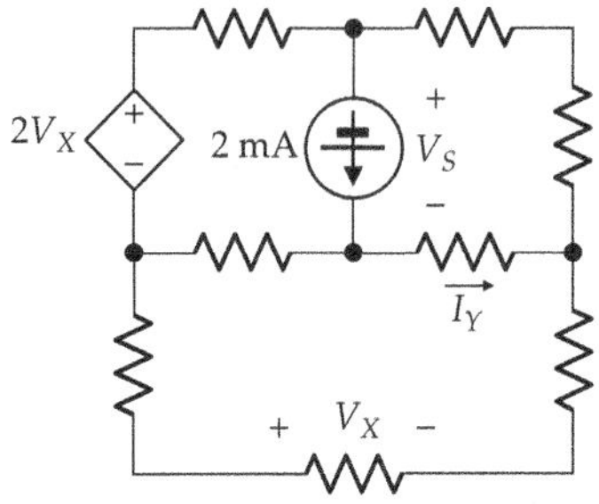

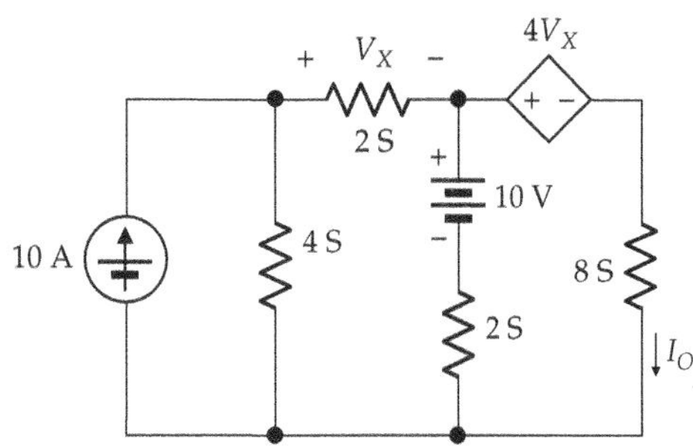

FIGURE P6.43

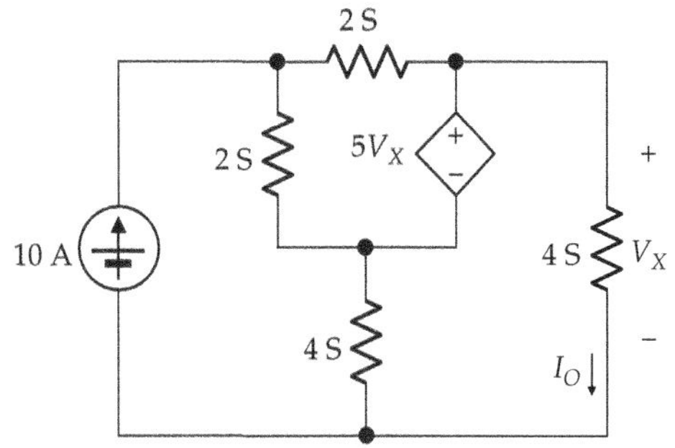

FIGURE P6.44

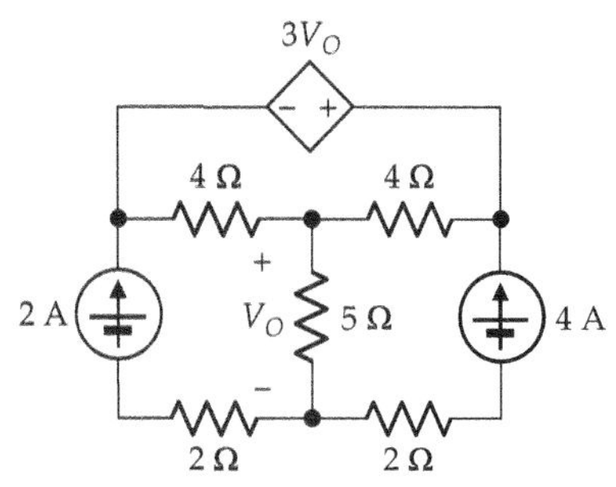

FIGURE P6.45

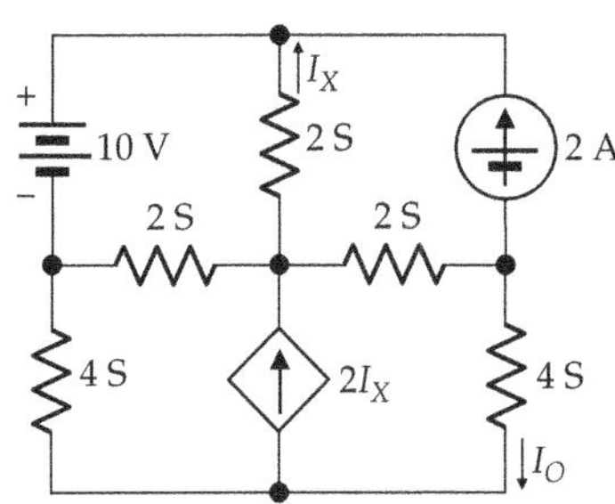

FIGURE P6.46

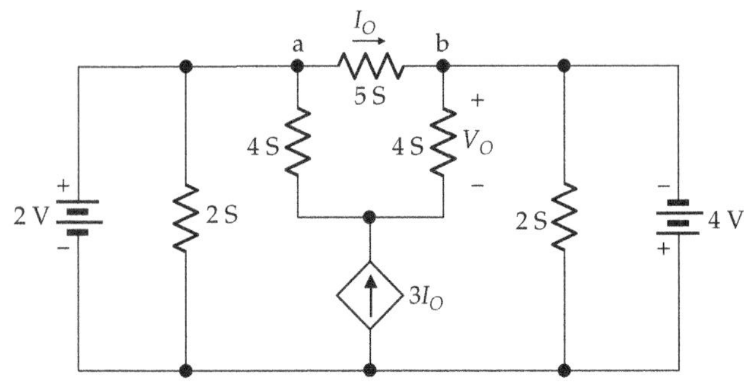

FIGURE P6.47

P6.48 Determine V_O in Figure P6.48.

Ans. −14.25 V.

P6.49 Determine V_O in Figure P6.49.

Ans. 15.51 V.

P6.50 Determine the power delivered or absorbed by each independent source in Figure P6.50.

Ans. 10 A source delivers 800 W, 20 V source 0 W.

P6.51 Determine I_X and V_Y in Figure P6.51.

Ans. −0.5 A, −10 V.

P6.52 Determine V_O in Figure P6.52, assuming all resistances are 2 Ω.

Ans. −8.67 V.

FIGURE P6.48

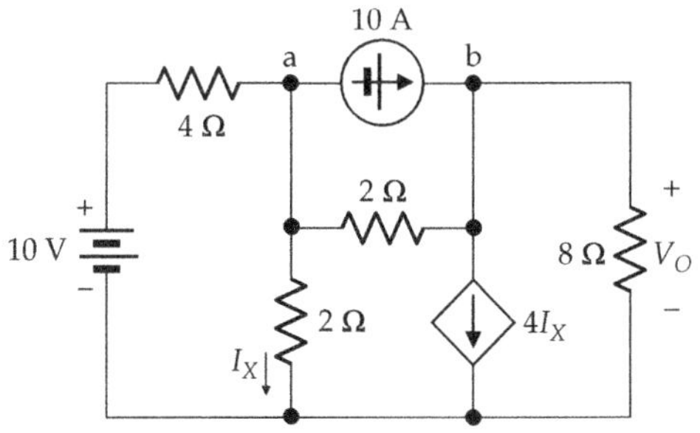

FIGURE P6.49

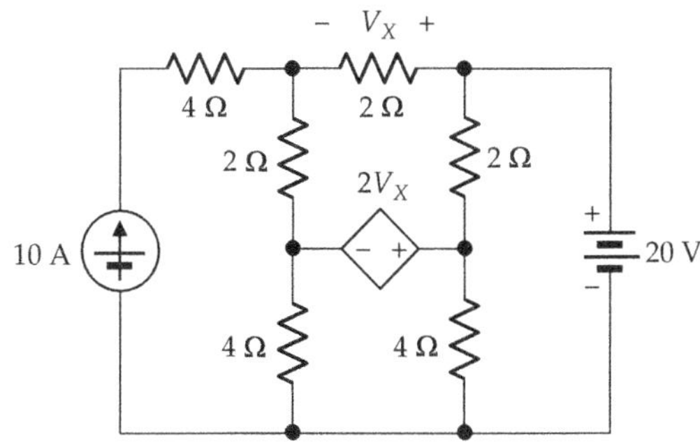

FIGURE P6.50

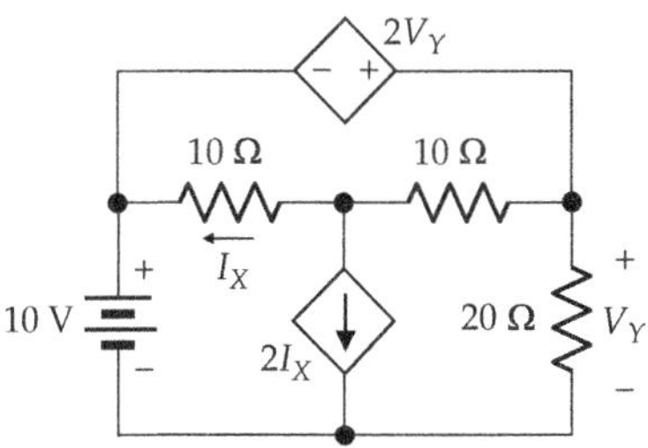

FIGURE P6.51

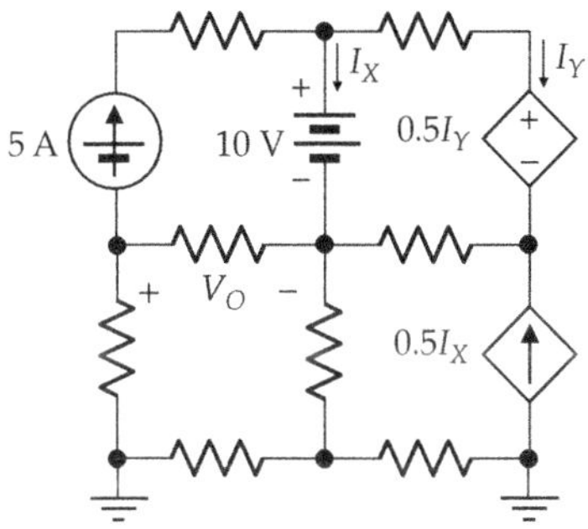

FIGURE P6.52

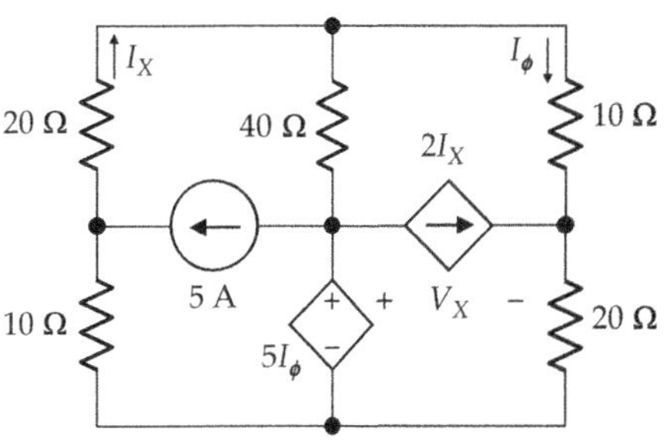

FIGURE P6.53

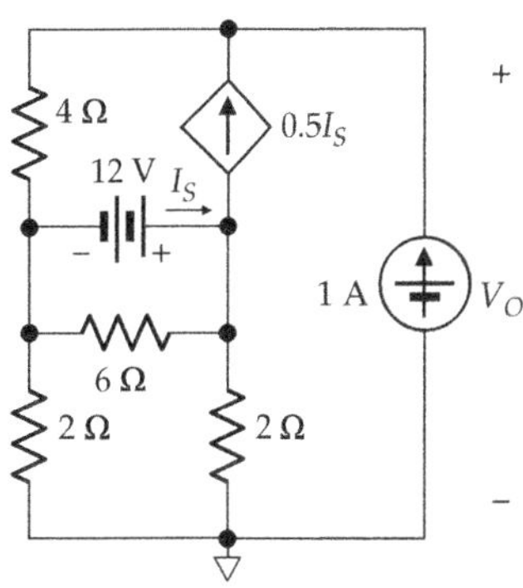

FIGURE P6.54

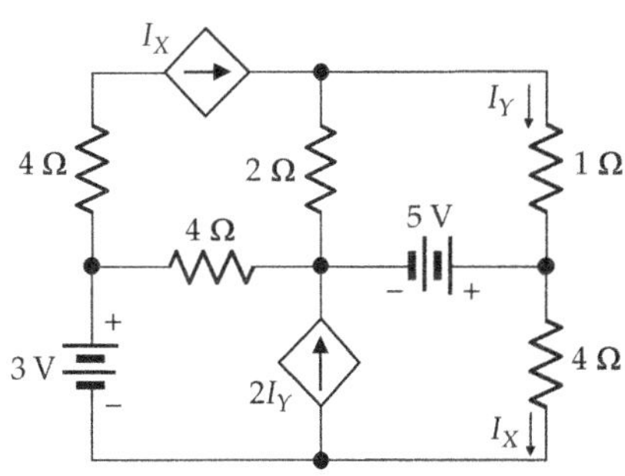

FIGURE P6.55

P6.53 Determine V_X in Figure P6.53.
Ans. −28.57 V.

P6.54 Determine V_O in Figure P6.54.
Ans. 21 V.

P6.55 Determine I_X and I_Y in Figure P6.55.
Ans. $I_X = 4\text{A}$ and $I_Y = 1\text{A}$

7

Capacitors, Inductors, and Duality

Objective and Overview

This chapter considers the fundamental properties of capacitors and inductors and introduces the very useful concept of duality in electric circuits.

The preceding chapters dealt with the analysis of resistive circuits only, in the dc steady state. Before discussing circuits in the sinusoidal steady state in the following chapters, the present chapter introduces some fundamental considerations on capacitors and inductors, including the capacitance and inductance of prototypical devices, voltage–current relations, stored energy, and the equivalent capacitance and inductance of capacitors and inductors in series or in parallel. Before discussing inductors, some basic notions on electromagnetism are considered.

The chapter concludes with an introduction of the very useful concept of duality in electric circuits, starting with duality of the relations for capacitors and inductors. Duality can be a useful aid in analyzing electric circuits involving energy storage elements.

7.1 Voltage–Current Relation of a Capacitor

Concept: *The fundamental attribute of a capacitor is its ability to store energy in the electric field resulting from separated positive and negative electric charges.*

The ability of a capacitor to store electric energy underlies its use in electric circuits to account for the electric energy stored in the electromagnetic field, as discussed in Section 1.9.

Figure 7.1a shows a prototypical capacitor consisting of two parallel metal plates, d m apart, each having an area of A m^2. The space between the plates is filled with a dielectric material of **permittivity**, or **dielectric constant**, ε farads/m. If the capacitor is momentarily connected to a source of voltage v, it acquires a charge $+q$ on the plate connected to the positive terminal of the source and an equal and opposite charge $-q$ on the plate connected to the negative terminal (Figure 7.1b). In an **ideal capacitor**, the charge on the capacitor is directly proportional to the voltage across it:

$$q = Cv \tag{7.1}$$

where C is the capacitance of the capacitor. C is a positive constant for an ideal capacitor, so q and v are linearly related. If q is in coulombs and v is in volts, C is in farads (F). Moreover, there is no power dissipation in an ideal capacitor. This implies that the insulation of the dielectric is perfect, that is, there is no leakage of charge between the plates of the capacitor, through the dielectric, so that the initial charge of an ideal, isolated capacitor is retained indefinitely.

The capacitance of the parallel-plate capacitor shown in Figure 7.1 can be readily calculated from elementary principles of electrostatics, with some simplifying assumptions. It will be assumed that the electric field ξ everywhere between the plates is normal to the surface of the plates and given by v/d V/m. This neglects fringe effects, which cause the electric field to curve outwards near the edges of the plates. If a cylinder of cross-sectional area ΔA m^2 is imagined to be oriented so that its sides are normal to the upper plate and extend above

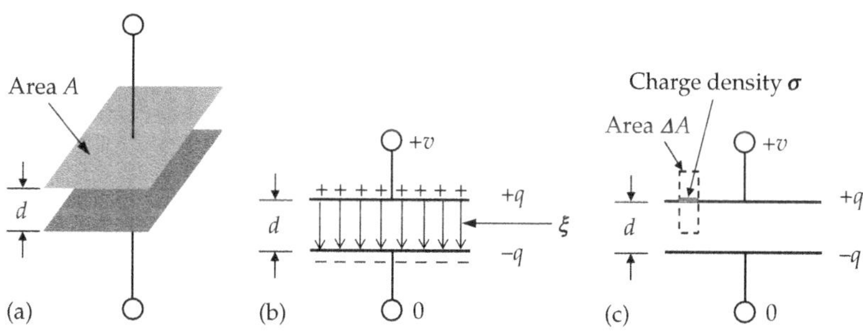

FIGURE 7.1
(a) Prototypical, parallel-plate capacitor, (b) charge q and electric field ξ of the capacitor, and (c) derivation of the expression for capacitance.

and below this plate, without reaching the lower plate (Figure 7.1c), then the charge enclosed by the cylinder may be denoted as $\sigma\Delta A$, where σ is the charge density q/A, assumed to be uniform all over the plate. From Gauss's law of electrostatics, the electric displacement D is equal to σ, where $D = \varepsilon\xi$. Substituting $D = q/A$ and $\xi = v/d$, in this relation,

$$q = \frac{\varepsilon A}{d}v \tag{7.2}$$

Comparing with Equation 7.1, C for the parallel-plate capacitor is

$$C = \frac{\varepsilon A}{d} \tag{7.3}$$

In general, capacitors can take many different forms, depending on the nature of the system and its geometry, which dictate how the electric field varies throughout the system. As exemplified by Equation 7.3, it is generally true that capacitance increases (1) with the permittivity of the dielectric between the conductive surfaces, (2) with the area of these surfaces, and (3) as the separation between these surfaces decreases. Note that, from Equation 7.3, the units of ε are farads/meter.

The charge q on the capacitor is related to the current i through the capacitor, by the defining relation for current as $i(t) = dq/dt$, or

$$q(t) = \int idt = \int_0^t idt + Q_0 \tag{7.4}$$

where Q_0 is any initial charge on the capacitor at $t = 0$. Note that the indefinite integral in Equation 7.4 does not explicitly involve Q_0, which is in fact the constant of integration in this case. On the other hand, Q_0 appears explicitly as the constant of integration in the expression for q in terms of the definite integral from $t = 0$ to any time t. From Equations 7.1 and 7.4,

$$v(t) = \frac{q}{C} = \frac{1}{C}\int idt = \frac{1}{C}\int_0^t idt + V_0 \tag{7.5}$$

where $V_0 = Q_0/C$ is the initial voltage across the capacitor at $t = 0$. Differentiating both sides of Equation 7.5 with respect to time and multiplying by C,

$$i(t) = C\frac{dv}{dt} \tag{7.6}$$

Note that Equation 7.5 is more general than Equation 7.6 in that it explicitly involves V_0.

7.1.1 Sign Convention

It should be noted that a definite sign convention is implied in writing Equations 7.5 and 7.6 with a positive sign on the RHS. The positive sign applies when the capacitor current i is assigned in the direction of a voltage drop v across the capacitor. But if i is assigned in the direction of a voltage rise v, then a negative sign should be used. This change of sign can be associated with the passive sign convention (Section 1.7) but will be justified next by the inherently positive value of C and by the flow of charge in a capacitor due to a change in voltage across the capacitor.

Suppose that i is in the direction of a voltage drop v, as in Figure 7.2. Since C is a positive quantity for a physical capacitor, then the value of i must be positive when the value of dv/dt is positive in Equation 7.6. If $i > 0$, then positive charges can be considered to flow into the positively charged plate (Figure 7.2a), thereby increasing the charge on the capacitor and hence the voltage across the capacitor, which means $dv/dt > 0$. With both i and dv/dt positive, then for C to be positive, a positive sign should be used in Equations 7.5 and 7.6.

On the other hand, if $i < 0$, positive charges can be considered to flow out of the positively charged plate (Figure 7.2b). The charge on the capacitor, and hence the voltage across it, will decrease, which means $dv/dt < 0$. With both i and dv/dt negative, C is again positive in Equations 7.5 and 7.6. This justifies writing these equations with a positive sign when i is in the direction of a voltage drop v.

Whereas for an ideal resistor, the assigned positive directions of v and i are always such that i is in the direction of a voltage drop v, this does not apply to a capacitor. The reason is that a resistor that obeys Ohm's law always dissipates power. On the other hand, a capacitor may actually absorb power when it is charging, or it may actually deliver power when it is discharging.

In Figure 7.2, for v positive as shown, a positive i gives a positive $p = vi$ for power absorbed, as argued in Example 1.4. If power is actually being delivered,

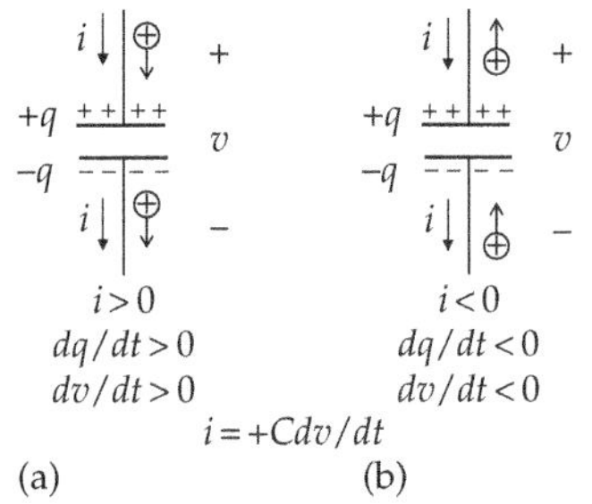

FIGURE 7.2
Capacitor current assigned in the direction of a voltage drop across the capacitor. The capacitor voltage is increasing in (a) and decreasing in (b).

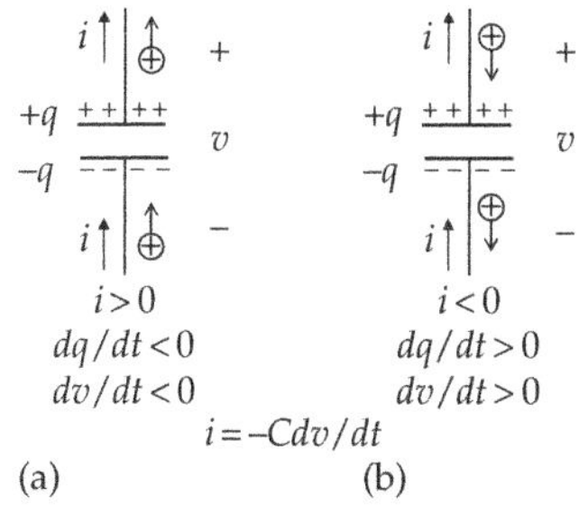

$i > 0$ $dq/dt < 0$ $dv/dt < 0$ (a)

$i < 0$ $dq/dt > 0$ $dv/dt > 0$ (b)

$i = -Cdv/dt$

FIGURE 7.3
Capacitor current assigned in the direction of a voltage rise across the capacitor. The capacitor voltage is increasing in (a) and decreasing in (b).

then the value of vi in Figure 7.2 is negative. To obtain a positive value for power delivered, as is convenient in some cases, the assigned positive directions of v and i in Figure 7.1 must be reversed relative to one another, as in Figure 7.3, where the assigned positive direction of i is reversed. But now, a positive i implies that positive charges are flowing out of the positively charged plate, which means that v is decreasing, so that $dv/dt < 0$ (Figure 7.3a). To have a positive value of C in the voltage–current relation, Equations 7.5 and 7.6 must be written with a negative sign as

$$v(t) = -\frac{1}{C}\int idt \tag{7.7}$$

and

$$i(t) = -C\frac{dv}{dt} \tag{7.8}$$

If i in Figure 7.3b has a negative value, then charges are flowing into the positively charged plate, v is increasing, so that $dv/dt > 0$. With i negative and dv/dt positive, C is again positive in Equations 7.7 and 7.8.

It should be noted that since the insulation in an ideal capacitor is perfect, and a perfect insulator does not allow movement of charge through it, the current in a capacitor cannot be a conduction current due to movement of charge from one plate of the capacitor to the other, through the insulation. Rather, the capacitor current is a **displacement current**. A simple interpretation of this current is to consider that the current i flowing into the positive terminal of the capacitor increases the positive charge on the positive plate of the capacitor by an amount δq in a time δt, where $\delta q = i\delta t$. The positive charge δq "displaces", through electrostatic repulsion, an equal positive charge δq from the negative plate. The displaced positive charge δq flows out of the negative terminal, thereby completing the current path through the capacitor, without any conduction current actually flowing between the plates of the capacitor. At the end of the time interval δt, the charge on the positive plate of the capacitor is $(+q + \delta q)$ and that on the negative pate is $(-q - \delta q)$.

7.1.2 Steady Capacitor Voltage

By a steady capacitor voltage is meant a voltage across the capacitor that is *constant for a specified duration.* In contrast, a dc voltage is, strictly speaking and from a signal analysis viewpoint, a steady voltage over all time, from $t = -\infty$ to $t = +\infty$. An example of a steady capacitor voltage is the voltage from $t = 1$ s to $t = 2$ s in Figure 7.5. A steady current is defined in a similar manner as a current that is constant for a specified duration.

When the capacitor voltage is a steady voltage V_{SD}, the charge on the capacitor does not change with time, so that $i(t) = CdV_{SD}/dt = 0$, in accordance with Equation 7.6. In other words, *the capacitor acts as an open circuit when the voltage across the capacitor is steady*, as illustrated in Figure 7.4a. The same is true, of course, of a dc voltage, V_{DC}, since $dV_{DC}/dt = 0$. A capacitor is said to "block" current due to a dc voltage and is often used in electronic circuits for this purpose.

However, it must be noted that a steady current can be forced through a capacitor by an ideal, steady current source I_{SD} (Figure 7.4b). Assuming the capacitor is initially uncharged, then from Equation 7.5,

$$v(t) = \frac{1}{C}\int_0^t I_{SD}dt = \frac{I_{SD}}{C}t \tag{7.9}$$

The capacitor voltage increases linearly with time, at a rate I_{SD}/C, during the interval in which the capacitor current is steady and therefore constant with respect to time (Figure 7.4b). Conversely, a capacitor voltage that increases linearly with time produces a steady capacitor

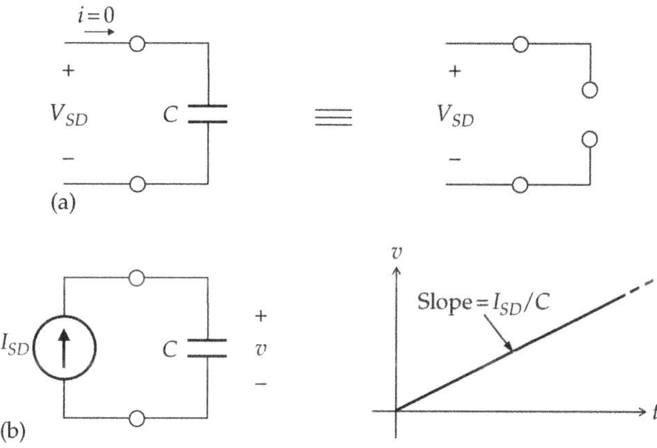

FIGURE 7.4
(a) Capacitor as an open circuit when a steady voltage is applied; (b) capacitor voltage increasing with time due to a steady applied current.

current. Thus, it follows from Figure 7.4b that if a voltage $v = I_{SD}t/C$ is applied to a capacitor, $i(t) = Cdv/dt = I_{SD}$.

7.1.3 Stored Energy

The energy stored in the electric field of a capacitor is equal to the work done in separating the charges $+q$ and $-q$, in opposition to the force of attraction between them. If a capacitor is being charged, and referring to Figure 7.2, the power input to the capacitor at any instant of time is $p = vi = vdq/dt$. The energy w absorbed by the capacitor in building up the charge on the capacitor from 0 to a value q is the time integral of p, in accordance with Equation 1.9. Assuming that the voltage is applied at $t = 0$ with $q = 0$ for $t \leq 0$, the energy supplied over the interval from $t = 0$ to t is

$$w(t) = \int_0^t pdt = \int_0^t v\frac{dq}{dt}dt = \int_0^q vdq \tag{7.10}$$

where the integration is from $t = 0$, when $q = 0$, to time t, when the charge is q. From Equation 7.1, $dq = Cdv$. Substituting in Equation 7.10,

$$w(t) = C\int_0^v vdv = \frac{1}{2}Cv^2 = \frac{1}{2}\frac{q^2}{C} = \frac{1}{2}qv \tag{7.11}$$

Primal Exercise 7.1

Determine the capacitance of a parallel-plate capacitor having plates of 4 × 3 cm, separated by 0.5 cm filled with an insulator having a relative permittivity $5000\varepsilon_0$, where $\varepsilon_0 = 8.85 \times 10^{-12}$ F/m is the permittivity of free space.

Ans. 10.62 nF.

Primal Exercise 7.2

An ideal, parallel-plate capacitor of 10 μF is charged to 1 V. If the separation between the parallel plates is doubled, determine the voltage across the capacitor. Compare the stored energies before and after the distance is doubled and explain the difference.

Ans. 2 V. Stored energy is doubled from 5 J to 10 J because of work done against attraction between plates.

Primal Exercise 7.3

If the voltage across a 1 μF capacitor is 2cos$t$ V, express the capacitor energy as a function of time and determine the difference between the maximum and minimum energy stored in the capacitor.

Ans. $w(t) = (\cos 2t - 1)$ μJ, 2 μJ.

Example 7.1: Capacitor Response to a Trapezoidal Voltage

The voltage shown in Figure 7.5 is applied to a 1 μF capacitor that is initially uncharged. Determine as a function of time (a) the charge on the capacitor, (b) the capacitor current, (c) the power absorbed by the capacitor, and (d) the energy absorbed by the capacitor. Assume the assigned positive directions of Figure 7.2.

Solution:

(a) Since various quantities are required as a function of time, v should be expressed analytically as a function of time. Thus, $v(t) = 5t$ V $0 \leq t \leq 1$ s, and $v(t) = 5$ V, $1 \leq t \leq 2$ s; $v(t)$ in the range $2 \leq t \leq 2.5$ s can be conveniently derived as follows: the slope of v with respect to t is $-5/0.5 = -10$ V/s in this range. The function $v(t) = -10t$ has the same slope but passes through the origin. The required function is displaced by 2.5 s to the right, so its equation is obtained by substituting $(t - 2.5)$ for t, giving: $v(t) = -10(t - 2.5)$ V, $2 \leq t \leq 2.5$ s. As a check, $v(t) = 5$ V when $t = 2$ s, and $v(t) = 0$ when $t = 2.5$ s.

From Equation 7.1, $q = Cv = 10^{-6} \times v$ coulombs = v microcoulombs (μC). It follows that $q(t) = 5t$ μC, $0 \leq t \leq 1$ s; $q(t) = 5$ μC, $1 \leq t \leq 2$ s; $q(t) = 10(2.5 - t)$ μC, $2 \leq t \leq 2.5$ s; and $q(t) = 0$, $t \geq 2.5$ s (Figure 7.6a).

(b) $i(t) = dq/dt$; hence, $i(t) = 5$ μC/s = 5 μA, $0 \leq t \leq 1$ s; $i(t) = 0$, $1 \leq t \leq 2$ s during the time that v is steady; $i(t) = -10$ μA, $2 \leq t \leq 2.5$ s; and $i(t) = 0$, $t \geq 2.5$ s (Figure 7.6b). Note that since q starts and ends with zero, the positive area of the i vs. t graph is equal in magnitude to the negative area.

(c) $p = vi$; multiplying v and i over the various time ranges gives $p(t) = 25t$ μW, $0 \leq t \leq 1$ s; $p(t) = 0$, $1 \leq t \leq 2$ s; $p(t) = -100(2.5 - t)$, $2 \leq t \leq 2.5$ s; and $p(t) = 0$, $t \geq 2.5$ s (Figure 7.6c). Note that the capacitor charges and absorbs power during the interval $0 \leq t \leq 1$ s while v and i are positive. No power is absorbed or delivered during the interval $1 \leq t \leq 2$ s when $i = 0$. During the interval $2 \leq t \leq 2.5$ s, $i < 0$, the capacitor discharges and delivers power. At $t = 2.5$ s, $q = 0$, and the capacitor is fully discharged. This means that all the energy absorbed during the interval $0 \leq t \leq 1$ s is returned

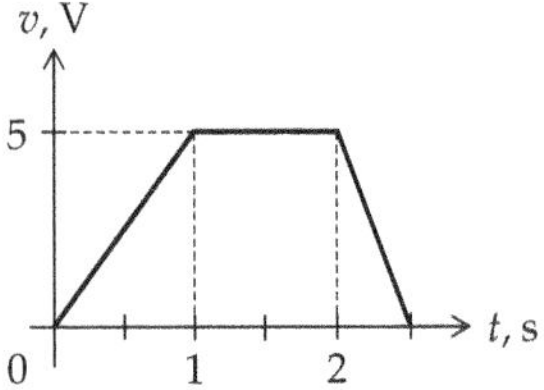

FIGURE 7.5
Figure for Example 7.1.

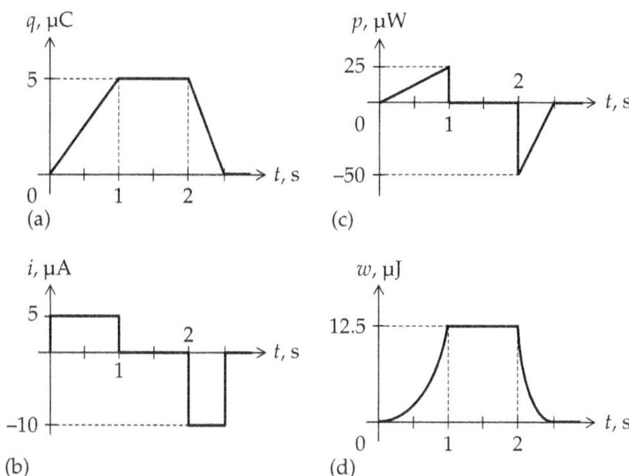

FIGURE 7.6
Figure for Example 7.1.

to the supply during the interval $2 \le t \le 2.5$ s. Since energy is the area under the power curve, the positive area of the triangle extending from $t = 0$ to $t = 1$ s, which is 12.5 μJ, must be equal in magnitude to the negative area extending from $t = 2$ to $t = 2.5$ s, which is in fact −12.5 μJ.

(d) Energy as a function of time is obtained as $w(t) = \int_0^t p dt$. Thus, $w(t) = \int_0^t 25dt = 12.5t^2, 0 \le t \le 1$ s. During the interval $1 \le t \le 2$ s, when no power is absorbed or delivered, the energy stored stays at 12.5 μJ. During the interval $2 \le t \le 2.5$ s, $w(t) = \int_2^t -100(2.5-t)dt + 12.5 = 50(t^2 - 5t + 6.25) =$ $50(t-2.5)^2$; $w(t) = 0$ at $t = 2.5$ s, and since $p(t) = 0$ for $t \ge 2.5$ s, $w(t) = 0$ for $t \ge 2.5$ s. The variation of w with t is shown in Figure 7.6d. Note that since the voltage returns to zero for $t \ge 2.5$ s, $w = 0$.

Simulation: The circuit is entered as in Figure 7.7. The capacitor is entered from the ANALOG_P library, because this capacitor has terminals 1 and 2 marked for convenience in determining unambiguously the assigned positive directions of capacitor voltage and current. The convention in PSpice is that positive current flows from terminal 1 to terminal 2 through any circuit element and that positive voltage is that of terminal 1 with respect to terminal 2.

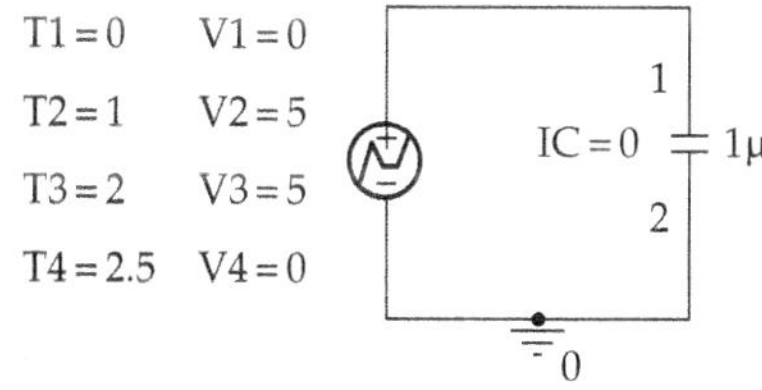

FIGURE 7.7
Figure for Example 7.1.

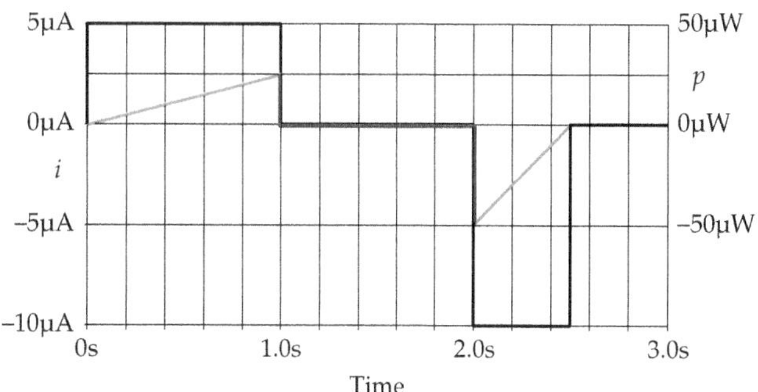

FIGURE 7.8
Figure for Example 7.1.

When using time-domain analysis, as in this case, initial conditions of energy storage elements should be specified, even if they are zero. To do so, display the Property Editor spreadsheet for the capacitor and enter 0 in the IC column. To display this entry, click on the Display button and select 'Name and Value' under 'Display Format'. The source is VPWL (piecewise-linear voltage source) from the source library, with the breakpoints entered as voltage values at the corresponding times in the Property Editor spreadsheet for the source (Appendix C). They are displayed in the same manner as the initial conditions for the capacitor just described. In the Simulation Settings, 'Analysis type' is 'Time Domain (Transient)', 'Run to time' is 3 s, 'Start Saving Data After' is 0, and 'Maximum Step size' is 0.5 m. After the simulation is run, then in order to display the current and power dissipated to appropriate vertical scales, select Trace/Add Trace in the Schematic1 window and enter 10*I(C1:1),W(C1). The time variation of capacitor current and dissipated power is displayed, as in Figure 7.8.

Primal Exercise 7.4

Repeat Example 7.1 assuming the voltage increases linearly from 0 to 5 V in 0.5 s, stays at 5 V for 1 s, then decreases linearly to zero in 1 s.

Ans. $0 < t < 0.5$ s: $q(t) = 10t$ μC, $i(t) = 10$ μA, $p(t) = 100t$ μW, $w(t) = 50t^2$ μJ; $0.5 < t < 1.5$ s: $q(t) = 5$ μC, $i(t) = 0$, $p(t) = 0$, $w(t) = 12.5$ μJ; $1.5 < t < 2.5$ s: $q(t) = -5t + 12.5$ μC, $i(t) = -5$ μA, $p(t) = 25t - 62.5$ μW, $w(t) = 12.5t^2 - 62.5t + 78.125$ μJ.

Example 7.2: Capacitive Circuit in dc State

It is required to determine V_L in Figure 7.9a under dc conditions and the total energy stored in the two capacitors.

Solution:

Under dc conditions, the two capacitors act as open circuits (Figure 7.9b). No current flows in the 2 μF

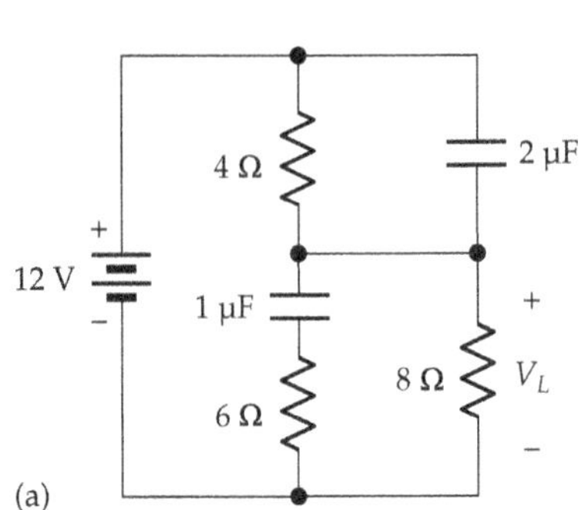

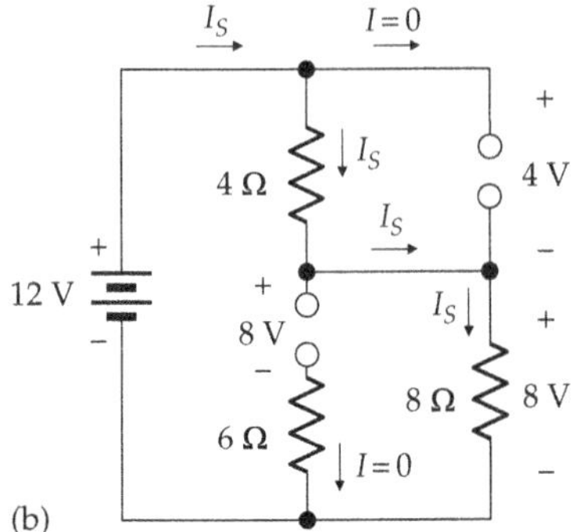

FIGURE 7.9
Figure for Example 7.2.

capacitor, nor in the 1 μF capacitor, nor in the 6 Ω resistor. The same current I_S flows in the 4 and 8 Ω resistors. Hence, these resistors are in series, so that voltage division applies. It follows that $V_L = \frac{8}{8+4} \times 12 = 8$ V, and $I_S = 12/12 = 1$ A.

Since no current flows in the 6 Ω resistor, the voltage across the 1 μF capacitor is 8 V, the same as V_L. The voltage across the 2 μF capacitor is the same as that across the 4 Ω resistor. From voltage division, this is $\frac{4}{8+4} \times 12 = 4$ V. As a check, KVL in the loop consisting of the voltage source and the 4 and 8 Ω resistors is 12 − 8 = 4 V. The energy stored in the 2 μF capacitor is $\frac{1}{2}CV^2 = \frac{1}{2} \times 2 \times (4)^2 = 16$ μJ, and the energy stored in the 1 μF capacitor is $\frac{1}{2}CV^2 = \frac{1}{2} \times 1 \times (8)^2 = 32$ μJ. The total stored energy is 48 μJ.

Simulation: The circuit is entered as in Figure 7.10. After selecting 'Bias Point' under 'Analysis type' in the Simulation Settings and running the simulation, pressing the V and I buttons displays the voltages and current shown.

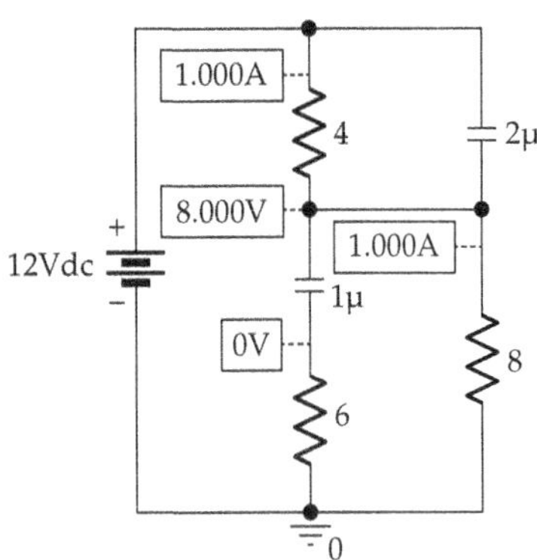

FIGURE 7.10
Figure for Example 7.2.

Problem-Solving Tip

- A capacitor is replaced by an open circuit under dc or steady conditions.

Primal Exercise 7.5

Repeat Example 7.2 assuming the source voltage is 18 V and interchanging the locations of the 4 Ω and 8 Ω resistors.

Ans. $V_L = 6$ V; $W = 162$ μJ.

7.2 Voltage–Current Relation of an Inductor

Before considering the *v–i* relation for the ideal inductor, some basic notions of electromagnetism will be reviewed.

7.2.1 Magnetic Fields and Related Quantities

A permanent magnet is surrounded by a magnetic field, in which a force is exerted on a magnetic object such as a compass needle, which is itself a small permanent magnet. The magnetic field can be conveniently visualized by means of imaginary lines referred to as **magnetic field lines**, as illustrated in Figure 7.11 for a bar magnet. Magnetic field lines are considered to be directed from the north pole of the magnet to its south pole, as indicated by the arrows on the lines. At any point along a field line, the direction of the tangent to the field line is the direction of the magnetic field at that point.

Magnetic field lines not only give the direction of the magnetic field but their density is indicative of the strength of the magnetic field, where density refers to the number of lines per unit area oriented normal to the direction of the lines.

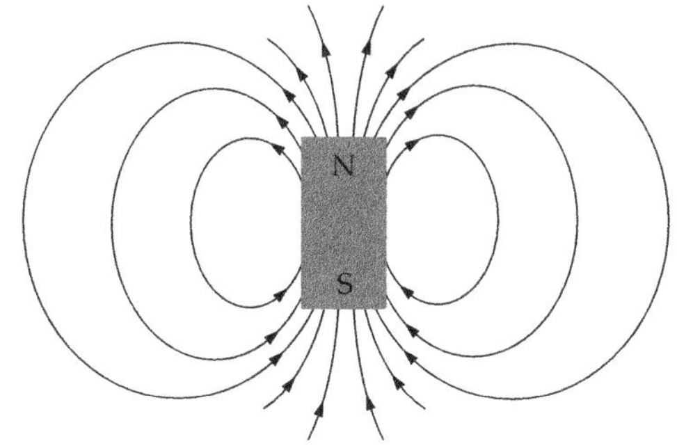

FIGURE 7.11
Magnetic field lines around a permanent magnet.

The totality of magnetic field lines emanating from the north pole of a magnet, or converging onto its south pole, is the **magnetic flux** of the magnet. The stronger the magnet, the larger the flux, the higher is the density of the magnetic field lines, and the stronger is the magnetic field. The unit of magnetic flux in SI units is the weber (Wb).

A magnetic field is associated with moving electric charges. The magnetic field lines for a long and straight current-carrying wire form concentric circles around the wire, in any plane normal to the wire (Figure 7.12). The direction of the magnetic field is related to that of the current by the right-hand rule: If the wire is gripped with the right hand, with the thumb extended in the direction of the current, the fingers point in the direction of the magnetic field. A basic quantity of the magnetic field is the **magnetic field strength** vector $\vec{H}$, related to current by **Ampere's circuital law**:

$$\oint H_l dl = I_{enc} \quad \text{or} \quad \oint \vec{H} \cdot d\vec{l} = I_{enc} \tag{7.12}$$

where the circle around the integral signifies integration around a closed path, dl is an increment of distance along this closed path, and I_{enc} is the current enclosed in traversing the closed path once. $\vec{H}$ is a vector having magnitude and direction. In the first form of the integral in Equation 7.12, H_l is the component of $\vec{H}$ in the direction of $d\vec{l}$. Alternatively, Ampere's law can be expressed in the form of the second integral in Equation 7.12, as the scalar product of the two vectors $\vec{H}$ and $d\vec{l}$. It is seen from Equation 7.12 that the unit of H in SI units is amperes/meter (A/m).

In the case of a long, straight wire carrying a dc current, $\vec{H}$ is of constant magnitude H around a circle of radius r centered on the wire in a plane normal to the wire (Figure 7.12a), and $\vec{H}$ is directed tangentially at every point along the circumference of the circle. The line integral over a circle of radius r is simply $2\pi rH$ in this case, and $I_{enc} = I$. Substituting in Equation 7.12,

$$H = \frac{I}{2\pi r} \tag{7.13}$$

Another basic quantity of the magnetic field is the **magnetic flux density**, $\vec{B}$, that is, flux per unit area. $\vec{B}$ is related to $\vec{H}$ as

$$\vec{B} = \mu_r \mu_0 \vec{H} \tag{7.14}$$

where μ_0 is the permeability of free space ($4\pi \times 10^{-7}$ N/A^2 or henries/m, the henry being the unit of inductance, as explained in Section 7.4) and μ_r is the relative permeability of the medium with respect to that of free space; that is, μ_r is the ratio of the permeability of the given medium to that of free space. Permeability is an important magnetic property of the medium, as discussed later. For the current-carrying conductor of Figure 7.12a, the flux density is the flux per unit area in the plane of the wire (Figure 7.12b). It is denoted by B_n, where the subscript 'n' emphasizes that the flux density is normal to the plane of the wire. Being flux density, the unit of B is the tesla (T), equivalent to 1 Wb/m^2.

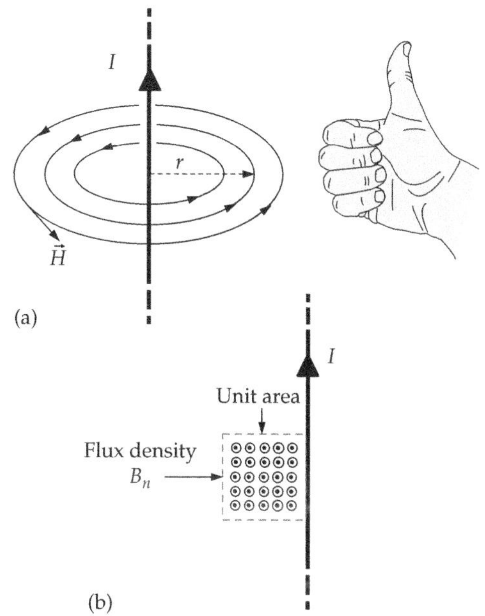

FIGURE 7.12
(a) Magnetic field of a long, straight wire, the directions of the current and the field being related by the right-hand rule and (b) magnetic flux density indicated as density of lines per unit cross-sectional area oriented normal to the direction of the field.

An essential difference between H and B is that H is directly related to current, independently of the magnetic properties of the medium surrounding the current-carrying conductor. B, on the other hand, is directly related to voltage, through Faraday's law (Equation 7.18, discussed later), independently of the magnetic properties of the medium. This implies that B and H must be related by the magnetic properties of the medium, as in Equation 7.14. Evidently, the stronger the magnetic field, the stronger is the magnetic field strength, H, and the larger is the magnetic flux density, B. The magnetic properties of a medium arise from the magnetic fields generated internally in the material in response to the applied magnetic field. In materials of high permeability, the magnetic effects due to the material play a dominant role in the magnetic behavior of the material.

7.2.2 Magnetic Flux Linkage

Consider a single turn of wire through which magnetic flux passes (Figure 7.13a). The magnetic flux linking the turn is simply the magnetic flux passing through the area enclosed by the turn, in the direction normal to the plane of the turn. This flux is the integral, over the area of the turn, of B_n, the component of magnetic flux density B that is normal to the plane of the turn, multiplied by 1 for a single turn:

$$\lambda = \left(\int_A B_n dA \right) \times 1 = \phi_n \times 1 \tag{7.15}$$

In Equation 7.15, the integral in brackets is ϕ_n, the normal component of the flux through the turn; ϕ_n multiplied by '1' for a single turn is the **magnetic flux linkage** λ. With the unit of ϕ_n being a weber (Wb), the unit of magnetic flux linkage is the weber-turn (Wb-T).

For two turns of thin wire that are closely packed (Figure 7.13b), the same magnetic flux may be assumed to pass through each turn, and in the same direction, so that ϕ_n is the same for each turn. The flux linkage of the two turns is ϕ_n multiplied by '2' for two turns, that is, $\lambda = \phi_n \times 2$. In effect, having two turns doubles the area through which flux passes, which doubles the flux linkage.

In the case of a coil of one or more turns, the relative directions of current in the coil and magnetic flux in the coil can be determined by the right-hand rule, which in this case can be alternatively expressed as follows: wrap the fingers of the right hand around the coil, with the fingers pointing in the direction of current in the turns of the coil. The extended thumb will point in the direction of flux in the coil (Figure 7.13c). Note that the right-hand rule of Figure 7.12 can still be applied, but one must be careful as to whether the flux whose direction is being determined is inside or outside the coil. Thus, current in the front part of the coil flows from left to right. If the thumb of the right hand points in the direction of this current, the fingers will be *inside* the coil and will point *upwards, in the direction of flux inside the coil.* However, the current in the rear part of the coil flows from right to left. If the thumb of the right hand points in the direction of this current, the fingers will be *outside* the coil and will point *downwards, in the direction of flux in the return path outside the coil.*

The definition of λ can be extended to a coil of N turns. However, when the coil is in a medium of small relative permeability, as in air or other nonmagnetic, or weakly magnetic material, not all the flux lines will pass through all the turns. Some of the flux lines will loop back around some of the turns of the coil without passing through all the turns of the coil, as illustrated in Figure 7.14a. λ can be formally defined as

$$\lambda = \sum_{j=1}^{N} \phi_{nj} \tag{7.16}$$

where ϕ_{nj}, the normal component of flux through the jth turn, is summed over all the N turns. In other words, *flux linkage is the total flux through all the turns of the coil in the direction normal to the plane of each turn.*

A convenient conceptualization is to define an effective flux ϕ_{eff} that passes through every turn of the coil, in the direction normal to the plane of each turn, as in Figure 7.14b, and which when multiplied by N gives the true value of flux linkage as in Equation 7.16. Thus,

$$\lambda = \phi_{eff} \times N \tag{7.17}$$

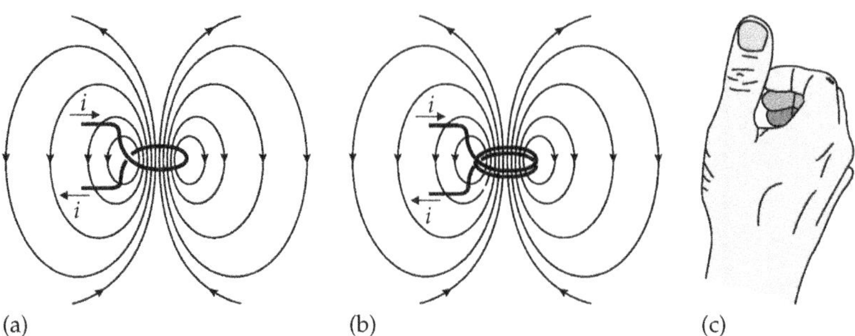

FIGURE 7.13
Magnetic flux linking a single turn (a) and two closely-packed turns (b), and (c) the directions of the current in the turn and the flux through the turn are related by another form of the right-hand rule.

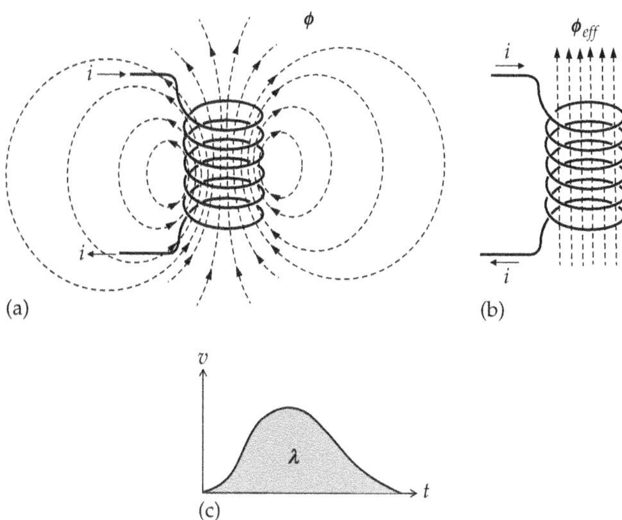

FIGURE 7.14
(a) Magnetic field lines of a multi-turn coil in a medium of low magnetic permeability, (b) effective flux of the solenoid, and (c) flux linkage of the solenoid.

The importance of magnetic flux linkage stems from the fact that **Faraday's law** of electromagnetic induction is expressed in terms of flux linkage as

$$|v(t)| = \left|\frac{d\lambda}{dt}\right| = N\left|\frac{d\phi_{eff}}{dt}\right| \tag{7.18}$$

In Equation 7.18, λ is time varying and could be positive or negative. $v(t)$ is the voltage induced in the coil. The magnitude symbol is used on both sides of Equation 7.18 to emphasize that it is the magnitude of $v(t)$ that is given by the equation. The polarity of $v(t)$ is determined by **Lenz's law**, according to which the polarity of the induced voltage is such that it opposes the change in the magnetic flux linkage that induces the voltage. Lenz's law will be illustrated in the discussion on induced voltage in the following section.

According to Equation 7.18, an alternative unit of magnetic flux linkage is the volt-second (Vs), which is also used in connection with voltages applied to coils. Note that Equation 7.18 makes ϕ_{eff} a well-defined quantity that can be determined from experimental measurements. For if a time-varying voltage is applied to a coil, it is possible, from measurement of this voltage, the coil current, and the coil resistance, to determine v across the ideal inductor component of the coil (Figure 7.14c). According to Equation 7.18, λ is the integral of this voltage with respect to time and is the area under the curve. ϕ_{eff} is then this area divided by the number of turns of the coil.

An important case that arises in practice, and which is conveniently simple to consider, is that of a coil wound around a toroidal core of high permeability (Figure 7.15a).

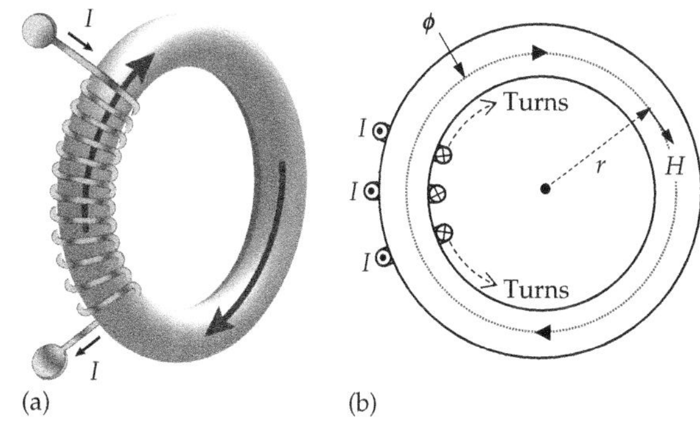

FIGURE 7.15
(a) Coil wound on a toroidal core of high magnetic permeability and (b) section through the toroidal core showing the magnetic field in the core.

The circular shape of the core, which naturally conforms to the circular path of the flux in the core, together with the high permeability, ensure that *the same flux in the core passes through every turn of the coil in a direction normal to the plane of each turn.* ϕ_{eff} for the flux in the core is in this case the same as the actual flux in the core.

To express Ampere's circuital law in the case of a toroidal core, consider a plane that is normal to the axis of the toroid and which cuts the toroid into two halves, one of which is shown in the cross section in Figure 7.15b. If we follow a path in this plane around a circle of radius r inside the toroid and centered on the axis of the toroid, we find that the current I in each of the N turns of the coil crosses the inside of the circle in the same direction in all the turns, as illustrated in Figure 7.15b for three turns. As the circle is traversed once, the current I is enclosed N times, once for each turn of the coil, so that I_{enc} in Equation 7.12 is NI. H is everywhere tangential to a given circle of radius r (Figure 7.15b) and is considered to have the same value at each point on the circumference of the circle. This follows from the fact that the flux is the same through any transverse cross section of the core that is normal to the tangent to the circle at any point on the circumference of this circle. B and hence H are therefore the same at any point on the circumference of the given circle of radius r. From Equation 7.12,

$$2\pi rH = NI \quad \text{or} \quad H = \frac{NI}{2\pi r} \tag{7.19}$$

Primal Exercise 7.6

The area of a 500-turn coil is 50 cm^2. Determine the voltage induced in the coil by a magnetic field of flux density that increases at a constant rate from 0 to 5×10^{-4} Wb/m^2 in 0.1 s.

Ans. 12.5 mV.

7.2.3 Inductance

Concept: *The fundamental attribute of an inductor is its ability to store energy in the magnetic field resulting from current flow in a conductor.*

As mentioned in Chapter 1, an ideal inductor accounts for the magnetic energy of the electromagnetic field. The circuit parameter associated with this energy storage ability is the inductance, defined as follows:

Definition: *The inductance of a coil is the flux linkage of the coil per unit current in the coil.*

According to this definition, the inductance L of a current-carrying coil is

$$L = \frac{\lambda}{i} \quad \text{or} \quad \lambda = Li \tag{7.20}$$

In an ideal inductor, L is a positive constant, so that λ and i are linearly related, and there is no power dissipation. With L positive, λ and i in Equation 7.20 must have the same sign, which means that a positive direction for λ should be associated with the positive direction for i. If i has a negative value, λ also has a negative value. This is analogous to the relation between q and v for a capacitor.

If the coil is in air or in a low-permeability medium, the effective flux can be used in the expression for flux linkage, as in Equation 7.18, to give

$$L = \frac{\lambda}{i} = N\frac{\phi_{eff}}{i} \tag{7.21}$$

A simple expression for the inductance of the coil in Figure 7.15 can be readily derived, based on two simplifying assumptions: (1) that the current-carrying wire is tightly wound around the core and that the wire diameter and thickness of insulation around the wire are sufficiently small so that the flux outside the core is negligible; in other words, all of the coil flux is confined to the core, and (2) that the diameter of the core is small compared with the mean diameter, a, of the toroid, so that H can be assumed constant across the transverse cross section of the core. It follows from Equation 7.19 that $H = NI/\pi a$, independently of r. From Equation 7.14, $B = \mu H$, where $\mu = \mu_r\mu_0$ (Equation 7.14). Hence,

$$B = \frac{\mu}{\pi a}NI \tag{7.22}$$

The flux in the core is $\phi = BA$, where A is the cross-sectional area of the core. The flux linkage is

$$\lambda = N\phi = \frac{\mu A}{\pi a}N^2 I \tag{7.23}$$

Dividing by the current gives the inductance of the coil as

$$L = \frac{\mu A}{\pi a}N^2 \tag{7.24}$$

Note that inductance is proportional to permeability and to the square of the number of turns in the coil.

7.2.4 Voltage–Current Relation

Differentiating both sides of Equation 7.20 with respect to time and using Faraday's law, $|v| = |d\lambda/dt|$ (Equation 7.18),

$$v(t) = L\frac{di}{dt} \tag{7.25}$$

where the magnitude designation has been dropped because of a definite sign convention associated with Equation 7.25, as will be explained shortly. When v is in volts, i in amperes, and t in seconds, L is in henries (H). Integrating both sides gives the equivalent i–v relation:

$$i(t) = \frac{1}{L}\int v dt = \frac{1}{L}\int_0^t v dt + I_0 \tag{7.26}$$

As mentioned in connection with Equation 7.6, Equation 7.26 is more general than Equation 7.25 in that it explicitly involves the initial inductor current at $t = 0$.

Since L is a positive quantity, a definite sign convention is implied in writing Equations 7.25 and 7.26 with a positive sign on the RHS, in that v must be positive when di/dt is positive. It will be argued next that in order to satisfy this requirement, the assigned positive direction of i and the polarity of v must be such that i is in the direction of a voltage drop v, as in Figure 7.16a. The justification can be explained with reference to Figure 7.16b and will be illustrated by a simulation of the circuit, assuming $L = 0.4$ H and $R_{src} = 2\ \Omega$. From KVL,

$$v_{SRC} - R_{src}i - v = 0 \tag{7.27}$$

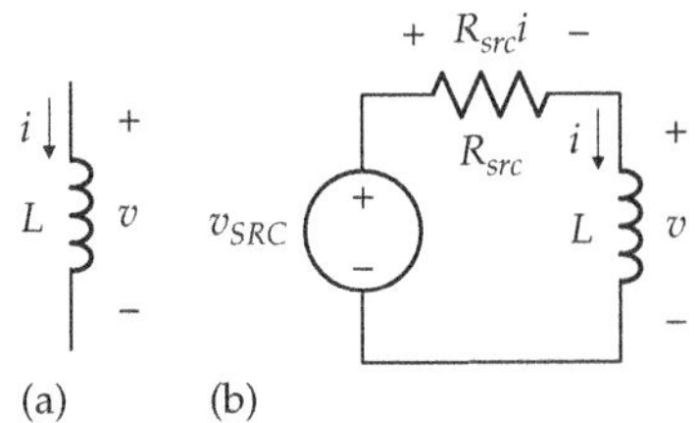

FIGURE 7.16
Assigned positive directions associated with a positive sign in Equation 7.25. (a) Current assigned in the direction of a voltage drop across the coil and (b) circuit for determining the polarity of the induced voltage.

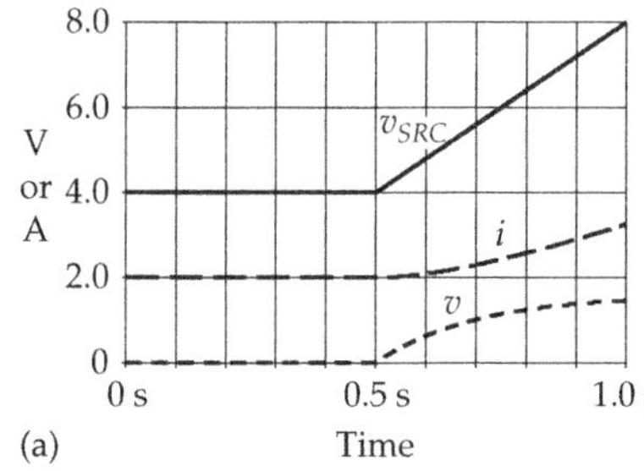

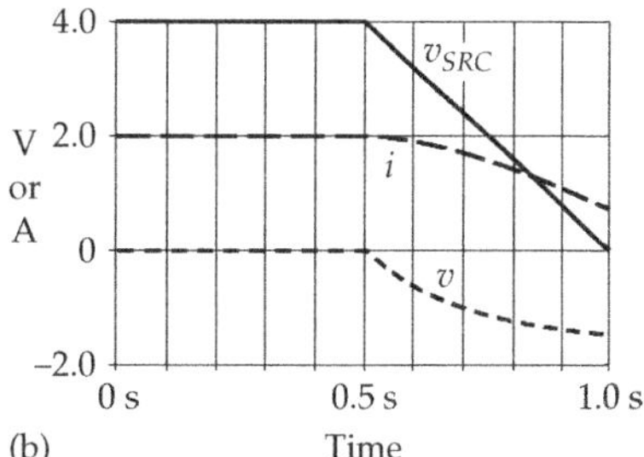

FIGURE 7.17
Illustration of Lenz's law in the circuit of Figure 7.16. Variation of i and v when v_{SRC} increases (a) or decreases (b) in Figure 7.16.

or

$$i = (v_{SRC} - v)/R_{src} \tag{7.28}$$

Let v_{SRC} be a steady voltage of 4 V, initially and up to t = 0.5 s (Figure 7.17a). When v_{SRC} has been steady for a sufficiently long time, i does not change with time so that $v = Ldi/dt = 0$. It follows from Equation 7.28 that i is steady at v_{SRC}/R_{src} = 2 A (Figure 7.17a).

When v_{SRC} increases with time, starting at t = 0.5 s, it is to be expected from Figure 7.16b and from Equation 7.28 that i, the current in the circuit, will also increase with v_{SRC}, so that $di/dt > 0$. According to Lenz's law, v opposes the increase in i by having a positive value that *subtracts* from v_{SRC} in Equation 7.28, as illustrated by the lower trace in Figure 7.17a. With di/dt and v, both positive, Equation 7.25 should have a positive sign so that L is positive.

On the other hand, if v_{SRC} decreases with time, as in Figure 7.17b for $t > 0.5$ s, i will decrease with time, in accordance with Equation 7.28, so that $di/dt < 0$. To oppose this decrease, in accordance with Lenz's law, the value of v becomes negative, as in the lower trace, which *adds* to v_{SRC} in Equation 7.28 and therefore opposes the decrease in i. It is seen that in both cases of i increasing or i decreasing with time, assigning i in the direction of a voltage drop v across the coil ensures that L has a positive value while satisfying Lenz's law. This justifies having a positive sign on the RHS of Equations 7.25 and 7.26.

As mentioned previously, an inductor is an energy storage element that can store magnetic energy and can deliver the stored energy to the rest of the circuit. Having i in the direction of a voltage drop v means that power absorbed by the inductor is positive and power delivered is negative. To make the power delivered positive, as is convenient in some cases, the relative directions of i and v should be reversed, as in Figure 7.18a, in which i is in the direction of a voltage rise v. It will be shown that in order to have a positive value of L under these conditions, the v–i relations for the inductor must be written with a negative sign:

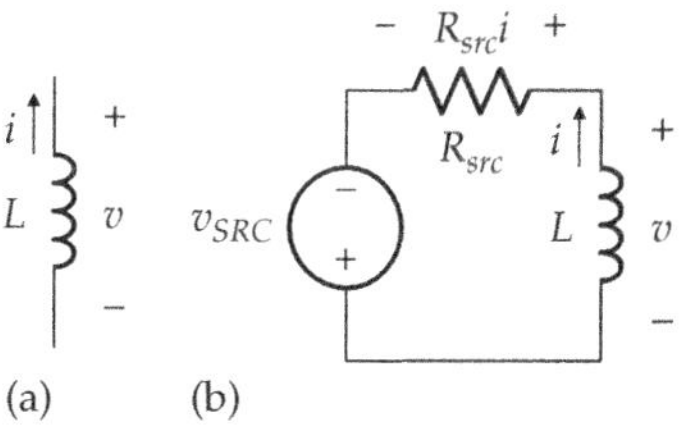

FIGURE 7.18
Assigned positive directions associated with a negative sign in Equation 7.29. (a) Current assigned in the direction of a voltage rise across the coil and (b) circuit for determining the polarity of the induced voltage.

$$v(t) = -L\frac{di}{dt} \tag{7.29}$$

and

$$i(t) = -\frac{1}{L}\int v dt = -\frac{1}{L}\int_0^t v dt + I_0 \tag{7.30}$$

The negative sign can be justified with reference to Figure 7.18b, in which the polarity of v_{SRC} is reversed in order to reverse the direction of i. From KVL,

$$-v_{SRC} + R_{src}i - v = 0 \tag{7.31}$$

or

$$i = \frac{v_{SRC} + v}{R_{src}} \tag{7.32}$$

When v_{SRC} decreases with time, i will also decrease with time in accordance with Equation 7.32, so that $di/dt < 0$. v opposes the decrease in i by having a positive value that *adds* to v_{SRC} in Equation 7.32, as

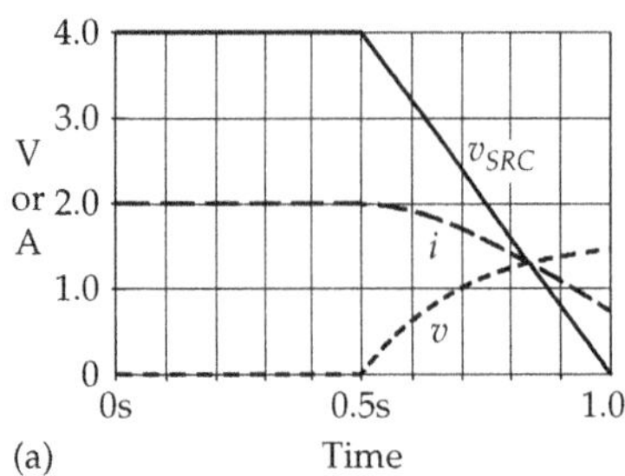

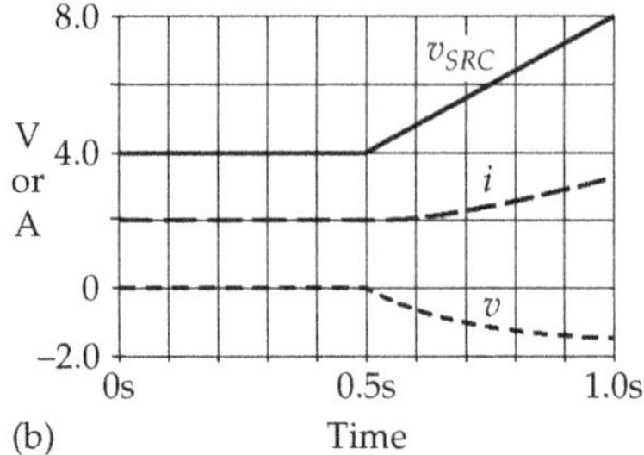

FIGURE 7.19
Illustration of Lenz's law in the circuit of Figure 7.18. Variation of i and v when v_{SRC} decreases (a) or increases (b) in Figure 7.18.

illustrated by the lower trace in Figure 7.19a. With di/dt negative and v positive, the negative sign in Equation 7.29 ensures that L is positive. On the other hand, if v_{SRC} increases with time, i will increase with time, in accordance with Equation 7.32, so that $di/dt > 0$. To oppose this increase, the value of v becomes negative, as in the lower trace in Figure 7.19b, which *subtracts* from v_{SRC} in Equation 7.32. The negative sign in Equation 7.29 again ensures that L is positive. It is seen that in both cases of i increasing or i decreasing with time, assigning i in the direction of a voltage rise across the coil ensures that L has a positive value while satisfying Lenz's law. This justifies writing Equations 7.29 and 7.30 with a negative sign on the RHS.

In both cases of Figures 7.16 and 7.18, the following concept applies:

Concept: *If the current through an inductor increases with time, this increase is opposed, in accordance with Lenz's law, by an induced voltage in the inductor that is a voltage drop in the direction of current. Conversely, if the current through an inductor decreases with time, this decrease is opposed by an induced voltage in the inductor that is a voltage rise in the direction of current.*

In the case of both capacitors and inductors, the following sign convention applies, consistent with the passive sign convention (Section 1.7):

Sign Convention: *If the assigned positive direction of current through a capacitor or an inductor is in the direction of a voltage drop across the circuit element, the voltage–current relations for the capacitor or the inductor are written with a positive sign. On the other hand, if the assigned positive direction of current through the capacitor or inductor is in the direction of a voltage rise across the circuit element, the voltage–current relations for the capacitor or the inductor are written with a negative sign. The foregoing ensures that C and L are positive quantities irrespective of the assignment of the positive direction of current through the circuit element and the positive polarity of voltage across the circuit element.*

7.2.5 Steady Inductor Current

When the inductor current is a steady current I_{SD}, the flux linkage in the inductor does not change with time, $v(t) = LdI_{SD}/dt = 0$, in accordance with Equation 7.25. In other words, *the inductor acts as a short circuit when the inductor current is steady,* as illustrated in Figure 7.20a.

However, it should be noted that if an ideal steady voltage source V_{SD} is applied across the inductor, with no initial flux in the inductor, it follows from Equation 7.26 that

$$i(t) = \frac{1}{L}\int_0^t V_{SD}dt = \frac{V_{SD}}{L}t \tag{7.33}$$

The inductor current increases linearly with time at a rate V_{SD}/L, as illustrated in Figure 7.20b. Conversely, an inductor current that increases linearly with time produces a steady inductor voltage. Thus, it follows from Figure 7.20b that if the inductor current is $V_{SD}t/L$, the inductor voltage is $v = Ldi/dt = V_{SD}$.

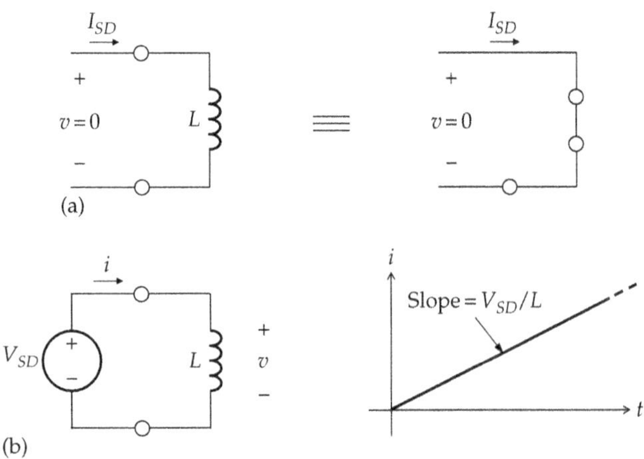

FIGURE 7.20
(a) Inductor as a short circuit when a steady current flows. (b) Inductor current increasing with time due to a steady applied voltage.

7.2.6 Stored Energy

The energy stored in the magnetic field of an inductor is equal to the work done in establishing the flux in the inductor against the induced voltage, which opposes the increase in inductor current during the establishment of the flux. Consider that the assigned direction of i through an inductor is that of a voltage drop v across L, as in Figure 7.16a. The instantaneous power input to the inductor is $p = vi$. Assuming that the voltage is applied at $t = 0$ with $i = 0$ and $\lambda = 0$ for $t < 0$, the energy supplied over the interval from $t = 0$ to t is

$$w(t) = \int_0^t pdt = \int_0^t vidt \quad (7.34)$$

Substituting $d\lambda = vdt$ from Faraday's law and $d\lambda = Ldi$ from Equation 7.18, Equation 7.34 becomes

$$w(t) = \int_0^t id\lambda = L\int_0^t idi = \frac{1}{2}Li^2 = \frac{1}{2}\frac{\lambda^2}{L} = \frac{1}{2}\lambda i \quad (7.35)$$

It should be kept in mind that electric energy in an electric circuit is represented by energy stored in capacitors. It is a function of voltages across these capacitors and represents potential energy of current carriers with respect to an arbitrary zero reference. On the other hand, magnetic energy in an electric circuit is represented by energy stored in inductors. It is a function of currents through these inductors and represents kinetic energy of current carriers.

Primal Exercise 7.7

Determine the inductance of a coil of 500 turns on a toroidal core, as in Figure 7.15, having a mean diameter of 10 cm, a cross-sectional area of 0.8 cm², and $\mu_r = 2000$.

Ans. 1.6 H.

Example 7.3: Inductor Response to Trapezoidal Current

The current shown in Figure 7.21 is applied to a 1 μH inductor of zero initial flux. Determine as a function of time (a) the flux linkage of the inductor, (b) the inductor voltage, (c) the power absorbed by the inductor, (d) and the energy absorbed by the inductor. Assume the assigned positive directions of Figure 7.16.

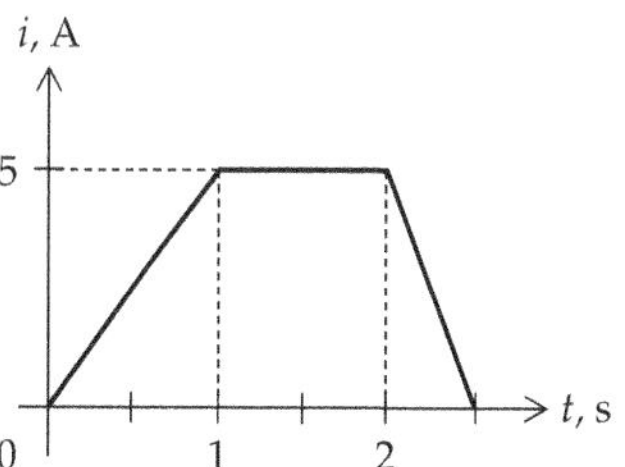

FIGURE 7.21
Figure for Example 7.3.

Solution:

(a) The first step is to express the current as a function of time, as explained in Example 7.1. Thus, $i(t) = 5t$ A, $0 \le t \le 1$ s; $i(t) = 5$ A, $1 \le t \le 2$ s; and $i(t) = -10(t - 2.5)$ A $= 10(2.5 - t)$, $2 \le t \le 2.5$ s.
   From Equation 7.20, $\lambda = Li = 10^{-6} \times i$ Vs $= i$ μVs. It follows that $\lambda(t) = 5t$ μVs, $0 \le t \le 1$ s; $\lambda(t) = 5$ μVs, $1 \le t \le 2$ s; $\lambda(t) = 10(2.5 - t)$ μVs, $2 \le t \le 2.5$ s; and $\lambda(t) = 0$, $t \ge 2.5$ s (Figure 7.22a).

(b) $v(t) = d\lambda/dt$; hence, $v(t) = 5$ μVs/s $= 5$ μV, $0 \le t \le 1$ s; $v(t) = 0$, $1 \le t \le 2$ s during the time that i is steady; $v(t) = -10$ μV, $\le t \le 2.5$ s; and $v(t) = 0$, $t \ge 2.5$ s (Figure 7.22b). Note that since λ starts and ends with zero, the positive area of the v vs. t graph is equal in magnitude to the negative area.

(c) $p = vi$; multiplying v and i over the various time ranges gives $p(t) = 25t$ μW, $0 \le t \le 1$ s; $p(t) = 0$, $1 \le t \le 2$ s; $p(t) = -100(2.5 - t)$, $2 \le t \le 2.5$ s; and $p(t) = 0$, $t \ge 2.5$ s (Figure 7.22c). Note that the flux linkage of the inductor increases and the inductor absorbs power during the interval $0 \le t \le 1$ s while v and i are positive. No power is absorbed or delivered during the interval $1 \le t \le 2$ s when $v = 0$. During the interval $2 \le t \le 2.5$ s, $i < 0$, the flux linkage decreases, and the inductor delivers power. At $t = 2.5$ s, $\lambda = 0$.

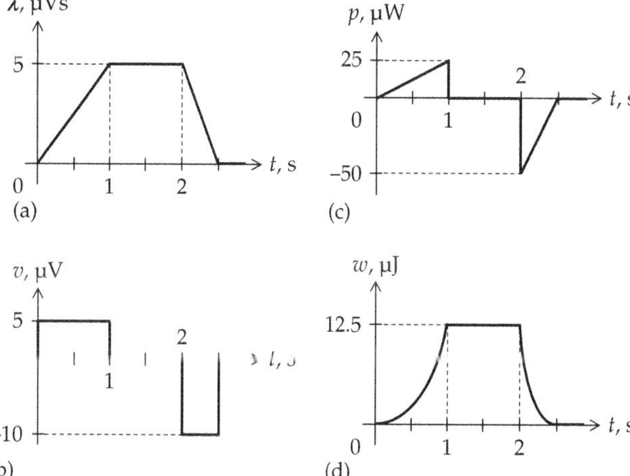

FIGURE 7.22
Figure for Example 7.3.

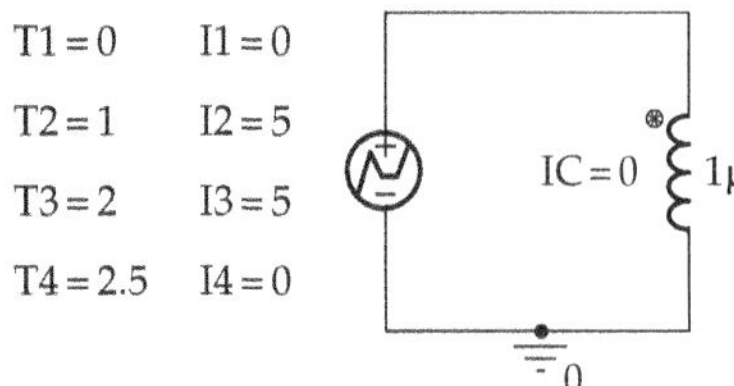

FIGURE 7.23
Figure for Example 7.3.

This means that all the energy absorbed during the interval $0 \le t \le 1$ s is returned to the supply during the interval $2 \le t \le 2.5$ s. Since energy is the area under the power curve, the positive area of the triangle extending from $t = 0$ to $t = 1$ s, which is 12.5 µJ, must be equal in magnitude to the negative area extending from $t = 2$ to $t = 2.5$ s, which is in fact −12.5 µJ.

(d) The energy as a function of time is obtained as $w(t) = \int_0^t p dt$. Thus, $w(t) = \int_0^t 25 dt = 12.5t^2$, $0 \le t \le 1$ s. During the interval $1 \le t \le 2$ s, when no power is absorbed or delivered, the energy stored stays at 12.5 µJ. During the interval $2 \le t \le 2.5$ s, $w(t) = \int_2^t -100(2.5-t)dt + 12.5 = 50(t^2 - 5t + 6.25) = 50(t - 2.5)^2$; $w = 0$ at $t = 2.5$ s, and since $p(t) = 0$ for $t \ge 2.5$ s, $w(t) = 0$ for $t \ge 2.5$ s. The variation of $w(t)$ with t is shown in Figure 7.22d. Note that since the current returns to zero for $t \ge 2.5$ s, $w = 0$.

Simulation: The circuit is entered as illustrated in Figure 7.23. The inductor is entered from the Analog library and has a marked terminal indicating that current entering this terminal is considered positive. Zero initial condition is entered as explained in Example 7.1. In the Simulation Settings, 'Analysis type' is 'Time Domain (Transient)', 'Run to time' is 3s, 'Start Saving Data After' is 0, and 'Maximum Step size' is 0.5m. After the simulation is run, then in order to display the current and power dissipated to appropriate vertical scales, select Trace/Add Trace in the Schematic1 window and enter 10*V(L1:1),W(L1). The time variation of inductor voltage and dissipated power is displayed, as in Figure 7.24, which is similar to Figure 7.8.

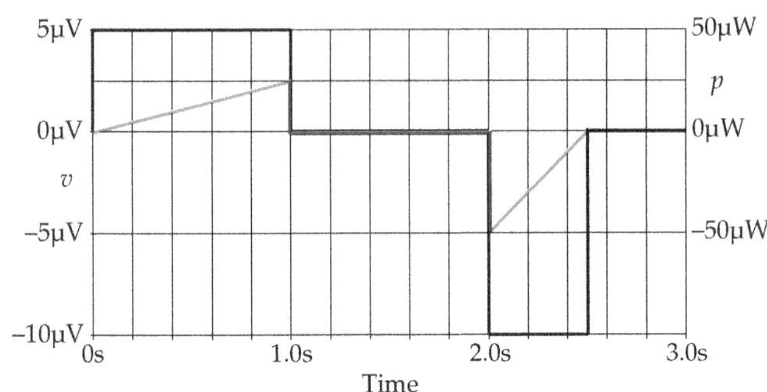

FIGURE 7.24
Figure for Example 7.3.

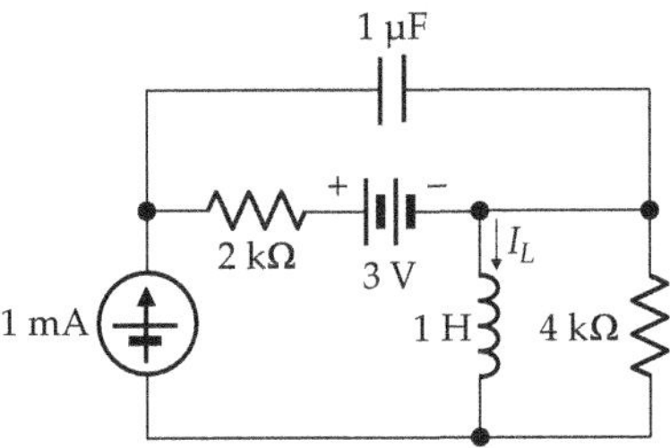

FIGURE 7.25
Figure for Primal Exercise 7.9.

Primal Exercise 7.8

Repeat Example 7.3 assuming the current increases linearly from 0 to 5 A in 0.5 s, stays at 5 A for 1 s, then decreases linearly to zero in 1 s.

Ans. $0 < t < 0.5$ s: $\lambda(t) = 10t$ µWb-T, $v(t) = 10$ µV, $p(t) = 100t$ µW, $w(t) = 50t^2$ µJ; $0.5 < t < 1.5$ s: $\lambda(t) = 5$ µWb-T, $i(t) = 0$, $p(t) = 0$, $w(t) = 12.5$ µJ; $1.5 < t < 2.5$ s: $\lambda(t) = -5t + 12.5$ µWb-T, $v(t) = -5$ µV, $p(t) = 25t - 62.5$ µW, $w(t) = 12.5t^2 - 62.5t + 78.125$ µJ.

Primal Exercise 7.9

Determine the energy stored in the circuit of Figure 7.25 under dc conditions.

Ans. 13 µJ.

7.3 Series and Parallel Connections of Initially Uncharged Capacitors

Ohm's law (Equation 2.2) and the v–i relation for a capacitor (Equation 7.6) are

$$i = Gv \quad \text{and} \quad i(t) = C\frac{dv}{dt} \tag{7.36}$$

Comparing these relations, it is seen that v is multiplied by G in the case of a resistor, whereas a function of v (its derivative) is multiplied by C in the case of a capacitor. This makes the derivations of the equivalent series and parallel elements analogous in both cases, which results in capacitances combining in series and in parallel in the same manner as conductances.

7.3.1 Series Connection of Initially Uncharged Capacitors

Figure 7.26a shows a series connection of 3 capacitors to which is applied a test current i_T that is an arbitrary

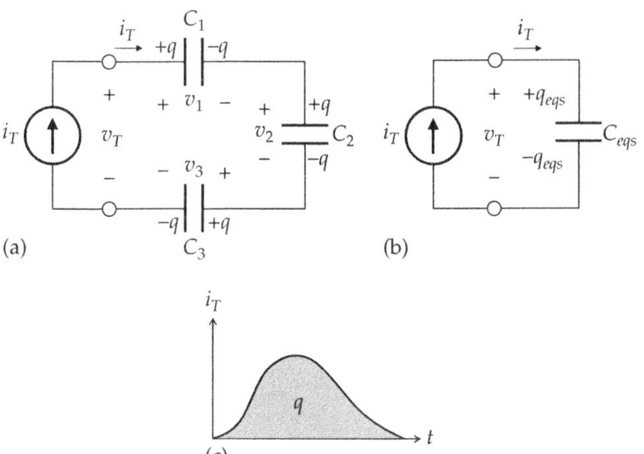

FIGURE 7.26
Series connection of capacitors (a), equivalent series capacitor (b), and capacitor current and charge (c).

function of time (Figure 7.26c). The area under the i_T vs. t curve represents the charge that is applied by the source. Considering i_T as the flow of positive charge, i_T adds a positive charge $+q$ to the left plate of C_1 and removes a positive charge $+q$ from the left plate of C_3, leaving a charge of $-q$ on this plate. Through electrostatic charge induction, charge is redistributed in the three capacitors so that each capacitor will have a charge of $+q$ on one plate and $-q$ on the opposite plate. With no initial charges on the capacitors, and no external connections to the plates of C_1 and C_2 that are connected together, nor to the plates of C_2 and C_3 that are connected together, the charges on the plates that are connected together add to zero, before and after i_T is applied. This is in accordance with conservation of charge.

If the voltage across the series combination is v_T, then according to KVL,

$$v_T = v_1 + v_2 + v_3 \tag{7.37}$$

The voltage across any capacitor is of the form

$$v(t) = \frac{1}{C}\int_0^t i_T dt + v(0) \tag{7.38}$$

Assuming zero initial charge, $v(0) = 0$. Substituting for each voltage in Equation 7.37 and factoring out the time integral of current,

$$v_T(t) = \left(\frac{1}{C_1} + \frac{1}{C_2} + \frac{1}{C_3}\right)\int_0^t i_T dt = \frac{1}{C_{eqs}}\int_0^t i_T dt \tag{7.39}$$

The bracketed sum defines an equivalent series capacitor that has the same v_T, i_T, and hence $\int_0^t i_T dt$ (Figure 7.26b). It follows from Equation 7.39 that

$$\frac{1}{C_{eqs}} = \frac{1}{C_1} + \frac{1}{C_2} + \frac{1}{C_3} \tag{7.40}$$

It is seen that *the reciprocals of capacitances in series add, just like conductances*. It is to be expected, therefore, that voltage divides between capacitors in series as in the case of conductances. The voltage across any of the capacitors, say, C_1, is

$$v_1(t) = \frac{1}{C_1}\int_0^t i_T dt, \quad \text{with } v_1(0) = 0,\ t \le 0 \tag{7.41}$$

Dividing Equation 7.41 by Equation 7.39,

$$\frac{v_1}{v_T} = \frac{1/C_1}{1/C_1 + 1/C_2 + 1/C_3} \tag{7.42}$$

Similarly, repeating the same procedure for v_2 and v_3,

$$\frac{v_2}{v_T} = \frac{1/C_2}{1/C_1 + 1/C_2 + 1/C_3} \quad \text{and} \quad \frac{v_3}{v_T} = \frac{1/C_3}{1/C_1 + 1/C_2 + 1/C_3} \tag{7.43}$$

It follows from Equations 7.42 and 7.43 that

$$\frac{v_1}{v_2} = \frac{C_2}{C_1}, \quad \frac{v_1}{v_3} = \frac{C_3}{C_1}, \quad \frac{v_2}{v_3} = \frac{C_3}{C_2} \tag{7.44}$$

The voltages across any two of the capacitors in series, without initial charges, are in inverse ratio to the capacitances.

The same i_T applied to C_{eqs} results in the same charge $q(t) = \int_0^t i_T dt$ on this capacitor, as on each of the series-connected capacitors (Figure 7.26c), so that

$$q = q_{eqs} \tag{7.45}$$

If Equation 7.37 is multiplied by $q_{eqs}/2$ or by the equal quantity $q/2$,

$$\frac{1}{2}qv_1 + \frac{1}{2}qv_2 + \frac{1}{2}qv_3 = \frac{1}{2}q_{eqs}v_T \tag{7.46}$$

In words, the energy stored in C_{eqs} is the same as the total energy stored in the three capacitors. This is to be expected, because v_T and i_T are the same, so the power input and energy are the same, in accordance with conservation of energy.

Although derived for the case of three capacitors, the preceding results can be generalized to any number of initially uncharged capacitors in series. In particular, if n identical capacitors, each of capacitance C, are connected in series, $C_{eqs} = C/n$ and the voltage across each capacitor is v_T/n. Moreover, C_{eqs} is smaller than the smallest of the series-connected capacitances, as in the case of conductances in series.

7.3.2 Parallel Connection of Initially Uncharged Capacitors

Consider a parallel connection of 3 capacitors across which a time-varying test voltage v_T is applied (Figure 7.27a). If the test current is i_T, the capacitor currents are, from KCL,

$$i_T = i_1 + i_2 + i_3 \tag{7.47}$$

The current through any capacitor is of the form $i(t) = Cdv_T/dt$. Substituting for each current in Equation 7.47 and factoring out the voltage derivative,

$$I_T(t) = (C_1 + C_2 + C_3)\frac{dv_T}{dt} = C_{eqp}\frac{dv_T}{dt} \tag{7.48}$$

The bracketed sum defines an equivalent parallel capacitor that has the same i_T, v_T, and hence dv_T/dt (Figure 7.27b). Thus,

$$C_{eqp} = C_1 + C_2 + C_3 \tag{7.49}$$

It is seen that *capacitances in parallel add, just like conductances*. It is to be expected, therefore, that current divides between capacitors in parallel as in the case of conductances. The current through any of the capacitors, say, C_1, is

$$i_1(t) = C_1\frac{dv_T}{dt} \tag{7.50}$$

Dividing Equation 7.50 by Equation 7.48,

$$\frac{i_1}{i_T} = \frac{C_1}{C_1 + C_2 + C_3} \tag{7.51}$$

Similarly,

$$\frac{i_2}{i_T} = \frac{C_2}{C_1 + C_2 + C_3} \quad \text{and} \quad \frac{i_3}{i_T} = \frac{C_3}{C_1 + C_2 + C_3} \tag{7.52}$$

It follows from Equations 7.51 and 7.52 that

$$\frac{i_1}{i_2} = \frac{C_1}{C_2}, \quad \frac{i_1}{i_3} = \frac{C_1}{C_3}, \quad \frac{i_2}{i_3} = \frac{C_2}{C_3} \tag{7.53}$$

The charge on any capacitor, say, C_1, is

$$q_1(t) = \int_0^t i_1 dt \qquad q_1 = 0, t \le 0 \tag{7.54}$$

Since v_T is the same for the paralleled capacitors, and $q = Cv_T$ for each capacitor, the charge on each capacitor is proportional to its capacitance. Hence, each current in Equations 7.51 through 7.53 can be replaced by the corresponding charge.

If both sides of Equation 7.49 are multiplied by the common voltage v_T, and substituting the relation $q = Cv$,

$$q_{eqp} = q_1 + q_2 + q_3 \tag{7.55}$$

That is, the total charge on the upper node in Figure 7.27a is the sum of the charges on all the capacitors and is the same as the charge on the equivalent parallel capacitor. This is to be expected, since the same current i_T delivers the same charge to the three capacitors in parallel as to C_{eqp}.

If both sides of Equation 7.55 are multiplied by $v_T/2$,

$$\frac{1}{2}q_1v_T + \frac{1}{2}q_2v_T + \frac{1}{2}q_3v_T = \frac{1}{2}q_{eqp}v_T \tag{7.56}$$

In words, the energy stored in C_{eqp} is the sum of the energies stored in the individual capacitors.

Although derived for the case of three capacitors, the preceding results can be generalized to any number of capacitors in parallel. In particular, if n identical

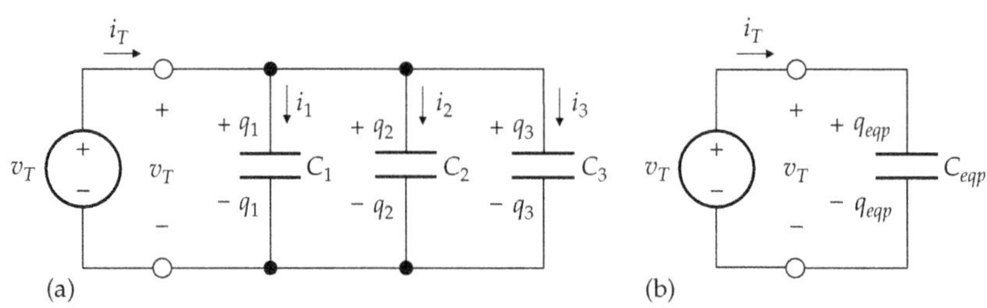

FIGURE 7.27
Parallel connection of capacitors (a) and equivalent parallel capacitor (b).

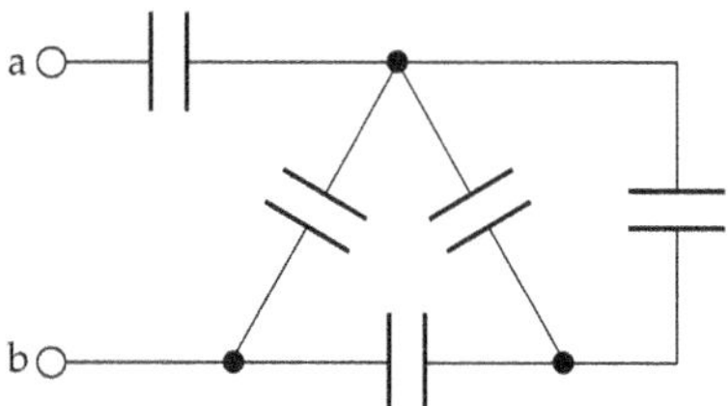

FIGURE 7.28
Figure for Primal Exercise 7.10.

capacitors, each of capacitance C, are connected in parallel, $C_{eqp} = nC$ and the charge on each capacitor is $q = q_{eqp}/n$. Moreover, C_{eqp} is larger than the largest of the parallel-connected capacitances, as in the case of conductances.

Summary: *In a series connection of initially uncharged capacitors, (i) the reciprocal of the equivalent series capacitance is the sum of the reciprocals of the individual capacitances, (ii) voltages divide inversely as the capacitances, and (iii) the charges on each of the capacitors and on the equivalent series capacitor are equal.*

In a parallel connection of initially uncharged capacitors, (i) the equivalent parallel capacitance is the sum of the individual capacitances, (ii) currents divide directly as the capacitances, and (iii) the charge on the equivalent parallel capacitance is the sum of the charges on the paralleled capacitances.

In both the series and parallel connections, the energy in the equivalent capacitor is the sum of the energies in the individual capacitors.

Primal Exercise 7.10

Determine the equivalent capacitance between terminals 'ab' in Figure 7.28, assuming all capacitances are 1 F.

Ans. 5/8 F.

7.4 Series and Parallel Connections of Initially Uncharged Inductors

Ohm's law (Equation 2.1) and the v–i relation for an inductor (Equation 7.25) are

$$v = Ri \quad \text{and} \quad v(t) = L\frac{di}{dt} \tag{7.57}$$

Comparing these relations, it is seen that i is multiplied by R in the case of a resistor, whereas a function of i (its derivative) is multiplied by L in the case of an inductor. This makes the derivations of the equivalent series and parallel elements analogous in both cases, which results in inductances combining in series and in parallel in the same manner as resistances.

7.4.1 Series Connection of Initially Uncharged Inductors

Consider a series connection of 3 inductors to which a time-varying test current i_T is applied (Figure 7.29a). If the test voltage is v_T, KVL gives in terms of the voltages across the inductors

$$v_T = v_1 + v_2 + v_3 \tag{7.58}$$

The voltage across any inductor is of the form $v(t) = Ldi_T/dt$. Substituting in Equation 7.58 for the voltage across each inductor and factoring out the di_T/dt term,

$$v_T(t) = (L_1 + L_2 + L_3)\frac{di_T}{dt} = L_{eqs}\frac{di_T}{dt} \tag{7.59}$$

The bracketed sum defines an equivalent series inductor that has the same v_T, i_T and hence di_T/dt (Figure 7.29b). Thus,

$$L_{eqs} = L_1 + L_2 + L_3 \tag{7.60}$$

It is seen that *inductances in series add, just like resistances*. To determine how the voltage divides between the inductors, we note that the voltage across any of the inductors, say, L_1, is $v_1(t) = L_1 di_T/dt$. Dividing this equation by Equation 7.59,

$$\frac{v_1}{v_T} = \frac{L_1}{L_1 + L_2 + L_3} \tag{7.61}$$

Similarly,

$$\frac{v_2}{v_T} = \frac{L_2}{L_1 + L_2 + L_3} \quad \text{and} \quad \frac{v_3}{v_T} = \frac{L_3}{L_1 + L_2 + L_3} \tag{7.62}$$

It follows from Equations 7.61 and 7.62 that

$$\frac{v_1}{v_2} = \frac{L_1}{L_2}, \quad \frac{v_1}{v_3} = \frac{L_1}{L_3}, \quad \frac{v_2}{v_3} = \frac{L_2}{L_3} \tag{7.63}$$

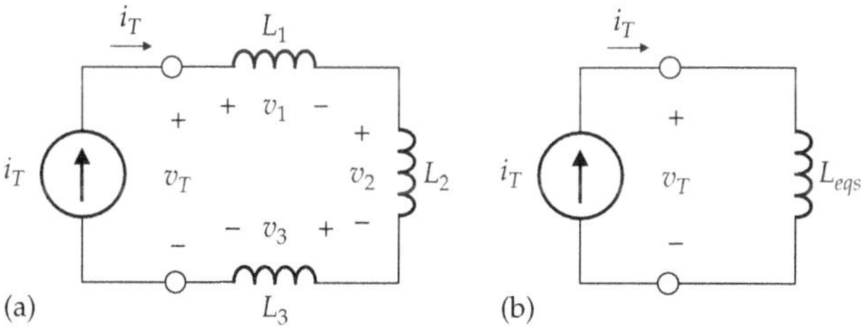

FIGURE 7.29
Series connection of inductors (a) and equivalent series inductor (b).

It is seen from Equations 7.63 that voltages across inductors in series divide in the same ratio as the inductances.

The flux linkage of any inductor, say, L_1, is

$$\lambda_1(t) = \int_0^t v_1 dt \quad \lambda_1(t) = 0, t \leq 0 \tag{7.64}$$

Since i_T is the same for the series-connected inductors, and $\lambda = Li_T$ for each inductor, the flux linkage of each inductor is proportional to its inductance. Hence, each voltage in Equations 7.61 through 7.63 can be replaced by the corresponding flux linkage.

If both sides of Equation 7.60 are multiplied by i_T and the relation $\lambda = Li$ substituted for each element, the resulting equation is

$$\lambda_{eqs} = \lambda_1 + \lambda_2 + \lambda_3 \tag{7.65}$$

The flux linkage in the equivalent series inductor is therefore the sum of the flux linkages in the individual inductors. This follows from the fact that the voltage across L_{eqs} is the sum of the voltages across the individual inductors, so that the time integral of the voltage across L_{eqs}, which is the flux linkage in L_{eqs}, is the sum of the integrals of the voltages across the individual inductors.

If both sides of Equation 7.65 are multiplied by $i_T/2$,

$$\frac{1}{2}\lambda_1 i_T + \frac{1}{2}\lambda_2 i_T + \frac{1}{2}\lambda_3 i_T = \frac{1}{2}\lambda_{eqs} i_T \tag{7.66}$$

In words, the energy stored in L_{eqs} is the same as the total energy stored in the three capacitors.

Although derived for the case of three inductors, the preceding results can be generalized to any number of inductors in series. In particular, if n identical inductors, each of inductance L, are connected in series, $L_{eqs} = nL$ and the voltage v across each inductor is v_T/n, where v_T is the total voltage across the series combination. Moreover, L_{eqs} is larger than the largest of the series-connected inductances, as in the case of resistances.

7.4.2 Parallel Connection of Initially Uncharged Inductors

Consider a parallel connection of 3 inductors across which is applied a test voltage v_T that is an arbitrary function of time (Figure 7.30c). The area under the v_T vs. t curve represents flux linkage that is established by the source. If the test current is i_T, it follows from KCL that

$$I_T = i_1 + i_2 + i_3 \tag{7.67}$$

The current through any inductor is $i(t) = \frac{1}{L}\int_0^t v_T dt$, assuming no initial current, and hence no initial flux linkage in the inductors. Substituting in Equation 7.67 for the current in each inductor and factoring out the $\int_0^t v_T dt$ term,

$$i_T(t) = \left(\frac{1}{L_1} + \frac{1}{L_2} + \frac{1}{L_3}\right)\int_0^t v_T dt = \frac{1}{L_{eqp}}\int_0^t v_T dt \tag{7.68}$$

The bracketed sum defines an equivalent parallel inductor that would have the same i_T, v_T, and hence $\int_0^t v_T dt$ (Figure 7.30b). Thus,

$$\frac{1}{L_{eqp}} = \frac{1}{L_1} + \frac{1}{L_2} + \frac{1}{L_3} \tag{7.69}$$

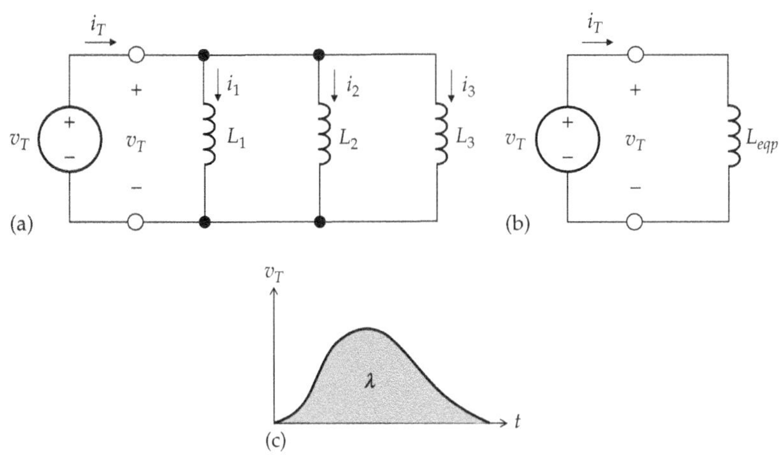

FIGURE 7.30
Parallel connection of inductors (a), equivalent parallel inductor (b), and inductor voltage and flux linkage (c).

It is seen that *the reciprocals of inductances in parallel add, just like resistances.* To determine how the current divides between inductors in parallel, we note that the current through any of the inductors, say, L_1, is $i_1(t) = \frac{1}{L_1}\int_0^t v_T dt$. Dividing by Equation 7.68,

$$\frac{i_1}{i_T} = \frac{1/L_1}{1/L_1 + 1/L_2 + 1/L_3} \quad (7.70)$$

Similarly,

$$\frac{i_2}{i_T} = \frac{1/L_2}{1/L_1 + 1/L_2 + 1/L_3} \text{ and } \frac{i_3}{i_T} = \frac{1/L_3}{1/L_1 + 1/L_2 + 1/L_3} \quad (7.71)$$

Applying Equations 7.70 and 7.71 to any two currents,

$$\frac{i_1}{i_2} = \frac{L_2}{L_1} \quad \frac{i_1}{i_3} = \frac{L_3}{L_1} \quad \frac{i_2}{i_3} = \frac{L_3}{L_2} \quad (7.72)$$

From Faraday's law in the form $\lambda(t) = \int_0^t v_T dt$ (Figure 7.30c), with no initial current in the inductors, and since v_P is the same for all inductors in parallel,

$$\lambda_1 = \lambda_2 = \lambda_3 = \lambda_{eqp} \quad (7.73)$$

As v_T builds up between 0 and t, the same flux linkage is established in all the paralleled inductors and in λ_{eqp}.

Multiplying both sides of Equation 7.67 by the common $\lambda/2$,

$$\frac{1}{2}\lambda_1 i_1 + \frac{1}{2}\lambda_2 i_2 + \frac{1}{2}\lambda_3 i_3 = \frac{1}{2}\lambda_{eqp} i_P \quad (7.74)$$

The energy stored in L_{eqp} is therefore the sum of the energies stored in the individual inductors.

Although derived for the case of three inductors, the results of this section can be generalized to any number of inductors in parallel. In particular, if n identical inductors, each of inductance L, are connected in parallel, $L_{eqp} = L/n$ and the current i in each inductor is $i = i_T/n$, where i_T is the current through the parallel combination. Moreover, L_{eqp} is smaller than the smallest of the parallel connected inductances, as in the case of resistors.

Summary: *In a series connection of initially uncharged inductors, (i) the equivalent series inductance is the sum of the individual inductances, (ii) voltages divide directly as the inductances, and (iii) the flux linkage of the equivalent series inductance is the sum of the flux linkages of the individual inductors.*

In a parallel connection of initially uncharged inductors, (i) the reciprocal of the equivalent parallel inductance is the sum of the reciprocals of the individual inductances, (ii) currents divide inversely as the inductances, and (iii) the flux linkages of each of the inductors and of the equivalent parallel inductance are equal.

In both the series and parallel connections, the energy in the equivalent inductor is the sum of the energies in the individual inductors.

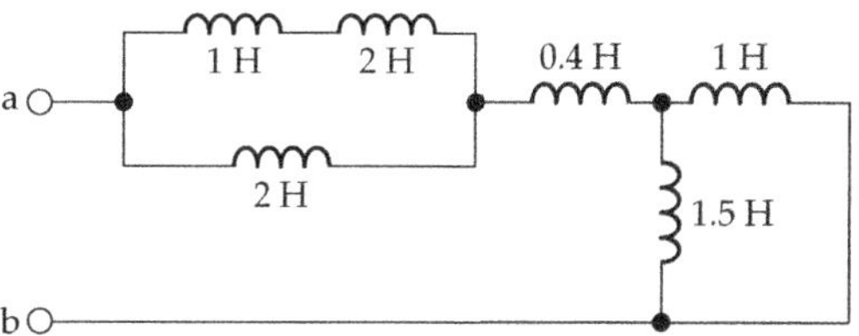

FIGURE 7.31
Figure for Primal Exercise 7.11.

Primal Exercise 7.11

Determine the equivalent inductance between terminals 'ab' in Figure 7.31.

Ans. 2.2 H.

7.5 Duality

Duality in electric circuits is an important and useful concept that facilitates circuit analysis in some cases and provides a better understanding of the comparative behavior of inductive and capacitive circuits.

The basis of duality is the sameness of expressions when voltage and current, as well as some circuit parameters, are interchanged in v–i relations. Consider the two expressions of Ohm's law and the v–i relations of capacitors and inductors:

$$v = Ri \qquad i = Gv \quad (7.75)$$

$$v(t) = L\frac{di}{dt} \qquad i(t) = C\frac{dv}{dt} \quad (7.76)$$

The two expressions of Ohm's law (Equations 7.75) are of the same form. If v and i are interchanged in the first expression ($v = Ri$), and R is replaced by G, the second expression ($i = Gv$) is obtained. Conversely, if i and v are interchanged in the second expression, and G is replaced by R, the first expression is obtained. The two expressions of Ohm's law become dual relations if the quantities being interchanged also have the same numerical values. For example, in the case of a 10 Ω resistor carrying a current of 2 A, Ohm's law, in the form $v = Ri$, is 20 (V) = (10 Ω) × (2 A). Its dual relation, in the form $i = Gv$, is 20 (A) = (10 S) × (2 V). In these relations, 20 V, 10 Ω, and 2 A in $v = Ri$ are replaced, respectively, by 20 A, 10 S, and 2 V in $i = Gv$. This expression now refers to a resistor of 10 S conductance, across which 2 V are applied.

Similarly, Equations 7.76 are of the same form, and one reduces to the other when v and i, as well as L and C, are interchanged. If 10 V are applied across a 2 H inductor, the v–i relation is 10 (V) = (2 H) × (di/dt A/s). The dual relation is 10 (A) = (2 F) × (dv/dt V/s) and represents a 2 F capacitor to which a 10 A current source is applied. In these dual relations, di/dt and dv/dt would have the same numerical values.

Formally, dual relations can be defined as follows:

Definition: *Two v–i relations are duals if one relation reduces to the other when the following are interchanged: (i) v and i, AND (ii) R and G, as well as L and C. Moreover, the quantities interchanged should have the same numerical values.*

The quantities involved in these interchanges are referred to as *dual quantities*. Thus, v and i are dual circuit variables, R and G are dual circuit parameters. A number of duality relationships are listed in Table 7.1. Dual circuit variables are not only the primary circuit variables of voltage and current. Thus, for a capacitor, $q = Cv$, whereas for an inductor $\lambda = Li$, and the relation $i = dq/dt$ is the dual of Faraday's law, $v = d\lambda/dt$. Since v is the dual of i and C is the dual of L, then q is the dual of λ. Independent voltage sources and independent current sources are duals. In the case of dependent sources, duality applies to the controlling variable as well. Thus, a CCVS and a VCCS are duals, as are a CCCS and a VCVS. KVL, according to which the algebraic sum of voltages around a mesh or a loop is zero, is the dual of KCL, according to which the algebraic sum of currents at a node is zero. It follows that the dual of a node is a mesh or a loop. A short circuit and an open circuit are duals, since the current in an open circuit is zero, whereas the voltage across a short circuit is zero. Similarly, a series connection and a parallel connection are duals, since in a series connection all the elements have the same current, whereas in a parallel connection all the elements have the same voltage. Consequently, voltage division and current division are duals.

TABLE 7.1

Dual Circuit Entities

Dual Circuit Variables and Quantities	
v	i
q	λ
Dual circuit parameters	
R	G
C	L
Dual relations	
Resistor: $v = Ri$	Resistor: $i = Gv$
Capacitor: $i = \dfrac{dq}{dt}$	Inductor: $v = \dfrac{d\lambda}{dt}$
$q = Cv, \quad i = C\dfrac{dv}{dt}$	$\lambda = Li, \quad v = L\dfrac{di}{dt}$
KVL: $\Sigma v = 0$ algebraically around a loop	KCL: $\Sigma i = 0$ algebraically at a node
Voltage division: voltages across series-connected resistors divide in proportion to resistances	Current division: currents in paralleled resistors divide in proportion to conductances
Dual sources	
Ideal voltage source: v_{SRC} is specified for all i	Ideal current source: i_{SRC} is specified for all v
Voltage-controlled voltage source: $v_{SRC} = \alpha v_\phi$	Current-controlled current source: $i_{SRC} = \alpha i_\phi$
Current-controlled voltage source: $v_{SRC} = \beta i_\phi$	Voltage-controlled current source: $i_{SRC} = \beta v_\phi$
Dual circuit connections	
Open circuit or open switch: $i = 0$ for all v	Short circuit or closed switch: $v = 0$ for all i
Series connection: same current flows in circuit elements	Parallel connection: same voltage appears across circuit elements
Mesh: voltages add algebraically	Node: currents add algebraically

In fact, duality extends to whole circuits as well. Thus,

Definition: *Two circuits are duals if the node-voltage equations of one circuit are the dual relations of the mesh-current equations of the other circuit.*

Duality can be used to derive a relation from its dual, or to check relations by comparing them with their duals, or to enhance understanding of the behavior of circuits by comparing them with that of their duals. Once a circuit is analyzed, its dual circuit is automatically analyzed at the same time.

We have already encountered the following examples of duality in the present chapter, as listed in Table 7.2:

1. Bearing in mind that L and C are dual circuit parameters, and that the series and parallel connections are dual connections (Table 7.1), it follows that the series connection of capacitors (Equation 7.40) and the parallel connection of inductors (Equation 7.69) are dual relations,

TABLE 7.2

Examples of Duality

$\dfrac{1}{C_{eqs}} = \dfrac{1}{C_1} + \dfrac{1}{C_2} + \dfrac{1}{C_3}$	$\dfrac{1}{L_{eqp}} = \dfrac{1}{L_1} + \dfrac{1}{L_2} + \dfrac{1}{L_3}$
$C_{eqp} = C_1 + C_2 + C_3$	$L_{eqs} = L_1 + L_2 + L_3$
Capacitor current is zero when capacitor voltage is steady, the capacitor acting like an open circuit	Inductor voltage is zero when the inductor current is steady, the inductor behaving as a short circuit
I_{SRC} applied to a capacitor produces a voltage $(I_{SRC}/C)t$	V_{SRC} applied to an inductor produces a current $(V_{SRC}/L)t$
Example 7.1: Voltage variation, capacitance value, variation of q and i	Example 7.2: Current variation, inductance value, variation of λ and v

as are the parallel connection of capacitors (Equation 7.49) and the series connection of inductors (Equation 7.60). That is, these relations are of the same form, and one can be obtained from the other by interchanging L and C having the same value.

2. The capacitor current is zero when the capacitor voltage is steady, the capacitor acting like an open circuit (Figure 7.4a). From duality, the inductor voltage is zero when the inductor current is steady, the inductor acting as a short circuit (Figure 7.20a). The dual entities that are interchanged are voltage and current, inductance and capacitance, and open circuit and short circuit.
3. A dc current source I_{SRC} applied to a capacitor produces a voltage $(I_{SRC}/C)t$ (Figure 7.4b). From duality, A dc voltage source V_{SRC} applied to an inductor produces a current $(V_{SRC}/L)t$ (Figure 7.20b). Voltage and current are interchanged, as are inductance and capacitance.
4. The voltage variation in Example 7.1 (Figure 7.5) and the capacitance value are the same as the current variation in Example 7.3 (Figure 7.21) and the inductance value. The variations of q and i in Example 7.1 (Figure 7.6) are the same as the variations of λ and v in Example 7.3 (Figure 7.22). Voltage and current are interchanged, as are charge and flux linkage.

Duality is further illustrated in the problems on capacitor and inductor relations at the end of this chapter.

Example 7.4: Dual of Loaded Linear-Output Voltage Source

It is required to derive the dual of the voltage source and load shown in Figure 7.32a.

Solution:

From KVL,

$$0.5I_L + 5.5I_L = 12\text{ V} \tag{7.77}$$

where the coefficients multiplying I_L are resistances. From Equation 7.77, $I_L = 2$ A. In the dual circuit equation V_L replaces I_L so that

$$0.5V_L + 5.5V_L = 12\text{ A} \tag{7.78}$$

where the coefficients multiplying V_L are conductances. Equation 7.78 is KCL for a circuit consisting of a 12 A source in parallel with a source conductance of 0.5 S and a load conductance of 5.5 S, as shown in Figure 7.32b. From Equation 7.78, $V_L = 2$ V. The following should be noted concerning these dual circuits:

1. Each element in the series circuit of Figure 7.32a is replaced by its dual circuit element, of the same value, in the parallel circuit of Figure 7.32b. Thus, the 12 A source is the dual of the 12 V source, the 0.5 S source conductance is the dual of the 0.5 Ω source resistance, and the 5.5 S load conductance is the dual of the 5.5 Ω load resistance.
2. The voltage and current values for each of the dual circuit elements are interchanged. Thus, for the voltage source, the voltage is 12 V and the current is 2 A, whereas for the current source, the current is 12 A and the voltage is 2 V. For the source resistance, the current is 2 A and the voltage is 0.5 × 2 = 1 V, whereas for the source conductance, the voltage is 2 V and the current is 0.5 × 2 = 1 A. For the load resistance, the current is 2 A and the voltage is 5.5 × 2 = 11 V, whereas for the load conductance, the voltage is 2 V and the current is 5.5 × 2 = 11 A.
3. Since voltage and current are interchanged for each circuit element, their product remains the same, which means that the power delivered or absorbed by each circuit element is the same as that delivered or absorbed by its dual circuit element. Both the voltage source and the current source deliver 12 × 2 = 24 W. Both the source resistance and source conductance dissipate 1 × 2 = 2 W, and both the load resistance and load conductance dissipate 11 × 2 = 22 W.
4. The dual circuit is of the same *form* as the equivalent current source (Example 3.5), but the values of

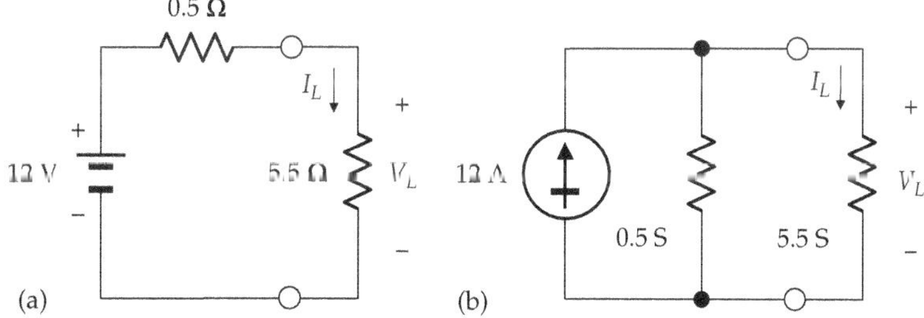

FIGURE 7.32
Figure for Example 7.4.

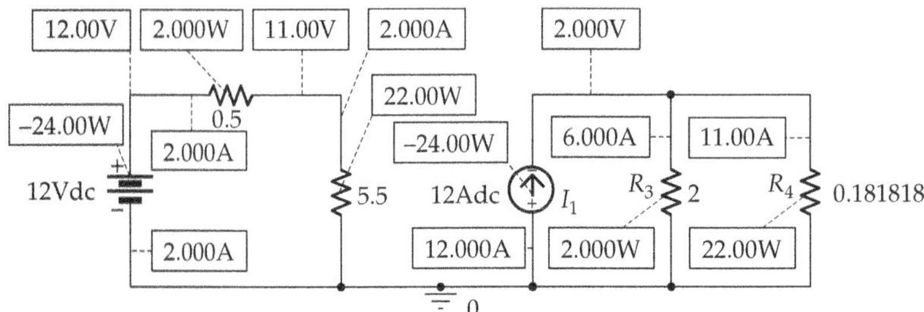

FIGURE 7.33
Figure for Example 7.4.

the circuit elements are *different*. This is because the criterion for source equivalence is that the v–i relation is the same at the source terminals, whereas in the dual circuit, v and i at these terminals are interchanged. In both cases, however, the same power is delivered to the load.

Simulation: The circuit is entered as in Figure 7.33. After selecting 'Bias Point' under 'Analysis type' in the Simulation Settings and running the simulation, pressing the pressing the I, V, and W buttons displays the currents, voltages, and powers, respectively, (Figure 7.33). The voltages and currents are duals in both cases. The same power of 24 W is delivered by both ideal source elements, the same power of 2 W is dissipated in the source resistances, and the same 22 W of power is delivered to the load.

A general procedure for deriving the dual of a planar circuit is presented in Appendix 7A. By a planar circuit is meant a circuit that can be drawn in two dimensions without any obligatory crossover connections. However, it is sufficient for our purposes to derive the dual of a circuit from duality between series and parallel connections. This is illustrated by the following example.

Example 7.5: Dual of Capacitive Voltage Divider

It is required to derive the dual of the circuit of Figure 7.9a and to determine its response under dc conditions.

Solution:

The circuit of Figure 7.9a, reproduced in Figure 7.34, consists of a 12 V source in series with two branches that are shown encircled by dashed ovals. The dual circuit consists of a 12 A source in parallel with two branches, each of which is the dual of the corresponding branch in the circuit of Figure 7.34a. The dual of the branch consisting of the 4 Ω resistor in parallel with 2 μF is a branch consisting of a 4 S resistor in series with a 2 μH inductor, as illustrated in Figure 7.34. Similarly, the dual of the branch consisting of 8 Ω in parallel with a series combination of 2 μF and 6 Ω is a branch consisting of 8 S in series with a parallel combination of 6 S and 1 μH (Figure 7.34). The circuit becomes as shown in Figure 7.35a.

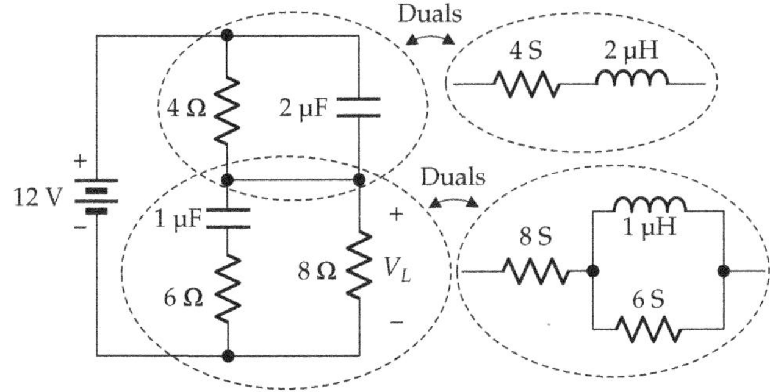

FIGURE 7.34
Figure for Example 7.5.

Under dc conditions, the inductors act as short circuits (Figure 7.35b). The 12 A current divides into 4 A in the 4 S resistor and 8 A in the 8 S resistor in accordance with current division between conductances (Equations 3.19 and 3.20). The voltage across each of these resistors, as well as the current source, is 4 A/4 S = 8 A/8 S = 1 V.

Duality between the two circuits is exemplified by the fact that the voltage and current values for each of the dual circuit elements are interchanged. This is summarized in Table 7.3. The first column lists all the corresponding dual elements in the two circuits. The second and third columns list the voltages and currents in the capacitive circuit, as shown in Figure 7.9b, whereas the last two columns list the currents and voltages in the inductive circuit of Figure 7.35b. For example, the 12 V source in the capacitive circuit has a voltage of 12 V and a current of 1 A. The dual 12 A source in the inductive circuit has a current of 12 A and a voltage of 1 V. The 4 Ω resistor in the capacitive circuit has a voltage of 4 V and a current of 1 A. Its 4 S dual in the inductive circuit has a current of 4 A and a voltage of 1 V. The 1 μF capacitor in the capacitive circuit acts as an open circuit ($I = 0$) with 8 V across it. Its 1 μH dual in the inductive circuit acts as a short circuit ($V = 0$) with 8 A through it, similarly for the other elements.

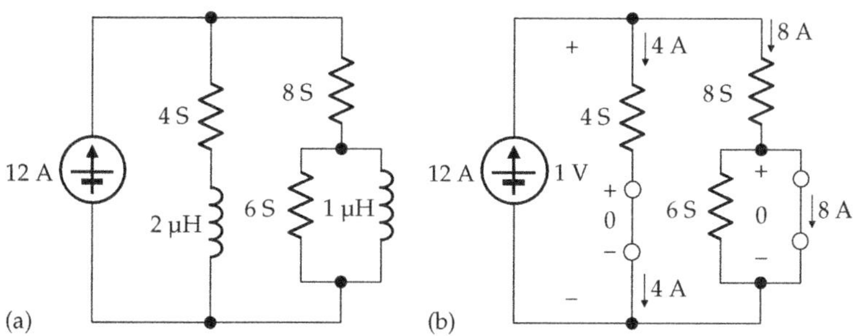

FIGURE 7.35
Figure for Example 7.5.

TABLE 7.3
Voltages and Currents in Dual Circuits

	Capacitive Circuit		Inductive Circuit	
Element	***V*, V**	***I*, A**	***I*, A**	***V*, V**
12 V/12 A sources	12	1	12	1
4 Ω/4 S resistors	4	1	4	1
6 Ω/6 S resistors	0	0	0	0
8 Ω/8 S resistors	8	1	8	1
1 μF/1 μH elements	8	0	8	0
2 μF/2 μH elements	4	0	4	0

The circuit of Figure 7.9b under dc conditions has one mesh. Its KVL equation is $4 + 8 = 12$. The circuit of Figure 7.35b is a two-essential-node circuit having the KCL equation $4 + 8 = 12$.

Simulation: The circuit is entered as in Figure 7.36. After selecting 'Bias Point' under 'Analysis type' in the Simulation Settings and running the simulation, pressing the V and I buttons displays the voltages and current shown.

Problem-Solving Tip

- An inductor is replaced by a short circuit under dc or steady conditions.

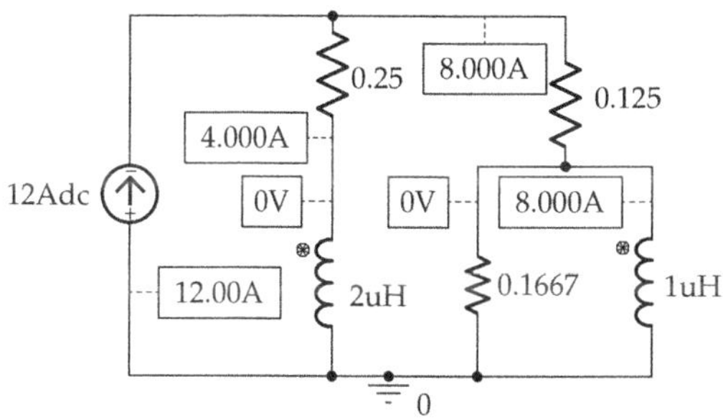

FIGURE 7.36
Figure for Example 7.5.

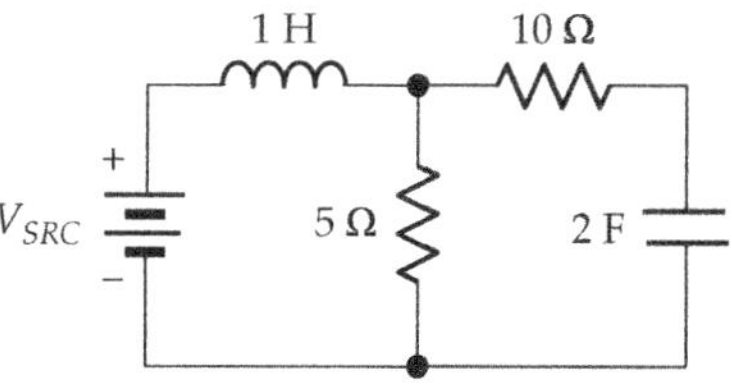

FIGURE 7.37
Figure for Primal Exercise 7.13.

Primal Exercise 7.12

Verify that the power delivered by the independent source is the same as that absorbed by the resistors in both the circuit of Figure 7.34 and its dual.

Primal Exercise 7.13

If the power dissipated in the 5 Ω resistor in Figure 7.37 is 1 W, determine the power dissipated in the dual of the 5 Ω resistor in the circuit that is the dual of the given circuit.

Ans. 1 W.

Exercise 7.14

Derive the dual circuit of Figure 7.35a by applying the procedure of Appendix 7A.

Learning Checklist: What Should Be Learned from This Chapter

- The fundamental attribute of a capacitor is its ability to store energy in the electric field resulting from separated positive and negative electric charges.
- If the assigned positive directions of current and voltage are such that the current through

a capacitor (or inductor) is in the direction of a voltage drop across the capacitor (or inductor), the voltage–current relation for the capacitor (or inductor) is written with a positive sign. On the other hand, if the assigned positive directions are such that the current through the capacitor (or inductor) is in the direction of a voltage rise across the capacitor (or inductor), the voltage–current relation for the capacitor (or inductor) is written with a negative sign. The foregoing ensures that C (or L) is a positive quantity irrespective of the assignment of the positive direction of capacitor (or inductor) current and the positive polarity of capacitor (or inductor) voltage.

- A capacitor acts as an open circuit when the voltage across the capacitor is steady, whereas an inductor acts as a short circuit when the inductor current is steady.
- The energy stored in a capacitor can be expressed as $\frac{1}{2}qv$, or $\frac{1}{2}Cv^2$, or $\frac{1}{2}q^2/C$.
- Magnetic field strength H is directly related to current by ampere's circuital law, $\oint H_l dl = I_{enc}$, independently of the magnetic properties of the medium surrounding the current-carrying conductor. Magnetic flux density B is directly related to voltage by Faraday's law, independently of the magnetic properties of the medium. B and H are related by the permeability of the medium: $B = \mu_r\mu_0 H$.
- The magnetic flux linkage λ of a coil is the sum, over all the turns of the coil, of the magnetic flux through each turn in the direction normal to the plane of the turn.
    1. When the same flux does not link all the turns of a coil, an effective flux can be defined that if multiplied by the number of turns gives the true flux linkage of the coil.
    2. In a toroidal core of high permeability, the same flux in the core passes through every turn of the coil in a direction normal to the plane of each turn.
- Faraday's law may be expressed as $|v| = |d\lambda/dt|$, where $|v|$ is the magnitude of the voltage induced by the time-varying flux linkage λ. The polarity of v is determined by Lenz's law, according to which the polarity of the induced voltage is such that it opposes the change in the magnetic flux linkage that induces the voltage.
- The fundamental attribute of an inductor is its ability to store energy in the magnetic field resulting from current flow in a conductor.
- The inductance of a coil is the flux linkage of the coil per unit current in the coil.
- If the current through an inductor increases with time, this increase is opposed, in accordance with Lenz's law, by an induced voltage in the inductor that is a voltage drop in the direction of current. Conversely, if the current through an inductor decreases with time, this decrease is opposed by an induced voltage in the inductor that is a voltage rise in the direction of current.
- In a series connection of initially uncharged capacitors, (1) the reciprocal of the equivalent series capacitance is the sum of the reciprocals of the individual capacitances, (2) voltages divide inversely as the capacitances, and (3) the charges on each of the capacitors and on the equivalent series capacitor are equal.
- In a parallel connection of initially uncharged capacitors, (1) the equivalent parallel capacitance is the sum of the individual capacitances, (2) currents divide directly as the capacitances, and (3) the charge on the equivalent parallel capacitance is the sum of the charges on the paralleled capacitances.
- In both the series and parallel connections, the energy in the equivalent capacitor is the sum of the energies in the individual capacitors.
- In a series connection of initially uncharged inductors, (1) the equivalent series inductance is the sum of the individual inductances, (2) voltages divide directly as the inductances, and (3) the flux linkage of the equivalent series inductance is the sum of the flux linkages of the individual inductors.
- In a parallel connection of initially uncharged inductors, (1) the reciprocal of the equivalent parallel inductance is the sum of the reciprocals of the individual inductances, (2) currents divide inversely as the inductances, and (3) the flux linkages of each of the inductors and of the equivalent parallel inductance are equal.
- In both the series and parallel connections, the energy in the equivalent inductor is the sum of the energies in the individual inductors.
- Two v–i relations are duals if one relation reduces to the other when the following are interchanged: (1) v and i, AND (2) R and G, as well as L and C. Moreover, the quantities being interchanged should have the same numerical values.
- Two circuits are duals if the node-voltage equations of one circuit are the dual relations of the mesh-current equations of the other circuit.

Problem-Solving Tips

1. Under steady dc or steady conditions, a capacitor is replaced by an open circuit and an inductor by a short circuit

Appendix 7A: Derivation of the Dual of a Planar Circuit

The procedure is based on the mesh-current equations in the given circuit being the dual relations of the node-voltage equations of the dual circuit to be derived. However, instead of writing the mesh-current and node-voltage equations, a graphical procedure can be followed. This procedure will be described using the circuit of Figure 7.38. It involves the following steps, illustrated in Figure 7.39:

1. Inside each mesh, place a node of the same number as the mesh and place a node outside the circuit (node 4 in Figure 7.39). This node is identified as the reference node in the dual circuit. For convenience, this node can have multiple representations outside the circuit, all of these being connected together and considered to be the same node.
2. The nodes are connected together by crossing all elements in the circuit. Whenever an element is crossed by a connection, the dual circuit element is placed in that connection. Consider node 2 for example. This node is connected to the reference node by crossing the 1 Ω resistor and the 3 H inductor. The dual of 1 Ω resistor, which is 1 S resistor, is placed in the first connection, and the dual of the 3 H inductor, which is a 3 F capacitor, is placed in the second connection. Node 2 is also connected to node 1 by crossing the 2 Ω resistor and placing its dual, a 2 S resistor in this connection. Node 2 is also connected to node 3 by crossing the 2 F capacitor and placing its dual, a 2 H inductor in this connection.
3. If a voltage source introduces a voltage rise in a given mesh when traversed in the clockwise direction of the mesh current, the direction of the dual current source is that of current entering the corresponding node. Conversely, if the voltage source introduces a voltage drop in a given mesh when traversed in the clockwise direction of the mesh current, the direction of the dual current source is that of current leaving the corresponding node. Thus, the 5 V source in mesh 1 in Figure 7.38 introduces voltage rise in mesh 1 in the direction of i_1. Its dual, a 5 A source, causes current to enter node 1 in Figure 7.39.
4. If a current source is in the direction of a mesh current, its dual voltage source will make the voltage of the node having the same number

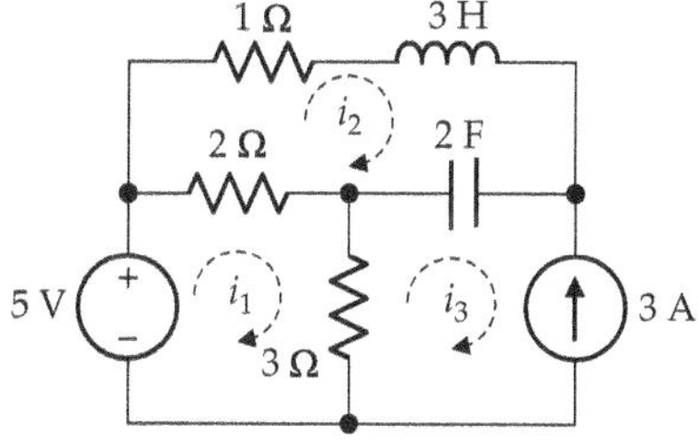

FIGURE 7.38
Derivation of mesh-current equations.

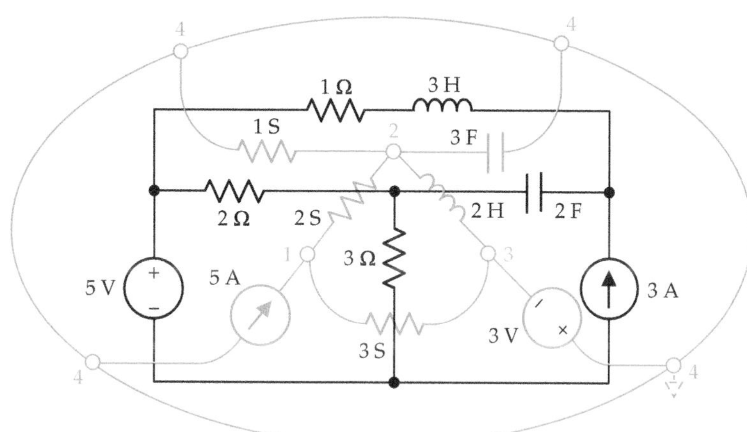

FIGURE 7.39
Derivation of dual of circuit of Figure 7.38.

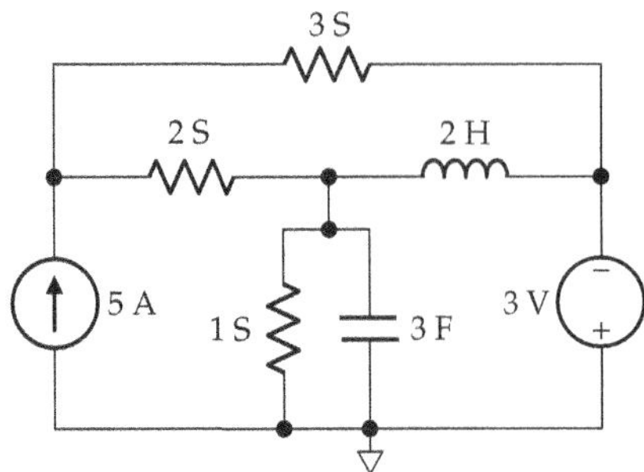

FIGURE 7.40
Dual circuit redrawn.

as the mesh more positive, and conversely. In Figure 7.38, the direction of the 3 A source is opposite that of i_3. Its dual is a 3 V source between node 3 and the reference node, and having a polarity that will make the voltage of node 3 negative with respect to that of the reference node.

5. Dependent sources are treated in the same manner as independent sources, the controlling variables being duals.
6. After deriving the dual circuit, this circuit can be redrawn for clarity, as illustrated in Figure 7.40.

Problems

Verify solutions by PSpice Simulation whenever feasible.

Capacitor Relations

P7.1 A parallel-plate capacitor consists of two plates, each being 5 cm^2, separated by 1 mm (Figure P7.1). Half the space between the plates, in the vertical direction, is filled by a dielectric of relative permittivity 10. Determine the capacitance, neglecting edge effects. Show that the capacitance can be considered as that of two capacitances in series.

Ans. 40.23 pF, equivalent to 44.25 pF in series with 442.5 pF.

P7.2 A parallel-plate capacitor consists of two plates, each being 5 cm square, separated by 1 mm (Figure P7.2), as in the preceding problem. Half the space between the plates, in the horizontal direction, is filled by a dielectric

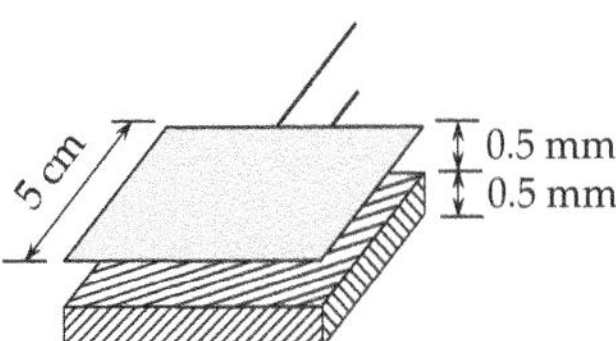

FIGURE P7.1

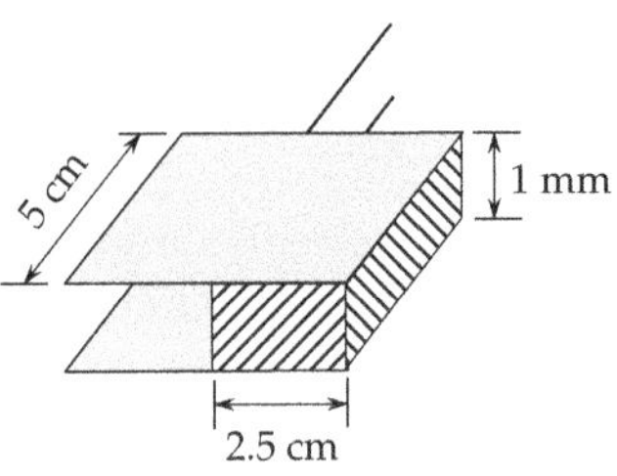

FIGURE P7.2

of relative permittivity 10. Determine the capacitance, neglecting edge effects. Show that the capacitance can be considered as that of two capacitances in parallel.

Ans. 121.7 pF, equivalent to 11.06 pF in parallel with 110.63 pF.

P7.3 A capacitor consists of two thin concentric cylinders l m long, the space between the two cylinders being filled with a dielectric of permittivity ε F/m. The radii of the inner and outer cylinders are a m and b m, respectively. Determine the capacitance, neglecting end effects. (*Hint*: Consider the voltage across a cylindrical shell of radius r and thickness dr, then integrate from a to b).

Ans. $\dfrac{2\pi\varepsilon l}{\ln(b/a)}$.

P7.4 A series of current pulses of 10 mA amplitude and 2 ms duration are applied to an initially uncharged, 5 μF capacitor. How many pulses are required to charge the capacitor to 20 V?

Ans. 5 pulses.

P7.5 A current pulse of amplitude 100 μA and 200 ms duration is applied at $t = 0$ to a 2 μF capacitor. Express the capacitor voltage as a function of time, assuming (a) the capacitor is initially uncharged and (b) the capacitor is initially charged to −10 V.

Ans. (a) $v(t) = 50t$ V, $0 \le t \le 200$ ms and $v(t) = 10$ V for $t \ge 200$ ms; (b) $v(t) = 50t - 10$ V, $0 \le t \le 200$ ms and $v(t) = 0$ for $t \ge 200$ ms.

P7.6 The current in a 1 μF capacitor is shown in Figure P7.6 as a function of time. Determine the total energy stored in the capacitor.

Ans. 200 μJ.

P7.7 The current in a 2 μF capacitor is shown in Figure P7.7 as a function of time. Determine the charge on the capacitor at $t = 3.4$ s.

Ans. 5.6 C.

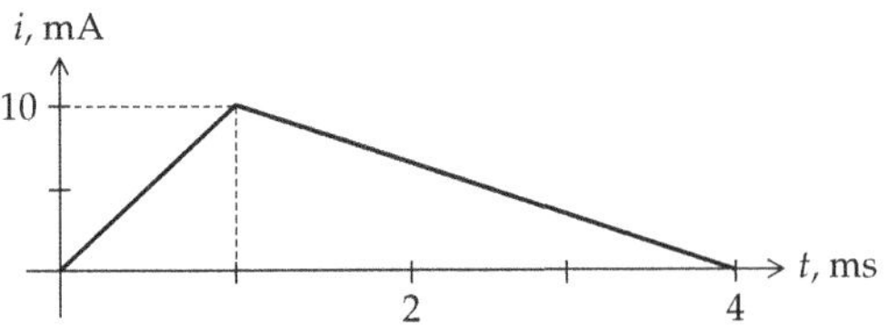

FIGURE P7.6

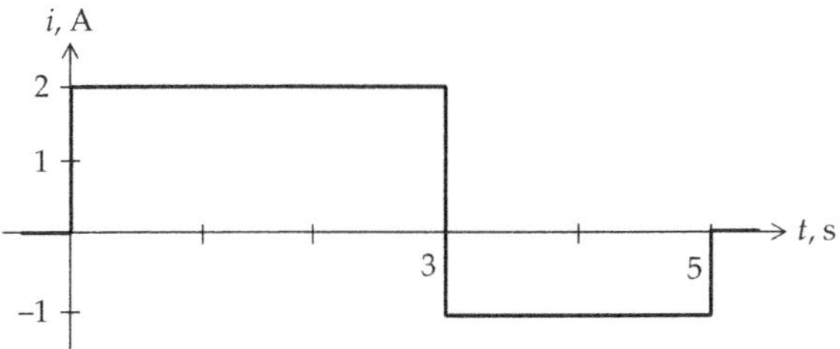

FIGURE P7.7

P7.8 The voltage waveform shown in Figure P7.8 is applied to a 1 μF capacitor. Determine the maximum value of the current through the capacitor.

Ans. 4 μA.

P7.9 The charge on a 0.1 μF capacitor varies with time as shown in Figure P7.9. Determine the average voltage across the capacitor.

Ans. 10 V.

P7.10 When the switch is closed in Figure P7.10, a current i flows that charges the capacitor. After a sufficiently long time, the capacitor is fully charged to 2 V. Determine, when the capacitor is fully charged, (a) the energy stored in the capacitor and (b) the total energy delivered by the battery, by considering the total charge delivered by the battery.

Ans. (a) 4 μJ; (b) 8 μJ.

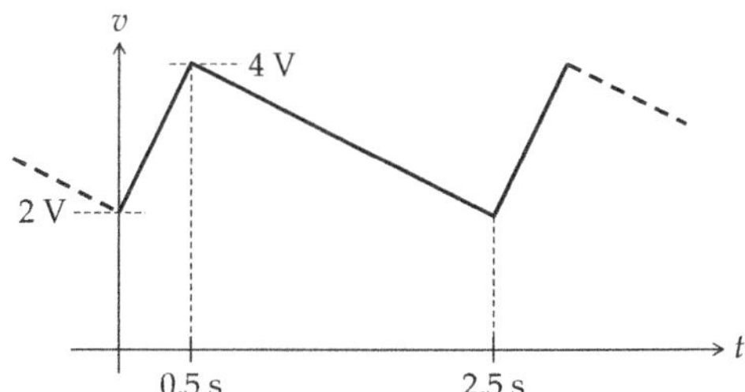

FIGURE P7.8

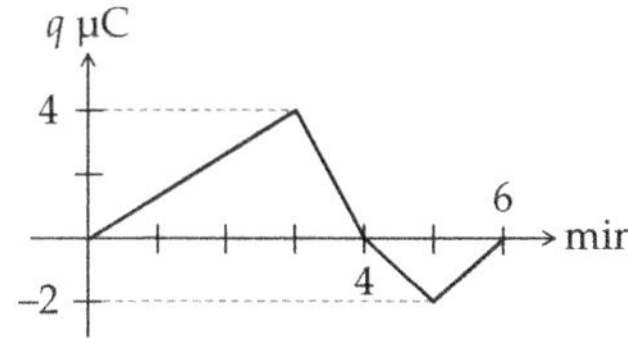

FIGURE P7.9

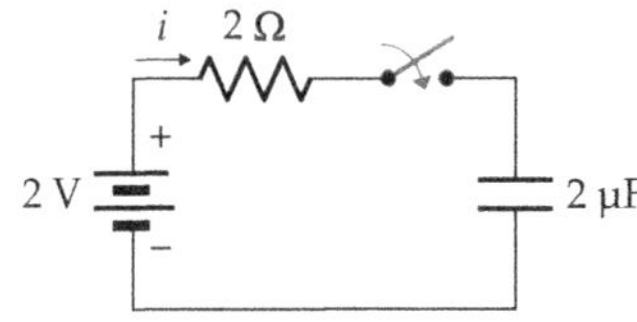

FIGURE P7.10

P7.11 The voltage applied to an initially uncharged 5 μF capacitor is $10te^{-5t}$ V, where t is in ms. Derive the expressions, as functions of time, for the energy stored in the capacitor, the capacitor current, and the instantaneous power input to the capacitor.

Ans. $w(t) = 250t^2e^{-10t}$ μJ, $i(t) = 50\left(e^{-5t} - 5te^{-5t}\right)$ mA, $p(t) = 500\left(te^{-10t} - 5t^2e^{-10t}\right)$ mW, t is in ms.

P7.12 The current applied to an initially uncharged 5 μF capacitor is $10e^{-5t}$ mA, where t is in ms. Derive the expressions, as functions of time, for the capacitor voltage, the energy stored in the capacitor, and the instantaneous power input to the capacitor.

Ans. $v(t) = 0.4\left(1 - e^{-5t}\right)$ V, $w(t) = 0.4\left(1 - e^{5t}\right)^2$ μJ, $p(t) = 4e^{-5t}\left(1 - e^{-5t}\right)$ mW, t is in ms.

P7.13 The voltage applied to an initially uncharged 0.1 μF capacitor is the first half-cycle of the waveform 10sin500t V, where t is in seconds. Derive the expressions, as functions of time, for the capacitor current, the energy stored in the capacitor, and the instantaneous power input to the capacitor. Sketch the time variation of these quantities.

Ans. $i(t)$ = 0.5cos(0.5t) mA, $w(t)$ = 5sin²(0.5t) μJ, $p(t)$ = 2.5sin$t$ mW, all for $0 \leq t \leq 2\pi$ ms, and are zero for $t > 2\pi$ ms.

P7.14 The current applied to an initially uncharged 0.1 μF capacitor is the first half-cycle of the waveform 10sin500t mA, where t is in seconds. Derive the expressions, as functions of time, for the capacitor voltage, the energy stored in the capacitor, and the instantaneous power input to the capacitor. Sketch the time variation of these quantities.

Ans. $v(t)$ = 200(1 – cos0.5t) V, $w(t)$ = 2(1 – cos0.5t)² mJ, $p(t)$ = 2sin0.5t – sin$t$ W, all for $0 \leq t \leq 2\pi$ ms, and are zero for $t > 2\pi$ ms.

P7.15 The triangular voltage pulse of Figure P7.15 is applied to an initially uncharged 0.1 μF capacitor. Plot as functions of time (a) the charge on the capacitor, (b) the energy stored in the capacitor, and (c) the instantaneous power input to the capacitor.

Ans. (a) $q(t) = \frac{t}{60}$ μC, $0 \leq t \leq 60$ s;

$q(t) = -\frac{t}{180} + \frac{4}{3}$ μC, $60 \leq t \leq 240$ s $60 \leq t \leq 240$ s;

$q(t) = 0$, $t \geq 240$ s; (b) $w(t) = \frac{1}{2}qv = \frac{t^2}{720}$ μJ, $0 \leq t \leq 60$ s;

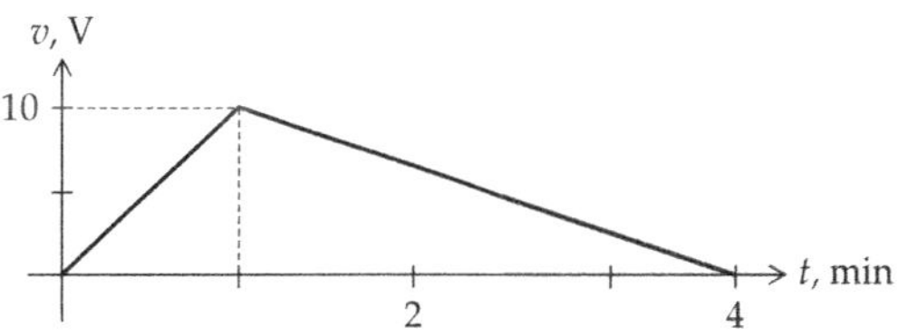

FIGURE P7.15

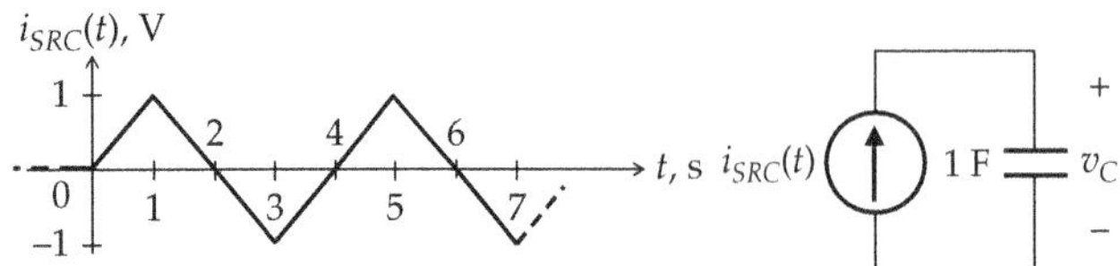

FIGURE P7.16

$$w(t)=\frac{1}{180}\left(\frac{t^2}{36}-\frac{40t}{3}+1600\right)\mu\text{J},\ 60\le t\le 240\text{ s};$$

$$w(t)=0,\quad t\ge 240\text{ s; (c) } p(t)=\frac{t}{360}\ \mu\text{W},\quad 0\le t\le 60\text{ s};$$

$$p(t)=\frac{1}{180}\left(\frac{t}{18}-\frac{40}{3}\right)\mu\text{W},\ 60\le t\le 240\text{ s};\ p(t)=0,\ t\ge 240\text{ s}.$$

P7.16 Determine $v_C(t)$ at $t = 7$ s in Figure P7.16, assuming an initially uncharged capacitor.

Ans. 0.5 V.

P7.17 The current waveform of Figure P7.17 is applied to an initially uncharged 0.5 μF capacitor. (a) Derive expressions, as functions of time, for the voltage across the capacitor during the time intervals: $0 \le t \le 10$ μs, $10 \le t \le 40$ μs, $40 \le t \le 60$ μs, $60 \le t \le 80$ μs, and $t > 80$ μs. (b) What is the charge on the capacitor at $t = 10$ μs and at $t = 50$ μs? Check the final value of voltage against the final charge. (c) What is the energy stored in the capacitor at $t = 80$ μs? (d) How do the expressions for the voltage across the capacitor derived in (a) above change if the capacitor was initially charged to 0.5 V?

Ans. (a) $v(t)=1.5t^2$ mV, $0\le t\le 10$ μs; $v(t)=-t^2+50t-250$ mV, $10\le t\le 40$ μs; $v(t)=-30t+1350$ mV, $40\le t\le 60$ μs; $v(t)=0.75t^2-120t+4050$ mV, $60\le t\le 80$ μs, $v(t)=-750$ mV, $t\ge 80$ μs. (b) 75 nC at $t = 10$ μs, and −75 nC at $t = 10$ μs. (c) 0.14 μJ; (d) all voltages are increased by 0.5 V.

P7.18 Determine the energy stored in the capacitor in Figure P7.18, assuming a dc steady state.

Ans. 4 J.

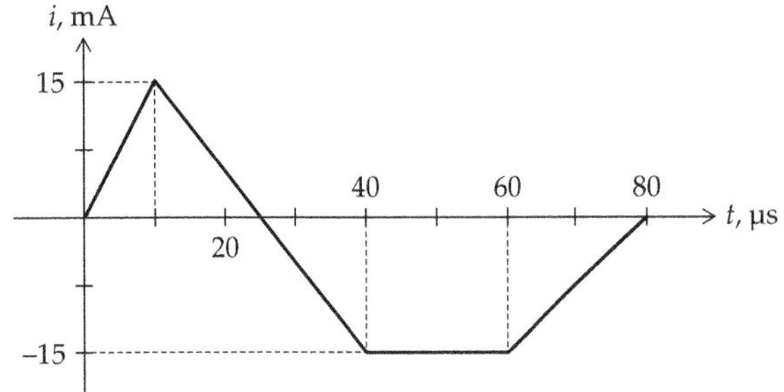

FIGURE P7.17

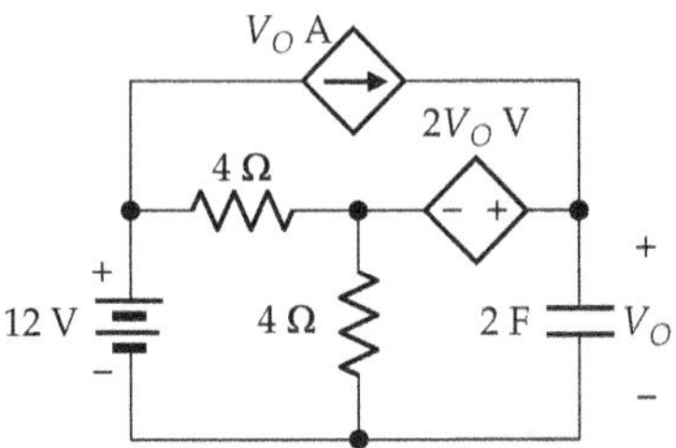

FIGURE P7.18

Inductor Relations

Note the duality with the corresponding problems on capacitor relations.

P7.19 A current of 10 mA through a coil of 100 turns results in a flux of 10^{-6} Wb in the coil. Assuming that all the flux links all the turns, determine the inductance of the coil.

Ans. 10 mH.

P7.20 A series of voltage pulses of 10 mV amplitude and 2 ms duration are applied to an initially uncharged 5 μH inductor. How many pulses are required to bring the inductor current to 20 A? (Dual of Problem P7.4).

Ans. 5 pulses.

P7.21 A voltage pulse of amplitude 100 μV and 200 ms duration is applied at $t = 0$ to a 2 μH inductor. Express the inductor current as a function of time, assuming (a) the inductor is initially uncharged and (b) the inductor current is initially −10 A. (Dual of Problem P7.5).

Ans. (a) $i(t) = 50t$ A, $0 \le t \le 200$ ms and $i(t) = 10$ A for $t \ge 200$ ms; (b) $i(t) = 50t - 10$ A, $0 \le t \le 200$ ms and $i(t) = 0$ for $t \ge 200$ ms.

P7.22 The voltage across a 1 μH inductor is shown in Figure P7.22 as a function of time. Determine the total energy stored in the inductor. (Dual of Problem P7.6).

Ans. 200 μJ.

P7.23 The voltage across a 2 μH inductor is shown in Figure P7.23 as a function of time. Determine the flux linkage in the inductor at $t = 3.4$ s. (Dual of Problem P7.7).

Ans. 5.6 Wb-T.

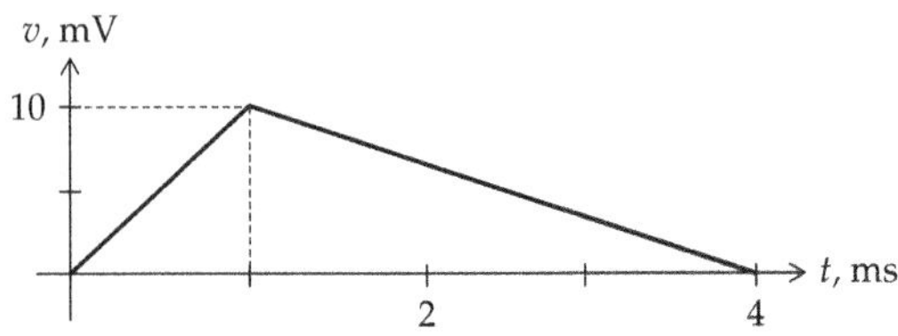

FIGURE P7.22

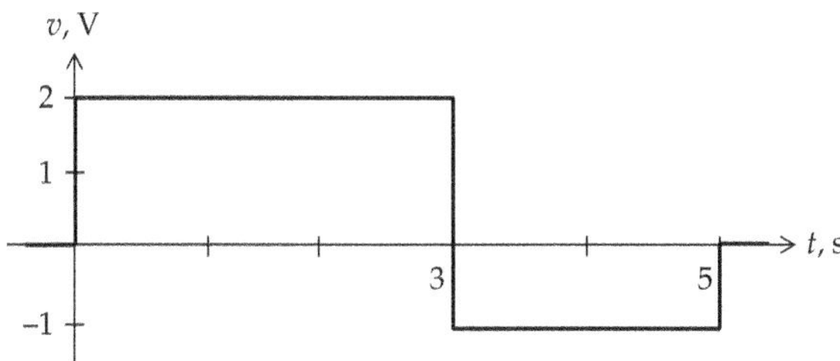

FIGURE P7.23

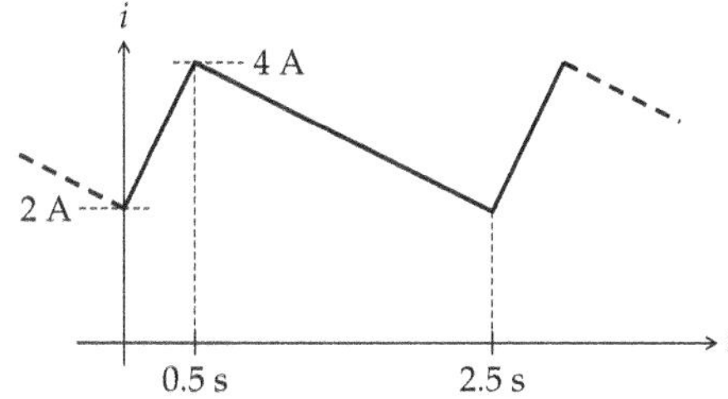

FIGURE P7.24

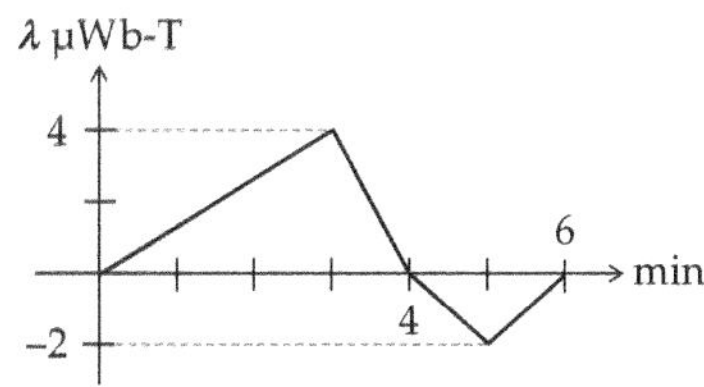

FIGURE P7.25

P7.24 The current waveform shown in Figure P7.24 is applied to a 1 μH inductor. Determine the maximum value of the voltage across the inductor. (Dual of Problem P7.8).

Ans. 4 μV.

P7.25 The flux linkage in a 0.1 μH inductor varies with time as shown in Figure P7.25. Determine the average current through the inductor. (Dual of Problem P7.9).

Ans. 10 A.

P7.26 When the switch is opened in Figure P7.26, a voltage is applied that establishes flux linkage in the inductor. After a sufficiently long time, the inductor is fully charged and the 2 A current flows in the inductor. Determine, when the inductor is fully charged, (a) the energy stored in the inductor and (b) the total energy delivered by the current source after the switch is opened. (Dual of Problem P7.10).

Ans. (a) 4 μJ; (b) 8 μJ.

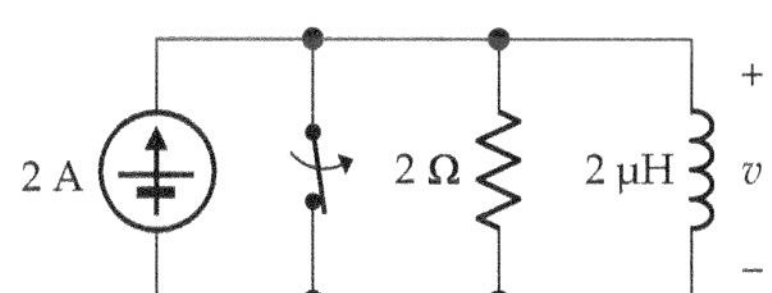

FIGURE P7.26

P7.27 The current applied to an initially uncharged 5 μH inductor is $10te^{-5t}$ A, where t is in ms. Derive the expressions, as functions of time, for the energy stored in the inductor, the inductor voltage, and the instantaneous power input to the inductor. (Dual of Problem P7.11).

Ans. $w(t) = 250t^2e^{-10t}\,\mu\text{J}$, $v(t) = 50\left(e^{-5t} - 5te^{-5t}\right)\text{mV}$, $p(t) = 500\left(te^{-10t} - 5t^2e^{-10t}\right)\text{mW}$, t is in ms.

P7.28 The voltage applied to an initially uncharged 5 μH inductor is $10e^{-5t}$ mV, where t is in ms. Derive the expressions, as functions of time, for the inductor current, the energy stored in the inductor, and the instantaneous power input to the inductor. (Dual of Problem P7.12).

Ans. $i(t) = 0.4\left(1 - e^{-5t}\right)\text{A}$, $w(t) = 0.4\left(1 - e^{5t}\right)^2\,\mu\text{J}$, $p(t) = 4e^{-5t}\left(1 - e^{-5t}\right)\text{mW}$, t is in ms.

P7.29 The current applied to an initially uncharged 0.1 μH inductor is the first half-cycle of the waveform $10\sin 500t$ A, where t is in seconds. Derive the expressions, as functions of time, for the inductor voltage, the energy stored in the inductor, and the instantaneous power input to the inductor. Sketch the time variation of these quantities. (Dual of Problem P7.13).

Ans. $v(t) = 0.5\cos(0.5t)$ mV, $w(t) = 5\sin^2(0.5t)$ μJ, $p(t) = 2.5\sin t$ mW, all for $0 \le t \le 2\pi$ ms, and are zero for $t > 2\pi$ ms.

P7.30 The voltage applied to an initially uncharged 0.1 μH inductor is the first half-cycle of the waveform $10\sin 500t$ mV, where t is in seconds. Derive the expressions, as functions of time, for the inductor current, the energy stored in the inductor, and the instantaneous power input to the inductor. Sketch the time variation of these quantities. (Dual of Problem P7.14).

Ans. $i(t) = 200(1 - \cos 0.5t)$ A, $w(t) = 2(1 - \cos 0.5t)^2$ mJ, $p(t) = 2\sin 0.5t - \sin t$ W, all for $0 \le t \le 2\pi$ ms, and are zero for $t > 2\pi$ ms.

P7.31 The voltage v_L shown in Figure P7.31 is applied to the initially uncharged inductor. Determine the value of t at which $i_L = -0.5$ A.

Ans. 3.75 s.

P7.32 The triangular current pulse of Figure P7.32 is applied to an initially uncharged 0.1 μH inductor. Plot as a function of time (a) the flux linkage in the inductor,

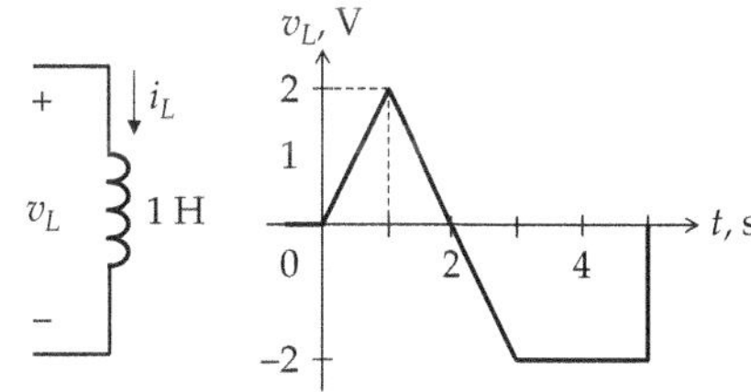

FIGURE P7.31

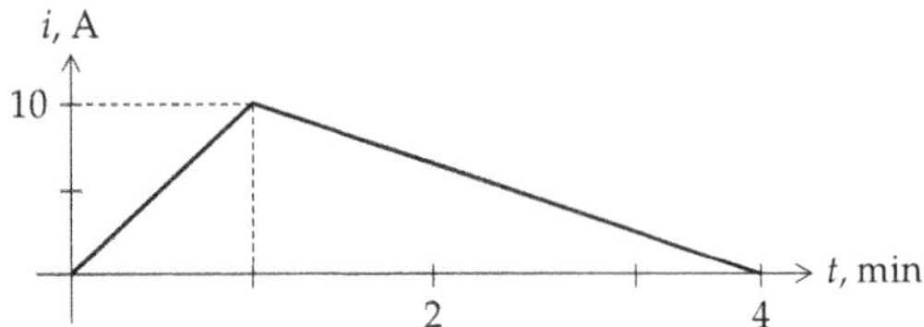

FIGURE P7.32

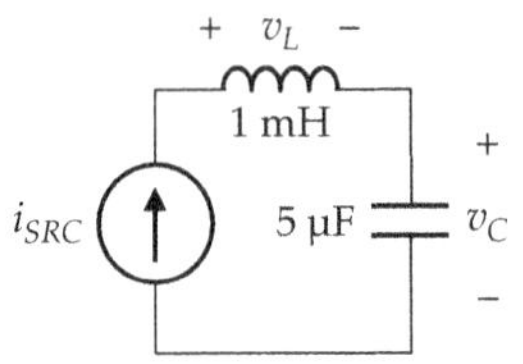

(b) the energy stored in the inductor, and (c) the instantaneous power input to the inductor. (Dual of Problem P7.15).

Ans. (a) $\lambda(t)=\frac{t}{60}$ μWb-T, $0\le t\le 60$ s; $\lambda(t)=-\frac{t}{180}+\frac{4}{3}$ μWb-T, $60\le t\le 240$ s; $\lambda(t)=0$, $t\ge 240$ s. (b) $w(t)=\frac{1}{2}qv=\frac{t^2}{720}$ μJ, $0\le t\le 60$ s; $w(t)=\frac{1}{180}\left(\frac{t^2}{36}-\frac{40t}{3}+1600\right)$ μJ, $60\le t\le 240$ s; $w(t)=0$, $t\ge 240$ s. (c) $p(t)=\frac{t}{360}$ μW, $0\le t\le 60$ s; $p(t)=\frac{1}{180}\left(\frac{t}{18}-\frac{40}{3}\right)$ μW, $60\le t\le 240$ s; $p(t)=0$, $t\ge 240$ s.

P7.33 The voltage waveform in Figure P7.33 is applied to an initially uncharged 0.5 μH inductor. (a) Derive expressions for the inductor current during the time intervals: $0\le t\le 10$ μs, $10\le t\le 40$ μs, $40\le t\le 60$ μs, $60\le t\le 80$ μs, and $t>80$ μs. (b) What is the flux linkage in the inductor at $t=10$ μs and at $t=50$ μs? Check the final value of current against the final flux linkage. (c) What is the energy stored in the inductor at $t=80$ μs? (d) How do the expressions for the current through the inductor derived in (a) above change if the inductor current was initially 0.5 A? (Dual of Problem P7.17).

Ans. (a) $i(t)=1.5t^2$ mA, $0\le t\le 10$ μs;
$i(t)=-t^2+50t-250$ mA, $10\le t\le 40$ μs;
$i(t)=-30t+1350$ mA, $40\le t\le 60$ μs;
$i(t)=0.75t^2-120t+4050$ mA, $60\le t\le 80$ μs,
$v(t)=-750$ mA, $t\ge 80$ μs. (b) 75 nWb-T at $t=10$ μs, and −75 nWb-T at $t=10$ μs; (c) 0.14 μJ; (d) all currents are increased by 0.5 A.

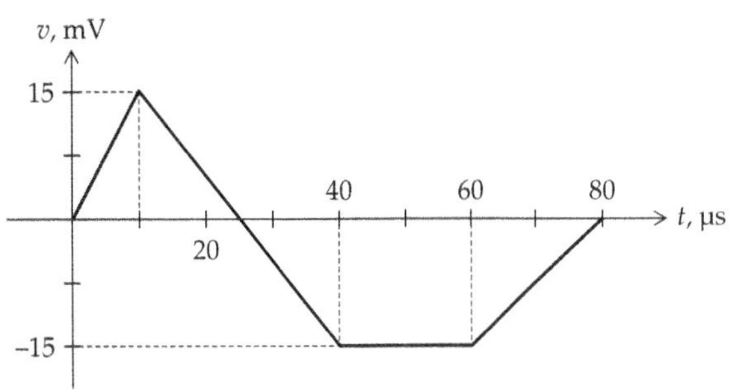

FIGURE P7.33

FIGURE P7.34

P7.34 Given $i_{SRC}(t)=0.1t$ A, $t\ge 0$ s, in Figure P7.34, with no energy stored in the circuit for $t\le 0$. Determine t at which $v_L=v_C$.

Ans. 0.1 ms.

P7.35 Given $i_{SRC}(t)$ shown in Figure P7.35, with zero initial energy storage for $t\le 0$. (a) Express v_L and v_C as functions of time and sketch this variation. (b) Determine (i) the instantaneous power absorbed by L and C as a function of time, and (ii) the energy stored by L and C at $t=2$ s.

Ans. (a) $v_L(t)=2$ mV, $0<t<1$ s, and $v_L(t)=0$, $t>1$ s; $v_C(t)=t^2$ kV, $0\le t\le 1$ s, and $v_C(t)=(2t-1)$ kV, $t\ge 1$ s. (b) (i) $p_L(t)=4t$ mW, $0<t<1$ s, and $p_L(t)=0$, $t>1$ s; $p_C(t)=2t^3$ kW, $0\le t\le 1$ s; $p_C(t)=4t-2$ kW, $t\ge 1$ s; (ii) 4.5 kJ.

P7.36 Determine V_X in Figure P7.36, assuming a dc steady state.

Ans. 1.6 V.

P7.37 Determine I_X in Figure P7.37, assuming a dc steady state.

Ans. 1 A.

P7.38 Determine I_X in Figure P7.38, assuming a dc steady state, all resistances are in ohms, all inductances are 1 H, and all capacitances are 1 F.

Ans. $75/14 = 5.36$ A.

P7.39 Determine the energy stored in the inductor in Figure P7.39, assuming a dc steady state.

Ans. 0.5 J.

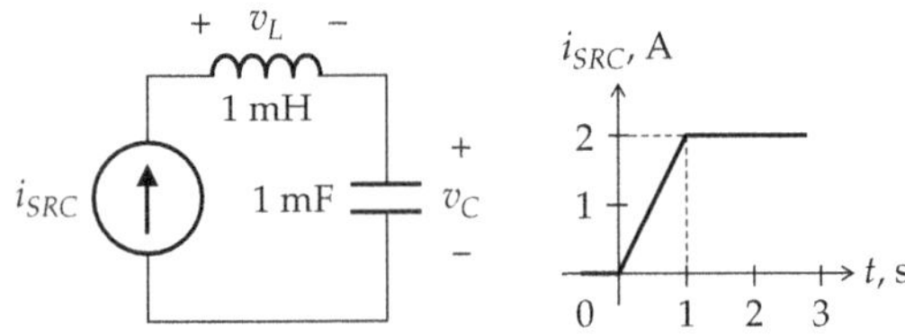

FIGURE P7.35

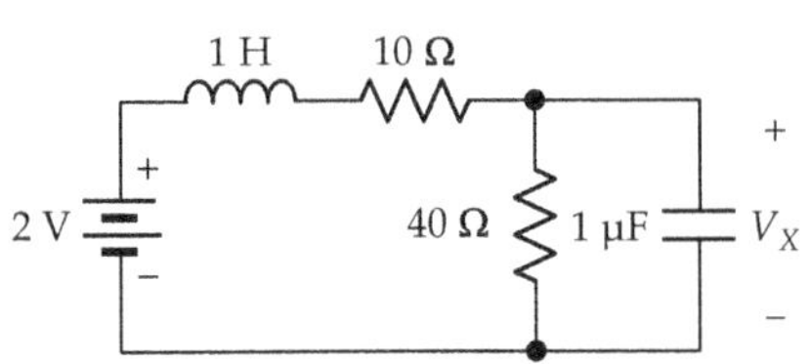

FIGURE P7.36

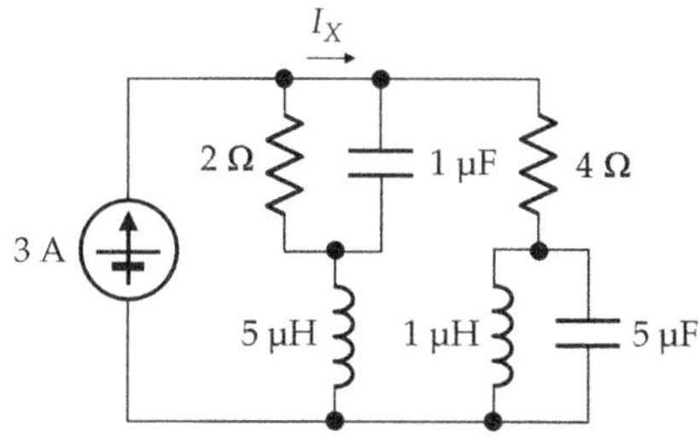

FIGURE P7.37

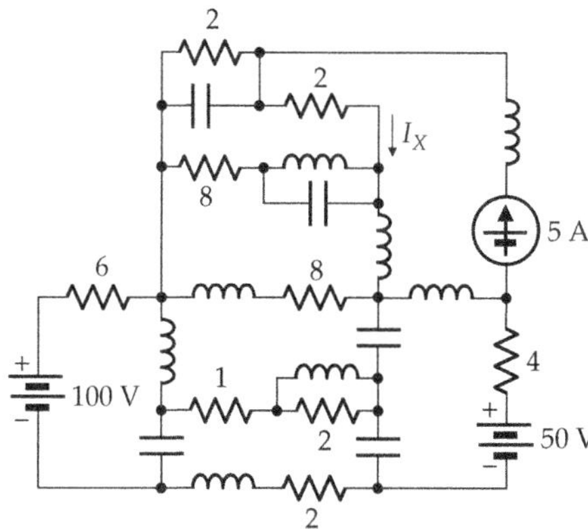

FIGURE P7.38

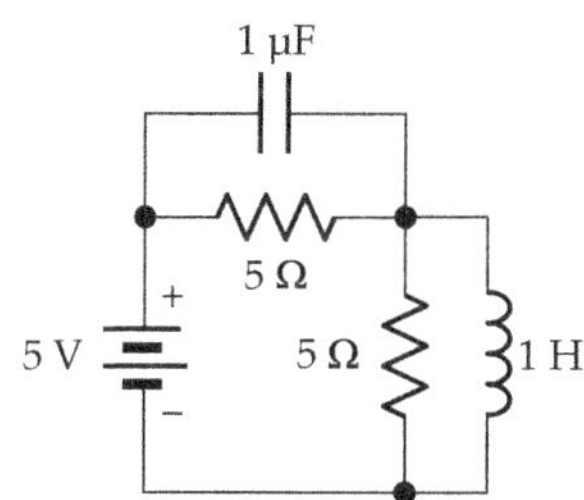

FIGURE P7.39

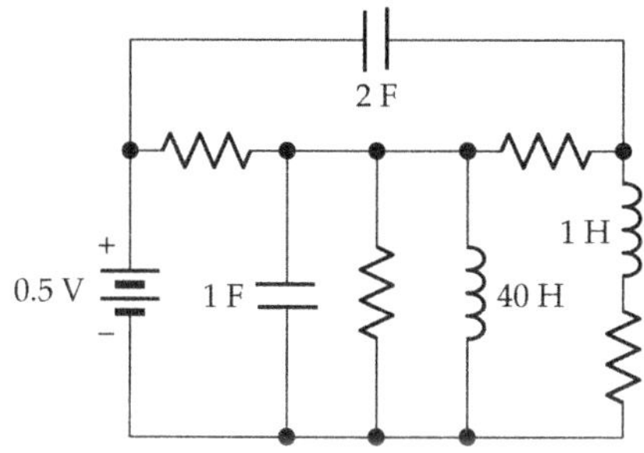

FIGURE P7.40

P7.40 Determine the energy stored in the circuit in Figure P7.40 in the dc state, assuming all resistances are 5 Ω.

Ans. 0.45 J.

P7.41 Determine C in Figure P7.41 so that the same energy is stored in C as in the inductor under dc conditions.

Ans. 0.5 mF.

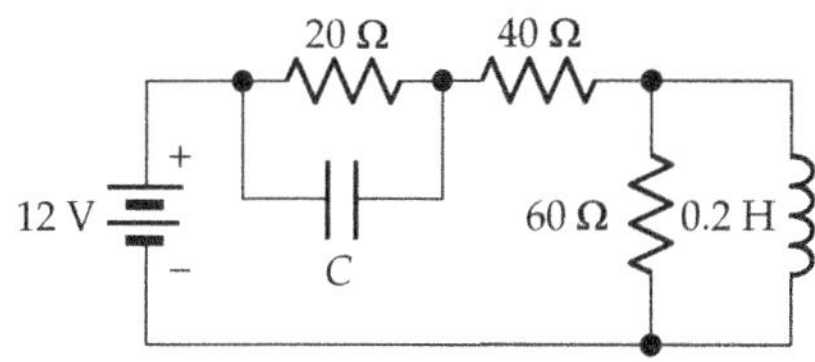

FIGURE P7.41

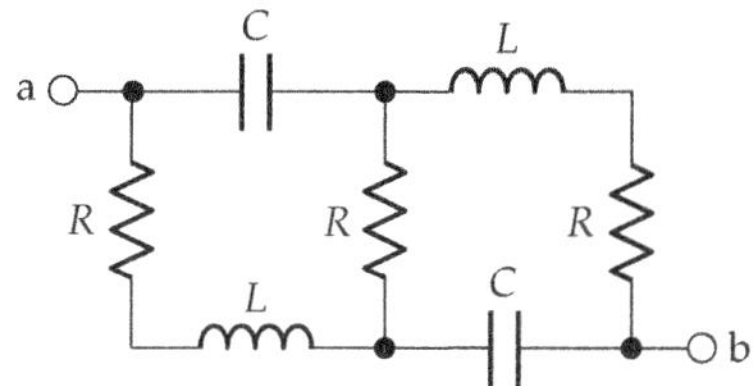

FIGURE P7.42

P7.42 Derive the relation between *R*, *L*, and *C* in Figure P7.42 so that the same energy is stored in *L* and in *C* under dc conditions.

Ans. $L = 4CR^2$.

Series and Parallel Connections of Capacitors and Inductors

P7.43 (a) Determine the equivalent capacitance between terminals 'ab' in Figure P7.43 if all the capacitances are 1 F. (b) Determine the equivalent inductance between terminals 'ab', assuming that all capacitors in Figure P7.43 are replaced by inductors of 1 H.

Ans. (a) 2/3 F; (b) 3/2 H.

P7.44 (a) Determine the equivalent capacitance between terminals 'ab' in Figure P7.44 if all the capacitances are 1 F. (b) Determine the equivalent inductance between terminals 'ab', assuming that all capacitors in Figure P7.44 are replaced by inductors of 1 H.

Ans. (a) $\frac{15}{41} = 0.37$ F; (b) $\frac{41}{15} = 2.73$ H.

P7.45 (a) Determine the equivalent inductance between terminals 'ab' in Figure P7.45 if all the inductances are 1 H. (b) Determine the equivalent capacitance between

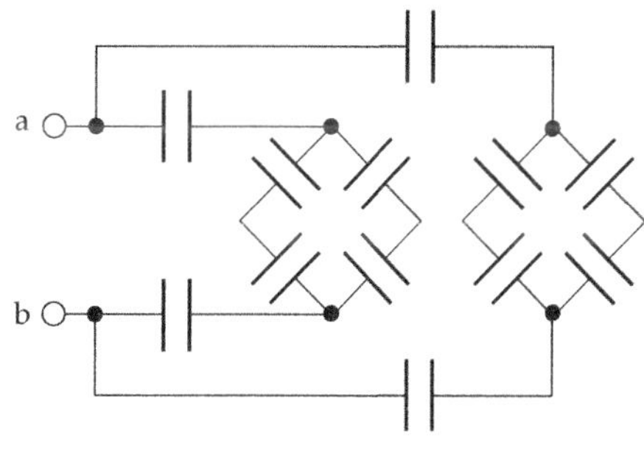

FIGURE P7.43

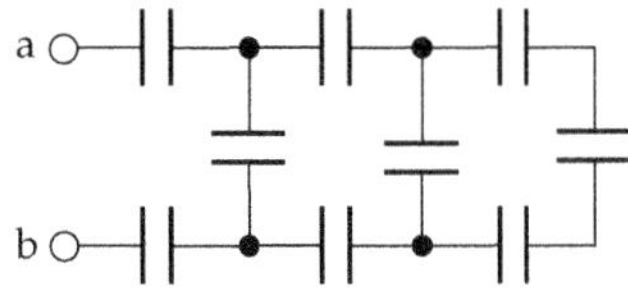

FIGURE P7.44

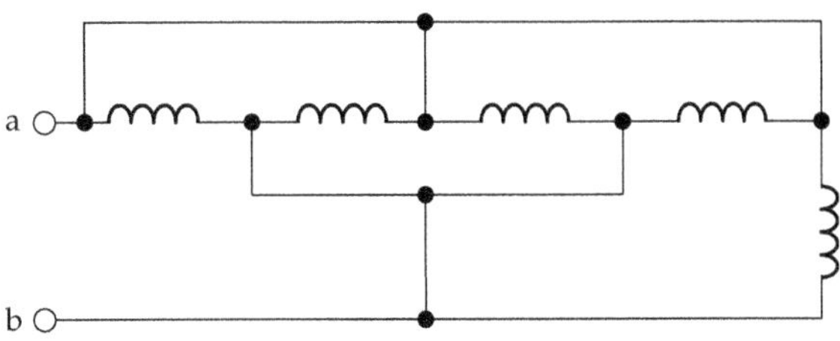

FIGURE P7.45

terminals 'ab', assuming that all inductors in Figure P7.45 are replaced by capacitors of 1 F.

Ans. (a) 1/5 H; (b) 5 F.

P7.46 (a) Determine the equivalent inductance between terminals 'ab' in Figure P7.46. (b) Determine the equivalent capacitance between terminals 'ab', assuming that all inductors in Figure P7.46 are replaced by capacitors of the same value in mF.

Ans. (a) 18 mH; (b) 850/7 = 121.4 mF.

P7.47 (a) Determine the equivalent inductance between terminals 'ab' in Figure P7.47, assuming all inductances are 2 μH (b) Determine the equivalent capacitance between terminals 'ab', assuming that all inductors in Figure P7.47 are replaced by capacitors of the same value in μF.

Ans. (a) 3/4 μH; (b) 16/3 μF.

P7.48 Repeat Exercise 7.10 replacing capacitances by inductances of 0.5 H (Figure P7.48).

Ans. 0.8 H.

P7.49 Repeat Exercise 7.11 replacing inductances by capacitances of equal value in farads (Figure P7.49).

Ans. 40/131 = 0.31 F.

P7.50 Ten 1 μF capacitors are charged in parallel to 1000 V. Determine (a) C_{eqp} and (b) the energy stored in the equivalent parallel capacitor. (c) Verify that the energy

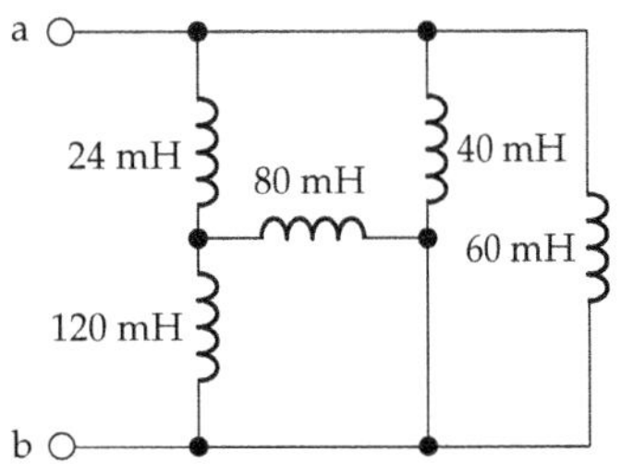

FIGURE P7.46

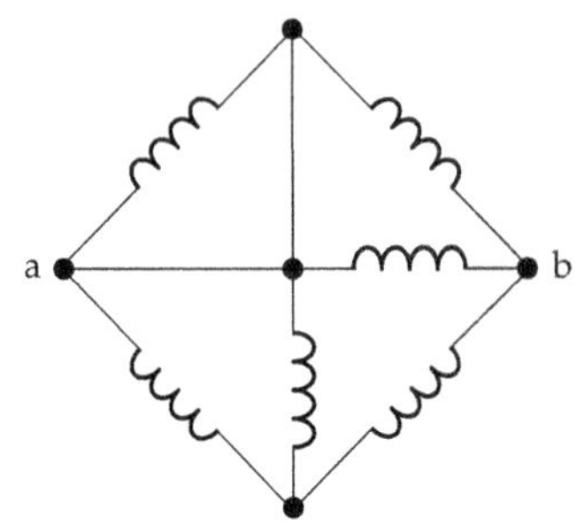

FIGURE P7.47

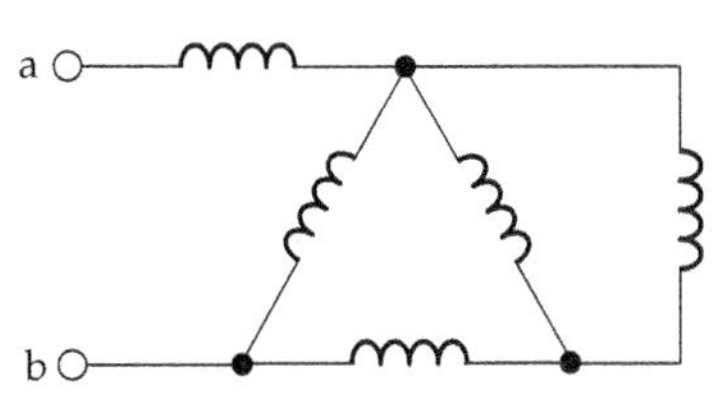

FIGURE P7.48

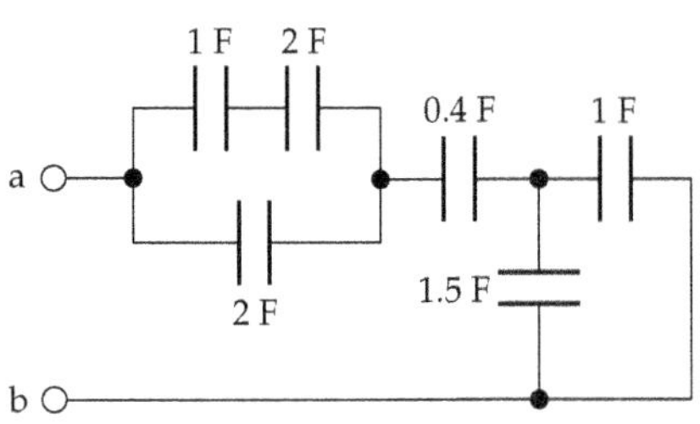

FIGURE P7.49

stored in this capacitor is the sum of the energies stored in the individual capacitors. After the capacitors are fully charged, they are connected in series. Determine (d) the voltage across the equivalent series capacitor and (e) the energy stored in this capacitor. (f) Deduce C_{eqs} for the charged capacitors. Is it the same as that for initially uncharged capacitors?

Ans. (a) 10 μF; (b) 5 J; (c) 0.5 J is stored per capacitor; (d) 10,000 V; (e) 5 J, (f) 0.1 μF, yes.

P7.51 All inductors in Figure P7.51 have an inductance L and all capacitors have a capacitance C. Determine L_{eq} in series with C_{eq} that are equivalent to the combination of L and C shown between terminals 'a' and 'b'.

Ans. $L_{eq} = 11\,L/6$, $C_{eq} = 3C/7$.

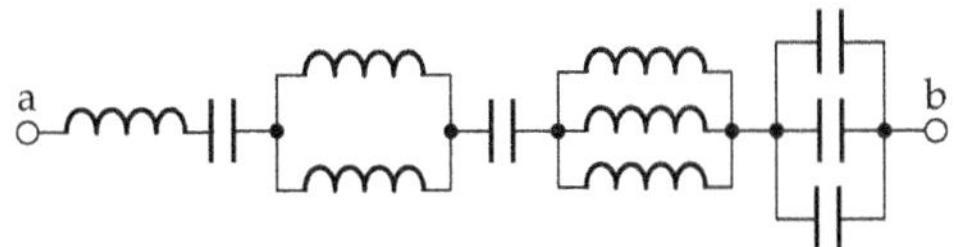

FIGURE P7.51

Duality

P7.52 Given the circuit of Figure P7.52, (a) derive the dual circuit, (b) compare the resistance or conductance seen by the independent source in each circuit, and (c) compare the power delivered or absorbed by each circuit element in the two circuits.

Ans. (a) 18 A source in parallel with 9, 12, and 15 S; (b) 36 Ω, 36 S; (c) sources deliver 9 W, 9 Ω/9 S resistors dissipate 2.25 W, 12 Ω/12 S resistors dissipate 3 W, and 15 Ω/15 S resistors dissipate 3.75 W.

P7.53 Given the circuit of Figure P7.53, (a) derive the dual circuit, (b) compare the resistance or conductance seen by the independent source in each circuit, and (c) compare the power delivered or absorbed by each circuit element in the two circuits.

Ans. (a) 12 A source in parallel with 2 S and with a series combination of 6 and 122; (b) 6 Ω, 6 S; (c) sources deliver 24 W, the 2 Ω/2 S resistors dissipate 8 W, the 6 Ω/6 S resistors dissipate 32/3 W, and the 12 Ω/12 S resistors dissipate 16/3 W.

P7.54 Given the circuit of Figure P7.54, (a) derive the dual circuit, (b) compare the resistance or conductance seen by the independent source in each circuit, and (c) compare the power delivered or absorbed by each circuit element in the two circuits.

Ans.(a) 30 A source in parallel with 10 S, 20 S and a series combination of 10 S and a CCVS; (b) 60 Ω, 60 S; (c) independent sources deliver 15 W, 10 Ω/10 S resistors on the LHS dissipate 2.5 W, 20 Ω/20 S resistors dissipate 5 W, 10 Ω/10 S resistors on the RHS dissipate 22.5 W, dependent sources deliver 15 W.

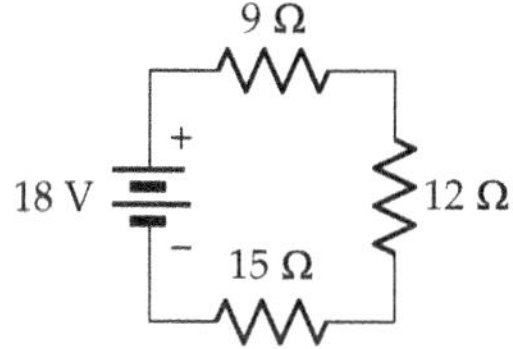

FIGURE P7.52

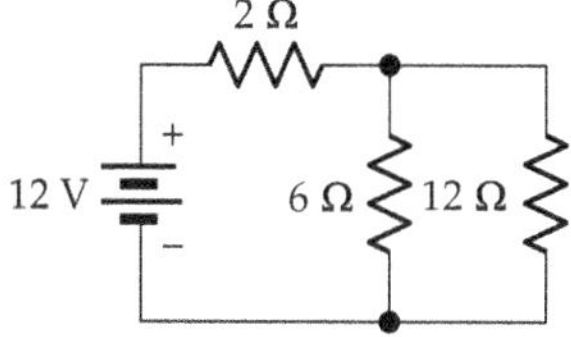

FIGURE P7.53

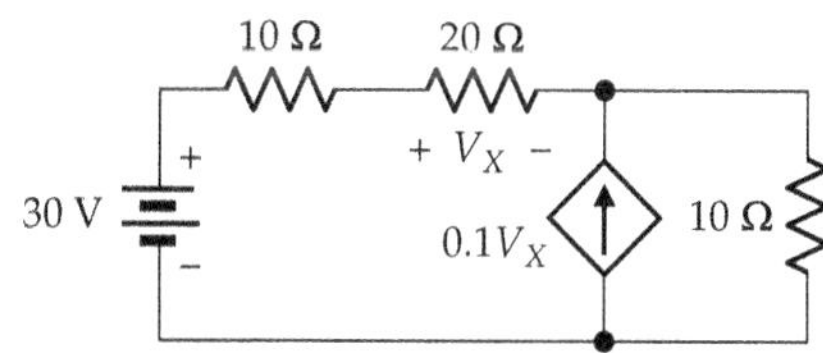

FIGURE P7.54

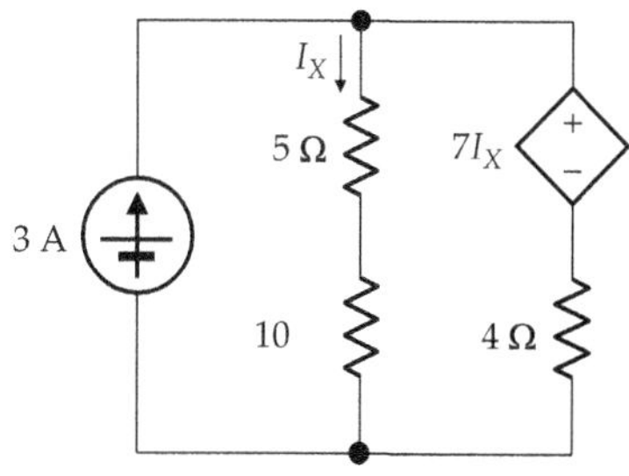

FIGURE P7.55

P7.55 Given the circuit of Figure P7.55, (a) derive the dual circuit, (b) compare the resistance or conductance seen by the independent source in each circuit, and (c) compare the power delivered or absorbed by each circuit element in the two circuits.

Ans. (a) 3 V source in series with a parallel combination of 5 and 10 S and a parallel combination of 4 S and a VCVS; (b) 5 Ω, 5 S; (c) independent sources deliver 45 W, 5 Ω/5 S resistors dissipate 5 W, 10 Ω/10 S resistors dissipate 10 W, 4 Ω/4 S resistors in parallel with dependent source dissipate 16 W, dependent sources absorb 14 W.

P7.56 (a) Derive the dual of the circuit of Figure P7.56; (b) represent the two circuits in the dc steady state; (c) compare voltage division in the given circuit with current division in the dual circuit; (d) compare the power delivered or absorbed by each circuit element in the two circuits.

Ans. Current source in parallel with 10 S, 10 μF, and a series connection of 20 S and 2 μH; (b) 15 V source applied to 10 Ω in series with 20 Ω, 15 A source applied to 10 S in parallel with 20 S; (c) 15 V divides in proportion to resistances, 15 A divides in proportion to conductances; (d) sources deliver 7.5 W, 10 Ω/10 S resistors dissipate 2.5 W, 20 Ω/20 S resistors dissipate 5 W.

P7.57 (a) Derive the dual of the circuit of Figure P7.57; (b) compare R_{in} of the dual circuit with R_{in} in Figure P7.57.

Ans. (b) R_{in} in Figure P7.57 is 10 Ω, R_{in} of dual circuit is 0.1 Ω.

P7.58 (a) Derive the mesh-current equations using the mesh currents shown in Figure P7.58, (b) deduce the node-voltage equations of the dual circuit, and (c) derive the circuit that will give these node-voltage equations with respect to a specified reference node.

Ans. Y-connection of 1, 2, and 3 mF inscribed in a Δ-connection of 1, 2, and 3 A current sources.

P7.59 Verify that the mesh-current equations of the circuit of Figure 7.38 are the dual relations of the node-voltage equations of Figure 7.40.

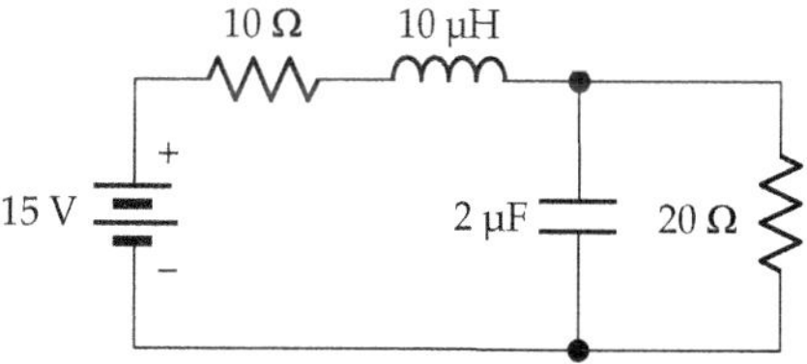

FIGURE P7.56

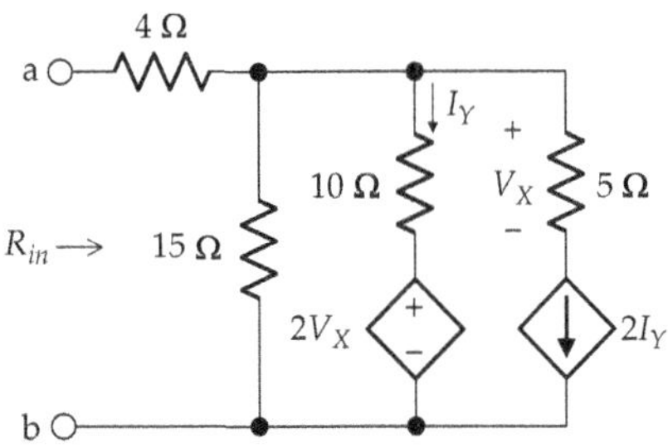

FIGURE P7.57

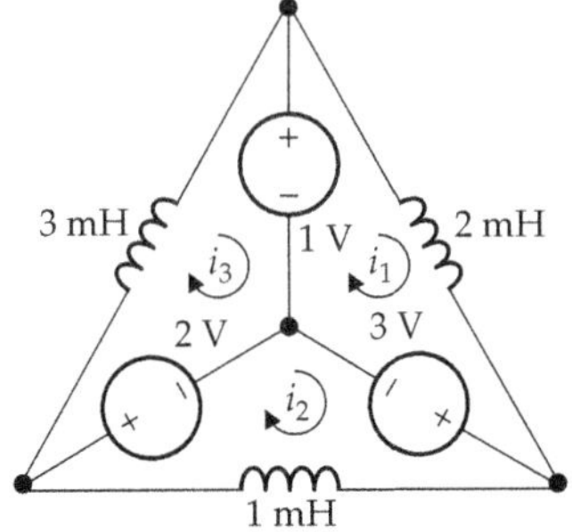

FIGURE P7.58

8

Sinusoidal Steady State

Objective and Overview

The chapter introduces phasor analysis, a powerful and very useful methodology that extends to the sinusoidal steady state all the concepts, theorems, and procedures applied to resistive circuits under dc conditions. The key underlying concept is the transformation of linear differential equations describing the behavior of circuits involving capacitors and inductors under steady-state sinusoidal conditions to algebraic equations involving the imaginary unit *j*. Appendix D provides an introduction to complex quantities.

The chapter begins with highlighting some fundamental properties of the sinusoidal function and responses to sinusoidal excitation, followed by a discussion of phasors and their properties. The *v–i* phasor relations for resistors, capacitors, and inductors are derived, leading to the concept of impedance. The representation of circuits in the frequency domain is then considered, including several illustrative examples. The chapter ends with a discussion of phasor diagrams.

8.1 The Sinusoidal Function

A voltage or a current that varies sinusoidally with time can be represented as

$$y(t) = Y_m \cos(\omega t + \theta) \tag{8.1}$$

where

Y_m is the amplitude of the sinusoidal function
ω is its angular frequency in rad/s
θ is its phase angle (Figure 8.1)

Note that the argument ($\omega t + \theta$) of a trigonometric function is in radians, which is dimensionless, as is ωt, but it is customary to express θ in degrees.

The time interval between successive repetitions of the same full range of values of y is the **period** T. The period is conveniently taken as the interval between successive maxima, or successive minima, or every other zero crossing, as illustrated in Figure 8.1 for successive zero crossings from positive to negative values of $y(t)$.

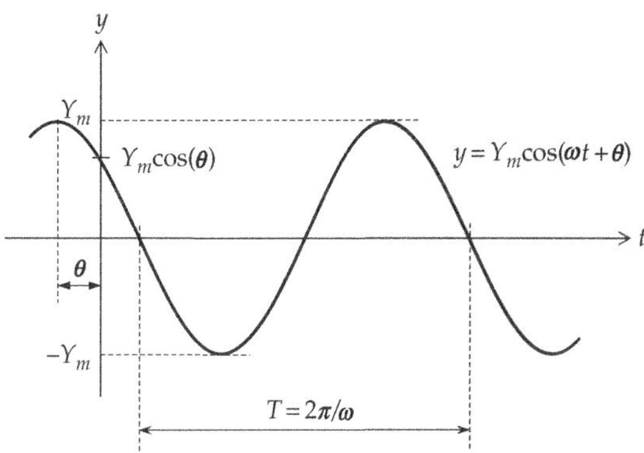

FIGURE 8.1
The sinusoidal function.

The full range of values of the function over a period is a **cycle**. The frequency f of repetitions of the function is

$$f = \frac{1}{T} = \frac{\omega}{2\pi} \tag{8.2}$$

where

T is in seconds
f is in cycles per second, or hertz (Hz)
ω is in rad/s

Voltages and currents that vary sinusoidally with time are designated in circuit analysis as **ac quantities**, where ac stands for 'alternating current'.

As ω decreases, T increases (Equation 8.2), the sinusoid becoming "flatter." If $\omega = 0$, the sinusoid of Figure 8.1 becomes a dc quantity of magnitude $Y_m\cos\theta$.

The sinusoidal function is a periodic function, that is, a function consisting of cycles that repeat at a given frequency. However, it is the only periodic function that is composed of a single frequency. All other periodic functions are composed of more than one frequency, in accordance with Fourier's theorem (Chapter 16).

An important property of the sinusoidal function is embodied in the following concept:

Concept: *When a sinusoidal excitation is applied to an LTI circuit, all the steady-state currents and voltages in the circuit are sinusoidal functions of the same frequency as the excitation, but which generally differ in amplitude and phase angle.*

The reason for this is that voltages and currents in an LTI circuit are subject to linear operations such as (1) scaling or multiplication by a constant, as in Ohm's law, voltage or current division, (2) addition or subtraction, as in KCL or KVL, and (3) differentiation and integration, as in the v–i relations of inductors and capacitors. When sinusoidal functions are subjected to any of these linear operations, they remain sinusoidal functions of the same frequency. The amplitudes and phase angles of the resulting sinusoidal functions will differ from those of the original functions, but no new frequencies are produced. It is assumed in this chapter that if more than a single source of excitation is applied to a given circuit, all the sources have the same frequency. Excitation by sources of more than one frequency is considered in Chapter 16.

Another important property of the sinusoidal function, which is used in this chapter, is embodied in Euler's formula:

$$Y_m e^{j\omega t} = Y_m \cos\omega t + jY_m \sin\omega t \tag{8.3}$$

where

$j = \sqrt{-1}$ is the *imaginary unit*

Y_m is the amplitude of the sinusoidal function

$Y_m e^{j\omega t}$ is a complex quantity, whose **real component** is $Y_m\cos\omega t$ and whose **imaginary component** is $jY_m\sin\omega t$. The term $Y_m\sin\omega t$, without j, is the **imaginary part**. The real part and the real component are one and the same.

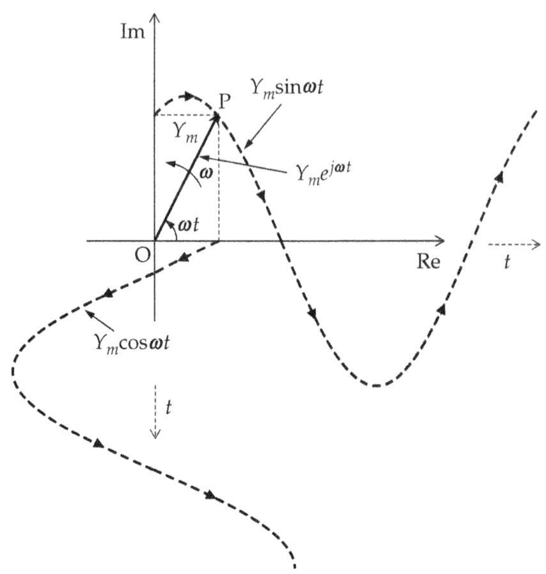

FIGURE 8.2
A rotating position vector on an Argand diagram.

Equation 8.3 can be plotted in the complex plane or Argand diagram, where the real part is plotted along the horizontal, or real, axis, and the imaginary part is plotted along the vertical, or imaginary, axis (Figure 8.2). $Y_m e^{j\omega t}$ is plotted in *polar form*, as a position vector OP, that is a vector drawn from the origin, whose magnitude is Y_m and whose angle with respect to the positive horizontal axis is ωt. This angle increases with time, so that the vector OP rotates counterclockwise (CCW) about the origin at an angular frequency ω rad/s. In *rectangular form*, $Y_m\cos\omega t$ is the projection of $Y_m e^{j\omega t}$ on the real axis, and $Y_m\sin\omega t$ is its projection on the imaginary axis. As time progresses, the lengths of these projections vary sinusoidally with time, as illustrated by the dashed extensions in Figure 8.2. As a vector, $\text{OP} = Y_m e^{j\omega t} = Y_m\cos\omega t + jY_m\sin\omega t$ as in Equation 8.3.

Analysis of the sinusoidal steady state is the derivation of the currents and voltages in a circuit after a sinusoidal excitation has been applied for a sufficiently long time. When any periodic excitation is suddenly applied to a circuit, there is generally an initial "transient," or temporary, response that dies out with time, as explained in later chapters. After this transient dies out, a steady state remains, in which the currents and voltages repeat according to some pattern that depends on the time variation of the excitation.

Analysis of the sinusoidal steady state is of fundamental importance for the following reasons:

1. The sinusoidal steady state is used in several practical cases, as in the frequency responses of systems (Chapter 14) and in power system analysis (Chapters 17 and 25). Electric power is efficiently generated, transmitted, and distributed as sinusoidal voltages and currents.
2. The methodology used in steady-state sinusoidal analysis is based on expressing sinusoidal voltages and currents as phasors and the v–i relations in terms of impedance. This is extremely convenient in that it extends to the sinusoidal steady state all the concepts, theorems, and procedures that apply to resistive circuits under dc conditions.
3. The representation of sinusoidal voltages and currents as phasors and the v–i relations in terms of impedance are later applied to the analysis of frequency responses of circuits (Chapter 14), the responses to periodic inputs (Chapter 16), power calculations (Chapter 17), three-phase systems (Chapter 25), and the Fourier transform (Chapter 22). Moreover, the phasor representation can be readily generalized to the most powerful method for analyzing linear systems, namely, that of the Laplace transform (Chapter 21).

Primal Exercise 8.1

Given $y(t) = 15\cos(200\pi t + 30°)$, determine (a) the duration of 10 cycles and (b) the time of occurrence of the first zero of the 11th cycle.

Ans. (a) 100 ms; (b) 101.67 ms.

8.2 Responses to Sinusoidal Excitation

8.2.1 Excitation in Trigonometric Form

Consider the circuit of Figure 8.3a in which a current source $i_{SRC}(t) = I_m\cos\omega t$ A is applied to a series RL circuit. It is desired to determine the voltage v across the series combination. Note that a tilde symbol (~) is added to the source to denote a steady-state sinusoidal excitation.

From KVL, $v = v_R + v_L$, where $v_R = Ri_{SRC}$ and $v_L = Ldi_{SRC}/dt$. Substituting for v_R and v_L,

$$v(t) = Ri_{SRC} + L\frac{di_{SRC}}{dt} \tag{8.4}$$

Substituting, $i_{SRC}(t) = I_m\cos\omega t$,

$$v(t) = RI_m\cos\omega t - \omega LI_m\sin\omega t \tag{8.5}$$

To combine the sine and cosine terms into a single function, the RHS is multiplied and divided by $\sqrt{R^2+\omega^2L^2}$ to give

$$v(t)=I_m\sqrt{R^2+\omega^2L^2}\left(\frac{R}{\sqrt{R^2+\omega^2L^2}}\cos\omega t - \frac{\omega L}{\sqrt{R^2+\omega^2L^2}}\sin\omega t\right) \tag{8.6}$$

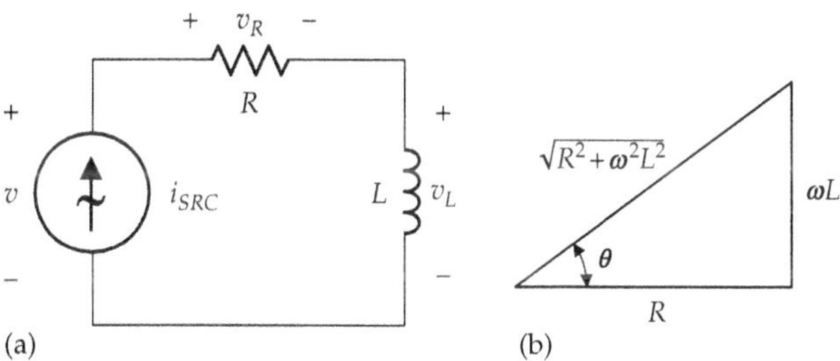

FIGURE 8.3
(a) Series RL circuit in the sinusoidal steady state and (b) angle θ equal to $\tan^{-1}\omega L/R$.

We now define an angle whose cosine is $R/\sqrt{R^2+\omega^2L^2}$ and whose sine is $\omega L/\sqrt{R^2+\omega^2L^2}$ (Figure 8.3b). Equation 8.6 becomes

$$\begin{aligned} v(t) &= I_m\sqrt{R^2+\omega^2L^2}\left(\cos\theta\cos\omega t - \sin\theta\sin\omega t\right) \\ &= I_m\sqrt{R^2+\omega^2L^2}\left(\cos(\omega t+\theta)\right) = V_m\cos(\omega t+\theta) \end{aligned} \tag{8.7}$$

where $V_m = I_m\sqrt{R^2+\omega^2L^2}$ is the amplitude of v. It is seen that this procedure for obtaining v is rather complicated, even in this simple case, as it involves writing KVL using a time derivative, followed by manipulation of trigonometric functions. In contrast, V in a similar dc case (Figure 8.4) can be written from Ohm's law as

$$V = (R_1+R_2)I_{SRC} \tag{8.8}$$

A pertinent question, therefore, is how to reduce the derivation of voltage in Figure 8.3a to be much like that in Figure 8.4. We will outline the two major steps involved in achieving this objective, before examining these steps and their implications:

1. The first step is to apply the excitation as a complex sinusoidal function $I_me^{j\omega t}$ rather than $I_m\cos\omega t$, so that $v_R(t) = RI_me^{j\omega t}$, and $v_L(t) = Ldi/dt = j\omega LI_me^{j\omega t}$. Note that differentiation has been reduced simply to multiplication by $j\omega$. KVL in Figure 8.3 gives

$$v(t) = (R + j\omega L)I_me^{j\omega t} \tag{8.9}$$

2. The second step is to set $t = 0$ in Equation 8.9, which makes the exponential term equal to unity, as this leads naturally to phasor notation and representation in the frequency domain. The RHS of Equation 8.9 becomes $(R + j\omega L)I_m$ and is of the same form as Equation 8.8 in that the voltage is the product of the amplitude of current and a complex quantity $(R + j\omega L)$, that is a constant for a given circuit and frequency of

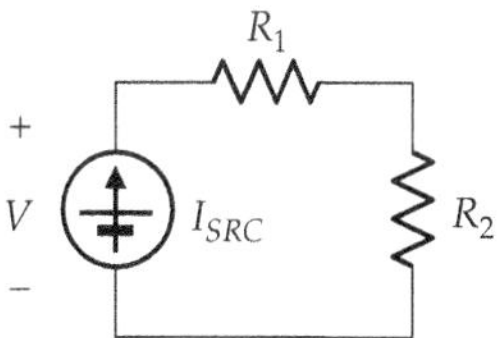

FIGURE 8.4
Voltage across two resistors in the dc state.

excitation. Moreover, $(R + j\omega L)$ in polar coordinates is $\sqrt{R^2+\omega^2L^2}e^{j\theta}$, where θ is the same angle in Figure 8.3b. Substituting in Equation 8.9, with $t = 0$, the RHS becomes

$$I_m\sqrt{R^2+\omega^2L^2}e^{j\theta} = V_m e^{j\theta} \tag{8.10}$$

where $V_m = I_m\sqrt{R^2+\omega^2L^2}$ is the amplitude of v, as in Equation 8.7. Equation 8.10 is in fact the precursor to phasor notation.

These two steps will be elaborated and justified in the following subsections, leading to the objective of applying "Ohm's law" to the sinusoidal steady state.

8.2.2 Complex Sinusoidal Excitation

A complex sinusoidal excitation is not physically realizable, but is an important mathematical construct that can be interpreted in terms of physically realizable trigonometric excitations. From Euler's formula,

$$I_m e^{j\omega t} = I_m\cos\omega t + jI_m\sin\omega t \tag{8.11}$$

It is seen from Equation 8.11 that a complex sinusoidal excitation is a combination of a real trigonometric excitation, $I_m\cos\omega t$, and an imaginary trigonometric excitation whose imaginary part is a real trigonometric excitation $I_m\sin\omega t$. If $(R + j\omega L)$ is expressed in polar coordinates as $\sqrt{R^2+\omega^2L^2}e^{j\theta}$, as in Equation 8.10. Equation 8.9 becomes

$$v(t) = I_m(R+j\omega L)e^{j\omega t} = I_m\sqrt{R^2+\omega^2L^2}e^{j(\omega t+\theta)} = V_m e^{j(\omega t+\theta)} \tag{8.12}$$

Equation 8.12 is the complex sinusoidal voltage response to the complex sinusoidal current excitation $I_m e^{j\omega t}$. Applying Euler's formula

$$v(t) = I_m\sqrt{R^2+\omega^2L^2}\left(\cos(\omega t+\theta)\right) + jI_m\sqrt{R^2+\omega^2L^2}\left(\sin(\omega t+\theta)\right) \tag{8.13}$$

It is seen that the real component of v is the same as that derived directly from an excitation $I_m\cos\omega t$ (Equation 8.7). This raises the question as to whether the imaginary part of the response in Equation 8.13 is the response to an excitation $i_{SRC} = I_m\sin\omega t$. We will show that this is indeed the case.

The most direct way of doing this is to express $I_m\sin\omega t$ as $I_m\cos(\omega t - 90°)$. This replaces ωt by $(\omega t - 90°)$ in Equations 8.5 through 8.7. Equation 8.7 becomes

$$v(t) = I_m\sqrt{R^2+\omega^2L^2}\left(\cos(\omega t-90°+\theta)\right) = I_m\sqrt{R^2+\omega^2L^2}\left(\sin(\omega t+\theta)\right) \tag{8.14}$$

Multiplying $I_m\sin\omega t$ by the constant j, as in Equation 8.11 multiplies the response in Equation 8.14 by j, which gives the imaginary component of v in Equation 8.13.

What we have shown, therefore, is that applying the excitation $I_m e^{j\omega t}$ is equivalent to applying the two excitations $I_m\cos\omega t$ and $I_m\sin\omega t$ at the same time, but with each of these excitations being independent of the other, as if it were applied alone. This underlies an important concept, namely,

Concept: *When a complex sinusoidal excitation $Y_m e^{j\omega t}$ is applied to an LTI circuit, the response is a complex sinusoidal function whose real part is the response to the real part of the excitation, $Y_m\cos\omega t$, applied alone, and whose imaginary part is the response to the imaginary part of the excitation, $Y_m\sin\omega t$, applied alone.*

This is illustrated in Figure 8.5, where the excitation $V_m e^{j\omega t}$ applied to an LTI circuit is the sum of two excitations $V_m\cos\omega t$ and $jV_m\sin\omega t$. Any voltage or current response in the circuit due to $V_m e^{j\omega t}$ is, by superposition, the sum of two responses: (1) the response due to a source equal to the real part of the excitation, $V_m\cos\omega t$ acting alone, with the source $jV_m\sin\omega t$ set to zero and (2) the response due to a source equal to the imaginary part of the excitation, $jV_m\sin\omega t$ acting alone, with the source $V_m\cos\omega t$ set to zero.

The reason for this separation of the responses is that in linear operations, the real and imaginary parts of a complex quantity do not mix; they retain their respective identities. Thus, if $x = (a + jb)$ and $y = (c + jd)$, then $x + y = (a + b) + j(c + d)$, where the real part of $(x + y)$ is the sum of the two real parts $(a + b)$, and the imaginary part is the sum of the two imaginary parts $(c + d)$. The same is true of other linear operations that are encountered in electric circuits, such as differentiation or integration. On the other hand, multiplying x and y, which is a nonlinear operation, gives $xy = (ab - cd) + j(bc + ad)$, where the real and imaginary parts of xy are mixtures of the real and imaginary parts of x and y.

It follows that if the response to an excitation $K\cos\omega t$ or $K\sin\omega t$ is required, a complex sinusoidal excitation $Ke^{j\omega t}$ can be applied and the real and imaginary parts of the response will be the responses to $K\cos\omega t$ and $K\sin\omega t$,

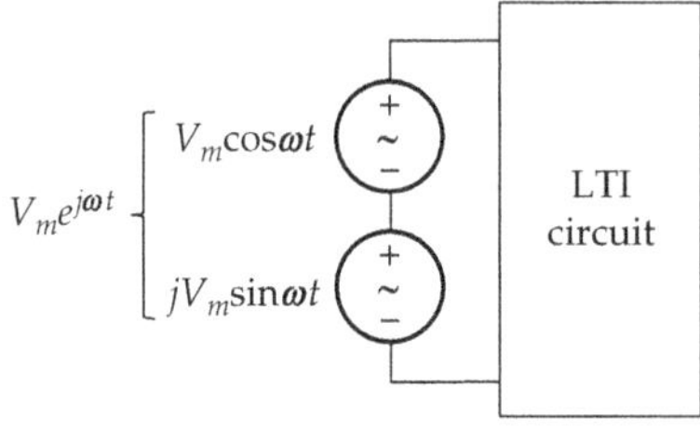

FIGURE 8.5
Complex sinusoidal excitation applied to an LTI circuit.

TABLE 8.1

Complex Sinusoidal Excitation and Responses

Time Function	Complex Form	Real Part	Imaginary Part
Excitation	$I_m e^{j\omega t}$	$I_m \cos\omega t$	$I_m \sin\omega t$
Response	$V_m e^{j(\omega t+\theta)}$	$V_m \cos(\omega t+\theta)$	$V_m \sin(\omega t+\theta)$

respectively. This is summarized in Table 8.1, where the excitation is assumed to be a current and the response a voltage.

Primal Exercise 8.2

Given $y(t) = 5\cos(\omega t + 30°) + 12\sin(\omega t + 30°)$, express $y(t)$ as a single cosine function, as a single sine function, and in terms of complex exponentials.

Ans. $13\cos(\omega t - 37.4°)$, $13\sin(\omega t + 52.6°)$, $(2.5 - j6)e^{j(\omega t+30°)} + (2.5 + j6)e^{-j(\omega t+30°)}$.

8.3 Phasors

8.3.1 Phasor Notation

The complex excitation $I_m e^{j\omega t}$ and the complex response $V_m e^{j(\omega t+\theta)}$ (Equation 8.12) can be plotted as position vectors on an Argand diagram (Figure 8.6a). Both vectors rotate CCW at an angular speed ω, retaining the phase angle θ between them as they rotate. The angular frequency ω is the same because, as emphasized earlier, all the voltages and current responses in an LTI circuit have the same frequency as the excitation. The information of interest in Figure 8.6a is the magnitudes of the two vectors and their relative phase angle. It follows that no significant information is lost by freezing the rotation of the vectors at a particular value of t or taking a snapshot of the rotating vectors at a time t, which can be conveniently taken as $t = 0$. $I_m e^{j\omega t}$ becomes $I_m e^0 = I_m$, and $V_m e^{j(\omega t+\theta)}$ becomes $V_m e^{j\theta}$, as in Figure 8.6b. Such complex quantities are denoted as **phasors**. Thus,

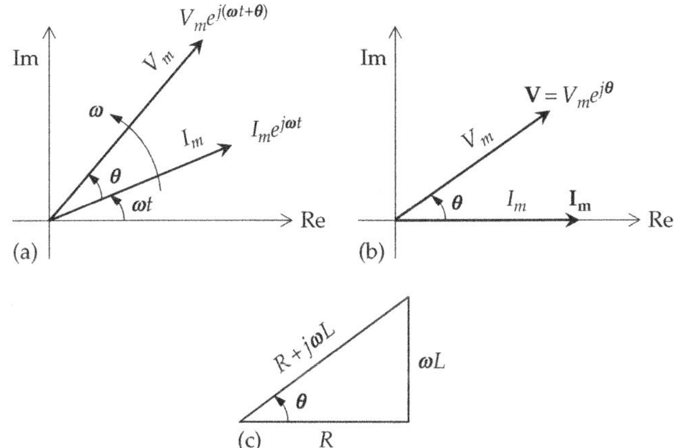

FIGURE 8.6
(a) Phasors rotating at the same angular velocity; (b) rotation frozen at $t = 0$; (c) $R + j\omega L$ drawn on a different Argand diagram.

Definition: *A phasor is a quantity such as $V_m e^{j\theta}$ representing a complex sinusoidal function of time, but with the time variation suppressed.*

Phasors are written in boldface and expressed as a magnitude and phase angle:

$$\mathbf{V} = V_m e^{j\theta} = V_m \angle\theta \tag{8.15}$$

It should be emphasized that phasors drawn on a given Argand diagram are implicitly complex sinusoidal functions of time rotating at the same angular frequency. Phasors that are implicitly rotating at different frequencies cannot, and should not, be drawn on the same Argand diagram. Nor should complex quantities, such as $(R + j\omega L)$, be drawn on the same Argand diagram as phasors (Figure 8.6c).

8.3.2 Properties of Phasors

Being position vectors, phasors have magnitude and direction, just like vectors, but with the additional property that their horizontal component is a real quantity, and their vertical component is an imaginary quantity. They can therefore be scaled, added, and subtracted like vectors.

Thus, multiplying a phasor by a real quantity K multiplies its magnitude by K, without changing its phase angle. Two phasors $\mathbf{Y}_1 = Y_1\angle\theta_1$ and $\mathbf{Y}_2 = Y_2\angle\theta_2$ can be added by drawing $\mathbf{Y}_2$ such that its origin lies at the tip of $\mathbf{Y}_1$ (Figure 8.7a). The sum $\mathbf{Y}_1 + \mathbf{Y}_2$ is the phasor whose origin is that of $\mathbf{Y}_1$ and whose tip is that of $\mathbf{Y}_2$. The sum of $\mathbf{Y}_1$ and $\mathbf{Y}_2$ may be also obtained by applying the "parallelogram rule", as in Figure 8.7a. The real part of $\mathbf{Y}_1 + \mathbf{Y}_2$ is $Y_1\cos\theta_1 + Y_2\cos\theta_2$, and its imaginary part is $Y_1\sin\theta_1 + Y_2\sin\theta_2$. It follows that the magnitude of $\mathbf{Y}_1 + \mathbf{Y}_2$ is

$$\left|\mathbf{Y}_1 + \mathbf{Y}_2\right| = \sqrt{Y_1^2 + Y_2^2 + 2Y_1Y_2\cos(\theta_2 - \theta_1)} \tag{8.16}$$

The phase angle of $\mathbf{Y}_1 + \mathbf{Y}_2$ is

$$\angle(\mathbf{Y}_1 + \mathbf{Y}_2) = \tan^{-1}\frac{Y_1\sin\theta_1 + Y_2\sin\theta_2}{Y_1\cos\theta_1 + Y_2\cos\theta_2} \tag{8.17}$$

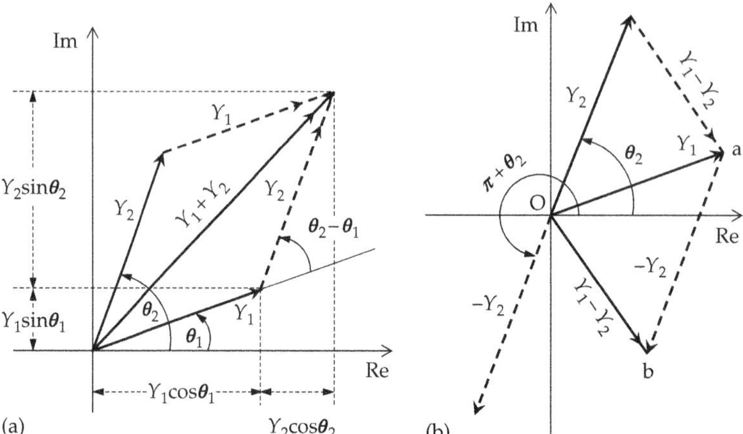

FIGURE 8.7
Phasor addition (a) and subtraction (b).

The phasor difference $\mathbf{Y}_1 - \mathbf{Y}_2$ is obtained by adding $\mathbf{Y}_1$ and $-\mathbf{Y}_2$, where $-\mathbf{Y}_2$ is a phasor of the same magnitude as $\mathbf{Y}_2$ but having a phase angle $(\theta_2 + \pi)$ (Figure 8.7b). Alternatively, the phasor $\mathbf{Y}_1 - \mathbf{Y}_2$ may be obtained as the phasor whose origin lies at the tip of $\mathbf{Y}_2$ and whose tip lies at the tip of $\mathbf{Y}_1$. Then $\mathbf{Y}_1 = \mathbf{Y}_2 + (\mathbf{Y}_1 - \mathbf{Y}_2)$. The phasor can be translated to the origin, without changing its magnitude and direction, so as to become a position vector. A phasor $\mathbf{Y} = Y\angle\theta$ (Figure 8.8a) can be multiplied by a complex quantity $A\angle\alpha$ (Figure 8.8b)

$$Ye^{j\theta} \times Ae^{j\alpha} = AYe^{j(\theta+\alpha)} \tag{8.18}$$

The product is a phasor of magnitude AY and phase angle $(\theta + \alpha)$ (Figure 8.8a).

A phasor $\mathbf{Y} = Y\angle\theta$ can be divided by a complex quantity $A\angle\alpha$:

$$\frac{Ye^{j\theta}}{Ae^{j\alpha}} = \frac{Y}{A}e^{j(\theta-\alpha)} \tag{8.19}$$

The quotient is a phasor of magnitude Y/A and phase angle $(\theta - \alpha)$ (Figure 8.8c).

A special case is multiplication and division by j. In the complex plane, j is an imaginary quantity of unit magnitude and a phase angle of $\pi/2$, that is,

$$j = 1 \times e^{j\frac{\pi}{2}} = \cos\frac{\pi}{2} + j\sin\frac{\pi}{2} \tag{8.20}$$

Multiplying a phasor by j rotates the phasor through an angle $\pi/2$ CCW without changing its magnitude (Figure 8.9a). Dividing a phasor by j, or multiplying it by $-j$, since $1/j = j/j^2 = -j$, rotates the phasor through an angle $\pi/2$ clockwise without changing its magnitude (Figure 8.9b).

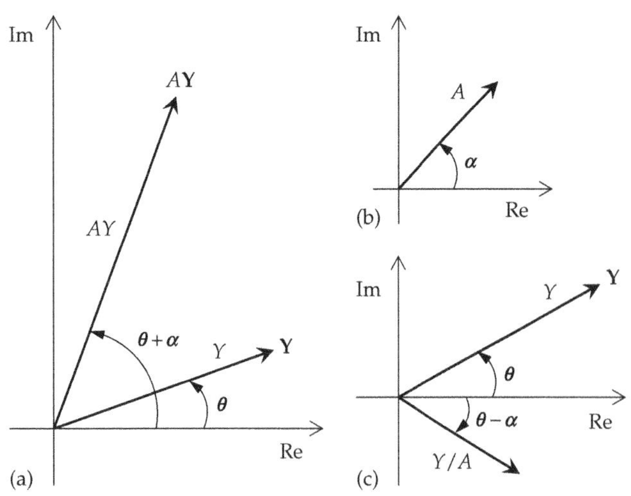

FIGURE 8.8
(a) Phasor **Y** multiplied by the complex number A shown in (b) and (c) **Y** divided by A.

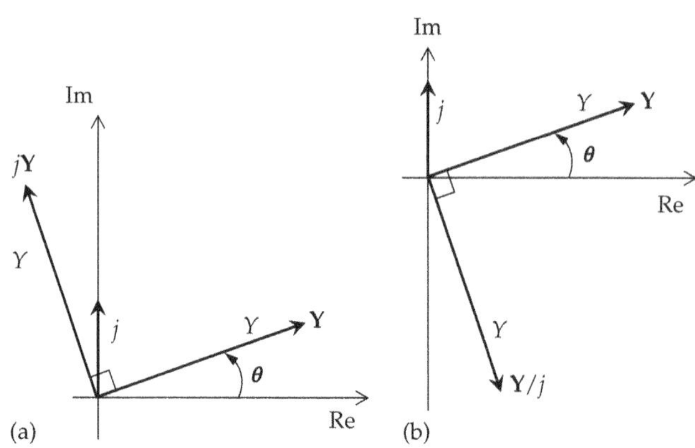

FIGURE 8.9
Phasor multiplication by j (a) and division by j (b).

Primal Exercise 8.3

Determine the product and quotient of $(12+j5)$ and $(3+j4)$ by working in rectangular coordinates and in polar coordinates.

Ans. *Product*: $16+j63=65\angle 75.8°$; *quotient*: $2.24-j1.32=2.6\angle -30.5°$.

Example 8.1: Manipulation of Complex Quantities

It is required to determine the magnitude, phase, real, and imaginary parts of $Y=(a+jb)/(c+jd)$, where a, b, c, and d are positive constants.

Solution:

Let us rationalize Y, that is, make its denominator real, by multiplying numerator and denominator by the complex conjugate of the denominator, $c-jd$. Thus,

$$Y=\frac{a+jb}{c+jd}\times\frac{c-jd}{c-jd}=\frac{ac+bd}{c^2+d^2}+j\frac{bc-ad}{c^2+d^2} \tag{8.21}$$

The real part of Y is $\frac{ac+bd}{c^2+d^2}$, its imaginary part is $\frac{bc-ad}{c^2+d^2}$.

The magnitude of Y is obtained as the square root of the sum of the squares of the real and imaginary parts. An easier way is to convert the numerator and denominator to polar coordinates. Thus, $Y=\frac{\sqrt{a^2+b^2}e^{j\theta_1}}{\sqrt{c^2+d^2}e^{j\theta_2}}=\frac{\sqrt{a^2+b^2}}{\sqrt{c^2+d^2}e}e^{j(\theta_1-\theta_2)}$. The magnitude of Y is therefore $\sqrt{a^2+b^2}/\sqrt{c^2+d^2}$. The phase angle of Y is $(\theta_1-\theta_2)=\tan^{-1}(b/a)-\tan^{-1}(d/c)$, that is, the phase angle of Y is the phase angle of the numerator minus that of the denominator. The following should be noted:

(a) Whereas the magnitude of Y is the magnitude of the numerator divided by that of the denominator, the real part of Y is *not* the real part of the numerator divided by that of the denominator. Nor is the imaginary part of Y the imaginary part of the numerator divided by that of the denominator.

(b) In determining the phase angle from the real and imaginary parts, the actual signs of these parts must be retained without change, as illustrated in Figure 8.10. Otherwise, the angle will be incorrect. Thus, assuming a and b are positive constants, $\tan^{-1}(b/a)$ is an angle θ in the first quadrant. However, $\tan^{-1}(-b/-a)$ is the ratio of two negative components, $-b$ and $-a$, so that the angle is $(\pi+\theta)$ in the third quadrant. Similarly, $\tan^{-1}(-b/a)$ is an angle $-\theta$ in the fourth quadrant, whereas $\tan^{-1}(b/-a)$ is an angle $(\pi-\theta)$ in the second quadrant.

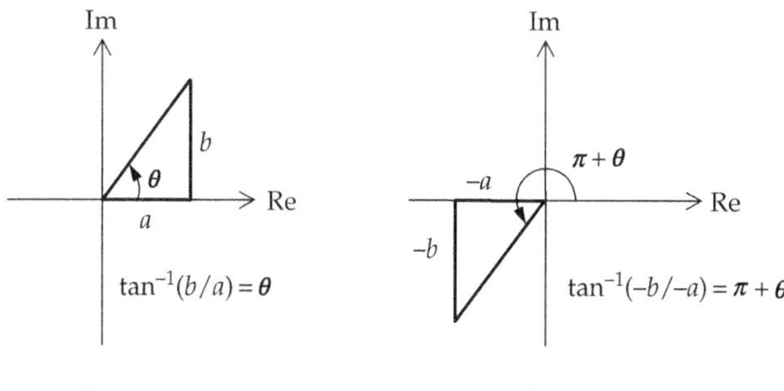

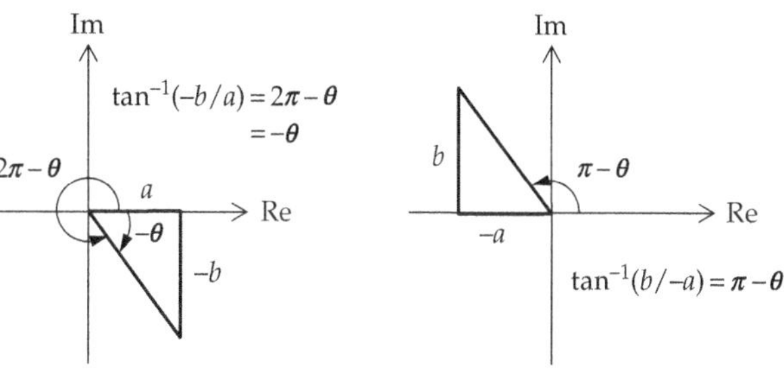

FIGURE 8.10
Figure for Example 8.1.

Problem-Solving Tip

- Complex quantities are conveniently added or subtracted in rectangular form and are conveniently multiplied or divided in polar form.

Exercise 8.4

Show that the square root of the sum of the squares of the real and imaginary parts of the RHS of Equation 8.21 reduces to $\sqrt{a^2+b^2}/\sqrt{c^2+d^2}$.

Primal Exercise 8.5

The phasor $(1-j)$ is rotated 90° clockwise. Represent the rotated phasor in the time domain as the imaginary component of the complex sinusoidal function of time, assuming an angular frequency ω rad/s.

Ans. $-\cos\omega t-\sin\omega t$.

Primal Exercise 8.6

Given $v_1(t)=12\sin(\omega t+60°)$ V and $v_2(t)=-6\cos(\omega t+30°)$ V. Express $v_1(t)$ as a cosine function, convert $v_1(t)$ and $v_2(t)$ to phasors, and draw them on a phasor diagram.

Ans. $v_1(t)=12\cos(\omega t-30°)$ V and $v_2(t)=6\cos(\omega t+210°)$ V, $\mathbf{V}_1=12\angle -30°$ V, $\mathbf{V}_2=6\angle 210°$ V (Figure 8.11).

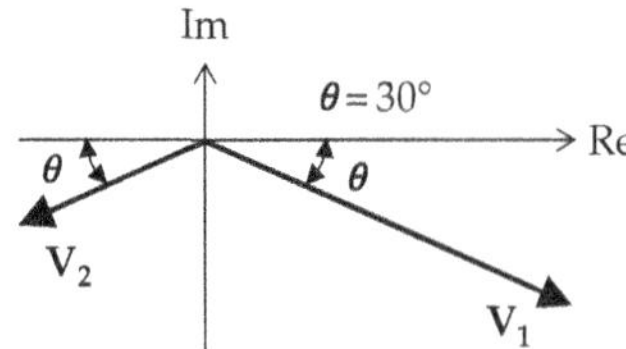

FIGURE 8.11
Figure for Primal Exercise 8.6.

Primal Exercise 8.7

Given the phasors $\mathbf{A} = 4\angle 45°$ and $\mathbf{B} = 2\angle -45°$, determine the phasor $\mathbf{A}'/\mathbf{B}'$, where $\mathbf{A}'$ is the phasor $\mathbf{A}$ multiplied by $(1 + j)$ and $\mathbf{B}'$ is the phasor $\mathbf{B}$ divided by $(1 - j)$.

Ans. $j4$.

8.4 Phasor Relations of Circuit Elements

Having defined phasors, the next step is to express the v–i relations of the three passive circuit elements in phasor notation.

8.4.1 Phasor Relations for a Resistor

If the current through a resistor is $I_m e^{j\omega t}$ A, the voltage across the resistor is $RI_m e^{j\omega t}$ V. In phasor notation $\mathbf{I} = I_m\angle 0°$ A, and $\mathbf{V} = RI_m\angle 0°$ V, (Figure 8.12a), or

$$\mathbf{V} = R\mathbf{I} \quad \text{or} \quad \mathbf{I} = G\mathbf{V} \tag{8.22}$$

According to the interpretation of complex sinusoidal excitation, a current $I_m\cos\omega t$ A produces a voltage $RI_m\cos\omega t$ V, and a current $I_m\sin\omega t$ V produces a voltage $RI_m\sin\omega t$ V. The voltage and current are in phase (Figure 8.12b).

If $i(t) = I_m\cos\omega t$ and $v(t) = RI_m\cos\omega t$, the instantaneous power dissipated by the resistor, based on the assigned positive directions of Figure 8.12a, is

$$p(t) = vi = RI_m^2\cos^2\omega t = \frac{RI_m^2}{2}(1+\cos 2\omega t) \tag{8.23}$$

p, plotted in Figure 8.12b, is never negative because a resistor does not deliver power. From Equation 8.23, p has an average component $P = RI_m^2/2$ and a sinusoidal component of zero average, amplitude $RI_m^2/2$, and frequency 2ω. The frequency of p is twice that of v or i because p is positive when v and i are both positive and both negative.

Formally, the average power P is obtained by integrating p over a period, which gives the energy dissipated over the period, and dividing by the period:

$$P = \frac{1}{T}\int_0^T p\,dt = \frac{1}{2\pi}\int_0^{2\pi}\frac{RI_m^2}{2}(1+\cos 2\omega t)\,d(\omega t)$$
$$= \frac{RI_m^2}{4\pi}\left[(\omega t) + \frac{1}{2}\sin 2\omega t\right]_0^{2\pi} = \frac{RI_m^2}{2} \tag{8.24}$$

Note that $I_m^2/2$ is the average, or mean, of the square of the current $i^2(t) = I_m^2\cos^2(\omega t)$, since this average is given by

$$\frac{1}{T}\int_0^T i^2 dt = \frac{1}{2\pi}\int_0^{2\pi} I_m^2\cos^2\omega t\,d(\omega t)$$
$$= \frac{I_m^2}{2\pi}\int_0^{2\pi}\frac{1}{2}(1+\cos 2\omega t)\,d(\omega t) = \frac{I_m^2}{2} \tag{8.25}$$

The square root of this mean is the **root-mean-square**, or rms value of i, which is denoted as I_{rms}. Thus,

$$I_{rms} = \sqrt{\frac{I_m^2}{2}} = \frac{I_m}{\sqrt{2}} \tag{8.26}$$

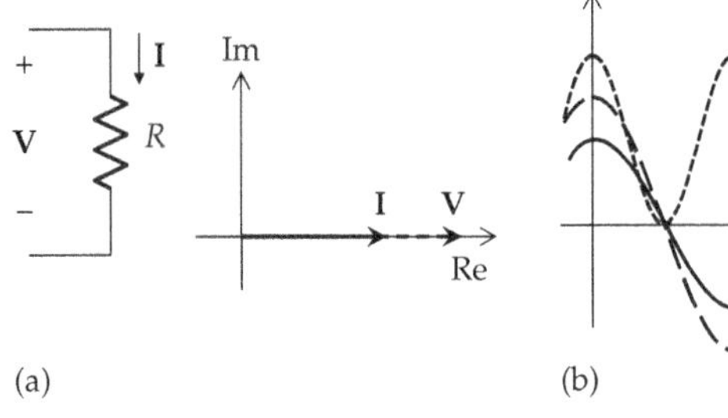

(a)

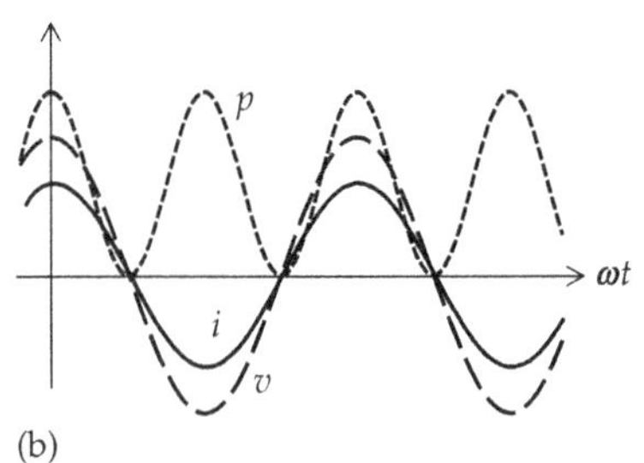

(b)

FIGURE 8.12
Sinusoidal responses of ideal resistors. (a) **V** and **I** for an ideal resistor shown as phasors and (b) v, i, and p for an ideal resistor in the time domain.

Substituting in Equation 8.24,

$$P = RI_{\text{rms}}^2 \quad (8.27)$$

Had we considered $i(t) = I_m\sin\omega t$ instead of $i(t) = I_m\cos\omega t$, only the sign of $\cos 2\omega t$ changes in preceding equations, but since the integral of this term averages to zero over the period, P and I_{rms} are unchanged. Moreover, if we consider $v(t) = V_m\cos\omega t$, where $V_m = RI_m$, the same procedure yields

$$P = \frac{V_m^2}{2R}, \quad V_{\text{rms}} = \frac{V_m}{\sqrt{2}}, \quad P = \frac{V_{\text{rms}}^2}{R} \quad (8.28)$$

It follows from Equations 8.24 and 8.28 that

$$P = \frac{RI_m^2}{2} = \frac{(RI_m)I_m}{2} = \frac{V_m I_m}{2} = \frac{V_m}{\sqrt{2}}\frac{I_m}{\sqrt{2}} = V_{\text{rms}}I_{\text{rms}} \quad (8.29)$$

Since the power dissipated in R by a dc current I is $P = VI = RI^2 = V^2/R$, it follows from Equations 8.27 through 8.29 that:

Concept: *A current of rms value I_{rms}, or a voltage of rms value V_{rms}, dissipate the same average power in a given resistor as a dc current, or a dc voltage, of the same value. The average power dissipated in the sinusoidal steady state is, like dc power, a real quantity that is independent of time, frequency, and phase angle.*

Using rms values of sinusoidal currents and voltages for power calculations results in expressions of the same form as under dc conditions. The rms value is also known as the **effective** value.

It should be emphasized that the *rms value of a phasor current or voltage is a real number that is $1/\sqrt{2}$ of the amplitude of the current or voltage, independently of the time, frequency, and phase angle*. For example, if $\mathbf{I} = (3 + j4) = 5\angle 53.1°$ A, so that $i(t) = 5\cos(\omega t + 53.1°)$ expressed as a cosine function, then $I_{\text{rms}} = 5/\sqrt{2}$ A.

8.4.2 Phasor Relations for a Capacitor

If the current through a capacitor is $I_m e^{j\omega t}$ A, the voltage across the capacitor is

$$v(t) = \frac{1}{C}\int i dt = \frac{1}{C}\int I_m e^{j\omega t} dt = \frac{I_m}{j\omega C}e^{j\omega t} = \frac{I_m}{\omega C}e^{j(\omega t - \pi/2)} + K \quad (8.30)$$

where K is the constant of integration that shifts the voltage in the vertical direction and defines its average value. When a purely sinusoidal current of zero average is applied to an uncharged capacitor, no net charge is deposited on the capacitor over a period, and the resulting steady-state voltage is sinusoidal, of zero average, so that $K = 0$.

In phasor notation, $\mathbf{I} = I_m\angle 0°$ A, and $\mathbf{V} = (I_m/j\omega C)\angle 0° = (I_m/\omega C)\angle -90°$ V, or

$$\mathbf{V} = \frac{1}{j\omega C}\mathbf{I} \quad \text{or} \quad \mathbf{I} = j\omega C\,\mathbf{V} \quad (8.31)$$

(Figure 8.13a). The magnitude of $\mathbf{V}$ is $(1/\omega C)$ times that of $\mathbf{I}$, and the phase angle of $\mathbf{V}$ is $-90°$, when the phase angle of $\mathbf{I}$ is zero (Figure 8.13a).

According to the interpretation of complex sinusoidal excitation, a current $I_m\cos\omega t$ A in a capacitor produces a voltage $(I_m/\omega C)\cos(\omega t - \pi/2) = (I_m/\omega C)\sin\omega t$ across the capacitor, and a current $I_m\sin\omega t$ A produces a voltage $(I_m/\omega C)\sin(\omega t - \pi/2) = -(I_m/\omega C)\cos\omega t$. The voltage *lags* the current by 90°, or the current *leads* the voltage by 90° (Figure 8.13). This can be ascertained by comparing the times of occurrence of the positive peaks of the waveforms of v and i in Figure 8.13b that are closest to one another. Since the peak of v occurs later in time than the nearest peak of i by a quarter of a period, v lags i by 90°.

The instantaneous power p absorbed by the capacitor in Figure 8.13b is

$$p(t) = vi = \left(\frac{I_m^2}{\omega C}\right)\sin\omega t\cos\omega t = \left(\frac{I_m^2}{2\omega C}\right)\sin 2\omega t \quad (8.32)$$

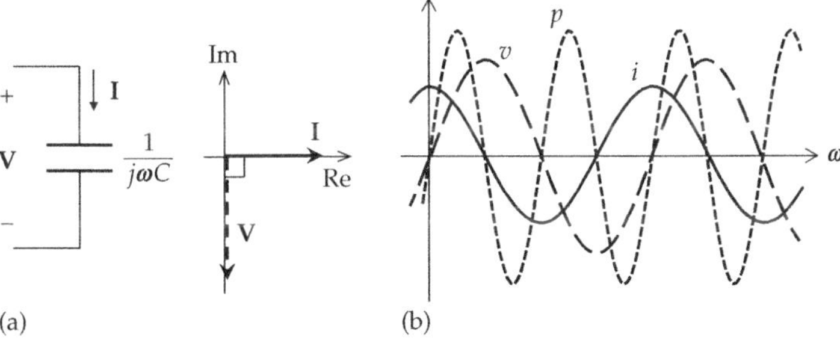

FIGURE 8.13
Sinusoidal responses of ideal capacitors. (a) $\mathbf{V}$ and $\mathbf{I}$ for an ideal capacitor shown as phasors and (b) v, i, and p for an ideal capacitor in the time domain.

and

$$P = \frac{1}{2\pi}\int_0^{2\pi}\left(\frac{I_m^2}{2\omega C}\right)\sin\omega 2td(\omega t) = 0 \qquad (8.33)$$

The average power P is zero because an ideal capacitor does not dissipate power. The power absorbed ($p > 0$) and stored as electric energy during a positive half-cycle of p is equal and opposite to the power delivered ($p < 0$) during a negative half-cycle of p, when the stored energy is returned to the supply. p varies at twice the frequency of v and i, as in the case of an ideal resistor and for the same reason previously mentioned. p is negative when either v or i is negative.

8.4.3 Phasor Relations for an Inductor

If the current through an inductor current is $I_m e^{j\omega t}$ A, the voltage across the inductor is

$$v(t) = L\frac{di}{dt} = j\omega L I_m e^{j\omega t} = \omega L I_m e^{j(\omega t + \pi/2)} \qquad (8.34)$$

In phasor notation, $\mathbf{I} = I_m\angle 0°$ A, and $\mathbf{V} = j\omega L I_m\angle 0° = \omega L I_m\angle 90°$ V, or

$$\mathbf{V} = j\omega L\mathbf{I} \quad \text{or} \quad \mathbf{I} = \frac{1}{j\omega L}\mathbf{V} \qquad (8.35)$$

The magnitude of **V** is ωL times that of **I**, and the phase angle of **V** is 90°, when the phase angle of **I** is zero (Figure 8.14a).

According to the interpretation of complex sinusoidal excitation, a current $I_m\cos\omega t$ A in the inductor produces a voltage $\omega L I_m\cos(\omega t + 90°) = -\omega L I_m\sin\omega t$ across the inductor, and a current $I_m\sin\omega t$ A produces a voltage $\omega L I_m\sin(\omega t + 90°) = \omega L I_m\cos\omega t$. The voltage *leads* the current by 90° or the current *lags* the voltage by 90° (Figure 8.13b). This may be ascertained by comparing the times of occurrence of the positive peaks of the waveforms of v and I that are closest to one another. Since the peak of v occurs earlier in time than the nearest peak of i by a quarter of a period, v leads i by 90°.

The instantaneous power p absorbed by the inductor in Figure 8.14b is

$$p(t) = vi = -\omega L I_m^2\sin\omega t\cos\omega t = -\left(\frac{\omega L I_m^2}{2}\right)\sin 2\omega t \qquad (8.36)$$

and

$$P = \frac{1}{2\pi}\int_0^{2\pi}\left(\frac{\omega L I_m^2}{2}\right)\sin 2\omega t d(\omega t) = 0 \qquad (8.37)$$

The average power P is zero because an ideal inductor does not dissipate power. The power absorbed ($p > 0$) and stored as magnetic energy during a positive half-cycle of p is equal and opposite to the power delivered ($p < 0$) during a negative half-cycle of p, when the stored energy is returned to the supply. p varies at twice the frequency of v and i, as in the case of an ideal capacitor and for the same reason previously mentioned.

From the preceding discussion, the corresponding v–i relations of capacitors and inductors in the time domain and in phasor notation are as follows:

$$\text{For a capacitor, } v(t) = \frac{1}{C}\int i dt \text{ and } \mathbf{V} = \frac{1}{j\omega C}\mathbf{I} \qquad (8.38)$$

$$\text{For an inductor, } v(t) = L\frac{di}{dt} \text{ and } \mathbf{V} = j\omega L\mathbf{I} \qquad (8.39)$$

It is seen that the following concept applies:

Concept: *In phasor notation, differentiation in time is expressed as multiplication by jω, and integration in time is expressed as division by jω. Thus, differential and integral relations are transformed to algebraic relations in jω for steady-state sinusoidal analysis* only.

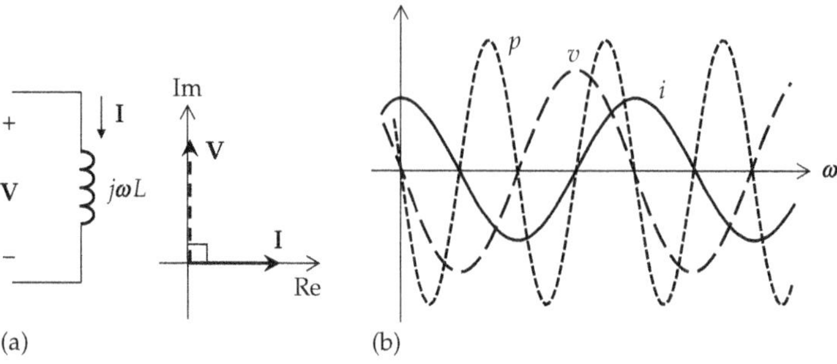

FIGURE 8.14
Sinusoidal responses of ideal inductors. (a) **V** and **I** for an ideal inductor shown as phasors and (b) v, i, and p for an ideal inductor in the time domain.

This concept in fact underlies the usefulness of phasor notation for steady-state sinusoidal analysis.

Another important observation is that v and i are in phase for an ideal resistor but are in phase quadrature for ideal capacitors and inductors. This is because an ideal resistor only dissipates power, which means that v and i are in phase. It does not store energy that must be returned later to the supply. If no power is delivered by an ideal resistor, p is never negative, so that v and i must be in phase.

On the other hand, ideal capacitors and inductors do not dissipate power, which means that the average power P over a period is zero. For this to be the case, v and i must be in phase quadrature, which means that if one is a cosine function, the other is a sine function, so that their product averages to zero over a period.

Concept: *The sinusoidal voltage and current for an ideal resistor are in phase because such a resistor is purely dissipative. They are in phase quadrature for ideal energy storage elements because these elements are nondissipative.*

Primal Exercise 8.8

The voltage applied to a 10 Ω resistor is $80\angle 35°$ V peak, the frequency being 50 Hz. Determine (a) the expression for the current in the time domain, assuming the voltage is a cosine function; and (b) the power dissipated in the resistor.

Ans. (a) $8\cos(100\pi t + 35°)$ A; (b) 320 W.

Primal Exercise 8.9

The current through a series combination of a 10 mH inductor and a 50 μF capacitor is $15\angle -75°$ mA rms. Determine the voltage across the inductor, and the voltage across the capacitor, (a) as phasors, (b) as functions of time, assuming that the frequency is 200 Hz and the time variation is a sine function.

Ans. (a) $\mathbf{V}_L = 60\pi\angle 15°$ mV rms, $\mathbf{V}_C = (750/\pi)\angle -165°$ mV rms; (b) $v_L(t) = 60\pi\sqrt{2}\sin(400\pi t + 15°)$ mV rms, $v_C(t) = (750\sqrt{2}/\pi)$ $\sin(400\pi t - 165°)$ mV rms.

Primal Exercise 8.10

A current $i(t) = 4\cos 5t$ A is applied to a series combination of 4 Ω resistor, 1 H inductor, and 0.1 F capacitor. (a) Express the voltage across the series combination as a phasor; (b) determine the average power dissipated in the resistor, using Equations 8.27 through 8.29.

Ans. (a) $\mathbf{V} = 16 + j12 = 20\angle 39.9°$ mV; (b) 32 W.

8.5 Impedance and Reactance

When considering opposition to the flow of current in a circuit under sinusoidal steady-state conditions, it is necessary to include, in addition to resistance, the effects of energy storage elements. This is because both the build-up of voltage across a capacitor and the induced voltage in an inductor oppose an increase in current through these elements by being voltage drops in the direction of current, as in the case of a resistor. Consider a part of a circuit or subcircuit that does not contain independent sources, for reasons that will be explained shortly. Let $\mathbf{V} = V_m\angle\theta_v$ be a voltage phasor across terminals 'ab' of the subcircuit, and let $\mathbf{I} = I_m\angle\theta_i$ be the current through the subcircuit in the direction of the voltage drop $\mathbf{V}$ (Figure 8.15). Then,

Definition: *Impedance Z is the ratio of the voltage phasor $V_m\angle\theta_v$ across a subcircuit to the current phasor $I_m\angle\theta_I$ flowing through the subcircuit in the direction of a voltage drop. The subcircuit could be a single R, L, or C, or a combination of these elements. The subcircuit should not contain independent sources but could contain dependent sources, as long as the controlling variables of these sources are all within the subcircuit.*

$$Z = \frac{\mathbf{V}}{\mathbf{I}} = \frac{V_m\angle\theta_v}{I_m\angle\theta_i} = \frac{V_m}{I_m}\angle\theta \tag{8.40}$$

where $\theta = \theta_v - \theta_i$ (Figure 8.16a). When $\mathbf{V}$ is in volts and $\mathbf{I}$ is in amperes, Z is in ohms. Since Z is in general complex, it can be expressed as

$$Z = R + jX \tag{8.41}$$

(Figure 8.16b). It should be emphasized that although Z is in general complex, it is not a phasor, because it is not a complex sinusoidal function of time in which the time variation has been suppressed. In fact, the exponential time variation cancels out in Equation 8.40. Z is therefore drawn on a separate diagram from that of $\mathbf{V}$ and $\mathbf{I}$. Moreover, the reason independent sources are excluded from the subcircuit in the definition of impedance is that in order to make the concept of impedance useful,

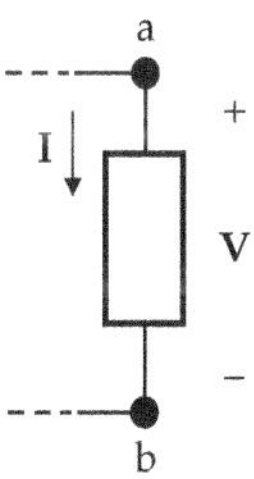

FIGURE 8.15
Definition of impedance.

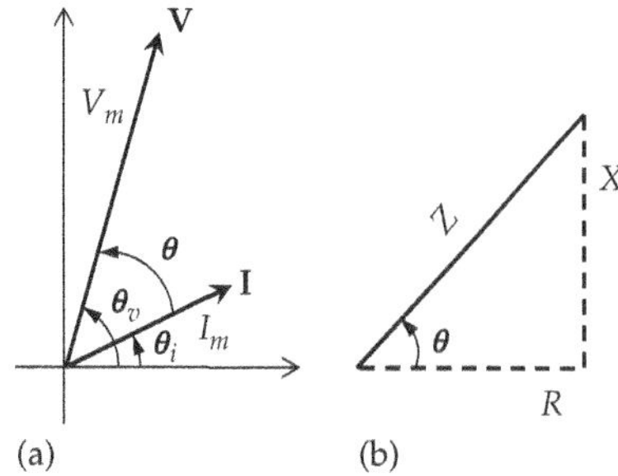

FIGURE 8.16
(a) Phasor diagram showing **V** and **I** and (b) impedance diagram.

impedance should depend only on the values of the passive circuit elements and their configuration, independently of any applied excitation due to independent sources. Dependent sources can be included, as long as their controlling voltages or currents are within the subcircuit. The effect of the dependent sources in this case is to alter values of *R*, *L*, and *C*.

In the case of an ideal resistor, **V** and **I** are in phase (Equation 8.22), which means that *Z* is real and equal to *R*. It follows that *R* in Equation 8.41 is the usual resistance that we have considered in previous chapters, and that for an ideal resistor, $X = 0$. *X*, the imaginary part of *Z* is the **reactance** and its unit is the ohm, just like resistance and impedance.

For an ideal capacitor, Equation 8.31 gives $Z = \mathbf{V}/\mathbf{I} = -j/\omega C$ so that $X = -1/\omega C$. For an ideal inductor, Equation 8.35 gives $Z = \mathbf{V}/\mathbf{I} = j\omega L$, so that $X = j\omega L$. For both ideal energy storage elements, $R = 0$, since these elements are nondissipative. The reactance represents the opposition to current flow due to energy storage elements.

It should be emphasized that when $X = 0$, **V** and **I** are in phase. When $X \neq 0$, *Z* is complex and **V** and **I** are out of phase. Hence,

Concept: *In a purely resistive circuit, all sinusoidal voltages and currents are in phase and their amplitude do not depend on frequency. Energy-storage elements possess reactance, which causes frequency-dependent phase differences between sinusoidal voltages and currents in the circuit and, in addition, makes the amplitudes of these sinusoidal voltages and currents depend on frequency.*

The following should be noted concerning reactance:

1. In any expression that involves impedance, reactance is always multiplied by *j*. For an ideal capacitor, $X = -1/\omega C$ and $Z = -j/\omega C$. For an ideal inductor, $X = \omega L$ and $Z = j\omega L$.
2. Since *j* appears in the **V**–**I** relations as a result of differentiating or integrating expressions involving $e^{j\omega t}$, *j* is always associated with ω in the expression for impedance. Hence, reactance is always a function of frequency.

The reciprocal of impedance is the **admittance** *Y*:

$$\frac{\mathbf{I}}{\mathbf{V}} = \frac{1}{Z} = Y = G + jB \tag{8.42}$$

where *B* is the **susceptance**. *B* and *Y* are in siemens, like *G*.

For an ideal resistor, $\mathbf{I}/\mathbf{V} = G = 1/R$ (Equation 8.15), so that $B = 0$, and *G* is the usual conductance. For an ideal capacitor, $Y = \mathbf{I}/\mathbf{V} = j\omega C$ (Equation 8.31), so that $G = 0$ and $B = j\omega C$. For an ideal inductor, $Y = \mathbf{I}/\mathbf{V} = 1/j\omega L = -1/\omega L$ (Equation 8.35), so that $G = 0$, and $B = -1/\omega L$. Table 8.2 lists the circuit properties of the three circuit elements. It is seen that under dc conditions $\omega = 0$, which makes the reactance of a capacitor infinite and its susceptance zero. This means that when the voltage across the capacitor is a dc voltage, the current is zero. The capacitor behaves as an open circuit, as argued in Section 7.1. Similarly, when $\omega = 0$, the reactance of an inductor is zero and its susceptance is infinite. This means that when the inductor current is dc, the voltage across the inductor is zero. The inductor behaves as a short circuit, as argued in Section 7.2.

Note that whereas $G = 1/R$ for an ideal resistor, $B = -1/X$ for an ideal capacitor or inductor, and not $+1/X$. This is because conversion from reactance to susceptance *must proceed through impedance and admittance*, which introduces a minus sign in the reciprocal of the imaginary component. Thus, $Z = jX$, $Y = 1/Z = 1/jX = -j/X$. But $Y = jB$, so that $B = -1/X$.

Since $\mathbf{V}/\mathbf{I} = Z = R + jX$ and *R* is the resistance, $R + jX$ can be interpreted as *R* in series with *jX*, as in Figure 8.17a. For if the current through the series combination is **I**, then it follows from KVL that $\mathbf{V} = R\mathbf{I} + jX\mathbf{I}$. Dividing by **I** gives $\mathbf{V}/\mathbf{I} = Z = R + jX$. Similarly, $\mathbf{I}/\mathbf{V} = Y = G + jB$ can be interpreted as *G* in parallel with *jB*, as in Figure 8.17b. For if the voltage across the parallel combination is **V**, then it follows from KCL that $\mathbf{I} = G\mathbf{V} + jB\mathbf{V}$. Dividing by **V** gives $\mathbf{I}/\mathbf{V} = Y = G + jB$.

TABLE 8.2
Circuit Properties of Circuit Elements

Circuit Property	Resistor	Inductor	Capacitor
Reactance (*X*)	0	ωL	$\frac{-1}{\omega C}$
Impedance (*Z*)	*R*	$j\omega L$	$\frac{1}{j\omega C} = \frac{-j}{\omega C}$
Susceptance (*B*)	0	$\frac{-1}{\omega L}$	ωC
Admittance (*Y*)	$G = \frac{1}{R}$	$\frac{1}{j\omega L} = \frac{-j}{\omega L}$	$j\omega C$

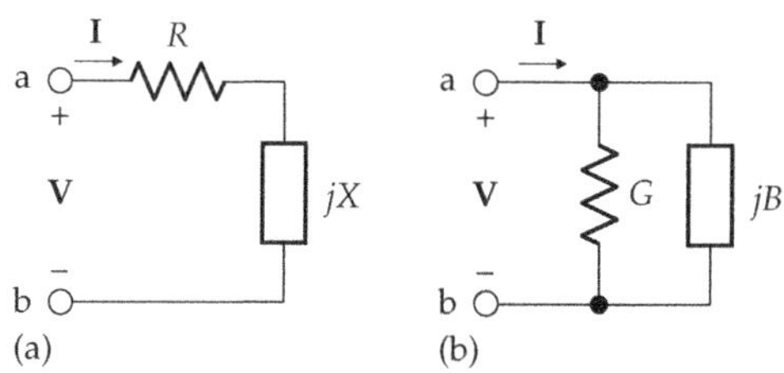

FIGURE 8.17
(a) Impedances in series and (b) admittances in parallel.

Primal Exercise 8.11

(a) Determine the reactance and susceptance of (i) the inductor and (ii) the capacitor of Primal Exercise 8.9; (b) determine the impedance and admittance of the inductor and capacitor (i) in series, (ii) in parallel.

Ans. (a) (i) 12.57 Ω, −0.0796 S, (ii) −15.92 Ω, 0.0628 S; (b) (i) − j3.349 Ω, j0.299 S, (ii) j59.72 Ω, −j0.168 S.

Exercise 8.12

Argue that (i) the negative sign of the reactance of a capacitor reflects the fact that the voltage across a capacitor lags the current through the capacitor by 90° in the time domain, and (ii) the magnitude of the reactance of a capacitor decreases with frequency because $i = Cdv/dt = dq/dt$. Apply a similar argument to the reactance of an inductor.

Example 8.2: Equivalent Parallel Impedance

Given an impedance $Z_S = R_S + j\omega L_S$, represented as R_S in series with L_S. It is required to determine R_P and L_P of the equivalent parallel combination (Figure 8.18).

Solution:

Since the two circuits are equivalent between terminals 'ab', they must have the same **V** and **I** at these terminals, so that $\frac{\mathbf{I}}{\mathbf{V}} = Y_P = \frac{1}{Z_S} = \frac{1}{R_S + j\omega L_S}$. Rationalizing this fraction by multiplying the numerator and denominator by the complex conjugate of the denominator, so as to make the denominator real,

$$Y_P = \frac{1}{R_S + j\omega L_S} \times \frac{R_S - j\omega L_S}{R_S - j\omega L_S} = \frac{R_S - j\omega L_S}{R_S^2 + \omega^2 L_S^2} = G_P + jB_P \quad (8.43)$$

Equating real and imaginary parts,

$$G_P = \frac{R_S}{R_S^2 + \omega^2 L_S^2} \quad \text{and} \quad B_P = -\frac{\omega L_S}{R_S^2 + \omega^2 L_S^2} \quad (8.44)$$

From Table 8.2, $G_P = 1/R_P$ and $B_P = -1/\omega L_P$. It follows that

$$R_P = \frac{R_S^2 + \omega^2 L_S^2}{R_S} = R_S + \frac{\omega^2 L_S^2}{R_S} \quad (8.45)$$

and

$$L_P = \frac{R_S^2 + \omega^2 L_S^2}{\omega^2 L_S} = L_S + \frac{R_S^2}{\omega^2 L_S} \quad (8.46)$$

It should be noted that if R_S and L_S are constants, such as 50 Ω and 10 mH, R_P and L_P are frequency dependent. They can only be considered constant at a specified frequency. Hence, the paralleled resistor and inductor of Figure 8.18b cannot be considered ideal circuit elements in this case.

It is instructive to verify some limiting cases:

1. $R_S = 0$, as for an ideal inductor. This makes $R_P \to \infty$, that is, an open circuit, and $L_P = L_S$. The same pure inductance appears between terminals 'ab' in both cases.
2. $L_S = 0$, as for an ideal resistor. This makes $L_P \to \infty$, that is, an open circuit, and $R_P = R_S$. The same pure resistance appears between terminals 'ab' in both cases.
3. $\omega = 0$, that is, dc conditions. Since L_S is assumed constant, $\omega L_S = 0$, which makes $R_S = R_P$ in Equation 8.45 and $L_P \to \infty$ in Equation 8.46. The resistance between terminals 'ab' is the same.
4. $\omega \to \infty$, which makes $\omega L_S \to \infty$ and $Z_S \to \infty$. Under these conditions, $R_P \to \infty$, and $L_P = L_S$. However, although the inductance is finite, the reactance $\omega L_P = \omega L_S$ tends to infinity. This makes $Z_P \to \infty$, so that terminals 'ab' are open-circuited in both cases.

Problem-Solving Tip

- Conversion of reactance, or reactance combined with resistance, to susceptance, or susceptance combined with conductance, must proceed through the intermediate, defining step of impedance as the reciprocal of admittance.

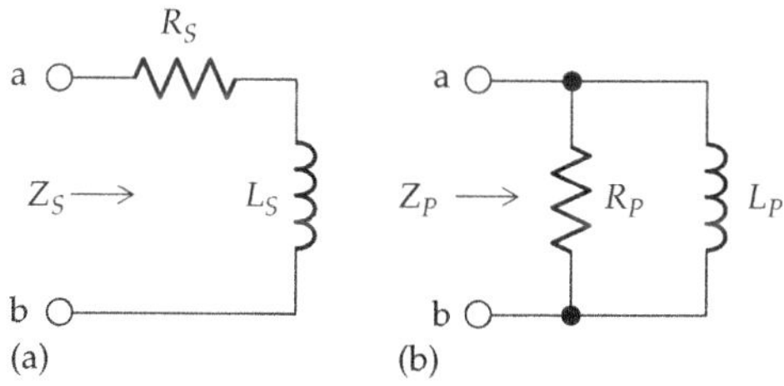

FIGURE 8.18
Figure for Example 8.2.

Primal Exercise 8.13

Given an inductor of 50 mH inductance in series with a 10 Ω resistance, determine (a) the impedance of the inductor and (b) the equivalent parallel resistance and inductance (Figure 8.18b), assuming a frequency of 1 krad/s.

Ans. (a) $10 + j0.05\omega$ Ω; (b) $R_p = 260$ Ω, $L_p = 52$ mH.

Primal Exercise 8.14

A 50 Ω resistor is connected in parallel with a 20 μF capacitor. Determine the resistance and reactance of the parallel combination at a frequency of 1 krad/s.

Ans. 25 Ω, $-j25$ Ω.

8.6 Governing Equations

The stated objective of sinusoidal steady-state analysis is to emulate dc analysis. In preceding chapters, circuit analysis under dc conditions was shown to be based on KCL, KVL, and Ohm's law. In phasor notation for voltages and currents in the sinusoidal steady state, time does not explicitly appear in the expressions for voltages and currents, just as in the dc case. This is because a steady state is assumed, so that only relative amplitudes and phases are relevant, and these are independent of time. Voltages and currents of circuit elements are related to impedance and its reciprocal, admittance, in the same way as Ohm's law relates dc voltages and currents of resistors to resistance and conductance. It remains to show, formally, that KCL and KVL apply in phasor notation.

If a current i_1 enters a node and currents i_2 and i_3 leave this node, KCL gives at any instant of time:

$$i_1 = i_2 + i_3 \tag{8.47}$$

where i_1, i_2, and i_3 are instantaneous values. If these currents vary sinusoidally with time, then at any time t, Equation 8.47 can be expressed as.

$$I_{m1}\cos(\omega t+\theta_1) = I_{m2}\cos(\omega t+\theta_2) + I_{m3}\cos(\omega t+\theta_3) \tag{8.48}$$

where the amplitudes and phases of the currents are such that Equation 8.48 is satisfied. Since Equation 8.48 must be satisfied at any time t, it is satisfied a quarter of a period later, that is, when t is replaced by $(t + \pi/2\omega)$:

$$I_{m1}\cos(\omega t+\theta_1+\pi/2) = I_{m2}\cos(\omega t+\theta_2+\pi/2) + I_{m3}\cos(\omega t+\theta_3+\pi/2) \tag{8.49}$$

or

$$I_{m1}\sin(\omega t+\theta_1) = I_{m2}\sin(\omega t+\theta_2) + I_{m3}\sin(\omega t+\theta_3) \tag{8.50}$$

Adding Equations 8.48 through 8.50 multiplied by j expresses the currents as complex sinusoidal functions:

$$I_{m1}e^{j(\omega t+\theta_1)} = I_{m2}e^{j(\omega t+\theta_2)} + I_{m3}e^{j(\omega t+\theta_3)} \tag{8.51}$$

Dropping the time variation leads to phasor notation:

$$\mathbf{I}_1 = \mathbf{I}_2 + \mathbf{I}_3 \tag{8.52}$$

It is seen that KCL applies to phasor currents. An exactly analogous argument shows that KVL too applies to phasor voltages.

Since all the concepts, theorems, and procedures discussed for the dc case are based on KCL, KVL, and Ohm's law, and since the counterparts of these relations are satisfied in the sinusoidal steady state using phasors and impedances, the inevitable conclusion is

Concept: *All circuit relations, concepts, theorems, and procedures that apply to resistive circuits under dc conditions apply to the sinusoidal steady state, with voltages and currents represented as phasors and impedances of circuit elements replacing resistance.*

Specifically, this applies to all the circuit equivalence relations of Chapter 3, the circuit theorems of Chapter 4, the circuit simplification procedures of Chapter 5, and the circuit equations of Chapter 6. In fact, the dc state can be considered as a special case of the sinusoidal steady state with the frequency set to zero, so that inductors are replaced by short circuits and capacitors by open circuits.

Example 8.3 illustrates voltage and current division using impedances.

Example 8.3: Voltage Divider in Sinusoidal Steady State

Given the circuit of Figure 8.19, it is required to determine $\mathbf{V}_L$, $\mathbf{I}_L$, $\mathbf{I}_1$, and $\mathbf{I}_2$, assuming $\omega = 100$ rad/s.

Solution:

The reactance of the capacitor is $-\dfrac{1}{\omega C} = -\dfrac{1}{100\times 10^{-3}} = -10$ Ω; $Z_2 = 10 - j10$ Ω. The reactance of the inductor is $\omega L = 100\times 0.2 = 20$ Ω; $Z_L = 20 + j20$ Ω. Z_2 in parallel with Z_L is $\dfrac{(20+j20)(10-j10)}{(20+j20)+(10-j10)} = \dfrac{20(1+j)(1-j)}{2(1+j)+(1-j)} = \dfrac{40}{3+j} = \dfrac{40(3-j)}{(3+j)(3-j)} = 4(3-j)$Ω.

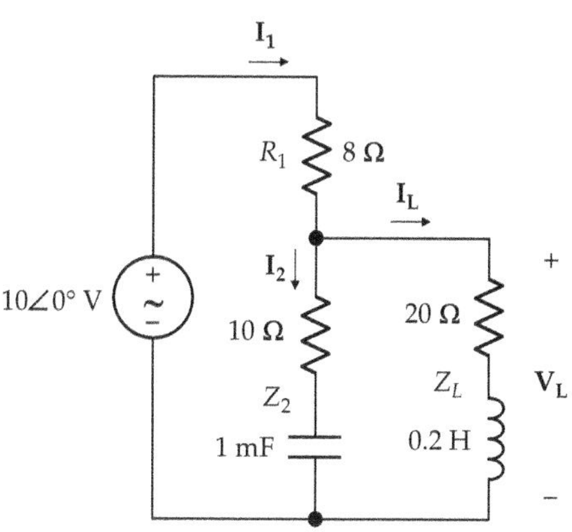

FIGURE 8.19
Figure for Example 8.3.

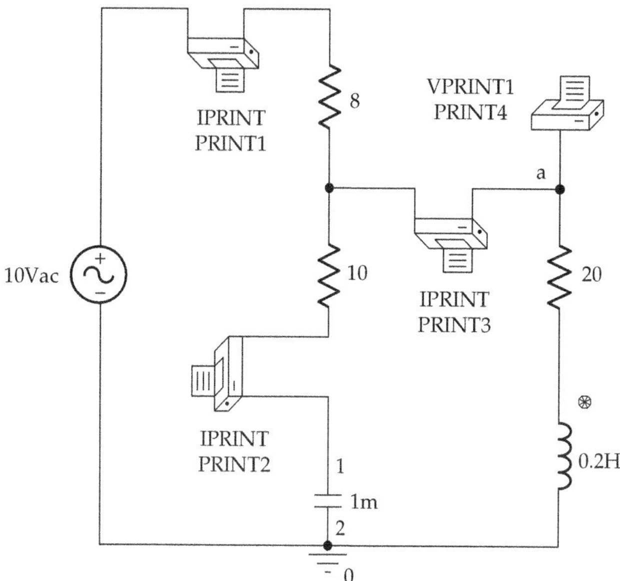

FIGURE 8.20
Figure for Example 8.3.

Hence,

$$\mathbf{I}_1 = \frac{10\angle 0^\circ}{8+4(3-j)} = \frac{10\angle 0^\circ}{20-j4} = \frac{10\angle 0^\circ}{20.4\angle -11.3^\circ}$$

$$= \frac{10\angle 0^\circ}{20.4\angle -11.3^\circ} = 0.49\angle 11.3^\circ \text{ A.}$$

From voltage division,

$$V_L = \frac{Z_L||Z_2}{R_1+Z_L||Z_2} \times \mathbf{V}_{\mathbf{SRC}} = \frac{4(3-j)}{8+4(3-j)} 10\angle 0^\circ$$

$$= \frac{3-j}{5-j} 10\angle 0^\circ = \frac{\sqrt{10}\angle -18.4^\circ}{\sqrt{26}\angle -11.3^\circ} 10\angle 0^\circ = 6.2\angle -7.1^\circ \text{ V.}$$

From current division,

$$\mathbf{I}_\mathrm{L} = \frac{10-10j}{(20+j20)+(10-10j)} \mathbf{I}_1 = \frac{1-j}{3+j} \times \frac{5}{10-j2} = \frac{5(1-j)}{4(8+j)}$$

$$= \frac{5(1-j)(8-j)}{4(8+j)(8-j)} = \frac{7-j9}{52} = 0.22\angle -52.1^\circ \text{ A.}$$

From KCL, $\mathbf{I}_2 = \mathbf{I}_1 - \mathbf{I}_\mathrm{L} = \dfrac{5}{2(5-j)} - \dfrac{5(1-j)}{4(8+j)} = \dfrac{5(3+j2)}{41-j3}$

$$= \frac{117+j91}{338} = 0.44\angle 37.9^\circ \text{ A.}$$

Simulation: The circuit is entered as in Figure 8.20. For the sinusoidal steady state, source VAC is used at a single frequency. A magnitude is entered, and a phase, if required, can be set in the Property Editor spreadsheet. Note that the value entered for VAC is considered by PSpice as an rms value for determining power. But for determining voltages or currents, the value entered could be interpreted either as a peak value or an rms value.

Printers from the SPECIAL library can be used for measuring voltages and currents. A current printer is inserted in series with one or more circuit elements through which the current in question passes. The positive direction of current is that entering the unmarked terminal of the current printer and leaving the terminal marked with a minus sign. A one-terminal voltage printer is used for measuring voltages with respect to ground, and a two-terminal voltage printer is used for measuring voltages between two nodes, neither of which is grounded. For AC measurements, a Y must be entered under AC in the Property Editor spreadsheet of each printer. The printer reading is in rectangular coordinates if a Y is entered under REAL and IMAG, and the reading is in polar coordinates if a Y is entered under MAG and PHASE.

In the Simulation Settings, 'AC Sweep/Noise' is chosen under 'Analysis type'. For sinusoidal steady-state analysis, a single frequency is used. Under AC Sweep Type, either 'Linear' or 'Logarithmic' may be selected. The frequency is entered in Hz and is $100/2\pi = 15.915$ Hz in this case. The same frequency is entered for 'Start Frequency' and 'End Frequency', and 1 is entered for 'Total Points'. After the simulation is run, the printer readings are available at the end of 'Simulation Output File'. This file can be viewed by selecting View/Output File at the top of the SCHEMATIC1 page, or by double clicking on the third icon from the top in the left margin of the SCHEMATIC1 page, labeled 'View Simulation Output File'. The printer readings are as follows:

FREQ	IM(V_PRINT1)	IP(V_PRINT1)
1.592E+01	4.903E−01	1.131E + 01
FREQ	IM(V_PRINT2)	IP(V_PRINT2)
1.592E+01	4.385E−01	3.788E + 01
FREQ	IM(V_PRINT3)	IP(V_PRINT3)
1.592E+01	2.193E−01	−5.213E+01
FREQ	VM(a)	VP(a)
1.592E+01	6.202E+00	−7.125E+00

The current printers are identified by their number, and the voltage printer by the label of the node to which the printer is connected. If the node is not labeled, the reference is to the pin number of the node. This number can be read by pointing the cursor at the node in the circuit entered. IM and IP refer to current magnitude and current phase; similarly for VM and VP. A value such as 4.903E−01 denotes 4.903×10^{-1}. It is seen that printer values agree with those calculated.

An alternative to using printers for reading values of currents and voltages is to use the 'Evaluate Measurement' feature of PSpice. In the SCHEMATIC1 page, select 'Trace/Evaluate Measurement'. Two windows are displayed. Under 'Functions or Macros' in the right-hand window, choose 'Analog Operators and Functions'. To read magnitudes, select the voltage or current required from the left-hand window labeled 'Simulation Output Variables'. To read phase angles, select P(), then select the variable required. For example, to read the voltage of the node labeled 'a', V(a) and P(V(a)) are selected. The values 6.20174 and −7.12502 are displayed in a window in the SCHEMATIC1 page. To read the values of $\mathbf{I}_1$, $\mathbf{I}_2$, and $\mathbf{I}_L$, the variables are −I(V1), I(C1), and I(L1), respectively.

Primal Exercise 8.15

Determine Z_{eq} in Figure 8.21 as the sum of a resistance and a reactance.

Ans. 29.2 − *j*14.4 Ω.

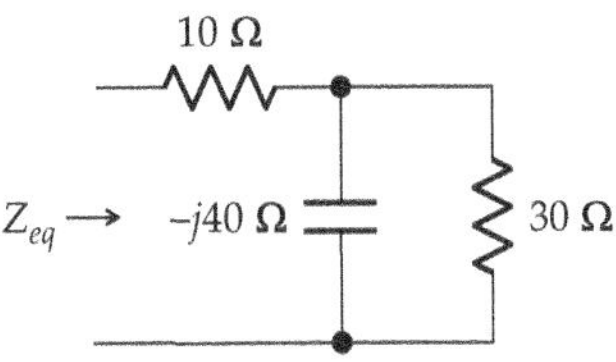

FIGURE 8.21
Figure for Primal Exercise 8.15.

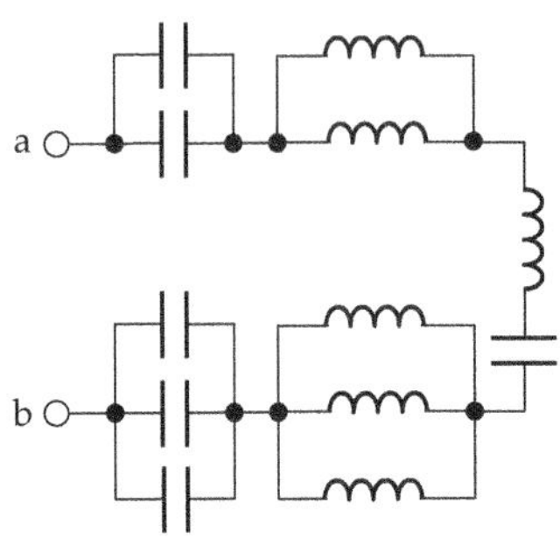

FIGURE 8.22
Figure for Primal Exercise 8.16.

Primal Exercise 8.16

Determine the impedance between terminals 'ab' in Figure 8.22, assuming all capacitances are 1 F, all inductances are 1 H, and $\omega = 1$ rad/s.

Ans. 0.

8.7 Representation in the Frequency Domain

Let us return to Figure 8.3 that we started with in order to illustrate the general procedure for analyzing the sinusoidal steady state in the same manner as a dc state. The source $I_m \cos\omega t$ is converted to a phasor $\mathbf{I}_{SRC} = I_m \angle 0°$, and R and L are expressed as impedances, R and $j\omega L$, respectively, in series with the source. The circuit in terms of phasors and impedances or admittances is now said to be in the *frequency domain*, as illustrated in Figure 8.23.

Direct application of KVL gives

$$\mathbf{V} = R\mathbf{I}_{SRC} + j\omega L\mathbf{I}_{SRC} = (R + j\omega L)\mathbf{I}_{SRC} = Z\mathbf{I}_{SRC} \tag{8.53}$$

This is an algebraic equation involving the phasors and the complex impedance $Z = R + j\omega L$. The relation $\mathbf{V} = Z\mathbf{I}_{SRC}$ is exactly analogous to Ohm's law in the dc case, bearing in mind that whereas a dc voltage or a dc current has only a magnitude and sign, an ac voltage or current has an amplitude and phase.

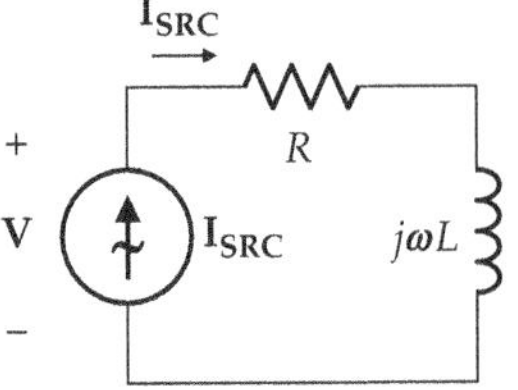

FIGURE 8.23
Series *RL* circuit in the frequency domain.

Expressing Z and $\mathbf{I_{SRC}}$ in polar coordinates, Equation 8.53 becomes

$$\mathbf{V} = \left(\sqrt{R^2+\omega^2L^2}\angle\theta\right)I_m\angle 0^\circ = I_m\sqrt{R^2+\omega^2L^2}\angle\theta \quad (8.54)$$

$\mathbf{V}$ can now be converted back to the time domain as a cosine function, since i_{SRC} is a cosine function. This gives

$$v = I_m\sqrt{R^2+\omega^2L^2}\left(\cos(\omega t+\theta)\right) \quad (8.55)$$

which is identical to Equation 8.7.

The procedure for deriving the sinusoidal steady-state response can be generalized and summarized as follows:

1. *The sinusoidal excitations are expressed as phasors. A single excitation $Y_m\cos(\omega t+\theta)$, or $Y_m\sin(\omega t+\theta)$, can be expressed as $Y_m\angle\theta$. In the case of several excitations, some of which are cosine functions and some are sine functions, they should all be converted to either cosine or sine functions before expressing them as phasors, taking into account the relative phase angles of cosine and sine functions.*
2. *Inductance and capacitance are expressed as impedance or admittance, as appropriate.*
3. *The circuit is now in the frequency domain and can be analyzed by any of the methods discussed in previous chapters for the dc case.*
4. *After obtaining the desired response, this response can be converted back to the time domain as a cosine function, if the time functions that were originally expressed as phasors were cosine functions, or converted as a sine function if the time functions that were originally expressed as phasors were sine functions.*

Primal Exercise 8.17

Given $v_{SRC}(t) = 10\cos t$ V in Figure 8.24, determine both as phasors and as functions of time: (a) $i(t)$; (b) $v_R(t)$; (c) $v_C(t)$; and (d) $v_L(t)$.

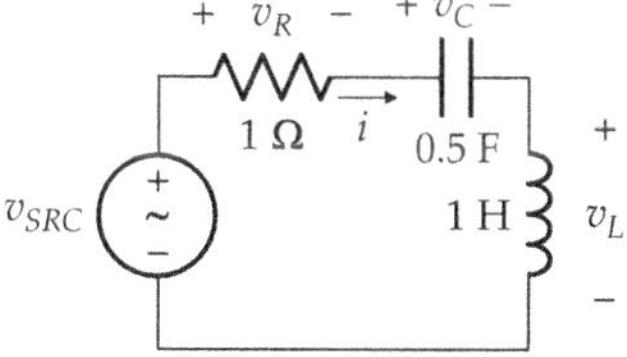

FIGURE 8.24
Figure for Primal Exercise 8.17.

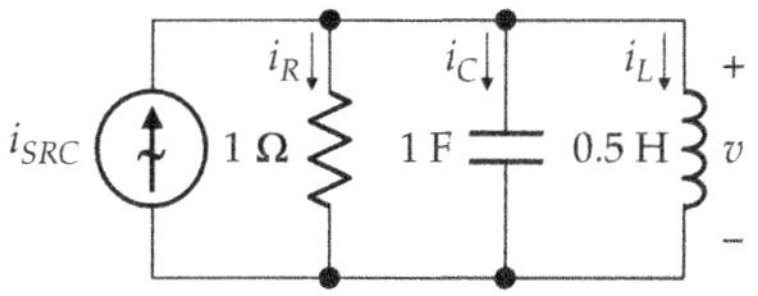

FIGURE 8.25
Figure for Primal Exercise 8.18.

Ans. (a) $\mathbf{I} = 5(1+j) = 5\sqrt{2}\angle 45^\circ$A, $i(t) = 5\sqrt{2}\cos(t+45^\circ) = 5(\cos t - \sin t)$A; (b) $\mathbf{V_R} = 5(1+j) = 5\sqrt{2}\angle 45^\circ$ V, $v_R(t) = 5\sqrt{2}\cos(t+45^\circ) = 5(\cos t - \sin t)$ V; (c) $\mathbf{V_C} = 10(1-j) = 10\sqrt{2}\angle -45^\circ$ V, $v_C(t) = 10\sqrt{2}\cos(t-45^\circ) = 10(\cos t + \sin t)$V, (d) $\mathbf{V_L} = 5(-1+j) = 5\sqrt{2}\angle 135^\circ$ V, $v_L(t) = 5\sqrt{2}\cos(t+135^\circ) = -5(\sin t + \cos t)$ V.

Primal Exercise 8.18

Given $i_{SRC}(t) = 10\sin t$ A in Figure 8.25, determine both as phasors and as functions of time: (a) $v(t)$; (b) $i_R(t)$; (c) $i_C(t)$; and (d) $i_L(t)$.

Ans. (a) $\mathbf{V} = 5(1+j) = 5\sqrt{2}\angle 45^\circ$ V, $v(t) = 5\sqrt{2}\sin(t+45^\circ) = 5(\sin t + \cos t)$ V; (b) $\mathbf{I_R} = 5(1+j) = 5\sqrt{2}\angle 45^\circ$ A, $i_R(t) = 5\sqrt{2}\sin(t+45^\circ) = 5(\sin t + \cos t)$ A; (c) $\mathbf{I_C} = 5(-1+j) = 5\sqrt{2}\angle 135^\circ$A, $i_C(t) = 5\sqrt{2}\sin(t+135^\circ) = 5(-\sin t + \cos t)$ A; (d) $\mathbf{I_L} = 10(1-j) = 10\sqrt{2}\angle -45^\circ$A, $i_L(t) = 10\sqrt{2}\sin(t-45^\circ) = 10(\sin t - \cos t)$ A.

Example 8.4: Norton's Equivalent Circuit in Sinusoidal Steady State

It is required to determine Norton's equivalent circuit (NEC) seen by the 20 Ω resistor between terminals 'ab' in Figure 8.26, assuming $v_{SRC1} = 200\sin(5\times10^4 t)$ V and $v_{SRC2} = 100\cos(5\times10^4 t)$ V.

Solution:

The circuit is first represented in the frequency domain. Since one source voltage is a cosine function and the other source voltage is a sine function, they should both be expressed as sine functions or cosine functions. As a cosine function, $v_{SRC1} = 200\cos(5\times10^4 t - 90^\circ)$ V. As phasors, $\mathbf{V_{SRC1}} = 200\angle -90^\circ = -j200$ V, and $\mathbf{V_{SRC2}} = 100\angle 0^\circ = 100$ V; $\omega = 5\times10^4$ rad/s, $j\omega L = j5\times10^4\times0.4\times10^{-3} = j20$ Ω, $\frac{-j}{\omega C} = \frac{-j}{5\times10^4\times10^{-6}} = -j20$ Ω. The circuit in the frequency domain is shown in Figure 8.27 with terminals 'ab' short circuited.

$\mathbf{I_{SC}}$ can be determined by superposition. With $\mathbf{V_{SRC1}}$ applied alone, and $\mathbf{V_{SRC2}}$ replaced by a short circuit

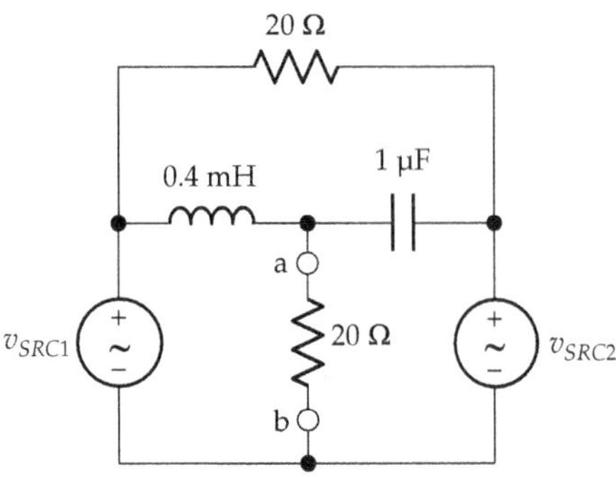

FIGURE 8.26
Figure for Example 8.4.

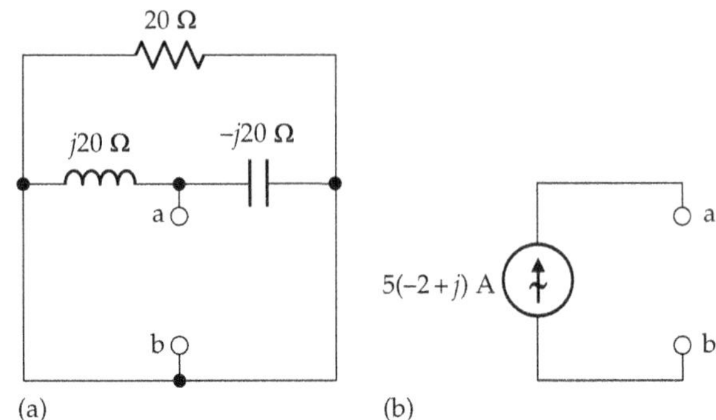

FIGURE 8.29
Figure for Example 8.4.

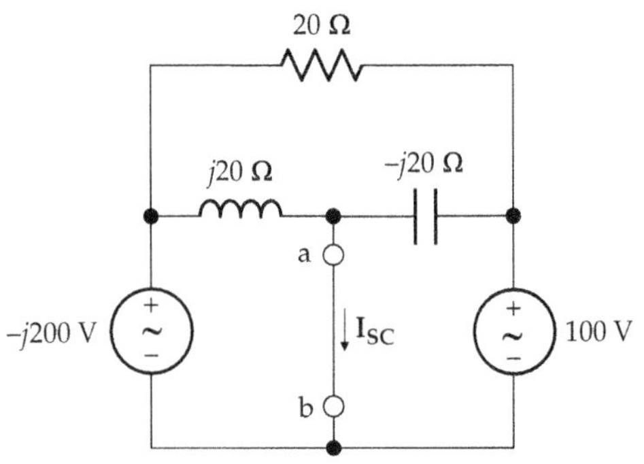

FIGURE 8.27
Figure for Example 8.4.

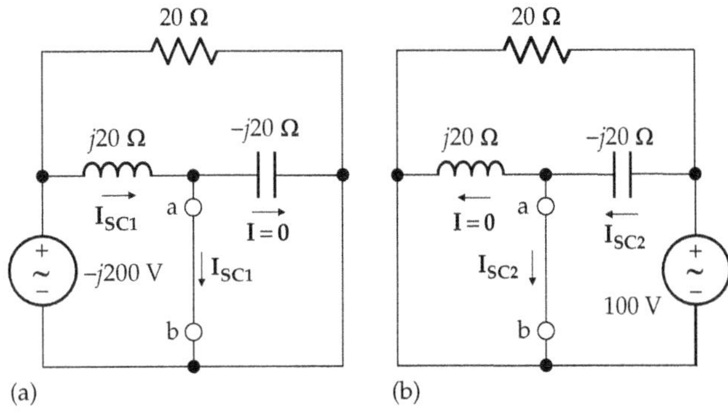

FIGURE 8.28
Figure for Example 8.4.

(Figure 8.28a), the capacitive reactance is short-circuited, and the current through it is zero. Hence, $\mathbf{I}_{SC1} = -j200/j20 = -10$ A. With $\mathbf{V}_{SRC2}$ applied alone and $\mathbf{V}_{SRC1}$ replaced by a short circuit (Figure 8.28b), the inductive reactance is short-circuited, and the current through it is zero. Hence, $\mathbf{I}_{SC2} = 100/(-j20) = j5$ A. It follows that $\mathbf{I}_{SC} = \mathbf{I}_N = -10 + j5 = 5(-2 + j)$ A.

If both sources are replaced by short circuits, the $j20\ \Omega$ and the $-j20\ \Omega$ appear in parallel across terminals 'ab' (Figure 8.29a). The 20 Ω resistor is short-circuited, does not carry any current, and can be removed. The impedance between terminals 'ab' is then $Z_{Th} = (j20)(-j20)/(j20 - j20) \to \infty$. Alternatively, the admittance of the inductor is $-j/20$ S, and the admittance of the capacitor is $j/20$ S. The admittance Y_{Th} between terminals 'ab' is zero. NEC is an ideal current source $5(-2 + j)$ A (Figure 8.29b). Since $Z_{Th} \to \infty$, TEC does not exist.

Simulation: Unlike the dc case, TEC or NEC cannot be obtained in a single simulation because ac quantities have both amplitude and phase. The circuit for determining Norton's current is entered as in Figure 8.30a. Because the current printer has zero resistance, a voltage-source-inductor loop is formed that is not allowed in PSpice, because the total resistance in the loop under dc conditions is zero, and PSpice may be required, in general, to derive currents under both dc and ac conditions. This loop can be broken by inserting a 1u resistance as shown, which is too small to affect the result. In the Simulation Settings, $f = 5\times10^4/2\pi = 7957.75$ Hz is entered for 'Start Frequency' and 'End Frequency', and a 1 for 'Total Points'. After the simulation is run, the printer readings are as follows:

FREQ	IR(V_PRINT1)	II(V_PRINT1)
7.958E+03	−1.000E+01	5.000E+00
7.958E+03	2.500E−09	3.576E−08

where the first reading is that of $\mathbf{I}_N$ in Figure 8.28a. The second reading is that of the current of the test voltage source in Figure 8.30b that is applied for determining Y_N. IR is the real part of the current and II is the imaginary part. The first reading agrees with the calculated value, and the second reading is insignificantly small, which is interpreted as zero.

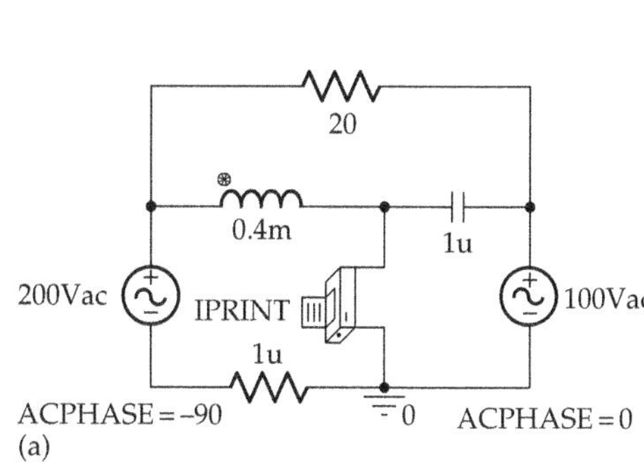

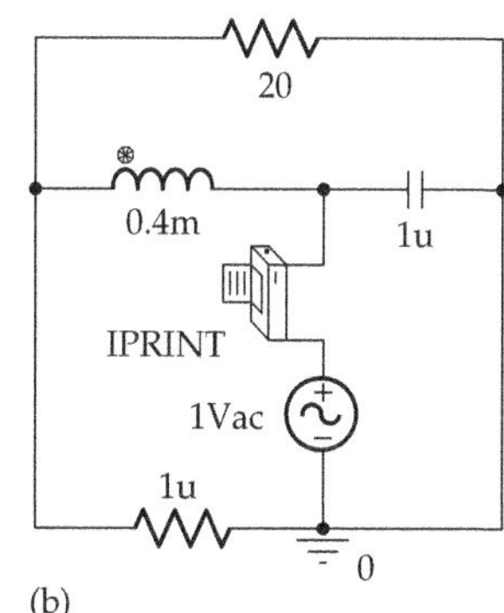

FIGURE 8.30
Figure for Example 8.4.

Example 8.5: Node-Voltage Analysis in Sinusoidal Steady State

It is required to determine v_O in Figure 8.31 assuming $v_{SRC} = 100\cos(10^3 - 30°)$ V.

Solution: $\mathbf{V}_{SRC} = 100\angle{-30°}$ V; $j\omega L = j10^3 \times 20 \times 10^{-3} = j20\ \Omega$, $1/j\omega C = 1/(10^3 \times 40 \times 10^{-6}) = -j25\ \Omega$.

The circuit will be analyzed by the node-voltage method. The voltage source in series with the 40 Ω resistance is transformed to a current source $\mathbf{I}_{SRC} = 2.5\angle{-30°}$ A in parallel with 40 Ω. In rectangular coordinates, $\mathbf{I}_{SRC} = 2.5(\cos 30° - j\sin 30°) = 1.25(\sqrt{3} - j)$. The circuit in the frequency domain becomes as shown in Figure 8.32. $\mathbf{I}_O = \mathbf{V}_2/20$. Bearing in mind that the coefficients of the node voltages are admittances when using phasor analysis, the node-voltage equation for the essential node on the left is

$$\left(\frac{1}{40} + \frac{1}{-j25}\right)\mathbf{V}_1 - \frac{1}{-j25}\mathbf{V}_2 = \frac{2}{20}\mathbf{V}_2 + 1.25\sqrt{3} - j1.25.$$

or

$$\left(\frac{1}{40} + \frac{j}{25}\right)\mathbf{V}_1 - \left(\frac{1}{10} + \frac{j}{25}\right)\mathbf{V}_2 = 1.25\sqrt{3} - j1.25.$$

The node-voltage equation for the essential node on the right is

$$-\frac{1}{-j25}\mathbf{V}_1 + \left(\frac{1}{-j25} + \frac{1}{20} + \frac{1}{10 + j20}\right)\mathbf{V}_2 = 0.$$

Solving for $\mathbf{V}_2$ gives $\mathbf{V}_2 = 5.021 + j27.65$ V.

Hence,

$$\mathbf{V}_O = \frac{10}{10 + j20}\mathbf{V}_2 = 12.1 + j3.52 \text{ V}.$$

Simulation: The circuit is entered as in Figure 8.33, where the phase angle is entered in the spreadsheet for the source. Note that as an alternative to reversing the

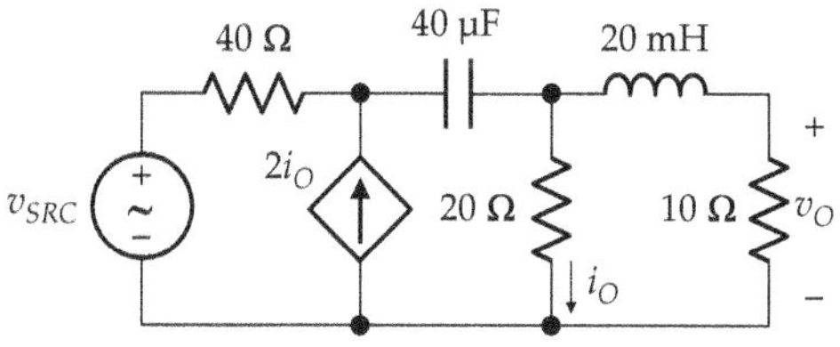

FIGURE 8.31
Figure for Example 8.5.

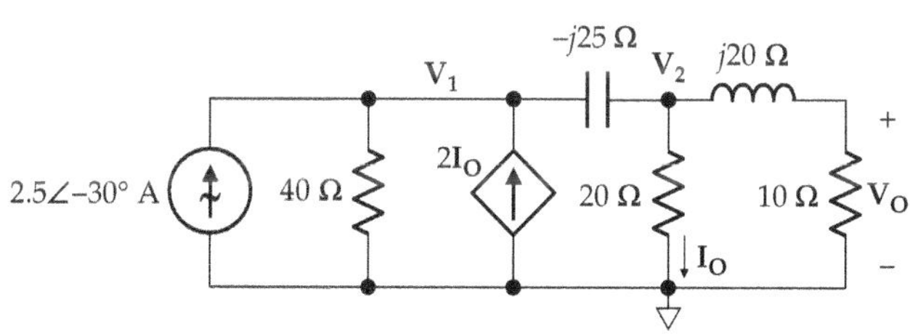

FIGURE 8.32
Figure for Example 8.5.

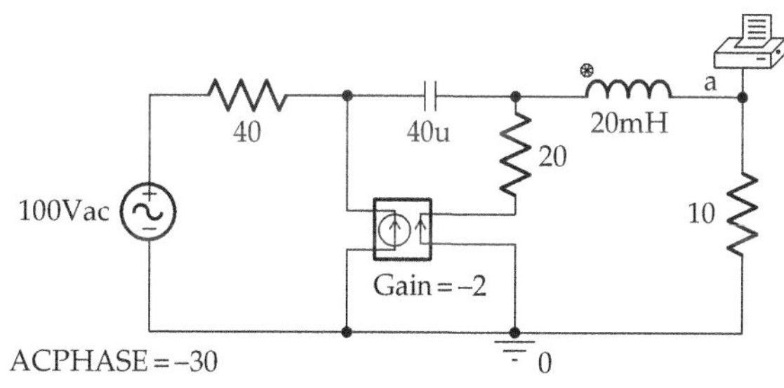

FIGURE 8.33
Figure for Example 8.5.

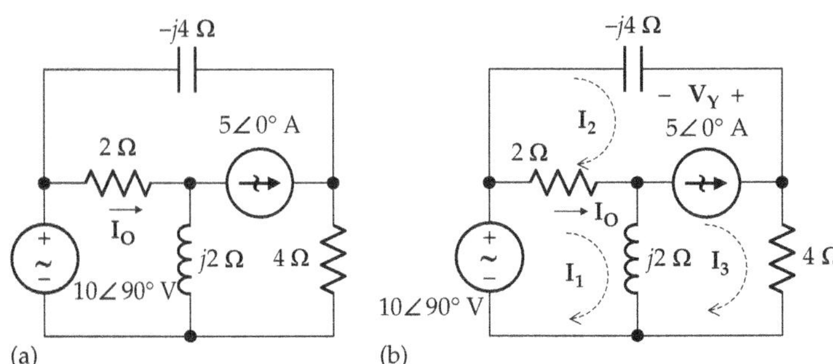

FIGURE 8.34
Figure for Example 8.6.

direction of $\mathbf{I}_O$ through the dependent source, the sign of the gain is made negative. In the Simulation Settings, $f = 10^3/2\pi = 1591.55$ Hz is entered for 'Start Frequency' and 'End Frequency', and 1 for 'Total Points'. The output node is labeled 'a'. After the simulation is run, the printer readings are as follows:

FREQ	VR(A)	VI(A)
1.592E+03	12.06E+00	3.522E+00

Example 8.6: Mesh-Current Analysis in Sinusoidal Steady State

It is required to determine $\mathbf{I}_O$ in Figure 8.34a.

Solution:

The circuit is given in the frequency domain. $\mathbf{I}_O$ will be determined by the mesh-current method, the mesh-current assignments being as in Figure 8.34b. Because of the nontransformable current source in meshes 2 and 3, a voltage $\mathbf{V}_Y$ is arbitrarily assigned across this source to allow writing the mesh-current equations for these meshes. The mesh-current equations are $(2 + j2)\mathbf{I}_1 - 2\mathbf{I}_2 - j2\mathbf{I}_3 = j10$; $-2\mathbf{I}_1 + (2 - j4)\mathbf{I}_2 = -\mathbf{V}_y$; and $-j2\mathbf{I}_1 + (4 + j2)\mathbf{I}_3 = \mathbf{V}_y$. Adding the last two equations to eliminate $\mathbf{V}_Y$, the resulting equation is $-(2 + j2)\mathbf{I}_1 + (2 - j4)\mathbf{I}_2 + (4 + j2)\mathbf{I}_3 = 0$. The current source gives $\mathbf{I}_3 - \mathbf{I}_2 = 5$. From the solution of the equations, $\mathbf{I}_1 = 1.25 + j3.75$ and $\mathbf{I}_2 = -3.75 - j1.25$, so that $\mathbf{I}_O = \mathbf{I}_1 - \mathbf{I}_2 = 5 + j5$ A.

Simulation: The circuit is entered as in Figure 8.35. Since the reactances are specified in the circuit, but inductance and capacitance values have to be entered in PSpice, any convenient value of ω can be assumed and L and C calculated accordingly. Thus, if $\omega = 1$ rad/s, then $L = X/\omega = 2$ H, and $C = -1/\omega X = 0.25$ F, as entered in the circuit. In the Simulation Settings, $f = 1/2\pi = 0.159155$ Hz is entered for 'Start Frequency' and 'End Frequency', and 1 for 'Total Points'. After the simulation is run, the printer readings are as follows:

FREQ	IR(V_PRINT1)	II(V_PRINT1)
1.592E−01	5.000E+00	5.000E+00

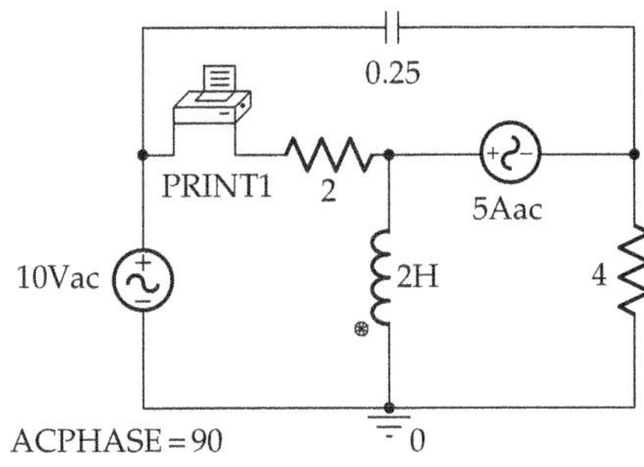

FIGURE 8.35
Figure for Example 8.6.

⋆8.8 Phasor Diagrams

Phasor diagrams showing various voltage and current phasors in a given circuit are useful for illustrating the interrelations between the various variables involved, particularly when some circuit parameter is varied, as illustrated by the following example.

Example 8.7: Phasor Diagram

Given the circuit of Figure 8.36. It is required to determine: (a) the value of R at which $\mathbf{V}_O$ is 90° out of phase with respect to v_I, and (b) how $\mathbf{V}_O$ changes as R is varied from 0 to infinity, assuming that no current is drawn at the output.

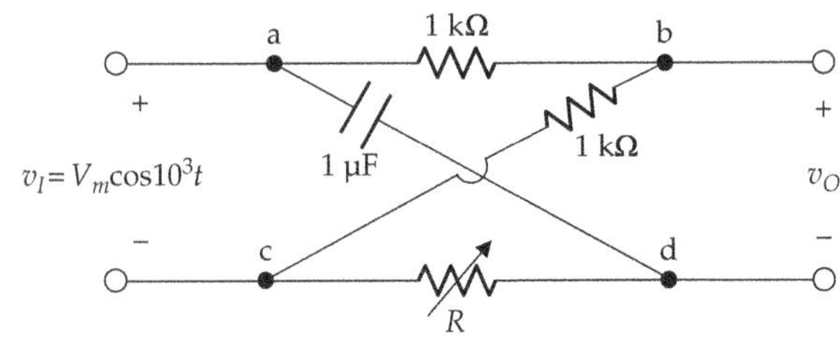

FIGURE 8.36
Figure for Example 8.7.

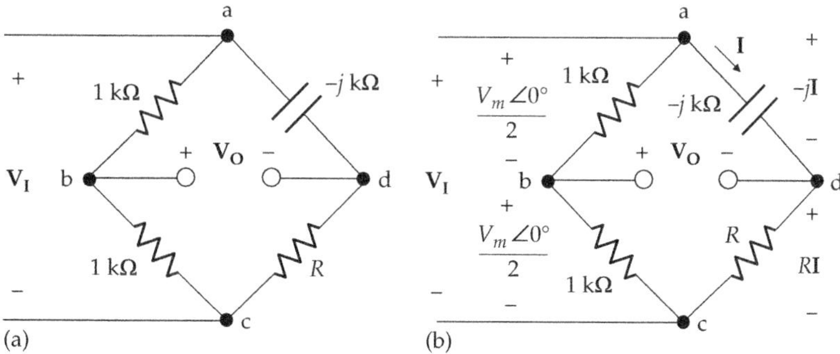

FIGURE 8.37
Figure for Example 8.7.

Solution:

(a) $-j/\omega C = -j/(10^3 \times 10^{-6}) = -j\,\text{k}\Omega$. The lattice configuration of Figure 8.36 can be redrawn as a bridge circuit for easier visualization as was done in Example 2.6. The resulting circuit is shown in Figure 8.37a in the frequency domain, with $\mathbf{V}_\text{I}$ applied between nodes 'a' and 'c', and $\mathbf{V}_\text{O}$ taken between nodes 'b' and 'd'. $\mathbf{V}_\text{I} = V_m\angle 0°$. By voltage division, $\mathbf{V}_{\text{bc}} = (V_m/2)\angle 0°$ and $\mathbf{V}_{\text{dc}} = V_m\angle 0° \times R/(R-j)$. It follows that.

$$\mathbf{V}_\text{O} = \mathbf{V}_{\text{bc}} - \mathbf{V}_{\text{dc}} = \frac{V_m\angle 0°}{2} - \frac{V_m\angle 0° R}{R-j} = \frac{V_m\angle 0°}{2}\left(1 - \frac{2R}{R-j}\right)$$
$$= \frac{V_m\angle 0°}{2}\frac{-R-j}{R-j} \tag{8.56}$$

The magnitude of $\mathbf{V}_\text{O}$ is given by

$$|\mathbf{V}_\text{O}| = \frac{V_m}{2}\frac{\sqrt{R^2+1}}{\sqrt{R^2+1}} = \frac{V_m}{2}. \tag{8.57}$$

It is seen that $|\mathbf{V}_\text{O}|$ is $V_m/2$, independently of R. To ascertain the correct value of the phase angle, the numerator and denominator in Equation 8.56 are drawn as position vectors in Figure 8.38. It is seen that $\angle V_\text{O} = \beta$, where

$$-\beta + (\pi + \theta) = 2\pi + \theta$$
$$\text{or,}\quad \beta = -\pi + 2\theta \tag{8.58}$$

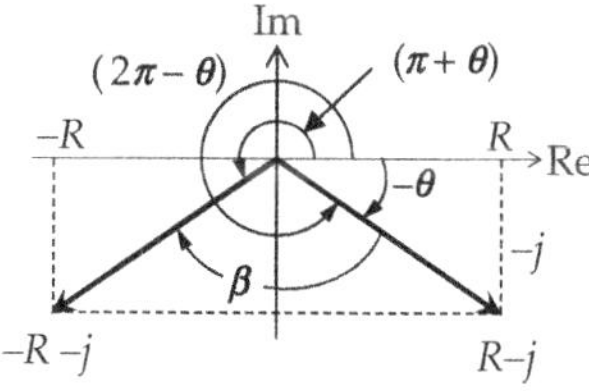

FIGURE 8.38
Figure for Example 8.7.

where $\theta = \tan^{-1}(1/R)$ and has a positive value that is less than 90°. $\angle\mathbf{V}_\text{O}$ is therefore negative, so that $\mathbf{V}_\text{O}$ lags $\mathbf{V}_\text{I}$ (Figure 8.40). To have $\mathbf{V}_\text{O}$ 90° out of phase with $\mathbf{V}_\text{I}$, $\angle\mathbf{V}_\text{O} = -90°$. Substituting in Equation 8.58, $2\theta = -90° + 180° = 90°$, so that $\theta = 45° = \tan^{-1}(1/R)$. This gives $R = 1\ \text{k}\Omega$.

(b) When $R = 0$, it follows from Equation 8.56 that

$$\mathbf{V}_\text{O} = \frac{V_m}{2}\left(\frac{-j}{-j}\right) = \frac{V_m}{2}\angle 0° \tag{8.59}$$

In Figure 8.39a, node 'd' is connected to 'c' when $R = 0$ which makes $\mathbf{V}_\text{O} = \mathbf{V}_{\text{bc}} = V_m\angle 0°/2$. To determine $\mathbf{V}_\text{O}$ as $R \to \infty$, the numerator and denominator of Equation 8.56 are first divided by R and then R made to approach infinity, so that

$$\mathbf{V}_\text{O} = \frac{V_m}{2}\left(\frac{-1-j/R}{1-j/R}\right) \xrightarrow{R\to\infty} -\frac{V_m}{2} = \frac{V_m}{2}\angle 180° \tag{8.60}$$

In Figure 8.39b, $\mathbf{I} = 0$ when $R \to \infty$, which makes nodes 'a' and 'd' at the same voltage, $\mathbf{V}_\text{O} = \mathbf{V}_{\text{ba}} = -V_m\angle 0°/2$, and $\angle\mathbf{V}_\text{O} = 180°$.

From KVL around the outer loop in Figure 8.37b, $\mathbf{V}_\text{I} = V_m\angle 0° = R\mathbf{I} - j\mathbf{I}$, where $X = -1\ \text{k}\Omega$, $\mathbf{I}$ is in mA, R is in kΩ, and voltages are in volts. The phasor $V_m\angle 0°$ is drawn in Figure 8.40 as OP. The phasor $-j\mathbf{I}$ lags the phasor $R\mathbf{I}$ by 90°, yet their phasor sum must always be $V_m\angle 0°$. It follows from the geometric properties of a circle that point Q joining these two phasors lies on the perimeter of a semicircle of diameter V_m (Figure 8.40). From KVL around the path 'cbdc' in Figure 8.37, $\mathbf{V}_\text{O} = (V_m/2) - R\mathbf{I}$. Hence, if a phasor $-R\mathbf{I}$ is drawn in Figure 8.40 from point T at the tip of the phasor $(V_m/2)\angle 0°$, $\mathbf{V}_\text{O}$ is the phasor from the origin O to S at the tip of $-R_\text{I}$.

When $R = 0$, phasors $R\mathbf{I}$ and $-R\mathbf{I}$ are zero, and $-j\mathbf{I} = V_m\angle 0°$. Q coincides with O, and S coincides with T, so that $\mathbf{V}_\text{O} = V_m/2$, as deduced from Figure 8.39a. When $R \to \infty$, $\mathbf{I} = 0$, $-j\mathbf{I} = 0$, and

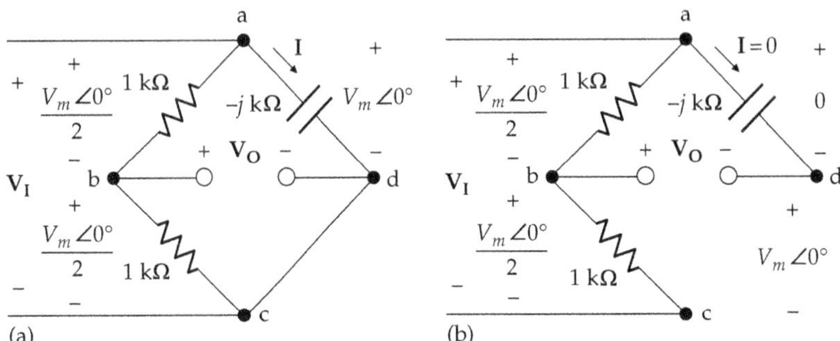

FIGURE 8.39
Figure for Example 8.7.

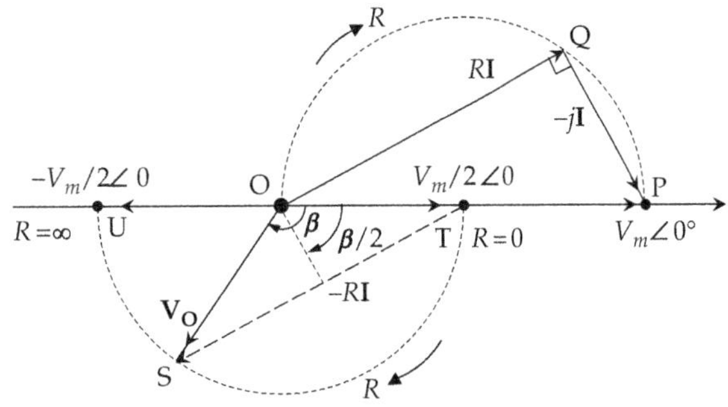

FIGURE 8.40
Figure for Example 8.7.

$R\mathbf{I} = V_m\angle 0°$ (Figure 8.40) so that Q coincides with P and S coincides with U. $\mathbf{V_O} = (-V_m/2)\angle 0°$, again as deduced from Figure 8.39b. As R increases from zero, Q moves clockwise around the perimeter of its semicircle. Likewise, S moves clockwise around the perimeter of a semicircle, and β, the phase angle of $\mathbf{V_O}$, varies between 0° and −180°.

It is seen from the geometry that $\sin(\beta/2) = -\dfrac{RI/2}{V_m/2}$, where $I = \dfrac{V}{\sqrt{R^2+X^2}} = \dfrac{V}{\sqrt{R^2+1}}$. Substituting for I, $\sin(\beta/2) = -R/\sqrt{R^2+1}$. When $R = 1$, $\sin(\beta/2) = -1/\sqrt{2}$, and $\beta/2 = -45°$. It follows that $\mathbf{V_O}$ lags $\mathbf{V_I}$ by 90°, as argued previously.

The circuit may be used to shift the phase of the output with respect to the input, over the range 0° to −180° without altering the magnitude.

Learning Checklist: What Should Be Learned from This Chapter

- The sinusoidal function is the only periodic function that is composed of a single frequency.
- When a sinusoidal excitation is applied to an LTI circuit, all the steady-state currents and voltages in the circuit are sinusoidal functions of the same frequency as the excitation, but which generally differ in amplitude and phase angle.
- According to Euler's formula, $Y_m e^{j\omega t} = Y_m \cos\omega t + Y_m \sin\omega t$, where $Y_m e^{j\omega t}$ is a position vector in the complex plane rotating CCW at an angular frequency ω rad/s, and whose projection along the real axis is $Y_m\cos\omega t$ and along the imaginary axis is $Y_m\cos\omega t$.
- When a complex sinusoidal excitation $Y_m e^{j\omega t}$ is applied to an LTI circuit, the response is a complex sinusoidal function whose real part is the response to the real part of the excitation, $Y_m\cos\omega t$, applied alone, and whose imaginary part is the response to the imaginary part of the excitation, $Y_m\sin\omega t$, applied alone.
- A phasor is a quantity such as $V_m e^{j\theta}$ representing a complex sinusoidal function of time, but with the time variation suppressed. Phasors have the same geometric properties as vectors but have real and imaginary components. They can be added and subtracted like vectors.
- The product of a phasor $\mathbf{Y} = Y\angle\theta$ and a complex quantity $A\angle\alpha$ is $AYe^{j(\theta+\alpha)}$. The quotient of $\mathbf{Y}$ and $\mathbf{A}$ is $(Y/A)e^{j(\theta-\alpha)}$.
- j in the complex plane has a magnitude of unity and a phase angle of $\pi/2$. Multiplying a phasor by j rotates the phasor through an angle $\pi/2$ counterclockwise without changing its magnitude; dividing a phasor by j, or multiplying it by $-j$, rotates the phasor through an angle $\pi/2$ clockwise without changing its magnitude.
- In phasor notation, $\mathbf{V} = R\mathbf{I}$, $\mathbf{V} = \mathbf{I}/j\omega C$, $\mathbf{V} = j\omega L\mathbf{I}$. The sinusoidal voltage and current for an ideal resistor are in phase because such a resistor is purely dissipative. They are in phase quadrature for ideal energy storage elements because

these elements are nondissipative. **I** lags **V** in an inductor and leads **V** in a capacitor.

- Differentiation in time is replaced by multiplication by $j\omega$ and integration in time is replaced by division by $j\omega$. Thus, differential and integral relations are transformed to algebraic relations in $j\omega$ for steady-state sinusoidal analysis *only*.
- The instantaneous power p in resistors, inductors, and capacitors under sinusoidal steady state conditions varies at twice the frequency of the voltage or current, because $p = vi$ is positive when v and i are both positive and both negative. p is negative when either v or i is negative.
- The instantaneous power p is never negative for an ideal resistor because such a resistor is purely dissipative. In ideal energy storage elements, p is of zero average over a period because these elements are nondissipative. They absorb power from the rest of the circuit during the positive half-cycle of p and return the same power to the rest of the circuit during the negative half-cycle of p.
- A current of rms value I_{rms}, or a voltage of rms value V_{rms}, dissipates the same power in a given resistor as a dc current, or a dc voltage, of the same value.
- The average power dissipated in the sinusoidal steady state is, like dc power, a real quantity that is independent of time.
- Impedance Z is the ratio of the voltage phasor $V_m\angle\theta_v$ at the terminals of a subcircuit, to the current phasor $I_m\angle\theta_i$ through the subcircuit in the direction of the voltage drop $V_m\angle\theta_v$. The real part of impedance is resistance and its imaginary part is reactance.
- Reactance is due to energy storage elements. Whereas sinusoidal voltages and currents are all in phase in a purely resistive circuit, energy storage elements introduce phase differences between sinusoidal voltages and currents.
- All circuit relations, concepts, theorems, and procedures that apply to resistive circuits under dc conditions apply to the sinusoidal steady state, with voltages and currents represented as phasors and impedances of circuit elements replacing resistance.
- The procedure for deriving the sinusoidal steady-state response can be summarized as follows:
  1. The sinusoidal excitations are expressed as phasors. A single excitation $Y_m\cos(\omega t + \theta)$, or $Y_m\sin(\omega t + \theta)$, can be expressed as $Y_m\angle\theta$. In the case of several excitations, some of which are cosine functions and some are sine functions, they should all be converted to either cosine or sine functions before expressing them as phasors.
  2. Inductance and capacitance are expressed as impedance or admittance, as appropriate.
  3. The circuit is now in the frequency domain and can be analyzed by any of the methods discussed in previous chapters for the dc case.
  4. After obtaining the desired response, this response can be converted back to the time domain as a cosine function, if the time functions that were originally expressed as phasors were cosine functions or converted as a sine function if the time functions that were originally expressed as phasors were sine functions.
- Phasor diagrams showing various voltage and current phasors in a given circuit are useful for illustrating the interrelations between the various variables involved, particularly when some circuit parameter is varied.

Problem-Solving Tips

1. Complex quantities are conveniently added or subtracted in rectangular form and are conveniently multiplied or divided in polar form.
2. Conversion of reactance, or reactance combined with resistance, to susceptance, or susceptance combined with conductance, must proceed through the intermediate step of impedance as the reciprocal of admittance.

Appendix 8A: ac Bridges

ac bridges can be used in the same manner as the dc Wheatstone bridge (Appendix 5A) to determine the value of a quantity from a condition of bridge balance, with the same advantages previously mentioned for a dc bridge.

An example of an ac bridge is the Maxwell L/C bridge, depicted in Figure 8.41 in the frequency domain. The bridge can be used for measuring the inductance and resistance of a coil in terms of known resistance and capacitance values. The condition for bridge balance can be derived in the same manner as for the dc bridge by determining the voltages $\mathbf{V}_{bc}$ and $\mathbf{V}_{dc}$ in terms of $\mathbf{V}_I$,

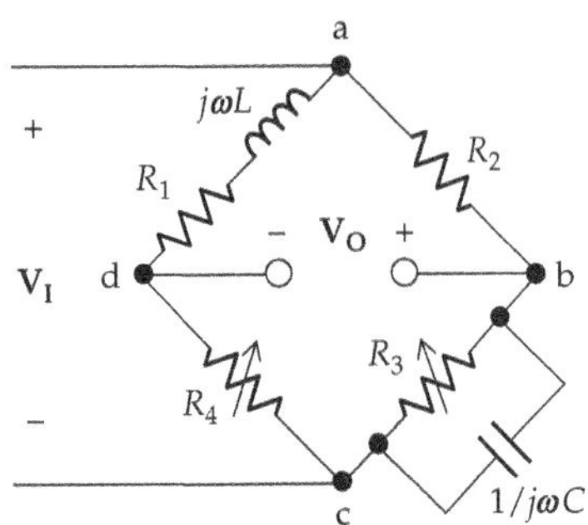

FIGURE 8.41
ac bridge.

from voltage division, and then equating $\mathbf{V}_{bc}$ to $\mathbf{V}_{dc}$. Before doing so, we note that the parallel impedance of the *RC* branch is

$$Z_p = \frac{R_3/j\omega C}{R_3 + 1/j\omega C} = \frac{R_3}{1 + j\omega CR_3} \tag{8.61}$$

It follows from voltage division that

$$\mathbf{V}_{bc} = \frac{Z_p}{R_2 + Z_p}\mathbf{V}_I \quad \text{and} \quad \mathbf{V}_{dc} = \frac{R_4}{R_4 + R_1 + j\omega L}\mathbf{V}_I \tag{8.62}$$

At bridge balance, $\mathbf{V}_O = 0$, so that $\mathbf{V}_{bc} = \mathbf{V}_{dc}$. Equating the two expressions from Equation 8.60 and simplifying,

$$R_1R_3 + j\omega LR_3 = R_2R_4 + j\omega CR_2R_3R_4 \tag{8.63}$$

For Equation 8.63 to be satisfied, the real parts alone must be equal and the imaginary parts alone must be equal. Equating the real parts gives the coil resistance as

$$R_1 = \frac{R_2R_4}{R_3} \tag{8.64}$$

Note that $R_1R_3 = R_2R_4$ is also the dc balance condition, when the inductor acts as a short circuit and the capacitor as an open circuit. Equating the imaginary parts gives the coil inductance as

$$L = CR_2R_4 \tag{8.65}$$

Having two conditions for ac bridge balance is due to ac quantities having both amplitude and phase or real and imaginary parts.

Problems

Apply ISDEPIC and verify solutions by PSpice simulation whenever feasible.

Sinusoidal Functions and Phasors

P8.1 Given $v(t) = 20\cos(\omega t - 45°)$ V and $i(t) = 10\sin(\omega t - 80°)$ A, determine which variable leads the other and by what angle.

Ans. $v(t)$ leads $i(t)$ by 125°.

P8.2 Given the sinusoidal time function $v(t)$ of Figure P8.2, express $v(t)$ as a function of time and as a phasor.

Ans. $v(t) = 10\cos(200\pi t + 72.54°)$ V, $\mathbf{V} = 10\angle 72.54°$ V.

P8.3 Given $i(t) = 2.5\cos(\omega t + 30°)$ A, determine the rms value of *i* and the power dissipated in an 8 Ω resistor.

Ans. 1.77 A, 25 W.

P8.4 Given a phasor $\mathbf{A} = A\angle\alpha$ in Figure P8.4, express the phasor 'Ob' in terms of **A** and *j*, assuming that the phasors 'Oa' and 'ab' have a magnitude *A*.

Ans. $j(1 + j)\mathbf{A}$.

P8.5 Given $\mathbf{A} = 10\angle 15°$, $\mathbf{B} = 20\angle 120°$, and $\mathbf{C} = 5\angle -45°$, determine the phasors resulting from the following operations: (a) $\mathbf{A} + \mathbf{B} + \mathbf{C}$, (b) $\mathbf{A} - \mathbf{B} + \mathbf{C}$, (c) $\mathbf{A} + \mathbf{B} - \mathbf{C}$, and (d) $\mathbf{A} - \mathbf{B} - \mathbf{C}$. Express the result in rectangular and polar forms.

Ans. (a) $3 + j16.37$, $16.68\angle 78.96°$; (b) $23.19 - j18.27$, $29.52\angle -38.22°$; (c) $-3.88 + j23.44$, $23.76\angle 99.39°$; (d) $16.12 - j11.20$, $19.63\angle -34.78°$.

P8.6 Given $\mathbf{A} = 3 + j5$, $\mathbf{B} = 10 - j8$, and $\mathbf{C} = j12$, determine the phasors resulting from the following operations: (a) **ABC**; (b) $(\mathbf{AB})/\mathbf{C}$; (c) $(\mathbf{A}/\mathbf{B})\mathbf{C}$; and (d) $\mathbf{A}/\mathbf{B}/\mathbf{C}$. Express the result in rectangular and polar forms.

Ans. (a) $-312 + j840$, $896.1\angle 110.4°$; (b) $2.17 - j5.83$, $6.22\angle -69.62°$; (c) $-5.4146 - j0.7317$, $5.46\angle 187.7°$; (d) $0.0376 + j0.0051$, $0.0379\angle 7.696°$.

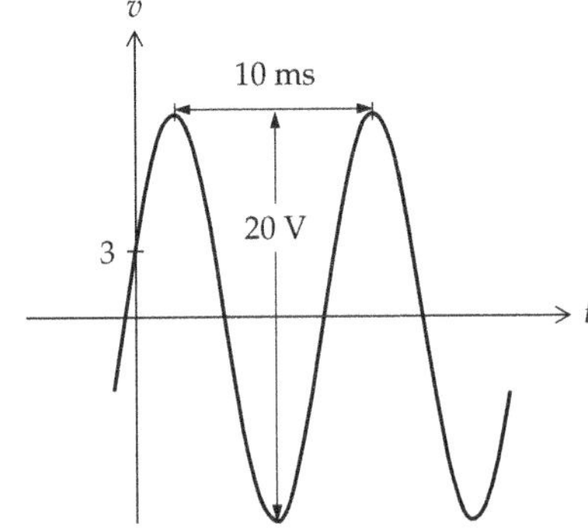

FIGURE P8.2

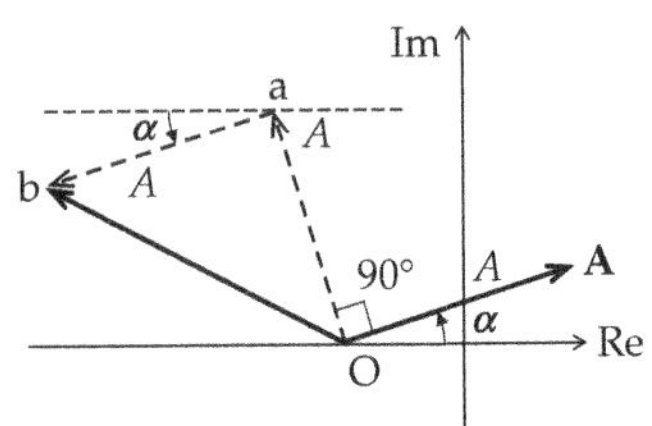

FIGURE P8.4

P8.7 Given $\mathbf{A} = 5 + j10$, determine the phasor that is **A** raised to the fourth power.

Ans. $15{,}625\angle{-106.3°}$ or $-4{,}375 - j15{,}000$.

P8.8 Given $\mathbf{A} = 24 + j32$, determine the phasor that is the cube root of **A**.

Ans. $3.42\angle 17.71°, 3.42\angle 137.71°, 3.42\angle 257.71°$.

P8.9 Using phasors, determine the steady-state y that satisfies the differential equation:

$$6\frac{dy}{dt} + 3y + 2\int_0^t ydt = 5\cos 4t + 10\sin 4t.$$

Express y as a cosine time function (Hint: express the RHS as a phasor).

Ans. $0.472\cos(4t - 146.2°)$.

P8.10 Given $v_1(t) = 2\cos\omega t - \sin\omega t$ V; $v_2(t) = \sqrt{2}\cos(\omega t - 135°)$ V. (a) Derive $v(t) = v_1(t) + v_2(t)$ as a single trigonometric function; (b) Express $v_1(t)$, $v_2(t)$, and $v(t)$ as phasors and draw them on a phasor diagram. Verify that $\mathbf{V} = \mathbf{V}_1 + \mathbf{V}_2$.

Ans. (a) $v(t) = \cos\omega t$ V; (b) $\mathbf{V}_1 = \sqrt{5}\angle 26.6°$ V; $\mathbf{V}_2 = \sqrt{2}\angle{-135°}$ V; $\mathbf{V} = 1\angle 0°$ V.

P8.11 Verify conservation of power in Example 8.7.

Impedance and Admittance

P8.12 A coil has a resistance of 10 Ω and an inductance L. When connected to a 100 V rms, 60 Hz supply, the magnitude of the coil current is 5 A rms. Determine L.

Ans. 45.9 mH.

P8.13 Given an impedance $0.1(4 + j3)$ Ω, determine the susceptance.

Ans. $B = -1.2$ S.

P8.14 $v(t)$ in Figure P8.14 is the voltage between two terminals of a given circuit, and $i(t)$ is the current entering these terminals in the direction of a voltage drop v. Determine the impedance looking into these terminals.

Ans. $2\angle{-45°}$ Ω.

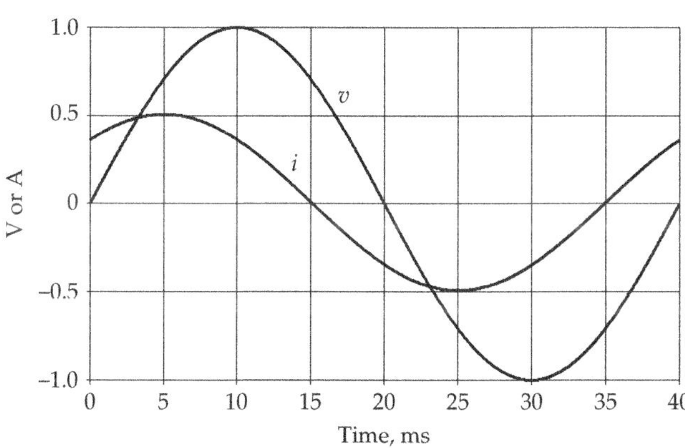

FIGURE P8.14

P8.15 A susceptance of −1 S is connected in series with an admittance $(3 + j4)$ S. Determine the reactance of the series combination.

Ans. 0.84 Ω.

P8.16 A capacitor of impedance Z_C is connected in parallel with a load of $(300 + j450)$ Ω. Determine Z_C so that the equivalent load is purely resistive.

Ans. $-j650$ Ω.

P8.17 Given a 40 μH inductor and a 100 nF capacitor: (a) At what frequency is the impedance of the series combination zero? (b) At what frequency is the admittance of the series combination zero? (c) At what frequency is the admittance of the parallel combination zero? (d) At what frequency is the impedance of the parallel combination zero?

Ans. (a) and (c) 500 krad/s; (b) and (d) 0 and ∞.

P8.18 (a) Determine the impedance seen by the source in Figure P8.18; (b) derive i as a phasor and in the time domain, assuming $v_{SRC}(t) = 60\cos(500t + 30°)$ V.

Ans. (a) $(30 - j30)$ Ω; (b) $\sqrt{2}\angle 75°$, $\sqrt{2}\cos(500t + 75°)$ V.

P8.19 Determine the frequency at which the impedance looking into terminals 'ab' in Figure P8.19 is purely resistive and specify this impedance.

Ans. 20 krad/s, 12 Ω.

P8.20 Determine the frequency at which the voltage across the 10 Ω resistor in Figure P8.20 is a maximum and specify the impedance seen by the source at this frequency.

Ans. $\omega = 2$ krad/s, 15 Ω.

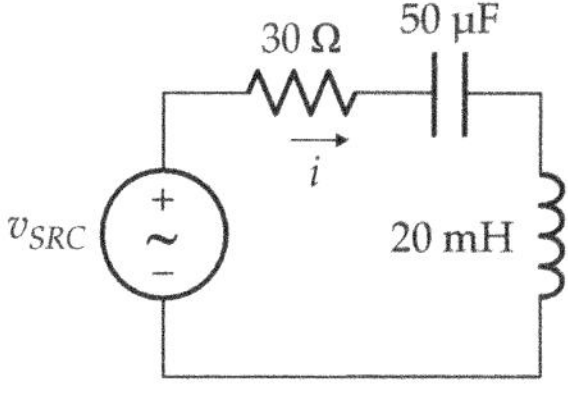

FIGURE P8.18

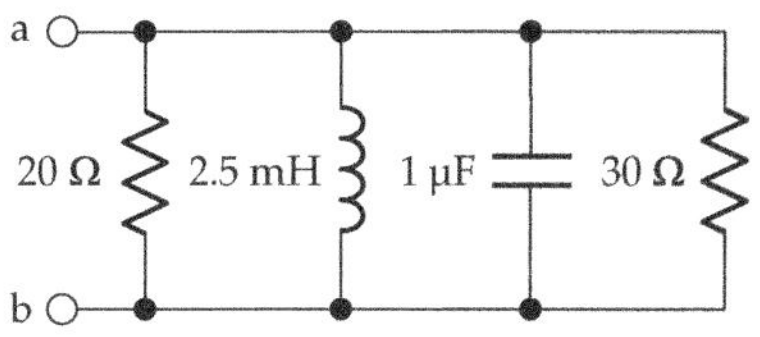

FIGURE P8.19

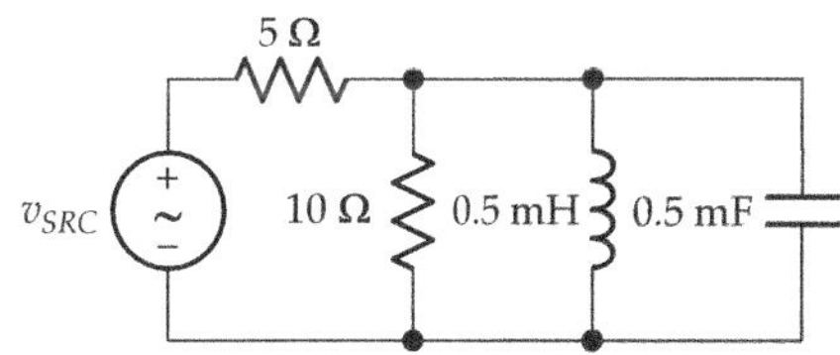

FIGURE P8.20

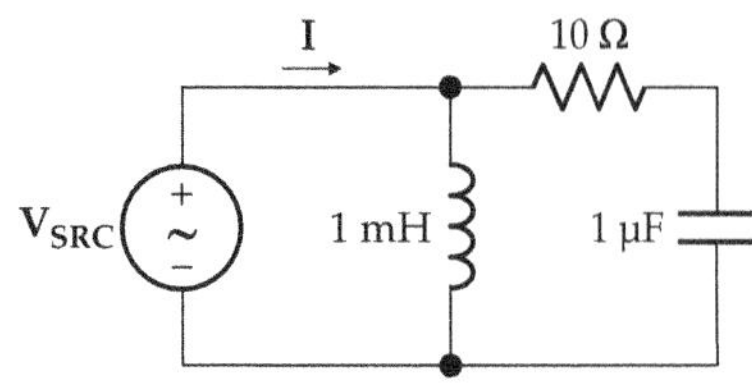

FIGURE P8.21

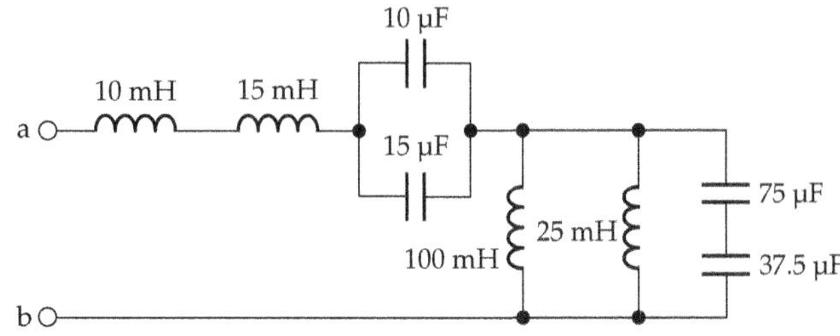

FIGURE P8.24

P8.21 Determine the frequency at which the source voltage and current in Figure P8.21 are in phase.

Ans. 100/3 krad/s.

P8.22 Determine Z_{in} in Figure P8.22.

Ans. $-j$ Ω.

P8.23 Determine Z_{in} in Figure P8.23, assuming all capacitances are 1/3 F, all inductances are 1 H, and ω = 1 rad/s. Apply (a) star-delta transformation; and (b) symmetry considerations.

Ans. $-j$ Ω.

P8.24 Determine the impedance and admittance between terminals 'ab' in Figure P8.24 at ω = 1 krad/s.

Ans. j25 Ω; $-j$40 mS.

P8.25 All the inductances in Figure P8.25 are j10 Ω, all the resistances are 20 Ω, and all the capacitances are $-j$12 Ω. Determine the impedance between terminals 'ab'.

Ans. 27 + j9 Ω.

P8.26 Determine Z_{in} in Figure P8.26 if Z = (1 + j) Ω.

Ans. j Ω.

P8.27 Determine Z_{in} in Figure P8.27.

Ans. j Ω.

FIGURE P8.25

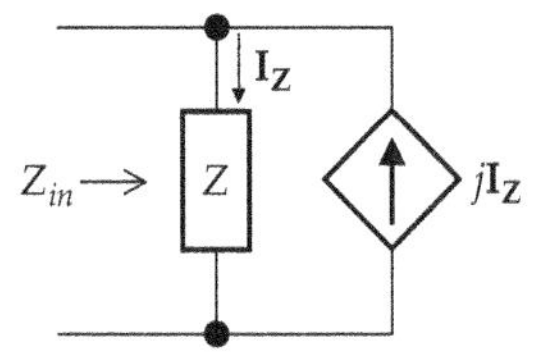

FIGURE P8.26

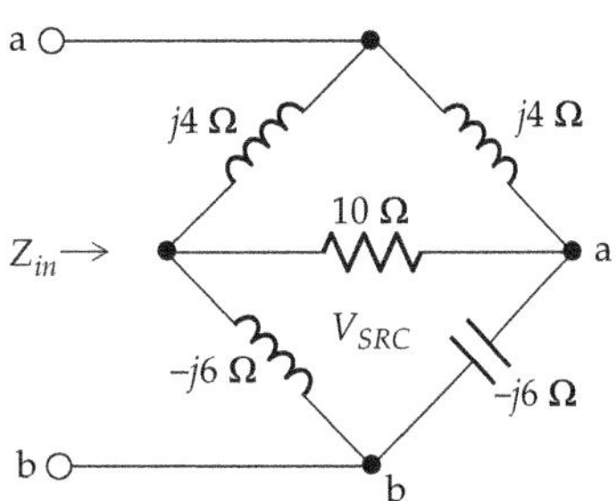

FIGURE P8.22

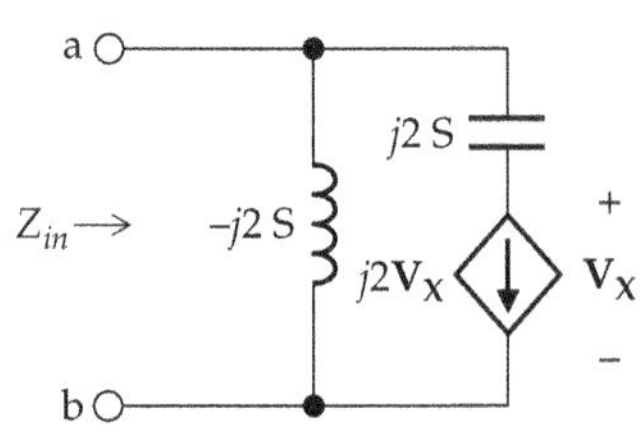

FIGURE P8.27

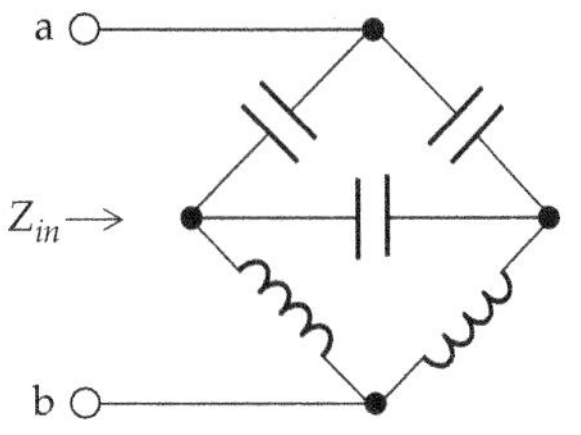

FIGURE P8.23

P8.28 Determine Z_{in} in Figure P8.28, assuming $\sigma = 1/(2 - j2)$ S.

Ans. j2 Ω.

P8.29 (a) Express the impedance looking into terminals 'ab' in Figure P8.29 in terms of ω; (b) determine ω so that Z_{in} is purely resistive.

Ans. (a) $j\omega \times 10^{-3} + \dfrac{40}{1 + j1.25 \times \omega \times 10^{-3}}$; (b) 5.6 krad/s.

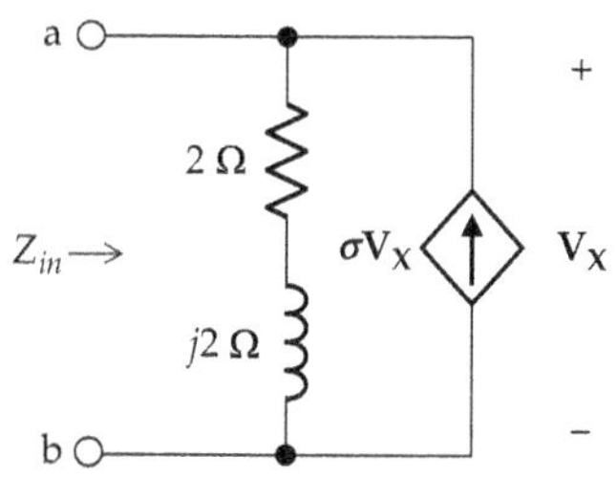

FIGURE P8.28

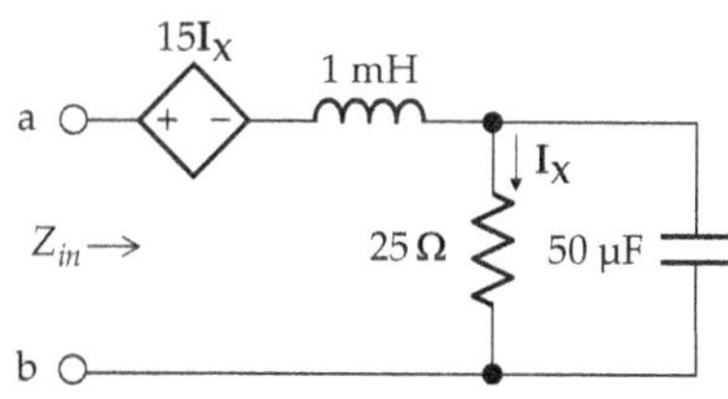

FIGURE P8.29

P8.30 Determine Y_{in} looking into terminals 'ab' in Figure P8.30.

Ans. 0.5(1 – *j*) S.

P8.31 Determine α in Figure P8.31 so that the impedance looking into terminals 'ab' is purely resistive.

Ans. $\alpha = 1\ \Omega$.

P8.32 Determine *Z* in Figure P8.32 so that the Thevenin's impedance between terminals 'ab' is 1 Ω.

Ans. 0.8 – *j*1.4 Ω.

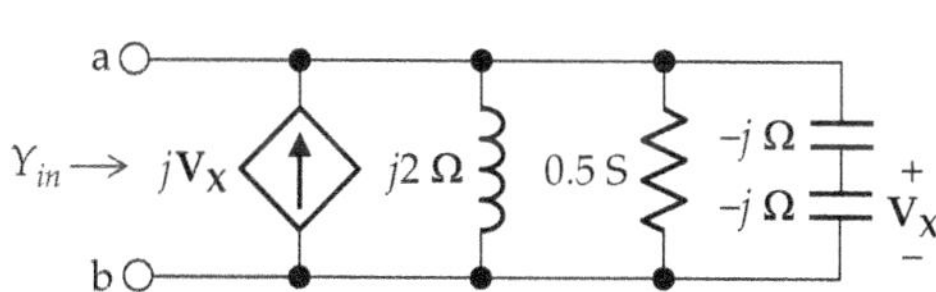

FIGURE P8.30

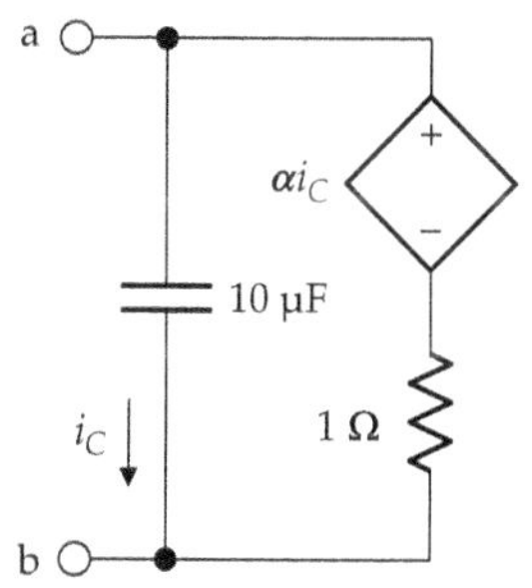

FIGURE P8.31

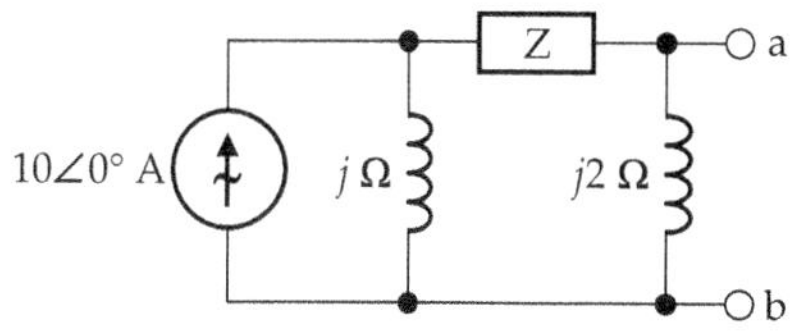

FIGURE P8.32

P8.33 Determine *Z* in Figure P8.33 given that $\mathbf{I_S} = 27.9\angle 57.8°$ A.

Ans. 5.00∠30.0°.

P8.34 Determine Z_{Th} looking into terminals 'ab' in Figure P8.34.

Ans. –*j*20 Ω.

P8.35 Determine Z_{Th} between terminals 'ab' in Figure P8.35.

Ans. 8(4 + *j*) Ω.

P8.36 Determine Z_{Th} looking into terminals 'ab' in Figure P8.36.

Ans. –*j* Ω.

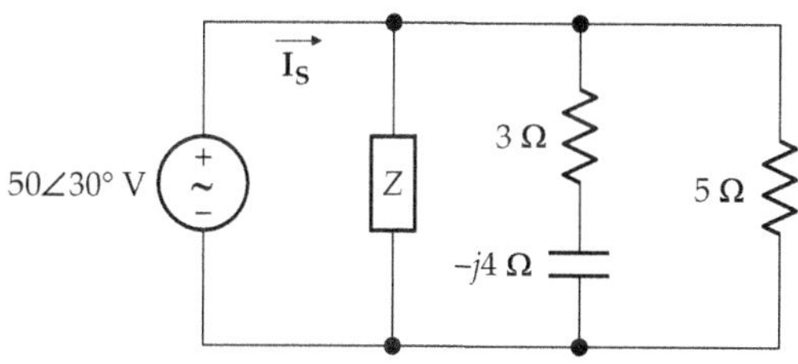

FIGURE P8.33

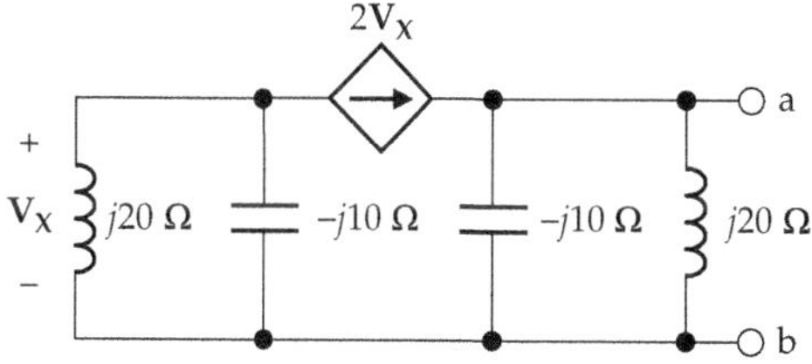

FIGURE P8.34

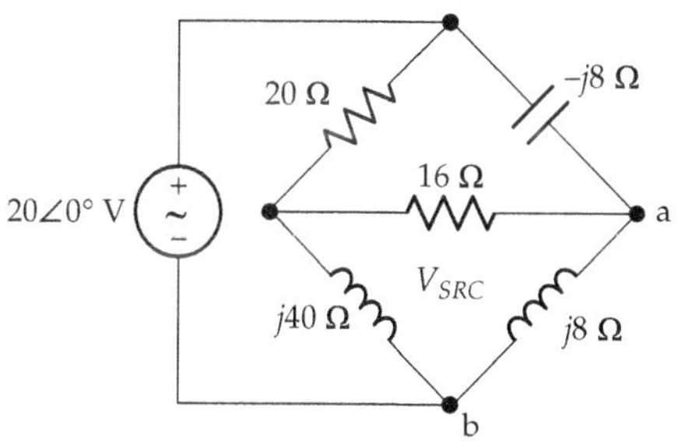

FIGURE P8.35

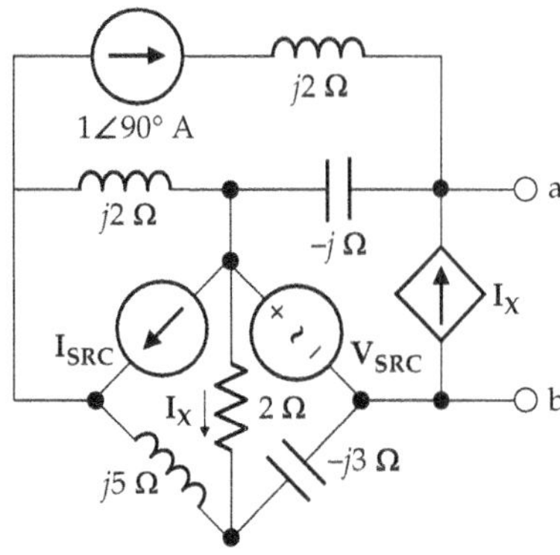

FIGURE P8.36

General Sinusoidal Steady-State Analysis

P8.37 Determine $\mathbf{I}_L$ in Figure P8.37, assuming $\omega = 100$ rad/s.

Ans. $1\angle{-90°}$ A.

P8.38 Determine $\mathbf{I}_C$ in Figure P8.38, assuming $\omega = 2$ krad/s.

Ans. $5\angle 135°$ A.

P8.39 Determine $i_S(t)$ in Figure P8.39, assuming $v_{SRC}(t) = 4\cos 100t$ V.

Ans. $2\cos 100t$ A.

P8.40 Determine $\mathbf{I}_1$, $\mathbf{I}_2$, $\mathbf{I}_3$, $\mathbf{V}_1$, and $\mathbf{V}_2$ in Figure P8.40.

Ans. $\mathbf{I}_1 = 2.50\angle 29.04°$ A; $\mathbf{I}_2 = 2.71\angle{-11.57°}$ A; $\mathbf{I}_3 = 1.82\angle 105.0°$ A; $\mathbf{V}_1 = 16.26\angle 78.43°$ V; $\mathbf{V}_2 = 7.28\angle 15°$ V.

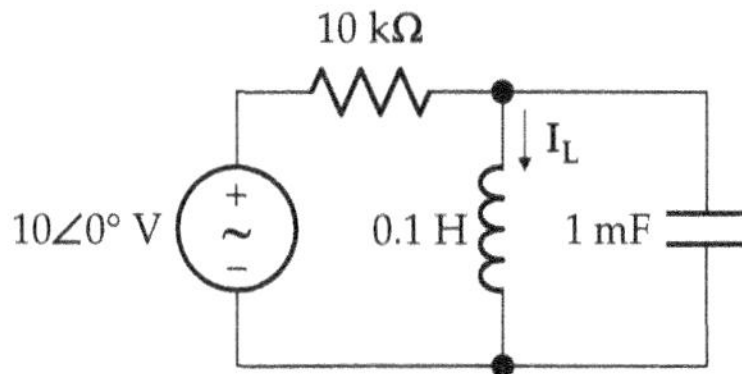

FIGURE P8.37

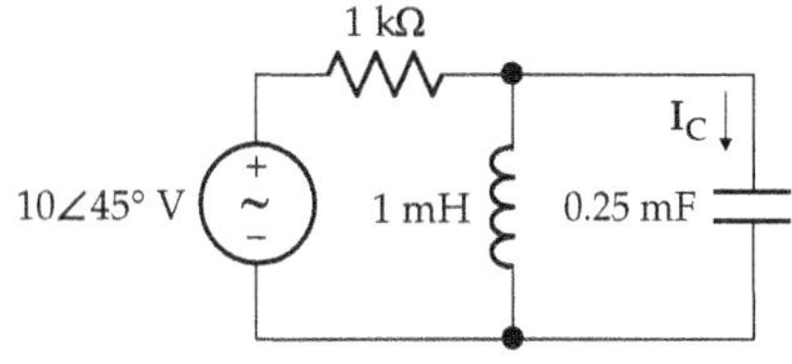

FIGURE P8.38

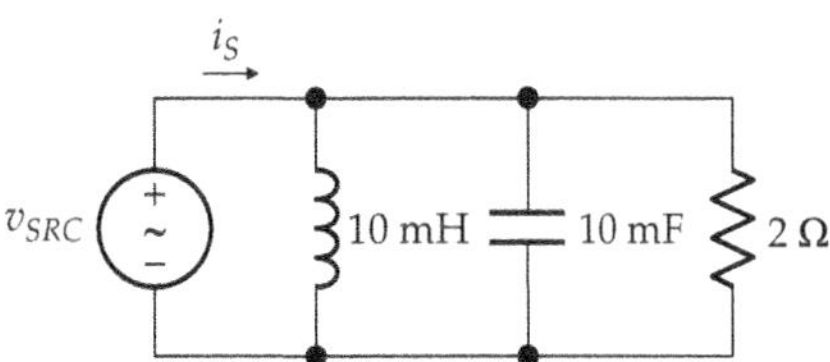

FIGURE P8.39

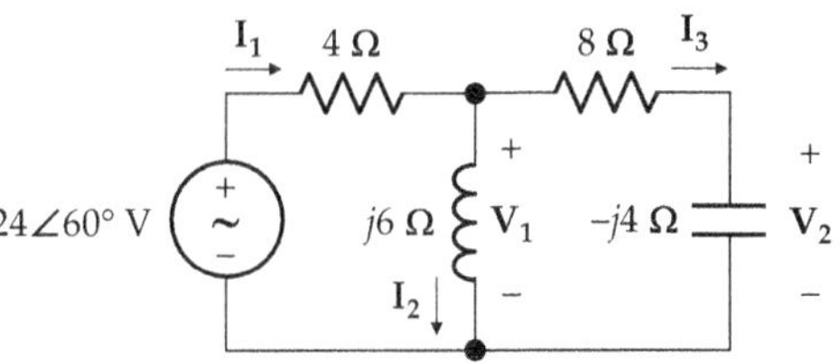

FIGURE P8.40

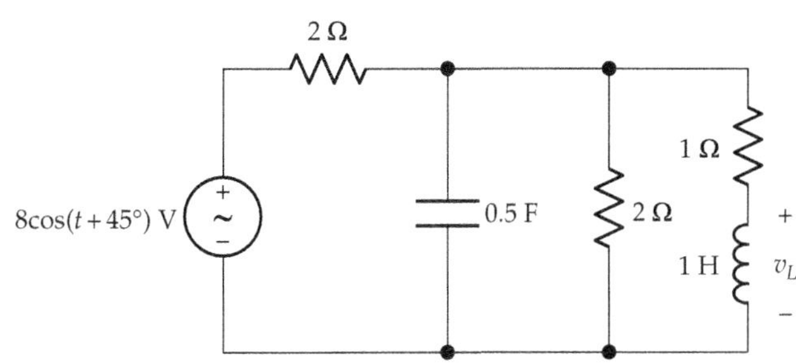

FIGURE P8.41

P8.41 Determine v_L in Figure P8.41 as a phasor and in the time domain.

Ans. $\mathbf{V}_L = j\frac{4}{3}\sqrt{2}$ V, $v_L(t) = 1.89\cos(t + 90°)$ V.

P8.42 Determine $\mathbf{I}_L$ and $\mathbf{V}_C$ in Figure P8.42, assuming the supply frequency is 1 krad/s.

Ans. $0.5 - j4$ A, $200 - j20$ V.

P8.43 Determine $\mathbf{V}_X$ and $\mathbf{I}_L$ in Figure P8.43 and the total power dissipated in the circuit, assuming the supply frequency is 1 krad/s and the source voltage is a peak value.

Ans. $2(1 + j5)$ V, $0.2\angle 0°$ A; 1.5 W.

P8.44 Given $i(t) = 8\sqrt{2}\cos(2500\pi t - 45°)$ A and $i_1(t) = 2\cos(2500\pi t)$ A in Figure P8.44. Determine: (a) $v(t)$ and $i_2(t)$ in the time and frequency domains; (b) Z, if composed of: (i) two series elements, or (ii) two parallel elements.

Ans. (a) $v(t) = 150.4\cos(2500\pi t - 57.86)$ V, $\mathbf{V} = 150.2\angle{-57.87°}$ V, $i_2(t) = 3.04\cos(2500\pi t - 110.48°)$ A, $\mathbf{I}_2 = 3.04\angle{-110.48°}$ A; (b) (i) $R = 15.97\ \Omega$ and $C = 19.96\ \mu$F, (ii) $G = 0.054$ S and $C = 2.75\ \mu$F.

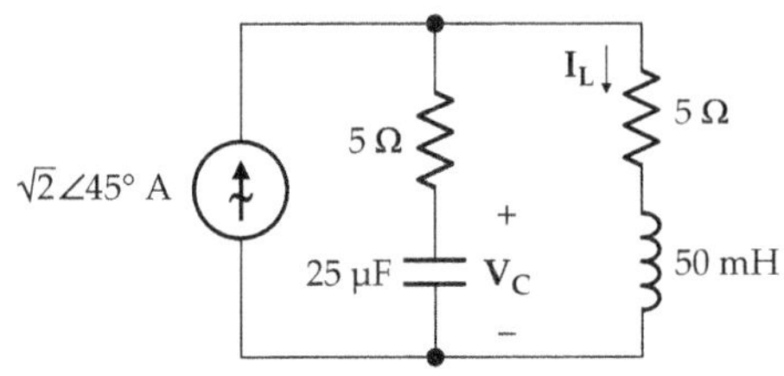

FIGURE P8.42

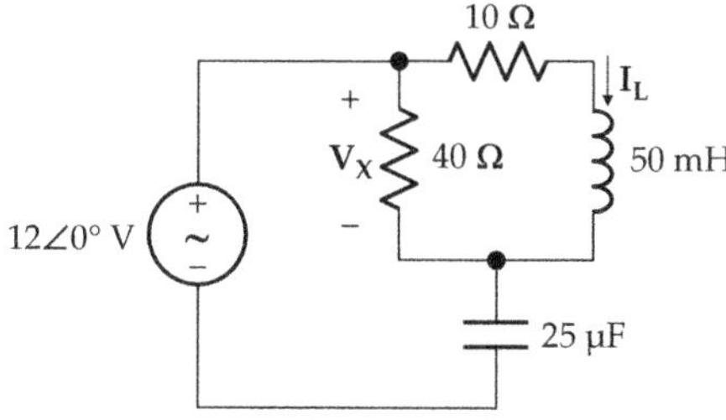

FIGURE P8.43

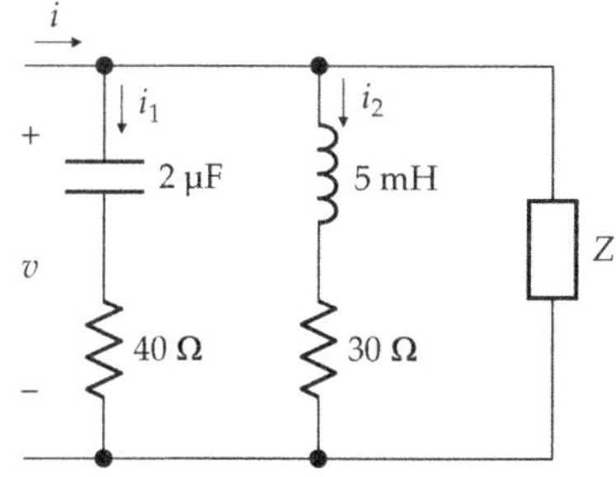

FIGURE P8.44

P8.45 Given that the voltage across the current source in Figure P8.45 is $v_S(t)=\sqrt{2}\cos(1000t-45°)$ V. Determine i_{SRC} as a phasor in rectangular coordinates.

Ans. $\mathbf{I}_{\mathbf{SRC}} = -j2$ mA.

P8.46 Determine ω, assuming $v_{SRC}(t)$ = $2\cos\omega t$ and $v_O(t)$ = $2\sin\omega t$ in Figure P8.46.

Ans. 2 rad/s.

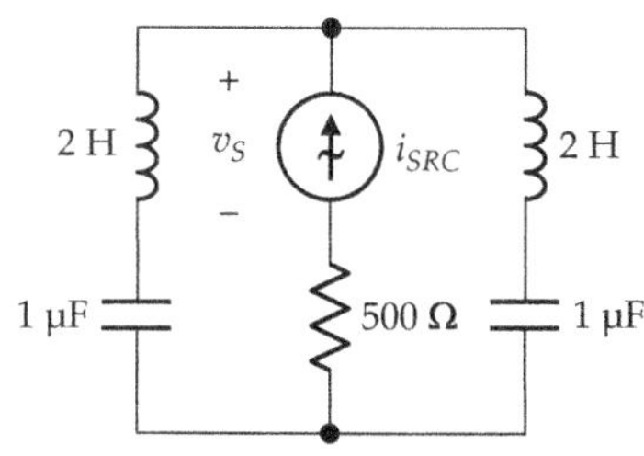

FIGURE P8.45

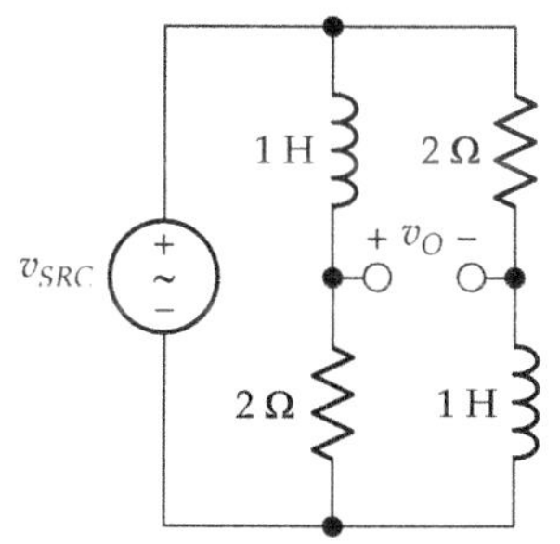

FIGURE P8.46

P8.47 Determine, in Figure P8.47, the maximum magnitude of v_L and its phase angle at this magnitude, given $v_{SRC}(t) = 2\cos\omega t$ V, and ω assuming any value between 0 and ∞.

Ans. $|v_L|_{max} = 2$ V, $\theta_L = 90°$.

P8.48 Determine $\mathbf{V}_{\mathbf{X}}$ in Figure P8.48.

Ans. $10\sqrt{2}\angle 15°$ V.

P8.49 Determine C in Figure P8.49 so that v_O has the same magnitude as v_I but lags it by 90°, assuming ω = 400 rad/s.

Ans. 5 μF.

P8.50 Determine the power dissipated in R_1 in Figure P8.50, assuming $\omega = 1$ krad/s.

Ans. 0.1 W.

P8.51 $|\mathbf{I}|$ in Figure P8.51 remains the same irrespective of whether the switch is open or closed. Show that under these conditions $2\omega^2LC = 1$.

P8.52 Determine $\mathbf{I}_{\mathbf{S}}$ in Figure P8.52.

Ans. 0.5 A.

P8.53 Determine i_C as a phasor in Figure P8.53, given v_{SRC} = $\sin 2t$ V, and assuming $\cos 2t$ to have zero phase angle.

Ans. $\frac{1}{\sqrt{2}}\angle -45°$ A.

P8.54 Determine $\mathbf{I}_1$ and $\mathbf{I}_2$ in Figure P8.54.

Ans. $\mathbf{I}_1 = -0.052 - j0.62$ A; $\mathbf{I}_2 = -0.47 + j1.91$ A

P8.55 Determine $\mathbf{V}_{\mathbf{O}}$ in Figure P8.55.

Ans. $5(\sqrt{2}+1) = 12.07$ V.

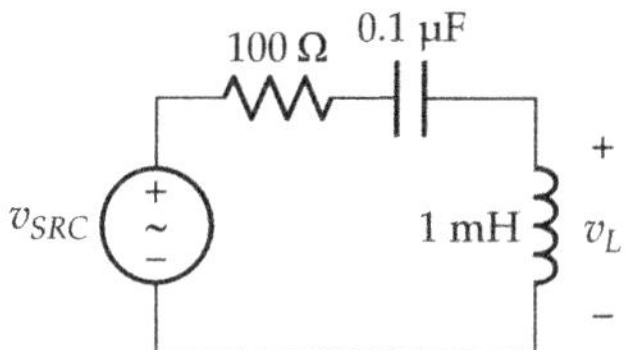

FIGURE P8.47

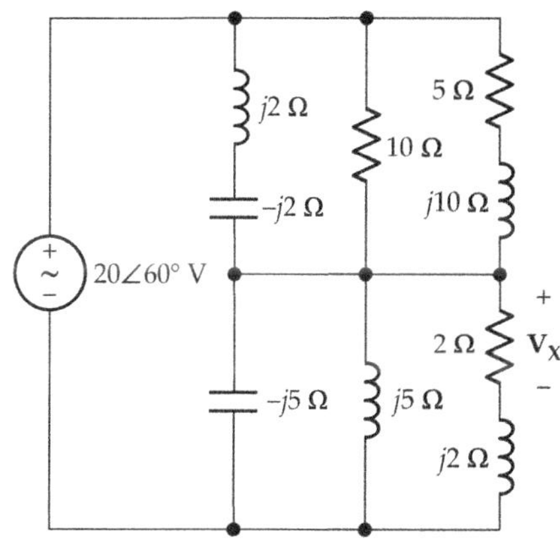

FIGURE P8.48

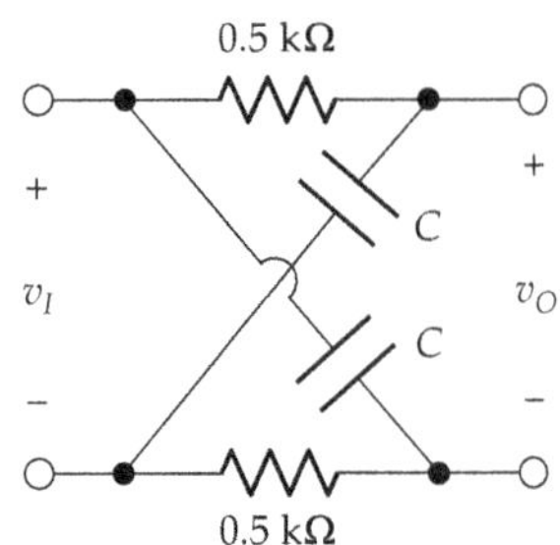

FIGURE P8.49

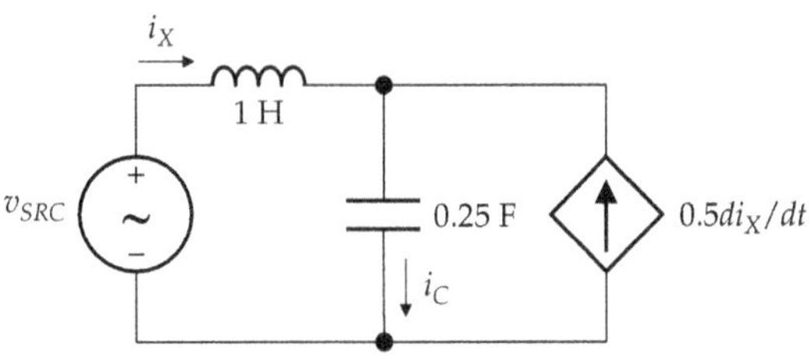

FIGURE P8.53

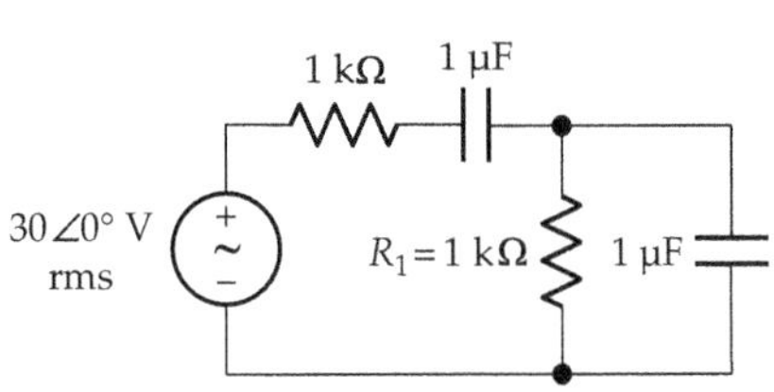

FIGURE P8.50

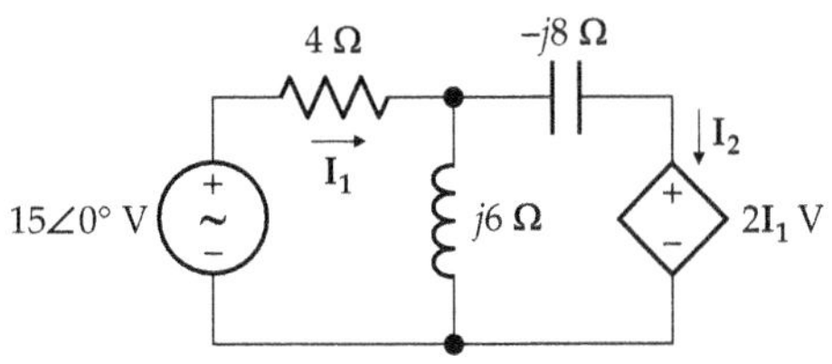

FIGURE P8.54

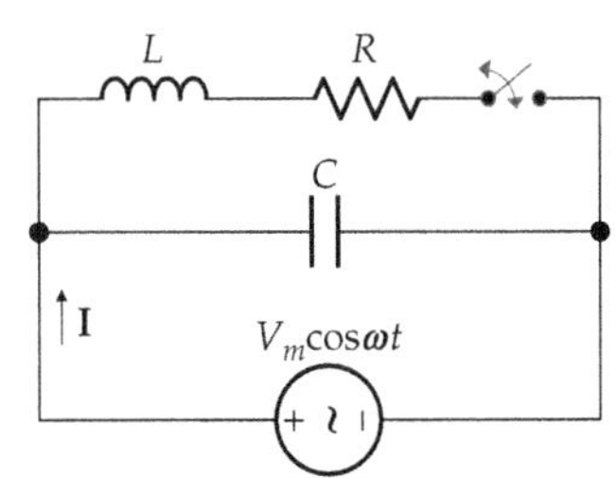

FIGURE P8.51

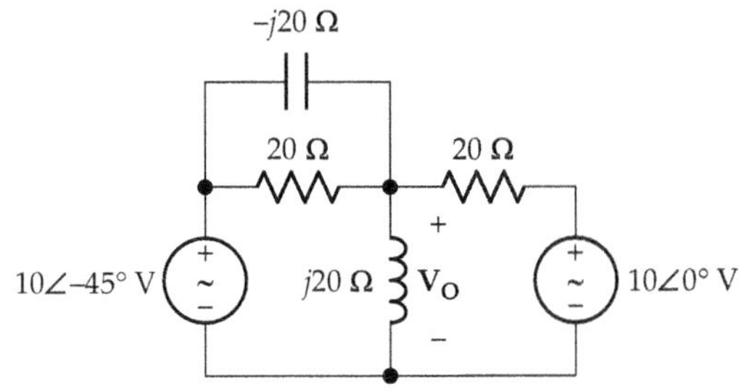

FIGURE P8.55

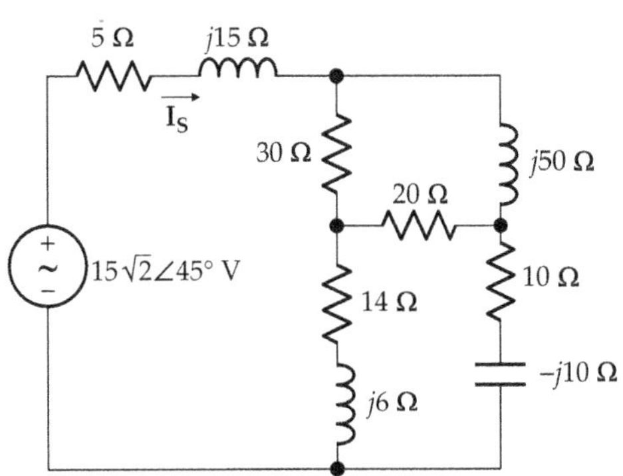

FIGURE P8.52

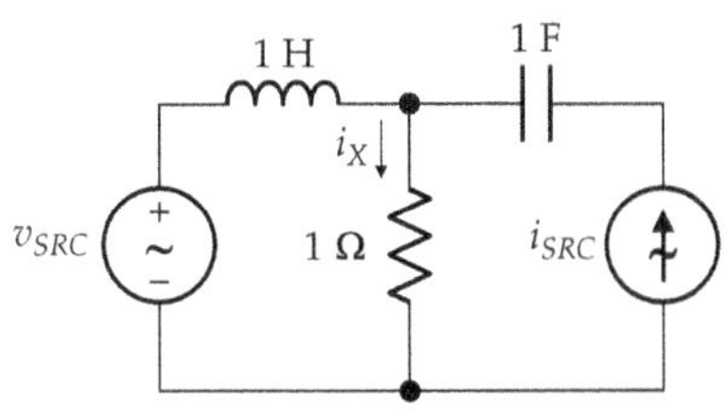

FIGURE P8.56

P8.56 Determine (a) i_X in Figure P8.56 given $v_{SRC}(t) = \cos t$ V and $i_{SRC}(t) = \sin 2t$ V and (b) the power dissipated in the resistor.

Ans. (a) $0.71\cos(t - 45°) + 0.89\sin(2t + 26.57°)$ A; 0.65 W.

P8.57 Determine $v_O(t)$ in Figure P8.57, assuming $\omega = 10^4$ rad/s.

Ans. $10.54\cos(\omega t - 63.43°)$ V.

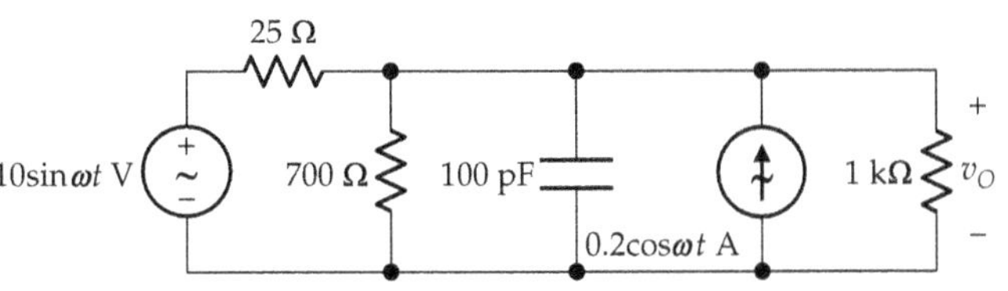

FIGURE P8.57

P8.58 Determine $\mathbf{I}_X$ in Figure P8.58.

Ans. $j3$ A

P8.59 Determine R in Figure P8.59, given $\mathbf{I} = 0$.

Ans. 1 Ω.

P8.60 Determine the net average power delivered in Figure P8.60, assuming $\mathbf{V}_{SRC} = 10\angle 0°$ V.

Ans. 2 W.

P8.61 Determine $\mathbf{I}$ so that $\mathbf{V}_O = 0$ in Figure P8.61, assuming that $X = 1$ Ω.

Ans. $4\angle 0°$ A.

P8.62 Determine $i_C(t)$ in Figure P8.62, assuming $v_{SRC}(t) = \sin(1000t + 30°)$ V.

Ans. $\sin(1000t + 120°)$ mA.

P8.63 Determine Z in Figure P8.63.

Ans. $68 + j24$ Ω.

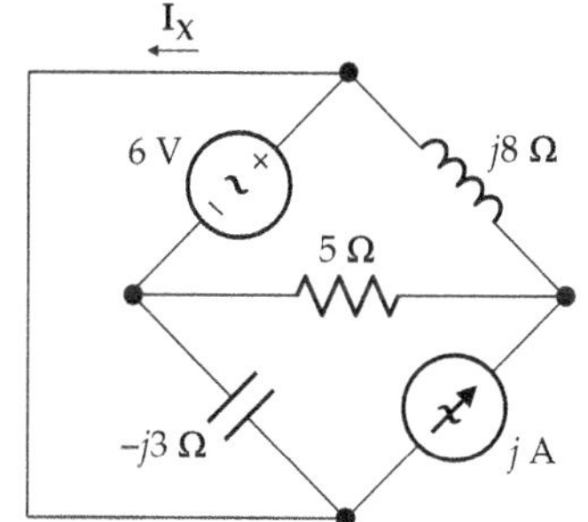

FIGURE P8.58

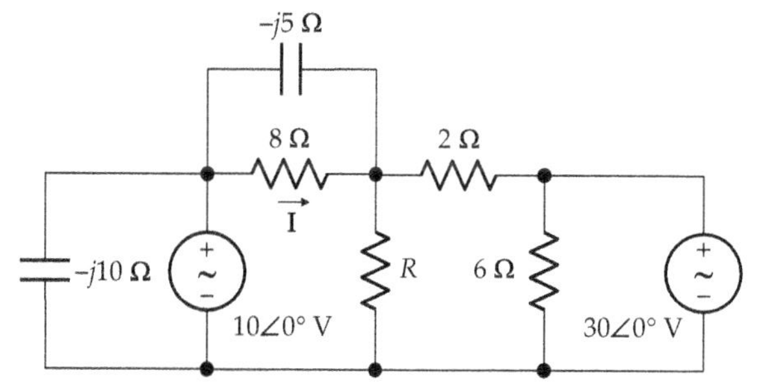

FIGURE P8.59

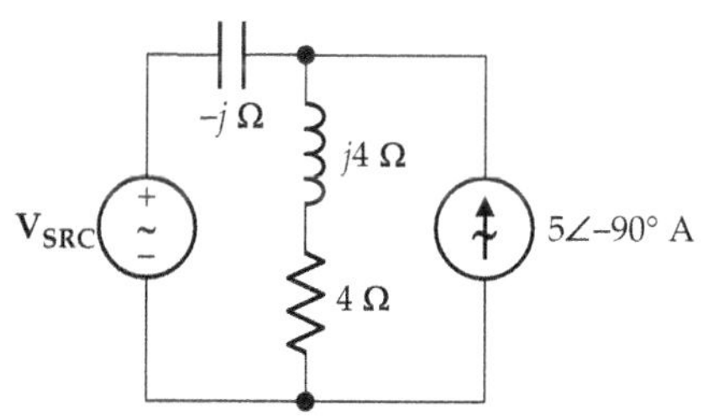

FIGURE P8.60

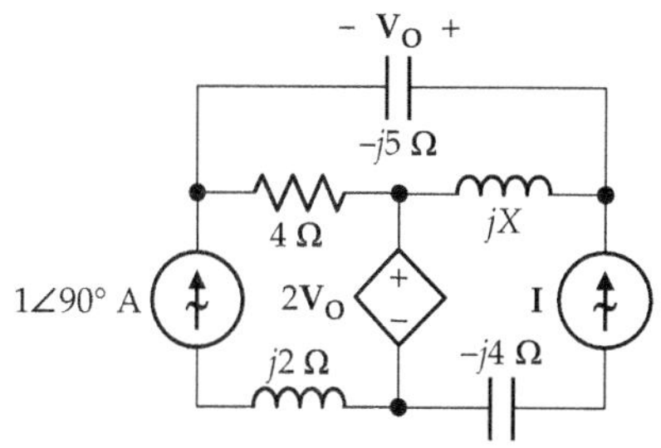

FIGURE P8.61

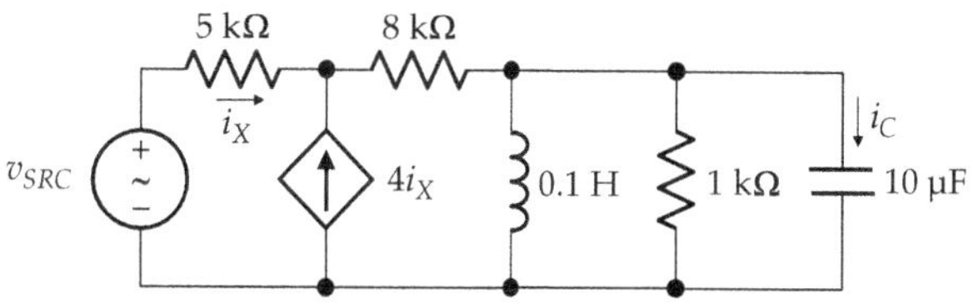

FIGURE P8.62

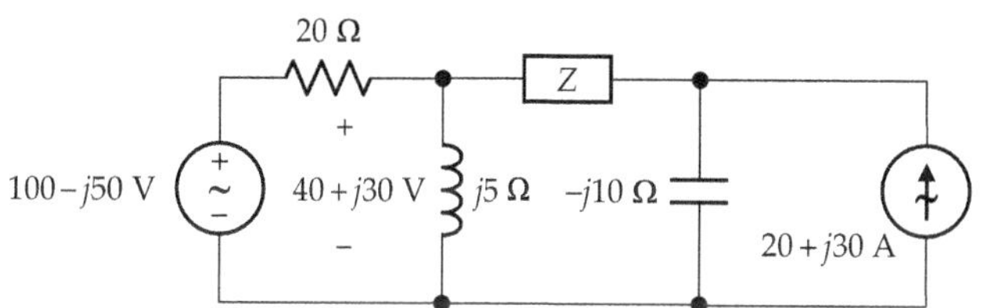

FIGURE P8.63

P8.64 Determine $\mathbf{I}_O$ in Figure P8.64.

Ans. $11.57\angle 89.63°$ A.

P8.65 Determine $\mathbf{V}_C$ and $\mathbf{I}_L$ in Figure P8.65

Ans. $\mathbf{V}_C = -30 - j90$ V, $\mathbf{I}_L = 8 - j6$ A.

P8.66 Determine $\mathbf{V}_O$ in Figure P8.66.

Ans. $90 + j10$ V.

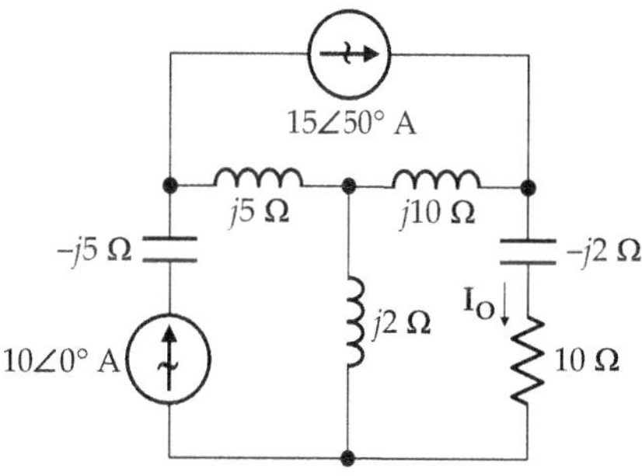

FIGURE P8.64

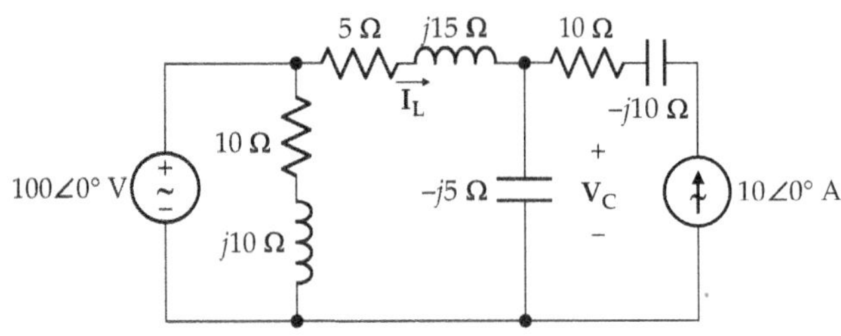

FIGURE P8.65

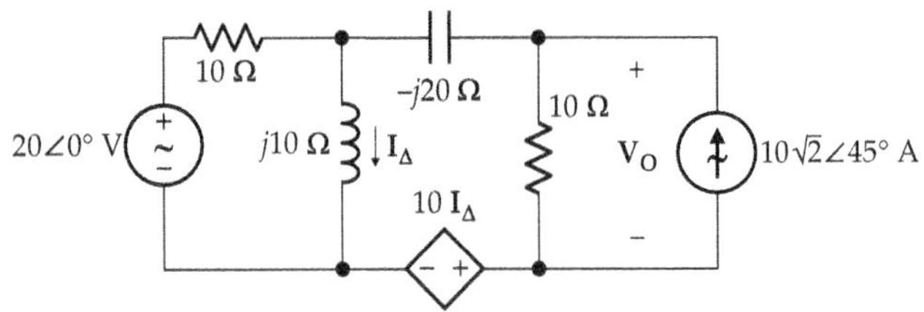

FIGURE P8.66

P8.67 Given $i_{SRC}(t) = \cos 10^8 t$ A in Figure P8.67, determine: (a) $\mathbf{V}_Y$ and $\mathbf{I}_X$; (b) $v_O(t)$.

Ans. (a) $\mathbf{V}_Y = 60 - j20$ V, $\mathbf{I}_X = 0.2 + j0.6$ A; (b) $\mathbf{V}_O = 60 - j120$ V; $v_O(t) = 60\sqrt{5}\cos(\omega t - 63.4°)$ V.

P8.68 Determine $\mathbf{V}_{ab}$ in Figure P8.68, assuming all impedances are in ohms.

Ans. $-741 + j494$ V.

P8.69 Determine $i_O(t)$ in Figure P8.69, given $v_{SRC}(t) = 10\cos(3000t)$ V.

Ans. $-1 + 2\cos(3000t)$ A.

P8.70 Determine the power dissipated in the 4 Ω resistor in Figure P8.70 by each independent source acting alone, given that $v_{SRC}(t) = 10\cos(10^3 t + 30°)$ V.

Ans. 2.25 W by dc source and 8 W by ac source.

P8.71 Determine the average power dissipated in the 1 Ω resistor in Figure P8.71, given that $v_{SRC1}(t) = 2\cos t$ V and $v_{SRC2}(t) = 2\sin t$ V.

Ans. 4 W.

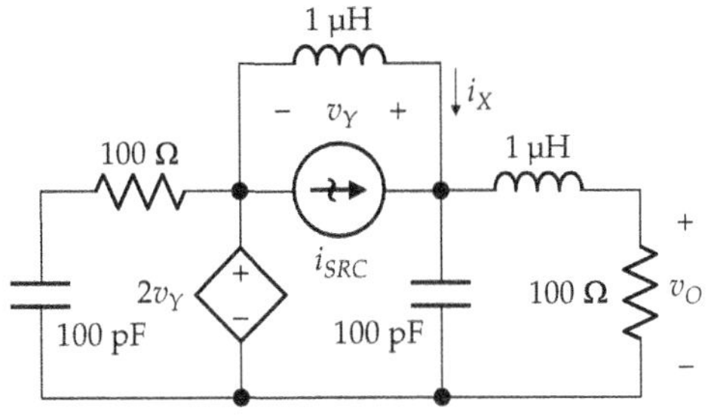

FIGURE P8.67

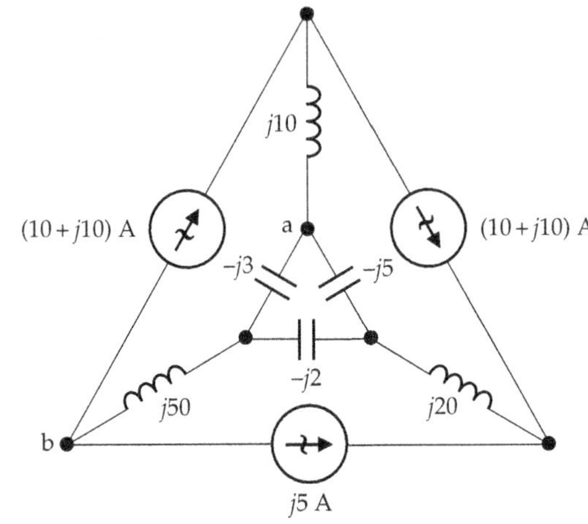

FIGURE P8.68

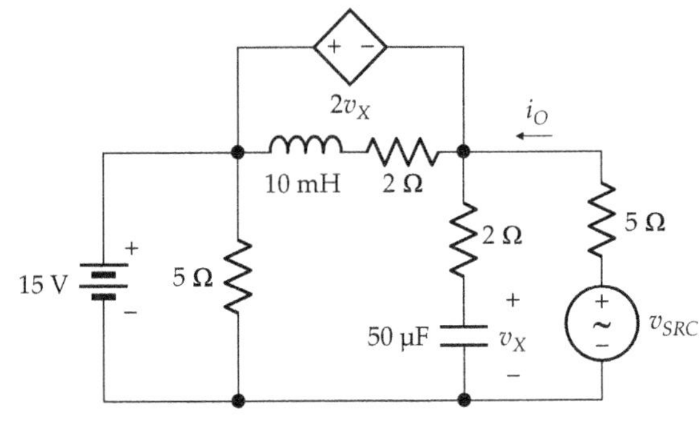

FIGURE P8.69

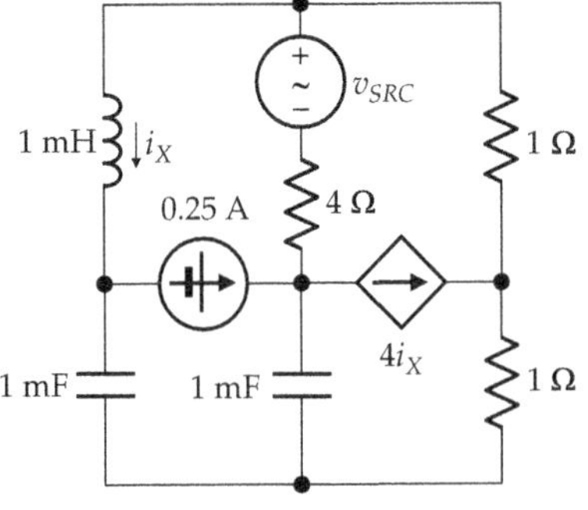

FIGURE P8.70

P8.72 Determine $i_1(t)$ in Figure P8.72 and the average power dissipated in the 2 Ω resistor.

Ans. $i_1(t) = 0.8 + \sqrt{2}\cos(100t - 45°)$ A; 3.28 W.

P8.73 Determine $\mathbf{I}_X$ in Figure P8.73.

Ans. $-j$ A.

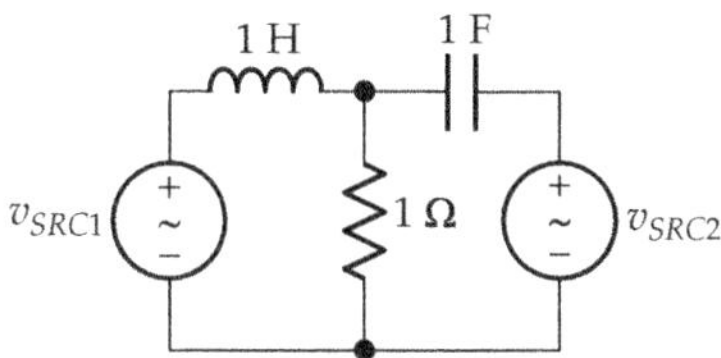

FIGURE P8.71

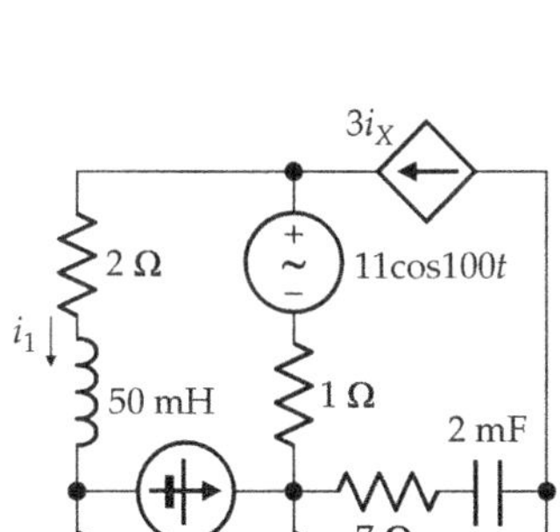

FIGURE P8.72

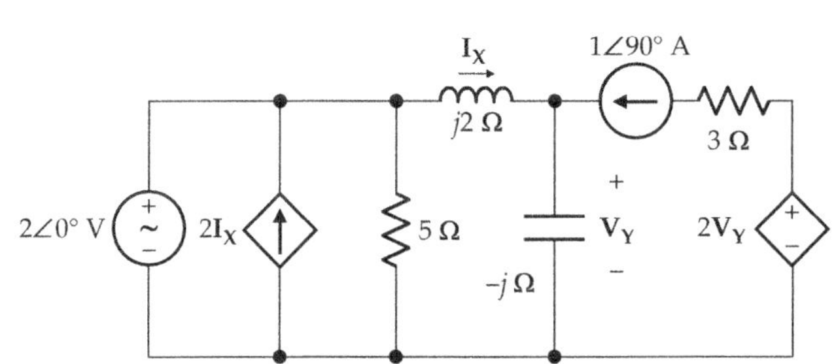

FIGURE P8.73

TEC and NEC

P8.74 Derive NEC looking into terminals 'ab' in Figure P8.74.
Ans. $\mathbf{I_N} = \mathbf{I_{ab}} = j3$ A, $Z_N = j10/3\ \Omega$.

P8.75 Derive TEC looking into terminals 'ab' in Figure P8.75.
Ans. $\mathbf{V_{Th}} = \mathbf{V_{ab}} = (-50 + j25)$ V, $Z_{Th} = (5 + j10)\ \Omega$.

P8.76 Derive TEC looking into terminals 'ab' in Figure P8.76.
Ans. $\mathbf{V_{Th}} = \mathbf{V_{ab}} = 57.35\angle -55.0°$ V, $Z_{Th} = (0.470 - j6.711)\ \Omega$.

P8.77 Derive TEC looking into terminals 'ab' in Figure P8.77.
Ans. $v_{Th}(t) = v_{ab} = 5\cos(10t)$ V in series with 0.1 Ω and 0.18 H.

P8.78 Derive NEC looking into terminals 'ab' in Figure P8.78.
Ans. $\mathbf{I_N} = \mathbf{I_{ab}} = 50\angle 0°$ A, $Z_N = (2 + j)\ \Omega$.

P8.79 Derive NEC looking into terminals 'ab' in Figure P8.79.
Ans. $\mathbf{I_N} = \mathbf{I_{ab}} = 1\angle 0°$ A, $Y_N = j2$ S.

P8.80 Derive TEC looking into terminals 'ab' in Figure P8.80.
Ans. $\mathbf{V_{Th}} = \mathbf{V_{ab}} = 12 + j6$ V, $Z_{Th} = 20\ \Omega$.

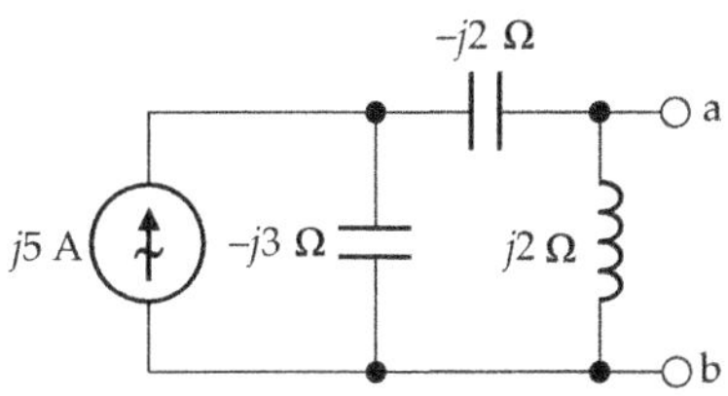

FIGURE P8.74

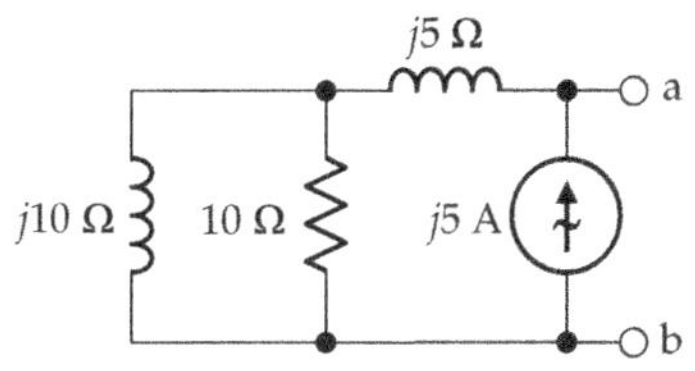

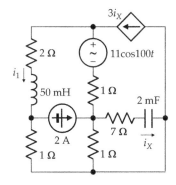

FIGURE P8.75

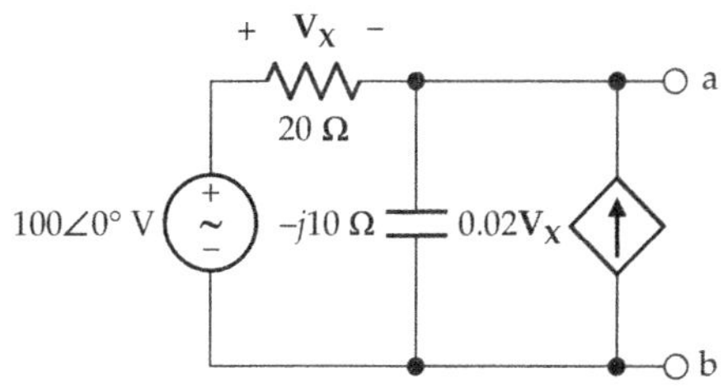

FIGURE P8.76

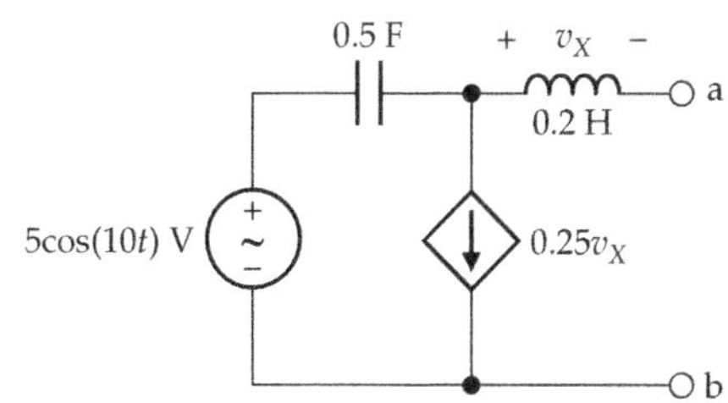

FIGURE P8.77

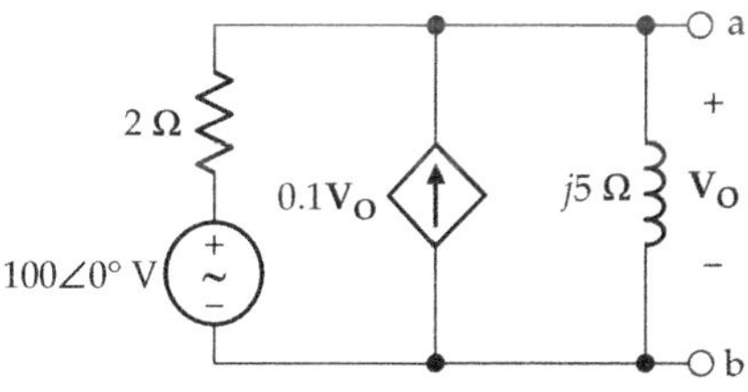

FIGURE P8.78

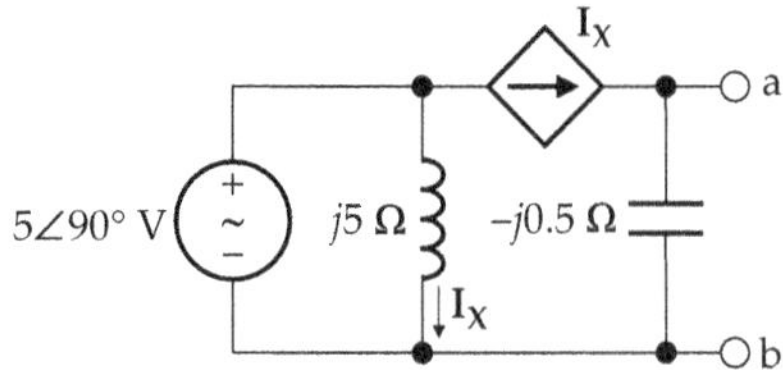

FIGURE P8.79

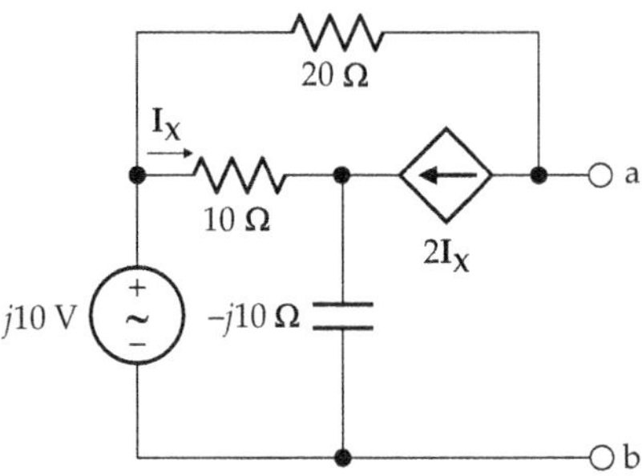

FIGURE P8.80

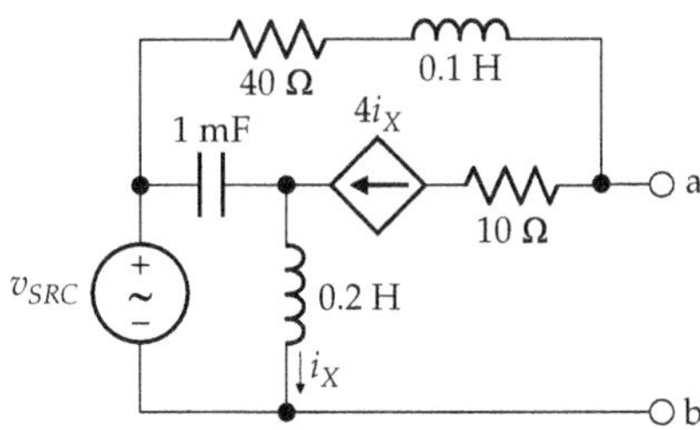

FIGURE P8.81

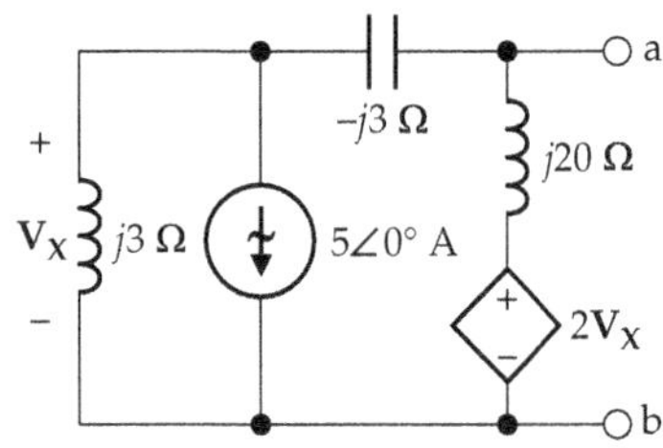

FIGURE P8.82

P8.81 Determine TEC looking into terminals 'ab' in Figure P8.81, assuming $v_{SRC}(t) = 10\cos 100t$ V.

Ans. $\mathbf{V_{Th}} = \mathbf{V_{ab}} = 2 + j32$ V; $Z_{Th} = 40 + j10\ \Omega$.

P8.82 Derive TEC looking into terminals 'ab' in Figure P8.82.

Ans. $\mathbf{V_{Th}} = \mathbf{V_{ab}} = -j15$ V, $Z_{Th} = 0$.

P8.83 Determine Z in Figure P8.83 so that $\mathbf{V_O}$ is in phase with the source voltage.

Ans. $j2.5\ \Omega$.

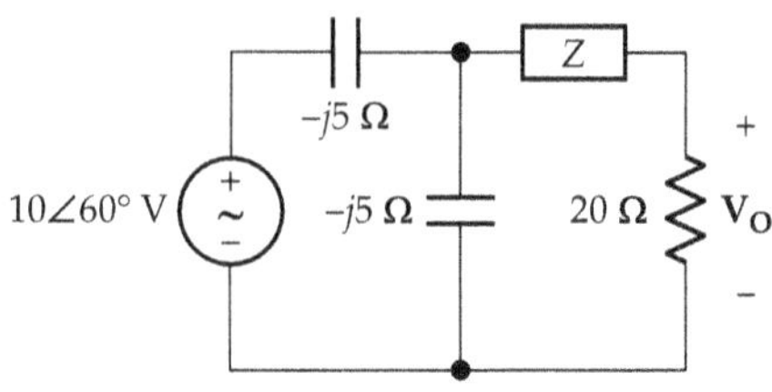

FIGURE P8.83

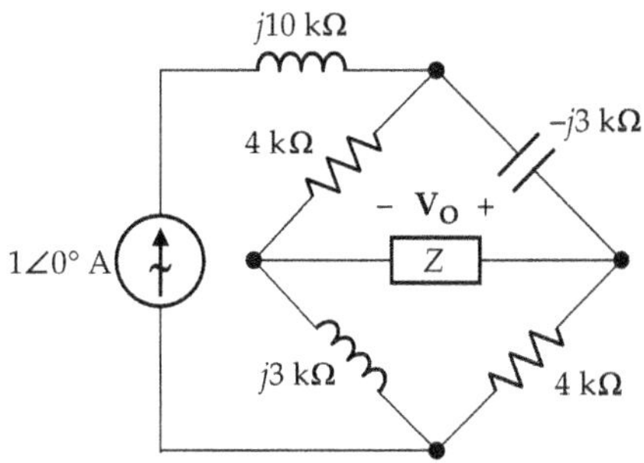

FIGURE P8.84

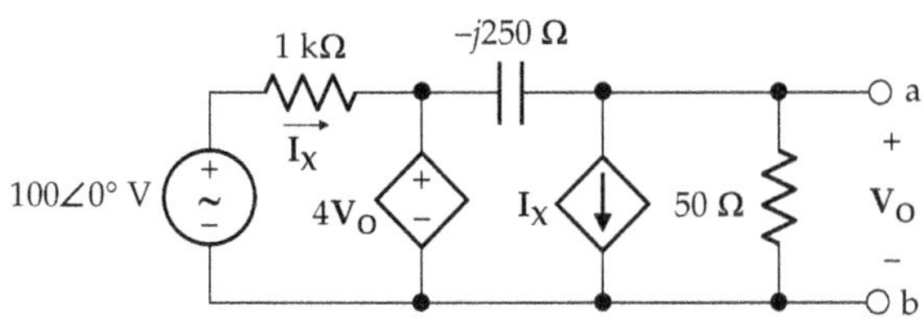

FIGURE P8.85

P8.84 Determine Z in Figure P8.84 so that $\mathbf{V_O} = 1\angle -90°$ V.

Ans. $-25(8 + j7)/113\ \Omega$.

P8.85 Derive NEC looking into terminals 'ab' in Figure P8.85.

Ans. $\mathbf{I_N} = \mathbf{I_{ab}} = -0.1$ A, $Y_N = (4 - j3)/250$ S.

P8.86 Derive NEC looking into terminals 'ab' in Figure P8.86. Ans. An ideal current source of $5/\sqrt{2}$ A.

Node-Voltage and Mesh-Current Methods

P8.87 (a) Given $i_{SRC}(t) = 10\cos 1000t$ mA in Figure P8.87, represent the circuit in the frequency domain; (b) derive the voltages indicated as phasors, using the node-voltage method; and (c) draw these voltages on a phasor diagram.

Ans. $\mathbf{V_1} = -500$ V, $\mathbf{V_2} = 400 + j200 = 447.2\angle 26.57°$ V, $\mathbf{V_S} = 900 + j200 = 922.0\angle 12.53°$ V, $\mathbf{V_C} = -400 - j200 = 447.2\angle 153.43°$ V, $\mathbf{V_L} = 800 + j400 = 894.4\angle 26.57°$ V.

P8.88 Determine $i_O(t)$ in Figure P8.88 using the mesh-current method, given $v_{SRC1}(t) = 10\cos(10^4t + 45°)$ V and $v_{SRC2}(t) = 10\cos(10^4t - 45°)$ V.

Ans. $0.133\cos(\omega t + 130.1°)$ A.

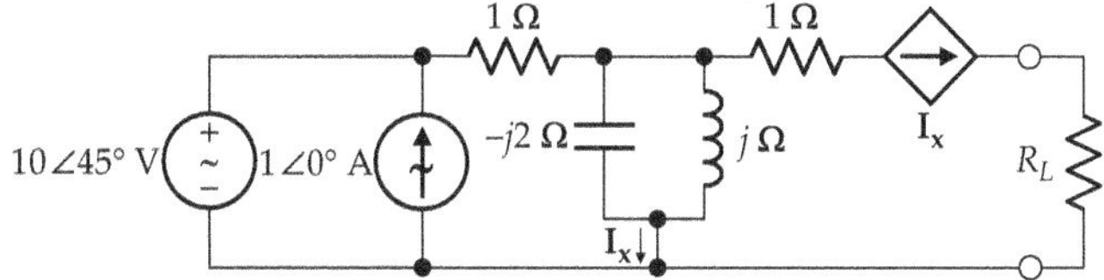

FIGURE P8.86

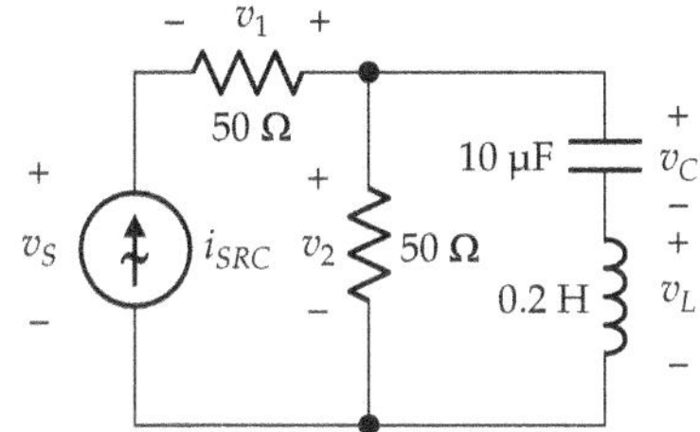

FIGURE P8.87

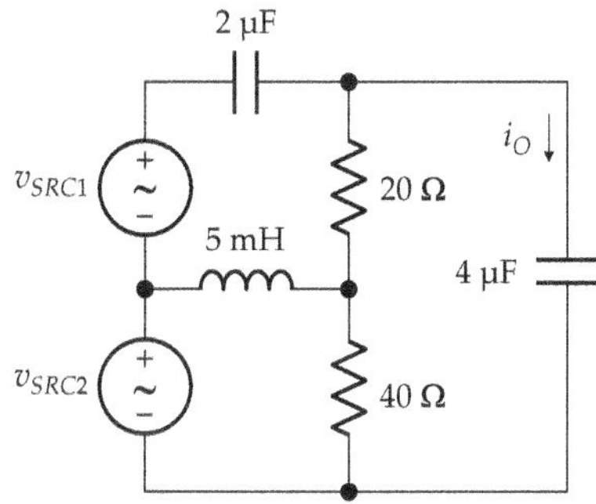

FIGURE P8.88

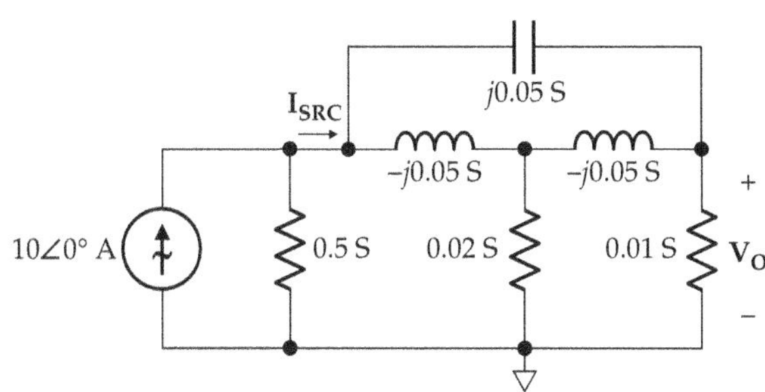

FIGURE P8.89

P8.89 Determine $\mathbf{I}_{SRC}$ and $\mathbf{V}_O$ in Figure P8.89 using the node-voltage method.

Ans. $\mathbf{V}_O = 13.4 + j11.7$ V, $\mathbf{I}_{SRC} = 0.469 + j0.164$ A.

P8.90 Determine $\mathbf{V}_O$ in Figure P8.90 using the mesh-current method.

Ans. $\mathbf{V}_O = 13.98 - j2.851$ V.

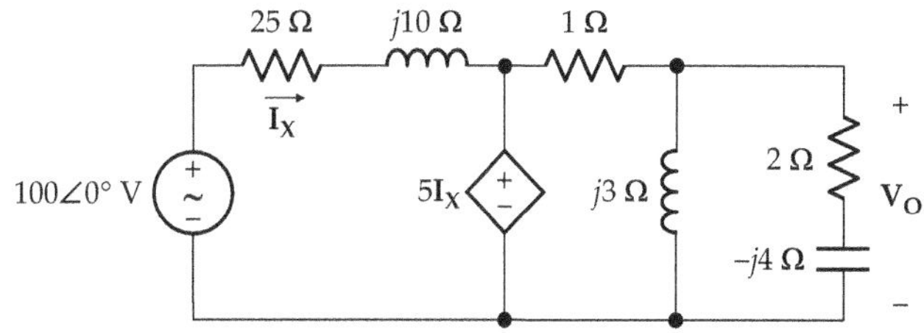

FIGURE P8.90

P8.91 Determine $\mathbf{V}_O$ in Figure P8.91 using the node-voltage method.

Ans. $\mathbf{V}_O = 4.88 - j20.0$ V.

P8.92 Determine $\mathbf{I}_O$ in Figure P8.92 using the mesh-current method.

Ans. $\mathbf{I}_O = 5 + j5$ A.

P8.93 Determine $v_C(t)$ in Figure P8.93 using the node-voltage method.

Ans. $12.1\sin(2000t + 6.86°)$ V.

P8.94 Determine $\mathbf{I}_O$ in Figure P8.94 using the mesh-current method.

Ans. $\mathbf{I}_O = 5 - j9$ A.

P8.95 Determine $\mathbf{V}_O$ in Figure P8.95 using the node-voltage method.

Ans. $\mathbf{V}_O = 1.82$ V.

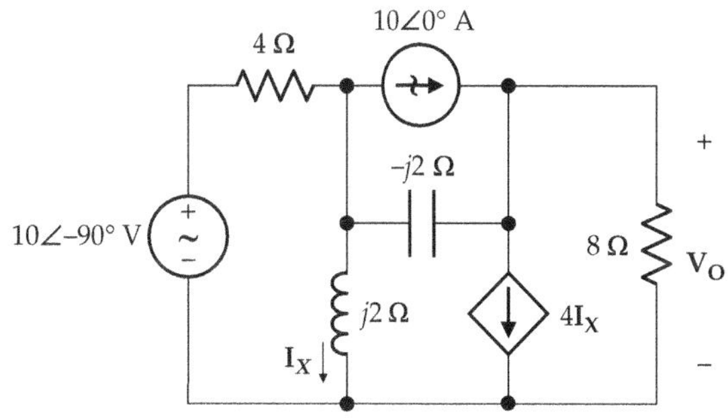

FIGURE P8.91

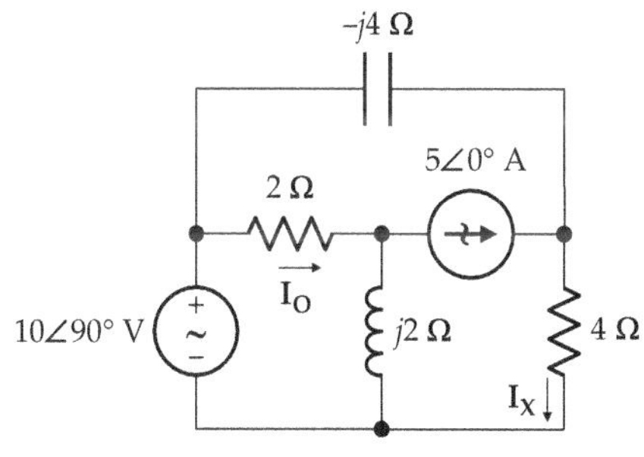

FIGURE P8.92

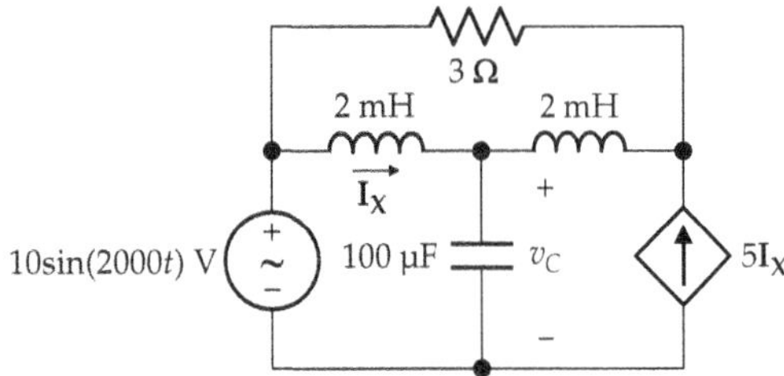

FIGURE P8.93

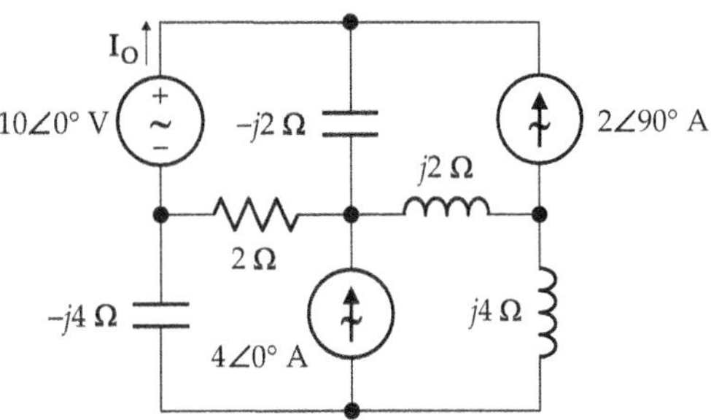

FIGURE P8.94

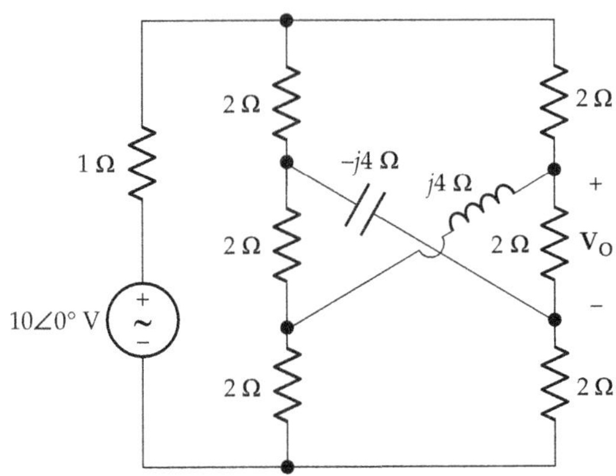

FIGURE P8.95

Probing Further

P8.96 Consider the ac bridge of Appendix 8A. Show that if the bridge is balanced and C is varied around balance, the phase angle changes by 180° in going through balance. Illustrate this using a phasor diagram. Note that the phase reversal around balance is true of all ac bridges.

9

Linear Transformer

Objective and Overview

This chapter introduces magnetic coupling between coils, whereby a time-varying magnetic field of a current-carrying coil induces a voltage in a nearby coil. Magnetic coupling is the basis for transformer action and plays an essential role in many types of devices, circuits, and systems.

The chapter begins with the basics of magnetic coupling, including the dot convention and the definitions of mutual inductance and the coupling coefficient. The general method for analyzing circuits involving mutual inductance is discussed, with emphasis on determining the polarity of the induced voltage. The T-equivalent circuit is presented, which eliminates the magnetic coupling and which can be used in analyzing most cases that are of practical interest.

9.1 Magnetic Coupling

When a current-carrying coil is in proximity to another coil, some of the flux due to the current-carrying coil links the other coil. If this flux is time varying, it induces a voltage in the other coil in accordance with Faraday's law. The two coils are said to be magnetically coupled. This is illustrated in Figure 9.1a for two coils in a medium of low permeability and in Figure 9.1b for two coils wound on a toroidal core of high permeability.

Suppose that coil 2 in Figure 9.1a and b is open-circuited. The flux linkage of coil 2 due to current in coil 1 is denoted by λ_{21}, where the first number in the subscript refers to the coil linked by the flux and the second number refers to the coil from which the flux originates. In the case of coils in a medium of low permeability (Figure 9.1a), an effective flux ϕ_{21eff} is defined, as discussed in connection with Figure 7.14, as the flux in coil 2 due to current in coil 1, which, if multiplied by N_2, the number of turns in coil 2, gives the true value of flux linkage in coil 2. That is, $\lambda_{21} = N_2\phi_{21eff}$.

In the case of coils on a high-permeability core (Figure 9.1b), and as explained in connection with Figure 7.15, ϕ_{21eff} is the same as the flux in the core ϕ_{21}. This same flux links every turn of the coil, so that $\lambda_{21} = N_2\phi_{21}$.

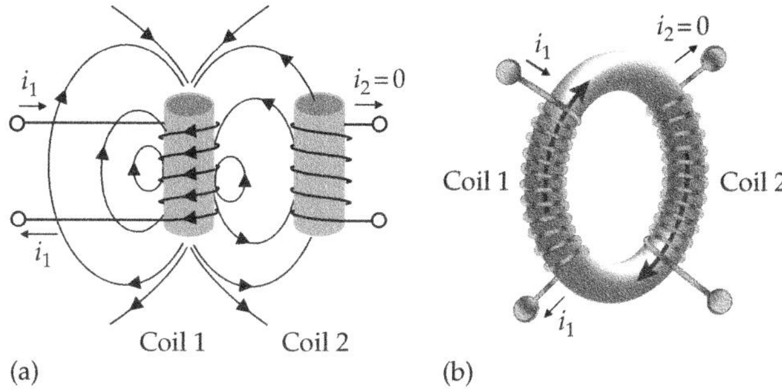

FIGURE 9.1
Magnetically coupled coils in air (a) or through a high-permeability toroidal core (b).

In both cases, Faraday's law can be expressed as

$$|v_{21}| = \left|\frac{d\lambda_{21}}{dt}\right| \tag{9.1}$$

where $|v_{21}|$ is the magnitude of the voltage induced in coil 2 due to current in coil 1. Thus, λ_{21} is well defined in terms of the time integral of v_{21}.

The polarity of v_{21} is determined by Lenz's law, just like the polarity of v_1 (Section 7.2), and could be deduced from the direction of current that flows if a resistor is connected between the coil terminals. However, an added consideration in this case is the relative sense of winding of the two coils, as will be explained with the aid of Figure 9.2. Note that the assigned positive polarity of v_{21} in Figure 9.2 is that of terminal 2 being positive with respect to terminal 2′.

In Figure 9.2a, the flux due to i_1 is directed downward in coil 1, in accordance with the right-hand rule (Figure 7.13), and is directed upward in coil 2. Assume that this flux is increasing with time; then according to Lenz's law, the polarity of the voltage induced in coil 2 is such that if current flows in coil 2 as a result of this induced voltage being applied across a resistor connected between terminals 2 and 2′, the flux in coil 2 due to this current is downward, so as to oppose the increasing flux in coil 2 due to i_1. According to the right-hand rule, i_2 flows out of terminal 2 of coil 2 so that the flux due to i_2 is downward (Figure 9.2b). Terminal 2 of coil 2 is then positive with respect to terminal 2′ to cause the flow of i_2 in the required direction. The induced voltage is of

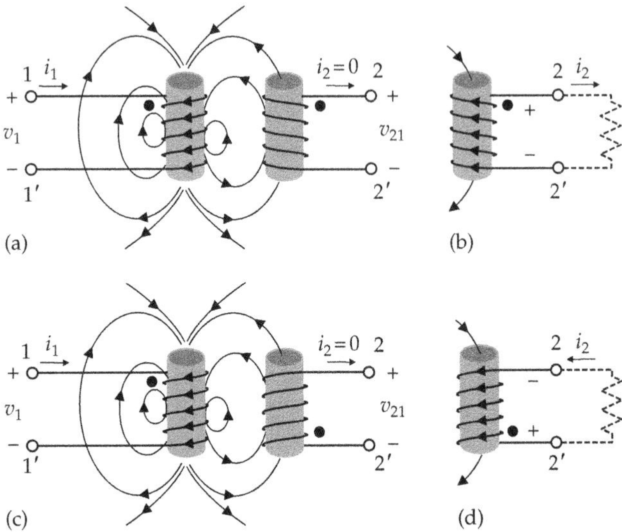

FIGURE 9.2
For the relative sense of winding in (a) the polarity of the induced voltage is as in (b), and for the relative sense of winding in (c) the polarity of the induced voltage is as in (d).

the same polarity as the assigned positive direction of v_{21} in Figure 9.2a, so the numerical value of v_{21} is positive.

Now consider Figure 9.2c, where the sense of winding of coil 2 has been reversed, while keeping the same sense of winding in coil 1. Thus, in proceeding from terminal 2 to terminal 2′ in Figure 9.2a, coil 2 is wound in the counterclockwise sense, when looking downward at the coil. On the other hand, coil 2 in Figure 9.2c is wound in the opposite sense, that is, clockwise, when looking downward at the coil, while proceeding from terminal 2 to terminal 2′. In order to have a downward flux in coil 2, i_2 must enter terminal 2. The induced voltage in coil 2 is now such that terminal 2′ is positive with respect to terminal 2, as shown in Figure 9.2d, so that the induced voltage applied across a resistor connected between terminals 2 and 2′ causes a flow of current in the required direction. This is opposite that of the assigned positive polarity of v_{21} in Figure 9.2c, so v_{21} will now have a negative numerical value.

9.1.1 Dot Convention

Having to show the sense of winding in a circuit diagram is rather awkward. It is avoided by using the **dot convention**. The basis for this convention can be readily understood with reference to Figure 9.2. Assuming that i_1 is increasing with time, the polarity of v_1 is such that terminal 1 is positive with respect to terminal 1′ so as to oppose the increase in current, as explained in Section 7.2. It was argued in connection with Figure 9.2a and b that when the flux linking coil 2 is increasing with time, terminal 2 is positive with respect to terminal 2′. *According to the dot convention, terminals that go positive together, or go negative together, on different coils are marked with a dot.* Thus, in Figure 9.2a and b, terminals 1 and 2 are marked with a dot as shown, or alternatively, terminals 1′ and 2′ could be so marked.

In Figure 9.2c and d, it was argued that when the flux linking coil 2 increases with time, terminal 2′ is positive with respect to terminal 2. According to the dot convention, therefore, terminals 1 and 2′ are marked with a dot as shown, or alternatively, terminals 1′ and 2 could be so marked.

The dot markings are further illustrated in Figure 9.3a and b for two coils coupled through a core of high permeability. Assuming i_1 increases with time, the polarity of v_1 is a voltage drop in the direction of i_1, as shown. In Figure 9.3a, the flux ϕ_{21} in the core due to i_1 is clockwise and increases with time. With the sense of winding of coil 2 as indicated, the induced voltage in coil 2 makes terminal 2 positive with respect to terminal 2′. This is because the current i_2 due to this induced voltage applied across a resistor connected between terminals 2 and 2′ leaves at terminal 2, thereby causing a counterclockwise flux ϕ_{12} that opposes ϕ_{21}. Hence, terminals 1 and 2 go positive together and are therefore marked with dots, as shown.

In Figure 9.3b, the sense of winding of coil 2 is reversed compared to coil 1. To have a flux ϕ_{12} that opposes ϕ_{21}, i_2 flows into terminal 2, which means that the induced voltage is such that terminal 2′ is positive with respect to terminal 2, so that if this voltage is applied across a resistor connected between terminals 2 and 2′, the resulting current is in the required direction. Terminals 1 and 2′ now go positive together and are therefore marked with dots.

It is seen that once terminals are marked with dots, there is no need to show the relative sense of winding

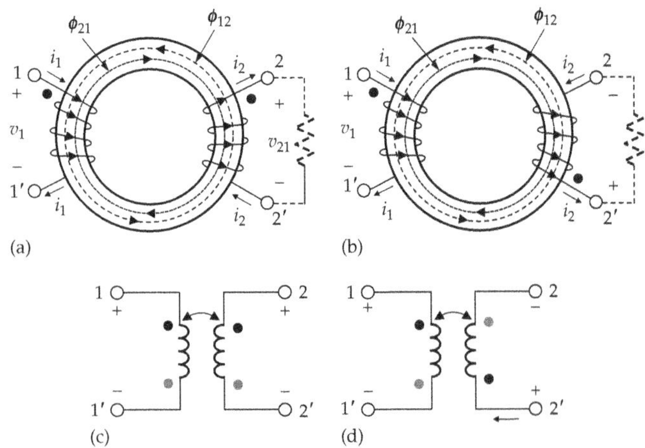

FIGURE 9.3
Dot marking to account for relative sense of winding. (a) Voltage polarities and directions of currents and fluxes when these are increasing with time, (b) same as (a) but with a reversed relative sense of winding of coil 2, and (c) and (d) show the circuit representation of the two coupled coils in (a) and (b), respectively.

of the coils. Magnetically coupled coils are drawn as in Figure 9.3c and d, with one terminal of each coil marked with a dot, according to the relative sense of winding. The fact that the two coils are magnetically coupled is indicated by an arc with two arrowheads at each end pointing to the two coils. It should be emphasized that there are only two relative senses of winding of the two coils, illustrated by the black dots in Figure 9.3c and d. The gray dots in each figure indicate the same relative sense of winding as the black dots in the same figure. Changing the relative sense of winding will henceforth be referred to as "reversal of the dot markings" from one configuration to the other, as between Figure 9.3c and d.

Although the argument used in Figures 9.2 and 9.3 concerning dot markings was based on the assumption that i_1 and its flux increase with time, the same conclusion is reached if these quantities were decreasing with time, rather than increasing. Figure 9.4a, for example, has the same sense of winding of both coils as in Figure 9.3a. If i_1 decreases with time, the induced voltage in coil 1 is now a voltage rise in the direction of i_1, as explained in Section 7.2. ϕ_{21}, and the flux linking coil 2 due to current in coil 1 is still clockwise around the core but decreases with time. According to Lenz's law, ϕ_{12}, the flux linking coil 1 due to current in coil 2 opposes the decrease of ϕ_{21} by adding to ϕ_{21} in the clockwise direction. i_2 now enters at terminal 2, so that the induced voltage reverses polarity, compared with Figure 9.3a in order to reverse the direction of i_2. But the coil terminals marked with dots now go negative together, so that the dot markings are still the same. A similar argument applies to Figure 9.4b in which the relative sense of winding is the same as in Figure 9.3b.

The current directions in Figure 9.4 provide an alternative interpretation of the dot markings, based on the directions of currents in the coils, relative to the dot markings, as opposed to the polarities of induced voltages, relative to the dot markings. This alternative interpretation is useful when writing the mesh-current equations or KVL involving magnetically coupled coils. The argument is as follows: In Figures 9.2 and 9.3, when i_1 enters at the dot-marked terminal and increases with time, which means that ϕ_{21} also increases with time, i_2 leaves the dot-marked terminal, so that its flux, ϕ_{12}, is in a direction that opposes ϕ_{21}. It follows that if i_2 enters the dot-marked terminal, instead of leaving this terminal, then ϕ_{12} is in the same direction as ϕ_{21}. This is illustrated in Figure 9.4 for i_1 decreasing with time. However, the direction of flux produced by a current in a coil is independent of whether the current is increasing or decreasing with time. Hence, the dots can also be interpreted on the basis that *if the currents of the two coils both enter, or both leave, the dotted terminals, the fluxes due to these currents are in the same direction in both coils*. In summary, the dot convention applies as follows:

Dot convention: *When the flux linking magnetically coupled coils changes with time, the polarities of the induced voltages are the same at the dot-marked terminals of each coil; that is, the dot-marked terminals will all become positive, or negative, together with respect to the unmarked terminal. As a corollary, fluxes due to currents both entering, or both leaving, the dot-marked terminals in each coil are in the same direction in both coils.*

Primal Exercise 9.1

Two magnetically coupled coils are connected as in Figure 9.5, where a voltage-indicating device is connected between terminals 22′ of coil 2. It is found that when the switch is closed, terminal 2′ goes positive with respect to terminal 2. (a) Specify the dot markings of the

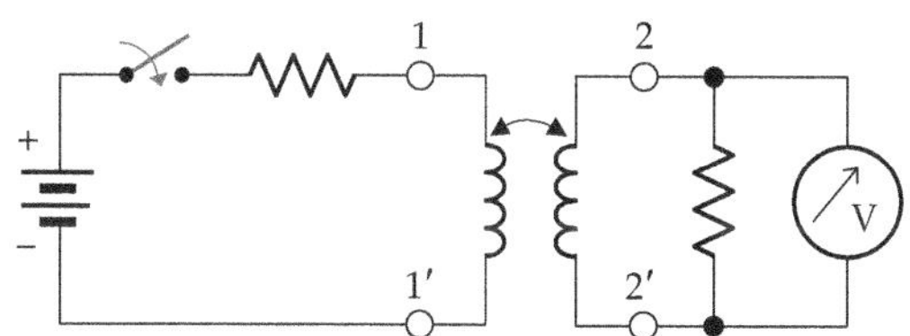

FIGURE 9.5
Figure for Primal Exercise 9.1.

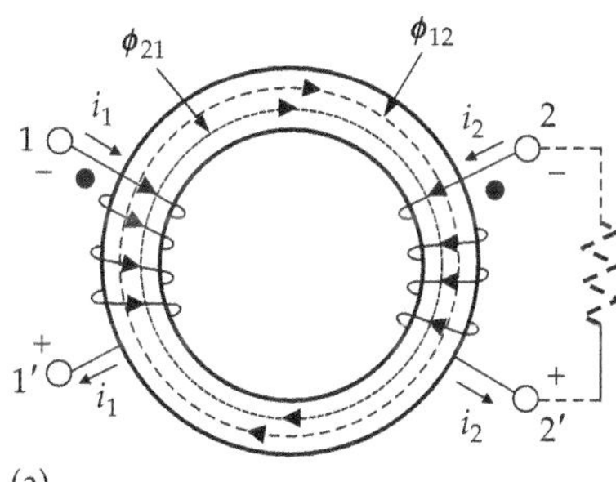

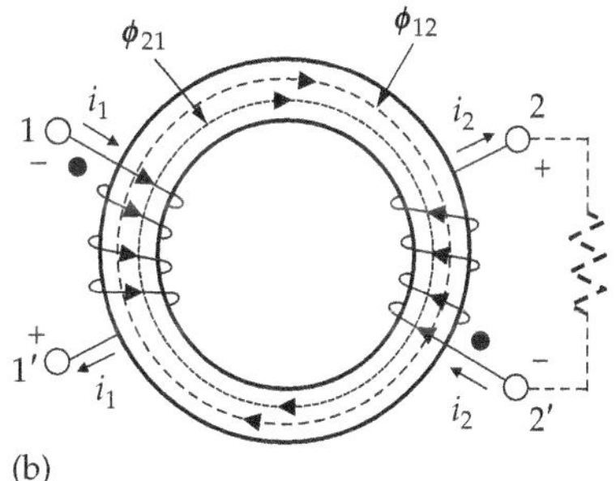

FIGURE 9.4
Relation of dot markings to current directions. (a) and (b) are the same coils in Figure 9.2a and b, respectively, but with currents and fluxes decreasing with time.

coils. (b) Which of the two terminals 2 or 2′ will go positive with respect to the other when the switch is opened?

Ans. (a) Terminals 1 and 2′ are marked with dots; (b) terminal 2 goes positive with respect to 2′.

Primal Exercise 9.2

The steady-state current in coil 2 is indicated in Figure 9.6. (a) Specify the dot markings of the coils based on the assigned positive directions of voltages across the coils; (b) verify the dot markings based on the directions of currents according to Lenz's law; (c) consider the source current in the time domain and verify the dot markings when the source current is increasing with time or is decreasing with time.

Ans. (a) Terminals 1 and 2′ are marked with dots; (b) current in coil 2 flows in the direction that opposes the flux due to the source current.

Primal Exercise 9.3

Two coils are wound on a high-permeability toroidal core of 2 cm^2 cross-sectional area, with coil 2 open-circuited (Figure 9.7). If the magnetic flux density B in the core, in the clockwise sense, due to current in coil 1, is decreasing at a constant rate of 0.1 Wb/cm^2/s, determine the magnitude and sign of v_2, assuming coil 2 has 10 turns.

Ans. −2 V.

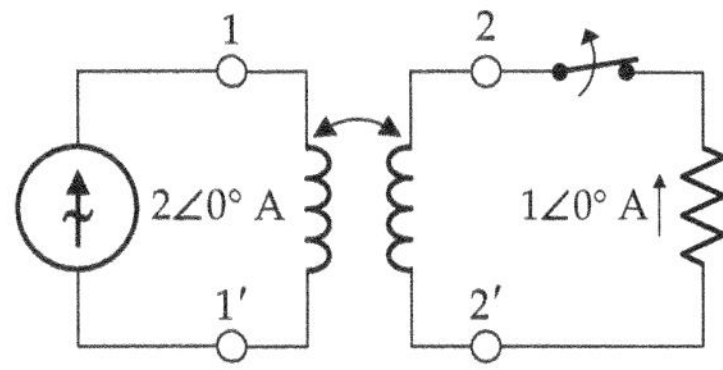

FIGURE 9.6
Figure for Primal Exercise 9.2.

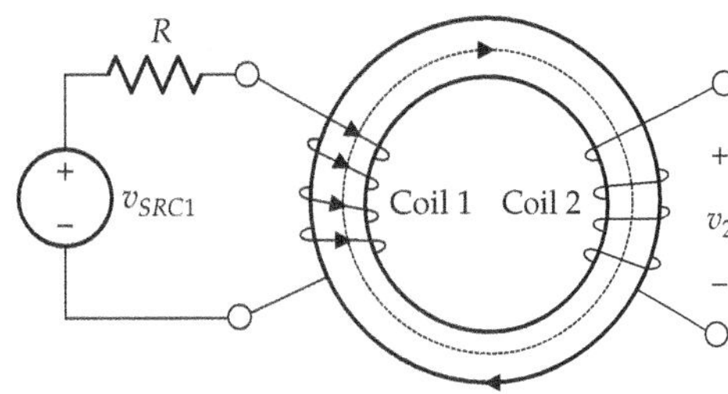

FIGURE 9.7
Figure for Primal Exercise 9.3.

9.2 Mutual Inductance

Faraday's law, in the form of Equation 9.1, expresses the magnitude of the voltage v_{21} induced in coil 2 by the time-varying flux linkage λ_{21} in coil 2 due to current in coil 1. If the coils are in a medium of low permeability, $\lambda_{21} = N_2\phi_{21eff}$. In the case of coils on a core of high permeability, $\lambda_{21} = N_2\phi_{21}$, where ϕ_{21} is in this case the core flux that links both coils. Since λ_{21} is due to i_1, it is desirable to express λ_{21} in terms of i_1. This is done by defining a quantity M_{21} as λ_{21} per unit current in coil 1. Thus,

$$M_{21} = \frac{\lambda_{21}}{i_1} \tag{9.2}$$

It follows that $\lambda_{21} = M_{21}i_1$. In a linear system, λ_{21} is directly proportional to i_1, so M_{21} is a constant. Faraday's law (Equation 9.1) becomes

$$|v_{21}| = \left|\frac{d\lambda_{21}}{dt}\right| = M_{21}\left|\frac{di_1}{dt}\right| \tag{9.3}$$

If coil 2 carries current, with coil 1 open-circuited, the same argument applies. Flux linkage λ_{12} is established in coil 1 due to i_2, where $\lambda_{12} = N_1\phi_{12eff}$ in the case of coils in a medium of low permeability and $\lambda_{12} = N_1\phi_{12}$ in the case of coils on a core of high permeability, where ϕ_{12} in this case is the core flux that links both coils. To express λ_{12} as a function of i_2, a quantity M_{12} is defined as λ_{12} per unit current in coil 2:

$$M_{12} = \frac{\lambda_{12}}{i_2} \tag{9.4}$$

where M_{12} is constant in a linear system. The magnitude of the voltage v_{12} induced in coil 1 is

$$|v_{12}| = \left|\frac{d\lambda_{12}}{dt}\right| = M_{12}\left|\frac{di_2}{dt}\right| \tag{9.5}$$

Equations 9.4 and 9.5 are the same as Equations 9.2 and 9.3, respectively, but with the subscripts '1' and '2' interchanged.

It is shown in Appendix 9A, based on conservation of energy, that

$$M_{12} = M_{21} = M \tag{9.6}$$

M is the **mutual inductance** between the two coils and is a constant in a linear system. It follows from Equations 9.2 and 9.4 that mutual inductance is defined as follows:

Definition: *The mutual inductance of two magnetically coupled coils is the flux linkage in one coil per unit current in the other coil. It is independent of which coil carries the current.*

That the flux linkage per unit current does not depend on which of two magnetically coupled coils carries the current is a consequence of M_{12} and M_{21} being equal. The definition of mutual inductance is seen to be a generalization of the definition of the inductance of a single coil, this being the flux linkage in the coil per unit current in the coil itself. In the case of two magnetically coupled coils, the mutual inductance is the flux linkage in one coil per unit current in the other coil. To distinguish the two types of inductance when both are present, the inductance L of a coil associated with current in the coil itself is referred to as the **self-inductance**. Just as L is an inherently positive quantity, so is M.

9.2.1 Coupling Coefficient

When two coils are magnetically coupled, not all the flux that links one coil links the other coil. This is particularly true for two coils in a nonmagnetic medium (Figure 9.8a). The flux due to i_1 in coil 1 extends in three dimensions around this coil but only a relatively small fraction of this three-dimensional flux links coil 2. Only this flux is shown in Figure 9.8a and in its two-dimensional representation in Figure 9.8b. In contrast, when the coils are coupled through a core of high permeability (Figure 9.9a), practically all the flux originating from one coil links the other coil.

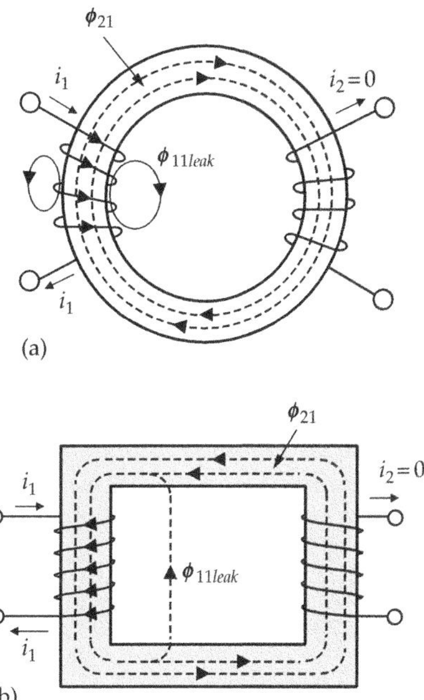

FIGURE 9.9
(a) Leakage flux of a current-carrying coil wound on a high-permeability core of toroidal shape (a) and rectangular shape (b).

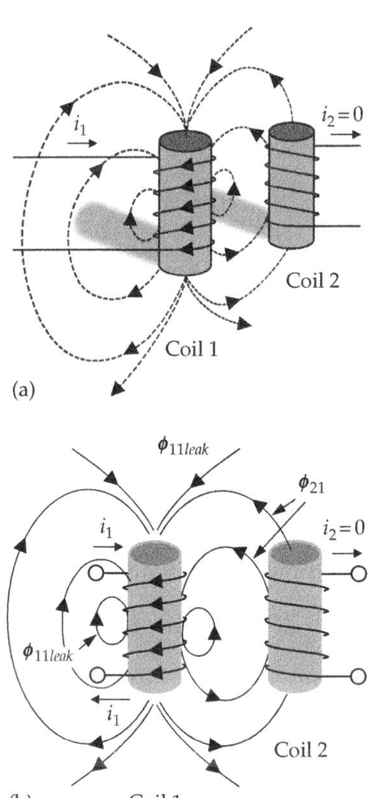

FIGURE 9.8
Flux due to one coil linking another coil in air. (a) The flux due to current in coil 1 extends in three dimensions in air around the coil, but only a small fraction of this flux links coil 2 (b).

The flux that links a current-carrying coil, but not another coil coupled to it, is the **leakage flux** of the current-carrying coil. It is seen that the leakage flux is relatively large in the case of coupling through a medium of low permeability (Figure 9.8a) and is relatively small for coupling through a core of high permeability (Figure 9.9a), where the leakage flux of coil 1 is denoted as ϕ_{11leak}. Even if the permeability of the core is high, the extent of the leakage flux depends on the geometry of the core and the coils. In the case of a rectangular core, for example, a leakage flux can occur between two opposite limbs of the core, as illustrated diagrammatically in Figure 9.9b; the lower the relative permeability of the core, the more significant is this leakage flux. A toroidal core, on the other hand, naturally conforms in shape to the flux path, so that this type of leakage flux is eliminated. However, flux leakage can still occur in the space between the winding and the core, as illustrated diagrammatically in Figure 9.9a. It is not possible to eliminate this space entirely by tightly winding the coil around the core, because of the finite thickness of the required insulation around the conducting wire of the coil and because of the finite diameter of the wire itself. Both of these factors imply that some leakage flux inevitably exists outside the core, including flux within the conducting wire itself.

For simplicity, and without loss of generality, the following discussion specifically refers to coils wound on a core of high permeability, as in Figure 9.9a. The degree of magnetic coupling between the two coils in Figure 9.9a is indicated by the coupling coefficient. From superposition, since the system is assumed to be linear, the total flux ϕ_{11} in coil 1 carrying a current i_1 is

$$\phi_{11} = \phi_{21} + \phi_{11eff(leak)} \tag{9.7}$$

where ϕ_{21} is the flux in coil 2 due to current in coil 1, which also links coil 1. When coil 2 does not carry current, ϕ_{21} is the flux in the core (Figure 9.9a). $\phi_{11eff(leak)}$ is the effective leakage flux of coil 1, which when multiplied by N_1 gives the true value of the flux linkage λ_{11leak} due to the leakage flux of coil 1. $\phi_{11eff(leak)}$ is used because all the leakage flux of coil 1 does not, in general, link all the turns of coil 1, as diagrammatically illustrated in Figures 9.1a and 9.2a.

The self-inductance of coil 1 is

$$L_1 = \frac{\lambda_{11}}{i_1} = \frac{N_1\left(\phi_{21} + \phi_{11eff(leak)}\right)}{i_1} \tag{9.8}$$

By definition, the mutual inductance M is the flux linkage in coil 2 per unit current in coil 1. From Equation 9.2, with $M_{21} = M$ and $\lambda_{21} = N_2\phi_{21}$,

$$M = \frac{\lambda_{21}}{i_1} = \frac{N_2\phi_{21}}{i_1} \tag{9.9}$$

Dividing Equation 9.9 by Equation 9.8 eliminates i_1:

$$\frac{M}{L_1} = \frac{N_2}{N_1}\frac{\phi_{21}}{\phi_{21} + \phi_{11eff(leak)}} \tag{9.10}$$

Similarly, if only coil 2 carries current,

$$\phi_{22} = \phi_{12} + \phi_{22eff(leak)} \tag{9.11}$$

where ϕ_{12} is the flux in coil 1 due to current in coil 2, which also links coil 2. When coil 1 does not carry current, ϕ_{12} is the flux in the core. $\phi_{22eff(leak)}$ is the effective leakage flux of coil 2, which when multiplied by N_2 gives the true value of the flux linkage λ_{22leak} of the leakage flux of coil 2.

The self-inductance of coil 2 is

$$L_2 = \frac{\lambda_{22}}{i_2} = \frac{N_2\left(\phi_{12} + \phi_{22eff(leak)}\right)}{i_2} \tag{9.12}$$

From Equation 9.2, with $M_{12} = M$ and $\lambda_{12} = N_1\phi_{12}$,

$$M = \frac{\lambda_{12}}{i_2} = \frac{N_1\phi_{12}}{i_2} \tag{9.13}$$

Dividing Equation 9.13 by Equation 9.12 eliminates i_2:

$$\frac{M}{L_2} = \frac{N_1}{N_2}\frac{\phi_{12}}{\phi_{12} + \phi_{22eff(leak)}} \tag{9.14}$$

Multiplying Equations 9.10 and 9.14 eliminates N_1 and N_2:

$$\frac{M^2}{L_1L_2} = k^2 = \frac{\phi_{21}}{\phi_{21} + \phi_{11eff(leak)}} \times \frac{\phi_{12}}{\phi_{12} + \phi_{22eff(leak)}} \tag{9.15}$$

where $k = M/\sqrt{L_1L_2}$ is the **coupling coefficient** and is a measure of how well the two coils are coupled together through the core. The first fraction, $\phi_{21}/(\phi_{21} + \phi_{11eff(leak)})$, is a measure of how well coil 1 is coupled to the core. This fraction is zero if there is no coupling ($\phi_{21} = 0$) and is unity if the coupling is perfect ($\phi_{11eff(leak)} = 0$). Similarly, the second fraction is a measure of how well coil 2 is coupled to the core. The product of the two fractions, k^2, is therefore a measure of how well the two coils are coupled together through the core. k assumes values between 0 and 1, where $k = 0$ denotes no coupling between the two coils. This occurs when $M = 0$, so that the core flux ϕ_{21} or ϕ_{12} is zero. $k = 1$ denotes perfect coupling, when neither coil has any leakage flux. The larger the value of k, the more "tightly" the coils are said to be coupled together.

For given L_1 and L_2, the largest value of M occurs when $k = 1$, and is $\sqrt{L_1L_2}$, the geometric mean of L_1 and L_2. The value of M is in this case intermediate between the values of L_1 and L_2. Clearly, k is inherently a positive quantity. However, PSpice allows a negative value of k in order to reverse the dot markings of the two coils, without changing their connections, as illustrated in Example 9.1.

Primal Exercise 9.4

Coil 1 of 100 turns and Coil 2 of 200 turns are wound on a high-permeability core. When coil 1 current is 2 A, with coil 2 open-circuited, the flux in the core is 3 mWb. Determine (a) the self-inductance of coil 1, assuming perfect coupling, and (b) the mutual inductance between the two coils.

Ans. (a) 150 mH; (b) 300 mH.

Primal Exercise 9.5

Two magnetically coupled coils have self-inductances of 2 and 8 mH, respectively, and a mutual inductance of 3 mH. Determine the coefficient of coupling between the two coils.

Ans. $k = 0.75$.

Primal Exercise 9.6

Given two identical coils wound on a core of high permeability, each of inductance 1 H, the coefficient of coupling being 0.5, determine the flux linkage in one coil when this coil is open-circuited and the current in the other coil is 1 A.

Ans. 0.5 Wb-T.

Primal Exercise 9.7

If $i_1(t) = I_m \cos\omega t$ A and the mutual inductance between coils 1 and 2 is M, (a) determine the magnitude of the voltage induced in coil 2. (b) Does this voltage depend on the current in coil 2?

Ans. (a) $\omega M I_m \sin\omega t$; (b) no.

Primal Exercise 9.8

Can the mutual inductance between two magnetically coupled coils be equal to the self-inductance of either coil?

Ans. If $L_1 = L_2 = L$, and $k = 1$, then $M = L$. If $L_1 \neq L_2$, M can be equal to the smaller of L_1 and L_2; it cannot be equal to the larger of L_1 and L_2, as this makes $k > 1$.

9.3 Linear Transformer

Definition: *A transformer consists of two or more coils that are magnetically coupled relatively tightly. In a linear transformer, permeability is constant, so that B and H, or λ and i, are linearly related.*

The magnetically coupled coils considered in Sections 9.1 and 9.2 are examples of a linear transformer for values of k that are not too small, say, larger than 0.1. *Linearity implies superposition of all magnetic variables, including fluxes.*

The general method of analyzing a circuit that includes a linear transformer is to apply KVL in the meshes that include the magnetically coupled coils. In writing the KVL equations, the assignment of positive directions of currents in the coils is either arbitrary or is dictated by convenience in analyzing the circuit. However, the assignment of positive directions of currents affects the sign of the Mdi/dt terms that account for the magnetic coupling. The question that we will address next is how to determine the sign of the Mdi/dt term.

Recall that in Section 7.2, the polarity of the induced voltage across a coil was argued from Lenz's law. Figure 7.26b is reproduced in Figure 9.10a, with the variables and parameters relabeled to conform to the present discussion on the two-coil, linear transformer. Note that v_{SRC1} and R_{src1} can represent, in general, TEC of the circuit connected to coil 1. From KVL,

$$v_{SRC1} - R_{src1} i_1 - v_{11} = 0 \tag{9.16}$$

or

$$i_1 = \frac{(v_{SRC1} - v_{11})}{R_{src1}} \tag{9.17}$$

The magnitude of v_{11} is, from Faraday's law,

$$|v_{11}| = \left|\frac{d\lambda_{21}}{dt}\right| = L_1 \left|\frac{di_1}{dt}\right| \tag{9.18}$$

As for the polarity of v_{11}, it was argued from Lenz's law that if, for the sake of argument, v_{SRC1} and i_1 increase with time, v_{11} opposes the increase in i_1 by having a positive value that subtracts from v_{SRC1} in Equation 9.17. The coil in Figure 9.10a is shown in Figure 9.10b as coil 1 of a linear transformer, with coil 2 open-circuited to begin with. The flux in the core due to i_1 is indicated as ϕ_{21} in the clockwise sense, in accordance with the right-hand rule.

Consider next that coil 1 is open-circuited and that coil 2 is connected to a circuit represented by its TEC consisting of v_{SRC2} and R_{src2}. If the current i_2 in coil 2 enters at terminal 2 as shown in Figure 9.10c, it produces a flux ϕ_{12} in the core in the clockwise sense, the same as ϕ_{21} in Figure 9.10b. As ϕ_{12} changes with time, the magnitude of the voltage v_{12} induced in coil 1 is given by Faraday's law as $|v_{12}| = |d\lambda_{12}/dt|$. Using Equation 9.13,

$$|v_{12}| = \left|\frac{d\lambda_{12}}{dt}\right| = M\left|\frac{di_2}{dt}\right| \tag{9.19}$$

What about the polarity of v_{12}? It is to be expected that since ϕ_{12} in Figure 9.10c is in the same direction as ϕ_{21} in Figure 9.10b, v_{12} is of the same polarity as v_{11}. According to Lenz's law, an increase in the flux in the core in the clockwise direction, whether due to an increase in ϕ_{21} or an increase in ϕ_{12}, results in an induced voltage of the same polarity. In other words, the polarity of the

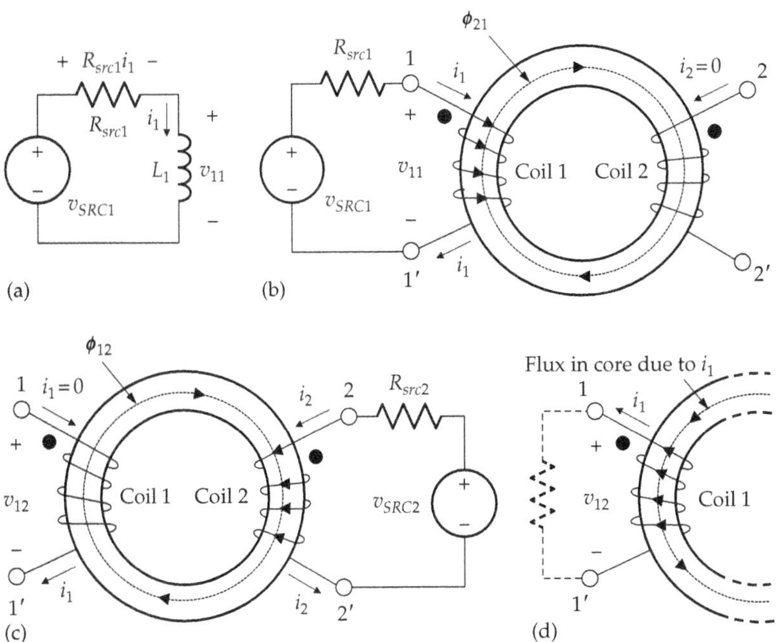

FIGURE 9.10
Polarities of induced voltages. (a) Excitation applied to a coil in series with a resistance, (b) the coil being magnetically coupled to another coil, (c) excitation applied to coil 2, with coil 1 open circuited, and (d) flux in the core if i_1 is allowed to flow under the influence of the induced voltage v_{12}.

voltage induced in coil 1 depends on whether the flux in the core is increasing or decreasing in the same direction in the core, irrespective of whether this flux is due to i_1 or i_2. This can be ascertained by imagining a resistor connected across coil 1, as in Figure 9.10d. If ϕ_{12} is increasing, a current i_1 will flow out of terminal 1, producing a flux in the core in the counterclockwise sense, as shown in Figure 9.10d, in opposition to ϕ_{12}. For i_1 to flow out of terminal 1, v_{12} should have the same polarity as v_{11}.

If both i_1 and i_2 are present at the same time, superposition applies. The total flux in the core is $(\phi_{21} + \phi_{12})$. The total induced voltage in coil 1 is $(L_1di_1/dt + Mdi_2/dt)$, where the sign of the Mdi_2/dt term is the same as that of the L_1di_1/dt, since v_{12} is of the same polarity as v_{11}, as argued in the preceding paragraph. KVL for the mesh that includes the source and coil 1 (Equation 9.16) becomes:

$$v_{SRC1} - R_{src}i_1 - \left(L_1\frac{di_1}{dt} + M\frac{di_2}{dt}\right) = 0 \qquad (9.20)$$

If the assigned positive direction of i_2 is reversed (Figure 9.11a), the direction of ϕ_{12} reverses, becoming counterclockwise. If, for the sake of argument, ϕ_{12} increases with time, then according to Lenz's law, v_{12} reverses polarity (Figure 9.11b), so that if i_1 flows due to v_{12}, it will produce a flux in the core in the clockwise sense so as to oppose the increase in ϕ_{12}. If both i_1 and i_2 are present at the same time, superposition applies. The total flux in the core is now $(\phi_{21} - \phi_{12})$. The total induced voltage in coil 1 is $(L_1di_1/dt - Mdi_2/dt)$, where the sign of the Mdi_2/dt term is now opposite that of the L_1di_1/dt, since the polarity of v_{12} is opposite that of v_{11}. KVL for the mesh that includes the source and coil 1 (Equation 9.16) becomes

$$v_{SRC1} - R_{src}i_1 - \left(L_1\frac{di_1}{dt} - M\frac{di_2}{dt}\right) = 0 \qquad (9.21)$$

Suppose, next, that the sense of winding of coil 2 is reversed, so that the dot marking on this coil is reversed (Figure 9.11c), while keeping the same assigned positive direction of i_2 as in Figure 9.11a. ϕ_{12} is in the clockwise sense, as in Figure 9.10c, and the same argument applies. If both i_1 and i_2 are present, ϕ_{12} adds to ϕ_{21}, so that Equation 9.20 applies. If the assigned positive direction of i_2 is reversed, while keeping the same dot markings as in Figure 9.11c, ϕ_{12} is now counterclockwise, opposing ϕ_{21} so that Equation 9.21 applies.

The foregoing is summarized by noting that if the assigned positive directions of i_1 and i_2 and the dot markings are such that *the fluxes in the core due to i_1 and i_2 are additive, the sign of the Mdi_2/dt term is the same as that of the L_1di_1/dt term* (Equation 9.20). On the other hand, if the assigned positive directions of i_1 and i_2 and the dot markings are such that *the fluxes in the core due to i_1 and i_2 are in opposition, the sign of the Mdi_2/dt term is opposite that of the L_1di_1/dt term* (Equation 9.21). Moreover, if coils

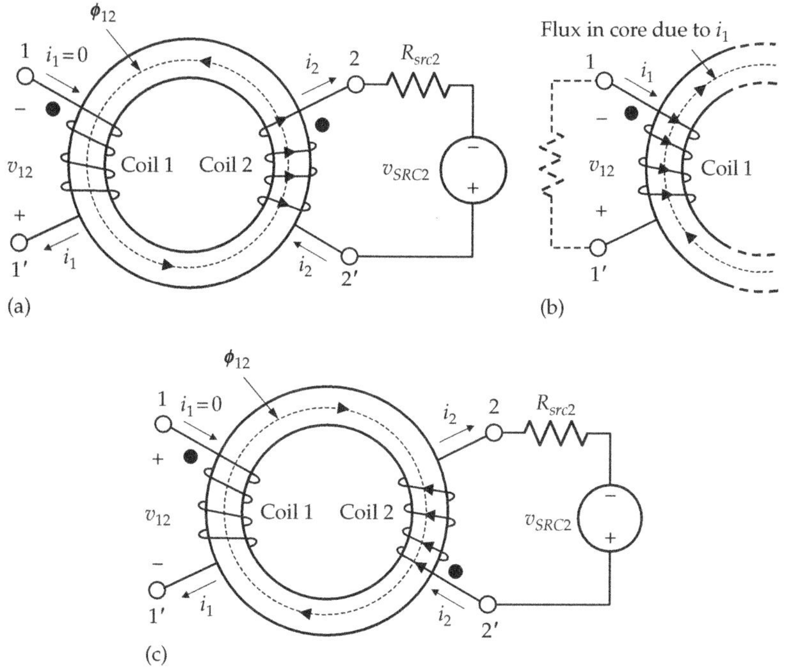

FIGURE 9.11
(a) Excitation applied to coil 2 with the reversed direction of i_2, (b) flux in the core if i_1 is allowed to flow under the influence of the induced voltage v_{12}, and (c) reversed sense of winding of coil 2.

1 and 2 are interchanged, the same argument applies to the polarities of the voltages induced in coil 2 due to i_1 and i_2. The subscripts 1 and 2 are simply interchanged in the preceding relations.

According to the dot convention, the fluxes due to i_1 and i_2 are in the same direction in the core if i_1 and i_2 both enter, or both leave, the dot-marked terminals of their respective coils. Otherwise, the fluxes in the core will be in opposition. The sign of the Mdi/dt term can therefore be determined in a given circuit as follows:

Sign of *Mdi/dt* term: *If the assigned positive directions of currents in the two coils are such that these currents both flow in, or both flow out, at the dot-marked terminals, the sign of the mutual inductance term (Mdi_1/dt, or Mdi_2/dt) for either coil is the same as that of the self-inductance term for that coil (L_1di_1/dt, or L_2di_2/dt, respectively). Otherwise, the sign of the mutual inductance term for either coil is opposite that of the self-inductance term for that coil.*

Example 9.1: Equivalent Inductances of Series-Connected Coupled Coils

Suppose that the two coils of Figure 9.10 are connected in series, as in Figure 9.12, where terminal 1′ of coil 1 is connected to terminal 2 of coil 2 in (a) and to terminal 2′ of coil 2 in (b). These connections reverse the dot markings, assuming the two coils are wound in the same sense in both cases. It is required to determine the induced voltages and the total inductance in each case.

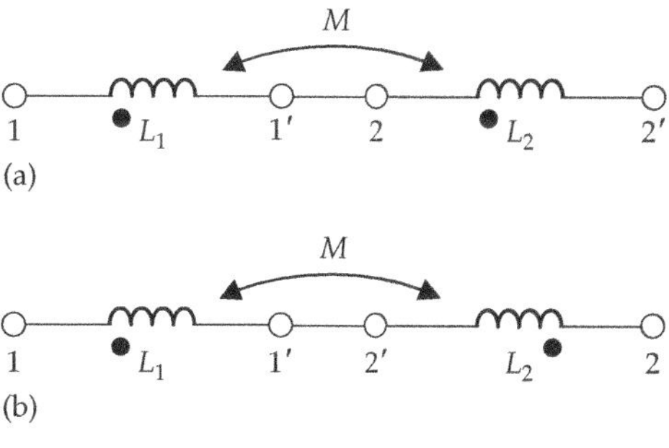

FIGURE 9.12
Figure for Example 9.1.

Solution:

(a) When a current i flows in the coils, there is an induced voltage in each coil due to the self-inductance of the coil, irrespective of whether or not the two coils are magnetically coupled. Assuming that $di/dt > 0$, the induced voltages L_1di/dt and L_2di/dt will have the polarities shown in Figure 9.13a, as voltage drops in the direction of current. Because of the magnetic coupling, i flowing in coil 1 induces a voltage Mdi/dt in coil 2, and i flowing in coil 2 induces a voltage Mdi/dt

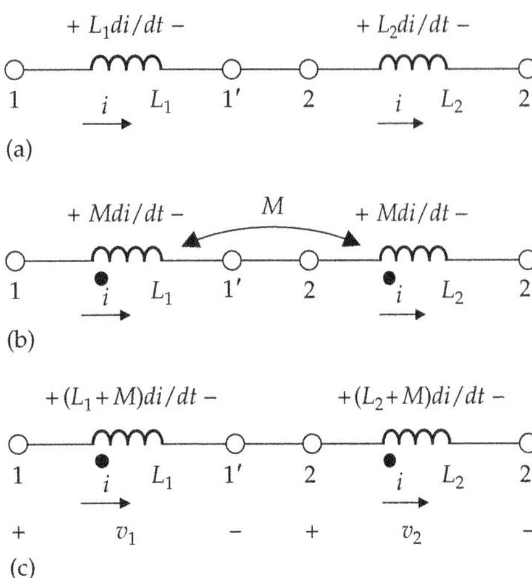

FIGURE 9.13
Figure for Example 9.1.

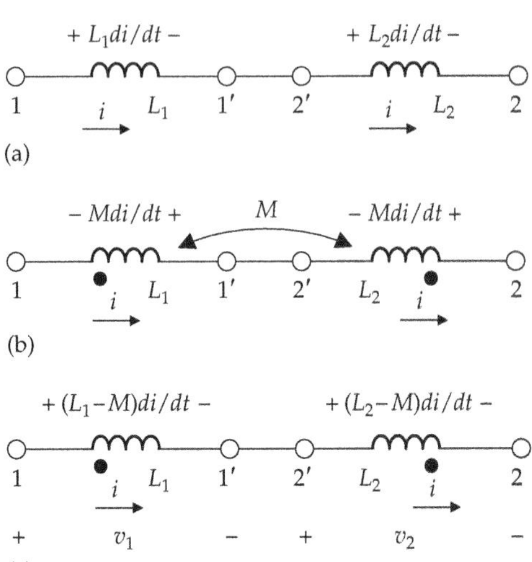

FIGURE 9.14
Figure for Example 9.1.

in coil 1. i enters the two coils at the dot-marked terminals in Figure 9.10, so that the flux due to i in either coil adds to the flux due to i in the other coil. Thus, ϕ_{12} and ϕ_{21} in Figure 9.10 add to one another. As argued previously, the sign of the Mdi/dt term is the same as the Ldi/dt term in each coil, as illustrated in Figure 9.13b. The total induced voltages v_1 and v_2 in coils 1 and 2 are therefore

$$\text{Coil 1:}\quad v_1 = L_1\frac{di}{dt} + M\frac{di}{dt} = (L_1 + M)\frac{di}{dt} \tag{9.22}$$

$$\text{Coil 2:}\quad v_2 = L_2\frac{di}{dt} + M\frac{di}{dt} = (L_2 + M)\frac{di}{dt} \tag{9.23}$$

as shown in Figure 9.13c. The effective inductance of coil 1 is therefore $(L_1 + M)$ and that of coil 2 is $(L_2 + M)$. The total voltage across the two coils is

$$v_1 + v_2 = (L_1 + L_2 + 2M)\frac{di}{dt} \tag{9.24}$$

The equivalent inductance is therefore

$$L_{eq} = L_1 + L_2 + 2M \tag{9.25}$$

(b) In the case of Figure 9.12b, i induces voltages L_1di/dt and L_2di/dt as voltage drops in the direction of i as before (Figure 9.14a). But i now enters coil 1 at the dot-marked terminal and coil 2 at the unmarked terminal, which means that the flux due to i in either coil opposes the flux due to i in the other coil. Thus, ϕ_{12} and ϕ_{21} are in opposition, as in Figure 9.11, and the sign of the Mdi/dt term is now opposite that of the Ldi/dt term in each coil, as illustrated in Figure 9.14b. The total induced voltages v_1 and v_2 in coils 1 and 2 are

$$\text{Coil 1:}\quad v_1 = L_1\frac{di}{dt} - M\frac{di}{dt} = (L_1 - M)\frac{di}{dt} \tag{9.26}$$

$$\text{Coil 2:}\quad v_2 = L_2\frac{di}{dt} - M\frac{di}{dt} = (L_2 - M)\frac{di}{dt} \tag{9.27}$$

as shown in Figure 9.14c. The total voltage across the two coils is

$$v_1 + v_2 = (L_1 + L_2 - 2M)\frac{di}{dt} \tag{9.28}$$

The equivalent inductance is therefore

$$L_{eq} = L_1 + L_2 - 2M \tag{9.29}$$

The effective inductances of coils 1 and 2 are $(L_1 - M)$ and $(L_2 - M)$, respectively. The following should be noted for the case of Figure 9.14:

1. If the value of M lies between L_1 and L_2, then one of the inductances, $(L_1 - M)$ or $(L_2 - M)$, will be negative. A negative inductance will have a negative reactance, just like capacitance, but with a magnitude that is directly proportional to frequency.
2. If $k = 1$ and $L_1 = L_2 = L$, then $M = L$, and $L_{eq} = 0$. The two inductances in series behave as a short circuit, because the induced voltage in each coil due to the magnetic coupling cancels out the induced voltage due to the self-inductance.

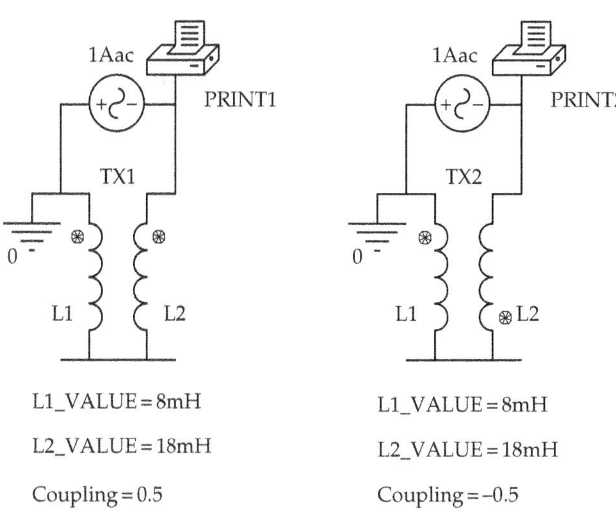

FIGURE 9.15
Figure for Example 9.1.

Simulation: It is required to verify Equations 9.25 and 9.29 for two magnetically coupled coils having inductances of 8 and 18 mH, with $k = 0.5$. The magnetically coupled coils are entered as the part XFRM_LINEAR from the ANALOG library. As entered (Figure 9.15), the coil on the LHS is L1, that on the RHS is L2, and the default dot markings are at the upper terminals of the two coils. The coil designations and the dot markings have been added in Figure 9.15 for clarity. The values of the inductances and the coupling coefficient are entered in the Property Editor spreadsheet. The coils are connected in series by joining the two lower terminals together. The left part of Figure 9.15 corresponds to Figure 9.14. To reverse the connections to one of the coils, PSpice allows changing the sign of the coupling coefficient instead of changing the wiring between the coils. With the coupling coefficient set to –0.5 in the right part of Figure 9.15, the connection corresponds to that of Figure 9.13.

To determine L_{eq}, a IAC source is used at a frequency of 1 rad/s. The voltage across the source is then numerically equal to the impedance L_{eq}, read by the voltage printer. In the Simulation Settings, 'Analysis type' is 'AC Sweep/Noise', 'Start Frequency' and 'End Frequency' are 0.159155, and 'Points/Decade' is 1. After the simulation is run, PRINT1 reads 0.014 V at a phase angle of 90°, corresponding to $L_{eq} = 14$ mH, and PRINT2 reads 0.038 V at a phase angle of 90°, corresponding to $L_{eq} = 38$ mH. Alternatively, the voltage magnitude and phase can be read using 'Evaluate Measurements' and selecting 'Analog Operators and Functions', as explained in Example 8.3.

An alternative method of implementing magnetic coupling between coils in PSpice, using the part K_Linear, is presented in Example 9.2.

Primal Exercise 9.9

Two magnetically coupled coils have self-inductances of $L_1 = 8$ mH and $L_2 = 18$ mH, with $k = 5/6$. Determine L_{eq}, and the effective inductance of each coil (a) when the coils are connected in series so as to give the larger L_{eq} and (b) when the connections to one of the coils are reversed.

Ans. $M = 10$ mH; (a) 46 mH, 18 mH, 28 mH; (b) 6 mH, −2 mH, 8 mH.

Primal Exercise 9.10

Given two magnetically coupled coils and an inductance measuring instrument, how would you determine the mutual inductance and mark the terminals of the coils with dots?

Ans. By measuring the total inductance, before and after reversal of connections between the coils.

Example 9.2: Analysis of Circuit Having Coupled Coils

Given the circuit of Figure 9.16 in which $v_{SRC}(t) = 100\cos 800t$ V and $k = 0.25$, it is required to determine v_O in the steady state.

Solution:

$\omega L_1 = 800 \times 10^{-2} = 8\ \Omega$; $\omega L_2 = 800 \times 4 \times 10^{-2} = 32\ \Omega$; $\omega M = k\sqrt{(\omega L_1)(\omega L_2)} = 0.25\sqrt{8 \times 32}\ \Omega = 4\ \Omega$; and $\dfrac{1}{\omega C} = \dfrac{1}{800 \times 0.25 \times 10^{-3}} = 5\ \Omega$. The circuit in the frequency domain is shown in Figure 9.17.

In writing the mesh-current equation for mesh 1, the total voltage drop in this mesh due to $\mathbf{I}_1$ equals, as usual, $\mathbf{I}_1$ multiplied by the total self-impedance of this mesh, that is, $(10 + j8 - j5)\mathbf{I}_1$. $\mathbf{I}_2$ introduces as usual a voltage rise of $Z_c\mathbf{I}_2$ in mesh 1, where $Z_c = -j5\ \Omega$ is the

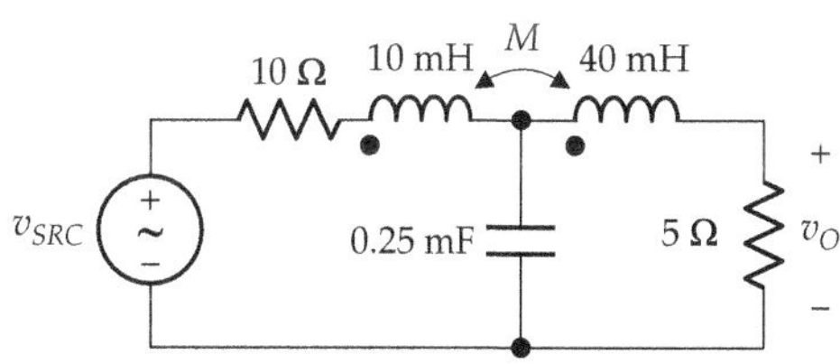

FIGURE 9.16
Figure for Example 9.2.

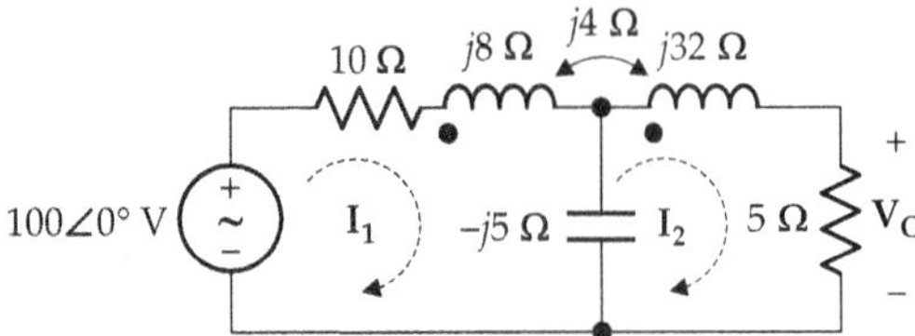

FIGURE 9.17
Figure for Example 9.2.

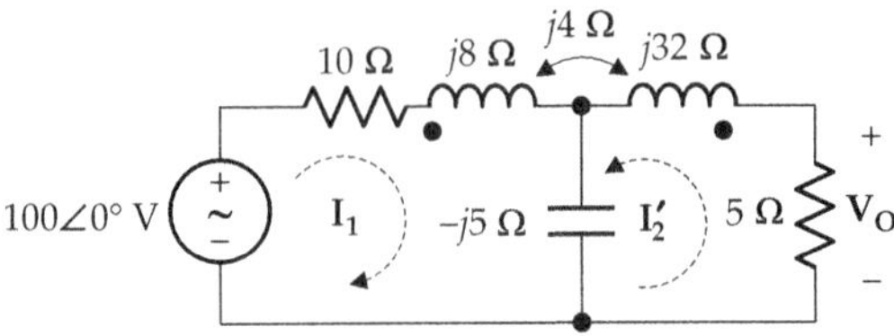

FIGURE 9.19
Figure for Example 9.2.

common impedance between meshes 1 and 2. As a voltage rise in mesh 1, the sign of this term is negative, so that this term becomes $-(-j5)\mathbf{I}_2$. In addition, there is a $j\omega M\mathbf{I}_2$ term due to the magnetic coupling between the coils in the two meshes. Since both $\mathbf{I}_1$ and $\mathbf{I}_2$ enter at the dotted terminals, the sign of the $j\omega M\mathbf{I}_2$ term is positive, the same as that of $j\omega L_1\mathbf{I}_1$ term in the equation of mesh 1, which is $+j8\mathbf{I}_1$. The mesh-current equation for mesh 1 is therefore

$$(10+j8-j5)\mathbf{I}_1-(-j4-j5)\mathbf{I}_2=100 \qquad (9.30)$$

Considering mesh 2, the self-impedance of the mesh is $(5 + j32 - j5)$. The current $\mathbf{I}_1$ introduces a mutual impedance term $-(-j5)\mathbf{I}_1$ in the mesh and a $j\omega M\mathbf{I}_1$ term whose sign is positive, the same as that of the $j\omega L_2\mathbf{I}_2$ term in the equation of mesh 2, which is $+j32\mathbf{I}_2$. The mesh-current equation for mesh 2 is

$$-(-j4-j5)\mathbf{I}_1+(5+j32-j5)\mathbf{I}_2=0 \qquad (9.31)$$

Solving for $\mathbf{I}_2$ gives $\mathbf{I}_2 = -3.0636 - j0.5375$ A, so that $\mathbf{V}_O = 5\mathbf{I}_2 = -15.3 - j2.69 = 15.55\angle{-170.0}°$ V, or $v_O(t) = 15.55\cos(800t - 170.0°)$ V.

If the assigned positive direction of $\mathbf{I}_2$ in Figure 9.17 is reversed and this current is denoted by $\mathbf{I}_2'$ (Figure 9.18), then $\mathbf{I}_2'$ flows into the unmarked terminal of coil 2. The sign of the mutual inductance term is now opposite that of the self-inductance term for each coil. The $Z_c\mathbf{I}_2'$ term in the mutual impedance between the two meshes becomes positive since the voltage drop due to $\mathbf{I}_2'$ flowing in Z_c is also a voltage drop in mesh 1. Equations 9.30 and 9.31 become

$$(10+j8-j5)\mathbf{I}_1+(-j4-j5)\mathbf{I}_2'=100 \qquad (9.32)$$

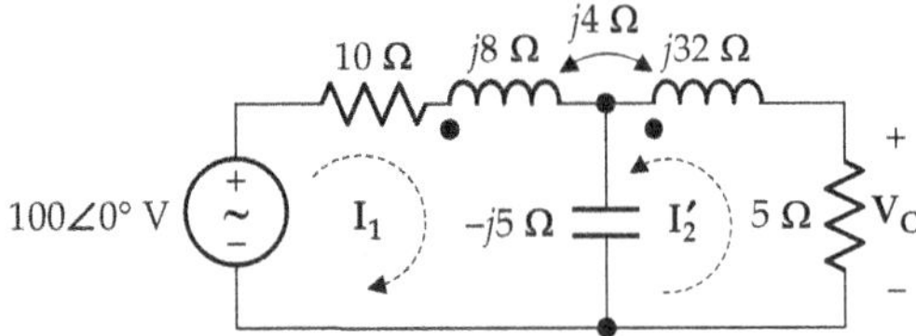

FIGURE 9.18
Figure for Example 9.2.

and

$$+(-j4-j5)\mathbf{I}_1+(5+j32-j5)\mathbf{I}_2'=0 \qquad (9.33)$$

Now $\mathbf{I}_2' = -\mathbf{I}_2$ and $\mathrm{V_O} = -R_2\mathbf{I}_2' = R_2\mathbf{I}_2$ as before.

If the dot marking on coil 2 is reversed (Figure 9.19), with $\mathbf{I}_2'$ still counterclockwise, then both currents now enter at the dot-marked terminals, and the sign of the mutual inductance term is again the same as that of the self-inductance terms. Equations 9.32 and 9.33 become

$$(10+j8-j5)\mathbf{I}_1+(j4-j5)\mathbf{I}_2'=100 \qquad (9.34)$$

and

$$(j4-j5)\mathbf{I}_1+(5+j32-j5)\mathbf{I}_2'=0 \qquad (9.35)$$

Simulation: The circuit of Figure 9.16 is entered as in Figure 9.20. In the Simulation Settings, 'Analysis type' is 'AC Sweep/Noise', 'Start Frequency' and 'End Frequency' are 127.324 ($800/2\pi$), and 'Points/Decade' is 1. After the simulation is run, $\mathbf{V}_O$ magnitude is read as 15.55 V and its phase angle as −170.0°, using Evaluate Measurements and selecting Analog Operators and Functions.

PSpice provides another method of implementing magnetic coupling between coils that can be generalized to coupling between more than two coils. The circuit is entered as in Figure 9.21, which is the given circuit of Figure 9.16, assuming no magnetic coupling between

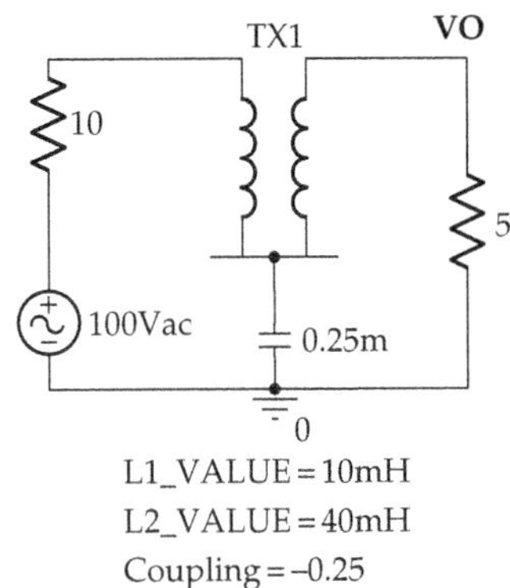

FIGURE 9.20
Figure for Example 9.2.

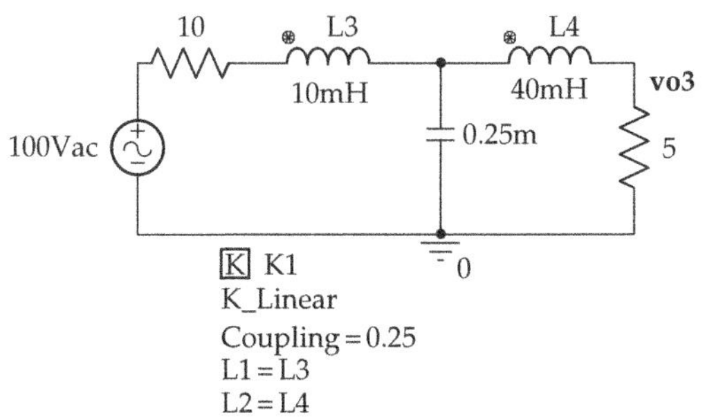

FIGURE 9.21
Figure for Example 9.2.

the coils. The two coils are arbitrarily labeled as L3 and L4, with the mark on the coils corresponding to the dots in Figure 9.16. The part K_Linear from the ANALOG library is entered on the schematic. When K is double-clicked, the Property Editor spreadsheet is displayed. Change the coupling coefficient to 0.25 and enter L3 as the value for L1 and L4 as the value for L2, as displayed in Figure 9.21. The simulation is run as before, and vo3 is read as having a magnitude of 15.55 V and a phase angle of −170.0°.

If, say, three magnetically coupled coils are to be coupled, the coils are entered as if there is no coupling between them, and with the coils oriented so that the marked coil terminals correspond to the dotted terminals in the circuit. The part K_Linear is entered three times, labeled as K1, K2, and K3. The coil part numbers, two at a time, are entered as L1 and L2 values in the spreadsheets, together with the corresponding coupling coefficients, and the simulation run as usual.

Problem-Solving Tip

- In deriving KVL or mesh-current equations, particularly in more complicated cases involving magnetic coupling between more than two coils, it is often helpful to derive these equations first in the absence of magnetic coupling and then systematically add the terms due to this coupling one at a time.

Primal Exercise 9.11

Determine v_O in Figure 9.19 and verify with PSpice simulation.

Ans. $1.745\angle 174.0°$ V.

Primal Exercise 9.12

Determine $\mathbf{V}_O$ in Figure 9.22.

Ans. $0.5(1 + j)$ V.

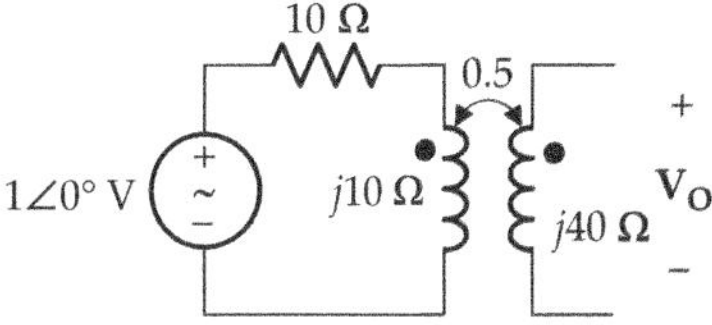

FIGURE 9.22
Figure for Primal Exercise 9.12.

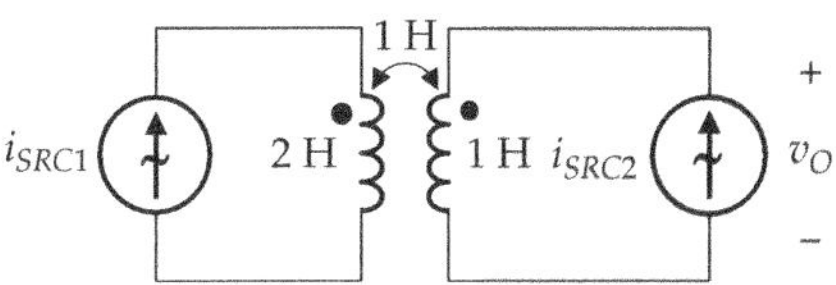

FIGURE 9.23
Figure for Primal Exercise 9.13.

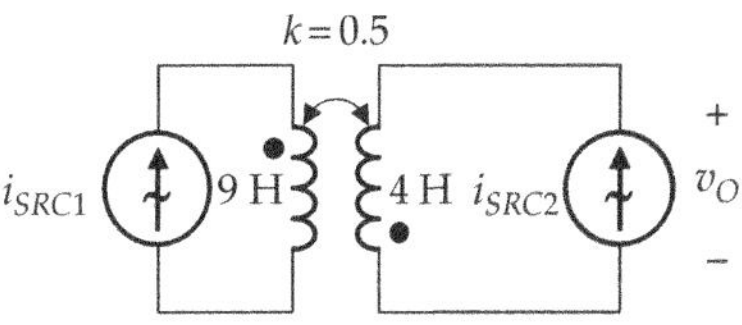

FIGURE 9.24
Figure for Primal Exercise 9.14.

Primal Exercise 9.13

Determine $\mathbf{V}_O$ in Figure 9.23 assuming $i_{SRC1}(t) = 2\cos 2t$ A and $i_{SRC2}(t) = \sin 2t$ A and considering the cosine function to have a zero phase angle.

Ans. $2 + j4$ V.

Primal Exercise 9.14

Determine $v_O(t)$ in Figure 9.24 as a cosine function, given that $i_{SRC1}(t) = \cos 10t$ A and $i_{SRC2}(t) = \sin 10t$ A.

Ans. $50\cos(10t - 36.9°)$ V.

Example 9.3: Input Impedance of Coupled Coils

It is required to determine the smallest input impedance seen by the current source $\mathbf{I}_1$ in Figure 9.25a as R is varied between zero and infinity.

Solution:

The problem will be solved step-by-step in order to clearly illustrate some of the concepts involved:

Step 1: If $R \to \infty$, coil 2 on the RHS is open-circuited. No current flows in this coil, so there is no induced voltage in coil 1 due to current in coil 2, but a voltage is induced in coil 2 by $\mathbf{I}_1$. With $\mathbf{I}_1$ entering the dot-marked terminal

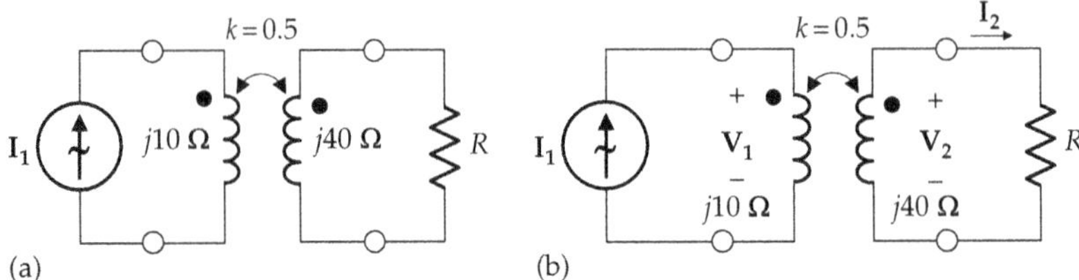

FIGURE 9.25
Figure for Example 9.3.

on coil 1 and $\mathbf{V}_1$ assigned a positive polarity as a voltage drop in the direction of $\mathbf{I}_1$, then $\mathbf{V}_1 = j\omega L_1 I_1$. The source sees an input impedance $Z_{in} = \mathbf{V}_1/\mathbf{I}_1 = j\omega L_1 = j10\ \Omega$.

Step 2: With $\mathbf{V}_1$ and $\mathbf{V}_2$ assigned the same positive polarities with respect to the dot markings (Figure 9.25b), the voltage induced in coil 2 due to $\mathbf{I}_1$ is $\mathbf{V}_2 = j\omega M\mathbf{I}_1$, independently of any $\mathbf{I}_2$.

Step 3: If R is connected, a current $\mathbf{I}_2$ flows in coil 2 under the influence of $\mathbf{V}_2$. According to Lenz's law, the direction of this current is such that it opposes the flux due to $\mathbf{I}_1$. Since $\mathbf{I}_1$ enters coil 1 at the dot-marked terminal, $\mathbf{I}_2$ is conveniently assigned a positive direction as leaving coil 2 at the dot-marked terminal (Figure 9.25b).

Step 4: KVL on the side of coil 1 is

$$\mathbf{V}_1 = j\omega L_1\mathbf{I}_1 - j\omega M\mathbf{I}_2 \tag{9.36}$$

whereas KVL on the side of coil 2 is

$$\mathbf{V}_2 = j\omega M\mathbf{I}_1 = \left(R + j\omega L_2\right)\mathbf{I}_2 \tag{9.37}$$

where $\mathbf{V}_2 = j\omega M\mathbf{I}_1$ is the induced voltage that drives current in coil 2. In Equation 9.37, the flow of $\mathbf{I}_2$ is opposed by the voltage drop $R\mathbf{I}_2$ and by the induced voltage $j\omega L_2\mathbf{I}_2$ due to $\mathbf{I}_2$ flowing in the inductor L_2, in accordance with Lenz's Law. Note that the sign of the $j\omega M$ term is opposite that of the $j\omega L$ term for both coils when these terms are on the same side of the equation. This is because the source current enters the dot-marked terminal and the current in coil 2 leaves at the dot-marked terminal.

Step 5: $\mathbf{V}_1$, and hence Z_{in}, has its smallest value when $\mathbf{I}_2$ has its largest value, since $\mathbf{I}_2$ opposes the flux due to $\mathbf{I}_1$. The largest $\mathbf{I}_2$ results in the smallest flux in the core and hence the smallest $\mathbf{V}_1$ and Z_{in}. This also follows from Equation 9.36 since the $j\omega M\mathbf{I}_2$ term subtracts from the $j\omega L_1\mathbf{I}_1$ term to give $\mathbf{V}_1$.

Step 6: From Equation 9.37, $\mathbf{I}_2 = j\omega M\mathbf{I}_1/(R + j\omega L_2)$. The only variable in this expression for $\mathbf{I}_2$ is R, and $\mathbf{I}_2$ is largest when R is smallest, that is, when $R = 0$.

Step 7: When $R = 0$, $\mathbf{I}_2 = (M/L_2)\mathbf{I}_1$. Substituting in Equation 9.36, $\mathbf{V}_1 = (j\omega L_1 - j\omega M^2/L_2)\mathbf{I}_1$, This gives $Z_{in} = \mathbf{V}_1/\mathbf{I}_1 = j\omega L_1 - j\omega^2 M^2/\omega L_2$ where $\omega M = 0.5\sqrt{10\times 40} = 10\ \Omega$. Substituting numerical values, $Z_{in} = j10 - j100/40 = j7.5\ \Omega$.

9.4 T-Equivalent Circuit

A very useful circuit is the T-equivalent circuit for two magnetically coupled coils. Because this circuit has three terminals, one being common to input and output, it applies when the two coils also have a common terminal, or can be connected in this manner, without changing the branch currents and voltages in the circuit, which is almost always the case in practice.

There are two forms of the T-equivalent circuit depending on the relative dot markings. To derive these circuits, we note, first, that two magnetically coupled coils of zero resistance are completely specified, in circuit terms, by three parameters: L_1, L_2, and M. This implies that three independent measurements are needed to fully describe the two coils and establish equivalence with another circuit.

Consider two magnetically coupled coils as in Figure 9.26a, with a time-varying current i_1 in coil 1, and coil 2 open-circuited. Let i_1, v_1, and v_2 be measured under these conditions. Since coil 2 does not carry current, the relation between v_1 and i_1 is that for a single coil that is not affected by current in the other coil, so that v_1 and i_1 are related by the v–i relation of an ideal inductor:

$$v_1 = L_1\frac{di_1}{dt} \tag{9.38}$$

where a positive sign is used because the assigned positive directions are those of i_1 in the direction of a voltage drop v_1 (Section 7.2). The voltage induced in the open-circuited coil 2 is, from Equation 9.3, with M substituted for M_{21},

$$v_2 = M\frac{di_1}{dt} \tag{9.39}$$

If $di_1/dt > 0$, terminal 1 is positive with respect to terminal 1′. According to the dot markings, terminal 2 is positive with respect to terminal 2′. Hence, Equation 9.39 is written with a positive sign, with the positive direction of v_2 assigned as in Figure 9.26a. Both v_1 and v_2 have positive values when $di_1/dt > 0$ and negative values when $di_1/dt < 0$.

Now consider the T-circuit of Figure 9.26b, with i_1 entering terminal 1 and terminal 2 open-circuited. i_1 flows

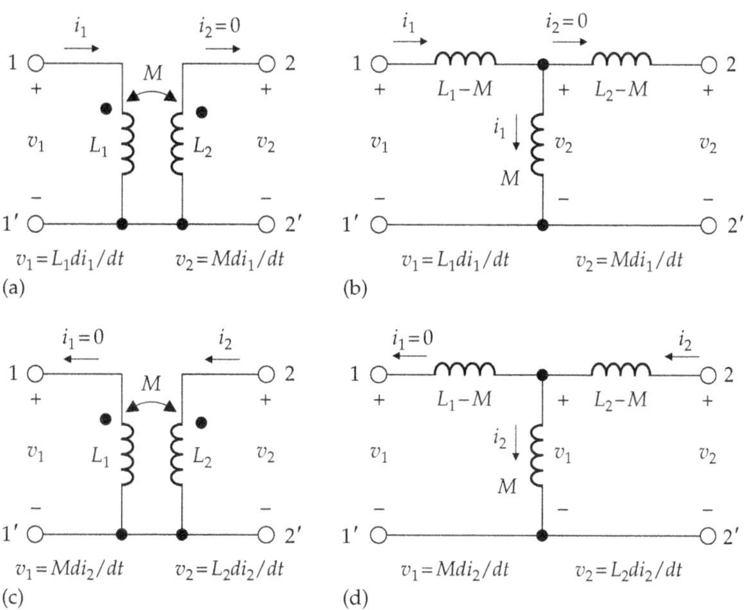

FIGURE 9.26
Equivalence between two coupled coils having the dot markings shown and a T-circuit. (a) Excitation is applied to coil 1, with coil 2 open circuited, (b) same conditions applied to T-circuit as in (a), (c) excitation applied to coil 2, with coil 1 open circuited, and (d) same conditions applied to T-circuit as in (c).

through two inductors in series, $(L_1 - M)$ and M, whose total inductance is $L_1 - M + M = L_1$, so that v_1 is related to i_1 by Equation 9.38. Because no current flows in $(L_2 - M)$, v_2 appears across M and is given by Equation 9.39.

Suppose that coil 2 has a time-varying current i_2, with coil 1 open-circuited (Figure 9.26c). Let i_2, v_2, and v_1 be measured under these conditions. The same aforementioned argument can be applied to deduce that

$$v_2 = L_2\frac{di_2}{dt} \tag{9.40}$$

and

$$v_1 = M\frac{di_2}{dt} \tag{9.41}$$

in both Figure 9.26c and d. Thus, it was shown that the circuit of Figure 9.26b or d reproduced four v–i relations at both the input and output terminals of the circuit of Figure 9.26a or c under open-circuit conditions. Note that equivalence between two circuits applies under the same conditions for both circuits. That is, in order to establish equivalence, the same terminals in the two circuits could be open circuited, or short circuited, or have the same resistance connected between them. In this case, an open circuit is the simplest case to consider for establishing equivalence. Hence, it can be concluded that the T-equivalent circuit of Figure 9.26b or d is equivalent to the two coupled coils with the dot markings indicated in Figure 9.26a.

Consider, next, the coupled coils of Figure 9.27a, having reversed dot markings. It is to be expected that reversing the dot markings changes the sign of the M term, so that the shunt branch is $-M$ and the series branches are $(L_1 + M)$ and $(L_2 + M)$. We will show that this is indeed the case.

In Figure 9.27a, v_1 is given by Equation 9.38, as before, and $v_2' = Mdi_1/dt$, in accordance with the dot markings. However, $v_2 = -v_2'$, so that

$$v_2 = -M\frac{di_1}{dt} \tag{9.42}$$

If $di_1/dt > 0$, and since M is a positive number, then to have v_2 negative in the T-equivalent circuit of Figure 9.27b, it is necessary that the inductance of the shunt branch be $-M$, instead of $+M$. The series inductances are now $(L_1 + M)$ and $(L_2 + M)$, so that the total series inductance through which i_1 flows is $L_1 + M - M = L_1$ and Equation 9.38 again applies.

If coil 2 carries a time-varying current i_2, with coil 1 open-circuited (Figure 9.27c), v_2 is given by Equation 9.40, as before. The induced voltage in coil 1 is $v_1' = Mdi_1/dt$ and its polarity will follow the dot markings, so that $v_1 = -v_1'$ (Figure 9.27c). The circuit of Figure 9.27d again satisfies the v–i relations under these conditions. The T-equivalent circuit of Figure 9.27b or d is thus equivalent to the two coupled coils with the dot markings as in Figure 9.27a or c, respectively. That the shunt inductance is now $-M$ is quite acceptable, as the circuit is only meant to reproduce the v–i relations for the two coupled coils and not to provide a physical correspondence with the two magnetically coupled coils. A more physically based equivalent

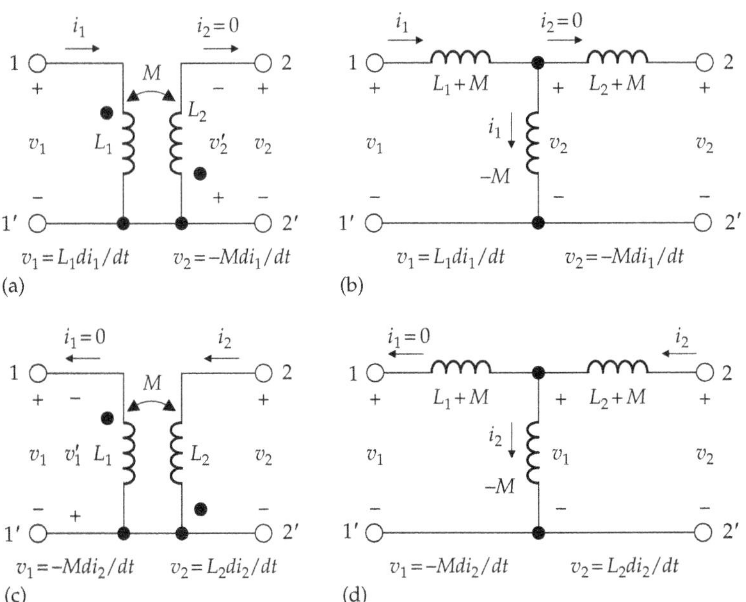

FIGURE 9.27
Equivalence between two coupled coils having the dot markings shown and a T-circuit. (a) Excitation is applied to coil 1, with coil 2 open circuited, (b) same conditions applied to T-circuit as in (a), (c) excitation applied to coil 2, with coil 1 open circuited, and (d) same conditions applied to T-circuit as in (c).

circuit of the two coupled coils is presented in Section 10.5. The T-equivalent circuit is useful in that it eliminates the magnetic coupling between the two coils, which allows application of any of the conventional methods of circuit analysis. If magnetic coupling is retained, analysis is generally limited to applying KVL or the mesh-current method.

It should be emphasized that the choice between the T-equivalent circuits of Figures 9.26 and 9.27 is dictated solely by the relative dot markings of the two coils and is *independent of the assigned positive directions of currents in the two coils or the polarities of the voltages across these coils.* Changing the assigned positive direction of the current in one of the coils, or changing the assigned positive polarity of the voltage across the coil, does not change the values of the inductances.

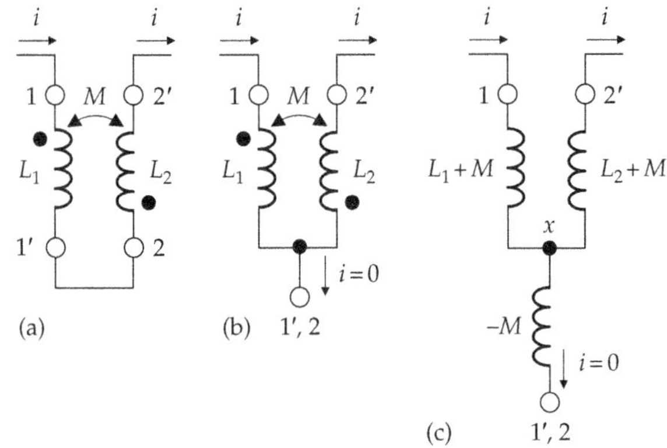

FIGURE 9.28
Figure for Example 9.4.

Example 9.4: Equivalent Inductance Using T-Equivalent Circuit

It is required to obtain L_{eq} of the series-connected, magnetically coupled coils of Example 9.1 using the T-equivalent circuit.

Solution:

The two coils in Figure 9.12a are rotated so as to bring them in parallel, for easier visualization, as shown in Figure 9.28a. Terminals 1′ and 2 are connected together and could be redrawn as in Figure 9.28b to make the connection between the two coils "internal" to the circuit. The coupled coils can now be replaced terminal for terminal by the T-equivalent circuit, as in Figure 9.28c, the three terminals being 1, 2′, and 1′–2 joined together. It is seen from the dot markings of the two coils that the appropriate T-equivalent circuit is that of Figure 9.27 having inductances $(L_1 + M)$, $(L_2 + M)$, and $-M$. The $-M$ branch is open-circuited, so that the inductance that appears between the outer terminals 1 and 2′ is $L_{eq} = L_1 + L_2 + 2M$, as in Equation 9.31. Note that terminal 'x' in Figure 9.28c is internal to the T-equivalent circuit and does not exist in the original series connection of the two coils. Since no current flows in the $-M$ branch, this branch could be replaced by a short circuit. The induced voltage between terminals 1 and 1′–2, or between terminal 1 and node 'x' is $(L_1 + M)di/dt$, so that the effective

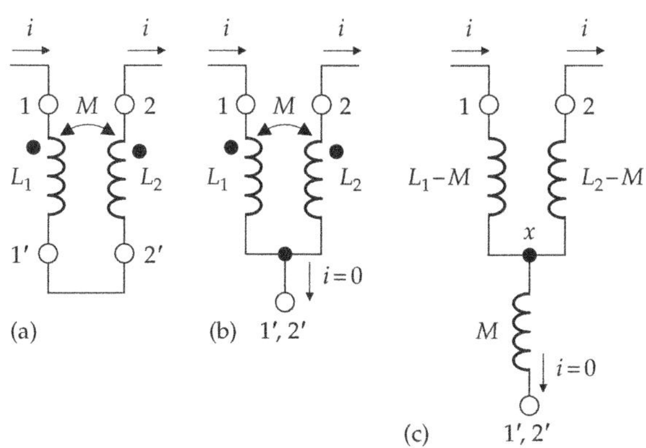

FIGURE 9.29
Figure for Example 9.5.

inductance of coil 1 is ($L_1 + M$) as in Figure 9.13. Similarly, the effective inductance of coil 2 is ($L_2 + M$).

The same procedure applied to the two coils in Figure 9.12b gives the T-equivalent circuit in Figure 9.29c, which is now that of Figure 9.26. The inductance that appears between the outer terminals 1 and 2′ is $L_{eq} = L_1 + L_2 - 2M$, as in Equation 9.35. The effective inductance of coil 1 is ($L_1 - M$), whereas the effective inductance of coil 2 is ($L_2 - M$).

Example 9.5: Circuit Analysis Using T-Equivalent Circuit

It is required to analyze the circuit in Figure 9.16, Example 9.2, using the T-equivalent circuit.

Solution:

The two coils in Figure 9.16 are rotated so as to bring them in parallel, for easier visualization, as shown in Figure 9.30a. They are then replaced, terminal for terminal, by the appropriate T-equivalent circuit, which is that of Figure 9.27. The resulting circuit is that of Figure 9.30b. The mesh-current equations are

$$(10 + j8 - j5)\mathbf{I}_1 - (-j4 - j5)\mathbf{I}_2 = 100$$

$$-(-j4 - j5)\mathbf{I}_1 + (5 + j32 - j5)\mathbf{I}_2 = 0$$

which are identical to Equations 9.30 and 9.31, respectively.

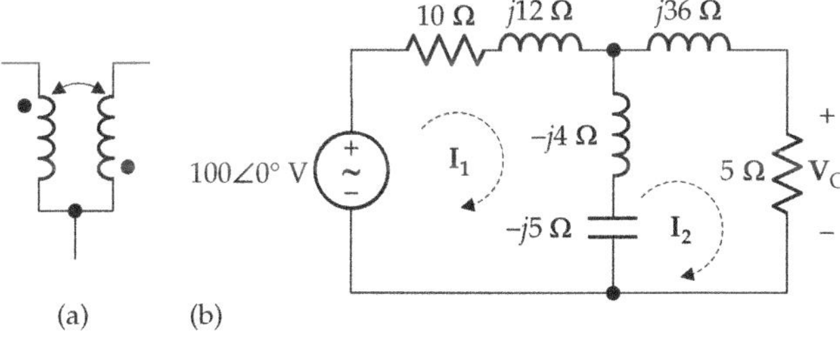

FIGURE 9.30
Figure for Example 9.5.

Problem-Solving Tip

- It is generally advantageous to replace magnetically coupled coils having a common connection by the appropriate T-equivalent circuit.

Learning Checklist: What Should Be Learned from This Chapter

- In magnetically coupled coils, the polarity of the induced voltage is determined by Lenz's law, as in the case of a single coil, but taking into consideration the relative sense of winding of the coils.
- The relative sense of winding of coils on the same core is accounted for by the dot convention.
- According to the dot convention, the polarities of the induced voltages are the same at the dot-marked terminals of each coil; that is, the dot-marked terminals will all become positive, or negative, together with respect to the unmarked terminal. As a corollary, fluxes due to currents entering, or leaving, the dot-marked terminals in each coil are in the same direction in both coils.
- The mutual inductance of two magnetically coupled coils is the flux linkage in one coil per unit current in the other coil. It is independent of which coil carries the current.
- When two coils are magnetically coupled together, there is some leakage flux of each coil that links one coil but not the other coil. The leakage flux is relatively large when the magnetic medium between the two coils is of low permeability and is relatively small when the coils are wound on a toroidal core of low permeability.
- The coupling coefficient $k = M/\sqrt{L_1L_2}$ is a measure of how well the two coils are coupled together through the core. It assumes positive values between 0 and 1, where $k = 0$ denotes no coupling between the two coils and $k = 1$ denotes perfect coupling.
- A transformer consists of two or more coils that are magnetically coupled relatively tightly. In a linear transformer, permeability is constant, so that B and H, or λ and i, are linearly related, and superposition applies to magnetic variables including fluxes.
- In writing KVL or mesh-current equations in circuits involving magnetically coupled coils, the sign of the term involving M depends on

both the dot markings of coil terminals and the assigned positive directions of currents in the coils. If the assigned positive directions of currents are such that these currents both flow in, or both flow out, at the dot-marked terminals, the sign of the mutual inductance term (Mdi_1/dt, or Mdi_2/dt) for either coil is the same as that of the self-inductance term for that coil (L_1di_1/dt or L_2di_2/dt, *respectively*). Otherwise, the sign of the mutual inductance term for either coil is opposite that of the self-inductance term for that coil.

- The T-equivalent circuit is a useful three-terminal circuit that can be used to replace, terminal for terminal, two magnetically coupled coils having a common terminal. It eliminates magnetic coupling between the coils, which allows application of any of the conventional methods of circuit analysis.
- There are two forms of the T-equivalent circuit, depending on the relative dot markings of the coil terminals, irrespective of the assigned positive directions of currents in the coils or the polarities of the voltages across these coils.

Problem-Solving Tips

1. In deriving KVL or mesh-current equations, particularly in more complicated cases involving magnetic coupling between more than two coils, it is often helpful to derive these equations first in the absence of magnetic coupling and then systematically add the terms due to this coupling one at a time.
2. It is generally advantageous to replace magnetically coupled coils having a common connection by the appropriate T-equivalent circuit.

Appendix 9A: Energy Stored in Magnetically Coupled Coils

Consider two magnetically coupled coils carrying steady currents I_1 and I_2. It is required to determine the energy expended in establishing these currents, starting from zero. It is convenient to assume that I_1 and I_2 are established by variable current sources in two steps: (1) i_1 is first increased from zero to I_1 with $I_2 = 0$ and (2) i_2 is then increased from zero to I_2 with $i_1 = I_1$.

The first step is illustrated in Figure 9.31a. While i_1 is increasing, the induced voltage $v_1 = L_1\dfrac{di_1}{dt}$ in coil 1 opposes the increase in i_1, in accordance with Lenz's law, by being a voltage drop across L_1 in the direction of i_1.

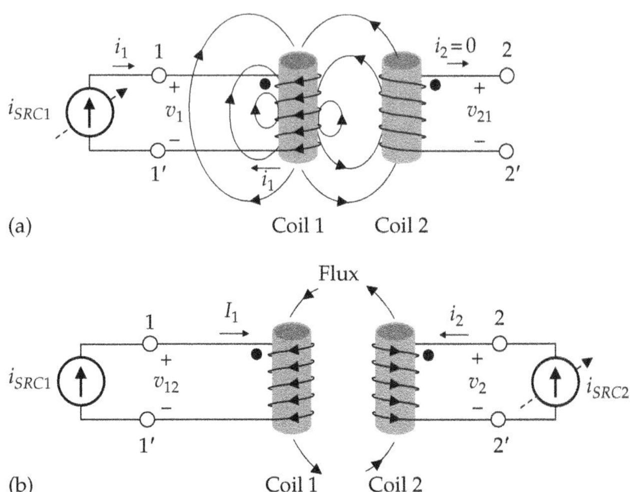

FIGURE 9.31
Energy stored in magnetic field due to coupled coils. (a) Build-up of current in coil 1 from 0 to I_1 while coil 2 is open circuited and (b) build-up of current in coil 2 from 0 to I_2 while the current in coil 1 is I_1.

This voltage is concurrently a voltage rise across the current source i_{SRC1}, so that the total energy w_{11} delivered by the source is

$$w_{11} = \int_0^t L_1\frac{di_1}{dt}i_1dt = L_1\int_0^{I_1} i_1di_1 = \frac{1}{2}L_1I_1^2 \tag{9.43}$$

The second step is illustrated in Figure 9.31b. With a constant current $i_1 = I_1$, the only voltage induced in coil 2 is that due to increasing i_2, and the total energy supplied by the current source i_{SRC2} in establishing I_2 is $\frac{1}{2}L_2I_2^2$, by analogy with Equation 9.43. However, as i_2 increases, it induces a voltage $v_{12} = M_{12}di_2/dt$ in coil 1. The sense of winding of coil 2 and the direction of i_2 are such that the flux associated with i_2 is also downward in coil 1, the same as the flux in coil 1 due to i_1. The effect of increasing i_2 is therefore the same as that of increasing i_1, so that v_{12} is of the same polarity as v_1 in Figure 9.31a and opposes the current in coil 1. The current source i_{SRC1} has therefore to deliver additional energy to maintain I_1. This energy is

$$w_{12} = \int_0^t v_{12}I_1dt = \int_0^t M_{12}\frac{di_2}{dt}I_1dt = M_{12}I_1\int_0^{I_2} di_2 = M_{12}I_1I_2 \tag{9.44}$$

The total energy expended in establishing I_1 and I_2 is

$$w_1 = \frac{1}{2}L_1I_1^2 + \frac{1}{2}L_2I_2^2 + M_{12}I_1I_2 \tag{9.45}$$

Suppose that I_1 and I_2 are established in the reverse order, that is, first, I_2 is established with $i_1 = 0$, and then

i_1 is increased to I_1 with $i_2 = I_2$. Following this same argument, the total energy expended in establishing I_1 and I_2 is

$$w_2 = \frac{1}{2}L_1I_1^2 + \frac{1}{2}L_2I_2^2 + M_{21}I_1I_2 \quad (9.46)$$

w_1 and w_2 must be equal, because in a lossless, linear system, the total energy expended must depend only on the final values of I_1 and I_2 and not on the time course of i_1 and i_2 that led to these values. Otherwise, it would be possible, at least in principle, to extract energy from the system at no energy cost, in violation of conservation of energy. Suppose, for example, that $w_1 < w_2$. Then I_1 and I_2 may be established from zero with expenditure of energy w_1. In principle, I_1 and I_2 may be reduced to zero by reversing the steps that led to w_2, with recovery of more energy than was expended, which is clearly impossible. Equating w_1 and w_2, it follows that

$$M_{12} = M_{21} = M \quad (9.47)$$

If either the polarity of i_{SRC2}, or the sense of winding of coil 2, is reversed in Figure 9.31, the flux due to i_2 becomes upward in coil 1. The polarity of v_{12} is reversed and becomes a voltage drop across the current source i_{SRC1}. Energy is therefore returned to the source and the sign of the energy term involving M becomes negative in Equations 9.45 and 9.46. M, however, is *always* a positive quantity.

Since I_1 and I_2 are arbitrary values, they might just as well be replaced by instantaneous values i_1 and i_2. The energy stored in the magnetic field in building up the currents in two magnetically coupled coils to i_1 and i_2, starting from zero, may therefore be expressed in general as

$$w = \frac{1}{2}L_1i_1^2 + \frac{1}{2}L_2i_2^2 \pm Mi_1i_2 \quad (9.48)$$

If the system is nonlinear and involves hysteresis (Section 10.5), the time course according to which i_1 and i_2 reach their final values does affect the energy expended, but the energy difference is dissipated as heat in the magnetic material.

Problems

Apply ISDEPIC and verify solutions by PSpice simulation whenever feasible.

Magnetically Coupled Coils

P9.1 Two coils are wound on a high-permeability core (Figure P9.1). Coil 1 has 1000 turns and carries a current $i_1 = 1$ A. Coil 2 has 500 turns. Determine the magnitude and direction of the current in coil 2 so that the net flux in the core is zero.

Ans. 2 A, entering terminal 2′.

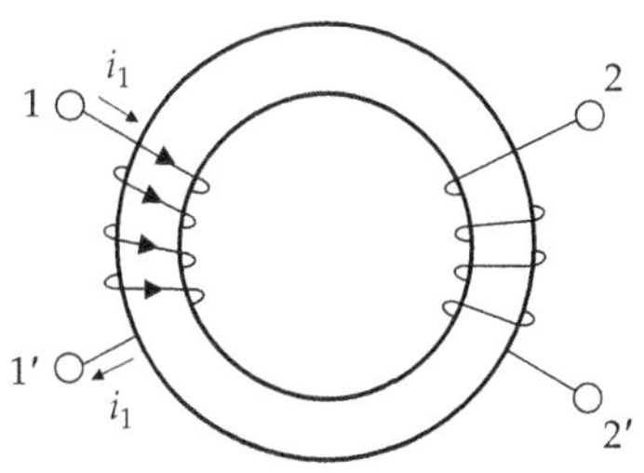

FIGURE P9.1

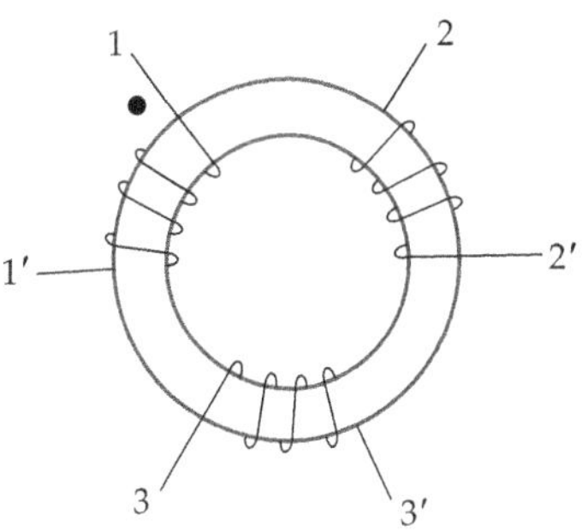

FIGURE P9.2

P9.2 The terminal of one coil in Figure P9.2 is marked with a dot. (a) Mark one terminal of the other coils with a dot; (b) connect the coils in series for maximum total inductance; (c) determine the inductances of coils 2 and 3, assuming $L_1 = 40$ mH; $M_{12} = M_{23} = 20$ mH, and $k_{12} = k_{23} = 1/\sqrt{2}$; (d) determine M_{13}, assuming $k_{13} = 0.5$.

Ans. (a) Marked terminals, 2′ and 3; (b) terminals 2′ and 3′ connected together, as are terminals 3 and 1′; (c) $L_2 = 20$ mH, $L_3 = 40$ mH; (d) $M_{13} = 20$ mH.

P9.3 Two coils are magnetically coupled through a core of high permeability. A current of 0.5 A in coil 1, with coil 2 open-circuited, results in a flux of 0.1 Wb in the core, whereas a current of 0.25 A in coil 2, with coil 1 open-circuited, results in a flux of 0.2 Wb in the core. If coil 1 has 100 turns, determine the number of turns of coil 2.

Ans. 400 turns.

P9.4 Two coils L_1 and L_2, having 1000 turns and 500 turns, respectively, are wound on a core of high permeability. When current is applied to one coil, with the other coil open-circuited, the effective leakage flux of either coil is 5% of the core flux. Determine the coefficient of coupling between the two coils.

Ans. 0.95.

P9.5 Two coils are wound on a core of high permeability. Coil 1 has 100 turns and coil 2 has 400 turns. A current of 1 A in coil 1, with coil 2 open-circuited, results in a core flux of 0.5 mWb. Determine the magnitude of the core flux resulting from a current of 0.8 A in coil 2, with coil 1 open-circuited.

Ans. 1.6 mWb.

P9.6 Two coils are tightly coupled to a high-permeability core, so that the leakage flux is negligibly small. If coil 1 has 100 turns and an inductance of 10 mH and the mutual inductance is 12.5 mH, determine (a) the number of turns of coil 2 and (b) the inductance of coil 2.

Ans. (a) 125; (b) 15.625 mH.

P9.7 Two coils having $N_1 = 1000$ turns and $N_2 = 510$ turns are coupled through a high-permeability core. The inductance of coil 1 is 1 mH and the mutual inductance is 0.5 mH. Determine the ratio of $\phi_{11eff(leak)}$ to ϕ_{21}.

Ans. 0.02.

P9.8 Two identical, magnetically coupled coils are connected in series. When the same current is passed though both coils, but with the connections of one coil reversed, the magnetic stored energy is multiplied by a factor of 2. Determine the coefficient of coupling of the coils.

Ans. 1/3.

P9.9 Two magnetically coupled coils have $k = 0.5$. When connected in series, the total inductance is 80 mH. When the connection to one of the coils is reversed, the total inductance is 40 mH. Determine the inductances of the two coils.

Ans. 52.36 and 7.64 mH.

P9.10 Given two coils, one having a self-inductance of 1 mH, the other a self-inductance of 4 mH. They are magnetically coupled and connected in series so that the fluxes due to the current in the two coils add to one another. In the T-equivalent circuit representing the two coupled coils, the magnetic energy that depends on the self-inductance of 4 mH is 2.2 times the magnetic energy that depends on the self-inductance of 1 mH. Determine the coefficient of coupling.

Ans. 0.75.

P9.11 Two coils are coupled through a high-permeability core. When $i_1 = 4$ A, $\phi_{11eff(leak)} = 0.1$ mWb and $\phi_{21} = 0.4$ mWb. When $i_2 = 3$ A, $\phi_{12} = 0.6$ mWb. If $N_2 = 1000$ turns and $L_2 = 400$ mH, determine N_1, L_1, M, $\phi_{22eff(leak)}$, and the total energy stored in the magnetic circuit when both coils carry the aforementioned currents, and the fluxes due to these currents are additive.

Ans. $N_1 = 500$ turns, $L_1 = 62.5$ mH, $M = 100$ mH, $\phi_{22eff(leak)} = 0.6$ mWb, energy is 3.5 J.

P9.12 Two coils having $N_1 = 800$ turns and $N_2 = 500$ turns are coupled through a high-permeability core. A current i_1 in coil 1 results in $\phi_{11eff(leak)} = 500$ μWb and $\phi_{21} = 400$ μWb, whereas a current $2i_1$ in coil 2 results in $\phi_{22eff(leak)} = 1400$ μWb. Determine (a) ϕ_{12} resulting from $2i_1$ in coil 2; (b) the coefficient of coupling; (c) the mutual inductance, assuming that the permeance of the core is 50 nWb/A-turn; and (d) the inductance of each coil.

Ans. (a) 500 μWb; (b) 0.342; (c) 20 mH; (d) $L_1 = 72$ mH and $L_2 = 47.50$ mH.

P9.13 Given two magnetically coupled coils. If the current in one coil is $10\sin 1000t$ mA, the voltage induced in the other coil is $32\cos 1000t$ V. When the two coils are connected in series, the largest measured inductance is 16.4 H. If the inductance of one coil is 3.6 H, determine the inductance of the other coil and the coefficient of coupling.

Ans. 6.4 H, 2/3.

P9.14 Determine the ratio v_1/v in Figure P9.14.

Ans. 0.7.

P9.15 Determine L_{eq} in Figure P9.15.

Ans. 4 H.

P9.16 Determine L_{eq} in Figure P9.16.

Ans. 0.

P9.17 Given $i_1 = 3$ A in Figure P9.17, with $di_1/dt = -0.2$ A/s at a given instant of time, determine v_{cd} at this instant, assuming $k = 0.7$.

Ans. 0.28 V.

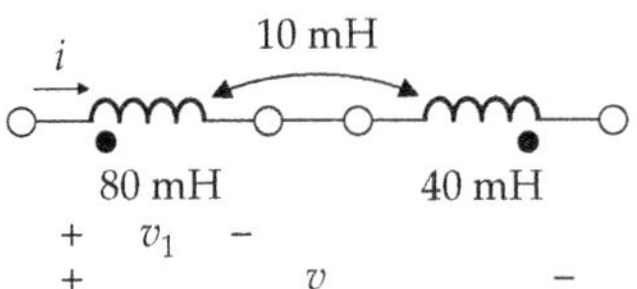

FIGURE P9.14

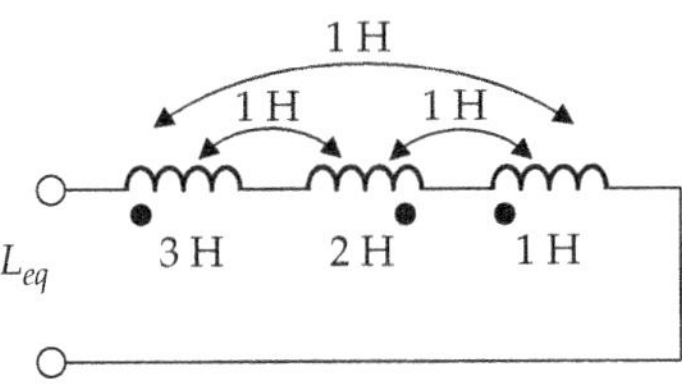

FIGURE P9.15

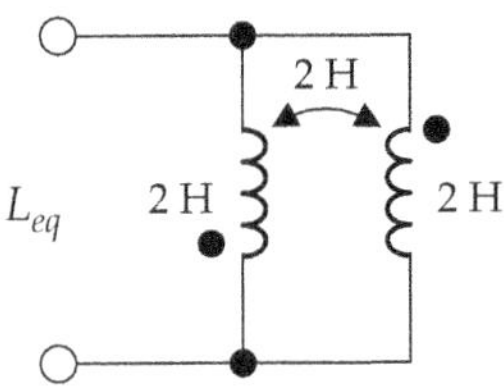

FIGURE P9.16

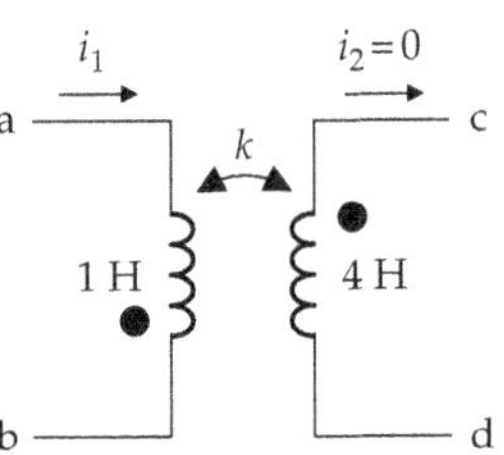

FIGURE P9.17

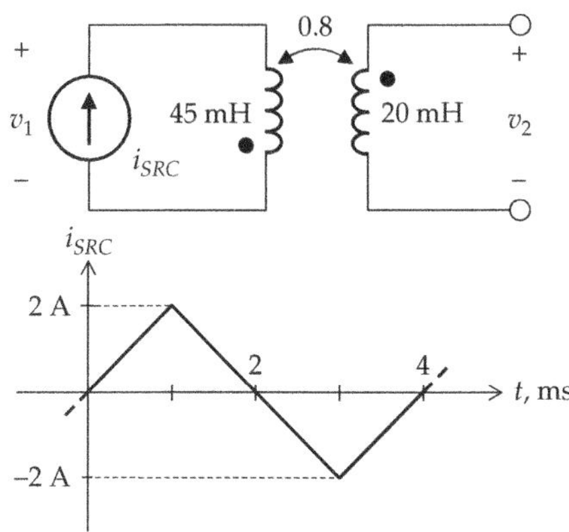

FIGURE P9.18

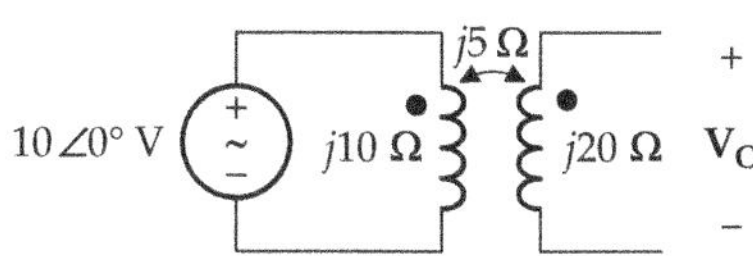

FIGURE P9.19

P9.18 Given the coupled coils of Figure P9.18, with i_{SRC} being the triangular waveform shown, sketch the waveforms of $v_1(t)$ and $v_2(t)$.

Ans. $v_1(t)$ and $v_2(t)$ are square waveforms of amplitudes 90 V and 48 V, respectively.

P9.19 Determine $\mathbf{V}_O$ in Figure P9.19.

Ans. $5\angle 0°$ V.

P9.20 Determine $v_1(t)$ and $v_2(t)$ in Figure P9.20, assuming $i_1(t) = (64t + 50)$ A and $i_2(t) = 15t$ A.

Ans. $v_1(t) = 335$ V; $v_2(t) = 109$ V.

P9.21 If $I_1 = 2$ A in Figure P9.21, determine the value of I_2 that minimizes the stored energy.

Ans. 2/3 A.

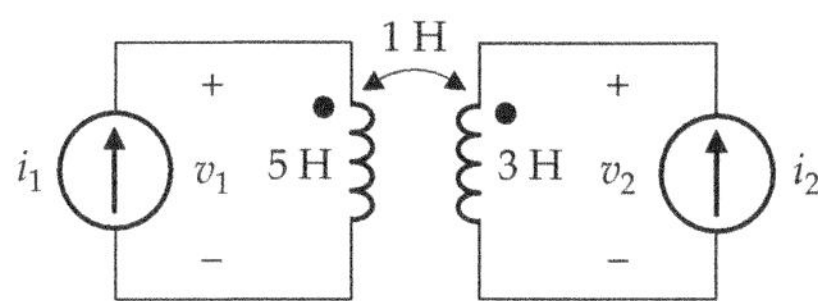

FIGURE P9.20

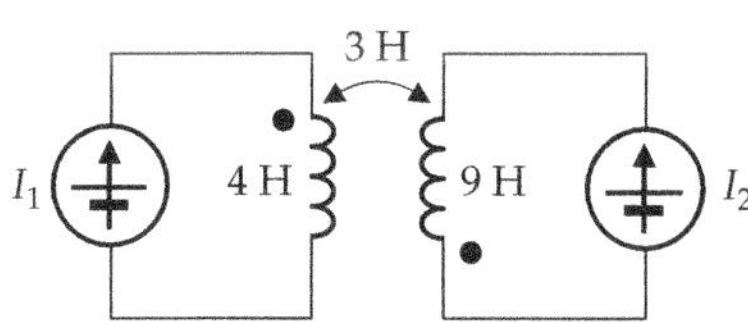

FIGURE P9.21

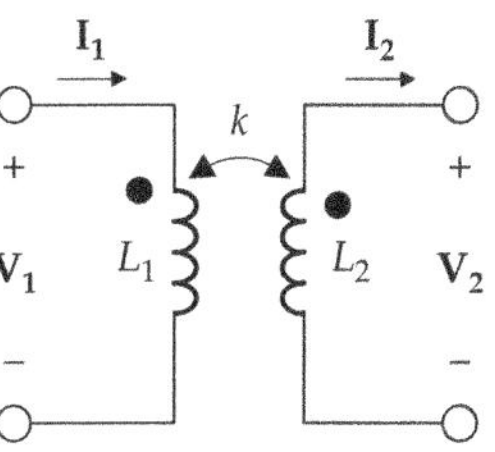

FIGURE P9.22

P9.22 The open-circuit ($\mathbf{I}_2 = 0$) voltage ratio of the linear transformer in Figure P9.22 is $\mathbf{V}_2/\mathbf{V}_1 = 0.25$, and the short-circuit ($\mathbf{V}_2 = 0$) current ratio is $\mathbf{I}_2/\mathbf{I}_1 = 1$. If the same coils are perfectly coupled, the mutual inductance is 8 H. Determine L_1, L_2, and k for the given coils.

Ans. $L_1 = 16$ H, $L_2 = 4$ H. $k = 0.5$.

Linear Transformer Circuits

P9.23 Determine L_{eq} in Figure P9.23.

Ans. 6 μH.

P9.24 Determine Z_{in} in Figure P9.24.

Ans. Infinite.

P9.25 Determine Z_{in} in Figure P9.25.

Ans. 0.

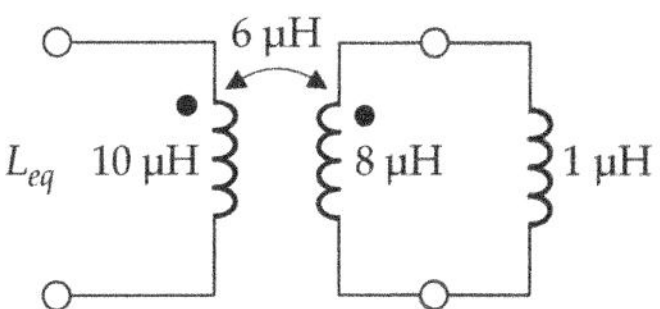

FIGURE P9.23

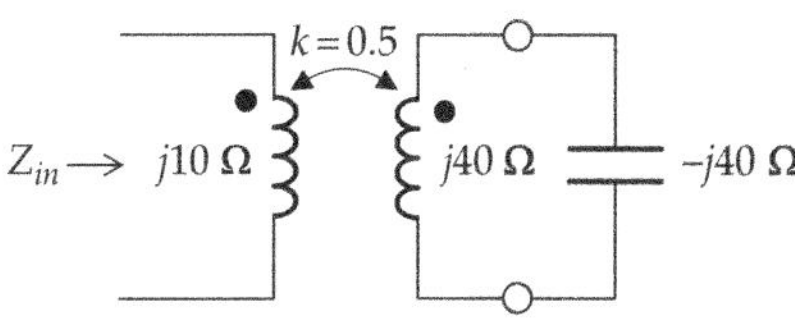

FIGURE P9.24

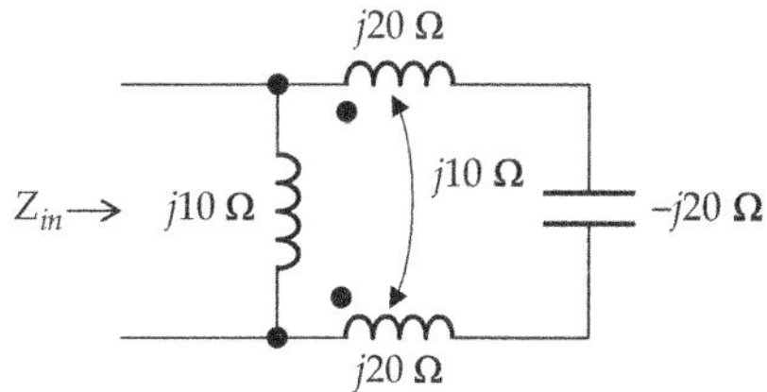

FIGURE P9.25

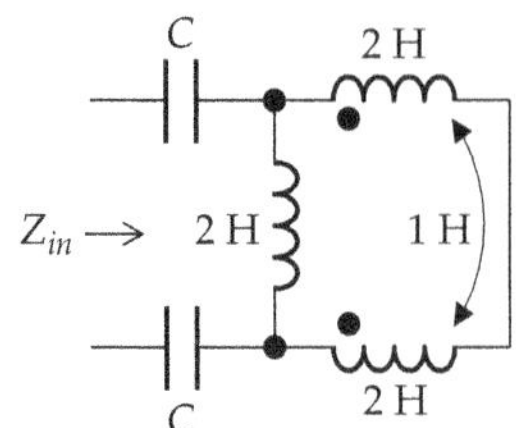

FIGURE P9.26

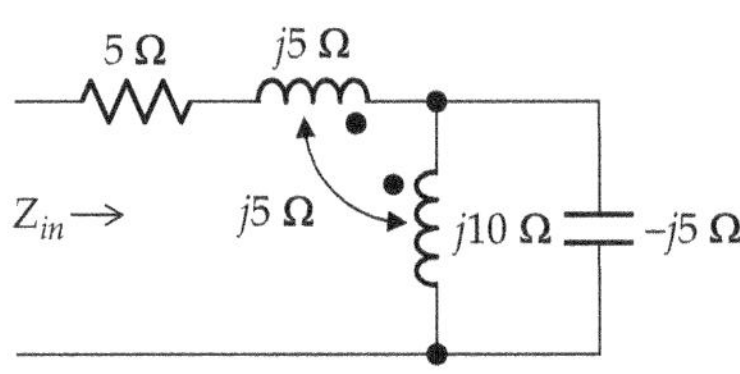

FIGURE P9.27

P9.26 Determine Z_{in}, assuming $\omega = 10$ rad/s and $C = 20$ mF in Figure P9.26.

Ans. 0.

P9.27 Determine Z_{in} in Figure P9.27.

Ans. 5 Ω.

P9.28 Determine Z_{in} in Figure P9.28, assuming $\omega = 10^6$ rad/s.

Ans. $-j162$ Ω.

P9.29 By substituting the T-equivalent circuit of either Figure 9.26b or Figure 9.26d for the linear transformer in Figure P9.29, show that the input impedance $\mathbf{V}_1/\mathbf{I}_1$ can be expressed as $Z_{in} = \dfrac{\mathbf{V}_1}{\mathbf{I}_1} = j\omega L_1 + \dfrac{\omega^2 M^2}{j\omega L_2 + Z_L}$. Note that because M is squared in this expression, Z_{in} is independent of the dot markings.

P9.30 Using the result of Problem P9.29, determine Z_{in} in Figure P9.30, assuming $\omega = 1$ krad/s.

Ans. (a) $25.6 + j40$ Ω.

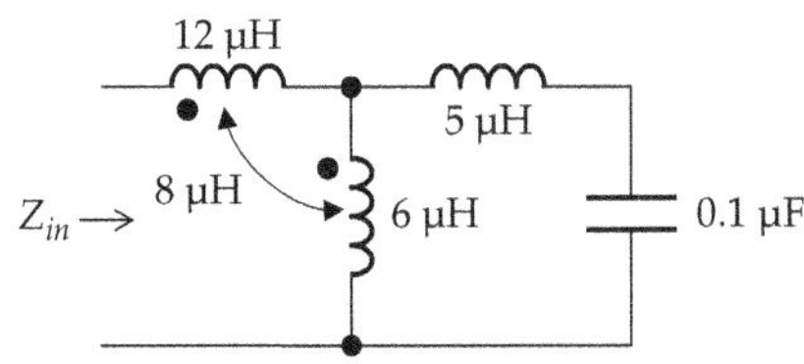

FIGURE P9.28

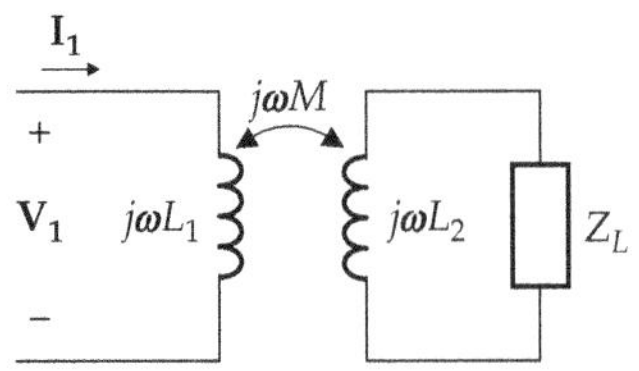

FIGURE P9.29

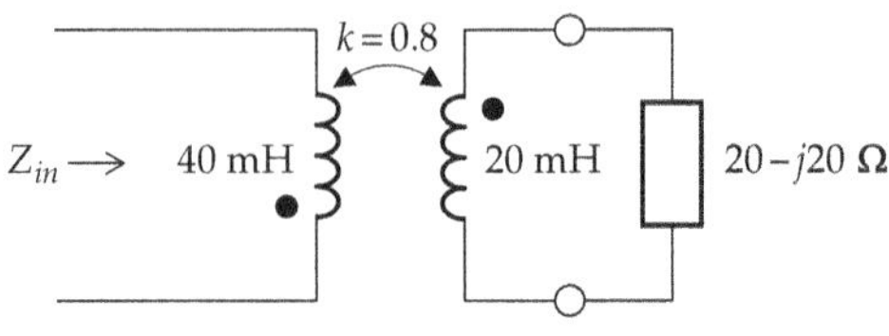

FIGURE P9.30

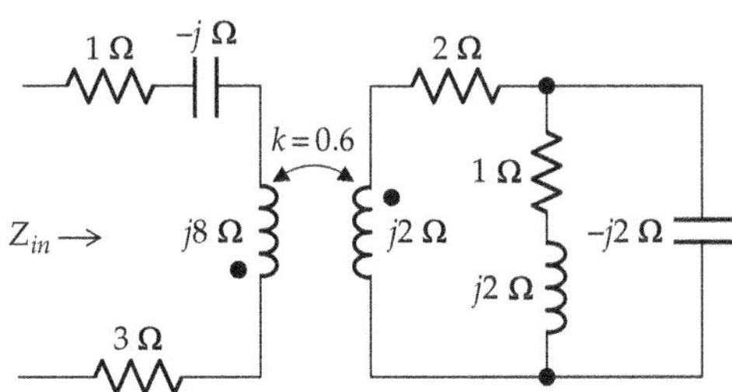

FIGURE P9.31

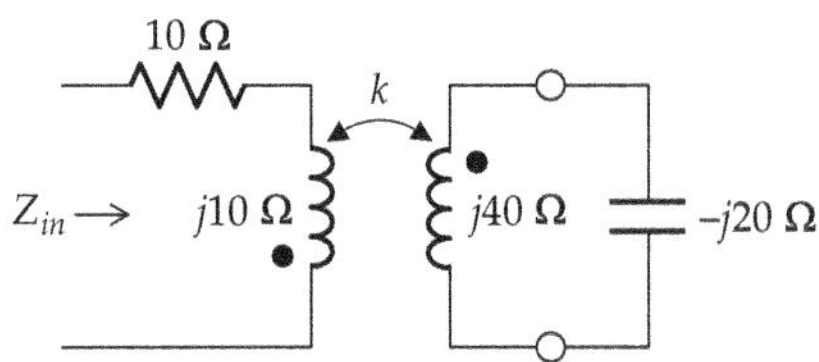

FIGURE P9.32

P9.31 Using the result of Problem P9.29, determine Z_{in} in Figure P9.31.

Ans. (a) $4.96 + j7$ Ω.

P9.32 Using the result of Problem P9.29, determine k in Figure P9.32 so that Z_{in} is purely resistive.

Ans. 0.71.

P9.33 Using the result of Problem P9.29, determine k in Figure P9.33 so that (a) the input impedance Z_{in} is purely resistive and (b) $\mathbf{I}_1$ is maximum.

Ans. (a) $k = 1$; (b) $k = 0$.

P9.34 Determine Z_{in} in Figure P9.34, assuming $Z = 3 + j4$ Ω.

Ans. $j2.75$ Ω.

P9.35 Determine M in Figure P9.35 so that $i = 0$.

Ans. 1 H.

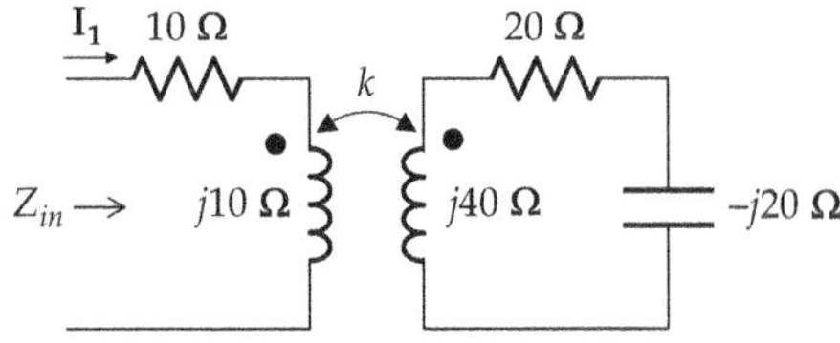

FIGURE P9.33

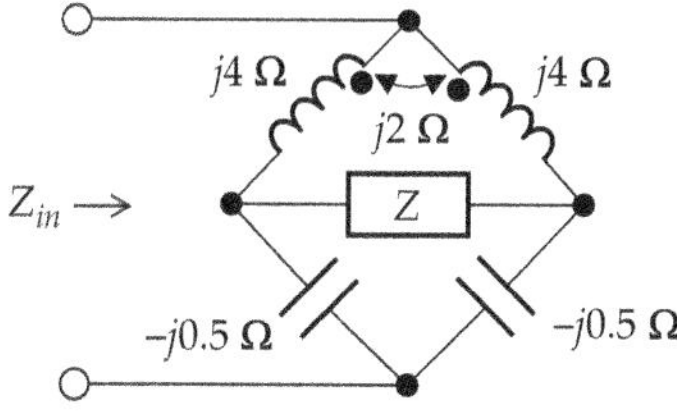

FIGURE P9.34

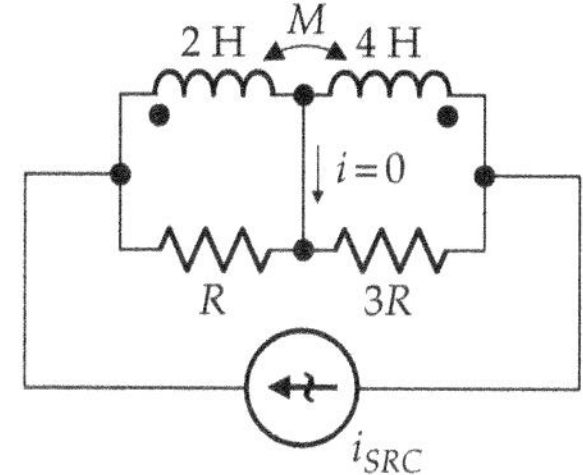

FIGURE P9.35

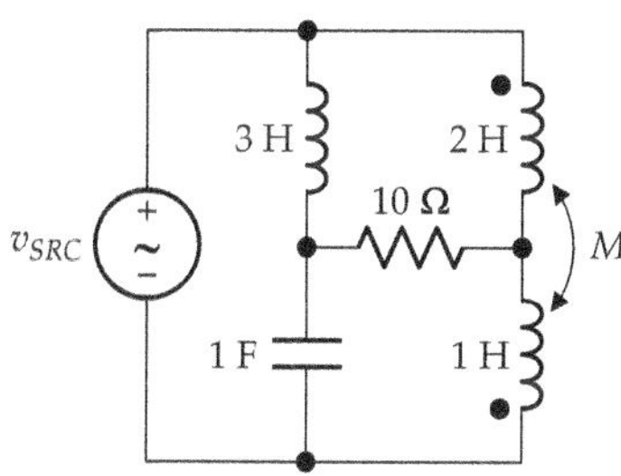

FIGURE P9.36

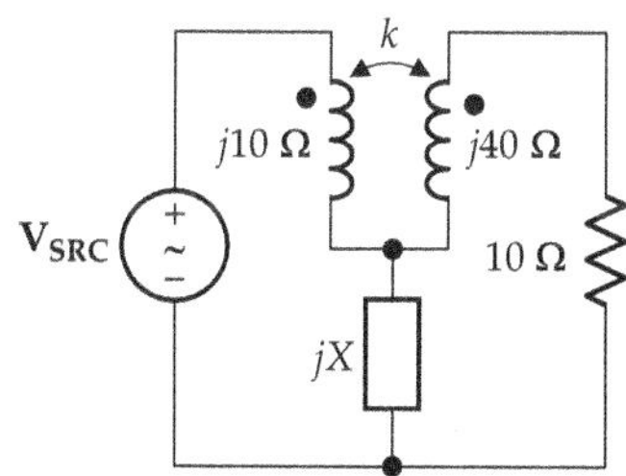

FIGURE P9.37

P9.36 Determine M in Figure P9.36 so that no current flows in the 10 Ω resistor, assuming $\omega = 1$ rad/s.

Ans. 1.25 H.

P9.37 Given $v_{SRC}(t) = 6\cos\omega t$ V and $k = 0.9$ in Figure P9.37, determine X so that no power is dissipated in the circuit.

Ans. −18 Ω.

P9.38 Determine the frequency at which the current i in Figure P9.38 has the same magnitude when the connections of one coil are reversed.

Ans. $\sqrt{10/3}$ rad/s.

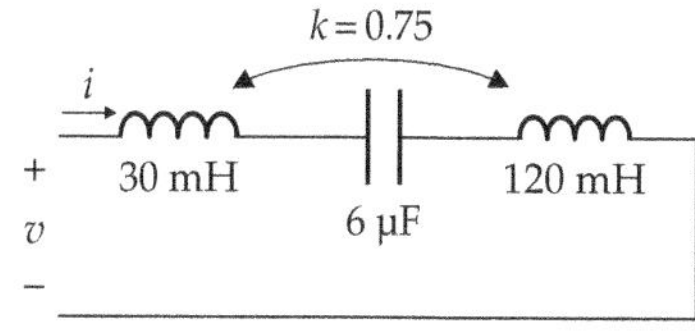

FIGURE P9.38

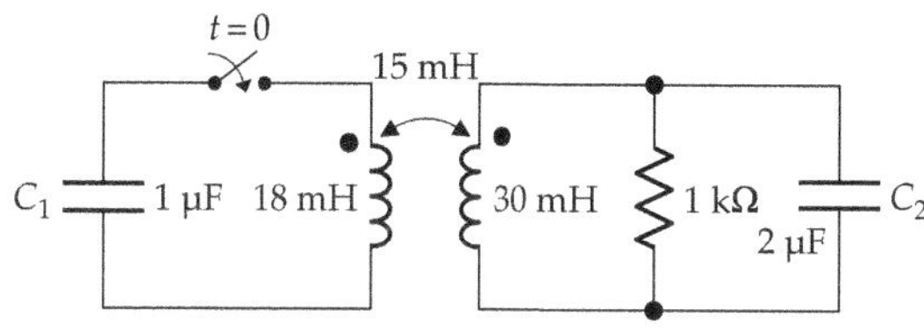

FIGURE P9.39

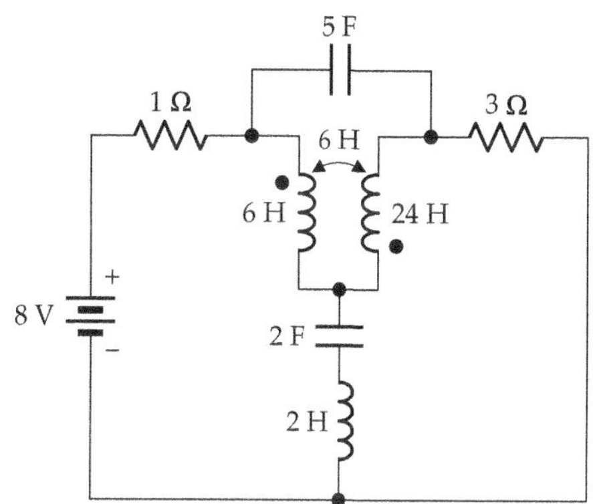

FIGURE P9.40

P9.39 C_1 in Figure P9.39 is initially charged to 6 V, and C_2 is uncharged. The switch is closed at $t = 0$. Calculate the total energy dissipated in the resistor.

Ans. 18 μJ.

P9.40 Determine the total energy stored in the circuit of Figure P9.40 in the dc steady state.

Ans. 120 J.

P9.41 Determine the stored energy in the circuit of Figure P9.41 in the dc steady state, assuming $M = 1$ H.

Ans. 14 J.

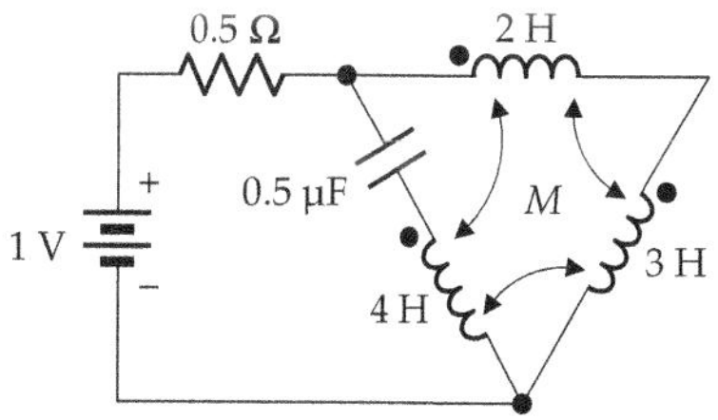

FIGURE P9.41

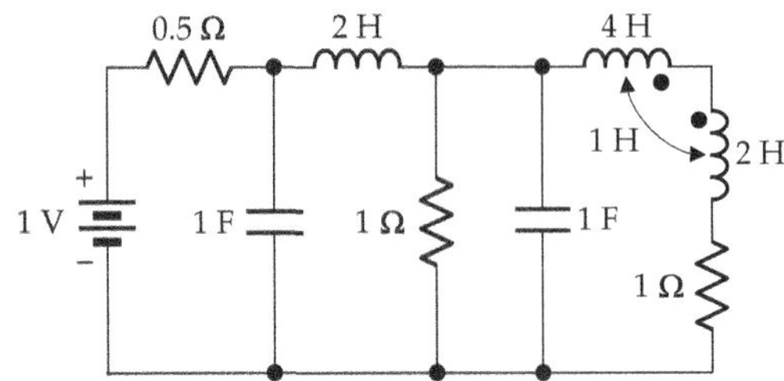

FIGURE P9.42

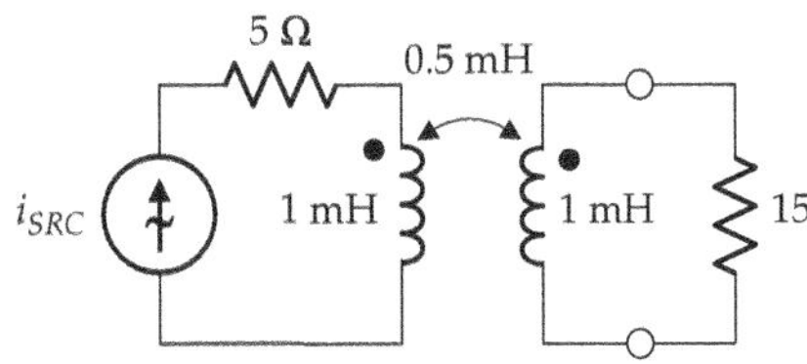

FIGURE P9.43

P9.42 Determine the energy stored in the circuit of Figure P9.42 in the dc steady state.

Ans. 1.75 J.

P9.43 Given $i_{SRC}(t) = 20\cos(10^4 t + 60°)$ A in Figure P9.43, determine the magnetic stored energy at $t = 0.1\pi$ ms.

Ans. 65.3 mJ.

P9.44 Using Equation 9.48, show that if two coils are perfectly coupled, the stored magnetic energy is $w = \frac{1}{2}\left(\sqrt{L_1}i_1 \pm \sqrt{L_2}i_2\right)^2$.

P9.45 Determine $\mathbf{I_S}$ in Figure P9.45.

Ans. $5\sqrt{2}\angle -45°$ A.

P9.46 Determine $\mathbf{V_O}$ in Figure P9.46.

Ans. $5.37\angle 3.43°$ V.

P9.47 Determine TEC looking into terminals 'ab' in Figure P9.47, assuming $v_{SRC}(t) = 2\cos(10^3 t - 60°)$ V and $k = 0.75$.

Ans. $\mathbf{V_{Th}} = \mathbf{V_{ab}} = 6\angle 120°$ V, $Z_{Th} = -j0.5$ Ω.

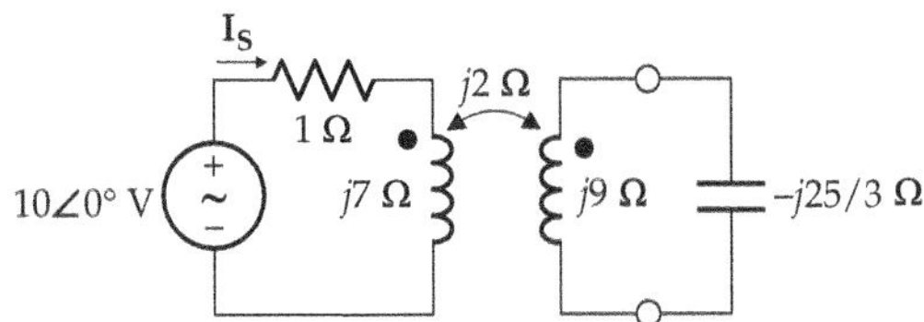

FIGURE P9.45

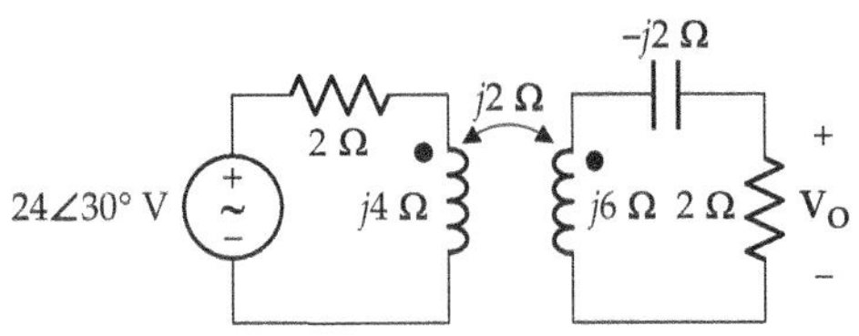

FIGURE P9.46

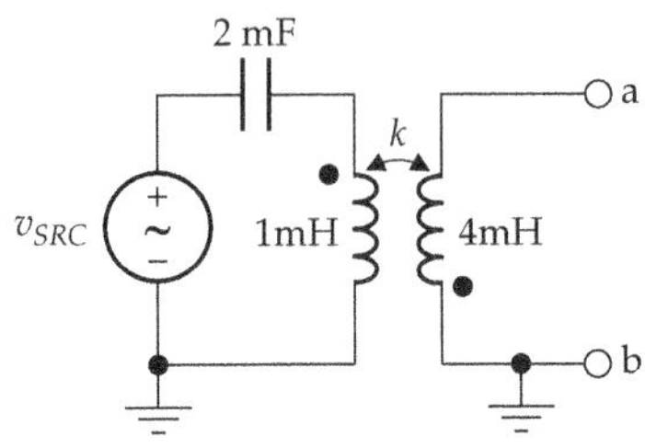

FIGURE P9.47

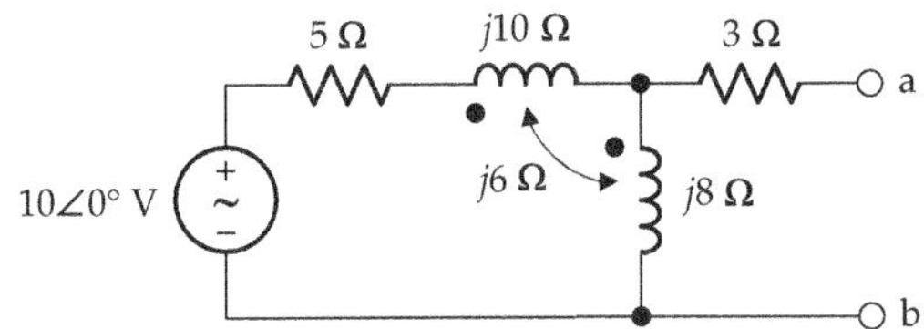

FIGURE P9.48

P9.48 Derive TEC between terminals 'ab' in Figure P9.48.

Ans. $\mathbf{V_{Th}} = \mathbf{V_{ab}} = \frac{28}{37}(6+j) = 4.54 + j0.76$ V, $Z_{Th} = \frac{751 + j304}{185} = 4.06 + j1.64$ Ω.

P9.49 Derive TEC looking into terminal 'ab' in Figure P9.49.

Ans. $\mathbf{V_{Th}} = \mathbf{V_{ab}} = 80/3$ V, $Z_{Th} = j200/9$ Ω.

P9.50 Derive TEC between terminals 'ab' in Figure P9.50.

Ans. $\mathbf{V_{Th}} = \mathbf{V_{ab}} = 60(1 + j)$ V, $Z_{Th} = 3(3 + j17)$ Ω.

P9.51 Determine k_X in Figure P9.51 so that no power is delivered or absorbed by v_{SRC2}, given that $v_{SRC1}(t) = 10\cos 10t$ V, $v_{SRC2}(t) = 10\sin 10t$ V, and $k = 0.05$.

Ans. $k_X = 1$.

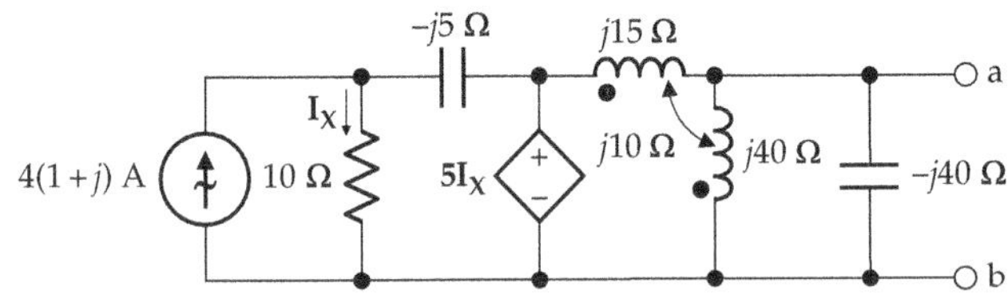

FIGURE P9.49

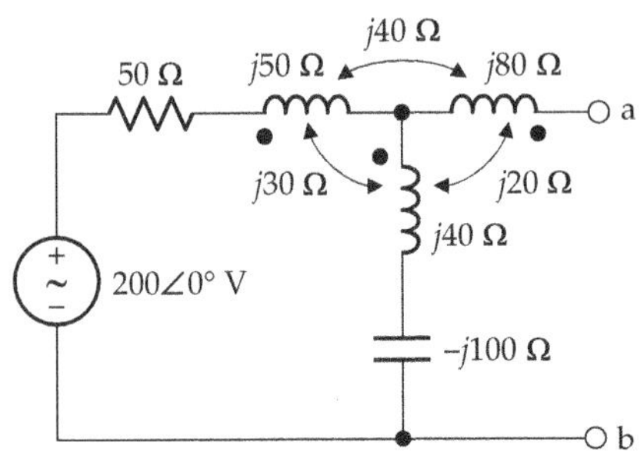

FIGURE P9.50

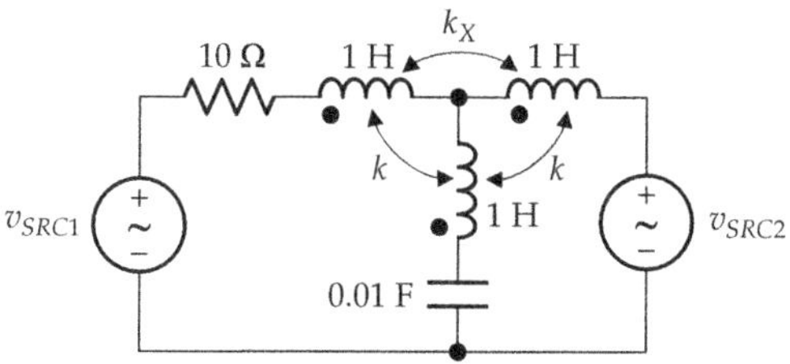

FIGURE P9.51

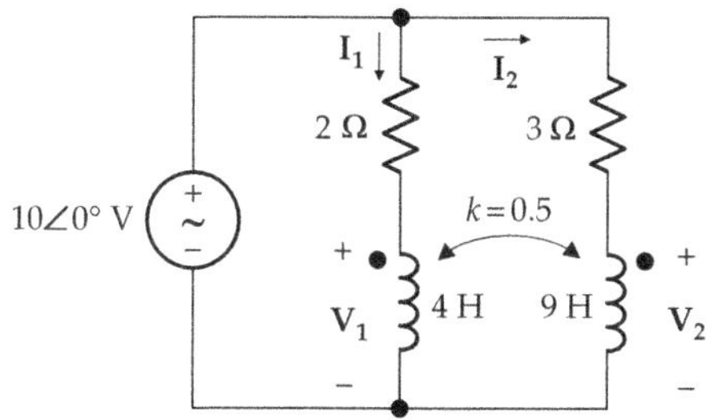

FIGURE P9.52

P9.52 Determine $\mathbf{I_1}$ and $\mathbf{I_2}$ in Figure P9.52, assuming $\omega = 1$ rad/s.

Ans. $\mathbf{I_1} = 0.873 - j1.61$ A, $\mathbf{I_2} = 0.0895 - j0.604$ A.

P9.53 Determine v_O in Figure P9.53 given that $v_{SRC}(t) = 2\cos 10^3 t$ V.

Ans. $-0.6\sqrt{2}\cos\left(10^3 t + 45°\right)$ V.

P9.54 Determine $i_O(t)$ in Figure P9.54 given that $v_{SRC}(t) = 200\sin(10^3 t)$ V.

Ans. $-0.0105\cos(10^3 t)$ A.

P9.55 Determine $\mathbf{I_X}$ in Figure P9.55.

Ans. $-0.0158 - j0.0123$ A.

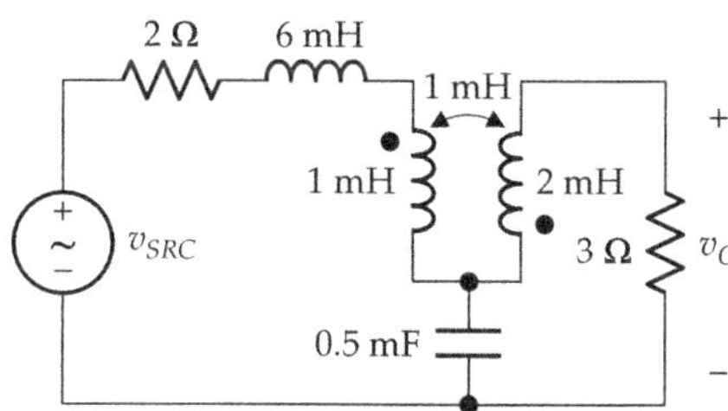

FIGURE P9.53

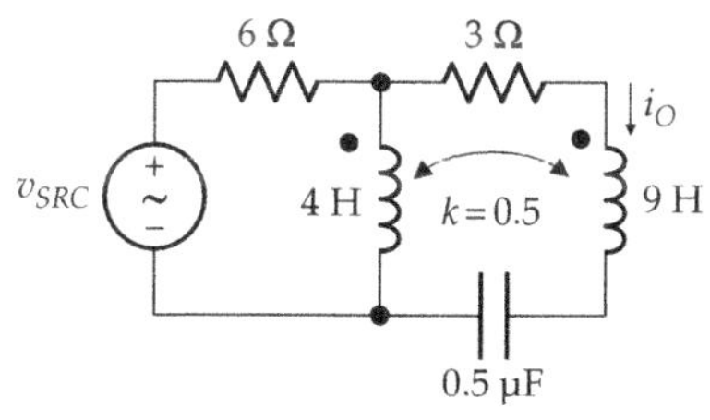

FIGURE P9.54

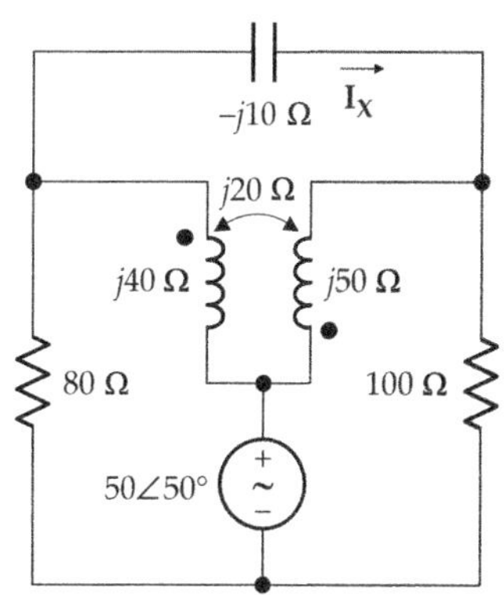

FIGURE P9.55

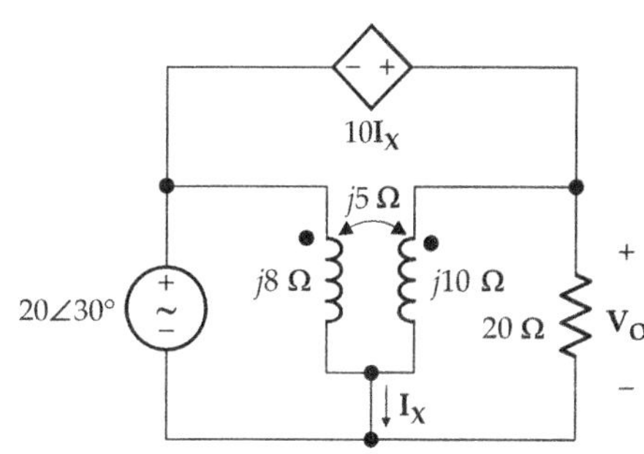

FIGURE P9.56

P9.56 Determine $\mathbf{V_O}$ in Figure P9.56.

Ans. $17.94 - j15.53$ V.

P9.57 Derive TEC looking into terminals 'ab' in Figure P9.57, given that $v_{SRC}(t) = 4\sqrt{2}\cos(t + 45°)$ V. Represent Thevenin's voltage in the time domain and express Z_{Th} in rectangular coordinates.

Ans. $\mathbf{V_{Th}} = \mathbf{V_{ab}} = -2\sin t$ V, $-0.25 + j0.25$ Ω.

P9.58 Determine $i_O(t)$ in Figure P9.58, given that $v_{SRC}(t) = 50\cos 500t$ V.

Ans. $3.33\cos(500t - 7.87°)$ A.

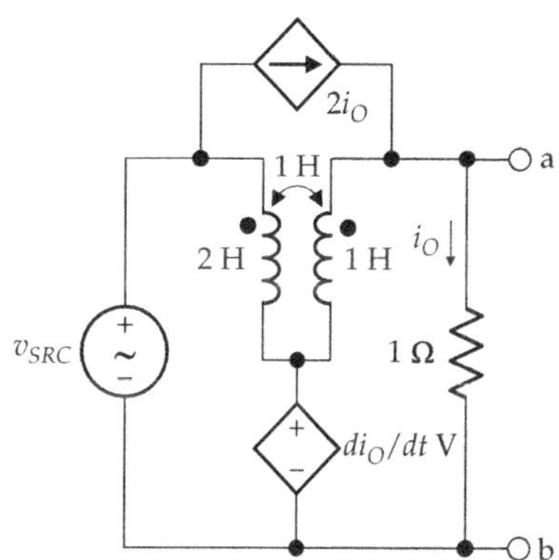

FIGURE P9.57

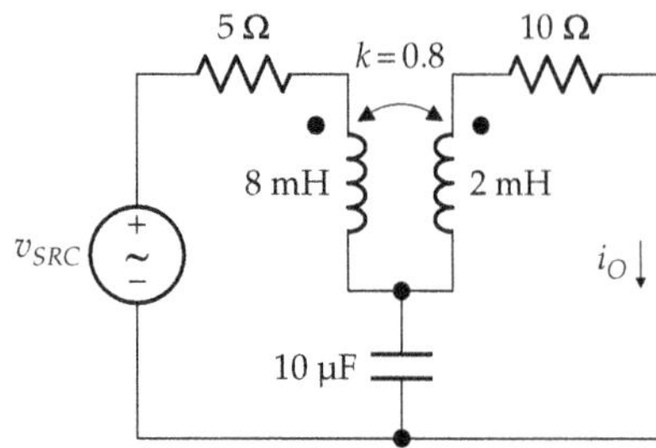

FIGURE P9.58

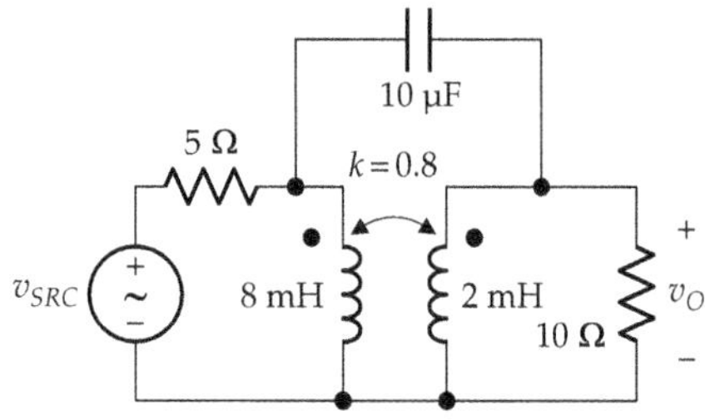

FIGURE P9.59

P9.59 Determine $v_O(t)$ in Figure P9.59 given that $v_{SRC}(t) = 50\cos 500t$ V.

Ans. 12.1cos(500*t* + 46.7°) V.

P9.60 Determine $\mathbf{V_O}$ and $\mathbf{V_S}$ in Figure P9.60

Ans. $\mathbf{V_O} = \frac{8\sqrt{2}}{3}\angle -90°$, $\mathbf{V_S} = \frac{8}{3}\angle 45°$ V.

P9.61 Determine the mesh currents in Figure P9.61.

Ans. $\mathbf{I_1} = 0.47\angle 13.5°$ A, $\mathbf{I_2} = 0.64\angle 104.1°$ A, $\mathbf{I_3} = 0.59\angle 67.7°$ A.

P9.62 Determine $\mathbf{I_O}$ in Figure P9.62.

Ans. 0.367 + *j*9.92 A.

P9.63 Determine $\mathbf{V_X}$ and $\mathbf{V_Y}$ in Figure P9.63, assuming *f* = 50 Hz.

Ans. $\mathbf{V_X} = 45.3\cos(100\pi t + 25.3°)$ V, $\mathbf{V_Y} = 16.89\cos(100\pi t - 143.6°)$ V.

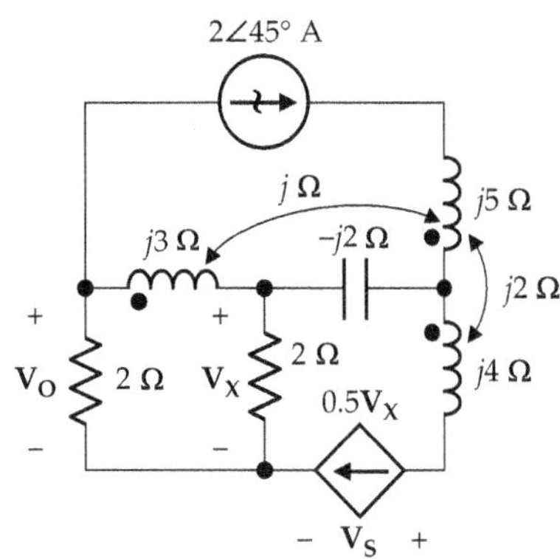

FIGURE P9.60

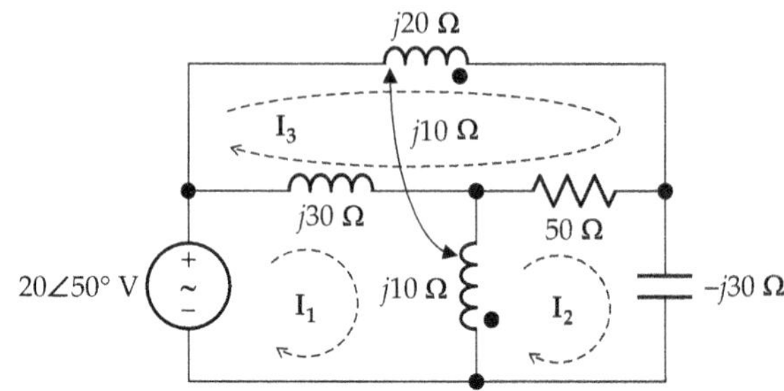

FIGURE P9.61

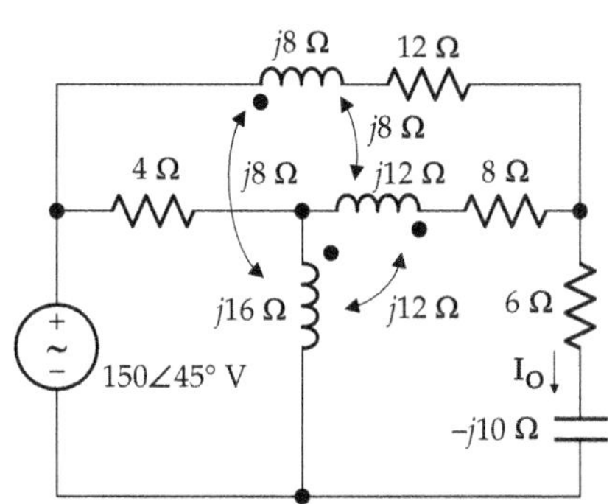

FIGURE P9.62

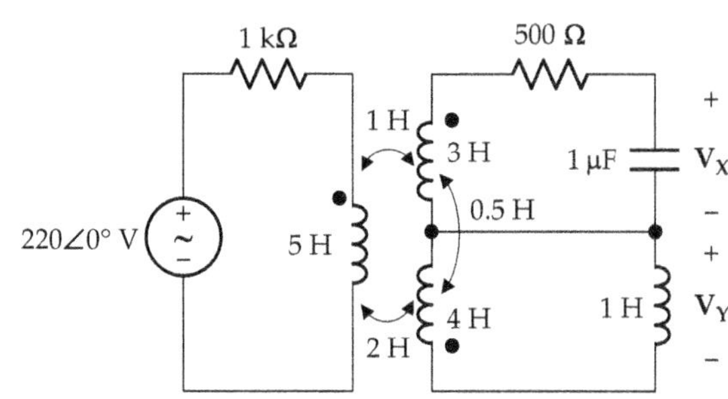

FIGURE P9.63

P9.64 For the circuit of Figure P9.64, (a) derive the mesh-current equations and (b) determine $\mathbf{V_{ab}}$.

Ans. $\mathbf{V_{ab}} = j10$ V.

P9.65 Determine *R*, *ω*, and *k* in Figure P9.65 so that V_{Th} looking into terminals 'ab' is $10\sqrt{2}\angle 0°$ and Z_{Th} is purely resistive.

Ans. *R* = 20 Ω, *ω* = 18 rad/s, *k* = 0.5, Z_{Th} = 10 Ω.

P9.66 Derive Thevenin's equivalent circuit between terminals 'ab' in Figure P9.66, assuming that $v_{SRC}(t) = 5\cos\omega t$ V.

Ans. $\mathbf{V_{Th}} = 5\cos\omega t$ V, $Z_{Th} = 0$.

P9.67 Show that at bridge balance ($v_O = 0$) in Figure P9.67, $M = \frac{L_1 R_4}{R_2 + R_4}$ and $M = C_2 R_1 R_4$.

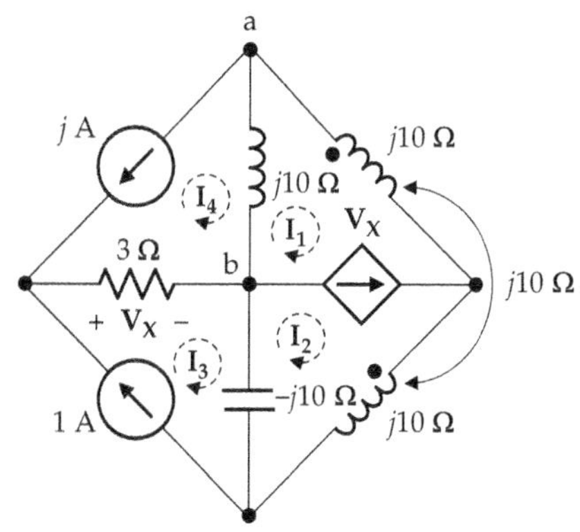

FIGURE P9.64

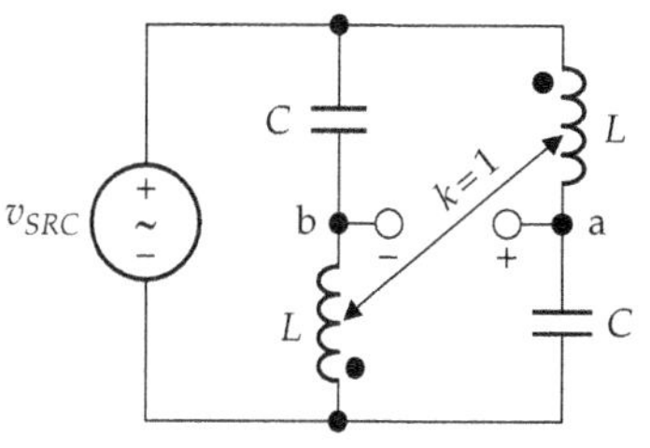

FIGURE P9.66

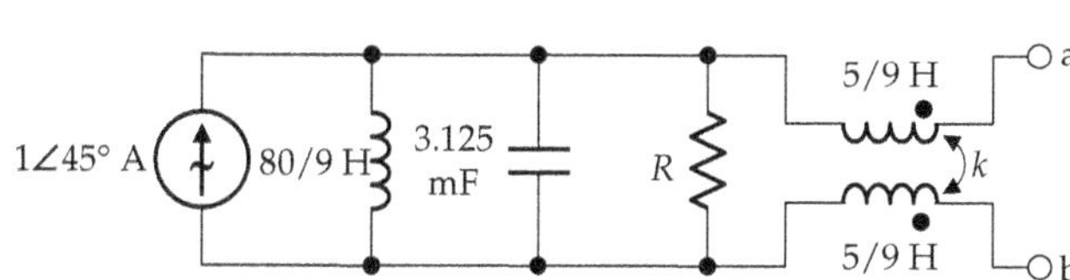

FIGURE P9.65

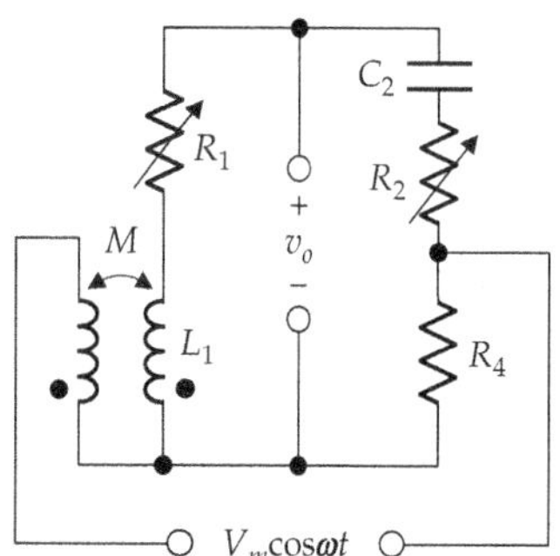

FIGURE P9.67

10

Ideal Transformers

Objective and Overview

This chapter considers a special case of the linear transformer, namely, the ideal transformer, which is of considerable practical and theoretical importance.

The chapter begins by highlighting some basic properties of magnetic circuits in order to better appreciate the properties of the ideal transformer. These properties are then specified, and their implications carefully considered, leading to the circuit attributes of the ideal transformer. Reflection of circuits from one side of the transformer to the other is discussed as an aid to analyzing circuits involving ideal transformers.

A special case of the ideal transformer, namely, the ideal autotransformer is presented. The chapter ends with transformer imperfections, their implications, and how they are dealt with in practical transformers.

This chapter builds on some of the basic concepts discussed in Section 7.2 and in Chapter 9.

10.1 Magnetic Circuit

A useful analogy exists between magnetic circuits and electric circuits. This analogy may be illustrated by considering a toroid of conducting material surrounded by an insulating medium, as in Figure 10.1a. A magnetic field exists inside the toroid, perpendicular to the plane of the toroid and directed into the plane of the paper. If this magnetic field changes with time, it induces a voltage in the toroid, in accordance with Faraday's law, which causes an electric current to flow in the toroid, as diagrammatically illustrated in Figure 10.1a. The direction of current is clockwise, in accordance with the right-hand rule.

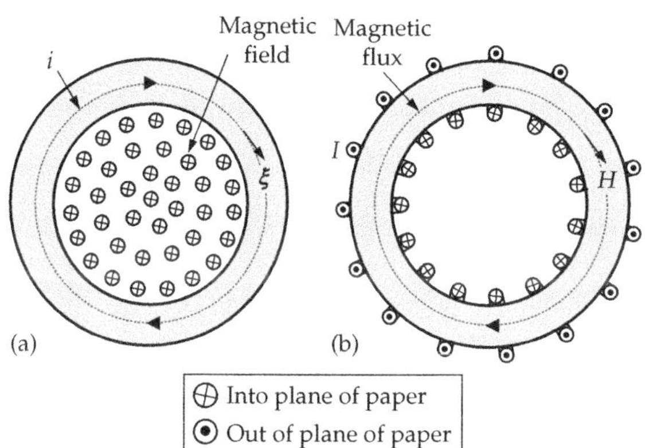

FIGURE 10.1
(a) Electric circuit; (b) magnetic circuit.

Consider next a toroid of magnetic material around which a coil is wound, as in Figure 10.1b. When a current I flows in the coil, a magnetic field is established in the toroid, as diagrammatically illustrated in the figure. If the current flows out of the plane of the paper around the outer circle of the toroid and into the plane of the paper around the inner circle, the magnetic field lines, or lines of magnetic flux, in the toroid are directed clockwise, in accordance with the right-hand rule.

The electrically conducting toroid in Figure 10.1a, in which a current flows under the influence of the induced voltage, is a simple electric circuit. The toroid of magnetic material in Figure 10.1b, in which a magnetic flux is established by the coil current, is a simple magnetic circuit. These two circuits are analogous in the following respects:

1. Both the lines of current flow in Figure 10.1a and the lines of magnetic flux in Figure 10.1b form closed loops, as is always the case. Current lines form closed loops because of conservation of charge, as explained in Section 1.2. Electric charge cannot be created or destroyed, which means that there are no sources or sinks of electric charge, so this charge can only circulate in closed loops. Magnetic flux lines form closed loops because magnetic monopoles, that is, isolated north or south poles, do not exist. The reason is that all magnetic fields are generated by circulating currents, as in Figure 10.1b, even at the atomic level. Magnetic flux is thus analogous to electric current:

$$\text{Electric current} \leftrightarrow \text{Magnetic flux} \qquad (10.1)$$

2. Current flows in the toroid of Figure 10.1a under the influence of the induced voltage, which in this case is more aptly described as an induced

emf (Section 2.4), since it drives the current in the toroid. This induced emf is related to the electric field ξ in the toroid by the integral relation:

$$\oint \xi dl = \text{Induced voltage or emf} \tag{10.2}$$

where ξ is tangential to the circular lines of current flow and the line integral is around the closed path of current in the toroid.

Magnetic flux in Figure 10.1b is established by the current in the coil wound around the toroid. Ampere's circuital law (Equation 7.12) gives the relation between magnetic field strength H in the toroid and the current producing the magnetic field as

$$\oint Hdl = NI \tag{10.3}$$

where

H is tangential to the circular flux lines and the line integral is around the closed path of magnetic flux lines in the toroid

N is the number of turns of the coil

Equations 10.2 and 10.3 are of the same form. The line integral on the LHS of Equation 10.2 is that of the electric field associated with the induced voltage, and the RHS of the equation is the driving voltage or emf. The LHS of Equation 10.3 is the line integral of the magnetic field strength associated with current in the coil, and the RHS of this equation is NI. By analogy with the RHS of Equation 10.2, NI can be considered as the "driving force" for the magnetic flux. NI is therefore referred to as the **magnetomotive force** (mmf), analogous to the emf. Thus,

$$\text{emf} \leftrightarrow \text{mmf} \tag{10.4}$$

Once the analogs of current and voltage are identified, the analogs of resistance and conductance readily follow, as indicated in Table 10.1. Resistance is voltage/current, so its magnetic circuit analog is mmf/(magnetic flux) and is termed **reluctance** (R). The magnetic circuit analog of conductance is flux/mmf and is the **permeance** (P). It follows that the analog of Ohm's law in the form of $v = i/G$ is

TABLE 10.1

Electric Circuit Analogy

Electric Circuit	Magnetic Circuit
Current	Flux (ϕ)
emf (V)	mmf (Ni)
Resistance	Reluctance (mmf/flux)
Conductance	Permeance (flux/mmf)
Conductivity	Permeability

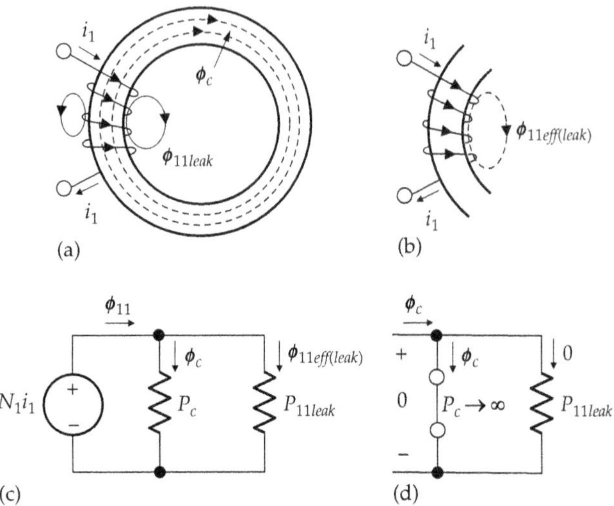

FIGURE 10.2
Coil wound on a toroidal core, (a) magnetic flux, (b) effective leakage flux, (c) equivalent electric circuit, (d) core of infinite permeability.

$$\text{mmf} = \frac{\text{Flux}}{\text{Permeance}} \tag{10.5}$$

The analogy between magnetic and electric circuits can be illustrated by considering a coil wound on a toroidal core, taking into account the leakage flux (Section 9.2), as in Figure 10.2a. It is assumed, for simplicity, as in the calculation of inductance of the coil in Figure 7.15, that the diameter of the core is small compared with the mean diameter, a, of the toroid. H_c in the core can then be considered constant across a transverse cross section of the core and is given by Ampere's circuital law as (Equation 7.19)

$$H_c = \frac{N_1 i_1}{\pi a} \tag{10.6}$$

where

πa is the length of path over which H_c is integrated

N_1 is the number of turns of coil 1

The magnetic flux density in the core, B_c, is

$$B_c = \mu_c H_c = \frac{\mu_c}{\pi a} N_1 i_1 \tag{10.7}$$

where $\mu_c = \mu_{rc}\mu_0$ is the permeability of the core. ϕ_c, the flux in the core, is

$$\phi_c = B_c A_c = \frac{\mu_c A_c}{\pi a} N_1 i_1 \tag{10.8}$$

where A_c is the cross-sectional area of the core.

The permeance of the core is the ratio of ϕ_c to the mmf, $N_1 i_1$. Thus,

$$P_c = \frac{\phi_c}{N_1 i_1} = \frac{\mu_c A_c}{\pi a} \tag{10.9}$$

The following should be noted concerning Equation 10.9:

1. The permeance depends only on the geometry and magnetic properties of the magnetic circuit and is independent of the coil current or the number of turns of the coil. In contrast, inductance is proportional to the square of the number of turns (Equation 7.14).
2. Equation 10.9 is analogous to Equation 3.25, in the form

$$G = \frac{\sigma A}{L} \tag{10.10}$$

with L identified with the length of path πa.

3. Comparing Equations 10.9 and 10.10, it is seen that permeability is analogous to conductivity. The higher the permeability of a magnetic material, the larger the flux per unit mmf is, just as the higher the conductivity of an electrical conductor, the larger the current per unit applied voltage is.

What about the leakage path? From superposition, the total flux ϕ_{11} in the coil is (Equation 9.7)

$$\phi_{11} = \phi_c + \phi_{11eff(leak)} \tag{10.11}$$

where $\phi_{11eff(leak)}$ is the effective leakage flux that, when multiplied by N_1, gives the true value of leakage flux linkage of coil 1. Since $\phi_{11eff(leak)}$, by definition, links all the turns of the coil, then if a closed path is followed along the path of $\phi_{11eff(leak)}$ in the space between the coil and the core and back, as illustrated in Figure 10.2b, the enclosed current is $N_1 i_1$, just as for a path involving the core flux ϕ_c. In other words, *the same mmf* $N_1 i_1$ *acts on the two paths of the fluxes,* ϕ_c *and* $\phi_{11eff(leak)}$.

Dividing both sides of Equation 10.11 by $N_1 i_1$ converts the equation to one involving permeances. Thus,

$$P_1 = P_c + P_{11eff(leak)} \tag{10.12}$$

where

P_c is given by Equation 10.9

$P_{11eff(leak)}$ is the effective permeance of the leakage path

By analogy to Equation 10.9, $P_{11eff(leak)}$ is zero when the area of the leakage path is zero, that is, when the coil conductor is very thin and is tightly wound around the core.

The electric circuit analog, illustrated in Figure 10.2c, is a driving voltage $N_1 i_1$ applied to two conductances in parallel. Equation 10.11 is KCL at either essential node. Ohm's law applied to two conductances G_1 and G_2 in parallel gives

$$v = \frac{i_1}{G_1} = \frac{i_2}{G_2} = \frac{i}{G_1 + G_2} \tag{10.13}$$

where

i_1 and i_2 are the currents in G_1 and G_2, respectively

$i = i_1 + i_2$

v is the voltage across the parallel combination

By analogy, it follows from Figure 10.2c that

$$N_1 i_1 = \frac{\phi_c}{P_c} = \frac{\phi_{11eff(leak)}}{P_{11eff(leak)}} = \frac{\phi_{11}}{P_c + P_{11eff(leak)}} \tag{10.14}$$

The following deductions from Equation 10.14 are of interest for the discussion in the next section on the ideal transformer:

1. For a given flux in the core (ϕ_c), the larger the permeability of the core (μ_c), the larger P_c is and the smaller the required mmf, $N_1 i_1$, is. In the limit, if $\mu_c \to \infty$, then $P_c \to \infty$, and with ϕ_c finite, $N_1 i_1 \to 0$ (Figure 10.2d). In terms of the electric circuit analogy, a finite current ϕ_c flowing in a short circuit ($P_c \to \infty$) requires a zero driving voltage ($N_1 i_1 = 0$).
2. It follows from Equation 10.14 that

$$\frac{\phi_{11eff(leak)}}{\phi_c} = \frac{P_{11eff(leak)}}{P_c} \quad \text{or} \quad \phi_{11eff(leak)} = \frac{P_{11eff(leak)}}{P_c}\phi_c \tag{10.15}$$

As in the case of current division between two conductances in parallel, the ratio of the fluxes in the two parallel paths is the same as the ratio of the permeances.

If $P_c \rightarrow \infty$, then for a given ϕ_c, $\phi_{11eff(leak)} \rightarrow 0$. This is a consequence of the voltage across the short circuit being zero, so that the current through any conductance in parallel with the short circuit is zero (Figure 10.2d).

If both sides of Equation 10.11 are multiplied by N_1,

$$\lambda_{11} = N_1\phi_{11} = N_1\phi_c + N_1\phi_{11eff(leak)} \tag{10.16}$$

If both sides of the flux linkage Equation 10.16 are divided by i_1, the inductances replace the flux linkages, which gives

$$L_1 = L_c + L_{11leak} \tag{10.17}$$

where L_c is given by Equation 7.24. It should be noted that the permeances of the core and the leakage path are in *parallel*, since the same mmf acts on both. However, the inductances of the two paths are in *series*, because the same time-varying current i_1 induces voltages due to the changing core flux and leakage flux. These induced voltages add together in the coil to give the total voltage induced in the coil. Thus, if both sides of Equation 10.17 are multiplied by di/dt,

$$L_1\frac{di_1}{dt} = L_c\frac{di_1}{dt} + L_{11leak}\frac{di_1}{dt} \tag{10.18}$$

where the term on the LHS is the total voltage induced by the two fluxes.

Primal Exercise 10.1

If i_1 = 2 A in Figure 10.2a results in ϕ_c = 0.1 Wb and $\phi_{11eff(leak)}$ = 5 mWb, determine P_c and $P_{11eff(leak)}$, assuming N_1 = 100 turns.

Ans. P_c = 0.5 mWb/A-turn; $P_{11eff(leak)}$ = 25 μWb/A-turn.

Example 10.1: Equivalent Electric Circuit of a Shell-Type Core

Given a coil wound on the central limb of a shell-type magnetic core, illustrated in Figure 10.3a, it is required to derive the analog electric circuit, neglecting leakage flux.

Solution:

The electric circuit analog is shown in Figure 10.3b. An mmf Ni is applied by the coil to a circuit consisting of the permeance of the central limb P_c in series with two conductances P_s in parallel, where P_s is the permeance of each of the two side limbs. If the flux in the central limb is ϕ_c, the flux in each side limb is $\phi_s = \phi_c/2$. The equivalent permeance P_{eq} in series with the source Ni is $P_{eq} = (P_c \times 2P_s)/(P_c + 2P_s) = (P_cP_s)/(P_c/2 + P_s)$, and $\phi_c = Ni/P_{eq}$.

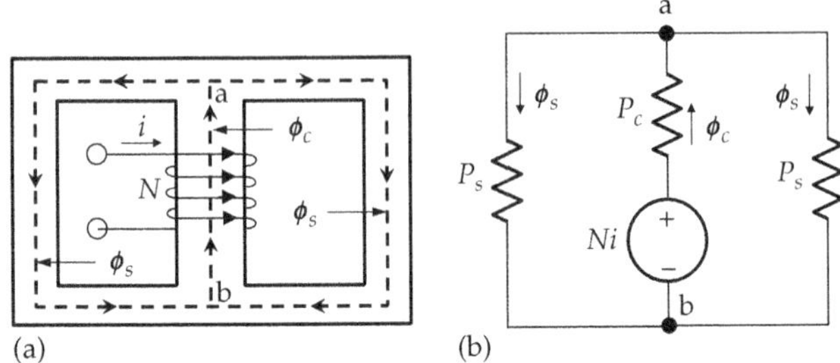

FIGURE 10.3
Figure for Example 10.1.

10.2 Ideal Transformer

10.2.1 Definition

Recall that ideal resistors, capacitors, and inductors were defined as possessing certain properties. We will similarly define an ideal transformer as an element having the following properties:

1. No power losses in the windings or in the core. No power loss in the windings implies zero resistance of the coils, and no power loss in the core implies certain properties of the core material, discussed in Section 10.5.
2. No energy stored in the electric field, as was assumed in the case of ideal resistors and inductors.
3. Perfect magnetic coupling to the core, which means no leakage path and no leakage flux.
4. Infinite core permeability μ_c. Since inductance and permeance are both directly proportional to permeability, the inductances of the coils, and the permeance of the core are also infinite.

Consider an ideal transformer consisting of two coils, as in Figure 10.4a. A time-varying voltage v_1 is applied to coil 1, coil 2 being open-circuited. With zero leakage flux, in accordance with item 3 of the definition, $\phi_{11} = \phi_c$, the flux in the core (Equation 10.11). The voltage induced in coil 1 is therefore entirely due to ϕ_c, the flux in the core. Moreover, with zero coil resistance and zero leakage flux, the terminal voltage v_1 is the same as the voltage induced in the coil by ϕ_c. It follows From Faraday's law that.

$$|v_1| = N_1\left|\frac{d\phi_c}{dt}\right| \quad \text{or} \quad |\phi_c| = \frac{1}{N_1}\left|\int v_1 dt\right| \tag{10.19}$$

According to Equation 10.19, $|\phi_c|$ for the ideal transformer in Figure 10.4 depends entirely on the number

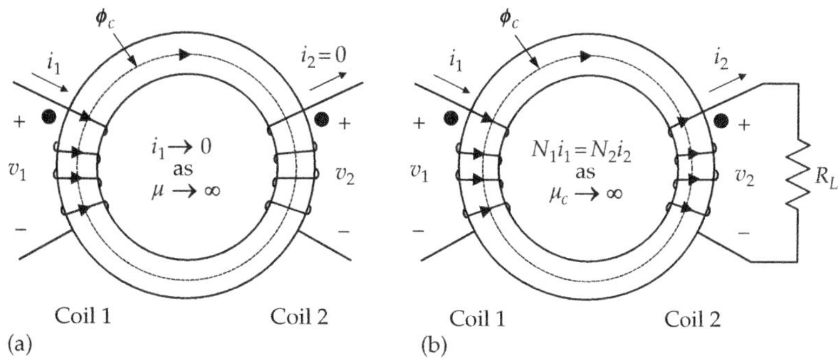

FIGURE 10.4
Voltages and currents of an ideal transformer. A two-coil ideal transformer with coil 2 open circuited (a) and carrying current (b).

of turns N_1 and on the magnitude and time variation of the applied voltage v_1. It does not depend on the permeability of the core nor on any current that may be flowing in the coil. For example, if $v_1 = 10\cos 100t$ V, with t in s and $N_1 = 1000$ turns, then ϕ_c is uniquely determined by v_1 and N_1 as $0.1\sin 100t$ mWb, irrespective of μ_c and i_1.

What is then the effect of infinite μ_c? The effect is on the current i_1. From the definition of inductance in terms of flux linkage,

$$i_1 = \frac{\lambda_1}{L_1} = \frac{N_1\phi_c}{L_1} \tag{10.20}$$

If ϕ_c is finite, as determined by v_1 and N_1, then $i_1 \to 0$ as $\mu_c \to \infty$, and hence $L_1 \to \infty$. This has two important implications:

1. *The mmf $N_1 i_1$ on the core is zero*. This is in accordance with Equation 10.5, for if the permeance is infinite, with the flux finite, the mmf is zero. It was argued in connection with the analog electric circuit of Figure 10.2d that having finite flux and infinite permeance is analogous to a finite current (the flux) flowing through a short circuit (infinite permeance), the resulting voltage (the mmf) being zero.
2. The power input into the transformer $v_1 i_1 = 0$. Hence, no energy is expended in establishing a finite ϕ_c in a core of infinite permeance, which means that no magnetic energy is stored in the core. With no leakage flux, and no *electric* energy stored in the transformer (assumption 2), this means that *no energy of any kind is stored in the ideal transformer*.

The time-varying ϕ_c induces a voltage v_2 in coil 2, whose magnitude is given by Faraday's law: $|v_2| = N_2|d\phi_c/dt|$. From Equation 10.19, $|v_1| = N_1|d\phi_c/dt|$. Dividing this by the former relation for $|v_2|$ gives

$$\frac{|v_1|}{|v_2|} = \frac{N_1}{N_2} \tag{10.21}$$

Equation 10.21 can be interpreted on the basis that ϕ_c is common to both coils of the ideal transformer of Figure 10.4a, which means that $|d\phi_c/dt|$ is the same for the two coils. In general, $|d\phi_c/dt| = |v|/N$, where $|v|/N$ is the *magnitude of the induced voltage per turn* in any coil wound on the core. Applying this equality to the two coils,

$$\left|\frac{d\phi_c}{dt}\right| = \frac{|v_1|}{N_1} = \frac{|v_2|}{N_2} \tag{10.22}$$

which is the same as Equation 10.21. Moreover, as in the case of coil 1, the terminal voltage v_2 is the same as the voltage induced in coil 2 by ϕ_c, even in the presence of coil current, because of the assumptions of zero leakage flux and zero resistance.

It should be emphasized that Equation 10.22 in terms of the volts induced per turn applies to any number of coils of the ideal transformer. In summary, the following concept applies:

Concept: *In an ideal transformer, the magnitude of the volts per turn induced by the flux in the core is the same for all coils of the ideal transformer. Because there is no leakage flux and no resistance, the induced voltage in any coil is the same as the terminal voltage of the coil. The sign of the terminal voltage in any given coil is determined by the assigned positive polarity of the voltage and the dot markings of the coils.*

In Figure 10.4a, the dot markings on the coils are indicated. According to the dot convention, the dot-marked terminals of both coils go positive together, or negative together. With the terminal voltages v_1 and v_2 assigned as shown, v_1 and v_2 in Figure 10.4a will always have the same sign. The magnitude designations in Equation 10.21 can be dropped, so that this equation can be written *in this case* as

$$\frac{v_1}{v_2} = \frac{N_1}{N_2} \tag{10.23}$$

If the assigned positive polarities of v_1 and v_2 do not conform to the dot markings, Equation 10.23 is written with a negative sign, as will be elaborated later.

Next, consider that coil 2 is connected to a resistor R_L (Figure 10.4b), so that i_2 flows as indicated. What is i_1 in this case? As explained earlier, ϕ_c depends only on v_1 and N_1 and not on any other variables such as i_1 or i_2. Moreover, when μ_c is infinite, the net mmf acting on the core is zero. Otherwise, a finite mmf multiplied by infinite permeance results in infinite flux, and hence infinite induced voltage, which is contrary to the assumption of a finite applied voltage. It follows that in the presence of i_2, i_1 *must be such that the net mmf acting on the core is zero*. Since permeance is a positive quantity, it is seen from Equation 10.5 that mmf and flux have the same sign or are in the same sense. That is, if the flux is clockwise, for example, then the mmf also acts clockwise. In Figure 10.4b, i_1 enters at the dot-marked terminal of coil 1, whereas i_2 leaves at the dot-marked terminal of coil 2. According to the dot convention, i_1 and i_2 produce fluxes in opposite directions in the core and therefore their mmfs are in opposition. To have a net mmf of zero acting on the core, $N_1i_1 - N_2i_2 = 0$, which gives

$$\frac{i_1}{i_2} = \frac{N_2}{N_1} \tag{10.24}$$

This underlies an important concept:

Concept: *In an ideal transformer, the net mmf acting on the core due to currents in all the coils of the ideal transformer must be zero. The sign, or sense, of the mmf due to current in a given coil depends on the assigned positive direction of current and on the dot markings.*

If Equations 10.23 and 10.24 are multiplied together,

$$v_1i_1 = v_2i_2 \tag{10.25}$$

In other words, the instantaneous power input and output are equal in an ideal transformer. This is to be expected, since an ideal transformer *neither dissipates nor stores energy*.

Figure 10.5 indicates the voltage and current ratios of an ideal transformer for different combinations of dot markings and assigned positive directions of currents and polarities of voltages. Note that the symbol for an ideal transformer has two parallel lines between the two coils. To justify the signs for the voltage and current ratios, consider Figure 10.5d, for example. The assigned positive direction of v_1 is as shown. v_1/N_1, the induced volts per turn in coil 1, is considered positive, with the positively assigned terminal being the terminal marked with a dot. The voltage induced in coil 2 is $N_2 \times$ (induced volts per turn), or $v_2' = (v_1/N_1)N_2$, and is positive because of the same dot markings with respect to v_1. According to the assigned positive polarity of v_2, $v_2 = -v_2'$. This makes $v_1/v_2 = -N_1/N_2$, as indicated. This is to be expected from the dot convention and the assigned positive polarities of v_1 and v_2.

The assigned positive directions of currents in Figure 10.5d are such that both i_1 and i_2 enter at the dot-marked terminals. The mmf's therefore add, so that for zero mmf in the core, $N_1i_1 + N_2i_2 = 0$. This gives $i_1/i_2 = -N_2/N_1$, as indicated. Note that when the dot markings are reversed, for the same assigned positive directions of voltage and current, the sign of the turns ratio also reverses in the expressions of v_1/v_2 and i_1/i_2. This is the case in Figure 10.5b compared to Figure 10.5d, and in Figure 10.5a compared to Figure 10.5c.

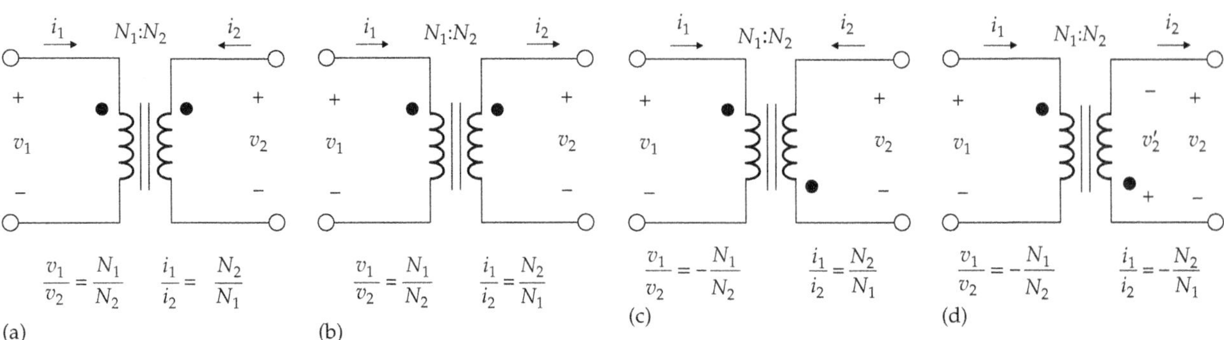

FIGURE 10.5
Voltage and current ratios of an ideal transformer for different combinations of dot markings and assigned positive directions of currents and polarities of voltages, as shown in (a) to (d).

Finally, it should be noted from the expression for inductance of the coils (Equation 7.24) that the ratio of the inductances of coils 1 and 2 in Figure 10.4 is

$$\frac{L_1}{L_2} = \left(\frac{N_1}{N_2}\right)^2 \qquad (10.26)$$

Although L_1 and L_2 are both infinite for an ideal transformer, their ratio is finite, because it is independent of μ_c and is equal to the square of the turns ratio.

In summary, the circuit implications of the properties of an ideal two-coil transformer are as follows:

1. $$\frac{v_1}{v_2} = \pm\frac{N_1}{N_2} \qquad (10.27)$$

   where the sign is positive if the assigned positive polarities of the voltages conform to the dot markings of the two coils and is negative otherwise. Equation 10.27 expresses the fundamental concept that the magnitude of the induced volts per turn is the same for the coils. It is the voltage equation of an ideal two-coil transformer.

2. $$\frac{i_1}{i_2} = \pm\frac{N_2}{N_1} \qquad (10.28)$$

   where the sign is negative if the assigned positive directions of currents are such that both currents flow into, or out of, the dotted terminals and is positive otherwise. Equation 10.28 expresses the fundamental concept that the net mmf on the core is zero. It is the current equation of an ideal two-coil transformer.

3. The magnitude of the current ratio is the reciprocal of that of the voltage ratio.
4. At any instant of time, the input power and output power are equal.
5. The inductances of the coils are infinite, but their ratio is finite and equals the square of the turns ratio.

Primal Exercise 10.2

Given $v_{SRC}(t) = 10\cos(100\pi t)$ V in Figure 10.6, determine v_2, i_2, and i_1, assuming (a) the dot markings shown and (b) reversed dot markings.

Ans. (a) $v_2(t) = 100\cos(100\pi t)$ V, $i_2(t) = 100\cos(100\pi t)$ mA, $i_1(t) = \cos(100\pi t)$ A.

(b) $v_2(t) = -100\cos(100\pi t)$ V, $i_2(t) = -100\cos(100\pi t)$ mA, $i_1 = \cos(100\pi t)$ A.

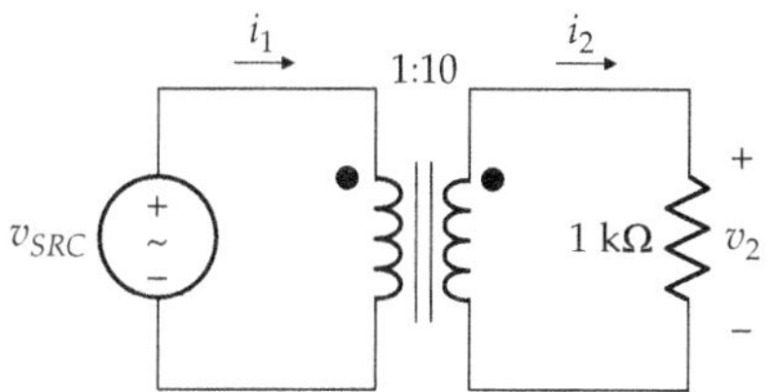

FIGURE 10.6
Figure for Primal Exercise 10.2.

10.2.2 Phasor Relations

The behavior of an ideal transformer can be further illustrated using phasor relations. Figure 10.7a shows an ideal transformer connected to a load impedance Z_L, with the corresponding phasor diagram depicted in Figure 10.7b. Figure 10.7c is a flow diagram of the causal relationships. A voltage $\mathbf{V}_1$ applied to coil 1 establishes a flux $\boldsymbol{\phi}_c$ in the core such that $\phi_c = \frac{1}{j\omega N_1}\mathbf{V}_1$, where integration with respect to time is equivalent to division by $j\omega$ in phasor notation. Note that $\boldsymbol{\phi}_c$ depends on $\mathbf{V}_1$, $\mathbf{N}_1$, and ω. Because of division by j, $\boldsymbol{\phi}_c$ lags $\mathbf{V}_1$ by 90°. The phase angle of $\mathbf{V}_1$ is arbitrarily taken as 90°, which makes the phase angle of $\boldsymbol{\phi}_c$ zero. $\boldsymbol{\phi}_c$ induces a voltage $\mathbf{V}_2$ in coil 2 such that $\mathbf{V}_2 = +j\omega N_2\phi_c$, assuming the assigned positive polarities and dot markings shown. Since $\mathbf{V}_2$ leads $\boldsymbol{\phi}_c$ by 90°, because of multiplication by j, it is in phase with $\mathbf{V}_1$, in accordance with the dot markings. $\mathbf{V}_2$ results in a current $\mathbf{I}_2$ that lags $\mathbf{V}_2$ by an angle θ, assuming Z_L is inductive. In order to have zero mmf in the core, a current $\mathbf{I}_1$ flows such that $N_1\mathbf{I}_1 - N_2\mathbf{I}_2 = 0$.

In a transformer having only one coil connected to a source of excitation, it is usual to refer to this coil as the **primary winding** and to all other coils as **secondary windings**. In Figure 10.7a, $\mathbf{V}_1$ is across the primary winding and Z_L is connected to the secondary winding. If $N_2 > N_1$, so that $|v_2| > |v_1|$, the transformer is a

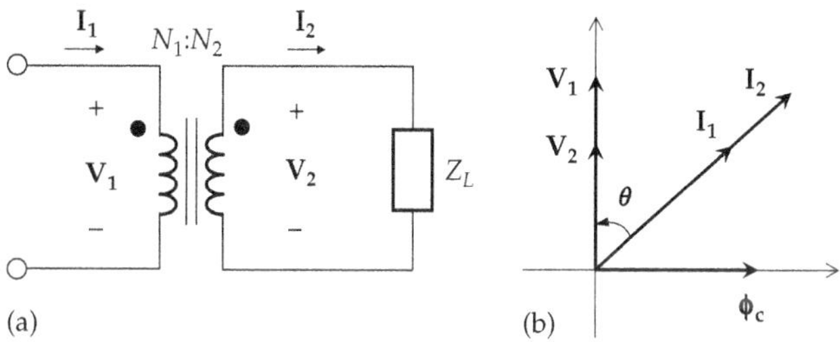

FIGURE 10.7
Phasor relations for an ideal transformer. (a) Two-winding transformer in the frequency domain, (b) phasor relations showing voltage and currents, and (c) flow diagram of causal relations.

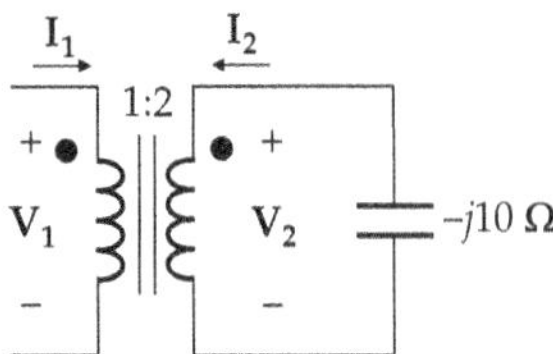

FIGURE 10.8
Figure for Primal Exercise 10.3.

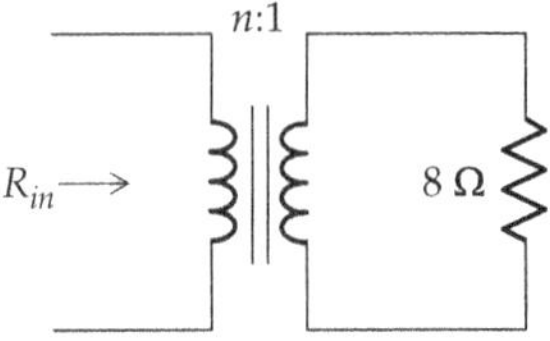

FIGURE 10.9
Figure for Primal Exercise 10.5.

step-up transformer, since the magnitude of the voltage of the secondary winding is larger than that of the primary winding. Conversely, if $N_2 < N_1$, the transformer is a **step-down transformer**.

Primal Exercise 10.3

Determine the phase angles of the phasor currents $\mathbf{I}_1$ and $\mathbf{I}_2$ in Figure 10.8 using the voltages as reference.

Ans. The phase angle of $\mathbf{I}_1$ is +90°, that of $\mathbf{I}_2$ is −90°.

Exercise 10.4

Consider the phasor relation $\mathbf{V}_1 = j\omega N_1 \mathbf{\Phi}_c$ (Figure 10.7). Show that this can be expressed as $V_{1rms} = 4.44 f N_1 \phi_{cpeak}$, where V_{1rms} is the rms value of $\mathbf{V}_1$ and ϕ_{cpeak} is the peak value of $\mathbf{\Phi}_c$.

10.2.3 Reflection of Impedance

The impedance Z_{Lp} that appears across the terminals of the primary winding in Figure 10.7a due to Z_L connected across the secondary winding is described as Z_L *reflected to the primary side*. This impedance is $\mathbf{V}_1/\mathbf{I}_1$ and its relation to Z_L can be readily determined from the voltage and current ratios. For the dot markings and assigned positive polarities of voltages and directions of currents in Figure 10.7a, $\mathbf{V}_1/\mathbf{V}_2 = N_1/N_2$ and $\mathbf{I}_2/\mathbf{I}_1 = N_1/N_2$. Multiplying these together,

$$\frac{\mathbf{V}_1}{\mathbf{I}_1}\frac{\mathbf{I}_2}{\mathbf{V}_2} = \left(\frac{N_1}{N_2}\right)^2 \tag{10.29}$$

Substituting $\mathbf{V}_2/\mathbf{I}_2 = Z_L$ and multiplying both sides of the equation by Z_L,

$$Z_{Lp} = \frac{\mathbf{V}_1}{\mathbf{I}_1} = Z_L\left(\frac{N_1}{N_2}\right)^2 \tag{10.30}$$

Equation 10.30 is *valid for any combination of dot markings and assignment of positive directions of currents and polarities of voltages*. Thus, reversing the dot markings alone changes the sign of the turns ratio in both the voltage and current ratios, as was pointed out in connection with Figure 10.5, but the product remains positive when the negative sign is squared. Reversal of either the assigned positive polarity of $\mathbf{V}_2$, or the assigned positive direction of $\mathbf{I}_2$, makes the turns ratio negative for the voltages or currents, respectively, but makes $\mathbf{V}_2 = -Z_L\mathbf{I}_2$, so that Equation 10.30 still applies. Similar considerations apply if either the assigned positive polarity of $\mathbf{V}_1$ or the assigned positive direction of $\mathbf{I}_1$ is reversed. Impedance never changes sign as it is reflected from one side of an ideal transformer to the other, because this violates conservation of energy (Problem P10.54).

It follows from Equation 10.30 that *a short circuit* (Z_L = 0) *is reflected as a short circuit and an open circuit* ($Z_L \to \infty$) *is reflected as an open circuit*.

Primal Exercise 10.5

Determine the turns ratio of the ideal transformer in Figure 10.9 so that the resistance reflected to the primary side is 200 Ω, irrespective of the dot markings.

Ans. 5:1.

Example 10.2: Reflection of Impedance

It is required to determine $\mathbf{V}_O$ in the circuit of Figure 10.10a.

Solution:

The first step in the general, basic analysis of circuits involving an ideal transformer is to assign voltages and currents to the ideal transformer. However, only two variables need be assigned, which can be conveniently chosen as the voltage on the side of smaller number of turns, and the current on the side of larger number of turns, in order to avoid working with fractions. In Figure 10.10b, the chosen voltage variable is the primary voltage $\mathbf{V}_1$, which makes the secondary voltage $3\mathbf{V}_1$. The assigned positive directions of the voltages are in accordance with the dot markings, in order to avoid dealing with negative voltages. The chosen current variable is the secondary current $\mathbf{I}_2$, which makes the primary current $3\mathbf{I}_2$. To avoid working with negative currents, the directions of the currents are chosen so

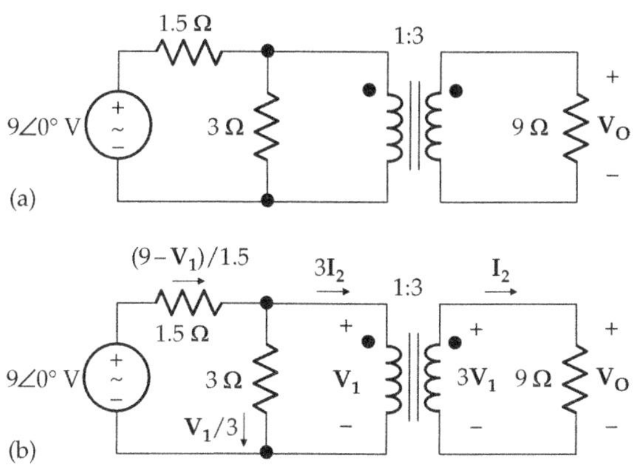

FIGURE 10.10
Figure for Example 10.2.

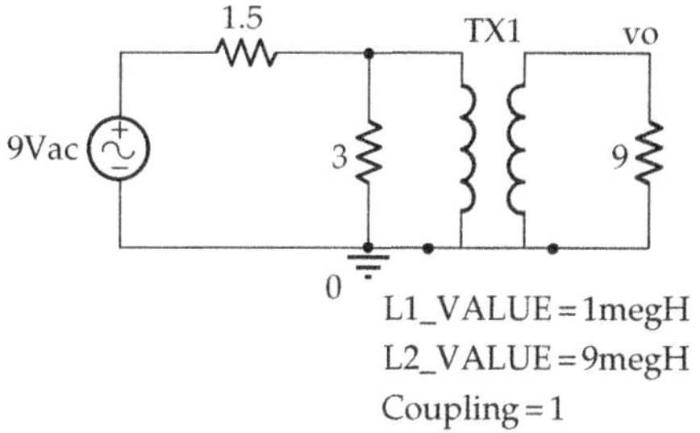

FIGURE 10.12
Figure for Example 10.2.

that the current ratio is positive. This means that if one current is assigned as entering the dot-marked terminal on one side, the other current is assigned as leaving the dot-marked terminal on the other side.

Since there are two unknowns, $\mathbf{V}_1$ and $\mathbf{I}_2$, two equations involving these variables need to be written and solved. Once $\mathbf{V}_1$ is determined, $\mathbf{V}_O$ follows as $3\mathbf{V}_1$. One equation involving $\mathbf{V}_1$ and $\mathbf{I}_2$ is provided by Ohm's law for the load resistance: $3\mathbf{V}_1 = 9\mathbf{I}_2$, or

$$\mathbf{V}_1 = 3\mathbf{I}_2 \tag{10.31}$$

The circuit on the primary side is a two-essential-node circuit, so that KCL at the upper node gives

$$\frac{9-\mathbf{V}_1}{1.5} = \frac{\mathbf{V}_1}{3} + 3\mathbf{I}_2 \tag{10.32}$$

Simplifying and collecting terms,

$$\mathbf{V}_1 + 3\mathbf{I}_2 = 6 \tag{10.33}$$

Solving Equations 10.31 and 10.33 gives $\mathbf{V}_1 = 3$ V, so that $\mathbf{V}_O = 9\angle 0°$ V, the same as the source voltage in this case.

An alternative method is to reflect the 9 Ω resistance to the primary side, in accordance with Equation 10.30, which gives $R_{Lp} = 9(1/3)^2 = 1$ Ω (Figure 10.11a). This 1 Ω resistance appears in parallel with the 3 Ω resistance, giving a resistance of 0.75 Ω. It follows from voltage division that $\mathbf{V}_1 = 9(0.75/2.25) = 3$ V (Figure 10.11b) and $\mathbf{V}_O = 9\angle 0°$ V, as before.

Simulation: The educational version of PSpice does not have a part for the ideal transformer. But the ideal transformer can be entered as a linear transformer, using the part XFRM_LINEAR from the ANALOG library. In accordance with the properties of the ideal transformer, the inductances of the two coils are made very large but in the ratio of the square of the number of turns. The coupling coefficient is unity and its sign is positive in accordance with the dot markings in Figure 10.10. The circuit is entered as in Figure 10.12, where the value of L_1 is set at 1 MH and that of L_2 as 9 MH. Note that 10^6 henries is entered in PSpice as 1 megH, and not 1MH, because PSpice is case insensitive and would interpret 1MH as 1 mH. In the Simulation Settings, 'Analysis type' is 'AC Sweep/Noise'; 0.159155 is entered for the 'Start Frequency' and for the 'End Frequency', corresponding to $\omega = 1$ rad/s; and 1 is entered for 'Points/Decade'. After the simulation is run, $\mathbf{V}_O$ magnitude is read as 9 V and its phase as essentially zero, using Evaluate Measurements and selecting Analog Operators and Functions. Note that an ideal transformer could also be entered in the same manner but using the part K_Linear, as in Example 10.5.

Problem-Solving Tips

- In analyzing a circuit using an ideal transformer, assign one voltage variable on one side of the transformer, usually the side of smaller number of

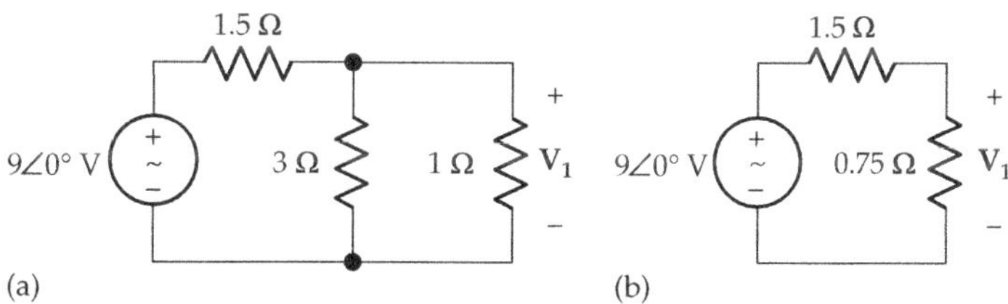

FIGURE 10.11
Figure for Example 10.2.

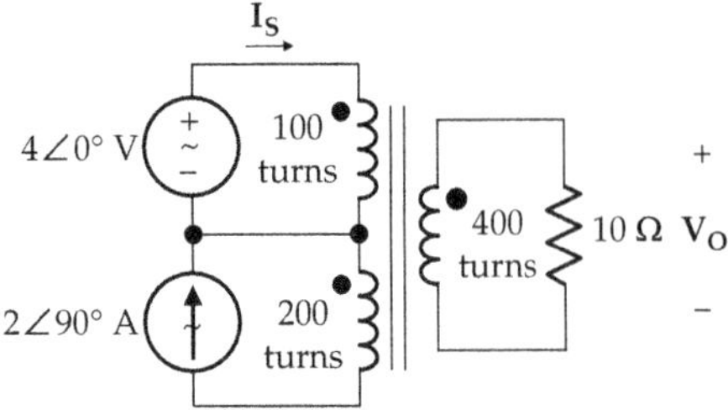

FIGURE 10.13
Figure for Primal Exercise 10.6.

turns, and one current variable on the other side, and express the other voltage and current of the transformer in terms of the assigned variables and the turns ratio.

- The polarities and directions of the transformer voltages and currents are assigned in accordance with the dot markings so as to avoid negative transformer voltages and currents on either side of the transformer.

Primal Exercise 10.6

Determine $\mathbf{V_O}$ and $\mathbf{I_S}$ in Figure 10.13.
Ans. $\mathbf{V_O} = 16\angle 0°$ V, $\mathbf{I_S} = 6.4 - j4$ A.

10.2.4 Applications of Transformers

After having discussed the basics of ideal transformers, it should be pointed out that the behavior of most types of practical transformers closely approximates that of ideal transformers, just as the behavior of most types of practical capacitors closely approximates that of ideal capacitors.

Transformers are widely used for a variety of purposes. They are used for stepping supply voltages up or down, for measurement of ac currents and voltages, for impedance matching, for generating polyphase supplies, and in rectifier circuits, inverters, coupled tuned amplifiers, oscillators, and a host of electronic circuits. A combination transformer/inductor, or ballast, is commonly used with fluorescent lamps.

10.3 Reflection of Circuits

Reflection applies not only to impedances but to whole circuits as well, which effectively removes the ideal transformer from the circuit. This is analogous to the removal of magnetic coupling between two coils by the T-equivalent circuit of a linear transformer. Consider, for example, Figure 10.14a having an ideal transformer of turns ratio 1:*a*, that is, N_1 turns on winding 1 and aN_1 turns on winding 2, where N_1 need not be specified. Since excitation is applied to both windings, we will not refer to the two sides as primary and secondary, but as side 1 (on the left) and side 2 (on the right). It follows that $\mathbf{I}_1 = a\mathbf{I}_2$, $\mathbf{V}_2 = a\mathbf{V}_1$, and $\mathbf{V}_L = \mathbf{V}_{S2} + \mathbf{V}_2$. KCL at node 'n' is

$$\mathbf{I}_2 + \mathbf{I}_{S2} = \mathbf{I}_L = \frac{\mathbf{V}_2 + \mathbf{V}_{S2}}{Z_L} \tag{10.34}$$

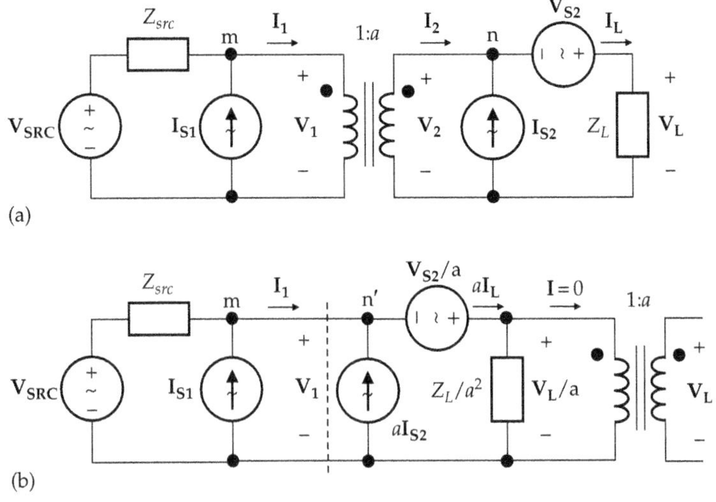

FIGURE 10.14
(a) Sources and impedances on both sides of an ideal transformer and (b) reflection of circuit from side 2 to side 1.

Multiplying Equation 10.34 by a,

$$a\mathbf{I}_2 + a\mathbf{I}_{S2} = a\mathbf{I}_L = \frac{a(\mathbf{V}_2 + \mathbf{V}_{S2})}{Z_L} \tag{10.35}$$

Dividing the numerator and denominator on the RHS by a^2 and replacing $a\mathbf{I}_2$ by $\mathbf{I}_1$ and $\mathbf{V}_2/a$ by $\mathbf{V}_1$, Equation 10.35 becomes

$$\mathbf{I}_1 + a\mathbf{I}_{S2} = a\mathbf{I}_L = \frac{\mathbf{V}_1 + \mathbf{V}_{S2}/a}{Z_L/a^2} \tag{10.36}$$

Equation 10.36 is KCL at node n′ to the right of the dashed line in Figure 10.14b. The circuit to the right of this line is the circuit connected to side 2 in Figure 10.14a but reflected to side 1. In reflecting voltages from one side to another, it is easy to remember that, according to the induced volts per turn concept, voltages are divided by the number of turns on the side of origin and multiplied by the number of turns on the destination side. Impedances are multiplied by the square of the factor by which voltages are multiplied, and currents are divided by the voltage factor. Thus, the voltage source is reflected as $\mathbf{V}_{S2}/a$, the impedance as Z_L/a^2, and the current source as $a\mathbf{I}_{S2}$. Winding 1 of the ideal transformer is now across the reflected Z_L/a^2 element, with winding 2 open-circuited. The current of winding 1 is zero, so that the ideal transformer is redundant and can be removed from the circuit.

Alternatively, the circuit on side 1 can be reflected to side 2. To show this, we note that KCL at node 'm' in Figure 10.15a is

$$\mathbf{I}_1 - \mathbf{I}_{S1} = \frac{\mathbf{V}_{SRC} - \mathbf{V}_1}{Z_{src}} \tag{10.37}$$

Dividing both sides by a,

$$\frac{\mathbf{I}_1}{\mathbf{a}} - \frac{\mathbf{I}_{S1}}{\mathbf{a}} = \frac{\mathbf{V}_{SRC} - \mathbf{V}_1}{aZ_{src}} \tag{10.38}$$

Multiplying the numerator and denominator on the RHS by a and replacing $\mathbf{I}_1/\mathbf{a}$ by $\mathbf{I}_2$ and $a\mathbf{V}_1$ by $\mathbf{V}_2$, Equation 10.38 becomes

$$\mathbf{I}_2 - \frac{\mathbf{I}_{S1}}{a} = \frac{a\mathbf{V}_{SRC} - \mathbf{V}_2}{a^2 Z_L} \tag{10.39}$$

Equation 10.39 is KCL at node m′ to the left of the dashed line in Figure 10.15b. The circuit to the left of this line is the circuit connected to side1 in Figure 10.15a but reflected to side 2. In accordance with the induced volts per turn concept, as explained previously, voltages are divided by the number of turns on the side of origin and multiplied by the number of turns on the destination side, currents are divided by this factor, and impedances are multiplied by the square of this factor. Thus, the voltage source is reflected as $a\mathbf{V}_{SRC}$, the impedance as a^2Z_{src}, and the current source as $\mathbf{I}_{S1}/a$. The ideal transformer now appears across the reflected source, with winding 1 open-circuited. The current of winding 2 is zero, so that the ideal transformer is redundant and can be removed from the circuit.

Suppose that the dot markings are reversed, keeping the same assigned positive directions of currents and voltages (Figure 10.16a). Equation 10.35 remains unchanged. But now $\mathbf{I}_1 = -a\mathbf{I}_2$ and $\mathbf{V}_2 = -a\mathbf{V}_1$. Substituting in Equation 10.35,

$$-\mathbf{I}_1 + a\mathbf{I}_{S2} = a\mathbf{I}_L = \frac{-a^2\mathbf{V}_1 + a\mathbf{V}_{s2}}{Z_L} \tag{10.40}$$

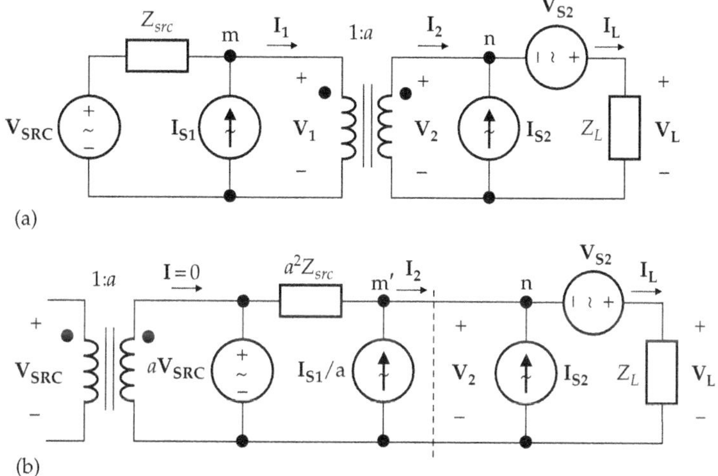

FIGURE 10.15
(a) Source and impedance on both sides of an ideal transformer, as in Figure 10.14 and (b) reflection of circuit from side 1 to side 2.

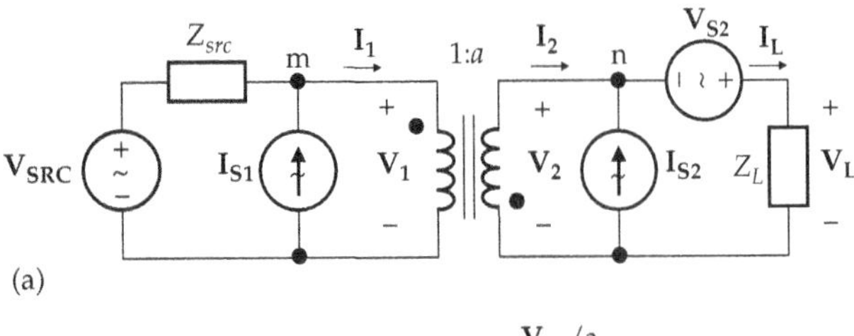

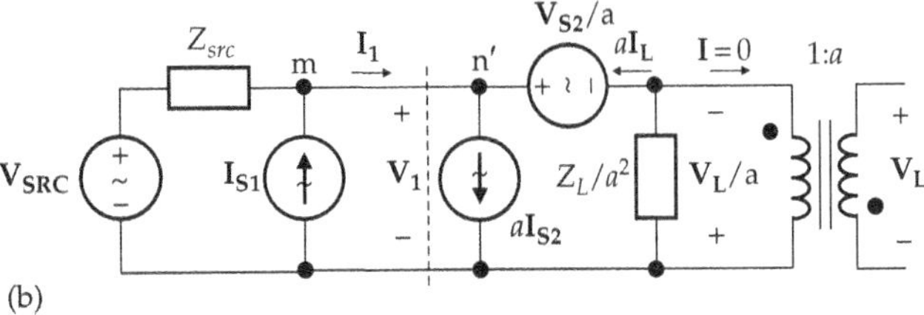

FIGURE 10.16
(a) Source and impedance on both sides of an ideal transformer, as in Figure 10.14 but with reversed dot markings and (b) reflection of circuit from side 2 to side 1.

Dividing the RHS of Equation 10.40 by a^2, rearranging, and changing signs,

$$\mathbf{I}_1 - a\mathbf{I}_{S2} = -a\mathbf{I}_L = \frac{\mathbf{V}_1 - \mathbf{V}_{S2}/a}{Z_L/a^2} \tag{10.41}$$

Compared to Equation 10.36, it is seen that the sign of a is reversed in Equation 10.41. This equation is KCL at node n′ in Figure 10.16b. Comparing this figure with Figure 10.14b, it is seen that reversing the dot markings reverses the directions of all the currents and voltages when reflected from side 2 to side 1.

The procedure for reflecting a circuit from one side of an ideal transformer to the other side can be summarized as follows:

1. In reflecting from the side of smaller number of turns to the side of larger number of turns, voltages are multiplied by the larger-than-unity turns ratio, impedances are multiplied by the square of this turns ratio, currents are divided by this turns ratio, and admittances are divided by the square of this turns ratio. Thus, voltages and impedances become larger, whereas currents and admittances become smaller. The converse applies when reflecting from a side of larger number of turns to the side of smaller number of turns: voltages and impedances become smaller, whereas currents and admittances become larger.
2. When the dot markings on the transformer windings are aligned horizontally, polarities or directions of source voltages and currents, required voltages or currents, and controlling voltages and currents are unchanged. When the dot markings are aligned diagonally, the polarities or directions of these voltages or currents are reversed.
3. In reflecting a circuit from one side of the transformer to the other, the order of the circuit elements must be retained.

The procedure for reflecting a circuit from one side of an ideal transformer to the other is illustrated by the following examples.

Example 10.3: Reflection of Circuit

It is required to determine v_O in Figure 10.17 by reflecting the circuit on either side of the transformer to the other side, assuming $\omega = 500$ krad/s.

Solution:

$1/j\omega C = 1/j(0.5 \times 10^6 \times 0.5 \times 10^{-6}) = -j4\ \Omega$. The voltage phasor representing the sinωt function is arbitrarily considered to have a reference phase angle of zero, which means that the phase angle of the current phasor representing the cosωt function is 90°, since cosωt = sin(ωt + 90°). The current phasor is therefore represented as $j2$ A. The circuit in the frequency domain is shown in Figure 10.18.

If the circuit on side 2 is reflected to side 1, the circuit becomes as in Figure 10.19a. Since the number of turns on side 1 is twice that of side 2, the current source is divided by 2, to become j A; the resistance is multiplied by 4, to become 8 Ω; and $\mathbf{V}_O$ across the 2 Ω resistance is multiplied by 2, to become $2\mathbf{V}_O$. The 6 V source in series with 4 Ω is transformed to a current source of 3/2 A in parallel with 4 Ω. The two current sources are replaced by a single current source of (3/2 + j) A, and the 4 Ω resistance is combined with the 8 Ω resistance in parallel

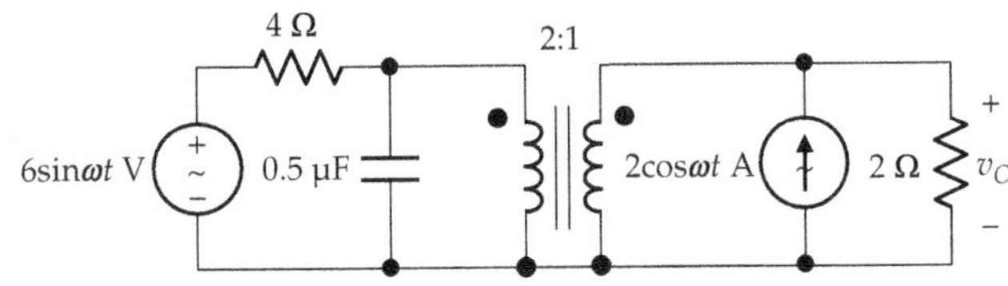

FIGURE 10.17
Figure for Example 10.3.

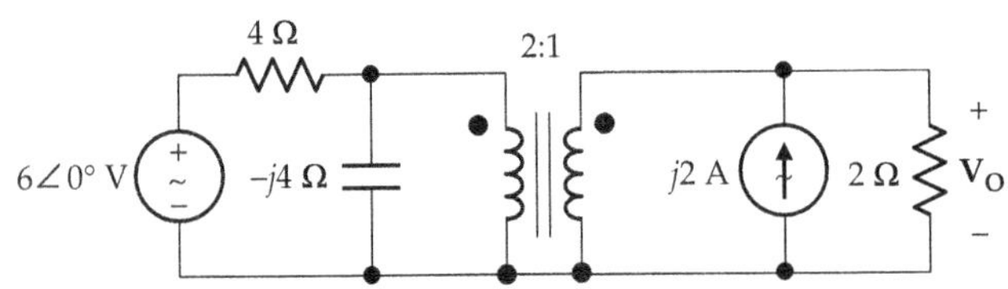

FIGURE 10.18
Figure for Example 10.3.

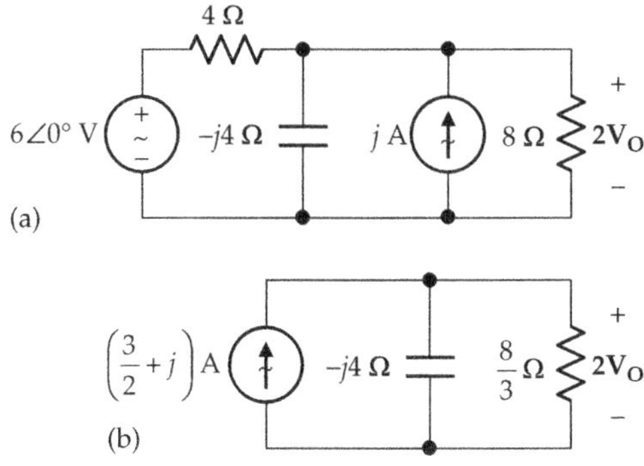

FIGURE 10.19
Figure for Example 10.3.

to give 8/3 Ω. The circuit becomes as in Figure 10.19b. It follows that $2\mathbf{V_O} = \left(\frac{3}{2}+j\right)\left(\frac{8}{3}\|(-j4)\right)$. Simplifying this expression gives $\mathbf{V_O} = 2$ V.

If the circuit on side 1 is reflected to side 2, the circuit becomes as in Figure 10.20a. Since the number of turns on side 2 is half that on side 1, the voltage source is divided by 2, to become 3 V, and the resistance and reactance are divided by 4, to become 1 Ω and $-j$ Ω. The 3 V source in series with 1 Ω is transformed to a current source of 3 A in parallel with 1 Ω. The two current sources are replaced by a single current source of $(3 + j2)$ A, and the 1 Ω resistance is combined with the 2 Ω resistance in parallel to give 2/3 Ω. The circuit becomes as in Figure 10.20b. It follows that $\mathbf{V_O} = (3+j2)\left(\frac{2}{3}\|(-j)\right)$. Simplifying this expression gives $\mathbf{V_O} = 2$ V.

If the dot markings on the transformer are reversed, and the circuit on side 2 is reflected to side 1, the circuit becomes as in Figure 10.21. Proceeding as for Figure 10.19 gives $-2\mathbf{V_O} = \left(\frac{3}{2}-j\right)\left(\frac{8}{3}\|(-j4)\right)$ or $\mathbf{V_O} = 2\angle 112.6°$ V. If the circuit on side 1 is reflected to side 2, the circuit becomes

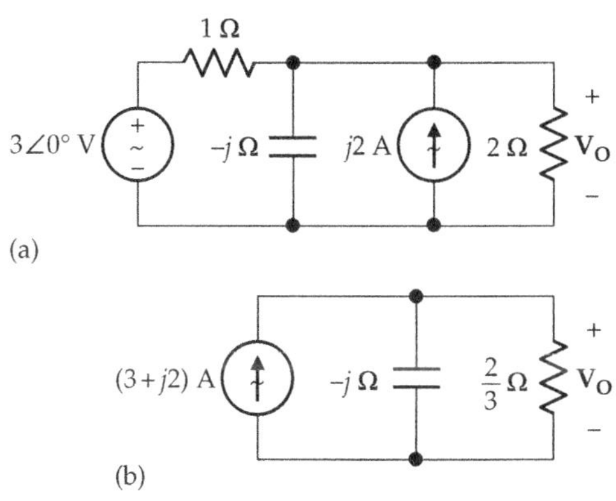

FIGURE 10.20
Figure for Example 10.3.

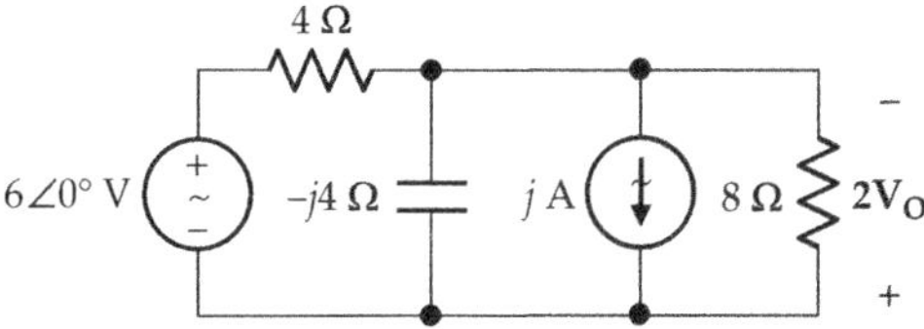

FIGURE 10.21
Figure for Example 10.3.

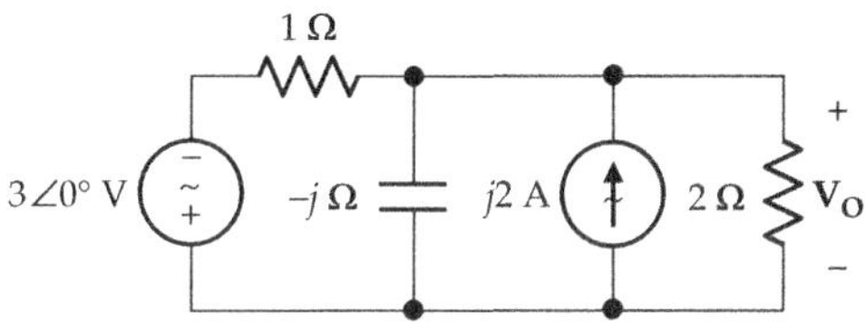

FIGURE 10.22
Figure for Example 10.3.

as in Figure 10.22. Proceeding as for Figure 10.20 gives $\mathbf{V_O} = (-3+j2)\left(\frac{2}{3}\|(-j)\right) = 2\angle 112.6°$ V.

The following example illustrates reflection involving a dependent source.

Example 10.4: Reflection of Circuit with Dependent Source

It is required to determine $\mathbf{V_O}$ in Figure 10.23 by reflecting the circuit on the primary side of the transformer to the secondary side.

Solution:

Following the procedure outlined previously, the voltage source, the dependent current source, and its controlling current are reflected to the secondary side, with reversed polarity, because of the placement of the dots. The magnitude of the source voltage is divided by 2 and the resistances are divided by 4. The magnitudes of the current source and its controlling current are multiplied by 2. Setting $2\mathbf{I_X} = \mathbf{I_Y}$, the dependent source current is related to its controlling current as on the primary side

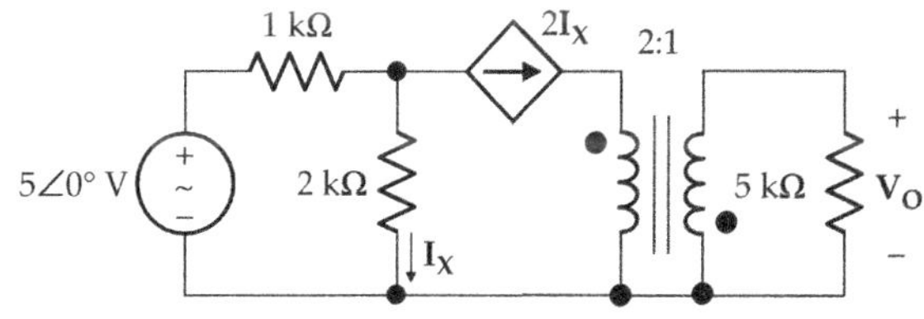

FIGURE 10.23
Figure for Example 10.4.

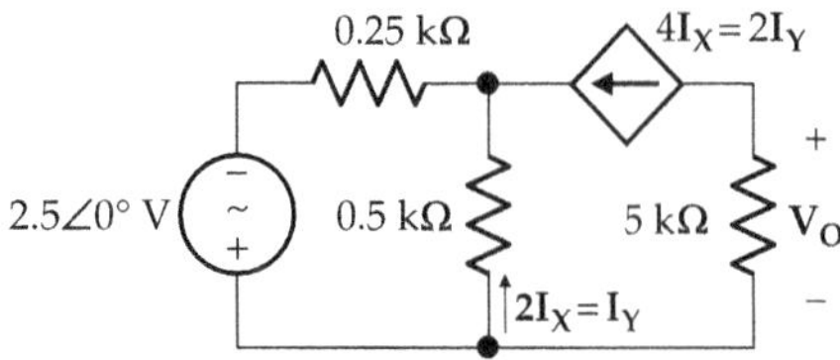

FIGURE 10.24
Figure for Example 10.4.

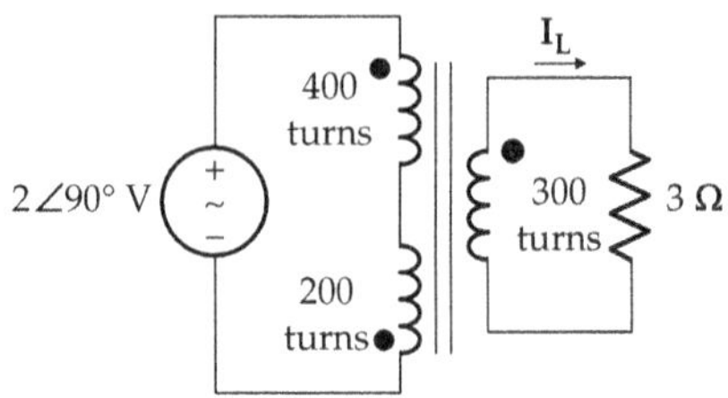

FIGURE 10.26
Figure for Primal Exercise 10.7.

but with reversed polarities (Figure 10.24). It follows from KCL that $2\mathbf{I}_Y + \mathbf{I}_Y = (2.5 - 0.5\mathbf{I}_Y)/0.25$, which gives $\mathbf{I}_Y = 2$ mA. Hence, $\mathbf{V}_O = -5 \times 2\mathbf{I}_Y = -20$ V.

It should be emphasized that the previously discussed simple reflection of circuits applies *only* in the case of two windings and in the absence of other coupling between the circuits connected to each winding. These circuits could be coupled by a subcircuit bridging the two sides, as in Figure 10.25a, or by a series-connected subcircuit as in Figure 10.25b. The subcircuit could be just a simple impedance.

Primal Exercise 10.7

Determine $\mathbf{I}_L$ in Figure 10.26 (a) by applying the volts/turn concept, and (b) noting, from the net zero mmf equation, that the effective number of turns of the primary winding is 200 turns and reflecting the source to the secondary side.

Ans. $1\angle 90°$ A.

10.4 Ideal Autotransformer

The ideal autotransformer is a special type of ideal transformer in which the primary and secondary windings are connected together in a particular way. To appreciate the idea behind an autotransformer, consider for the sake of argument an ideal, step-up ($N_2 > N_1$), two-winding transformer having the assigned positive polarities of voltages and directions of currents, as in Figure 10.27a. The output voltage v_2 is that appearing solely across the secondary winding. As an alternative, it is possible to have a smaller induced voltage v_2' added to the primary voltage v_1, so that $v_2 = v_1 + v_2'$, as illustrated in Figure 10.27b. Such an ideal transformer is referred to as an **ideal autotransformer**.

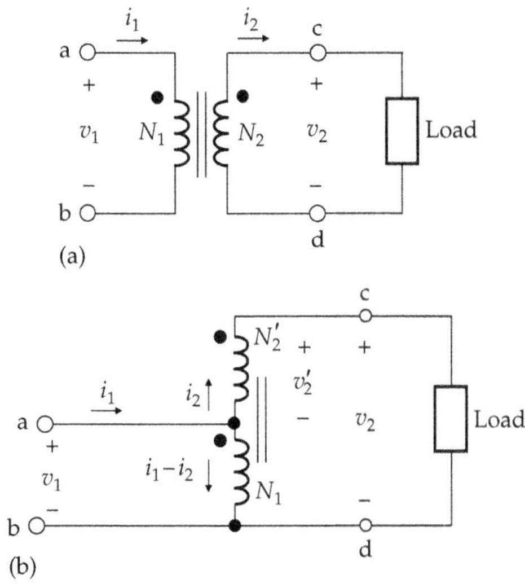

FIGURE 10.27
(a) A step-up two-winding transformer ($v_2 > v_1$) and (b) alternative connection of windings for the same v_1 and v_2.

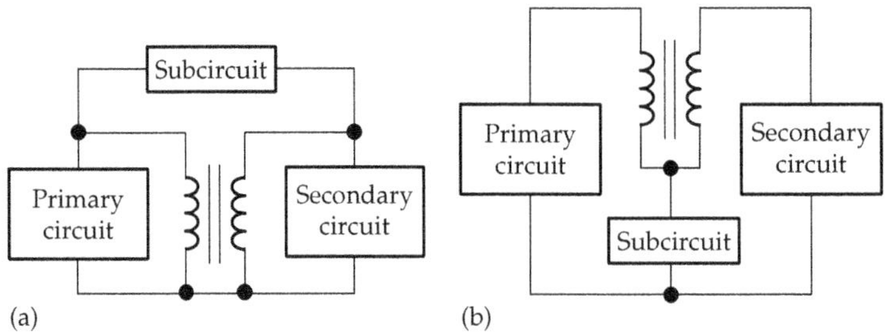

FIGURE 10.25
Circuits on the primary and secondary sides of an ideal transformer are coupled by a subcircuit bridging the two circuits (a), or connected in series with the transformer windings (b).

The principal advantage of an autotransformer is that the added voltage v_2', which is smaller than v_2, requires a winding having a number of turns N_2' that is less than N_2. This makes the autotransformer smaller, lighter, and less expensive than a two-winding transformer having the same input and output voltages and currents. In addition, the two windings having the number of turns N_1 and N_2' could be a single winding with a bare part, devoid of insulation. Terminal 'a' in Figure 10.27b is then connected to a slider over this bare part of the winding, resulting in an autotransformer of *variable turns ratio.*

If terminals 'c' and 'd' in Figure 10.27b are connected to the source and terminals 'a' and 'b' are connected to the load, a step-down autotransformer is obtained, in which the output voltage is smaller in magnitude than the input voltage.

The principal disadvantage of the autotransformer is that the input and output sides are directly connected together. That is, there is a conductive pathway between input and output, unlike a two-winding transformer. Thus, if a battery is connected between terminals 'a' and 'd' of the ideal, two-winding transformer of Figure 10.27a, no current flows (Figure 10.28). This is because the insulation between the two windings is assumed to be perfect in an ideal transformer, just like the insulation between the two plates of an ideal capacitor. Perfect insulation has infinite resistance, so that the dc current I_B is V_B divided by an infinite resistance between the two windings, which gives $I_B = 0$. In other words, there is no conductive pathway between the two windings. The two windings are said to be **electrically isolated** from one another. On the other hand, if a battery is connected between terminals 'a' and 'd' of the autotransformer of Figure 10.27b, the battery is short-circuited by the zero dc resistance of the primary winding between terminals 'a' and 'bd'. Electrical isolation is sometimes required in practice to safeguard against electric shock, or for other reasons, such as having different grounded nodes on the two sides of a two-winding transformer. An autotransformer cannot be used in these cases.

The voltage and current relations for the autotransformer follow directly from those of the two-winding transformer. The magnitude of the induced volts per turn in the primary winding is v_1/N_1. Hence, $v_2' = (v_1/N_1)N_2'$. For the dot markings in Figure 10.27b, $v_2 = v_1 + v_2'$. Dividing this relation by v_1 and substituting $v_2'/v_1 = N_2'/N_1$ from the preceding relation give

$$\frac{v_2}{v_1} = \frac{v_1 + v_2'}{v_1} = 1 + \frac{N_2'}{N_1} = \frac{N_1 + N_2'}{N_1} \quad (10.42)$$

Setting the net mmf to zero in Figure 10.27b, $N_1(i_1 - i_2) = N_2' i_2$. Dividing by i_2 and simplifying,

$$\frac{i_1}{i_2} = 1 + \frac{N_2'}{N_1} = \frac{N_1 + N_2'}{N_1} \quad (10.43)$$

Equations 10.42 and 10.43 are equivalent, respectively, to the voltage and current ratios of a two-winding transformer of turns ratio $(N_1 + N_2'): N_1$, since N_2' effectively adds to N_1 to give the total number of turns on one side of the autotransformer.

If the dot markings are reversed, keeping the same assigned positive polarities of voltages and directions of currents, then $v_2 = v_1 - v_2'$, and the turns ratio N_2'/N_1 changes sign in Equations 10.42 and 10.43 (Exercise 10.8).

Note that, in principle, any two-winding transformer can be connected as an autotransformer to give an effectively larger, or smaller, turns ratio.

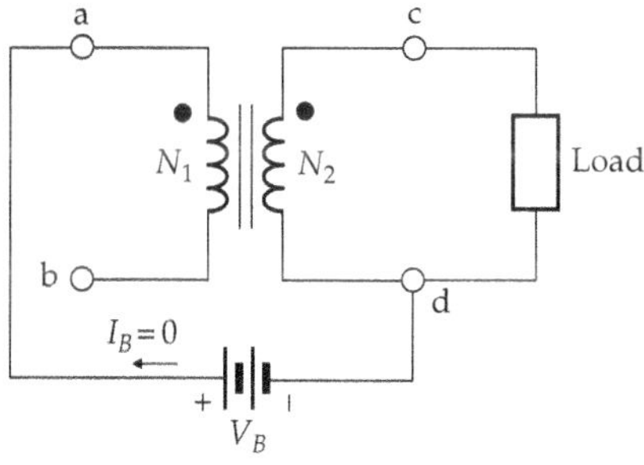

FIGURE 10.28
Electrical isolation in a two-winding transformer.

Exercise 10.8

Show that if the dot marking on either winding of an autotransformer is reversed, the turns ratio N_2'/N_1 changes sign.

Primal Exercise 10.9

(a) Determine v_2/v_1 and i_1/i_2 when (i) $N_1 = 200$ turns and $N_2' = 100$ turns, with the dot markings as in Figure 10.27b, (ii) $N_2' = 100$ turns with the relative dot markings of Figure 10.27b reversed, and (iii) $N_2' = 300$ turns with the relative dot markings of Figure 10.27b reversed, as in (2). (b) Does it make any difference in (ii) and (iii) in which winding the dots in Figure 10.27b are reversed?

Ans. (a) Both ratios are 1.5 in (i), 0.5 in (ii), and −0.5 in (iii); (b) no.

Example 10.5: Three-Winding Ideal Transformer

Given a three-winding ideal transformer with loads Z_2 and Z_3 connected as shown in Figure 10.29, it is required to determine the input impedance $\mathbf{V}_1/\mathbf{I}_1$.

Solution:

The transformer can be analyzed using the two fundamental concepts of the ideal transformer: (1) equality of the magnitude of the induced volts per turn in all

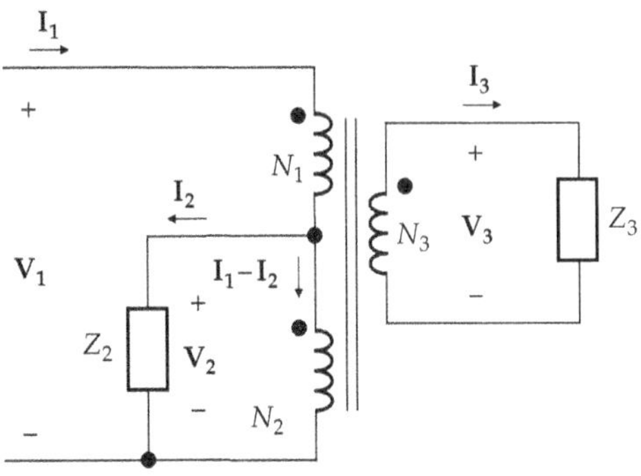

FIGURE 10.29
Figure for Example 10.5.

the windings and (2) net zero mmf on the core. $\mathbf{V}_1$ is the applied voltage across $(N_1 + N_2)$ turns and is also the induced voltage in these turns, because of the absence of flux leakage and resistance in an ideal transformer. The magnitude of the induced volts per turn is $V_1/(N_1 + N_2)$. The magnitudes of the induced voltages and terminal voltages of the other windings are $V_2 = [V_1/(N_1 + N_2)]N_2$ and $V_3 = [V_1/(N_1 + N_2)]N_3$. The assigned positive polarities of $\mathbf{V}_1$, $\mathbf{V}_2$, and $\mathbf{V}_3$ are in accordance with the dot markings on the three windings. It follows that the voltage ratios have positive signs: $\frac{\mathbf{V}_2}{\mathbf{V}_1} = \frac{N_2}{N_1 + N_2}$ and $\frac{\mathbf{V}_3}{\mathbf{V}_1} = \frac{N_3}{N_1 + N_2}$.

Consider next the mmf acting on the core. The currents $\mathbf{I}_1$ and $(\mathbf{I}_1 - \mathbf{I}_2)$ enter at the dotted terminals of their respective windings. Their mmfs are in the same sense and may be assigned a positive sign. On the other hand, $\mathbf{I}_3$ leaves its winding at the dotted terminals, so its mmf is assigned a negative sign. Since the net mmf in the core must be zero, $N_1\mathbf{I}_1 + N_2(\mathbf{I}_1 - \mathbf{I}_2) - N_3\mathbf{I}_3 = 0$.

Moreover, $\mathbf{I}_2 = (1/Z_2)\mathbf{V}_2$ and $\mathbf{I}_3 = (1/Z_3)\mathbf{V}_3$. Eliminating $\mathbf{I}_2$, $\mathbf{I}_3$, $\mathbf{V}_2$, and $\mathbf{V}_3$ from the preceding equations gives

$$\frac{\mathbf{V}_1}{\mathbf{I}_1} = \frac{(N_1 + N_2)^2}{N_2^2/Z_2 + N_3^2/Z_3} \quad (10.44)$$

If $Z_2 \to \infty$, the impedance reflected to the primary side is $Z_3(N_1 + N_2)^2/N_3^2$, as for a two-winding transformer of turns ratio $(N_1 + N_2)/N_3$. If $Z_3 \to \infty$, the reflected impedance at the input is $Z_2(N_1 + N_2)^2/N_2^2$, as for an autotransformer of turns N_1 and N_2, equivalent to a two-winding transformer of number of turns $(N_1 + N_2)$ and N_2. These reflected impedances in parallel give the input impedance of Equation 10.44. This is equivalent to applying each of Z_2 and Z_3, one at a time, and then paralleling the resulting impedances in a manner reminiscent of superposition. It should be emphasized that superposition does not apply, in general, to impedances. However, in some cases such as this, the substitution theorem allows replacing Z_2 and Z_3 by current sources to which superposition can be applied (Problem 10.55).

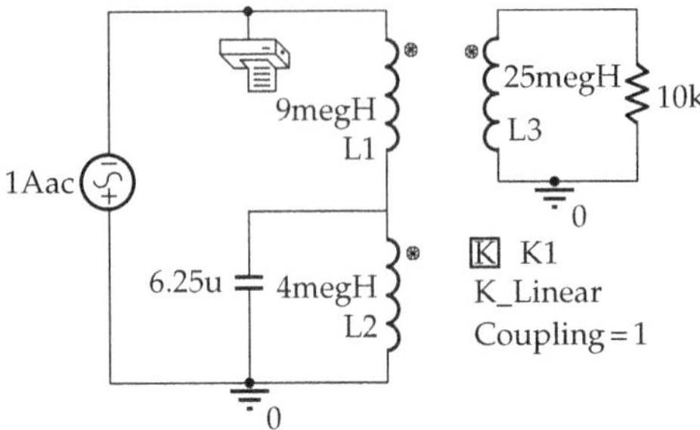

FIGURE 10.30
Figure for Example 10.5.

Simulation: The circuit is simulated using the following numerical values: N_1 = 3000 turns, N_2 = 2000 turns, N_3 = 5000 turns, Z_3 = 10 kΩ, and Z_2 is represented by a 6.25 μF capacitor, with ω = 100 rad/s, so that $Z_2 = -j1600$ Ω. The circuit is entered as in Figure 10.30. The three-winding ideal transformer is simulated using the part K_Linear from the Analog Library. The entries made in the Property Editor spreadsheet for this part are L1 for L1, L2 for L2, and L3 for L3. The values entered for L1, L2, and L3 in the Display Properties window are as indicated in the figure. These inductance values are in the ratio of the square of the number of turns and are large enough to simulate the infinite inductances of the windings of an ideal transformer. A 1 A current source is applied so that voltage read by the voltage printer is numerically equal to the input impedance required. In the Simulation Settings, 'Analysis type' is 'AC Sweep/Noise'; 15.9155 is entered for the Start Frequency and for the End Frequency, corresponding to ω = 100 rad/s; and 1 is entered for Points/Decade. After the simulation is run, the printer readings are 5.000E+03 for the real part and −5.000E+03 for the imaginary part, in accordance with Equation 10.44.

Problem-Solving Tip

- Voltage and current ratios in an ideal transformer, including an autotransformer, having any number of windings can always be determined from (1) the equality of the magnitude of the induced volts per turn in all windings and (2) a net zero mmf acting on the core, taking into account in both cases the dot markings at the terminals of the windings.

10.5 Transformer Imperfections

Practical transformers depart from the ideal in the following respects:

1. Finite resistance of the windings. This may be accounted for, at least at low frequencies, by adding appropriate resistances at the terminals of the windings.

2. Core losses, as described later in this section. These can be accounted for, in an approximate manner, by a resistance connected across the primary winding.
3. Energy stored in the electric field associated with voltage differences between windings and between the turns of each winding. This energy can be theoretically represented by distributed capacitance but can be accounted for, in an approximate manner, by lumped capacitances connected across the terminals of each winding and between windings (Problem P10.58).
4. Finite inductance of the windings and finite leakage flux. These will be considered in what follows because of their decisive effect on the frequency range over which a transformer can operate satisfactorily.

A transformer that is ideal except for finite inductance of the windings and finite leakage flux is in effect a linear transformer consisting of two coupled, lossless coils, of self-inductances L_1 and L_2 and a coupling coefficient that is less than unity. The transformer is shown represented in the frequency domain in Figure 10.31, connected to a source $\mathbf{V}_{SRC}$ on the primary side and to a load Z_L on the secondary side. The governing KVL equations are

$$j\omega L_1\mathbf{I}_1 - j\omega M\mathbf{I}_2 = \mathbf{V}_{SRC} \tag{10.45}$$

and

$$-j\omega M\mathbf{I}_1 + \left(Z_L + j\omega L_2\right)\mathbf{I}_2 = 0 \tag{10.46}$$

where the sign of the M term is opposite that of the L_1 and L_2 terms because $\mathbf{I}_1$ enters the dot-marked terminal of coil 1, whereas $\mathbf{I}_2$ leaves the dot-marked terminal of coil 2, so that their fluxes are in opposition. Substituting for $\mathbf{I}_2$ from Equation 10.46 in Equation 10.45 and dividing by $\mathbf{I}_1$,

$$\frac{\mathbf{V}_{SRC}}{\mathbf{I}_1} = \frac{j\omega L_1 Z_L + \omega^2\left(M^2 - L_1L_2\right)}{Z_L + j\omega L_2} \tag{10.47}$$

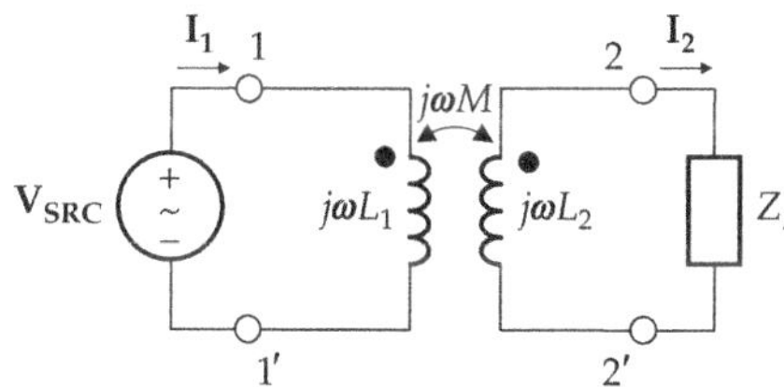

FIGURE 10.31
Ideal transformer having finite coil inductances and nonzero leakage flux.

$\mathbf{V}_{SRC}/\mathbf{I}_1$ is the impedance seen by the source and is the same as that derived in Problem P9.29, using the T-equivalent circuit (Exercise 10.11). If $k^2L_1L_2$ is substituted for M^2, this impedance becomes at the two extreme values of $Z_L = 0$ and $Z_L \to \infty$

$$Z_L = 0: \quad \frac{\mathbf{V}_{SRC}}{\mathbf{I}_1} = \frac{\omega^2\left(M^2 - L_1L_2\right)}{j\omega L_2} = \frac{\omega^2 L_1L_2\left(k^2 - 1\right)}{j\omega L_2}$$

$$= j\omega L_1\left(1 - k^2\right) \tag{10.48}$$

$$Z_L \to \infty: \quad \frac{\mathbf{V}_{SRC}}{\mathbf{I}_1} = j\omega L_1 \tag{10.49}$$

Equation 10.49 is to be expected, for if the secondary winding is open-circuited, the impedance seen by the source is simply that of the primary winding.

Any circuit that represents the effect of finite inductances of windings and finite leakage fluxes must satisfy Equations 10.45 through 10.49 under the same conditions, such as open-circuited secondary, short-circuited secondary, and perfect or imperfect coupling.

10.5.1 Finite Inductance of Windings

We will consider the coils to be perfectly coupled to begin with, that is, $M^2 = L_1L_2$. Substituting $M^2 = L_1L_2$ in Equation 10.47, multiplying the numerator and denominator by L_2, and rearranging L_1 and L_2 in the numerator, the RHS of this equation becomes

$$\frac{\mathbf{V}_{SRC}}{\mathbf{I}_1} = \frac{j\omega L_1 Z_L}{Z_L + j\omega L_2} = \frac{L_1}{L_2}\frac{j\omega L_2 Z_L}{Z_L + j\omega L_2} = \frac{L_1}{L_2}\left(j\omega L_2 \| Z_L\right) \tag{10.50}$$

The interpretation of Equation 10.50 is that $\mathbf{V}_{SRC}/\mathbf{I}_1$, the impedance seen by the source, is the parallel impedance of L_2 and Z_L on the secondary side, reflected to the primary side of an ideal transformer of inductance ratio $L_1{:}L_2$, as shown in Figure 10.32a. This is in accordance with impedance reflection (Equation 10.30), bearing in mind that the ratio $L_1{:}L_2$ is the square of the turns ratio (Equation 10.26). The impedance $j\omega L_2$, when reflected to the primary side, becomes $j\omega L_2(L_1/L_2) = j\omega L_1$, as in Figure 10.32b. As a check, if $Z_L \to \infty$, $\mathbf{V}_{SRC}/\mathbf{I}_1 = j\omega L_1$, as in Equation 10.49. Note that the finite inductance of the two windings is accounted for either by L_2 across the secondary winding or by L_1 across the primary winding, *but not both*.

10.5.2 Finite Leakage Flux

The next step is to modify the circuit of Figure 10.32b so as to include the effect of leakage flux. Since equivalence must apply for any value of Z_L, we can conveniently assume the special case of $Z_L = 0$, in which case $\mathbf{V}_{SRC}/\mathbf{I}_1$ is given by Equation 10.48. When terminals 22′

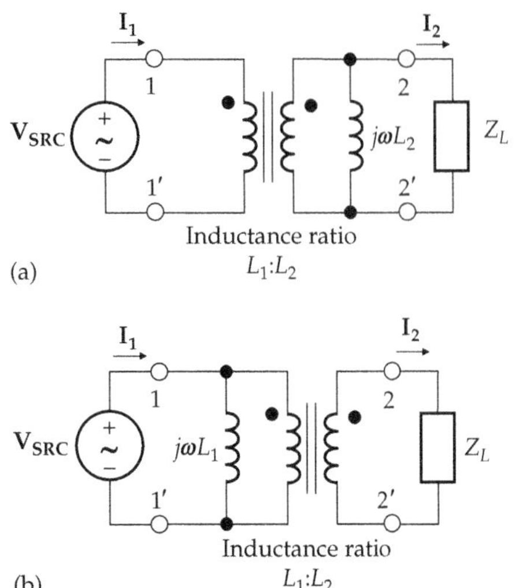

FIGURE 10.32
Finite inductance of windings is accounted for by connecting an inductance L_2 across the secondary winding (a), or an inductance L_1 across the primary winding (b), with the ideal transformer having an inductance ratio L_1:L_2.

in Figure 10.32b are short-circuited by having $Z_L = 0$, the short circuit is reflected to the primary side, which short-circuits $j\omega L_1$. To satisfy Equation 10.48, a series impedance $j\omega L_1\left(1-k^2\right)$ should be inserted as shown in Figure 10.33a. However, when terminals 22′ are open-circuited, Equation 10.49 must be satisfied. This will be the case if the shunt impedance is $j\omega L_1 k^2$ instead of $j\omega L_1$, so that the impedance seen by the source is $j\omega L_1\left(1-k^2\right)+j\omega L_1 k^2 = j\omega L_1$.

The condition for finite inductance of windings must still be satisfied. This condition now requires that $j\omega L_2$ be reflected as $j\omega k^2 L_1$. To satisfy this requirement the

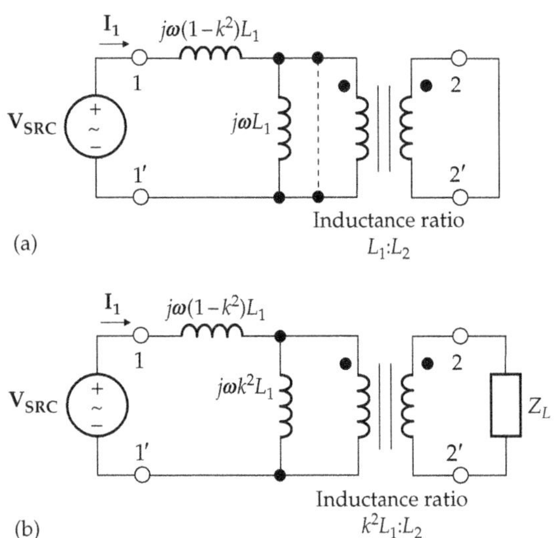

FIGURE 10.33
Accounting for finite leakage flux with the secondary winding short circuited (a) and including the finite reactance of the windings (b).

inductance ratio is modified to k^2L_1:L_2, as in Figure 10.33b. The circuit in this figure is then the equivalent circuit of a transformer that is ideal except for finite inductances of the windings and finite leakage flux.

That the circuit of Figure 10.33b satisfies Equations 10.45 and 10.46 is verified in Example 10.6.

★Example 10.6: Verification of Transformer Equivalent Circuit

It is required to verify that the equivalent circuit of Figure 10.33b satisfies Equations 10.45 through 10.47.

Solution:

On the primary side in Figure 10.34, $\mathbf{V}_1' = j\omega k^2 L_1\left(\mathbf{I}_1-\mathbf{I}_1'\right)$, and on the secondary side, $\mathbf{V}_2 = Z_L\mathbf{I}_2$. From the voltage relations for the ideal transformer, $\mathbf{V}_2/\mathbf{V}_1' = \sqrt{L_2}/\left(k\sqrt{L_1}\right)$, or $k\sqrt{L_1}\mathbf{V}_2 = \mathbf{V}_1'\sqrt{L_2}$. Substituting for $\mathbf{V}_1'$ and $\mathbf{V}_2$ in terms of the currents gives $k\sqrt{L_1}Z_L\mathbf{I}_2 = j\omega k^2 L_1\left(\mathbf{I}_1-\mathbf{I}_1'\right)\sqrt{L_2}$, or

$$Z_L\mathbf{I}_2 = j\omega k\sqrt{L_1L_2}\left(\mathbf{I}_1-\mathbf{I}_1'\right) = j\omega M\mathbf{I}_1 - j\omega M\mathbf{I}_1' \quad (10.51)$$

From the current relations for the ideal transformer,

$$k\sqrt{L_1}\mathbf{I}_1' = \sqrt{L_2}\mathbf{I}_2 \quad (10.52)$$

Multiplying both sides of Equation 10.52 by $j\omega\sqrt{L_2}$ gives $j\omega M\mathbf{I}_1' = j\omega L_2\mathbf{I}_2$. Substituting for $j\omega M\mathbf{I}_1'$ in Equation 10.51 and rearranging this equation gives Equation 10.46.

On the primary side in Figure 10.34, $\mathbf{V}_{\mathrm{SRC}} = j\omega(1-k^2)L_1\mathbf{I}_1 + j\omega k^2 L_1\left(\mathbf{I}_1-\mathbf{I}_1'\right) = j\omega L_1\mathbf{I}_1 - j\omega k^2 L_1\mathbf{I}_1'$. Multiplying both sides of Equation 10.52 by $j\omega k\sqrt{L_1}$ gives $j\omega k^2 L_1\mathbf{I}_1' = j\omega M\mathbf{I}_2$. Substituting for $j\omega k^2 L_1\mathbf{I}_1'$ in the equation for $\mathbf{V}_{\mathrm{SRC}}$ and rearranging this equation give Equation 10.45.

We can also show that the circuit of Figure 10.33b satisfies Equation 10.47 by first expressing this equation as

$$\frac{\mathbf{V}_{\mathrm{SRC}}}{\mathbf{I}_1} = \frac{j\omega L_1 Z_L + \omega^2\left(M^2 - L_1L_2\right)}{Z_L + j\omega L_2} = \frac{j\omega L_1 Z_L + \left(j\omega L_1\right)\left(j\omega L_2\right)\left(1-k^2\right)}{Z_L + j\omega L_2} \quad (10.53)$$

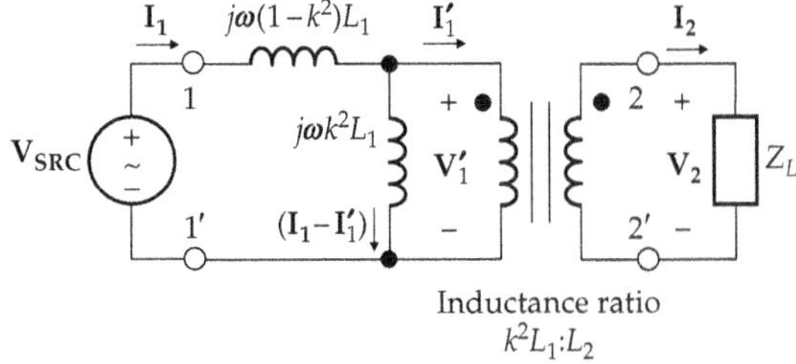

FIGURE 10.34
Figure for Example 10.6.

The numerator can be expressed in the following form:

$$\frac{\mathbf{V_{SRC}}}{\mathbf{I}_1} = \frac{j\omega L_1\left(1-k^2\right)\left(Z_L + j\omega L_2\right) + j\omega k^2 L_1 Z_L}{Z_L + j\omega L_2} \quad (10.54)$$

Dividing the numerator by the denominator,

$$\frac{\mathbf{V_{SRC}}}{\mathbf{I}_1} = j\omega L_1\left(1-k^2\right) + \frac{j\omega k^2 L_1 Z_L}{Z_L + j\omega L_2} \quad (10.55)$$

According to Equation 10.55, the input impedance is $j\omega L_1(1 - k^2)$ in series with an impedance represented by the second term. The numerator and denominator of this term are multiplied by k^2L_1/L_2 to give

$$\frac{j\omega k^2 L_1 Z_L}{Z_L + j\omega L_2} = \frac{\left(j\omega k^2 L_1\right)\left(Z_L k^2 L_1/L_2\right)}{\left(Z_L k^2 L_1/L_2\right) + j\omega k^2 L_1} \quad (10.56)$$

The RHS of Equation 10.56 represents an impedance $j\omega k^2 L_1$ in parallel with Z_L reflected to the primary side using an inductance ratio k^2L_1/L_2, in accordance with Figure 10.33b.

Primal Exercise 10.10

Determine the input impedance $\mathbf{V_{SRC}}/\mathbf{I_S}$ in Figure 10.35, assuming $L_1 = 0.1$ H, $L_2 = 0.2$ H, and $\omega = 10$ kHz and perfect coupling by (a) considering L_2 across the secondary winding, (b) considering L_1 across the primary winding, and (c) applying Equation 10.47. Compare with the input impedance of an ideal transformer of inductance ratio L_1:L_2.

Ans. (a), (b), and (c) 447.2∠26.6°; 500 Ω.

Exercise 10.11

Show that the input impedance (Equation 10.47) can be expressed as

$$\frac{\mathbf{V_{SRC}}}{\mathbf{I}_1} = j\omega L_1 + \frac{\omega^2 M^2}{Z_L + j\omega L_2} \quad (10.57)$$

as derived in Problem P9.29, from the T-equivalent circuit.

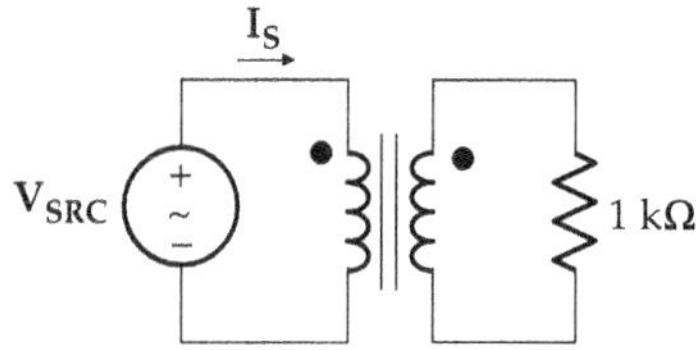

FIGURE 10.35
Figure for Primal Exercise 10.10.

10.5.3 Frequency Range

For simplicity and to illustrate clearly the basic ideas involved, we will assume resistive load and source impedances. Consider a load R_L connected to a source of voltage $\mathbf{V_{SRC}}$ and impedance R_{src} through a transformer represented by the equivalent circuit of Figure 10.33b. Let R_L' and V_L' be, respectively, the load impedance and voltage referred to the primary side. The circuit becomes that of Figure 10.36a. At a low enough frequency, the reactance $\omega k^2 L_1$ in parallel with R_L' becomes small compared to R_L' so that $\left(R_L' \| j\omega k^2 L_1\right) \cong j\omega k^2 L_1$. The circuit reduces to that of Figure 10.36b, where,

$$\mathbf{V}_L' = \frac{j\omega k^2 L_1}{R_{src} + j\omega L_1}\mathbf{V_{SRC}} \quad (10.58)$$

According to Equation 10.58, V_L' progressively decreases with frequency. At dc ($\omega = 0$) the transformer is inoperative, since the core flux is not time varying and no voltage is induced in the secondary winding. Note that the preceding argument still applies if at low frequencies $j\omega(1 -)k^2L_1$ is much smaller than R_{src}.

At a high enough frequency, the shunting effect of the impedance $j\omega k^2 L_1$ on R_L' becomes negligible, and the circuit reduces to that of Figure 10.36c. It follows that

$$\mathbf{V}_L' = \frac{R_L'}{R_L' + R_{src} + j\omega\left(1-k^2\right)L_1}\mathbf{V_{SRC}} \quad (10.59)$$

The load voltage V_L' progressively decreases as the frequency increases. Note that the preceding argument

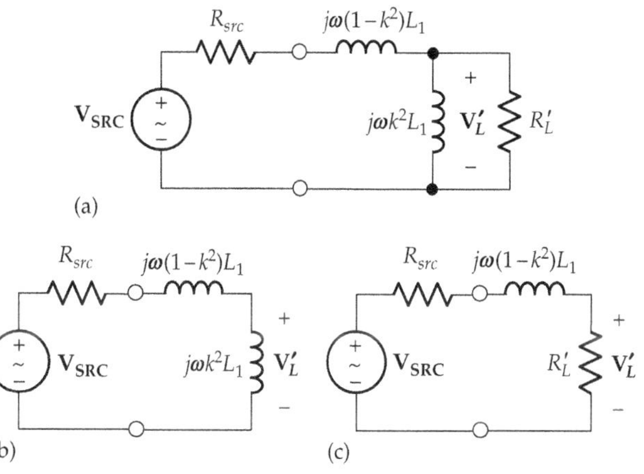

FIGURE 10.36
(a) Transformer equivalent circuit with resistive load referred to the primary side. The equivalent circuit at low frequencies (b) and at high frequencies (c).

still applies if at high frequencies $j\omega(1 -)k^2L_1$ is much larger than R_{src}. It follows that

Concept: *The performance of a nonideal transformer is limited at low frequencies by the reactance of the windings and at high frequencies by the leakage reactance.*

It is seen from the preceding discussion that the low and high frequencies at which a transformer is no longer useful depends on the source and load impedances. But it remains generally true that transformers intended for low frequencies must have large inductances of windings, which requires a large number of turns and a large cross-sectional area of the core, for a given core permeability (Equation 7.24). The transformer becomes large in size and more expensive. On the other hand, high-frequency transformers, such as pulse transformers, can be of small size but must have a low leakage flux. They commonly have a toroidal core in order to reduce the leakage flux.

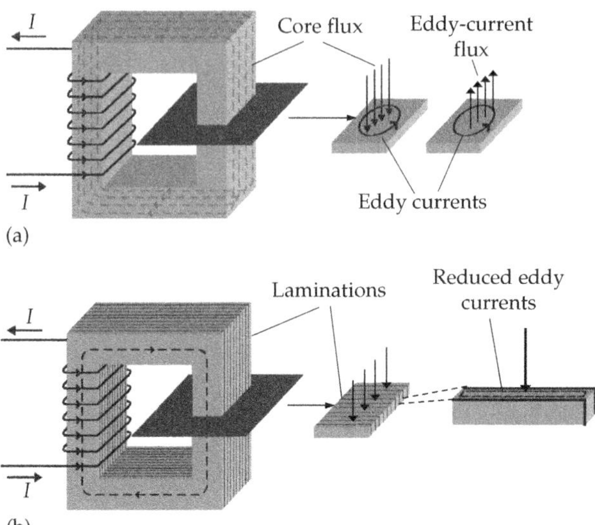

FIGURE 10.37
(a) Induced eddy currents in a conducting core and (b) effect of laminations on eddy currents.

10.5.4 Core Losses

These are of two types: eddy-current losses and hysteresis losses. Eddy-current losses occur in any core made of electrically conducting material. We have so far neglected the effect of the time-varying magnetic flux on the core itself. This time-varying flux induces a voltage around any closed path in the core that encloses some or all of the core flux. If the core material is electrically nonconducting, the induced voltage does not cause the flow of current in the core, and there is no effect on transformer behavior. However, it is impractical to have transformer cores made of perfectly nonconducting material, which means that the time-varying flux in the core will result in current in the core.

The case of a core of electrically conducting material is illustrated in Figure 10.37a, which shows a coil wound on such a core. A cross section of the core, shown enlarged, will have a time-varying core flux that is normal to the cross section. Any closed path around the flux lines in this cross section will have a voltage induced in it in accordance with Faraday's law. Since the core is conducting, this voltage will produce a current in this closed path, known as an **eddy current**, similar to the eddies produced in a turbulent flow of water. There are innumerable closed paths in an innumerable number of cross sections around the core, with the resulting eddy currents merging into a definite pattern of current flow in the core that depends on the shape of both the core and its cross section as well as the frequency of the applied voltage.

The eddy currents have two undesirable effects: (1) as in any current flow through a conductor, power is dissipated because of resistance, which results in heating, and (2) in accordance with Lenz's law, the eddy currents produce a flux, indicated in Figure 10.37a as eddy-current flux, which opposes the inducing flux. The resulting flux in the core is the difference between the flux due to the winding in the absence of eddy currents and that due to the eddy currents. The second undesirable effect of eddy currents is therefore to reduce the flux in the core.

The eddy currents are reduced in one of two ways. In high-frequency transformers, such as small toroidal transformers, the core is made of ceramic-type materials, known as **ferrites**, composed of oxides of iron and other metals such as nickel, zinc, and manganese. These have high resistivity, which, for a given voltage induced by flux in a loop in the core, practically eliminates eddy currents. At low frequencies, ferrite cores are impractical because of transformer size. Eddy currents are reduced in low-frequency transformers by assembling the core using thin laminations. These are insulated from one another and stacked together so that the flux is in a direction parallel to the plane of the laminations (Figure 10.37b). The induced currents are confined by the insulation to within each lamination. This effectively reduces the cross-sectional area of the loop that encloses flux and which can give rise to current flow, as illustrated in the figure. The induced voltage and hence the eddy currents are thereby greatly reduced. Laminations are usually made of iron alloys such as silicon steel, in which the added silicon increases the resistivity. The insulation is usually in the form of a thin layer of varnish.

Iron, steel, nickel, cobalt, and their alloys are **ferromagnetic** materials that have the desired characteristic of high permeability. However, this permeability

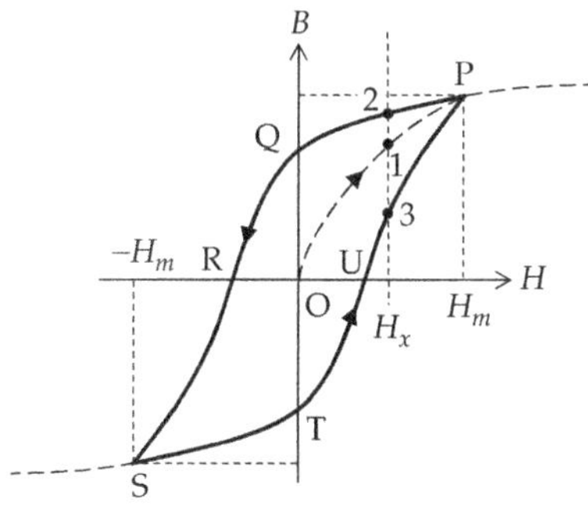

FIGURE 10.38
Hysteresis loss.

is not constant, so that the B–H relation is nonlinear and also exhibits another type of nonlinearity in the form of **hysteresis**. In general, hysteresis arises when an effect lags behind its cause, as can happen in mechanical systems due to friction. As a result of hysteresis, the state of a system depends on its previous history, that is, the manner in which this state is reached. To illustrate hysteresis, consider that H in an initially unmagnetized ferromagnetic specimen is increased from zero. B increases along a curve such as OP in Figure 10.38 that flattens out at large values of H. This flattening of the curve B–H curve at high values of H is described as **magnetic saturation**. If, at point P, H is reduced, B lags behind H. When H is reduced to zero at Q, B retains the value of the positive intercept on the B-axis. To reduce B to zero at R requires a negative H. Making H more negative still brings the operating point to S. If H is now increased back to H_m, B changes along the lower part of the curve, STUP. It is seen that at a particular value H_x, for example, B can take on different values corresponding to points 1, 2, or 3, or intermediate values, depending on how H_x is reached. The loop that is traced by a cyclic variation in H is a hysteresis loop. With H proportional to current and B proportional to voltage, the area enclosed by the hysteresis loop represents power dissipated in the core and not returned to the supply. The power dissipation also appears as heat in the core.

Hysteresis is put to good use in permanent magnets, where the residual magnetism of such a magnet is due to the intercept OQ on the vertical axis of the B–H hysteresis curve. The intercept OQ is referred to as the **remanence**, whereas the intercept OR is the **coercivity**.

Both eddy-current loss and hysteresis loss are a function of the magnitude of the core flux ϕ_c. The voltage across the branch $j\omega k^2L_1$ in Figure 10.33b can be considered proportional to ϕ_c. Core losses are sometimes accounted for by adding, in parallel with this branch, a resistance R_c whose value is such that the same power is dissipated in this resistance as in the core.

10.5.5 Construction of Small Inductors and Transformers

Small transformers at high frequencies commonly have toroidal ferrite cores, as illustrated in Figure 10.39a. Small inductors and transformers at power and audio frequencies have laminated cores that are usually rectangular in shape or of the shell type. Figure 10.39b is an outside view of a small, two-winding transformer having a shell-type core. Figure 10.39c is a cutaway view of a shell-type transformer.

The windings are wound on formers of rectangular or preferably square cross section that are fitted around one or both sides of a rectangular core (Figure 10.37) or around the central limb of the shell-type core. The formers are commonly made from plastic or impregnated cardboard. They are not shown in Figure 10.39 for a clearer view of the windings. The shell-type core is compact and rugged and naturally conforms to the two-sided distribution of the magnetic field of the windings (Figure 10.3).

Both primary and secondary windings are wound using insulated wire and are usually mounted one on top of the other, with the lower-voltage winding closer to the core, to reduce the stress on the insulation, since the core is usually connected to ground. Having the primary and secondary windings on separate limbs of a rectangular core (Figure 9.9) markedly reduces the coupling between them but reduces the capacitance between the two windings and practically eliminates the possibility of a direct short circuit between the two windings in the event of insulation breakdown. This type of construction is seldom used except for special applications, such as safety isolation transformers where it is essential to have the primary and secondary sides isolated even in the event of insulation breakdown. Alternatively, isolation between primary and secondary windings could be achieved by having a grounded metallic screen between the two windings. The screen also prevents coupling between primary and secondary windings through the interwinding capacitance. This is illustrated diagrammatically in Figure 10.40, where a grounded screen is interposed between the two windings. Each winding now has a capacitance to ground but not directly to the other winding.

Learning Checklist: What Should Be Learned from This Chapter

- According to the analogy between magnetic and electric circuits, magnetic flux is analogous to electric current; mmf, which is equal to the ampere-turns in Amperes circuital law,

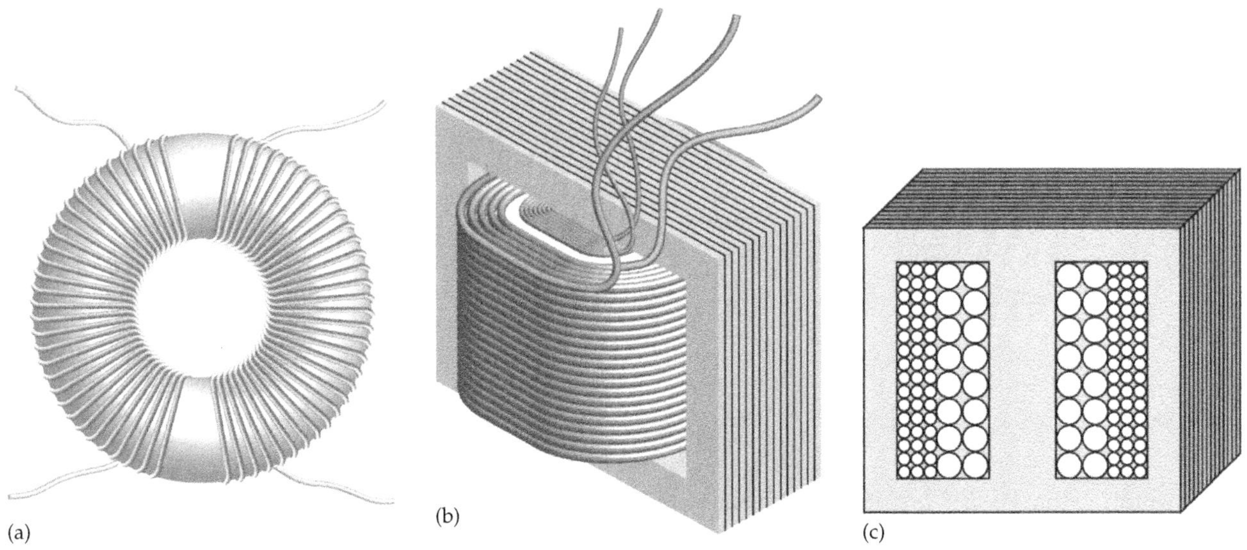

FIGURE 10.39
Construction of small transformers. (a) Transformer having a toroidal core, (b) transformer having a shell type core, shown in section (c).

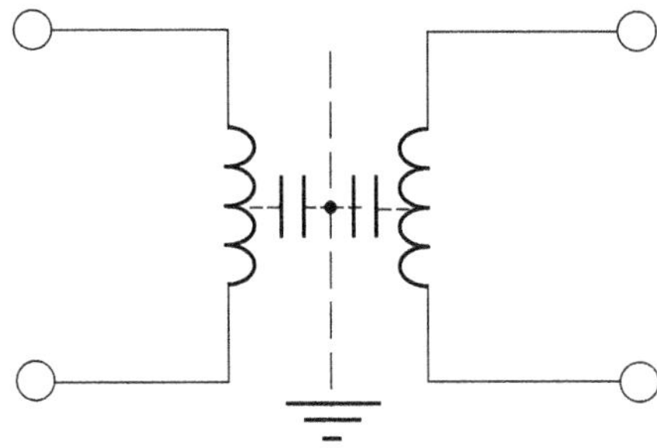

FIGURE 10.40
Grounded metallic screens between transformer windings.

is analogous to a driving voltage. Permeance and permeability are analogous to conductivity and conductance, respectively. The analog of Ohm's law in the form $v = i/G$ is mmf = flux/permeance.

- Permeance depends only on the geometry and magnetic properties of the magnetic circuit and is independent of the coil current or the number of turns of the coil.
- For a coil wound on a magnetic core, the following applies:
  1. The total permeance of the magnetic circuit is the sum of the permeance of the core and the leakage path, with the same mmf acting on both the core and the leakage path. The electric circuit analog is an emf acting on two conductances in parallel.
  2. If the core permeability is infinite, the core permeance and inductance are infinite. Zero mmf is required to establish a finite flux in a core of infinite permeability. The zero mmf will not drive any flux in the leakage path that is in parallel with the core path.
  3. The permeances of the core and the leakage path are in parallel because the same mmf acts on both. The inductances of the core and the leakage path are in series because the same time-varying coil current induces voltages due to the changing core flux and leakage flux that add together in the coil.
- The defining properties of an ideal transformer and their implications are as follows:
  1. No power losses, implies zero coil resistance and no eddy currents or hysteresis in the core.
  2. No electric stored energy, implies neglecting the electric field between the turns of a transformer coil and between different coils of the transformer.
  3. Perfect coupling to the core, that is, zero leakage flux, implies, together with zero resistance of the coils, that the terminal voltage of any given coil is the same as the voltage induced in the coil by the core flux.
  4. Infinite core permeability, implies zero mmf on the core, and no magnetic energy stored in the transformer.

- The circuit implications of the properties of an ideal two-coil transformer are as follows:
  1. Transformer voltage equation: $\frac{v_1}{v_2} = \pm\frac{N_1}{N_2}$, where the sign is positive if the assigned positive polarities of the voltages conform to the dot markings of the two coils and is negative otherwise. The voltage equation expresses the fundamental concept that the magnitude of the induced volts per turn is the same for all the coils of an ideal transformer
  2. Transformer current equation: $\frac{i_1}{i_2} = \pm\frac{N_2}{N_1}$, where the sign is negative if the assigned positive directions of currents are such that both currents flow into, or out of, the dotted terminals and is positive otherwise. The current equation expresses the fundamental concept that the net mmf on the core is zero.
  3. At any instant of time, the input power and output power are equal.
  4. The inductances of the coils are infinite, but their ratio is finite, because it is independent of μ_c and is equal to the square of the turns ratio.
  5. Impedance can be reflected from one side of an ideal transformer to the other by multiplying by the square of the ratio by which the voltage is multiplied. The reflected impedance is independent of the dot markings. Consequently, a short circuit ($Z_L = 0$) is reflected as a short circuit, and an open circuit ($Z_L \rightarrow \infty$) is reflected as an open circuit.
- In the case of an ideal transformer having two windings that are not externally coupled, as through a bridge-connected or a series-connected subcircuit, the circuits on either side of the transformer can be reflected to the other side. The procedure for reflecting a circuit from one side of an ideal transformer to the other side can be summarized as follows:
  1. In reflecting from the side of smaller number of turns to the side of larger number of turns, voltages are multiplied by the larger-than-unity turns ratio, impedances are multiplied by the square of this turns ratio, currents are divided by this turns ratio, and admittances are divided by the square of this turns ratio. Thus, voltages and impedances become larger, whereas currents and admittances become smaller. The converse applies when reflecting from a side of larger number of turns to the side of smaller number of turns: voltages and impedances become smaller, whereas currents and admittances become larger.
  2. When the dot markings on the transformer windings are aligned horizontally, polarities or directions of source voltages and currents, required voltages or currents, and controlling voltages and currents are unchanged. When the dot markings are aligned diagonally, the polarities or directions of these voltages or currents are reversed.
  3. In reflecting a circuit from one side of the transformer to the other, the order of the circuit elements must be preserved.
- In an ideal autotransformer, a second winding is connected so that its voltage adds to, or subtracts from, the voltage of the primary winding. The result is a smaller total number of turns, but a loss of electrical isolation between the circuits on the two sides of the transformer.
- The ideal autotransformer is analyzed in exactly the same way as a conventional two-winding transformer, based on the same magnitude of the induced volts per turn in each winding, and zero net mmf in the core, while taking into account the dot markings at the terminal of each winding.
- The finite inductance of the windings and the finite leakage flux in a nonideal transformer can be accounted for by adding appropriate series and shunt inductances to an ideal transformer of modified turns ratio. The performance of a nonideal transformer is limited at low frequencies by the reactance of the windings and at high frequencies by the leakage reactance.
- There are two types of core losses in a nonideal transformer: eddy-current loss and hysteresis loss. Both of these losses result is power dissipation in the core that results in heating of the core.
  1. Eddy-current loss arises because of the currents induced in an electrically conducting core, with resultant power dissipation and reduction of flux in the core. It is mitigated at low frequencies by laminating the core and at high frequencies by using a core of metallic oxides that have high resistivity.

2. Hysteresis is exhibited by ferromagnetic materials due to a lag of B behind H. As a result, a hysteresis loop is formed when H is varied in a cyclic manner. The area of the loop represents a power loss that results in heating of the core. Moreover, a given value of, say, H can be associated with a range of values of B, depending on how this value of H is reached.

Problem-Solving Tips

1. In analyzing a circuit using an ideal transformer, assign one voltage variable on one side of the transformer, usually the side of smaller number of turns, and one current variable on the other side. The other voltage and current of the transformer are expressed in terms of the assigned variables and the turns ratio.
2. The polarities and directions of the transformer voltages and currents are assigned in accordance with the dot markings so as to avoid negative transformer voltages and currents on either side of the transformer.
3. In reflecting from a side of smaller number of turns to the other side, voltages and impedances become larger, whereas currents and admittances become smaller. The converse applies when reflecting from a side of larger number of turns to the other side: voltages and impedances become smaller, whereas currents and admittances become larger.
4. Voltage and current ratios in an ideal transformer, including an autotransformer, having any number of windings can always be determined from (a) the equality of the magnitude of the induced volts per turn in all windings and (b) a net zero mmf acting on the core, taking into account in both cases the dot markings at the terminals of the winding.

Problems

Apply ISDEPIC and verify solutions by PSpice simulation whenever feasible.

Ideal Transformer

P10.1 Determine $\mathbf{V_O}$ in Figure P10.1 using the induced volts per turn concept and taking into account the dot markings.

Ans. −36 V.

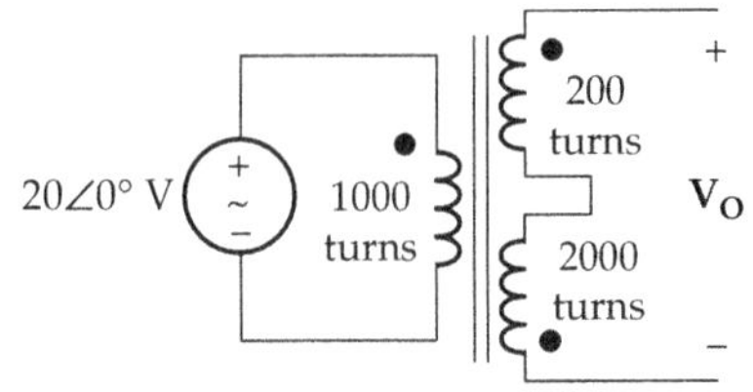

20∠0° V, 1000 turns, 200 turns, 2000 turns, $\mathbf{V_O}$

FIGURE P10.1

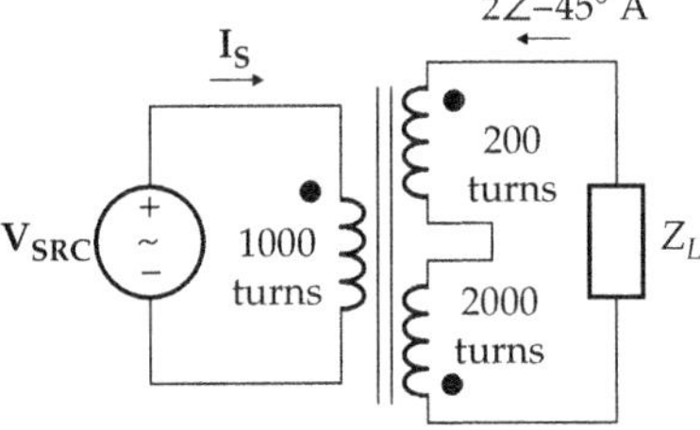

FIGURE P10.2

P10.2 Determine $\mathbf{I_S}$ in Figure P10.2, using the net mmf concept and taking into account the dot markings.

Ans. 3.6∠−45° A.

P10.3 Determine $\mathbf{I_O}$ in Figure P10.3.

Ans. −*j*10 A.

P10.4 Determine $v_O(t)$ in Figure P10.4.

Ans. 5cos(400*t* − 53.1°) V.

P10.5 Determine $\mathbf{V_O}$ and $\mathbf{I_S}$ in Figure P10.5.

Ans. $\sqrt{10}\angle -108.4°$ V, $\dfrac{1}{2\sqrt{10}}\angle -18.4°$ A.

P10.6 Determine $v_O(t)$ in Figure P10.6, assuming $v_{SRC}(t)$ = 50sinωt V and $\omega = 10^5$ rad/s.

Ans. 19.71sin(ωt − 9.8°) V.

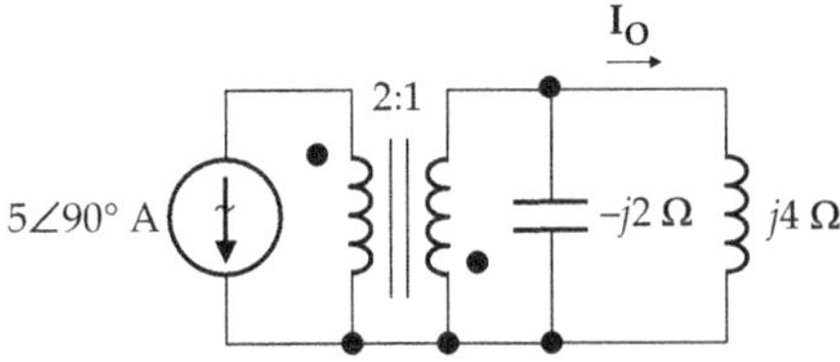

FIGURE P10.3

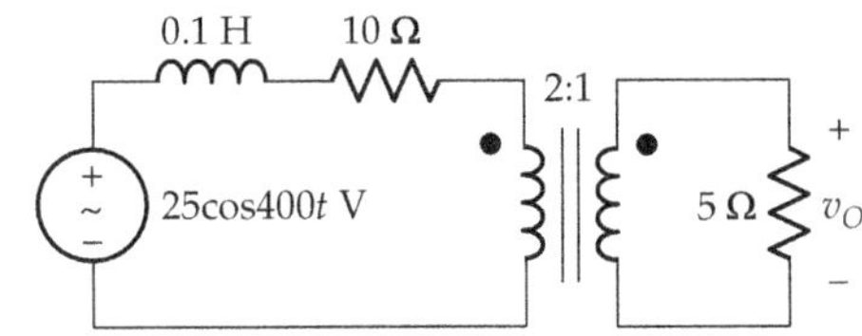

FIGURE P10.4

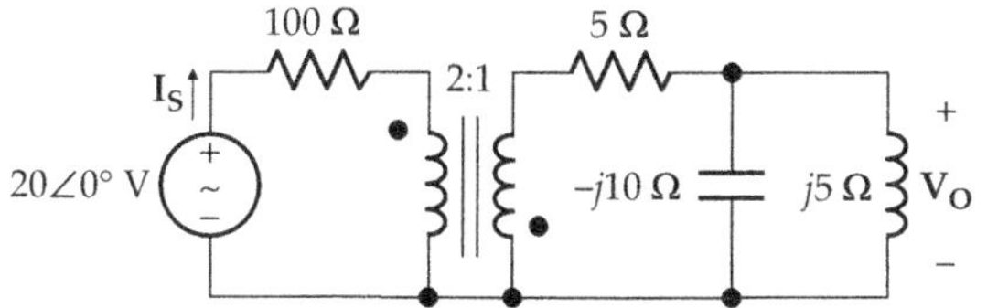

FIGURE P10.5

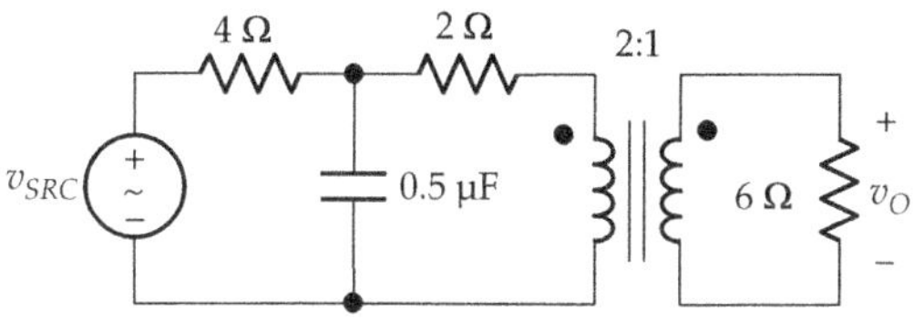

FIGURE P10.6

P10.7 Determine $\mathbf{I_O}$ in Figure P10.7.

Ans. $9.33\angle{-193.2^\circ}$ A.

P10.8 Determine the impedance seen by the source in Figure P10.8.

Ans. $2.5(3 - j)$ Ω.

P10.9 Determine $\mathbf{V_O}$ in Figure P10.9.

Ans. $-14.4 + j19.2$ V.

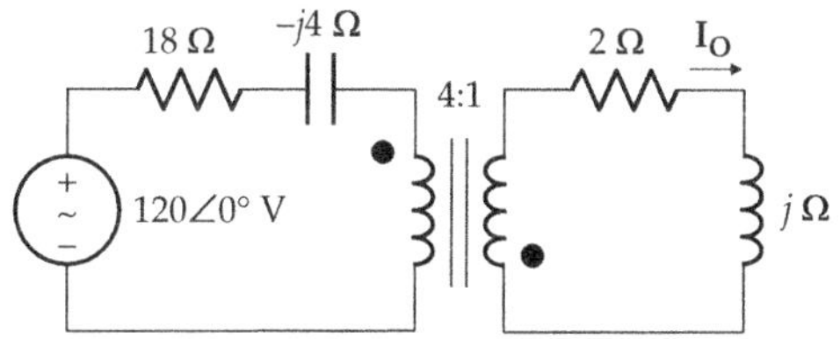

FIGURE P10.7

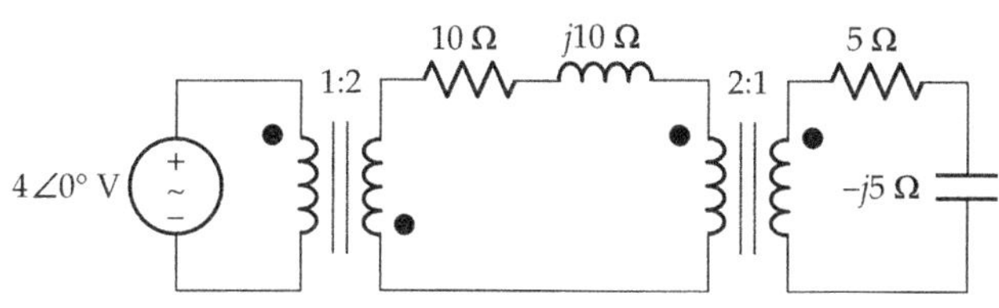

FIGURE P10.8

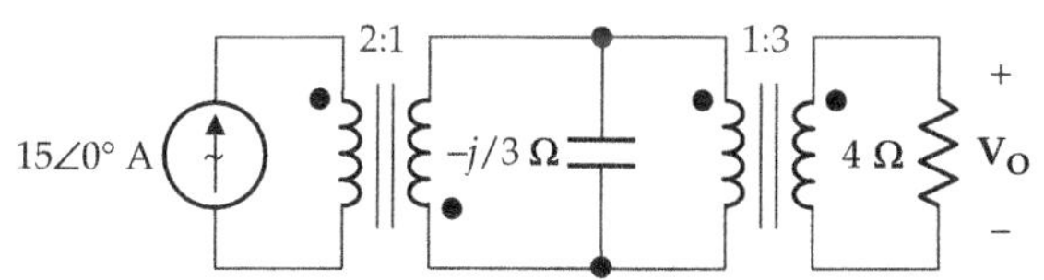

FIGURE P10.9

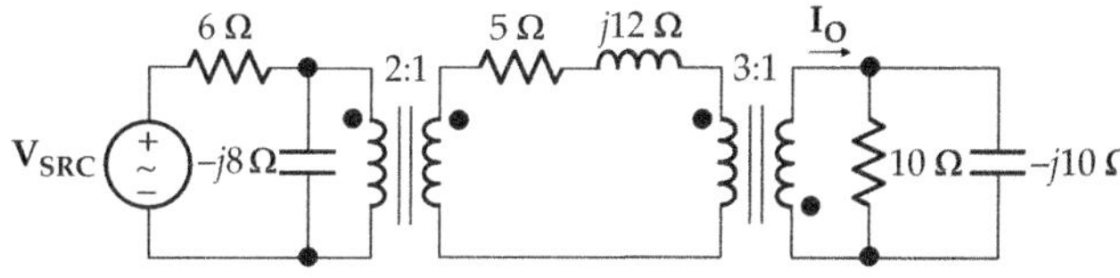

FIGURE P10.10

P10.10 Determine $\mathbf{I_O}$ in Figure P10.10, given that $v_{SRC}(t) = 100\sin(100\pi t)$ V.

Ans. $1.96\cos(100\pi t + 86.6^\circ)$ A.

P10.11 Determine $\mathbf{V_O}$ in Figure P10.11.

Ans. $2.58\angle 180^\circ$ V.

P10.12 Derive TEC looking into terminals 'ab' in Figure P10.12.

Ans. $\mathbf{V_{Th}} = \mathbf{V_{ab}} = 40\angle 0^\circ$ V, $Z_{Th} = 50$ Ω.

P10.13 Derive TEC looking into terminals 'ab' in Figure P10.13.

Ans. $\mathbf{V_{Th}} = \mathbf{V_{ab}} = -3\angle 0^\circ$ V, $Z_{Th} = 32$ Ω.

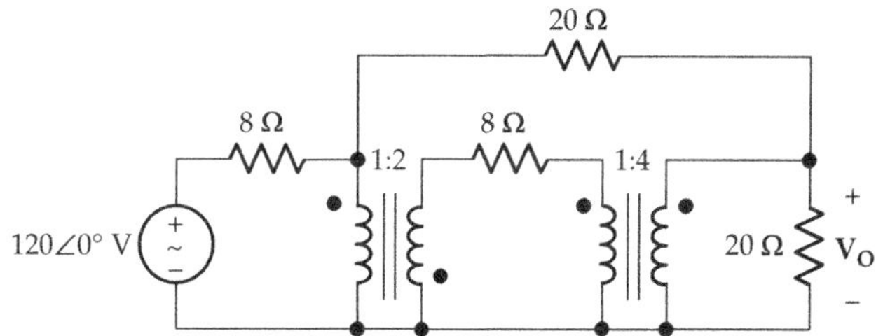

FIGURE P10.11

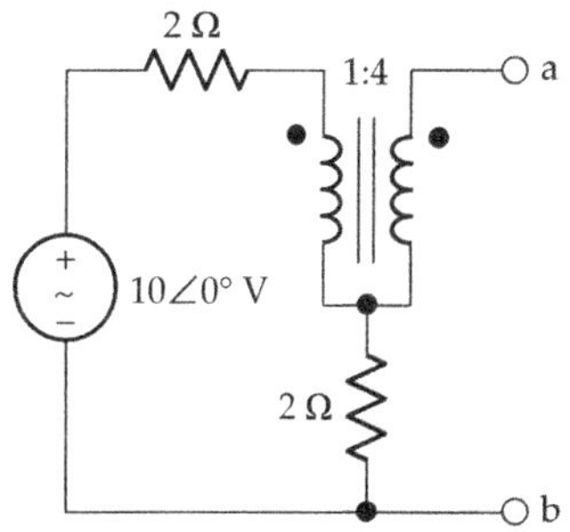

FIGURE P10.12

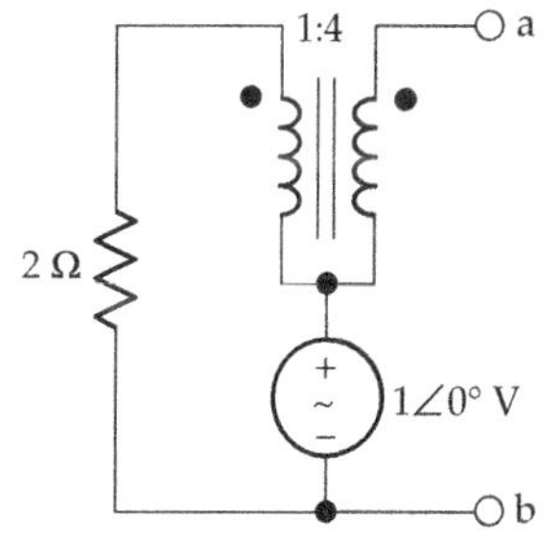

FIGURE P10.13

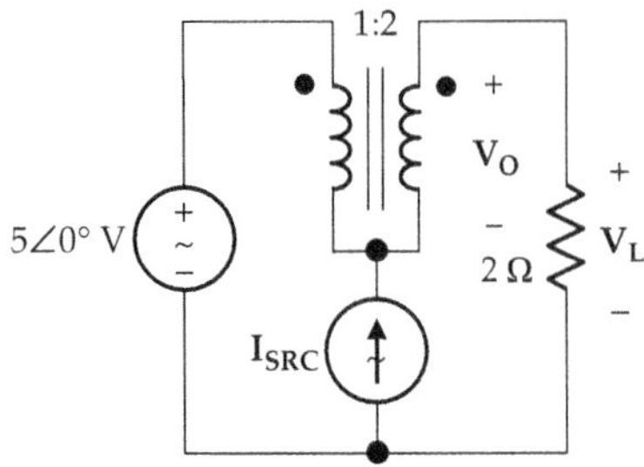

FIGURE P10.14

P10.14 Determine in Figure P10.14 (a) $\mathbf{V_L}$ if $\mathbf{I_{SRC}} = 1\angle 0°$ A and (b) $\mathbf{V_O}$ if $\mathbf{I_{SRC}}$ is such that $\mathbf{V_L} = 7\angle 0°$ V.

Ans. (a) $2\angle 180°$ V; (b) $4\angle 0°$ V.

P10.15 Show that $Z_{in} = a^2 Z_L + (1 \pm a)^2 Z_c$ in Figure P10.15, where the plus or minus sign is used depending on the relative dot markings of the two windings.

P10.16 Derive TEC looking into terminals 'ab' in Figure P10.16.

Ans. $\mathbf{V_{Th}} = \mathbf{V_{ab}} = -\frac{160}{197}(14 + j)$ V in series with $\frac{20}{197}(183 - j)$ Ω.

P10.17 Determine Z_{Th} looking into terminals 'ab' in Figure P10.17.

Ans. $j4$ Ω.

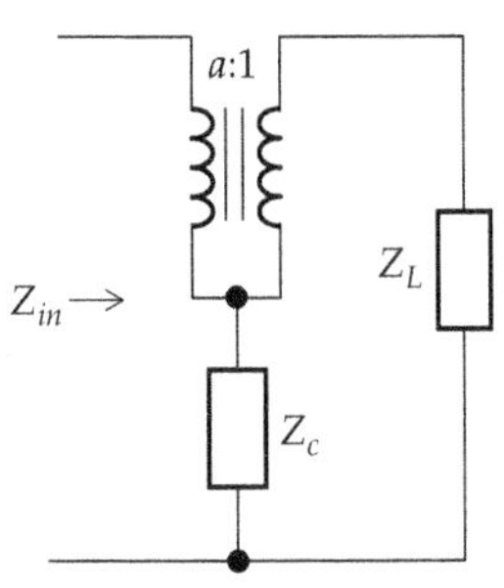

FIGURE P10.15

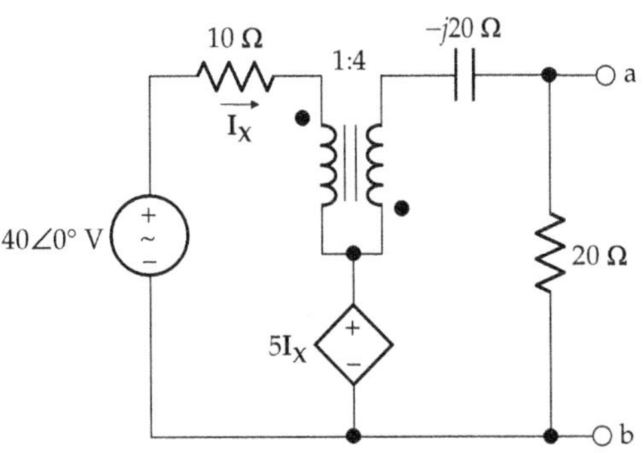

FIGURE P10.16

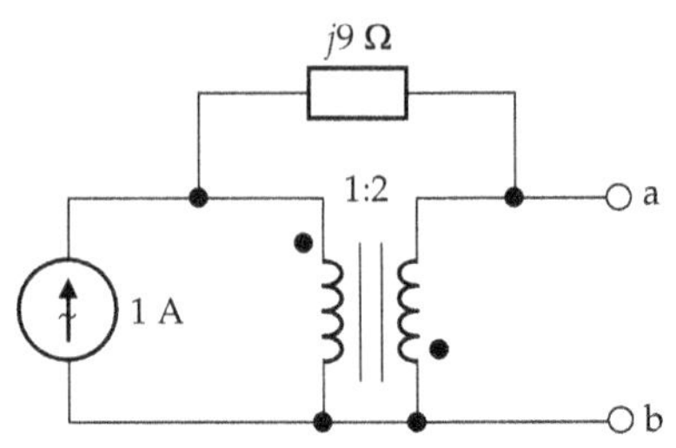

FIGURE P10.17

P10.18 Determine Z_X in Figure P10.18 so that $\mathbf{V_O} = 0$.

Ans. $Z_X = -j\omega La$.

P10.19 Determine the power dissipated in the 30 Ω resistor in Figure P10.19 given that $v_{SRC}(t) = 30\cos(10t + 15°)$ V.

Ans. 15 W.

P10.20 Determine $\mathbf{V_L}$ in Figure P10.20.

Ans. 2 V.

P10.21 Derive TEC looking into terminal 'ab' in Figure P10.21.

Ans. $\mathbf{V_{Th}} = \mathbf{V_{ab}} = 15\angle 0°$ V, $Z_{Th} = 10$ Ω.

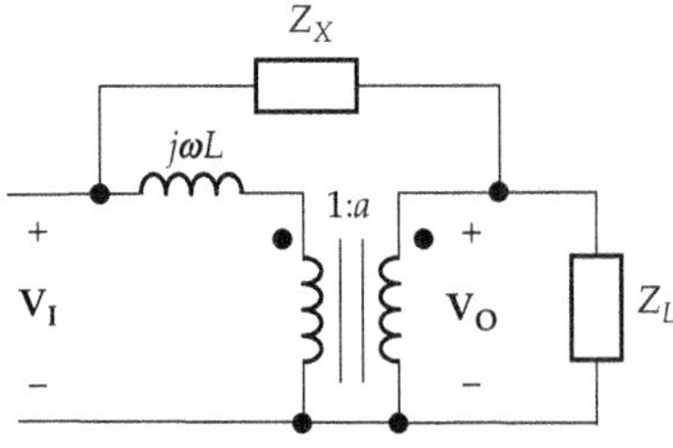

FIGURE P10.18

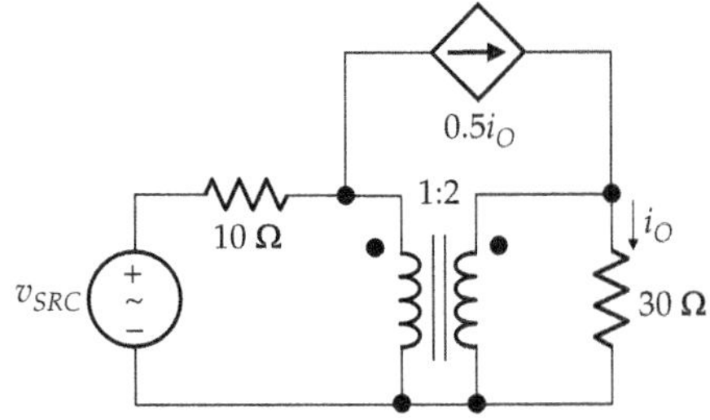

FIGURE P10.19

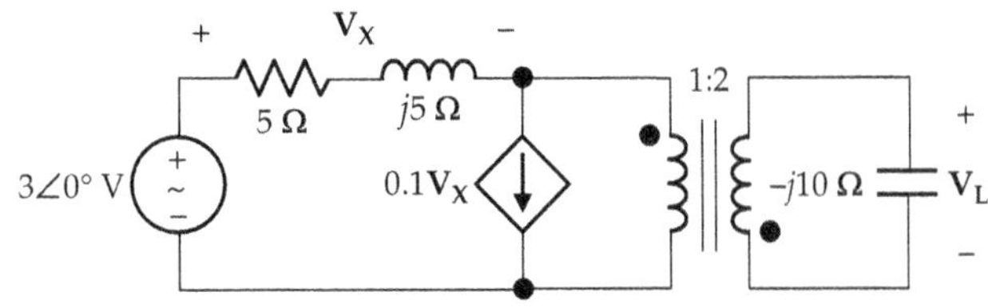

FIGURE P10.20

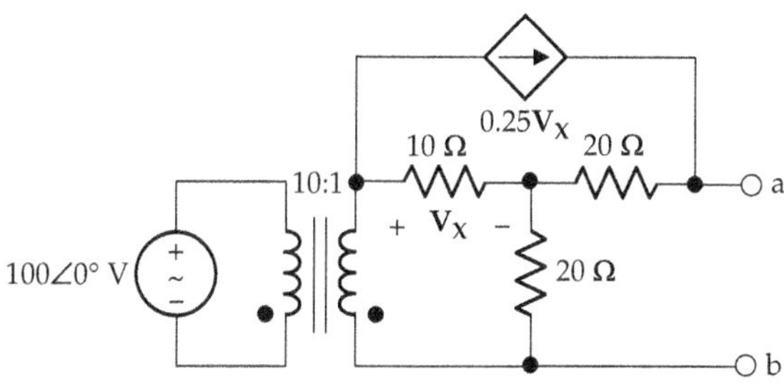

FIGURE P10.21

P10.22 Derive TEC looking into terminal 'ab' in Figure P10.22.

Ans. $\mathbf{V_{Th}} = \mathbf{V_{ab}} = 6\angle 0°$ V, $Z_{Th} = -j5\ \Omega$.

P10.23 Determine $v_O(t)$ in Figure P10.23, assuming $v_{SRC}(t) = 2\sqrt{2}\cos(10t - 10°)$ V.

Ans. $3\cos(10t + 125°)$ V.

P10.24 Derive TEC between terminals 'ab' in Figure P10.24.

Ans. $\mathbf{V_{Th}} = \mathbf{V_{ab}} = -10$ V, $Z_{Th} = 2.5\ \Omega$.

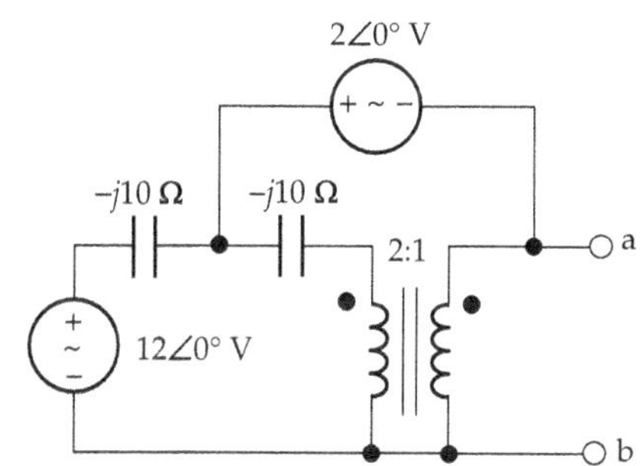

FIGURE P10.22

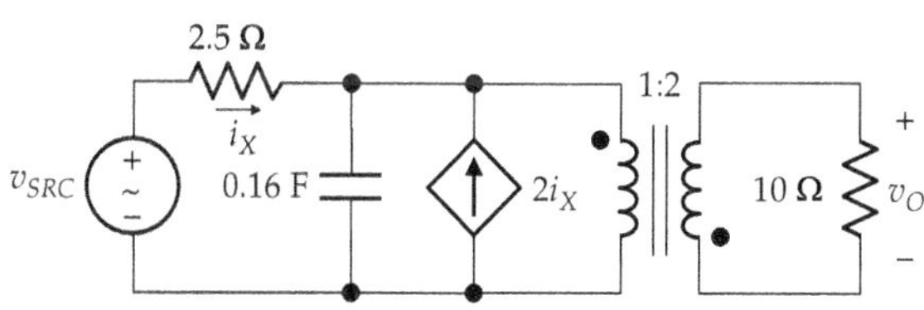

FIGURE P10.23

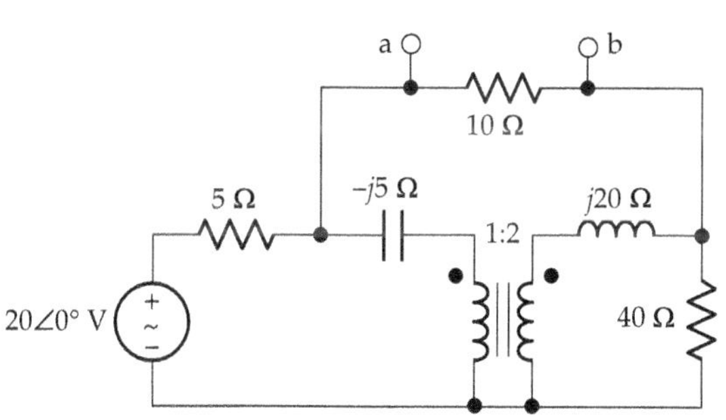

FIGURE P10.24

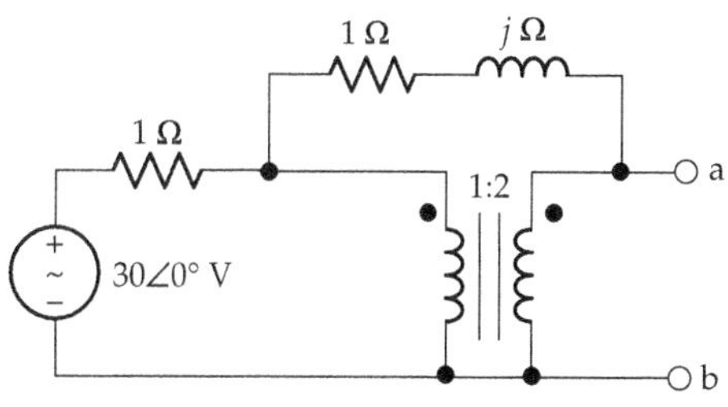

FIGURE P10.25

P10.25 Derive TEC between terminals 'ab' in Figure P10.25.

Ans. $\mathbf{V_{Th}} = \mathbf{V_{ab}} = (36 + j12)$ V, $Z_{Th} = (2.4 + j0.8)\ \Omega$.

P10.26 Determine $\mathbf{I_X}$ in Figure P10.26.

Ans. 2 A.

P10.27 Determine $i_O(t)$ and $i_1(t)$ in Figure P10.27, given that $v_{SRC}(t) = 5\sqrt{2}\cos(2t - 45°)$ V and $i_{SRC}(t) = 0.5\sin 2t$ A.

Ans. $i_O(t) = 0.5\sin 2t$ A, $i_1(t) = -0.5\sin 2t$ A.

P10.28 Derive TEC looking into terminals 'ab' in Figure P10.28.

Ans. $\mathbf{V_{Th}} = \mathbf{V_{ab}} = 2\angle 0°$ V, 7/6 Ω.

P10.29 Determine $\mathbf{I_S}$ in Figure P10.29.

Ans. $2\angle 0°$ A.

P10.30 Determine Z_{in} in Figure P10.30.

Ans. $-j2\ \Omega$.

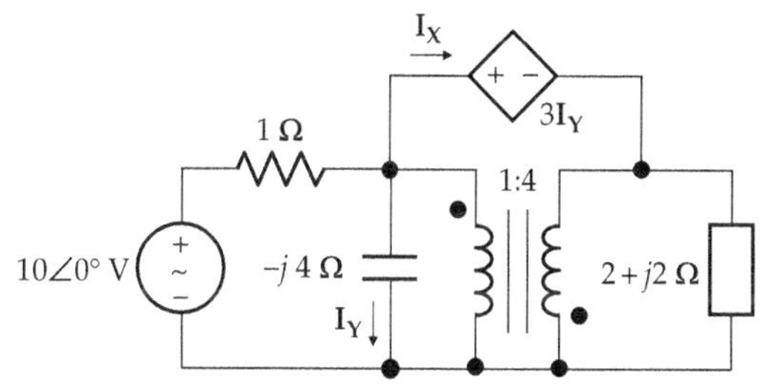

FIGURE P10.26

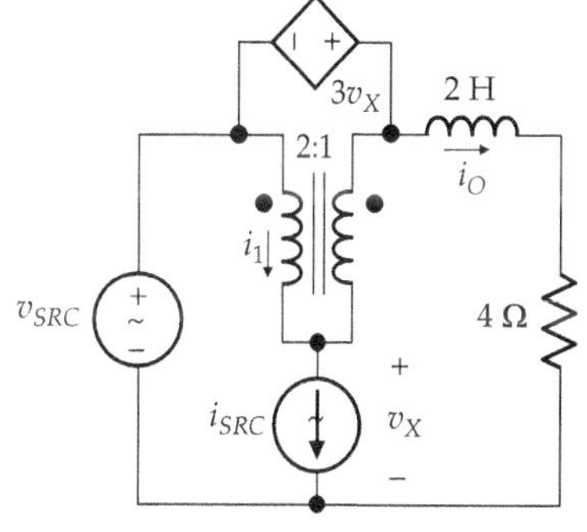

FIGURE P10.27

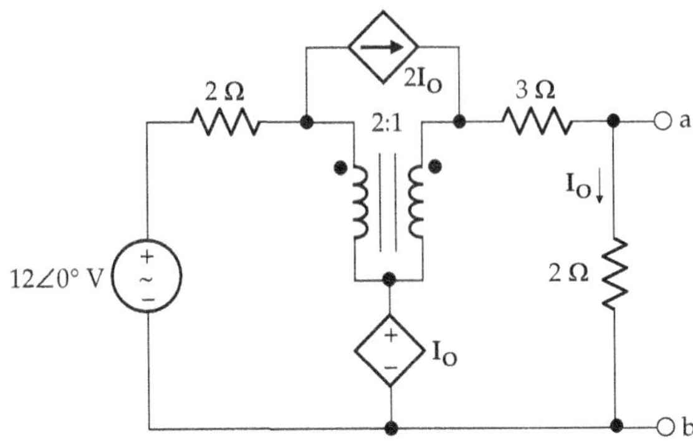

FIGURE P10.28

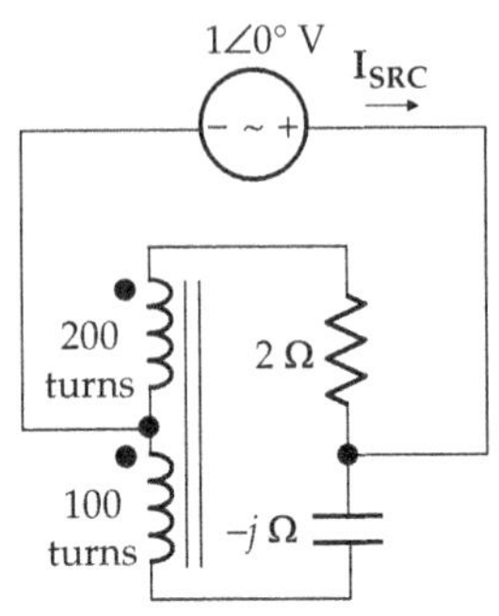

FIGURE P10.31

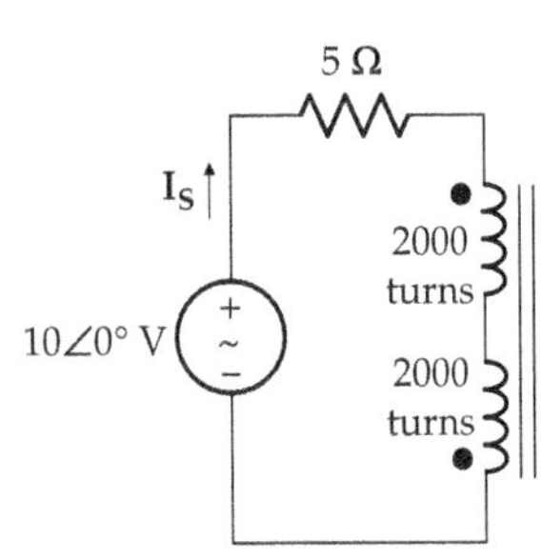

FIGURE P10.29

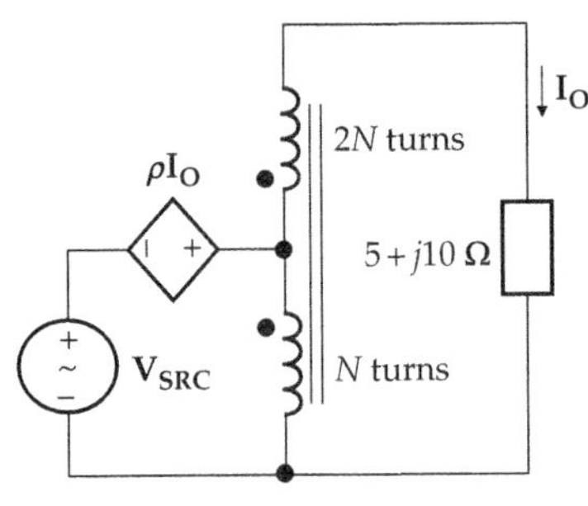

FIGURE P10.32

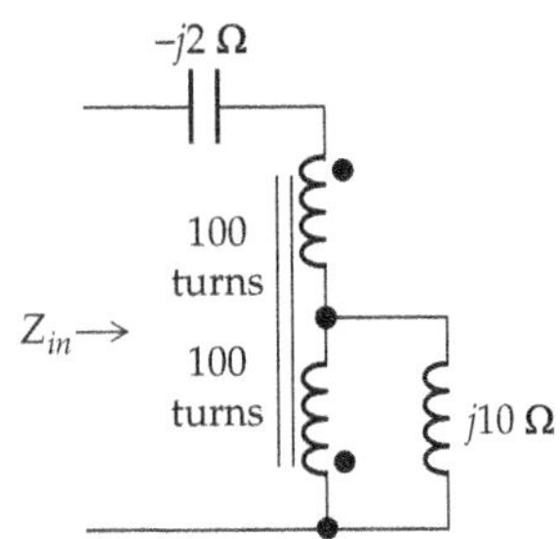

FIGURE P10.30

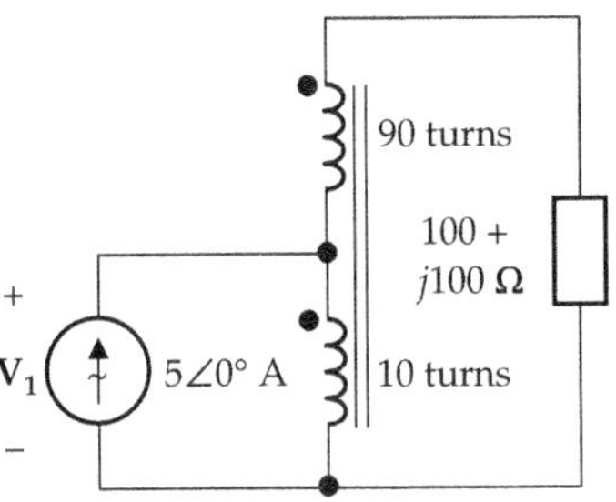

FIGURE P10.33

P10.31 Determine $\mathbf{I}_{SRC}$ in Figure P10.31.

Ans. $0.9(1 + j2)$ A.

P10.32 Determine ρ in Figure P10.32 so that the impedance seen by the independent source is purely reactive, and specify this reactance.

Ans. $\rho = -5$, and $X = 10\ \Omega$.

P10.33 Determine $\mathbf{V}_1$ in Figure P10.33.

Ans. $5(1 + j)$ V.

P10.34 Determine Z_{in} looking into terminals 'ab' in Figure P10.34.

Ans. $10.92\angle{-36.12°} = 8.818 - j6.435\ \Omega$.

P10.35 Determine $\mathbf{I}_1$, $\mathbf{I}_2$, $\mathbf{I}_o$, $\mathbf{V}_1$, and $\mathbf{V}_2$ in Figure P10.35.

Ans. $\mathbf{I}_1 = 1.851 - j0.584$ A, $\mathbf{I}_2 = -1.481 + j0.467$ A, $\mathbf{I}_O = 0.37 - j0.117$ A.

$\mathbf{V}_1 = 0.772 - j0.651$ V, $\mathbf{V}_2 = 3.86 - j3.25$ V.

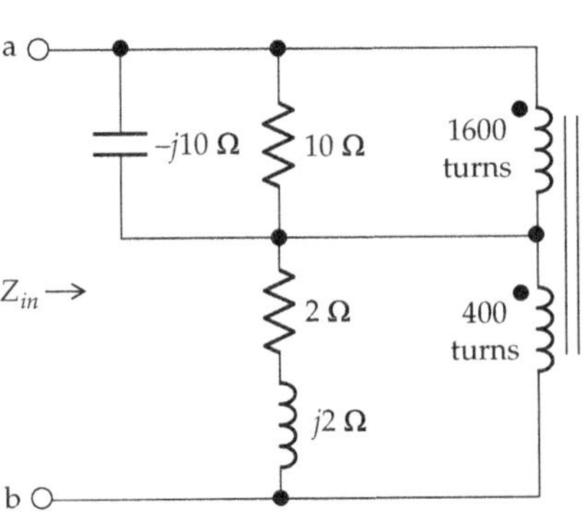

FIGURE P10.34

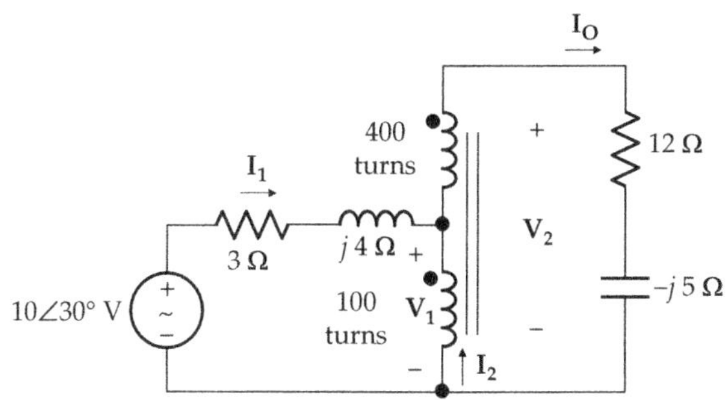

FIGURE P10.35

P10.36 Determine the average power dissipated in the 2 Ω resistor in Figure P10.36.

Ans. 1 W.

P10.37 Derive TEC looking into terminals 'ab' in Figure P10.37.

Ans. $\mathbf{V}_{\mathbf{Th}} = \mathbf{V}_{\mathbf{ab}} = 10(1 - j)$ V, $Z_{Th} = 100(1 - j)$ Ω.

P10.38 Determine $i_S(t)$ in Figure P10.38, assuming that $v_{SRC}(t) = 20\cos 1000t$ V.

Ans. $\sqrt{2}\cos 1000t$ A.

P10.39 Determine $\mathbf{V}_O$ in Figure P10.39.

Ans. −400∠0° V.

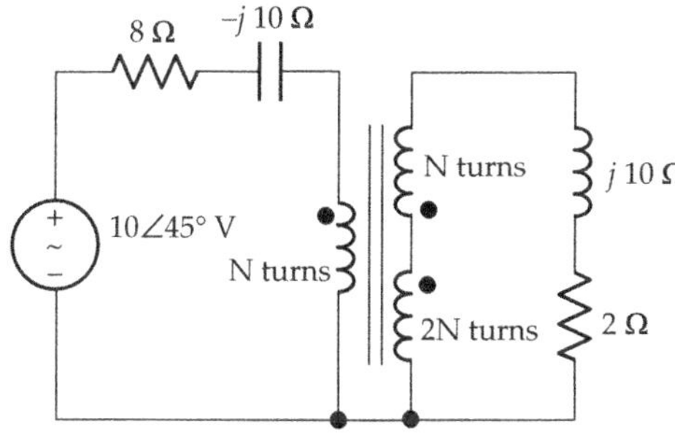

FIGURE P10.36

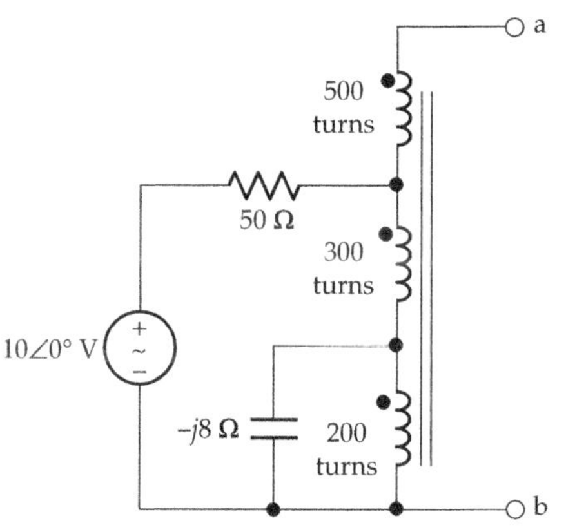

FIGURE P10.37

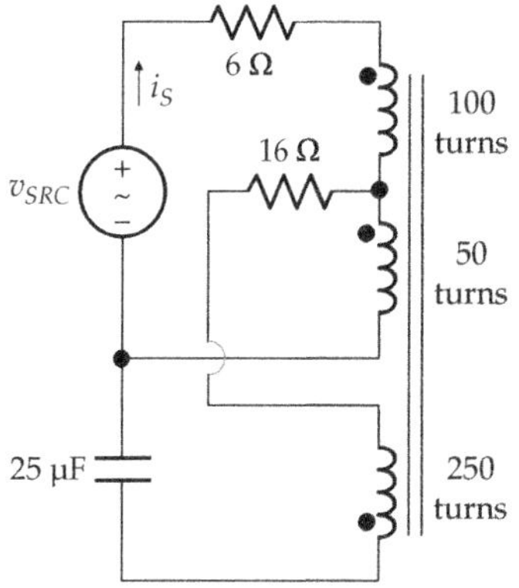

FIGURE P10.38

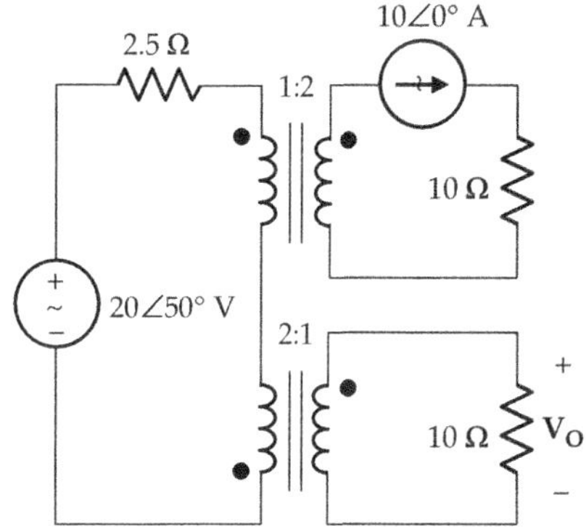

FIGURE P10.39

P10.40 Determine Z_{in} in Figure P10.40.

Ans. 10 Ω.

P10.41 Derive TEC looking into terminal 'ab' in Figure P10.41.

Ans. $\mathbf{V}_{\mathbf{Th}} = \mathbf{V}_{\mathbf{ab}} = (-64 + j48)$ V, $Z_{Th} = \frac{96}{5}(4 - j3)$ Ω.

P10.42 Derive TEC looking into terminal 'ab' in Figure P10.42.

Ans. $\mathbf{V}_{\mathbf{Th}} = \mathbf{V}_{\mathbf{ab}}$ (−26.1 + j 9.65) V in series with (15.44 + j1.69) Ω.

P10.43 Determine $\mathbf{I}_X$ in Figure P10.43.

Ans. 7.5 A.

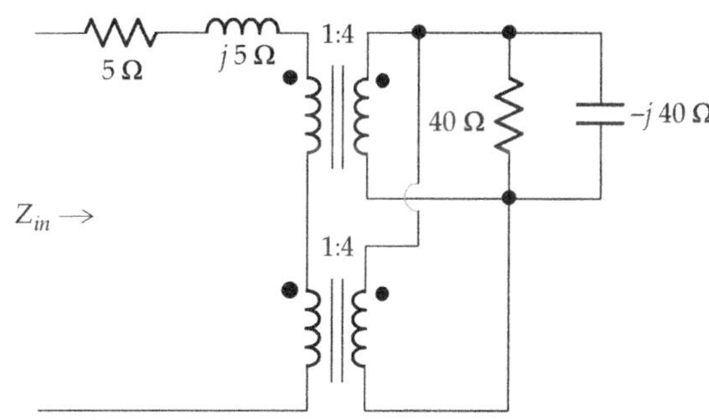

FIGURE P10.40

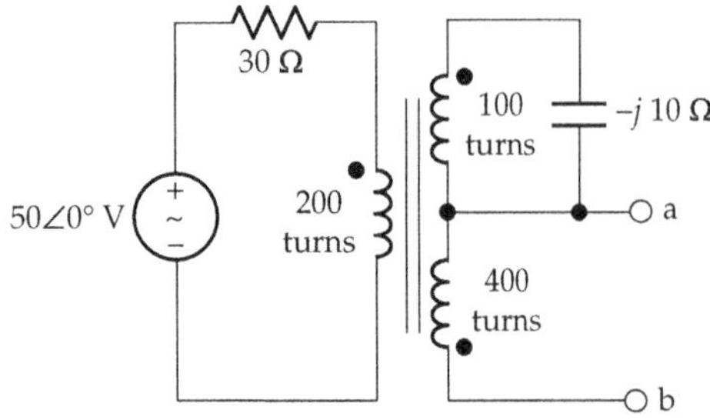

FIGURE P10.41

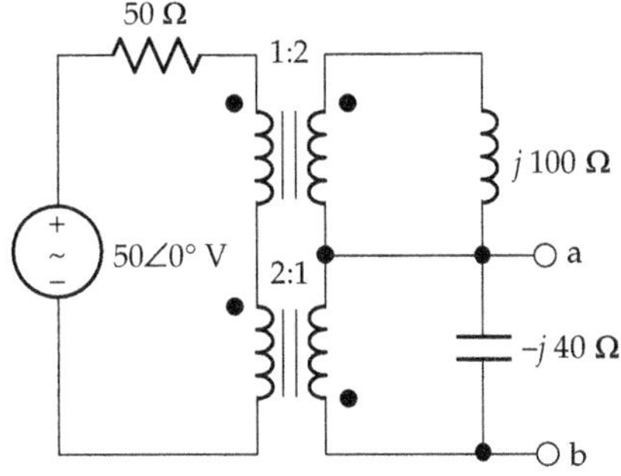

FIGURE P10.42

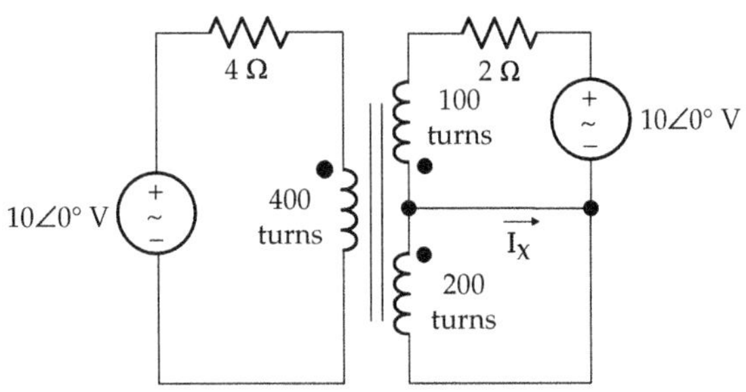

FIGURE P10.43

P10.44 Determine Z_{Th} looking into terminals 'ab' in Figure P10.44 assuming (a) the number of turns $N_1 = N_2$, (b) $N_1 \neq N_2$, and (c) reversed dot markings, irrespective of the relation between N_1 and N_2. Note that in this case, that of a series ideal transformer, changing the dot markings gives a different Z_{Th} when $N_1 = N_2$.

Ans. (a) Z; (b) and (c) ∞.

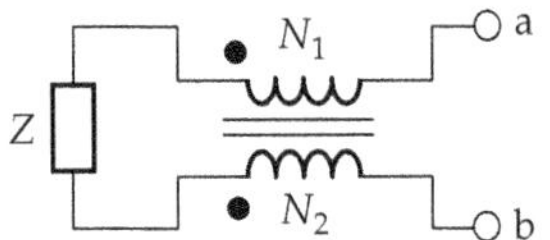

FIGURE P10.44

Linear and Ideal Transformers

P10.45 Determine $\mathbf{V}_{Th} = \mathbf{V}_{ab}$. In Figure P10.45.

Ans. $2(-1 + j)$ V.

P10.46 Derive TEC looking into terminals 'ab' in Figure P10.46, assuming $v_{SRC}(t) = 10\cos(10^3t + 45°)$.

Ans. $\mathbf{V}_{Th} = \mathbf{V}_{ab} = 20\angle 45°$ V, $Z_{Th} = (20 + j40)\ \Omega$.

P10.47 Determine k in Figure P10.47 so that no current flows in Z_X.

Ans. 0.4.

P10.48 Determine X in Figure P10.48 so no current flows in the 5 Ω resistor.

Ans. 2.5 Ω.

P10.49 Determine L_{eq} in Figure P10.49.

Ans. 6 H.

P10.50 Determine L_{eq} in Figure P10.50.

Ans. 4 H.

P10.51 Derive TEC looking into terminals 'ab' in Figure P10.51, where $\mathbf{V}_{Th} = \mathbf{V}_{ab}$.

Ans. $\mathbf{V}_{Th} = \mathbf{V}_{ab} = 7\angle -90°$ V, $Z_{Th} = j100\ \Omega$.

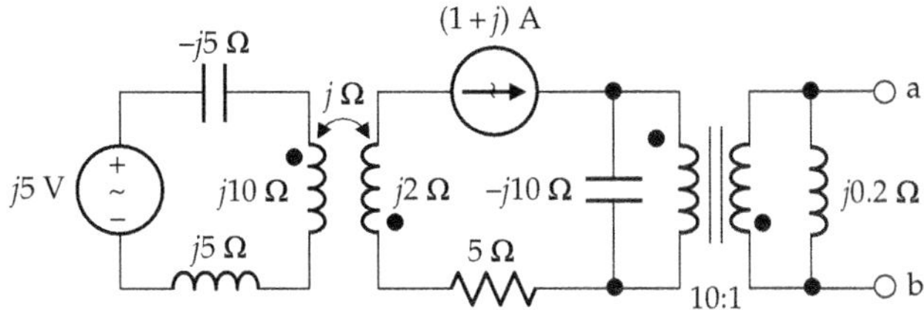

FIGURE P10.45

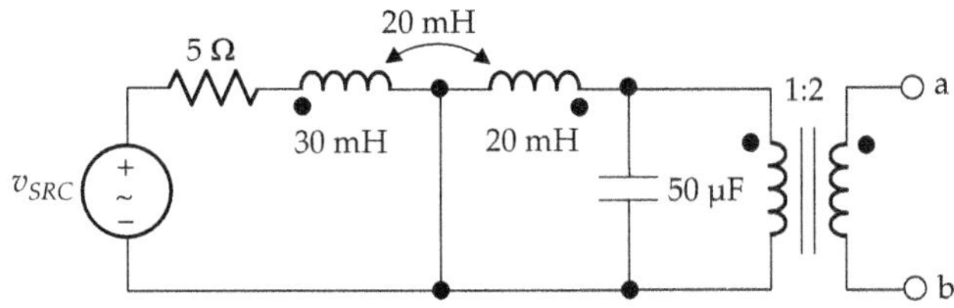

FIGURE P10.46

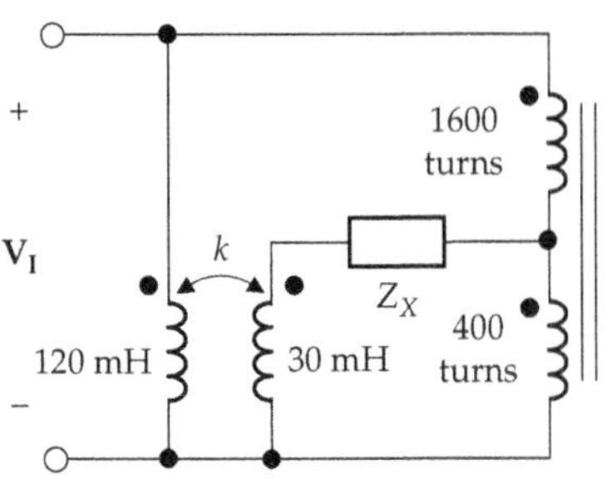

FIGURE P10.47

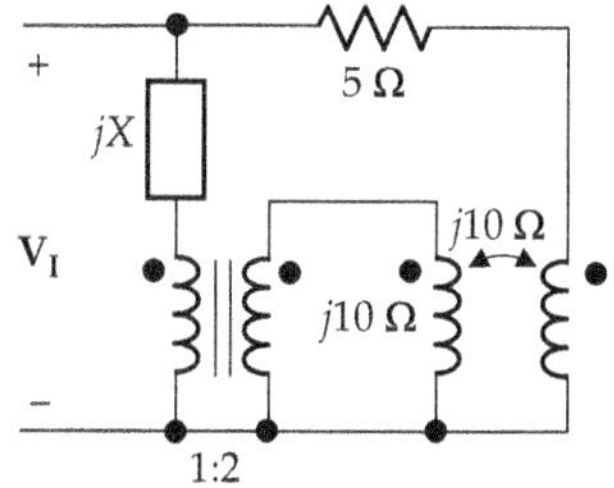

FIGURE P10.48

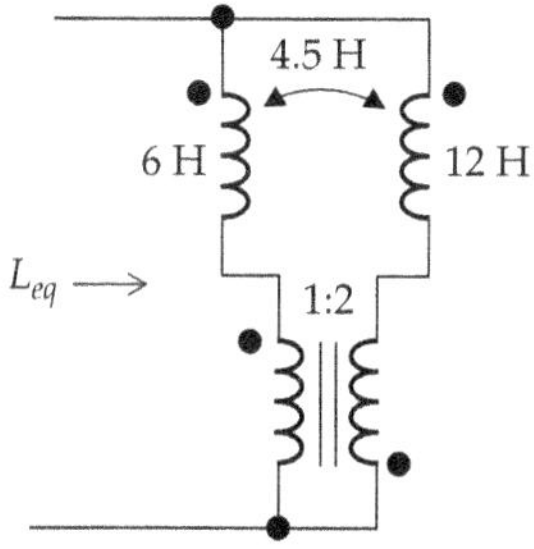

FIGURE P10.49

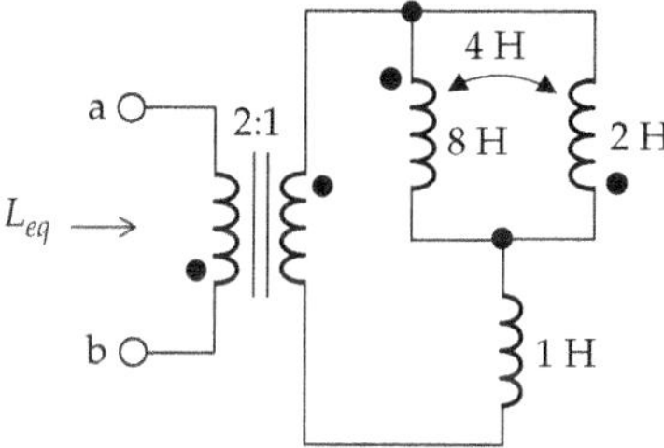

FIGURE P10.50

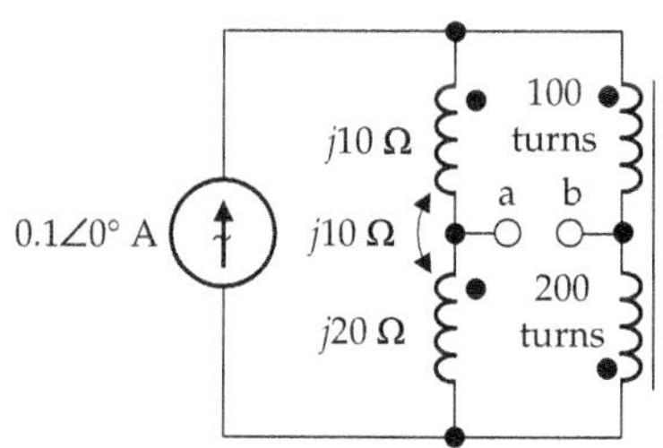

FIGURE P10.51

P10.52 Determine a in Figure P10.52 so that $Y_{in} = 0$, assuming $\omega = 1$ Mrad/s

Ans. 2.

Probing Further

P10.53 Consider the definition of an ideal transformer. Argue that the voltage equation (Equation 10.23) is satisfied by assuming zero coil resistance, no coil currents, and infinite core permeability, even in the presence of a leakage path. Then argue that in the presence of current in the coils, the equation is not satisfied unless it is assumed that the coupling is perfect, that is, there is no leakage path and hence no leakage flux.

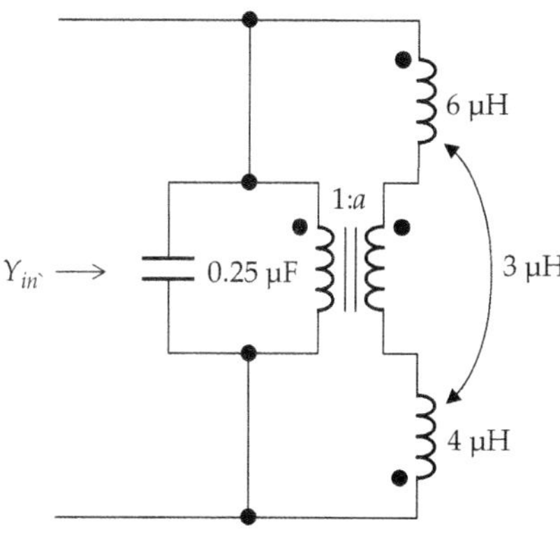

FIGURE P10.52

P10.54 The circuit of Figure P10.54 is purely passive, no energy being generated. The circuit remains passive if R is reflected as a^2R to the left side. However, if R is reflected as $-a^2R$, argue that energy is delivered to R_A, which violates conservation of energy.

P10.55 Referring to Example 10.5, replace Z_2, by a current source $\mathbf{I}_2 = \mathbf{V}_2/Z_2$ and Z_3 by a current source $\mathbf{I}_3 = \mathbf{V}_3/Z_3$. Determine $\mathbf{I}_1$ by superposition, applying $\mathbf{I}_2$ and $\mathbf{I}_3$ one at a time, then substituting for $\mathbf{V}_2$ and $\mathbf{V}_3$ in terms of $\mathbf{V}_1$ so as to obtain Equation 10.44.

P10.56 Show that by first reflecting the shunt reactance and then the series reactance in Figure 10.33b, the transformer equivalent circuit can be represented as in Figure P10.56.

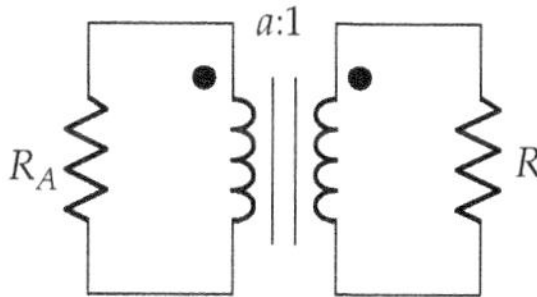

FIGURE P10.54

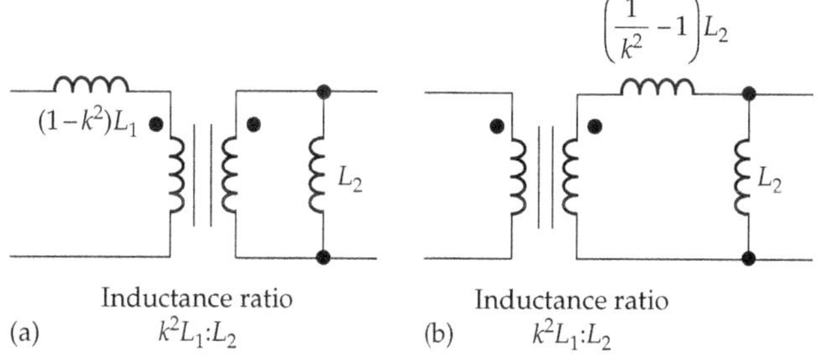

FIGURE P10.56

P10.57 Show that, after the leakage flux is added in Figure 10.33a, Equation 10.49 can be satisfied by the circuit of Figure P10.57. In order to satisfy the relation $\mathbf{V}_2 = j\omega M\mathbf{I}_1$ with the secondary open-circuited, the inductance ratio should be $L_1{:}k^2L_2$, leading to an alternative form of the transformer equivalent circuit.

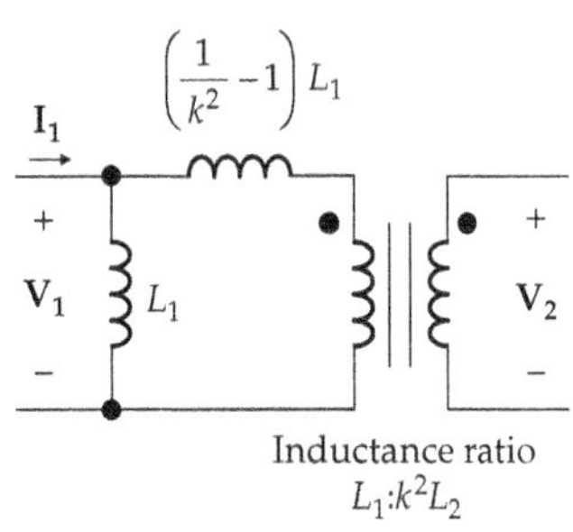

FIGURE P10.57

P10.58 In Figure P10.58a, three lumped capacitors are added in the high-frequency equivalent circuit of the transformer to account for the capacitances of the windings as well as the interwinding capacitance C_i. Show that C_i can be replaced by the capacitances shown in Figure P10.58b.

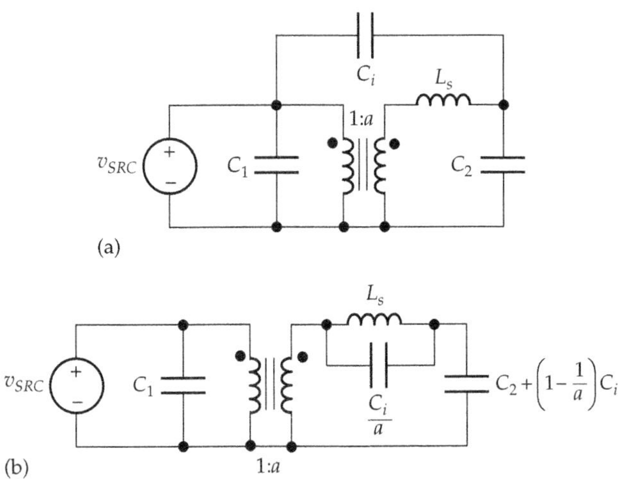

FIGURE P10.58

11

Basic Responses of First-Order Circuits

Objective and Overview

This chapter examines the basic responses of *RC* and *RL* circuits to initial energy storage or to dc excitation.

The first part of the chapter is concerned with the discharging of a capacitor or an inductor through a resistor and the charging of these energy storage elements when associated with a resistor and connected to a dc source. In discussing the circuit responses, duality is extensively illustrated and physical interpretations are strongly emphasized.

First-order circuits are later generalized to include more than one resistor and more than one capacitor or inductor as well as dependent sources. A general procedure is formulated for analyzing first-order circuits in response to initial energy storage or to dc excitation without having to derive and solve the differential equation. The chapter ends with an explanation of the role of the transient component of circuit responses.

11.1 Capacitor Discharge

Consider a capacitor C connected to a battery of voltage V_{SRC} and to a resistor R_C through a changeover switch, as illustrated in Figure 11.1a. When the switch is in position 'a', the capacitor is charged to the battery voltage V_{SRC} and stores electric energy $(1/2)CV_{SRC}^2$. Suppose that the switch is moved at $t = 0$ to position 'b' so that the capacitor and resistor in parallel are isolated from the battery (Figure 11.1b). In all switching operations, it is tacitly assumed that switching occurs instantaneously.

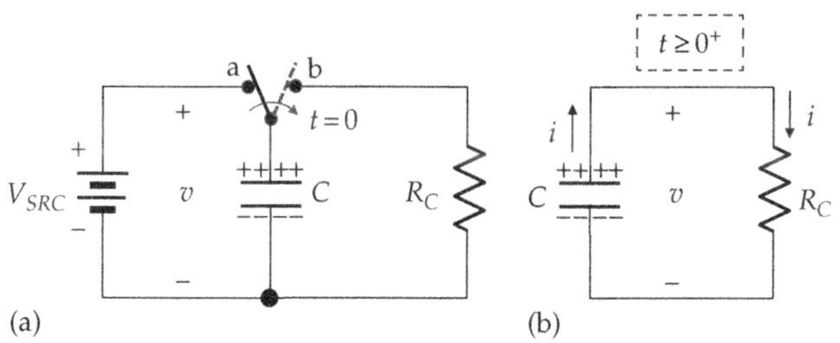

FIGURE 11.1
(a) Capacitor is charged when the switch is in position 'a' and begins to discharge when the switch is moved to position 'b' at $t = 0$ and (b) capacitor discharge for $t \geq 0^+$.

If the switch is moved at $t = 0$, then $t = 0^-$ denotes the instant of time just before the switch is moved, whereas $t = 0^+$ denotes the instant of time just after the switch is moved. These two instances of time are considered to be infinitesimally separated. It should be noted in this connection that if a function is continuous at $t = 0$, that is, it does not have a step, or discontinuity at $t = 0$, the value of the function is the same at $t = 0$, $t = 0^-$, or $t = 0^+$.

In Figure 11.1a, the voltage across the capacitor, just before the switch is moved, is $v(0^-) = V_{SRC}$. What is $v(0^+)$ just after the switch is moved? The answer is that it is still V_{SRC} because of the following *universal* concept:

Concept: *Energy in general, including stored energy, cannot be changed instantaneously by any physically realizable means, as this would require infinite driving forces, that is, infinite voltages or currents in the case of energy storage elements in electric circuits. Such infinite values are not physically realizable.*

Mathematically, this concept is embodied in the derivative/integral form of the v–i relation of ideal energy storage circuit elements. Considering the derivative form of the v–i relation of a capacitor, $i = Cdv/dt$, an instantaneous change in v implies that $dv/dt \to \infty$. This would require an infinite i, which is not physically realizable (but refer to Example 11.7 for more on this). Alternatively, it can be argued that $dv = idt/C$. As $dt \to 0$ at the instant of switching, then as long as i remains finite, $dv \to 0$, which means that the change in the capacitor charge, and hence v, is zero at the instant of switching. It follows that in Figure 11.1, v at $t = 0^+$ is the same as $t = 0^-$, although i starts to flow at $t = 0^+$. It is like suddenly opening an outlet valve at the bottom of a vessel containing some water. The outflow, analogous to i, in Figure 11.1b jumps instantly from zero to some finite value. However, at the instant of opening the outlet valve, water has not yet flowed out of the vessel so that the amount of water in the tank, analogous to the stored charge, does not change at this instant.

Similarly, for an inductor, $di = vdt/L$. As $dt \to 0$ at the instant of switching, then as long as v remains finite, $di \to 0$, which means that the change in i is zero at the instant of switching.

Although $v(0^+) = V_{SRC}$ in Figure 11.1b, v does not remain constant at this value for $t \geq 0^+$. This is because

R_C is now connected across C, resulting in a current $i(0^+) = V_{SRC}/R_C$. This current carries positive charge from the upper plate of the capacitor through the resistor to the lower plate, thereby neutralizing some of the negative charge on this plate. The capacitor charge, and hence voltage, therefore decreases with time, as long as i continues to flow. As $t \to \infty$, the charge on the capacitor is reduced to zero, that is, the capacitor is fully discharged, so that v and i become zero. It is required to determine how v and i vary with time for $t \geq 0^+$.

If we choose i as the variable for analyzing the circuit in Figure 11.1b, then we are in effect considering the circuit to be a series circuit, in which KCL is automatically satisfied and KVL has to be expressed in terms of the v–i relations of C and R_C. On the other hand, if we chose v as the variable for analyzing the circuit, we are in effect considering the circuit to be a parallel circuit, in which KVL is automatically satisfied and KCL has to be expressed in terms of the v–i relations of C and R_C. Choosing the latter case, we have, for the resistor, $i = v/R_C$, whereas for the capacitor, $i = -Cdv/dt$, where the minus sign applies because i through the capacitor is in the direction of a voltage rise v across the capacitor (Equation 7.7). KCL gives

$$\frac{v}{R_C} = -C\frac{dv}{dt} \quad \text{or} \quad v + R_C C\frac{dv}{dt} = 0 \tag{11.1}$$

Dividing by $R_C C$ and rearranging,

$$\frac{dv}{dt} + \frac{1}{R_C C}v = 0 \quad \text{or} \quad \frac{dv}{dt} + \frac{v}{\tau_C} = 0, \quad t \geq 0^+ \tag{11.2}$$

where $\tau_C = R_C C$ is the **time constant**.

Mathematically, the responses of lumped-parameter, linear time-invariant (LTI) circuits are governed by linear, ordinary, differential equations with constant coefficients, whose order depends on the number of distinct energy storage elements in the circuit, as will be clarified later. Equation 11.2 is an example of such a differential equation of the first order, since the highest order of the derivative of v is the first derivative. For the sake of brevity, we will henceforth refer only to the order of circuit differential equations, without the other attributes. Circuits having a single energy storage element are referred to as first order because their responses obey first-order differential equations.

The variation of v with time is given by the solution of Equation 11.2. Although the solution can be readily obtained by straightforward integration after separation of variables (Exercise 11.1), let us digress a little and consider a more general approach that will be first applied to the first-order differential equation

$$\frac{dy}{dt} + \frac{y}{\tau} = 0, \quad t \geq 0^+ \tag{11.3}$$

where y is a variable that can be a current or a voltage. A differential equation with the variable and its derivatives on the LHS and zero on the RHS, as in Equation 11.3, is referred to as the **homogeneous differential equation**. The general solution of a homogeneous, linear, ordinary, differential equation of any order, but having constant coefficients, is the sum of exponentials of the form Ae^{st}, where A and s are constants that depend on the circuit and on the initial conditions, and where the number of exponential terms in the sum is equal to the order of the differential equation. Since Equation 11.3 is of the first order, its general solution is $y(t) = Ae^{st}$. Substituting in Equation 11.3 and collecting terms,

$$Ae^{st}\left(s + \frac{1}{\tau}\right) = 0 \tag{11.4}$$

To satisfy Equation 11.4, either $A = 0$ or the terms in brackets must add to zero. If $A = 0$, then $y = 0$, and the response is zero. This is a trivial solution that contradicts the assumption of initial energy storage in the circuit. It follows that in order to satisfy the differential equation,

$$s + \frac{1}{\tau} = 0 \tag{11.5}$$

This equation in s is the **characteristic equation** of the differential equation. It gives the value of s as $-1/\tau$, which shows that s is a constant that depends only on the circuit parameters, which are R and C in the case of the RC circuit of Figure 11.1. The general solution is then

$$y(t) = Ae^{-t/\tau}, \quad t \geq 0^+ \tag{11.6}$$

A is an arbitrary constant that corresponds to the constant of integration, since, formally, the solution of a first-order differential equation involves a single integration. The value of A can be determined from initial conditions. Substituting $t = 0^+$ in Equation 11.6 gives $y(0^+) = A$. This initial value may also be denoted as y_0. We conclude, therefore, that the solution to Equation 11.3 with $y = y_0$ at $t = 0^+$ is

$$y(t) = y_0 e^{-t/\tau}, \quad t \geq 0^+ \tag{11.7}$$

Equation 11.7 is a decaying exponential function. It is plotted in Figure 11.2 in normalized form, where the horizontal axis represents $x = t/\tau$ and the vertical axis represents $e^{-x} = y/y_0$. Note that e^{-x} is continuous at $x = 0$ so that its value is unity at $t = 0^-$ or $t = 0^+$. At $x = 1$ or $t = \tau$, the response decreases to $1/e$ or approximately 36.8% of its initial value. Another interpretation of τ is based on the magnitude of the slope at $x = 0$. Thus,

$$\frac{de^{-x}}{dx} = -e^{-x} \quad \text{and} \quad \left|\frac{de^{-x}}{dx}\right| = 1 \quad \text{at } x = 0 \tag{11.8}$$

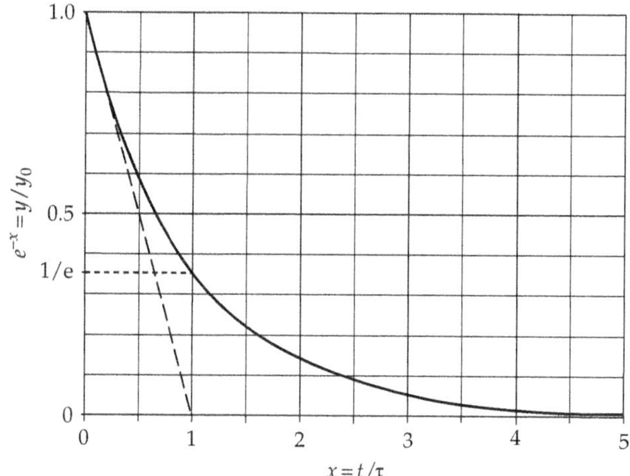

FIGURE 11.2
Decaying exponential response.

In other words, the magnitude of the slope of the tangent to e^{-x} at $x = 0$ is unity. Since the value of $e^{-x} = y/y_0 = 1$ at $x = 0$, the tangent intersects the horizontal axis at $x = 1$ or $t = \tau$. The interpretation is that if the response decreases linearly at the initial, maximum rate, instead of exponentially, the response is reduced to zero at $t = \tau$.

Note that if a variable decays exponentially with a time constant τ, as in the case of $y(t)$ in Equation 11.7, then $y(t + \tau) = y(t)/e$, which means that for every time interval τ, $y(t)$ is reduced at the end of the interval to $1/e$ of its value at the beginning of the interval. After five time constants, the response decays to $1/e^5 = 0.0067$ or 0.67% of its initial value. This is close enough to zero for most practical purposes.

The time constant is an important parameter of a first-order response in that it is indicative of the speed of response of the circuit. The larger the time constant, the longer it would take the response to reach a given fraction of its initial value; conversely, the smaller the time constant, the faster is this fraction reached. This is illustrated in Figure 11.3, which shows a simulation of the three exponentials corresponding to time constants of 1, 0.5, and 0.25 s. Note that in each case, the tangent to the curve at $t = 0$ intersects the time axis at $t = \tau$.

Concept: *The time constant of a first-order circuit is a measure of its speed of response. The larger the time constant, the slower or more 'sluggish' the response is and conversely.*

Comparing Equations 11.2 and 11.3, the variation of v with time is obtained from Equation 11.7 by replacing y with v, y_0 with V_0, and τ with $\tau_C = RC$, which gives

$$v(t) = V_0 e^{-t/\tau_C} \quad t \geq 0^+ \tag{11.9}$$

and

$$i(t) = \frac{v}{R_C} = I_0 e^{-t/\tau_C}, \quad t \geq 0^+ \tag{11.10}$$

where
$V_0 = v(0^+) = V_{SRC}$
$I_0 = V_0/R_C$

Both v and i are decaying exponentials as in Figure 11.2, since they are related by Ohm's law for R_C. As $t \to \infty$, v and i go to zero. The charge carried by the current completely discharges the capacitor (Primal Exercise 11.2), and the energy initially stored in the capacitor is dissipated in the resistor (Problem P11.57).

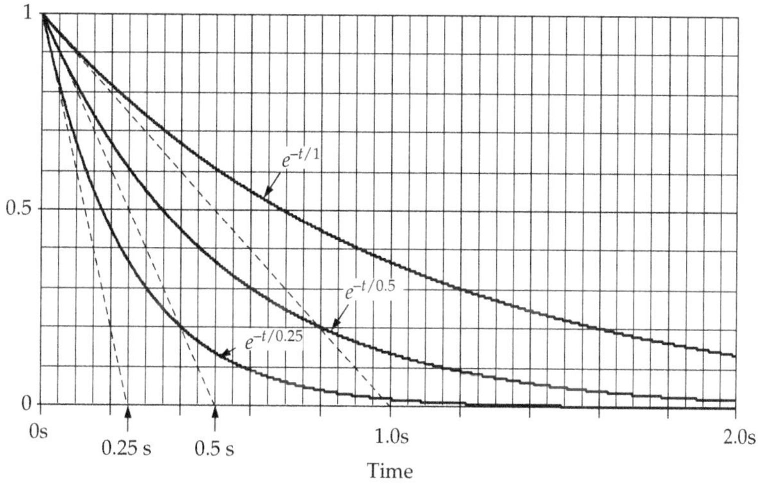

FIGURE 11.3
Effect of time constant on rate of decay.

It may be wondered why v and i decay exponentially with time rather than linearly. The answer lies in the rate at which v decreases. A linear decrease implies that v decreases at a constant rate, irrespective of the value of v at any given time. In fact, however, v does not decrease at a constant rate. From the v–i relation for the capacitor, the rate of decrease of v, which is $-dv/dt$, is

$$-\frac{dv}{dt}=\frac{i}{C} \tag{11.11}$$

That is, the decrease in v is proportional to i, because with the assigned positive directions of v and i as in Figure 11.1b, the decrease in v is due to the discharging of the capacitor by i. But with R_C connected across the capacitor, v and i are related by Ohm's law, $v = R_C i$. Substituting in Equation 11.11,

$$-\frac{dv}{dt}=\frac{v}{R_C C} \tag{11.12}$$

As v decreases, therefore, the rate of decrease of v also decreases. In other words, v decreases at a progressively reduced rate as the capacitor discharges. Mathematically, when the rate of decrease of a variable at any instant t is proportional to the value of the variable at that instant, the variable decays exponentially.

The system response that arises solely from the initial energy storage, with no other inputs, is the **natural response** of the system.

Exercise 11.1

Rearrange Equation 11.3 in the form of $dy/y = -(1/\tau)dt$. Integrate both sides of this equation, including a constant of integration. Express the equation in exponential form and show that it reduces to Equation 11.6.

Primal Exercise 11.2

Integrate Equation 11.10 to show that the charge carried by the current is equal to that initially stored in the capacitor.

Primal Exercise 11.3

A 0.5 F capacitor is initially charged to 5 V and is paralleled at $t = 0$ with a 1 Ω resistor. Derive the expression for the voltage across the capacitor as a function of time for $t \geq 0^+$.

Ans. $5e^{-2t}$ V, t is in s.

Primal Exercise 11.4

A 1 μF capacitor is initially charged to 1 V and disconnected from the charging source at $t = 0$. If the capacitor voltage drops to 0.9 V after 100 hours (h), determine (a) the time constant and (b) the resistance R_p that is effectively in parallel with the capacitor. This resistance is that of the dielectric between the plates of the capacitor.

Ans. (a) 949.1 h, (b) 3.42×10^{12} Ω.

Example 11.1: Capacitor Discharge

A 5 mF capacitor charged to 3 V is paralleled at $t = 0$ with a 2 Ω resistor. Another 2 Ω resistor is connected in parallel with the combination at $t = 5$ ms (Figure 11.4). It is required to determine v for $t \geq 0^+$.

Solution:

$0^+ \leq t \leq 5$ ms: from Equation 11.9, $v(t) = V_0 e^{-t/\tau_{C1}} = 3e^{-t/10}$ V, where $\tau_{C_1} = R_C C = 2\times5\times10^{-3} \equiv 10$ ms; t in the exponent is in ms because τ is in ms. v decays exponentially from 3 V toward zero with a time constant of 10 ms. At $t = 5$ ms, $v = 3e^{-0.5} = 1.82$ V.

$t \geq 5$ ms: the resistance in parallel with C is 2||2 = 1 Ω and $\tau_{C_2} = 1\times5\times10^{-3} \equiv 5$ ms. Applying Equation 11.9, $v(t) = 1.82e^{-(t-5)/5}$ V, where at $t = 5$ ms, the exponent is zero and $v = 1.82$ V.

Simulation: The circuit is entered as in Figure 11.5. The capacitor having marked terminals is entered from the ANALOG_P library and oriented so that the voltage of terminal 1 is positive with respect to ground. The initial

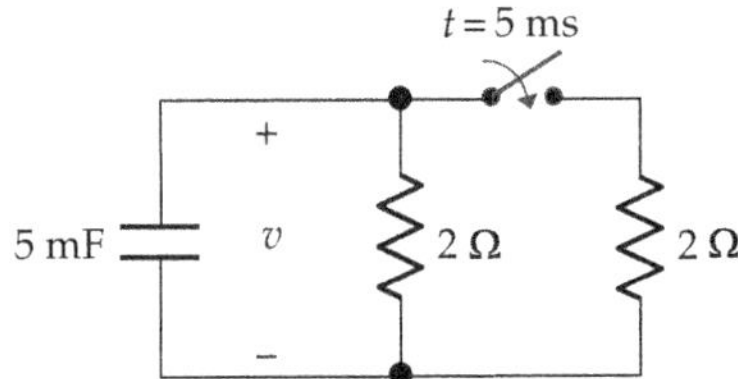

FIGURE 11.4
Figure for Example 11.1.

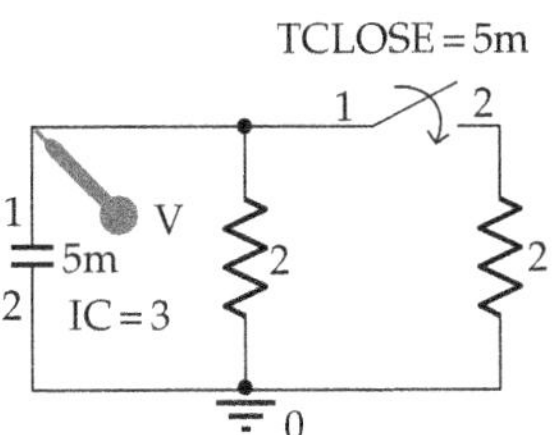

FIGURE 11.5
Figure for Example 11.1.

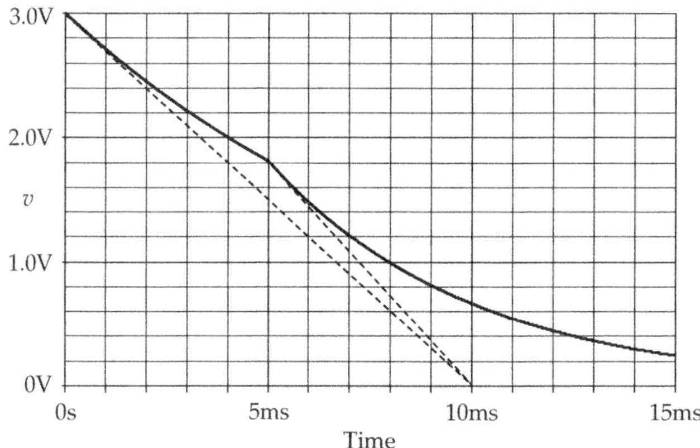

FIGURE 11.6
Figure for Example 11.1.

condition of 3 V is entered in the Property Editor spreadsheet. A normally open switch is set to close at $t = 5$ ms. In the Simulation Settings, 'Analysis type' is 'Time Domain (Transient),' 'Run to time' is 15m, 'Start Saving Data After' is 0, and 'Maximum Step size' is 1u. After the simulation is run, the plot shown in Figure 11.6 is displayed and consists of two exponentials: the first decaying from $v = 5$ V at $t = 0$ to $v = 1.82$ V at $t = 5$ ms, with a time constant of 10 ms, and the second decaying from $v = 1.82$ V at $t = 5$ ms toward zero, with a time constant of 5 ms. Note that the tangent at the start of the first exponential intersects the time axis at $\tau_{C_1} = 10$ ms and the tangent to the second exponential at $t_0 = 5$ ms intersects the time axis at $t_0 + \tau_{C_2} = 5 + 5 = 10$ ms, which is equal to τ_{C_1} in this case.

Problem-Solving Tip

- In the exponent t/τ, t naturally has the same units as τ. If τ is in seconds, then t is in seconds, and if τ is in ms, then t is in ms. If t is in seconds and is to be expressed in ms, while keeping the same value of the exponent, then the new t in ms should be multiplied by s/ms or 10^{-3}. Another way of looking at this is that t in ms is a larger number than t in s, so t in ms in the exponent must be multiplied by 10^{-3} to maintain the same numerical value of the exponent as when t is in s.

11.2 Capacitor Charging

Consider next the case of an initially uncharged capacitor connected through a resistor and a normally open switch to a battery of voltage V_{SRC} (Figure 11.7a). The switch is closed at $t = 0$. At $t = 0^-$, the capacitor voltage is zero, by assumption, and no energy is stored in the capacitor. At $t = 0^+$, the stored energy and hence the voltage remain zero, because of the previously mentioned fundamental concept that the stored energy cannot be changed instantaneously by any physically realizable means. Terminal 'b' of the resistor is therefore at 0 V, with respect to ground at $t = 0^+$, whereas terminal 'a' is connected to the battery and is at a voltage V_{SRC} with respect to ground. The voltage across R_C is V_{SRC}, causing a current V_{SRC}/R_C to flow through the resistor and the capacitor at $t = 0^+$. Note that the uncharged capacitor acts as a *short circuit* at the instant of switching, since the voltage across it is zero at this instant, while the capacitor current is V_{SRC}/R_C.

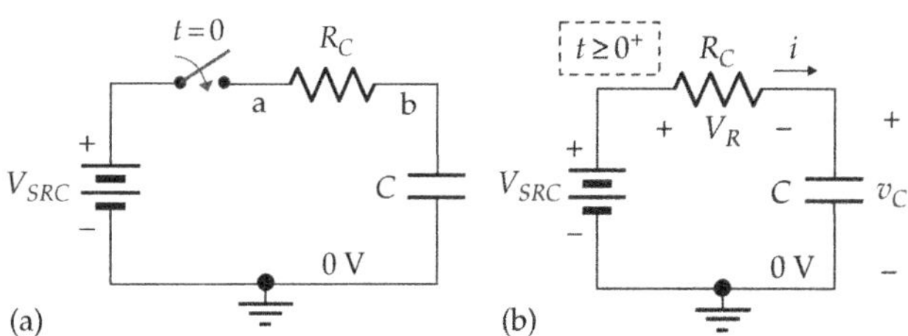

FIGURE 11.7
(a) Capacitor begins to charge when the switch is closed at $t = 0$ and (b) capacitor charging for $t \geq 0^+$.

The flow of current through the resistor charges the capacitor, increasing v_C, the voltage across the capacitor (Figure 11.7b). However, as v_C increases, the voltage across the resistor decreases, because $v_C + v_R = V_{SRC}$ is constant so that the current i decreases. Eventually, the capacitor charges to V_{SRC}, and i drops to zero. The capacitor now acts as an *open circuit* in the steady state. It is required to determine how v_C and i vary with time for $t \geq 0^+$.

Since the circuit is a series circuit, KCL is automatically satisfied. At any time t, KVL in Figure 11.7b gives $v_C + v_R = V_{SRC}$. Substituting for v_C in terms of the integral form of the v–i relation for the capacitor and for v_R in terms of Om's law,

$$\frac{1}{C}\int idt + R_C i = V_{SRC}, \quad t \geq 0^+ \tag{11.13}$$

Note that the v–i relation for the capacitor is written with a positive sign since i is in the direction of a voltage drop v_C. If both sides of Equation 11.13 are differentiated with respect to time, the RHS becomes zero, since V_{SRC} is a constant. Dividing the LHS by R and substituting $\tau_C = R_C C$, Equation 11.13 becomes

$$\frac{di}{dt} + \frac{i}{\tau_C} = 0, \quad t \geq 0^+ \tag{11.14}$$

Equation 11.14 is of the same form as Equation 11.3. Its solution is given by Equation 11.7 by replacing y with i, y_0 with $I_0 = V_{SRC}/R_C$, and τ with τ_C, which gives

$$i(t) = I_0 e^{-t/\tau_C}, \quad t \geq 0^+ \tag{11.15}$$

i therefore decays exponentially, as in Figure 11.2, from an initial value of $I_0 = V_{SRC}/R_C$ at $t = 0^+$ to zero as $t \to \infty$, as the capacitor becomes fully charged.

The time variation of v_C is obtained from the v–i relation of the capacitor as

$$v_C(t) = \frac{1}{C}\int_{0^+}^{t} idt + 0, \quad t \geq 0^+ \tag{11.16}$$

where $v_C = 0$ at the lower limit of integration, $t = 0^+$. Substituting for i from Equation 11.15 and integrating,

$$v_C(t) = \frac{V_{SRC}}{\tau_C}\int_{0^+}^{t} e^{-t/\tau_C}dt = V_{SRC}\left[-e^{-t/\tau_C}\right]_{0^+}^{t}$$

or

$$v_C(t) = V_{SRC}\left(1 - e^{-t/\tau_C}\right) = V_{CF}\left(1 - e^{-t/\tau_C}\right), \quad t \geq 0^+ \tag{11.17}$$

where $V_{CF} = V_{SRC} = v_C(\infty)$ is the final value of v_C and $v_C(0^+) = 0$. Equation 11.17 is plotted in Figure 11.8 in normalized form as v_C/V_{CF} vs. t/τ. The response is a "saturating" exponential that starts from zero and approaches unity as $t \to \infty$. At $t = \tau$, the response is $(1 - 1/e)$ or approximately 63.2% of its final value. Another interpretation of τ is that if the response increases linearly at its maximum initial value, it will reach the final value at $t = \tau$. This can be demonstrated by differentiating Equation 11.17, which gives

$$\frac{d(v/V_{CF})}{d(t/\tau)} = e^{-t/\tau} = 1 \quad \text{at } t = 0^+. \tag{11.18}$$

This means that $v/V_{CF} = 1$ at $t/\tau = 1$ or $v = V_{CF}$ at $t = \tau$.

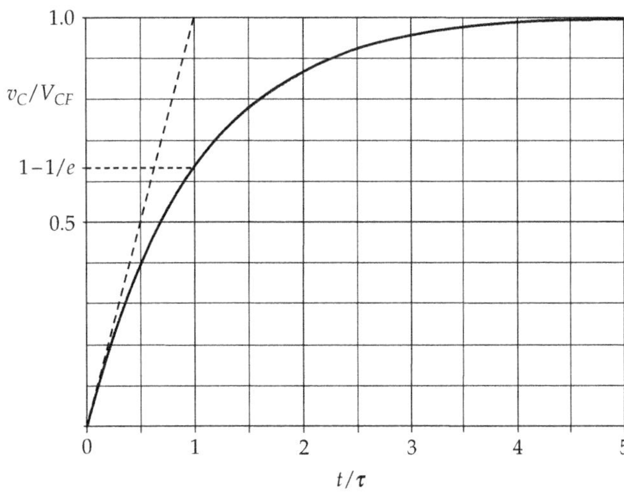

FIGURE 11.8
Saturating exponential response.

The exponential variation of v_C again follows from the argument that as v_C increases, its rate of increase is reduced as a linear function of v_C itself. Thus, the rate of increase of v_C is $dv_C/dt = i/C$. But $i = (V_{SRC} - v_C)/R_C$ so that as v_C increases, i decreases and the rate of increase of v_C also decreases as a linear function of v_C.

Note that as $t \to \infty$, the energy dissipated in the resistor is equal to that stored in the capacitor (Problem P11.58).

11.2.1 Charging with Initial Energy Storage

Suppose the capacitor in Figure 11.7 is initially charged to a voltage $V_0 < V_{SRC}$. The initial value of the current is $(V_{SRC} - V_0)/R_C$ instead of V_{SRC}/R_C for an initially uncharged capacitor. Equation 11.15 becomes

$$i(t) = \frac{(V_{SRC} - V_0)}{R_C}e^{-t/\tau_C} = I_0e^{-t/\tau_C}, \quad t \geq 0^+ \tag{11.19}$$

It is seen from this equation that $i(0^+) = I_0 = (V_{SRC} - V_0)/R_C$ and $i(\infty) \to 0$, as required. Equation 11.16 for v_C becomes

$$\begin{aligned} v_C(t) &= \frac{1}{C}\int_{0^+}^{t} idt + V_0 = (V_{SRC} - V_0)\left[-e^{-t/\tau_C}\right]_{0^+}^{t} + V_0 \\ &= V_{SRC} + (V_0 - V_{SRC})e^{-t/\tau_C} \\ &= V_{CF} + (V_0 - V_{CF})e^{-t/\tau_C}, \quad t \geq 0^+ \end{aligned} \tag{11.20}$$

Thus, $v_C(0^+) = V_0$ and $v_C(\infty) = V_{CF}$, as required.

It is instructive to derive Equations 11.19 and 11.20 from superposition, since the circuit is LTI. According to the principle of superposition, the total response is the algebraic sum of the response to the excitation applied to an uncharged capacitor and the response due to a charged capacitor with the excitation set to zero. Considering i first, Equation 11.15 applies to the response when the capacitor is charged from an initially uncharged state. When the capacitor is initially charged and the battery is replaced by a short circuit, the capacitor discharges, as in Figure 11.1b. The discharge current is given by Equation 11.10, but the direction of the discharge current is opposite to that of i in Figure 11.7b. Hence, the current in Equation 11.10, with $I_0 = V_0/R_C$, should be subtracted from that given by Equation 11.15. This gives

$$i(t) = \frac{V_{SRC}}{R_C}e^{-t/\tau_C} - \frac{V_0}{R_C}e^{-t/\tau_C} = \frac{(V_{SRC} - V_0)}{R_C}e^{-t/\tau_C}, \quad t \geq 0^+$$

which is the same as Equation 11.19. Similarly, v_C in this case is the sum of v_C given by Equation 11.17 for the charging of a capacitor from an initially uncharged state and v given by Equation 11.9 for capacitor discharge. Thus,

$$v_C(t) = V_{SRC}\left(1 - e^{-t/\tau_C}\right) + V_0 e^{-t/\tau_C} = V_{SRC} + \left(V_0 - V_{SRC}\right)e^{-t/\tau_C},$$
$$t \geq 0^+$$

which is the same as Equation 11.20. This is a fundamental and very useful property of LTI circuits that is expressed by the following concept:

Concept: *The responses in an LTI circuit, with initial energy storage, to an applied excitation can be derived as the algebraic sum of two responses: (i) the response to the applied excitation acting alone, with zero initial energy storage, and (ii) the response to the initial energy storage acting alone, with zero applied excitation.*

Capacitor charging through a current source is discussed in Example 11.2.

Primal Exercise 11.5

Assume that in Figure 11.7 $C = 1\ \mu F$. Determine R_C such that $v_C(10\ \mu s) = V_{SRC}/2$.

Ans. 14.4 Ω.

Primal Exercise 11.6

A 1 μF capacitor that is initially charged to 6 V is connected at $t = 0$ to a 12 V battery through a 10 kΩ resistor, as in Figure 11.7. Determine (a) the time constant and (b) the expressions for capacitor current and voltage for $t \geq 0^+$.

Ans. (a) 10 ms; (b) $i(t) = 0.6e^{-0.1t}$ mA, $v_C(t) = 12 - 6e^{-0.1t}$ V, t is in ms.

Primal Exercise 11.7

Repeat Primal Exercise 11.6 assuming the initial voltage on the capacitor is 18 V.

Ans. (a) 10 ms; (b) $i(t) = -0.6e^{-0.1t}$ mA, $v_C(t) = 12 + 6e^{-0.1t}$ V, where t is in ms.

Example 11.2: Capacitor Charging by Current Source

A dc current source I_{SRC} is applied at $t = 0$ to an uncharged capacitor C in parallel with a resistor R_C (Figure 11.9a). It is required to determine how v across the parallel combination, i_R and i_C, vary with time for $t \geq 0^+$ (Figure 11.9b).

Solution:

At $t = 0^-$, the stored energy in C is zero, since the voltage across C is zero, because of the closed switch. At $t = 0^+$, this energy and hence v remain zero. $v = 0$ means $i_R = 0$ so that all of I_{SRC} initially flows in C. C acts as a short circuit at this instant since the voltage across it is zero, while the current through it is I_{SRC}. C begins to charge, increasing v. As v increases, i_R increases in accordance with Ohm's law. This reduces i_C because $i_C + i_R = I_{SRC}$ is constant. Eventually, C charges fully, i_C drops to zero, and all of I_{SRC} flows through R_C. C acts as an open circuit, which means that the final value of v is $V_{CF} = R_C I_{SRC}$.

To analyze the circuit, we note that KVL is automatically satisfied in Figure 11.9b, because of the parallel connection. KCL gives $i_C + i_R = I_{SRC}$. Substituting $i_C = Cdv/dt$ and $i_R = v/R_C$,

$$C\frac{dv}{dt} + \frac{v}{R_C} = I_{SRC}, \quad t \geq 0^+ \qquad (11.21)$$

Dividing by C and substituting $\tau_C = R_C C$,

$$\frac{dv}{dt} + \frac{v}{\tau_C} = \frac{I_{SRC}}{C}, \quad t \geq 0^+ \qquad (11.22)$$

Equation 11.22 differs in form from Equation 11.14 in that the RHS is not zero. In other words, Equation 11.22 is a **nonhomogeneous differential equation**. The solution

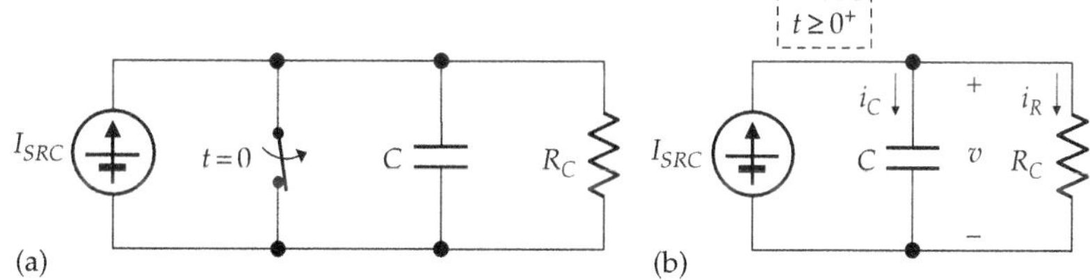

FIGURE 11.9
Figure for Example 11.2.

to this equation is discussed in detail in Section 11.5, where it is argued that the solution is the sum of two components: (1) v that is the solution to the homogeneous differential equation and (2) any v that satisfies Equation 11.22. In particular, as $t \to \infty$, steady conditions prevail so that $dv/dt = 0$ and $V_{CF} = (\tau_C/C)I_{SRC} = R_C I_{SRC}$, as argued previously. The first component of the solution is given by Equation 11.6, with an arbitrary constant A. It follows that the complete solution is

$$v(t) = Ae^{-t/\tau_C} + R_C I_{SRC}, \quad t \geq 0^+ \tag{11.23}$$

A is determined from the initial condition that $v(0^+) = 0$, which gives $A = -RI_{SRC}$. The complete solution is then

$$v(t) = R_C I_{SRC}\left(1 - e^{-t/\tau_C}\right) = V_{CF}\left(1 - e^{-t/\tau_C}\right), \quad t \geq 0^+ \tag{11.24}$$

This is a saturating exponential of zero initial value and a final value of $R_C I_{SRC}$, as required.

How will v change with time if C is charged to a voltage V_0 at $t = 0^-$ in Figure 11.9b? At $t = 0^-$, the energy stored in the capacitor is $(1/2)CV_0^2$. At $t = 0^+$, this energy, and hence V_0, remains the same, since the stored energy cannot be changed instantaneously by any physically realizable means. At this instant, $i_R = V_0/R_C$ so that $i_C = I_{SRC} - V_0/R_C$. As $t \to \infty$, C charges fully, $i_C = 0$, I_{SRC} flows though R_C, and the final voltage is $R_C I_{SRC}$, irrespective of V_0. The analytical solution is obtained from the general solution (Equation 11.23) but substituting the initial condition that $v = V_0$ at $t = 0$. This gives $A = V_0 - R_C I_{SRC}$ so that

$$v(t) = \left(V_0 - R_C I_{SRC}\right)e^{-t/\tau_C} + R_C I_{SRC}$$
$$= V_{CF} + \left(V_0 - V_{CF}\right)e^{-t/\tau_C}, \quad t \geq 0^+ \tag{11.25}$$

where $V_{CF} = R_C I_{SRC}$. The resistor current is

$$i_R(t) = \frac{v}{R_C} = \left(\frac{V_0}{R_C} - I_{SRC}\right)e^{-t/\tau_C} + I_{SRC}$$
$$= I_{RF} + \left(I_{R0} - I_{RF}\right)e^{-t/\tau_C}, \quad t \geq 0^+ \tag{11.26}$$

where $I_{R0} = i_R(0^+) = V_0/R_C$ and $I_{RF} = i_R(\infty) = I_{SRC}$. The capacitor current is

$$i_C(t) = C\frac{dv}{dt} = -\frac{C}{\tau_C}\left(V_0 - R_C I_{SRC}\right)e^{-t/\tau_C} = \left(I_{SRC} - I_{R0}\right)e^{-t/\tau_C}$$
$$= I_{C0}e^{-t/\tau_C}, \quad t \geq 0^+ \tag{11.27}$$

where

$I_{C0} = i_C(0^+) = I_{SRC} - V_0/R_C$
$i_C(\infty) = 0$

The current values at the two limits are in agreement with those argued previously. Note that $i_C + i_R = I_{SRC}$ for all t.

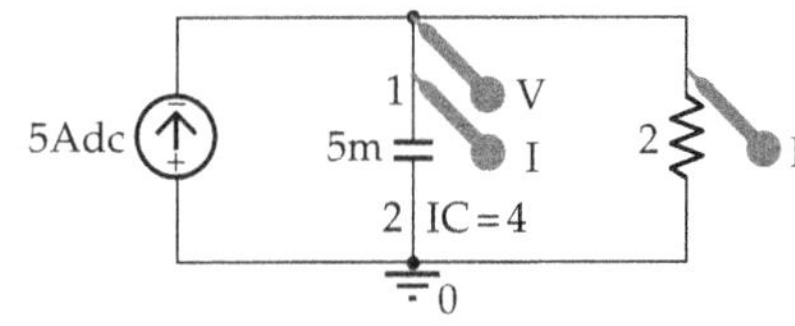

FIGURE 11.10
Figure for Example 11.2.

Equations 11.25 through 11.27 can be derived from superposition. Thus, with the excitation applied alone, v is given by Equation 11.24. When the initial energy storage is applied alone, with I_{SRC} in Figure 11.9b set to zero, that is, the ideal source replaced by an open circuit, v is given by Equation 11.9. Adding the two responses gives Equation 11.25, bearing in mind that $V_{CF} = R_C I_{SRC}$. Similar considerations apply to i_R and i_C (Exercise 11.9).

If $C = 5$ mF, $R_C = 2\ \Omega$, $I_{SRC} = 5$ A, and $V_0 = 4$ V, then $R_C C = 10$ ms, $V_{CF} = 10$ V, $I_{R0} = 2$ A, and

$$v(t) = 10 - 6e^{-t/10}\ \text{V}, \quad t \geq 0^+\ \text{ms} \tag{11.28}$$

$$i_R(t) = 5 - 3e^{-t/10}\ \text{A}, \quad t \geq 0^+\ \text{ms} \tag{11.29}$$

and

$$i_R(t) = 3e^{-t/10}\ \text{A}, \quad t \geq 0^+\ \text{ms} \tag{11.30}$$

Simulation: The circuit is entered as in Figure 11.10. A 5 A, IDC source is applied without a switch, since PSpice considers a dc source applied at $t = 0$, at the beginning of the simulation, as if the dc source is applied through a switch at $t = 0$.

In the Simulation Settings, 'Analysis type' is 'Time Domain (Transient),' 'Run to time' is 30m, 'Start Saving Data After' is 0, and 'Maximum Step size' is 1u. After the simulation is run, the plots shown in Figure 11.11 are displayed, which shows the time variation of v, i_R, and i_C in accordance with Equations 11.28 through 11.30. Note that in the three plots, a tangent at $t = 0$ intersects the asymptote of the plot at $t = \tau_C$.

Problem-Solving Tip

- Never apply initial conditions except in the complete solution, as in Equation 11.23.

Exercise 11.8

Transform the current source I_{SRC} in parallel with R_C to a voltage source $R_C I_{SRC} = V_{CF}$ in series with R_C. Verify that v and i_C are given by the same expressions.

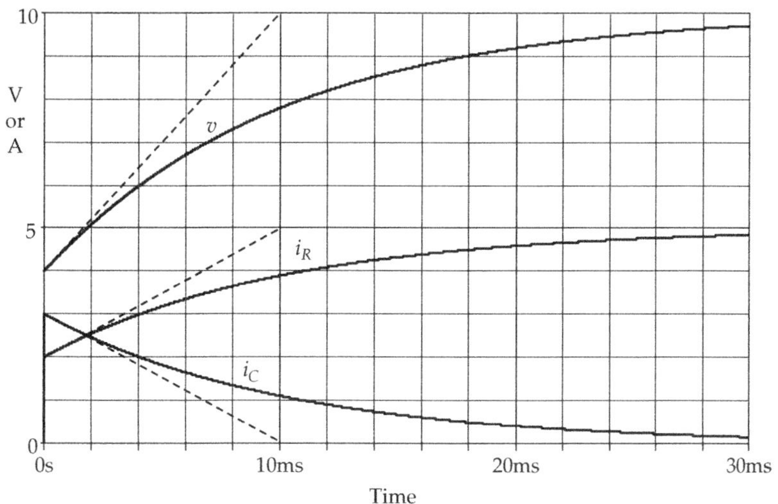

FIGURE 11.11
Figure for Example 11.2.

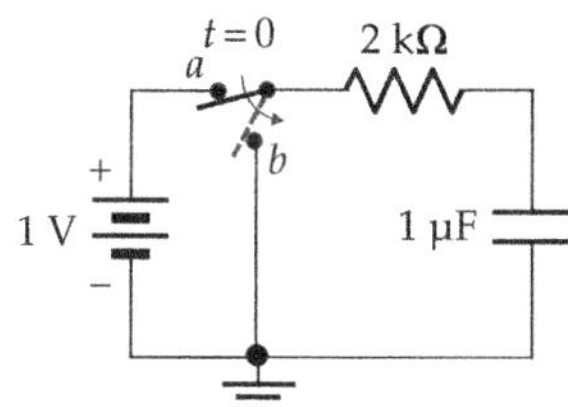

FIGURE 11.12
Figure for Primal Exercise 11.10.

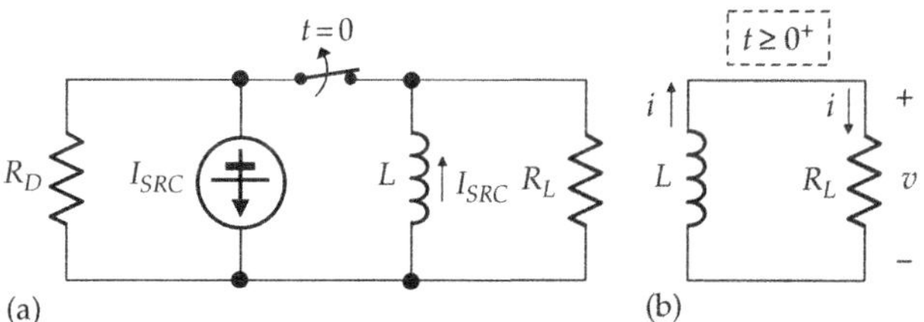

FIGURE 11.13
(a) Inductor is charged when the switch is opened at $t = 0$ and (b) inductor discharge for $t \geq 0^+$.

Exercise 11.9

Derive Equations 11.26 and 11.27 for i_R and i_C using superposition.

Primal Exercise 11.10

The switch is moved from position 'a' to position 'b' in Figure 11.12 at $t = 0$, after being in position 'a' for a long time. The switch is moved back to position 'a' after 1 ms. Determine v_C at $t = 2$ ms.

Ans. $1 + e^{-1} - e^{-1/2}$ V.

11.3 Inductor Discharge

In the circuit of Figure 11.13a, a dc current source is connected to R_L and L through a normally closed switch that is opened at $t = 0$. The switch is assumed to have been closed for a long time so that at $t = 0^-$, steady conditions prevail. This means that the inductor acts as a short circuit so that the current I_{SRC} flows through the inductor, with zero voltage across the paralleled circuit elements. R_D ensures that after the switch is opened, the ideal current source is not left open-circuited, as this would result in a theoretically infinite voltage across the source (Section 3.6).

At $t = 0^-$, the current in the inductor is I_{SRC}, as in Figure 11.13a, and the energy stored in the inductor is $(1/2)LI_{SRC}^2$. At $t = 0^+$, the inductor current remains I_{SRC} because the stored energy cannot be changed instantaneously by any physically realizable means. The circuit reduces to that of Figure 11.13b for $t \geq 0^+$, with the positive directions of i and v assigned as shown and with i having an initial value $I_0 = I_{SRC}$ at $t = 0^+$. It is required to determine how i and v vary with time for $t \geq 0^+$.

That i decreases from its initial value of I_{SRC} is evident from energy considerations. i flowing through R_L dissipates energy, which can only come from the energy stored in the inductor. As energy continues to be dissipated in R_L, i decreases until all the initially stored energy is dissipated so that i and v eventually become zero.

Choosing i as the variable for analysis is tantamount to considering the circuit of Figure 11.13b as a series circuit, so KCL is automatically satisfied. In terms of i, the voltage drop across the resistor is R_Li in the direction of i, whereas the voltage drop across the inductor is Ldi/dt also in the direction of i. From KVL, the sum of these voltage drops is zero, that is, $R_Li + Ldi/dt = 0$. Dividing by L,

$$\frac{di}{dt}+\frac{R_L}{L}i=0 \quad \text{or} \quad \frac{di}{dt}+\frac{i}{\tau_L}=0, \quad t\geq 0^+ \qquad (11.31)$$

where $\tau_L = L/R_L$ is the **time constant** and the initial conditions are $i(0^+) = I_0 = I_{SRC}$ and $v = R_LI_{SRC}$ at $t = 0^+$. Note that Equation 11.31 is the dual relation of Equation 11.2. It can be derived from Equation 11.2 by replacing v with i and $\tau_C = R_CC$ with $\tau_L = L/R_L$, where L is the dual of C and $G_L = 1/R_L$ is the dual of R_C. The series circuit of Figure 11.13b is the dual of the parallel circuit of Figure 11.1b.

From duality or from the general solution (Equation 11.7), the solution to Equation 11.31 is

$$i(t)=I_0e^{-t/\tau_L}, \quad t\geq 0^+ \qquad (11.32)$$

and

$$v(t)=R_Li=V_0e^{-t/\tau_L}, \quad t\geq 0^+ \qquad (11.33)$$

where
$i(0^+) = I_0 = I_{SRC}$
$v(0^+) = V_0 = R_LI_{SRC}$

$i(\infty) = v(\infty) = 0$, as the energy initially stored in the inductor is dissipated in the resistor (Problem P11.57). Both the current and the voltage decay exponentially.

It is instructive to interpret the decay of i and v in terms of flux linkage. Suppose that i_{SRC} in Figure 11.14a increases, during $t \leq 0^-$, while the switch is closed, from zero to a steady value $I_0 = I_{SRC}$ in some manner that need not be specified. The induced voltage v' opposes the buildup of current through the inductor by being a voltage drop in the direction of current through the inductor (Figure 11.14a). Hence, for $t \leq 0^-$, v' varies with time in some arbitrary manner, as illustrated in Figure 11.14b. λ in the inductor is equal to the area under the v' vs. t graph. Eventually at $t = 0^-$, the circuit is in a steady state, the current in the inductor is I_0, $v' = 0$, and the flux linkage $\lambda_0 = LI_{SRC}$ is the area under the graph of v vs. t for $t \leq 0^-$.

At $t = 0^+$, $v = R_LI_0$. Since the polarity of v across the inductor is now opposite to that of v' during the establishment of I_0, the time integral of v, $t \geq 0^+$, is a flux linkage $\int vdt$ that subtracts from λ_0. At any time t, λ is (Figure 11.14c).

$$\lambda(t)=\lambda_0-\int_{0^+}^{t}vdt=LI_0-R_LI_0\int_{0^+}^{t}e^{-t/\tau_L}dt=LI_0e^{-t/\tau_L}, \quad t\geq 0^+ \qquad (11.34)$$

At $t = 0^+$, $\lambda = \lambda_0 = LI_0$. As $t \to \infty$, $\lambda = 0$. Dividing $\lambda(t)$ in Equation 11.34 by L gives $i(t)$ in Equation 11.32. Note that λ is the dual of q, the charge on the capacitor in Figure 11.1b.

The rate of decrease of i is

$$-\frac{di}{dt}=-\frac{v}{L}=\frac{R_Li}{L} \qquad (11.35)$$

Since the rate of decrease of i is proportional to i, the decay of i is exponential. v, being proportional to i, will also decay exponentially.

It must not be construed, because the initial current I_0, due to flow of electric charges, eventually decays to zero that charge is not being conserved. Recall that a current source excites a circuit by imparting kinetic energy to electric charges, and that current is the *rate* of flow of electric charge. Hence, as the inductor discharges, the charges loose kinetic energy, without any loss of charge.

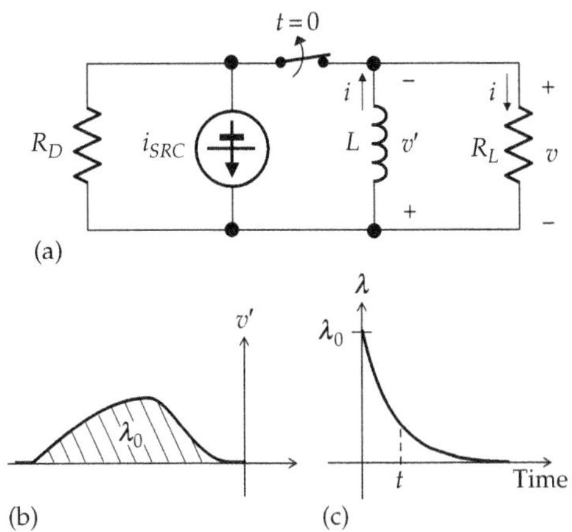

FIGURE 11.14
(a) Charging of the inductor while switch is closed, (b) arbitrarily assumed variation of voltage across the inductor while charging, and (c) decay of flux linkage of the inductor for $t \geq 0^+$.

Primal Exercise 11.11

A 0.5 H inductor having an initial current of 5 A is paralleled at $t = 0$ with a 2 Ω resistor. Derive the expression for the inductor current as a function of time for $t \geq 0^+$.

Ans. $5e^{-4t}$ A, t is in s.

Example 11.3: Inductor Discharge

A 5 mH inductor having an initial current of 3 A is connected at $t = 0$ in series with a 0.5 Ω resistor and a closed switch. Another 0.5 Ω resistor is added in series with the combination at $t = 5$ ms (Figure 11.15) by opening the switch. It is requited to determine the inductor current for $t \geq 0^+$.

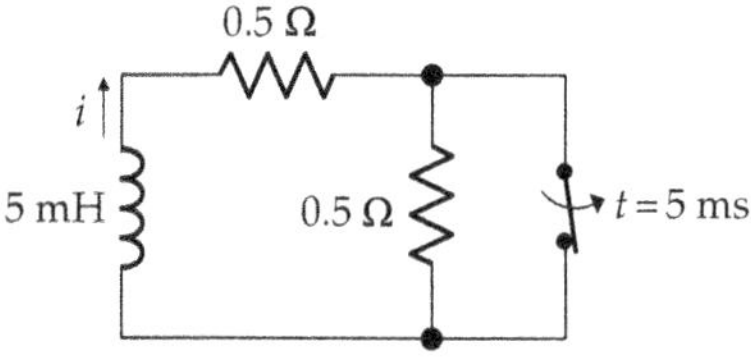

FIGURE 11.15
Figure for Example 11.3.

Solution:

The circuit in Figure 11.15 is the dual of that in Figure 11.4. The 5 mH inductor is the dual of the 5 mF capacitor. The dual of the 2 Ω resistor in Figure 11.4 is a 2 S resistor or a 0.5 Ω resistor, as in Figure 11.15. For $t \geq 5$ ms, the two resistors are in parallel with the 5 mF capacitor in Figure 11.4 and in series with the 5 mH inductor in Figure 11.14. The normally open switch in Figure 11.4 is the dual of the normally closed switch as in Figure 11.15.

$0^+ \leq t \leq 5$ ms: From Equation 11.32, $i(t) = I_0 e^{-t/\tau_{L1}} = 3e^{-t/10}$ A, where $\tau_{L1} = L/R_L = 5 \times 10^{-3}/0.5 \equiv 10$ ms. i decays exponentially from 3 A with a time constant of 10 ms. At $t = 5$ ms, $i = 3e^{-0.5} = 1.82$ A.

$t \geq 5$ ms: The resistance in series with L is 1 Ω and $\tau_{L2} = 5 \times 10^3/1 \equiv 5$ ms. Equation 11.32 applies with $I_0 = 1.82$ A and $\tau_2 = 5$ ms. Thus, $i = 1.82e^{-(t-5)/5}$ A, where at $t = 5$ ms, the exponent is zero and $i = 1.82$ V.

Simulation: The circuit is entered as in Figure 11.16. The inductor is oriented so that positive inductor current flows upward. The initial condition of 3 A is entered in the Property Editor spreadsheet. A normally closed switch is set to open at $t = 5$ ms. In the Simulation Settings, 'Analysis type' is 'Time Domain (Transient),' 'Run to time' is 15m, 'Start Saving Data After' is 0, and 'Maximum Step size' is 1u. After the simulation is run, the same plot shown in Figure 11.6 is displayed, with v V replaced by i A.

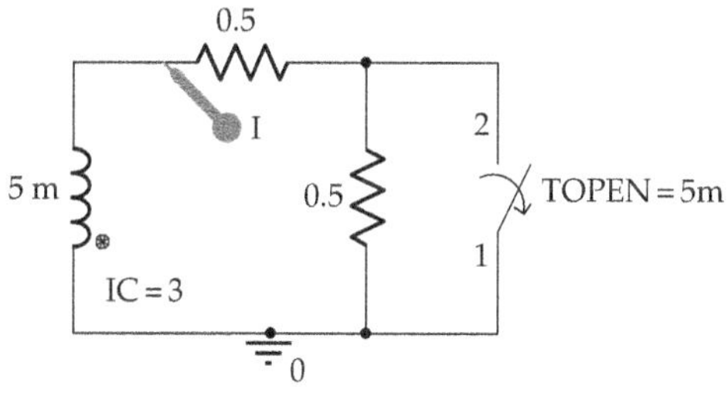

FIGURE 11.16
Figure for Example 11.3.

11.4 Inductor Charging

In Figure 11.17, a dc current source I_{SRC} is connected in parallel with L and R_L. For $t \leq 0^-$, I_{SRC} flows through a normally closed switch connected across the parallel combination. The current in R_L and L and the voltage across the combination are all zero. The switch is opened at $t = 0$.

At $t = 0^+$, the energy in the inductor, and hence the inductor current, remains zero. This means that I_{SRC} flows through R_L so that $v = R_L I_{SRC}$ at this instant. Note that the inductor acts as an open circuit at the instant of switching, since its current is zero, while the voltage v across it jumps from 0 to $R_L I_{SRC}$. The time integral of v, the flux linkage λ, increases with time, as does the inductor current $i_L = \lambda/L$. But as i_L increases, i_R decreases, because $i_L + i_R = I_{SRC}$ is constant, which means that v decreases. Eventually as $t \to \infty$, v, and hence i_R, goes to zero. All of I_{SRC} flows through L, and the inductor behaves as a short circuit in the final steady state. It is required to determine how v and i_L vary with time.

With KVL automatically satisfied by the assignment of v across the parallel elements, KCL gives $i_L + i_R = I_{SRC}$. Substituting for i_L in terms of the integral form of the v–i relation for the inductor and for i_R in terms of Ohm's law,

$$\frac{1}{L}\int v dt + \frac{v}{R_L} = I_{SRC}, \quad t \geq 0^+ \tag{11.36}$$

If both sides of Equation 11.36 are differentiated with respect to time, the RHS becomes zero, since I_{SRC} is a constant. Multiplying the LHS by R_L and substituting $\tau_L = L/R_L$, Equation 11.36 becomes

$$\frac{dv}{dt} + \frac{v}{\tau_L} = 0, \quad t \geq 0^+ \tag{11.37}$$

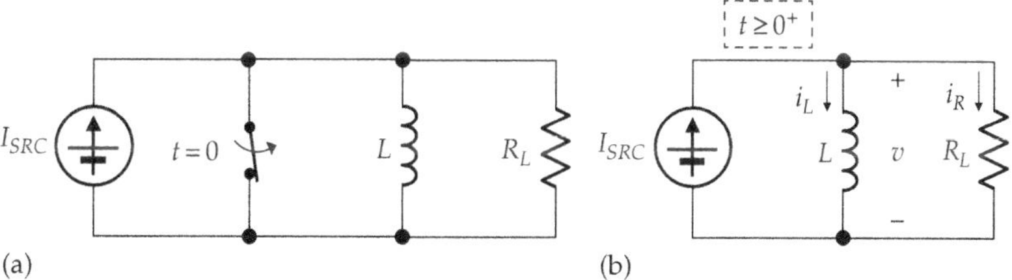

FIGURE 11.17
(a) Inductor begins to charge when the switch is opened at $t = 0$ and (b) inductor charging for $t \geq 0^+$.

The solution to Equation 11.37 is given by Equation 11.7 with v replacing y; y_0 replaced by the initial value of v at $t = 0$, which is $R_L I_{SRC}$; and τ_L replacing τ. Thus,

$$v(t) = V_0 e^{-t/\tau_L}, \quad t \geq 0^+ \tag{11.38}$$

where $V_0 = R_L I_{SRC}$; v therefore decays exponentially, as in Figure 11.2, from an initial value of V_0 to zero as $t \to \infty$, as the inductor becomes fully charged.

The time variation of i_L is obtained from the v–i relation of the inductor as

$$i_L(t) = \frac{1}{L}\int_{0^+}^{t} v dt + 0, \quad t \geq 0^+ \tag{11.39}$$

where $i_L(0^+) = 0$ at $t = 0^+$. Substituting for v from Equation 11.38 and then integrating it,

$$i_L(t) = \frac{I_{SRC}}{\tau_L}\int_{0^+}^{t} e^{-t/\tau_L} dt = I_{SRC}\left[-e^{-t/\tau_L}\right]_{0^+}^{t}$$

or

$$i_L(t) = I_{LF}\left(1 - e^{-t/\tau_L}\right) \quad t \geq 0^+ \tag{11.40}$$

where $i_L(\infty) = I_{LF} = I_{SRC}$. Equation 11.40 satisfies the initial condition that $i_L(0^+) = 0$ and represents a saturating exponential, as in Figure 11.8.

The exponential variation again follows from the argument that the rate of change of the variables at a given instant is proportional to the variable itself at that instant. Considering v, its rate of decrease is, from Equation 11.37, $-dv/dt = v/\tau_L$, which is proportional to v. Since v changes exponentially, so will the currents.

What if in Figure 11.17 the inductor had an initial current I_0? In this case the initial value of the voltage is $R_L(I_{SRC} - I_0)$ instead of $R_L I_{SRC}$. Equation 11.38 becomes

$$v(t) = R_L\left(I_{SRC} - I_0\right)e^{-t/\tau_L} = V_0 e^{-t/\tau_L}, \quad t \geq 0^+ \tag{11.41}$$

It is seen from this equation that $v(0^+) = V_0$ and $v(\infty) = 0$. Substituting for $v(t)$ in Equation 11.39,

$$i_L(t) = \frac{1}{L}\int_{0^+}^{t} v dt + I_0 = \left(I_{SRC} - I_0\right)\left[-e^{-t/\tau_L}\right]_{0^+}^{t} + I_0$$
$$= I_{LF} + \left(I_0 - I_{LF}\right)e^{-t/\tau_L}, \quad t \geq 0^+ \tag{11.42}$$

where $I_{LF} = i(\infty) = I_{SRC}$ and $i(0^+) = I_0$, as required.

Equations 11.41 and 11.42 can be derived from superposition, since the circuit is LTI. Considering v, Equation 11.38 applies to the response when the inductor is initially uncharged. Figure 11.13b applies to the current through the inductor acting alone, with the current source replaced by an open circuit. However, the direction of the discharge current through R_L is opposite to i_R in Figure 11.17b so that the response given by Equation 11.33 should be subtracted from that given by Equation 11.38. This gives.

$$v(t) = R_L I_{SRC} e^{-t/\tau_L} - R_L I_0 e^{-t/\tau_L} = R_L\left(I_{SRC} - I_0\right)e^{-t/\tau_L}, \quad t \geq 0^+$$

which is the same as Equation 11.41. Similarly, i_L in this case is the sum of i_L given by Equation 11.40 and i given by Equation 11.32. Thus,

$$i_L(t) = I_{LF}\left(1 - e^{-t/\tau_L}\right) + I_0 e^{-t/\tau_L} = I_{LF} + \left(I_0 - I_{LF}\right)e^{-t/\tau_L} \quad t \geq 0^+$$

which is the same as Equation 11.42.

As to be expected, duality applies. The circuit in Figure 11.17b is the dual of that in Figure 11.7b, with L replacing C, $G_L = 1/R_L$ in Figure 11.17b replacing R_C in Figure 11.7b, I_{SRC} replacing V_{SRC}, I_0 replacing V_0, v replacing i, i_L replacing v_C, and λ replacing q. All the relations derived in this section can be obtained from those in Section 11.2 by applying these replacements. For example, Equation 11.41 becomes, in terms of G_L,

$$v(t) = \frac{\left(I_{SRC} - I_0\right)}{G_L} e^{-t/\tau_L}, \quad t \geq 0^+ \tag{11.43}$$

Making the aforementioned replacements gives

$$i(t) = \frac{\left(V_{SRC} - V_0\right)}{R_C} e^{-t/\tau_C}, \quad t \geq 0^+$$

which is the same as Equation 11.19, where τ in Equation 11.43 is $G_L L$, the dual of $R_C C$ in Equation 11.19.

Inductor charging through a voltage source is discussed in Example 11.4.

Primal Exercise 11.12

A 1 μH inductor having an initial current of 6 A is connected at $t = 0$ to a 12 A source through a 10 kS resistor, as in Figure 11.17. Determine (a) the time constant and (b) the expressions for inductor current and voltage for $t \geq 0$.

Ans. (a) 10 ms; (b) $v = 0.6e^{-0.1t}$ mV, $i_L(t) = 12 - 6e^{-0.1t}$ A, t is in ms. Note that the numerical values make this exercise the dual of that Exercise 11.6.

Example 11.4: Inductor Charging by Voltage Source

A battery voltage V_{SRC} is applied at $t = 0$ to an uncharged inductor L in series with a resistor R_L (Figure 11.18a). It is required to determine how i, v_R, and v_L vary with time (Figure 11.18b).

Solution:

At $t = 0^-$, the stored energy in the uncharged L is zero, by assumption. At $t = 0^+$, this energy remains zero. This implies that $i = 0$ and $v_R = 0$ so that all of V_{SRC} initially appears across L, which therefore acts as an open circuit. The time integral of v_L, which is the flux linkage λ, increases with time, as does $i = \lambda/L$. As i increases, v_R increases in accordance with Ohm's law. This reduces v_L, because $v_L + v_R = V_{SRC}$ is constant. Eventually, L charges fully, v_L drops to zero, and all of V_{SRC} appears across R_L, with L acting as a short circuit. The final value of i is therefore $i(\infty) = I_F = V_{SRC}/R_L$.

From KVL, $v_L + v_R = V_{SRC}$. Substituting $v_L = Ldi/dt$ and $v_R = R_L i$,

$$L\frac{di}{dt} + R_L i = V_{SRC}, \quad t \geq 0^+ \tag{11.44}$$

Dividing by L and substituting $\tau_L = L/R_L$,

$$\frac{di}{dt} + \frac{i}{\tau_L} = \frac{V_{SRC}}{L}, \quad t \geq 0^+ \tag{11.45}$$

As in Example 11.2, the general solution is the sum of two components: (1) i that is the solution of the homogeneous differential equation and (2) any i that satisfies the nonhomogeneous differential Equation 11.45. In particular, as $t \to \infty$, steady state conditions prevail so that $di/dt = 0$ and $I_F = (\tau/L)V_S = V_{SRC}/R_L$ as argued previously. The first component is given by Equation 11.6, with an arbitrary constant A. It follows that the general solution is

$$i(t) = Ae^{-t/\tau_L} + V_{SRC}/R_L, \quad t \geq 0^+ \tag{11.46}$$

A is determined from the initial condition that $i(0^+) = 0$, which gives $A = -V_{SRC}/R_L$. The general solution is then

$$i(t) = (V_{SRC}/R_L)(1 - e^{-t/\tau_L}) = I_F(1 - e^{-t/\tau_L}), \quad t \geq 0^+ \tag{11.47}$$

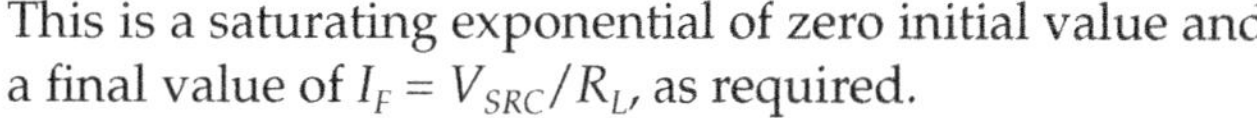

This is a saturating exponential of zero initial value and a final value of $I_F = V_{SRC}/R_L$, as required.

Suppose that i has an initial value I_0 at $t = 0^-$ so that the energy stored in the inductor is $(1/2)LI_0^2$ at $t = 0^-$. At $t = 0^+$, this stored energy, and hence I_0, remains the same, since the stored energy cannot be changed instantaneously by any physically realizable means. At this instant, $V_{R0} = R_L I_0$ so that $v_L = V_{SRC} - R_L I_0$. Eventually, L charges fully, $v_L = 0$, V_{SRC} appears across R_L, and the final current is $I_F = V_{SRC}/R_L$, irrespective of I_0. The analytical solution is obtained from the general solution (Equation 11.46) by substituting the initial condition that $i(0^+) = I_0$. This gives $A = I_0 - V_{SRC}/R_L$ so that

$$\begin{aligned} i(t) &= (I_0 - V_{SRC}/R_L)e^{-t/\tau_L} + V_{SRC}/R_L \\ &= I_F + (I_0 - I_F)e^{-t/\tau_L}, \quad t \geq 0^+ \end{aligned} \tag{11.48}$$

where $I_F = V_{SRC}/R_L$. The voltages are

$$\begin{aligned} v_R(t) = R_L i &= (V_{R0} - V_{SRC})e^{-t/\tau_L} + V_{SRC} \\ &= V_{SRC} + (V_{R0} - V_{SRC})e^{-t/\tau_L}, \quad t \geq 0^+ \end{aligned} \tag{11.49}$$

and

$$\begin{aligned} v_L(t) = L\frac{di}{dt} &= -\frac{L}{\tau_L}(I_0 - V_{SRC}/R_L)e^{-t/\tau_L} \\ &= (V_{SRC} - V_{R0})e^{-t/\tau_L}, \quad t \geq 0^+ \end{aligned} \tag{11.50}$$

where $v_R(0^+) = V_{R0} = R_L I_0$, $v_L(0^+) = V_{SRC} - V_{R0}$, $v_R(\infty) = V_{SRC}$, and $v_L(\infty) = 0$, as argued previously. Note that $v_L + v_R = V_{SRC}$ at all t.

Again, i, v_R, and v_L can be derived from superposition (Exercise 11.14).

If $L = 5$ mH, $R_L = 0.5\ \Omega$, $V_{SRC} = 5$ V, and $I_0 = 4$ A, then $L/R_L = 10$ ms, $I_F = 10$ A, $V_{R0} = 2$ V, and

$$i(t) = 10 - 6e^{-t/10}\ \text{A}, \quad t \geq 0^+\ \text{ms} \tag{11.51}$$

$$v_R(t) = 5 - 3e^{-t/10}\ \text{V}, \quad t \geq 0^+\ \text{ms} \tag{11.52}$$

and

$$v_L(t) = 3e^{-t/10}\ \text{V}, \quad t \geq 0^+\ \text{ms} \tag{11.53}$$

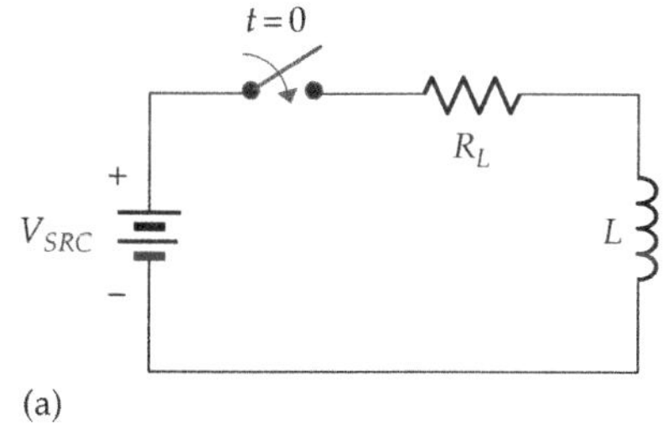

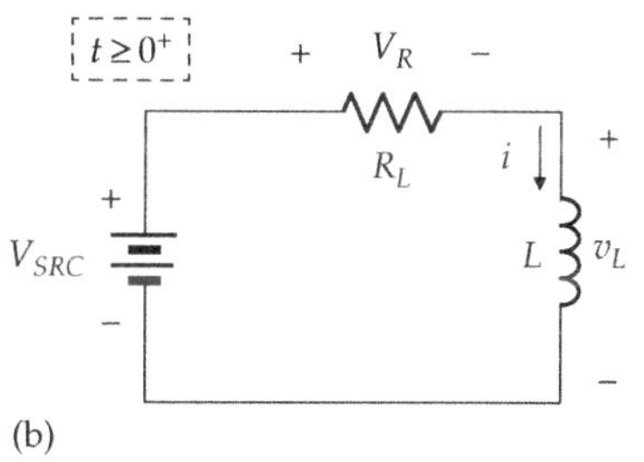

FIGURE 11.18
Figure for Example 11.4.

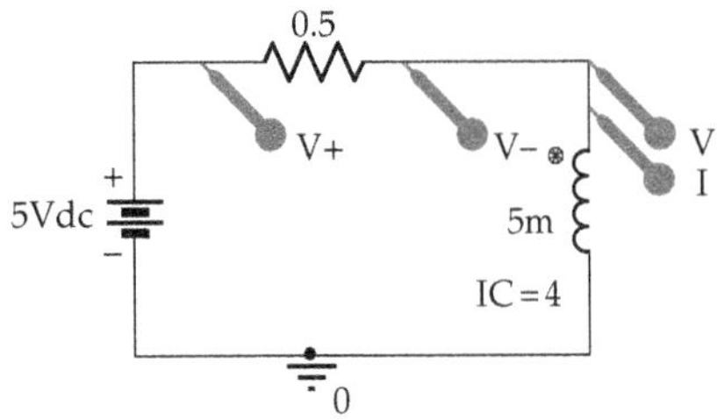

FIGURE 11.19
Figure for Example 11.4.

Duality should be noted. The circuit in Figure 11.18 is the dual of that in Figure 11.9 when R in one circuit is replaced by G in the other circuit. Equations 11.44 through 11.53 are the duals of Equations 11.21 through 11.30 when the following are interchanged: V_{SRC} and I_{SRC}, v and i, v_R and i_R, v_L and i_C, L and C, and R_C and $G_L = 1/R_L$. The numerical values in this example are the duals of those in Example 11.2.

Simulation: The circuit is entered as in Figure 11.19. The source VDC applies a voltage of 5 V at $t = 0$, at the start of the simulation, so no switch is needed. A differential marker is used to display the voltage across the 0.5 Ω resistor.

In the Simulation Settings, 'Analysis type' is 'Time Domain (Transient),' 'Run to time' is 30 m, 'Start Saving Data After' is 0, and 'Maximum Step size' is 1u. After the simulation is run, the plot is as shown in Figure 11.1 for the dual quantities, that is, i replaces v, v_R replaces i_R, and v_L replaces i_C.

Problem-Solving Tip

- Always check the correctness of a derived response for current or voltage as a function of time by determining the values at $t = 0$ and as $t \to \infty$ and by satisfying KVL and KCL.
- Duality can be helpful in checking the correctness of assumptions and expressions.

Exercise 11.13

Transform the voltage source V_{SRC} in series with R_L in Figure 11.18 to a current source $I_{SRC} = V_{SRC}/R_L$ in parallel with R_L. Verify that v_L and the inductor current are given by the same expressions.

Exercise 11.14

Derive i, v_R, and v_L (Equations 11.48 through 11.50) using superposition.

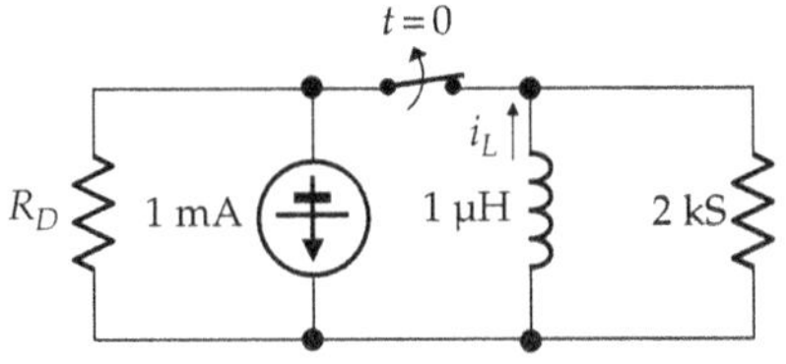

FIGURE 11.20
Figure for Primal Exercise 11.15.

Primal Exercise 11.15

The switch in Figure 11.20 is opened at $t = 0$, after being closed for a long time. The switch is closed again after 1 ms. Determine i_L at $t = 2$ ms. Assume that the conductance of R_D is much less than 2 kS so it can be neglected. Note that this is the dual of Primal Exercise 11.10.

Ans. $1 + e^{-1} - e^{-1/2}$ A.

11.5 Generalized First-Order Circuits

The first-order circuits discussed so far were basic, prototypical, LTI circuits consisting of a single resistor and a single energy storage element. First-order circuits, however, can be more complex, consisting of more than a single resistor, capacitor, or inductor, and can include dependent sources. Nevertheless, for a circuit to be first order, *it must be an LTI circuit that is reducible to one of the prototypical circuits that have a single ideal capacitor or a single ideal inductor, in combination with a single resistor.* For example, the circuit of Figure 11.21 includes two capacitors that cannot be combined into a single ideal capacitor. The circuit therefore is not first order. But in the absence of the 4 Ω resistor, the two capacitors combine into a single 1.25 F capacitor in series with a 1 Ω resistor, which is a first-order circuit. It may be argued that the 4 Ω resistor and 0.25 F capacitor can be converted to an equivalent resistor in parallel with an equivalent capacitor that can then be combined with the 1 F capacitor, resulting in a first-order circuit. However, the equivalent resistance and capacitance are functions of frequency (Exercise 11.16). The equivalent parallel

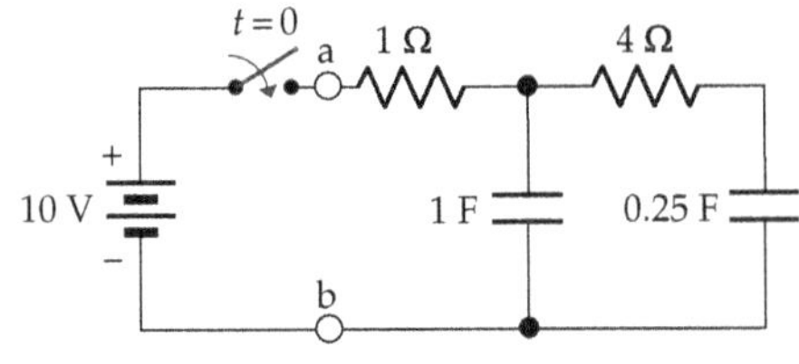

FIGURE 11.21
Second-order circuit that cannot be reduced to a first-order circuit.

capacitor cannot therefore be considered as the ideal capacitor of a first-order circuit, since the capacitance of an ideal capacitor is a constant that is independent of frequency.

Exercise 11.16

Determine the equivalent parallel resistance and capacitance of the 4 Ω resistor and 0.25 F capacitor (Figure 11.21).

Ans. $G_{eqp} = \frac{1}{4}\frac{\omega^2}{\omega^2+1}\,\Omega$ and $C_{eqp} = \frac{1}{4}\frac{1}{\omega^2+1}$ F.

11.5.1 Generalized Response

Any natural response of a first-order circuit, or the response to a dc excitation, obeys a first-order differential equation of the form

$$\frac{dy}{dt} + \frac{y}{\tau} = K, \quad t \geq 0^+ \tag{11.54}$$

where
- y is the voltage or current under consideration
- K is zero for a natural response or a constant in the case of a dc excitation. Thus, $K = 0$ in Equations 11.2 and 11.31, whereas $K = I_{SRC}/C$ in Equation 11.22 and $K = V_{SRC}/L$ in Equation 11.45.

The general solution of Equation 11.54 is the sum of two components:

1. A component that is the solution of the homogeneous differential equation. This component is of the form of Equation 11.7, including an arbitrary constant of integration. It is termed the **complementary function** in mathematics and the **transient response** in circuit analysis, because it decays with time in a stable circuit. The time course of the transient is that of the natural response, and the role of the transient is explained in Section 11.6.
2. A component that satisfies the nonhomogeneous differential equation. This component is the **particular integral** in mathematics and the final, **steady-state response** in circuit analysis. As its name implies, it is the value of the variable as $t \to \infty$. When K is a constant, the steady-state, final value Y_F is obtained from Equation 11.54 by setting $dy/dt = 0$, which gives $Y_F = K\tau$.

The complete solution is the sum of these two components:

$$y = Ae^{-t/\tau} + Y_F, \quad t \geq 0^+ \tag{11.55}$$

That Equation 11.55 is the general solution of Equation 11.54 is evidenced by (1) satisfying Equation 11.54, as can be readily verified by substitution, and (2) having an arbitrary constant of integration, A, which can be found from initial conditions in the circuit. Thus, if $y = Y_0$ at $t = 0$, it follows from Equation 11.55 that

$$A = Y_0 - Y_F \tag{11.56}$$

Substituting for A in Equation 11.55,

$$y(t) = Y_F + (Y_0 - Y_F)e^{-t/\tau}, \quad t \geq 0^+ \tag{11.57}$$

It must be emphasized that the initial conditions are applied to the complete solution (Equation 11.55) and *not to the transient component of the solution alone.*

All the currents and voltages derived in the preceding sections are of the form of Equation 11.57, with various values for Y_0, Y_F, and τ.

The attractive feature of Equation 11.57 is that it gives $y(t)$ without having to derive and solve the differential equation for y. It suffices to determine Y_0, Y_F, and τ, which can be done directly from the circuit, as discussed in what follows. It must also be emphasized that Equation 11.57 applies only when the RHS of Equation 11.54 is zero or a constant, that is, the desired response is a natural response, with zero applied excitation, or is the response to a dc excitation. *Equation 11.57 does not apply when the excitation is time varying.*

The homogeneous differential equation (Equation 11.54 with $K = 0$) is a very basic attribute of any circuit. The following features of this equation and their implications should be carefully noted:

1. Independent sources appear in the nonhomogeneous differential equation, but not in the homogeneous differential equation, which implies that *the homogeneous differential equation can just as well be derived with independent sources set to zero.* Thus, the homogeneous differential equation of v in Figure 11.9, which is Equation 11.22 with zero on the RHS, can be derived from Figure 11.9 with I_{SRC} set to zero. Similarly, the homogeneous differential equation of i in Figure 11.18, which is Equation 11.45 with zero on the RHS, can be derived from Figure 11.18 with V_{SRC} set to zero (Exercise 11.17).
2. In the standard form of the homogeneous differential equation of a first-order circuit

(Equation 11.54), the coefficient of the dy/dt term is unity and the coefficient of the y term is the reciprocal of the time constant. A first-order homogeneous differential equation is therefore completely determined by the time constant τ. In turn, $\tau = CR$ or L/R, irrespective of independent sources, which means that τ can be determined with independent sources set to zero. Accordingly, τ is the same for the charging of a capacitor from a voltage source V_{SRC} (Equation 11.19) as for its discharging, with $V_{SRC} = 0$ (Equation 11.10), similarly for an inductor.

3. The solution of the homogeneous differential equation contains the arbitrary constant and the exponent in terms of τ. As explained in Section 8.1, all voltages and currents in an LTI circuit are related by linear operations. All these operations do not affect the exponent in the expressions for the voltages and currents. It follows from the preceding two items '1' and '2' that all voltages and currents have the same τ and therefore obey the same homogeneous differential equation. They differ only in the values of the arbitrary constant and of the final steady state.

Exercise 11.17

Verify that the homogeneous differential equation (Equation 11.22) can be derived from Figure 11.9 with $I_{SRC} = 0$ and that the homogeneous differential equation (Equation 11.45) can be derived from Figure 11.18 with $V_{SRC} = 0$.

11.5.2 Determining Initial and Final Values

Y_0 and Y_F can be determined by any of the circuit techniques discussed so far in this book, bearing in mind the following basic considerations:

1. In the steady state, inductors act as short circuits and capacitors as open circuits.
2. When a sudden change is made at $t = 0$ in a circuit containing energy storage elements, the voltage across a capacitor and the current through an inductor remain the same at $t = 0^+$, because the stored energy cannot be changed instantaneously by any physically realizable means.
3. The voltages and currents at $t = 0^+$ are the initial values used in determining the arbitrary constant in the general solution of the differential equation for a given voltage or current.
4. However, *the current through a capacitor and the voltage across an inductor are not directly related to stored energy and can therefore change instantaneously to satisfy KCL or KVL.*
5. When energy storage elements are charged through a sudden change, an *uncharged* capacitor acts as a short circuit, and an *uncharged* inductor acts as an open circuit.
6. If the capacitor or inductor is initially charged, the preceding procedure is applied assuming no initial energy storage. The effect of the initial energy storage, acting alone, can then be added algebraically by superposition.

Note that Y_0 and Y_F in the circuits considered in this chapter are either zero or dc values of current or voltage. Such values are independent of L and C, since under dc conditions, inductors act as short circuits and capacitors as open circuits.

11.5.3 Effect of Sources on Time Constant

A physical interpretation can be given for τ not being affected by independent dc sources: during charging or discharging of energy storage elements, currents and voltages change with time. If the current through an ideal, independent dc voltage source varies by Δi_{SRC}, the source voltage does not vary, by definition of an ideal, independent voltage source. This also follows from the fact that the source resistance of an ideal voltage source is zero so that $\Delta v_{SRC} = 0 \times \Delta i_{SRC} = 0$. Since Δi_{SRC} produces no change in voltage across the dc voltage source, the source appears as a short circuit as far as the changing current is concerned. Similarly, if the voltage across an ideal, independent dc current source changes by Δv_{SRC}, the source current does not change. Hence, the source appears as an open circuit as far as the changing voltage is concerned. The effective time constant is therefore determined with ideal, independent voltage sources replaced by short circuits and ideal, independent current sources by open circuits.

The time constant is formally derived as the resistance seen by the energy storage element in a first-order circuit. This is illustrated in Figure 11.22a, in which the energy

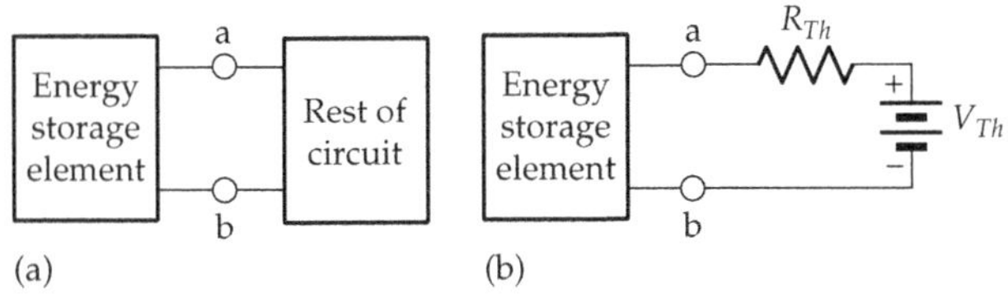

FIGURE 11.22
Energy storage element is separated from the rest of the circuit (a) and which can be represented by its TEC (b).

storage element is shown, separated from the rest of the circuit at terminals 'ab' of the energy storage element. The rest of the circuit can be represented by its Thevenin's equivalent circuit (TEC) as seen by the energy storage element (Figure 11.22b). Since independent sources can be set to zero when determining the time constant, $V_{Th} = 0$. *The resistance that combines with the circuit parameter of the energy storage element to determine the time constant is* R_{Th}. That is, the time constant is $R_{Th}C$ in the case of a capacitor and L/R_{Th} in the case of an inductor.

The foregoing can be illustrated by the circuit of Figure 11.23, where C is assumed to have an initial voltage V_0 at $t = 0^-$. It is required to determine v and i_C for $t \geq 0^+$, after the switch is closed at $t = 0$. The circuit can be reduced to a prototypical, series RC circuit by deriving TEC seen by the capacitor. This will be done in two easy steps. The first is to derive TEC for the circuit to the left of R_p in Figure 11.23, as shown in Figure 11.24a. When terminals 'ab' are open-circuited, the current

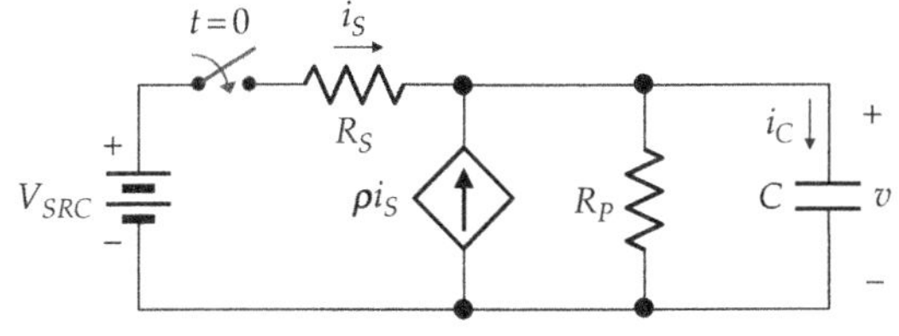

FIGURE 11.23
Circuit for determining the resistance seen by energy storage element.

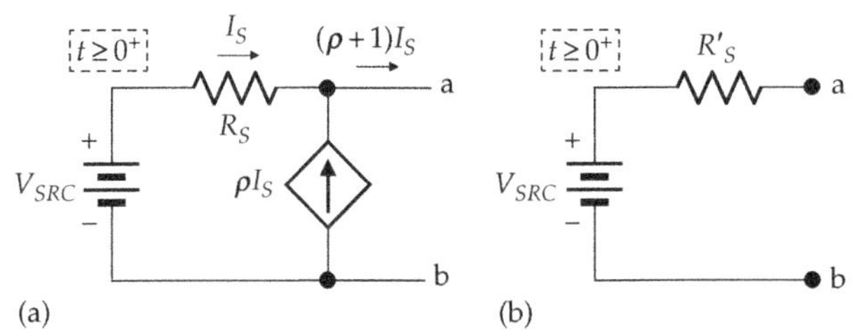

FIGURE 11.24
(a) Circuit to the left of R_p in Figure 11.23 for $t \geq 0^+$ and (b) intermediate TEC looking into terminals 'ab'.

$(\rho + 1)I_S = 0$, which means that $I_S = 0$, since ρ can have, in general , any value. The current source is replaced by an open circuit and no current flows in R_S. It follows that $V_{Th} = V_{SRC}$. When terminals 'ab' are short-circuited, the current that flows in the short circuit is $I_{SC} = (\rho + 1)I_S$, where $I_S = V_{SRC}/R_S$. This gives $I_{SC} = (\rho + 1)V_{SRC}/R_S$ and $R_{Th} = R'_S = R_S/(\rho + 1)$. TEC is shown in Figure 11.24b.

The second step is to add R_P, as in Figure 11.25a. If V_{SRC} is set to zero, the new R_{Th} is $R''_S = (R'_S \parallel R_P)$, and the new V_{Th} is

$$V_{Th} = \frac{R_P}{R'_S + R_P} V_{SRC} = \frac{R'_S R_P}{R'_S(R'_S + R_P)} V_{SRC} = \frac{R''_S}{R'_S} V_{SRC} \quad (11.58)$$

The final TEC, with C connected, is shown in Figure 11.25b. $v(0^+) = V_0$ and $v(\infty) = V_F = (R''_S/R'_S)V_{SRC}$. When the voltage source is set to zero, the resistance seen by C is R''_S and $\tau = CR''_S$. It follows that

$$v(t) = V_F + (V_0 - V_F)e^{-t/\tau}, \quad t \geq 0^+ \quad (11.59)$$

The final value of i_C is zero, when C is fully charged. The initial value of i_C is

$$I_{C0} = \frac{(R''_S/R'_S)V_{SRC} - V_0}{R''_S} = \frac{V_{SRC}}{R'_S} - \frac{V_0}{R''_S} \quad t \geq 0^+ \quad (11.60)$$

and

$$i_C(t) = I_{C0}e^{-t/\tau}, \quad t \geq 0^+ \quad (11.61)$$

The following should be noted:

1. V_F and I_{C0} are functions of both the independent and dependent sources, as well as the resistances, but are not a function of C.
2. $\tau = CR''_S$ is a function of the resistances and the dependent source but not the independent source.

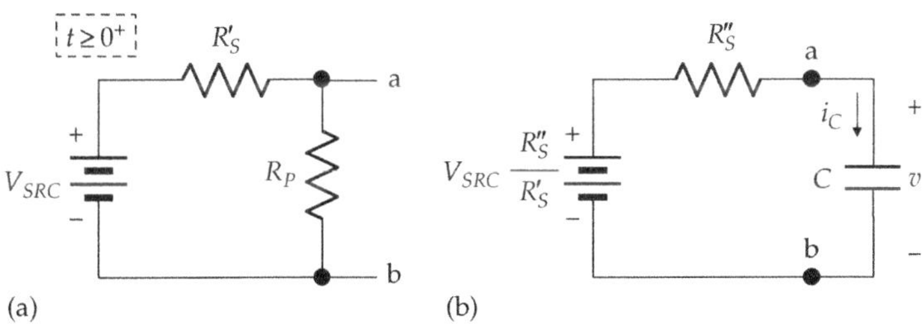

FIGURE 11.25
Determining Thevenin's resistance. (a) Intermediate TEC with R_p connected and (b) capacitor connected to the final TEC.

In summary, the following concepts apply to first-order circuits:

Concepts:

1. *Any natural response or response to dc excitation in a first-order circuit is of the general form*

$$y(t) = Y_F + (Y_0 - Y_F)e^{-t/\tau}, \quad t \geq 0^+$$

where
Y_0 *is the initial value at* $t = 0^+$, *after the change in the circuit takes place*
Y_F *is the final, steady-state value*
τ *is the time constant*

Y_0, Y_F, *and* τ *can be determined directly from the circuit. It is important to note that* Y_0 *for any given voltage or current in the circuit is determined by the constancy of capacitor voltages and inductor currents. One cannot assume that variables other than these remain constant at the instant of switching. This is illustrated by Example 11.6.*

2. τ *can be determined from Thevenin's resistance seen by C or L and is the same for all currents and voltages in the circuit.*
3. *Both independent and dependent sources affect the initial and steady-state, final values of currents or voltages in a first-order circuit. Dependent sources, but not independent sources, affect the time constant so that independent sources can be set to zero when determining the time constant.*
4. *Initial and steady-state, final values of currents and voltages are independent of C or L.*

Exercise 11.18

Verify that the same expressions for v and i_C in Figure 11.23 can be obtained by solving the differential equations satisfied by these variables.

11.5.4 Effective Values of Circuit Elements

If a first-order circuit contains more than one resistor and more than one capacitor or inductor, with or without dependent sources, then the circuit should be reducible to a single effective resistor and a single effective capacitor or inductor, with or without an independent source. Otherwise, the circuit is not first order, as in the case of the circuit of Figure 11.21.

In Figure 11.26a, for example, the three inductors can be combined into a single inductor, L_{eff}, and the three resistors can be combined into a single resistor R_{eff}, where

$$L_{eff} = L_1 \| (L_2 + L_3) \quad \text{and} \quad R_{eff} = R_1 \| (R_2 + R_3) \tag{11.62}$$

The circuit reduces to a prototypical circuit consisting of a voltage source applied to a series combination of L_{eff} and R_{eff} (Figure 11.26b).

As demonstrated in previous chapters, dependent sources generally alter the coefficients multiplying currents and voltages in circuit equations. They will generally affect the effective values of resistance, capacitance, and inductance and hence initial values, final values, and the time constant.

Primal Exercise 11.19

(a) Assume that in Figure 11.26, L_1 = 5 μH, L_2 = 3 μH, L_3 = 2 μH, R_1 = 8 kΩ, R_2 = 6 kΩ, and R_3 = 2 kΩ. Determine L_{eff}, R_{eff}, and τ_{eff}. (b) Repeat (a) if the inductors are replaced by capacitors having the same numerical values in microfarads.

Ans. (a) L_{eff} = 2.5 μH, R_{eff} = 4 kΩ, τ_{eff} = 0.625 ns; (b) C_{eff} = 6.2 μF, R_{eff} = 4 kΩ, τ_{eff} = 24.8 ms.

Example 11.5: Analysis of First-Order Circuit

The switch in Figure 11.27 is closed at $t = 0$ after being open for a long time. It is required to determine $i_L(t)$ for $t \geq 0^+$.

Solution:

$i_L(t)$ will be determined from its initial value, final, steady-state value, and the time constant. The initial

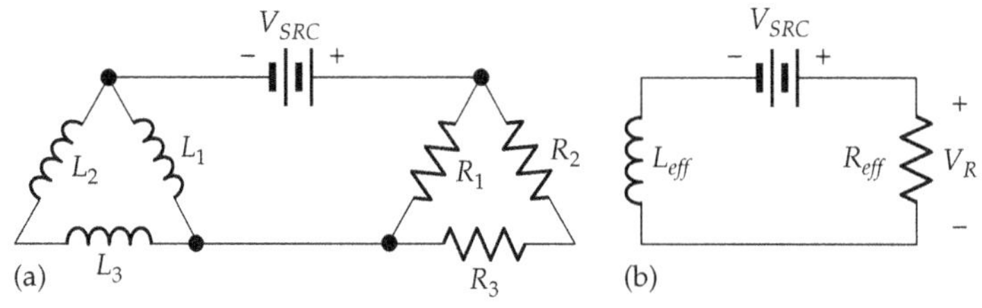

FIGURE 11.26
Inductances and resistances in the circuit shown in (a) are replaced by the effective inductance and resistance in (b).

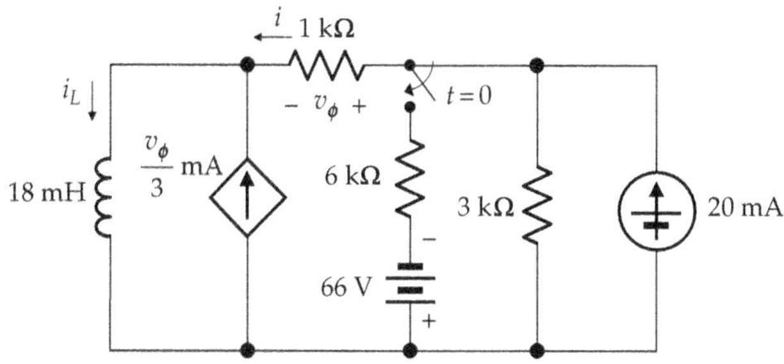

FIGURE 11.27
Figure for Example 11.5.

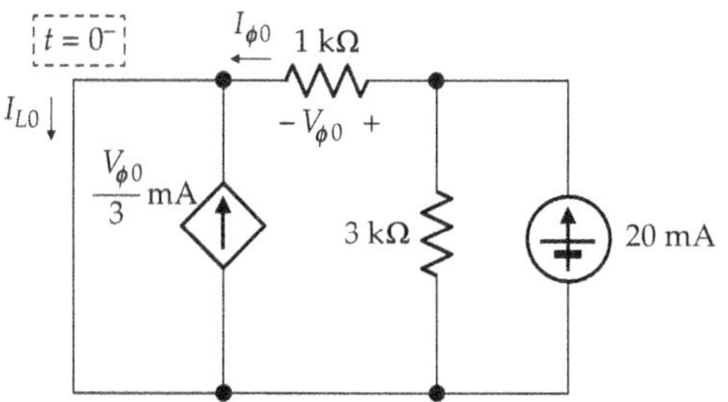

FIGURE 11.28
Figure for Example 11.5.

value I_{L0} is that which applies at $t = 0^+$. But because the stored energy cannot change instantaneously, $i_L(0^+) = i_L(0^-)$, where $i_L(0^-) = I_{L0}$ is the steady-state value with the switch open, which means that the inductor acts as a short circuit (Figure 11.28). From current division, $I_{\phi 0} = 20 \times 3/4 = 15$ mA. $I_{L0} = I_{\phi 0} + V_{\phi 0}/3$, where $V_{\phi 0} = (I_{\phi 0}$ mA$) \times (1$ kΩ). It follows that $I_{L0} = (4/3)I_{\phi 0} = 20$ mA.

When the switch is closed, the 66 V source and the 6 kΩ resistor are added to the circuit. This combination can be transformed to its equivalent current source, the circuit to the right of the 1 kΩ resistor becoming as shown in Figure 11.29. The two current sources can be combined into a 9 mA current source, directed upward, in parallel with 6||3 kΩ, which is 2 kΩ. In the steady state, the inductor is again a short circuit so that the circuit becomes as in Figure 11.28 but with the new source values (Figure 11.30). From current division, $I_{\phi F} = 9 \times 2/3 = 6$ mA. This gives $I_{LF} = (4/3)I_{\phi F} = 8$ mA.

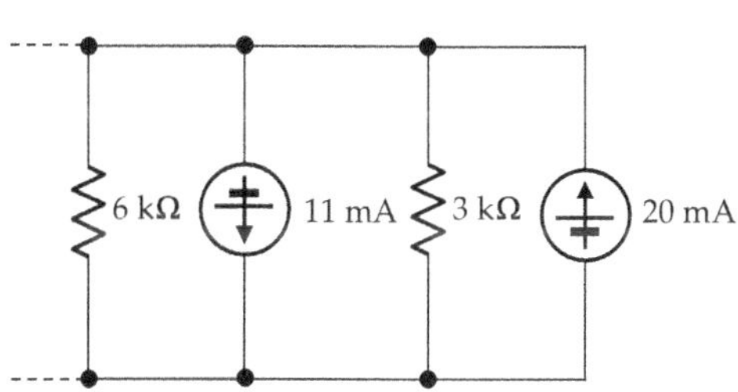

FIGURE 11.29
Figure for Example 11.5.

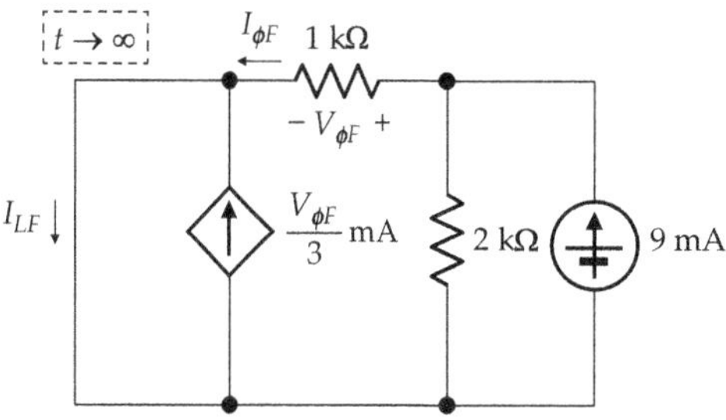

FIGURE 11.30
Figure for Example 11.5.

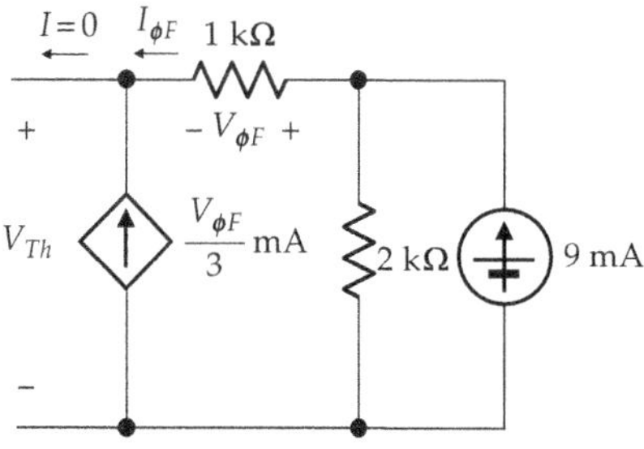

FIGURE 11.31
Figure for Example 11.5.

To determine the time constant, we note that the resistance seen by the inductor is Thevenin's resistance to the right of the inductor terminals. The short-circuit current between these terminals is $I_{LF} = 8$ mA, as has just been determined. When the inductor is replaced by an open circuit (Figure 11.31), $V_{\phi F}/1 + V_{\phi F}/3 = 0$ so that $V_{\phi F} = 0$ and $I_{\phi F} = 0$. It follows that V_{Th} is the voltage that appears across the 2 kΩ resistor, which is (9 mA) × (2 kΩ) = 18 V. $R_{Th} = 18/8 = 2.25$ kΩ. The time constant is $\tau = L/R = 18$ mH/(2.25 kΩ) $= 18 \times 10^{-3}/(2.25 \times 10^3) = 8 \times 10^{-6}$ s ≡ 8 μs. From Equation 11.57, $i_L(t) = 8 + (20-8)e^{-t/8} = 8 + 12e^{-t/8}$ mA, $t \geq 0^+$, where t is in μs.

Simulation: The circuit is entered as in Figure 11.32. When the simulation is started at $t = 0$, with zero initial current in the inductor, PSpice applies the 20 mA current source at $t = 0$, so sufficient time must be allowed for the circuit to reach a steady state before the switch is closed.

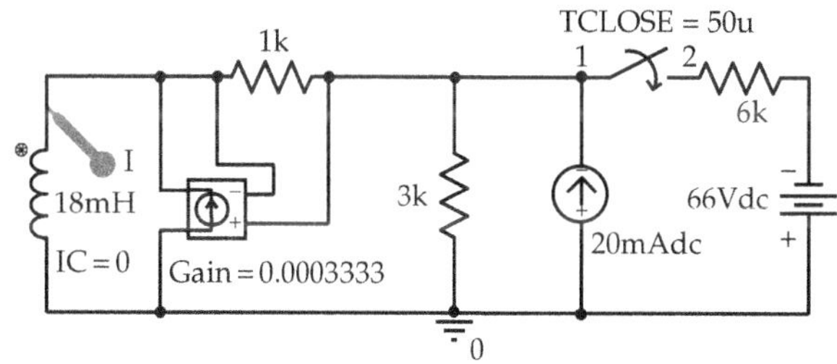

FIGURE 11.32
Figure for Example 11.5.

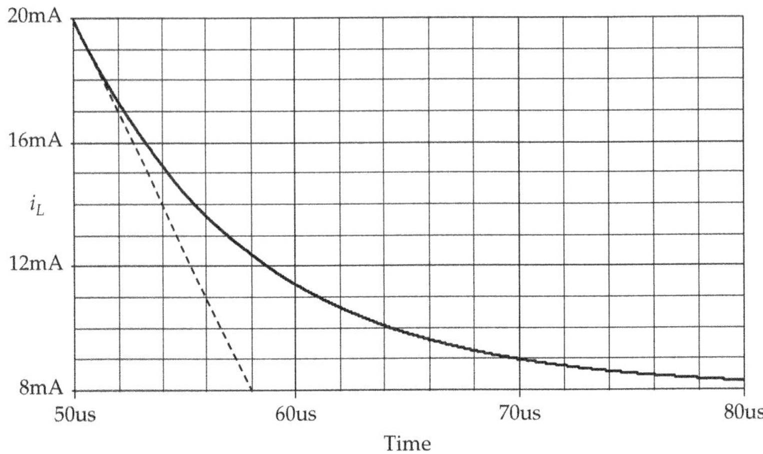

FIGURE 11.33
Figure for Example 11.5.

In Figure 11.32, the switch is closed at $t = 50$ μs. Since we are not interested in i_L prior to closing the switch, 50u is entered for 'Start saving data after' in the Simulation Settings. The other entries in the Simulation Settings are 'Time Domain (Transient)' for 'Analysis type,' 80u for 'Run to time,' and 1n for 'Maximum Step size'. A default time of 1 μs is allowed in PSpice for the switch to close, which means that closure will actually occur at 51 μs. To have closure occur more nearly at 50 μs, the default value for TTRAN is changed to 1n in the Property Editor spreadsheet of the switch. After the simulation is run, the plot for i_L is displayed as in Figure 11.33.

Alternatively, the initial steady-state current in the inductor can be calculated and entered as the IC value. Or zero is entered for 'Start saving data after' and both the buildup and decay of i_L displayed. This would also allow confirmation that a steady state is reached before the switch is closed.

Example 11.6: Jump in Voltage upon Switching

The switch is closed at $t = 0$ in Figure 11.34 after being open for a long time. It is required to determine $v_X(t)$ for $t \geq 0^+$.

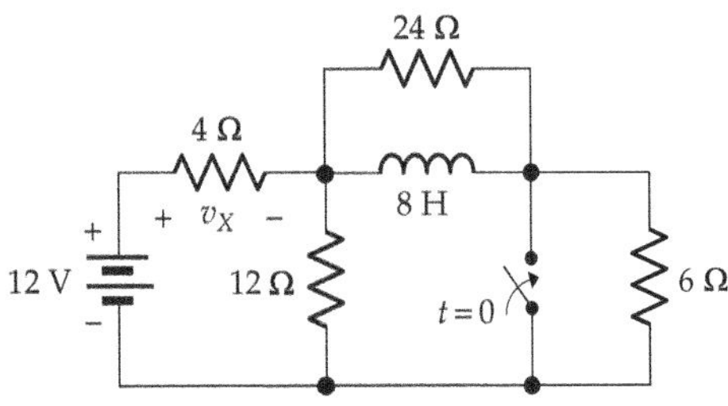

FIGURE 11.34
Figure for Example 11.6.

Solution:

When the switch has been open for a long time, the inductor acts as a short circuit. The circuit reduces to that shown in Figure 11.35. $(12\|6) = 4\ \Omega$. The current in the 4 Ω resistor is $12/(4+4) = 1.5$ A, and $i_L(0^-) = I_{L0} = 1.5 \times 12/18 = 1$ A; $v_X(0^-) = 4 \times 1.5 = 6$ V.

At $t = 0^+$, with the switch closed, the circuit becomes as in Figure 11.36. The 24 Ω resistor is in parallel with the 12 Ω resistor so that $(12\|24) = 8\ \Omega$; $i(0^+) = I_{L0} = 1$ A, from the constancy of stored energy. From KCL, $v_X/4 = (12 - v_X)/8 + 1$, which gives $v_X(0^+) = 20/3$ V. Note that v_X jumps at $t = 0$ from 6 V to 20/3 V. It is the latter value that should be used as the initial value of v_X for $t \geq 0^+$, based on the constancy of i_L. As $t \to \infty$, the inductor acts as a short circuit, and $V_{XF} = 12$ V. The resistance seen by the inductor when the 12 V source is set to zero

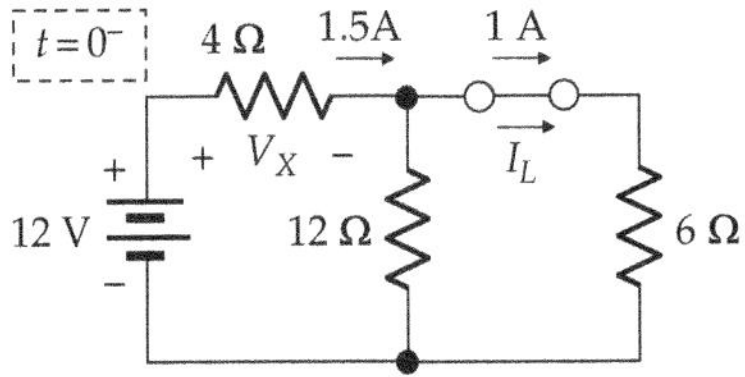

FIGURE 11.35
Figure for Example 11.6.

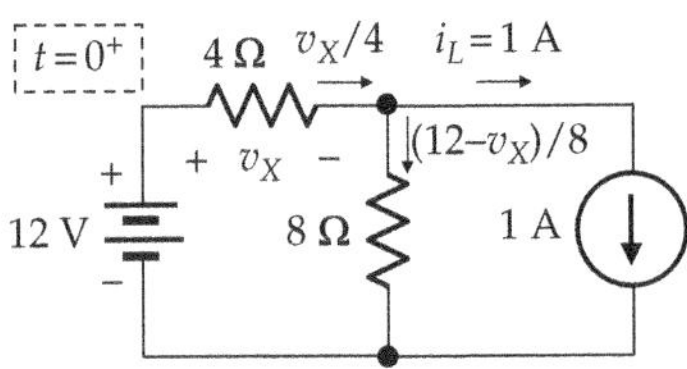

FIGURE 11.36
Figure for Example 11.6.

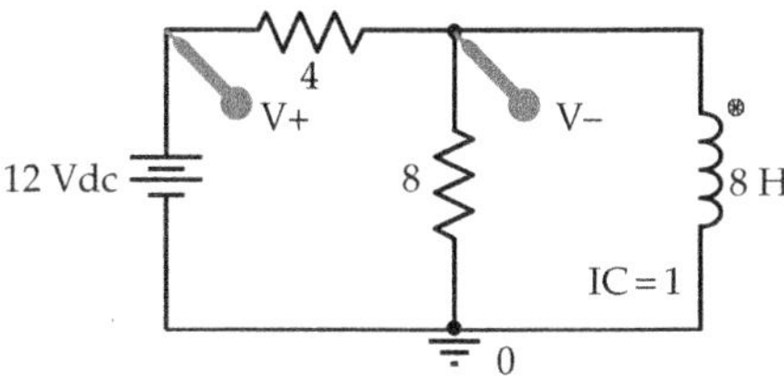

FIGURE 11.37
Figure for Example 11.6.

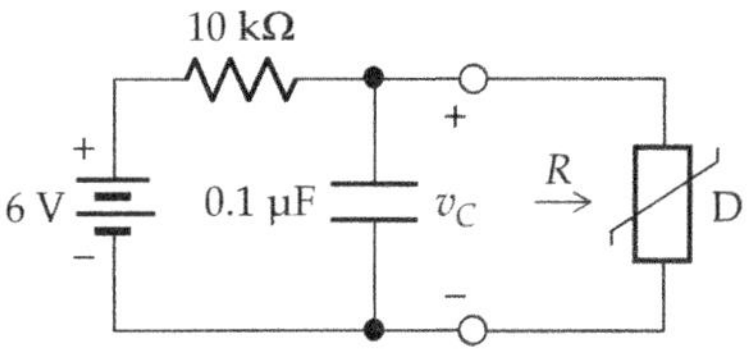

FIGURE 11.39
Figure for Example 11.7.

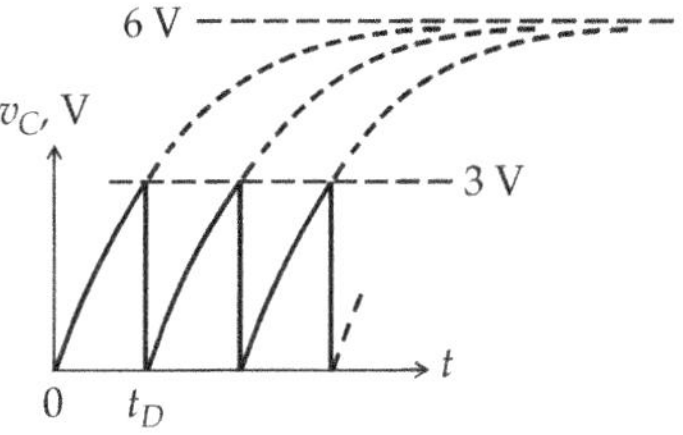

FIGURE 11.40
Figure for Example 11.7.

is 4||12||24 = 8/3 Ω. Hence, $\tau = 8/(8/3) = 3$ s. It follows that $v_X(t) = 12 + (20/3 - 12)e^{-t/3} = 12 - (16/3)e^{-t/3}$ V.

Simulation: The circuit is entered as in Figure 11.37. In the Simulation Settings, 'Analysis type' is 'Time Domain (Transient),' 'Run to time' is 10s, 'Start Saving Data After' is 0, and 'Maximum Step size' is 2m. After the simulation is run, the plot is as shown in Figure 11.38. v_X jumps at $t = 0$ from 6 to 6.6667 V, and $\tau = 3$ s.

*Example 11.7: Analysis of Repetitive Response

A nonlinear "threshold" device D is connected across the capacitor in Figure 11.39. The device changes its resistance R abruptly once the increasing voltage v_C reaches a certain level or threshold. It is assumed that R is infinite for $0 < v_C < 3$ V, while the capacitor is charging. When $v_C = 3$ V, R becomes zero, instantly discharging the capacitor, and immediately becomes infinite again when the capacitor is discharged. The capacitor starts charging again, and the cycle is repeated. v_C is therefore a periodic waveform that repeats at a certain frequency, as illustrated by the solid-line in Figure 11.40. It is required to determine the frequency of oscillation. This type of oscillator, based on a nonlinear element, such as D, that repetitively charges and discharges an energy storage element is referred to as a **relaxation oscillator**.

Solution:

Let $t = 0$ be the time when the capacitor is fully discharged and R becomes infinite. For $t \geq 0$, the capacitor charges toward 6 V with $\tau = 10\times10^3\times0.1\times10^{-6} = 10^{-3} \equiv 1$ ms. Hence, $v_C(t) = 6(1 - e^{-t})$ V, where t is in ms and $v_c \leq 3$ V. When $v_c = 3$ V, $1 - e^{-t_D} = 3/6 = 0.5$. This gives $e^{-t_D} = 0.5$, or $t_D = \ln 2$ ms. Hence, $f_D = 1/\ln 2 = 1.44$ kHz.

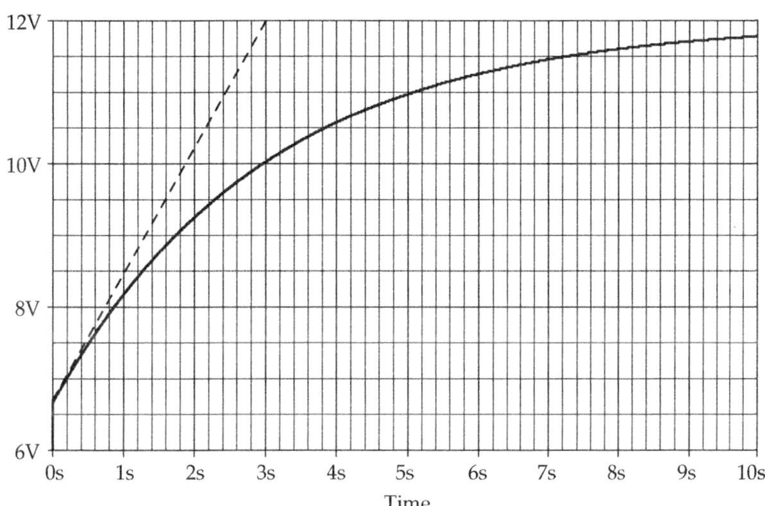

FIGURE 11.38
Figure for Example 11.6.

This problem illustrates some interesting and important points concerning the analysis and behavior of electric circuits:

1. It was repeatedly stressed in this chapter that the voltage across a capacitor cannot be changed instantaneously by any physically realizable means. Yet it is assumed in Figure 11.40 that device D instantly discharges C. Does this mean that the instantaneous discharge depicted in Figure 11.40 is not physically realizable? The answer is yes. Theoretically, C can be discharged instantaneously by a current *impulse*, which is discussed in detail in Chapter 18. The impulse function is not physically realizable, being a purely mathematical construct that is nevertheless extremely useful and important in the analysis of signals and systems, as demonstrated in Chapters 18 through 23.
2. If the assumption of instantaneous discharge of C is not physically realizable, are we justified in considering it? The answer is emphatically yes. The reason is that such assumptions provide an easy way of determining the *limiting behavior* of the circuit. The frequency f_C derived earlier is the highest theoretically possible frequency of oscillation for the circuit of Figure 11.39, for the assumed values of battery voltage, R, C, and threshold level. In practice, the resistance of device D when it discharges the capacitor is small and nonlinear. But it can only prolong the time of discharge and hence reduce f_C. Even analyzing the circuit assuming that R is small and linear complicates the analysis considerably (Problem P11.59). Assuming $R = 0$ when R discharges C greatly simplifies the analysis and provides a first approximation to the behavior of the circuit. Such an assumption is an extremely useful first step in designing a practical circuit, so this approach is extensively used in engineering design. The analysis can be subsequently refined as in Problem P11.59, leading to a final design that is usually completed through simulation, taking into account such practical considerations as the behavior of device D according to its manufacturer's data sheet, manufacturing tolerances of values of R and C, the effect of the drop in battery voltage with use, the effect of temperature variation, etc.
3. It should be carefully noted that in Figure 11.40, C charges toward a steady-state value of 6 V, the battery voltage, and not to 3 V, the threshold level. Fundamentally, the reason is that physical systems "operating in real time" are "causal". Operation in real time means that the system response unfolds for the first time as the current time progresses. This is in contrast to a recorded signal that was captured at an earlier time and is being replayed later. At any instant of the recording, the future response is already available. For example, a football game that is being watched "live" is being watched in real time. The action in the game unfolds for the first time as the current time progresses. On the other hand, when a recorded game is watched, any future outcome, such as the final score, is available at any time. By causal is meant that the system cannot anticipate what is going to happen in the future. This is true when one is watching the football game in real time. One cannot tell for sure what might happen next. In the same manner, the capacitor in Figure 11.40 cannot anticipate that it is going to be discharged when v_C reaches 3 V. An associated attribute of this behavior is that the system is "memoryless" in the sense that even if it is discharged at 3 V in one cycle, it does not remember this in the next cycle. The consequence of being causal and memoryless is that the capacitor charges in every cycle in normal fashion toward a final value of 6 V.
4. How about the nature of device D? D could be a simple neon lamp that acts as an open circuit at low voltages, without emitting light. When the voltage across the lamp reaches a certain threshold level, usually about few tens of volts, the gas in the lamp breaks down, an arc is struck between the two electrodes of the lamp, light is emitted by the arc, and the resistance across the lamp falls to a low value, thereby discharging the capacitor. When v_C falls to a low enough level, the arc is extinguished, and the lamp reverts to the open-circuit state. The result is a "light flasher" at a suitable frequency (Problems P11.33 and P11.59). Alternatively, D could be a threshold electronic circuit that produces some signal when its threshold is reached. If a switch is inserted in series with R, the circuit can act as a simple "timer" that produces a signal when the threshold is reached at a preset time after the switch is closed. The time can be varied by varying R or C (Problem P11.52).
5. A variation of the circuit of Figure 11.39 is a photoflash unit in which C is connected through a changeover switch, as in Figure 11.41. In one position of the switch, C is charged to V_{SRC}. When the switch is moved, the low-resistance flash lamp is connected across C. The large discharge current rapidly discharges C and produces an intense flash.

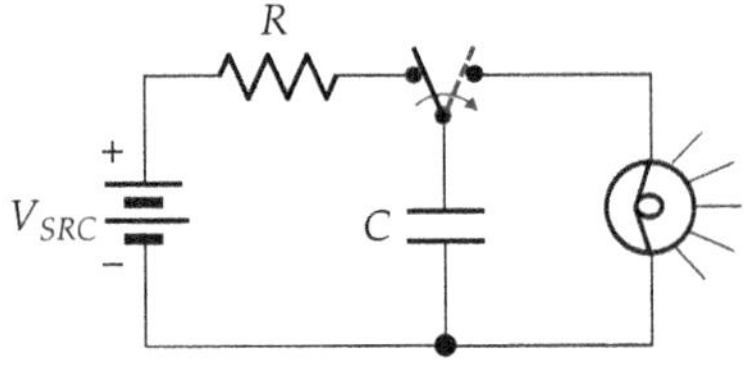

FIGURE 11.41
Figure for Example 11.7.

11.6 Role of Transient

It was emphasized throughout this chapter that stored energy cannot be changed instantaneously by any physically realizable means. Since the energy stored in a capacitor is directly related to the voltage across the capacitor and the energy stored in an inductor is directly related to the current though the inductor, capacitor voltages and inductor currents cannot be changed instantaneously by any physically realizable means. So what if the steady-state, final values of these circuit variables are different from the initial values, as they usually are? Evidently, there should be a smooth transition from the initial values to the steady-state, final values. This smooth transition is accomplished by the transient response.

Consider, for example, the case of a capacitor of initial voltage V_0 being charged by a battery of voltage V_{SRC} through a resistor R (Figure 11.42a). After the switch is closed and as $t \to \infty$, the capacitor acts as an open circuit in the final steady state, so the final voltage across the capacitor is V_{SRC}. The voltage across the capacitor at any time $t \geq 0^+$ is, from Equation 11.57,

$$v(t) = V_{SRC} + (V_0 - V_{SRC})e^{-t/\tau}, \quad t \geq 0^+ \qquad (11.63)$$

$v(0^+) = V_0$ and $v(\infty) = V_F = V_{SRC}$. The transition from V_0 to V_{SRC} is accomplished by the transient term, which is the second term on the RHS of Equation 11.63 and which contains the exponential. This term is shown in the lower trace of Figure 11.42b, assuming for the sake of argument that $V_0 < V_{SRC}$. When added to V_{SRC}, the total response is obtained, starting from V_0 at $t = 0^+$ and approaching V_{SRC} as $t \to \infty$. The time course of the transient is determined by the time constant in a first-order circuit. In the special case of $V_0 = V_{SRC}$, there is no transient response, because the final state is assumed from the very beginning at $t = 0^+$.

The same role is played by the transient response in the cases of inductor charging as well as the discharging of a capacitor or an inductor. In the case of a capacitor discharging through the resistor, for example, the final voltage across the capacitor is zero. The total response of capacitor voltage is a transient response that takes this voltage from its initial value to its final value of zero. The time-varying components of all the responses discussed in Sections 11.1 through 11.4 are transient responses that take the circuit from the initial to the final state. In fact, this role of the transient response is perfectly general:

Concept: *The transient response in a given circuit response provides a smooth transition from the initial value of the response to its steady-state, final value. This transition from the initial value to the final value cannot occur instantaneously because energy stored in energy storage elements cannot be changed instantaneously by any physically realizable means. The time course of this transition in a first-order circuit is determined by the time constant.*

Evidently, there is no transient in purely resistive circuits so that both resistor voltages and currents can change instantly from initial to final values. It may be wondered if this is a violation of the general principle enunciated in Section 11.1 that energy values cannot be changed instantaneously by any physically realizable means. The answer is, of course, no. It must be recognized that when resistor voltages or currents change instantaneously, it is the *power dissipated*, p, that changes instantaneously, not the energy dissipated, w. These two quantities are related by $p = dw/dt$ or $dw = pdt$. If p is finite and $dt \to 0$, as between $t = 0^-$ and $t = 0^+$, then $dw = 0$, that is, the *energy dissipated* does not change. The exception of infinite p implies infinite resistor voltages or currents, which are not physically realizable. Hence, the general principle of constancy of energy values at the instant of change, under physically realizable conditions, is not violated.

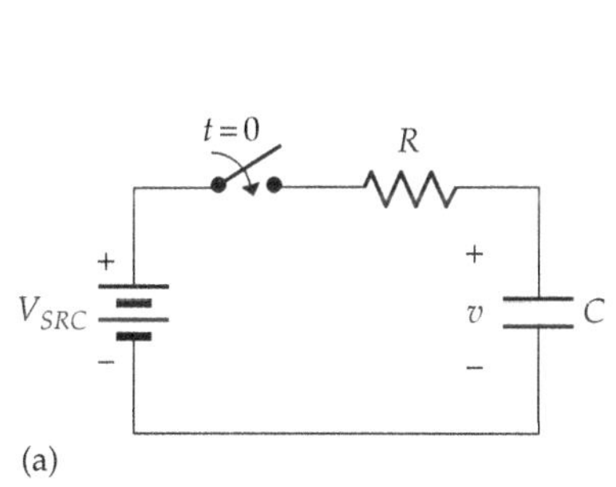

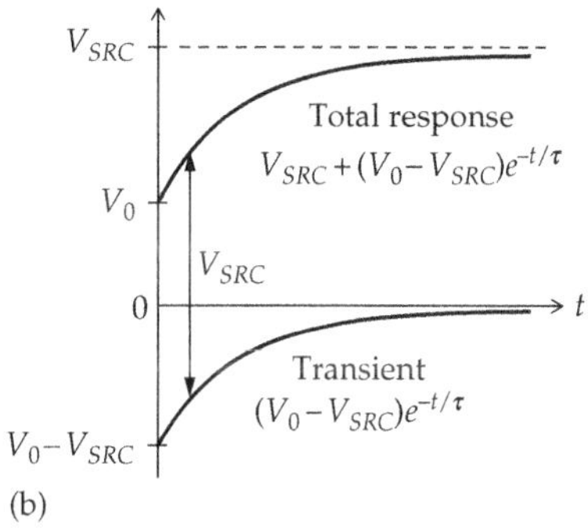

FIGURE 11.42
(a) Charging of a capacitor through a resistor and (b) the transient response and the total response.

Learning Checklist: What Should Be Learned from This Chapter

- Energy, in general, including stored energy, cannot be changed instantaneously by any physically realizable means, as this would require infinite driving forces, that is, infinite voltages or currents in the case of energy storage elements in electric circuits. Such infinite values are not physically realizable.
- The behavior of a first-order circuit is governed by a first-order, linear, ordinary, differential equation with constant coefficients. The general solution to this equation contains an arbitrary constant that is determined from the initial value at $t = 0^+$ of the variable in the differential equation. Once this substitution is made, the response for $t \geq 0^+$ is obtained.
- The time constant of a first-order circuit is a measure of its speed of response. The larger the time constant, the slower or more "sluggish" the response is and conversely.
- The time constant is RC for a capacitive circuit and L/R for an inductive circuit.
- When energy storage elements are charged through a sudden change, an uncharged capacitor acts as a short circuit and an uncharged inductor acts as an open circuit.
- The responses in an LTI circuit, with initial energy storage, to an applied excitation can be derived as the algebraic sum of two responses: (1) the response to the applied excitation acting alone, with zero initial energy storage, and (2) the response to the initial energy storage acting alone, with zero applied excitation.
- The general solution of a linear, ordinary, differential equation is the sum of two components: (1) a transient term, obtained from the solution of the homogeneous differential equation, and (2) a steady-state term that is the solution of the nonhomogeneous differential equation as $t \to \infty$.
- The natural response of a first-order circuit, or the response to a dc excitation, is of the form $y = Y_F + (Y_0 - Y_F)e^{-t/\tau}$, $t \geq 0^+$, where Y_F is the steady-state, final value of y and Y_0 is its initial value. These values and the time constant τ can be determined directly from the circuit.
- The following are important features of the homogeneous differential equation of a first-order circuit:
  1. Independent sources do not appear in the homogeneous differential equation, which means that the equation can be derived with independent sources set to zero.
  2. The homogeneous differential equation is completely determined by the time constant τ.
  3. The same homogeneous differential equation applies to all voltage and current responses in the circuit. These responses have the same time constant, so they differ only in the values of the arbitrary constant and of the final and steady state.
- The voltages and currents at $t = 0^+$ are the initial values used in determining the arbitrary constant in the general solution of the differential equation for a given voltage or current.
- Although the voltage across a capacitor and the current through an inductor do not change at $t = 0$, because the stored energy cannot be changed instantaneously by any physically realizable means, other voltages or currents in the circuit, including the current through a capacitor and the voltage across an inductor, are not directly related to stored energy and can therefore change instantaneously to satisfy KCL or KVL.
- Both independent and dependent sources affect the initial and steady-state, final values of currents or voltages in a first-order circuit. Dependent sources, but not independent sources, affect the time constant so that independent sources can be set to zero when determining the time constant.
- The time constant can be determined from Thevenin's resistance seen by C or L, when all independent sources are set to zero.
- Initial and steady-state, final values of currents and voltages are independent of C or L, because in the steady state, inductors act as short circuits and capacitors as open circuits.
- If a first-order circuit contains more than one resistor and more than one capacitor or inductor, with or without dependent sources, then the circuit should be reducible to a single effective resistor and a single effective capacitor or inductor, with or without an independent source.
- The transient response in a given circuit response provides a smooth transition from the initial value of the response to its steady-state, final value. This transition from the initial value to the final value cannot occur instantaneously because energy stored in energy storage elements cannot be changed instantaneously by any physically realizable means. The time course of the transient response in a first-order circuit is determined by the time constant.

Problem-Solving Tips

1. When the exponent t/τ is first formed, t naturally has the same units as τ. If τ is in seconds, then t is in seconds, and if τ is in ms, then t is in ms. If t is τ is in seconds and is to be expressed in ms, while keeping the same value of the exponent, then the new t in ms should be multiplied by s/ms or 10^{-3}. Another way of looking at this is that t in ms is a larger number than t in s, so t in ms in the exponent must be multiplied by 10^{-3} to maintain the same value as when t is in s.
2. Never apply initial conditions except to the complete solution, and not to the transient solution alone.
3. Always check the correctness of a derived response for current or voltage as a function of time by determining the values at $t = 0$ and as $t \to \infty$.
4. Duality can be helpful in checking the correctness of expressions.

Problems

Apply ISDEPIC and verify solutions by PSpice simulation whenever feasible.

Single Energy Storage Elements

P11.1 $v(t)$ and $i(t)$ in Figure P11.1 are given by $v(t) = 50e^{-10t}$ V and $i(t) = 20e^{-10t}$ mA, $t \geq 0^+$s. Determine τ, R_C, C, $q(0^+)$.

Ans. 0.1 s, 2.5 kΩ, 40 μF, 2 mC.

P11.2 $v(t)$ and $i(t)$ in Figure P11.2 are given by $v(t) = 40e^{-100t}$ V and $i(t) = 10e^{-100t}$ A, $t \geq 0^+$s. Determine τ, R_L, L, $\lambda(0^+)$.

Ans. 10 ms, 4 Ω, 40 mH, 0.4 Wb-T.

P11.3 $I_{L0} = 2$ A in Figure P11.3 at $t = 0$. Just before the switch is opened at $t = 1$ s, $i_L(1^-) = I_{L0}/2$ A. Determine $i_L(t)$ at $t = 2$ s.

Ans. 0.25 A.

P11.4 $v(t)$ in a first-order circuit is governed by the differential equation $2(dv/dt) + (2/3)v = 10$, $t \geq 0^+$s, with $v(0^+) = 3$ V. Determine τ, V_F, and $v(t)$.

Ans. 3 s, 15 V, $v(t) = 15 - 12e^{-t/3}$ V.

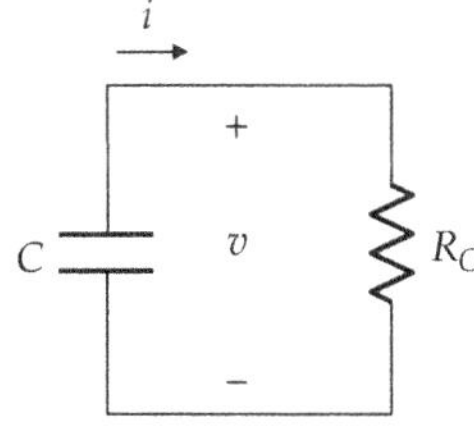

FIGURE P11.1

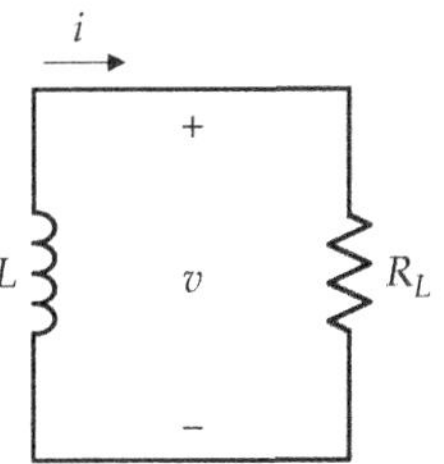

FIGURE P11.2

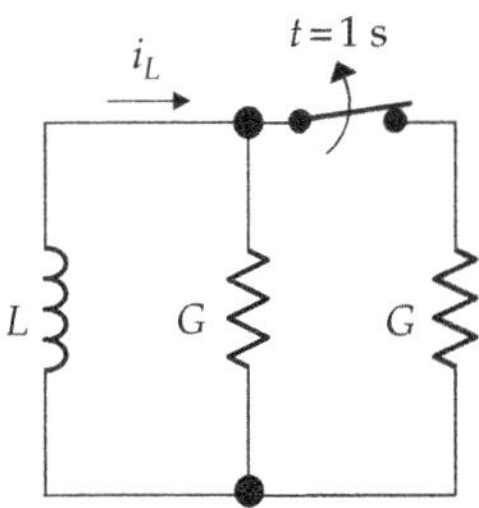

FIGURE P11.3

P11.5 The switch in Figure P11.5 is moved from position 'a' to position 'b' after being in position 'a' for a long time. Determine $v(t)$, $t \geq 0^+$.

Ans. $v(t) = 20e^{-125t}$ V, t is in s.

P11.6 The switch in Figure P11.6 is moved to position 'b' at $t = 0$ after being in position 'a' for a long time. Determine, for $t \geq 0^+$, (a) $v_C(t)$, (b) $v_O(t)$, (c) $i_O(t)$, and (d) the total energy dissipated in the 60 kΩ resistor as $t \to \infty$.

Ans. (a) $v_C(t) = 100e^{-25t}$ V, t is in s; (b) $v_O(t) = 60e^{-25t}$ V; (c) $i_O(t) = e^{-25t}$ mA. (d) 1.2 mJ.

P11.7 The switch in Figure P11.7 is opened at $t = 0$ after being closed for a long time. Determine $v_O(t)$ for $t \geq 0^+$.

Ans. $v_O(t) = -40e^{-100t}$ V, t is in s.

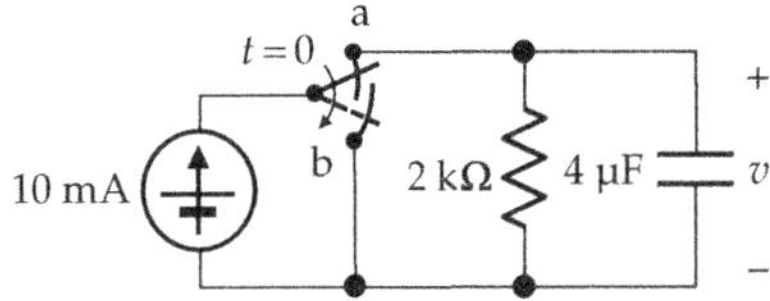

FIGURE P11.5

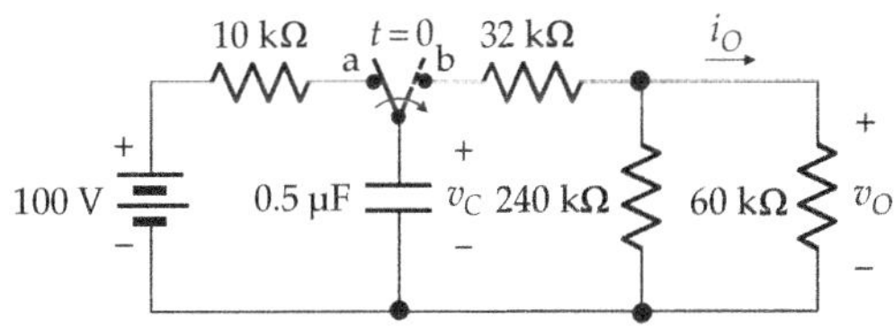

FIGURE P11.6

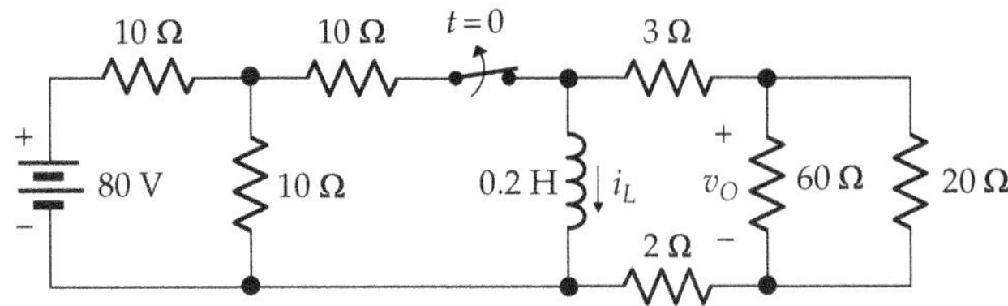

FIGURE P11.7

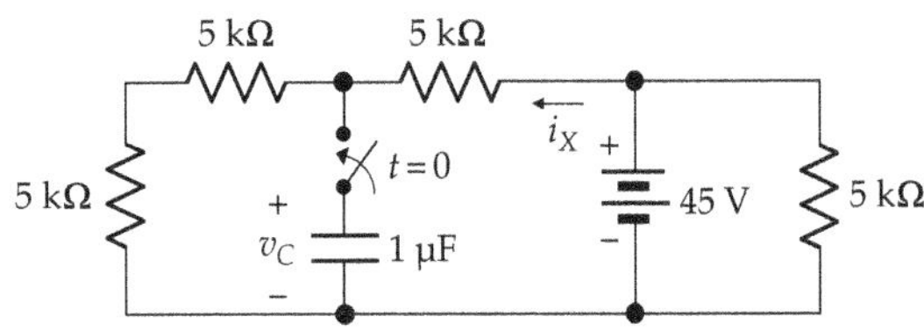

FIGURE P11.8

P11.8 The switch in Figure P11.8 is closed at $t = 0$, the capacitor being initially uncharged. Determine, for $t \geq 0^+$, (a) $v_C(t)$ and (b) $i_X(t)$.

Ans. (a) $v_C(t) = 30\left(1 - e^{-0.3t}\right)$V, t is in ms; (b) $i_X(t) = 3 + 6e^{-0.3t}$ mA.

P11.9 (a) The switch in Figure P11.9 is closed at $t = 0$ after being opened for a long time. Determine $v_O(t)$ for $t \geq 0^+$. (b) The switch opens again at $t = 10$ s. Determine $v_O(t)$ for $t \geq 10$ s.

Ans. (a) $v_O(t) = 2.25 + 0.75e^{-t/15}$ V, $0 \leq t \leq 10$ s;
(b) $v_O(t) = 3 - 0.37e^{-0.05(t-10)}$ V, $t \geq 10$ s.

P11.10 The switch in Figure P11.10 is closed at $t = 0$ after being in the open position for a long time. Determine, for $t \geq 0^+$, (a) $i_O(t)$ and (b) $v_O(t)$.

Ans. (a) $i_O(t) = 20 - 15e^{-12.5t}$ A; (b) $v_O(t) = 15e^{-12.5t}$ V, t is in s.

P11.11 The switch in Figure P11.11 is opened at $t = 0$ after being closed for a long time. Determine $i_X(t)$, $t \geq 0^+$.

Ans. $i_X(t) = e^{-5t}$ mA, t is in s.

P11.12 Both switches in Figure P11.12 are opened at $t = 0$ after being closed for a long time. Determine, for $t \geq 0^+$, (a) $v_C(t)$ and (b) the time it takes to dissipate 75% of the energy initially stored in the capacitor.

Ans. (a) $v_C(t) = -102e^{-25t}$ V, t is in s; (b) 27.73 ms.

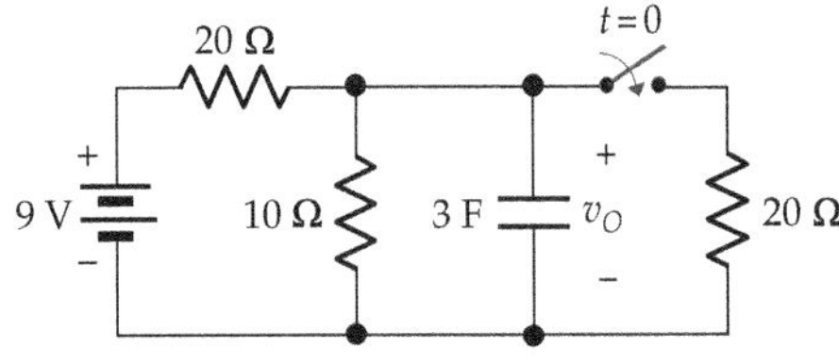

FIGURE P11.9

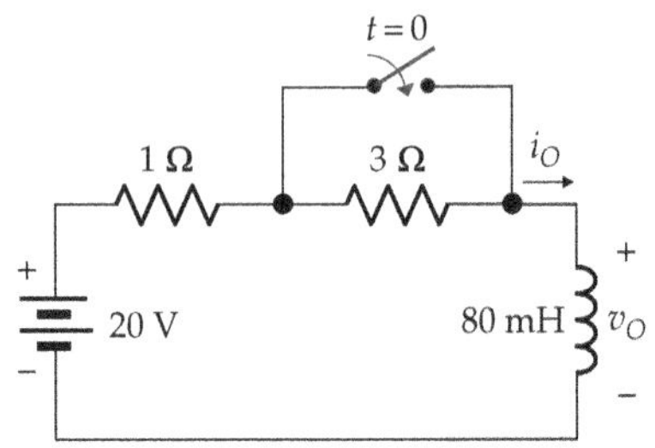

FIGURE P11.10

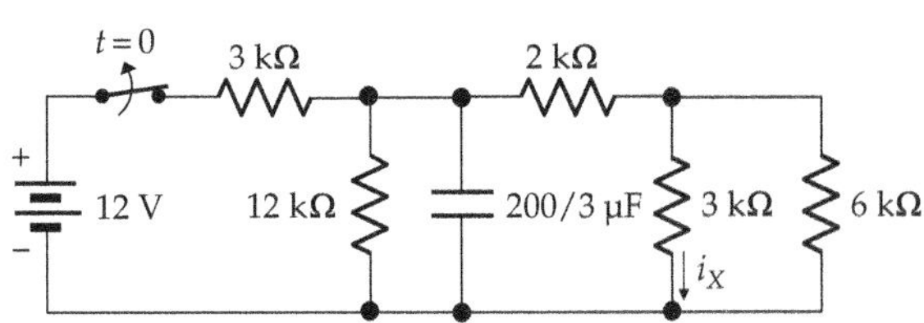

FIGURE P11.11

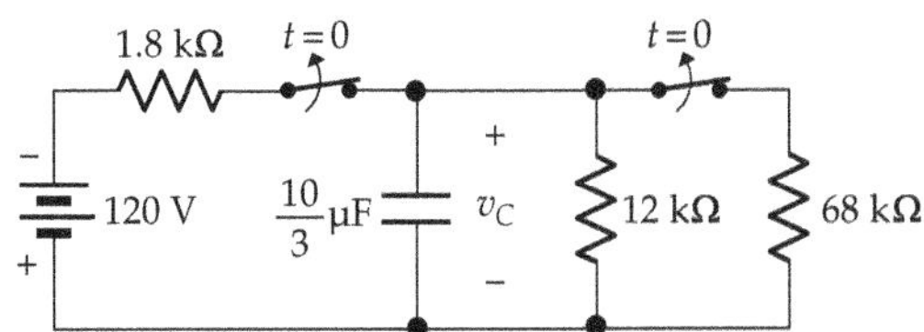

FIGURE P11.12

P11.13 Both switches in Figure P11.13 have been closed for a long time. The first switch opens at $t = 0$ and the second switch opens at $t = 35$ ms. Determine (a) $i_L(t)$ for $0 \leq t \leq 35$ ms, (b) $i_L(t)$ for $t \geq 35$ ms, and (c) the percentage of the energy initially stored in the inductor that is dissipated in the 18 Ω resistor.

Ans. (a) $i_L(t) = 6e^{-t/25}$ A, t is in ms; (b) $i_L(t) = 1.48e^{-3(t-35)/50}$ A; (c) 31.31%.

P11.14 The switch in Figure P11.14 is moved to position 'b' at $t = 0$, after being in position 'a' for a long time. Determine $v_C(t)$ for $t \geq 0^+$.

Ans. $v_C(t) = -20 - 10e^{-10t}$ V, t is in s.

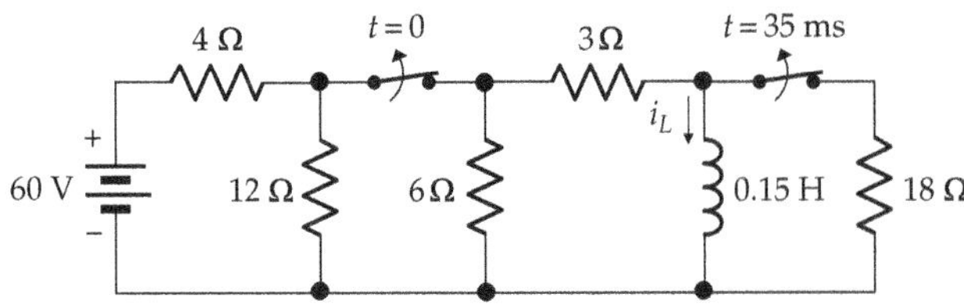

FIGURE P11.13

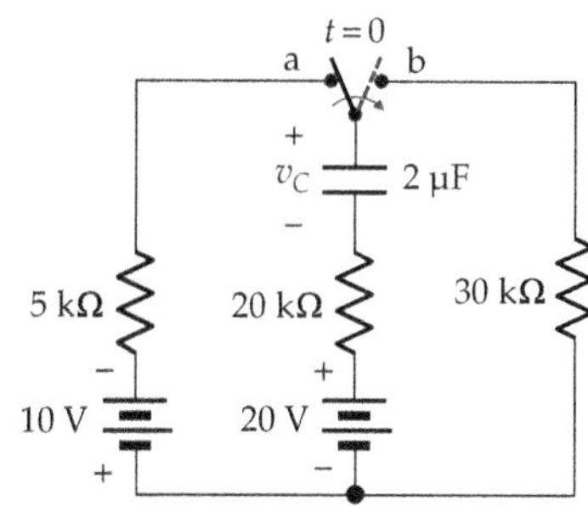

FIGURE P11.14

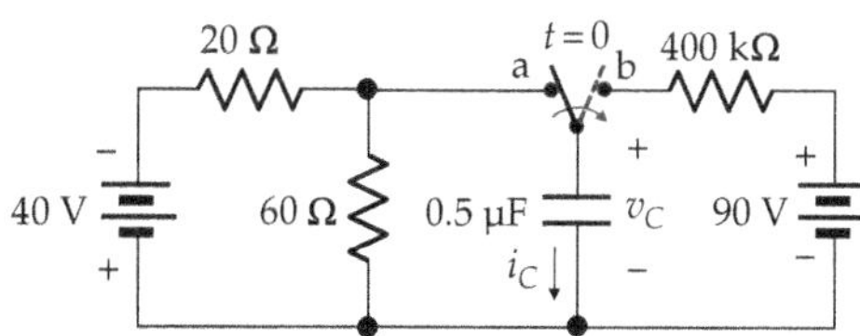

FIGURE P11.15

P11.15 The switch in Figure P11.15 is moved to position 'b' at $t = 0$ after being in position 'a' for a long time. Determine, for $t \geq 0^+$, (a) $v_C(t)$ and (b) $i_C(t)$ from the initial and final values and from the v–i relation for the capacitor.

Ans. (a) $v_C(t) = 90 - 120e^{-5t}$ V, t is in s; (b) $i_C(t) = 0.3e^{-5t}$ mA.

P11.16 The switch in Figure P11.16 is moved to position 'b' at $t = 0$ after being in position 'a' for a long time. Determine, for $t \geq 0^+$, (a) $v_c(t)$ and (b) $i_c(t)$ from the initial and final values and from the v–i relation for the capacitor.

Ans. (a) $v_c(t) = -60 + 90e^{-100t}$ V, t is in s; (b) $i_c(t) = -225e^{-100t}$ mA.

P11.17 Switch S_1 in Figure P11.17 is closed at $t = 0$, with the capacitor initially uncharged. Switch S_2 is closed at $t = 30$ ms, and switch S_3 is closed at $t = 50$ ms. Determine $v_C(t)$ for all $t \geq 0^+$.

Ans. $v_C(t) = 40\left(1 - e^{-t/10}\right)$ V, $0 \leq t \leq 30$ ms, $v_C(t) = 20 + (18)e^{-(t-30)/2}$ V, $t \geq 30$ ms.

P11.18 The switch in Figure P11.18 is moved to position 'b' at $t = 0$ after being in position 'a' for a long time. Determine, for $t \geq 0^+$, (a) $v_O(t)$ and (b) $i_O(t)$.

Ans. (a) $v_O(t) = 40e^{-10t}$ V, t is in s; (b) $i_O(t) = 12 - 20e^{-10t}$ A.

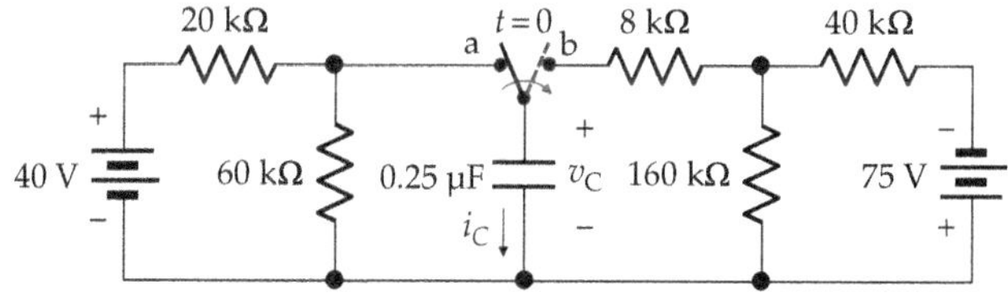

FIGURE P11.16

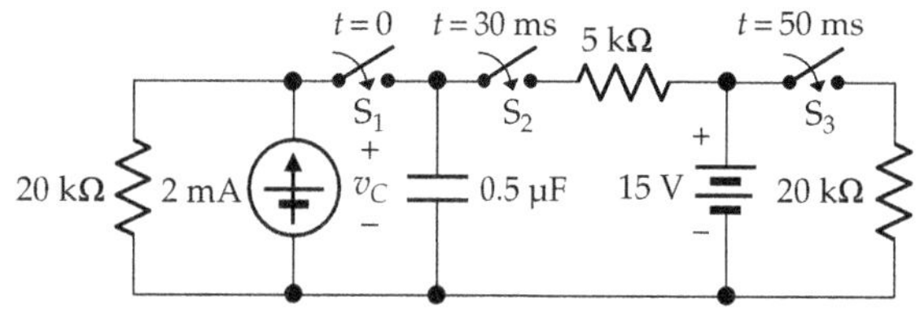

FIGURE P11.17

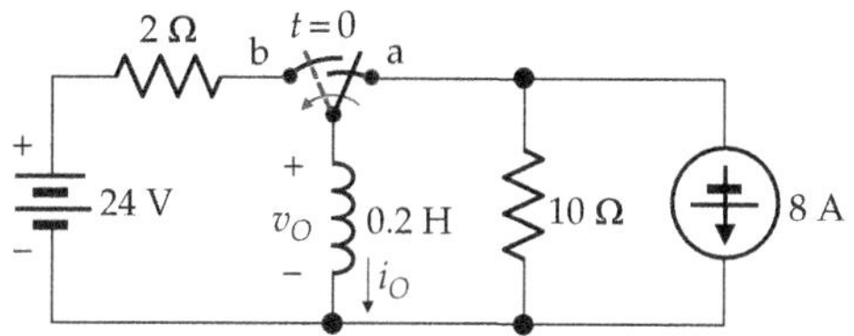

FIGURE P11.18

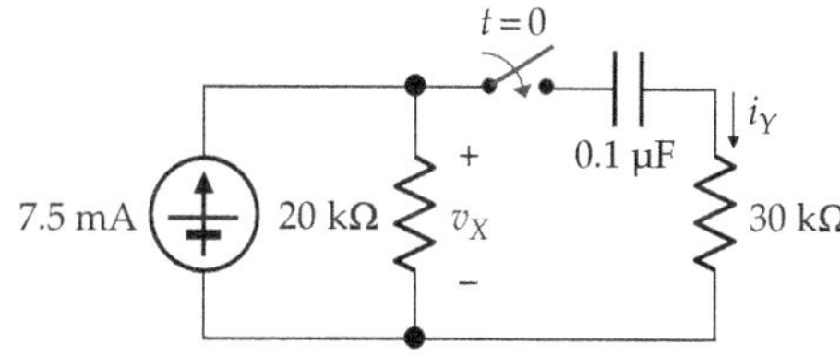

FIGURE P11.19

P11.19 The switch in Figure P11.19 is closed at $t = 0^+$, with the capacitor initially uncharged. Determine, for $t \geq 0^+$, (a) $i_Y(t)$ and (b) $v_X(t)$.

Ans. (a) $i_Y(t) = 3e^{-200t}$ mA, t is in s; (b) $v_X(t) = 150 - 60e^{-200t}$ V.

P11.20 The capacitor in Figure P11.20 was initially uncharged and the switch was in position 'a'. At $t = 0$, the switch is moved to position 'b'. Determine for $t \geq 0^+$ (a) $v_C(t)$, (b) $i_C(t)$, (c) the energy delivered by the 12 V battery as $t \to \infty$, (d) the energy absorbed by the 6 V battery as $t \to \infty$, and (e) the energy dissipated in the resistor as $t \to \infty$; verify this by integrating the power dissipated by i_C from $t = 0^+$ to $t \to \infty$. (f) If after a long time, $t' = 0$, the switch is moved to position 'c', determine $v_C(t)$ and $i_C(t)$ for $t' \geq 0^+$, the energy delivered by the

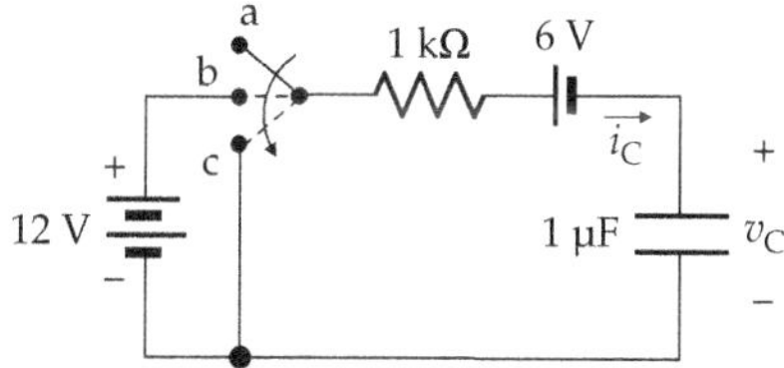

FIGURE P11.20

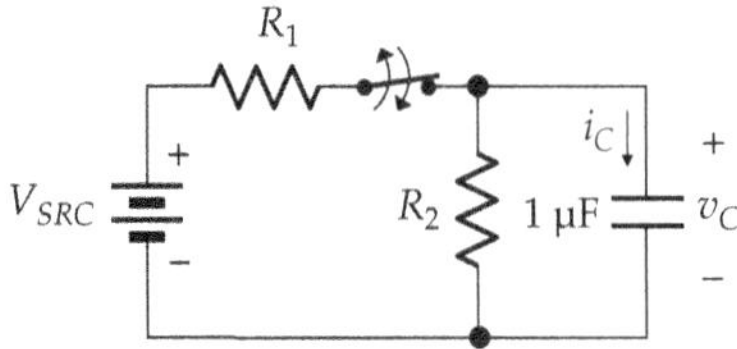

FIGURE P11.21

6 V battery, the energy gained or lost by the capacitor, and the energy dissipated in the resistor.

Ans. (a) $v_C(t) = 6(1 - e^{-t})$ V, t is in ms; (b) $i_C(t) = 6e^{-t}$ mA; (c) 72 μJ; (d) 36 μJ; (e) 18 μJ; (f) $v_C(t) = -6 + 12e^{-t'}$ V, $i_C(t) = -12e^{-t'}$, 72 μJ, 0, 72 μJ.

P11.21 It is given in Figure P11.21 that (i) the time constant with the switch open is twice the time constant with the switch closed, and (ii) if the switch is closed with an initial voltage V_{c0} = 1 V across the capacitor, the initial value of the capacitor current is I_{c0} = 1 mA and the final value of the capacitor voltage is $v_C(\infty)$ = 2 V. (a) Determine V_{SRC}, R_1, and R_2. (b) If the switch is opened at t = 0, with V_{c0} = 1 V, determine $v_C(t)$, $t \geq 0^+$. (c) If the switch is then closed at t = 1 ms, determine $v_C(t)$, $t \geq 1$ ms.

Ans. (a) V_{SRC} = 4 V, $R_1 = R_2$ = 2 kΩ; (b) $v_C(t) = e^{-0.5t}$ V, $0 \leq t \leq$ ms; (c) $v_C(t) = 2 + (e^{-0.5} - 2)e^{-(t-1)}$ V, $t \geq 1$ ms.

P11.22 Both switches in Figure P11.22 have been open for a long time. S_1 is closed at t = 0 and S_2 at t = 1 s. Determine $v_C(t)$ for $t \geq 1$ s.

Ans. $v_C(t) = 8 - 19.21e^{-35(t-1)/12}$ V, t is in s.

P11.23 Switch S_1 in Figure P11.23 has been in position a for a long time, with switch S_2 closed. At t = 0, S_1 is moved to position b and remains in this position. S_2 is opened at t = 50 μs and closed again at t = 100 μs. Determine v_C for (a) $0 \leq t \leq 50$ μs, (b) $50 \leq t \leq 100$ μs, and (c) $t \geq 100$ μs.

Ans. (a) $\frac{500}{7}e^{-17t/1400}$ V, t is in μs; (b) $38.9e^{-(t-50)/140}$ V; (c) $27.2e^{-17(t-100)/1400}$ V.

P11.24 The switch in Figure P11.24 is moved at t = 0 to position 'b' after being in position 'a' for a long time. Determine $v_C(t)$ for $t \geq 0^+$ and the time at which $v_C(t)$ = 0.

Ans. $v_C(t) = -50 + 95e^{-t/30}$ V, t is in ms; v_C = 0 at $t = 30\ln\left(\frac{95}{50}\right) = 19.3$ ms.

P11.25 The switch in Figure P11.25 is moved to position 'b' at t = 0, after being in position 'a' for a long time. Determine, for $t \geq 0^+$, (a) $v_L(t)$ and (b) $i_L(t)$. A make-before-break switch is used to avoid open-circuiting the source.

Ans. (a) $v_L(t) = 5e^{-2.5t}$ V, t is in s; (b) $i_L(t) = 5 - e^{-2.5t}$ A.

P11.26 The switch in Figure P11.26 is moved to position 'b' at t = 0 after being in position 'a' for a long time. Determine $v_C(t)$ for $t \geq 0^+$.

Ans. $v_C(t) = \frac{160}{7}(1 - 2e^{-7t})$ V, t is in s.

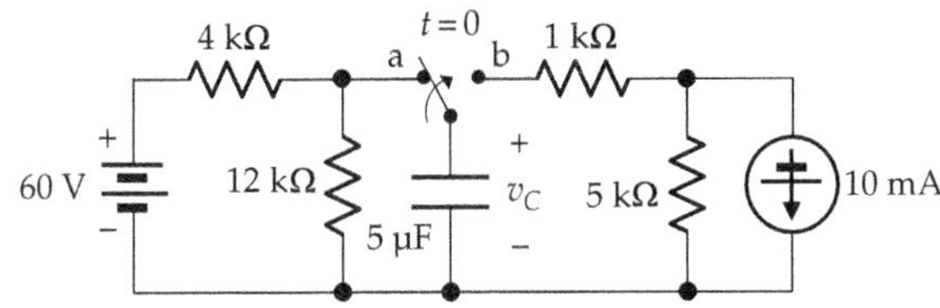

FIGURE P11.24

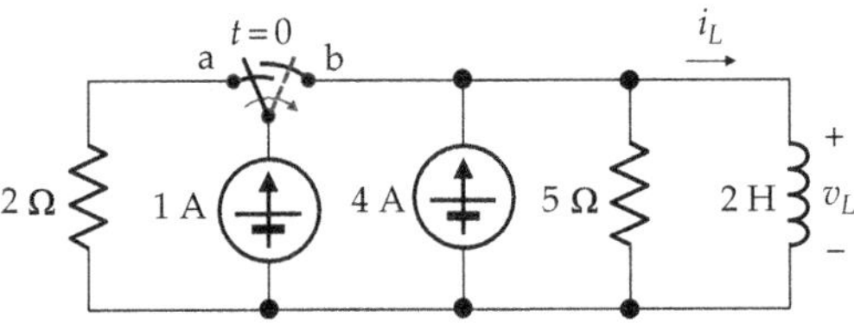

FIGURE P11.25

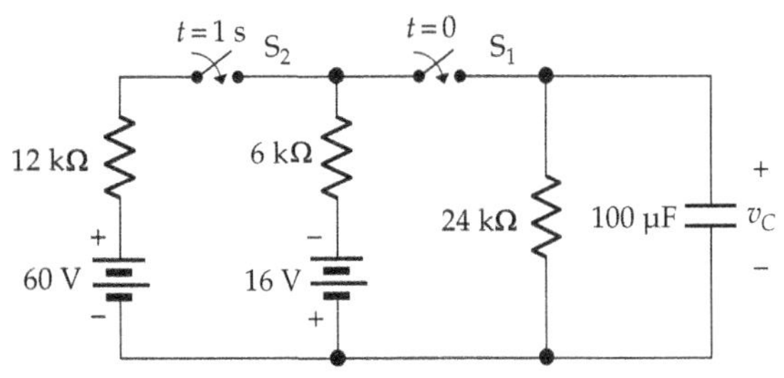

FIGURE P11.22

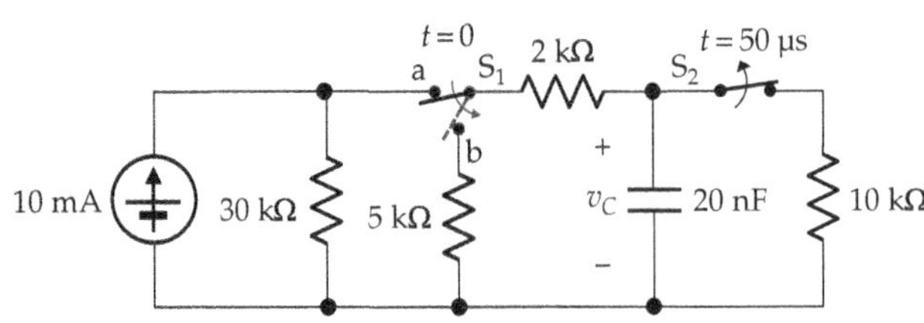

FIGURE P11.23

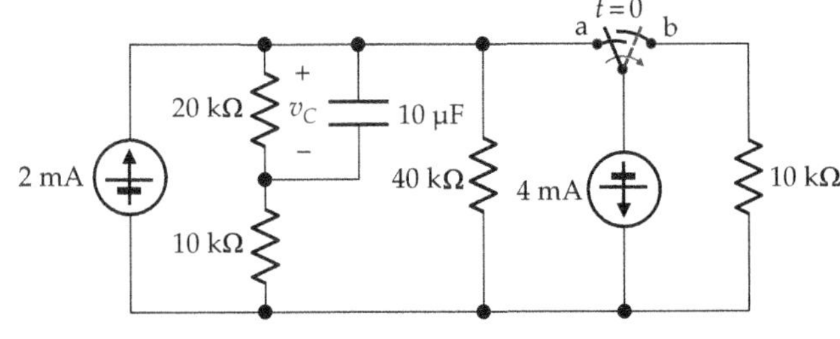

FIGURE P11.26

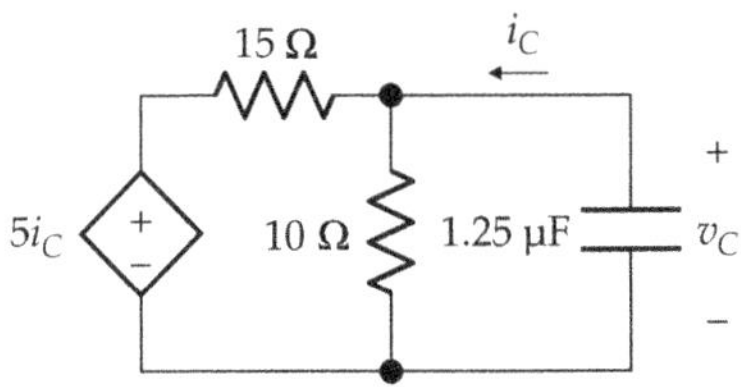

FIGURE P11.27

P11.27 The capacitor in Figure P11.27 is charged to 10 V at $t = 0$. Determine $v_C(t)$ for $t \geq 0$.

Ans. $v_C(t) = 10e^{-t/10}$ V, t is in ms.

P11.28 The voltage across the inductor in Figure P11.28 is 10 V at $t = 0$. Determine $v_L(t)$ for $t \geq 0^+$.

Ans. $v_L(t) = 10e^{-40t/3}$ V, t is in μs.

P11.29 The switch in Figure P11.29 is opened at $t = 0$ after being closed for a long time. Determine $i_L(t)$ for $t \geq 0^+$.

Ans. $i_L(t) = 2e^{-8t}$ V, t is in s.

P11.30 The switch in Figure P11.30 is moved to position 'b' after being in position 'a' for a long time. Determine $v_O(t)$ for $t \geq 0^+$.

Ans. $v_O(t) = \dfrac{1680}{101} + \dfrac{168}{1010}e^{-101t/16}$ V, t is in s.

P11.31 The switch in Figure P11.31 is closed at $t = 0$ after being open for a long time. Determine $v_O(t)$ for $t \geq 0^+$.

Ans. $v_O(t) = 24 + 48e^{-t/0.12}$ V, t is in s.

P11.32 The switch in Figure P11.32 is opened at $t = 0$ after having been closed for a long time. Determine $i_X(t)$ and $v_X(t)$, $t \geq 0^+$.

Ans. $i_X(t) = 4e^{-1.5t}$ mA; $v_X(t) = -2e^{-1.5t}$ V.

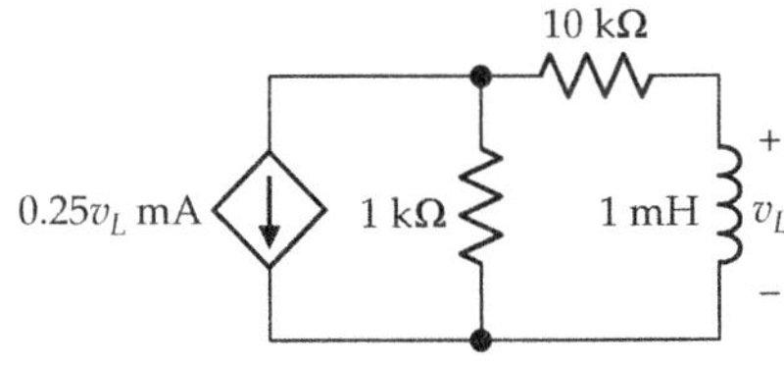

FIGURE P11.28

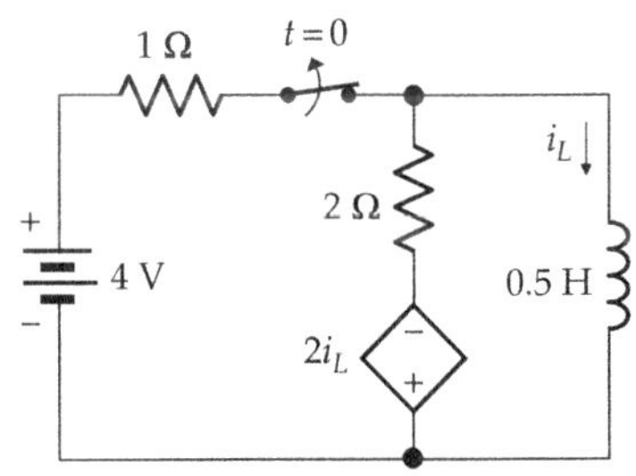

FIGURE P11.29

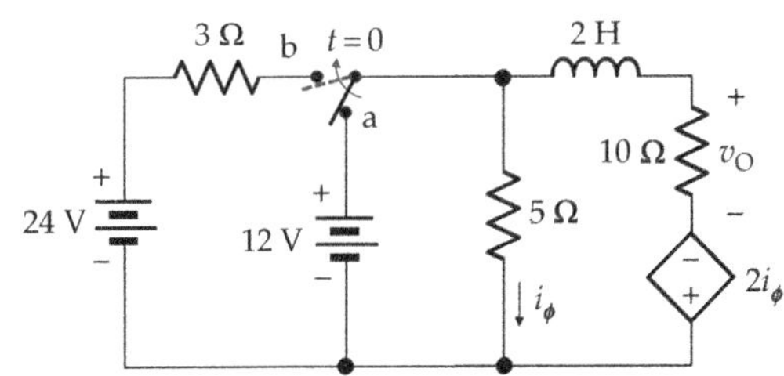

FIGURE P11.30

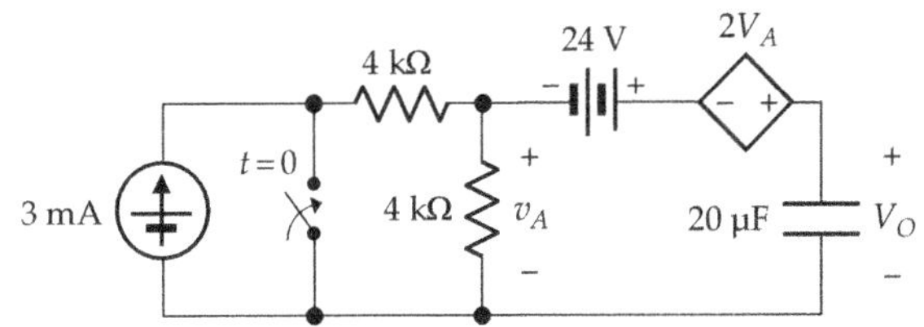

FIGURE P11.31

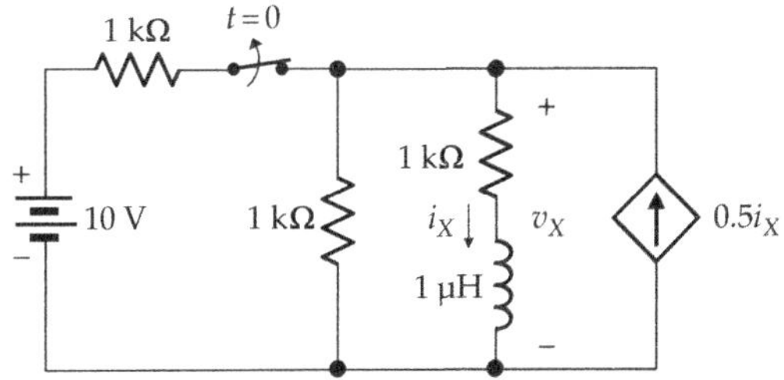

FIGURE P11.32

Multiple Energy Storage Elements

P11.33 The switch in Figure P11.33 is closed at $t = 0$, with the 1 μF capacitor charged to 10 V and the 4 μF capacitor uncharged. Determine $i(t)$, for $t \geq 0^+$.

Ans. $i(t) = 8e^{-t}$mA, where t is in ms.

P11.34 The switch in Figure P11.34 is closed at $t = 0$, with both capacitors charged to 6 V each. Determine, for $t \geq 0^+$, (a) $i(t)$, (b) $v_1(t)$, and (c) $v_2(t)$.

Ans. (a) $i(t) = 3e^{-t/4}$ mA, t is in ms; (b) $v_1(t) = 10 - 4e^{-t/4}$ V; (c) $v_2(t) = 8 - 2e^{-t/4}$ V.

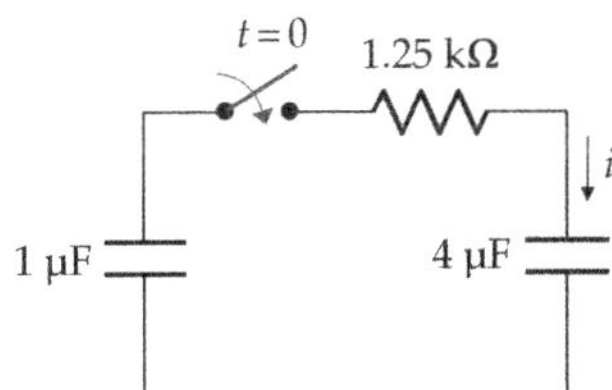

FIGURE P11.33

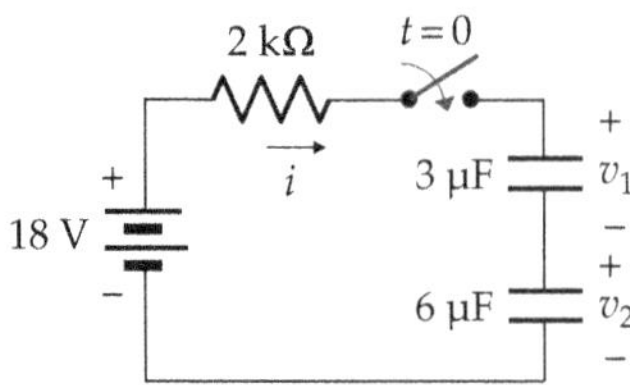

FIGURE P11.34

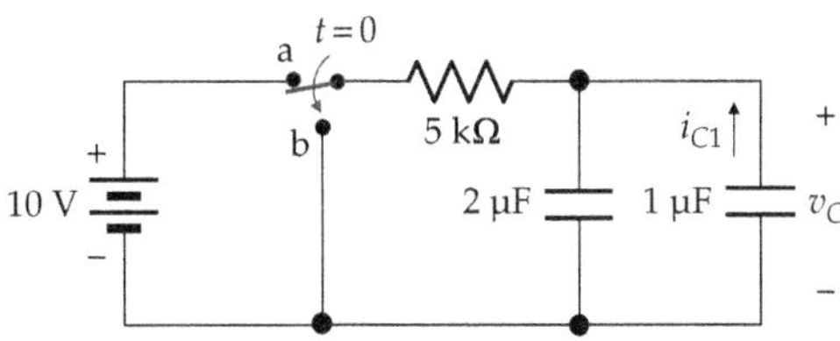

FIGURE P11.35

P11.35 The switch in Figure P11.35 has been in position 'a' for a long time. It is moved to position 'b' at $t = 0$. Determine (a) $v_C(0^+)$, (b) V_{CF}, (c) τ, (d) $v_C(t)$ for $t \geq 0^+$, and (e) $i_{c1}(t)$ for $t \geq 0^+$.

Ans. (a) 10 V; (b) 0; (c) 15 ms; (d) $v(t)=10e^{-t/15}$ V, t is in ms; (e) $i_{C1}(t)=\frac{2}{3}e^{-t/15}$ mA.

P11.36 The switch in Figure P11.36 is moved to position 'b' at $t = 0$ after being in position 'a' for a long time. Determine $i_X(t)$, $t \geq 0^+$.

Ans. $i_X(t)=\frac{1}{3}e^{-t/1.2}$ mA, t is in ms.

P11.37 The switch in Figure P11.37 is moved to position 'b' at $t = 0$ after having been in position 'a' for a long time. Determine $v_O(t)$ for $t \geq 0^+$.

Ans. $v_O(t)=-80e^{-5t}$ V, t is in s.

P11.38 Both switches in Figure P11.38 are opened at $t = 0$ after being closed for a long time. Determine $i_O(t)$ for $t \geq 0^+$.

Ans. $i_O(t)=10e^{-5t}$ mA, t is in ms.

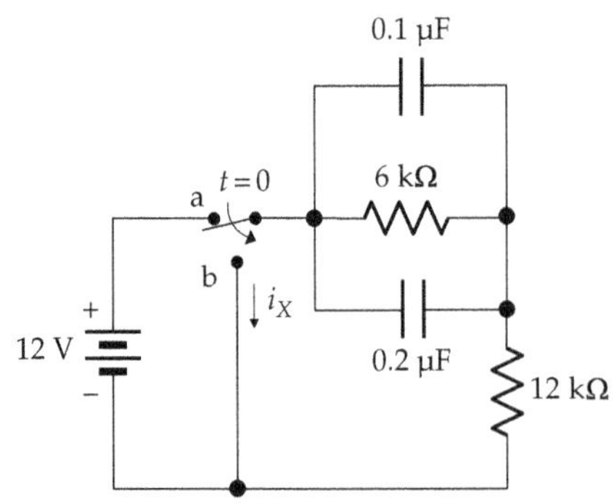

FIGURE P11.36

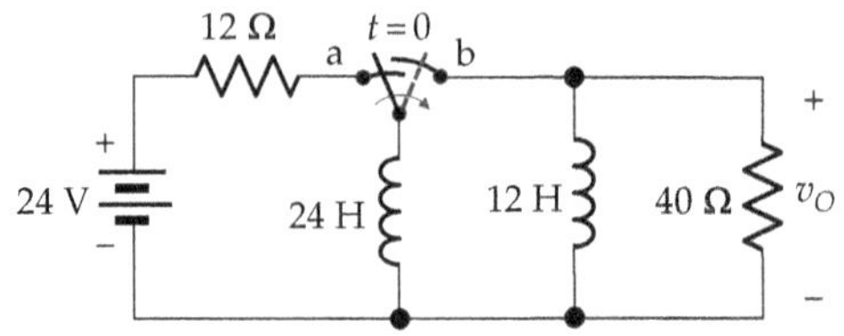

FIGURE P11.37

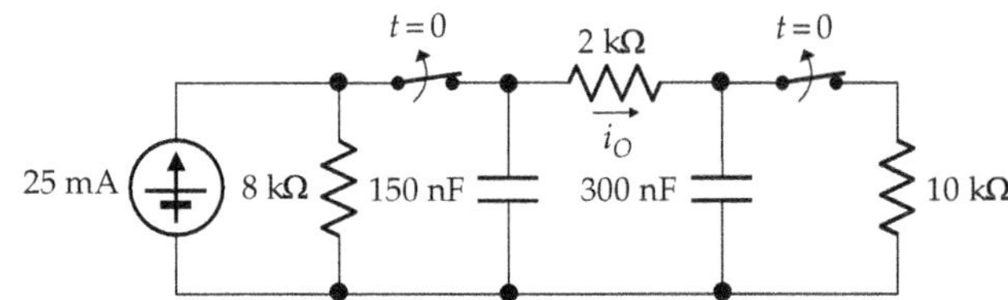

FIGURE P11.38

P11.39 The switch in Figure P11.39 is closed at $t = 0$ after being open for a long time. Determine $i_X(t)$, for $t \geq 0$.

Ans. $-\frac{1}{6}e^{-\frac{t}{0.72}}$ mA, t is in ms.

P11.40 The switch in Figure P11.40 is initially in position 'a', with the capacitors uncharged. At $t = 0$, the switch is moved to position 'b'. When $v_C = 10$ V, the switch is moved to position 'c'. Determine $v_C(t)$ for the time when the switch (a) is in position 'b' and (b) after it was moved to position 'c'.

Ans. (a) $v_C(t)=14.86\left(1-e^{-t/0.74}\right)$V, t is in ms; (b) $v_C(t)=0.94+9.06e^{(t-0.83)/4.76}$ V.

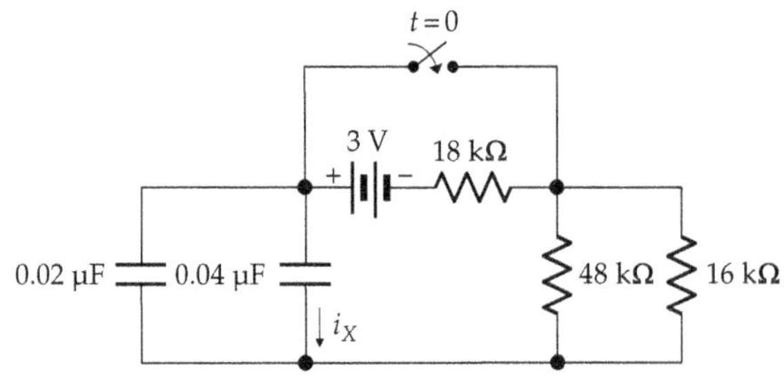

FIGURE P11.39

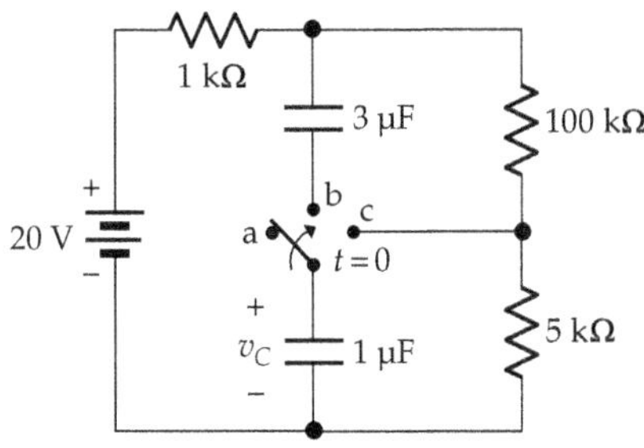

FIGURE P11.40

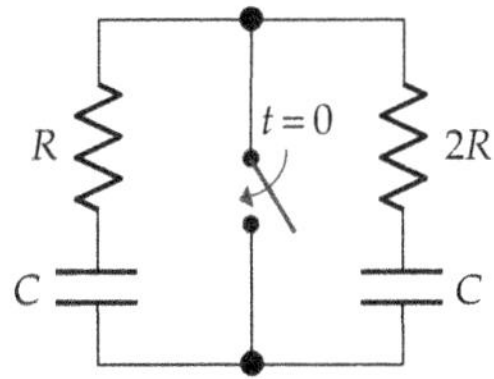

FIGURE P11.41

P11.41 $RC = 1$ ms in Figure P11.41. Before the switch is closed, the circuit is in a steady state, with a total energy storage of 1 J. If the switch is closed at $t = 0$, determine the total energy stored at $t = 1$ ms.

Ans. $0.5(e^{-2} + e^{-1}) = 0.25$ J.

P11.42 The switch in Figure P11.42 is moved at $t = 0$ from position 'a' to position 'b' after being in position 'a' for a long time. Determine, for $t \geq 0^+$, (a) $v(t)$ and (b) energy dissipated in the resistor.

Ans. (a) $v(t) = 20e^{-t/20}$ V, t is in μs; (b) 2 μJ.

P11.43 The switch in Figure P11.43 is closed at $t = 0$ after being open for a long time. Determine, for $t \geq 0^+$, (a) $v(t)$ and (b) the energy dissipated in the resistor.

Ans. (a) $v(t) = 6\left(1 - e^{-10t}\right)$ V, t is in s; (b) 0.18 J.

P11.44 The switch in Figure P11.44 is moved at $t = 0$ from position 'a' to position 'b' after being in position 'a' for a long time. Determine, for $t \geq 0^+$, (a) $v(t)$ and (b) the energy dissipated in the resistor.

Ans. (a) $v(t) = -10e^{-t/5}$ V, t is in ms; (b) 12.5 mJ.

P11.45 The switch in Figure P11.45 has been closed for a long time. It is opened at $t = 0$ and is closed at $t = 1$ ms. Determine $i_1(t)$ and $v_O(t)$, $t \geq 0^+$.

Ans. $0 \leq t \leq 1$ ms, $i_1(t) = 0.2e^{-3t}$ A, $v_O(t) = 30e^{-3t}$ V; $t \geq 1$ ms, $i_1(t) = 0.2 - 0.19e^{-2.5(t-1)}$ A, $v_O(t) = -23.75e^{-2.5(t-1)}$ V.

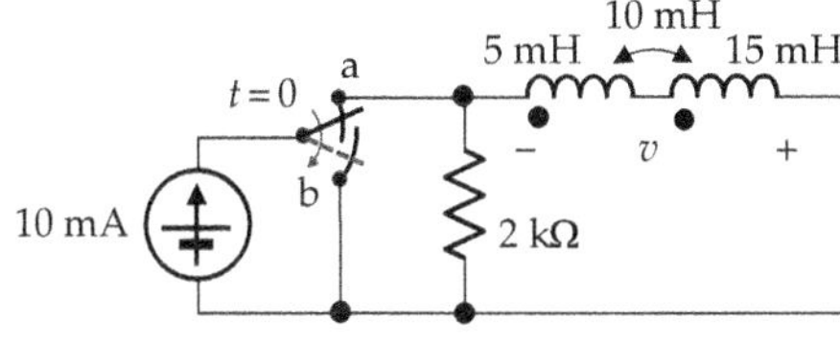

FIGURE P11.42

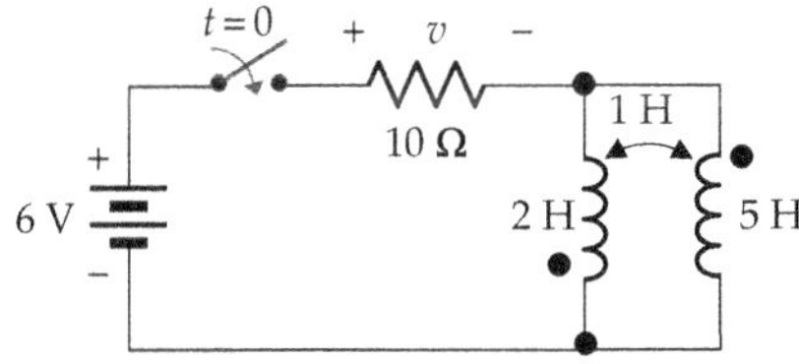

FIGURE P11.43

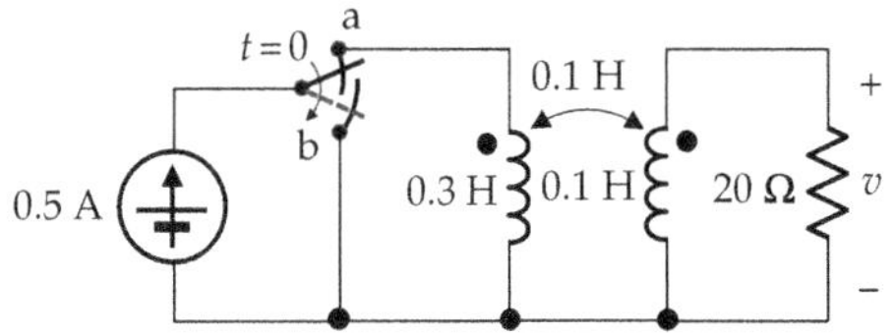

FIGURE P11.44

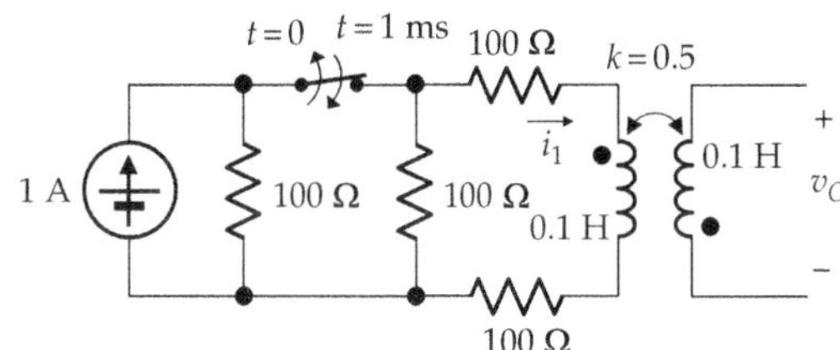

FIGURE P11.45

P11.46 The switch in Figure P11.46 is moved from position 'a' to position 'b' at $t = 0$ and back to position 'a' at $t = 1$ s. Determine $v(t)$, $t \geq 0^+$.

Ans. $0 \leq t \leq 1$ s, $v(t) = -4e^{-t/2}$ V; $t \geq 1$ s, $v(t) = 4\left(1 - e^{-0.5}\right) e^{-0.5(t-1)}$ V.

P11.47 The switch in Figure P11.47 is closed at $t = 0$ after being open for a long time. Determine $v(t)$, $t \geq 0^+$.

Ans. $v(t) = 10e^{-0.5t}$ V.

P11.48 The switch in Figure P11.48 is moved at $t = 0$ from position 'a' to position 'b', after being in position 'a' for a long time. Determine for $t \geq 0^+$ (a) $i_O(t)$ and (b) $v_O(t)$.

Ans. (a) $i_O(t) = -16 + 28e^{-5t}$ A; (b) $v_O(t) = -56e^{-5t}$ V, t is in s.

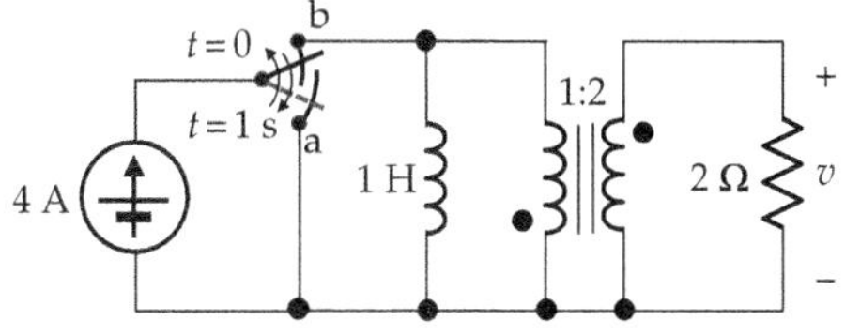

FIGURE P11.46

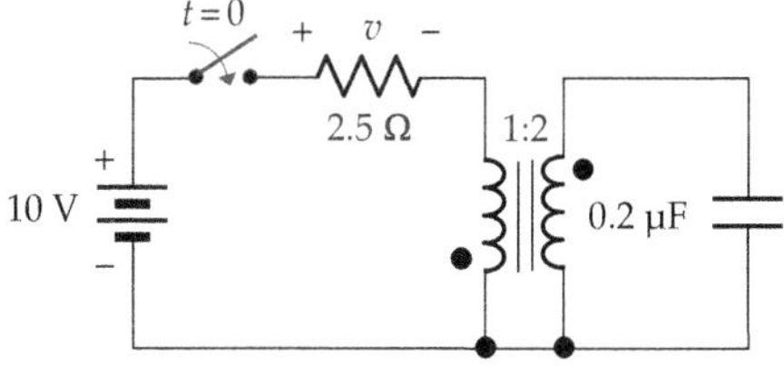

FIGURE P11.47

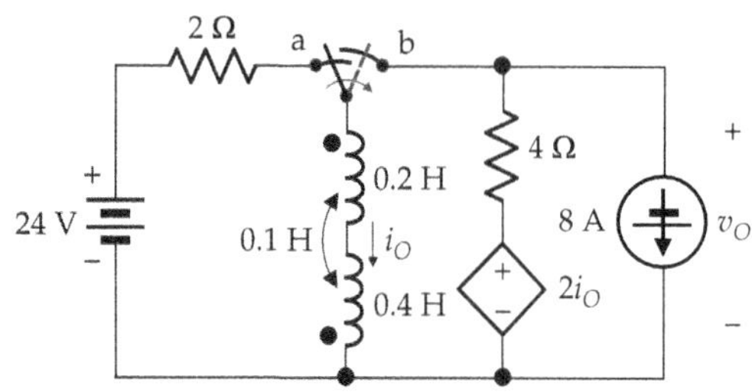

FIGURE P11.48

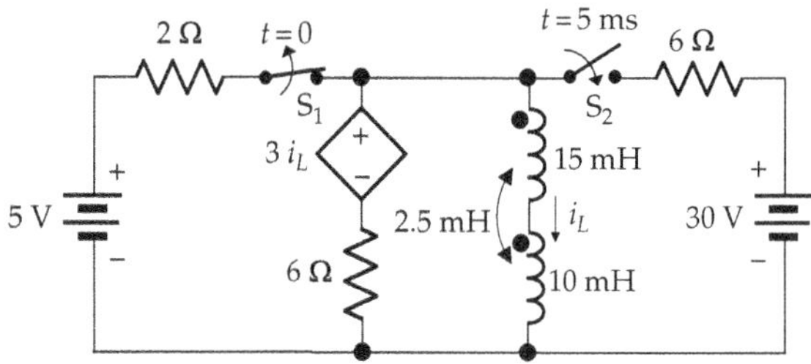

FIGURE P11.49

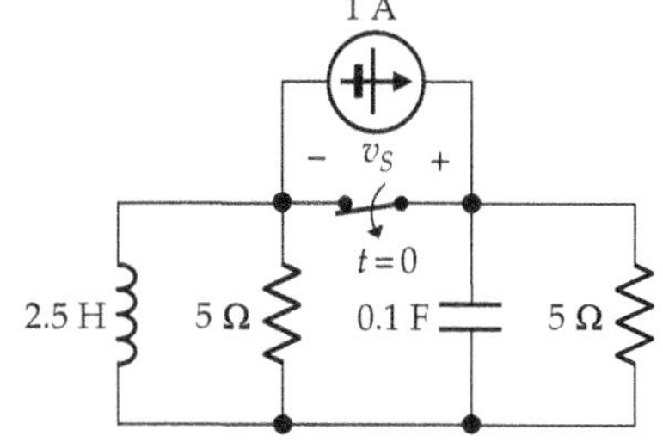

FIGURE P11.50

P11.49 Switch S_1 in Figure P11.49 is opened at $t = 0$ after being closed for a long time. Switch S_2 is closed at $t = 5$ ms. Determine $i_L(t)$ at $t = 20$ ms.

Ans. 6.69 A.

P11.50 The switch in Figure P11.50 is opened at $t = 0$ with no energy initially stored in the circuit. Determine $v_S(t)$, $t \geq 0^+$.

Ans. 5 V.

Design Problems

P11.51 A 1 μF capacitor in a cardiac defibrillator is charged to 2 kV during operation. When the defibrillator is switched off, the capacitor should discharge to 20 V within 1 s. What should be the effective resistance in parallel with the capacitor?

Ans. < 217 kΩ.

P11.52 Figure P11.52 illustrates the basic circuit of a timer. Timing starts when the switch is closed, with the capacitor discharged, and ends when the threshold device D is activated by a voltage of 5 V across the capacitor.

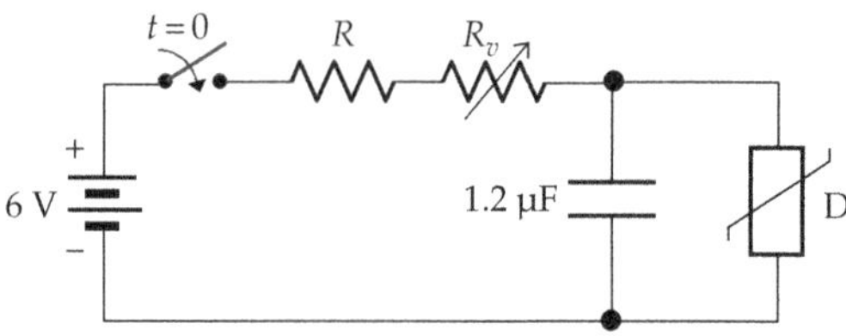

FIGURE P11.52

The timing interval is to be varied between 1 and 5 s by means of a variable resistor R_v in series with a fixed resistor R. Select suitable values of R and R_v.

Ans. $R = 465$ kΩ; $R_v = 1.86$ MΩ.

P11.53 It is required to design a neon lamp flasher to flash for 1 s every 5 s (P11.59). Assume $V_{SRC} = 100$ V, $V_{max} = 80$ V, $V_{min} = 10$ V, and $C = 1$ μF. Determine R_1 and R_2. (*Hint*: Determine R' from T_{on} by successive approximation using a spreadsheet program.)

Ans. $R_1 = 2.659$ MΩ; $R_2 = 261.1$ kΩ.

P11.54 An RC circuit is to be used to integrate a 6 V step (Problem P11.60). Determine RC if the error at $T = 10$ s is not to exceed 5% and derive the value of the integrated voltage.

Ans. 96.71 s, 0.589 V.

P11.55 The circuit of P11.61 is to be used to differentiate a pulse 100 μs wide. Determine the maximum RC that will give at the end of the pulse an output that does not exceed 1% of the pulse height.

Ans. 21.71 μs.

P11.56 An electronic voltage measuring instrument has an input impedance of 5 MΩ in parallel with 90 pF and a maximum input voltage of 100 V (Figure P11.56). Specify R and C for an attenuator probe that allows measurement of voltages up to 1000 V without phase shift.

Ans. 45 MΩ, 10 pF.

Probing Further

P11.57 Show that the energy dissipated in R_C in Figure 11.1b is $w_R(t) = \frac{1}{2}CV_0^2\left(1 - e^{-2t/\tau_C}\right)$, $t \geq 0^+$, which indicates that as $t \to \infty$, the energy initially stored in the capacitor is dissipated in the resistor. Show that analogous considerations apply in the case of inductor discharge in Figure 11.13b.

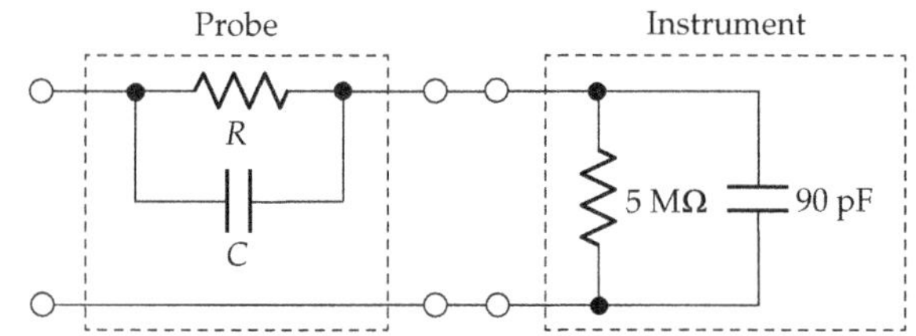

FIGURE P11.56

P11.58 Note that in capacitor charging (Figure 11.7b), the energy delivered by the battery for $t \geq 0^+$ is $\int_{0^+}^{t} V_{SRC} i dt = V_{SRC} q$, where q is the charge delivered by the battery and stored in the capacitor. Deduce that as $t \to \infty$, the energy dissipated in the resistor is equal to that stored in the capacitor. Show that analogous considerations apply in the case of inductor charging in Figure 11.17b.

P11.59 Figure P11.59 illustrates a neon lamp flasher. The neon lamp is assumed to turn on when $v = V_{max}$ and to turn off when v falls to V_{min}. The lamp acts as an open circuit when turned off and has a negligible resistance when turned on. Show that $T_{off} = R_1 C \ln \frac{V_{SRC} - V_{min}}{V_{SRC} - V_{max}}$ and $T_{on} = R'C \ln \frac{V_{max} - V'_{SRC}}{V_{min} - V'_{SRC}}$, where T_{on} is the flash duration, T_{off} is the duration of the off interval, $R' = (R_1 \| R_2)$, and $V'_{SRC} = V_{SRC} R_2 / (R_1 + R_2)$.

P11.60 The switch in Figure P11.60 is closed at $t = 0$, with the capacitor initially uncharged. For $t \leq 0^-$, $V_S = 0$ and $v_C = 0$, whereas for $t \geq 0^+$, $V_S = V_{SRC}$. If $v_C << V_S$, $i \cong V_S/R$ and $v_C \cong \frac{1}{C}\int_0^t i dt \cong \frac{V_S}{RC} t = V_S t'$, where t' is a normalized, dimensionless time variable. It is seen that the voltage across the capacitor is proportional to the time integral of the applied voltage step V_S so that the circuit acts as an integrator at low output voltages. The expression for $v_C(t)$ is $v_C = V_S\left(1 - e^{-t/\tau}\right)$. Expand $\left(1 - e^{-t/\tau}\right)$ to show that for an integration interval T, the deviation from the true value T/τ of the integral of v_C/V_S is $\left[-\frac{1}{2}\left(\frac{T}{\tau}\right)^2 + \frac{1}{3!}\left(\frac{T}{\tau}\right)^3 - \frac{1}{4!}\left(\frac{T}{\tau}\right)^4 + \cdots\right]$, and the percentage error is this deviation divided by T/τ and multiplied by 100. For a given T, the error decreases with increasing τ, but for a given V_S, the integral $v_C = V_S T/\tau$ decreases as τ is increased. The magnitude of v_C could be increased using an operation amplifier (Chapter 13).

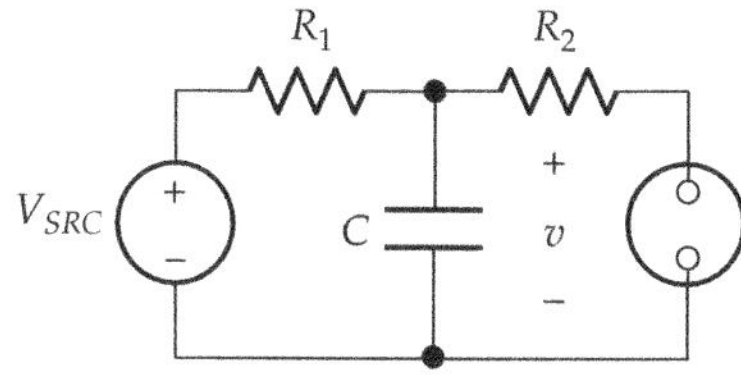

FIGURE P11.59

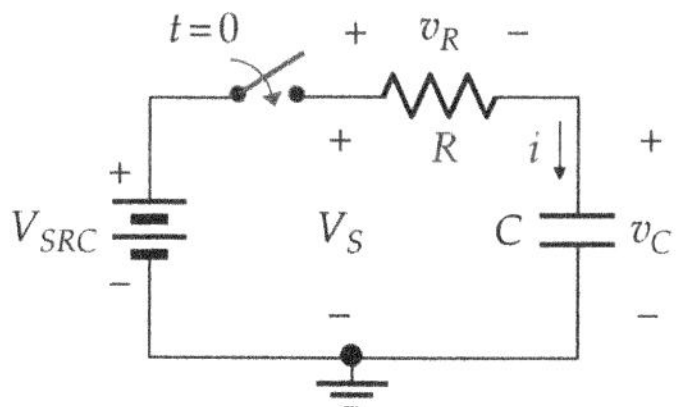

FIGURE P11.60

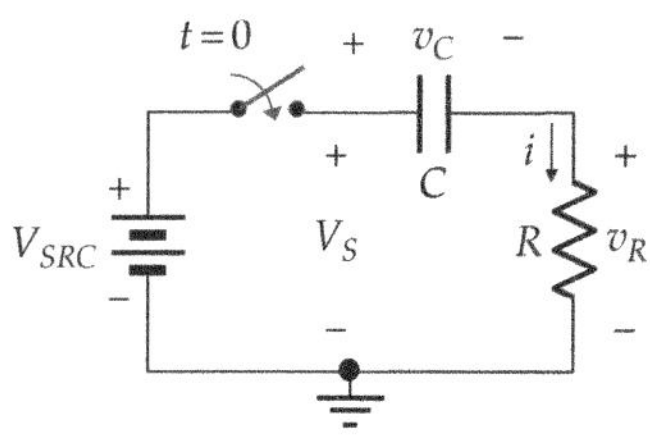

FIGURE P11.61

P11.61 The switch in Figure P11.61 is closed at $t = 0$, with the capacitor initially uncharged. For $t \leq 0^-$, $V_S = 0$ and $v_C = 0$, whereas for $t \geq 0^+$, $V_S = V_{SRC}$. If $v_R << V_S$, $i \cong CdV_S/dt$ and $v_R = RCdV_S/dt$. In other words, the voltage across the resistor is proportional to the time derivative of the applied voltage step V_S so that the circuit acts as a differentiator at low output voltages. For an ideal differentiator, $v_R = 0$ at $t = 0^+$. In the circuit in Figure P11.61, $v_R \cong 0$ at $t \cong 5\tau$. Hence, to approximate a good differentiator, τ should be small, which means the output is small.

P11.62 The coil of an electromagnetically operated switch or relay is sometimes controlled by an electronic transistor switch, as illustrated in Figure P11.62a, where the circuit connected to the relay coil is represented by its TEC. (a) If the relay coil has an inductance 0.5 H and the transistor switch interrupts a relay current of 1 A in 1 ms, determine the magnitude and polarity of the voltage that appears across the relay coil and the transistor as it interrupts the current, neglecting the voltage drop in the resistances. (b) To protect the transistor from this voltage, a diode (Figure P11.62b) is connected across terminals 'ab' of the coil. The diode presents an open circuit when terminal 'a' is positive with respect to terminal 'b', during normal operation, but when the voltage polarity reverses, the diode conducts, virtually short-circuiting the coil terminals 'ab'. The main limitation is a small delay in de-energizing the relay. Determine this delay if the relay de-energizes when the current drops to 0.2 A, assuming a relay coil resistance of 10 Ω.

Ans. (a) 500 V across the coil, with terminal 'b' positive with respect to terminal 'a', (500 + V_{Th}) across the transistor; (b) 80.5 ms.

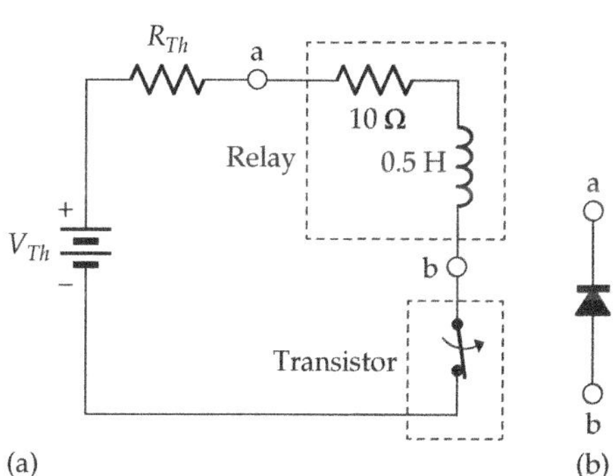

FIGURE P11.62

12

Basic Responses of Second-Order Circuits

Objective and Overview

This chapter extends the discussion of Chapter 11 to prototypical, second-order circuits consisting of a resistor, capacitor, and inductor, connected in series or in parallel. The main objective is to present the three basic types of responses that are possible with second-order circuits, namely, overdamped, critically damped, and underdamped responses, and to explore the main characteristic features of these responses.

The natural responses of the prototypical series and parallel circuits are discussed in the first part of the chapter. These natural responses are due to initial energy storage in capacitors or inductors, in the absence of any other excitation. The second part of the chapter is concerned with the responses of the prototypical circuits to applied, steady excitations. Appendix 12A outlines the procedure for deriving the natural responses of more general second-order circuits and the responses of these circuits to applied dc excitations. Responses of second-order circuits to more general forms of excitation are discussed in Part II of the book.

12.1 Natural Responses of Series *RLC* Circuit

A second-order LTI circuit is so called because its responses obey a second-order, linear, ordinary differential equation with constant coefficients. Second-order circuits discussed in this chapter are reducible to a combination of a resistor, a capacitor, and an inductor, connected in series or in parallel.

Figure 12.1a illustrates an *RLC* circuit in which the switch is opened at $t = 0$. It is assumed that the switch has been closed for a sufficiently long time so that steady-state conditions prevail before the switch is opened. Under these steady-state conditions, the inductor acts as a short circuit, so that the source current I_{SRC} flows through the inductor and the voltage across the capacitor is zero. With the switch open, the circuit becomes a series *RLC* circuit, as in Figure 12.1b. The source current is diverted through R_D, so that the ideal current source is not left open-circuited. It is required to determine how the current i in the circuit

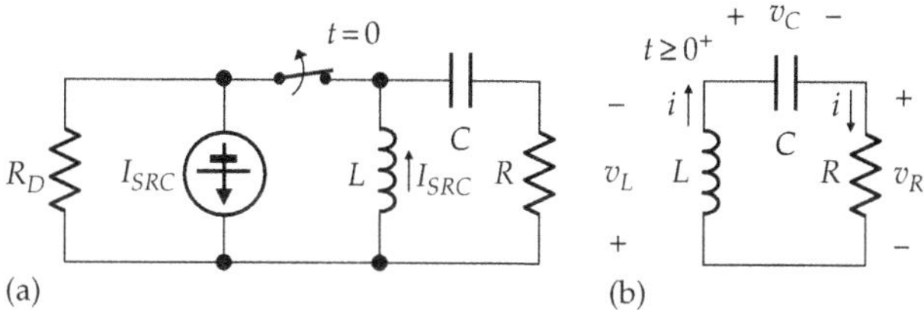

FIGURE 12.1
(a) Inductor is charged when the switch has been in the closed position for a long time and (b) series *RLC* circuit having an initially charged inductor.

of Figure 12.1b and all the voltages in the circuit vary with time for $t \geq 0^+$.

KCL is automatically satisfied in the series circuit. From KVL, $v_R + v_C + v_L = 0$. Substituting the v–i relations for the circuit elements,

$$Ri + \frac{1}{C}\int i dt + L\frac{di}{dt} = 0, \quad t \geq 0^+ \tag{12.1}$$

Differentiating, dividing by L, and rearranging,

$$\frac{d^2i}{dt^2} + \frac{R}{L}\frac{di}{dt} + \frac{i}{LC} = 0, \quad t \geq 0^+ \tag{12.2}$$

Equation 12.2 is a homogeneous differential equation in i. It is of the general form

$$\frac{d^2y}{dt^2} + 2\alpha\frac{dy}{dt} + \omega_0^2 y = 0, \quad t \geq 0^+ \tag{12.3}$$

where y is a voltage or current variable and α and ω_0^2 are given by

$$\alpha = \frac{R}{2L} \quad \text{and} \quad \omega_0 = \frac{1}{\sqrt{LC}} \tag{12.4}$$

Note that α and ω_0 depend on the circuit parameters R, L, and C only. Moreover, since the first term, d^2i/dt^2, in Equation 12.3. has the units of y divided by t^2, the other two terms must have the same units. This means that α and ω_0 are in radians/s. For reasons that will become clear later, α is referred to as the **damping factor** and ω_0 as the **resonant frequency**.

As mentioned in connection with Equation 11.3, the general solution of a homogeneous, linear, ordinary, differential equation of any order, but having constant coefficients, is the sum of exponentials of the form Ae^{st}, where s and A are constants for a given circuit and initial conditions, and the number of exponential terms in the sum is equal to the order of the equation. Substituting in Equation 12.3 and collecting terms,

$$Ae^{st}\left(s^2 + 2\alpha s + \omega_0^2\right) = 0 \tag{12.5}$$

In order to satisfy Equation 12.5 for all t, either $A = 0$ or the bracketed terms must sum to zero. If $A = 0$, all the responses in the circuit will be zero. The solution is trivial and contradicts the assumption of nonzero initial energy storage. Hence,

$$s^2 + 2\alpha s + \omega_0^2 = 0 \tag{12.6}$$

Equation 12.6 is the characteristic equation of the differential Equation 12.3, similar to Equation 11.4, for a first-order homogeneous differential equation. It is of fundamental importance in the analysis of circuits and systems, because it governs the basic natural response of the circuit or system. The roots s_1 and s_2 of Equation 12.6 are

$$s_1 = -\alpha + \sqrt{\alpha^2 - \omega_0^2} \quad \text{and} \quad s_2 = -\alpha - \sqrt{\alpha^2 - \omega_0^2} \tag{12.7}$$

where s_1 and s_2 depend on R, L, and C only, since α and ω_0 depend only on R, L, and C (Equation 12.4). s_1 and s_2 are related by the following equations:

$$s_1 + s_2 = -2\alpha = -\frac{R}{L}, \quad s_1 - s_2 = 2\sqrt{\alpha^2 - \omega_0^2}, \quad \text{and} \quad s_1 s_2 = \omega_0^2 \tag{12.8}$$

The general solution of the homogeneous differential equation is then of the form

$$y(t) = Ae^{s_1 t} + Be^{s_2 t}, \quad t \geq 0^+ \tag{12.9}$$

where A and B are arbitrary constants that depend, in general, on the initial conditions as well as s_1 and s_2. Note that since Equation 12.3 is second order and formally involves two integrations, the general solution should have two arbitrary constants, as in Equation 12.9. Fundamentally, the initial conditions are the values of the energy-related variables at $t = 0^+$, that is, inductor current and capacitor voltage, which do not change between $t = 0^-$ and $t = 0^+$. Mathematically, and for convenience, the initial conditions applied are those of the variable under consideration and its first derivative at $t = 0^+$.

It is evident from Equation 12.7 that there are three cases to consider, corresponding to $\alpha > \omega_0$, $\alpha < \omega_0$, and $\alpha = \omega_0$. These three conditions result in three different types of natural responses of second-order circuits, as will be discussed in the following subsections. Before doing so, however, we will make some generalizations based on the discussion in the preceding chapter on first-order circuits:

1. As $t \to \infty$, all responses in Figure 12.1 will go to zero, because stored energy will eventually be dissipated in the resistor, which means that all steady-state responses will be zero. The solution to Equation 12.3 is therefore a purely transient response.
2. The second-order homogeneous differential equation is completely specified by α and ω_0 and hence by the exponents s_1 and s_2. Linear operations in LTI circuits do not change the exponents s_1 and s_2. Hence, as discussed in Section 11.5, the following concept applies:

Concept: *All the circuit variables in a given second-order circuit obey the same homogeneous differential equation. The same characteristic equation applies to all the circuit variables, which means that all these variables have the same s_1, s_2, α, and ω_0 and, hence, the same form of the natural response.*

It follows that the natural responses in Figure 12.1 have the general form

$$i(t) = A_i e^{s_1 t} + B_i e^{s_2 t}, \quad t \geq 0^+ \tag{12.10}$$

$$v_C(t) = A_C e^{s_1 t} + B_C e^{s_2 t}, \quad t \geq 0^+ \tag{12.11}$$

$$v_L(t) = A_L e^{s_1 t} + B_L e^{s_2 t}, \quad t \geq 0^+ \tag{12.12}$$

where the exponents are the same but the arbitrary constants are, in general, different, and $s_1 \neq s_2$. All the responses vanish as $t \to \infty$, because s_1 and s_2 are either negative real or have negative real parts (Equation 12.7). $v_R = Ri$ is directly proportional to i.

Since the same form of the general solution applies to all the circuit variables in a given circuit, and in cases where the responses to more than one variable are required, it will be demonstrated that in a series circuit, it is convenient to derive first the general solution for v_C, the voltage across the capacitor. This has two advantages: (1) the general responses for i, or v_R, and v_L, are obtained by successive differentiation. This is more convenient than having to derive some responses through integration by parts, as can happen if a variable other than v_C is considered first. (2) It is easy to verify the voltage responses by checking that they satisfy KVL.

Exercise 12.1

Verify, by applying the relations between v_C and i, and between v_L and i, in Equation 12.1, that v_C and v_L obey the same homogeneous differential equation in i (Equation 12.2).

12.1.1 Overdamped Responses

The natural responses will be illustrated by considering a numerical example having $R = 500\ \Omega$, $C = 12.5\ \mu F$, $L = 0.5$ H, and $I_0 = 3$ A. Hence, $\alpha = R/2L = 500$ rad/s and $\omega_0 = 1/\sqrt{LC} = 400$ rad/s. With $\alpha > \omega_0$. It follows that $s_1 = -500 + \sqrt{(500)^2 - (400)^2} = -200$ rad/s and $s_2 = -500 - \sqrt{(500)^2 - (400)^2} = -800$ rad/s. Considering v_C, as explained earlier, and substituting in Equation 12.11

$$v_C(t) = Ae^{-200t} + Be^{-800t}, \quad t \geq 0^+ \qquad (12.13)$$

where the subscripts have been dropped from the arbitrary constants A and B for convenience. Note that, according to Equation 12.13, A and B have the same units as $v_C(t)$. A and B are determined from the initial conditions. In the circuit of Figure 12.1, the stored energy just before the switch is opened at $t = 0^-$ is zero in the capacitor and $(1/2)LI_0^2$ in the inductor. It follows from the constancy of stored energy in the capacitor at the instant of switching that $v_C(0^+) = 0$. Substituting in Equation 12.13,

$$A + B = 0 \qquad (12.14)$$

or $A = -B$. To apply the second initial condition, Equation 12.13 is differentiated to give

$$\frac{dv_C(t)}{dt} = -200Ae^{-200t} - 800Be^{-800t}, \quad t \geq 0^+ \qquad (12.15)$$

$i(0^-) = I_0 = 3$ A and does change at $t = 0^+$ because the stored energy in the inductor does not change at the instant of switching. It follows that $(dv_C/dt)_{0+} = i(0^+)/C = 3/C = 3/(12.5 \times 10^{-6}) = 24 \times 10^4$ V/s. With $A = -B$, Equation 12.15 gives $600A = 24 \times 10^4$, or $A = 400$ V $= -B$. Substituting in Equation 12.13,

$$v_C(t) = 400\left(e^{-200t} - e^{-800t}\right)\text{V}, \quad t \geq 0^+ \qquad (12.16)$$

Once v_C is obtained, all the other responses readily follow. Thus,

$$i(t) = C\frac{dv_C}{dt} = 12.5 \times 10^{-6} \times 400\left(-200e^{-200t} + 800e^{-800t}\right)$$

or

$$i(t) = -e^{-200t} + 4e^{-800t}\ \text{A}, \quad t \geq 0^+ \qquad (12.17)$$

$$v_R(t) = Ri = -500e^{-200t} + 2000e^{-800t}\ \text{V}, \quad t \geq 0^+ \qquad (12.18)$$

$$v_L(t) = L\frac{di}{dt} = 0.5\left(200e^{-200t} - 3200e^{-800t}\right)$$

or

$$v_L(t) = 100e^{-200t} - 1600e^{-800t}\ \text{V}, \quad t \geq 0^+ \qquad (12.19)$$

It is important to note the following checks on the expressions for the responses: (1) $v_C(0^+) = 0$, $i(0^+) = 3$ A, $v_R(0^+) = 3 \times 500 = 1500$ V, and $v_L(0^+) = -1500$ V, in accordance with KVL at $t = 0^+$, as illustrated in Figure 12.2; (2) as $t \to \infty$, $i(\infty) = v_C(\infty) = v_L(\infty) = 0$, since all responses become zero when the energy initially stored in the inductor is dissipated in the resistor; and (3) $v_R + v_C + v_L = 0$ for all t, as required by KVL.

Note that constancy of stored energy in C at the instant of switching requires that $v_C(0^+) = 0$, but the capacitor current can jump instantaneously from zero at $t = 0^-$ to $I_0 = 3$ A at $t = 0^+$ in order to satisfy KCL. Similarly, constancy of stored energy in the inductor at the instant of switching requires that the $i(0^+) = I_0$, but the inductor voltage can jump from 0 to $-RI_0$ in order to satisfy KVL.

Figure 12.3 shows the variation with time, for $t \geq 0^+$, of i, v_C, and more conveniently, $v_L' = -v_L$, rather than v_L. A positive i charges the capacitor, which stores electric energy. Both the energy stored in the capacitor and that dissipated in the resistor can only come from the energy initially stored in the inductor. This means that i must decrease with time and eventually go to zero. However, the decrease in i is not monotonic, as in a first-order circuit. The capacitor continues to charge while i is positive but decreasing. When $i = 0$, $v_C = v_L'$, and v_C is at a maximum, since $dv_C/dt = i/C = 0$ at this instant, as indicated by the dashed line on the left in Figure 12.3. Beyond this time, the capacitor discharges and drives i in the negative direction. i then reaches a minimum before decaying to zero. Note that since $v_L' = -Ldi/dt$, then $v_L' > 0$ when i is decreasing, $v_L' < 0$ when i is increasing, and $v_L' = 0$

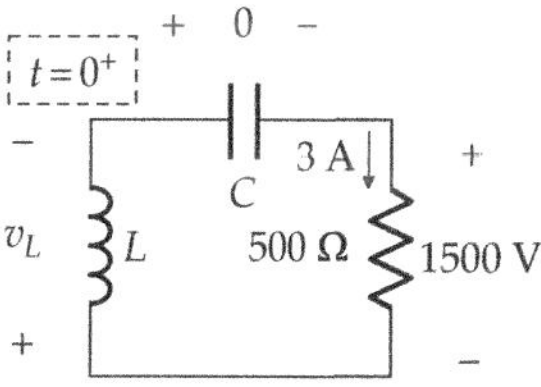

FIGURE 12.2
KVL at $t = 0^+$ in the circuit of Figure 12.1.

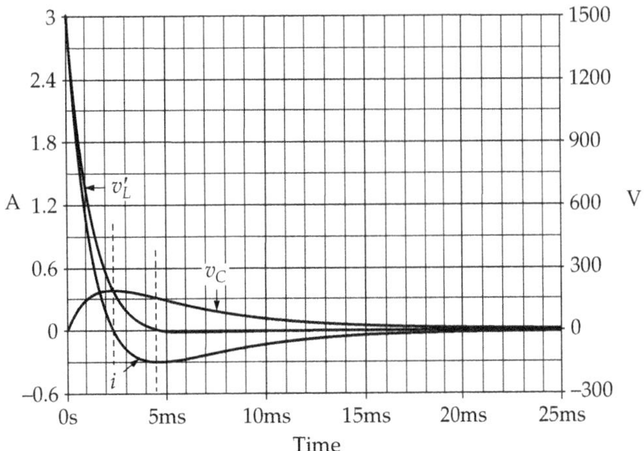

FIGURE 12.3
Overdamped, natural responses of series *RLC* circuit.

when i is minimum. Eventually, all the energy initially stored in the inductor is dissipated in the resistor, and all the circuit responses decay to zero. The responses of Figure 12.3 are described as **overdamped**.

What is the effect of increasing R on the overdamped response? Suppose that R is increased from 500 to 1500 Ω. This gives $s_1 = -1500 + \sqrt{(1500)^2 - (400)^2} = -54.3$ rad/s and $s_2 = -1500 - \sqrt{(1500)^2 - (400)^2} = 2946$ rad/s. It is seen that as R increases, the magnitude of s_1 decreases, whereas the magnitude of s_2 increases, so that e^{s_1t} decreases at a slower rate, whereas e^{s_2t} decreases at a faster rate. The overall effect can be argued from $v_L(0^+)$. With $I_0 = 3$ A, as before, and $R = 1500$ Ω, $v_L(0^+) = -4500$ V/s instead −1500 V/s. But $v_L(0^+) = (L)(di/dt)_{0^+}$. Thus, as R increases, $v_L(0^+)$ becomes more negative, and i decreases at a faster rate. This is illustrated in Figure 12.4 for $R = 1500$ Ω and $R = 500$ Ω. The latter is the same as in Figure 12.3 but on an expanded time scale.

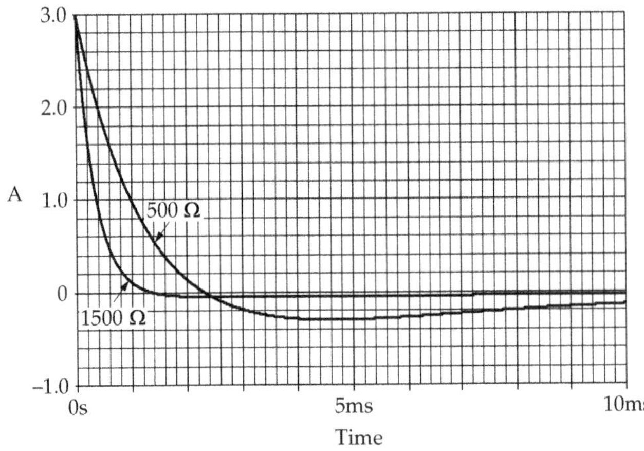

FIGURE 12.4
Effect of resistance on the overdamped current response in a series *RLC* circuit.

Primal Exercise 12.2

Determine (a) α, ω_0, s_1, and s_2 for the series *RLC* circuit in Figure 12.1b with $R = 2.6$ kΩ, $L = 0.1$ mH, and $C = 0.4$ nF, and (b) $i(t)$, $t \geq 0^+$, if $i(0^+) = 24$ mA and $v_C(0^+) = 0$.

Ans. (a) $\alpha = 13$ Mrad/s, $\omega_0 = 5$ Mrad/s, $s_1 = -1$ Mrad/s, $s_2 = -25$ Mrad/s. (b) $i(t) = -e^{-t} + 25e^{-25t}$ mA, t is in μs.

12.1.2 Underdamped Responses

Consider next the case when R is reduced to 100 Ω. This makes $\alpha = 100$ rad/s, which is less than $\omega_0 = 400$ rad/s. Under these conditions, s_1 and s_2, as given by Equation 12.7, become complex conjugates and can be expressed as

$$s_1 = -\alpha + j\omega_d \quad \text{and} \quad s_2 = -\alpha - j\omega_d \tag{12.20}$$

where

$$\omega_d = \sqrt{\omega_0^2 - \alpha^2} \tag{12.21}$$

and is referred to as the **damped natural frequency**. Substituting the numerical values, $\omega_d = \sqrt{(400)^2 - (100)^2} = 100\sqrt{15} = 387.3$ rad/s. Equation 12.11 becomes

$$v_C(t) = A_C e^{-100(1-j\sqrt{15})t} + B_C e^{-100(1+j\sqrt{15})t}, \quad t \geq 0^+ \tag{12.22}$$

where A_C and B_C are in general complex constants. If $e^{j100\sqrt{15}t} = \cos 100\sqrt{15}t + j\sin 100\sqrt{15}t$ and $e^{-j100\sqrt{15}t} = \cos 100\sqrt{15}t - j\sin 100\sqrt{15}t$ are substituted in Equation 12.22, the general solution of Equation 12.22 can be expressed as

$$v_C(t) = e^{-100t}\left(A\cos 100\sqrt{15}t + B\sin 100\sqrt{15}t\right), \quad t \geq 0^+ \tag{12.23}$$

where $A = (A_C + B_C)$ and $B = j(A_C - B_C)$ are new arbitrary constants. Evidently, A and B are real, since $v_C(t)$ must be a real function of time. The general solution of Equation 12.23 is more convenient to work with than Equation 12.22 because it does not involve any complex quantities. The initial condition that $v_C(0^+) = 0$ gives $A = 0$, and Equation 12.23 becomes

$$v_C(t) = Be^{-100t}\sin 100\sqrt{15}t, \quad t \geq 0^+ \tag{12.24}$$

To apply the second initial condition, Equation 12.24 is differentiated to give

$$\frac{dv_C}{dt} = B\left(-100e^{-100t}\sin 100\sqrt{15}t + 100\sqrt{15}e^{-100t}\cos 100\sqrt{15}t\right),\quad t \geq 0^+ \qquad (12.25)$$

Substituting $(dv_C/dt)_{0+} = 24\times10^4$, as was done in connection with Equation 12.15, gives $B = 2400/\sqrt{15}$ V. Hence,

$$v_C(t) = \frac{2400}{\sqrt{15}}e^{-100t}\sin 100\sqrt{15}t \text{ V},\quad t \geq 0^+ \qquad (12.26)$$

It follows from Equation 12.26 that

$$i(t) = C\frac{dv_C}{dt} = 12.5\times10^{-6}\times \frac{2400}{\sqrt{15}}\left(-100e^{-100t}\sin 100\sqrt{15}t + 100\sqrt{15}e^{-100t}\cos 100\sqrt{15}t\right)$$

or

$$i(t) = 3e^{-100t}\left(\cos 100\sqrt{15}t - \frac{1}{\sqrt{15}}\sin 100\sqrt{15}t\right) \text{ A},\quad t \geq 0^+ \qquad (12.27)$$

$$v_R(t) = 300e^{-100t}\left(\cos 100\sqrt{15}t - \frac{1}{\sqrt{15}}\sin 100\sqrt{15}t\right) \text{ V},\quad t \geq 0^+ \qquad (12.28)$$

$$v_L(t) = L\frac{di}{dt} = 1.5\left[-100e^{-100t}\left(\cos 100\sqrt{15}t - \frac{1}{\sqrt{15}}\sin 100\sqrt{15}t\right) + e^{-100t}\left(-100\sqrt{15}\sin 100\sqrt{15}t - 100\cos 100\sqrt{15}t\right)\right]$$

or

$$v_L(t) = -150e^{-100t}\left[2\cos 100\sqrt{15}t + \frac{14}{\sqrt{15}}\sin 100\sqrt{15}t\right] \text{ V},\quad t \geq 0^+ \qquad (12.29)$$

The same checks apply as for Equations 12.16 through 12.19.

i, $v_L' = -v_L$ and v_C are plotted in Figure 12.5. The responses are damped sinusoids, that is, sinusoidal functions whose amplitudes decrement with time because of the exponential term $e^{-\alpha t}$ that multiplies the sinusoids. That is why α is termed the damping factor. The responses of Figure 12.5 are described as **underdamped**.

When α is large, the amplitudes attenuate rapidly, conversely when α is small. This is illustrated in Figure 12.6 for i and two values of R. i decrements more slowly when $R = 20\ \Omega$ and $\alpha = 20$ rad/s than when R and α have larger values.

Because of the marked oscillations, there is an appreciable alternation of stored energy between the inductor and the capacitor. Thus, when $i = 0$, energy is stored only in the capacitor, but when $v_C = 0$, energy is stored only in the inductor.

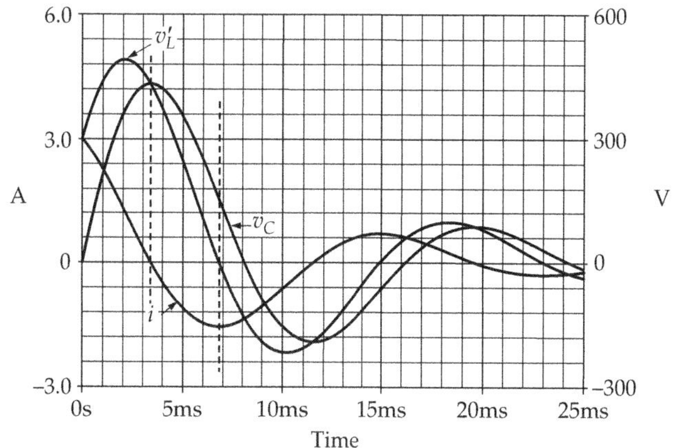

FIGURE 12.5
Underdamped, natural responses of series RLC circuit.

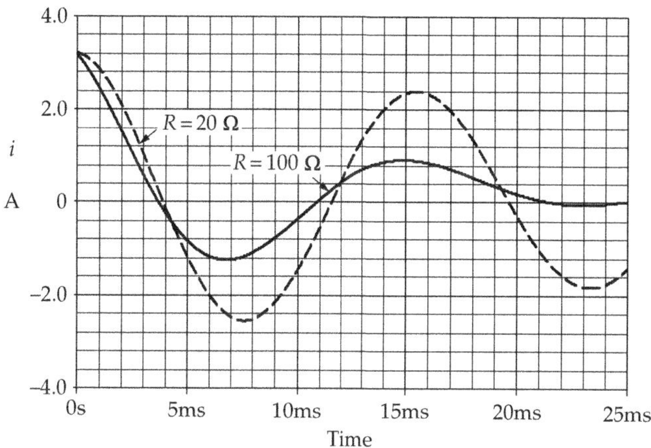

FIGURE 12.6
Effect of resistance on the underdamped current response in a series RLC circuit.

Primal Exercise 12.3

Determine (a) α, ω_0, and ω_d for the series RLC circuit of Figure 12.1b, with $R = 0.6$ kΩ, $L = 0.1$ mH, and $C = 0.4$ nF, and (b) $i(t)$, $t \geq 0^+$, if $i(0^+) = 24$ mA and $v_C(0^+) = 0$.

Ans. (a) $\alpha = 3$ Mrad/s, $\omega_0 = 5$ Mrad/s, $\omega_d = 4$ Mrad/s; (b) $i(t) = 6e^{-3t}(4\cos 4t - 3\sin 4t)$ mA, t is in μs.

12.1.3 Critically Damped Responses

The third possibility, when $\alpha = \omega_0$, makes $s_1 = s_2 = -\omega_0 = -400$ rad/s. The value of R is then $R = 2\omega_0 L = 400\ \Omega$. The responses in this case are described as **critically damped**.

Equation 12.11 becomes $v_C(t) = (A_C + B_C)e^{-\omega_0 t}$, having only one arbitrary constant ($A_C + B_C$), which means that it is not the general solution of the homogeneous differential equation. The general solution in this case, with only one exponential $e^{-\omega_0 t}$, is

$$v_C(t) = Ae^{-\omega_0 t} + Bte^{-\omega_0 t}, \quad t \geq 0^+ \qquad (12.30)$$

where the subscripts have been dropped from A and B for convenience. That Equation 12.30 satisfies Equation 12.10 can be readily verified by substitution. The initial condition that $v_C(0^+) = 0$ makes $A = 0$.

To apply the second initial condition, Equation 12.30 is differentiated to give, with $A = 0$,

$$\frac{dv_C}{dt} = Be^{-\omega_0 t} - B\omega_0 te^{-\omega_0 t}, \quad t \geq 0^+ \qquad (12.31)$$

Substituting $(dv_C/dt)_{0+} = 24 \times 10^4$ V/s, as was done in connection with Equation 12.15, gives $B = 24 \times 10^4$ V/s. Hence,

$$v_C(t) = 24 \times 10^4 te^{-400t}\ \text{V}, \quad t \geq 0^+ \qquad (12.32)$$

It follows from Equation 12.32 that

$$i(t) = C\frac{dv_C}{dt} = 12.5 \times 10^{-6} \times 24 \times 10^4 \left(e^{-400t} - 400te^{-400t}\right)$$

or

$$i(t) = 3e^{-400t}(1 - 400t)\ \text{A}, \quad t \geq 0^+ \qquad (12.33)$$

$$v_R(t) = 1200e^{-400t}(1 - 400t)\ \text{V}, \quad t \geq 0^+ \qquad (12.34)$$

$$v_L(t) = L\frac{di}{dt} = 1.5\left[-400e^{-400t}(1 - 400t) - 400e^{-400t}\right]$$

or

$$v_L(t) = L\frac{di}{dt} = -600e^{-400t}[2 - 400t], \quad t \geq 0^+ \qquad (12.35)$$

The same checks apply as for Equations 12.16 through 12.19.

i, $v'_L = -v_L$ and v_C are plotted in Figure 12.7. It is seen that the responses are qualitatively similar to the overdamped responses of Figure 12.3, although the mathematical expressions are of quite different form.

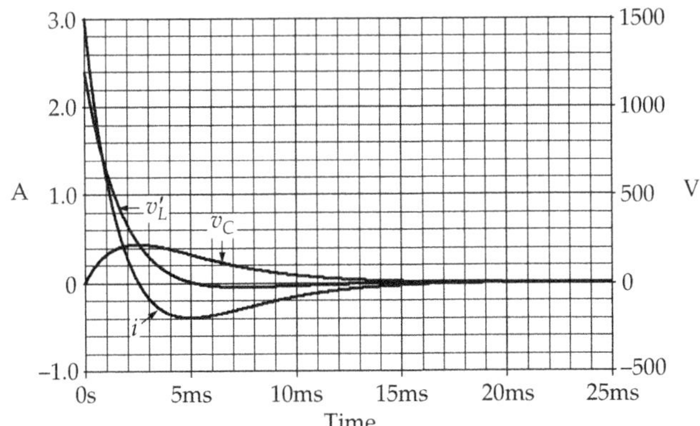

FIGURE 12.7
Critically damped, natural responses of series *RLC* circuit.

Summary:

1. *When R is large enough so that $\alpha > \omega_0$, s_1 and s_2 are negative real and distinct. The responses are overdamped.*
2. *When R is such that $\alpha = \omega_0$, then $s_1 = s_2 = -\omega_0$. The responses are critically damped.*
3. *When R is small enough so that $\alpha < \omega_0$, s_1 and s_2 are complex, the responses are damped sinusoids. The responses are underdamped.*

Primal Exercise 12.4

If the series *RLC* circuit of Figure 12.1b has $L = 0.1$ mH, and $C = 0.4$ nF, determine (a) R for critical damping and (b) $i(t)$, $t \geq 0^+$, if $i(0^+) = 24$ mA and $v_C(0^+) = 0$.

Ans. (a) $R = 1\ \text{k}\Omega$; (b) $i(t) = 24e^{-5t}(1 - 5t)$ mA, t is in μs.

What if initial energy was stored in the capacitor rather than in the inductor? As emphasized earlier, this does not change α and ω_0, and hence s_1 and s_2, because these depend only on R, L, and C, and not on the initial conditions. The type of response, whether overdamped, underdamped, or critically damped, is the same, because the type of response depends only on the relative magnitudes of α and ω_0. But the arbitrary constants change with the initial conditions, which changes the coefficients of the terms in the responses. Example 12.1 considers the cases of initial energy storage in the capacitor alone, and in both the capacitor and inductor. The following important concept is illustrated:

Concept: *The natural responses due to initial energy storage in both the inductor and capacitor can be obtained by superposition, that is, by adding algebraically the responses due to energy storage in the inductor alone and in the capacitor alone.*

Example 12.1: Natural Responses of Series *RLC* Circuit with Initial Energy Storage

Determine $i(t)$, $v_R(t)$, $v_C(t)$, and $v_L(t)$ in the circuit of Figure 12.1b if at $t = 0$, (a) the capacitor has an initial voltage $V_0 = -1200$ V and the inductor has zero initial current and (b) the capacitor has an initial voltage $V_0 = -1200$ V and the inductor has an initial current of $i(0^-) = 3$ A.

Solution:

(a) The circuit is illustrated in Figure 12.8a, where the capacitor is charged by the battery to $V_0 = V_{SRC}$ and is connected via the changeover switch at $t = 0$ to be in series with a resistor R and an uncharged inductor L. The circuit becomes as in Figure 12.8b for $t \geq 0^+$. To avoid unnecessary repetition, only the case of overdamped responses, with $R = 500\ \Omega$, will be considered.

As determined previously for the overdamped case, $\alpha = 500$ rad/s, $\omega_0 = 400$ rad/s, $s_1 = -200$ rad/s, and $s_2 = -800$ rad/s. The general solution of Equation 12.13 therefore applies, expressed as

$$v_C(t) = Ae^{-200t} + Be^{-800t}, \quad t \geq 0^+$$

At $t = 0^-$, the energy stored in the capacitor is $(1/2)CV_0^2$ and no energy is stored in the inductor. At $t = 0^+$, these energies stay the same, so that $v_C(0^+) = V_0$ and $i(0^+) = 0$. This gives

$$A + B = -1200 \tag{12.36}$$

To apply the second initial condition, Equation 12.13 is differentiated to give

$$\frac{dv_C}{dt} = -200Ae^{-200t} - 800Be^{-800t} \tag{12.37}$$

$(dv_C/dt)_{0+} = i(0^+)/C = 0$, so that $A = -4B$. Substituting in Equation 12.36 gives $A = -1600$ V and $B = 400$ V. It follows that

$$v_C(t) = -400\left(4e^{-200t} - e^{-800t}\right) \text{ V}, \quad t \geq 0^+ \tag{12.38}$$

$$i(t) = C\frac{dv_C}{dt} = 12.5 \times 10^{-6} \times 400\left(800e^{-200t} - 800e^{-800t}\right)$$

or

$$i(t) = 4\left(e^{-200t} - e^{-800t}\right) \text{ A}, \quad t \geq 0^+ \tag{12.39}$$

$$v_L(t) = 0.5\frac{di}{dt} = -400\left(e^{-200t} - 4e^{-800t}\right) \text{ V}, \quad t \geq 0^+ \tag{12.40}$$

As a check, (1) $v_C(0^+) = -1200$ V, $i(0^+) = 0$, $v_L(0^+) = 1200$ V; (2) as $t \to \infty$, $v_C(\infty) = i(\infty) = v_L(\infty) = 0$, since all responses become zero when the energy initially stored in the capacitor is dissipated in the resistor; and (3) $v_R + v_C + v_L = 0$ for all t, as required by KVL.

i, $v_C' = -v_C$, and v_L are plotted in Figure 12.9. At $t = 0^+$, $v_C' = 1200$ V, which means that the upper plate in Figure 12.8b is positively charged with respect to the lower plate. The capacitor begins to discharge, driving i in the positive direction, so i increases. As i varies with time, it induces a voltage v_L in the inductor. While i is increasing, its increase is opposed by a positive v_L. i increases at a decreasing rate because of the drop in v_C' as the capacitor discharges, and because the increase in i is opposed by a positive v_L, added to the voltage drop across the resistor. v_L decreases with the decreasing rate of increase of i. i reaches a maximum, at which time $v_L = Ldi/dt = 0$. i then decreases towards zero, resulting in a negative v_L that opposes the decrease in i. Eventually, all the energy initially stored in the capacitor is dissipated in the resistor.

(b) The general solution of Equation 12.13 applies, that is,

$$v_C(t) = Ae^{-200t} + Be^{-800t}, \quad t \geq 0^+$$

At $t = 0^-$, the energy stored in the capacitor is $(1/2)CV_0^2$ and the energy stored in the inductor is $(1/2)LI_0^2$. At $t = 0^+$, these energies stay the same. This means that at $v_C(0^+) = V_0 = -1200$ V and $i(0^+) = I_0 = 3$ A. The first condition gives, as before, $A + B = -1200$. From Equation 12.37,

$$i(t) = C\frac{dv_C}{dt} = -12.5 \times 10^{-6} \times 200\left(Ae^{-200t} + 4Be^{-800t}\right) \tag{12.41}$$

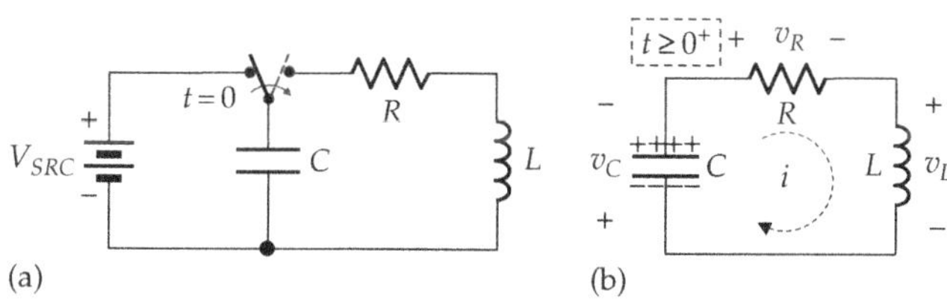

FIGURE 12.8
Figure for Example 12.1.

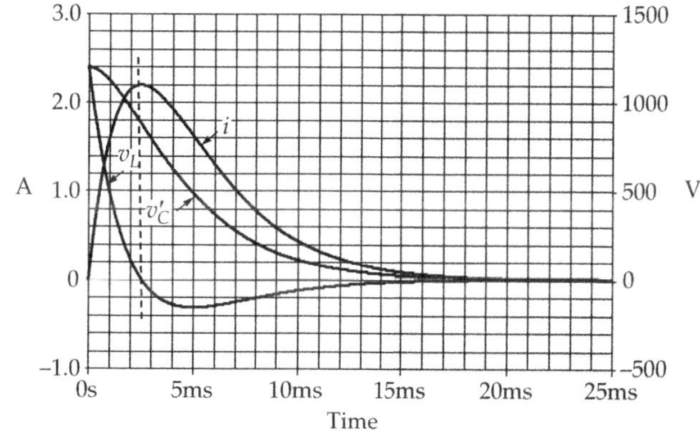

FIGURE 12.9
Figure for Example 12.1.

At $t = 0^+$, Equation 12.41 reduces to $A + 4B = -1200$. It follows that $A = -1200$ V and $B = 0$, so that

$$v_C(t) = -1200e^{-200t} \text{ V}, \quad t \geq 0^+ \tag{12.42}$$

$$i(t) = C\frac{dv_C}{dt} = 12.5\times10^{-6}\times1200\times200e^{-200t}$$

or

$$i(t) = 3e^{-200t} \text{ A}, \quad t \geq 0^+ \tag{12.43}$$

$$v_R(t) = 1500e^{-200t} \text{ V}, \quad t \geq 0^+ \tag{12.44}$$

$$v_L(t) = 0.5\frac{di}{dt} = -300e^{-200t} \text{ V}, \quad t \geq 0^+ \tag{12.45}$$

It should be noted that the initial conditions in this example have been deliberately chosen so as to give $B = 0$. It can be shown that the condition for $B = 0$ is that $I_0 = s_1CV_0$, which is satisfied in this case (Problem P12.63). This condition is tantamount to having the capacitor initially charged to V_O and discharging with a time constant $1/s_1$. $v_C(t)$ is then given by $-V_0e^{-s_1t}$. The initial value of the discharge current is s_1CV_0. If this happens to be equal to I_0, the initial current in the circuit, as is the case in this example, then $B = 0$. This eliminates the term involving the exponential in s_2, leaving only the term involving the exponential in s_1. All the responses will have a single time constant, as for a first-order circuit.

As a check, (1) $v_C(0^+) = -1200$ V, $i(0^+) = 3$ A, $v_L(0^+) = -300$ V; (2) as $t \to \infty$, $v_C(\infty) = i(\infty) = v_L(\infty) = 0$, since all responses become zero when the energy initially stored in the capacitor and in the inductor is dissipated in the resistor; and (3) $v_R + v_C + v_L = 0$ for all t, as required by KVL.

The responses are shown in Figure 12.10. Being simple exponentials, the tangents to the curves at $t = 0$ intersect the time axis at $t = 1/200$ s, or 5 ms.

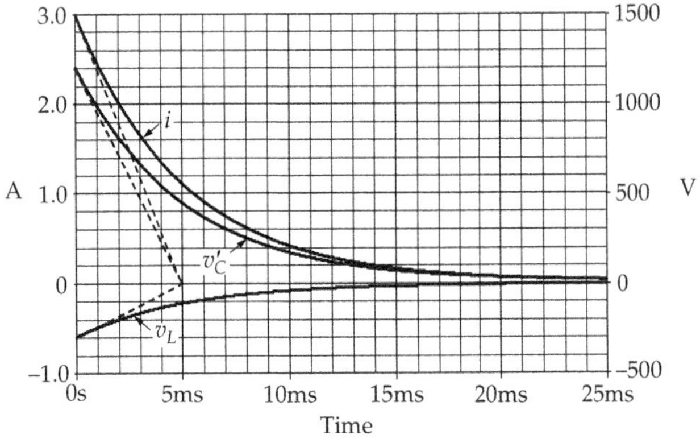

FIGURE 12.10
Figure for Example 12.1.

TABLE 12.1
Superposition of Responses due to Stored Energy

Quantity	Energy Stored in Inductor	Energy Stored in Capacitor	Energy Stored in Inductor and Capacitor
i, A	$-e^{-200t} + 4e^{-800t}$	$4\left(e^{-200t} - e^{-800t}\right)$	$3e^{-200t}$
v_L, V	$100e^{-200t} - 1600e^{-800t}$	$-400\left(e^{-200t} - 4e^{-800t}\right)$	$-300e^{-200t}$
v_C, V	$400\left(e^{-200t} - e^{-800t}\right)$	$-400\left(4e^{-200t} - e^{-800t}\right)$	$-1200e^{-200t}$

It is important to note that superposition applies. Table 12.1 lists the overdamped responses due to stored energy in the inductor alone, in the capacitor alone, and in both the inductor and the capacitor. Superposition applies, of course, to the underdamped and critically damped responses as well.

Primal Exercise 12.5

Repeat Example 12.1 for the cases of underdamped and critically damped responses.

Ans. Critically damped responses: $v_C(t) = -1200\left(e^{-400t} + 200te^{-400t}\right)$ V, $i(t) = 3\left(e^{-400t} + 400te^{-400t}\right)$ A, $v_L(t) = -24\times10^4te^{-400t}$ V.

Underdamped responses: $v_C(t) = -1200e^{-100t}\times\left(\cos100\sqrt{15}t - \left(1/\sqrt{15}\right)\sin100\sqrt{15}t\right)$ V, $i(t) = 1.5e^{-100t}\times\left(2\cos100\sqrt{15}t + \left(14/\sqrt{15}\right)\sin100\sqrt{15}t\right)$ A,

$$v_L(t) = 300e^{-100t}\left(3\cos100\sqrt{15}t - \left(11/\sqrt{15}\right)\sin100\sqrt{15}t\right) \text{ V.}$$

Exercise 12.6

Verify that in all the responses of the series *RLC* derived so far, the time integral of the current is equal to the initial charge on the capacitor, and the time integral of the voltage across the inductor is equal to the initial flux linkage of the inductor.

Primal Exercise 12.7

Both switches in Figure 12.11 are moved at $t = 0$, the initial stored energies being 12.5 J in the capacitor and 0.5 J in the inductor. Determine $v_L(0^+)$.

Ans. 8 V.

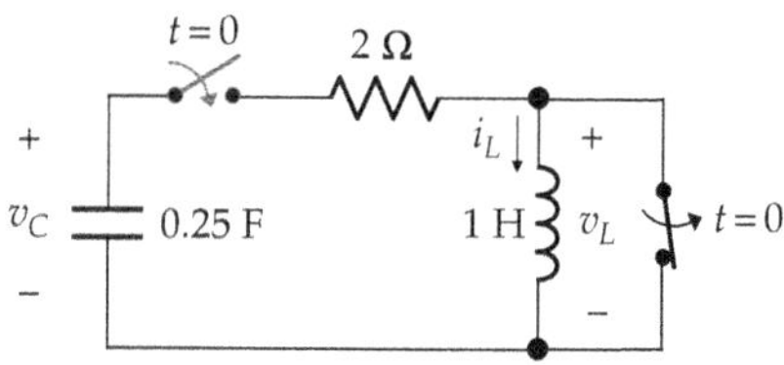

FIGURE 12.11
Figure for Primal Exercise 12.7.

Primal Exercise 12.8

Determine whether the response of the circuit in Figure 12.11 for $t \geq 0^+$ is overdamped, critically damped, or underdamped.

Ans. Underdamped.

Primal Exercise 12.9

In a series *RLC* circuit, the natural voltage response across one of the elements is $2e^{-4t}\cos 3t$ V, t is in s. If R is divided by 2, what would be the values of α and ω_d?

Ans. $\alpha = 2$ rad/s, $\omega_d = \sqrt{21}$ rad/s.

Table 12.2 lists, for reference purposes, the responses due to energy storage in each of the capacitor or inductor alone. These responses are derived by following the same procedure as in the preceding numerical cases, and in Example 12.1, but retaining the general symbols (Exercise 12.10).

Exercise 12.10

Derive the general expressions listed in Table 12.2.

Exercise 12.11

Verify that the numerical, overdamped responses i, v_R, v_L, and v_C derived in this section, for both the cases of initial inductor current and initial capacitor voltage, are in accordance with the general expressions listed in Table 12.2.

12.1.4 Sustained Oscillations

The series *RLC* circuit is shown in Figure 12.12a but with $R = 0$. Assuming $L = 0.5$ H, $C = 12.5$ μF, and $I_0 = 3$ A, as

TABLE 12.2
Natural Responses of Series *RLC* Circuit, $t \geq 0^+$ (Assigned Positive Directions as in Figure 12.1b)

	Initial Inductor Current I_0	Initial Capacitor Voltage V_0
Over damped responses	$i = \frac{I_0}{(s_1 - s_2)}\left[s_1e^{s_1t} - s_2e^{s_2t}\right]$	$i = -\frac{V_0}{L(s_1 - s_2)}\left[e^{s_1t} - e^{s_2t}\right]$
	$v_R = \frac{RI_0}{(s_1 - s_2)}\left[s_1e^{s_1t} - s_2e^{s_2t}\right]$	$v_R = -\frac{2\alpha V_0}{(s_1 - s_2)}\left[e^{s_1t} - e^{s_2t}\right]$
	$v_L = \frac{LI_0}{(s_1 - s_2)}\left[s_1^2e^{s_1t} - s_2^2e^{s_2t}\right]$	$v_C = -\frac{V_0}{(s_1 - s_2)}\left[s_2e^{s_1t} - s_1e^{s_2t}\right]$
	$v_C = \frac{I_0}{C(s_1 - s_2)}\left[e^{s_1t} - e^{s_2t}\right]$	$v_L = -\frac{V_0}{(s_1 - s_2)}\left[s_1e^{s_1t} - s_2e^{s_2t}\right]$
Critically damped responses	$i = I_0e^{-\omega_0t}(1 - \omega_0t)$	$i = -\frac{V_0}{L}te^{-\omega_0t}$
	$v_R = RI_0e^{-\omega_0t}(1 - \omega_0t)$	$v_R = -2\omega_0V_0te^{-\omega_0t}$
	$v_L = -\omega_0LI_0e^{-\omega_0t}(2 - \omega_0t)$	$v_L = -V_0e^{-\omega_0t}(1 - \omega_0t)$
	$v_C = \frac{I_0}{C}te^{-\omega_0t}$	$v_C = V_0e^{-\omega_0t}(1 + \omega_0t)$
Under damped responses	$i = I_0e^{-\alpha t}\left(\cos\omega_dt - \frac{\alpha}{\omega_d}\sin\omega_dt\right)$	$i = -\frac{V_0e^{-\alpha t}}{\omega_dL}\sin\omega_dt$
	$v_R = RI_0e^{-\alpha t}\left(\cos\omega_dt - \frac{\alpha}{\omega_d}\sin\omega_dt\right)$	$v_R = -\frac{2\alpha V_0e^{-\alpha t}}{\omega_d}\sin\omega_dt$
	$v_L = -LI_0e^{-\alpha t}\left[2\alpha\cos\omega_dt + \left(\omega_d - \frac{\alpha^2}{\omega_d}\right)\sin\omega_dt\right]$	$v_L = -V_0e^{-\alpha t}\left(\cos\omega_dt - \frac{\alpha}{\omega_d}\sin\omega_dt\right)$
	$v_C = \frac{I_0}{\omega_dC}e^{-\alpha t}\sin\omega_dt$	$v_C = V_0e^{-\alpha t}\left(\cos\omega_dt + \frac{\alpha}{\omega_d}\sin\omega_dt\right)$

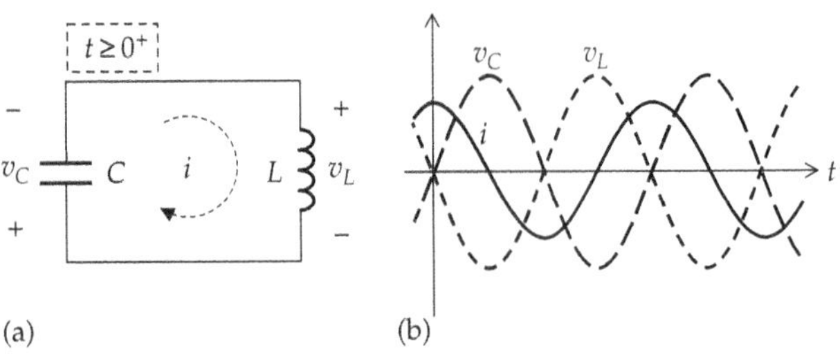

FIGURE 12.12
(a) Lossless LC circuit and (b) variations with time of i, v_L, and v_C.

in Section 12.1.2, this makes $\alpha = 0$ and $\omega_d = 400$ rad/s. Equation 12.23 becomes $v_C(t) = A\cos 400t + B\sin 400t$. Having $v(0^+) = 0$ makes $A = 0$; so that $v_C(t) = B\sin 400t$; B is determined from the second initial condition that $i(0^+) = C(dv_C/dt)_{0+} = 3$ A. This gives $12.5\times 10^{-6}\times 400B = 3$, or $B = 600$. It follows that.

$$v_C(t) = 600\sin 400t \text{ V}, \quad t \geq 0^+ \tag{12.46}$$

$$v_L(t) = -v_C(t) = -600\sin 400t \text{ V}, \quad t \geq 0^+ \tag{12.47}$$

$$i(t) = C\frac{dv_C}{dt} = 3\cos 400t \text{ A}, \quad t \geq 0^+ \tag{12.48}$$

The responses are nondecrementing, sustained oscillations (Figure 12.12b). As to be expected, v_L leads i by 90° and v_C lags i by 90°. With $R = 0$, there is no power dissipation in the circuit. Energy is continually exchanged between magnetic energy stored in the inductor and electric energy stored in the capacitor. When $v_C = v_L = 0$, the magnitude of i is maximum, and all the energy is stored in the inductor. The energy at this instant is $(1/2)Li^2 = 9/4 = 2.25$ J. When $i = 0$, the magnitude of v_C is maximum, and all the energy is stored in the capacitor. The energy at this instant is $(1/2)Cv^2 = (1/2)\times 12.5\times 10^{-6}\times (600)^2 = 2.25$ J. At intermediate times, energy is stored partly in the inductor and partly in the capacitor, but the total energy $w(t)$ at any instant is 2.25 J. Thus,

$$\begin{aligned} w(t) &= \frac{1}{2}Li^2 + \frac{1}{2}Cv_C^2 \\ &= \frac{1}{2}\left(0.5\times(3\cos 400t)^2 + 12.5\times 10^{-6}\times(600\sin 400t)^2\right) \\ &= \frac{1}{2}\left(4.5\cos^2 400t) + 4.5\sin^2 400t\right) = 2.25 \text{ J} \end{aligned}$$

Having $R = 0$ may seem an ideal case that cannot be realized in practice. But active devices, such as transistors, can be used to introduce a negative resistance that effectively reduces R to zero. In other words, the active device introduces sufficient energy to compensate for the power dissipated in the circuit, so that continuous oscillations are produced. This is the principle behind LC **oscillators**, which are commonly used in practice.

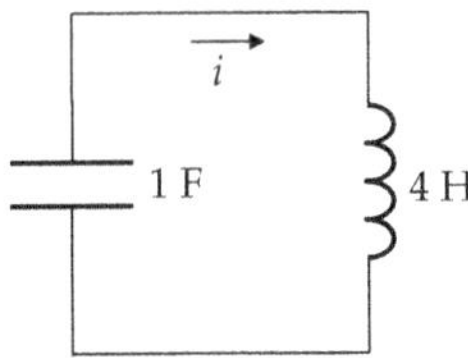

FIGURE 12.13
Figure for Primal Exercise 12.14.

Primal Exercise 12.12

If $L = 10$ μH, determine C that gives an oscillation frequency of 100 kHz.

Ans. 0.253 μF.

Primal Exercise 12.13

Assume that the circuit of Figure 12.12 started oscillating at 1 Mrad/s, with initial energies of 1.5 μJ in the inductor and 0.5 μJ in the capacitor. Determine L if the peak voltage across the circuit is 2 V.

Ans. 1 μH.

Primal Exercise 12.14

Given that $i = \sin\omega_0 t$ A in Figure 12.13, determine the smallest $t > 0$ at which the instantaneous energy stored in the capacitor (a) is a maximum, and (b) is equal to that stored in the inductor.

Ans. (a) $2\pi = 6.28$ s; (b) $\pi/2 = 1.57$ s.

12.2 Natural Response of Parallel *GCL* Circuit

The dual of the series RLC circuit considered in Section 12.1 is the parallel GCL circuit (Figure 12.14). For $t < 0^-$, the capacitor is charged to $V_0 = V_{SRC}$ and the inductor is uncharged (Figure 12.14a). At $t = 0^+$, the switch is moved, and the circuit becomes as in Figure 12.14b. Table 12.3 lists the dual quantities in both cases.

The initial conditions first considered in Section 12.1 were $i(0^+) = 3$ A and $v_C(0^+) = 0$ V. The corresponding initial conditions in the dual GCL circuit are, from Table 12.3, $v(0^+) = 3$ V and $i_L(0^+) = 0$. We will analyze this case for overdamped responses from first principles to illustrate that v and the currents in Figure 12.14 are the same as those derived from substituting the dual quantities of Table 12.3 in the equations for the series

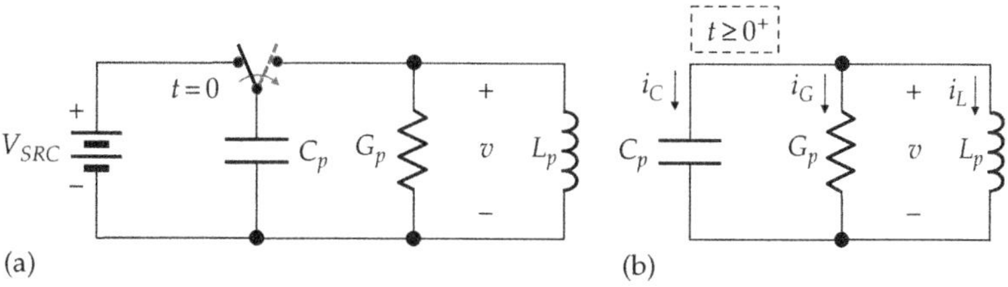

FIGURE 12.14
(a) Capacitor is charged when connected to the battery for $t \le 0^-$ and (b) parallel *GCL* circuit having an initially charged capacitor. Inductances and resistances in the circuit shown in (a) are replaced by the effective inductance and resistance in (b).

TABLE 12.3
Dual Quantities

Series *RLC*	R	L	C	i	v_R	v_C	v_L	$\alpha = R/2L$	$\omega_0 = 1/\sqrt{LC}$
Parallel *GCL*	G_p	C_p	L_p	v	i_G	i_L	i_C	$\alpha_p = G_p/2C_p$	$\omega_0 = 1/\sqrt{C_pL_p}$

RLC circuit. Just as it is convenient in the case of the series *RLC* circuit to derive the solution for v_C first, it is convenient in the case of the parallel *GCL* circuit to derive the solution for i_L, the dual of v_C, first.

From KCL, $i_G + i_C + i_L = 0$. The voltage v across the parallel combination is $v(t) = L_p di_L/dt$, and $i_G(t) = G_p v = G_pL_pdi_L/dt$; $i_C(t) = C_pdv/dt = L_pC_pd^2i_L/dt^2$. Substituting in the *KCL* equation,

$$G_pL_p\frac{di_L}{dt} + L_pC_p\frac{di_L}{dt} + i_L = 0, \quad t \ge 0^+ \qquad (12.49)$$

Dividing by L_pC_p and rearranging,

$$\frac{d^2i_L}{dt^2} + \frac{G_p}{C_p}\frac{di_L}{dt} + \frac{1}{C_pL_p}i_L = 0, \quad t \ge 0^+ \qquad (12.50)$$

This is represented in standard form as

$$\frac{d^2i_L}{dt^2} + 2\alpha_p\frac{di_L}{dt} + \omega_0^2 i_L = 0, \quad t \ge 0^+ \qquad (12.51)$$

where

$$\alpha_p = \frac{G_p}{2C_p} \quad \text{and} \quad \omega_0 = \frac{1}{\sqrt{C_pL_p}} \qquad (12.52)$$

Note that each of α and ω_0 has the same value in dual series and parallel circuits, because the numerical values of R and G_p are the same, as are the numerical values of L and C_p, and C and L_p. The expression for ω_0 is the same for both circuits in terms of the circuit parameters, being $1/\sqrt{\text{inductance} \times \text{capacitance}}$ in both circuits, *but the expression for α is different.*

Substituting $i_L = Ae^{st}$ in Equation 12.51,

$$Ae^{st}\left(s^2 + 2\alpha_p s + \omega_0^2\right) = 0 \qquad (12.53)$$

In order to satisfy Equation 12.53 for all t, and have nonzero responses, the bracketed terms must sum to zero:

$$s^2 + 2\alpha_p s + \omega_0^2 = 0 \qquad (12.54)$$

The roots s_1 and s_2 of Equation 12.54 are given by the same expressions as in Equation 12.7 and would have the same numerical values for dual circuits.

The general solution is of the form

$$i_L(t) = Ae^{s_1t} + Be^{s_2t}, \quad t \ge 0^+ \qquad (12.55)$$

Next, we will substitute numerical values in accordance with duality, namely, $G_p = 500$ S, $L_p = 12.5$ μH, $C_p = 0.5$ F, and $V_0 = 3$ V. This gives $\alpha_p = G_p/2C_p = 500$ rad/s, $\omega_0 = 1/\sqrt{C_pL_p} = 400$ rad/s. It follows that $s_1 = -200$ rad/s and $s_2 = -800$ rad/s. Substituting in Equation 12.55,

$$i_L(t) = Ae^{-200t} + Be^{-800t}, \quad t \ge 0^+ \qquad (12.56)$$

From the initial condition that $i_L(0^+) = 0$, $A = -B$. To apply the initial condition of $v(0^+) = L_p\left(di_L/dt\right)_{0^+}$, Equation 12.56 is differentiated and both sides multiplied by L_p to give

$$L_p\frac{di_L}{dt} = 12.5\times 10^{-6}\left(-200Ae^{-200t} - 800Be^{-800t}\right) = 3 \text{ V} \qquad (12.57)$$

At $t = 0^+$, this reduces to $A + 4B = -1200$. Substituting $A = -B$ gives $A = 400$ A and $B = -400$ A. It follows that

$$i_L(t) = 400\left(e^{-200t} - e^{-800t}\right) \text{ A}, \quad t \geq 0^+ \qquad (12.58)$$

$$v(t) = L_p \frac{di_L}{dt} = 12.5 \times 10^{-6} \times 400\left(-200e^{-200t} + 800e^{-800t}\right)$$

or

$$v(t) = -e^{-200t} + 4e^{-800t} \text{ V}, \quad t \geq 0^+ \qquad (12.59)$$

$$i_G(t) = -500e^{-200t} + 2000e^{-800t} \text{ A}, \quad t \geq 0^+ \qquad (12.60)$$

$$i_C(t) = C_p \frac{dv}{dt} = 0.5\left(200e^{-200t} - 3200e^{-800t}\right)$$

or

$$i_C(t) = 100e^{-200t} - 1600e^{-800t} \text{ A}, \quad t \geq 0^+ \qquad (12.61)$$

As a check, (1) $v(0^+) = 3$ V, $i_C(0^+) = -1500$ A, $i_G(0^+) = 1500$ A, in accordance with KCL; (2) as $t \to \infty$, $v(\infty) = i_L(\infty) = i_C(\infty) = 0$, since all responses become zero when the energy initially stored in the capacitor is dissipated in the resistor; and (3) $i_G + i_C + i_L = 0$ for all t, as required by KCL. Equations 12.58 through 12.61 are the dual relations of Equations 12.16 through 12.19, respectively.

The general expressions for the overdamped, critically damped, and underdamped natural responses of the parallel *GCL* circuit follow from those of the series *RLC* circuit listed in Table 12.2, but with dual quantities interchanged in accordance with Table 12.3.

Exercise 12.15

Derive from duality, using Tables 12.2 and 12.3, the responses of the parallel *GCL* circuit when initial energy is stored in the inductor alone.

Primal Exercise 12.16

Both switches in Figure 12.15 are moved at $t = 0$, the initial stored energies being 12.5 J in the inductor and 0.5 J in the capacitor. Determine i_C at $t = 0^+$. Note that this is the dual case of Primal Exercise 12.7.

Ans. 8 A.

Primal Exercise 12.17

The switch in Figure 12.16 is closed at $t = 0$, with an initial voltage $V_{C0} = 5$ V, and no energy stored in the inductor.

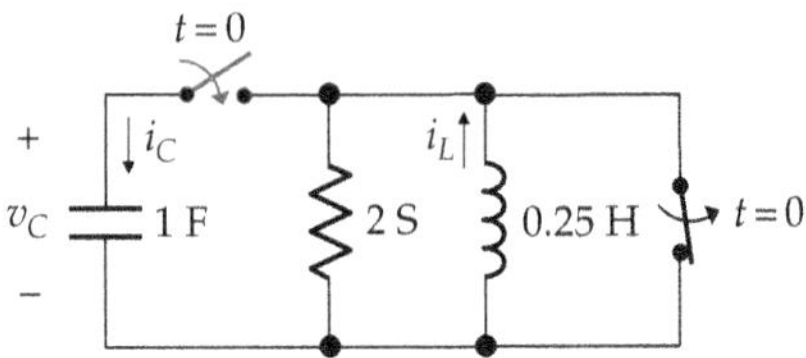

FIGURE 12.15
Figure for Primal Exercise 12.16.

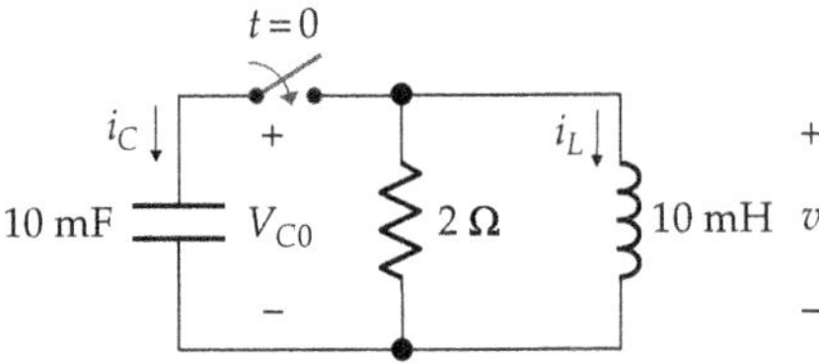

FIGURE 12.16
Figure for Primal Exercise 12.17.

(a) Determine $v(0^+)$, $i_L(0^+)$, and $i_C(0^+)$; (b) specify whether the response for $t \geq 0^+$ is overdamped, critically damped, or underdamped.

Ans. (a) $v(0^+)$ 5 V, $i_L(0^+) = 0$, $i_C(0^+) = -2.5$ A; (b) underdamped.

12.3 Charging of Series *RLC* Circuit

In Figure 12.17, a battery of voltage V_{SRC} is applied at $t = 0$ to a series *RLC* circuit with no initial energy storage. It is required to determine the circuit responses for $t \geq 0^+$.

From KVL after the switch is closed, $v_R + v_L + v_C = V_{SRC}$. As explained earlier, it is convenient in a series circuit to derive v_C first. We note that $v_R = Ri = RCdv_C/dt$ and that $v_L = Ldi/dt = LCd^2v_C/dt^2$. Substituting these relations in KVL,

$$RC\frac{dv_C}{dt} + LC\frac{d^2v_C}{dt^2} + v_C = V_{SRC}, \quad t \geq 0^+ \qquad (12.62)$$

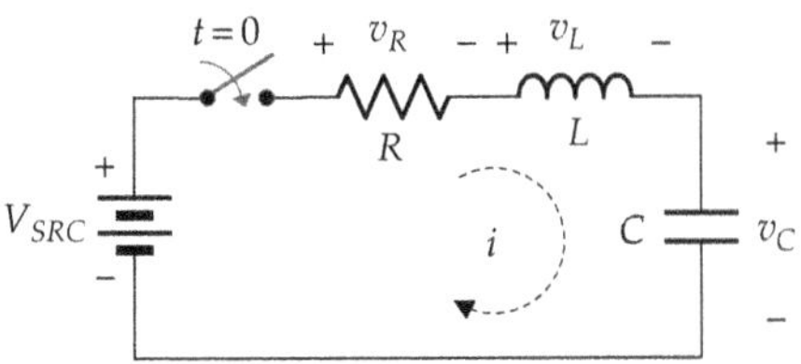

FIGURE 12.17
Charging of a series *RLC* circuit.

Dividing by LC, and rearranging,

$$\frac{d^2v_C}{dt^2}+\frac{R}{L}\frac{dv_C}{dt}+\frac{1}{LC}v_C=\frac{V_{SRC}}{LC},\quad t\geq 0^+ \quad (12.63)$$

In standard form, Equation 12.63 is

$$\frac{d^2v_C}{dt^2}+2\alpha\frac{dv_C}{dt}+\omega_0^2v_C=\frac{V_{SRC}}{LC},\quad t\geq 0^+ \quad (12.64)$$

The complete solution of Equation 12.64 is the sum of two components: (1) the transient component, which is the solution of the homogeneous differential equation, and (2) the steady-state component. The homogeneous differential equation is the same as Equation 12.10, so that the transient component is given by Equation 12.11. The steady-state component, obtained by setting $dv_C/dt = 0 = d^2v_C/dt^2$ in Equation 12.62, is $v_C = V_{SRC}$. It follows that the complete solution for v_C is

$$v_C(t)=Ae^{s_1t}+Be^{s_2t}+V_{SRC},\quad t\geq 0^+ \quad (12.65)$$

where s_1 and s_2 are given by Equation 12.7. Note that s_1 and s_2 either are negative real or have a negative real part, so that $v_C \rightarrow V_{SRC}$ as $t \rightarrow \infty$. It is clear from the circuit of Figure 12.17 that in the steady state after the switch is closed, the inductor acts as a short circuit and the capacitor as an open circuit, so that $i(\infty) = 0$ and $v_C(\infty) = V_{SRC}$.

The arbitrary constants A and B are determined by *applying the initial conditions to the complete solution*. The stored energy is zero at $t = 0^-$, so it remains zero at $t = 0^+$. This means that $i(0^+) = 0$, $v_C(0^+) = 0$, and $v_L(0^+) = V_{SRC}$ in order to satisfy KVL at this instant (Figure 12.18). Since, $v_C(0^+) = 0$, Equation 12.65 gives

$$A+B=-V_{SRC} \quad (12.66)$$

The second initial condition that $i(0^+) = 0$, which means that $(dv_C/dt)_{0+} = 0$, is applied by differentiating Equation 12.65 to obtain, at $t = 0^+$:

$$s_1A+s_2B=0 \quad (12.67)$$

Solving Equations 12.66 and 12.67 gives

$$A=\frac{s_2V_{SRC}}{s_1-s_2}\quad\text{and}\quad B=-\frac{s_1V_{SRC}}{s_1-s_2} \quad (12.68)$$

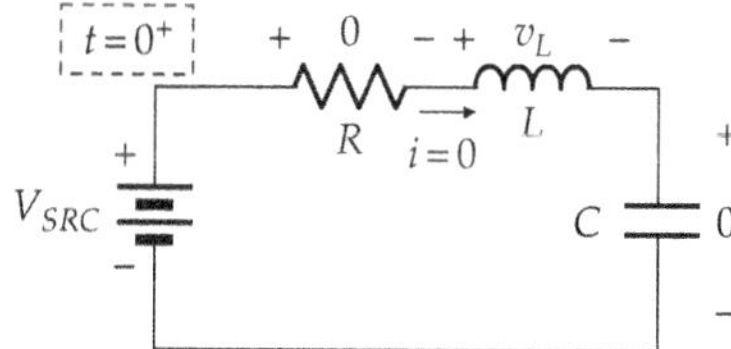

FIGURE 12.18
KVL at $t = 0^+$ in the circuit of Figure 12.17.

The following equations are derived:

$$v_C(t)=V_{SRC}\left[\frac{s_2e^{s_1t}}{s_1-s_2}-\frac{s_1e^{s_2t}}{s_1-s_2}+1\right],\quad t\geq 0^+ \quad (12.69)$$

$$i(t)=C\frac{dv_C}{dt}=CV_{SRC}\left[\frac{s_1s_2e^{s_1t}}{s_1-s_2}-\frac{s_1s_2e^{s_2t}}{s_1-s_2}\right]$$

or

$$i(t)=\frac{V_{SRC}}{L(s_1-s_2)}\left[e^{s_1t}-e^{s_2t}\right],\quad t\geq 0^+ \quad (12.70)$$

where the substitution $Cs_1s_2 = C\omega_0^2 = 1/L$ was made.

$$v_R(t)=Ri=\frac{2\alpha V_{SRC}}{(s_1-s_2)}\left[e^{s_1t}-e^{s_2t}\right],\quad t\geq 0^+ \quad (12.71)$$

where the substitution $R/L = 2\alpha$ was made.

$$v_L(t)=L\frac{di}{dt}=\frac{V_{SRC}}{(s_1-s_2)}\left[s_1e^{s_1t}-s_2e^{s_2t}\right],\quad t\geq 0^+ \quad (12.72)$$

Equations 12.69 through 12.72 apply to overdamped and underdamped responses. They do not apply to the critically damped case, because $s_1 = s_2$ gives 0/0 in the responses, which is indeterminate.

12.3.1 Underdamped Response

Rather than substitute $s_1 = -\alpha + j\omega_d$, $s_2 = -\alpha - j\omega_d$, $s_1 - s_2 = j2\omega_d$, and $\omega_0^2 = \omega_d^2 + \alpha^2$ in Equations 12.69 through 12.72, it is more convenient to start with the general solution for v_C being of the form

$$v_C(t)=e^{-\alpha t}\left(A'\cos\omega_dt+B'\sin\omega_dt\right)+V_{SRC},\quad t\geq 0^+ \quad (12.73)$$

Since $v_C = 0$ at $t = 0^+$, then $A' = -V_{SRC}$. B' is obtained from the initial condition that $i(0^+) = (dv_C/dt)_{0+} = 0$. To apply this condition, Equation 12.73 is differentiated and set to zero. But only the cosine terms need be considered in the derivative since the sine terms are zero at $t = 0^+$. Thus,

$$\frac{dv_C}{dt}=V_{SRC}\alpha e^{-\alpha t}\cos\omega_dt+\omega_de^{-\alpha t}B'\cos\omega_dt+X(t) \quad (12.74)$$

where $X(t)$ consists of sine terms that are zero at $t = 0$ and the substitution $A' = -V_{SRC}$ was made. Setting $t = 0^+$

in Equation 12.74 gives $B' = -(\alpha/\omega_d)V_{SRC}$. The following equations are derived:

$$v_C(t) = V_{SRC}\left[1 - e^{-\alpha t}\left(\cos\omega_d t + \frac{\alpha}{\omega_d}\sin\omega_d t\right)\right] \quad (12.75)$$

$$i(t) = C\frac{dv_C}{dt} = CV_{SRC}\left[\alpha e^{-\alpha t}\left(\cos\omega_d t + \frac{\alpha}{\omega_d}\sin\omega_d t\right)\right.$$

$$\left. - e^{-\alpha t}\left(-\omega_d \sin\omega_d t + \alpha\sin\omega_d t\right)\right]$$

or

$$i(t) = \frac{V_{SRC}}{\omega_d L}e^{-\alpha t}\sin\omega_d t, \quad t \geq 0^+ \quad (12.76)$$

where the substitution $C\left[\left(\alpha^2/\omega_d\right) + \omega_d\right] = C\left(\alpha^2 + \omega_d^2\right)/\omega_d = C\omega_0^2/\omega_d = 1/\omega_d L$ was made.

$$v_R(t) = \frac{2\alpha V_{SRC}}{\omega_d}e^{-\alpha t}\sin\omega_d t, \quad t \geq 0^+ \quad (12.77)$$

where the substitution $R/L = 2\alpha$ was made.

$$v_L(t) = L\frac{di}{dt} = V_{SRC}e^{-\alpha t}\left(\cos\omega_d t - \frac{\alpha}{\omega_d}\sin\omega_d t\right) \quad (12.78)$$

12.3.2 Critically Damped Response

In a critically damped circuit, $\alpha = \omega_0$, so that $s_1 = s_2 = -\omega_0$. The transient response is given by Equation 12.30. In the final steady state, the inductor acts as a short circuit and the capacitor as an open circuit, so that $i(\infty) = 0$, $v_L(\infty) = 0$, and $v_C(\infty) = V_{SRC}$. The complete solution for v_C is

$$v_C(t) = Ae^{-\omega_0 t} + Bte^{-\omega_0 t} + V_{SRC}, \quad t \geq 0^+ \quad (12.79)$$

Since $v_C = 0$ at $t = 0$, then $A = -V_{SRC}$. B is obtained from $i(0^+) = (dv_C/dt)_{0+} = 0$. Differentiating Equation 12.79, with $A = -V_{SRC}$, and setting $t = 0$, gives $B = -\omega_0 V_{SRC}$. The following equations are derived:

$$v_C(t) = V_{SRC}\left[1 - e^{-\omega_0 t} - \omega_0 te^{-\omega_0 t} +\right], \quad t \geq 0^+ \quad (12.80)$$

$$i(t) = C\frac{dv_C}{dt} = C\left[\omega_0 V_{SRC}e^{-\omega_0 t} - \omega_0 V_{SRC}e^{-\omega_0 t} + \omega_0^2 tV_{SRC}e^{-\omega_0 t}\right]$$

or

$$i(t) = \frac{V_{SRC}}{L}te^{-\omega_0 t}, \quad t \geq 0^+ \quad (12.81)$$

where $C\omega_0^2 = 1/L$ has been substituted.

$$v_R(t) = 2\omega_0 V_{SRC}te^{-\omega_0 t}, \quad t \geq 0^+ \quad (12.82)$$

where $R/L = 2\omega_0$ has been substituted.

$$v_L(t) = L\frac{di}{dt} = V_{SRC}e^{-\omega_0 t}\left(1 - \omega_0 t\right), \quad t \geq 0^+ \quad (12.83)$$

Note that Equations 12.69 through 12.72 for the overdamped response, Equations 12.75 through 12.78 for the underdamped response, and Equations 12.80 through 12.83 for the critically damped response all satisfy initial conditions, final steady-state conditions, and KVL: (1) $v_C(0^+) = 0$, $i(0^+) = 0$, $v_L(0^+) = V_{SRC}$; (2) $i(\infty) = 0$, $v_L(\infty) = 0$, $v_C(\infty) = V_{SRC}$; and (3) $v_R + v_L + v_C = V_{SRC}$, all in accordance with KVL.

The responses of the series *RLC* circuit during charging, with zero initial conditions, are summarized in the first column of Table 12.4.

12.3.3 Comparison of Responses

It is important to appreciate the differences in the step responses of the overdamped, critically damped, and underdamped circuits. In Figure 12.19, a voltage of 10 V is applied at $t = 0$ to a series *RLC* circuit having $L = 0.5$ H and $C = 12.5$ μF, as used in Section 12.1. Figures 12.20 and 12.21 show the variation of i and v_C for $t \geq 0^+$ and for $R = 100$, 200, 400, and 700 Ω. The responses are overdamped for $R = 700$ Ω, are critically damped for $R = 400$ Ω, and are underdamped for $R = 100$ and 200 Ω. The simulation procedure is described in Example 12.2.

All the current plots in Figure 12.20 have the same slope of (40 mA)/(2 ms) = 20 A/s at $t = 0$. This is because $(di/dt)_{0^+} = v_L(0^+)/L = V_{SRC}/L = 20$ A/s, irrespective of R. As to be expected, the overdamped response has the smallest peak current value, because of the largest resistance. This peak value increases for smaller values of resistance. The two underdamped responses are oscillatory, the oscillations becoming more marked as the resistance decreases.

The following characteristics of the responses of v_C shown in Figure 12.21 should be noted:

1. The overdamped response is the slowest of the responses to approach the steady-state value. This is to be expected, since v_C is proportional to the integral of the current, and the amplitude of the current decreases with increasing R (Figure 12.20). This means that the larger R, the smaller the area under the current curve is, at least up to t between 7 and 8 ms in Figure 12.20. The area under the i vs. t curve is the same for all R as $t \to \infty$, because the final value of v_C is the same

TABLE 12.4

Charging of Dual Circuits, $t \geq 0^+$, with Zero Initial Energy Storage (Assigned Positive Directions as in Figures 12.17 and 12.25)

	Series *RLC* Circuit	Parallel *GCL* Circuit
Overdamped responses	$i = \frac{V_{SRC}}{L(s_1 - s_2)}\left[e^{s_1 t} - e^{s_2 t}\right]$	$v = \frac{I_{SRC}}{C_p(s_1 - s_2)}\left[e^{s_1 t} - e^{s_2 t}\right]$
	$v_R = \frac{2\alpha V_{SRC}}{(s_1 - s_2)}\left[e^{s_1 t} - e^{s_2 t}\right]$	$i_G = \frac{2\alpha I_{SRC}}{(s_1 - s_2)}\left[e^{s_1 t} - e^{s_2 t}\right]$
	$v_L = \frac{V_{SRC}}{(s_1 - s_2)}\left[s_1 e^{s_1 t} - s_2 e^{s_2 t}\right]$	$i_C = \frac{I_{SRC}}{(s_1 - s_2)}\left[s_1 e^{s_1 t} - s_2 e^{s_2 t}\right]$
	$v_C = V_{SRC}\left[\frac{s_2}{s_1 - s_2}e^{s_1 t} - \frac{s_1}{s_1 - s_2}e^{s_2 t} + 1\right]$	$i_L = I_{SRC}\left[\frac{s_2}{s_1 - s_2}e^{s_1 t} - \frac{s_1}{s_1 - s_2}e^{s_2 t} + 1\right]$
Critically damped responses	$i = \frac{V_{SRC}}{L}te^{-\omega_0 t}$	$v = \frac{I_{SRC}}{C_p}te^{-\omega_0 t}$
	$v_R = 2\alpha V_{SRC}te^{-\omega_0 t}$	$i_G = 2\alpha V_{SRC}te^{-\omega_0 t}$
	$v_L = V_{SRC}e^{-\omega_0 t}(1 - \omega_0 t)$	$i_C = I_{SRC}e^{-\omega_0 t}(1 - \omega_0 t)$
	$v_C = V_{SRC}\left[1 - e^{-\omega_0 t} - \omega_0 te^{-\omega_0 t}\right]$	$i_L = I_{SRC}\left[1 - e^{-\omega_0 t} - \omega_0 te^{-\omega_0 t}\right]$
Underdamped responses	$i = \frac{V_{SRC}}{\omega_d L}e^{-\alpha t}\sin\omega_d t$	$v = \frac{I_{SRC}}{\omega_{dp}C_p}e^{-\alpha_p t}\sin\omega_{dp} t$
	$v_R = \frac{2\alpha V_{SRC}}{\omega_d}e^{-\alpha t}\sin\omega_d t$	$i_G = \frac{2\alpha I_{SRC}}{\omega_{dp}}e^{-\alpha_p t}\sin\omega_{dp} t$
	$v_L = V_{SRC}e^{-\alpha t}\left(\cos\omega_d t - \frac{\alpha}{\omega_d}\sin\omega_d t\right)$	$i_C = I_{SRC}e^{-\alpha_p t}\left(\cos\omega_{dp} t - \frac{\alpha_p}{\omega_{dp}}\sin\omega_{dp} t\right)$
	$v_C = V_{SRC}\left[1 - e^{-\alpha t}\left(\cos\omega_d t + \frac{\alpha}{\omega_d}\sin\omega_d t\right)\right]$	$i_L = I_{SRC}\left[1 - e^{-\alpha_p t}\left(\cos\omega_{dp} t + \frac{\alpha_p}{\omega_{dp}}\sin\omega_{dp} t\right)\right]$

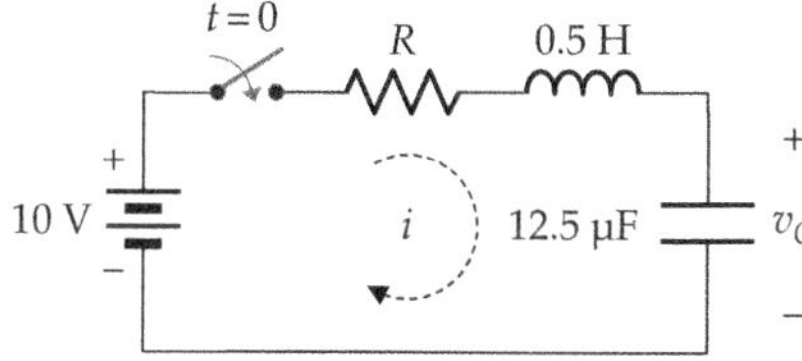

FIGURE 12.19
A voltage of 10 V applied to a series *RLC* circuit.

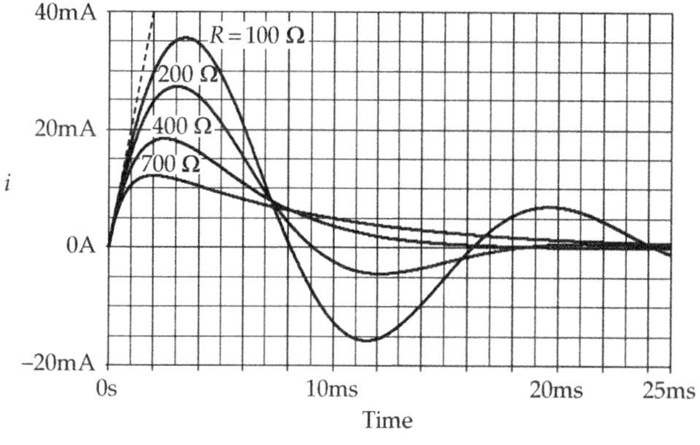

FIGURE 12.20
Current responses in the circuit of Figure 12.19 for various values of resistance.

2. The critically damped response is the fastest of the responses to approach the steady-state value *without overshoot*.
3. The underdamped response is the fastest to reach the steady-state value, but overshoots this value and oscillates about it at a frequency ω_d before settling to the final value. As the resistance decreases, the overshoot increases, and the oscillations become more pronounced. In the limit, as $R \to 0$, $\omega_d \to \omega_0$.

If we consider the 9 V level in Figure 12.21, for example, the overdamped response (700 Ω) reaches this level at 19.2 ms, the critically damped response at 9.72 ms, the underdamped response having $R = 200\ \Omega$ at 5.31 ms, and the underdamped response having $R = 100\ \Omega$ at 4.32 ms. However, the underdamped response can have unacceptable overshoot and undershoot.

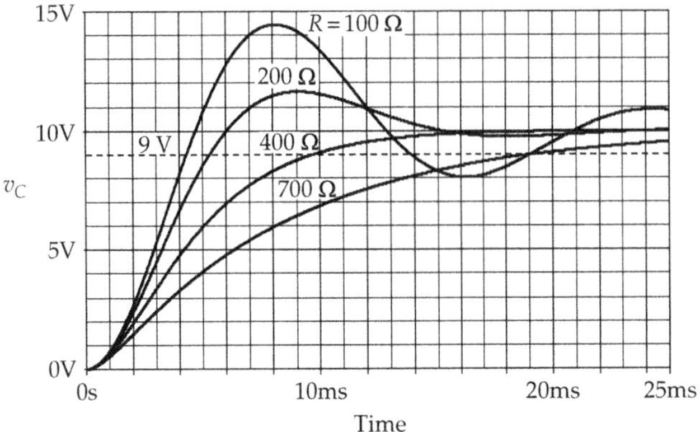

FIGURE 12.21
Capacitor voltage response in the circuit of Figure 12.19 for various values of resistance.

Example 12.2: Simulation of Step Responses of Series *RLC* Circuit

It is required to simulate the circuit in Figure 12.17 to obtain the variation of i and v_c for $t \geq 0^+$ and for $R = 100$, 200, 400, and 700 Ω.

Simulation: The schematic is entered as in Figure 12.22, with zero initial conditions (IC = 0) in the inductor and capacitor. PSpice automatically applies the 10 V VDC source at the start of the simulation ($t = 0$) without the need to connect a switch that closes at $t = 0$. In order to display a family of curves for various values of R, R is first declared as a global parameter, as follows: double-click on the default resistance value displayed, which invokes the 'Display Properties' window. In the 'Value' field, enter a chosen designation enclosed in curly brackets, which tells PSpice that this is a parameter and not a fixed value. In the present example, {*R*_val} is entered. The next step is to declare {*R*_val} a global parameter. Place the part 'PARAM' from the 'SPECIAL' library; this shows on the schematic as 'PARAMETERS:.' When this word is double-clicked, the Property Editor spreadsheet is displayed. Click on the 'New Column' button to display the 'Add New Column' dialog box. Enter R_val in the 'Name' field and any value, say, 100, in the 'Value' field.

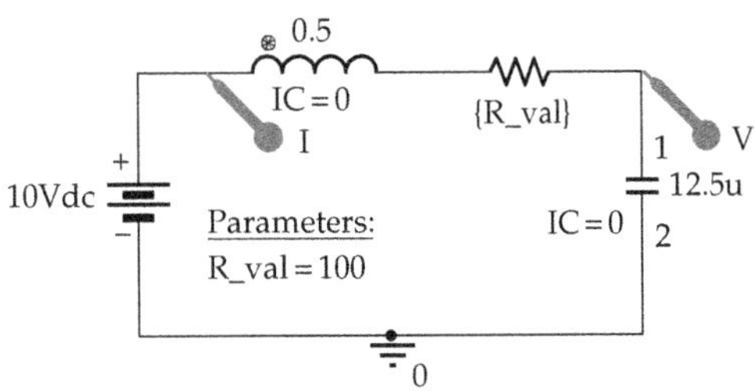

FIGURE 12.22
Figure for Example 12.2.

A new column R_val is added to the spreadsheet with the entry 100. To have this displayed on the schematic, click on the 'Display' button and choose 'Name' and 'Value' in the 'Display Properties' dialog box. R_val = 100 appears under 'PARAMETERS:,' as shown in Figure 12.22.

To run the simulation, select 'Time Domain (Transient)' in the Simulation. Select 'General Settings' under 'Options' and enter 25m for 'Run to time', 0 for 'Start saving data after', and 5u for 'Maximum step size.' Then select 'Parametric Sweep' under 'Options', choose 'Global parameter' under 'Sweep variable', and enter R_val for 'Parameter name'. In the 'Value list' under 'Sweep type', enter: 100,200,400,700. Apply a current marker to the marked terminal of the inductor, without the voltage marker at the capacitor terminal, and run the simulation. An available sections window is displayed that allows you to select plots for one or more values of R. This is useful for identifying which trace belongs to which value of R. When all traces are selected, the current plots of Figure 12.20 are displayed. When the simulation is run with the voltage marker applied at the capacitor terminal, without the current marker at the inductor terminal, the plots of Figure 12.21 are displayed.

Primal Exercise 12.18

An underdamped, series *RLC* circuit has $\omega_d = 30$ krad/s and $\alpha = 40$ krad/s. If $C = 1$ μF, determine R.

Ans. 32 Ω.

Primal Exercise 12.19

Given a critically damped, series *RLC* circuit having $L = 1$ mH and $R = 2$ kΩ. Determine C.

Ans. 1 nF.

Primal Exercise 12.20

The switch in Figure 12.23 is closed at $t = 0$ with no energy initially stored in the circuit. Determine the largest value of di/dt.

Ans. 2 A/ms at $t = 0$.

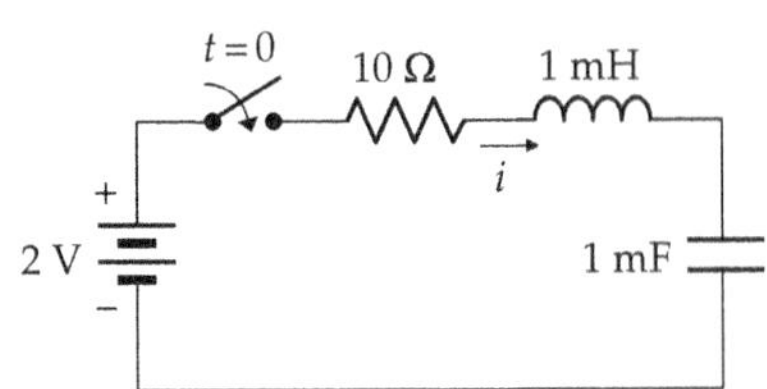

FIGURE 12.23
Figure for Primal Exercise 12.20.

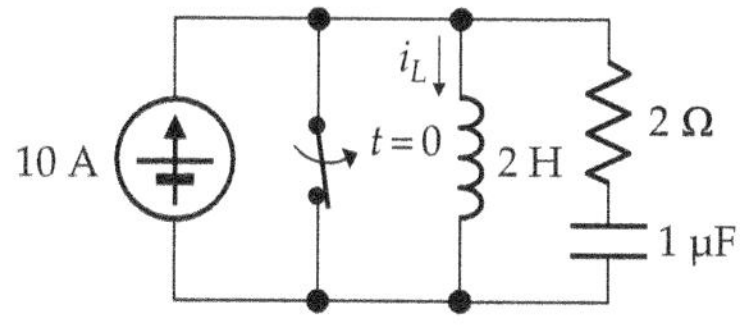

FIGURE 12.24
Figure for Primal Exercise 12.21.

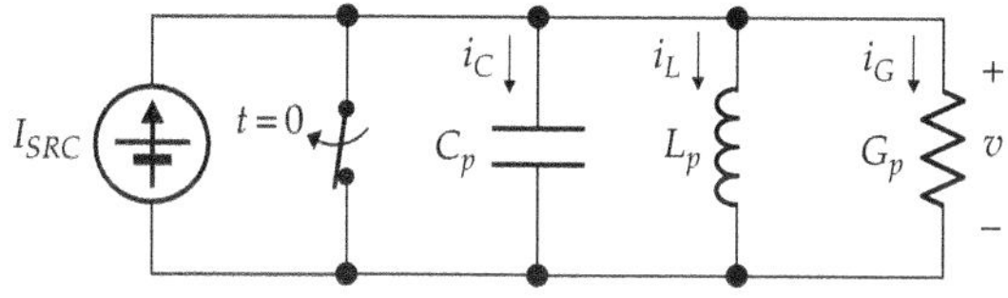

FIGURE 12.25
Charging of a parallel *GCL* circuit.

Primal Exercise 12.21

The switch in Figure 12.24 is opened at $t = 0$ with no initial energy storage in the circuit. Determine di_L/dt at $t = 0^+$.

Ans. 10 A/s.

12.3.4 Charging of Parallel *GCL* Circuit

The dual of the series *RLC* circuit charged by a dc voltage V_{SRC} is the parallel *GCL* circuit charged by a dc current I_{SRC} (Figure 12.25), where the dual quantities are those listed in Table 12.3 and with the value of I_{SRC} numerically equal to that of V_{SRC}. The responses are derived in a manner exactly analogous to that of the series *RLC* circuit for the three cases of overdamping, critical damping, and underdamping. These responses are summarized in the second column of Table 12.4, for zero initial conditions. The dual of the response i of the series *RLC* circuit in Figure 12.17 is v of the parallel *GCL* parallel circuit, whereas the dual response of v_C of the series *RLC* circuit in Figure 12.17 is i_L of the parallel *GCL* parallel circuit.

12.4 Procedure for Analyzing Prototypical Second-Order Circuits

All second-order circuits containing both inductors and capacitors that are considered in this chapter are reducible to the prototypical series *RLC* or parallel *GCL* circuits discussed in the preceding sections. An essential difference between the series and parallel configurations when deriving the circuit responses is the expression for α. The expression for ω_0 is $1/\sqrt{LC}$ in both cases. But $\alpha = R/2L$ in the case of the series circuit, whereas in the case of the parallel circuit,

$$\alpha_p = \frac{G_p}{2C_p} = \frac{1}{2R_pC_p} \tag{12.84}$$

where $R_p = 1/G_p$. Only when the series and parallel circuits are duals are the numerical values of α and α_p equal to one another. It is seen that, for given L and C, *increasing the resistance R in the series circuit increases α and the damping*, whereas *increasing the resistance R_p in the parallel circuit decreases α_p and the damping*.

In the absence of independent sources, differential equations governing circuit responses are homogeneous, which means that the responses are purely transient in nature, with zero responses in the final steady state. In the presence of independent sources:

Complete response = transient response + final, steady-state response

The transient response is the solution of the homogeneous differential equation, and the final, steady-state response is the solution of the nonhomogeneous differential equation, for $t \geq 0^+$, with all time derivatives set to zero.

It is important to bear in mind the following:

1. *When determining the transient response, independent sources are set to zero.* The argument is analogous to that made in Section 11.5, to justify that the time constant is not affected by independent sources. First, independent sources do not appear in the homogeneous differential equation; second, as voltages and currents change in the course of the transient responses, ideal voltage sources act as short circuits, and ideal current sources act as open circuits, as far as the changes are concerned.
2. Setting independent sources to zero also ascertains, in case of uncertainty, whether the circuit configuration is series or parallel and hence whether to use the expression for α or α_p.
3. The arbitrary constants are part of the transient response. However, they should be determined by *applying initial conditions to the complete response*.
4. Since all voltages and currents in a given circuit are governed by the same homogeneous differential equation, all the responses have the same ω_0, α, or α_p.

The preferred procedure for analyzing a prototypical, second-order *RLC* circuit is not based on deriving the differential equation for a particular circuit variable, as this complicates the analysis unnecessarily. The general procedure is as follows:

1. *Determine the initial values of the circuit variables of interest at $t = 0^-$ and $t = 0^+$ as well as their final, steady-state values.*
2. *Set independent source to zero in the circuit for $t \geq 0^+$ in order to ascertain whether the configuration is series or parallel and to derive the transient response.*
3. *Reduce the circuit to a prototypical series RLC or parallel GCL circuit, as appropriate. Effective values of circuit parameters may have to be derived, including the effects of any dependent sources.*
4. *Determine ω_0, α, or α_p, to ascertain the type of circuit damping, that is, whether the circuit is overdamped, underdamped, or critically damped.*
5. *Decide on the circuit variable whose response is to be derived first. It is generally more convenient to derive v_C first in a series circuit and i_L first in a parallel circuit.*
6. (a) *If the circuit is overdamped, any response of the circuit is of the form*

$$y(t) = Ae^{s_1 t} + Be^{s_2 t} + Y_F, \quad t \geq 0^+ \qquad (12.85)$$

where y is the circuit variable of interest and Y_F is the final, steady-state value of y, and is either zero or a constant. s_1 and s_2 are determined in accordance with Equation 12.7.

(b) *If the circuit is critically damped, any response is of the form*

$$y(t) = e^{-\omega_0 t}(A + Bt) + Y_F, \quad t \geq 0^+ \qquad (12.86)$$

Note that ω_0 is the same, whether the circuit is series or parallel.

(c) *If the circuit is underdamped, any response is of the form*

$$y(t) = e^{-\alpha t}(A\cos\omega_d t + B\sin\omega_d t) + Y_F, \quad t \geq 0^+ \qquad (12.87)$$

7. *The arbitrary constants A and B in Equations 12.85 through 12.87 are determined from $y(0^+)$ and $(dy/dt)_{0^+}$.*
8. *Once the response of the variable of interest is obtained, other circuit variables can be derived from the v–i relations of the circuit elements.*
9. *Always check that all responses satisfy initial conditions; final, steady-state conditions; and KCL or KVL, as may be appropriate.*

The procedure is illustrated by the following examples and by problems at the end of the chapter.

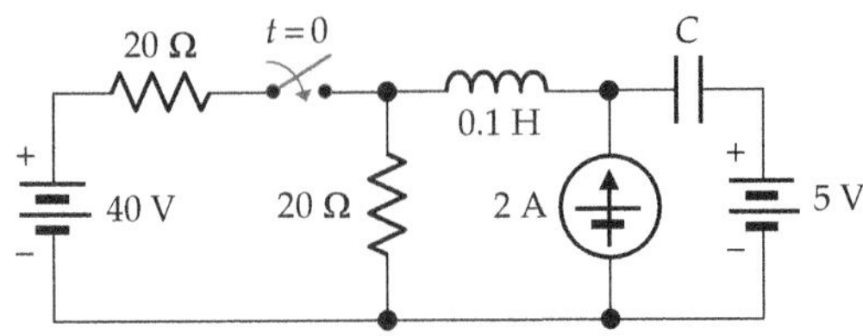

FIGURE 12.26
Figure for Primal Exercise 12.22.

Primal Exercise 12.22

The switch in Figure 12.26 is closed at $t = 0$ after being open for a long time. Determine whether the circuit responses are overdamped, underdamped, or critically damped, assuming that C is (a) 25 mF, (b) 4 mF, or (c) 1 mF.

Ans. (a) overdamped; (b) critically damped; (c) underdamped.

Example 12.3: Responses of Series *RLC* Circuit with Initial Energy Storage

The switch in Figure 12.27 is moved at $t = 0$ from position 'a' to position 'b' after being in position 'a' for a long time. The switch shown is a make-before-break type of switch, which makes contact with terminal 'b' before breaking with terminal 'a'. This ensures that the ideal current source is not open-circuited as the current is diverted from the *RLC* circuit to R_D. It is required to determine (a) R for critical damping and (b) the circuit responses for $t \geq 0^+$.

Solution:

At $t = 0^-$, after the switch has been in position 'a' for a long time, the inductor acts as a short circuit under steady conditions and the capacitor as an open circuit (Figure 12.28a). It follows that $I_{L0} = 0.05$ A and $V_{C0} = 0.05R$ V. For $t \geq 0^+$, the current source is diverted

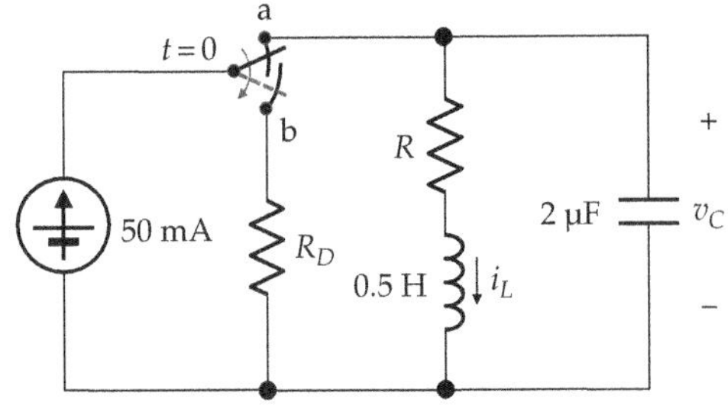

FIGURE 12.27
Figure for Example 12.3.

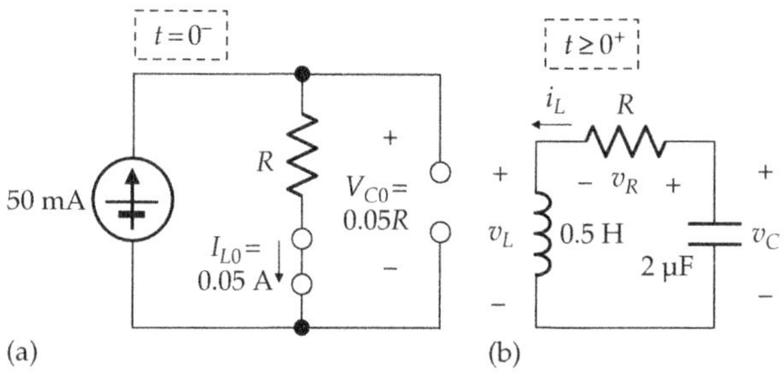

FIGURE 12.28
Figure for Example 12.3.

to R_D and isolated from the circuit, which becomes as in Figure 12.28b.

(a) For $t \geq 0^+$, the circuit is series RLC. $\omega_0 = 1/\sqrt{LC} = 1/\sqrt{0.5\times 2\times 10^{-6}} = 10^3$ rad/s. For critical damping, $\alpha = R/2L = \omega_0 = 10^3$ rad/s. Hence, $R = 2 \times 0.5 \times 10^3 = 1000\ \Omega$.

(b) Since the circuit is critically damped, all the circuit responses are of the form of Equation 12.86, with final, steady-state values equal to zero, because all initially stored energy is eventually dissipated in the resistor. Since all circuit responses are required, it is convenient to derive the response for v_C first. Hence,

$$v_C(t) = e^{-\omega_0 t}(A + Bt), \quad t \geq 0^+ \tag{12.88}$$

Substituting $v_C(0^+) = V_{C0} = 0.05 \times 1000 = 50$ V gives $A = 50$ V. To obtain B, Equation 12.88 is differentiated, the right-hand side multiplied by C and set equal to -0.05 at $t = 0^+$; the minus sign is used because i_L is in a direction of a voltage rise across C. The resulting equation is $2 \times 10^{-6}(-50 \times 10^3 + B) = -0.04$, which gives $B = 25 \times 10^3$ V/s. Hence,

$$v_C(t) = e^{-10^3 t}\left(50 + 25\times 10^3 t\right)\ \text{V}, \quad t \geq 0^+\ \text{s}$$

or

$$v_C(t) = e^{-t}(50 + 25t)\ \text{V}, \quad t \geq 0^+\ \text{ms} \tag{12.89}$$

$$i_L(t) = -C\frac{dv_C}{dt} = -2\times 10^{-6}\left[-10^3 e^{-10^3 t}\left(50 + 25\times 10^3 t\right) + e^{-10^3 t}\times 25\times 10^3\right]$$

or

$$i_L(t) = e^{-10^3 t}(0.05 + 50t)\ \text{A}, \quad t \geq 0^+\ \text{s}$$

or

$$i_L(t) = 50e^{-t}(1 + t)\ \text{mA}, \quad t \geq 0^+ \text{ms} \tag{12.90}$$

$$v_R(t) = i_L(t)\text{V, where } i_L(t) \text{ is in mA} \tag{12.91}$$

$$v_L(t) = L\frac{di_L}{dt} = 0.5\left[-10^3 e^{-10^3 t}(0.05 + 50t) + e^{-10^3 t}\times 50\right]$$

or

$$v_L(t) = -25\times 10^3 te^{-10^3 t}\ \text{V}, \quad t \geq 0^+\ \text{s}$$

or

$$v_L(t) = -25te^{-t}\ \text{V}, \quad t \geq 0^+\ \text{ms} \tag{12.92}$$

It can be readily verified that the expressions for the responses are the sum of the corresponding expressions in the two columns for critical damping in Table 12.2, bearing in mind that the relative positive polarity of v_C is opposite that assumed in the table. The responses satisfy the initial conditions, the final steady-state conditions, and KVL. Note that if t is in ms instead of s, the responses assume simpler forms.

Simulation: The circuit is entered as in Figure 12.29 with initial values as indicated, taking into account the polarity markings on L and C to correspond to the assigned positive directions in Figure 12.28b. In the Simulation Settings, 'Analysis type' is 'Time Domain (Transient),' 'Run to time' is 5m, 'start saving data after' is 0, and 'Maximum step size' is 1u. After the simulation is run, the plots of Figure 12.30 are displayed. v_C decreases monotonically as C discharges. i_L also decreases monotonically, but $di_L/dt = 0$ at $t = 0$ because $v_L = 0$ at $t = 0$. Since i_L is decreasing, v_L is always negative ($-v_L > 0$). v_L is bound to have a minimum (or $-v_L$ a maximum) because it starts at zero at $t = 0$ and ends at zero as $t \to \infty$, but is not zero in between. The minimum of v_L occurs at the point of inflection of i_L.

Problem-Solving Tips

- It is safer to analyze second-order circuits using the standard units of amperes, volts, and seconds. Conversion to more convenient units can be left to the end.

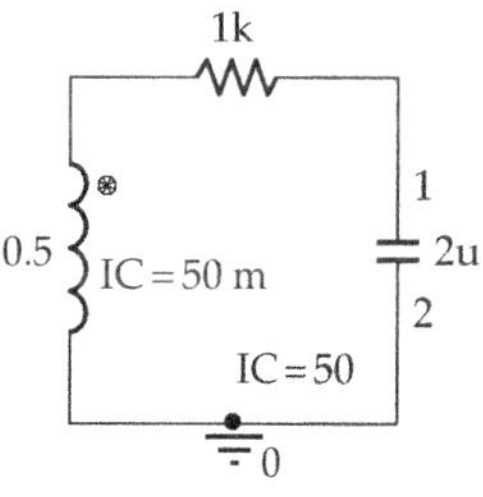

FIGURE 12.29
Figure for Example 12.3.

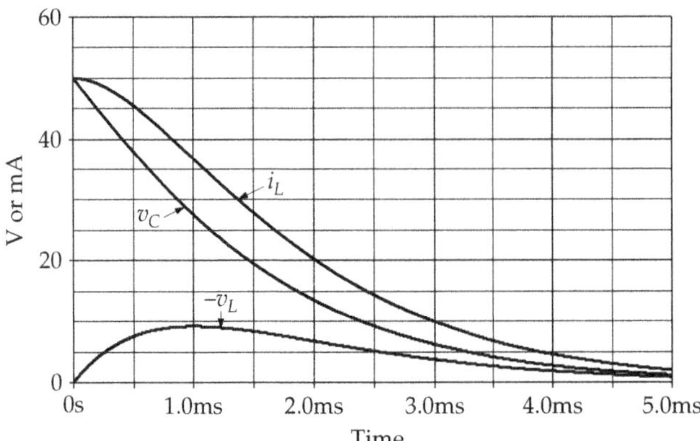

FIGURE 12.30
Figure for Example 12.3.

- It is often convenient and helpful to keep α, α_p, ω_0, ω_d, s_1, and s_2 as symbols and to substitute numerical values at the end of the analysis or wherever necessary in the course of the analysis. The same is sometimes true of the circuit parameters R, L, and C.
- When differentiating expressions containing arbitrary constants in order to apply initial conditions, ignore terms in t or $\sin\omega_0 t$ that are zero at $t = 0$.

Exercise 12.23

Verify that the responses of Example 12.3 satisfy the corresponding entries in Table 12.2.

Exercise 12.24

Verify that the minimum of v_L and the point of inflection of i_L in Example 12.3 both occur at $t = 1$ ms. Use the expressions with t in ms and ignore the multiplying coefficients.

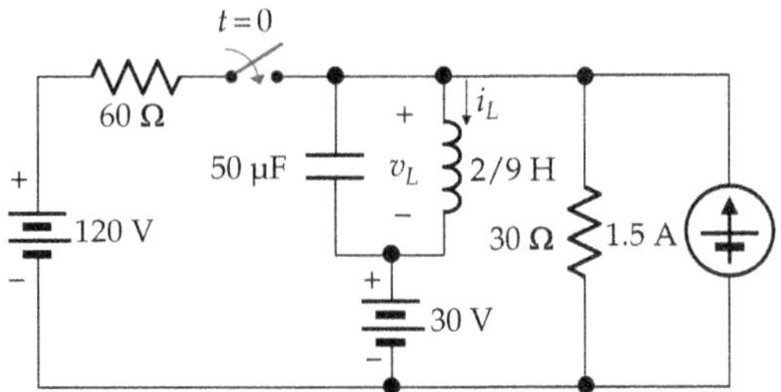

FIGURE 12.31
Figure for Example 12.4.

Example 12.4: Forced Responses of Parallel *GCL* Circuit with Initial Energy Storage

The switch in Figure 12.31 is closed at $t = 0$ after being open for a long time. It is required to determine $i_L(t)$ and $v_L(t)$ for $t \geq 0^+$.

Solution:

In the steady state at $t = 0^-$, the inductor acts as a short circuit and the capacitor as an open circuit, as shown in Figure 12.32a. Taking the lower node as reference, the voltage of the upper node 'a' is 30 V. The current flowing downward in the 30 Ω resistor is therefore 1 A. From KCL, $I_{L0} = 0.5$ A, and $V_{L0} = 0$.

In the final steady state, as $t \to \infty$, the inductor again acts as a short circuit and the capacitor as an open circuit, but with the 120 V source and the 60 Ω resistor added to the circuit, as in Figure 12.32b. V_a is 30 V, so that the current through the 60 Ω resistor is (120–30)/60 = 1.5 A; $i_L(\infty) = I_{LF} = 2$ A, and $v_L(\infty) = V_{LF} = 0$.

When the independent sources are set to zero in the circuit for $t \geq 0^+$ and the 60 and 30 Ω resistors are combined in parallel into a 20 Ω resistor, the circuit reduces to the parallel circuit of Figure 12.32c. $\omega_0 = 1/\sqrt{LC} = 1/\sqrt{(2/9)\times 50\times 10^{-6}} = 300$ rad/s. From Equation 12.84, $\alpha_p = 1/\left(2\times 20\times 50\times 10^{-6}\right) = 500$ rad/s. Since $\alpha_p > \omega_0$,

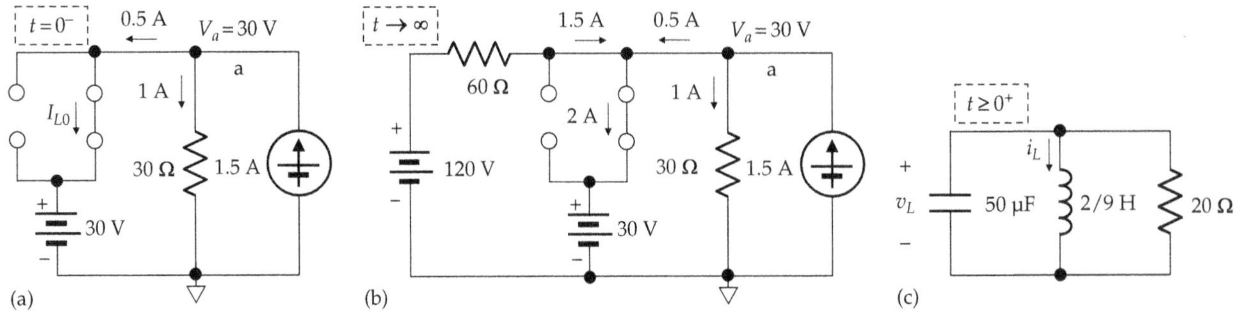

FIGURE 12.32
Figure for Example 12.4.

the circuit is overdamped. The roots of the characteristic equation are:

$s_1 = -500 + 100\sqrt{25-9} = -100$ rad/s and $s_2 = -5\times100 - 100\sqrt{25-9} = -900$ rad/s.

Since the circuit is a parallel circuit, it is convenient to derive first the response for i_L. The complete response for i_L is of the form of Equation 12.85:

$$i_L(t) = Ae^{s_1t} + Be^{s_2t} + 2\text{ A} \tag{12.93}$$

The initial conditions are $i_L(0^+) = 0.5$ A and $v_L(0^+) = L(di_L/dt)_{0+} = 0$, since the capacitor voltage is initially zero. The circuit at $t = 0^+$ is shown in Figure 12.33. The condition that $i_L(0^+) = 0.5$ A gives

$$A + B = -1.5 \tag{12.94}$$

To obtain a second equation in A and B, Equation 12.93 is differentiated and the right-hand side set equal to zero at $t = 0^+$. The resulting equation is

$$A + 9B = 0 \tag{12.95}$$

Solving Equations 12.94 and 12.95 gives $A = -1.6875$ A and $B = 0.1875$ A. Hence,

$$i_L(t) = -1.6875e^{-100t} + 0.1875e^{-900t} + 2\text{ A},\quad t \geq 0^+ \tag{12.96}$$

$$v_L(t) = L\frac{di_L}{dt} = \frac{2}{9}\left(168.75e^{-100t} - 168.75e^{-900t}\right)\text{ A}$$

or

$$v_L(t) = 37.5\left(e^{-100t} - e^{-900t}\right)\text{ V},\quad t \geq 0^+ \tag{12.97}$$

i_L and v_L satisfy the initial conditions and the final steady-state conditions.

Simulation: The circuit is entered as in Figure 12.34 with initial values as indicated, taking into account the polarity marking on L to correspond to the assigned positive direction of i_L in Figure 12.31. In the Simulation Settings, 'Analysis type' is 'Time Domain (Transient),' 'Run to time' is 50m, 'Start saving data after' is 0, and 'Maximum step size' is 1u. After the simulation is run, the plots of Figure 12.35 are displayed. i_L increases monotonically from its initial value of 0.5 A to its final value of 2 A, with $di_L/dt = 0$ at $t = 0$ because $V_{L0} = 0$ at $t = 0$. Since i_L is increasing, v is always positive. v is bound to have a maximum because it starts at zero at $t = 0$ and ends at zero as $t \to \infty$, but is not zero in between. The maximum of v_L occurs at the point of inflection of i_L, as in Example 12.3.

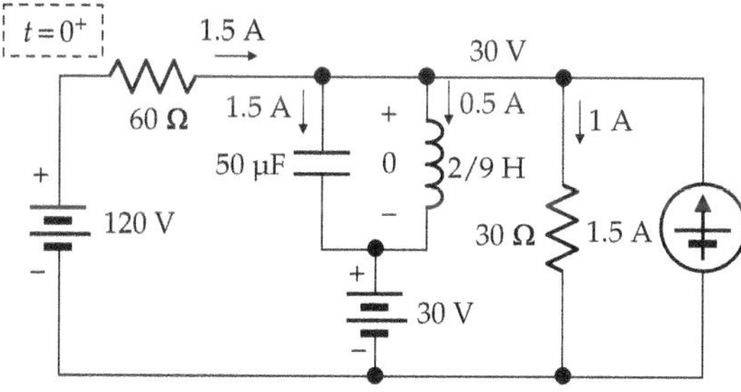

FIGURE 12.33
Figure for Example 12.4.

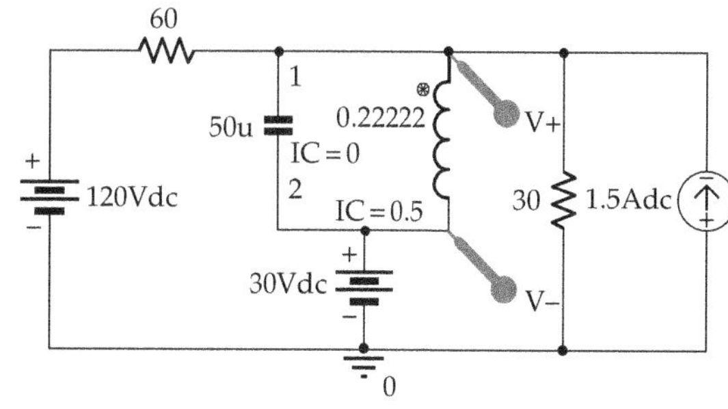

FIGURE 12.34
Figure for Example 12.4.

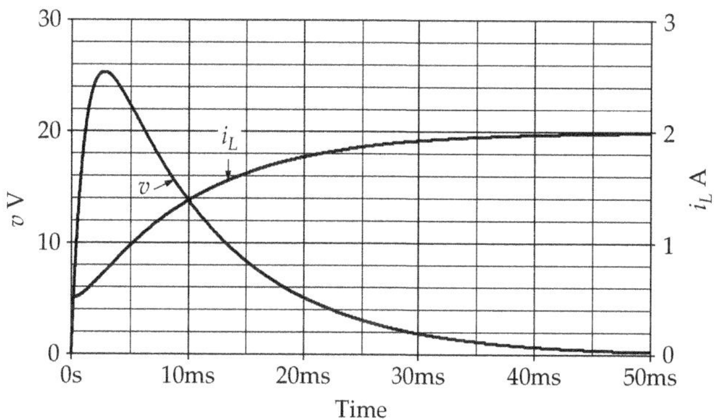

FIGURE 12.35
Figure for Example 12.4.

Exercise 12.25

Verify $v_C = v_L$ in Example 12.4 from the v–i relation of the capacitor.

Example 12.5: Forced Responses of Series *RLC* Circuit with Dependent Source and Initial Energy Storage

The switch in Figure 12.36 is closed at $t = 0$ after being open for a long time. It is required to determine the responses for $t \geq 0^+$, assuming an initial charge of 5 V on the capacitor.

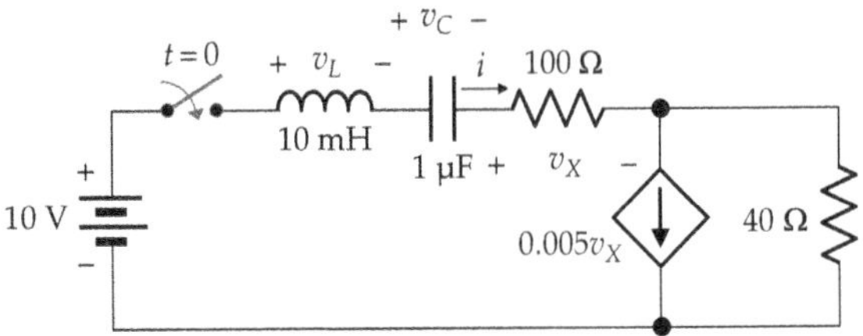

FIGURE 12.36
Figure for Example 12.5

Solution:

For $t \geq 0^+$, the circuit is a series *RLC* circuit but with an effective resistance because of the resistive circuit that includes a dependent source. The initial values at $t = 0^-$ are $i(0^-) = 0$, $v_L(0^-) = 0$, $v_C(0^-) = 5$ V, and $v_X(0^-) = 0$. At $t = 0^+$, i and v_C do not change, because the stored energy does not change at the instant of switching, so that $i(0^+) = 0$, $v_C(0^+) = 5$ V, $v_X(0^+) = 0$, the dependent current source is zero, and the voltage drop across the two resistors is zero. It follows that $v_L(0^+) = 5$ V in order to satisfy KVL. As $t \to \infty$, $i(\infty) \to 0$, since the capacitor acts as an open circuit, $v_L(\infty) \to 0$, since the inductor acts as a short circuit. $v_X(\infty) = 0$, the dependent current source is zero, and the voltage drop across the two resistors is zero. It follows that $v_C(\infty) = 10$ V to satisfy KVL.

The next step is to determine the effective series resistance. The general procedure is to apply KVL, KCL, and Ohm's law, as illustrated in Figure 12.37. The dependent source current is $0.005v_X = 0.005 \times 100i = 0.5i$. The current in the 40 Ω resistor is therefore $0.5i$. Hence, $v_R = 100i + 40 \times 0.5i = 120i$, so that the effective series resistance is $v_R/i = 120$ Ω.

Alternatively, the source absorption theorem can be applied in this case. The voltage across the 40 Ω resistor is $40(i - 0.005v_x) = 40(0.01v_x - 0.005v_x) = 40 \times 0.005v_x$. The dependent source is therefore equivalent to a resistance $40 \times 0.005v_x/0.005v_x = 40$ Ω. The effective resistance is 100 Ω in series with two 40 Ω resistances in parallel, which gives 120 Ω.

$\omega_0 = 1/\sqrt{LC} = 1/\sqrt{10^{-2} \times 10^{-6}} = 10^4$ rad/s. $\alpha = R/2L = 120/(2 \times 10^{-2}) = 6 \times 10^3$ rad/s. As $\alpha < \omega_0$, the circuit is underdamped, and $\omega_d = \sqrt{\omega_0^2 - \alpha^2} = 10^3\sqrt{100 - 36} = 8 \times 10^3$ rad/s. Since the circuit is a series circuit and all responses are required, it is convenient to derive first the response for v_C. The complete response for v_C is of the form of Equation 12.87:

$$v_C(t) = e^{-\alpha t}\left(A\cos\omega_d t + B\sin\omega_d t\right) + 10, \quad t \geq 0^+ \qquad (12.98)$$

From the initial condition $v_C(0^+) = 5$ V, it follows that $A = -5$ V. To obtain B, Equation 12.98 is differentiated and the right-hand side set equal to 0 at $t = 0^+$, since $i(0^+) = 0$. The resulting equation, with only the cosine terms considered is

$$\frac{dv_C}{dt} = 5\alpha e^{-\alpha t}\cos\omega_d t + \omega_d B e^{-\alpha t}\cos\omega_d t + X(t) \qquad (12.99)$$

where $X(t)$ consists of sine terms and is zero at $t = 0^+$. Equating the RHS to zero at $t = 0^+$ gives $B = -5\alpha/\omega_d$. It follows that

$$v_C(t) = -5e^{-\alpha t}\left(\cos\omega_d t + \frac{\alpha}{\omega_d}\sin\omega_d t\right) + 10, \ t \geq 0^+ \qquad (12.100)$$

$$i(t) = C\frac{dv_C}{dt} = -5C\left[-\alpha e^{-\alpha t}\left(\cos\omega_d t + \frac{\alpha}{\omega_d}\sin\omega_d t\right)\right.$$

$$\left. + e^{-\alpha t}\left(-\omega_d\sin\omega_d t + \alpha\cos\omega_d t\right)\right] = 5Ce^{-\alpha t}\left[\left(\frac{\alpha^2}{\omega_d} + \omega_d\right)\sin\omega_d t\right]$$

$$= 5Ce^{-\alpha t}\left[\left(\frac{\alpha^2 + \omega_d^2}{\omega_d}\right)\sin\omega_d t\right]$$

or

$$i(t) = \frac{5}{\omega_d L}e^{-\alpha t}\sin\omega_d t, \quad t \geq 0^+ \qquad (12.101)$$

$$v_L(t) = L\frac{di}{dt} = \frac{5}{\omega_d}e^{-\alpha t}\left(-\alpha\sin\omega_d t + \omega_d\cos\omega_d t\right)$$

or

$$v_L(t) = 5e^{-\alpha t}\left(\cos\omega_d t - \frac{\alpha}{\omega_d}\sin\omega_d t\right) \qquad (12.102)$$

Substituting numerical values and replacing 10^3t s by t ms,

$$v_C(t) = 5\left(2 - e^{-6t}\left(\cos 8t + 0.75\sin 8t\right)\right) \text{ V}, \quad t \geq 0^+ \text{ ms} \qquad (12.103)$$

$$i(t) = 62.5e^{-6t}\sin 8t \text{ mA}, \quad t \geq 0^+ \qquad (12.104)$$

$$v_L(t) = 5e^{-6t}\left(\cos 8t - 0.75\sin 8t\right) \text{ V}, \quad t \geq 0^+ \text{ ms} \qquad (12.105)$$

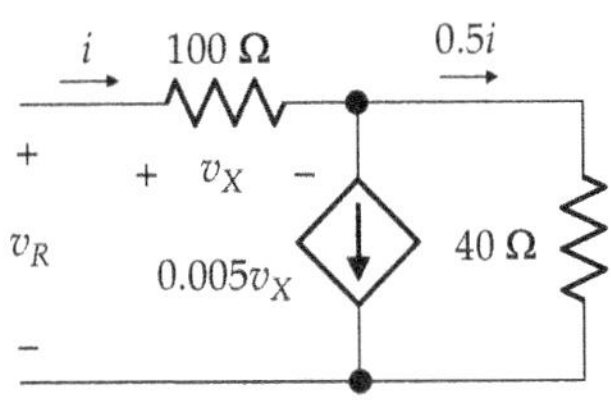

FIGURE 12.37
Figure for Example 12.5.

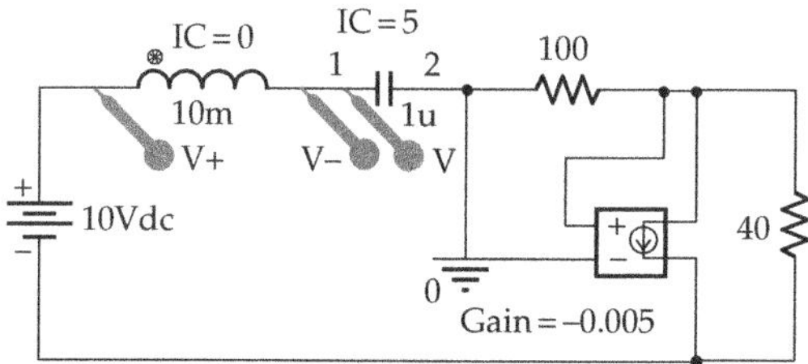

FIGURE 12.38
Figure for Example 12.5.

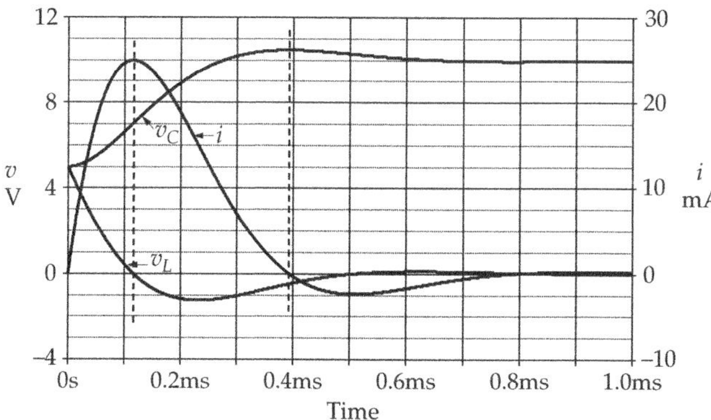

FIGURE 12.39
Figure for Example 12.5.

Simulation: The circuit is entered as in Figure 12.38 with initial values as indicated, taking into account the polarity marking on C to correspond to the assigned positive direction of i in Figure 12.36. In the Simulation Settings, 'Analysis type' is 'Time Domain (Transient),' 'Run to time' is 1m, 'Start saving data after' is 0, and 'Maximum step size' is 0.1u. After the simulation is run, the plots of Figure 12.39 are displayed. It is seen that the responses are only weakly oscillatory because the circuit is not heavily underdamped. i increases when the switch is closed, which increases the charge on the capacitor. As the capacitor voltage increases, the rate of increase of i decreases. Eventually, v_C reaches a maximum, and $i = 0$ at this instant. The capacitor then discharges slightly, and i goes negative. v_L is positive for increasing i, is negative for decreasing i, and is zero when i has a maximum or a minimum.

Exercise 12.25

Verify that (a) v_C in Example 12.5 is the same as that obtained from the v–i relation of the capacitor, (b) i is maximum when $v_L = 0$ and at the point of inflection of the v_C curve, and (c) the responses in Example 12.5 can be derived by superposition of the responses due to V_{SRC} acting alone, and V_{c0} acting alone.

Learning Checklist: What Should Be Learned from This Chapter

- The natural response of a prototypical second-order RLC circuit is governed by a second-order homogeneous differential equation: $\frac{d^2y}{dt^2} + 2\alpha\frac{dy}{dt} + \omega_0^2 y = 0$ where y is a voltage or a current, $\alpha = R/2L$ is the damping factor for the series circuit, and $\omega_0 = 1/\sqrt{LC}$ is the resonant frequency. These values determine the nature of the roots s_1 and s_2 of the characteristic equation of the circuit: $s^2 + 2\alpha s + \omega_0^2 = 0$, where s_1 and s_2 depend only on the circuit parameters.
- All the circuit variables in a given circuit obey the same homogeneous differential equation. The same characteristic equation applies to all the circuit variables, which means that all these variables have the same s_1, s_2, α, and ω_0 and, hence, the same form of the natural response.
- The natural responses of the second-order, series circuit for $t \geq 0^+$ are of the form $i(t) = A_i e^{s_1 t} + B_i e^{s_2 t}$, $v_C(t) = A_C e^{s_1 t} + B_C e^{s_2 t}$, and $v_L(t) = A_L e^{s_1 t} + B_L e^{s_2 t}$ where the exponents are the same but the arbitrary constants are, in general, different, and $s_1 \neq s_2$.
- For given L and C, the nature of the roots s_1 and s_2 depends on R:
  1. When R is large enough so that $\alpha > \omega_0$, s_1 and s_2 are negative real and distinct. The responses are said to be overdamped.
  2. When R is such that $\alpha = \omega_0$, then $s_1 = s_2 = -\omega_0$. The responses are said to be critically damped.
  3. When R is small enough so that $\alpha < \omega_0$, s_1 and s_2 are complex, the responses are damped sinusoids. The responses are said to be underdamped.
- The natural responses due to initial energy storage in both the inductor and capacitor can be obtained by superposition, that is, by adding algebraically the responses due to energy storage in the inductor acting alone and in the capacitor acting alone.
- The responses of series RLC circuits and parallel GCL circuits are related by duality considerations. The expression $\omega_0 = 1/\sqrt{LC}$ is the same in both circuits, but for the parallel circuit, $\alpha_p = \frac{G_p}{2C_p} = \frac{1}{2R_pC_p}$, where $R_p = 1/G_p$.
  1. Only when the series and parallel circuits are duals are the numerical values of α and α_p

equal to one another. It is seen that, for given L and C, increasing the resistance R in the series circuit increases α and the damping, whereas increasing the resistance R_p in the parallel circuit decreases α_p and the damping.

- In the charging of a series RLC circuit or a parallel GCL circuit:
  1. The overdamped response is the slowest of the responses to approach the steady-state value.
  2. The critically damped response is the fastest of the responses to approach the steady-state value without overshoot.
  3. The underdamped response is the fastest to reach the steady-state value, but overshoots this value and oscillates about it at a frequency ω_d before settling to the final value.
- In a second-order circuit, the transient response and circuit configuration, that is, whether series or parallel, can be determined after setting all independent sources to zero. This is because the transient response is determined by the homogeneous differential equation, which does not involve independent sources.
- The responses of a second-order LC circuit can be determined without having to derive the differential equation of the circuit by following a systematic procedure of general applicability:
  1. After ascertaining whether the configuration is series or parallel, the values of ω_0, α, or α_p, are derived to determine whether the circuit is overdamped, underdamped, or critically damped.
  2. If the responses of more than one circuit variable are required, it is convenient to derive v_C first in a series circuit and i_L first in a parallel circuit, since all other responses follow from the v–i relations of the circuit elements through successive differentiation.
  3. The general expressions for the response of any variable can be expressed in one of the following forms, for $t \geq 0^+$:

$$\text{Overdamped: } y(t) = Ae^{s_1 t} + Be^{s_2 t} + Y_F$$

$$\text{Critically damped: } y(t) = e^{-\omega_0 t}(A + Bt) + Y_F$$

$$\text{Underdamped: } y(t) = e^{-\alpha t}(A\cos\omega_d t + B\sin\omega_d t) + Y_F$$

  4. The arbitrary constants A and B are determined from $y(0^+)$ and $(dy/dt)_{0^+}$.

Problem-Solving Tips

1. It is safer to analyze second-order circuits using the standard units of amperes, volts, and seconds. Conversion to more convenient units can be left to the end.
2. It is often convenient and helpful to keep α, α_p, ω_0, ω_d, s_1, and s_2 as symbols and to substitute numerical values at the end of the analysis or wherever appropriate in the course of the analysis. The same is sometimes true of the circuit parameters R, L, and C.
3. When differentiating expressions containing arbitrary constants in order to apply initial conditions, ignore terms in t or $\sin\omega_0 t$ that are zero at $t = 0$.
4. In a series RLC circuit, it is generally convenient to derive the response v_C first and to use $i = Cdv_C/dt$ to find the second arbitrary constant.
5. In a parallel GCL circuit, it is generally convenient to derive the response i_L first and to use $v = Ldi_L/dt$ to find the second arbitrary constant.
6. It is important to always check that all responses satisfy initial conditions, final, steady-state conditions, and KCL or KVL, as may be appropriate.

Appendix 12A: More General Second-Order Circuits

The procedure described in Section 12.4 can be readily generalized to derive the natural and step responses of second-order circuits that are not reducible to the prototypical series RLC or parallel GCL. The procedure will be illustrated with reference to the circuit of Figure 12.40. It is assumed in this circuit that the switch is closed at $t = 0$, after being open for a long time, which means that no initial energy is stored in the circuit at $t = 0^-$.

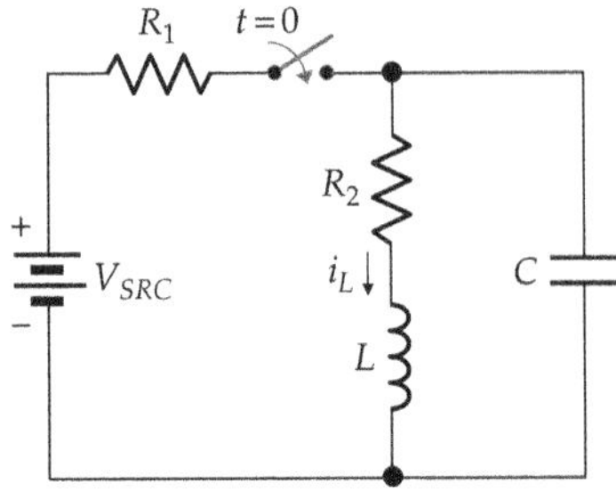

FIGURE 12.40
A more general RLC Circuit.

The general procedure is to set independent sources to zero and derive the homogeneous differential equation for a selected variable, for $t \geq 0^+$. The general solution of this equation is formulated, having two arbitrary constants and taking into account the final steady-state value of the selected variable. The values of these constants are obtained from the values of this variable and its first derivative at $t = 0^+$.

Suppose i_L is the required variable. In order to derive the homogeneous differential equation satisfied by i_L for $t \geq 0^+$, the independent source is set to zero, with the switch closed. The circuit becomes as shown in Figure 12.41. KCL at the upper node gives

$$\frac{v_C}{R_1} + i_L + C\frac{dv_C}{dt} = 0 \qquad (12.106)$$

From KVL around the mesh on the right,

$$L\frac{di_L}{dt} + R_2 i_L = v_C \qquad (12.107)$$

Substituting for v_C and dv_C/dt from Equation 12.107 in Equation 12.106 and dividing by LC gives the homogeneous differential equation satisfied by i_L as

$$\frac{d^2 i_L}{dt^2} + \left(\frac{R_2}{L} + \frac{1}{CR_1}\right)\frac{di_L}{dt} + \left(\frac{R_1 + R_2}{LCR_1}\right) i_L = 0 \qquad (12.108)$$

This can be put in standard form as

$$\frac{d^2 i_L}{dt^2} + 2\alpha\frac{di_L}{dt} + \omega_0^2 i_L = 0 \qquad (12.109)$$

where

$$\alpha = \frac{1}{2}\left(\frac{R_2}{L} + \frac{1}{CR_1}\right) \quad \text{and} \quad \omega_0^2 = \left(\frac{R_1 + R_2}{LCR_1}\right) \qquad (12.110)$$

Note that if $R_2 = 0$, α and ω_0 reduce to those for a parallel GCL circuit.

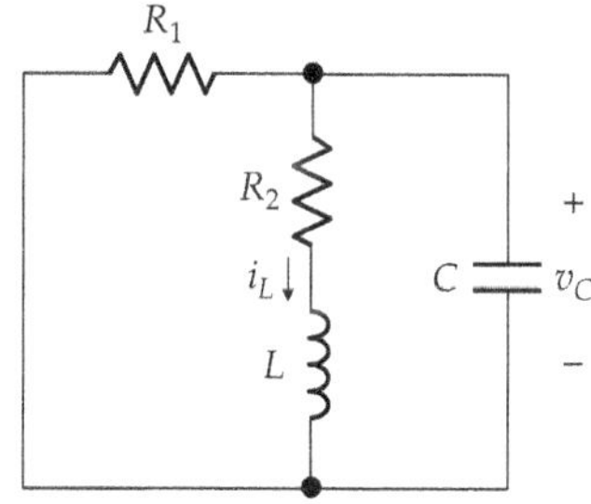

FIGURE 12.41
Circuit of Figure 12.40 with independent source set to zero.

Depending on the relative values of α and ω_0, the complete solution for i_L is given by Equations 12.85 through 12.87 as

$$i_L(t) = Ae^{s_1 t} + Be^{s_2 t} + I_{LF}, \quad \alpha > \omega_0 \qquad (12.111)$$

or

$$i_L(t) = e^{-\omega_0 t}(A + Bt) + I_{LF}, \quad \alpha = \omega_0 \qquad (12.112)$$

or

$$i_L(t) = e^{-\alpha t}(A\cos\omega_d t + B\sin\omega_d t) + I_{LF}, \quad \alpha < \omega_0 \qquad (12.113)$$

With zero initial energy storage, $i_L(0^+) = i_L(0^-) = 0$ and $v_C(0^+) = v_C(0^-) = 0$. To determine $(di_L/dt)_{0+}$, we note that with $i_L(0^+)$ and $v_C(0^+)$ both zero, the voltage across the inductor is zero, so that $(di_L/dt)_{0+} = 0$. V_{SRC} appears across R_1 at $t = 0^+$ and the current V_{SRC}/R_1 flows through C. The two initial conditions are used in Equation 12.111, 12.112, or 12.113 appropriately to determine A and B. I_{LF} is $V_{SRC}/(R_1 + R_2)$.

Problems

Apply ISDEPIC and verify solutions by PSpice simulation whenever feasible

Parameters and Initial Values of Second-Order Circuits

P12.1 The switch in Figure P12.1 is opened at $t = 0$. (a) Determine ω_0 and α; (b) deduce the type of damping of the responses for $t \geq 0$.

Ans. (a) $\omega_0 = 0.25$ rad/s, $\alpha_p = 0.025$ rad/s; (b) underdamped.

P12.2 The switch in Figure P12.2 is closed at $t = 0$. Choose R so that the responses are critically damped for $t \geq 0^+$.

Ans. 250 Ω.

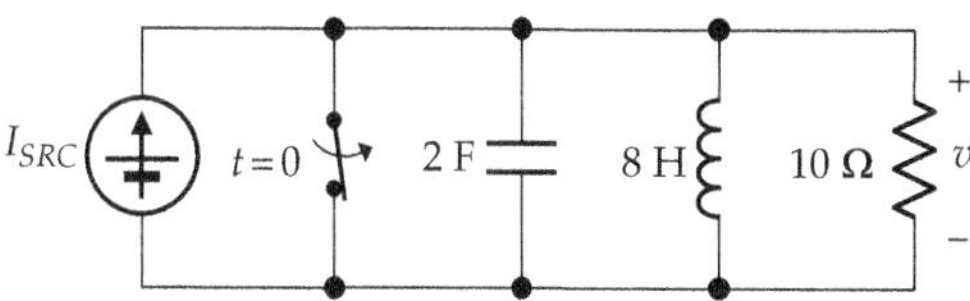

FIGURE P12.1

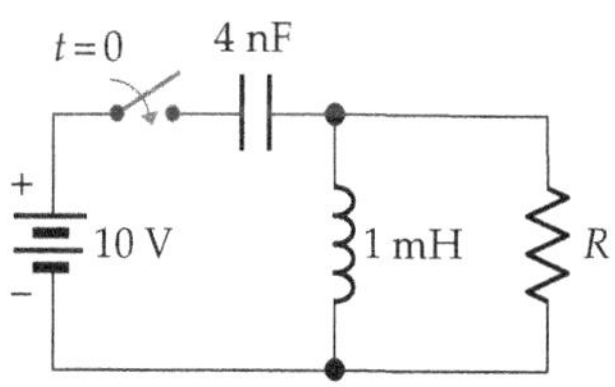

FIGURE P12.2

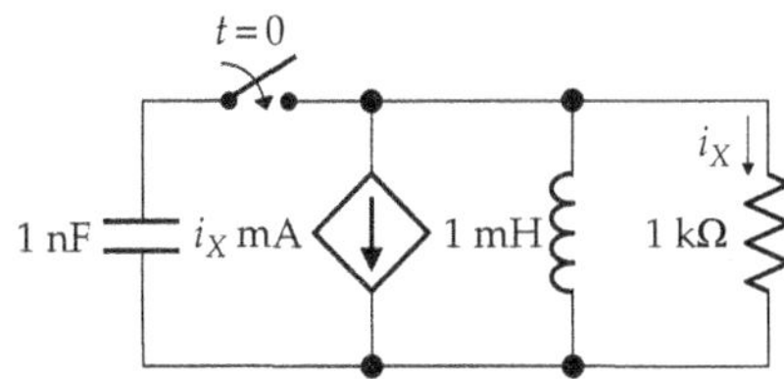

FIGURE P12.3

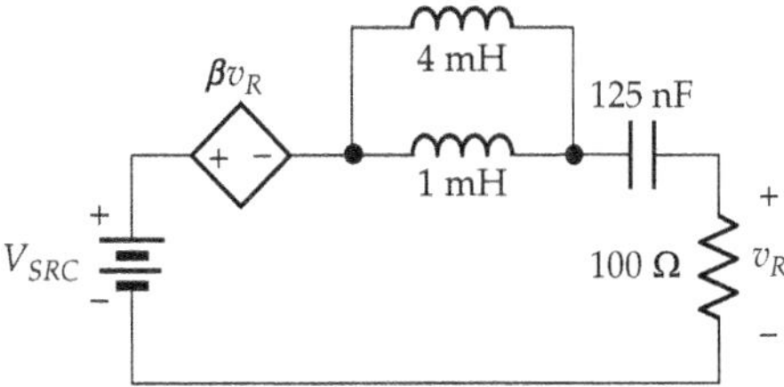

FIGURE P12.4

P12.3 The switch in Figure P12.3 is closed at t = 0, with the capacitor initially charged to 10 V. (a) Determine ω_0 and α; (b) deduce the type of damping of the responses for $t \geq 0$.

Ans. (a) ω_0 = 1 Mrad/s = α_p; (b) critically damped.

P12.4 Determine β in Figure P12.4 so that the response is critically damped by applying (i) the source absorption theorem and (ii) KVL and comparing coefficients with those of a prototypical series *RLC* circuit.

Ans. 0.6.

P12.5 Derive the dual of the circuit of Figure P12.4 and determine β following the corresponding procedures. Compare the behavior of the dual circuits.

Ans. Paralleled elements: I_{SRC}, 0.8 mF, 125 nH, 100 S, and a CCCS of $0.6i_C$. The two circuits have the same ω_0 and α, so dual variables have the same expressions.

P12.6 Determine (a) the effective inductance and effective capacitance in Figure P12.6 by applying (i) the source absorption theorem and (ii) KVL and comparing coefficients with those of a prototypical series *RLC* circuit, (b) ω_0 and α; and (c) whether the circuit is overdamped, underdamped, or critically damped.

Ans. (a) 40 mH, 0.25 μF; (b) ω_0 = 10 krad/s, α = 1.25 krad/s, underdamped.

P12.7 Derive the dual of the circuit of Figure P12.6 and analyze in the same manner.

Ans. Paralleled elements: I_{SRC}, 40 mF, 0.25 μH, 100 S, and a CCCS of $3i_C + 7i_L$.

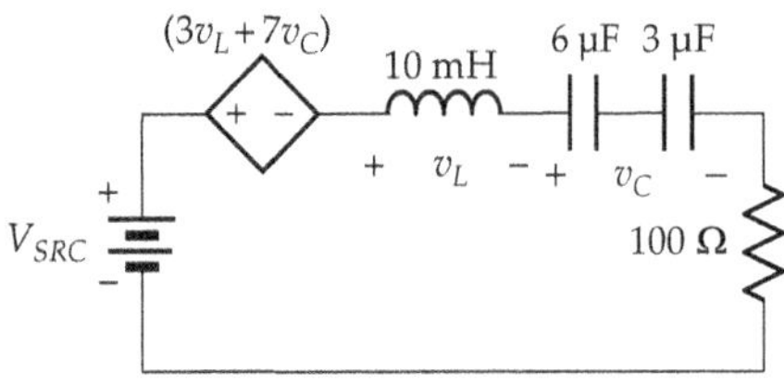

FIGURE P12.6

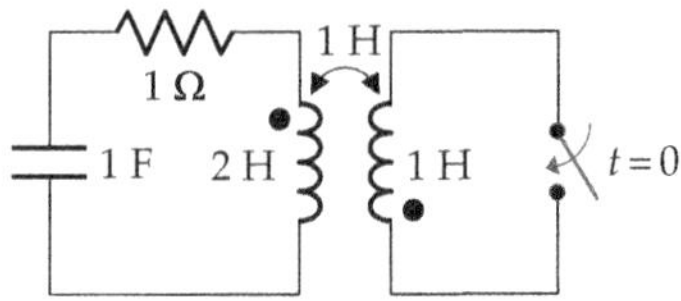

FIGURE P12.8

P12.8 The capacitor in Figure P12.8 is charged to 10 V at the instant the switch is closed at t = 0. Determine the damping coefficient α of the circuit for $t > 0^+$.

Ans. 0.5 rad/s.

P12.9 Given that at $t = 0^-$, $i_L = 0$ and i_R = 1 A in Figure P12.9, determine at $t = 0^+$ (a) $\frac{di_L}{dt}$ and (b) $\frac{di_R}{dt}$.

Ans. (a) −2.5 A/s; (b) −50 mA/s.

P12.10 The switch in Figure P12.10 is opened at t = 0 after being closed for a long time. (a) Determine i_L, v_C, dv_C/dt, and di_L/dt just after the switch is opened. (b) Is the circuit reducible to a prototypical series *LCR* circuit or a parallel *GCL* circuit?

Ans. (a) i_L = 1 mA, v_C = 6 V, dv_C/dt = −1 kV/s, di_L/dt = −5 A/s; (b) no.

P12.11 The switch in Figure P12.11 is opened at t = 0 after being closed for a long time. Determine i_C just after the switch is opened, given that V_{SRC} = 1 V.

Ans. 0.5 mA.

P12.12 Consider the circuit of Problem P12.2, reproduced in Figure P12.12. Determine v_C, i_L, dv_C/dt, and di_L/dt at $t = 0^+$, assuming the capacitor and inductor are initially uncharged and R = 250 Ω.

Ans. $v_C = 0$, $i_L = 0$, $\frac{dv_C}{dt} = -10^7$ V/s, $\frac{di_L}{dt} = 10^4$ A/s.

P12.13 The switch in Figure P12.13 is opened at t = 0, after being closed for a long time. Determine dv/dt at $t = 0^+$.

Ans. −1 V/ms.

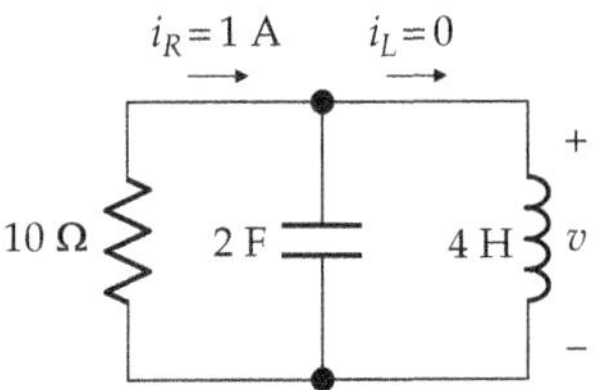

FIGURE P12.9

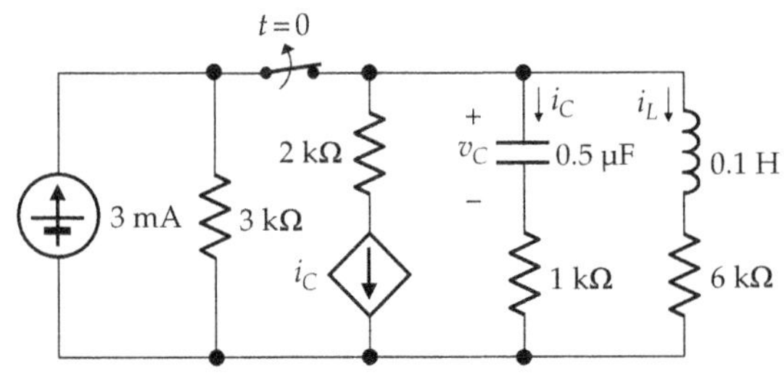

FIGURE P12.10

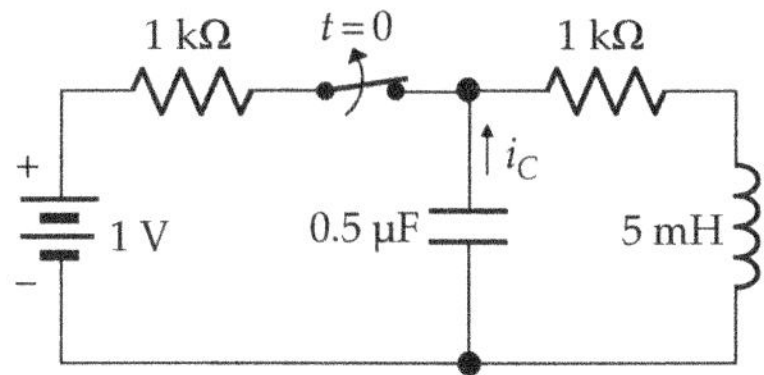

FIGURE P12.11

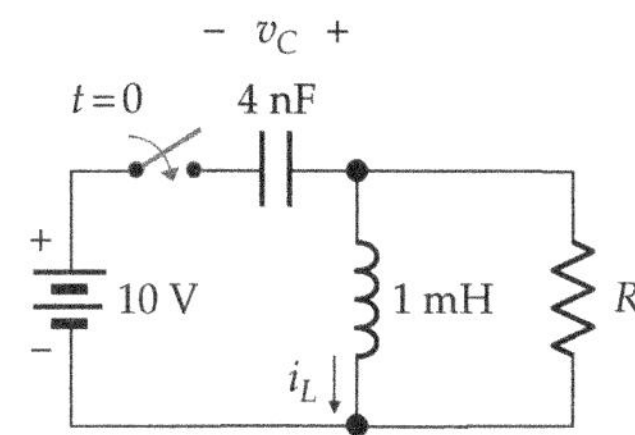

FIGURE P12.12

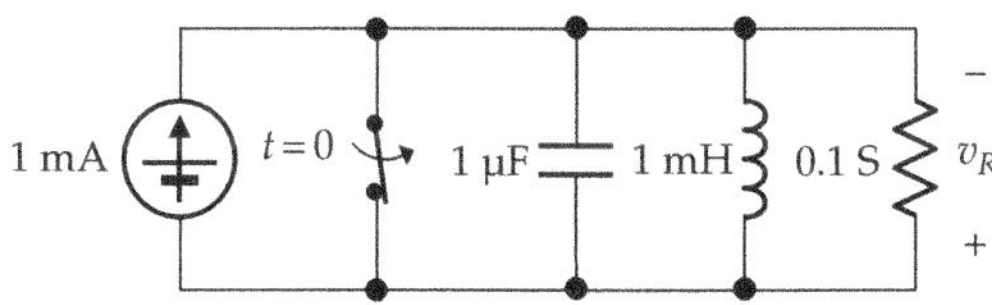

FIGURE P12.13

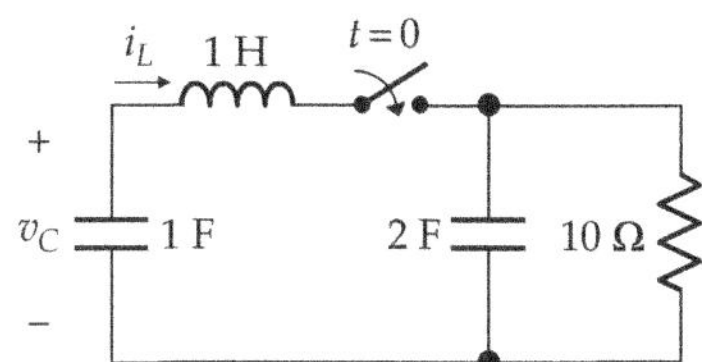

FIGURE P12.14

P12.14 The switch in Figure P12.14 is closed at $t = 0$, with the 1 F capacitor charged to 6 V, the inductor having a current of 4 A, whereas the 2 F capacitor is uncharged. (a) Determine the energy dissipated in the 10 Ω resistor from $t = 0$ to infinity. (b) Is the circuit reducible to a prototypical series *LCR* circuit or a parallel *GCL* circuit?

Ans. (a) 26 J; (b) no.

Natural Responses of Second-Order Circuits

Verify solutions with entries in Tables 12.2 through 12.4.

P12.15 The switch in Figure P12.15 is closed at $t = 0$, after being open for a long time. Determine *R*, given that $\omega_0 = 1$ rad/s and the circuit is critically damped.

Ans. 1 Ω.

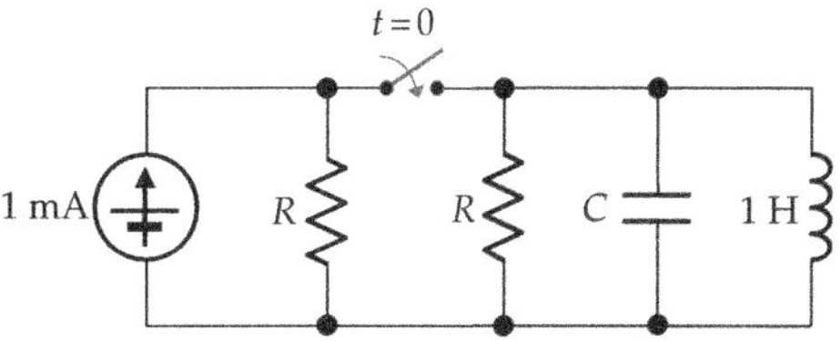

FIGURE P12.15

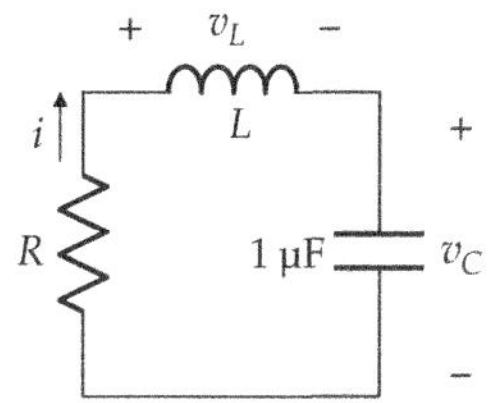

FIGURE P12.16

P12.16 Given that $i(t)=20e^{-400t}(1-400t)$ mA in Figure P12.16, *t* is in s, determine *L*, *R*, $v_L(t)$, and $v_C(t)$, assuming zero initial charge on the capacitor. Verify that $v_R + v_L + v_C = 0$.

Ans. $L = 6.25$ H, $R = 5$ kΩ, $v_L(t)=-50e^{-400t}(2-400t)\ V$, $v_C(t)=2\times10^4 te^{-400t}$ V.

P12.17 v in Figure P12.17 is known to be of the form: $v(t)=Ae^{-500t}+Bte^{-500t}$, *t* is in s, with $V_0 = 8$ V and $I_{L0} = -10$ mA. Determine (a) *R*, *C*, *A*, and *B*; and (b) $i_C(t)$ for $t \geq 0^+$.

Ans. (a) $R = 1$ kΩ, $C = 1$ μF, $A = 8$ V, $B = 6000$ V/s; (b) $i_C(t)=2e^{-500t}(1-1500t)$ mA.

P12.18 The switch in Figure P12.18 is moved from position 'a' to position 'b' at $t = 0$ after being in position 'a' for a long time. Determine $v_C(t)$ and $i_L(t)$ for $t \geq 0^+$.

Ans. $v_C(t)=\frac{-1}{4\sqrt{6}}\left(e^{-(5-2\sqrt{6})t}-e^{-(5-2\sqrt{6})t}\right)$ kV, $i_L(t)=\frac{1}{2\sqrt{6}}\left(\left(-5+2\sqrt{6}\right)e^{-(5-2\sqrt{6})t}+\left(5+2\sqrt{6}\right)e^{-(5-2\sqrt{6})t}\right)$ A, *t* is in ms.

P12.19 Given $v_C = 0$ and $i_L = -4$ A at $t = 0$ in Figure P12.19. Determine $v_C(t)$ for $t \geq 0^+$.

Ans. $v_C(t) = 10\sin 5t$ V.

P12.20 Given the charge on the capacitor in Figure P12.20 is $q(t) = \cos\omega_0 t$ C. Determine the magnitude of the

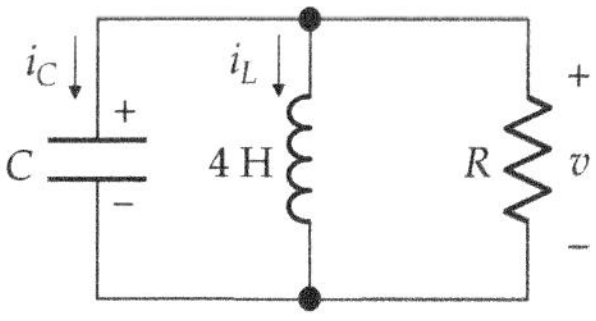

FIGURE P12.17

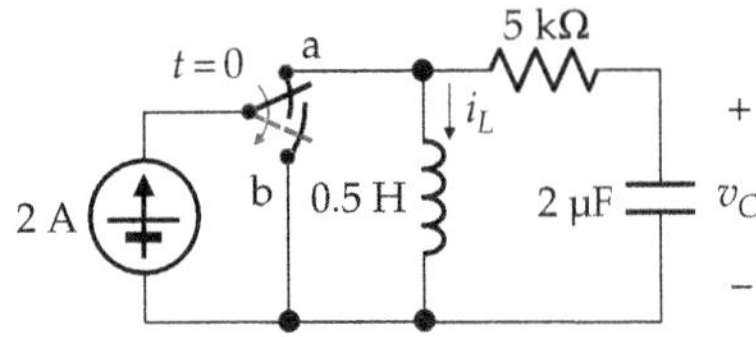

FIGURE P12.18

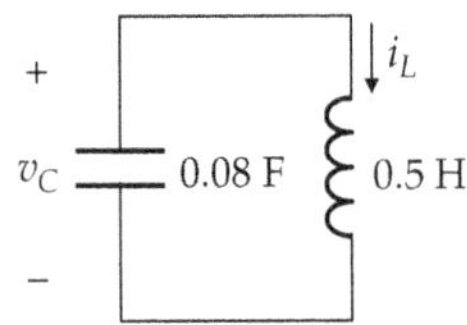

FIGURE P12.19

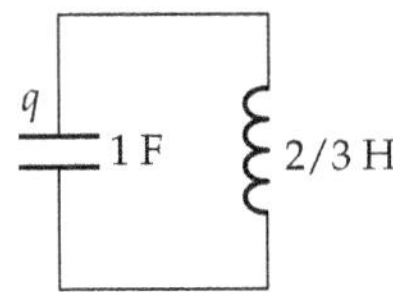

FIGURE P12.20

inductor current when the energy stored in the capacitor is one-third of its maximum value.

Ans. 1 A.

P12.21 The *LC* circuit in Figure P12.21 is undergoing continuous sinusoidal oscillations. At $t = 2$ s, the energy stored in the capacitor is 6 J and that stored in the inductor is 2 J. Determine the peak value of current *i* in the circuit.

Ans. $i_p = \sqrt{8} = 2.83$ A.

P12.22 Given that $v = 0$, $I_{10} = I_{20} = 2$ A at $t = 0$ in Figure P12.22, determine the total instantaneous energy in the inductors as a function of time.

Ans. $8\cos^2 t$ J.

P12.23 The switch in Figure P12.23 is opened at $t = 0$ after being closed for a long time, Determine i_R at $t = 1$ ms.

Ans. 0.271 A.

P12.24 The switch in Figure P12.24 is opened at $t = 0$ after being closed for a long time. Determine $v_O(t)$.

Ans. $v_O(t) = e^{-t}\left[\left(160+280/\sqrt{3}\right)e^{\sqrt{3}t/2}+\left(160-280/\sqrt{3}\right)e^{-\sqrt{3}t/2}\right]$ V

t is in ms.

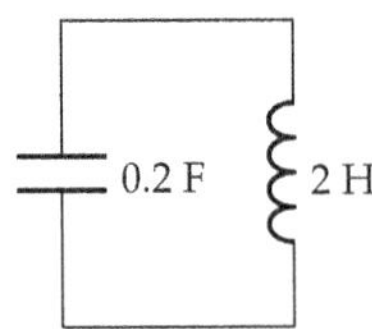

FIGURE P12.21

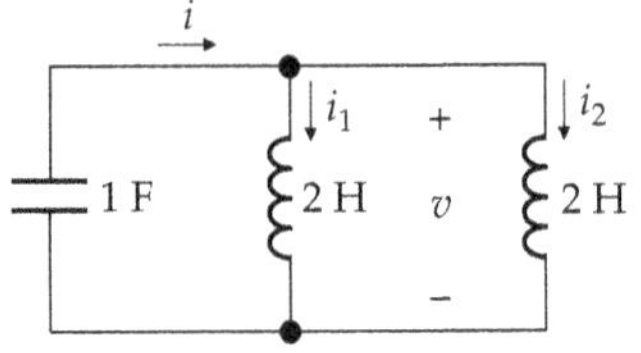

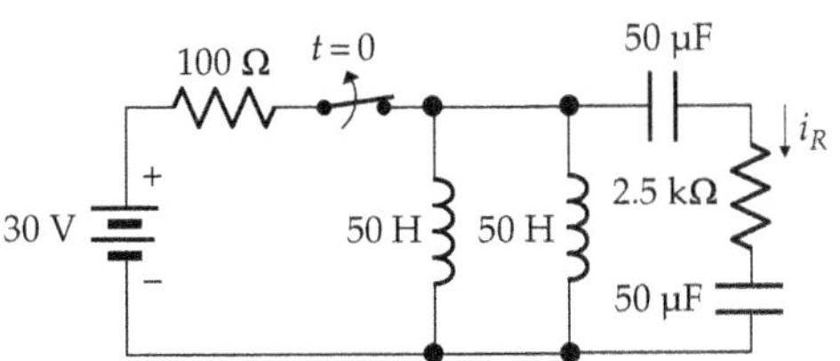

FIGURE P12.22

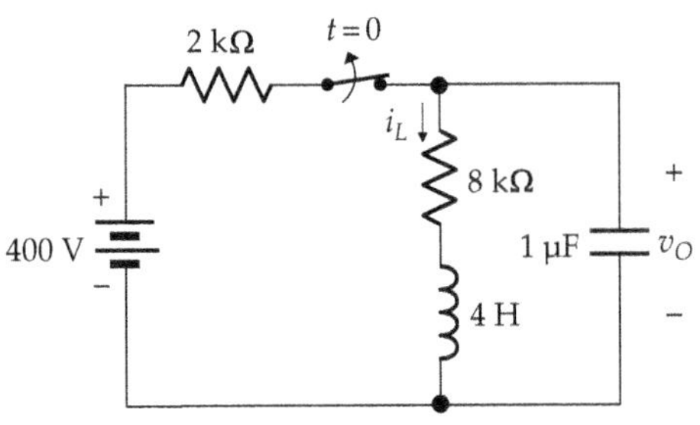

FIGURE P12.23

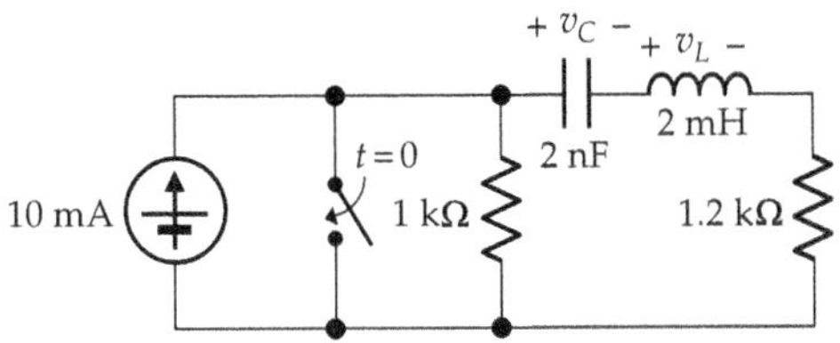

FIGURE P12.24

FIGURE P12.25

P12.25 The switch in Figure P12.25 is closed at $t = 0$ after being open for a long time. Determine $v_L(t)$ and $v_C(t)$ for $t \geq 0^+$.

Ans. $v_L(t) = -10e^{-0.3t}\left(\cos 0.4t - 0.75\sin 0.4t\right)$ V, $v_C(t) = 10e^{-0.3t}\left(\cos 0.4t + 0.75\sin 0.4t\right)$ V, *t* is in μs.

P12.26 The switch in Figure P12.26 is moved to position 'b' at $t = 0$ after being in position 'a' for a long time. Determine $i(t)$ and $v_O(t)$ for $t \geq 0^+$.

Ans. $i(t) = 20e^{-t}\sin t$ mA, $v_O(t) = 10e^{-t}\left(\cos t - \sin t\right)$ V, *t* is in ms.

P12.27 The switch in Figure P12.27, is moved to position 'b' at $t = 0$ after being in position a for a long time. (a) Choose *R* so that the response is critically damped;

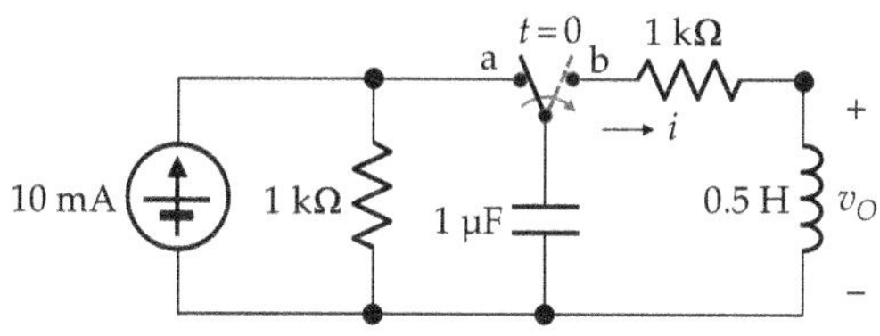

FIGURE P12.26

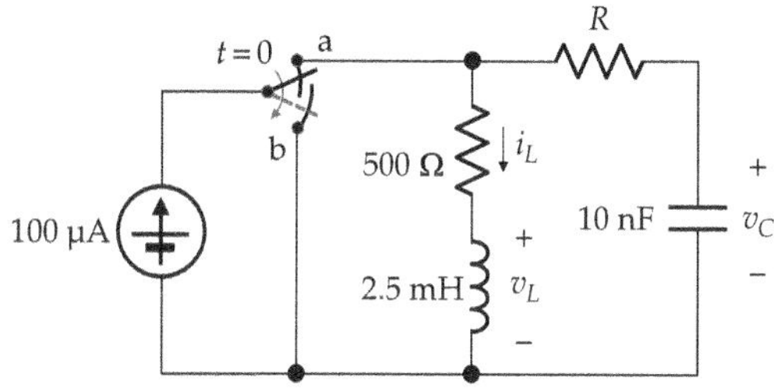

FIGURE P12.27

(b) determine the initial values $v_C(0^+)$, $i_L(0^+)$, and $v_L(0^+)$; (c) determine $v_C(t)$ and $i_L(t)$ for $t \geq 0^+$.

Ans. (a) 500 Ω; (b) $i_L(0^+) = 100\ \mu\text{A}$, $v_C(0^+) = 50\ \text{mV}$, $v_L(0^+) = -50\ \text{mV}$; (c) $v_C(t) = 50e^{-0.2t}\ \text{mV}, i_L(t) = 100e^{-0.2t}\ \mu\text{A}$, t is in μs.

P12.28 The switch in Figure P12.28 is moved from position 'a' to position 'b' at $t = 0$ after being in position 'a' for a long time. Determine $v_C(t)$ and $i_L(t)$ for $t \geq 0^+$.

Ans. $i_L(t) = e^{-5t}\left(4\cos 20t + \frac{1}{5}\sin 20t\right)\text{A}, v_C(t) = e^{-5t}$ $\left(6\cos 20t - 19.75\sin 20t\right)$ V, t is in s.

P12.29 The switch in Figure P12.29 is opened at $t = 0$ after being closed for a long time. Determine $v_L(t)$ for $t \geq 0^+$.

Ans. $v_L(t) = -\frac{40}{\sqrt{3}}e^{-0.5t}\sin 0.5\sqrt{3}t$ V, t is in ms.

P12.30 The switch in Figure P12.30 is closed at $t = 0$, with each of the capacitors charged to 10 V in the polarity indicated and with no initial energy storage in the inductor. Determine (a) R for critical damping; (b) $v_O(t)$ for $t \geq 0^+$.

Ans. (a) 5 Ω; (b) $v_O(t) = 10(1-t)e^{-t}$ V, t is in ms.

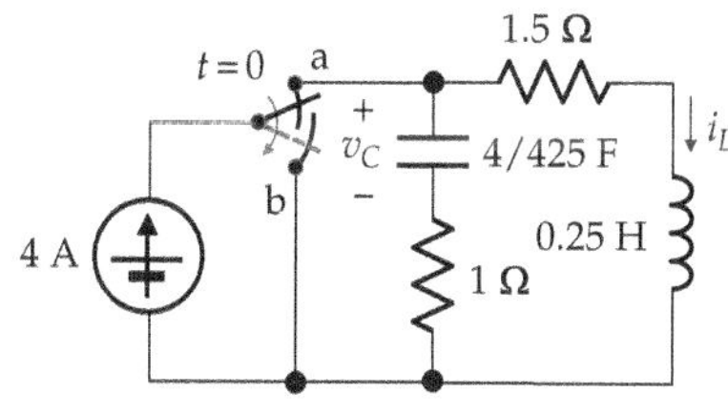

FIGURE P12.28

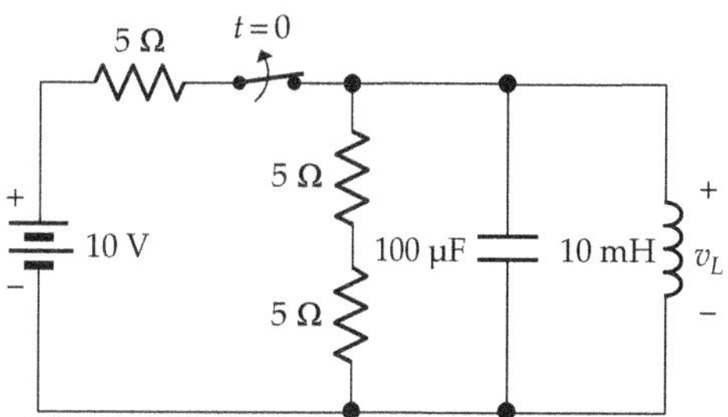

FIGURE P12.29

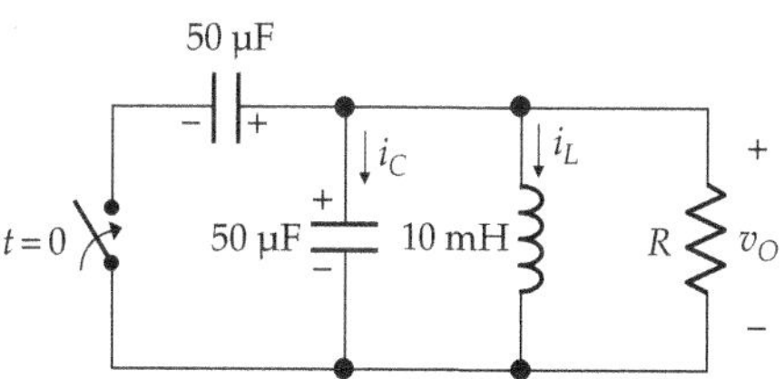

FIGURE P12.30

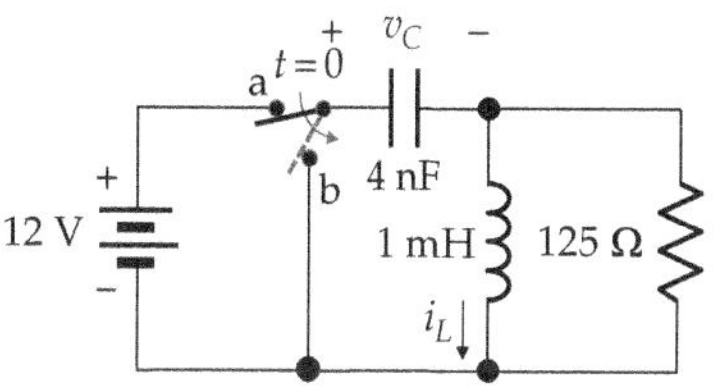

FIGURE P12.31

P12.31 The switch in Figure P12.31 is moved from position 'a' to position 'b' at $t = 0$ after being in position 'a' for a long time. Determine $v_C(t)$ and $i_L(t)$ for $t \geq 0^+$.

Ans. $i_L(t) = \frac{-12}{\sqrt{3}}\left(e^{-(1-\sqrt{3}/2)t} - e^{-(1+\sqrt{3}/2)t}\right)$ mA, $v_C(t) = \frac{12}{\sqrt{3}}\left(\left(-1+\frac{\sqrt{3}}{2}\right)e^{-(1-\sqrt{3}/2)t} + \left(1+\frac{\sqrt{3}}{2}\right)e^{-(1+\sqrt{3}/2)t}\right)$ V, t is in μs.

P12.32 The switch in Figure P12.32 is moved to position 'b' at $t = 0$ after being in position 'a' for a long time. (a) Choose G so that the response is critically damped; (b) determine the initial values $i_L(0^+)$, $v_C(0^+)$, and $i_C(0^+)$; (c) determine $i_L(t)$ and $v_C(t)$ for $t \geq 0^+$. Note that the circuit is the dual of that of problem P12.27.

Ans. (a) 500 S; (b) $v_C(0^+) = 100\ \mu\text{V}$, $i_L(0^+) = 50\ \text{mA}$, $i_C(0^+) = -50\ \text{mA}$; (c) $i_L(t) = 50e^{-0.2t}\ \text{mA}, v_C(t) = 100e^{-0.2t}\ \mu\text{V}$, t is in μs.

P12.33 The switch in Figure P12.33 is moved to position 'b' at $t = 0$ after being in position 'a' for a long time. Determine $v_O(t)$ for $t \geq 0^+$.

Ans. $v_O(t) = \frac{5e^{-t}}{\sqrt{3}}\left[-\left(1+\sqrt{3}/2\right)e^{\sqrt{3}t/2} + \left(1-\sqrt{3}/2\right)e^{-\sqrt{3}t/2}\right]$ V, t is in ms.

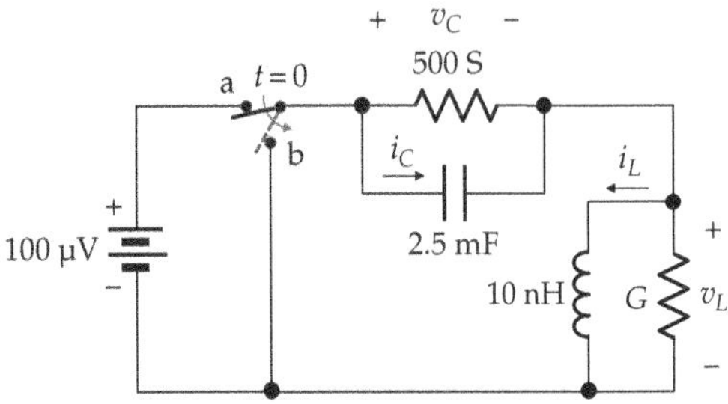

FIGURE P12.32

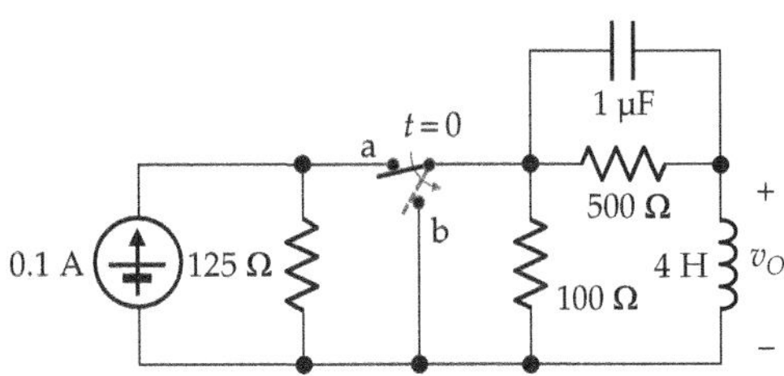

FIGURE P12.33

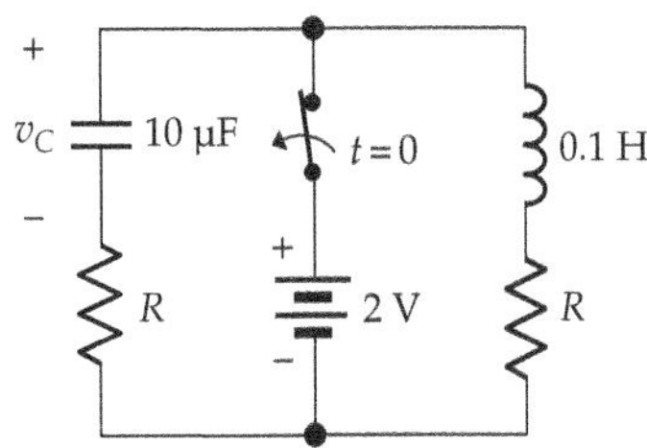

FIGURE P12.34

P12.34 The switch in Figure P12.34 is opened at $t = 0$ after being closed for a long time: (a) determine R for critical damping; using this value of R, determine (b) $v_C(t)$, $t \geq 0^+$, t in ms, and (c) the energy in µJ stored in the capacitor and inductor as functions of time.

Ans. (a) 100 Ω; (b) $V_C(t) = 2e^{-t}$ V; (c) $w_C(t) = 5V_{SRC}^2 e^{-2t}$, $\omega_L(t) = 5V_{SRC}^2 e^{-2t}$ µJ, $t \geq 0^+$, t in ms.

P12.35 The switch in Figure P12.35 is moved to position 'b' at $t = 0$, with zero initial energy storage. Determine (a) ρ for critical damping and (b) $v(t)$, $i_C(t)$, and $i_L(t)$ for $t \geq 0^+$.

Ans. (a) $\rho = 0$; (b) $v(t) = 100te^{-10t}$ V, $i_C(t) = 100(e^{-10t} - 10te^{-10t})$ mA, $i_L(t) = 100 - 100(e^{-10t} + 10te^{-10t})$ mA, t is in ms.

P12.36 The switch in Figure P12.36 is moved at $t = 0$ from position 'a' to position 'b' after being in position 'a' for a long time. Determine $v_O(t)$ for $t \geq 0^+$.

Ans. $v_O(t) = 10e^{-100t}(3\cos 300t - \sin 300t)$ V, t is in s.

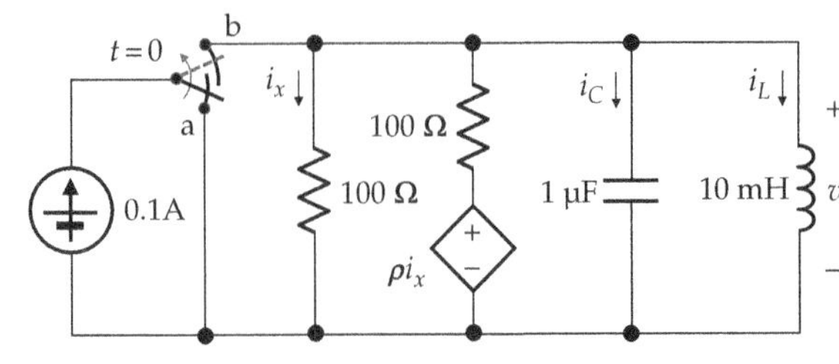

FIGURE P12.35

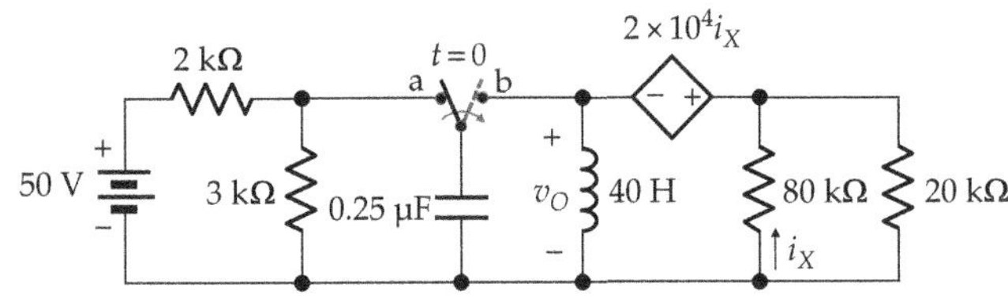

FIGURE P12.36

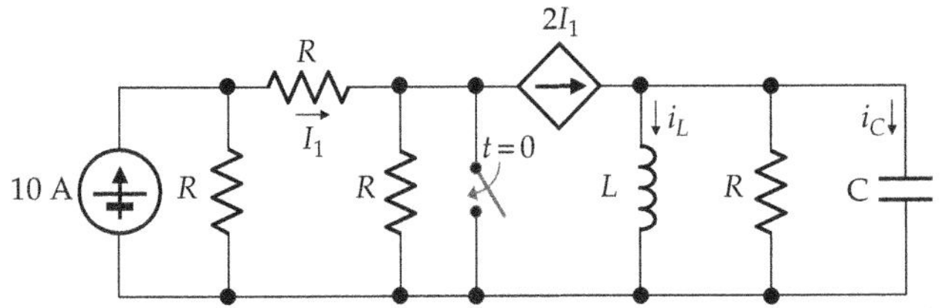

FIGURE P12.37

P12.37 The switch in Figure P12.37 is closed at $t = 0$ after being open for a long time. Determine (a) $i_L(0^-)$, $i_C(0^-)$, $i_L(0^+)$, and $i_C(0^+)$; and (b) $i_L(t)$ for $t \geq 0^+$, assuming $R = 1$ Ω, $L = 1$ H, and $C = 1$ F.

Ans. (a) $i_L(0^-) = 20$ A, $i_C(0^-) = 0$, $i_L(0^+) = 20$ A, $i_C(0^+) = -10$ A;

(b) $i_L(t) = 10\left(1 + e^{-0.5t}\left(\cos 0.5\sqrt{3}t + \frac{1}{\sqrt{3}}\sin 0.5\sqrt{3}t\right)\right)$ A.

P12.38 Switch S_1 in Figure P12.38 is moved at $t = 0$ from position 'a' to position 'b' after being in position 'a' for a long time. Switch S_2 is opened at $t = 0.2$ s. Determine $i_L(t)$, $t \geq 0^+$.

Ans. $i_L(t) = 2e^{-10t}$ A, $0 \leq t \leq 0.2$ s; $i_L(t) =$ $0.271e^{-6(t-0.2)}(\cos 8(t-0.2) - 0.75\sin 8(t-0.2))$ A, $t \geq 0.2^+$ s

Forced Responses of Second-Order Circuits

P12.39 The current in a second-order circuit is governed by the differential equation $\frac{d^2i}{dt^2} + 2\frac{di}{dt} + 4i = 12$, $t \geq 0^+$ with $i(0^+) = 0 = \left(\frac{di(t)}{dt}\right)_{0^+}$. Determine $i(t)$ for $t \geq 0^+$.

Ans. $3 - \sqrt{3}e^{-t}\left(\sqrt{3}\cos\sqrt{3}t + \sin\sqrt{3}t\right)$ A.

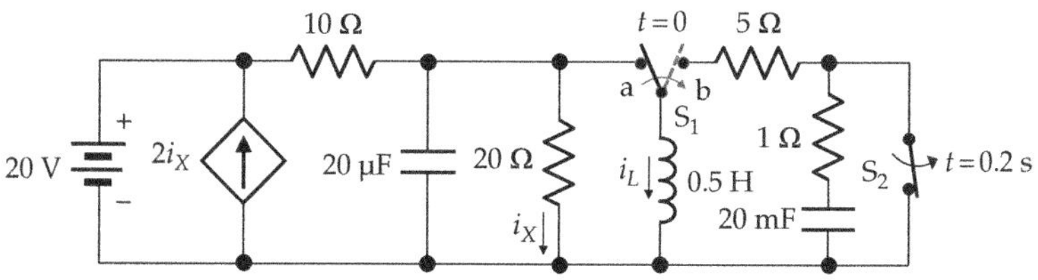

FIGURE P12.38

P12.40 The response of a series *RLC* circuit to a voltage that is suddenly applied at $t = 0$ is $v_C(t) = 20 - 10e^{-1000t} - 10e^{-4000t}$ V, $i_L(t) = 2e^{-1000t} + 8e^{-4000t}$ mA, $t \geq 0^+$. Determine *C*.

Ans. 0.2 μF.

P12.41 The switch in Figure P12.41 is closed at $t = 0$, with zero initial energy storage in the circuit. Determine $v_C(t)$ and $i_L(t)$ for $t \geq 0^+$.

Ans. $v_C(t) = 10\left(1 - e^{-0.1t}(1 + 0.1t)\right)$ V, $i_L(t) = 50\left(2 - e^{-0.1t}(2 + 0.1t)\right)$ mA, *t* is in μs.

P12.42 The switch in Figure P12.42 is moved at $t = 0$ from position 'a' to position 'b' after being in position 'a' for a long time, with the capacitor uncharged. Determine $v_L(t)$ for $t \geq 0^+$.

Ans. $v_L(t) = -2e^{-40t} - 8e^{-160t}$ V, *t* is in s.

P12.43 The switch in Figure P12.43 is closed at $t = 0$ after being open for a long time. Choose R_X for critical damping, and determine $v_O(t)$ and $v_C(t)$ for $t \geq 0^+$.

Ans. $R_X = 1.5$ kΩ; $v_O(t) = 1 - 0.5te^{-t}$ V, $v_C(t) = 1 - e^{-t} - te^{-t}$ V, *t* is in μs.

P12.44 The switch in Figure P12.44 is closed at $t = 0$ after being open for a long time. Determine $i_C(t)$ for $t \geq 0^+$.

Ans. $i_C(t) = \frac{8}{3}e^{-4t}\sin 3t$ A, *t* is in s.

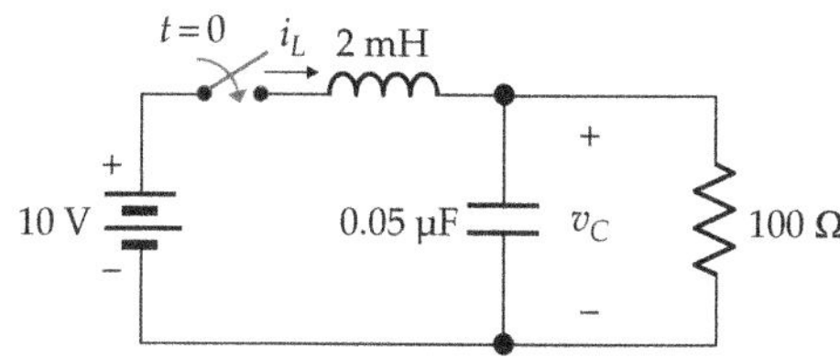

FIGURE P12.41

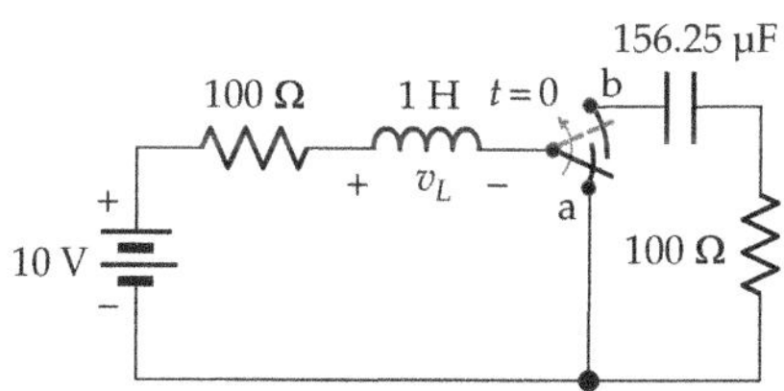

FIGURE P12.42

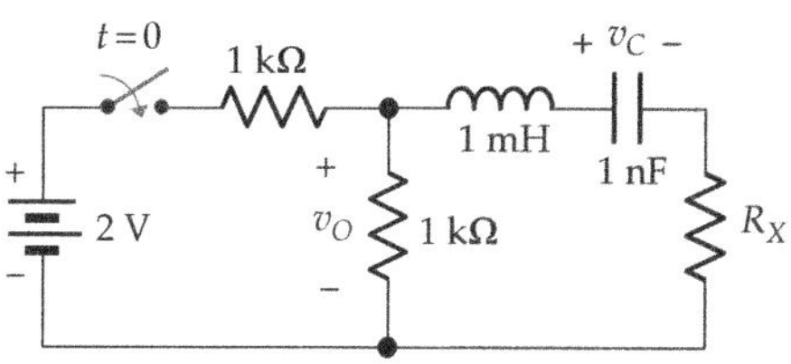

FIGURE P12.43

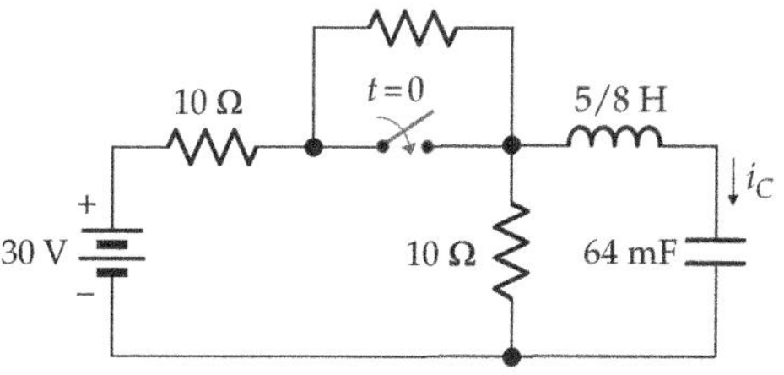

FIGURE P12.44

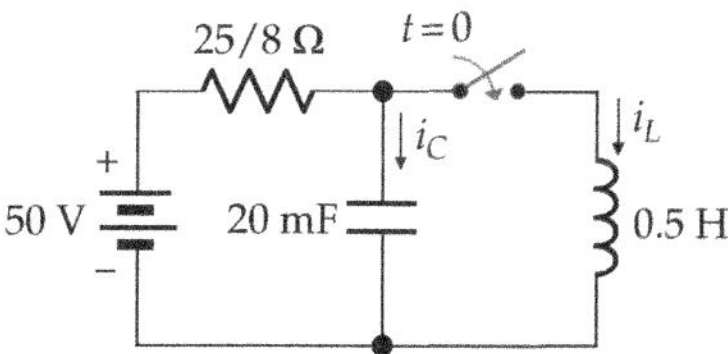

FIGURE P12.45

P12.45 The switch in Figure P12.45 is closed at $t = 0$ after being open for a long time. Determine $i_L(t)$ and $i_C(t)$ for $t \geq 0^+$.

Ans. $i_L(t) = -e^{-8t}\left(16\cos 6t + \frac{14}{3}\sin 6t\right) + 16$ A; $i_C(t) = -\frac{50}{3}e^{-8t}\sin 6t$ A, *t* is in s.

P12.46 The switch in Figure P12.46 is moved at $t = 0$ from position 'a' to position 'b' after being in position 'a' for a long time. Determine $v_C(t)$ and $v_L(t)$ for $t \geq 0^+$.

Ans. $v_C(t) = 36e^{-8t} - 16e^{-18t} + 20$ V, $v_L(t) = 16e^{-8t} - 36e^{-18t}$ V, *t* is in s.

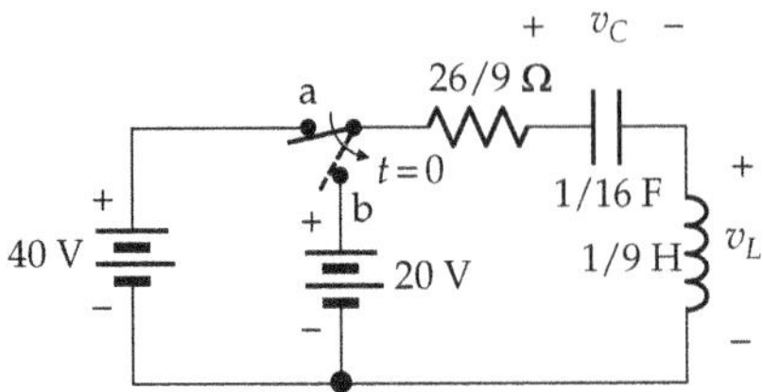

FIGURE P12.46

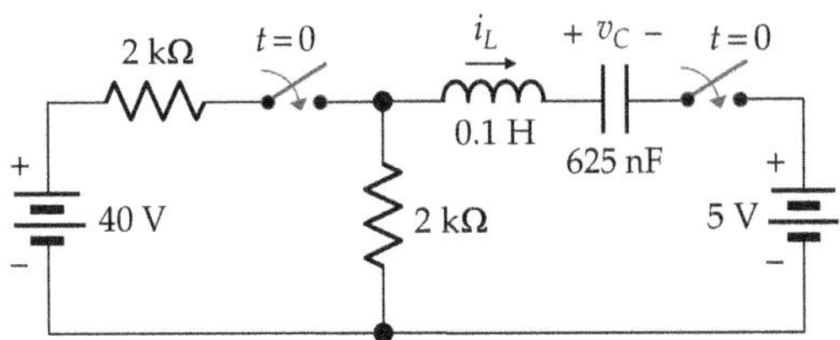

FIGURE P12.47

P12.47 Both switches in Figure P12.47 are closed at $t = 0$, with zero initial energy storage in the circuit. Determine $v_C(t)$ and $i_L(t)$ for $t \geq 0^+$.

Ans. $v_C(t) = -20e^{-2t} + 5e^{-8t} + 15$ V, $i_L(t) = 25\left(e^{-2t} - e^{-8t}\right)$ mA, t is in ms.

P12.48 The switch in Figure P12.48 is moved at $t = 0$ from position 'a' to position 'b' after being in position 'a' for a long time. Determine $v_C(t)$ and $i_L(t)$ for $t \geq 0^+$.

Ans. $v_C(t) = 20 + 22.5e^{-3t}\sin 4t + 30e^{-3t}\cos 4t$ V, $i_L(t) = -15e^{-3t}\sin 4t$ A, t is in s.

P12.49 The switch in Figure P12.49 is moved at $t = 0$ from position 'a' to position 'b' after being in position 'a' for a long time, with the capacitor initially uncharged. Determine $v_C(t)$ and $i_L(t)$ for $t \geq 0^+$.

Ans. $v_C(t) = 80 - 80e^{-10t} - 400te^{-10t}$ V, $i_L(t) = 2e^{-10t} + 20te^{-10t}$ A, t is in s.

P12.50 When the switch is opened at $t = 0$ in Figure P12.50, with no energy initially stored in the circuit, it is found that $v(t) = 2te^{-2t}$ V, $t \geq 0^+$ t is in s. Determine I_{SRC}.
Ans. 1 A.

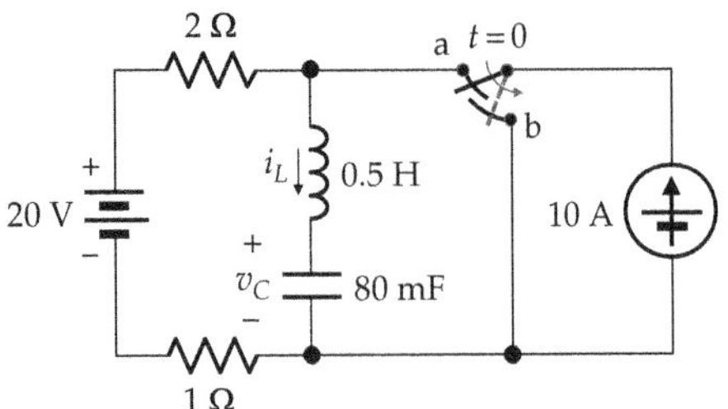

FIGURE P12.48

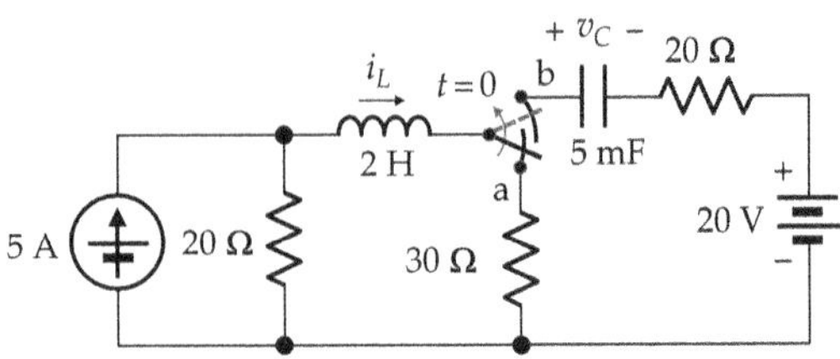

FIGURE P12.49

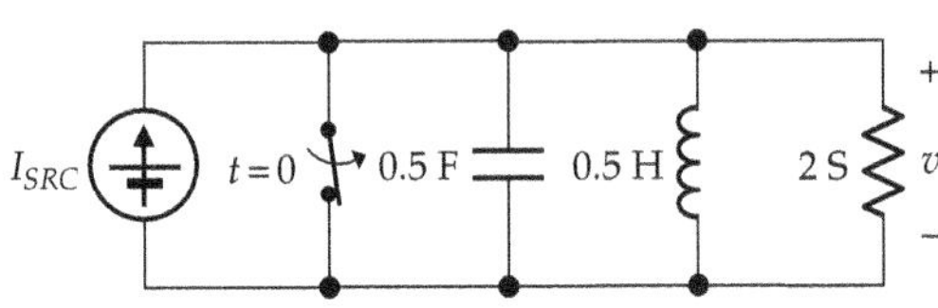

FIGURE P12.50

P12.51 The switch in Figure P12.51 is opened at $t = 0$ after being closed for a long time. Determine $i_L(t)$ and $i_C(t)$ for $t \geq 0^+$.

Ans. $i_L(t) = 2e^{-10t} + 20te^{-10t} + 2$ A, $i_C(t) = -2e^{-10t} + 20te^{-10t}$ A, t is in s.

P12.52 The switch in Figure P12.52 is moved at $t = 0$ from position 'a' to position 'b' after being in position 'a' for a long time. Determine $v_C(t)$ and $i_L(t)$ for $t \geq 0^+$.

Ans. $v_C(t) = -5e^{-60t}(4\cos 80t + 3\sin 80t) + 40$ V, $i_L(t) = 25e^{-60t}\sin 80t$ mA, t is in ms.

P12.53 The switch in Figure P12.53 is moved at $t = 0$ from position 'a' to position 'b' after being in position 'a' for a long time. Determine $v_C(t)$ and $i_L(t)$ for $t \geq 0^+$.

Ans. $v(t) = 100te^{-2.5t}$ V, $i_L(t) = -100e^{-2.5t} - 250te^{-2.5t} + 150$ mA, t is in ms.

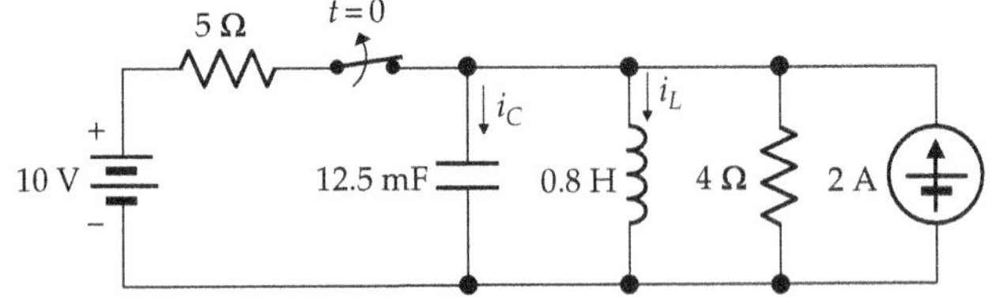

FIGURE P12.51

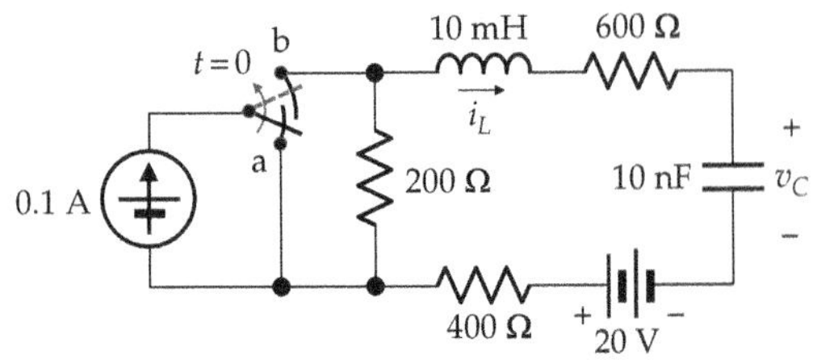

FIGURE P12.52

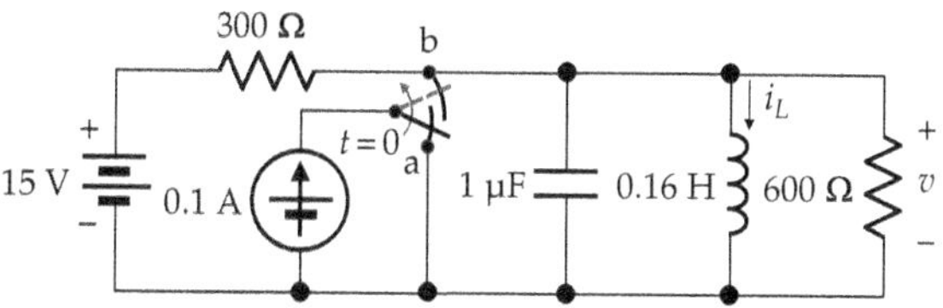

FIGURE P12.53

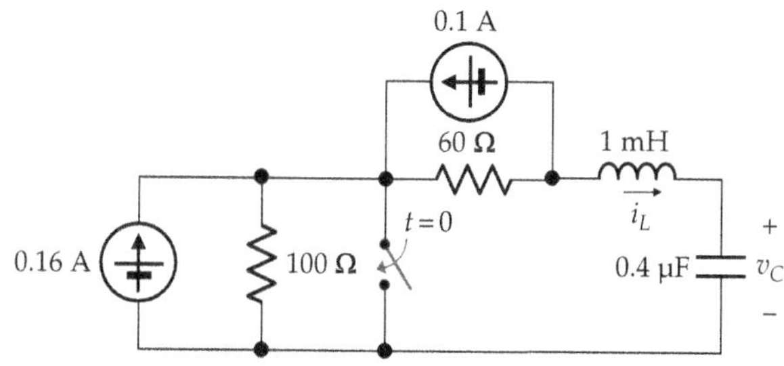

FIGURE P12.54

P12.54 The switch in Figure P12.54 is closed at $t = 0$ after being opened for a long time. Determine $v_C(t)$ and $i_L(t)$ for $t \geq 0^+$.

Ans. $v_C(t) = 4e^{-30t}(4\cos40t + 3\sin40t) - 6$ V, $i_L(t) = -0.4e^{-30t}\sin40t$ A, t is in ms.

P12.55 Both switches in Figure P12.55 are moved at $t = 0$ after being in their initial positions for a long time. Determine $v_C(t)$ and $i_L(t)$ for $t \geq 0^+$.

Ans. $v_C(t) = 10e^{-0.4t}(4\cos0.3t - 3\sin0.3t) + 40$ V, $i_L(t) = 10e^{-0.4t}\cos0.3t$ A, t is in s.

P12.56 The switch in Figure P12.56 is moved at $t = 0$ from position 'a' to position 'b' after being in position 'a' for a long time. Determine $v_C(t)$ and $i_L(t)$ for $t \geq 0^+$.

Ans. $v_C(t) = 9e^{-t} + e^{-9t} + 10$ V, $i_L(t) = 1.5e^{-t} + 1.5e^{-9t} + 3$ A, t is in s.

P12.57 The switch in Figure P12.57 is opened at $t = 0$ after being closed for a long time. Determine $v_C(t)$ and $i_L(t)$ for $t \geq 0^+$.

Ans. $v_C(t) = 10e^{-5t} + 10$ V, $i_L(t) = 5e^{-5t}$ A, t is in s.

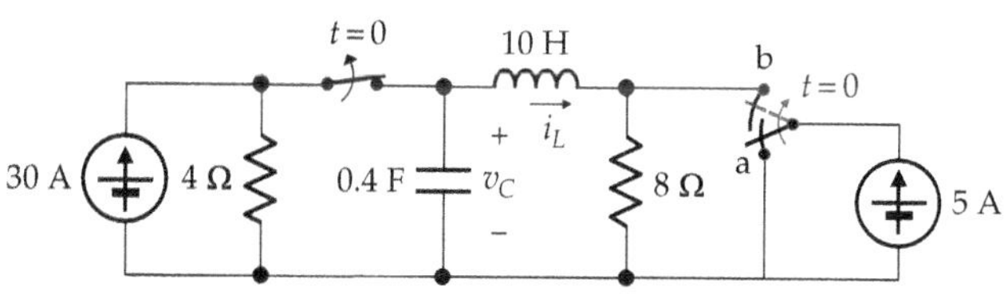

FIGURE P12.55

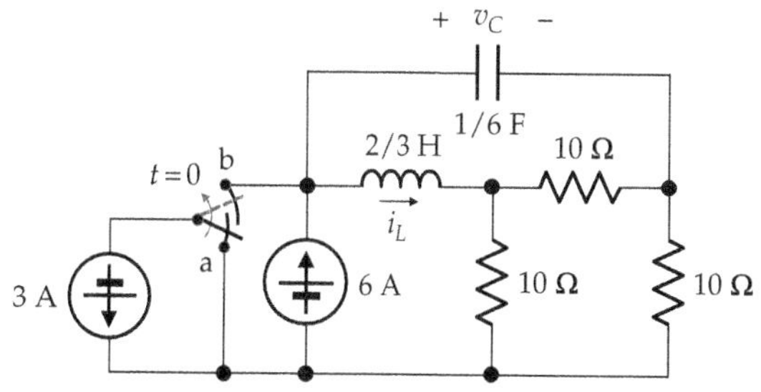

FIGURE P12.56

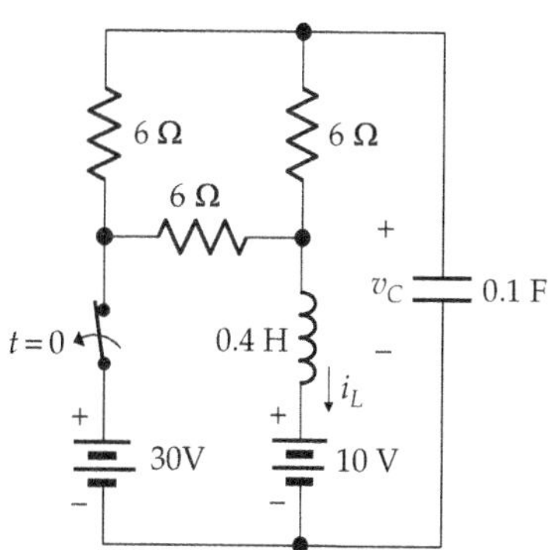

FIGURE P12.57

Design Problems

P12.58 A parallel *GCL* circuit has $C_p = 10$ nF and $L_p = 100$ μH. Determine R_p if the maximum percentage overshoot in the response i_L is not to exceed 5%, given that the maximum percentage overshoot is $100e^{-(\alpha_p/\omega_d)\pi}$ (Refer to Problem P12.62).

Ans. 72.45 Ω.

P12.59 Given a coil of 0.5 H inductance and 50 Ω resistance connected to a 5 V supply, as in Figure P12.59. It is required to generate a voltage $v_O \cong 2500$ V upon opening the switch at $t = 0$. Show that a value of C of 0.8 nF produces this voltage at the first maximum of an underdamped response across C.

P12.60 It is desired to generate a sequence of five pulses during the interval $0 \leq t \leq 6$ s using the circuit shown in Figure P12.60. D is a threshold device that generates an output pulse when its input voltage is 0.3 V and increasing, but not when decreasing. The second-order

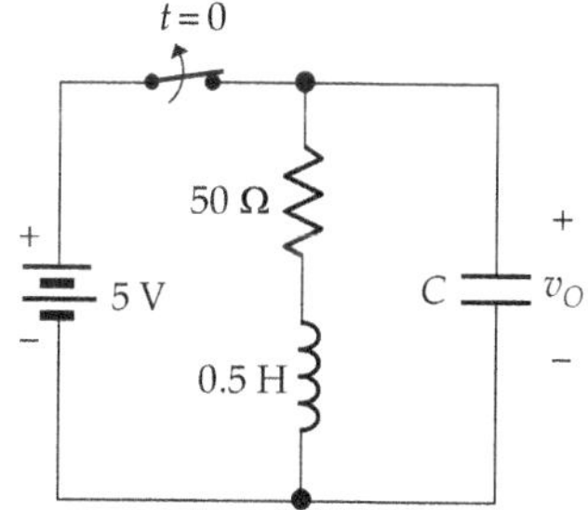

FIGURE P12.59

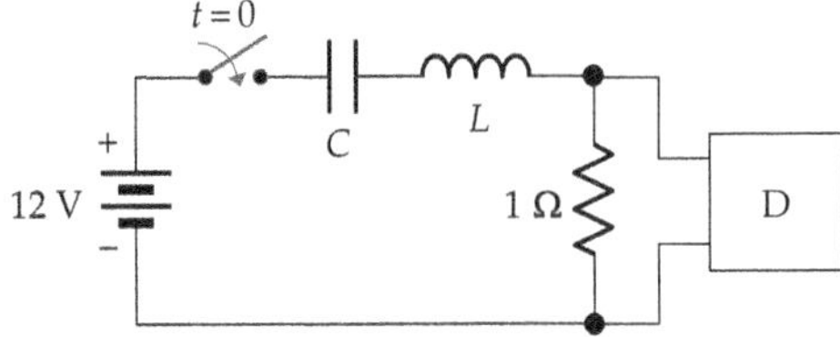

FIGURE P12.60

circuit is required to generate across the 1 Ω resistor a decaying sinusoidal voltage after the switch is closed at $t = 0$, so as to produce the required number of pulses. Show that suitable values of L and C are $L = 2.4$ H and $C = 18.8$ mF. Verify with a PSpice simulation. Use the results of Problem P12.62.

Probing Further

P12.61 Deduce from Table 12.2 for the natural response due to an initial current in the inductor that (a) for the overdamped response, $i = 0$ at $t = t_0 \dfrac{\ln(s_2/s_1)}{s_1 - s_2}$, and $v_L = 0$ at $t = 2\ t_0$, and (b) for the critically damped response, $i = 0$ at $t = 1/\omega_0$, and $v_L = 0$ at $t = 2/\omega_0$. Thus, the first zero of v_L occurs at a time that is twice that of the first zero of i.

P12.62 Consider the response v_C given by Equation 12.75. Show that (a) the maxima of the overshoots occur when $\omega_d t$ is an odd multiple of π, that is, at $t_{\max} = (2n+1)\dfrac{\pi}{\omega_d}$, $n = 0, 1, 2$, etc., and that the minima of the undershoots occur when $\omega_d t$ is an even multiple of π, that is, at $t_{\min} = (2n)\dfrac{\pi}{\omega_d}$, $n = 1, 2$, etc. (b) The values of the maxima are $v_{C\max} = V_{SRC}\left(1 + e^{-\alpha t_{\max}}\right)$ and the values of the minima are $v_{C\min} = V_{SRC}\left(1 + e^{-\alpha t_{\min}}\right)$. (c) The maximum overshoot beyond the steady-state value V_{SRC} is $V_{SRC} e^{-\frac{\alpha}{\omega_d}\pi}$ or $100 e^{-\frac{\alpha}{\omega_d}\pi}$ as a percentage of the steady-state value.

P12.63 Three cases were encountered in this chapter in which a series, second-order circuit has first-order responses. (a) Show that in Example 12.1, where the response is overdamped, $B = 0$ if $I_0 = s_1 C V_{C0}$. (b) Show that in Problem P12.33, where the response is critically damped, $B = 0$ if $I_0 = \omega_0 C V_{C0}$. (c) Show that in Problem P12.57, where the response is also critically damped but the final capacitor voltage is not zero, $B = 0$ if $I_0 = \omega_0 C(V_{C0} - V_{CF})$.

Part II

Topics in Circuit Analysis

13

Ideal Operational Amplifier

Objective and Overview

An operational amplifier, or op amp in short, is an electronic device that was commonly used in analog computers in the 1950s, before the advent of digital computers. Operational amplifiers are so called because they were at the heart of various building blocks that performed mathematical operations, such as addition, subtraction, differentiation, and integration. These building blocks were used to solve differential equations on analog computers. Nowadays, op amps of high performance and low cost are widely available in integrated-circuit (IC) form, which makes them an important building block in a variety of signal-processing applications. In electric circuits, op amps are used in active filters. They are therefore introduced in this chapter before passive filters are discussed in Chapter 14 and active filters in Chapter 15.

This chapter begins with some basic considerations on op amps, including the definition and properties of ideal and almost-ideal op amps. The very important concept of feedback is then introduced as a prelude to discussing the two basic op amp configurations: inverting and noninverting. These configurations are then considered in detail, including some of their variants, such as the unity-gain follower, adder, ideal integrator, and differentiator. The chapter ends with the analysis of the difference amplifier.

13.1 Basic Properties

Definition: *An op amp is a three-terminal, voltage-operated device whose output voltage is directly proportional to the difference between the voltages applied to its two input terminals.*

Figure 13.1 illustrates the symbol of an operational amplifier. One of the inputs is designated as a noninverting input, denoted by the (+) sign, whereas the other input is designated as an inverting input, denoted by the (−) sign. According to the preceding definition, the output voltage v_O is related as follows to the voltages v_P and v_N applied to the noninverting and inverting terminals, respectively:

$$v_O = A_v(v_P - v_N) = A_v\varepsilon \tag{13.1}$$

A_v is the **voltage gain** of the amplifier. The voltages v_O, v_P, and v_N are referenced with respect to a common ground. The effective input to the op amp is the voltage difference $(v_P - v_N)$, denoted by ε and referred to as the **differential input** to the amplifier. Because of the dependence of the output on the differential input, the op amp is an example of a **differential amplifier**.

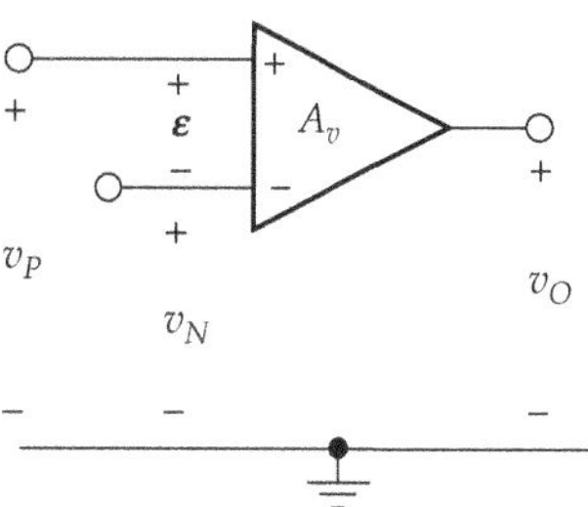

FIGURE 13.1
Symbol of operational amplifier.

It is seen from Equation 13.1 that the sign of v_O is the same as that v_P but is opposite that of v_N, hence the designation of v_P as a noninverting, or positive (P) input and v_N as an inverting, or negative input (N).

Definition: *An ideal op amp has the following properties:*

1. *The output is an ideal voltage source of voltage v_O given by*

$$v_O = A_v(v_P - v_N), \quad \textit{with} \quad A_v \to \infty \tag{13.2}$$

2. *Like an ideal voltage source, the amplifier has zero output resistance and can deliver any output voltage or current at any frequency.*
3. *Both inputs behave as open circuits.*
4. *The op amp is free from all imperfections, such as nonlinearities or any form of distortion of the output with respect to the input.*

Just as in the case of ideal circuit elements, the ideal op amp is a very useful abstraction that can be used in initial designs and which is commonly invoked to illustrate some important circuit concepts. Many practical op amps approach the ideal in several respects, so the concept is not as farfetched as it may seem.

13.1.1 Almost-Ideal Op Amp

In order to gain some insight into the basic properties of op amps, consider, to begin with, an op amp that is ideal

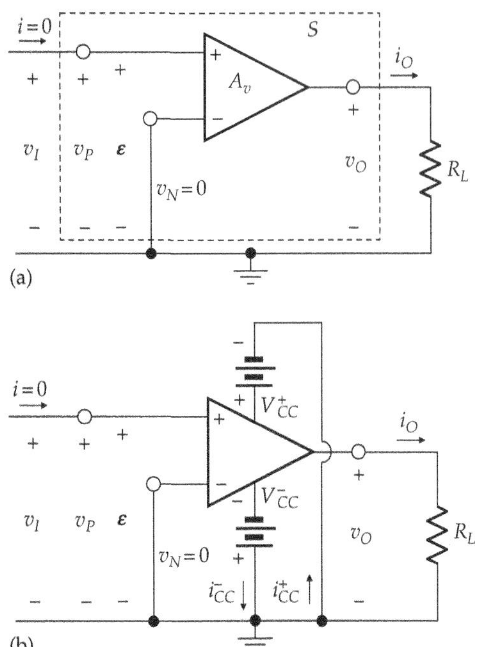

FIGURE 13.2
Operational amplifier driving a load R_L (a), with the dc bias supplies shown (b).

except for a finite A_v. An input v_I is applied to the noninverting input, with the inverting input grounded and a load resistance R_L connected to the output, as illustrated in Figure 13.2a. According to Equation 13.1, with $v_N = 0$ and $v_P = v_I = \varepsilon$,

$$v_O = A_v v_I = A_v \varepsilon \tag{13.3}$$

With v_O finite, a current $i_O = v_O/R_L$ flows through R_L in accordance with Ohm's law. According to the definition of an ideal op amp, the op amp inputs behave as open circuits, which means that the input current is zero (Figure 13.2a). If the op amp is enclosed by a surface S, then the current entering this surface is zero, whereas the current leaving it is $i_O \neq 0$, in apparent violation of KCL. Moreover, the power input to the op amp is zero, but the instantaneous power output is $v_O i_O \neq 0$, in apparent violation of conservation of power. Clearly, this cannot be the case, so what is missing?

Fundamentally, the characterizing attribute of any amplifying device, including an op amp, is its ability to *amplify power*; that is, at any instant of time, the power output of the device, due to a source applied to the input of the device, is larger than the power delivered by this source to the device. For this to happen, at least one external dc supply, referred to as a **bias supply**, must be connected to the device. When the current through this supply and the power delivered by it are taken into account, both KCL and conservation of energy are satisfied.

Op amps usually have two dc bias supplies, one that applies a positive voltage V_{CC}^+ to the op amp, and the other applies a negative voltage V_{CC}^-, as illustrated in Figure 13.2b. The magnitudes of these voltages are usually equal, but they need not be. The bias supplies are connected between specially provided pins of the IC op amp and ground. When these bias supplies are taken into account, then KCL gives at every instant

$$i_O + i_{CC}^- = i_{CC}^+ \tag{13.4}$$

From conservation of power, the instantaneous power delivered to the load is equal to the instantaneous power delivered by the power supplies, that is,

$$v_O i_O = V_{CC}^+ i_{CC}^+ + V_{CC}^- i_{CC}^- \tag{13.5}$$

If, for example, the input of the op amp is connected to a microphone and the output to a loudspeaker, then the audio-signal power delivered to the loudspeaker is derived from the bias supplies.

The next step is to consider the input–output characteristic of an op amp, that is, the v_O-ε relation, taking into account a finite A_v and the presence of bias supplies. It follows from Equation 13.3 that this relation is a straight line passing through the origin and having a slope A_v, assuming A_v is constant. This is illustrated in Figure 13.3a, from which it is seen that the line extends on both sides of the origin; that is, a positive input gives a positive output, whereas a negative input gives a negative output. If $A_v \to \infty$, the straight-line characteristic coincides with the vertical axis, since the ratio of v_O to ε goes to infinity. Theoretically, the straight-line characteristic having a finite A_v extends to infinity in both directions. In practice, the output cannot become more positive than V_{CC}^+, or more negative than V_{CC}^-, as illustrated by the horizontal dashed

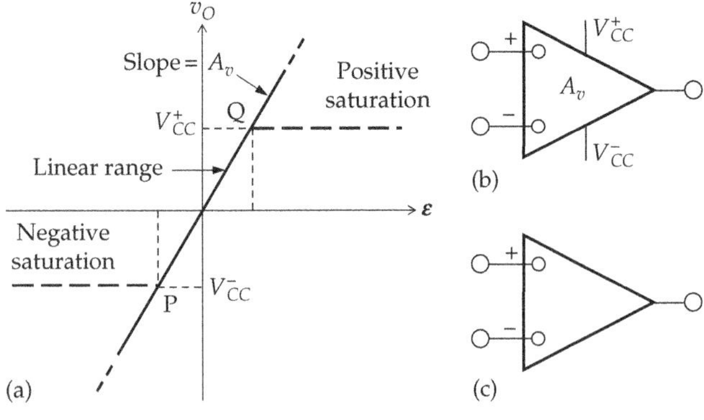

FIGURE 13.3
Input–output characteristic of an almost-ideal op amp (a) and its symbol (b); (c) symbol of ideal op amp.

lines in Figure 13.3a. The horizontal region where v_O is ideally equal to V_{CC}^+ is the region of *positive saturation*, since v_O is clamped at V_{CC}^+ in this region. The region of positive saturation extends from point 'Q', having $\varepsilon = V_{CC}^+/A_v$, to more positive values of ε. Similarly, the horizontal region, where v_O is ideally equal to V_{CC}^-, is the region of *negative saturation*. It extends from point P having $\varepsilon = -V_{CC}^-/A_v$, to more negative values of ε. The part of the characteristic between 'P' and 'Q' is the region of *linear operation* of the op amp, where v_O is directly proportional to ε. Note that the v_O-ε characteristic relates v_O to ε under *static conditions*, that is, a point on the plot is obtained by applying a dc differential input and observing the corresponding dc output under steady conditions.

In analyzing ideal op amp circuits, it is often convenient and instructive to assume a finite A_v to begin with, and then let $A_v \to \infty$. It is also required sometimes to assume finite values of V_{CC}^+ and V_{CC}^-, rather than the infinite values assumed for an ideal op amp. To avoid confusion over terminology, we will adopt the following definition of an **almost-ideal op amp**:

Definition: *In an almost-ideal op amp, either A_v is a finite constant or the bias supplies, V_{CC}^+ and V_{CC}^-, have finite values, or both. In all other respects, the almost-ideal op amp is ideal.*

The OPAMP part in PSpice is an almost-ideal op amp having default values of $A_v = 10^6$, $V_{CC}^+ = 15$ V, and $V_{CC}^- = -15$ V. However, these default values can be changed, if desired, in the Property Editor spreadsheet of the OPAMP part.

The symbol used in this book for an almost-ideal op amp is illustrated in Figure 13.3b, in which finite values of A_v, V_{CC}^+ and V_{CC}^-, are indicated. In contrast, these symbols are omitted from the symbol for an ideal op amp (Figure 13.3c). In both cases, the inputs are shown open circuited inside the symbol for emphasis.

It should be noted that in an almost-ideal op amp, and in accordance with Equation 13.1, v_O depends only on the differential input $\varepsilon = v_P - v_N$, independently of the absolute values of v_P and v_N; that is, adding any positive or negative voltage to both v_P or v_N does not affect v_O, as long as $v_P - v_N$ remains the same.

13.1.2 Equivalent Circuit

Over the linear operating region, corresponding to region PQ of the input–output characteristic in Figure 13.3a, the almost-ideal op amp can be represented on the output side by a VCVS of source voltage $A_v\varepsilon$, where ε is the differential input between the noninverting and inverting inputs, as illustrated in Figure 13.4.

As explained in the preceding subsection, an amplifying device, such as an op amp, delivers power in a circuit. It is therefore an *active* circuit element that is represented by an ideal source (Section 1.8). The source is a dependent source, since energy is derived from an electric source of energy, the dc bias supplies, rather than converted from a nonelectric source of energy, as is the case with independent sources (Section 2.6).

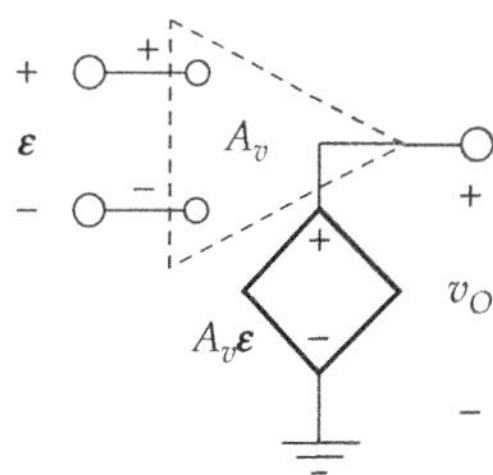

FIGURE 13.4
VCVS equivalent circuit of almost-ideal op amp.

V_{CC}^+ and V_{CC}^- do not appear in the linear equivalent circuit. The extent of the linear range does depend on V_{CC}^+ and V_{CC}^-, but within the linear operating range, over which the equivalent circuit is valid, v_O depends only on A_v and ε, and not on V_{CC}^+ and V_{CC}^-. In the saturation regions, the op amp output is represented by a battery of voltage V_{CC}^+ in the positive saturation region and a battery of voltage V_{CC}^- in the negative saturation region.

The equivalent circuit can be useful in analyzing circuits involving almost-ideal op amps of finite A_v. But when $A_v \to \infty$ and $\varepsilon \to 0$, in an ideal op amp, the equivalent circuit is no longer useful. Circuits involving ideal op amps are therefore analyzed using a different approach, as illustrated in the rest of the chapter.

Example 13.1: Simulation of Input–Output Characteristic of the Op Amp in PSpice

It is required to simulate the input–output characteristic of the op amp in PSpice.

Simulation: The circuit is entered as in Figure 13.5 using the part OPAMP from the ANALOG library, without changing the default values of $A_v = 10^6$, $V_{CC}^+ = 15$ V, and

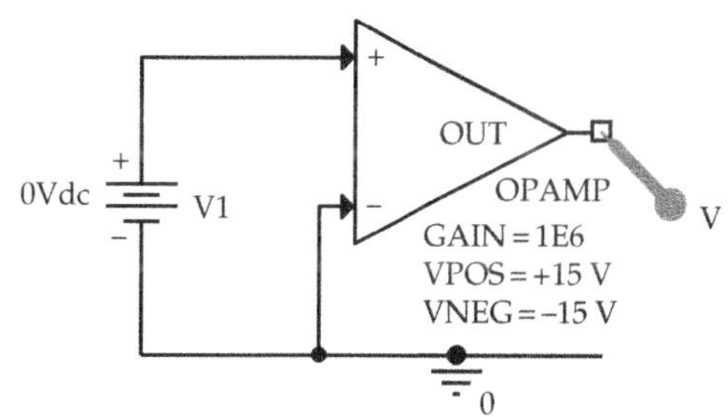

FIGURE 13.5
Figure for Example 13.1.

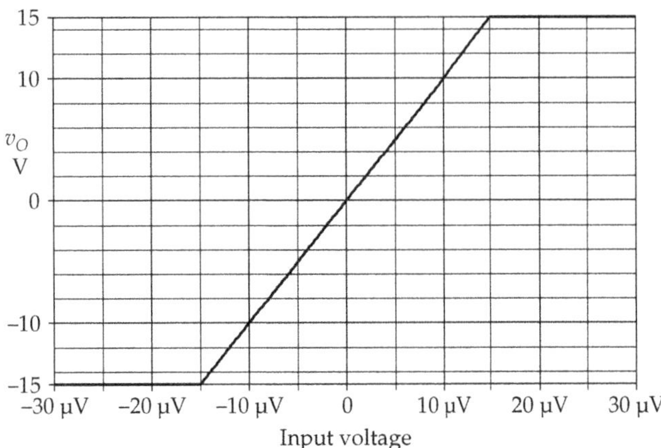

FIGURE 13.6
Figure for Example 13.1.

$V_{CC}^- = -15$ V, which are indicated in the figure. To run the simulation, select 'DC Sweep' under 'Analysis type' in the Simulation Settings dialog box. Under 'Sweep variable' select 'Voltage source' and enter V1 in the 'Name' field. Under 'Sweep type,' select 'Linear' and enter 30u for 'Start value,' −30u for 'End value,' and 0.1u for 'Increment.' The input–output characteristic is displayed as in Figure 13.6. It is seen that positive saturation is at +15 V, starting at an input voltage of 15 V/10^6 = 15 µV and extending in the positive direction. The negative saturation starts at −15V starting at an input voltage of −15 V/10^6 = −15 µV and extends in the negative direction.

Primal Exercise 13.1

Given an almost-ideal op amp having $A_v = 10^4$, $V_{CC}^+ = 20$ V, and $V_{CC}^- = -20$ V, determine (a) the breakpoints of the linear operating region, and the regions of positive saturation and of negative saturation and (b) the differential input voltage ε when the output voltage is 10 V; specify (c) the current inputs to the differential inputs under these conditions and (d) the source resistance at the output of the op amp.

Ans. (a) Linear operating range is from −2 mV to +2 mV; the positive saturation region extends for $\varepsilon \geq 2$ mV, and the negative saturation region extends for $\varepsilon \leq -2$ mV; (b) 1 mV; (c) 0 A; (d) 0 Ω.

13.2 Feedback

Feedback, as the word implies, refers to feeding part or all of the output of a device or system back to its input, as illustrated in Figure 13.7. The connection from the

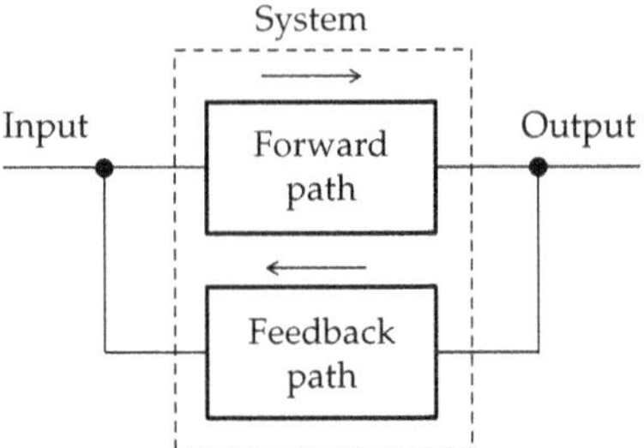

FIGURE 13.7
Feedback system.

output back to the input is the **feedback path**, whereas the connection from input to output through the system is the **forward path**. The two paths together form the **feedback loop**. Feedback plays a critical role in modifying the behavior of a circuit or system and is extensively employed in many types of systems, particularly electronic circuits and control systems.

Feedback is almost invariably present in op amp circuits. But since an op amp has two inputs, noninverting and inverting, feedback in op amp circuits could be to either of these inputs, exclusively or predominantly. In Figure 13.8a, the feedback path is from the output to the noninverting input of an almost-ideal op amp, whereas the inverting input is grounded, for simplicity. In the forward path, $v_O = A_v\varepsilon$ (Equation 13.1); from KVL on the input side, $\varepsilon = v_P - v_N = v_P$, since $v_N = 0$. The feedback connection makes $v_P = v_O$. If the output voltage v_O of the op amp tends to increase for some reason, as indicated by the upward arrow next to v_O at the extreme right of Figure 13.8b, v_P and ε will also increase by the same amount as v_O through the feedback path. The increase

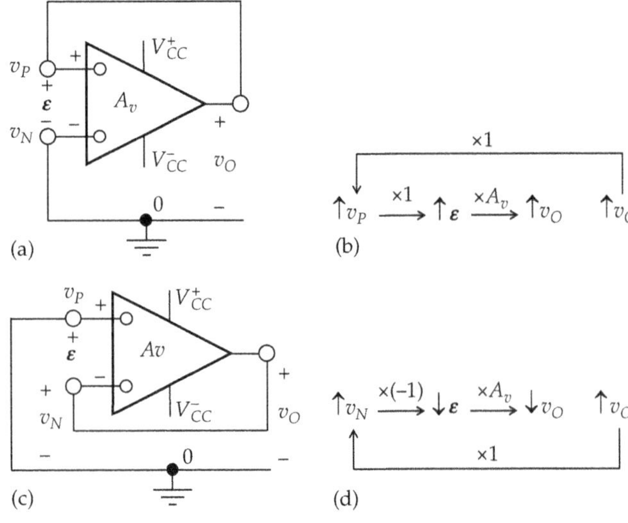

FIGURE 13.8
Positive feedback (a) and responses to an increase in v_O (b); negative feedback (c) and responses to an increase in v_O (d).

in ε will increase v_O by A_v times the initial increase in v_O, through the forward path. Thus, an initial change in v_O in a given direction results in a change in v_O in the *same direction as the initial change*. In other words, the feedback tends to reinforce the initial change. This type of feedback connection is described as **positive feedback**.

In Figure 13.8c, the feedback path is from the output to the inverting input, whereas the noninverting input is grounded for simplicity. In the forward path, $v_O = A_v\varepsilon$ as before; from KVL on the input side, $\varepsilon = v_P - v_N = -v_N$, since $v_P = 0$. The feedback connection makes $v_N = v_O$. If v_O tends to increase, as indicated by the upward arrow next to v_O at the extreme right of Figure 13.8d, v_N will increase, but since $\varepsilon = -v_N$, ε will increase in the negative direction through the feedback path, and v_O will also increase in the negative direction through the forward path. Thus, an initial change in v_O in a given direction results in a change in v_O in the *direction opposite that of the initial change*. In other words, the initial change is opposed. This type of feedback connection is described as **negative feedback**.

The stipulation that v_O tends to increase or decrease in the preceding argument should be justified. In practice, small changes in voltage or current always occur in any circuit, without any change in the applied inputs, because of "noise" due to random fluctuations in the motion of current carriers in any conductor (Section 2.1). The op amp output is therefore subject to small fluctuations due to noise in the op amp itself, as well as noise in the bias supplies. The input of the op amp is also subject to noise and to pickup of extraneous signals or interference from the surroundings through electric or magnetic coupling. It may be noted that whereas noise is always present in real circuits, it is absent in simulations, unless deliberately introduced.

The presence of positive or negative feedback has a decisive effect on the behavior of the op amp circuit. Consider, to begin with, the negative feedback circuit of Figure 13.8c but with an input voltage V_I applied between the noninverting input and ground as in Figure 13.9a. The output and input of the op amp are related in the steady state by the v_O-ε characteristic of the op amp. From KVL at the input, $V_I = \varepsilon + v_N = \varepsilon + v_O$, or $v_O = -\varepsilon + V_I$. As a plot of v_O vs. ε, this is the equation of a line of slope −1 and intercept V_I on both axes as in Figure 13.9b. The operating point is the point of intersection 'P' of this line with the v_O-ε characteristic of the op amp, since the two relations involving v_O and ε are satisfied at this point. 'P', whose coordinates are denoted as ε_0 and V_{O0}, is the only point of intersection between the two plots in this case. It is a stable operating point, or a point of *stable equilibrium*, in the sense that any tendency for the input or output to change from their values at 'P' produces a countereffect that restores these values to ε_0 and v_{O0}. For example, suppose that v_O *tends to increase* from V_{O0} by $+\Delta v_O$. This will also increase v_N by $+\Delta v_O$ because of the direct connection between input and output through the feedback path. However, an increase in v_N causes an equal decrease in ε, because $\varepsilon + v_N = V_I$, which is constant. The change $+\Delta v_O$ thus produces a change $\Delta\varepsilon = -\Delta v_O$. This change at the input in the negative direction will produce a change of $-A_v\Delta v_O$ at the output of the op amp that is in a direction opposite that of the initial $+\Delta v_O$. This means that v_O will move back toward its initial value V_{O0}. As v_O so moves, $\Delta v_O \rightarrow 0$, which restores v_O and ε to their original values at 'P', and no further change occurs. Similar considerations apply if the initial change in v_O is $-\Delta v_O$ instead of $+\Delta v_O$. 'P' is therefore a stable operating point. Moreover, if V_I is varied over an appropriate range, *the negative feedback connection allows a stable operating point anywhere in the linear region of the v_O-ε characteristic of the op amp.*

Consider next the positive feedback circuit in Figure 13.8a, but with an input voltage V_I applied between the inverting input and ground, as in Figure 13.10a. The output and input of the op amp are related in the steady state by the v_O-ε characteristic of the op amp, as always. From KVL at the input, $v_P = \varepsilon + V_I$, or $v_O = \varepsilon + V_I$. As a plot of v_O vs. ε, this is the equation of a line of slope +1 and intercept of magnitude V_I on both axes (Figure 13.10b). There are now three points of

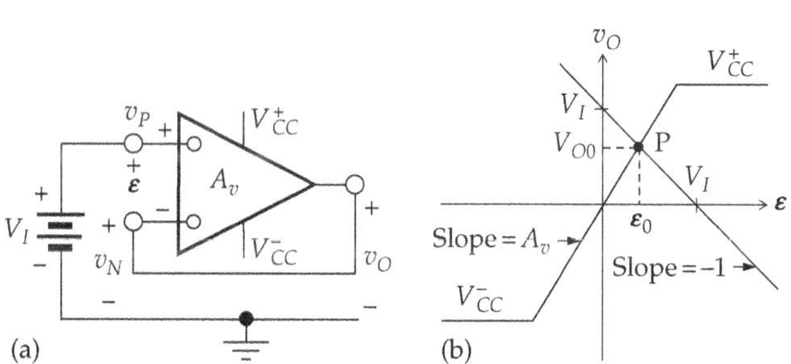

FIGURE 13.9
Stable operating point with negative feedback. (a) Op amp with negative feedback and (b) graphical construction for the operating point.

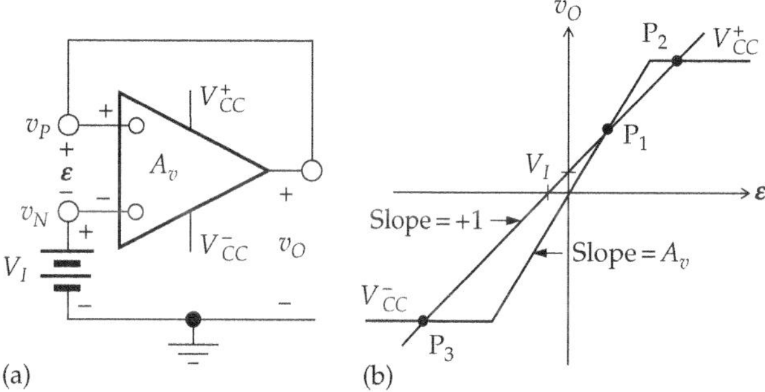

FIGURE 13.10
Unstable and stable operating points with positive feedback. (a) Op amp with positive feedback and (b) graphical construction for the operating point.

intersection at which the values of ε and v_O satisfy both the equation of the line and the v_O-ε characteristic of the op amp. We will argue that 'P$_1$' is an unstable operating point if $A_v > 1$ in Figure 13.10, as is almost invariably the case. Thus, if v_O tends to increase by $+\Delta v_O$, v_P also increases by $+\Delta v_O$ because of the direct connection between input and output through the feedback path. However, an increase in v_P causes an equal increase in ε, because $\varepsilon + V_I = v_P$. An increase $+\Delta v_O$ in ε at the input of the op amp will produce a change of $+A_v\Delta v_O$ at the output of the op amp. This new value of output is in turn fed back to the input and results in a new output $(A_v)^2\Delta v_O$, and so on. Here, we have to distinguish two cases: (1) $A_v < 1$, in which case the successive changes at the output and input of the op amp get progressively smaller, eventually reverting to their original values at 'P$_1$'. This means that 'P$_1$' is a stable operating point (Exercise 13.2). (2) $A_v > 1$, in which case the successive changes at the output and input of the op amp get progressively larger. v_O and ε will increase very rapidly until they reach the intersection point 'P$_2$' in the positive saturation region. 'P$_2$' is a stable operating point: v_O does not change because of positive saturation; any tendency for ε to change cannot be sustained because ε must remain constant to satisfy KVL at the input. Similarly, if initially v_O tends to decrease by Δv_O from its value at 'P$_1$', v_O and ε will decrease very rapidly until they reach the intersection point 'P$_3$' in the negative saturation region. 'P$_3$' is also a stable operating point because v_O must remain constant. It is seen that with positive feedback and $A_v > 1$, two stable states are possible, corresponding to intersection points 'P$_2$' and 'P$_3$' that are not within the linear operating region of the op amp. The op amp can be used in this case as a *switch*, where the two saturation states correspond to the two stable states of the switch: open or closed. Although switching can occur without positive feedback, by applied small positive and negative values of ε, positive feedback increases the switching speed and enhances performance.

Note that in Figure 13.10, the gain around the feedback loop is $A_v \times 1 = A_v$, where the "1" for the feedback path is due to the fact that all of v_O is applied to the noninverting input ($v_P = v_O$). In the general case, where only a fraction of the output is fed back to the input, the preceding argument applies to the gain around the feedback loop, rather than to the amplifier gain A_v alone. Thus, if $A_v = 2$ and only a quarter, say, of the output is applied to the noninverting input ($v_P = 0.25v_O$), the loop gain is $2 \times 0.25 = 0.5$. A change of Δv_O in the output is fed back as a change of $0.5\Delta v_O$. The positive feedback configuration is stable in this case, despite the fact that $A_v = 2$.

The use of negative feedback in amplifying-type op amp circuits is, in contrast to switching-type op amp circuits, imperative for two reasons: (1) it allows stable operation in the linear region of the v_O-ε characteristic of the op amp, and (2) It confers some desirable properties on the op amp circuit, as explained later. Hence, the following concept applies:

Concept: *Nonswitching op amp circuits invariably employ negative feedback, which necessitates some circuit connection between the output and the inverting input of the op amp.*

The negative feedback circuit connection between the output and the inverting input could be a direct connection, as in the case of the unity-gain amplifier (Figure 13.18), or the connection could be through a single circuit element or a combination of circuit elements.

It must not be assumed, however, that negative feedback systems are always stable. The preceding discussion assumes no delays, or phase shift, around the feedback loop. If the phase shift around the feedback loop is 180°, the feedback changes sign and becomes effectively positive. If the loop gain under these conditions exceeds unity, the negative feedback system will be unstable.

There are two basic op amp configurations that employ negative feedback, namely, the noninverting and the inverting configurations. In the noninverting configuration, the feedback signal is in series with the input signal, whereas in the inverting configuration, the feedback signal is effectively in shunt with the input signal. These configurations are discussed in the following two sections.

Exercise 13.2

Plot the v_O-ε characteristic of the op amp, with $A_v < 1$, and the line relating v_O and ε in accordance with KVL at the input with positive feedback (Figure 13.10a). Argue that the point of intersection is a stable operating point. What if $A_v = 1$?

Ans. If $A_v < 1$, the change in output is not sufficient to sustain the change in input. If $A_v = 1$, the two lines coincide over the linear range of the op amp characteristic. Theoretically, this is neutral equilibrium. In practice, A_v cannot be *exactly* 1.

Exercise 13.3

An argument similar to that used with positive feedback (Figure 13.10) is sometimes applied to negative feedback (Figure 13.9), namely, that if v_O changes by Δv_O, ε changes by $-\Delta v_O$, and v_O changes by $-A_v\Delta v_O$, then by $+(A_v)^2\Delta v_O$, $-(A_v)^3\Delta v_O$, and so on, which seemingly makes 'P' in Figure 13.9 unstable. What is wrong with this argument?

Ans. If v_O changes by $-A_v\Delta v_O$, then v_O moves to the other side of 'P', through zero. But as it approaches 'P', $\Delta v_O \to 0$. When $\Delta v_O = 0$, there is no tendency for v_O to move in the opposite direction, and v_O will stop at 'P'.

13.3 Noninverting Configuration

Figure 13.11 illustrates an almost-ideal op amp connected in the noninverting configuration. The input v_I is applied directly to the noninverting terminal, whereas the inverting terminal is connected to a voltage divider across the op amp output. This means that not all of v_O, but a fraction of it, is fed back to the inverting input. That is, $v_N = \beta v_O$, where β is the **feedback factor**. Because the inverting input, by definition of an ideal op amp, does not draw any current; it follows from simple voltage division that

$$v_N = \frac{R_r}{R_r + R_f} v_O \quad \text{or} \quad \beta = \frac{R_r}{R_r + R_f} \tag{13.6}$$

It is desired to derive the relation between the output voltage v_O and the input voltage v_I. The governing equations are (1) the v_O-ε characteristic of the op amp over the linear range, represented by the equivalent circuit, and (2) KVL at the input:

$$v_O = A_v\varepsilon = A_v(v_P - v_N) \quad \text{and} \quad v_P = \varepsilon + v_N \tag{13.7}$$

with $v_P = v_I$ and $v_N = \beta v_O$. Note that according to Equation 13.7, v_N subtracts from v_P to give ε. This is characteristic of a series connection in which voltages add algebraically (Section 2.9). Substituting for v_P and v_N, in terms of v_I and v_O, eliminating ε, and simplifying,

$$\frac{v_O}{v_I} = \frac{A_v}{1 + \beta A_v} \tag{13.8}$$

With A_v and β positive numbers, the sign of v_O is the same as that of v_I, which gives the noninverting configuration its name. The noninversion is basically because the input is applied to the noninverting input.

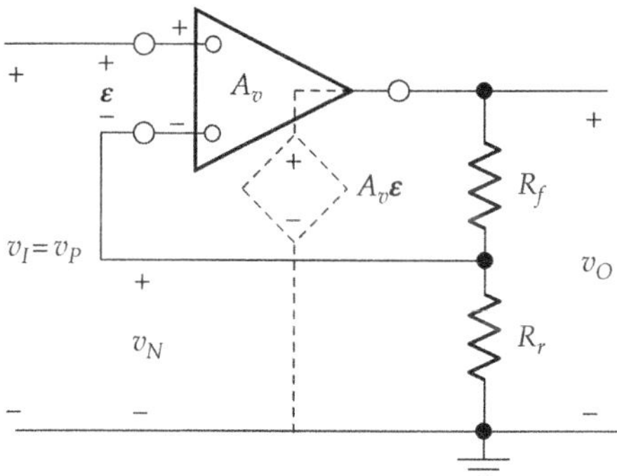

FIGURE 13.11
Noninverting configuration with op amp of finite gain.

Normally, $\beta A_v \gg 1$, so that

$$\frac{v_O}{v_I} \cong \frac{1}{\beta} = \frac{R_r + R_f}{R_r} = 1 + \frac{R_f}{R_r} \tag{13.9}$$

If, for example, $R_r = 1\,\text{k}\Omega$, $R_f = 4\,\text{k}\Omega$, then $\beta = 1/5$. If $A_v = 10^5$, which is typical for a general-purpose, low-frequency op amp, then $v_O/v_I = 10^5/(1 + (10^5)/5) = 4.99975 \cong 1/\beta$. It is seen from Equation 13.9 that the voltage gain v_O/v_I cannot be less than unity.

It may be wondered at this stage, and quite legitimately, as to the wisdom of starting with an op amp of high voltage gain, such as 10^5, and ending with an amplifier circuit having a much smaller voltage gain, say, 5 or 100. Are we throwing away voltage gain for nothing? If a certain value of gain is required for a given application, why not design the op amp for this value of gain and get done with it? For example, a stable and precise value of voltage gain of 1.125 is required in digital thermometers, so why not design op amps for this application having $A_v = 1.125$? The answer is that the gain A_v of the op amp cannot be precisely defined, because of manufacturing tolerances, that is, inevitable variations in the properties of components during the manufacturing process. These variations can be extremely difficult and costly to control to a high degree of precision. Moreover, the value of A_v is subject to variations during normal use of the op amp, due to various factors such as (1) changes in internal components because of environmental effects, such as temperature and humidity, or due to "aging," which is some variation in the properties of these components with time, because of some long-term physical or chemical changes, or (2) variations in bias voltages, as when batteries begin to run down. When an op amp of very large A_v is connected as in Figure 13.11, or in other negative feedback configurations as will be demonstrated later, the gain is no longer determined by A_v but by resistance values, as in Equation 13.9. Discrete-component resistors can be accurate to 0.01% and highly stable. IC resistors have a larger tolerance, typically between 5% and 20%, but the *ratio* of resistances, as in Equation 13.9, is at least an order of magnitude more precise than the values of individual resistances in the same integrated circuit. Consequently, when precision amplification is required, it is highly advantageous to trade off the high gain of an op amp for a precise and stable value of amplification.

The preceding discussion underscores a concept that is of great practical importance, utilizing negative feedback:

Concept: *By employing negative feedback, op amp circuits can be designed to trade off the high voltage gain of the op amp for some desirable characteristics of the circuit.*

High gain in IC op amps is quite inexpensive and can be advantageously traded off, not only for precision and stability of voltage gain, as illustrated by the noninverting configuration, but for other desirable characteristics such as reduced distortion, wider bandwidth, increased input impedance, or reduced output resistance.

For an ideal op amp, $A_v \rightarrow \infty$, and the voltage gain in Equation 13.9 is then

$$\frac{v_O}{v_I} = \frac{1}{\beta} = \frac{R_r + R_f}{R_r} \quad (13.10)$$

Equation 13.10 can be derived directly by applying a very important concept that greatly facilitates the analysis of ideal op amp circuits. Recall that in this case the equivalent op amp circuit is not applicable, because $A_v \rightarrow \infty$.

In Figure 13.11, $\varepsilon = v_O/A_v$, as is true, by definition, of all op amp circuits (Equation 13.1). If $A_v \rightarrow \infty$, then $\varepsilon \rightarrow 0$ for finite v_O. A zero differential input ε implies a **virtual short circuit** between the noninverting and inverting terminals of the op amp; that is, these two terminals are at the same voltage, without being physically connected together. This virtual short circuit is indicated by the dotted connection between the two input terminals in Figure 13.12. It is seen that one can immediately write $v_I = v_N$, with $v_N = \beta v_O$. Eliminating v_N gives $v_O = v_I/\beta$, as in Equation 13.10.

This concept can be readily generalized as follows:

Concept: *In a negative feedback connection employing an ideal op amp ($A_v \rightarrow \infty$), the differential input is vanishingly small for any finite output, so that the two inputs of the op amp are at the same voltage and can be considered to be virtually short-circuited to one another.*

It should be emphasized that the virtual short circuit is not a physical connection between the inverting and noninverting terminals. Although these terminals are maintained at the same voltage by the negative feedback, *no current flows through the virtual short circuit.* Resistors that are effectively in parallel because of a virtual short circuit *have the same voltage across them* as when they are physically in parallel. However, because a virtual short circuit does not pass any current, the *distribution of currents in the rest of the circuit is different in the two cases*, which means that the circuits behave differently altogether. The distribution of currents in the case of a virtual short circuit is illustrated in Example 13.9.

With $A_v \rightarrow \infty$, the output can be considered as an ideal, independent voltage source of voltage v_O, in accordance with the substitution theorem.

Example 13.2: Noninverting Configuration

It is required to determine V_O and I_O in Figure 13.13, assuming an ideal op amp.

Solution:

With the virtual short circuit between the noninverting and inverting terminals, $V_N = V_P = 6$ V. From voltage division, $V_N = V_O(2/3)$. It follows that $V_O = (3/2)V_N = 9$ V. Alternatively, it can be argued that since V_N appears across the 2 kΩ resistor, $I_2 = 6/2 = 3$ mA. I_2 flows through the 1 kΩ resistor, producing a voltage drop of 3 V across this resistor. It follows that $V_O = 3 + 6 = 9$ V. Note that as with an ideal voltage source, V_O is independent of I_O and hence does not depend on whatever is connected to the output of the ideal op amp.

From KVL around the mesh at the output of the op amp, the current in the upper 2 kΩ resistor is $I_1 = (9 - 5)/2 = 2$ mA. It follows from KCL at the output node that $I_O = 2 + 3 = 5$ mA.

Simulation: The circuit is entered as in Figure 13.14 using the part OPAMP from the ANALOG library, without changing the default values of A_v, V_{CC}^+, and V_{CC}^-. The 1u resistor added at the output of the op amp is simply to have PSpice display the output current of the op amp. To run the simulation, select 'Bias Point' under 'Analysis type' in the Simulation Settings dialog box. After the simulation is run, pressing the V and I buttons displays the values shown in Figure 13.15. Note that the 9 V output is less than the default V_{CC}^+ of 15 V, which confirms that the op amp operates in the linear region of its v_O-ε characteristic.

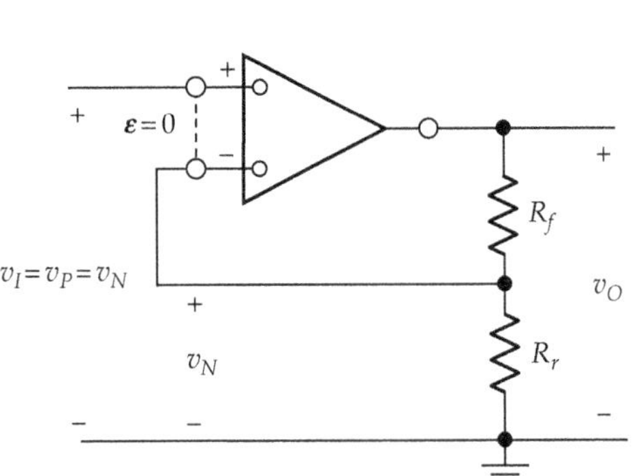

FIGURE 13.12
Noninverting configuration with ideal op amp.

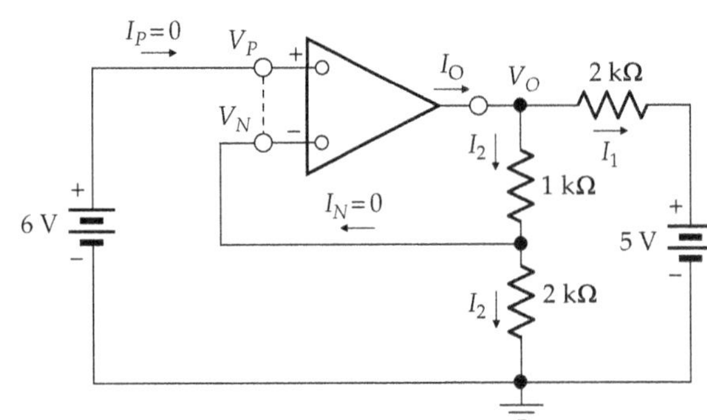

FIGURE 13.13
Figure for Example 13.2.

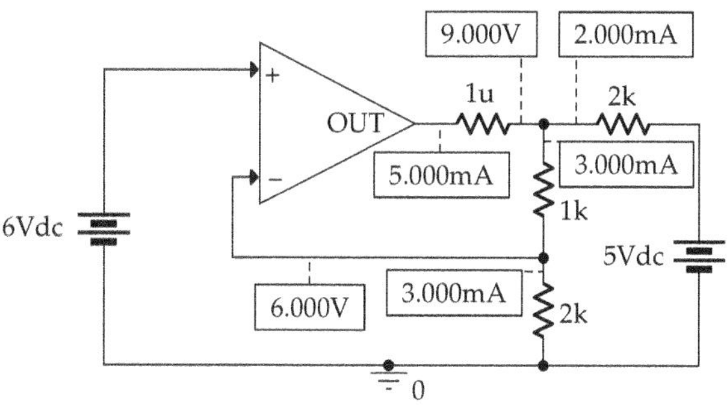

FIGURE 13.14
Figure for Example 13.2.

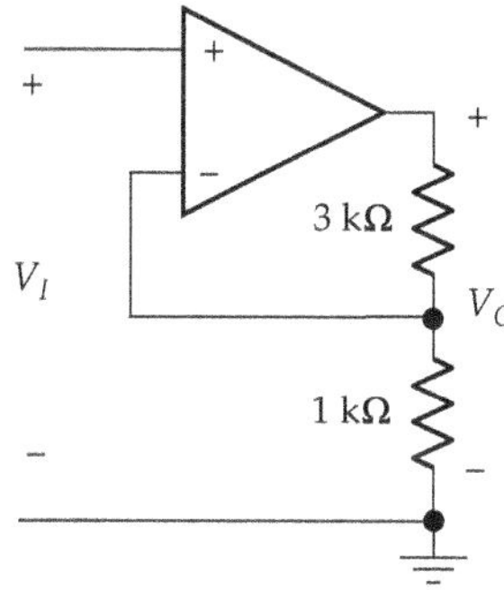

FIGURE 13.15
Figure for Primal Exercise 13.4.

Problem-Solving Tips

- In solving problems involving *ideal* op amps, always make use of the virtual short circuit between the two inputs.
- In solving problems involving almost-ideal op amps, always check that the op amp is operating in the linear region of its input–output characteristic.

Primal Exercise 13.4

Draw the circuit diagram of an amplifier having a gain of +4, using an ideal op amp and a 1 kΩ resistor connected to common ground.

Ans. The circuit is as shown in Figure 13.15.

Primal Exercise 13.5

Determine V_O in Figure 13.16 using the concept of the virtual short circuit between the input terminals of the op amp.

Ans. 4 V.

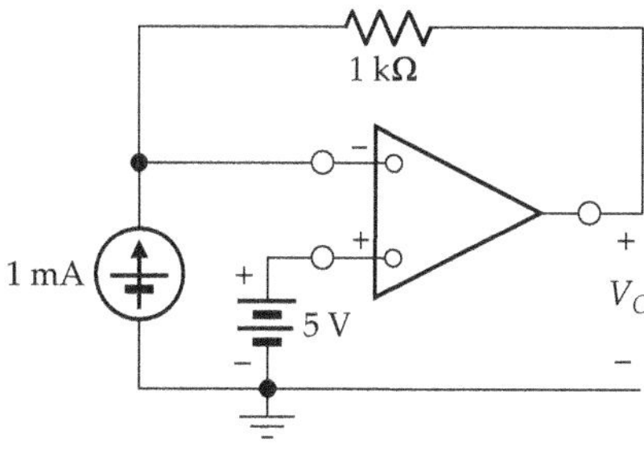

FIGURE 13.16
Figure for Primal Exercise 13.5.

Primal Exercise 13.6

Repeat Primal Exercise 13.5 with the 1 mA source replaced by a 5 kΩ resistor

Ans. 6 V.

Primal Exercise 13.7

Determine V_O in Figure 13.17

Ans. 3 V.

13.3.1 Unity-Gain Amplifier

An important special case of the noninverting configuration is the **unity-gain amplifier**, in which $R_f = 0$, so that the output is directly connected to the inverting input (Figure 13.18). R_r across the ideal voltage source output of the op amp becomes redundant and is omitted. $R_f = 0$ makes $\beta = 1$, so that Equation 13.8 becomes

$$\frac{v_O}{v_I} = \frac{A_v}{1 + A_v} \cong 1, \quad \text{if } A_v \gg 1 \tag{13.11}$$

In terms of the virtual short circuit in the case of an ideal amplifier, $v_{SRC} = v_P = v_N = v_O$, so that $v_O/v_{SRC} = 1$. The unity-gain amplifier is also known as a **voltage follower**,

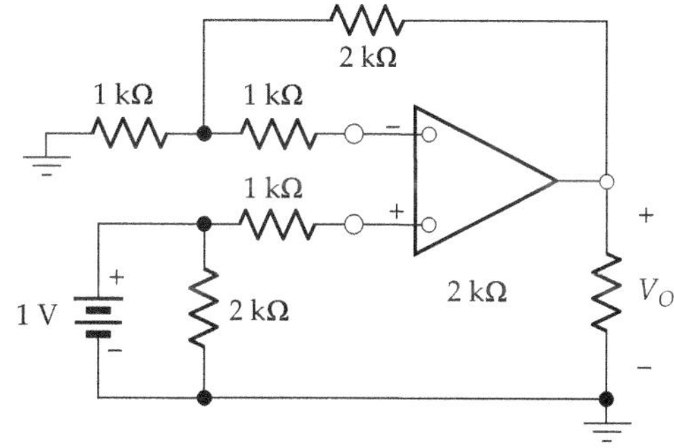

FIGURE 13.17
Figure for Primal Exercise 13.7.

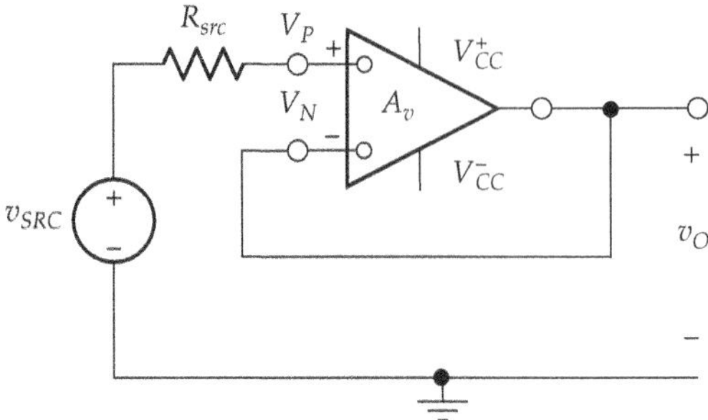

FIGURE 13.18
Unity-gain amplifier.

since the output voltage follows the input voltage because even in the case of an almost-ideal op amp, $v_O = v_N \cong v_P$. A special symbol is sometimes used for a unity-gain amplifier consisting of a triangle marked with '×1', as in Figure 13.19b.

The unity-gain amplifier is extremely useful for isolating a source from a load, as is desirable in many cases. Consider, for example, a linear-output voltage source connected to a load resistance (Figure 13.19a). From voltage division,

$$v_O = \frac{R_L}{R_L + R_{src}} v_{SRC} \tag{13.12}$$

It is seen that because of R_{src}, variations in R_L affect the load voltage v_O. Moreover, if R_L is comparable to or considerably smaller than R_{src}, then v_O is significantly less than v_{SRC} and the power delivered to the load is also significantly reduced. For example, if $V_{SRC} = 6$ V, $R_{src} = 1.1$ kΩ, and $R_L = 100$ Ω, then v_O is only $(0.1/1.2) \times 6 = 0.5$ V, and the power delivered to the load is $(0.5)^2/100 \equiv 2.5$ mW. Most of v_{SRC} appears across R_{src}, and most of the power supplied by the source is dissipated in R_{src}. A unity-gain amplifier can be used as in Figure 13.19b to isolate the source from the load. No current is drawn from the voltage source, so that this source is now ideally terminated by an open circuit. Variations in R_L do not affect the voltage at the source terminals, which remains at v_{SRC}. The output voltage of the unity-gain amplifier is v_{SRC}, which is applied to R_L, so $v_O = 6$ V using the same numerical values. The power delivered to the load is $(6)^2/100 \equiv 360$ mW, assuming the op amp can handle this power. Another advantage is that if the load is an impedance Z_L instead of R_L, then v_O is frequency dependent in Figure 13.19a, but not in Figure 13.19b. The op amp output in the latter case is virtually an ideal voltage source that maintains v_{SRC} irrespective of the output current.

Example 13.3: Unity-Gain Amplifier

It is required to determine V_O in Figure 13.20a, assuming an ideal op amp.

Solution:

Let the current in the upper branch be I_X, as in Figure 13.20b. Because of the virtual short circuit, $V_P = V_N$, so that $V_O = V_N = V_P$. This means that there is no voltage drop between the output terminal and the noninverting terminal. From KVL, $-I_X \times 1 + 10 = 0$, which gives $I_X = 10/1 = 10$ mA. Because of the open circuit at the op amp input terminals, I_X flows in the 2 kΩ resistor. It follows that $V_P = 2 \times 10 = 20$ V and $V_O = V_P = 20$ V.

Simulation: The circuit is entered as in Figure 13.21 using the part OPAMP from the ANALOG library. To maintain operation in the linear region of the input–output characteristic, V_{CC}^+ is changed to 25 V from its default value of 15 V. There is no need to change A_v nor V_{CC}^-, which only changes the level of the negative saturation region. To run the simulation, select 'Bias Point' under 'Analysis type' in the Simulation Settings dialog box. After the simulation is run, pressing the V and I buttons displays the values shown in Figure 13.21.

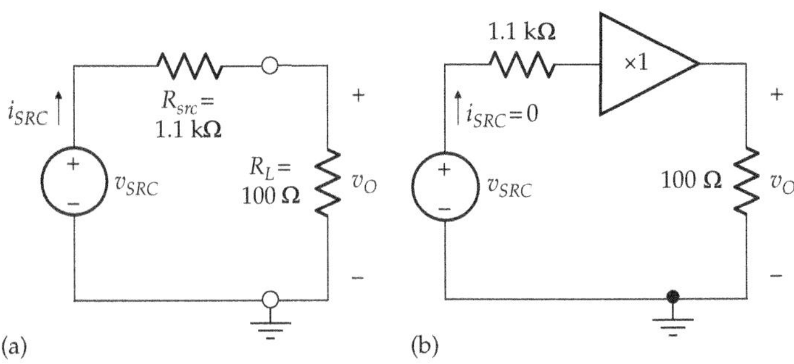

FIGURE 13.19
A load R_L connected directly to a linear-output voltage source (a) and through an isolating, unity-gain amplifier (b).

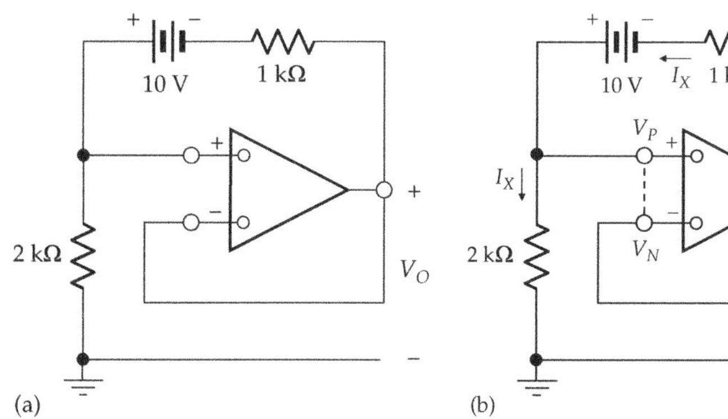

FIGURE 13.20
Figure for Example 13.3.

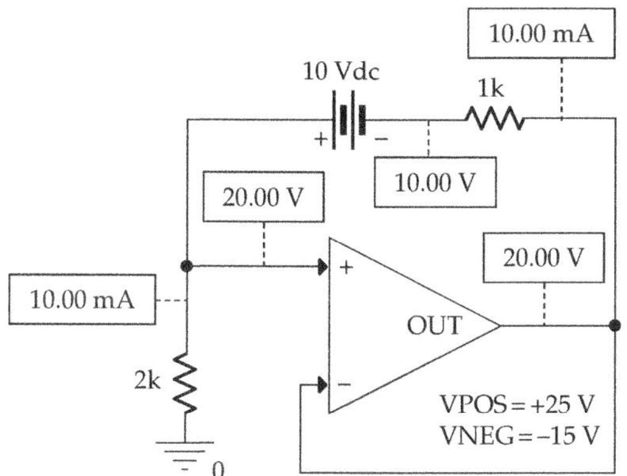

FIGURE 13.21
Figure for Example 13.3.

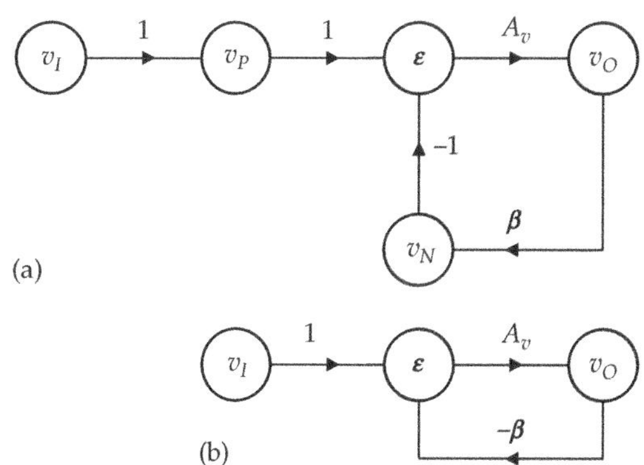

FIGURE 13.22
Figure for Example 13.4.

Exercise 13.8

Assume that the op amp in Figure 13.20 is almost-ideal, having a voltage gain A_v. Determine V_O as a function of A_v and verify that V_O is 20 V as $A_v \to \infty$.

Ans. $V_O = \dfrac{20}{1+3/A_v}$ V.

The following example illustrates the analysis of the noninverting configuration using **a signal-flow diagram**. Such diagrams can serve as useful aids in investigating the behavior of feedback systems.

*Example 13.4: Signal-Flow Diagram for Noninverting Configuration

It is required to analyze the noninverting configuration using a signal-flow diagram.

Analysis:

1. The variables of interest are identified and surrounded by circles. In the case of the noninverting configuration of Figure 13.11, the variables of interest are v_I. v_P, v_N, v_O, and ε, shown encircled in Figure 13.22a.
2. The circles are joined by lines with arrows pointing in the direction of signal flow.
3. In the noninverting configuration, the signal flow is from v_I (which equals v_P), via ε, to v_O in the forward path and then through v_N and back to ε in the feedback path.
4. The lines joining the encircled variables are labeled with multiplying factors in accordance with the relationship between the variables in the direction of signal flow. Thus, the line joining v_I to v_P is labeled with 1, since $v_P = v_I \times 1$, and the line joining ε to v_O is labeled with A_v because $v_O = \varepsilon \times A_v$. The relation $\varepsilon = v_P - v_N$ is represented by a label of 1 on the line joining v_P to ε and by a label of −1 on the line joining v_N to ε.

The following should be noted:

1. The convergence of two or more lines on an encircled variable signifies weighted addition or subtraction of the encircled variables of origin of the convergent lines.

2. The sign of the feedback is that of the product of the labels around the feedback loop. In Figure 13.22a, the feedback is from ε to v_O and back through v_N to ε. The product of the labels around the loop is $(-1) \times A_v$, which has a negative sign denoting a negative feedback.
3. The product of the labels in a signal path is the *gain* along the path. Thus, $1 \times A_v = A_v$ is the gain in the forward path, or the *forward gain*, from v_I to v_O, whereas $(-1) \times A_v \times \beta = -\beta A_v$ is the gain around the feedback loop, or the *loop gain*.

Evidently, the signal-flow diagram provides a graphical representation of the relations between the variables involved. The question therefore arises as to whether it is possible to derive directly from the diagram the relation between any two variables. Indeed, this is possible, using some well-defined rules. It suffices for our purposes to use the following simple rule:

Rule: *The value of a given variable equals the value of another variable upstream along the signal path, multiplied by the forward gain between these variables and divided by (1 – loop gain) of any feedback loop encountered anywhere along the path between the two variables, as long as the variables in the loop do not have additional inputs.*

(13.13)

In Figure 13.22a, for example, the forward gain from v_I to v_O is $1 \times A_v = A_v$, and the loop gain around the feedback loop is $-\beta A_v$. The variables in the loop, ε, v_O, and v_N do not have inputs additional to those under consideration. Hence, according to Rule 13.13, $v_O/v_I = A_v/(1 + \beta A_v)$, which is the same as Equation 13.8.

It is seen that signal-flow diagrams can be very convenient in that they can be derived directly from the circuit diagram and allow a simple formulation of the relations between the variables involved, without having to write and solve equations. Intermediate variables that are not of immediate interest can be omitted from the signal flow diagram, with the lines appropriately relabeled. For example, v_P can be omitted from Figure 13.22a, and the line from v_I to ε labeled $1 \times 1 = 1$. Similarly, v_N can be omitted, and the line from v_O to ε labeled $(-1) \times \beta = -\beta$ (Figure 13.22b). This simplifies the diagram without losing significant information.

Primal Exercise 13.9

Determine V_O in Figure 13.23.

Ans. 15 V.

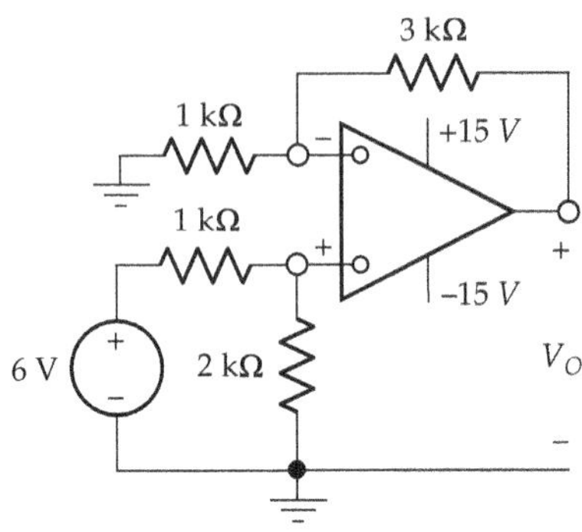

FIGURE 13.23
Figure for Primal Exercise 13.9.

13.4 Inverting Configuration

The basic inverting configuration is shown in Figure 13.24. The noninverting input is connected to common ground, and negative feedback is provided by the resistance R_f between the output terminal and the inverting terminal. The circuit will first be analyzed assuming an ideal op amp. As explained in connection with Figure 13.12, a virtual short circuit exists between the two input terminals of an ideal op amp in the presence of negative feedback. Because the noninverting terminal is grounded, the virtual short circuit becomes in this case a **virtual ground** at the inverting terminal. This is indicated by the dashed ground symbol at the inverting terminal in Figure 13.24.

The virtual ground is a most important feature of the inverting configuration, as will become clear in the next section. The virtual ground has two critical implications:

1. The inverting terminal is effectively at a ground voltage of zero. This means that the input current i_r is determined entirely by the circuit on the input

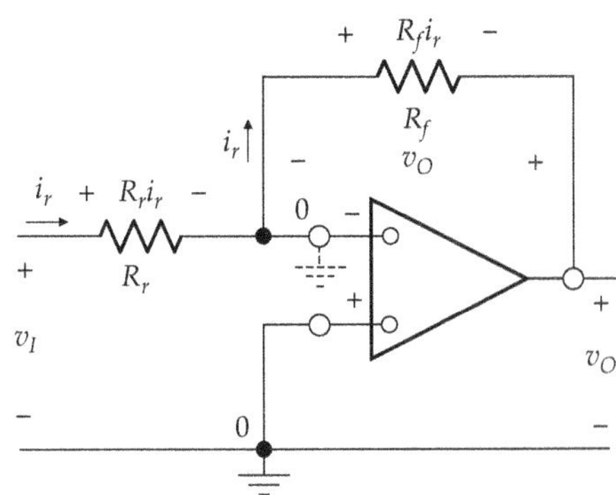

FIGURE 13.24
Inverting configuration using ideal op amp.

side of the op amp, independently of the circuit connected to the output of the op amp. Thus,

$$i_r = \frac{v_I}{R_r} \tag{13.14}$$

2. Because of the open circuit at the inverting terminal, i_r flows through R_f irrespective of R_f. Recall from the definition of an ideal current source that if a resistor is connected to an ideal current source, the current in the resistor is determined by the source irrespective of the resistance value. Since i_r is forced to flow through R_f irrespective of R_f, then R_f sees an ideal current source i_r.

Since the voltage at the inverting terminal is zero,

$$v_O = -R_f i_r \tag{13.15}$$

Eliminating i_r between the two preceding equations,

$$\frac{v_O}{v_I} = -\frac{R_f}{R_r} \tag{13.16}$$

It is seen that the sign of v_O is opposite that of v_I, which gives the inverting configuration its name. The inversion is basically because the input is applied to the inverting input, albeit through a circuit element R_r in this case.

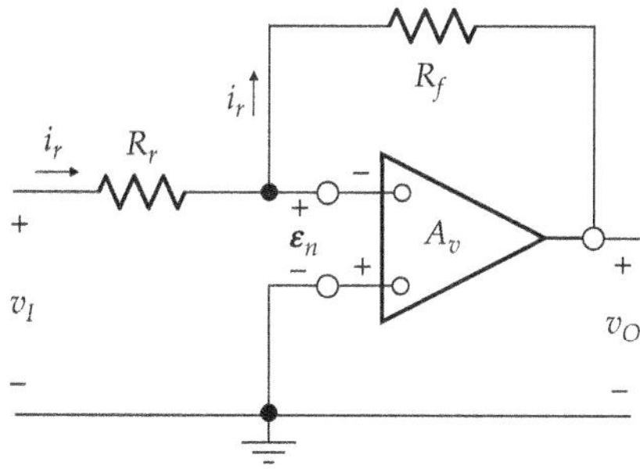

FIGURE 13.25
Inverting configuration using op amp of finite gain.

We will next analyze the inverting configuration assuming an almost-ideal op amp. Because the noninverting terminal is grounded, it is convenient to define $\varepsilon_n = -\varepsilon$ (Figure 13.25), so that $v_O = -A_v\varepsilon_n$ for the op amp. Since ε_n is a function of v_I and v_O, it is instructive to derive the v_O–v_I relation by superposition, rather than KVL and KCL (Exercise 13.20), as this provides better insight into the behavior of the circuit. To do so, v_I is applied alone with v_O set to zero (Figure 13.26a), then v_O is applied alone, with v_I set to zero (Figure 13.26b).

v_O should not be set to zero by short-circuiting the output of the op amp, as this would short-circuit an ideal voltage source, leading to an infinite current. But we can imagine that the op amp is modified to have $A_v = 0$ (Figure 13.26a), which gives $v_O = 0$, so that the output is effectively grounded. It follows from voltage division that

$$\varepsilon_{n1} = \frac{R_f}{R_r + R_f} v_I \tag{13.17}$$

To apply v_O with v_I set to zero, it is best to disconnect the op amp output and apply a voltage source v_O as in Figure 13.26b, where $v_O = -\varepsilon_{n2}A_v$. This would avoid connecting the ideal voltage source v_O in parallel with the voltage source $\varepsilon_{n2}A_v$ at the output of the op amp although theoretically, the two sources have the same voltage. It follows from Figure 13.26b, with $v_I = 0$, that

$$\varepsilon_{n2} = \frac{R_r}{R_r + R_f} v_O \tag{13.18}$$

Applying superposition,

$$\varepsilon_n = \varepsilon_{n1} + \varepsilon_{n2} = -\frac{v_O}{A_v} = \frac{R_f}{R_r + R_f} v_I + \frac{R_r}{R_r + R_f} v_O \tag{13.19}$$

Equation 13.19 can be rearranged to give

$$\frac{v_O}{v_I} = -\frac{R_f}{R_r + R_f} \frac{A_v}{1 + \beta A_v} \tag{13.20}$$

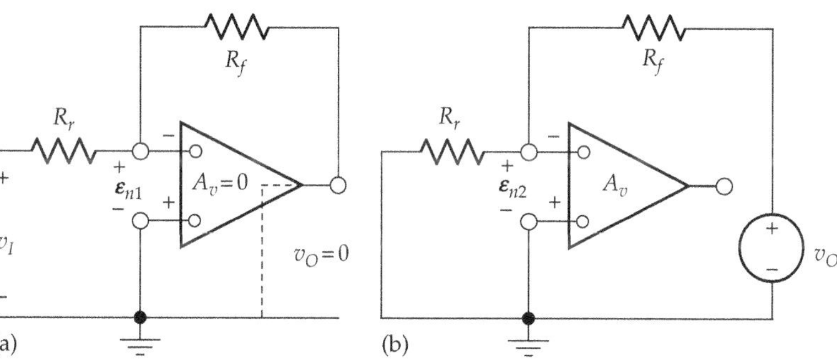

FIGURE 13.26
Application of superposition to the inverting configuration; v_o is set to zero in (a) and v_I is set to zero in (b).

where $\beta = R_r/(R_r + R_f)$, as for the noninverting configuration (Equation 13.6). If $A_v \to \infty$, Equation 13.20 becomes

$$\frac{v_O}{v_I} = -\frac{R_f}{R_r + R_f}\frac{1}{\beta} = -\frac{R_f}{R_r} \tag{13.21}$$

as in Equation 13.16. It should be noted that even in the case of an almost-ideal op amp, the inverting configuration is, for most practical purposes, a virtual ground. Even if A_v is relatively low at 10^4, and $v_O = 10$ V, say, then $\varepsilon_n = -10/10^4 \equiv -1$ mV. It is seen that the difference between the noninverting and inverting configurations, in terms of feedback, is that in the noninverting configuration, the feedback signal is effectively in series with the applied input and subtracts from it, so that $\varepsilon = v_I - v_N$ (Figure 13.11). Thus, the sign in the feedback path is negative, and the sign in the forward path is positive ($v_O = \varepsilon A_v$), so that the sign of the product around the feedback loop is negative, as it should be for a negative feedback system. In the inverting configuration, on the other hand, the feedback signal is effectively in shunt with the input, through R_f and R_r and adds to it (Figure 13.26 and Equation 13.19, where $\varepsilon_n = \varepsilon_{n1} + \varepsilon_{n2}$). The sign in the feedback path is positive, but the sign in the forward path is negative ($v_O = -\varepsilon_n A_v$), so that the sign of the product around the feedback loop is again negative, as it should be for a negative feedback system.

If $R_f = R_r$, then $v_O = -v_I$; the inverting configuration becomes a unity-gain inverter.

Equation 13.16 can be readily generalized to the sinusoidal steady state, where R_r and R_f are replaced by Z_r and Z_f, respectively, and all voltages and currents are considered as phasors (Figure 13.27). Because of the virtual ground at the inverting input of the op amp, $\mathbf{I}_r = \mathbf{V}_I/Z_r$ and flows through Z_f, resulting in $\mathbf{V}_O = -Z_f\mathbf{I}_r$. Eliminating $\mathbf{I}_r$ gives

$$\frac{\mathbf{V_O}}{\mathbf{V_I}} = -\frac{Z_f}{Z_r} \tag{13.22}$$

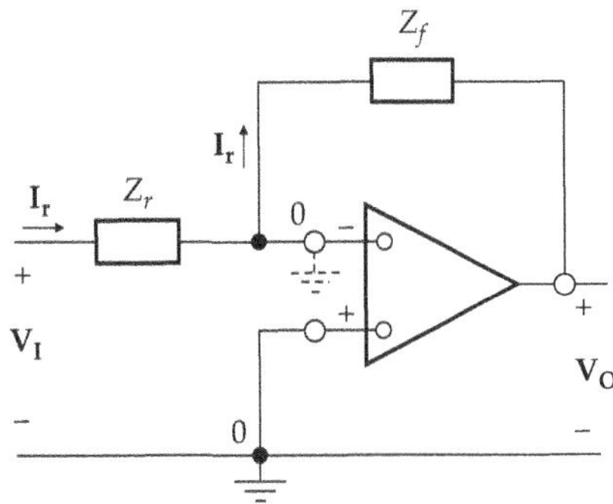

FIGURE 13.27
Inverting configuration having impedances.

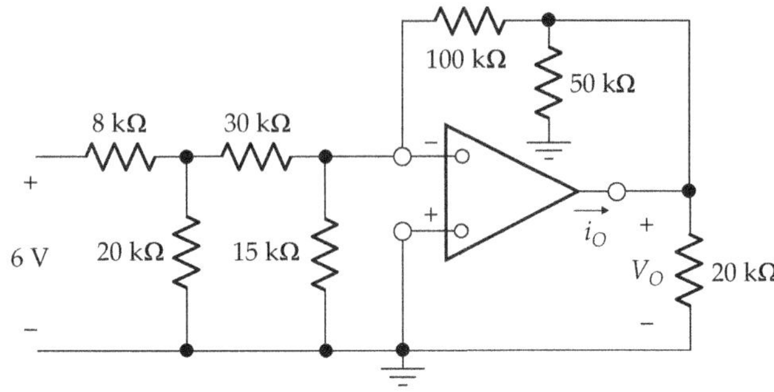

FIGURE 13.28
Figure for Example 13.5.

Exercise 13.10

Derive Equation 13.20 using the governing equations: $v_I - \varepsilon_n = R_r i_r$, $\varepsilon_n - v_O = R_f i_r$, and $v_O = -A_v\varepsilon_n$, eliminating i_r, and ε_n between these three equations.

Example 13.5: Inverting Configuration

It is required to determine v_O and i_O in Figure 13.28, assuming an ideal op amp.

Solution:

Because of the virtual ground (Figure 13.29), the current through the 15 kΩ resistor is zero, which makes the 30 kΩ resistor effectively in parallel with the 20 kΩ resistor, since they have the same voltage across them. The parallel resistance is (30) × (20)/50 = 12 kΩ. R_r = (8 + 12) = 20 kΩ, and the input current is 6/20 = 0.3 mA. By current division, the current in the 30 kΩ resistor is 0.3 × (20/50) = 0.12 mA. This current flows through the 100 kΩ resistor and produces a voltage drop of (100) × 0.12 = 12 V. It follows that $v_O = -12$ V.

The currents in the 50 and 20 kΩ connected to the output of the op amp are 0.24 and 0.6 mA, respectively, so that $-i_O = 0.12 + 0.24 + 0.6 = 0.96$ mA.

Simulation: The circuit is entered as in Figure 13.30 using the part OPAMP from the ANALOG library, without changing the default values. As in Figure 13.14, the 1u resistors are added in order to have PSpice display the currents. To run the simulation, select 'Bias Point' under 'Analysis type' in the Simulation Settings dialog box. After the simulation is run, pressing the V and I buttons displays the values shown in Figure 13.30. It is seen that, except for the voltage at the inverting terminal and the current through the 15 kΩ resistor, which are effectively zero, the indicated currents and voltages are the same as those assuming an ideal op amp. Note that $\varepsilon_n = -\dfrac{-12}{10} \equiv 12\,\mu$V.

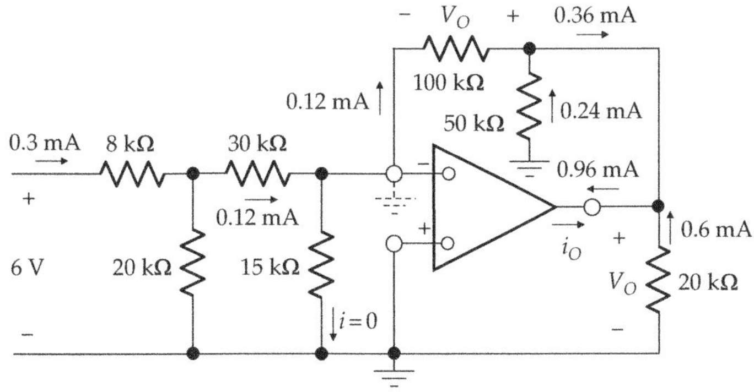

FIGURE 13.29
Figure for Example 13.5.

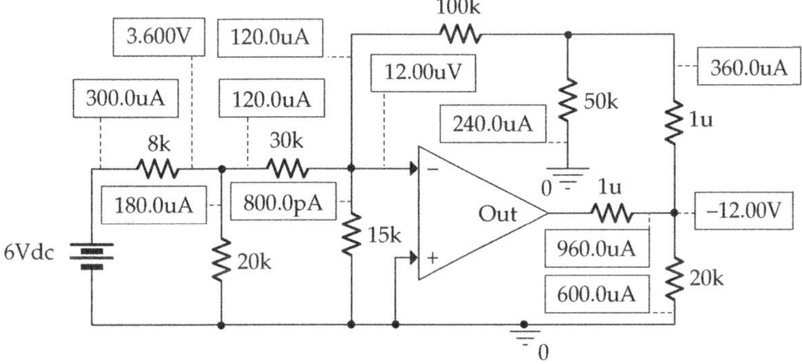

FIGURE 13.30
Figure for Example 13.5.

Problem-Solving Tips

- A combination of circuit elements connected across the voltage source output of an op amp does not affect the output voltage of the op amp but affects its output current.
- No current flows through a passive circuit element, or a combination of such elements, connected across a virtual short circuit at the op amp inputs.

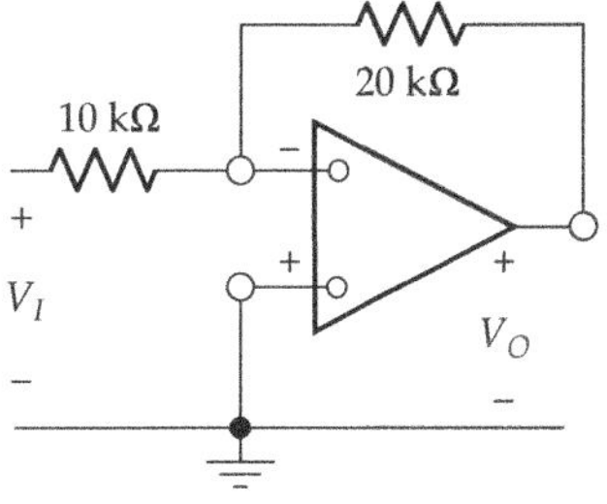

FIGURE 13.31
Figure for Primal Exercise 13.11.

Primal Exercise 13.11

Draw the circuit diagram of an amplifier having a gain of −2 and input resistance of 10 kΩ, using an ideal op amp.

Ans. The circuit is as shown in Figure 13.31.

Primal Exercise 13.12

Determine V_O in Figure 13.32 by two methods: (1) applying the concept of a virtual short circuit between the two input terminals and (2) applying one battery at a time and using superposition. Note that when the 4 V

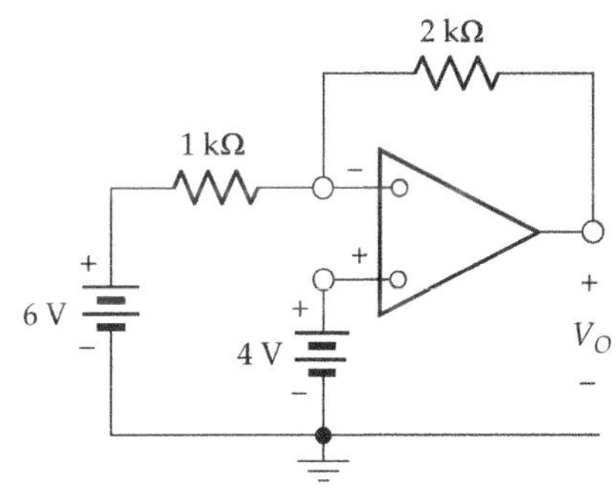

FIGURE 13.32
Figure for Primal Exercise 13.12.

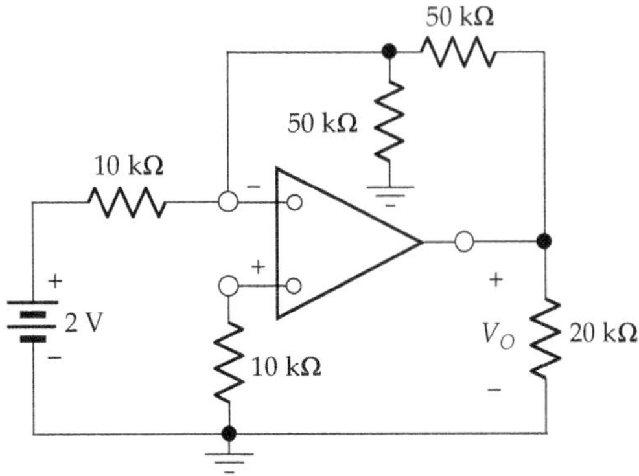

FIGURE 13.33
Figure for Primal Exercise 13.13.

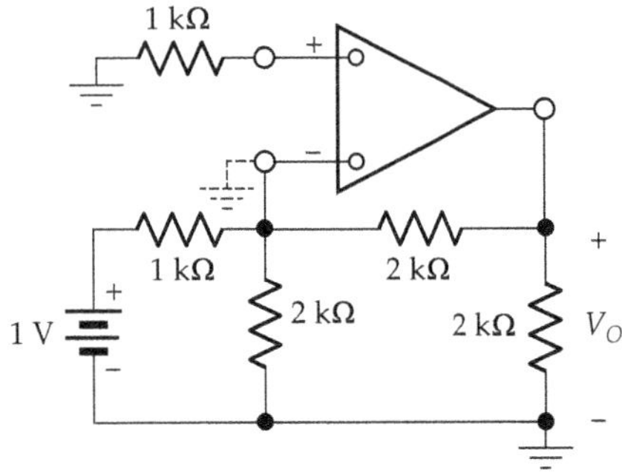

FIGURE 13.34
Figure for Primal Exercise 13.14.

battery is set to zero, the circuit is an inverting amplifier, whereas when the 6 V battery is set to zero, the circuit is a noninverting amplifier.

Ans. $V_O = 0$.

Primal Exercise 13.13

Determine V_O in Figure 13.33.

Ans. −10 V.

Primal Exercise 13.14

Determine V_O in Figure 13.34.

Ans. −2 V.

★Example 13.6: Signal-Flow Diagram for Inverting Configuration

It is required to analyze the inverting configuration using a signal-flow diagram.

Analysis: The signal-flow diagram can be derived directly from Figure 3.25. The main variables of interest are v_I, ε_n, and v_O (Figure 13.35). For the op amp, $v_O = -A_v\varepsilon_n$. If $v_O = 0$, it follows from Figure 13.26a that $\varepsilon_{n1} = R_f v_I/(R_r + R_f)$. When $v_I = 0$, it follows from Figure 13.26b that $\varepsilon_{n2} = R_r v_O/(R_r + R_f) = \beta v_O$. The loop gain is βA_v, the same as in the noninverting configuration (Figure 13.22b), but the negative sign is now associated with the forward path rather than the feedback path, as in the inverting configuration. Applying Rule 13.13 gives

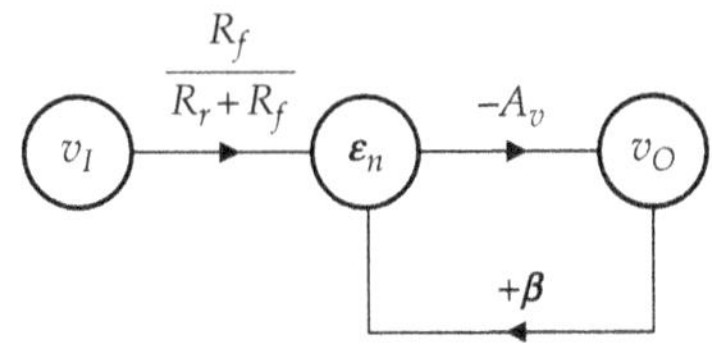

FIGURE 13.35
Figure for Example 13.6.

$$\frac{v_O}{v_I} = -\frac{R_f}{R_r + R_f}\frac{A_v}{1+\beta A_v} \tag{13.23}$$

which is the same as Equation 13.20.

13.5 Applications of the Inverting Configuration

13.5.1 Current-Source-to-Voltage-Source Converter

The circuit of Figure 13.36 converts a nonideal current source of source resistance R_{src}, connected to the input of an op amp in the inverting configuration, to an ideal voltage source at the output. Because of the virtual ground, no current flows in R_{src}. The current source is ideally terminated with a short circuit, so that all of i_{SRC} flows toward the inverting input and through R_f, resulting in an ideal voltage source output of $v_O = -R_f i_{SRC}$.

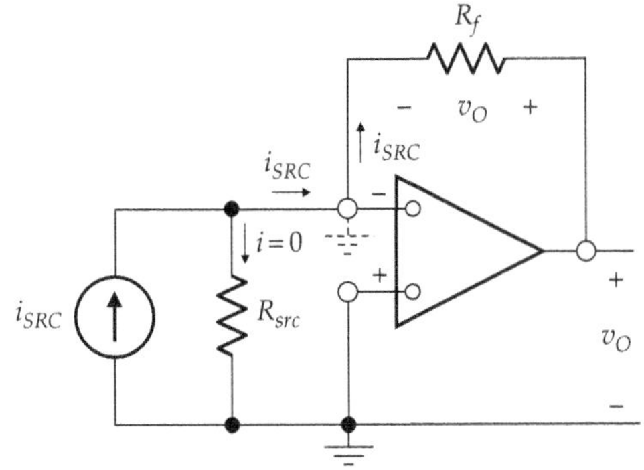

FIGURE 13.36
Current-source-to-voltage-source converter.

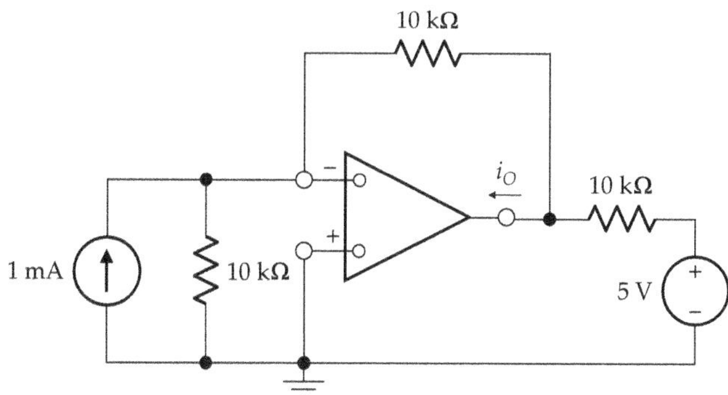

FIGURE 13.37
Figure for Primal Exercise 13.15.

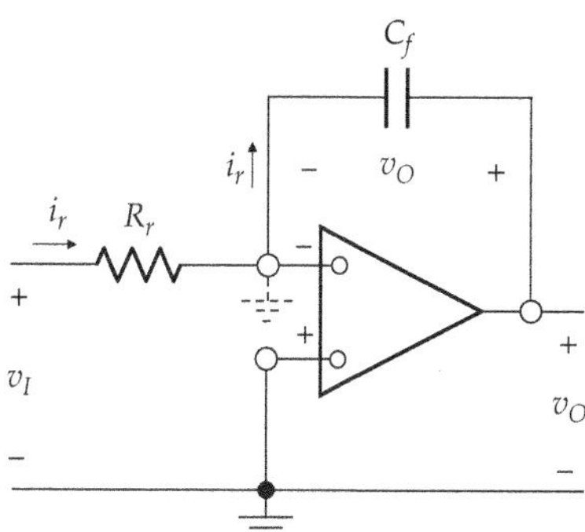

FIGURE 13.38
Ideal integrator.

Primal Exercise 13.15

Determine i_O in Figure 13.37 (Note that the circuit at the output is a two-essential-node circuit).

Ans. 2.5 mA.

13.5.2 Ideal Integrator

The feedback resistor R_f is replaced by a capacitor C_f in Figure 13.38; i_r remains equal to v_I/R, because of the virtual ground, and flows through C_f. The resulting output voltage is

$$v_O = -\frac{1}{C_f}\int_0^t \frac{v_I}{R_r}dt + v_O(0) \tag{13.24}$$

If $v_O(0) = 0$, v_O is $-1/R_rC_f$ times the time integral of v_I. Ideally, the circuit acts as a perfect, or ideal, integrator with a gain $-1/R_rC_f$.

In the frequency domain, the ratio of the output voltage to the input voltage of the ideal integrator is obtained from Equation 13.22 by substituting $Z_r = R_r$ and $Z_r = 1/j\omega C_f$:

$$\frac{\mathbf{V_O}}{\mathbf{V_I}} = -\frac{1}{j\omega C_f R_r} \tag{13.25}$$

Primal Exercise 13.16

Determine R_r in the ideal integrator for an integrator gain of −1, assuming $C_f = 1$ μF.

Ans. 1 MΩ.

13.5.3 Ideal Differentiator

It is evident from Figure 13.38 that if C_f is replaced by an inductor, the circuit becomes a differentiator. However, the use of inductors is avoided in practice as much as possible in favor of capacitors for several reasons: (1) inductors are bulky and expensive compared to capacitors, (2) practical capacitors are generally closer to the ideal than practical inductors, and (3) capacitors of relatively small capacitance can be incorporated in integrated circuits, whereas this is much less practical with inductors. Differentiation can be obtained by interchanging the resistor and capacitor in Figure 13.38, as illustrated in Figure 13.39.

Because of the virtual ground, $i_r = C_r dv_I/dt$; i_r flowing through R_f produces a voltage output given by

$$v_O = -R_f C_r \frac{dv_I}{dt} \tag{13.26}$$

Ideally, the circuit acts as a perfect, or ideal, differentiator with a gain of $-R_fC_r$.

In the frequency domain, the ratio of the output voltage to the input voltage of the ideal differentiator is obtained from Equation 13.22 by substituting $Z_r = -1/j\omega C_r$ and $Z_f = R_f$:

$$\frac{\mathbf{V_O}}{\mathbf{V_I}} = -j\omega C_r R_F \tag{13.27}$$

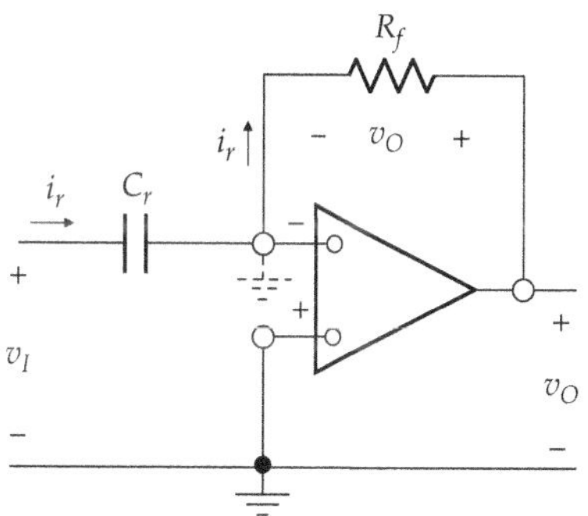

FIGURE 13.39
Ideal differentiator.

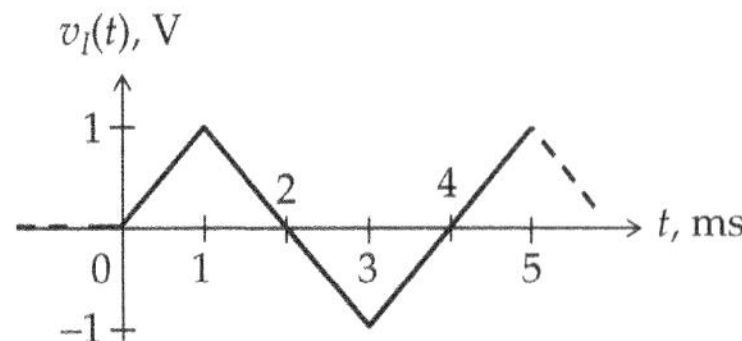

FIGURE 13.40
Figure for Primal Exercise 13.17.

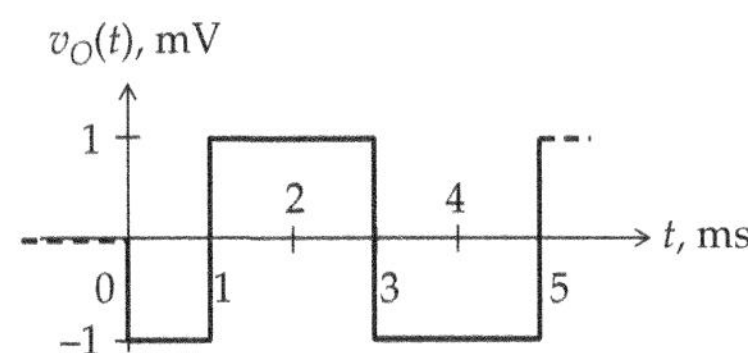

FIGURE 13.41
Figure for Primal Exercise 13.17.

Primal Exercise 13.17

Assume that the triangular $v_i(t)$ of Figure 13.40, starting at $t = 0$, is applied to the ideal differentiator of Figure 13.39, having $C_r = 0.1$ μF and $R_f = 10$ kΩ. Sketch $v_O(t)$.

Ans. $v_O(t)$ is shown in Figure 13.41.

13.5.4 Adder

The virtual ground can act as a summing point, leading to an adding circuit, as illustrated in Figure 13.42 for three inputs. Because of the virtual ground, $i_{r1} = v_{I1}/R_{r1}$, $i_{r2} = v_{I2}/R_{r2}$, and $i_{r3} = v_{I3}/R_{r3}$. The current i_r flowing toward the inverting terminal is the sum of the three currents, $i_r = i_{r1} + i_{r2} + i_{r3}$, and flows through R_f. The output is

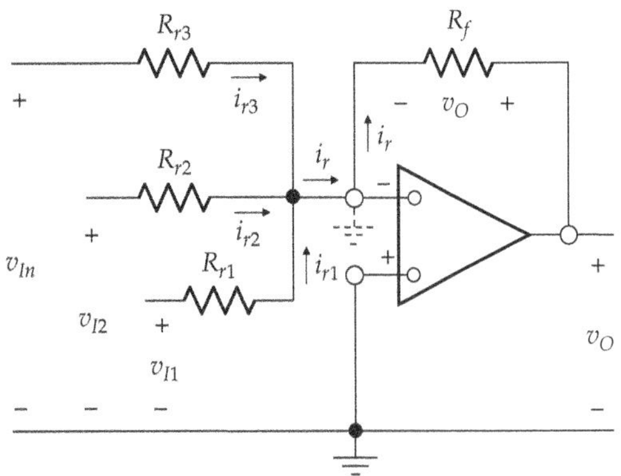

FIGURE 13.42
Adder.

$v_O = -R_f i_r = -R_f(i_{r1} + i_{r2} + ir_3)$. Substituting for the currents in terms of the voltage inputs,

$$v_O = -R_f\left(\frac{v_{I1}}{R_{r1}} + \frac{v_{I2}}{R_{r2}} + \frac{v_{I3}}{R_{r3}}\right) \tag{13.28}$$

In other words, the output is the weighted sum of the inputs, with sign inversion. If all the input resistances are equal to R_r,

$$v_O = -\frac{R_f}{R_r}\left(v_{I1} + v_{I2} + v_{I3}\right) \tag{13.29}$$

The output voltage is now the sum of the input voltages multiplied by the gain $-R_f/R_r$. An important feature of this adding circuit is that the virtual ground prevents interaction between the input circuits, thereby effectively isolating them from one another. A change in v_{I1} or R_{r1}, for example, affects only i_{r1} and has no effect on the other input circuits.

If subtraction of some inputs is required, these inputs can be inverted before being applied to the adder. Alternatively, a difference amplifier (Section 13.6) can be used.

Primal Exercise 13.18

Determine V_O in Figure 13.43.

Ans. −10 V.

Example 13.7: Noninverting Integrator

It is required to analyze the circuit of the noninverting integrator of Figure 13.44.

Analysis: The circuit is noteworthy in several respects: (1) the integrated output is not inverted, (2) the value of C_f can be magnified by a multiplying factor, and (3) the inverting input of the integrator is a virtual ground, rather than the noninverting input. This is because the op amp on the right is in the inverting configuration, having an input v_O and an output $v_{O2} = -(R_b/R_a)v_O$ (Figure 13.45). Due to this inversion in the feedback path, the forward path through the op amp should be noninverting, so as

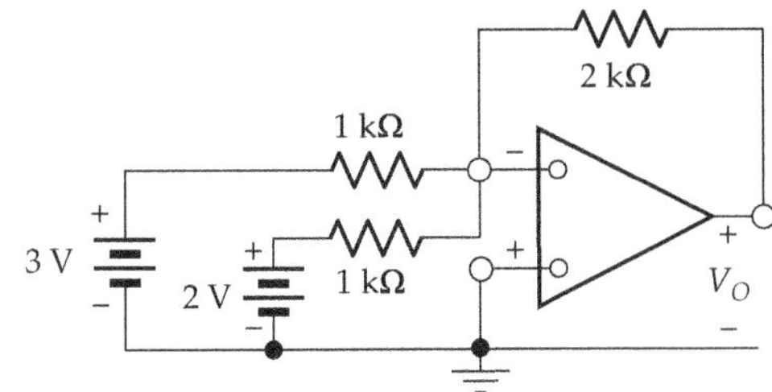

FIGURE 13.43
Figure for Primal Exercise 13.18.

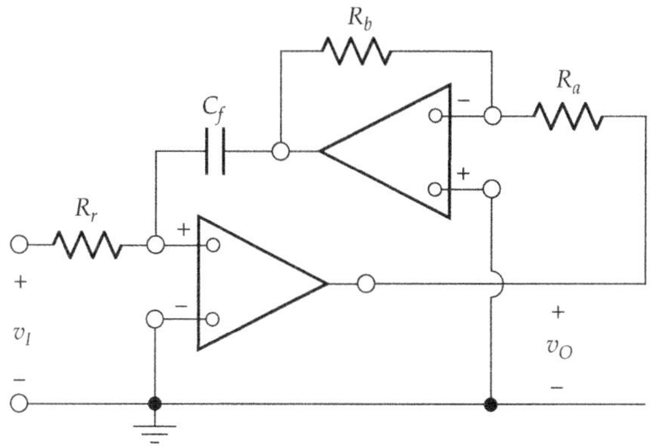

FIGURE 13.44
Figure for Example 13.7.

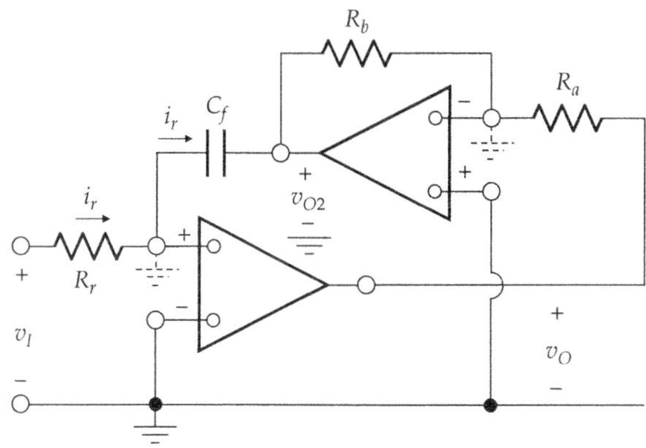

FIGURE 13.45
Figure for Example 13.7.

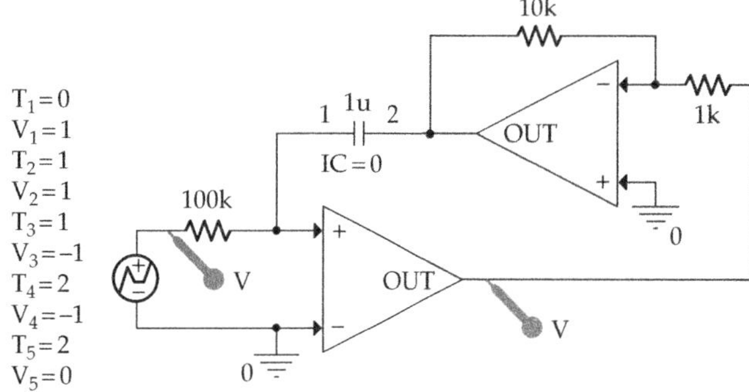

FIGURE 13.46
Figure for Example 13.7.

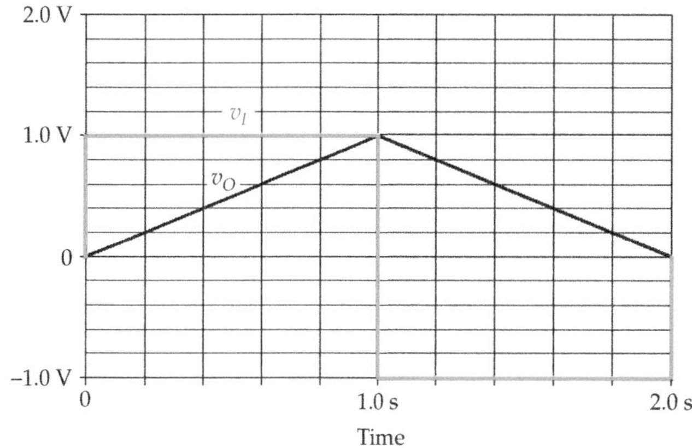

FIGURE 13.47
Figure for Example 13.7.

to have a negative feedback loop. The inverting terminal of the integrator op amp should therefore be grounded. For a finite v_O, the voltage at the noninverting input terminal of the op amp is vanishingly small, which means that this terminal is a virtual ground (Figure 13.45).

It is seen from Figure 13.45 that $v_{O2} = -(R_b/R_a)v_O$ and that $i_r = v_I/R_r$; i_r flowing through C_f produces a voltage drop $\frac{1}{C_f}\int_0^t i_r dt = \frac{1}{R_rC_f}\int_0^t v_I dt$ across C_f, assuming C_f is initially uncharged. This voltage drop is $-v_{O2}$, which gives

$$\frac{1}{R_rC_f}\int_0^t v_I dt = \left(R_b/R_a\right)v_O \quad \text{or} \quad v_O = \frac{1}{R_r\left(C_fR_b/R_a\right)}\int_0^t v_I dt \tag{13.30}$$

C_f is thus multiplied by R_b/R_a, the magnitude of the gain of the inverting amplifier, and v_O is not inverted with respect to v_I.

Simulation: Suppose that $R_r = 100$ kΩ, $C_f = 1$ μF, $R_b = 10$ kΩ, and $R_a = 1$ kΩ. The circuit is entered as in Figure 13.46 using the part OPAMP from the ANALOG library, without changing the default values. The source VPWL having the parameters indicated is a biphasic pulse v_I, as in Figure 13.47. To run the simulation, select 'Time Domain (Transient)' under 'Analysis type,' 2 s for 'Run to time,' 0 for 'Start saving data after,' and 1m for 'Maximum step size.' After the simulation is run, the input voltage and output voltage are displayed as in Figure 13.47. From Equation 13.30, the integrator gain is $1/(10^5 \times 10^{-6} \times (10/1)) = 1$. The integral of the biphasic input pulse of 1 V amplitude and 1 s duration of each phase is a triangular waveform of 1 V peak, as in Figure 13.47. Note that the output of the inverting amplifier of gain −10 is a triangular waveform of −10 V peak, so that the op amp operates in the linear region of its characteristic. Note also that the results of using the almost-ideal op amp of PSpice are not significantly different from those assuming an ideal op amp.

Primal Exercise 13.19

Determine V_O in Figure 13.48.

Ans. 1 V.

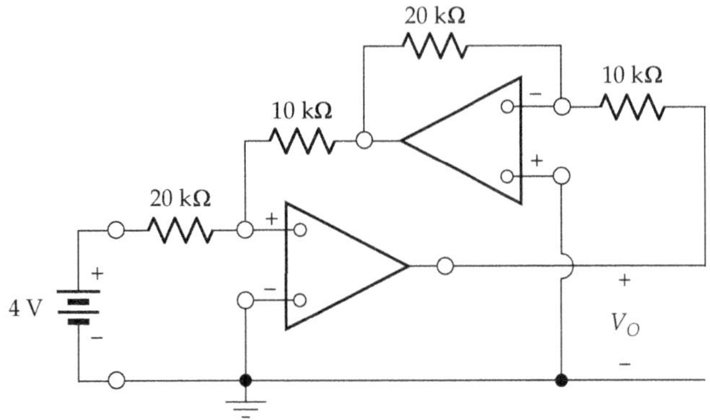

FIGURE 13.48
Figure for Primal Exercise 13.19.

13.6 Difference Amplifier

The basic circuit of the difference amplifier is shown in Figure 13.49. The input v_P at the noninverting input is simply the output of a voltage divider: $v_P = R_d v_{SRC1}/(R_c + R_d)$. The current flowing through R_a toward the inverting terminal is $(v_{SRC2} - v_N)/R_a$, and the current flowing through R_b away from the inverting terminal is $(v_N - v_O)/R_b$. From KCL, these currents are equal. Moreover, $v_P = v_N$ because of the virtual short circuit at the op amp inputs. Eliminating v_P and v_N between these equations and rearranging gives

$$v_O = -\frac{R_b}{R_a}v_{SRC2} + \frac{1+R_b/R_a}{1+R_c/R_d}v_{SRC1} \tag{13.31}$$

If $\frac{R_b}{R_a} = \frac{R_d}{R_c} = k$, Equation 13.31 reduces to

$$v_O = k\left(v_{SRC1} - v_{SRC2}\right) \tag{13.32}$$

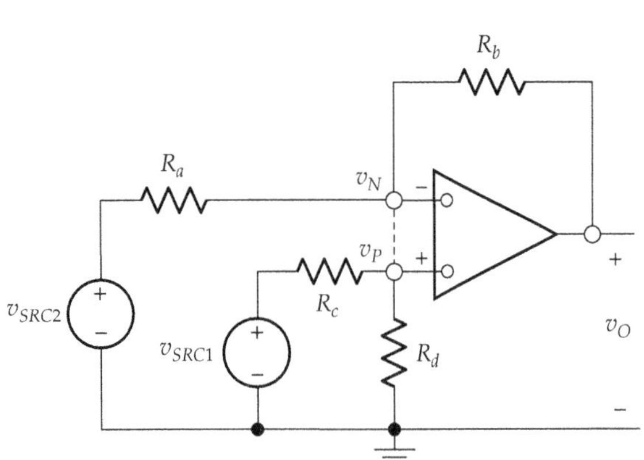

FIGURE 13.49
Difference amplifier.

The circuit can therefore be used for directly subtracting one signal from another.

The difference amplifier can be considered as a combination of the inverting and noninverting configurations. Accordingly, Equation 13.31 can be derived by superposition of v_{SRC1} and v_{SRC2} (Problem P13.42). Alternatively, Equation 13.31 can be derived by applying superposition of v_{SRC2} and v_O, as in Figure 13.26, to derive v_N, and then substituting $v_P = R_d v_{SRC1}/(R_c + R_d)$ for v_N.

Example 13.8: Instrumentation Amplifier

It is required to analyze the circuit of Figure 13.50 representing an instrumentation amplifier (IA).

Analysis: The amplifier consists of two stages, the first stage having two op amps in a noninverting configuration. The outputs of these op amps are combined so as to provide a differential output that is applied to a difference amplifier in the second stage. Since the input sources are applied to noninverting inputs, the input impedance is high. Moreover, as is shown in what follows, the overall gain is determined by the value of a single resistance R_1. The output v_O depends on the difference $(v_{SRC1} - v_{SRC2})$ between the inputs. This is advantageous, as outside interference that adds the same voltage to v_{SRC1} and v_{SRC2} is canceled out. For best performance, the resistors having the same subscript number should be closely matched in value.

The differential gain of the first stage can be very simply derived by noting that, because the inputs behave as open circuits, $v_{P1} = v_{SRC1}$ and $v_{P2} = v_{SRC2}$ (Figure 13.51). Because of the virtual short circuit at the inputs of an ideal op amp, $v_{N1} = v_{P1} = v_{SRC1}$ and $v_{N2} = v_{P2} = v_{SRC2}$.

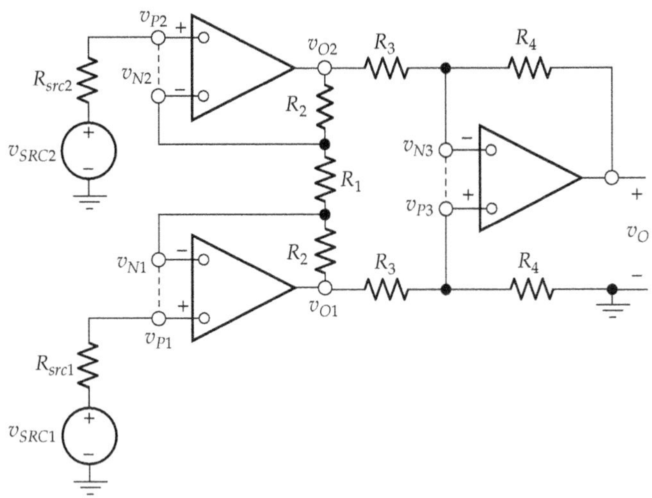

FIGURE 13.50
Figure for Example 13.8.

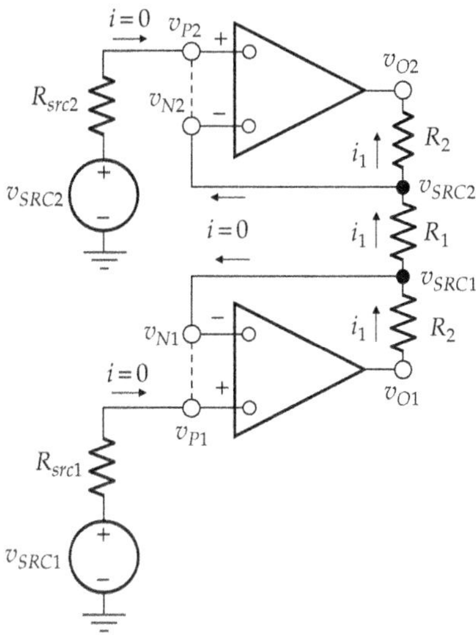

FIGURE 13.51
Figure for Example 13.8.

This means that the voltage ($v_{N1} - v_{N2}$) across R_1 equals the differential input voltage ($v_{SRC1} - v_{SRC2}$), so $i_1 = (v_{SRC1} - v_{SRC2})/R_1$. This same current flows in the two resistances R_2. Hence,

$$v_{O1} - v_{O2} = \frac{(v_{SRC1} - v_{SRC2})}{R_1}(R_1 + 2R_2) \qquad (13.33)$$

The differential gain of the first stage is therefore

$$\frac{v_{O1} - v_{O2}}{v_{SRC1} - v_{SRC2}} = 1 + 2\frac{R_2}{R_1} \qquad (13.34)$$

In order to obtain the overall gain, this differential gain has to be multiplied by that of the difference amplifier (Equation 13.32). Identifying k, v_{SRC1}, and v_{SRC2} in Equation 13.32, with their counterparts R_4/R_3, v_{O1}, and v_{O2}, respectively, in Figure 13.50, it follows that

$$\frac{v_O}{v_{O1} - v_{O2}} = \frac{R_4}{R_3} \qquad (13.35)$$

Multiplying Equations 13.34 and 13.35 gives

$$\frac{v_O}{v_{SRC1} - v_{SRC2}} = \left(1 + 2\frac{R_2}{R_1}\right)\frac{R_4}{R_3} \qquad (13.36)$$

It is seen that the gain can be varied by varying a single resistance R_1 that does not have be to be matched to any other resistance. Because of its many attractive features, the IA is available in IC form.

Simulation: Suppose that $R_1 = 2$ kΩ and all other resistances are 1 kΩ. From Equation 13.36, the voltage gain $v_O/(v_{SRC1} - v_{SRC2}) = 2$. The circuit is entered as in

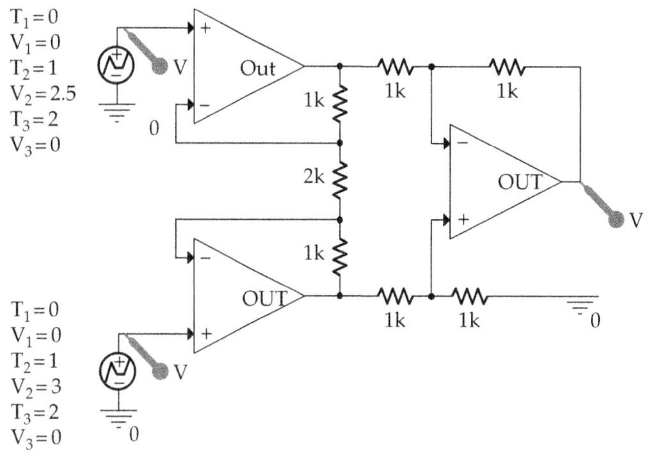

FIGURE 13.52
Figure for Example 13.8.

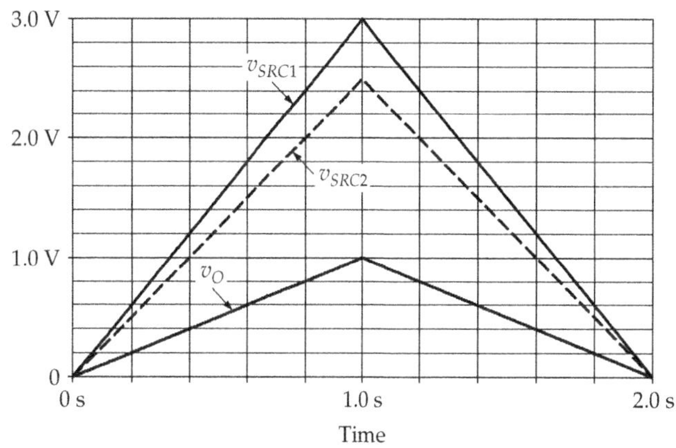

FIGURE 13.53
Figure for Example 13.8.

Figure 13.52 using the part OPAMP from the ANALOG library, without changing the default values. The VPWL sources, having the parameters indicated, are triangular waveforms (Figure 13.53). To run the simulation, select 'Time Domain (Transient)' under 'Analysis type,' 2s for 'Run to time,' 0 for 'Start saving data after,' and 1m for 'Maximum step size.' After the simulation is run, the input voltage and output voltage are displayed as in Figure 13.53, where v_{SRC1} and v_{SRC2} are triangular waveforms of amplitudes 3 and 2.5 V, respectively. The output voltage is a triangular waveform of amplitude 1 V.

Primal Exercise 13.20

Determine I_S in Figure 13.54.

Ans. –4 mA.

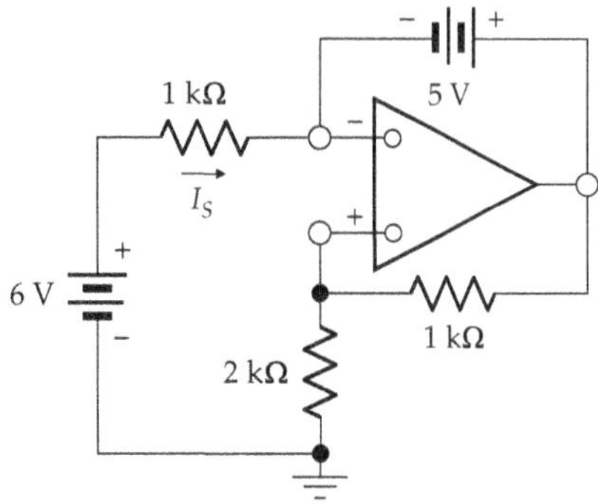

FIGURE 13.54
Figure for Primal Exercise 13.20.

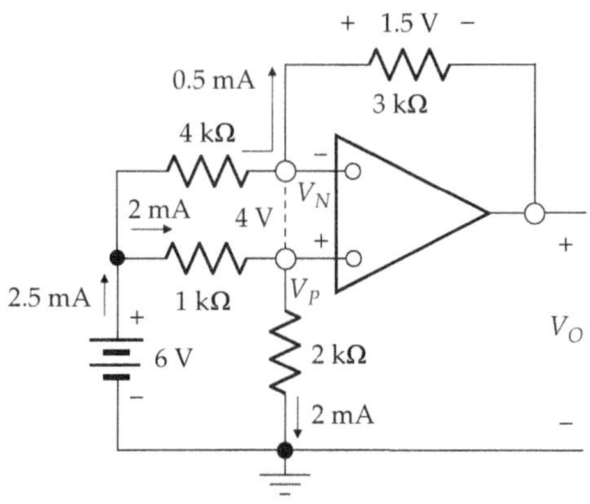

FIGURE 13.56
Figure for Example 13.9.

*Example 13.9: Current Distribution with Virtual Short Circuit

The objective of this example is to clarify the difference between having resistors physically in parallel and having them effectively in parallel through a virtual short-circuit, using the circuit of Figure 13.55.

Analysis: Since the op amp is assumed ideal, a virtual short circuit exists between the inverting and noninverting inputs (Figure 13.56). From voltage division, $V_P = 6 \times 2/3 = 4$ V, so that $V_N = V_P = 4$ V. Note that the 4 and 1 kΩ resistors are *effectively* in parallel: they have the same 2 V across them, resulting in currents of 0.5 and 2 mA through these resistors. However, when applying voltage division, the 1 kΩ resistor is considered to be in series with the 2 kΩ resistor, since no current flows through the virtual short circuit. The 0.5 mA current through the 4 kΩ resistor flows through the 3 kΩ resistor, producing a voltage drop of 1.5 V. It follows that $V_O = V_N - 1.5 = 2.5$ V.

What exactly is the difference between having the 1 kΩ and 4 kΩ resistors effectively in parallel through a virtual short circuit and having them actually in parallel through a physical short circuit, as in Figure 13.57?

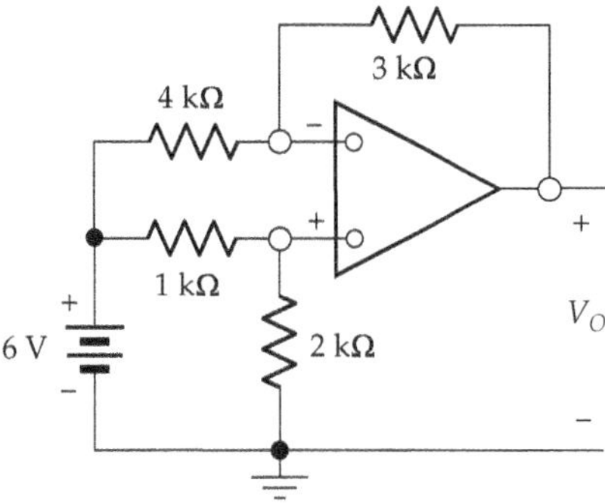

FIGURE 13.55
Figure for Example 13.9.

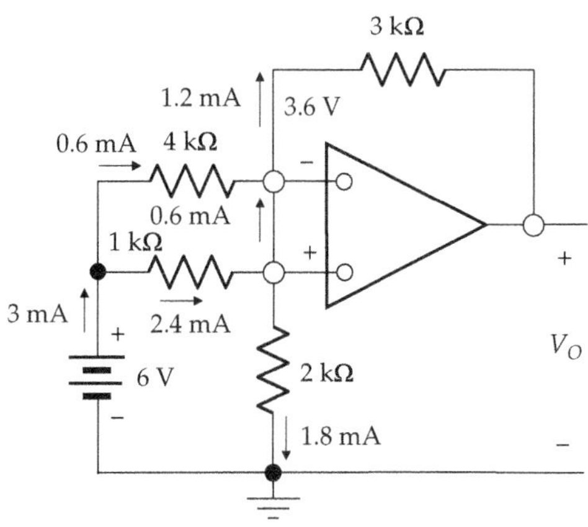

FIGURE 13.57
Figure for Example 13.9.

It is clear that the circuit is now a different circuit altogether. Since the inverting and noninverting inputs are connected together, the differential input is zero, so that $V_O = 0$. With $V_O = 0$, the 3 and 2 kΩ resistors are in parallel, having a parallel resistance of $3 \times 2/5 = 1.2$ kΩ. The parallel resistance of the 4 and 1 kΩ resistors is $4 \times 1/5 = 0.8$ kΩ. By voltage division, the voltage of the common op amp inputs is $6 \times 1.2/(1.2 + 0.8) = 3.6$ V. The currents in the circuit are shown in Figure 13.57. With $V_O = 0$, the current in the 3 kΩ resistor is 1.2 mA.

In comparing Figures 13.56 and 13.57, the following should be noted:

1. The voltage across the 4 and 1 kΩ resistors is the same 2 V in Figure 13.56 and the same 2.4 V in Figure 13.57, the corresponding currents in the two resistors being determined by Ohm's law in each case. In this respect, a physical parallel connection and an effective parallel connection behave in the same manner in that they have the same voltage across them.

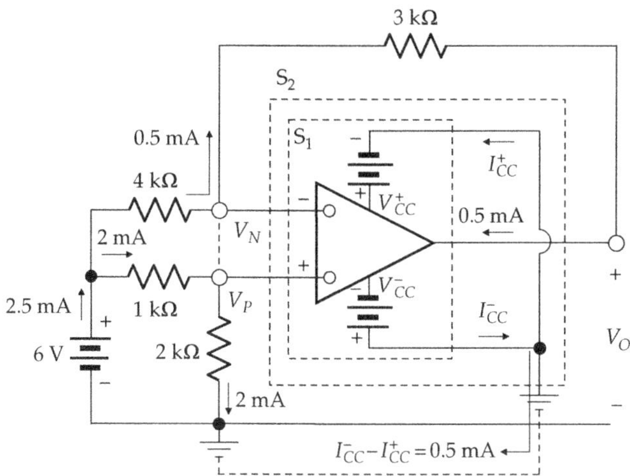

FIGURE 13.58
Figure for Example 13.9.

2. However, because a virtual short circuit does not pass current, unlike a physical short circuit, the *distribution of currents in the rest of the circuit is different in the two cases*. This means that the overall circuits are different in the two cases, resulting, in general, in different voltages and currents in each case. As noted earlier, the current in the 1 kΩ resistor in Figure 13.56 flows through the 2 kΩ resistor, and the current in the 4 kΩ resistor flows through the 3 kΩ resistor.

It is of interest in this connection to trace the flow of the 0.5 mA through the 3 kΩ resistor in Figure 13.56, into the output of the op amp, and back to the 6 V battery. This is illustrated in Figure 13.58, which shows the two V_{CC} supplies with their common connection grounded. It follows from applying KCL to surface S_1 that $I_{CC}^+ + 0.5 = I_{CC}^-$ or $I_{CC}^- - I_{CC}^+ = 0.5$. When the op amp output current is zero, $I_{CC}^+ = I_{CC}^-$. It follows from KCL at the ground of the V_{CC} supplies or from applying KCL to surface S_2 that the 0.5 mA current flows from the ground connections of the V_{CC} supplies back to the ground connection of the 6 V battery. The 0.5 mA current adds to the 2 mA current through the 2 kΩ resistor to give 2.5 mA through the battery.

13.7 Solving Problems on Operational Amplifiers

The ISDEPIC approach presented in Section 2.10 can be adapted to solving problems on op amps. The steps are as follows:

Step 1—Initialize: a. Mark on the circuit diagram all given values of circuit parameters, currents and voltages, as well as the unknowns to be determined. Mark the differential input ε in the case of almost-ideal op amps, and indicate virtual short circuits and virtual grounds in the case of ideal op amps. Mark op amp input currents and currents in virtual short circuits as zero, by means of an 'x'.

b. If the solution requires that a given value of current or voltage be satisfied, assume this value from the very beginning.

Step 2—Simplify: Try and reduce the circuit to a simpler form, if possible.

Step 3—Deduce: Determine any values of current or voltage that follow from immediate, direct application of Ohm's law, KCL, KVL, without introducing any additional unknowns. Look for the immediate consequences of voltage division or current division, as these are encountered often in op amp circuits. Pay particular attention to implications of virtual short circuits and virtual grounds in the case of ideal op amps. If this step does not provide the solution, go to Step 4. Steps 2 and 3 can sometimes be alternated.

Step 4—Explore: Consider the nodes and meshes in the circuit to see if KCL or KVL can be satisfied by assignment of a single unknown current or voltage AND if this unknown can then be directly determined from KCL or KVL. It is generally convenient to select as this single unknown, the output voltage of an op amp, or the nonzero voltage at the inputs of an ideal op amp that is not in the inverting configuration. Once this unknown is determined, other required variables can be derived from KCL, KVL, or Ohm's law. It is helpful in implementing this step to bear in mind the following:

a. The nodes that should be examined are particularly those at op amp inputs and virtual grounds. KCL at op amp output terminals is generally not helpful to begin with because the output current is not known.
b. Make use of the output–input relations for inverting and noninverting configurations based on ideal op amps.
c. Make use of the v_O-ε characteristic (equivalent circuit) in the case of almost-ideal op amps.
d. Avoid introducing additional unknowns unnecessarily.

Steps 5 and 6—Plan and Implement are usually unnecessary.

Step 7—Check your calculations and results. PSpice simulations are very helpful.

The preceding steps are illustrated by Example 13.10.

Example 13.10: Two-Stage Amplifier

It is required to determine V_O in Figure 13.59.

Solution: The implementation of the steps outlined is illustrated in Figure 13.60.

Step 1—Initialize: The circuit parameters and the required voltage V_O are indicated. The zero input currents are indicated by an 'x' at the inputs of the op amps. The virtual short circuit at the inputs of the first op amp, and the virtual ground at the inverting input of the second op amp are indicated and also marked with x to emphasize that no currents flow in these virtual connections.

Step 2—Simplify: The circuit is in a simple enough form.

Step 3—Deduce: Because of the virtual ground, the 3 kΩ and 6 kΩ resistors connected to the noninverting input of the first op amp are effectively in parallel, the parallel resistance being 2 kΩ. This parallel resistance, in series with the 1 kΩ resistance connected at the input of first op amp, is across the 9 V source. The current in the 1 kΩ resistor is therefore 3 mA. From current division, the currents in the 3 and 6 kΩ resistors are 2 mA and 1 mA, respectively. The voltage at the noninverting input of the first op amp is from Ohm's law or voltage division 6 V. Because of the virtual short circuit, this is the voltage at the inverting input of the first op amp and at the output of this op amp, by virtue of the unity-gain follower connection. The current in the 6 kΩ resistor connected to the inverting input of the second op amp is therefore 1 mA.

The total current entering the node at the inverting input of the second op amp is 1 mA + 2 mA = 3 mA. This current flows through the 3 kΩ resistor, producing a voltage drop of (3) × 3 = 9 V. Because of the virtual ground, $V_O = -9$ V.

Simulation: The circuit is entered as in Figure 13.61 using the part OPAMP from the ANALOG library, without changing the default values. To run the simulation, select Bias Point under 'Analysis type' in the Simulation Settings dialog box. After the simulation is run, pressing the V and I buttons displays the values shown in Figure 13.61. It is seen that, except for the 9 μV voltage at the inverting terminal of the second op amp, the indicated currents and voltages are the same as those assuming an ideal op amp.

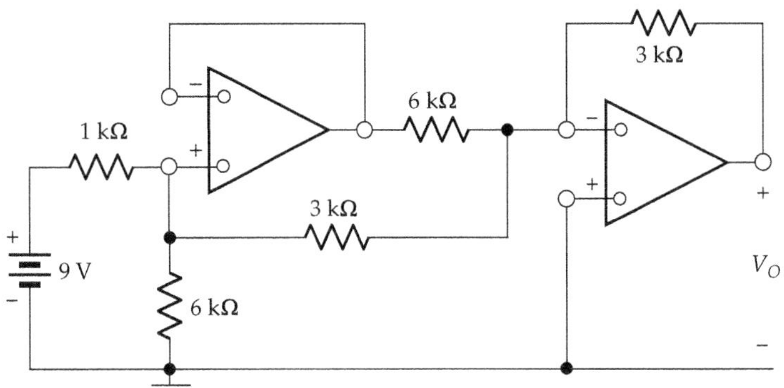

FIGURE 13.59
Figure for Example 13.10.

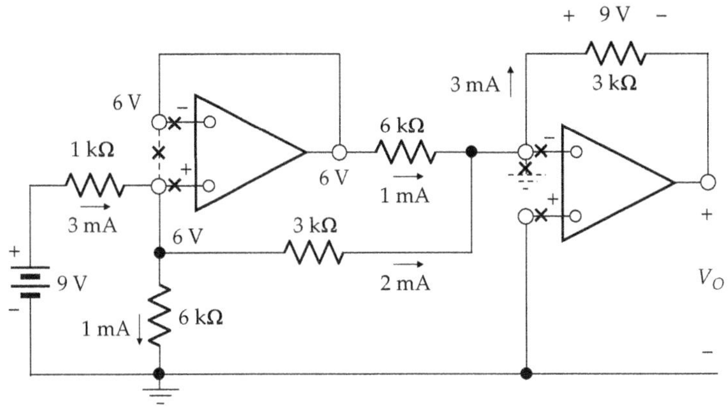

FIGURE 13.60
Figure for Example 13.10.

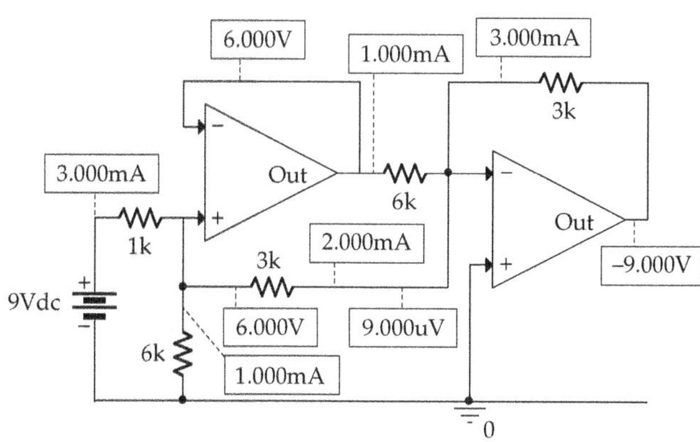

FIGURE 13.61
Figure for Example 13.10.

*Example 13.11: Simulation of Response of Practical Op Amp

This example has two objectives: (1) to compare the results of simulating a difference amplifier using the PSpice part of a popular, general-purpose op amp, the uA741, in order to show that the analytical results based on ideal op amps give the same results in this case and (2) examine the effects of tolerances of resistance values. These tolerances are deviations from nominal values due to inevitable variations in the materials and processes used in manufacturing the components. The tolerance, specified by the manufacturer, is expressed as a percentage of the nominal value. For example, the actual resistance of a 1 kΩ resistor of 2% tolerance can be anywhere in the range of 980–1020 Ω. Commonly used resistors have tolerances of 5% or 10%, but resistors of 0.1%, 0.25%, 0.5%, 1%, 2%, and 20% are available. The smaller the tolerance, the more expensive the resistor is.

Simulation: The circuit is entered as in Figure 13.62 using the uA741 part from the EVAL library. The only additions made to the part, as entered, are the bias supplies of ±15 V between pins 4 and 7 and ground. The output voltage can be readily calculated assuming an ideal op amp. The noninverting terminal (pin 3) is at 5 V. Because of the virtual short circuit, the inverting terminal (pin 2) is also at 5 V. The current flowing through the 10 kΩ resistor toward the inverting terminal is $(6 - 5)/10 = 0.1$ mA. This current flows through the 10 kΩ feedback resistor, producing a voltage drop of 1 V from inverting input to output. It follows that $V_O = 4$ V.

To run the simulation, with full sensitivity analysis, select 'Bias Point' under 'Analysis type' in the Simulation Settings dialog box, check the 'Perform Sensitivity analysis' box under 'Output File Options,' and enter V(U1:OUT) in the 'Output variable(s)' field. After the simulation is run, pressing the V and I buttons displays the values shown in Figure 13.62. Note that the simulated output voltage of 4.000 V agrees with the calculated value of 4 V to within three significant figures beyond the decimal point. Note also that the input currents of the op amp are not zero, but 81.59 and 81.91 nA. These are **input bias currents** of the op amp. Moreover, the voltages of the input terminals are 4.999 and 5.000 V. This difference is the algebraic sum of ε, the differential input voltage, and an inherent voltage difference referred to as the **input offset voltage** of the op amp.

It must be emphasized that deviations from the ideal can have a drastic effect on some types of op amp circuits. For example, the aforementioned input bias currents and offset voltage are themselves integrated by the capacitor of an ideal integrator, thereby driving the op amp to saturation after a long enough period, even with zero input. For this reason, the capacitor in a practical integrator is usually short-circuited by an electronic switch until the start of integration.

The results of the sensitivity analysis are tabulated in the output file for all component and parameter values,

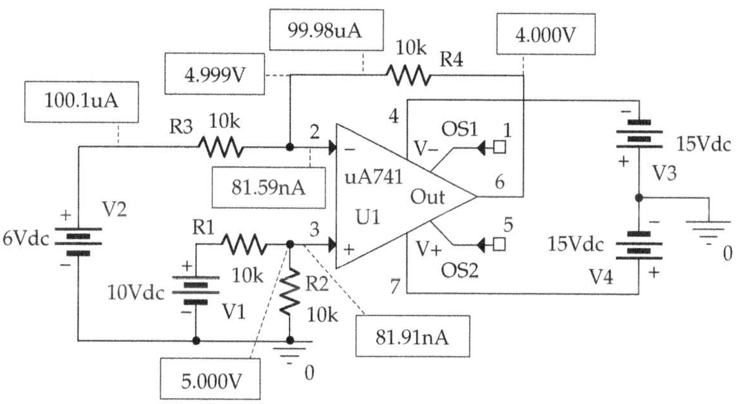

FIGURE 13.62
Figure for Example 13.11.

including those of the op amp. The following entries for the values of the sources and resistors have been selected from the table:

Element Name	Element Value	Element Sensitivity (Volts/Unit)	Normalized Sensitivity (Volts/Percent)
R_R1	1.000E+04	−5.000E−04	−5.000E−02
R_R2	1.000E+04	4.999E−04	4.999E−02
R_R3	1.000E+04	1.001E−04	1.001E−02
R_R4	1.000E+04	−9.997E−05	−9.997E−03
V_V1	1.000E+01	1.000E+00	1.000E−01
V_V2	6.000E+00	−1.000E+00	−6.000E−02
V_V3	1.500E+01	−3.598E−05	−5.397E−06
V_V4	1.500E+01	3.598E−05	5.397E−06

Considering R_1, for example, the third column indicates that an increase of 1 Ω in R_1 decreases (because of the minus sign) V_O by 5.000 × 10^{-4} V, conversely if R_1 deceases. The fourth column indicates that an increase of 1% in R_1 decreases V_O by 5.000 × 10^{-2} V. Similar interpretations apply to the other rows.

To examine the effect of the 2% tolerance, this tolerance is entered, as a percentage, in the Property Editor Spreadsheet of each of the four resistors, in the respective row or column. Select 'DC Sweep' under 'Analysis type.' With the 'Primary Sweep' box checked and 'Voltage source' selected as the 'Sweep Variable,' enter the name of the voltage source, V1 or V2, in the 'Name' field. Under 'Sweep type,' with 'Linear' selected and assuming V1 was chosen as the voltage source, enter 10 for 'Start value,' also 10 for 'End Value' and 1 for 'Increment.' Check the 'Monte Carlo/Worst Case' box under 'Options,' which automatically opens a new window. Select 'Worst Case/Sensitivity' and enter V(U1:OUT) in the 'Output variable' field. Press the 'More Settings' button, choose the 'the maximum value (MAX)' from the pull-down menu in the 'Find' field, and select 'Hi' under 'Worst-Case direction.' Press the 'OK' buttons and run the simulation. In the 'Available Sections' window, press the 'All' button and then the 'OK' button. The output file shows the results of a limited sensitivity analysis for the four resistors, indicating a maximum output voltage of 4.235 V, or 105.88% of nominal, for worst-case combination of tolerances.

To find the minimum voltage for the worst-case combination of tolerances, the preceding procedure is repeated, except that after pressing 'More Settings' button, choose 'the minimum value (MIN)' from the pull-down menu in the 'Find' field, and select 'Low' under 'Worst-Case direction.' After running the simulation, the output file shows a minimum output voltage of 3.7548 V, or 93.877% of nominal, for worst-case combination of tolerances.

Learning Checklist: What Should Be Learned from This Chapter

- An op amp is a three-terminal, voltage-operated device whose output voltage is directly proportional to the difference between the voltages applied to its two input terminals: $v_O = A_v(v_P - v_N) = A_v\varepsilon$, where A_v is the voltage gain of the amplifier and ε is the differential input.
- An ideal op amp has the following properties:
  1. The output is an ideal voltage source of voltage v_O given by

  $$v_O = A_v(v_P - v_N), \quad \text{with } A_v \to \infty$$

  2. Like an ideal voltage source, the amplifier has zero output resistance and can deliver any output voltage or current, at any frequency.
  3. Both inputs behave as open circuits.
  4. The op amp is free from all imperfections, such as nonlinearities or any form of distortion of the output with respect to the input.
- An op amp, like any amplifying device, fundamentally amplifies power. Some external dc bias supply is needed to provide the difference between the output power and the input power. Power is conserved and KCL is satisfied when the bias supplies are taken into account.
- If A_v is assumed finite and the bias supplies are taken into consideration, the input–output characteristic of the op amp can be divided into three regions: (1) a linear region of slope A_v centered at the origin; (2) a positive saturation region, where the output is clamped at V_{CC}^+ and which extends from an input V_{CC}^+/A_v to more positive inputs; and (3) a negative saturation region, where the output is clamped at V_{CC}^- and which extends from $-V_{CC}^-/A_v$ to more negative inputs.
- In an almost-ideal op amp, either A_v is a finite constant or the bias supplies, V_{CC}^+ and V_{CC}^-, have finite values or both. In all other respects the almost-ideal op amp is ideal.
- Over the linear operating range, an almost-ideal op amp can be represented by an ideal VCVS of voltage $A_v\varepsilon$.
- In a feedback system, part or all of the output is fed back to the input so as to modify the behavior of the system in some way. The feedback

is positive if a change in the output in a given direction results in a change in the output in the same direction, which tends to reinforce the initial change. The feedback is negative if a change in the output in a given direction results in a change in the output in the opposite direction, that is, the initial change is opposed.

- Positive feedback in an op amp circuit having a loop gain of more than 1 results in stable operation in the positive and negative saturation regions only. The op amp behaves in this case like a switch.
- Negative feedback allows stable operation anywhere in the linear region of the op amp characteristic.
- Nonswitching op amp circuits invariably employ negative feedback, which necessitates some circuit connection between the output and the inverting input of the op amp. The negative feedback allows trading off the high gain of the op amp for some desirable characteristics of the circuit, such as a precise and stable amplification that depends on resistance ratios only, irrespective of variations in A_v and other circuit parameters.
- In a negative feedback connection employing an ideal op amp ($A_v \rightarrow \infty$), the differential input is vanishingly small for any finite output, so that the two inputs of the op amp are at the same voltage and can be considered to be virtually short-circuited to one another.
- In the noninverting configuration using an ideal op amp, $\frac{v_O}{v_I} = 1 + \frac{R_f}{R_r}$, where R_f is the resistance of the feedback voltage divider that is connected directly to the output of the op amp.
- The unity-gain amplifier is a special case of the noninverting configuration. It can be advantageously used to isolate a nonideal voltage source from a load, so that the load sees an ideal voltage source output whose value is equal to the open-circuit voltage of the nonideal source.
- In the inverting configuration using an ideal op amp, $\frac{v_O}{v_I} = -\frac{R_f}{R_r}$, where R_f is the feedback resistance connected between the output of the op amp and the inverting input. This relation is generalized to $\frac{\mathbf{V_O}}{\mathbf{V_I}} = -\frac{Z_f}{Z_r}$ in phasor notation.
- The virtual short circuit between the noninverting and inverting terminals of the op amp becomes a virtual ground at the inverting terminal in the inverting configuration. The virtual ground effectively isolates the circuit on the input side of the op amp from the circuit connected to the output of the op amp, while forcing the input current to flow through the output circuit.
- The virtual ground is an important feature of the inverting configuration that allows a number of useful applications such as a current-source-to-voltage-source converter, an ideal integrator, an ideal differentiator, and an adder.
- A difference amplifier is essentially a combination of the inverting and noninverting configurations. It can be combined with a noninverting input stage to form a high-performance IA.
- The ISDEPIC approach can be advantageously adapted to solving problems on op amps.

Problem-Solving Tips

1. In solving problems involving ideal op amps, always make use of the virtual short circuit between the two inputs.
2. In solving problems involving almost-ideal op amps, always check that the op amp is operating in the linear region of its input–output characteristic.
3. A combination of circuit elements connected across the voltage source output of an op amp does not affect the output voltage of the op amp but affects its output current.
4. No current flows through a passive circuit element, or a combination of such elements, connected across a virtual short circuit at the op amp inputs.

Problems

Apply ISDEPIC and verify solutions by PSpice simulation, whenever feasible.

Single-Op Amp Circuits

P13.1 Determine V_O in Figure P13.1.
Ans. 8 V.

P13.2 Determine i_L in Figure P13.2. Note that i_L is independent of R_L, so that R_L sees a current source of value i_L.
Ans. 1 mA.

P13.3 Determine i_L in Figure P13.3.
Ans. −15 mA.

P13.4 Determine R_{in} in Figure P13.4.
Ans. 10 kΩ.

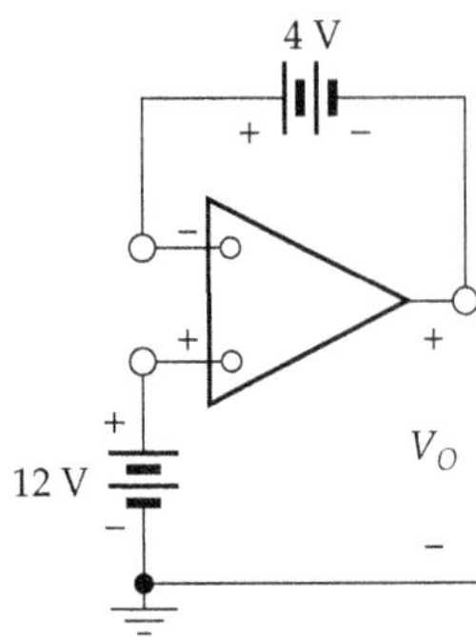

FIGURE P13.1

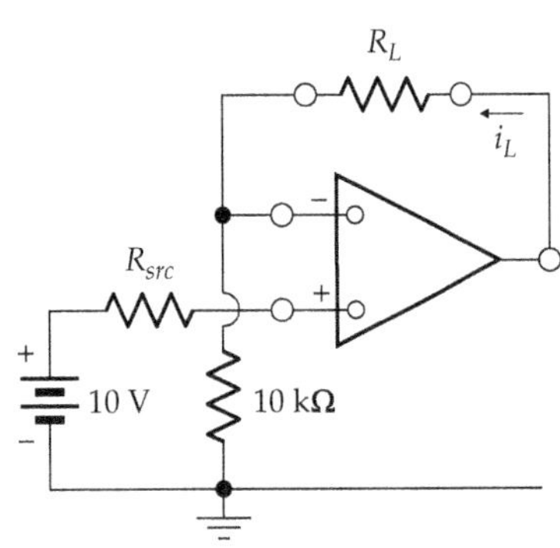

FIGURE P13.2

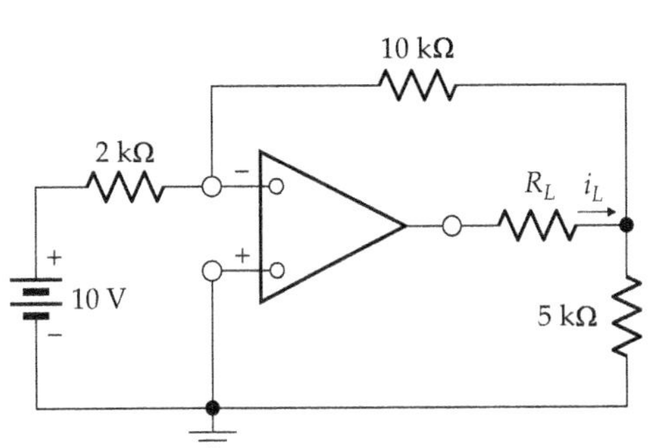

FIGURE P13.3

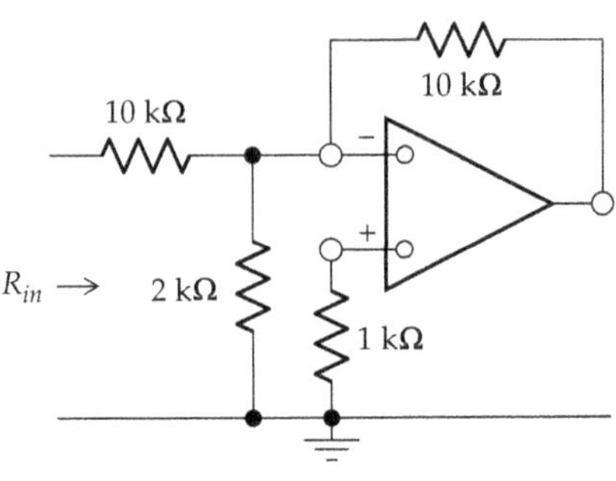

FIGURE P13.4

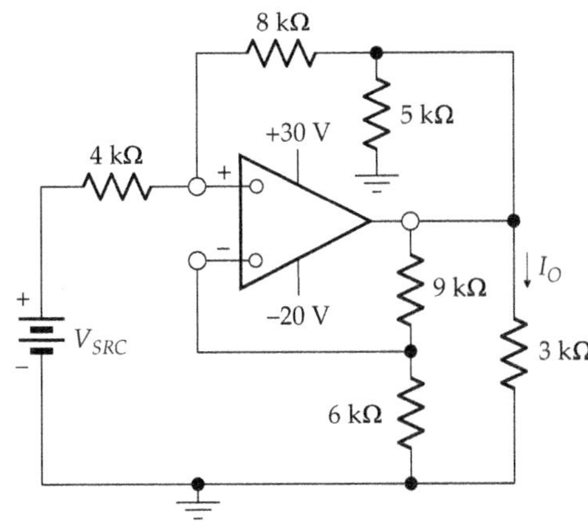

FIGURE P13.5

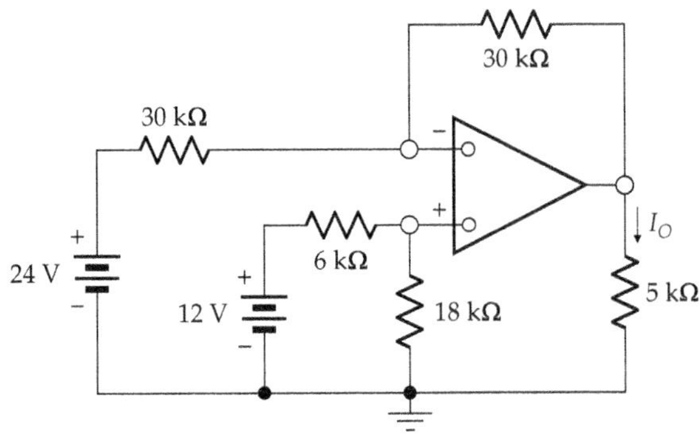

FIGURE P13.7

P13.5 Determine I_O in Figure P13.5 assuming $V_{SRC} = 2$ V.

Ans. 20/3 mA.

P13.6 Determine the range of V_{SRC} in the preceding problem that will maintain operation in the linear region of the input–output characteristic of the op amp.

Ans. $-2\text{ V} \le V_{SRC} \le 3$ V.

P13.7 Determine I_O in Figure P13.7.

Ans. −1.2 mA.

P13.8 Determine V_O in Figure P13.8.

Ans. 14 V.

P13.9 Given that $v_1(t) = \cos\omega t$ V in Figure P13.9 and $v_2(t) = \sin\omega t$ V, determine R so that $v_O(t) = -\cos(\pi/16)\cos(\omega t - \pi/16)$ V.

Ans. $\cot(\pi/16)$ Ω.

P13.10 The circuit of Figure P13.10 uses matched resistor pairs G_1, G_2, and G_3 to provide the average of the applied inputs. Show that $v_O = -(a_1v_1 + a_2v_2 + a_3v_3)$, where $a_k = \dfrac{G_k}{G_1 + G_2 + G_3}$, $k = 1, 2,$ or 3, and $a_1 + a_2 + a_3 = 1$.

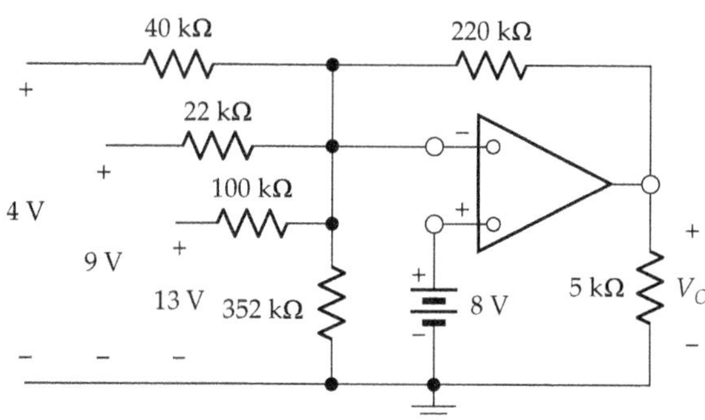

FIGURE P13.8

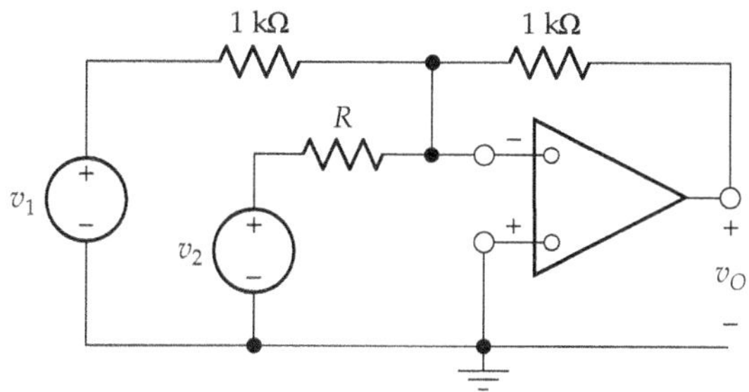

FIGURE P13.9

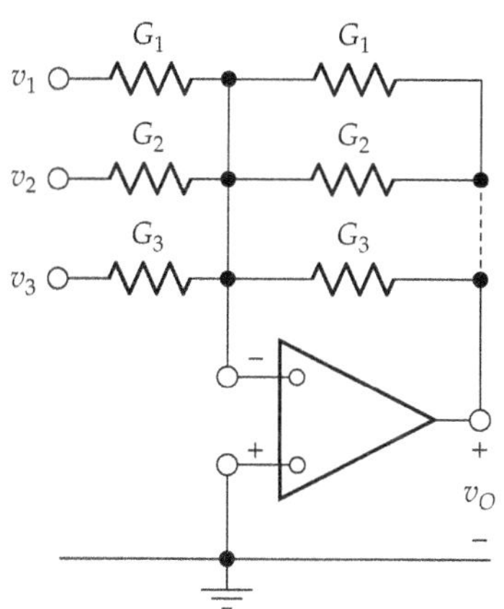

FIGURE P13.10

P13.11 An amplifier in the inverting configuration has R_r = 10 kΩ, R_f = 40 kΩ, and bias supplies of ±12 V. Sketch the amplifier output if the input is the waveform shown in Figure P13.11.

Ans. Output is $-4 \times v_I$, for $-3 \text{ V} \leq v_I \leq +3$ V, and is clipped at ±12 V for $|v_I| \geq 3$ V.

P13.12 If $v_i(t) = 4\sin 100t$ V in the amplifier of Problem P13.11, determine the time intervals in which the amplifier is in positive and negative saturation.

Ans. Negative saturation from 8.48 to 22.9 ms and positive saturation from 39.9 to 54.4 ms, during the first period.

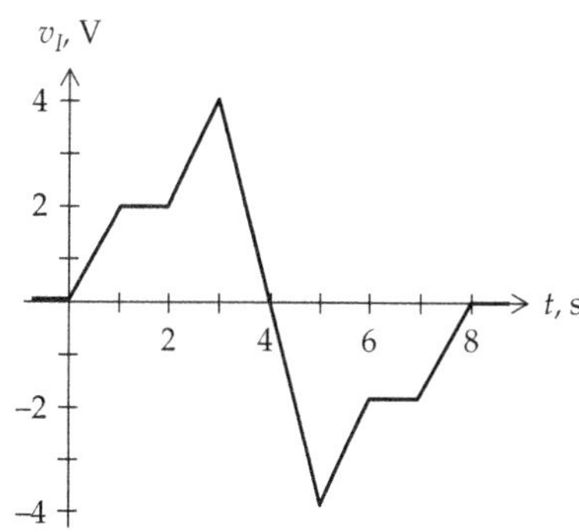

FIGURE P13.11

P13.13 Show that in Figure P13.13 $v_O = R_2\left(1+\frac{R_f}{R_1}\right)i_{SRC2} - R_f i_{SRC1}$.

P13.14 Determine $R > 0$ in Figure P13.14 so that the op amp just reaches saturation.

Ans. 28 kΩ.

P13.15 Determine v_O in Figure P13.15 if $v_{SRC}(t) = 4\cos 100\pi t$ V, $t \geq 0$.

Ans. $9\cos 100\pi t$ V.

P13.16 Determine V_O in Figure P13.16.

Ans. −4 V.

P13.17 Determine V_O in Figure P13.17.

Ans. 6 V.

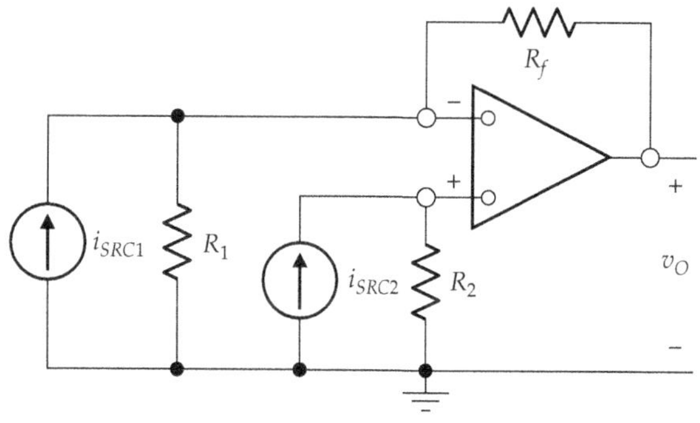

FIGURE P13.13

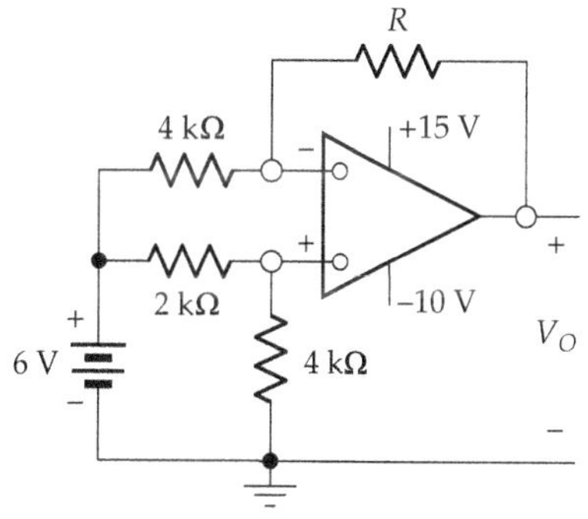

FIGURE P13.14

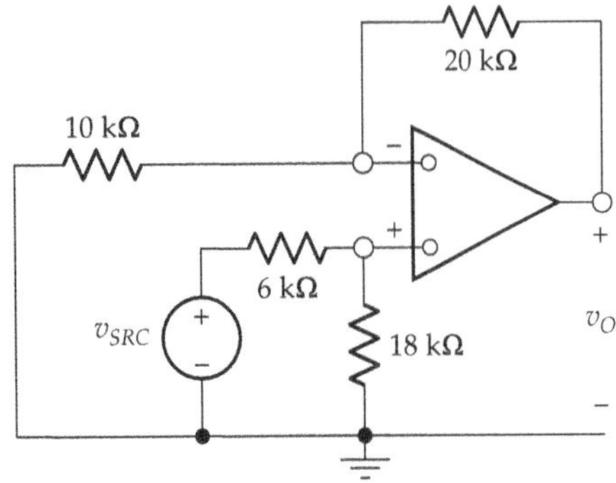

FIGURE P13.15

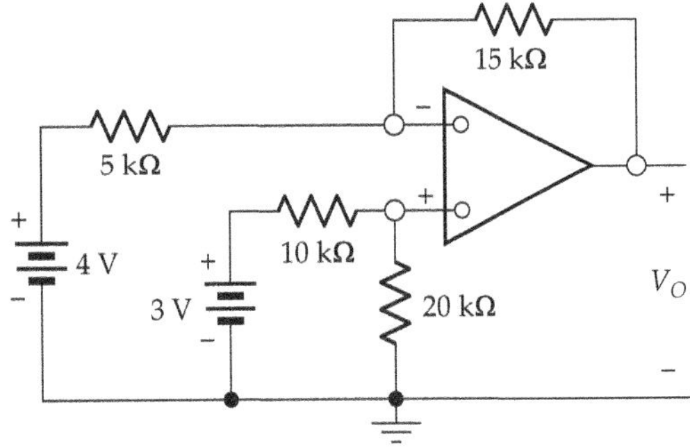

FIGURE P13.16

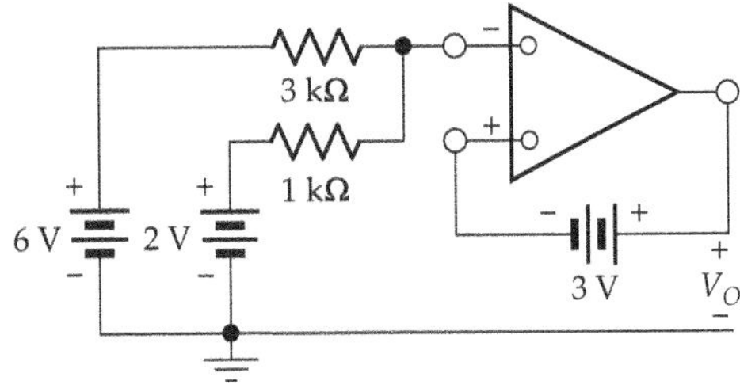

FIGURE P13.17

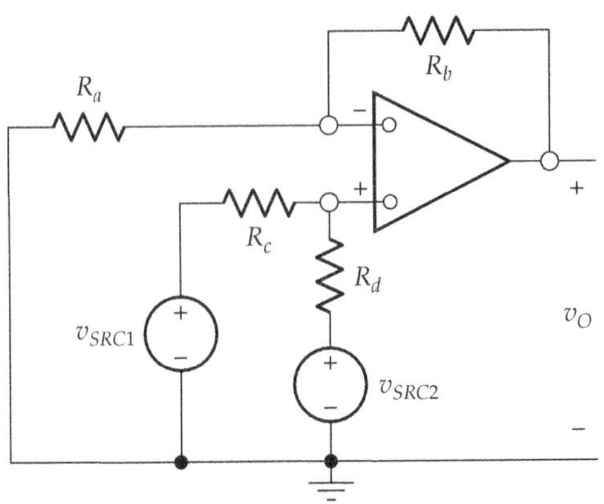

FIGURE P13.18

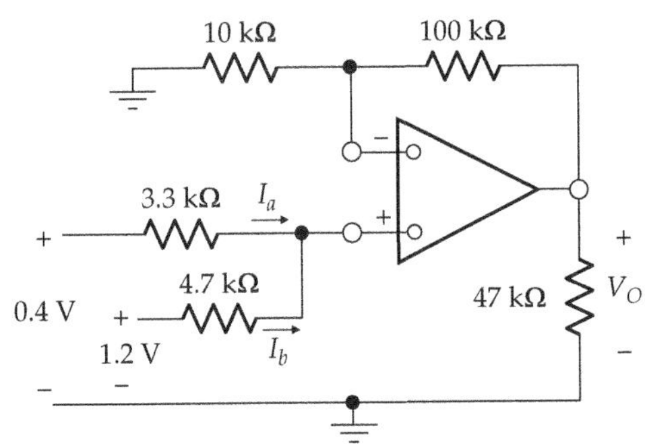

FIGURE P13.19

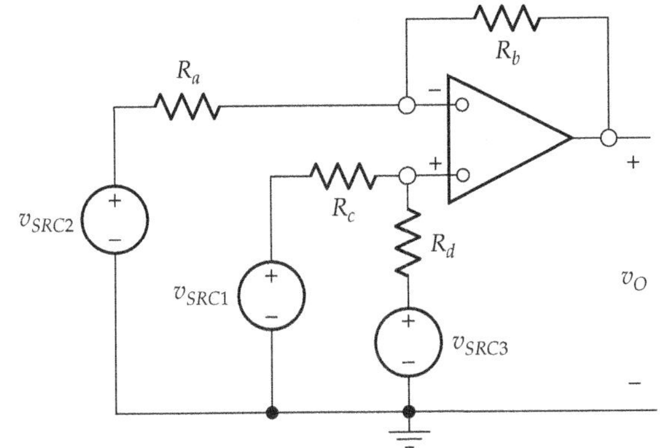

FIGURE P13.20

P13.18 Show that in Figure P13.18, $v_O = \frac{1}{R_c + R_d}\left(1 + \frac{R_b}{R_a}\right) \times (R_d v_{SRC1} + R_c v_{SRC2})$. Note that this circuit is a noninverting adder. However, changes in one input circuit affect the current in the other circuit.

P13.19 Determine in Figure P13.19 (a) I_a, (b) I_b, and (c) V_O.

Ans. (a) −0.1 mA; (b) 0.1 mA; (c) 8.03 V.

P13.20 Show that if v_{SRC3} is added to the difference amplifier of Figure 13.49, as in Figure P13.20, the output voltage v_O is given by $v_O = -\frac{R_b}{R_a} v_{SRC2} + \frac{R_d}{R_c + R_d} \frac{R_a + R_b}{R_a} v_{SRC1} + \frac{R_c}{R_c + R_d} \frac{R_a + R_b}{R_a} v_{SRC3}$.

P13.21 Figure P13.21 illustrates a circuit that may be used with a strain gauge bridge, where the resistance of one of the elements of the bridge changes by a fraction α. Show that $v_O = \frac{R}{R_o} \frac{\alpha}{(1+\alpha)(1+R_o/R)+1} V_{DC}$. Note that if $\frac{R_o}{R} \ll 1$ and $\alpha \ll 1$, then $v_O = \frac{\alpha R}{2R_o} V_{DC}$, and that the bridge output can be effectively doubled if R_o in the upper left arm of the bridge also varies as $R_o(1 + \alpha)$.

P13.22 Determine R_{in} in Figure P13.22.

Ans. 5 kΩ.

P13.23 Determine R_{in} in Figure P13.23.

Ans. 0.

P13.24 Show that in Figure P13.24, $i_L = (v_2 - v_1)/R_2$. Note that the load sees a current source whose value is determined by the differential input $(v_2 - v_1)$.

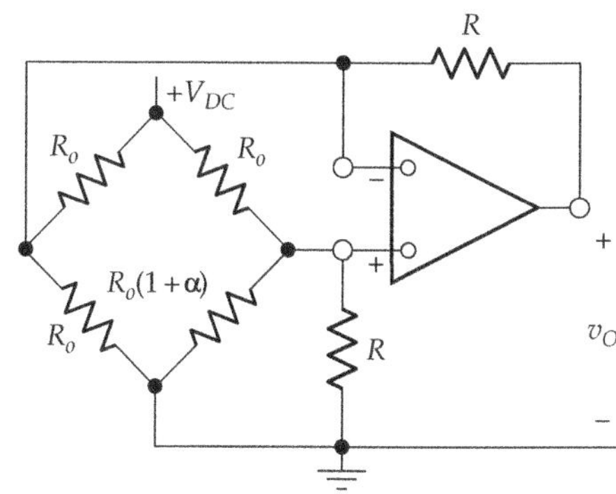

FIGURE P13.21

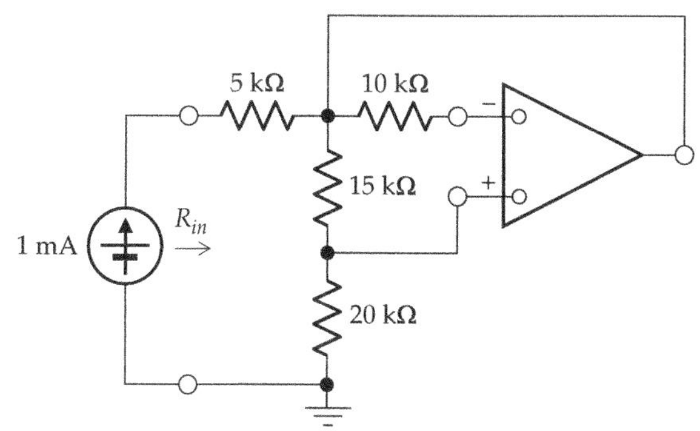

FIGURE P13.22

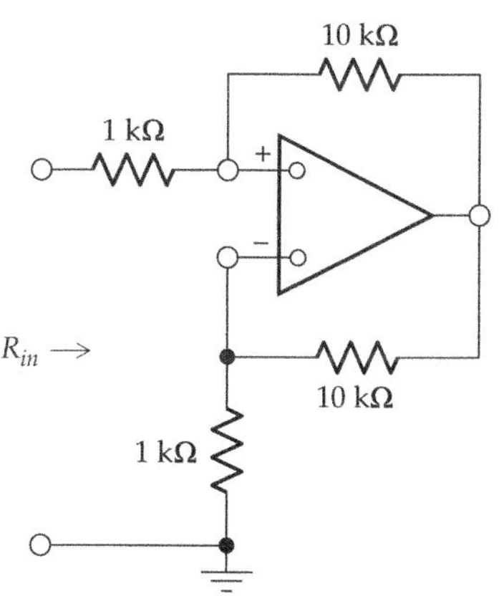

FIGURE P13.23

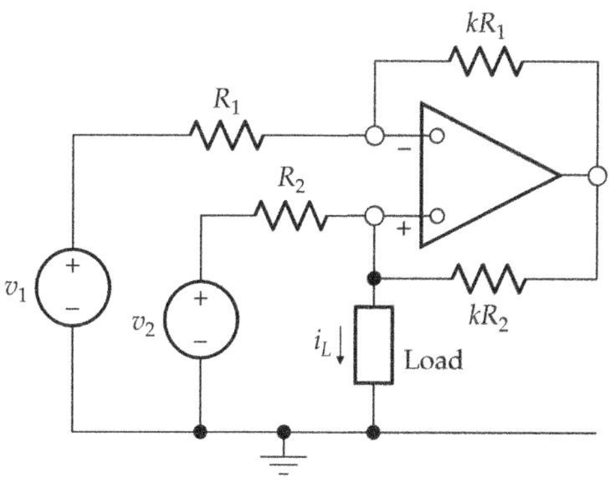

FIGURE P13.24

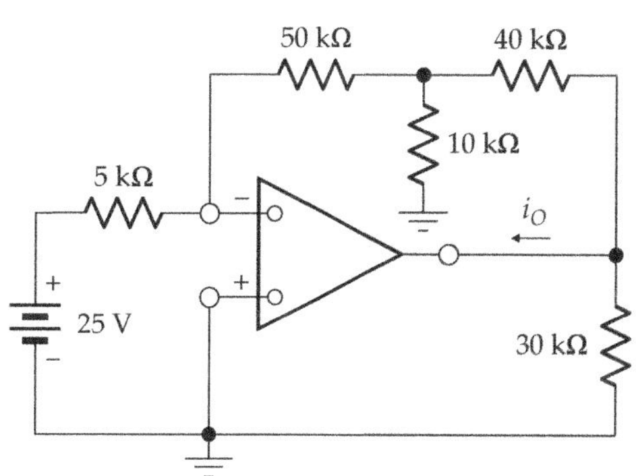

FIGURE P13.25

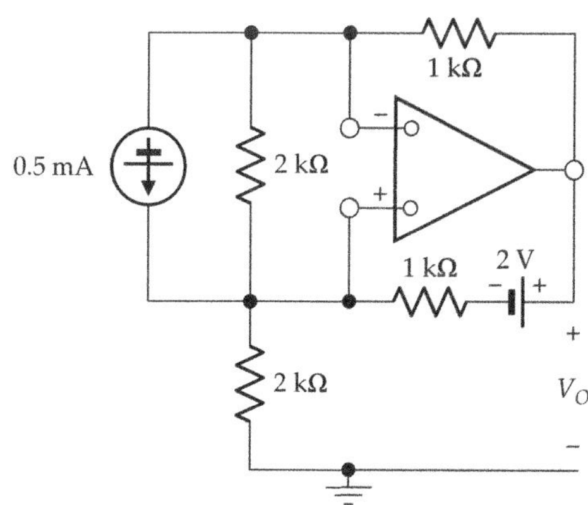

FIGURE P13.26

P13.25 Determine i_O in Figure P13.25.

Ans. 235/3 mA.

P13.26 Determine in Figure P13.26 (a) V_O and (b) power delivered or absorbed by each source.

Ans. (a) –1.5 V; (b) power delivered by the current source is zero, power delivered by voltage source is 3 mW.

P13.27 Determine V_O in Figure P13.27.

Ans. 1.5 V.

P13.28 Determine V_O in Figure P13.28.

Ans. 23 V.

P13.29 Determine V_O in Figure P13.29.

Ans. –72/37 V.

P13.30 Determine TEC looking into terminals 'ab' in Figure P13.30.

Ans. $V_{Th} = V_{ab} = 3$ V, $R_{Th} = 4/7\ \Omega$.

P13.31 Show that in Figure P13.31, $\dfrac{v_O}{v_1 - v_2} = -\dfrac{R_2}{R_1}\dfrac{1}{1+\dfrac{R_1}{2R_{\text{var}}}}$.

Note that this circuit is that of a difference amplifier having variable gain.

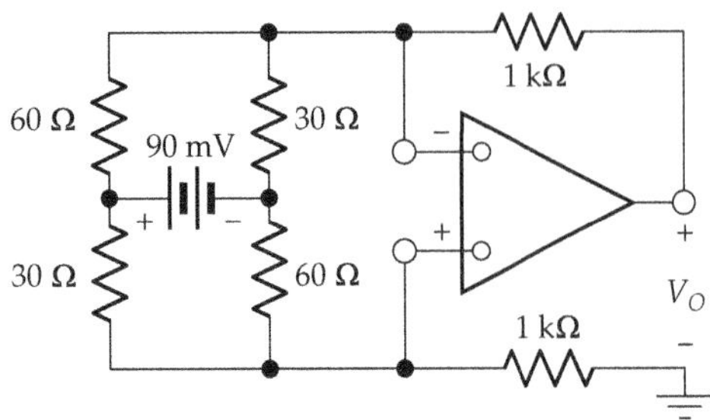

FIGURE P13.27

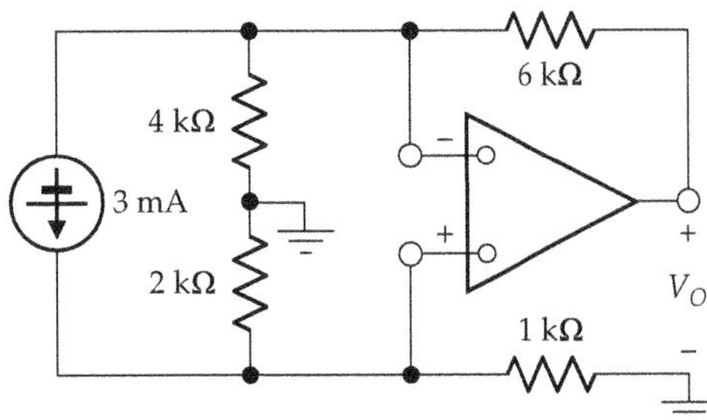

FIGURE P13.28

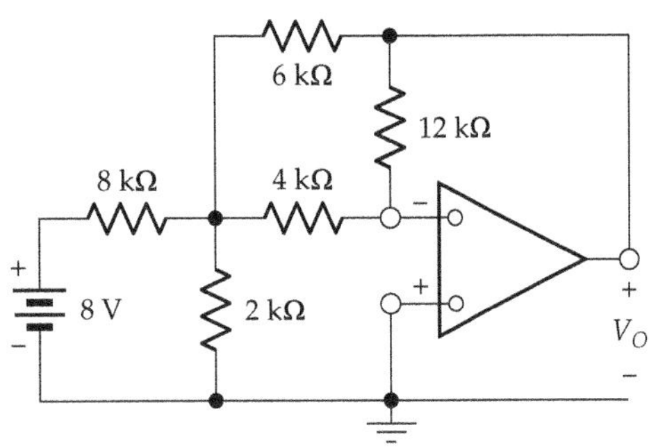

FIGURE P13.29

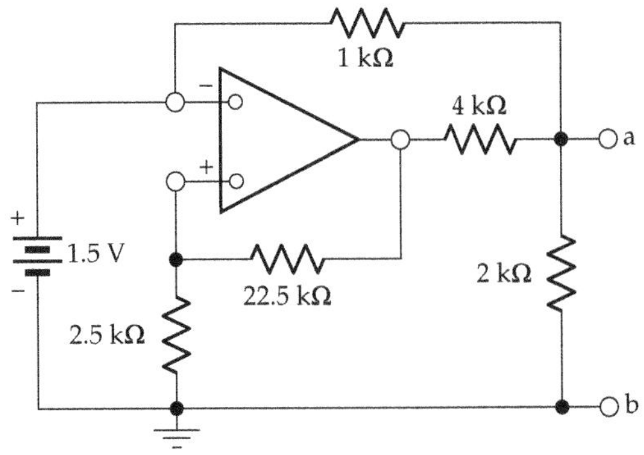

FIGURE P13.30

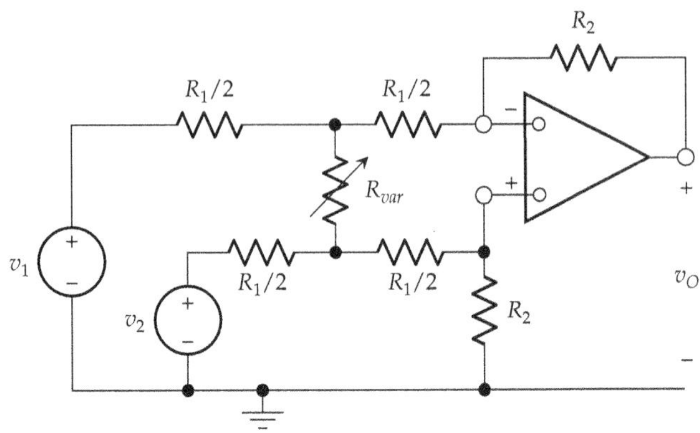

FIGURE P13.31

P13.32 Show that in Figure P13.32, $\frac{v_O}{v_1 - v_2} = \frac{R_2}{R_1}\left(1 + \frac{R_2}{2R_{\text{var}}}\right)$.

Note that this circuit is that of a difference amplifier having variable gain.

P13.33 Determine $v_O(t)$ in Figure P13.33, assuming $v_{SRC}(t) = 5\cos 100t$ V, ≥ 0.

Ans. $-\cos(100t - 36.9°)$ V.

P13.34 Determine $v_O(t)$ and $i_O(t)$ in Figure P13.34, assuming $v_{SRC}(t) = \sin 1000t$ V, $t \geq 0$.

Ans. (a) $v_O(t) = \sin(1000t + 90°)$ V; (b) $i_O(t) = \sqrt{2}\cos(1000 + 135°)$ mA.

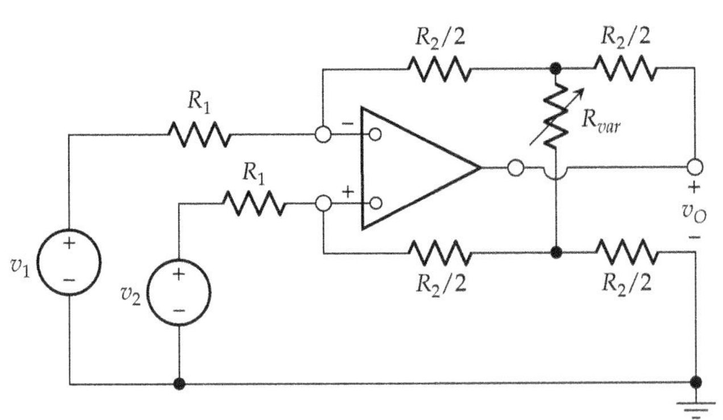

FIGURE P13.32

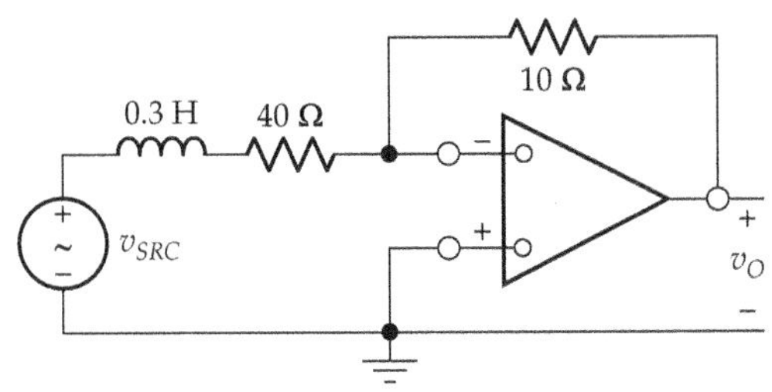

FIGURE P13.33

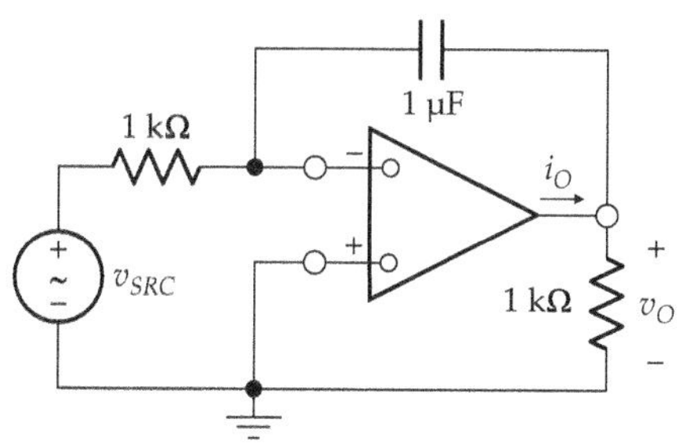

FIGURE P13.34

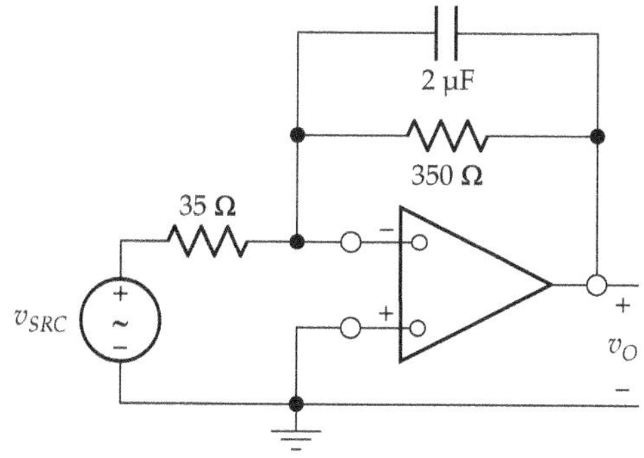

FIGURE P13.35

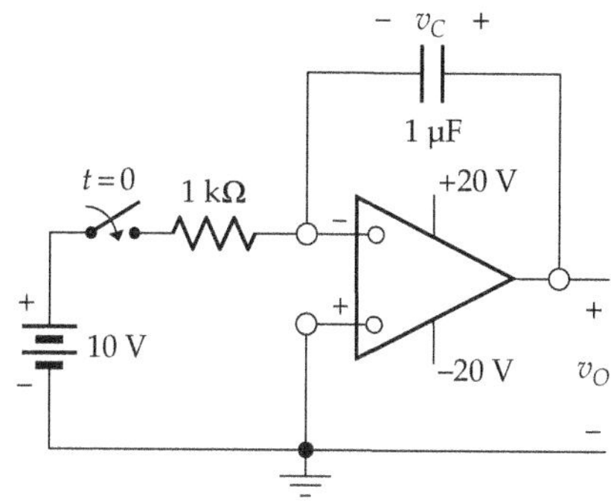

FIGURE P13.37

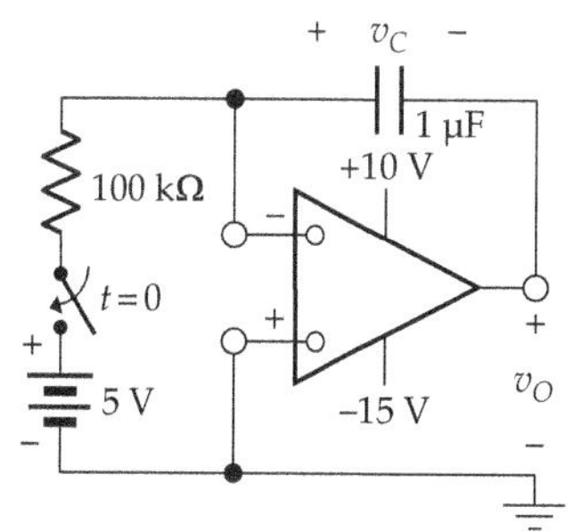

FIGURE P13.38

P13.35 Determine $v_O(t)$ in Figure P13.35 if $v_{SRC}(t) = 10\sqrt{2}\cos 10^4 t$ V, $t \geq 0$.

Ans. $20\cos(10^4 t + 98.1°)$ V.

P13.36 Determine $v_O(t)$ in Figure P13.36 if $v_{SRC1}(t) = \sin 10^3 t$ V and $v_{SRC2} = \cos 10^3 t$ V, $t \geq 0$.

Ans. $v_O(t) = \frac{\sqrt{2}}{2}\cos(10^3 t + 135°)$ V.

P13.37 The switch in Figure P13.37 is closed at $t = 0$, with $V_{CO} = 20$ V at $t = 0$. Determine t at which the op amp just reaches negative saturation.

Ans. 4 ms.

P13.38 The switch in Figure P13.38 is closed at $t = 0$, with the capacitor having an initial voltage $V_{CO} = 5$ V. Determine the time at which v_O reaches a steady value that does not change with time.

Ans. 200 ms.

P13.39 The switch in Figure P13.39 is closed at $t = 0$, with the capacitor initially uncharged. Determine v_O, given that $v_{SRC}(t) = 10\cos 100t$ V, $t \geq 0$.

Ans. $-(100t + 20\sin 100t)$ V.

P13.40 The switch in Figure P13.40 is closed at $t = 0$, with the inductor initially uncharged. Determine v_O, given that $v_{SRC}(t) = 5\sin 100t$ V.

Ans. $-2.5\cos 100t$ V.

P13.41 The circuit of Figure P13.41 provides double integration using a single op amp. Use phasor analysis to show that $\frac{V_o}{V_{src}} = -\frac{1}{(\omega\tau)^2}$, where $\tau = RC$.

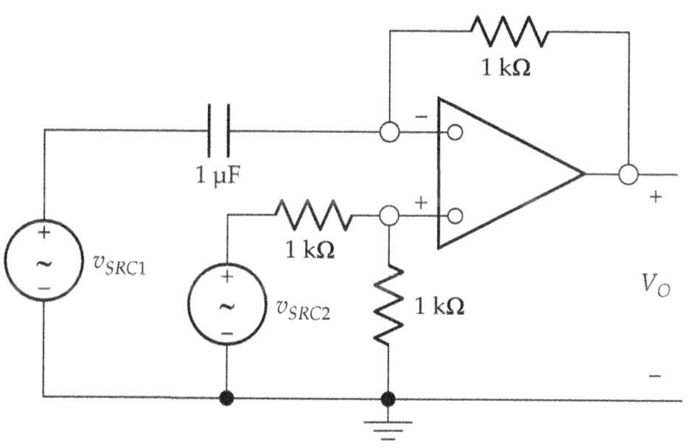

FIGURE P13.36

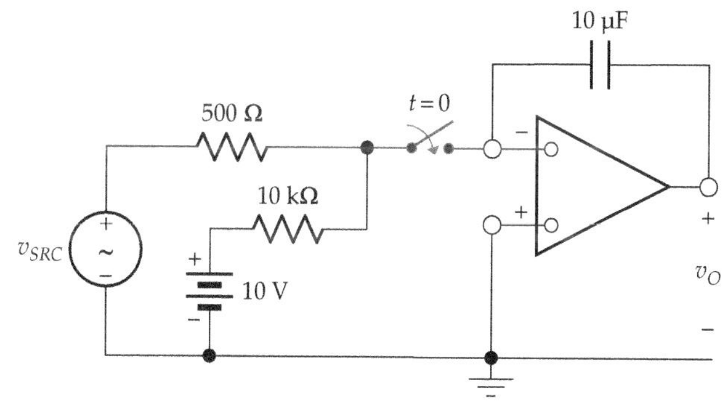

FIGURE P13.39

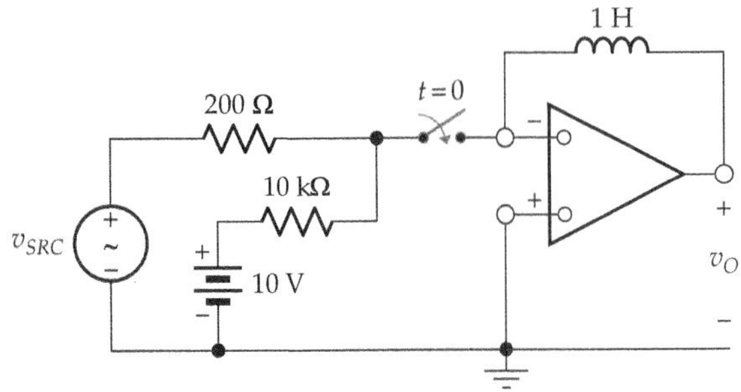

FIGURE P13.40

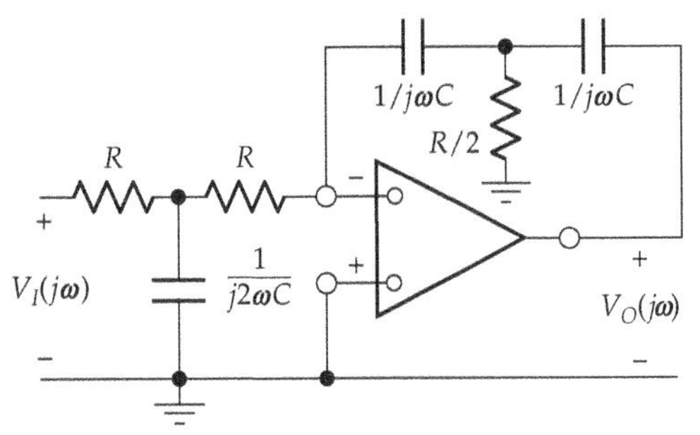

FIGURE P13.41

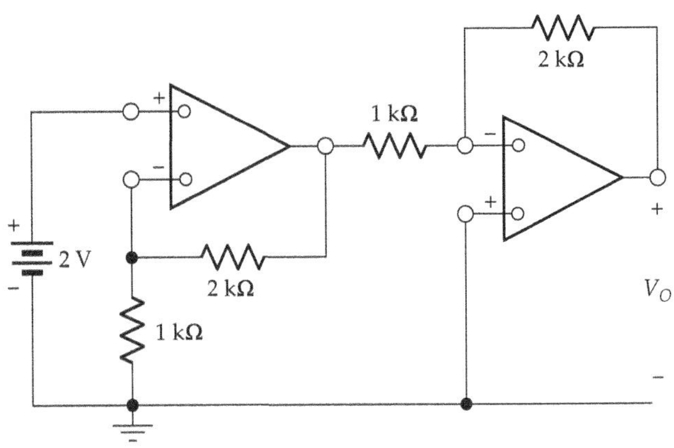

FIGURE P13.43

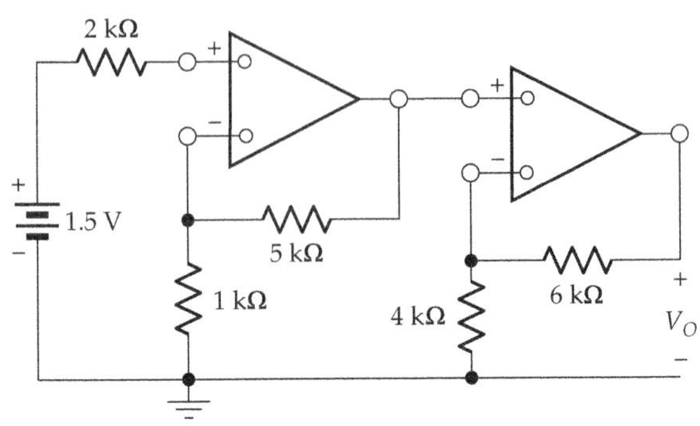

FIGURE P13.44

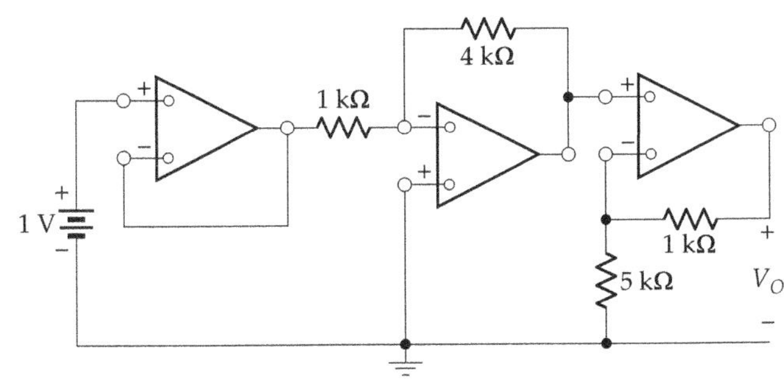

FIGURE P13.45

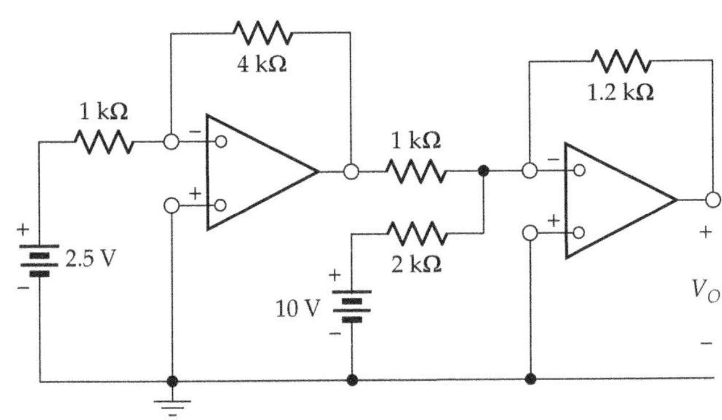

FIGURE P13.46

P13.42 Derive Equation 13.31 in two ways: (i) superposition of v_{SRC2} applied alone, with v_{SRC1} set to zero, and v_{SRC1} applied alone with v_{SRC2} set to zero and (ii) superposition of v_{SRC2} and v_O to obtain v_N and then substituting v_P in terms of v_{SRC1} for v_N.

Multi-Op Amp Circuits

P13.43 Determine V_O in Figure P13.43.

Ans. −12 V.

P13.44 Determine V_O in Figure P13.44.

Ans. +22.5 V.

P13.45 Determine V_O in Figure P13.45.

Ans. −4.8 V.

P13.46 Determine V_O in Figure P13.46.

Ans. 6 V.

P13.47 Determine V_O in Figure P13.47.

Ans. 18 V.

P13.48 Determine V_O in Figure P13.48.

Ans. 24 V.

P13.49 Determine V_O in Figure P13.49.

Ans. 5.5 V.

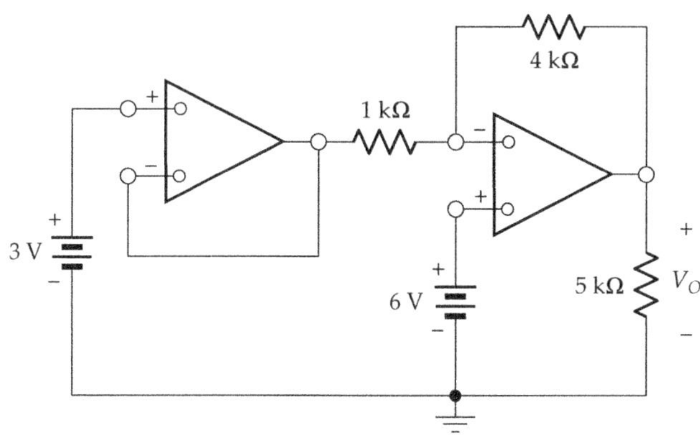

FIGURE P13.47

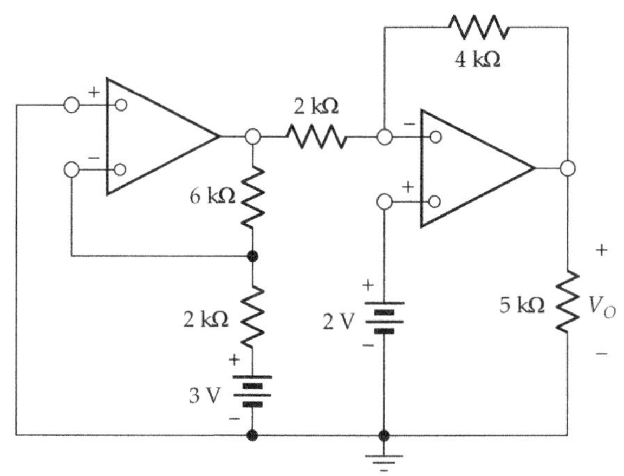

FIGURE P13.48

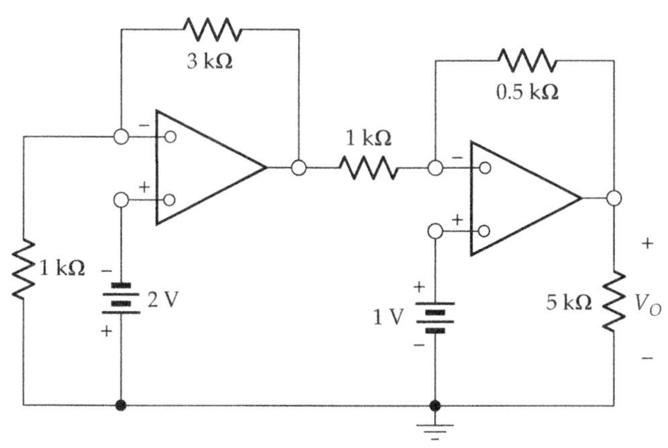

FIGURE P13.49

P13.50 Determine V_O in Figure P13.50.

Ans. 0.

P13.51 Determine V_O in Figure P13.51.

Ans. −6 V.

P13.52 Determine V_O in Figure P13.52.

Ans. −1.5 V.

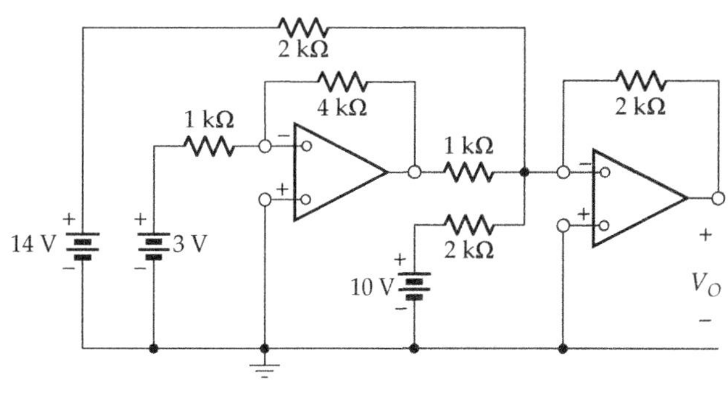

FIGURE P13.50

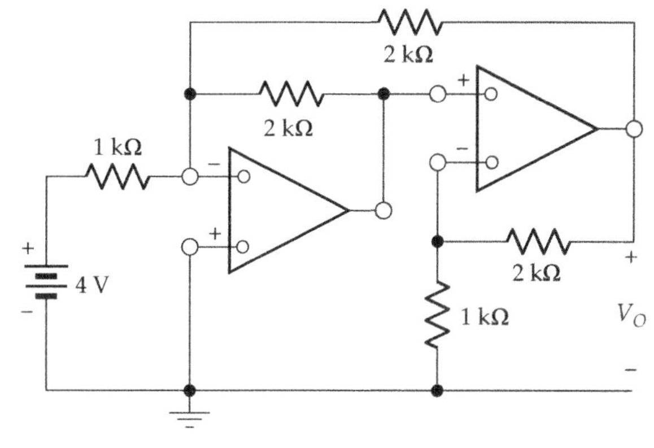

FIGURE P13.51

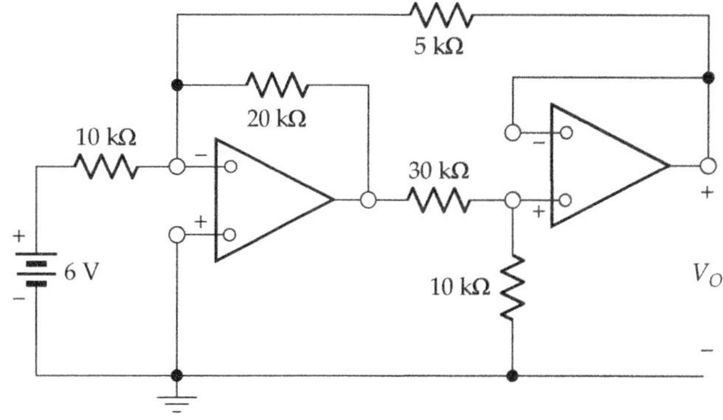

FIGURE P13.52

P13.53 Determine I_O in Figure P13.53.

Ans. 2.25 mA.

P13.54 Determine V_{O1} and I_O in Figure P13.54.

Ans. 1.36 V, −2.96 mA.

P13.55 Determine V_{O1} and V_{O2} in Figure P13.55.

Ans. V_{O1} = 15.85 V, V_{O2} = 13.6 V.

P13.56 Show that v_O in Figure P13.56 is given by $v_O = R_5\left(\dfrac{R_2}{R_1R_3}v_2 - \dfrac{v_1}{R_4}\right)$.

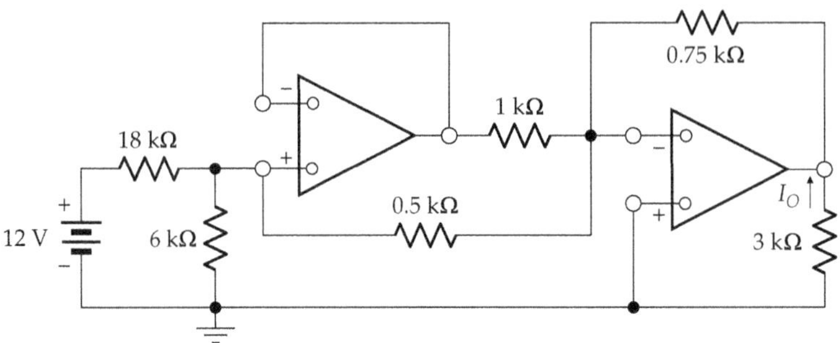

FIGURE P13.53

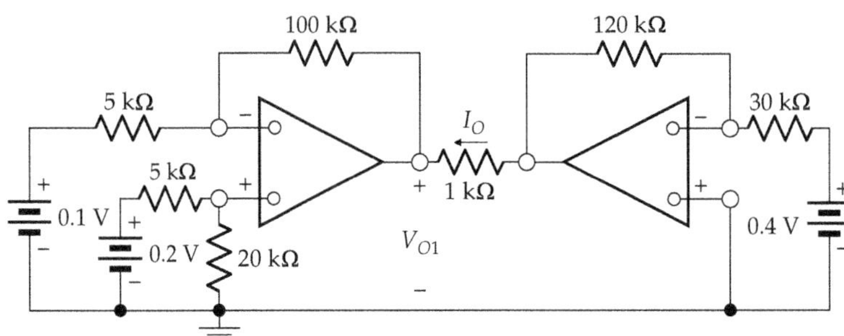

FIGURE P13.54

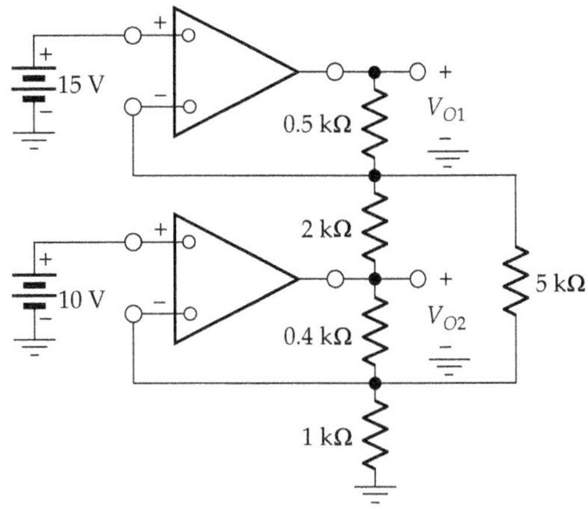

FIGURE P13.55

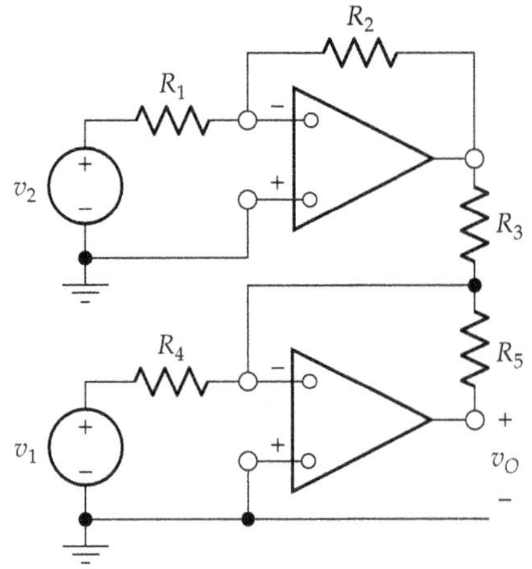

FIGURE P13.56

P13.57 Determine v_O/v_{SRC} in Figure P13.57.

Ans. 8.

P13.58 Determine (a) R in Figure P13.58 so that $I_S = 0$ and (b) maximum and minimum V_{SRC} for operation in the linear region.

Ans. 70 kΩ, ±1.5 V.

P13.59 Determine v_O in Figure P13.59 in terms of v_1 and v_2.

Ans. $v_O = 4v_1 - 6v_2$.

P13.60 Determine V_O in Figure P13.60.

Ans. 12.5 V.

P13.61 Determine V_O in Figure P13.61.

Ans. −20 V.

P13.62 Determine V_O in Figure P13.62.

Ans. 3 V.

P13.63 Determine V_O in Figure P13.63.

Ans. 10 V.

P13.64 Show that $C_{in} = C/\alpha$ in Figure P13.64, where α is the fraction of R_p that appears between nodes 'a' and 'b'.

P13.65 Show that in Figure P13.65, $Z_{in} = R^2/Z$. This means that if Z is an ideal capacitor, then $Z = 1/j\omega C$, and $Z_{in} = j\omega CR^2$. This means that an ideal inductor having $L = \omega CR^2$ appears across the input terminals. Conversely, if Z is an ideal inductor having $Z = j\omega L$, then $Z_{in} = R^2/j\omega L = 1/j(\omega L/R^2)$, which represents an

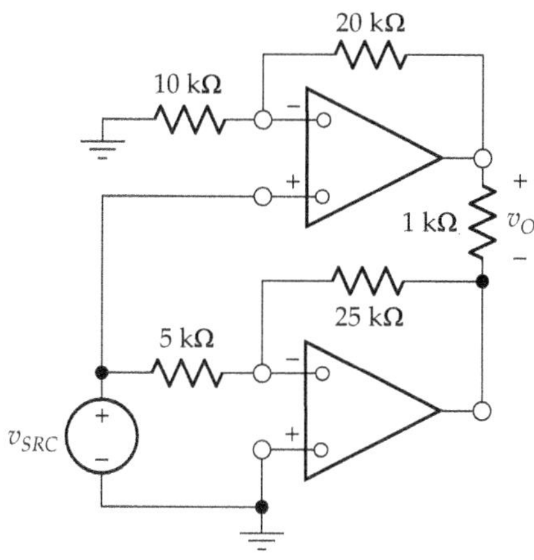

FIGURE P13.57

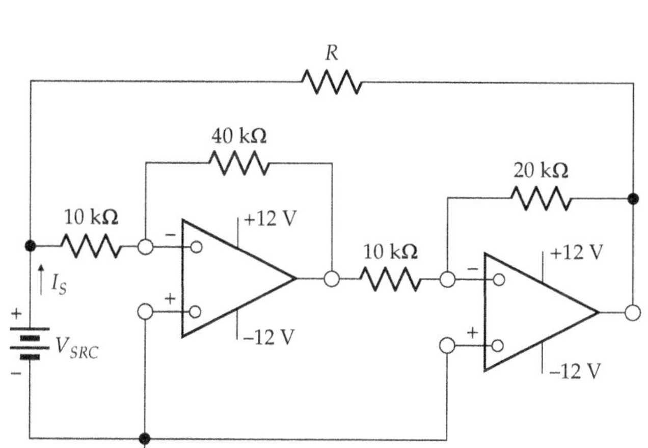

FIGURE P13.58

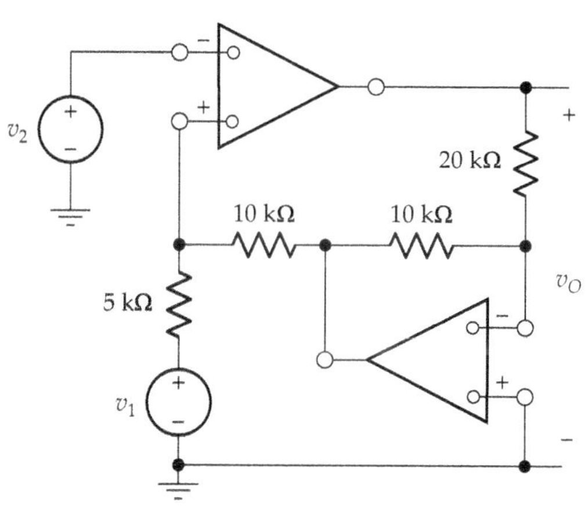

FIGURE P13.59

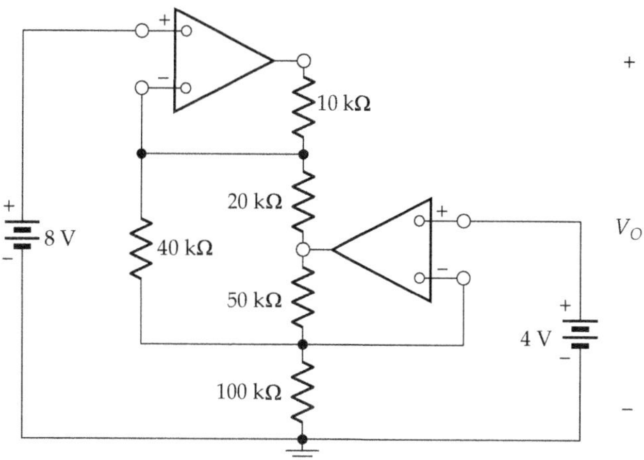

FIGURE P13.60

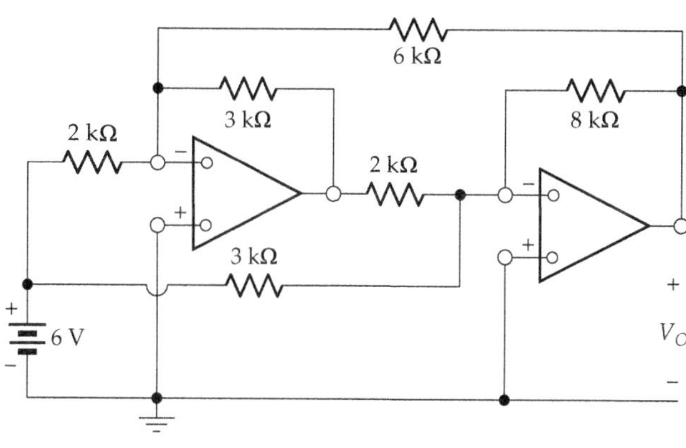

FIGURE P13.61

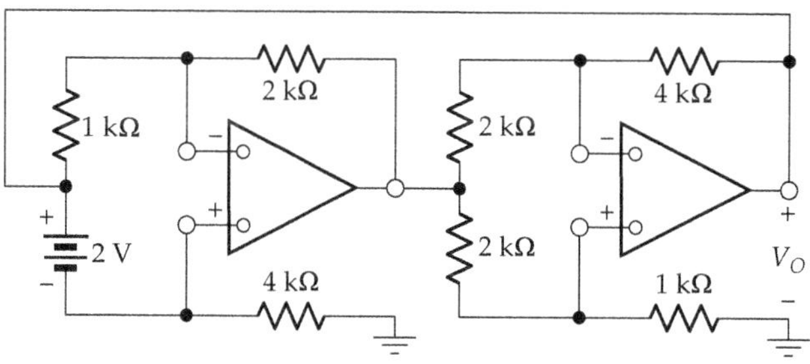

FIGURE P13.62

ideal capacitor having $C = L/R^2$. This type of circuit is known as a **gyrator**.

P13.66 Show that in Figure P13.66, $Z_{in} = R^2/Z$, as in Problem P13.65.

Design Problems

P13.67 Design an inverting adder, similar to the circuit of Figure 13.42, whose output is $v_O = -(4v_1 + 2v_2 - v_3 + 5v_4)$, where all inputs are in the range ±1 V. The current drawn from each source should not exceed 100 μA.

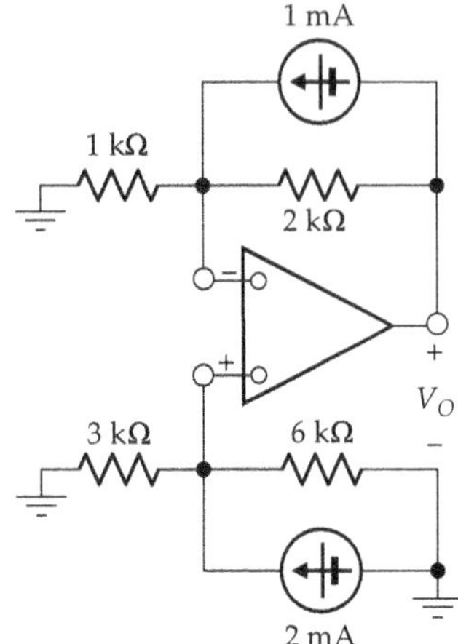

FIGURE P13.63

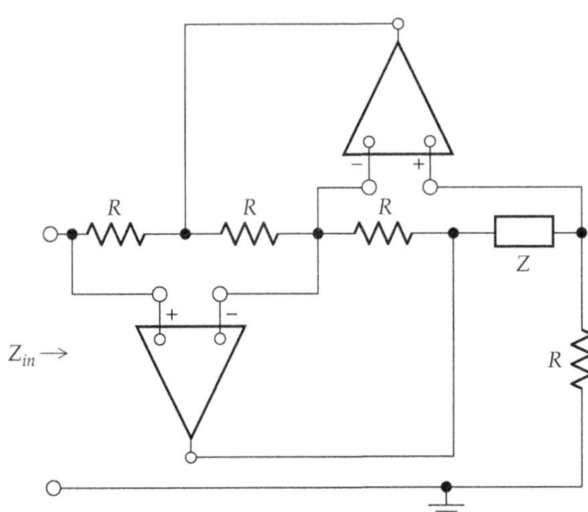

FIGURE P13.66

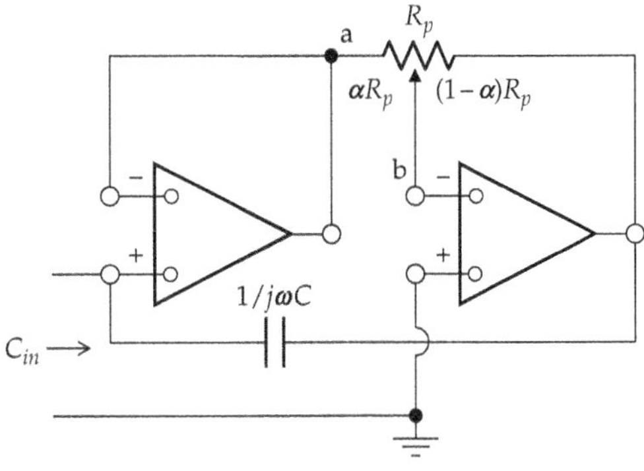

FIGURE P13.64

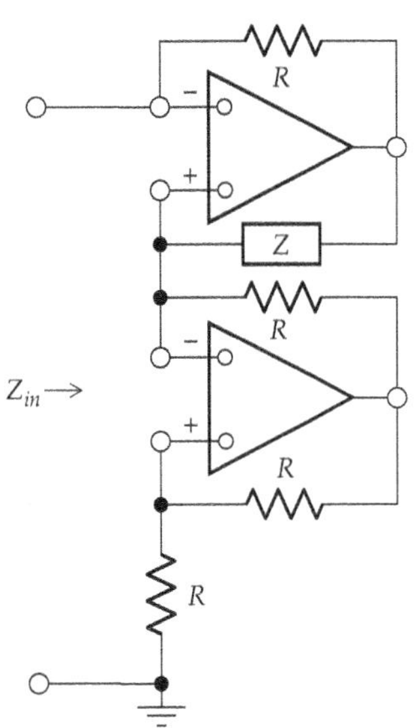

FIGURE P13.65

Specify the minimum magnitude of the bias supplies so as to avoid op amp saturation.

Ans. $R_1 = 12.5$ kΩ, $R_2 = 25$ kΩ, $R_3 = 50$ kΩ, $R_4 = 10$ kΩ, $R_f = 50$ kΩ; 12 V.

P13.68 Design a difference amplifier whose output is $v_O = 5(v_1 - v_2)$, such that each source sees an input resistance of at least 20 kΩ.

Ans. $R_a = R_c = 20$ kΩ, and $R_b = R_d = 100$ kΩ.

P13.69 Design a noninverting adder, similar to the circuit of Problem P13.18, whose output is $v_O = 3v_1 + v_2 + 2v_3$, using resistors in the range 10–100 kΩ.

Ans. $R_1 = 20$ kΩ, $R_2 = 60$ kΩ, $R_3 = 60$ kΩ, $R_f = 50$ kΩ, and $R_r = 10$ kΩ.

P13.70 Design a circuit, similar to that of Figure P13.24 that converts a 10 V source of 1 kΩ source resistance to a 1 mA ideal current source using only two resistance values.

Ans. 1 and 10 kΩ resistors.

P13.71 Design an averaging circuit, similar to that of Figure P13.10, whose output is $v_O = -(0.2v_1 + 0.3v_2 + 0.5v_3)$ using resistors in the range 10–60 kΩ.

Ans. $R_1 = 60$ kΩ, $R_2 = 40$ kΩ, $R_3 = 24$ kΩ.

P13.72 Design a precision amplifier to amplify inputs in the range of ±3 V, with a gain of 5% ± 1% using resistors of 10 and 40 kΩ. Select the largest suitable resistor tolerances required from the standard tolerances of 0.25%, 0.5%, 1%, and 2%.

Ans. 0.5% tolerance.

P13.73 The circuit of Figure P13.73 uses an *R*-2*R* ladder (Problem P5.32) to implement a 4 bit digital-to-analog converter (DAC). The 4 bit word can be expressed as $(b_3b_2b_1b_0)$, where each of the bits 'b' can be 1 or 0. When a given bit is zero, the corresponding transistor switch of the same number subscript is in the left position, whereas when the bit is 1, the switch is in the right position. Show that $V_O = -V_{ref}\dfrac{R_f}{16R}\left(8b_3 + 4b_2 + 2b_1 + b_0\right)$.

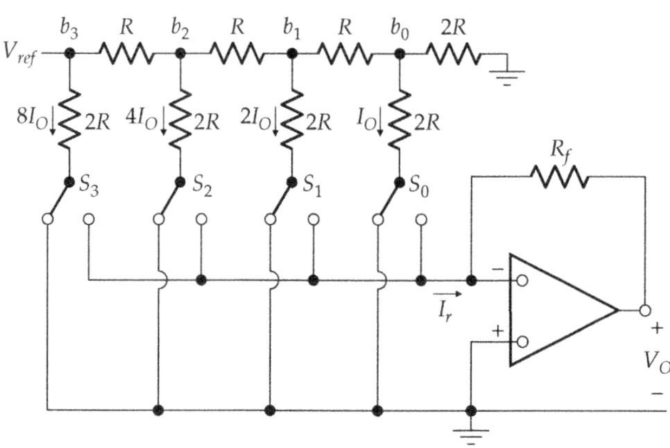

FIGURE P13.73

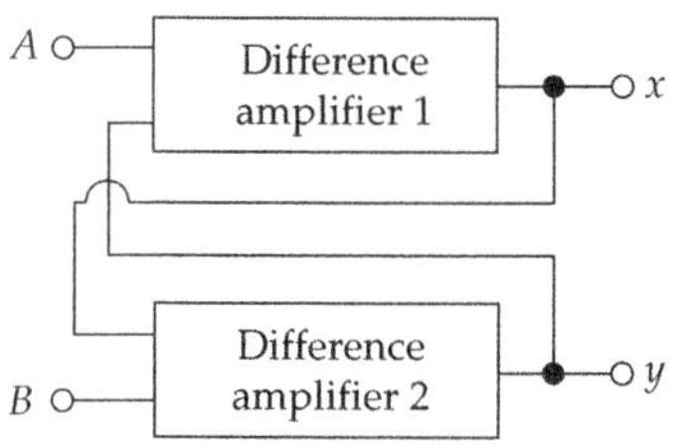

FIGURE P13.74

Design a DAC using a 5 V V_{ref}, with $R = 10$ kΩ that will give a maximum $|V_O| = 12$ V.

Ans. $R_f = 3.375$ kΩ.

P13.74 Consider the equations: $a_1x + a_2y = A$ and $b_1x + b_2y = B$. Express these as $x = \frac{1}{a_1}A - \frac{a_2}{a_1}y$ and $y = \frac{1}{b_1}B - \frac{b_2}{b_1}x$. Verify that the circuit of Figure P13.74 can be used to solve these equations. Design a circuit for solving the equations $2x + 5y = 20$ and $5x - 2y = 21$. Verify your circuit with PSpice simulation.

P13.75 Verify that the circuit of Figure P13.75, with $RC = 1$, can be used to solve the first-order differential equation: $\frac{dy}{dt} + \frac{1}{\tau}y = f(t)$, $t \geq 0$, with given initial value $y(0^+)$.

Design a circuit for solving the differential equation: $\frac{dy}{dt} + \frac{y}{2} = 10\sin 2t$, $t \geq 0$, with $y(0^+) = 1$ V.

Probing Further

P13.76 Consider the inverting configuration using an op amp that is ideal except for finite A_v and a finite differential input resistance R_{id} between the inverting and noninverting inputs of the op amp. Show that $\frac{v_O}{v_{SRC}} = -\frac{R_f}{R_r}\frac{1}{1+1/\beta' A_v}$, where $\frac{1}{\beta'} = 1 + \frac{R_f}{R_r \| R_{id}}$.

P13.77 Consider the inverting configuration connected to a load R_L and using an op amp that is ideal except for finite A_v and a finite output resistance R_o. Show that $\frac{v_O}{v_{SRC}} = -\frac{R_f}{R_r}\frac{1}{1+1/\beta A_v'}$, where $A_v' = -\frac{v_O}{\varepsilon_n}$ is the modified ratio of the output voltage to the differential input and is given by $A_v' = \frac{R_f \| R_L}{R_o + R_f \| R_L}A_v - \frac{R_o \| R_L}{R_f + R_o \| R_L} \cong \frac{R_L}{R_o + R_L}A_v$. (*Hint*: Apply the substitution theorem and superposition.)

P13.78 The circuit of Figure P13.78 is the basic circuit of a **charge amplifier** that may be used with a piezoelectric transducer. This consists of a special type of crystal, such as barium titanate, which, when subjected to a force between two opposite faces of the crystal, generates a voltage between these surfaces. Electrically, the crystal can be represented by the dashed rectangle, with v_{src}, C_{src}, and R_{src} representing, respectively, the voltage generated, and the capacitance and resistance appearing between the crystal faces.

A change in either v_{src} or C_{src} changes the charge on C_{src}. This charge is transmitted to C, thereby causing a change in the output voltage. Show that $\Delta v_O = -\frac{\Delta C_{src}}{C}v_{SRC} - \frac{C_{src}}{C}\Delta v_{SRC}$.

P13.79 Derive a circuit using two integrators, along the same lines as in Problem P13.75, to solve the second-order differential equation: $\frac{d^2y}{dt^2} + a\frac{dy}{dt} + by = f(t)$, $t \geq 0$, with given initial conditions for $y(0^+)$ and $(dy/dt)_{0^+}$.

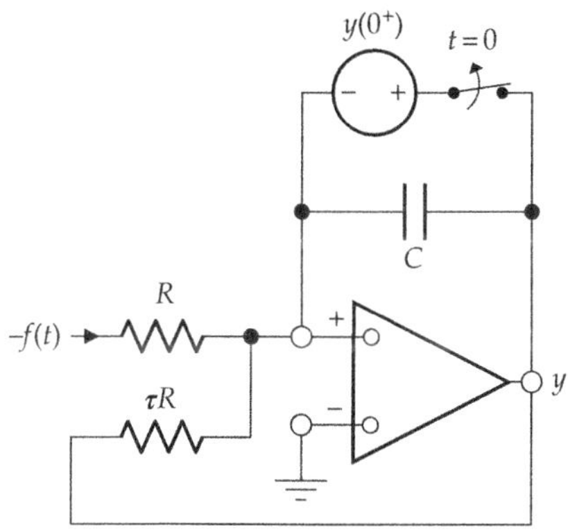

FIGURE P13.75

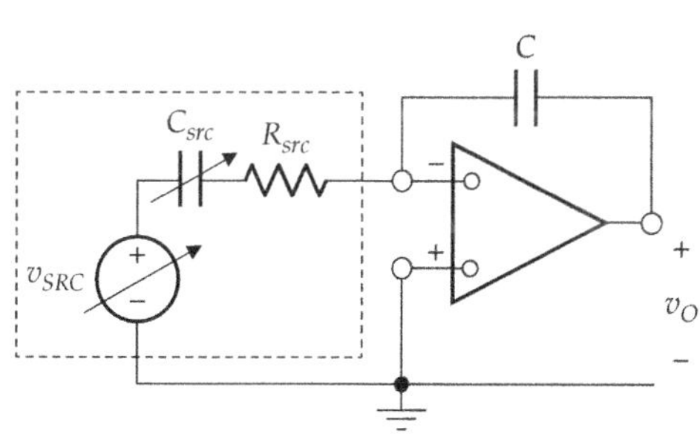

FIGURE P13.78

14

Frequency Responses

Objective and Overview

A frequency-selective circuit or filter is a circuit intended to provide an output that varies with frequency in a desired manner. Frequency-selective circuits are an important class of electric circuits that are extensively used in communications, control, and signal processing systems. Understanding the frequency response of such circuits is basic to designing frequency selective circuits that behave in a desired manner.

The study of frequency response is important for another reason. Energy storage elements are present in all but the simplest physical systems, whether electrical or nonelectrical in nature. These energy storage elements make the response frequency dependent. In particular, the response of practically all physical systems is reduced at high frequencies because of inherent physical processes that can be represented by shunt capacitance in the equivalent electric circuit. The responses of some systems show pronounced peaks, or resonance, at particular frequencies, which can have a major impact on system performance. The frequency response is therefore relevant to understanding the behavior of physical systems and for taking measures to improve their performance.

This chapter is concerned with the basic frequency responses of first-order and second-order circuits, with emphasis on understanding the fundamental concepts and physical interpretations behind these responses. The chapter begins with some basic assumptions in filter analysis, followed by presenting the responses of ideal filters. First-order filters are then discussed and Bode plots are introduced as a very useful tool for describing the frequency response. The four types of second-order responses are all analyzed in detail in terms of a series *RLC* circuit. Duality is invoked in deriving the responses to parallel first-order and second-order circuits. The chapter ends by examining some general features of frequency responses.

14.1 Analysis of Filters

The passive filters considered in this chapter consist, in general, of resistors, capacitors, inductors, and dependent sources. Independent source excitation is only present at the input terminals and the output is taken at a designated pair of terminals.

The following assumptions are made in the basic analysis of filters:

1. The input to the filter is a sinusoidal function of time having a fixed amplitude and phase angle, but a frequency that can be varied over a range that may extend from zero (dc) to a virtually infinite value.
2. The input is assumed to have been applied for a long time, so that the filter output is derived under sinusoidal, steady-state conditions.

It follows from these assumptions that phasor methods (Chapter 8) provide an appropriate tool for analyzing filters. The filter is represented in the frequency domain, and all currents and voltages are designated as phasors. The object of the analysis is to determine how the output of the filter varies in amplitude and phase as the frequency of the input is varied over a specified range, while keeping the amplitude and phase of the input constant.

Because variation with frequency is of primary interest, it is customary to represent phasor voltage and currents in a filter as functions of $j\omega$. This is illustrated in Figure 14.1, where the filter input and output are denoted as $V_I(j\omega)$ and $V_O(j\omega)$, respectively, bearing in mind that the filter input or output could be current rather than voltage. The frequency response of the filter is generally expressed as the ratio of the output voltage or current to the input voltage or current and is referred to as the **transfer function**, $H(j\omega)$, of the filter. Although the transfer function has a special meaning that will be elaborated in Section 22.3, in connection with the Laplace transform, we will, for present purposes, define the transfer function $H(j\omega)$ in the frequency domain as follows:

Definition: *The transfer function $H(j\omega)$ in the frequency domain is the ratio of the phasor of the designated output to the phasor of a single input applied to the filter.*

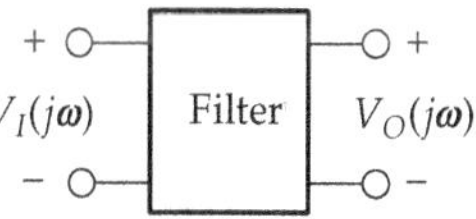

FIGURE 14.1
Filter input and output.

Thus, for the filter of Figure 14.1,

$$H(j\omega) = |H(j\omega)|\angle\theta = \frac{|V_O(j\omega)|\angle\theta_O}{|V_I(j\omega)|\angle\theta_I} \quad (14.1)$$

The frequency response of the filter is therefore expressed in terms of the magnitude $|H(j\omega)|$ of the transfer function and its phase angle θ. It should be emphasized that the transfer function expressed by Equation 14.1 is derived under two conditions: (1) the filter does not have any independent sources other than at the input and (2) the filter is in a sinusoidal steady state, which implies that any initial conditions in the circuit have died out.

14.2 Ideal Frequency Responses

Frequency selectivity arises because of the presence of energy storage elements, which could be capacitors or inductors. As explained in Section 8.4, these energy storage elements, by virtue of their reactance, make the voltages and currents in a circuit depend on frequency. It is this frequency dependence that underlies the frequency selectivity of filters, as will be demonstrated in the following sections.

There are four basic frequency responses of filters, as illustrated in Figure 14.2a through d, in terms of $|H(j\omega)|$ for the *ideal* cases. Figure 14.2a is the response of an ideal **low-pass filter**. As its name implies, all frequencies between 0 (dc) and a **cutoff frequency** ω_{cl} are passed through the filter without attenuation, that is, with $|H(j\omega)| = 1$, which makes the magnitude of the output equal to that of the input. This range of frequencies from zero to ω_{cl} is the **passband**. At frequencies above ω_{cl}, the output is zero, that is, $|H(j\omega)| = 0$. These frequencies constitute the **stopband**.

Figure 14.2b is that of an ideal **high-pass filter**. $|H(j\omega)| = 0$ over the stopband, from dc up to a cut-off frequency ω_{ch}, but $|H(j\omega)| = 1$ in the passband, for frequencies greater than ω_{ch}. The response of Figure 14.2c is that of an ideal **bandpass filter**. The passband, having $|H(j\omega)| = 1$, is for frequencies in the range ω_{cl} to ω_{ch}, whereas the stopband, having $|H(j\omega)| = 0$, extends over the lower and higher frequencies outside this range. Finally, the response of Figure 14.2d is that of an ideal **bandstop filter**. The stopband, having $|H(j\omega)| = 0$, is for frequencies in the range ω_{cl} to ω_{ch}, whereas the passband, having $|H(j\omega)| = 1$, extends over the lower and higher frequencies outside this range. It should be emphasized that the responses in Figure 14.2a through d are ideal and cannot be physically realized. However, they can be closely approached by various filter designs.

The four types of filter responses illustrated in Figure 14.2 are interrelated. For example, if a low-pass response of cutoff frequency ω_{cl} is subtracted from unity, a high-pass response of the same cutoff frequency is obtained and conversely (Figure 14.3a). Similarly, if a bandpass response is subtracted from unity, a bandstop response of the same cutoff frequencies is obtained and conversely (Figure 14.3b). In this sense, the low-pass response is the complement of the high-pass response with respect to unity and conversely, as will be demonstrated later for a first-order circuit. Similarly, the bandpass response is the complement of the bandstop response with respect to unity and conversely, as will be demonstrated later for a second-order circuit.

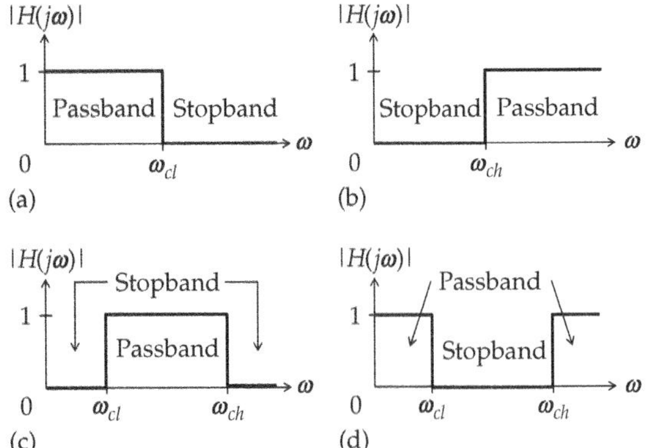

FIGURE 14.2
Ideal frequency responses. (a) Lowpass, (b) highpass, (c) bandpass, and (d) bandstop.

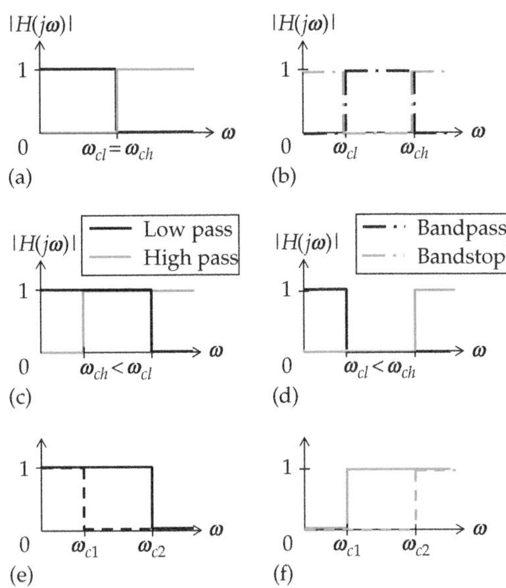

FIGURE 14.3
Interrelations between ideal frequency responses having the cutoff frequencies indicated. (a) Low pass or high pass response subtracted from unity, (b) bandpass or bandstop response subtracted from unity, (c) low pass and high pass response multiplied together, (d) low pass and high pass response added together, (e) low pass response subtracted from another low pass response, and (f) high pass response subtracted from another high pass response.

Consider next a low-pass response of cutoff frequency ω_{cl} and a high-pass response of cutoff frequency ω_{ch}, with $\omega_{ch} < \omega_{cl}$ (Figure 14.3c). If these two responses are multiplied together, the result is a bandpass response having a passband from ω_{ch} to ω_{cl}. This follows from the fact that for $\omega < \omega_{ch}$, the zero response of the high-pass filter when multiplied by the unity response of the low-pass filter gives a zero response. Similarly, for $\omega > \omega_{cl}$, the zero response of the low-pass filter when multiplied by the unity response of the high-pass filter gives a zero response. But for $\omega_{ch} < \omega < \omega_{cl}$, both responses are unity, and their product is unity.

If a low-pass response of cutoff frequency ω_{cl} is added to a high-pass response of cutoff frequency ω_{ch}, with $\omega_{cl} < \omega_{ch}$, the resulting response is bandstop having a stopband from ω_{cl} to ω_{ch} (Figure 14.3d). This is because the response is zero in the range $\omega_{cl} < \omega < \omega_{ch}$, where both responses are zero. Outside this range, one of the responses is unity, so their sum is unity.

Finally, a bandpass response results from subtracting a low-pass response of cutoff frequency ω_{c1} from another low-pass response of cutoff frequency $\omega_{c2} > \omega_{c1}$ (Figure 14.3e). Similarly, a bandpass response results from subtracting a high-pass response of cutoff frequency ω_{c2} from another high-pass response of cutoff frequency $\omega_{c1} < \omega_{c2}$ (Figure 14.3f).

The interrelation between the various frequency responses will be illustrated and utilized in this and the following chapter.

14.3 First-Order Responses

Prototypical first-order filters consist of a single resistor and a single energy storage element, which could be a capacitor or an inductor. They are referred to as first order because their responses obey a first-order, linear, ordinary differential equation with constant coefficients, as discussed in Chapter 11.

A prototypical, first-order, series capacitive filter is illustrated in Figure 14.4a in the frequency domain, where the input voltage $V_I(j\omega)$ is the source voltage $V_{SRC}(j\omega)$ and the output voltage $V_O(j\omega)$ is the voltage $V_C(j\omega)$ across the capacitor. From voltage division,

$$H_C(j\omega) = \frac{V_C(j\omega)}{V_{SRC}(j\omega)} = \frac{1/j\omega C}{R + 1/j\omega C} = \frac{1}{1 + j\omega CR} \tag{14.2}$$

or

$$\left|H_C(j\omega)\right| = \frac{\left|V_C(j\omega)\right|}{\left|V_{SRC}(j\omega)\right|} = \frac{1}{\sqrt{1+\omega^2C^2R^2}} \quad \text{and}$$

$$\angle H_C(j\omega) = -\tan^{-1}\omega CR \tag{14.3}$$

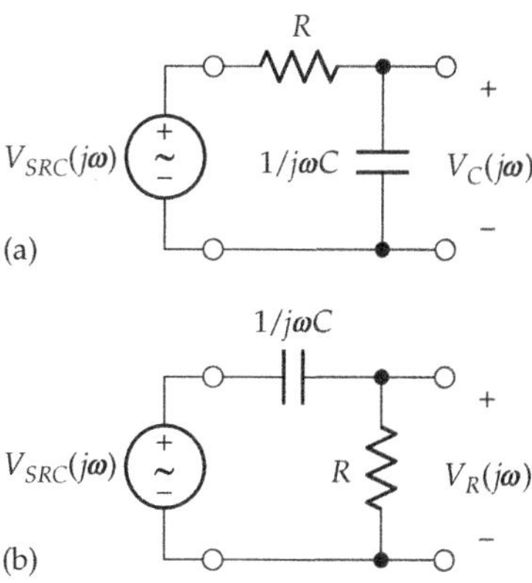

FIGURE 14.4
RC low-pass filter (a) and high-pass filter (b).

Assume that $V_{SRC}(j\omega)$ remains constant as its frequency is varied. When $\omega = 0$, the capacitor acts as an open circuit in the steady state, no current flows, $V_C(j\omega) = V_{SRC}(j\omega)$, and $H_C(j\omega) = 1$. As the frequency increases, $|V_C(j\omega)|$ decreases continuously, in accordance with Equation 14.3. When $\omega \to \infty$, the reactance of the capacitor approaches zero, that is, the capacitor acts as a short circuit, $V_C(j\omega) \to 0$, and $H_C(j\omega) \to 0$. As discussed in Section 14.2, the response is low pass, since high-frequency signals are attenuated, whereas low-frequency signals are transmitted with little attenuation.

If the output is taken across the resistor, as illustrated in Figure 14.4b for the same circuit but redrawn, then

$$H_R(j\omega) = \frac{V_R(j\omega)}{V_{SRC}(j\omega)} = \frac{R}{R + 1/j\omega C} = \frac{j\omega CR}{1 + j\omega CR} \tag{14.4}$$

$$\left|H_R(j\omega)\right| = \frac{\left|V_R(j\omega)\right|}{\left|V_{SRC}(j\omega)\right|} = \frac{\omega CR}{\sqrt{1+\omega^2C^2R^2}} \quad \text{and}$$

$$\angle H_C(j\omega) = 90^\circ - \tan^{-1}\omega CR \tag{14.5}$$

When $\omega = 0$, no current flows, and $V_R(j\omega) = 0$. As the frequency increases, $\left|V_R(j\omega)\right|$ increases continuously. When $\omega \to \infty$, the reactance of C approaches zero, and $V_R(j\omega) \to V_{SRC}(j\omega)$. The response is now high pass, since low-frequency signals are attenuated, whereas high-frequency signals are transmitted with little attenuation.

Similar responses are obtained from an *RL* circuit (Figure 14.5). Following the same procedure as in the case of the capacitive circuit, it is seen that

$$H_R(j\omega) = \frac{V_R(j\omega)}{V_{SRC}(j\omega)} = \frac{R}{R + j\omega L} = \frac{1}{1 + j\omega L/R} \tag{14.6}$$

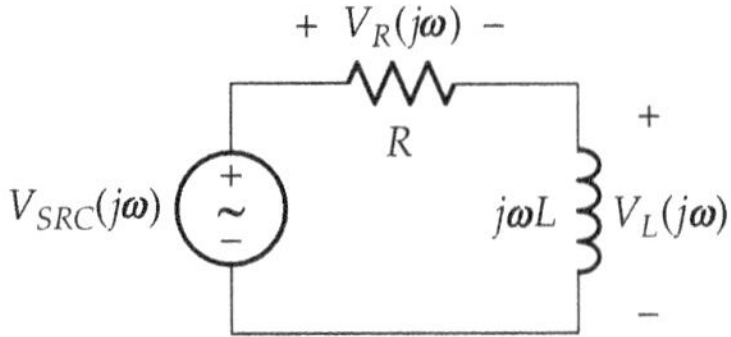

FIGURE 14.5
RL filter.

$$|H_R(j\omega)| = \frac{|V_R(j\omega)|}{|V_{SRC}(j\omega)|} = \frac{1}{\sqrt{1+\omega^2L^2/R^2}} \quad \text{and}$$

$$\angle H_R(j\omega) = -\tan^{-1}\omega L/R \tag{14.7}$$

The response across the resistor is now low pass, since at $\omega = 0$ the inductor acts as a short circuit, $V_R(j\omega) = V_{SRC}(j\omega)$, and $H_R(j\omega) = 1$. As $\omega \to \infty$, the inductor acts as an open circuit, $V_R(j\omega) \to 0$, and $H_R(j\omega) \to 0$.

If the response is taken across the inductor,

$$H_L(j\omega) = \frac{V_L(j\omega)}{V_{SRC}(j\omega)} = \frac{j\omega L}{R + j\omega L} \tag{14.8}$$

$$|H_L(j\omega)| = \frac{|V_L(j\omega)|}{|V_{SRC}(j\omega)|} = \frac{\omega L/R}{\sqrt{1+\omega^2L^2/R^2}} \quad \text{and}$$

$$\angle H_R(j\omega) = 90° - \tan^{-1}\omega L/R \tag{14.9}$$

The response across the inductor is high pass. At $\omega = 0$, the inductor acts as a short circuit so that $V_L(j\omega) = 0$, and $H_L(j\omega) = 0$. As $\omega \to \infty$, the inductor acts as an open circuit, $V_L(j\omega) \to V_{SRC}(j\omega)$, and $H_L(j\omega) \to 1$.

Since the basic first-order circuits of Figures 14.4 and 14.5 consist of two circuit elements, only two outputs are possible from each circuit, one output being low pass, the other high pass. This is a general property of first-order filters:

Concept: *The frequency response of a first-order filter is either low pass or high pass, the variation of the response with frequency being due to the frequency-dependent reactance of energy storage elements.*

There are no first-order bandpass or bandstop responses, as these responses require two transitions between a passband and a stopband. This cannot be achieved by a single energy storage element, as will be clarified later.

14.3.1 Parallel First-Order Filters

The parallel circuit duals of the first-order series circuits of Figures 14.4 and 14.5 are shown in Figure 14.6.

In Figure 14.6a and at $\omega = 0$, the inductor acts as a short circuit, and $I_{Lp}(j\omega) = I_{SRC}(j\omega)$. As $\omega \to \infty$, the inductor acts as an open circuit, and $I_{Gp}(j\omega) = I_{SRC}(j\omega)$. It is to be expected, therefore, that the response $I_{Lp}(j\omega)$ is low pass and the response $I_{Gp}(j\omega)$ is high pass. The analytical expressions can be derived from current division or from duality. The circuit of Figure 14.6a is the dual of that of Figure 14.4, the dual of $V_C(j\omega)$ being $I_{Lp}(j\omega)$ and the dual of $V_R(j\omega)$ being $I_{Gp}(j\omega)$. Replacing C by L_p and R by G_p in Equation 14.2,

$$H_{Lp}(j\omega) = \frac{I_{Lp}(j\omega)}{I_{SRC}(j\omega)} = \frac{1/j\omega L_p}{G_p + 1/j\omega L_p} = \frac{1}{1 + j\omega L_p G_p} \tag{14.10}$$

Equation 14.4 becomes

$$H_{Gp}(j\omega) = \frac{I_G(j\omega)}{I_{SRC}(j\omega)} = \frac{G_p}{G_p + 1/j\omega L_p} = \frac{j\omega L_p G_p}{1 + j\omega L_p G_p} \tag{14.11}$$

It is seen that the response $H_{Lp}(j\omega)$ is low pass and the response $H_{Gp}(j\omega)$ is high pass, as argued previously for a first-order circuit.

In Figure 14.6b and at $\omega = 0$, the capacitor acts as an open circuit, and $I_{Gp}(j\omega) = I_{SRC}(j\omega)$. As $\omega \to \infty$, the capacitor acts as a short circuit, and $I_{Cp}(j\omega) = I_{SRC}(j\omega)$. It is to be expected, therefore, that the response $I_{Gp}(j\omega)$ is low pass and the response $I_{Cp}(j\omega)$ is high pass. The analytical expressions can be derived from current division or from duality. The circuit of Figure 14.6b is the dual of that of Figure 14.5, the dual of $V_L(j\omega)$ being $I_{Cp}(j\omega)$ and the dual of $V_R(j\omega)$ being $I_{Gp}(j\omega)$. Replacing L by C_p and R by G_p in Equation 14.6,

$$H_{Gp}(j\omega) = \frac{I_{Gp}(j\omega)}{I_{SRC}(j\omega)} = \frac{G_p}{G_p + j\omega C_p} = \frac{1}{1 + j\omega C_p/G_p} \tag{14.12}$$

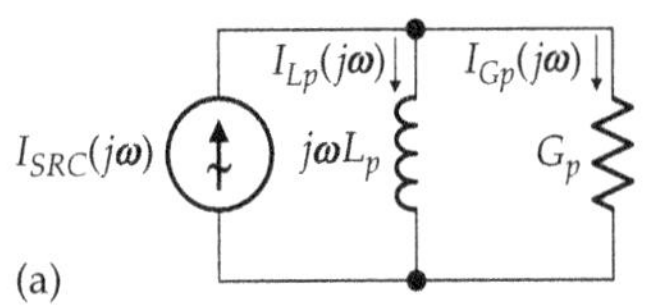

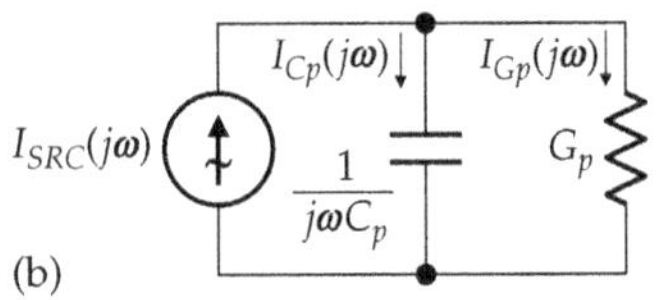

FIGURE 14.6
First-order, parallel *RL* filter (a) and *RC* filter (b).

Equation 14.8 becomes

$$H_{Cp}(j\omega) = \frac{I_{Cp}(j\omega)}{I_{SRC}(j\omega)} = \frac{j\omega C_p}{G_p + j\omega C_p} \quad (14.13)$$

It is seen that the response $H_{Gp}(j\omega)$ is low pass and the response $H_{Cp}(j\omega)$ is high pass, as argued previously for a first-order circuit.

It will be noted that in all the first-order responses considered, where the response across one element is low pass, the response across the other element is high pass, with these two responses being complementary with respect to the applied source. In Figure 14.4, for example, $\frac{V_C(j\omega)}{V_{SRC}(j\omega)} + \frac{V_R(j\omega)}{V_{SRC}(j\omega)} = 1$. Similar considerations apply to the inductive filter of Figure 14.5 and to the parallel circuits of Figure 14.6. This is in accordance with what was mentioned in Section 14.2 about the low-pass and high-pass responses being the complement of one another. In a first-order circuit, the sum of the low-pass and high-pass voltage or current responses is equal to the applied source excitation. The low-pass and high-pass responses are therefore the complement of one another with respect to the applied source excitation.

Although it is straightforward to plot the magnitudes and phase angles of the aforementioned responses as a function of ω, it is more useful and convenient, for reasons that will become clear later, to use logarithmic rather than linear plots. These plots are known as **Bode plots** after Hendrik W. Bode, the engineer who introduced them. We will therefore digress a little to discuss Bode plots.

Primal Exercise 14.1

Draw the circuit of simple first-order *RL* filter having the transfer function $\frac{1}{2(1 + j\omega/2)}$, where ω is in rad/s, using a 1 H inductor.

Ans. The circuit is shown in Figure 14.7.

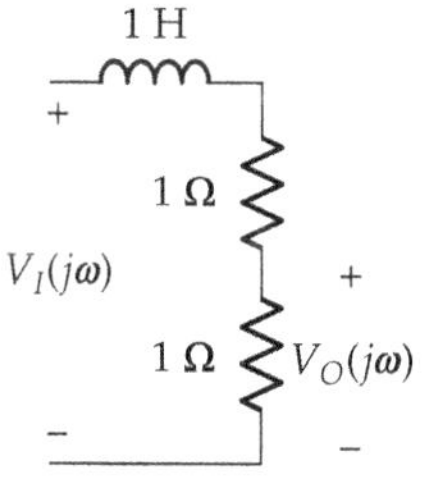

FIGURE 14.7
Figure for Primal Exercise 14.3.

14.4 Bode Plots

Consider by way of illustration the function $y = ax^n$. Taking logarithms to base 10 of both sides: $\log_{10}y = \log_{10}a + n\log_{10}x$. The plot of $\log_{10}y$ vs. $\log_{10}x$ is a line of slope n. *The coefficient 'a' shifts the line vertically, without affecting its slope.* If $y = 0.1x^2$, for example, and x assumes the values $x_1 = 10$, $x_2 = 100$, and $x_3 = 1000$, $\log_{10}x$ of these values of x are 1, 2, and 3, respectively. The corresponding values of y are $y_1 = 10$, $y_2 = 1000$, and $y_3 = 10^5$; $\log_{10}y$ of these values of y are 1, 3, and 5, respectively. The equation of the line is $\log_{10}y = \log_{10}0.1 + 2\log_{10}x = -1 + 2\log_{10}x$. This line, plotted in Figure 14.8, has a slope of 2, which is the power of x. The line $\log_{10}y = 2\log_{10}x$ is shifted downward by 1 because of the 0.1 multiplier.

It should be noted that *equal spacing along a logarithmic axis is equivalent to equal ratios of the variable involved.* Considering the horizontal axis for example, the intercepts of 'P$_1$', 'P$_2$', and 'P$_3$' on this axis are $\log_{10}x_1 = 1$, $\log_{10}x_2 = 2$, and $\log_{10}x_3 = 3$, respectively. These are equally spaced along the logarithmic axis, that is,

$$\log_{10}x_2 - \log_{10}x_1 = \log_{10}x_3 - \log_{10}x_2 = 1$$

corresponding to

$$\frac{x_2}{x_1} = \frac{100}{10} = 10 \quad \text{and} \quad \frac{x_3}{x_2} = \frac{1000}{100} = 10$$

That is, the equal spacing along the logarithmic horizontal axis is equivalent to equal ratios of the x variable, similarly for the vertical axis and the y variable.

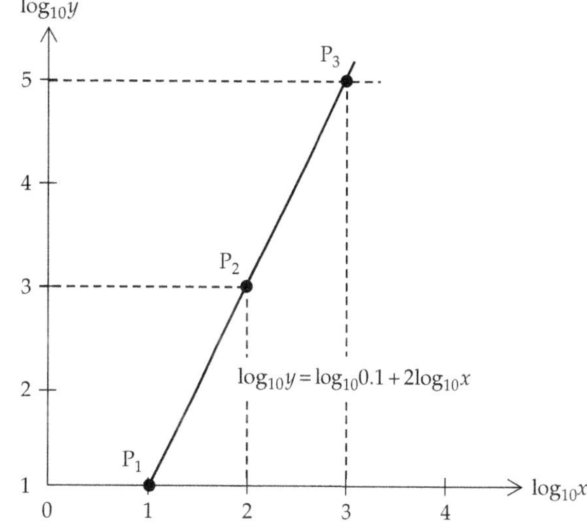

FIGURE 14.8
Logarithmic plot of $y = 0.1x^2$.

Figure 14.9 illustrates an alternative but a more convenient way of plotting $\log_{10}$ of variables. In Figure 14.8, the axes are labeled as $\log_{10}x$ or $\log_{10}y$ and are scaled *linearly* in terms of $\log_{10}x$ or $\log_{10}y$. In contrast, the axes in Figure 14.9 are labeled as x or y but are scaled *logarithmically as* $\log_{10}x$ *or* $\log_{10}y$. Thus, the markings on the x-axis are x values of 1, 10, 10^2, 10^3, etc. But the markings are equidistant along the x-axis, corresponding to a logarithmic scaling on this axis. Each interval between successive powers of ten is also scaled logarithmically. Thus, the interval 1–10 is shown expanded in Figure 14.9. If the length of this interval is denoted as 'mn', the marked value of 2, for example, is scaled as $\log_{10}2 = 0.30$ and is placed at a distance of 0.30 × 'mn' to the right of 'm'. Similarly, the marked value of 5 is scaled as $\log_{10}5 = 0.70$ and is placed at a distance of 0.70 × 'mn' to the right of 'm'. Clearly, this is a more convenient way of plotting logarithmic values, because the values of the variables are entered directly, without the need to take their logarithms. 'P_2', for example, is entered as (100, 1000) in Figure 14.9, rather than ($\log_{10}100$, $\log_{10}1000$) in Figure 14.8. The plot is, of course, the same straight line $\log_{10}y = \log_{10}0.1 + 2\log_{10}x$ in both cases. The slope of the line in Figure 14.9 is $(\log_{10}10^5 - \log_{10}10^3)/(\log_{10}10^3 - \log_{10}10^2) = \log_{10}100/\log_{10}10 = 2$, as in Figure 14.8.

In Bode magnitude plots, the horizontal axis is labeled in terms of ω but is scaled logarithmically as $\log_{10}\omega$. However, the vertical axis is labeled and scaled *linearly* as $20\log_{10}|H(j\omega)|$, so the plot is still log–log. In general, the expression $20\log_{10}B$, where B is the magnitude of a transfer function in terms of voltage or current, denotes the dB value of B, where dB is the abbreviation of **decibel**. This representation originated in the early days of telephony for specifying power loss in cascaded circuits, in which the overall power loss is the product of the ratios of the power outputs to the power inputs of the individual circuits. Using logarithms was convenient because (1) it meant that the logarithms of the power ratios of the individual circuits could be added to obtain the logarithm of the overall power loss of the cascaded circuits and (2) it allowed specifying a large range of power loss or gain as a much smaller range of the logarithm of these values. The decibel was defined as $10\log_{10}A$, where A is a power ratio. Since the power ratio for a given resistor is proportional to the square of the ratio of voltage or current, $20\log_{10}$ is used with the ratios of voltages and currents. Thus, voltage ratios of 10^3, 10^4, and 10^5 are expressed as 60 dB, 80 dB, and 100 dB, respectively. The advantages of using Bode plots for describing the responses of frequency-selective circuits will become clear from future discussions.

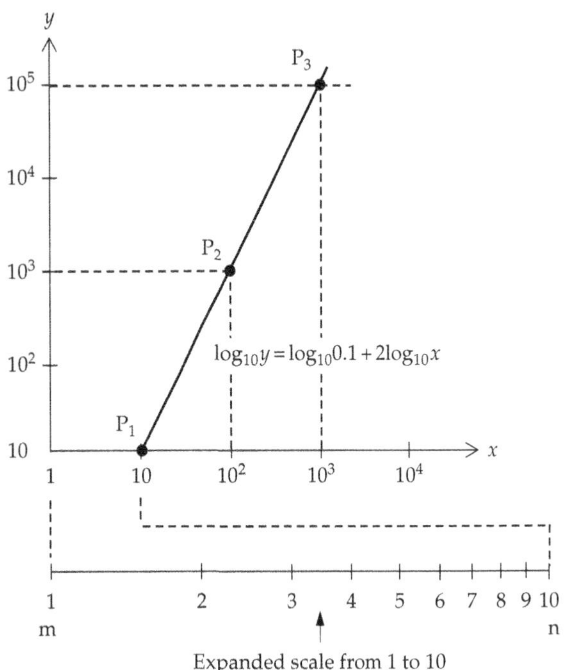

FIGURE 14.9
Alternative logarithmic plot of $y = 0.1x^2$.

Primal Exercise 14.2

Scale and mark the x-axis and y-axis logarithmically as in Figure 14.9, over a suitable range, plot the function $y = 0.2x^{1.5}$ by locating points on the line, and verify that the slope of the line is 1.5.

Primal Exercise 14.3

(a) Convert the following voltage ratios to dB: (i) 4, (ii) 1/4, (iii) 40; (b) convert the following dB values to voltage ratios: (i) 2 dB, (ii) −4 dB, (iii) 6 dB.

Ans. (a) (i) 12.04 dB, (ii) −12.04 dB, (iii) 32.04 dB; (b) (i) 1.259, (ii) 0.6310, (iii) 1.995.

14.4.1 Low-Pass Response

$H(j\omega)$ for the first-order low-pass response can be expressed in normalized form by replacing the time constants RC and L/R by $1/\omega_{cl}$ in all the low-pass transfer functions derived so far. All these transfer functions take the form

$$H(j\omega) = \frac{1}{1 + j\omega/\omega_{cl}} \tag{14.14}$$

where

$\omega_{cl} = 1/CR$ for the series RC circuit
$\omega_{cl} = R/L$ for the series RL circuit
$\omega_{cl} = 1/L_pG_p$ for the parallel GL circuit
$\omega_{cl} = G_p/C_p$ for the parallel GC circuit

The magnitude of the response is

$$\left|H(j\omega)\right| = \frac{1}{\sqrt{1+\left(\omega/\omega_{cl}\right)^2}} \tag{14.15}$$

It is desired to plot the magnitude of the response as dB vs. $\log_{10}\omega$. A convenient feature of Bode plots is that they can be approximated by straight-line asymptotes for limiting values of ω. To illustrate this feature, consider first the very low values of ω, for which $\omega/\omega_{cl} \ll 1$. Equation 14.15 reduces to

$$\left|H(j\omega)\right| = 1, \quad \text{as } \omega \to 0 \tag{14.16}$$

In terms of dB, taking $\log_{10}$ of both sides of Equation 14.16 and multiplying by 20, $20\log_{10}|H(j\omega)| = 20\log_{10}(1) = 0$ dB, which is the equation of the low-frequency asymptote, as $\omega \to 0$. It is, in fact, the horizontal axis or 0 dB line (Figure 14.10a).

For large values of ω, for which $\omega/\omega_{cl} \gg 1$, Equation 14.15 reduces to

$$\left|H(j\omega)\right| = \frac{1}{\sqrt{\left(\omega/\omega_{cl}\right)^2}} = \frac{\omega_{cl}}{\omega}, \quad \text{as } \omega \to \infty \tag{14.17}$$

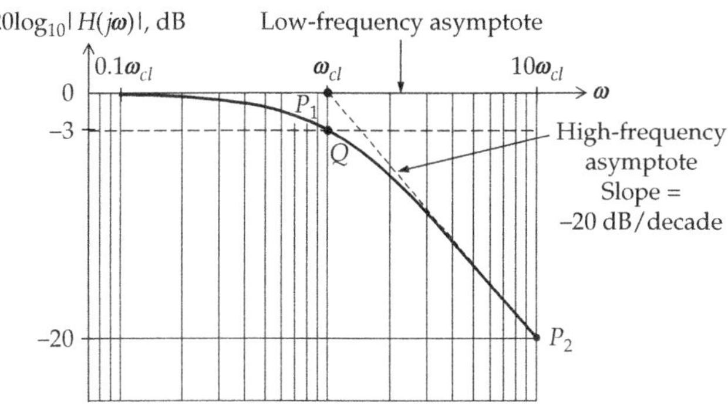

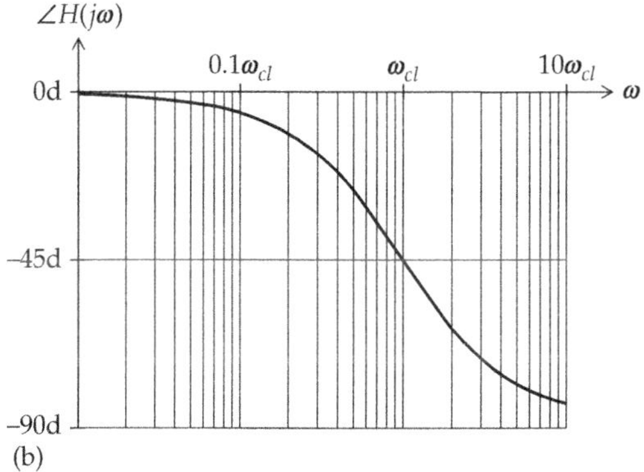

FIGURE 14.10
Bode magnitude plot (a) and Bode phase plot (b) of a first-order, low-pass response.

In terms of dB, taking $\log_{10}$ of both sides of Equation 14.17 and multiplying by 20,

$$20\log_{10}\left|H(j\omega)\right| = 20\log_{10}\omega_{cl} - 20\log_{10}\omega, \quad \text{as } \omega \to \infty \tag{14.18}$$

This is the equation of a straight line of the form $y = a - bx$, where y is $20\log_{10}|H(j\omega)|$, x is $\log_{10}\omega$, and a and b are constants. To draw this high-frequency asymptote, it is sufficient to identify a point on the line and its slope. It is seen that when $\omega = \omega_{cl}$, Equation 14.18 gives

$$20\log_{10}\left|H(j\omega_{cl})\right| = 20\log_{10}\omega_{cl} - 20\log_{10}\omega_{cl} = 0 \text{ dB} \tag{14.19}$$

This means that the high-frequency asymptote passes through the point (ω_{cl}, 0 dB). In other words, the high-frequency asymptote intersects the horizontal axis, that is, the 0 dB line, at $\omega = \omega_{cl}$, as indicated by point 'P_1' in Figure 14.10a.

To determine the slope of the line, the usual procedure can be followed by identifying the two points (x_1, y_1) and (x_2, y_2) on the line, the slope being $(y_2 - y_1)/(x_2 - x_1)$. We have already identified one point $P_1(x_1, y_1)$ as (ω_{cl}, 0 dB). To identify a second point 'P_2', consider $\omega = 10\omega_{cl}$. Substituting this in Equation 14.18,

$$20\log_{10}\left|H(j10\omega_{cl})\right| = 20\log_{10}\omega_{cl} - 20\log_{10}10\omega_{cl}$$
$$= -20\log_{10}10 = -20 \text{ dB} \tag{14.20}$$

It is seen that $(y_2 - y_1) = -20 - 0 = -20$ dB. As for $(x_2 - x_1)$, note that in moving from 'P_1' to 'P_2', the frequency is increased from ω_{cl} to $10\omega_{cl}$. In frequency terms, multiplying the frequency by 10 is described as moving up a *decade* of frequency, whereas dividing the frequency by 10 is moving down a decade of frequency. The slope of the high-frequency asymptote can therefore be specified as −20 dB/decade, as indicated in Figure 14.10a.

We have thus far identified the low-frequency and high-frequency asymptotes of the Bode magnitude plot of the first-order, low-pass response, but have not yet identified any points on the plot itself. A salient point on this plot is obtained by substituting $\omega = \omega_{cl}$ in Equation 14.15. This gives

$$\left|H(j\omega_{cl})\right| = \frac{1}{\sqrt{2}} \quad \text{and} \quad 20\log_{10}\left|H(j\omega_{cl})\right| = -3.0103 \cong -3\text{dB} \tag{14.21}$$

A point on the Bode magnitude plot is therefore 'Q' having coordinates (ω_{cl}, −3 dB), as in Figure 14.10a. Because of this, ω_{cl} is referred to as the **3 dB cutoff frequency**. A synonymous term is the **half-power frequency**, because if we imagine the input to be of constant amplitude but variable frequency and the output to be across a fixed resistor R, then at very low frequencies the output voltage may be V_O rms, and the average power dissipated in R is $P = V_O^2/R$. At $\omega = \omega_{cl}$, the output voltage drops to $V_O/\sqrt{2}$ rms, and the power becomes $P/2$; hence, the half-power designation. Because ω_{cl} is the intersection of the two asymptotes, it is also the **corner frequency**. However, it should be noted that whereas the half-power frequency is, by definition, the same as the 3 dB cutoff frequency, the corner frequency is not always the same as the 3 dB cutoff frequency, as will be clarified later.

With the asymptotes and the 3 dB cutoff point defined, the low-pass Bode magnitude plot of Figure 14.10a can be sketched with reasonable accuracy. The accuracy can be improved by noting the following: (1) The −3 dB at $\omega = \omega_{cl}$ represents the maximum deviation of the plot from the asymptotes; (2) at a frequency of $0.5\omega_{cl}$, $20\log_{10}\left|H(j\omega)\right| = 20\log_{10}\left(1/\sqrt{1.25}\right) = -0.969 \cong -1\,\text{dB}$, that is, 1 dB below the low-frequency asymptote; and (3) at a frequency of $2\omega_{cl}$, $20\log_{10}|H(j\omega)|$ is approximately 1 dB below the high-frequency asymptote (Exercise 14.8). Changing the frequency by a factor of 2 is described as moving through an *octave* of frequency.

As for the Bode phase plot, it follows from Equation 14.14 that the phase angle of $H(j\omega)$ is that of the numerator, which is zero, minus that of the denominator, which is arctan of the ratio of the imaginary part to the real part. Thus,

$$\angle H(j\omega) = -\tan^{-1}(\omega/\omega_{cl}) \tag{14.22}$$

When $\omega = 0, \angle H(j\omega) = 0$. At $\omega = \omega_{cl}, \angle H(j\omega) = -45°$, and as $\omega \to \infty$, $\angle H(j\omega) \to -90°$ (Figure 14.10b). At very low frequencies, $\angle H(j\omega) = -\tan^{-1}(\omega/\omega_{cl}) \cong -\omega/\omega_{cl}$.

14.4.2 High-Pass Response

$H(j\omega)$ for the first-order high-pass response may be expressed in normalized form by replacing the time constants RC, L/R, L_pG_p, and C_p/G_p by $1/\omega_{ch}$ in all the high-pass transfer functions derived so far. All these transfer functions take the form

$$H(j\omega) = \frac{j\omega/\omega_{ch}}{1 + j\omega/\omega_{ch}} = \frac{1}{1 + \omega_{ch}/j\omega} \tag{14.23}$$

Note that although ω_{cl} and ω_{ch} are given by the same expression, they refer to two different responses, one low pass, the other high pass (Example 14.1). It follows from Equation 14.23 that the magnitude and phase of the high-pass response can be expressed as

$$\left|H(j\omega)\right| = \frac{1}{\sqrt{1 + (\omega_{ch}/\omega)^2}} \tag{14.24}$$

and

$$\angle H(j\omega) = 90° - \tan^{-1}\left(\frac{\omega}{\omega_{ch}}\right) \tag{14.25}$$

In a manner exactly analogous to the low-pass response, the high-frequency asymptote is obtained by letting $\omega \to \infty$ in Equation 14.24. This gives $|H(j\omega)| = 1$ and $20\log_{10}1 = 0$ so that the high-frequency asymptote on the logarithmic plot is the horizontal axis or 0 dB line. As $\omega \to 0$, $|H(j\omega)| \to \omega/\omega_{ch}$. On the logarithmic plot, the low-frequency asymptote is therefore a line whose equation is

$$20\log_{10}\left|H(j\omega)\right| = -20\log_{10}\omega_{ch} + 20\log_{10}\omega, \quad \omega \to 0 \tag{14.26}$$

Following an argument similar to that used for the low-pass response, it is seen that the low-frequency asymptote is a line of slope of +20 dB/decade that intersects the horizontal axis, that is, the 0 dB line, at $\omega = \omega_{ch}$ (Figure 14.11a). This is also the 3 dB cutoff frequency or the half-power frequency. It is also the corner frequency in this case.

It follows from Equations 14.23 and 14.25 that the Bode phase plot of the high-pass response is that of the low-pass response shifted upward by 90°, because of the $j\omega$ term in the numerator of the transfer function (Figure 14.11b).

The following should be noted concerning first-order responses and the 3 dB cutoff frequencies:

1. In all the preceding first-order transfer functions, the reciprocals of the 3 dB cutoff frequencies are the same as the time constants defined in Chapter 11 for the corresponding circuits. Fundamentally, this is due to the relationship between circuit responses in the time domain and in the frequency domain, as exemplified by the Laplace transform (Chapter 21).
2. It was pointed out in Section 11.5, that the time constant could be derived from the resistance seen by the energy storage element when independent sources are set to zero.

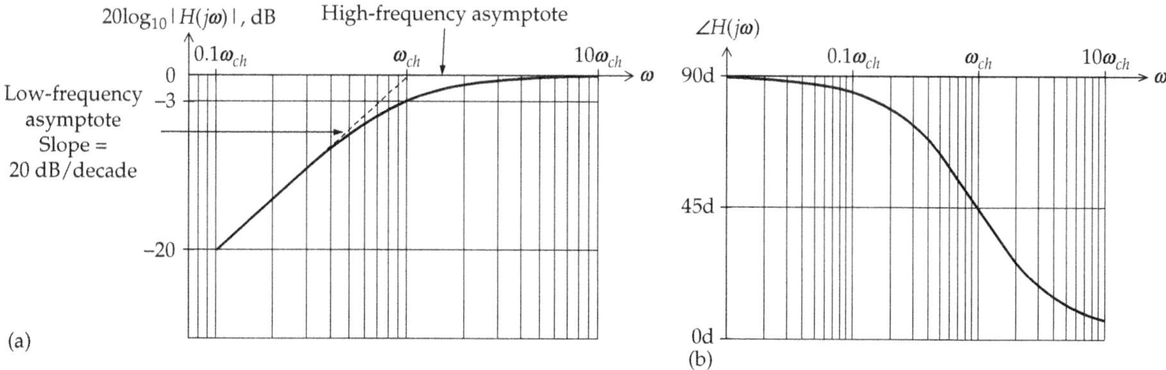

FIGURE 14.11
Bode magnitude plot (a) and Bode phase plot (b) of a first-order, high-pass response.

This resistance is in fact R_{Th} of TEC seen by the energy storage element. The same applies to finding the resistance involved in the 3 dB cutoff frequency.

3. The 3 dB cutoff frequency can also be determined from the transfer function in two other ways:
   a. From the transfer functions in standard form in terms of $j\omega$. In this form the denominator is (1 + $j\omega$ term), where the $j\omega$ term is $j\omega/\omega_{cl}$ for a low-pass response and is $\omega_{ch}/j\omega$ for a high-pass response. It is seen that ω_{cl} is the denominator of the term whose numerator is $j\omega$, and ω_{ch} is the numerator of the term whose denominator is $j\omega$.
   b. From the magnitude of the normalized, standard form of the transfer function by setting this magnitude equal to $1/\sqrt{2}$ and solving for ω. Note that the 3 dB cutoff frequency is defined as 3 dB below *the maximum* value of $20\log_{10}|H(j\omega)|$ in the passband. In the normalized form, the maximum value of $|H(j\omega)|$ is unity so that the maximum value of magnitude of $20\log_{10}|H(j\omega)|$ is 0 dB. If the transfer function is multiplied by a scalar K so that $|H(j\omega)| = K/\sqrt{1+(\omega/\omega_{cl})^2}$ in the case of the low-pass response, for example, then the maximum value of $|H(j\omega)|$ in the passband is K, and the value of this magnitude at the 3 dB cutoff frequency is $K/\sqrt{2}$ and not $1/\sqrt{2}$.
4. The high-pass response (Equations 14.23) can be derived from the low-pass response (Equation 14.14) by interchanging $j\omega$ and ω_c, where ω_c is the corresponding 3 dB cutoff frequency. This is a general principle that will be illustrated in other cases as well.

It is clear from the preceding discussion that first-order responses can only have one cutoff frequency, ω_{cl} or ω_{ch}. On the other hand, bandpass and bandstop responses require two cutoff frequencies, as illustrated in Figure 14.2c and d, because there are two transitions between the passband and the stopband, with each transition involving a cutoff frequency. Hence, there are no first-order bandpass and bandstop filters.

Primal Exercise 14.4

Consider the *RC* circuit of Figure 14.4. Express, in terms of the time constant τ_C, the transfer function when the output is taken across (a) *C* or (b) across *R*; (c) determine the 3 dB cutoff frequency when $R = 100\ \Omega$ and $C = 5\ \mu F$.

Ans. (a) $\dfrac{1}{1+j\omega\tau_C}$; (b) $\dfrac{j\omega\tau_C}{1+j\omega\tau_C}$; (c) 2 krad/s.

Primal Exercise 14.5

Consider the circuit that is the dual of that of Primal Exercise 14.1. (a) Determine the values of L_p, G_p, and the 3 dB cutoff frequency. Express, in terms of the time constant τ_L, the transfer function when the output is taken as the current through (b) L_p, or (c) G_p;

Ans. (a) 5 μH, 100 S, 2 krad/s; (b) $\dfrac{1}{1+j\omega\tau_L}$; (c) $\dfrac{j\omega\tau_L}{1+j\omega\tau_L}$.

Primal Exercise 14.6

Consider that the high-frequency asymptote of Figure 14.10a, in the form $H(j\omega) = \omega_{cl}/j\omega$, is extended to lower and higher frequencies. Determine the dB value at (a) $\omega = 0.1\omega_{cl}$ and (b) $100\omega_{cl}$. Do the same for the low-frequency asymptote of Figure 14.11a, in the form

$H(j\omega) = j\omega/\omega_{ch}$, at (a) $\omega = 0.01\omega_{cl}$ and (b) $10\omega_{cl}$. Interpret all these results in terms of the slopes of the asymptotes and decade changes of frequency.

Ans. High-frequency asymptote: (a) 20 dB; (b) −40 dB. Low-frequency asymptote: (a) −40 dB; (b) 20 dB.

Exercise 14.7

Show that a slope of 20 dB/decade is equivalent to 6 dB/octave.

Exercise 14.8

Show that (a) in a low-pass Bode magnitude plot, the response an octave higher than ω_{cl} is 1 dB below the high-frequency asymptote, and (b) in a high-pass Bode magnitude plot, the response an octave lower than ω_{cl} is 1 dB below the low-frequency asymptote.

Example 14.1: First-Order Responses

Given $R = 1$ kΩ and $C = 0.5$ μF in Figure 14.4. Determine L in Figure 14.5 for a corner frequency that is 10 times that of the RC filter, assuming a resistance of 2 kΩ. Obtain by simulation both the low-pass and the high-pass responses of the RL filter.

Solution:

$\omega_C = 1/RC = 1/(10^3 \times 0.5 \times 10^{-6}) = 1000/0.5 = 2$ krad/s. $\omega_L = 20$ krad/s $= R/L$. Hence, $L = (2\text{ k}\Omega)/(20\text{ krad/s}) = 0.1$ H.

Simulation: The circuit is entered as in Figure 14.12, using the source VAC of default magnitude 1 V. The resistor is entered as part 'r' from the ANALOG_P library. This resistor has its terminals marked 1 and 2 to facilitate plotting the responses. To run the simulation, select 'AC Sweep/Noise' for 'Analysis type' in the Simulation Settings dialog box. Under 'AC Sweep Type', select 'Logarithmic' and enter 100 for 'Start Frequency', 100k for 'End Frequency', and 1000 for 'Points/Decade'. After the simulation is run, select in the SCHEMATIC1 page Trace/Add Trace and then DB() from the right window and V(L1:1) or V1(L1) from the left window so that the entry in Trace expression appears as DB(V(L1:1)). The high-pass plot is displayed. To display the low-pass response and after selecting DB() as for the high-pass response, enter V(R1:1) – V(L1:1) or V1(R1) – V1(L1) inside the DB parentheses so that the entry in Trace expression appears as DB(V(R1:1) – V(L1:1)). The two plots appear as in Figure 14.13. To have a more expanded display along the vertical axis, select Plot/Axis Settings/Y-Axis/User Defined, and enter '−30' to '0'. The vertical scale now extends from −30 dB to 0.

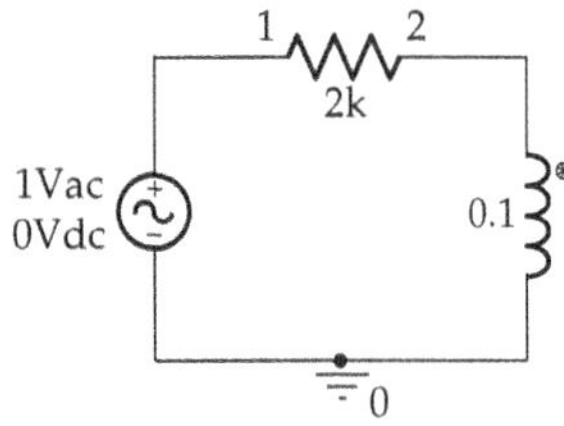

FIGURE 14.12
Figure for Example 14.1.

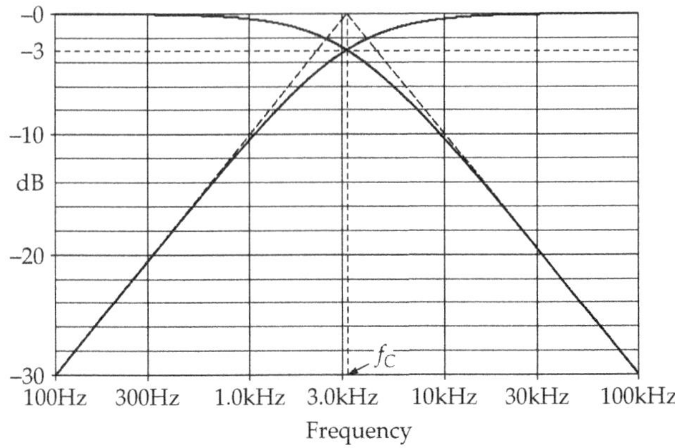

FIGURE 14.13
Figure for Example 14.1.

To display the low-frequency asymptote of the high-pass response, select Trace/Add Trace and enter in the Trace Expression field the RHS of Equation 14.26 as $-20 * \text{LOG10}(3183.1) + 20 * \text{LOG10}(\text{Frequency})$ where $f_{ch} = (20\text{ krad/s})/2\pi = 3183.1$ Hz. Note that LOG10() can be entered from the right window and 'Frequency' from the left window. 'Frequency' as entered is in Hz. To display the high-frequency asymptote of the low-pass response, the entry is 20*LOG10(3183.1) − 20*LOG10(Frequency), which is the negative of the low-frequency asymptote.

The 3 dB cutoff frequency can be read using the cursor search command. Click first on the Toggle cursor icon and then on the Cursor Search icon. In the Search Command window enter sle(max-3.01), as explained in Appendix C. The values 3.1801K and 3.1860K are displayed in the cursor window for the high-pass and low-pass responses, respectively. Alternatively, select Trace/Evaluate Measurement, then Cutoff_Highpass_3dB(1), and enter V(L1:1) in place of '1'. In the Measurement

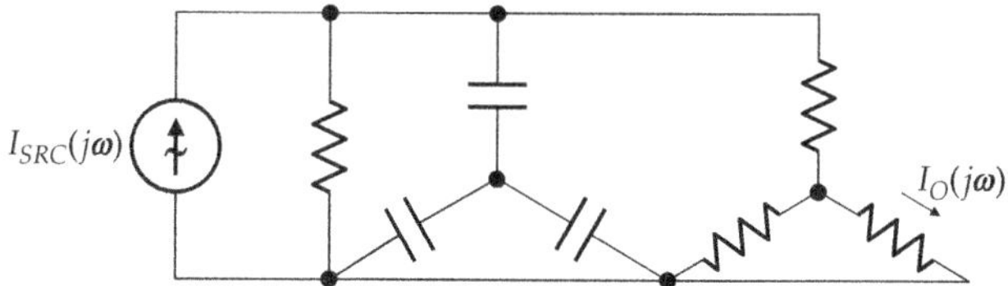

FIGURE 14.14
Figure for Primal Exercise 14.9.

Results window, 3.18743k is displayed. Similarly, choosing Cutoff_Highpass_3dB(V(R1:1) – V(L1:1)) displays 3.17869k in the Measurement Results window.

The following should be noted: (1) f_c is the same for both responses, since it is $L/(2\pi R)$ in the same circuit; (2) the two responses are symmetrical on a logarithmic scale with respect to a line through $f = f_c$, because ω/ω_{cl} (or f/f_{cl}) in the low-pass response is replaced by its reciprocal ω_{cl}/ω (or f_{cl}/f) in the high-pass response; and (3) both asymptotes intersect the 0 dB line at $f = f_c$. This is because the equations of the asymptotes are f/f_c and f_c/f. The two asymptotes intersect where $f/f_c = f_c/f$ or $f = f_c$. At this frequency, either ratio is unity, which corresponds to 0 dB.

Primal Exercise 14.9

Determine the maximum $|I_O(j\omega)|$ in Figure 14.14 and specify the frequency at which it occurs, assuming $|I_{SRC}(j\omega)|$ = 5 mA, all resistances are 4 kΩ, and all capacitances are 100 nF.

Ans. 1 mA at $\omega = 0$.

14.5 Second-Order Bandpass Response

As defined in Section 12.2, second-order LTI circuits are so called because their responses obey a second-order, linear, ordinary differential equation with constant coefficients. These circuits are of two general types: (1) circuits that are reducible to a series or a parallel combination of a resistor, an inductor, and a capacitor or (2) circuits that have resistors in combination with two capacitors or two inductors, where these energy storage elements cannot be combined into a single capacitor or inductor, as in Figure 11.21, for example.

In contrast to first-order frequency responses, which, as discussed in Section 14.3, are either low pass or high pass, second-order frequency responses can be any of the four basic types mentioned in Section 14.2. These responses will be discussed using the prototypical series *RLC* circuit illustrated in Figure 14.15. The sinusoidal source $V_{SRC}(j\omega)$ is assumed to be of constant magnitude

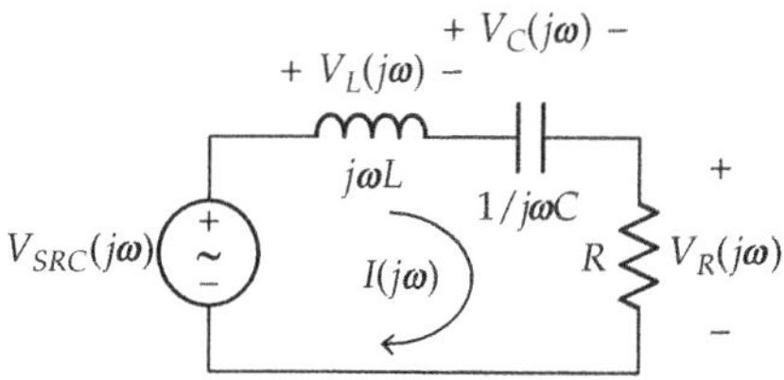

FIGURE 14.15
Prototypical series *RLC* circuit.

and phase angle, but of variable frequency. The current in the circuit and the voltages across the circuit elements are phasors.

The impedance seen by the source is the impedance of the three elements in series

$$Z = R + j\omega L + 1/j\omega C = R + jX = R + j(\omega L - 1/\omega C) \quad (14.27)$$

where the combined reactance X is $(\omega L - 1/\omega C)$. The current is

$$I(j\omega) = \frac{V_{SRC}(j\omega)}{Z} = \frac{V_{SRC}(j\omega)}{R + j(\omega L - 1/\omega C)} \quad (14.28)$$

A striking feature of the combined reactance X in Equation 14.28 is that at some frequency, the capacitive and inductive reactances are equal in magnitude but opposite in sign so that their sum $X = \omega L - 1/\omega C = 0$. The frequency ω_0 at which this occurs is given by

$$\omega_0 L = \frac{1}{\omega_0 C} \quad \text{or} \quad \omega_0^2 LC = 1 \quad \text{or} \quad \omega_0 = \frac{1}{\sqrt{LC}} \quad (14.29)$$

Thus, at the frequency ω_0, the series combination of L and C acts as a short circuit ($X = 0$), and $Z = R$, as illustrated in the impedance diagram of Figure 14.16a. At a frequency $\omega < \omega_0$, $\omega L < 1/\omega C$ so that the net reactance X is capacitive, with $|Z| = \sqrt{R^2 + X^2} > R$, and $\angle z < 0$ (Figure 14.16b). At a frequency $\omega > \omega_0$, $\omega L > 1/\omega C$ so that the net reactance X is inductive, with $|Z| = \sqrt{R^2 + X^2} > R$, and $\angle z > 0$ (Figure 14.16c). It is seen that at $\omega = \omega_0$, Z has a minimum value of R and is purely resistive. With a constant magnitude and phase angle of $V_{SCR}(j\omega)$, it follows from Equation 14.28 that at $\omega = \omega_0$, the current $I(j\omega_0)$ will be a maximum and in phase with $V_{SCR}(j\omega_0)$. At frequencies that are below or above ω_0, $|I(j\omega)|$ is reduced, with $I(j\omega)$ leading $V_{SCR}(j\omega)$ at frequencies below ω_0, since the net reactance is capacitive, and with $I(j\omega)$ lagging $V_{SCR}(j\omega)$ at frequencies above ω_0, when the net reactance is inductive.

Having examined qualitatively the behavior of the series *RLC* circuit of Figure 14.1, the next step is to investigate this behavior quantitatively. We will start by

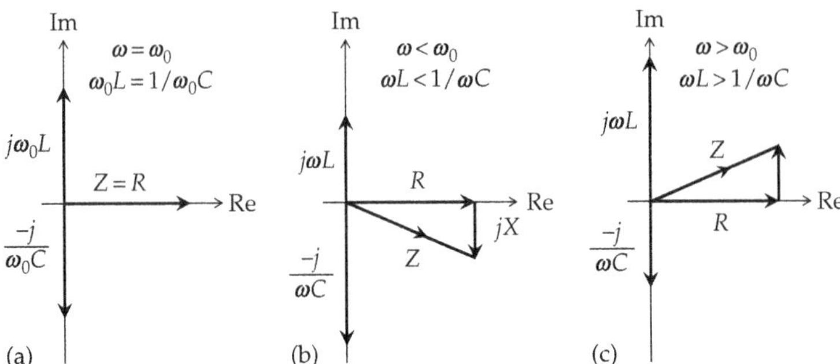

FIGURE 14.16
Impedance of series *RLC* circuit at resonance (a), at a frequency below (b), and at a frequency above (c), the resonant frequency.

considering the output voltage $V_R(j\omega)$ across the resistor. It follows from Ohm's law and Equation 14.1 or from voltage division that

$$H_{BP}(j\omega) = \frac{V_R(j\omega)}{V_{SRC}(j\omega)} = \frac{R}{j\omega L + R + 1/j\omega C} = \frac{j\omega CR}{1 - \omega^2 LC + j\omega CR} \tag{14.30}$$

$H_{BP}(j\omega)$ can be expressed in a more basic, standard form, using a quantity Q that is defined for the series *RLC* circuit as

$$Q = \frac{\omega_0 L}{R} = \frac{1}{\omega_0 CR} \tag{14.31}$$

Q is a dimensionless **quality factor**. Fundamentally, it depends on the ratio of the peak energy stored in L or C to the average energy dissipated in R during a cycle (Problem P14.67). For a given L and C, the smaller R, the larger Q is and conversely.

Substituting $CR = 1/\omega_0 Q$ (Equation 14.31) and $LC = 1/\omega_0^2$ (Equation 14.29) in Equation 14.30,

$$H_{BP}(j\omega) = \frac{j\omega/\omega_0 Q}{1 - \omega^2/\omega_0^2 + j\omega/\omega_0 Q} \tag{14.32}$$

Multiplying the numerator and denominator by $\omega_0 Q/j\omega$ and simplifying,

$$H_{BP}(j\omega) = \frac{1}{1 + jQ(\omega/\omega_0 - \omega_0/\omega)} \tag{14.33}$$

It follows that

$$|H_{BP}(j\omega)| = \frac{1}{\sqrt{1 + Q^2(\omega/\omega_0 - \omega_0/\omega)^2}} \tag{14.34}$$

and

$$\angle H_{BP}(j\omega) = -\tan^{-1} Q(\omega/\omega_0 - \omega_0/\omega) \tag{14.35}$$

The Bode magnitude and phase plots are shown in Figure 14.17a and b for $Q = 10$ and $Q = 5$.

It is seen from Equation 14.34 that when $\omega = \omega_0$, $|H_{BP}(j\omega_0)|$ has its maximum value of unity so that $20\log_{10}|H_{BP}(j\omega)| = 0$ dB, irrespective of Q. This is because the combination of L and C acts as a short circuit at this frequency so that $V_{SRC}(j\omega_0)$ appears across R. $|H_{BP}(j\omega)|$ decreases on either side of ω_0. At $\omega = 0$, no current flows, because C acts as an open circuit, so $V_R(j\omega) = 0$. Similarly, as $\omega \to \infty$, L acts as an open circuit, so $V_R(j\omega) \to 0$. The frequency response is limited to a range of frequencies and is therefore of the bandpass type (Figure 14.2c).

Note from Equation 14.34 that because of the term $(\omega/\omega_0 - \omega_0/\omega)^2$ in the denominator, $|H_{BP}(j\omega)|$ at $\omega/\omega_0 = k$ has the same as at $\omega/\omega_0 = 1/k$. On a logarithmic scale, $\omega = k\omega_0$ and $\omega = \omega_0/k$ are equidistant, on either side of ω_0, since $\log_{10}k\omega_0 = \log_{10}\omega_0 + \log_{10}k$ and $\log_{10}\omega_0/k = \log_{10}\omega_0 - \log_{10}k$. This means that $20\log_{10}|H_{BP}(j\omega)|$ is symmetrical about the vertical line through $\omega = \omega_0$, as in Figure 14.17a. Moreover, it follows from Equation 14.34 that as $\omega \to 0$,

$$|H_{BP}(j\omega)| \to \frac{1}{\sqrt{Q^2(-\omega_0/\omega)^2}} = \frac{1}{Q\omega_0}\omega \tag{14.36}$$

so that that the low-frequency asymptote of the Bode magnitude plot has a slope of +20 dB/decade, as for a first-order high-pass response. As $\omega \to \infty$,

$$|H_{BP}(j\omega)| \to \frac{1}{\sqrt{Q^2(\omega/\omega_0)^2}} = \frac{\omega_0}{Q}\frac{1}{\omega} \tag{14.37}$$

which means that the high-frequency asymptote of the Bode magnitude plot has a slope of −20 dB/decade, as for a first-order low-pass response. This is because as $\omega \to 0$, $\omega L \to 0$ so that the circuit reduces to a high-pass *RC* circuit. Similarly, as $\omega \to \infty$, $1/\omega C \to 0$ so that the circuit reduces to a low-pass *RL* circuit. Equating the RHS's

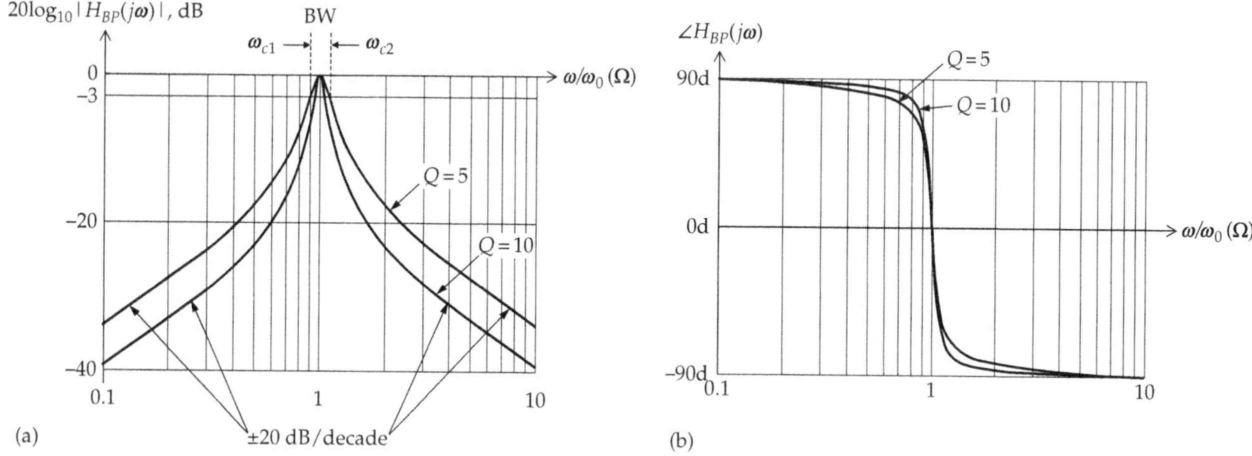

FIGURE 14.17
Bode magnitude plot (a) and Bode phase plot (b) of a second-order, bandpass response for two values of Q.

of Equations 14.36 and 14.37 gives $\omega = \omega_0$ as the point of intersection of the two asymptotes.

The phase angle is zero at $\omega = \omega_0$, since the reactances cancel out and $V_R(j\omega_0) = V_{SRC}(j\omega_0)$. When $\omega < \omega_0$, the capacitive reactance predominates (Figure 14.16b) so that the current, and hence $V_R(j\omega)$, leads $V_{SRC}(j\omega)$. As $\omega \to 0$, $\tan\angle H_{BP}(j\omega) \to +\infty$, and $\angle H_{BP}(j\omega) \to 90°$, as for an ideal capacitor. When $\omega > \omega_0$, the inductive reactance predominates (Figure 14.16c) so that the current, and hence $V_R(j\omega)$, lags $V_{SRC}(j\omega)$. As $\omega \to \infty$, $\tan\angle H_{BP}(j\omega) \to -\infty$, and $\angle H_{BP}(j\omega) \to -90°$, as for an ideal inductor.

At the half-power frequencies ω_{c1} and ω_{c2}, $\left|H_{BP}\left(j\omega_{c1,2}\right)\right| = 1/\sqrt{2}$, by definition. Substituting this in Equation 14.34 squaring both sides and cross multiplying gives

$$\frac{\omega_c}{\omega_0} - \frac{\omega_0}{\omega_c} = \pm\frac{1}{Q} \tag{14.38}$$

Solving for ω_c and retaining only the positive roots,

$$\omega_{c1} = \omega_0\left[\sqrt{1+(1/2Q)^2} - 1/2Q\right],$$

$$\omega_{c2} = \omega_0\left[\sqrt{1+(1/2Q)^2} + 1/2Q\right] \tag{14.39}$$

The 3 dB bandwidth (BW) is defined as $\omega_{c2} - \omega_{c1}$. It follows from Equation 14.39 that

$$\text{BW} = \omega_{c2} - \omega_{c1} = \frac{\omega_0}{Q} \quad \text{and} \quad \sqrt{\omega_{c1}\,\omega_{c2}} = \omega_0 \tag{14.40}$$

For a given ω_0, the larger the Q, the narrower the BW is, that is, the sharper the peak of the bandpass response is, as in Figure 14.17a. If the BW does not exceed 10% of the center frequency, corresponding to a Q of at least 10, the bandpass response is usually described as **narrowband**.

The pronounced peaking of the frequency response at some frequency ω_0 is described as **resonance**, ω_0 being the **resonant frequency**. As illustrated earlier, resonance results in a bandpass response that allows selection of a relatively narrowband of frequencies out of a wide range of frequencies that may be present. Because of this, a resonant circuit is also referred to as a **tuned circuit**. More elaborate, near-ideal bandpass responses are used in radio, television, and communication receivers to select or tune to particular stations or channels. Dynamical systems, in general, exhibit resonance, which could be quite pronounced if Q is high. A structure, such as a bridge, may resonate at some frequency. If excited at this frequency, the amplitude of vibration may be large enough to cause failure of the bridge.

Recall that the damping factor was defined in Equation 12.4, as $\alpha = R/2L$ for the series circuit. Comparing this to the definition of Q as $\omega_0 L/R$ (Equation 14.31), it is seen that

$$Q = \frac{\omega_0}{2\alpha} \tag{14.41}$$

The smaller the R, the lesser is the damping, the smaller is the α, and the larger is the Q. For a critically damped circuit, $\alpha = \omega_0$ so that $Q = 0.5$. $Q < 0.5$ for an overdamped circuit and $Q > 0.5$ for an underdamped circuit.

Example 14.2 illustrates a bandpass response of a second-order circuit involving two capacitors. It exemplifies some important concepts that are discussed after the example.

Primal Exercise 14.10

Consider the circuit of Figure 14.15, with $L = 0.1$ H, $C = 4$ nF, and $R = 1$ kΩ. Determine (a) BW (as R/L), (b) ω_0, (c) Q (as ω_0/BW), and (d) ω_{c1} and ω_{c2} (from BW and $\sqrt{\omega_{c1}\omega_{c2}} = \omega_0$).

Ans. (a) 10 krad/s; (b) 50 krad/s; (c) 5; (d) $\omega_{c2} = 55.25$ krad/s, $\omega_{c1} = 45.25$ krad/s.

Primal Exercise 14.11

A second-order bandpass filter has a maximum response at 10 kHz and a lower 3 dB cutoff frequency of 7.5 kHz. Determine the 3 dB BW.

Ans. 5.83 kHz.

Exercise 14.12

Interpret the relation $\sqrt{\omega_{c1}\omega_{c2}} = \omega_0$ on the basis that ω_{c1} and ω_{c2} are symmetrically located with respect to ω_0 when the frequency axis is logarithmic.

Ans. Symmetry with respect to ω_0 on a logarithmic frequency scale requires that $\omega_{c2}/\omega_0 = \omega_0/\omega_{c1}$.

Example 14.2: Second-Order Bandpass *RC* Circuit

It is required to derive the transfer function $H_{BP}(j\omega) = V_O(j\omega)/V_{SRC}(j\omega)$ and determine Q in Figure 14.18.

Solution:

$H_{BP}(j\omega)$ will be derived using voltage division. Since $\frac{V_O(j\omega)}{V_1(j\omega)} = \frac{R}{R + 1/j\omega C} = \frac{j\omega CR}{j\omega CR + 1}$, it is necessary to derive $V_1(j\omega)/V_{SRC}(j\omega)$. To do so, we note that the impedance of $1/j\omega C$ in parallel with $R + 1/j\omega C$ is $Z_p = \frac{(1/j\omega C)(R + 1/j\omega C)}{1/j\omega C + R + 1/j\omega C} = \frac{1}{j\omega C}\frac{R + 1/j\omega C}{R + 2/j\varpi C} = \frac{1}{j\omega C}\frac{1 + j\omega CR}{2 + j\omega CR}$.

It follows that $\frac{V_1(j\omega)}{V_{SRC}(j\omega)} = \frac{Z_p}{R + Z_p} = \frac{j\omega CR + 1}{1 - \omega^2C^2R^2 + 3j\omega CR}$. Multiplying $V_1(j\omega)/V_{SRC}(j\omega)$ by $V_O(j\omega)/V_1(j\omega)$ gives

$$H_{BP}(j\omega) = \frac{V_O(j\omega)}{V_{SRC}(j\omega)} = \frac{j\omega CR}{1 - \omega^2C^2R^2 + 3j\omega CR} \tag{14.42}$$

Dividing the numerator and denominator by $j\omega CR$ and taking 3 outside the denominator, Equation 14.42 can be expressed as

$$H_{BP}(j\omega) = \frac{1}{3}\frac{1}{1 + (j/3)(\omega/\omega_0 - \omega_0/\omega)} \tag{14.43}$$

where $\omega_0 = 1/CR$. Equation 14.43 is of exactly the same form as Equation 14.33, but with a multiplying factor of 1/3. This scalar multiplier does not affect the type of frequency response, which is therefore bandpass. At $\omega = \omega_0$, $H_{BP}(j\omega)$ has its maximum value of 1/3. Comparing the denominators of Equations 14.33 and 14.43, it follows that $Q = 1/3$.

Simulation: The circuit is entered as in Figure 14.19, assuming $C = 1$ μF and $R = 1000/2\pi = 159.155$ Ω, so that $\omega_0 = 1/RC = 2\pi/(10^3 \times 10^{-6}) = 2\pi \times 10^3$ rad/s ≡ 1 kHz. The details of the simulation are as previously explained in Example 14.1. Simulation Settings include a start frequency of 10 Hz, an end frequency of 100 kHz, and 1000 points/decade. After the simulation is run, select DB(V(R2:1)). $|H_{BP}(j\omega)|$ is displayed as in Figure 14.20. The maximum value at $\omega_0 = 1$ kHz is $20\log_{10}(1/3) = -9.54$ dB.

The low-frequency asymptote, as $\omega \to 0$, is $|H_{BP}(j\omega)| = (1/3)3\omega/\omega_0 = \omega/\omega_0$, and the high-frequency asymptote, as $\omega \to \infty$, is $|H_{BP}(j\omega)| = (1/3)3\omega_0/\omega = \omega_0/\omega$. The two asymptotes intersect at $\omega = \omega_0$, at the 0 dB line.

To display the low-frequency asymptote, select Trace/Add Trace and enter 20*LOG10(Frequency)-20*LOG10(1000) in the Trace expression field. To display the high-frequency asymptote, the entry is -20*LOG10(Frequency) + 20*LOG10(1000).

ω_0, ω_{c1}, and ω_{c2} can be read using the cursor. Click first on the Toggle cursor icon and then on the Cursor

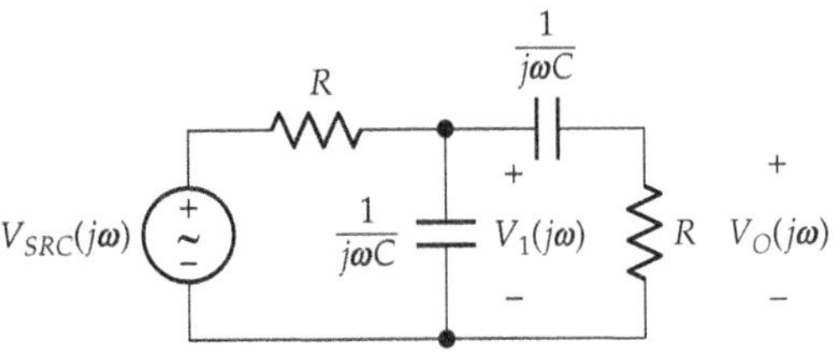

FIGURE 14.18
Figure for Example 14.2.

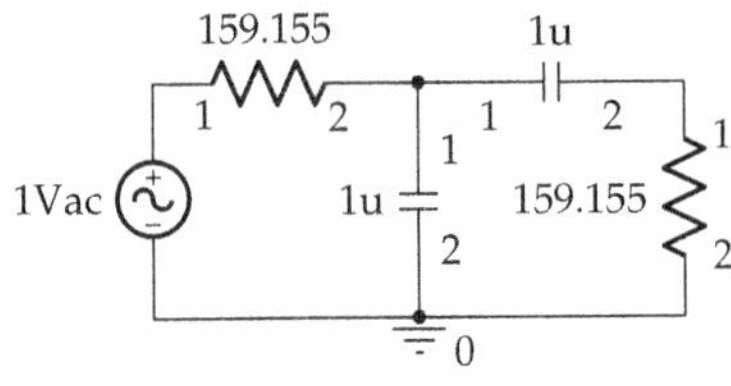

FIGURE 14.19
Figure for Example 14.2.

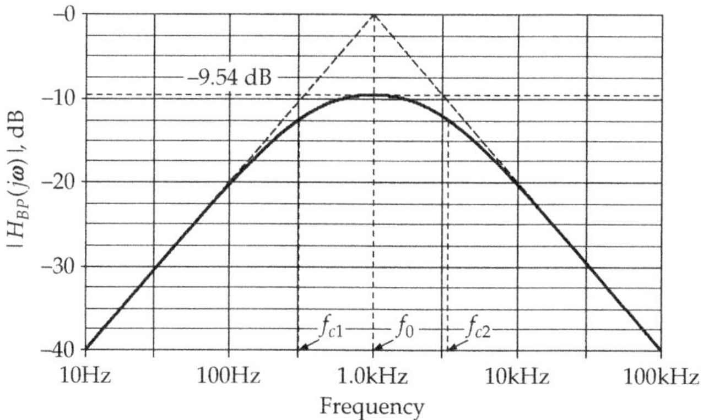

FIGURE 14.20
Figure for Example 14.2.

Max icon. ω_0 is read in the Cursor Window as 1.0000K. With the cursor at ω_0, enter the search command sle(max-3). Cursor 1 moves to ω_{c2} and its value is displayed in the Cursor Window as 3.2963K. Repeat the same command but with Cursor 2 selected in the Search Command window. Cursor 2 moves to ω_{c1} and its value is displayed in the Cursor Window as 303.374. The difference Y1-Y2 is displayed as 2.9929K. This gives a Q of $\omega_0/\text{BW} = 1/2.9929$, which is nominally 1/3.

The Bode phase plot (Figure 14.21) is obtained by selecting Trace/add Trace and then P() from the right window and V(R1:1) from the left window.

It will be noted that the first RC circuit whose output is $V_1(j\omega)$ in Figure 14.18 is a first-order low-pass filter, whereas the second RC circuit whose output is $V_O(j\omega)$ is a first-order high-pass filter, with both filters having the same values of R and C. The two filters are cascaded by having the output of the low-pass filter applied as the input to the high-pass filter so that the transfer functions $V_1(j\omega)/V_{SRC}(j\omega)$ and $V_O(j\omega)/V_1(j\omega)$ are multiplied by one another in order to derive the overall response (Equation 14.42). As explained in connection with Figure 14.3c, multiplication of a low-pass response and a high-pass response results in a bandpass response. Note that multiplication of the magnitudes of transfer functions is equivalent to adding them on the Bode magnitude plots.

A relevant question is, how would Q change if the two cascaded first-order filters are isolated from one another? In Figure 14.18, the high-pass filter 'loads' the low-pass filter, that is, it draws its input current from the output of the low-pass filter, thereby affecting the output voltage $V_1(j\omega)$. The two filters can be isolated by inserting a unity-gain amplifier between the two filters, as discussed in Section 13.3 and illustrated in Figure 14.22.

It follows from Equation 14.2 that $V_1(j\omega)/V_{SRC}(j\omega) = 1/(1 + j\omega CR)$, and it follows from Equation 14.4 that $V_O(j\omega)/V_1(j\omega) = j\omega CR/(1 + j\omega CR)$. The overall transfer function is

$$\frac{V_O(j\omega)}{V_{SRC}(j\omega)} = \frac{V_1(j\omega)}{V_{SRC}(j\omega)} \times \frac{V_O(j\omega)}{V_1(j\omega)} = \frac{j\omega CR}{(j\omega CR+1)^2}$$

$$= \frac{j\omega CR}{1-\omega^2C^2R^2+2j\omega CR} \tag{14.44}$$

Dividing the numerator and denominator by $j\omega CR$ and taking 2 outside the denominator, as was done with Equation 14.42, Equation 14.44 becomes

$$H_{BP}(j\omega) = \frac{1}{2}\frac{1}{1+(j/2)(\omega/\omega_0-\omega_0/\omega)} \tag{14.45}$$

where $\omega_0 = 1/CR$. It is seen that Q is increased by isolation from 1/3 to 1/2. This result is quite general and can be expressed as the following concept:

Concept: *Second-order circuits having two capacitors or two inductors cannot have a Q exceeding 0.5.*

Having $Q = 0.5$ means that the circuit is critically damped, as explained in connection with Equation 14.41. Hence, second-order circuits having two capacitors or two inductors are either overdamped or, at best, critically damped. They cannot be underdamped. In other words, *underdamped, second-order, passive circuits of* $Q > 0.5$ *must have a capacitor and an inductor.*

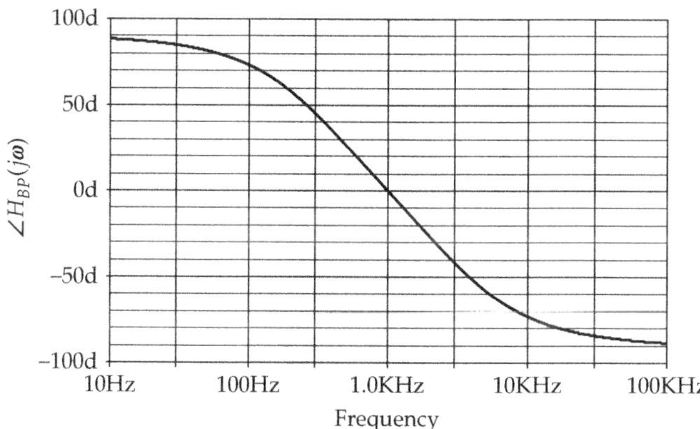

FIGURE 14.21
Figure for Example 14.2.

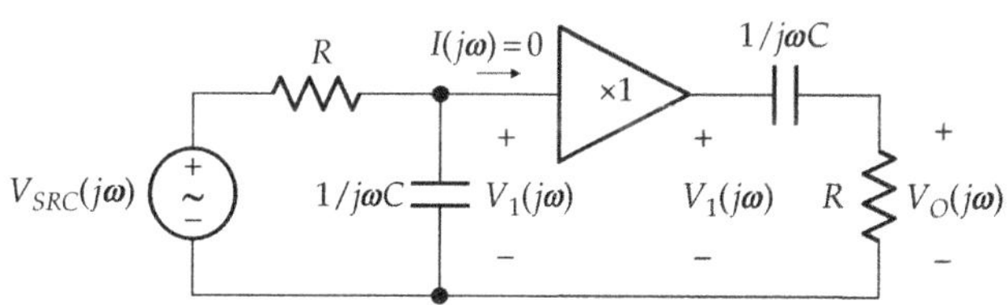

FIGURE 14.22
Cascaded first-order filters, with isolation.

14.6 Second-Order Bandstop Response

It was pointed out in Section 14.2 that the second-order bandpass and bandstop responses are complementary with respect to the applied input. This implies that if the voltage response across R is bandpass, the voltage response, which when added to the bandpass response equals the source voltage, must be bandstop. In other words, the response $V_{LC}(j\omega)$ across L and C together, as illustrated in Figure 14.23, must be bandstop.

That this is the case can be ascertained qualitatively before analyzing the circuit, as was done with the circuits considered so far. At $\omega = 0$, the capacitor acts as an open circuit, $I(j\omega) = 0$, and $V_{SRC}(j\omega)$ appears across the capacitor so that $V_{LC}(j\omega) = V_{SRC}(j\omega)$. As $\omega \to \infty$, the inductor acts as an open circuit, $I(j\omega) = 0$, and $V_{SRC}(j\omega)$ appears across the inductor so that, again, $V_{LC}(j\omega) = V_{SRC}(j\omega)$. At the resonant frequency ω_0, the series combination of L and C acts as a short circuit, $V_{LC}(j\omega) = 0$, and $I(j\omega) = V_{SRC}(j\omega)/R$. The nature of the response is therefore bandstop.

Analytically, $V_{LC}(j\omega) = V_L(j\omega) + V_C(j\omega)$, where, from voltage division,

$$H_{BS}(j\omega) = \frac{V_{LC}(j\omega)}{V_{SRC}(j\omega)} = \frac{j\omega L + 1/j\omega C}{j\omega L + R + 1/j\omega C} = \frac{1-\omega^2 LC}{1-\omega^2 LC + j\omega CR} \quad (14.46)$$

It is seen that the response is zero when $1 - \omega^2 LC = 0$, at the resonant frequency ω_0. The denominator is the same as that of Equation 14.30. Proceeding as for the bandpass response, that is, substituting $CR = 1/\omega_0 Q$ (Equation 14.31), $LC = 1/\omega_0^2$ (Equation 14.29) in Equation 14.46,

$$H_{BS}(j\omega) = \frac{1-\omega^2/\omega_0^2}{1-\omega^2/\omega_0^2 + j\omega/\omega_0 Q} \quad (14.47)$$

Multiplying the numerator and denominator by $\omega_0 Q/j\omega$ and simplifying,

$$H_{BS}(j\omega) = \frac{jQ(\omega/\omega_0 - \omega_0/\omega)}{1 + jQ(\omega/\omega_0 - \omega_0/\omega)} \quad (14.48)$$

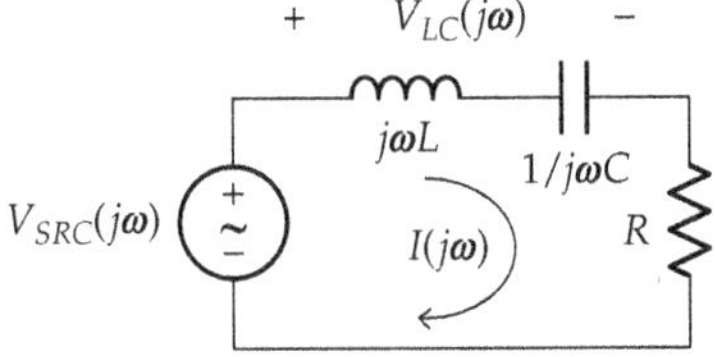

FIGURE 14.23
Second-order bandstop response across L and C.

It follows that

$$|H_{BS}(j\omega)| = \frac{|Q(\omega/\omega_0 - \omega_0/\omega)|}{\sqrt{1+Q^2(\omega/\omega_0 - \omega_0/\omega)^2}} \quad (14.49)$$

To obtain the phase angle, it is convenient to divide the numerator and denominator of Equation 14.48 by $jQ(\omega/\omega_0 - \omega_0/\omega)$ to give

$$\angle H_{BS}(j\omega) = \tan^{-1}\left[1/Q(\omega/\omega_0 - \omega_0/\omega)\right] \quad (14.50)$$

The amplitude and phase plots are shown in Figure 14.24a and b for Q of 5 and 10. $|H_{BS}(j\omega)|$ is plotted and not its logarithm, because $|H_{BS}(j\omega)| = 0$ at $\omega = \omega_0$, and $\log_{10} 0 \to -\infty$. For $\omega < \omega_0$, the net reactance is capacitive and can be represented as $1/\omega C_{eff} = (1/\omega C - \omega L)$. The circuit is equivalent to R in series with C_{eff}, and the response is akin to that of a low-pass RC filter. When ω is only slightly larger than zero, $1/\omega C_{eff}$ is large and $V_{LC}(j\omega) \cong V_{SRC}(j\omega)$ so that the magnitude of the response is near unity and the phase angle is small. When ω is slightly less than ω_0, $1/\omega C_{eff}$ is small, like $1/\omega C$ of a low-pass RC filter as $\omega \to \infty$. The response is small and the phase angle is near −90° (Figure 14.24).

For $\omega > \omega_0$, the net reactance is inductive and can be represented as $\omega L_{eff} = (\omega L - 1/\omega C)$. The circuit is equivalent to R in series with L_{eff}, and the response is akin to that of a high-pass RL filter. When ω is slightly larger than ω_0, ωL_{eff} is small, like ωL of the RL filter for $\omega \cong 0$. The response is small and the phase angle is near +90°. As $\omega \to \infty$, ωL_{eff} is large, the magnitude of the response is near unity, and the phase angle is small, like that of the high-pass RL filter.

At the half-power frequencies, $|H(j\omega)| = 1/\sqrt{2}$. Equation 14.49 reduces to Equation 14.38 for the bandpass case. The half-power frequencies are given by Equation 14.39 and the BW by Equation 14.40, the same as for the bandpass response.

Primal Exercise 14.13

Repeat Primal Exercise 14.10, but with the output taken across L and C.

Ans. Exactly the same as in Primal Exercise 14.10.

Primal Exercise 14.14

Determine the largest value of $|H(j\omega)|$ in Figure 14.25, ω_0, and Q.

Ans. 2/3; 10^4 rad/s; 50.

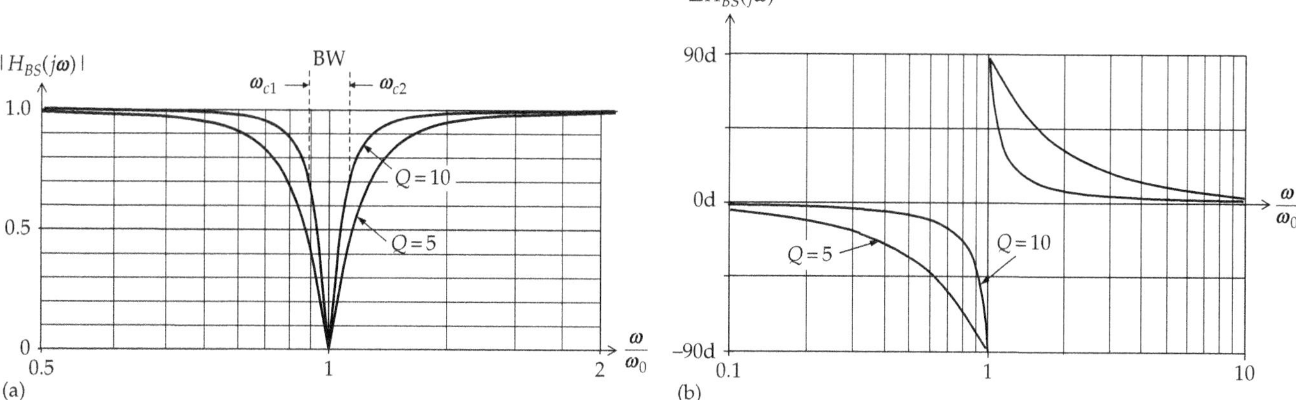

FIGURE 14.24
Bode magnitude plot (a) and Bode phase plot (b) of a second-order, bandstop response for two values of Q.

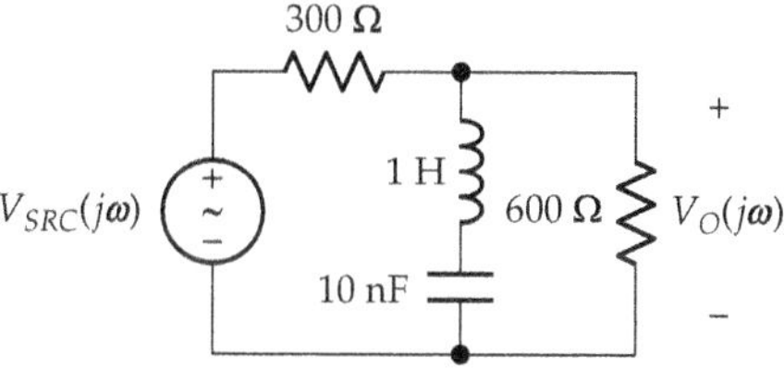

FIGURE 14.25
Figure for Primal Exercise 14.14.

14.7 Second-Order Low-Pass and High-Pass Responses

14.7.1 Low-Pass Response

Suppose that the response is taken across the capacitor in the series *RLC* circuit, as shown in Figure 14.26. At $\omega = 0$, the capacitor acts as an open circuit and the inductor as a short circuit, $I(j\omega) = 0$, and $V_C(j\omega) = V_{SRC}(j\omega)$. As $\omega \to \infty$, the inductor acts as an open circuit and the capacitor as a short circuit and $V_C(j\omega) = 0$. The response is clearly of a low-pass nature. It is generally true that when the response of a filter is taken across a capacitor, the response tends to zero as $\omega \to \infty$ so that the response is of a low-pass nature.

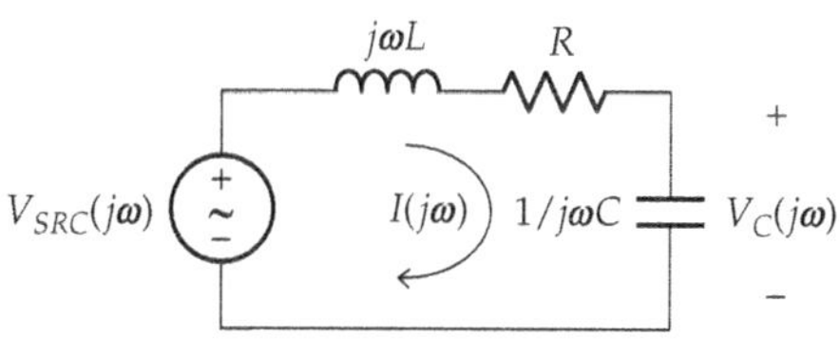

FIGURE 14.26
Second-order lowpass response across *C*.

However, at the resonant frequency, the net reactance is zero, and if *R* is small enough, the current at this frequency will be large. Thus, as $R \to 0$, $I(j\omega_0) \to \infty$. When this large current is multiplied by the impedance of the capacitor, $V_C(j\omega)$ can exceed $V_{SRC}(j\omega)$. In other words, the second-order low-pass response can exhibit some peaking, unlike the first-order low-pass response.

It follows from voltage division that

$$H_{LP}(j\omega) = \frac{V_C(j\omega)}{V_{SRC}(j\omega)} = \frac{1/j\omega C}{j\omega L + R + 1/j\omega C} = \frac{1}{1-\omega^2 LC + j\omega CR}$$

$$= \frac{1}{1-\omega^2/\omega_0^2 + j\omega/\omega_0 Q} \tag{14.51}$$

Note that the denominator of Equation 14.51 is the same as that for the bandpass response (Equation 14.30) and the bandstop response (Equation 14.46). The reason for this will be explained later.

It follows from Equation 14.51 that

$$|H_{LP}(j\omega)| = \frac{1}{\sqrt{\left(1-\left(\frac{\omega}{\omega_0}\right)^2\right)^2 + \left(\frac{\omega}{\omega_0 Q}\right)^2}} \tag{14.52}$$

and

$$\angle H_{LP}(j\omega) = -\tan^{-1}\frac{\left(\frac{\omega}{\omega_0 Q}\right)}{1-\left(\frac{\omega}{\omega_0}\right)^2} \tag{14.53}$$

Note that in Equation 14.53 the imaginary part, $\omega/\omega_0 Q$ in Equation 14.51, is always positive, but the real part,

$\left(1-(\omega/\omega_0)^2\right)$, can be positive or negative. This means that +tan^{-1} (imaginary part/real part) is an angle in the first quadrant when ω is small, and moves counter clockwise to the second quadrant as $\omega \to \infty$. Its negation, $\angle H_{LP}(j\omega)$, is in fourth or third quadrants, respectively.

We wish to determine if $|H_{LP}(j\omega)|$ has a maximum at some frequency, which means that the expression under the square root in Equation 14.52 has a minimum. To check this, it is convenient to denote ω/ω_0 by u so that the expression becomes

$$f(u) = \left(1-u^2\right)^2 + u^2/Q^2 \tag{14.54}$$

Deriving $df(u)/du$, setting it equal to zero, and retaining only positive nonzero values of u give

$$u_{\max} = \sqrt{1-\frac{1}{2Q^2}} \quad \text{or} \quad \omega_{\max} = \omega_0\sqrt{1-\frac{1}{2Q^2}} \tag{14.55}$$

Hence, $|H_{LP}(j\omega)|$ has a maximum at some nonzero value $\omega_{\max}$ if $Q > 1/\sqrt{2}$. That $|H_{LP}(j\omega)|$ has a maximum at $\omega = \omega_{\max}$ and not a minimum can be confirmed by substituting $\omega_{\max}/\omega_0$ in Equation 14.52, which gives

$$\left|H_{LP}(j\omega)\right|_{\max} = \frac{Q}{\sqrt{1-1/4Q^2}} \tag{14.56}$$

If $R = 0$, $Q \to \infty$, and $|H_{LP}(j\omega)|_{\max} \to \infty$ at $\omega = \omega_{\max}$, confirming that $|H_{LP}(j\omega)|$ has a maximum at this frequency. For R small but not zero, the response is highly peaked, as was argued qualitatively. Moreover, since the impedance of the capacitor is inversely proportional to frequency, maximum $V_C(j\omega) = I(j\omega)/j\omega C$ occurs not when $I(j\omega)$ is maximum, that is, at $\omega = \omega_0$, but at a slightly lower frequency, as given by Equation 14.55. It may be noted that in Equation 14.56, $|H_{LP}(j\omega)|_{\max} \to \infty$ when $Q^2 = 1/4$ or $Q = 0.5$. But $\omega_{\max}$ is an imaginary frequency for $Q < 1/\sqrt{2}$ (Equation 14.55), so no peaking in fact occurs for $Q \le 1/\sqrt{2}$.

It is of interest to determine the equations of the asymptotes of $|H_{LP}(j\omega)|$. As $\omega \to 0$, $|H_{LP}(j\omega)| \to 1$ (Equation 14.52), and $20\log_{10}|H_{LP}(j\omega)| \to 0$ dB. The low-frequency asymptote is therefore the 0 dB line, that is, the horizontal axis. As $\omega \to \infty$, the u^4 term in the denominator of Equation 14.52 dominates so that $|H_{LP}(j\omega)| \to (\omega_0/\omega)^2$. Taking the logarithm to base 10 and multiplying by 20, the equation of the high-frequency asymptote is

$$20\log_{10}\left|H_{LP}(j\omega)\right| = 40\log_{10}\omega_0 - 40\log_{10}\omega \tag{14.57}$$

It is seen that the high-frequency asymptote has a slope of −40 dB/decade. When $\omega = \omega_0$, the RHS of Equation 14.57 is zero, which means that asymptote intersects the 0 dB line at $\omega = \omega_0$. A larger magnitude of the slope at high frequencies means that these unwanted frequencies attenuate more rapidly. This is desirable and closer to the ideal low-pass characteristic of Figure 14.2a.

The Bode magnitude and phase plots are discussed in more detail in Example 14.3.

Example 14.3: Bode Plots of Second-Order Low-Pass Response

It is required to obtain the Bode magnitude and phase plots of the low-pass response of the series *RLC* circuit assuming $L = 0.5$ H, $C = 12.5$ μF, and $R = 10$, 100, 283, and 400 Ω.

Solution:

$\omega_0 = 1/\sqrt{0.5\times 12.5\times 10^{-6}} = 400$ rad/s, $2\omega_0 L = 400\ \Omega$, and the chosen values of R correspond to $Q = 20$ ($R = 10\ \Omega$), $Q = 2$ ($R = 100\ \Omega$), $Q = 1/\sqrt{2}$ ($R = 283\ \Omega$), and $Q = 0.5$ ($R = 400\ \Omega$).

Simulation: The circuit is entered as in Figure 14.27. To display multiple traces for different resistance values, double-click on the default resistance value displayed, which invokes the Display Properties window. In the Value field, enter a chosen designation enclosed in curly brackets, which tells PSpice that this is a parameter and not a fixed value. In the present example, {R_val} is entered. The next step is to declare {R_val} a global parameter. Place the part PARAM from the SPECIAL library; this shows on the schematic as PARAMETERS:. When this word is double-clicked, the Property Editor spreadsheet is displayed. Click on the New Column button to display the Add New Column dialog box. Enter R_val in the Name field and any value, say, 1k, in the Value field. A new column R_val is added to the spreadsheet with the entry 1k. To have this displayed on the schematic, click on the Display button and choose Name and Value in the Display Properties dialog box. R_val = 1k appears under PARAMETERS:, as shown in Figure 14.27.

To run the simulation, select AC Sweep/Noise in the Simulation Settings dialog box and enter a start frequency of 10 Hz, an end frequency of 1000 Hz, and

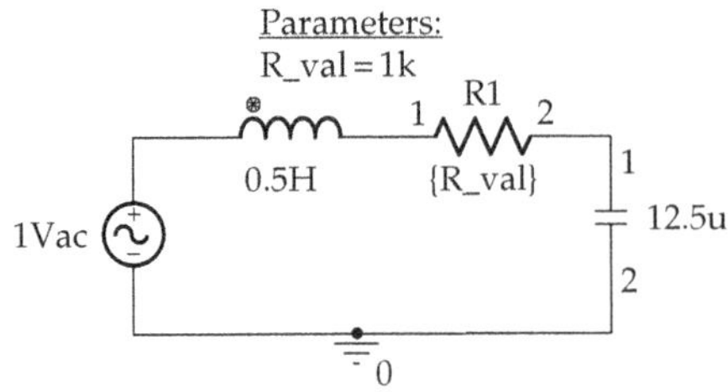

FIGURE 14.27
Figure for Example 14.3.

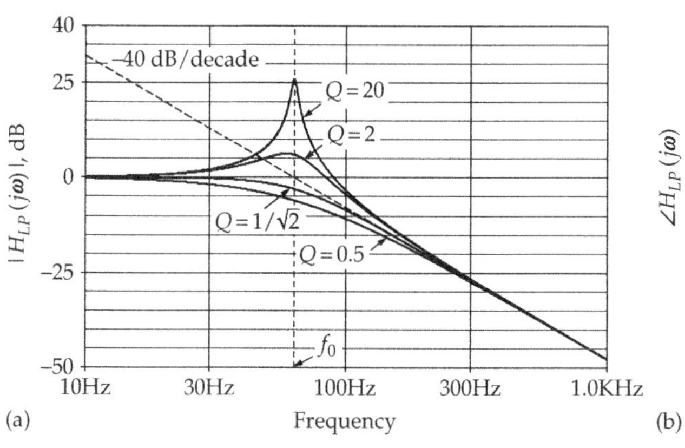

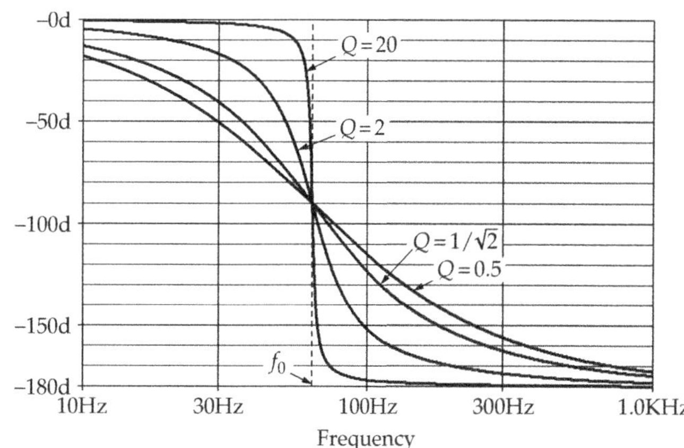

FIGURE 14.28
Figure for Example 14.3.

1000 points/decade. Then choose Parametric Sweep under Options, select Global parameter and enter R_val for Parameter name. Under Sweep type enter 10,100,283,400 in 'Value' list.

After the simulation is run, the magnitude plot for the four resistance values is displayed, as in Figure 14.28a. The high-frequency asymptote is added by selecting Trace/Add traces, then entering in the Trace Expression field 40*LOG10[63.662] – 40*LOG10[Frequency], where $f_0 = (400/2\pi) = 63.662$ Hz. Other details of the simulation are as in Example 14.2.

The cursor can be used to obtain quantitative information. To select a particular trace, such as that for Q = 20, for example, click on its colored marker just below the horizontal axis on the LHS. Press the Toggle cursor button and then the Cursor Max button. The cursor reads the max as 26.021 dB at 63.631 Hz. The calculated values from Equations 14.58 and 14.57 are 26.023 dB at 63.662 Hz.

The trace for $Q = 1/\sqrt{2}$ is of special interest as it represents a maximally flat or Butterworth response (Section 15.2). For larger values of Q, the response is peaked. The trace having Q = 0.5 represents critical damping. Note that the corner frequency, where the two asymptotes intersect, is at $f = f_0$. However, this is the 3 dB cutoff frequency only for the plot having $Q = 1/\sqrt{2}$, for substituting this value of Q in Equation 14.52 gives $\left|H_{LP}(j\omega)\right| = 1/\sqrt{2}$. The 3 dB cutoff frequency for all other plots is not at $f = f_0$. This illustrates that, in general, the 3 dB cutoff frequency is different from the corner frequency.

To obtain the phase plot, select Trace/Add Traces, then choose P( ) and enter V(C1:1). The phase plot is displayed as in Figure 14.28b.

It is instructive to interpret the phase response. For $0 < \omega < \omega_0$, the net series reactance is capacitive. For small ω/ω_0, this reactance is large, $V_C(j\omega) \cong V_{SRC}(j\omega)$, so $\angle H_{LP}(j\omega)$ is small and negative, in accordance with Equation 14.53, and as for a low-pass RC filter at $\omega \cong 0$. For ω/ω_0 slightly less than 1, the net capacitive reactance is small, $I(j\omega)$ is nearly in phase with $V_{SRC}(j\omega)$, and $V_C(j\omega)$ lags $I(j\omega)$ by almost 90°. $\angle H_{LP}(j\omega) \cong -90°$, like that of a low-pass RC filter as $\omega \to \infty$. From Equation 14.53 the argument of the arctan function is large and positive, so the angle is near 90° and its negation is near −90°. For $\omega > \omega_0$, the net series reactance is inductive. For ω/ω_0 slightly larger than 1, the net inductive reactance is small, $I(j\omega)$ lags $V_{SRC}(j\omega)$ by a small angle, and $V_C(j\omega)$ lags $I(j\omega)$ by 90°. $\angle H_{LP}(j\omega)$ is slightly more negative than −90°. In Equation 14.53 the real part in the denominator is negative, and the argument of the arctan function is large and negative, so the angle is slightly more 90°, in the second quadrant, and its negation is slightly more negative than −90°, in the third quadrant. As $\omega \to \infty$, the net inductive reactance is large, $I(j\omega)$ lags $V_{SRC}(j\omega)$ by almost 90° and $V_C(j\omega)$ lags $I(j\omega)$ by another 90°, so $\angle H_{LP}(j\omega)$ approaches −180° from the third quadrant.

Primal Exercise 14.15

Consider the series circuit of Primal Exercise 14.10 having L = 0.1 H, C = 4 nF, and R = 1 kΩ. Determine (a) Q, (b) $\left|H_{LP}(j\omega)\right|_{max}$ in dB, and (c) ω_{max}.

Ans. (a) 5; (b) 14.02 dB; (c) 49.50 kHz.

14.7.2 High-Pass Response

Suppose that the response is taken across the inductor in the series RLC circuit, as shown in Figure 14.29. At $\omega = 0$, the capacitor acts as an open circuit and the inductor as a short circuit, $I(j\omega) = 0$, and $V_L(j\omega) = 0$. As $\omega \to \infty$,

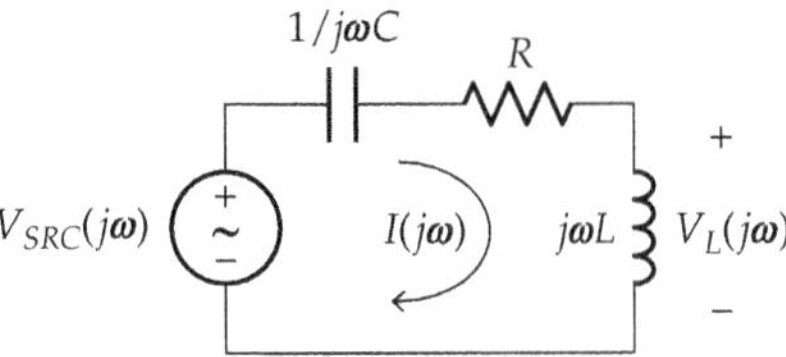

FIGURE 14.29
Second-order lowpass response across *L*.

the capacitor acts as a short circuit and the inductor as an open circuit and $V_C(j\omega) = V_{SRC}(j\omega)$. The response is clearly of a high-pass nature. It is generally true that when the response of a filter is taken across an inductor, the response tends to zero as $\omega \to 0$ so that the response is of a high-pass nature.

At the resonant frequency, the net reactance is zero, and if R is small enough, the current at this frequency will be large. Thus, as $R \to 0$, $I(j\omega_0) \to \infty$. When this large current is multiplied by the impedance of the inductor, $V_L(j\omega)$ can exceed $V_{SRC}(j\omega)$. In other words, the second-order high-pass response can exhibit some peaking, unlike the first-order high-pass response.

It follows from voltage division that

$$H_{HP}(j\omega) = \frac{V_L(j\omega)}{V_{SRC}(j\omega)} = \frac{j\omega L}{j\omega L + R + 1/j\omega C} = \frac{-\omega^2 LC}{1-\omega^2 LC + j\omega CR}$$

$$= \frac{-\omega^2/\omega_0^2}{1-\omega^2/\omega_0^2 + j\omega/\omega_0 Q} \tag{14.58}$$

The denominator of Equation 14.58 is the same as that of the second-order responses previously considered. Dividing the numerator and denominator of Equation 14.58 by $-\omega^2 LC$,

$$H_{HP}(j\omega) = \frac{1}{1-\omega_0^2/\omega^2 - j\omega_0/\omega Q} \tag{14.59}$$

Comparing Equations 14.59 and 14.51, it is seen that the high-pass and low-pass responses can be derived, one from the other, by interchanging $j\omega$ and ω_0, as was noted in connection with first-order responses.

It follows from Equation 14.59 that

$$\left|H_{HP}(j\omega)\right| = \frac{1}{\sqrt{\left(1-\left(\frac{\omega_0}{\omega}\right)^2\right)^2 + \left(\frac{\omega_0}{\omega Q}\right)^2}} \tag{14.60}$$

and

$$\angle H_{HP}(j\omega) = -\tan^{-1}\frac{\left(-\frac{\omega_0}{\omega Q}\right)}{1-\left(\frac{\omega_0}{\omega}\right)^2} \tag{14.61}$$

Note that in Equation 14.61 the imaginary part, $-\omega_0/\omega Q$, is always negative, but the real part, $\left(1-(\omega_0/\omega)^2\right)$, can be positive or negative. This means that +tan^{-1}(imaginary part/real part) is an angle in the third quadrant when ω is small, and moves counterclockwise to the fourth quadrant as $\omega \to \infty$. Its negation, $\angle H_{HP}(j\omega)$, is in the second or first quadrants, respectively.

If ω_0/ω is denoted by r, the expression under the square root in Equation 14.60 can be represented as

$$g(u) = \left(1-r^2\right)^2 + r^2/Q^2 \tag{14.62}$$

Equation 14.62 is of exactly the same form as Equation 14.54 with $r = \omega_0/\omega$ replacing $u = \omega/\omega_0$. The same conclusions reached for the low-pass response apply to the high-pass response, but with ω/ω_0 replaced by ω_0/ω. Thus, by analogy to Equation 14.55, $|H_{HP}(j\omega)|$ has a maximum at

$$r_{max} = \sqrt{1-1/2Q^2} \quad \text{or} \quad \omega_{max} = \omega_0/\sqrt{1-1/2Q^2} \tag{14.63}$$

The interpretation is that maximum $I(j\omega)$ occurs at $\omega = \omega_0$, when the net reactance is zero. However, $V_L(j\omega) = j\omega L I(j\omega)$ so that maximum $V_L(j\omega)$ occurs at a frequency slightly higher than ω_0, as indicated by Equation 14.63. The value of $|H_{HP}(j\omega)|_{max}$ is the same as that of $|H_{LP}(j\omega)|_{max}$ given by Equation 14.56. This is to be expected, for if $|H_{HP}(j\omega)|$ is the same function of ω_0/ω as $|H_{LP}(j\omega)|$ is a function of ω/ω_0, the high-pass and low-pass Bode magnitude plots for the same circuit are symmetrical with respect to the line $\omega = \omega_0$. Thus, at a frequency $\omega = k\omega_0$, where k is a constant, $\omega/\omega_0 = k$ and $\left|H_{LP}(j\omega)\right| = 1/\sqrt{\left(1-k^2\right)^2 + k^2/Q^2}$ (Equation 14.52). At a frequency $\omega_0/\omega = k$, $|H_{HP}(j\omega)|$ will have the same value (Equation 14.60). On a log scale, the frequencies $k\omega_0$ and ω_0/k are equidistant on either side of ω_0 so that the two plots are symmetrical with respect to the line $\omega = \omega_0$.

It can be readily shown, as was done for the low-pass response, that the low-frequency asymptote has a slope of +40 dB/decade and the high-frequency asymptote coincides with the horizontal axis. The Bode magnitude and phase plots are discussed in more detail in Example 14.4.

Example 14.4: Bode Plots of Second-Order High-Pass Response

It is required to obtain the Bode magnitude and phase plots of the high-pass response of the series *RLC* circuit assuming the same values of *R*, *L*, and *C* as in Example 14.3 (Figure 14.30).

Solution:

The values of ω_0 and Q are the same as in Example 14.3.

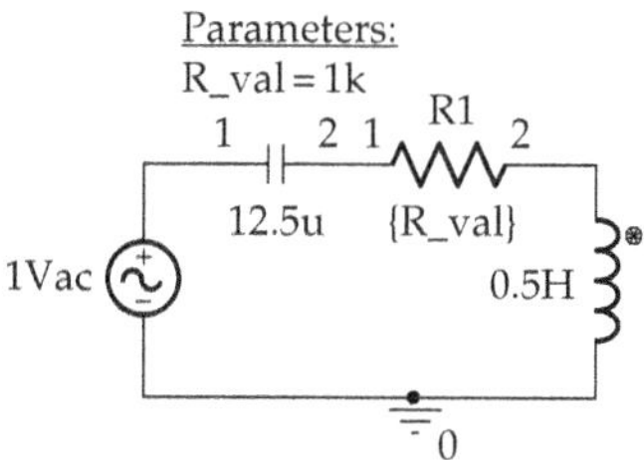

FIGURE 14.30
Figure for Example 14.4.

Simulation: The circuit is entered as in Figure 14.30; the simulation procedure is the same as in Example 14.3, but with the voltage across the inductor V(L1:1) selected for plotting.

After the simulation is run, the magnitude plot for the four resistance values is displayed, as in Figure 14.31a. The high-frequency asymptote is added by selecting Trace/Add traces and then entering in the Trace Expression field −40*LOG10[63.662]+ 40*LOG10[Frequency], where $f_0 = (400/2\pi) = 63.662$ Hz. Note the symmetry with respect to the line through f_0 between the low-pass response of Figure 14.28a and the high-pass response of Figure 14.31a.

It is instructive to interpret the phase response. For $0 < \omega < \omega_0$, the net series reactance is capacitive. For small ω, this reactance is large, $I(j\omega)$ leads $V_{SRC}(j\omega)$ by slightly less than 90° and $V_L(j\omega)$ leads $I(j\omega)$ by 90°, so $\angle H_{LP}(j\omega)$ is slightly less than 180°. In Equation 14.53, ω_0/ω is large, the imaginary part and the real part are negative. Their ratio is proportional to ω/ω_0 and is small. The angle given by the arctan function is in the third quadrant, slightly more than 180°. Its negation is in the second quadrant, slightly less than 180°. For ω slightly less than ω_0, the net capacitive reactance is small, $I(j\omega)$ leads $V_{SRC}(j\omega)$ by a small angle and $V_L(j\omega)$ leads $I(j\omega)$ by 90°, so $\angle H_{LP}(j\omega)$ is slightly more than 90°. In Equation 14.53, ω_0/ω is slightly larger than 1. The imaginary part and the real part are still negative but their ratio is positive and large. The angle is in the third quadrant, slightly less than 270°. Its negation is in the second quadrant, slightly more than 90°. For $\omega > \omega_0$, the net series reactance is inductive. For ω slightly larger than ω_0, the net inductive reactance is small, $I(j\omega)$ lags $V_{SRC}(j\omega)$ by a small angle and $V_L(j\omega)$ leads $I(j\omega)$ by 90°, so $\angle H_{LP}(j\omega)$ is slightly less than 90°. In Equation 14.53, ω_0/ω is slightly less than 1. The imaginary part is negative and the real part is positive. Their ratio is negative and large. The angle is in the fourth quadrant, slightly larger than 270°. Its negation is in the first quadrant, slightly less than 90°. As $\omega \to \infty$, the net inductive reactance is large, $V_L(j\omega) \cong V_{SRC}(j\omega)$, so $\angle H_{LP}(j\omega)$ approaches 0° in the first quadrant.

Primal Exercise 14.16

Compare the high-pass response of the series circuit of Primal Exercise 14.15 with its low-pass response considered in that exercise. Determine (a) Q, (b) $\left|H_{HP}(j\omega)\right|_{max}$ in dB, and (c) ω_{max}.

Ans. (a) 5, the same as for low-pass response; (b) 14.02 dB, the same as for low-pass response; (c) 50.51 kHz.

Primal Exercise 14.17

Determine the type of response $V_O(j\omega)/V_{SRC}(j\omega)$ in Figure 14.32 and its passband gain in dB.

Ans. High pass, −6 dB.

Primal Exercise 14.18

Determine the type of response $I_O(j\omega)/I_{SRC}(j\omega)$ in Figure 14.33 and its passband gain in dB.

Ans. Low pass, ≅ −8 dB.

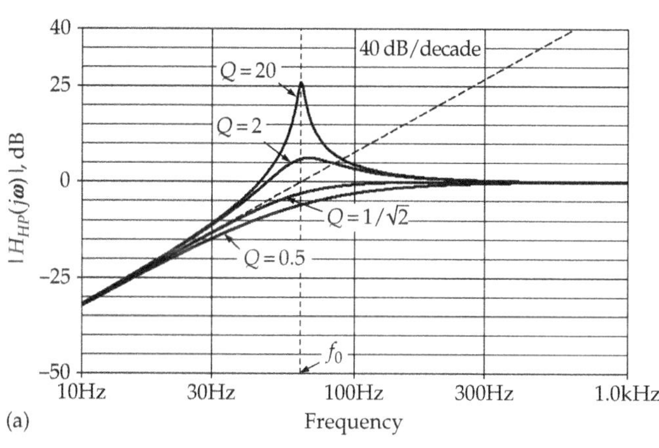

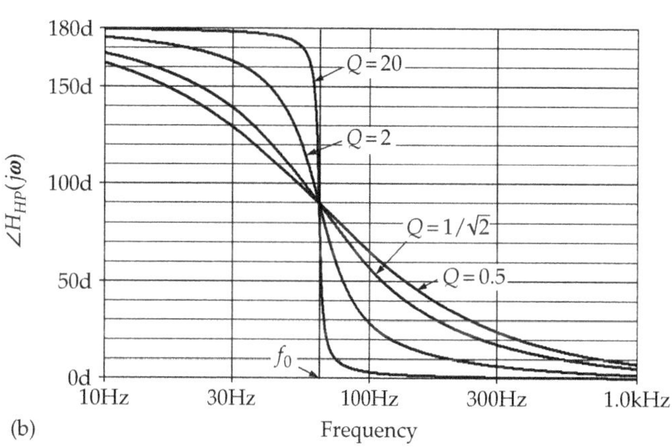

FIGURE 14.31
Figure for Example 14.4.

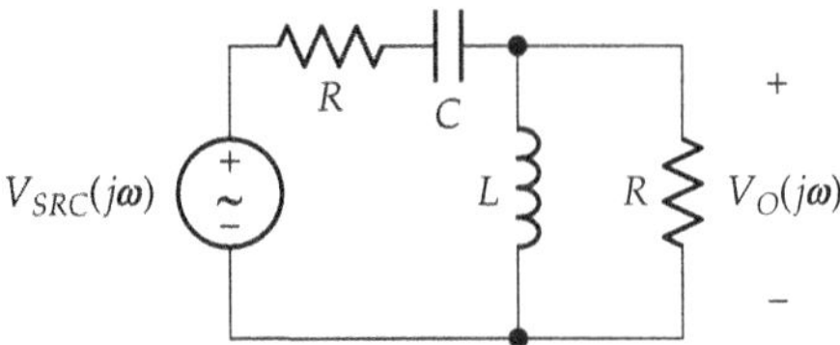

FIGURE 14.32
Figure for Primal Exercise 14.17.

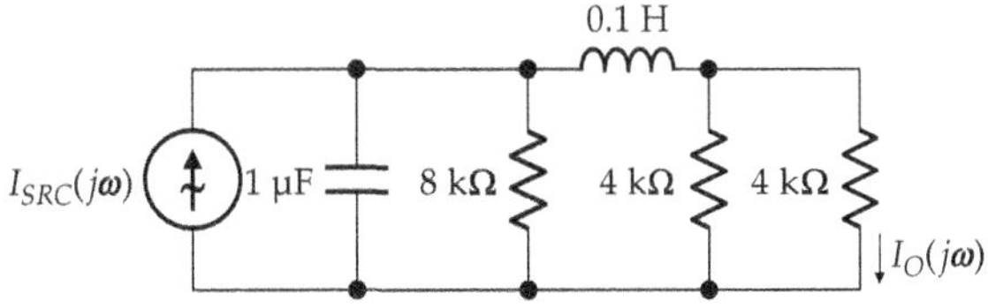

FIGURE 14.33
Figure for Primal Exercise 14.18.

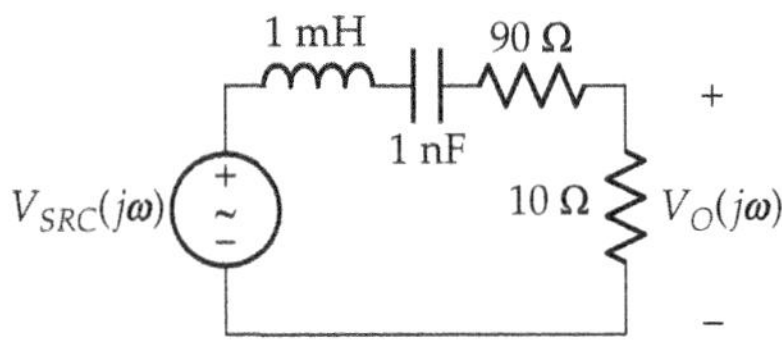

FIGURE 14.34
Figure for Primal Exercise 14.19

Primal Exercise 14.19

Determine the frequency at which the high-frequency asymptote of the response $V_O(j\omega)/V_{SRC}(j\omega)$ in Figure 14.34 intersects the 0 dB line.

Ans. 10 krad/s.

14.8 Parallel Circuit

The parallel *GCL* circuit excited by a current source (Figure 14.35) is the dual of the series *RLC* circuit of Figure 14.15, the dual quantities being the same as in Table 12.3. This table is reproduced as Table 14.1 in terms of the phasor variables, Q, ω_0, and BW.

ω_0 is clearly the same in both cases, since duality interchanges L and C, which leaves their product unaltered.

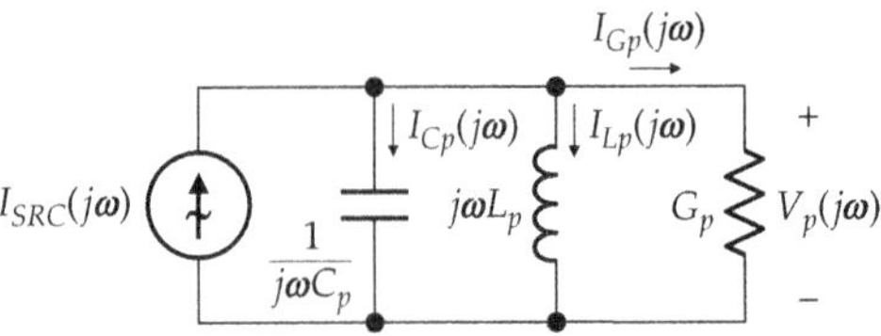

FIGURE 14.35
Prototypical parallel *GCL* circuit.

Q for the series circuit is given by Equation 14.31 as $Q = \omega_0 L/R = 1/\omega_0 CR$. Replacing quantities by their duals,

$$Q_p = \frac{\omega_0 C_p}{G_p} = \frac{1}{\omega_0 L_p G_p} \tag{14.64}$$

In terms of $R_p = 1/G_p$, Equation 14.64 becomes

$$Q_p = \omega_0 C_p R_p = \frac{R_p}{\omega_0 L_p} \tag{14.65}$$

Note that the dual of G_p is R in the series case, both having the same numerical values. But R_p is the reciprocal of G_p. It is also helpful to remember that the expressions for Q_p in Equation 14.65 are reciprocal in form to the expressions for Q in Equation 14.31. Thus, $\omega_0 C_p R_p$ is reciprocal in form to $1/\omega_0 CR$, and $R_p/\omega_0 L_p$ is reciprocal in form to $\omega_0 L/R$. Note that for a parallel resonant circuit, the larger R_p, the larger is Q_p.

It should be noted that Q, whether for a series or a parallel circuit, is determined only by circuit parameters, irrespective of any independent sources that may be present and their connections. Hence, the circuit configuration, whether series or parallel, can be ascertained, and the correct expression and value of Q derived, by setting independent sources to zero. This is the same as applied in Section 12.4, for ascertaining the circuit configuration and determining α.

With ω_0 the same, it is seen that ω_{c1}, ω_{c2}, and BW are given by the same expressions as Equations 14.39 and 14.40, but with Q replaced by Q_p:

$$\omega_{c1} = \omega_0\left[\sqrt{1+\left(1/2Q_p\right)^2} - 1/2Q_p\right],$$

$$\omega_{c2} = \omega_0\left[\sqrt{1+\left(1/2Q_p\right)^2} + 1/2Q_p\right] \tag{14.66}$$

TABLE 14.1
Dual Quantities

Series *RLC*	R	L	C	$V_R(j\omega)$	$V_L(j\omega)$	$V_C(j\omega)$	$I(j\omega)$	$Q = \omega_0 L/R = 1/\omega_0 CR$	$\omega_0 = 1/\sqrt{LC}$	$\text{BW} = \omega_0/Q$
Parallel *GCL*	G_p	C_p	L_p	$I_{Gp}(j\omega)$	$I_{Cp}(j\omega)$	$I_{Lp}(j\omega)$	$V_p(j\omega)$	$Q_p = \omega_0 C_p R_p = R_p/\omega_0 L_p$	$\omega_0 = 1/\sqrt{C_p L_p}$	$\text{BW} = \omega_0/Q_p$

TABLE 14.2

Dual Frequency Responses

Response	Transfer Functions: Series Circuit	Parallel Circuit
Bandpass	$\frac{V_R(j\omega)}{V_{SRC}(j\omega)} = \frac{j\omega/\omega_0 Q}{1-\omega^2/\omega_0^2 + j\omega/\omega_0 Q}$	$\frac{I_{Gp}(j\omega)}{I_{SRC}(j\omega)} = \frac{j\omega/\omega_0 Q_p}{1-\omega^2/\omega_0^2 + j\omega/\omega_0 Q_p}$
Bandstop	$\frac{V_{LC}(j\omega)}{V_{SRC}(j\omega)} = \frac{1-\omega^2/\omega_0^2}{1-\omega^2/\omega_0^2 + j\omega/\omega_0 Q}$	$\frac{I_{LpCp}(j\omega)}{I_{SRC}(j\omega)} = \frac{1-\omega^2/\omega_0^2}{1-\omega^2/\omega_0^2 + j\omega/\omega_0 Q_p}$
Lowpass	$\frac{V_C(j\omega)}{V_{SRC}(j\omega)} = \frac{1}{1-\omega^2/\omega_0^2 + j\omega/\omega_0 Q}$	$\frac{I_{Lp}(j\omega)}{I_{SRC}(j\omega)} = \frac{1}{1-\omega^2/\omega_0^2 + j\omega/\omega_0 Q_p}$
Highpass	$\frac{V_L(j\omega)}{V_{SRC}(j\omega)} = \frac{-\omega^2/\omega_0^2}{1-\omega^2/\omega_0^2 + j\omega/\omega_0 Q}$	$\frac{I_{Cp}(j\omega)}{I_{SRC}(j\omega)} = \frac{-\omega^2/\omega_0^2}{1-\omega^2/\omega_0^2 + j\omega/\omega_0 Q_p}$

and

$$\text{BW} = \omega_{c2} - \omega_{c1} = \frac{\omega_0}{Q_p} \quad \text{and} \quad \sqrt{\omega_{c1}\,\omega_{c2}} = \omega_0 \tag{14.67}$$

The frequency responses follow readily from duality. These are listed in Table 14.2 for the series *RLC* circuit, as has previously been derived, together with the dual *GCL* counterpart. It is seen that the only difference in the dual responses when expressed in this form is the replacement of Q of the series circuit by Q_p of the parallel circuit.

The responses of the parallel circuit can be readily argued qualitatively. Consider, for example, the response $I_{Gp}(j\omega)/I_{SRC}(j\omega)$ and its complement with respect to the source, $I_{LpCp}(j\omega)/I_{SRC}(j\omega) = (I_{Lp}(j\omega) + I_{Cp}(j\omega))/I_{SRC}(j\omega)$. The key to understanding the behavior of the circuit is to recognize that at the resonant frequency ω_0, the susceptances of the inductor and capacitor are equal in magnitude but opposite in sign so that their sum is zero, that is, $\omega_0 C - 1/\omega_0 L = 0$. In other words, the impedance at resonance of L_p in parallel with C_p is infinite; that is, the parallel combinations act as an open circuit. This follows also for the parallel impedance at resonance as

$$\left(j\omega_0 L_p \,\|\, \left(1/j\omega_0 C_p\right)\right) = \frac{j\omega_0 L_p / j\omega_0 C_p}{j\left(\omega_0 L_p - 1/\omega_0 C_p\right)} \to \infty \tag{14.68}$$

This is in contrast to the series connection of L and C, where the impedance at resonance is zero, so that the series combination acts as a short circuit.

If the parallel combination of L_p and C_p acts as an open circuit at resonance, then the whole of the source current flows through G_p so that the combined current through the inductor and capacitor is $I_{LpCp}(j\omega_0) = I_{Lp}(j\omega_0) + I_{Cp}(j\omega_0) = 0$, and $I_{Gp}(j\omega_0) = I_{SRC}(j\omega_0)$. At $\omega = 0$ the capacitor acts as an open circuit and the inductor as a short circuit. All of $I_{SRC}(j\omega)$ is diverted through L_p so that $I_{LpCp}(j\omega) = I_{SRC}(j\omega)$, and $I_{Gp}(j\omega_0) = 0$. If $\omega \to \infty$, the inductor acts as an open circuit and the capacitor as a short circuit. All of the $I_{SRC}(j\omega)$ is diverted through C_p so that, again, $I_{LpCp}(j\omega) = I_{SRC}(j\omega)$, and $I_{Gp}(j\omega_0) = 0$. It is seen that the response $I_{Gp}(j\omega)/I_{SRC}(j\omega)$ is bandpass, whereas the response $I_{LpCp}(j\omega)/I_{SRC}(j\omega)$ is bandstop, as expected from duality. Similarly, it follows that the response $I_{Cp}(j\omega)/I_{SRC}(j\omega)$ is high pass, whereas the response $I_{Lp}(j\omega)/I_{SRC}(j\omega)$ is low pass.

It may be noted that the circuit of Figure 14.35 is of special interest because it is the small-signal representation of a parallel-tuned circuit driven by a transistor amplifier. Examples 14.5 and 14.6 consider parallel *GCL* circuits. Example 14.6 illustrates that in determining the circuit configuration, whether series or parallel, independent sources are set to zero, as was done in Section 12.4. The same argument applies, namely, when voltages and currents in a circuit change, for any reason, including a variation in frequency; ideal, independent voltage sources act as short circuits; and ideal, independent current sources act as open circuits, as far as changes are concerned.

Primal Exercise 14.20

Consider the parallel *GCL* circuit that is the dual of the series circuit of Primal Exercise 14.15. Determine (a) L_p, C_p, G_p, R_p, (b) ω_0, (c) Q_p, and (d) BW; (e) compare ω_0, Q_p, and BW with those of the dual series circuit.

Ans. (a) L_p = 4 nH, C_p = 0.1 F, G_p = 1 kS, R_p =1 mΩ; (b) 50 krad/s; (c) 5; (d) 10 krad/s; (e) the same as those of the dual series circuit.

Exercise 14.21

It was demonstrated earlier that the impedance of a series combination of L and C is zero at resonance, whereas the admittance of a parallel combination of L and C is zero at resonance. Express this behavior in terms of duality.

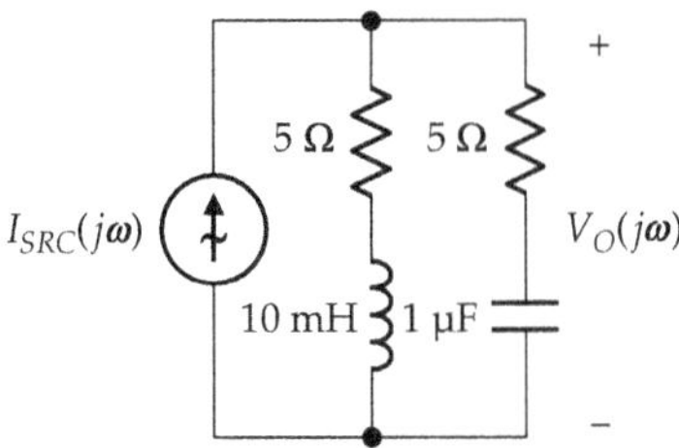

FIGURE 14.36
Figure for Primal Exercise 14.22.

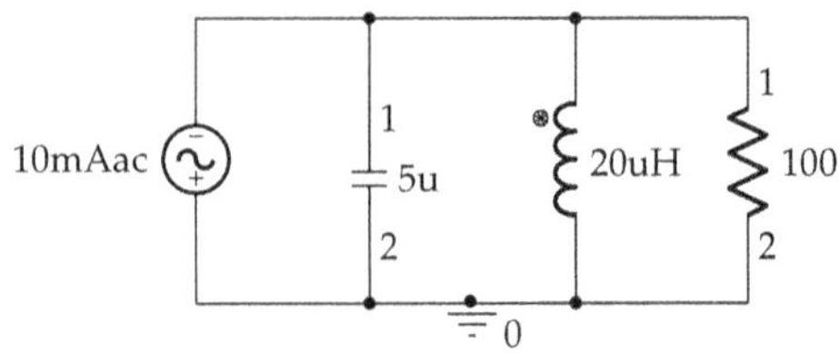

FIGURE 14.37
Figure for Example 14.5.

Primal Exercise 14.22

Determine Q of the response $V_O(j\omega)/I_{SRC}(j\omega)$ in Figure 14.36
Ans. 10.

Example 14.5: Parallel *GCL* Circuit

A parallel *GCL* circuit excited by a current source is given, as in Figure 14.35. If a current input of 10 mA peak is to produce a maximum response of 1 V peak at 100 krad/s, it is required to determine L_p and R_p to give a BW of 2 krad/s and to derive ω_{c1} and ω_{c2}. This example also illustrates using MATLAB to derive Bode plots.

Solution:

At resonance, the 10 mA current source sees only R_p, since L_p and C_p in parallel act as an open circuit. Hence, to produce a 1 V response at resonance, 1 V = $R_p \times 1$ mA, which gives $R_p = \dfrac{1}{0.01} = 100\ \Omega$.

If BW = 2 krad/s and ω_0 = 100 krad/s, it follows from Equation 14.67 that $Q_p = \dfrac{\omega_0}{\text{BW}} = 50$. With $Q_p = \omega_0 C_p R_p$, $C_p = \dfrac{Q_p}{\omega_0 R_p} \equiv 5\ \mu\text{F}$. Then $L_p = \dfrac{1}{\omega_0^2 C_p} = 20\ \mu\text{H}$.

From Equation 14.66, $\omega_{c1} = 100\left[\sqrt{1+10^{-4}} - 0.01\right] = 99.005$ krad/s and $\omega_{c2} = 100\left[\sqrt{1+10^{-4}} + 0.01\right] = 101.005$ krad/s. The geometric mean $\sqrt{\omega_{c1}\omega_{c2}}$ is 100 krad/s (Equation 14.40) and is very nearly equal to the arithmetic mean $(\omega_{c1} + \omega_{c2})/2$, because of the relatively large Q_p.

Simulation: The circuit is entered as in Figure 14.37, using the source IAC. The resistor and capacitor are entered from the ANALOG_P library. To run the simulation, select AC Sweep/Noise for the Analysis type in the Simulation Settings dialog box. Under AC Sweep Type select 'Logarithmic' and enter 100 for 'Start Frequency', 1meg for 'End Frequency', and 3000 for 'Points/Decade'. After the simulation is run, select in the SCHEMATIC1 page Trace/Add Trace and then DB() from the right window and V(R1:1) from the left window. The plot appears as in Figure 14.38a.

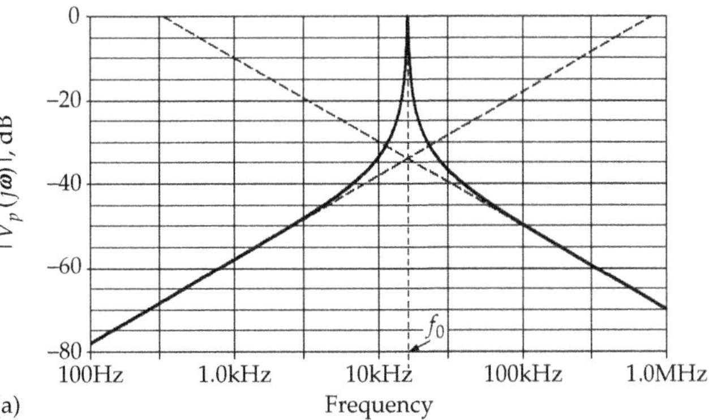

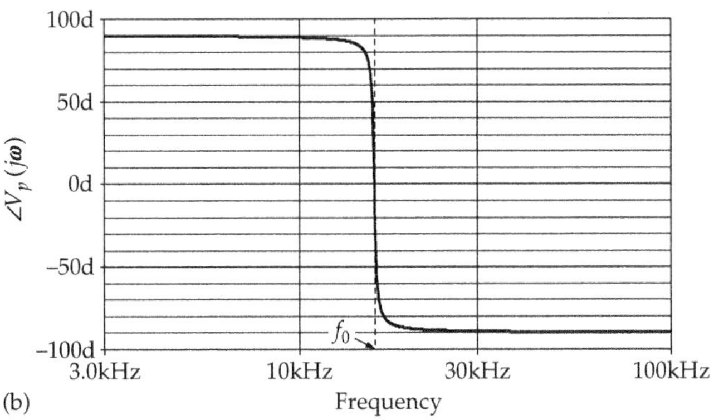

FIGURE 14.38
Figure for Example 14.5.

Note that this is a plot of $V_p(j\omega)$ in response to the 10 mA source and is not a plot of the transfer function $V_p(j\omega)/I_{SRC}(j\omega)$, which is obtained with IAC = 1 A. The analytical expression for $V_p(j\omega)$ is

$$\left|V_p(j\omega)\right| = \frac{1}{\sqrt{1+(50)^2(\omega/\omega_0 - \omega_0/\omega)^2}}\ \text{V} \tag{14.69}$$

At $\omega = \omega_0$, $|V_p(j\omega_0)|$ = 1 V, and $20\log_{10}|V_p(j\omega_0)| = 0$, as in Figure 14.38a. The low-frequency asymptote, as $\omega \to 0$, is $|V_p(j\omega)| = \omega/50\omega_0$, and the high-frequency asymptote, as $\omega \to \infty$, is $|V_p(j\omega)| = \omega_0/50\omega$. The two asymptotes intersect, where $\omega/50\omega_0 = \omega_0/50\omega$, which occurs at $\omega = \omega_0$.

To display the low-frequency asymptote, select Trace/Add Trace and enter –20*LOG10(15915.5) – 20*LOG10(50) + 20*LOG10(Frequency) in the Trace expression field, where f_0 = 15915.5 Hz. To display the high-frequency asymptote, the entry is 20*LOG10(15915.5) – 20*LOG10(50 – 20*LOG10(frequency). The two asymptotes intersect at $f = f_0$, as explained earlier.

f_0, f_{c1}, and f_{c2} can be read using the cursor. Click first on the Toggle cursor icon and then on the Cursor Max icon. f_0 is read in the Cursor Window as 15.922 K, compared to $10^5/(2\pi)$ = 15.916 kHz. With the cursor at f_0, enter the search command sle(max-3). Cursor 1 moves to f_{c2} and its value is displayed in the Cursor Window as 16.075 K, corresponding to 100.002 krad/s. Repeat the same command, but with Cursor 2 selected in the Search Command window. Cursor 2 moves to f_{c1} and its value is displayed in the Cursor Window as 15.757 K, corresponding to 99.004 krad/s. The difference Y2-Y1 is displayed as 318.000, corresponding to 1.998 krad/s. Alternatively, select Trace/Evaluate Measurement and then Bandwidth_Bandpass_3dB(1), and enter V(R1:1) in place of '1'. In the Measurement Results window, 318.11378 is displayed. The value of Q can be displayed by selecting Q_Bandpass(1,db_level) and entering V(R1:1) in place of '1' and 3 in place of db_level. The value of 50.00340 is displayed in the Measurement Results window.

The Bode phase plot can be derived by selecting Trace/add Trace and then P() from the right window and V(R1:1) from the left window. The plot appears as in Figure 14.38b.

Bode plots from MATLAB: The coefficients of $s = j\omega$ in the numerator and denominator of the transfer function are first entered as arrays in the order of decreasing powers of s, including the zero power for any constant term. The numerator in this case is $2 \times 10^3 s$ rad/s, so the coefficients are entered as

```
>>num = [2*10^3, 0]
```

The denominator is $s^2 + 2 \times 10^3 s + 10^{10}$, so the coefficients are entered as

```
>>den = [1, 2*10^3, 10^10]
```

The bode command is entered as

```
>>bode(num, den)
```

The Bode magnitude and phase plots are displayed as in Figure 14.39, after some editing for clarity.

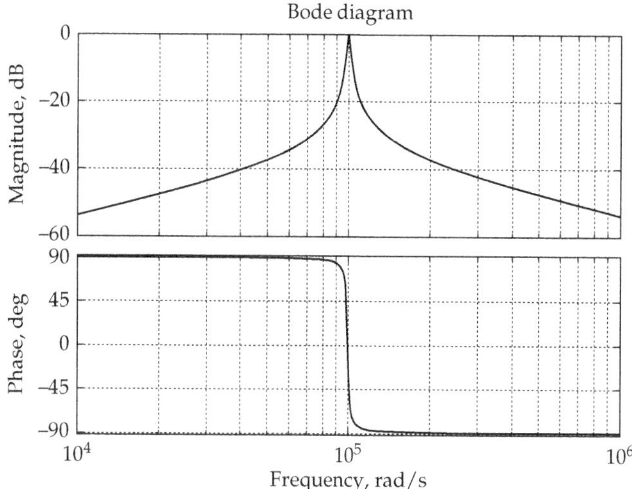

FIGURE 14.39
Figure for Example 14.5.

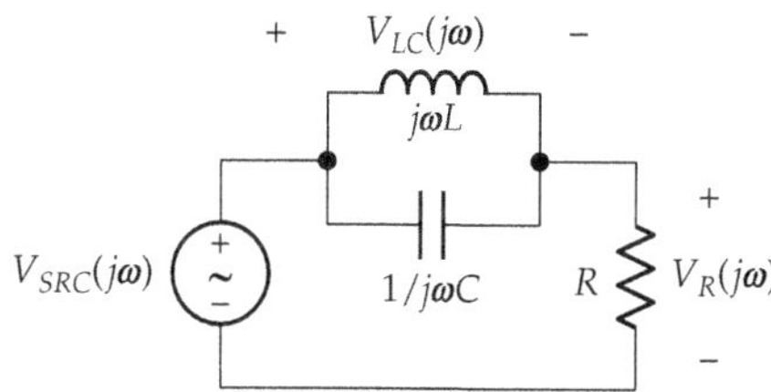

FIGURE 14.40
Figure for Example 14.6.

Example 14.6: Bandstop Response

(a) It is required to verify that the response $V_R(j\omega)$ of the circuit of Figure 14.40 is bandstop and that the response $V_{LC}(j\omega)$ is bandpass.

(b) If $R = 1\text{ k}\Omega$, determine L and C so that the half-power frequencies are 1000 and 1200 rad/s.

Solution:

(a) At the resonant frequency, L in parallel with C acts as an open circuit so that the current through R is zero, $V_R(j\omega) = 0$, and $V_{LC}(j\omega) = V_{SRC}(j\omega)$. At $\omega = 0$, the inductor acts as a short circuit so that $V_{LC}(j\omega) = 0$ and $V_R(j\omega) = V_{SRC}(j\omega)$. Similarly, as $\omega \to \infty$, the capacitor acts as a short, and again $V_{LC}(j\omega) = 0$ and $V_R(j\omega) = V_{SRC}(j\omega)$. It follows that the response across R is bandstop, whereas the response across L and C is bandpass. The transfer function is

$$H_{BS}(j\omega) = \frac{V_R(j\omega)}{V_{SRC}(j\omega)} = \frac{R}{R + \dfrac{j\omega L/j\omega C}{j\omega L + 1/j\omega C}} = \frac{R}{R + \dfrac{j\omega L}{1-\omega^2 LC}}$$

$$= \frac{R\left(1-\omega^2 LC\right)}{R\left(1-\omega^2 LC\right) + j\omega L} = \frac{1-\omega^2/\omega_0^2}{1-\omega^2/\omega_0^2 + \dfrac{j\omega}{\omega_0}\left(\dfrac{\omega_0 L}{R}\right)}$$

$$= \frac{1-\omega^2/\omega_0^2}{1-\omega^2/\omega_0^2 + \dfrac{j\omega}{\omega_0}\left(\dfrac{1}{Q_p}\right)} \quad (14.70)$$

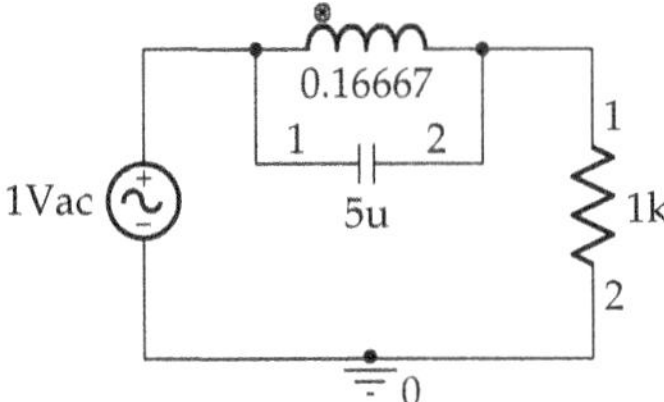

FIGURE 14.41
Figure for Example 14.6.

Equation 14.70 is of exactly the same form as Equation 14.47, with Q_p, replacing Q, where $Q_p = R/\omega_0 L$ is Q of the parallel circuit, as given by Equation 14.65, with R_p replaced by R. This confirms that the circuit configuration can be ascertained by setting the voltage source to zero, in which case the circuit reduces to a parallel GCL circuit.

(b) $\omega_0 = \sqrt{1000 \times 1200} = 1095.4$ rad/s. From Equation 14.65, $C = \dfrac{Q_p}{\omega_0 R} = \dfrac{1}{\text{BW} \times R} = \dfrac{1}{200 \times 1 \times 10^3} \equiv 5\,\mu\text{F}$; $L = 1/\omega_0^2 C = 1/6$ H. $Q_p = \omega_0 CR = 5.48$.

Simulation: The circuit is entered as in Figure 14.41, as in previous examples in this chapter. To run the simulation, select AC Sweep/Noise for the Analysis type in the Simulation Settings dialog box. Under AC Sweep, 10 is entered for 'Start Frequency', 1k for 'End Frequency', and 3000 for 'Points/Decade'. After the simulation is run, select in the SCHEMATIC1 page Trace/Add Trace and then DB() from the right window and V(R1:1) from the left window. The x-axis range is set from 30 Hz to 1 kHz using Plot/Axis Settings. The plot appears as in Figure 14.42. Cursor minimum gives f_0 = 174.350. Cursor command 'sle(0.70711)' gives f_{c2} = 190.984. Repeating this command, but selecting cursor 2, gives f_{c1} = 159.153. The calculated values of f_0, f_{c2}, and f_{c1} are 174.346, 190.986, and 159.155 Hz, respectively.

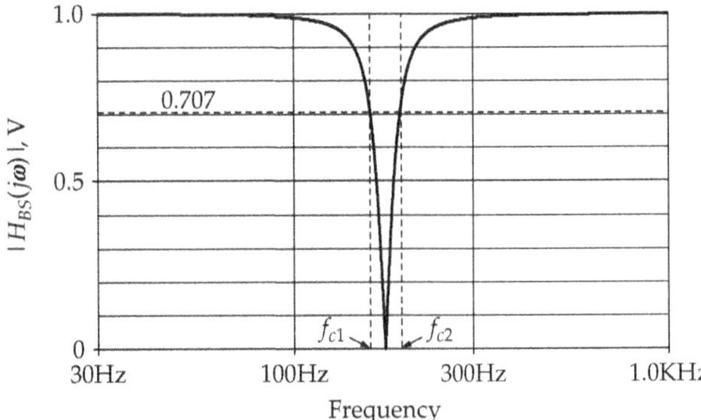

FIGURE 14.42
Figure for Example 14.6.

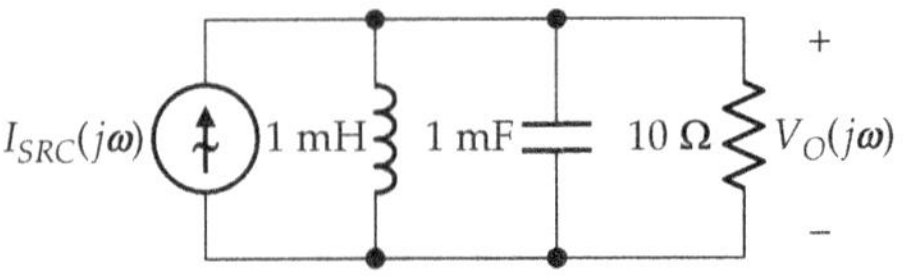

FIGURE 14.43
Figure for Primal Exercise 14.23.

Primal Exercise 14.23

Determine the type of response $V_O(j\omega)/I_{SRC}(j\omega)$, its maximum magnitude, and Q in Figure 14.43.

Ans. Response is bandpass, maximum response is 10 Ω; $Q = 10$.

14.9 Summary of Second-Order Responses

Prototypical, second-order responses are summarized in Table 14.3, normalized to a maximum response magnitude of 1 and with $j\omega$ replaced by s.

Although s, strictly speaking, is the complex frequency of the Laplace transform (Chapter 20), it can, in fact, be replaced by $j\omega$ for the sinusoidal steady state (Section 22.3), as is the case with frequency responses. Substituting $s = j\omega$ gives the same responses as in Table 14.2 multiplied by ω_0^2, with $s^2 = -\omega_0^2$.

The following should be noted concerning Table 14.3:

1. By definition, the prototypical responses in terms of s have a unity coefficient of the term representing the highest power of s in the denominator, which is s^2 in the case of second-order responses, as well as in the numerator, when s^2 occurs.
2. The frequency responses can, in general, be multiplied by a scalar K, without changing the nature of the response.
3. When the responses are normalized to a maximum response magnitude of 1, the numerator of the low-pass response must be ω_0^2, the constant term in the denominator, so that the response is 1 when $\omega = 0$ or $s = 0$. Similarly, the numerator of the high-pass response must be s^2 so that the response is 1 when $\omega \to \infty$ or $s \to \infty$. The numerator of the bandpass response must be $(\omega_0/Q)s$ so that the response is 1 when $s^2 = -\omega_0^2$.

TABLE 14.3

Prototypical, Second-Order Responses

Type of Response	Transfer Function
Lowpass	$\dfrac{\omega_0^2}{s^2+\dfrac{\omega_0}{Q}s+\omega_0^2}$
Bandpass	$\dfrac{\dfrac{\omega_0}{Q}s}{s^2+\dfrac{\omega_0}{Q}s+\omega_0^2}$
Bandstop	$\dfrac{s^2+\omega_0^2}{s^2+\dfrac{\omega_0}{Q}s+\omega_0^2}$
Highpass	$\dfrac{s^2}{s^2+\dfrac{\omega_0}{Q}s+\omega_0^2}$
Allpass	$\dfrac{s^2-\dfrac{\omega_0}{Q}s+\omega_0^2}{s^2+\dfrac{\omega_0}{Q}s+\omega_0^2}$

4. Each term in the numerator and denominator has the units of (rad/s)[2] so that the transfer function is dimensionless in term of ratios of voltages or currents.
5. The denominator is the same in all the responses. This is a consequence of the denominator being an expression of the characteristic equation of the differential equation (Equation 12.6) and the fact that the homogeneous differential equation is of the same form for all responses in the circuit.
6. The coefficient ω_0/Q of s in the denominator is the 3 dB BW of bandpass and bandstop responses. Q in the low-pass and high-pass responses is defined as in Table 14.1 for series and parallel circuits.
7. Being independent of the excitation, Q can, in all cases, be determined by setting independent sources to zero.
8. Whereas the maximum change in the phase angle is 90° in first-order responses, it is 180° in second-order responses, except the all-pass response. In all the types of fist-order and second-order circuits discussed in this chapter, the phase angle becomes more lagging as the frequency increases.
9. The all-pass response has a magnitude of unity at all frequencies but produces a phase shift of 360° as a function of frequency (Problem P14.38).
10. The responses are interrelated. For example, the all-pass response is the bandpass response subtracted from the bandstop response (Problem P14.38). The bandstop response is the sum of the low-pass and high-pass responses.
11. Interchanging s and ω_0 changes a low-pass response to a high-pass response and conversely. This is equivalent to interchanging inductors and capacitors. For if we consider $\omega_0^2=\dfrac{1}{L_kC_m}$, then $\dfrac{\omega_0^2}{s^2}=\dfrac{1}{(sL_k)(sC_m)}$. Interchanging L and C is equivalent to replacing the impedance sL_k by an impedance $1/sC_k$ and replacing sC_m by $1/sL_m$. The expression ω_0^2/s^2 becomes $\dfrac{1}{(1/sC_k)(1/sL_m)}=s^2L_mC_k=\dfrac{s^2}{\omega_0^2}$, where the expression for ω_0^2 becomes $1/L_mC_k$. It is seen that interchanging L and C is equivalent to interchanging ω_0^2/s^2 and s^2/ω_0^2, that is, interchanging s and ω_0.
12. The response of a circuit can be a combination of the prototypical responses of Table 14.3, as illustrated by the problems at the end of this chapter.

Primal Exercise 14.24

Specify the transfer function of a bandpass filter that has $\omega_0 = 5$ rad/s, $Q = 1$, and a maximum passband gain of 2.

Ans. $\dfrac{10s}{s^2+5s+25}$.

Primal Exercise 14.25

The transfer function of a filter is of the form $\dfrac{As}{s^2+As+B}$, where $s = j\omega$ and A and B depend on circuit parameters but not on ω. Determine B if the half-power frequencies are 20 and 45 krad/s.

Ans. 900 krad[2]/s[2].

Exercise 14.26

Based on items 1 and 3 of Section 14.9, derive the prototypical responses in terms of s for first-order filters. Compare to Equations 14.14 and 14.23. Note that interchanging s and the 3 dB cutoff frequency changes a low-pass response to a high-pass response, and conversely.

Ans. $L_P:\dfrac{\omega_{cl}}{s+\omega_{cl}}$; $H_P:\dfrac{s}{s+\omega_{ch}}$.

Learning Checklist: What Should Be Learned from This Chapter

- The transfer function $H(j\omega)$ in the frequency domain is the ratio of the phasor of the designated output to the phasor of a single input applied to the filter:
  1. As a ratio of two phasors, $H(j\omega)$ has a magnitude and phase angle, both of which are, in general, functions of frequency.
  2. $H(j\omega)$ is derived under steady-state sinusoidal conditions, with independent sources present only at the input of the filter.
- There are four basic frequency responses: low pass, high pass, bandpass, and bandstop. The range of frequencies over which there is little or no attenuation between input and output is the passband, whereas the range of frequencies over which there is considerable attenuation between input and output is the stopband.
- The basic frequency responses are interrelated in the sense that some responses can be derived from other responses through addition, subtraction, and multiplication or by complementing the normalized response with respect to unity.
- The frequency response of a first-order filter is either low pass or high pass, the variation of the response with frequency being due to the frequency-dependent reactance of energy storage elements.
- In logarithmic plots,
  1. The plot of $y = ax^n$ is a straight line of slope n; a shifts the line vertically, without affecting its slope.
  2. Equal spacing along a logarithmic axis is equivalent to equal ratios of the variable involved.
- In Bode magnitude plots, the horizontal axis is labeled in terms of ω but scaled logarithmically as $\log_{10}\omega$. The vertical axis is labeled and scaled *linearly* as $20\log_{10}|H(j\omega)|$ dB, which is the dB value of $|H(j\omega)|$. In Bode phase plots, the horizontal axis is the same as in the Bode magnitude plots, whereas the vertical axis is the phase shift in degrees.
- A characteristic feature of Bode plots is that they can be approximated over some frequency ranges by straight-line asymptotes. First-order, low-pass, and high-pass responses have a horizontal-line asymptote over the passband. The asymptote in the stopband is a line of slope −20 dB/decade in the case of a low-pass filter and of slope +20 dB/decade in the case of a high-pass filter.
- The points of intersection of asymptotes of Bode plots are corner frequencies. In the case of first-order filters, the corner frequency is also the 3 dB cutoff frequency, at which the magnitude of the transfer function is $1/\sqrt{2}$ of its limiting, maximum value in the passband. In all cases, the 3 dB cutoff frequency is synonymous with the half-power frequency, at which the power delivered to a resistive load at the output of the filter is one-half the maximum power delivered to the same load at the limiting maximum value in the passband.
- Second-order responses are more varied and can be low pass, high pass, bandpass, or bandstop. The first three types of response show pronounced peaking or resonance when the damping is small.
- A second-order response can be obtained from two independent energy storage elements of the same type, that is, elements that cannot be combined into a single element. However, Q of these circuits does not exceed 0.5.
- In a series *RLC* circuit, the sum of the inductive and capacitive reactances is zero at the resonant frequency ω_0 so that the series combination of L and C acts as a short circuit. In a parallel *GCL* circuit, the sum of the inductive and capacitive susceptances is zero at the resonant frequency ω_0 so that the parallel combination of L and C acts as an open circuit.
- A bandpass response can result from cascading a low-pass response and a high-pass response. A bandstop response can result from the summation of a low-pass and a high-pass response.
- The 3 dB BW of bandpass and bandstop circuits equals ω_0/Q. The higher the Q, the narrower the BW for a given ω_0, where Q depends on the power dissipated in the circuit, relative to the maximum energy store.
- $\omega_0 = 1/\sqrt{LC}$ for both series *RLC* circuits and parallel *GCL* circuits. In a series circuit, $Q = \omega_0 L/R = 1/\omega_0 CR$. In a parallel circuit, $Q_p = \omega_0 C_p R_p = R_p/\omega_0 L_p$.

- In all prototypical second-order circuits, the denominator is of the form $s^2(\omega_0/Q)s+\omega_0^2$. In the case of bandpass and bandstop responses, ω_0/Q is the BW.
- Interchanging L and C in a low-pass, second-order circuit changes the response to high pass and conversely.

Problem-Solving Tips

1. Always determine how a filter circuit behaves at $\omega = 0$, as $\omega \to \infty$, and at the resonant frequency in the case of second-order circuits.
2. In determining the phase response, retain the signs of the imaginary and real parts, whose ratio is the tangent of the phase angle, so as to locate the phase angle in the correct quadrant.

Problems

Verify solutions by PSpice simulation.

First-Order Responses

P14.1 Given the transfer function $H(j\omega)=\dfrac{10^7}{10^6+j\omega}$, sketch the magnitude Bode plot and determine the output voltage as a function of time, when the input voltage is $0.1\sin\omega t$, where (a) $\omega = 0.3\times10^6$ rad/s, (b) $\omega = 10^6$ rad/s, and (c) $\omega = 3\times10^6$ rad/s.

Ans. (a) $0.958\sin(0.3\times10^6t - 16.7°)$ V; (b) $0.707\sin(10^6t - 45°)$ V; (c) $0.316\sin(3\times10^6t - 71.6°)$ V.

P14.2 An RC low-pass filter is required having a 3 dB cutoff frequency of 500 Hz, using a 50 nF capacitor. Determine (a) R and (b) the transfer function, and specify where the output voltage is taken.

Ans. (a) 6366 Ω; (b) $3141.6/(3141.6 + j\omega)$, the output voltage being across the capacitor.

P14.3 Suppose that the filter in the preceding problem is loaded with a resistor having the same value of R. Determine (a) the 3 dB cutoff frequency in rad/s, (b) the transfer function, and (c) the passband gain at $\omega = 0$.

Ans. (a) 6283.2 rad/s; (b) $3141.6/(6283.2 + j\omega)$; (c) 1/2.

P14.4 Given a low-pass RL filter having $R = 1$ kΩ and $L = 20$ mH. Determine (a) the 3 dB cutoff frequency in krad/s and (b) the low-pass transfer function.

Ans. (a) 50 krad/s; (b) $5\times10^4/(5\times10^4 + j\omega)$.

P14.5 Determine, for the filter in Figure P14.5, (a) the transfer function, (b) the 3 dB cutoff frequency in rad/s, (c) the magnitude of the gain and the phase shift as a function of frequency, and (d) the maximum gain in the passband.

Ans. (a) $j3\omega/(1 + j9\omega)$; (b) 1/9 rad/s; (c) $3\omega/\sqrt{1+81\omega^2}$, $90° - \tan^{-1}(9\omega)$; (d) 1/3 as $\omega \to \infty$.

P14.6 Determine, for the filter in Figure P14.6, (a) the transfer function, (b) the 3 dB cutoff frequency in rad/s, (c) the magnitude of the gain and the phase shift as a function of frequency, and (d) the maximum gain in the passband.

Ans. (a) $j0.04\omega/(160 + j0.2\omega)$; (b) 800 rad/s; (c) $\omega/\sqrt{(800)^2+\omega^2}$, $90° - \tan^{-1}(\omega/800)$; (d) 0.2 as $\omega \to \infty$.

P14.7 Determine, for the filter in Figure P14.7, (a) the transfer function $V_O(j\omega)/I_{SRC}(j\omega)$, (b) the 3 dB cutoff frequency in krad/s, (c) the magnitude of the gain and the phase shift as a function of frequency, and (d) the maximum gain in the passband.

Ans. (a) $5\times10^3/(1+j\omega)$ V/A, ω in krad/s; (b) 1 krad/s; (c) $5\times10^3/\sqrt{1+\omega^2}$, $-\tan^{-1}\omega$; (d) 5×10^3 as $\omega \to 0$.

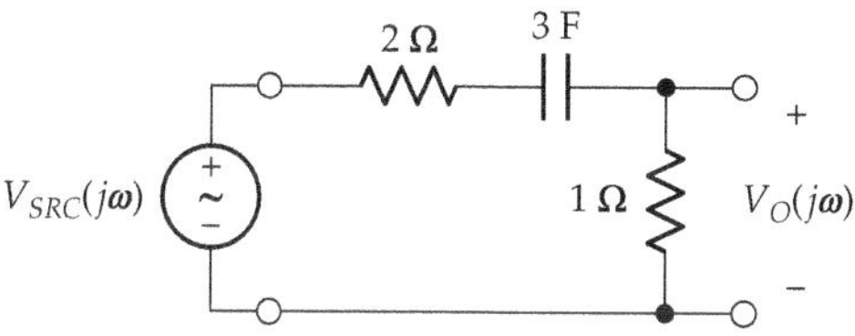

FIGURE P14.5

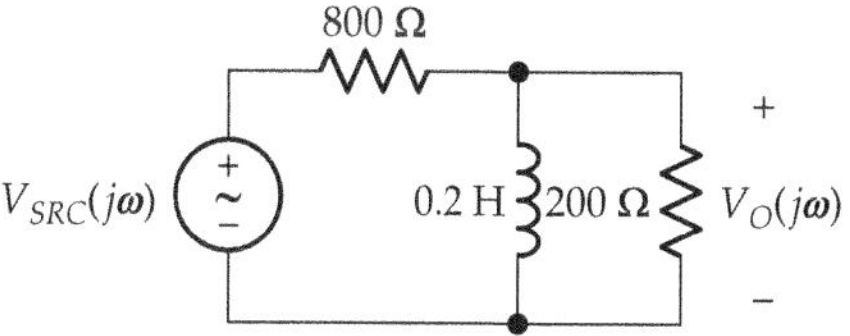

FIGURE P14.6

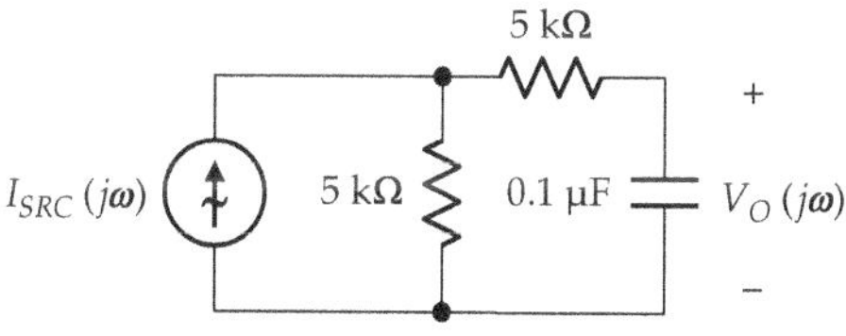

FIGURE P14.7

P14.8 Repeat the preceding problem, assuming $V_O(j\omega)$ is taken across the 5 kΩ resistor in series with the capacitor.

Ans. (a) $2.5 \times 10^3 j\omega/(1 + j\omega)$ V/A, ω in krad/s; (b) 1 krad/s; (c) $2.5\times10^3\omega/\sqrt{1+\omega^2}$, $90° - \tan^{-1}\omega$ (d) 2.5×10^3 as $\omega \to \infty$.

P14.9 Determine, for the filter in Figure P14.9, (a) the transfer function $V_O(j\omega)/V_{SRC}(j\omega)$, (b) the 3 dB cutoff frequency in Mrad/s, (c) the magnitude of the gain and the phase shift as a function of frequency, and (d) the maximum gain in the passband.

Ans. (a) $0.5/(1 + j40\omega)$, ω is in Mrad/s; (b) 25 krad/s; (c) $0.5/\sqrt{1+(40\omega)^2}$, $-\tan^{-1}(40\omega)$; (d) 0.5 as $\omega \to 0$.

P14.10 Repeat the preceding problem with $V_O(j\omega)$ taken across the 10 kΩ resistor.

Ans. (a) $j40\omega/(1 + j40\omega)$, ω is in Mrad/s; (b) 25 krad/s; (c) $40\omega/\sqrt{1+(40\omega)^2}$, $90° - \tan^{-1}(40\omega)$; (d) 1 as $\omega \to \infty$.

P14.11 Determine, for the filter in Figure P14.11, (a) the transfer function $V_O(j\omega)/V_{SRC}(j\omega)$, (b) the 3 dB cutoff frequency in krad/s, (c) the magnitude of the gain and the phase shift as a function of frequency, and (d) the maximum gain in the passband.

Ans. (a) $j2\omega/(1 + j1.8\omega)$, ω is in krad/s; (b) 5/9 krad/s; (c) $2\omega/\sqrt{1+(1.8\omega)^2}$, $90° - \tan^{-1}(1.8\omega)$; (d) 10/9 as $\omega \to \infty$.

P14.12 Repeat the preceding problem with the capacitor replaced by a 0.1 H inductor.

Ans. (a) $(10/9)/(1 + j\omega/180)$ V/A, ω is in krad/s; (b) 180 krad/s; (c) $(10/9)/\sqrt{1+(j\omega/180)^2}$, $-\tan^{-1}(j\omega/180)$; (d) 10/9 as $\omega \to 0$.

P14.13 (a) Determine Norton's equivalent circuit (NEC) seen by the resistor in Figure P14.13, (b) $I_O(j\omega)$, and (c) the 3 dB cutoff frequency.

Ans. (a) 2 A in parallel with 2/3 μF; (b) $2/(1 + j\omega/1500)$ A, ω is in rad/s; (c) 1.5 krad/s.

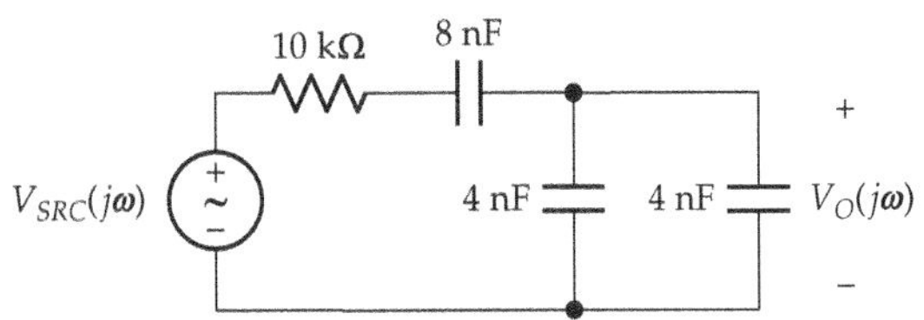

FIGURE P14.9

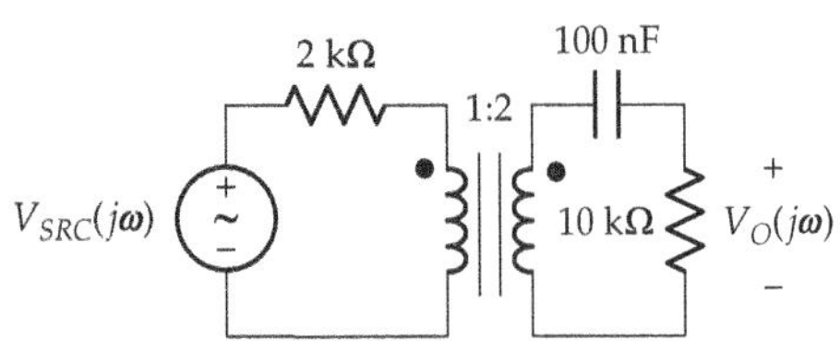

FIGURE P14.11

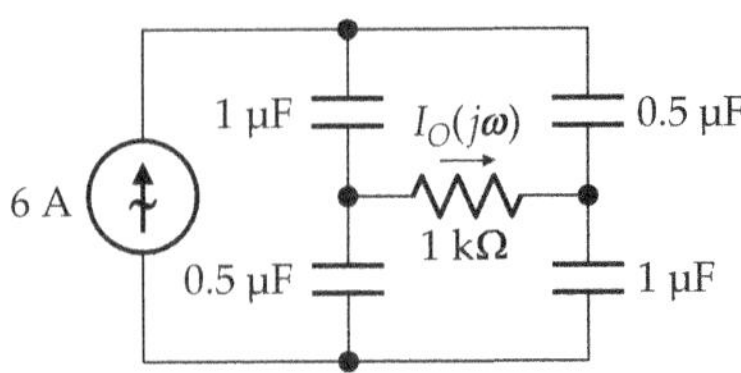

FIGURE P14.13

P14.14 Repeat the preceding problem assuming the capacitors are replaced by inductors, the μF units being replaced by H units.

Ans. (a) 2 A is parallel with 3/4 H; (b) $-j1.5\omega/(1000 + j0.75\omega)$ A, ω is in rad/s; (c) 4/3 krad/s.

P14.15 Determine, for the filter in Figure P14.15, (a) the transfer function $V_O(j\omega)/I_{SRC}(j\omega)$, (b) the 3 dB cutoff frequency in krad/s, (c) the magnitude of the gain and the phase angle as a function of frequency, and (d) the maximum gain in the passband.

Ans. (a) $2 \times 10^3/(1 + j0.09\omega)$ V/A, ω is in krad/s; (b) 100/9 krad/s; (c) $2\times10^3/\sqrt{1+(0.09\omega)^2}$, $-\tan^{-1}(0.09\omega)$; (d) 2×10^3 as $\omega \to 0$.

P14.16 Determine, for the filter in Figure P14.16, (a) the transfer function $V_O(j\omega)/V_{SRC}(j\omega)$, (b) the 3 dB cutoff frequency in krad/s, (c) the magnitude of the gain and the phase shift as a function of frequency, and (d) the maximum gain in the passband.

Ans. (a) $j5 \times 10^{-5}\omega/(1 + j3 \times 10^{-4}\omega)$, ω is in rad/s; (b) 10/3 krad/s; (c) $5\times10^{-5}\omega/\sqrt{1+(3\times10^{-4}\omega)^2}$, $90° - \tan^{-1}(3 \times 10^{-4}\omega)$; (d) 1/6 as $\omega \to \infty$.

P14.17 Repeat the preceding problem with the inductor replaced by a 50 nF capacitor.

Ans. (a) $(1/6)/(1 + j\omega 25/3)$, ω is in Mrad/s; (b) 120 krad/s; (c) $(1/6)/\sqrt{1+(25\omega/3)^2}$, $-\tan^{-1}(25\omega/3)$; (d) 1/6 as $\omega \to 0$.

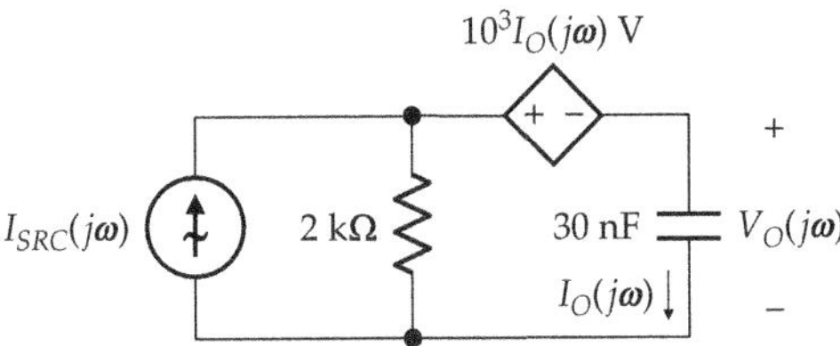

FIGURE P14.15

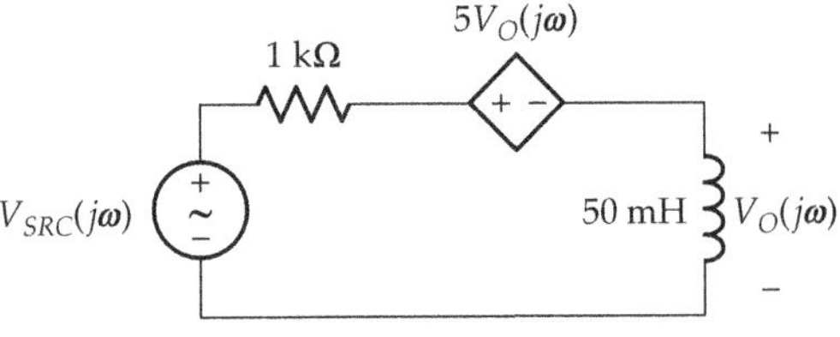

FIGURE P14.16

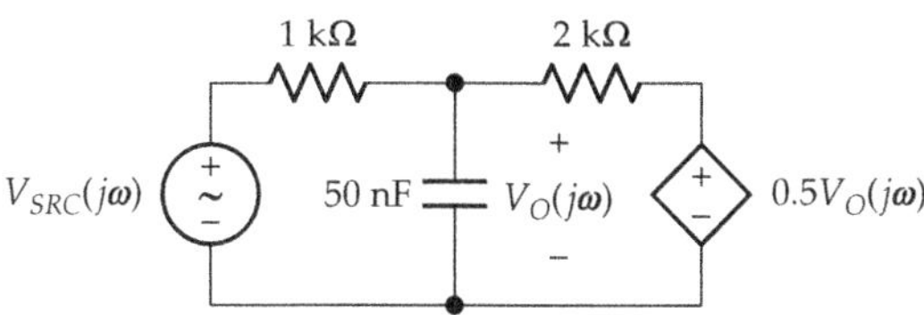

FIGURE P14.18

P14.18 Determine, for the filter in Figure P14.18, (a) the transfer function $V_O(j\omega)/V_{SRC}(j\omega)$, (b) the 3 dB cutoff frequency in krad/s, (c) the magnitude of the gain and the phase shift as a function of frequency, and (d) the maximum gain in the passband.

Ans. (a) $0.8/(1+j0.04\omega)$, ω is in krad/s; (b) 25 krad/s; (c) $0.8/\sqrt{1+(0.04\omega)^2}$, $-\tan^{-1}(0.04\omega)$; (d) 0.8 as $\omega \to 0$.

P14.19 Repeat the preceding problem with the capacitor replaced by a 50 mH inductor.

Ans. (a) $j0.05\omega/(1 + j\omega/16)$, ω is in krad/s; (b) 16 krad/s; (c) $0.05\omega/\sqrt{1+(\omega/16)^2}$, $90° - \tan^{-1}(\omega/16)$; (d) 0.8 as $\omega \to \infty$.

P14.20 Reduce the circuit of Figure P14.20 to a first-order circuit and specify the values of the circuit elements.

Ans. 5 Ω in series with 1/6 μF.

P14.21 Determine the transfer functions $V_C(j\omega)/I_{SRC}(j\omega)$ and $V_L(j\omega)/I_{SRC}(j\omega)$ in Figure P14.21. Note that each is a first-order transfer function that is independent of the parameters of the other subcircuit.

Ans. $R_1/(1 + j\omega CR_1)$ V/A, $-j\omega LR_2/(R_2 + j\omega L)$ V/A

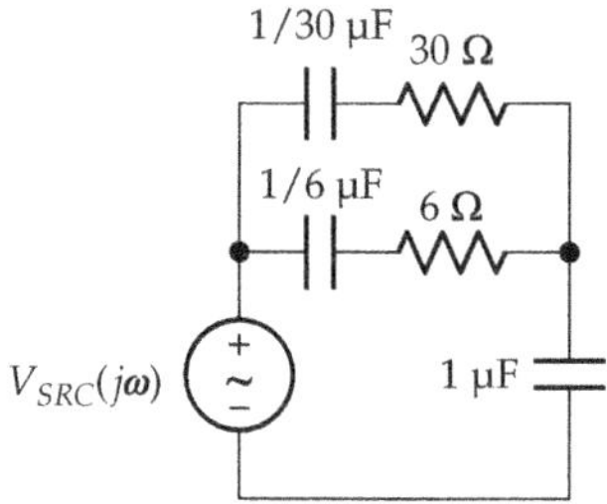

FIGURE P14.20

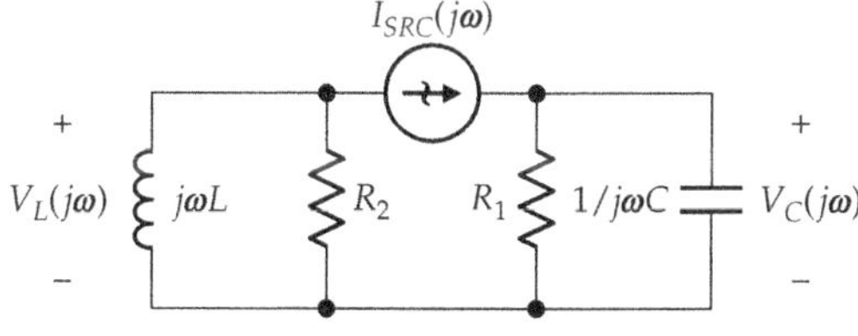

FIGURE P14.21

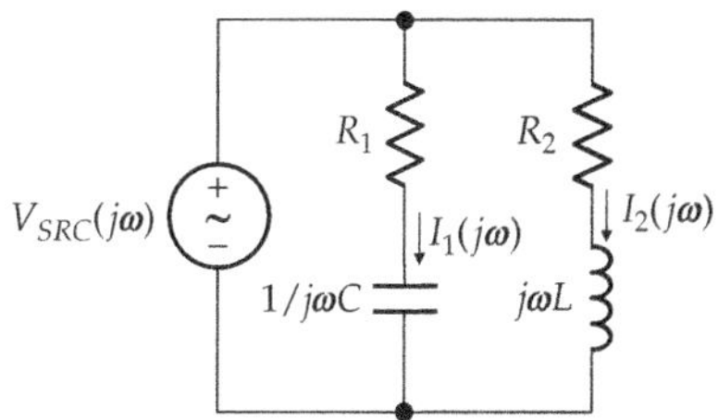

FIGURE P14.22

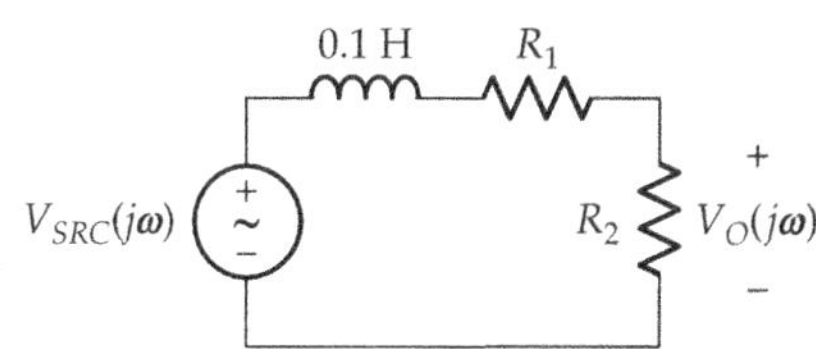

FIGURE P14.23

P14.22 Determine the transfer functions $I_1(j\omega)/V_{SRC}(j\omega)$ and $I_2(j\omega)/V_{SRC}(j\omega)$ in Figure P14.22. Note that each is a first-order transfer function that is independent of the parameters of the other subcircuit.

Ans. $j\omega C/(1 + j\omega CR_1)$ A/V, $1/(R_2 + j\omega L)$ A/V.

P14.23 The LF and HF asymptotes of the response $|V_O(j\omega)/V_{SRC}(j\omega)|$ in Figure P14.23 intersect at (1 krad/s, −12 dB). Determine R_2.

Ans. 25.12 Ω.

P14.24 A low-pass, first-order transfer function $H_1(j\omega)$ has a maximum gain in the passband of −2 dB and a corner frequency of 50 kHz. Another low-pass, first-order transfer function $H_2(j\omega)$ has a magnitude that is 8 times that of $H_1(j\omega)$ and a corner frequency of 1 kHz. Determine $|H_1(j\omega)H_2(j\omega)|$ in dB when it falls 3 dB below its maximum value at low frequencies.

Ans. 11 dB.

Second-Order Responses

P14.25 A series *RLC* circuit has R = 1 kΩ and half-power frequencies of 20 and 100 kHz. Determine L and C.

Ans. 1.99 mH, 0.251 μF.

P14.26 A series *RLC* circuit has a resonant frequency of 150 kHz and a BW of 75 kHz. Determine the half-power frequencies.

Ans. 117.1, 192.1 kHz.

P14.27 For the circuit of Figure P14.27, determine (a) the transfer function $V_O(s)/V_{SRC}(s)$, where $s = j\omega$, (b) ω_0, (c) Q, (d) BW, and (e) $v_O(t)$ if $v_{SRC}(t) = 450\cos(\omega_0 t)$ V.

Ans. (a) $\frac{4s}{s^2+5s+625}$, s is in krad/s; (b) 25 krad/s; (c) 5; (d) 5 krad/s; (e) $360\cos(\omega_0 t)$ V.

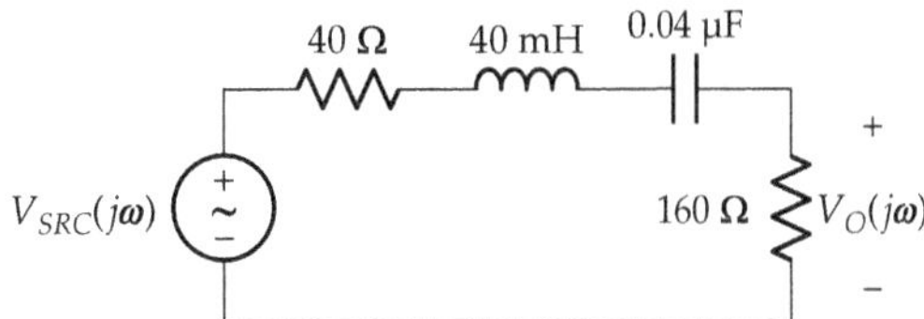

FIGURE P14.27

P14.28 Determine R_x in Figure P14.28 so that the BW does not exceed 750 rad/s.

Ans. $R_x \leq 500\ \Omega$.

P14.29 Determine in Figure P14.29 (a) the maximum value of the transfer function $|V_O(j\omega)/I_{SRC}(j\omega)|$; (b) Q and BW.

Ans. (a) 25 V/A; (b) 3.125, 3.2 krad/s.

P14.30 Determine R, L, and C in Figure P14.30 so that the maximum response is 1 V, ω_0 = 100 krad/s, and BW = 4 krad/s.

Ans. R = 10 kΩ, L = 4 mH, C = 25 nF.

P14.31 Determine ω_0 and Q in Figure P14.31.

Ans. 100 krad/s, 10.

P14.32 Determine ω_0 and Q in Figure P14.32.

Ans. 1.1 Mrad/s, 11.

P14.33 Determine R and L in Figure P14.33 so that the resonant frequency is 4 kHz and Q = 5.

Ans. 15.92 Ω, 3.17 mH.

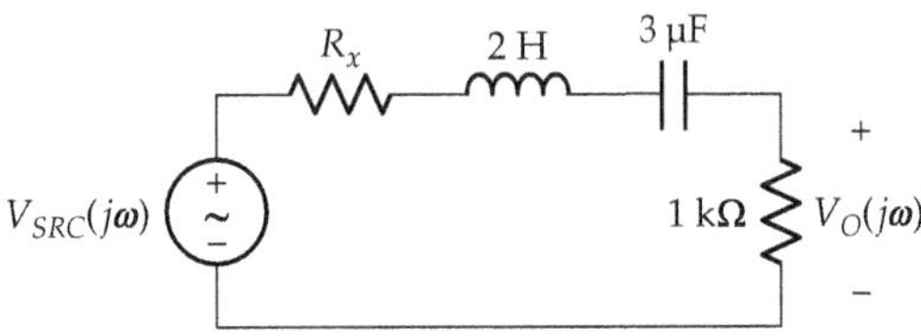

FIGURE P14.28

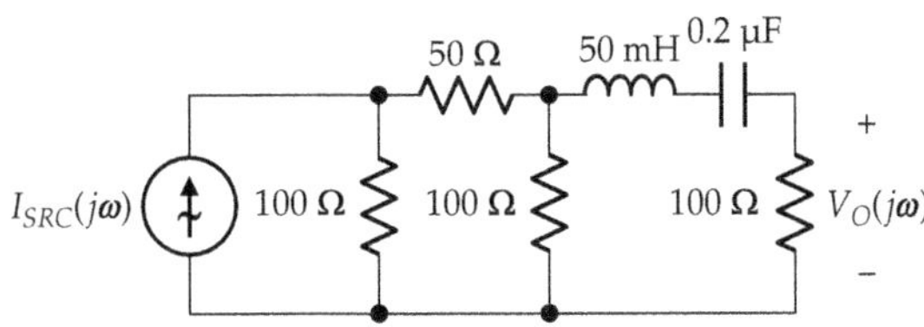

FIGURE P14.29

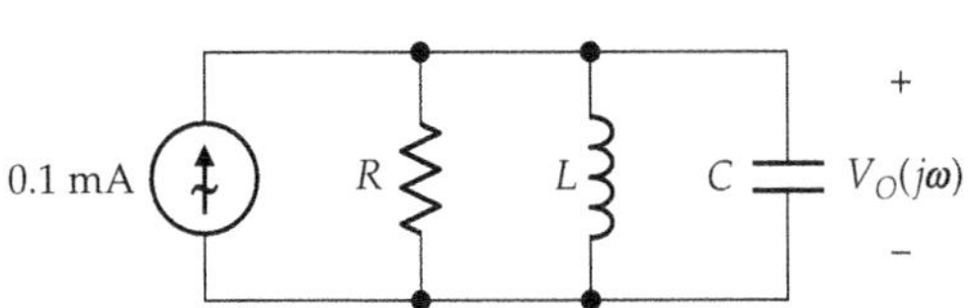

FIGURE P14.30

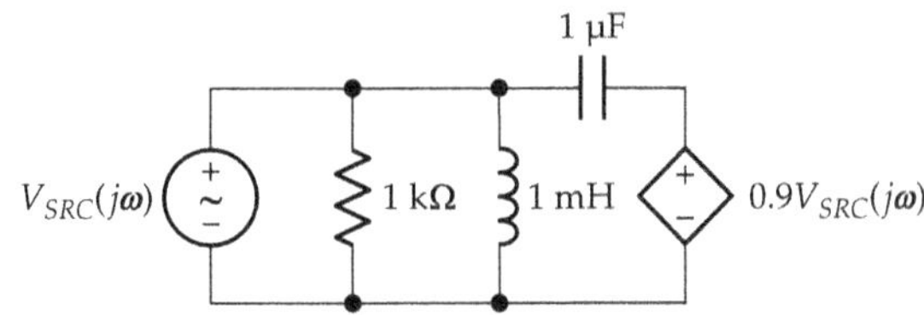

FIGURE P14.31

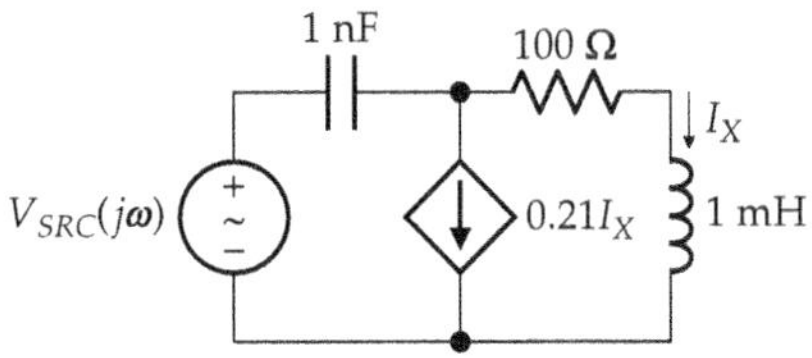

FIGURE P14.32

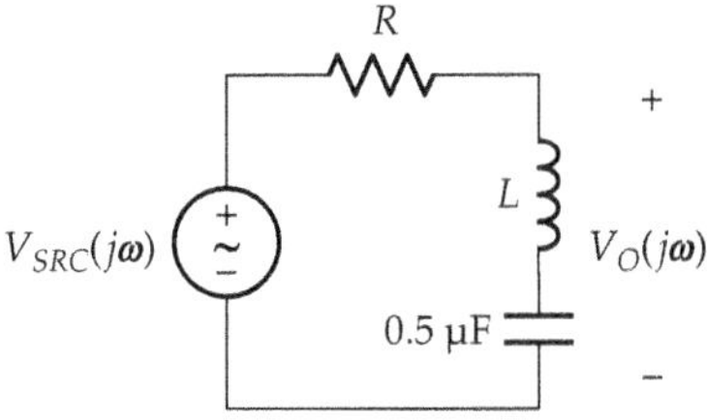

FIGURE P14.33

P14.34 Determine in Figure P14.34 (a) the minimum value of the transfer function $|V_O(j\omega)/V_{SRC}(j\omega)|$; (b) Q and BW.

Ans. (a) 1/15; (b) 80, 12.5 rad/s.

P14.35 Determine Q and BW in Figure P14.35.

Ans. 15, 10.61 kHz.

P14.36 If a 4 MΩ resistor is connected in parallel with L and C in Figure P14.36, determine the percentage change in BW.

Ans. BW increases by 25%.

P14.37 For the circuit of Figure P14.37, determine (a) ω_0, (b) Q, and (c) $V_O(j\omega_0)$ if $V_{SRC}(j\omega_0) = 1$ V.

Ans. (a) 10^4 rad/s; (b) 8; (c) 0.8 V.

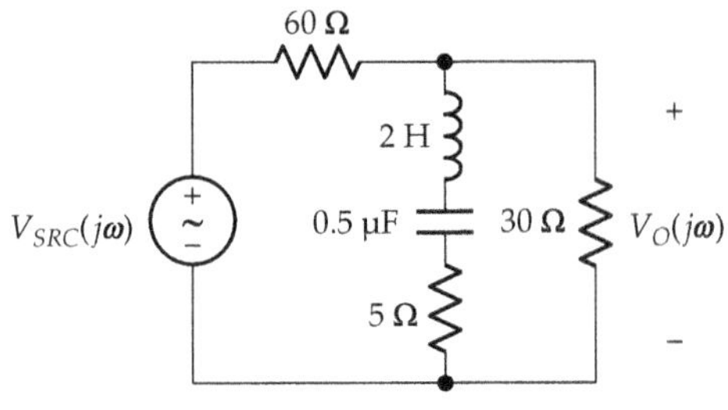

FIGURE P14.34

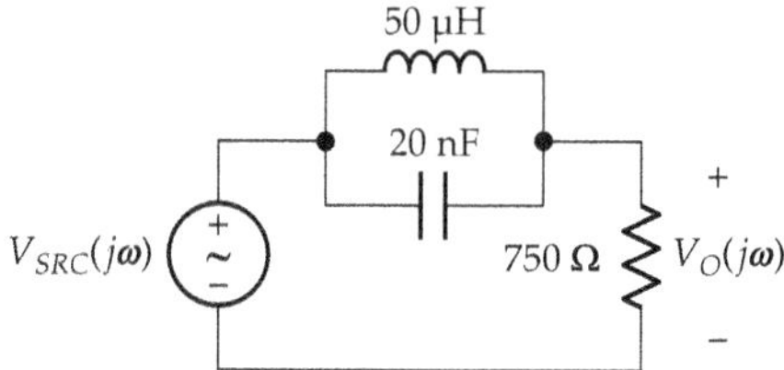

FIGURE P14.35

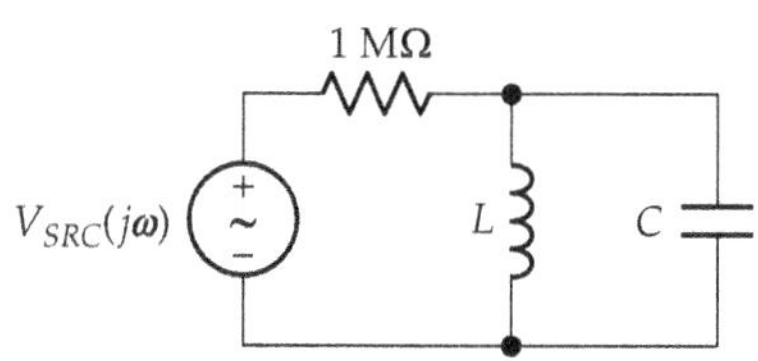

FIGURE P14.36

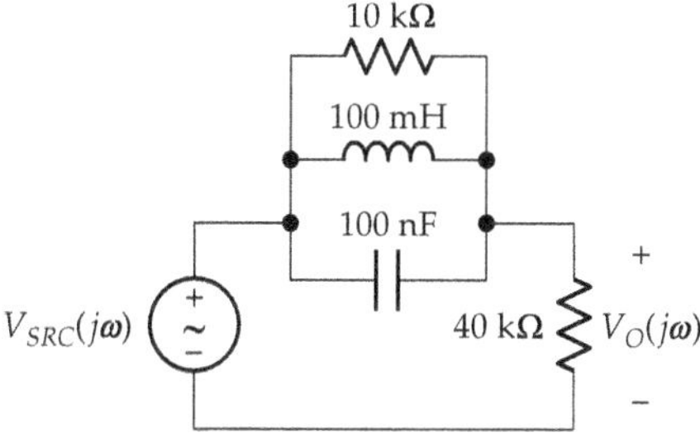

FIGURE P14.37

P14.38 (a) Show that the response $V_O(j\omega)/V_{SRC}(j\omega)$ in Figure P14.38 is an all-pass response. (b) Determine the frequency at which the phase shift is 180°, assuming $R = 10$ kΩ, $L = 1$ μH, and $C = 1$ μF.

Ans. 1 Mrad/s.

P14.39 Determine the frequency at which the response $V_O(j\omega)/V_I(j\omega)$ in Figure P14.39 is a maximum.

Ans. 10^4 rad/s.

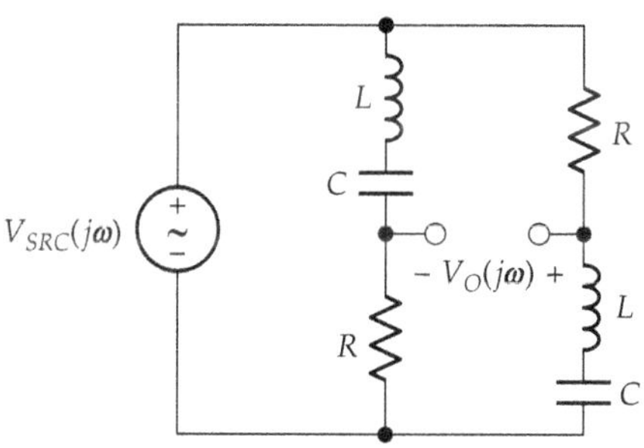

FIGURE P14.38

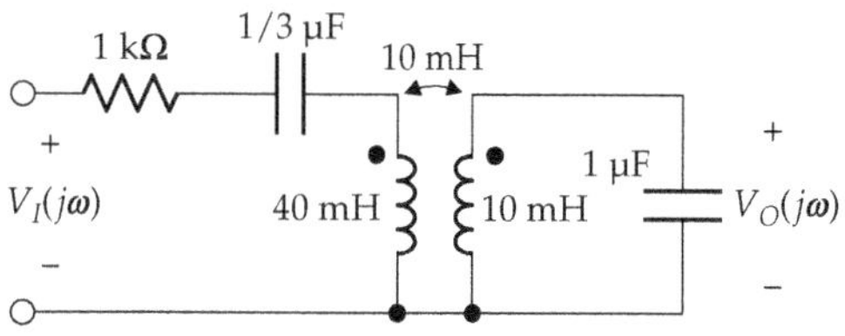

FIGURE P14.39

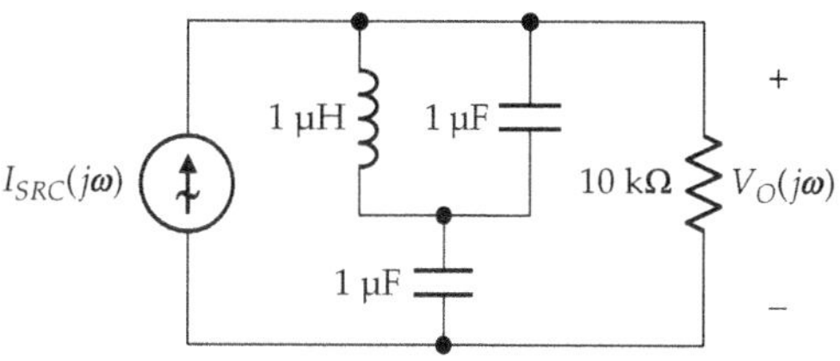

FIGURE P14.40

P14.40 Determine the frequency at which $V_O(j\omega) = 0$ in Figure P14.40.

Ans. $1/\sqrt{2}$ Mrad/s.

P14.41 Determine L_1 and L_2 in Figure P14.41 so that the magnitude of the transfer function $V_O(j\omega)/V_I(j\omega)$ is unity at 1 Mrad/s and zero at 0.5 Mrad/s.

Ans. $L_1 = 0.1$ mH, $L_2 = 0.3$ mH.

P14.42 Determine (a) the nature of the response in Figure P14.42, (b) ω_0, and (c) BW.

Ans. (a) Bandpass; (b) 1 krad/s; (c) 3 krad/s.

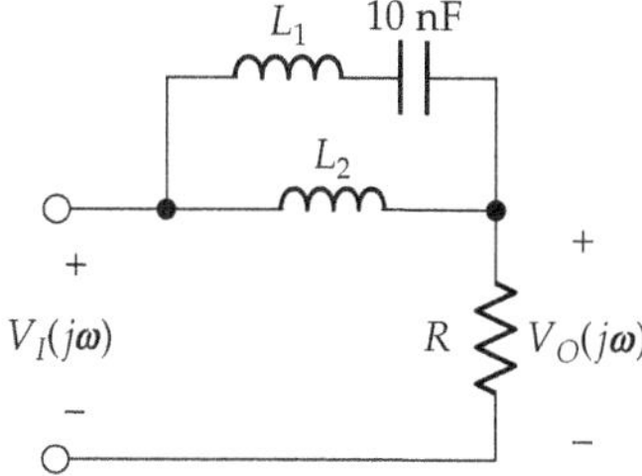

FIGURE P14.41

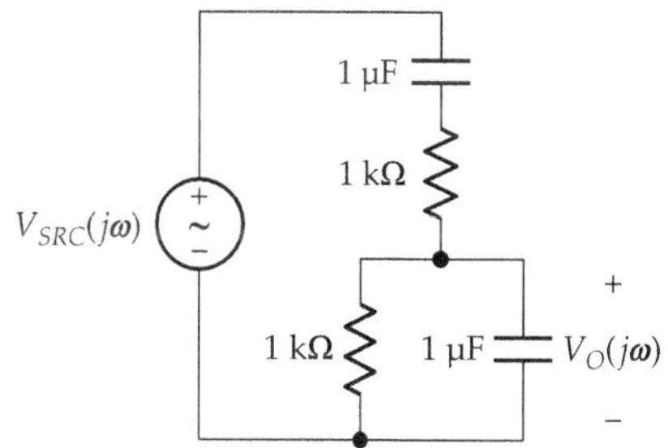

FIGURE P14.42

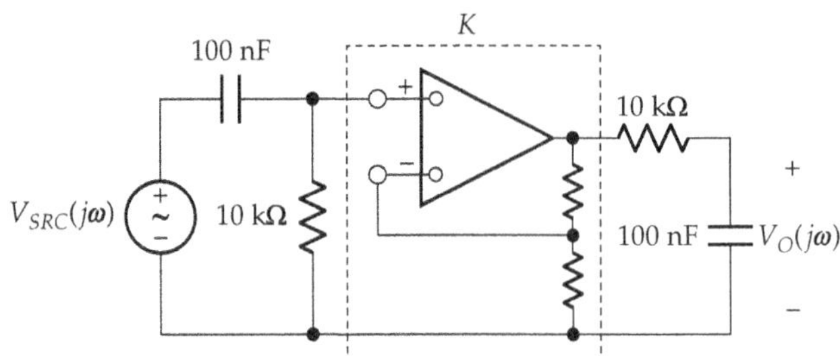

FIGURE P14.43

P14.43 Determine K, the gain of the op amp stage in Figure P14.43 so that the maximum magnitude of the response $V_O(j\omega)/V_{SRC}(j\omega)$ is 20 dB.

Ans. 20.

P14.44 (a) Show that the response $V_O(j\omega)/V_{SRC}(j\omega)$ in Figure P14.44 is the product of a first-order low-pass response and a first-order high-pass response.
(b) Choose R_f and C_f so that the two first-order responses have the same 3 dB cutoff frequency, and the maximum magnitude of the overall response is 0 dB, assuming $R = 2$ kΩ and $C = 0.5$ μF.
(c) Specify Q of the overall response.

Ans. (b) $R_f = 8$ kΩ, $C_f = 0.125$ μF; (c) 0.5.

P14.45 Given the circuit of Figure P14.45, where k is a positive constant. (a) Derive the transfer function $V_O(s)/V_{SRC}(s)$; (b) specify the type of response and derive the expression for Q; (c) determine Q when (i) $k = 1$ and (ii) k is very large. Explain the difference in Q in the two cases.

Ans. (a) $\dfrac{V_O(s)}{V_{SRC}(s)} = \dfrac{1}{C^2R^2}\dfrac{1}{s^2 + \dfrac{s}{CR}(2+1/k) + \dfrac{1}{C^2R^2}}$; (b) low pass, $Q = \dfrac{1}{2+1/k}$; (c) (i) 1/3; (ii) 0.5 because the second circuit does not load the first.

P14.46 Given the circuit of Figure P14.46, where k is a positive constant, (a) derive the transfer function $I_O(s)/I_{SRC}(s)$; (b) specify the type of response and derive the expression for Q; (c) determine Q when (i) $k = 1$ and (ii) k is very large. Explain the difference in Q in the two cases.

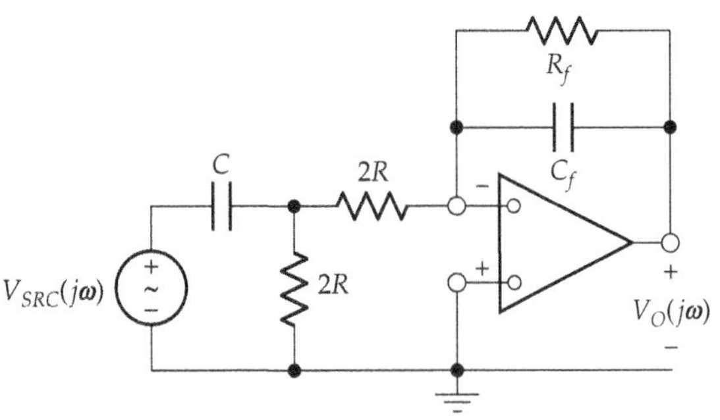

FIGURE P14.44

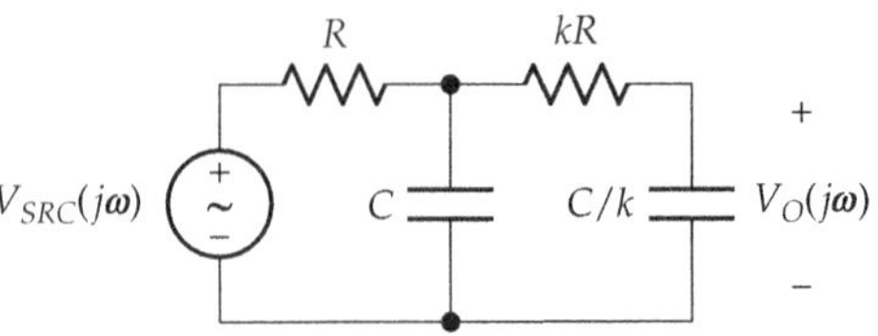

FIGURE P14.45

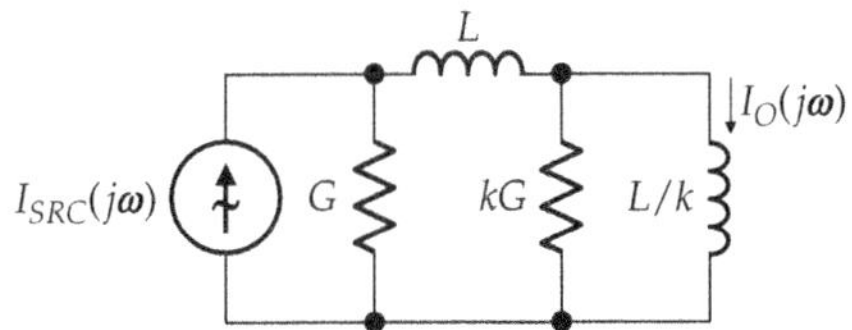

FIGURE P14.46

Note that the circuit is the dual of that of the preceding problem.

Ans. (a) $\dfrac{I_O(s)}{I_{SRC}(s)} = \dfrac{1}{L^2G^2}\dfrac{1}{s^2 + \dfrac{s}{LG}(2+1/k) + \dfrac{1}{L^2G^2}}$; (b) low pass, $Q = \dfrac{1}{2+1/k}$; (c) (i) 1/3; (b) 0.5 because the second circuit does not load the first.

P14.47 (a) Derive $H(j\omega) = V_O(j\omega)/V_{SRC}(j\omega)$ in Figure P14.47 and specify the type of response; (b) determine the maximum value of $H(j\omega)$ and the frequency at which it occurs; (c) determine the 3 dB BW; (d) derive the expressions for the low-frequency and high-frequency asymptotes and specify their slopes in dB/decade.

Ans. (a) $\dfrac{1}{3}\dfrac{300s}{s^2 + 300s + 2\times 10^4}$, bandpass; (b) 1/3 at $\omega_0 = 100\sqrt{2}$ rad/s; (c) 300 rad/s; (d) $|H(j\omega)| = \omega/200$, of slope +20 dB/decade, and $|H(j\omega)| = 100/\omega$, of slope −20 dB/decade.

P14.48 Determine the 3 dB BW of the response $V_O(j\omega)/I_{SRC}(j\omega)$ in Figure P14.48.

Ans. 5 rad/s.

P14.49 Determine the maximum magnitude of $V_O(j\omega)$ in Figure P14.49 and the phase angle at this magnitude.

Ans. $1\angle 180°$ V.

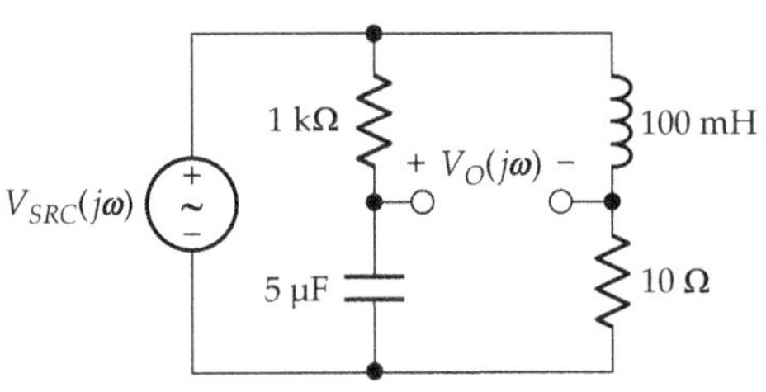

FIGURE P14.47

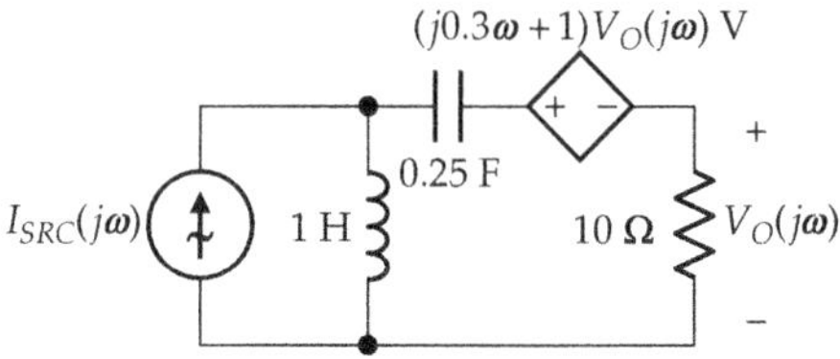

FIGURE P14.48

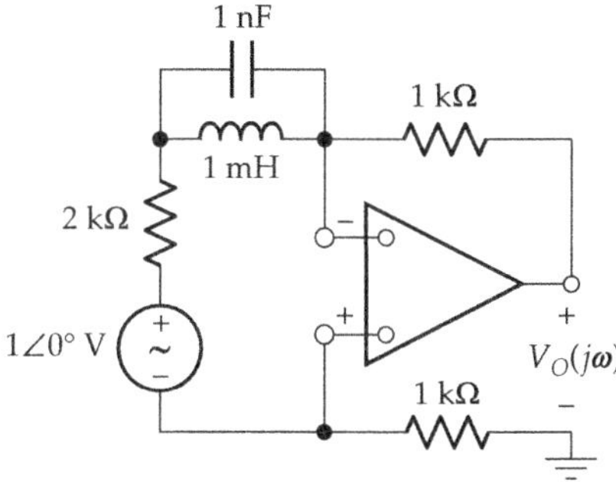

FIGURE P14.49

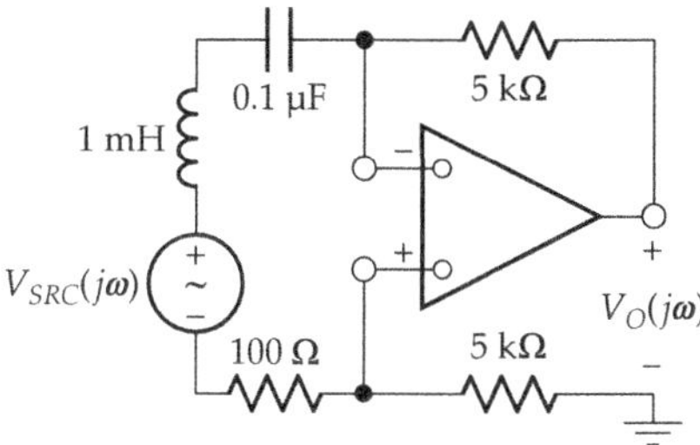

FIGURE P14.50

P14.50 Determine the maximum magnitude of the transfer function $V_O(j\omega)/V_{SRC}(j\omega)$ in dB in Figure P14.50.
Ans. 40 dB.

P14.51 $C = 2/3$ nF in Figure P14.51. (a) Derive NEC looking into terminals 'ab'; (b) derive the transfer function $V_O(s)/V_{SRC}(s)$, expressing it in standard form in terms of $s = j\omega$. (c) From the transfer function, determine the maximum gain, ω_0, and Q.

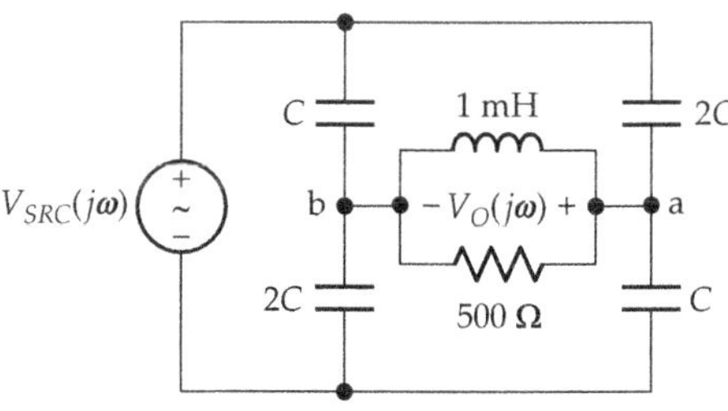

FIGURE P14.51

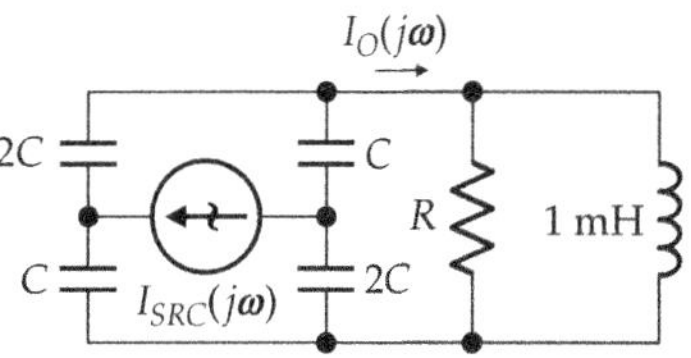

FIGURE P14.52

Ans. (a) Norton's capacitance: 1 nF, $\mathbf{I_{ab}} = \mathbf{I_N} = (1/3)j\omega V_{SRC}(j\omega)$; (b) $\dfrac{1}{3}\dfrac{s^2}{s^2+\left(2\times10^6\right)s+10^{12}}$; (c) 1/3, 1 Mrad/s, 0.5.

P14.52 (a) Determine R in Figure P14.52 so that $Q = 10$, assuming $C = 75$ nF; (b) derive the transfer function $I_O(j\omega)/I_{SRC}(j\omega)$; (c) characterize the response in terms of the basic responses of second-order circuits; (d) determine the gain in dB as $\omega \to 0$ and at $\omega = \omega_0$.

Ans. (a) 1 kΩ; (b) $\dfrac{1}{3}\dfrac{10^4 s+10^{10}}{s^2+10^4 s+10^{10}}$; (c) the response is the sum of a bandpass and a low-pass response; (d) −9.54 and 10.5 dB.

Transfer Functions

P14.53 Given $H(s)=\dfrac{10^6}{s^2+1600s+10^6}$, determine the maximum gain in dB at $\omega = 10^3$ rad/s.

Ans. −4.1 dB.

P14.54 Given $H(s)=\dfrac{35s}{s^2+50s+3600}$, determine the maximum of $|H(s)|$.

Ans. 0.7.

P14.55 Given $H(s)=\dfrac{s^2+3600}{s^2+50s+3600}$, determine the smaller 3 dB cutoff frequency.

Ans. 40 rad/s.

P14.56 Given $H(s)=\dfrac{100s}{s^2+100s+10^8}$, determine Q and BW.

Ans. 100, 100 rad/s.

P14.57 Given the transfer function $\dfrac{10s}{s^2+100s+\omega_0^2}$, determine Q if the product of the two 3 dB cutoff frequencies is 10^6 (rad/s)^2.

Ans. 10.

P14.58 Given $H(s)=\dfrac{4s^2+1000s+100}{5s^2+500s+125}$, determine the maximum response in dB.

Ans. 6 dB.

P14.59 Determine the transfer function whose asymptotic Bode magnitude plot is shown in Figure P14.59.

Ans. $\dfrac{2\times10^4 s}{(s+10)(s+200)}$.

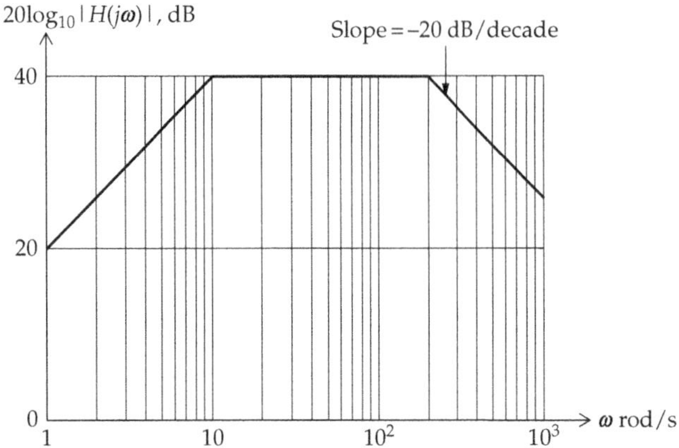

FIGURE P14.59

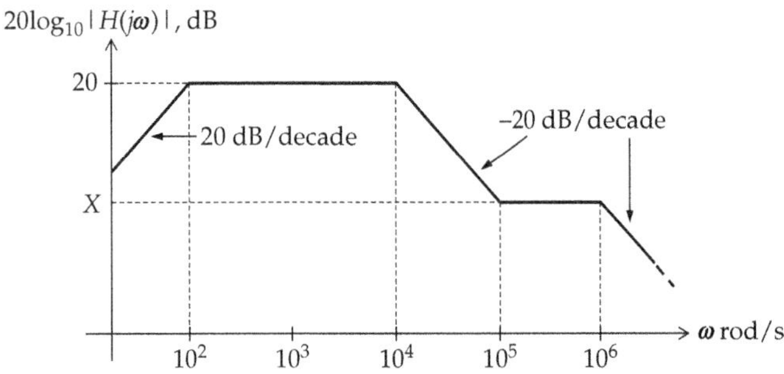

FIGURE P14.60

P14.60 Given the asymptotic Bode magnitude plot of Figure P14.60, determine (a) the X dB level and (b) the transfer function represented by this asymptotic plot.

Ans. (a) 0 dB; (b) $\dfrac{s\left(s+10^5\right)\times 10^6}{\left(s+100\right)\left(s+10^4\right)\left(s+10^6\right)}$.

P14.61 Determine the transfer function whose asymptotic Bode magnitude plot is shown in Figure P14.61.

Ans. $\dfrac{s\left(s+25\right)}{25\left(s+2\right)}$.

Design Problems

P14.62 A high-pass filter is to be used to reduce the drift, that is, the slow variation with time, of a biological signal applied to an electronic instrument having an input resistance of 1 MΩ and negligible input capacitance. If a low-frequency signal of 0.25 Hz is not to be attenuated by more than 3 dB, determine the smallest capacitance that must be connected at the input of the instrument.

Ans. 0.64 μF.

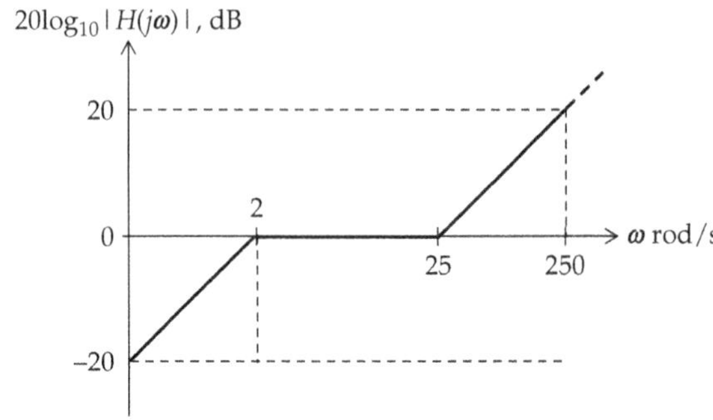

FIGURE P14.61

P14.63 In high-quality sound reproduction, different loudspeakers handle different ranges of audio frequencies. A *woofer* loudspeaker reproduces lower audio frequencies, up to about 3 kHz, whereas a *tweeter* loudspeaker reproduces audio frequencies in the range 3 kHz to about 20 kHz. The output of the audio amplifier is applied to the loudspeakers through a **crossover circuit** that directs the appropriate range of frequencies to each loudspeaker. In its simplest form, this circuit consists of first-order low-pass and high-pass filters that feed the woofer and tweeter, respectively, as in Figure P14.63. The responses of the two filters cross at the **crossover frequency** f_c, which is the −3 dB frequency of each filter. Determine L and C, assuming $f_c = 2.5$ kHz and that each loudspeaker presents a pure resistance of 8 Ω.

Ans. Approximately 0.5 mH and 8 μF.

P14.64 Determine L and C in Figure P14.64 so as to have a bandpass filter having a resonant frequency of 2 kHz and a BW of 500 Hz.

Ans. 4.97 mH, 1.27 μF.

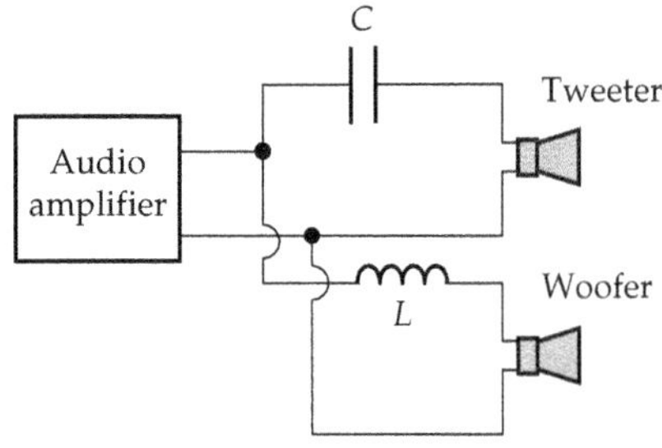

FIGURE P14.63

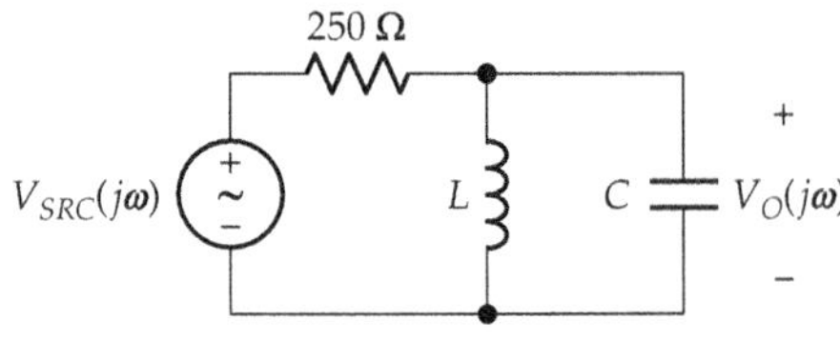

FIGURE P14.64

P14.65 A bandpass, series *RLC* circuit is required having $\omega_0 = 10^6$ rad/s, BW = 10^4 rad/s, $L = 1$ mH, and a maximum gain of 0.5. Implement the circuit using 5 Ω resistors, 2 mH inductors, and 2 nF capacitors.

Ans. The circuit consists of a series combination of two 2 mH inductors in parallel, two 2 nF capacitors in series, and two 5 Ω resistors in series, across one of which the output is taken.

P14.66 In touch-tone telephone dialing, the push buttons are arranged in four rows and three columns, with buttons in each row arranged to generate one of four frequencies in a low-frequency group and buttons in each column arranged to generate one of three frequencies in a high-frequency group, as illustrated in Figure P14.66. In this way, each button is identified by one of seven combinations of a low frequency in the low-frequency group and a high frequency in the high-frequency group. Pressing the number 5 button, for example, generates tones of frequencies 770 and 1336 Hz. The frequencies for each button are discriminated from neighboring circuits by bandpass filters. In the case of the number 5 button, the center frequency of the low-frequency bandpass filter is 770 Hz, whereas its 3 dB cutoff frequencies are 697 and 852 Hz, which are the center frequencies for the bandpass filters for the neighboring frequency on each side. Similarly, the center frequency of the high-frequency bandpass filter is 1336 Hz, whereas its 3 dB cutoff frequencies are 1209 and 1477 Hz. Determine *L* and *C* for a series *RLC* circuit that discriminates the frequencies in the low-frequency group, considering the outermost frequencies of this group to be 3 dB cutoff frequencies, and assuming *R* in standard telephone circuits is 600 Ω.

Ans. $L = 0.39$ H, $C = 0.1$ μF.

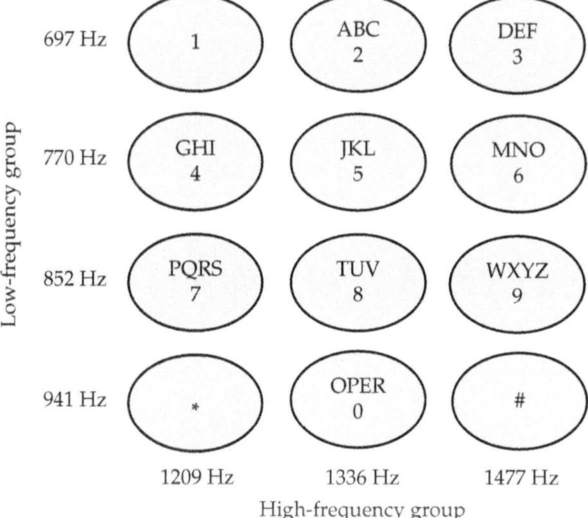

FIGURE P14.66

Probing Further

P14.67 Consider the series *RLC* circuit of Figure 14.15. Let the current at resonance be $i = I_m\cos\omega_0 t$ A. (a) Show that the peak energy stored in the inductor, $w_P = \frac{L}{2}I_m^2$, is the same as the peak energy $\frac{1}{2\omega_0^2 C}I_m^2$ stored in the capacitor; (b) show that the average energy dissipated per cycle in the resistor is $\frac{2\pi}{\omega_0}\left(\frac{1}{2}RI_m^2\right)$ and the average energy dissipated per radian of the cycle is $w_D = \frac{1}{\omega_0}\left(\frac{1}{2}RI_m^2\right)$; (c) deduce that $Q = w_P/w_D$.

P14.68 Show that the peaked, second-order low-pass response $\left(Q < 1/\sqrt{2}\right)$ crosses the 0 dB axis $(y = 1)$ at a frequency $\omega_0\sqrt{2\left(1 - 1/\left(2Q^2\right)\right)} = \omega_{max}\sqrt{2}$.

P14.69 Show that the second-order low-pass response at ω_0 is $H_{LP}(j\omega_0) = Q$.

P14.70 Show that for the low-pass response, the half-power frequency is given by $\omega_{1/2} = \omega_0\sqrt{\sqrt{\left(1 - 1/\left(2Q^2\right)\right)^2 + 1} + \left(1 - 1/\left(2Q^2\right)\right)}$. Verify this result for the cases of critical damping ($Q = 0.5$) and maximally flat response $\left(Q = 1/\sqrt{2}\right)$.

P14.71 Show that if the low-pass response is peaked, the −3 dB frequencies with respect to the peak are given by $\omega_{c1} = \omega_0\sqrt{\left(1 - 1/\left(2Q^2\right)\right) - (1/Q)\sqrt{1 - 1/\left(4Q^2\right)}}$ and $\omega_{c2} = \omega_0\sqrt{\left(1 - 1/\left(2Q^2\right)\right) + (1/Q)\sqrt{1 - 1/\left(4Q^2\right)}}$. Verify that if Q is large, the BW is approximately ω_0/Q, as for the bandpass response.

P14.72 Consider a plot of the phase angle of Equation 14.53 as a function of $\log_{10}u$. Show that the slope at $\omega = \omega_0$ is $-4.6Q$ rad/decade. Deduce that a line of this slope intersects the 0° line at $u_1 = 10^{-\frac{\pi}{9.2}Q}$, $u_1 = 10^{-\frac{\pi}{9.2}Q}$ and intersects the −180° line at $u_2 = 10^{\frac{\pi}{9.2}Q}$, $u_2 = 10^{\frac{\pi}{9.2Q}}$. Note that such a line may be used to approximate the Bode phase plot.

P14.73 Consider the relation $\omega_0 = 1/\sqrt{LC}$. Show that $\frac{d\omega_0}{\omega_0} = -\frac{1}{2}\frac{dC}{C}$. Note that this means that a 1% increase in C deceases ω_0 by 0.5%.

15

Butterworth and Active Filters

Objective and Overview

Having considered the basic first-order and second-order frequency responses in Chapter 14, the present chapter focuses on two types of filters that are of considerable practical importance, namely, Butterworth and active filters.

Butterworth filters characteristically have a maximally flat response in the passband, which makes them particularly useful as low-pass and high-pass filters. Active filters, as their name implies, incorporate an active element, usually in the form of an operational amplifier, previously discussed in Chapter 13. These filters are also commonly used, because the op amp provides many advantages, including the convenience of using only capacitors and resistors, with a Q that is not limited to 0.5 or less, as is the case with passive RC circuits.

This chapter begins by explaining the procedure of scaling, which focuses on designing a particular type of filter in normalized form. The nature of Butterworth filters is then explained and examples are given of the design of such filters. First-order active filters are introduced and their responses analyzed. Second-order active filters of various types, using a single op amp in the noninverting or inverting configuration, are then presented. The chapter ends with a discussion of a second-order universal active filter that can implement any of the second-order responses.

15.1 Scaling

Scaling provides a convenient, generalized, and systematic procedure for implementing a given type of passive or active filter. The type of filter under consideration is first designed in a normalized form, based on a 3 dB cutoff frequency ω_c, or a center frequency ω_0, of 1 rad/s, as well as any conditions imposed by the nature of the filter. Scaling is then applied to implement the filter in the form having the desired ω_c or ω_0 and any desired constraints, such as the value of a capacitance, inductance, or resistance.

Consider, for example, a normalized second-order bandpass filter having $\omega_0 = 1$ rad/s. To have $\omega_0 = 1$ rad/s, it is convenient to assume normalized values of $L = 1$ H and $C = 1$ F, since these values give $\omega_0 = 1/\sqrt{LC} = 1$ rad/s. The normalized value of R cannot also be 1, because this would constrain $Q = \omega_0 L/R$ for a series circuit, or $Q_p = \omega_0 CR$ for a parallel circuit, to be 1. Hence, the value of R in the normalized filter depends on the desired Q. If $Q = 10$, for example, then $R = 0.1\ \Omega$ for the series circuit. Based on these normalized values, scaling can be applied to derive a filter having any desired ω_0, such as 100 krad/s, and any particular value of L, C, or R, such as $L = 1$ mH.

It is convenient to consider that scaling is applied in two steps: (1) multiplying the magnitudes of the impedances in the circuit by a magnitude scaling factor, without changing the frequency, and (2) multiplying the frequency by a frequency scaling factor, without changing the magnitudes of the impedances. When both impedance magnitudes and frequency are to be changed, both scaling factors are applied.

In scaling impedance magnitudes, the object is to multiply these magnitudes by a positive, real scale factor k_m, which could be less than or greater than unity, without changing the frequency. To implement this, R is multiplied by k_m. L is also multiplied by k_m so that the impedance magnitude ωL is multiplied by k_m, with ω unchanged. C must be divided by k_m so that the impedance magnitude $1/\omega C$ is multiplied by k_m, with ω unchanged. Thus,

$$R'_m = k_m R; \quad L'_m = k_m L; \quad \text{and} \quad C'_m = C/k_m \qquad (15.1)$$

where unprimed parameters denote the initial normalized values and primed parameters having an m subscript denote the scaled values in accordance with impedance magnitude scaling.

Note that cutoff frequencies such as R/L or $1/CR$, or center frequencies such as $1/\sqrt{LC}$, are not affected by impedance magnitude scaling, since k_m cancels out from these expressions. This justifies considering scaling of impedance magnitudes independently of changing the frequency.

In frequency scaling, frequencies are changed without affecting magnitudes of impedances. Since resistance is not a function of frequency, it is not affected by frequency scaling. If the frequency is multiplied by a positive, real scale factor k_f, without changing the magnitude of the impedance ωL, then L must be divided by k_f. Similarly, if the frequency is multiplied by k_f, without changing the magnitude of the impedance $1/\omega C$, then C must also be divided by k_f. If $\omega' = k_f\omega$ is the scaled frequency, then in frequency scaling alone,

$$\omega' = k_f\omega, \quad R_f' = R, \quad L_f' = L/k_f, \quad \text{and} \quad C_f' = C/k_f \qquad (15.2)$$

where unprimed parameters denote the initial normalized values and primed parameters having an f subscript denote the scaled values in accordance with frequency scaling.

If both magnitude and frequency scaling are applied, then the scaling coefficients for each of the parameters in Equations 15.1 and 15.2 are multiplied together to give

$$R' = k_m R, \quad \omega' = k_f\omega, \quad L' = \frac{k_m}{k_f}L, \quad \text{and} \quad C' = \frac{1}{k_m k_f}C \qquad (15.3)$$

where unprimed parameters denote the initial normalized values and primed parameters without a subscript denote the scaled values in accordance with both impedance magnitude scaling and frequency scaling.

The bandwidth BW, being a frequency, scales like ω. $Q = \omega_0/\text{BW} = \omega_0 L/R$, or $Q_p = \omega_0 C_p R_p$, are *unchanged* by frequency or magnitude scaling, since they are dimensionless. Thus,

$$(\text{BW})' = k_f \times \text{BW}, Q' = Q, \quad \text{and} \quad Q' = Q_p \qquad (15.4)$$

Consider, for example, the normalized, series RLC, bandpass filter having $\omega_0 = 1$ rad/s, $L = 1$ H, $C = 1$ F, $Q = 10$, in which case $R = \omega_0 L/Q = 0.1\ \Omega$, as explained previously. Suppose that a filter of this type is required to have $\omega_0 = 100$ krad/s and $L = 1$ mH. Applying frequency scaling to ω_0 gives $k_f = (100\text{ krad/s})/(1\text{ rad/s}) = 10^5$. To determine k_m when both impedance magnitude and frequency scaling are used, Equation 15.3 is applied to L and L'. Thus, $k_m = k_f L'/L = (10^5 \times (1\text{ mH})/(1\text{ H}) = 100$. It follows from Equations 15.3 that $R' = 100 \times (0.1\ \Omega) = 10\ \Omega$ and $C' = (1\text{ F})/(10^5 \times 100) = 0.1\ \mu\text{F}$. It is seen that $\omega_0 = 1/\sqrt{10^{-3} \times 10^{-7}} = 100$ krad/s, and $Q = \omega_0 L/R = 10^5 \times 10^{-3}/10 = 10$, or $Q = 1/\omega_0 CR = 1/(10^5 \times 10^{-7} \times 10) = 10$, as required.

Note that scaling applies to filters of any order and is not restricted to normalized filters but can be applied to any filter when it is desired to change a parameter value or frequency, as illustrated by Exercises 15.3 and 15.4.

Primal Exercise 15.1

Given a first-order, normalized RL filter having $R = 1\ \Omega$ and $L = 1$ H, determine the frequency and magnitude scaling factors to have a 3 dB cutoff frequency of 10 krad/s and an inductance of 1 mH.

Ans. $k_f = 10^4$, $k_m = 10$.

Primal Exercise 15.2

Scale the normalized second-order series RLC bandpass filter having $L = 1$ H, $C = 1$ F, and $Q = 10$ to have a bandwidth of 5 krad/s, using a 50 nF capacitor.

Ans. $L' = 8$ mH, $R' = 40\ \Omega$.

Primal Exercise 15.3

Consider a series RLC filter having $R = 40\ \Omega$, $L = 8$ mH, and $C = 50$ nF, which makes $\omega_0 = 50$ kHz, $Q = 10$, and BW = 5 kHz. Suppose that the filter is to be modified to have $R = 1$ kΩ and $\omega_0 = 100$ kHz. Determine k_f, k_m, the scaled values of L' and C' and verify that Q remains the same.

Ans. $k_f = 2$, $k_m = 25$, $L' = 0.1$ H, $C' = 1$ nF.

Primal Exercise 15.4

Consider the filter of Primal Example 15.3 having $R = 40\ \Omega$, $L = 8$ mH, and $C = 50$ nF, which makes $\omega_0 = 50$ kHz, $Q = 10$, and BW = 5 kHz. Suppose that the filter is to be modified to have $\omega_0 = 10$ kHz, $L = 20$ mH, and $Q = 5$. Determine k_m, k_f, the scaled values of R' and C' and verify that the required values are obtained. Note that the frequency should first be scaled, which keeps Q the same, and then R changed to change Q.

Ans. $k_f = 0.2$, $k_m = 0.5$, $C' = 0.5\ \mu$F, $R = 40\ \Omega$.

15.2 Butterworth Response

The Butterworth response will be illustrated, to begin with, using the second-order low-pass response. Recall from Section 14.2 that an ideal low-pass filter has unity gain in the passband, zero gain in the stopband, with an infinitely sharp transition, or roll-off, between the two bands. The Bode magnitude plot is as illustrated in Figure 15.1a. Referring to Figure 14.28a, consider the Bode magnitude plots for $Q = 1/\sqrt{2}$ and $Q = 0.5$, reproduced in Figure 15.1b. The plot for $Q = 1/\sqrt{2}$ is maximally flat in the sense that, if $Q > 1/\sqrt{2}$, the plot rises to a maximum above the 0 dB line, which means that the response is not flat in the passband. On the other hand, if $Q < 1/\sqrt{2}$, the plot begins to fall below the 0 dB at smaller values of ω/ω_0 than in the case of $Q = 1/\sqrt{2}$. This again means that the response is not as flat in the passband as in the case of $Q = 1/\sqrt{2}$. We wish to investigate the reason for this behavior and hence deduce the form of maximally flat responses for filters of order greater than second order.

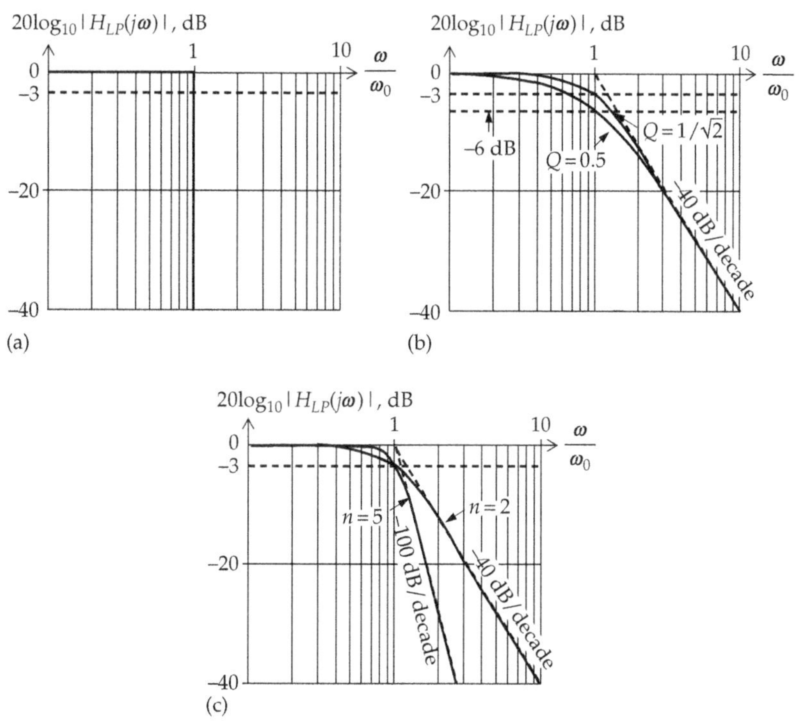

FIGURE 15.1
(a) Ideal low-pass filter response; (b) second-order Butterworth response and second-order low-pass response having $Q = 0.5$; (c) second-order and fifth-order Butterworth responses.

From Table 14.3, the prototypical, second-order low-pass response is

$$H_{LP2}(j\omega) = \frac{\omega_0^2}{s^2 + (\omega_0/Q)s + \omega_0^2} = \frac{1}{-\omega^2/\omega_0^2 + j\omega/\omega_0 Q + 1} \quad (15.5)$$

where $j\omega$ is substituted for s and the numerator and denominator are divided by ω_0^2. It follows that

$$|H_{LP2}(j\omega)| = \frac{1}{\sqrt{(1-\omega^2/\omega_0^2)^2 + (\omega/\omega_0 Q)^2}} = \frac{1}{\sqrt{(1-u^2)^2 + (u/Q)^2}} = \frac{1}{\sqrt{y}} \quad (15.6)$$

where $u = \omega/\omega_0$, and

$$y = (1-u^2)^2 + \left(\frac{u}{Q}\right)^2 = u^4 + u^2\left(\frac{1}{Q^2} - 2\right) + 1 \quad (15.7)$$

as in Equation 14.54.

It follows from Equation 15.7 that, for the two values of Q in Figure 15.1b,

$$Q = 0.5 \quad y_1 = u^4 + 2u^2 + 1 \quad (15.8)$$

$$Q = 1/\sqrt{2} \quad y_2 = u^4 + 1 \quad (15.9)$$

Table 15.1 lists several values of u in the passband, between 0 and 1, in 0.1 increments, and the corresponding values of u^2, $2u^2$, u^4, y_1, and y_2. The key observation is that the increase in y_1 with u, and hence the drop in $|H_{LP}(j\omega)|$ with ω, is dominated in the range $0 \le u \le 1$ by the term $2u^2$ and not by the term u^4. This is because as a fraction, u^4 is smaller than u^2, particularly for small value of u. As a result, y_2 increases less rapidly with u than y_1, resulting in a flatter response in the passband for $Q = 1/\sqrt{2}$, compared to the response for $Q = 0.5$.

It is seen from Equations 15.6 through 15.9 that for the maximally flat, second-order response having $Q = 1/\sqrt{2}$,

$$|H_{LP2}(j\omega)| = \frac{1}{\sqrt{1+u^4}} = \frac{1}{\sqrt{1+(u^2)^2}} \quad (15.10)$$

The same argument can be extended to a response of higher order n, in which case the power of ω in the transfer function is n and becomes a power of $2n$ in the magnitude response when the real part is squared. This is the case with the second-order response, since the power of

TABLE 15.1

Second-Order Butterworth Response

u	0.1	0.2	0.3	0.4	0.5	0.6	07	0.8	0.9
u^2	0.01	0.04	0.09	0.16	0.25	0.36	0.49	0.64	0.81
$2u^2$	0.02	0.08	0.18	0.32	0.5	0.72	0.98	1.28	1.62
u^4	0.0001	0.0016	0.0081	0.0256	0.0625	0.1296	0.2401	0.4096	0.6561
y_1	1.0201	1.0816	1.1881	1.3456	1.5625	1.8496	2.2201	2.6896	3.2761
y_2	1.0001	1.0016	1.0081	1.0256	1.0625	1.1296	1.2401	1.4096	1.6561

ω is 2 in Equation 15.5 and becomes 4, as u^4, in Equation 15.10. The response of order n is then

$$\left|H_{LP2}(j\omega)\right| = \frac{1}{\sqrt{1+\left(u^n\right)^2}} = \frac{1}{\sqrt{1+u^{2n}}} \tag{15.11}$$

Such a response has three important features:

1. When $n = 2$ and $u = 0.3$, for example, $y_2 = (1 + u^4) = 1.0081$ (Table 15.1), and $\left|H_{LP2}(j\omega)\right| = 1/\sqrt{1.0081} = 0.996$. On the other hand, when $n = 5$, for example, $(1 + u^{10})$ has the same value of 1.0081 if $u = 0.618$, or more than twice the frequency of 0.3. This means that the larger the n, the larger the frequency is in the passband at which the response is at a certain value less than unity. In other words, *the larger n, the "flatter" the response is in the passband.*
2. The high-frequency asymptote, as $u \to \infty$, is $20\log_{10}|H_{LP2}(j\omega)| = -20\log_{10}(u^n) = -20n\log_{10}(u)$. That is, the slope is $20n$ dB/decade. As n increases, not only is the response flatter in the passband, but *the transition between the passband and the stopband is also sharper.* The response becomes closer to the ideal.
3. When $u = 1$, then $u^{2n} = 1$ for all n, $20\log_{10}\left(1/\sqrt{2}\right) = -3$ dB. That is, *all the responses have an attenuation of −3 dB at u = 1, or $\omega = \omega_0$, regardless of n.* Hence, ω_0 is the 3 dB cutoff frequency or the half-power frequency for all n. It is also the corner frequency, because the asymptote $-20n\log_{10}(u)$ is zero at $u = 1$. In other words, the high-frequency asymptote intersects the 0 dB line at $\omega = \omega_0$. Because of this, ω_0 can be replaced by ω_c, the symbol for a cutoff frequency. The responses for $n = 2$ and $n = 5$ are shown in Figure 15.1c.

Higher-order responses of the form of Equation 15.11 are indeed possible and are known as **Butterworth responses**. They are based on the Butterworth polynomials given in Table 15.2 for orders up to eight.

TABLE 15.2

Normalized Butterworth Polynomials of Order n

n	Factors of Polynomial $B_n(s)$
1	$(s + 1)$
2	$(s^2 + 1.414s + 1)$, where $\sqrt{2} = 1.414$
3	$(s + 1)(s^2 + s + 1)$
4	$(s^2 + 0.765s + 1)(s^2 + 1.848s + 1)$
5	$(s + 1)(s^2 + 0.618s + 1)(s^2 + 1.618s + 1)$
6	$(s^2 + 0.518s + 1)(s^2 + 1.414s + 1)(s^2 + 1.932s + 1)$
7	$(s + 1)(s^2 + 0.445s + 1)(s^2 + 1.247s + 1)(s^2 + 1.802s + 1)$
8	$(s^2 + 0.390s + 1)(s^2 + 1.111s + 1)(s^2 + 1.663s + 1)(s^2 + 1.962s + 1)$

These polynomials are in normalized form, corresponding to a cutoff frequency of 1 rad/s. The polynomials of order greater than two are expressed as products of second-order and first-order power functions, which are appropriate for implementation as a cascade of first-order and second-order filters, as will be clarified shortly.

To show that these polynomials do give the required responses, consider, the first-order Butterworth polynomial, $B_1(s)$. Substituting $s = j\omega$, $B_1(j\omega) = 1 + j\omega$, and

$$\left|B_1(j\omega)\right| = \sqrt{1+\omega^2} \tag{15.12}$$

which is the same as in the denominator of Equation 15.11, with $n = 1$ and $u = \omega$, since ω_0 is normalized to 1.

Substituting $s = j\omega$ in the expression for the second-order polynomials gives

$$B_2(j\omega) = 1-\omega^2 + j\omega\sqrt{2} \tag{15.13}$$

and

$$\left|B_2(j\omega)\right| = \sqrt{\left(1-\omega^2\right)^2 + 2\omega^2} = \sqrt{1+\omega^4} \tag{15.14}$$

as in the denominator of Equation 15.10, or Equation 15.11 with $n = 2$.

Expanding the third-order polynomial and substituting $s = j\omega$,

$$B_3(j\omega) = s^3 + 2s^2 + 2s + 1 = 1 - 2\omega^2 + j\omega(2 - \omega^2) \quad (15.15)$$

and

$$|B_3(j\omega)| = \sqrt{(1 - 2\omega^2)^2 + \omega^2(2 - \omega^2)^2} = \sqrt{1 + \omega^6} \quad (15.16)$$

as in the denominator Equation 15.11, with $n = 3$.

In summary, the Butterworth polynomials embody the following concept:

Concept: *In the expression for the magnitude of a Butterworth polynomial of order n, the only term in ω is a term raised to the power 2n.*

How are these polynomials used in low-pass and high-pass filters? To answer this question, we will consider first-order and second-order responses only, because, in practice, higher-order filters are implemented as a cascade of such filters. For a first-order response ($n = 1$), the magnitude of the normalized, low-pass response is given by Equation 15.11, with $\omega_0 = 1$ and $u = \omega$, as

$$|H_{LP1}(j\omega)| = \frac{1}{\sqrt{1 + \omega^2}} \quad (15.17)$$

corresponding to a transfer function

$$H_{LP1}(j\omega) = \frac{1}{1 + j\omega} \quad (15.18)$$

which is the same as Equation 14.14, with $\omega_{cl} = 1$ rad/s. That is, an ordinary first-order response is in fact a Butterworth response of the first order.

For a second-order response ($n = 2$), the magnitude of the normalized, low-pass response is given by Equation 15.11, with $\omega_0 = 1$ and $u = \omega$, is

$$|H_{LP2}(j\omega)| = \frac{1}{\sqrt{1 + \omega^4}} \quad (15.19)$$

The denominator of Equation 15.19 is $|B_2(j\omega)|$ of Equation 15.14, which is the magnitude of $B_2(j\omega)$ of Equation 15.13, corresponding to $B_2(s) = s^2 + \sqrt{2}s + 1$. It follows that the denominator of $H_{LP2}(s)$ for a normalized, second-order, Butterworth low-pass response is $B_2(s)$. Thus,

$$H_{LP2}(s) = \frac{1}{s^2 + \sqrt{2}s + 1} \quad (15.20)$$

It follows from Table 15.2 *that the normalized Butterworth polynomial of the second order is the denominator of the normalized Butterworth low-pass response.*

As for a high-pass response, recall from Table 14.3, that the transformation from a low-pass to a high-pass filter involves replacing s/ω_0 by ω_0/s or replacing $j\omega$ by $1/j\omega$ in the case of a normalized response having $\omega_0 = 1$ rad/s. Thus, Equation 15.18 becomes

$$H_{HP1}(j\omega) = \frac{1}{1 + 1/j\omega} = \frac{j\omega}{1 + j\omega} \quad (15.21)$$

and

$$|H_{HP1}(j\omega)| = \frac{1}{\sqrt{1 + (1/\omega)^2}} \quad (15.22)$$

corresponding to Equations 14.23 and 14.24, with $\omega_{ch} = 1$ rad/s.

Replacing ω by $1/\omega$ in Equation 15.19,

$$|H_{HP2}(j\omega)| = \frac{1}{\sqrt{1 + (1/\omega)^4}} = \frac{\omega^2}{\sqrt{1 + \omega^4}} \quad (15.23)$$

corresponding to the transfer function

$$H_{LP2}(s) = \frac{s^2}{s^2 + \sqrt{2}s + 1} \quad (15.24)$$

which is also obtained by replacing s by $1/s$ in Equation 15.20. The denominator of Equation 15.24 is the normalized Butterworth polynomial of the second order. This transfer function is the same as that of the prototypical high-pass response of Table 14.3, with $\omega_0 = 1$ rad/s and $Q = 1/\sqrt{2}$.

The preceding arguments can be generalized in the following concepts:

Concepts:

1. *In a* ***normalized*** *Butterworth, low-pass response of order n, the Butterworth polynomial of Table 15.2 is the denominator of the transfer function, the numerator of the transfer function being unity.*
2. *In a* ***normalized*** *Butterworth, high-pass response of order n, (a) the Butterworth polynomial of Table 15.2, with s replaced by 1/s, and the numerator of the transfer function being unity, is the denominator of the transfer function, or (b) the Butterworth polynomial of Table 15.2 is the denominator of the transfer function, the numerator of the transfer function being* s^n.

What about bandpass and bandstop filters? Bandpass filters having a flat, or nearly flat response, are very desirable in many applications. Such responses can be derived using a more complicated transformation than that between low-pass and high-pass responses. In practice, the required response is commonly obtained using magnetically coupled, second-order tuned circuits in cascaded amplifier stages, or by "stagger-tuning" second-order tuned circuits in cascaded amplifier stages, that is, by having slightly different resonant frequencies of these circuits. Bandstop circuits normally have narrowband responses in order to reject a particular frequency, such as the 50 Hz power frequency, so that bandstop Butterworth responses are not normally required.

Although a high-order low-pass or high-pass Butterworth response can approach the ideal filter response, other types of filters are more efficient in achieving the same sharp transition between the passband and the stopband but with a lower-order filter, which means having fewer components in the circuit. However, this is achieved by allowing some "ripple," or oscillation, in the magnitude of the response in the passband, as in Chebyshev filters, or in both the passband and the stopband, as in elliptic filters.

Primal Exercise 15.5

The low-frequency asymptote of a high-pass Butterworth response of maximum gain 9 dB has a slope of 80 dB/decade and intersects the 9 dB line at a frequency of 1 kHz. Determine the frequency at which the gain is 6 dB.

Ans. 1 kHz.

Primal Exercise 15.6

Given the transfer function $\frac{s^2}{s^2+\sqrt{2}s+1}$ for a normalized second-order Butterworth high-pass filter, determine the transfer function of the filter when the frequency is scaled to 10 rad/s.

Ans. $\frac{s^2}{s^2+10\sqrt{2}s+100}$.

Exercise 15.7

A fourth-order Butterworth high-pass filter has a passband gain of 10 and a 3 dB cutoff frequency of 10 kHz. Determine the frequency at which the gain is 3×10^{-6}.

Ans. 234.0 Hz.

Example 15.1: Second-Order and Third-Order Butterworth Low-Pass Filters

It is required to design a low-pass Butterworth filter having a cutoff frequency of 100 krad/s, using 50 nF capacitors and to implement the filter as (a) a second-order filter and (b) a third-order filter, both filters having a maximum gain of unity in the passband. The transfer functions are to be derived and the filters implemented using series *RLC* circuits.

Solution:

(a) From Table 15.2, $B_2(s)=\left(s^2+\sqrt{2}s+1\right)$, which means that $Q=1/\sqrt{2}$ and $R=\omega_0 L/Q=1\times1\times\sqrt{2}$ in the normalized response. To scale the frequency from 1 rad/s to 100 krad/s requires $k_f = 10^5$. To scale C from 1 F to 50 nF, with frequency scaling, requires k_m = (1 F)/($k_f \times$ 50 nF) = $1/(10^5 \times 50 \times 10^{-9}) = 200$ (Equation 15.3). It follows that $L' = (k_m/k_f) \times$ (1 H) = 200×10^{-5} = 2 mH, and $R' = 200\sqrt{2} = 282.8\ \Omega$. The circuit is implemented as in Figure 15.2, the output being taken across the capacitor. The transfer function is, from Table 14.9.1, with $\omega_0 = 10^5$ rad/s and $Q=1/\sqrt{2}$:

$$H_{LP2}=\frac{10^{10}}{s^2+10^5\sqrt{2}s+10^{10}} \tag{15.25}$$

Note that whereas R' of the required filter depends on k_m, and L' depends on both k_m and k_f, the transfer function of the required filter (Equation 15.25) depends only on k_f. This is because the units of each term in the numerator and denominator of the voltage/voltage or current/current transfer function is $(\text{rad/s})^2$ (Table 14.3), which implies that these terms are affected by frequency scaling only. This means that the required response of Equation 15.25 can be derived from the normalized response of Equation 15.20 by applying frequency scaling only. To do so, we note that the cutoff frequency is 1 rad/s in the normalized response and 10^5 rad/s in the required response. Hence, to derive the transfer function of the required response from that of the normalized response, frequency in the normalized transfer function (Equation 15.20) should be divided by $k_f = 10^5$ to give

$$H_{LP2}(s)=\frac{1}{\left(s/10^5\right)^2+\sqrt{2}\left(s/10^5\right)+1} \tag{15.26}$$

Note that s in Equation 15.26 corresponds to the scaled frequency, that is, the normalized frequency multiplied by k_f to give a cutoff frequency of 100 krad/s. Hence, s must be divided by 10^5 to

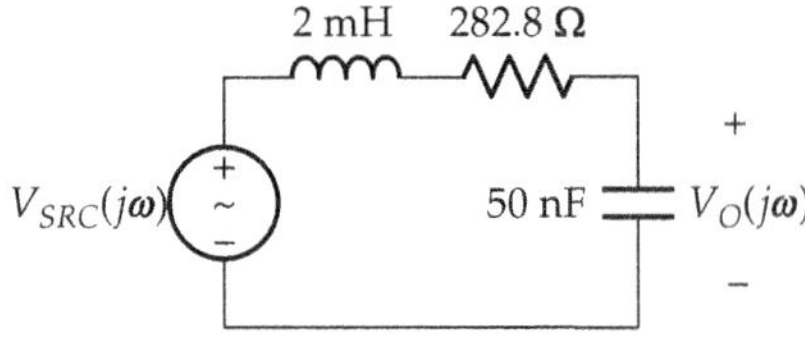

FIGURE 15.2
Figure for Example 15.1.

correspond to a normalized cutoff frequency of 1 rad/s, when substituting in Equation 15.20.

Multiplying the numerator and denominator of Equation 15.26 by 10^{10} gives the transfer function of Equation 15.25. To verify that k_m cancels out in the transfer function, we note that, when both magnitude and frequency scaling are applied,

$$\omega_0' = \frac{1}{\sqrt{L'C'}} = \frac{1}{\sqrt{\frac{k_m}{k_f}L \times \frac{1}{k_m k_f}C}} = \frac{k_f}{\sqrt{LC}} = k_f \omega_0 \quad (15.27)$$

k_m cancels out in the term ω_0^2 and in the term ω_0/Q, with Q unaffected by scaling.

(b) From Table 15.2, $B_3(s) = (s + 1)(s^2 + s + 1)$, which is a cascade of a first-order circuit and a second-order circuit. The normalized first-order circuit has $\omega_0 = 1$ rad/s, which means $C = 1$ F and $R = 1\ \Omega$. As in (a) $k_f = 10^5$ and $k_m = 200$, so that $C' = 50$ nF and $R' = 200\ \Omega$, which gives $\omega_{cl} = 100$ krad/s. The circuit is shown in Figure 15.3.

For the second-order circuit, k_f, k_m, and L' are the same. However, the coefficient of s in the bracketed second-order term in $B_3(s)$ is unity, and not $\sqrt{2}$, as in $B_2(s)$. This means that $\omega_0/Q = 1$, or $Q = 1$, since $\omega_0 = 1$ rad/s in the normalized response. Hence, $R = 1/\omega_0 CQ = 1\ \Omega$, so that $R' = 200 \times 1 = 200\ \Omega$. The overall transfer function is the product of low-pass transfer functions of the first and second orders:

$$H_{LP3}(s) = \frac{10^5}{s + 10^5} \times \frac{10^{10}}{s^2 + 10^5 s + 10^{10}} \quad (15.28)$$

Multiplying the open-circuit transfer functions corresponds to cascading the circuits, with isolation, as discussed at the end of this section. The circuit is as shown in Figure 15.3, where isolation is provided by a unity-gain amplifier. The second-order circuit is shown preceding the first-order circuit. The same transfer function is of course obtained if the first-order circuit precedes the second-order circuit. In practice, the circuit that has the larger input impedance is generally preferred as the input circuit so as to minimize loading of the input source.

Simulation: Only the third-order circuit will be simulated, since the second-order circuits were simulated

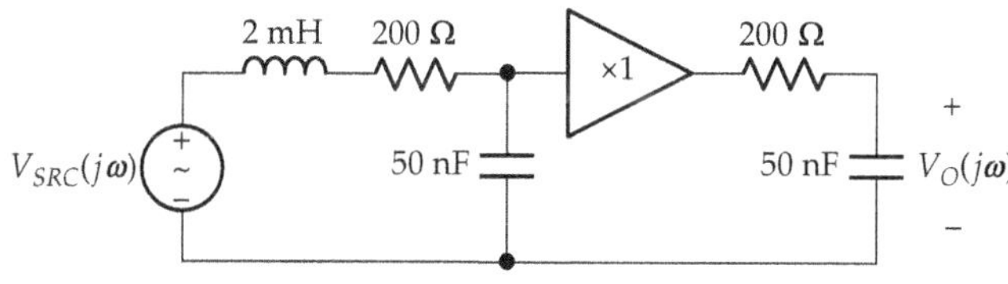

FIGURE 15.3
Figure for Example 15.1.

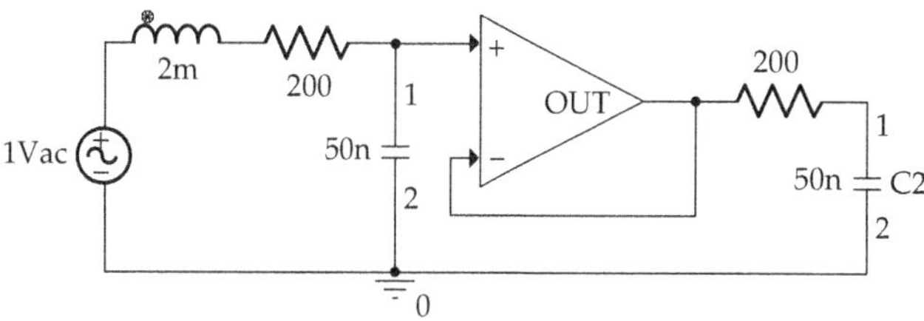

FIGURE 15.4
Figure for Example 15.1.

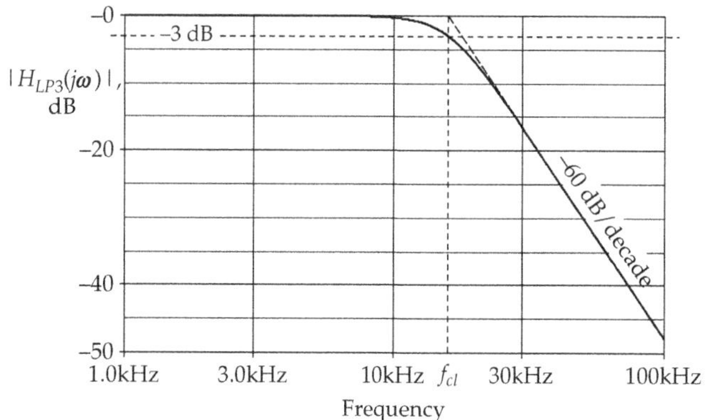

FIGURE 15.5
Figure for Example 15.1.

in Chapter 14. The circuit is entered as in Figure 15.4. Simulation Settings are a start frequency of 1k, an end frequency of 100k, and 3000 points/decade. After the simulation is run, select DB(V(C2:1)). $|H_{LP3}(j\omega)|$ is displayed as in Figure 15.5. Using cursor search, the 3 dB cutoff frequency is read as 15.903K, the calculated value being 15.916 kHz. The high-frequency asymptote is entered as 60*LOG10(15915.5)-60*LOG10(Frequency). The phase shift is displayed by entering P(V(C2:1)) and varies between 0° at low frequencies to −270° at high frequencies. By entering sxval(15.9155k), the phase shift at the 3 dB cutoff frequency is read as −135.000. This is to be expected, since the phase shift at ω_0 is −45° for the first-order filter (Figure 14.10b) and is −90° for the second-order filter (Figure 14.28b).

Example 15.2: Second-Order and Third-Order Butterworth High-Pass Filters

It is required to design a high-pass Butterworth filter having a cutoff frequency of 100 krad/s, using 50 nF capacitors and to implement the filter as (a) a second-order filter and (b) a third-order filter. Derive the transfer functions and implement the filters using series *RLC* circuits.

Solution:

(a) The values of L' and R' are the same as those determined in part (a) of Example 15.1, namely, 2 mH and 282.8 Ω. The circuit is implemented as in Figure 15.6, the output being taken across

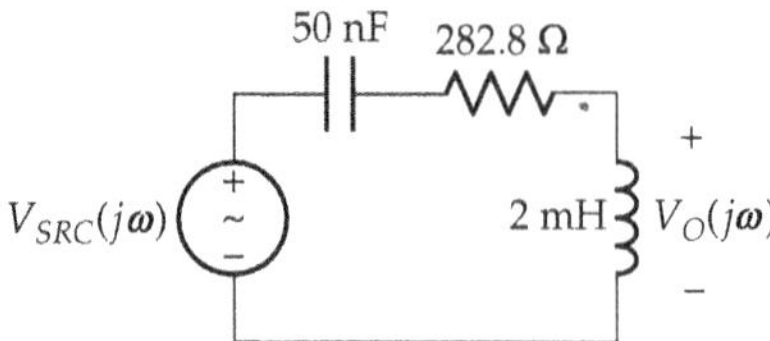

FIGURE 15.6
Figure for Example 15.2.

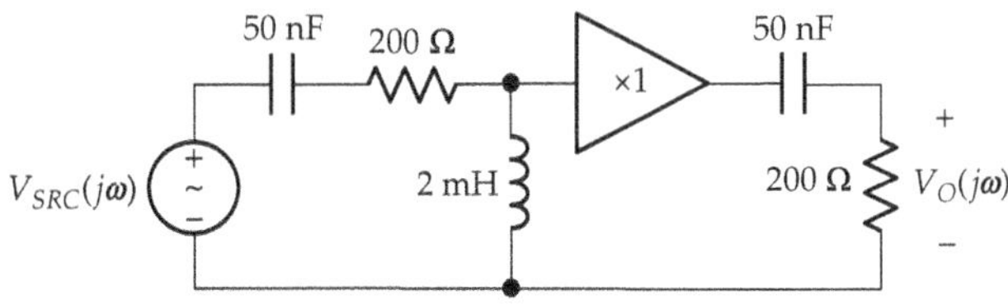

FIGURE 15.7
Figure for Example 15.2.

the inductor. The transfer function is, from Table 14.9.1, with $\omega_0 = 10^5$ rad/s and $Q = 1/\sqrt{2}$:

$$H_{HP2} = \frac{s^2}{s^2 + 10^5\sqrt{2}s + 10^{10}} \tag{15.29}$$

As argued in connection with Equation 15.26, the response of Equation 15.29 can be derived from the normalized Butterworth high response by replacing s with $s/10^5$.

(b) The values of the circuit components of the first-order and second-order filters are the same as those determined in part (b) of Example 15.1. The circuit is as shown in Figure 15.7, the transfer function being the product of the open-circuit high-pass transfer functions of the first- and second-order responses:

$$H_{HP3}(s) = \frac{s}{s + 10^5} \times \frac{s^2}{s^2 + 10^5 s + 10^{10}} \tag{15.30}$$

Simulation: The circuit is entered as in Figure 15.8. Simulation Settings are a start frequency of 1k, an end frequency of 100k, and 3000 points/decade. After the simulation is run, select DB(V(R2:1)). $|H_{HP3}(j\omega)|$ is displayed as

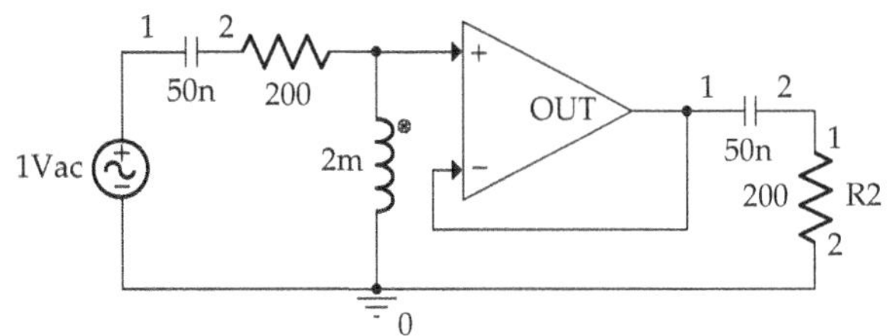

FIGURE 15.8
Figure for Example 15.2.

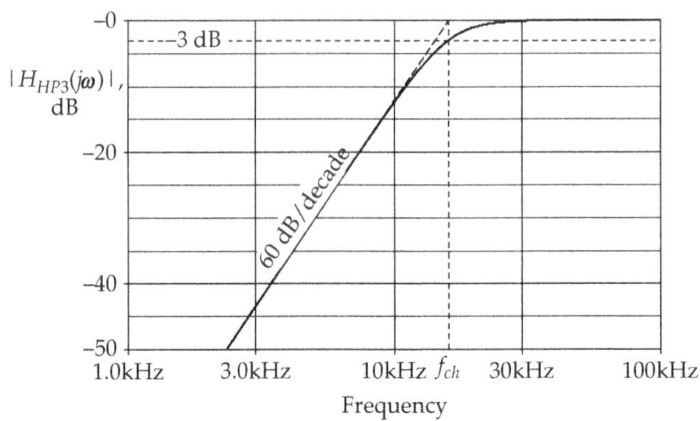

FIGURE 15.9
Figure for Example 15.2.

in Figure 15.9. Using cursor search, the 3 dB cutoff frequency is read as 15.903K, the calculated value being 15.916 kHz. The low-frequency asymptote is entered as –60*LOG10(15915.5)+60*LOG10(Frequency). The phase shift is displayed by entering P(V(R2:1)) and varies between 270° at low frequencies to 0° at high frequencies. By entering sxval(15.9155k), the phase shift at the 3 dB cutoff frequency is read as 135.000. This is to be expected, since the phase shift at ω_0 is 45° for the first-order filter (Figure 14.11b), and is 90° for the second-order filter (Figure 14.31b).

Primal Exercise 15.8

Consider three, identical, low-pass, first-order filters that are cascaded with isolation. (a) Determine the ratio of the response at the 3 dB cutoff frequency of a single filter to that at very low frequencies; (b) compare with ratio of the responses for a third-order Butterworth low-pass filter having the same 3 dB cutoff frequency as the single filter in (a). Express the ratio in terms of dB and as a decimal fraction.

Ans. (a) –9 dB, $(0.707)^3 = 0.354$; (b) –3 dB, 0.707.

Primal Exercise 15.9

Determine the gain of a Butterworth low-pass filter of order n at a frequency that is a decade higher than the cutoff frequency.

Ans. Very nearly $-20n$ dB.

Exercise 15.10

In the time domain, is a second-order Butterworth response, overdamped, critically damped, or underdamped?

Ans. Slightly underdamped, because $Q = 0.71 > 0.5$ for critical damping.

15.2.1 Product of Transfer Functions

It is opportune at this stage to emphasize how the product of transfer functions is implemented. Consider two circuits 'N_1' and 'N_2' having open-circuit transfer functions $H_1(s)$ and $H_2(s)$, as in Figure 15.10a and b. What happens when these two circuits are cascaded, as in Figure 15.10c? The overall transfer function $V'_{O2}(s)/V_{I1}(s)$ can be expressed as

$$H(s)=\frac{V'_{O2}(s)}{V_{I1}(s)}=\frac{V'_{O1}(s)}{V_{I1}(s)}\times\frac{V'_{O2}(s)}{V'_{O1}(s)}=H'_1(s)H_2(s) \quad (15.31)$$

where, in cascading, the output voltage $V'_{O1}(s)$ of 'N_1' is applied as the input voltage of 'N_2'. Note that $H_2(s)$ is the same, because 'N_2' is open-circuited in Figure 15.10b and c. However, $V'_{O1}(s)$ in Figure 15.10c is not the same as $V_{O1}(s)$ in Figure 15.10a because the current I_{12} that flows between the two cascaded circuits alters, in general, the output voltage of 'N_1'. This means that the transfer function $H(s)$ should be derived by analyzing circuits 'N_1' and 'N_2' together. In some cases, however, the transfer functions of the two individual circuits can be simply multiplied together to obtain the overall transfer function, which is the case when the second circuit has no effect on the first. We wish to clarify under what conditions this occurs. There are three general cases to consider:

1. When I_{12} is zero or insignificantly small. I_{12} is zero if an isolating amplifier, of infinite input impedance, is interposed between the two circuits, as in Figure 15.7. I_{12} is insignificantly small when the input impedance of 'N_2', that is, $V'_{O1}(s)/I_{12}(s)$ is large compared with the output impedance of 'N_1', which is Thevenin's impedance looking into the output terminals of 'N_1', with the input source set to zero. This case was considered in Problem P14.45.

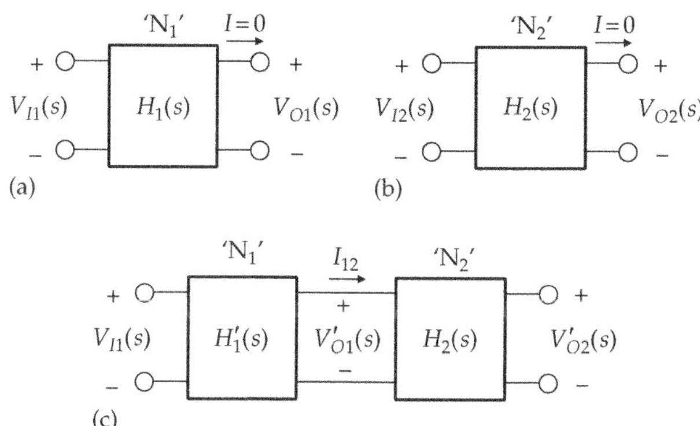

FIGURE 15.10
Product of transfer functions. Two circuits (a) and (b) cascaded in (c).

2. When the output of 'N_1' is an ideal voltage source, as when this output is that of an ideal op amp. In this case the open-circuit output voltage of 'N_1' is the same as in the presence of a finite I_{12} because, by definition of an ideal voltage source, the source voltage is independent of the source current. Many such cases will be encountered later on, as in the case of the broadband filter (Figure 15.14).
3. When 'N_2' input is a virtual ground. In this case, $I_{12}(s)/V_{I1}(s)$ is determined by 'N_1' alone, and $V_{O2}(s)/I_{12}(s)$ is determined by 'N_2' alone. The overall transfer function is then the product of these two individual transfer functions. This is the case when 'N_2' is an op amp in the inverting configuration.

Another point that should be kept in mind is that when transfer functions are multiplied in the frequency domain, their magnitudes are multiplied together and their phase angles are added together. Thus,

$$\begin{aligned}H(j\omega)&=H_1(j\omega)H_2(j\omega)=\left|H_1(j\omega)\right|e^{j\theta_1}\times\left|H_2(j\omega)\right|e^{j\theta_2}\\&=\left|H_1(j\omega)\right|\left|H_2(j\omega)\right|e^{j(\theta_1+\theta_2)}\end{aligned} \quad (15.32)$$

On the Bode magnitude plot, the Bode magnitude plots of $|H_1(j\omega)|$ and $|H_2(j\omega)|$ add together to give the Bode magnitude plot of the overall transfer function. Thus, $|H(j\omega)| = |H_1(j\omega)||H_2(j\omega)|$, so that

$$20\log_{10}\left|H(j\omega)\right|=20\log_{10}\left|H_1(j\omega)\right|+20\log_{10}\left|H_2(j\omega)\right| \quad (15.33)$$

15.3 First-Order Active Filters

Op amps may be used with *RC* circuits to construct active filters. *RC*, rather than *RL*, circuits are used for this purpose, because capacitors are less bulky and expensive than inductors and are closer to an ideal circuit element. Moreover, inductors are more awkward to implement in integrated circuits than capacitors. Op amps in active filters can be used to provide gain, isolation, load drive, and a higher Q for improved performance. Q of Butterworth filters exceeds 0.5, which means that such filters cannot be realized using *RC* circuits alone, since the maximum Q of these filters is 0.5. Active filters are extensively used for frequencies up to a few megahertz, limited by the frequency response of op amps. For higher frequencies, passive filters are generally used. The simplest forms of active filters are the first-order filters discussed in this section.

15.3.1 Low-Pass Filter

The transfer function of the ideal integrator is $H_{int}(s) = -1/sC_fR_r$ (Equation 13.25). The Bode magnitude plot is $-20\log_{10}(\omega C_fR_r)$, which is a straight line of slope −20 dB/decade that intersects the 0 dB line at $\omega = 1/C_fR_r$, as this gives $\log_{10}(1) = 0$ (Figure 15.11a). The gain is high at low frequencies because of the reactance $1/\omega C_f$ of the capacitor, which makes $Z_f/Z_r \to \infty$ as $\omega \to 0$ (Equation 13.22). In order to have a low-pass characteristic, the response must become flat at low frequencies, as indicated by the thick horizontal line in Figure 15.11a, which coincides with the 0 dB line for $\log_{10}\omega C_fR_r < 1$. This can be achieved by connecting a resistance R_f in parallel with C_f (Figure 15.11b), so that at low frequencies, as the reactance of the capacitance becomes very large, the gain of the amplifier becomes essentially independent of frequency. The transfer function is

$$H_{LP1}(s) = -\frac{R_f \| (1/sC_f)}{R_r} = -\frac{R_f}{R_r}\frac{\omega_{cl}}{s+\omega_{cl}} \tag{15.34}$$

where $\omega_{cl} = 1/C_fR_f$. The transfer function $H_{LP1}(s)$ is that of a first-order low-pass response having a cutoff frequency ω_{cl} and a gain of $-R_f/R_r$ as $s \to 0$.

Primal Exercise 15.11

A filter of the type of Figure 15.11b is required to have a passband gain of −10 and a 3 dB cutoff frequency of 1 krad/s, using a 0.1 μF capacitor. Determine R_r and R_f.

Ans. $R_r = 1\ \text{k}\Omega$, $R_f = 10\ \text{k}\Omega$.

15.3.2 High-Pass Filter

Interchanging resistors and capacitors in the low-pass filter results in a high-pass filter (Exercise 15.13). However, the passband gain in this case is given by the ratio of two capacitances, which is not as convenient to control precisely as the ratio of two resistances.

Alternatively, a high-pass filter using two resistors and a capacitor can be derived from the ideal differentiator, whose transfer function is $H_{dif}(s) = -sC_rR_f$ (Equation 13.27). The Bode magnitude plot is $+20\log_{10}(\omega C_rR_f)$, which is a straight line of slope + 20 dB/decade that intersects the 0 dB line at $\omega = 1/C_rR_f$ (Figure 15.12a). The gain is R_f divided by $(1/\omega C_r)$ and is high at high frequencies because of the small reactance $1/\omega C_r$ of the capacitor, since $Z_f/Z_r \to \infty$ as $\omega \to \infty$ (Equation 13.22). In order to have a high-pass characteristic, the response must become flat at high frequencies, as indicated by the thick horizontal line in Figure 15.12a, which coincides with the 0 dB line for $\log_{10}\omega C_fR_r > 1$. This can be achieved by connecting a resistance R_r in series with C_r (Figure 15.12b), so that at high frequencies, as the reactance of the capacitance becomes very small, the gain of the amplifier becomes essentially independent of frequency. The transfer function is

$$H_{HP1}(s) = -\frac{R_f}{R_r + 1/sC_r} = -\frac{R_f}{R_r}\frac{s}{s+\omega_{ch}} \tag{15.35}$$

where $\omega_{ch} = 1/C_rR_r$. The transfer function of Equation 15.35 is that a first-order high-pass response having a cutoff frequency ω_{ch} and a gain of $-R_f/R_r$ as $s \to \infty$.

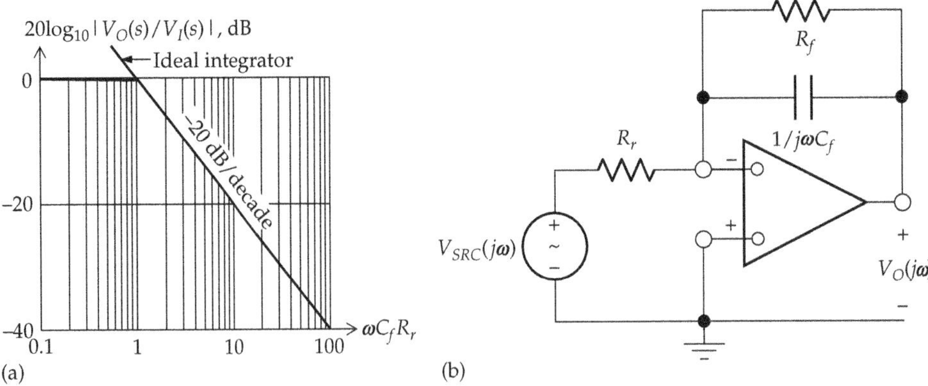

FIGURE 15.11
(a) Bode magnitude plot of first-order low-pass response derived from that of an ideal integrator; (b) circuit implementation using an op amp in the inverting configuration.

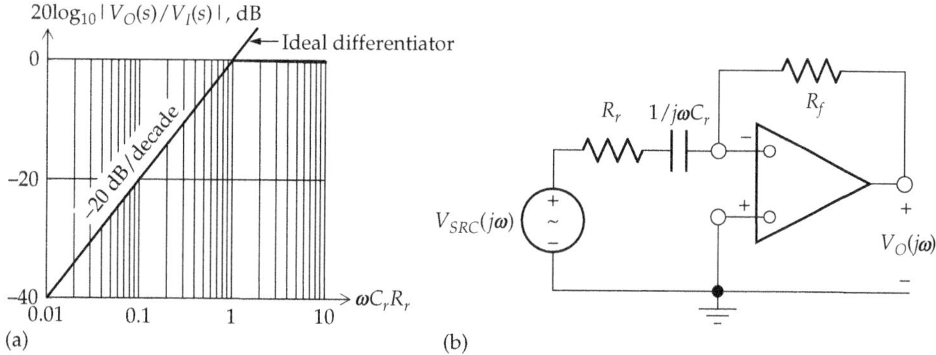

FIGURE 15.12
(a) Bode magnitude plot of first-order high-pass response derived from that of an ideal differentiator; (b) circuit implementation using an op amp in the inverting configuration.

Primal Exercise 15.12

A filter of the type of Figure 15.12 is required to have a passband gain of −10, a 3 dB cutoff frequency of 10 krad/s, and an input impedance of magnitude 10 kΩ at 20 krad/s. Determine C_r, R_r and R_f.

Ans. $C_r = 5\sqrt{5}$ nF, $R_r = 20/\sqrt{5}$ kΩ, $R_f = 200/\sqrt{5}$ kΩ.

Exercise 15.13

Show that a high-pass filter can be derived from the low-pass filter in Figure 15.11b by (a) replacing C_f by an inductor or (b) by replacing R_r by a capacitor, which amounts to interchanging resistors and capacitors in the circuit. Derive the transfer functions and the passband gain in each case.

Ans. (a) $-\dfrac{R_f}{R_r}\dfrac{s}{s+\omega_{ch}}$, where $\omega_{ch} = R_f/L_f$, passband gain: $-R_f/R_r$; (b) $-\dfrac{C_r}{C_f}\dfrac{s}{s+\omega_{ch}}$, where $\omega_{ch} = 1/C_fR_f$, passband gain: $-C_r/C_f$.

Primal Exercise 15.14

Specify the type of response $V_O(j\omega)/V_{SRC}(j\omega)$ in Figure 15.13.

Ans. Ideal integrator.

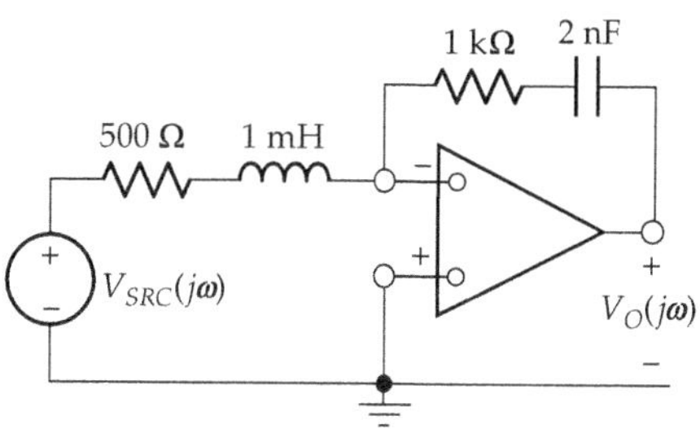

FIGURE 15.13
Figure for Primal Exercise 15.14.

As explained in Section 14.2, a second-order bandpass response can be obtained by multiplying low-pass and high-pass responses. This is conveniently accomplished using active filters. When the upper 3 dB cutoff frequency of the resulting bandpass response is much larger than the lower 3 dB cutoff frequency, the filter is referred to as **broadband**. The design of such a filter is particularly simple, as illustrated by the following Example.

Example 15.3: Broadband Bandpass Filter

It is required to design a broadband bandpass filter in the audio frequency range 20 Hz to 15 kHz, having a passband gain of 5, using 50 nF capacitors and op amps in the inverting configuration.

Solution:

The overall transfer function is the product of the low-pass and high-pass transfer functions given by Equations 15.34 and 15.35, respectively:

$$H_{BP2}(s) = \left[-\frac{R_{fl}}{R_{rl}}\frac{\omega_{c2}}{s+\omega_{c2}}\right]\left[-\frac{R_{fh}}{R_{rh}}\frac{s}{s+\omega_{c1}}\right]$$

$$= K\frac{1}{1+s/\omega_{c2}}\frac{1}{1+\omega_{c1}/s} \tag{15.36}$$

or

$$H_{BP2}(s) = K\frac{\omega_{c2}s}{s^2+(\omega_{c1}+\omega_{c2})s+\omega_{c1}\omega_{c2}} \tag{15.37}$$

where the l and h in the subscripts of the resistors refer to the low-pass and high-pass filters, respectively, $K = R_{fl}R_{fh}/R_{rl}R_{rh}$, $\omega_{c1} = 1/C_{rh}R_{rh}$ is the 3 dB cutoff frequency of the high-pass filter, and $\omega_{c2} = 1/C_{rl}R_{rl}$ is the 3 dB cutoff frequency of the low-pass filter, with $\omega_{c1} < \omega_{c2}$. Equation 15.37 is of the form of a bandpass response (Table 14.3), where $\omega_0 = \sqrt{\omega_{c1}\omega_{c2}}$ is the center frequency of the passband.

$|H_{BP2}(j\omega)|$ is the product of the magnitudes of each of the fractions in Equation 15.36. Thus,

$$|H_{BP2}(j\omega)| = \frac{K}{\sqrt{(1+\omega^2/\omega_{c2}^2)(1+\omega_{c1}^2/\omega^2)}} \tag{15.38}$$

For a broadband filter, $\omega_{c2} \gg \omega_{c1}$. In this case, $\omega_{c2}/\omega_{c1} = (2\pi \times 15{,}000)/(2\pi \times 20)$, or $\omega_{c2} = 750\omega_{c1}$. If $\omega < \omega_{c2}/10$, say, that is, less than $2\pi \times 1.5$ krad/s in this case, then $\omega^2/\omega_{c2}^2 < 0.01$. The term ω^2/ω_{c2}^2 under the square root can therefore be neglected for frequencies in the vicinity of ω_{c1}, since this term has very little effect on the response.

$$|H_{BP2}(j\omega)| \cong \frac{K}{\sqrt{(1+\omega_{c1}^2/\omega^2)}} \tag{15.39}$$

The filter behaves as a high-pass filter of 3 dB cutoff frequency ω_{c1} that is hardly affected by ω_{c2}. Similarly, if $\omega > 10\omega_{c1}$, say, which is larger than $2\pi \times 100$ rad/s, in this case, then $\omega_{c1}^2/\omega^2 < 0.01$, and

$$|H_{BP2}(j\omega)| \cong \frac{K}{\sqrt{(1+\omega^2/\omega_{c2}^2)}} \tag{15.40}$$

The filter behaves as a low-pass filter of 3 dB cutoff frequency ω_{c2} that is hardly affected by ω_{c1}. In other words, as long as ω_{c2} is at least 10 times ω_{c1}, each filter has a negligible effect on the other, so that *the two filters can be designed separately*.

For the high-pass filter, $R_{rh} = 1/(\omega_{c1}C_{rh}) = 1/(2\pi \times 20 \times 50 \times 10^{-9}) \cong 160$ kΩ. For the low-pass filter, $R_{fl} = 1/(\omega_{c2}C_{fl}) = 1/(2\pi \times 15{,}000 \times 50 \times 10^{-9}) \cong 210$ Ω. $K = R_{fl}R_{fh}/R_{rl}R_{rh} = 5$. We may choose $R_{fh} = 400$ kΩ, so $R_{rl} = 105$ Ω. The filter circuit is shown in Figure 15.14. As mentioned in connection with Figure 15.3, it is more advantageous that the high-pass filter precedes the low-pass filter in this case in order to have a higher input impedance.

It should be noted that according to the statement of the problem, $\omega_{c1} = 2\pi \times 20$ rad/s and $\omega_{c2} = 2\pi \times 15$ krad/s, which gives a 3 dB bandwidth of $(\omega_{c2} - \omega_{c1})$. But the cascading of the low-pass and high-pass filters results in a bandpass bandwidth of $(\omega_{c2} + \omega_{c1})$, in accordance with Equation 15.36, in which the coefficient of s in the denominator is $(\omega_{c2} + \omega_{c1})$. This wider bandwidth is due to the effect of one filter on the other, which has been neglected in the preceding solution for the sake of a much simplified design procedure. When $\omega_{c2} \gg \omega_{c1}$, the resulting error is small. It is in this case $((f_{c2}+f_{c1}) - (f_{c2}-f_{c1}))/(f_{c2}-f_{c1}) = 2f_{c1}/(f_{c2}-f_{c1}) = 40$ Hz/(15 kHz – 20 Hz). This evaluates to about 0.3% and is in fact erring on the right side, as a slightly wider bandwidth is generally more desirable. This discrepancy is quite acceptable, considering that, in practice, the tolerance on the values of commonly used components is at least ±2% for resistors and ±5% for capacitors.

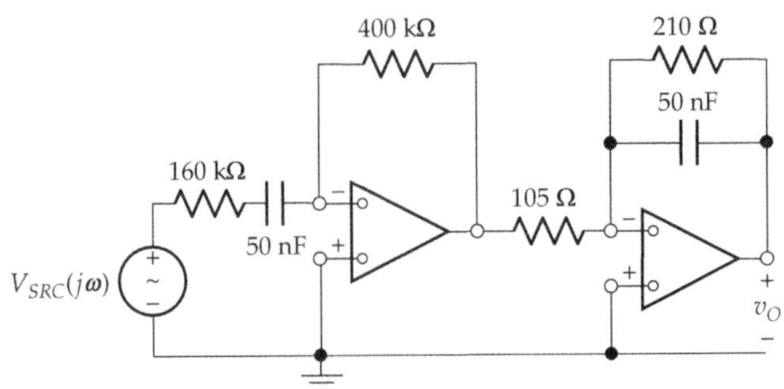

FIGURE 15.14
Figure for Example 15.3.

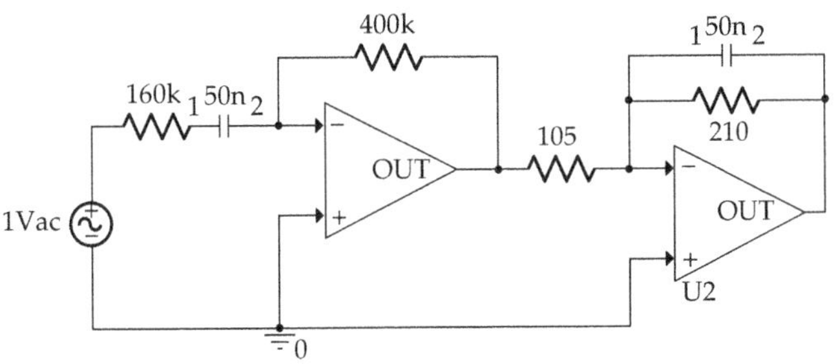

FIGURE 15.15
Figure for Example 15.3.

Simulation: The circuit is entered as in Figure 15.15. Simulation Settings are a start frequency of 1 Hz, an end frequency of 100 kHz, and 2000 points/decade. After the simulation is run, select DB(V(U2:OUT)). $|H_{BP3}(j\omega)|$ is displayed as in Figure 15.16. Using cursor max, the maximum gain in the passband is read as 13.968 dB at 551.442 Hz, compared to $20\log_{10}(5) = 13.979$ dB at $f_0 = \sqrt{20 * 15{,}000} = 547.72$ Hz. Cursor search gives the lower 3 dB cutoff frequency as 19.889 Hz and the upper 3 dB cutoff frequency as 15.161 kHz. The equation of the low-frequency asymptote is 20*LOG10(5) –20*LOG10(20)+20*LOG10(Frequency), whereas the equation of the high-frequency asymptote is 20*LOG10(5) +20*LOG10(15E3)–20*LOG10(Frequency).

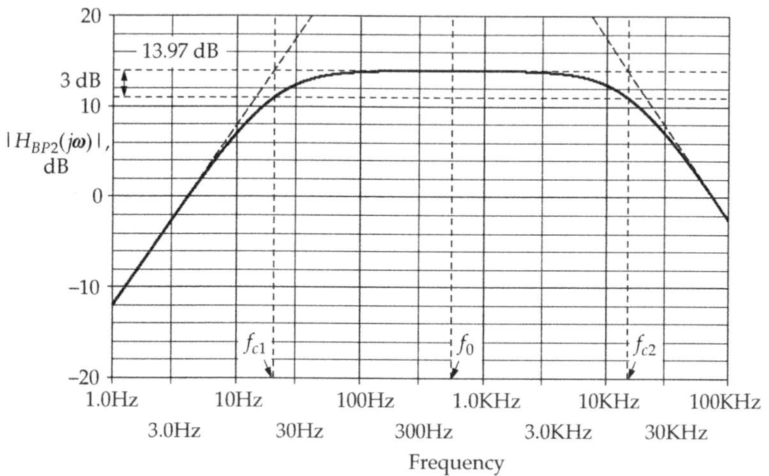

FIGURE 15.16
Figure for Example 15.3.

15.4 Noninverting Second-Order Active Filters

15.4.1 High-Pass Filter

A high-pass noninverting filter is illustrated in Figure 15.17. Using the node-voltage method, for example, the transfer function can be shown to be

$$\frac{V_O(j\omega)}{V_{SRC}(j\omega)} = A\frac{s^2}{s^2 + s\left[\frac{1}{R_2}\left(\frac{1}{C_1}+\frac{1}{C_2}\right)+\frac{1}{C_1R_1}(1-A)\right]+\omega_{ch}^2} \quad (15.41)$$

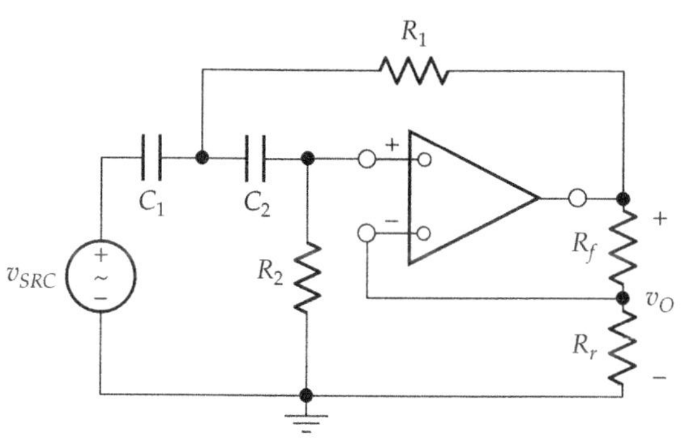

FIGURE 15.17
Noninverting second-order high-pass filter.

where

$$A = \left(1+\frac{R_f}{R_r}\right)$$

$$\omega_{ch} = 1/\sqrt{C_1C_2R_1R_2}$$

$$Q = \frac{\sqrt{C_1C_2R_1R_2}}{(C_1+C_2)R_1 + C_2R_2(1-A)}$$

$$s = j\omega$$

As $\omega \rightarrow \infty$, the capacitors act as short circuits. R_2 is connected across v_{SRC} and is redundant as far as v_O is concerned. R_1 is connected between v_{SRC} and the ideal voltage source output of the ideal op amp. It is also redundant because the current through it only affects the currents in these ideal sources and nothing else in the circuit. With C_1 and C_2 replaced by short circuits and R_1 and R_2 removed, the circuit reduces to a noninverting amplifier of gain A.

As $\omega \rightarrow 0$, the capacitors act as open circuits. The amplifier is isolated from the input source, its input is grounded, through R_2, so that the output is zero. The response is therefore high pass.

To facilitate the design, A is set to 1, that is, the amplifier is connected as a unity-gain amplifier. To normalize the response, it is convenient to set $C_1 = C_2 = 1$ F, so that the coefficient of s, and hence Q, in Equation 15.41 is determined by R_2 only. To have $\omega_{ch} = 1$ rad/s, $R_1R_2 = 1$. Equation 15.41 becomes in normalized form:

$$\frac{V_O(j\omega)}{V_{SRC}(j\omega)} = \frac{s^2}{s^2+\frac{2}{R_2}s+1}, \quad \text{with } R_1R_2 = 1 \quad (15.42)$$

Example 15.4: Second-Order Noninverting High-Pass Butterworth Filter

It is required to design a unity-gain, second-order, Butterworth high-pass filter having a 3 dB cutoff frequency of 500 Hz, using 0.5 μF capacitors.

Solution:

For a second-order Butterworth normalized response, the coefficient of s in the dominator of Equation 15.42 should be $\sqrt{2}$ (Table 15.2). This gives $R_2 = 2/\sqrt{2} = \sqrt{2}\ \Omega$, so that $R_1 = 1/\sqrt{2}\ \Omega$. To move the cutoff frequency from 1 rad/s to $2\pi \times 500$ Hz requires a frequency scale factor $k_f = 1000\pi$. To use capacitors of 0.5 μF requires a magnitude scale factor $k_m = \dfrac{1}{1000\pi \times 0.5 \times 10^{-6}} = \dfrac{2000}{\pi}$ (Equation 15.3). Resistances are multiplied by k_m, so that $R_1 = \dfrac{1000\sqrt{2}}{\pi} = 450.16\ \Omega$ and $R_2 = \dfrac{2000\sqrt{2}}{\pi} = 900.32\ \Omega$.

Simulation: The circuit is entered as in Figure 15.18. Simulation Settings are a start frequency of 10 Hz, an end frequency of 10 kHz, and 2000 points/decade. After the simulation is run, select DB(V(U1:OUT)). $|H_{HP2}(j\omega)|$ is displayed as in Figure 15.19. Using cursor search gives the 3 dB cutoff frequency as 500.591 Hz. The equation of the low-frequency asymptote is: −20*LOG10(500)+20*LOG10(Frequency).

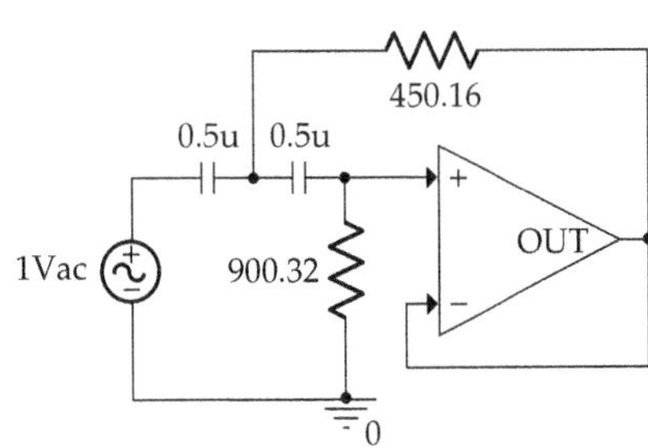

FIGURE 15.18
Figure for Example 15.4.

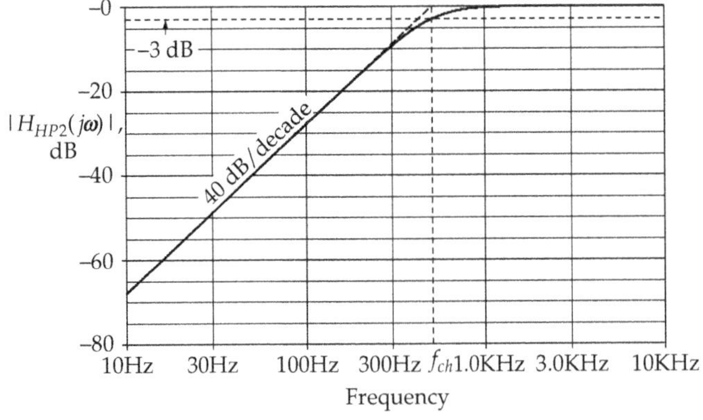

FIGURE 15.19
Figure for Example 15.4.

15.4.2 Low-Pass Filter

It was explained in connection with Table 14.3 that a low-pass filter can be derived from the high-pass filter or, conversely, by interchanging s and ω_0. In the case of the active *RC* filters under consideration, this corresponds to interchanging s and ω_c, the 3 dB cutoff frequency, by replacing every resistor of the filter circuit, that is, excluding R_r and R_f of the op amp, by a capacitor, and conversely. In the transfer function, this replaces every R_m by $1/sC_m$, and every $1/sC_k$ by R_k, or every C_k by $1/sR_k$. With $\omega_c = \dfrac{1}{C_k R_m}$, this changes $\dfrac{\omega_c}{s} = \dfrac{1}{sC_k R_m}$ to $R_k(sC_m) = \dfrac{s}{\omega_c}$, where the expression for ω_c becomes $\dfrac{1}{C_m R_k}$.

It is seen that making these replacements interchanges ω_c/s and s/ω_c, that is, interchanges s and ω_c. The circuit in Figure 15.17 becomes that in Figure 15.20, where R_1 in Equation 15.41 is replaced by $1/sC_1$, R_2 by $1/sC_2$, C_1 by $1/sR_1$, and C_2 by $1/sR_2$. After the resulting equation is simplified, the transfer function becomes

$$\frac{V_O(j\omega)}{V_{SRC}(j\omega)} = \frac{A}{C_1C_2R_1R_2} \frac{1}{s^2 + s\left[\dfrac{1}{C_1}\left(\dfrac{1}{R_1} + \dfrac{1}{R_2}\right) + \dfrac{1}{C_2R_2}(1-A)\right] + \omega_{cl}^2} \tag{15.43}$$

where

$$\omega_{cl} = 1/\sqrt{C_1C_2R_1R_2}$$

$$Q = \frac{\sqrt{C_1C_2R_1R_2}}{C_2(R_1+R_2) + C_1R_1(1-A)}$$

As $\omega \to 0$, the capacitors act as open circuits. R_1 and R_2 do not carry any current because they are connected

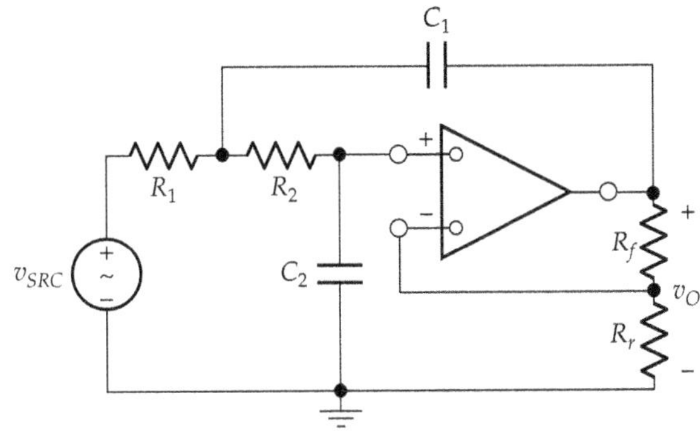

FIGURE 15.20
Noninverting second-order low-pass filter.

to open circuits. The circuit reduces to a noninverting configuration of gain A. As $\omega \to \infty$, the capacitors act as short circuits. C_2 effectively grounds the input of the op amp, so the output is zero. The response is low pass.

To facilitate the design, the amplifier is connected as a unity-gain amplifier, so that $A = 1$. The response is conveniently normalized by setting $R_1 = R_2 = 1\ \Omega$, so that the coefficient of s, and hence Q is determined by C_1 only. To have $\omega_{cl} = 1$ rad/s, $C_1C_2 = 1$. Equation 15.43 becomes

$$\frac{V_O(j\omega)}{V_{SRC}(j\omega)} = \frac{1}{s^2 + \frac{2}{C_1}s + 1}, \quad \text{with } C_1C_2 = 1 \quad (15.44)$$

Example 15.5: Third-Order Noninverting Butterworth Low-Pass Filter

It is required to design a third-order, Butterworth low-pass filter based on the circuit in Figure 15.20, having a gain of 3 and a 3 dB cutoff frequency of 2000 rad/s, using 1 kΩ resistors in the filter sections.

Solution:

Referring to Table 15.2, the normalized third-order Butterworth polynomial is $(s^2 + s + 1)(s + 1)$. Equating the unity coefficient of s in $s^2 + s + 1$ to $2/C_1$ in Equation 15.44 gives $C_1 = 2$ F, so that $C_2 = 0.5$ F in the normalized filter. To move the cutoff frequency to 2000 rad/s requires $k_f = 2000$. To use 1 kΩ resistors requires $k_m = 1000$. It follows that $C_1 = (2\text{ F})/(1000 \times 2.000) = 1\ \mu$F and $C_2 = C_1/4 = 0.25\ \mu$F.

The normalized first-order filter has $R = 1\ \Omega$, $C = 1$ F, and $\omega_{cl} = 1$ rad/s. It has, therefore, the same k_f and k_m as the second-order filter. Hence, $C = (1\text{ F})/(1000 \times 2.000) = 0.5\ \mu$F. If the first-order filter is implemented using an active filter (Figure 15.11b), the overall gain is negative, which requires an additional inverting amplifier stage to obtain a positive overall gain. Alternatively, a simple RC low-pass filter can be used, followed by a noninverting amplifier having a gain of three. The amplifier isolates the RC filter from the load so as to have the required transfer function. The filter circuit is shown in Figure 15.21. The first-order filter could precede the second-order filter in this case.

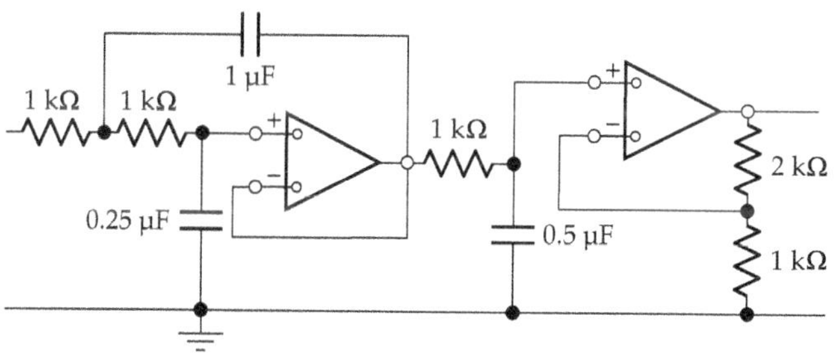

FIGURE 15.21
Figure for Example 15.5.

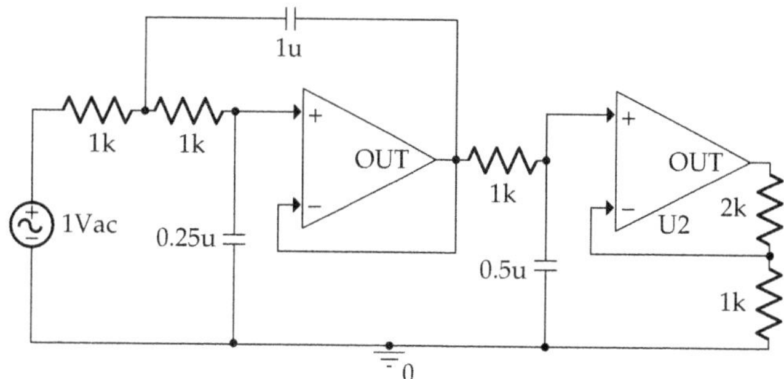

FIGURE 15.22
Figure for Example 15.5.

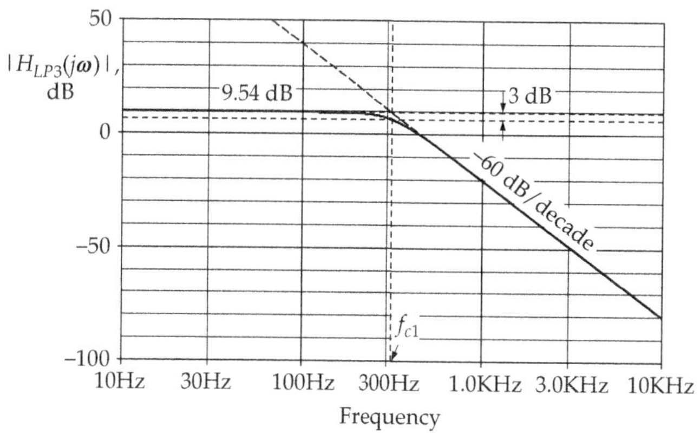

FIGURE 15.23
Figure for Example 15.5.

Simulation: The circuit is entered as in Figure 15.22. Simulation Settings are a start frequency of 10 Hz, an end frequency of 10 kHz, and 2000 points/decade. After the simulation is run, select DB(V(U2:OUT)). $|H_{LP3}(j\omega)|$ is displayed as in Figure 15.23. The passband gain is read from the cursor as 9.542 dB, which equals $20\log_{10}(3)$. Cursor search gives the 3 dB cutoff frequency as 318.058 Hz, compared to a calculated value of 318.31 Hz. The equation of the high-frequency asymptote is 20*LOG10(3)+20*LOG10(318.31)−20*LOG10(Frequency).

15.4.3 Bandpass Filter

A bandpass filter is shown in Figure 15.24, whose transfer function is

$$\frac{V_O(j\omega)}{V_{SRC}(j\omega)} = \frac{\frac{A}{C_1R_3}s}{s^2 + s\left[\frac{1}{(R_1 \| R_3)}\left(\frac{1}{C_1} + \frac{1}{C_2}\right) + \frac{1}{C_2}\left(\frac{1}{R_2} - \frac{A}{R_1}\right)\right] + \omega_0^2} \quad (15.45)$$

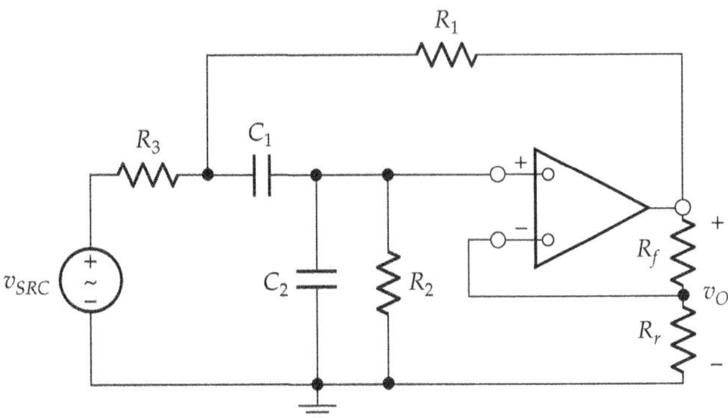

FIGURE 15.24
Noninverting second-order bandpass filter.

where $\omega_0 = 1/\sqrt{C_1C_2(R_1 \| R_3)R_2}$. Since ω_0/Q is the coefficient of s in the denominator, $Q = \dfrac{1/\omega_0}{R_2(C_1+C_2)+C_1(R_1 \| R_3)[1-(AR_2/R_1)]}$. The maximum gain is when $\omega = \omega_0$ so that $s^2 + \omega_0^2 = 0$ in the denominator; s cancels out from the numerator and denominator, which gives the maximum gain as $K = (A/C_1R_3)/(\omega_0/Q) = QA/(\omega_0C_1R_3)$.

That the response is bandpass can be readily checked. As $\omega \to 0$, C_1 isolates the input of the op amp from the source and the output is zero. As $\omega \to \infty$, C_2 grounds the input of the op amp, and the output is again zero. This implies that the nonzero output is a maximum at some intermediate value, so the response is bandpass.

It will be noted that normalizing Equation 15.45 is not as simple as in the low-pass and high-pass filters. A filter of this type is considered in Problem P15.43. An inverting bandpass filter that is easier to design is presented in the next section.

15.5 Inverting Second-Order Active Filters

15.5.1 Bandpass Filter

An inverting second-order bandpass filter is illustrated in Figure 15.25, whose transfer function is

$$\frac{V_O(j\omega)}{V_{SRC}(j\omega)} = -\frac{1}{R_2C_2}\frac{s}{s^2 + \dfrac{1}{R_1}\left(\dfrac{1}{C_1}+\dfrac{1}{C_2}\right)s + \omega_0^2} \qquad (15.46)$$

where $\omega_0 = 1/\sqrt{C_1C_2R_1(R_2 \| R_3)}$, $\dfrac{\omega_0}{Q} = \dfrac{1}{R_1}\left(\dfrac{1}{C_1}+\dfrac{1}{C_2}\right)$ is the 3 dB bandwidth. The gain at the center frequency, obtained by setting $s = j\omega_0$, is $K = -\dfrac{R_1C_1}{R_2(C_1+C_2)}$.

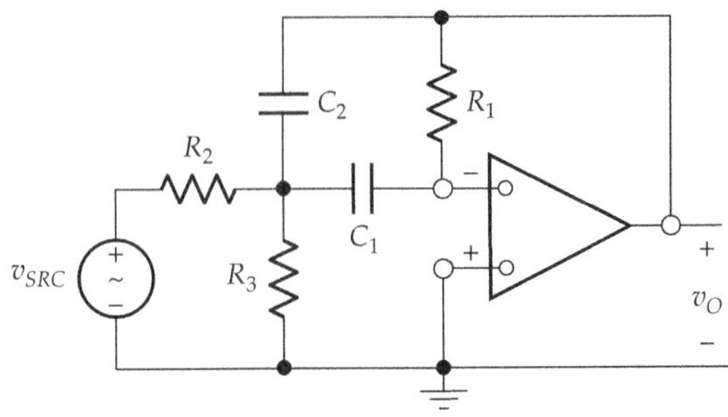

FIGURE 15.25
Inverting second-order bandpass filter.

That the response in Figure 15.25 is bandpass can be readily ascertained. As $\omega \to 0$, the capacitors act as open circuits, no current passes through R_1, the output terminal is at virtual ground, and the output voltage is therefore zero. As $\omega \to \infty$, the capacitors act as short circuits, thereby connecting the output terminal directly to virtual ground and resulting in zero output voltage. The nonzero output voltage must have a maximum magnitude at some intermediate value, so the response is bandpass.

If the filter response is normalized by having $C_1 = C_2 = 1$ F and $\omega_o = 1$ rad/s, Equation 15.46 becomes

$$\frac{V_O(j\omega)}{V_{SRC}(j\omega)} = -\frac{1}{R_2}\frac{s}{s^2 + \dfrac{2}{R_1}s + 1}, \quad \text{with } R_1(R_2 \| R_3) = 1 \qquad (15.47)$$

where $\dfrac{1}{Q} = \text{BW} = \dfrac{2}{R_1}$ and $K = -\dfrac{Q}{R_2}$. It follows that $R_1 = 2Q$, $R_2 = -Q/K$, and $R_3 = \dfrac{Q}{2Q^2+K}$. Once the center frequency ω_o, the passband gain K, and BW or Q are specified, R_1, R_2, and R_3 can be determined, as illustrated by the following example.

Example 15.6: Second-Order Inverting Bandpass Filter

It is required to design a bandpass filter having a center frequency of the passband of 20 krad/s, $Q = 10$, and a passband gain of −4, using 0.05 μF capacitors.

Solution:

Using the relations derived in connection with Equation 15.47, $R_1 = 2Q = 20\ \Omega$, $R_2 = \dfrac{10}{4} = 2.5\ \Omega$, and $R_3 = \dfrac{10}{2\times(10)^2-4} = \dfrac{10}{196}\ \Omega$. The frequency scale factor k_f is 20,000. To have $C = 0.05$ μF requires an impedance magnitude scale factor $k_m = \dfrac{10^6}{20{,}000\times 0.05} = 1{,}000$. Hence,

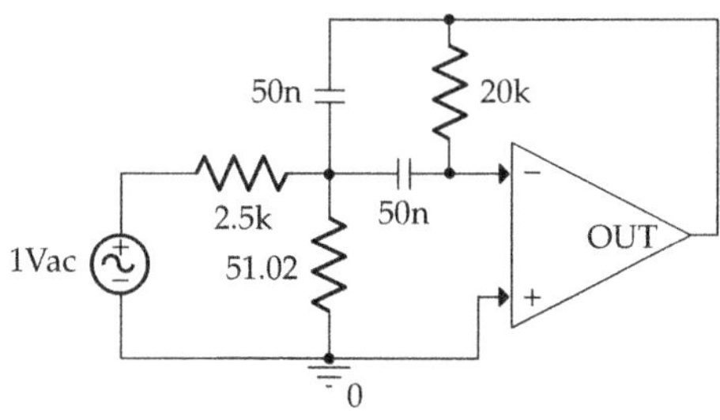

FIGURE 15.26
Figure for Example 15.6.

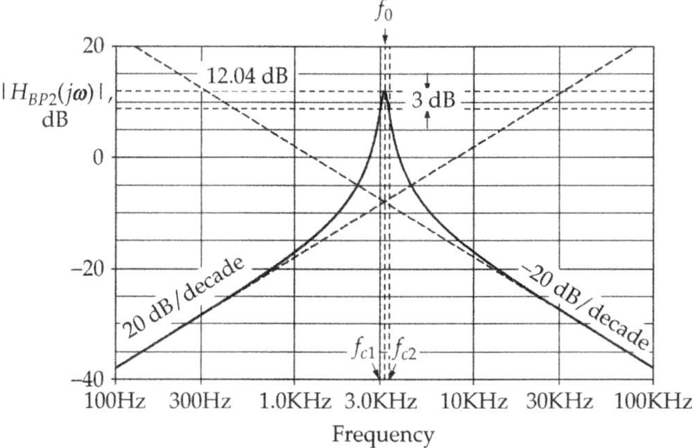

FIGURE 15.27
Figure for Example 15.6.

$R_1 = (20\ \Omega) \times 1000 = 20\ \text{k}\Omega$, $R_2 = (2.5\ \Omega) \times 1000 = 2.5\ \text{k}\Omega$, and $R_3 = \frac{10}{196} \times 1000 = 51.02\ \Omega$. The bandwidth is $\frac{\omega_o}{Q} = \frac{2}{R_1} k_f = 2000\ \text{rad/s}$.

Simulation: The circuit is entered as in Figure 15.26. Simulation Settings are a start frequency of 100 Hz, an end frequency of 100 kHz, and 2000 points/decade. After the simulation is run, select DB(V(U1:OUT)). $|H_{BP2}(j\omega)|$ is displayed as in Figure 15.27. The maximum gain is read from cursor maximum as 12.038 dB at 3.1842K, or 20.007 krad/s, compared to $20\log_{10}(4) = 12.041$ dB at 20 krad/s. Cursor search gives the upper 3 dB cutoff frequency f_{c2} as 3.3459K and the lower 3 dB cutoff frequency f_{c1} as 3.0283K. This gives BW = 317.6 Hz, or 1995.5 rad/s, compared to a calculated value of 2000 rad/s, and a Q of 10.026.

From Equation 15.46, $|H_{BP2}(j\omega)| = \omega/\left(R_2C_2\omega_0^2\right)$ as $\omega \to 0$. The equation of the low-frequency asymptote is

$$20\log_{10} f - 20\log_{10}\left(R_2C_2\omega_0^2/2\pi\right) = 20\log_{10} f - 20\log_{10}\left(2.5\times10^3 \times 5\times10^{-8} \times \left(20\times10^3\right)^2/2\pi\right) = 20\log_{10}(f) - 20\log_{10}(7057.7)$$

As $\omega \to \infty$, $|H_{BP2}(j\omega)| \to 1/(\omega C_2R_2)$. The equation of the high-frequency asymptote is $20\log_{10}(1/2\pi C_2R_2) - 20\log_{10}(f) = 20\log_{10}(1273.2) - 20\log_{10}(f)$.

15.5.2 High-Pass Filter

If in Figure 15.25, R_2 is replaced by a capacitor C_3, the transfer function of Equation 15.46 becomes

$$\frac{V_O(j\omega)}{V_{SRC}(j\omega)} = -\frac{C_3}{C_2}\frac{s^2}{s^2 + \frac{1}{R_1}\left(\frac{1}{C_1} + \frac{1}{C_2} + \frac{C_3}{C_1C_2}\right)s + \omega_{ch}^2} \quad (15.48)$$

where

$$\omega_{ch} = 1/\sqrt{C_1C_2R_1R_3}$$

$$Q = \omega_{ch}R_1C_1C_2/(C_1 + C_2 + C_3)$$

Equation 15.48 is the transfer function of a high-pass response, the circuit being as in Figure 15.28.

As $\omega \to 0$, the capacitors act as open circuits, no current passes through R_1, the output terminal is at virtual ground, and the output voltage is therefore zero, as in Figure 15.25. As $\omega \to \infty$, the reactances of the capacitors become very small and the transfer function Z_f/Z_r approaches $-C_3/C_2$, as in Equation 15.48.

The response can be normalized by setting $C_1 = C_2 = C_3 = 1$ F, $\omega_{ch} = 1$ rad/s, so that $R_1R_3 = 1$, and the maximum magnitude of the gain becomes unity. The transfer function of the normalized response becomes

$$\frac{V_O(j\omega)}{V_{SRC}(j\omega)} = -\frac{s^2}{s^2 + \frac{3}{R_1}s + \omega_{ch}^2} \quad R_1R_3 = 1 \quad (15.49)$$

where $R_1 = 3Q$. The design and simulation of the circuit are left as a problem (P15.45).

15.5.3 Low-Pass Filter

Replacing capacitors by resistors, and conversely, result in an inverting low-pass filter as in Figure 15.29, where

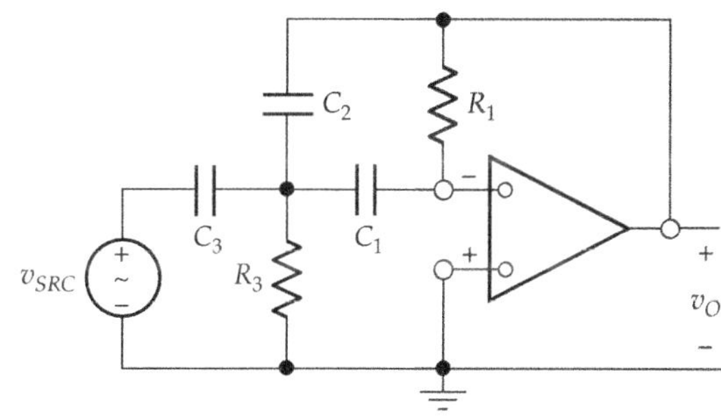

FIGURE 15.28
Inverting second-order high-pass filter.

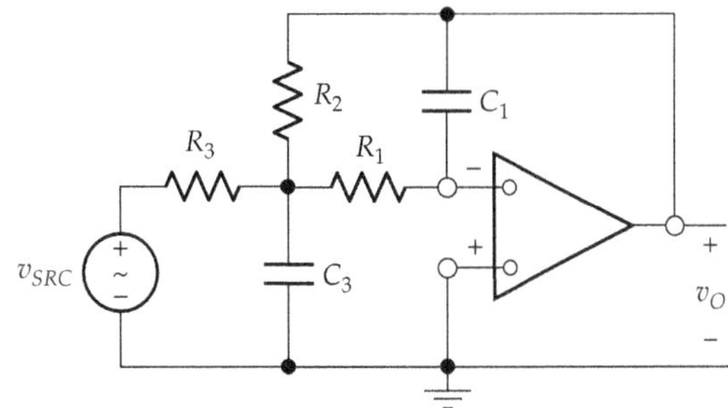

FIGURE 15.29
Inverting second-order low-pass filter.

the same subscript is used for the elements that replace one another. The transfer function is

$$\frac{V_O(j\omega)}{V_{SRC}(j\omega)} = -\frac{1}{C_1C_3R_1R_3}\frac{1}{s^2 + \frac{1}{C_3}\left(\frac{1}{R_1} + \frac{1}{R_2} + \frac{1}{R_3}\right)s + \omega_{cl}^2} \quad (15.50)$$

where

$$\omega_{cl} = 1/\sqrt{C_1C_3R_1R_2}$$

$$Q = \omega_{cl}C_3\left(R_1 \| R_2 \| R_3\right)$$

As $\omega \to 0$, the capacitors act as open circuits. No current flows in R_1, and the circuit reduces to an inverting amplifier having a transfer function $-R_2/R_3$. As $\omega \to \infty$, the capacitors act as short circuits, with C_3 short-circuiting the input to the op amp, so that the output voltage is zero.

The response can be normalized by setting $R_1 = R_2 = R_3 = 1\ \Omega$, $\omega_{ch} = 1$ rad/s, so that $C_1C_3 = 1$, and the maximum magnitude of the gain becomes unity. The transfer function of the normalized response becomes

$$\frac{V_O(j\omega)}{V_{SRC}(j\omega)} = -\frac{s^2}{s^2 + \frac{3}{C_3}s + \omega_{ch}^2}, \quad C_1C_3 = 1 \quad (15.51)$$

where $C_3 = 3Q$. The design and simulation of the circuit are left as a problem (P15.46).

15.6 Universal Filter

A universal filter is illustrated in Figure 15.30. The filter has three inputs and is capable of providing any of the five, second-order filter responses, depending on how

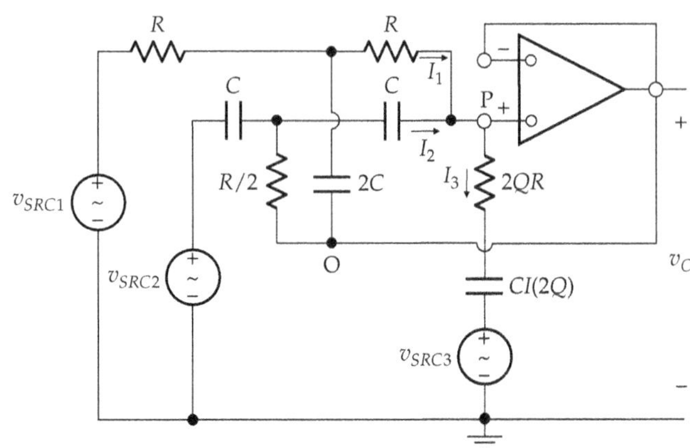

FIGURE 15.30
Universal filter.

the three inputs are applied. The output of the filter as a function of all three inputs is

$$V_O(j\omega) = \frac{s^2V_{SRC2}(j\omega) + (\omega_0/Q)sV_{SRC3}(j\omega) + \omega_0^2V_{SRC1}(j\omega)}{s^2 + (\omega_0/Q)s + \omega_0^2} \quad (15.52)$$

where $\omega_0 = 1/RC$ and Q is a designated value. The input combinations for the different responses are indicated in Table 15.3. To obtain a low-pass response, for example, the input is applied as $V_{SRC1}(j\omega)$, with the other two inputs connected to ground.

Equation 15.52 can be derived by the node-voltage method; voltage variables are assigned to the essential nodes at 'P', 'O', and the essential nodes between the two capacitors and the two resistors. KCL is derived at each of these nodes and the resulting simultaneous equations solved. It is simpler and more instructive to derive Equation 15.52 using the substitution theorem (Section 4.4).

Since the amplifier is unity-gain, the voltage at the noninverting input 'P' of the op amp is the same as the output voltage $V_O(j\omega)$ at node 'O'. The op amp can therefore be replaced by an ideal independent voltage source $V_O(j\omega)$ between node 'O' and ground, and the branch connected between node 'P' and ground can also be replaced by this source, in accordance with the substitution theorem.

TABLE 15.3
Input Combinations and Responses of Universal Filter

$V_{SRC1}(j\omega)$	$V_{SRC2}(j\omega)$	$V_{SRC3}(j\omega)$	Response
$V_{SRC}(j\omega)$	0	0	Low pass
0	$V_{SRC}(j\omega)$	0	High pass
0	0	$V_{SRC}(j\omega)$	Bandpass
$V_{SRC}(j\omega)$	$V_{SRC}(j\omega)$	0	Bandstop
$V_{SRC}(j\omega)$	$V_{SRC}(j\omega)$	$-V_{SRC}(j\omega)$	Allpass

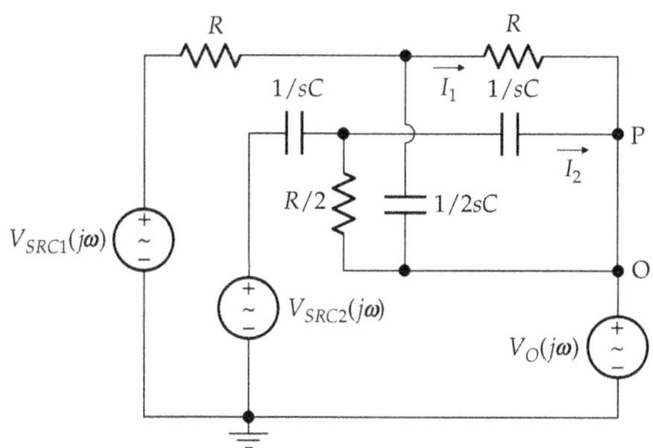

FIGURE 15.31
Derivation of the response of the universal filter.

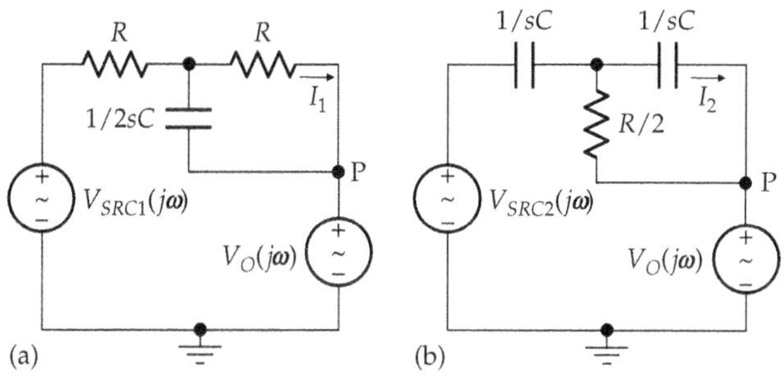

FIGURE 15.32
Output currents of T-circuits of the universal filter. (a) Output current of the upper T-circuit in Figure 15.31 and (b) output current of the lower T-circuit in Figure 15.31.

Since nodes 'P' and 'O' are at the same voltage $V_O(j\omega)$ with respect to ground, they can be connected together. The circuit in the frequency domain becomes as in Figure 15.31. We wish to determine the currents I_1 and I_2. The circuit for I_1 is shown in Figure 15.32a.

The voltage drop across R on the RHS is RI_1, which is also the voltage across the capacitor. The capacitor current is therefore $2sCRI_1$. The total current is $I_1(1 + 2sCR)$. From KVL, $V_{SRC1}(j\omega) = RI_1(1 + 2sCR) + RI_1 + V_O(j\omega)$. It follows that

$$I_1 = \frac{1}{2R}\frac{\omega_0}{s+\omega_0}\left(V_{SRC1}(j\omega) - V_O(j\omega)\right) \tag{15.53}$$

where $\omega_0 = 1/RC$. The circuit for determining the current I_2 is shown in Figure 15.32b. I_2 may be determined in the same way as I_1 in Figure 15.32a. Alternatively, it may be noted that if R and $1/sC$ in Figure 15.32a are replaced by $1/sC$ and R, respectively, the circuit of Figure 15.32b is obtained. Making these replacements in Equation 15.53 gives

$$I_2 = \frac{C}{2}\frac{s^2}{s+\omega_0}\left(V_{SRC2}(j\omega) - V_O(j\omega)\right) \tag{15.54}$$

Referring to Figure 15.30, I_3 is given by

$$I_3 = \frac{V_O(j\omega) - V_{SRC3}(j\omega)}{2QR + 2Q/sC} \tag{15.55}$$

From KCL, $I_3 = I_1 + I_2$. Substituting for the currents in this equation, collecting terms, and simplifying give Equation 15.52.

Example 15.7: Notch Filter

It is required to configure the universal filter as a bandstop filter having a center frequency of 1 krad/s and a Q of 50 using 0.1 μF capacitors.

Solution:

$R = 1/(\omega_0 C) = 1/(10^3 \times 10^{-7}) = 10$ kΩ. For a Q of 50, $2RQ = 20 \times 10^3 \times 50 = 1$ MΩ, and $C/(2Q) = 10^{-7}/100 = 1$ nF.

Simulation: The circuit is entered as in Figure 15.33. Simulation Settings are a start frequency of 100 Hz,

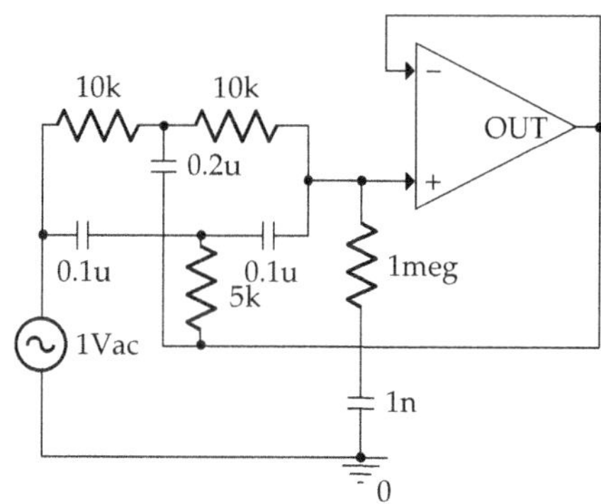

FIGURE 15.33
Figure for Example 15.7.

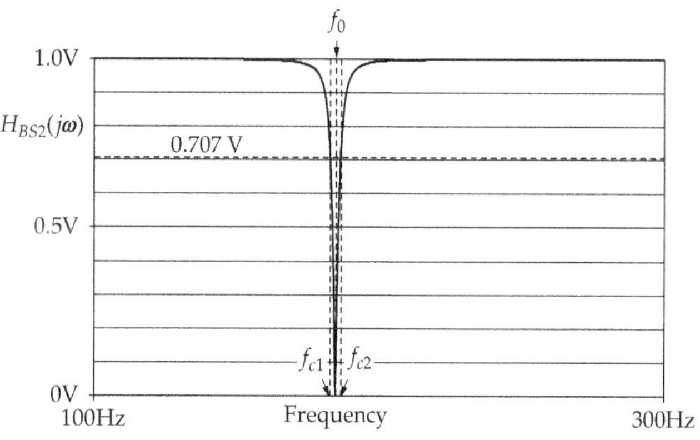

FIGURE 15.34
Figure for Example 15.7.

an end frequency of 300 Hz, and 5000 points/decade. After the simulation is run, select DV(U1:OUT). $|H_{BP2}(j\omega)|$ is displayed as in Figure 15.34. The minimum gain of zero is read from cursor minimum as at 159.148, compared to a calculated value of 159.155 Hz. Cursor search gives the upper 3 dB cutoff frequency f_{c2} as 160.755 and the lower 3 dB cutoff frequency f_{c1} as 157.571. This gives BW = 3.184 Hz, corresponding to a Q of 159.148/3.184 = 49.98.

Learning Checklist: What Should Be Learned from This Chapter

- Scaling is a convenient procedure based on designing a certain type of filter in normalized form and then scaling the values of these normalized filters as may be required for any particular implementation of this type of filter.
    1. For a normalized first-order filter, the values chosen are $\omega_c = 1$ rad/s, $R = 1\ \Omega$, $C = 1$ F, and $L = 1$ H.
    2. For a normalized second-order filter, the values chosen are $\omega_c = 1$ rad/s, $C = 1$ F, and $L = 1$ H. The normalized R is chosen to give the required value of Q.
- The scaled values are $\omega' = k_f\omega$, $R' = k_mR$, $L' = \frac{k_m}{k_f}L$, and $C' = \frac{1}{k_mk_f}C$ where the frequency scaling factor k_f is determined by the desired ω_c, ω_0, or BW, and the magnitude scaling factor k_m is determined by the desired value of some circuit parameter.
    1. BW scales like ω, Q is unaffected by scaling.
- Scaling can be applied to any filter when it is desired to change a parameter value or frequency.
- A Butterworth low-pass or high-pass response is a maximally flat response in the passband. The reason for this is that in the expression for the magnitude of a Butterworth polynomial of order n, the only term in ω is a term raised to the power $2n$, so that the change in the attenuation with frequency in the passband is as small as possible.
- In a normalized Butterworth, low-pass response of order n, the normalized Butterworth polynomial is the denominator of the transfer function, the numerator of the transfer function being unity. The magnitude of the normalized, Butterworth low-pass response of order n is of the form $1/\sqrt{1+u^{2n}}$, where $u = \omega/\omega_c$.
- In a normalized Butterworth, high-pass response of order n, the normalized Butterworth polynomial is the denominator of the transfer function, the numerator of the transfer function being s^n. The magnitude of the normalized, Butterworth high-pass response of order n is of the form $u^n/\sqrt{1+u^{2n}}$.
- All Butterworth low-pass or high-pass responses have an attenuation of −3 dB with respect to the passband gain at $\omega = \omega_c$, irrespective of n. ω_c is both the 3 dB cutoff and the corner frequency. The magnitude of the slope of the asymptote of the response is $20n$ dB/decade.
- Op amps in active filters provide gain, isolation, load drive, and high Q.
- Op amps are used with RC circuits to surpass the limit of $Q = 0.5$ for passive RC circuits and implement active, Butterworth RC filters having $Q > 0.5$.
- A first-order low-pass active filter can be derived from an ideal integrator by connecting a resistor in parallel with the capacitor in order to limit the gain at low frequencies.
- A first-order high-pass active filter can be derived from an ideal differentiator by connecting a resistor in series with the capacitor in order to limit the gain at high frequencies.
- A second-order, broadband, bandpass filter can be implemented by cascading a first-order low-pass active with a first-order high-pass active filter. The two filters can be designed separately, as long as the 3 dB cutoff frequencies are at least a decade of frequency apart.
- Second-order low-pass, high-pass, bandpass, and bandstop active filters having any desired Q may be constructed using operational amplifiers in the noninverting or inverting configuration.

Problem-Solving Tips

1. Scaling is convenient to use in filter design. In applying scaling, remember that, in general, values of L and C are scaled by both frequency and magnitude scale factors, that bandwidth scales like frequency, and that Q is invariant with scaling.
2. Always determine how an active filter circuit behaves at $\omega \to 0$, as $\omega \to \infty$.
3. In all types of filters (low pass, high pass, bandpass RLC, and bandstop RLC), it is generally possible to determine the gains of interest from

the circuit, without deriving the transfer function, by examining the circuits at the frequency extremes ($\omega \to 0$ and $\omega \to \infty$) or at the resonant frequency ω_0. However, in the case of bandpass RC circuits, the maximum gain can only be derived, in general, from the transfer function.

4. The 3 dB cutoff frequency of a first-order filter can be obtained from R_{Th} seen by the energy storage element without having to derive the transfer function.

Problems

Verify solutions by PSpice Simulation.

Scaling and Passive Butterworth Filters

P15.1 Given a transfer function of a series RLC circuit as $\frac{10s}{s^2+0.1s+1}$, normalized to have $\omega_0 = 1$ rad/s, $C = 1$ F, $L = 1$ H, and $R = 0.1\ \Omega$, determine L, C, R, and Q if (a) ω_0 is scaled to 1 Mrad/s without magnitude scaling and (b) if ω_0 is scaled to 1 Mrad/s and C is scaled to 100 nF.

Ans. (a) 1 μH, 1 μF, 0.1 Ω, $Q = 10$; (b) 10 μH, 100 nF, 1 Ω, $Q = 10$.

P15.2 Given the magnitude response $|H(j\omega)| = \frac{\omega^3}{\sqrt{1+\omega^6}}$, specify the type of response.

Ans. Third-order, Butterworth high pass.

P15.3 Given a normalized, series RLC circuit with $L = 1$ H and $C = 1$ F, select R so as to have a second-order, low-pass, normalized Butterworth response when the output is taken across C. Scale the parameters so as to have $\omega_0 = 10$ krad/s and $C = 100$ nF.

Ans. $L' = 0.1$ H, $R' = 1000\sqrt{2}\ \Omega$.

P15.4 Repeat Problem P15.3 considering the filter to be high pass instead of low pass.

Ans. $L' = 0.1$ H, $R' = 1000\sqrt{2}\ \Omega$.

P15.5 Given a parallel GCL circuit with $L_p = 10$ mH, $C_p = 4$ μF, (a) select R_p so as to have a second-order, low-pass, normalized Butterworth response when the output is taken as the current through L_p and (b) select L_p and R_p so as to have $\omega_0 = 100$ krad/s and $C_p = 10$ nF.

Ans. (a) $R_p = 25\sqrt{2}\ \Omega$; (b) $L_p = 10$ mH, $R_p = 500\sqrt{2}\ \Omega$.

P15.6 (a) Derive the transfer function of the circuit in Figure P15.6; (b) determine L and C so as to have a second-order, normalized Butterworth low-pass response, with $R_1 = 2\ \Omega$ and $R_2 = 1\ \Omega$; (c) scale the parameters so as to have $R_2 = 10$ kΩ and a 3 dB cutoff frequency $f_0 = 10$ kHz.

Ans. (a) $\frac{V_O(j\omega)}{V_{SRC}(j\omega)} = \frac{R_2}{LCR_1}\frac{1}{s^2+s\left(\frac{1}{CR_1}+\frac{R_2}{L}\right)+\frac{1}{LC}\frac{R_1+R_2}{R_1}}$;

(b) $LC = \frac{3}{2}$, with $C = \frac{\sqrt{6}}{4}\left(\sqrt{3}\pm1\right)$ F and $L = \frac{\sqrt{6}}{2}\left(\sqrt{3}\mp1\right)$H;

(c) $R_1' = 20$ kΩ, $C' = \frac{5}{2\pi}\left(\sqrt{3}\pm1\right)$ nF, $L' = \frac{1}{2\pi}\left(\sqrt{3}\mp1\right)$ H.

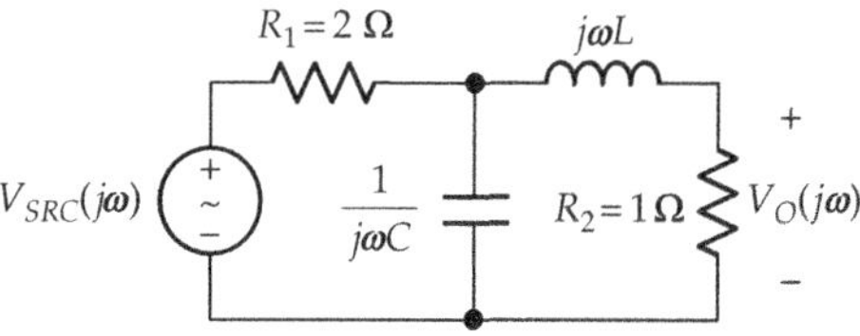

FIGURE P15.6

P15.7 (a) Derive the transfer function of the circuit in Figure P15.7; (b) determine L and C so as to have a third-order, normalized Butterworth high-pass response, with 1 Ω resistors; (c) scale the parameters so as to have resistance values of 2 kΩ and a cutoff frequency $f_0 = 10$ kHz.

Ans. (a) $\frac{V_O(j\omega)}{V_{SRC}(j\omega)} = \frac{s^3LC^2}{2s^3LC^2+s^2\left(2LC+C^2\right)+2sC+1}$;

(b) $C = 1$ F and $L = 0.5$ H; (c) $L' = 15.9$ mH, $C' = 7.96$ nF.

P15.8 (a) Derive the transfer function $V_O(j\omega)/V_{SRC}(j\omega)$ in Figure P15.8; (b) select R so as to have a normalized Butterworth high-pass filter; (c) scale R, L, and C so as to have $\omega_0 = 1$ krad/s, using a 1 mH inductor.

Ans. (a) $\frac{V_O(j\omega)}{V_{SRC}(j\omega)} = \frac{0.25s^2}{s^2+0.125Rs+1}$; (b) $8\sqrt{2}\ \Omega$; (c) $C' = 0.25$ mF, $R' = 4\sqrt{2}\ \Omega$.

P15.9 Determine R and L in Figure P15.9 for a second-order, Butterworth high-pass filter having $\omega_0 = 10$ krad/s, using a 0.1 μF capacitor

Ans. $L_1' = 10$ mH, $R' = 100\sqrt{2}\ \Omega$.

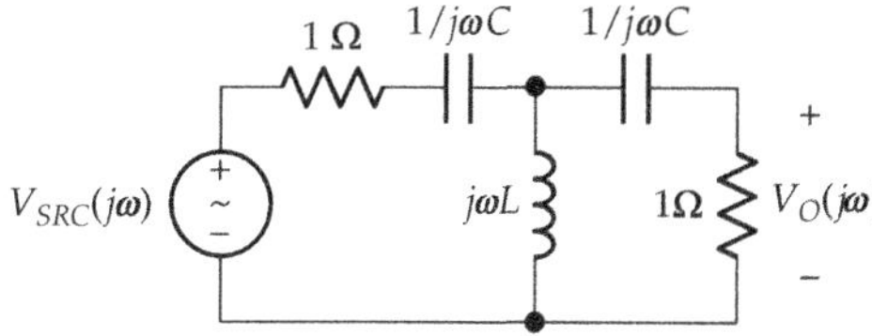

FIGURE P15.7

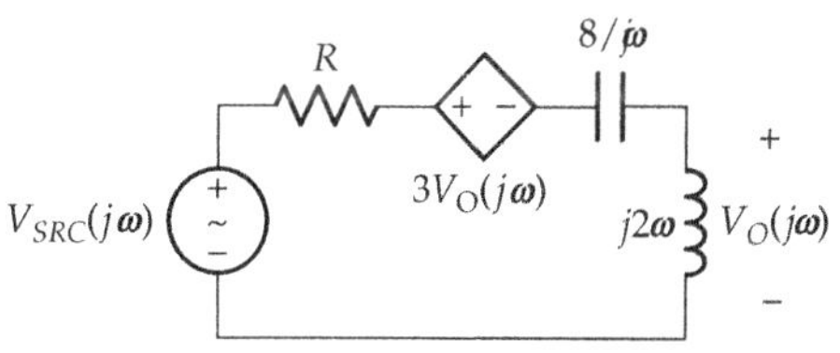

FIGURE P15.8

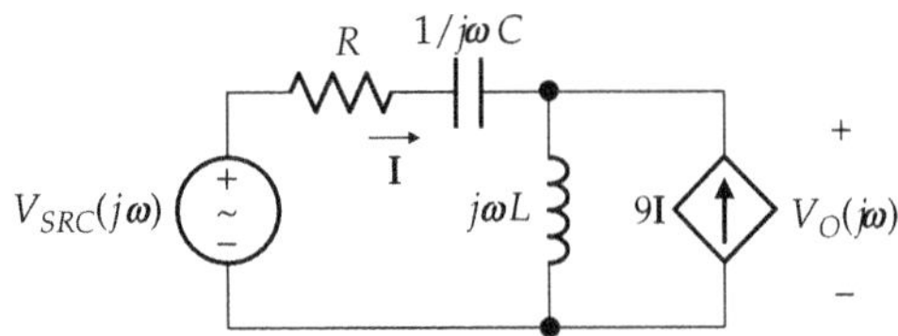

FIGURE P15.9

First-Order Active Filters

P15.10 Derive the transfer function of a first-order low-pass filter having a passband gain of 0.5 and a corner frequency of 1 kHz.

Ans. $\dfrac{1000\pi}{s+2000\pi}$.

P15.11 Derive the transfer function of a first-order high-pass filter having a passband gain of 10 and a corner frequency of 100 rad/s.

Ans. $\dfrac{10s}{s+100}$.

P15.12 Given a first-order, low-pass active filter having $R_r = 20\ \Omega$, $R_f = 800\ \Omega$, and $\omega_{cl} = 200$ rad/s, determine R_r and R_f that will make $\omega_{cl} = 1$ krad/s, without changing the gain or capacitor value.

Ans. $R_f = 160\ \Omega$, $R_r = 4\ \Omega$.

P15.13 Derive the transfer function of the circuit of Figure P15.13 and determine the gain as $\omega \to 0$ and as $\omega \to \infty$. Sketch the Bode plots when (i) $C_rR_r > 10C_fR_f$, (ii) $C_fR_f > 10C_rR_r$.

Ans. $-\dfrac{R_f}{R_r}\dfrac{1+sC_rR_r}{1+sC_fR_f}, -\dfrac{R_f}{R_r}, -\dfrac{C_r}{C_f}$.

P15.14 Determine the passband gain and the 3 dB cutoff frequency of the filter shown in Figure P15.14.

Ans. Gain = 1, $\omega_{cl} = 250$ rad/s.

P15.15 Determine, for the circuit of Figure P15.15, (a) the frequency of maximum gain, (b) K for a maximum gain of 20 dB, and (c) the bandwidth.

Ans. (a) 2 rad/s; (b) $K = 50$; (c) 5 rad/s.

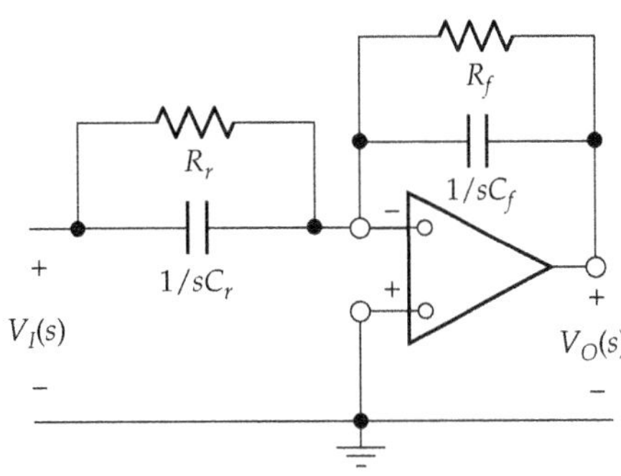

FIGURE P15.13

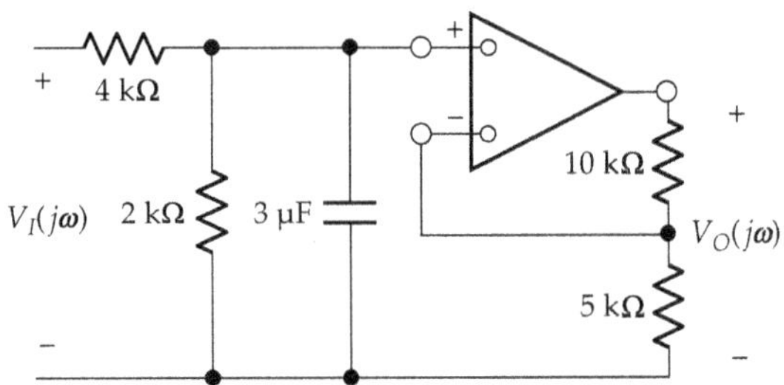

FIGURE P15.14

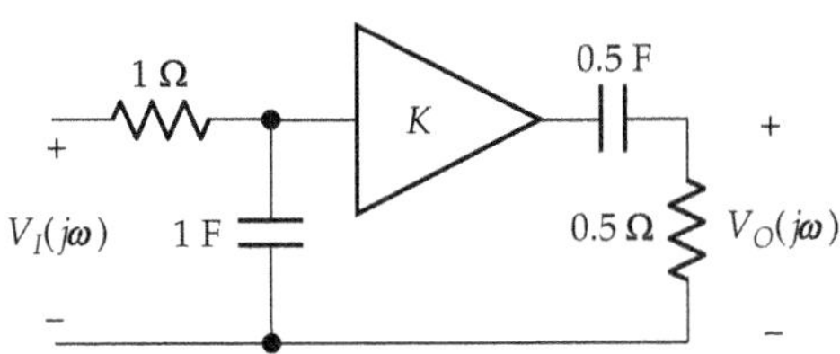

FIGURE P15.15

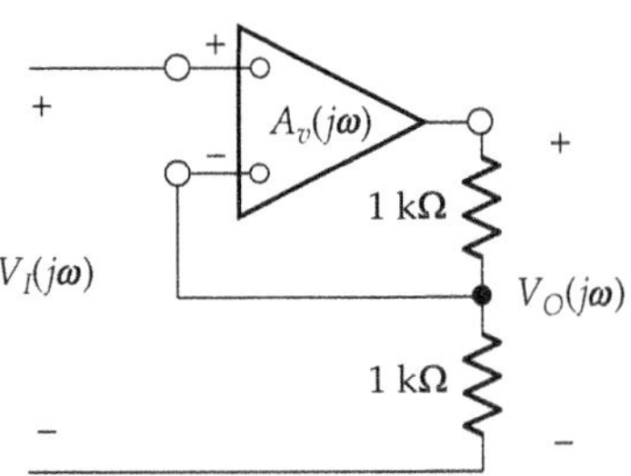

FIGURE P15.16

P15.16 The op amp in Figure P15.16 is ideal except that its gain is frequency dependent and given by $A_v(j\omega) = \dfrac{10^5}{1+j\omega}$.

Determine the 3 dB cutoff frequency of the response $|V_O(j\omega)/V_I(j\omega)|$.

Ans. 50.0 krad/s.

P15.17 If a resistance R is connected in series with the input of the circuit in Figure P13.64, a first-order low-pass filter is obtained having a variable cutoff frequency and a magnified capacitance value. Determine the 3 dB cutoff frequency if $R = 10\ \text{k}\Omega$, $C = 100$ nF, and $\alpha = 0.2$.

Ans. 200 rad/s.

P15.18 (a) Derive the transfer function $V_O(j\omega)/V_I(j\omega)$ in Figure P15.18; (b) determine the low-frequency and high-frequency asymptotes of the Bode plot, the corner frequency, and the gain at this frequency in dB; (c) sketch the Bode magnitude and phase plots.

Ans. (a) $2 + j\omega$; (b) LF asymptote: $20\log_{10}|H(j\omega)| = 6$ dB, HF asymptote: $20\log_{10}|H(j\omega)| = 20\log_{10}\omega$, $\omega_c = 2$ rad/s, gain at this frequency: 9 dB.

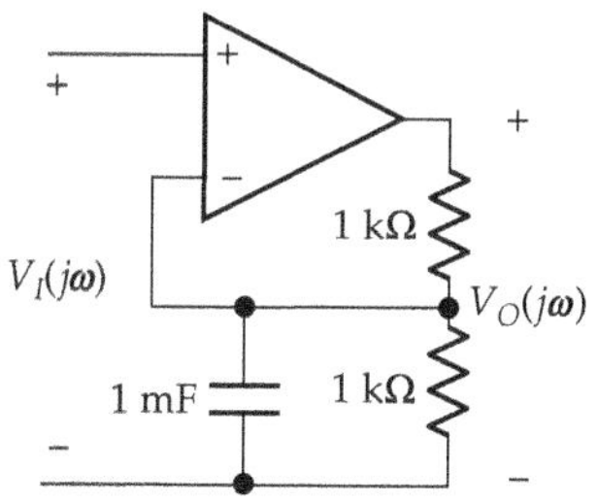

FIGURE P15.18

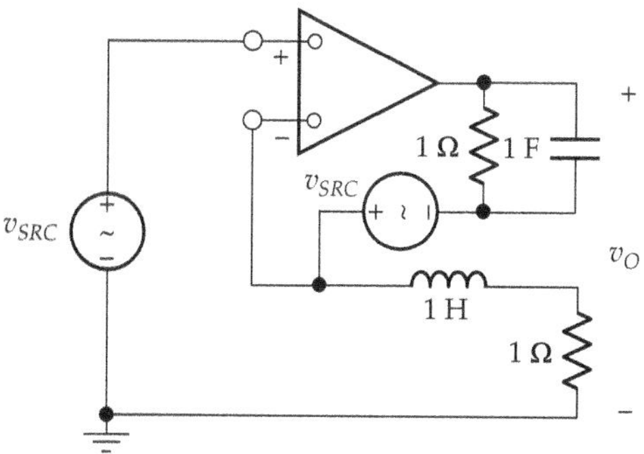

FIGURE P15.19

P15.19 Derive the transfer function $V_O(j\omega)/V_{SRC}(j\omega)$ in Figure P15.19.

Ans. $\dfrac{1}{s^2+2s+1}$.

P15.20 (a) Sketch, on the same graph, the Bode plots of $|V_{O1}(j\omega)/V_{SRC}(j\omega)|$ and $|V_{O2}(j\omega)/V_{O1}(j\omega)|$ in Figure P15.20; (b) determine (i) the maximum dB value of $|V_{O2}(j\omega)/V_{SRC}(j\omega)|$, (ii) the frequency at which this maximum occurs, and (iii) the dB value of $|V_{O2}(j\omega)/V_{SRC}(j\omega)|$ a decade above and a decade below the frequency of maximum $|V_{O2}(j\omega)/V_{SRC}(j\omega)|$. Note that this is a broadband filter composed of two isolated passive filters.

Ans. (b) (i) −6 dB, (ii) 1 krad/s, (iii) −9 dB.

Higher-Order Active Filters

P15.21 Specify the response $\mathbf{V}_O/\mathbf{V}_I$ of the filter circuit of Figure P15.21 and determine the passband gain. Note that this is the same as the high-pass filter of Figure 15.17 with the capacitors replaced by inductors.

Ans. Second order, low pass of passband gain of 3.

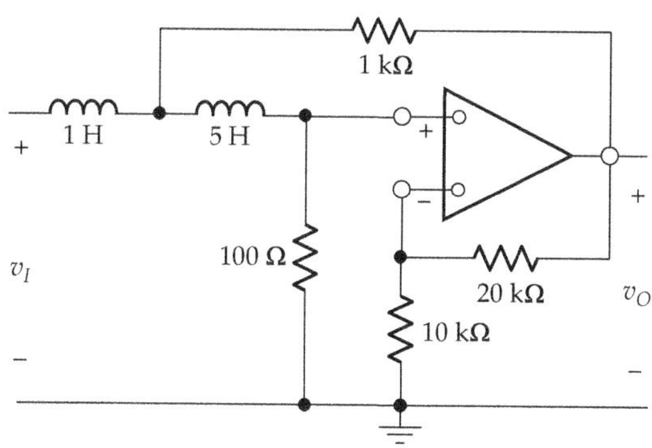

FIGURE P15.21

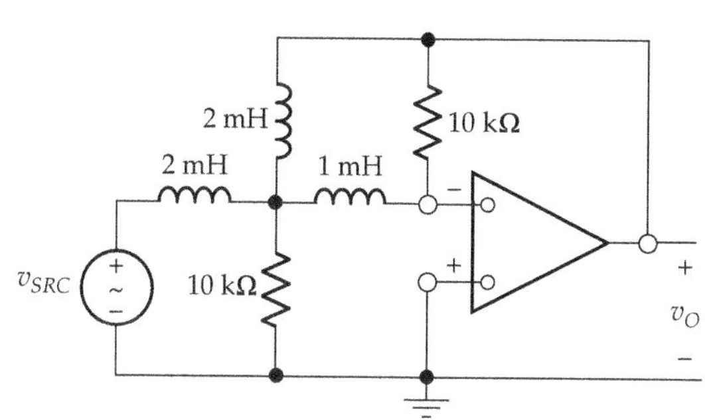

FIGURE P15.22

P15.22 Specify the response $\mathbf{V}_O/\mathbf{V}_I$ of the filter circuit of Figure P15.22 and determine the passband gain. Note that this is the same as the high-pass filter of Figure 15.28 with the capacitors replaced by inductors.

Ans. Second order, low pass of passband gain of −1.

P15.23 Show that the transfer function of the circuit in Figure P15.23 is $-\dfrac{1}{sC_fR_r}\dfrac{sC_fR_f}{1+sC_fR_f}\dfrac{sC_rR_r}{1+sC_rR_r}$. Note that if $C_fR_f = C_rR_r = \tau$, then the transfer function becomes $-\dfrac{1}{sC_fR_r}\left(\dfrac{s\tau}{1+s\tau}\right)^2 V_{src}$. Deduce that the circuit behaves as a bandpass filter for frequencies in the neighborhood of $\omega = 1/\tau$, as a differentiator at low frequencies and as an integrator at high frequencies.

P15.24 Derive the transfer function of the filter shown in Figure P15.24.

Ans. $\dfrac{-25{,}000s}{(s+5{,}000)(s+10{,}000)}$.

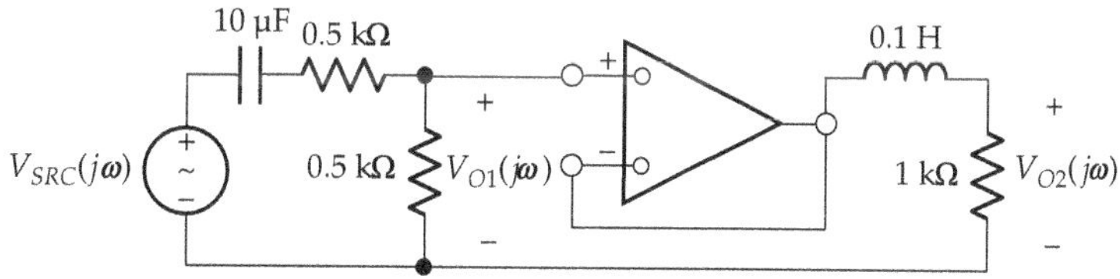

FIGURE P15.20

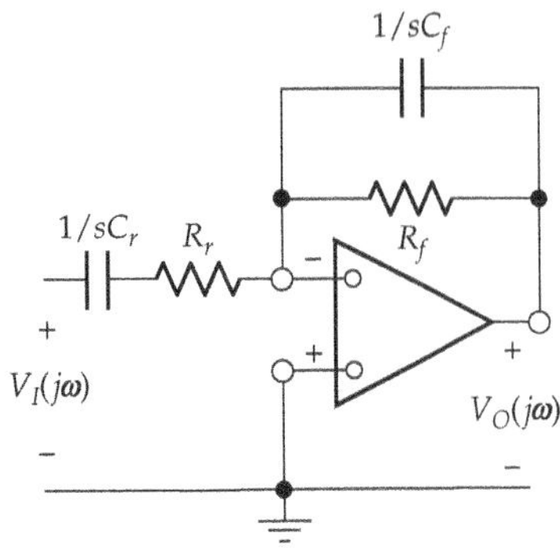

FIGURE P15.23

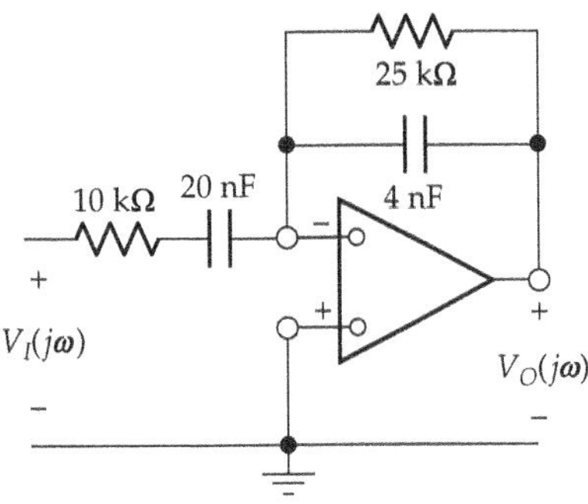

FIGURE P15.24

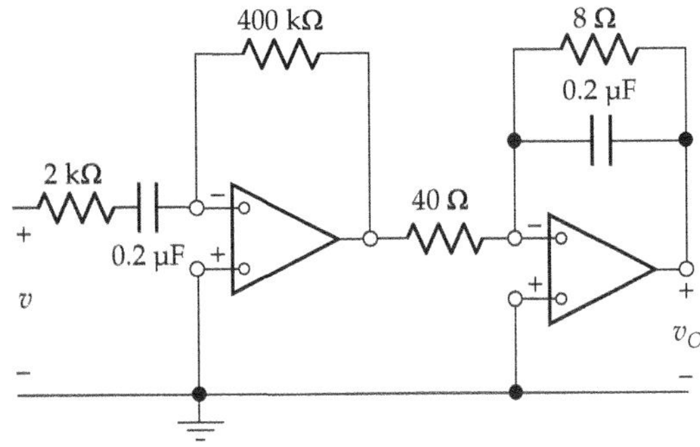

FIGURE P15.25

P15.25 Determine the center frequency of the filter shown in Figure P15.25.

Ans. 39.53 krad/s.

P15.26 (a) Derive the transfer function of the filter shown in Figure P15.26. (b) Determine the maximum gain. (c) Determine the bandwidth.

Ans. (a) $\dfrac{10^5 s}{s^2 + 2\times10^4 s + 10^8}$; (b) 5; (c) 20 krad/s.

P15.27 (a) Derive the transfer function of the filter shown in Figure P15.27. (b) Determine the center frequency, the maximum gain, Q, BW, and the 3 dB cutoff frequencies. (c) If the circuit is to be used as a broadband filter, determine (i) the capacitor values that give a 3 dB

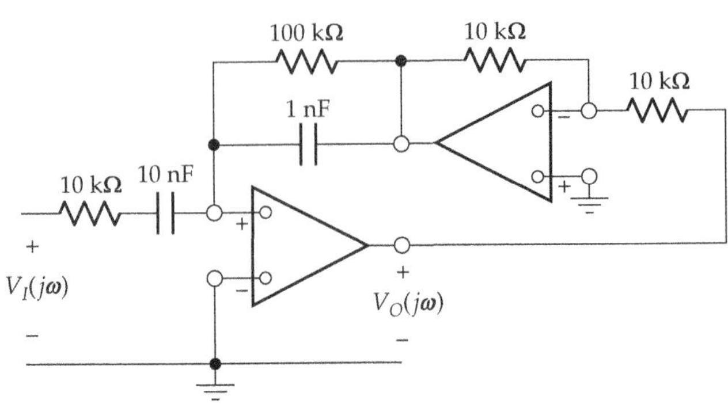

FIGURE P15.26

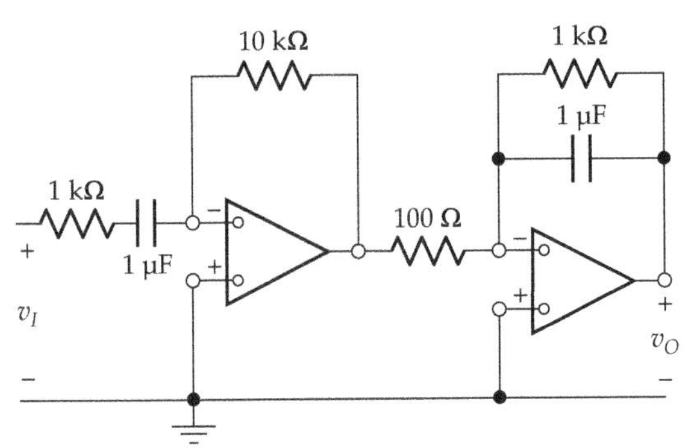

FIGURE P15.27

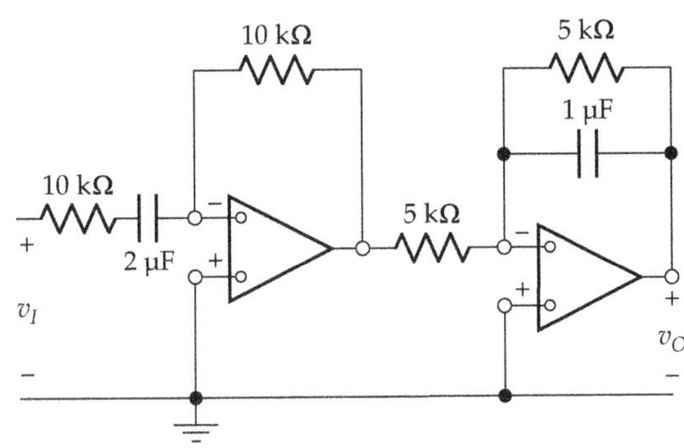

FIGURE P15.28

bandwidth from 100 rad/s to 1 Mrad/s and (ii) the maximum gain in dB.

Ans. (a) $\dfrac{10^5 s}{s^2 + 2\times10^3 s + 10^6}$; (b) $\omega_0 = 1$ krad/s, 50, $Q = 0.5$, BW = 2 krad/s, $\omega_c = \left(\sqrt{2} \pm 1\right)$ krad/s; (c) capacitance of high-pass filter: 10 μF, capacitance of low-pass filter: 1 nF, 40.0 dB.

P15.28 Determine the maximum gain of the filter shown in Figure P15.28 and the frequency at which it occurs.

Ans. 0.8 at 100 rad/s.

P15.29 Given two first-order filters of transfer functions $-\frac{200}{s+200}$ and $-\frac{s}{s+50}$. The inputs of these filters are paralleled together and the outputs are added in an inverting adder of unity gain. (a) Derive the transfer function and characterize the resulting response; (b) determine the maximum gain and the frequency at which it occurs.

Ans. (a) $\frac{s^2+400s+10{,}000}{s^2+250s+10{,}000}$, sum of bandpass and bandstop responses; (b) 1.6 at 100 rad/s.

P15.30 (a) Derive the transfer function $V_O(j\omega)/V_I(j\omega)$ in Figure P15.30; (b) characterize the response; (c) determine the maximum gain, the minimum gain, both in dB, and the frequencies at which they occur.

Ans. (a) $10\frac{s^2+250s+125\times10^5}{s^2+100125s+125\times10^5}$; (b) the response is a combination of second-order bandpass and second-order bandstop responses; (c) maximum gain is 20 dB and occurs as $\omega\to 0$ or $\omega\to\infty$; minimum gain is −32 dB and occurs at 3.54 krad/s.

P15.31 Given the filter of Figure P15.31, (a) derive the transfer function and (b) determine A, if $Q = 0.8$.

Ans. (a) $\frac{A(s^2+\omega_0^2)}{s^2+2(2-A)\omega_0 s+\omega_0^2}$, where $\omega_0 = 1/CR$; (b) 1.375.

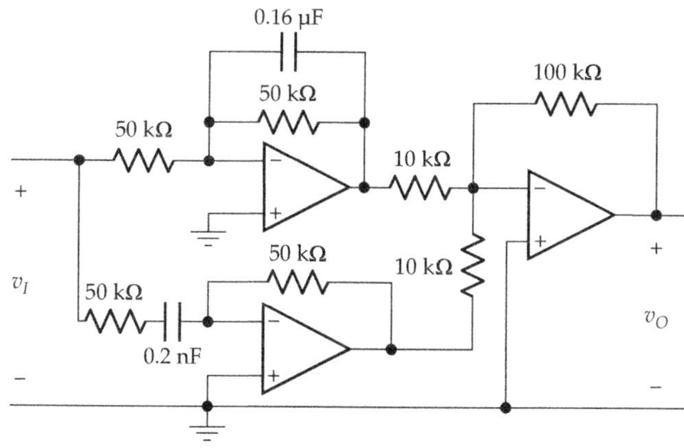

FIGURE P15.30

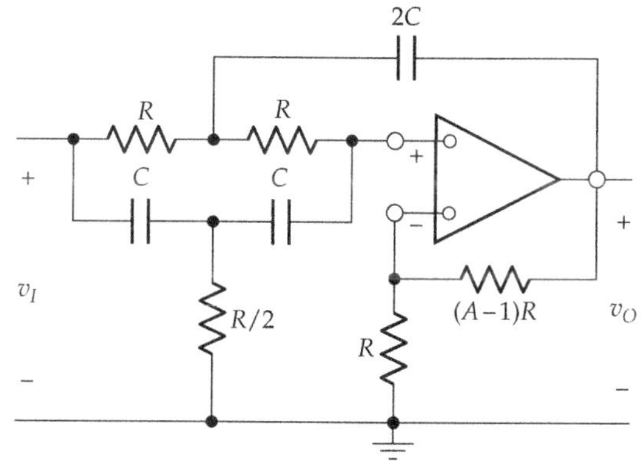

FIGURE P15.31

P15.32 The circuit of Figure P15.32 provides an adjustable response in which the output could lag or lead the input. Show that the transfer function is $-\frac{R_{p2}}{R_{p1}}\frac{1+s\tau_1}{1+s\tau_{p1}}\frac{1+s\tau_{p2}}{1+s\tau_2}$, where $\tau_1 = C_1R_1$, $\tau_2 = C_2R_2$, $\tau_{p1} = C_1\left[R_1 + R_{p1}\alpha_1(1-\alpha_1)\right]$, and $\tau_{p2} = C_2\left[R_2 + R_{p2}\alpha_2(1-\alpha_2)\right]$.

P15.33 Two circuits are cascaded, with isolation (Figure P15.33). $H_1(s)$, is the transfer function of a first-order, normalized, Butterworth low-pass filter of passband gain of 2. The overall open-circuit transfer function $H(s) = V_O(s)/V_I(s)$ is that of a second-order, normalized, Butterworth low-pass response of passband gain of 1/2. (a) Derive $H_2(s)$; (b) sketch the magnitude Bode plots of $H_1(j\omega)$ and $H(j\omega)$; (c) determine $|H_2(j\omega)|$ and evaluate $|H_2(j1)|$ in dB; (d) derive the equations of the low-frequency and high-frequency asymptotes of $|H_2(j\omega)|$ and evaluate the 3 dB cutoff frequency

Ans. (a) $\frac{1}{4}\frac{s+1}{s^2+\sqrt{2}s+1}$; (c) $\left|H_2(j\omega)\right| = \frac{1}{4}\frac{\sqrt{1+\omega^2}}{\sqrt{1+\omega^4}}$, $|H_2(j1)| = 20\log_{10}(1/4) = -12$ dB; (d) LF asymptote: $20\log_{10}(1/4) = -12$ dB, HF asymptote: as $\omega\to\infty$ is $-12\text{ dB} - 20\log_{10}\omega$, $\omega_c = 1.55$ rad/s.

P15.34 Three filters, each having the response $s/(s+1)$ are cascaded with isolation. (a) Characterize the response; (b) determine the half-power frequency; (c) if each filter is implemented as an active filter having 1 kΩ resistors, determine the value of the required capacitor; (d) using scaling, determine the value of the capacitor if the half-power frequency is

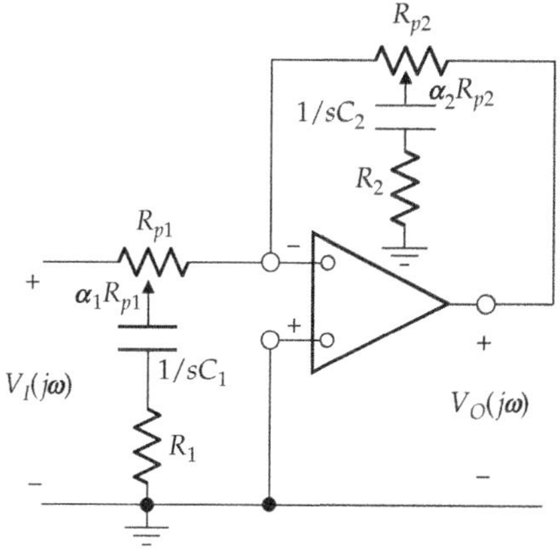

FIGURE P15.32

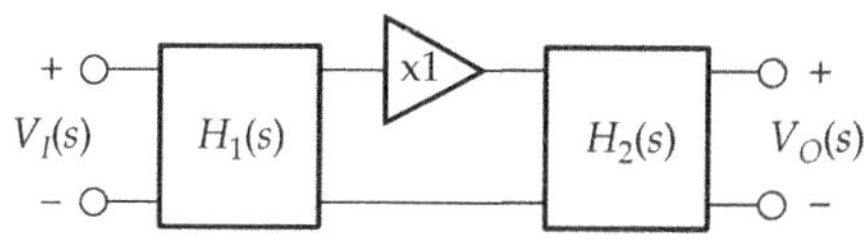

FIGURE P15.33

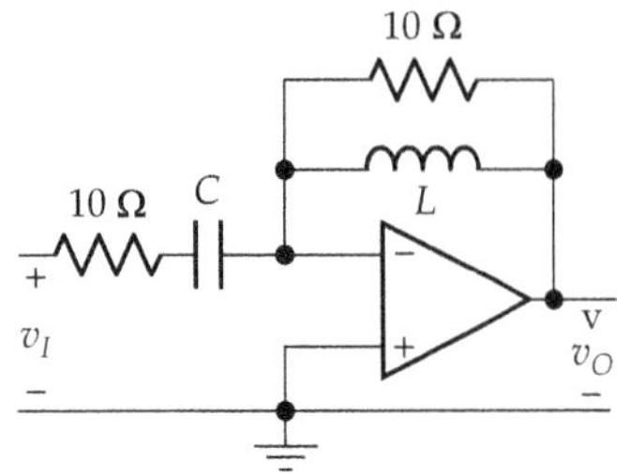

FIGURE P15.35

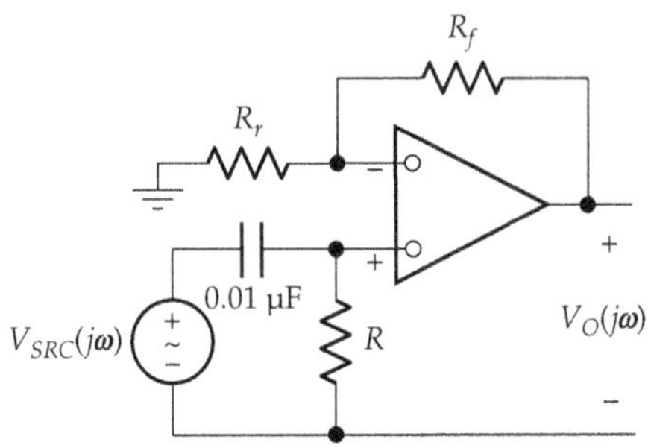

FIGURE P15.38

multiplied by a factor of 1000 and the resistances are reduced to 500 Ω.

Ans. (a) Third-order high pass; (b) $\frac{1}{\sqrt{\sqrt[3]{2}-1}} = 1.96$ rad/s; (c) 1 mF; (d) 2 μF.

P15.35 (a) Determine L and C in Figure P15.35 so that the denominator of the transfer function $V_o(j\omega)/V_I(j\omega)$ is the product of two equal factors having a root at $\omega = -10$ rad/s; (b) using these values of L and C, determine the 3 dB cutoff frequency of the filter response; (c) what type of damping would the response in the time domain have?

Ans. (a) $L = 1$ H, $C = 0.01$ F; (b) 15.54 rad/s; (c) critical damping.

Design Problems

P15.36 It is required to design a second-order, high-pass Butterworth filter that has a 3 dB cutoff frequency of 10 krad/s, using a 10 mH inductor. Determine R, C, and Q, if the filter is implemented as (a) a series RLC circuit and (b) a parallel GCL circuit.

Ans. (a) $C = 1$ μF, $R = 100\sqrt{2}$ Ω, $Q = 1/\sqrt{2}$; (b) $C = 1$ μF, $R = 100/\sqrt{2}$ Ω, $Q = 1/\sqrt{2}$.

P15.37 (a) Design a first-order, low-pass active filter having $\omega_{cl} = 500$ rad/s and a gain of −10 using a 0.1 μF capacitor; (b) implement the filter as in Figure P15.37 with a gain of +10, using $R_r = R$.

Ans. $R_f = 20$ kΩ, $R_r = 2$ kΩ; (b) $R = 20$ kΩ, $R_f = 180$ kΩ.

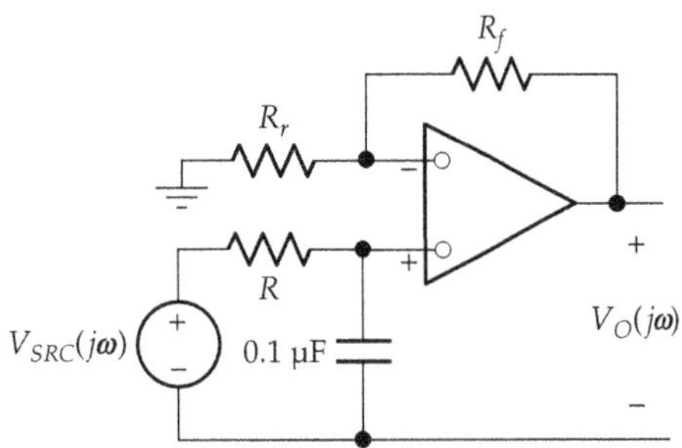

FIGURE P15.37

P15.38 (a) Design a first-order, high-pass active filter having $\omega_{cl} = 10$ krad/s and a magnitude of gain of 33 dB using a capacitor of 0.01 μF, as in Figure P15.38 with $R_r = R_i$; (b) implement the filter using $R_r = 10$ kΩ.

Ans. $R_r = 10$ kΩ, $R_f = 446.7$ kΩ; $R_f = 446.7$ kΩ.

P15.39 Design a broadband bandpass filter having 3 dB cutoff frequencies of 100 Hz and 10 kHz and a passband gain of 2 using 0.2 μF capacitors. The filter is to have a very high input impedance.

Ans. Low-pass filter: $R_f = 80.0$ Ω, $R_r = 40$ Ω; high-pass filter: $R_f = R_r = 7958$ Ω. First stage is a unity-gain amplifier.

P15.40 Design a broadband bandpass filter to meet the following specification: (a) 3 dB cutoff frequencies of 10 Hz and 20 kHz, (b) a 1 nF capacitor is to be used in the low-pass filter and a 100 nF capacitor in the high-pass filter, (c) minimum impedance of 10 kΩ resistive, and (d) overall gain of −10.

Ans. First stage, amplifier of gain −10 and $R_r = 10$ kΩ; second stage, high-pass filter, $R_r = 159.2$ kΩ, $R_f = 200$ kΩ; third stage, low-pass filter, $R_r = 10$ kΩ, $R_f = 7.958$ kΩ.

P15.41 Design a broadband bandpass filter to meet the following specification: (a) 3 dB cutoff frequencies of 10 Hz and 20 kHz, (b) a passband gain of 2, (c) a purely resistive input impedance of 1 kΩ, (d) a slope of low-frequency and high-frequency asymptotes of ±20 dB/decade, (e) 1 μF capacitors are to be used.

Ans. First stage, low-pass filter: $R_r = 1$ kΩ, $R_f = 8$ Ω; second stage, high-pass filter, $R_r = 16$ kΩ, $R_f = 4$ MΩ.

P15.42 It is required to design a third-order high-pass Butterworth filter using a first-order high-pass filter cascaded with a second-order high-pass, noninverting filter of the type shown in Figure 15.17, reproduced in Figure P15.42 using a unity-gain amplifier. The filter should have a gain of 20 dB and a 3 dB cutoff frequency

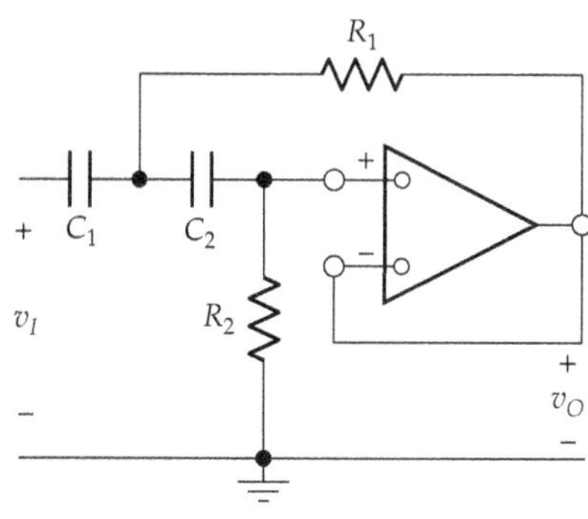

FIGURE P15.42

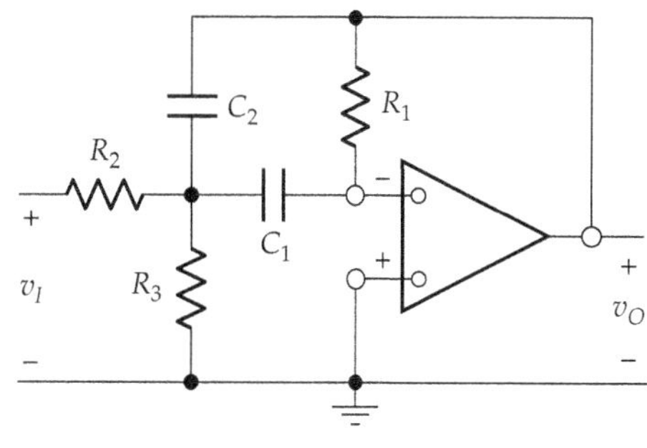

FIGURE P15.44

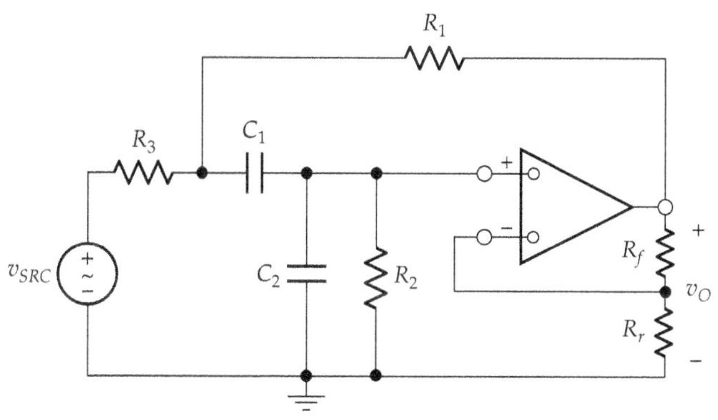

FIGURE P15.43

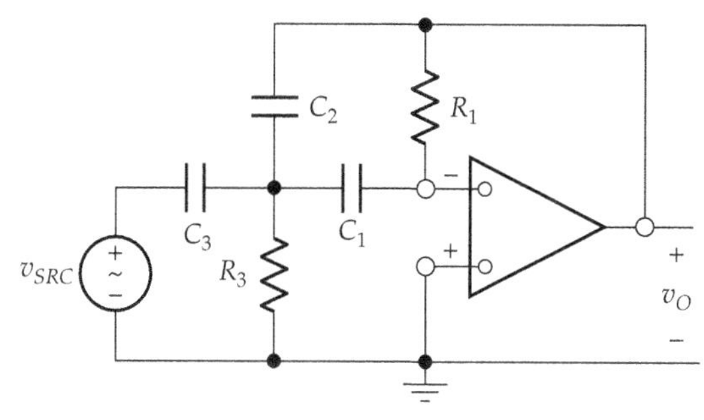

FIGURE P15.45

of 10 krad/s, using 0.1 μF capacitors. Determine the required values of resistances.

Ans. $R_1 = 500\ \Omega$, $R_2 = 2\ \text{k}\Omega$; first-order filter, $R_r = 1\ \text{k}\Omega$, $R_f = 10\ \text{k}\Omega$.

P15.43 A noninverting bandpass filter of the type shown in Figure 15.24 is required having a center frequency of 100 kHz and a Q of 20. The filter is to have 1 kΩ resistors and capacitors of equal value C. Determine C, A, and the maximum gain.

Ans. $C = 10/\left(\pi\sqrt{2}\right) = 2.25\ \text{nF}$, $A = 5 - \dfrac{\sqrt{2}}{20} = 4.93$,

maximum gain $= \dfrac{QA}{\sqrt{2}} \cong 70 = 37\ \text{dB}$

P15.44 It is required to design a second-order bandpass filter of the type shown in Figure 15.25, reproduced in Figure P15.43. The filter should have a gain of −10, Q = 10, ω_0 = 500 rad/s, using 1 μF capacitors. Determine the required values of resistances.

Ans. $R_1 = 40\ \text{k}\Omega$, $R_2 = 2\ \text{k}\Omega$, $R_3 = 105.3\ \Omega$ (Figure P15.44).

P15.45 It is required to design a unity-gain, inverting, second-order, Butterworth high-pass filter of the type shown in Figure 15.28, reproduced in Figure P15.45. The filter should have a 3 dB cutoff frequency of 500 Hz, using 0.5 μF capacitors, as in Example 15.4. Determine the values of R_1 and R_3.

Ans. $R_1 = \dfrac{6000}{\pi\sqrt{2}} \cong 1350\ \Omega$ and $R_3 = \dfrac{2000\sqrt{2}}{3\pi} \cong 300\ \Omega$.

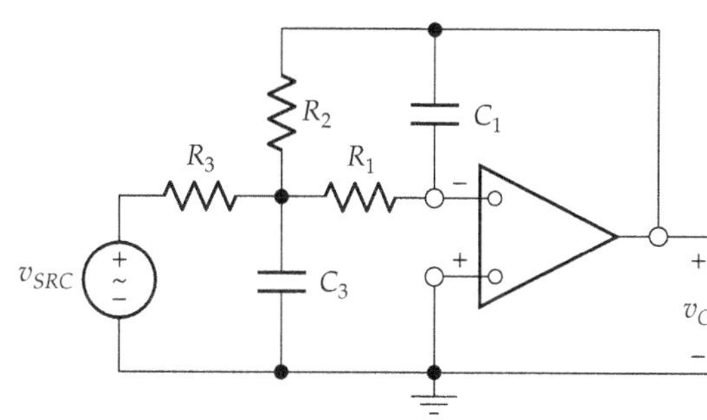

FIGURE P15.46

P15.46 It is required to design a unity-gain, inverting, second-order, Butterworth low-pass filter of the type shown in Figure 15.29, reproduced in Figure P15.46. The filter should have a 3 dB cutoff frequency of 500 Hz, using

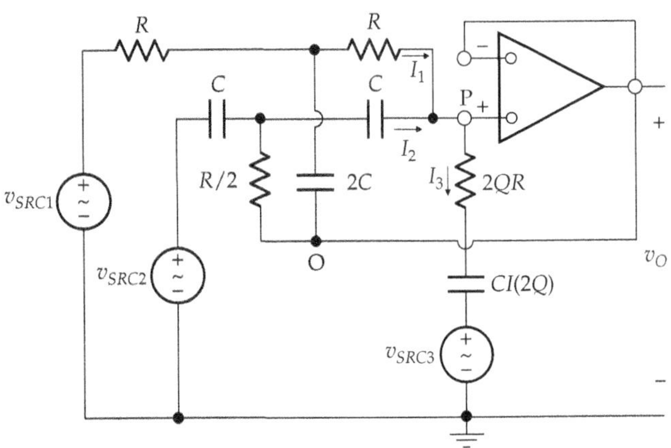

FIGURE P15.47

0.5 μF capacitors, as in Example 15.4. Determine the values of R_1 and R_3.

Ans. $C_1 = \dfrac{100\sqrt{2}}{3\pi} = 15.00$ nF and $C_3 = \dfrac{300}{\pi\sqrt{2}} = 67.52$ nF.

P15.47 It is required to design a bandstop filter based on the universal filter in Figure 15.30, reproduced in Figure P15.47. The filter should reject the power frequency of 50 Hz with a Q of 100, using $C = 1$ μF.

Ans. $R = \dfrac{10}{\pi} = 3.183$ kΩ, $2RQ = 636.6$ kΩ, $C/2Q = 5$ nF.

P15.48 A Butterworth low-pass filter is required having an attenuation of 1 dB at 100 Hz and at least 50 dB at 1 kHz. (a) Specify the order of the filter; (b) determine the cutoff frequency of the filter.

Ans. (a) 3; (b) 787 rad/s.

16

Responses to Periodic Inputs

Objective and Overview

Chapter 8 was concerned with the sinusoidal steady state in which all the sources applied to a given circuit are of the same frequency. This case can be readily generalized to inputs that are nonsinusoidal but periodic. The basis for this generalization is Fourier's theorem, according to which a periodic signal of a given frequency can be expressed, in general, as an infinite series of sine and cosine functions whose frequencies are integral multiples of the frequency of the periodic signal. The amplitudes of the higher-frequency sinusoids decrease fairly rapidly in most cases so that the periodic signal can be represented in practice by a finite sum of sinusoids.

Once a periodic signal is represented as a series expansion of sinusoids of different frequencies, the response of a linear time-invariant (LTI) circuit to the periodic signal is the sum of the responses to the individual frequency components. The different frequencies do not interact in an LTI circuit so that each of these responses can be obtained using phasor analysis, as was done in Chapter 8.

Circuit responses to periodic signals are of considerable practical interest, because these signals are very common. The steady state of any linear or nonlinear circuit, other than the dc steady state, is periodic. Thus, the outputs of free-running oscillators, the time bases of TV and computer displays, and continuous vibrations of all kinds are periodic signals.

The Fourier series expansion (FSE) of periodic functions can be generalized to nonperiodic functions by means of the Fourier transform (Chapter 23).

This chapter begins by presenting Fourier's theorem and deducing how the coefficients of the sine and cosine terms of the FSE can be obtained. The derivation of the FSE is significantly simplified if the given periodic function possesses some symmetry properties or if it's the sum, the product, the derivative, or the integral of functions whose FSE is already known. These simplifications are discussed in considerable detail. This is followed by considering circuit responses to periodic signals and the average power involved, leading to the definition and derivation of the root-mean-square (rms) value of a periodic voltage or current.

16.1 Fourier Series

Figure 16.1 illustrates a periodic function $f(t)$ of arbitrary waveform. Its period T is defined, like that of a sinusoidal function, as the time interval between successive repetitions of the same full range of values of the periodic function.

Mathematically, the defining property of any periodic function $f(t)$ of period T is

$$f(t) = f(t + nT) \tag{16.1}$$

for any t within the period T, where n is any positive or negative integer. This means that the function repeats every period T. Thus, In Figure 16.1, $f(t_0)$ has the same value as $f(t_0 + T)$, $f(t_0 - T)$, and so on for any integral multiple of T. The sinusoidal function discussed in Chapter 8 is an example of a periodic function so that the same definitions of cycle, frequency, and angular frequency for a sinusoidal function (Equation 8.2) apply to periodic functions in general. In particular, the reciprocal of T is the **fundamental frequency** of the periodic waveform.

From a mathematical viewpoint, a periodic time function extends over all time, from $-\infty$ to $+\infty$. In practice, a function can be assumed periodic if it has been in a steady state for a time interval that is large compared to the period, say, 10 times. Periodic functions need not be time functions. They can be spatial functions, in which case the period is the **wavelength**.

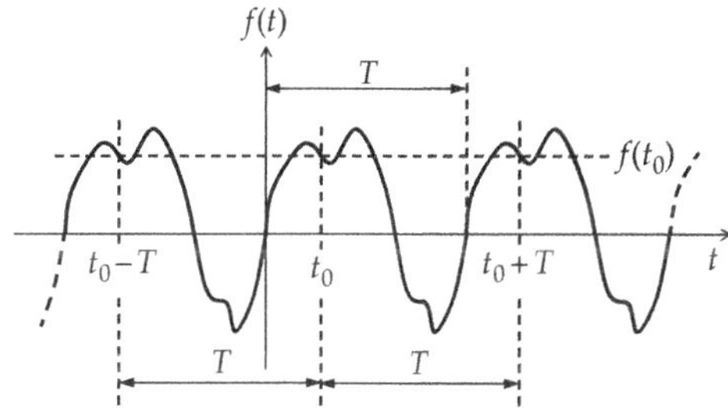

FIGURE 16.1
Periodic function.

A remarkable theorem on periodic functions is Fourier's theorem:

Fourier's Theorem: *A periodic function f(t) can be expressed, in general, as an infinite series of cosine and sine functions*:

$$f(t) = a_0 + \sum_{n=1}^{\infty} \left(a_n \cos n\omega_0 t + b_n \sin n\omega_0 t\right) \quad (16.2)$$

where

n is a positive integer

a_0, a_n, and b_n are constants, known as *Fourier coefficients*, that depend on $f(t)$

The component having $n = 1$ is the **fundamental** and is represented by the sum of two terms, $a_1\cos\omega_0 t + b_1\sin\omega_0 t$, where $\omega_0 = 2\pi f = 2\pi/T$. The component having $n > 1$ is the nth **harmonic** and is represented by the sum of two terms, $a_n\cos n\omega_0 t + b_n \sin n\omega_0 t$. The series expression of Equation 16.2 is the **Fourier series expansion** (FSE) of $f(t)$.

To appreciate the idea behind Fourier's theorem, consider the periodic triangular waveform of Figure 16.2 having a unit amplitude and a period of 4 s. The FSE of the negation of this waveform is derived in Example 16.5. The FSE of the waveform of Figure 16.2 is

$$f_{tr}(t) = \frac{8}{\pi^2}\left(\cos\omega_0 t + \frac{1}{9}\cos 3\omega_0 t + \frac{1}{25}\cos 5\omega_0 t + \cdots\right) \quad (16.3)$$

Comparing Equations 16.2 and 16.3, it is seen that no sine terms are present in Equation 16.3 and the cosine terms are nonzero for odd values of n only. This is generally true of many FSEs, that is, some values of a_n and b_n in Equation 16.2 may be zero in some cases. The first term in the FSE in Equation 16.3, which is $(8/\pi^2)\cos\omega_0 t$, is shown in Figure 16.3a for the first period, 0–4 s, and is a rough, first approximation to the triangular waveform. Adding the second term in the FSE, $(8/9\pi^2)\cos 3\omega_0 t$, which is the third harmonic, gives a better approximation, as in Figure 16.3b. Adding the fifth harmonic, the third term in the FSE, and higher harmonics in the FSE further improve the approximation.

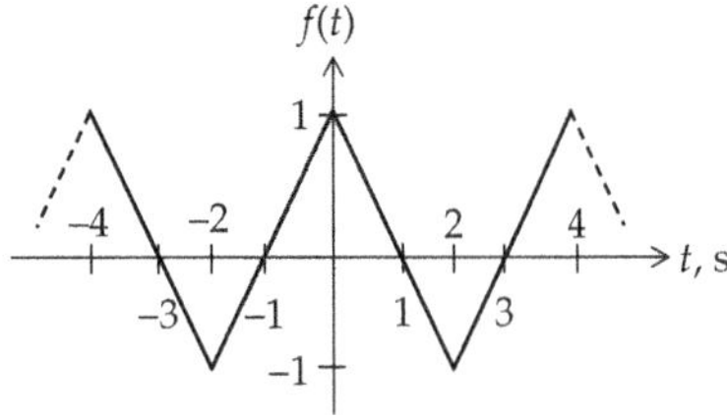

FIGURE 16.2
Triangular periodic waveform.

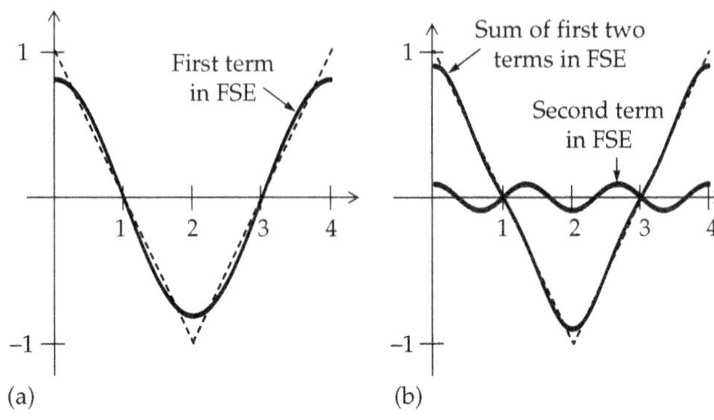

FIGURE 16.3
(a) A first approximation of a triangular waveform by a sinusoidal waveform and (b) adding a third-harmonic gives a better approximation.

Although periodic functions encountered in practice can be usually expressed as FSEs, it is mathematically of interest to determine if Fourier's theorem applies to any arbitrary periodic function. **Dirichlet's conditions**, which are *sufficient* to ensure that the FSE of a given periodic function $f(t)$ converges to $f(t)$ for any t, may be stated as follows:

1. The integral $\int_{t_0}^{t_0+T} |f(t)|\,dt$ exists, that is, the integral is finite, for any arbitrary t_0.
2. $f(t)$ is single valued over a period, with only a finite number of finite discontinuities, maxima, or minima.

Combining the sine and cosine terms for the same n in Equation 16.2, the FSE becomes

$$f(t) = c_0 + \sum_{n=1}^{\infty} c_n \cos(n\omega_0 t + \theta_n) \quad (16.4)$$

$$\text{where } c_0 = a_0,\ c_n = \sqrt{a_n^2 + b_n^2},\ \text{and } \theta_n = -\tan^{-1}\frac{b_n}{a_n} \quad (16.5)$$

16.2 Fourier Analysis

The object of Fourier analysis is to derive the Fourier coefficients of a given FSE. Some integral trigonometric relations needed for this purpose can be summarized as follows:

Summary: *Given the four functions $\cos m\omega_0 t$, $\sin m\omega_0 t$, $\cos n\omega_0 t$, and $\sin n\omega_0 t$, where m and n are nonzero integers, the integral of the product of any two of these functions over the period of the fundamental, $T = 2\pi/\omega_0$, is zero, except for the products $\cos^2 n\omega_0 t$ and $\sin^2 n\omega_0 t$, having $m = n$, in which case the integral is $T/2$.*

Thus,

$$\int_0^T \cos n\omega_0 t \sin m\omega_0 t dt = 0, \quad \text{for all } n \text{ and } m \tag{16.6}$$

$$\int_0^T \cos n\omega_0 t \cos m\omega_0 t dt = 0 = \int_0^T \sin n\omega_0 t \sin m\omega_0 t dt,$$

$$\text{for } n \neq m \tag{16.7}$$

$$\int_0^T \cos^2 n\omega_0 t dt = \frac{T}{2} = \int_0^T \sin^2 n\omega_0 t dt \quad n \neq 0 \tag{16.8}$$

Equations 16.6 through 16.8 are proved next, using basic trigonometric relations (Appendix B). The proof is based on the following basic concept:

Concept: *The integral of the kth harmonic, $\sin k\omega_0 t$ or $\cos k\omega_0 t$, where k is a nonzero integer, is zero over the period of the fundamental component, $T = 2\pi/\omega_0$.*

In Figure 16.4, the integral of the fundamental component $\sin\omega_0 t$ is clearly zero over a period, such as that from $t = 0$ to T, since the area of a positive half-cycle is equal in magnitude but opposite in sign to the area of a negative half-cycle. The integral of $0.5\sin 3\omega_0 t$ between $t = 0$ and T is also zero because three full periods are included in this interval. In general,

$$\int_0^{T=2\pi/\omega_0} A\sin k\omega_0 t dt = \frac{A}{k\omega_0}\left[-\cos k\omega_0 t\right]_0^{2\pi/\omega_0}$$

$$= \frac{A}{k\omega_0}\left[-\cos 2k\pi + \cos 0\right]$$

$$= \frac{A}{k\omega_0}\left[-1+1\right] = 0, \quad \text{for all nonzero, integer } k \tag{16.9}$$

When $k = 0$, $A\sin k\omega_0 t = 0$, and its integral over a period is zero. In a similar manner,

$$\int_0^{T=2\pi/\omega_0} A\cos k\omega_0 t dt = 0, \quad \text{for all nonzero, integer } k \tag{16.10}$$

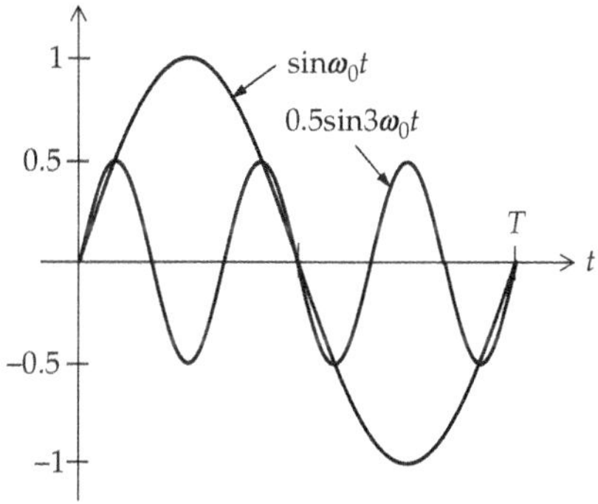

FIGURE 16.4
Fundamental and third harmonic over a period of the fundamental.

When $k = 0$, $A\cos k\omega_0 t = 1$, and its integral over a period is $AT = 2A\pi/\omega_0$.

Consider the trigonometric identities

$$\sin(\alpha+\beta) = \sin\alpha\cos\beta + \cos\alpha\sin\beta \tag{16.11}$$

and

$$\sin(\alpha-\beta) = \sin\alpha\cos\beta - \cos\alpha\sin\beta \tag{16.12}$$

If Equations 16.11 and 16.12 are added together,

$$\sin\alpha\cos\beta = \frac{1}{2}\left(\sin(\alpha+\beta) + \sin(\alpha-\beta)\right) \tag{16.13}$$

Identifying β with $n\omega_0 t$ and α with $m\omega_0 t$, the integral in Equation 16.6 becomes

$$\int_0^T \cos n\omega_0 t \sin m\omega_0 t\, dt$$

$$= \frac{1}{2}\left[\int_0^T \sin(m+n)\omega_0 t dt + \int_0^T \sin(m-n)\omega_0 t dt\right] \tag{16.14}$$

Since m and n are integers, $(m + n)$ or $(m - n)$ is either a nonzero integer or a zero ($m = n$). The nonzero integer can be identified with k in Equation 16.9 so that the integrals on the RHS of (Equation 16.14) are zero. When $m = n$, the integrand of the second integral on the RHS of Equation 16.14 is zero, and the first integral is zero, so that the integrals on the RHS of Equation 16.14 are again zero. This proves Equation 16.6.

Consider next the trigonometric identities

$$\cos(\alpha+\beta) = \cos\alpha\cos\beta - \sin\alpha\sin\beta \tag{16.15}$$

and

$$\cos(\alpha-\beta) = \cos\alpha\cos\beta + \sin\alpha\sin\beta \tag{16.16}$$

If Equations 16.15 and 16.16 are added together,

$$\cos\alpha\cos\beta = \frac{1}{2}\left(\cos(\alpha+\beta) + \cos(\alpha-\beta)\right) \tag{16.17}$$

Identifying β with $n\omega_0 t$ and α with $m\omega_0 t$, the first integral in Equation 16.7 becomes

$$\int_0^T \cos n\omega_0 t \cos m\omega_0 t\, dt$$

$$= \frac{1}{2}\left[\int_0^T \cos(m+n)\omega_0 t dt + \int_0^T \cos(m-n)\omega_0 t dt\right] \tag{16.18}$$

As long as $m \neq n$, $(m + n)$ and $(m - n)$ are nonzero integers, and the integrals are zero (Equation 16.10). If $m = n$, the first integral on the RHS of Equation 16.18 is zero, and the second integral is T, so that the RHS of Equation 16.18 is $T/2$. This proves Equations 16.7 and 16.8 for the cosine functions. The proof for the sine functions follows along the same lines, starting with subtracting Equation 16.15 from Equation 16.16, which gives the difference between two cosine functions rather than their sum.

Returning to Fourier analysis and the determination of Fourier coefficients, we note that a_0 in the FSE is determined by integrating both sides of Equation 16.2 over a period:

$$\int_0^T f(t)dt = \int_0^T a_0\,dt + \sum_{n=1}^{\infty}\int_0^T a_n \cos n\omega_0 t\,dt + \sum_{n=1}^{\infty}\int_0^T b_n \sin n\omega_0 t\,dt$$

$$= a_0 T + 0 + 0$$

where the second and third integrals on the RHS are zero in accordance with Equations 16.9 and 16.10. It follows that

$$a_0 = \frac{1}{T}\int_{t_0}^{t_0+T} f(t)dt \qquad (16.19)$$

a_0 is therefore the average of $f(t)$ over a period. It is the *dc component* of $f(t)$, whereas the cosine and sine terms are the *ac component*.

We wish to determine next a_n and b_n, where n refers to a particular positive integer in the range 1 to ∞, such as a_3 or b_5. But since it is common practice to also use n as an indexing positive integer in the range 1 to ∞, as in Equation 16.2, it is necessary to distinguish between the indexing positive integer in the range 1–∞ and a particular value in this range because both will occur in the same expression. We will denote the indexing integer in the FSE of Equation 16.2 by k and let n denote the order of the harmonic for which a_n and b_n are to be determined.

If both sides of Equation 16.2 are multiplied by $\cos n\omega_0 t$ and are integrated over a period and invoking Equations 16.6 through 16.8, we obtain

$$\int_0^T f(t)\cos n\omega_0 t\,dt = \int_0^T a_0 \cos n\omega_0 t\,dt$$

$$+\sum_{k=1}^{\infty}\int_0^T a_k \cos k\omega_0 t \cos n\omega_0 t\,dt + \sum_{k=1}^{\infty}\int_0^T b_k \sin k\omega_0 t \cos n\omega_0 t\,dt$$

$$= 0 + \sum_{k=1}^{\infty}\int_0^T \frac{a_k}{2}\left[\cos(k+n)\omega_0 t + \cos(k-n)\omega_0 t\right]dt$$

$$+\sum_{k=1}^{\infty}\int_0^T \frac{b_k}{2}\left[\sin(k+n)\omega_0 t + \sin(k-n)\omega_0 t\right]dt$$

$$= 0 + \frac{T}{2}a_n + 0$$

where the only nonzero integral is that of $\cos(k - n)$ for $k = n$, which results in an integral $(T/2)a_n$.

This gives

$$a_n = \frac{2}{T}\int_0^T f(t)\cos n\omega_0 t\,dt \qquad (16.20)$$

Note that setting $n = 0$ in Equation 16.20 gives twice the value of a_0 (Equation 16.19). Hence, a_0 cannot, in general, be obtained from a_n by setting $n = 0$.

To determine b_n, we multiply both sides of Equation 16.2 by $\sin n\omega_0 t$, integrate over a period, and invoke Equations 16.6 through 16.8 to obtain

$$\int_0^T f(t)\sin n\omega_0 t\,dt = \int_0^T a_0 \sin n\omega_0 t\,dt$$

$$+\sum_{k=1}^{\infty}\int_0^T a_k \cos k\omega_0 t \sin n\omega_0 t\,dt + \sum_{k=1}^{\infty}\int_0^T b_k \sin k\omega_0 t \sin n\omega_0 t\,dt$$

$$= 0 + \sum_{k=1}^{\infty}\int_0^T \frac{a_k}{2}\left[\sin(k+n)\omega_0 t - \sin(k-n)\omega_0 t\,dt\right]$$

$$+\sum_{k=1}^{\infty}\int_{t_0}^T \frac{b_k}{2}\left[\cos(k-n)\omega_0 t - \cos(k-n)\omega_0 t\right]dt$$

$$= 0 + 0 + \frac{T}{2}b_n$$

for $k = n$, as in the preceding case. This gives

$$b_n = \frac{2}{T}\int_0^T f(t)\sin n\omega_0 t\,dt, \qquad (16.21)$$

The derivation of the Fourier coefficients can be summarized as follows:

Summary: *a_0 is the average of $f(t)$ over a period, a_n is twice the average of $f(t)\cos n\omega_0 t$ over a period, and b_n is twice the average of $f(t)\sin n\omega_0 t$ over a period.*

Primal Exercise 16.1

Express the following functions as FSEs and determine the period by applying the definition of periodicity ($f(t) = f(t + nT)$): (a) $\sin 2t \cos 4t$ and (b) $\sin 2t + \sin 5t + \sin 7t$. Note that the fundamental frequency is the largest common factor of the frequencies of the individual components.

Ans. (a) $-0.5\sin 2t + 0.5\sin 6t$, π; (b) periodic of period 2π.

Example 16.1: Fourier Analysis of Sawtooth Waveform

It is required to derive the Fourier coefficients of the sawtooth waveform of Figure 16.5.

Solution:

During the interval $0 < t < T$, $f_{st}(t) = \frac{A}{T}t$; $a_0 = \frac{1}{T}\int_0^T \frac{A}{T}t\,dt = \frac{A}{T^2}\left[\frac{t^2}{2}\right]_0^T = \frac{A}{2}$; this is readily checked, since the period is a triangle of area $AT/2$, and the average over a period is the area divided by the period or $A/2$. An easier way of obtaining the average value in many cases is to determine by how much the horizontal axis should be moved so that the area enclosed by the function above the new horizontal axis is the same as the area enclosed by the function below this axis. It is seen that in Figure 16.5, if the horizontal axis is moved in the positive direction (i.e., upwards) by $+A/2$, as in Figure 16.6a, as much positive area is enclosed by the function as negative area. The average value is therefore $+A/2$.

$$a_n = \frac{2}{T}\int_0^T \frac{A}{T} t\cos n\omega_0\, dt = \frac{2A}{T^2}\int_0^T t\cos n\omega_0 t\, dt$$

Integrating by parts or using the Table of Integrals (Appendix B),

$$\begin{aligned} a_n &= \frac{2A}{T^2}\left[\frac{1}{n^2\omega_0^2}\cos n\omega_0 t + \frac{t}{n\omega_0}\sin n\omega_0 t\right]_0^{2\pi/\omega_0} \\ &= \frac{2A}{T^2}\left[\frac{1}{n^2\omega_0^2}\cos 2n\pi + \frac{2\pi}{n\omega_0^2}\sin 2n\pi - \frac{1}{n^2\omega_0^2}\cos 0 - 0\right] \\ &= [1+0-1-0] = 0 \text{ for odd or even integer values of } n \end{aligned}$$

The FSE does not have any cosine terms for reasons that will be explained later.

$$\begin{aligned} b_n &= \frac{2}{T}\int_0^T \frac{A}{T} t\sin n\omega_0\, dt \\ &= \frac{2A}{T^2}\int_0^T t\sin n\omega_0 t\, dt \\ &= \frac{2A}{T^2}\left[\frac{1}{n^2\omega_0^2}\sin n\omega_0 t - \frac{t}{n\omega_0}\cos n\omega_0 t\right]_0^{2\pi/\omega_0} \\ &= \frac{2A}{T^2}\left[\frac{1}{n^2\omega_0^2}\sin 2n\pi - \frac{2\pi}{n\omega_0^2}\cos 2n\pi - 0 + 0\right] \\ &= \frac{2A}{T^2}\left[-\frac{2\pi}{n\omega_0^2}\right] = -\frac{A}{\pi n}, \end{aligned}$$

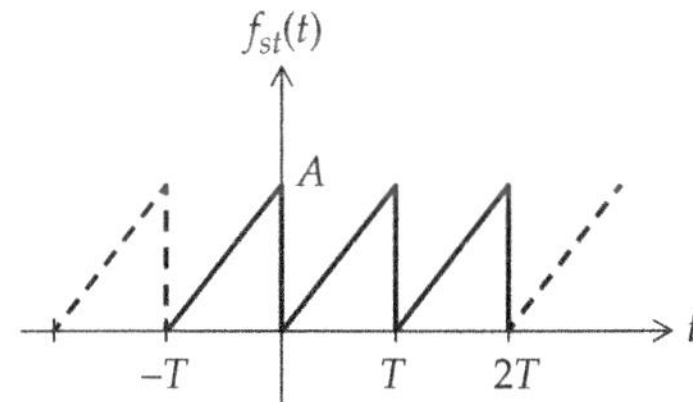

FIGURE 16.5
Figure for Example 16.1.

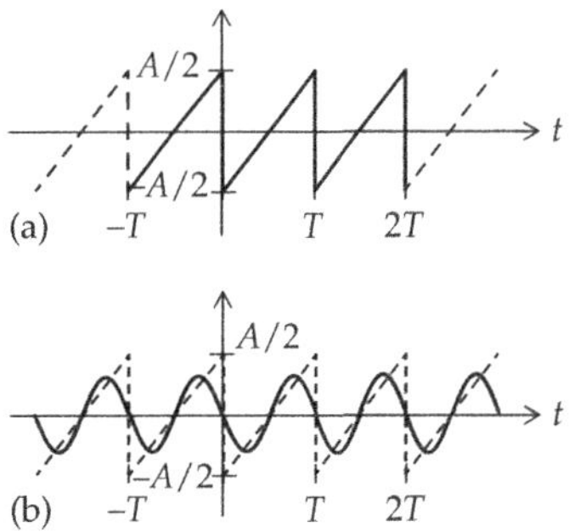

FIGURE 16.6
(a) Sawtooth waveform of zero average, and is approximated by a negative sinusoid in (b).

where $\omega_0 T = 2\pi$. The trigonometric form of $f_{st}(t)$ is therefore

$$\begin{aligned} f_{st}(t) &= \frac{A}{2} - \frac{A}{\pi}\sum_{n=1}^{\infty}\frac{\sin n\omega_0 t}{n} \\ &= \frac{A}{2} - \frac{A}{\pi}\left[\sin\omega_0 t + \frac{\sin 2\omega_0 t}{2} + \frac{\sin 3\omega_0 t}{3} + \cdots\right] \end{aligned} \tag{16.22}$$

The FSE of Equation 16.22 makes sense. For example, if the dc component $A/2$ is subtracted from the FSE, the ac component, of zero average, is as shown in Figure 16.6a. The sinusoidal function that roughly approximates the waveform of Figure 16.6a is $-B\sin\omega_0 t$, where B is some appropriate amplitude. This agrees with the fundamental component in the FSE. The additional harmonics, all being negative sinusoids, improve the approximation.

At the points of discontinuity, where $f(t)$ jumps between 0 and A, $t = kT$, where $k = 0, \pm 1, \pm 2, \pm 3, \ldots$. The sinusoidal terms are of the form $\sin(n\omega_0 \times kT) = \sin(2\pi nk)$, where $\omega_0 T = 2\pi$. Since nk is an integer, $\sin(2\pi nk) = 0$. It follows that all the sinusoidal terms vanish at $t = kT$ and $f(t) = A/2$, which is the average value of $f(t)$ at the two ends of the discontinuity. This is a general feature of the FSE.

Problem-Solving Tips

- Remember that $\omega_0 T = 2\pi$.
- When the period has a simple geometric form, a_0 can be easily determined by dividing the area over the period by the period T or by shifting the horizontal axis so as to obtain a zero average. The average value is then equal to the shift of the horizontal axis. If the shift is upwards, the average value is positive, and conversely.
- Whenever feasible, check if the form of the FSE makes sense as a rough approximation to the given periodic function.

Primal Exercise 16.2

Determine ω_0 for the function of Figure 16.5 if (a) $T = 1$ s and (b) $T = \pi$ s.

Ans. (a) 2π rad/s; (b) 2 rad/s.

16.2.1 Exponential Form

The FSE can also be conveniently expressed in exponential form. For this purpose, $\cos n\omega_0 t$ and $\sin n\omega_0 t$ in Equation 16.2 are replaced by their exponential forms, and terms having the same exponent are grouped together. Thus,

$$f(t) = a_0 + \sum_{n=1}^{\infty}\left[a_n\left(\frac{e^{jn\omega_0 t} + e^{-jn\omega_0 t}}{2}\right) + b_n\left(\frac{e^{jn\omega_0 t} - e^{-jn\omega_0 t}}{2j}\right)\right]$$
$$= a_0 + \sum_{n=1}^{\infty}\left[\left(\frac{a_n - jb_n}{2}\right)e^{jn\omega_0 t} + \left(\frac{a_n + jb_n}{2}\right)e^{-jn\omega_0 t}\right] \tag{16.23}$$

Let $C_n = (a_n - jb_n)/2$. Substituting for a_n and b_n from Equations 16.20 and 16.21, C_n can be expressed as

$$C_n = \frac{1}{T}\int_0^T f(t)(\cos n\omega_0 t - j\sin n\omega_0 t)dt \tag{16.24}$$

Using Euler's formula (Equation 8.3), we obtain

$$C_n = \frac{1}{T}\int_0^T f(t)e^{-jn\omega_0 t}dt \tag{16.25}$$

It follows that for $n = 0$,

$$C_0 = \frac{1}{T}\int_0^T f(t)dt = a_0 \tag{16.26}$$

Note that unlike a_n (Equation 16.20), C_0 can be obtained from C_n by setting $n = 0$ in the integrand in Equation 16.25.

If the complex conjugate of C_n is denoted by $C_n^* = (a_n + jb_n)/2$, it can be readily shown that C_n^* is simply derived from C_n by changing the sign of n, as this changes the sign of the sine component but not the cosine component. Thus, substituting for a_n and b_n from Equations 16.20 and 16.21 and using Euler's formula,

$$C_n^* = \frac{1}{T}\int_0^T f(t)(\cos n\omega_0 t + j\sin n\omega_0 t)dt$$
$$= \frac{1}{T}\int_0^T f(t)e^{jn\omega_0 t}dt = C_{-n} \tag{16.27}$$

Equation 16.23 can be then expressed as

$$f(t) = C_0 + \sum_{n=1}^{\infty}C_n e^{jn\omega_0 t} + \sum_{n=1}^{\infty}C_{-n}e^{-jn\omega_0 t} \tag{16.28}$$

Note that the two summations on the RHS of Equation 16.28 represent the ac component and have a zero average.

If the index values of n in the last term on the RHS are changed to negative integers from -1 to $-\infty$, C_{-n} becomes C_n, and the sign of the exponent in the last term on the RHS becomes positive. Equation 16.28 can now be written as

$$f(t) = C_0 + \sum_{n=1}^{\infty}C_n e^{jn\omega_0 t} + \sum_{n=-1}^{-\infty}C_n e^{jn\omega_0 t} \tag{16.29}$$

Equation 16.29 can be expressed more compactly as

$$f(t) = \sum_{n=-\infty}^{\infty}C_n e^{jn\omega_0 t} \tag{16.30}$$

where $n = 0$ gives the first term in Equation 16.29, positive values of n give the middle term, and negative values of n give the last term. It should be noted, however, that whereas Equation 16.30 is the compact mathematical form, Equation 16.29 is the more practical form. This is because in some cases, as in Example 16.2, deriving C_n using Equation 16.25 and then setting $n = 0$ do not give a finite value for C_0. In such cases, C_0 is derived directly from the average value of $f(t)$ and used in Equation 16.29.

The relationships between C_n, a_n, and b_n readily follow from the definition of C_n. Thus,

$$a_n = 2\mathrm{Re}(C_n) \quad \text{and} \quad b_n = -2\mathrm{Im}(C_n) \tag{16.31}$$

$$|C_n| = \left|\frac{a_n - jb_n}{2}\right| = \frac{\sqrt{a_n^2 + b_n^2}}{2} = \frac{c_n}{2} \quad \text{and} \quad \angle C_n = -\tan^{-1}\frac{b_n}{a_n} = \theta_n \tag{16.32}$$

where c_n and θ_n are given by the relations of Equation 16.5.

Compared to the trigonometric form, the exponential form is advantageous for several reasons: (1) it is generally easier to apply Equation 16.25 to determine C_n rather than using Equations 16.20 and 16.21 to determine a_n and b_n, (2) deriving C_n gives both a_n and b_n simultaneously, and (3) some important properties of $f(t)$, such as its frequency spectrum, are expressed directly in terms of C_n, as discussed next.

16.2.2 Frequency Spectrum

The plots of $|C_n|$ and θ_n against frequency are, respectively, the **amplitude spectrum** and the **phase spectrum** of $f(t)$. They both constitute the **frequency spectrum** of $f(t)$. Because frequencies in the FSE have discrete values only, the frequency spectrum of a periodic function is a *line spectrum* that consists of a series of lines at $\omega = n\omega_0$, where $n = 0$, ± 1, ± 2, ± 3, … (Figure 16.7). Since $C_{-n} = C_n^*$, it is seen that $|C_n| = |C_{-n}|$ and $\angle C_n = -\tan^{-1}(b_n/a_n) = -\angle C_{-n}$, because C_n

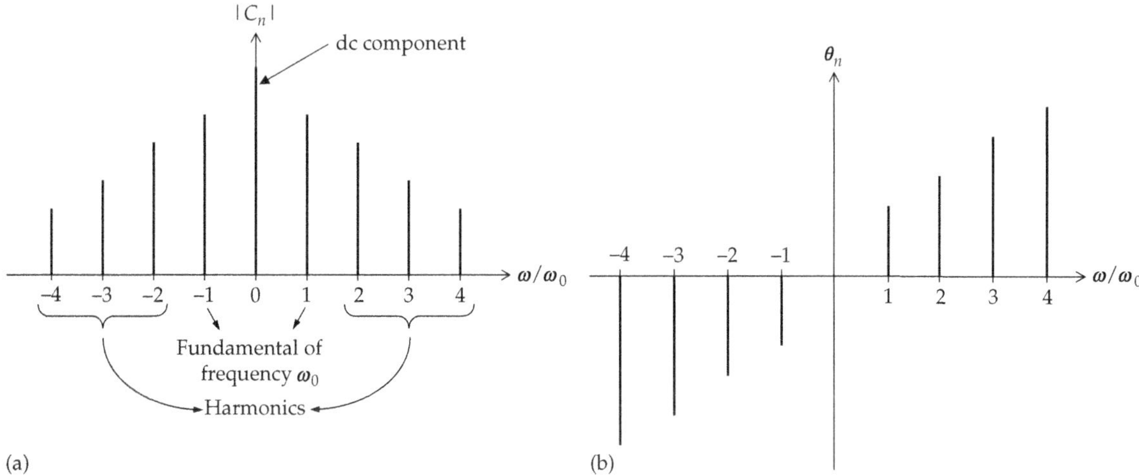

FIGURE 16.7
(a) Amplitude spectrum; (b) frequency spectrum.

and its complex conjugate have the same magnitude but phase angles of opposite sign. The amplitude spectrum is therefore an even function, that is, it is symmetrical about the vertical axis (Figure 16.7a). Note that $|C_0|$ is the magnitude of the dc component and $|C_n|$ is half the amplitude of the fundamental component ($n = 1$) or the harmonics ($n > 1$), in accordance with Equation 16.32. The phase spectrum, on the other hand, is an odd function (Figure 16.7b). However, when C_n is purely real ($b_n = 0$), then $\theta_n = 0°$, when $C_n > 0$, or $\theta_n = 180°$, when $C_n < 0$. In these cases, the negation of 0° or 180° is the same angle of 0° or 180°.

It should be noted that ω and ω_0 *are positive quantities*. Negative values of ω/ω_0 in Figure 16.7 arise because of negative values of n that multiply a positive value of ω_0 to give a negative value, $-n\omega_0$. So negative values of $n\omega_0$ are *not* negative frequencies, which have no physical meaning. Mathematically, these negative values combine with their positive counterparts in complex exponentials (Equation 16.30) to give real cosine and sine terms, because all physical voltages and currents are real and cannot have imaginary components.

The exponential form of the FSE for the sawtooth waveform is derived in Example 16.2, together with its frequency spectrum. This example also demonstrates an important and useful technique, namely, that of deriving the FSE of a different but related version of a given $f(t)$ from the already derived FSE of $f(t)$.

Example 16.2: Exponential Fourier Series of Sawtooth Waveform

It is required to derive the exponential Fourier coefficients of the sawtooth waveform of Figure 16.5 and plot its amplitude and phase spectra.

Solution:

$C_n = \frac{1}{T}\int_0^T \frac{A}{T} te^{-jn\omega_0 t}\,dt$. Integrating by parts (Appendix B), with $\omega_0 T = 2\pi$,

$$C_n = -\frac{A}{T^2 n^2 \omega_0^2}\left[-jn\omega_0 te^{-jn\omega_0 t} - e^{-jn\omega_0 t}\right]_0^{2\pi/\omega_0}$$

$$= -\frac{A}{4\pi^2 n^2}\left[-j2n\pi\right] = j\frac{A}{2\pi n} \quad (16.33)$$

C_n is imaginary, which means that $a_n = 0$ (Equation 16.31). Note that the average value of $f_{st}(t)$ is $A/2$ and cannot be obtained by setting $n = 0$ in Equation 16.33. This is because if $n = 0$, then it should be substituted in the expression for C_n before integration. Otherwise, the integration would mean dividing by zero, which invalidates the result.

From Equation 16.29, the exponential form of $f_{st}(t)$ is

$$f_{st}(t) = \frac{A}{2} + \frac{A}{2\pi}\sum_{\substack{n=-\infty \\ n\neq 0}}^{\infty} \frac{j}{n} e^{jn\omega_0 t} \quad (16.34)$$

From Equation 16.33, $|C_n| = A/2\pi n$. The amplitude spectrum consists of a line of height $A/2$ at $\omega = 0$ and lines of height $A/(2\pi n)$ at $n\omega/\omega_0$, where n is a positive or a negative integer (Figure 16.8a). The height of $A/(2\pi n)$ is one-half the amplitude of the fundamental or the nth harmonic. The phase angle of C_n is +90° for n, a positive integer, and −90° for n, a negative integer (Figure 16.8b).

C_n can be obtained using MATLAB's int(E,t,a,b) command. Ignoring for the moment A/T^2, the integral

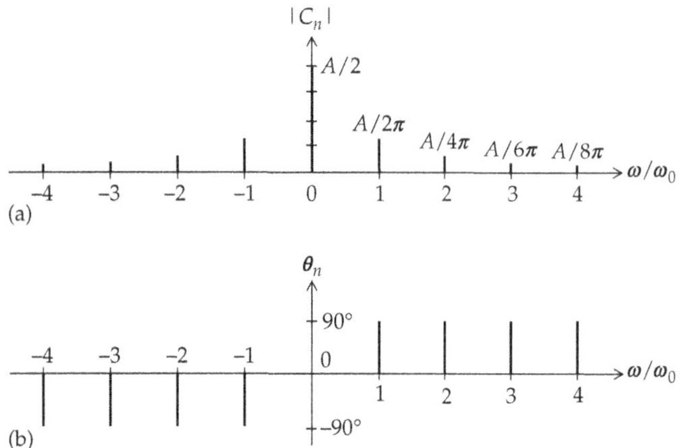

FIGURE 16.8
Amplitude spectrum (a) and phase spectrum (b) of the sawtooth waveform.

$\int_0^T te^{-jn\omega_0 t}dt$ can be evaluated by entering the following code:

```
syms t n w
int(t*exp(-j*n*w*t),t,0,2*pi/w)
```

where w denotes ω_0. MATLAB returns:

```
-1/(n^2*w^2)+((1/exp(2*pi*n*i))*
(1+2*pi*n*i))/(n^2*w^2)
```

which is

$$-\frac{1}{n^2\omega_0^2}+\frac{1}{e^{+j2\pi n}}\frac{1+j2n\pi}{n^2\omega_0^2}=\frac{j2n\pi}{n^2\omega_0^2}=\frac{j2\pi}{n\omega_0^2}$$

since $e^{j2\pi n}=1$ for all n. Multiplying by $A/T^2 = A\omega_0^2/4\pi^2$ gives $C_n = j\dfrac{A}{2\pi n}$, as in Equation 16.33.

If the function $f_{st}(t)$ of Figure 16.5 is negated (Figure 16.9a), then shifted upward by A, it becomes the "reversed sawtooth" waveform $f_{str}(t)$ of Figure 16.9b. The FSE of $f_{str}(t)$ is obtained by adding A to the negation of the RHS of Equation 16.34:

$$f_{str}(t)=\frac{A}{2}-\frac{A}{\pi}\sum_{n=1}^{\infty}\frac{\sin n\omega_0 t}{n} \tag{16.35}$$

Equation 16.35 also follows from $f_{st}(t) + f_{str}(t) = A$.

C_0' of $f_{str}(t)$ is $A/2$ and its C_n' is $-j\dfrac{A}{2\pi n}$. The amplitude spectrum is unchanged, but the phase spectrum is negated.

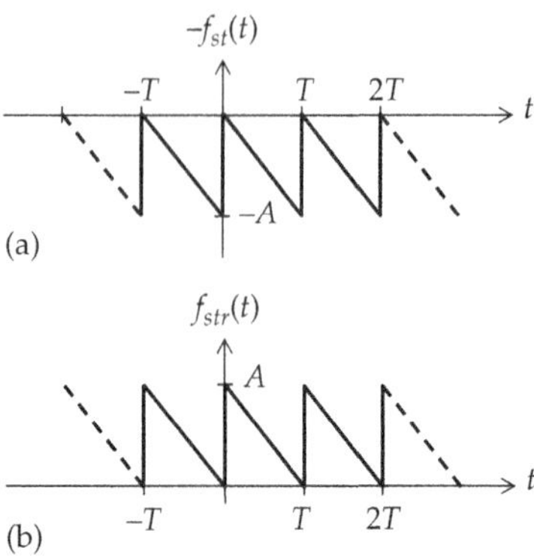

FIGURE 16.9
Sawtooth waveform of Figure 16.5 negated (a), and shifted upward by A (b).

Primal Exercise 16.3

Derive the Fourier coefficients of the reversed sawtooth waveform of Figure 16.9b by applying Equations 16.19 through 16.21.

Example 16.3: Fourier Analysis of Rectangular Waveform

It is required to derive the Fourier coefficients of the rectangular pulse train $f_{pt}(t)$ illustrated in Figure 16.10 and plot its amplitude and phase spectra.

Solution:

Since $f_{pt}(t)$ is an even function, it is convenient to take a period that is symmetrical about the vertical axis, as shown. Hence, Equation 16.25 becomes

$$C_n=\frac{1}{T}\int_{-T/2}^{T/2}f(t)e^{-jn\omega_0 t}dt$$

$$=\frac{1}{T}\int_{-\tau/2}^{\tau/2}Ae^{-jn\omega_0 t}dt=\frac{A}{T}\left[\frac{e^{-jn\omega_0 t}}{-jn\omega_0}\right]_{-\tau/2}^{\tau/2}$$

$$=\frac{A}{n\omega_0 T}\left(\frac{e^{jn\omega_0\tau/2}-e^{-jn\omega_0\tau/2}}{j}\right)$$

$$=\frac{2A}{n\omega_0 T}\sin(n\omega_0\tau/2)=A\frac{\tau}{T}\frac{\sin(n\omega_0\tau/2)}{(n\omega_0\tau/2)}$$

$$=A\frac{\tau}{T}\text{sinc}(n\omega_0\tau/2) \tag{16.36}$$

where $\text{sinc}(x)=(\sin x)/x$. The sinc($x$) function, illustrated in Figure 16.11, is an important function in signal analysis. sinc(x) = 0, when sin(x) = 0, that is, when $x = n\pi$, where n is a positive or negative nonzero integer. When $x \rightarrow 0$,

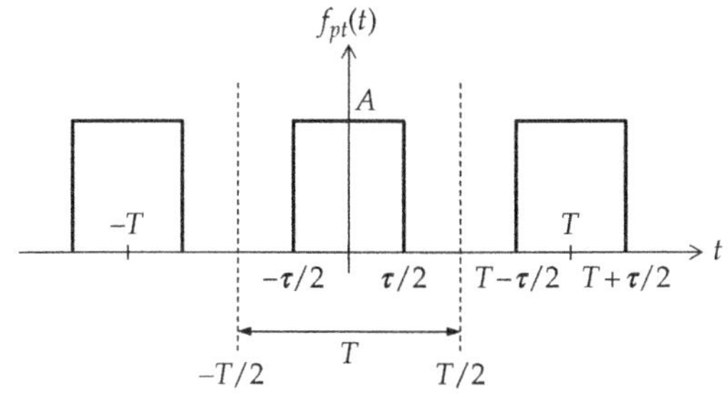

FIGURE 16.10
Figure for Example 16.3.

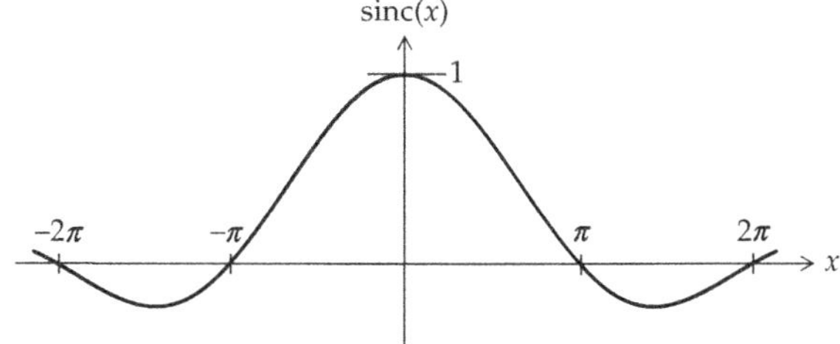

FIGURE 16.11
Figure for Example 16.3.

sinc(0) → 0/0, which is indeterminate. However, according to L'Hopital's rule (Appendix A),

$$\underset{x\to 0}{\text{limit}}\frac{\sin x}{x}=\underset{x\to 0}{\text{limit}}\frac{d(\sin x)dx}{d(x)/dx}=\underset{x\to 0}{\text{limit}}\frac{\cos x}{1}=1 \tag{16.37}$$

It follows that sinc(0) = 1, as in Figure 16.11. For $0<x<\pi$, $\sin(x)>0$, and $\text{sinc}(x)>0$. For $\pi<x<2\pi$, $\sin(x)<0$, and sinc(x) < 0. For $x > 0$, sinc(x) therefore has alternating positive and negative lobes between successive zero crossings at integral multiples of π. But since sinc(x) is a sine function divided by its argument, the magnitude of sinc(x) decreases as the magnitude of x increases. sinc(x) is an even function, because when x changes sign, sin(x) also changes sign, since $\sin(-x) = -\sin(x)$.

Returning to Equation 16.36, we note that since C_n is real, $b_n = 0$, and the FSE of $f_{pt}(t)$ function does not have any sine terms. The reason for this is that the function is even, as explained in the next section. The average value of $f_{pt}(t)$ is $A\tau/T$. However, it can be obtained in this case by setting $n = 0$ in Equation 16.36, because in this case, the numerator of the integral is zero when $n = 0$, so L'Hopital's rule can be applied and gives a finite result.

If we set $\alpha = \tau/T$ and replace $\omega_0 T$ by 2π,

$$C_n=\alpha A\,\text{sinc}(\alpha n\pi)=\frac{A}{n\pi}\sin(\alpha n\pi) \tag{16.38}$$

and

$$f_{pt}(t)=\alpha A\sum_{n=-\infty}^{\infty}\text{sinc}(\alpha n\pi)e^{jn\omega_0 t} \tag{16.39}$$

To express the FSE in trigonometric form, we substitute $a_0=C_0=\alpha A$, $a_n = 2C_n$, and $b_n = 0$ in Equation 16.2:

$$f_{pt}(t)=\alpha A+\frac{2A}{\pi}\left[\sin\alpha\pi\cos\omega_0 t+\frac{\sin 2\alpha\pi}{2}\cos 2\omega_0 t +\frac{\sin 3\alpha\pi}{3}\cos 3\omega_0 t+\cdots\right] \tag{16.40}$$

The amplitude spectrum is easier to visualize if a particular value of α is selected, say, $\alpha = 1/5$. From Equation 16.38, $|C_n| = (A/5)|\text{sinc}(n\pi/5)|$. It is seen that $|C_0| = A/5$, and the lines of the amplitude spectrum are bounded by the magnitude of the sinc function for continuous n, as illustrated in Figure 16.12a. The line spectra occur at integer values of $n = \omega/\omega_0$. The amplitude is zero for values of n that are integral multiples of 5. The phase spectrum is shown in Figure 16.12b. Since C_n is real, its phase angle is zero when $C_n > 0$ and is 180° when $C_n < 0$. The phase angle is zero when $n = 0$, since $C_0 = A/5$ is positive and is not defined, strictly speaking, when $C_n = 0$, as when $n = \pm 5$, because the phase angle can be zero or 180° when the magnitude is zero.

If α is small, it is seen from Equation 16.40 that the fundamental and harmonics all have essentially the same amplitude $2\alpha A$, since for small x, $\sin x \cong x$. This is an

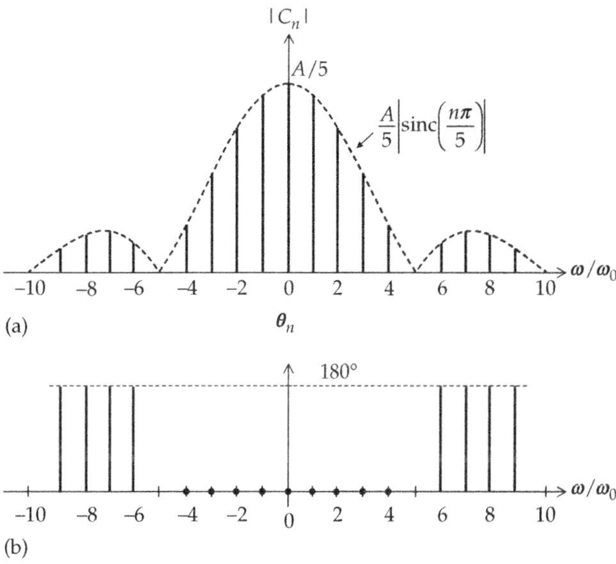

FIGURE 16.12
Figure for Example 16.3.

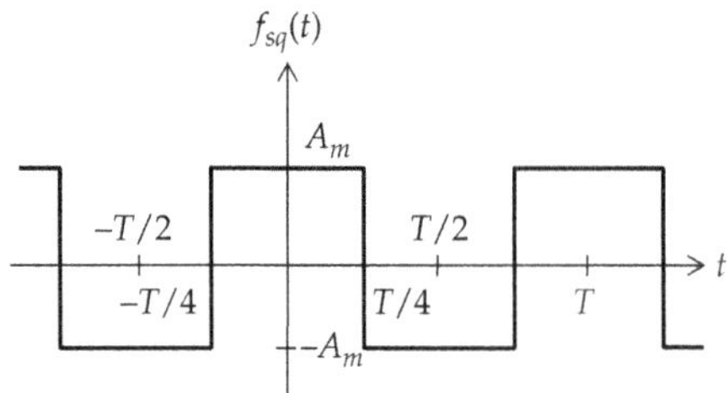

FIGURE 16.13
Figure for Example 16.3.

important result in signal analysis, according to which, the narrower the pulses, the more significant the higher harmonics are, relative to the fundamental.

To determine C_n using MATLAB's int(E,t,a,b) command, we enter the following:

```
syms t n w a.
int(exp(-j*n*w*t),t,-a/2,a/2)
```

where
w is ω_0
a is τ

MATLAB returns: (2*sin((a*n*w)/2))/(n*w), which is $\frac{2\sin(\tau n\omega_0/2)}{n\omega_0}$. Substituting $\omega_0/2 = \pi/T$ and $\alpha = \tau/T$, this becomes $\frac{T\sin(\alpha n\pi)}{n\pi}$. Multiplying by A/T gives C_n as in Equation 16.38.

We can deduce from Equation 16.40 the FSE $f_{sq}(t)$ of a square wave of amplitude A_m and zero average value (Figure 16.13). To do so, we set $\alpha = 1/2$ and $A_m = A/2$ and remove the dc value from $A_{pt}(t)$ (Equation 16.40). When n is even, $\sin(n\pi/2) = 0$, whereas for odd n, $\sin(n\pi/2)$ is alternately +1 and −1. This gives

$$f_{sq}(t) = \frac{4A_m}{\pi}\left[\cos\omega_0 t - \frac{1}{3}\cos 3\omega_0 t + \frac{1}{5}\cos 5\omega_0 t - \frac{1}{7}\cos 7\omega_0 t + \cdots\right] \quad (16.41)$$

Note that the cosine function is a first rough approximation to an even square wave.

Exercise 16.4

Derive the Fourier coefficients of the rectangular waveform of Figure 16.10 by applying Equations 16.19 through 16.21.

Exercise 16.5

Determine A and $\alpha = \tau/T$ in Figure 16.10 so that $C_0 = 1/2$ and $C_1 = 1/\pi$.

Ans. $A = 1$, $\alpha = 1/2$.

16.2.3 Translation in Time

The exponential form is convenient for determining the effect of translation in time. If a periodic waveform $f(t)$ is delayed by t_d, it becomes $f(t - t_d)$ with respect to the same time origin. Replacing t by $(t - t_d)$ in Equation 16.30,

$$f(t-t_d) = \sum_{n=-\infty}^{\infty} C_n e^{jn\omega_0(t-t_d)} = \sum_{n=-\infty}^{\infty}\left[C_n e^{-jn\omega_0 t_d}\right]e^{jn\omega_0 t} \quad (16.42)$$

The effect is to replace C_n by $C_n e^{-jn\omega_0 t_d}$. The magnitude of C_n, and hence the amplitude spectrum, remains unchanged because $|e^{-jn\omega_0 t_d}| = |\cos n\omega_0 t - j\sin n\omega_0 t|$ $\sqrt{\cos^2 n\omega_0 t + \sin^2 n\omega_0 t} = 1$. However, the new phase angle θ_n' is

$$\theta_n' = \theta_n - n\omega_0 t_d \quad (16.43)$$

where θ_n is the phase angle of C_n. Conversely, if the function is advanced by t_a, C_n is replaced by $C_n e^{+jn\omega_0 t_a}$; the phase angle θ_n is increased by $n\omega_0 t_a$. Note that a change in θ_n (Equation 16.32) implies a change of $n\omega_0 t_d$ in the phase angle of each term of the FSE (Equation 16.4).

Example 16.4: Translation in Time and Fourier Analysis of Square Wave

It is required to derive the FSE of the square wave (Figure 16.12) when delayed or advanced by $T/4$.

Solution:

Since $t_d = T/4$, $n\omega_0 t_d = n\omega_0 T/4 = n\pi/2$. When a function is delayed in time, a given part of the waveform will occur later in time, which means that the function is shifted to the right. Thus, consider the transition that occurs at $t = -T/4$ in Figure 16.13 from $-A_m$ to $+A_m$. When the function is delayed by $T/4$, this same transition will occur at $t = 0$. In effect, the function is shifted by $T/4$ to the right (Figure 16.14a). Conversely, if the function is advanced in time by $T/4$, the same transition now occurs at $t = -T/2$. In effect, the function is shifted by $T/4$ to the left (Figure 16.14b).

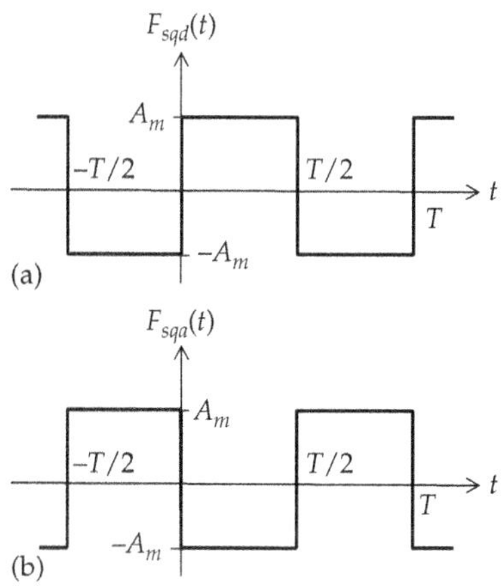

FIGURE 16.14
Figure for Example 16.4.

When the square wave function of Figure 16.12 is delayed by $T/4$ (Figure 16.14a), the phase angle of each of the terms in Equation 16.41 is decreased by $n\pi/2$:

$$f_{sqd}(t) = \frac{4A_m}{\pi}\left[\cos\left(\omega_0 t - \frac{\pi}{2}\right) - \frac{1}{3}\cos\left(3\omega_0 t - \frac{3\pi}{2}\right) + \frac{1}{5}\cos\left(5\omega_0 t - \frac{5\pi}{2}\right) - \frac{1}{7}\cos\left(7\omega_0 t - \frac{7\pi}{2}\right) + \cdots\right]$$

or

$$f_{sqd}(t) = \frac{4A_m}{\pi}\left[\sin\omega_0 t + \frac{1}{3}\sin 3\omega_0 t + \frac{1}{5}\sin 5\omega_0 t + \cdots\right] \quad (16.44)$$

If the square wave of Figure 16.12 is advanced by $T/4$ (Figure 16.14b), it becomes the negation of Figure 16.14a, so that its FSE is the negation of Equation 16.44:

$$f_{sqa}(t) = -\frac{4A_m}{\pi}\left[\sin\omega_0 t + \frac{1}{3}\sin 3\omega_0 t + \frac{1}{5}\sin 5\omega_0 t + \cdots\right] \quad (16.45)$$

As argued in connection with Figure 16.6, $+\sin\omega_0 t$ is qualitatively a first approximation in shape to the square wave of Figure 16.14a, whereas $-\sin\omega_0 t$ is qualitatively a first approximation in shape to the square wave of Figure 16.14b.

Simulation: PSpice can be used to (1) obtain a file printout of the amplitude and phase of each frequency component of a periodic waveform and (2) display the amplitude spectrum of a periodic waveform. Periodic sources are available in PSpice, such as VPULSE and VSIN from the SOURCE library, which can be modified using parts from the analog behavioral module (ABM) Library. For example, a VSIN source followed by an ABS block from the ABM Library can be used to generate a full-wave rectified waveform (Figure 16.27b). VPULSE can be used to generate practically any periodic waveform of rectangular, triangular, or trapezoidal pulses by proper choice of pulse durations and rise and fall times. More general periodic waveforms consisting of straight-line segments can be generated by the source VPWL_RE_FOREVER (Example 16.8).

The square waveform of Figure 16.14a is simulated, assuming $A_m = 5$ V and $T = 2$ s. VPULSE (Appendix C) is used to obtain a pulse train of 10 V peak-to-peak amplitude and zero average value, as in Figure 16.15. The output is labeled vo using net alias. To run the simulation, Time Domain (Transient) is selected for Analysis type, 6 is entered for 'Run to time', 0 for 'Start saving data after', and 0.1m for 'Maximum step size'. In the same Simulation Settings window, select Output File Options. In the new Transient File Output Options window, select Perform Fourier Analysis, and enter 0.5 for 'Center Frequency', 9 for 'Number of Harmonics', and V(vo) for 'Output variables'. After the simulation is run, selecting Trace/Add Trace and then V(vo) displays vo as a function of time. Selecting Trace/Fourier displays the amplitudes of the Fourier components as a function of frequency. Expand the x-axis by selecting Plot/Axis settings and define an x-axis range of 0–5 Hz. The plot of Figure 16.16 is displayed. To label the first peak, for example, enter the cursor command sxv(0.5). After the cursor moves to the peak, press the Mark Label icon. The peak value is displayed as 6.3663 corresponding to a calculated value of $4A_m/\pi = 20/\pi = 6.3662$. The other peaks are inversely proportional to the order of the harmonic, in accordance with Equation 16.44. Note that the amplitude of the Fourier components displayed by PSpice is $c_n = 2|C_n|$ (Equation 16.32).

To view the printout, select View Simulation Output File from the Schematic1 page and scroll down to the Fourier analysis part. The results tabulated below are displayed. The dc component is listed separately at the top and is insignificant. The first column is the

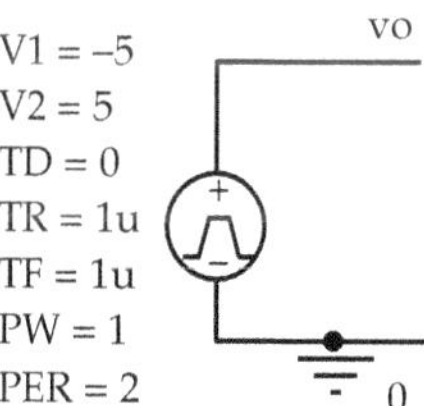

FIGURE 16.15
Figure for Example 16.4.

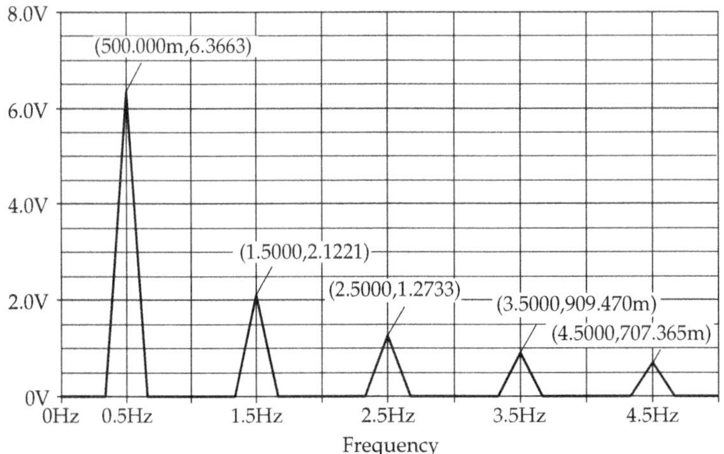

FIGURE 16.16
Figure for Example 16.4.

harmonic number, with the harmonic number 1 being the fundamental. The second column is the frequency in Hz of the harmonics. The third column is the magnitude of the harmonics. It is seen that the even harmonics are nominally zero. The magnitudes of the odd harmonics are in accordance with those in Figure 16.16. The fourth column is the normalized magnitude of the harmonics with respect to the fundamental. The fifth column is the phase angle of the harmonics, assuming a sine function, as in Equation 16.44, so that the phase angles of the odd harmonics are nominally zero. The last column is the normalized phase angle with respect to the fundamental. The total percentage harmonic component at the end of the table is a measure of the total harmonic content of the signal.

Primal Exercise 16.6

Given the periodic function $f(t)$ in Figure 16.17, determine the dc component of $f(t)$.

Ans. 1.

Primal Exercise 16.7

Given the periodic function $f(t)$ in Figure 16.18, determine the dc component of $f(t)$ in two ways: (a) from the area enclosed by $f(t)$ and (b) by shifting the horizontal axis so that the areas above and below the axis are equal. Note how much this method is easier in this case.

Ans. 4.

Fourier Components of Transient Response V(VO)
dc Component = 3.500184E−04

Harmonic No.	Frequency, Hz	Fourier Component	Normalized Component	Phase, deg	Normalized Phase, deg
1	5.000E−01	6.366E+00	1.000E+00	−2.070E−02	0.000E+00
2	1.000E+00	7.000E−04	1.100E−04	8.996E+01	9.001E+01
3	1.500E+00	2.122E+00	3.333E−01	−6.210E−02	−4.850E−10
4	2.000E+00	7.000E−04	1.100E−04	8.993E+01	9.001E+01
5	2.500E+00	1.273E+00	2.000E−01	−1.035E−01	−2.425E−09
6	3.000E+00	7.000E−04	1.100E−04	8.989E+01	9.002E+01
7	3.500E+00	9.095E−01	1.429E−01	−1.449E−01	−6.789E−09
8	4.000E+00	7.000E−04	1.100E−04	8.986E+01	9.002E+01
9	4.500E+00	7.074E−01	1.111E−01	−1.863E−01	−1.455E−08

Total Harmonic Distortion = 4.287948E+01 percent

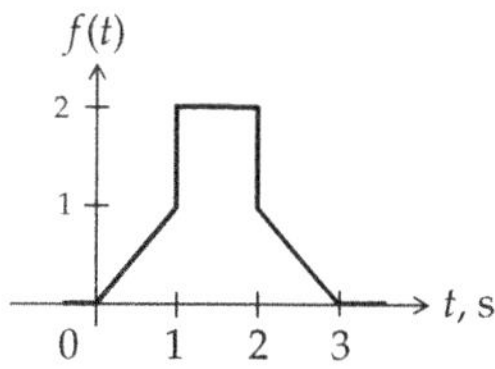

FIGURE 16.17
Figure for Primal Exercise 16.6.

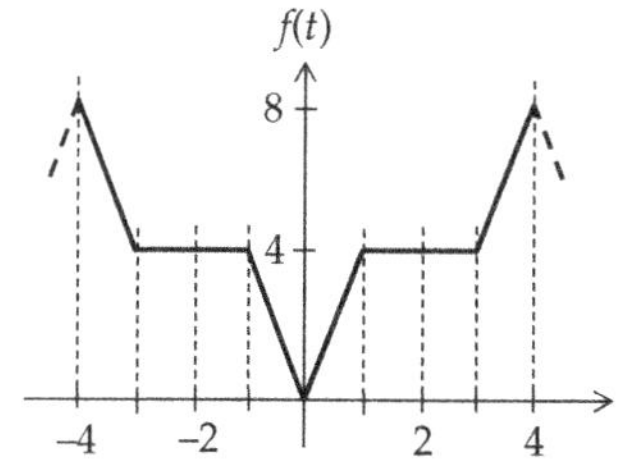

FIGURE 16.18
Figure for Primal Exercise 16.7.

Primal Exercise 16.8

Derive (a) the Fourier coefficients of the square waveforms of Equations 16.44 and 16.45 by applying Equations 16.19 through 16.21, and (b) the exponential form of the square waves $f_{sq}(t)$, $f_{sqd}(t)$, and $f_{sqa}(t)$.

Ans. (b) $f_{sq}(t) = \frac{2A_m}{\pi}\sum_{\substack{n=-\infty \\ n\neq 0}}^{n=\infty}\frac{1}{n}\sin\left(\frac{n\pi}{2}\right)e^{jn\omega_0 t}$, $f_{sqd}(t)$

$= -\frac{j2A_m}{\pi}\sum_{\substack{n=-\infty \\ n\neq 0}}^{n=\infty}\frac{1}{n}e^{jn\omega_0 t}$, $f_{sqa}(t) = \frac{j2A_m}{\pi}\sum_{\substack{n=-\infty \\ n\neq 0}}^{n=\infty}\frac{1}{n}e^{jn\omega_0 t}$,

where n is odd.

Exercise 16.9

Derive and compare the amplitude and phase spectra of the square waves of Equations 16.41, 16.44, and 16.45.

Ans. The amplitude spectrum is $|C_n| = \frac{2A_m}{|n|\pi}$, where n is odd, for the three cases. The phase spectrum of $f_{sq}(t)$ is zero for $n = 1, 5, 9, 13$, etc., and is 180° for $n = 3, 7, 11, 15$, etc. The phase spectrum of $f_{sqd}(t)$ is –90° for positive n. The phase spectrum of $f_{sqa}(t)$ is 90° for positive n. In all cases, the amplitude spectrum is an even function and the phase spectrum is an odd function.

16.3 Symmetry Properties of Fourier Series

16.3.1 Even-Function Symmetry

Consider an even function such as

$$f_{even}(t) = 2\cos\omega_0 t + 5\cos 2\omega_0 t + 10\cos 3\omega_0 t \quad (16.46)$$

Such a function is even, that is, $f_{even}(t) = f_{even}(-t)$, because all of its components are even. If an odd function, such as $4\sin\omega_0 t$, is added to the even components, then the function is no longer even, since $4\sin\omega_0(-t) = -4\sin\omega_0(t)$. This leads to the following general concept:

Concept: *The FSE of an even periodic function does not contain any sine terms; its Fourier coefficients can be evaluated over half a period.*

When the function is even, a period of the function is centered about the vertical axis, from $-T/2$ to $+T/2$. Since the FSE of an even periodic function does not contain any sine terms, $b_n = 0$ and $C_n = a_n/2$ is real. It follows that

$$C_n = \frac{1}{T}\int_{-T/2}^{T/2} f(t)e^{-jn\omega_0 t}dt = \frac{a_n}{2}$$

$$= \frac{1}{T}\left[\int_{-T/2}^{0} f(t)\cos n\omega_0 t dt + \int_{0}^{T/2} f(t)\cos n\omega_0 t dt\right] \quad (16.47)$$

If we substitute $t = -t'$ in the first integral in brackets, this integral becomes $\int_{T/2}^{0} f(-t')\cos n\omega_0(-t')(-dt') = \int_{0}^{T/2} f(-t')\cos n\omega_0(-t')dt'$. Changing the dummy integration variable back to t and invoking the property of an even function that $f(t) = f(-t)$, with the cosine function itself an even function, the integral becomes $\int_{0}^{T/2} f(t)\cos n\omega_0 dt$, the same as the second integral. Hence,

$$C_n = \frac{2}{T}\int_{0}^{T/2} f(t)\cos n\omega_0 t dt = \frac{2}{T}\text{Re}\left[\int_{0}^{T/2} f(t)e^{-jn\omega_0 t}dt\right] \quad (16.48)$$

The dc component is obtained from the first integral in Equation 16.48 by setting $n = 0$. With $a_n = 2C_n$, it follows that

$$a_0 = \frac{2}{T}\int_{0}^{T/2} f(t)dt$$

$$a_n = \frac{4}{T}\int_{0}^{T/2} f(t)\cos n\omega_0 t dt, \text{ and } b_n = 0 \text{ for all } n \quad (16.49)$$

Note that an even function can have a dc component and still remain even, because the dc component, being a constant, has even symmetry. Examples of even functions are the rectangular pulse train (Figure 16.10) and the square waveform of Figure 16.13. The corresponding FSEs (Equations 16.40 and 16.41) do not have any sine terms.

16.3.2 Odd-Function Symmetry

In a manner exactly analogous to that of even-function symmetry, it can be argued that an odd function can only contain odd terms, without any even terms. This implies that the FSE of an odd function consists of sine terms only without any cosine terms or a dc component. Thus, the following concept applies:

Concept: *The FSE of an odd periodic function does not contain an average term nor any cosine terms; its Fourier coefficients can be evaluated over half a period.*

When the function is odd, a period of the function extends from $-T/2$ to $+T/2$. Since the FSE of an odd periodic function does not contain any cosine terms, $a_n = 0$ and $C_n = -jb_n/2$ is imaginary. It follows that

$$C_n = \frac{1}{T}\int_{-T/2}^{T/2} f(t)e^{-jn\omega_0 t}dt = -j\frac{b_n}{2}$$

$$= -\frac{j}{T}\left[\int_{-T/2}^{0} f(t)\sin n\omega t dt + \int_{0}^{T/2} f(t)\sin n\omega_0 t dt\right] \tag{16.50}$$

If we substitute $t = -t'$ in the first integral in brackets, this integral becomes $\int_{T/2}^{0} f(-t')\sin n\omega_0(-t')(-dt') = \int_{0}^{T/2} f(-t')\sin n\omega_0(-t')dt'$. Changing the dummy integration variable back to t and invoking the property of an odd function that $f(t) = -f(-t)$, with the sine function itself an odd function, the integral becomes $\int_{0}^{T/2} f(t)\sin n\omega_0 dt$, the same as the second integral. Hence, for the odd function,

$$C_0 = 0, \text{ and } C_n = -\frac{2j}{T}\int_{0}^{T/2} f(t)\sin n\omega_0 t dt$$

$$= \frac{2}{T}\text{Im}\left[\int_{0}^{T/2} f(t)e^{-jn\omega t}dt\right] \tag{16.51}$$

or

$$a_0 = 0 = a_n \text{ for all } n \text{ and } b_n = \frac{4}{T}\int_{0}^{T/2} f(t)\sin n\omega_0 t dt \tag{16.52}$$

Note that the term between square brackets in Equation 16.51 is the imaginary component, including j and the sign.

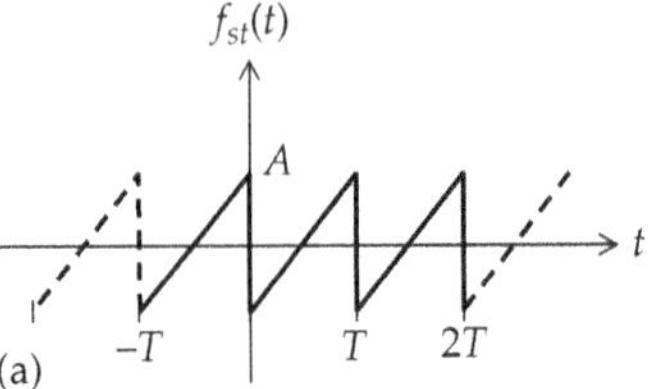

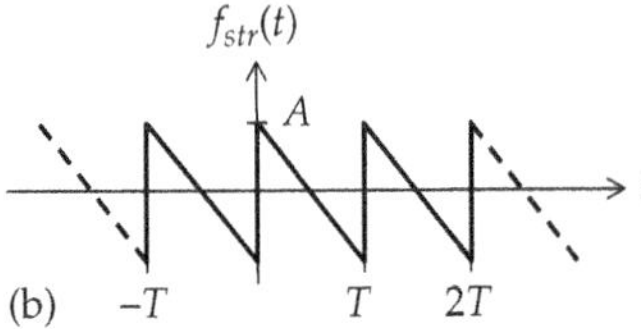

FIGURE 16.19
dc component removed from the sawtooth waveform (a), and from the reversed sawtooth waveform (b).

An example of an odd function is the square waves of Figure 16.14. The FSEs of these functions (Equations 16.44 and 16.45) consist of sine terms only.

Note that a periodic function that appears to be neither odd nor even can become odd when the dc component is removed. An example is the sawtooth waveforms of Figures 16.5 and 16.9b. If the dc component $A/2$ is subtracted, the functions become odd, as illustrated in Figure 16.19. The FSEs (Equations 16.22 and 16.35) have a dc component and sine terms only, corresponding to an odd ac component. In such cases, Equations 16.51 and 16.52 should be applied to the function after the dc component is removed.

16.3.3 Half-Wave Symmetry

A periodic function $f(t)$ possesses half-wave symmetry if

$$f(t) = -f(t+T/2) \text{ or } f(t) = -f(t-T/2) \quad 0 \le t \le T/2 \tag{16.53}$$

The two forms of the definition in Equation 16.53 are identical, since the second form is obtained by subtracting T from the argument of the first form, in accordance with the definition of a periodic function (Equation 16.1). An example of a half-wave symmetric waveform is shown in Figure 16.20, where increasing or decreasing t by $T/2$ negates the value of the function, in accordance with Equation 16.53. Geometrically, half-wave symmetry means that if either half-cycle or half-wave is displaced horizontally by $T/2$ toward the other half, so that the two half-waves are aligned vertically, the half-waves are symmetrical with respect to the horizontal axis.

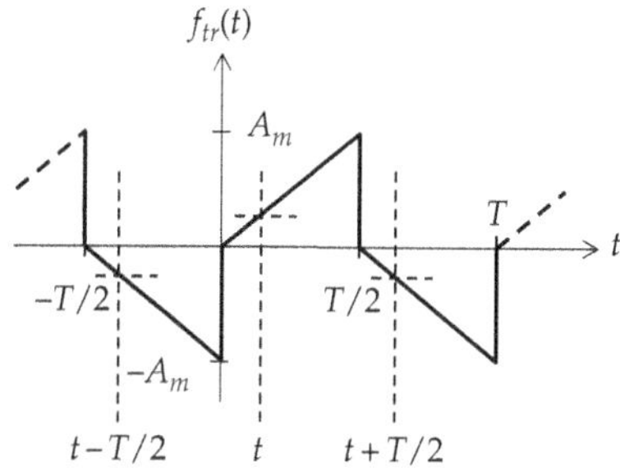

FIGURE 16.20
Half-wave symmetric periodic function.

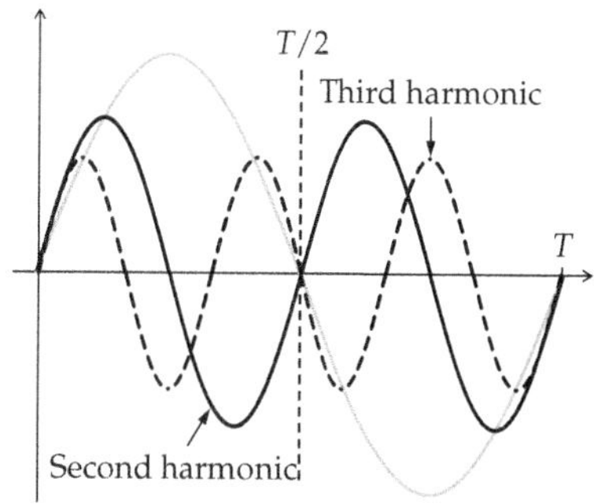

FIGURE 16.21
Half-wave symmetry of odd and even harmonics.

Concept: *The FSE of a half-wave symmetric periodic function does not contain an average term or any even harmonics; its Fourier coefficients can be evaluated over half a period. Thus,*

$$C_n = \frac{2}{T}\int_0^{T/2} f(t)e^{-jn\omega_0 t}dt, \quad \text{for } n \text{ odd}$$

$$\text{and } C_n = 0, \quad \text{for } n \text{ even or zero} \tag{16.54}$$

To prove this property, we express C_n as

$$C_n = \frac{1}{T}\int_0^{T/2} f(t)e^{-jn\omega_0 t}dt + \frac{1}{T}\int_{T/2}^{T} f(t)e^{-jn\omega_0 t}dt \tag{16.55}$$

Substituting $t' = t - T/2$, the second integral becomes $\frac{1}{T}\int_0^{T/2} f(t'+T/2)e^{-jn\omega_0(t'+T/2)}dt'$. Changing the dummy variable t' back to t, invoking the half-wave symmetry property, and substituting $n\omega_0\frac{T}{2} = n\pi$, the integral becomes $\frac{1}{T}\int_0^{T/2} -f(t)e^{-jn\omega_0 t}e^{-jn\pi}dt$. Equation 16.55 reduces to

$$C_n = \frac{1}{T}\int_0^{T/2} f(t)e^{-jn\omega_0 t}dt - \frac{1}{T}\int_0^{T/2} f(t)e^{-jn\omega_0 t}e^{-jn\pi}dt \tag{16.56}$$

It is seen that for $n = 0$ or even, $e^{-jn\pi} = \cos n\pi - j\sin n\pi = 1$, so that $C_n = 0$. But for n odd, $e^{-jn\pi} = -1$. Equation 16.54 then follow.

In terms of the coefficients a_n and b_n of the trigonometric form

$$a_0 = 0, \quad a_n = 0 = b_n \quad \text{for } n \text{ even}$$

and

$$a_n = \frac{4}{T}\int_0^{T/2} f(t)\cos n\omega_0 t\,dt,$$

$$b_n = \frac{4}{T}\int_0^{T/2} f(t)\sin n\omega_0 t\,dt \quad \text{for } n \text{ odd} \tag{16.57}$$

The reason, of course, why half-wave symmetric waveforms have odd harmonics only is that odd harmonics, such as the third harmonic in Figure 16.21, are half-wave symmetric. Thus, if the half-wave of the third harmonic between $T/2$ and T in Figure 16.21 is shifted by $T/2$ to the left, it becomes symmetrical to the half-wave of the third harmonic between 0 and $T/2$ with respect to the horizontal axis. On the other hand, even harmonics are not half-wave symmetric. Thus, if the half-wave of the second harmonic between $T/2$ and T in Figure 16.21 is shifted to the left by $T/2$, it coincides with the half-wave of the second harmonic between $T/2$ and T. In fact, if when the half-wave of a periodic waveform between $T/2$ and T is shifted to the left by $T/2$ it coincides with the half-wave between 0 and $T/2$, then the FSE of the periodic waveforms contains even harmonics only. In effect, the period of such a waveform is $T/2$ rather than T. An example of such a waveform is the full-wave rectified waveform discussed later in Example 16.6.

It should be noted that if a dc component, which is even, is added to a half-wave symmetric periodic waveform, the half-wave symmetric property is destroyed, but the ac component still does not contain any even harmonics. In this case, Equations 16.54 and 16.57 should be applied to the function after the dc component is removed.

16.3.4 Quarter-Wave Symmetry

A half-wave symmetric function that is odd or even is also symmetrical about a vertical line through the middle of its positive or negative half-cycles. Consider, for example, the function of Figure 16.22a. The function is

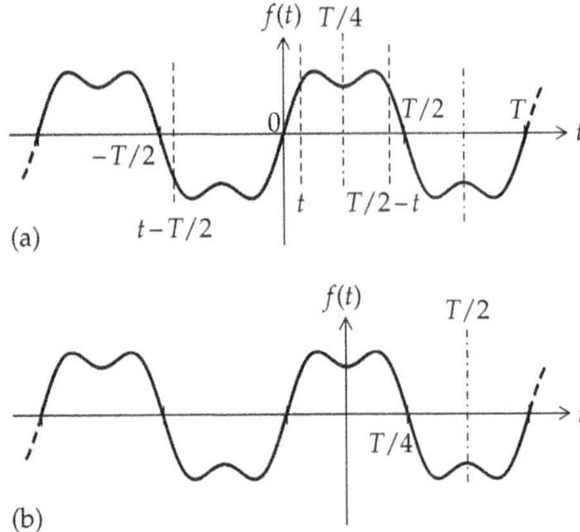

FIGURE 16.22
Quarter-wave symmetric periodic waveform.

half-wave symmetric, for if the negative half-cycle from 0 to $-T/2$ is shifted to the right by $T/2$, the shifted half-wave and the positive half-wave from 0 to $T/2$ are symmetrical with respect to the horizontal axis. If instead of being shifted to the right by $T/2$, the negative half-wave is folded around the vertical axis, so that the point at $-T/2$ coincides with the point at $T/2$, the two half-waves are again symmetrical with respect to the horizontal axis. This implies that each of the positive and negative half-cycles is symmetrical about a vertical line through its middle. Note that the half-wave symmetric function of Figure 16.20 does not have this property.

Formally, this can be proved as follows, considering an odd function to begin with: from the definition of half-wave symmetry (Equation 16.53), $f(t) = -f(t - T/2)$, $0 \leq t \leq T/2$. From the definition of an odd function, $f(t) = -f(-t)$, which means that negating the argument negates the function. Negating the argument of $-f(t - T/2)$ and the function, it becomes $f(T/2 - t)$, so that

$$f(t) = f(T/2 - t) \quad 0 \leq t \leq T/2 \tag{16.58}$$

Equation 16.58 implies that $f(t)$ is symmetrical about a vertical line through the middle of the positive half-cycle, at $t = T/4$, as illustrated in Figure 16.22a. From half-wave symmetry, $f(t)$ is also symmetrical about the vertical line through the middle of the negative half-cycle.

If the odd function of Figure 16.22a is shifted by $T/4$ to the left, it becomes the even function of Figure 16.22b. The symmetry about a vertical line through the middle of the positive or negative half-cycles is retained. A half-wave symmetric function that is also symmetrical about a vertical line through the middle of its positive or negative half-cycles is said to possess **quarter-wave symmetry**. The square waves of Figures 16.15 and 16.16 are examples of quarter-wave symmetric waveforms.

Since a quarter-wave symmetric waveform is also half-wave symmetric, the FSE of an odd, quarter-wave symmetric function consists of odd sine terms only, so that

$$a_0 = 0, \qquad a_n = 0, \qquad \text{for all } n, \qquad b_n = 0, \qquad \text{for even } n$$

From Equation 16.57,

$$b_n = \frac{4}{T}\int_0^{T/2} f(t)\sin(n\omega_0 t)dt = \frac{4}{T}\int_0^{T/4} f(t)\sin(n\omega_0 t)dt + \frac{4}{T}\int_{T/4}^{T/2} f(t)\sin(n\omega_0 t)dt, \quad n \text{ odd} \tag{16.59}$$

Substituting $t' = T/2 - t$, the second integral becomes

$$\frac{4}{T}\int_{T/4}^{T/2} f(t)\sin(n\omega_0 t)dt = \frac{4}{T}\int_{T/4}^{0} f\left(\frac{T}{2} - t'\right)\sin n\omega_0\left(\frac{T}{2} - t'\right)(-dt') \tag{16.60}$$

Because each half-cycle of $f(t)$ is symmetrical about its midline, $f(T/2 - t') = f(t')$; $\sin(n\omega_0 T/2 - n\omega_0 t') = \sin(n\pi - n\omega_0 t') = \sin(n\omega_0 t')$ when n is odd. Equation 16.59 becomes

$$b_n = \frac{4}{T}\int_0^{T/4} f(t)\sin(n\omega_0 t)dt + \frac{4}{T}\int_0^{T/4} f(t')\sin(n\omega_0 t')dt', \quad n \text{ odd} \tag{16.61}$$

Replacing the dummy integration variable t' with t, the second integral becomes identical with the first, so that

$$b_n = \frac{8}{T}\int_0^{T/4} f(t)\sin n\omega_0 dt, \quad n \text{ odd} \tag{16.62}$$

In other words, b_n need only be evaluated over a quarter period, from $t = 0$ to $t = T/4$. This is, in fact, because both $f(t)$ and $\sin n\omega_0 t$, with n odd, are symmetrical about the middle of the half-cycle from $t = 0$ to $t = T/2$.

Similarly, the FSE of an even, quarter-wave symmetric function consists of odd cosine terms only, so that

$$a_0 = 0, \quad b_n = 0 \quad \text{for all } n, \quad a_n = 0 \quad \text{for even } n$$

Moreover, a_n for odd n need be evaluated over a quarter period only:

$$a_n = \frac{8}{T}\int_0^{T/4} f(t)\cos n\omega_0 t dt \quad n \text{ odd} \tag{16.63}$$

TABLE 16.1
Summary of Symmetry Properties of Periodic Functions

Type of Symmetry			b_n	a_n	a_0
Neither odd nor even			$\frac{2}{T}\int_0^T f(t)\sin n\omega_0 t\,dt$	$\frac{2}{T}\int_0^T f(t)\cos n\omega_0 t\,dt$	$\frac{1}{T}\int_0^T f(t)dt$
Even			0	$\frac{4}{T}\int_0^{T/2} f(t)\cos n\omega_0 t\,dt$	$\frac{2}{T}\int_0^{T/2} f(t)dt$
Odd			$\frac{4}{T}\int_0^{T/2} f(t)\sin n\omega_0 t\,dt$	0	0
Half-wave symmetry	Neither odd nor even		$\frac{4}{T}\int_0^{T/2} f(t)\sin n\omega_0 t\,dt$ n odd, 0 for n even	$\frac{4}{T}\int_0^{T/2} f(t)\cos n\omega_0 t\,dt$ n odd, 0 for n even	0
	Quarter-wave symmetry	Neither odd nor even	$\frac{8}{T}\int_0^{T/4} f(t)\sin n\omega_0 t\,dt$ n odd, 0 for n even	$\frac{8}{T}\int_0^{T/4} f(t)\cos n\omega_0 t\,dt$ n odd, 0 for n even	0
		Even	0	$\frac{8}{T}\int_0^{T/4} f(t)\cos n\omega_0 t\,dt$ n odd, 0 for n even	0
		Odd	$\frac{8}{T}\int_0^{T/4} f(t)\sin n\omega_0 t\,dt$ n odd, 0 for n even	0	0

This can be proved in the same way as for b_n in the case of an odd, quarter-wave symmetric function. Again, this is because both $f(t)$ and $\cos n\omega_0 t$, with n odd, are symmetrical about the middle of the half-cycle from $t = -T/4$ to $t = T/4$.

Symmetry properties are summarized in Table 16.1.

Primal Exercise 16.10

(a) Determine the average value of the periodic function $f(t)$ in Figure 16.23 and (b) specify all the symmetry properties of the ac component of $f(t)$.

Ans. (a) 1.5; (b) ac component is even, half-wave symmetric, and quarter-wave symmetric.

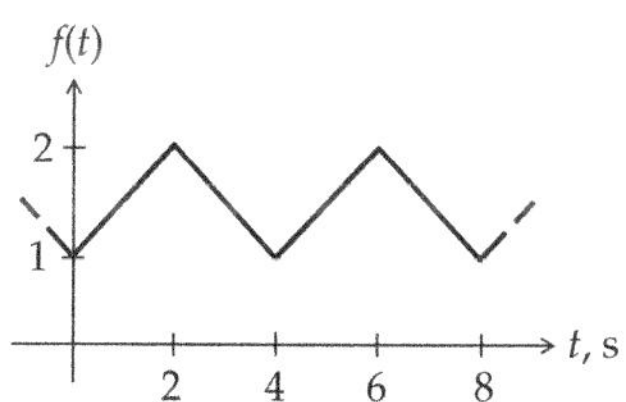

FIGURE 16.23
Figure for Primal Exercise 16.10.

Example 16.5: Fourier Analysis of Triangular Waveform

It is required to determine the FSE of the triangular waveform of Figure 16.24.

Solution:

The function has zero average, is even, and possesses half-wave symmetry. It is also quarter-wave symmetric. Its FSE must contain odd cosine terms only. Over the interval $0 \leq t \leq T/4$, $f_{tr}(t) = (4A_m/T)(t - T/4)$. It follows from Equation 16.63 that

$$a_n = \frac{8}{T}\int_0^{T/4} 4\frac{A_m}{T}(t - T/4)\cos n\omega_0 t\,dt$$

$$= \frac{32A_m}{T^2}\mathrm{Re}\left[\int_0^{T/4}\left(t - \frac{T}{4}\right)e^{-jn\omega_0 t}\,dt\right] \quad (16.64)$$

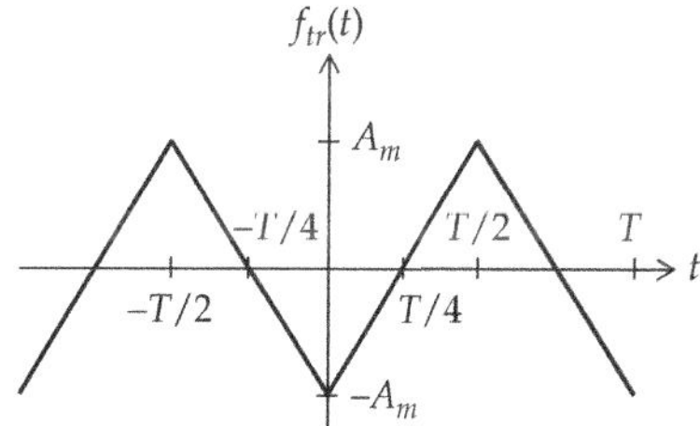

FIGURE 16.24
Figure for Example 16.5.

where using the exponential form makes the integration by parts somewhat simpler. It follows that

$$a_n = \frac{32A_m}{T^2}\text{Re}\left[\frac{te^{-jn\omega_0 t}}{-jn\omega_0} + \frac{e^{-jn\omega_0 t}}{(n\omega_0)^2} - \frac{T}{4}\frac{e^{-jn\omega_0 t}}{(-jn\omega_0)}\right]_0^{T/4}$$

$$= -\frac{32A_m}{T^2(n\omega_0)^2} = -\frac{8A_m}{\pi^2 n^2}.$$

In evaluating this expression, *even values of n should not be used*, because in applying Equation 16.64, we have already restricted n to be odd on account of half-wave symmetry. Hence,

$$a_n = -\frac{32A_m}{T^2(n\omega_0)^2} = -\frac{8A_m}{\pi^2 n^2}, \quad \text{where } n \text{ is odd} \quad (16.65)$$

The FSE of $f(t)$ is therefore

$$f_{tr}(t) = -\frac{8A_m}{\pi^2}\left(\cos\omega_0 t + \frac{1}{9}\cos 3\omega_0 t + \frac{1}{25}\cos 5\omega_0 t + \cdots\right) \quad (16.66)$$

Problem-Solving Tip

- When the Fourier coefficients a_n and b_n are evaluated using expressions that restrict the values of n, such as n being odd or even, only these restricted values should be used in deriving the individual coefficients a_n and b_n.

Exercise 16.11

Derive the FSE of the triangular waveform of Figure 16.24 when advanced by $T/4$, so that the origin is at the midpoint of the side of positive slope.

Ans. $f_{tr}(t) = \frac{8A_m}{\pi^2}\sum_n^{\infty}\left[\frac{1}{n^2}\sin\left(\frac{n\pi}{2}\right)\right]\sin n\omega_0 t$, where n is odd.

Primal Exercise 16.12

Consider the periodic function of period 8 s in Figure 16.25. (a) Determine a_0; (b) remove the dc component and specify whether or not the ac component is half-wave symmetric by translating the negative half-cycle by 4 s so as to align it vertically with the positive half-wave; (c) specify whether the ac component is odd or even or quarter-wave symmetric; (d) specify the expression for determining a_n.

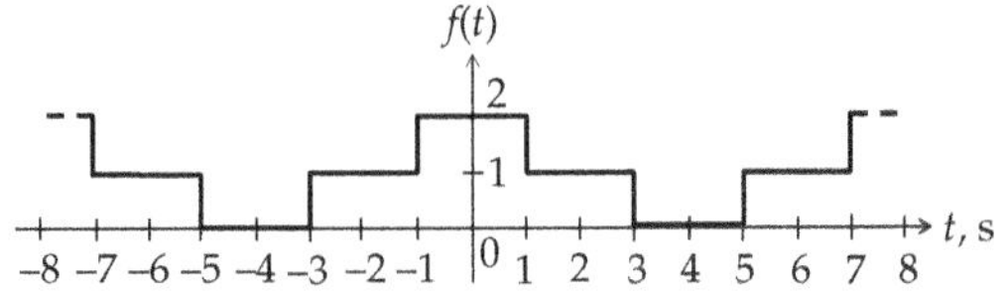

FIGURE 16.25
Figure for Primal Exercise 16.12.

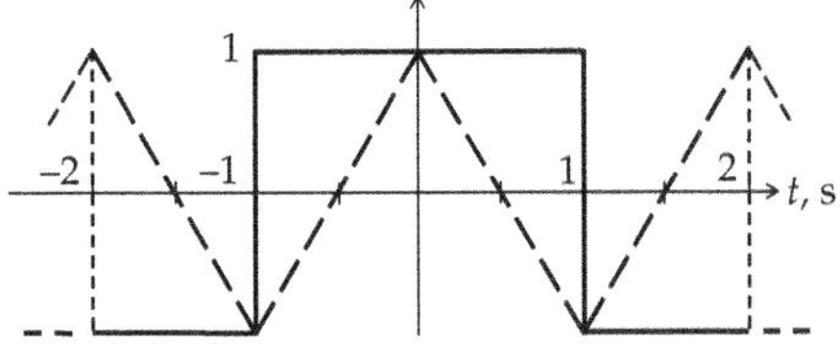

FIGURE 16.26
Figure for Primal Exercise 16.13.

Ans. (a) 1; (b) ac component is half-wave symmetric; (c) ac component is even and therefore quarter-wave symmetric; (d) $a_n = \int_0^1 \cos(n\pi t/4)dt$, n odd.

Primal Exercise 16.13

$f_1(t)$ and $f_2(t)$ are, respectively, the square and triangular waveforms shown in Figure 16.26, each having an amplitude of 1 unit and zero average value. (a) Determine the component of $(f_1(t) + f_2(t))$ having a frequency of $3\pi/2$ rad/s. (b) Are $f_1(t)$ and $f_2(t)$ quarter-wave symmetric? (c) Is the sum $(f_1(t) + f_2(t))$ half-wave symmetric?

Ans. (a) $-0.42\cos 1.5\pi t$; (b) yes; (c) no.

16.4 Derivation of FSEs from Those of Other Functions

16.4.1 Addition/Subtraction/Multiplication

Concept: *The FSEs of some functions can be derived from FSEs of other functions having the same period, through addition, subtraction, or multiplication.*

This is illustrated by the following example.

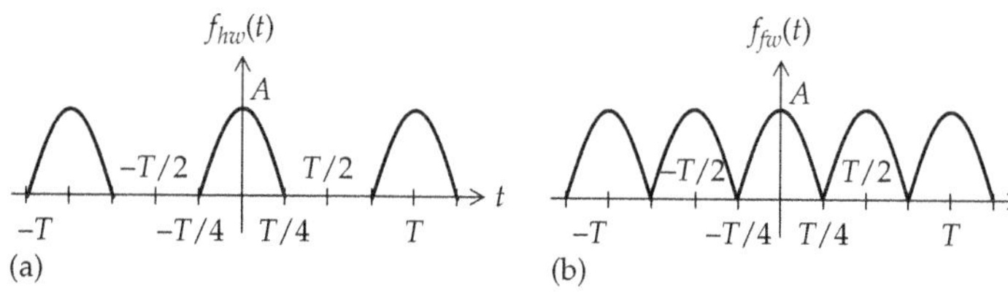

FIGURE 16.27
Figure for Example 16.6.

Example 16.6: Half-Wave and Full-Wave Rectified Waveforms

It is required to determine the FSE of (a) the half-wave rectified waveform of Figure 16.27a and (b) the full-wave rectified waveform of Figure 16.27b.

Solution:

(a) The given half-wave rectified waveform can be considered to be the product of a cosine function of amplitude A and a square pulse train of unity amplitude, both functions having the same period T (Figure 16.28). The FSE of the pulse train is that of Equation 16.40, with $A = 1$ and $\alpha = 1/2$. The FSE of the cosine function is the function itself. Hence,

$$f_{hw}(t) = A\cos\omega_0 t\left\{\frac{1}{2} + \frac{2}{\pi}\left[\cos\omega_0 t - \frac{1}{3}\cos 3\omega_0 t + \frac{1}{5}\cos 5\omega_0 t - \frac{1}{7}\cos 7\omega_0 t + \cdots\right]\right\} \quad (16.67)$$

Multiplying each term of the FSE of the pulse train by $\cos\omega_0 t$ and using the trigonometric identity $\cos\alpha\cos\beta = (1/2)[\cos(\alpha - \beta) + \cos(\alpha + \beta)]$,

$$f_{hw}(t) = \frac{A}{2}\cos\omega_0 t + \frac{A}{\pi}\left(\cos(0) + \cos 2\omega_0 t - \frac{1}{3}\cos 2\omega_0 t - \frac{1}{3}\cos 4\omega_0 t + \frac{1}{5}\cos 4\omega_0 t + \frac{1}{5}\cos 6\omega_0 t - \frac{1}{7}\cos 6\omega_0 t - \frac{1}{7}\cos 8\omega_0 t + \cdots\right)$$

$$= \frac{A}{\pi} + \frac{A}{2}\cos\omega_0 t + \frac{2A}{\pi}\left(\frac{1}{3}\cos 2\omega_0 t - \frac{1}{15}\cos 4\omega_0 t + \frac{1}{35}\cos 6\omega_0 t + \cdots + \frac{(-1)^{n+1}}{4n^2 - 1}\cos 2n\omega_0 t + \cdots\right), \quad n = 1, 2, 3, \ldots \quad (16.68)$$

The FSE contains a dc component A/π, a fundamental component $A/2$, and even harmonics as cosine terms, as to be expected for an even function.

(b) The FSE of the full-wave rectified waveform of Figure 16.27b can be derived by considering it as the sum of a half-wave rectified waveform of amplitude $2A$ and the function $-A\cos\omega_0 t$ (Figure 16.29). During the interval $-T/4 \leq t \leq T/4$, the negative half-cycle of $-A\cos\omega_0 t$ subtracts from the positive half-cycle of the half-wave rectified waveform having twice the magnitude, resulting in a positive

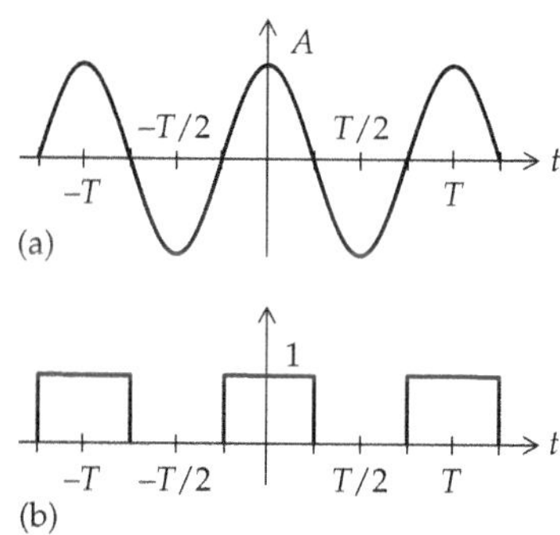

FIGURE 16.28
Figure for Example 16.6.

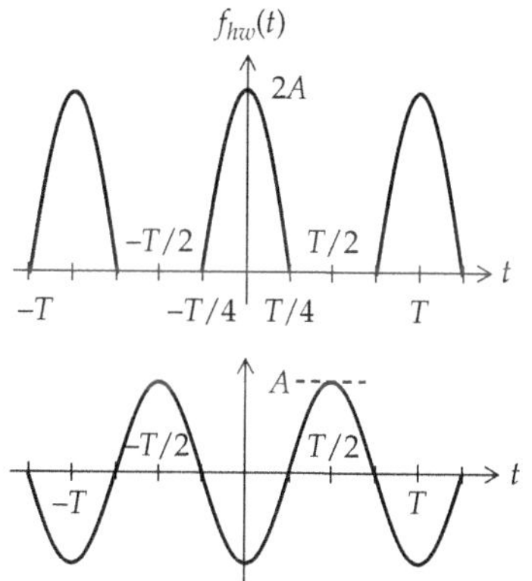

FIGURE 16.29
Figure for Example 16.6.

half-cycle of amplitude A, as in Figure 16.27. During the interval $T/4 \leq t \leq 3T/4$, the positive half-cycle of $-A\cos\omega_0 t$ simply fills the empty half-cycles of the half-wave rectified waveform, as required in the full-wave rectified waveform. The same is true for the other half-cycles in Figure 16.29.

From Equation 16.68, with A replaced by $2A$,

$$f_{hw}(t) = \frac{2A}{\pi} + A\cos\omega_0 t + \frac{4A}{\pi}\left(\frac{1}{3}\cos 2\omega_0 t - \frac{1}{15}\cos 4\omega_0 t + \frac{1}{35}\cos 6\omega_0 t + \cdots + \frac{(-1)^{n+1}}{4n^2-1}\cos 2n\omega_0 t + \cdots\right), \quad n = 1,2,3,\ldots \tag{16.69}$$

Adding $-A\cos\omega_0 t$ cancels the $A\cos\omega_0 t$ term and gives the FSE for full-wave rectified waveform:

$$f_{fw}(t) = \frac{2A}{\pi} + \frac{4A}{\pi}\left(\frac{1}{3}\cos 2\omega_0 t - \frac{1}{15}\cos 4\omega_0 t + \frac{1}{35}\cos 6\omega_0 t - \cdots + \frac{(-1)^{n+1}}{4n^2-1}\cos 2n\omega_0 t + \cdots\right), \quad n = 1,2,3,\ldots \tag{16.70}$$

where $n = 1$ refers to the first term, $n = 2$ refers to the second term, etc.

Note that ω_0 in Equation 16.70 is that of $\cos\omega_0 t$, which is also the fundamental frequency of the half-wave rectified waveform. The fundamental frequency of the full-wave rectified waveform is in fact $2\omega_0$. If this is denoted by ω_0', Equation 16.70 becomes

$$f_{fw}(t) = \frac{2A}{\pi} + \frac{4A}{\pi}\left(\frac{1}{3}\cos\omega_0' t - \frac{1}{15}\cos 2\omega_0' t + \frac{1}{35}\cos 3\omega_0' t - \cdots + \frac{(-1)^{n+1}}{4n^2-1}\cos n\omega_0' t + \cdots\right), \quad n = 1,\ 2,\ 3,\ \ldots, \tag{16.71}$$

The FSE now contains both odd and even harmonics of the fundamental frequency ω_0'. The full-wave rectified waveform can also be derived in alternative ways (Problem P16.30).

Simulation: The half-wave and full-wave rectified waveforms are generated from a sinusoidal waveform of 2 V amplitude and 50 Hz frequency. The VSIN source is entered as in Figure 16.30, with a phase angle of 90° so as to obtain a cosine function as in Figure 16.27. The half-wave rectified waveform is obtained by applying the sinusoidal voltage to a voltage limiter, available in PSpice from ABM library as the part LIMIT. The default settings of the limits are 0 and 10, which means that only voltages in the range 0 to 10 V will be allowed through the limiter. The negative half-cycles of the sinusoid are therefore removed and only the positive half-cycles will appear at the output of the limiter. This output is labeled 'hw' in Figure 16.30.

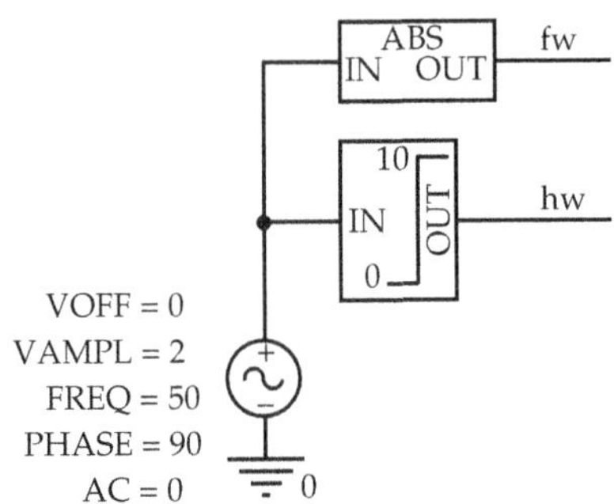

FIGURE 16.30
Figure for Example 16.6.

The full-wave rectified waveform is obtained by applying the sinusoidal voltage to the ABS part from the ABM library. The output of this module is the absolute value of the input, so that the negative half-cycles of the sinusoid are "inverted" and become positive half-cycles, as required for the full-wave rectified waveform. The output of the ABS module is labeled 'fw' in Figure 16.30.

The Fourier analysis is performed as described in Example 16.4. Time Domain (Transient) is selected for Analysis type, and 60m is entered for 'Run to time', 0 for 'Start saving data after', and 10u for 'Maximum step size'. In the new Transient File Output Options window, under 'Perform Fourier Analysis', 50 is entered for 'Center Frequency', 6 for 'Number of Harmonics', and 'V(hw), V(fw)' for 'Output variables'. After the simulation is run, select Trace/Fourier and then Trace/Add V(hw) and Trace/Add Trace V(fw), and expand the x-axis by selecting Plot/Axis settings and define an x-axis range of 0 to 350 Hz. The plot of Figure 16.31 is displayed. Considering the amplitude of the Fourier components for the half-wave rectified waveform, shown dashed, the magnitudes of the successive peaks correspond to the calculated magnitudes of the terms in Equation 16.68 of 0.6366, 0.4244, 0.08488, and 0.03638, all in volts. The magnitudes of the successive peaks of the full-wave rectified waveform correspond to the calculated magnitudes of the terms in Equation 16.70 of 1.2732, 0.8488, 0.1698, and 0.07276, all in volts. The output file gives the same magnitude values. The positive nonzero harmonics have a phase angle of nominally +90°, whereas the negative nonzero harmonics have a phase angle of nominally −90°. These ±90° phase angles give the cosine terms in the FSEs, as in the source VSIN.

Primal Exercise 16.14

Derive the FSE of the half-wave and full-wave rectified waveforms by applying Equations 16.19 to 16.21.

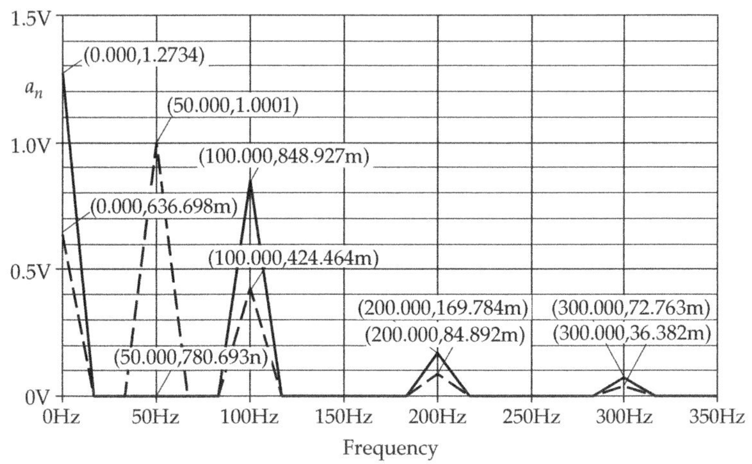

FIGURE 16.31
Figure for Example 16.6.

Primal Exercise 16.15

Consider a periodic half-sinusoid $f(t)$ of period $T > \pi/2$, defined as $f(t) = \cos t$, $-\pi/2 \leq t \leq \pi/2$, and $f(t) = 0$ for the remainder of the period. (a) Determine a_0, a_n, and b_n of the FSE of $f(t)$, in terms of n and ω_0, as applicable and (b) verify that the FSE reduces to Equations 16.68 and 16.71 in the case of a half-wave waveform and a full-wave waveform, respectively.

Ans. (a) $b_n = 0$, $a_0 = \dfrac{\omega_0}{\pi}$, $a_n = -\dfrac{2\omega_0}{\pi}\dfrac{1}{n^2\omega_0^2 - 1}\cos n\dfrac{\omega_0\pi}{2}$; (b) $\omega_0 = 1$ rad/s for the half-wave rectified waveform, and $\omega_0 = 2$ rad/s for the full-wave rectified waveform.

Primal Exercise 16.16

Derive the FSE of the half-wave rectified and full-wave rectified waveforms when delayed by a quarter period of the supply so that they start as $\sin\omega_0 t$ functions.

Ans. Half-wave rectified waveform: $f(t) = \dfrac{A}{\pi} + \dfrac{A}{2}\sin\omega_0 t - \dfrac{2A}{\pi}\sum_{n=1}^{\infty}\dfrac{\cos 2n\omega_0 t}{4n^2 - 1}$ where ω_0 is the supply frequency. Full-wave rectified waveform: $f(t) = \dfrac{2A}{\pi} - \dfrac{4A}{\pi}\sum_{n=1}^{\infty}\dfrac{\cos n\omega_0' t}{4n^2 - 1}$ where $\omega_0' = 2\omega_0$.

Primal Exercise 16.17

Determine (a) C_n for the full-wave rectified waveform in Figure 16.32a and (b) C_1 for the modified full-wave rectified waveform in Figure 16.32b.

Ans. (a) $C_n = (-1)^{n+1}\dfrac{2}{\pi}\dfrac{1}{4n^2 - 1}$, $n = 1, 2, 3\ldots$; (b) 0.085.

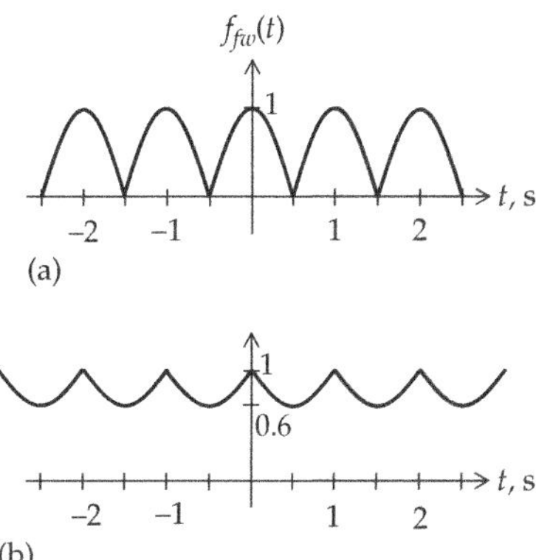

FIGURE 16.32
Figure for Primal Exercise 16.17.

16.4.2 Differentiation/Integration

Concept: *The FSE of a given periodic function can be differentiated or integrated term by term. The result is the FSE of a periodic function that is the derivative or integral of the given function, except that integrating the dc component destroys the periodicity of the function.*

This follows quite simply from differentiating or integrating both sides of the FSE. When a periodic function having a dc component is differentiated, the dc component vanishes and the resulting function is periodic with zero average. But when a periodic function having a dc component is integrated, the integral of the dc component increases linearly with time, which destroys the periodicity of the function, although the integrated ac component is still periodic.

Example 16.7: Derivation of FSE through Integration

It is required to obtain the FSE of the triangular waveform of Figure 16.24 as the integral of the delayed square waveform of Figure 16.14a.

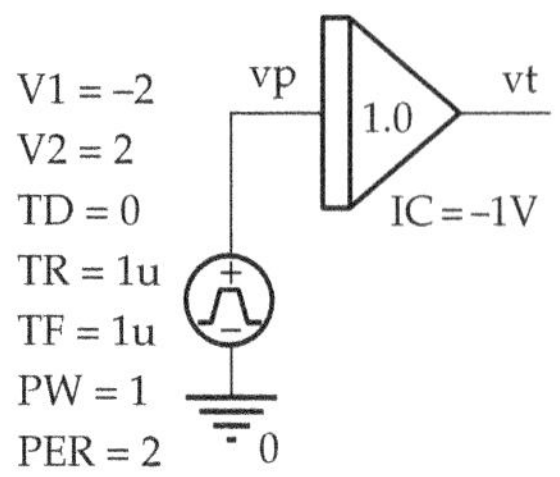

V1 = −2
V2 = 2
TD = 0
TR = 1u
TF = 1u
PW = 1
PER = 2
vp
1.0
vt
IC = −1V
0

FIGURE 16.33
Figure for Example 16.7.

Solution:

Consider the delayed square waveform of Figure 16.14a, whose FSE is given by Equation 16.44. Integrating this FSE gives

$$\int f_{sqd}(t)dt$$

$$= -\frac{4A_m}{\pi\omega_0}\left[\cos\omega_0 t + \frac{1}{9}\cos 3\omega_0 t + \frac{1}{25}\cos 5\omega_0 t + \cdots\right] + C' \quad (16.72)$$

where the constant of integration C' is the average value of the function. The RHS of Equation 16.72, with $C' = 0$, is identical to $f_{tr}(t)$ (Equation 16.66), bearing in mind that the peak-to-peak amplitude of the triangular wave equals the area under one half-cycle of the square wave. Thus,

$$2A_{mtr} = A_{msq} \times (T/2) \quad (16.73)$$

Hence, $\dfrac{4A_{msq}}{\pi\omega_0} = \dfrac{4}{\pi\omega_0} \times \dfrac{4A_{mtr}}{T} = \dfrac{8A_{mtr}}{\pi^2}$ as in Equation 16.66.

Simulation: The simulation is based on the square wave of Figure 16.14a having $A_m = 2$ V, with zero average, and $T = 2$ s. The source VPULSE is used having the parameters shown in Figure 16.33. The source voltage is applied as the input to integrating module INTEG from the ABM library. If the waveform of Figure 16.13 is integrated starting at $t = 0$, the resulting triangular waveform is always positive, that is, its 'sits' on the horizontal axis. To have a zero average, an initial value of $-A_{mtr}$ should be used in the integration, where A_{mtr} is, from Equation 16.73, $2 \times (2/2)/2 = 1$. Hence, −1 is entered as the IC for the integration, as shown in Figure 16.33. The resulting square and triangular waveforms are shown in Figure 16.34.

The Fourier analysis is performed as described in the preceding examples, the amplitude of the harmonic components being shown in Figure 16.35. The magnitudes of the successive peaks of the square waveform, shown dashed, correspond to the calculated magnitudes of the terms in Equation 16.44 of 2.5465, 0.8488, 0.5093, 0.3638, and 0.2829. The magnitudes of the successive peaks of the triangular waveform correspond to the calculated magnitudes of the terms in Equation 16.66 of 0.8106 V, 90.06 mV, 32.42 mV, 16.54 mV, and 10.10 mV. Note that the harmonics of the triangular waveform attenuate much more rapidly than those of the square waveform because they vary as $1/n^2$ rather than $1/n$.

Example 16.8: Fourier Simulation of Piece-Wise Linear Waveforms

It is required to obtain the FSE of the periodic waveform of Figure 16.36 analytically and by simulation, to demonstrate the simulation of triangular, trapezoidal, and other piecewise-linear, periodic waveforms.

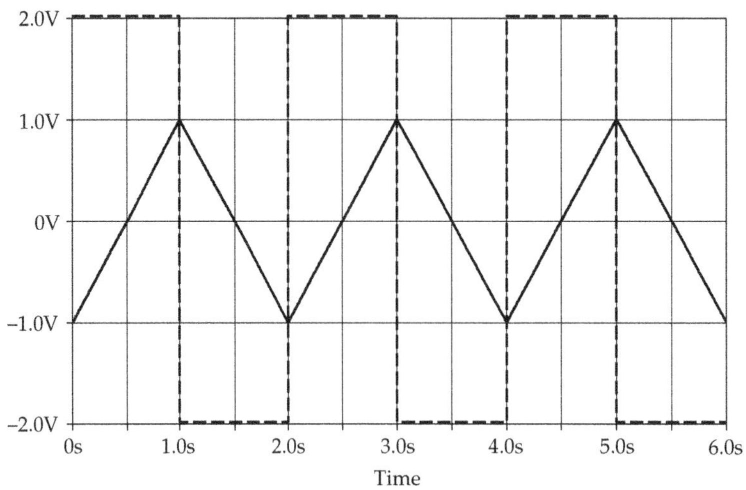

FIGURE 16.34
Figure for Example 16.7.

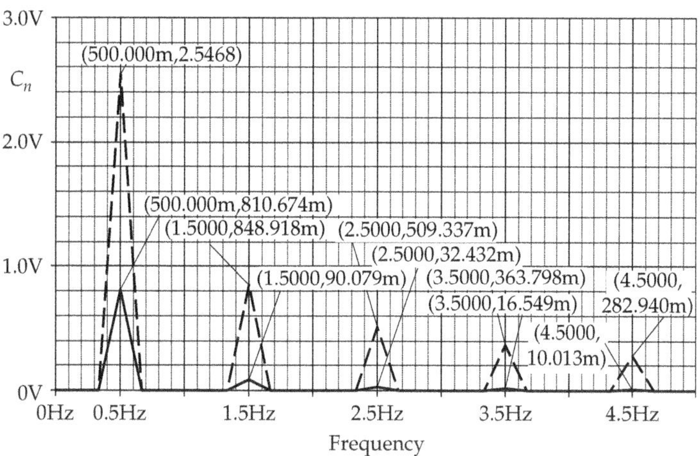

FIGURE 16.35
Figure for Example 16.7.

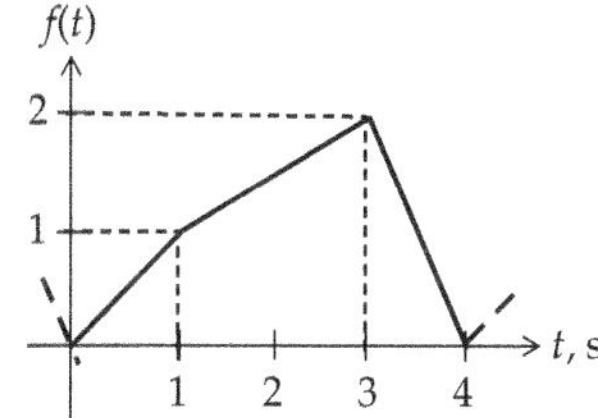

FIGURE 16.36
Figure for Example 16.8.

Solution:

The equations of the three line segments are $f(t) = t$, $0 \leq t \leq 1$; $f(t) = 0.5(t + 1)$, $1 \leq t \leq 3$; and $f(t) = 2(-t + 4)$, $3 \leq t \leq 4$. It follows that

$$C_0 = \frac{1}{4}\left[\int_0^1 t\,dt + \int_1^3 0.5(t+1)dt + \int_3^4 2(-t+4)dt\right]$$

$$= \frac{1}{4}\left[\frac{t^2}{2}\right]_0^1 + \frac{1}{8}\left[\frac{t^2}{2} + t\right]_1^3 + \frac{1}{2}\left[-\frac{t^2}{2} + 4t\right]_3^4 = 1.125;$$

$$C_n = \frac{1}{4}\left[\int_0^1 te^{-jn\omega_0 t}dt + \int_1^3 0.5(t+1)^{-jn\omega_0 t}dt + \int_3^4 2(-t+4)^{-jn\omega_0 t}dt\right]$$

$$= \frac{1}{4}\left[e^{-jn\omega_0 t}\left(\frac{jt}{n\omega_0} + \frac{1}{n^2\omega_0^2}\right)\right]_0^1 + \frac{1}{8}\left[e^{-jn\omega_0 t}\left(\frac{jt}{n\omega_0} + \frac{1}{n^2\omega_0^2} + \frac{j}{n\omega_0}\right)\right]_1^3$$

$$+ \frac{1}{2}\left[e^{-jn\omega_0 t}\left(-\frac{jt}{n\omega_0} - \frac{1}{n^2\omega_0^2} + \frac{j4}{n\omega_0}\right)\right]_3^4$$

$$= \frac{1}{8n^2\omega_0^2}\left(-2 + e^{-jn\omega_0} + 5e^{-j3n\omega_0} - 4e^{-j4n\omega_0}\right).$$

As a check, all terms in $n\omega_0$ should cancel out, because the harmonics should decrease as $1/n^2$, since the function is continuous but its first derivative is not, as discussed in Section 16.5.

Substituting $\omega_0 = 2\pi/4 = \pi/2$,

$$C_n = \frac{1}{8n^2\omega_0^2}\left(-2 + e^{-jn\pi/2} + 5e^{-j3n\pi/2} - 4e^{-j2n\pi}\right)$$

$$= \frac{1}{2n^2\pi^2}\left(-6 + e^{-jn\pi/2} + 5e^{-j3n\pi/2}\right)$$

$$= \frac{1}{2n^2\pi^2}$$

$$\times\left[-6 + \cos n\pi/2 + 5\cos 3n\pi/2 - j\left(\sin n\pi/2 + 5\sin 3n\pi/2\right)\right].$$

It follows that $a_n = \frac{1}{n^2\pi^2}\left(-6 + \cos n\pi/2 + 5\cos 3n\pi/2\right) = -6/n^2\pi^2$ for n odd.

$b_n = \frac{1}{n^2\pi^2}\left(\sin n\pi/2 + 5\sin 3n\pi/2\right) = 0$ for n even. Substituting values of n, the following Fourier coefficients are obtained:

For $n = 1$, $a_1 = -6/\pi^2$, $b_1 = -4/\pi^2$, $c_1 = 2\sqrt{13}/\pi^2 = 0.7306$, $\theta_1 = \tan^{-1}(-2/-3) = -146.3°$.

For $n = 2$, $a_2 = -3/\pi^2$, $b_2 = 0$, $c_2 = 3/\pi^2 = 0.3040$, $\theta_2 = 0$.

For $n = 3$, $a_3 = -2/3\pi^2$, $b_3 = 4/9\pi^2$, $c_3 = 2\sqrt{13}/9\pi^2 = 0.0812$, $\theta_3 = \tan^{-1}(2/-3) = 146.3°$.

For $n = 4$, $a_4 = 0$, $b_4 = 0$, $c_4 = 0$, θ_4 is undefined.

For $n = 5$, $a_5 = -6/25\pi^2$, $b_5 = -4/25\pi^2$, $c_5 = 2\sqrt{13}/25\pi^2 = 0.0292$, $\theta_5 = \tan^{-1}(-2/-3) = -146.3°$.

For $n = 6$, $a_6 = -1/3\pi^2$, $b_6 = 0$, $c_6 = 1/3\pi^2 = 0.0338$, $\theta_6 = 0$.

For $n = 7$, $a_7 = -6/49\pi^2$, $b_7 = 4/49\pi^2$, $c_7 = 2\sqrt{13}/49\pi^2 = 0.0149$, $\theta_7 = \tan^{-1}(2/-3) = 146.3°$. Note that the sixth harmonic is of larger amplitude than the fifth harmonic.

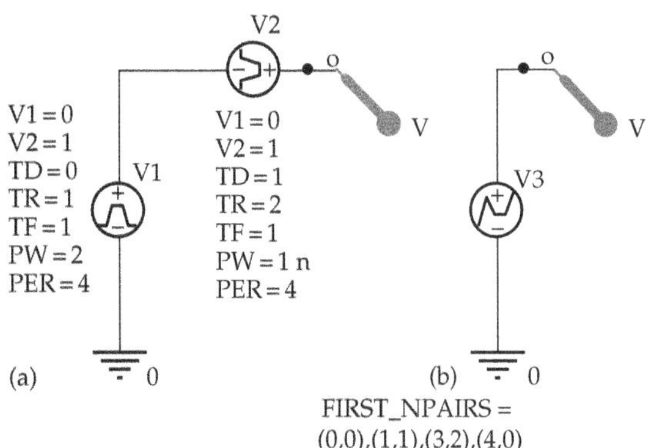

FIGURE 16.37
Figure for Example 16.8.

Simulation: The periodic waveform is generated in two ways. The first, illustrated in Figure 16.37a, utilizes two VPULSE sources having the parameters indicated. V1 generates a periodic, trapezoidal waveform, where the trapezoid in each period has a short base of 2 s, a long base of 4 s, a time-to-rise of 1 s, and a time-to-fall of 1 s. V2 generates a periodic, triangular waveform that is delayed by 1 s with respect to the trapezoid and has a base of 3 s, a height of 1 s, a time-to-rise of 2 s, and a time-to-fall of 1 s. Adding these two waveforms results in the periodic waveform of Figure 16.36.

The second method, illustrated in Figure 16.37b, uses the periodic, piecewise-linear source VPWL_RE_FOREVER from the source library. The source parameters are entered in the FIRST_NPAIRS field as coordinate pairs for each of the four corners, as indicated in Figure 16.37b.

The simulation is performed as described in the previous examples. The plot of the amplitudes of the Fourier components is shown in Figure 16.38 and is in agreement with the calculated values.

16.5 Concluding Remarks on FSEs

16.5.1 Rate of Attenuation of Harmonics

FSEs are theoretically infinite series. In practice, a periodic waveform is truncated, that is, approximated by a finite number of terms of the FSE, to any desired degree of accuracy. The more rapidly the magnitudes of the harmonics decrease with the order of the harmonic, the fewer are the number of terms of the FSE that have to be included to obtain a given degree of accuracy. The rate of attenuation of harmonics is related to the degree of "smoothness" of the function:

Concept: *The smoother the function, the more rapidly the harmonics decrease in magnitude.*

To apply this concept, it is necessary to have some measure of the smoothness of a function. One such measure is obtained by successively differentiating the function to determine the order of the derivative that first becomes discontinuous, that is, shows a jump in value during every period. In the case of a square wave, for example, the function itself is discontinuous, twice in every period. The derivative of order 0, that is, the function itself is said to be discontinuous. The sawtooth waveform is also discontinuous, once every period. A triangular waveform such as that of Figure 16.24 is continuous at the corners of the triangle, but the slope suddenly changes at these corners. In other words, the first derivative of the triangular waveform is a square wave and is discontinuous. Similarly, the half-wave and full-wave rectified waveforms are continuous but their first derivatives are not. In such cases, the first derivative or derivative of order 1 is discontinuous.

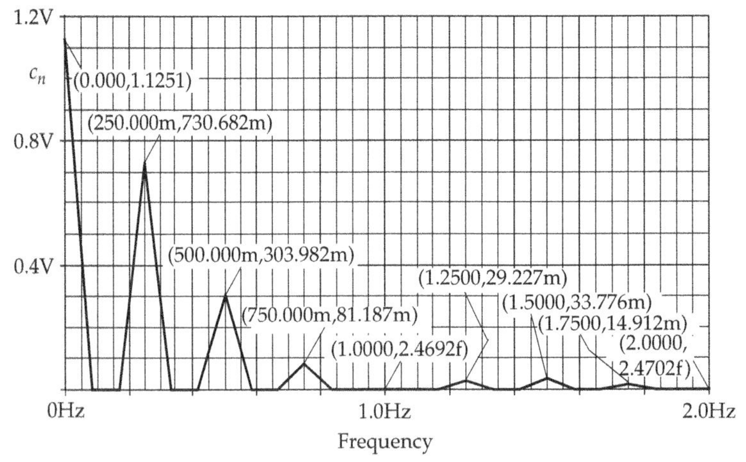

FIGURE 16.38
Figure for Example 16.8.

It can be shown that if the mth derivative of a periodic function is discontinuous, with all the derivatives of lower order being continuous, the magnitudes of the harmonics decrease approximately as $1/n^{m+1}$, $m=0, 1, 2, \ldots$, the derivative of order 0 being the function itself. Thus, in the case of the square wave, the function is discontinuous, so that $m = 0$, and the harmonics decrease as $1/n$. In the case of a triangular waveform, the first derivative is discontinuous, $m = 1$, and the nth harmonic is proportional to $1/n^2$. Whereas the aforementioned statement holds exactly for the square and triangular waves, it holds approximately for other functions. For example, the relative magnitudes of the harmonics of the pulse train decrease as $\sin(an\pi)/n$ (Equation 16.40), so the $1/n$ is multiplied by a sine function. In the case of the half-wave and full-wave rectified waveforms (Equations 16.68 and 16.70), the magnitude decreases as $1/(4n^2 - 1)$. This property is useful for checking the FSE of a given periodic function.

16.5.2 Application to Nonperiodic Functions

A nonperiodic function that satisfies Dirichlet's conditions over a given interval can be represented over this interval by a Fourier series. It is convenient to choose this interval to be half a period of a periodic function and to assume that the periodic function is odd or even. If the periodic function is considered odd, its FSE will consist of sine terms only, whereas if the periodic function is considered even, its FSE will consist of cosine terms only. This is illustrated by Problem P16.19.

16.5.3 Shifting Horizontal and Vertical Axes

Shifting the time axis up or down affects the dc component of the FSE without affecting the ac component. Shifting the vertical axis to the right or left (1) affects odd–even symmetry, without affecting half-wave or quarter-wave symmetry, that is, sine or cosine terms may be introduced in the FSE or removed from it, but if the periodic function is half-wave symmetric, only odd harmonics will be present, and (2) modifies the phase angles of the ac components, without affecting their magnitudes and without affecting the dc component.

16.6 Circuit Responses to Periodic Functions

Concept: *The steady-state response of an LTI circuit to a periodic signal is the sum of the responses to each component acting alone.*

This follows readily from the principle of superposition (Section 5.1), which applies to all LTI systems. Consider a general periodic function $v_I(t)$ of the form of Equation 16.4:

$$v_I(t) = V_{I0} + V_{I1}\cos(\omega_0 t + \theta_{I1}) + V_{I2}\cos(2\omega_0 t + \theta_{I2}) + \cdots V_{In}\cos(n\omega_0 t + \theta_{In}) + \cdots \quad (16.74)$$

If $v_I(t)$ is applied to an LTI circuit (Figure 16.39), $v_I(t)$ can be represented as the series connection of ideal voltage sources, each source representing a term in the FSE. From superposition, the output is the sum of components, each component being due to one source acting alone, with all the other sources set to zero, that is, replaced by short circuits.

Since the ac components of the input are sinusoids, the sinusoidal steady-state output due to each of these components can be determined by phasor analysis. Suppose v_I of Equation 16.74 is applied to the circuit of Figure 16.40a. Figure 16.40b shows the circuit in the frequency domain for the nth harmonic, where $\mathbf{V_{In}}$ is the input phasor $V_{In}\angle\theta_{In}$, $\mathbf{V_{On}}$ is the output phasor $V_{On}\angle\theta_{On}$, and the impedance of the capacitor at the frequency $n\omega_0$ is $1/jn\omega_0 C_2$. From voltage division,

$$\frac{\mathbf{V_{On}}}{\mathbf{V_{In}}} = \frac{R_2 \| (1/jn\omega_0 C_2)}{R_1 + R_2 \| (1/jn\omega_0 C_2)} = \frac{R_2}{R_1 + R_2} \frac{1}{1 + jn\omega_0 C_2 (R_1 \| R_2)} \quad (16.75)$$

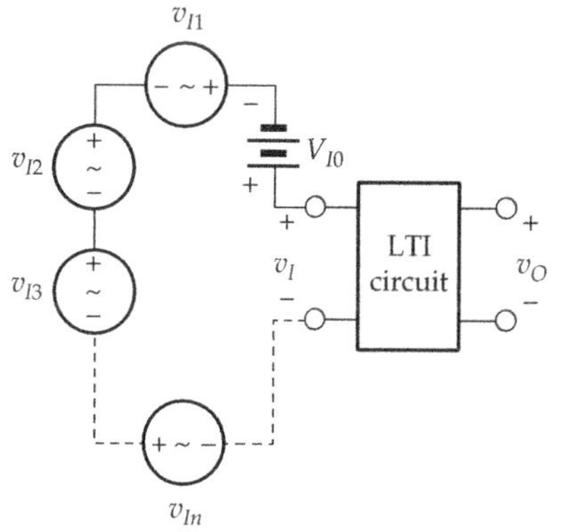

FIGURE 16.39
Periodic function applied to an LTI circuit.

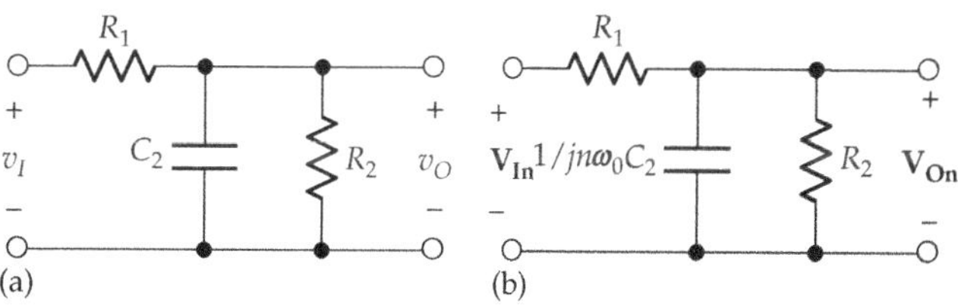

FIGURE 16.40
(a) Periodic waveform applied to an RC circuit and (b) the circuit in the frequency domain.

The magnitude and phase angle of the nth harmonic in the output are

$$V_{On} = \frac{R_2}{R_1 + R_2}\frac{V_{In}}{\sqrt{1+\left[n\omega_0 C_2\left(R_1 \parallel R_2\right)\right]^2}}, \quad (16.76)$$

$$\theta_{On} = \theta_{In} - \tan^{-1} n\omega_0 C_2\left(R_1 \parallel R_2\right)$$

For $\omega_0 = 0$, the circuit is a resistive voltage divider, and the dc component is

$$V_{O0} = \frac{R_2}{R_1 + R_2}V_{I0} \quad (16.77)$$

By superposition, the output is the sum of the dc and ac components given by Equations 16.76 and 16.77 for $n = 1, 2, 3, \ldots$. In the time domain,

$$v_O(t) = V_{O0} + V_{O1}\cos(\omega_0 t + \theta_{O1}) + V_{O2}\cos(2\omega_0 t + \theta_{O2}) + \cdots V_{On}\cos(n\omega_0 t + \theta_{On}) + \cdots \quad (16.78)$$

Example 16.9: Response of *RC* Circuit to a Periodic Input

It is required to determine the output voltage $v_O(t)$ in Figure 16.41 when the input voltage is the square waveform of Figure 16.14a, with $A_m = 5$ V, $T = 2$ ms, $R = 1$ kΩ, and $C = 1$ μF.

Solution:

The FSE of the input voltage is given by Equation 16.44, with $A_m = 5$ V. According to Equation 16.76, with $R_2 \to \infty$, $R_1 = R$, and $C_2 = C$, the nth harmonic of the output has a magnitude that is $1/\sqrt{1+(n\omega_0 CR)^2} = 1/\sqrt{1+n^2\pi^2}$ times that of the corresponding input component and lags this component by $\tan^{-1} n\pi$, where $CR = 1$ ms and $\omega_0 = 2\pi \times 0.5 = \pi$ krad/s. The FSE of the output can therefore be expressed as

$$v_O(t) = \frac{20}{\pi}\sum_{n=1,3,5,\ldots}^{\infty}\frac{\sin(n\omega_0 t - \beta_n)}{n\sqrt{1+(n\pi)^2}}, \quad \beta = \tan^{-1} n\pi \quad (16.79)$$

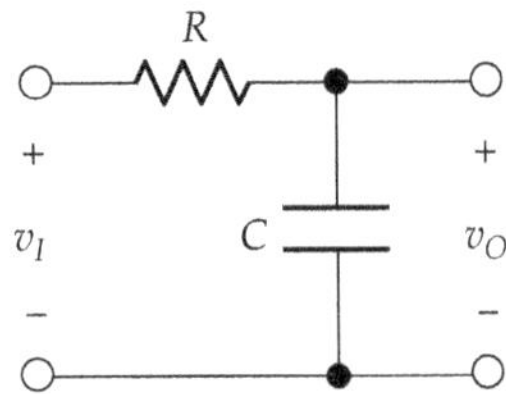

FIGURE 16.41
Figure for Example 16.9.

In general, the output waveform can be approximated to any desired degree of accuracy by adding a sufficient number of terms from the FSE of Equation 16.79. However, it is possible in this case to obtain an "exact" expression for the output waveform in the steady state as the repetitive transient charging and discharging of the capacitor through a resistor. The procedure is as follows.

In the steady state, let the minimum $v_O(t)$ be V_{Omin} at an arbitrary time $t = 0$ and at $t = T$, and let the maximum $v_O(t)$ be V_{Omax} at $t = T/2$ (Figure 16.42). During the interval $0 \le t \le T/2$, the capacitor charges from an initial value V_{Omin} toward a final value of +5 V. From Equation 11.57,

$$v_O(t) = 5 + (V_{Omin} - 5)e^{-t/RC}, \quad 0 \le t \le T/2 \quad (16.80)$$

At $t = T/2 = 1$ ms and with $RC = 1$ ms, $v_O = V_{Omax}$, and Equation 16.80 gives

$$V_{Omax} = 5 + (V_{Omin} - 5)e^{-1}, \quad 0 \le t \le T/2 \quad (16.81)$$

During the interval $T/2 \le t \le T$, the capacitor discharges from an initial value V_{Omax} toward a final value of −5 V. From Equation 11.57,

$$v_O(t) = -5 + (V_{Omax} + 5)e^{-(t-T/2)/RC}, \quad T/2 \le t \le T \quad (16.82)$$

At $t = T = 2$ ms and with $RC = 1$ ms, $v_O = V_{Omin}$, and Equation 16.82 gives

$$V_{Omin} = -5 + (V_{Omax} + 5)e^{-1}, \quad 0 \le t \le T/2 \quad (16.83)$$

Solving Equations 16.81 and 16.83 gives $V_{Omax} = -V_{Omin} = 5(e-1)/(e+1) = 2.3106$ V.

Note that if $(t - T/2)$ in Equation 16.82 is replaced by t, the half-wave between $T/2$ and T is shifted by $T/2$ to the left. With $V_{Omax} = V_{Omin}$, the RHS of Equation 16.82 is the negation of the RHS of Equation 16.80, which implies that $v_O(t)$ is half-wave symmetric.

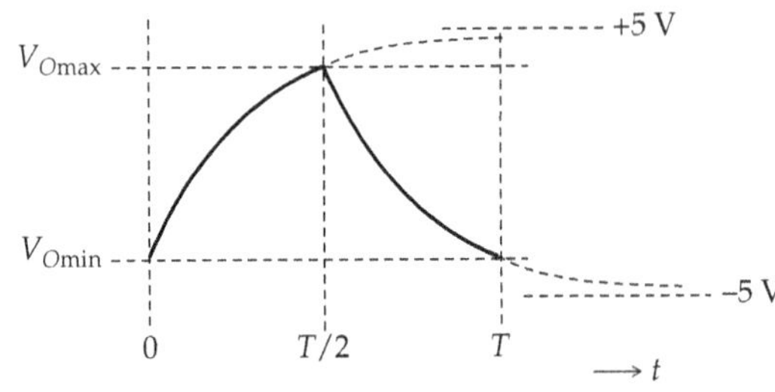

FIGURE 16.42
Figure for Example 16.9.

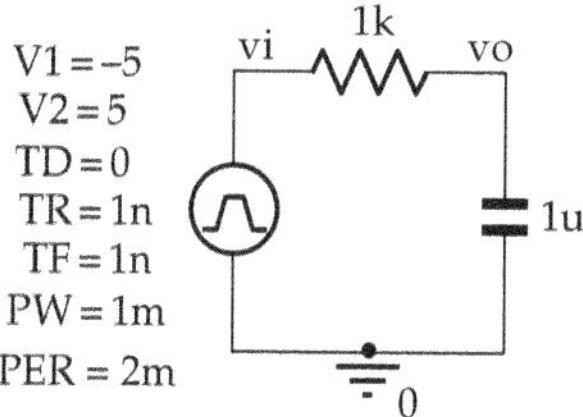

FIGURE 16.43
Figure for Example 16.9.

Simulation: The simulation is based on the square wave of Figure 16.14a having $A_m = 5$ V, with zero average, and $T = 2$ ms. The source VPULSE is used having the parameters shown in Figure 16.43. The input and output waveforms shown in Figure 16.44 are obtained by entering in the simulation profile for Time Domain (Transient) analysis 20 m for 'Run to time', 1u for 'Maximum step size', and 10 m for 'Start saving data after' in order to allow time for the initial transient to die down and the circuit to approach a steady state. v_{Omax} and v_{Omin} are read from the simulation as 2.3115 and 2.3053 V, respectively, in agreement with the calculated values.

The Fourier analysis is performed as described in the preceding examples, the amplitudes of the Fourier components being shown in Figure 16.45. The input voltage is the same waveform as in Example 16.4 and has the same amplitudes (Figure 16.15). The amplitudes of the successive peaks of the output waveform are those given by Equation 16.79, the calculated magnitudes for the successive peaks being 1.9310 V, 0.2239 V, 80.893 mV, 41.313 mV, and 25.002 mV.

The output file gives the following table for the output voltage:

Fourier Components of Transient Response V(VO)
dc Component = 4.999902E−03

Harmonic No.	Frequency, Hz	Fourier Component	Normalized Component	Phase, deg	Normalized Phase, deg
1	5.000E+02	1.931E+00	1.000E+00	−7.252E+01	0.000E+00
2	1.000E+03	1.572E−03	8.140E−04	8.687E+00	1.537E+02
3	1.500E+03	2.239E−01	1.160E−01	−8.447E+01	1.331E+02
4	2.000E+03	7.933E−04	4.108E−04	3.836E+00	2.939E+02
5	2.500E+03	8.089E−02	4.189E−02	−8.724E+01	2.754E+02
6	3.000E+03	5.298E−04	2.744E−04	1.967E+00	4.371E+02
7	3.500E+03	4.131E−02	2.139E−02	−8.863E+01	4.190E+02
8	4.000E+03	3.976E−04	2.059E−04	8.508E−01	5.810E+02
9	4.500E+03	2.500E−02	1.295E−02	−8.955E+01	5.631E+02
10	5.000E+03	3.181E−04	1.648E−04	3.580E−02	7.252E+02

Total Harmonic Distortion = 1.258014E+01 percent

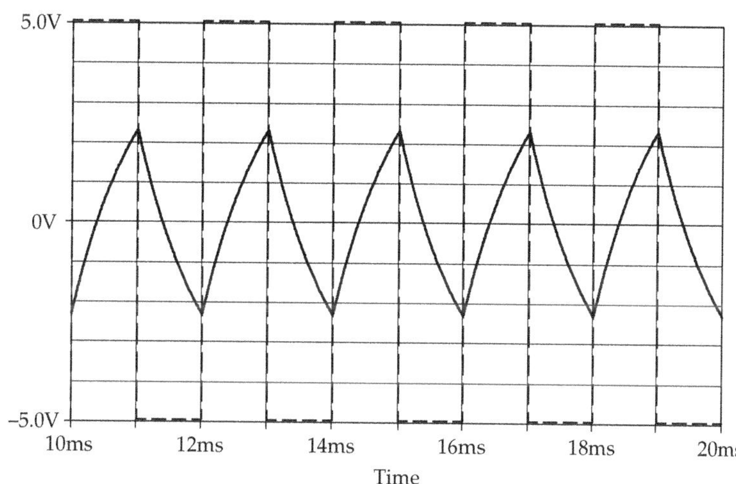

FIGURE 16.44
Figure for Example 16.9.

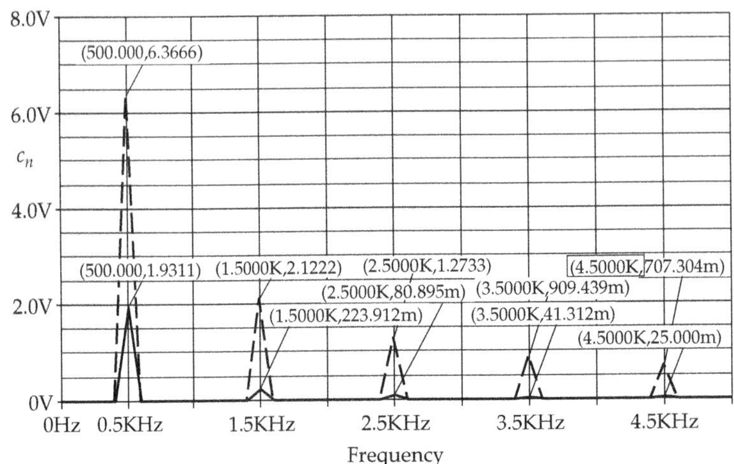

FIGURE 16.45
Figure for Example 16.9.

From Equation 16.79, the calculated values of the phase angle $-\beta_n$ for the odd ac components are −72.34°, −83.94°, −86.36°, −87.40°, and −87.97°, respectively, in near agreement with the values in the table. Note that the phase angles in the fifth column of the table of Fourier components in PSpice are for a sine function and not a cosine function.

Exercise 16.18

Show that C_n of $v_O(t)$ of Example 16.9 evaluated over the interval $0 \le t \le T/2$ is $\frac{2V_m}{n\pi}\left(\frac{-j}{1+jn\omega_0 CR}\right)$ for n odd and zero for n even, in agreement with Equation 16.79.

16.7 Average Power and rms Values

Concept: *In an LTI circuit, components of different frequencies do not interact, and the total average power is the sum of the average powers due to each component acting alone.*

To justify this, consider a periodic input voltage $v_I(t)$ of the form

$$v_I(t) = V_0 + \sum_{n=1}^{\infty} V_n \cos(n\omega_0 t + \theta_{vn}) \quad (16.84)$$

Let this voltage be applied to two terminals of an LTI circuit. The input current $i_I(t)$ at these terminals in the steady state is also periodic, of the same frequency, and can be expressed as

$$i_I(t) = I_0 + \sum_{n=1}^{\infty} I_n \cos(n\omega_0 t + \theta_{in}) \quad (16.85)$$

Assuming $i_I(t)$ to be in the direction of a voltage drop $v_I(t)$, the instantaneous power input to the circuit is

$$p_I = v_I i_I \quad (16.86)$$

The average power input to the circuit is

$$P = \frac{1}{T}\int_0^T v_I\, i_I\, dt \quad (16.87)$$

To illustrate the basic concepts involved, assume that the FSEs of $v_I(t)$ and $i_I(t)$ consist of a dc component and two ac components. Thus,

$$v_I(t) = V_0 + V_1\cos(\omega_0 t + \theta_{v1}) + V_2\cos(2\omega_0 t + \theta_{v2}) \quad (16.88)$$

and

$$i_I(t) = I_0 + I_1\cos(\omega_0 t + \theta_{i1}) + I_2\cos(2\omega_0 t + \theta_{i2}) \quad (16.89)$$

The instantaneous power is

$$\begin{aligned} p_I(t) &= v_I(t)i_I(t) \\ &= V_0I_0 + V_1I_1\cos(\omega_0 t + \theta_{v1})\cos(\omega_0 t + \theta_{i1}) \\ &\quad + V_2I_2\cos(2\omega_0 t + \theta_{v2})(2\omega_0 t + \theta_{i2}) \\ &\quad + V_0I_1\cos(\omega_0 t + \theta_{i1}) + V_0I_2\cos(2\omega_0 t + \theta_{i2}) \\ &\quad + V_1I_0\cos(\omega_0 t + \theta_{v1}) \\ &\quad + V_1I_2\cos(\omega_0 t + \theta_{v1})\cos(2\omega_0 t + \theta_{i2}) + V_2I_0\cos(2\omega_0 t + \theta_{v2}) \\ &\quad + V_2I_1\cos(2\omega_0 t + \theta_{v2})\cos(\omega_0 t + \theta_{i1}) \end{aligned} \quad (16.90)$$

Each of the first three terms in Equation 16.91 is the product of a voltage and a current of the same frequency: dc, fundamental, and second harmonic. All the remaining terms are products of terms of different frequencies. The four terms that contain the product of a dc term and an ac component, namely, $V_0I_1\cos(\omega_0t + \theta_{i1})$, $V_0I_2\cos(2\omega_0t + \theta_{i2})$, $V_1I_0\cos(\omega_0t + \theta_{v1})$, and $V_2I_0\cos(2\omega_0t + \theta_{v2})$, integrate to zero over a period. For example,

$$\int_0^T V_0 I_1\cos(\omega_0t+\theta_{i1})dt = \frac{V_0 I_1}{\omega_0}\left[\sin(\omega_0t+\theta_{i1})\right]_0^{2\pi/\omega_0}$$

$$= \frac{V_0I_1}{\omega_0}\left[\sin(2\pi+\theta_{i1})-\sin(\theta_{i1})\right]=0 \tag{16.91}$$

The two product terms of an ac component of one frequency and an ac component of another frequency, namely, $V_1I_2\cos(\omega_0t+\theta_{v1})\cos(2\omega_0t+\theta_{i2})$ and $V_2I_1\cos(2\omega_0t+\theta_{v2})\cos(\omega_0t+\theta_{i1})$, also integrate to zero over a period. Thus, using the trigonometric identity $\cos\alpha\cos\beta = (1/2)[\cos(\alpha - \beta) + \cos(\alpha + \beta)]$, the term $V_1I_2\cos(\omega_0t + \theta_{v1})\cos(2\omega_0t + \theta_{i2}) = (1/2)[\cos(\omega_0t + \theta_{i2} - \theta_{v1}) + \cos(3\omega_0t + \theta_{i2} + \theta_{v1})]$. Each of these terms integrates to zero over a period, as in Equation 16.91.

Only the three terms in Equation 16.90, each of which is the product of a voltage and a current of the same frequency, do not integrate to zero over a period. Let us evaluate the average power associated with each of these frequencies. The average power associated with the dc terms is

$$P_0 = \frac{1}{T}\int_0^T V_0I_0\,dt = V_0I_0 \tag{16.92}$$

Using the aforementioned trigonometric identity, the average power associated with the fundamental frequency is

$$P_1 = \frac{1}{T}\int_0^T V_1I_1\cos(\omega_0t+\theta_{v1})\cos(\omega_0t+\theta_{i1})dt$$

$$= \frac{V_1 I_1}{2T}\int_0^T\left[\cos(\theta_{v1}-\theta_{i1})+\cos(2\omega_0t+\theta_{v1}+\theta_{i1})\right]dt$$

$$= \frac{V_1 I_1}{2}\cos(\theta_{v1}-\theta_{i1}) \tag{16.93}$$

where the second term in the integrand evaluates to zero as in Equation 16.91.

Similarly, the average power associated with the second harmonic is

$$P_2 = \frac{1}{T}\int_0^T V_2I_2\cos(2\omega_0t+\theta_{v2})\cos(2\omega_0t+\theta_{i2})dt$$

$$= \frac{V_2 I_2}{2T}\int_0^T\left[\cos(\theta_{v2}-\theta_{i2})+\cos(4\omega_0t+\theta_{v2}+\theta_{i2})\right]dt$$

$$= \frac{V_2 I_2}{2}\cos(\theta_{v2}-\theta_{i2}) \tag{16.94}$$

It follows that the total average power is

$$P = V_0I_0 + \frac{V_1I_1}{2}\cos(\theta_{v1}-\theta_{i1})+\frac{V_2I_2}{2}\cos(\theta_{v2}-\theta_{i2})$$

$$= V_0I_0 + V_{1\text{rms}}I_{1\text{rms}}\cos(\theta_{v1}-\theta_{i1})+V_{2\text{rms}}I_{2\text{rms}}\cos(\theta_{v2}-\theta_{i2}) \tag{16.95}$$

where a sinusoidal voltage or current of amplitude Y_n has an rms value of $Y_n/\sqrt{2}$ (Section 8.3).

This result can be readily generalized to the case of $v_l(t)$ and $i_l(t)$ given by Equations 16.84 and 16.85, respectively. The same procedure leads to the conclusion that

$$P = V_0I_0 + \sum_{n=1}^{\infty}\frac{V_nI_n}{2}\cos(\theta_{vn}-\theta_{in})$$

$$= V_0I_0 + \sum_{n=1}^{\infty}V_{n\text{rms}}I_{n\text{rms}}\cos(\theta_{vn}-\theta_{in}) \tag{16.96}$$

It is seen from Equation 16.96 that the average power is due only to components of voltage and current of the same frequency. The average power of each frequency component is given by the product of the rms voltage, the rms current, and $\cos(\theta_{vn} - \theta_{in})$, the cosine of the phase difference between the voltage and current sinusoids of the same frequency. Multiplying by $\cos(\theta_{vn} - \theta_{in})$ ensures that the average power is due the components of voltage and current that are in phase. Thus, if the voltage and current of a certain frequency component $n\omega_0$ are in phase, as in the case of the voltage across a resistor and the current through it, $\theta_{vn} = \theta_{in}$, $\cos(\theta_{vn} - \theta_{in}) = 1$, and average power dissipated is $V_{n\text{rms}}I_{n\text{rms}}$, in accordance with Equation 8.29. On the other hand, if the voltage and current are 90° out of phase, as in the case of the voltage across a capacitor or inductor and the current through it, $\theta_{vn} - \theta_{in} = \pm 90°$, $\cos(\theta_{vn} - \theta_{in}) = 0$, and the average power is zero, in accordance with Equations 8.33 and 8.37.

It should be noted that the instantaneous power $p(t)$ is a function of all the ac components of the signal, including the components of different frequencies and the components of frequency $2n\omega_0$, such as the $4\omega_0t$ term in Equation 16.94. However, *only components of the same frequency contribute to the average power.* Thus, in Figure 16.46, for example, the instantaneous power p varies with time in a manner that depends on all the terms present in the product of voltage and current. At any instant t_1, the instantaneous power has a value p_1 that depends on the variation of $p(t)$ with t. The average power P is such that over a period T, the area above the line P is the same as the area below the line P. This average power depends only on the voltage and current components of the same frequency.

It should be noted that when more than one independent source of the *same* frequency is applied to an LTI

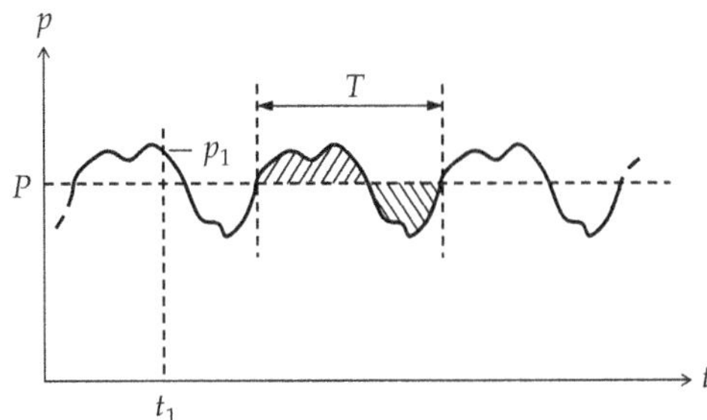

FIGURE 16.46
Instantaneous and average power.

circuit, the voltages and currents due to the individual sources interact with one another and are superposed, but the average powers due to these sources do not superpose, because the sum of squares of voltages or currents is not equal to the square of the sum, as explained in Section 5.1. However, when the various sources are of different frequencies, the voltages and currents due to the individual sources do not interact, that is, each response retains its individual identity, so that the average powers due to the sources of different frequency can be added together, in accordance with Equation 16.96. Superposition applies in this case to voltages, currents, and the average power.

16.7.1 rms Value

It is of interest to determine the rms value of a periodic waveform $f(t)$. By definition, F_{rms}, the rms value of $f(t)$, is given by

$$F_{rms} = \sqrt{\frac{1}{T}\int_0^T \left[f(t)\right]^2 dt} \qquad (16.97)$$

In other words, as its name implies, the rms value is obtained by squaring the given function, evaluating the mean or average of the square over a period, and then taking the square root of the mean. Assuming $f(t)$ to be given by Equation 16.4,

$$F_{rms}^2 = \frac{1}{T}\int_0^T \left[c_0 + \sum_{n=1}^{\infty} c_n\cos\left(n\omega_0 t + \theta_n\right)\right]^2 dt \qquad (16.98)$$

When the sum between the square brackets is squared, it will contain terms that are the products of different frequencies as well as squared terms. For example, if the series is $c_0 + c_1\cos(\omega_0 t + \theta_1) + c_2\cos(2\omega_0 t + \theta_2)$, the square is

$$c_0^2 + c_1^2\cos^2\left(\omega_0 t + \theta_1\right) + c_2^2\cos^2\left(2\omega_0 t + \theta_2\right) + 2c_0c_1\cos\left(\omega_0 t + \theta_1\right) + 2c_0c_2\cos\left(2\omega_0 t + \theta_2\right) + 2c_1c_2\cos\left(\omega_0 t + \theta_1\right)\cos\left(2\omega_0 t + \theta_2\right) \qquad (16.99)$$

As explained in connection with Equation 16.90, the last three terms in Equation 16.99 integrate to zero over a period T, leaving only the squared terms. Hence, when the terms between the square brackets in Equation 16.98 are squared, only the squared individual terms need be retained. Equation 16.98 can then be expressed as

$$F_{rms}^2 = \frac{1}{T}\int_{t_0}^{t_0+T}\left[c_0^2 + \sum_{n=1}^{\infty} c_n^2\cos^2\left(n\omega_0 t + \theta_n\right)\right]dt \qquad (16.100)$$

Using the trigonometric identity, $\cos^2\alpha = (1/2)(1 + \cos 2\alpha)$ (Appendix B), Equation 16.100 becomes

$$F_{rms}^2 = c_0^2 + \frac{1}{T}\int_0^{tT}\sum_{n=1}^{\infty}\frac{c_n^2}{2}\left[1 + \cos 2\left(n\omega_0 t + \theta_n\right)\right]dt \qquad (16.101)$$

The terms $\cos 2(n\omega_0 t + \theta_n)$ integrate to zero over a period, so that

$$F_{rms}^2 = c_0^2 + \sum_{n=1}^{\infty}\frac{c_n^2}{2} \qquad (16.102)$$

or

$$F_{rms} = \sqrt{c_0^2 + \sum_{n=1}^{\infty}\frac{c_n^2}{2}} \qquad (16.103)$$

Recall that c_n is the amplitude of the nth harmonic, so $c_n^2/2$ is the square of the rms value of this harmonic. By definition, the dc value is the same as the rms value. The summation in Equation 16.103 is over all the ac components of the FSE of the periodic function. According to Equation 16.103, the following concept applies:

Concept: *The rms value of a periodic function is the square root of the sum of the squares of the rms values of the individual components of the FSE.*

In terms of a and b coefficients (Equation 16.5), Equation 16.103 can be expressed as

$$F_{rms} = \sqrt{a_0^2 + \sum_{n=1}^{\infty}\frac{\left(a_n^2 + b_n^2\right)}{2}} \qquad (16.104)$$

If a periodic voltage $v(t)$ is applied across a resistor R, the rms current component I_{nrms} corresponding to a voltage component V_{nrms} is $I_{nrms} = V_{nrms}/R$, with $\theta_{in} = \theta_{vn}$. Substituting for I_{nrms} in Equation 16.96 and using Equation 16.102,

$$P = \frac{1}{R}\left[V_0^2 + \sum_{n=1}^{\infty} V_{nrms}^2\right] = \frac{V_{rms}^2}{R} \qquad (16.105)$$

where V_{rms} is the rms value of the periodic voltage v. When a single voltage of rms value V_{rms} is applied to a resistor R, the average power dissipated in R is V_{rms}^2/R. When a periodic voltage of rms value V_{rms} is applied to a resistor R, the average power dissipated in R is given by exactly the same expression, V_{rms}^2/R, in accordance with Equation 16.105, where V_{rms} is given by Equation 16.104.

Similarly, in terms of current,

$$P = R\left[I_0^2 + \sum_{n=1}^{\infty} I_{nrms}^2\right] = I_{rms}^2 R \qquad (16.106)$$

where I_{rms} is the rms value of the periodic current through R.

If a periodic waveform $f(t)$ is expressed analytically, it is usually much simpler to determine its rms value from direct application of Equation 16.97 rather than from its Fourier coefficients (Equation 16.98). Consider, for example, the half-wave rectified waveform of Figure 16.27a given by $A\cos\omega_0 t$, $-\pi/2 \le \omega_0 t \le \pi/2$, and zero over the rest of the period. The mean of its square over a period, which is the square of the rms value, is given by

$$(\text{HWR})_{rms}^2 = \frac{A^2}{2\pi}\int_{-\pi/2}^{\pi/2} \cos^2(\omega_0 t)\,d(\omega_0 t) = \frac{A^2}{4} \qquad (16.107)$$

The rms value is therefore $A/2$. The dc component is A/π (Equation 16.68). According to Equation 16.103, the rms value of a periodic function is the square root of the sum of the square of the dc component and the square of the rms value of the ac component of the periodic function. It follows that

$$\frac{A}{2} = \sqrt{\left(\frac{A}{\pi}\right)^2 + (\text{HWR})_{ac\text{-rms}}^2} \qquad (16.108)$$

where $(\text{HWR})_{ac\text{-rms}}$ is the rms value of the ac component alone. Hence,

$$(\text{HWR})_{ac\text{-}rms} = \frac{A}{2}\sqrt{1-\left(\frac{2}{\pi}\right)^2} = 0.39\text{ A} \qquad (16.109)$$

Example 16.10 rms Value of Triangular Waveform

It is required to determine the rms value of the periodic triangular waveform $f(t)$ shown in Figure 16.47 and deduce the rms value of the ac component.

Solution:

For $0 \le t \le \tau$, $f(t) = (A/\tau)t$. The square of this waveform is $(A/\tau)^2 t^2$ and the area under the curve is

$$\int_0^{\tau}\left(\frac{A}{\tau}\right)^2 t^2 dt = \left(\frac{A}{\tau}\right)^2 \frac{\tau^3}{3} = \frac{A^2\tau}{3} \qquad (16.110)$$

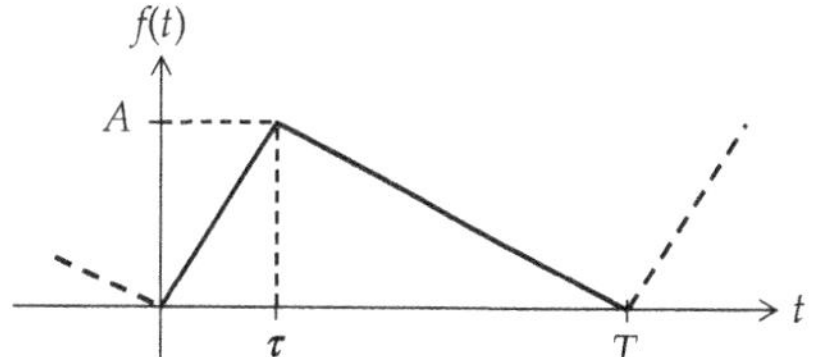

FIGURE 16.47
Figure for Example 16.10.

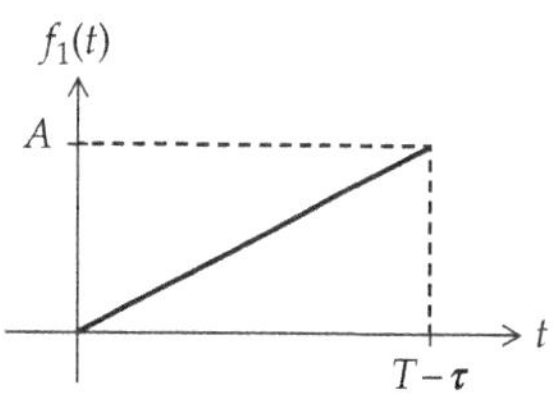

FIGURE 16.48
Figure for Example 16.10.

For $\tau \le t \le T$, the area under the curve of the squared function can be more easily determined by noting that this area is the same as that for $f_1(t)$ in Figure 16.48. By analogy with Equation 16.110, the area under the square function is $A^2(T-\tau)/3$, $\tau \le t \le T$. The total squared area for one period of $f(t)$ is the sum $A^2\tau/3 + A^2(T-\tau)/3 = A^2T/3$. The mean is $A^2/3$, and the rms value is $A/\sqrt{3}$. Because it is independent of τ, this result applies to any triangular waveform that varies between 0 and A and repeats continuously without interruption. It applies, for example, to a sawtooth waveform having $\tau = 0$, or $\tau = T$.

Since the variation of $f(t)$ over a period consists of straight-line segments, the dc or average value of $f(t)$ is half the amplitude, which is $A/2$. If the rms value of the ac component is denoted by $F_{ac\text{-rms}}$, then $\frac{A}{\sqrt{3}} = \sqrt{\left(\frac{A}{2}\right)^2 + F_{ac\text{-rms}}^2}$. This gives $F_{ac\text{-rms}} = A\sqrt{\frac{1}{3}-\frac{1}{4}} = \frac{A}{\sqrt{12}} = \frac{A}{2\sqrt{3}}$. That is, the rms value of the ac component of the FSE of the triangular function is one-half the rms value of the function.

Primal Exercise 16.19

(a) By inspection, what is the rms value of the full-wave rectified waveform of Figure 16.27b? (b) Determine the rms value of the periodic waveform shown in Figure 16.49.

Ans. (a) $A/\sqrt{2}$, the same as a sinusoid of amplitude A, because the squared function is the same; (b) $\frac{A}{2}\sqrt{\frac{13}{6}}$.

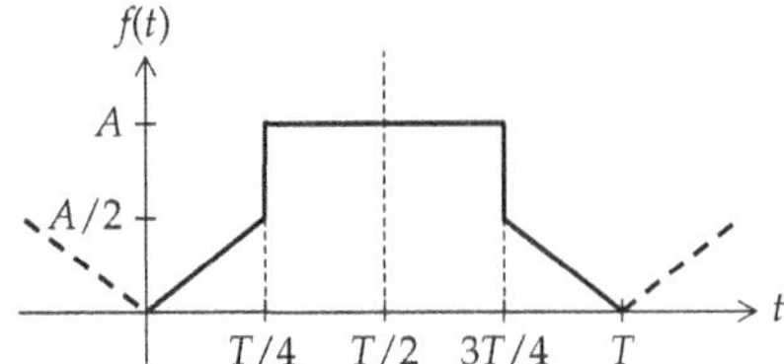

FIGURE 16.49
Figure for Primal Exercise 16.19.

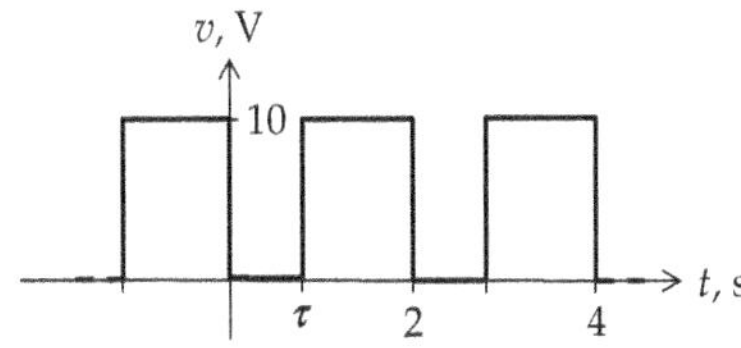

FIGURE 16.50
Figure for Primal Exercise 16.20.

Primal Exercise 16.20

Determine τ in Figure 16.50 if the rms value of the periodic voltage shown is 5 V.

Ans. 1.5 s.

Primal Exercise 16.21

Determine A and τ in Figure 16.51 so that the average value of $f(t)$ is 1 and its rms value is 3.

Ans. $A = 9$, $\tau = 2/9$ s.

Primal Exercise 16.22

$f(t)$ in Figure 16.52 is a periodic waveform of period 4 s, its ac component is quarter-wave symmetric, with $f(t) = 6 + kt^2$, $0 < t \leq 1$, where k is such that the rms value of $f(t)$ is 8. Determine the rms value of the *ac component* of $|f(t)|$ when delayed by 1 s.

Ans. 5.92.

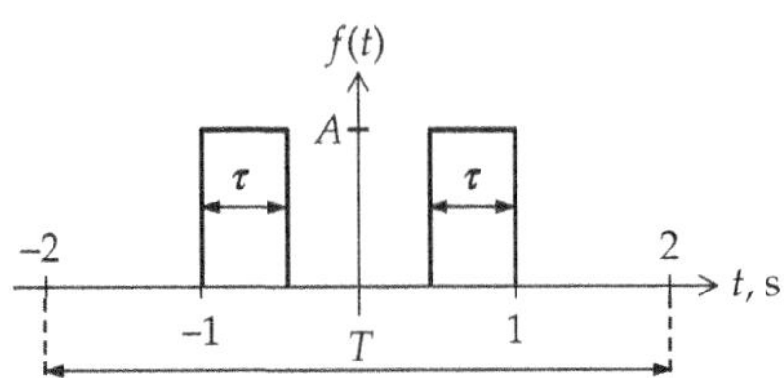

FIGURE 16.51
Figure for Primal Exercise 16.21.

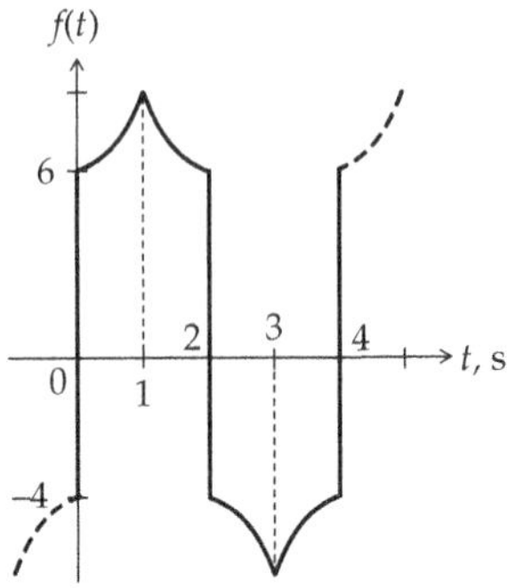

FIGURE 16.52
Figure for Primal Exercise 16.22.

Learning Checklist: What Should Be Learned from This Chapter

- A periodic function $f(t)$ is defined by the relation $f(t) = f(t + nT)$ for any t within the period T, where n is any positive or negative nonzero integer.
  1. The period T is defined, like that of a sinusoidal function, as the time interval between successive repetitions of the same full range of values of the periodic function. The reciprocal of T is the fundamental frequency of the periodic waveform, f_0 Hz. It follows that $\omega_0 T = 2\pi$.
- Given the four functions $\cos m\omega_0 t$, $\sin m\omega_0 t$, $\cos n\omega_0 t$, and $\sin n\omega_0 t$, where m and n are integers, the integral of the product of any two of these functions over a period $T = 2\pi/\omega_0$ is zero, except the products $\cos^2 n\omega_0 t$ and $\sin^2 n\omega_0 t$, having $m = n$, in which case the integral is $T/2$. The zero value of the integral of the product of different frequencies is based on the fact that the average of any harmonic over the period of the fundamental is zero.
- According to Fourier's theorem, a periodic function $f(t)$ can be expressed, in general, as an infinite series of cosine and sine functions in an FSE: $f(t) = a_0 + \sum_{n=1}^{\infty}\left(a_n \cos n\omega_0 t + b_n \sin n\omega_0 t\right)$, where n is a positive integer and a_0, a_n, and b_n are constants, known as Fourier coefficients, that depend on $f(t)$.
  1. The sum of the two terms $(a_n \cos n\omega_0 t + b_n \sin n\omega_0 t)$ represents the nth harmonic, where $n = 1$ represents the fundamental.
  2. a_0 is the average or dc component of $f(t)$: $a_0 = \frac{1}{T}\int_{t_0}^{t_0+T} f(t)dt$.

3. The cosine and sine terms of the FSE are the ac component of $f(t)$, where a_n and b_n are given by $a_n = \frac{2}{T}\int_0^T f(t)\cos n\omega_0 t\,dt$, which is twice the average of $f(t)\cos n\omega t$ over a period, and $b_n = \frac{2}{T}\int_0^T f(t)\sin n\omega_0 t\,dt$, which is twice the average of $f(t)\sin n\omega t$ over a period.

- The FSE can be expressed in the alternative trigonometric form as $f(t) = c_0 + \sum_{n=1}^{\infty} c_n \cos(n\omega_0 t + \theta_n)$, where $c_0 = a_0$, $c_n = \sqrt{a_n^2 + b_n^2}$, and $\theta_n = \tan^{-1}\frac{b_n}{a_n}$.
- The FSE can be expressed in exponential form as $f(t) = C_0 + \sum_{n=1}^{\infty} C_n e^{jn\omega_0 t} + \sum_{n=-1}^{-\infty} C_n e^{jn\omega_0 t} = \sum_{n=-\infty}^{\infty} C_n e^{jn\omega_0 t}$, where $C_0 = a_0$, and $C_n = \frac{1}{T}\int_0^T f(t)e^{-jn\omega_0 t}$ $dt = \frac{a_n - jb_n}{2}$.
- The plot of $|C_n|$ against $n\omega_0$ is the amplitude spectrum of $f(t)$ and is an even function of $n\omega_0$. The plot of $\angle C_n$ against $n\omega_0$ is the phase spectrum of $f(t)$ and is an odd function of $n\omega_0$.
- If a periodic waveform $f(t)$ is delayed by t_d, θ_n is reduced by $n\omega_0 t_d = 2\pi n(t_d/T)$. Conversely, if $f(t)$ is advanced by t_d, θ_n is increased by $n\omega_0 t_d = 2\pi n(t_a/T)$.
- The FSE of an even periodic function does not contain any sine terms; its Fourier coefficients can be evaluated over half a period:

$$a_0 = \frac{2}{T}\int_0^{T/2} f(t)dt,$$

$$a_n = \frac{4}{T}\int_0^{T/2} f(t)\cos n\omega_0 t\,dt,$$

and

$$b_n = 0 \quad \text{for all } n$$

- The FSE of an odd periodic function does not contain an average term nor any cosine terms; its Fourier coefficients can be evaluated over half a period:

$$a_0 = 0 = a_n \quad \text{for all } n, \quad \text{and}$$

$$b_n = \frac{4}{T}\int_0^{T/2} f(t)\sin n\omega_0 t\,dt$$

1. The ac component of a periodic function may be odd, but the periodic function itself may be neither odd nor even because of a dc component.

- In a half-wave symmetric waveform, if either half-cycle or half-wave is displaced horizontally by $T/2$ toward the other half, the two half-waves are aligned vertically and are symmetrical with respect to the horizontal axis. Mathematically, $f(t) = -f(t+T/2)$ or $f(t) = -f(t-T/2)$, $0 \le t \le T/2$.
  1. The FSE of a half-wave symmetric periodic function does not contain an average term or any even harmonics; its Fourier coefficients can be evaluated over half a period: $a_0 = 0, a_n = 0 = b_n$ for n even, and $a_n = \frac{4}{T}\int_0^{T/2} f(t)\cos n\omega_0 t\,dt$, $b_n = \frac{4}{T}\int_0^{T/2} f(t)\sin n\omega_0 t\,dt$, for n odd.
  2. The ac component of a periodic function may have odd harmonics only, but the periodic function itself may not be half-wave symmetric because of a dc component.
- A half-wave symmetric function that is also symmetrical about a vertical line through the middle of its positive or negative half-cycles is said to possess quarter-wave symmetry. Such a function can be made either odd or even by shifting the function in time.
  1. If the function is odd, $b_n = \frac{8}{T}\int_0^{T/4} f(t)\sin n\omega_0 dt$ for n odd, $b_n = 0$ for n even, $a_0 = 0$, and $a_n = 0$ for all n.
  2. If the function is even, $a_n = \frac{8}{T}\int_0^{T/4} f(t)\cos n\omega_0 dt$ for n odd, $a_n = 0$ for n even, $a_0 = 0$, and $b_n = 0$, for all n.
  3. If the function is neither odd nor even, a_n and b_n are nonzero for n odd and are given by the preceding expressions.
- The FSE of some periodic waveforms can be derived from those of other periodic waveforms through addition, subtraction, multiplication, differentiation, or integration.
- The FSE of a given periodic function can be differentiated or integrated term by term. The result is the FSE of a periodic function that is the derivative or integral of the given function, except that integrating the dc component destroys the periodicity of the function.
- The smoother the function, the more rapidly the harmonics decrease in magnitude. Consequently, if the mth derivative of a periodic function is discontinuous, with all the derivatives of lower order being continuous, the magnitudes of the harmonics decrease

approximately as $1/n^{m+1}$, $m=0,1,2,\ldots$, the derivative of order 0 being the function itself.

- Shifting the time axis up or down affects the dc component of the FSE without affecting the ac component. Shifting the vertical axis to the right or left (1) affects odd–even symmetry, without affecting half-wave or quarter-wave symmetry, and (2) modifies the phase angles of the ac components, without affecting their magnitudes and without affecting the dc component.
- The steady-state response of an LTI circuit to a periodic signal is the sum of the responses to each component acting alone.
- In an LTI circuit, components of different frequencies do not interact, and the total average power is the sum of the average powers due to each component acting alone.
- The rms value of a periodic function is the square root of the sum of the squares of the rms values of the individual components, bearing in mind that a dc value is the same as the rms value.
    1. If a periodic waveform $f(t)$ is expressed analytically, it is usually much simpler to determine its rms value from the definition of this value, that is, by squaring the function, deriving the average of this square, and taking the square root.

Problem-Solving Tips

1. Remember that $\omega_0 T = 2\pi$.
2. When the period has a simple geometric form, a_0 can be easily determined by dividing the area over the period by the period T or by determining the shift of the horizontal axis that would result in the area enclosed by the function above the new horizontal axis is the same as the area enclosed by the function below this axis.
3. Whenever feasible, check if the form of the FSE makes sense as a rough approximation to the given periodic function.
4. When the Fourier coefficients a_n and b_n are evaluated using expressions that restrict the values of n, such as n being odd or even, only these restricted values should be used in deriving the individual coefficients a_n and b_n.
5. The first step in Fourier analysis is to ascertain if the function possesses any type of symmetry. If the function has a nonzero average, this average should be removed to see if the function becomes odd or half-wave or quarter-wave symmetric.

Problems

Verify solutions by PSpice simulation.

Fourier Analysis

P16.1 Verify the following properties of odd and even functions:

1. The product of two odd functions is an even function.
2. The product of an odd function and an even function is an odd function.
3. The product of two even functions is an even function.
4. The sum or difference of two odd functions is an odd function.
5. The sum or difference of two even functions is an even function.
6. The sum or difference of an odd function and an even function is neither odd nor even.

P16.2 Determine the period of each of the following functions: (a) $f(t) = 5 + 10\cos 100\pi t + 5\cos 200\pi t + 2\cos 400\pi t$; (b) $g(t) = \dfrac{\cos 100\pi t \sin 300\pi t - \sin 100\pi t \cos 300\pi t}{\sin 200\pi t}$.

Ans. (a) 20 ms; (b) function is not periodic.

P16.3 Given the two functions (a) $f(t) = \cos(100\pi t)\sin(200\pi t)$ (b) $g(t) = \cos^2(100\pi t)\sin^2(200\pi t)$, show that $f(t)$ and $g(t)$ are periodic, determine their periods, and derive their FSEs.

Ans. (a) 20 ms, $0.5\sin(100\pi t) + 0.5\sin(300\pi t)$; (b) 10 ms, $0.25 + 0.125\cos(200\pi t) - 0.25\cos(400\pi t) - 0.125\cos(600\pi t)$.

P16.4 Determine a_0, a_5, and b_5 for the periodic function shown in Figure P16.4. Obtain a_5 by applying both Equations 16.20 and 16.63. Note that the latter applies to the ac component only.

Ans. $a_0 = 1.5$, $a_5 = 1.15$, $b_5 = 0$.

P16.5 The current through a 1 μF capacitor is $2\cos^2(100\pi t)$ mA, where t is in s. Determine the period of the voltage across the capacitor.

Ans. Voltage is nonperiodic.

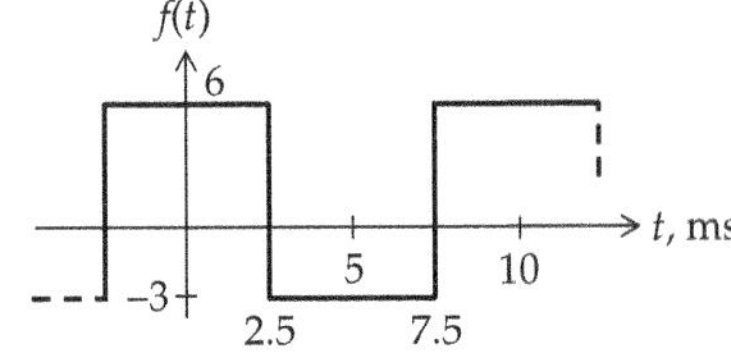

FIGURE P16.4

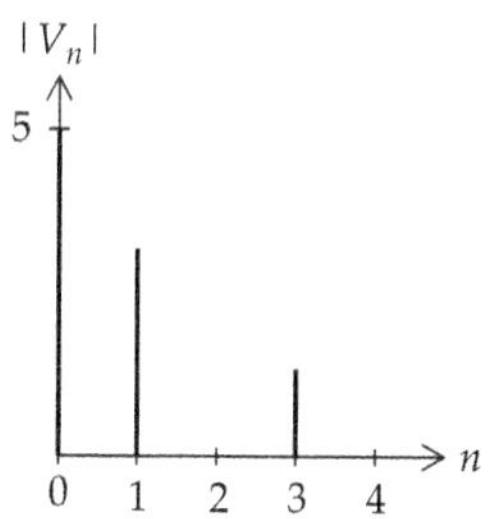

FIGURE P16.6

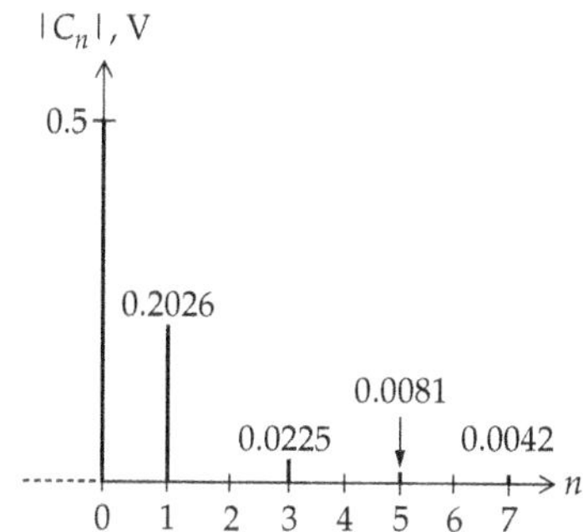

FIGURE P16.7

P16.6 A periodic voltage is represented by the expansion $v(t)=\sum_{n=-\infty}^{\infty} V_n e^{j200\pi nt}$ and has the V_n magnitude spectrum shown in Figure P16.6. Determine the average value and the fundamental frequency of $v(t)$.

Ans. 5, 100 Hz.

P16.7 Verify that the amplitude spectrum shown in Figure P16.7 is the FSE of the following function: $f(t)=\frac{1}{2}+\frac{4}{\pi^2}\cos\omega_0 t+\frac{4}{(3\pi)^2}\cos 3\omega_0 t+\frac{4}{(5\pi)^2}\cos 5\omega_0 t+\cdots$.

What can be said about the discontinuities of the function?

Ans. Function is continuous, its first derivative is discontinuous.

P16.8 Given $f(t) = 2\cos(100\pi t) + 3\cos(300\pi t) + 6\cos(500\pi t) + 9\sin(300\pi t)$, determine the coefficients of the exponential FSE of $f(t)$.

Ans. $C_1 = C_{-1} = 1$, $C_3 = 1.5 - j4.5$, $C_{-3} = 1.5 + j4.5$, $C_5 = C_{-5} = 3$.

P16.9 Specify the type of symmetry of the periodic function shown in Figure P16.9 and characterize the coefficients a_n and b_n of its FSE.

Ans. Quarter-wave symmetric and odd; $a_n = 0$ for all n, $b_n = 0$ for even n and nonzero for odd n.

P16.10 Given the periodic function $f(t)$ shown in Figure P16.10, determine the amplitude of the third harmonic.

Ans. $\frac{80}{3\pi\sqrt{2}}$.

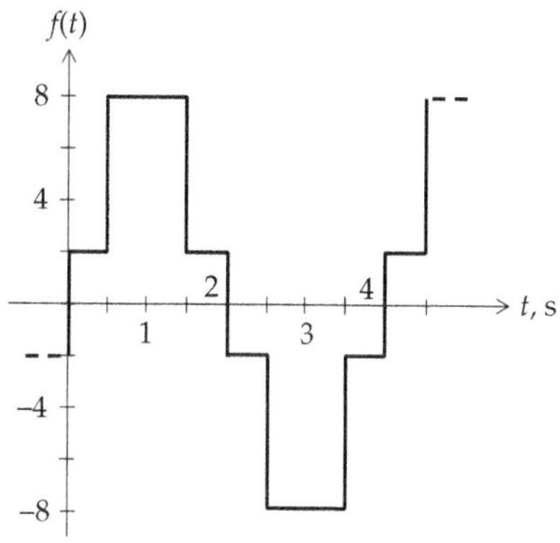

FIGURE P16.9

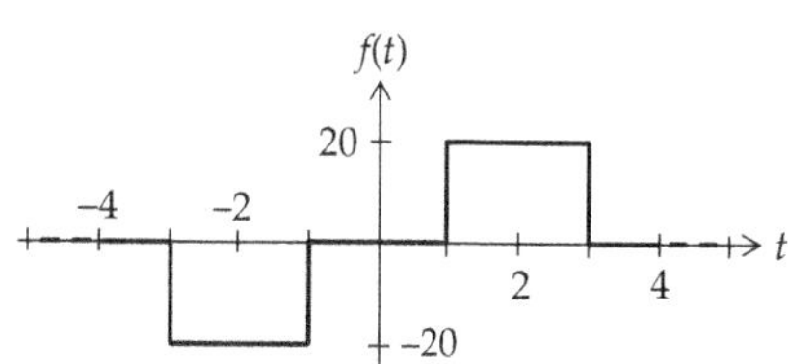

FIGURE P16.10

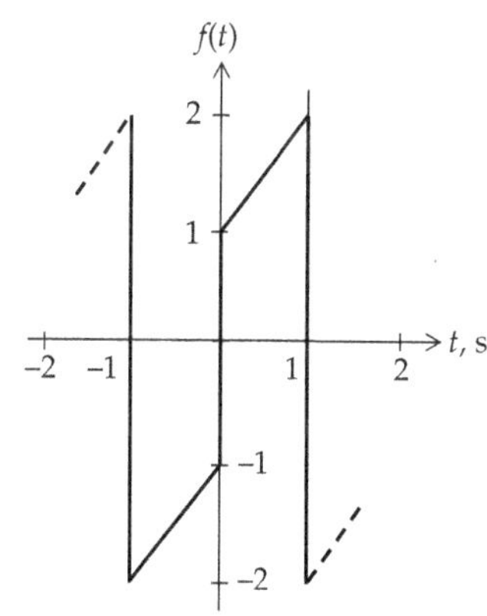

FIGURE P16.12

P16.11 Two periodic functions of period 6 s are defined by $f(t) = -t$, $-3 \le t \le 0$, and $f(t) = t$, $0 \le t \le 3$; $g(t) = 1$, $0 < t < 3$, and $g(t) = -1$, $3 < t < 6$. Determine the ratio of the amplitude of the third harmonic in $f(t)$ to that in $g(t)$.

Ans. $1/\pi$.

P16.12 Derive the FSE of the function shown in Figure P16.12.

Ans. $a_0 = 0 = a_n$ for all n. $b_n = \frac{2}{\pi}\left(\frac{1}{n}\left(1+2(-1)^{n+1}\right)\right)$.

P16.13 Determine a_5 for the periodic voltage shown in Figure P16.13.

Ans. 0.

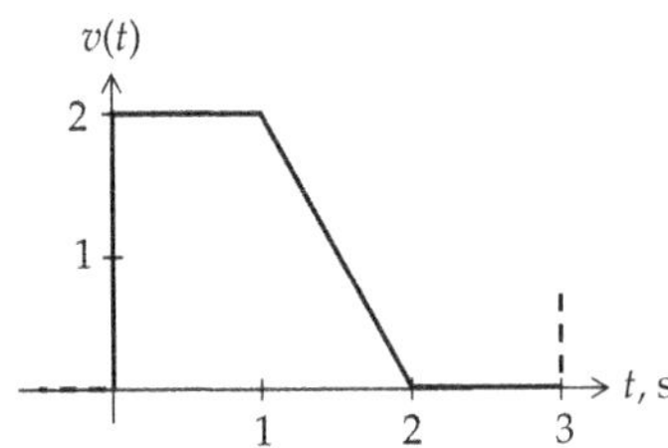

FIGURE P16.13

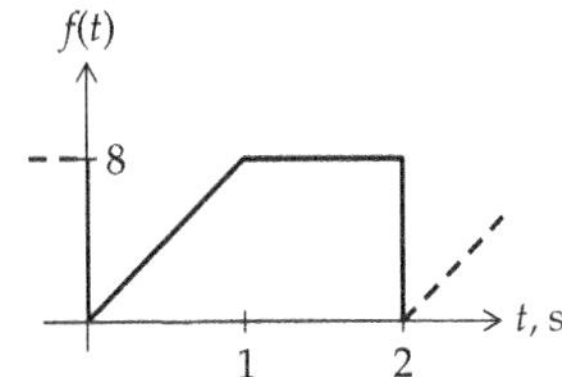

FIGURE P16.14

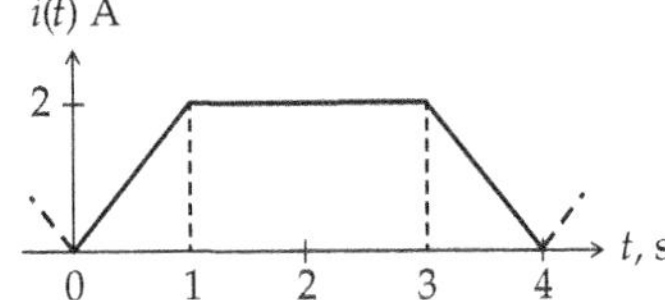

FIGURE P16.16

P16.14 Derive the FSE expansion of the periodic function shown in Figure P16.14.

Ans. $f(t)=6-\sum_{n=1,3,5,\ldots}^{\infty}\frac{16}{\pi^2 n^2}\cos n\pi t-\sum_{n=1,2,3\ldots}^{\infty}\frac{8}{\pi n}\sin n\pi t$

P16.15 Derive the FSE of a periodic trapezoidal waveform defined over a period 0–2 as $f(t) = t + 2$, $0 < t < 1$, and $f(t) = 0$ elsewhere in the period.

Ans. $C_0 = 2.5$, $C_n = \frac{1}{2}\left[\frac{e^{-jn\pi}-1}{n^2\pi^2}+j\frac{3e^{-jn\pi}-2}{n\pi}\right]$.

P16.16 (a) Derive the first three terms of the trigonometric FSE of the periodic function $i(t)$ A shown in Figure P16.16; (b) repeat (a) when $i(t)$ when advanced or delayed by 2 s.

Ans. (a) $i(t)=1.5-\frac{4}{\pi^2}(2\cos\pi t/2+\cos\pi t)$ A; (b) $1.5+\frac{4}{\pi^2}(2\cos\pi t/2-\cos\pi t)$A.

P16.17 Given a full-wave rectified waveform of period T as shown in Figure 16.27b, except that because of dissymmetry in the rectifier circuit, the half-sinusoids are not all of the same amplitude but alternate with amplitudes of 12 and 10 V. Derive the FSE, assuming that the sinusoid centered at the origin has a an amplitude of 12 V.

Ans. $\frac{22}{\pi}+\cos\omega_0 t+\frac{44}{3\pi}\cos 2\omega_0 t-\frac{44}{15}\cos 4\omega_0 t+\frac{44}{35}\cos 6\omega_0 t$

$+\cdots+\frac{(-1)^{n+1}44}{4n^2-1}\cos 2n\omega_0 t+\cdots \quad n=1,2,3,\ldots$

P16.18 Derive the FSE of the function shown in Figure 16.20.

Ans. $C_n=-\frac{2A_m}{n^2\pi^2}-j\frac{A_m}{n\pi}$

P16.19 A function is defined over half a period by $e^t, 0<t<1$. Derive the FSE if the function is (a) even, (b) odd.

Ans. (a) $a_0 = e - 1$, $a_n=\frac{2(e-1)}{1+n^2\pi^2}$ for even n, and $a_n=-\frac{2(e+1)}{1+n^2\pi^2}$ for odd n.

(b) $a_0=0, b_n=-\frac{2n\pi(e-1)}{1+n^2\pi^2}$ for even n, and $b_n=\frac{2n\pi(e+1)}{1+n^2\pi^2}$ for odd n.

P16.20 Determine the magnitude and phase angle of the fundamental component of the periodic function $f(t)$ shown in Figure P16.20, where $f(t) = \sin(2\pi t)$, for $0 \le t \le 0.25$ s, and $f(t) = 0.5\sin(2\pi t)$, for $0.25 \le t \le 0.5$ s.

Ans. 0.75, zero phase angle of the sine term.

P16.21 The periodic function shown in Figure P16.21 is described by $f(t) = 3 + \sin t$, $0 \le t \le \pi$, and $f(t) = -2 - \sin t$, $\pi \le t \le 2\pi$. Determine the average value of $f(t)$ and the fundamental component.

Ans. $a_0 = 1.317$ V, $a_1 = 0$, $b_1 = 3.183$ V.

P16.22 v in Figure P16.22 is a periodic function of period π described by $v(t) = 10\sin t$, $0 \le t < \pi/2$, and $v(t) = 0$, $\pi/2 < t \le \pi$. Determine (a) the average value of $v(t)$,

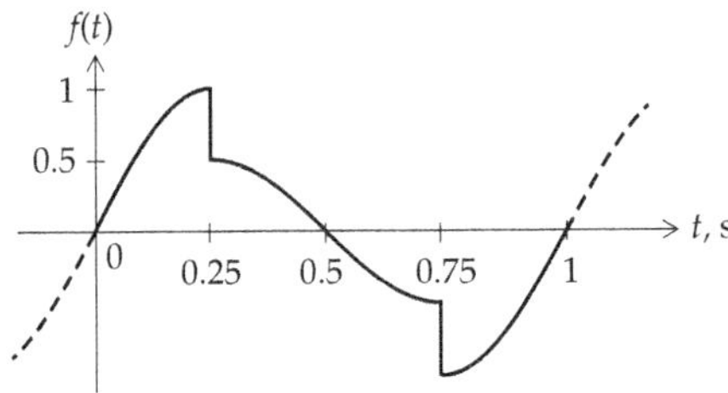

FIGURE P16.20

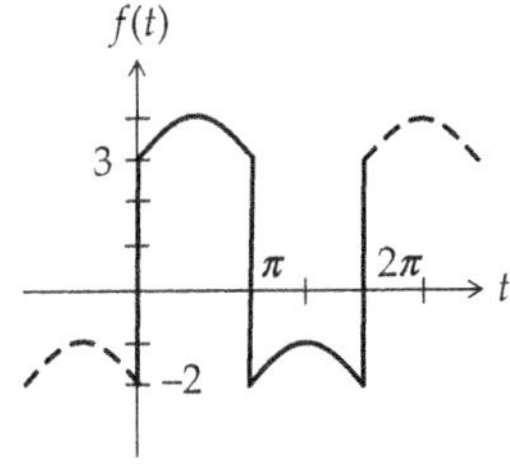

FIGURE P16.21

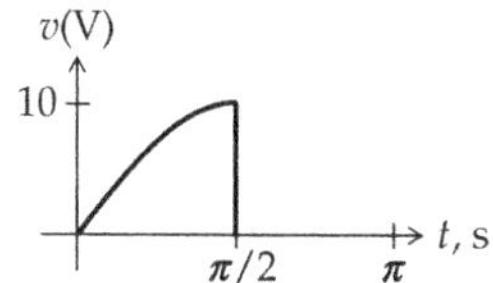

FIGURE P16.22

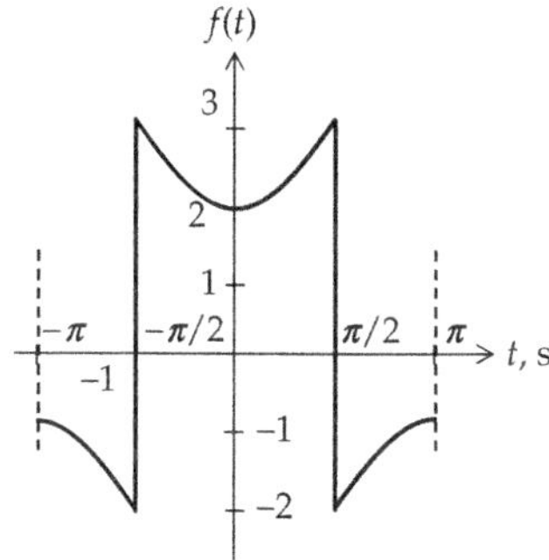

FIGURE P16.23

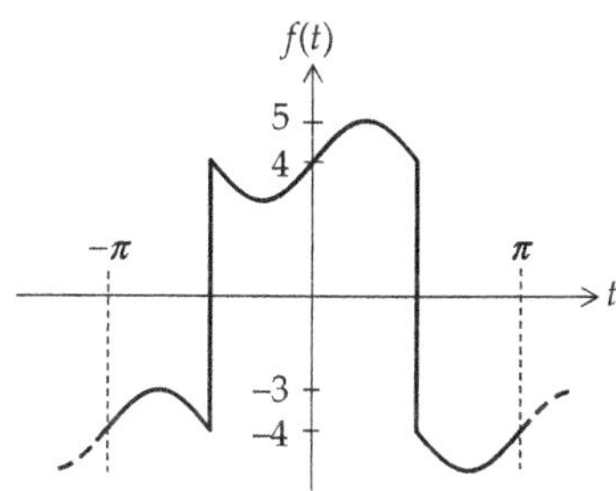

FIGURE P16.24

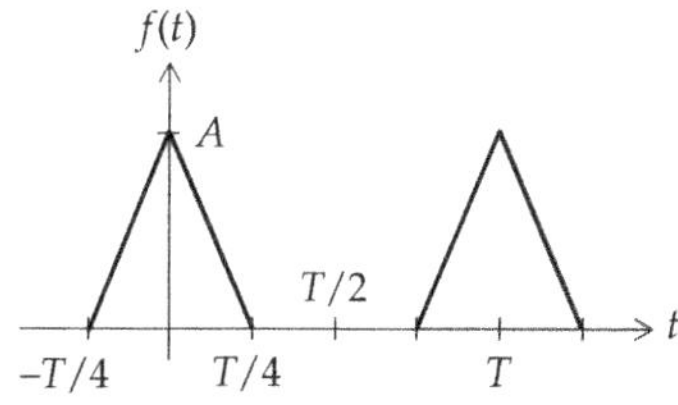

FIGURE P16.25

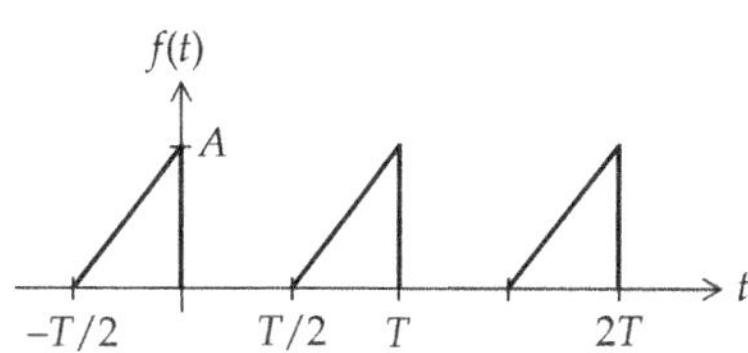

FIGURE P16.26

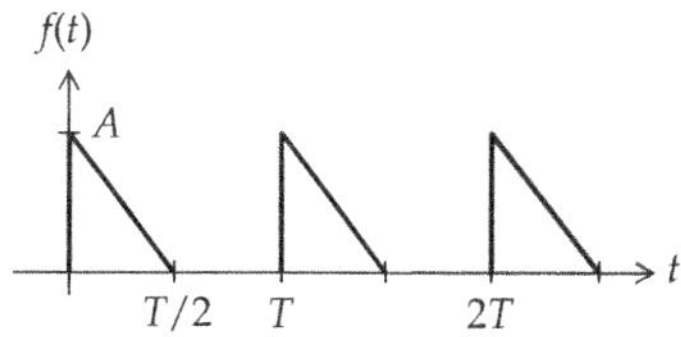

FIGURE P16.27

(b) a_3 and b_3, (c) the rms value of the ac component of v and (d) power dissipated in a 10 Ω resistor to which v is applied.

Ans. (a) $10/\pi$ V; (b) $a_3 = -10/\pi$, $b_3 = 10/\pi$; (c) 3.86 V; (d) 2.5 W.

P16.23 Derive the FSE of the periodic function shown in Figure P16.23, defined as

$f(t) = \cos(t + \pi) - 2$, $-\pi < t < -\pi/2$

$f(t) = -\cos t + 3$, $-\pi/2 < t < +\pi/2$

$f(t) = \cos(t - \pi) - 2$, $\pi/2 < t < \pi$

Ans. $a_0 = 0.5$, $a_n = 0$ for even n, and $a_n = \dfrac{10}{n\pi}\sin\dfrac{n\pi}{2}$ for odd n.

P16.24 (a) Derive the first three terms of the FSE of $f(t)$ in Figure P16.24, where $f(t) = 4 + \sin 2t$, $-\pi/2 < t < \pi/2$ and $f(t) = -4 + \sin 2t$, $-\pi \le t < -\pi/2$ and $\pi/2 < t \le \pi$; (b) determine these terms when $f(t)$ is advanced by a quarter period.

Ans. (a) $(16/\pi)\cos t + \sin 2t - (16/3\pi)\cos 3t$; (b) $-(16/\pi)\sin t - \sin 2t - (16/3\pi)\sin 3t$.

P16.25 (a) Derive the FSE of the waveform of Figure P16.25 by direct evaluation. (b) Show that if it is added to a delayed and negated version, the result agrees with Equation 16.66. (c) Indicate how the FSE can be obtained as the product of a rectangular pulse train of unit height (Figure 16.8) and a triangular waveform derived from that of Figure 16.22.

Ans. $a_0 = A/4, a_n = \dfrac{16A}{T^2n^2\omega_0^2}\left[1-\cos\dfrac{n\pi}{2}\right] = \dfrac{4A}{\pi^2n^2}\left(1-\cos\dfrac{n\pi}{2}\right)$, $n = 1, 2, 3, \ldots$.

P16.26 (a) Derive the FSE of the waveform of Figure P16.26 by direct evaluation. (b) Indicate how the FSE can be obtained as the product of a rectangular pulse train of unit height (Figure 16.8) and a sawtooth waveform derived from that of Figure 16.5.

Ans. $C_0 = A/4, C_n = \dfrac{A}{2\pi^2n^2}(1-\cos n\pi) + j\dfrac{A}{2\pi n}$

P16.27 Derive the FSE of the waveform of Figure P16.27 in two ways: (a) directly and (b) as the product of a rectangular pulse train of unit height (Figure 16.9) and a reversed sawtooth waveform derived from that of Figure 16.8b.

Ans. $C_0 = A/4, C_n = \dfrac{A}{2\pi^2n^2}(1-\cos n\pi) - j\dfrac{A}{2\pi n}$.

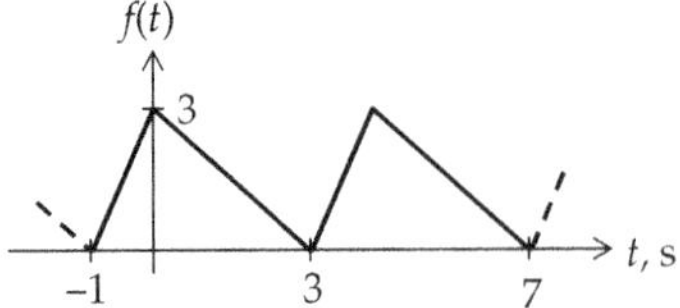

FIGURE P16.29

P16.28 (a) Show that if the FSE of Problem P16.26 is combined with the negated FSE of Problem P16.27, the FSE of the sawtooth waveform of Figure 16.5 is obtained. (b) Derive the FSE of Problem P16.26 from that of Problem P16.25 and that of a rectangular pulse train of amplitude A and period T.

P16.29 Derive the FSE of the waveform of Figure P16.29 in two ways: (a) direct evaluation of coefficients and (b) from that of its derivative.

Ans. $C_0 = 6$, $C_n = \dfrac{1}{n^2\omega_0^2}\left(1-\cos\left(\dfrac{n\pi}{2}\right)-j\sin\left(\dfrac{n\pi}{2}\right)\right)$.

P16.30 Obtain the FSE of a full-wave rectified waveform in two ways: (a) as the sum of two half-wave rectified waveforms, with one waveform shifted by half a period with respect to the other waveform, and (b) as the product of a square wave of zero average and $\cos\omega_0 t$.

P16.31 Derive the FSE of the waveform of Figure P16.31 in two ways: (a) direct evaluation of coefficients and (b) as the sum of two shifted rectangular pulse trains.

Ans. $a_0 = 1/4$, $a_n = \dfrac{\sin(n\pi/2)}{n\pi} = \dfrac{(-1)^{(n+3)/2}}{n\pi}$ for odd n and zero for even n, $b_n = \dfrac{9}{n\pi}$ for odd n, and $\dfrac{1+(-1)^{(n+2)/2}}{n\pi}$ for even n.

P16.32 A periodic function of period 1 s is defined as $f(t) = 2t^3 - \dfrac{t}{2}$, $-\dfrac{1}{2} \le t \le \dfrac{1}{2}$. Determine how the magnitudes of the harmonics vary with the order n of the harmonic?

Ans. As $1/n^3$.

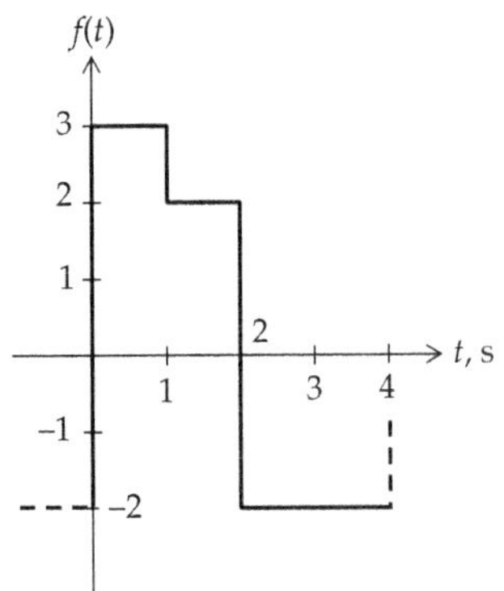

FIGURE P16.31

P16.33 Express the function of Figure 16.47 as a Fourier series in both exponential and trigonometric forms, given that $A = 3$ units, $T = 4$ s, and $\tau = T/4$. Verify that the harmonics decrease at a rate of $1/n^2$ and that the function becomes odd if advanced by 0.5 s and the dc value removed.

Ans. $f(t)=1.5+\dfrac{4}{\pi^2}\sum_{\substack{n=-\infty\\n\neq 0}}^{\infty}\dfrac{1}{n^2}\left(\cos\left(\dfrac{n\pi}{2}\right)-1-j\sin\left(\dfrac{n\pi}{2}\right)\right)$; $a_0 = 1.5$, $a_n = \dfrac{8}{\pi^2 n^2}\left(\cos\left(\dfrac{n\pi}{2}\right)-1\right) = -\dfrac{8}{\pi^2 n^2}$, $n = 1, 3, 5, 7$, etc., $a_n = -\dfrac{16}{\pi^2 n^2}$, $n = 2, 6, 10, 14$, etc., and $a_n = 0$, $n = 4, 8, 12, 16$, etc., $b_n = \dfrac{8}{n^2\pi^2}\sin\left(\dfrac{n\pi}{2}\right) = 0$ for even n, $b_n = \dfrac{8}{\pi^2 n^2}1$, $n = 1, 5, 9, 13$, etc., and $b_n = -\dfrac{8}{\pi^2 n^2}$, $n = 3, 7, 11, 15$, etc.

P16.34 Derive the FSE of the periodic triangular waveform of Figure P16.34. Show that it can be made that of an even function or an odd function by a shift in time.

Ans. $f(t) = \sum_{n=1}^{\infty}\dfrac{16}{\pi^2 n^2}\cos\dfrac{n\pi}{4}(t-1)$.

P16.35 Given a half-wave symmetric function $f(t)$, half a period of which is shown in Figure P16.35, determine the coefficients of the FSE up to and including the fifth harmonic, neglecting higher harmonics, and assuming that the function can be approximated by $f(\pi/6) = 0.6$, $f(\pi/3) = 1.9$, $f(\pi/2) = 3.3$, $f(2\pi/3) = 4$, and $f(5\pi/6) = 3.3$.

Ans. $a_1 = -1.1024$, $a_3 = 0.6460$, $a_5 = 0.4564$, $b_1 = 3.4065$, $b_3 = 0.2$, $b_5 = 0.0935$.

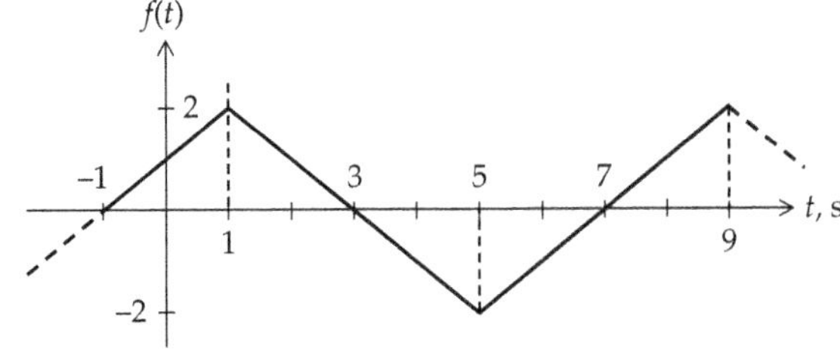

FIGURE P16.34

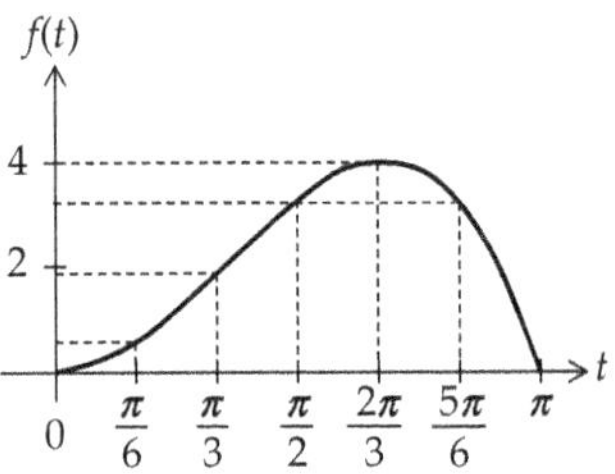

FIGURE P16.35

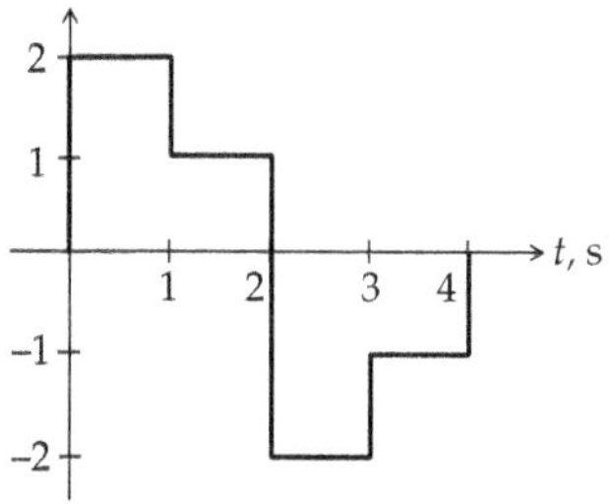

FIGURE P16.36

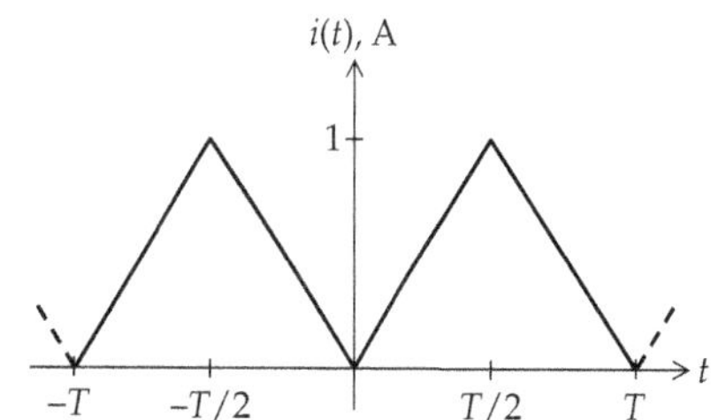

FIGURE P16.41

P16.36 The first period of the periodic function $f(t)$ is shown in Figure P16.36. Determine (a) C_0, C_1, C_2, and C_3; (b) derive the amplitude and phase angle of the third harmonic component of $f(t)$.

Ans. (a) $C_0 = 0$, $C_1 = \frac{1}{\pi} - j\frac{3}{\pi}$, $C_2 = 0$, $C_3 = -\frac{1}{3\pi} - j\frac{1}{\pi}$; (b) $0.67A\cos(3\omega_0 t + \alpha)$, where $\alpha = \tan^{-1}(-3) = 108.34°$.

P16.37 $v_{SRC1}(t)$ and $v_{SRC2}(t)$ are the periodic functions shown in Figure P16.37. Determine τ so that $v_O(t)$ is periodic.

Ans. 1.25 s.

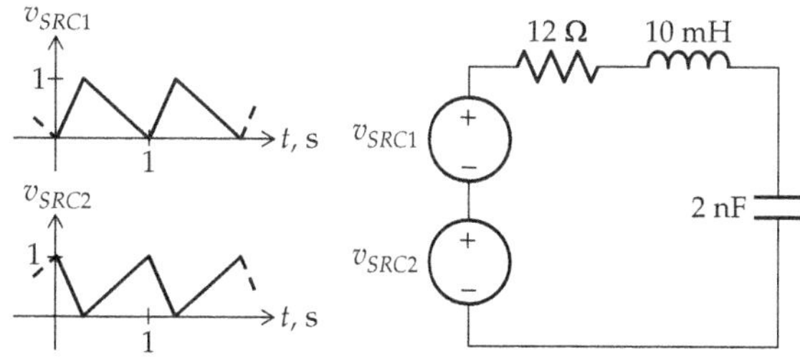

FIGURE P16.43

Responses to Periodic Inputs

P16.38 A voltage $v(t) = \cos\omega_0 t + 2\sin 2\omega_0 t$ is applied to a circuit that gives an output $v_O = v + v^2$. Determine the FSE of the output.

Ans. $2.5 + \sqrt{5}\cos(\omega_0 t - 63.4°) + 2.06\cos(2\omega_0 t - 76.0°) + 2\sin 3\omega_0 t - 2\cos 4\omega_0 t$.

P16.39 The output v_O of a nonlinear network is related to the input v_i by the power relation: $v_O = 5v_i + 0.5v_i^2 + 0.1v_i^3$. If $v_i(t) = 4\sin 2000\pi t$, determine the magnitude of each term of the Fourier expansion of the output. Does the output possess half-wave symmetry? Why not?

Ans. $v_O(t) = 4 + 24.8\sin 2000\pi t - 4\cos 4000\pi t - 1.6\sin 6000\pi t$ V; no, because of the dc and the second harmonic cosine term.

P16.40 A current $i = i_1 + i_2$, where $i_1(t) = 10\cos(100\pi t)$ A and $i_2(t) = \cos(300\pi t)$ A, is applied to an impedance $(3 - j4)$ Ω at the frequency of 100π rad/s. Determine the ratio $|\mathbf{V}_2|/|\mathbf{V}_1|$, where $\mathbf{V}_1$ and $\mathbf{V}_2$ are the voltages across the impedance due to i_1 and i_2, respectively.

Ans. 0.066.

P16.41 The current through a 1 H inductor is the periodic triangular waveform of Figure P16.41, with $T = 1/2\pi$. Determine the amplitude of the fundamental component of the voltage across the inductor.

Ans. 16 V.

P16.42 The current waveform of Figure P16.41 is applied to a 2 Ω resistor in parallel with a very large capacitor. Determine the voltage across the capacitor.

Ans. 1 Vdc.

P16.43 The periodic voltages $v_{SRC1}(t)$ and $v_{SRC2}(t)$ are applied to the circuit shown in Figure P16.43. Determine the voltage across the capacitor.

Ans. 1 Vdc.

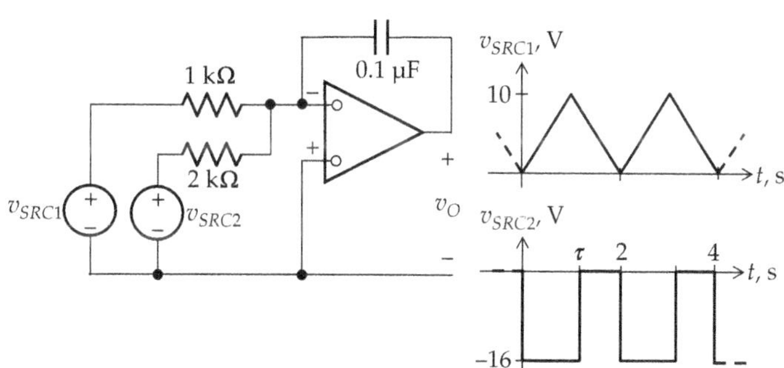

FIGURE P16.37

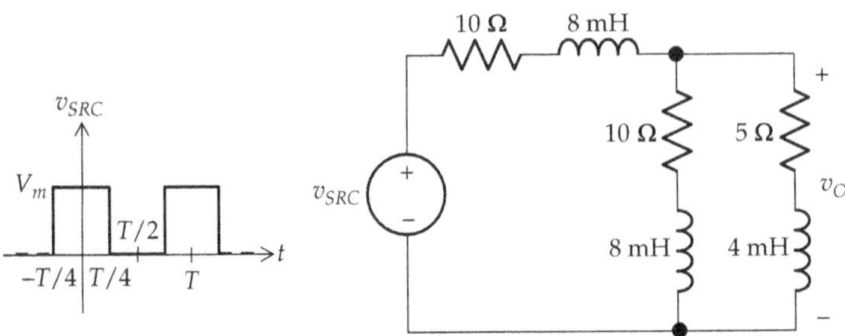

FIGURE P16.44

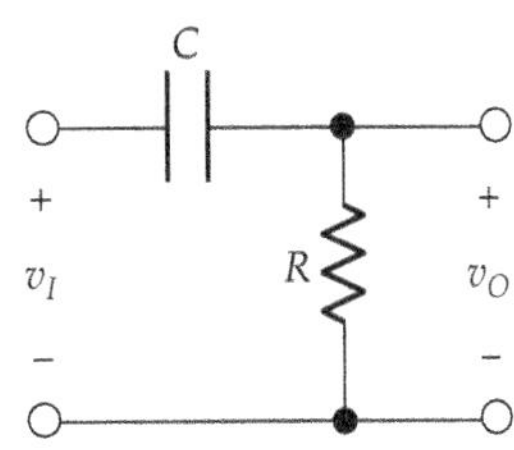

FIGURE P16.45

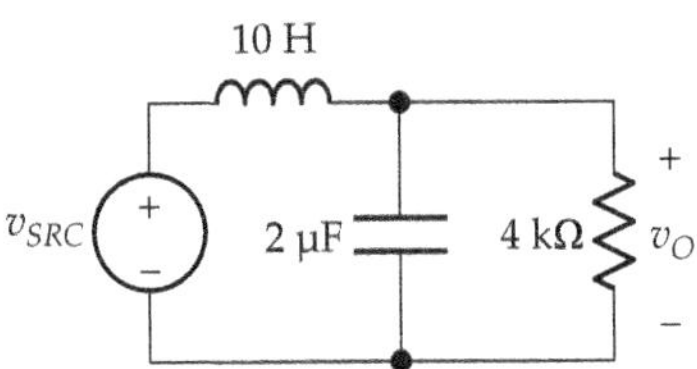

FIGURE P16.46

P16.44 The periodic voltage $v_{SRC}(t)$ is applied to the circuit shown in Figure P16.44. Determine $v_O(t)$.

Ans. $v_O(t) = \dfrac{V_m}{8} + \dfrac{V_m}{2\pi}\sum_{n=1}^{\infty}\dfrac{\sin(n\pi/2)}{n}\cos n\omega_0 t.$

P16.45 The triangular pulse train of Figure 16.22 is applied as the input $v_I(t)$ to the circuit of Figure P16.45. Determine $v_O(t)$. Show that if $\omega CR \ll 1$, $v_O \to CR\dfrac{dv_I}{dt}$, and v_O becomes the square pulse train of Figure 16.13 with an appropriate amplitude.

Ans. $$v_O(t) = \frac{8A_m}{\pi^2}\left[\frac{\omega_0 CR}{\sqrt{1+\omega_0^2C^2R^2}}\sin(\omega_0 t - \alpha_1) + \frac{\omega_0 CR}{3\sqrt{1+9\omega_0^2C^2R^2}}\sin(3\omega_0 t - \alpha_3) + \frac{\omega_0 CR}{5\sqrt{1+25\omega_0^2C^2R^2}}\sin(5\omega_0 t - \alpha_5) + \cdots\right],$$ where $\tan\alpha_n = n\omega_0 CR$.

P16.46 $v_{SRC}(t)$ in Figure P16.46 is the full-wave rectified waveform of Figure 16.27b having $T = 1/50$ s and an amplitude of 50 V. The purpose of the LC filter is to attenuate the ac components of $v_{SRC}(t)$, leaving a near-dc voltage across the 4 kΩ load. Determine the first four nonzero terms in the FSE of v_O and compare with those of $v_{SRC}(t)$.

Ans. The relative rates of attenuation of the first three ac terms are 0.14, 0.033, and 0.014, respectively.

P16.47 Repeat Problem P16.46 assuming that $v_{SRC}(t)$ is a square wave 50 V peak to peak, 25 V average value, and $T = 1/50$ s. Compare with the results of Problem P16.46.

Ans. The relative rates of attenuation of the first three ac terms are 0.80, 0.059, and 0.02, respectively.

P16.48 The sawtooth waveform of Figure 16.5 is applied to the circuit of Figure P16.48, where C is a very large capacitor. Determine the first five terms in the FSE of $v_O(t)$, assuming $A = 5$ V and $\omega = 10^6$ rad/s. What is the effect of C?

Ans. The magnitudes and phase angles are, respectively, 16 mV, 90.38°; 1.8 mV, 90.13°; 0.74 mV, 90.08°, 0.41 mV 90.06°; and 0.26 m, 90.05° to block the dc component.

P16.49 A triangular current source having the waveform of Figure 16.24, with $A_m = 10$ mA and $\omega = 10^5$ rad/s, is applied to the circuit of Figure P16.49. Determine the first three terms in the FSE of v_O.

Ans. $v_O(t) = 85.1\cos(\omega_0 t - 85.8°) + 3.03\cos(3\omega_0 t - 89.7°) + 0.651\cos(5\omega_0 t - 89.8°)$ V.

P16.50 $v_{SRC}(t)$ in Figure P16.50 is the triangular waveform of Figure 16.24. Determine the first three terms of $v_O(t)$, assuming $\omega = 1$ rad/s and $A_m = 10$ V.

Ans. $v_O(t) = 4.37\cos(\omega_0 t + 158.5°) + 0.34\cos(3\omega_0 t + 130.2°) + 0.085\cos(5\omega_0 t + 116.9°)$ V.

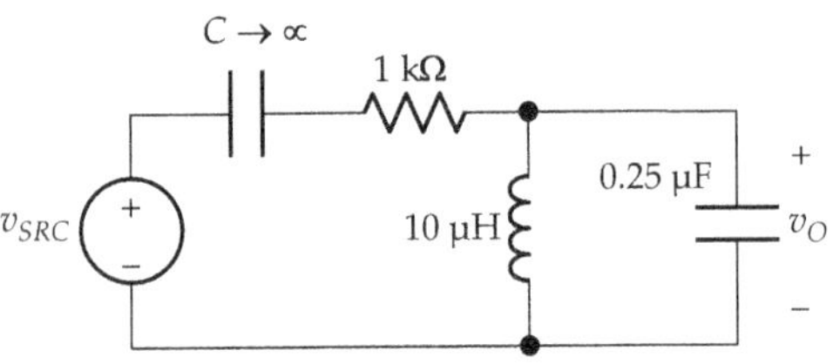

FIGURE P16.48

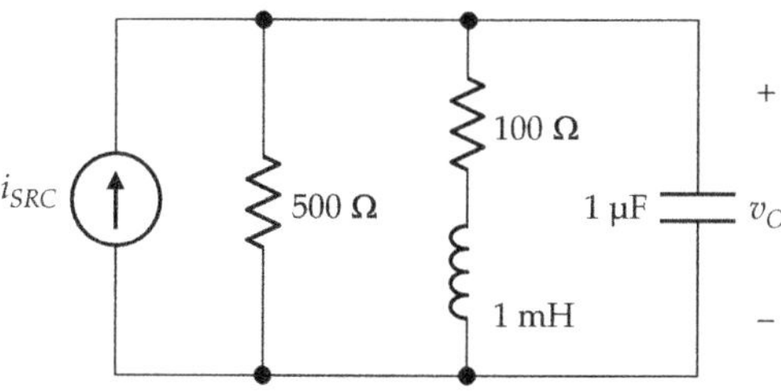

FIGURE P16.49

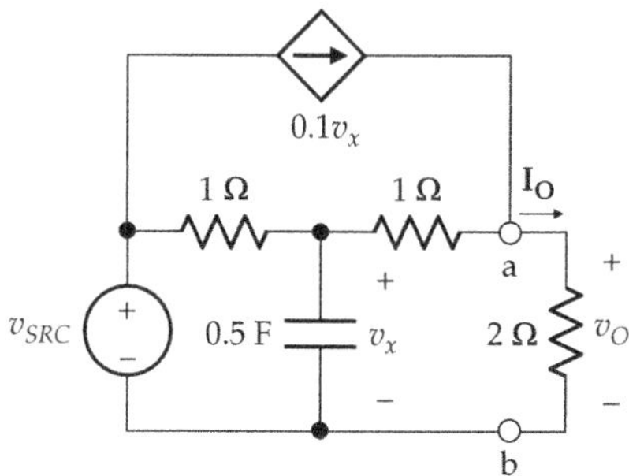

FIGURE P16.50

Power and rms Values

P16.51 Determine the average power dissipated in a resistance of 2 Ω if a voltage of $32\sin(3t)\cos^2(t/2)$ V is applied across the resistor.

Ans. 96 W.

P16.52 Given $f(t) = 5 + 10\cos 100\pi t + 5\cos 200\pi t + 2\cos 400\pi t$, (a) determine the rms value of $f(t)$; (b) if $f(t)$ is an approximation of a periodic voltage that dissipates 90 W in a 1 Ω resistor, determine the % error in real power involved in the approximation.

Ans. (a) 9.46 V; (b) 0.56%.

P16.53 Given $f(t) = 20\cos(10^3t + 90°) + 4\cos(3 \times 10^3t - 90°) + \cos(5 \times 10^3t + 90°) + 0.2\cos(7 \times 10^3t - 90°)$, (a) determine the rms value of the signal; (b) specify whether or not the signal is even or odd and whether or not it has half-wave symmetry. Repeat (a) and (b) for the same signal but with the phase angles of the components negated.

Ans. (a) 14.44; (b) odd and half-wave symmetric; (a) and (b) remain the same after the phase angles are negated.

P16.54 The voltage across two given terminals of a circuit is $v(t) = 5\sin(t) + 10\sin(3t)$ V. The current input at these terminals, in the direction of a voltage drop, is $i(t) = 7\sin(2t) + 50\sin(8t)$ A. Determine the average power absorbed by the circuit.

Ans. 0.

P16.55 The voltage across two given terminals of a circuit is $v(t) = 4\sin t + 5\cos 2t + 10\sin 4t$ V. The current input at these terminals, in the direction of the voltage drop, is $i(t) = 6\sin t + 8\cos 5t + 12\sin 8t$ A. Determine (a) the rms value of v, (b) the rms value of i, and (c) the average power delivered to the network at the given terminals.

Ans. (a) 8.40 V; (b) 11.04 A; (c) 12 W.

P16.56 The voltage across two given terminals of a circuit is $v(t) = 15 + 400\cos(500t) + 100\sin(1500t)$ V. The current input at these terminals, in the direction of the voltage drop, is $i(t) = 2 + 5\sin(500t + 60°) + 3\cos(1500t - 15°)$ A. Determine (a) the average power delivered to the circuit, (b) the rms value of v, and (c) the rms value of i.

Ans. (a) 934.85 W; (b) 291.93 V; (c) 4.58 A.

P16.57 A full-wave rectified waveform of 5 V amplitude and a frequency of 1 kHz is applied to a 5 Ω resistor in series with a parallel combination of another 5 Ω resistor and a 1 F capacitor. Determine the average power dissipated in the circuit.

Ans. 1.487 W.

P16.58 The voltage across two given terminals of a circuit is $v(t) = 2 + 2\cos(1000t) + \cos(2000t)$ V. The current at these terminals, in the direction of a voltage rise, is $i(t) = 1 + \sin(1000t) + 0.5\sin(2000t)$ A. Determine the average power delivered or absorbed by the circuit.

Ans. 2 W delivered.

P16.59 A voltage $(5\sin 2t - 6)$ V is applied to a 5 Ω resistor. Determine the average power dissipated in the resistor.

Ans. 9.7 W.

P16.60 The voltage applied across a 1 Ω resistor is expressed in exponential form of an FSE as $e^{j100\pi t} + (2+j4)e^{j200\pi t} + (3+j9)e^{j300\pi t}$ V, $n > 0$. Determine the energy dissipated in the resistor in the interval 1/50 to 2/50 s.

Ans. 4.44 J.

P16.61 A voltage $5\sin\omega_0 t$ V applied to a given resistor dissipates 5 W. Determine the power dissipated in the same resistor when the applied voltage is (a) a full-wave voltage and (b) a half-wave rectified voltage, both having the same amplitude of 5 V.

Ans. (a) 5 W; (b) 2.5 W.

P16.62 The voltage applied across a 5 Ω resistor is $v(t) = 1 - \sum_{n=1}^{\infty} \frac{\cos 500nt}{n^2}$ V. Determine the power dissipated in the resistor using the first four nonzero terms.

Ans. 0.31 W.

P16.63 The current through a coil having a resistance of 1Ω and an inductance of 10 mH consists of a fundamental and a third harmonic. The rms current through the coil is 5 A and the rms voltage cross the coil is 20 V. If the frequency of the fundamental is 300 rad/s, what are the rms values of the fundamental and third harmonic components of the coil current and voltage?

Ans. $I_1 = 2.5\sqrt{11/3}$ A, $I_3 = I_3 = 2.5/\sqrt{3}$ A, $V_1 = 2.5\sqrt{110/3}$ V, and $V_3 = 2.5\sqrt{82/3}$ V.

P16.64 A voltage $v(t) = 100 + 50\sin(500t) + 25\sin(1500t)$ V is applied across a coil having an inductance of 20 mH and a resistance of 5 Ω. Determine (a) the coil current $i(t)$ and (b) the average power dissipated in the coil.

Ans. (a) $i(t) = 20 + 4.47\sin(500t - 63.4°) + 0.822\sin(1500t - 80.54°)$ A; (b) Approx. 2052.0 W.

P16.65 A periodic waveform of 240 V rms has 20% 3rd harmonic content, 5% 5th harmonic content, and 2% 7th harmonic content, all the percentages being with respect to the fundamental. Determine the rms values of the 3rd and 7th harmonics.

Ans. 47 V and 4.7 V.

P16.66 A current $i(t) = 2 + 3\sin\omega_0 t - 2\sin 2\omega_0 t$ A is applied to a 3 Ω resistor in parallel with a 1 H inductor. Determine (a) the average power dissipated in the circuit and (b) the change in the average power when $i(t)$ is advanced or delayed by 2 s.

Ans. (a) 6.04 W; (b) no change.

P16.67 Each source in Figure P16.67 is $15\cos(10t)$ V, and the power dissipated in the middle resistor is 50 W. Determine the power dissipated in this resistor if the frequency of one of the sources is doubled.

Ans. 25 W.

P16.68 Determine the power dissipated in the circuit of Figure P16.68 if $v_{SRC1}(t) = 10\cos t$ and $v_{SRC2}(t) = 10\cos 3t$.

Ans. 50.7 W.

P16.69 Determine the rms value of $i(t)$ in Figure P16.69.

Ans. 11.87 A.

P16.70 A voltage $v(t) = 400\sqrt{2}\sin(\omega t + \theta_1) + 180\sqrt{2}\sin(3\omega t + \theta_3)$ V is applied to a series RLC circuit having $R = 60$ Ω. At the frequency of the third harmonic, the reactances of L and C have equal magnitudes, and the ratio L/C is 900 Ω². Determine the rms current in the circuit.

Ans. 5 A.

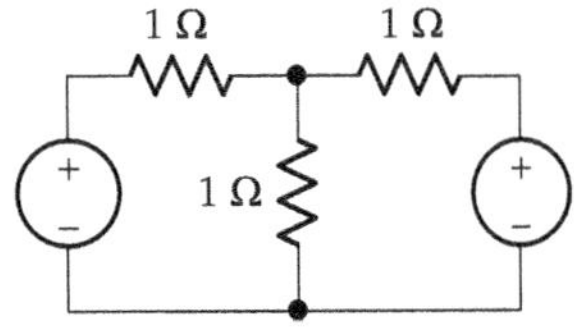

FIGURE P16.67

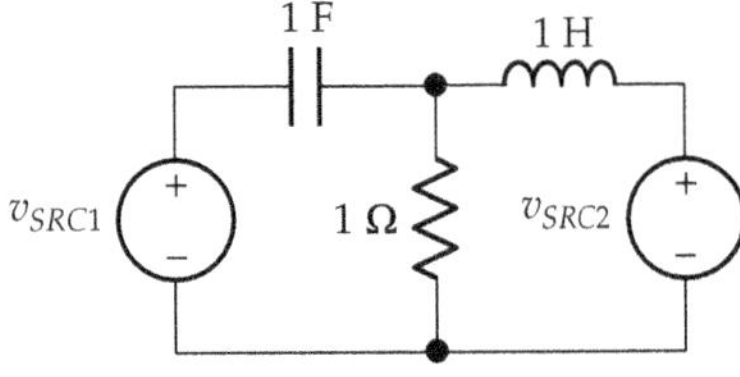

FIGURE P16.68

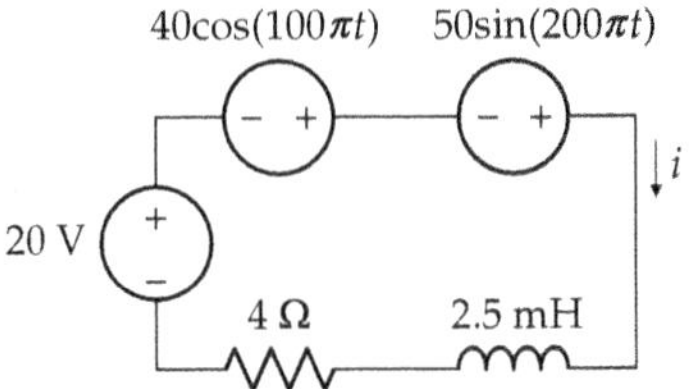

FIGURE P16.69

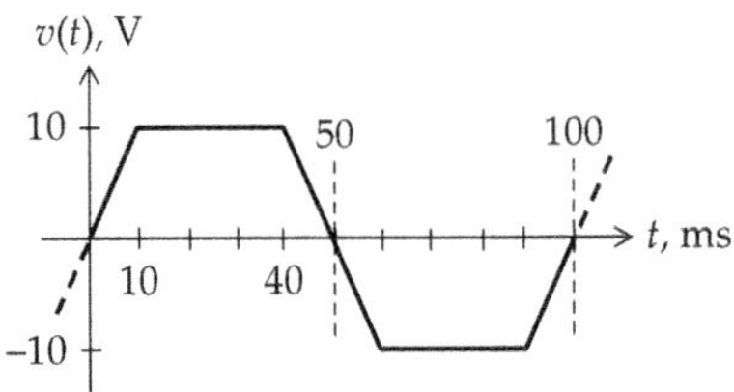

FIGURE P16.71

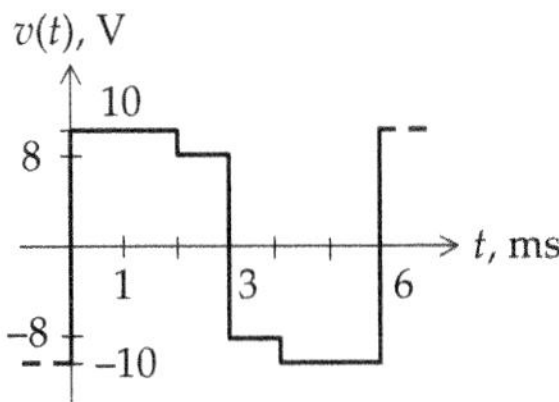

FIGURE P16.72

P16.71 Determine the rms value of the voltage waveform shown in Figure P16.71.

Ans. $2\sqrt{55/3}$ V.

P16.72 The periodic voltage shown in Figure P16.72 is applied across a 10 Ω resistor. Determine the average power dissipated in the resistor.

Ans. 8.8 W.

P16.73 The periodic current whose first period is shown in Figure P16.73 is applied to a 5 Ω resistor. Determine the average power dissipated in the resistor.

Ans. 10 W.

P16.74 Determine the rms value of the periodic function shown in Figure P16.74.

Ans. 1.29.

P16.75 Determine the rms value of the voltage shown in Figure P16.75 over the time interval (0, 5) s.

Ans. 0.86.

P16.76 $f_2(t)$ in Figure P16.76 is the same function $f_1(t)$ lowered by 1 unit. Which function has the larger rms value?

Ans. $f_1(t)$.

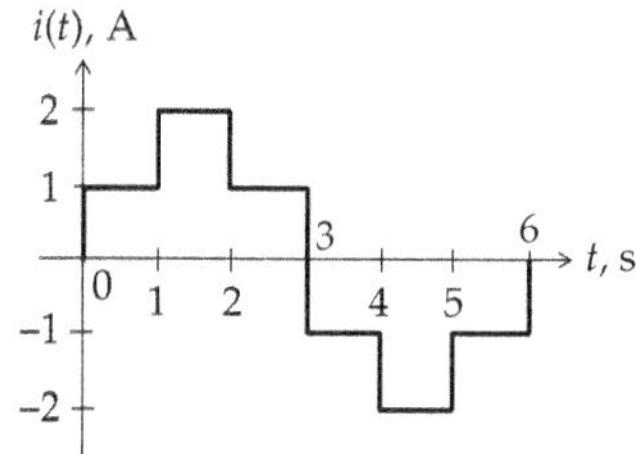

FIGURE P16.73

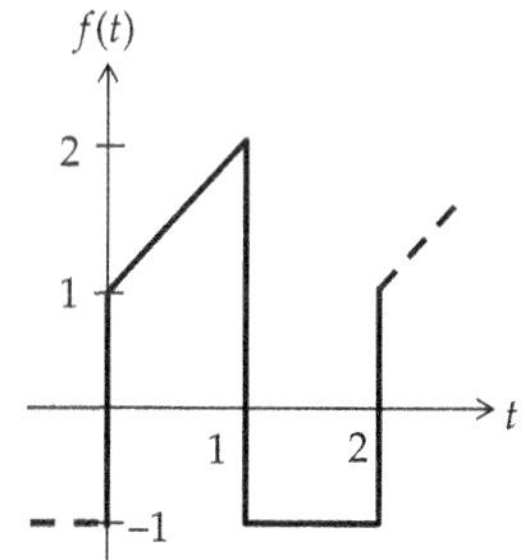

FIGURE P16.74

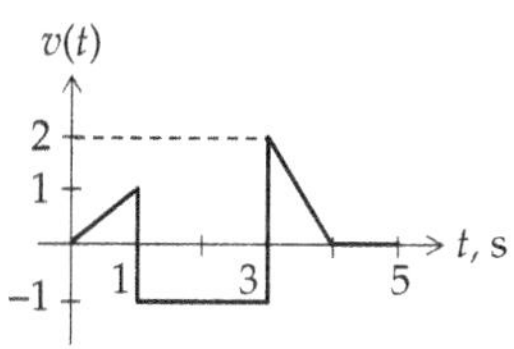

FIGURE P16.75

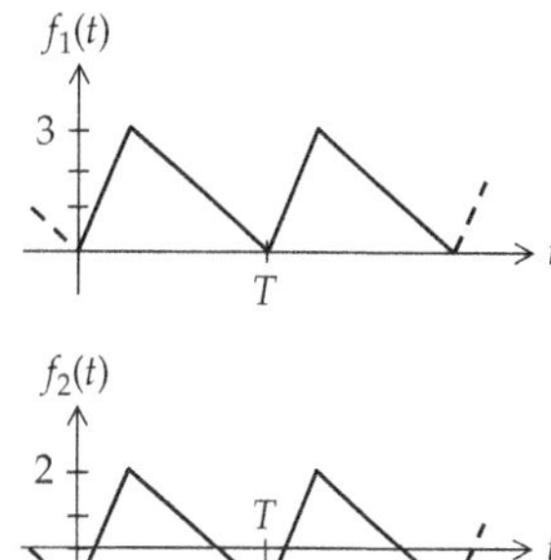

FIGURE P16.76

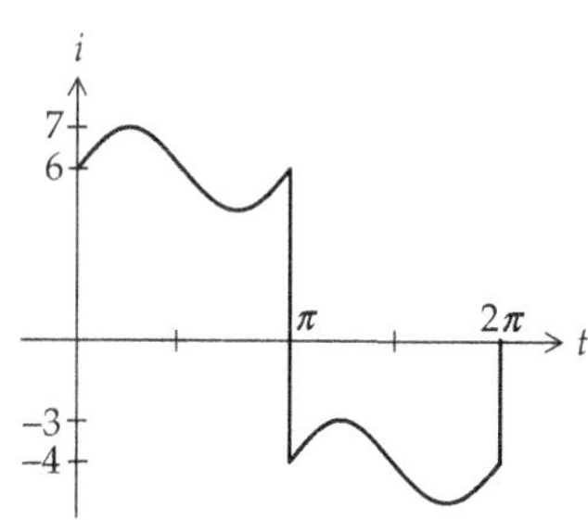

FIGURE P16.77

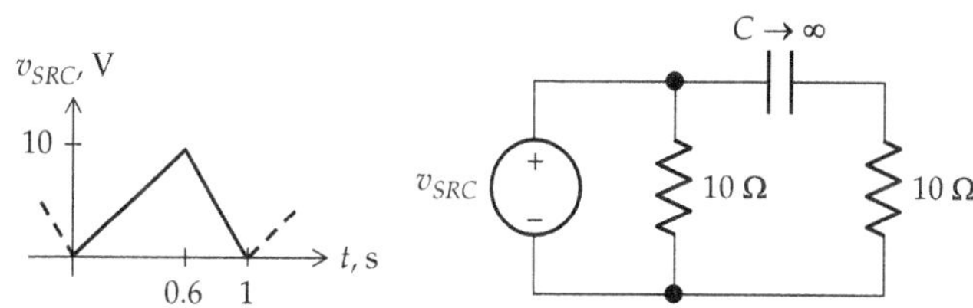

FIGURE P16.78

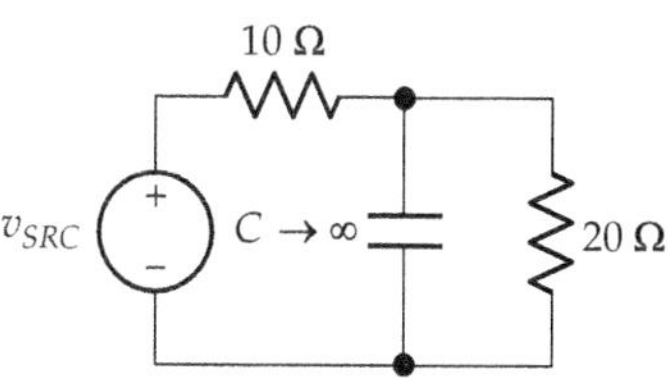

FIGURE P16.79

P16.77 The periodic current shown in Figure P16.77 is described over a period as

$i(t) = 6 + \sin 2t$ A, $0 \le t \le \pi$

$= -4 + \sin 2(t - \pi)$ A, $\pi \le t \le 2\pi$.

Determine the rms value of $i(t)$.

Ans. 5.15 A.

P16.78 The periodic voltage v_{SRC} is applied as shown in Figure P16.78. Determine the average power dissipated in the circuit.

Ans. 4.17 W

P16.79 The half-wave rectified waveform of Figure 16.27a, with $A = 10$ V and $f = 50$ Hz, is applied to the circuit of Figure P16.79. Determine the average power dissipated in the circuit.

Ans. 1.83 W.

P16.80 A periodic voltage having the waveform of Figure P16.25, with $A = 8$ V and $T = 1$ s, is applied to a coil having a resistance of 4 Ω and an extremely large inductance. Determine the average power dissipated in the circuit.

Ans. 1 W.

Design Problems

P16.81 $v_I(t)$ in Figure P16.81 is a rectangular pulse train (Figure 16.9) of 1 V amplitude, with $T = 10$ ms and $\tau = 2$ ms. Select the largest R and smallest C that will satisfy the following requirements: (i) the dc component in v_O should not be attenuated by more than 10% of that in v_I and (ii) the magnitude of the fundamental in v_O should not exceed 2% of the dc component.

Ans. $R = 100$ kΩ, $C = 1.5$ μF.

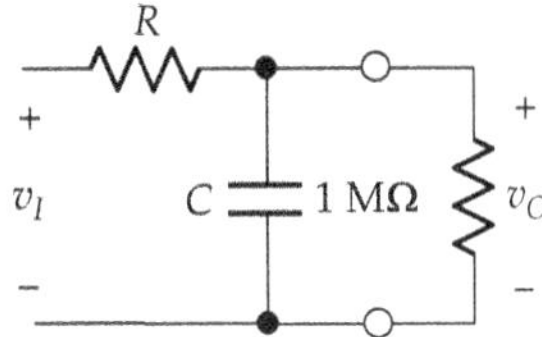

FIGURE P16.81

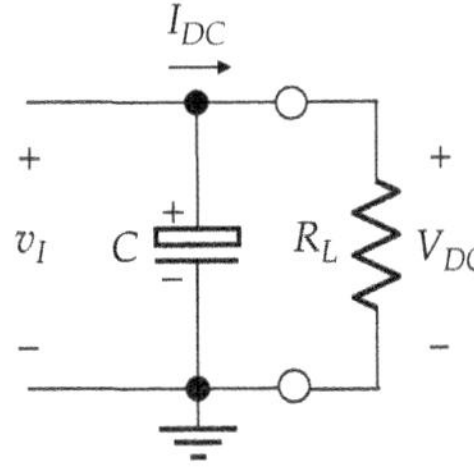

FIGURE P16.82

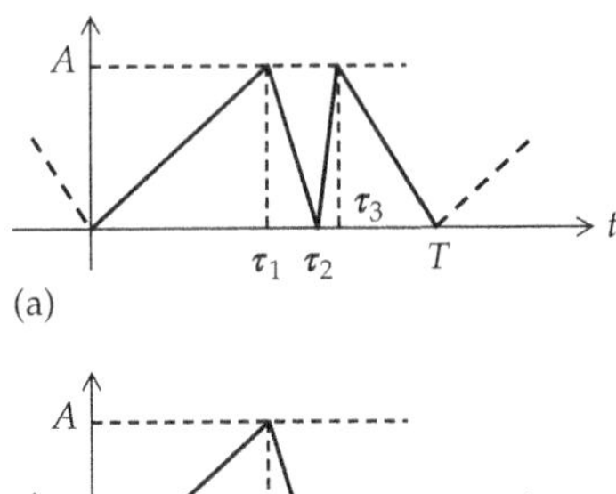

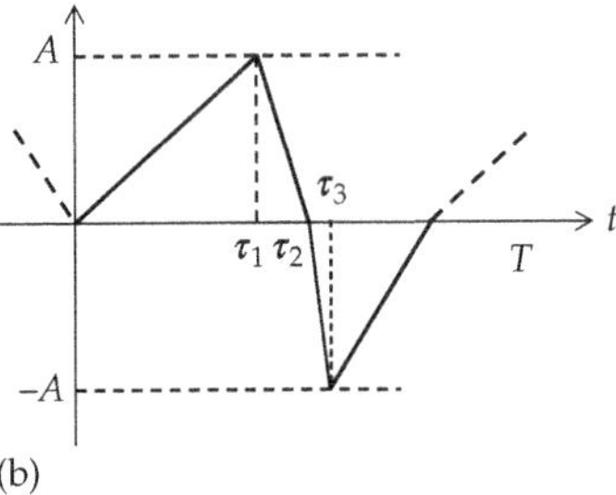

FIGURE P16.85

P16.82 Figure P16.82 illustrates a capacitor-input filter that is commonly used to obtain a nominally dc voltage from a full-wave rectified waveform (Figure 16.27b) $v_I(t)$. C is an electrolytic capacitor, which has a large capacitance, and is polarized by the dc voltage across it. In the approximate analysis of the filter, the governing equations are $V_{DC} = V_m - \frac{v_r}{2}$, $v_r = \frac{I_{DC}}{2fC}$, with $V_{DC} = R_L I_{DC}$, where V_m is the peak voltage of the input, v_r is the peak-to-peak ripple in the load voltage, and f is the frequency of the supply from which v_I is derived. It is required to have $V_{DC} = 12$ V across a load resistance $R_L = 20\ \Omega$, with $f = 50$ Hz, and v_r not exceeding 5% of V_{DC}. Determine (a) V_m, (b) C, and (c) the rms value of the ac component of v_r, assuming v_r has a nearly triangular waveform.

Ans. (a) 12.3 V; (b) 10,000 μF; (c) $100\sqrt{3} = 173$ mV.

P16.83 An LC filter (Figure P16.46) is to be used to supply a 100 Ω resistive load from a full-wave rectified waveform (Figure 16.27b) of 15 V amplitude derived from a 50 Hz supply. Determine C if $L = 0.1$ H and the amplitude of the first ac component of v_O is not to exceed about 3.5% of the dc component.

Ans. 508 μF.

Probing Further

P16.84 Consider the square pulse train of Figure 16.14a whose FSE is given by Equation 16.44. Assume that $A_m = 1$ and $\omega_0 = 1$ rad/s. Using an appropriate computer program, plot the FSE expansion using an increasing number of harmonics. Observe that with the ninth harmonic, the waveform begins to look like a square wave. Note, however, that no matter how many harmonics you add (necessarily a finite number), there always remains a small overshoot and some damped oscillations following the discontinuities. This is referred to as **Gibb's phenomenon**.

P16.85 Generalize the result of Example 16.10 to the triangular waveform of Figure P16.85a having two triangles per period of equal amplitude A but arbitrary values of τ_1, τ_2, and τ_3 within a period, where $\tau_1 \le \tau_2 \le \tau_3$. Show that the rms value is the same if one of the triangles is negative-going but of the same amplitude (Figure P16.85b). Deduce that the rms value of a function that alternates linearly between $+A$ and $-A$ a number of times is $A/\sqrt{3}$.

Ans. rms value remains $A/\sqrt{3}$.

17

Real, Reactive, and Complex Power

Objective and Overview

Power calculations are generally straightforward but tedious because they involve complex voltages and currents. They are considerably facilitated by utilizing the concept of complex power. The real part of complex power is real, or average, power, whereas its imaginary part is the reactive power associated with energy storage elements. The usefulness of complex power stems from its conservation, which implies that real and reactive power can be summed branch by branch in any given circuit.

Energy storage elements alternately store energy and return this energy to the supply. The reactance of energy storage elements places added current and voltage burdens on the power system, which necessitates sometimes adding capacitors that counteract the normally inductive reactances of power systems. Another important practical consideration is when a source inherently has a relatively large source impedance. It is of interest in such cases to determine the load impedance that results in maximum power transfer to the load.

The chapter begins by discussing instantaneous power in resistors, inductors, and capacitors, followed by the more general case of a circuit that is a combination of these elements. This serves to define real power and reactive power and explain their nature. Complex power is then defined, its conservation justified in detail, and its usefulness in circuit analysis illustrated in power factor correction that counteracts the inductive reactive power in a circuit.

The rest of the chapter addresses the problem of maximum power transfer, first in the case of purely resistive circuits, followed by the more general case in terms of impedances and admittances of a source and a load. The conditions of maximum power transfer from a source to a load are examined, first without constraints and then subject to some constraints that may be encountered under certain conditions.

17.1 Instantaneous and Real Power

17.1.1 Resistor

Consider a voltage $v(t) = V_m\cos(\omega t + \theta)$ applied to a resistor R (Figure 17.1a). The current through the resistor is $i(t) = I_m\cos(\omega t + \theta)$, where $I_m = V_m/R$. The instantaneous power dissipated in the resistor at any time t is

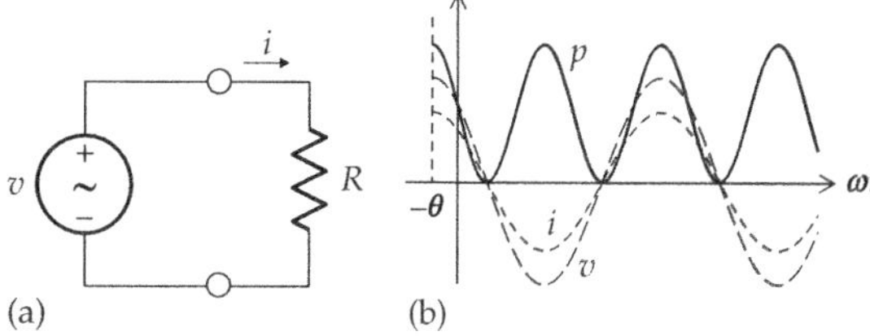

FIGURE 17.1
Instantaneous power for an ideal resistor. (a) Resistor connected to a sinusoidal source and (b) variation of voltage, current, and power with time.

$$p(t) = v(t)i(t) = V_mI_m\cos^2(\omega t + \theta) = \frac{V_mI_m}{2}\left[1 + \cos 2(\omega t + \theta)\right] \quad (17.1)$$

as shown in Figure 17.1b. The instantaneous power varies at twice the supply frequency because power is proportional to the square of v or i, so that the variation of power with time is the same during the positive half-cycles of v and i as during the negative half-cycles. p is never negative, since the resistor does not return power to the supply. If p is averaged over a full period, $2\pi/\omega$, the cosine term averages to zero, so that the average of p over a period is

$$P = \frac{V_mI_m}{2} = \frac{V_m}{\sqrt{2}}\frac{I_m}{\sqrt{2}} = V_{rms}I_{rms} \quad (17.2)$$

Equation 17.2 agrees with what was derived in Section 8.3. P is referred to as the **real**, or **average**, power. It represents electric energy that is converted to another form of energy, which in this case is heat if R represents a heating element. In general, real power could represent conversion to another form of energy, if R, for example, represents a lamp. The expression "real power" will be used in this chapter rather than "average power" to emphasize that real power is the real part of complex power.

Equation 17.1 is expressed in terms of P as

$$p(t) = P\left[1 + \cos 2(\omega t + \theta)\right] \quad (17.3)$$

It is seen that P is both the magnitude of the real power and the amplitude of the alternating component of the instantaneous power of frequency 2ω rad/s. According to Equation 17.3, the maximum value of p is $2P$.

Primal Exercise 17.1

A current of $5\cos 100\pi t$ A flows through a 4 Ω resistor. Determine (a) the real power dissipated in the resistor and (b) the amplitude of the alternating component of the instantaneous power and its frequency.

Ans. (a) 50 W; (b) 50 W, 200π rad/s.

17.1.2 Inductor

If the voltage $v(t) = V_m\cos(\omega t + \theta)$ is applied across an inductor L (Figure 17.2a), the current through the inductor is $i(t) = I_m\cos(\omega t + \theta - 90°) = I_m\sin(\omega t + \theta)$, where $I_m = V_m/\omega L$ (Section 8.3). The instantaneous power absorbed by the inductor at any time t is

$$p(t) = v(t)i(t) = V_mI_m\cos(\omega t+\theta)\sin(\omega t+\theta)$$

$$= \frac{V_mI_m}{2}\sin 2(\omega t+\theta) \quad (17.4)$$

as shown in Figure 17.2b. When p is positive, the inductor actually absorbs power and stores it as energy in the magnetic field. When p is negative, the inductor delivers power, thereby returning the stored energy to the source. The real power is zero, so that as much power flows in one direction as in the opposite direction and no power is dissipated. p varies at twice the supply frequency because it is the product of v and i, so that p is positive when v and i are both positive or both negative, and p is negative otherwise.

By analogy with the case of the resistor, Equation 17.4 is expressed as

$$p(t) = Q_L\sin 2(\omega t+\theta) \quad (17.5)$$

$$Q_L = \frac{V_mI_m}{2} = \frac{V_m}{\sqrt{2}}\frac{I_m}{\sqrt{2}} = V_{rms}I_{rms} \quad (17.6)$$

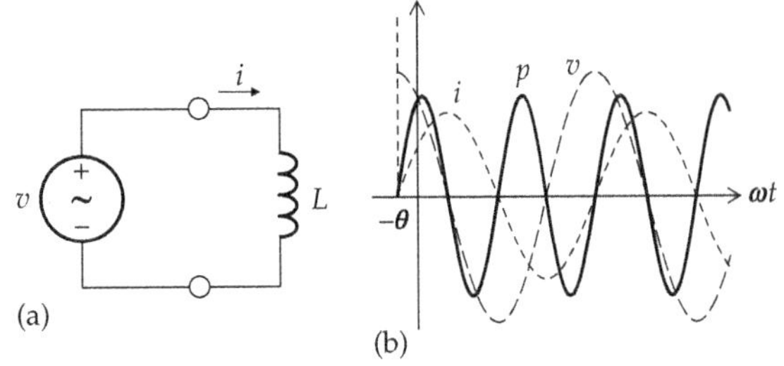

FIGURE 17.2
Instantaneous power for an ideal inductor. (a) Inductor connected to a sinusoidal source and (b) variation of voltage, current, and power with time.

Q_L is referred to as the **reactive power** in the case of an inductor. It is the amplitude of the power that is alternately absorbed in the inductor and returned to the supply. In this sense, reactive power is confined to the electric circuit and is not available outside the circuit. Because of its different nature, the unit of Q is not a watt but a volt-ampere reactive, abbreviated as VAR.

17.1.3 Capacitor

Similarly, if the voltage $v(t) = V_m\cos(\omega t + \theta)$ is applied across a capacitor C (Figure 17.3a), the current through the capacitor is $i(t) = I_m\cos(\omega t + \theta + 90°) = -I_m\sin(\omega t + \theta)$, where $I_m = \omega CV_m$ (Section 8.3). The instantaneous power delivered to the capacitor at any time t is

$$p(t) = v(t)i(t) = -V_mI_m\cos(\omega t+\theta)\sin(\omega t+\theta)$$

$$= -\frac{V_mI_m}{2}\sin 2(\omega t+\theta) \quad (17.7)$$

as shown in Figure 17.3b. When p is positive, the capacitor actually absorbs power and stores it as energy in the electric field. When p is negative, the capacitor delivers power, thereby returning the stored energy to the source. The real power is zero, so that as much power flows in one direction as in the opposite direction and no power is dissipated. p varies at twice the supply frequency, as explained for the case of an inductor.

Similar to the case of the inductor, Equation 17.7 can be expressed as

$$p(t) = Q_C\sin 2(\omega t+\theta) \quad (17.8)$$

$$Q_C = -\frac{V_mI_m}{2} = -\frac{V_m}{\sqrt{2}}\frac{I_m}{\sqrt{2}} = -V_{rms}I_{rms} \quad (17.9)$$

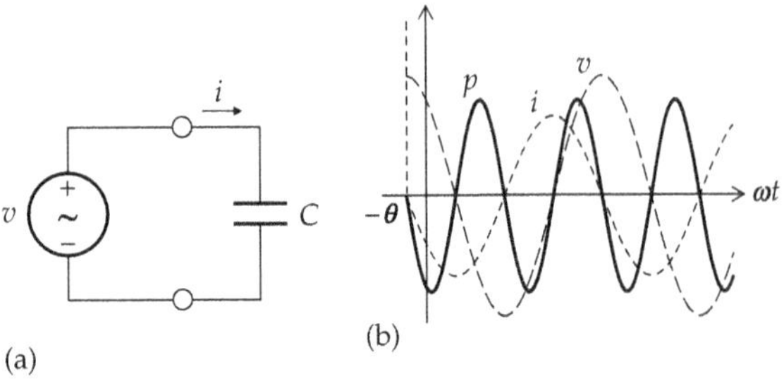

FIGURE 17.3
Instantaneous power for an ideal capacitor. (a) Capacitor connected to a sinusoidal source and (b) variation of voltage, current, and power with time.

where Q_C is again a reactive power equal to the amplitude of the power that is alternately stored in the capacitor and returned to the source. Note that the reactive power is positive for an inductor and negative for a capacitor.

Primal Exercise 17.2

A current $2\cos\omega t$ flows through a 1 mH inductor in series with a 1 mF capacitor. Determine Q for (a) the inductor and (b) the capacitor, assuming ω = 2000 rad/s; (c) calculate the series reactance and verify that Q for this reactance is the algebraic sum of the reactive power in (a) and (b).

Ans. (a) 4 VAR; (b) −1 VAR; (c) series reactance is 1.5 Ω inductive, Q = 3 VAR.

Primal Exercise 17.3

Repeat Primal Exercise 17.2 assuming ω = 500 rad/s.

Ans. (a) 1 VAR; (b) −4 VAR; (c) series reactance is 1.5 Ω capacitive, Q = −3 VAR.

Primal Exercise 17.4

(a) Determine the voltages across the inductor and capacitor $v_L(t)$ and $v_C(t)$ in Primal Exercise 17.2; (b) verify that the phasor sum $\mathbf{V}_L + \mathbf{V}_C$ is the phasor representation of $v_L(t) + v_C(t)$.

Ans. (a) $-4\sin 2000t$ V, $\sin 2000t$ V.

17.1.4 General Case

We will now generalize the preceding discussion to a circuit that consists of resistors as well as inductors and capacitors, in the sinusoidal steady state. Let v be the voltage applied to the terminals of such a circuit 'N', and let i be the current entering these terminals in the direction of the voltage drop v (Figure 17.4a). The instantaneous power delivered to the circuit is

$$p = vi \tag{17.10}$$

In general, the power p is partly dissipated in the resistors of the circuit and partly stored in the energy storage elements. If $v(t) = V_m\cos(\omega t + \theta_v)$ and $i(t) = I_m\cos(\omega t + \theta_i)$

$$p(t) = V_m I_m \cos(\omega t + \theta_v)\cos(\omega t + \theta_i) \tag{17.11}$$

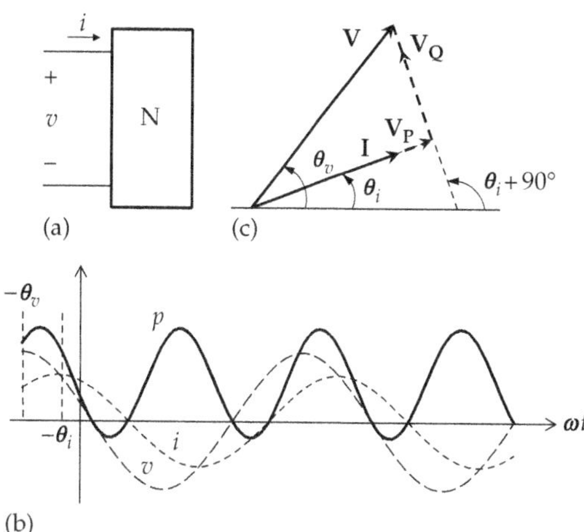

FIGURE 17.4
Instantaneous power for an LTI circuit. (a) Sinusoidal excitation applied to a circuit 'N', (b) variation of voltage, current, and power with time, and (c) voltage phasor diagram resolved into two components.

as shown in Figure 17.4b. If the circuit is inductive, $\theta_v > \theta_i$, that is, the current lags the voltage, as assumed in Figure 17.4, for the sake of illustration. On the other hand, if the circuit is capacitive, $\theta_i > \theta_v$, that is, the current leads the voltage. Since the circuit contains, in general, both resistors and energy storage elements, p has a nonzero average, because of power dissipation in the resistors, and p is negative over part of the cycle because of the power returned to the supply by the energy storage elements, as in Figures 17.2 and 17.3.

Figure 17.4c illustrates the phasor diagram for v and i as phasors: $\mathbf{V} = V_m\angle\theta_v$ and $\mathbf{I} = I_m\angle\theta_i$, with $\theta_v > \theta_i$, as assumed. Let us resolve $\mathbf{V}$ into two components: a component $\mathbf{V_P}$ that is in phase with $\mathbf{I}$ and a component $\mathbf{V_Q}$ that is in phase quadrature with $\mathbf{I}$. The magnitude of $\mathbf{V_P}$ is V_m, the magnitude of $\mathbf{V}$ multiplied by $\cos(\theta_v - \theta_i)$, and its phase angle is θ_i, the same as that of $\mathbf{I}$. In the time domain, $\mathbf{V_P}$ is

$$v_P(t) = \left[V_m\cos(\theta_v - \theta_i)\right]\cos(\omega t + 0_i)] \tag{17.12}$$

The magnitude of $\mathbf{V_Q}$ is V_m multiplied by $\sin(\theta_v - \theta_i)$, and its phase angle is θ_i, + 90° (Figure 17.4c). In the time domain, $\mathbf{V_Q}$ is

$$\begin{aligned} v_Q(t) &= [V_m\sin(\theta_v - \theta_i)]\cos(\omega t + \theta_i + 90°) \\ &= -[V_m\sin(\theta_v - \theta_i)]\sin(\omega t + \theta_i) \end{aligned} \tag{17.13}$$

Multiplying v_P from Equation 17.12 by i and denoting the product as p_P,

$$\begin{aligned} p_P(t) &= v_P i = V_m I_m \cos(\theta_v - \theta_i)\cos^2(\omega t + \theta_i) \\ &= \frac{V_m I_m}{2}\cos(\theta_v - \theta_i)\left[1 + \cos 2(\omega t + \theta_i)\right] \end{aligned} \tag{17.14}$$

Since v_P is the component of voltage that is in phase with i, p_P represents the instantaneous power dissipated in the resistances in the circuit. If p_P is averaged over a full period, the $\cos2(\omega t + \theta_i)$ term averages to zero, leaving the average of p_P as the constant term $(V_mI_m/2)$ $\cos(\theta_v - \theta_i)$. This is the real power P in the general case, so that Equation 17.14 can then be written as

$$p_P(t) = P\left[1+\cos2(\omega t+\theta_i)\right] \tag{17.15}$$

where

$$P = \frac{V_mI_m}{2}\cos(\theta_v - \theta_i) = V_{rms}I_{rms}\cos(\theta_v - \theta_i) \tag{17.16}$$

It is seen that Equation 17.15 is of the same form as Equation 17.3 and that when the circuit 'N' is purely resistive, $\theta_v = \theta_I$, and P in Equation 17.16 reduces to P in Equation 17.2. On the other hand, $\theta_v - \theta_I = +90°$ for a purely inductive circuit and $\theta_v - \theta_I = -90°$ for a purely capacitive circuit. In either case, $P = 0$, since no power is dissipated in a purely reactive circuit. It follows that the general expression for real power is given by Equation 17.16 and involves $\cos(\theta_v - \theta_I)$, which accounts for the component of voltage that is in phase with the current.

Multiplying v_Q from Equation 17.13 by i and denoting the product as p_Q,

$$\begin{aligned} p_Q(t) = v_Qi &= -V_mI_m\sin(\theta_v-\theta_i)\sin(\omega t+\theta_i)\cos(\omega t+\theta_i) \\ &= -\frac{V_mI_m}{2}\sin(\theta_v-\theta_i)\sin2(\omega t+\theta_i) \\ &= \frac{V_mI_m}{2}\sin(\theta_v-\theta_i)\cos\left[2(\omega t+\theta_i)+90°\right] \\ &= Q\cos\left[2(\omega t+\theta_i)+90°\right] \end{aligned} \tag{17.17}$$

where

$$Q = \frac{V_mI_m}{2}\sin(\theta_v - \theta_i) = V_{rms}I_{rms}\sin(\theta_v - \theta_i) \tag{17.18}$$

Again, Equation 17.18 is of the same form as Equations 17.5 and 17.8. For a purely inductive circuit, $\theta_v - \theta_i = 90°$, and Equation 17.18 reduces to Equation 17.6, whereas for a purely capacitive circuit, $\theta_v - \theta_i = -90°$, and Equation 17.18 reduces to Equation 17.9. For a purely resistive circuit, $\theta_v = \theta_I$, so $Q = 0$, since a resistive circuit does not store electric or magnetic energy. It is seen that the general expression for Q is given by Equation 17.18 and involves $\sin(\theta_v - \theta_I)$, which accounts for the component of voltage that is in phase quadrature with the current. Again, Q is positive for an inductive reactance ($\theta_v > \theta_I$) and is negative for a capacitive reactance ($\theta_v < \theta_I$).

The preceding results can be summarized as follows:

Summary: *When v is the voltage across the terminals of a circuit consisting of resistors, inductors, and capacitors and i is the current entering these terminals in the direction of a voltage drop v, the instantaneous power $p = vi$ is pulsating at a frequency that is twice that of v and i. The real power dissipated in the resistors is P, the average of the product of i and the component of v that is in phase with I; $P = V_{rms}I_{rms}$ $\cos(\theta_v - \theta_i)$, where P is also the amplitude of the alternating component of the product of i and the component of v that is in phase with i. The reactive power, representing the energy that is alternately stored in the energy storage elements and returned to the rest of the circuit, is the product of i and the component of v that is in phase quadrature with i. The reactive power is purely alternating, is of zero average, and has an amplitude of $Q = V_{rms}I_{rms}\sin(\theta_v - \theta_i)$. When the circuit is inductive, $\theta_v > \theta_I$, so Q is positive. When the circuit is capacitive, $\theta_v < \theta_I$, so Q is negative.*

From conservation of power, the power input at every instant is the algebraic sum of the power dissipated in the resistances and the power stored as energy in energy storage elements at that instant. In other words,

$$p = p_P + p_Q \tag{17.19}$$

This is illustrated in Figure 17.5, where p_P has an average P and an alternating component $P\cos2(\omega t + \theta_i)$. p_Q is purely alternating and of amplitude Q and leads the alternating component of p_P by 90°, assuming the circuit is inductive. At any instant of time t_1, p is the algebraic sum of p_P and p_Q.

If both sides of Equation 17.19 are divided by i,

$$v = v_P + v_Q \tag{17.20}$$

In other words, at any instant of time, v is the algebraic sum of v_P and v_Q. In the frequency domain, $\mathbf{V}$ is the phasor sum of $\mathbf{V_P}$ and $\mathbf{V_Q}$ (Figure 17.4c).

Example 17.1 illustrates real and reactive power in an RL circuit.

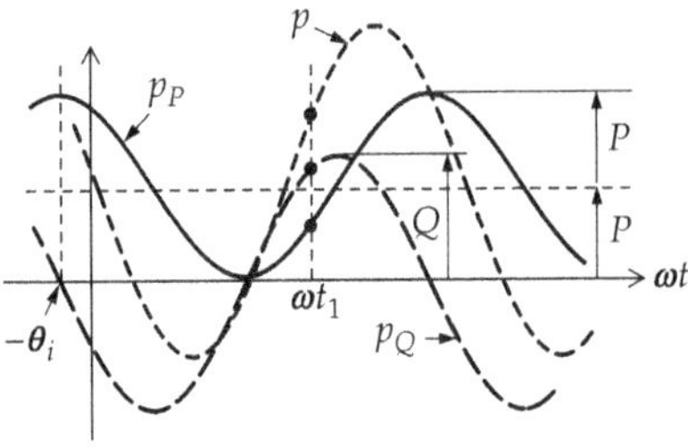

FIGURE 17.5
Instantaneous, real, and reactive power in the time domain.

Exercise 17.5

Show that Equations 17.19 and 17.20 are satisfied by applying trigonometric identities to (a) ($v_P + v_Q$), where v_P and v_Q are given by Equations 17.12 and 17.13, respectively, and (b) ($p_P + p_Q$), where p_P and p_Q are given by Equations 17.14 and 17.17, respectively.

Exercise 17.6

Given that $p_P(t) = P[1 + \cos2(\omega t + \theta_i)]$ and $p_Q(t) = -Q\sin2(\omega t + \theta_i)$, show that the maximum and minimum values of the instantaneous power p are given by

$$p_{max} = P + \sqrt{P^2 + Q^2} \quad \text{and} \quad p_{min} = P - \sqrt{P^2 + Q^2} \quad (17.21)$$

Example 17.1: Real and Reactive Power in an Inductive Circuit

Consider a voltage $v(t) = 100\cos(1000t + 30°)$ V applied to a 30 Ω resistor in series with a 40 mH inductor (Figure 17.6a). It is required to examine the real and reactive powers in the circuit.

Solution:

$\omega L = 1000 \times 0.04 = 40$ Ω. In terms of the phasor voltage **V** and the phasor current **I**, $\mathbf{I} = \frac{\mathbf{V}}{Z} = \frac{100\angle 30°}{30 + j40} = 2\angle\theta_i$ where $\theta_i = 30° - \tan^{-1}(4/3) = -23.13°$ (Figure 17.6b). In the time domain, $i(t) = 2\cos(1000t - 23.13°)$ A, and $v_R(t) = Ri = 60\cos(1000t - 23.13°)$ V, where $v_R(t)$ is the component of v in phase with i (v_P in Equation 17.12). $v_L(t) = Ldi/dt = -80\sin(1000t - 23.13°)$ V, $= 80\cos(1000t - 23.3° + 90°)$ V and is the component of v that leads i by 90° (v_Q in Equation 17.13). Note that $v = v_R + v_L$ (Exercise 17.7). As phasors, $\mathbf{V_R} = 60\angle\theta_i$ V and $\mathbf{V_L} = 80\angle(\theta_i + 90°)$ V, so that $\mathbf{V_R} + \mathbf{V_R} = \mathbf{V}$, as in Figure 17.6b.

It should be carefully noted that the real power dissipated in the resistor is, from Equation 17.2, $P = V_{Rrms} \times I_{rms} = (60 \times 2)/2 = 60$ W *in terms of the voltage across the resistor and current through it*. But *in terms of the voltage and current at the terminals of the circuit*, $P = V_{rms} \times I_{rms} \cos(\theta_v - \theta_i)$, where $(\theta_v - \theta_i) = \tan^{-1}(4/3)$, so that $\cos(\theta_v - \theta_i) = 3/5$ and $P = (100 \times 2/2) \times (3/5) = 60$ W, as it should be. The instantaneous power dissipated in the resistor is, from Equation 17.15: $60[1 + \cos2(\omega t - 23.13°)$, the maximum instantaneous power dissipation being 120 W.

From Equation 17.18, and *in terms of the voltage and current at the terminals of the circuit*, $Q = V_{rms}I_{rms}\sin(\theta_v - \theta_i) = (100 \times 2/2) \times (4/5) = 80$ VAR, where $\sin(\theta_v - \theta_i) = 4/5$, since $\tan(\theta_v - \theta_i) = 4/3$. *In terms of the current and voltage across the inductor*, $Q = V_{Lrms}I_{rms} = (80 \times 2/2) = 80$ VAR, as it should be. From Equation 17.18, $v_Qi(t) = 80\cos\left[2(1000t - 23.13°) + 90°\,\text{VAR}\right]$.

It is instructive to determine Q from the instantaneous energy stored in the inductor. This energy is at any instant: $w(t) = (1/2)Li^2 = (1/2)LI_m^2\cos^2(\omega t + \theta_i)$. The instantaneous reactive power in the inductor p_L is dw/dt. It follows that $p_L(t) = -(1/2)\left(2\omega LI_m^2\cos(\omega t + \theta_i)\sin(\omega t + \theta_i)\right) = -(1/2)\omega LI_m^2\sin2(\omega t + \theta_i) = (1/2)\omega LI_m^2\cos\left[2(\omega t + \theta_i) + 90°\right]$. The amplitude $(1/2)\omega LI_m^2$ is $Q = (1/2) \times 1000 \times 0.04 \times 4 = 80$ VAR.

Problem-Solving Tip

- When working with power, it is convenient to express voltages and currents as rms values.

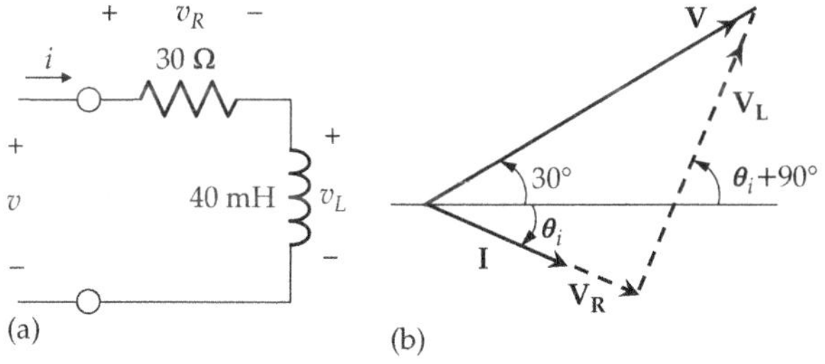

FIGURE 17.6
Figure for Example 17.1.

Exercise 17.7

Show that $v = v_R + v_L$ in Example 17.1.

Primal Exercise 17.8

Repeat Example 17.1 with the inductor replaced by a 25 μF capacitor.

Ans. $P = 60$ W, $Q = -80$ VAR.

17.2 Complex Power

17.2.1 Complex Power Triangle

P and Q are the amplitudes of purely alternating quantities, namely, $P\cos2(\omega t + \theta_i)$ and $Q\cos[2(\omega t + \theta_i) + 90°]$, respectively (Equations 17.15 and 17.18). Since these components are sinusoidal functions of time having the same frequency 2ω, they can be represented as phasors on an Argand diagram. As real power, **P** is drawn with zero phase angle, along the real axis, and **Q** is drawn with a phase angle of 90°, for an inductive reactance

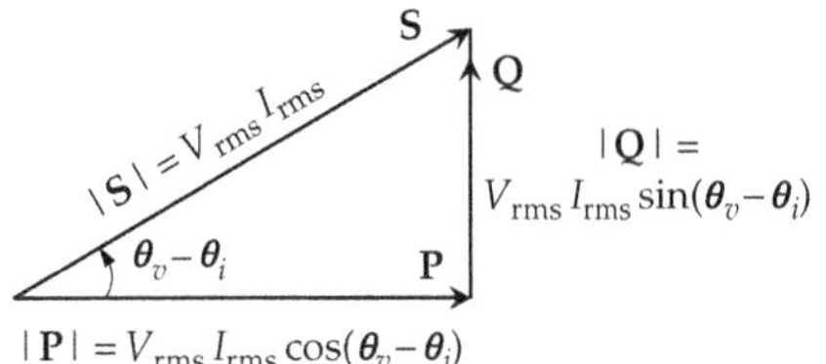

FIGURE 17.7
Complex power triangle.

(Figure 17.7), and with a phase angle of −90° for a capacitive reactance. The phasor sum is

$$\mathbf{S} = \mathbf{P} + \mathbf{Q} = |\mathbf{P}| + j|\mathbf{Q}| \tag{17.22}$$

where $|\mathbf{P}| = P = V_{rms}I_{rms}\cos(\theta_v - \theta_i)$ and $|\mathbf{Q}| = Q = V_{rms}I_{rms}\sin(\theta_v - \theta_i)$. **S** is the **complex power** having a magnitude $|\mathbf{S}| = S = \sqrt{P^2 + Q^2} = V_{rms}I_{rms}$ and a phase angle of $(\theta_v - \theta_i)$. The magnitude S is referred to as the **apparent power**. The unit of S is the volt-ampere (VA). Although **P**, **Q**, and **S** are of the nature of phasors, they are not commonly referred to as such and are not drawn on the same diagram as voltage and current phasors, because they have twice the frequency.

It follows from Figure 17.7 that **S** can be expressed as

$$\mathbf{S} = V_{rms}I_{rms}\angle(\theta_v - \theta_i) = V_{rms}\angle\theta_v \times I_{rms}\angle -\theta_i = \mathbf{V}_{rms}\mathbf{I}^*_{rms} \tag{17.23}$$

where $\mathbf{I}^*_{rms}$ is the conjugate of $\mathbf{I}_{rms}$, having a magnitude I_{rms} and phase angle $-\theta_i$. Multiplication of $\mathbf{V}_{rms}$ by $\mathbf{I}^*_{rms}$ in Equation 17.23 is necessary in order to have the phase angle of S equal to $(\theta_v - \theta_i)$. Note that it is usually more convenient in power calculations to express magnitudes of voltages and currents as rms values.

Having defined complex power and expressed it in terms of voltage and current in Equation 17.23, the next step is to express complex power in terms of current and impedance. To do so, it is convenient to consider a circuit 'N' consisting of resistors, capacitors, and inductors. Let the input impedance Z looking into the terminals of the circuit be represented as a series combination of resistance R and reactance X (Figure 17.8a), so that $Z = R + jX$. Let $\mathbf{V}_{rms} = V_{rms}\angle\theta_v$ be the voltage at the input terminals of 'N' and $\mathbf{I}_{rms} = I_{rms}\angle\theta_i$ be the input current in the direction of the voltage drop $\mathbf{V}_{rms}$. It follows that $Z = \mathbf{V}_{rms}/\mathbf{I}_{rms}$, so that the phase angle of Z is $(\theta_v - \theta_i)$. Moreover, substituting $\mathbf{V}_{rms} = Z\mathbf{I}_{rms}$ in Equation 17.23,

$$\mathbf{S} = Z\mathbf{I}_{rms}\mathbf{I}^*_{rms} = ZI_{rms}\angle\theta_i \times I_{rms}\angle -\theta_i = ZI^2_{rms} = RI^2_{rms} + jXI^2_{rms} \tag{17.24}$$

where the product of $\mathbf{I}_{rms}$ and its complex conjugate is the real quantity equal to the square of the magnitude I_{rms}. It follows from Equation 17.24 that

$$P = RI^2_{rms} \quad \text{and} \quad Q = XI^2_{rms} \tag{17.25}$$

The impedance triangle is shown in Figure 17.8b. Multiplying the sides of the triangle by I^2_{rms} gives $RI^2_{rms} = P$ for the base, $XI^2_{rms} = Q$ for the height, and $|Z|I^2_{rms} = S$ for the hypotenuse, in accordance with Equations 17.24 and 17.25. It is seen that the complex power triangle of Figure 17.7 is simply a scaled version of the impedance triangle, the scaling factor being I^2_{rms}. Moreover, $\mathbf{V}_{rms} = \mathbf{V}_R + \mathbf{V}_X = R\mathbf{I}_{rms} + jX\mathbf{I}_{rms}$, where $\mathbf{V}_R = R\mathbf{I}_{rms}$ is the component of $\mathbf{V}_{rms}$ in phase with $\mathbf{I}_{rms}$ and $\mathbf{V}_X = jX\mathbf{I}_{rms}$ is the component of $\mathbf{V}_{rms}$ in phase quadrature with $\mathbf{I}_{rms}$.

Complex power can equally well be expressed in terms of voltage and admittance, rather than current and impedance. To do so, we consider the input admittance $Y = G + jB$ at the input terminals of circuit 'N', rather than the input impedance Z (Figure 17.9). It follows that

$$\frac{\mathbf{I}_{rms}}{\mathbf{V}_{rms}} = Y = \frac{1}{|Z|\angle(\theta_v - \theta_i)} = |Y|\angle -(\theta_v - \theta_i) = G + jB \tag{17.26}$$

The admittance diagram is shown in Figure 17.9b, where B is negative for the assumed inductive reactance $(\theta_v > \theta_i)$. **S** may be expressed as

$$\mathbf{S} = \mathbf{V}_{rms}\mathbf{I}^*_{rms} = \mathbf{V}_{rms}(Y\mathbf{V}_{rms})^* = \mathbf{V}_{rms}Y^*\mathbf{V}^*_{rms}$$

$$= Y^*V^2_{rms} = \frac{1}{Z^*}V^2_{rms} \tag{17.27}$$

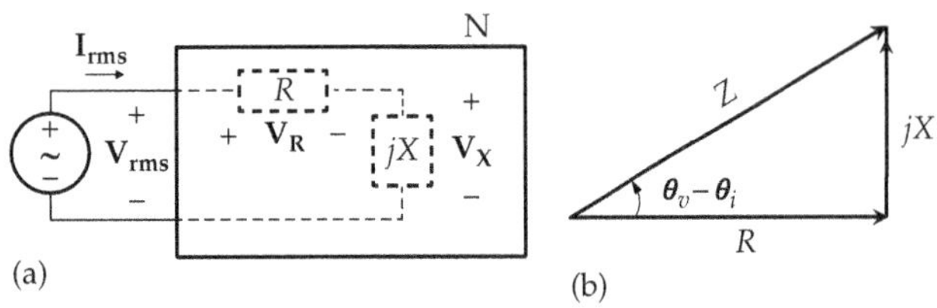

FIGURE 17.8
Complex power in a series circuit. (a) Circuit 'N' represented by a resistance and reactance in series and (b) impedance triangle for circuit 'N'.

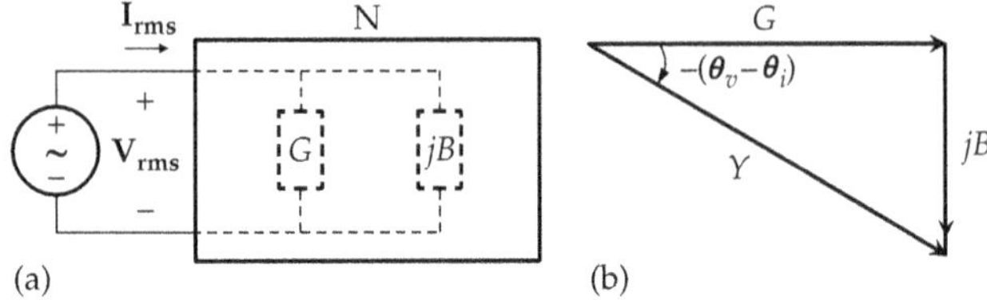

FIGURE 17.9
Complex power in a parallel circuit. (a) Circuit 'N' represented by a conductance and susceptance in parallel and (b) admittance triangle for circuit 'N'.

TABLE 17.1

Complex Power Relations

	$\mathbf{S} = P + jQ = \mathbf{VI}^*_{rms}$	
	Series Connection	**Parallel Connection**
	$Z = R + jX$	$Y = G + jB$
S	ZI^2_{rms}	$Y^*V^2_{rms} = V^2_{rms}/Z^*$
P	RI^2_{rms}	GV^2_{rms}
Q	XI^2_{rms}	$-BV^2_{rms}$

where the conjugate complex of the product of two complex quantities is equal to the product of the complex conjugate of each quantity. Thus, if the two complex quantities are expressed as $Ae^{j\alpha}$ and $Be^{j\beta}$, their complex conjugates are $Ae^{-j\alpha}$ and $Be^{-j\beta}$. Their product is $ABe^{j(\alpha+\beta)}$, whose complex conjugate $ABe^{-j(\alpha+\beta)}$ is the product of $Ae^{-j\alpha}$ and $Be^{-j\beta}$. Substituting $Y^* = G - jB$,

$$\mathbf{S} = (G - jB)V^2_{rms} = GV^2_{rms} - jBV^2_{rms} \tag{17.28}$$

which gives

$$P = GV^2_{rms} \quad \text{and} \quad Q = -BV^2_{rms} \tag{17.29}$$

The complex power triangle of Figure 17.7 is thus a scaled version of the admittance triangle, with B inverted, and a scaling factor of V^2_{rms}. Complex power relations are summarized in Table 17.1.

Primal Exercise 17.9

(a) Determine **P**, $\mathbf{Q_L}$, $\mathbf{Q_C}$, and **S** delivered by the source and absorbed by the load impedance in Figure 17.10; verify that **S** = **VI***; and (b) represent **P**, $\mathbf{Q_L}$, $\mathbf{Q_C}$, and **S** on a phasor diagram.

Ans. (a) $\mathbf{P} = 1\angle 0°$ W, $\mathbf{Q_L} = 0.5\angle 90°$ VAR, $\mathbf{Q_C} = 1\angle -90°$ VAR, $\mathbf{S} = 0.5\sqrt{5}\angle -26.6°$ VA.

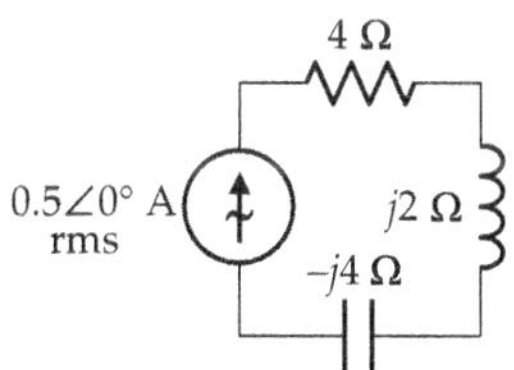

FIGURE 17.10
Figure for Primal Exercise 17.9.

Primal Exercise 17.10

Repeat Primal Exercise 17.9 for the circuit that is the dual of that of Figure 17.10. Note that the circuit of Figure 17.10 is capacitive, but its dual is inductive.

Ans. (a) $\mathbf{P} = 1\angle 0°$ W, $\mathbf{Q_C} = 0.5\angle -90°$ VAR, $\mathbf{Q_L} = 1\angle +90°$ VAR, $\mathbf{S} = 0.5\sqrt{5}\angle 26.6°$ VA.

Primal Exercise 17.11

Determine the reactive power absorbed by the inductor and the capacitor in Figure 17.11.

Ans. $Q_L = 5$ VAR, $Q_C = -10$ VAR.

Primal Exercise 17.12

Determine C in Figure 17.12 if the complex power delivered by the source is $-j5$ kVA, with $\omega = 1$ krad/s.

Ans. 1.5 mF.

17.2.2 Conservation of Complex Power

We will argue in this subsection that complex power is conserved because its real and imaginary parts, P and Q, are each conserved in any given circuit. Before doing so, let us clarify what exactly is meant by conservation of power in a circuit. Recall from the discussion in Chapters 1 and 2 that conservation of power in a circuit under dc conditions means that the power delivered by sources in the circuit is equal to the power dissipated in the resistors in the circuit at every instant. This power is real power.

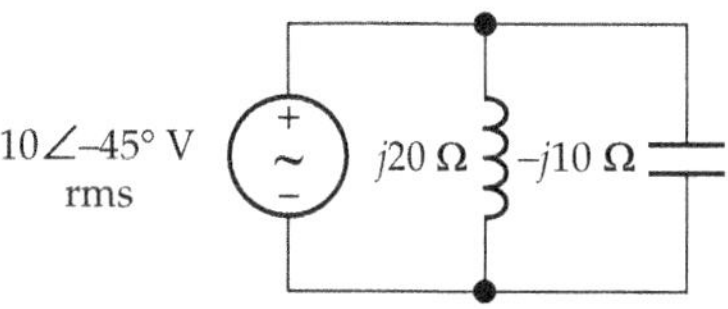

FIGURE 17.11
Figure for Primal Exercise 17.11.

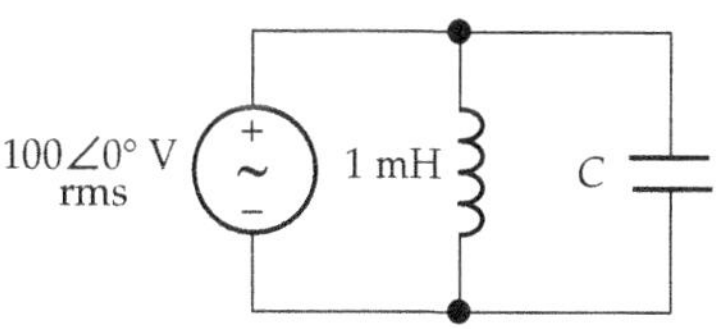

FIGURE 17.12
Figure for Primal Exercise 17.12.

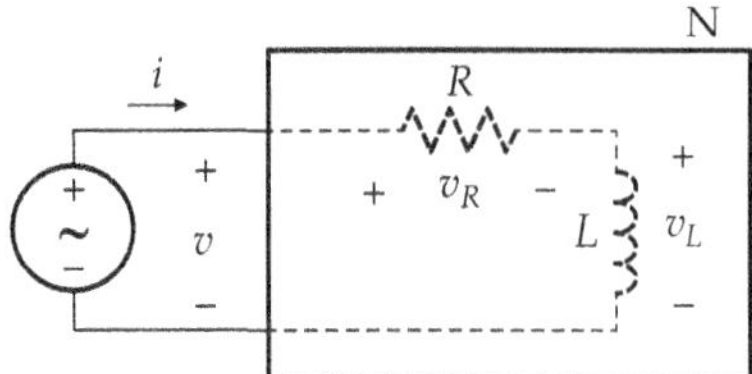

FIGURE 17.13
Conservation of power in the time domain.

The circuit of Figure 17.8 is illustrated in Figure 17.13 in the time domain, where the reactance X is represented by an inductance L for the purpose of illustration. Conservation of power means that at every instant, the power delivered by the source is equal to that absorbed by the load, so that power is conserved in the circuit as a whole at every instant. Since instantaneous power in the sinusoidal steady state is given by Equation 17.19, we can express conservation of power at every instant as

$$p_{PS} + p_{QS} = p_{PL} + p_{QL} \tag{17.30}$$

The LHS in Equation 17.30 is the instantaneous power delivered by the source, and the RHS is the instantaneous power absorbed by the load, both in accordance with Equation 17.19. The question is whether (1) Equation 17.30 necessitates that $p_{PS} = p_{PL}$ and $p_{QS} = p_{QL}$ or (2) that p_{PS} can be different from p_{PL} and p_{QS} different from p_{QL} as long as Equation 17.30 is satisfied. In the latter case, suppose that, for the sake of argument, $p_{PS} > p_{PL}$ and $p_{QS} < p_{QL}$. This means that more real power p_{PS} is supplied by the source than is absorbed by the load and that the load converts some of this real power to reactive power p_{QL} so as to satisfy Equation 17.30. However, this is physically impossible, because real and reactive powers are very different in nature. Real power is power expended in a load, such as that dissipated in a resistor, whereas reactive power is power associated with the magnetic energy stored in inductors and the electric energy stored in capacitors. Real power is useful power that can be converted to other forms of energy, such as heat, light, and mechanical energy, and that is available outside the circuit. Reactive power is power that flows back and forth between the source and energy storage elements, with zero average, and is not available outside the circuit. It must be concluded, therefore, that in satisfying Equation 17.30, the real powers p_{PS} and p_{PL} must be equal and the reactive powers p_{QS} and p_{QL} must be equal. Mathematically, if the averages of both sides of Equation 17.30 are taken over a period, then

$$\frac{1}{T}\int_0^T p_{PS} + \frac{1}{T}\int_0^T p_{QS} = \frac{1}{T}\int_0^T p_{PL} + \frac{1}{T}\int_0^T p_{QL} \tag{17.31}$$

Since the averages of the reactive powers p_{QS} and p_{QL} are zero, the averages of p_{PS} and p_{PL} are equal. But the average of real power is also the amplitude of its alternating component (Equation 17.15). This means that if the averages of p_{PS} and p_{PL} are equal, then $p_{PS} = p_{PL}$, and in turn $p_{QS} = p_{QL}$.

The same conclusion can be reached in the frequency domain, with some additional inferences. Consider that $\mathbf{V_{SRC}}$ rms is applied as in Figure 17.14a to a resistance R_1 alone. The load absorbs a real power $P_1 = |\mathbf{V_{SRC}}|^2/R_1$, and the source delivers an equal real power $P_{S1} = P_1$. Suppose an inductance L is added in parallel with R_1 (Figure 17.14b). P_1 does not change because it depends on $|\mathbf{V_{SRC}}|$, which has not changed. The inductor absorbs a reactive power $Q_L = -B_L|\mathbf{V_{SRC}}|^2 = |\mathbf{V_{SRC}}|^2/\omega L$. The source must supply a reactive power $Q_{SL} = Q_L$,

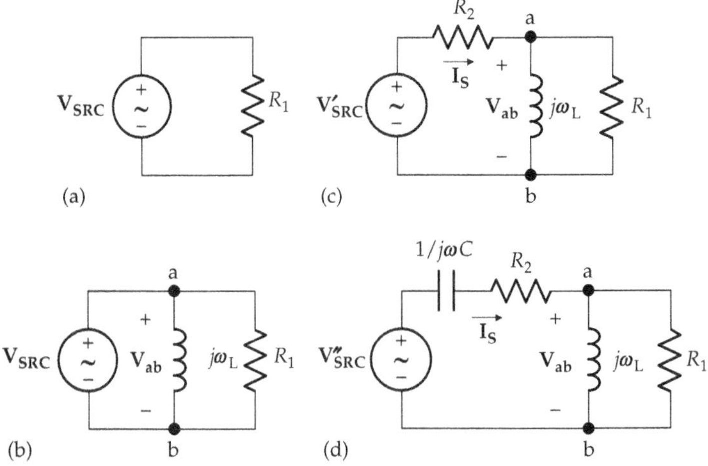

FIGURE 17.14
Conservation of complex power. (a) Resistor connected to a sinusoidal voltage source, (b) addition of a parallel inductor, (c) addition of a series resistor, keeping $\mathbf{V_{ab}}$ as in (b), and (d) addition of a series capacitor, keeping $\mathbf{V_{ab}}$ and $\mathbf{I_S}$ as in (c).

in accordance with conservation of power at every instant. It is seen that

$$P_{S1} + Q_{SL} = P_1 + Q_L \quad (17.32)$$

Equation 17.32 is the counterpart of Equation 17.30 in the frequency domain.

Next, consider that a resistance R_2 is added in series with R_1 and L (Figure 17.4c) and that $\mathbf{V}_{SRC}$ is changed to $\mathbf{V}'_{SRC}$, such that $\mathbf{V}_{ab} = \mathbf{V}_{SRC}$ remains the same, that is, $\mathbf{V}'_{SRC} = R_2\mathbf{I}_S + \mathbf{V}_{ab}$. P_1 and Q_L do not change, since $\mathbf{V}_{ab}$ has not changed, but the load connected to the source now absorbs an additional real power $P_2 = |\mathbf{I}_S|^2R_2$. The source must therefore supply an additional real power $P_{S2} = P_2$, in accordance with conservation of power at every instant. Equation 17.32 now becomes

$$P_{S1} + Q_{S1} + P_{S2} = P_1 + Q_L + P_2 \quad (17.33)$$

Let a capacitance C be added in series with R_2 (Figure 17.4d) and let $\mathbf{V}'_{SRC}$ be changed to $\mathbf{V}''_{SRC}$ such that $\mathbf{V}_{ab}$ and $\mathbf{I}_S$ remain the same, that is, $\mathbf{V}''_{SRC} = \mathbf{V}_{ab} + (R_2 + 1/j\omega C)\mathbf{I}_S$. P_1, Q_L, and P_2 do not change, since $\mathbf{V}_{ab}$ and $\mathbf{I}_S$ have not changed. The load connected to the source now absorbs an additional reactive power $Q_C = X_C|\mathbf{I}_S|^2 = -|\mathbf{I}_S|^2/\omega C$. The source must deliver an additional reactive power $Q_{SC} = Q_C$, in accordance with conservation of power at every instant, so that

$$P_S + Q_S = P_1 + Q_L + P_2 + Q_C \quad (17.34)$$

where $P_S = P_{S1} + P_{S2} = P_1 + P_2$, and $Q_S = Q_{SL} + Q_{SC} = Q_L + Q_C$.

According to Equation 17.34, real power P and reactive power Q are conserved, as in Equation 17.30 in the time domain. In addition, the preceding argument illustrates a very useful feature of complex power analysis, namely, that P and Q can each be added branch by branch across the circuit. Thus, as one proceeds from right to left in Figure 17.14d, P_1 of R_1 is added to Q_L of the inductor, to P_2 of R_2, and to Q_C of the capacitor to obtain the complex power absorbed by these elements. By conservation of power, this is also the power delivered by the source.

Since P and Q are separately conserved in any given circuit, their complex sum, $\mathbf{S} = P + jQ$, must also be conserved. In summary,

Summary: *In any circuit in the sinusoidal steady state, reactive power is conserved in the circuit as a whole, because each energy storage element alternately absorbs and returns to the supply the same energy during each half period of the supply. Real power is conserved in the circuit as a whole, because it is an average power, and the average of reactive power is zero, so that the sources in the circuit must supply this average power. With each of* **P** *and* **Q** *conserved, their sum* **S** *is conserved.*

Concept: *Real and reactive powers can each be summed branch by branch in a given circuit. The total sum of each, over all branches of the circuit, including sources, is zero, in accordance with conservation of real and reactive powers.*

Note that in summing real and reactive powers, real power and inductive, reactive power are considered positive if absorbed by any given passive circuit element or subcircuit, whereas capacitive, reactive power is considered negative if absorbed. As always, power delivered is the negative of power absorbed.

Example 17.2: Application of Complex Power

Two loads 'L_1' and 'L_2' are connected across a $1000\angle 0°$ V rms supply (Figure 17.15a). 'L_1' absorbs real power of 40 kW and reactive, inductive power of 30 kVAR, whereas 'L_2' absorbs real power of 80 kW and reactive, inductive power of 60 kVAR. The loads are fed through a power line having a resistance of 0.1 Ω and a reactance of 0.5 Ω. It is required to determine the voltage $\mathbf{V}_{SRC}$.

Solution:

The real powers of 'L_1' and 'L_2' are added together to give 120 kW, and their reactive powers are added together to give 90 kVAR lagging, that is, inductive. The complex power triangle at terminals 'ab' is shown in Figure 17.15b. Thus, $\mathbf{S} = 120{,}000 + j90{,}000 = 1{,}000\angle 0° \times \mathbf{I}_S^*$, which gives $\mathbf{I}_S^* = \dfrac{120{,}000 + j90{,}000}{1{,}000} = 150\angle 36.9°$ A. Hence, $\mathbf{I}_S = 150\angle -36.9°$A.

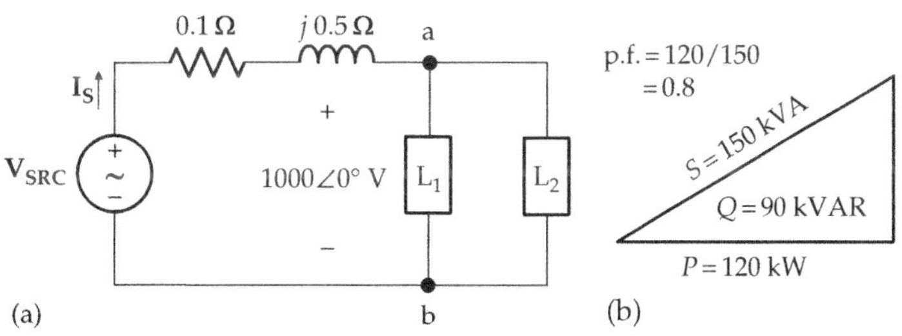

FIGURE 17.15
Figure for Example 17.2.

The real power absorbed by the 0.1 Ω resistance is $0.1\times(150)^2 = 2.25$ kW, and the reactive power absorbed by the 0.5 Ω reactance is $0.5\times(150)^2 = 11.25$ kVAR. The total real power at the inputs of the supply terminals is 122.25 kW, and the total reactive power is 101.25 kVAR. The complex power at these terminals is therefore $\mathbf{S} = 122{,}250 + j101{,}250 = 158{,}734\angle 39.6° = \mathbf{V_{SRC}}\angle\theta_V \times \mathbf{I}^*_{\mathbf{SRC}} = V_{SRC}\angle\theta_v \times 150\angle 36.9°$. Hence, $\mathbf{V_{SRC}} = 1058\angle 2.7°$ V.

The real power delivered by the source is 122.25 kW. It is the sum of the real power dissipated in the line resistance and in loads 'L_1' and 'L_2'. The reactive power delivered by the source is 101.25 kVAR. It is also the sum of the inductive reactive power in the line reactance and in the two loads. Real and reactive power are conserved in the system as a whole.

Problem-Solving Tip

- In a series circuit, it is convenient to derive powers as $P = RI^2_{rms} = R|\mathbf{I}_{rms}|^2$ and $Q = XI^2_{rms} = X|\mathbf{I}_{rms}|^2$, whereas in a parallel circuit it is convenient to derive powers as $P = GV^2_{rms} = G|\mathbf{V}_{rms}|^2$ and $Q = -BV^2_{rms} = -B|\mathbf{V}_{rms}|^2$. Note that only magnitudes are involved in these relations and not phase angles.

17.3 Power Factor Correction

Concept: *Reactive power, although it averages to zero over a cycle, generally increases the voltage and current requirements of a load.*

A series reactance increases the voltage needed for delivering a required real power to a load. This is seen in Figure 17.8a, where the real power delivered to the load is $P = RI^2_{rms}$. In the presence of a series reactance X, a larger voltage across the series combination is required for a given I_{rms} and hence a given P. Similarly, a parallel reactance increases the current needed for delivering a required real power to a load. In Figure 17.9a, the real power delivered to the load is $P = GV^2_{rms}$. In the presence of the susceptance B, a larger current through the parallel combination is required for a given I_{rms} and hence a given P. In the presence of both series and parallel reactive elements, as is generally the case with real-life loads, both a larger voltage and a larger current are required for delivering a given real power to the load. This places an additional burden on the power system supplying the load. A larger current requires a greater current-carrying capacity of the supply conductors, and a larger voltage requires a higher degree of insulation of these conductors.

The contribution of the reactive elements relative to the resistive, or power-consuming, elements is indicated by the phase angle $(\theta_v - \theta_i)$, where $\cos(\theta_v - \theta_i)$ is the **power factor** (abbreviated p.f.) and $\sin(\theta_v - \theta_i)$ is the **reactive factor**. For a purely resistive load, the p.f. is unity. Since the p.f. is the same for a positive $(\theta_v - \theta_i)$ as for a negative $(\theta_v - \theta_i)$, these two cases are distinguished by adding the attribute "lagging" or "leading," respectively. For example, for a purely inductive load, $\theta_v - \theta_i = 90°$, and the p.f. is zero lagging, whereas for a purely capacitive load, $\theta_v - \theta_i = -90°$, and the p.f. is zero leading. In practice, ac loads generally have a lagging p.f., mainly due to ac motors as well as the inductances associated with transformers and ballasts of fluorescent lights. The p.f. may be as low as 0.8 lagging or less, particularly during motor starting. A low p.f. is undesirable because of the additional current and voltage burdens placed on the supply, as explained previously. In the case of large loads, the additional costs involved can be quite considerable, so that some measures are taken to improve the p.f.. This **power factor correction** is achieved by adding capacitive reactance to counteract the inductive reactance of the load. The reactive power due to inductance is positive, whereas the reactive power due to capacitance is negative. By adding the proper value of capacitance, the net reactive power can be reduced to zero, thereby achieving a unity power factor.

Example 17.3: Power Factor Correction

Assuming a supply frequency of 50 Hz, determine the capacitance that must be added in parallel with the loads of Figure 17.15 of Example 17.2 so as to make the p.f. unity at the load terminals 'ab'. Determine the effect of this capacitance on the supply current and voltage.

Solution:

The reactive power at terminals 'ab' was found in Example 17.2 to be 90 kVAR. The reactive power of the added capacitor must be −90 kVAR. The value of capacitance is determined using Equation 17.29, $Q = -BV^2_{rms}$, where $Q = -90$ kVAR, $B = 2\pi fC$, and $V_{rms} = 1000$ V. This gives $C = 286.5$ μF.

The total reactive power at terminals 'ab' is now zero. Hence, $S = 120{,}000 = 1000\angle 0° \times \mathbf{I}^*_S$. This gives $\mathbf{I}^*_S = 120\angle 0°$ A. The real power absorbed by the 0.1 Ω resistance is $0.1\times(120)^2 = 1.44$ kW, and the reactive power absorbed by the 0.5 Ω reactance is $0.5\times(120)^2 = 7.2$ kVAR. The total real power at the source terminals is 121.44 kW and the total reactive power is 17.2 kVAR. The complex power at these terminals is therefore $\mathbf{S} = 121{,}440 + j7{,}200 = 121{,}653\angle 3.4° = \mathbf{V_{SRC}}\angle\theta_V \times \mathbf{I}^*_S = V_S\angle\theta_v \times 120\angle 0°$. Hence, $\mathbf{V_{SRC}} = 1014\angle 3.4°$ V rms.

The p.f. at the load was initially 120/150 = 0.8. By correcting it to 1, the source current was reduced from 150 to

120 A and the source voltage was reduced by a relatively small amount in this case, from 1058 to 1014 V rms, due to the reduced voltage drop across the line impedance.

Problem-Solving Tip

- The relation $\mathbf{S} = \mathbf{V}_{rms}\mathbf{I}_{rms}^*$ is in terms of magnitudes $|\mathbf{S}| = |\mathbf{V}_{rms}||\mathbf{I}_{rms}|$, or $S = V_{rms}I_{rms}$, independent of phase angles.

Primal Exercise 17.13

The load 'L' in Figure 17.16 absorbs 30 kW and 9 kVAR inductive. C is such that the p.f. is unity. Determine (a) $\mathbf{I}_S$ and (b) C, assuming the frequency is 1 krad/s.

Ans. (a) $300\angle 0°$ A or $100\angle 0°$ A; (b) 100 μF, 0.9 mF.

Primal Exercise 17.14

A load absorbs $10(1 - j)$ kVA. It is required to change the p.f. to unity. Determine the reactance that must be added (a) in series with the load, assuming a load current of 10 Arms, or (b) in parallel with the load, assuming a load voltage of 100 V rms.

Ans. (a) +100 Ω; (b) +1 Ω.

Exercise 17.15

Determine (a) the capacitance that must be added in parallel with the loads of Example 17.2 so as to make the p.f. unity at the source terminals, assuming a frequency of 50 Hz and (b) the new values of source current and voltage.

Ans. (a) 309.5 μF; (b) $120.22\angle 3.45°$A; $1010.2\angle 3.45°$V.

17.3.1 Power Measurements

Since $|\mathbf{S}| = |\mathbf{V}_{rms}\mathbf{I}_{rms}^*| = |\mathbf{V}_{rms}||\mathbf{I}_{rms}|$, the apparent power $|\mathbf{S}|$ at any location is determined by measuring $|\mathbf{V}_{rms}|$ and $|\mathbf{I}_{rms}|$ at that location by means of an ac voltmeter and an ac ammeter, respectively. To measure the real power, a wattmeter

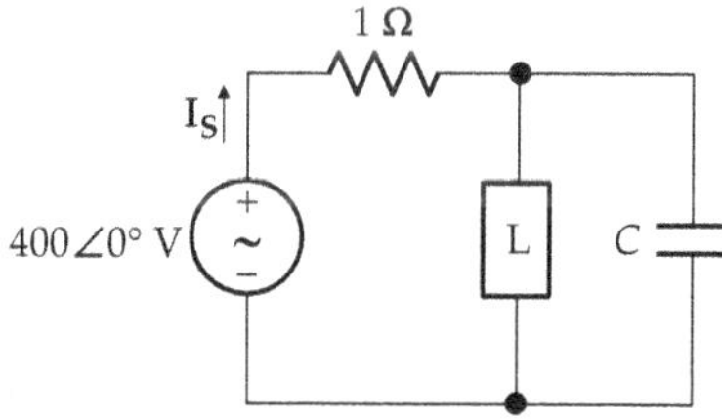

FIGURE 17.16
Figure for Primal Exercise 17.13.

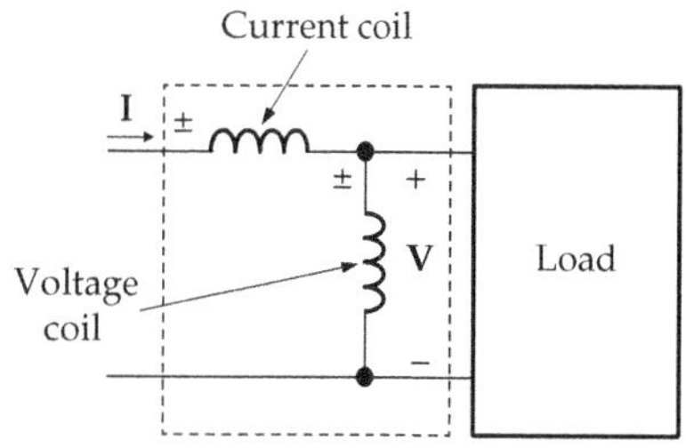

FIGURE 17.17
Wattmeter connection.

is used. The wattmeter has two coils, a current coil and a voltage coil with ± polarity markings. The current coil is connected like an ammeter, so that the load current flows through it. The voltage coil is connected across the load terminals, like a voltmeter. When the polarities of the two coils with respect to **I** and **V** are as indicated in Figure 17.17, the wattmeter reads the power absorbed, $P = V_{rms}I_{rms}\cos(\theta_v - \theta_i)$. If the polarity of either coil is reversed, the wattmeter gives a negative indication. Knowing P and $|\mathbf{S}|$, Q and the p.f. readily follow from the complex power triangle. Ideally, the voltage coil draws no current and the voltage across the current coil is zero.

Primal Exercise 17.16

In Example 17.2, a wattmeter is used to measure the total real power delivered to the combined load. If the voltage terminal of the wattmeter is connected through a 20:1 voltage step-down transformer, and the current coil is connected through a 10:1 current step-down transformer, with normal polarities maintained, determine the wattmeter reading. If the voltage and current at the load terminals are measured to be 1000 V and 150 A, respectively, determine the power factor of the combined load.

Ans. 600 W; 0.8.

17.4 Maximum Power Transfer

A matter of practical importance is to transfer maximum power to a load from a given source of specified open-circuit voltage and source impedance, under steady-state conditions. Before considering the general case, we will start with the purely resistive case.

17.4.1 Purely Resistive Circuit

In Figure 17.18, a dc source having an open-circuit voltage V_{SRC} and a source resistance R_{src} supplies a resistive load R_L. The ideal voltage source and source

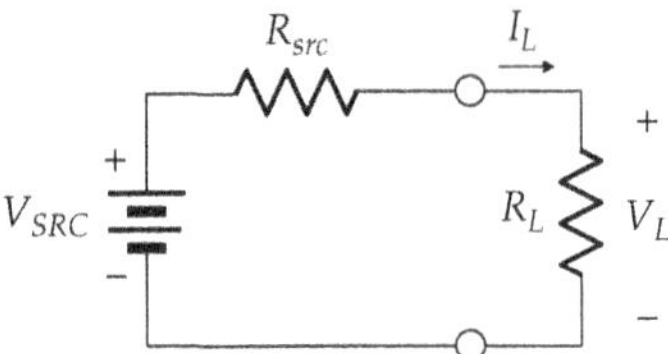

FIGURE 17.18
Maximum power transfer in a resistive circuit.

resistance could represent, in general, TEC of a circuit connected to R_L.

The load current is $I_L = V_{SRC}/(R_{src} + R_L)$, and the power transferred to R_L is

$$P_L = R_L\left(\frac{V_{SRC}}{R_{src} + R_L}\right)^2 = \frac{R_L}{(R_{src} + R_L)^2}V_{SRC}^2 \quad (17.35)$$

With V_{SRC} and R_{src} constant, it is required to determine R_L that maximizes P_L. If dP_L/dR_L is derived and set to zero, it is found that P_L is maximum for R_L given by

$$R_{Lm} = R_{src} \quad (17.36)$$

When Equation 17.36 is satisfied, the source and load resistances are said to be **matched**. The voltage across R_L is $V_{Lm} = V_{SRC}/2$, the current is $I_{Lm} = V_{SRC}/2R_{src}$, and the power transferred to the load is $P_{Lm} = V_{SRC}^2/4R_{Lm}$. Figure 17.19 shows the source characteristic and load line, as previously derived in connection with Figure 3.29. The power $V_{Lm}I_{Lm}$ transferred to the load is represented by the area of the rectangle OAQB. This area is a maximum when Equation 17.36 is satisfied. Any other load such as R_L' or R_L'' results in a rectangle of smaller area.

Figure 17.20 illustrates the various power relations in the circuit under conditions of maximum power transfer. The source characteristic and load line are the same

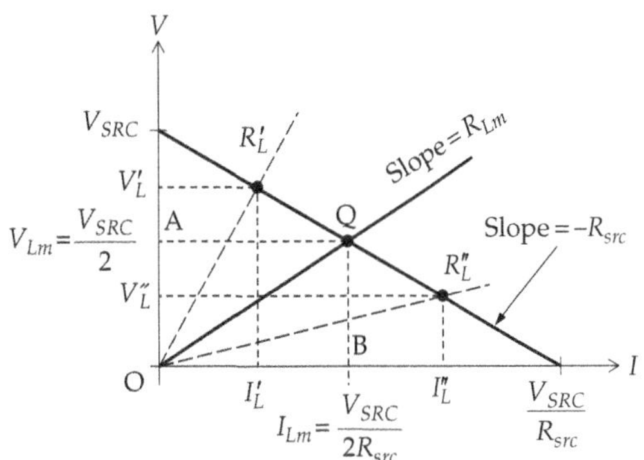

FIGURE 17.19
Graphical interpretation of maximum power transfer.

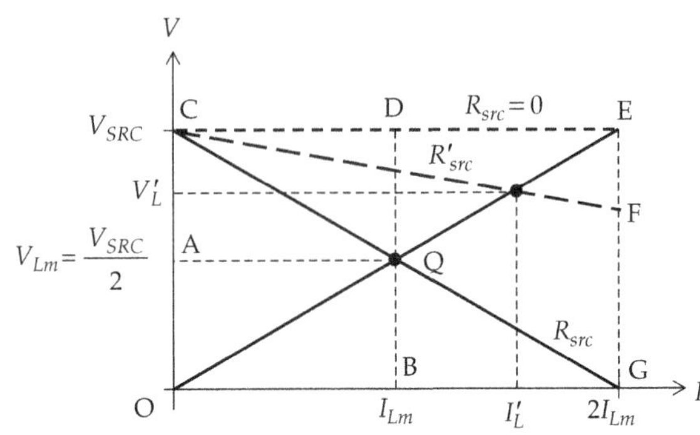

FIGURE 17.20
Power relations under conditions of maximum power transfer.

as in Figure 17.19, and the power delivered to the load is the area of the rectangle OAQB, as in Figure 17.19. The power dissipated in R_{src} is $V_{SRC}^2/4R_{Lm}$, the same as that transferred to the load, since the resistances and currents are equal. It is represented by the area of the rectangle ACDQ. The total power delivered by the ideal voltage source v_{SRC} is $V_{SRC}^2/2R_{Lm}$, represented by the area of the rectangle OCDB.

It should be emphasized that the condition of maximum power transfer applies to the case where V_{SRC} and R_{src} are kept constant and R_L is varied. If R_L is kept constant and R_{src} is varied, the results are quite different. It is seen from Figure 17.20 that if R_L is kept equal to R_{Lm}, the load line OE remains the same. If R_{src} is reduced to R_{src}', the source characteristic is represented by the line CF. The load current increases to $I_L' > I_{Lm}$, the load voltage increases to $V_L' > V_{Lm}$, and the power transferred to the load is represented by the area $V_L'I_L'$ of the rectangle, which is larger than the area of the rectangle OAQB. In the limit, if $R_{src} = 0$, the source characteristic becomes the horizontal line CE. All of V_{SRC} is applied to R_{Lm}, $I_L = V_{SRC}/R_{Lm}$, and the power transferred is V_{SRC}^2/R_{Lm}, corresponding to the area of the rectangle OCEG. Hence, given a source having a specified open-circuit voltage V_{SRC}, maximum power is transferred to the load when $R_{src} = 0$.

When the source has a small internal resistance, the condition for maximum power transfer is generally not of practical interest. For example, in the case of a 12 V battery having a source resistance of 0.04 Ω, maximum power transfer occurs when the load is unrealistically small at 0.04 Ω. The battery current under these conditions is 12/0.08 = 150 A, which will most likely damage the battery. The normal battery load is much larger than 0.04 Ω. On the other hand, consider a transistor amplifier having an output resistance of 100 Ω and required to drive a loudspeaker of nominal resistance 8 Ω. Maximum power transfer is of practical interest in this case and is usually implemented in practice using an audio transformer.

Primal Exercise 17.17

Consider a load connected to a source having an open-circuit voltage of 12 V and a source resistance of 50 Ω. Determine (a) the load resistance for maximum power transfer, (b) the power delivered to the load, and (c) the power dissipated in the source resistance.

Ans. (a) 50 Ω, (b) 0.72 W; (c) 0.72 W.

Example 17.4: Maximum Power Transfer in Resistive Circuit

It is required to determine the value of R_L that should be connected between terminals 'ab' in Figure 17.21a for maximum power transfer.

Solution:

We need only determine R_{Th} looking into terminals 'ab'. R_{Lm} for maximum power transfer is then equal to R_{Th}. To determine R_{Th}, we apply a test source V_T between terminals 'ab' with the 4 V source set to zero, and determine I_T (Figure 17.21b). It is seen that $I_b \times 1 = -10^{-4}V_T$. On the output side, KCL gives $20 \times 10^{-4}V_T + I_T = V_T/100$. It follows that $V_T/I_T = R_{Th} = R_{Lm} = 125$ kΩ.

Simulation: We can determine R_{Th} using dc sweep as was done in Example 4.1. Instead we will demonstrate PSpice's parametric sweep feature that allows designating the value of a circuit component as a global parameter, which can then be varied continuously or for some discrete values. The circuit is entered as shown in Figure 17.22, with a power probe added to R3. To use parametric sweep with R3, double click on the default resistance value displayed, which invokes the Display Properties window. In the Value field, enter a chosen designation enclosed in curly brackets, which tells PSpice that this is a parameter. In the present example, {*R*_val} is entered. The next step is to declare {*R*_val} a global parameter. Place the part PARAM from the SPECIAL library; this shows on the schematic as *PARAMETERS*: When this word is double-clicked, the Property Editor spreadsheet is displayed. Click on the New Column button to display the Add New Column dialog box. Enter R_val in the Name field and any value, say 1k, in the Value field. A new column R_val is added to the spreadsheet with the entry 1k. To have this displayed on the schematic, click on the Display button and choose Name and Value in the Display Properties dialog box. R_val = 1k appears under *PARAMETERS*, as shown in Figure 17.22.

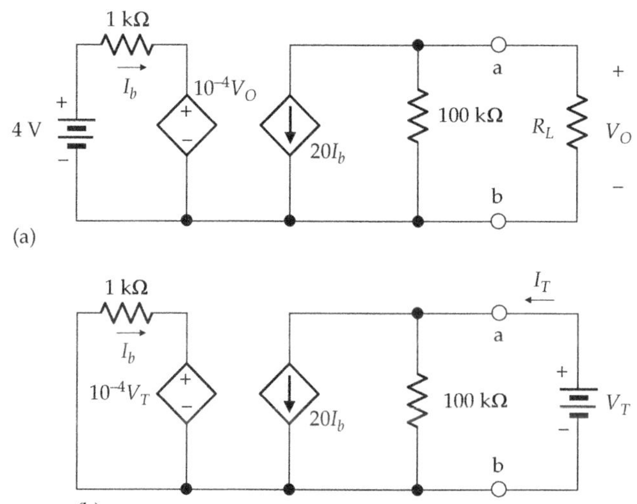

FIGURE 17.21
Figure for Example 17.4.

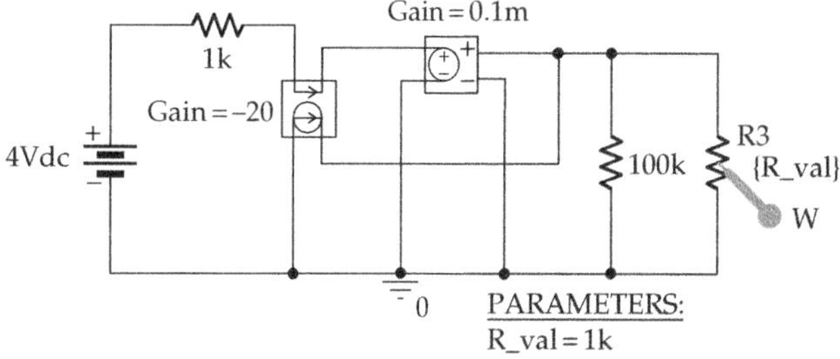

FIGURE 17.22
Figure for Example 17.4.

To run the simulation, select DC Sweep in the Simulation Settings dialog box. Under Sweep variable, choose Global parameter and enter R_val for 'Parameter name'. Under Sweep type choose Linear and enter 50 k for 'Start value', 250 k for 'End value', and 100 for 'Increment'. It may be necessary to try different ranges of values of R_val before identifying the range that shows a maximum. When PSpice is run, the plot of Figure 17.23 is displayed. Click on the cursor button and select Trace/Cursor/Max. The cursor reading gives 125.000K, 200.000. Thus, $R_{L\max} = 125$ kΩ, and the maximum power transferred is 200 W. If the power probe is not used, the plot of Figure 17.23 is displayed by adding the trace W[R3] to the schematic page.

It should be noted that parametric sweep can also be implemented under ac conditions using the source VAC instead of VDC and AC Sweep instead of DC Sweep.

Problem-Solving Tip

- When only the condition for maximum power transferred is required, without the value of this power, it is only necessary to determine R_{Th}, or Z_{Th}, or Y_N, as the case may be, without determining V_{Th}.

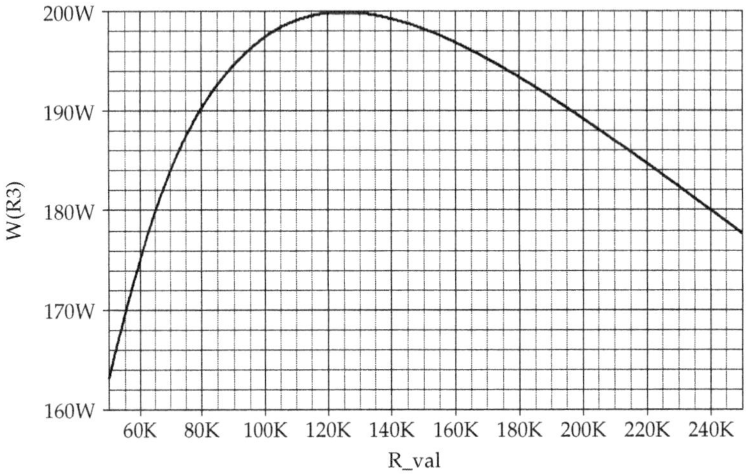

FIGURE 17.23
Figure for Example 17.4.

Exercise 17.18

Verify by analysis that the maximum power transferred in Example 17.4 is 200 W.

Exercise 17.19

The total power delivered by the source V_{SRC} varies with R_L in accordance with the relation $P_{SRC} = V_{SRC}I_L = V_{SRC}^2/(R_{src} + R_L)$. Because this power varies with R_L, maximum power dissipation in R_L does *not* in general coincide with minimum power dissipation in R_{src}, and conversely. Determine R_L and the power dissipated in R_L and R_{src} when the power dissipated in R_{src} is (a) maximum and (b) minimum.

Ans. (a) $R_L = 0, P_L = 0, P_{src} = V_{src}^2/R_{src}$; (b) $R_L = \infty$, $P_L = 0$, $P_{src} = 0$.

17.4.2 Source and Load Impedances

The condition for maximum power transfer will be generalized next to the case of source or load impedances under steady-state sinusoidal conditions. Let $\mathbf{V_{SRC}}$ and $Z_{src} = R_{src} + jX_{src}$ represent, in general, TEC as seen from terminals 'ab' of a given circuit connected to a load $Z_L = R_L + jX_L$ (Figure 17.24). The current phasor $\mathbf{I_L}$ is given by

$$\mathbf{I_L} = \frac{1}{Z_{src} + Z_L}\mathbf{V_{SRC}} = \frac{V_{src}\angle\theta_{src}}{(R_{src} + R_L) + j(X_{src} + X_L)} \quad (17.37)$$

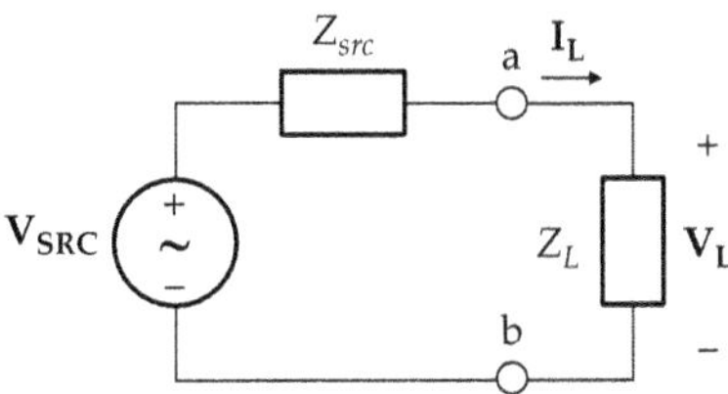

FIGURE 17.24
Maximum power transfer in terms of impedances.

Assuming that $\mathbf{V_{SRC}}$ and hence $\mathbf{I_L}$ are expressed as rms values, the power transferred to R_L is

$$P_L = R_L|\mathbf{I_L}|^2 = V_{src}^2\frac{R_L}{(R_{src} + R_L)^2 + (X_{src} + X_L)^2} \quad (17.38)$$

We wish to determine the condition that maximizes P_L, assuming that V_{src}, R_{src}, and X_{src} are fixed, whereas R_L and X_L are variable. If R_L and X_L can be varied independently, without restriction, it is clear that, with R_L fixed, P_L is maximum for X_L given by

$$X_{Lm} = -X_{src} \quad (17.39)$$

With this condition satisfied, Equation 17.38 reduces to Equation 17.35. P_L is maximum when Equation 17.36 is satisfied, that is, $R_{Lm} = R_{src}$. Combining this condition with that of Equation 17.39 gives the condition for maximum power transfer as

$$Z_{Lm} = Z_{src}^* \quad (17.40)$$

where Z_{src}^* is the complex conjugate of Z_{src}. When Equation 17.40 is satisfied, X_L is equal in magnitude but opposite in sign to X_{src}, so that the two reactances cancel out. With the additional condition that $R_{Lm} = R_{src}$, maximum power is transferred to R_L.

What would be the condition for maximum power transfer if R_L and X_L can again be varied independently, but their range of variation is restricted, so that Equation 17.40 cannot be satisfied? It is clear from Equation 17.38 that under these conditions, with R_L fixed, P_L is maximum when $(X_{src} + X_L)$ is as small as possible. With $(X_{src} + X_L)$ considered constant, the condition for maximum power transfer can be determined by deriving dP_L/dR_L from Equation 17.38 and setting it to zero. This gives

$$R_{Lm} = \sqrt{R_{src}^2 + (X_L + X_{src})^2} \tag{17.41}$$

If R_L cannot be made equal to this value, then a value of R_L as close to it as possible will give maximum power transfer.

Another case of interest arises from the use of a transformer with Z_L. The load impedance $Z_L = |Z_L|\angle\theta_L$ reflected to the primary side is ideally $Z_{Lp} = |Z_{Lp}|\angle\theta_L = (N_p/N_s)^2|Z_L|\angle\theta_L$, where N_p and N_s are the number of turns of the primary and secondary windings, respectively. It follows that $|Z_{Lp}| = (N_p/N_s)^2|Z_L|$. The effect of an ideal transformer is to vary $|Z_L|$ by the square of the turns ratio, $(N_p/N_s)^2$, while the phase angle remains constant. To derive the condition for maximum power transfer under these conditions, when $|Z_L|$ is varied in Figure 17.24, we substitute in Equation 17.38 $R_L = |Z_L|\cos\theta_L$, $X_L = |Z_L|\sin\theta_L$, $R_{src} = |Z_{src}|\cos\theta_{src}$, and $X_{src} = |Z_{src}|\sin\theta_{src}$ to obtain

$$P_L = V_{src}^2 \frac{|Z_L|\cos\theta_L}{(|Z_L|\cos\theta_L + |Z_{src}|\cos\theta_{src})^2 + (|Z_L|\sin\theta_L + |Z_{src}|\sin\theta_{src})^2} \tag{17.42}$$

where the only variable on the RHS is $|Z_L|$. Deriving $\frac{dP_L}{d|Z_L|}$ and setting it equal to zero gives for the condition of maximum power transfer

$$|Z_{Lm}| = |Z_{src}| \tag{17.43}$$

The various conditions for maximum power transfer are summarized in Table 17.2. Note that Equation 17.43 is consistent with Equation 17.40.

Primal Exercise 17.20

Determine R_L and C in Figure 17.25 for maximum power transfer to R_L, assuming R_L and C can be varied independently without restriction, with $\omega = 100$ krad/s.

Ans. $R_L = 100\ \Omega$, $C = 1$ nF.

TABLE 17.2

Conditions for Maximum Power Transfer

Allowed Variation	Condition for Maximum Power Transfer
R_L and X_L can be varied independently without restriction.	$R_{Lm} = R_{src}$ and $X_{Lm} = -X_{src}$
R_L is fixed but X_L can be varied.	$X_{Lm} = -X_{src}$
X_L is fixed but R_L can be varied.	$R_{Lm} = \sqrt{R_{src}^2 + (X_L + X_{src})^2}$
R_L and X_L can be varied independently over a restricted range.	X_{Lm} as close to $-X_{src}$ as possible, R_{Lm} as close to $\sqrt{R_{src}^2 + (X_L + X_{src})^2}$ as possible
$\lvert Z_L\rvert$ can be varied, while $\angle\theta_L$ is constant.	$\lvert Z_{Lm}\rvert = \lvert Z_{src}\rvert$

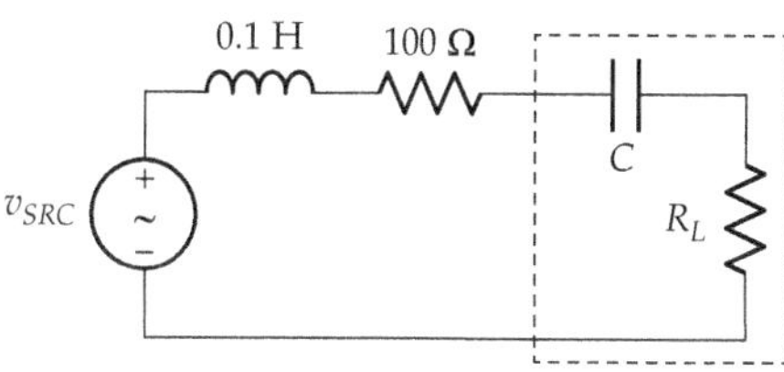

FIGURE 17.25
Figure for Primal Exercise 17.20.

Primal Exercise 17.21

Determine R_L in Figure 17.26 for maximum power transfer to it and calculate this power.

Ans. 8 Ω, 12.5 W.

Primal Exercise 17.22

If X in Figure 17.27 can be varied between 5 and 10 Ω, determine the value of X that results in maximum power transfer to the load and calculate this power, assuming $v_{SRC} = 5\sin(200t + 45°)$ V.

Ans. 10 Ω, 0.52 W.

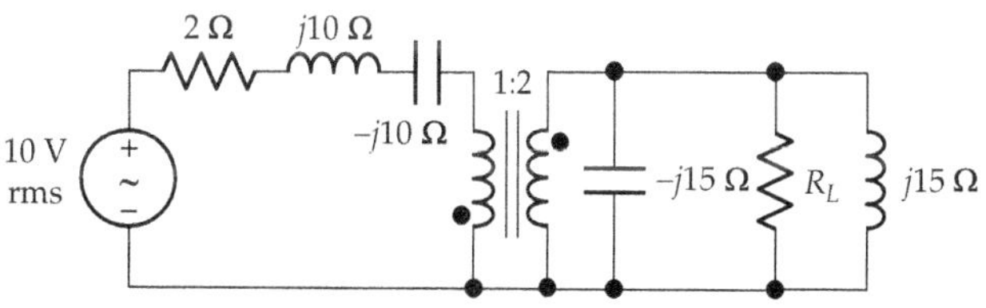

FIGURE 17.26
Figure for Primal Exercise 17.21.

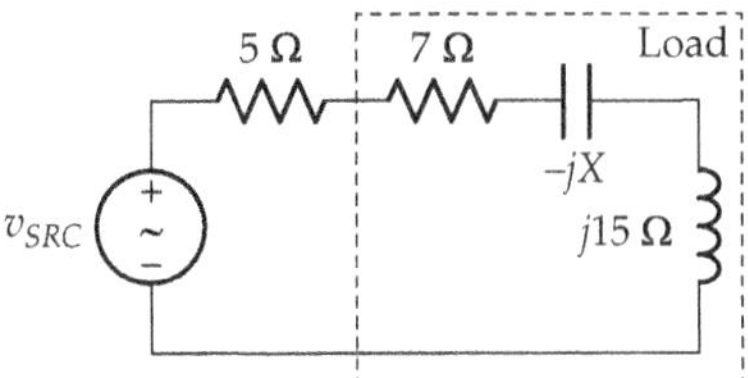

FIGURE 17.27
Figure for Primal Exercise 17.22.

Primal Exercise 17.23

A voltage source having a source impedance of $(15 + j60)\ \Omega$ is connected to a load $(R_L + jX_L)$, where R_L and the magnitude of X_L can each be varied independently in the range 10–50 Ω. Determine R_L for maximum power transfer to the load.

Ans. 18.03 Ω.

Primal Exercise 17.24

Determine a of the transformer in Figure 17.28 so that maximum power is transferred to the inductive load, and calculate this power. Note that there are no dot markings on the transformer because these are irrelevant to the condition for maximum power transfer.

Ans. 3; 5 W.

Example 17.5: Maximum Power Transfer in Terms of Impedances

Given the circuit of Figure 17.29, it is required to determine the condition for maximum power transfer, and the power transferred to R_L, under the following conditions:

1. Both R_L and X_L are variable independently.
2. R_L is fixed at 35 Ω and X_L is variable.
3. X_L is fixed at −50 Ω and R_L is variable.
4. R_L is fixed at 35 Ω, X_L is fixed at −50 Ω, and the transformer can be selected to have any desired turns ratio.

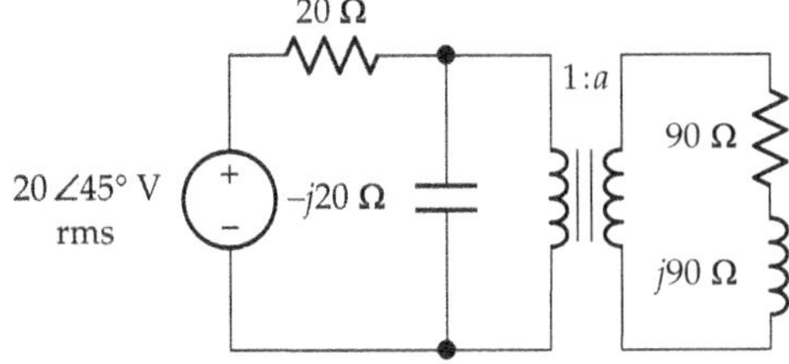

FIGURE 17.28
Figure for Primal Exercise 17.24.

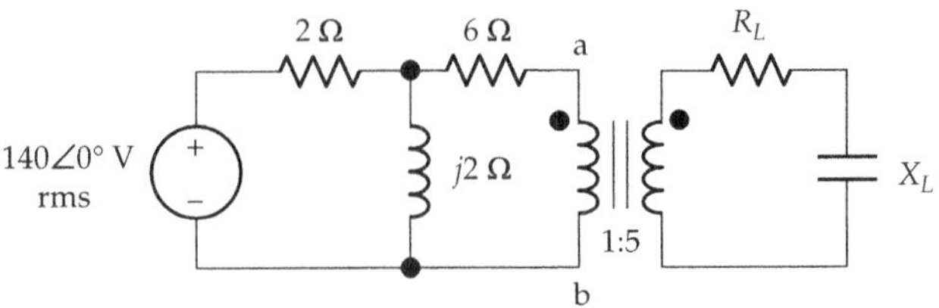

FIGURE 17.29
Figure for Example 17.5.

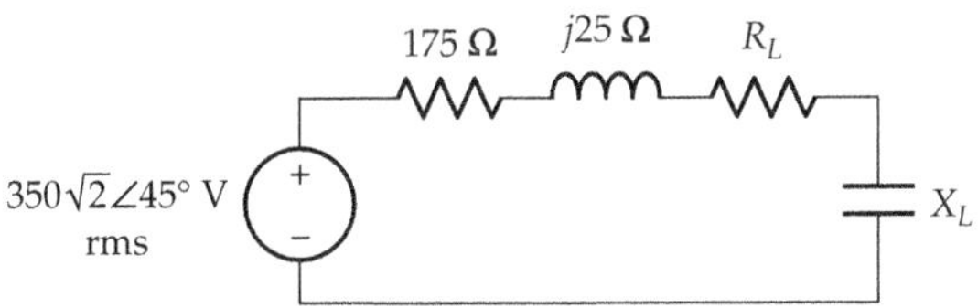

FIGURE 17.30
Figure for Example 17.5.

Solution:

The first step is to derive TEC looking into terminals 'ab' toward the source and reflect TEC to the secondary side.

When terminals 'ab' are open-circuited, it follows from voltage division that $\mathbf{V}_{\text{Th}} = 140\angle 0^\circ \dfrac{j2}{2+j2} = 70(1+j)$ V rms. With the source set to zero, the impedance looking into terminals 'ab' is $Z_{Th} = 6 + \dfrac{j2\times 2}{2+j2} = 7 + j\Omega$. Reflecting TEC to the secondary side, it becomes a voltage source of $70(1+j)\times 5 = 350(1+j) = 350\sqrt{2}\angle 45^\circ$ V rms in series with $(7 + j) \times 25 = 175 + j25\ \Omega$ (Figure 17.30).

(a) If both R_L and X_L can be varied independently, the condition for maximum power transfer is given by Equation 17.40: $R_{Lm} = 175\ \Omega$ and $X_{Lm} = -25\ \Omega$. The power transferred to R_{Lm} is $\dfrac{\left|350\sqrt{2}\right|^2}{4\times 175} = 350$ W.

(b) If $R_L = 35\ \Omega$ and X_L is variable, maximum power transfer occurs when $X_{Lm} = -25\ \Omega$ as in (a). The power transferred to R_L is $\dfrac{\left(350\sqrt{2}\right)^2\times 35}{(175+35)^2} = 194.4$ W.

(c) If $X_L = -50\ \Omega$ and R_L is variable, the condition for maximum power transfer is given by Equation 17.41: $R_{Lm} = \sqrt{(175)^2 + (25-50)^2} = 176.8\,\Omega$. The magnitude of the current is $\dfrac{350\sqrt{2}}{\sqrt{(175+176.8)^2+(25)^2}} =$ 1.4 A rms, and the power transferred to R_{Lm} is $(1.4)^2\times 176.8 = 348.2$ W.

(d) If R_L is fixed at 35 Ω and X_L is fixed at −50 Ω, $|Z_L| = \sqrt{(35)^2 + (50)^2} = 61.0\,\Omega$. The magnitude of Z_{Th} is $\sqrt{49+1} = 7.1\,\Omega$. The turns ratio, instead of 1:5, should be $1:\sqrt{\frac{61.0}{7.1}}$, or 1:2.94, which is nearly 1:3. TEC reflected to the secondary side will be a source of magnitude $70\sqrt{2} \times 3 \cong 300$ V rms in series with a resistance of $7 \times (3)^2 = 63\,\Omega$ and a reactance of $1 \times (3)^2 = 9\,\Omega$. The magnitude of the current is $\frac{300}{\sqrt{(35+63)^2 + (9-50)^2}} = 2.8$ Arms, and the power transferred to R_L is $(2.8)^2 \times 35 = 274.4$ W.

Note that the power transferred in (a), when both R_L and X_L can be varied independently, is larger than in all other cases.

Problem-Solving Tip

- The maximum power transferred to a load when R_L and X_L of a series-connected load can be varied independently is larger than the maximum power transferred when R_L and X_L are constrained in some manner. The largest maximum power transferred is $|V_{Thrms}|^2/4R_{Th}$, where R_{Lm} of the load for maximum power transfer equals R_{Th}.

17.4.3 Admittance Relations

Let $\mathbf{I_{SRC}}$ and Y_{src} represent, in general, NEC as seen from terminals 'ab' of a given circuit connected to a load $Y_L = G_L + jB_L$, as in Figure 17.31. The circuit is the dual in form of that of Figure 17.24. The voltage phasor $\mathbf{V_L}$ is

$$\mathbf{V_L} = \frac{1}{Y_{src} + Y_L}\mathbf{I_{SRC}} = \frac{I_{src}\angle\theta_{src}}{(G_{src} + G_L) + j(B_{src} + B_L)} \tag{17.44}$$

Assuming that $\mathbf{V_L}$ is expressed as an rms value, the power transferred to G_L is

$$P_L = G_L|\mathbf{V_L}|^2 = I_{src}^2 \frac{G_L}{(G_{src} + G_L)^2 + (B_{src} + B_L)^2} \tag{17.45}$$

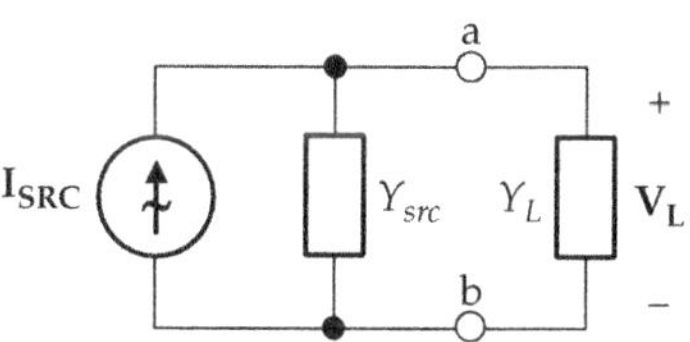

FIGURE 17.31
Maximum power transfer in terms of admittances.

Equation 17.45 is the dual of Equation 17.38. The conditions for maximum power transfer are those of Figure 17.24 with admittances, conductances, and susceptances replacing impedances, resistances, and reactances, respectively:

$$Y_{Lm} = Y_{src}^* \tag{17.46}$$

$$G_{Lm} = \sqrt{G_{src}^2 + (B_L + B_{src})^2} \tag{17.47}$$

$$|Y_{Lm}| = |Y_{src}| \tag{17.48}$$

Example 17.6: Maximum Power Transfer in Terms of Admittances

Given the circuit of Figure 17.32, it is required to determine the condition for maximum power transfer, and the power transferred to R_L, assuming the following:

(a) Both R_L and X are variable independently.
(b) Both R_L and X are variable independently but only over the range 30–50 Ω each.
(c) If R_L is fixed at 30 Ω and X is variable, what is the condition for minimum power dissipated in R_{src} and how much is this power?

Solution:

(a) Let us convert the source and its impedance to its NEC. Norton's source current is $I_N = \frac{30\angle 0°}{10 + j10} = \frac{30\angle 0°}{10\sqrt{2}\angle 45°} = 1.5\sqrt{2}\angle -45°$ A rms. Norton's admittance is $Y_N = \frac{1}{10 + j10} = \frac{10 - j10}{200} = \frac{1}{20} - \frac{j}{20}$ S. The circuit becomes as shown in Figure 17.33.

According to Equation 17.46, maximum power is transferred to the load when $B_m = -B_N$ and $G_{Lm} = G_N$. The first condition gives $-\frac{1}{20} - \frac{1}{20} + \frac{1}{X} = 0$,

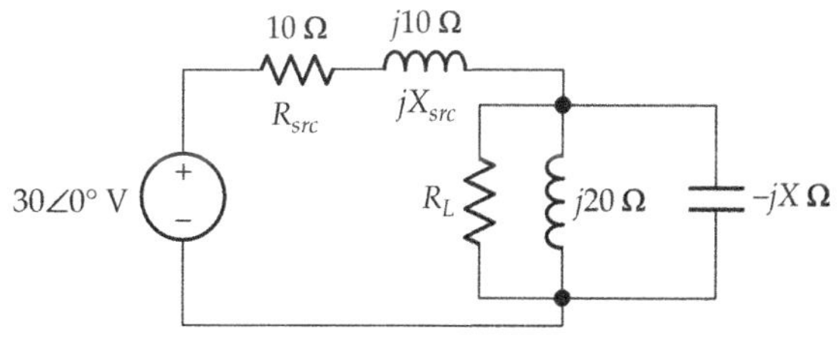

FIGURE 17.32
Figure for Example 17.6.

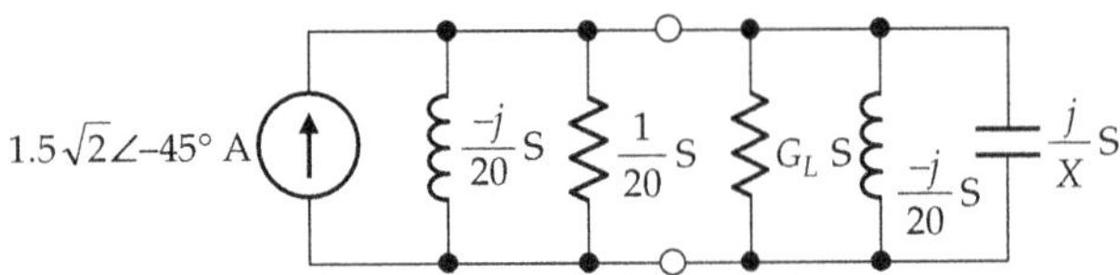

FIGURE 17.33
Figure for Example 17.6.

from which $X_m = 10\ \Omega$. The second condition gives $G_L = G_{src} = \frac{1}{20}$ S, or $R_L = 20\ \Omega$. Under these conditions, the current in G_L is $\frac{I_N}{2} = \frac{1.5}{\sqrt{2}}$, and the power transferred to G_L is $\left(\frac{1.5}{\sqrt{2}}\right)^2 \frac{1}{G_{Lm}} = \frac{45}{2} = 22.5$ W.

(b) The total susceptance is $-\frac{1}{20} - \frac{1}{20} + \frac{1}{X} = -\frac{1}{10} + \frac{1}{X}$. For maximum power transfer, this should be as close to zero as possible, which means that X has its smallest positive value, which is 30 Ω. Under these conditions the total susceptance is $\left(-\frac{1}{10} + \frac{1}{30}\right) = -\frac{1}{15}$. G_{Lm} for maximum power transfer is then given by Equation 17.47 as $G_{Lm} = \sqrt{(1/20)^2 + (1/15)^2} = 1/12$ S. The value of G_L nearest to this is 1/30 S. Maximum power is transferred under these conditions when R_L and X are 30 Ω each.

To find the maximum power transferred, we note that the circuit reduces under these conditions to that shown in Figure 17.34. The admittance Y_x of the combination is $\left(-\frac{j}{15} + \frac{1}{20} + \frac{1}{30}\right) = \left(-\frac{j}{15} + \frac{1}{12}\right) = \frac{5 - j4}{60} = \frac{\sqrt{41}}{60}\angle{-\tan^{-1}0.8}$. The voltage $\mathbf{V_x} = \frac{1}{Y_x}\mathbf{I_N}$, so, $|\mathbf{V_x}| = \frac{1}{|Y_x|}|\mathbf{I_N}| = 1.5\sqrt{2}\frac{60}{\sqrt{41}}$ and the power transferred to G_L is $G_L|\mathbf{V_x}|^2 = 13.2$ W.

(c) If R_L is fixed and X is variable, minimum power is dissipated in R_{src} when the current in this resistor is a minimum. With the source reactance fixed, the current in R_{src} is a minimum when the impedance of the parallel combination is a maximum,

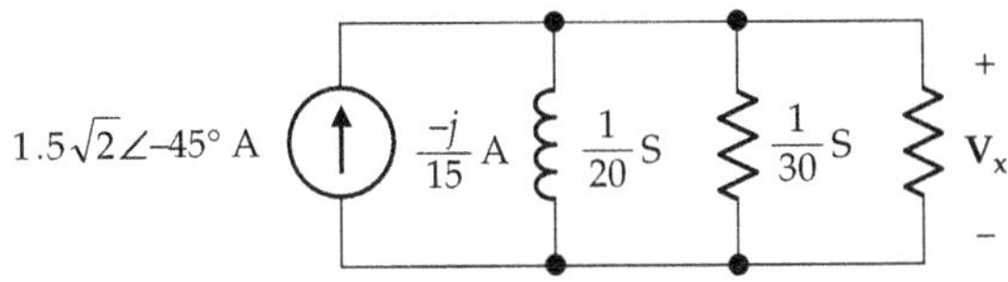

FIGURE 17.34
Figure for Example 17.6.

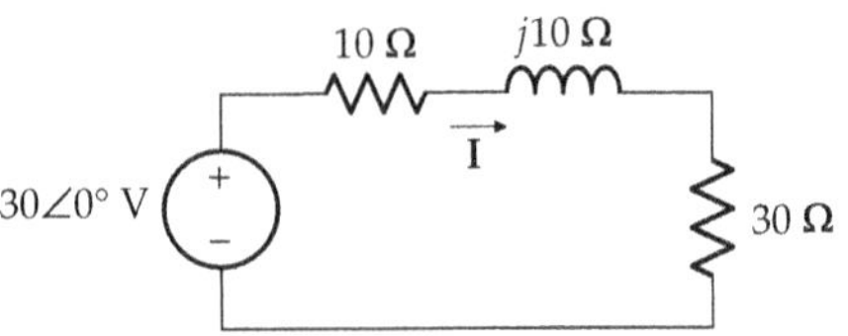

FIGURE 17.35
Figure for Example 17.6.

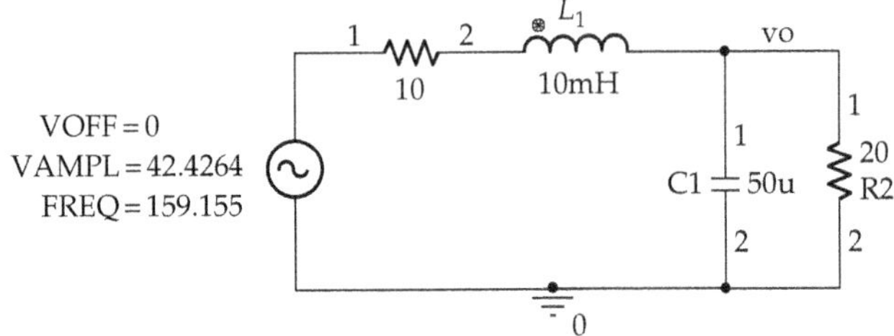

FIGURE 17.36
Figure for Example 17.6.

that is, when the admittance Y_L is a minimum. This occurs when $-\frac{1}{20} + \frac{1}{X} = 0$ or $X = 20\ \Omega$. The circuit reduces to that of Figure 17.35. The current is $\mathbf{I} = \frac{30\angle 0^\circ}{10 + 30 + j10} = \frac{3}{4 + j}$, and $|\mathbf{I}| = \frac{3}{\sqrt{17}}$ A. The power dissipated in R_{src} is $R_{src}|\mathbf{I}|^2 = \frac{9}{17} \times 10 = 5.3$ W.

Simulation: The schematic is shown in Figure 17.36, using the source VSIN, in order to display variation with respect to time. The amplitude is set as $30\sqrt{2} = 42.4264$ V and the frequency as $\omega = 1$ krad/s ($f = 159.155$ Hz). The source inductance is 10 mH to give a reactance of 10 Ω. Under conditions of maximum power transfer, as determined in (a), the reactance of the load is 20 Ω in parallel with −10 Ω, which gives a load reactance of −20 Ω, corresponding to a capacitance of 50 μF at 1 krad/s. Note that the load impedance is a resistance of 20 Ω in parallel with a capacitive reactance of 20 Ω, which gives a phase angle of $\theta_v - \theta_i = -45°$, the current leading the voltage by 45°. The simulation is run using Time Domain (Transient) analysis, with a 'Run to time' of 20 m, a 'Start saving data after' of 10 m, to allow for the initial transient to die down, and a 'Maximum step size' of 1u. The instantaneous power delivered to the load is displayed by selecting Trace/Add trace and entering V(vo)*I(L1), since the current through the inductor is the same as that through the R2-C1 load. The real and reactive powers are displayed by entering V(vo)*I(R2) and V(vo)*I(C1), respectively. The power plots appear in Figure 17.37.

Using the cursor, the peak of p_P is 45.000, which gives $P = 22.5$ W, as determined in (a) for the maximum power

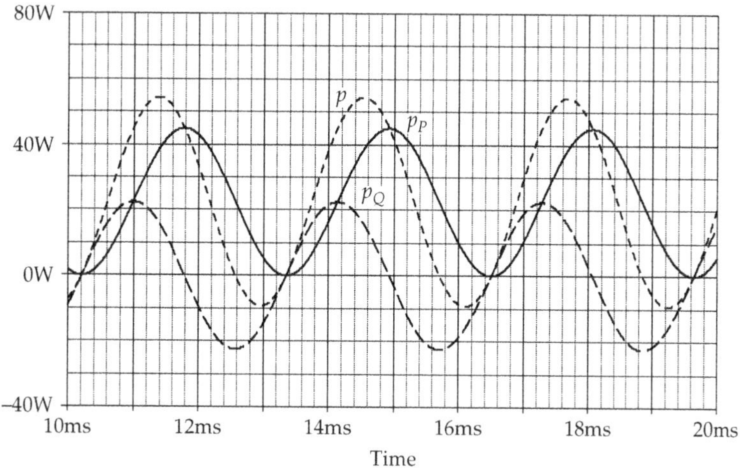

FIGURE 17.37
Figure for Example 17.6.

transferred to the load. Since the phase angle of the load is 45°, Q is also 22.5 VAR. Cursor max gives $p_Q = 22.499$. The maximum of p is read as 54.320 and the minimum as −9.3198, corresponding, respectively, to calculated values of $P+\sqrt{P^2+Q^2}$ and $P-\sqrt{P^2+Q^2}$ (Equation 17.21).

Problem-Solving Tip

- The maximum power transferred to a load when G_L and B_L of a parallel-connected load can be varied independently is larger than the maximum power transferred when G_L and B_L are constrained in some manner. The largest maximum power transferred is $|I_{Nrms}|^2/4G_N$, where G_{Lm} of the load for maximum power transfer equals G_N.

Learning Checklist: What Should Be Learned from This Chapter

- When v is the instantaneous voltage across the terminals of a circuit consisting of resistors, inductors, and capacitors and i is the instantaneous current entering these terminals in the direction of a voltage drop v, the instantaneous power absorbed is $p = vi$ and is pulsating at a frequency that is twice that of v and i. The real power dissipated in the resistors is $P = V_{rms}I_{rms} \times \cos(\theta_v - \theta_i)$, where P is also the amplitude of the alternating component of the product of i and the component of v that is in phase with i. The reactive power, representing the energy that is alternately stored in the energy storage elements and returned to the rest of the circuit is the product of i and the component of v that is in phase quadrature with i. The reactive power is purely alternating, of zero average, and has an amplitude of $Q = V_{rms}I_{rms}\sin(\theta_v - \theta_i)$. When the circuit is inductive, $\theta_v > \theta_I$, so Q is positive. When the circuit is capacitive, $\theta_v < \theta_I$, so Q is negative.
- Complex power **S** is the phasor sum of **P** and **Q**: **S** = **P** + **Q**. It is $= \mathbf{V}_{rms}\mathbf{I}^*_{rms}$, where $\mathbf{I}_{rms}$ is in the direction of a voltage drop $\mathbf{V}_{rms}$.
- The complex power absorbed by an impedance Z is ZI^2_{rms}, where the real power dissipated in R is $P = RI^2_{rms}$ and the reactive power absorbed by X is $Q = XI^2_{rms}$.
- The complex power absorbed by an admittance Y is $Y^*V^2_{rms}$, where the real power dissipated is $P = GV^2_{rms}$ and the reactive power absorbed by B is $Q = -BV^2_{rms}$.
- In any circuit in the sinusoidal steady state, reactive power is conserved in the circuit as a whole, because each energy storage element alternately absorbs and returns the same energy during each half period of the supply. Real power is conserved in the circuit as a whole, because it is an average power, and the average of reactive power is zero, so that the sources in the circuit must supply this average power. With **P** and **Q** conserved, their phasor sum **S** is conserved.
- Real, reactive, and complex powers can each be summed branch by branch in a given circuit, with the total sum of each, over all branches of the circuit, equal to zero.
  1. In summing real and reactive powers, real power and inductive, reactive power are considered positive if absorbed by any given passive circuit element or subcircuit,

whereas capacitive, reactive power is considered negative if absorbed. Power delivered is the negative of power absorbed.

- Reactive power, although it averages to zero over a cycle, generally increases the voltage and current requirements of a load.
- $\cos(\theta_v - \theta_i)$ is known as the power factor (p.f.). It is the cosine of the angle between impedance and resistance in the impedance triangle, or between complex power and real power in the complex power triangle. The p.f. is unity for a pure resistance, is zero lagging for a purely inductive reactance, and is zero leading for a purely capacitive reactance.
- The p.f. can be brought to unity by adding capacitance to counteract the normally inductive reactance of the load.
- If a source of open-circuit voltage $\mathbf{V}_{SRC}$ and source resistance R_{src} is connected to a load R_L, maximum power is transferred from the source to the load when $R_{Lm} = R_{src}$. The maximum power transferred is $(V_{SRC})^2/4R_{Lm}$.
- If a source of open-circuit voltage $\mathbf{V}_{SRC}$ and source impedance Z_{src} is connected to a load $Z_L = R_L + jX_L$ and R_L and X_L can be varied independently without restriction, maximum power is transferred from the source to the load when $Z_{Lm} = Z_{src}^*$, where Z_{src}^* is the conjugate of Z_{src}. If R_L and X_L can be varied independently, but only over a restricted range, maximum power is transferred when $X_L + X_{src}$ is as small as possible and R_L is as close as possible to $\sqrt{R_{src}^2 + (X_L + X_{src})^2}$. If $|Z_L|$ can be varied, as when a transformer is used, maximum power is transferred when $|Z_{Lm}| = |Z_{src}|$. Similar relations apply if admittances are used instead of impedances.
  1. Similar relations apply for admittances, where resistance is replaced by conductance and reactance is replaced by susceptance.

Problem-Solving Tips

1. When working with power, it is convenient to express voltages and currents as rms values.
2. In a series circuit, it is convenient to derive powers as $P = RI_{rms}^2 = R|\mathbf{I}_{rms}|^2 =$ and $Q = XI_{rms}^2 = X|\mathbf{I}_{rms}|^2$, whereas in a parallel circuit it is convenient to derive powers as $P = GV_{rms}^2 = G|\mathbf{V}_{rms}|^2$ and $Q = -BV_{rms}^2 = -B|\mathbf{V}_{rms}|^2$. Note that magnitudes are involved in these relations and not phase angles.
3. The relation $\mathbf{S} = \mathbf{V}_{rms}\mathbf{I}_{rms}^*$ is in terms of magnitudes $|\mathbf{S}| = |\mathbf{V}_{rms}||\mathbf{I}_{rms}|$, or $S = V_{rms}I_{rms}$, independent of phase angles.
4. When only the condition for maximum power transferred is required, without the value of this power, it is only necessary to determine R_{Th}, or Z_{Th}, or Y_N, as the case may be, without determining V_{Th}.
5. The maximum power transferred to a load when R_L and X_L of a series-connected load can be varied independently is larger than the maximum power transferred when R_L and X_L are constrained in some manner. The largest maximum power transferred is $|V_{Thrms}|^2/4R_{Th}$, where R_{Lm} of the load for maximum power transfer equals R_{Th}.
6. The maximum power transferred to a load when G_L and B_L of a parallel-connected load can be varied independently is larger than the maximum power transferred when G_L and B_L are constrained in some manner. The largest maximum power transferred is $|I_{Nrms}|^2/4G_N$, where G_{Lm} of the load for maximum power transfer equals G_N.

Problems

Verify solutions by PSpice simulation.

Complex Power

P17.1 If $v(t) = 80\cos(100t)$ V and $i(t) = -30\sin(100t - 30°)$ A, determine (a) the instantaneous power and (b) the average power.

Ans. (a) $600 + 1200\cos(200t + 60°)$ W; (b) 600 W.

P17.2 Determine the impedance of a load that absorbs 20 kVAR at 0.6 p.f. lagging when the current through the load has a magnitude of 50 Arms.

Ans. $6 + j8\ \Omega$.

P17.3 The conjugate of the complex power delivered by a current source is $400 - j400$ VA. If the source current is $10\angle 45°$ rms, determine the rms voltage across the source.

Ans. $40\sqrt{2}\angle 90°$V.

P17.4 A load that absorbs a complex power of $5 + j10$ VA is connected in parallel with a load that absorbs 20 W at a lagging power factor 0.8. Determine the phase difference between the voltage across the parallel combination and the current through it.

Ans. Voltage leads current by 45°.

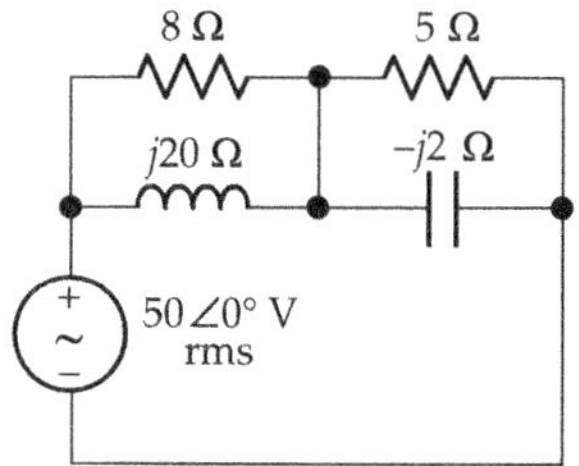

FIGURE P17.7

P17.5 An impedance 4 + *j*4 Ω is connected in parallel with an impedance of 12 + *j*6 Ω. If the input reactive power is 1000 VAR, determine the real power absorbed.

Ans. 1210.5 W.

P17.6 Two impedances 9.8∠−78° Ω and 18.5∠21.8° Ω are connected in parallel and the combination is connected in series with an impedance 5∠53° Ω. If the circuit is connected across a 100 Vrms source, determine the real power delivered by the source.

Ans. 980.0 W.

P17.7 The capacitor in the circuit of Figure P17.7 absorbs −200 VAR. Determine the power dissipated in the 5 Ω resistor.

Ans. 80 W.

P17.8 A load absorbs a complex power of 20 + *j*15 kVA at a voltage of 200∠50° Vrms and a frequency of 60 Hz. Determine the value of the capacitance that should be added across the load to correct the power factor to unity.

Ans. 994.7 μF.

P17.9 An impedance 2 + *j*4 Ω is connected in parallel with a resistance *R*. Determine *R* so that the p.f. of the combination is 0.9 lagging.

Ans. 3.2 Ω.

P17.10 Two inductive loads of 0.88 kW, 0.8 p.f., and 1.32 kW, 0.6 p.f., are connected across a 220 Vrms, 50 Hz supply. (a) Calculate the total complex power of the loads and the supply current. (b) Determine the capacitance that should be connected in parallel with the loads to bring the p.f. to (i) unity and (ii) 0.9 lagging.

Ans. (a) 2.2 + *j*2.42 kVA, 14.87∠−47.7° A; (b) (i) 159 μF, (ii) 89 μF.

P17.11 An electric motor draws 100 kW at 0.8 p.f. lagging from a 240 Vrms, 60 Hz supply. The motor is in parallel with another load of 0.1 + *j*0.4 Ω. Determine the value of the capacitance that should be added in parallel with the combination to raise the power factor to 0.95 lagging.

Ans. 7.67 mF.

P17.12 A capacitor of −*j*30 Ω is connected in parallel with a load that absorbs 1200 W at 0.8 p.f. lagging and a voltage of 300 Vrms magnitude. The parallel combination is supplied through a line of 0.5 Ω resistance. Determine the power dissipated in this resistance.

Ans. 32.5 W.

P17.13 An inductor of *j*5 Ω is connected in parallel with an impedance of 5 − *j*5 Ω. If the current to the parallel combination is 1∠0° Arms, determine the total reactive power absorbed.

Ans. 5 VAR.

P17.14 A load that absorbs 160 W at 0.8 p.f. lagging at a voltage of 200∠0° Vrms is connected in parallel with another load that absorbs 320 VAR at 0.6 p.f. lagging and with a capacitor chosen for unity p.f. across the whole combination. Determine the current drawn by this combination.

Ans. 2∠0° Arms.

P17.15 A load that absorbs 15 kVA at 0.6 p.f. lagging is connected in parallel with a load that absorbs 4.8 kW at 0.8 p.f. leading, across a source of voltage 200∠0° Vrms and 50 Hz frequency. Determine the capacitance that should be connected in parallel with the loads so as to have a minimum magnitude of current supplied by the source.

Ans. 0.67 mF.

P17.16 Three loads are supplied in parallel at 240 Vrms, 50 Hz. 'L_1' absorbs 240 W at 0.6 p.f. lagging, 'L_2' absorbs 200 VAR at 0.5 p.f. lagging, and 'L_3' absorbs 100 VA at 0 p.f. leading. Determine (a) the total apparent power, (b) the p.f., (c) the magnitude of supply current, (d) the parallel capacitance that raises the p.f. to unity, and (e) the resulting magnitude of supply current.

Ans. (a) 550.2 VA; (b) 0.646 lagging; (c) 2.29 A; (d) 23.2 μF; (e) 1.48 A.

P17.17 Three loads are supplied in parallel at 500 Vrms. 'L_1' absorbs 12 kW at 0.6 p.f. lagging, 'L_2' absorbs 15 kW at unity p.f., and 'L_3' absorbs 6 kVAR at 0.8 p.f. leading. The line has a resistance of 1 Ω and negligible reactance. Determine (a) the rms magnitude of the source voltage, (b) the combined power factor of the load, and (c) the percentage of the real power delivered by the source that is absorbed by the loads.

Ans. (a) 570.3∠−2° V; (b) 0.962; (c) 86.85%.

P17.18 Determine the reading of the wattmeter in Figure P17.18.

Ans. 211.1 W.

P17.19 R_1, L, R_2, and C in Figure P17.19 are unknown. Load 1 absorbs a complex power of 100∠45° VA, and load 2 absorbs a complex power of 50∠−45° VA. Determine R_1.

Ans. $250\sqrt{2}$ Ω.

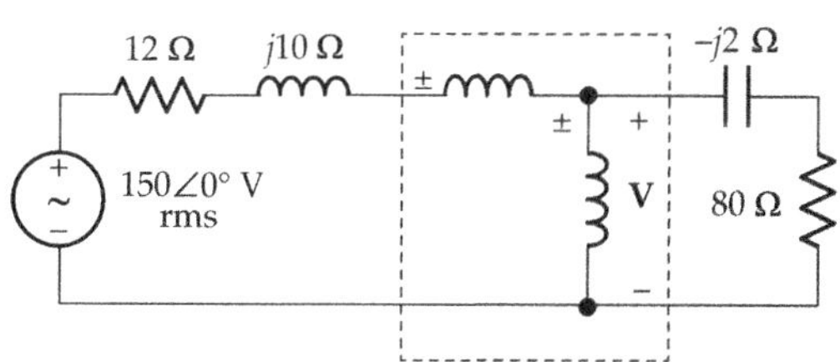

FIGURE P17.18

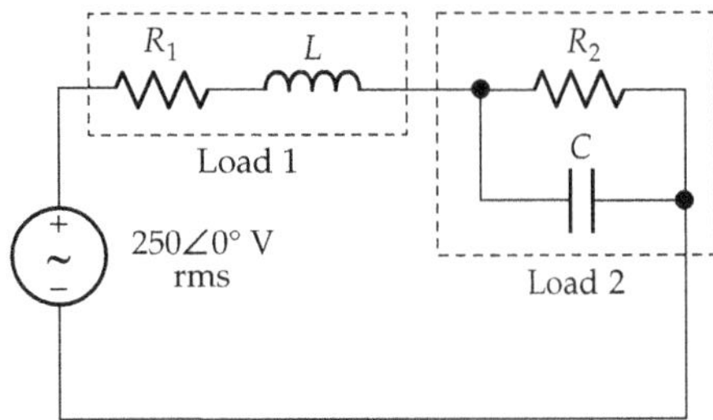

FIGURE P17.19

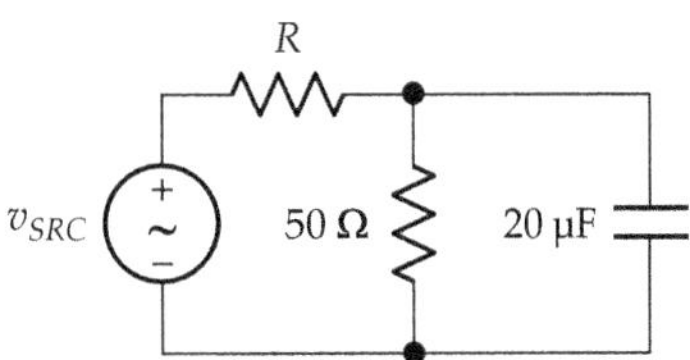

FIGURE P17.20

P17.20 Determine R and the rms magnitude of $\mathbf{V}_{SRC}$ in Figure P17.20, given that each resistor absorbs 2 W and $\omega = 1000$ rad/s.

Ans. 25 Ω, $5\sqrt{10}\angle 18.43°$ V.

P17.21 Determine the complex power delivered by the independent current source in Figure P17.21 and verify that it equals the complex power absorbed in the rest of the circuit.

Ans. $-7.5 + j10$ VA.

P17.22 Determine C in Figure P17.22 if the capacitor absorbs 5 VAR and the frequency is 50 Hz. Derive the power absorbed by C from conservation of power in the circuit.

Ans. 49.1 µF.

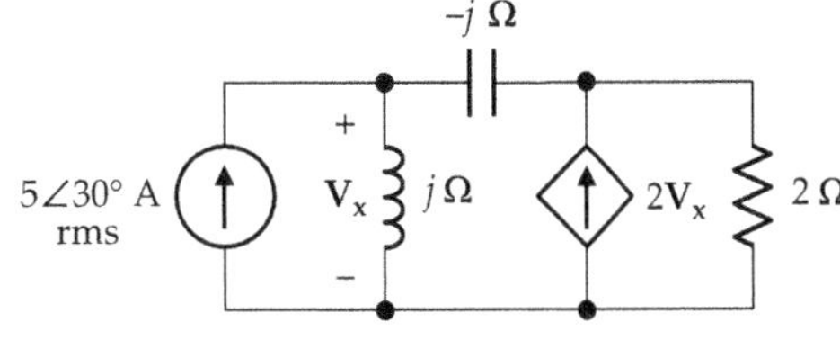

FIGURE P17.21

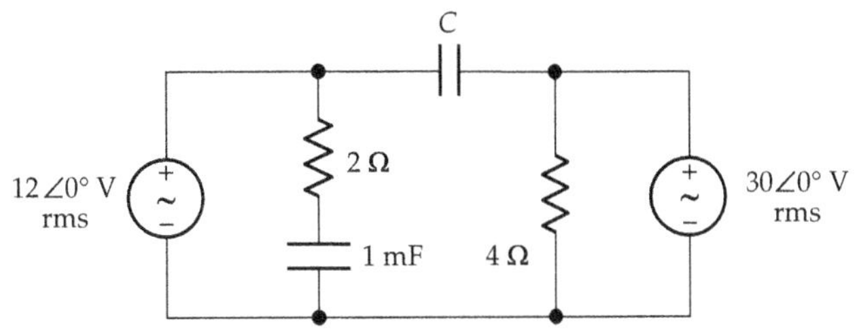

FIGURE P17.22

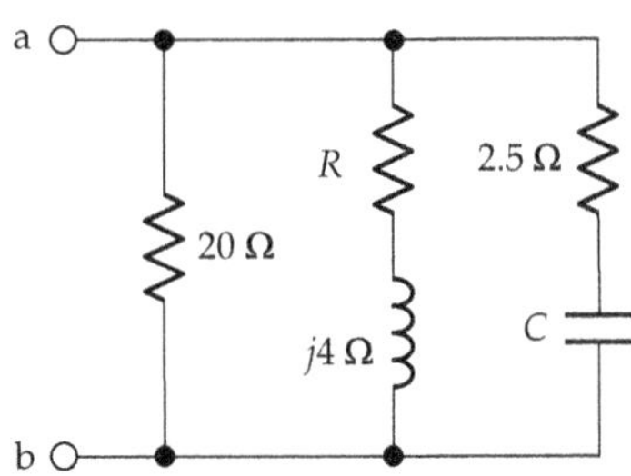

FIGURE P17.23

P17.23 Given that the complex power absorbed by the inductive branch in Figure P17.23 is $12 + j16$ VA, determine C so that the power factor at terminals ab is unity, assuming $\omega = 1$ rad/s.

Ans. 0.2 F, or 0.8 F.

P17.24 Determine the reactance that must be placed in parallel with terminals 'ab' in Figure P17.24 so that the power factor is unity at these terminals.

Ans. $-j10/9$ Ω.

P17.25 Given that the load 'L' in Figure P17.25 absorbs 100 kW at 0.8 p.f. lagging, determine C so that the p.f. at the source terminals is unity.

Ans. $7/(8\pi) = 0.28$ mF.

P17.26 ω, L, and C in Figure P17.26 are such that the power factor seen by the source is unity. Determine the power factor seen by the source if ω is doubled and C is halved, assuming $I_C = 5$ Arms.

Ans. 0.37 leading.

P17.27 Determine the complex power delivered by $\mathbf{V}_{SRC1}$ and $\mathbf{V}_{SRC2}$ in Figure P17.27 given that 'L_1' absorbs 4 kW at a power factor of 0.6 lagging, 'L_2' absorbs 3 kW at a power factor of 0.6 leading, and the complex power absorbed

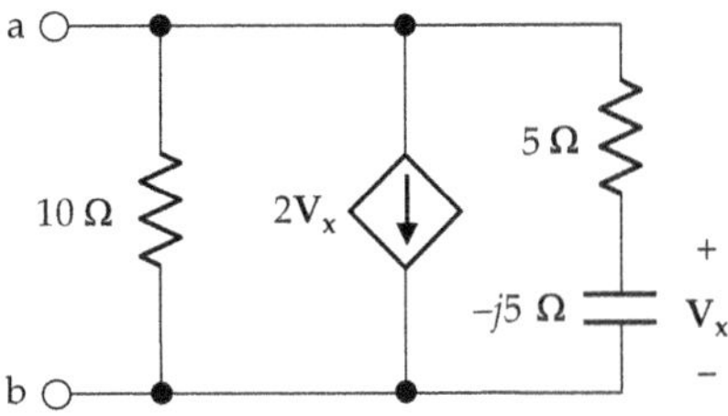

FIGURE P17.24

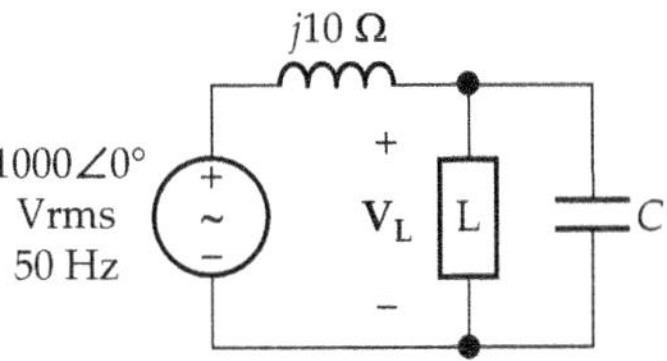

FIGURE P17.25

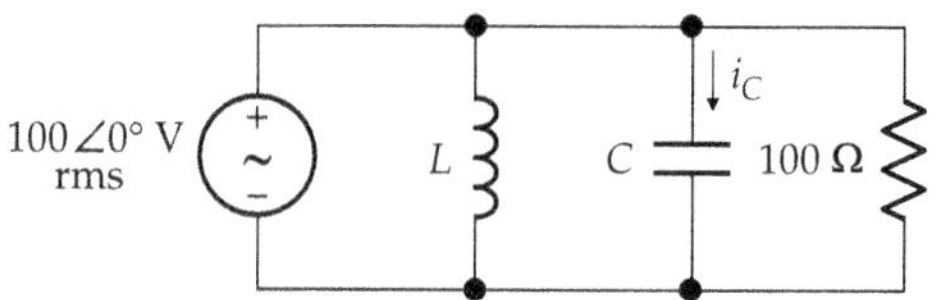

FIGURE P17.26

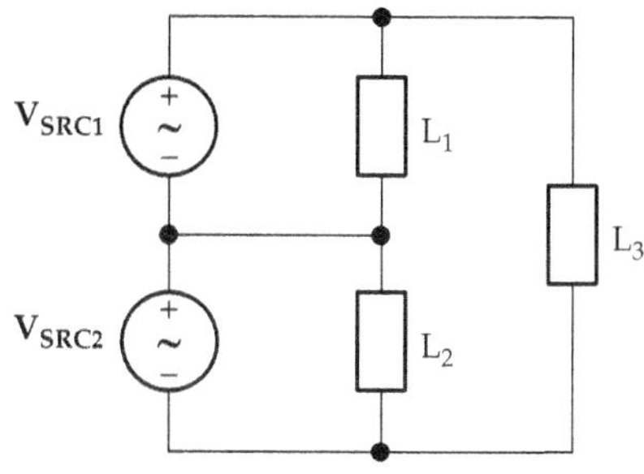

FIGURE P17.27

by 'L_3' is 12 + j5 kVA. Assume that $\mathbf{V}_{SRC1} = 400\angle 0°$ Vrms and $\mathbf{V}_{SRC2} = 400\angle 90°$ Vrms.

Ans. $\mathbf{V}_{SRC1}$ delivers 12.5 + j1.83 kVA and $\mathbf{V}_{SRC2}$ delivers 6.5 + j4.5 kVA.

P17.28 In Figure P17.28, determine the instantaneous power, the real power, and reactive power delivered by the source, given that $v_{SRC} = 10\cos 10^6 t$.

Ans. $\mathbf{S} = 25 + j100$ VA, so $P = 25$ W and $Q = 100$ VAR.

P17.29 Determine the total complex power delivered by the two sources in Figure P17.29 if $v_{SRC1} = 5\cos(2t + 45°)$ V and $v_{SRC2} = 5\cos 2t$ V.

Ans. 2.93 + j1.46 VA.

P17.30 Determine the complex power delivered by each source in Figure P17.30.

Ans. Power delivered by voltage source = j7.5 VA. Power delivered by current source = 0.5 − j8 VA.

P17.31 (a) A load 'L_1' consisting of a 12 kΩ resistor in series with a 40 H inductor is connected in parallel with a load 'L_2' consisting of a 75 nF capacitor in parallel with a resistor having a conductance of 40 μS, the frequency being 400 rad/s (Figure P17.31). If the 40 μS resistor dissipates a real power 4 kW, (i) determine the total complex power absorbed by 'L_1' and 'L_2', (ii) draw the power phasor diagrams for the two loads, and (iii) calculate the p.f. of the combined load of 'L_1' and 'L_2'.

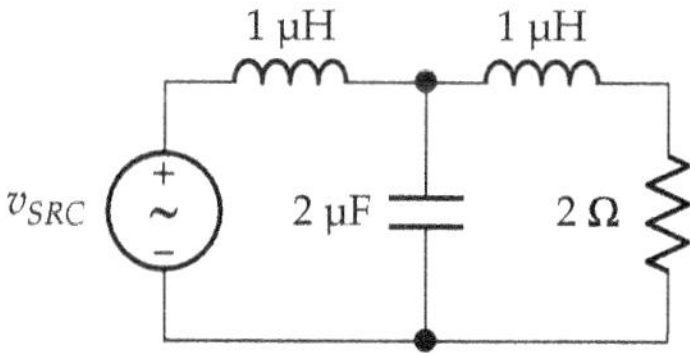

FIGURE P17.28

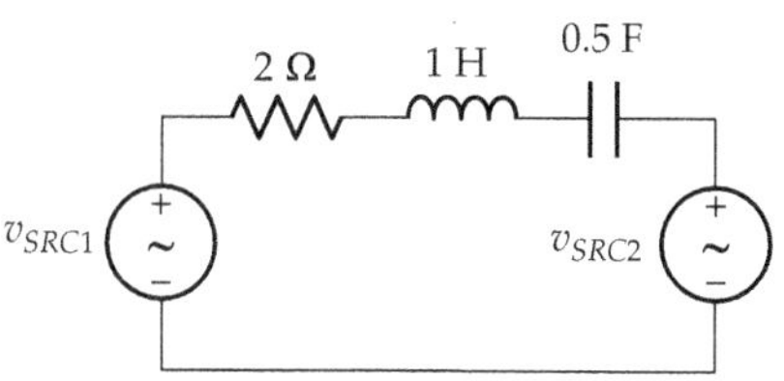

FIGURE P17.29

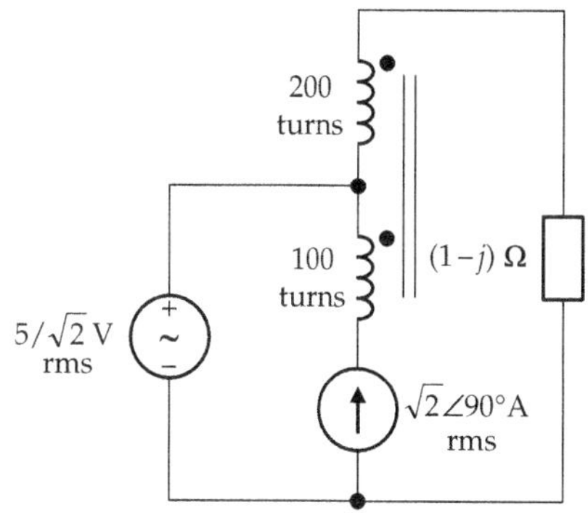

FIGURE P17.30

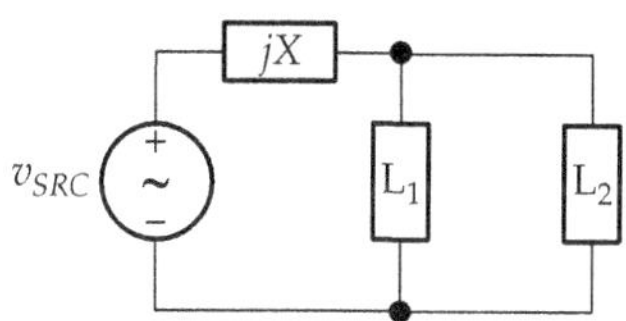

FIGURE P17.31

(b) The voltage across 'L_1' and 'L_2' is kept the same as in (a), but a reactance X is connected between 'L_1' and 'L_2' and the source, as shown. Determine (i) the value of X that results in a p.f. of unity at the source and (ii) the rms value of the source voltage.

Ans. (a) (i) 7000 + j1000 VA, (iii) 0.99; (b) (i) −2000 Ω, (ii) 9900 V rms.

Maximum Power Transfer

P17.32 Determine R_L in Figure P17.32 that will absorb maximum power and calculate this power.

Ans. 10 Ω, 62.5 W.

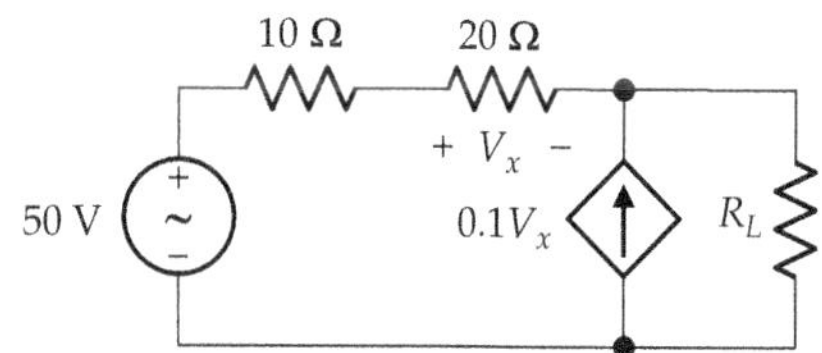

FIGURE P17.32

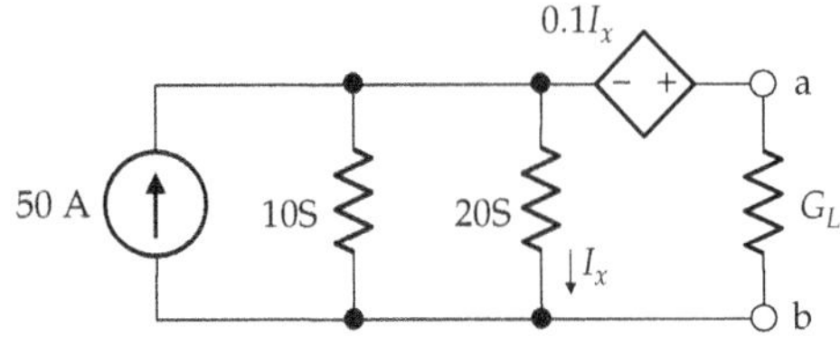

FIGURE P17.33

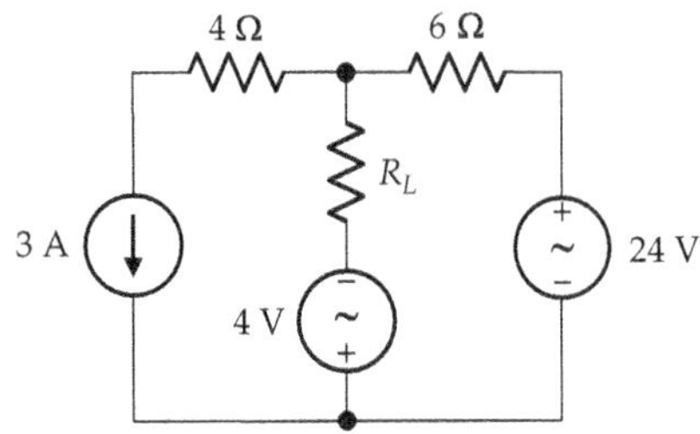

FIGURE P17.34

P17.33 Determine G_L in Figure P17.33 for maximum power transfer and calculate this power.

Ans. 0.1 Ω, 62.5 W.

P17.34 R_L in Figure P17.34 is restricted to the range 1–5 Ω. Determine the value of R_L that results in maximum power transfer to it and calculate the value of this power.

Ans. 5 Ω, 4.13 W.

P17.35 Determine R_L in Figure P17.35 for maximum power and calculate this power.

Ans. 9 Ω, 13.44 W.

P17.36 Determine R_L in Figure P17.36 for maximum power transfer and calculate this power.

Ans. 10/3 Ω, 40.83 W.

P17.37 Determine R_L in Figure P17.37 that satisfies the maximum power condition.

Ans. 15 Ω.

P17.38 Determine a of the transformer in Figure P17.38 so that maximum power is transferred to the 200 Ω load, and calculate this power.

Ans. a = 2, 2 W.

P17.39 Determine R_L in Figure P17.39 for maximum power transfer and calculate this power.

Ans. 16 Ω, 25 W.

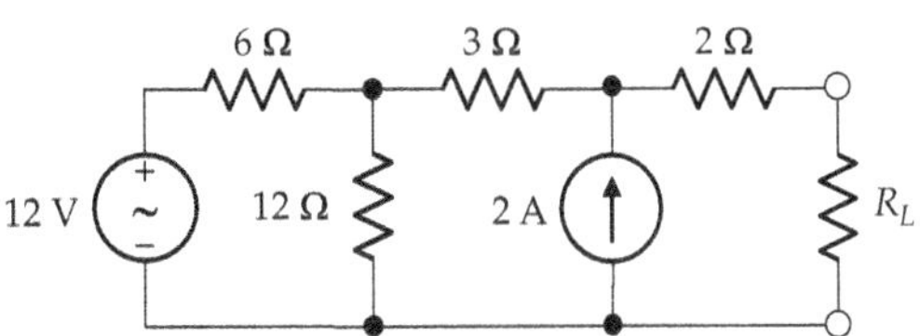

FIGURE P17.35

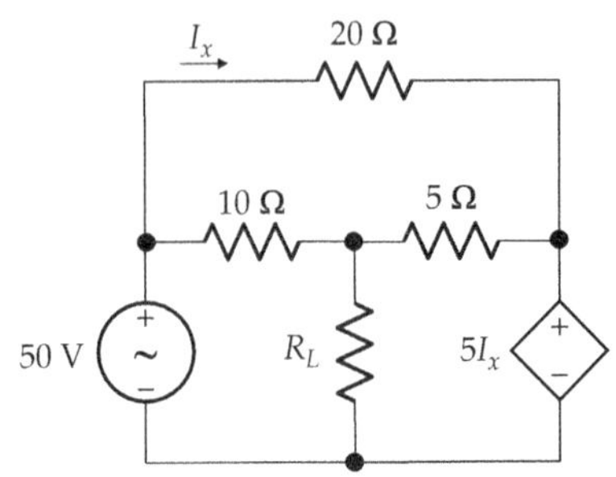

FIGURE P17.36

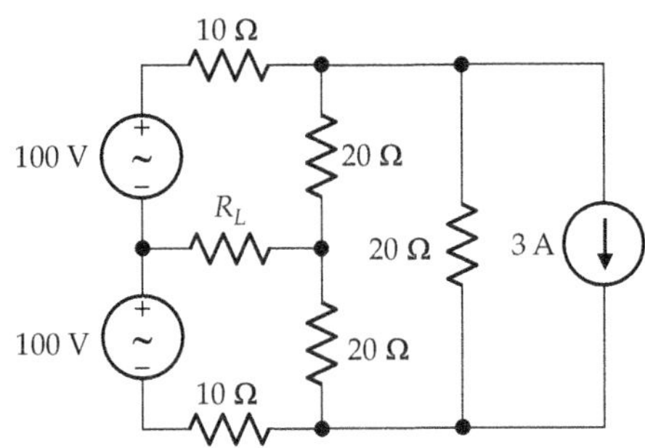

FIGURE P17.37

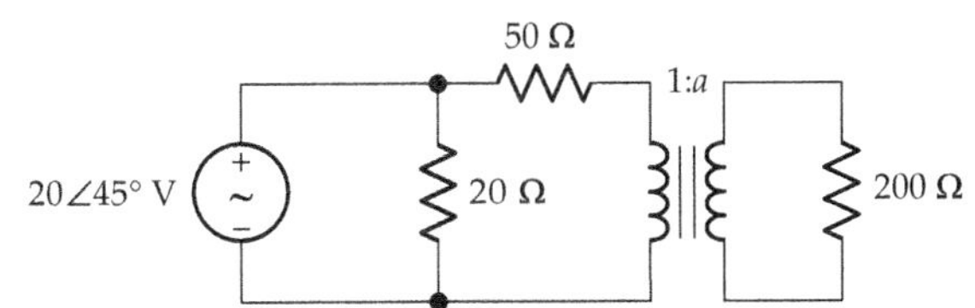

FIGURE P17.38

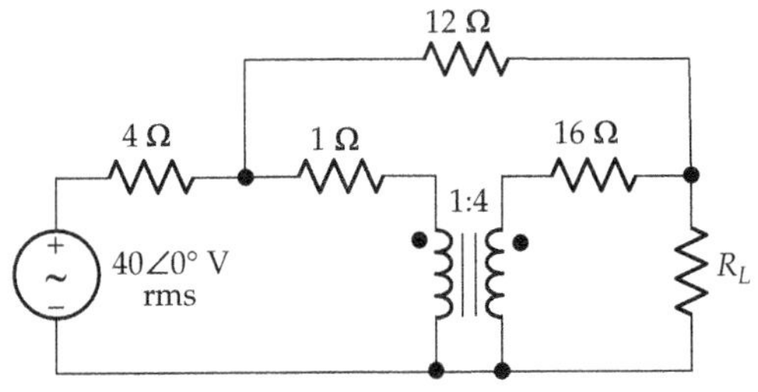

FIGURE P17.39

P17.40 Determine R_L in Figure P17.40 for maximum power transfer and calculate this power.

Ans. 18 Ω, 8/9 W.

P17.41 Determine the frequency at which maximum power is dissipated in the 10 Ω resistor in Figure P17.41.

Ans. 1 rad/s.

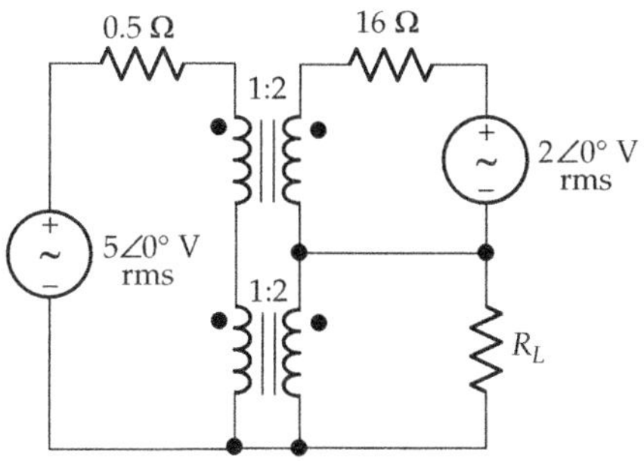

FIGURE P17.40

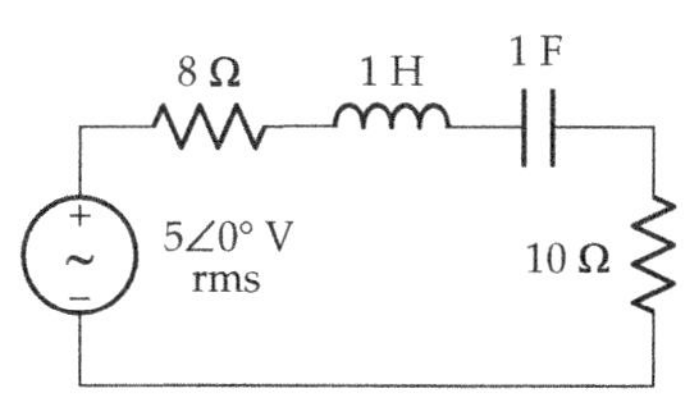

FIGURE P17.41

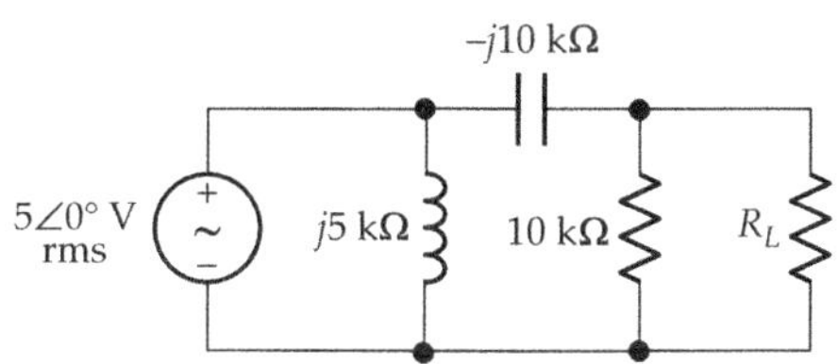

FIGURE P17.42

P17.42 Determine R_L in Figure P17.42 for maximum power transfer.

Ans. $5\sqrt{2}$ Ω.

P17.43 Determine Z_L in Figure P17.43 for maximum power transfer, assuming $i_{SRC} = 3\cos 5000t$ A.

Ans. 20 – *j*10 Ω.

P17.44 Determine Z_L in Figure P17.44 for maximum power transfer and calculate this power.

Ans. 8.123 – *j*3.785 Ω, 549.2 W.

P17.45 Determine Z_L in Figure P17.45 for maximum power transfer and calculate this power.

Ans. 5 – *j*5 Ω, 180 W.

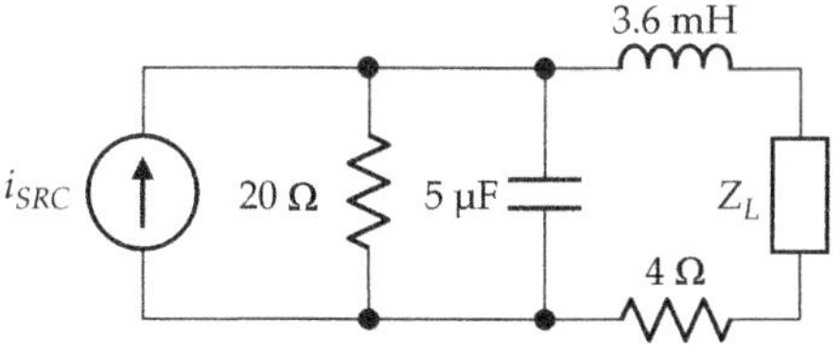

FIGURE P17.43

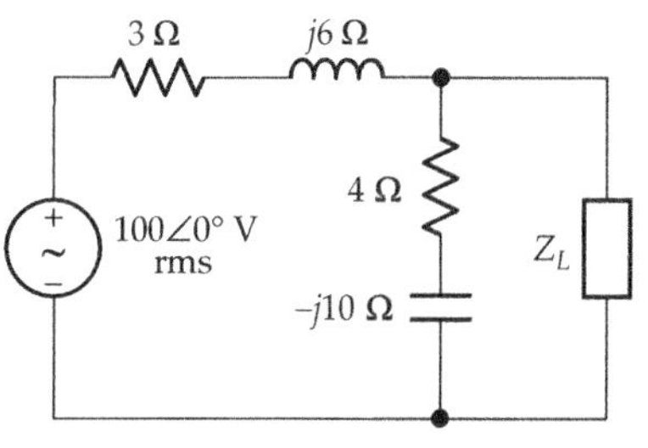

FIGURE P17.44

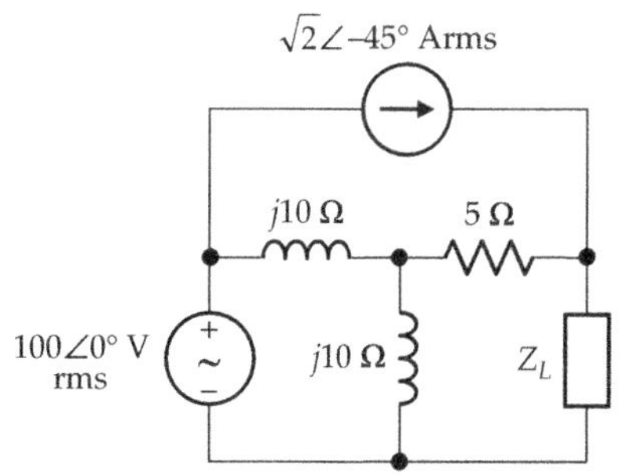

FIGURE P17.45

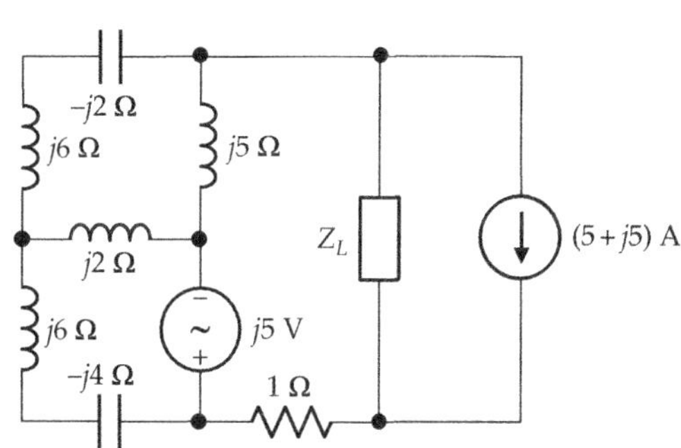

FIGURE P17.46

P17.46 Determine Z_L in Figure P17.46 for maximum power transfer.

Ans. 1 – *j*2.5 Ω.

P17.47 Determine Z_L in Figure P17.47 for maximum power transfer.

Ans. 3 + *j* Ω.

P17.48 Determine Z_L in Figure P17.48 for maximum power transfer and calculate this power.

Ans. 0.8 – *j*0.4 Ω, 7.81 W.

P17.49 Determine R_L in Figure P17.49 for maximum power transfer and calculate this power.

Ans. 6.00 Ω, 4.67 W.

P17.50 Determine R_L in Figure P17.50 for maximum power transfer.

Ans. $1/\sqrt{5}$ Ω.

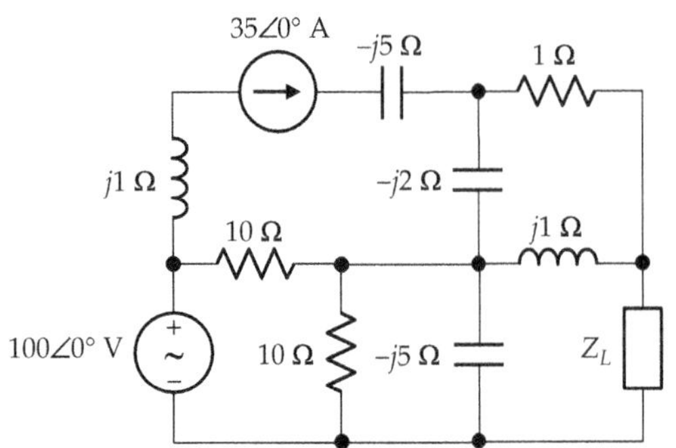

FIGURE P17.47

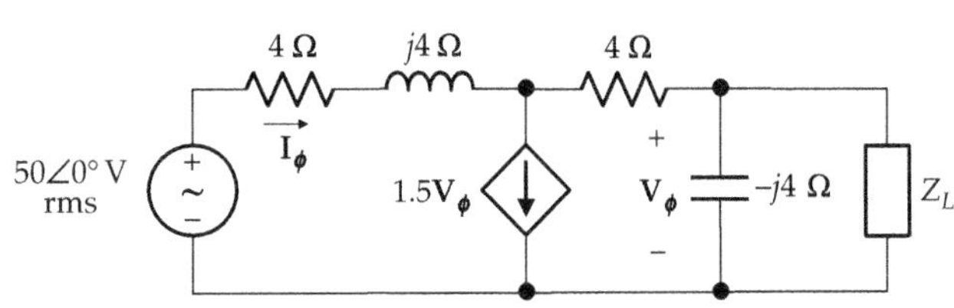

FIGURE P17.48

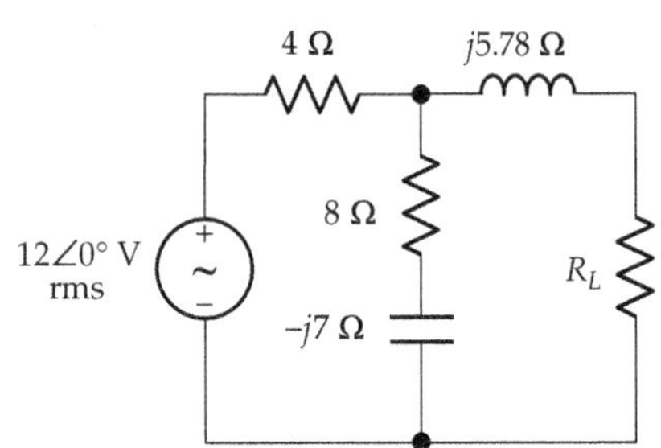

FIGURE P17.49

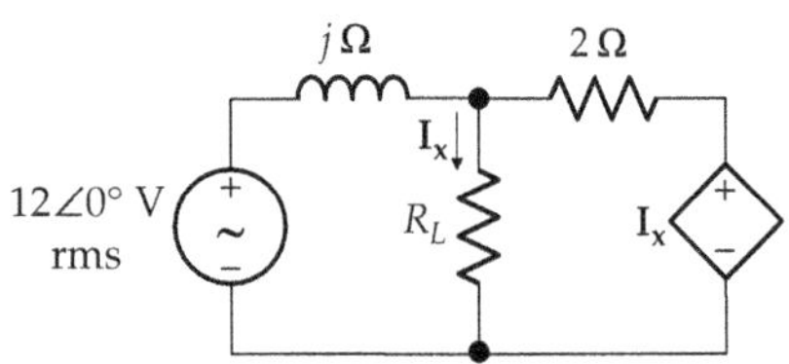

FIGURE P17.50

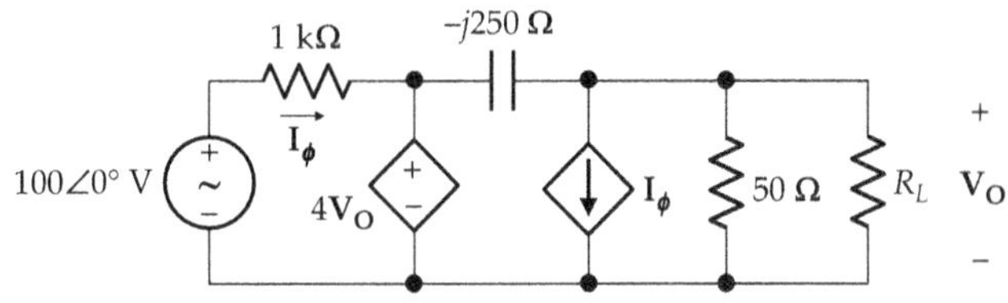

FIGURE P17.51

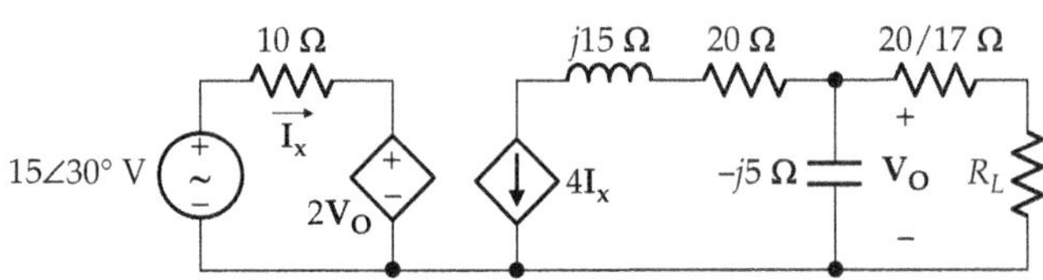

FIGURE P17.52

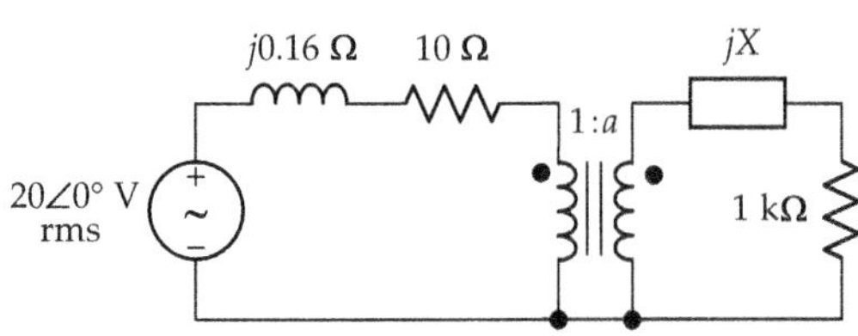

FIGURE P17.53

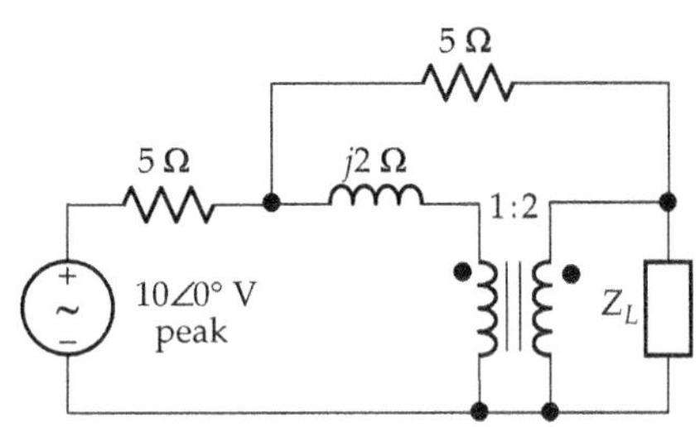

FIGURE P17.54

P17.51 Determine R_L in Figure P17.51 for maximum power transfer and calculate this power.

Ans. 50 Ω, 0.139 W.

P17.52 Determine R_L in Figure P17.52 for maximum power transfer and calculate this power.

Ans. 5/17 Ω, 90 W.

P17.53 Determine a and X in Figure P17.53 so that maximum power is transferred to the 1 kΩ resistor.

Ans. $a = 10$, $X = -16$ Ω.

P17.54 Determine Z_L in Figure P17.54 for maximum power transfer and calculate this power.

Ans. 10 Ω, 1.25 W.

P17.55 Determine R_L in Figure P17.55 so that it absorbs maximum power and calculate this power.

Ans. 160/7 Ω, 20/7 W.

P17.56 Determine ωM in Figure P17.56 so that maximum power is absorbed by the 20 Ω resistor and calculate this power.

Ans. $\omega M = 20\sqrt{2}$ Ω, 478 W.

P17.57 Determine R_L in Figure P17.57 for maximum power absorption in it and calculate this power.

Ans. 4.38 Ω, 1.25 W.

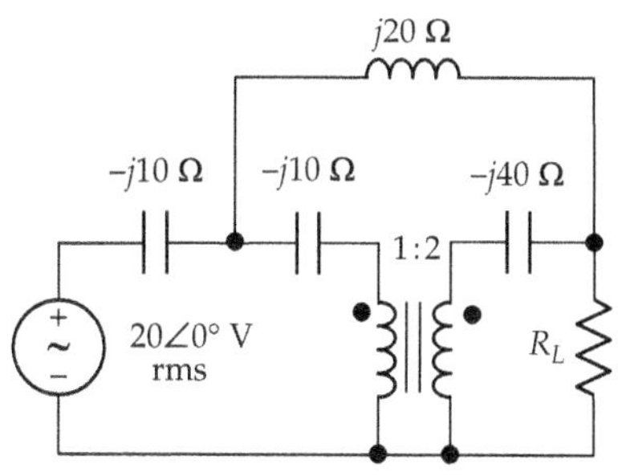

FIGURE P17.55

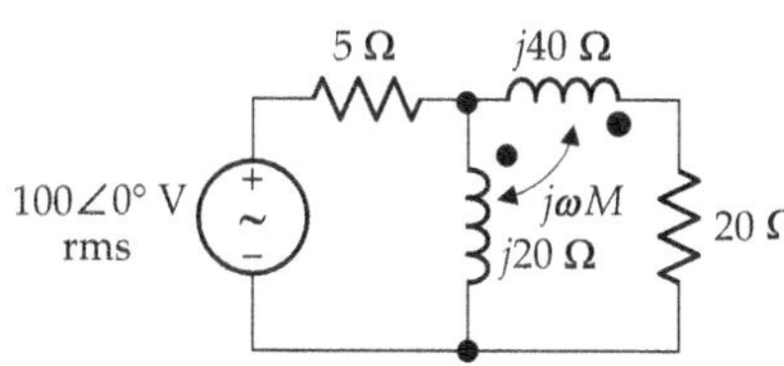

FIGURE P17.56

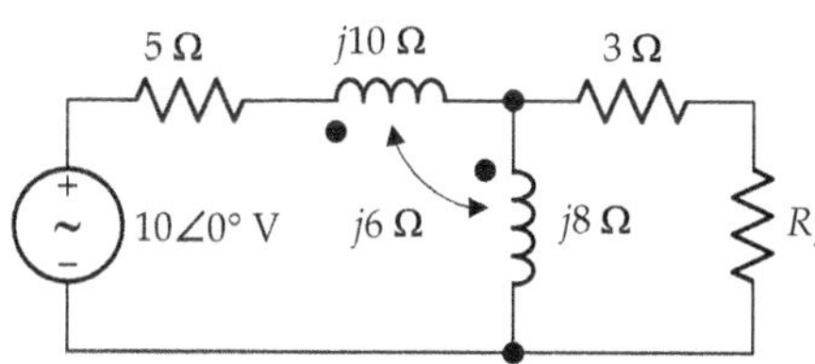

FIGURE P17.57

P17.58 Determine a and R_L in Figure P17.58 so that maximum power is transferred to R_L.

Ans. $a = 2$, 8 Ω.

P17.59 Determine Z_L in Figure P17.59 for maximum power transfer to it and calculate this power.

Ans. $2.4 - j0.8$ Ω, 150 W.

P17.60 Determine a in Figure P17.60 so that maximum power is absorbed by the 20 Ω resistor and calculate this power.

Ans. $a = 2.54$, 86.21 W.

P17.61 Determine X and R_L in Figure P17.61 for maximum power transfer to R_L and calculate this power.

Ans. $R_L = 10$ Ω, $X = -10$ Ω, 0.8 W

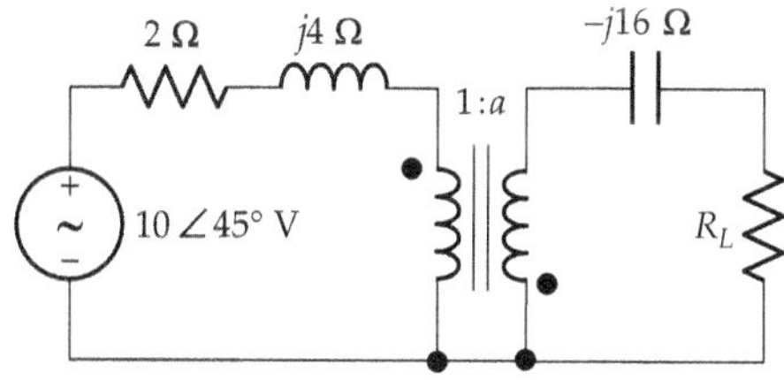

FIGURE P17.58

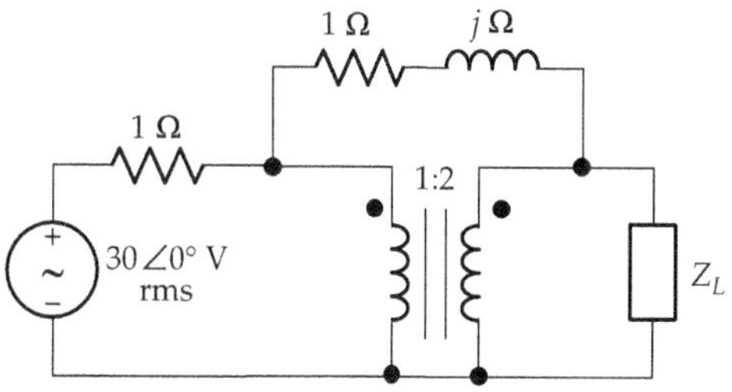

FIGURE P17.59

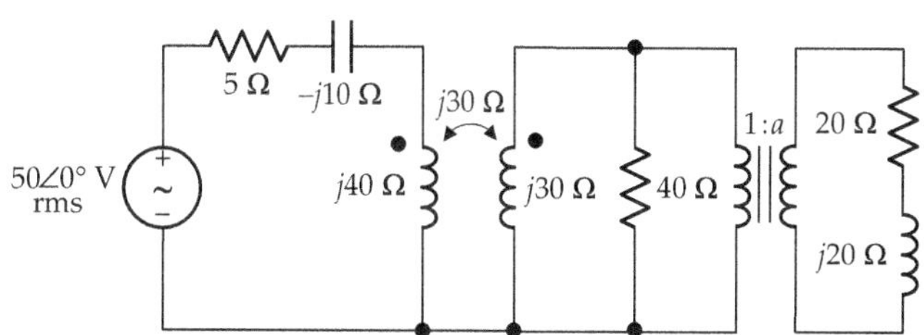

FIGURE P17.60

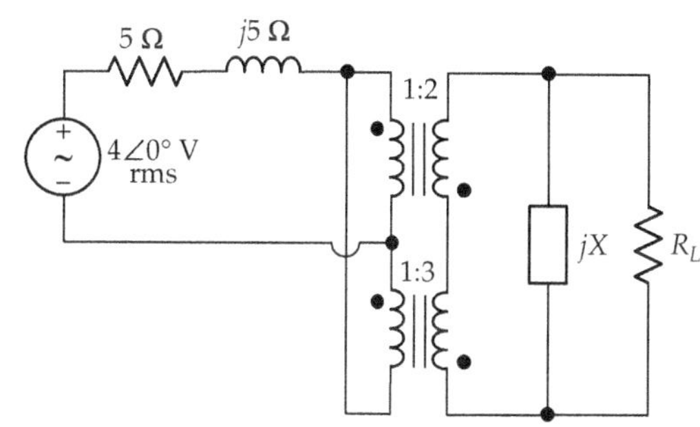

FIGURE P17.61

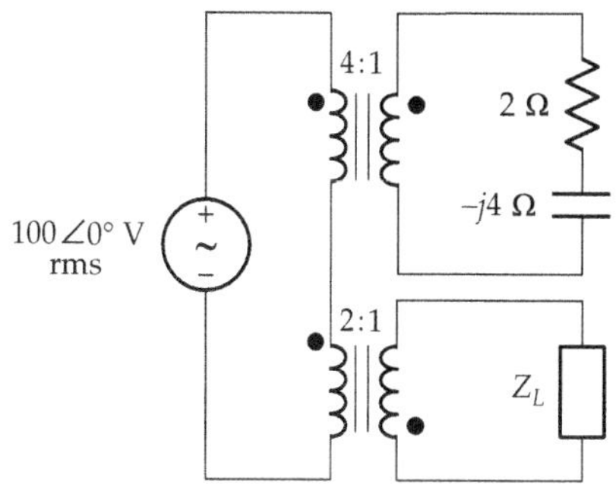

FIGURE P17.62

P17.62 Determine Z_L in Figure P17.62 for maximum power transfer and calculate this power.

Ans. $8(1 + j2)$ Ω, 78.125 W.

P17.63 (a) Determine Z_L in Figure P17.63 that results in maximum power absorption in Z_L and calculate this power. (b) If Z_L consists of a resistor in parallel with a capacitor,

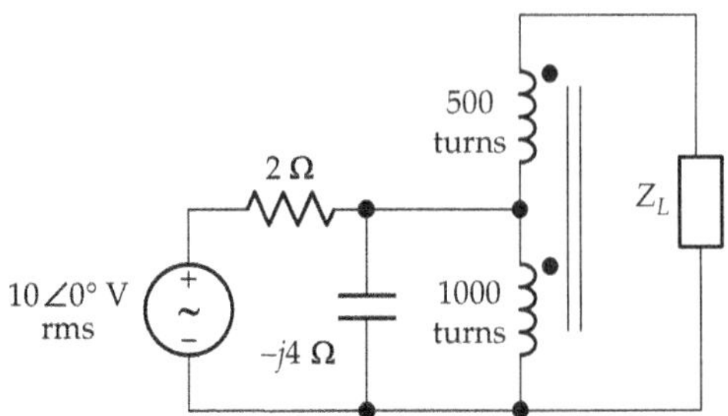

FIGURE P17.63

determine the values of these elements that will result in maximum power absorption in Z_L, and calculate this power.

Ans. (a) 3.6 + j1.8 Ω, 12.5 W; (b) C = 0, $R_L = 9/\sqrt{5}\,\Omega$, 11.8 W.

P17.64 (a) Determine Y_L in Figure P17.64 that results in maximum power absorption in Y_L and the value of this power. (b) If Y_L consists of a resistor in parallel with an inductor, what values of these elements will result in maximum power absorption in Y_L, and how much is this power?

Ans. (a) 3.6 + j1.8 S; 12.5 W; (b) $L \to \infty$, $G_L = 9/\sqrt{5}\,\Omega$, 11.8 W.

P17.65 (a) Determine the turns ratio a in Figure P17.65 so that maximum power is transferred to the 10 Ω resistor and calculate this power, assuming X = 30 Ω. (b) Assuming a = 2 and R_x = 10 Ω, determine X that results in maximum power absorption in R_x and calculate this power. (c) Assuming a = 2 and X = 15 Ω, determine R_x that results in maximum power absorption in this resistor and calculate this power. (d) Assuming a = 2 and R_x and X are variables, determine R_x and X that will result in maximum power absorption in R_x and calculate this power.

Ans. (a) 2.34, 1.18 W; (b) 80/29 Ω, 3.19 W; (c) R_x = 25.96 Ω, 3.53 W; (d) R_x = 664/29 Ω, X = 80/29 Ω, 3.77 W.

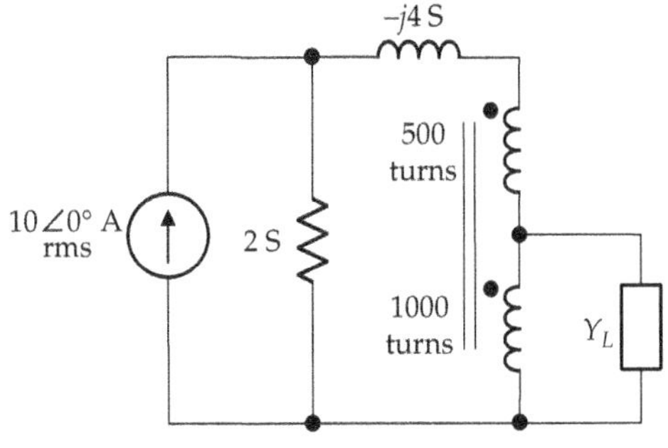

FIGURE P17.64

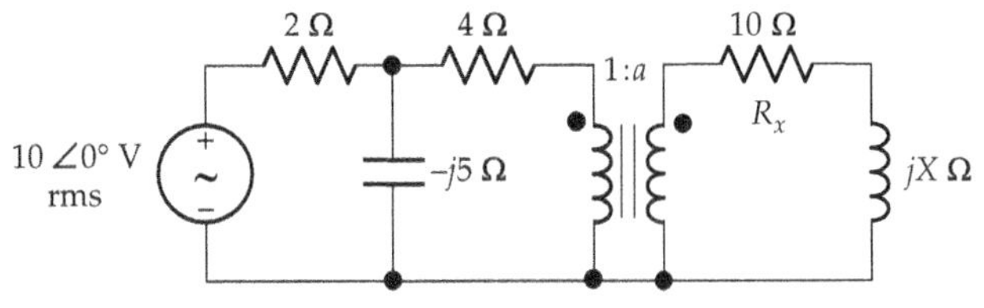

FIGURE P17.65

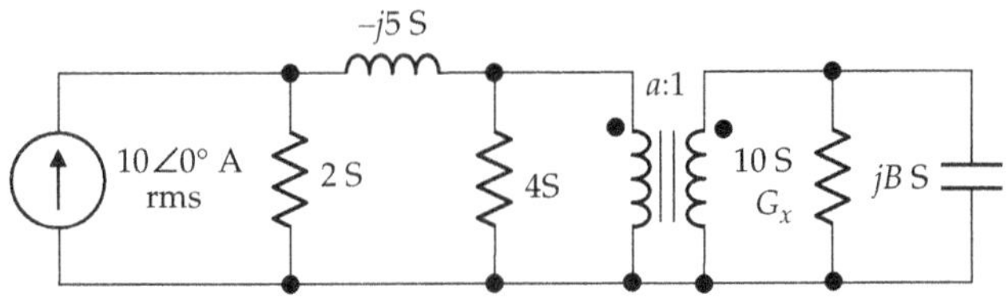

FIGURE P17.66

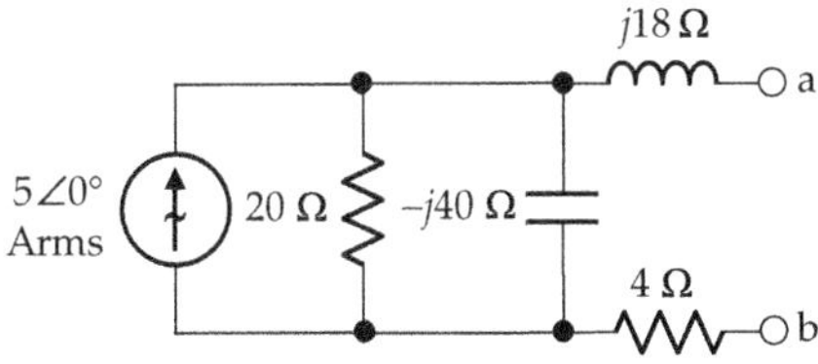

FIGURE P17.67

P17.66 (a) Determine the turns ratio a in Figure P17.66 so that maximum power is transferred to the 10 S resistor and calculate this power, assuming B = 30 S. (b) Assuming a = 2 and G_x = 10 S, determine B that results in maximum power absorption in G_x and calculate this power. (c) Assuming a = 2 and B = 15 S, determine G_x that results in maximum power absorption in this resistor and calculate this power. (d) Assuming a = 2 and G_x and B are variables, determine G_x and B that will result in maximum power absorption in G_x and calculate this power.

Ans. (a) 2.34, 1.18 W; (b) 80/29 S, 3.19 W; (c) G_x = 25.96 Ω, 3.53 W; (d) G_x = 664/29 Ω, B = 80/29 Ω, 3.77 W.

P17.67 A load $R_L + jX_L$ is connected between terminals 'ab' in Figure P17.67, where R_L can be varied between 0 and 10 Ω and X_L can be varied between ±5 Ω. Determine (a) the values of R_L and X_L that result in maximum power dissipated in R_L and (b) the complex power absorbed by the load under these conditions.

Ans. (a) X_{Lm} = −5 Ω, R_{Lm} = 10 Ω; (b) $\mathbf{S_L}$ = 86.5 − j43.2 VA.

Design Problems

P17.68 An amplifier having an output resistance of 75 Ω drives a load of 300 Ω resistance and 1.5 mH inductance at 1 MHz. Determine the transformer ratio and the capacitor that must be inserted in series with the amplifier output for maximum power transfer to the load.

Ans. Transformer ratio 1:2, C = 2 nF.

P17.69 An amplifier having an open-circuit voltage of 4 Vrms and an output resistance of 100 Ω supplies a load of 25 Ω and 0.1 H inductance at 400 Hz. A capacitor is connected in series with the output of the amplifier in order to block the dc voltage at the output of the amplifier. Select the capacitance so that maximum power is dissipated in the load, and determine this power.

Ans. 1.58 μF, 25.6 mW.

P17.70 A generator supplies 400 kW at 0.8 pF lagging at the current rating of the generator. Determine the largest resistive load that can be added to the generator, without exceeding the current rating of the generator, if the power factor is increased to unity, and calculate the capacitive kVAR required.

Ans. 500 kW, −300 kVAR.

18

Responses to Step and Impulse Inputs

Objective and Overview

The present chapter is concerned with responses to step and impulse inputs, which are the most basic examples of "sudden changes" applied to circuits. Impulse responses are fundamental to signals and systems analysis, including convolution and the Laplace transform, to be discussed in subsequent chapters.

The impulse response is introduced as a limiting case of a pulse response of a capacitor, after which the fundamental attributes of an impulse are presented. This is followed by a discussion of the step and impulse responses of single capacitors and series and parallel combinations of a capacitor and a resistor. The same procedure is applied to the step and impulse responses of single inductors and series and parallel combinations of an inductor and a resistor. These cases are used to illustrate some important concepts concerning the responses of capacitors and inductors to sudden changes.

The chapter ends with a generalization of the concepts presented to circuits consisting of capacitors, inductors, and resistors.

18.1 Capacitor Response to Current Pulse

Consider a step current I_{SRC} applied at $t = 0$ to a capacitor that is initially uncharged, as illustrated in Figure 18.1a. The current step implies that $i_{SRC} = 0$ for $t < 0$, and $i_{SRC} = I_{SRC}$, for $t > 0$, so that the source current suddenly changes at $t = 0$ from 0 to I_{SRC}. The voltage v across the capacitor is

$$v(t) = \frac{1}{C}\int_0^t I_{SRC}dt + 0 = \frac{I_{SRC}t}{C}, \quad t \geq 0 \qquad (18.1)$$

The voltage across the capacitor increases linearly with time at a rate I_{SRC}/C as shown in Figure 18.1a.

Suppose a current pulse of amplitude I_{SRC} that starts at $t = 0$ and ends at $t = a$ is applied to the uncharged capacitor (Figure 18.1b). Initially, for $t < a$, the current pulse is indistinguishable from the current step of the same amplitude, which causes the voltage across the capacitor to increase linearly at the same rate I_{SRC}/C as before. At $t = a$, $v = aI_{SRC}/C$, since the charge delivered to the capacitor from $t = 0$ to $t = a$ is aI_{SRC}, the area under the current pulse. The voltage is $1/C$ times this charge, that

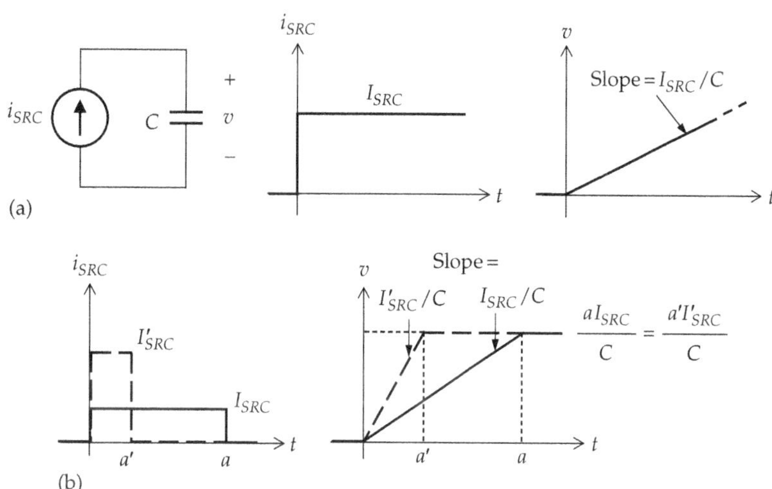

FIGURE 18.1
Capacitor response to current step (a) and to current pulses of different durations but of the same area (b).

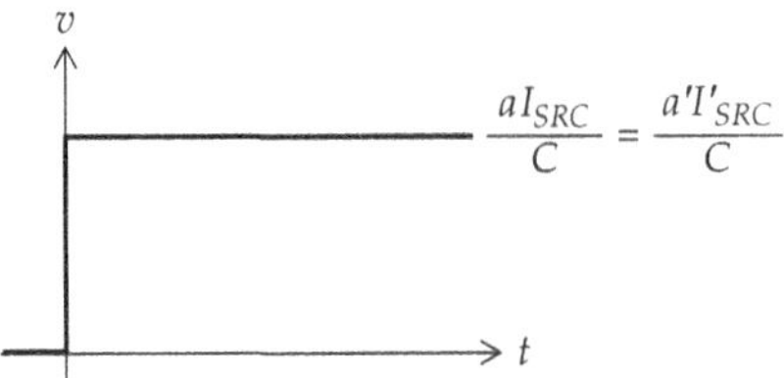

FIGURE 18.2
Step change in capacitor voltage.

is, aI_{SRC}/C. Just after $t = a$, the pulse ends, so that i_{SRC} becomes zero, and the ideal current source acts as an open circuit. Since the capacitor is assumed ideal, the charge is maintained for $t > a$, and v remains at its value of aI_{SRC}/C at $t = a$.

Let the amplitude of the pulse be increased to I'_{SRC} and its duration reduced to a' (Figure 18.1b), while keeping the area of the pulse the same, that is, $a'I'_{SRC} = aI_{SRC}$ = charge delivered at the end of current pulse. For $t < a'$, $dv/dt = I'_{SRC}/C > I_{SRC}/C$. The change in v is steeper, but the final value of v remains the same because the charge transferred is the same. Thus, from Equation 18.1, the final value of v is $\left(a'I'_{SRC}/C\right) = \left(aI_{SRC}/C\right)$ but is reached at a faster rate.

The question arises as to what shape of pulse will cause a step or instantaneous change in voltage across the capacitor, from 0 to aI_{SRC}/C at $t = 0$ (Figure 18.2). This means that $a' \to 0$, so that the pulse width becomes infinitesimal. But if the current pulse is of infinitesimal width, yet must deliver the same finite amount of charge aI_{SRC} or $a'I'_{SRC}$, then the pulse amplitude must tend to infinity so as to have a finite area. In other words, $i = Cdv/dt \to \infty$ at $t = 0$, since dv/dt, the slope of the step function in Figure 18.2, is infinite at $t = 0$. We can also consider the pulses in Figure 18.1b to be of duration $a\Delta$ and height I_{SRC}/Δ, so that the pulse area is aI_{SRC}, independently of Δ. In the case of the pulse of width a and height I_{SRC}, $\Delta = 1$, whereas in the case of the pulse of width a' and height I'_{SRC}, with $a'I'_{SRC} = aI_{SRC}$, $\Delta = a'/a$.

If we now let $\Delta \to 0$, the current pulse of Figure 18.3 becomes of infinitesimal width and of infinite height, as required to produce a step change in the voltage across the capacitor at $t = 0$ in Figure 18.2. The area or charge remains aI_{SRC} so that the amplitude of the voltage step across the capacitor is aI_{SRC}/C. The pulse now becomes an **impulse function** or a **Dirac delta function**. Qualitatively, such a function is a pulse of infinitesimal duration and infinite amplitude but having a finite area or **strength**, which is aI_{SRC} in this case. The impulse function will be considered in more detail in the following section.

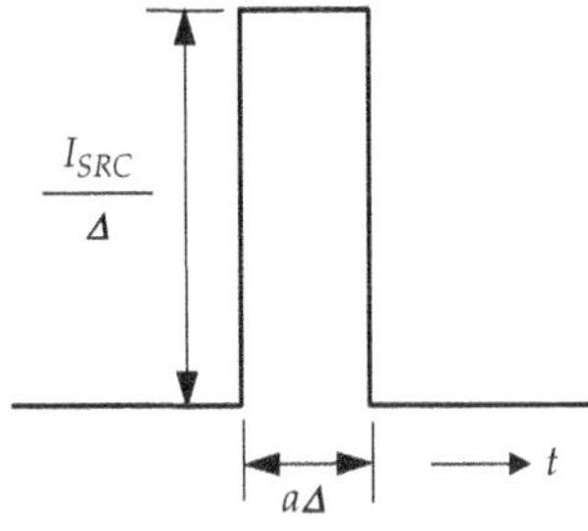

FIGURE 18.3
Current impulse.

Primal Exercise 18.1

Determine the capacitor voltage as a function of time if a current of 5 mA is applied for 10 ms to a 10 μF, uncharged capacitor.

Ans. v rises to 5 V at 10 ms and remains at this value thereafter.

18.2 The Impulse Function

The impulse function is of fundamental importance in the theory of signals and systems, as will become apparent from future discussions. Mathematically, the impulse function is not a function in the ordinary sense but is a *singular, generalized function* that has some rather peculiar properties. Because the amplitude of the impulse goes to infinity, care must be exercised in dealing with infinities. Mathematically, this is avoided, where necessary, by considering the finite area of the impulse, instead of its infinite amplitude, as will be illustrated in future discussions.

A unit impulse function at the origin is denoted by $\delta(t)$ and is formally defined such that

$$\int_{-\infty}^{+\infty} \delta(t)dt = 1, \quad \text{with } \delta(t) = 0 \quad \text{for } t \neq 0 \qquad (18.2)$$

According to Equation 18.2, the area or strength of $\delta(t)$ is unity; hence, its designation as a unit impulse. Moreover, $\delta(t)$ is zero everywhere except at $t = 0$.

Figure 18.4 helps provide a "feel" for $\delta(t)$ to clarify its variation with respect to time and its relation to another important function, namely, the unit step function at the origin, $u(t)$. In this figure, the interval between $t = 0^-$ and $t = 0^+$ is greatly expanded, where $t = 0^-$ is the time just less than 0, and $t = 0^+$ is the time just greater than 0 (Figure 18.4). The following should be noted concerning $\delta(t)$:

1. $\delta(t) = 0$, for $t \leq 0^-$ and $\delta(t) = 0$, for $t \geq 0^+$, so that $\delta(t) \neq 0$ *only* in the interval between $t = 0^-$ and $t = 0^+$. In words, $\delta(t)$ is zero for *t less than or equal*

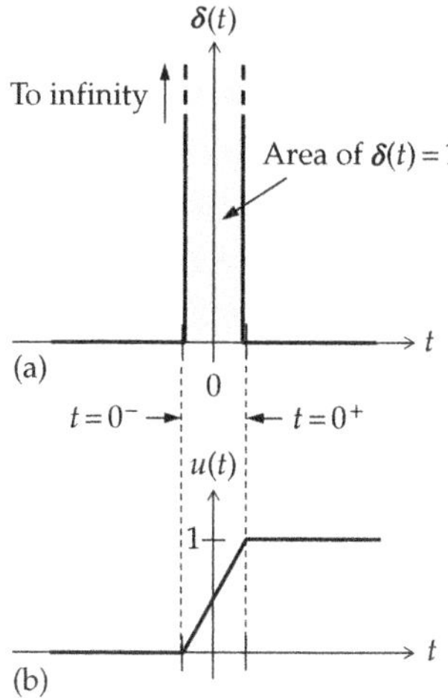

FIGURE 18.4
Impulse function (a) and its relation to a step function (b).

to 0^- and is also zero for t *greater than or equal to* 0^+, bearing in mind that $t = 0^-$ and $t = 0^+$ are infinitesimally separated. In other words, $\delta(t)$ is concentrated at $t = 0$. This is not as alien a concept as may appear at first sight, for it is customary to consider charge or mass to be concentrated at a point. This means that a finite charge or mass extends over an infinitesimal distance only. The mass/unit distance or charge/unit distance is therefore an impulse at the point where the mass or charge is located. What is being considered here is an extension of this concept to voltages or currents that occur at a "point" in time.

2. The time integral of $\delta(t)$ is the unit step function $u(t)$ (Figure 18.4b), also known as the **Heaviside step function**. From Equation 18.2 and Figure 18.4,

$$\int_{-\infty}^{t} \delta(t)dt = \int_{0^-}^{t} \delta(t)dt = 0, \quad t \le 0^- \qquad (18.3)$$

where the lower limit has been changed from $-\infty$ to 0^-, without changing the value of the integral, because, by definition (Equation 18.1), $\delta(t) = 0$ between $t = -\infty$ and $t = 0^-$, so that no area is enclosed in this interval. Extending the upper limit to $t \ge 0^+$ would include the area of the unit impulse, so that the value of the integral becomes unity, again by definition of the unit impulse. Thus,

$$\int_{-\infty}^{t} \delta(t)dt = \int_{0^-}^{t} \delta(t)dt = 1, \quad t \ge 0^+ \qquad (18.4)$$

t in Equation 18.4 can be extended to $+\infty$, as in Equation 18.2, without changing the value of the integral, because $\delta(t) = 0$ between $t = 0^+$ and $t = \infty$, so that no additional area is included in this interval. Equations 18.3 and 18.4 define $u(t)$, the unit step function at the origin as

$$u(t) = 0, \text{ for } t \le 0^- \text{ and } u(t) = 1, \text{ for } t \ge 0^+ \quad (18.5)$$

In other words, $u(0^-) = 0$, and $u(0^+) = 1$. Equations 18.3 and 18.4 involving two values of the integral of $\delta(t)$, one for $t \le 0^-$ and one for $t \ge 0^+$ can be combined into a single expression using the definition of $u(t)$ from Equation 18.5:

$$\int_{-\infty}^{t} \delta(t)dt = u(t), \text{ for all } t \qquad (18.6)$$

$u(t)$ changes from 0 to 1 between $t = 0^-$ and $t = 0^+$, as illustrated in Figure 18.4b. Since $t = 0^-$ and $t = 0^+$ are infinitesimally separated, $u(t)$ is, strictly speaking, undefined at $t = 0$, where it changes "instantaneously" from 0 to 1. However, $u(t)$ can be interpreted as having a value of 1/2 at $t = 0$ (Problem P18.49). This is reminiscent of the Fourier series expansion for a periodic function having a step discontinuity, in which case, the FSE gives the average value of the function at the two ends of the discontinuity (Example 16.1).

3. Differentiating both sides of Equation 18.6 with respect to time,

$$\delta(t) = \frac{du(t)}{dt} \qquad (18.7)$$

This makes sense intuitively, as the derivative of $u(t)$ would be zero, for $t \le 0^-$ and for $t \ge 0^+$, where $u(t)$ does not change with time and would tend to infinity at $t = 0$ where $u(t)$ changes instantaneously. Note that, strictly speaking, when a function is discontinuous at a certain value of time, as is $u(t)$ at $t = 0$, the derivative at this instant is not defined, because it tends to infinity, and infinity is not precisely defined. However, $\delta(t)$ is a special type of infinity, of infinitesimal duration. Equations 18.6 and 18.7 define the relation between $\delta(t)$, the unit impulse at the origin and $u(t)$, the unit step at the origin.

In order to avoid any possible confusion, it should be carefully noted that at $t = 0^-$, $\delta(t)$ *has not yet occurred*, and $u(t)$ *is still zero*. At $t = 0^+$, $\delta(t)$ *is over*, and $u(t) = 1$. In other words, $\delta(t)$ and the change in $u(t)$ *are both confined to between* $t = 0^-$ and $t = 0^+$.

Conventionally, and in terms of $t = 0$, rather than $t = 0^-$ and $t = 0^+$ as in Equation 18.5, $u(t)$ is sometimes defined as

$$u(t)=0, \text{ for } t<0 \text{ and } u(t)=1, \text{ for } t \geq 0 \quad (18.8)$$

It should be noted, however, that this definition can lead to ambiguities and inconsistencies in some cases (Problem P18.49). In analyzing circuits involving a sudden change, such as step or impulse inputs, or switching operations, it is more advantageous and much clearer to consider the state of the circuit just before and just after the sudden change, that is, at $t = 0^-$ and $t = 0^+$ for a sudden change at $t = 0$. The definition of $u(t)$ according to Equations 18.5 through 18.7 is much better suited for such cases and is used exclusively in this book.

4. As implied in Figure 18.4a, $\delta(t)$ is an even function of time, that is, $\delta(t) = \delta(-t)$. This can be easily proved using the defining integral (Equation 18.2) and substituting $t = -t'$ in $\delta(t)$, in dt, and in the integration limits. The integral becomes

$$\int_{-\infty}^{+\infty}\delta(t)dt=\int_{+\infty}^{-\infty}\delta(-t')(-dt')=\int_{-\infty}^{+\infty}\delta(-t')dt'=$$

$$\int_{-\infty}^{+\infty}\delta(-t)dt \quad (18.9)$$

The negative sign of dt' in the integrand in the second integral from the left is removed by interchanging the limits of integration. t' is replaced by t in the last integral, because, in a definite integral, the limits are defined, so that the integration variable is a dummy variable that can be designated by any symbol, without changing the value of the integral.

5. For consistency in all the relations involving $\delta(t)$ and $u(t)$, particularly those of the Laplace transform and its properties (Chapter 21), $u(t)$ is considered dimensionless, so that $\delta(t)$ has the dimensions of t^{-1}, that is, per unit time. This is consistent with Equation 18.4 and means that the integral in Equation 18.2 is dimensionless.

6. It follows that a current impulse can be expressed as $q\delta(t)$ A, where q is the strength, or area, of the current impulse. Since this area is in ampere-second, or coulombs, it represents charge. With $\delta(t)$ having the dimensions of t^{-1}, time cancels out, so that the unit of $q\delta(t)$ is the ampere. Similarly, a voltage impulse can be expressed as $\lambda\delta(t)$ V, where λ is the strength or area of the voltage impulse. Since this area is in volt-second, or weber-turn, it represents flux linkage. With $\delta(t)$ having the dimensions of t^{-1}, time cancels out, so that the unit of $\lambda\delta(t)$ is the volt. Unless otherwise specified, impulses will be denoted as $K\delta(t)$, with K interpreted as appropriate for the case under consideration.

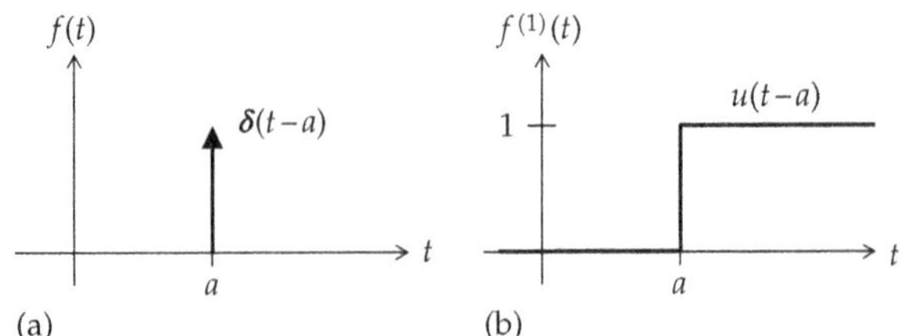

FIGURE 18.5
Delayed impulse (a) and delayed step (b).

7. $\delta(t - a)$ is a unit impulse $\delta(t)$ that is delayed by a, so it occurs at $t = a$, between $t = a^-$ and $t = a^+$, as illustrated in Figure 18.5a. Equation 18.3 becomes

$$\int_{-\infty}^{t}\delta(t-a)dt=\int_{a^-}^{t}\delta(t-a)dt=0, \quad t\leq a^- \quad (18.10)$$

since $\delta(t - a) = 0$ for $t \leq a^-$. Equation 18.4 becomes

$$\int_{-\infty}^{t}\delta(t-a)dt=\int_{a^-}^{t}\delta(t-a)dt=1, \quad t\geq a^+ \quad (18.11)$$

since the impulse of unit area occurs between $t = a^-$ and $t = a^+$. Following the same argument as for Equations 18.3 and 18.4, Equations 18.10 and 18.11 can be combined in a single expression involving $u(t - a)$:

$$\int_{-\infty}^{t}\delta(t-a)dt=u(t-a), \quad \text{for all } t \quad (18.12)$$

where $u(t - a)$ is the unit step function $u(t)$ that is delayed by a, so that it jumps from zero to unity at $t = a$ (Figure 18.5b). Differentiating both sides of Equation 18.12

$$\delta(t-a)=\frac{du(t-a)}{dt} \quad (18.13)$$

An important property of $\delta(t)$ is presented in Example 18.1. Other properties of $\delta(t)$ are considered in problems at the end of the chapter.

Example 18.1: Sampling of Function by Impulse

It is required to prove the following property of the impulse function:

$$f(t)\delta(t-a)=f(a)\delta(t-a) \quad \text{and} \quad f(t)\delta(t-a)=0 \quad \text{if } f(a)=0$$

where $f(t)$ is continuous at $t = a$.

Solution:

Figure 18.6 illustrates a unit impulse $\delta(t-a)$ at $t = a$ and a function $f(t)$ that is continuous at $t = a$. Continuity of $f(t)$ at $t = a$ implies that $f(a^-) = f(a) = f(a^+)$. It may be argued, at first sight, that multiplying the infinite $\delta(t - a)$ by a finite value of $f(t)$ is still infinity. The impulse function, however, is a special type of infinity that is of infinitesimal duration. The proper procedure is to consider, first, the integral of the product $f(t)\delta(t - a)$, then apply the definition of the impulse and the condition of continuity of $f(t)$ at $t = a$. The integral of the product $f(t)\delta(t - a)$ can be expanded as

$$\int_{-\infty}^{t} f(t)\delta(t-a)dt = \int_{-\infty}^{a^-} f(t)\delta(t-a)dt + \int_{a^-}^{a^+} f(t)\delta(t-a)dt + \int_{a^+}^{t} f(a)\delta(t-a)dt \quad (18.14)$$

Since $\delta(t - a)$ is zero for $t \le a^-$ and for $t \ge a^+$, it follows that

$$\int_{-\infty}^{t} f(t)\delta(t-a)dt = 0 \quad \text{for } t \le a^-$$

and

$$\int_{-\infty}^{t} f(t)\delta(t-a)dt = \int_{a^-}^{a^+} f(t)\delta(t-a)dt \quad \text{for } t \ge a^- \quad (18.15)$$

Because the function is continuous at $t = a$, $f(t)$ in the interval between $t = a^-$ and $t = a^+$ is $f(a)$. This is a constant and can be taken outside the integral. From the definition of the impulse at $t = a$, the integral of $\delta(t - a)$ between $t = a^-$ and $t = a^+$ is unity. Thus,

$$\int_{-\infty}^{t} f(t)\delta(t-a)dt = 0 \quad \text{for } t \le a^-$$

and

$$\int_{-\infty}^{t} f(t)\delta(t-a)dt = f(a)\int_{a^-}^{a^+} \delta(t-a)dt = f(a) \quad \text{for } t \ge a^+ \quad (18.16)$$

As was done with Equation 18.12, the two conditions in Equation 18.16 can be combined into a single expression in terms of the step function at $t = a$:

$$\int_{-\infty}^{t} f(t)\delta(t-a)dt = f(a)u(t-a) \quad (18.17)$$

Differentiating both sides of Equation 18.17 and using Equation 18.13,

$$f(t)\delta(t-a) = f(a)\frac{d}{dt}\big(u(t-a)\big) = f(a)\delta(t-a) \quad (18.18)$$

If $f(a) = 0$, then it must not be inferred from Equation 18.18 that multiplying zero by infinity is indeterminate. If $f(a) = 0$, the RHS of Equation 18.17 is zero. Differentiating both sides of this equation gives $f(t)\delta(t - a) = 0$ if $f(a) = 0$. Again, this emphasizes that in dealing with impulses, the finite area of the impulse should be invoked wherever necessary to obtain mathematically correct results. Intuitively, it seems reasonable that since the impulse $\delta(t - a)$ occurs at the point $t = a$, multiplying $\delta(t - a)$ by $f(a)$ multiplies the strength of the impulse by $f(a)$.

Because multiplying $f(t)$ by $\delta(t - a)$ returns the value $f(a)$, the impulse is said to *sample* the function at $t = a$. Note that if $f(t)$ has a step discontinuity at $t = a$, the value sampled is the mean of $f(a^-)$ and $f(a^+)$ (Problem P18.49).

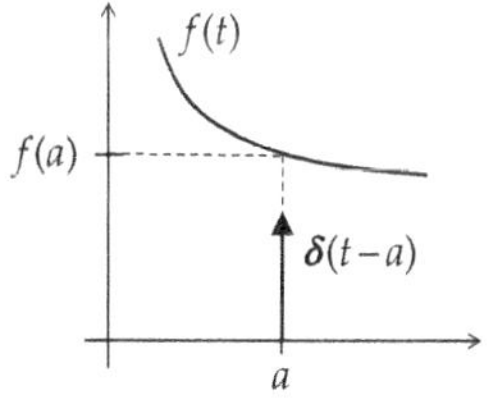

FIGURE 18.6
Figure for Example 18.1.

Primal Exercise 18.2

Determine (a) $5\delta(t)\sin t$ and (b) $5\delta(t)\cos t$.

Ans. (a) 0; (b) $5\delta(t)$.

Primal Exercise 18.3

The response of an LTI circuit to $\delta(t)$ is $u(t - 1) - u(t - 3)$. Determine the response of the circuit to the delayed impulse $\delta(t - 2)$.

Ans. $u(t - 3) - u(t - 5)$.

18.3 Responses of Capacitive Circuits to Step and Impulse Inputs

18.3.1 Single Capacitor

The fact that the v–i relation for a capacitor involves differentiation or integration with respect to time underlies an important characteristic of the capacitor. Consider the v–i relation of a capacitor to be rearranged as $dv = (1/C)idt$. Integrating this relation between two limits t_1 and t_2 gives

$$\int_{v_1}^{v_2} dv = \frac{1}{C}\int_{t_1}^{t_2} idt \tag{18.19}$$

where $v(t_2) = v_2$ and $v(t_1) = v_1$. It follows that

$$v(t_2)-v(t_1) = \frac{1}{C}\int_{t_1}^{t_2} idt \tag{18.20}$$

Let $t_2 \to t_1$, that is, $t_2 - t_1 = \Delta t \to 0$. We will consider the current integral on the RHS of Equation 18.20 as an area, assuming first that i is finite and continuous at $t = t_1$, as illustrated in Figure 18.7a. For small Δt, the area under the curve from $t = t_1$ to $t = t_2$ is that of a trapezoid of width Δt and a finite mean height. This area tends to zero as $\Delta t \to 0$, so that $v(t_2) = v(t_1)$.

If i is finite and has a finite discontinuity at $t = t_1$ (Figure 18.7b), then the mean height of the trapezoid of width Δt is finite, so that the area of the trapezoid tends to zero as $\Delta t \to 0$. Again, Equation 18.20 gives $v(t_2) = v(t_1)$.

It is seen that as long as i is finite and continuous at $t = t_1$, or is finite and has a finite discontinuity at $t = t_1$, the integral on the RHS of Equation 18.20 tends to zero as $\Delta t \to 0$, which makes $v(t_2) = v(t_1)$. In other words, the capacitor voltage does not change instantaneously due to a finite change in capacitor current.

On the other hand, if a current impulse $i = q\delta(t - t_1)$ occurs at $t = t_1$, then

$$v(t_2)-v(t_1) = \frac{1}{C}\int_{t_1}^{t_2} idt = \frac{1}{C}\int_{t_1}^{t_2} q\delta(t-t_1)dt = \frac{1}{C}\int_{t_1^-}^{t_1^+} q\delta(t-t_1)dt = \frac{q}{C} \tag{18.21}$$

This underlies an important concept,

Concept: *The voltage across a capacitor cannot be changed instantaneously except by a current impulse, which deposits on the capacitor a charge equal to the strength of the impulse.*

A current impulse of $q\delta(t)$ A has a strength or area of q coulombs. When such an impulse flows through a capacitor of C farads, it deposits q coulombs on the capacitor instantaneously, between $t = 0^-$ and $t = 0^+$, which changes the voltage across the capacitor instantaneously by q/C V. Physically, an instantaneous change in the charge or voltage of a capacitor is an instantaneous change in the stored energy. As emphasized in earlier chapters, stored energy cannot be changed instantaneously by any finite "force" or any physically realizable means. This does not contradict the effect of the impulse because the infinite amplitude of the impulse and its infinitesimal duration are not physically realizable. The impulse, after all, is a mathematical construct that is, nevertheless, extremely useful for theoretical purposes in that it readily leads to important conclusions and concepts concerning the behavior of linear systems, including circuits. Practically, an impulse can be approximated by a pulse whose area is equal to that of the impulse, and whose duration is small compared to the shortest response time of the system, as will be illustrated by PSpice simulations.

It should be carefully noted that whereas the voltage across a capacitor cannot be changed instantaneously except by a current impulse, *the current through a capacitor*

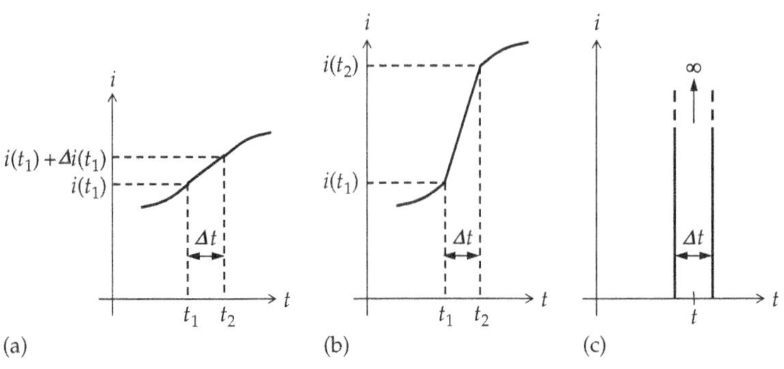

FIGURE 18.7
Integrals of a continuous function (a), or a function including a step discontinuity (b), or an impulse (c).

can change instantaneously, even in the absence of any impulse. This is because the energy stored in the capacitor is not directly related to current, but to charge or voltage.

The responses of a capacitor to current or voltage steps or impulses can be readily derived from the v–i relation of a capacitor. These responses, illustrated in Figure 18.8, are discussed in what follows, assuming the capacitor is initially uncharged. Any initial charge on the capacitor can, in principle, be added to these responses algebraically, as appropriate and in accordance with superposition. Units of v, i, R, and C are not explicitly specified but are assumed to be consistent.

Case 1: $i_{SRC}(t) = Ku(t)$ (Figure 18.8a)

$$v(t)=\frac{1}{C}\int_{0^-}^{t} idt=\frac{K}{C}\int_{0^-}^{0^+} u(t)dt+\frac{K}{C}\int_{0^+}^{t} u(t)dt=0+\frac{K}{C}t,\ t\geq 0^+.$$

The integral of the step function between 0^- and 0^+ is zero. Although $u(t) = 1$ at $t = 0^+$, the area involved in the integration is $1 \times (0^+ - 0^-) = 1 \times \Delta t = 0$ as $\Delta t \to 0$. The integral of $u(t)$ for $t \geq 0^+$ is the ramp function t. The response is shown in Figure 18.8a. $v(t)$ is continuous at $t = 0$, that is, $v(0^-) = 0$ and $v(0^+) = 0$, because the change in i is finite, as was argued in connection with Figure 18.7b. But the slope of $v(t)$ is discontinuous at $t = 0$.

Case 2: $i_{SRC}(t) = \delta u(t)$ (Figure 18.8b)

$v(t)=\frac{1}{C}\int_{0^-}^{t} idt=\frac{K}{C}\int_{0^-}^{0^+}\delta(t)dt=\frac{K}{C}u(t)$. As explained previously, the current impulse changes the capacitor voltage instantaneously from zero to K/C. The voltage response will therefore be $(K/C)u(t)$ (Figure 18.8b). Note that

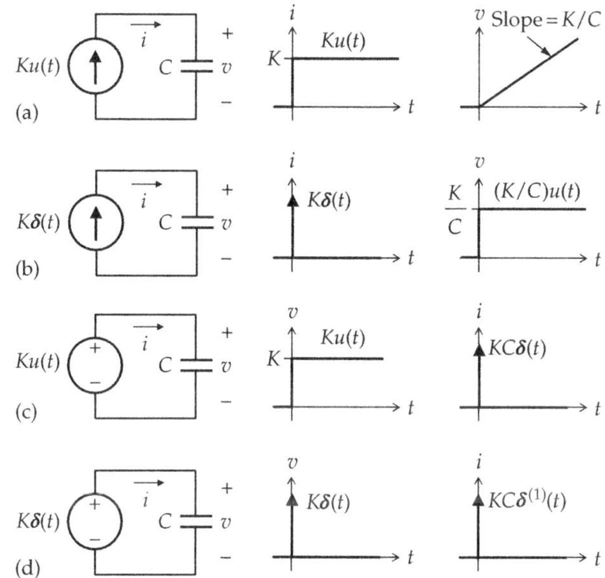

FIGURE 18.8
Capacitor responses to a current step (a), to a current impulse (b), to a voltage step (c), and to a voltage impulse (d).

since an impulse function is the time derivative of the step function (Equation 18.8), the current input of Figure 18.8b is the time derivative of that of Figure 18.8a. Because the voltage–current relation for a capacitor is linear, the voltage response of Figure 18.8b is also the time derivative of that of Figure 18.8a.

Case 3: $v_{SRC}(t) = Ku(t)$ (Figure 18.8c)

$i(t)=C\frac{dv}{dt}=KC\delta(t)$. With the voltage across the capacitor being a step function at the origin, the capacitor current is an impulse at the origin of strength KC (Figure 18.8c). This is the charge deposited by the current impulse on the capacitor, so that the voltage across the capacitor changes instantaneously from 0 at $t = 0^-$ to K at $t = 0^+$, as required by the step function.

Case 4: $v_{SRC}(t) = K\delta(t)$ (Figure 18.8d)

$i(t)=C\frac{dv}{dt}=KC\delta^{(1)}(t)$, where $\delta^{(1)}(t)$ is the derivative of the unit impulse (Figure 18.8d). When the applied voltage is an impulse the voltage first increases toward infinity and eventually returns to zero at $t = 0^+$. By this time, there is no charge on the capacitor. The derivative of the impulse is biphasic: it increases in the positive direction on the rising phase of the impulse, then goes negative in mirror fashion on the negative phase of the impulse. The positive part of the $\delta^{(1)}(t)$ current deposits charge on the capacitor, and the negative part removes this charge, so that by $t = 0^+$, the voltage is zero, and there is no net charge on the capacitor, in accordance with v being zero at $t = 0^+$. Because of its biphasic nature $\delta^{(1)}(t)$ is also known as the **unit doublet**.

Before ending this discussion, the following argument that was used in the preceding discussion should be emphasized, namely, that

$$\int_{t_1}^{t_2} ydt \to 0 \quad \text{as } t_2 \to t_1, \text{if } y \text{ remains finite during the interval } t_1 \text{ to } t_2 \text{ even though it may have a finite jump or step at } t = t_1 \tag{18.22}$$

This argument will be invoked on many occasions in future discussions. A helpful analogy is that of the water vessel invoked in Section 11.1. Consider a water tap emptying into a vessel, where water flow from the tap is analogous to current and the volume of water in the vessel is analogous to charge. If the water tap is suddenly turned on at $t = 0$, causing a finite flow of water into the vessel, then at $t = 0^+$, water has not yet accumulated in the vessel. In other words, the time integral of water flow, from $t = 0^-$, just before the tap is turned on, and $t = 0^+$, just after the tap is turned on, is zero.

A consequence of Equation 18.22 is that a finite quantity added to or subtracted from an impulse does not affect the impulse. Thus,

$$\delta(t) \pm y = \delta(t), \text{ if } y \text{ remains finite during the impulse} \quad (18.23)$$

To prove this, the LHS is integrated over the duration of the impulse and Equation 18.22 applied. Thus,

$$\int_{0^-}^{0^+} \delta dt \pm \int_{0^-}^{0^+} y dt = \int_{0^-}^{0^+} \delta dt \pm 0 \quad (18.24)$$

It is seen from Equation 18.24 that the integral of y between $t = 0^-$ and $t = 0^+$ is zero, so that the impulse is not affected.

18.3.2 *RC* Circuit

There are two basic circuits to consider: (1) a series *RC* circuit excited by a voltage step or impulse and (2) a parallel *RC* circuit excited by a current step or impulse. When a series *RC* circuit is excited by an ideal current source, including a current step or impulse source, the resistor is redundant as far as the capacitor voltage is concerned, because the capacitor current is determined by the ideal current source, independently of the resistor. The capacitor voltage will then be as in Figure 18.8a or b. Similarly, when a parallel *RC* circuit is excited by an ideal voltage source, including a voltage step or impulse source, the resistor is redundant as far as the capacitor current is concerned, because the capacitor voltage is determined by the ideal voltage source, independently of the resistor. The capacitor current will then be as in Figure 18.8c or d.

The responses of interest are summarized in Figure 18.9. These responses can be derived from the generalized response of first-order circuits, as given by Equation 11.57, based on the initial and final values of the voltage or current and the time constant. The final value can be readily determined from steady-state conditions. The resistance seen by the capacitor is Thevenin's resistance, which is R. Hence, $\tau = RC$ in all cases. It remains, therefore, to specify the initial value of voltage or current at $t = 0^+$, just after the sudden change due to the step or impulse. It is assumed as in Figure 18.8 that the capacitor is initially uncharged. Physical interpretation of circuit behavior will be emphasized throughout.

Case 1: $v_{SRC}(t) = Ku(t)$ (Figure 18.9a)

At $t = 0^-$, both i and v are zero, by assumption. The energy stored in the capacitor is zero at $t = 0^-$.

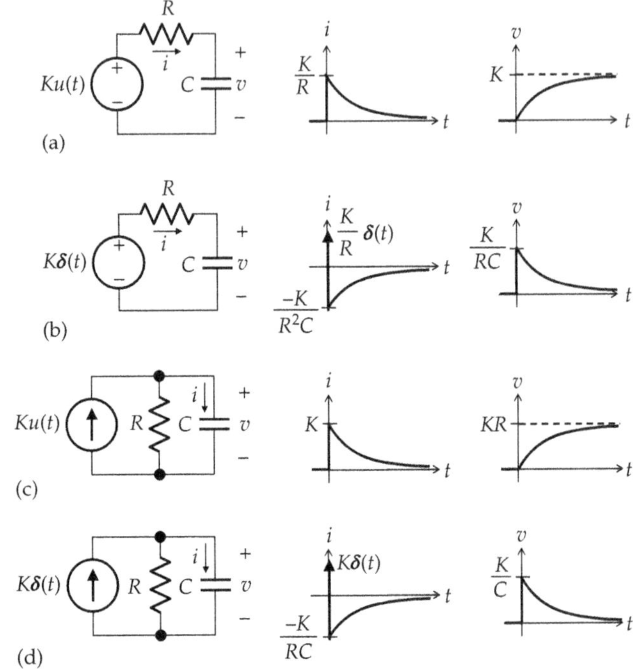

FIGURE 18.9
Responses of *RC* circuit to a voltage step (a), to a voltage impulse (b), to a current step (c), and to a current impulse (d).

The capacitor voltage is not forced to change at $t = 0$, for, according to KVL, $Ku(t) = Ri + v$. When $u(t)$ changes at $t = 0$, i can change so as to satisfy KVL, with v remaining zero, as required by the constancy of stored energy. Hence, $v(0^+) = 0$. In other words, *the uncharged capacitor acts as a short circuit* during the interval $t = 0^-$ *to* $t = 0^+$, *in response to the voltage step*. This means that $i(0^+) = K/R$ in order to satisfy KVL at $t = 0^+$. As $t \to \infty$, the capacitor is fully charged, so that $i(\infty) = 0$ and $v(\infty) = K$. With the initial values $i(0^+)$ and $v(0^+)$ determined, as well as the final values, it follows from Equation 11.57 that

$$i(t) = \frac{K}{R}e^{-t/RC} \quad \text{and} \quad v(t) = K\left(1 - e^{-t/RC}\right), \quad t \geq 0^+ \quad (18.25)$$

The time variations of $i(t)$ and $v(t)$ are illustrated in Figure 18.9a.

Case 2: $v_{SRC}(t) = K\delta(t)$ (Figure 18.9b)

During the rising phase of the impulse, the impulse appears like a step of large magnitude. It would be expected, therefore, that the uncharged capacitor would initially act as a short circuit, as in Case 1. If the capacitor acts as a short circuit, the voltage impulse appears across R, resulting in a current impulse $i = K\delta(t)/R$. The current impulse flowing through the capacitor deposits a charge K/R on the capacitor at $t = 0^+$, so that $v(0^+) = K/RC$. This means that, unlike a true short circuit, $v(0^+) \neq 0$. The fact that v increases from 0 to K/RC during the impulse does not invalidate the argument that a full voltage impulse appears across R, as argued in connection with Equation 18.24, when a finite quantity is added to, or subtracted from, an impulse.

At $t = 0^+$, when the impulse is over, the ideal voltage source acts as a short circuit, and the circuit reduces to a capacitor that is charged to K/RC and having a resistor R across it. The capacitor begins to discharge, thereby reversing the direction of i. The initial value of i at $t = 0^+$ is $-K/R^2C$. This can also be determined from KVL, which is

$$v(t)+Ri(t)=K\delta(t), \quad \text{for all } t \qquad (18.26)$$

At $t = 0^+$, $v(0^+) = K/RC$, and the impulse is over, so that the RHS of Equation 18.26 is zero. This gives $i(0^+) = -v(0^+)/R = -K/R^2C$.

As $t \to \infty$, both v and i tend to zero. It follows from the discussion of Section 11.1 that

$$i(t)=-\frac{K}{R^2C}e^{-t/RC} \quad \text{and} \quad v(t)=\frac{K}{RC}e^{-t/RC}, \quad t \geq 0^+ \qquad (18.27)$$

The time variations of $i(t)$ and $v(t)$ are illustrated in Figure 18.9b.

Case 3: $i_{SRC}(t) = Ku(t)$ (Figure 18.9c)

The responses in this case can be derived from those of Case 1 by transforming the $Ku(t)$ voltage source in series with R to a $Ku(t)/R$ current source in parallel with R. Multiplying i and v of Figure 18.9a by R gives the required responses as illustrated in Figure 18.9c. Nevertheless, it is instructive to derive these responses from first principles.

At $t = 0^-$, both i and v are zero, by assumption, and the energy stored in the capacitor is zero. The capacitor voltage is not forced to change at $t = 0$, for, according to KCL, $Ku(t) = v/R + i$. When $u(t)$ changes at $t = 0$, i can change so as to satisfy KCL, with v remaining zero, as required by the constancy of stored energy. Hence, $v(0^+) = 0$. The uncharged capacitor acts as a short circuit during the interval $t = 0^-$ to $t = 0^+$, in response to the current step. This means that $i(0^+) = K$ in order to satisfy KCL at $t = 0^+$.

As $t \to \infty$, the capacitor is fully charged, so that $i(\infty) = 0$ and $v(\infty) = KR$. Using Equation 11.57,

$$i(t)=Ke^{-t/RC} \quad \text{and} \quad v(t)=KR\left(1-e^{-t/RC}\right), \quad t \geq 0^+ \qquad (18.28)$$

The time variations of $i(t)$ and $v(t)$ are illustrated in Figure 18.9c.

Case 4: $i_{SRC} = K\delta(t)$ (Figure 18.9d)

The responses in this case can be derived from those of Case 2 by transforming the $K\delta(t)$ voltage source in series with R to a $K\delta(t)/R$ current source in parallel with R. Multiplying i and v of Figure 18.9b by R gives the response in this case, as illustrated in Figure 18.9d. Nevertheless, it is instructive to derive these responses from first principles.

During the rising phase of the impulse, the impulse appears like a step of large magnitude. It would be expected, therefore, that the capacitor would initially act as a short circuit. This means that the applied current impulse flows through the capacitor, thereby depositing a charge K on the capacitor and increasing v to K/C at $t = 0^+$. Although the capacitor acts initially as a short circuit in response to the current impulse, the capacitor voltage at the end of the impulse is finite, unlike that of a true short circuit. As v increases from 0 to K/C at $t = 0^+$ during the impulse, the current in R increases from 0 to K/RC at $t = 0^+$. This does not invalidate the argument that a full current impulse flows through the capacitor, as argued in connection with Equation 18.24, when a finite quantity is added to, or subtracted from, an impulse.

At $t = 0^+$, when the impulse is over, the ideal current source acts as an open circuit and the circuit reduces to a capacitor that is charged to K/C and having a resistor R across it. The capacitor begins to discharge, thereby reversing the direction of i. The initial value of i at $t = 0^+$ is $-K/RC$. This can also be determined from KCL, which is

$$i+v/R=K\delta(t), \quad \text{for all } t \qquad (18.29)$$

At $t = 0^+$, $v(0^+) = K/C$, and the impulse is over, so that the RHS of Equation 18.29 is zero. This gives $i(0^+) = -v(0^+) = -K/RC$.

As $t \to \infty$, both v and i tend to zero. It follows from the discussion of Section 11.1 that

$$i(t)=-\frac{K}{RC}e^{-t/RC} \quad \text{and} \quad v(t)=\frac{K}{C}e^{-t/RC}, \quad t \geq 0^+ \qquad (18.30)$$

The time variations of $i(t)$ and $v(t)$ are illustrated in Figure 18.9d.

18.3.3 Summary of Responses of Capacitive Circuits

The responses discussed in this section can be summarized by the following important concepts:

1. *The voltage across a capacitor does not change instantaneously unless it is forced to change in order to satisfy KVL. On the other hand, the capacitor current can change instantaneously in order to satisfy KVL and KCL.*
2. *If the voltage across a capacitor is forced to change instantaneously, this change is accomplished by a current impulse through the capacitor.*

In Figure 18.8c, the capacitor voltage is forced to change instantaneously to become equal to K, in accordance with KVL, the capacitor current being an impulse $KC\delta(t)$. In Figure 18.9a and c the capacitor voltage is not forced to change instantaneously by the applied excitation. The capacitor current jumps instantaneously to satisfy KVL in Figure 18.9a and KCL in Figure 18.9c.

3. *When a voltage or current* step *is applied to an RC circuit, and the capacitor voltage is not forced to change, an* uncharged *capacitor acts as a short circuit at the time the step is applied. Since the voltage of an uncharged capacitor is zero before the step is applied, it will remain zero at just after the step, because the stored energy does not change instantaneously. As for a short circuit, the capacitor voltage is zero but the capacitor current can have any finite value.*

   In Figure 18.9a, the capacitor acts as a short circuit in response to the voltage step, so that at $t = 0^+$, $v = 0$ and $i = K/R$. Similarly in Figure 18.9c, the capacitor acts as a short circuit in response to the current step, so that at $t = 0^+$, $v = 0$ and the source current K passes through the capacitor.

4. *If in the preceding case the capacitor is initially charged, the capacitor voltage does not change just after the step, because the stored energy does not change instantaneously if no current impulse flows through the capacitor. Superposition can be applied to determine the currents and voltages in the rest of the circuit, but it is generally easier to determine these values by replacing the charged capacitors by batteries of equal voltage, in accordance with the substitution theorem.*

5. *When a voltage or current* impulse *is applied to an RC circuit, an* uncharged *capacitor* initially *acts as a short circuit in response to the impulse, which determines the path followed by the current impulse, or where in the circuit the voltage impulse appears. Just after the impulse, the capacitor voltage is not zero, as in a short circuit, but is equal to the charge deposited by the current impulse divided by the capacitance.*

   The justification is that the rising phase of an impulse is indistinguishable to begin with from a step of large amplitude. The capacitor will therefore initially act as a short circuit. But if a current impulse flows through the capacitor, then the capacitor voltage at $t = 0^+$ is not zero, as in the case of a short circuit, but is the charge deposited by the impulse divided by the capacitance. Thus, in Figure 18.9d, the capacitor acts initially as a short circuit, so that the current impulse flows through the capacitor, depositing a charge K coulombs and producing a capacitor voltage K/C at $t = 0^+$. In Figure 18.9b, the capacitor acts initially as a short circuit, so that the voltage impulse appears across R and results in a current impulse $(K/R)\delta(t)$ through the capacitor. The current impulse deposits a charge K/R, the resulting voltage being K/RC.

6. *If in the preceding case the capacitor is initially charged, superposition can be applied. The capacitor is first considered uncharged, and the preceding concepts applied. The charge or voltage due to initially stored energy is then added algebraically to the charge or voltage due to the impulse.*

It should be kept in mind that the current and voltage values at $t = 0^+$ are the initial values for determining current and voltage as functions of time by solving the differential equation involved.

Example 18.2: Responses of a Capacitive Circuit to Voltage Step and Impulse

Given the circuit of Figure 18.10, (a) determine $v_1(0^+)$, $v_2(0^+)$, $i_1(0^+)$, and $i_2(0^+)$ assuming $v_{SRC}(t) = 2u(t)$ V and the capacitors to be initially uncharged. (b) Repeat (a) assuming initial voltages of the capacitors at $t = 0$ of $V_{10} = 4$ V and $V_{20} = 6$ V, and determine how KVL and KCL are satisfied at $t = 0^+$. (c) Repeat (a) assuming $v_{SRC}(t) = 4\delta(t)$ mV. (d) Repeat (b) assuming $v_{SRC}(t) = 4\delta(t)$ mV.

Solution:

KVL in Figure 18.10 is

$$2i_1 + v_1 + v_2 = v_{SRC}, \quad \text{for all } t \tag{18.31}$$

Equation 18.31 applies irrespective of the nature of v_{SRC}. When v_{SRC} is a voltage step, v_1 and v_2 are not forced to change at $t = 0^+$, because the term in i_1 can change to satisfy KVL. The stored energy and hence the capacitor

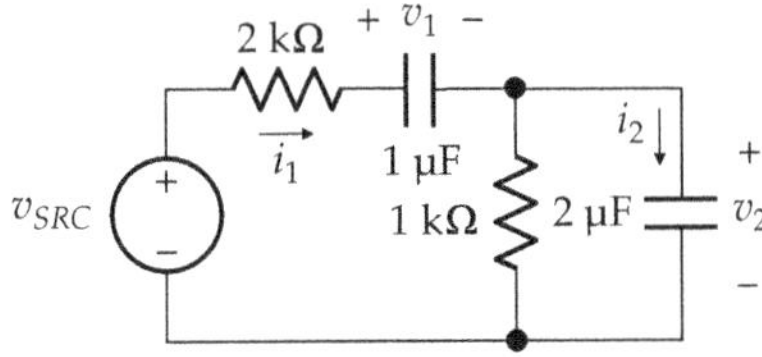

FIGURE 18.10
Figure for Example 18.2.

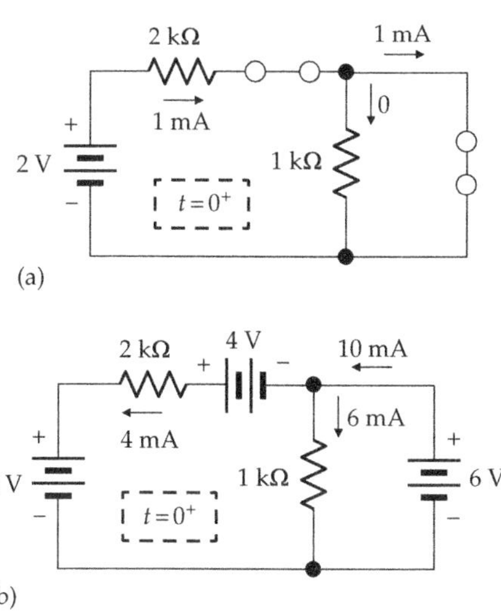

FIGURE 18.11
Figure for Example 18.2.

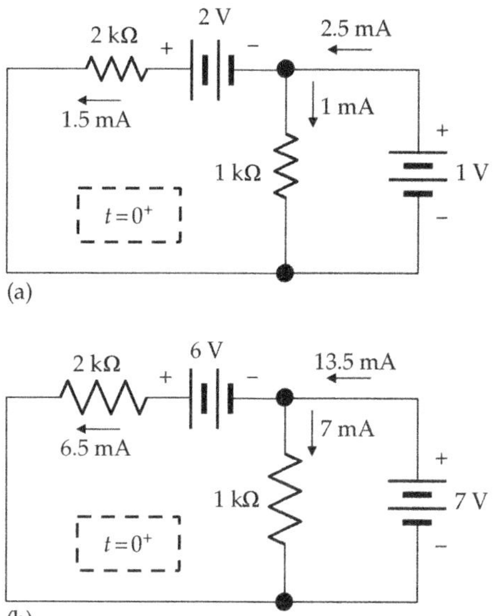

FIGURE 18.12
Figure for Example 18.2.

voltages remain the same, at $t = 0^-$ and at $t = 0^+$, irrespective of the initial charges on the capacitors.

(a) When the step is applied, the capacitors act as short circuits (Figure 18.11a). It follows that $v_1(0^+) = 0$ and $v_2(0^+) = 0$. Moreover, $i_1(0^+) = 2\text{ V}/2\text{ k}\Omega = 1$ mA. Since $v_2(0^+) = 0$ the current in the 1 kΩ is zero, so that $i_2(0^+) = i_1(0^+) = 1$ mA.

(b) As argued in connection with Equation 18.31, v_1 and v_2 are not forced to change at $t = 0$. Hence, $v_1(0^+) = V_{10} = 4$ V and $v_2(0^+) = V_{20} = 6$ V. At $t = 0^+$, the circuit becomes as shown in Figure 18.11b, where $v_{SRC} = 2$ V and the capacitors have been replaced by ideal dc voltage sources in accordance with the substitution theorem (Section 4.4). It follows from Equation 18.31 that $2i_1(0^+) + 4 + 6 = 2$, which gives $i_1(0^+) = -4$ mA. The current in the 1 kΩ resistor is 6 mA in the direction shown, so that $i_2(0^+) = -(6 + 4) = -10$ mA in order to satisfy KCL. Note that superposition can also be applied, considering the capacitors to be initially uncharged, and then adding to $i_1(0^+)$ and $i_2(0^+)$ the components due to the initial voltages acting alone. However, it is simpler in this case to apply KVL to the circuit of Figure 18.11b. It should be kept in mind that replacing the 1 and 2 μF capacitors by batteries of 4 and 6 V, respectively, applies at $t = 0^+$ only, and at no other time.

(c) When the impulse is applied, the capacitors initially act as short circuits as in Figure 18.11a. The voltage impulse will therefore appear across the 2 kΩ resistor and cause a current impulse of $4\delta(t)$ mV/2 kΩ = $2\delta(t)$ μA through both capacitors. The current impulse deposits a charge of 2 μC on the plate of the 1 μF capacitor connected to the 2 kΩ resistor and removes 2 μC from the plate of the 2 μF capacitor connected to the negative source terminal. This leaves each of the two capacitors with 2 μC, so that $v_1(0^+) = 2$ V and $v_2(0^+) = 1$ V (Figure 18.12a). At $t = 0^+$, the impulse is over and the voltage source acts as a short circuit; KVL around the outer loop gives $i_1(0^+) = -1.5$ mA. With 1 mA flowing in the 1 kΩ resistor, KCL gives $i_2(0^+) = -2.5$ mA.

(d) When the capacitors are initially charged, the current impulse can be considered first to deposit 2 μC on uncharged capacitors. This charge adds to the charges already present. The total charge on the 1 μF capacitor is 4 + 2 = 6 μC, which gives $v_1(0^+) = 6$ V, whereas the total charge on the 2 μF capacitor is 12 + 2 = 14 μC, and $v_2(0^+) = 7$ V (Figure 18.12b). It follows from KVL around the outer loop that $i_1(0^+) = -6.5$ mA. With 7 mA flowing in the 1 kΩ resistor, KCL gives $i_2(0^+) = -13.5$ mA.

Simulation: The circuit for simulation of the voltage step is shown in Figure 18.13a. PSpice applies VDC at the start of the simulation like a voltage step. Initial voltages across the capacitors are assumed as given.

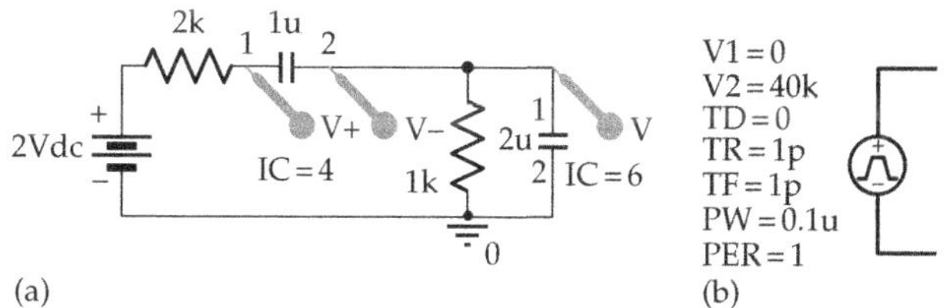

FIGURE 18.13
Figure for Example 18.2.

Applying time domain analysis, cursor readings give $v_1(0^+) = 4.0000$ V and $v_2(0^+) = 6.000$ V. To read the currents using the cursor, it is necessary to remove the switching transient at $t = 0$. In the SCHEMATIC1 page, select 'Plot/Axis Settings/X-Axis/User Defined', and enter 1 us as the beginning of the trace, instead of zero. The cursors, when toggled, read I(C1) = $i_1(0^+) = -3.9995$ m and I(C2) = $i_2(0^+) = -9.999$ m.

To simulate the impulse, the battery is replaced by VPULSE with the parameters entered as shown in Figure 18.13b. Because the time response of the circuit is of the order of kilo-ohms × microfarads, or milliseconds, the voltage impulse is simulated by a pulse of 0.1 μs duration and 40 kV amplitude, which makes the area 4 mV as required. The initial values can be read using the cursor by selecting a user-defined time axis starting at 0.101 μs, after the impulse is over. The readings are $v_1(0^+) = 5.9988$ V, $v_2(0^+) = 6.9990$ V, $i_1(0^+) = -6.4989$ m, and $i_2(0^+) = -13.498$ m.

Primal Exercise 18.4

Determine $v_C(0^+)$ and $i_C(0^+)$ in Figure 18.14, assuming $v_{SRC}(t) = 6u(t)$, with (a) the capacitor initially uncharged; (b) the capacitor initially charged to 2 V. (c) and (d) repeat (a) and (b) assuming $v_{SRC}(t) = 6\delta(t)$.

Ans. (a) $v_C(0^+) = 0$ V, $i_C(0^+) = 3$ A; (b) $v_C(0^+) = 2$ V, $i_C(0^+) = 2$ A; (c) $v_C(0^+) = 1.5$ V, $i_C(0^+) = -0.75$ A; (d) $v_C(0^+) = 3.5$ V, $i_C(0^+) = -1.75$ A.

Primal Exercise 18.5

Determine $v_C(0^+)$ and $i_C(0^+)$ in Figure 18.15.

Ans. $v_C(0^+) = 3$ V, $i_C(0^+) = -1.5$ A.

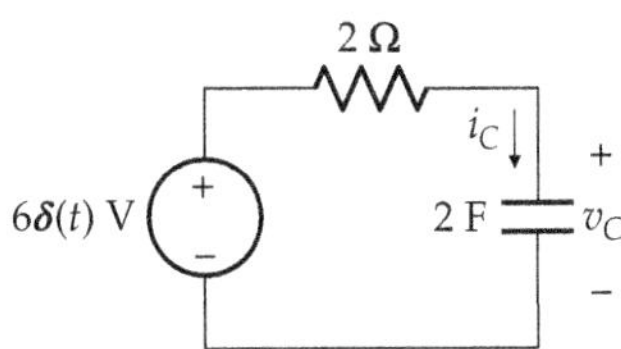

FIGURE 18.14
Figure for Primal Exercise 18.4.

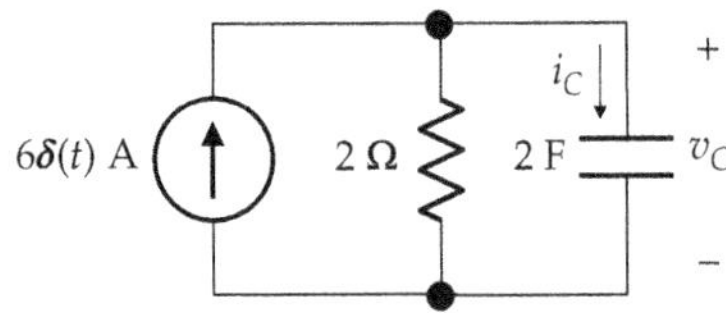

FIGURE 18.15
Figure for Primal Exercise 18.5.

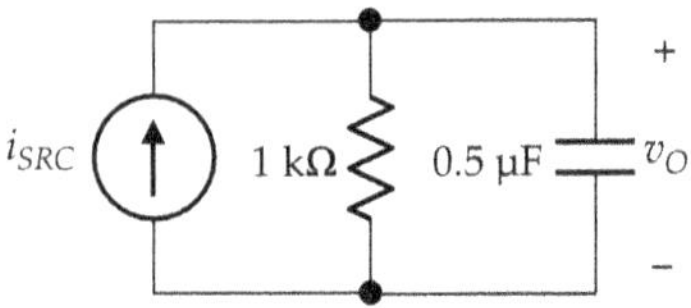

FIGURE 18.16
Figure for Primal Exercise 18.6.

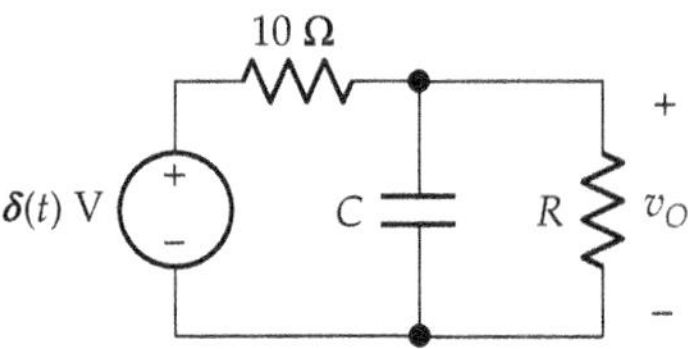

FIGURE 18.17
Figure for Primal Exercise 18.7.

Primal Exercise 18.6

Determine $v_O(t)$ in Figure 18.16 given $i_{SRC}(t) = 10\delta(t)$ μA + $5u(t)$ mA, with the capacitor initially uncharged. Note that the current impulse establishes the initial charge on the capacitor at $t = 0^+$.

Ans. $v_O(t) = 5 + 15e^{-2t}$ V, t is in ms.

Primal Exercise 18.7

Determine R and C in Figure 18.17 given that $v_O(t) = 0.1e^{-0.2t}u(t)$V, t is in s, in response to a unit impulse, $\delta(t)$.

Ans. $C = 1$ F, $R = 10$ Ω.

18.4 Inductor Response to Voltage Pulse

Consider a step voltage V_{SRC} applied at $t = 0$ to an inductor that is initially uncharged, as illustrated in Figure 18.18a. The voltage step implies that $V_{SRC} = 0$ for $t < 0$, and $v_{SRC} = V_{SRC}$, for $t > 0$, so that the source voltage suddenly changes at $t = 0$ from 0 to V_{SCR}. The current i through the inductor is

$$i(t) = \frac{1}{L}\int_0^t V_{SRC}dt + 0 = \frac{V_{SRC}t}{L} \tag{18.32}$$

The inductor current increases linearly with time at a rate V_{SRC}/L as shown.

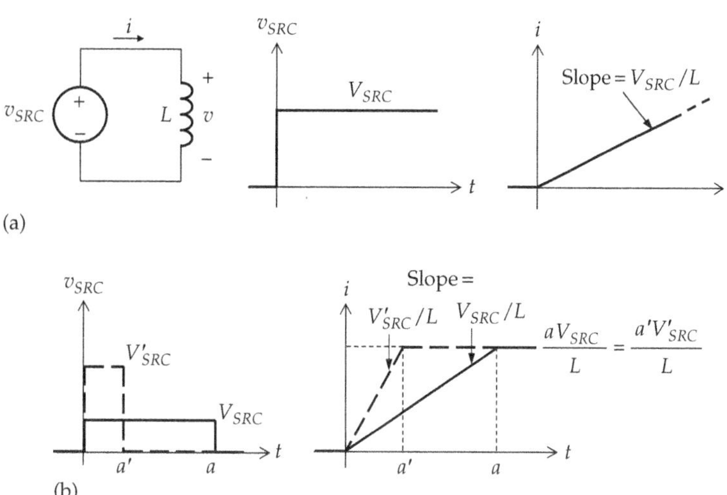

FIGURE 18.18
Inductor response to voltage step (a) and to voltage pulses of different durations but of the same area (b).

Suppose that a voltage pulse of amplitude V_{SRC} that starts at $t = 0$ and ends at $t = a$ is applied to the uncharged inductor (Figure 18.18b). Initially, for $t < a$, the voltage pulse is indistinguishable from the voltage step of the same amplitude, which causes the inductor current to increase linearly at the same rate V_{SRC}/L as before. At $t = a$, $i = aV_{SRC}/L$, since the flux linkage established in the inductor from $t = 0$ and $t = a$ is aV_{SRC}, the area under the voltage pulse. The current is $1/L$ times this flux linkage, that is, aV_{SRC}/L. Just after $t = a$, the pulse ends, so that v_{SRC} becomes zero, and the ideal voltage source acts as a short circuit. Since the inductor is assumed ideal, the current is maintained for $t > a$ at its value of aV_{SRC}/L at $t = a$.

Let the amplitude of the pulse be increased to V'_{SRC} and its duration reduced to a' (Figure 18.18b), while keeping the area of the pulse the same, that is, $a'V'_{SRC} = aV_{SRC}$ = flux linkage established by the voltage pulse. For $t < a'$, $di/dt = (V'_{SRC}/L) > (V_{SRC}/L)$. The change in i is steeper, but the final value of i remains the same because the flux linkage established is the same. Thus, from Equation 18.32, the final value of v is $(a'V'_{SRC}/L) = (aV_{SRC}/L)$ but is reached at a faster rate.

Note the similarity, because of duality, between Figures 18.1 and 18.32, whereby v replaces i and L replaces C. As in the case of the capacitor considered in Section 18.1, if the width of the voltage pulse becomes infinitesimally small, while the area of the pulse remains at aV_{SRC}, then the voltage amplitude becomes infinitely large, resulting in a voltage impulse of strength aV_{SRC} Vs. The resulting change in current becomes a step function, so that the current changes from 0 to aV_{SRC}/L instantaneously at $t = 0$.

Primal Exercise 18.8

Determine the inductor current as a function of time if a voltage of 5 mV is applied for 10 ms to a 10 μH inductor.

Ans. i rises to 5 A at 10 ms and remains at this value thereafter 0.5.

18.5 Responses of Inductive Circuits to Step and Impulse Inputs

This section closely parallels Section 18.3 in accordance with duality between capacitive and inductive circuits.

18.5.1 Single Inductor

Consider the v–i relation of an inductor to be rearranged as $di = (1/L)vdt$. Integrating this relation between two limits t_1 and t_2 gives

$$\int_{i_1}^{i_2} di = \frac{1}{L}\int_{t_1}^{t_2} vdt \tag{18.33}$$

where $i(t_2) = i_2$, and $i(t_1) = i_1$. It follows that

$$i(t_2) - i(t_1) = \frac{1}{L}\int_{t_1}^{t_2} vdt \tag{18.34}$$

As argued in Section 18.3, if v is finite and continuous at $t = t_1$, or is finite and has a finite discontinuity at $t = t_1$, the integral on the RHS of Equation 18.34 tends to zero as $t_2 \to t_1$, which makes $i(t_2) = i(t_1)$. In other words, the inductor current does not change instantaneously due to a finite change in inductor voltage.

On the other hand, if a voltage impulse of strength λ occurs at $t = t_1$, then

$$i(t_2)-i(t_1) = \frac{1}{L}\int_{t_1}^{t_2} v dt = \frac{1}{L}\int_{t_1}^{t_2} \lambda\delta(t-t_1)dt = \frac{1}{L}\int_{t_1^-}^{t_1^+} \lambda\delta(t-t_1)dt = \frac{\lambda}{L} \quad (18.35)$$

This underlies an important concept,

Concept: *The current through an inductor cannot be changed instantaneously except by a voltage impulse, which establishes in the inductor a flux linkage equal to the strength of the impulse.*

A voltage impulse of $\lambda\delta(t)$ V has a strength or area of λ Wb-T. When such an impulse is applied across an inductor of L henries, it establishes a flux linkage λ Wb-T in the inductor instantaneously, between $t = 0^-$ and $t = 0^+$, which changes the inductor current instantaneously by λ/L A. Physically, an instantaneous change in the flux linkage or current of an inductor is an instantaneous change in the stored energy. As argued previously, stored energy cannot be changed instantaneously by any finite 'force'. But a voltage impulse of infinite amplitude and infinitesimal duration can change the energy stored in the inductor instantaneously, and hence the inductor current.

Whereas the current through an inductor cannot be changed instantaneously except by a voltage impulse, *the voltage across an inductor can change instantaneously*, even in the absence of any impulse. This is because the energy stored in the inductor is not directly related to voltage, but to flux linkage or current.

The responses of an inductor to current or voltage steps or impulses can be readily derived from the v–i relation of an inductor, as illustrated in Figure 18.19, assuming the inductor is initially uncharged. Any initial current in the inductor can, in principle, be added algebraically to these responses, as appropriate and in accordance with superposition. Units of v, i, R, and L are not explicitly specified but are assumed to be consistent.

Case 1: $v_{SRC}(t) = Ku(t)$ (Figure 18.19a)

$$i(t)=\frac{1}{L}\int_{0^-}^{t} vdt = \frac{K}{L}\int_{0^-}^{0^+} u(t)dt + \frac{K}{L}\int_{0^+}^{t} u(t)dt = 0 + \frac{K}{L}t, t \geq 0^+.$$

i is continuous at $t = 0$, but its slope is discontinuous. Moreover, $i(0^-) = 0$ and $i(0^+) = 0$, because the change in v is finite.

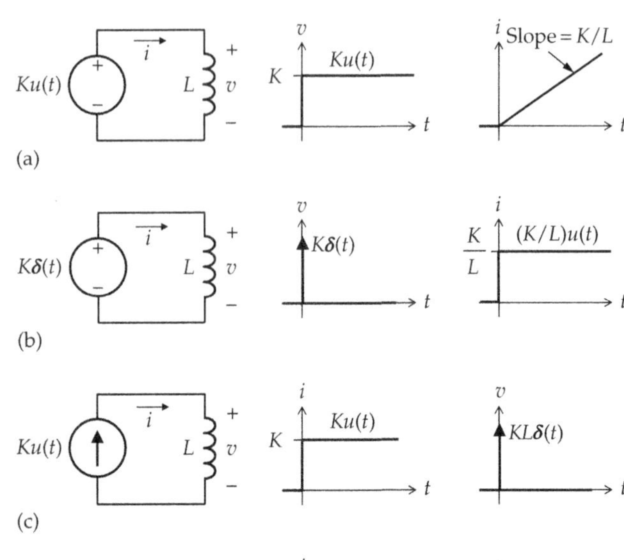

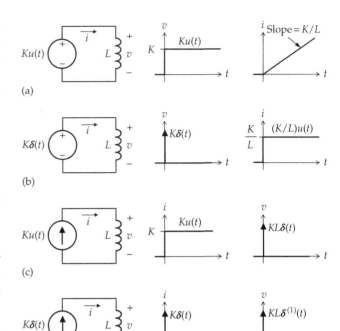

FIGURE 18.19
Inductor responses to a voltage step (a), to a voltage impulse (b), to a current step (c), and to a current impulse (d).

Case 2: $v_{SRC}(t) = K\delta(t)$ (Figure 18.19b)

$i(t)=\frac{1}{L}\int_{0^-}^{t} vdt=\frac{K}{L}\int_{0^-}^{0^+} \delta(t)dt = \frac{K}{L}u(t)$. The voltage impulse changes the inductor current instantaneously from zero to K/L. The current response will therefore be $u(t)$ (Figure 18.19b). Since the voltage input of Figure 18.19b is the time derivative of that of Figure 18.19a, the current response of Figure 18.19b is also the time derivative of that of Figure 18.19a.

Case 3: $i_{SRC}(t) = Ku(t)$ (Figure 18.19c)

$v(t)=L\frac{di}{dt}=KL\delta(t)$, that is, an impulse at the origin of strength KL (Figure 18.19c). The flux linkage established by the voltage impulse in the inductor is KL and the current through the inductor is K, as required by the step function.

Case 4: $i_{SRC} = Ku(t)$ (Figure 18.19d)

$v(t)=L\frac{di}{dt}=KL\delta^{(1)}(t)$, where $\delta^{(1)}(t)$ is the derivative of the unit impulse. The positive part of the $\delta^{(1)}(t)$ voltage establishes flux linkage in the inductor, and the negative part abolishes this flux linkage, so that by $t = 0^+$, the current is zero, and there is no net flux linkage in the inductor, in accordance with i being zero at $t = 0^+$.

18.5.2 *RL* Circuit

There are two basic circuits to consider: (1) a series *RL* circuit excited by a voltage step or impulse and (2) a parallel

RL circuit excited by a current step or impulse. When a series *RL* circuit is excited by an ideal current source, such as a current step or impulse source, the resistor is redundant as far as the inductor voltage is concerned, because the inductor current is determined by the ideal current source, independently of the resistor. The inductor voltage will then be as in Figure 18.19c or d. Similarly, when a parallel *RL* circuit is excited by an ideal voltage source, including a voltage step or impulse source, the resistor is redundant as far the inductor current is concerned, because the inductor voltage is determined by the ideal voltage source, independently of the resistor. The inductor current will then be as in Figure 18.19a or b.

The responses of interest are summarized in Figure 18.20. These responses can be derived from the generalized response of first-order circuits, as given by Equation 11.57, based on the initial and final values of the voltage or current and the time constant. The final value can be readily determined from steady-state conditions. The resistance seen by the inductor is Thevenin's resistance, which is R. Hence, $\tau = L/R$ in all cases. It remains, therefore, to specify the initial value of voltage or current at $t = 0^+$, just after the sudden change due to the step or impulse. It is assumed as in Figure 18.19 that the inductor is initially uncharged. The physical interpretation of circuit behavior will be emphasized throughout.

Case 1: $i_{SRC} = Ku(t)$ (Figure 18.20a)

At $t = 0^-$, both i and v are zero, by assumption. The energy stored in the inductor is zero at $t = 0^-$. The inductor current is not forced to change at $t = 0$, for, according to KCL, $Ku(t) = v/R + i$. When $u(t)$ changes at $t = 0$, v can change to satisfy KCL, with i remaining zero, as required by constancy of stored energy. Hence, $i(0^+) = 0$. In other words, *the uncharged inductor acts as an open circuit* during the interval $t = 0^-$ *to* $t = 0^+$, *in response to the current step*. The applied current step flows through R *at* $t = 0^-$, which means that v jumps from 0 at $t = 0^-$ to KR at $t = 0^+$ in order to satisfy KVL. As $t \to \infty$, the inductor acts as a short circuit and all the source current passes through the inductor, so that $v(\infty) = 0$ and $i(\infty) = K$. With the initial and final values of $i(0^+)$ and $v(0^+)$ determined, it follows from Equation 11.57 that

$$v(t) = KRe^{-Rt/L} \quad \text{and} \quad i(t) = K\left(1 - e^{-Rt/L}\right), \quad t \geq 0^+ \qquad (18.36)$$

The time variations of $i(t)$ and $v(t)$ are illustrated in Figure 18.20a. Note that the dual of R in the capacitive case is G in the inductive case, which has been replaced by its reciprocal R.

Case 2: $v_{SRC}(t) = K\delta(t)$ (Figure 18.20b)

During the rising phase of the impulse, the impulse appears like a step of large magnitude. It would be

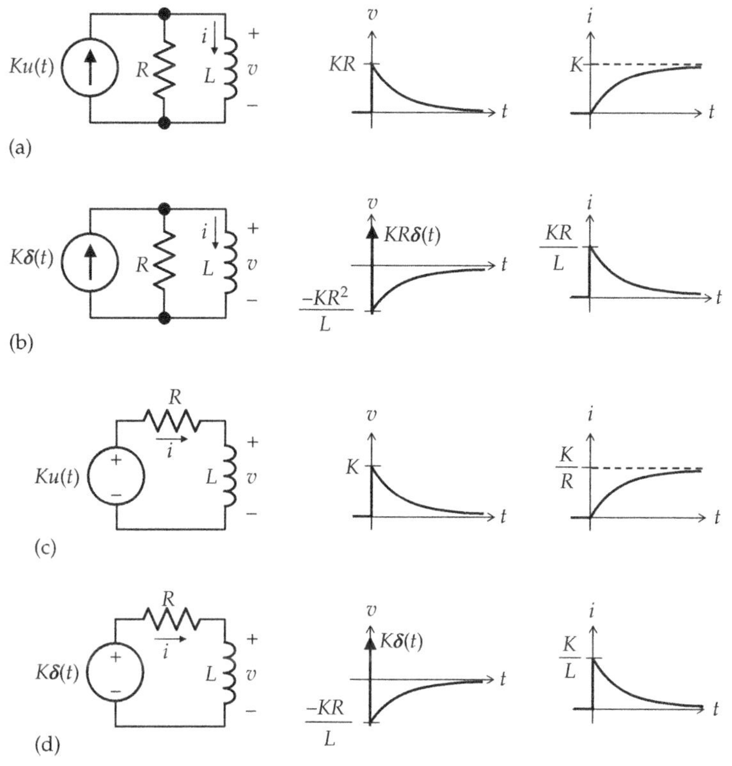

FIGURE 18.20
Responses of *RL* circuit to a current step (a), to a current impulse (b), to a voltage step (c), and to a voltage impulse (d).

expected, therefore, that the inductor would act as an open circuit. In this case, the current impulse flows through R, resulting in a voltage impulse $v = KR\delta(t)$. The voltage impulse establishes a flux linkage KR in the inductor at $t = 0^+$, so that $i(0^+) = KR/L$. This means that, unlike a true open circuit, $i(0^+) \neq 0$. The fact that i increases from 0 to KR/L during the impulse does not invalidate the argument that the full current impulse flows through R, as argued in connection with Equation 18.24, when a finite quantity is added to, or subtracted from, an impulse.

At $t = 0^+$, when the impulse is over, the ideal current source acts as an open circuit and the circuit reduces to an inductor having an initial current of KR/L and a resistor R across it. The inductor begins to discharge, with a reversal of the polarity of v. The initial value of v at $t = 0^+$ is $-KR^2/L$. This can also be determined from KCL, which is

$$i + v/R = K\delta(t), \quad \text{for all } t \tag{18.37}$$

At $t = 0^+$, $i(0^+) = KR/C$, and the impulse is over, so that the RHS of Equation 18.37 is zero. This gives $v(0^+) = -Ri(0^+) = -KR^2/L$.

As $t \to \infty$, both v and i tend to zero. It follows from the discussion of Section 11.3 that

$$v(t) = -\frac{KR^2}{L}e^{-Rt/L} \quad \text{and} \quad i(t) = \frac{KR}{L}e^{-Rt/L}, \quad t \geq 0^+ \tag{18.38}$$

The time variations of $i(t)$ and $v(t)$ are illustrated in Figure 18.20b.

Case 3: $v_{SRC}(t) = Ku(t)$ (Figure 18.20c)

The responses in this case can be derived from those of Case 1 by transforming the $Ku(t)$ current source in parallel with R to a $KRu(t)$ voltage source in series with R. Dividing i and v of Figure 18.20a by R gives the response in this case, as illustrated in Figure 18.20c. Nevertheless, it is instructive to derive these responses from first principles.

At $t = 0^-$, both i and v are zero, by assumption, and the energy stored in the inductor is zero. The inductor current is not forced to change at $t = 0$, for according to KVL, $Ku(t) = Ri + v$. When $u(t)$ changes at $t = 0$, v can change so as to satisfy KVL, with i remaining zero, as required by the constancy of stored energy. Hence, $i(0^+) = 0$. The uncharged inductor acts as an open circuit during the interval $t = 0^-$ to $t = 0^+$, in response to the voltage step. This means that $v(0^+) = K$ in order to satisfy KVL at $t = 0^+$.

As $t \to \infty$, the inductor acts as a short circuit, so that $v(\infty) = 0$ and $i(\infty) = K/R$. Using Equation 11.57,

$$v(t) = Ke^{-Rt/L} \quad \text{and} \quad i(t) = \frac{K}{R}\left(1 - e^{-t/RC}\right), \quad t \geq 0^+ \tag{18.39}$$

The time variations of $i(t)$ and $v(t)$ are illustrated in Figure 18.20c.

Case 4: $v_{SRC}(t) = K\delta(t)$ (Figure 18.20d)

The responses in this case can be derived from those of Case 2 by transforming the $K\delta(t)$ current source in parallel with R to a $KR\delta(t)$ voltage source in series with R. Dividing i and v of Figure 18.20b by R gives the response in this case, as illustrated in Figure 18.20d. Nevertheless, it is instructive to derive these responses from first principles.

During the rising phase of the impulse, the impulse appears like a step of large magnitude. It would be expected, therefore, that the inductor would act as an open circuit. This means that the applied voltage impulse appears across the inductor, thereby establishing a flux linkage K in the inductor and increasing i to K/L at $t = 0^+$. Although the inductor acts initially as an open circuit in response to the voltage impulse, the inductor current at the end of the impulse is finite, unlike that of a true open circuit. As i increases from 0 to K/L at $t = 0^+$ during the impulse, the voltage across R increases from 0 to KR/L at $t = 0^+$. This does not invalidate the argument that a full voltage impulse appears across the inductor, as argued in connection with Equation 18.24, when a finite quantity is added to or subtracted from an impulse.

At $t = 0^+$, when the impulse is over, the ideal voltage source acts as a short circuit and the circuit reduces to an inductor having an initial current K/L and a resistor R across it. The inductor begins to discharge, thereby reversing the direction of v. The initial value of v at $t = 0^+$ is $-KR/L$. This can also be determined from KCL, which is

$$v + Ri = K\delta(t), \quad \text{for all } t \tag{18.40}$$

At $t = 0^+$, $i(0^+) = K/L$, and the impulse is over, so that the RHS of Equation 18.40 is zero. This gives $v(0^+) = -Ri(0^+) = -KR/L$.

As $t \to \infty$, both v and i tend to zero. It follows from the discussion of Section 11.3 that

$$v(t) = -\frac{KR}{L}e^{-Rt/L} \quad \text{and} \quad i(t) = \frac{K}{L}e^{-Rt/L}, \quad t \geq 0^+ \tag{18.41}$$

The time variations of $i(t)$ and $v(t)$ are illustrated in Figure 18.20d.

18.5.3 Summary of Responses of Inductive Circuits

The responses discussed in this section can be summarized by the following important concepts:

1. *The current through an inductor does not change instantaneously unless it is forced to change in order to satisfy KCL. On the other hand, the inductor*

voltage can change instantaneously in order to satisfy KVL and KCL.

2. *If the current through an inductor is forced to change instantaneously, this change is accomplished by a voltage impulse across the inductor.*

   In Figure 18.19c, the inductor current is forced to change instantaneously to become equal to K, in accordance with KCL, the inductor voltage being an impulse $KL\delta(t)$. In Figure 18.20a and c, the inductor current is not forced to change instantaneously by the applied excitation. The inductor voltage jumps instantaneously to satisfy KCL in Figure 18.20a and KVL in Figure 18.20c.

3. *When a voltage or current* step *is applied to an RL circuit, and the inductor current is not forced to change, an* uncharged *inductor acts as an open circuit at the time the step is applied. Since the current of an uncharged inductor is zero before the step is applied, it will remain zero just after the step, because the stored energy does not change instantaneously. As for an open circuit, the inductor current is zero but the inductor voltage can have any finite value.*

   In Figure 18.20a, the inductor acts as an open circuit from $t = 0^-$ to $t = 0^+$, so that at $t = 0^+$, $i = 0$, and $v = KR$. Similarly in Figure 18.20c, the inductor acts as an open circuit from $t = 0^-$ to $t = 0^+$, so that at $t = 0^+$, $i = 0$, and the source voltage K appears across the inductor.

4. *If in the preceding case the inductor is initially charged, the inductor current does not change just after the step, because the stored energy does not change instantaneously if no voltage impulse is applied across the inductor. Superposition can be applied to determine the currents and voltages in the rest of the circuit, but it is generally easier to determine these values by replacing the charged inductors by dc sources of equal current, in accordance with the substitution theorem.*

5. *When a voltage or current* impulse *is applied to an RL circuit, an* uncharged *inductor* initially *acts as an open circuit in response to the impulse, which determines where in the circuit the voltage impulse appears, or the path followed by the current impulse. Just after the impulse, the current through the inductor is not zero but is equal to the flux linkage established by the voltage impulse divided by the inductance.*

   The justification is that the rising phase of an impulse is indistinguishable to begin with from a step of large amplitude. The inductor will therefore initially act as an open circuit. But if a voltage impulse appears across the inductor, then the inductor current at $t = 0^+$ is not zero, as in the case of an open circuit, but is the flux linkage established by the impulse divided by the inductance. Thus, in Figure 18.20d, the inductor acts initially as an open circuit, so that the voltage impulse appears across the inductor, establishing a flux linkage K and producing an inductor current K/L at $t = 0^+$. In Figure 18.20b, the inductor acts initially as an open circuit, so that the current impulse flows through R and results in a voltage impulse $(KR)\delta(t)$ across the inductor. The voltage impulse establishes a flux linkage KR, the resulting current being KR/L.

6. *If in the preceding case the inductor is initially charged, superposition can be applied. The inductor is first considered uncharged, and the preceding concepts applied. The flux linkage or current due to initially stored energy is then added algebraically to the flux linkage or current due to the impulse.*

It should be kept in mind that the current and voltage values at $t = 0^+$ are the initial values for determining current and voltage as functions of time by solving the differential equation involved.

Example 18.3: Responses of an Inductive Circuit to Current Step and Impulse

Consider the circuit of Figure 18.21, which is of the form of the dual of Figure 18.10, but with different component values. (a) Assuming $i_{SRC}(t) = 2u(t)$ mA and the inductors to be initially uncharged, determine $i_1(0^+)$, $i_2(0^+)$, $v_1(0^+)$, and $v_2(0^+)$. (b) Repeat (a) assuming initial currents in the inductors at $t = 0$ of $I_{10} = 3$ mA and $I_{20} = 4$ mA, and determine how KVL and KCL are satisfied at $t = 0^+$. (c) Repeat (a) assuming $i_{SRC}(t) = 2\delta(t)$ μA. (d) Repeat (b) assuming $i_{SRC}(t) = 2\delta(t)$ μA.

Solution:

KCL in Figure 18.21 is

$$v_1/2 + i_1 + i_2 = i_{SRC}, \quad \text{for all } t \tag{18.42}$$

Equation 18.42 applies irrespective of the nature of i_{SRC}. When i_{SRC} is a current step, i_1 and i_2 are not forced to

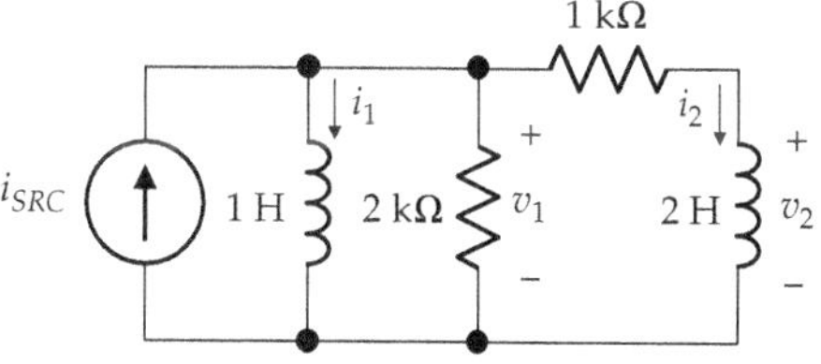

FIGURE 18.21
Figure for Example 18.3.

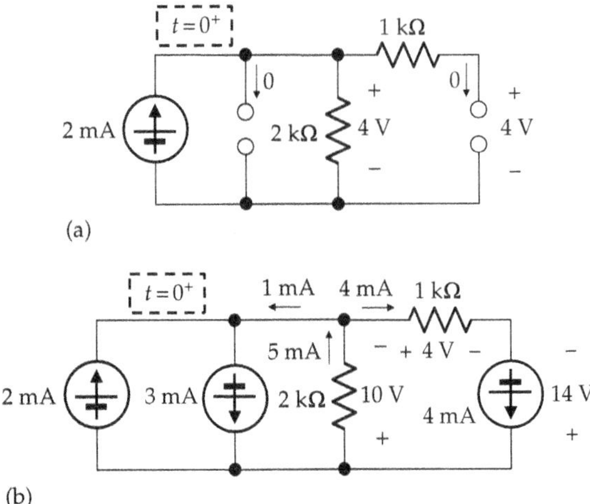

FIGURE 18.22
Figure for Example 18.3.

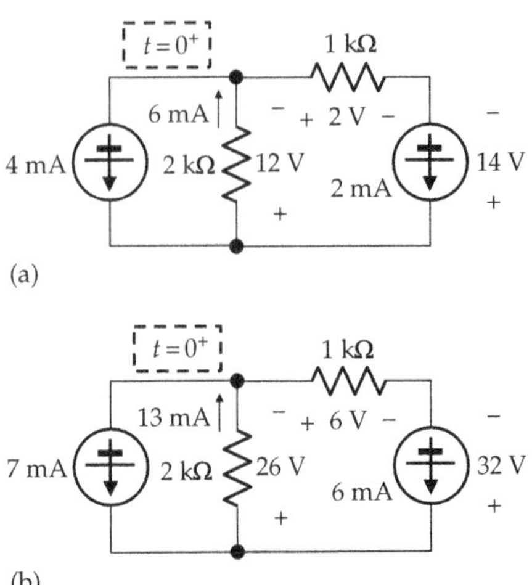

FIGURE 18.23
Figure for Example 18.3.

change at $t = 0^+$, because the term in v_1 can change to satisfy KCL. The stored energy and hence the inductor currents remain the same, at $t = 0^-$ and at $t = 0^+$, irrespective of the initial currents in the inductors.

(a) When the step is applied, the inductors act as open circuits (Figure 18.22a). It follows that $i_1(0^+) = 0$ and $i_2(0^+) = 0$. Moreover, $v_1(0^+) = 2$ mA × 2 kΩ = 4 V. Since $i_2(0^+) = 0$ the voltage across the 1 kΩ is zero, so that $v_2(0^+) = v_1(0^+) = 4$ V.

(b) As argued in connection with Equation 18.42, i_1 and i_2 are not forced to change. Hence, $i_1(0^+) = 3$ mA and $i_2(0^+) = I_{20} = 4$ mA. At $t = 0^+$, the circuit becomes as shown in Figure 18.22b, where $i_{SRC} = 2$ mA and the inductors have been replaced by ideal dc current sources in accordance with the substitution theorem (Section 4.4). It follows from Equation 18.42 that $2 = 3 + v_1(0^+)/2 + 4$, which gives $v_1(0^+) = -10$ V. The voltage across the 1 kΩ resistor is 4 V of the polarity shown, so that $v_2(0^+) = -(10 + 4) = -14$ V in order to satisfy KVL. Note that superposition can also be applied, considering the inductors to be initially uncharged, and then adding to $v_1(0^+)$ and $v_2(0^+)$ the components due to the initial currents acting alone. However, it is simpler in this case to apply KCL to the circuit of Figure 18.22b. It should be kept in mind that replacing the 1 and 2 H inductors by current sources of 3 and 4 A, respectively, applies at $t = 0^+$ only, and at no other time.

(c) When the impulse is applied, the inductors initially act as open circuits, as in Figure 18.22a. The current impulse will therefore flow through the 2 kΩ resistor and cause a voltage impulse of $2\delta(t)$ µA × 2 kΩ = $4\delta(t)$ mV. The voltage impulse establishes a flux linkage of 4 mWb-T in both inductors. This produces initial currents $i_1(0^+) = 4$ mA and $i_2(0^+) = 2$ mA (Figure 18.23a). At $t = 0^+$, the impulse is over, and the current source acts as an open circuit. From KCL, the current in the 2 kΩ resistor is 6 mA, and $v_1 = -12$ V. KVL around the mesh on the RHS gives $v_2 = -14$ V.

(d) When the inductors are initially charged, the voltage impulse still establishes a flux linkage of 4 mWb-T in each inductor, which adds to the flux linkages already present. The total flux linkage in the 1 H inductor becomes 4 + 3 = 7 mWb-T, and $i_1(0^+) = 7$ mA, whereas the total flux linkage in the 2 H inductor becomes 4 + 8 = 12 mWb-T, and $i_2(0^+) = 6$ mA (Figure 18.23b). The current in the 2 kΩ resistor is 13 mA, and $v_1 = -26$ V. KVL around the mesh on the RHS gives $v_2 = -32$ V.

Simulation: The circuit for simulation of the current step is shown in Figure 18.24a. PSpice applies IDC at the start of the simulation like a current step. Initial currents through the inductors are assumed. Applying time domain analysis, cursor readings give $i_1(0^+) = 3.0000$ mA, $i_2(0^+) = 4.000$ mA, $v_1(0^+) = -10.0000$ V, and $v_2(0^+) = -14.0000$ V.

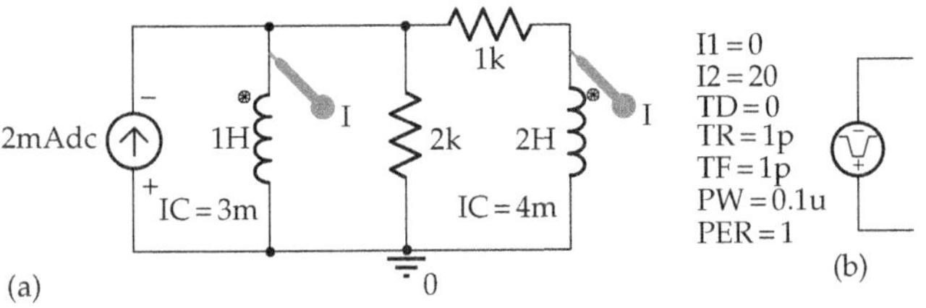

FIGURE 18.24
Figure for Example 18.3.

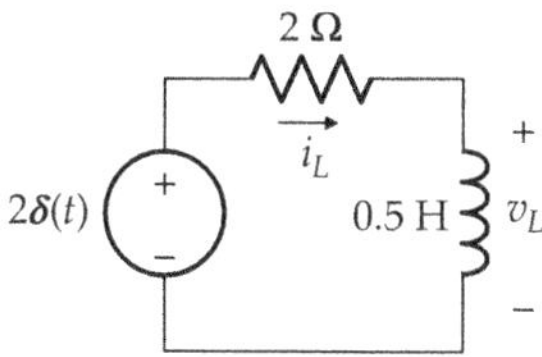

FIGURE 18.25
Figure for Primal Exercise 18.9.

The impulse is simulated by IPULSE with the parameters entered as shown in Figure 18.24b. The initial values can be read using the cursor by selecting a user-defined time axis starting at 0.101 μs, after the impulse is over. The readings are $i_1(0^+)$ = 6.9966 m, $i_2(0^+)$ = 5.9981 m, $v_1(0^+)$ = −25.989, and $v_2(0^+)$ = −31.987.

Primal Exercise 18.9

Determine $i_L(0^+)$ and $v_L(0^+)$ in Figure 18.25.
Ans. $i_L(0^+)$ = 4 A, $v_L(0^+)$ = −8 V.

Primal Exercise 18.10

Determine $i_L(0^+)$ and $v_L(0^+)$ in Figure 18.26.
Ans. $i_L(0^+)$ = 8 A, $v_L(0^+)$ = −16 V.

Primal Exercise 18.11

Determine $i(t)$ in Figure 18.27 given $v_{SRC} = 2\delta(t) + 2u(t)$ A.
Ans. $i(t)$ = 2 A.

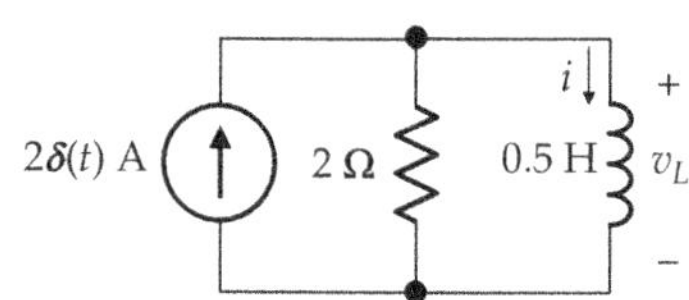

FIGURE 18.26
Figure for Primal Exercise 18.10.

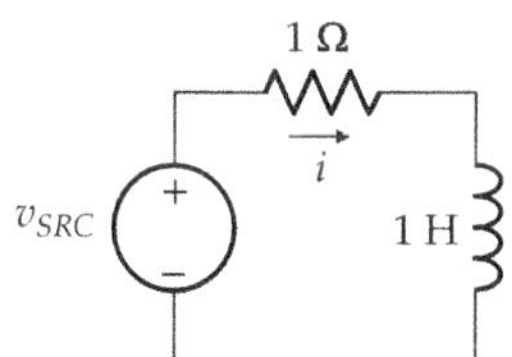

FIGURE 18.27
Figure for Primal Exercise 18.11.

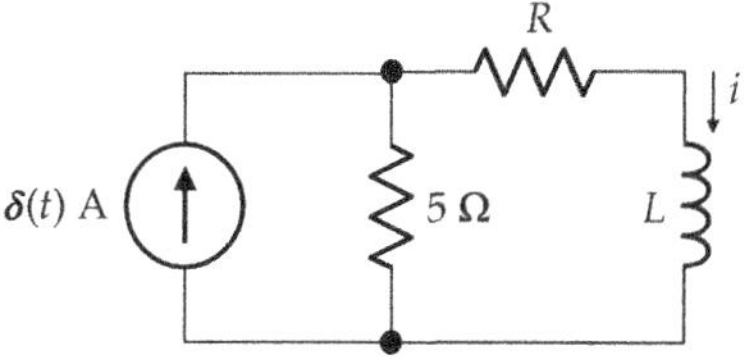

FIGURE 18.28
Figure for Primal Exercise 18.12

Primal Exercise 18.12

Determine R and L in Figure 18.28 given that $i(t) = 0.5e^{-t}u(t)\text{A}$ t is in s, in response to a unit impulse, $\delta(t)$.
Ans. L = 10 H, R = 5 Ω.

18.6 Responses of *RLC* Circuits to Step and Impulse Inputs

When step or impulse inputs are applied to an *RLC* circuit, capacitors and inductors act, respectively, as summarized at the ends of Sections 18.3 and 18.5. This behavior is illustrated in Examples 18.4 and 18.5.

Example 18.4: Responses of *RLC* Circuit to Voltage Step and Impulse

Consider the circuit of Figure 18.29, with an initial voltage of 2 V on the capacitor and an initial current of 2 mA in the inductor, determine (a) $i_1(0^+)$, $i_2(0^+)$, $v_1(0^+)$, and $v_2(0^+)$, assuming $v_{SRC}(t) = 9u(t)$ V, and verify that KVL and KCL are satisfied. (b) Repeat (a) assuming $v_{SRC}(t) = 6\delta(t)$ mV.

Solution:

(a) KVL is

$$v_{SRC} = 2i_1 + v_1 + v_2 \tag{18.43}$$

and KCL is

$$i_1 = i_2 + v_2/1 \tag{18.44}$$

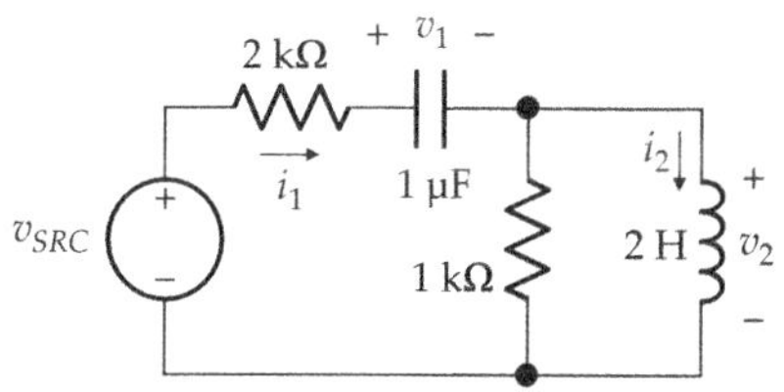

FIGURE 18.29
Figure for Example 18.4.

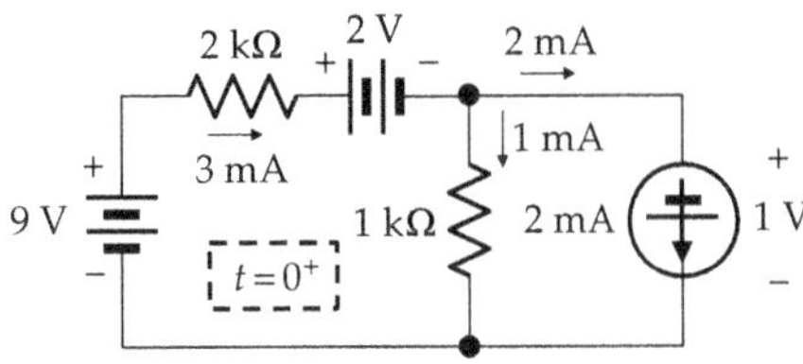

FIGURE 18.30
Figure for Example 18.4.

Neither v_1 nor i_2 are forced to change at $t = 0^+$ in order to satisfy KVL or KCL because KVL can be satisfied by changes in i_1 or v_2 in Equation 18.43, and KCL can be satisfied by changes in i_1 in Equation 18.44. It follows that $v_1(0^+) = 2$ V, and $i_2(0^+) = 2$ mA. At $t = 0^+$, the capacitor can be replaced by a 2 V battery and the inductor by a 2 mA dc current source. If the current in the 1 kΩ resistor is denoted as $v_2/1$ mA (Figure 18.30), then $i_1 = v_2/1 + 2$ mA, and KVL around the mesh on the LHS at $t = 0^+$ is $9 = 2(v_2/1 + 2) + 2 + v_2$, which gives $v_2(0^+) = 1$ V and $i_1(0^+) = 3$ mA. The currents and voltages at $t = 0^+$ will be as shown in Figure 18.30.

(b) In the interval $0^- < t < 0^+$, the capacitor acts as a short circuit and the inductor as an open circuit (Figure 18.31a). The $6\delta(t)$ mV voltage impulse produces a $2\delta(t)$ μA current impulse, which in turn produces a voltage impulse of $2\delta(t)$ mV across the inductor. The $2\delta(t)$ μA current impulse deposits a charge of 2 μC on the capacitor, which adds to the initial charge of 2 μC, so that the charge at $t = 0^+$ is 4 μC and $v_1(0^+) = 4$ V (Figure 18.31b). The voltage impulse establishes a flux linkage of 2 mWb-T in the inductor, which adds to the initial 4 mWb-T, so that the flux linkage at $t = 0^+$ is 6 mWb-T and $i_2(0^+) = 3$ mA. $i_1(0^+) = 3 + v_2/1$ mA, and KVL in the mesh on the LHS is $2(3 + v_2/1) + 4 + v_2 = 0$. This gives $v_2(0^+) = -10/3$ V and $i_1(0^+) = -1/3$ mA, as in Figure 18.31b.

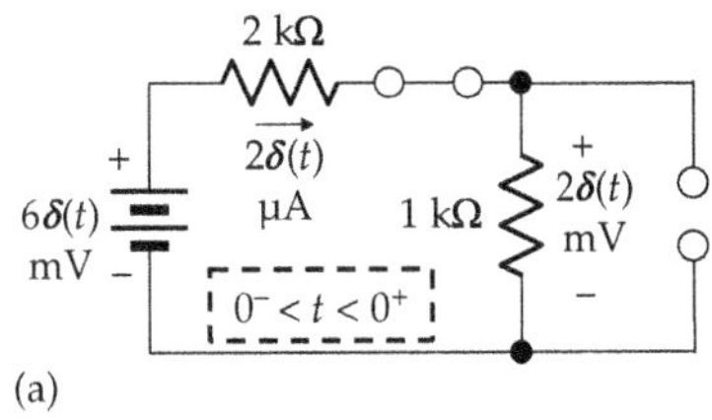

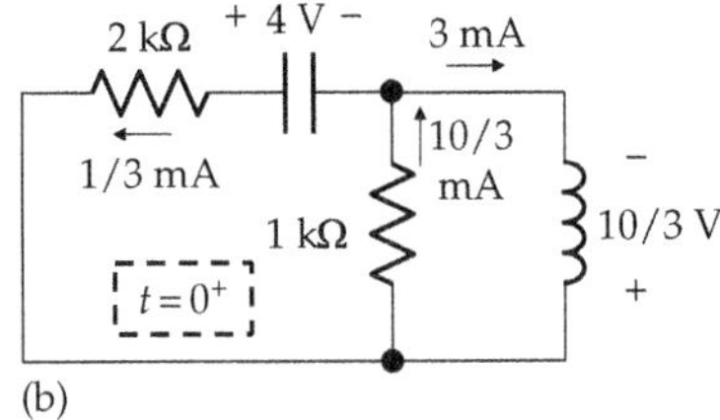

FIGURE 18.31
Figure for Example 18.4.

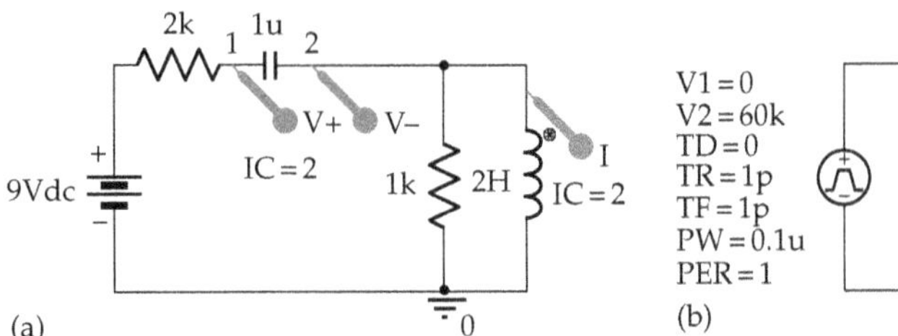

FIGURE 18.32
Figure for Example 18.4.

Simulation: The circuit for simulation of the voltage step is shown in Figure 18.32a. Applying time domain analysis, cursor readings give $i_1(0^+) = 3.0000$ mA, $i_2(0^+) = 2.000$ mA, $v_1(0^+) = 2.0000$ V, and $v_2(0^+) = 1.0000$ V.

The impulse is simulated by VPULSE having the parameters entered as shown in Figure 18.32b. The initial values can be read using the cursor by selecting a user-defined time axis starting at 0.101 μs, after the impulse is over. The readings are $i_1(0^+) = -333.268$u, $i_2(0^+) = 2.9995$ m, $v_1(0^+) = 3.9993$, and $v_2(0^+) = -3.3328$.

Example 18.5: Responses of *RLC* Circuit to Current Step and Impulse

Consider the circuit of Figure 18.33, with an initial voltage of 2 V on the capacitor and an initial current of 2 mA in the inductor, determine (a) $i_1(0^+)$, $i_2(0^+)$, $v_1(0^+)$, and $v_2(0^+)$, assuming $i_{SRC}(t) = 6u(t)$ mA, and verify that KVL and KCL are satisfied. (b) Repeat (a) assuming $i_{SRC}(t) = 6\delta(t)$ μA.

Solution:

(a) KCL is

$$i_{SRC} = i_1 + v_1/2 + i_2 \tag{18.45}$$

and KVL around the mesh on the RHS is

$$v_1 = 1 \times i_2 + v_2 \tag{18.46}$$

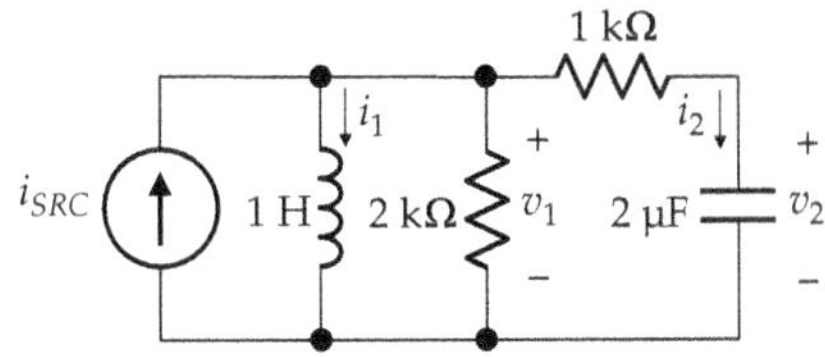

FIGURE 18.33
Figure for Example 18.5.

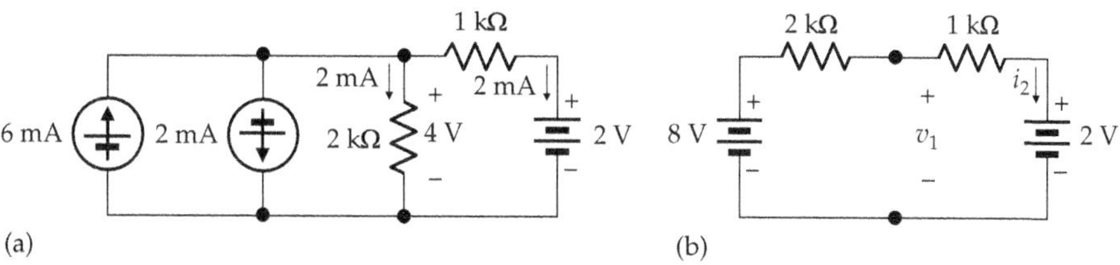

FIGURE 18.34
Figure for Example 18.5.

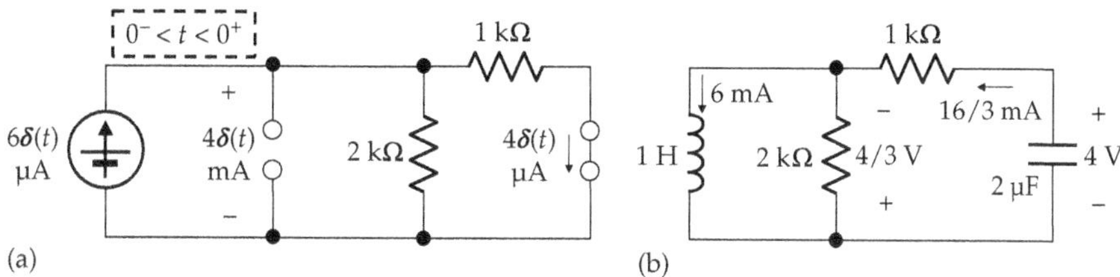

FIGURE 18.35
Figure for Example 18.5.

Neither i_1 nor v_2 are forced to change at $t = 0^+$ in order to satisfy KVL or KCL, because KCL and KVL can be satisfied by changes in v_1 or i_2. It follows that $i_1(0^+) = 2$ mA, and $v_2(0^+) = 2$ V. At $t = 0^+$, the inductor can be replaced by a 2 mA current source and the capacitor by a 2 V battery, as shown in Figure 18.34a. $v_1(0^+)$ and $i_2(0^+)$ can be determined by first combining the two current sources into a single 4 mA current source directed upwards, and then transforming this source, in combination with the 2 kΩ resistor in parallel with it, to an equivalent voltage source of 8 V in series with a 2 kΩ resistor (Figure 18.34b). From KVL, $8 = 3i_2(0^+) + 2$, which gives $i_2(0^+) = 2$ mA. It follows that $v_1(0^+) = 2 + 2 \times 1 = 4$ V.

(b) In the interval $0^- < t < 0^+$, the capacitor acts as a short circuit and the inductor as an open circuit (Figure 18.35a). The resistance of the two resistors in parallel is 2/3 kΩ. The $6\delta(t)$ μA current impulse produces a $6 \times 2/3\delta(t) = 4\delta(t)$ mV voltage impulse, which in turn produces a current impulse of $4/1\delta(t) = 4\delta(t)$ μA through the capacitor. The $4\delta(t)$ μA current impulse deposits a charge of 4 μC on the capacitor, which adds to the initial charge of 4 μC, so that the charge at $t = 0^+$ is 8 μC and $v_2(0^+) = 4$ V (Figure 18.35b). The voltage impulse establishes a flux linkage of 4 mWb-T in the inductor, which adds to the initial 2 mWb-T, so that the flux linkage at $t = 0^+$ is 6 mWb-T and $i_2(0^+) = 6$ mA. KCL in terms of $v_1(0^+)$ is $i_1(0^+) = 6 = -v_1(0^+)/2 - [v_1(0^+) - 4]/1$, which gives $v_1(0^+) = -4/3$ V and $i_2(0^+) = -16/3$ mA, as in Figure 18.35b.

Simulation: The circuit for simulation of the voltage step is shown in Figure 18.36a. Applying time domain analysis, cursor readings give $i_1(0^+) = 2.0000$ mA, $i_2(0^+) = 2.000$ mA, $v_1(0^+) = 4.0000$ V, and $v_2(0^+) = 2.0000$ V.

The impulse is simulated by IPULSE with the parameters entered as shown in Figure 18.36b. The initial values can be read using the cursor by selecting a user-defined time axis starting at 0.101 μs, after the impulse is over. The readings are $i_1(0^+) = 5.9987$ m, $i_2(0^+) = -5.3322$ m, $v_1(0^+) = -1.3330$, and $v_2(0^+) = 3.9992$.

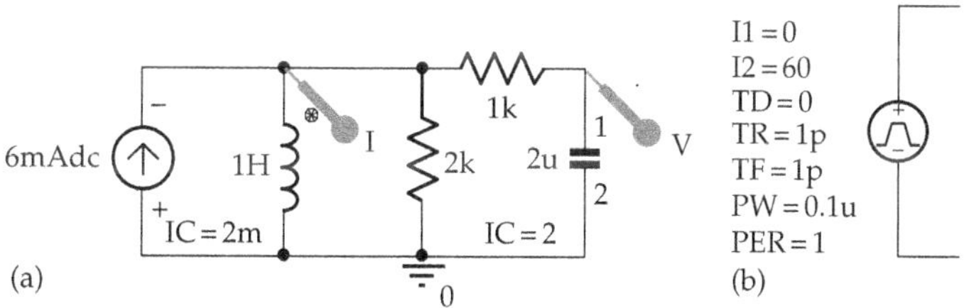

FIGURE 18.36
Figure for Example 18.5.

Primal Exercise 18.13

Determine $i_L(0^+)$, $i_C(0^+)$, $v_L(0^+)$, and $v_C(0^+)$ in Figure 18.37, assuming $i_L(0^-) = 1$ mA and $v_C(0^-) = 3$ V, if (a) $i_{SRC}(t) = 1.5u(t)$ mA; (b) $i_{SRC}(t) = \delta(t)$ μA.

Ans. (a) $i_L(0^+) = 1$ mA, $i_C(0^+) = 0.5$ mA; $v_L(0^+) = 4$ V, $v_C(0^+) = 3$ V; (b) $i_L(0^+) = 1.5$ mA, $i_C(0^+) = -1.5$ mA; $v_L(0^+) = 1$ V, $v_C(0^+) = 4$ V.

Primal Exercise 18.14

Determine R in Figure 18.38 so that the response $v_O(t)$ to $v_{SRC}(t) = \delta(t)$ is critically damped.

Ans. 1 Ω.

Primal Exercise 18.15

Determine $v_R(t)$, $v_O(t)$, $i_C(t)$, and $i_L(t)$, in Figure 18.39.

Ans. $v_R(t) = \delta(t)$ V, $v_O(t) = 1000\cos 10^3 t u(t)$ V, $i_C(t) = \delta(t) - 1000\sin 10^3 t u(t)$ A, $I_L(t) = 1000\sin 10^3 t u(t)$ A.

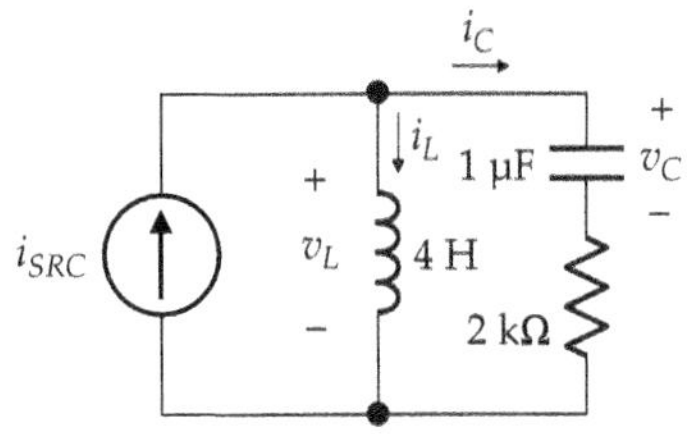

FIGURE 18.37
Figure for Primal Exercise 18.13.

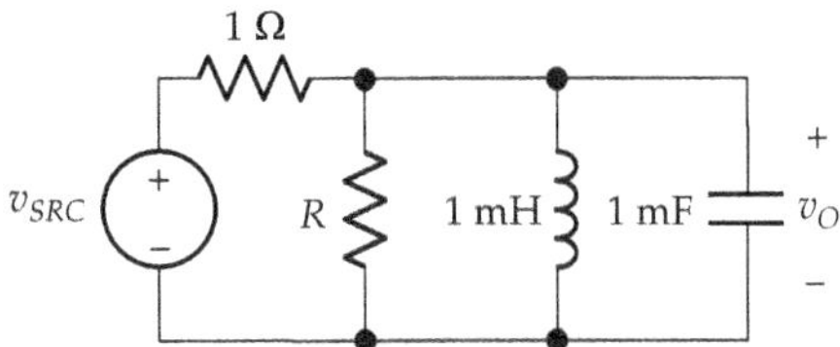

FIGURE 18.38
Figure for Primal Exercise 18.14.

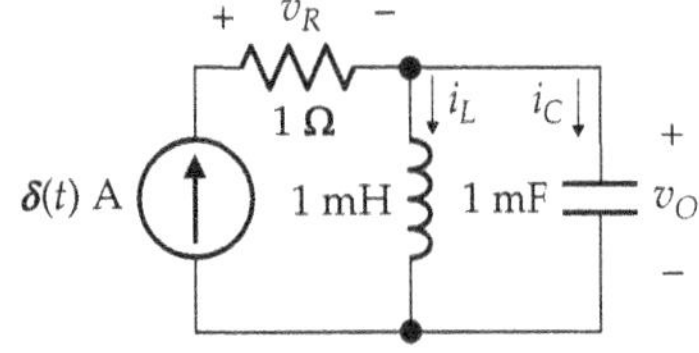

FIGURE 18.39
Figure for Primal Exercise 18.15.

Learning Checklist: What Should Be Learned from This Chapter

- The impulse function has infinite magnitude, infinitesimal duration, and a finite area, designated as the strength of the impulse.
  1. A unit impulse at the origin is defined as $\int_{-\infty}^{+\infty} \delta(t)dt = 1$, with $\delta(t) = 0$ for $t \neq 0$.
  2. The impulse at the origin occurs between $t = 0^-$ and $t = 0^+$. At $t = 0^-$, the impulse has not yet started ($\delta(t) = 0$), and the impulse is over at $t = 0^+$.
- $\delta(t)$ is related to $u(t)$, the unit step at the origin, by $\int_{-\infty}^{t} \delta(t)dt = u(t)$, for all t, or by $\delta(t) = du(t)/dt$.
  1. The unit step function is dimensionless, whereas the unit impulse function has the units of t^{-1}.
- The voltage across a capacitor cannot be changed instantaneously except by a current impulse, which deposits on the capacitor a charge equal to the strength of the impulse.
- The current through an inductor cannot be changed instantaneously except by a voltage impulse, which establishes in the inductor a flux linkage equal to the strength of the impulse.
- The voltage across a capacitor does not change instantaneously unless it is forced to change in order to satisfy KVL. On the other hand, the capacitor current can change instantaneously in order to satisfy KVL and KCL.
- The current through an inductor does not change instantaneously unless it is forced to change in order to satisfy KCL. On the other hand, the inductor voltage can change instantaneously in order to satisfy KVL and KCL.
- When a voltage or current step is applied to an RC circuit, an uncharged capacitor acts as a short circuit at the time the step is applied. Since the voltage of an uncharged capacitor is zero before the step is applied, it will remain zero at just after the step, because the stored energy does not change instantaneously. As for a short circuit, the capacitor voltage is zero but the capacitor current can have any finite value.
  1. If in the preceding case the capacitor is initially charged, the capacitor voltage does not change just after the step, because the

stored energy does not change instantaneously if no current impulse flows through the capacitor. Superposition can be applied to determine the currents and voltages in the rest of the circuit, but it is generally easier to determine these values by replacing the charged capacitors by batteries of equal voltage, in accordance with the substitution theorem.

- When a voltage or current step is applied to an *RL* circuit, an uncharged inductor acts as an open circuit at the time the step is applied. Since the current of an uncharged inductor is zero before the step is applied, it will remain zero just after the step. As for an open circuit, the inductor current is zero but the inductor voltage can have any finite value.
    1. If in the preceding case the inductor is initially charged, the inductor current does not change just after the step, because the stored energy does not change instantaneously if no voltage impulse is applied across the inductor. Superposition can be applied to determine the currents and voltages in the rest of the circuit, but it is generally easier to determine these values by replacing the charged inductors by dc sources of equal current, in accordance with the substitution theorem.
- When a voltage or current impulse is applied to an *RC* circuit, an uncharged capacitor initially acts as a short circuit in response to the impulse, which determines the path followed by the current impulse, or where in the circuit the voltage impulse appears. Just after the impulse, the capacitor voltage is not zero, as in a short circuit, but is equal to the charge deposited by the current impulse divided by the capacitance,
    1. If in the preceding case the capacitor is initially charged, superposition can be applied. The capacitor is first considered uncharged, and the preceding concepts applied. The charge or voltage due to initially stored energy is then added algebraically to the charge or voltage due to the impulse.
- When a voltage or current impulse is applied to an *RL* circuit, an uncharged inductor initially acts as an open circuit in response to the impulse, which determines where in the circuit the voltage impulse appears, or the path followed by the current impulse. Just after the impulse, the current through the inductor is not zero but is equal to the flux linkage established by the voltage impulse divided by the inductance.
    1. If in the preceding case the inductor is initially charged, superposition can be applied. The inductor is first considered uncharged, and the preceding concepts applied. The flux linkage or current due to initially stored energy is then added algebraically to the flux linkage or current due to the impulse.
- In an *RLC* circuit, and in response to step or impulse inputs, capacitors and inductors act as in *RC* and *RL* circuits, respectively.
- Current and voltage values at $t = 0^+$ are the initial values for determining current and voltage as functions of time.

Problem-Solving Tips

1. Always check if at $t = 0^+$, capacitor voltages or inductor currents are forced to change in order to satisfy KVL and KCL.
2. In the case of impulse inputs, ascertain the passage of impulses in the circuit, with capacitors acting as short circuits and inductors as open circuits.

Problems

Verify solutions by PSpice simulation.

Impulse Function

P18.1 The voltage drop across a device is $u(t)$ V and the current through it, in the direction of voltage drop, is $\delta(t)$A. Determine the total energy delivered to the device.

Ans. 1/2 J.

P18.2 The voltage drop across a device is $\delta(t-2)$V and the current through it, in the direction of voltage drop, is $5t$ A. Determine the instantaneous power and energy delivered to the device.

Ans. $10\delta(t-2)$ W, 10 J.

P18.3 The voltage drop across a device is $\delta(t-1)$ V and the current through it, in the direction of voltage drop, is $10\left(1-e^{0.5t}\right)$A. Determine the instantaneous power and energy delivered to the device.

Ans. $10\left(1-e^{0.5}\right)\delta(t-1)$ W, −6.49 J.

P18.4 Evaluate the following integrals involving impulse functions:

(a) $\int_{-\infty}^{\infty} 10e^t \sin 2\pi t\, \delta(t-0.75)dt$;

(b) $\int_{-\infty}^{\infty} \left[4\delta(t)+\cos 2\pi t\delta(t-1)+2t^2\delta(t-2)\right]dt$.

(c) $\int_{-\infty}^{\infty} \frac{j\omega\delta(\omega)}{3+j\omega}d\omega$

Ans. (a) −21.17; (b) 13; (c) 0.

P18.5 Show that $\int_{-\infty}^{\infty} \delta(at)dt = \frac{1}{|a|}$. (*Hint*: Change the time variable to $t' = at$. Note that Equation 18.2 applies only when t in the argument of the impulse function is multiplied by unity).

P18.6 Evaluate

(a) $\int_{-\infty}^{\infty} 24\delta(1-12t)\cos 4\pi t dt$. (*Hint*: Apply the procedure of P18.5).

(b) $\int_{-\infty}^{\infty} (\cos t)\delta\left(2t-\frac{\pi}{2}\right)dt$

Ans. (a) 1; (b) $1/2\sqrt{2}$.

P18.7 Using integration by parts, show that

(a) $\int_{-\infty}^{\infty} f(t)\delta^{(1)}(t-a)dt = -f^{(1)}(a)$, where the (1) superscript denotes the first derivative.

P18.8 Show that Equations 18.7 and 18.8 are invariant with respect to the unit of time.

First-Order Capacitive Circuits

P18.9 Determine i_S and v_L at $t = 0^+$ in Figure P18.9, given $v_{SRC}(t) = 0.1\delta(t)$ V and assuming that the capacitor is initially uncharged.

Ans. $v_L(0^+) = 2$ V, $i_S(0^+) = -40$ A.

P18.10 Determine i_S, v_1, and v_2 at $t = 0^+$ in Figure P18.10, given $v_{SRC}(t) = 10\delta(t)$ V and assuming that the capacitors (a) are initially uncharged and (b) have initial voltages $V_{10} = V_{20} = 2$ V.

Ans. (a) $v_1(0^+) = 5/3$ V, $v_2(0^+) = 10/3$ V, $i_S(0^+) = -5$ mA; (b) 11/3 V, 16/3 V, $i_S(0^+) = -9$ mA.

P18.11 Determine C in Figure P18.11, given that $v_O(0^+) = 0.5$ V in response to $v_{SRC}(t) = \delta(t)$ V.

Ans. 0.2 F.

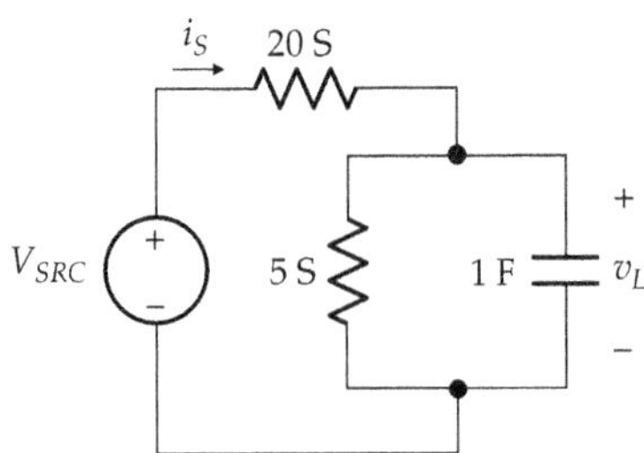

FIGURE P18.9

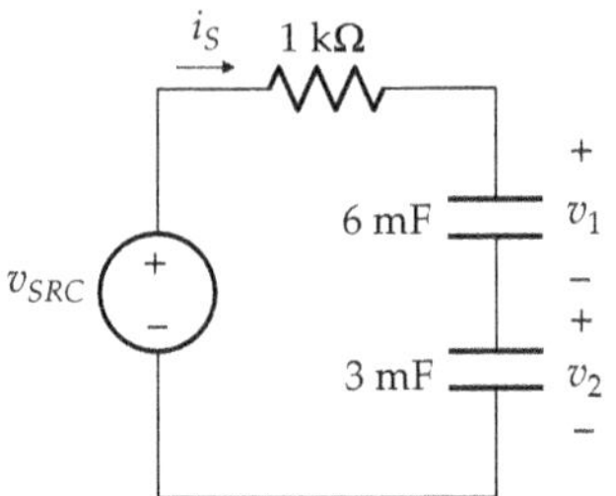

FIGURE P18.10

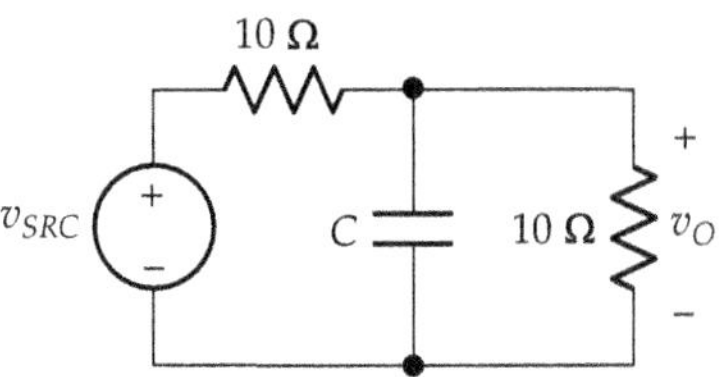

FIGURE P18.11

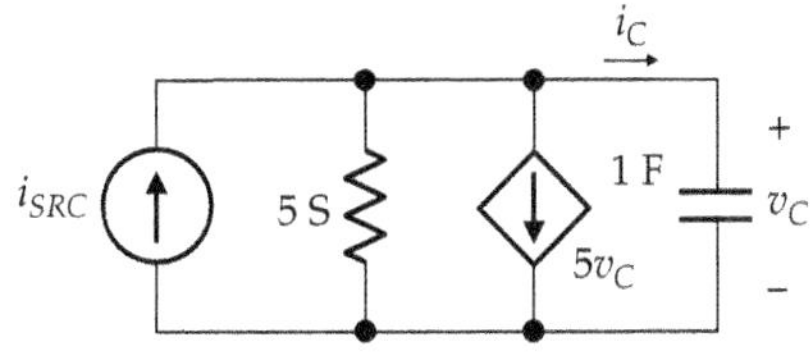

FIGURE P18.12

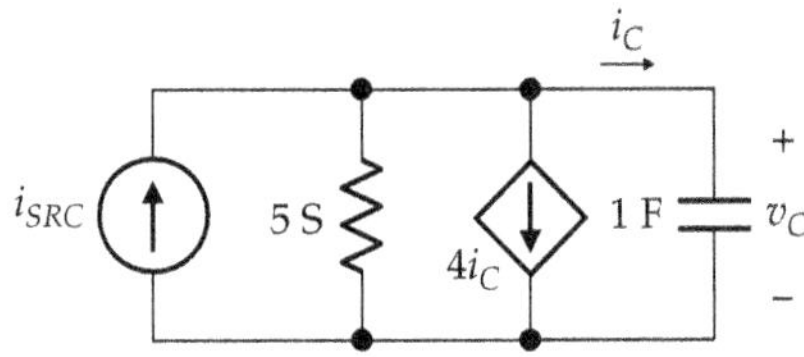

FIGURE P18.13

P18.12 Determine i_C at $t = 0^+$ in Figure P18.12, given $i_{SRC}(t) = \delta(t)$ A and assuming that the capacitor is initially uncharged.

Ans. −10 A.

P18.13 Determine i_C at $t = 0^+$ in Figure P18.13, given $i_{SRC}(t) = 10\delta(t)$ A and assuming that the capacitor is initially uncharged.

Ans. −2 A.

P18.14 Determine v_1 and v_2 for $t = 0^+$ in Figure P18.14, given $i_{SRC}(t) = 4\delta(t)$ A and assuming initial voltages $v_1 = 1$ V and $v_2 = 1$ V.

Ans. $v_1 = 5/3$ V, $v_2 = 4/3$ V.

P18.15 Determine $v_O(t)$ in Figure P18.15 under the following conditions: (a) $v_{SRC}(t) = \delta(t)$ V, with the capacitor initially uncharged; (b) $v_{SRC}(t) = u(t)$ V, with the capacitor initially uncharged (determine the response both directly and from the impulse response); (c) $v_{SRC}(t)$ is

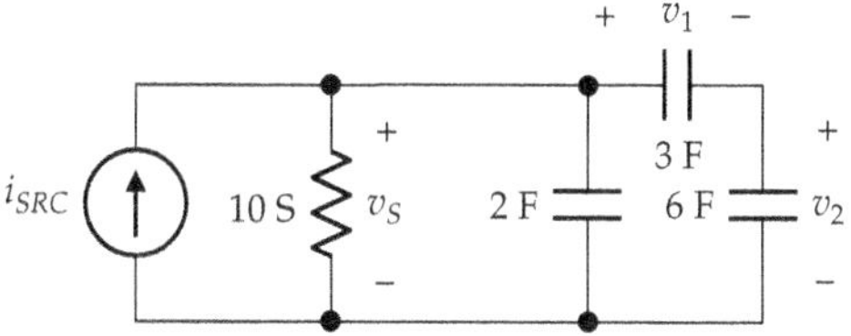

FIGURE P18.14

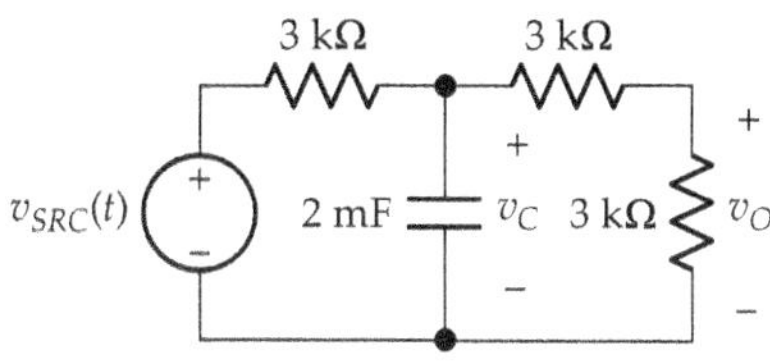

FIGURE P18.15

a pulse of 1 V amplitude and of duration from $t = 0$ to $t = 1$ s, with the capacitor initially uncharged; and (d) $v_{SRC}(t) = 15u(t)$, with the capacitor initially charged to 1 V.

Ans. (a) $(1/12)e^{-0.25t}$ V; (b) $(1/3)\left(1-e^{-0.25t}\right)$ V; (c) $(1/3)\left(e^{-0.25(t-1)}-e^{-0.25t}\right)$ V; (d) $5-4.5^{-0.25t}$, t is in s.

P18.16 Determine in Figure P18.16: (a) v_O at $t = 0^+$; (b) $i_C(t)$ for $t \geq 0^+$, assuming zero initial energy storage.

Ans. (a) 1 V; (b) $-\dfrac{e^{-t/8}}{8}$ mA, $t \geq 0^+$, t is in s.

P18.17 Determine $v_C(0^+)$ in Figure P18.17, assuming the capacitor is initially uncharged.

Ans. 20 V.

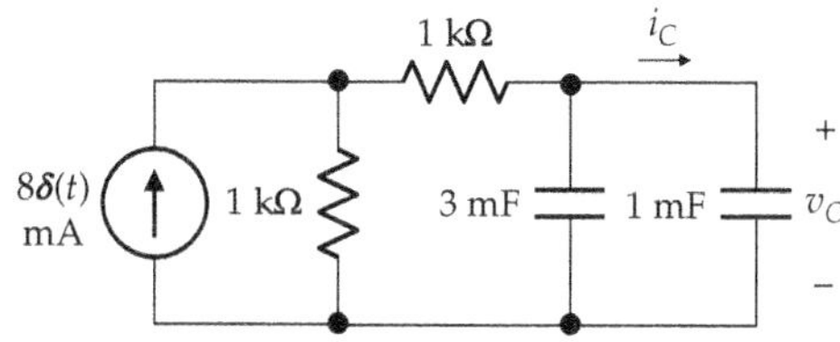

FIGURE P18.16

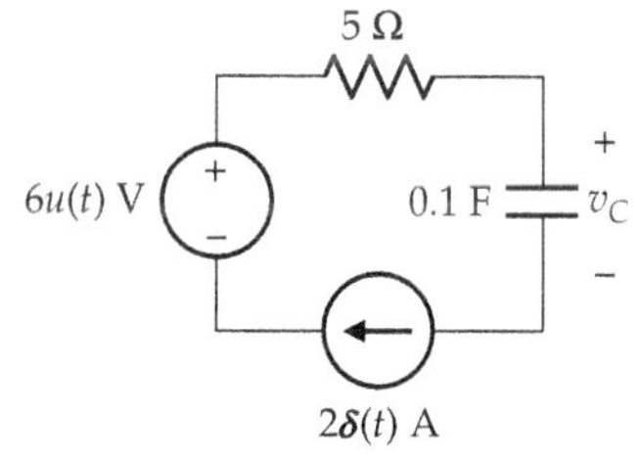

FIGURE P18.17

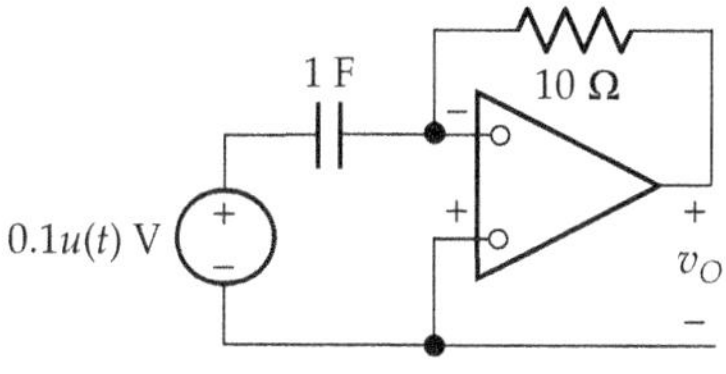

FIGURE P18.18

P18.18 Determine $v_O(t)$ in Figure P18.18.

Ans. $-\delta(t)$ V.

P18.19 Determine $v_O(t)$ in Figure P18.19, assuming an initial voltage of 2 V on C_1 and an initial voltage $V_{20} = 2$ V on C_2.

Ans. $v_O(t) = 2e^{-t/4}$ V, t is in ms.

P18.20 The switch is closed in Figure P18.20 at $t = 0$, with the capacitor initially uncharged. Derive the expression for $i_S(t)$, $t \geq 0^-$.

Ans. $CV_{SRC}(1-\alpha)\delta(t) + V_{SRC}u(t)/R$.

P18.21 Determine dv_C/dt at $t = 0^+$ in Figure P18.21, assuming no initial energy storage.

Ans. 2 V/s.

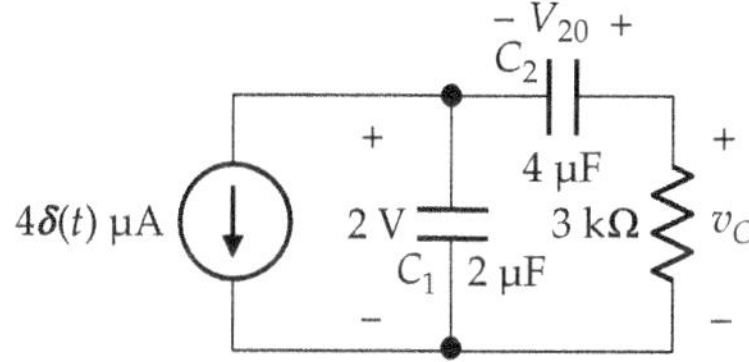

FIGURE P18.19

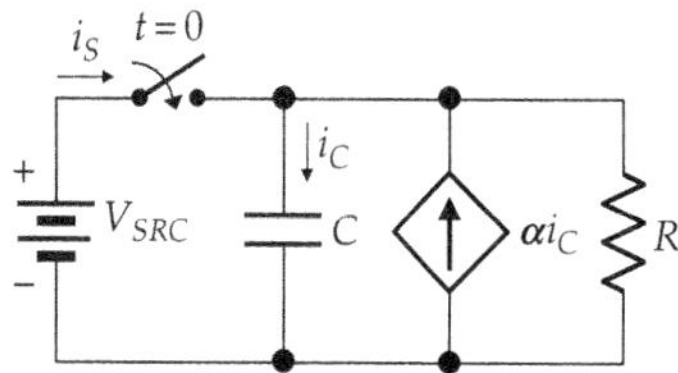

FIGURE P18.20

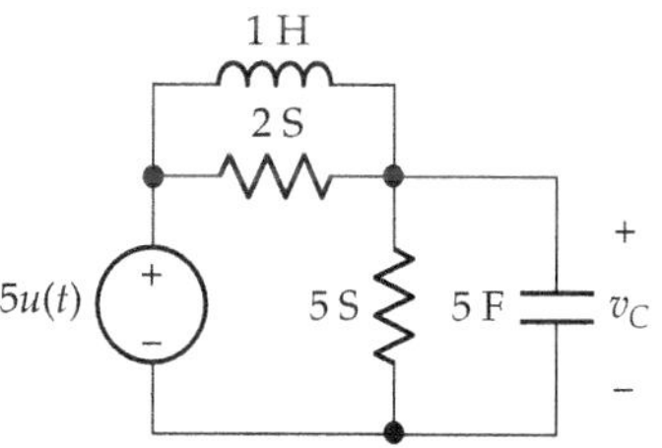

FIGURE P18.21

First-Order Inductive Circuits

P18.22 Determine v_S and i_L at $t = 0^+$ in Figure P18.22, given $i_{SRC}(t) = 0.1\delta(t)$ A and assuming the inductor is initially uncharged.

Ans. $i_L(0^+) = 2$ A, $v_S(0^+) = -40$ V.

P18.23 Determine v_S, i_1, and i_2 at $t = 0^+$ in Figure P18.23, given $i_{SRC}(t) = \delta(t)$ μA and assuming that the inductors (a) are initially uncharged and (b) have initial currents $I_{10} = I_{20} = 2$ mA.

Ans. (a) $i_1(0^+) = 5/3$ mA, $i_2(0^+) = 10/3$ mA, $v_S(0^+) = -50$ V; (b) 11/3 mA, 16/3 mA, $v_S(0^+) = -90$ V.

P18.24 Determine $i_{L1}(t)$, $i_{L2}(t)$, and $i_S(t)$, $t \geq 0^+$, in Figure P18.24 assuming the inductors are initially uncharged.

Ans. $i_{L1}(t) = 2e^{-t}$ A, $i_{L2}(t) = 2$ A, and $i_S(t) = -2(1 + e^{-t})$ A, t is in μs.

P18.25 Determine v_L at $t = 0^+$ in Figure P18.25, given $v_{SRC}(t) = 10\delta(t)$ V and assuming that the inductor is initially uncharged.

Ans. −2 V.

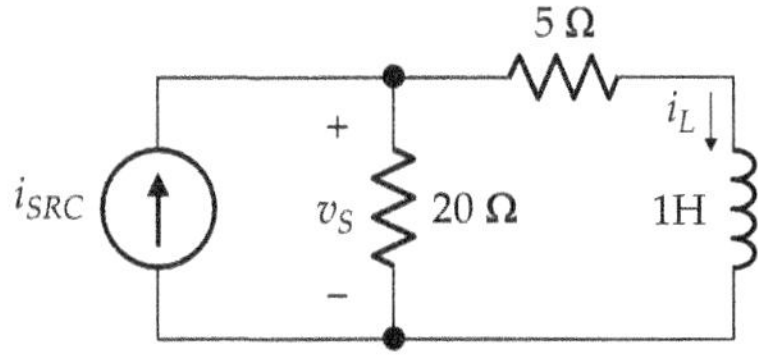

FIGURE P18.22

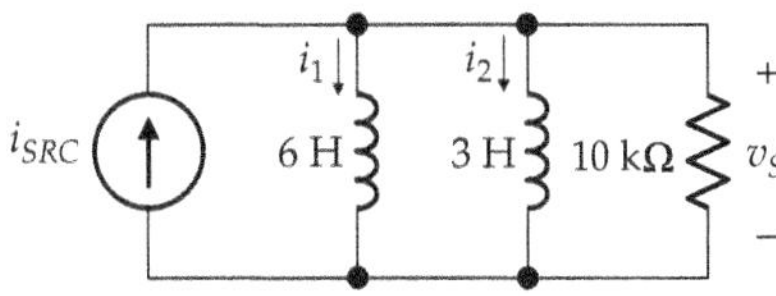

FIGURE P18.23

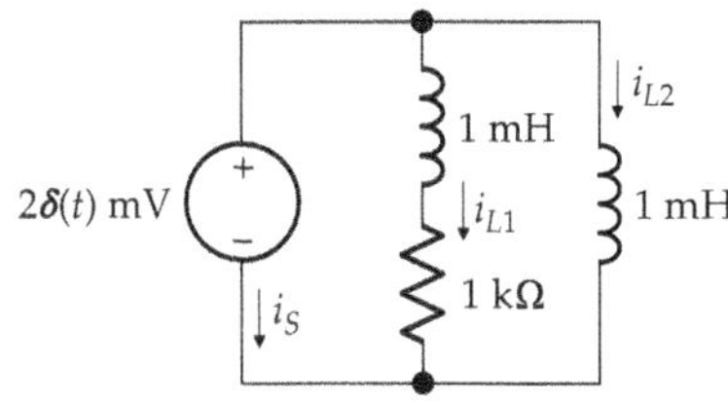

FIGURE P18.24

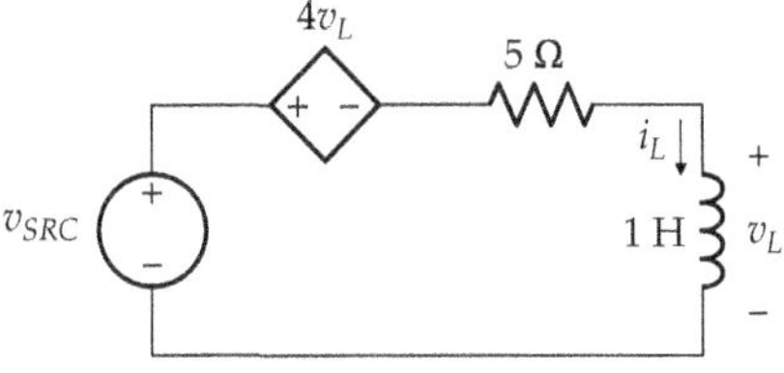

FIGURE P18.25

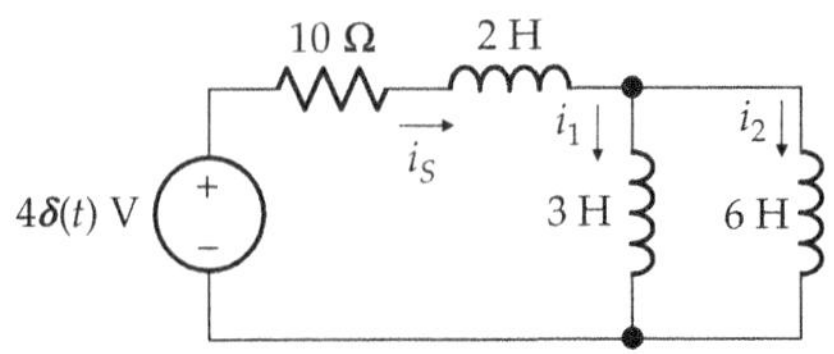

FIGURE P18.26

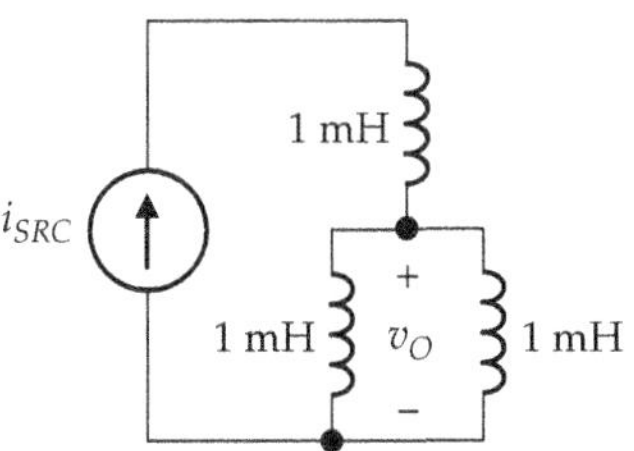

FIGURE P18.27

P18.26 Determine i_1 and i_2 for $t = 0^+$ in Figure P18.26 assuming initial currents; $i_1 = 1$ A and $i_2 = 1$ A.

Ans. $i_1 = 5/3$ A, $i_2 = 4/3$ A.

P18.27 Determine $i_{SRC}(t)$ in Figure P18.27 given that v_O is an impulse at the origin of strength 1 mVs, and assuming the inductors are initially uncharged.

Ans. $2u(t)$ A.

P18.28 The initial currents in L_1 and L_4 are shown in Figure P18.28, those in L_2 and L_3 are not shown. Determine $i_3(0^+)$,

Ans. 3 A.

P18.29 Determine $i_O(t)$ in Figure P18.29 assuming the inductor current to be initially to 6 mA.

Ans. $5 - 2e^{-2t}$ A, t is in μs.

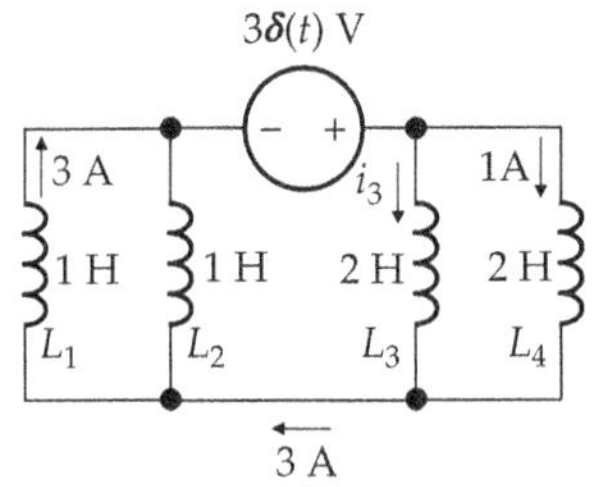

FIGURE P18.28

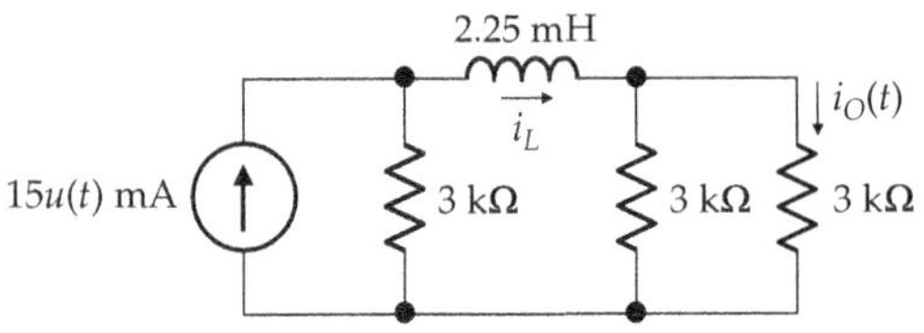

FIGURE P18.29

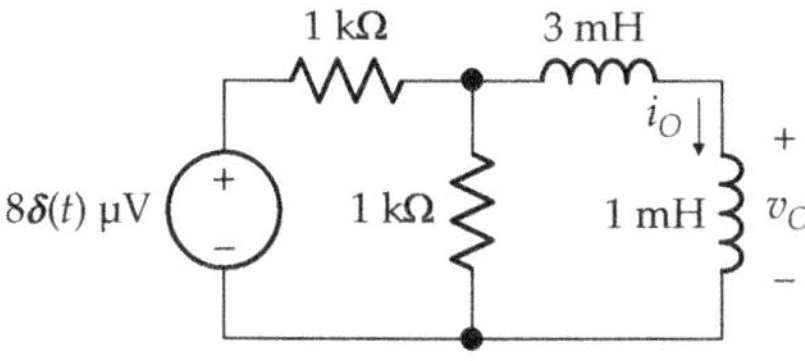

FIGURE P18.30

P18.30 Determine in Figure P18.30: (a) i_O at $t = 0^+$; (b) $v_O(t)$ for $t \geq 0^+$, assuming no initial energy storage.

Ans. (a) 1 mA; (b) $-\frac{e^{-t/8}}{8}$ V, $t \geq 0^+$, t is in μs.

P18.31 The switch is opened in Figure P18.31 at $t = 0$, with no initial energy in the inductor. Derive the expression for $v_S(t)$, $t \geq 0^-$.

Ans. $RI_{SRC}u(t) + LI_{SRC}(1 - \alpha)\delta(t)$.

P18.32 Determine $v_O(t)$ in Figure P18.32 assuming no initial energy storage.

Ans. $v_O(t) = 4e^{-t/2}u(t)$ V, t is in s.

Second-Order *RLC* Circuits

P18.33 Determine i_L and v_C at $t = 0^+$ in Figure P18.33, given $i_{SRC}(t) = 20\delta(t)$ μA, and assuming the inductor and capacitor are initially uncharged.

Ans. $v_C(0^+) = 10$ V, $i_L(0^+) = 0$.

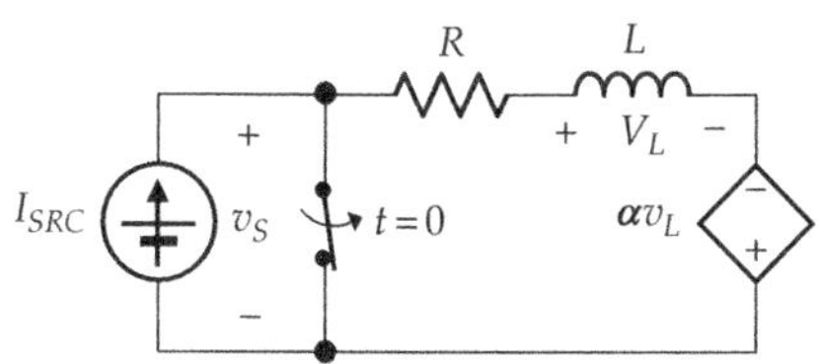

FIGURE P18.31

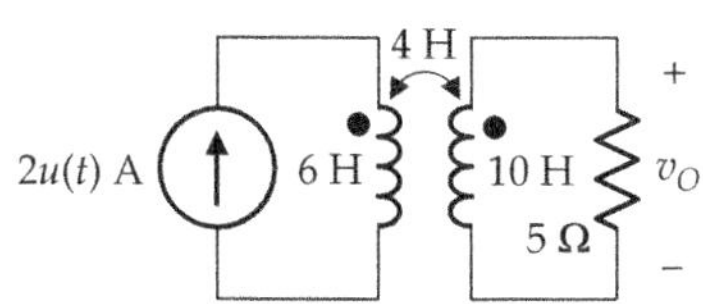

FIGURE P18.32

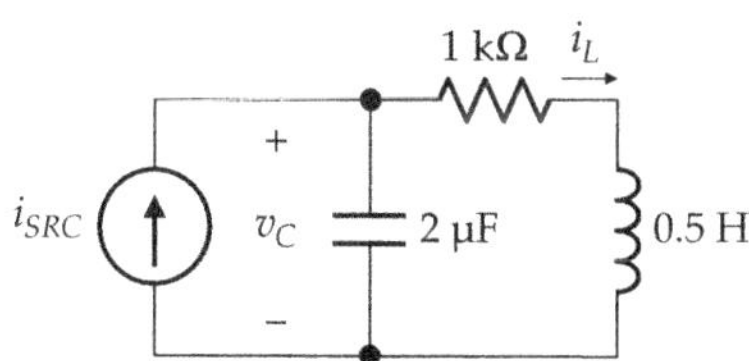

FIGURE P18.33

P18.34 Determine i_L and v_C at $t = 0^+$ in Figure P18.34, given $v_{SRC}(t) = 20\delta(t)$ μV, and assuming the inductor and capacitor are initially uncharged.

Ans. $v_C(0^+) = 0$, $i_L(0^+) = 10$ A.

P18.35 Determine i_L and v_C at $t = 0^+$ in Figure P18.35, given $i_{SRC}(t) = 20\delta(t)$ μA, and assuming the inductor and capacitor are initially uncharged.

Ans. $v_C(0^+) = 10$ V, $i_L(0^+) = 40$ mA.

P18.36 Determine i_L and v_C at $t = 0^+$ in Figure P18.36, given $v_{SRC}(t) = 10\delta(t)$ μV, and assuming the inductor and capacitor are initially uncharged.

Ans. $v_C(0^+) = -10$ V, $i_L(0^+) = 10$ mA.

P18.37 Determine v_S at $t = 0^+$ in Figure P18.37, given $i_{SRC}(t) = 20\delta(t)$ A, and assuming the inductor and capacitor are initially uncharged.

Ans. $v_S(0^+) = 20$ V.

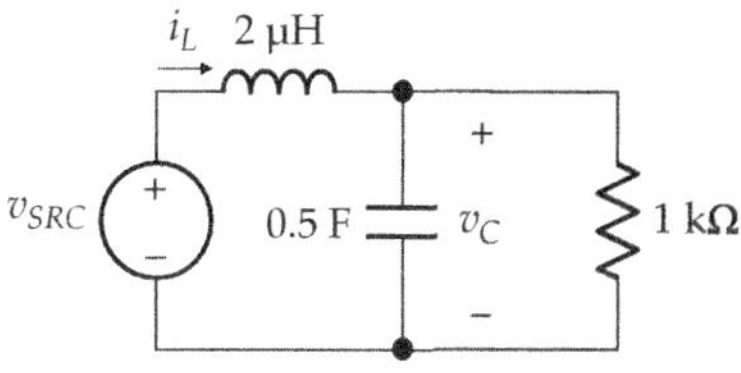

FIGURE P18.34

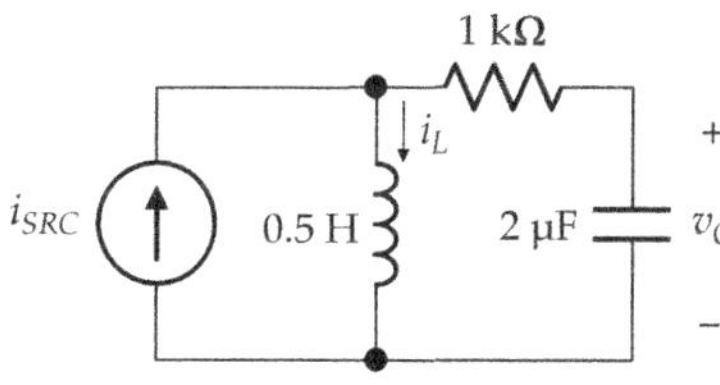

FIGURE P18.35

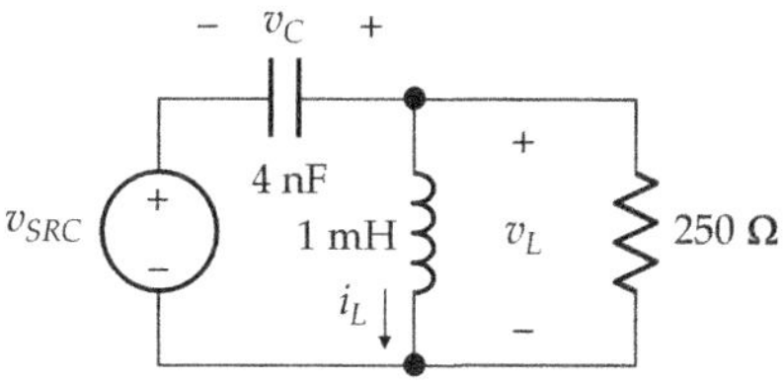

FIGURE P18.36

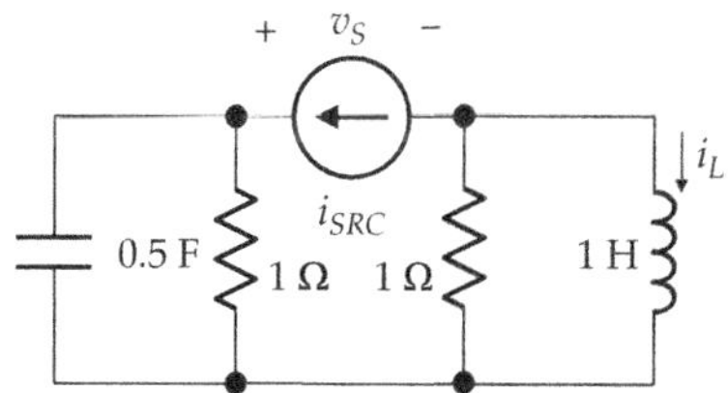

FIGURE P18.37

P18.38 Determine di_L/dt at $t = 0^+$ in Figure P18.38, given $i_{SRC}(t) = 5u(t)$ A, assuming the inductor and capacitor are initially uncharged.

Ans. 2 A/s.

P18.39 Determine i_L and v_C at $t = 0^+$ in Figure P18.39, given $i_{SRC}(t) = \delta(t)$ mA, and assuming the inductor and capacitor are initially uncharged.

Ans. $i_L(0^+) = 2$ A, $v_C(0^+) = 0$ V.

P18.40 Determine i_L, i_C, i_R, and v at $t = 0^+$ in Figure P18.40, given $i_{SRC}(t) = 0.1u(t)$ A, and assuming the inductor and capacitor are initially uncharged.

Ans. $v(0^+) = 0$, $i_L(0^+) = 0$, $i_C(0^+) = 0.1$ A, $i_R(0^+) = 0$.

P18.41 Determine $i_C(0^+)$ in Figure P18.41, assuming no initial current in the inductor and an initial $V_0 = 1$ V.

Ans. 1 A.

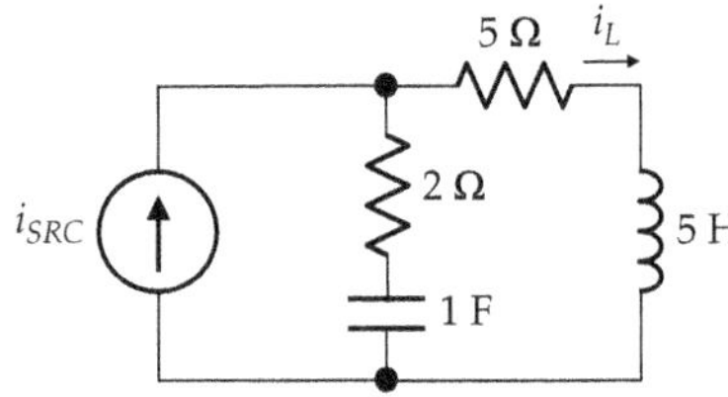

FIGURE P18.38

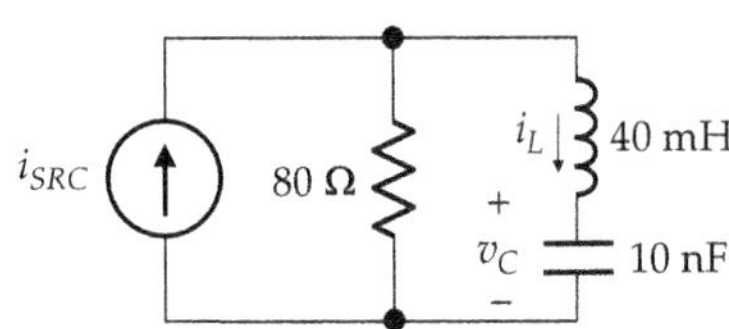

FIGURE P18.39

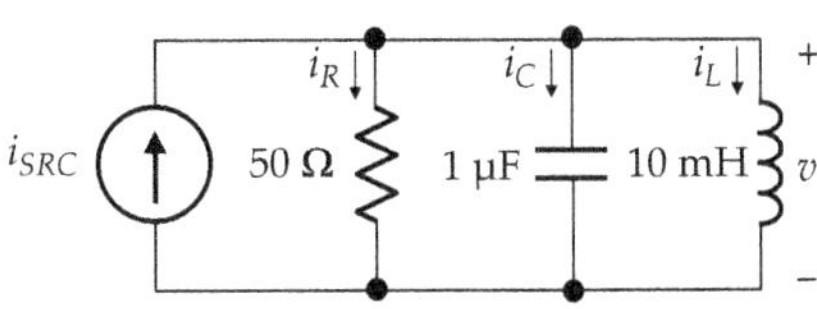

FIGURE P18.40

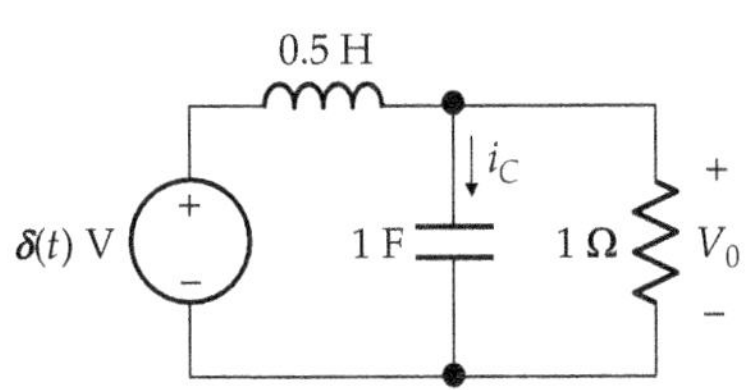

FIGURE P18.41

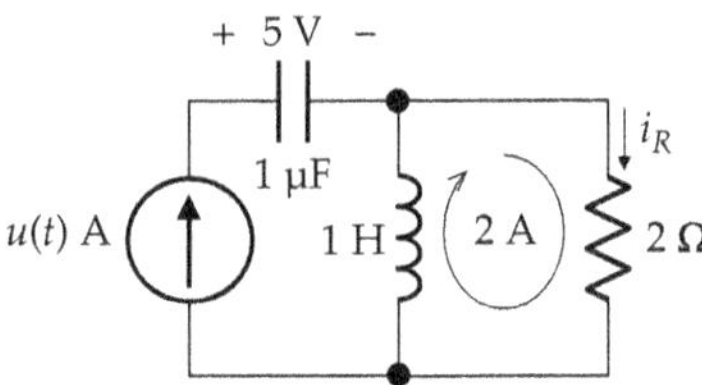

FIGURE P18.42

P18.42 Determine $i_R(t)$, $t \geq 0^+$ in Figure P18.42, assuming at $t = 0^-$ an initial voltage of 5 V across the capacitor and an initial circulating current of 2 A.

Ans. $3e^{-2t}$ A, t is in s.

P18.43 Determine v_L, v_C, i, and v_x at $t = 0^+$ in Figure P18.43, given $v_{SRC}(t) = 0.1u(t)$ V, and assuming $\rho = 8$ mA/V, and the inductor and capacitor are initially uncharged.

Ans. $i(0^+) = 0$, $v_x(0^+) = 0$, $v_C(0^+) = 0$, $v_L(0^+) = 0.1$ V.

P18.44 Determine v_O, v_L, i_L, and v_x at $t = 0^+$ in Figure P18.44, given $i_{SRC}(t) = 10u(t)$ A, and assuming the capacitor is initially uncharged and the inductor has an initial current of 0.1 A.

Ans. $v_x(0^+) = 0$, $v_O(0^+) = 0$, $i_L(0^+) = 0.1$ A, $v_L(0^+) = -0.1$ V.

P18.45 Repeat Problem P13.44 assuming the inductor is initially uncharged and the voltage across the capacitor is initially $v_X = 2$ V.

Ans. $v_x(0^+) = 2$ V, $v_O(0^+) = 1$ V, $i_L = 0$, $v_L(0^+) = 1$ V.

P18.46 The initial voltage on the capacitor in Figure P18.46 is $V_{10} = 6$ V, and the initial current in the inductor is $I_{20} = 2$ A. Determine $i_1(0^+)$, $i_2(0^+)$, $v_1(0^+)$, and $v_2(0^+)$.

Ans. $i_1(0^+) = -1$ A, $i_2(0^+) = 4$ A, $v_1(0^+) = 10$ V, $v_2(0^+) = -5$ V.

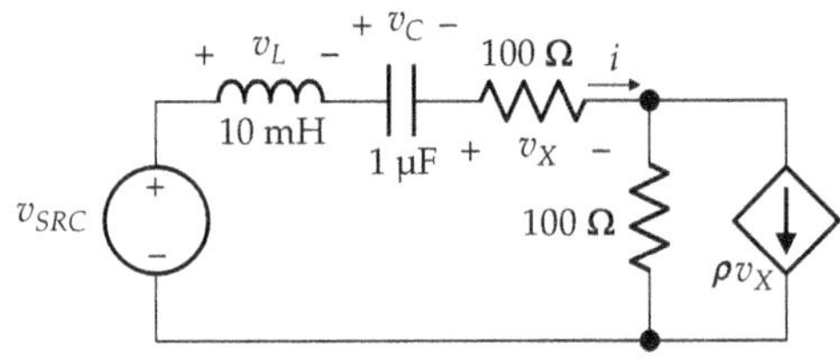

FIGURE P18.43

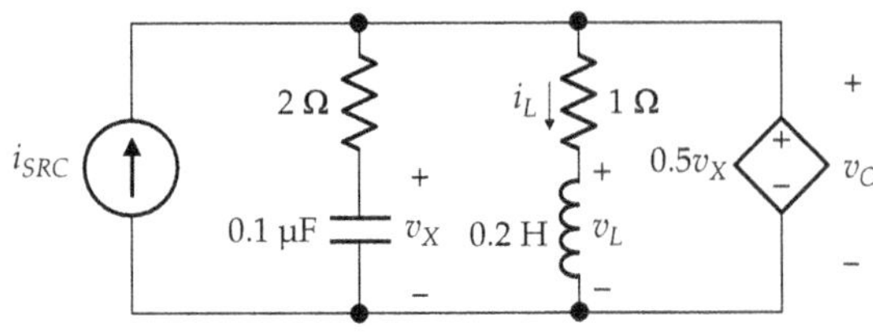

FIGURE P18.44

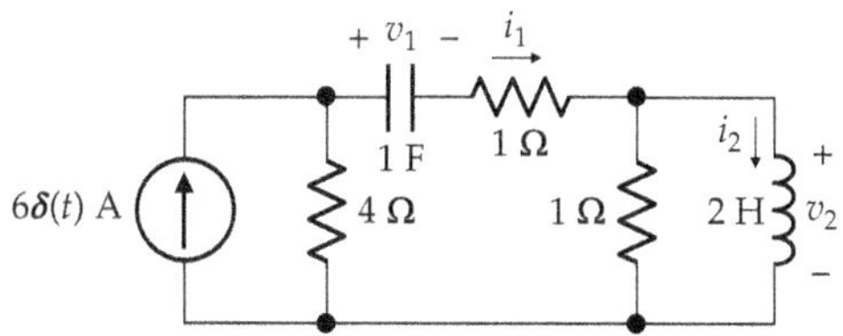

FIGURE P18.46

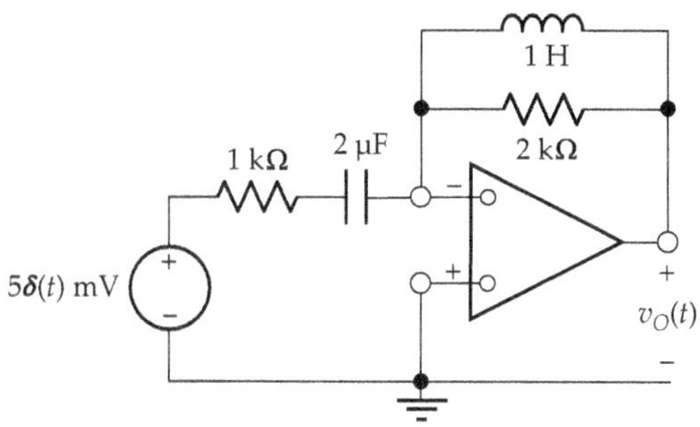

FIGURE P18.47

P18.47 Determine $v_O(0^+)$ in Figure P18.47, assuming no initial energy storage.

Ans. 25 V.

P18.48 Determine $i_L(0^+)$ and $v_C(0^+)$ in Figure P18.48, assuming no initial energy storage.

Ans. $i_L(0^+) = 2$ A; $v_C(0^+) = 2$ V.

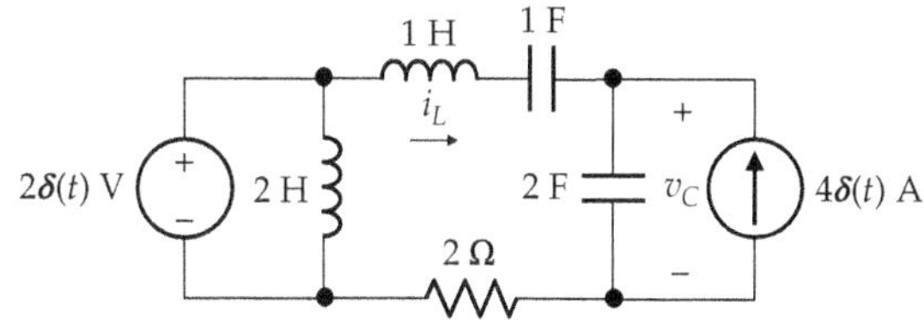

FIGURE P18.48

Probing Further

P18.49 (a) Show that $\int_{-\infty}^{\infty} u(t)\delta(t)dt = \frac{1}{2}$, using integration by parts. Note that this implies that $u(t) = 1/2$ at $t = 0$, for if this value is taken outside the integral, $\frac{1}{2}\int_{0^-}^{0^+} \delta(t)dt = \frac{1}{2}$, which is correct. (b) Consider that a function $f(t)$ has a step discontinuity K at $t = a$, where a is a constant. Express $f(t)$ as $f'(t) + Ku(t - a)$, where $f'(t)$ is continuous at $t = a$. Using the integral in (a), show that $f(t)\delta(t-a) = \frac{1}{2}\left[f(a^-) + f(a^+)\right]\delta(t-a)$. Note that according to the definition of Equation 18.8, $\int_{-\infty}^{\infty} u(t)\delta(t)dt = 1$, so that if a function $f(t)$ has a step discontinuity K at $t = a$, the value sampled by an impulse at $t = a$, is $f(t)\delta(t - a) = f(a^+)\delta(t - a)$, irrespective of K, which does not make sense.

19

Switched Circuits with Initial Energy Storage

Objective and Overview

Chapter 18 was concerned with circuit responses to the basic step and impulse inputs. The present chapter extends the analysis to responses resulting from switching operations in circuits having initial energy storage in capacitors and inductors, which involve instantaneous redistribution of charge in capacitors and flux linkage in inductors. These instantaneous redistributions are necessarily effected by current or voltage impulses. The discussion illustrates some important and useful concepts.

The chapter begins with the series and parallel connections of capacitors having initial charges and then considers the dual case of series and parallel connections of inductors having initial currents. It is shown that the analysis of several types of circuits is greatly facilitated by applying the concept of equivalent capacitance and inductance. As to be expected from the linearity of the circuits under consideration, superposition can be applied to obtain the total charges of capacitors, and the total flux linkages of inductors, due to applied excitation in the presence of initial energy storage. The discussion illustrates how charge is conserved in capacitive circuits and how flux linkage is conserved in inductive circuits. The effect of adding resistance is examined and it is demonstrated that energy can be trapped in a circuit even in the presence of resistors.

19.1 Series and Parallel Connections of Capacitors with Initial Charges

When capacitors having initial stored energy are connected in series or in parallel, with or without additional external input, superposition applies because the circuit is LTI. In order to determine the redistribution of capacitor charges or voltages, it is usually convenient to apply the concept of equivalent series or parallel capacitance, discussed in Section 7.3. It may be wondered, however, if the values of these equivalent capacitors under conditions of no initial charges apply in the presence of initial charges. The answer is embodied in the following concept:

Concept: *When capacitors in series or in parallel have initial charges, the equivalent series or parallel capacitance is the same as in the absence of initial charges.*

This conclusion follows quite simply from the fact that the capacitance of an ideal capacitor is a positive constant (Equation 7.1) that is determined by the geometry of the system and the dielectric properties of the medium in which the electric field exists, as exemplified by Equation 7.3. The capacitance is independent of capacitor charge or voltage. If the capacitance depends on the charge, then the system is not linear any more. It follows that C_{eqs} and C_{eqp} are given by Equations 7.40 and 7.49, respectively, in the presence of initial charges on the capacitances.

19.1.1 Capacitors in Parallel

To investigate the behavior of capacitors in parallel, consider the case of a 2 μF capacitor C_1 having an initial charge $q_1(0) = 16$ μC connected at $t = 0$ to a 6 μF capacitor C_2 having an initial charge $q_2(0) = 24$ μC (Figure 19.1a). The initial voltages of C_1 and C_2 are $16/2 = 8$ V and $24/6 = 4$ V, respectively.

When connected together, the voltages across C_1 and C_2 are forced to change instantaneously to become equal, in accordance with KVL. Since the change is instantaneous, it can only be accomplished by a current impulse that transfers charge from the capacitor of larger initial voltage (C_1 having a voltage of 8 V) to the capacitor of lower initial voltage (C_2 having a voltage of 4 V). This reduces the voltage of C_1 and increases the voltage of C_2 so as to

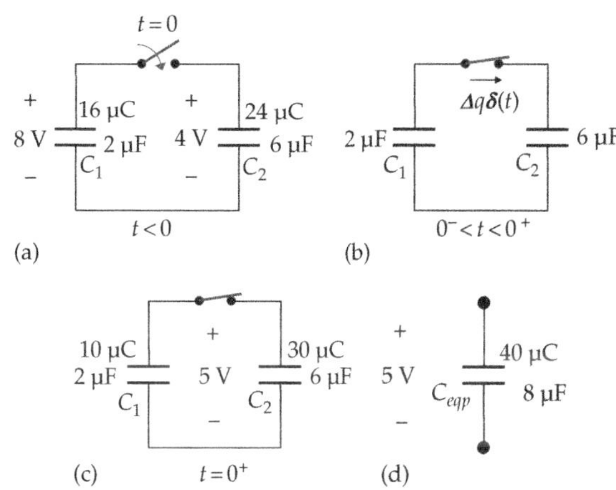

FIGURE 19.1
Paralleling of capacitors and equivalent parallel capacitor. Charged capacitor before switch is closed (a), during redistribution of charge (b), and just after the charge is redistributed (c); (d) equivalent parallel capacitor.

equalize the voltage across the two capacitors. Evidently, the final voltage will be intermediate between the initial voltages of the two capacitors. The current impulse is shown as $\Delta q\delta t$ in Figure 19.1b flowing from C_1 to C_2 between $t = 0^-$ and $t = 0^+$. Note that charge flows in the direction of a voltage drop, from the capacitor having the higher initial voltage to the capacitor having the lower initial voltage although C_1 is of smaller capacitance than C_2 and has the smaller charge.

The total initial charge of 40 μC becomes distributed over the two capacitors in parallel, having a combined capacitance of 8 μF. The final voltage will therefore be 40/8 = 5 V. The final charges on C_1 and C_2 are 10 μC and 30 μC, respectively (Figure 19.1c). C_1 has lost 6 μC and C_2 has gained 6 μC. The strength of the impulse is the charge transferred, so that Δq = 6 μC. Since the charge lost by C_1 is the charge gained by C_2, charge is conserved between the two capacitors. This must be the case, because charge is confined to the capacitor plates that are connected together and cannot flow anywhere else.

The preceding argument can be formalized by using the concept of the equivalent parallel capacitor, C_{eqp}, whose value is (2 + 6) = 8 μF. Recall from Chapter 3 that equivalence of two circuits between a given pair of terminals means that the two circuits have the same voltage–current relation between these terminals. It can be argued that the voltage across C_{eqp} is the same as that of the parallel combination of C_1 and C_2. If we consider the initial charges on C_1 and C_2 in Figure 19.1c as resulting from some currents $i_1(t)$ and $i_2(t)$, respectively, then the total current of C_1 and C_2 in parallel (Figure 19.1c) is $i_1(t) + i_2(t)$. By equivalence, the charge on C_{eqp} must result from a current $i_1(t) + i_2(t)$. Since q_1 is the time integral of $i_1(t)$ and q_2 is the time integral of $i_2(t)$, then the charge on C_{eqp} is the sum of the two integrals, that is, the sum of q_1 and q_2, which is 40 μC (Figure 19.1d). The voltage of C_{eqp} is therefore 40/8 = 5 V, which, by equivalence, must also be the voltage across the parallel combination of C_1 and C_2.

It is of interest to compare the initial and final energies stored in the two capacitors. The initial stored energy is

$$w_0 = \frac{1}{2}(16\,\mu\text{C})\times(8\,\text{V}) + \frac{1}{2}(24\,\mu\text{C})\times(4\,\text{V}) = 112\ \mu\text{J} \quad (19.1)$$

and the final energy is

$$w_f = \frac{1}{2}(40\,\mu\text{C})\times(5\,\text{V}) = 100\ \mu\text{J} \quad (19.2)$$

As a result of redistribution of charge, 12 μJ are lost. That the final energy is less than the initial energy does not violate conservation of energy. Mathematically, this discrepancy is manifested as the integral of $p = Ri^2$ over the duration of the impulse, where R is the ideally zero resistance of the connection between the two capacitors and i^2 is the square of an impulse. The product Ri^2 is indeterminate, but not zero. Physically, energy is dissipated in some residual resistance of the connection, which is inevitably present, and in electromagnetic radiation due to charge acceleration. If the capacitors are connected by a resistor, it can be readily shown that the power dissipated in the resistor accounts for the difference in the initial and final energies of the capacitors.

It may be concluded, therefore, that when two or more capacitors having initial energy storage are paralleled together, charge is redistributed among the capacitors by current impulses so as to equalize the voltage across them. If a current pulse is applied to the parallel combination, charge is added to each capacitor, while keeping a common voltage across them. The distribution of charge among the capacitors is conveniently determined using the equivalent parallel capacitor, as illustrated in the following example.

Example 19.1: Paralleling of Initially Charged Capacitors

Given three capacitors C_1 = 2 μF having a charge of 24 μC, C_2 = 4 μF having a charge 24 μC, and C_3 = 6 μF having a charge of 12 μC, it is required to determine (a) the voltage across the capacitors when connected in parallel and (b) the voltage and charge on each capacitor after the application of a 9 mA, 4 ms, pulse to the paralleled capacitors.

Solution:

(a) The initial charge and voltage on each capacitor are shown in Figure 19.2a. When connected in parallel, C_{eqp} = 12 μF (Figure 19.2b). From conservation of charge, the charge on C_{eqp} is the sum of the charges on the individual capacitors, which is 60 μC. The voltage of C_{eqp} is 60/12 = 5 V. From equivalence, this must be the voltage across the parallel combination. It follows that the charges on C_1, C_2, and C_3 are, respectively, 2 × 5 = 10 μC, 4 × 5 = 20 μC, and 6 × 5 = 30 μC (Figure 19.2c). The total charge is 60 μC, as it should be.

(b) The 9 mA pulse of 4 ms duration delivers a charge of 9 × 4 = 36 μC. Applied to C_{eqp}, the total charge is now 96 μC, and the voltage is 96/12 = 8 V (Figure 19.3a). From equivalence, the voltage across the parallel combination is also 8 V, which means that the charges on C_1, C_2, and C_3 are, respectively, 2 × 8 = 16 μC, 4 × 8 = 32 μC, and 6 × 8 = 48 μC (Figure 19.3b). The total charge is 96 μC, as it should be.

Simulation: The circuit for part (a) is entered as in Figure 19.4a. PSpice requires that every node should have a path of finite or zero resistance to ground, which means that nodes isolated from ground by ideal capacitors or current sources are not allowed and will give a "floating" node error. In circuit terminology, **floating**

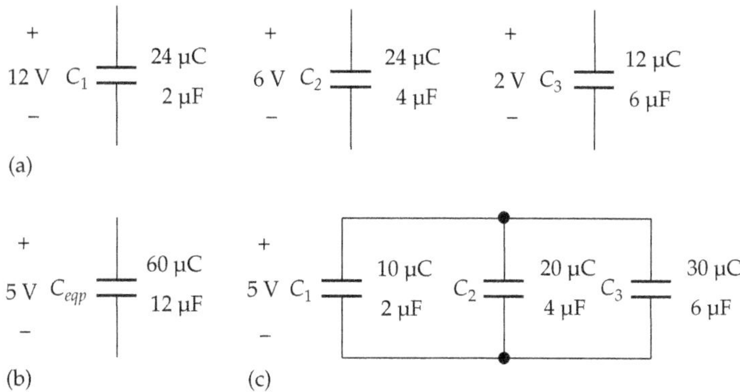

FIGURE 19.2
Figure for Example 19.1.

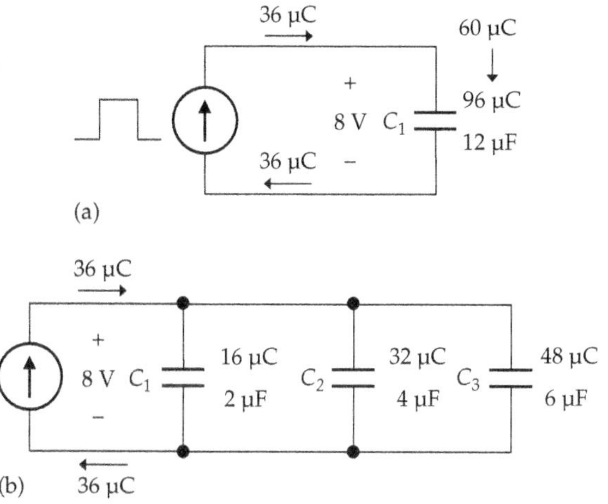

FIGURE 19.3
Figure for Example 19.1.

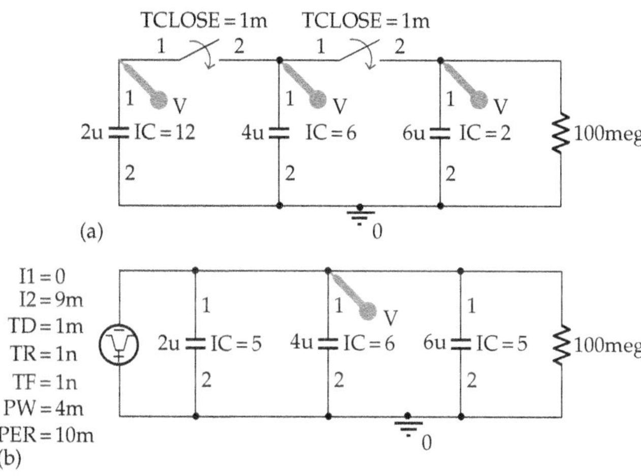

FIGURE 19.4
Figure for Example 19.1.

denotes isolation from ground. A 100 MΩ resistor connected between the upper node and ground in Figure 19.4 avoids the "floating" of the upper node and is sufficiently large so as not to significantly affect the results. The initial voltages on the capacitors are entered as explained previously. Two normally open switches have been added as the Sw_tClose part from the EVAL library. The desired closure time of the switch is set at 1 ms in the Property Editor spreadsheet of the switch so as to show both the initial and final voltages of the capacitors. Voltage markers have been added to indicate voltages. In the simulation profile, Time domain (Transient) analysis is chosen, 2 ms entered for 'Run to time', 0 entered for 'Start saving data after', and 0.5 μs entered for 'Maximum step size'. After the simulation is run, the graph in Figure 19.5 is displayed showing the initial voltages of 2, 6, and 12 V, and the final voltage of 5 V, as determined previously.

The circuit for part (b) is entered as illustrated in Figure 19.4b. The current source is IPULSE from the SOURCE library. The parameters of the source are more fully explained in Appendix C. I1 is the initial current level, which is zero; I2 is the higher level of current, which is the 9 mA magnitude; and TD is the time delay, after $t = 0$, when the pulse is applied. This is set to 1 ms in the simulation so as to display the initial conditions, TR is the rise time of the pulse; TF is its fall time; PW is the pulse width, which is 4 ms; and PER is the period of the pulse. Since only a single pulse is required, the period is set larger than the duration of the simulation. TR and TF are set at 1 ns, which is small compared to the pulse width of 4 ms, so that the pulse is "square" rather than trapezoidal. In the simulation profile, Time domain (Transient) analysis is chosen, 6m entered for 'Run to time', 0 entered for 'Start saving data after', and 0.5u entered for 'Maximum step size'. After the simulation is run, the graph of Figure 19.6 is displayed. The voltage increases linearly during the pulse from 5 to 8 V.

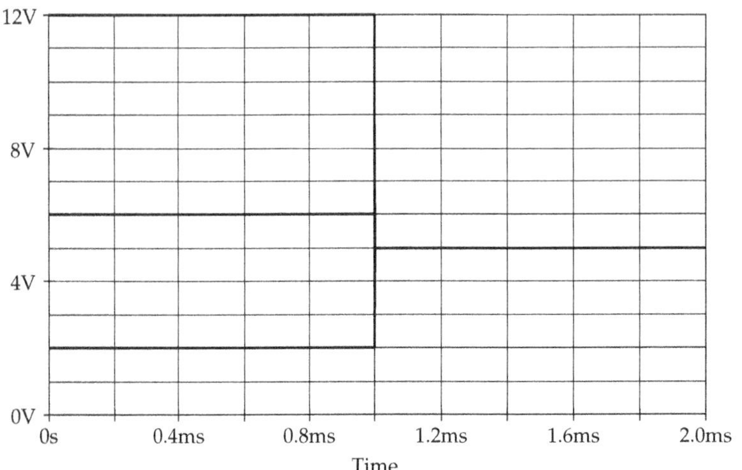

FIGURE 19.5
Figure for Example 19.1.

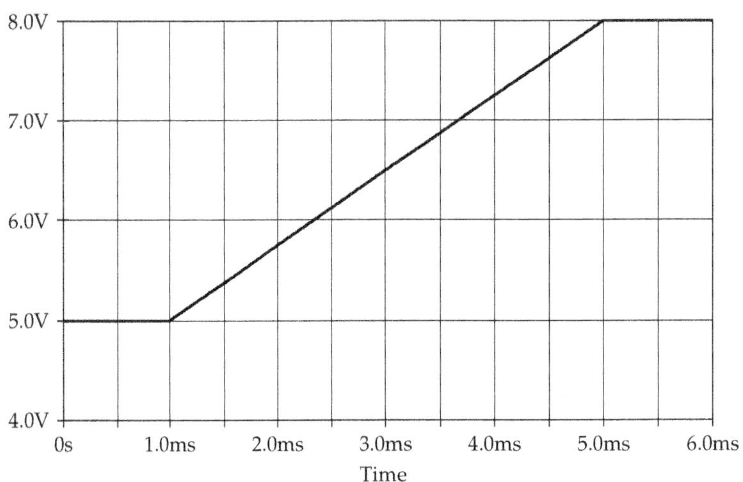

FIGURE 19.6
Figure for Example 19.1.

Primal Exercise 19.1

Assume that in Figure 19.1a C_1 is initially charged to 2 V and C_2 is uncharged. Determine the common voltage after the switch is closed.

Ans. 0.5 V.

Primal Exercise 19.2

Repeat Example 19.1 assuming a charge of 12 μC on C_1, 12 μC on C_2, and 24 μC on C_3.

Ans. (a) 4 V; (b) 7 V, $q_1 = 14$ μC, $q_2 = 28$ μC, $q_3 = 42$ μC.

19.1.2 Capacitors in Series

When considering capacitors connected in series and having initial stored energy, it is of interest to determine the voltages across the individual capacitors in the following cases:

1. A current pulse is applied to the series combination.
2. A steady voltage is applied to the series combination.
3. The two terminals of the series combination are connected together.

These cases will be considered in the following three examples. It will be seen that the concept of the equivalent series capacitance is very useful, particularly in the last two cases.

Example 19.2: Charged Capacitors in Series

A current pulse of 3 A amplitude and 4 s duration is applied to three capacitors in series: C_1 = 6 F, C_2 = 3 F, and C_3 = 2 F (Figure 19.7a). It is required to determine the voltages across the capacitors (a) when the capacitors are initially uncharged and (b) when C_1, C_2, and C_3 have initial charges of 12, 6, and, 2 C, respectively.

Solution:

(a) Overall, the current pulse can be considered to deposit a charge of +12 C on the top plate of C_1 and to remove +12 C from the bottom plate of C_3, leaving a charge of −12 C on this plate. There will be induced charges in the remaining plates of the capacitors (Figure 19.7b) so that each capacitor will have a charge of 12 C on its top plate and −12 C on its bottom plate. There is zero net charge (+12 and −12 C) on the capacitor plates that are connected together. The voltages on the capacitors are v_1 = 12/6 = 2 V, v_2 = 12/3 = 4 V, and v_3 = 12/2 = 6 V. The total voltage across the three capacitors is 12 V.

C_{eqs} can be evaluated by considering two capacitors at a time. The series capacitance of the 6 F and 3 F capacitors is (6 × 3)/(6 + 3) = 2 F. The series capacitance of this and the 2 F capacitor is 1 F (Figure 19.7). The 3 A current pulse deposits 12 C of charge at t = 4 s, the voltage across C_{eqs} being 12/1 = 12 V, the same as that across the series combination.

(b) When the capacitors have arbitrarily assigned initial charges (Figure 19.8a), the charge on C_{eqs} is clearly not the same as the individual charge of any of the three capacitors, because these individual charges are all different. But because of equivalence, the v–i relation is the same, even with i = 0, so that the voltage across C_{eqs} must be the same as the voltage across the series combination. The voltages of the individual capacitors are v_{10} = 12/6 = 2 V, v_{20} = 6/3 = 2 V, and v_{30} = 2/2 = 1 V. The total voltage is 5 V, which must be the voltage across C_{eqs}. It follows that the charge of C_{eqs} is 1 × 5 = 5 C. That is, the charge on C_{eqs} must be consistent with the value of this capacitor and the voltage across it. The total energy stored in the three capacitors is no longer the same as that stored in C_{eqs}, because the three capacitors are charged individually and arbitrarily.

After the current pulse is applied to the three capacitors, the resulting charges and voltages of the three capacitors can be determined from superposition, since the system is linear. This implies that the charge on each capacitor due to the current pulse simply adds to the initial charge already present on that capacitor. The current pulse deposits 12 C on C_{eqs} at t = 4 s, increasing the charge on this capacitor to 17 C and the voltage to 17/1 = 17 V. The charge on each of the three capacitors must increase by the same amount as that on C_{eqs}, since the same current is applied for the same duration. The final charges are q_{1f} = 12 + 12 = 24 C, q_{2f} = 6 + 12 = 18 C, and q_{3f} = 2 + 12 = 14 C (Figure 19.7b). The voltages on the three capacitors are v_{1f} = 24/6 = 4 V, v_{2f} = 18/3 = 6 V, and v_{3f} = 14/2 = 7 V, the total being 17 V, as for C_{eqs}.

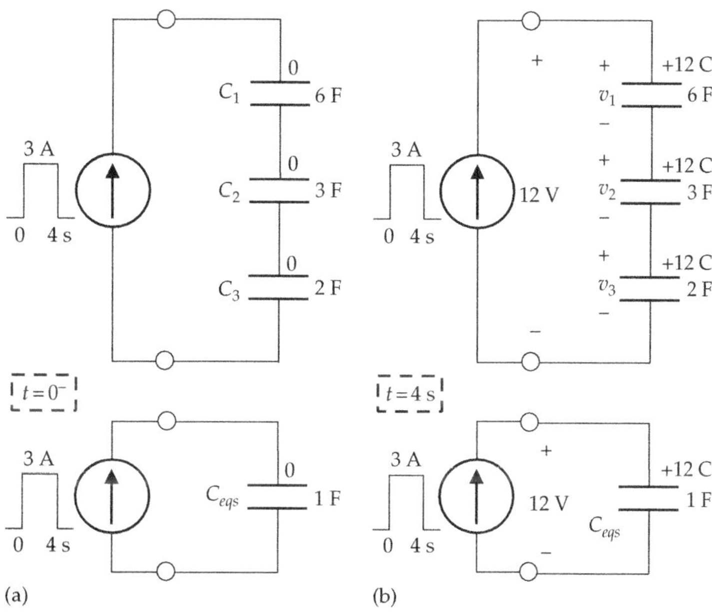

FIGURE 19.7
Figure for Example 19.2.

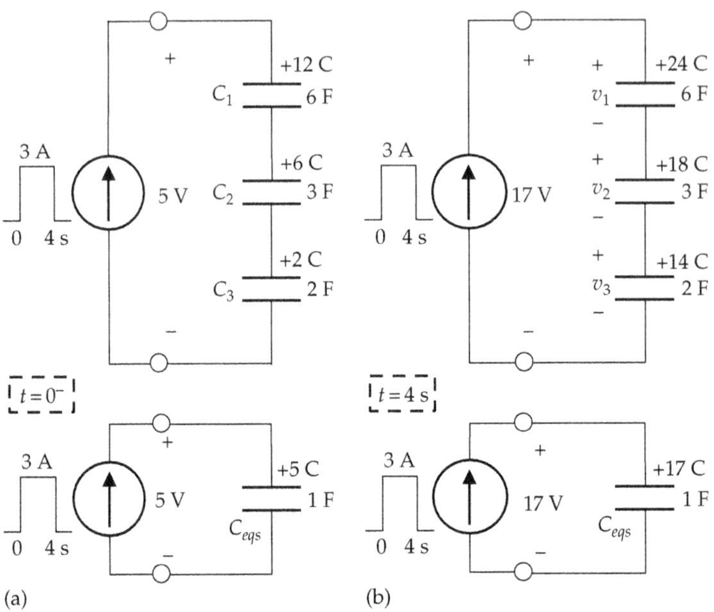

FIGURE 19.8
Figure for Example 19.2.

TABLE 19.1
Capacitor Charges and Voltages

	$C_1 = 6$ F		$C_2 = 3$ F		$C_3 = 2$ F		$C_{eqs} = 1$ F	
	q_1, C	v_1, V	q_2, C	v_2, V	q_3, C	v_3, V	Q_{eqs}, V	V_{eqs}, V
No initial storage	12	2	12	4	12	6	12	12
Initial conditions	12	2	6	2	2	1	5	5
Final ($t = 4$ s)	24	4	18	6	14	7	17	17

The charges and voltages for C_1, C_2, C_3, and C_{eqs} are listed in Table 19.1. The "No initial storage" row refers to the values found in (a) prior to the 4 s pulse. The row that follows lists the initial conditions. The last row indicates the charges and voltages at $t = 4$ s, these being the sum of the corresponding values in the two preceding rows.

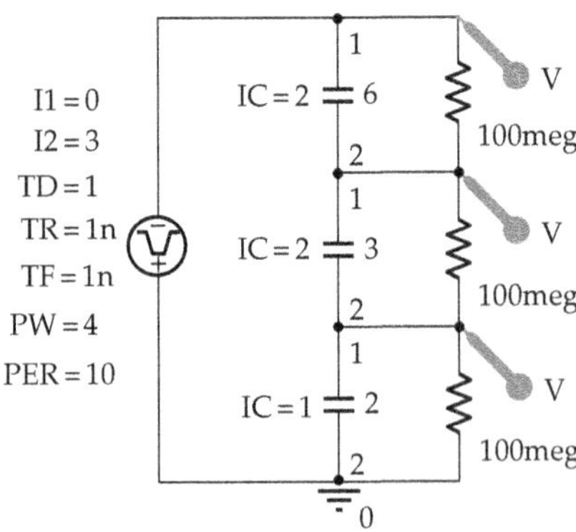

FIGURE 19.9
Figure for Example 19.2.

Simulation: The circuit is entered as illustrated in Figure 19.9. 100 MΩ resistors are added in order to avoid a floating node error in the simulation, and the current source is entered as explained in the preceding example. The initial voltages on the capacitors are added as described previously. In the simulation profile, Time domain (Transient) analysis is chosen, 6 entered for 'Run to time', 0 entered for 'Start saving data after', and 0.5 m entered for 'Maximum step size'. After the simulation is run, the graph in Figure 19.10 is displayed showing the initial and final voltages at the nodes with voltage markers. The initial voltages, with respect to ground, are (i) 1 V, the voltage across C_3; (ii) 3 V, the sum of the voltages across C_2 and C_3; and (iii) 5 V, the voltage across the three capacitors. These voltages increase linearly during the pulse and reach, in the same order, 7, 13, and 17 V, corresponding, at $t = 4$ s, to $v_3 = 7$ V, $v_2 = 6$ V, and $v_1 = 4$ V, as in Figure 19.8b.

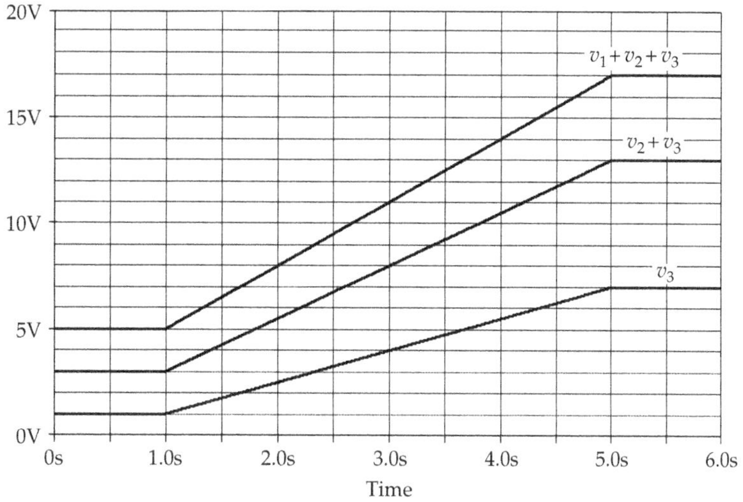

FIGURE 19.10
Figure for Example 19.2.

Primal Exercise 19.3

A 1A pulse of 2 μs duration is passed through three capacitors of 0.2, 0.8, and 0.04 μF connected in series. The capacitors were initially uncharged. Determine, at the end of the pulse, (a) the charge of each capacitor and the voltage across it and (b) the capacitance of the equivalent capacitor, its charge, and the voltage across it.

Ans. (a) 2 μC; 10, 2.5, and 50 V, respectively; (b) 0.032 μF, 2 μC; 62.5 V.

Example 19.3: Steady Voltage Applied to Capacitors in Series

Let a steady voltage of 15 V be applied at $t = 0$ to the series-connected capacitors in Figure 19.11, where C_1 = 6 μF, C_2 = 3 μF, and C_3 = 2 μF. It is required to determine the charges and voltages on the capacitors at $t \geq 0^+$: (a) with zero initial energy storage and (b) with the initial voltages on the capacitors as $v_{10} = 2$ V, $v_{20} = 2$ V, and $v_{30} = 1$ V.

Solution:

(a) The capacitance of 6 μF in series with 3 μF is 2 μF, and the series capacitance of this and the 2 μF capacitor

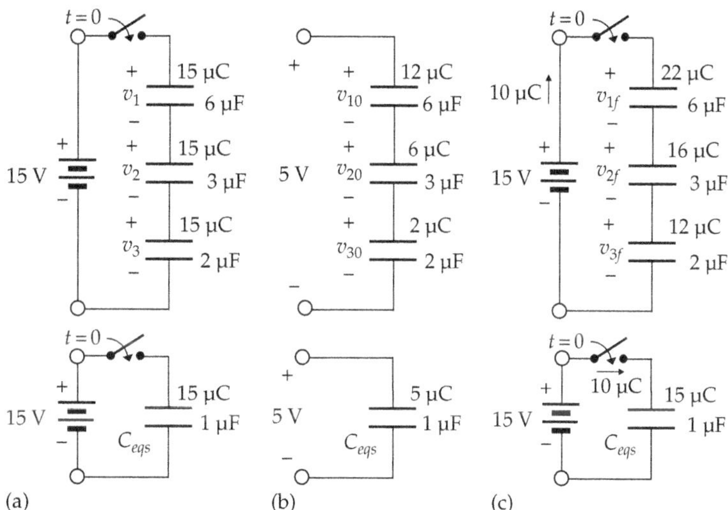

FIGURE 19.11
Figure for Example 19.3.

is $C_{eqs} = 1$ μF. When 15 V are applied to the series combination and to C_{eqs}, a current impulse flows that will instantaneously charge C_{eqs} to 15 V by depositing a charge of 15 μC on C_{eqs} (Figure 19.11a). By circuit equivalence, this same charge is applied to the three capacitors. Since the capacitors have zero initial energy storage, the charge on each of the three capacitors will be the same 15 μC (Equation 7.45). It follows that $v_1 = 15/6 = 2.5$ V, $v_2 = 15/3 = 5$ V, and $v_3 = 15/2 = 7.5$ V. The total voltage is 15 V, as it should be. Note how the charge was determined first from C_{eqs} and deduced for the individual capacitors.

(b) When the capacitors are initially charged, the total voltage is 5 V, which, by circuit equivalence, is also the voltage on C_{eqs}. The charge on C_{eqs} will be (5 V) × (1 μF) = 5 μC (Figure 19.11b). The applied 15 V will bring the charge on C_{eqs} to 15 μC, which must be due to a flow of charge of 10 μC from the applied source. By circuit equivalence, this same charge must flow through the three series capacitors, as illustrated in Figure 19.11c, increasing the charge on each capacitor by 10 μC. Hence, the final charges and voltages will be as follows:

$$q_{1f} = 2 \times 6 + 10 = 22\ \mu\text{C} \quad \text{and} \quad v_{1f} = 11/3\ \text{V}$$

$$q_{2f} = 2 \times 3 + 10 = 16\ \mu\text{C} \quad \text{and} \quad v_{2f} = 16/3\ \text{V}$$

$$q_{3f} = 1 \times 2 + 10 = 12\ \mu\text{C} \quad \text{and} \quad v_{3f} = 12/2 = 6\ \text{V}$$

It is seen that $v_{1f} + v_{2f} + v_{3f} = (11/3) + (16/3) + 6 = 15$ V as it should be.

Simulation: The circuit is entered as in Figure 19.12. 100 MΩ resistors and initial voltages on the capacitors have been added as explained previously. A normally open switch is added and set to close at 1 m.

In the simulation profile, Time domain (Transient) analysis is chosen, 2 m entered for 'Run to time', 0 entered for 'Start saving data after', and 0.5 μs entered for 'Maximum step size'. After the simulation is run, the graph in Figure 19.13 is displayed showing the initial and final voltages at the nodes with voltage markers. The initial voltages, with respect to ground, are read using the cursor as 5.000, 3.000, and 1.000 V corresponding to $v_{10} = 5-3 = 2$ V, $v_{20} = 3 - 1 = 2$ V, and $v_{30} = 1$ V. The final voltages, with respect to ground, are read using the cursor as 6,000, 11,333, and 15,000 V corresponding to $v_{1f} = 15-34/3 = 11/3$ V, $v_{2f} = (34/3) - 6 = 16/3$ V, and $v_{3f} = 6$ V.

The preceding two examples illustrate the following concept:

Concept: *When an excitation is applied to capacitors in series or in parallel having initial stored energy, the charge due to the applied excitation adds algebraically to the charge initially stored in each capacitor, in accordance with superposition.*

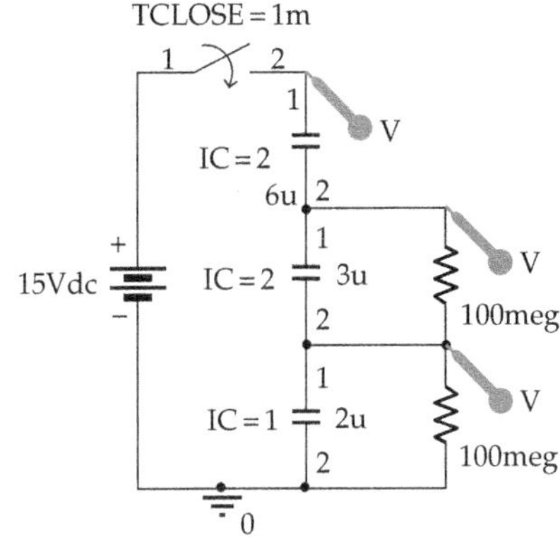

FIGURE 19.12
Figure for Example 19.3.

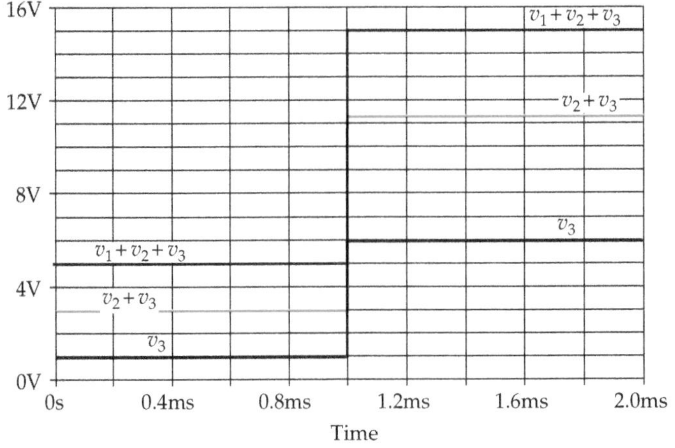

FIGURE 19.13
Figure for Example 19.3.

Primal Exercise 19.4

Repeat Example 19.3 assuming the initial charges on the capacitors are q_{10} = 18 μC, q_{20} = 12 μC, and q_{30} = 10 μC.

Ans. (a) As in Example 19.3; (b) q_{1f} = 21 μC, q_{2f} = 15 μC, q_{3f} = 13 μC.

Example 19.4: Ring Connection of Capacitors

Consider three capacitors having initial charges and voltages as shown in Figure 19.14 and connected in series with a switch that is closed at t = 0. It is required to determine the final charges and voltages on the three capacitors.

Solution:

C_{eqs} can be determined by considering the capacitors two at time. Thus, the two 8 μF capacitors C_1 and C_2 in series give a capacitance of 4 μF. This, in series with the 4 μF capacitor C_3, gives C_{eqs} = 2 μF. The voltage on C_{eqs} is the voltage drop across the switch, which is $V_{aa'}$ = 2 + 10 − 4 = 8 V. It follows that the charge on C_{eqs} is (8 V) × (2 μF) = 16 μC (Figure 19.15a). When the switch is closed, C_{eqs} completely discharges, causing the flow of 16 μC in the clockwise direction (Figure 19.15b). By circuit equivalence, 16 μC will also flow in the clockwise direction in the original circuit when the switch is closed in this circuit (Figure 19.16). As a result, C_1 is completely discharged. The 16 μC add to the charge on the top plate of C_2, which becomes +48 μC. The charge on the negatively charged plate of C_3 will be −40 + 16 = −24 μC (Figure 19.16). The final voltages on the capacitors will be v_{1f} = 0, v_{2f} = 48/8 = 6 V, and v_{3f} = 24/4 = 6 V. The total voltage around the mesh is zero, in accordance with KVL.

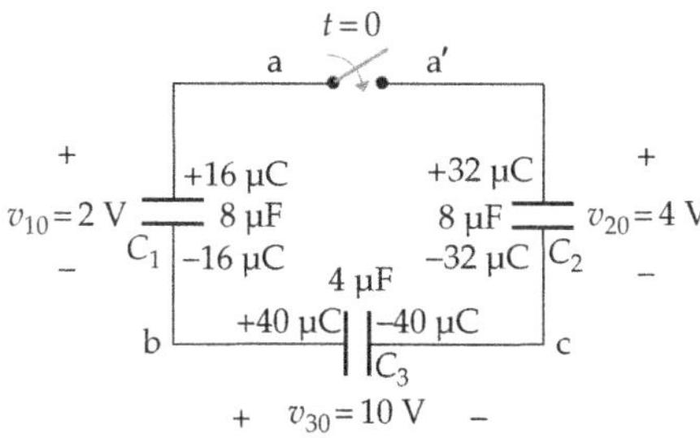

FIGURE 19.14
Figure for Example 19.4.

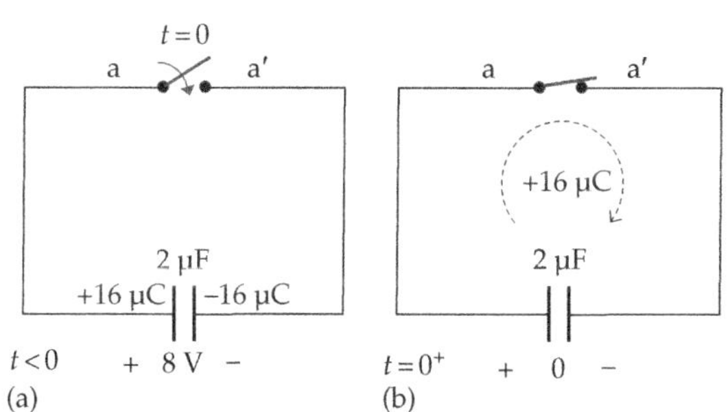

FIGURE 19.15
Figure for Example 19.4.

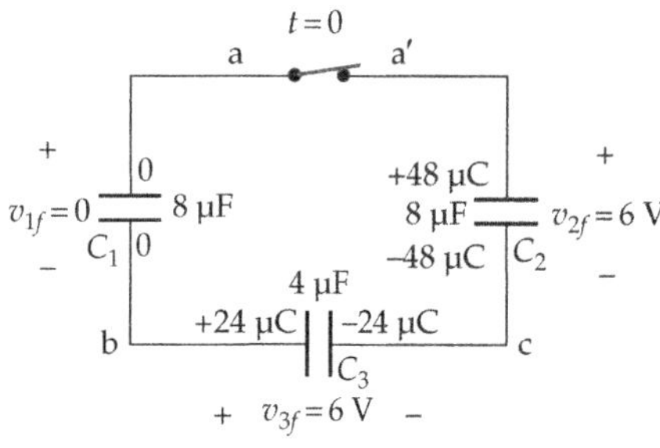

FIGURE 19.16
Figure for Example 19.4.

It is seen from Figures 19.14 and 19.16 that the charges at the capacitor nodes, before and after the switch is closed, are as follows:

'a' and a': 48 μC, 'b': 24 μC, 'c': −72 μC

Charge is therefore conserved at each of the nodes. Evidently, this is because the charge can only be redistributed between the capacitor plates connected to a given node and cannot flow elsewhere.

It should be noted that when the switch connected between terminals 'a' and a' is closed, charges are redistributed instantaneously through current impulses. As explained in the case of parallel connection of capacitors, energy is lost. Thus, the energy stored in Figure 19.14 is 0.5 × 2 × 16 + 0.5 × 10 × 40 + 0.5 × 4 × 32 = 280 μJ. The energy stored in Figure 19.16 is 0.5 × 6 × 24 + 0.5 × 6 × 48 = 216 μJ.

Simulation: The circuit is entered as in Figure 19.17. Two large 100 MΩ resistances are added, in order to avoid a

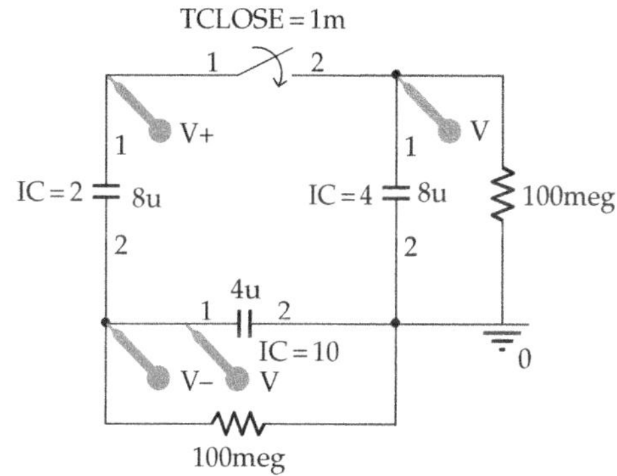

FIGURE 19.17
Figure for Example 19.4.

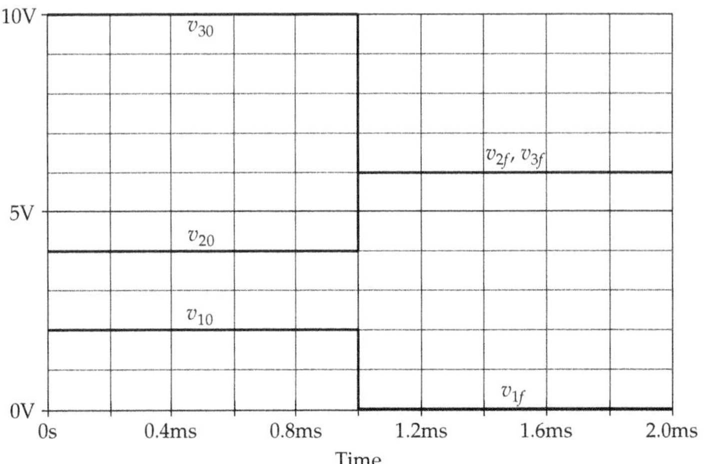

FIGURE 19.18
Figure for Example 19.4.

floating node error in the simulation, as explained previously. The initial voltages on the capacitors are entered and a normally open switch is added and set to close at 1 m. Capacitor voltages are indicated using a differential voltage marker for C_1 and voltage markers for C_2 and C_3. In the simulation profile, Time domain (Transient) analysis is chosen, 2 m entered for 'Run to time', 0 entered for 'Start saving data after', 0 entered for 'Start saving data after', and 0.5u entered for 'Maximum step size'. After the simulation is run, the graph in Figure 19.18 is displayed showing the initial and final voltages across the capacitors.

Problem-Solving Tip

- Equivalent series or parallel capacitors can be conveniently used to facilitate the solution of problems involving capacitors.

Example 19.4 illustrates the following concept concerning conservation of charge:

Concept: *Charge is conserved at any node in a circuit at the instant of a sudden change as long as no current impulses are applied from an external source to the node at the instant of the sudden change.*

Conservation of charge is in accordance with KCL.

Primal Exercise 19.5

Repeat Example 19.4 using initial values of 20, 24, and 30 μC on C_1, C_2, and C_3, respectively.

Ans. $q_{1f} = 6$ μC, $v_{1f} = 3/4$ V, $q_{2f} = 38$ μC, $v_{2f} = 19/4$ V, $q_{3f} = 16$ μC, $v_{3f} = 4$ V.

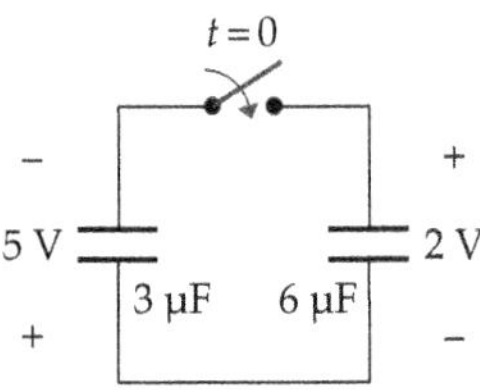

FIGURE 19.19
Figure for Primal Exercise 19.6.

Primal Exercise 19.6

The switch in Figure 19.19 is closed at $t = 0$, with the initial voltages across the capacitors as shown. Determine the voltage after the switch is closed, considering the two capacitors to be (a) in series or (b) in parallel.

Ans. 1/3 V, the upper node being negative with respect to the lower node.

19.2 Series and Parallel Connections of Inductors with Initial Currents

The case of inductors with initial currents parallels closely that of capacitors discussed in the preceding section, as to be expected from duality. Thus,

Concept: *When inductors in series or in parallel have initial currents, the equivalent series and parallel inductances are the same as when there are no initial currents.*

This conclusion follows from the fact that the inductance of an ideal inductor is a positive constant that depends on the geometry of the system, on the number of turns of the coil, and on the magnetic properties of the medium in which the magnetic field exists, as exemplified in Equation 7.24. The inductance is independent of the current in the inductor. If the inductance depends on the current, then the system is not linear any more. It follows that L_{eqs} and L_{eqp} are given by Equations 7.60 and 7.69, Chapter 3, respectively, in the presence of initial currents in the inductors.

The following examples are essentially the duals of those considered for the case of capacitors.

19.2.1 Inductors in Series

To investigate the behavior of inductors in series, suppose that a 2 μH inductor L_1 having a current $i_1(0) = 8$ A is connected at $t = 0$ to a 6 μH inductor L_2 having a current $i_2(0) = 4$ A by simultaneously moving the two switches, as shown in Figure 19.20. It is required to determine the final current in each inductor and in the equivalent inductor and to compare the initial and final energies stored in L_1 and L_2.

Figure 19.20 illustrates how the initial currents may be established in practice prior to connecting the inductors in series. The switches are of the make-before-break type, that is, they make contact with the final contact before breaking with the initial contact. This ensures that the inductor current is not interrupted during switching. Interrupting the inductor current during switching means that the inductor current becomes zero in a very short time interval, which implies a large di/dt. This, in turn, gives rise to a large Ldi/dt voltage across the switch contacts as they interrupt the current. The large voltage causes breakdown of the air dielectric between the contacts, leading to an arc discharge between the contacts. The high current density of the arc can damage the contacts due to localized burning. This kind of precaution is always necessary when interrupting currents through inductors. The resistors across the current sources ensure that these sources in Figure 19.20 are not left open circuited after the switches move to their new positions.

The initial flux linkages in L_1 and L_2 are $\lambda_1(0) = (2\ \mu\text{H})(8\ \text{A}) = 16\ \mu\text{Vs}$ and $\lambda_2(0) = (6\ \mu\text{H})(4\ \text{A}) = 24\ \mu\text{Vs}$, respectively (Figure 19.21a). When connected together, the currents

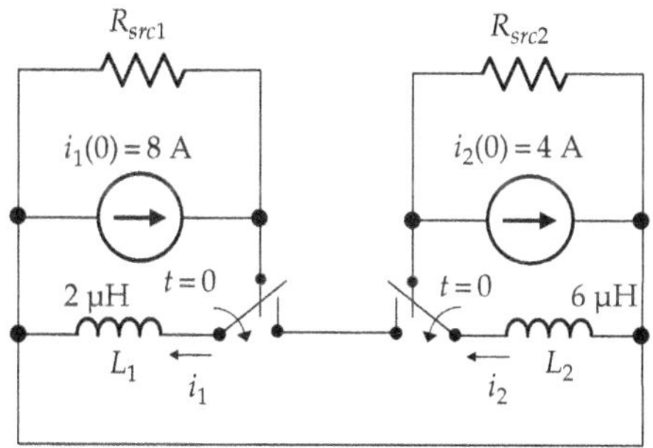

FIGURE 19.20
Series connection of inductors.

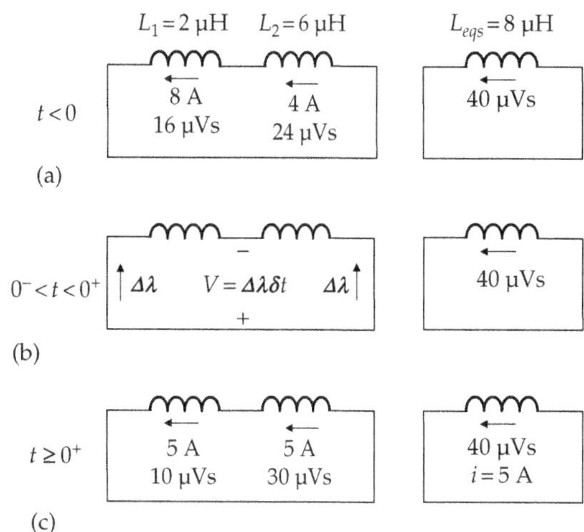

FIGURE 19.21
Inductors in series and equivalent series inductor. (a) Initial currents in inductors, (b) polarity of voltage impulse that equalized the currents, and (c) the currents after equalization.

through L_1 and L_2 are forced to change instantaneously to become equal, in accordance with KCL. Since the change is instantaneous, it can only be accomplished by a voltage impulse that reduces the flux linkage of the inductor having the larger initial current (L_1 of current 8 A) and increases the flux linkage of the inductor having the smaller initial current (L_2 of current 4 A), so as to equalize the current through the two inductors. Evidently, the final current will be intermediate between the initial currents of the two inductors. The voltage impulse, shown as $\Delta\lambda\delta t$ in Figure 19.21b between $t = 0^-$ and $t = 0^+$, is of the polarity indicated. It adds $\Delta\lambda$ to the flux linkage of the inductor having the larger inductance and flux linkage, but smaller current, and subtracts $\Delta\lambda$ from the flux linkage of the inductor having the smaller inductance and flux linkage, but larger current. The total flux linkage remains the same.

The total initial flux linkage of 40 μVs becomes distributed over the two inductors in series, having a combined inductance of 8 μH. The final current will therefore be 40/8 = 5 A. The final flux linkages of L_1 and L_2 are 10 and 30 μVs, respectively (Figure 19.21c). The flux linkage in L_1 has decreased by 6 μVs, whereas the flux linkage of L_2 has increased by 6 μVs. The strength of the impulse is the flux linkage transferred, so that $\Delta\lambda = 6$ μVs. Since the flux linkage lost by L_1 is the flux linkage gained by L_2, flux linkage is conserved around the mesh formed by the two inductors. Note that in going around the mesh, say in the counterclockwise direction, the flux linkages of L_1 and L_2, as well as $\Delta\lambda$ that adds to the flux linkage of L_2, are all in the same direction and can all be considered to have the same positive sign. On the other hand, $\Delta\lambda$ that subtracts from the flux linkage of L_2 is in the opposite direction and will therefore have a negative sign.

The preceding argument can be formalized by using the concept of the equivalent series inductor, L_{eqs}, whose value is (2 + 6) = 8 μH. From the concept of equivalence, the current through L_{eqs} is the same as that of the series combination of L_1 and L_2. If we consider the initial flux linkages of L_1 and L_2 as resulting from some voltages $v_1(t)$ and $v_2(t)$, respectively, then the total voltage of L_1 and L_2 in series is $v_1(t) + v_2(t)$. By equivalence, the flux linkage of L_{eqs} must result from a voltage $v_1(t) + v_2(t)$. Since the flux linkage of L_1 is the time integral of $v_1(t)$ and the flux linkage of L_2 is the time integral of $v_2(t)$, then the flux linkage of L_{eqs} is the sum of the two integrals, that is, the sum of the two flux linkages, which is 40 μVs (Figure 19.1). The current through L_{eqs} is therefore 40/8 = 5 A, which by equivalence, must also be the current through the series combination of L_1 and L_2.

The initial energy stored in the two inductors is $(1/2)L_1(i_1(0))^2 + (1/2)L_2(i_2(0))^2 = (1/2)\times 2\times 64 + (1/2)\times 6\times 16 = 112\,\mu\text{J}$. The final energy stored in the two inductors is $(1/2)(L_1+L_2)(i_f)^2 = (1/2)\times 8\times 25 = 100\,\mu\text{J}$. As a result of readjustment of current by the voltage impulse, 12 μJ are lost.

As explained for the case of two paralleled capacitors in connection with Figure 19.1, the discrepancy between the initial and final values of energy does not violate conservation of energy. The discrepancy is attributed to an impulse of voltage that appears across the two inductors, involving an infinite voltage due to the impulse applied across an infinite parallel resistance. In the presence of such a resistance, it can be readily shown that the power dissipated in the resistor accounts for the difference in the initial and final energies of the inductors.

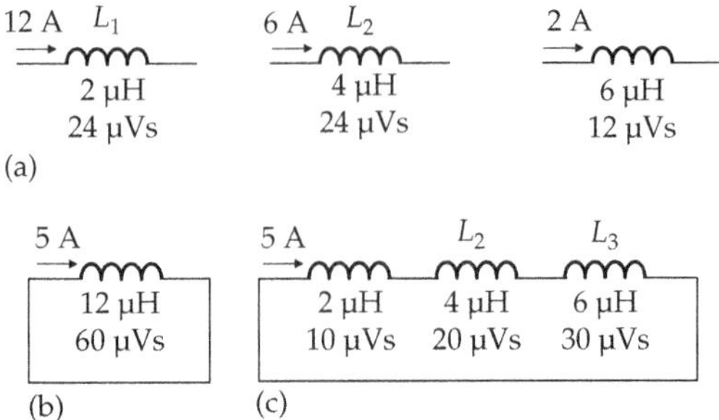

FIGURE 19.22
Figure for Example 19.5.

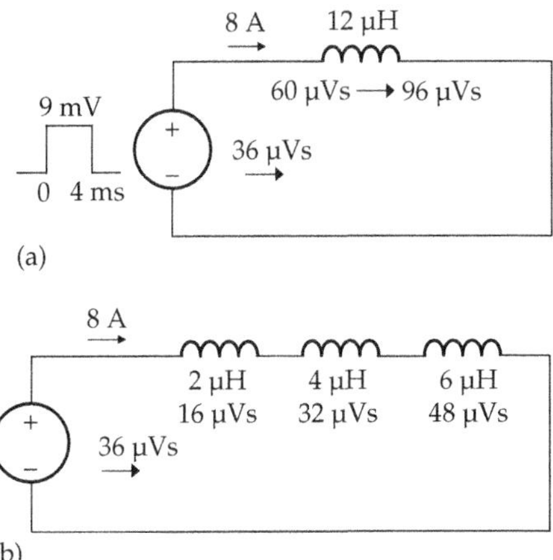

FIGURE 19.23
Figure for Example 19.5.

Example 19.5: Series Connection of Charged Inductors

Given three inductors, L_1 = 2 μH having a current of 12 A, L_2 = 4 μH having a current of 6 A, and L_3 = 6 μH having a current of 2 A, it is required to determine (a) the current through the inductors when connected in series and (b) the current and flux linkage for each inductor after the application of a 9 mV, 4 ms, pulse to the series combination of inductors.

Solution:

(a) The initial current and flux linkage for each inductor are shown in Figure 19.22a. When connected in series, L_{eqs} = 12 μH (Figure 19.22b). From equality of voltages across the series combination and across L_{eqs}, the flux linkage of L_{eqs} is the sum of the flux linkages on the individual inductors, which is 60 μVs. The current of L_{eqs} is 60/12 = 5 A. From equivalence, this must be the current through the series combination. It follows that the flux linkages of L_1, L_2, and L_3 are, respectively, 2 × 5 = 10 μVs, 4 × 5 = 20 μVs, and 6 × 5 = 30 μVs (Figure 19.22c). The total flux linkage is 60 μVs, as it should be.

(b) The 9 mV pulse of 4 ms duration establishes a flux linkage of 9 × 4 = 36 μVs of the same sign as that of L_{eqs}. The total flux linkage is now 96 μVs, and the current is 96/12 = 8 A (Figure 19.23a). From equivalence, the current through the series combination is also 8 A, which means that the flux linkages of L_1, L_2, and L_3 are, respectively, 2 × 8 = 16 μVs, 4 × 8 = 32 μVs, and 6 × 8 = 48 μVs (Figure 19.23b). The total flux linkage is 96 μVs, as it should be.

Simulation: The circuit for part (a) is entered as illustrated in Figure 19.24a. The following should be noted concerning this schematic: (i) In order to provide a path for the initial current in each inductor, normally closed switches are connected in parallel rather than in series. The switches are set to open at t = 1 ms, thereby connecting the conductors in series. (ii) PSpice does not allow loops of zero resistance, which can arise from loops of inductors, or voltage sources, or both. The 1 nΩ resistor in Figure 19.24a avoids such a loop when the switches open. When the switches are closed, a zero resistance is avoided in the other two meshes by the resistances of the closed switches.

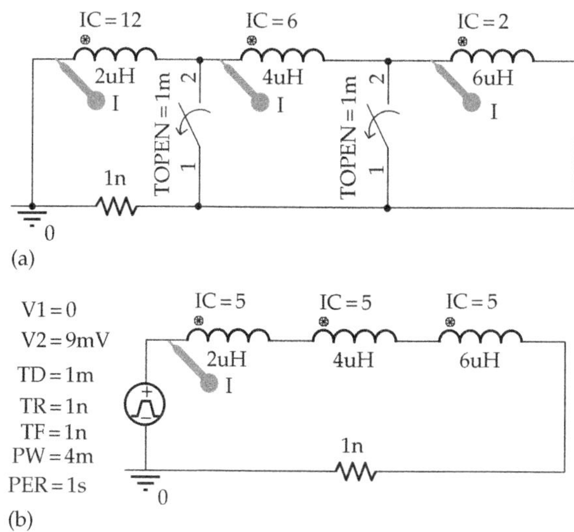

FIGURE 19.24
Figure for Example 19.5.

The 1 nΩ resistance is too small to significantly affect the results. (iii) The default resistance of a closed switch is 0.01 Ω, which is too large for the simulation involving pure inductances. This resistance is reduced to 1 nΩ in the Property Editor spreadsheet of the switch.

Current markers are used to indicate the currents. Time domain (Transient) analysis is chosen, 2 ms is entered for 'Run to time', 0 for 'Start saving data after', and 0.5 μs for 'Maximum step size'. After the simulation is run, the graph in Figure 19.25 is displayed showing the initial currents of 2, 6, and 12 A, and the final current of 5 A, as determined previously.

The circuit for part (b) is entered as illustrated in Figure 19.24b. The voltage source is VPULSE from the source library. The parameters of the source are entered as explained in Example 19.1. Time domain (Transient) analysis is chosen, 6m is entered for 'Run to time', 0 for 'Start saving data after', and 0.5u for 'Maximum step size'. After the simulation is run, the graph in Figure 19.26 is displayed. The current increases linearly during the pulse from 5 to 8 A.

Primal Exercise 19.7

Repeat Example 19.5 assuming a flux linkage of 12 μVs on L_1, 12 μVs on L_2, and 24 μVs on L_3.

Ans. (a) 4 A; (b) 7 A, $\lambda_1 = 14$ μVs, $\lambda_2 = 28$ μVs, $\lambda_3 = 42$ μVs.

---

*19.2.2 Inductors in Parallel

The behavior of inductors in parallel is examined in the next three examples, which are the duals of Examples 19.2 through 19.4 for capacitors.

Example 19.6: Charged Inductors in Parallel

A voltage pulse of 3 V amplitude and 4 s duration is applied to three inductors in parallel: $L_1 = 6$ H, $L_2 = 3$ H, and $L_3 = 2$ H (Figure 19.27a). It is required to determine the currents in the inductors (a) when the inductors are initially uncharged and (b) when L_1, L_2, and L_3 have initial flux linkages of 12, 6, and 2 Vs, respectively.

Solution:

(a) The voltage pulse establishes a flux linkage of +12 Vs, or Wb-turn, in each of the inductors and in L_{eqp}. The currents in the inductors are $i_1 = 12/6 = 2$ A,

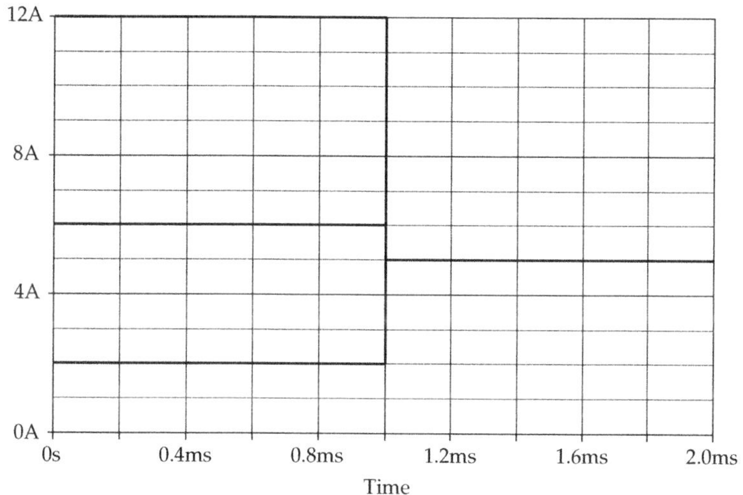

FIGURE 19.25
Figure for Example 19.5.

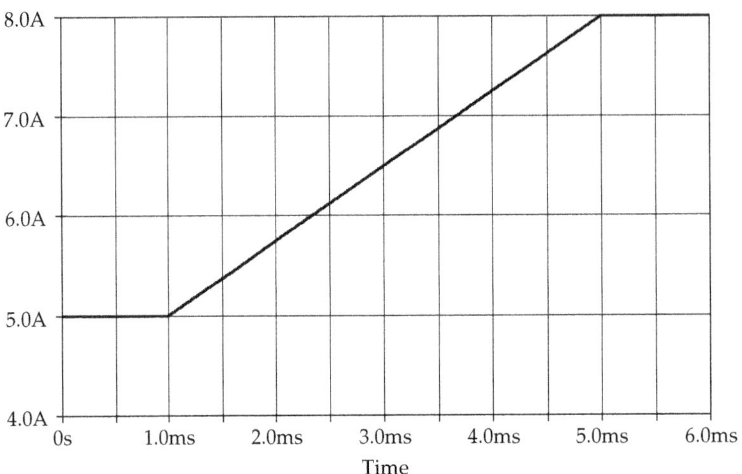

FIGURE 19.26
Figure for Example 19.5.

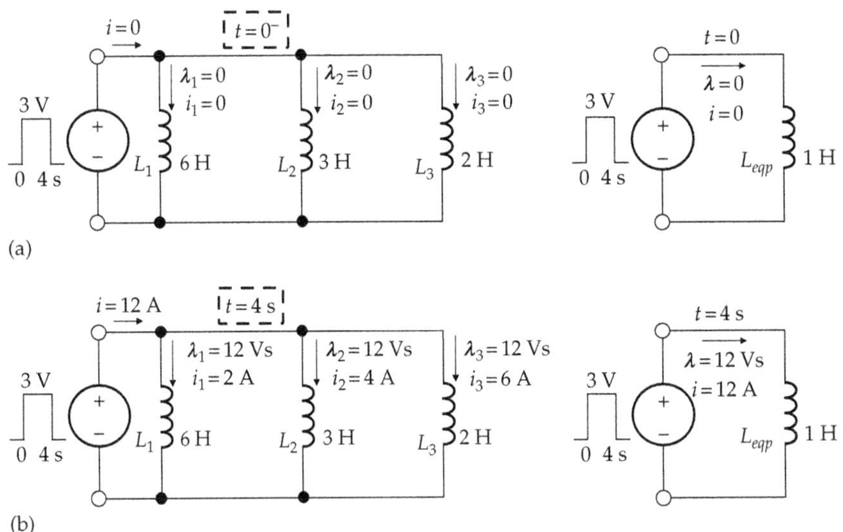

FIGURE 19.27
Figure for Example 19.6.

$i_2 = 12/3 = 4$ A, and $i_3 = 12/2 = 6$ A. The total current through the three inductors is 12 A (Figure 19.27b). L_{eqp} can be evaluated by considering two inductors at a time. The parallel inductance of the 6 and 3 H inductors is $(6 \times 3)/(6 + 3) = 2$ H. The parallel inductance of this and the 2 H inductor is 1 H. The 3 V voltage establishes 12 Vs of flux linkage at $t = 4$ s, the current through L_{eqp} being $12/1 = 12$ A, the same as that through the parallel combination (Figure 19.27b).

(b) When the voltage is applied to the three inductors, with initial flux linkages (Figure 19.28a), the flux linkage in L_{eqp} is clearly not the same as the individual flux linkage in any of the three inductors, since these individual flux linkages are all different. But because of equivalence, the current in L_{eqp} must be the same as that in the paralleled inductors. The currents in the individual inductors are $i_{10} = 12/6 = 2$ A, $i_{20} = 6/3 = 2$ A, and $i_{30} = 2/2 = 1$ A. The total current is 5 A, which must be the current in L_{eqp}. It follows that the flux linkage in L_{eqp} is $1 \times 5 = 5$ Vs (Figure 19.28a).

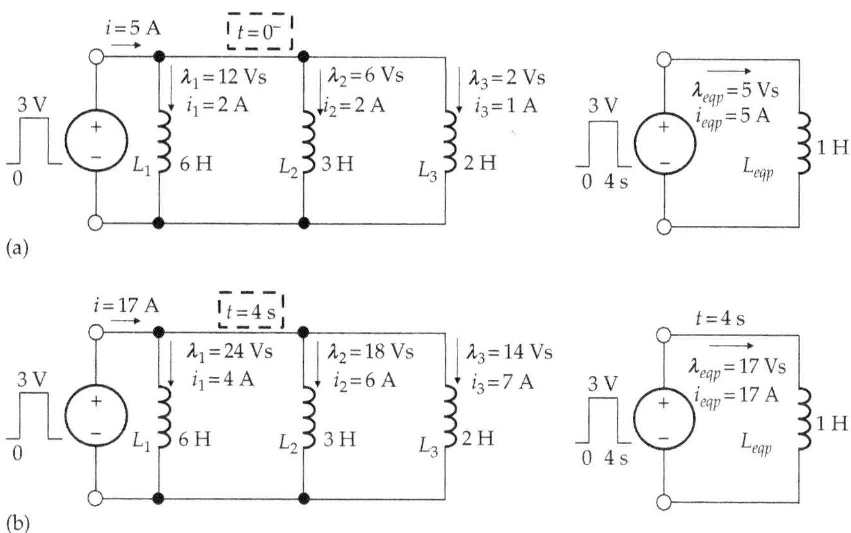

FIGURE 19.28
Figure for Example 19.6.

TABLE 19.2
Inductor Flux Linkages and Currents

	L_1		L_2		L_3		L_{eqp}	
	λ_1, Vs	**i_1, A**	**λ_2, Vs**	**i_2, A**	**λ_3, Vs**	**i_3, A**	**λ_{eqp}, Vs**	**i_{eqp}, A**
No initial storage	12	2	12	4	12	6	12	12
Initial conditions	12	2	6	2	2	1	5	5
Final (t = 4 s)	24	4	18	6	14	7	17	17

The voltage pulse of 3 V amplitude and 4 s duration adds a flux linkage of 12 Vs to the flux linkages present in L_{eqp} and in each of the three paralleled inductors. λ_{eqp} becomes 17 Vs and i_{eqp} = 17 A (Figure 19.28b). The application of superposition to the individual inductors is illustrated in Table 19.2. The "No initial storage" row refers to the values found previously (Figure 19.27b). The row that follows lists the initial conditions. The last row indicates the flux linkages and currents at t = 4 s, after the voltage pulse is over. The values in this row are the sums of the corresponding values in the two preceding rows. The initial and final values are also shown in Figure 19.28. The final flux linkages are $\lambda_{1f} = 12 + 12 = 24$ Vs, $\lambda_{2f} = 6 + 12 = 18$ Vs, $\lambda_{3f} = 2 + 12 = 14$ Vs, and $\lambda_{eqp} = 2 + 12 = 17$ Vs (Figure 19.28b). The currents in the three inductors are $i_{1f} = 24/6 = 4$ A, $i_{2f} = 18/3 = 6$ A, and $i_{3f} = 14/2 = 7$ A, the total being 17 A, as for L_{eqp}. Note from Table 19.2 that superposition can be applied to flux linkages as well as currents.

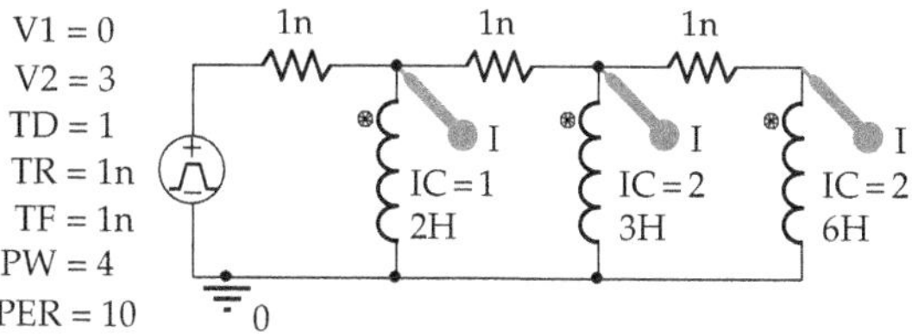

FIGURE 19.29
Figure for Example 19.6.

Simulation: The circuit is entered as in Figure 19.29. 1 nΩ resistors are added in order to avoid short-circuit loops, and the voltage source is entered as explained in the preceding example. The initial currents in the inductors are added as described previously in the Property Editor spreadsheet. In the simulation profile, Time domain (Transient) analysis is chosen, 6 is entered for 'Run to time', 0 for 'Start saving data after', and 0.5 m for 'Maximum step size'. After the simulation is run,

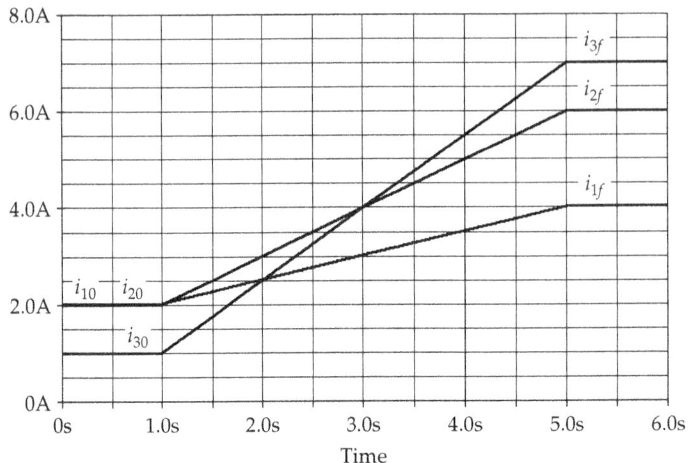

FIGURE 19.30
Figure for Example 19.6.

the graph displayed in Figure 19.30 shows the initial and final currents in the three inductors. The currents increase linearly from their initial values to their final values during the application of the 4 s pulse.

Primal Exercise 19.8

A 1 V pulse of 2 μs duration is applied to three inductors of 0.2, 0.8, and 0.04 μH connected in parallel. The inductors were initially uncharged. Determine, at the end of the pulse, (a) the flux linkage of each inductor and the current through it and (b) the inductance of the equivalent inductor, its flux linkage, and current.

Ans. (a) 2 μVs; 10, 2.5, and 50 A, respectively; (b) 0.032 μH, 2 μVs, 62.5 A.

Example 19.7: Steady Current Applied to Inductors in Parallel

Let a steady current of 15 A be applied to the parallel-connected inductors in Figure 19.31, by opening the switch at $t = 0$, where $L_1 = 6$ μH, $L_2 = 3$ μH, and $L_3 = 2$ μH. It is required to determine the flux linkages and currents of the inductors: (a) with zero initial energy storage and (b) with the initial currents on the inductors of $i_{10} = 2$ A, $i_{20} = 2$ A, and $i_{30} = 1$ A.

Solution:

(a) The inductance of 6 μH in parallel with 3 μH is 2 μH, and the parallel inductance of this and the 2 μH capacitor is $L_{eqp} = 1$ μH. When 15 A are applied to the parallel combination and to L_{eqp}, a voltage impulse

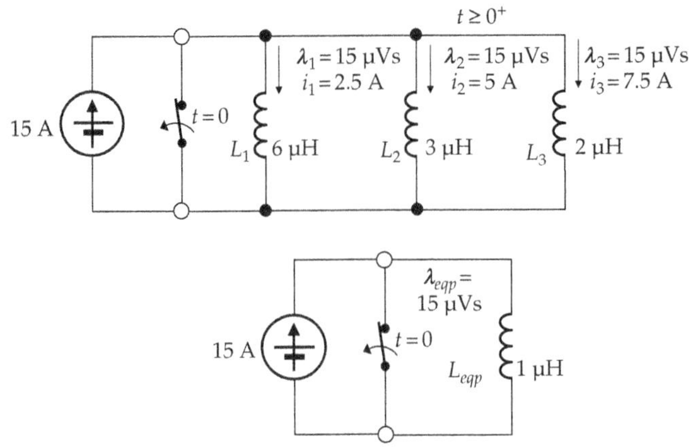

FIGURE 19.31
Figure for Example 19.7.

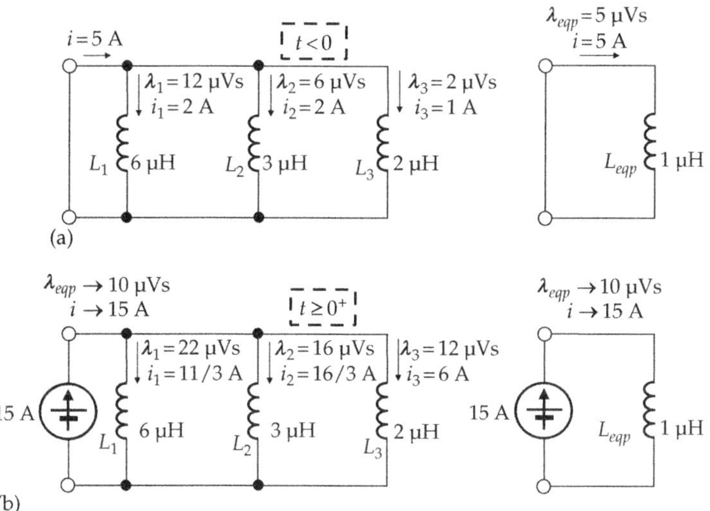

FIGURE 19.32
Figure for Example 19.7.

occurs that instantaneously establishes a flux linkage of 15 μVs in L_{eqp} (Figure 19.31). From circuit equivalence, this same voltage impulse is applied to the parallel combination of inductors. Since these inductors have zero initial energy storage, the flux linkage on each of the three inductors will be the same as on L_{eqp}. It follows that $i_1 = 15/6 = 2.5$ A, $i_2 = 15/3 = 5$ A, and $i_3 = 15/2 = 7.5$ A. The total current is 15 A, as it should be.

(b) When the inductors are initially charged, the total current is 5 A, which, by circuit equivalence, is also the current in L_{eqp}. The flux linkage of L_{eqp} will be (5 A) × (1 μH) = 5 μVs (Figure 19.32a). The applied 15 A will bring the flux linkage to 15 μVs, which means an increase of flux linkage of 10 μVs from the applied source. By circuit equivalence, this same integral of voltage is applied to the three inductors, which increases the flux linkage of each inductor by 10 μVs. Hence, the final flux linkages and currents will be as follows (Figure 19.32b):

$$\lambda_{1f} = 2 \times 6 + 10 = 22\ \mu\text{Vs} \quad \text{and} \quad i_{1f} = 22/6 = 11/3\ \text{A}$$

$$\lambda_{2f} = 2 \times 3 + 10 = 16\ \mu\text{Vs} \quad \text{and} \quad i_{2f} = 16/3\ \text{A}$$

$$\lambda_{3f} = 1 \times 2 + 10 = 12\ \mu\text{Vs} \quad \text{and} \quad i_{3f} = 12/2 = 6\ \text{A}$$

It is seen that $i_{1f} + i_{2f} + i_{3f} = (11/3) + (16/3) + 6 = 15$ A as it should be.

Simulation: The circuit is entered as illustrated in Figure 19.33. The opening time of the normally closed switch is set at 1 ms and its resistance when closed (RCLOSED)

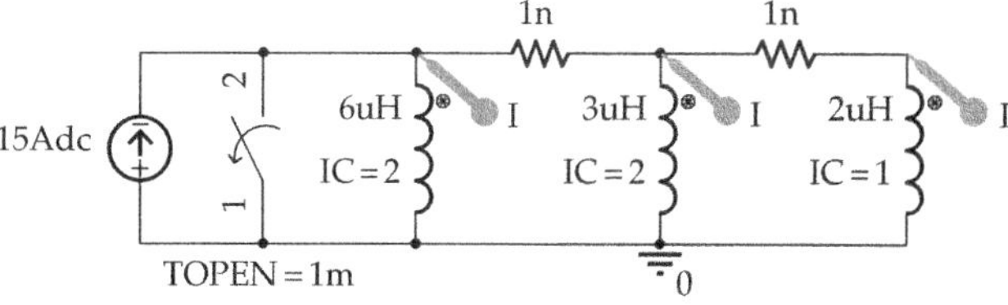

FIGURE 19.33
Figure for Example 19.7.

is set to 1 nΩ. Time domain (Transient) analysis is chosen, 2 ms is entered for 'Run to time', 0 for 'Start saving data after', and 0.5 μs for 'Maximum step size'. After the simulation is run, the graph in Figure 19.34 is displayed showing the initial and final currents for each inductor, using current markers. The final currents are read using the cursor as 3.6667 A, 5.3333 A, and 5.9971 A corresponding to $i_{1f} = 11/3$ A, $i_{2f} = 16/3$ A, and $i_{3f} = 6$ A.

Problem-Solving Tip

- Equivalent series or parallel inductors can be conveniently used to facilitate the solution of problems involving inductors.

The preceding two examples illustrate the following concept:

Concept: *When an excitation is applied to inductors in series or in parallel and having initial stored energy, the flux linkage due to the applied excitation adds algebraically to the flux linkage initially stored in each inductor.*

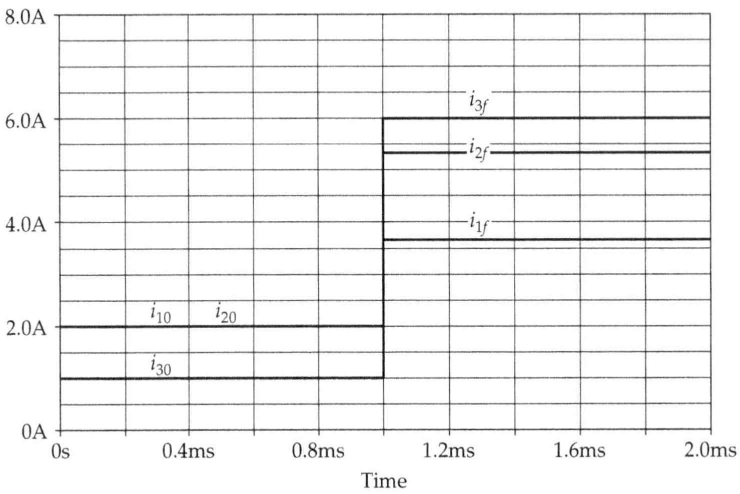

FIGURE 19.34
Figure for Example 19.7.

Primal Exercise 19.9

Repeat Example 19.7 assuming the initial flux linkages of the inductors are λ_{10} = 18 μVs, λ_{20} = 12 μVs, and λ_{20} = 10 μVs.

Ans. (a) As in Example 19.7; (b) λ_{1f} = 21 μVs, λ_{2f} = 15 μVs, λ_{3f} = 13 μVs.

Example 19.8: Redistribution of Currents in Paralleled Inductors

Given three inductors having initial flux linkages and currents as shown in Figure 19.35 and are paralleled with a switch that is opened at $t = 0$. It is required to determine the final currents and flux linkages of the three inductors. This example is the dual of Example 19.4. Because there are two voltage rises and one voltage drop in Figure 19.14, two currents are shown entering the upper node in Figure 19.35 and one current leaving the node.

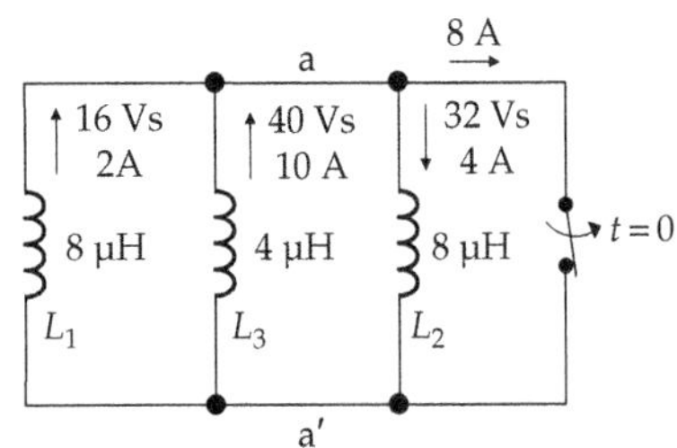

FIGURE 19.35
Figure for Example 19.8.

Solution:

L_{eqp} can be determined by considering the inductors two at time. Thus, the two 8 μH inductors in parallel give an inductance of 4 μH. This, in parallel with 4 μH, gives L_{eqp} = 2 μH. The current in L_{eqp} is the current through the switch, in which 2 + 10 − 4 = 8 A. It follows that the flux linkage of L_{eqp} is (8 A) × (2 μH) = 16 Vs (Figure 19.36a). When the switch is opened, L_{eqp} completely discharges, which is equivalent to applying 16 Vs in the opposite direction (Figure 19.36b) by means of a voltage impulse of strength 16 Vs, the polarity of the voltage impulse being node 'a' positive with respect to node a′. By circuit equivalence, the same voltage impulse appears between the corresponding nodes in the original circuit. The resulting flux linkage of 16 Vs opposes the flux linkages in L_1 and L_3 and adds to the flux linkage in L_2. It follows that $\lambda_1 = 0$, λ_3 = 24 Vs, and λ_2 = 48 Vs, so that $i_1 = 0$, i_2 = 6 A, and i_3 = 6 A (Figure 19.37). Note that KCL is satisfied.

Flux linkage is conserved around the three closed paths formed by any two inductors. The flux linkages

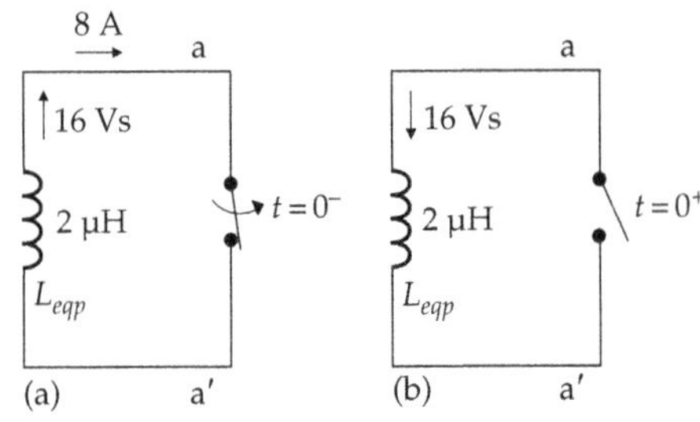

FIGURE 19.36
Figure for Example 19.8.

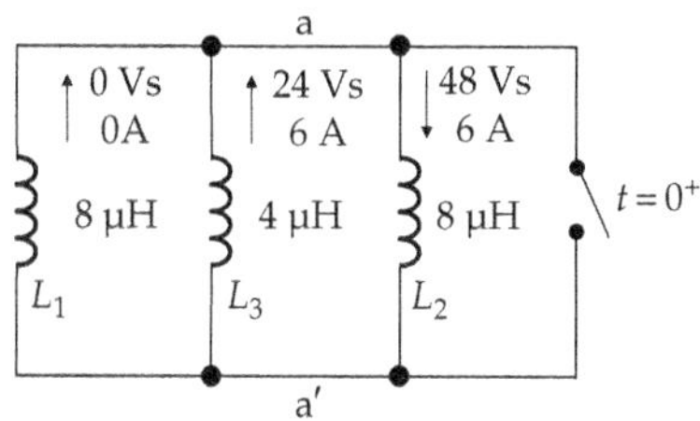

FIGURE 19.37
Figure for Example 19.8.

in the clockwise direction around the closed paths are as follows:

$L_1L_3: 16-40=-24$ Vs at $t=0^-$ and $0-24=-24$ Vs at $t=0^+$.

$L_2L_3: 40+32=72$ Vs at $t=0^-$ and $24+48=72$ Vs at $t=0^+$.

$L_1L_2: 16+32=48$ Vs at $t=0^-$ and $0+48=48$ Vs at $t=0^+$.

Note that in going clockwise around a closed path, flux linkage in the same direction is taken as positive, whereas flux linkage in the opposite direction is taken as negative.

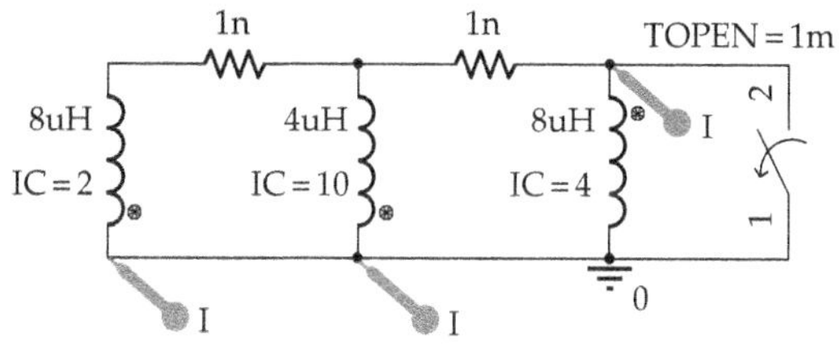

FIGURE 19.38
Figure for Example 19.8.

Simulation: The circuit is entered as illustrated in Figure 19.38. The orientation of the inductors and the placement of the current markers are such that positive current values are indicated. The opening time of the normally closed switch is set at 1 ms and its resistance when closed (RCLOSED) is set to 1 nΩ. Time domain (Transient) analysis is chosen, 2 ms is entered for 'Run to time', 0 for 'Start saving data after', and 0.5u for 'Maximum step size'. After the simulation is run, the graph in Figure 19.39 is displayed showing the initial and final currents for each inductor.

Example 19.8 illustrates the following concept concerning conservation of flux linkage:

Concept: *Flux linkage is conserved around any closed path in a circuit at the instant of a sudden change as long as no voltage impulses are applied from an external source in the closed path at the instant of the sudden change.*

Conservation of flux linkage is in accordance with KVL.

Primal Exercise 19.10

Repeat Example 19.8 using initial values of 20, 24, and 30 μVs on L_1, L_2, and L_3, respectively.

Ans. $\lambda_{1f}=6$ μVs, $i_{1f}=3/4$ A, $\lambda_{2f}=38$ μVs, $i_{2f}=19/4$ A, $\lambda_{3f}=16$ μVs, $i_{3f}=4$ A.

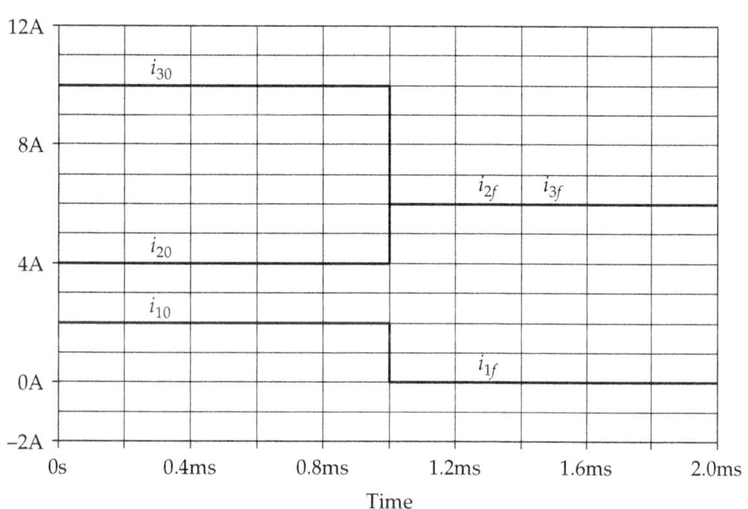

FIGURE 19.39
Figure for Example 19.8.

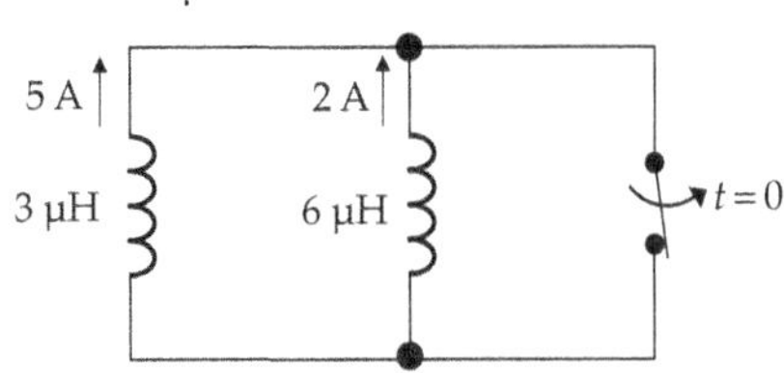

FIGURE 19.40
Figure for Primal Exercise 19.11.

Primal Exercise 19.11

The switch in Figure 19.40 is opened at $t = 0$, with the initial currents in the inductors as shown. Determine the current after the switch is opened, considering the two inductors to be (a) in series or (b) in parallel.

Ans. 1/3 A clockwise.

19.3 Switched Circuits

By switched circuits are meant circuits in which the movement of one or more switches changes the circuit configuration, as by adding or removing circuit elements or by modifying the applied excitation in some specified manner. It is to be expected that a switching operation, being another form of a sudden change in a circuit, is subject to the same basic considerations discussed in this chapter and the preceding chapter, such as the following:

1. A capacitor voltage does not change at the instant of switching unless forced to by a current impulse in order to satisfy KVL. Similarly, an inductor current does not change at the instant of switching unless forced to by a voltage impulse in order to satisfy KCL.
2. KCL and KVL must be satisfied at all times, before switching, just after switching, and in the steady state.

These basic concepts allow the determination of the initial conditions, just after switching, that are required for deriving the complete solution following the switching operation. This is illustrated by the following examples of switching in first-order and second-order circuits. Examples 19.9 and 19.10 illustrate impulsive readjustment in the presence of a resistor, whereas Examples 19.11 and 19.12 illustrate the "trapping" of stored energy in a circuit.

Adding an inductor to the capacitive circuits considered in this chapter, or a capacitor to the inductor circuits, does not introduce any important concepts beyond those of Chapter 12. The equivalent series or parallel capacitor or inductor will have an initial stored energy, after the initial switching operation, as in the circuits of Chapter 12 and can be analyzed in the same manner discussed in that chapter. More general switched circuits can be analyzed using the Laplace transform (Chapter 22).

Example 19.9: Switched Parallel *RC* Circuit

In the circuit in Figure 19.41, capacitor C_1 is initially charged to $V_{10} = 10$ V and C_2 is uncharged. The switch is closed at $t = 0$. It is required to determine the variation with time of v, the voltage across the capacitors, for $t \geq 0^+$.

Solution:

Just after the switch is closed, the voltages across C_1 and C_2 are forced to change instantaneously from their initial values to a common voltage, by a current impulse, as in the case of the two capacitors in Figure 19.1. At $t = 0^-$, v and i_R are both zero. At $t = 0^+$, v and hence i_R jump by a finite amount due to redistribution of charge between the capacitors. As argued in connection with Equation 18.22, the time integral of i_R, which is the charge leaked through the resistor, is zero between $t = 0^-$ and at $t = 0^+$, because v and hence i_R are finite during this interval. It follows that *charge is conserved at the instant of switching*, as it flows from C_1 to C_2.

The initial charge on C_1 is $10 \times 3 = 30$ C, and $C_{eqp} = 5$ F. v at $t = 0^+$ is $30/5 = 6$ V and $i_R = 6$ A. As $t \to \infty$, both capacitors are completely discharged and v_F and i_{RF} are both zero. $\tau = RC = 5$ s. It follows from Equation 11.57 that

$$v(t) = 6e^{-0.2t}\,\text{V}, \quad t \geq 0^+ \tag{19.3}$$

It is of interest to examine the capacitor currents. To take into consideration the switching interval, v can be expressed for $t \geq 0^-$ as

$$v(t) = 6e^{-0.2t}u(t)\text{V}, \tag{19.4}$$

where $v = 0$ for $t \leq 0^-$ and is given by Equation 19.3 for $t \geq 0^+$. The current i_2 is

$$i_2(t) = C_2\frac{dv}{dt} = 2\left[6e^{-0.2t}\delta(t) - 1.2e^{-0.2t}u(t)\right], \quad t \geq 0^- \tag{19.5}$$

$$= 12\delta(t) - 2.4e^{-0.2t}u(t)\,\text{A}, \quad t \geq 0^- \tag{19.6}$$

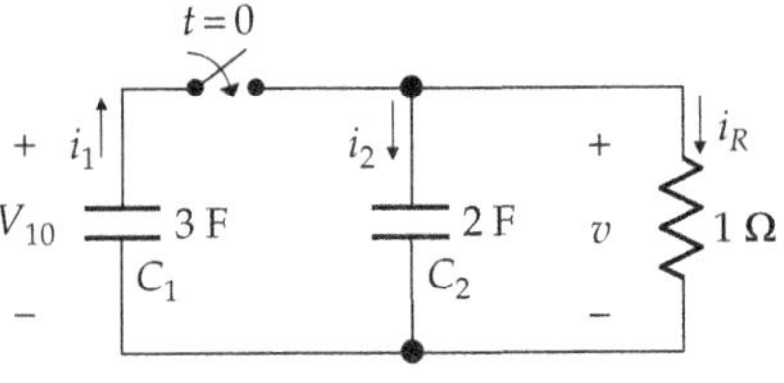

FIGURE 19.41
Figure for Example 19.9.

where the first term between brackets in Equation 19.5 evaluates to $6\delta(t)$ at $t = 0$. Similarly, the voltage across C_1 having the same polarity as v can be expressed as

$$v_1(t) = 10 - \left(10 - 6e^{-0.2t}\right)u(t)\ \text{V}, \tag{19.7}$$

so that $v_1 = 10$ V for $t \le 0^-$, and v_1 is given by Equation 19.3 for $t \ge 0^+$. The current i_1 is

$$i_1(t) = -C_1 \frac{dv_1}{dt}$$

$$= -3\left[-\left(10 - 6e^{-0.2t}\right)\delta(t) - 1.2e^{-0.2t}u(t)\right],\quad t \ge 0^- \tag{19.8}$$

$$= 12\delta(t) + 3.6e^{-0.2t}u(t)\ \text{A},\quad t \ge 0^- \tag{19.9}$$

It is seen that a current impulse flows from C_1 to C_2, transferring a charge of 12 C, so that the charges at $t = 0^+$ are $30 - 12 = 18$ C on C_1 and 12 C on C_2, the resulting voltage being 6 V on the two capacitors. KCL is $i_1 - i_2 = i_R$, so that the current impulse cancels out from i_R, leaving $i_R = v/R$, for $t \ge 0^+$, where v is given by Equation 19.3.

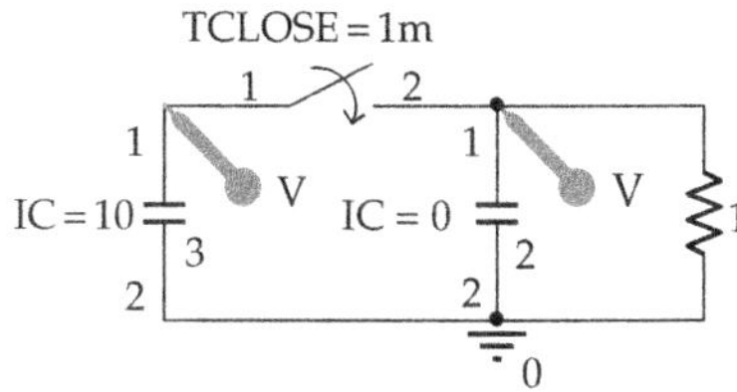

FIGURE 19.42
Figure for Example 19.9.

Simulation: The circuit is entered as in Figure 19.42. A normally open switch is set to close at 1 ms so as to show the initial voltages across the capacitors, which are in turn entered as 10 V for C_1 and 0 for C_2. 'Time domain (Transient)' analysis is chosen, 15 ms is entered for 'Run to time', 0 for 'Start saving data after', and 1 μs for 'Maximum step size'. After the simulation is run, the graph displayed in Figure 19.43 shows the initial values of voltages across C_1 and C_2 as 10 V and 0, respectively. When the switch is closed, both voltages change to 6 V and decay exponentially thereafter with a time constant of 5 ms. The tangent to the exponential at $t = 1$ ms is entered as −1200 * Time + 7.2.

*Example 19.10: Switched Series *RL* Circuit

The circuit in Figure 19.44 is the dual of that in Figure 19.41. Inductor L_1 has an initial current of $I_{10} = 10$ A and L_2 is uncharged. The switch is closed at $t = 0$. It is required to determine the variation with time of i, the current through the inductors for $t \ge 0^+$.

Solution:

Just after the switch is closed, the currents through L_1 and L_2 are forced to change instantaneously from their initial values to a common value, by a voltage impulse, as in the case of the two inductors in Figure 19.20. At $t = 0^-$, i and v_R are both zero. At $t = 0^+$, i and hence v_R jump by a finite amount. In the interval $0^- \le t \le 0^+$, when a voltage impulse appears across the inductors, v_R is finite and will not affect this impulse (Equation 18.24). The reduction in flux linkage of L_1 will therefore be equal to the increase in flux linkage of L_2, so that flux linkage is conserved at the instant of switching.

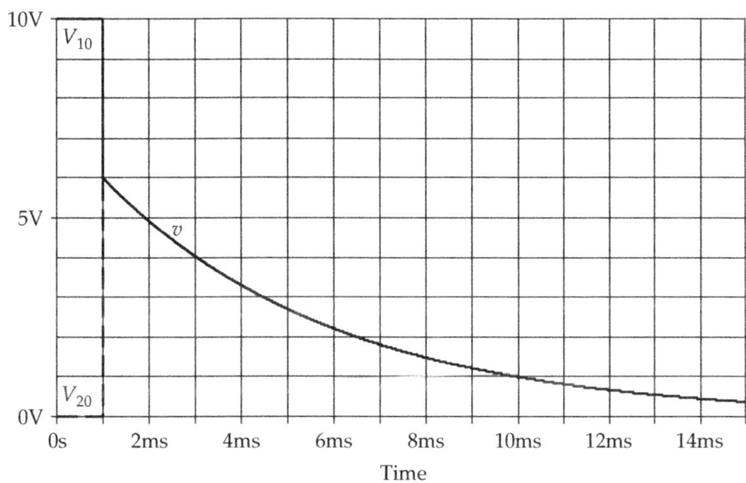

FIGURE 19.43
Figure for Example 19.9.

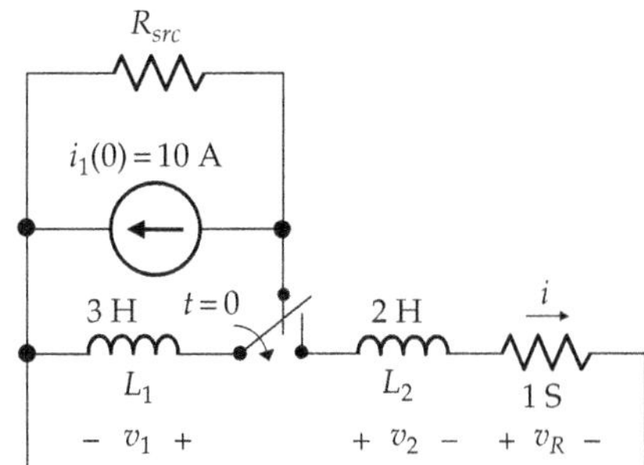

FIGURE 19.44
Figure for Example 19.10.

The initial flux linkage of L_1 is 10 × 3 = 30 Vs, and L_{eqs} = 5 H. i at $t = 0^+$ is 30/5 = 6 A and v_R = 6 V. As $t \to \infty$, both inductors are completely discharged and i_f and v_{Rf} are both zero. It follows from Equation 11.57, that

$$i(t) = 6e^{-0.2t} \text{ A}, \quad t \geq 0^+ \tag{19.10}$$

It is of interest to examine the inductor voltages. To take into consideration the switching interval, i can be expressed as

$$i_2(t) = 6e^{-0.2t}u(t) \text{ V}, \tag{19.11}$$

where $i_2 = 0$ for $t \leq 0^-$, and i_2 is given by Equation 19.10 for $t \geq 0^+$. The voltage v_2 is

$$v_2(t) = L_2\frac{di_2}{dt} = 2\left[6e^{-0.2t}\delta(t) - 1.2e^{-0.2t}u(t)\right], \quad t \geq 0^- \tag{19.12}$$

$$= 12\delta(t) - 2.4e^{-0.2t}u(t) \text{ V}, \tag{19.13}$$

where the first term between brackets in Equation 19.12 evaluates to $6\delta(t)$. Similarly, the current through L_1 having the same polarity as i can be expressed as

$$i_1(t) = 10 - \left(10 - 6e^{-0.2t}\right)u(t) \text{ A}, \tag{19.14}$$

so that i_1 = 10 A for $t \leq 0^-$, and i_1 is given by Equation 19.10 for $t \geq 0^+$. v_1 is

$$v_1(t) = -L_1\frac{di_1}{dt}$$

$$= -3\left[-\left(10 - 6e^{-0.2t}\right)\delta(t) - 1.2e^{-0.2t}u(t)\right], \quad t \geq 0^- \tag{19.15}$$

$$= 12\delta(t) + 3.6e^{-0.2t}u(t) \text{ V} \tag{19.16}$$

It is seen that a voltage impulse of 12 Vs appears across L_1 and L_2, reducing the flux linkage in L_1 at $t = 0^+$ to 30 – 12 = 18 Vs and establishing a flux linkage of 12 Vs in L_2. The resulting current is 6 A in both inductors. KVL is $v_1 - v_2 = v_R$, so that the voltage impulse cancels out from v_R, leaving $v_R = i/G$, for $t \geq 0^+$, where i is given by Equation 19.10.

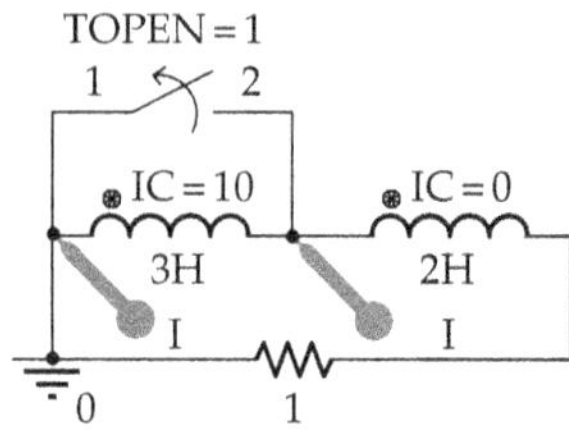

FIGURE 19.45
Figure for Example 19.10.

Simulation: The circuit is entered as in Figure 19.45. A normally closed switch is set to open at 1 s so as to connect the two inductors in series. The initial currents in the inductors are entered as 10 A for L_1 and 0 for L_2. Time domain (Transient) analysis is chosen, 15 s is entered for 'Run to time', 0 for 'Start saving data after', and 1 ms for 'Maximum step size'. After the simulation is run, the graph displayed is the same as that in Figure 19.43 but with the vertical axis representing current instead of voltage. When the switch is opened, the currents in the two inductors change to 6 A and decay exponentially thereafter with a time constant of 5 s.

The following concept is illustrated by Examples 19.9 and 19.10:

Concept: *A finite current does not add to, or subtract from, charge at a node during the infinitesimal interval just before and just after a switching operation. Similarly, a finite voltage does not add to, or subtract from, flux linkage in a closed path during the interval just before and just after a switching operation.*

Thus, charge is conserved at the nodes in Figure 19.41 at the instant of switching. Similarly, flux linkage is conserved in the mesh in Figure 19.44 at the instant of switching.

Example 19.11: Switched Series *RC* Circuit

Given two capacitors C_1 = 3 F and C_2 = 2 F charged to 10 V each and connected at t = 0 in series with a 2 Ω resistor (Figure 19.46a), it is required to determine i as a function of time and the final values of v_1 and v_2.

Solution:

C_{eqs} = (3 × 2)/(3 + 2) = 1.2 F. The initial voltage of C_{eqs} is the sum of the initial voltages across the two capacitors, that is, 20 V. The initial charge on C_{eqs} is q = 1.2 × 20 = 24 C.

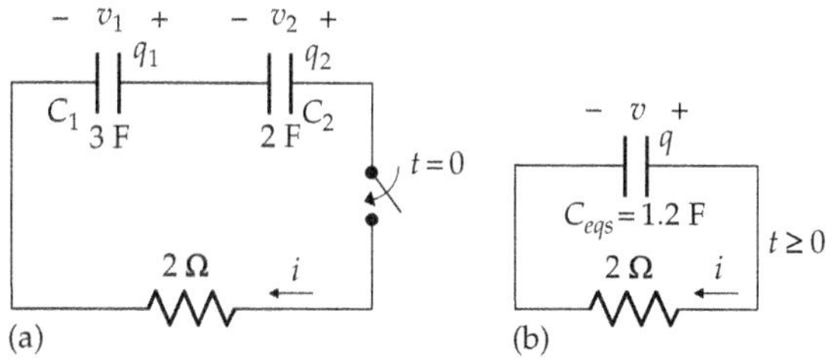

FIGURE 19.46
Figure for Example 19.11.

The circuit for $t \geq 0$ is shown in Figure 19.46b. As $t \to \infty$, C_{eqs} will completely discharge, and $i = 0$. The initial value of i is 10 A, and the time constant is $2 \times 1.2 = 2.4$ s. It follows that i is given by

$$i(t) = 10e^{-t/2.4}\text{ A}, \quad t \geq 0, t \text{ in s} \qquad (19.17)$$

The charge $q = 24$ C is moved by i through the circuit in the clockwise direction. The initial charge on C_1 is 30 C. When 24 C are moved clockwise, the residual charge on C_1 is 6 C, and the final value of v_1 is $v_{1f} = 6/3 = 2$ V. The initial charge on C_2 is 20 C. When 24 C are moved clockwise, the residual charge on C_2 is −4 C and the final value of v_2 is $v_{2f} = -4/2 = -2$ V. The voltage across the resistor is 2 − 2 = 0, since i goes to zero. Some residual charge therefore remains trapped in the capacitors.

With the initial and final values of v_1 and v_2 known, it follows that

$$v_1(t) = 2 + 8e^{-t/2.4}\text{V}, \quad \text{and}$$

$$v_2(t) = -2 + 12e^{-t/2.4}\text{V} \quad t \geq 0, t \text{ in s} \qquad (19.18)$$

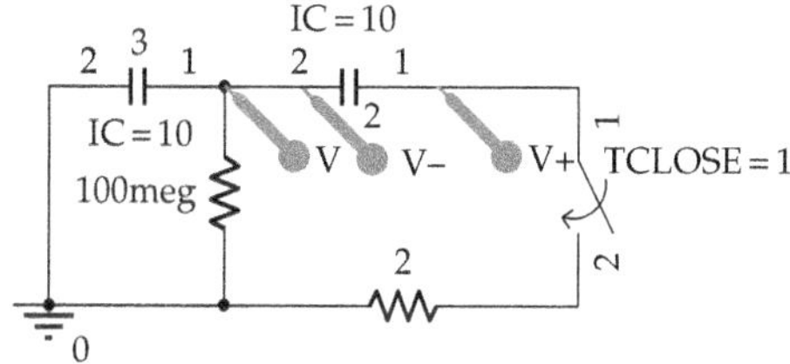

FIGURE 19.47
Figure for Example 19.11.

Simulation: The circuit is entered as in Figure 19.47. A normally open switch is set to close at 1 s so as to display the initial voltages across the capacitors, which are in turn entered as 10 V for C_1 and C_2. A 100 meg resistor is added to avoid a floating node between the capacitors. Time domain (Transient) analysis is chosen, 10 s is entered for 'Run to time', 0 for 'Start saving data after', and 1 ms for 'Maximum step size'. After the simulation is run, the graph displayed in Figure 19.48 is showing the initial values of voltages across C_1 and C_2. When the switch is closed, both voltages decrease exponentially with a time constant of 2.4 s, the final voltages being 2 for v_1 and −2 for v_2.

*Example 19.12: Switched Parallel *RL* Circuit

The circuit in Figure 19.49a is the dual of that in Figure 19.46a. Both inductors have an initial current of 10 A. The switch is opened at $t = 0$. It is required to determine v as a function of time and the final values of i_1 and i_2.

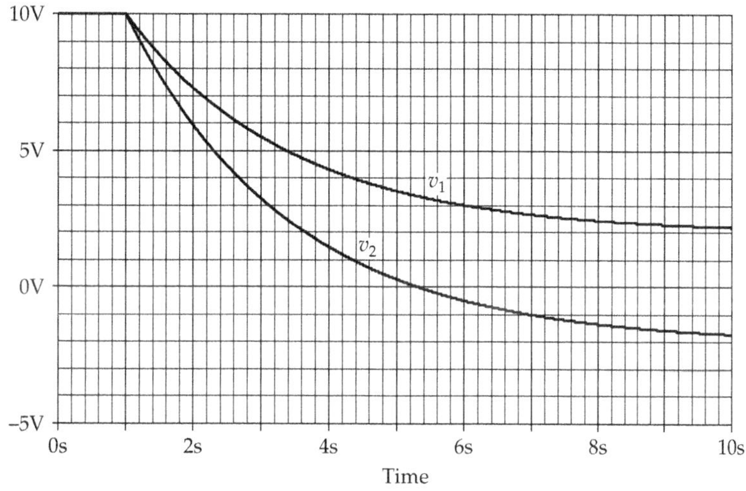

FIGURE 19.48
Figure for Example 19.11.

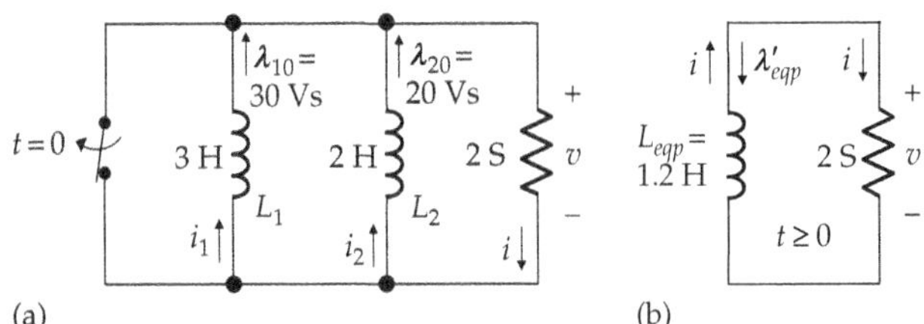

FIGURE 19.49
Figure for Example 19.12.

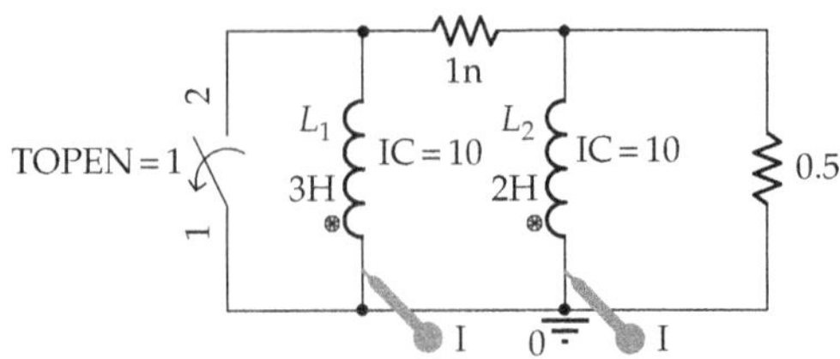

FIGURE 19.51
Figure for Example 19.12.

Solution:

$L_{eqp} = (3 \times 2)/(3 + 2) = 1.2$ H (Figure 19.49b). The initial current of L_{eqp} is the sum of the initial currents of the two inductors, that is, 20 A. The initial flux linkage of L_{eqp} is $\lambda_{eqp} = 1.2 \times 20 = 24$ Vs. As $t \to \infty$, L_{eqp} will completely discharge, and $v = 0$. Initial value of v is $i/G = 10$ V, and the time constant is $2 \times 1.2 = 2.4$ s. It follows that v is given by

$$v(t) = 10e^{-t/2.4}\ \text{V}, \quad t \geq 0^+ \qquad (19.19)$$

The initial flux linkage λ_{eqp} of 24 Vs is reduced to zero as $t \to \infty$. This can be considered to be due to an opposing flux linkage $\lambda'_{eqp} = -\lambda_{eqp}$ that arises from the time integral of v (Figure 19.49b). By equivalence, this λ'_{eqp} also appears across L_1 and L_2 and subtracts from the initial flux linkages. The final flux linkages are $\lambda_{1f} = 6$ Vs in L_1, directed upward, and $\lambda_{2t} = 4$ Vs in L_2, directed downward. A current of 2 A therefore circulates in the two inductors (Figure 19.50), representing a residual flux linkage that remains trapped in the inductors.

With the initial and final values of i_1 and i_2 known, it follows that

$$i_1(t) = 2 + 8e^{-t/2.4}\ \text{A}, \quad \text{and}$$

$$i_2(t) = -2 + 12e^{-t/2.4}\ \text{V}, \quad t \geq 0 \qquad (19.20)$$

Simulation: The circuit is entered as in Figure 19.51. A normally closed switch is set to open at 1 s so as to display the initial currents through the inductors, which are in turn entered as 10 A for L_1 and L_2. A 1 nΩ resistor is added to avoid a zero resistance inductor loop. Time domain (Transient) analysis is chosen, 10 s is entered for 'Run to time', 0 for 'Start saving data after', and 1 ms for 'Maximum step size'. After the simulation is run, the graph displayed is the same as that in Figure 19.48 but with the vertical axis representing current instead of voltage. When the switch is opened, the currents in the two inductors decrease exponentially with a time constant of 2.4 s, the final currents being 2 for i_1 and −2 for i_2,

The following concept is illustrated by Examples 19.11 and 19.12:

Concept: *Stored energy can be trapped in capacitors or in inductors in the final, steady state when initial charges on capacitors, or initial flux linkages in inductors, are not completely neutralized, even in the presence of resistors, while KCL, KVL, and Ohm's law are satisfied.*

Primal Exercise 19.12

The switch in Figure 19.52 is closed at $t = 0$, with the inductor initially uncharged. Determine $v_O(0^+)$, $v_L(0^+)$, and $i_L(t)$, $t \geq 0^+$.

Ans. $v_O(0^+) = 0$, $v_L(0^+) = 4$ V, $i_L(t) = \left(1 - e^{-4t}\right)$, $t \geq 0^+$ s.

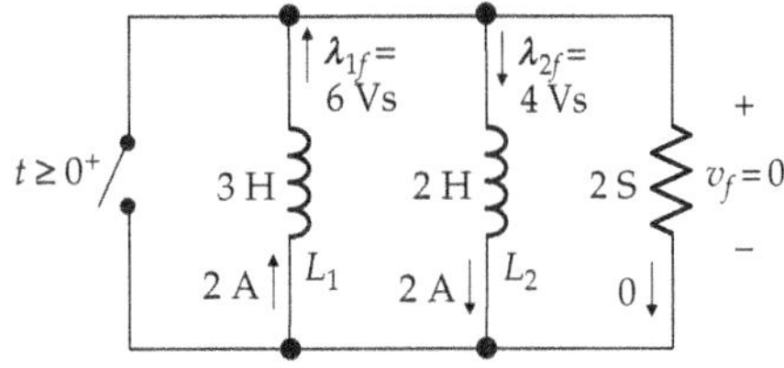

FIGURE 19.50
Figure for Example 19.12.

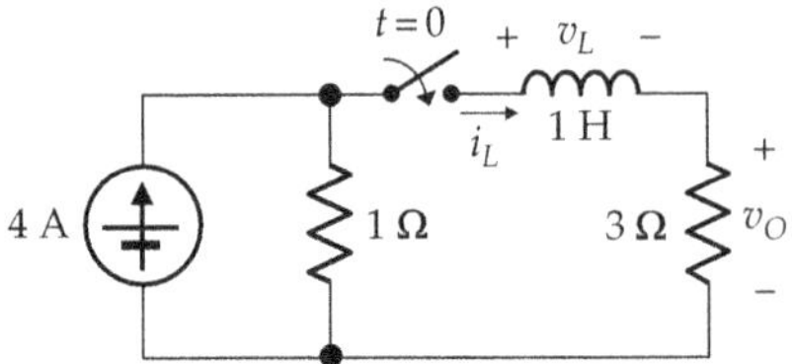

FIGURE 19.52
Figure for Primal Exercise 19.12.

Learning Checklist: What Should Be Learned from This Chapter

- When capacitors in series or in parallel have initial charges, the equivalent series or parallel capacitance is the same as in the absence of initial charges.
- When inductors in series or in parallel have initial currents, the equivalent series and parallel inductances are the same as when there are no initial currents.
- When capacitors having different voltages are paralleled, charge is instantaneously redistributed by current impulses, so as to have a common voltage across the capacitors. This voltage can be readily determined from conservation of charge and the equivalent parallel capacitance.
- When inductors having different currents are connected in series, flux linkage is instantaneously redistributed by voltage impulses, so as to have the same current through the inductors. This current can be readily determined from conservation of flux linkage and the equivalent series inductance.
- When an excitation is applied to capacitors in series or in parallel having initial stored energy, the charge due to the applied excitation adds algebraically to the charge initially stored in each capacitor, in accordance with superposition. The charge added and the final voltages across the capacitors are most easily determined using the equivalent series or parallel capacitance, as applicable.
- When an excitation is applied to inductors in series or in parallel having initial stored energy, the flux linkage due to the applied excitation adds algebraically to the flux linkage initially present in each inductor, in accordance with superposition. The flux linkage added, and the final currents in the inductors are most easily determined using the equivalent series or parallel inductance, as applicable.
- Charge is conserved at any node in a circuit at the instant of a sudden change as long as no current impulses are applied from an external source to the node at the instant of the sudden change.
- Flux linkage is conserved around any closed path in a circuit at the instant of a sudden change as long as no voltage impulses are applied from an external source in the closed path at the instant of the sudden change.
- There is an apparent loss of energy in impulsive redistribution of charges in capacitive circuits or in the impulsive redistribution of flux linkages in inductive circuits, due to indeterminate power dissipation by a current impulse flowing through zero resistance or a voltage impulse appearing across zero conductance.
- In a switching operation, a capacitor voltage does not change at the instant of switching unless forced to by a current impulse. Similarly, an inductor current does not change at the instant of switching unless forced to by a voltage impulse. KCL and KVL must be satisfied at all times, before switching, just after switching, and in the steady state.
- A finite current does not add to, or subtract from, charge at a node during the infinitesimal interval just before and just after a switching operation. Similarly, a finite voltage does not add to, or subtract from, flux linkage in a closed path during the infinitesimal interval just before and just after a switching operation.
- Stored energy can be trapped in capacitors or in inductors in the final, steady state when initial charges on capacitors, or initial flux linkages in inductors, are not completely neutralized, even in the presence of resistors, while KCL, KVL, and Ohm's law are satisfied.

Problem-Solving Tips

1. Equivalent series or parallel capacitors can be conveniently used to facilitate the solution of problems involving capacitors.
2. Equivalent series or parallel inductors can be conveniently used to facilitate the solution of problems involving inductors.

Problems

Verify solutions by PSpice simulation.

Capacitive Circuits

P19.1 The switch in Figure P19.1 is closed at $t = 0$, with C_1 initially charged to $V_{10} = 2$ V. Determine the initial voltage V_{20} of C_2 that will make each of C_1 and C_2 completely discharge as $t \to \infty$.

Ans. 3 V.

P19.2 The switch in Figure P19.2 is closed at $t = 0$ after being open for a long time. Determine $v_O(t)$, $t \geq 0^+$, assuming $V_{SRC} = 8$ V.

Ans. 4 V.

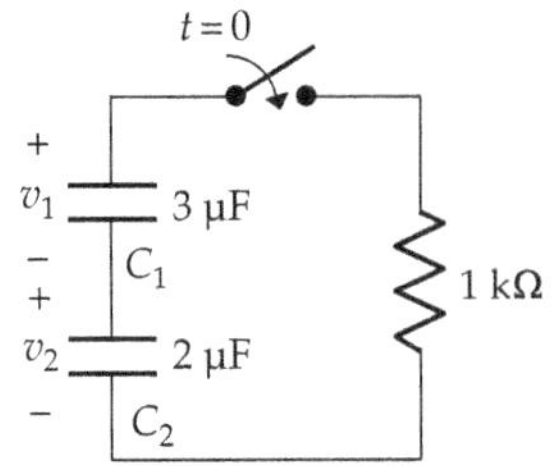

FIGURE P19.1

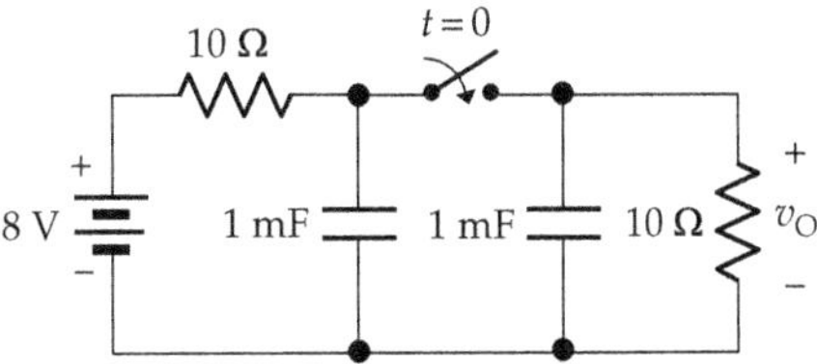

FIGURE P19.2

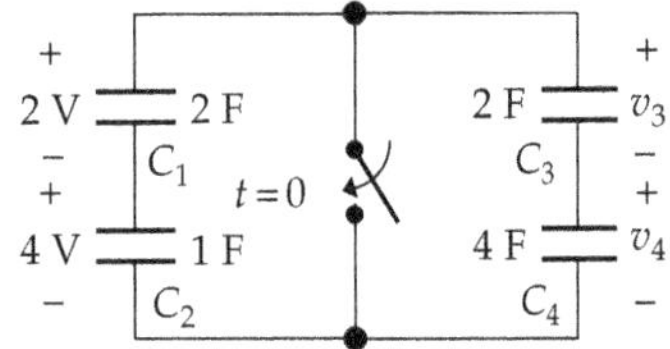

FIGURE P19.3

P19.3 The switch in Figure P19.3 is closed at $t = 0$, with initial voltages on C_1 and C_2 and with $q_4(0^-) = 4$ C. Initial voltage on C_3 is not shown. Determine (a) the charge through the switch when it closes and (b) the final value of v_4.

Ans. (a) 12 C; (b) −1 V.

P19.4 C_2 in Figure P19.4 is initially uncharged and the switch has been in position 'a' for a long time. The switch is moved to position 'b' at $t = 0$. Determine $v_{C1}(t)$ for $t \geq 0^+$.

Ans. $10\left(0.4 + 0.6e^{-t/3}\right)$, t is in ms.

P19.5 The switches in Figure P19.5 open at $t = 0$ after being closed for a long time. Determine the final values of v_{C1} and v_{C2}.

Ans. $v_1 = 4$ V, $v_2 = -4$ V.

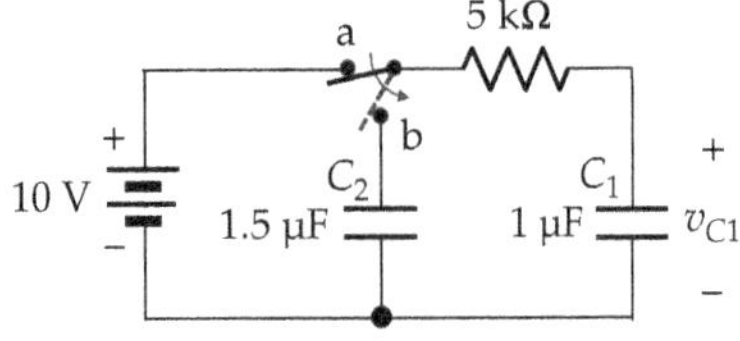

FIGURE P19.4

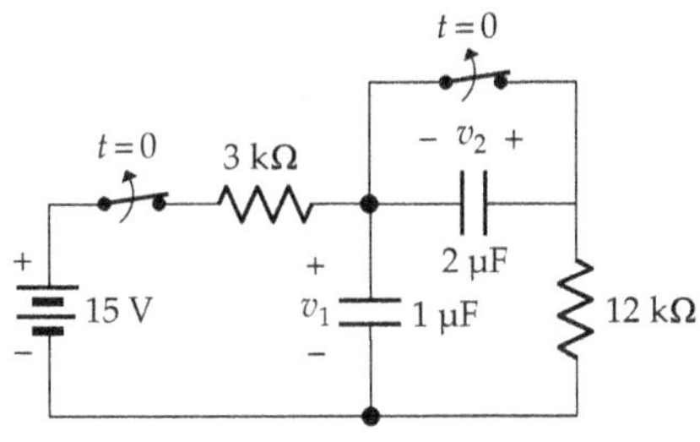

FIGURE P19.5

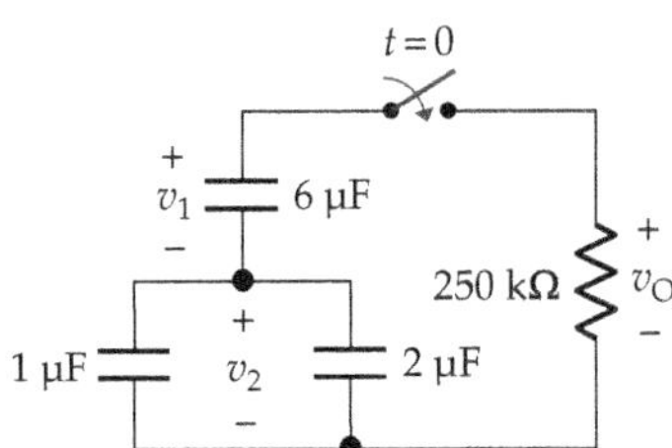

FIGURE P19.6

P19.6 The capacitors in Figure P19.6 have initial voltages $V_{10} = -5$ V and $V_{20} = 30$ V. If the switch is closed at $t = 0$, determine (a) $v_1(t)$, $v_2(t)$, and $v_O(t)$ for $t \geq 0^+$ and (b) the energy trapped in the circuit.

Ans. (a) $v_1(t) = -\dfrac{40}{3} + \dfrac{25}{3}e^{-2t}$ V, $v_2(t) = \dfrac{40}{3} + \dfrac{50}{3}e^{-2t}$ V, $v_O(t) = 25e^{-2t}$ V, $t \geq 0^+$, t is in s; (b) 0.8 mJ.

P19.7 The switch in Figure P19.7 is closed at $t = 0$, the initial values of the capacitor voltages being $V_{10} = 1$ V, $V_{20} = 0.2$ V, and $V_{30} = 1.2$ V. Determine (a) $i_O(t)$ for $t \geq 0^+$ and (b) the final values of v_1, v_2, and v_3.

Ans. $i_O(t) = 12e^{-21t/80}$ A, $t \geq 0^+$, t is in ms, $v_{1f} = 43/35$ V, $v_{2f} = 3/35$ V, $v_{3f} = 8/7$ V.

P19.8 The capacitors in Figure P19.8 were initially charged as shown. When the switch is closed at $t = 0$, the current i_ϕ is found to be $0.18e^{-t}$ mA where t is in ms. Determine (a) $v_\phi(t)$ for $t \geq 0^+$, (b) R, and (c) the final voltages across the capacitors.

Ans. (a) $75e^{-t}$ V; (b) 1250/3 kΩ; (c) −24 V across the 60/13 nF, 24 V across the parallel combination.

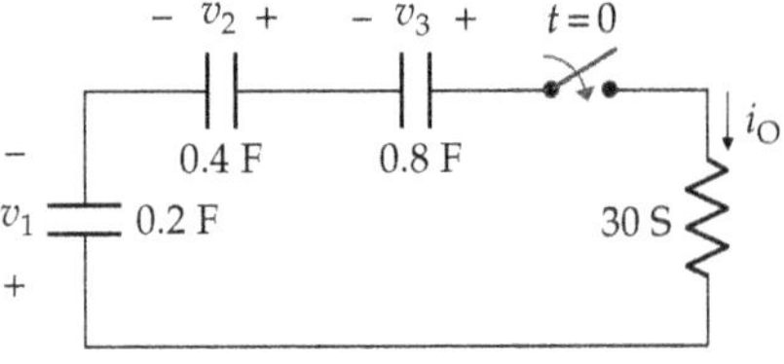

FIGURE P19.7

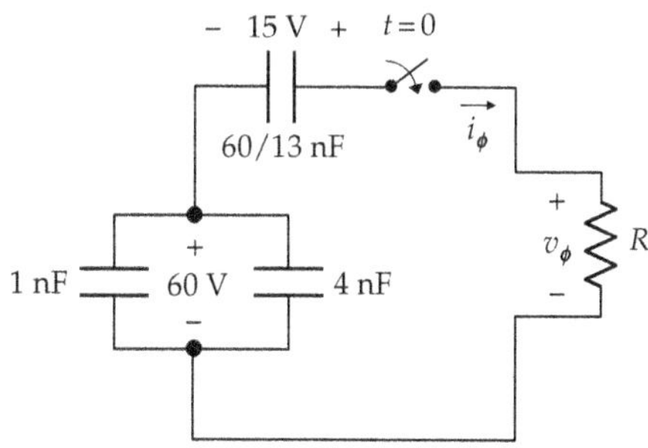

FIGURE P19.8

P19.9 The initial voltages on the capacitors in Figure P19.9, before the switch is closed are, 2 V on C_1, 6 V on C_2, and 0 V on C_3. The switch is closed at $t = 0$. Determine (a) the charges on the three capacitors at $t = 0^+$; (b) $v_R(t)$, $t \geq 0^+$; and (c) the final voltages on the capacitors.

Ans. (a) $q_1(0^+) = 0$, $q_2(0^+) = 8$ μC, $q_3(0^+) = 4$ μC; (b) $v_R(t) = 4e^{-0.5t}$ V, t is in ms; (c) $v_{1F} = -2$ V, $v_{2F} = +2$ V, $v_{3F} = 0$.

P19.10 The switch in Figure P19.10 is closed at $t = 0$, with C_1 initially charged to $V_{10} = 9$ V, and C_2 and C_3 uncharged. Determine (a) $v_1(0^+)$, (b) $i(t)$ for $t \geq 0^+$, and (c) the voltage across C_3 as $t \to \infty$, by evaluating the charge deposited on C_3 by i as $t \to \infty$.

Ans. (a) 6 V; (b) $6e^{-t/2}$ mA where t is in ms; (c) 4 V.

P19.11 Given, in Figure P19.11, $C_1 = 8$ F, initially charged to $V_{10} = 6$ V; $C_2 = 6$ F, initially charged to $V_{20} = 4$ V; and $C_3 = 3$ F, initially charged to $V_{30} = 2$ V, the switch is closed at $t = 0$. Determine the final values of v_2 and v_3.

Ans. $v_{2f} = 2$ V; $v_{3f} = -2$ V.

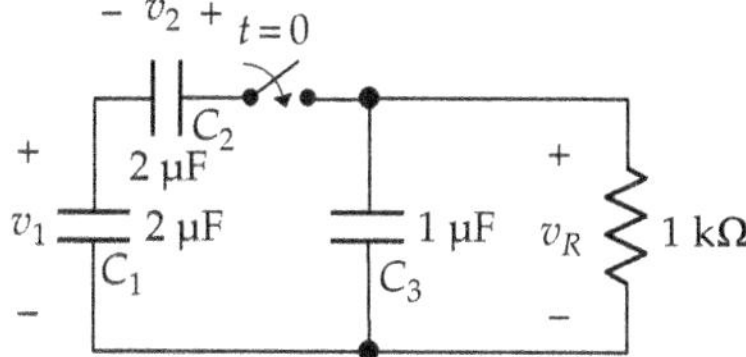

FIGURE P19.9

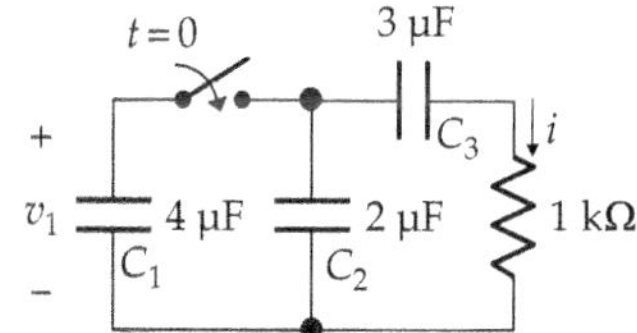

FIGURE P19.10

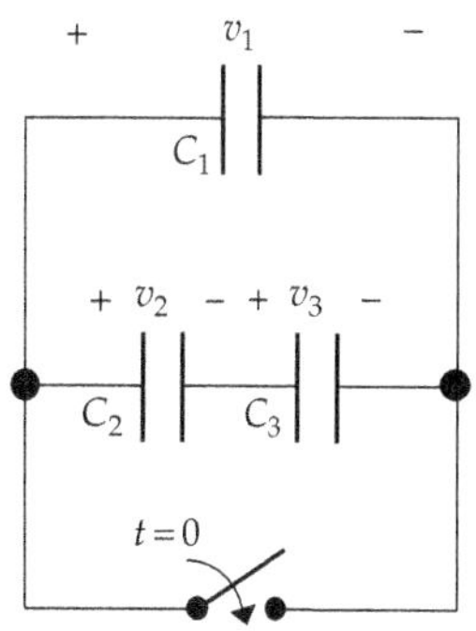

FIGURE P19.11

Inductive Circuits

P19.12 The switch in Figure P19.12 is closed at $t = 0$. Determine $i(t)$ for $t \geq 0^+$ if (a) the initial currents in the inductors are as indicated in Figure P19.12 and (b) if each initial current is 5 A pointing from left to right.

Ans. (a) $1.6e^{-2t}$ A; (b) $-e^{-2t}$ A.

P19.13 The switch in Figure P19.13 is opened at $t = 0$, with an initial current of 10 A in the 2 H inductor. Determine the initial current I_{20} in the 4 H inductor so that both inductors completely discharge as $t \to \infty$.

Ans. 5 A.

P19.14 The switch in Figure P19.14 is moved to position 'b' at $t = 0$ after being in position 'a' for a long time. Determine the final values of i_1 and i_2.

Ans. $i_1 = 4$ A, $i_2 = -4$ A.

P19.15 The inductors in Figure P19.15 have initial currents $I_{10} = 30$ A and $I_{20} = -5$ A. If the switch is opened at $t = 0$, determine (a) $i_1(t)$, $i_2(t)$, and $i_O(t)$ for $t \geq 0^+$ and (b) the energy trapped in the circuit.

Ans. (a) $i_1(t) = \frac{40}{3} + \frac{50}{3}e^{-2t}$ A, $i_2(t) = -\frac{40}{3} + \frac{25}{3}e^{-2t}$ A, $i_O(t) = 25e^{-2t}$ A, $t \geq 0^+$, t is in s;

(b) 0.8 kJ.

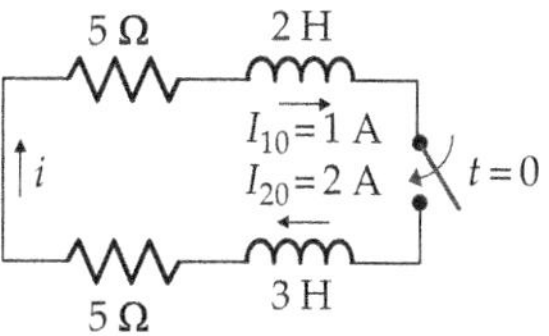

FIGURE P19.12

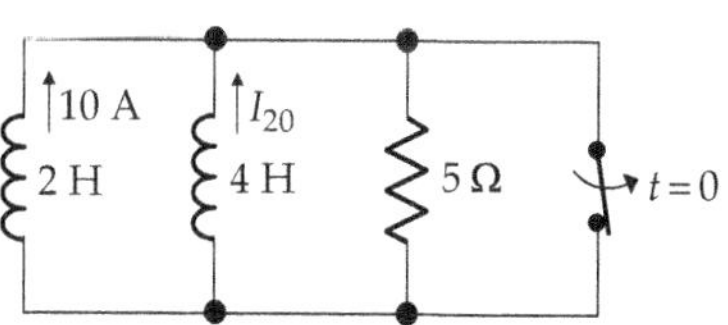

FIGURE P19.13

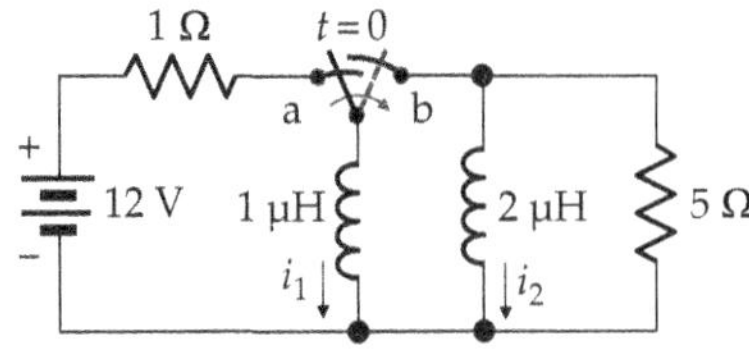

FIGURE P19.14

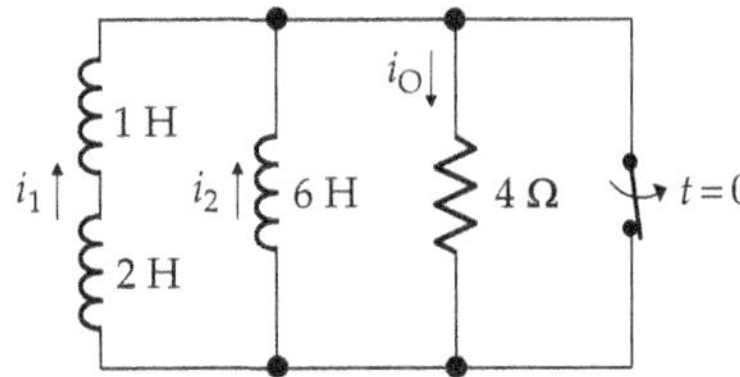

FIGURE P19.15

P19.16 The three inductors in Figure P19.16 have the initial currents shown. The switch is opened at $t=0$. Determine $v_O(t)$ for $t \geq 0$ and the final currents in the inductors.

Ans. $v_O(t)=12e^{-21t/80}$ V, $t \geq 0^+$, t is in ms, $i_{1f}=43/35$ A, $i_{2f}=3/35$ A, $v_{3f}=8/7$ A.

P19.17 The inductors in Figure P19.17 have initial currents as shown. When the switch is opened at $t=0$, the voltage $v_\phi(t)$ is found to be $50e^{-2.5t}$ V, where t is in s. Determine (a) $i_\phi(t)$ for $t \geq 0^+$, (b) R, and (c) the final currents in the inductors.

Ans. (a) $i_\phi(t)=e^{-2.5t}$ A, $t \geq 0^+$, t is in s; (b) 50 Ω; (c) 0.

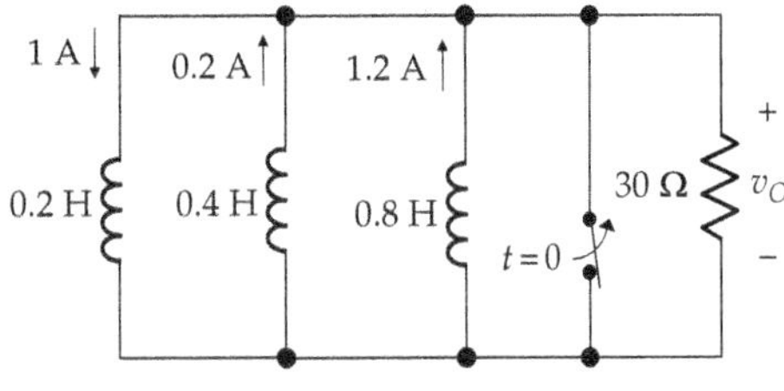

FIGURE P19.16

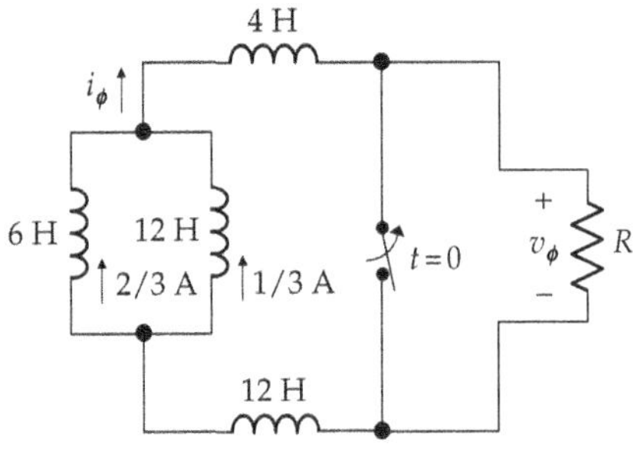

FIGURE P19.17

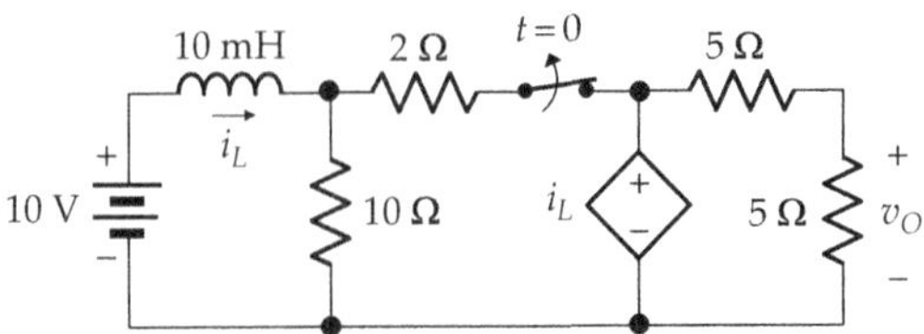

FIGURE P19.19

P19.18 Derive and analyze the circuit that is the dual of that of Problem P19.9.

Ans. (a) $\lambda_1(0^+)=0$, $\lambda_2(0^+)=8$ μVs, $\lambda_3(0^+)=4$ μVs; (b) $i_G(t)=4e^{-0.5t}$ A, t is in ms; (c) $i_{1F}=-2$ A, $i_{2F}=+2$ A, $i_{3F}=0$.

P19.19 The switch in Figure P19.19 is opened at $t=0$ after being closed for a long time. Determine $I_L(0^+)$ and $v_O(t)$.

Ans. $i_L(0^+)=4$ A, $v_O(t)=0.5(1+3e^{-t})$V, $t \geq 0^+$ms.

RLC Circuits

P19.20 The switch in Figure P19.20 is closed at $t=0$, with $i_L(0^-)=0$ and $v_C(0^-)=2$ V. Assuming $\rho=3$ kΩ, determine (a) $i_L(0^+)$ and $v_O(0^+)$ and (b) $v_O(t)$.

Ans. (a) $i_L(0^+)=6$ mA; $v_O(0^+)=-6$ V; (b) $v_O(t)=-6e^{-t}$ $t \geq 0^+$ μs.

P19.21 The switch in Figure P19.21 is opened at $t=0$, after being closed for a long time, and just after the impulse is over at $t=0^+$. Determine (a) $i_L(0^-)$, $i_L(0^+)$, $v_O(0^-)$, and $v_O(0^+)$ and (b) $v_O(t)$ assuming the value of R for critical damping.

Ans. (a) $i_L(0^-)=0$, $v_O(0^-)=20$ V, $i_L(0^+)=10$ A, $v_O(0^+)=20$ V; $v_O(t)=10e^{-t}(1+2te^{-t})+10$ V, $t \geq 0^+$ms.

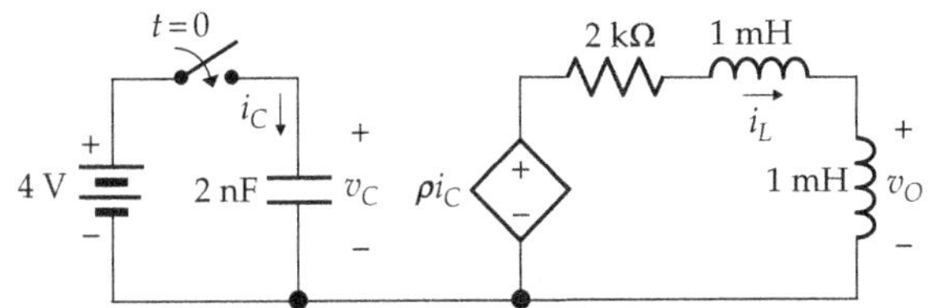

FIGURE P19.20

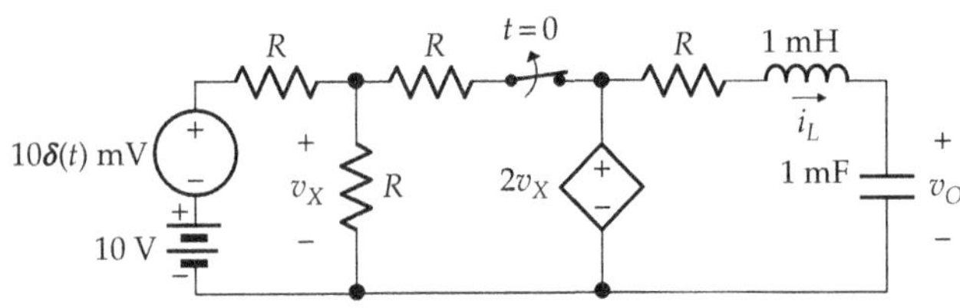

FIGURE P19.21

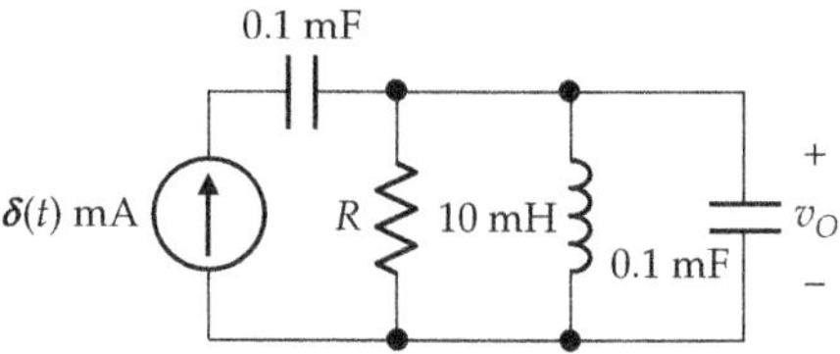

FIGURE P19.22

P19.22 Determine $v_O(t)$ in Figure P19.22 assuming zero initial energy storage and a critically damped response.

Ans. $v_O(t) = 10e^{-t}(1-t)$ V, $t \geq 0^+$ ms.

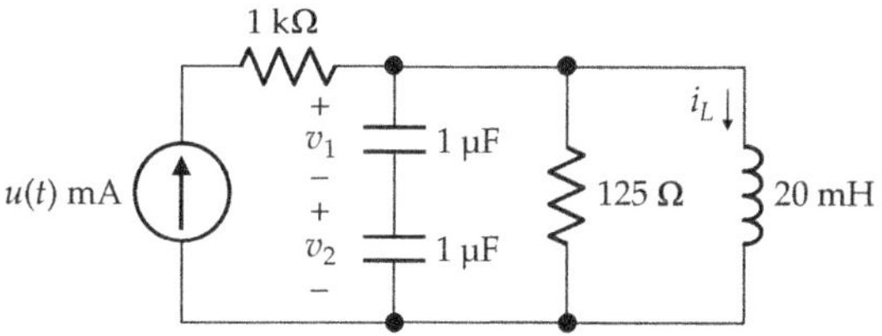

FIGURE P19.23

P19.23 Determine $v_1(t)$ and $v_2(t)$, $t \geq 0^+$, in Figure P19.23, assuming $i_L(0^-) = 0$, $v_1(0^-) = 0$ and $v_2(0^-) = 1$ V.

Ans. $v_1(t) = 0.5e^{-8t}(\cos 6t - \sin 6t) - 0.5$ V, $v_2(t) = 0.5e^{-8t}(\cos 6t - \sin 6t) + 0.5$ V.

20

Convolution

Objective and Overview

This chapter introduces convolution, which is a fundamental operation in the time domain that gives the response of a linear time-invariant (LTI) system to an arbitrary input, based solely on its impulse response, without any knowledge about the system or its constituents.

The chapter begins by explaining two operations that are commonly encountered in convolution, namely, shifting a function in time and folding a function around the vertical axis. The convolution integral is then introduced and interpreted graphically, which leads to a general procedure for deriving the convolution integral based on the graphical interpretation. Some basic properties of the convolution operation are then presented, followed by some important special cases of convolution, namely, convolution of staircase functions and convolution with impulse and step functions. The chapter ends with a summary of some general properties of the convolution integral, illustrated with additional examples.

20.1 Shifting in Time and Folding

Before considering convolution, some basics concerning shifting of functions in time and folding around the vertical axis are discussed.

20.1.1 Shifting in Time

Consider the function, $h(\lambda)$, where λ is a time variable, as shown in the middle trace in Figure 20.1a. It is assumed for simplicity and for reasons that will become clear later that $h(\lambda) = 0$ for $\lambda < 0^-$ and decreases like an exponential for $\lambda > 0^+$. Suppose that the function is shifted to the right by a, where a is a positive number, such as 5 units of time. The origin, where $\lambda = 0$, is unchanged so that the vertical edge of the function is shifted from the origin to $\lambda = a$. The magnitude of the function at a particular value of λ, say $\lambda = \lambda_1 > 0$, is $h(\lambda_1)$, where λ in the function $h(\lambda)$ is replaced by λ_1. The value of the shifted function should be the same at $\lambda = (\lambda_1 + a)$ as at $\lambda = \lambda_1$. For this to be the case, *λ in the shifted function should be decreased by a*, that is, λ is replaced by $(\lambda - a)$. The shifted function becomes $h(\lambda - a)$, so that at $\lambda = (\lambda_1 + a)$, $h(\lambda - a) = h((\lambda_1 + a) - a) = h(\lambda_1)$, the same value as for the unshifted function, as it should be.

Similarly, if the function is shifted by $-a$ to the left, where a is a positive number, the vertical edge of the function is moved from the origin to $\lambda = -a$ (Figure 20.1a). The value $h(\lambda_1)$ of the unshifted function should occur at $\lambda = (\lambda_1 - a)$ for the shifted function. *λ in the shifted function should therefore be increased by a*, that is, λ is replaced by $(\lambda + a)$. The shifted function becomes $h(\lambda + a)$, so that at $\lambda = (\lambda_1 - a)$, $h(\lambda + a) = h((\lambda_1 - a) + a) = h(\lambda_1)$, the same value as for the unshifted function, as it should be.

If the function is shifted by a variable amount denoted by the symbol t, which could be positive or negative, the same expression

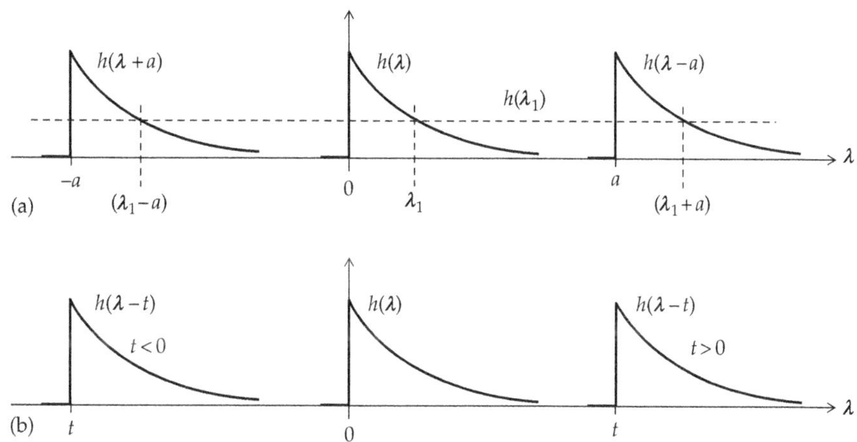

FIGURE 20.1
Shifting in time. (a) Function $f(\lambda)$ shifted to the right or left by a positive number a, and (b) $f(\lambda)$ shifted to the right or left by a variable t.

$h(\lambda - t)$ applies in terms of t as for a shift to the right by a positive quantity t (Figure 20.1b). If t has a positive value a, the expression for the shifted function is $h(\lambda - a)$, as in Figure 20.1a. If t has a negative value, $t = -a$, then the expression for the shifted function is $h(\lambda + a)$, again as in Figure 20.1a.

20.1.2 Folding around the Vertical Axis

Next, consider the function $h(-\lambda)$ (Figure 20.2a). This is the same function $h(\lambda)$ of Figure 20.2a but with λ replaced by $-\lambda$. It is thus the mirror image of the function $h(\lambda)$ with respect to the vertical axis. It is also described as $h(\lambda)$ folded around the vertical axis. Shifting the function $h(-\lambda)$ to the right or left is subject to the same rule illustrated in Figure 20.1a, namely, in shifting to the right by a with respect to the same origin, λ in the shifted function is decreased by a, whereas in shifting to the left by a, λ in the shifted function is increased by a, where a is a positive number. Thus, if the function $h(-\lambda)$ is shifted to the right by a, it becomes $h(-(\lambda - a)) = h(a - \lambda)$ (Figure 20.2b). Note that *the shift is applied to λ in the argument of the given function and not to the whole argument $-\lambda$*. Thus, the argument $-\lambda$ becomes $(a - \lambda)$ and not $(-\lambda - a)$. Similarly, if the function $h(-\lambda)$ is shifted by a variable amount denoted by the symbol t, which could be positive or negative, the folded function becomes $h(t - \lambda)$. In other words, *λ in the function is replaced by $(t - \lambda)$ when the function is folded around the vertical axis and shifted by a variable t.*

Primal Exercise 20.1

Derive the expressions for the following functions when first folded around the vertical axis and then shifted by t: (a) $u(\lambda)$, (b) $u(\lambda - 3)$, (c) $\delta(\lambda)$, (d) $\lambda u(\lambda)$, (e) $(\lambda - 2)u(\lambda - 2)$, and (f) $\cos \pi\lambda$. Note that an even function is unaltered by folding around the vertical axis. When folded and shifted, the argument of the function could be $(t - \lambda)$ or $(\lambda - t)$.

Ans. (a) $u(-\lambda)$, $u(t - \lambda)$; (b) $u(-\lambda - 3)$, $u(t - \lambda - 3)$; (c) $\delta(-\lambda) = \delta(\lambda)$, $\delta(t - \lambda) = \delta(\lambda - t)$; (d) $-\lambda u(-\lambda)$, $(t - \lambda)u(t - \lambda)$; (e) $(-\lambda - 2)u(-\lambda - 2)$, $(t - \lambda - 2)u(t - \lambda - 2)$; (f) $\cos(-\pi\lambda) = \cos(\pi\lambda)$, $\cos(\pi(t - \lambda)) = \cos(\pi(\lambda - t))$.

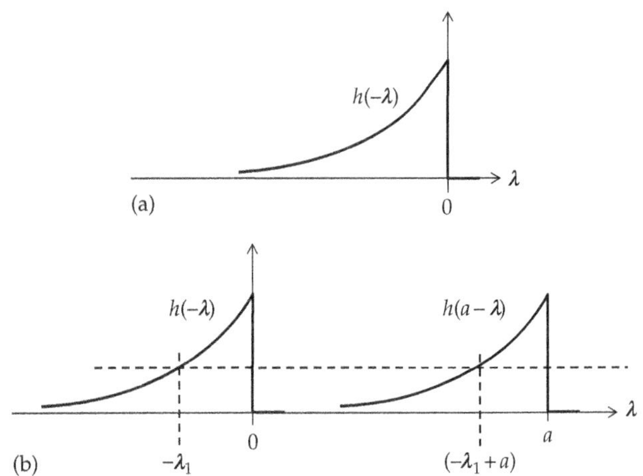

FIGURE 20.2
Function $f(\lambda)$ folded around the vertical axis (a) and shifted to the right by a positive number a (b).

20.2 Convolution Integral

To illustrate the meaning and significance of convolution in circuit analysis, consider an LTI circuit or system whose response $h(t)$ to a unit impulse $\delta(t)$ (Figure 20.3a) can be determined in some manner, either analytically, or by simulation, or experimentally by applying a large, brief input of a duration that is much shorter than the smallest time constant, or response time, of the system, as was done in the simulation examples in preceding chapters. The question is, knowing $h(t)$, can one determine the response $y(t)$ to an arbitrary input $x(t)$ (Figure 20.3b)? According to convolution theory, $y(t)$ can indeed be determined as

$$y(t) = \int_{-\infty}^{\infty} x(\lambda)h(t - \lambda)d\lambda \tag{20.1}$$

where λ is an arbitrary integration variable having the dimensions of time. The RHS of Equation 20.1 is a definite integral that evaluates to a function of t alone.

Equation 20.1 defines a convolution operation, which is represented as

$$y(t) = x(t) * h(t) \tag{20.2}$$

In words, $x(t)$ is said to be convolved with $h(t)$. The key concept behind convolution is that the input $x(t)$ can be considered as a series of very narrow pulses, each of which approximates an impulse. Since the system is linear, these impulse responses can be summed together to give $y(t)$.

The first step in deriving the convolution integral, therefore, is to approximate $x(t)$ by a series of pulses of equal width, $\Delta\lambda$, as illustrated in Figure 20.4a, where the approximation can be as close as desired by making $\Delta\lambda$ sufficiently small. The ith pulse starts at $t = \lambda_i$ and is of duration $\Delta\lambda$ and height $x(\lambda_i)$. As $\Delta\lambda$ is made very

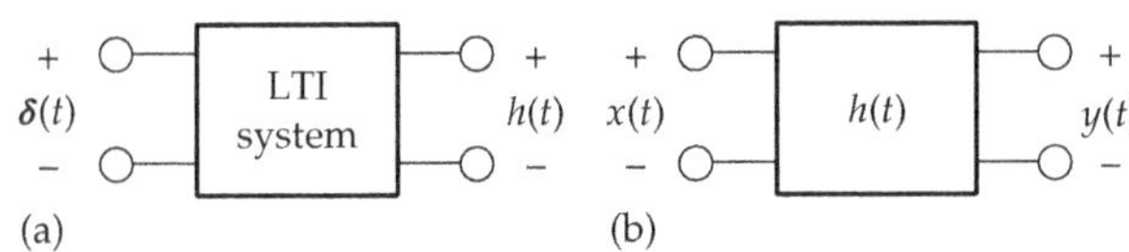

FIGURE 20.3
Interpretation of convolution integral. Knowing the response $h(t)$ of an LTI system to $\delta(t)$ (a), the response $y(t)$ of the system to an arbitrary input $x(t)$ (b) can be determined by convolution.

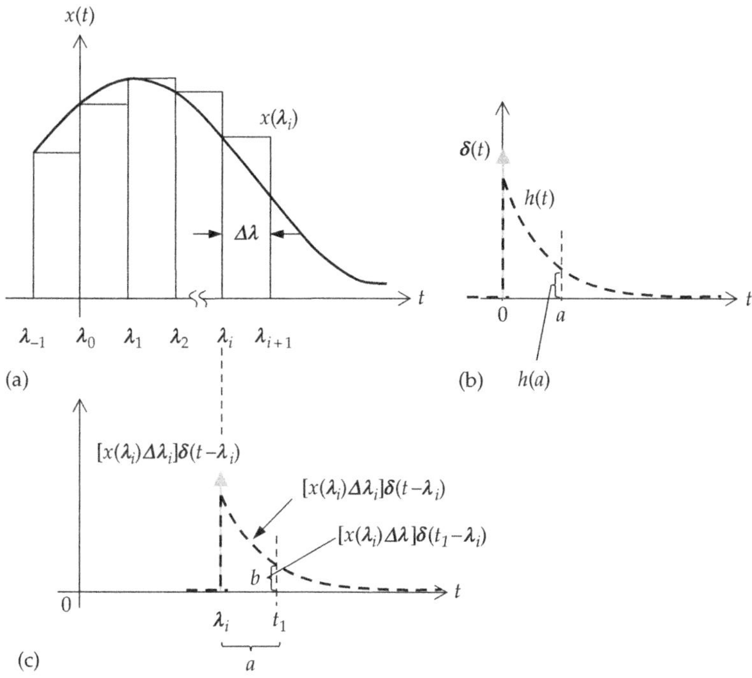

FIGURE 20.4
Graphical derivation of convolution integral. (a) A function $x(t)$ approximated by a series of narrow pulses, (b) the response $h(t)$ of the system to $\delta(t)$, and (c) response of the system to the narrow pulse at $t = \lambda_i$ *in* (a).

small, the response to the ith pulse can be considered to be the same as the response to an impulse at $t = \lambda_i$ whose strength is the same as the area of the pulse, which is $x(\lambda_i)\Delta\lambda$. Let the response to $\delta(t)$, a unit impulse at the origin, be $h(t)$, as illustrated in Figure 20.4b. Note that since $\delta(t)$ occurs at the origin, the response $h(t)$ is zero for $t < 0^-$, since there is no response to the impulse before it occurs. The response to the unit impulse at $t = a$ is $h(a)$. A unit impulse that is delayed by λ_i is $\delta(t - \lambda_i)$. If the strength of the impulse is $[x(\lambda_i)\Delta\lambda]$, rather unity, the impulse is represented as $[x(\lambda_i)\Delta\lambda]\delta(t - \lambda_i)$, and the response to the impulse for $t \geq \lambda_i$ is $y(t) = [x(\lambda_i)\Delta\lambda]h(t - \lambda_i)$, where $(t - \lambda_i)$ is the interval of time t after the beginning of the impulse response, as illustrated in Figure 20.4c. At any particular instant of time $t = t_1$, where $t_1 - \lambda_I = a$, for example, the response is $h(a)$ as in Figure 20.4b, but multiplied by the strength of the impulse. That is, the response is $[x(\lambda_i)\Delta\lambda]h(t_1 - \lambda_i)$, represented by the ordinate b in Figure 20.4c. Since the system is LTI, superposition applies, so that the total response due to all the pulses λ_i that occur up to t_1 is the sum of the responses to all these individual pulses. It follows that

$$y(t_1) = \sum_{\lambda_i=-\infty}^{\lambda_i=t_1} x(\lambda_i)h(t_1 - \lambda_i)\Delta\lambda \tag{20.3}$$

In the limit, as $\Delta\lambda \rightarrow d\lambda$, the summation becomes an integration. Since λ_i is arbitrary, the subscript can be dropped, replacing λ_i by λ. Similarly, t_1 can be replaced by t. The lower limit of the integral is $-\infty$ and the upper limit is t. The summation becomes the convolution integral:

$$y(t) = \int_{-\infty}^{t} x(\lambda)h(t - \lambda)d\lambda \tag{20.4}$$

Mathematically, the integral of Equation 20.4 represents an infinite sum, over a continuum of time, of responses to impulses, each having an infinitesimal strength. Practically, the summation of Equation 20.3 is the sum of responses to pulses whose duration is finite but small compared with the smallest time constant of the circuit or the reciprocal of the highest frequency in the circuit's natural response. Note that in changing the summation to integration, the primary time variable becomes λ instead of t, the latter becoming a constant with respect to the integration.

The integral of Equation 20.4 is the same as that of Equation 20.1, except for the upper limit. It is seen from Figure 20.4b that if λ_i, the time at which the impulse occurs, exceeds t_1, the time at which response is required, then the response is zero. This is because a physical system operating in real time does not respond to an impulse before this impulse occurs. Operation in real time means that the system response unfolds for the first time as current time progresses. This is in contrast to a recorded signal that was captured at an earlier time and is being replayed later. At any instant of the recording, the future response is already available. Physical systems do not anticipate the occurrence of an input before it occurs and cannot therefore respond to such an input, as explained in Example 11.7. If the response is zero for $\lambda > t$, then the upper limit in Equation 20.4 could just as well be infinity rather than t.

In summary, the following concept applies:

Concept: *The response of an LTI circuit to an arbitrary input can be considered as the superposition of responses to sufficiently narrow pulses, each having an amplitude determined by the input. These narrow pulses can be approximated by impulses, which naturally leads to a convolution integral involving the input function and the impulse response of the circuit.*

It is convenient, though by no means essential, to assume that $x(t) = 0$, for $t < 0^-$, which means that the lower limit of integration in Equation 20.4 can be taken as 0^- instead of $-\infty$. This is usually done because of the link between convolution and the one-sided Laplace transform, which has 0^- as the lower limit of integration (Section 21.1). Having the lower limit as 0^- rather than 0 would include an impulse at the origin. With the lower limit taken as 0^-, Equation 20.4 reduces to the form of the convolution integral that is conventionally applied in circuit analysis, namely,

$$y(t) = \int_{0^-}^{t} x(\lambda)h(t-\lambda)d\lambda \qquad (20.5)$$

If the integrand does not include an impulse at the origin, then the lower limit of integration can be unambiguously taken as zero in Equation 20.5.

It should be emphasized that according to Equation 20.5, the given function $x(t)$ is now $x(\lambda)$, a function of the time variable λ rather than t. The integration is with respect to this new time variable λ, so that t is a constant as far as the integration with respect to λ is concerned. t can be a particular numerical value, in which case Equation 20.5 evaluates $y(t)$ at that particular value of t. More generally, t can assume any value within a specified range that depends on the function under consideration, in which case $y(t)$ is obtained as a function of t over the given range of t.

20.2.1 Graphical Interpretation

We wish to interpret the convolution integral graphically, as this leads to a very useful procedure for deriving the convolution integral. We note first of all that, in general, an integral represents an area. Equation 20.5 can therefore be considered as the area under the product of two functions: (1) $x(\lambda)$, which is simply the given function $x(t)$ with the time variable t replaced by another time variable λ, and (2) $h(t - \lambda)$. But what is $h(t - \lambda)$? According to Figure 20.2, $h(t - \lambda)$ can be considered to result from three steps: (1) t in the impulse response $h(t)$ is replaced by λ to give $h(\lambda)$, as in $x(\lambda)$; (2) $h(\lambda)$ is folded around the vertical axis, resulting in $h(-\lambda)$, as in Figure 20.2a; and (3) $h(-\lambda)$ is shifted by t to the right, to give $h(t - \lambda)$, as in Figure 20.2b. These steps are illustrated in Figure 20.5a. The product $x(\lambda)h(t - \lambda)$ is shown in Figure 20.5b. According to Equation 20.5, the value of the convolution integral at t is the area under the product from $\lambda = 0$ to $\lambda = t$, shown shaded in Figure 20.5b. As t is changed, the position of the shifted function $h(t - \lambda)$ changes, the product function changes, and so does the area under the product from $\lambda = 0$ to $\lambda = t$.

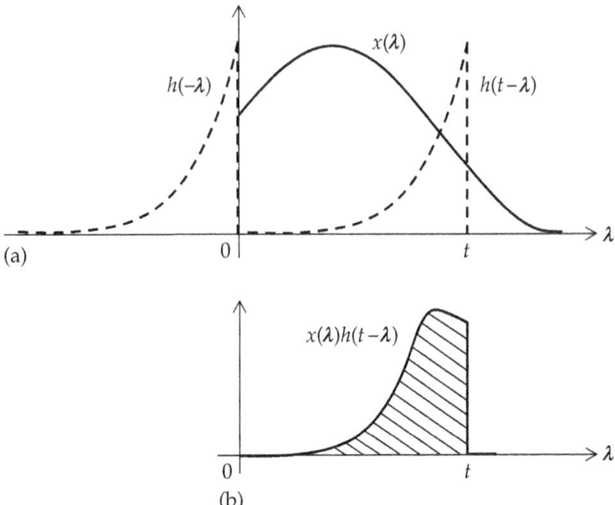

FIGURE 20.5
Graphical interpretation of convolution integral. (a) Impulse response folded around the vertical axis and shifted to the right by t and (b) the product of $x(\lambda)$ and $h(t - \lambda)$.

20.2.2 Procedure Based on Graphical Interpretation

Figure 20.5b forms the basis for the graphical evaluation of the convolution integral. The following steps are involved:

1. *Express $h(t)$ and $x(t)$ as function of λ.*
2. *Fold the impulse response around the vertical axis, that is, draw it backward as $h(-\lambda)$.*
3. *Shift $h(-\lambda)$ by t to obtain $h(t - \lambda)$.*
4. *Determine the area under the product $x(\lambda)h(t - \lambda)$ over the given range of t. The result is $y(t)$ for this range of t.*
5. *Repeat steps 3 and 4 for various values of t to obtain $y(t)$ over the whole range of t.*

The preceding procedure is illustrated by Examples 20.1 and 20.2, which also demonstrate, in principle, how convolution can be applied to obtain a circuit response due to an excitation that varies arbitrarily with time.

Example 20.1: Response of *RL* Circuit to a Rectangular Voltage Pulse

Consider the series *RL* circuit of Figure 20.6a, with $R = 1\ \Omega$, $L = 1$ H, zero initial energy storage, and $v_{SRC}(t)$ as the pulse shown in Figure 20.6b. It is required to determine $i(t)$ as $v_{SRC}(t)*h(t)$, where $h(t)$ is the current response to a unit voltage impulse.

Solution:

$h(t)$ is given by Equation 18.5, with $K = 1$, $L = 1$, and $\tau = 1$. Replacing t by λ, the impulse response becomes (Figure 20.6c)

$$h(\lambda) = e^{-\lambda}\ \text{A, for } \lambda \geq 0^+,$$

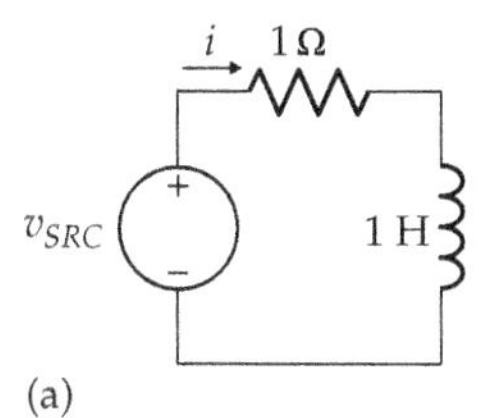

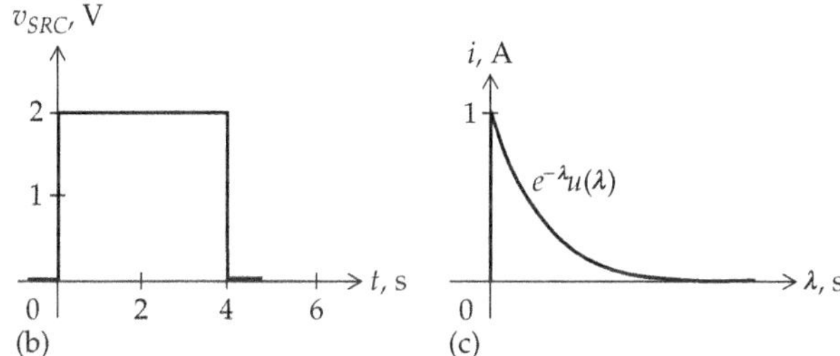

FIGURE 20.6
Figure for Example 20.1.

and

$$h(\lambda) = 0, \quad \text{for } \lambda \leq 0^- \tag{20.6}$$

The two different expressions of $h(\lambda)$ for $\lambda \leq 0^-$ and $\lambda \geq 0^+$ can be combined for analytical purposes into a single expression by using the unit step function $u(\lambda)$, as defined by Equation 18.5:

$$h(\lambda) = e^{-\lambda}u(\lambda)\ \text{A} \tag{20.7}$$

The next step is to fold $h(\lambda)$ around the vertical axis, as explained in connection with Figure 20.2a. The function becomes

$$h(-\lambda) = e^{\lambda}u(-\lambda) \tag{20.8}$$

as shown in Figure 20.7a. $h(-\lambda)$ does not overlap $v_{SRC}(\lambda)$, so that the product $v_{SRC}(\lambda)h(-\lambda)$ is zero for $t = 0$. When the folded function $h(-\lambda)$ is shifted by t to the right, it becomes

$$h(t-\lambda) = e^{-(t-\lambda)}u(t-\lambda) \tag{20.9}$$

as explained in connection with Figure 20.2b and shown in Figure 20.7b.

The next step is to form the product $v_{SRC}(\lambda)h(t-\lambda)$ and determine the area under the product for $t > 0$. Whereas the expression for $h(t-\lambda)u(t-\lambda)$ is the same for any $t > 0$, $v_{SRC}(\lambda) = 2$, for $0 < t < 4$ s, and $v_{SRC}(\lambda) = 0$, for $t > 4$ s. It follows that the range of the shift t must be divided into two, in accordance with the time variation of $v_{SRC}(\lambda)$. The convolution integral for the first range of t is

$$i'(t) = v_{SRC}(\lambda) * h(\lambda) = \int_0^t 2e^{-(t-\lambda)}u(t-\lambda)d\lambda\ \text{A} \tag{20.10}$$

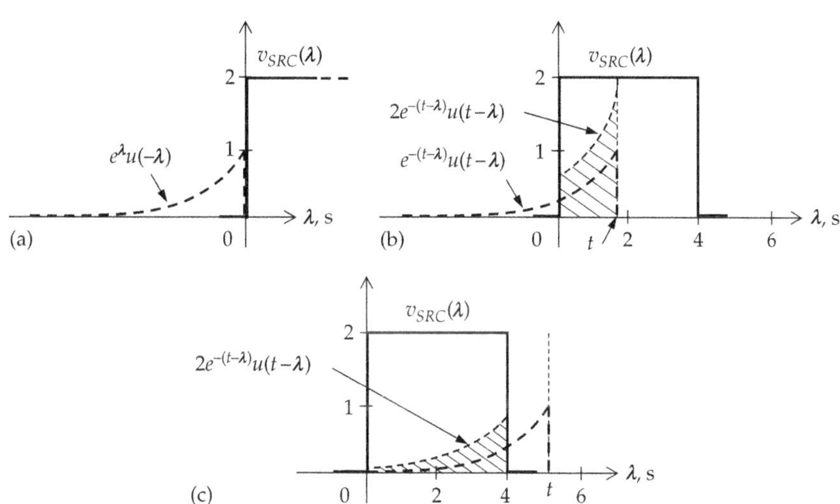

FIGURE 20.7
Figure for Example 20.1.

The following should be noted concerning this integral:

1. The integrand does not have any impulses and is finite over the range of t from $t = 0$ to $t = 4$. It follows that the lower limit of integration could be 0^- or 0^+, and the upper limit of integration could be 4^- or 4^+, without affecting the value of the integral, since the product of a finite quantity and an infinitesimal interval is zero. The range of integration can therefore be considered as $0 \leq t \leq 4$ s. Ranges of t will henceforth be considered inclusive of the lower and upper limits unambiguously, without specifying (limit)$^-$ and (limit)$^+$ when no impulses are included in the integrand. In the presence of impulses, the ranges should be specified more carefully.
2. By definition of $u(t)$, $u(0^+) = 1$ and $u(0^-) = 0$ (Equation 18.5). It follows that over the range of integration from $\lambda = 0$ to λ just less than t, $u(t - \lambda) = 1$, whereas outside the integration range, when λ just exceeds t, $u(t - \lambda) = 0$. This means that $u(t - \lambda)$ in Equation 20.10 is redundant and can be ignored. It is helpful to remember that $u(x) = 1$ for x positive and $u(x) = 0$ for x negative. $u(0)$ is finite between $x = 0^-$ and $x = 0^+$. The integral of a finite integrand between infinitesimal limits is zero. So *when the integrand is finite, it can be assumed that $u(x) = 0$ for x negative and $u(x) = 1$ for x zero or positive.*

Equation 20.10 becomes

$$i'(t) = v_{SRC}(\lambda) * h(\lambda) = \int_0^t 2e^{-(t-\lambda)} d\lambda = 2\left(1 - e^{-t}\right) \text{ A}, \quad 0 \leq t \leq 4 \text{ s} \quad (20.11)$$

The area under the product from $\lambda = 0$ to $\lambda = t$, $0 \leq t \leq 4$ s, is shown shaded in Figure 20.7b.

When t is increased beyond 4 s, the product of the two functions is zero beyond the edge of the pulse at $t = 4$ s. The upper limit of integration is 4, so that the convolution integral becomes

$$i''(t) = v_{SRC}(\lambda) * h(\lambda) = \int_0^4 2e^{-(t-\lambda)} d\lambda = 2\left(e^{-(t-4)} - e^{-t}\right) \text{A}, \quad t \geq 4 \text{ s} \quad (20.12)$$

The complete response $i(t)$ is $i(t) = i'(t)$, $0 \leq t \leq 4$ s, and $i(t) = i''(t)$ for $t \geq 4$ s, as shown by the simulation plot of Figure 20.10. The following should be noted concerning Equations 20.11 and 20.12:

1. The integrand is the same in both cases but the upper limit is different, which makes the *integrals* different.
2. $i'(t)$ and $i''(t)$ have the same value at $t = 4$ s, although the voltage pulse is, strictly speaking, undefined at $t = 4$ s. Mathematically, this is because the integral of a finite integrand between two infinitesimally separated limits is zero (Equation 18.22), so that the integral is the same at $t = 4^-$ s as at $t = 4^+$ s. This justifies the inclusive limits, $t \leq 4$ s and $t \geq 4$ s. The lower inclusive limit $0 \leq t$ in Equation 20.11 is similarly justified, in the absence of an impulse at the origin.

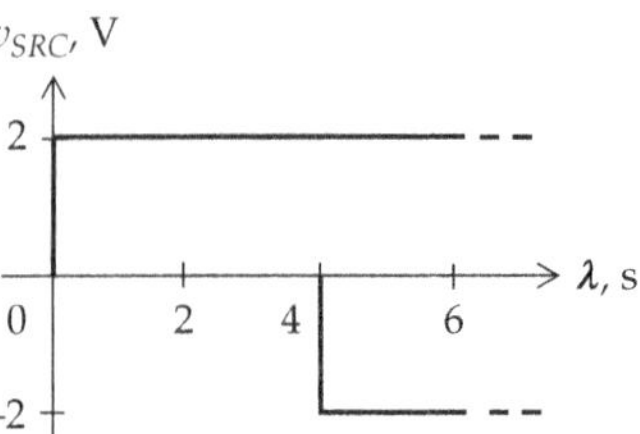

FIGURE 20.8
Figure for Example 20.1.

It is instructive to derive $i(t)$ as the response to two step functions: $2u(t)$, starting at the origin, and $-2u(t - 4)$, which is a negative, delayed step stating at $t = 4$ s (Figure 20.8). When these two steps are added, the result is a pulse of 2 V amplitude and 4 s duration (Figure 20.6b). The response to $2u(t)$ is obtained from Equation 11.57, with an initial value $i(0^+) = 0$ and a final value $i_F = 2$ A and $\tau = 1$ s. This gives Equation 20.11 as the response for $0 \leq t \leq 4$ s. The response to the negative step is obtained by noting that since the circuit is LTI, *time invariance implies that delaying the input simply delays the response by the same interval.* The response to the negative delayed step is therefore $-2\left(1 - e^{-(t-4)}\right)$ A. By superposition, the response for $t \geq 4$ s is the sum of this response and that given by Equation 20.11. Adding these two responses gives the response of Equation 20.12.

Simulation: The circuit is entered as in Figure 20.9. The voltage source is the pulse source VPULSE having the parameters shown. In the simulation profile, Time domain (Transient) analysis is selected, 10s is entered for 'Run to time', 0 for 'Start saving data after', and 1ms for 'Maximum step size'. After the simulation is run, the graph of Figure 20.10 is displayed, showing the currents during each of the integration intervals, as given by Equations 20.11 and 20.12. The source voltage has been added for convenience.

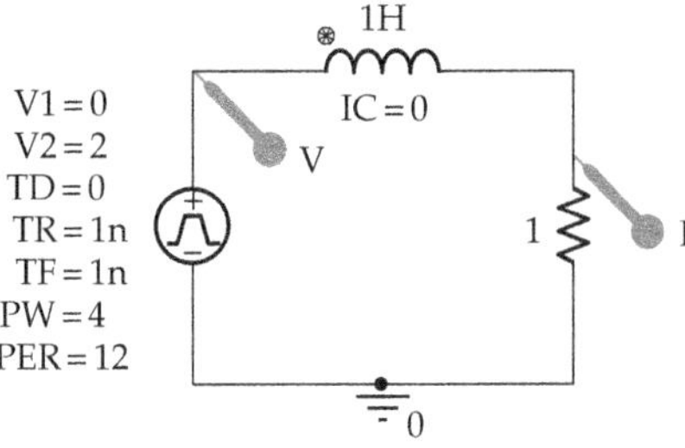

FIGURE 20.9
Figure for Example 20.1.

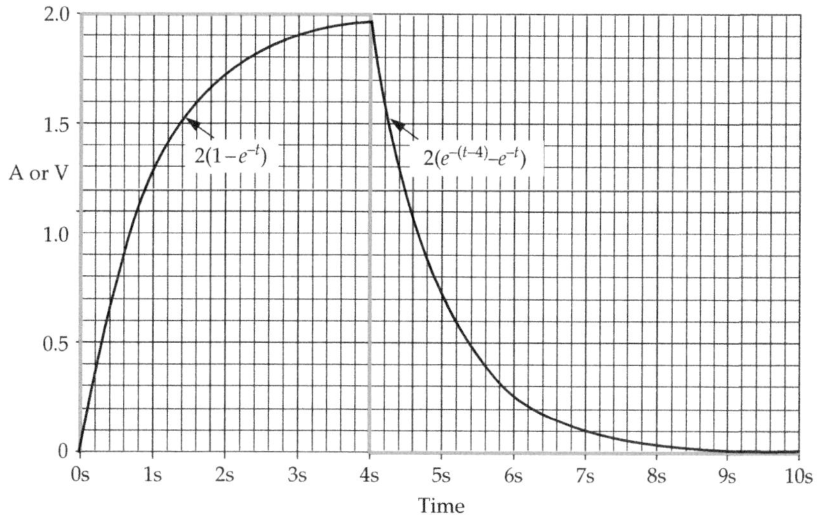

FIGURE 20.10
Figure for Example 20.1.

Example 20.2: Response of *RL* Circuit to a Trapezoidal Voltage Pulse

Consider the same series *RL* circuit of Example 20.1 but with $v_{SRC}(t)$ as in Figure 20.11. It is required to determine $i(t)$ as $v_{SRC}(t)*h(t)$.

Solution:

$h(t)$ is the same as before, and the folded impulse response is given by Equation 20.8 and shown in Figure 20.12a.

The next step is to form the product function $v_{SRC}(\lambda)h(t-\lambda)u(t-\lambda)$ and determine the area under the product for different values of the shift t. Considering the first range of t, the shifted function $h(t-\lambda)\,u(t-\lambda)$ is shown in Figure 20.12b, having the vertical edge located at $\lambda = t$. The product $v_{SRC}(\lambda)h(t-\lambda)u(t-\lambda)$ is $2h(t-\lambda)u(t-\lambda)$. Denoting the convolution integral $y(t)$ in this range by i', it follows that i' is the area under the product $2h(t-\lambda)u(t-\lambda)$ from $\lambda = 0$ to $\lambda = t$. This area is shown shaded in Figure 20.12b and is obtained analytically as

$$i'(t) = \int_0^t 2e^{-(t-\lambda)}d\lambda = 2\left(1-e^{-t}\right)\text{ A},\quad 0 \le t \le 4\text{ s} \qquad (20.13)$$

as in Equation 20.11. For a shift t beyond 4 s and up to $t = 6$ s, the folded and shifted impulse response $h(t-\lambda)u(t-\lambda)$ is as shown in Figure 20.13. This function is multiplied by $v_{SRC}(\lambda) = (6-\lambda)$ over the interval $4 \le t \le 6$ s. *Note that although t is restricted to the range between 4 and 6 s, the convolution integral is with respect to λ and starts from λ = 0, in accordance with Equation 20.15.* The convolution integral is the shaded area in Figure 20.13a. It is denoted by i'' over the given range of t and is given by

$$i''(t) = \int_0^4 2e^{-(t-\lambda)}d\lambda + \int_4^t (6-\lambda)e^{-(t-\lambda)}d\lambda =$$

$$7 - t - 2e^{-t} - e^{-(t-4)}\text{ A},\quad 4 \le t \le 6\text{ s} \qquad (20.14)$$

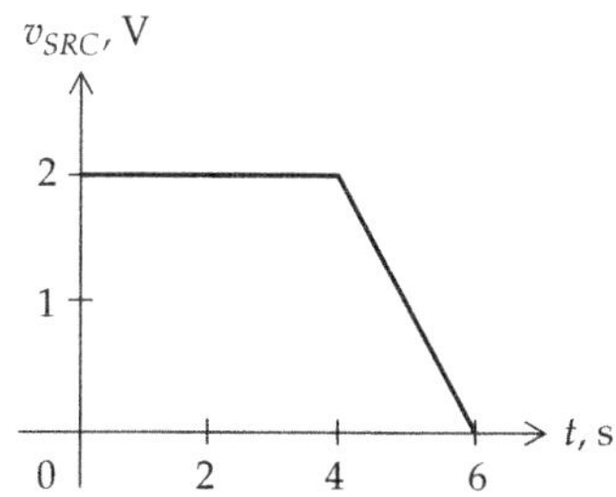

FIGURE 20.11
Figure for Example 20.2.

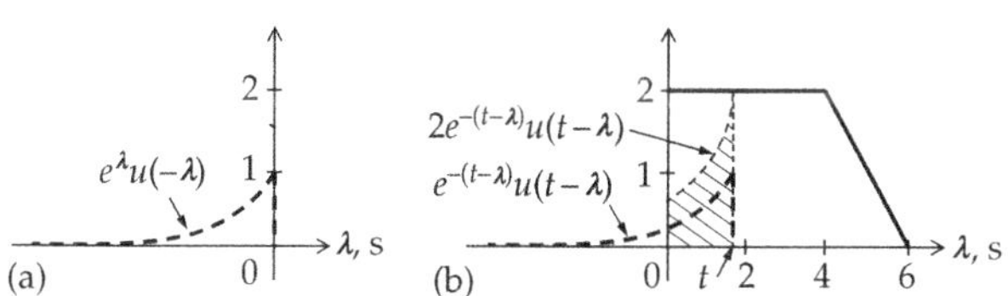

FIGURE 20.12
Figure for Example 20.2.

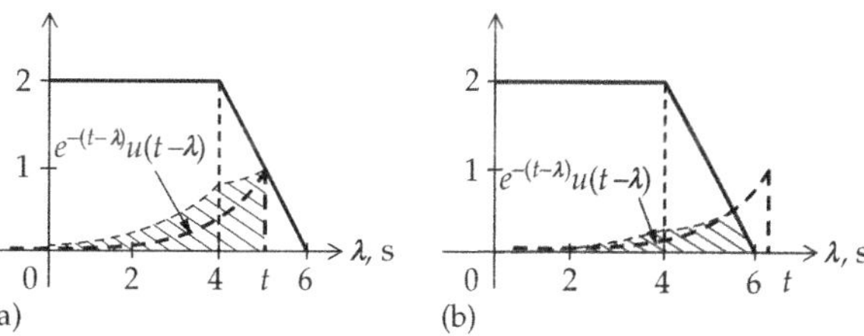

FIGURE 20.13
Figure for Example 20.2.

For $t \geq 6$ s the folded and shifted impulse response $h(t-\lambda)u(t-\lambda)$ is as shown in Figure 20.13b. This function is multiplied by $v_{SRC}(\lambda)$, which is 0 for $t \geq 6$ s. The convolution integral is the shaded area in Figure 20.13b and extends from $\lambda = 0$ to $\lambda = 6$. It is denoted by i''' over the given range of t and is given by

$$i'''(t)' = \int_0^4 2e^{-(t-\lambda)}d\lambda + \int_4^6 (6-\lambda)e^{-(t-\lambda)}d\lambda$$
$$= -2e^{-t} - e^{-(t-4)} + e^{-(t-6)}\text{A}, \quad t \geq 6\text{ s} \qquad (20.15)$$

The complete response i is the convolution integral i as a function of t. It is given by i', i'', and i''' over the respective time intervals, as shown in the simulation plot of Figure 20.15. Note that i and its first derivative are continuous at the breakpoints $t = 4$ s and $t = 6$ s. The continuity of i follows from the fact that the current through the inductor is not being forced to change. The continuity of di/dt follows from Kirchhoff's voltage law: $v_{SRC}(t) = Ri + Ldi/dt$. Since v_{SRC} and i are continuous at the breakpoints, then di/dt is continuous at these points.

Note that Equations 20.13 and 20.14 give the same value at $t = 4$ s. Similarly, Equations 20.14 and 20.15 give the same value at $t = 6$ s. This serves as a check on the convolution integral.

Simulation: The circuit is entered as illustrated in Figure 20.14. The voltage source is the piecewise-linear VPWL having the parameters shown. In the simulation profile, Time domain (Transient) analysis is selected, 10 s is entered for 'Run to time', 0 for 'Start saving data after', and 0.5s for 'Maximum step size'. After the simulation is run, the graph of Figure 20.15 is displayed showing the source voltage and the three currents during each of the integration intervals.

Problem-Solving Tip

- A simple check on convolution is that the convolution integral should have the same value at the common limits of consecutive ranges of t.

Primal Exercise 20.2

If $h(t)$ and $x(t)$ are as shown in Figure 20.16, determine the value of $y(t) = f(t)*g(t)$ at $t = 1.5$ s.

Ans. −0.5.

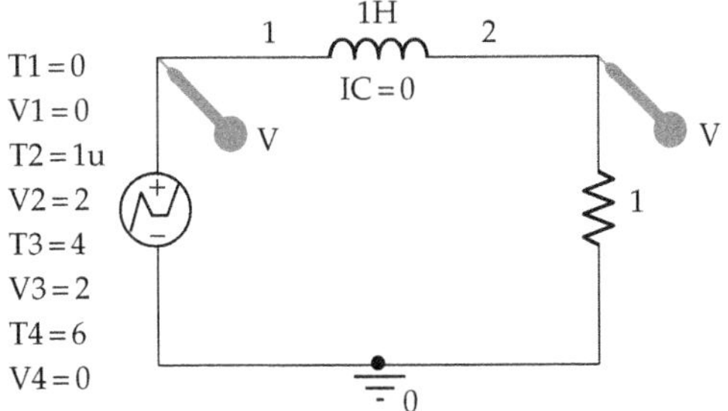

FIGURE 20.14
Figure for Example 20.2.

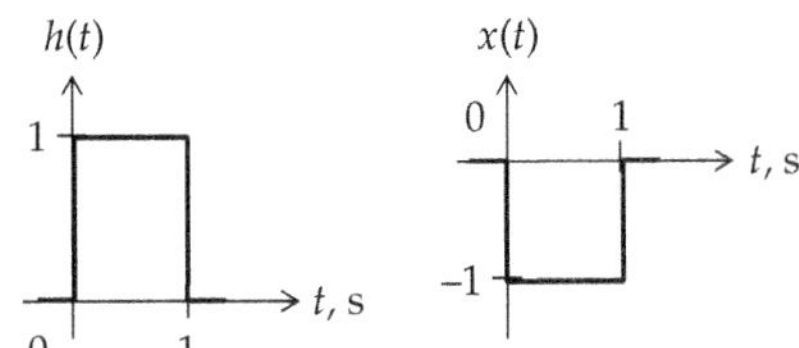

FIGURE 20.16
Figure for Primal Exercise 20.2.

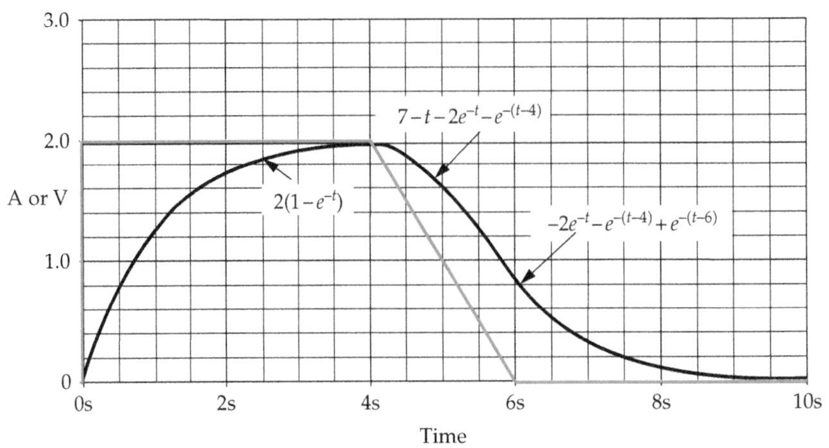

FIGURE 20.15
Figure for Example 20.2.

20.3 Operational Properties of Convolution

The convolution integral is not restricted to the interpretation of Figure 20.3 in terms of the impulse response. It is in fact encountered in other contexts, such as probability theory. In such cases, the convolution integral is to be regarded simply as a mathematical operation on two functions of time, which may be denoted as $f(t)$ and $g(t)$, which need not be interpreted in terms of the input to a system and its impulse response. This is illustrated by Example 20.3 and is used in the following discussion on the operational properties of convolution.

20.3.1 Commutative Property

$$f(t)*g(t)=g(t)*f(t) \quad (20.16)$$

The convolution integral is *commutative*, which means that the two functions in the integral can be interchanged. In terms of Equation 20.10, this means

$$y(t)=\int_{0^-}^{t} f(\lambda)g(t-\lambda)d\lambda=\int_{0^-}^{t} g(\lambda)f(t-\lambda)d\lambda \quad (20.17)$$

The second integral in Equation 20.17 can be derived from the first by substituting $u = t - \lambda$ in the first integral, which gives

$$y(t)=\int_{t}^{0^-} f(t-u)g(u)(-du)=\int_{0^-}^{t} f(t-u)g(u)du \quad (20.18)$$

Replacing the dummy integration variable u by λ gives the second integral in Equation 20.17.

As a consequence of the commutative property, the roles of the impulse response $h(t)$ and the applied input $x(t)$ can be interchanged without affecting the convolution integral.

Primal Exercise 20.3

Repeat Example 20.1 considering the rectangular waveform being the impulse response and the exponential function being the applied input.

20.3.2 Distributive Property

$$x(t)*\left[f(t)+g(t)\right]=x(t)*f(t)+x(t)*g(t) \quad (20.19)$$

The proof of this property readily follows from the distributive property of integration. Thus,

$$x(t)*\left[f(t)+g(t)\right]=\int_{0^-}^{t} x(t)\left[f(t-\lambda)+g(t-\lambda)\right]d\lambda$$

$$=\int_{0^-}^{t} x(t)f(t-\lambda)d\lambda+\int_{0^-}^{t} x(t)g(t-\lambda)d\lambda$$

$$=x(t)*f(t)+x(t)*g(t).$$

We have in fact implicitly used the distributive properties in Example 20.2 when $h(t)$ was convolved with $x(t)$ considered as the sum of functions over different ranges of t.

20.3.3 Associative Property

$$x(t)*\left[f(t)*g(t)\right]=\left[x(t)*f(t)\right]*g(t) \quad (20.20)$$

To prove this property, we note that $f(t)*g(t)=\int_{0^-}^{t} f(t)g(t-\lambda)d\lambda$. Hence, using the commutative property,

$$x(t)*\left[f(t)*g(t)\right]=\int_{0^-}^{t}\left[\int_{0^-}^{t} f(t)g(t-\lambda)d\lambda\right]x(t-\sigma)d\sigma \quad (20.21)$$

where σ is a dummy integration variable. Equation 20.21 can be expressed as

$$\int_{0^-}^{t}\int_{0^-}^{t} f(t)g(t-\lambda)x(t-\sigma)d\lambda d\sigma \quad (20.22)$$

Similarly, $x(t)*f(t)=\int_{0^-}^{t} f(t)x(t-\sigma)d\sigma$, and

$$\left[x(t)*f(t)\right]*g(t)=\int_{0^-}^{t}\left[\int_{0^-}^{t} f(t)x(t-\sigma)d\sigma\right]g(t-\lambda)d\lambda$$

$$=\int_{0^-}^{t}\int_{0^-}^{t} f(t)g(t-\lambda)x(t-\sigma)d\lambda d\sigma$$

as in Equation 20.22.

20.3.4 Invariance with Inverse Integration and Differentiation

According to this property,

$$f(t)*g(t)=f^{(n)}(t)*g^{(-n)}(t)=f^{(-n)}(t)*g^{(n)}(t) \quad (20.23)$$

where

n is a positive integer

superscript (n) denotes the nth derivative

superscript $(-n)$ denotes the nth integral

Equation 20.23 can be easily proved using the Laplace transform for functions that are zero for $t < 0$ (Equation 21.98). Its general proof in the time domain for functions that are not zero for $t < 0$ will not be given here. This property is useful in deriving the convolution integral for functions involving linear segments and step discontinuities, as illustrated by Example 20.5.

Example 20.3: Convolution of Ramp and Cosine Functions

Convolve the two functions $f(t) = tu(t)$ and $g(t) = \cos t u(t)$.

Solution:

Since the two functions are zero for $t < 0$, the lower limit of integration is zero and the upper limit is t. As there is no impulse at the origin, a lower limit of 0 is the same as 0^-. The first step is to replace t by λ as the integration variable, because t will be the variable in the resulting expression for the convolution integral. The functions become $f(\lambda) = \lambda u(\lambda)$ and $g(\lambda) = \cos\lambda u(\lambda)$. The next step is to replace λ by $(t - \lambda)$ in one of the functions. It does not matter in which function this replacement is made because of the commutative property of the convolution integral. We will choose for illustration the replacement of λ by $(t - \lambda)$ in the cosine function. The convolution integral becomes

$$y(t) = \int_0^t \lambda\cos(t-\lambda)d\lambda = \int_0^t \lambda(\cos t\cos\lambda + \sin t\sin\lambda)d\lambda$$
$$= \cos t\int_0^t \lambda\cos\lambda d\lambda + \sin t\int_0^t \lambda\sin\lambda d\lambda \qquad (20.24)$$

Both integrals can be evaluated by integration by parts. This is facilitated by performing the integration by parts on the complex exponent $e^{j\lambda}$. Thus,

$$\int_0^t \lambda e^{j\lambda}d\lambda = \left[\frac{\lambda}{j}e^{j\lambda}\right]_0^t - \frac{1}{j}\int_0^t e^{j\lambda}d\lambda = \left[\frac{\lambda}{j}e^{j\lambda}\right]_0^t + \left[e^{j\lambda}\right]_0^t$$
$$= -jte^{jt} - 0 + e^{jt} - 1 = -jt(\cos t + j\sin t) + (\cos t + j\sin t) - 1$$
$$= (\cos t + t\sin t - 1) + j(\sin t - t\cos t) \qquad (20.25)$$

The real part is the integral of $\lambda\cos\lambda$, whereas the imaginary part is the integral of $\lambda\sin\lambda$. It follows that

$$y(t) = \cos t(\cos t + t\sin t - 1) + \sin t(\sin t - t\cos t)$$
$$= \cos^2 t + t\sin t\cos t - \cos t + \sin^2 t - t\sin t\cos t$$

or

$$y(t) = (1 - \cos t)u(t) \qquad (20.26)$$

Exercise 20.4

Repeat Example 20.3 by replacing λ by $(t - \lambda)$ in $\lambda u(\lambda)$ rather than in $\cos\lambda$ and verify that the result is the same as Equation 20.26.

Primal Exercise 20.5

Convolve (a) $(1-e^{-t})u(t)$ and $e^{-2t}u(t)$; (b) $tu(t)$ and $tu(t)$.

Ans. (a) $\left(\frac{1}{2} - e^{-t} + \frac{e^{-2t}}{2}\right)u(t)$; (b) $(t^3/6)u(t)$.

20.4 Special Cases of Convolution

20.4.1 Convolution of Staircase Functions

A function of finite magnitude and duration can in general be approximated by a series of steps at regular intervals of time (Figure 20.17), the smaller the interval, the better is the approximation. Such a stepped function is a **staircase function**. The convolution integral of two staircase functions is a piecewise-linear function consisting of a series of straight line segments, as will be demonstrated in this section. Analytically, the convolution of two such functions reduces to multiplication of two polynomials and can be readily performed using MATLAB's 'conv' command, as illustrated by Example 20.4. This can be useful in practice, as when the functions are derived experimentally and cannot be expressed analytically.

Example 20.4: Convolution of Staircase Functions

It is required to convolve $f(t)$ and $g(t)$ of Figure 20.18. Such functions, which consist essentially of a series of rectangles, can be used to illustrate the convolution of staircase functions. Convolving rectangular functions is particularly simple, as is demonstrated in this example.

Solution:

The usual procedure based on the graphical interpretation is applied for the sake of illustration. The functions

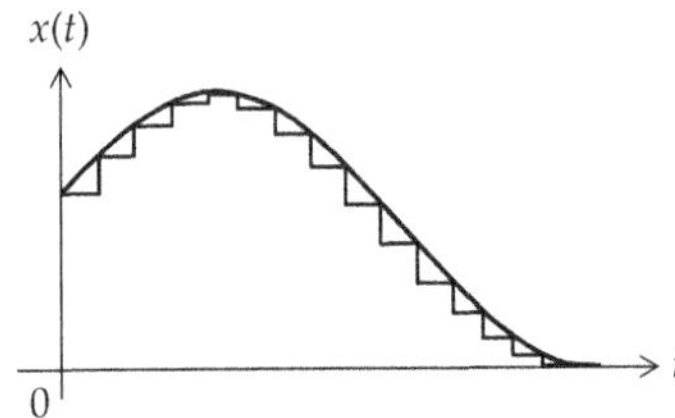

FIGURE 20.17
Staircase approximation to a function.

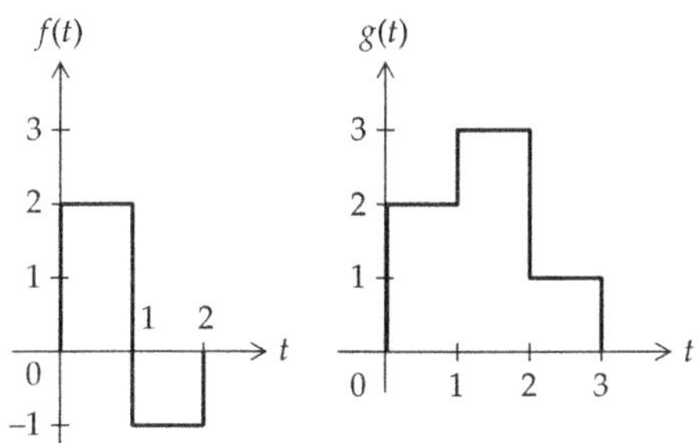

FIGURE 20.18
Figure for Example 20.4.

are first expressed in terms of λ. Either $f(\lambda)$ or $g(\lambda)$ may be folded and shifted by t. $f(\lambda)$ will be folded and shifted in this example. It is generally advisable to fold and shift the simpler function, having the fewer number of distinct ranges of values, or to fold and shift a function that starts at a negative time.

Because of the discrete steps in the two functions, there are several ranges of t to consider:

(a) $0 \le t \le 1$: Figure 20.19a shows $f(-\lambda)$ shifted to the right by t. Over the range $0 \le t \le 1$, the magnitude of $f(-\lambda)$ is 2 and that of $g(\lambda)$ is also 2. The area under the product is $4t$, which is linear in t and varies from 0 at $t = 0$ to 4 at $t = 1$. Because of this linearity, no formal integration need be performed. $y(t)$ over this range is a line that joins the origin, when $t = 0$ and $y(0) = 0$, to the point $y(1) = 4$ at $t = 1$, as shown in Figure 20.22. Hence, *the value of the area under the product need only be determined for the two end values of t in the given range*. When $t = 0$, there is no overlap between $f(-\lambda)$ and $g(\lambda)$, and the product is zero. When $t = 1$, Figure 20.19b applies, and the area under the product is $4 \times 1 = 4$.

(b) $1 \le t \le 2$: The linear variation of area with the shift of t applies no matter how many rectangles are involved. Linearity applies to every pair of rectangles that are being multiplied together, and the sum of the linear functions is still linear in t. The situation when $1 \le t \le 2$ is shown in Figure 20.20a. There are three rectangular areas involved in this case: (i) for the rectangles between 0 and $(t - 1)$, the area is $(-1) \times 2 \times (t - 1) = 2 - 2t$; (ii) for the rectangles between $(t - 1)$ and 1, the area is $2 \times 2 \times (1 - (t - 1)) = 4(2 - t) = 8 - 4t$; and (iii) for the rectangles between 1 and t, the area is $2 \times 3 \times (t - 1) = 6t - 6$. Each of these areas is linear in t. When added together, the sum is 4, which happens to be independent of t in this case. Thus, when $t = 1$, Figure 20.19b applies, and the area is 4. When $t = 2$, the area is $(-1) \times 2 + 2 \times 3 = 4$ (Figure 20.20b). It follows that $y(t)$ remains at 4 from $t = 1$ to $t = 2$, as shown in Figure 20.22. Thus, the area need only be calculated at the two boundary values of t, that is, at $t = 1$ and $t = 2$ in this case. In what follows, only the areas at the boundary values of t will be calculated.

(c) $2 \le t \le 3$: When $t = 3$, Figure 20.20c applies, and the total area under the product is $(-1) \times 3 \times 1 + 1 \times 2 \times 1 = -1$. $y(t)$ changes from 4 at $t = 2$ to -1 at $t = 3$ (Figure 20.22).

(d) $3 \le t \le 4$: When $t = 4$, Figure 20.21a applies, and the total area under the product is $(-1) \times 1 \times 1 = -1$. Hence, $y(t)$ remains at -1 from $t = 3$ to $t = 4$ (Figure 20.22).

(e) $4 \le t \le 5$: When $t = 5$, Figure 20.21b applies and the product becomes zero, because there is no overlap between the two functions.

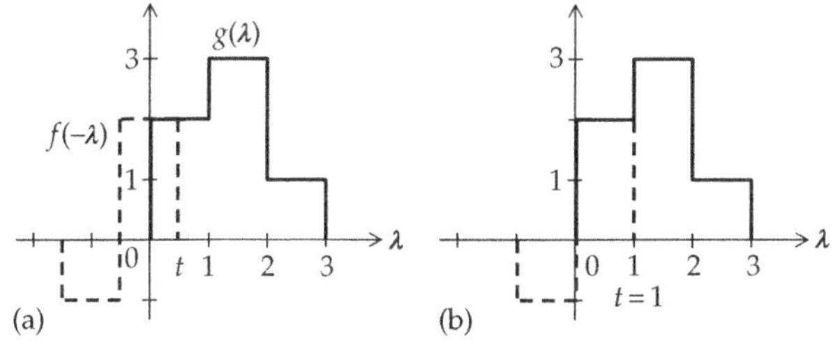

FIGURE 20.19
Figure for Example 20.4.

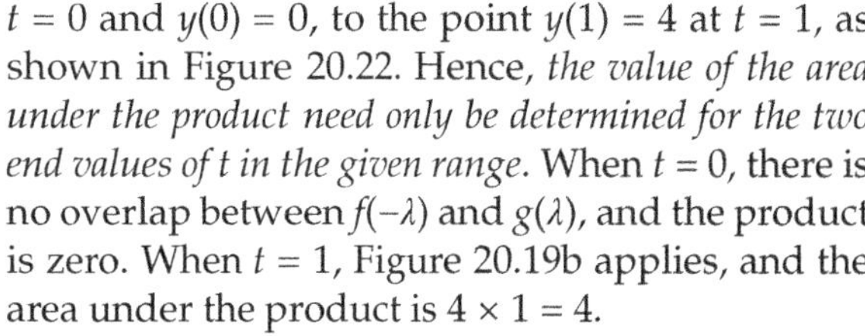

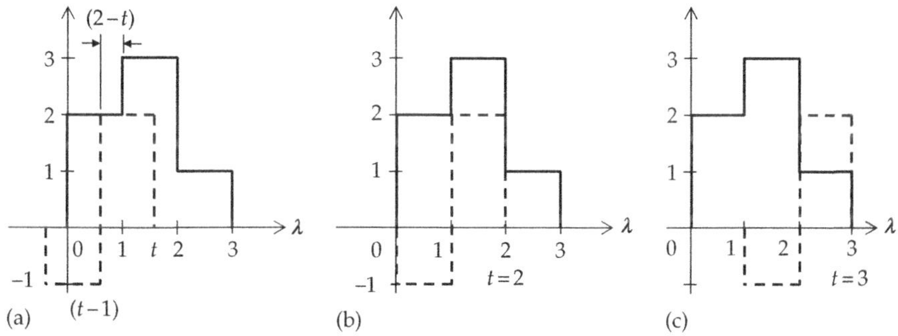

FIGURE 20.20
Figure for Example 20.4.

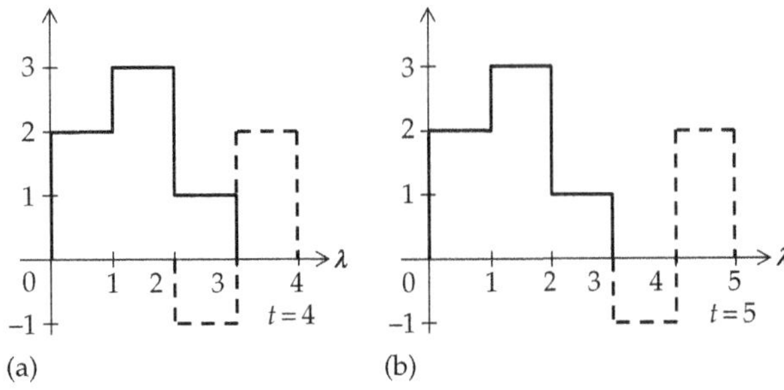

FIGURE 20.21
Figure for Example 20.4.

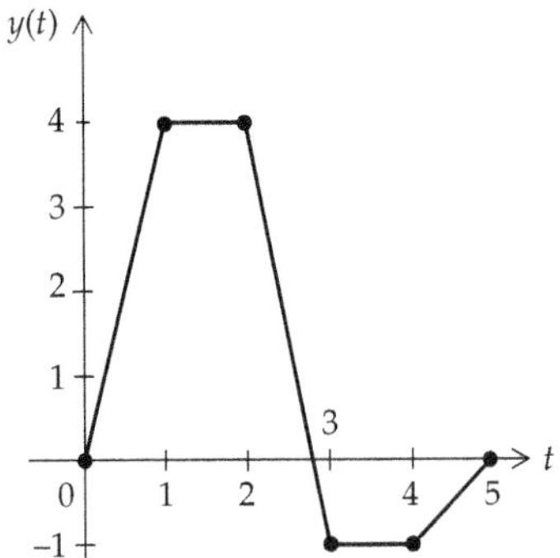

FIGURE 20.22
Figure for Example 20.4.

The convolution integral $y(t)$ is shown in Figure 20.22.

To use MATLAB for convolving two staircase functions, two polynomials representing $f(t)$ and $g(t)$ are formed. The independent variable of the polynomial is arbitrary but will be kept as t, since this is the argument of $y(t)$. However, capital letters are used for the polynomials in order to distinguish the polynomial from the given function. The coefficients of the powers of t in a polynomial are equal to the various levels of the respective staircase function, in the order of decreasing powers t, with the last nonzero level in the positive direction of the time variable t of the function being the constant term of the polynomial. Thus, since $f(t)$ in Figure 20.18 consists of two levels, the first level is the coefficient of t in the polynomial in t and the second level is the constant term. The polynomial is written as $F(t) = 2t - 1$. $G(t)$ consists of three levels: the first level is the coefficient of t^2, the second level is the coefficient of t, and the third level is the constant term. The polynomial is written as $G(t) = 2t^2 + 3t + 1$. The product of the two polynomials will be denoted by $B(t)$, the breakpoint polynomial, given by

$$B(t) = F(t)G(t) = (2t-1)(2t^2+3t+1) = 4t^3 + 4t^2 - t - 1$$

It is seen that the coefficients of $B(t)$ are the values of $y(t)$ at the successive nonzero breakpoints between the first and last zero values of $y(t)$. Note that $B(t)$ is not the same as $y(t)$, the convolution integral. $B(t)$ is just a polynomial whose coefficients give the nonzero breakpoints of $y(t)$.

The multiplication of large polynomials is facilitated by MATLAB's 'conv' command. If we enter the coefficients of $F(x)$ and $G(x)$ as arrays,

```
>> F = [2 -1]
>> G = [2 3 1]
```

followed by the command:

```
>> Y = conv(F, G)
```

MATLAB returns

```
Y =
4    4    -1    -1
```

corresponding to the values of $y(t)$ at the successive nonzero breakpoints, between the first and last zero values of $y(t)$.

When deriving the convolution integral as the product of two polynomials, the following should be noted:

(a) It is assumed that unit time intervals are used. When the time intervals are not unity, the intervals should be scaled to unity, but the values of the function should also be scaled so as to keep the same area under the product of the functions (see Problem P20.26).
(b) It is most convenient to assume that both staircase functions start at $t = 0$. If that is not the case, the functions are advanced or delayed so as to make them start at $t = 0$, then the convolution integral is shifted back by an amount that is opposite that of the *total net* shift of the two functions. This is illustrated by problems at the end of the chapter.

Problem-Solving Tip

- It is generally advisable to fold and shift the simpler function, having the fewer number of distinct ranges of values, or to fold and shift a function that starts at negative time.

20.4.2 Convolution with Impulse Function

It is required to convolve a function $x(t)u(t)$ with $\delta(t)$, a unit impulse at the origin (Figure 20.23a). When t is replaced by λ and the impulse function is folded around

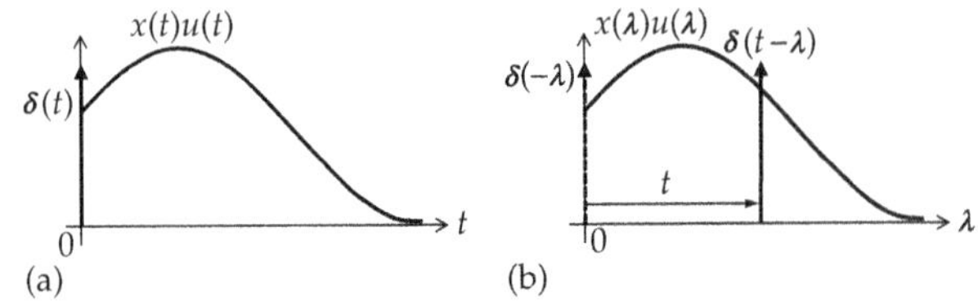

FIGURE 20.23
Convolution with impulse at origin. (a) A function $x(t)u(t)$ is to be convolved with $\delta(t)$ and (b) $\delta(\lambda)$ folded around the vertical axis and shifted by t to form the convolution integral.

the vertical axis, the resulting function $\delta(-\lambda)$ remains an impulse at the origin because $\delta(t)$ is an even function (Equation 18.8). When shifted by t to the right (Figure 20.23b), it becomes $\delta(t-\lambda)$. This is the same as $\delta(\lambda-t)$, because of the evenness of the impulse function. The convolution integral becomes

$$x(t)*\delta(t)=\int_{0^-}^{t}x(\lambda)u(\lambda)\delta(\lambda-t)d\lambda,\quad t>0 \tag{20.27}$$

It is assumed that $x(\lambda)$ does not have an impulse at the origin, so that the lower limit of integration could be unambiguously 0, as explained in connection with Equation 20.5. The integrand is nonzero over the duration of the impulse from $\lambda = t^-$ to $\lambda = t^+$. Assuming that the function is continuous at $\lambda = t$, this value is substituted for λ in $x(\lambda)u(\lambda)$, as usual in evaluating expressions involving impulses. The term $x(t)$ is a constant as far as the integration with respect to λ is concerned and can be taken outside the integral. Since $u(\lambda) = 1$ over the range of integration, with t positive, $u(\lambda)$ is redundant in the integrand. Equation 20.27 becomes

$$x(t)*\delta(t)=\int_{t^-}^{t^+}x(\lambda)\delta(\lambda-t)d\lambda=$$

$$x(t)\int_{t^-}^{t^+}\delta(\lambda-t)d\lambda=x(t),\quad t>0 \tag{20.28}$$

where the integral of a unit impulse over the duration of the impulse is, by definition, unity. The restriction $t > 0$ is removed by multiplying the function by $u(t)$. Equation 20.28 becomes

$$x(t)*\delta(t)=x(t)u(t) \tag{20.29}$$

It is seen that convolution of a function with a unit impulse at the origin is the function itself, for $t > 0$, assuming the function is continuous over the given range of t.

In terms of the graphical interpretation of Figure 20.23, recall that multiplying a continuous function $x(\lambda)$ by an impulse at time $\lambda = t$ makes the impulse "sample" the function at this instant, that is, return the value of the function $x(t)$ as the strength of the impulse $x(t)\delta(\lambda-t)$ at time t. The area under this product, over the duration of the impulse, is $x(t)$. Hence, as t is increased from zero, the function is "swept" by the impulse, resulting in $x(t)$ over the range of the given function.

Equation 20.29 can be interpreted in terms of the commutative property of convolution. Thus, $\delta(t)$ can be considered as an input and $x(t)$ as the response to $\delta(t)$. Equation 20.29 is simply an expression of the fact that if the input is an impulse $\delta(t)$, the response is evidently $x(t)$.

To appreciate the significance of convolution with an impulse, suppose that the impulse is delayed by $a > 0$,

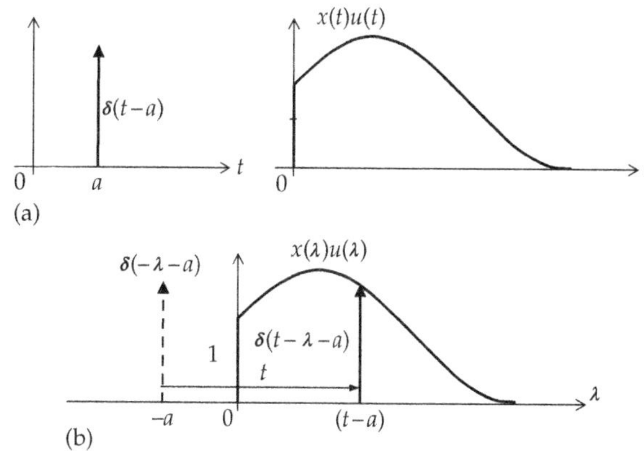

FIGURE 20.24
Convolution with delayed impulse. (a) A function $x(t)u(t)$ is to be convolved with $\delta(t-a)$, and (b) $\delta(\lambda-a)$ folded around the vertical axis and shifted by t to form the convolution integral.

while $x(t)$ remains as it was (Figure 20.24a). Replacing t by λ, the functions become $x(\lambda)$ and $\delta(\lambda-a)$. When the impulse function is folded and shifted to the right by t, λ is replaced by $(t-\lambda)$ so that the impulse function becomes $\delta(t-\lambda-a)$ (Figure 20.24b). Since the impulse function is even, the argument can be negated, without changing the value of the function. Thus, $\delta(t-\lambda-a) = \delta(\lambda-t+a) = \delta(\lambda-(t-a))$. The convolution integral becomes

$$x(t)*\delta(t-a)=\int_{0^-}^{t}x(\lambda)u(\lambda)\delta(\lambda-(t-a))d\lambda,\quad t>a \tag{20.30}$$

The integrand is nonzero over the duration of the impulse, from $\lambda = (t-a)^-$ to $\lambda = (t-a)^+$. Assuming that the function is continuous at $\lambda = t-a$, this value is substituted for λ in the expression $x(\lambda)u(\lambda)$, as usual in evaluating expressions involving impulses, to give $x(t-a)u(t-a)$; $x(t-a)$ is a constant as far as integration with respect to λ is concerned and can be taken outside the integral. $u(t-a) = 1$ for $t > a$. Equation 20.30 becomes

$$x(t)*\delta(t-a)=x(t-a)\int_{(t-a)^-}^{(t-a)^+}\delta(\lambda-(t-a))d\lambda=x(t-a),\ t>a \tag{20.31}$$

The restriction $t > a$ is removed by multiplying the function by $u(t-a)$. Equation 20.31 becomes

$$x(t)*\delta(t-a)=x(t-a)u(t-a) \tag{20.32}$$

It is seen that convolution of a function with a delayed impulse delays the function by the delay of the impulse. This result can be interpreted along the same lines as in Figure 20.23. The difference is that the given function is not swept until t just exceeds a, the delay of the impulse. This means that the function that is returned by the convolution integral is the given function delayed by a.

The next step is to generalize convolution with an impulse to the case where the impulse is delayed by $a > 0$ and the function $x(t)$ is delayed by $b > 0$. The delayed function is considered as $x(t - b)u(t - b)$ and not as $x(t - b)u(t)$, which in general is a different function altogether. To appreciate the difference, consider the function $x(t)u(t)$, with $x(t) = t$, as in Figure 20.25a. The extension of $x(t) = t$ to $t < 0$ is shown in dashed line. If $x(t)$ is delayed by $b > 0$ and multiplied by $u(t)$, the resulting function is shown as the solid line in Figure 20.25b. On the other hand, if the delayed function $x(t - b)$ is multiplied by the delayed step function $u(t - b)$, then the function of Figure 20.25a is delayed as is by b, as shown in Figure 20.25c. In other words, $x(t - b)u(t - b) = 0$ for $t < b$ and is $x(t - b)$ for $t > b$, just as the function $x(t)u(t) = 0$ for $t < 0$ and is $x(t)$ for $t > 0$.

The function $x(t)u(t)$ of Figure 20.23a is shown as a delayed function $x(t - b)u(t - b)$ in Figure 20.26a, together with the delayed impulse $\delta(t - a)$. When t is replaced by λ, and the impulse is folded around the vertical axis and shifted by t to the right, λ is replaced by $(t - \lambda)$ so that the impulse becomes $\delta(t - \lambda - a)$ (Figure 20.26b). Since the impulse function is even, the argument can be negated, without changing the value of the function. Thus, $\delta(t - \lambda - a) = \delta(\lambda - t + a) = \delta(\lambda - (t - a))$. The convolution integral becomes

$$x(t-b)*\delta(t-a)=$$
$$\int_{0^-}^{t} x(\lambda-b)u(\lambda-b)\delta(\lambda-(t-a))d\lambda, \quad t>(b+a) \quad (20.33)$$

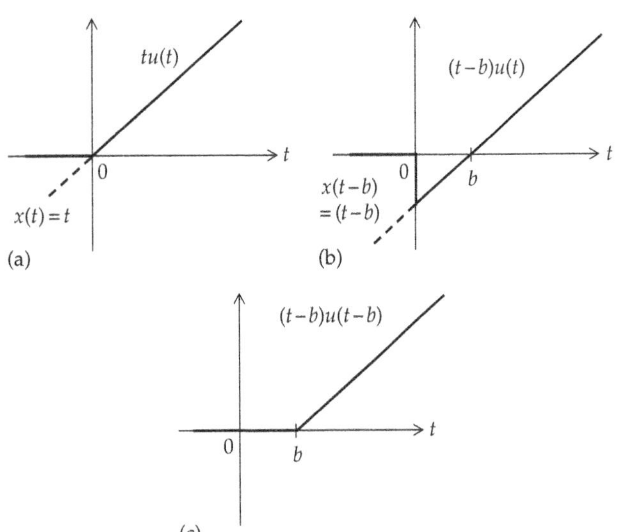

FIGURE 20.25
Delaying a function. (a) The functions $x(t) = t$ and $x(t) = tu(t)$, (b) the function $(t - b)u(t)$, and (c) the function $(t - b)u(t - b)$.

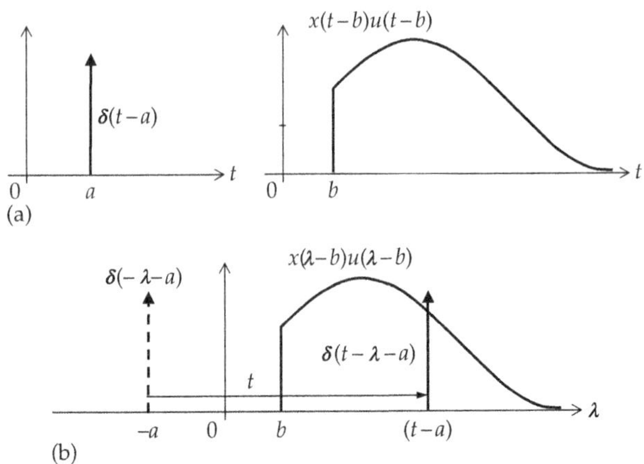

FIGURE 20.26
Convolution of delayed function with delayed impulse. (a) A delayed function $x(t - b)u(t - b)$ is to be convolved with $\delta(t - a)$, and (b) $\delta(\lambda - a)$ folded around the vertical axis and shifted by t to form the convolution integral.

There is no overlap if $t < (b + a)$. The integrand is nonzero over the duration of the impulse, from $\lambda = (t - a)^-$ to $\lambda = (t - a)^+$. Assuming that the function is continuous at $\lambda = t - a$, this value is substituted for λ in the expression $x(\lambda - b)u(\lambda - b)$, as usual in evaluating expressions involving impulses. The terms $x(\lambda - b)u(\lambda - b)$ in the integrand become $x(t - a - b)u(t - a - b)$. The term $x(t - a - b)$ is a constant as far as integration with respect to λ is concerned and can be taken outside the integral. The step function $u(t - a - b)$ is redundant since its argument is positive for $t > (b + a)$. Equation 20.33 becomes

$$x(t-b)*\delta(t-a)=$$
$$x(t-a-b)\int_{(t-a)^-}^{(t-a)^+} \delta(\lambda-(t-a))d\lambda = x(t-b-a), \quad t>(b+a) \quad (20.34)$$

The restriction $t > (b + a)$ is removed by multiplying the function by $u(t - b - a)$. Equation 20.34 becomes

$$x(t-b)*\delta(t-a) = x(t-a-b)u(t-a-b) \quad (20.35)$$

In effect, as t increases beyond $(a + b)$, the impulse sweeps over the $x(\lambda - b)u(\lambda - b)$ function, returning the values of the function $x(t - b)$, but after a further delay a.

Concept: *Convolving a function with a unit impulse delays the function by the delay of the unit impulse, assuming the function to be continuous over given range of the shift t.*

Note that whereas multiplication of a function by a unit impulse samples the function at the time of occurrence of the impulse (Example 18.1), convolving the function with a unit impulse delays the function by the delay of the impulse. If the unit impulse occurs at the origin, the function is unaltered.

20.4.3 Convolution with Step Function

It is required to convolve a function $x(t)u(t)$ with $u(t)$, a unit step function at the origin. If t is replaced by λ, and $u(\lambda)$ is folded around the vertical axis, the functions become as shown in Figure 20.27. When $u(-\lambda)$ is shifted to the right by t, it becomes $u(t-\lambda)$. The product $x(\lambda)u(\lambda)u(t-\lambda)$ is the part of the function $x(\lambda)$ between 0 and t. The area under the product, which is the convolution integral, is the shaded area. It is seen that this area is simply the integral of $x(\lambda)$ between 0 and t. Thus,

$$x(t)*u(t)=\int_0^t x(\lambda)d\lambda \tag{20.36}$$

Suppose that $u(t)$ is delayed by $a>0$, which is the same as shifting $u(t)$ to the right by a, so it becomes $u(t-a)$ (Figure 20.28a). Suppose that the function $x(t)$ is delayed by $b>0$ (Figure 20.28a) to become $x(t-b)u(t-b)$, as in Figure 20.28a. When t is replaced by λ, and the step function $u(\lambda-a)$ is folded around the vertical axis and shifted by t to the right, λ is replaced by $(t-\lambda)$ so that the step function becomes $u(t-\lambda-a)$. It is seen from Figure 20.28b that if $t<(a+b)$, there is no overlap between the functions $u(t-\lambda-a)$ and $x(\lambda-b)u(\lambda-b)$, so that the product is zero. The product is nonzero only for $t>(a+b)$. The convolution integral is the shaded area under the product and is given by

$$x(t-b)*u(t-a)=\int_b^{t-a} x(\lambda-b)u(\lambda-b)d\lambda,\quad t>(a+b) \tag{20.37}$$

$u(\lambda-b)$ in the integrand is redundant, because it is zero for $\lambda<b$, and the lower limit of integration is b, anyway. With $t>(a+b)$, it follows that $t-a>b$, so that the step function is unity over the range of integration. The restriction that $t>(a+b)$ can be accounted for by multiplying the integral by $u(t-a-b)$. Equation 20.37 becomes

$$x(t-b)*u(t-a)=\left[\int_b^{t-a} x(\lambda-b)d\lambda\right]u(t-b-a) \tag{20.38}$$

The following concept applies:

Concept: *Convolving a function with a unit step is equivalent to integrating the function and delaying it by the delay of the unit step. The range of integration is from the start of the function to a value of the time variable λ that is t minus the delay of the unit step.*

If the unit step occurs at the origin, the function is integrated up to t, as in Equation 20.36. Example 20.5 illustrates convolution with step and impulse functions, as well as Equation 20.23.

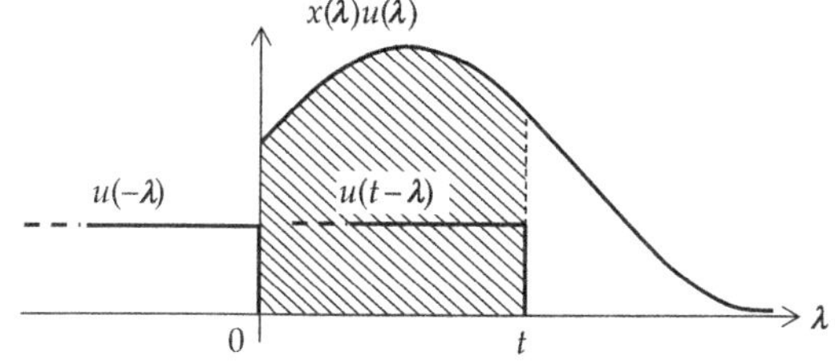

FIGURE 20.27
Convolution with step at origin.

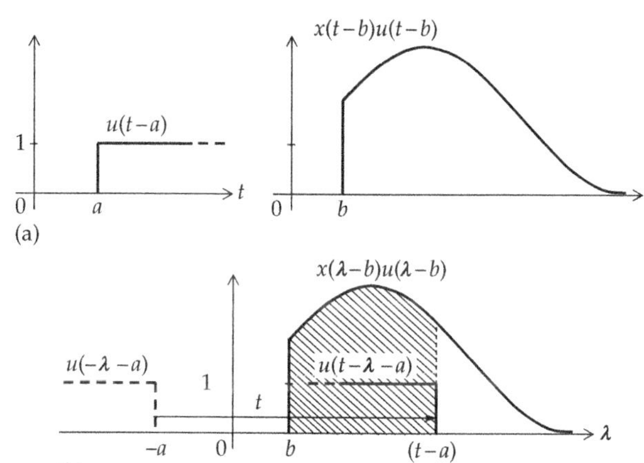

FIGURE 20.28
Convolution of delayed function with delayed step.

Primal Exercise 20.6

Determine (a) $y(t)=u(t)*u(t)$; (b) $y(t)=u(t)*\delta(t)$, considering the convolution as being either with a unit step at the origin or with a unit impulse at the origin; and (c) $y(t)=u(t)*tu(t)$.

Ans. (a) $tu(t)$; (b) $u(t)$; (c) $(1/2)t^2u(t)$.

Primal Exercise 20.7

Determine the convolution of $f(t)$ shown in Figure 20.29 with $\delta(t-0.5)$, where t is in s.

Ans. $f(t)$ is delayed by 0.5 s.

Primal Exercise 20.8

Determine the value of the convolution integral at $t=0$ when $f(t)$ in Figure 20.30 is convolved with the impulses $2(\delta(t+1)+\delta(t-1))$.

Ans. 8.

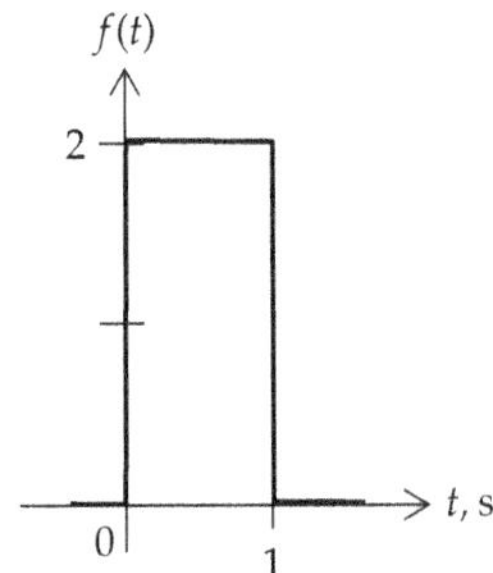

FIGURE 20.29
Figure for Primal Exercise 20.7.

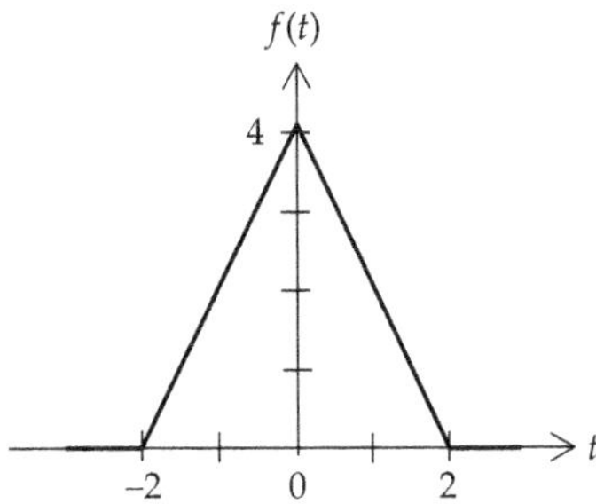

FIGURE 20.30
Figure for Primal Exercise 20.8.

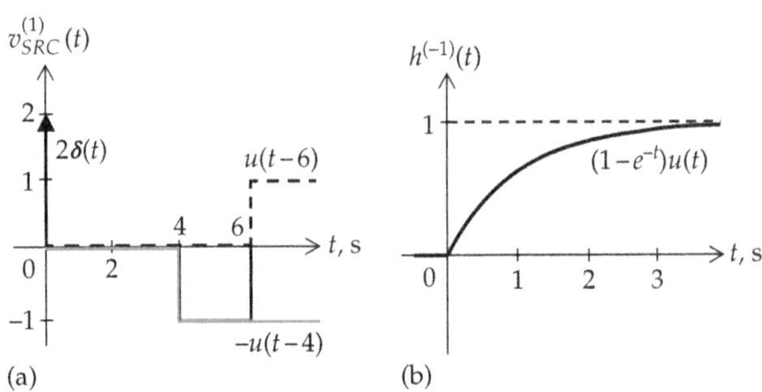

FIGURE 20.31
Figure for Example 20.5.

Example 20.5: Convolution with Derivative of Trapezoidal Function

It is required to evaluate the convolution integral of Example 20.2 using Equation 20.23 with $n = 1$.

Solution:

Either $v_{SRC}(t)$ or $h(t)$ can be identified with $f(t)$ or $g(t)$. Identifying $v_{SRC}(t)$ with $f(t)$, the derivative of $v_{SRC}(t)$ is as shown in Figure 20.31a. $v_{SRC}(t) = 0, t < 0$, with a step jump to $v_{SRC}(t) = 2$ at $t = 0$, whose derivative is the impulse $2\delta(t)$. The derivative $v_{SRC}^{(1)}(t) = 0$, $0 < t < 2$, $v_{SRC}^{(1)}(t) = -1$, $4 < t < 6$ s, and $v_{SRC}^{(1)}(t) = 0$, $t > 6$ s. The rectangular function in the interval $4 < t < 6$ s can be considered as the sum of two step functions $-u(t - 4)$ and $u(t - 6)$, shown dashed in Figure 20.31a. The integral of the response to the unit impulse is (Figure 20.31b)

$$h^{(-1)}(t) = \int_0^t e^{-t} dt = (1 - e^{-t})u(t) \tag{20.39}$$

For the interval between 0 and 4 s, $h^{(-1)}(t)$ is convolved only with the impulse $2\delta(t)$, that is,

$$i'(t) = h^{(-1)}(t) * 2\delta(t) = 2(1 - e^{-t}) \quad 0 \le t \le 4\text{ s} \tag{20.40}$$

which is identical with Equation 20.13 over the interval $0 \le t \le 4$ s.

Convolution with the negative step is obtained from Equation 20.38 with $a = 4$ and $b = 0$. This gives

$$\left[(1 - e^{-t})u(t)\right] * (-u(t - 4))$$

$$= -\int_0^{t-4} (1 - e^{-\lambda}) d\lambda = -t + 5 - e^{-(t-4)}, \quad 4 \le t \le 6\text{ s} \tag{20.41}$$

To obtain the convolution integral for the interval from 4 to 6 s, the convolution with the step $-u(t - 4)$ is *added* to the convolution integral with the impulse, because the convolution integral starts at $t = 0^-$ (Equation 20.5). Adding Equations 20.40 and 20.41 gives Equation 20.14.

Convolution with the positive step is obtained from Equation 20.38 with $a = 6$ and $b = 0$. This gives

$$\left[(1 - e^{-t})u(t)\right] * (u(t - 6))$$

$$= \int_0^{t-6} (1 - e^{-\lambda}) d\lambda = t - 7 + e^{-(t-6)}, \quad t \ge 6\text{ s} \tag{20.42}$$

To obtain the convolution integral for $t \ge 6$ s, Equation 20.42 should be added to Equations 20.40 and 20.41, which gives Equation 20.15.

*20.4.4 Implications of Impulse Response

Convolution with impulse and step functions can be interpreted in terms of system memory and distortion. Consider Figure 20.32a, which is essentially the same as Figure 20.4. Recall that in evaluating the convolution integral at time t, a narrow-pulse constituent of the input such as $x(\lambda_i)$ at $\lambda = \lambda_i$, shown in gray in Figure 20.32a, is interpreted as an impulse, and its contribution to the output at t is evaluated by multiplying $x(\lambda_i)$ by the value of the impulse response $h(t - \lambda_i)$, at a time $(t - \lambda_i)$ after the start of the impulse response. Time t can be interpreted as the "present instant" at which the convolution integral, or output, is desired. Pulses occurring at $\lambda > t$ are "future" inputs that have not yet occurred. As pointed out earlier, the contributions of these pulses to

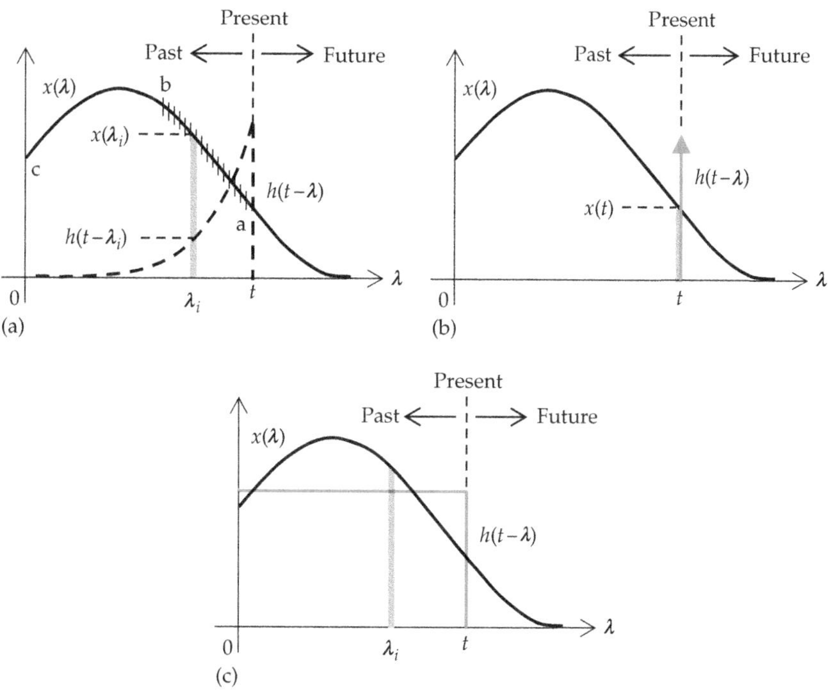

FIGURE 20.32
Memory and distortion.

the present input are excluded by taking t as the upper limit of the convolution integral, on the basis that physical systems operating in real time are causal, that is, they do not anticipate, and hence do not respond, to an input before it occurs.

Pulses occurring at $\lambda < t$ are "past" inputs whose contributions are summed by the integration to obtain the total response at time t. The contributions of these pulses are weighted by the impulse response $h(t - \lambda_i)$. It is seen from Figure 20.32a that this weighting depends on the shape of the impulse response. Because of the exponential decay of the impulse response shown, pulses occurring over part 'ab', shown crossed in Figure 20.32a, have more weight than pulses occurring earlier in time, over the part 'bc'. The contributions of pulses occurring at $\lambda < t$ can be interpreted as a kind of "memory" possessed by the circuit, in the sense that past inputs are not "forgotten", since they affect the output at the present time t. If the impulse response is itself an impulse (Figure 20.32b), the circuit is "memoryless" in the sense that the present output depends only on the present input. On the other hand, a circuit whose impulse response is a step function (Figure 20.32c) has "perfect memory" in the sense that past inputs are not only remembered, but their contributions to the present input are equally weighted.

It may be wondered as to the kind of circuit that is memoryless or has perfect memory. If we consider a unit impulse of current $\delta(t)$ applied to a resistor R, the voltage response is $h(t) = R\delta(t)$, as in Figure 20.32b. An ideal resistor is therefore memoryless, and by extension, purely resistive circuits are memoryless as well. On the other hand, if a unit impulse of current $\delta(t)$ is applied to a capacitor C, the voltage response is $h(t) = (1/C)\int_0^t \delta(t)dt = u(t)/C$, which is a step function, as shown in Figure 20.32c. An ideal capacitor therefore has perfect memory, and by extension, purely capacitive circuits have perfect memory as well.

Another implication of the impulse response is that if this response is an impulse at the origin, the output is the same as the input, but scaled by the strength of the impulse (Equation 20.29). The output is therefore an undistorted form of the input, where distortion refers to deviation of the output waveform from that of the input. If the impulse response is a delayed impulse, the output is also undistorted but is delayed with respect to the input (Equation 20.32). In the frequency domain, a distortionless delay is produced by circuits having a magnitude response that is independent of frequency and a phase shift that is directly proportional to frequency (Equation 23.32). On the other hand, if the impulse response is a step function, the output waveform is highly distorted by the integration of the input (Equation 20.38).

In conclusion, the larger the deviation of the impulse response of a circuit from an impulse, the more substantial is the circuit memory and the larger is the distortion of the output, compared to the input.

20.5 Some General Properties of the Convolution Integral

We will summarize in what follows some general properties of the convolution integral that can serve as a guide to its derivation and a check on its correctness.

Property 1: *If both functions being convolved start with finite or zero values, the convolution integral starts with a zero value, at a value of t when overlap between the two functions is just about to begin.*

Thus, the two functions convolved in Examples 20.1, 20.2, and 20.4 both start with finite values at $t = 0^+$. The convolution integrals $y(t)$ in Figures 20.10, 20.14, and 20.22 all start at the origin. This follows from the fact that if one function is folded, without shift ($t = 0$), it extends from the origin in the negative direction of λ, whereas the other function extends from the origin in the positive direction of λ. The two functions do not overlap for $t < 0$, but overlap for $t > 0$. Hence, $y(t)$ also starts at the origin. In Example 20.5, $v_{SRC}^{(1)}(t)$ has an impulse at the origin, whereas $h^{(-1)}(t)$ starts with a zero value at the origin. $y(t)$ starts at the origin (Figure 20.31) because the product of an impulse and zero is zero (Example 18.1). Had $h^{(-1)}(t)$ started with a finite value, then $y(t)$ would have started with a finite value.

Property 2: *If both functions being convolved end with a value of zero, the convolution integral also ends with a value zero, at a finite value of t when there is no longer any overlap between the two functions.*

In Example 20.4, both functions end with a zero value. $y(t)$ ends with a zero value. In Examples 20.1 and 20.2, $h(t)$ theoretically extends to infinity, so that $y(t)$ also extends to infinity.

Property 3: *If both functions being convolved are of finite duration, the duration of the convolution integral is the sum of the durations of the two functions.*

Thus, $f(t)$ and $g(t)$ in Figure 20.18 are of durations of two and three units, respectively. The duration of the convolution integral in Figure 20.22 is five units. The justification follows from Figure 20.21b. The overlap between the two functions just ends, and $y(t)$ becomes zero, for a value of t that is the sum of the durations of the two functions.

Property 4: *Delaying one or both functions being convolved delays the start of the convolution integral by an interval that is the sum of the delays of the two functions.*

In Equation 20.35, the convolution integral is delayed by the sum of the delays of $x(t)$ and the impulse function, until overlap just begins. Similarly, in Equation 20.38, the convolution integral is delayed by the sum of the delays of $x(t)$ and the step function.

Examples 20.6 and 20.8 illustrate convolution with a function that starts at a finite negative time. In such cases, the negative value of t at which a function starts is effectively a "negative delay". Moreover, the time-invariant property of LTI systems can be invoked in these cases, as illustrated by Example 20.6.

Example 20.6: Convolution of Pulse with Itself

$f(t)$ in Figure 20.33a is usually denoted as $A\text{rect}(t/\tau)$. It is required to derive the convolution integral of $f(t)$ when convolved with itself.

Solution:

Method 1: When t is replaced by λ, and $f(\lambda)$ is folded around the vertical axis, its waveform does not change because of its even symmetry. The product is $[A\text{rect}(t/\tau)]^2$ and the area under the product is $A^2\tau$. Evidently, this is the maximum value of the convolution integral, for any shift to the right or left reduces the area under the product.

Because $f(t)$ starts with a finite value at $t = -\tau/2$, the convolution integral must start at zero, with no overlap. In order for overlap to just begin, the folded function should be shifted to the left by τ (Figure 20.33b), that is, $t = -\tau$. For $-\tau \le t \le 0$, the shifted function is shown in Figure 20.34a. The convolution function $y(t)$ is reduced from its maximum value of $A^2\tau$ by the product area A^2t. The area under the product becomes $A^2(t + \tau)$, $-\tau \le t \le 0$, which is the equation of a straight line of slope 1 and which intersects the time axis at $t = -\tau$. Thus, $y(t) = A^2\tau$ for $t = 0$ and decreases linearly to zero at $t = -\tau$ (Figure 20.35).

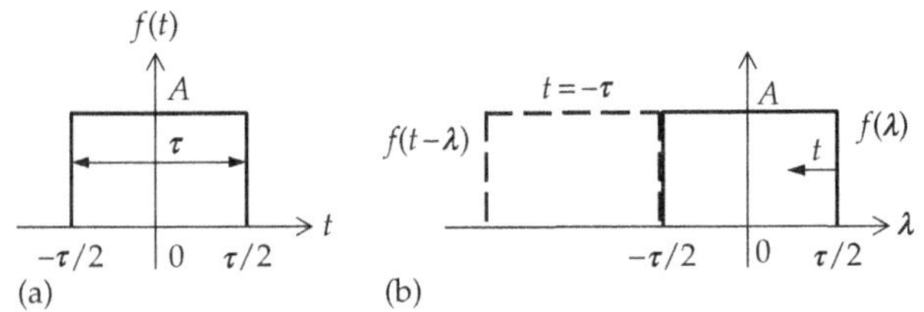

FIGURE 20.33
Figure for Example 20.6.

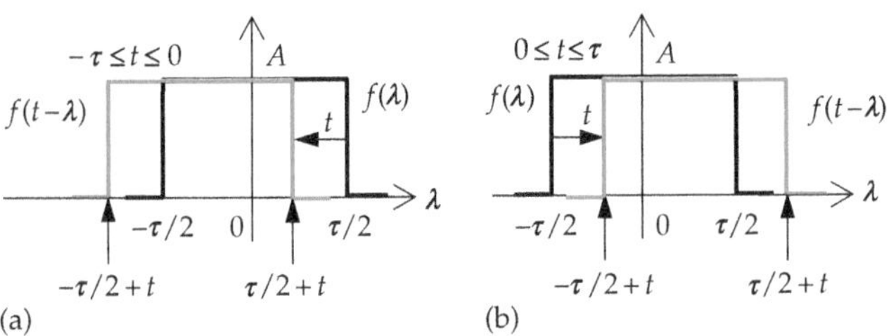

FIGURE 20.34
Figure for Example 20.6.

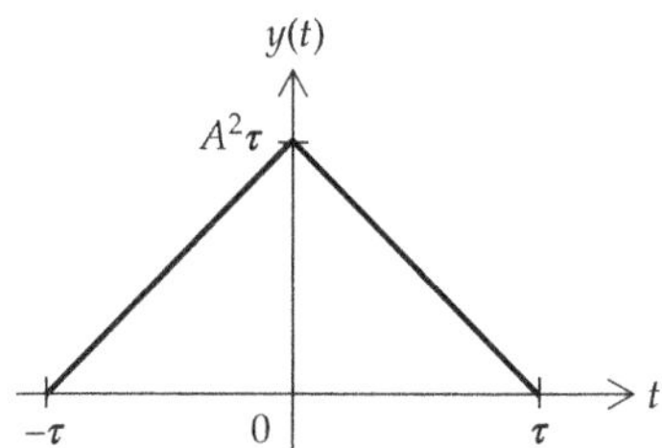

FIGURE 20.35
Figure for Example 20.6.

Note that the leading edge of the folded and shifted function is at $\tau/2 - (-t) = \tau/2 + t$, $-\tau \le t \le 0$, and its trailing edge is at $\tau/2 + t - \tau = -\tau/2 + t$. The convolution integral is $y_1(t) = A^2\int_{-\tau/2}^{\tau/2+t} d\lambda = A^2(t+\tau), -\tau \le t \le 0$, as before.

If the folded function is shifted to the right by t, the area under the product is also reduced until there is no longer any overlap at $t = \tau$ (Figure 20.34b). For $0 \le t \le \tau$, the maximum area is reduced by A^2t, so that $y(t) = A^2(-t + \tau)$, which is the equation of a straight line of slope − 1 and which intersects the time axis at $t = \tau$. It is seen that $y(t) = A^2\tau$ for $t = 0$ and decreases linearly to zero at $t = \tau$ (Figure 20.35). The leading edge of the folded and shifted function is at $\tau/2 + t$, $0 \le t \le \tau$, and its trailing edge is at $\tau/2 + t - \tau = -\tau/2 + t$, the same as for negative t, since t is a variable that can assume positive or negative values, so that as a symbol t can be treated like a positive number. The convolution integral is $y_1(t) = A^2\int_{-\tau/2+t}^{\tau/2} d\lambda = A^2(-t+\tau), 0 \le t \le \tau$, as before.

Method 2: Let $f(t)$ be shifted by $\tau/2$ to the right, so that it starts at the origin (Figure 20.36a). When t is replaced by λ, and $f(\lambda)$ is folded around the vertical axis, the right-hand vertical edge of $f(-\lambda)$ coincides with the vertical axis, so that no overlap occurs at $t = 0$ between $f(\lambda)$ and $f(-\lambda)$. The convolution integral $y(t)$ starts from zero at $t = 0$. When the folded function is shifted by t to the right, $0 \le t \le \tau$ (Figure 20.36b), the area under the product is A^2t and reaches a maximum value of $A^2\tau$ at $t = \tau$. For $\tau \le t \le 2\tau$ (Figure 20.36c), the area under the product is $y(t) = A^2(\tau - (t - \tau)) = A^2(2\tau - t)$. $y(t)$ decreases linearly from $A^2\tau$ at $t = \tau$ to zero at $t = 2\tau$ (Figure 20.36d).

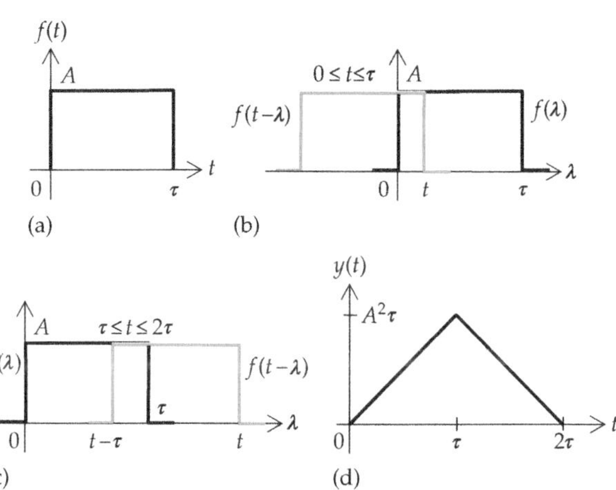

FIGURE 20.36
Figure for Example 20.6.

Since the system is an LTI, then shifting the input by a certain time interval shifts the output by the same time interval. If $f(t)$ in Figure 20.36a is shifted to the left by $\tau/2$, it becomes as in Figure 20.33a. But since the function is convolved with itself, this means that two input functions are each shifted to the left by $\tau/2$. The output is shifted to the left by τ. If $y(t)$ in Figure 20.36d is shifted to the left by $2(\tau/2) = \tau$, it becomes as in Figure 20.35.

An important consideration in convolution based on the graphical procedure is identifying the consecutive ranges of t of the convolution integral. In general, a distinct range of t of the convolution integral is defined by the following:

1. The integrands of the convolution integral are the same over the given range of t, but are different from those of the preceding range of t or the following range of t, where an integrand is the product, over the range, of the unshifted function and the folded and shifted function
2. The integrands of the convolution integral are the same as those of a preceding range of t, or a following range of t, but the integration limits are different.

As mentioned previously, a useful test for the correctness of the ranges of t is to verify that the convolution integral has the same value for values of t that are common between successive ranges of t. The following two examples illustrate these points.

Example 20.7: Convolution of Pulse with Two Half-Sinusoids

It is required to derive the convolution integral of the functions $f(t)$ and $g(t)$ in Figure 20.37, where $g(t) = 2\sin\pi t$, $0 \le t \le 1$ s, and $g(t) = -2\sin\pi t$, $1 \le t \le 2$ s.

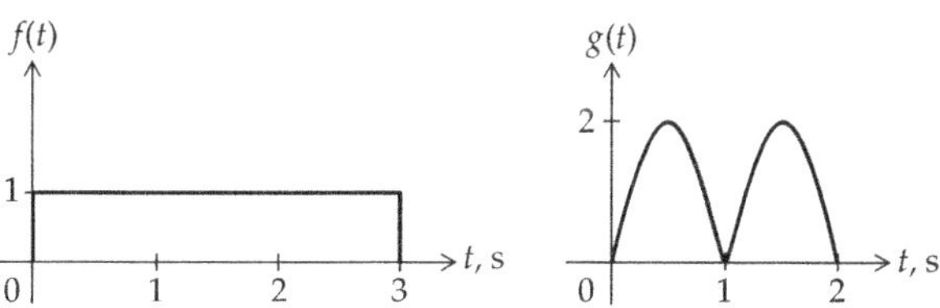

FIGURE 20.37
Figure for Example 20.7.

Solution:

After replacing t in the given functions by λ, $f(\lambda)$ is folded and shifted to the right by t. The first range of t is $0 \le t \le 1$ s (Figure 20.38a). Over this range, the product of $f(t - \lambda)$ and $g(\lambda)$ has the same expression, but this expression is different from that for $t > 1$. The convolution integral is

$$y_1(t)=\int_0^t 2\sin\pi\lambda\, d\lambda=\frac{2}{\pi}\left[-\cos\pi\lambda\right]_0^t=\frac{2}{\pi}(1-\cos\pi t),\ 0\le t\le 1\text{ s} \tag{20.43}$$

The end values are $y_1(0) = 0$ and $y_1(1) = 4/\pi$. As $f(t - \lambda)$ is shifted further to the right, the next range of t is $1 \le t \le 2$ s (Figure 20.38b), over which the product of $f(t - \lambda)$ and $g(\lambda)$ has the same expression, but this expression is different from that for $t < 1$, because of the negative sine function and the added product with the first sine function. The convolution integral is

$$y_2(t)=\int_0^1 2\sin\pi\lambda\, d\lambda+\int_1^t -2\sin\pi\lambda\, d\lambda=\frac{2}{\pi}\left[-\cos\pi\lambda\right]_0^1 +\frac{2}{\pi}\left[\cos\pi\lambda\right]_1^t=\frac{2}{\pi}(3+\cos\pi t),\quad 1\le t\le 2\text{ s} \tag{20.44}$$

The end values are $y_2(1) = 4/\pi$ and $y_2(2) = 8/\pi$. As $f(t - \lambda)$ is shifted further to the right, the next range of t is $2 \le t \le 3$ s (Figure 20.39a). The convolution integral is

$$y_3(t)=\int_0^1 2\sin\pi\lambda\, d\lambda+\int_1^2 -2\sin\pi\lambda\, d\lambda=\frac{2}{\pi}\left[-\cos\pi\lambda\right]_0^1 +\frac{2}{\pi}\left[\cos\pi\lambda\right]_1^2=\frac{8}{\pi},\quad 2\le t\le 3\text{ s} \tag{20.45}$$

$y_3(t)$ is independent of t over this range. Note that the integrands in Equations 20.44 and 20.45 are the same, but the upper integration limit of the second integral is different. This gives a different convolution integral and hence defines a distinct range of t.

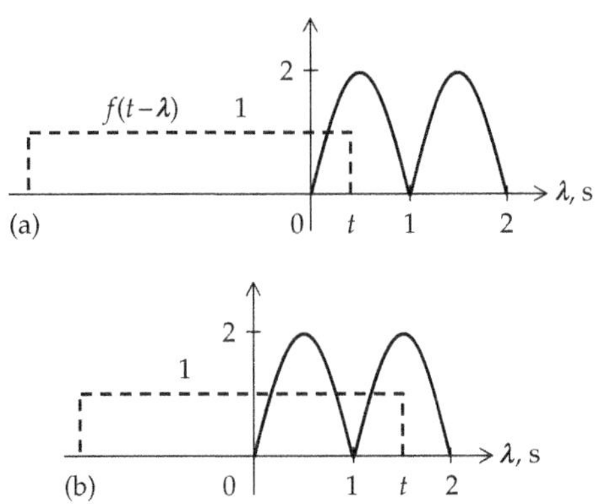

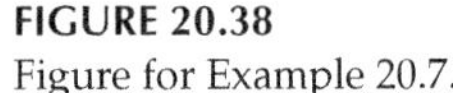

FIGURE 20.38
Figure for Example 20.7.

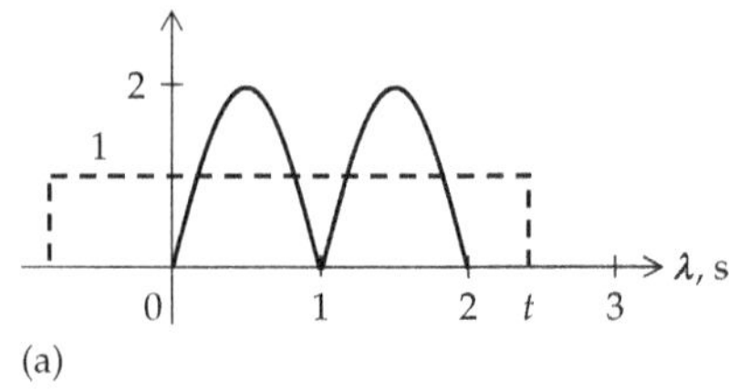

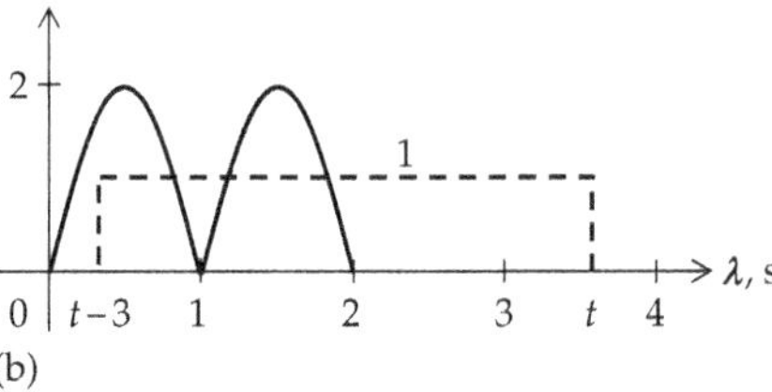

FIGURE 20.39
Figure for Example 20.7.

As $f(t - \lambda)$ is shifted further to the right, the next range of t is that for which the trailing edge of $f(t - \lambda)$, at $(t - 3)$, is between 0 and 1 (Figure 20.39b), that is, $3 \le t \le 4$ s. Over this range, the product of $f(t - \lambda)$ and $g(\lambda)$ has the same integrands as in Equation 20.45 but a different lower limit of the first integral. The convolution integral is

$$y_4(t)=\int_{t-3}^1 2\sin\pi\lambda\, d\lambda+\int_1^2 -2\sin\pi\lambda\, d\lambda=\frac{2}{\pi}\left[-\cos\pi\lambda\right]_{t-3}^1 +\frac{2}{\pi}\left[\cos\pi\lambda\right]_1^2=\frac{2}{\pi}(3-\cos\pi t),\quad 3\le t\le 4\text{ s} \tag{20.46}$$

The end values are $y_4(3) = 8/\pi$ and $y_4(4) = 4/\pi$. As $f(t - \lambda)$ is shifted further to the right, the next range of t is that for which the trailing edge of $f(t - \lambda)$, at $(t - 3)$, is between 1 and 2 (Figure 20.40a), that is, $4 \le t \le 5$ s. Over this range, the product of $f(t - \lambda)$ and $g(\lambda)$ has the same

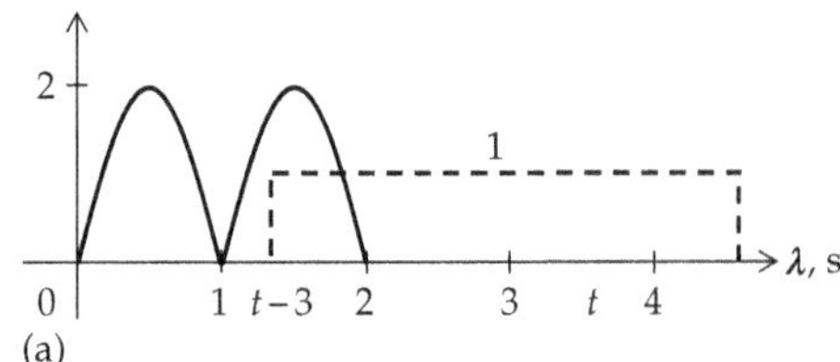

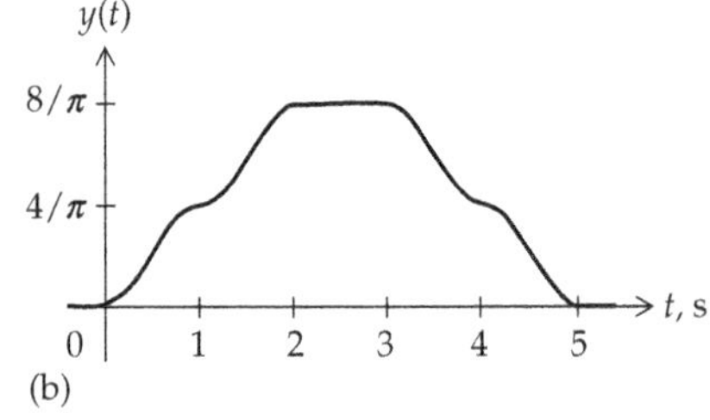

FIGURE 20.40
Figure for Example 20.7.

expression, but this expression is different from that for $t < 4$. The convolution integral is

$$y_5(t) = \int_{t-3}^{2} -2\sin\pi\lambda d\lambda = \frac{2}{\pi}\left[\cos\pi\lambda\right]_{t-3}^{2} = \frac{2}{\pi}(1+\cos\pi t),$$

$$4 \le t \le 5\text{ s} \qquad (20.47)$$

The end values are $y_5(4) = 4/\pi$ and $y_5(5) = 0$. For $t \ge 5$, the there is no overlap and $y(t) = 0$. The convolution integral $y(t)$ is shown in Figure 20.40b.

In convolving functions starting at a finite negative time, specifying the different ranges of t of the convolution integral is considerably facilitated by identifying a part of the function whose abscissa has the same value as the shift in t. This is illustrated by the following example.

Example 20.8: Convolution of Triangular and Biphasic Pulses

It is required to derive the convolution integral of the functions $f(t)$ and $g(t)$ shown in Figure 20.41.

Solution:

$g(t)$ extends into negative time. When t is replaced by λ, and $g(\lambda)$ is folded around the vertical axis, its waveform does not change because of its even symmetry (Figure 20.42a). For $t = 0$, that is, without any shift of $g(-\lambda)$, the product $g(-\lambda)f(\lambda) \neq 0$, and the convolution integral $y(0)$ is the shaded area under the product, which is 0.5. This is not the start of the convolution integral, because when the two functions being convolved start with zero or finite values, the convolution integral should start with a zero value, corresponding to no overlap between the unshifted function and the other function that has been folded. These functions just overlap if $g(-\lambda)$ is shifted to the left by 1 s (Figure 20.42b). Note that the λ-coordinate of the vertex of the shifted function $g(t - \lambda)$ is equal to the shift: with no shift ($t = 0$), the vertex is on the vertical axis ($\lambda = 0$), whereas with a shift of $t = -1$, the λ-coordinate of the vertex is -1. Hence, the amount of shift is the same

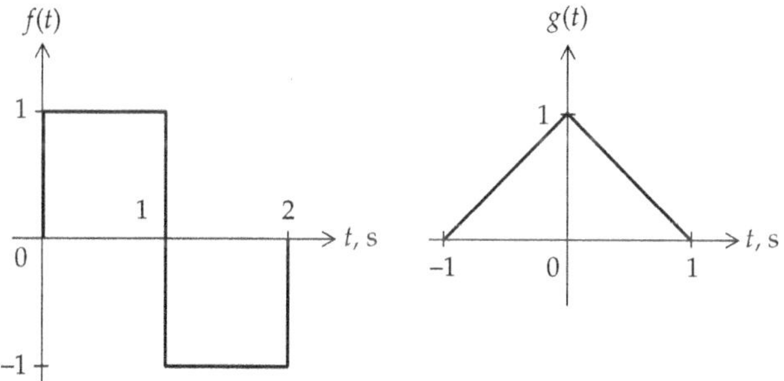

FIGURE 20.41
Figure for Example 20.8.

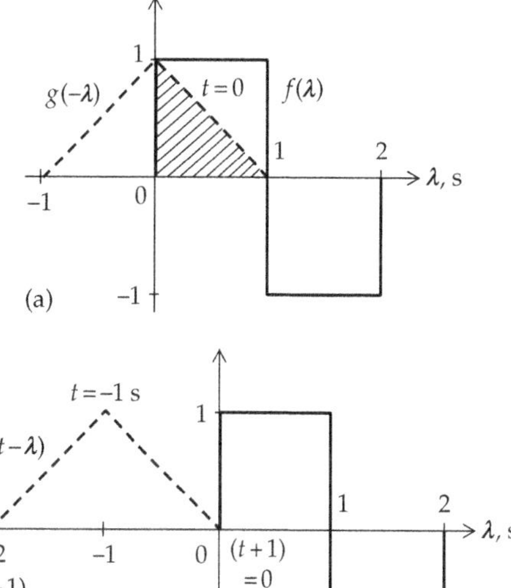

FIGURE 20.42
Figure for Example 20.8.

as the λ-coordinate of the vertex. The leading endpoint of $g(t - \lambda)$ is at $\lambda = (t + 1)$ and its trailing endpoint is at $\lambda = (t - 1)$. Note that in terms of t as a symbol, the coordinates of the vertex and the two endpoints are the same for any shift of t, positive or negative.

The convolution integral $y(t)$ is obtained by increasing t over consecutive ranges until there is no overlap between $f(\lambda)$ and $g(t - \lambda)$. As $g(t - \lambda)$ in Figure 20.42b is shifted to the right, the first range of t is when the vertex is between $\lambda = -1$ and $\lambda = 0$ or when the leading endpoint, $t + 1$, is between $\lambda = 0$ and $\lambda = 1$ (Figure 20.43a). That is, $0 \le t + 1 \le 1$ or $-1 \le t \le 0$. Over this range, $f(\lambda) = 1$

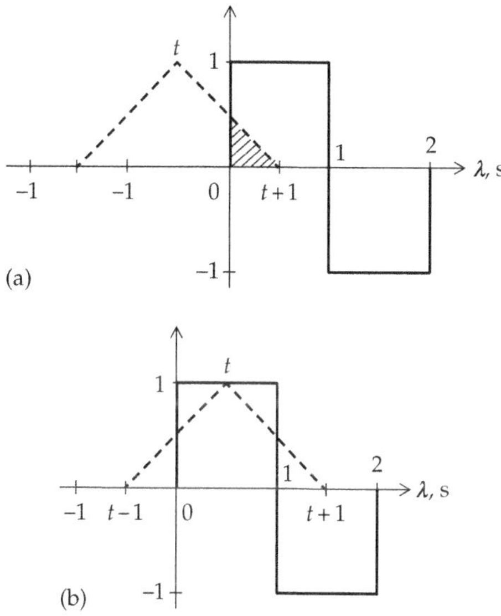

FIGURE 20.43
Figure for Example 20.8.

TABLE 20.1
Convolution Integral

		$y(t)$ at Range Limits	
Component	**Range**	**Low End**	**High End**
$y_1(t) = 0.5t^2 + t + 0.5$	$-1 \le t \le 0$ s	0	0.5
$y_2(t) = -1.5t^2 + t + 0.5$	$0 \le t \le 1$ s	0.5	0
$y_3(t) = 1.5t^2 - 5t + 3.5$	$1 \le t \le 2$ s	0	−0.5
$y_4(t) = -0.5t^2 + 3t - 4.5$	$2 \le t \le 3$ s	−0.5	0
$y(t) = 0$	$t \ge 3$ s	—	—

and $g(t - \lambda) = (-\lambda + t + 1)$, the equation of the negatively sloping line. This equation is most easily obtained by noting that the slope of the line is −1, and its ordinate is zero at $\lambda = (t + 1)$. The product has the same expression over this range, but a different expression for $(t + 1) > 1$ or $t > 0$. The convolution integral $y_1(t)$ over this range is

$$y_1(t) = \int_0^{t+1} (-\lambda + t + 1) \times 1 d\lambda = \left[-\frac{\lambda^2}{2} + \lambda t + \lambda \right]_0^{t+1}$$
$$= \frac{t^2}{2} + t + \frac{1}{2}, \quad 1 \le t \le 0 \text{ s} \tag{20.48}$$

This is in fact the area $(t + 1)^2/2$ of a right isosceles triangle having a side $(t + 1)$. The end values are $y_1(-1) = 0$ and $y_1(0) = 0.5$ (Table 20.1).

As $g(t - \lambda)$ is shifted further to the right, the next range of t is that when the vertex is between $\lambda = 0$ and $\lambda = 1$ or the leading endpoint, $t + 1$, is between $\lambda = 1$ and $\lambda = 2$ (Figure 20.43b). That is, $1 \le t + 1 \le 2$ or $0 \le t \le 1$. Over this range, the product of $f(\lambda)$ and $g(t - \lambda)$ has the same expression, but this expression is different from that for $(t + 1) < 1$ or for $(t + 1) > 2$. The equation of the positively sloping line is $g(t - \lambda) = (\lambda - t + 1)$. The slope of the line is +1 and its ordinate is zero at $\lambda = (t - 1)$. The convolution integral is

$$y_2(t) = \int_0^t (\lambda - t + 1) \times 1 d\lambda + \int_t^1 (-\lambda + t + 1) \times 1 d\lambda$$
$$+ \int_1^{t+1} (-\lambda + t + 1)(-1) d\lambda = -\frac{t^2}{2} + t - \frac{t^2}{2} + \frac{1}{2} - \frac{t^2}{2}$$
$$= -\frac{3}{2}t^2 + t + \frac{1}{2}, \quad 0 \le t \le 1 \text{ s} \tag{20.49}$$

Note that, geometrically, the area under the product is the algebraic sum of areas of the three figures: (i) a trapezoid of width t and of sides 1 and $(1 - t)$ (this area is $t(2 - t)/2 = t - t^2/2$); (ii) a trapezoid of width $(1 - t)$ and of sides 1 and $(1 - (1 - t)) = t$ (this area is $(1 + t)(1 - t)/2 = 1/2 - t^2/2$); and (iii) a right isosceles triangle of side t and area $t^2/2$. Adding the positive areas of the trapezoids to the negative area of the triangle gives the same area as in Equation 20.49. The end values are $y_2(0) = 0.5$ and $y_2(1) = 0$ (Table 20.1). It should also be noted that the equations of the two lines in terms of t as a symbol are the same for all ranges of t.

As $g(t - \lambda)$ is shifted further to the right, the next range of t is that when the vertex is between $\lambda = 1$ and $\lambda = 2$ or the trailing endpoint, $t - 1$, is between $\lambda = 0$ and $\lambda = 1$ (Figure 20.44a). That is, $0 \le t - 1 \le 1$ or $1 \le t \le 2$. Over this range, the product of $f(\lambda)$ and $g(t - \lambda)$ has the same expression, but this expression is different from that for $(t + 1) < 2$ or for $(t + 1) > 3$. The convolution integral is

$$y_3(t) = \int_{t-1}^1 (\lambda - t + 1) \times 1 d\lambda + \int_1^t (\lambda - t + 1) \times (-1) d\lambda$$
$$+ \int_t^2 (-\lambda + t + 1) \times (-1) d\lambda = \frac{3}{2}t^2 - 5t + \frac{7}{2}, \quad 1 \le t \le 2 \text{ s} \tag{20.50}$$

As in the preceding range, the product area can be considered the algebraic sum of three areas: (i) the area $(2 - t)^2/2 = 2 - 2t + t^2/2$ of a right isosceles triangle, (ii) the area $-(t - 1)(3 - t)/2 = 3/2 - 2t + t^2/2$ of the trapezoid on the left, and (iii) the area $-t(2 - t)/2 = -t + t^2/2$ of the trapezoid on the right. The sum of these areas is the same as in Equation 20.50. The end values are $y_3(1) = 0$ and $y_3(2) = -0.5$ (Table 20.1).

As $g(t - \lambda)$ is shifted further to the right, the next range of t is that when the vertex is between $\lambda = 2$ and $\lambda = 3$ or the trailing endpoint, $t - 1$, is between $\lambda = 1$ and $\lambda = 2$ (Figure 20.44b). That is, $1 \le t - 1 \le 2$ or $2 \le t \le 3$.

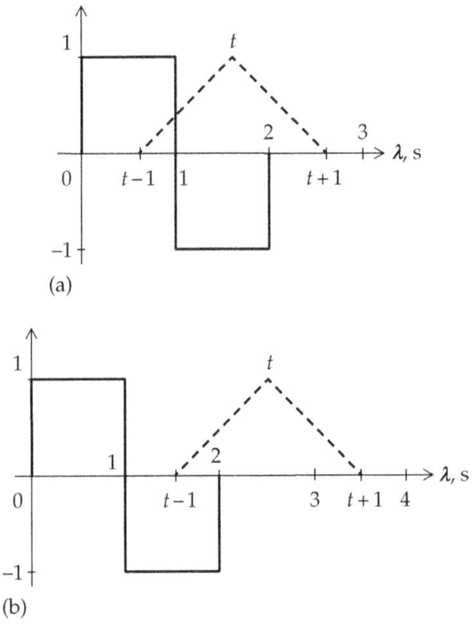

FIGURE 20.44
Figure for Example 20.8.

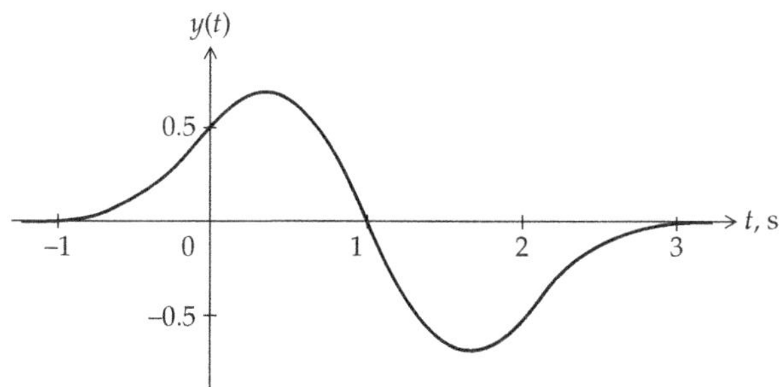

FIGURE 20.45
Figure for Example 20.8.

Over this range, the product of $f(\lambda)$ and $g(t-\lambda)$ has the same expression, but this expression is different from that for $(t-1)<1$ or for $(t-1)>2$, where there is no more overlap. The convolution integral is

$$y_4(t)=\int_{t-1}^{2}(\lambda-t+1)\times(-1)d\lambda=-\frac{t^2}{2}+3t-\frac{9}{2},\quad 2\le t\le 3\text{ s} \tag{20.51}$$

which is the area $(3-t)^2/2$ of a right isosceles triangle of side $(3-t)$. The end values are $y_4(2)=-0.5$ and $y_4(3)=0$ (Table 20.1). Beyond $t=3$ s, there is no overlap, so that $y(t)=0$. The convolution integral is plotted in Figure 20.45.

Problem-Solving Tip

- After a function is folded, and if the convolution integral is not zero at $t=0$, it is advisable to first shift the function using a positive value of t, as this will easily define the coordinates on the λ-axis of the salient features of the folded function, such as leading or trailing edges or endpoints. The expressions of these coordinates in terms of t remain the same for all values of t, positive or negative.

Exercise 20.9

Apply time invariance to Example 20.8 by shifting $g(t)$ 1 s to the right so that it starts at $t=0$. Derive the convolution integral and then shift it back 1 s to the left and compare with $y(t)$ of Figure 20.45.

Exercise 20.10

Assume that $g(t)$ in Figure 20.41 is as in Figure 20.46. When $g(\lambda)$ is folded around the vertical axis, and shifted by t to the right, specify the λ-coordinate of the vertex in terms of t so that it gives the correct value of λ when $t=0$. Note that this expression applies for any shift t.

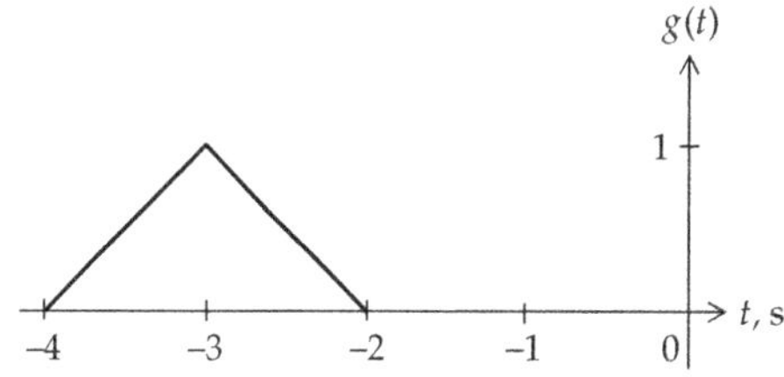

FIGURE 20.46
Figure for Primal Exercise 20.10.

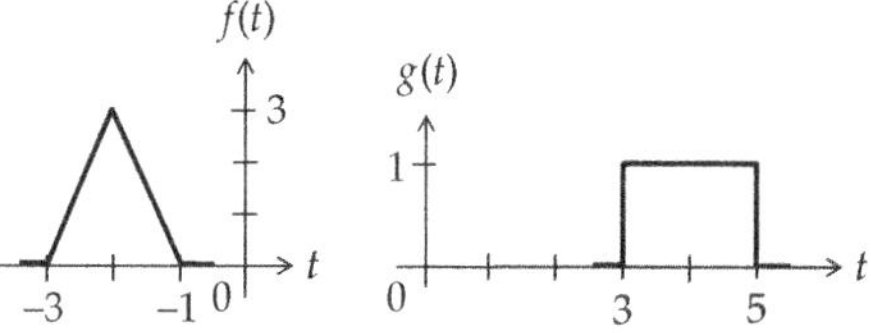

FIGURE 20.47
Figure for Primal Exercise 20.11.

Ans. $\lambda=t+3$, so that the vertex is at $\lambda=3$ when there is no shift ($t=0$) and $\lambda=-1$ when the triangle is shifted 4 units to the left ($t=-4$) so there is no overlap at the start of the convolution integral.

Primal Exercise 20.11

Determine the maximum value of the convolution integral $f(t)*g(t)$ in Figure 20.47 and the time at which it occurs.

Ans. 3 units at $t=2$ units.

Learning Checklist: What Should Be Learned from This Chapter

- If a function is shifted to the right along the x-axis by a positive number a, the x-axis variable in the function is replaced by $(x-a)$. Conversely, if a function is shifted to the left along the x-axis by a positive number a, the x-axis variable in the function is replaced by $(x+a)$. If the function is shifted by a variable amount denoted by a symbol such as t, which could have positive or negative values, the x-axis variable in the function is replaced by $(x-t)$.
- If a function is folded around the y-axis, the x-axis variable in the function is negated. If the folded function is shifted along the x-axis, the same aforementioned rules for shifting apply to *only* the x-axis variable in the function, not including any minus sign of this variable.

- The response of an LTI circuit to an arbitrary input can be considered as the superposition of responses to sufficiently narrow pulses, each having an amplitude determined by the input. These narrow pulses can be approximated by impulses, which naturally leads to a convolution integral involving the input function and the impulse response of the circuit.
- The convolution integral that is conventionally applied in circuit analysis is $y(t)=\int_{0^-}^{t} x(\lambda)h(t-\lambda)d\lambda$, where t is constant with respect to the integration and the integral gives the value of the convolution function at t.
- The procedure for graphical derivation of the convolution integral is as follows:
  1. Express $h(t)$ and $x(t)$ as function of λ.
  2. Fold the impulse response around the vertical axis, that is, draw it backward as $h(-\lambda)$.
  3. Shift $h(-\lambda)$ by t to obtain $h(t-\lambda)$.
  4. Determine the area under the product $x(\lambda)$ $h(t-\lambda)$ over the given range of t. The result is $y(t)$ for this range of t.
  5. Repeat steps 3 and 4 for various values of t to obtain $y(t)$ over the whole range of t.
- The convolution integral is not restricted to the interpretation in terms of the impulse response but can be regarded simply as a mathematical operation on two functions of time, which may be denoted as $f(t)$ and $g(t)$, which need not be interpreted in terms of the input to a system and its impulse response.
- The convolution integral is commutative, distributive, associative, and invariant with integration of one function a number of times and differentiation of the other function the same number of times.
- Practical functions may be approximated by staircase functions. The convolution integral of two staircase functions is a piecewise-linear function consisting of a series of straight line segments. The breakpoints between these segments occur at the discrete values of the shift of the folded function. The value of the convolution integral at each breakpoint is equal to the area under the product of the unshifted function and the shifted function, for each discrete shift of the folded function. Graphically, this area need only be calculated at the boundary values of the discrete shifts, and not at intermediate values of the shifts.
  1. Analytically, convolution of staircase functions reduces to multiplication of two polynomials representing the functions being convolved. The coefficients of the terms of a given polynomial are equal to the various levels of the respective staircase approximation of the function, in the order of decreasing power of the assigned variable of the polynomial, with the last nonzero level in the positive direction of the time variable t of the function being the constant term of the polynomial. The coefficients of the terms of the polynomial product of the two polynomials are the values of the convolution integral at the successive breakpoints between the first and last zero values of the convolution integral.
- Convolving a function with a unit impulse delays the function by the delay of the unit impulse, assuming the function to be continuous over the given range of the shift t.
- Convolving a function with a unit step is equivalent to integrating the function and delaying it by the delay of the unit step. The range of integration is from the start of the function to a value of the time variable λ that is t minus the delay of the unit step.
- The larger the deviation of the impulse response of a circuit from an impulse, the more substantial is the circuit memory and the larger is the distortion of the output, compared to the input.
- The convolution integral of two functions has the following properties:
  1. If both functions being convolved start with finite or zero values, the convolution integral starts with a zero value, at a value of t when there is no overlap between the two functions.
  2. If both functions being convolved end with a value of zero, the convolution integral also ends with a value zero, at a finite value of t when there is no longer any overlap between the two functions.
  3. If both functions being convolved are of finite duration, the duration of the convolution integral is the sum of the durations of the two functions.
  4. Delaying one or both functions being convolved delays the start of the convolution integral by an interval that is the sum of the delays of the two functions.
- In general, a distinct range of t of the convolution integral is defined by the following:
  1. The integrands of the convolution integral are the same over the given range, but are different from those of a preceding range of t or a following range of t, where an integrand is

the product, over the range, of the unshifted function and the folded and shifted function

- The integrands of the convolution integral are the same as those of a preceding range of t, or a following range of t, but the integration limits are different

Problem-Solving Tips

1. A simple check on convolution is that the convolution integral should have the same value at the limits of overlapping ranges of t.
2. It is generally advisable to fold and shift the simpler function, having the fewer number of distinct ranges of values, or to fold and shift a function that starts at negative time.
3. After a function is folded, and if the convolution integral is not zero at $t = 0$, it is advisable to first shift the function using a positive value of t, as this will easily define the coordinates on the λ-axis of the salient features of the folded function, such as leading or trailing edges or endpoints. The expressions of these coordinates in terms of t remain the same for all values of t, positive or negative.

Problems

Verify solutions by PSpice simulation.

Analytical Convolution

P20.1 Convolve $2u(t)$ and $\sin t\cos t$.

Ans. $0.5(1 - \cos 2t)$.

P20.2 Evaluate $\delta(t + 1)*\delta(t)*\delta(t - 1)$.

Ans. $\delta(t)$.

P20.3 Convolve e^{-t} with $\left(1-e^{-2t}\right)$.

Ans. $1+e^{-2t}-2e^{-t}$.

P20.4 Evaluate the following convolution operations: (a) $\sin\omega t * \cos\omega t$, (b) $\cos\omega t * \cos\omega t$, and (c) $\sinh at * \sin\omega t$. (Note that $\sinh x = -j\sin jx$.)

Ans. (a) $\frac{t}{2}\sin\omega t$; (b) $\frac{t}{2}\cos\omega t+\frac{\sin\omega t}{2\omega}$; (c) $\frac{\omega\sinh at - a\sin\omega t}{\omega^2+a^2}$.

P20.5 The impulse response of a circuit is $h(t) = 4u(t) - 2u(t - 5)$. Determine the circuit output when an excitation $x(t) = 2\delta(t - 1) - 3\delta(t - 3)$ is applied with zero initial conditions. Calculate the output at $t = 7$ s.

Ans. $8u(t - 1) - 12u(t - 3) - 4u(t - 6) + 6u(t - 8)$; -8.

P20.6 Given $f(t) = \cos t$, $-\pi/2 \le t \le +\pi/2$, and $f(t) = 0$ elsewhere. Determine the maximum value of the convolution integral when $f(t)$ is convolved with itself.

Ans. $\pi/2$.

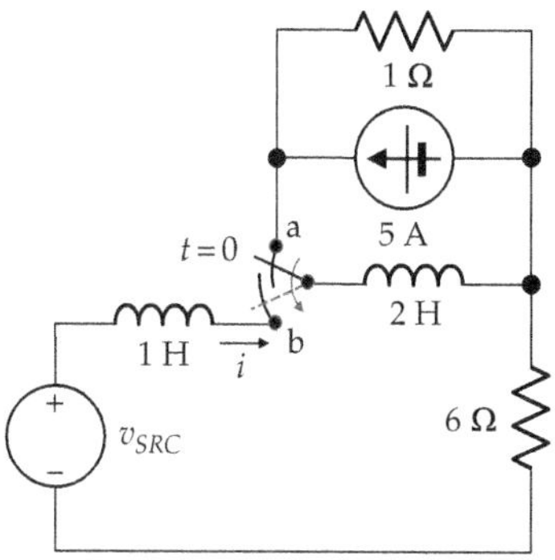

FIGURE P20.9

P20.7 Evaluate $\left(\sqrt{t-2}\right)u(t-2)*\delta(t-4)$ at $t = 7$ s.

Ans. 1.

P20.8 Evaluate $t^2u(t)*\cos tu(t)$.

Ans. $2(t - \sin t)$.

P20.9 The switch in Figure P20.9 is moved at $t = 0$ from position 'a' to position 'b' after being in position 'a' for a long time. Determine i for $t \ge 0$, assuming $v_{SRC} = 2te^{-3t}u(t)$ V.

Ans. $\left(4e^{-2t}-(2/3)e^{-3t}-(2/3)te^{-3t}\right)u(t)$ A.

P20.10 An input defined as $v_I = (t + 1)$ V, $-1 \le t \le 0$, $v_I = (-t + 1)$ V, $0 \le t \le 1$, and $v_I = 0$ elsewhere is applied to a circuit having the impulse response $h(t)=e^{-t}u(t)$. Determine the output for all t.

Ans. $t+e^{-(t+1)}$, $-1 \le t \le 0$, $e^{-(t+1)}-t+2-2e^{-t}$, $0 \le t \le 1$, and $e^{-(t+1)}-2e^{-t}+e^{-(t-1)}$, $t \ge 1$.

P20.11 An input voltage e^{-at} is applied from $t = -\infty$ to R and C in series. Obtain an expression for the output voltage across C in terms of a convolution integral taking the lower limit as $t = -\infty$. Show that the response at $t = 0$ is (a) finite if $a < 1/CR$ and (b) infinite if $a \ge 1/CR$, where $s = -1/CR$ is the pole of the circuit transfer function.

P20.12 The response of a linear, time-invariant circuit to a unit step input, $u(t)$, is $\left(1-e^{-t}\right)u(t)$. Determine the response to an input $e^{-t}u(t)$.

Ans. $te^{-t}u(t)$.

Graphical Convolution

P20.13 Given $h(t)$ and $x(t)$ in Figure P20.13, determine $h(t)*x(t)$ at $t = 8.1$ s.

Ans. 160.

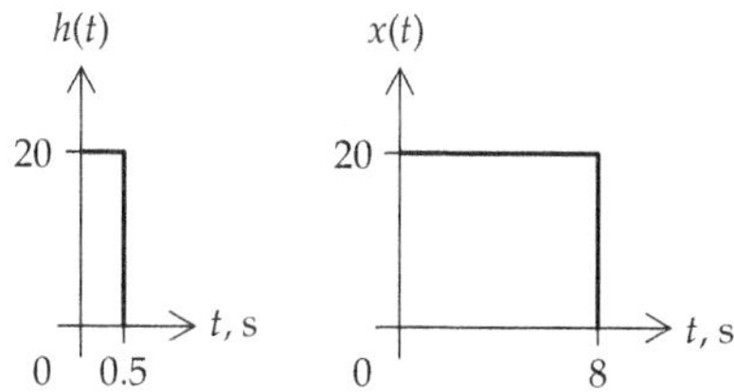

FIGURE P20.13

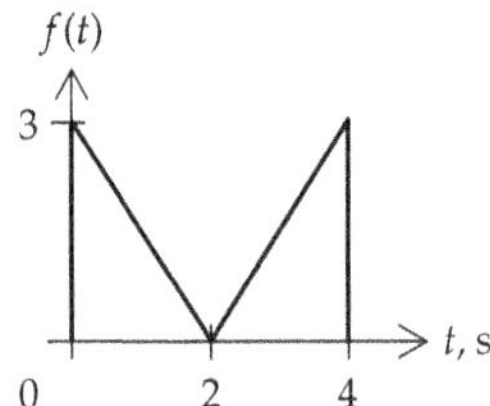

FIGURE P20.14

P20.14 Given $f(t)$ in Figure P20.14, determine $f(t)*2u(t)$ at $t = 5$ s.

Ans. 12.

P20.15 Given $h(t)$ and $x(t)$ in Figure P20.15, determine $h(t)*x(t)$ at $t = 6$ s.

Ans. 108.

P20.16 Given the function $4e^{-t}u(t)$ and $f(t)$ in Figure P20.16, determine $h(t)*x(t)$ at $t = 3$ s. Verify the result by direct integration.

Ans. $8\left(e^{-1} - e^{-3}\right)$.

P20.17 Determine $y(0)$, where $y(t) = f(t)*g(t)$ and $f(t)$ is the semicircle shown in Figure P20.17.

Ans. $4 - \pi$.

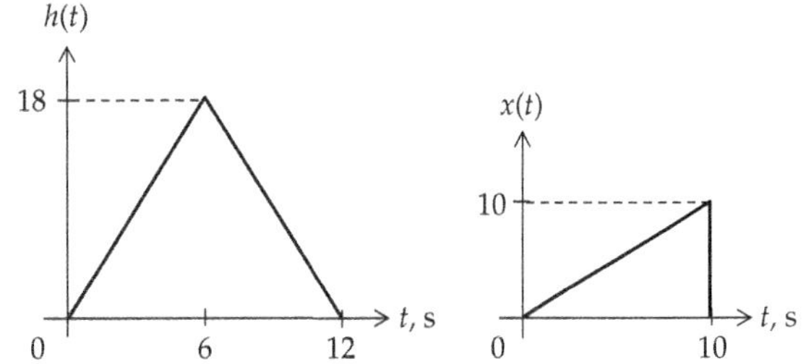

FIGURE P20.15

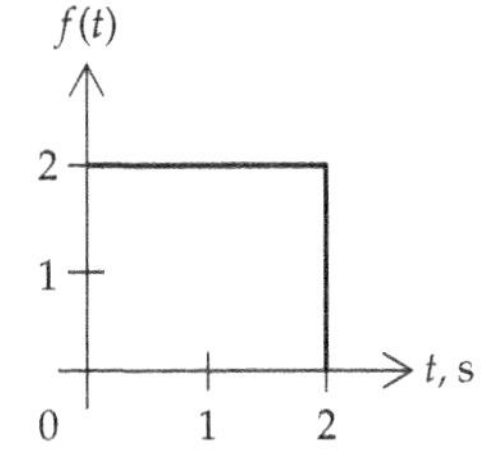

FIGURE P20.16

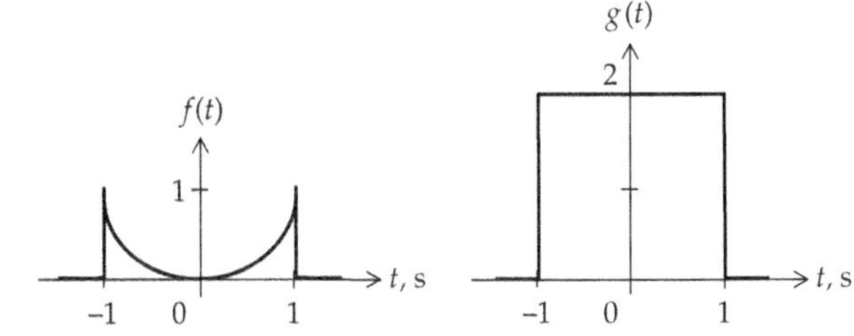

FIGURE P20.17

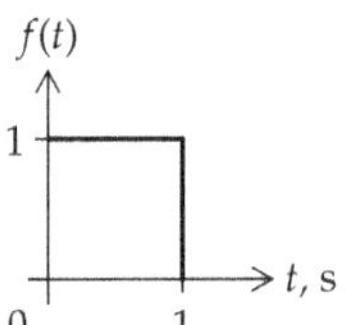

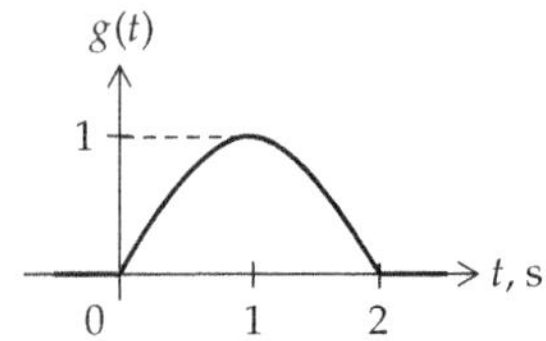

FIGURE P20.18

P20.18 Given $f(t)$ and $g(t)$ as in Figure P20.18, where $g(t) = \sin(\pi t/2), 0 \le t \le 2$, and $g(t) = 0$ elsewhere, determine $f(t)*g(t)$ for all t. Verify the result by direct integration.

Ans. $y(t) = \dfrac{2}{\pi}\left(1 - \cos(\pi t/2)\right), \quad 0 \le t \le 1;$

$y(t) = \dfrac{2}{\pi}\left(\sin(\pi t/2) - \cos(\pi t/2)\right), \quad 1 \le t \le 2;$

$y(t) = \dfrac{2}{\pi}\left(1 + \sin(\pi t/2)\right), \quad 2 \le t \le 3; y(t) = 0, \quad t \ge 3$

P20.19 Given $f(t)$ and $g(t)$ as in Figure P20.19, evaluate $f(t)*g(t)$ for all t.

Ans. $y(t) = 2t - 2$.

P20.20 Given $f(t)$ and $g(t)$ as in Figure P20.20, evaluate $f(t)*g(t)$ for all t. Verify the result using multiplication of polynomials. Determine the value of the convolution integral at $t = 2.5$ s.

Ans. $B(t) = 6t^2 + 7t - 3$; 2.

P20.21 Given $f(t)$ and $g(t)$ as in Figure P20.21, evaluate $f(t)*g(t)$ for all t. Verify the result using multiplication of polynomials.

Ans. $y(3) = 0, y(4) = 6, y(5) = 4, y(6) = 2, y(7) = 4, y(8) = 0$.

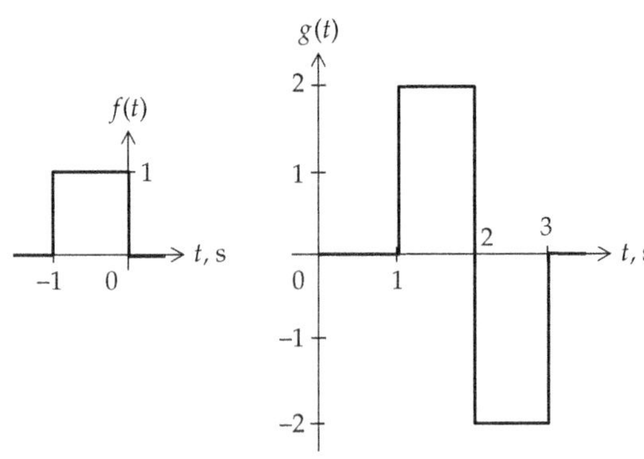

FIGURE P20.19

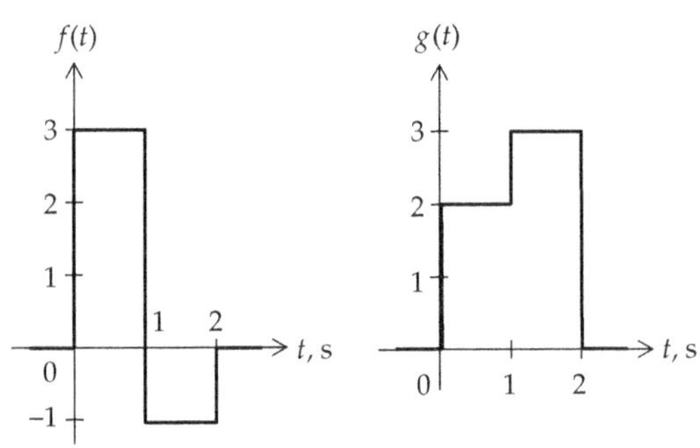

FIGURE P20.20

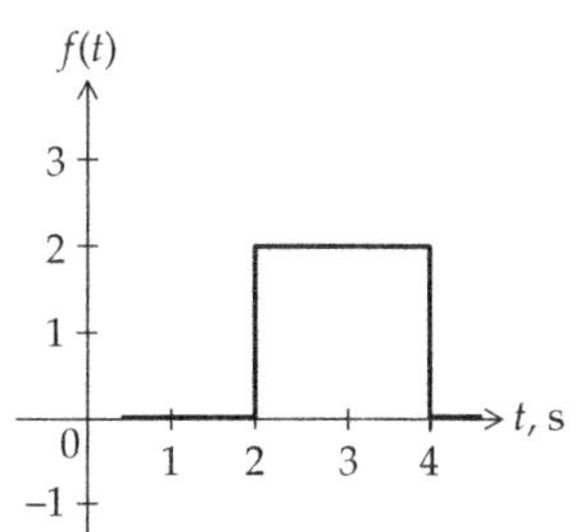

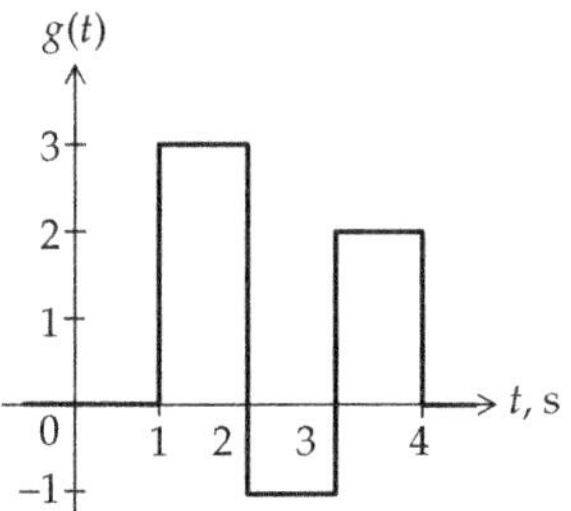

FIGURE P20.21

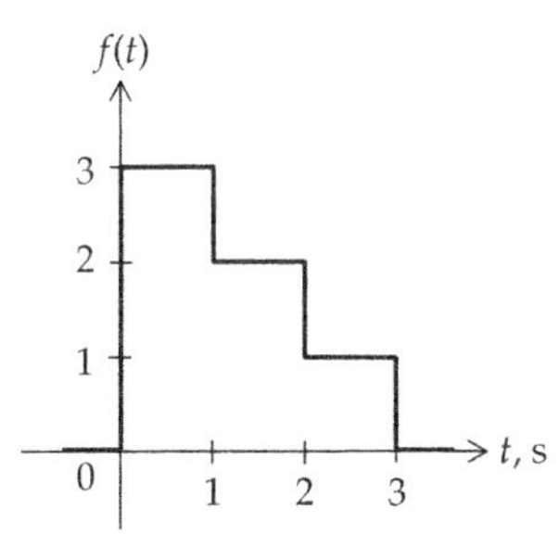

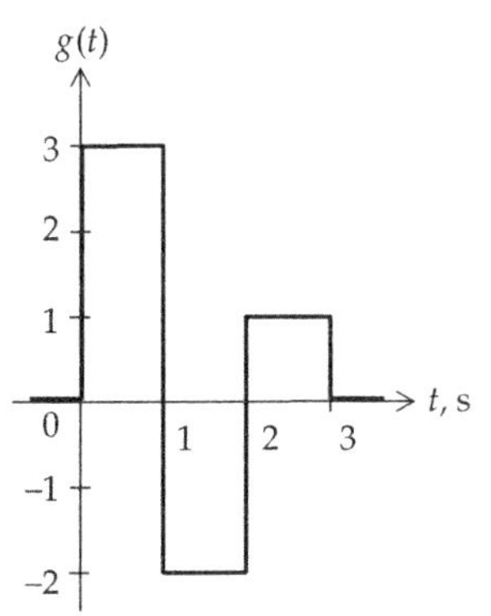

FIGURE P20.22

P20.22 Given $f(t)$ and $g(t)$ as in Figure P20.22, evaluate $f(t)*g(t)$ for all t. Verify the result using multiplication of polynomials.

Ans. $B(t) = 9t^4 + 2t^2 + 1$.

P20.23 Given $f(t)$ and $g(t)$ as in Figure P20.23, evaluate $f(t)*g(t)$ for all t. Verify the result using multiplication of polynomials.

Ans. $B(t) = t^4 + 2t^2 + 9$.

P20.24 Given $f(t)$ and $g(t)$ as in Figure P20.24, where $g(t) = \sin\pi(t - 1)$, $1 \leq t \leq 2$, and $g(t) = 0$ elsewhere, determine $f(t)*g(t)$ for all t. Verify by convolving with step functions.

Ans. $y(t) = \frac{1}{\pi}(1 + \cos\pi t)$, $1 \leq t \leq 2$; $y(t) = \frac{2}{\pi}$, $2 \leq t \leq 3$; $y(t) = \frac{1}{\pi}(1 - \cos\pi t)$, $3 \leq t \leq 4$; $y(t) = 0$, $t \geq 4$.

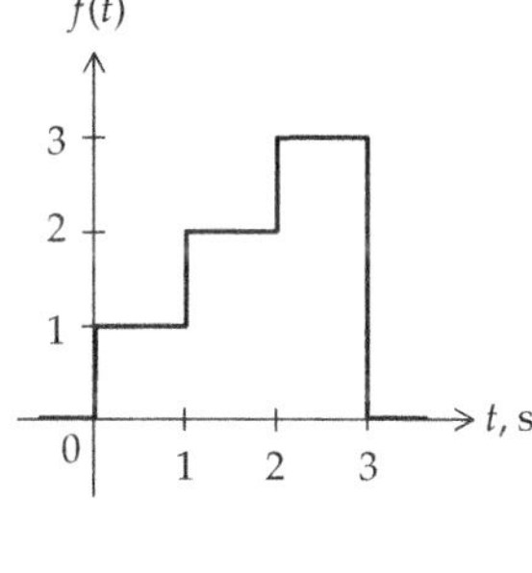

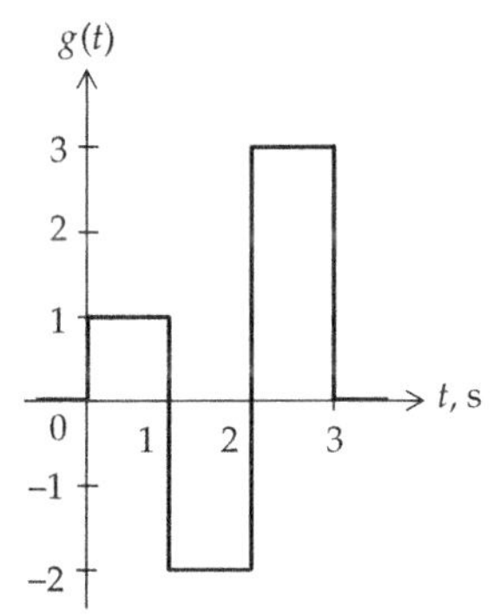

FIGURE P20.23

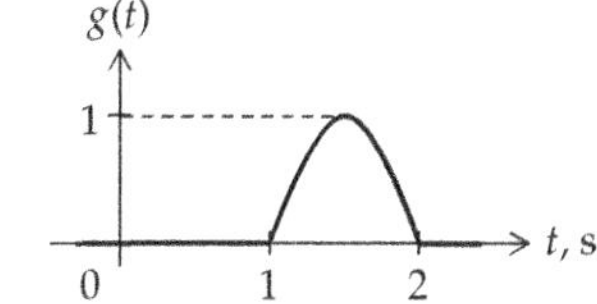

FIGURE P20.24

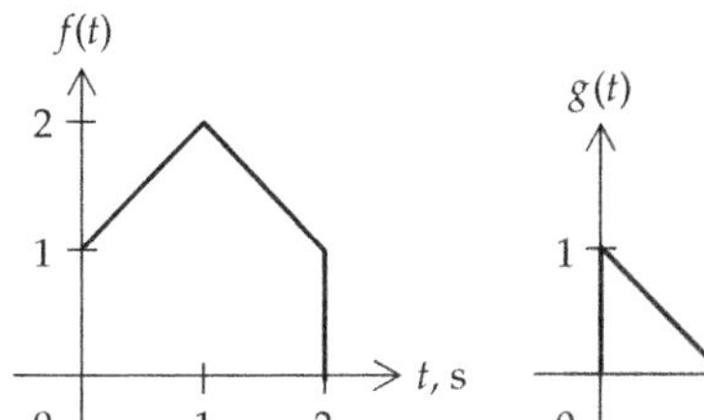

FIGURE P20.25

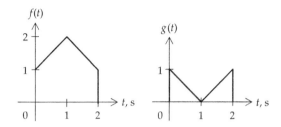

P20.25 Given $f(t)$ and $g(t)$ in Figure P20.25, determine $f(t)*g(t)$.

Ans. $y(t) = -t^3/6 + t$, $0 \leq t \leq 1$;
$y(t) = t^3/2 - 2t^2 + 3t - 2/3$, $1 \leq t \leq 2$;
$y(t) = -t^3/2 + 4t^2 - 11t + 34/3$, $y(t) = 0$ for $t \geq 4$.2
$\leq t \leq 3$; $y(t) = t^3/6 - 2t^2 + 7t - 20/3$, $3 \leq t \leq 4$.

P20.26 Given $f(t)$ in Figure P20.26, determine the convolution integral when $f(t)$ is convolved with itself. Verify the result analytically, by convolution of step functions and by product of polynomials.

Ans. $y(t) = (t + \tau/2)u(t + \tau/2) - 2tu(t) + (t - \tau/2)u(t - \tau/2)$.

P20.27 Given $f(t)$ and $g(t)$ in Figure P20.27, (a) evaluate $f(t)*g(t)$; (b) if the impulse response of a circuit is $4e^{-t}u(t)$, determine the response of the circuit to $f(t)$.

Ans. (a) $y(t) = (t + 3)u(t + 3) - 2tu(t) + (t - 3)u(t - 3)$; (b) $y(t) = \left(1 - e^{-(t+2)}\right)u(t + 2) - \left(1 - e^{-(t-2)}\right)u(t - 1)$.

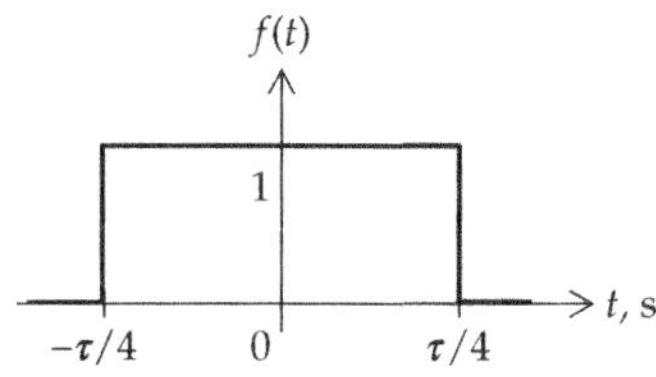

FIGURE P20.26

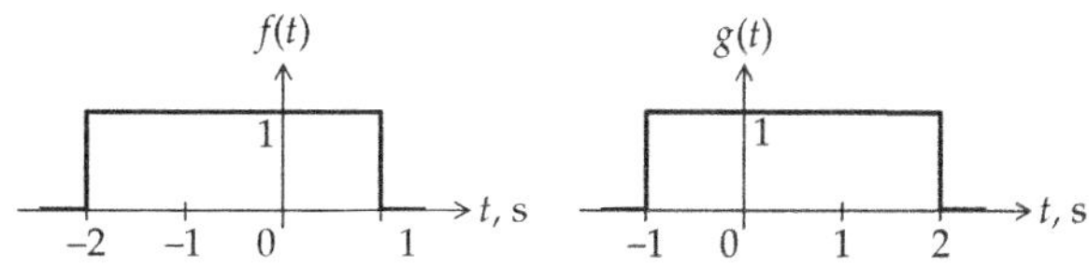

FIGURE P20.27

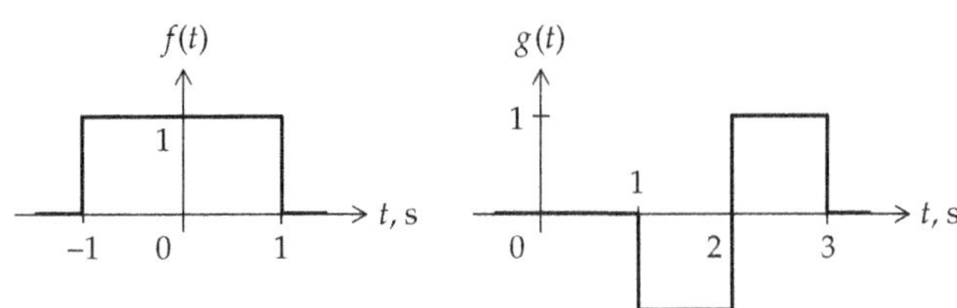

FIGURE P20.28

P20.28 Given $f(t)$ and $g(t)$ in Figure P20.28, evaluate $f(t)*g(t)$ for all t.

Ans. $y(t) = -t,\ 0 \leq t \leq 1$; $y(t) = (t-2),\ 1 \leq t \leq 3$; $y(t) = -(t-4),\ 3 \leq t \leq 4$; $y(t) = 0,\ t \geq 4$.

P20.29 Verify the commutative property for the functions of Example 20.2 by folding and shifting v_{SRC} instead of $h(t)$.

P20.30 Derive the convolution integral in Example 20.7 by considering $f(t)$ as the sum of step functions.

P20.31 Derive the convolution integral in Example 20.7 by differentiating $f(t)$ and integrating $g(t)$.

P20.32 Convolve $f(t) = \sin\pi t,\ 0 \leq t \leq \pi$, and $g(t)$ in Figure P20.32.

Ans. $\frac{1}{\pi}(1-\cos\pi t),\ 0 \leq t \leq 1;\ -\frac{2\cos\pi t}{\pi},\ 1 \leq t \leq 2;$

$-\frac{1}{\pi}(1+\cos\pi t),\ 2 \leq t \leq 3;\ 0,\ t \geq 3.$

P20.33 Convolve $f(t)$ and $g(t)$ in Figure P20.33.

Ans. $\frac{(t+1)^2}{2},\ -1 \leq t \leq 0;\ -\frac{t^2}{2}+2t+\frac{1}{2},\ 0 \leq t \leq 1;$

$-\frac{t^2}{2}+\frac{5}{2},\ 1 \leq t \leq 2;\ \frac{(3-t)^2}{2},\ 2 \leq t \leq 3;\ 0,\ t \geq 3.$

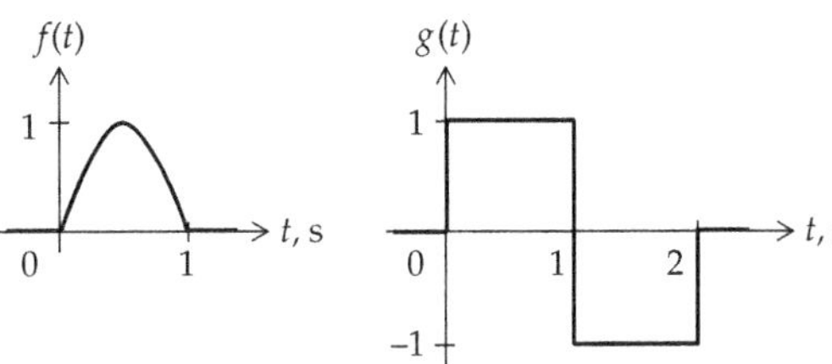

FIGURE P20.32

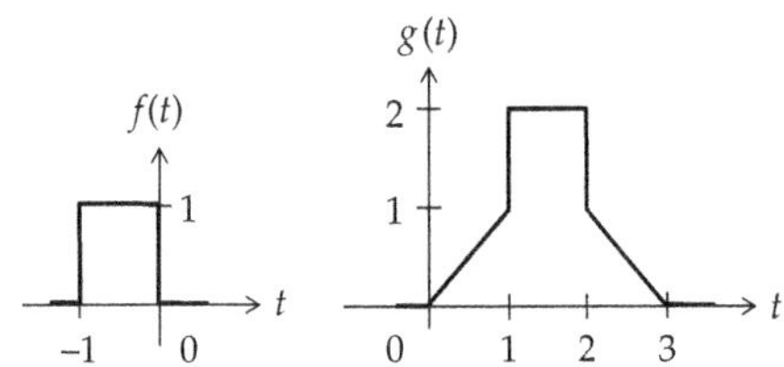

FIGURE P20.33

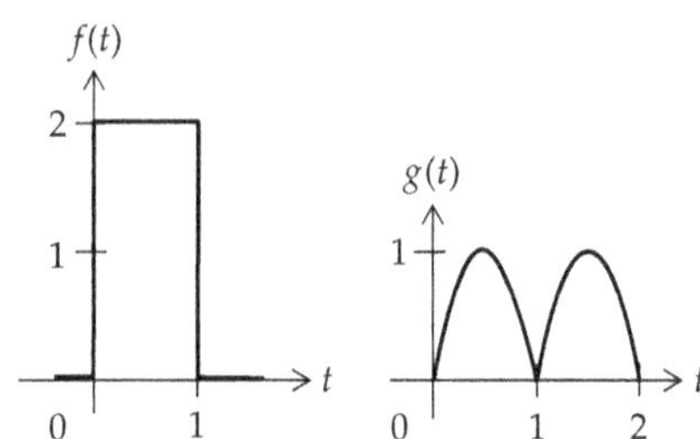

FIGURE P20.34

P20.34 Convolve $f(t)$ and $g(t)$ in Figure P20.34, where $f(t)$ is a pulse of 2 units amplitude and 1 unit duration and $g(t)$ consists of two sinusoidal half-cycles represented by $g(t) = \sin\pi t,\ 0 \leq t \leq 1$, and $g(t) = -\sin\pi t,\ 1 \leq t \leq 2$.

Ans. $y_1(t) = \frac{2}{\pi}(1-\cos\pi t),\ 0 \leq t \leq 1;\ y_2(t) = \frac{4}{\pi},\ 1 \leq t \leq 2;$

$y_3(t) = \frac{2}{\pi}[1+\cos\pi t],\ 2 \leq t \leq 3;\ y_4(t) = 0,\ t \geq 3.$

P20.35 Convolve $f(t)$ and $g(t)$ in Figure P20.35.

Ans. $y_1(t) = \frac{t^2}{2}+t+\frac{1}{2},\ -1 \leq t \leq 0;\ y_2(t) = -t^2+\frac{1}{2},\ 0 \leq t \leq 1;$

$y_3(t) = \frac{t^2}{2}-t,\ 1 \leq t \leq 2;\ y_4(t) = 0,\ t \geq 2.$

P20.36 Convolve $f(t)$ and $g(t)$ in Figure P20.36. Sketch $y(t)$.

Ans. $y_1(t) = u(t+4),\ t \leq -3;\ y_2(t) = 2t+7,\ -3 \leq t \leq -1;$ $y_3(t) = 5,\ -1 \leq t \leq -1;\ y_4(t) = -2t+7,\ 1 \leq t \leq 3;\ y_5(t) = u(4-t),\ t \geq 3.$

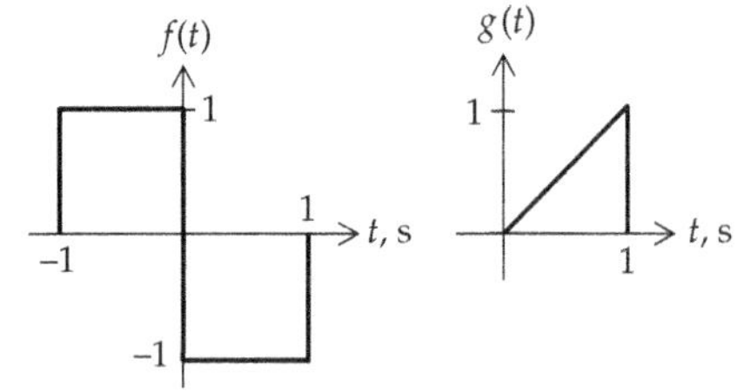

FIGURE P20.35

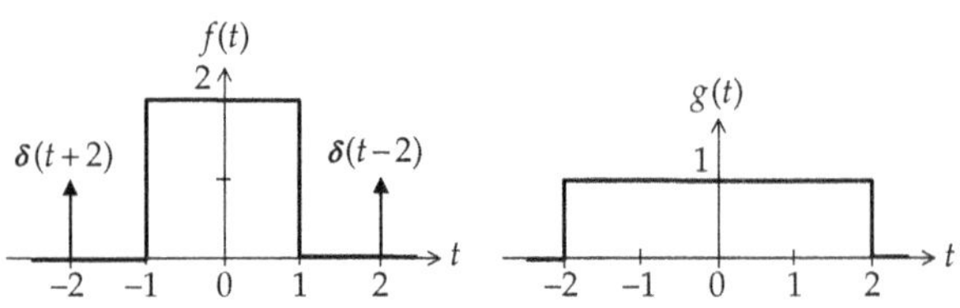

FIGURE P20.36

21

Properties of the Laplace Transform

Objective and Overview

This chapter and the next are concerned with the Laplace transform (LT), which provides an extremely powerful tool for analyzing linear systems. Recall that the power and usefulness of the phasor approach is in transforming linear, ordinary differential equations with constant coefficients to algebraic equations for analyzing the sinusoidal steady state only. The LT extends this approach by transforming linear, ordinary differential equations also to algebraic equations, but for deriving the complete response, that is, steady-state plus transient. Moreover, the excitation is not limited to sinusoidal excitation but could be any arbitrary excitation that has an LT.

The present chapter first introduces the LT before presenting its most important properties. This is followed by a discussion of the solution of linear, ordinary differential equations using the LT, and the inversion of the LT to obtain the time function. The power of the LT method is extended by some useful theorems, which are presented and applied at the end of the chapter. Of special interest are the initial-value theorem and the convolution theorem. The latter theorem provides the important link between the convolution integral and the LT.

21.1 General

Before defining the LT, the notion of complex frequency is explained. A complex frequency s can be expressed as the sum of a real component σ and an imaginary component $j\omega$, that is, $s = \sigma + j\omega$. An s-plane can be postulated as an Argand diagram whose real axis is σ and whose imaginary axis is $j\omega$, as illustrated in Figure 21.1. The point 'P', for example, having coordinates −2 and 3, represents a complex frequency $s = -2 + j3$ rad/s.

We have encountered such a frequency in deriving the roots of an underdamped, second-order circuit as $s_1 = -\alpha + j\omega_d$ and $s_2 = -\alpha - j\omega_d$ (Equation 12.20). Both s_1 and s_2 are complex frequencies, where ω_d is a physical frequency of sinusoidal functions, such as $\sin\omega_d t$ and $\cos\omega_d t$. On the other hand, α, although having the units of frequency, is not a physical frequency in the same

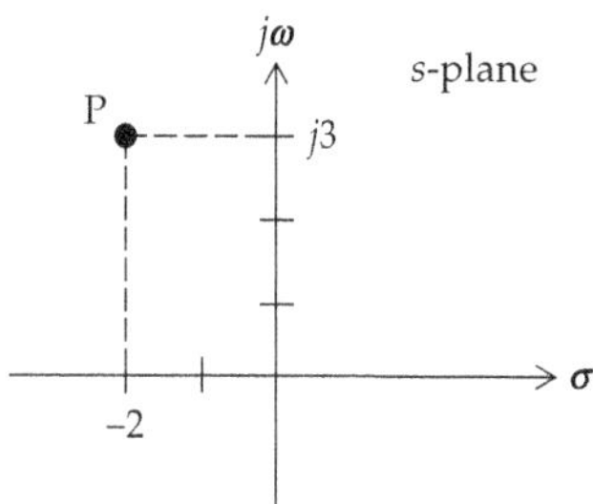

FIGURE 21.1
A complex frequency in the s-plane.

sense as ω_d. Rather, it appears in circuit responses in the exponent of the term $e^{-\alpha t}$. Similarly, in the general expression $s = \sigma + j\omega$, ω is a physical frequency of sinusoidal functions and σ is a quantity that appears in the exponents of exponential terms.

Formally, the LT is a transformation from the time domain to the s-domain, which means that a function of time, $f(t)$, becomes a function of s, according to some defining mathematical relation. Given an arbitrary function $f(t)$ defined over all time, the general, or two-sided, LT of $f(t)$ is defined as

$$\text{Two-sided LT of } f(t) = \int_{-\infty}^{+\infty} f(t)e^{-st}dt \quad (21.1)$$

The integral, assuming it exists, is clearly a function of s alone. Let the interval from $-\infty$ to $+\infty$ be divided into two intervals, one from $-\infty$ to 0^-, the other from 0^- to $+\infty$:

$$\text{Two-sided LT of } f(t) = \int_{-\infty}^{0^-} f(t)e^{-st}dt + \int_{0^-}^{+\infty} f(t)e^{-st}dt \quad (21.2)$$

In circuit analysis, and for reasons that will become clear in the following chapter, a one-sided LT is defined as the second integral in Equation 21.2, considering the first integral to be zero. This is tantamount to ignoring values of $f(t)$, for $t < 0^-$.

The single-sided LT will henceforth be denoted as $\mathcal{L}\{f(t)\}$ or $F(s)$. From the preceding discussion,

$$\mathcal{L}\{f(t)\} = F(s) = \int_{0^-}^{\infty} f(t)e^{-st}dt, \quad t \geq 0^- \tag{21.3}$$

Concept: *In the one-sided LT F(s), values of f(t) for t < 0⁻ are ignored.*

Specifying the lower limit of integration in Equation 21.3 as 0^- includes in the integration any impulse at the origin. This is desirable, because such an impulse is commonly encountered in circuit analysis.

Let the interval of integration in Equation 21.3 be divided as follows:

$$\int_{0^-}^{\infty} f(t)e^{-st}dt = \int_{0^-}^{0^+} f(t)e^{-st}dt + \int_{0^+}^{\infty} f(t)e^{-st}dt \tag{21.4}$$

According to Equation 18.22, the integral from 0^- to 0^+ is zero as long as $f(t)$ remains finite during this interval, as when $f(t)$ does not have an impulse at the origin. Hence, Equation 21.4 becomes *for a function f(t) that does not have an impulse at the origin*

$$F(s) = \int_{0^-}^{\infty} f(t)e^{-st}dt = \int_{0^+}^{\infty} f(t)e^{-st}dt \tag{21.5}$$

In other words, if a function does not have an impulse at the origin, although it may have a finite step at the origin, the lower limit in the integral for the LT could just as well be 0^+ instead of 0^-. In this case, the LT of $f(t)$ is the same as the LT of $f(t)u(t)$, irrespective of the lower limit of integration being 0^- or 0^+ or 0. That is, as long as $f(t)$ does not have an impulse at the origin, the integral of $f(t)u(t)$ is zero between the two infinitesimal limits 0^- and 0^+. Hence, for a function that does not have an impulse at the origin,

$$F(s) = \int_{0^-}^{\infty} f(t)e^{-st}dt = \int_{0^+}^{\infty} f(t)e^{-st}dt$$
$$= \int_{0^-}^{\infty} f(t)u(t)e^{-st}dt = \int_{0^+}^{\infty} f(t)u(t)e^{-st}dt \tag{21.6}$$

Some useful LTs can be readily derived from the basic definition of Equation 21.5. Thus,

$$\mathcal{L}\{u(t)\} = \int_{0^-}^{\infty} u(t)e^{-st}dt = \int_{0^+}^{\infty} e^{-st}dt = \frac{1}{s}\left[-e^{-st}\right]_{0^+}^{\infty} = \frac{1}{s} \tag{21.7}$$

If K is a constant, which by definition is independent of time,

$$\mathcal{L}\{K\} = \int_{0^+}^{\infty} Ke^{-st}dt = \frac{K}{s}\left[-e^{-st}\right]_{0^+}^{\infty} = \frac{K}{s} \tag{21.8}$$

Note that since K does not have an impulse at the origin, its LT is the same as that of $Ku(t)$.

If $f(t) = e^{-at}$,

$$\mathcal{L}\{e^{-at}\} = \int_{0^-}^{\infty} e^{-at}e^{-st}dt = \frac{1}{s+a}\left[-e^{-(s+a)t}\right]_{0^-}^{\infty} = \frac{1}{s+a} \tag{21.9}$$

To evaluate $\mathcal{L}\{\delta(t)\}$, we note that $\delta(t) = 0$ for $t \leq 0^-$ and $\delta(t) = 0$ for $t \geq 0^+$. Hence,

$$\mathcal{L}\{\delta(t)\} = \int_{0^-}^{\infty} \delta(t)e^{-st}dt = \int_{0^-}^{0^+} \delta(t)e^{-st}dt$$
$$= \int_{0^-}^{0^+} \delta(t) \times 1dt + 0 = 1 \tag{21.10}$$

The following should be noted concerning the LT:

1. Mathematically, sufficient conditions for the existence of the LT are (a) convergence of the defining integral of the LT and (b) the function is piecewise continuous on a finite interval of time for $t \geq 0$, which means that $f(t)e^{-st}$ is integrable over any such interval of time. Functions of interest in circuit analysis normally possess LTs.
2. Theoretically, σ can be chosen to be greater than some constant c so that multiplying $f(t)$ by e^{-ct} ensures convergence of the integral, that is, gives a finite value of the integral. However, some functions, such as t^t or e^{t^n}, increase so rapidly with t that $t^t e^{-\sigma t}$ and $e^{t^n} e^{-\sigma t}$ do not tend to zero as $t \to \infty$, no matter how large is σ. Such functions do not have LTs, but generally, are not of practical interest.
3. The inverse LT (ILT) transforms the function $F(s)$ back to $f(t)$ according to the following integral relation:

$$\mathcal{L}^{-1}\{F(s)\} = f(t) = \frac{1}{2\pi j}\int_{\sigma-j\infty}^{\sigma+j\infty} F(s)e^{st}ds \tag{21.11}$$

   Because this is integration in the complex plane, and requires some specialized procedures, it is seldom used in practice. Other methods, discussed later, are used for inverting the LT.
4. The LT of a given function, as defined by Equation 21.5, is unique. Conversely, if two functions have the same LT, these functions cannot differ over any time interval of finite length,

although they may differ at isolated points. Functions having finite values at isolated points in time are not of practical interest.

5. Because $F(s)$ is derived through integration with respect to time, its units are those of $f(t)$ multiplied by the unit of time used in the integration.

The LT is, in general, derived in the following ways:

1. Direct evaluation of the integral, as in Equations 21.7 through 21.10.
2. From known LTs of functions, using the properties of the LT discussed in Section 21.2.
3. Lookup in tables of LT pairs. Extensive tables are available of the LTs of many types of functions and their inverse functions.
4. Use of MATLAB's 'laplace' command. For example, if we enter

```
>>syms a t
>>laplace(exp(-a*t))
MATLAB returns: 1/(s+a)
```

21.2 Operational Properties of the Laplace Transform

Multiplication by a constant. If $\mathcal{L}\{f(t)\}=F(s)$, then

$$\mathcal{L}\{Kf(t)\}=KF(s) \tag{21.12}$$

Proof: $\mathcal{L}\{Kf(t)\}=\int_{0^-}^{\infty}Kf(t)e^{-st}dt=K\int_{0^-}^{\infty}f(t)e^{-st}dt=KF(s)$.

Thus, the LT of $K\delta(t)$ is K, and the LT of Ke^{-at} is $\dfrac{K}{s+a}$.

Addition/subtraction. If $\mathcal{L}\{f_1(t)\}=F_1(s)$, and $\mathcal{L}\{f_2(t)\}=F_2(s)$, then

$$\mathcal{L}\{f_1(t)\pm f_2(t)\}=F_1(s)\pm F_2(s) \tag{21.13}$$

Proof: $\mathcal{L}\{f_1(t)\pm f_2(t)\}=\int_{0^-}^{\infty}\left[f_1(t)\pm f_2(t)\right]e^{-st}dt=\int_{0^-}^{\infty}f_1(t)e^{-st}dt\pm\int_{0^-}^{\infty}f_2(t)e^{-st}dt=F_1(s)\pm F_2(s)$. We can use this property to derive the LTs of $\cos\omega t$ and $\sin\omega t$ using Equation 21.9. Thus,

$$\mathcal{L}\{\cos\omega t\}=\mathcal{L}\left\{\frac{e^{j\omega t}+e^{-j\omega t}}{2}\right\}=\frac{1}{2}\left[\frac{1}{s-j\omega}+\frac{1}{s+j\omega}\right]=\frac{s}{s^2+\omega^2} \tag{21.14}$$

$$\mathcal{L}\{\sin\omega t\}=\mathcal{L}\left\{\frac{e^{j\omega t}-e^{-j\omega t}}{2j}\right\}=\frac{1}{2j}\left[\frac{1}{s-j\omega}-\frac{1}{s+j\omega}\right]=\frac{\omega}{s^2+\omega^2} \tag{21.15}$$

Time scaling. If $\mathcal{L}\{f(t)\}=F(s)$, then

$$\mathcal{L}\{f(at)\}=\frac{1}{a}F\left(\frac{s}{a}\right),\quad \text{where } a>0 \tag{21.16}$$

Proof: $\mathcal{L}\{f(at)\}=\int_{0^-}^{\infty}f(at)e^{-st}dt$. Changing the variable of integration to $x=at$, $\mathcal{L}\{f(at)\}=\frac{1}{a}\int_{0^-}^{\infty}f(x)e^{-\frac{s}{a}x}dx=\frac{1}{a}F\left(\frac{s}{a}\right)$. The restriction $a>0$ ensures that $f(at)$ is defined only for $t>0$.

Primal Exercise 21.1

(a) Apply time scaling to the LT of $e^{-t}u(t)$ to show that the LT of $e^{-at}u(t)$ is $1/(s+a)$ and (b) derive the LT of the following functions: (i) $\cosh at=(e^{at}+e^{-at})/2$, (ii) $2\delta(t)-4+5e^{-3t}$.

Ans. (b) (i) $s/(s^2-a^2)$, (ii) $2-4/s+5/(s+3)$.

Integration in time. If $\mathcal{L}\{f(t)\}=F(s)$, then

$$\mathcal{L}\left\{\int_{0^-}^{t}f(x)dx\right\}=\frac{1}{s}F(s) \tag{21.17}$$

Proof: $\mathcal{L}\left\{\int_{0^-}^{t}f(x)dx\right\}=\int_{0^-}^{\infty}\left[\int_{0^-}^{t}f(x)dx\right]e^{-st}dt$. Let $\int_{0^-}^{t}f(x)dx=u$ and $e^{-st}dt=dv$. Integrating by parts (AppendixB): $\int_{0^-}^{\infty}\left[\int_{0^-}^{t}f(x)dx\right]e^{-st}dt=\left[-\frac{e^{-st}}{s}\int_{0^-}^{t}f(x)dx\right]_{t=0^-}^{t=\infty}+\frac{1}{s}\int_{0^-}^{t}f(t)e^{-st}dt$. The first term vanishes at both limits, and the second term is $\frac{1}{s}F(s)$.

Equation 21.17 can be generalized to

$$\mathcal{L}\left\{f^{(-n)}(t)\right\}=\frac{1}{s^n}F(s) \tag{21.18}$$

where $f^{(-n)}$ is the nth integral of $f(t)$.

If $g(t)=\int f(t)dt=\int_{0^-}^{t}f(x)dx+g(0^-)$, then using Equations 21.7 and 21.18

$$\mathcal{L}\left\{\int f(t)dt\right\}=\frac{1}{s}F(s)+\frac{1}{s}g(0^-) \tag{21.19}$$

Example 21.1: Laplace Transform of Powers of t

It is required to prove that

$$\mathcal{L}\{t^n u(t)\} = \frac{n!}{s^{n+1}} \quad (21.20)$$

Solution:

We will successively apply the integration-in-time property, starting with $\mathcal{L}\{\delta(t)\}=1$ (Equation 21.10):

$$\mathcal{L}\{u(t)\} = \mathcal{L}\left\{\int_{0^-}^{t} \delta(x)dx\right\} = \frac{1}{s} \quad (21.21)$$

and $\mathcal{L}\{tu(t)\} = \mathcal{L}\left\{\int_{0^-}^{t} u(x)dx\right\} = \frac{1}{s}\cdot\frac{1}{s} = \frac{1}{s^2}$, $\mathcal{L}\{t^2u(t)\}$ $= 2\mathcal{L}\left\{\int_{0^-}^{t} xdx\right\} = 2\cdot\frac{1}{s}\cdot\frac{1}{s^2} = \frac{2}{s^3}$, $\mathcal{L}\{t^3u(t)\} = 3\mathcal{L}\left\{\int_{0^-}^{t} x^2dx\right\}$ $= 3\cdot\frac{1}{s}\cdot\frac{2}{s^3} = \frac{3}{s^4}$, and so on.

Primal Exercise 21.2

Verify that $\mathcal{L}\left\{\int_{0^-}^{t} xdx\right\}$ according to the integration-in-time property is the same as that obtained by first integrating x, and then taking the LT.

Differentiation in time. If $\mathcal{L}\{f(t)\}=F(s)$, then

$$\mathcal{L}\left\{\frac{df(t)}{dt}\right\} = sF(s) - f(0^-) \quad (21.22)$$

and in general,

$$\mathcal{L}\left\{\frac{df^n(t)}{dt^n}\right\} = s^nF(s) - s^{n-1}f(0^-)$$
$$-s^{n-2}f^{(1)}(0^-) - \cdots - f^{(n-1)}(0^-) \quad (21.23)$$

where $f^{(m)}(0^-)$ is the mth derivative of $f(t)$ evaluated at $t = 0^-$.

Proof: $\mathcal{L}\left\{\frac{df(t)}{dt}\right\} = \int_{0^-}^{\infty} \frac{df(t)}{dt}e^{-st}dt = \left[f(t)e^{-st}\right]_{0^-}^{\infty}$ $-\int_{0^-}^{\infty}(-s)f(t)e^{-st}dt = sF(s) - f(0^-)$. Note that $f(t)e^{-st}$ evaluates to zero as $t\to\infty$. Otherwise, the integral of Equation 21.5 does not converge, and $f(t)$ will not have an LT $F(s)$, as assumed.

Similarly, if $g(t) = \frac{df(t)}{dt}$, then $\frac{dg(t)}{dt} = \frac{d^2f(t)}{dt^2}$. Equation 21.22 gives $\mathcal{L}\left\{\frac{dg(t)}{dt}\right\} = s\mathcal{L}\{g(t)\} - g(0^-)$. But $\mathcal{L}\{g(t)\} = \mathcal{L}\left\{\frac{df(t)}{dt}\right\} =$ $sF(s) - f(0^-)$, and $g(0^-)=f^{(1)}(0^-)$. Substituting term by term, $\mathcal{L}\left\{\frac{d^2f(t)}{dt^2}\right\} = \mathcal{L}\left\{\frac{dg(t)}{dt}\right\} = s\left(sF(s) - f(0^-)\right) - g(0^-) =$ $s^2F(s) - sf(0^-) - f^{(1)}(0^-)$. Successive application of the differentiation property leads to Equation 21.23.

According to the differentiation-in-time property,

$$\mathcal{L}\left\{\frac{d\delta(t)}{dt}\right\} = s,\quad \mathcal{L}\left\{\frac{d^2\delta(t)}{dt^2}\right\} = s^2,\quad \text{and}\quad \mathcal{L}\left\{\frac{d\delta^n(t)}{dt^n}\right\} = s^n \quad (21.24)$$

To illustrate the differentiation-in-time property, consider the differentiation of a constant, K. Evidently, $dK/dt = 0$. Let us apply the differentiation-in-time property to see how the same result is obtained. The LT of K is the LT of $Ku(t)$, which is K/s (Equation 21.8). According to the differentiation-in-time property, the LT of dK/dt is

$$\mathcal{L}\left\{\frac{dK}{dt}\right\} = s\mathcal{L}\{Ku(t)\} - K\big|_{t=0^-} = s\frac{K}{s} - K = 0 \quad (21.25)$$

where K, being independent of time, is K at $t = 0^-$. Since the LT of dK/dt is zero, dK/dt must be zero, as expected. Note that $f(0^-)$ is that of the function as given, which is K in this case, and not that of $f(t)u(t)$, which is zero.

Primal Exercise 21.3

Verify that $\mathcal{L}\left\{d\frac{e^{-t}}{dt}\right\}$ according to the differentiation-in-time property is the same as that obtained by first differentiating e^{-t}, and then taking the LT.

Example 21.2: Laplace Transforms of sine and cosine Functions

It is required to verify that the LTs of $\sin\omega t$ and $\cos\omega t$, as given by Equations 21.14 and 21.15, satisfy the integration-in-time and differentiation-in-time properties.

Solution:

(a) Consider the integral of the sine function:

$$\int_{0^-}^{t} \sin\omega t dt = \frac{1}{\omega}\left[-\cos\omega t\right]_{0^-}^{t} = \frac{1}{\omega}\left[1 - \cos\omega t\right] \quad (21.26)$$

According to the integration-in-time property (Equation 21.17) and using Equation 21.15, the LT of the LHS of Equation 21.26 is $\omega/s(s^2+\omega^2)$. We have to show that this is equal to the LT of the RHS of Equation 21.26. This LT is

$$\mathcal{L}\left\{\frac{1}{\omega}\left[1-\cos\omega t\right]\right\}=\frac{1}{\omega}\left[\frac{1}{s}-\frac{s}{s^2+\omega^2}\right]=\frac{\omega}{s\left(s^2+\omega^2\right)} \quad (21.27)$$

which gives the same expression, as required.

The derivative of $\sin\omega t$ is

$$\frac{d\left(\sin\omega t\right)}{dt}=\omega\cos\omega t \quad (21.28)$$

According to the differentiation-in-time property (Equation 21.22) and using Equation 21.15, the LT of the LHS of Equation 21.28 is $s\omega/(s^2+\omega^2)-0$, since the value of $\sin\omega t$ at $t = 0^-$ is zero. From Equation 21.14, the LT of the RHS of Equation 21.28 is also $s\omega/(s^2+\omega^2)$.

(b) Consider the integral of the cosine function:

$$\int_{0^-}^{t}\cos\omega t dt=\frac{1}{\omega}\left[\sin\omega t\right]_{0^-}^{t}=\frac{1}{\omega}\left[\sin\omega t\right] \quad (21.29)$$

According to the integration-in-time property (Equation 21.17) and using Equation 21.14, the LT of the LHS of Equation 21.29 is $s/s(s^2+\omega^2)=1/(s^2+\omega^2)$. From Equation 21.15, the LT of the RHS of Equation 21.29 is also $\omega/\omega(s^2+\omega^2)=1/(s^2+\omega^2)$.

The derivative of $\cos\omega t$ is

$$\frac{d\left(\cos\omega t\right)}{dt}=-\omega\sin\omega t \quad (21.30)$$

According to the differentiation-in-time property and using Equation 21.14, with $\cos\omega t$ at $t = 1$ at $t = 0^-$, the LT of the LHS of Equation 21.30 is

$$\frac{s\times s}{s^2+\omega^2}-1=-\frac{\omega^2}{s^2+\omega^2} \quad (21.31)$$

The RHS of Equation 21.31 is indeed the LT of the RHS of Equation 21.30.

It should be emphasized that, whereas the LTs of $f(t)$ and $f(t)u(t)$ are the same, when $f(t)$ does not have an impulse at the origin, the function being differentiated in the differentiation-in-time property is $f(t)$ and not $f(t)u(t)$. Thus, the derivative of $(\cos\omega t)u(t)$ is $(\cos\omega t)\delta(t) - \omega(\sin\omega t)u(t) = \delta(t) - \omega(\sin\omega t)u(t)$, which has an impulse at the origin, because of the sudden jump of $(\cos\omega t)u(t)$ at the origin from 0 at $t = 0^-$ to unity at $t = 0^+$. The LT of this derivative is $1 - \omega^2/(s^2 + \omega^2) = s^2/(s^2 + \omega^2)$, whereas the LT of the derivative of $(\cos\omega t)$ is $-\omega^2/(s^2 + \omega^2)$. Nevertheless, the differentiation-in-time property also applies to $(\cos\omega t)u(t)$. For according to this property,

$$\mathcal{L}\left\{d\left[\left(\cos\omega t\right)u(t)\right]dt\right\}=s\mathcal{L}\left\{\left(\cos\omega t\right)u(t)\right\}-\left[\left(\cos\omega t\right)u(t)\right]_{t=0^-}$$

$$=s^2/\left(s^2+\omega^2\right),\text{ as before, since }\left[\left(\cos\omega t\right)u(t)\right]_{t=0^-}=0.$$

Translation in s-domain. If $\mathcal{L}\{f(t)\}=F(s)$, then

$$\mathcal{L}\left\{e^{-at}f(t)\right\}=F(s+a) \quad (21.32)$$

Proof: $\mathcal{L}\left\{e^{-at}f(t)\right\}=\int_{0^-}^{\infty}e^{-at}f(t)e^{-st}dt=\int_{0^-}^{\infty}f(t)e^{-(s+a)t}dt = F(s + a)$.

Equation 21.32 allows easy derivation of LTs of functions multiplied by e^{-at}. For example, $\mathcal{L}\left\{e^{-at}u(t)\right\}=\frac{1}{s+a}$ (Equation 21.9), given that $\mathcal{L}\left\{u(t)\right\}=\frac{1}{s}$ (Equation 21.7). It also follows that

$$\mathcal{L}\left\{e^{-at}\cos\omega t\right\}=\frac{s+a}{\left(s+a\right)^2+\omega^2} \quad (21.33)$$

and

$$\mathcal{L}\left\{e^{-at}\sin\omega t\right\}=\frac{\omega}{\left(s+a\right)^2+\omega^2} \quad (21.34)$$

Translation in time. If a unit impulse $\delta(t)$ at the origin is translated in time by a positive constant a, to become $\delta(t - a)$ in positive time, its LT is

$$\mathcal{L}\left\{\delta(t-a)\right\}=\int_{0^-}^{\infty}\delta(t-a)e^{-st}dt \quad (21.35)$$

Substituting $t'=t-a$, Equation 21.35 becomes

$$\mathcal{L}\left\{\delta(t-a)\right\}=\int_{-a^-}^{\infty}\delta(t')e^{-s(t'+a)}dt'=e^{-as}\int_{-a^-}^{\infty}\delta(t')e^{-st'}dt' \quad (21.36)$$

Changing the dummy variable t' to t and noting that the integrand is nonzero only over the range 0^- to 0^+, it follows that

$$\mathcal{L}\left\{\delta(t-a)\right\}=e^{-as}\int_{0^-}^{0^+}\delta(t)e^{-st}dt=e^{-as} \quad (21.37)$$

Consider next a function $f(t)u(t)$ that is translated in time by a positive constant a, to become $f(t-a)u(t-a)$, as was explained in connection with Figure 20.25. The LT of $f(t-a)u(t-a)$ is

$$\mathcal{L}\{f(t-a)u(t-a)\}=\int_{0^-}^{\infty} f(t-a)u(t-a)e^{-st}dt \quad (21.38)$$

Because of the presence of $u(t-a)$, the integrand is zero for $t \leq a^-$ and is $f(t-a)$ for $t \geq a^+$. Assume to begin with that $f(t)$ does not have an impulse at the origin, so that $f(t-a)$ does not have an impulse at $t=a$. The integral of Equation 21.38 is therefore the same if the lower limit is a^- or a^+. Taking the lower limit as a^+ and substituting $t'=t-a$:

$$\mathcal{L}\{f(t-a)u(t-a)\}=\int_{0^+}^{\infty} f(t')e^{-as}e^{-t's}dt'$$

$$=e^{-as}\int_{0^+}^{\infty} f(t)e^{-st}dt=e^{-as}F(s) \quad (21.39)$$

where the dummy variable of integration was changed back from t' to t, and e^{-as} was taken outside the integration sign because it is a constant as far as integration with respect to t' is concerned. If $f(t)$ has an impulse $K\delta(t)$ at the origin, then it could be expressed as $f(t)=K\delta(t)+f'(t)$, where $f'(t)$ does not have an impulse at the origin. The shifted function becomes $f(t-a)u(t-a)=K\delta(t-a)+f'(t-a)u(t-a)$. From Equations 21.37 and 21.39, the LT of the RHS of this equation is $Ke^{-as}+e^{-as}F'(s)=e^{-as}(K+F'(s))=e^{-as}F(s)$, where $F'(s)$ is the LT of $f'(t)$. It follows that

$$\mathcal{L}\{f(t-a)u(t-a)\}=e^{-as}F(s) \quad (21.40)$$

irrespective of whether or not $f(t)$ has an impulse at the origin. Note the similarity to the Fourier series, where a delay t_d multiplies C_n by $e^{-jn\omega_0 t_d}$ (Equation 16.42).

As a simple example, consider the rectangular pulse of Figure 21.2. The pulse can be expressed as $f(t)=Ku(t)-Ku(t-a)$. Since $\mathcal{L}\{Ku(t)\}=\frac{K}{s}$ and $\mathcal{L}\{Ku(t-a)\}=\frac{K}{s}e^{-as}$, it follows that

$$\mathcal{L}\{f(t)\}=\frac{K}{s}-\frac{K}{s}e^{-as}=\frac{K}{s}\left(1-e^{-as}\right) \quad (21.41)$$

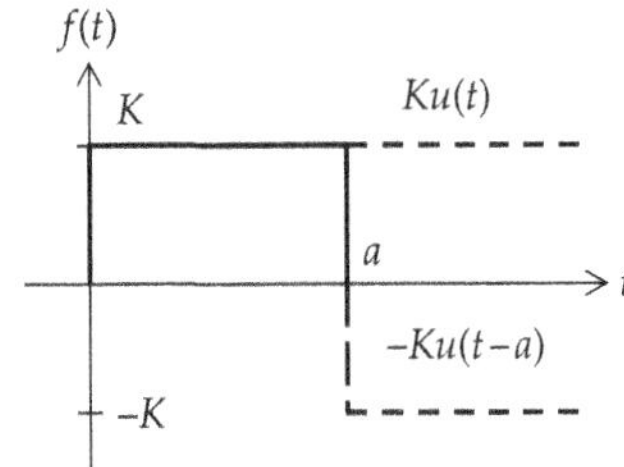

FIGURE 21.2
A rectangular pulse as the sum of step functions.

The translation-in-time property can be used to determine the LT of a periodic function $f(t)$ of period T that starts at $t=0$, given that the LT of the first period is $G(s)$. Since periods after the first are successively translated by T, it follows that

$$\mathcal{L}\{f(t)\}=G(s)\left[1+e^{-sT}+e^{-2sT}+\cdots\right]=\frac{G(s)}{1-e^{-sT}} \quad (21.42)$$

A useful generalization of the translation-in-the-time-domain property is when $f(t)$ is delayed by b and the step function is delayed by a, where a and b are positive constants with $b \geq a$. The LT becomes $\mathcal{L}\{f(t-b)(u(t-a))\}=\int_{0^-}^{\infty} f(t-b)u(t-a)e^{-st}dt$. Proceeding as in Equation 21.39 and substituting $t'=t-a$, it follows that

$$\mathcal{L}\{f(t-b)u(t-a)\}=e^{-as}\mathcal{L}\{f(t-b+a)\}, \quad b \geq a \quad (21.43)$$

irrespective of whether or not $f(t)$ has an impulse at the origin. The restriction $b \geq a$ ensures that $f(t-b+a)$ is not shifted to negative time, which is excluded by the single-sided LT.

Multiplication by *t*. If $\mathcal{L}\{f(t)\}=F(s)$, then

$$\mathcal{L}\{tf(t)\}=-\frac{dF(s)}{ds} \quad (21.44)$$

and

$$\mathcal{L}\{t^n f(t)\}=(-1)^n\frac{d^nF(s)}{ds^n} \quad (21.45)$$

Proof: $\mathcal{L}\{tf(t)\}=\int_{0^-}^{\infty} tf(t)e^{-st}dt=-\int_{0^-}^{\infty} f(t)\frac{d}{ds}\left(e^{-st}\right)dt=-\frac{d}{ds}\int_{0^-}^{\infty} f(t)e^{-st}dt=-\frac{dF(s)}{ds}$.

Hence, $\mathcal{L}\{t\cdot tf(t)\}=-\frac{d}{ds}\left[\mathcal{L}\{tf(t)\}\right]=\frac{d^2F(s)}{ds^2}$, and so on.

We can deduce that

$$\mathcal{L}\{t\cos\omega t\}=-\frac{d}{ds}\left[\frac{s}{s^2+\omega^2}\right]=\frac{s^2-\omega^2}{\left(s^2+\omega^2\right)^2} \quad (21.46)$$

$$\mathcal{L}\{t\sin\omega t\}=-\frac{d}{ds}\left[\frac{\omega}{s^2+\omega^2}\right]=\frac{2\omega s}{\left(s^2+\omega^2\right)^2} \quad (21.47)$$

Primal Exercise 21.4

Derive the LT of $2u(t)+3\delta(t-1)$.

Ans. $2/s+3e^{-s}$.

Primal Exercise 21.5

Derive the LT of a half-sinusoid defined by $f(t) = \sin t$, $0 \le t \le \pi$, and $f(t) = 0$, $t \ge \pi$, using (a) translation in time, by adding $\sin(t-\pi)u(t-\pi)$ to $\sin t u(t)$, and (b) direct evaluation, considering $\sin t = \text{Im}(e^{jt})$. Note that in (a) $\sin t$ is added to $\sin(t-\pi)u(t-\pi)$ and not to $\sin(t-\pi)$.

Ans. $\dfrac{1+e^{-\pi s}}{s^2+1}$.

Primal Exercise 21.6

Verify that the LT of te^{-t} is the same if multiplication by t is applied to e^{-t} or translation in the s-domain is applied to t.

Primal Exercise 21.7

Derive the LT of $tu(t-2)$

Ans. $\dfrac{e^{-2s}}{s^2} + \dfrac{2e^{-2s}}{s}$.

Division by t. If $\mathcal{L}\{f(t)\} = F(s)$, then, assuming that $\dfrac{f(t)}{t}$ has an LT,

$$\mathcal{L}\left\{\frac{f(t)}{t}\right\} = \int_s^\infty F(s)ds \tag{21.48}$$

Proof: $\int_s^\infty F(s)ds = \int_s^\infty \left[\int_{0^-}^\infty f(t)e^{-st}dt\right]ds$. Reversing the order of integration,

$$\int_s^\infty F(s)ds = \int_{0^-}^\infty\left[\int_s^\infty f(t)e^{-st}ds\right]dt = \int_{0^-}^\infty\left[-\frac{f(t)}{t}e^{-st}\right]_s^\infty dt$$
$$= \int_{0^-}^\infty\left[\frac{f(t)}{t}e^{-st}\right]dt.$$

For example, $\mathcal{L}\{u(t)\} = \mathcal{L}\left\{\dfrac{tu(t)}{t}\right\} = \int_s^\infty \dfrac{1}{s^2}ds = \dfrac{1}{s}$.

The reader may have noted that in Section 21.1 we referred to the region of convergence in the s-plane, which assured convergence of the integral (Equation 21.3), yet no reference to this region of convergence was made in subsequent discussions of the LT. In fact, the validity of the unilateral LT is not restricted to the region of convergence but extends to the whole of the s-plane except at the poles (Section 21.3), where the LT becomes infinite.

Table 21.1 summarizes the basic properties of the LT.

TABLE 21.1

Basic Properties of the LT $\mathcal{L}\{f(t)\} = F(s)$

$\mathcal{L}\{Kf(t)\}$	$KF(s)$
$\mathcal{L}\{f_1(t) \pm f_2(t)\}$	$F_1(s) \pm F_2(s)$
$\mathcal{L}\{f(at)\}$, $a>0$	$\frac{1}{a}F\left(\frac{s}{a}\right)$
$\mathcal{L}\left\{\int_{0^-}^t f(x)dx\right\}$	$\frac{1}{s}F(s)$
$\mathcal{L}\left\{\int_{0^-}^t f(x)dx + f(0^-)\right\}$	$\frac{1}{s}F(s) + \frac{1}{s}f(0^-)$
$\mathcal{L}\left\{\frac{df(t)}{dt}\right\}$	$sF(s) - f(0^-)$
$\mathcal{L}\left\{\frac{df^n(t)}{dt^n}\right\}$	$s^nF(s) - s^{n-1}f(0^-) - s^{n-2}f^{(1)}(0^-) - \cdots - f^{(n-1)}(0^-)$
$\mathcal{L}\{e^{-at}f(t)\}$	$F(s+a)$
$\mathcal{L}\{tf(t)\}$	$-\frac{dF(s)}{ds}$
$\mathcal{L}\{t^nf(t)\}$	$(-1)^n\frac{d^nF(s)}{ds^n}$
$\mathcal{L}\left\{\frac{f(t)}{t}\right\}$	$\int_s^\infty F(u)du$
$\mathcal{L}\{f(t-a)u(t-a)\}$	$e^{-as}F(s)$
$\mathcal{L}\{f(t-b)u(t-a)\}$	$e^{-as}\mathcal{L}\{f(t-b+a)\}$
$\mathcal{L}\{f(t)\}$, where $f(t)$ is periodic of period T	$\frac{G(s)}{1-e^{-sT}}$, where $G(s)$ is LT of first period
$\mathcal{L}\{f(t)*g(t)\}$	$F(s)G(s)$

Example 21.3: Laplace Transform of a Sawtooth Pulse

As an illustrative example of the properties of the LT and of graphical manipulation of basic waveforms, we will consider the derivation of the LT of the single sawtooth pulse of Figure 21.3 by four methods.

Solution:

Method 1: The LT can be evaluated by direct integration:

$$F(s) = \int_{0^-}^T \frac{A}{T}te^{-st}dt = \frac{A}{T}\left[-\frac{t}{s}e^{-st}\right]_{0^-}^T + \frac{A}{sT}\int_{0^-}^T e^{-st}dt$$
$$= \frac{A}{T}\left[-\frac{t}{s}e^{-st}\right]_{0^-}^T + \frac{A}{sT}\left[-\frac{1}{s}e^{-st}\right]_{0^-}^T$$
$$= \frac{A}{T}\left(-\frac{T}{s}e^{-Ts} - \frac{1}{s^2}e^{-Ts} + \frac{1}{s^2}\right) = A\left(-\frac{e^{-Ts}}{s} + \frac{1-e^{-Ts}}{Ts^2}\right) \tag{21.49}$$

Method 2: The sawtooth pulse, $f(t)$, can be considered as the sum of three components: (1) A ramp function

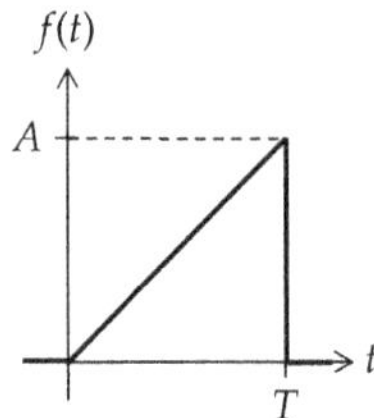

FIGURE 21.3
Figure for Example 21.3.

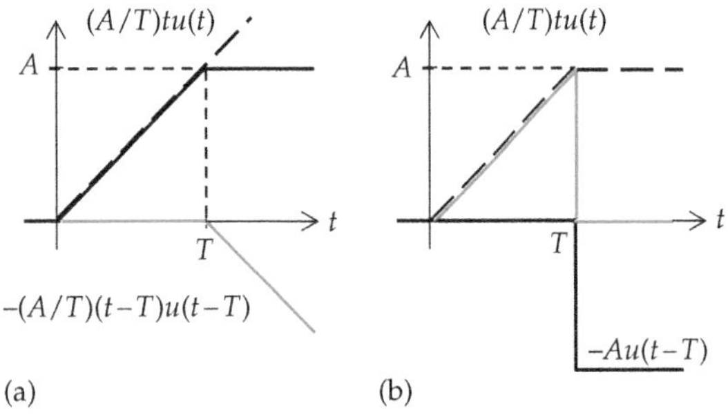

FIGURE 21.4
Figure for Example 21.3.

$(A/T)tu(t)$, as shown by the line having the long dashes in Figure 21.4a; (2) a delayed and negated ramp function $-(A/T)(t-T)u(t-T)$ having the same magnitude of slope as the first component, and illustrated by the blue line in Figure 21.4a. The delayed ramp, before negation is $(A/T)(t-T)u(t-T)$, as explained in connection with Figure 20.25. When these two components are added, the opposing slopes cancel one another for $t \geq T$, resulting in the solid line plot in Figure 21.4a; (3) to obtain the sawtooth from this function, a third component, the negated and delayed step function, $-Au(t-T)$ should be added. It follows that

$$f(t) = \frac{A}{T}tu(t) - \frac{A}{T}(t-T)u(t-T) - Au(t-T) \quad (21.50)$$

The LT of Equation 21.50 is

$$F(s) = \frac{A}{T}\frac{1}{s^2} - \frac{A}{T}\frac{e^{-Ts}}{s^2} - \frac{Ae^{-Ts}}{s} = A\left[-\frac{e^{-Ts}}{s} + \frac{1-e^{-Ts}}{Ts^2}\right] \quad (21.51)$$

Method 3: Consider the rectangular pulse of amplitude A/T and duration T (Figure 21.5a). The integral of this pulse increases linearly with a slope A/T over the range $0 \leq t \leq T$, and equals A at $t = T$. The integral remains at this value for $t \geq T$ (Figure 21.5b). The LT of the rectangular pulse of Figure 21.5a is given by Equation 21.36 for Figure 21.2, with A/T replacing K and T replacing a, to give $(A/T)(1-e^{-Ts})/s$. Dividing this by s gives the LT of the integral function. When the delayed step $-Au(t-T)$,

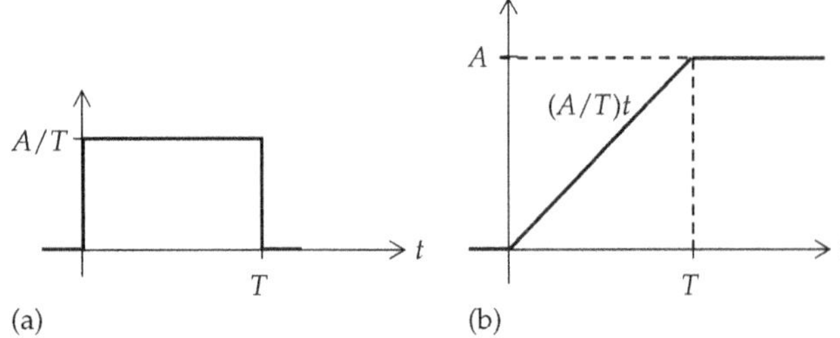

FIGURE 21.5
Figure for Example 21.3.

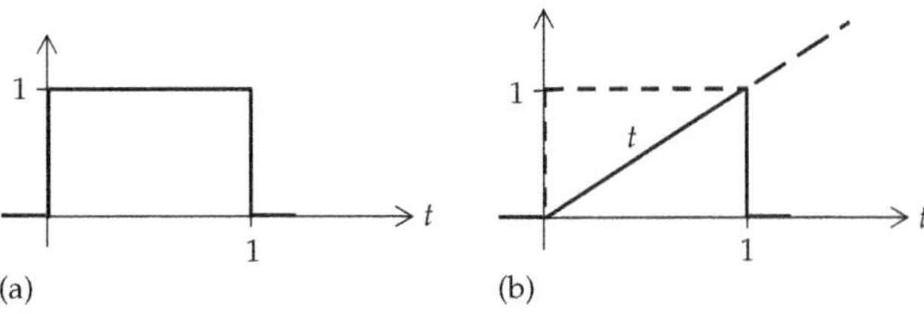

FIGURE 21.6
Figure for Example 21.3.

whose LT is $-Ae^{-Ts}/s$, is added to the integral function, the LT of the given pulse is obtained. The sum of the LT of the integrated function and that of the delayed step gives the same result as in Equation 21.51.

As a variation on this method, the derivative of $f(t)$ is the rectangular pulse of Figure 21.5a plus an impulse $-A\delta(t - T)$. The LT of the combination is $(A/T)(1-e^{-Ts})/s - Ae^{-Ts}$. Dividing by s to obtain the LT of the integral gives the same result.

Method 4: Consider a rectangular pulse of unit amplitude and unit duration (Figure 21.6a). If a ramp function $x(t) = t$ is multiplied by this unit rectangular pulse, the product is a sawtooth pulse of unit amplitude and unit duration (Figure 21.6b). The LT of the unit rectangular pulse is $(1-e^{-s})/s$ (Equation 21.41). Multiplication by t is equivalent to differentiating this with respect to s and changing the sign (Equation 21.44). This gives $(1-e^{-s}-se^{-s})/s^2$. If the sawtooth pulse is of height 1 and duration T, then the ramp function becomes t/T, so that the value is unity at $t = T$ rather than $t = 1$. This is equivalent to time scaling (Equation 21.16) with $a = 1/T$. The LT and s are multiplied by T, which gives $(1-e^{-sT}-sTe^{-sT})/Ts^2$. Multiplying this by A gives Equation 21.51.

21.3 Solution of Linear, Ordinary Differential Equations

To illustrate the application of the LT to the solution of linear, ordinary differential equations with constant coefficients, consider, for simplicity and without loss of generality, a second-order differential equation of the form

$$a\frac{d^2y(t)}{dt^2}+b\frac{dy(t)}{dt}+cy(t)=f(t) \tag{21.52}$$

where $f(t)$, referred to as the **forcing function**, is some arbitrary function that has an LT. Taking the LT of both sides, making use of the differentiation-in-time property:

$$as^2Y(s)-ay^{(1)}(0^-)-asy(0^-)+bsY(s)-by(0^-)$$
$$+cY(s)=F(s) \tag{21.53}$$

where $Y(s)$ and $F(s)$ are the LTs of $y(t)$ and $f(t)$, respectively. Solving for $Y(s)$,

$$Y(s)=\frac{(as+b)y(0^-)+ay^{(1)}(0^-)}{as^2+bs+c}+\frac{F(s)}{as^2+bs+c} \tag{21.54}$$

The following concept should be noted in connection with Equation 21.54 that is an important feature of the LT method:

Concept: *A linear, ordinary differential equation with constant coefficients is transformed by the LT to an algebraic equation in powers of s that can be solved for the LT of the variable of the equation Y(s), as in any algebraic equation. When this is done, the initial conditions $y(0^-)$ and $y^{(1)}(0^-)$ appear like applied inputs.*

Formally, the solution to the differential equation, $y(t)$ can be expressed as

$$y(t)=\mathcal{L}^{-1}\left\{\frac{(as+b)y(0^-)+ay^{(1)}(0^-)}{as^2+bs+c}\right\}+\mathcal{L}^{-1}\left\{\frac{F(s)}{as^2+bs+c}\right\} \tag{21.55}$$

where $\mathcal{L}^{-1}$ denotes the ILT.

The first term on the RHS of Equation 21.55 is the response $y(t)$ for nonzero initial conditions, with no forcing function. It is the **natural response** or **zero-input response**. The second term is the response $y(t)$ to the forcing function alone. It is the steady-state response, the **forced response,** or **zero-state response**. Because the system is linear, the total response is the sum of the natural and forced responses, the initial conditions being treated as inputs at $t=0$ (Section 22.1). The denominator as^2+bs+c is analogous, term by term to the LHS of Equation 21.52 with the first and second derivatives replaced by s and s^2, respectively. The equation $as^2+bs+c=0$ is the **characteristic equation** previously derived from the homogenous differential equation (Equation 12.6).

Concept: *The characteristic equation of a linear differential equation is a polynomial in s obtained by taking the LT of the equation with zero forcing function (the homogeneous equation), zero initial conditions, and assuming a nonzero value of the variable of the equation. The LHS of the characteristic equation appears in the denominator of the LT of all the responses derived from the differential equation.*

Thus, if we take the LT of Equation 21.52, with $f(t)=0$ and zero initial conditions, we obtain $(as^2+bs+c)Y(s)=0$. In general, $Y(s)\neq 0$, so that $as^2+bs+c=0$ is the characteristic equation.

21.3.1 Inverse Laplace Transform

The ILT can be obtained in one of the following ways:

1. Lookup in tables of LT pairs, as previously mentioned.
2. Numerical inversion of the LT.
3. The LTs of responses of LTI circuits having lumped circuit parameters are rational functions of s. These can be expressed, at least in part, as proper rational functions. Such functions can be expanded as partial fractions, which are then inverted term by term.
4. Use of MATLAB's 'ilaplace' command. For example, if we enter:

```
>> syms s
>> ilaplace (1/s^5)
MATLAB returns: 1/24*t^4
```

For convenience in determining ILTs, Table 21.2 summarizes the LTs of some commonly encountered functions in circuit analysis, based on the properties of the LT discussed in the previous section.

21.3.2 Partial Fraction Expansion

Consider an LT $F(s)$ that is a rational function of s, that is, a ratio of two polynomials in s:

$$F(s)=\frac{a_ms^m+a_{m-1}s^{m-1}+\cdots+a_1s+a_0}{b_ns^n+b_{n-1}s^{n-1}+\cdots+b_1s+b_0} \tag{21.56}$$

where the a's and b's are real coefficients, and m and n are integers. $F(s)$ can be expressed in terms of factors of the numerator and denominator as

$$F(s)=K\frac{(s+z_1)(s+z_2)\ldots(s+z_m)}{(s+p_1)(s+p_2)\ldots(s+p_n)} \tag{21.57}$$

TABLE 21.2

LT Pairs

f(t)	*F(s)*	Reference Equation
$\delta(t)$	1	(21.10)
$\delta(t-a)$	e^{-as}	(21.37)
$\delta^{(n)}(t)$	s^n	(21.24)
$u(t)$	$\frac{1}{s}$	(21.7)
$tu(t)$	$\frac{1}{s^2}$	(21.20)
$t^n u(t)$	$\frac{n!}{s^{(n+1)}}$	(21.20)
$e^{-at}u(t)$	$\frac{1}{s+a}$	(21.9)
$te^{-at}u(t)$	$\frac{1}{(s+a)^2}$	(21.20) and (21.32)
$t^n e^{-at}u(t)$	$\frac{n!}{(s+a)^{(n+1)}}$	(21.20) and (21.32)
$\sin\omega t u(t)$	$\frac{\omega}{s^2+\omega^2}$	(21.15)
$\cos\omega t u(t)$	$\frac{s}{s^2+\omega^2}$	(21.14)
$\sin(\omega t+\theta)u(t)$	$\frac{s\sin\theta+\omega\cos\theta}{s^2+\omega^2}$	(21.14) and (21.15)
$\cos(\omega t+\theta)u(t)$	$\frac{s\cos\theta-\omega\sin\theta}{s^2+\omega^2}$	(21.14) and (21.15)
$e^{-at}\sin\omega t u(t)$	$\frac{\omega}{(s+a)^2+\omega^2}$	(21.34)
$e^{-at}\cos\omega t u(t)$	$\frac{s+a}{(s+a)^2+\omega^2}$	(21.33)
$t\sin\omega t u(t)$	$\frac{2\omega s}{(s^2+\omega^2)^2}$	(21.47)
$t\cos\omega t u(t)$	$\frac{s^2-\omega^2}{(s^2+\omega^2)^2}$	(21.46)
$te^{-at}\sin\omega t u(t)$	$\frac{2\omega(s+a)}{((s+a)^2+\omega^2)^2}$	(21.32) and (21.47)
$te^{-at}\cos\omega t u(t)$	$\frac{(s+a)^2-\omega^2}{((s+a)^2+\omega^2)^2}$	(21.32) and (21.46)

where $K = a_m/b_n$. The values $-z_1, -z_2, \ldots, -z_m$ are **zeros** of $F(s)$ since $F(s) = 0$ when s assumes any of these values. The values $-p_1, -p_2, \ldots, -p_n$ are **poles** of $F(s)$, since $F(s) \to \infty$ when s assumes any of these values. In general zeros and poles may be complex, in which case they must occur in complex conjugate pairs, because the a's and b's are real in the LTs of physical systems. Note that $Y(s)$, derived from the differential equation of a circuit response (Equation 21.54), is a rational function.

In general, $m > n$ in Equation 21.56, so by dividing the numerator by the denominator, $F(s)$ can be expressed as

$$F(s) = k_{m-n}s^{m-n} + k_{m-n-1}s^{m-n-1} + \cdots + k_1 s + k_0 + \frac{c_r s^r + c_{r-1}s^{r-1} + \cdots + c_1 s + c_0}{b_n s^n + b_{n-1}s^{n-1} + \cdots + b_1 s + b_0} \quad (21.58)$$

where $r < n$. This makes the ratio of polynomials on the RHS a *proper* rational function. The inverse transform of k_0 is $k_0\delta(t)$, that of $k_1 s$ is $k_1\delta^{(1)}(t)$, and that of $k_{m-n}s^{m-n}$ is $k_{m-n}\delta^{(m-n)}(t)$ (Table 21.2). It remains to determine the ILT of the proper rational function, which we will denote as $\frac{N(s)}{D(s)}$.

Most textbooks discuss methods for deriving the partial fraction expansion (PFE) of proper rational functions. These methods are based on examples in which the coefficients of s in $N(s)$ and $D(s)$ are relatively small whole numbers, and the power of s in $D(s)$ is usually 2 or 3. Such examples, however, are rather artificial, and the aforementioned methods become impractical in more realistic cases where the coefficients of s are not small whole numbers and the power of s in the denominator exceeds 3. In practice, the ILT is obtained in these cases using MATLAB's 'ilaplace' command, without deriving the PFE, as illustrated by Example 21.4. Moreover, this command can be used irrespective of whether or not the rational function is proper. Nevertheless, the PFE is of theoretical importance. We will therefore consider in the following paragraphs some aspects of the PFE that are important for future discussions.

Consider the proper rational function:

$$F(s) = \frac{N(s)}{(s+p_1)(s+p_2)\ldots(s+p_n)} \quad (21.59)$$

where $N(s)$ is a polynomial of power less than n and all the poles have different, real values. $F(s)$ can be expressed in partial fraction form as

$$F(s) = \frac{K_1}{s+p_1} + \frac{K_2}{s+p_2} + \cdots + \frac{K_n}{s+p_n} \quad (21.60)$$

where the K's are constant coefficients known as the **residues** of $F(s)$. An easy way of determining any of these coefficients, say, K_1, is to multiply both sides of Equation 21.60 by the corresponding denominator, $(s+p_1)$ in this case, and set $s = -p_1$. The RHS of Equation 21.60 reduces simply to K_1. The LHS becomes $\left.\frac{N(s)}{(s+p_2)\ldots(s+p_n)}\right|_{s=-p_1}$, which can be readily evaluated to give K_1. This is the **residue method** for determining the PFE. For example, consider $F(s) = \frac{2s^2+8s+7}{(s+1)(2s+3)(s+2)}$. The PFE is of the form

$$F(s) = \frac{2s^2+8s+7}{(s+1)(2s+3)(s+2)} = \frac{K_1}{s+1} + \frac{K_2}{2s+3} + \frac{K_3}{s+2} \quad (21.61)$$

To determine K_1, both sides of Equation 21.61 are multiplied by $(s + 1)$. The resulting equation is

$$\frac{2s^2+8s+7}{(2s+3)(s+2)} = K_1 + \frac{K_2(s+1)}{2s+3} + \frac{K_3(s+1)}{s+2} \quad (21.62)$$

Substituting $s = -1$, the LHS of Equation 21.62 gives 1, and the RHS reduces to K_1, so that $K_1 = 1$. Similarly, multiplying both sides of Equation 21.61 by $(2s + 3)$ and setting $s = -3/2$ gives $K_2 = 2$, whereas multiplying both sides of Equation 21.61 by $(s + 2)$ and substituting $s = -2$ gives $K_3 = -1$. It follows that the PFE of $F(s)$ is

$$F(s) = \frac{2s^2+8s+7}{(s+1)(2s+3)(s+2)} = \frac{1}{s+1} + \frac{2}{2s+3} - \frac{1}{s+2} \quad (21.63)$$

The ILT of $F(s)$ is

$$f(t) = \left[e^{-t} + e^{-1.5t} - e^{-2t}\right]u(t) \quad (21.64)$$

Note that the ILT of terms in the PFE other than those involving $\delta(t)$ and $u(t)$, is automatically multiplied by $u(t)$, because the time function is assumed to be zero for $t \leq 0^-$.

Consider, next, the proper rational function $F(s) = \frac{3s^2+8s+4}{(s+1)^3}$. To see how $F(s)$ can be represented by a PFE, we express the numerator in terms of powers of $(s + 1)$. To do so, the numerator is written as $k_1(s + 1)^2 + k_2(s + 1) + k_3$. This is set equal to $3s^2 + 8s + 4$, and k_1, k_2, and k_3 are determined by comparing coefficients of the terms having the same power of s and the constant term. It follows that $k_1 = 3$, $k_2 = 2$, and $k_3 = -1$. Hence,

$$F(s) = \frac{3s^2+8s+4}{(s+1)^3} = \frac{3(s+1)^2+2(s+1)-1}{(s+1)^3}$$

$$= \frac{3}{(s+1)} + \frac{2}{(s+1)^2} - \frac{1}{(s+1)^3} \quad (21.65)$$

Similar reasoning leads to the conclusion that if the denominator of $F(s)$ consists of the repeated root $(s+p)^n$, the PFE of $F(s)$ is of the general form

$$F(s) = \frac{N(s)}{(s+p)^n} = \frac{K_1}{s+p} + \frac{K_2}{(s+p)^2} + \cdots + \frac{K_{n-1}}{(s+p)^{n-1}} + \frac{K_n}{(s+p)^n} \quad (21.66)$$

The residue method can only be used to determine K_n by multiplying both sides of Equation 21.66 by $(s + p)^n$ and setting $s = -p$. Thus, if both sides are multiplied by $(s + p)^{n-1}$, for example, in order to determine K_{n-1}, the LHS becomes $N(s)/(s + p)$, and the term in K_n on the RHS becomes $K_n/(s + p)$. Setting $s = -p$ makes both sides infinite. The same is true if both sides are multiplied by $(s + p)$ raised to any of the smaller powers in the denominator. The residues K_1 to K_{n-1} can be determined analytically either through differentiation or by equating the coefficients of the same power of s, as well as the constant term, on both sides of the equation (Examples 21.4 and 21.6, respectively). It should be noted that whereas the poles in Equation 21.60 are simple poles, $-p$ in Equation 21.66 is referred to as a **multiple pole of order n**.

The roots considered so far are real, but the same considerations apply if the roots are complex, except that, since the coefficients in $N(s)$ and $D(s)$ are real, complex poles must occur in conjugate pairs and their residues must also be conjugate pairs. We will illustrate working with simple complex poles in Example 21.4, but as mentioned earlier, the ILT can be derived using MATLAB's 'ilaplace' command without determining the residues, and irrespective of whether the poles are real or complex, single or multiple.

Example 21.4: Inverse Laplace Transform

It is required to determine the ILT of $F(s) = \frac{25}{s(s^2+2s+5)}$ using MATLAB and by PFE.

Solution:

We invoke MATLAB's 'ilaplace' command by entering

```
>> syms s
>> ilaplace (25/(s*(s^2+2*s+5)))
MATLAB returns: 5 - (5*(cos(2*t) +
  sin(2*t)/2))/exp(t)
```

This can be rearranged as

$$f(t) = 5 - 5e^{-t}\left[\cos 2t + \frac{1}{2}\sin 2t\right], \quad t \geq 0 \quad (21.67)$$

To determine the ILT by using the PFE, we note that $s^2 + 2s + 5 = (s + 1 + j2)(s + 1 - j2)$, so that the roots are complex. The PFE is therefore

$$F(s) = \frac{25}{s\left(s^2 + 2s + 5\right)} = \frac{K_1}{s} + \frac{K_2}{s+1+j2} + \frac{K_2^*}{s+1-j2} \quad (21.68)$$

where the residues K_2 and K_2^* are complex conjugates because their respective poles are complex conjugates. This ensures that the coefficients in the numerator of $F(s)$ are real quantities, as they should be for any physical system. K_1 is determined in the usual manner by multiplying both sides of Equation 21.68 by s and setting $s = 0$. This gives $K_1 = 5$. K_2 is determined in the same manner by first multiplying both sides of Equation 21.68 by $(s + 1 + j2)$. Thus,

$$(s+1+j2) \times \frac{25}{s(s+1+j2)(s+1-j2)} = \frac{25}{s(s+1-j2)}$$

$$= \frac{K_1}{s}(s+1+j2) + K_2 + \frac{K_2^*}{s+1-j2}(s+1+j2) \quad (21.69)$$

Substituting $s = -1 - j2$, and simplifying gives $K_2 = -\left(\frac{5}{2} + j\frac{5}{4}\right)$, so that $K_2^* = -\left(\frac{5}{2} - j\frac{5}{4}\right)$. The PFE becomes

$$F(s) = \frac{25}{s\left(s^2 + 2s + 5\right)} = \frac{5}{s} - \left(\frac{5}{2} + j\frac{5}{4}\right)\frac{1}{s+1+j2}$$

$$-\left(\frac{5}{2} - j\frac{5}{4}\right)\frac{1}{s+1-j2} \quad (21.70)$$

Referring to Table 21.2, the ILT is

$$f(t) = 5u(t) - \left(\frac{5}{2} + j\frac{5}{4}\right)e^{-(t+j2)}u(t)$$

$$-\left(\frac{5}{2} - j\frac{5}{4}\right)e^{-(t-j2)}u(t) \quad (21.71)$$

Simplifying and expressing the complex exponents in terms of sinusoidal functions gives Equation 21.67.

It is seen that working with complex numbers is rather awkward and error prone. This can be avoided by expressing Equation 21.68 as

$$F(s) = \frac{25}{s\left(s^2 + 2s + 5\right)} = \frac{k_1}{s} + \frac{k_2 s + k_3}{s^2 + 2s + 5} \quad (21.72)$$

Note that in the PFE, if the denominator is a term for a simple pole, real or complex, the numerator is a constant and equal to a residue of the function, as in Equations 21.61 and 21.68. If the denominator is a term for a multiple pole p^n, the PFE includes all terms having in the denominator $(s + p)$ raised to all integral powers from 1 to n, the numerators of these terms being constants, as in Equation 21.66. However, when the two terms involving simple, complex conjugate poles are combined, the denominator is quadratic in s, and the numerator is, in general, linear in s, as it should be in a proper rational function. The numerator is therefore expressed as $k_2 s + k_3$ in the second term on the RHS in Equation 21.72. k_1 is determined as before by multiplying both sides of Equation 21.72 by s and setting $s = 0$, which gives $k_1 = 5$; k_2 and k_3 are obtained by combining the two terms on the RHS of Equation 21.72 into a single term having a common denominator. The coefficients of terms having s raised to the same integral power in the numerator on both sides of the equation are then compared, as well as the constant term. With $k_1 = 5$, Equation 21.72 is expressed as

$$F(s) = \frac{25}{s\left(s^2 + 2s + 5\right)} = \frac{(k_2 + 5)s^2 + (10 + k_3)s + 25}{s\left(s^2 + 2s + 5\right)} \quad (21.73)$$

To have equal numerators on both sides of this equation, we must have $k_2 = -5$ and $k_3 = -10$. Equation 21.72 becomes

$$F(s) = \frac{25}{s\left(s^2 + 2s + 5\right)} = \frac{5}{s} - \frac{5s + 10}{s^2 + 2s + 5} \quad (21.74)$$

The next step is to complete the square of the quadratic in the denominator on the RHS. Thus, $s^2 + 2s + 5 = [(s + 1)^2 + 2^2]$. The numerator is expressed in terms of the squared term in s in the denominator, that resulted from completing the square. This term is $(s + 1)$. Thus, $5s + 10 = 5(s + 1) + 5$. $F(s)$ becomes

$$F(s) = \frac{5}{s} - \frac{5s + 10}{s^2 + 2s + 5} = \frac{5}{s} - \frac{5(s+1)}{(s+1)^2 + 2^2} - \frac{5}{2}\frac{2}{(s+1)^2 + 2^2} \quad (21.75)$$

Referring to Table 21.2, the ILT of the first term on the RHS is $5u(t)$, that of the second term is $-5e^{-t}\cos(2t)u(t)$, and that of the third term is $-\frac{5}{2}e^{-t}\sin(2t)u(t)$. Adding these terms gives Equation 21.67.

Problem-Solving Tip

- Working with complex quantities can be avoided by combining complex conjugate poles in the PFE into a single term. The numerator of this term is linear in s when its denominator is quadratic in s and the complex poles are simple.

Primal Exercise 21.8

Derive the ILT of the following functions: (a) $\frac{5}{s^2+3s+2}$; (b) $\frac{3}{(s+1)^2}$; (c) $\frac{2s+3}{s^2+4s+6}$.

Ans. (a) $5(e^{-t}-e^{-2t})$; (b) $3e^{-t}(1-t)$; (c) $e^{-2t}\left(2\cos\sqrt{2}t-\frac{1}{\sqrt{2}}\sin\sqrt{2}t\right)$.

21.4 Theorems on the Laplace Transform

21.4.1 Final-Value Theorem

This is a useful theorem that gives the final value of a function of time from its LT without having to invert the transform:

If $\mathcal{L}\{f(t)\} = F(s)$, where all the poles of F(s) have negative real parts, except for a simple pole at the origin, if such a pole exists, then

$$\lim_{t\to\infty} f(t) = \lim_{s\to 0} sF(s) \tag{21.76}$$

***Proof*:** If $F(s)$ has only a simple pole at the origin, its PFE can be expressed as

$$F(s) = p(s) + \frac{K_1}{s} + \cdots + \frac{K_r}{(s+a_r)^m} + \cdots \tag{21.77}$$

where

- $p(s)$ is a polynomial in s
- the pole $-a_r$ is negative real, of order m, or complex of order m but having a negative real part
- m is an integer equal to or larger than 1

Note that $-a_r$ cannot be a purely imaginary pole, like that of the LT of $\cos\omega t$ and $\sin\omega t$ because such poles are on the imaginary axis, with no negative real part. They are excluded in the statement of the final-value theorem. Multiplying both sides of Equation 21.77 by s and setting $s = 0$ gives $F(s) = K_1$. It follows from this equation that $\lim_{s\to 0} sF(s) = K_1$.

The ILT of the RHS of Equation 21.77 is

$$f(t) = (\text{Impulse function and its derivatives}) + K_1 + (\text{Terms multiplied by } e^{-a_r t}) \tag{21.78}$$

The impulse function and its derivatives are over at $t=0^+$. When $-a_r$ is real, the ILT of the term $K_r/(s + a_r)^m$ is, from Table 21.2, of the form $e^{-a_r t}$ multiplied by an integral power of t. When $-a_r$ is complex, equal to $-\sigma_r \pm j\omega_r$, where σ_r and ω_r are positive numbers, the ILT is of the form $e^{-a_r t}$ multiplied, in general, by terms containing $\cos\omega_r t$ and $\sin\omega_r t$, which in turn may be multiplied by t raised to some power. In all these cases the exponential multiplier $e^{-a_r t}$ approaches zero as $t \to \infty$, so that these terms vanish. It follows that $f(t) \to K_1$ as $t \to \infty$. Note that a multiple pole at the origin makes $sF(s) \to \infty$ for $s = 0$ and is not in accordance with the statement of the final-value theorem. In this case $f(t) \to \infty$ as $t \to \infty$ (Primal Exercise 21.9). However, the equal limits at infinity cannot be considered as satisfying Equation 21.76.

21.4.2 Initial-Value Theorem

If $\mathcal{L}\{f(t)\} = F(s)$ where F(s) is a proper rational function, then

$$\lim_{t\to 0^+} f(t) = \lim_{s\to\infty} sF(s) \tag{21.79}$$

***Proof*:** $\mathcal{L}\left\{\frac{df(t)}{dt}\right\} = sF(s) - f(0^-)$. The LT of the derivative of $f(t)$ is $\int_{0^-}^{\infty}\frac{df(t)}{dt}e^{-st}dt = \int_{0^-}^{0^+}\frac{df(t)}{dt}1dt + \int_{0^+}^{\infty}\frac{df(t)}{dt}e^{-st}dt$. If we take the limit as $s \to \infty$, the last integral vanishes because of the term e^{-st}. Since $F(s)$ is a proper rational function, $f(t)$ does not have any impulses or their derivatives at the origin, and $\int_{0^-}^{0^+}\frac{df(t)}{dt}1dt = f(0^+) - f(0^-)$. It follows that $\lim_{s\to\infty} sF(s) = f(0^-) + \lim_{s\to\infty}\mathcal{L}\left\{\frac{df(t)}{dt}\right\} = f(0^-) + f(0^+) - f(0^-) = f(0^+)$.

If $F(s)$ is not a proper rational function, $f(t)$ has impulses or their derivatives at the origin. But these are over at $t = 0^+$. The correct value of $f(0^+)$ is obtained by applying the initial-value theorem to the proper rational fraction part of $F(s)$.

The initial- and final-value theorems are useful for checking the LT for a given variable, because it is often obvious from the nature of the problem what the initial and final values of the variable are. Moreover, the initial-value theorem is particularly useful in the case of impulsive readjustments at $t = 0$ when the voltage across a capacitor, or the current through an inductor, is forced to change instantaneously. The initial-value theorem gives the correct values at $t = 0^+$ after the impulsive readjustments are over, as will be demonstrated in the next chapter.

Example 21.5: Final-Value and Initial-Value Theorems

Apply the final-value and initial-value theorems to the following: (a) $\frac{3s+10}{s^2+7s+12}$; (b) $\frac{2s+5}{s^2+4}$; (c) $3\frac{s^3-6s^2+15s+40}{s(s^2+4s+5)}$. Compare to the limiting values of the ILT.

Solution:

(a) $F(s)=\dfrac{3s+10}{s^2+7s+12}=\dfrac{3s+10}{(s+3)(s+4)}$. The conditions for applying the final-value theorem are satisfied. Multiplying by s and setting $s = 0$ gives $sF(s) = 0$, so that $f(\infty) = 0$. The conditions for applying the initial-value theorem are also satisfied. Multiplying by s and letting $s \to \infty$ gives $sF(s) = 3$, so that $f(0^+) = 3$.

To derive the ILT, $F(s)$ is expressed as a PFE:

$$F(s)=\frac{3s+10}{(s+3)(s+4)}=\frac{k_1}{s+3}+\frac{k_2}{s+4} \tag{21.80}$$

Multiplying both sides by $(s + 3)$ and setting $s = -3$ gives $k_1 = 1$. Multiplying both sides by $(s + 4)$ and setting $s = -4$ gives $k_2 = 2$. It follows that $f(t)=e^{-3t}+2e^{-4t}$. It is seen that $f(0^+) = 3$ and $f(\infty) = 0$.

(b) Multiplying by s and setting $s = 0$ gives $sF(s) = 0$, so that $f(\infty) = 0$. This, however, is not correct, because the conditions for applying the final-value theorem are not satisfied since there are two complex poles on the imaginary axis at $\pm j2$, without having negative real parts. The conditions for the initial-value theorem are satisfied. Multiplying by s and letting $s \to \infty$ gives $sF(s) = 2$, so that $f(0^+) = 2$.

The ILT is

$$f(t)=2\cos 2t+\frac{5}{2}\sin 2t \tag{21.81}$$

It is seen that $f(0^+) = 2$, in accordance with the initial-value theorem, but as $t \to \infty$, $f(t)$ alternates between two finite values.

(c) The conditions for applying the final-value theorem are satisfied: (i) the pole at the origin is simple, and (ii) the complex poles are at $-2 \pm j$ have negative real parts. Multiplying by s and setting $s = 0$ gives $sF(s) = 24$, so that $f(\infty) = 24$. To apply the initial-value theorem, we note that the function is not a proper rational function. If multiplied by s and $s \to \infty$, $F(s) \to \infty$. To obtain the value at $t = 0^+$, the numerator is divided by the denominator to give a proper rational function. This gives

$$F(s)=3\frac{s^3-6s^2+15s+40}{s(s^2+4s+5)}=3\left(1+\frac{-10s^2+10s+40}{s(s^2+4s+5)}\right) \tag{21.82}$$

Considering the proper fraction, multiplying it by s, and letting $s \to \infty$ gives $3 \times (-10) = -30$. Hence, $f(0^+) = -30$.

To derive $f(t)$, the proper fraction is expressed as a PFE:

$$\frac{-10s^2+10s+40}{s(s^2+4s+5)}=\frac{k_1}{s}+\frac{k_2 s+k_3}{s^2+4s+5} \tag{21.83}$$

Multiplying both sides by s and setting $s = 0$ gives $k_1 = 8$. Multiplying out on the RHS and collecting terms in the numerator,

$$\frac{-10s^2+10s+40}{s(s^2+4s+5)}=\frac{(8+k_2)s^2+(32+k_3)s+40}{s(s^2+4s+5)} \tag{21.84}$$

Comparing coefficients on both sides, $k_2 = -18$ and $k_3 = -42$. It follows that

$$\frac{-10s^2+10s+40}{s(s^2+4s+5)}=\frac{8}{s}-\frac{18s+22}{s^2+4s+5} \tag{21.85}$$

We next complete the square in the quadratic as $(s + 2)^2 + 1$, and express the numerator as $18s + 22 = 18(s + 2) - 10$. Hence, $F(s)$ becomes

$$F(s)=3\left(1+\frac{8}{s}-\frac{18(s+2)}{(s+2)^2+1}+\frac{10}{(s+2)^2+1}\right) \tag{21.86}$$

Taking the ILT,

$$f(t)=3\left(\delta(t)+8u(t)-18e^{-2t}\cos t+10e^{-2t}\sin t\right) \tag{21.87}$$

It is seen that $f(0^+) = 3(0 + 8 - 18 + 0) = -30$ in accordance with the initial-value theorem, and $f(\infty) = 24$, in accordance with the final-value theorem.

Primal Exercise 21.9

Given (a) $F(s)=\dfrac{s+2}{s(s+1)(s+3)}$ and (b) $F(s)=\dfrac{s+2}{s^2(s+1)(s-3)}$, determine $f(0^+)$ and $f(\infty)$ and verify by deriving $f(t)$.

Ans. (a) $f(0^+)=0$, $f(\infty)=2/3$, $f(t)=\frac{2}{3}u(t)-\frac{1}{9}u(t)-\frac{e^{-t}}{2}u(t)-\frac{e^{-3t}}{6}u(t)$; (b) $f(0^+) = 0$, final-value theorem does not apply, because of a double pole at the origin. $f(t)=-\frac{2}{3}tu(t)+\frac{1}{9}u(t)-\frac{e^{-t}}{4}u(t)+\frac{5e^{3t}}{36}u(t)$. Both limits are infinite.

21.4.3 Convolution Theorem

If $F(s) = \mathcal{L}\{f(t)\}$ *and* $G(s) = \mathcal{L}\{g(t)\}$, *then*

$$F(s)G(s) = \mathcal{L}\{f(t) * g(t)\} \tag{21.88}$$

Proof: We start with $F(s) = \int_{0^-}^{\infty} f(\lambda)e^{-s\lambda}d\lambda$, where λ is considered the time variable in the definition of the LT. Since $G(s)$, being a function of s, is a constant as far as integration with respect to λ is concerned, $F(s)G(s)$ can be written as

$$F(s)G(s) = \int_{0^-}^{\infty} G(s)e^{-s\lambda}f(\lambda)d\lambda \tag{21.89}$$

$G(s)e^{-s\lambda}$ in the integrand on the RHS of Equation 21.89 can be expressed in terms of the time-shift property of the LT as

$$G(s)e^{-s\lambda} = \mathcal{L}\{g(t-\lambda)u(t-\lambda)\} = \int_{0^-}^{\infty} g(t-\lambda)u(t-\lambda)e^{-s\lambda}\,dt \tag{21.90}$$

Substituting for $G(s)e^{-s\lambda}$ in Equation 21.89,

$$F(s)G(s) = \int_{0^-}^{\infty}\left[\int_{0^-}^{\infty} g(t-\lambda)u(t-\lambda)e^{-s\lambda}\,dt\right]f(\lambda)d\lambda \tag{21.91}$$

$f(\lambda)$ and $e^{-s\lambda}$ are constants as far as integration with respect to t is concerned. Hence, $f(\lambda)$ could be moved inside the inner integral and $e^{-s\lambda}$ could be moved outside it. Thus,

$$F(s)G(s) = \int_{0^-}^{\infty}\left[\int_{0^-}^{\infty} f(\lambda)g(t-\lambda)u(t-\lambda)dt\right]e^{-s\lambda}d\lambda \tag{21.92}$$

Interchanging the order of integration by interchanging $d\lambda$ and dt,

$$F(s)G(s) = \int_{0^-}^{\infty}\left[\int_{0^-}^{\infty} f(\lambda)g(t-\lambda)u(t-\lambda)d\lambda\right]e^{-s\lambda}dt \tag{21.93}$$

$u(t-\lambda)$ in the inner integral ensures that $g(t-\lambda)u(t-\lambda)=0$ for $t<\lambda$ or $\lambda>t$. Hence, it can be omitted if the upper limit of integration in the inner integral is made t instead of infinity.

$$F(s)G(s) = \int_{0^-}^{\infty}\left[\int_{0^-}^{t} f(\lambda)g(t-\lambda)d\lambda\right]e^{-s\lambda}dt \tag{21.94}$$

The inner integral is $f(t) * g(t)$, and the outer integral is its LT, which proves the convolution theorem.

The convolution theorem allows the derivation of the convolution integral of two functions of time, other than by direct evaluation or using the graphical method, as illustrated by Example 21.6.

The convolution theorem can be used to prove the invariance of convolution with inverse integration and differentiation (Equation 20.23) for functions that are zero for $t < 0^-$. Thus, if $F(s)$ is the LT of $f(t)$ that is zero for $t < 0^-$, then the LT of $f^{(n)}(t)$, the nth derivative of $f(t)$, is $s^nF(s)$ (Equation 21.23). The LT of $f^{(-n)}(t)$, the nth integral of $f(t)$, is $F(s)/s^n$ (Equation 21.17). Similarly for a function $g(t)$ that is zero for $t < 0$. But,

$$F(s)G(s) = \left(\left[s^nF(s)\right]\left[\frac{G(s)}{s^n}\right]\right) = \left(\left[\frac{F(s)}{s^n}\right]\left[s^nG(s)\right]\right) \tag{21.95}$$

Applying the convolution theorem,

$$\mathcal{L}\{f(t) * g(t)\} = \mathcal{L}\{f^{(n)}(t) * g^{(-n)}(t)\} = \mathcal{L}\{f^{(-n)}(t) * g^{(n)}(t)\} \tag{21.96}$$

Taking the ILT proves Equation 20.23.

Example 21.6: Convolution of Two Functions

Derive the convolution integral of $f(t) = te^{-t}u(t)$ and $g(t) = (\cos t - 3\sin t)u(t)$.

Solution:

$F(s) = \dfrac{1}{(s+1)^2}$ and $G(s) = \dfrac{s-3}{s^2+1}$. The product can be expressed as

$$F(s)G(s) = \frac{s-3}{(s+1)^2(s^2+1)} = \frac{K_1}{(s+1)} + \frac{K_2}{(s+1)^2} + \frac{K_3s+K_4}{s^2+1} \tag{21.97}$$

where the numerator of the last term is expressed as a linear function of s, as in Equation 21.72. K_2 is determined, as usual, by multiplying both sides by $(s+1)^2$ and setting $s = -1$; K_1, K_3, and K_4 can be determined by combining the terms on the RHS in a single term having a common denominator and comparing coefficients of the numerator on both sides of the equation, as was done in connection with Equation 21.73. Instead, it is instructive to determine K_1 in an alternative manner that can generally be applied in PFEs having multiple poles of any order. Multiplying both sides by $(s+1)^2$

$$\frac{s-3}{(s^2+1)} = K_1(s+1) + K_2 + X(s)(s+1)^2 \tag{21.98}$$

where the function $(K_3s + K_4)/(s^2 + 1)$ has been denoted as $X(s)$. Setting $s = -1$ gives $K_2 = -2$. If both sides of Equation 21.98 are differentiated with respect to s, the resulting equation is

$$\frac{-s^2+6s+1}{(s^2+1)^2} = K_1 + 2X(s)(s+1) + X^{(1)}(s)(s+1)^2 \quad (21.99)$$

Setting $s = -1$ gives $K_1 = -1.5$. Equation 21.97 becomes

$$F(s)G(s) = \frac{s-3}{(s+1)^2(s^2+1)} = -\frac{1.5}{(s+1)} - \frac{2}{(s+1)^2} + \frac{K_3s+K_4}{s^2+1} \quad (21.100)$$

Combining the terms on the RHS in a single term gives

$$F(s)G(s) = \frac{s-3}{(s+1)^2(s^2+1)} = \frac{(-1.5+K_3)s^3 + (-1.5-2+2K_3+K_4)s^2}{(s+1)^2(s^2+1)} + \frac{(-1.5+K_3+2K_4)s + (-1.5-2-K_4)}{(s+1)^2(s^2+1)} \quad (21.101)$$

Setting the coefficient of s^3 to zero gives $K_3 = 1.5$, and setting the constant term to −3 gives $K_4 = 0.5$. As a check, the coefficient of s^2 is zero, and the coefficient of s is 1. It follows that

$$F(s)G(s) = \frac{s-3}{(s+1)^2(s^2+1)} = -\frac{1.5}{(s+1)} - \frac{2}{(s+1)^2} + \frac{1.5s+0.5}{s^2+1} \quad (21.102)$$

The ILT of $F(s)G(s)$ gives

$$f(t)*g(t) = -(1.5+2t)e^{-t}u(t) + 1.5\cos tu(t) + 0.5\sin tu(t) \quad (21.103)$$

Primal Exercise 21.10

The impulse response of a circuit is $h(t) = \delta^{(1)}(t) + 5\delta(t)$. If the input to the circuit is $x(t) = (1 - e^{-2t})u(t)$, determine the response of the circuit as $t \to \infty$.

Ans. 5.

Example 21.7: Application of Convolution Theorem

A voltage $v_{SRC}(t)$ that is varying in an arbitrary and unknown manner is applied at $t = 0$ to the circuit of Figure 21.7, with zero initial conditions. If the voltage v_1 across C_1 is known experimentally and cannot be expressed analytically, it is required to determine $i_{SRC}(t)$.

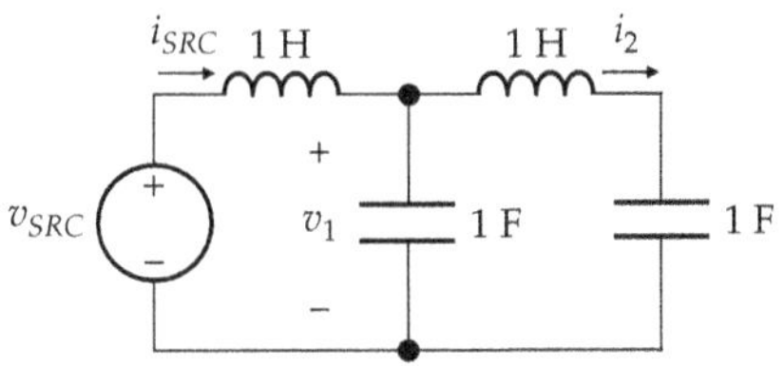

FIGURE 21.7
Figure for Example 21.7.

Solution:

From KVL around the mesh on the RHS,

$$1 \times \frac{di_2}{dt} + \frac{1}{1}\int i_2 dt = v_1(t) \quad (21.104)$$

Moreover,

$$v_1(t) = \frac{1}{1}\int (i_{SRC} - i_2)dt \quad (21.105)$$

Taking the LT of Equations 21.104 and 21.105 gives

$$sI_2(s) + \frac{I_2(s)}{s} = V_1(s), \quad V_1(s) = \frac{I_{SRC}(s) - I_2(s)}{s} \quad (21.106)$$

Eliminating $I_2(s)$ between these two equations yields

$$I_{SRC}(s) = \frac{s(s^2+2)}{s^2+1}V_1(s) = X(s)V_1(s) \quad (21.107)$$

where $X(s) = s(s^2 + 2)/(s^2 + 1)$. According to the convolution theorem,

$$I_{SRC}(s) = X(s)V_1(s) = \mathcal{L}\{x(t)*v_1(t)\} \quad (21.108)$$

Taking the ILT of Equation 21.108,

$$i_{SRC}(t) = x(t)*v_1(t) \quad (21.109)$$

where $x(t)$ is the ILT of $X(s)$ and can be derived by first dividing the numerator of $X(s)$ by its denominator, which gives

$$X(s) = \frac{s(s^2+2)}{s^2+1} = s + \frac{s}{s^2+1} \quad (21.110)$$

The ILT transform is

$$x(t) = \delta^{(1)}(t) + \cos t \quad (21.111)$$

It follows that

$$i_{SRC}(t) = \cos t * v_1(t) + \delta^{(1)}(t) * v_1(t) \quad (21.112)$$

Since $v_1(t)$ is experimentally derived, it can be represented by a staircase function (Section 20.4). $\cos t$ can also be represented as a staircase function, so that $\cos t * v_1(t)$ can be obtained as the product of two polynomials.

From the invariance of convolution with inverse integration and differentiation (Section 20.3),

$$\delta^{(1)}(t) * v_1(t) = \delta(t) * v_1^{(1)}(t) = v_1^{(1)}(t) \quad (21.113)$$

where $v_1^{(1)}(t)$ is the first derivative of $v_1(t)$ and can also be obtained from the staircase approximation or by other numerical methods.

Learning Checklist: What Should Be Learned from This Chapter

- In the one-sided LT, $F(s) = \int_{0^-}^{\infty} f(t)e^{-st}dt$, values of $f(t)$ for $t < 0^-$ are ignored.
- For a function $f(t)$ that does not have an impulse at the origin, the lower limit in the one-sided LT could be 0^- or 0^+ or 0: $F(s) = \int_{0^-}^{\infty} f(t)e^{-st}dt = \int_{0^+}^{\infty} f(t)e^{-st}dt$. In this case, the LT of $f(t)$ is the same as the LT of $f(t)u(t)$, irrespective of the lower limit of integration being 0^- or 0^+ or 0.
- The LT transform has many properties that can be used for the derivation of the LTs of other functions. These properties include multiplication by a constant, addition/subtraction, time scaling, integration, differentiation, translation in the s-domain, translation in time, multiplication by t, and division by t. These properties are listed in Table 21.1.
- A linear differential equation is transformed by the LT to an algebraic equation in powers of s that can be solved for the LT of the variable of the equation $Y(s)$, as in any algebraic equation. When this is done, the initial conditions $y(0^-)$ and $y^{(1)}(0^-)$ appear like applied inputs.
- The characteristic equation of a linear differential equation is a polynomial in s obtained by taking the LT of the equation with zero forcing function (the homogeneous equation), zero initial conditions, and assuming a nonzero value of the variable of the equation. The LHS of the characteristic equation appears in all the responses derived from the differential equation.
- The LTs of responses of LTI circuits having lumped circuit parameters are rational functions of s. These can be expressed, at least in part, as proper rational functions. Such functions can be expanded as partial fractions, which are then inverted term by term.
- The zeros of $F(s)$ are the roots of the numerator of $F(s)$ so that $F(s) = 0$, when s equals any of these roots. The poles of $F(s)$ are the roots of the denominator of $F(s)$ so that $F(s) \to \infty$ when s equals any of these roots.
- When the poles are simple and real, the PFE is of the form $F(s) = \frac{K_1}{s+p_1} + \frac{K_2}{s+p_2} + \cdots + \frac{K_n}{s+p_n}$, where K_r is the residue of pole p_r, $r = 1, 2, \ldots, n$. K_r can be readily determined by the residue method, that is, by multiplying both sides of the equation by $(s + p_r)$ and setting $s = -p_r$.
- When the poles are complex, they occur in complex conjugate pairs, and their residues are also complex conjugates so as to have real coefficients of the polynomials in the numerator and denominator of $F(s)$. Working with complex quantities is avoided by the following procedure:
  1. The two terms involving complex conjugate poles are combined, resulting in a term that is quadratic in s in the denominator, and, in general, linear in s in the numerator.
  2. The coefficients of the linear function in the numerator are determined by comparing coefficients of terms having s raised to the same integral power, as well as the constant term, on both sides of the equation for the PFE.
  3. The square is completed in the quadratic of the denominator, and the numerator is expressed in terms of the squared term in s in the denominator, after completing the square.
  4. The resulting expression in s is inverted in terms of cosine and sine functions of time, multiplied by an exponential function of time.
- The PFE of a multiple pole or order n is of the form $F(s) = \frac{N(s)}{(s+p)^n} = \frac{K_1}{s+p} + \frac{K_2}{(s+p)^2} + \cdots + \frac{K_{n-1}}{(s+p)^{n-1}} + \frac{K_n}{(s+p)^n}$. Only K_n can be determined by the residue method; the other residues are determined either by successive differentiation or by multiplying out to have a common denominator and comparing coefficients of various powers of s, as well as the constant term, on both sides of the equation for the PFE.
- *Final-value theorem*: If all the poles of $F(s)$ have negative real parts, except for a simple

pole at the origin, if such a pole exists, then $\lim_{t\to\infty} f(t) = \lim_{s\to 0} sF(s)$. The final-value theorem cannot be applied if $F(s)$ has a multiple pole at the origin or has poles on the imaginary axis.

- *Initial-value theorem*: If $F(s)$ is a proper rational function, then $\lim_{t\to 0^+} f(t) = \lim_{s\to\infty} sF(s)$. If $F(s)$ is not a proper rational function, the numerator is divided by the denominator, and the initial-value theorem is applied to the rational fraction part to obtain $f(0^+)$. In case of impulsive readjustment in a circuit at $t = 0$, the initial-value theorem gives the responses just after the impulse is over at $t = 0^+$.
- The convolution theorem provides the link between the convolution integral and the LT. It states that $F(s)G(s) = \mathcal{L}\{f(t) * g(t)\}$. The convolution theorem allows an alternative derivation of the convolution integral of two functions of time.

Problem-Solving Tips

1. In applying the differentiation-in-time property, the function that is differentiated is $f(t)$, and not $f(t)u(t)$. It follows that $f(0^-)$ in Equation 21.22 is the value of $f(t)$ at $t = 0^-$ and not the value of $f(t)u(t)$ at $t = 0^-$, which is zero.
2. Working with complex quantities can be avoided by combining complex conjugate poles in the PFE into a single term. The numerator of this term is, in general, linear in s when the complex poles are simple.
3. The initial- and final-value theorems can provide a useful check on the LT of a given response, using the limiting values of this response.
4. In inverting the LT, MATLAB's 'ilaplace' command can be used irrespective of whether $F(s)$ is a proper rational function or not, and irrespective of whether the poles are real or complex, single or multiple.

Appendix 21A: Simplification of Rational Functions of *s*

It is required to simplify rational functions involving s multiplied by integer powers of 10. Consider the function $F(s) = \dfrac{s+10^3}{10^{-3}s^2 + 5s + 6\times 10^3}$. If we substitute $s = 10^3 s'$ and divide the numerator and denominator by 10^3, we obtain a more convenient function $F'(s') = \dfrac{s'+1}{(s')^2 + 5s' + 6} = \dfrac{2}{s'+3} - \dfrac{1}{s'+2}$. Clearly, if s is in rad/s, then s' is in krad/s. This is because if s' is in krad/s, its numerical value is 10^{-3} times that of s in rad/s. Hence s' should be multiplied by 10^3 to have the same numerical value as s.

However, replacing s by $10^3 s'$ does not change the units of $F'(s')$, which remain s, the same as $F(s)$. But when inverting $F'(s')$ the unit of time should be in ms, since s' is in krad/s. This means that when inverting $F'(s')$, the ILT should be multiplied by 10^3 to convert it from s to ms. Thus, the ILT of $F'(s')$ is $f(t) = 10^3(2e^{-3t'} - e^{-2t'})$, with t' in ms. This can be readily verified. Thus,

$$F(s) = \frac{10^3\left(s+10^3\right)}{s^2 + 5\times 10^3 s + 6\times 10^6} = 10^3\left(\frac{2}{s+3\times 10^3} - \frac{1}{s+2\times 10^3}\right)$$

The ILT is $f(t) = 10^3(2e^{-3\times 10^3 t} - e^{-2\times 10^3 t})$, which is the same result, with t in s.

A more formal argument for multiplying the ILT by 10^3 is based on the ILT $f(t) = \dfrac{1}{2\pi j}\displaystyle\int_{\sigma-j\infty}^{\sigma+j\infty} F(s)e^{st}ds$ (Equation 21.11). If we substitute $s = 10^3 s'$, ignoring the change in the limits at infinity, which remain at infinity, we obtain $f(t) = \dfrac{10^3}{2\pi j}\displaystyle\int_{\sigma-j\infty}^{\sigma+j\infty} F'(s')e^{10^3 s't}ds' = \dfrac{10^3}{2\pi j}\displaystyle\int_{\sigma-j\infty}^{\sigma+j\infty} F'(s')e^{s't'}ds'$ $= 10^3 f'(t')$, where $f'(t')$ is the inverse transform of $F'(s')$ and t' is in ms. Hence, $f(t) = 2\times 10^3 e^{-3t'} - 10^3 e^{-2t'}$.

Note that in applying the initial- and final-value theorems, when $F'(s')$ is multiplied by s' and the limit obtained as s' tends to infinity or zero, the result must also be multiplied by 10^3 in order to obtain the correct limiting value of $f(t)$.

Similar considerations apply if s is replaced by $10^6 s'$, where s' is in Mrad/s. The ILT should then be multiplied by 10^6.

Problems

Laplace Transform of Analytical Functions

P21.1 Determine the LT of $f(t) = (2t + 3t^2)^2$.

Ans. $\dfrac{8s^2 + 72s + 216}{s^5}$.

P21.2 Determine the LT of $-2e^{-3(t-4)}u(t-4)$.

Ans. $\dfrac{-2e^{-4s}}{s+3}$.

P21.3 Determine the LTs of the following functions: (a) $\left(\dfrac{e^{at} + e^{-at}}{2}\right)u(t)$; (b) $\left(\dfrac{e^{at} - e^{-at}}{2}\right)u(t)$.

Ans. (a) $\dfrac{s}{s^2 - a^2}$; (b) $\dfrac{a}{s^2 - a^2}$.

P21.4 Determine the LT of the following functions: (a) $(t-1)\,u\,(t-2)$; (b) $(t-2)u(t-1)$. Verify the results by expressing each function as the sum of a delayed ramp function and a delayed step function and deriving their LTs.

Ans. (a) $e^{-2s}\left(\frac{1}{s^2}+\frac{1}{s}\right)$; (b) $e^{-s}\left(\frac{1}{s^2}-\frac{1}{s}\right)$.

P21.5 Determine the LTs of the following functions: (a) $2e^{-t}u(t-1)$; (b) $\cos(4t-1)u(t)$.

Ans. (a) $2\frac{e^{-(s+2)}}{s+2}$; (b) $\frac{s\cos(1)+4\sin(1)}{s^2+16}$.

P21.6 Determine the LT of $te^{-t}\cosh 4t\,u(t)$.

Ans. $\frac{s^2+2s+17}{\left(s^2+2s-15\right)^2}$

P21.7 Determine the LT of $e^{-t}\sqrt{t}u(t)$ given that the LT of $e^{-t}/\sqrt{t}$ is $\frac{\sqrt{\pi}}{\sqrt{s+1}}$.

Ans. $\frac{\sqrt{\pi}}{2(s+1)^{3/2}}$.

P21.8 Using the division by t property and referring to a table of integrals (Appendix B), show that $\mathcal{L}\left\{\frac{\sin 5t}{t}u(t)\right\}=\frac{\pi}{2}-\tan^{-1}\left(\frac{s}{5}\right)$.

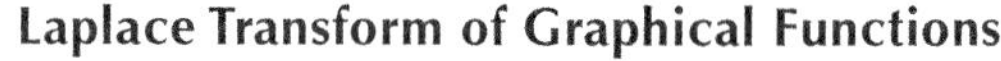

Laplace Transform of Graphical Functions

P21.9 Determine the LT of $f(t)$ in Figure P21.9.

Ans. $\frac{1}{s}\left(1+e^{-2s}\right)$.

P21.10 Determine the LT of $f(t)$ in Figure P21.10.

Ans. $\frac{5\left(e^{-2s}+e^{-4s}+e^{-6s}-3e^{-8s}\right)}{s}$.

P21.11 Determine the LT of the single triangular pulse of Figure P21.11.

Ans. $\frac{A}{\alpha T}\frac{1}{s^2}-\frac{A}{\alpha(1-\alpha)T}\frac{e^{-\alpha Ts}}{s^2}+\frac{A}{(1-\alpha)T}\frac{e^{-Ts}}{s^2}$.

P21.12 Determine the LT of $f(t)$ in Figure P21.12.

Ans. $-\frac{2}{s}e^{-s}+\frac{1}{2s^2}\left(2-e^{-s}-e^{-3s}\right)$.

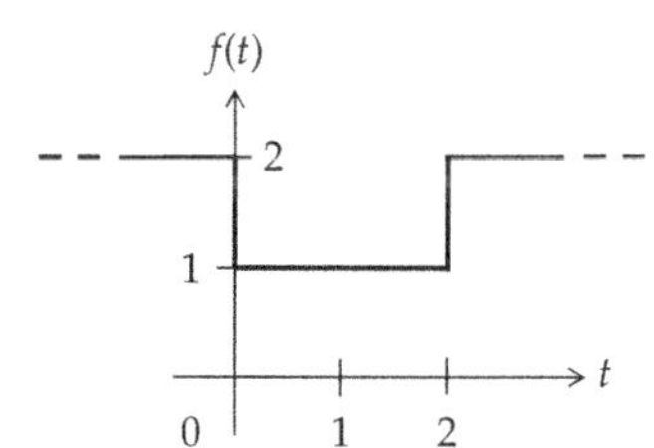

FIGURE P21.9

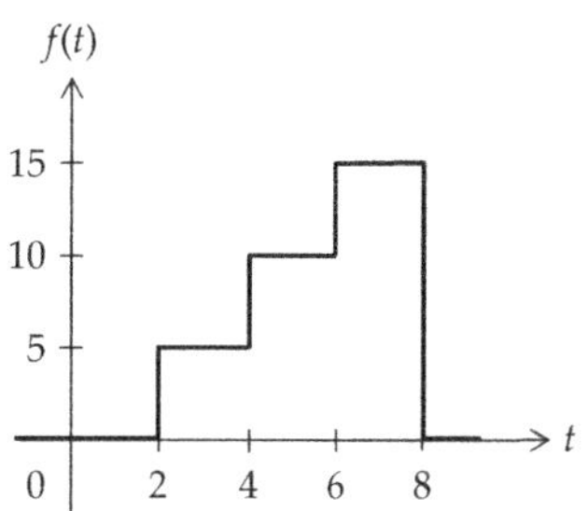

FIGURE P21.10

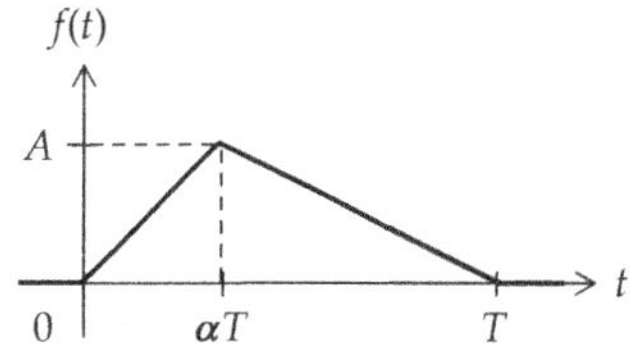

FIGURE P21.11

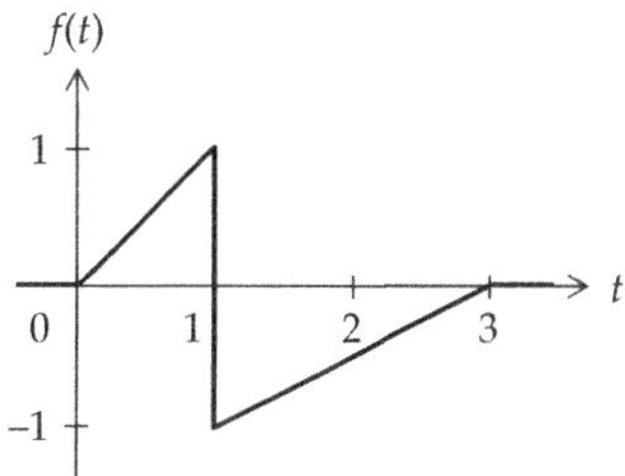

FIGURE P21.12

P21.13 Determine the LT of (a) $f(t)$ in Figure P21.13; (b) $f^{(1)}(t)$. Note that dividing the LT of $f^{(1)}(t)$ by s does not give back the LT of $f(t)$, but the LT of $[f(t)+u(t)]$ having a value of zero for $t<0$. This is because integration of a function adds an arbitrary constant to the function. In this case the arbitrary constant makes the function zero for $t<0$.

Ans. (a) $\frac{2}{s^2}-\left(\frac{1}{s}+\frac{3}{s^2}\right)e^{-s}+\frac{1}{s^2}e^{-2s}$; (b) $1+\frac{2}{s}-\frac{3}{s}e^{-s}-e^{-s}+\frac{1}{s}e^{-2s}$.

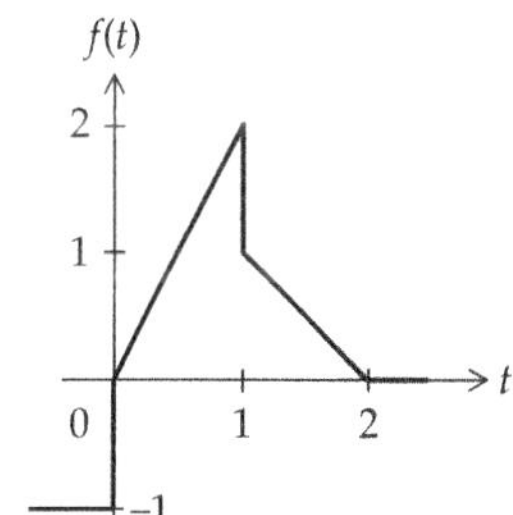

FIGURE P21.13

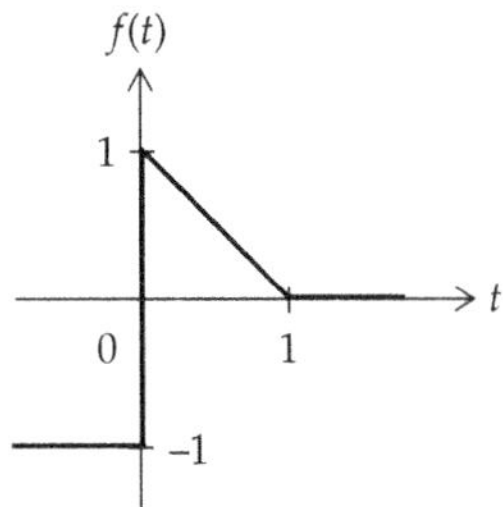

FIGURE P21.14

P21.14 Determine the LT of the time integral of $f(t)$ in Figure P21.14.

Ans. $\frac{1}{s^2} - \frac{1}{s^3} + \frac{1}{s^3}e^{-s}$.

P21.15 Determine the LT of the derivative of $f(t)$ in Figure P21.15.

Ans. $\frac{1 - 2e^{-s} + (1+s)e^{-3s}}{s}$.

P21.16 Determine the LT of the derivative of $f(t)$ in Figure P21.16.

Ans. $-1 + \left(e^{-s} - e^{-2s}\right) + \frac{1}{s}\left(1 - e^{-s} - e^{-2s} + e^{-3s}\right)$.

P21.17 Determine the LT of the derivative of $f(t)$ in Figure P21.17.

Ans. $2\left(1 - \frac{1}{s}\right) + \frac{2e^{-2s}}{s}\left(2 - e^{-s}\right)$.

P21.18 Determine the LT of the derivative of $f(t)$ in Figure P21.18. Check by deriving $F(s)$ and applying the differentiation-in-time property.

Ans. $\frac{1}{s}\left(1 - e^{-s} - 2se^{-s}\right)$.

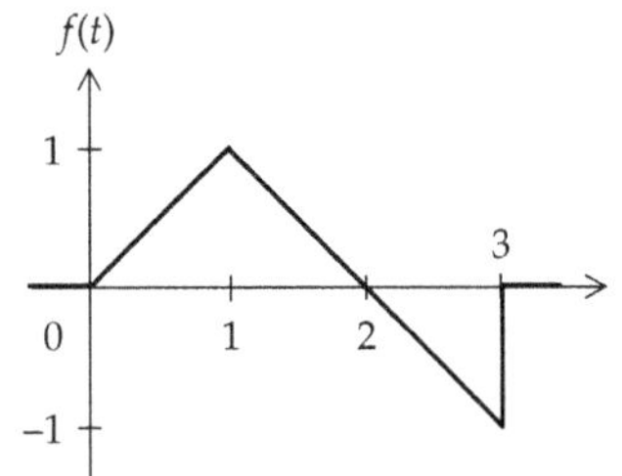

FIGURE P21.15

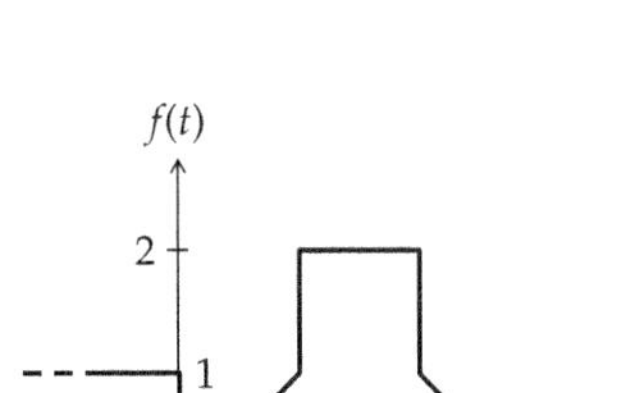
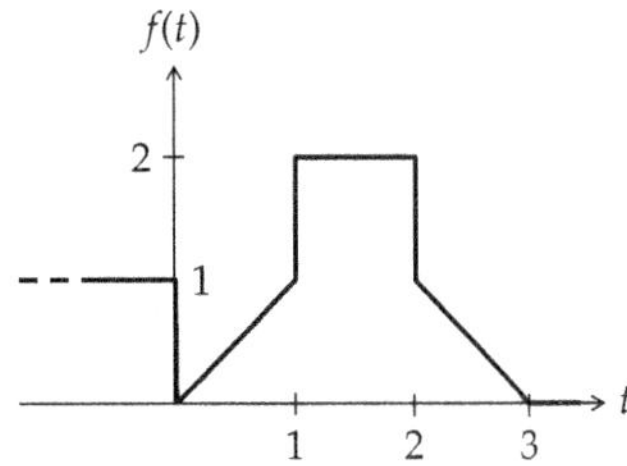

FIGURE P21.16

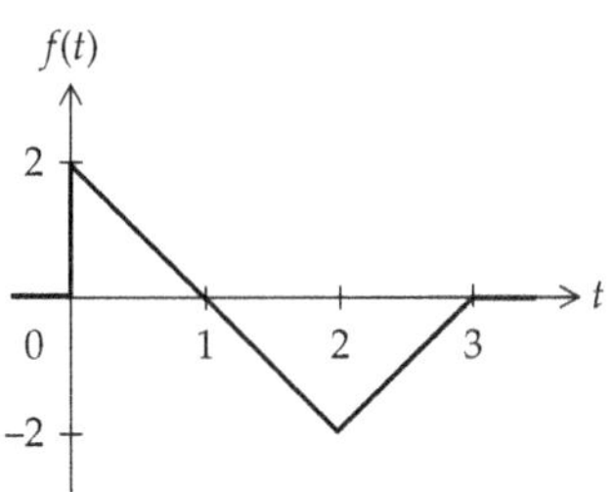

FIGURE P21.17

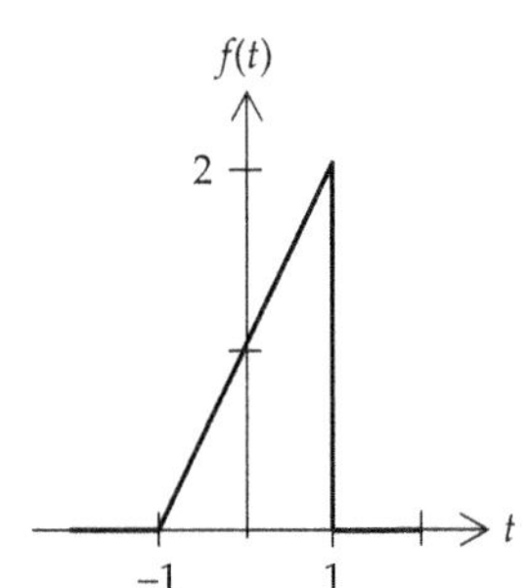

FIGURE P21.18

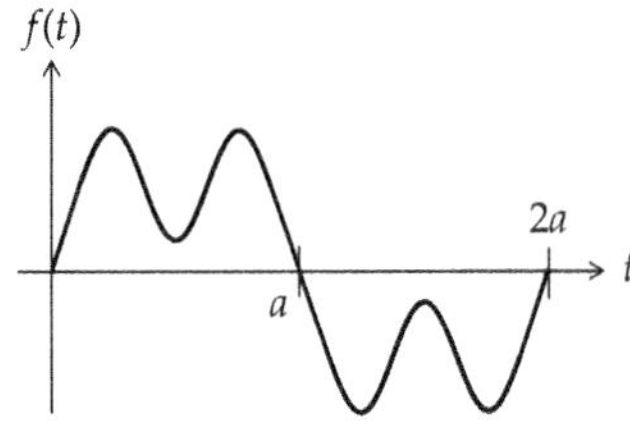

FIGURE P21.19

P21.19 Figure P21.19 shows two identical consecutive pulses each of duration a, the second pulse being inverted with respect to the first. If $F(s)$ is the LT of the two pulses shown, determine, in terms of $F(s)$, the LT of $f(t)$ shifted to the left by a.

Ans. $\frac{-F(s)}{1 - e^{-as}}$.

Laplace Transform of Periodic Functions

P21.20 From the LT of a single rectangular pulse (Equation 21.41), deduce that the LT of the square wave of Figure P21.20 can be expressed as $\frac{A_m}{s}\frac{\left(1 - e^{-sT/2}\right)}{\left(1 + e^{-sT/2}\right)}$.

P21.21 From the LT of a single sawtooth (Equation 21.51), deduce that the LT of the sawtooth waveform of Figure P21.21 is $\frac{A}{T}\left[\frac{1}{s^2} - \frac{Te^{-Ts}}{s\left(1 - e^{-Ts}\right)}\right]$.

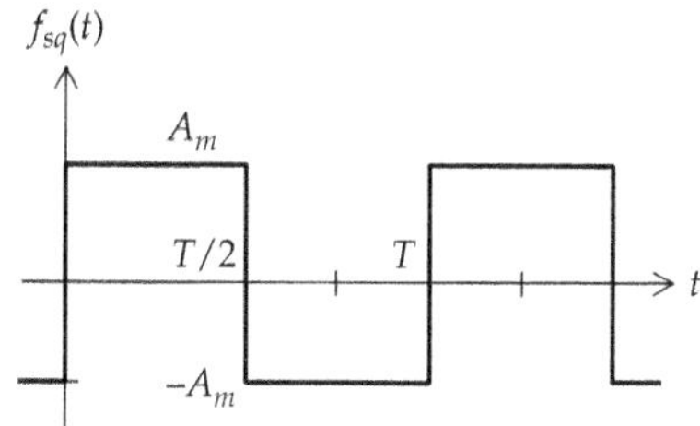

FIGURE P21.20

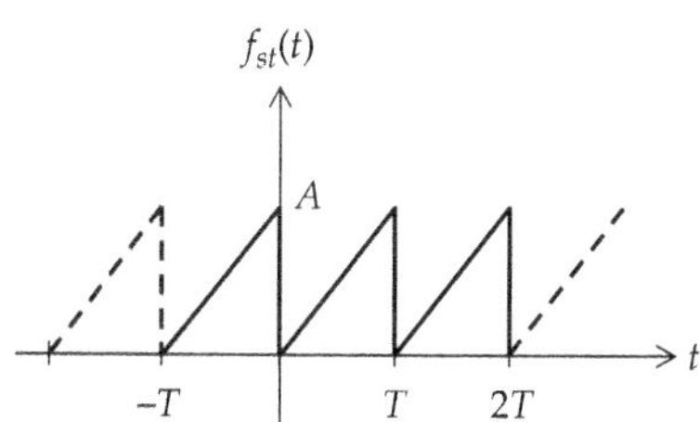

FIGURE P21.21

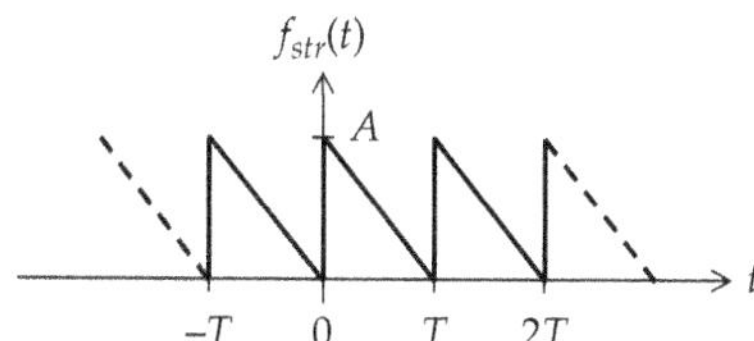

FIGURE P21.22

P21.22 Given the periodic train of reversed sawtooth pulses of Figure P21.22. Show that the LT of the first period between 0 and T is $\frac{A}{s} - \frac{A}{Ts^2}\left(1 - e^{-Ts}\right)$. Deduce that the LT of the inverted sawtooth waveform of Figure P21.22 is $\frac{A}{T}\left[-\frac{1}{s^2} + \frac{Te^{-Ts}}{s\left(1 - e^{-Ts}\right)}\right]$.

P21.23 Determine the LT of (a) the impulse train $A[\delta(t) + \delta(t-T) + \delta(t-2T) + \cdots]$ and (b) a negative impulse train $-A[\delta(t-T/2) + \delta(t-3T/2) + \delta(t-5T/2) + \cdots]$.

From the sum of these two LTs, deduce the LT of the square wave of Figure P21.20.

Ans. (a) $\frac{A}{\left(1-e^{-sT}\right)}$; (b) $-\frac{Ae^{-sT/2}}{\left(1-e^{-sT}\right)}$.

P21.24 If $f(t) = (\cos\pi t/2)u(t)$ and $g(t)$ is an infinite train of unit impulses of period 1, as shown in Figure P21.24, determine the LT of the product $f(t) \times g(t)$.

Ans. $\frac{1}{1+e^{-2s}}$.

P21.25 Using the translation-in-time property, show that the LT of a single half sinusoid described by $f(t) = A_m\sin\omega t$, $0 \leq t \leq \pi/\omega$, and $f(t) = 0$, elsewhere, is $\frac{\omega A_m}{s^2+\omega^2}\left(1 + e^{-s\pi/\omega}\right)$. Verify the result by direct integration using integration by parts. Deduce that the LT of a full-wave rectified waveform of amplitude A_m can be expressed as $\frac{\omega A_m}{s^2+\omega^2}\coth\left(\frac{\pi}{2\omega}s\right)$.

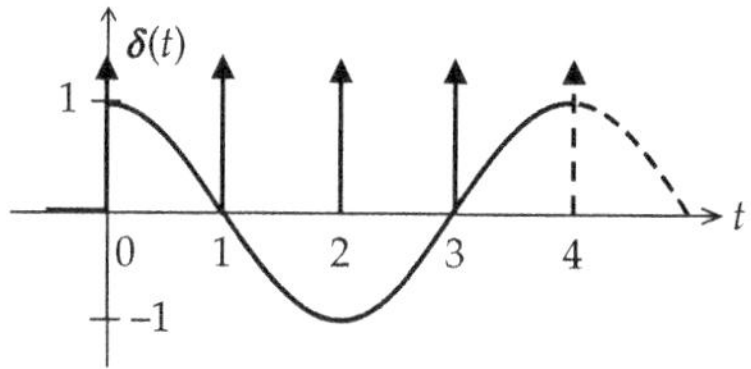

FIGURE P21.24

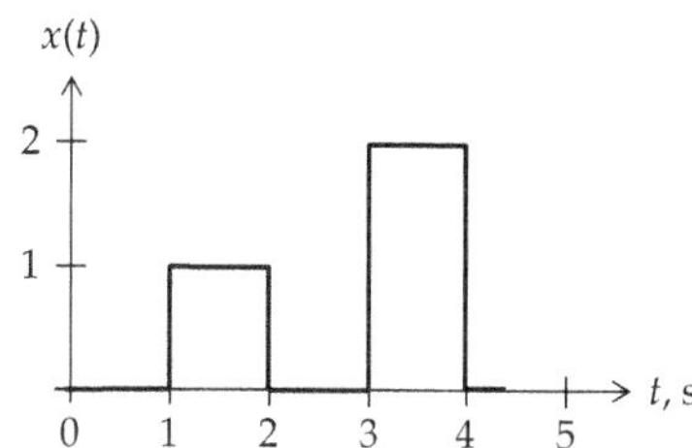

FIGURE P21.26

P21.26 If the LT of the response of an LTI circuit to $\delta(t)$ is $1/(s+1)$, determine the response to the input $x(t)$ shown in Figure P21.26.

Ans. $(1-e^{-(t-1)})u(t-1) - (1-e^{-(t-2)})u(t-2) + 2(1+e^{-(t-3)})$ $u(t-3) - 2(1-e^{-(t-4)})u(t-4)$.

Inverse Laplace Transform

P21.27 Determine the ILT of $F(s) = \frac{8s^2+4s+6}{(2s+3)(s+4)}$.

Ans. $4\delta(t) + (3.6e^{-1.5t} - 23.6e^{-4t})u(t)$.

P21.28 Determine the ILT of $F(s) = \frac{10\left(3s^2+4s+4\right)}{s(s+2)^2}$.

Ans. $(10 + 20e^{-2t} - 40te^{-2t})u(t)$.

P21.29 Determine the ILT of $F(s) = \frac{(s+3)}{(s+4)^2}$.

Ans. $e^{-4t}(1-t)u(t)$.

P21.30 Determine the ILT of $F(s) = \frac{s+1}{s^2+6s+9}$.

Ans. $e^{-3t}(1-2t)u(t)$.

P21.31 Determine the ILT of $F(s) = \frac{s}{s^2+4s+20}$.

Ans. $e^{-2t}(\cos 4t - 0.5\sin 4t)u(t)$.

P21.32 Determine the ILT of $F(s) = \frac{s^2+s+2}{s^2+4}$.

Ans. $\delta(t) + (\cos 2t - \sin 2t)u(t)$.

P21.33 Determine the ILT of $F(s)=\dfrac{40+2(s^2+4s+5)^2}{(s^2+4s+5)^2}$.

Ans. $2\delta(t)+20e^{-2t}(\sin t-t\cos t)u(t)$.

P21.34 Determine the ILT of $F(s)=\dfrac{s\sin\phi+\omega\cos\phi}{s^2+\omega^2}$.

Ans. $\sin(\omega t+\phi)u(t)$.

P21.35 If $F(s)=\dfrac{1+e^{-3s}}{s^2+1}$, determine $f(t)$ at $t=2$ s.

Ans. $f(t)=\sin t u(t)+\sin(t-3)u(t-3)=\sin 2$.

P21.36 If $F(s)=\dfrac{1+e^{-s}}{(s+1)^2}$, determine $f(t)$ at $t=2$ s.

Ans. $f(t)=te^{-t}u(t)+(t-1)e^{-(t-1)}u(t-1)=2e^{-2}+e^{-1}=0.64$.

P21.37 Determine the ILT of $F(s)=\dfrac{s+3}{s^2-s-2}$.

Ans. $(5/3)e^{2t}u(t)-(2/3)e^{-t}u(t)$.

P21.38 Determine the ILT of $F(s)=\dfrac{1}{(s-1)^3}+\dfrac{1}{s^2+2s-8}$.

Ans. $(1/2)t^2e^tu(t)+(1/6)e^{2t}u(t)-(1/6)e^{-4t}u(t)$.

Theorems on the Laplace Transform

P21.39 If $F(s)=\dfrac{8s^3+89s^2+311s+300}{s(s+2)(s^2+8s+15)}$, determine $f(t)$ as $t\to\infty$.

Ans. 10.

P21.40 If $F(s)=\dfrac{s+2}{s^2+2s-3}$, determine $f(t)$ as $t\to\infty$.

Ans. Infinite; final-value theorem does not apply.

P21.41 If $F(s)=\dfrac{4s+3}{s^2-2s+2}$, determine $f(t)$ as $t\to\infty$.

Ans. Infinite; final-value theorem does not apply.

P21.42 If $F(s)=\dfrac{2(s+2)}{(s+1)(s^2+1)}$, determine $f(t)$ as $t\to\infty$.

Ans. The final-value theorem does not apply because of the poles at $\pm j$. These poles result in a $\cos(t+\theta)$ term in $f(t)$, which makes $f(t)$ oscillate between two finite values as $t\to\infty$.

P21.43 If $F(s)=\dfrac{2s^2+1}{s^2+4s+1}$, determine $f(0^+)$.

Ans. –8.

P21.44 If $F(s)=\dfrac{s^3-6s^2+15s+50}{s(s^2+4s+5)}$, determine $f(0^+)$.

Ans. 10.

P21.45 A function $x(t)$ when convolved with the function $(1-e^{-2t})u(t)$ gives the function $(1-e^{-2t}-2te^{-t})u(t)$. Determine $x(t)$.

Ans. $te^{-t}u(t)$.

P21.46 Determine the value of the convolution integral $y(t)$ at t = 0.5 s, where $y(t)=f(t)*g(t)$, with $f(t)=\sin t$, and $g(t)=2\delta(t)+\delta^{(2)}(t)$.

Ans. $\delta(t)+\sin t=\sin 0.5$.

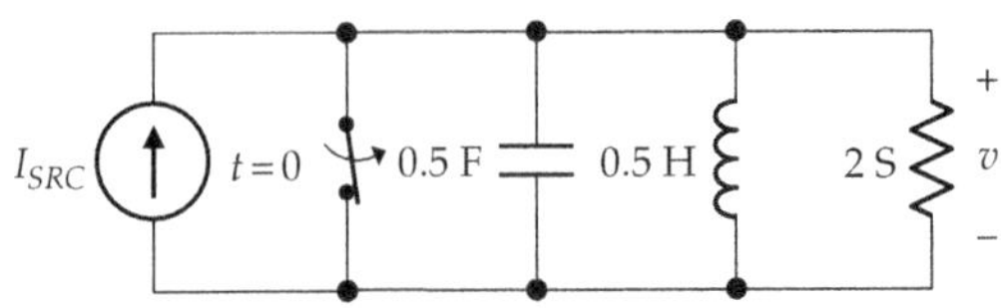

FIGURE P21.48

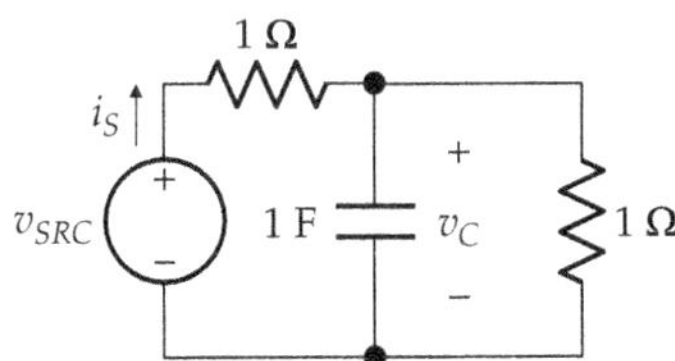

FIGURE P21.49

P21.47 Invert $\dfrac{1}{s(s^2+1)}$ using the convolution theorem.

Ans. $(1-\cos t)u(t)$.

P21.48 When the switch is opened in Figure P21.48 at t = 0, after being closed for a long time, it is found that $V(s)=\dfrac{2}{(s+4)^2}$. Determine I_{SRC}. (*Hint*: Use the differentiation-in-time property and the initial-value theorem.)

Ans. 1 A.

P21.49 Determine $i_S(t)$ in Figure P21.49, assuming $v_{SRC}(t)=e^{-t}(t+\cos t)u(t)$ V and an initial voltage $V_{C0}=1$ V.

Ans. $i_S(t)=e^{-t}(1-\sin t)u(t)$ A.

Miscellaneous

P21.50 Use the time scaling property to show that $\mathcal{L}\{\delta(at)\}=\dfrac{1}{a}$. Verify by direct evaluation, with a change of the variable of integration so as to have the impulse function at the origin in the standard form, with unity coefficient of t.

P21.51 The instantaneous power in a circuit is expressed in the s-domain as $P(s)=\dfrac{s+48}{(s+1)(s+2)}$. Determine the energy delivered to the circuit between t = 0 and any arbitrary t.

Ans. $24-47e^{-t}+23e^{-2t}$.

P21.52 Given the differential equation, $x^{(2)}(t)+2x^{(1)}(t)+x(t)=e^{j2t}$, with zero initial conditions at t = 0, determine $x(t)$ for $t\geq 0$. Note that the response to the cosine function can be obtained from that to the sine function by straight differentiation, whereas the converse is not true. Explain why. (*Hint*: Recall that the applied signals are $(\sin 2t)u(t)$ and $(\cos 2t)u(t)$.)

Ans. $\left[-\dfrac{3+j4}{25}e^{+j2t}-\dfrac{1-j2}{5}te^{-t}+\dfrac{3+j4}{25}e^{-t}\right]u(t)$.

22

Laplace Transform in Circuit Analysis

Objective and Overview

After having considered the basic properties of the Laplace transform (LT) in Chapter 21, the application of the LT to circuit analysis is presented in this chapter. The discussion naturally begins with the representation of circuit elements in the *s*-domain. Because the LT of a given function excludes values of the function for $t < 0^-$, special consideration must be given to initial conditions in energy storage elements. This is followed by a discussion of the general procedure for analyzing circuits in the *s*-domain, including switching circuits, since the LT transform offers some unique advantages in these cases. The LT transform of circuit responses naturally leads to the concept of transfer function, which allows some important conclusions in terms of impulse response, stability, and sinusoidal steady-state response. The chapter ends with interpretations of poles and zeros in the *s*-domain and the responses of first-order and second-order circuits.

22.1 Representation of Circuit Elements in the *s*-Domain

In order to fully utilize the power of the LT approach, the circuit should be analyzed entirely in the *s*-domain. It is necessary for this purpose to be able to represent circuit elements in the *s*-domain, including any initial conditions of energy storage elements. Once this is done, the conventional circuit analysis techniques that were applied to circuits in the frequency domain can be carried over to the *s*-domain, with currents and voltages represented by their LTs. We will discuss in this section how resistors, capacitors, inductors, and linear transformers are represented in the *s*-domain.

22.1.1 Resistor

Taking the LT of both sides of Ohm's law $v = Ri$,

$$V(s) = RI(s) \tag{22.1}$$

It is seen that in the *s*-domain, where the resistor voltage and current are $V(s)$ and $I(s)$, respectively, the resistor is simply represented by its resistance R.

22.1.2 Capacitor

Consider an uncharged capacitor to begin with. The v–i relation is

$$v(t) = \frac{1}{C}\int_{0^+}^{t} idt + 0, \quad t \geq 0^+ \tag{22.2}$$

where i is in the direction of a voltage drop v across the capacitor (Figure 22.1a). It is assumed that $i(t)$ does not include an impulse at $t = 0$. Hence, i remains finite during the interval from $t = 0^-$ to $t = 0^+$, so that the lower limit of integration could just as well be 0^+ instead of 0^-. Taking the LT of both sides,

$$V(s) = \frac{I(s)}{sC}, \quad V_0 = 0 \tag{22.3}$$

Recall that in the sinusoidal steady state, where any initial voltage would have died out as part of the natural response, $V(j\omega) = I(j\omega)/j\omega C$. In Equation 22.3, $j\omega$

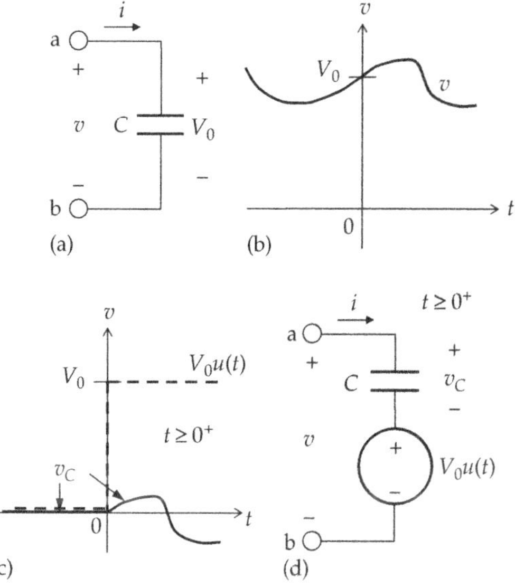

FIGURE 22.1
Initial voltage across a capacitor as a voltage step at the origin. (a) Assigned capacitor voltage and current, (b) variation of capacitor voltage with time, (c) variation of capacitor voltage in (b) for $t \geq 0$ considered as the sum of a step voltage $V_0u(t)$ and a voltage v_C across an uncharged capacitor, and (d) circuit implementation of $V_0u(t)$ and v_C in (c).

is replaced by s for an initially uncharged capacitor. $1/sC$ can therefore be considered as the *impedance* of an uncharged capacitor in the s-domain.

Suppose that the voltage across the capacitor is changing arbitrarily with time (Figure 22.1b). Equation 22.2 becomes

$$v(t) = \frac{1}{C}\int_{0^+}^{t} idt + V_0, \quad t \geq 0^+ \tag{22.4}$$

where V_0 is the initial voltage across the capacitor at $t = 0$. Taking the LT of both sides of Equation 22.4,

$$V(s) = \frac{I(s)}{sC} + \frac{V_0}{s} = V_C(s) + \frac{V_0}{s} \tag{22.5}$$

Comparing Equation 22.5 with Equation 22.3, it is seen that $V_C(s) = I(s)/sC$ is the LT of the voltage across an uncharged capacitor. V_0/s is the LT of $V_0u(t)$. The interpretation of Equation 22.5 is that because the single-sided LT excludes values of v for $t < 0^-$, and since $V_C(s) = I(s)/sC$ is the LT of the voltage across an uncharged capacitor, then in order to account for the initial voltage across the capacitor, it is necessary to add a voltage step $V_0u(t)$, whose LT is V_0/s. In other words, v is divided into two components (Figure 22.1c): (1) a voltage v_C across an initially uncharged capacitor and (2) a step function $V_0u(t)$. In circuit terms, these two components are represented as in Figure 22.1d. The step voltage establishes the initial voltage V_0 at $t = 0^+$ between terminals 'ab'. Thereafter, for $t \geq 0^+$, the voltage v_C across the initially uncharged capacitor changes in accordance with i, added to the initial voltage V_0. Taking the LT of the sum $v = (v_C + V_0u(t))$ gives Equation 22.5. The circuit in the s-domain is shown in Figure 22.2a.

The ideal voltage source V_0/s in series with the uncharged capacitor of impedance $1/sC$ (Figure 22.2a) can be transformed to a current source $(V_0/s)/(1/sC) = CV_0$ in parallel with the same uncharged capacitor, as in Figure 22.2b, where CV_0 is the LT of an impulse of strength CV_0. Alternatively, this representation of the initial capacitor voltage can be derived from the LT of the alternative form of the v–i relation of the capacitor $i = Cdv/dt$. This gives

$$I(s) = sCV(s) - CV_0 \tag{22.6}$$

Figure 22.2b is the representation of the circuit of Figure 22.2c in the s-domain. In this figure, C is initially uncharged. The capacitor provides a short-circuit path for the impulse $CV_0\delta(t)$, which deposits a charge CV_0 on the uncharged capacitor, resulting in the initial voltage V_0 at $t = 0^+$. Thereafter, for $t \geq 0^+$, the current source behaves as an open circuit, so that i flows through C and v changes accordingly.

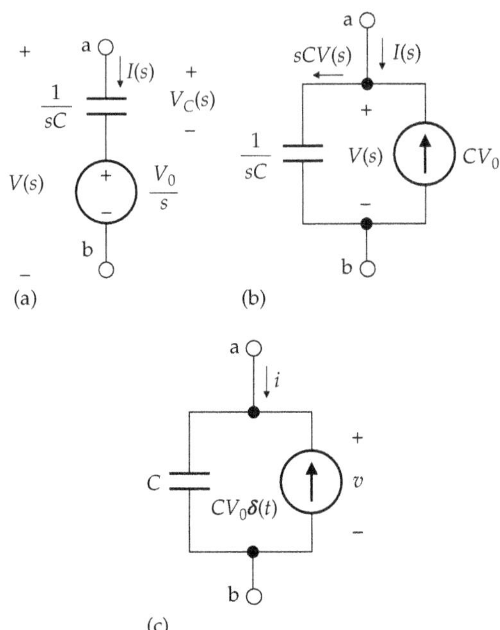

FIGURE 22.2
s-domain representation of capacitor with initial voltage. (a) Representation of the circuit of Figure 22.1d in the s domain, (b) voltage source in (a) is transformed to its equivalent current source, and (c) representation in the time domain of the circuit in (b).

It was assumed in the preceding that i does not include an impulse at $t = 0$. If i did include such an impulse, then by superposition, the impulse simply deposits additional charge on the capacitor between $t = 0^-$ and $t = 0^+$.

Note that whereas in the time domain, the relation $i = Cdv/dt$ is incomplete in the sense that it does not explicitly involve V_0, as noted in connection with Equation 7.6, its transformation to the s-domain does involve V_0 in accordance with the differentiation-in-time property of the LT.

How does one obtain $Q(s)$, the LT of the charge on the capacitor in the s-domain in the circuits of Figure 22.2a and b? In both cases, $Q(s) = CV(s)$, as follows from taking the LT of both sides of the equation $q(t) = Cv(t)$. However, in Figure 22.2a, the charge on the ideal capacitor alone is the charge accumulated only for $t > 0^+$, to which must be added the initial charge at $t = 0^+$. Thus, $v(t) = v_C(t) + V_0$ (Equation 22.4), so that $q(t) = C\left(v_C(t) + V_0\right)$, $t \geq 0^+$. Taking the LT of both sides gives $Q(s)$ as $Q(s) = C(V_C(s) + V_0/s) = CV(s)$. Hence, in determining $Q(s)$ from Figure 22.2a, the voltage step $V_0u(t)$ must be included as part of the voltage across the capacitor.

On the other hand, the impulse in Figure 21.1b deposits the initial charge CV_0 and is over at $t = 0^+$. The charge on the ideal capacitor at any time $t \geq 0^+$ is therefore the true charge. This is reflected in having $V_C(s)$ the same $V(s)$ in Figure 22.2b.

22.1.3 Inductor

Consider an uncharged inductor to begin with. The v–i relation can be expressed in the form

$$i(t) = \frac{1}{L}\int_{0+}^{t} v dt + 0, \quad t \geq 0^+ \tag{22.7}$$

where i is in the direction of a voltage drop v across the inductor (Figure 22.3a). Again, it is assumed that $v(t)$ does not include an impulse at $t = 0$. Taking the LT of both sides,

$$I(s) = \frac{V(s)}{sL} \quad \text{or} \quad V(s) = sLI(s) \quad I_0 = 0 \tag{22.8}$$

Comparing with the frequency-domain representation of $V(j\omega) = j\omega LI(j\omega)$, it follows that sL is the *impedance* of an uncharged inductor in the s-domain.

If the inductor has an initial current I_0 at $t = 0^-$, Equation 22.7 becomes

$$i(t) = \frac{1}{L}\int_{0^+}^{t} v dt + I_0, \quad t \geq 0^- \tag{22.9}$$

Taking the LT of both sides of Equation 22.9,

$$I(s) = \frac{1}{sL}V(s) + \frac{I_0}{s} \tag{22.10}$$

Comparing Equation 22.10 with Equation 22.8, it is seen that $V(s)/sL$ is the LT of the current through an uncharged inductor. I_0/s is the LT of $I_0u(t)$. The interpretation of Equation 22.10 is similar to that of the case of the capacitor. Since $V(s)/sL$ is the LT of the current through an uncharged inductor, the initial current I_0 is established between terminals 'ab' at $t = 0^+$ by a current step $I_0u(t)$ in parallel with an ideal uncharged inductor. For $t \geq 0^+$, v between terminals 'ab' varies according to the current i_L through the uncharged inductor, added to the initial current I_0 (Figure 22.3b). The LT of $I_0u(t)$ is I_0/s, so that the

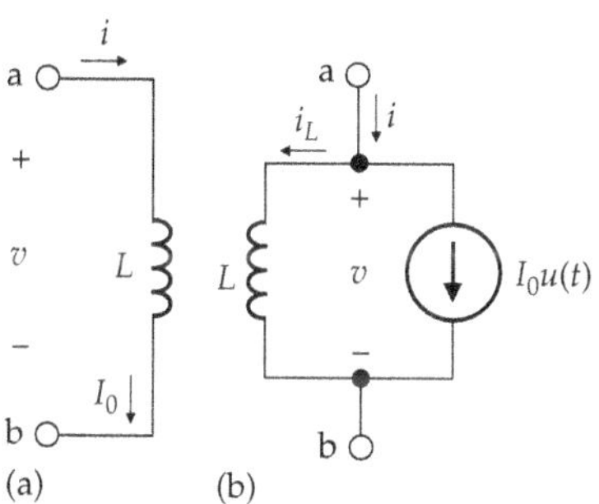

FIGURE 22.3
(a) Assigned inductor voltage and current and (b) initial current through the inductor accounted for by a current step at the origin.

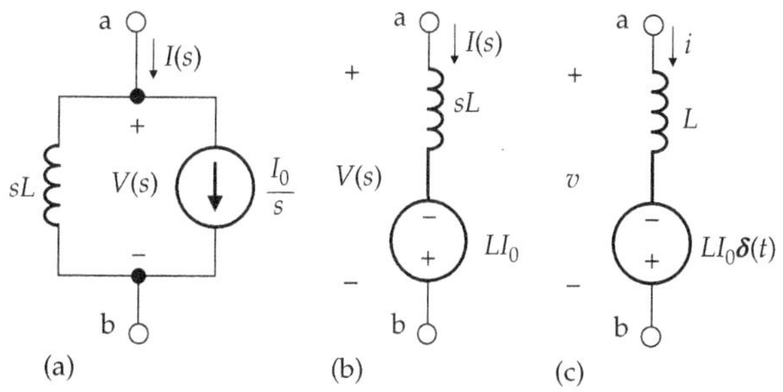

FIGURE 22.4
s-domain representation of inductor having initial current. (a) Representation of the circuit of Figure 22.3b in the s domain, (b) current source in (a) is transformed to its equivalent voltage source, and (c) representation in the time domain of the circuit in (b).

circuit of Figure 22.3b is represented in the s-domain as in Figure 22.4a, in accordance with Equation 22.10.

The ideal current source I_0/s in parallel with the impedance sL in Figure 22.4a can be transformed to an equivalent voltage source $(I_0/s) \times sL = LI_0$ in series with the impedance sL. The ideal voltage source in Figure 22.4b is an impulse $LI_0\delta(t)$ in the time domain that is a voltage rise in the direction of current through the inductor (Figure 22.4c). The impulse, acting through the circuit connected to the inductor, establishes an initial current I_0 in the inductor between $t = 0^-$ and $t = 0^+$, with the inductor acting as on open circuit. Thereafter, for $t \geq 0^+$, the voltage source behaves as a short circuit, and i changes with the current through the inductor, added to the initial current I_0.

The representation of R, L, and C in the s-domain can be summarized as follows:

Summary: *In the s-domain, the impedance of a resistor is R, that of an uncharged capacitor is 1/sC, and that of an uncharged inductor is sL. To account for initial energy storage, while considering the values of all responses to be zero at* $t = 0^-$*, in accordance with the LT, step or impulse, ideal, independent sources are added in series or in parallel with the ideal energy storage element so as to provide the required initial value at* $t = 0^+$.

The fact that sources are added to account for initial conditions at the terminals of the energy storage element in the s-domain implies the following concept:

Concept: *The sources that are added to account for initial conditions in the energy storage elements in the s-domain become an integral part of the representation of the energy storage element in the s-domain.*

This is indicated in Figures 22.2 and 22.4, where the LT $V(s)$ of the terminal voltage across the energy storage element includes the voltage of the series-connected voltage source (Figures 22.2a and 22.4b). The LT $I(s)$ of the terminal current through the energy storage element includes the current of the parallel-connected current source (Figures 22.2b and 22.4a).

An important implication is that these independent sources can be treated just like any sources of excitation, so that they can be included in superposition, for example. This is in accordance with the fundamental concept that in an LTI circuit the responses to initial conditions can be added, by superposition, to the responses due to applied inputs acting alone. This fundamental concept is further exemplified by Equation 21.55, which indicates that initial conditions can be treated as inputs at $t = 0$.

Example 22.1: *RL* Circuit with Initial Current in Inductor

A voltage $V_{SRC}u(t)$ is applied at $t = 0$ to a series RL circuit in which the inductor has an initial current I_0 (Figure 22.5a). It is desired to derive the expressions for the current in the circuit and the voltage across the inductor as functions of time.

Solution:

The s-domain representation of the circuit is shown in Figure 22.5b. The series-source representation is clearly more convenient to use in this case than the parallel-source representation. From KVL,

$$RI(s) + sLI(s) - LI_0 = \frac{V_{SRC}}{s} \tag{22.11}$$

This gives

$$I(s) = \frac{V_{SRC}}{s(R + sL)} + \frac{LI_0}{R + sL} \tag{22.12}$$

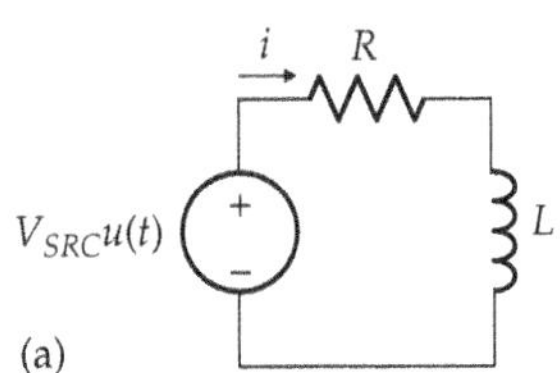

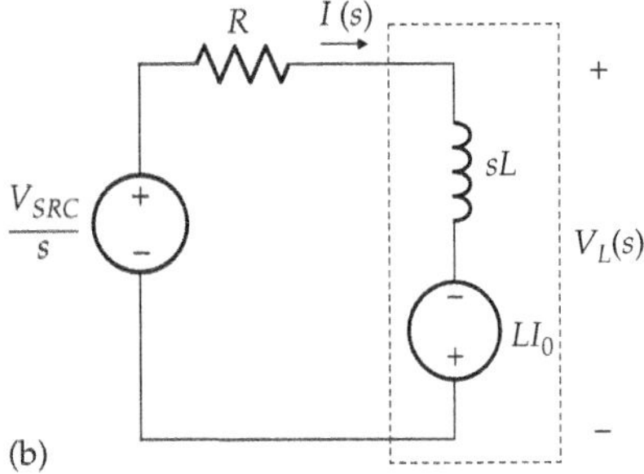

FIGURE 22.5
Figure for Example 22.1.

Note that Equation 22.12 follows from superposition by applying one source at a time, with the other source set to zero. The equation can be rearranged as

$$I(s) = \frac{1}{L}\frac{V_{SRC} + sLI_0}{s(s + 1/\tau)} \tag{22.13}$$

where $\tau = L/R$ is the time constant. Equation 22.13 can be expressed as a PFE:

$$I(s) = \frac{V_{SRC}}{sR} + \frac{I_0 - V_{SRC}/R}{s + 1/\tau} \tag{22.14}$$

Taking the ILT of Equation 22.14,

$$i(t) = \frac{V_{SRC}}{R} + \left(I_0 - \frac{V_{SRC}}{R}\right)e^{-t/\tau} \tag{22.15}$$

Equation 22.15 is of the same form as that of Equation 11.57, derived in terms of initial value, final value, and the time constant.

The voltage across the inductor in the s-domain is

$$V_L(s) = sLI(s) - LI_0 = \frac{V_{SCR} - RI_0}{s + 1/\tau} \tag{22.16}$$

Taking the ILT,

$$v_L(t) = (V_{SRC} - RI_0)e^{-t/\tau} \tag{22.17}$$

Again, this result follows from Equation 11.57. The final value of the voltage across the inductor is zero. With an initial current I_0 in the circuit, the initial value of voltage across the inductor is $(V_{SRC} - RI_0)$.

Problem-Solving Tip

- Always include the sources that account for initial conditions in energy storage elements as an integral part of these elements when analyzing the circuit.

Primal Exercise 22.1

Given a series RLC circuit with initial current I_0 in the inductor and V_0 in the capacitor, as shown in Figure 22.6, represent the circuit in the s-domain and derive the expression for $I(s)$.

Ans. $I(s) = \dfrac{V_1(s) + LI_0 - V_0/s}{sL + R + 1/sC}$.

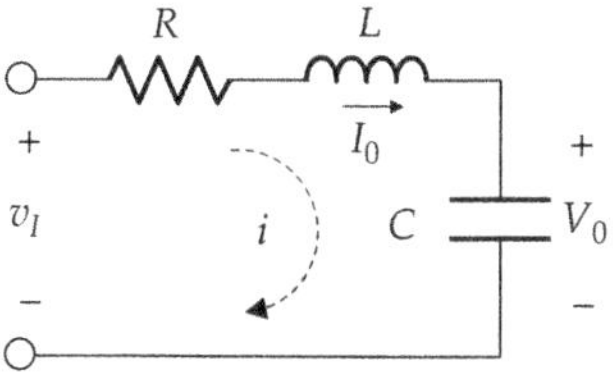

FIGURE 22.6
Figure for Primal Exercise 22.1.

Exercise 22.2

Interpret $\lambda(s)$ in Figure 22.4a and b in the same manner as was done for $Q(s)$ in Figure 22.2a and b. Is this in accordance with duality?

Ans. Yes.

22.1.4 Magnetically Coupled Coils

The mesh-current equations of the basic circuit of Figure 22.7a are

$$L_1\frac{di_1}{dt} - M\frac{di_2}{dt} = v_1$$

$$-M\frac{di_1}{dt} + L_2\frac{di_2}{dt} + v_2 = 0 \qquad (22.18)$$

Taking the LT and including initial values I_{10} and I_{20},

$$sL_1(s) - L_1I_{10} - sMI_2(s) + MI_{20} = V_1(s)$$

$$-sM(s)I_1(s) + MI_{10} + sLI_2(s) - L_2I_{20} + V_2(s) = 0 \qquad (22.19)$$

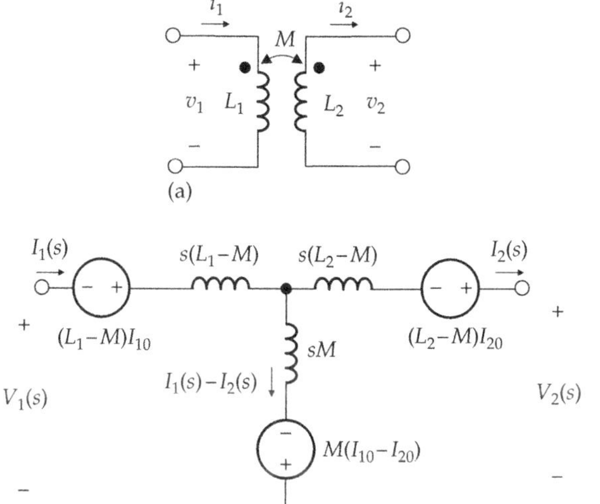

FIGURE 22.7
(a) Assigned voltages and currents of magnetically coupled coils and (b) s-domain representation of the coils in (a) having initial currents.

Any s-domain representation of the coupled coils must satisfy Equations 22.19. It can be readily verified that the T-circuit of Figure 22.7b satisfies these equations and is in accordance with the voltage source representation of the initial current through an inductor (Figure 22.4b). Thus, the s-domain impedances are the inductances of the T-equivalent circuit of Figure 9.26b multiplied by s. The initial current on the input side is I_{10} and is associated with an inductance $(L_1 - M)$, so that the source voltage accounting for the initial conditions in this inductor is a voltage rise $(L_1 - M)I_{10}$ in the direction of current. Similarly on the output side. The initial current in the shunt branch is $(I_{10} - I_{20})$ downward and the source voltage is a voltage rise $M(I_{10} - I_{20})$ in the direction of the current.

Primal Exercise 22.3

Represent the initial current in each of the three inductors in the circuit of Figure 22.7b by a parallel current source, as in Figure 22.4a. Assume that the dot markings on one coil are reversed from those of Figure 22.7a.

22.2 Solution of Circuit Problems in the s-Domain

It was emphasized earlier that in circuit analysis, the fundamental relations that must be satisfied are KCL, KVL, and the v–i relations of the circuit elements. When KCL is satisfied by instantaneous currents at a given node, then by taking LTs, KCL will also be satisfied in the s-domain. Similarly for voltage rises and drops that satisfy KVL. In the preceding section, we have seen how the v–i relations of circuit elements are represented in the s-domain, including initial values. The following concept applies:

Concept: *All circuit laws and techniques discussed previously for the frequency domain using phasor notation, apply equally well to the s-domain. These include series and parallel combinations of s-domain impedances and admittances; Y-Δ transformation; node-voltage, mesh-current, and loop-current methods of analysis; TEC and NEC; and superposition.*

We have already applied in problems of the preceding sections, KVL, superposition, and source transformation in the s-domain.

The first step in the general procedure for analyzing circuits using the LT method is to transform the circuit to the s-domain, representing passive circuit elements by their s-domain impedances, initial values of voltages across capacitors and currents though inductors by appropriate sources, and independent and dependent voltage or current sources by their LTs. The circuit is then analyzed by any of the techniques described previously for phasor analysis, and the LT of the desired circuit response is derived. The essential feature of the LT method is that the circuit equations are now algebraic in s, just as they were algebraic in $j\omega$ in the case of phasor analysis. However, the ILT gives the complete response in the time domain, both transient and steady state, to any arbitrary excitation that has an LT.

Example 22.2: Circuit Response to a Unit Voltage Impulse

It is required to determine $I_1(s)$, $I_2(s)$, and $I_o(s)$ in Figure 22.8 in response to a unit voltage impulse, assuming zero initial conditions, and to interpret the behavior of the circuit.

Solution:

Considering $I_1(s)$ and $I_2(s)$ to be mesh currents, the mesh-current equations may be written as

$$(s+1)I_1(s)-I_2(s)=V_{src}(s) \quad (22.20)$$

$$-I_1(s)+(1+1/s)I_2(s)=-\rho I_o(s) \quad (22.21)$$

where $I_o(s)=I_1(s)-I_2(s)$. Solving these equations,

$$I_1(s)=\frac{s(1-\rho)+1}{s^2(1-\rho)+s+1}V_{src}(s) \quad (22.22)$$

$$I_2(s)=\frac{s(1-\rho)}{s^2(1-\rho)+s+1}V_{src}(s) \quad (22.23)$$

$$I_o(s)=\frac{1}{s^2(1-\rho)+s+1}V_{src}(s) \quad (22.24)$$

It follows from the initial-value theorem, with $V_{src}(s) = 1$, that

$$i_1(0^+)=1\,\text{A},\quad i_2(0^+)=1\,\text{A},\quad \text{and}\quad i_O(0^+)=0 \quad (22.25)$$

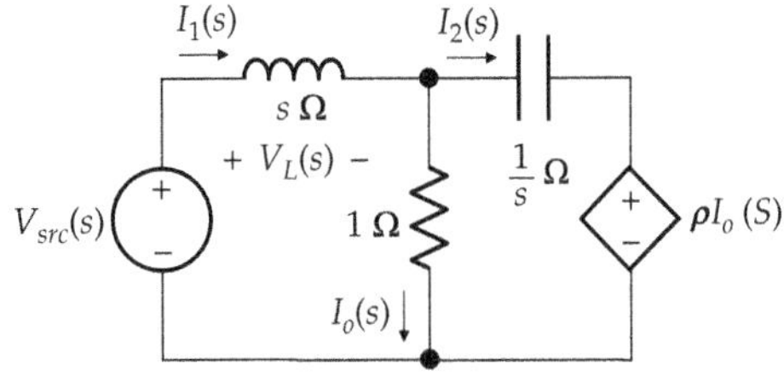

FIGURE 22.8
Figure for Example 22.2.

These results can be interpreted on the basis that when the voltage impulse is applied, the uncharged inductor initially behaves as an open circuit, as discussed in Section 18.5. The impulse appears across the inductor, and the inductor current jumps to $1/L = 1$ A, so that $i_1(0^+) = 1$ A. Since the voltage across the uncharged capacitor is not being forced to change by any current impulse, it remains zero. This means that the voltage across the resistor is the same as that across the dependent source, that is, $i_O(0^+)\times 1=\rho i_O(0^+)$ or

$$i_O(0^+)[\rho-1]=0 \quad (22.26)$$

If $\rho\neq 1$, Equation 22.26 can only be satisfied by having $i_O(0^+) = 0$, which makes $i_2(0^+) = i_1(0^+)$, in accordance with Equation 22.25. The voltage across the inductor is $V_L(s)=sI_1(s)=\frac{s^2(1-\rho)+s}{s^2(1-\rho)+s+1}=1-\frac{1}{s^2(1-\rho)+s+1}$. The 1 term accounts for the impulse and the term $\frac{1}{s^2(1-\rho)+s+1}$ is $V_L(s)$ after the impulse is over. Applying the initial-value theorem to this term gives $v_L(0^+)=0$. This also follows from the circuit, for at $t = 0^+$, the voltage source behaves as a short circuit, so that $v_L(0^+) = -1 \times i_O(0^+) = 0$.

If $\rho=1$, the voltage across the resistor is the same as that across the dependent source, so $i_2 = 0$ for all t. Hence, $I_1(s) = I_O(s) = 1/(s + 1)$, and $i_1(t)=i_O(t)=e^{-t}$, as for R in series with L.

It will be seen that the LT automatically gives all the correct responses of the circuit.

Exercise 22.4

Derive Equation 22.24 for $I_o(s)$ using TEC.

22.2.1 Switching

As mentioned in connection with the initial-value theorem (Section 21.4), the LT method is particularly useful for solving switching problems in which impulsive readjustment takes place at the time of switching because capacitor voltages or inductor currents are forced to change at this instant. The general procedure is to represent the circuit in the s-domain, as usual, *after* switching occurs, that is, $t\geq 0^+$, but including the initial conditions for voltages across capacitors, and currents through inductors, just *before* switching takes place ($t = 0^-$). The initial-value theorem gives any desired circuit response at $t = 0^+$. When no impulsive readjustments are involved at $t = 0$, the initial conditions for voltages across capacitors and currents through inductors are of course the same at $t = 0^-$ as at $t = 0^+$.

Example 22.3: Switched Responses of Capacitors in Parallel with a Resistor

In the circuit of Figure 22.9a, C_1 has an initial voltage $V_{10} = 50$ V, and C_2 is uncharged. If the switch is closed at $t = 0$, determine the voltage across the parallel combination and the currents in the circuit as functions of time. This example is the same as Example 19.9, but with different numerical values.

Solution:

The circuit for $t \geq 0^+$, after the switch is closed, is shown in Figure 22.9b, indicating v and the currents in the three branches. This circuit is shown in the s-domain in Figure 22.10, but with the initial conditions for $t = 0^-$, before the switch is closed. The initial condition in the 0.3 F capacitor is represented by the parallel, current-impulse source, since the circuit is a parallel circuit. The source is $CV_0 = 0.3 \times 50 = 15$ As or C. $V(s)$ is $I_1(s)$ multiplied by $1/0.2s$ Ω in parallel with 2 Ω:

$$V(s) = \frac{2/0.2s}{2 + 1/0.2s} I_1(s) \tag{22.27}$$

From KCL,

$$I_1(s) = 15 - 0.3sV(s) \tag{22.28}$$

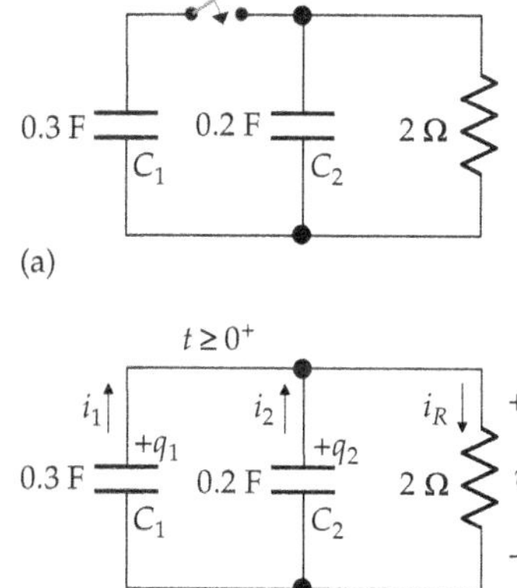

FIGURE 22.9
Figure for Example 22.3.

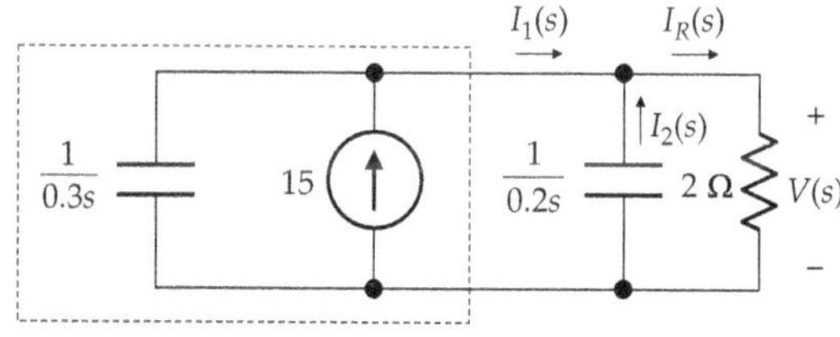

FIGURE 22.10
Figure for Example 22.3.

Equations 22.27 and 22.28 are solved for $V(s)$ and $I_1(s)$; $I_2(s)$ is obtained as $-0.2sV(s)$ and $I_R(s)$ as $I_1(s) + I_2(s)$. These functions and their ILTs are

$$V(s) = \frac{30}{s+1}; \quad v(t) = 30e^{-t}u(t) \text{ V}, \quad \text{for all } t \tag{22.29}$$

$$I_1(s) = \frac{6s+15}{s+1} = 6 + \frac{9}{s+1}; \; i_1(t) = 6\delta(t) + 9e^{-t}u(t) \text{ A, for all } t \tag{22.30}$$

$$I_2(s) = -0.2sV(s) = -\frac{6s}{s+1} = -6 + \frac{6}{s+1};$$

$$i_2(t) = -6\delta(t) + 6e^{-t}u(t) \text{ A}, \quad \text{for all } t \tag{22.31}$$

$$I_R(s) = I_1(s) + I_2(s) = \frac{15}{s+1}; \quad i_R(t) = 15e^{-t}u(t) \text{ A}, \quad \text{for all } t \tag{22.32}$$

Physically, the initial charge on C_1 is 15 C. When the switch is closed, this charge is distributed between C_1 and C_2 so as to equalize the voltage across them. The parallel capacitance is 0.5 F and the common voltage at $t = 0^+$ is $v(0^+) = 15/0.5 = 30$ V, as in Equation 22.29. Charge is conserved at $t = 0^+$, bearing on mind that no charge is transferred by i_R at this instant, as was explained in Example 19.9. The charge on C_1 at $t = 0^+$ is $0.3 \times 30 = 9$ C, and the charge on C_2 at $t = 0^+$ is $0.2 \times 30 = 6$ C. Hence, 6 C are transferred between $t = 0^-$ and $t = 0^+$ by a current impulse $6\delta(t)$ A, as indicated in the expressions for $i_1(t)$ and $i_2(t)$ in Equations 22.30 and 22.31. $v(t)$ decays exponentially from its initial value of 30 V at $t = 0^+$ with a time constant $2 \times 0.5 = 1$ s, as indicated in Equation 22.29. It follows that $i_R = i_1 + i_2$, for $t \geq 0^+$, in accordance with Equations 22.30 and 22.31. Clearly, $i_R = v/2$, as expected.

Let us examine next what information can be derived from the initial- and final-value theorems of the LT. Multiplying the LTs of the voltage and currents by s and setting $s = 0$ makes all the final values zero, as expected. Multiplying $V(s)$ and $I_R(s)$ by s and letting $s \to \infty$ gives $v(0^+) = 30$ V and $i_R(0^+) = 15$ A, also as expected. To apply the initial-value theorem to $I_1(s)$ and $I_2(s)$, the numerator is divided by the denominator to obtain the LT of an impulse and a proper rational function. Because the impulse is over at $t = 0^+$, it does not contribute to $i_1(0^+)$ and $i_2(0^+)$, so that these values are determined by the proper rational function part of the LT. Multiplying these by s and letting $s \to \infty$ gives the correct values of $i_1(0^+) = 9$ A and $i_2(0^+) = 6$ A. This is an interesting feature of the LT in that when using the initial values at $t = 0^-$ in a circuit that applies for $t \geq 0^+$ and that involves impulsive readjustment between $t = 0^-$ and $t = 0^+$, the initial-value theorem gives correct values at $t = 0^+$ for the circuit variables having an impulse.

The following observations are of interest:

1. $i_1(t)$ and $i_2(t)$ are given by Equations 22.30 and 22.31 for all t. Considering $i_1(t)$, it is seen that $i_1(t) = 0$ for $t \leq 0^-$. Between $t = 0^-$ and $t = 0^+$, $i_1(t) = 6\delta(t)$ A, since the term $9e^{-t}u(t)$ is finite during this interval and does not affect the impulse. For $t \geq 0^+$, the impulse is over and $i_1(t) = 9e^{-t}u(t)$ A. $i_2(t)$ is similarly interpreted.
2. $I_1(s)$, the LT of the current through C_1, is from Figure 22.10, given by $I_1(s) = 15 - 0.3sV(s) = (6s + 15)/(s + 1)$, as in Equation 22.30. The 15 As current source, being part of the representation of the initial conditions of C_1, must be included as part of the capacitor in the s-domain when evaluating the capacitor current.
3. Can i_1 and i_2 be determined from the v–i relation $i = Cdv/dt$ of the capacitors? To do so, one must have an expression for capacitor voltage that applies to either capacitor for all t. Since the capacitors have different initial conditions, they would have different voltage expressions. v in Figure 22.9b is the voltage across all the paralleled elements for $t \geq 0^+$. In Equation 22.29, the expression for v is multiplied by $u(t)$, which means that $v = 0$ at $t \leq 0^-$ and $v = 30$ V at $t = 0^+$. This expression for v is therefore appropriate for v_2, the voltage across C_2, which has these initial conditions. Differentiating both sides of Equation 22.29, considering the RHS as the product of two variables, gives $-C_2dv/dt = -0.2(30e^{-t}\delta(t) - 30e^{-t}u(t)) = -6\delta(t) + 6e^{-t}u(t)$, as in Equation 22.31.

   As for v_1, the voltage across C_1, this voltage is 50 V for $t \leq 0^-$, and $v_1(t) = 30e^{-t}u(t)$ V for $t \geq 0^+$. These two conditions can be combined as

   $$v_1(t) = 50 + \left(-50 + 30e^{-t}\right)u(t), \quad \text{for all } t \tag{22.33}$$

   Differentiating Equation 22.33 with respect to t and multiplying by $-C_1$ gives $i_1(t)$ in Equation 22.30. Note that $V_1(s)$ from Equation 22.33 is the same as $V(s)$ in Equation 22.29, because the single-sided LT is the same if the lower limit of integration is $t = 0^-$ or $t = 0^+$ when the function does not have an impulse at the origin. Although $i_1(t)$ has an impulse at the origin, $v_1(t)$ has a step at the origin.
4. $Q_2(s) = C_2V(s) = 6/(s + 1)$ is the LT of the charge on C_2; $q_2(t) = 6e^{-t}u(t)$ C, with $q_2(0^+) = 6$ C, as follows also from the initial-value theorem. $q_1(t) = C_1v_1(t)$, where $v_1(t)$ is given by Equation 22.33. $Q_1(s) = C_1V_1(s) = C_1V(s) = 9/(s + 1)$, with $q_1(0^+) = 9$ C, as follows also from the initial-value theorem.

Problem-Solving Tip

- In switching problems that involve impulsive readjustments at the instant of switching, the initial values in energy storage elements just before switching should be used in the s-domain representation of the circuit after switching.

Primal Exercise 22.5

Given $C_1 = C_2 = 1$ F, $R = 0.5\ \Omega$, and $V_0 = 10$ V in Example 22.3, determine $i_1(t)$, $v(t)$, $i_2(t)$, $q_1(t)$, and $q_2(t)$.

Ans. $i_1(t) = 5\delta(t) + 5e^{-t}u(t)$ A; $v(t) = 5e^{-t}u(t)$ V; $i_2(t) = -5\delta(t) - 5e^{-t}u(t)$ A, $q_1(t) = q_2(t) = 5e^{-t}u(t)$ C.

Example 22.4: Switched Response of an *RC* Circuit

In Figure 22.11, the switch is moved to position 'b' at $t = 0$ after being in position 'a' for a long time. The switch is moved to position 'a' at $t = 1$ s and back to position 'b' at $t = 2$ s. Determine $v_O(t)$ during the time intervals $0 \leq t \leq 1$ s, $1 \leq t \leq 2$ s, and $t \geq 2$ s, given that $v_{SRC}(t) = 15\sin t u(t)$ V.

Solution:

At $t = 0^-$, $v_O(0^-) = 5$ V, from voltage division, with the capacitor fully charged. During the interval $0 \leq t \leq 1$ s, the circuit in the s-domain is as shown in Figure 22.12, with $I_0 = CV_0 = 5$ As, and $V_{src}(s) = 15/(s^2 + 1)$. From superposition and PFE,

$$v_o(s) = \frac{5}{s+2} + \frac{15}{(s+2)(s^2+1)} = \frac{5s^2+20}{(s+2)(s^2+1)} = \frac{8}{s+2} - \frac{3(s-2)}{s^2+1} \tag{22.34}$$

Hence,

$$v_O(t) = 8e^{-2t} - 3\cos t + 6\sin t, \quad 0 \leq t \leq 1\text{ s} \tag{22.35}$$

At $t = 1$ s, $v_O(1) = 4.51$ V. During the interval $1 \leq t \leq 2$ s, the circuit in the s-domain is as shown in Figure 22.12, with $I_0 = 4.51$ As and $V_{src}(s) = 10/s$. From superposition and PFE,

$$v_o(s) = \frac{4.51}{s+2} + \frac{10}{s(s+2)} = \frac{4.51s+10}{s(s+2)} = \frac{5}{s} - \frac{0.49}{s+2} \tag{22.36}$$

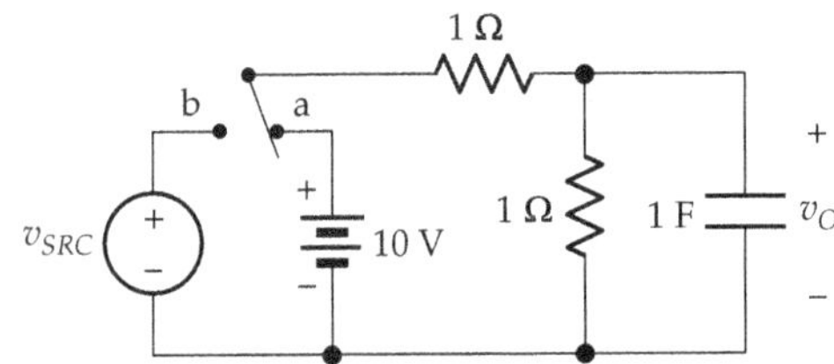

FIGURE 22.11
Figure for Example 22.4.

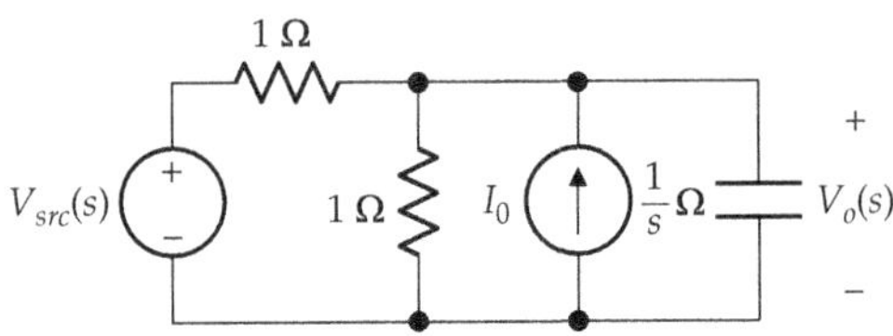

FIGURE 22.12
Figure for Example 22.4.

It should be noted that when we consider the LT of the 10 V source to be $10/s$, it is implicitly assumed that zero time is the instant when this source is applied at $t = 1$ s. In other words, when we take the ILT of Equation 22.36, it is with respect to a time variable $t' = t - 1$. Thus,

$$v_O(t') = 5 - 0.49e^{-2t'}, \quad 0 \le t' \le 1\text{ s} \tag{22.37}$$

or

$$v_O(t) = 5 - 0.49e^{-2(t-1)}, \quad 1 \le t \le 2\text{ s} \tag{22.38}$$

At $t = 2$ s, $v_O(t) = 4.93$ V. For $t \ge 2$ s, the circuit in the s-domain is as shown in Figure 22.12, with $I_0 = 4.93$ As, and $V_{src}(s) = 15/(s^2 + 1)$. From superposition and PFE,

$$v_o(s) = \frac{4.93}{s+2} + \frac{15}{(s+2)(s^2+1)} = \frac{4.93s^2 + 19.93}{(s+2)(s^2+1)}$$

$$= \frac{7.93}{s+2} - \frac{3(s-2)}{s^2+1} \tag{22.39}$$

As explained in connection with switching at $t = 1$ s, the ILT is with respect to $t'' = t - 2$. Hence,

$$v_O(t'') = 7.93e^{-2t''} - 3\cos t'' + 6\sin t'', \quad t'' \ge 1\text{ s} \tag{22.40}$$

or

$$v_O(t) = 7.93e^{-2(t-2)} - 3\cos(t-2) + 6\sin t(t-2), \quad t \ge 2\text{ s} \tag{22.41}$$

Problem-Solving Tip

- In delayed switching, the LT implicitly assumes a time origin at $t = 0$.

22.3 Transfer Function

Definition: *When a single excitation is applied to a given circuit having no initial energy storage, the transfer function H(s) is the ratio of the LT Y(s) of a designated response to the LT X(s) of the applied excitation:*

$$H(s) = \frac{Y(s)}{X(s)} \tag{22.42}$$

In Example 22.2, $I_1(s)/V_{src}(s)$, $I_2(s)/V_{src}(s)$, and $I_1(s)/V_{src}(s)$ are all examples of transfer functions in the same circuit. Being the ratio of a response to an excitation, the transfer function is independent of the nature of the excitation. It depends on the circuit, on where in the circuit the excitation is applied, and on which voltage or current in the circuit is the designated response.

A special case of interest is when the applied excitation $x(t)$ is $\delta(t)$, a unit impulse at the origin. $X(s) = 1$, so that $Y(s) = H(s)$ in the s-domain. By definition $y(t) = h(t)$, the response to the unit impulse. It follows that the transfer function $H(s)$ and the impulse response $h(t)$ are related through a Laplace transformation and its inverse. Thus,

$$\mathcal{L}\{h(t)\} = H(s) \quad \text{and} \quad \mathcal{L}^{-1}\{H(s)\} = h(t) \tag{22.43}$$

Concept: *The LT of the response to a unit impulse of excitation is the transfer function, and the ILT of the transfer function is the response in the time domain to a unit impulse of excitation.*

This concept can be used to relate circuit responses in the s- and time domains. Let the only excitation applied to a circuit without initial energy storage be $v_1(t)$ (Figure 22.13a). According to the definition of convolution (Section 20.1),

$$v_2(t) = v_1(t) * h(t) \tag{22.44}$$

Taking the LT of both sides,

$$V_2(s) = \mathcal{L}\{v_1(t) * h(t)\} \tag{22.45}$$

Using the convolution theorem (Equation 21.91),

$$V_2(s) = V_1(s)H(s) \tag{22.46}$$

which is in accordance with the definition of the transfer function (Equation 22.42).

If two circuits are cascaded (Figure 22.13c), then using Equation 22.46

$$V_2'(s) = V_1(s)H_1'(s) \quad \text{and} \quad V_3(s) = V_2'(s)H_2(s), \tag{22.47}$$

where $V_2'(s)$ and hence $H_1'(s)$ are generally different from the open-circuit transfer function $H(s)$ (Figure 22.13b), because of the loading effect of the second circuit on the

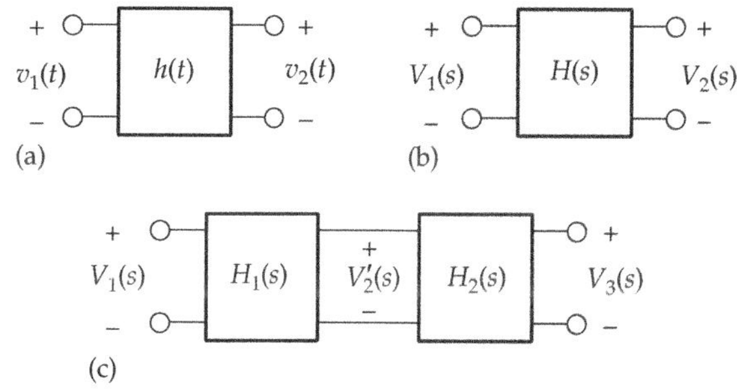

FIGURE 22.13
Cascading of circuits in the time and s domains. (a) A circuit in the time domain having an impulse response $h(t)$, (b) the same circuit as (a) in the s-domain, and (c) two cascaded circuits in the s-domain.

first circuit, as explained in connection with Figure 15.10. It follows from Equation 22.47 that

$$\frac{V_3(s)}{V_1(s)} = \frac{V_3(s)}{V_2'(s)} \times \frac{V_2'(s)}{V_1(s)} = H_2(s)H_1'(s) \tag{22.48}$$

That is, as noted earlier, when circuits are cascaded, the overall transfer function is the product of the individual transfer functions of the two circuits. This does not apply in the time domain, where the outputs of the individual circuits are related to their inputs by convolution:

$$v_2'(t) = v_1(t) * h_1'(t), \quad v_3(t) = v_2'(t) * h_2(t), \quad \text{and}$$
$$v_3(t) = v_1(t) * h_1'(t) * h_2(t) \tag{22.49}$$

This illustrates a great convenience of working with transfer functions.

Another important property of the transfer function is its relation to the characteristic equation:

Concept: *The poles of the transfer function are the roots of the characteristic equation.*

Recall from Section 12.1 that all the responses in a circuit obey the same homogeneous differential equation, so that the same characteristic equation applies to all the variables in a given circuit. It was pointed out in Section 21.3, that the characteristic equation is the LT of the homogeneous equation, assuming zero initial conditions, as in the definition of the transfer function. The LHS of the characteristic equation appears in the denominator of all the responses in a given circuit, as in Equation 21.55, and as was demonstrated in the LT of circuit responses derived in this chapter. In other words, the LHS of the characteristic equation appears as the denominator of the transfer function for a given response in a circuit. It follows that the roots of the characteristic equation are the poles of the transfer function.

22.3.1 Stability

Definition: *A circuit is stable if the response to a unit impulse tends to zero as $t \to \infty$. The circuit is unstable if the response to a unit impulse increases without limit, that is, is unbounded as $t \to \infty$. The circuit is marginally stable, or metastable, if as $t \to \infty$, the response to a unit impulse does not approach zero but remains bounded.*

Recall that the response in the time domain to a unit impulse is the ILT, $h(t)$, of the transfer function $H(s)$ (Equation 22.43), whose poles are the roots of the characteristic equation. Circuit stability is therefore determined by the location of the poles of the transfer function in the s-plane. Specifically, a distinction is made between (1) poles located on the imaginary axis, (2) poles located in the open left half of the s-plane, that is, the left half of the s-plane, excluding the imaginary axis, and (3) poles located in the open right half of the s-plane, excluding the imaginary axis.

Concept: *If all the poles of the transfer function lie in the open left half of the s-plane, the circuit is stable. If at least one pole lies in the open right half of the s-plane, the circuit is unstable. If all the poles of the transfer function are simple poles on the imaginary axis, the circuit is metastable. If at least one pole on the imaginary axis is a multiple pole, the circuit is unstable.*

This statement can be justified by considering the ILTs of terms in the PFE of $H(s)$. Consider terms of the form $K/(s + p_r)$ or $K/(s + p_r)^n$, which denote, respectively, a simple or a multiple pole $-p_r$. If located in the open left half of the s-plane, the simple pole is generally of the form $-p_r = -\alpha_r - j\omega_r$, with $-\alpha_r < 0$. Complex poles must occur in complex conjugate pairs in order to have real coefficients of $H(s)$. The ILT of a real pole is $Ke^{-\alpha_r t}u(t)$, if the pole is simple, and $Kt^{n-1}e^{-\alpha_r t}u(t)/(n-1)!$, if the pole is of order n (Table 21.2). If the pole is complex, these ILTs are multiplied by a cosine term of frequency ω_r and a phase angle that depends on the residues of the poles. In all these cases, these terms vanish as $t \to \infty$ because of the presence of the $e^{-\alpha_r t}$ term, so that the circuit is stable. On the other hand, if the pole lies in the open right half of the s-plane, $-\alpha_r > 0$, and $e^{-\alpha_r t} \to \infty$ as $t \to \infty$. The impulse response is unbounded and the circuit is unstable.

As for poles on the imaginary axis, a simple pole at the origin results in a term K/s in the transfer function, whose ILT is $Ku(t)$. In the absence of a pole in the open right half of the s-plane, the impulse response $Ku(t)$ is bounded, and the circuit is metastable. Similarly, a pair of simple complex poles on the imaginary axis at $\pm j\omega_r$ result in a term $(K_1 s + K_2)/(s^2 + \omega_r^2)$ in the transfer function, whose ILT is $K\cos(\omega_r t + \theta)$ where θ depends on K_1 and K_2. The impulse response remains bounded as $t \to \infty$, and the circuit is metastable. On the other hand, if the poles on the imaginary axis are multiple poles, the ILTs will contain t or powers of t (Table 21.2). The impulse response is unbounded as $t \to \infty$, and the circuit is unstable.

A constant term, a term in s, or terms in powers of s in the transfer function have ILTs that are, respectively, an impulse, its first derivative, and its higher derivatives. The ILTs of all these terms are zero at $t = 0^+$. In the absence of a pole in the open right half of the s-plane or of a multiple pole on the imaginary axis, the impulse response is zero as $t \to \infty$, and the circuit is stable.

The following should be noted concerning the location of poles in the s-plane:

1. Poles on the imaginary axis have $\alpha_r = 0$. When these are the only poles present in the transfer function, the circuit is lossless and consists of ideal inductors and capacitors only. Poles in the left half of

the s-plane arise because of the presence of resistances in the circuit, which make $\alpha_r > 0$. The circuit is dissipative. By analogy, poles in the right half of the s-plane can be considered to be due to a negative resistance, which supplies energy to the circuit and causes an unbounded response.

2. A circuit with a simple pole at the origin is metastable, as explained previously. However, if the circuit is excited at the frequency of the pole, the *response is unbounded*. Consider, for example, a capacitor C. Its transfer function in terms of $V(s)/I(s)$ is the impedance $1/sC$, which denotes a simple pole at the origin. A dc source is of zero frequency, so that the frequency of the source is also at the origin in the s-plane. If C is excited by a dc current source I_{dc}, whose LT is I_{dc}/s, then $V(s) = I_{dc}/s^2C$; $v(t) = (I_{dc}/C)tu(t)$ and is unbounded. Example 22.5 discusses the case of a lossless LC circuit that is excited at the frequency of its poles on the imaginary axis.

Example 22.5: Responses of LC Circuit

Given a parallel LC circuit (Figure 22.14a), with zero initial conditions, excited by a current source, it is required to determine (a) the transfer function $V_o(s)/I_{src}(s)$, (b) v_O in response to a unit current impulse, and (c) v_O in response to a sinusoidal excitation of the same frequency as the poles.

Solution:

(a) It follows from the circuit in the s-domain (Figure 22.14b) that $V_o(s) = \dfrac{sL(1/sC)}{sL+1/sC} I_{src}(s)$, so that

$$H(s) = \frac{V_o(s)}{I_{src}(s)} = \frac{1}{C}\frac{s}{s^2+\omega_0^2} \tag{22.50}$$

where $\omega_0^2 = 1/LC$. $H(s)$ has a pair of simple, conjugate poles on the imaginary axis at $s = \pm j\omega_0$.

(b) If $i_{SRC}(t) = \delta(t)$:

$$h(t) = v_O(t) = \mathcal{L}^{-1}\{H(s)\} = \frac{1}{C}\cos\omega_0 t \tag{22.51}$$

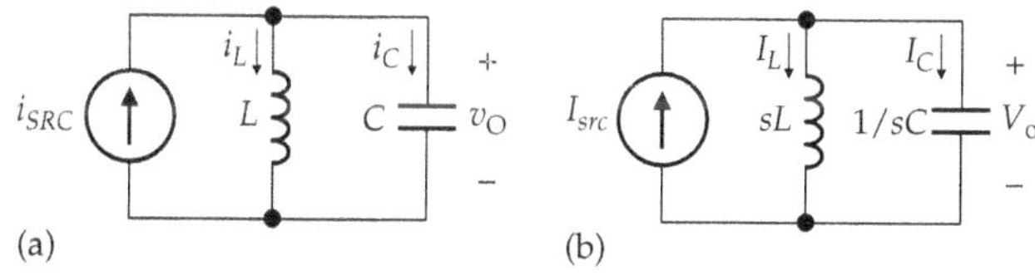

FIGURE 22.14
Figure for Example 22.5.

This result is readily interpreted. The unit impulse instantaneously charges the capacitor to a voltage $1/C$ V, the energy in the capacitor being initially $(1/2)Cv_O^2(0) = (1/2C)$. The circuit then continuously oscillates at a frequency ω_0, the amplitude of v_O being $1/C$. The oscillations are sustained, as described in Section 12.1, for the same circuit in response to an initial current in the inductor. $i_L(t) = \dfrac{1}{L}\displaystyle\int_0^t v_O(t)dt = \omega_0 \sin\omega_0 t$. The current in the inductor in Figure 22.14a has its largest magnitude when $\omega_0 t = n\pi/2$, where n is an integer. At these instants of time $v_O = 0$, so no energy is stored in the capacitor. The energy stored in the inductor is then $\dfrac{1}{2}LI_m^2 = \dfrac{1}{2}L\dfrac{1}{LC} = \dfrac{1}{2C}$, which is the same as that initially stored in the capacitor. It is seen that the energy continuously oscillates between electric energy stored in the capacitor, at $\omega_0 t = n\pi$, and magnetic energy stored in the inductor, at $\omega_0 t = n\pi/2$. At intermediate times, energy is stored in both the inductor and the capacitor, the total energy being $1/2C$, as required by conservation of energy. Since the amplitude of oscillation is bounded, the circuit is metastable.

(c) If $i_{SRC}(t) = A\cos\omega t$, where $\omega \neq \omega_0$, $I_{src}(s) = A\dfrac{s}{s^2+\omega^2}$, and

$$V_o(s) = \frac{A}{C}\frac{s}{s^2+\omega_0^2}\frac{s}{s^2+\omega^2} \tag{22.52}$$

Because the numerator and each term in the denominator is in s^2, s^2 can be replaced by a variable x, so that the PFE can be expressed as

$$V_o(s) = \frac{A}{C}\left(\frac{x}{(x+\omega_0^2)(x+\omega^2)}\right) = \frac{A}{C}\left(\frac{K_1}{x+\omega_0^2} + \frac{K_2}{x+\omega^2}\right) \tag{22.53}$$

Multiplying both sides by $(x+\omega_0^2)$ and substituting $x = -\omega_0^2$ gives $K_1 = \omega_0^2/(\omega_0^2-\omega^2)$. Multiplying both sides by $(x+\omega^2)$ and substituting $x = -\omega^2$ gives $K_2 = -\omega^2/(\omega_0^2-\omega^2)$. The PFE of $V_O(s)$ is

$$V_o(s) = \frac{A}{C(\omega_0^2-\omega^2)}\left[\frac{\omega_0^2}{s^2+\omega_0^2} - \frac{\omega^2}{s^2+\omega^2}\right] \tag{22.54}$$

as can be readily verified. The ILT is

$$v_O(t) = \frac{A}{C(\omega_0^2-\omega^2)}\left[\omega_0\sin\omega_0 t - \omega\sin\omega t\right] \tag{22.55}$$

The voltage is thus a combination of the applied signal and the natural oscillation of the circuit. The response remains bounded.

If $i_{SRC}(t) = A\cos\omega_0 t$, then

$$V_o(s) = \frac{A}{C}\frac{s^2}{\left(s^2+\omega_0^2\right)^2} \quad (22.56)$$

Replacing s^2 by a variable x, as was done in connection with Equation 22.53, the PFE can be expressed as

$$\frac{x}{\left(x+\omega_0^2\right)^2} = \frac{K_1}{x+\omega_0^2} + \frac{K_2}{\left(x+\omega_0^2\right)^2} \quad (22.57)$$

Multiplying both sides by $\left(x+\omega_0^2\right)^2$ and substituting $x = -\omega_0^2$ gives $K_2 = -\omega_0^2$. Multiplying both sides by $\left(x+\omega_0^2\right)^2$, Equation 22.57 becomes $x = K_1\left(x+\omega_0^2\right) + K_2$. Comparing coefficients, with $K_2 = -\omega_0^2$, gives $K_1 = 1$. Hence,

$$\frac{s^2}{\left(s^2+\omega_0^2\right)^2} = \frac{1}{s^2+\omega_0^2} - \frac{\omega_0^2}{\left(s^2+\omega_0^2\right)^2} \quad (22.58)$$

To put this in a form which can be readily inverted, $s^2/\left(s^2+\omega_0^2\right)^2$ is added to both sides and the two sides are divided by two. The PFE of $V_o(s)$ becomes

$$V_o(s) = \frac{A}{2C}\left[\frac{1}{s^2+\omega_0^2} + \frac{s^2-\omega_0^2}{\left(s^2+\omega_0^2\right)^2}\right] \quad (22.59)$$

as can be readily verified. The ILT is (Table 21.2)

$$v_O(t) = \frac{A}{2C}\left[\frac{1}{\omega_0}\sin\omega_0 t + t\cos\omega_0 t\right] \quad (22.60)$$

The response is now unbounded, because the frequency of excitation is the same as that of the poles of the circuit. Note that in this case, it is not possible to solve the problem using phasors, because the circuit never reaches a steady state, as required by phasor analysis.

If two identical LC circuits are cascaded with isolation, so that the second circuit does not load the first, the transfer function is of the form $s^2/\left(s^2+\omega_0^2\right)^2$ (Equation 22.56). The impulse response is now unbounded because of the double pole on the imaginary axis. An impulse excites the first circuit into continuous oscillation, which provides excitation to the second circuit at its resonant frequency. This is illustrated by the following simulation.

Simulation: The Schematic is entered as in Figure 22.15, where the first circuit, having L = 1 H and C = 0.1 mF is excited by a current impulse of strength $10^4 \times 10^{-7} = 10^{-3}$ As. The second LC circuit is identical with the first circuit and is excited by the output of the first circuit though a VCCS. In this way, the first circuit is not loaded by the second circuit, and this circuit is excited at the frequency of the pole. The negative sign of the gain of the dependent source is to compensate for the reversed connection of the controlling voltage of this source. In the simulation profile, Time Domain (Transient) is chosen as the Analysis type, 1s is entered for 'Run to time', 0 for 'Start saving data after', and 0.1m for 'Maximum step size'. After the simulation is run, the waveforms of Figure 22.16 are displayed, where the dashed sinusoid is the voltage across the first LC circuit. The voltage jumps to 10 V at the end of the impulse and continues as a cosine function of 10 V amplitude, as indicated by the voltage scale on the RHS of the graph. This is in accordance with the impulse flowing through the capacitor and depositing a charge of 10^{-3} C, which results in a voltage of $10^{-3}/10^{-4}$ = 10 V. The measured period is 62.617 ms, corresponding to a frequency of 15.97 Hz, compared to a calculated value of 15.92 Hz. As indicated by the voltage marker, the output of the second circuit is unbounded, starting from zero and reaching 43.3 kV at 1 s. It should be noted that in order to have the two voltages on the same display, the output of the first circuit is selected as V(C1:1)*2E03.

22.3.2 Sinusoidal Steady-State Response

The transfer function applies to a circuit having no initial energy storage. In the sinusoidal steady state, any initial stored energy would have died down by the time the steady state is established. Moreover, the frequencies of sinusoidal functions are represented on the imaginary axis of the s-plane, where $s = j\omega$. It should be possible therefore to derive the sinusoidal steady-state response from the transfer function by substituting $s = j\omega$. It is shown in this section how this can be accomplished.

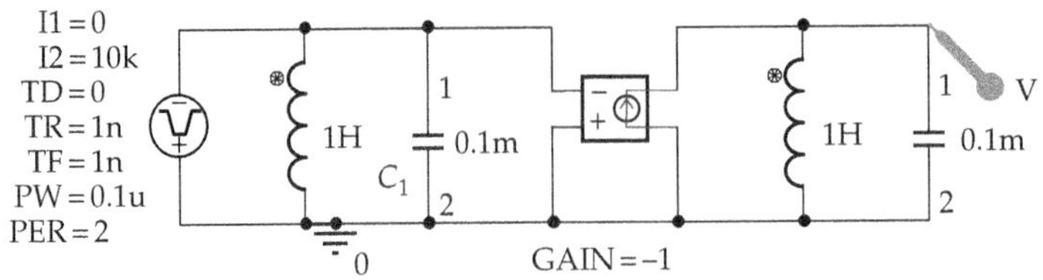

FIGURE 22.15
Figure for Example 22.5.

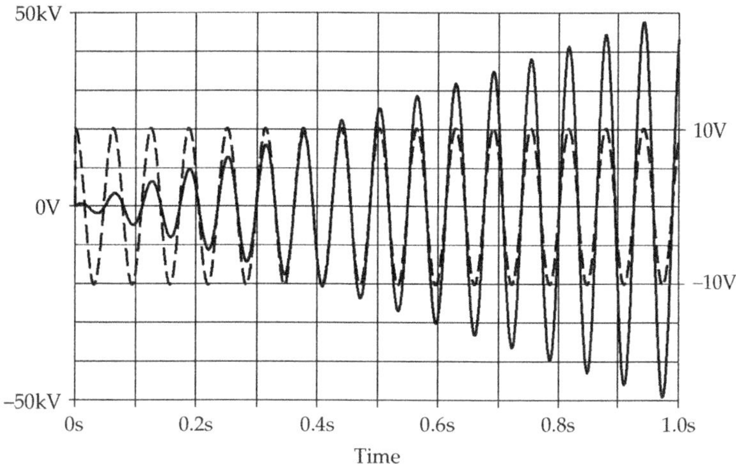

FIGURE 22.16
Figure for Example 22.5.

Consider a stable circuit to which a sinusoidal excitation $x(t) = X_m\cos\omega t$ is applied. Let the transfer function of the circuit be a proper rational function $H(s)$. If the transfer function is not a proper rational function, then $H(s)$ refers to the proper rational fraction part, after dividing the numerator by the denominator. This is justified by the fact that the quotient is terms in s or powers of s, representing an impulse and its higher derivatives, which are zero at $t = 0^+$. However, the response $Y(s)$ of the circuit contains, in general, transients for $t > 0^+$. $H(s)$ can be expressed as

$$H(s)=\frac{Y(s)}{X(s)}=\frac{N(s)}{D(s)} \tag{22.61}$$

where $N(s)$ and $D(s)$ are, respectively, the numerator and denominator polynomials of $H(s)$, with $N(s)/D(s)$ being a proper rational function. Substituting $X(s) = X_m s/(s^2 + \omega^2)$ in Equation 22.61,

$$Y(s)=X_m\frac{s}{s^2+\omega^2}\frac{N(s)}{D(s)} \tag{22.62}$$

This can be expressed as a PFE in the form

$$Y(s)=X_m\frac{s}{(s-j\omega)(s+j\omega)}\frac{N(s)}{D(s)}=\frac{K_1}{s-j\omega}+\frac{K_1^*}{s+j\omega}+U(s) \tag{22.63}$$

where $U(s)$ represents terms whose denominators involve the roots of $D(s)$. To determine K_1, we proceed in the usual manner by multiplying both sides by $(s - j\omega)$ and substituting $s = j\omega$. All terms on the RHS, except K_1 will go to zero, including the terms $(s - j\omega)U(s)$. This is because the circuit is assumed to be stable, which means that there are no poles at $s = \pm j\omega$. Such poles will make the circuit metastable, and if excited at this frequency, the response will be unbounded, as discussed earlier. It follows from Equation 22.63, with $H(s)$ replacing $N(s)/D(s)$, and substituting $s = j\omega$ after cancelling out $(s - j\omega)$ from the numerator and denominator, that

$$K_1=\frac{X_m s(s-j\omega)H(s)}{(s-j\omega)(s+j\omega)}=X_m\frac{j\omega}{j2\omega}H(j\omega)=\frac{X_m}{2}H(j\omega)$$
$$=\frac{X_m}{2}\left|H(j\omega)\right|e^{j\phi} \tag{22.64}$$

where $H(j\omega)$ has been expressed in terms of its magnitude and phase angle as $\left|H(j\omega)\right|e^{i\phi}$. Hence,

$$K_1^*=\frac{X_m}{2}\left|H(j\omega)\right|e^{-j\phi} \tag{22.65}$$

The two terms in K_1 and its conjugate can be combined as

$$\frac{K_1}{s-j\omega}+\frac{K_1^*}{s+j\omega}=\frac{X_m}{2}\left|H(j\omega)\right|\left(\frac{e^{j\phi}}{s-j\omega}+\frac{e^{-j\phi}}{s+j\omega}\right) \tag{22.66}$$

Equation 22.63 becomes

$$Y(s)=\frac{X_m}{2}\left|H(j\omega)\right|\left(\frac{e^{j\phi}}{s-j\omega}+\frac{e^{-j\phi}}{s+j\omega}\right)+U(s) \tag{22.67}$$

Since the circuit is assumed to be stable, all the poles of $U(s)$ have negative real parts, so that their ILTs approach zero in the steady state, as $t \to \infty$. Ignoring $U(s)$, the steady-state $Y(s)$, denoted as $Y_{ss}(s)$ is

$$Y_{ss}(s)=\frac{X_m}{2}\left|H(j\omega)\right|\left(\frac{e^{j\phi}}{s-j\omega}+\frac{e^{-j\phi}}{s+j\omega}\right) \tag{22.68}$$

Note that the final-value theorem (Section 21.4) cannot be applied to Equation 22.67 to find the steady-state value of $Y(s)$ because of the poles $\pm j\omega$ on the imaginary axis. Using Equation 21.9, the ILT of $Y_{SS}(s)$ is

$$y_{SS}(t)=\frac{X_m}{2}|H(j\omega)|\left(e^{j(\omega t+\phi)}+e^{-j(\omega t+\phi)}\right)=X_m|H(j\omega)|\cos(\omega t+\phi) \tag{22.69}$$

where
$|H(j\omega)|$ is the magnitude of $H(s)$ after substituting $s = j\omega$
ϕ is the phase of $H(j\omega)$

It is seen that the steady-state sinusoidal response can be obtained from the transfer function, in accordance with the following concept:

Concept: *The sinusoidal steady-state response is obtained from a transfer function that is a proper rational function by substituting $s = j\omega$, multiplying the magnitude of the excitation by $|H(j\omega)|$ and adding the phase angle of $H(j\omega)$ to that of the excitation.*

Consider, for example, an excitation $5\cos(4t + 30°)$ V, having $\omega = 4$ rad/s, applied to circuit whose transfer function is

$$H(s)=\frac{100(s-2)}{(s+3)(s^2+2s+5)} \tag{22.70}$$

Substituting $s = j4$, the transfer function becomes

$$H(j4)=\frac{100(j4-2)}{(j4+3)(-16+j8+5)} \tag{22.71}$$

It follows that

$$|H(j4)|=\frac{100\sqrt{(16+4)}}{\left(\sqrt{16+9}\right)\left(\sqrt{(121+64)}\right)}=6.58 \tag{22.72}$$

and

$$\begin{aligned}\angle H(j4)&=\tan^{-1}(4/-2)-\tan^{-1}(4/3)-\tan^{-1}(8/-11)\\&=116.6°-53.1°-144.0°=-80.5°\end{aligned} \tag{22.73}$$

where account has been taken of the signs of the real and imaginary components, as explained in connection with Figure 8.10. The response to $5\cos(4t + 30°)$ V is therefore $5 \times 6.58\cos(4t + 30° - 80.5°) = 32.9\cos(4t - 50.5°)$ V.

The magnitude and phase angle of $H(j4)$ in Equations 22.72 and 22.73 are of course the same as those evaluated in the conventional manner. The denominator of $H(s) = s^3 + 5s^2 + 11s + 15 \equiv (15 - 5\omega^2) + j\omega(11 - \omega^2)$, which gives for $\omega = 4$, $H(j4)=100\frac{-2+j4}{-65-j20}=6.58\angle(116.6°-197.1)=6.58\angle(-80.5°)$.

The factors of $H(j\omega)$ of Equation 22.71 can be interpreted geometrically in the s-plane. It is more instructive for this purpose to factor the quadratic expression in the denominator of Equation 22.70 into its complex roots. Thus, $s^2 + 2s + 5 = (s + 1 + j2)(s + 1 - j2)$, so that $H(s)$ becomes

$$H(s)=\frac{100(s-2)}{(s+3)(s+1+j2)(s+1-j2)} \tag{22.74}$$

$H(s)$ has a zero at $s = 2$ a pair of complex conjugate poles p_1 and p_2 at $s = -1 - j2$ and $s = -1 + j2$, respectively, and a pole p_3 at $s = -3$ (Figure 22.17). When $s = j\omega$ is substituted, $H(j\omega)$ becomes

$$\begin{aligned}H(j\omega)&=\frac{100(j\omega-2)}{(j\omega+3)(j\omega+1+j2)(j\omega+1-j2)}\\&=\frac{100(-2+j\omega)}{(3+j\omega)\left[1+j(\omega+2)\right]\left[1+j(\omega-2)\right]}\end{aligned} \tag{22.75}$$

$s = j\omega$ is represented by point 'Q' at a distance $+\omega$ on the imaginary axis. Each of the factors on the RHS of Equation 22.75 can be represented as a vector (Figure 22.17). Thus, $(-2 + j\omega)$ is the vector zQ. It is the sum of the vector zO, which is −2, and the vector $j\omega$. Similarly, $p_3Q = p_3O + OQ = 3 + j\omega$; $p_2Q = 1 + j(\omega - 2)$; and $p_1Q = 1 + j(\omega + 2)$. These vectors can be evaluated in terms of magnitude and phase in the Argand diagram in the same way as phasors. Thus, if $\omega=4$, $zQ = 2\sqrt{5}\angle 116.6°$, $p_1Q = \sqrt{37}\angle 80.5°$, $p_2Q = \sqrt{5}\angle 63.4°$, and $p_3Q=5\angle 53.1°$. Combining these terms gives the same result as in Equations 22.72 and 22.73.

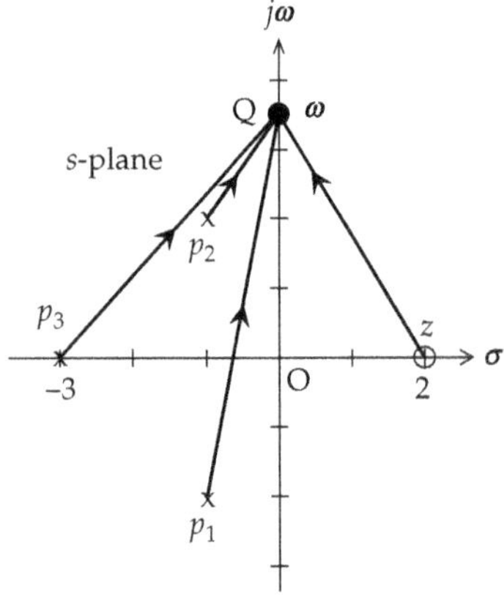

FIGURE 22.17
Interpretation of transfer function in the s-plane.

It should be kept in mind that the response to a dc voltage or current excitation is equivalent to a sinusoidal excitation of zero frequency. The dc response can therefore be obtained by multiplying the dc excitation by $H(0)$, that is, the transfer function $H(s)$ with s set to zero (see Primal Exercise 22.8).

22.3.3 Interpretation of Zeros and Poles

It is seen from the preceding discussion that physically realizable frequencies in the form of trigonometric functions such as $\cos\omega t$ or $\sin\omega t$ are those that lie on the imaginary axis of the s-plane, with the origin representing $\omega = 0$, or dc conditions. A frequency ω is represented on the imaginary axis by two points, $+j\omega$ and $-j\omega$ corresponding to $e^{j\omega t}$ and $e^{-j\omega t}$. These combine to form a trigonometric function as explained in Section 16.2. Zeros of the form $\left(s^2+\omega_z^2\right)$ lie on the imaginary axis at $s=\pm j\omega_z$ and make the transfer function equal to zero at a frequency ω_z. Such a zero was encountered in the bandstop response (Table 14.3). Similarly, poles of the form $\left(s^2+\omega_p^2\right)$ theoretically make the value of the function infinite at a frequency ω_p, as in the case of resonance in a circuit of zero resistance.

22.4 Interpretations of Circuit Responses in the s-Domain

22.4.1 Natural Responses of First-Order Circuits

Figure 22.18 shows an RL circuit in the s-domain, assuming zero initial conditions. It follows from KVL that $V_{src}(s) = RI(s) + sLI(s)$, which gives

$$\frac{I(s)}{V_{src}(s)}=\frac{1}{R+sL}=\frac{1}{L}\frac{1}{s+1/\tau} \tag{22.76}$$

where $\tau = L/R$ is the time constant. If the applied excitation is a voltage impulse of strength K Vs, then $V_{src}(s) = K$, and the impulse establishes an initial current K/L in the inductor at $t = 0^+$, resulting in a natural response for $t \geq 0^+$. Substituting for $V_{src}(s)$ in Equation 22.76 gives

$$I(s)=\frac{K}{L}\frac{1}{s+1/\tau} \tag{22.77}$$

Once $I(s)$ is known, $V_R(s)$ and $V_L(s)$ readily follow. Thus,

$$V_R(s)=RI(s)=\frac{K}{\tau}\frac{1}{s+1/\tau} \tag{22.78}$$

and

$$V_L(s)=sLI(s)=K\frac{s}{s+1/\tau} \tag{22.79}$$

It will be noted that all the responses $I(s)$, $V_R(s)$, and $V_L(s)$ have a pole at $s = -1/\tau$, which is also the root of the characteristic equation, in accordance with the discussion at the beginning of the preceding section. Thus, the differential equation with zero forcing function is $L\frac{di}{dt}+Ri=0$. Taking the LT, with zero initial conditions, gives $(sL + R)I(s) = 0$. The characteristic equation is therefore $sL + R = 0$, whose root is $s = -1/\tau$.

Concept: *All the responses of a stable first-order circuit in the s-domain are characterized by a pole on the negative real axis whose magnitude is the reciprocal of the time constant.*

This is illustrated in Figure 22.19. As $R \rightarrow 0$, $1/\tau \rightarrow 0$, and the pole approaches the origin in the s-plane. The transfer function $I(s)/V_{src}(s)$ reduces to $1/sL$, the admittance of an uncharged inductor. As R increases, the pole moves along the negative real axis, away from the origin.

Primal Exercise 22.6

If $L = 1$ H in the circuit of Figure 22.18, determine how the pole moves along the negative real axis as R varies between 1 and 100 Ω.

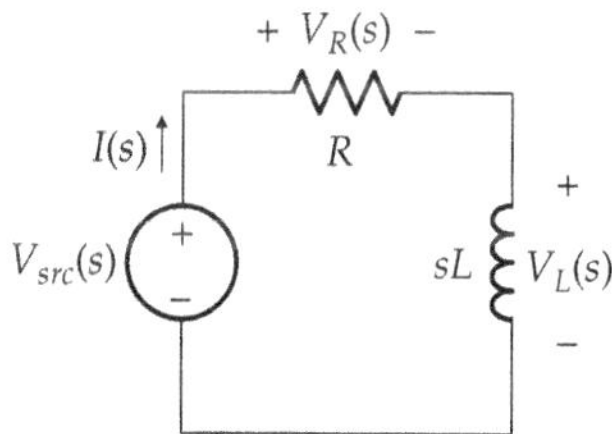

FIGURE 22.18
RL circuit in the s-domain.

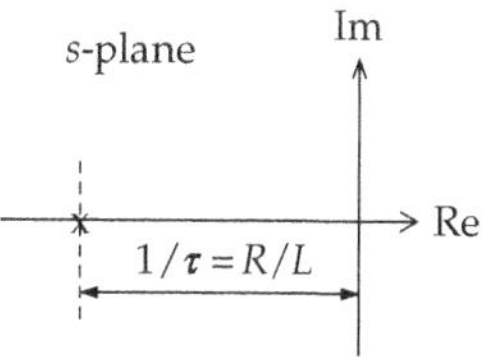

FIGURE 22.19
Pole location of a stable first-order circuit.

Example 22.6: Transfer Function of an *RC* Circuit

Consider the *RC* circuit shown in Figure 22.20. It is desired to derive and interpret the transfer function $I(s)/V_{src}(s)$, assuming no initial energy storage.

Solution:

It follows from Figure 22.20 that

$$H_I(s) = \frac{I(s)}{V_{src}(s)} = \frac{1}{R + 1/sC} = \frac{1}{R}\frac{s}{s + 1/\tau} \quad (22.80)$$

where $\tau = RC$. $H_I(s)$, being an admittance, has a zero at $s = 0$ and a pole at $s = -1/\tau$. The zero occurs at a frequency $\omega = 0$ which corresponds to dc conditions. Under dc conditions, the capacitor behaves as an open circuit, so that the admittance is zero. At any frequency $s = j\omega$, the admittance is $H_I(j\omega) = 1/(R + 1/j\omega C)$, as obtained using phasor analysis. Note that impedance and admittance in terms of $j\omega$ are defined under sinusoidal steady-state conditions. The pole is on the negative real axis. Formally, $H_I(s) \to \infty$ at $s = -1/\tau$, but this does not occur at any physically realizable frequency, because such frequencies are only represented on the imaginary axis of the *s*-plane.

If the excitation is a voltage step $Ku(t)$, $V_{src}(s) = K/s$. Substituting for $V_{src}(s)$ in Equation 22.80

$$I(s) = \frac{K}{R}\frac{1}{s + 1/\tau} \quad (22.81)$$

The ILT is

$$i(t) = \frac{K}{R}e^{-t/\tau}u(t) \quad (22.82)$$

This is the same response discussed for case (a) in Figure 18.9. The final-value theorem gives $i(t) = 0$, as $t \to \infty$, and the initial-value theorem gives $i(0^+) = K/R$.

If the excitation is a voltage impulse of strength *K* Vs, $V_{src}(s) = K$. Substituting for $V_{src}(s)$ in Equation 22.80,

$$I(s) = \frac{K}{R}\frac{s}{s + 1/\tau} = \frac{K}{R}\left(1 - \frac{1/\tau}{s + 1/\tau}\right) = \frac{K}{R} - \frac{K}{CR^2}\frac{1}{s + 1/\tau} \quad (22.83)$$

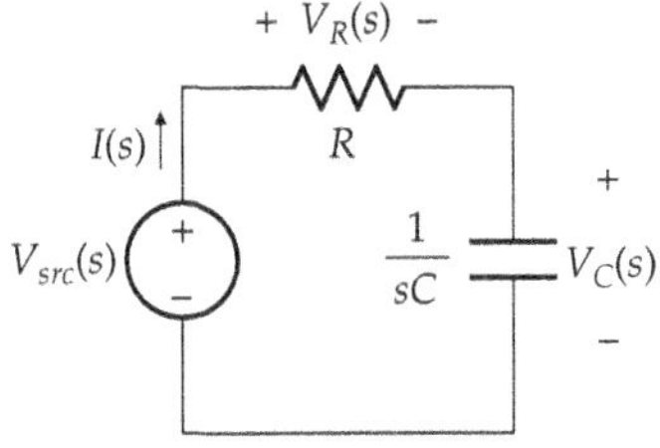

FIGURE 22.20
Figure for Example 22.6.

The ILT is

$$i(t) = \frac{K}{R}\delta(t) - \frac{K}{CR^2}e^{-t/\tau}u(t) \quad (22.84)$$

This is the same response discussed for case (b) in Figure 18.9. The final-value theorem gives $i(t) = 0$, as $t \to \infty$, and the initial-value theorem applied to the rational function part gives $i(0^+) = -K/CR^2$.

Substituting $s = j\omega$,

$$H_I(j\omega) = \frac{j\omega C}{1 + j\omega CR} = \frac{1}{\sqrt{R^2 + 1/\omega^2C^2}}\angle(90° - \phi),$$

$$\phi = \tan^{-1}(\omega CR) \quad (22.85)$$

If $v(t) = V_m\cos(\omega t + \theta)$, then

$$i(t) = \frac{V_m}{\sqrt{R^2 + 1/\omega^2C^2}}\cos(\omega t + \theta - \phi + 90°)$$

$$= -\frac{V_m}{\sqrt{R^2 + 1/\omega^2C^2}}\sin(\omega t + \theta - \phi), \quad \phi = \tan^{-1}(\omega CR) \quad (22.86)$$

as obtained from phasor analysis.

Simulation: This example illustrates the direct simulation of a transfer function using the part named LAPLACE from the PSpice Analog Behavioral Module (ABM) library. When this part is entered, a block is displayed having the default transfer function $1/(s + 1)$, but the numerator and denominator can be changed to the desired polynomial in *s* (Figure 22.21a). The transfer function $s/(s + 1)$ corresponds to $K = 1$ Vs, $C = 1$ F, and $R = 1\ \Omega$ in Equation 22.83. To invert the transfer function, a unit impulse function is applied at the IN terminal and the output is derived at the OUT terminal. In the simulation profile, Time Domain (Transient) is chosen as the Analysis type, 5s is entered for 'Run to time', 0 for 'Start saving data after', and 0.5m for 'Maximum step size'. After the simulation is run, the setting of the *x*-axis is changed by choosing Plot/Axis Settings/X-axis, then selecting User Defined and replacing the lower value of 0 by 10ms, in order to hide the initial transient. The plot of Figure 22.22 is displayed corresponding to the response after the impulse in Equation 22.84. Note that the output is interpreted as a voltage, but the transfer function is that of a current, so that the ordinates in Figure 22.22 are in fact in amperes.

To display the steady-state sinusoidal response, the schematic of Figure 22.21b is entered, representing an arbitrarily chosen input of 10cos*t* V and the same transfer function. The simulation profile is changed to AC Sweep (Noise), with start and end frequencies of 0.159155, and 1 point per decade. After the simulation is run, the voltage printer readings in the output file are a magnitude of 7.071 V and a phase angle of 45°, in accordance with Equation 22.86.

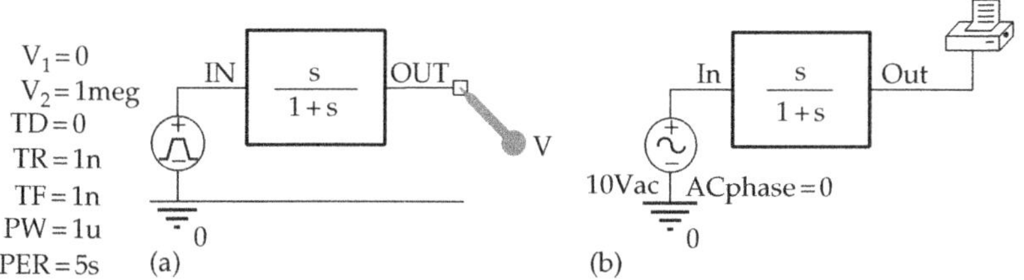

FIGURE 22.21
Figure for Example 22.6.

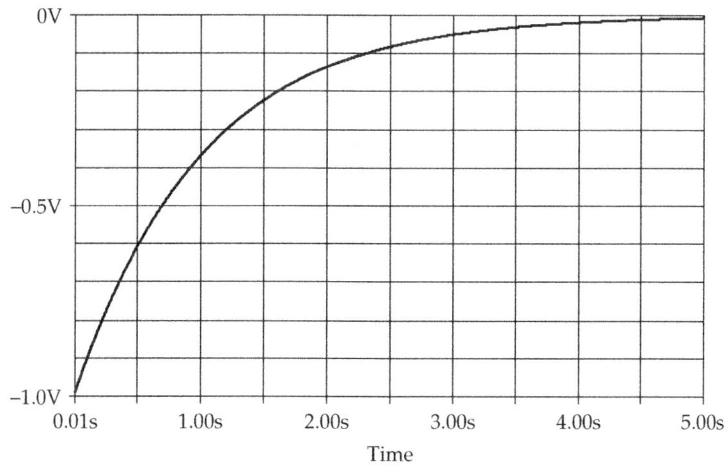

FIGURE 22.22
Figure for Example 22.6.

Exercise 22.7

Derive and interpret the transfer function $V_C(s)/V_{src}(s)$ in the circuit of Figure 22.20.

Ans. $\dfrac{V_C(s)}{V_{src}(s)} = \dfrac{1}{\tau}\dfrac{1}{s+1/\tau}$.

Primal Exercise 22.8

The transfer function of a circuit is $F(s) = 500/(s+100)$. If the steady-state output of the circuit is $v_O(t) = 4 + 10\cos(100t + 45°)$ V, determine the input $v_I(t)$.

Ans. $0.8 - 2\sqrt{2}\sin 100t$ V.

22.4.2 Natural Responses of Second-Order Circuits

Consider the series *RLC* in the *s*-domain, with zero initial conditions (Figure 22.23). The transfer function $I(s)/V_{src}(s)$ is

$$H(s) = \frac{I(s)}{V_{src}(s)} = \frac{1}{R + sL + 1/sC}$$
$$= \frac{1}{L}\frac{s}{s^2 + sR/L + 1/LC} = \frac{1}{L}\frac{s}{s^2 + 2\alpha s + \omega_0^2} \quad (22.87)$$

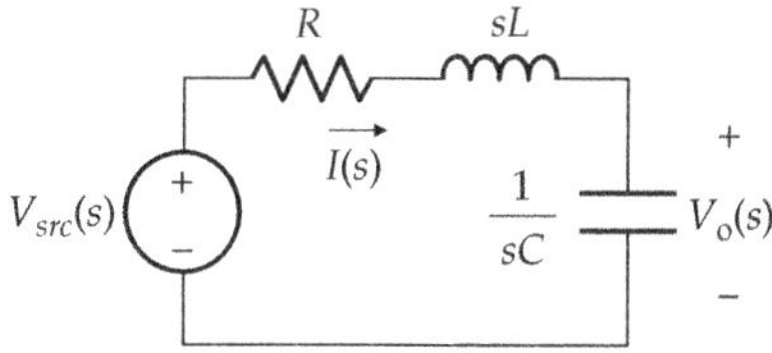

FIGURE 22.23
Series *RLC* circuit in the *s*-domain.

and $\omega_0 = 1/\sqrt{LC}$. The denominator of Equation 22.87 is the LHS of the characteristic equation of the linear differential equation derived earlier (Equation 12.6). The poles are the roots of the characteristic equation $s_1 = -\alpha + \sqrt{\alpha^2 - \omega_0^2}$ and $s_2 = -\alpha - \sqrt{\alpha^2 - \omega_0^2}$ (Equation 12.7), which, as explained in Section 12.1, determine the type of response. Thus, for the overdamped response, the poles s_1 and s_2 are on the negative real axis (Figure 22.24a). As *R* decreases, the poles move closer together on the real axis, until they coalesce into a double pole at $s = -\omega_0$, corresponding to critical damping ($\alpha = \omega_0$ as in Figure 22.24b). With further decrease in *R*, the poles become complex conjugates lying on the semicircle of radius ω_0. This is because the real part of the poles is α and their imaginary part is ω_d, where $\alpha^2 + \omega_d^2 = \omega_0^2$, and ω_0 is the radius of

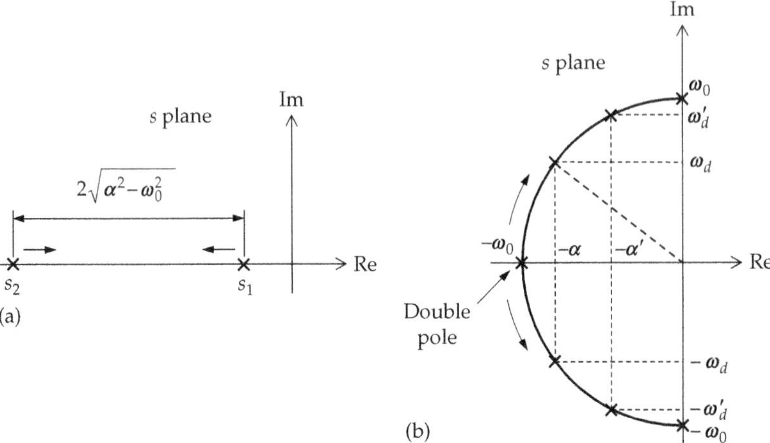

FIGURE 22.24
(a) Location of poles for an overdamped response and (b) location of poles for critically damped and underdamped responses.

a semicircle. The response is now underdamped. As R becomes smaller still, the poles move closer to the imaginary axis. In the limiting case of $R = 0$, $\alpha = 0$, and the poles become purely imaginary at $\pm j\omega_0$.

Concept: *The poles of an overdamped second-order circuit lie on the negative real axis of the s-plane. A critically damped circuit has a double pole on the negative real axis, at a distance ω_0 from the origin. The poles of an underdamped circuit occur in complex conjugate pairs in the left half of the s-plane, on a semicircle of radius ω_0 centered on the origin.*

Example 22.7: Transfer Function of an *RLC* Circuit

Consider the *RLC* circuit shown in Figure 22.25. It is desired to derive and interpret the transfer function $I(s)/V_{src}(s)$. This is the same bandstop circuit analyzed in Example 14.6.

Solution:

The combined impedance of L and C is $sL/(s^2LC+1)$. It follows that $\frac{I(s)}{V_{src}(s)} = \frac{1}{R+sL/(s^2LC+1)}$, which simplifies to

$$H(s) = \frac{I(s)}{V_{src}(s)} = \frac{1}{R}\frac{s^2+\omega_0^2}{s^2+2\alpha_p s+\omega_0^2} \tag{22.88}$$

where

$\alpha_p = 1/2CR$
$\omega_0 = 1/\sqrt{LC}$

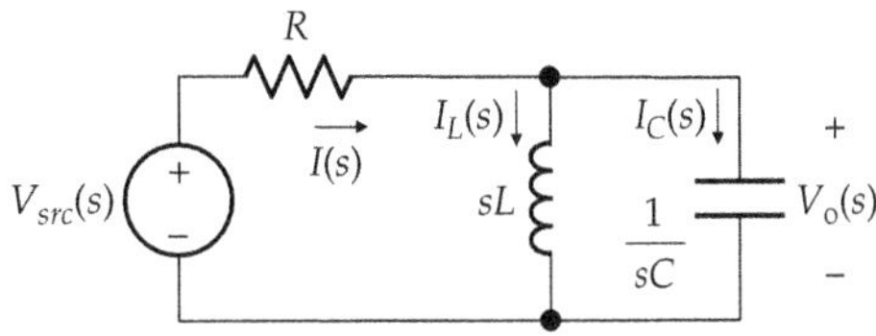

FIGURE 22.25
Figure for Example 22.7.

$H(s)$ has zeros at $s = \pm j\omega_0$ and has the poles of a parallel *GCL* circuit, since the circuit reduces to a parallel circuit when the source is set to zero. The zeros occur when the *LC* branch is in parallel resonance, so that its impedance is infinite, or its admittance is zero. This makes $I(s)$ zero and $V_o(s) = V_{src}(s)$.

If the source is sinusoidal, $v_{SRC}(t) = V_m\cos(\omega t+\theta)$:

$$I(s) = \frac{V_m}{R}\frac{s^2+\omega_0^2}{s^2+2\alpha_p s+\omega_0^2}\left[\frac{s\cos\theta}{s^2+\omega^2} - \frac{\omega\sin\theta}{s^2+\omega^2}\right] \tag{22.89}$$

When $\omega = \omega_0$, $s^2+\omega_0^2$ cancels out from the numerator and denominator, so that

$$I(s) = \frac{V_m}{R}\frac{s\cos\theta-\omega_0\sin\theta}{s^2+2\alpha_p s+\omega_0^2} \tag{22.90}$$

Since the poles of the denominator have negative real parts, it follows from the final-value theorem that $I_{SS}(s) = \lim_{t\to\infty} i(t) = \lim_{s\to 0} sI(s) = 0$, because of parallel resonance, as argued earlier. Applying the initial-value theorem to $I(s)$ in Equation 22.89 or Equation 22.90 gives $i(0^+) = V_m\cos\theta/R$. This is to be expected since $v_{SRC}(0) = V_m\cos\theta$ and C behaves as a short circuit in response to this step input at $t = 0$, irrespective of whether or not ω equals ω_0.

It should be noted that whereas in the steady state $I_{ss}(s) = 0$ when $\omega = \omega_0$, this is not true of $I_L(s)$ and $I_C(s)$. Thus, using Equation 22.90,

$$I_L(s) = \frac{1/sC}{sL+1/sC}I(s) = \frac{V_m}{RLC}\frac{s\cos\theta-\omega_0\sin\theta}{s^2+2\alpha_p s+\omega_0^2}\frac{1}{s^2+\omega_0^2} \tag{22.91}$$

From the initial-value theorem, $i_L(0^+) = \lim_{s\to\infty} sI_L(s) = 0$ since the inductor behaves as an open circuit at $t = 0$. The steady-state value $I_{Lss}(s)$ cannot be obtained by applying the final-value theorem to Equation 22.91 because of the poles $\pm j\omega_0$ on the imaginary axis. However, the only terms of interest in the PFE of $I_L(s)$ are $K/(s - j\omega_0)$ and $K^*/(s + j\omega_0)$, since the other terms vanish as $t \to \infty$. To determine K, we multiply the RHS of Equation 22.91 by $(s - j\omega_0)$ and set $s = j\omega_0$. This gives $K = \frac{V_m}{2\omega_0 L}(\sin\theta - j\cos\theta)$ and $K^* = \frac{V_m}{2\omega_0 L}(\sin\theta + j\cos\theta)$. Substituting and simplifying,

$$I_{Lss}(s) = \frac{V_m}{\omega_0 L}\frac{\omega_0\cos\theta + s\sin\theta}{s^2 + \omega_0^2} \tag{22.92}$$

and

$$i_{LSS}(t) = \frac{V_m}{\omega_0 L}\sin(\omega_0 t + \theta) \tag{22.93}$$

Similarly,

$$I_C(s) = \frac{sL}{sL + 1/sC}I(s) = \frac{V_m}{R}\frac{s^3\cos\theta - s^2\omega_0\sin\theta}{s^2 + 2\alpha_p s + \omega_0^2}\frac{1}{s^2 + \omega_0^2} \tag{22.94}$$

From the initial-value theorem, $i_C(0^+) = \lim_{s\to\infty} sI_C(s) = V_m\cos\theta/R$, since C behaves as a short circuit at $t = 0^+$. To determine $i_{CSS}(t)$, we proceed in the same manner as I_{LSS} by considering the terms $K'/(s - j\omega_0)$ and $K'^*/(s + j\omega_0)$. Multiplying the RHS of Equation 22.94 by $(s - j\omega_0)$ and setting $s = j\omega_0$ gives $K' = -\frac{CV_m}{2}(\sin\theta - j\cos\theta)$ and $K'^* = -\frac{CV_m}{2}(\sin\theta + j\cos\theta)$. Substituting and simplifying,

$$I_{Css}(s) = -\omega_0 CV_m\frac{\omega_0\cos\theta + s\sin\theta}{s^2 + \omega_0^2} \tag{22.95}$$

and

$$i_{CSS}(t) = -\omega_0 CV_m\sin(\omega_0 t + \theta) \tag{22.96}$$

To check the values of $i_{LSS}(t)$ and $i_{CSS}(t)$, we note that at $\omega = \omega_0$, $i_{SS} = 0$, and $v_O(t) = V_m\cos(\omega_0 t + \theta)$. It follows that $i_{LSS}(t) = \frac{1}{L}\int v_O(t)dt = \frac{V_m}{\omega_0 L}\sin(\omega_0 t + \theta)$, ignoring the constant of integration in the steady state, which agrees with Equation 22.93. It also follows that $i_{CSS}(t) = Cdv_O(t)/dt = -\omega_0 CV_m\sin(\omega_0 t + \theta)$, in agreement with Equation 22.96. Moreover, $i_{LSS}(t) + i_{CSS}(t) = 0$. In other words, the current $\omega_0 CV m\sin(\omega_0 t + \theta)$ circulates in the mesh composed of the inductor and capacitor.

Exercise 22.9

Show that when $\omega \neq \omega_0$, the steady-state sinusoidal response $I_{SS}(s)$ in Example 22.7 is the same as that obtained from phasor analysis.

Ans. $H(j\omega) = \frac{1}{R}\frac{\omega_0^2 - \omega^2}{\omega_0^2 - \omega^2 + j\omega/CR}$.

Exercise 22.10

Derive $I(s)$, $I_L(s)$, and $I_C(s)$ in Example 22.7 when v_{SRC} is a unit impulse. Interpret the result in terms of the initial- and final-value theorems.

Ans. $I(s) = \frac{1}{R}\left[1 - \frac{2\alpha_p s}{s^2 + 2\alpha_p s + \omega_0^2}\right]$; $I_L(s) = \frac{\omega_0^2}{R}\frac{1}{s^2 + 2\alpha_p s + \omega_0^2}$; $I_C(s) = \frac{1}{R}\left[1 - \frac{2\alpha_p s + \omega_0^2}{s^2 + 2\alpha_p s + \omega_0^2}\right]$.

Learning Checklist: What Should Be Learned from This Chapter

- In the *s*-domain, the impedance of a resistor is R, that of an uncharged capacitor is $1/sC$, and that of an uncharged inductor is sL. To account for initial energy storage, while considering the values of all responses to be zero at $t = 0^-$, in accordance with the single-sided LT, step or impulse, ideal, independent sources are added in series or in parallel with the ideal energy storage element so as to provide the required initial value at $t = 0^+$.
- The sources that are added to account for initial conditions in the energy storage elements in the *s*-domain become an integral part of the representation of the energy storage element in the *s*-domain.
- The *s*-domain representation of the linear transformer is based on the T-equivalent circuit. The ideal voltage source representation of the initial current in an inductor is applied to the inductors in the series and shunt branches of the T-equivalent circuit.
- All circuit laws and techniques, discussed previously for the frequency domain using phasor notation, apply equally well to the *s*-domain. These include series and parallel combinations of *s*-domain impedances and admittances; node-voltage, mesh-current, and loop-current methods of analysis; TEC and NEC; and superposition and Y-Δ transformation.

1. In the *s*-domain, passive circuit elements are represented by their *s*-domain impedances,

with initial values of voltages across capacitors and currents though inductors accounted for by appropriate sources. Independent and dependent voltage or current sources are expressed in terms of their LTs.

2. The circuit is then analyzed by any of the techniques described previously for phasor analysis, and the LT of the desired circuit response is derived by solving algebraic equations for the LT of the desired responses.
3. The ILT of the desired response gives the complete response in the time domain, both transient and steady state, to any arbitrary excitation that has an LT.

- In switching problems that involve impulsive readjustments at the instant of switching, initial values in energy storage elements just before switching are used in the *s*-domain representation of the circuit after switching. The initial-value theorem gives circuit responses just after switching.
- When a single excitation is applied to a given circuit having no initial energy storage, the transfer function $H(s)$ is the ratio of the LT $Y(s)$ of a designated response to the LT $X(s)$ of the applied excitation.
- The LT of the response to a unit impulse of excitation is the transfer function, and the ILT of the transfer function is the response in the time domain to a unit impulse of excitation.
- The poles of the transfer function are the roots of the characteristic.
- A circuit is stable if the response to a unit impulse tends to zero as $t \to \infty$. The circuit is unstable if the response to a unit impulse increases without limit, that is, is unbounded as $t \to \infty$. The circuit is marginally stable, or metastable, if as $t \to \infty$, the response to a unit impulse does not approach zero but remains bounded.
- If all the poles of the transfer function lie in the open left half of the *s*-plane, the circuit is stable. If at least one pole lies in the open right half of the *s*-plane, the circuit is unstable. If all the poles of the transfer function are simple poles on the imaginary axis, the circuit is metastable. If at least one pole on the imaginary axis is a multiple pole, the circuit is unstable.
- If a metastable circuit is excited at the frequency of its poles on the imaginary axis, the response in unbounded.
- The sinusoidal steady-state response is obtained from a transfer function that is a proper rational function by substituting $s = j\omega$, multiplying the magnitude of the excitation by $|H(j\omega)|$ and adding the phase angle of $H(j\omega)$ to that of the excitation. As ω is varied, *s* moves along the imaginary axis.
- All the responses of a stable first-order circuit in the *s*-domain are characterized by a pole on the negative real axis whose magnitude is the reciprocal of the time constant.
- The poles of an overdamped second-order circuit lie on the negative real axis of the *s*-plane. A critically damped circuit has a double pole on the real axis. The poles of an underdamped circuit occur in complex conjugate pairs at a distance ω_0 from the origin.

Problem-Solving Tips

1. Always include the sources that account for initial conditions in energy storage elements as an integral part of these elements when analyzing the circuit.
2. In delayed switching, the LT implicitly assumes a time origin at $t = 0$.

Problems

Verify solutions by PSpice simulation.

Transfer Function

P22.1 If the impulse response of a given circuit is $e^{-2t}u(t)$, determine the input that will produce an output $\delta(t)$.

Ans. $\delta^{(1)}(t)+2\delta(t)$.

P22.2 When an input $\delta(t - 1) - \delta(t - 2)$ is applied to an LTI circuit, the output is $\delta(t - 3)$. Determine the impulse response, $h(t)$, of the circuit.

Ans. $\sum_{n=2}^{\infty}\delta(t-n)$.

P22.3 Determine *L* and *C* in Figure P22.3 given that the transfer function $V_2(s)/V_1(s)$ has a pole at $-100 + j700$ rad/s.

Ans. 2.5 H, 0.8 μF.

P22.4 Determine the impulse response $h(t)$ if the transfer function is $H(s)=\dfrac{s^2-4s-4}{s^4+8s^2+16}$.

Ans. $t\cos 2tu(t) - t\sin 2tu(t)$.

P22.5 Determine the steady-state sinusoidal response corresponding to the transfer function of the preceding problem when the excitation frequency is 4 rad/s.

Ans. Response in unbounded.

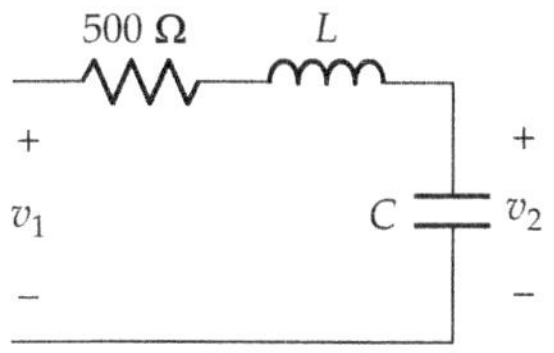

FIGURE P22.3

P22.6 The current in a circuit is governed by the differential equation $\frac{d^2i}{dt^2} + 2\frac{di}{dt} + i = 0$, with $i(0^+) = 1$ and $di/dt = 0$ at $t = 0^+$. Determine $i(t)$.

Ans. $e^{-t}(1+t)$ A.

P22.7 Two identical circuits are cascaded, without the second circuit loading the first. If the overall transfer function of the cascade is $4/(s+1)^2$, determine the impulse response of each circuit.

Ans. $\pm 2e^{-t}$.

P22.8 If an input $u(t)$ is applied to the first circuit of the preceding problem, determine the output of the second circuit as function of time.

Ans. $-4te^{-t} - 4e^{-t} + 4$.

P22.9 Two circuits are cascaded, without the second circuit loading the first. If the impulse response of the first circuit is $2\delta^{(1)}(t)$, and the impulse response of the second circuit is $e^{-t}\sin 2t$, determine the impulse response of the cascade.

Ans. $2e^{-t}(2\cos 2t - \sin 2t)$.

P22.10 Determine the impulse response of the cascaded circuits in Figure P22.10, assuming $RC = 1$.

Ans. $te^{-t}u(t)$.

P22.11 Given $\frac{V_o(s)}{V_i(s)} = \frac{10/s}{0.4s + 4 + 10/s}$ and $v_i(t) = 10\big(u(t) - u(t-0.1)\big)$, determine $v_O(t)$ for $t \geq 0.1$.

Ans. $-10e^{-5t}(5t+1) + 10e^{-5(t-1)}(5t+0.5)$.

P22.12 The response of a circuit to an input $v_I(t) = -\delta(t) + e^{-t}\delta^{(2)}(t)$ is $v_O(t) = 2\big(\delta(t) - \delta(t-1)\big)$. Determine the impulse response of the circuit.

Ans. $u(t) - e^{-2t}u(t) - u(t-1) + e^{-2(t-1)}u(t-1)$.

P22.13 The impulse response of a circuit is $u(t)$. Determine the steady-state response to the sinusoidal input $5\cos 2t$.

Ans. $(5/2)\sin 2t$.

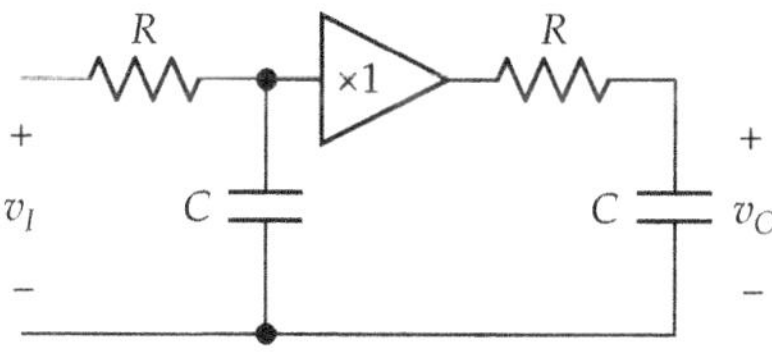

FIGURE P22.10

P22.14 The transfer function of a given circuit is $(s+3)/(s^2+2s+5)$. Determine (a) the response of the circuit in the time domain to a step of $5u(t)$ and (b) the sinusoidal steady-state response of the circuit to an excitation $2\cos t$.

Ans. (a) $(3 - 3e^{-t}\cos 2t + e^{-t}\sin 2t)u(t)$; (b) $\sqrt{2}\cos(t - 8.13^\circ)$.

P22.15 A source of unknown voltage is applied to the circuit shown in Figure P22.15. Determine $i(t)$ given that $i_C(t) = (1 + t)u(t)$ A.

Ans. $(1 + t + t^2/2)u(t)$ A.

P22.16 An unknown voltage $v_{SRC}(t)$ is applied to the circuit of Figure P22.16. If $v_C(t) = 5e^{-4t}$, determine $i_{SRC}(t)$ and $v_{SRC}(t)$.

Ans. $i_{SRC}(t) = 2.5\delta(t) - 5e^{-4t}$; $v_{SRC}(t) = 2.5\delta(t)$.

P22.17 An unknown voltage $v_{SRC}(t)$ is applied to the circuit of Figure P22.17. If $v_O(t) = 2 - 3e^{-t}$, determine $i_C(t)$.

Ans. $i_C(t) = 2\sqrt{2}\sin t/\sqrt{2} - \cos t/\sqrt{2}$ A.

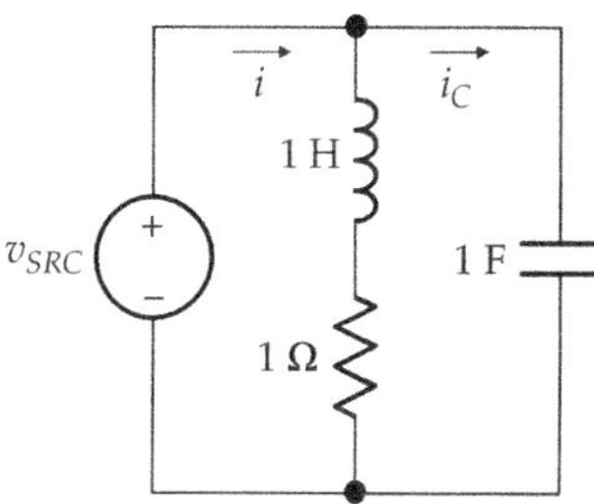

FIGURE P22.15

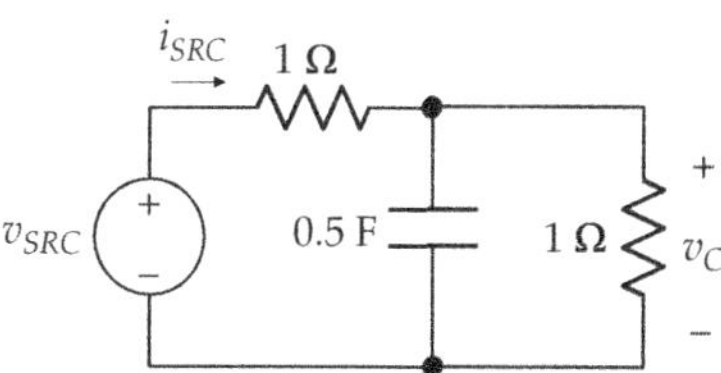

FIGURE P22.16

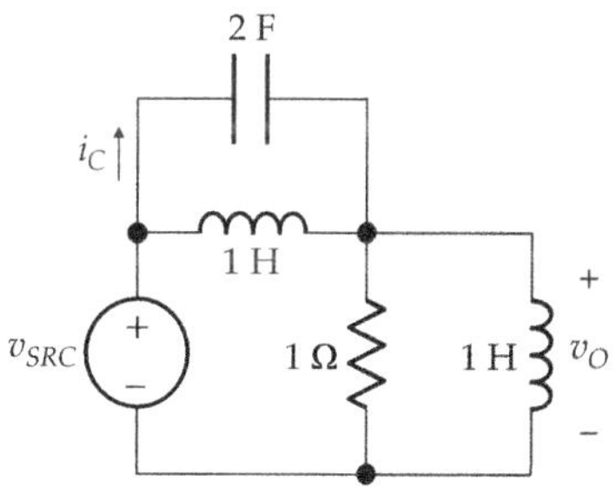

FIGURE P22.17

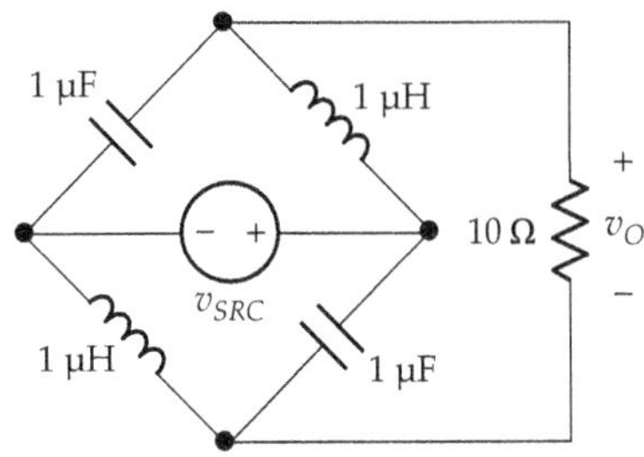

FIGURE P22.18

P22.18 Derive the transfer function $V_o(s)/V_{src}(s)$ in Figure P22.18.

Ans. $\dfrac{1-(s')^2}{(s')^2+0.2s'+1}$, s' is in Mrad/s.

P22.19 Determine the transfer function $V_o(s)/I_i(s)$ in Figure P22.19 assuming $Z_1 = 1+s$, $Z_2 = 1+1/s$, $Y_1 = \dfrac{s}{2+s}$, and $Y_2 = 3s$.

Ans. $\dfrac{1}{6s+7}$.

P22.20 Determine the transfer function $V_o(s)/I_i(s)$ in Figure P22.20.

Ans. $\dfrac{2s(s+1)}{2s^2+2s+1}$.

P22.21 Determine the transfer function $I_o(s)/V_{src}(s)$ in Figure P22.21.

Ans. $\dfrac{1}{5(s+5)}$.

P22.22 Determine the transfer function $I_o(s)/I_{src}(s)$ in Figure P22.22.

Ans. $\dfrac{s(s+7)}{6s^2+17s+20}$.

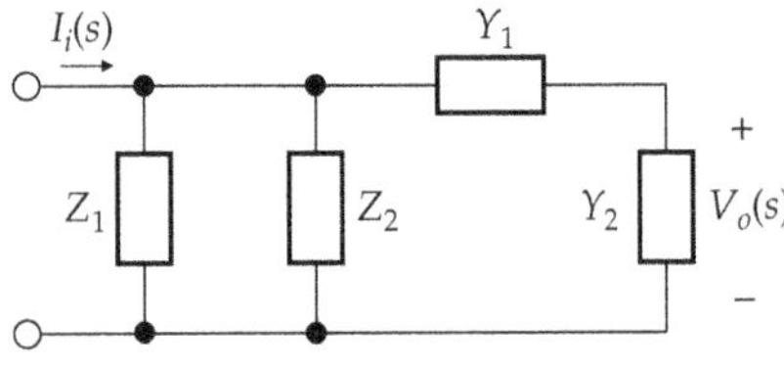

FIGURE P22.19

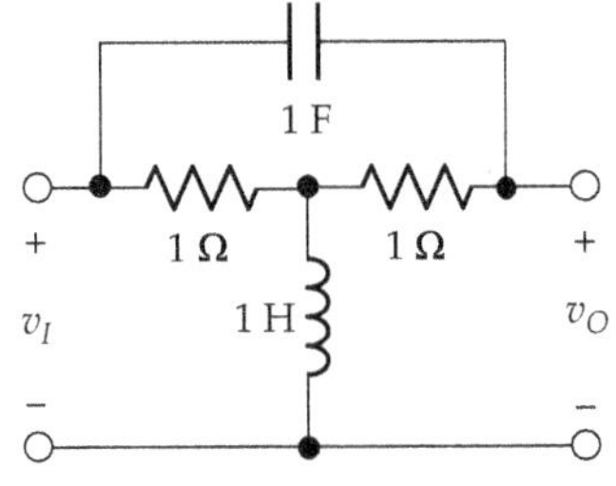

FIGURE P22.20

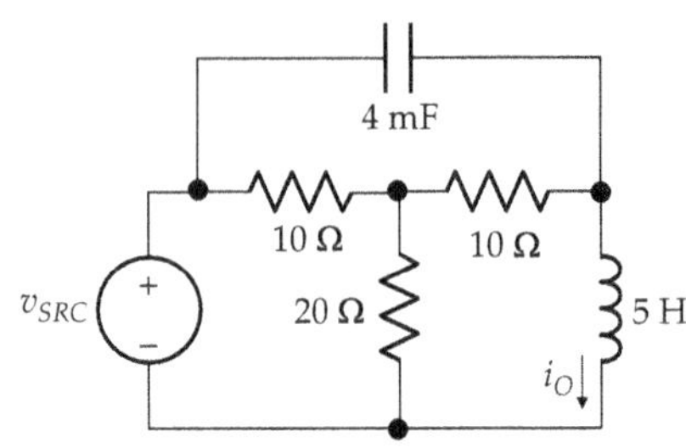

FIGURE P22.21

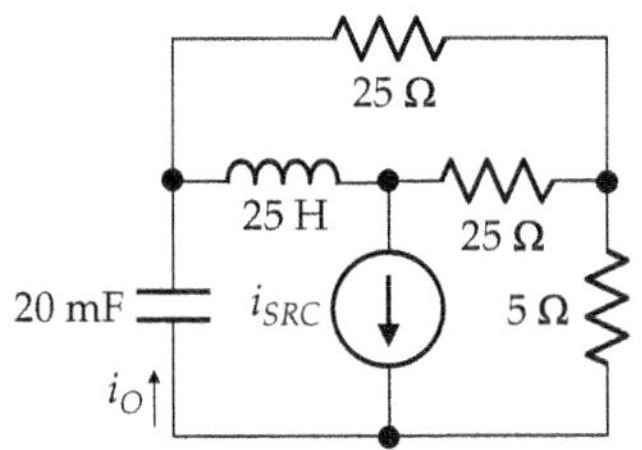

FIGURE P22.22

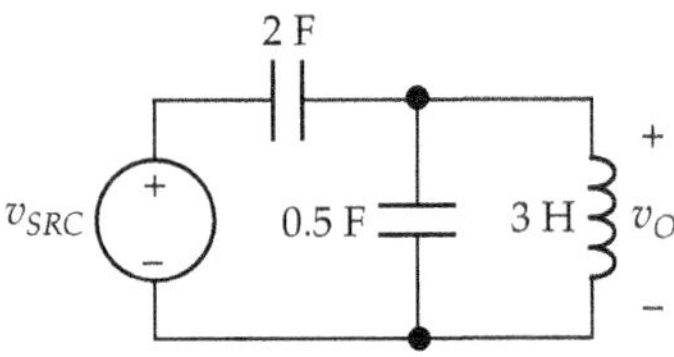

FIGURE P22.23

P22.23 Determine the poles and zeros of the transfer function $V_o(s)/V_{src}(s)$ in Figure P22.23.

Ans. Double zero at the origin, poles at $\pm j0.37$ rad/s.

P22.24 Determine $V_L(s)$ in Figure P22.24, assuming that the voltage sources are applied at t = 0, with zero initial energy storage in the capacitors.

Ans. $\dfrac{2s}{2s^2+2s+1}$.

P22.25 The transfer function of an LTI circuit is $\dfrac{10}{1+s/10^3}$.

Determine the steady-state response when the input is $(1 + \cos 1000t)$ V.

Ans. $10+\dfrac{10}{\sqrt{2}}\cos\left(1000t-\dfrac{\pi}{4}\right)$ V.

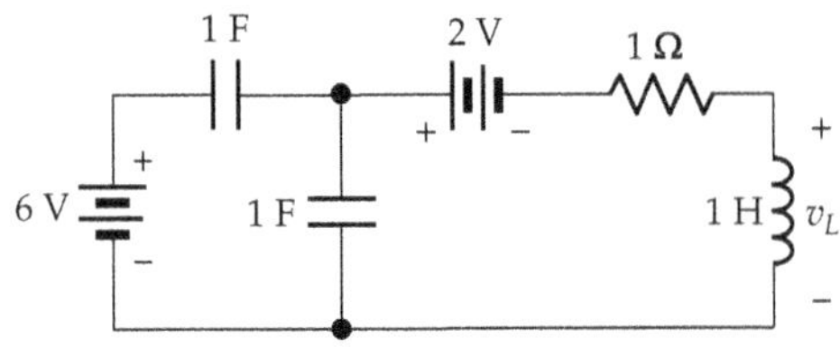

FIGURE P22.24

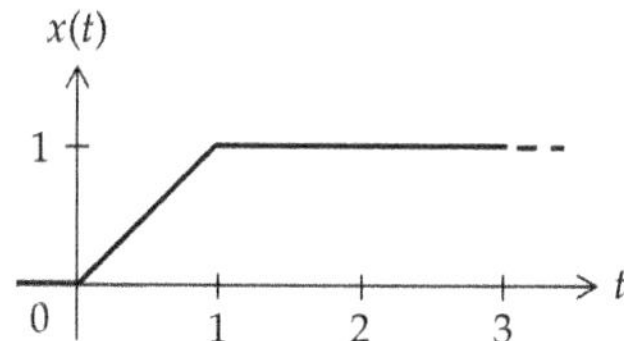

FIGURE P22.26

P22.26 The step response of an LTI circuit is $\left(1-e^{-t}\right)u(t)$. Determine (a) the response to the input $x(t)$ shown in Figure P22.26 and (b) 10sin$t$.

Ans. (a) $\left(t-1+e^{-t}\right)u(t)-\left(t-2+e^{-(t-1)}\right)u(t-1)$

(b) $\frac{10}{\sqrt{2}}\sin\left(t-\frac{\pi}{4}\right)$.

P22.27 The response of a given circuit to a unit voltage impulse is $\left(e^{t}-e^{-t}\right)'$ V. Determine the response to a sawtooth pulse defined by $v(t) = t$ V, $0 \le t \le 1$ s, and $v(t) = 0$, for $t < 0$, and $t > 1$ s.

Ans. $\left(e^{t}-e^{-t}-2t\right)u(t)-\left(e^{t-1}-e^{-t+1}-2t+2\right)u(t-1)$ V; $\left(e^{t}-e^{-t}-2t\right)u(t)-2\left(e^{(t-1)}-2t\right)u(t-1)$ V.

P22.28 When an input $240u(t)$ is applied to a circuit, the response is $100e^{-800t}u(t)$. Determine the steady-state response to a sinusoidal input 5cos800t.

Ans. 0.295cos(800t + 45°).

First-Order Circuits

P22.29 The switch in Figure P22.29 is opened at $t = 0$, after being closed for a long time. Determine $v_O(t)$ and the pole location of $V_o(s)$.

Ans. $v_O(t) = 5e^{-t}$ V where t is in ms. The pole is located at $s = -1$ krad/s.

P22.30 The switch in Figure P22.30 is closed at $t = 0$, after being open for a long time. Determine $I(s)$.

Ans. $\frac{5}{6}\frac{43s+52}{s(11s+14)}$.

P22.31 The switch in Figure P22.31 is moved to position 'b' at $t = 0$, after being in position 'a' for a long time. Determine $I_s(s)$.

Ans. $10+\frac{10}{5s+2}$.

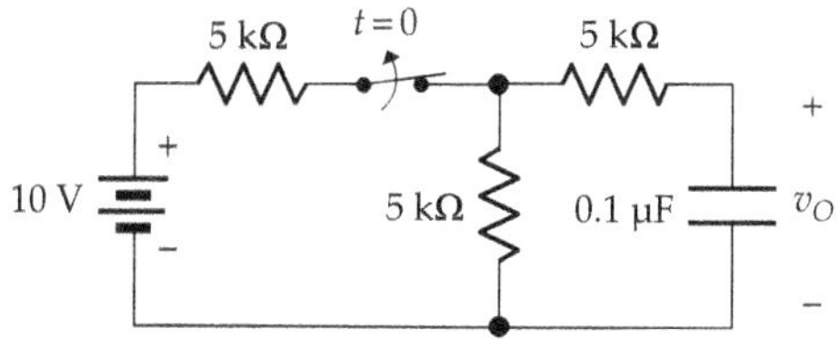

FIGURE P22.29

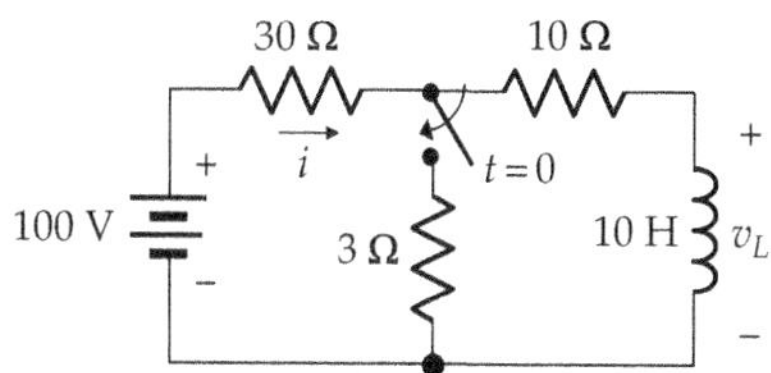

FIGURE P22.30

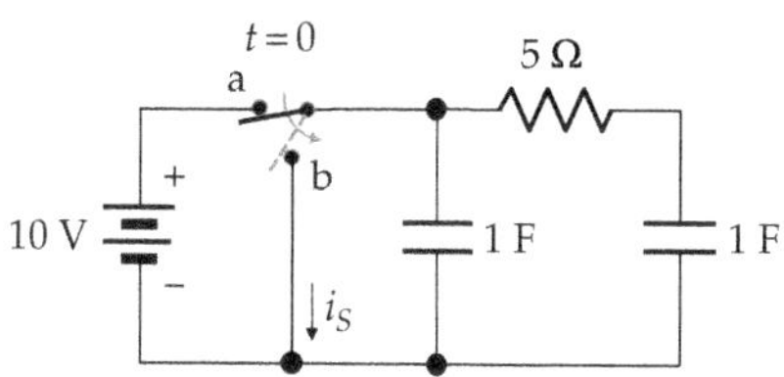

FIGURE P22.31

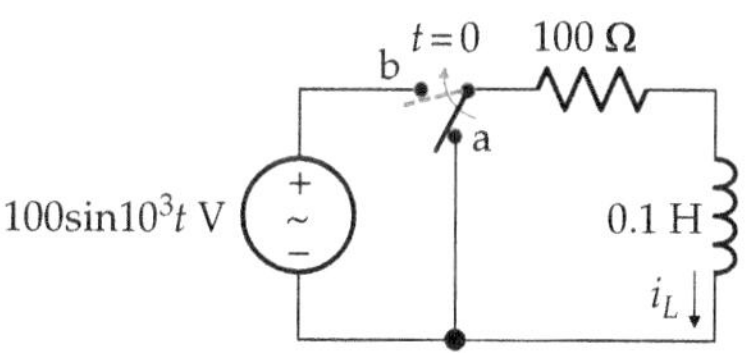

FIGURE P22.32

P22.32 The switch in Figure P22.32 is moved to position 'b' at $t = 0$, after being in position 'a' for a long time. Determine $I_L(s)$, given $i_L(0^+) = 0.5$ A.

Ans. $\frac{0.5s^2+1.5\times10^6}{\left(s^2+10^6\right)\left(s+10^3\right)}$.

P22.33 Both switches in Figure P22.33 are opened at $t = 0$ after being closed for a long time. Determine $i_O(t)$.

Ans. $10e^{-5t}$ mA, t is in ms.

P22.34 The switch in Figure P22.34 is moved to position 'b' at $t = 0$, after being in position 'a' for a long time. Determine $i_O(t)$ and the pole location of $I_o(s)$.

Ans. $i_O(t)=(10/3)e^{-t/90}$ mA, t is in μs. The pole of $I_o(s)$ is at −1/90 Mrad/s.

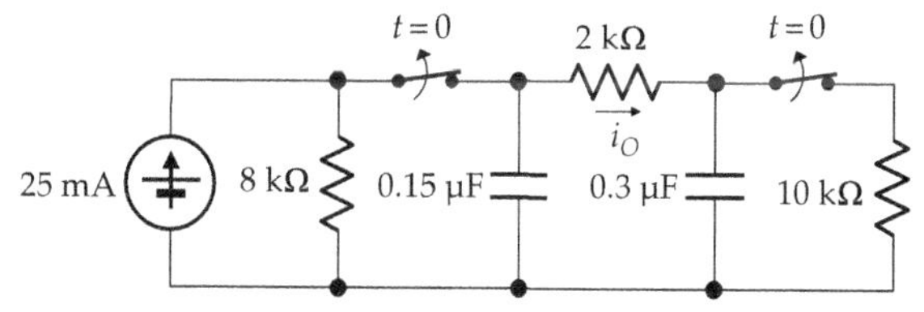

FIGURE P22.33

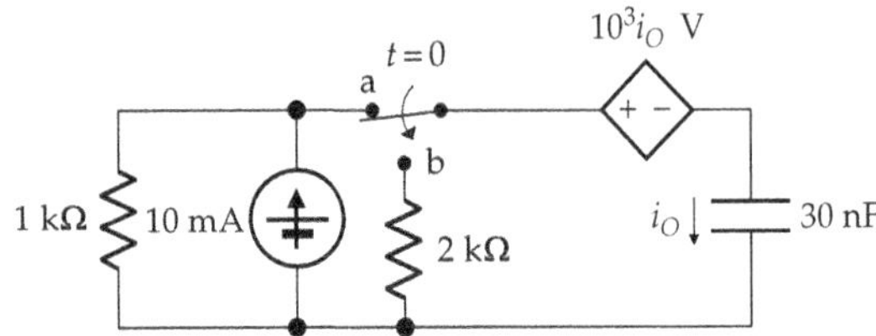

FIGURE P22.34

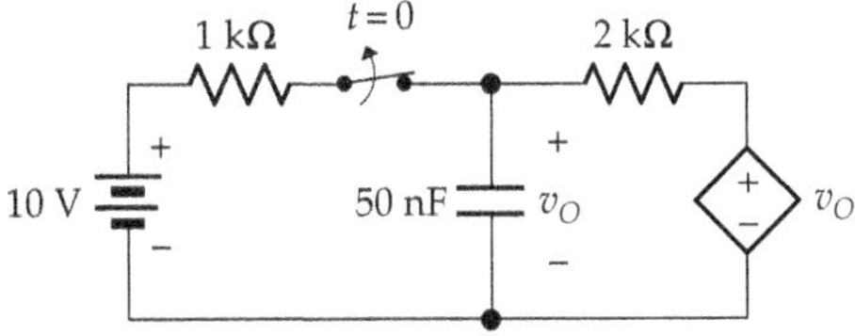

FIGURE P22.35

P22.35 The switch in Figure P22.35 is opened at t = 0, after being closed for a long time. Determine $v_O(t)$ and the pole location of $V_o(s)$.

Ans. $v_O(t)$ = 10 V. The pole of V_o is at the origin.

P22.36 The switch in Figure P22.36, is moved to position 'b' at t = 0, after being in position 'a' for a long time. Determine $v_O(t)$ and the pole location of $V_o(s)$.

Ans. $v_O(t) = 0.5e^{-t/3}$ V, where t is in ms. The pole of $V_o(s)$ is at −1000/3 rad/s.

P22.37 The switch in Figure P22.37 is moved to position 'b' at t = 0, after being in position 'a' for a long time Determine $v_O(t)$ and the pole location of $V_o(s)$.

Ans. $v_O(t) = 900/19 - (710/19)e^{-38t/3}$ V, where t is in ms. The poles of $V_o(s)$ are at zero and 38/3 krad/s.

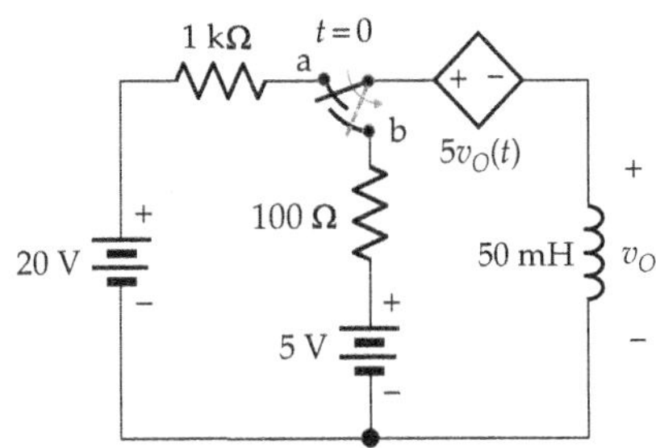

FIGURE P22.36

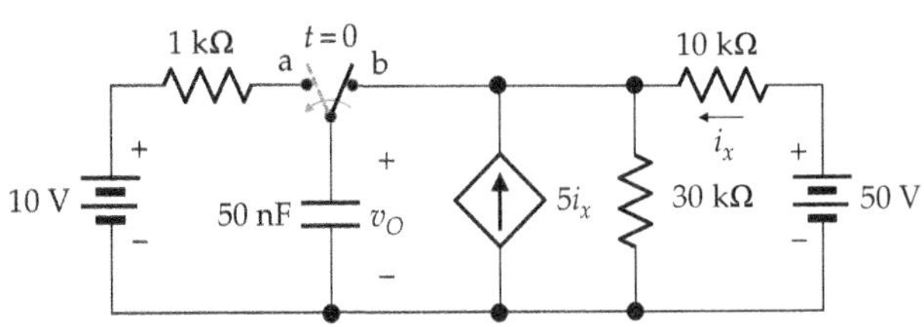

FIGURE P22.37

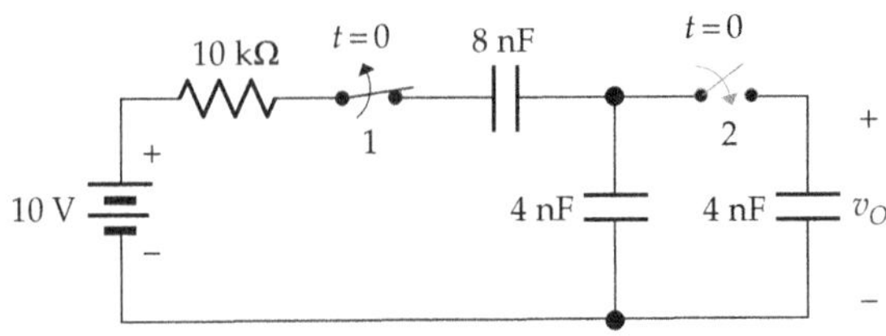

FIGURE P22.38

P22.38 Switch 1 in Figure P22.38 has been closed for a long time, with the capacitors initially uncharged when the switch was first closed. At t = 0, switch 1 is opened and switch 2 is closed. Determine $v_O(t)$ and the pole location of $V_o(s)$.

Ans. $v_O(t)$ = 10/3 V. The pole of $V_o(s)$ is at the origin.

P22.39 The switch in Figure P22.39 is closed at $t = 0^-$, just before the impulse is applied, with i_2 = 0. Determine $I_1(s)$ and $I_2(s)$.

Ans. $I_1(s) = \dfrac{2s+1}{s(s+1)}$, $I_2(s) = \dfrac{1}{2s}$.

P22.40 The switch in Figure P22.40 is opened at t = 0 after being closed for a long time. Determine $I(s)$, $V_{L1}(s)$, $V_{L2}(s)$, and the corresponding ILTs.

Ans. $I(s)=3\dfrac{s+2}{s(s+1)}$ A, $V_{L1}(s)=-\dfrac{6s}{s+1}$ V, $V_{L2}(s)=6\dfrac{s+2}{s+1}$ V,

$i(t)=\left(6-3e^{-t}\right)u(t)$ A, $v_{L1}(t)=-6\delta(t)+6e^{-t}u(t)$ V,

$v_{L2}(t)=6\delta(t)+6e^{-t}u(t)$ V.

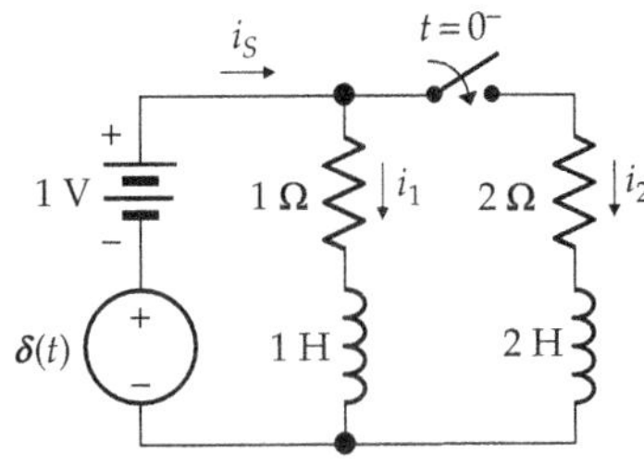

FIGURE P22.39

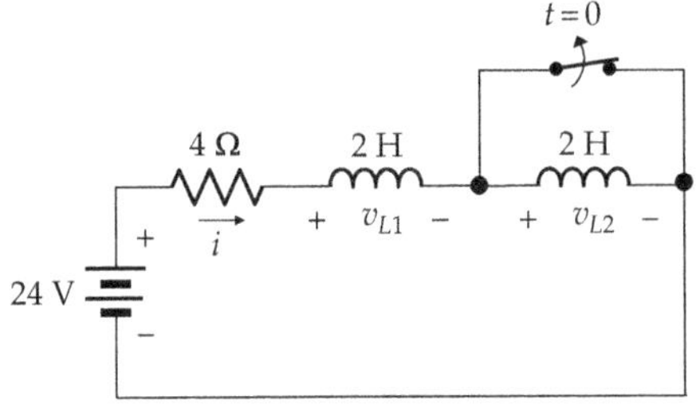

FIGURE P22.40

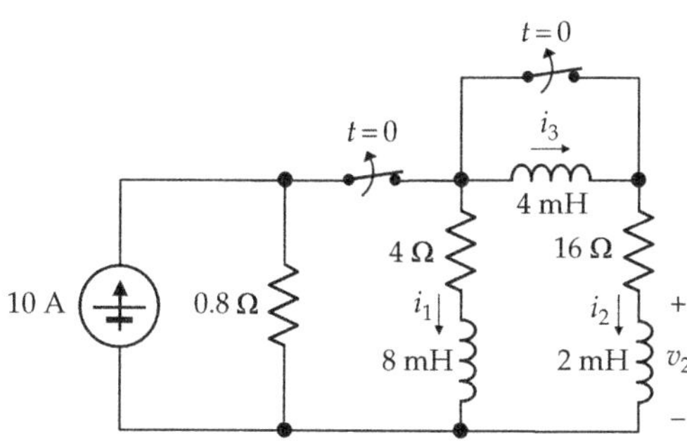

FIGURE P22.41

P22.41 Both switches in Figure P22.41 are opened at $t = 0$ after being closed for a long time. Determine $I_2(s)$, $V_2(s)$, and the corresponding ILTs.

Ans. $I_2(s) = -6/(7s+10)$, $V_2(s) = -0.8-12s/(7s+10)$; $i_2(t) = -(6/7)e^{-10t/7}$ A; $v_2(t) = -(88/35)\delta(t) - (120/49)e^{-10t/7}$ V.

P22.42 The switch in Figure P22.42 is opened at $t = 0$ after being closed for a long time, with $i_2(0^-) = 1$ A. Determine $I_2(s)$, $V_1(s)$, and the corresponding ILTs.

Ans. $I_2(s) = -0.6/(s+0.5)$ A, $V_1(s) = -11.2-2.4/(s+0.5)$ V; $i_2(t) = -0.6e^{-0.5t}$ A, $v_1(t) = -11.2\delta(t) - 2.4e^{-0.5t}$ V.

P22.43 The switch in Figure P22.43 is opened at $t = 0$ after being closed for a long time. Determine $i(t)$.

Ans. $2u(t)$ A.

P22.44 $i_{SRC}(t) = e^{-t/\tau}$ A in Figure P22.44, where $\tau = RC$, with zero initial stored energy. Determine $v(t)$.

Ans. $v(t) = \dfrac{1}{C}te^{-t/\tau}$ V.

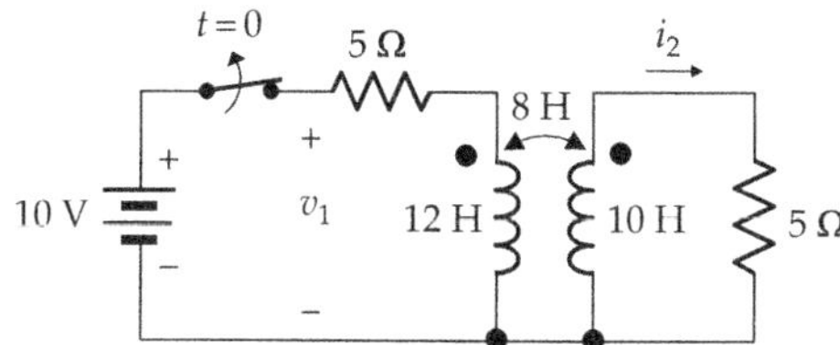

FIGURE P22.42

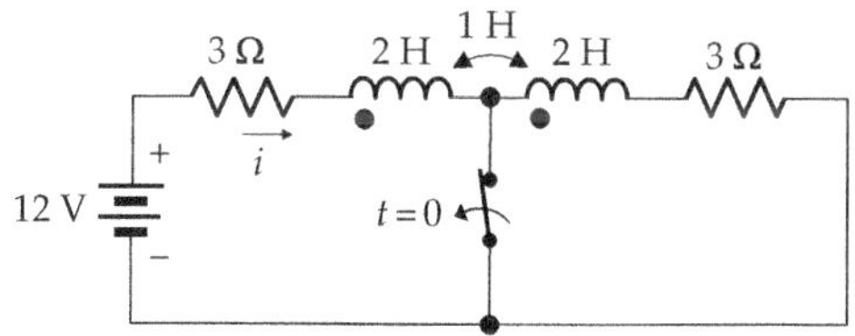

FIGURE P22.43

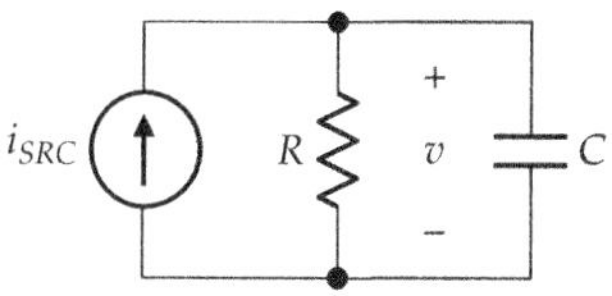

FIGURE P22.44

Second-Order Circuits

P22.45 Determine (a) the response v_O of the circuit in Figure P22.45 to $v_I = \delta(t)$ and (b) the steady-state response to a sinusoid having a unit amplitude and a frequency $\omega = 10^6$ rad/s.

Ans. (a) $\dfrac{10^6}{\sqrt{5}}\left[e^{-\left(\frac{3-\sqrt{5}}{2}\right)t} - e^{-\left(\frac{3+\sqrt{5}}{2}\right)t}\right]$ V, t is in μs;

(b) $|H(j\omega)| = 1/3$, and $\angle H(j\omega) = -90°$.

P22.46 Determine the frequency of the excitation $v_{SRC}(t) = V_m\cos\omega t$ at which the circuit of Figure P22.46 will have an unbounded response. Consider any convenient voltage as a typical response.

Ans. 1.265×10^5 rad/s.

P22.47 Show that the response $v_O(t)$ to the input $\delta(t)$ in Figure P22.47 is unbounded.

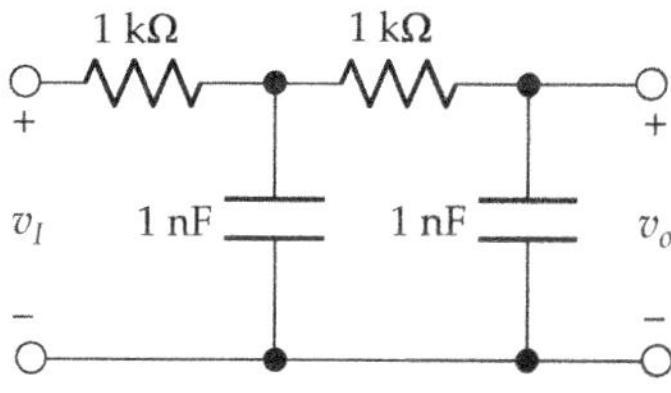

FIGURE P22.45

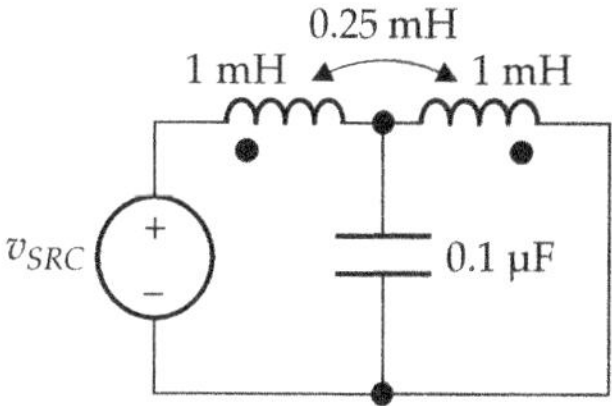

FIGURE P22.46

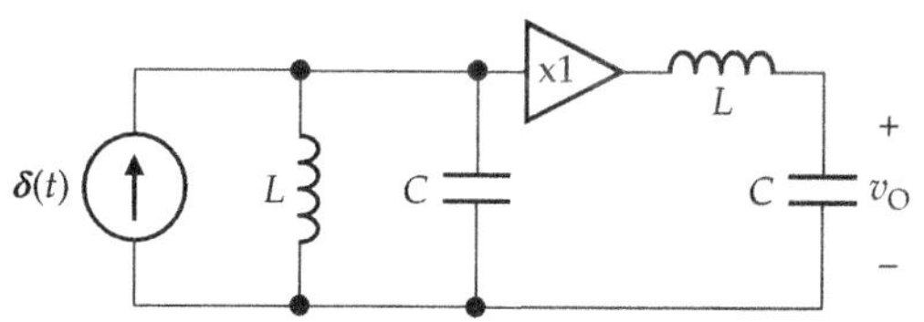

FIGURE P22.47

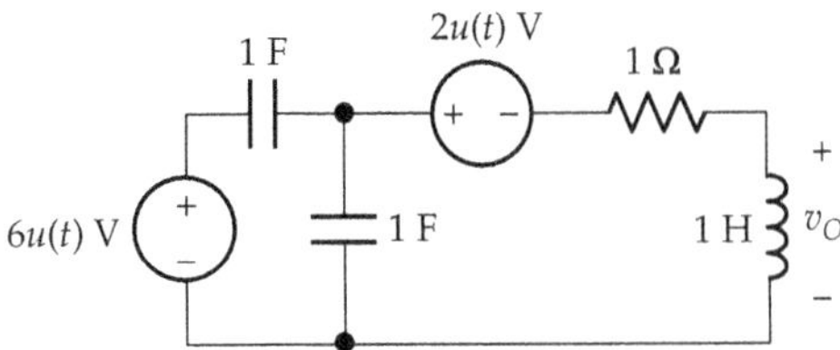

FIGURE P22.48

P22.48 Determine $V_o(s)$ and $v_O(t)$ for the circuit in Figure P22.48, assuming zero initial conditions.

Ans. $V_o(s)=s/(s^2+s+1/2)$, $v_O(t)=\sqrt{2}e^{-t/2}(\cos t/2+45°)$ V.

P22.49 Determine $v_O(t)$ for the circuit in Figure P22.49 if $v_{SRC}(t)=(6+3t)u(t)$, assuming zero initial conditions.

Ans. $v_O(t)=(1.5+4.5e^{-2t})u(t)$ V.

P22.50 (a) Determine the transfer function $V_o(s)/V_{src}(s)$ for the circuit in Figure P22.50, assuming zero initial conditions; (b) verify the initial-value and final-value theorems for unit impulse and step inputs; (c) derive the magnitude and phase shift of the transfer function under sinusoidal steady-state conditions.

Ans. (a) $H(s)=\dfrac{2s^2}{3s^2+11s+5}$;

(c) $|H(j\omega)|=\dfrac{2\omega^2}{\sqrt{(5-3\omega^2)^2+121\omega^2}}$, $\phi=180°-\tan^{-1}\left(\dfrac{11\omega}{5-3\omega^2}\right)$.

P22.51 Determine $V_o(s)$ in Figure P22.50 if $v_{SRC}(t)=te^{-t}u(t)$, assuming zero initial conditions. Use MATLAB to derive $v_O(t)$.

Ans. $V_o(s)=\dfrac{2s^2}{(s+1)^2(3s^2+11s+5)}$; $v_O(t)=\dfrac{2e^{-t}}{3}\left(\dfrac{1}{3}-t\right)-$ $0.552e^{-3.14t}+0.339e^{-0.532t}$ V

P22.52 Determine $V_o(s)$ in Figure P22.50 if $v_{SRC}(t)$ is a single rectangular pulse of 1 V amplitude and 1 s duration, assuming zero initial conditions. Use MATLAB to derive $v_O(t)$.

Ans. $V_o(s)=\dfrac{2s}{3s^2+11s+5}(1-e^{-s})$; $v_O(t)=v_{O1}(t)u(t)-v_{O1}(t-1)$ $u(t-1)$ V, where $v_{O1}(t)=0.803e^{-3.14t}-(-0.031e^{-0.532t})$ V.

P22.53 Determine $V_o(s)$ in Figure P22.50 if $v_{SRC}(t)$ is a single half-sinusoid of 1 V amplitude and $\omega=1$ rad/s, assuming zero initial conditions.

Ans. $V_o(s)=\dfrac{2s^2}{(s^2+1)(3s^2+11s+5)}(1+e^{-\pi s})$.

P22.54 Determine $V_o(s)$ in the circuit of Figure P22.54 if $v_{SRC}(t)=te^{-t}u(t)$, assuming zero initial conditions. Verify the initial-value theorem, the final-value theorem, and derive the steady-state sinusoidal response when $v_{SRC}(t)=\cos t$ V.

Ans. $V_o(s)=\dfrac{2s^2+s+2}{(s+1)(2s^2+2s+3)}$; $\dfrac{1}{\sqrt{5}}\cos(t+\phi)$ V,

$\phi=90°-\tan^{-1}(2)$.

P22.55 (a) Determine v_O in Figure P22.55 if $i_{SRC}(t)$ is a single rectangular pulse of 1 A amplitude and 1 s duration, (b) verify the initial-value theorem and the final-value theorem, and (c) derive the steady-state sinusoidal response when $i_{SRC}(t)=3\sin 2t$ A.

Ans. (a) $v_O(t)=\dfrac{1}{2}(1-e^{-t}(\cos t+\sin t))u(t)-$ $\dfrac{1}{2}(1-e^{-(t-1)}(\cos(t-1)+\sin(t-1)))u(t-1)$V;

(c) $\dfrac{3}{2\sqrt{5}}\sin(2t-116.6°)$V.

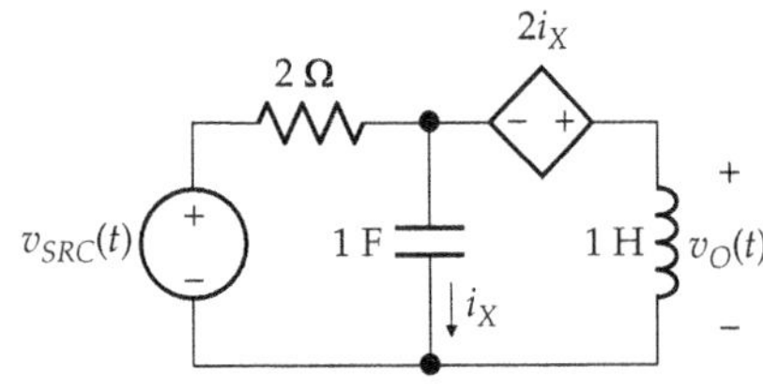

FIGURE P22.49

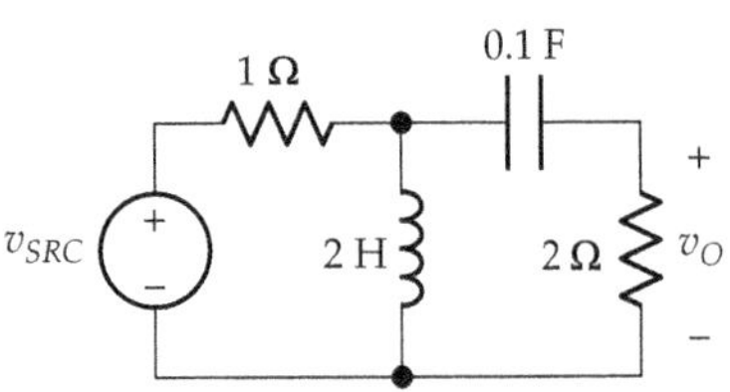

FIGURE P22.50

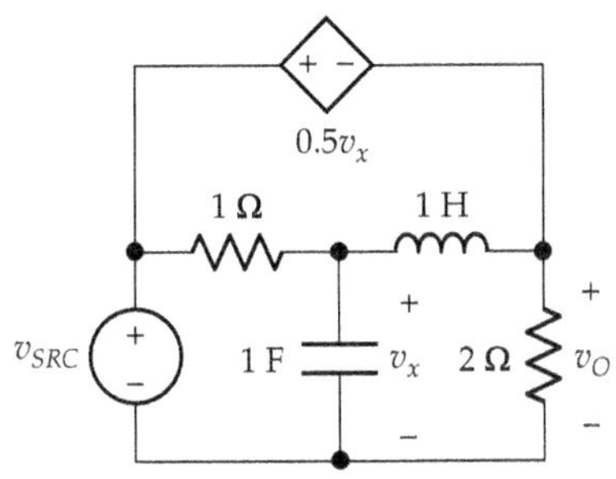

FIGURE P22.54

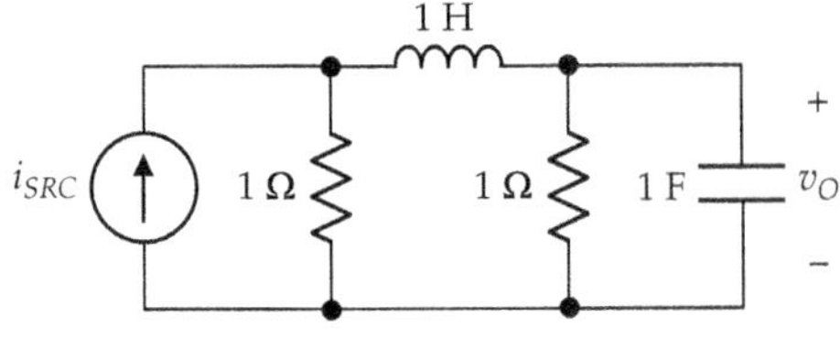

FIGURE P22.55

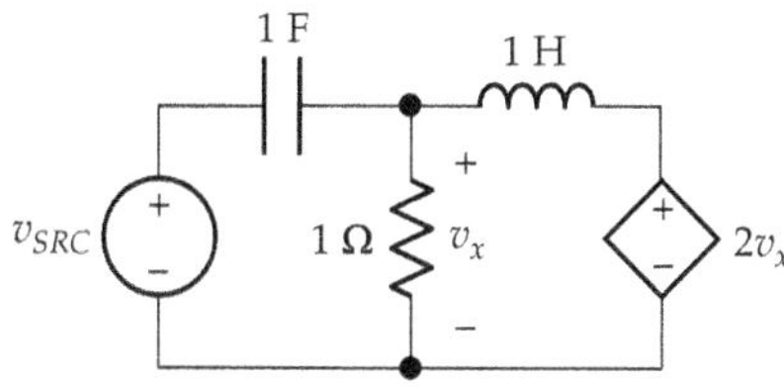

FIGURE P22.56

P22.56 (a) Determine $V_x(s)$ in Figure P22.56 if $v_{SRC}(t) = t\sin tu(t)$, (b) verify the initial-value theorem and the final-value theorem, and (c) derive the steady-state sinusoidal response when $v_{SRC}(t) = 5\cos t$ V.

Ans. (a) $V_x = \dfrac{2s^3}{(s^2+s-1)(s^2+1)^2}$ V; (c) $\sqrt{5}\cos(t+26.6°)$ V.

P22.57 Determine $v_O(t)$ in Figure P22.57 if the initial current in the inductor is 0.1 A and the initial voltage across the capacitor is zero.

Ans. $v_O(t) = 20 - \dfrac{262}{59}e^{-40t} - \dfrac{40}{177}e^{-2t/3}$ V.

P22.58 Determine $v_O(t)$ in Figure P22.57 if the initial current in the inductor is zero and the initial voltage across the capacitor is 12 V, having the polarity of a voltage rise in the direction of i_x.

Ans. $v_O(t) = 2 - \dfrac{518}{59}e^{-40t} + \dfrac{20}{177}e^{-2t/3}$ V.

P22.59 Determine $v_O(t)$ in Figure P22.59 if the initial current in the inductor is 0.2 A and the initial voltage across the capacitor is zero.

Ans. $v_O(t) = 20 - 20\cos 5\sqrt{2}t + \dfrac{39\sqrt{2}}{10}\sin 5\sqrt{2}t$ V.

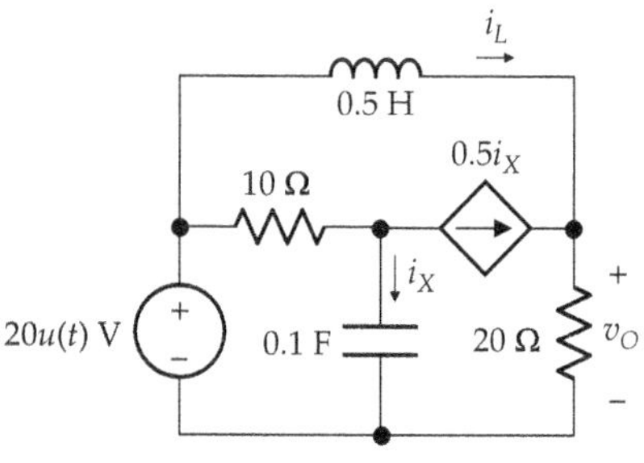

FIGURE P22.57

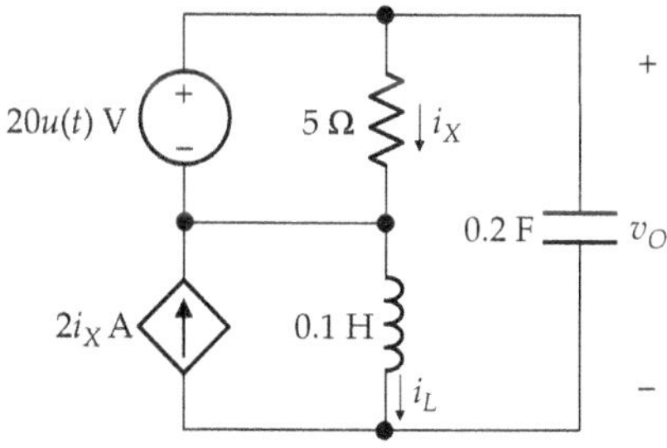

FIGURE P22.59

P22.60 Determine $v_O(t)$ in Figure P22.59 if the initial current in the inductor is zero and the initial voltage across the capacitor is 3 V and of the same polarity as v_O.

Ans. $v_O(t) = 20 - 17\cos 5\sqrt{2}t + 4\sqrt{2}\sin 5\sqrt{2}t$ V.

P22.61 The switch in Figure P22.61 is opened at $t = 0$ after being closed for a long time. Determine $I(s)$.

Ans. $I(s) = \dfrac{120(2s+1)}{60s^2 + 30s + 1}$.

P22.62 Determine $V_{Th}(s)$ looking into terminals 'ab' in Figure P22.62.

Ans. $V_{Th}(s) = V_{ab}(s) = \dfrac{12(s+4)}{s(s+6)}$.

P22.63 The circuit in Figure P22.63 is the s-domain representation of a parallel LC circuit with initial energy storage in L and C. Determine the initial values I_{L0} and V_{C0}.

Ans. $I_{L0} = -1$ A, $V_{C0} = 2$ V.

P22.64 Determine in Figure P22.64: (a) $v_O(t)$ if $v_{SRC}(t) = u(t) - u(t-1)$ V; (b) $v_{SRC}(t)$ if $v_O(t) = \left(-\dfrac{3}{2} + t + e^{-t} + \dfrac{e^{-2t}}{2}\right)u(t)$ V.

Ans. (a) $v_O(t) = (1 - e^{-t} - e^{-2t})u(t) - (1 - e^{-(t-1)} - e^{-2(t-1)})u(t-1)$ V; (b) $tu(t)$ V.

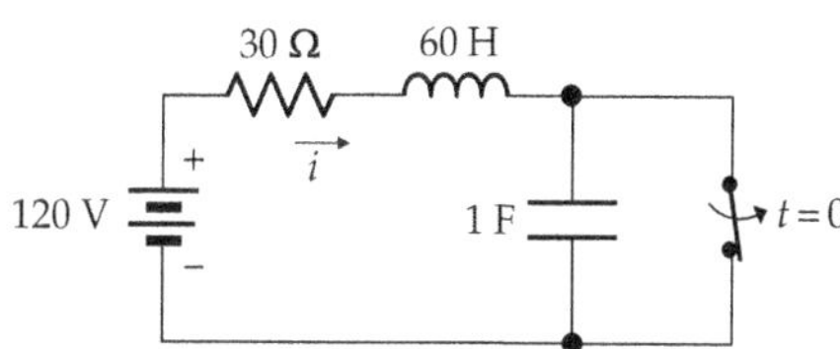

FIGURE P22.61

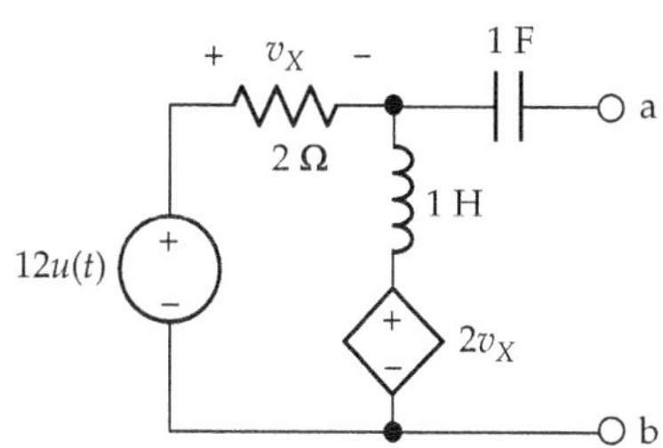

FIGURE P22.62

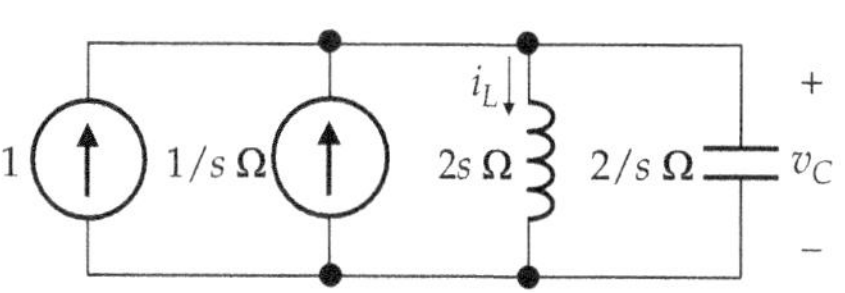

FIGURE P22.63

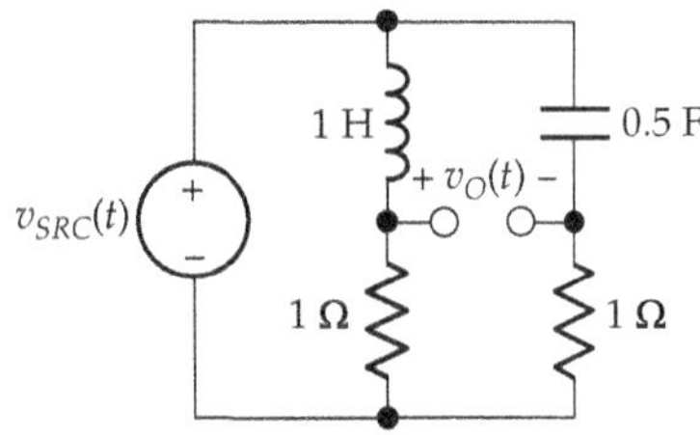

FIGURE P22.64

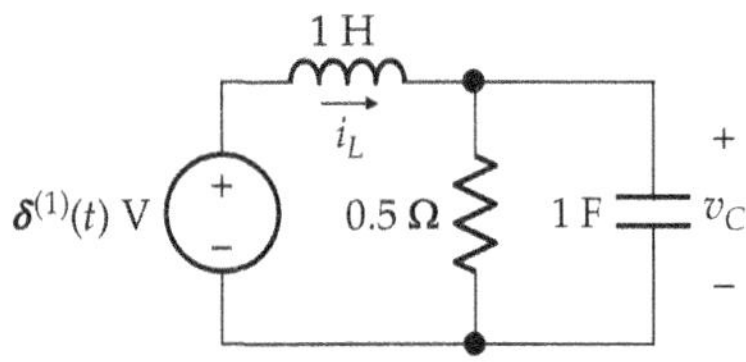

FIGURE P22.65

P22.65 Determine $i_L(t)$ in Figure P22.65, given that I_{L0} = 1 A and V_{C0} = 2 V.

Ans. $i_L(t) = \delta(t) + e^{-t}u(t) - 2te^{-t}u(t)$ A.

P22.66 The switch in Figure P22.66 is closed at t = 0, after being open for a long time. Determine $I_C(s)$.

Ans. $I_C(s) = \dfrac{s}{s^2 + 10^4 s + 25\times 10^6}$.

P22.67 The switch in Figure P22.67 is opened at t = 0, after being closed for a long time. Determine $i(t)$.

Ans. $i(t) = 2e^{-2t}(-2\cos t + \sin t)$ A.

P22.68 The switch in Figure P22.68 is opened at t = 0, after being closed for a long time. Determine $I(s)$.

Ans. $I(s) = \dfrac{5(s + 15\times 10^3)}{s^2 + 25\times 10^3 s + 10^8}$.

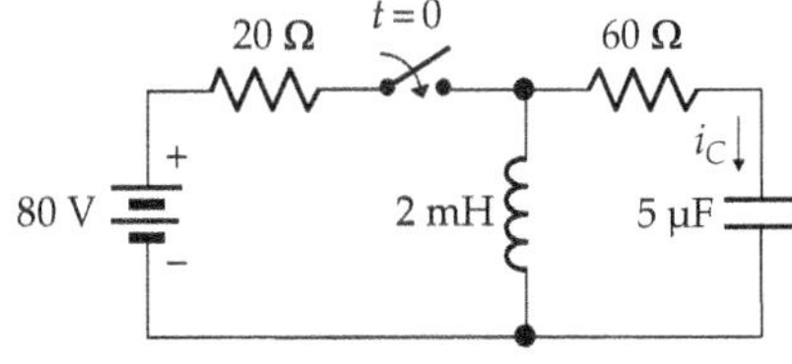

FIGURE P22.66

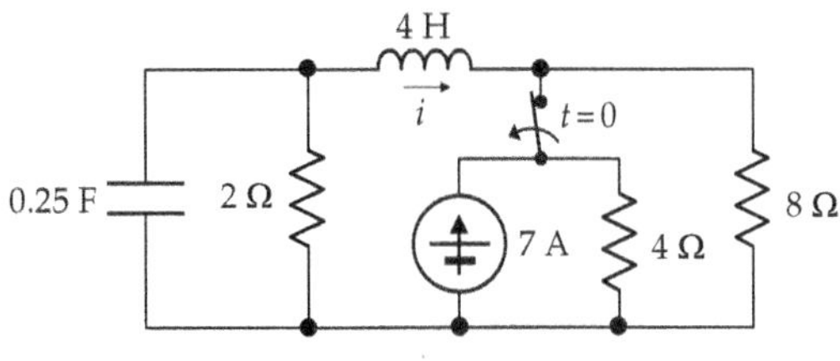

FIGURE P22.67

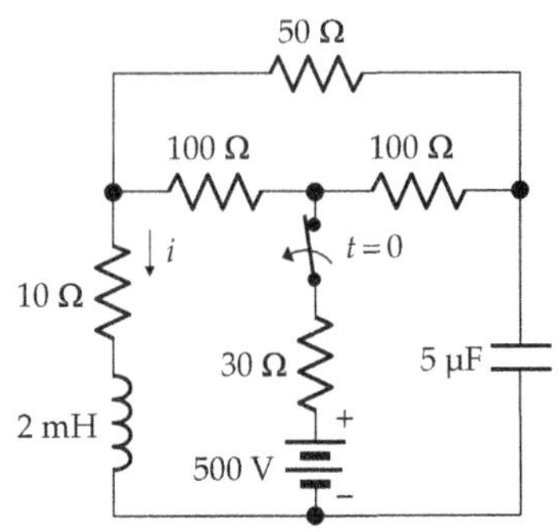

FIGURE P22.68

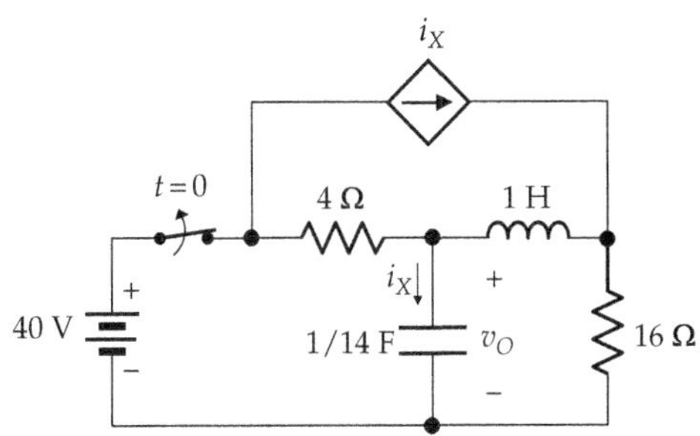

FIGURE P22.69

P22.69 The switch in Figure P22.69 is opened at t = 0, after being closed for a long time. Determine $v_O(t)$.

Ans. $v_O(t) = 35e^{-t}u(t) - 3e^{-7t}u(t)$ V.

P22.70 Determine $v_O(t)$ in Figure P22.70, assuming V_{C0} = 10 V and I_{L0} = 1 A

Ans. $10u(t)$ V.

P22.71 The switch in Figure P22.71 is moved at t = 0⁻ from position 'a' to position 'b', just before $u(t)$ and $\delta(t)$ are applied, after being in position 'a' for a long time. Determine $v_O(t)$.

Ans. $v_O(t) = u(t) + e^{-0.5t}\left(\cos\dfrac{\sqrt{3}}{2}t - \dfrac{1}{\sqrt{3}}\sin\dfrac{\sqrt{3}}{2}t\right)u(t)$ V.

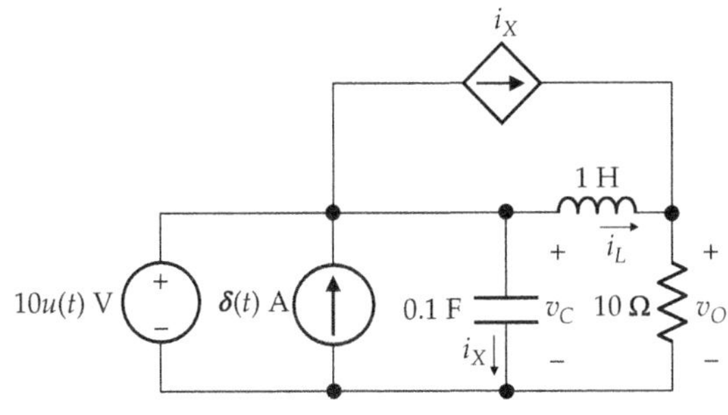

FIGURE P22.70

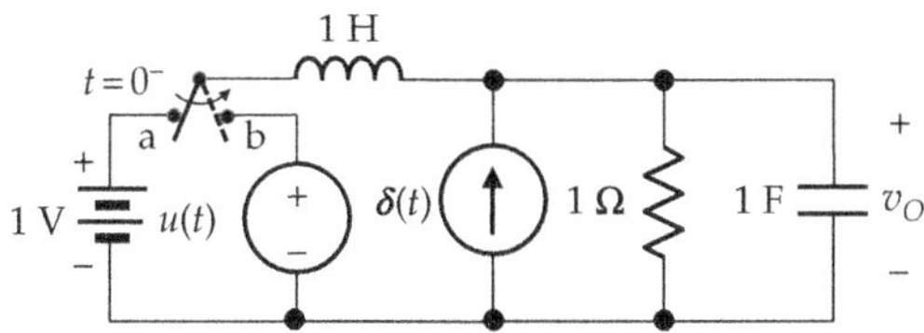

FIGURE P22.71

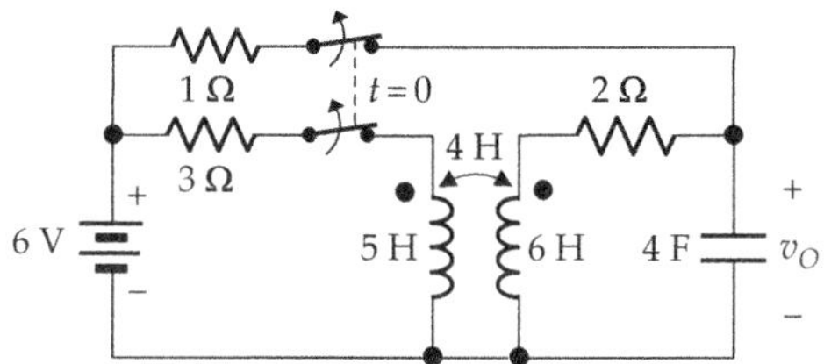

FIGURE P22.72

P22.72 The double switch in Figure P22.72 is opened at $t = 0$ after being closed for a long time. Determine $v_O(t)$.

Ans. $v_O(t) = 4e^{-t/6}\cos\frac{6}{\sqrt{2}}t - \sqrt{2}e^{-t/6}\sin\frac{6}{\sqrt{2}}t$ V.

P22.73 The switch in Figure P22.73 is opened at $t = 0$ after being closed for a long time. (a) Determine $V_C(s)$ from the circuit in the s-domain and (b) derive $v_C(t)$ for $t \geq 0^+$.

Ans. (a) $V_C(s) = \frac{s}{s^2 + 2s + 2}$; (b) $v_C(t) = e^{-t}(\cos t - \sin t)$.

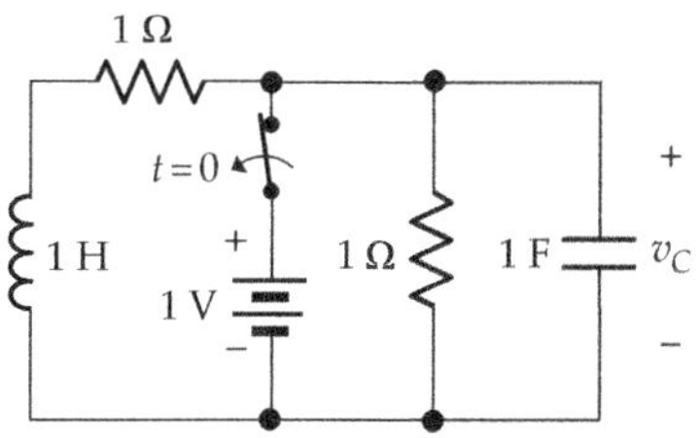

FIGURE P22.73

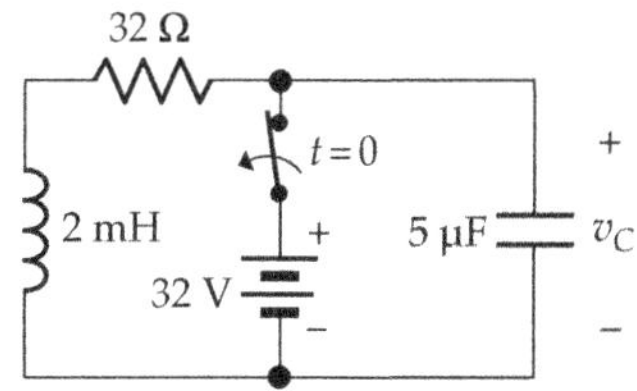

FIGURE P22.74

P22.74 The switch in Figure P22.74 is opened at $t = 0$ after being closed for a long time. Determine (a) derive $V_C(s)$; (b) $v_C(t)$, $t \geq 0^+$.

Ans. (a) $32\left(\frac{s + 9.75\times10^3}{s^2 + 16\times10^3 s + 10^8}\right)$; (b) $e^{-8t}\left(32\cos(6t) + (28/3)\sin(6t)\right)u(t)$ V.

23

Fourier Transform

Objective and Overview

Although the Laplace transform (LT) has several advantages over the Fourier transform (FT) in circuit analysis, the FT is fundamental to signal analysis. Being conceptually an extension of Fourier analysis to nonperiodic signals, it utilizes the same frequency-domain representation as phasor analysis. It shares many of the operational properties of the LT but has some unique and very useful properties that are explored in this chapter.

The FT provides a powerful tool for working in the frequency domain. This has many important applications in signal processing, communications, and control systems. The usefulness of FT techniques has been greatly enhanced by digital computation, based on a rapid and efficient algorithm known as the **fast Fourier transform** that computes the **discrete FT (DFT)**. This transform is an approximation to the FT that produces a finite set of discrete-frequency spectrum values from a finite set of discrete-time values.

The FT is first derived from the Fourier series expansion (FSE) and its salient characteristics are highlighted. Some general properties of the FT and their implications are then presented followed by the operational properties of the FT and their applications. Circuit applications of the FT are discussed and compared with those of the LT. This chapter ends with Parseval's theorem, which is concerned with energy in the frequency domain.

23.1 Derivation of the Fourier Transform

The FT can be derived from the exponential form of the FSE of a periodic waveform as the period becomes infinitely large. Consider as an example the FSE of the rectangular pulse train analyzed in Example 16.3. As the period of the function becomes infinitely large, the waveform reduces to a single pulse of amplitude A and duration τ, extending from $t=-\tau/2$ to $t=+\tau/2$, often denoted by $A\text{rect}(t/\tau)$, as mentioned earlier. C_n of the waveform is given by $C_n=(A\tau/T)\text{sinc}(n\omega_0\tau/2)$ (Equation 16.36). Since T will be made very large and ω_0 correspondingly very small, let us consider C_nT instead of C_n. Thus, from Equation 16.36

$$C_nT=\int_{-T/2}^{T/2} f(t)e^{-jn\omega_0 t}dt=A\tau\text{sinc}(n\omega_0\tau/2),\ n=0,\pm1,\pm2,\ldots \tag{23.1}$$

Figure 23.1a shows the plot of C_nT vs. $n\omega_0$ for the case of $T=5\tau$ illustrated in Example 16.3. The plot is

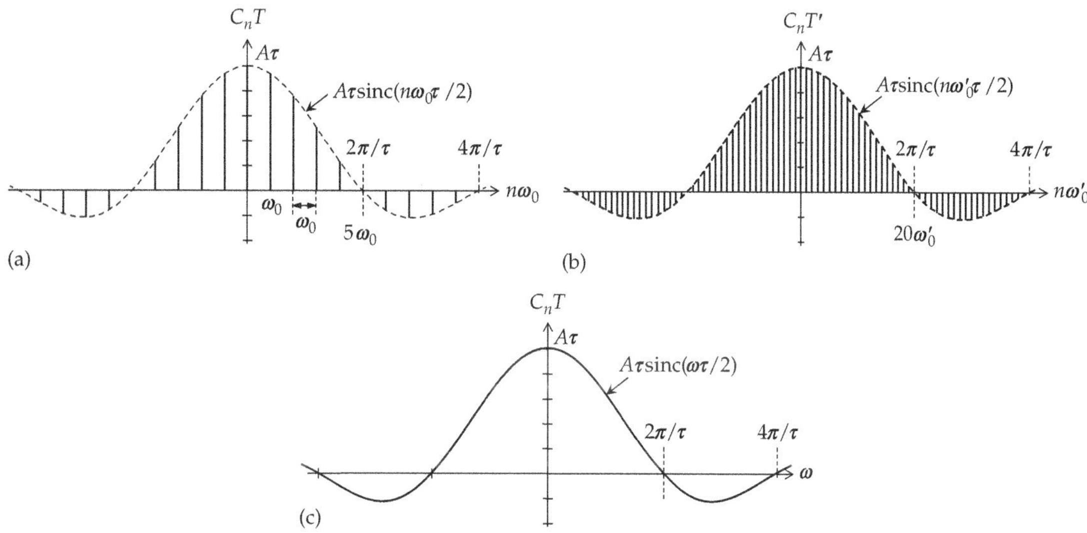

FIGURE 23.1
C_nT of a pulse train as the period is increased. (a) Combined amplitude and phase spectra of the rectangular pulse train of period T, whose pulse around the origin is $A\text{rect}(t/\tau)$, with $T=5\tau$, (b) the function in (a) when T is multiplied by 4, and (c) as $T\to\infty$, the function shown is the FT of $A\text{rect}(t/\tau)$.

a combination of the magnitude and phase spectra of Figure 16.12. C_n is shown positive when $\angle C_n = 0$ and negative when $\angle C_n = 180°$. The zero crossings occur at $n\omega_0\tau/2 = \pm m\pi$ or $n\omega_0 = \pm 2m\pi/\tau$, $m = 1, 2, 3, \ldots$. The zero crossings depend on τ, independently of T. When $\tau = T/5$, the first zero crossing for $n\omega_0 > 0$ occurs at $n\omega_0\tau/2 = \pi$, or $n\omega_0 = 2\pi/\tau = (2\pi)\times(5/T) = 10\pi/T = 10\pi/(2\pi/\omega_0) = 5\omega_0$, that is, after 5 intervals of ω_0.

If τ is kept constant while T is multiplied by 4, say, to give $T' = 4T$, then $\omega_0' = 2\pi/T'$ is divided by 4. The spectral lines are now closer together, being separated by ω_0' (Figure 23.1b), and the first zero crossings still occur at $\pm 2\pi/\tau = \pm(2\pi)\times(5/T) = \pm(2\pi)\times(5\times 4/T') = \pm 20\omega_0'$, that is, after twenty intervals of ω_0'. The general shape of the curve remains the same, because the zero crossings are independent of T.

As $T \to \infty$, the function becomes the aperiodic function $A\text{rect}(t/\tau)$, and the separation $\omega_0 = 2\pi/T$ between neighboring spectral lines becomes an infinitesimal $d\omega$. The abscissa $n\omega_0$ becomes ω, so that its infinitesimal variation is $d\omega$. Equation 23.1 becomes

$$C_nT = \int_{-\infty}^{\infty} f(t)e^{-j\omega t}dt = A\tau\,\text{sinc}\left(\frac{\omega\tau}{2}\right) = \frac{2A}{\omega}\sin\left(\frac{\omega\tau}{2}\right) \tag{23.2}$$

as illustrated in Figure 23.1c.

The integral C_nT in Equation 23.2 is the FT of the aperiodic time function $f(t)$ and represents a transformation from the time domain to the frequency domain. Thus,

$$\mathfrak{F}\{f(t)\} = F(\omega) = \int_{-\infty}^{\infty} f(t)e^{-j\omega t}dt \tag{23.3}$$

where $F(\omega)$ is the FT of $f(t)$.

The inverse transformation, from the frequency domain to the time domain, is obtained from the exponential form of the FSE: $f(t) = \sum_{n=-\infty}^{\infty} C_n e^{jn\omega_0 T} = \sum_{n=-\infty}^{\infty}(C_nT)e^{jn\omega_0 T}(1/T)$. In the limit, as $T \to \infty$, C_nT is replaced by $F(\omega)$ and $n\omega_0$ by ω. Now $(1/T) = (\omega_0/2\pi)$, where ω_0 is the separation between adjacent spectral lines (Figure 23.1a). As $T \to \infty$, this separation becomes $d\omega$, as noted earlier. Hence, $1/T$ is replaced by $d\omega/2\pi$, and the summation becomes an integration, which gives

$$\mathfrak{F}^{-1}\{F(\omega)\} = f(t) = \frac{1}{2\pi}\int_{-\infty}^{\infty} F(\omega)e^{j\omega t}d\omega \tag{23.4}$$

where $\mathfrak{F}^{-1}\{F(\omega)\}$ is the inverse FT (IFT) of $F(\omega)$.

The following should be noted concerning the Fourier transformation:

1. The FT can be interpreted along the same line as the FSE. The frequency spectrum of a periodic function is discrete, defined at frequencies $n\omega_0$, the amplitude of the nth harmonic being $2|C_n|$. On the other hand, as $T \to \infty$, $2|C_n| = 2\frac{|F(\omega)|}{T} = \frac{|F(\omega)|}{\pi}d\omega$ becomes infinitesimally small, and the frequency separation ω_0 also becomes infinitesimal; that is, the amplitudes of the sinusoids are now a continuous function of frequency instead of being defined as harmonic only at the discrete frequencies $n\omega_0$. The periodic function is the sum of a denumerably infinite number of sinusoids of finite amplitude, successively separated in frequency by ω_0. The aperiodic function can be viewed as the sum of a nondenumerably infinite number of sinusoids of infinitesimal amplitude and infinitesimal frequency separation. The amplitude of each of these components is $(|F(\omega)|/\pi)d\omega$ and its phase is that of $F(\omega)$. The distinction between a denumerably infinite and a nondenumerably infinite number of sinusoids is that a denumerably infinite number is, in principle, countable, as 1, 2, 3, … all the way to infinity. On the other hand, when the frequency is continuous, counting of sinusoids, even over a finite frequency range, is nondenumerably infinite. The counting becomes impossible, even in principle, because of infinitesimal frequency separation.
2. Sufficient conditions for the existence of the FT are as follows: (a) $\int_{-\infty}^{+\infty}|f(t)|dt$ is finite, that is, $|f(t)|$ is absolutely integrable, and (b) the function is piecewise continuous on a finite interval of time, as in the case of the LT, which means that the function is integrable over any finite interval. These conditions are *sufficient but not necessary*, so that functions for which $\int_{-\infty}^{+\infty}|f(t)|dt$ is infinite can have an FT. Examples are dc quantities, step functions, and sinusoidal functions. The FTs of these functions cannot be obtained directly using Equation 23.3 but are obtained in other ways, as illustrated later.
3. Compared to the one-sided LT of Equation 21.3, the FT is defined from $t = -\infty$ to $t = +\infty$, with s in the LT replaced by $j\omega$. The extension over all time means that values of functions for $t < 0$ are included in the FT, but initial conditions at $t = 0$ are not easily included. Moreover, because of the $e^{-\sigma t}$ in the LT integral, where σ is the real part of s, the LT integral converges for more functions than does the FT integral. As a consequence, functions such as $tu(t)$ have an LT but not an FT.

4. The importance of the FT in signal analysis stems from the fact that it transforms a time function extending over all time, from $t = -\infty$ to $t = +\infty$, to the frequency domain, thereby yielding a true representation of the frequency content of the function. Strictly speaking, a dc or a sinusoidal signal extends from $t = -\infty$ to $t = +\infty$. In order to portray the frequency content of such signals through a transformation from the time domain, the transformation should include all time, as in the FT.
5. Some signal processing applications involve the FT in terms of spatial coordinates in two or three dimensions rather than time. Equation 23.3 applies in this case for each spatial coordinate, with the spatial coordinate replacing t.
6. The FT can have a physical representation. For example, if an image on a transparent film is placed at the focal point of a convex lens and illuminated by coherent light, as from a laser, the image seen at the other focal point is the two-dimensional spatial FT of the image on the transparency. In fact, when monochromatic x-rays, that is, x-rays of a single frequency, are diffracted by a crystal, the diffraction pattern is the FT of the three-dimensional crystal structure for that particular angle of incidence of the x-rays.

We illustrate in what follows the derivation of the FT of some basic functions, starting with a unit impulse $\delta(t)$ at the origin. It follows from Equation 23.3 that

$$\mathfrak{F}\{\delta(t)\} = \int_{-\infty}^{\infty} \delta(t)e^{-j\omega t}dt = \int_{0^-}^{0^+} \delta(t)\times 1dt = 1 \quad (23.5)$$

as for the LT.

We next derive the FT of $f(t) = 1$, a constant. In this case $\int_{-\infty}^{+\infty} |f(t)|dt$ is infinite, so the FT cannot be obtained directly from Equation 23.3 but can be derived from Equation 23.5. According to this equation, the IFT of 1 is $\delta(t)$. Applying the inverse transform relation (Equation 23.4) to 1 and setting this equal to $\delta(t)$,

$$\mathfrak{F}^{-1}\{1\} = \frac{1}{2\pi}\int_{-\infty}^{\infty} 1\times e^{j\omega t}d\omega = \delta(t) \quad (23.6)$$

Interchanging the variables ω and t gives

$$\frac{1}{2\pi}\int_{-\infty}^{\infty} 1\times e^{j\omega t}dt = \delta(\omega) \quad (23.7)$$

Replacing ω by $-\omega$, with $\delta(-\omega) = \delta(\omega)$,

$$\int_{-\infty}^{\infty} 1\times e^{-j\omega t}dt = 2\pi\delta(\omega) \quad (23.8)$$

The LHS of Equation 23.8 is the FT of a constant equal to 1 that extends over all time. Hence,

$$\mathfrak{F}\{1\} = 2\pi\delta(\omega) \quad (23.9)$$

and is an impulse of strength 2π at $\omega = 0$ in the frequency domain. This is to be expected because the constant 1 can be considered as a dc signal that extends from $t = -\infty$ to $t = +\infty$. A dc signal is of zero frequency, so that its representation in the frequency domain should be at the point $\omega = 0$ only, which is an impulse. On the other hand, it was pointed out in connection with Equation 16.40, that if the pulse is narrow enough, fundamental and harmonics all have essentially the same amplitude. In the limit, an impulse at $t = 0$, being of infinitesimal duration, has a frequency spectrum extending over all frequencies, and with all these frequency components having the same relative amplitude. Its FT is therefore a constant extending from $-\infty$ to $+\infty$, as in Equation 23.5.

Let us derive next the FT of $f(t) = e^{j\omega_0 t}$. From Equation 23.3, $F(\omega) = \int_{-\infty}^{\infty}\left(e^{j\omega_0 t}\right)e^{-j\omega t}dt = \int_{-\infty}^{\infty} e^{-j(\omega-\omega_0)t}dt$. Comparing with Equation 23.8, it is seen that ω in the integral on the LHS of this equation has been replaced by $(\omega - \omega_0)$. Replacing ω on the RHS of Equation 23.8 by $(\omega - \omega_0)$ gives

$$\mathfrak{F}\left\{e^{j\omega_0 t}\right\} = 2\pi\delta(\omega - \omega_0) \quad (23.10)$$

Once we have the FT of $e^{j\omega_0 t}$, the FTs of $\cos\omega_0 t$ and $\sin\omega_0 t$ readily follow. Thus,

$$\mathfrak{F}\{\cos\omega_0 t\} = \mathfrak{F}\left\{\frac{e^{j\omega_0 t} + e^{-j\omega_0 t}}{2}\right\} = \pi\left[\delta(\omega + \omega_0) + \delta(\omega - \omega_0)\right] \quad (23.11)$$

The FT of $\cos\omega_0 t$ is therefore two impulse functions, one at $\omega = \omega_0$ and the other at $\omega = -\omega_0$, each of strength π. Since a function $\cos\omega_0 t$ that extends from $t = -\infty$ to $t = +\infty$ has a single frequency component ω_0, it is represented by impulses in the frequency domain at $\omega = \pm\omega_0$. As discussed in connection with the exponential form of Fourier series (Section 16.2), the positive and negative frequencies combine to give a real function $\cos\omega_0 t$. Similarly,

$$\mathfrak{F}\{\sin\omega_0 t\} = \mathfrak{F}\left\{\frac{e^{j\omega_0 t} - e^{-j\omega_0 t}}{2j}\right\} = j\pi\left[\delta(\omega+\omega_0) - \delta(\omega-\omega_0)\right] \quad (23.12)$$

In this case, the impulse function at $\omega = -\omega_0$ has a phase angle of $\pi/2$, whereas the impulse function at $\omega = \omega_0$ has a phase angle of $-\pi/2$. The sine function is thus 90° out of phase with the cosine function, as it should be. As ω_0 decreases, the two impulses of the FT of a cosine function come closer to the origin. When $\omega_0 = 0$, the two impulses add at the origin to give a single impulse of strength 2π, which is the FT of unity. In the time domain, as ω_0 decreases, $\cos\omega_0 t$ becomes "flatter," as illustrated in Figure 23.2a for $\cos\omega_0 t$ and for $\cos 0.25\omega_0 t$. As $\omega_0 \to 0$, $\cos\omega_0 t \to 1$, whose FT is 2π. In the case of $\sin\omega_0 t$, the two impulses are 180° out of phase, so they cancel out at the origin, when $\omega_0 = 0$, giving a zero FT. In the time domain, $\sin\omega_0 t$ becomes "flatter" as ω_0 decreases (Figure 23.2b). As $\omega_0 \to 0$, $\sin\omega_0 t \to 0$, so *the* FT is zero.

As in the case of the LT, MATLAB can be used to find the FT and IFT. For example, if we enter

```
>> syms t w
>> fourier(sin(2*t))
```

MATLAB returns

```
>> -pi*(dirac(w - 2) - dirac(w + 2))*i
```

where Dirac() denotes δ() and i denotes *j*. If we then enter

```
>> ifourier(-pi*(dirac(w - 2) - dirac(w + 2)))
```

MATLAB returns

```
>> (1/exp(2*x*i))/2 - exp(2*x*i)/2,
```

which is cos*x*.

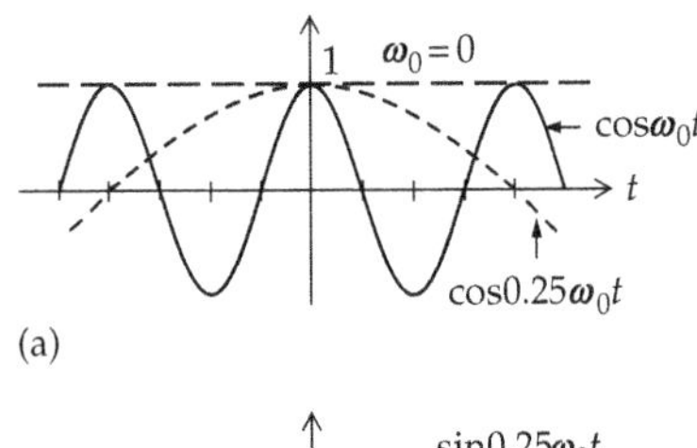

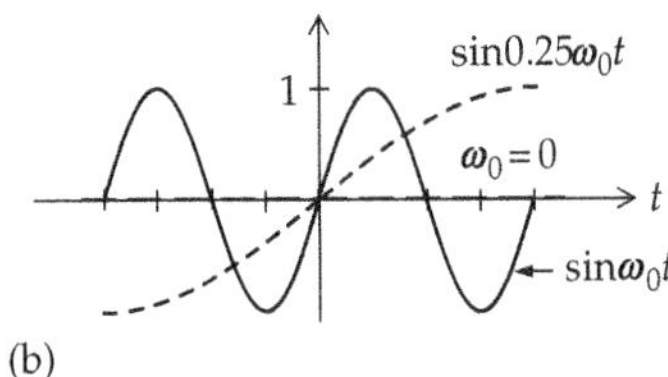

FIGURE 23.2
Cosine (a) and sine (b) functions as the frequency approaches zero.

Example 23.1: Fourier Transform of Biphasic Pulse

It is required to derive the FT of the function shown in Figure 23.3.

Solution:

$$\mathfrak{F}\{f(t)\} = -\int_{-\tau/2}^{0} Ae^{-j\omega t}dt + \int_{0}^{\tau/2} Ae^{-j\omega t}dt$$

$$= \frac{A}{j\omega}\left[e^{-j\omega t}\right]_{-\tau/2}^{0} - \frac{A}{j\omega}\left[e^{-j\omega t}\right]_{0}^{\tau/2}$$

$$= -j\left(\frac{2A}{\omega}\right)\left(1-\cos\frac{\omega\tau}{2}\right) = \left(\frac{2A}{j\omega}\right)\left(1-\cos\frac{\omega\tau}{2}\right) \quad (23.13)$$

Example 23.2: Inverse Fourier Transform of a Symmetrical Function

The FT of a function is shown in Figure 23.4. It is required to find *f*(*t*).

Solution:

$$\mathfrak{F}^{-1}\{F(\omega)\} = \frac{1}{2\pi}\int_{-2}^{-1} 2e^{j\omega t}d\omega + \frac{1}{2\pi}\int_{-1}^{1} e^{j\omega t}d\omega + \frac{1}{2\pi}\int_{1}^{2} 2e^{j\omega t}d\omega$$

$$= \frac{1}{\pi jt}\left[e^{j\omega t}\right]_{-2}^{-1} + \frac{1}{2\pi jt}\left[e^{j\omega t}\right]_{-1}^{1} + \frac{1}{\pi jt}\left[e^{j\omega t}\right]_{1}^{2}$$

$$= \frac{1}{\pi jt}\left[e^{-jt} - e^{-j2t} + e^{j2t} - e^{jt}\right] + \frac{1}{2\pi jt}\left[e^{jt} - e^{-jt}\right]$$

$$= \frac{2}{\pi t}\left[\sin 2t - \sin t\right] + \frac{1}{\pi t}\left[\sin t\right] = \frac{1}{\pi t}\left[2\sin 2t - \sin t\right].$$

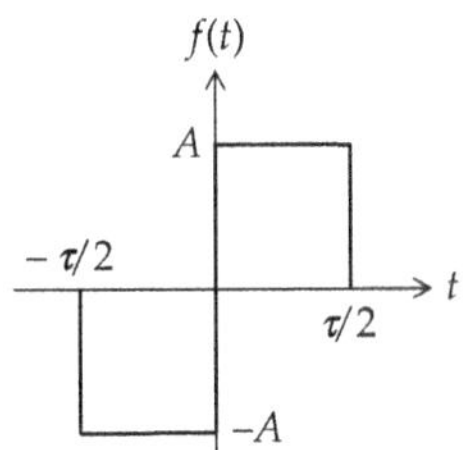

FIGURE 23.3
Figure for Example 23.1.

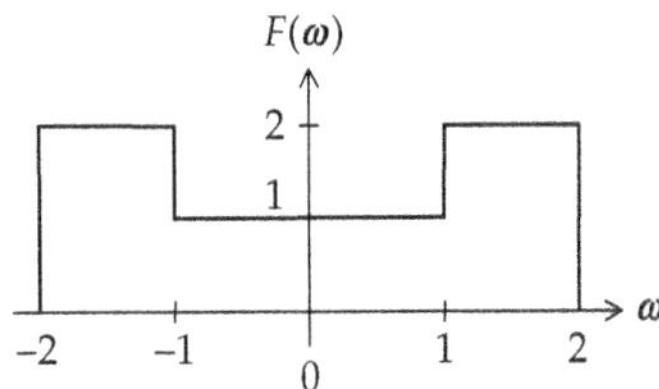

FIGURE 23.4
Figure for Example 23.2.

Primal Exercise 23.1

Determine the FT of $e^{-a|t|}$, $a > 0$. Note that this is the sum of $e^{-at}u(t)$, having $t > 0$, and $e^{at}u(-t)$, having $t < 0$. The two functions are symmetrical with respect to the vertical axis.

Ans. $$\mathcal{F}\left\{e^{-a|t|}\right\} = \frac{2a}{a^2+\omega^2} \tag{23.14}$$

23.2 Some General Properties of the Fourier Transform

23.2.1 Real and Imaginary Parts

It should be emphasized that, according to Equation 23.3, $F(\omega)$ is, in general, a complex quantity. In practice, $f(t)$ is always a real function of time, so that

$$F(\omega) = \int_{-\infty}^{\infty} f(t)e^{-j\omega t}dt = \int_{-\infty}^{\infty} f(t)\cos\omega t dt - j\int_{-\infty}^{\infty} f(t)\sin\omega t dt$$
$$= A(\omega) + jB(\omega) \tag{23.15}$$

where $A(\omega)$ and $B(\omega)$ are the real and imaginary parts, respectively, of $F(\omega)$, expressed as

$$A(\omega) = \int_{-\infty}^{\infty} f(t)\cos\omega t dt, \quad B(\omega) = -\int_{-\infty}^{\infty} f(t)\sin\omega t dt \tag{23.16}$$

The following may be deduced from the preceding relations:

Deductions:

1. *The real part of* $F(\omega)$ is even, since $A(\omega) = A(-\omega)$.
2. *The imaginary part of* $F(\omega)$ is odd, since $B(\omega) = -B(-\omega)$.
3. *The magnitude of* $F(\omega)$, which is $\sqrt{A^2(\omega)+B^2(\omega)}$, *is even.*
4. *The phase angle of* $F(\omega)$, which is $\tan^{-1}\left(B(\omega)/A(\omega)\right)$, *is odd.*
5. *Replacing* ω *by* $-\omega$ *gives the complex conjugate of* $F(\omega)$, *that is,* $F(-\omega) = F^*(\omega)$.
6. If $f(t)$ is even, $f(t)\cos\omega t$ is an even function of t, and $f(t)\sin\omega t$ is an odd function of t, which makes $B(\omega) = 0$. Hence, $F(\omega)$ is real and even, with $A(\omega) = 2\int_0^{\infty} f(t)\cos\omega t dt$.
7. If $f(t)$ is odd, $f(t)\cos\omega t$ is an odd function of t, and $f(t)\sin\omega t$ is an even function of t, which makes $A(\omega) = 0$. Hence, $F(\omega)$ is imaginary and odd, with $B(\omega) = -2\int_0^{\infty} f(t)\sin\omega t dt$.

For example, $\mathcal{F}\{\cos\omega_0 t\}$ is real and even (Equation 23.11), whereas $\mathcal{F}\{\sin\omega_0 t\}$ is imaginary and odd (Equation 23.12). Similarly, the FT of $\text{rect}(t/\tau)$ is real and even (Equation 23.2), and the FT of two antisymmetrical pulses is imaginary and odd (Equation 23.13).

23.2.2 Fourier Transform at Zero Frequency

If $\omega = 0$ in Equation 23.3,

$$F(0) = \int_{-\infty}^{\infty} f(t)dt \tag{23.17}$$

That is, the value of FT at $\omega = 0$ is the net positive area subtended by $f(t)$. This area is the average, or dc component, of $f(t)$ multiplied by the interval over which $f(t) \neq 0$.

It should be emphasized at this stage that considerable confusion concerning relations involving the FT can be avoided if a distinction is made between (1) "well-behaved" functions for which $\int_{-\infty}^{+\infty} |f(t)|dt$ is finite, which is one of the sufficient conditions for the existence of the FT, and (2) functions for which $\int_{-\infty}^{+\infty} |f(t)|dt$ is infinite, in which case $F(\omega)$, if it exists, cannot be obtained directly from Equation 23.3. In this case, the FT generally has impulses. Since Equation 23.17 is derived from Equation 23.3, the following should be noted:

1. If $\int_{-\infty}^{+\infty} |f(t)|dt$ is finite, if $f(t)$ is of finite duration, and if the dc component of $f(t)$ is zero, then $F(0) = 0$. This can serve as a useful check on the FTs of such functions. For example, $f(t)$ of Figure 23.3 has a finite $\int_{-\infty}^{+\infty} |f(t)|dt$, is of finite duration, and has zero dc. According to Equation 23.17, its $F(0)$ must be zero. Substituting $\omega = 0$ in Equation 23.13 gives $F(0) = 0/0$, which is indeterminate. However, according to L'Hopital's rule, $F(0)$ can be obtained by first differentiating the numerator and denominator of the function, separately with respect to ω, before substituting $\omega = 0$. Differentiating the numerator of Equation 23.13 with respect to ω gives $A\sin\omega\tau/2$, and differentiating the denominator with respect to ω gives j. Substituting $\omega = 0$ results in $F(0) = 0$, as expected.

2. On the other hand, if $\int_{-\infty}^{+\infty}|f(t)|dt$ is infinite, and even if the dc component of $f(t)$ is zero, it does not follow that $F(0) = 0$. For example, a signum function (Example 23.3) has an infinite $\int_{-\infty}^{+\infty}|f(t)|dt$, and zero dc component, yet $F(0) \neq 0$ (Equation 23.19). In the case of $\cos\omega_0 t$ and $\sin\omega_0 t$, $\int_{-\infty}^{+\infty}|f(t)|dt$ is also infinite. The dc component is zero because it equals a finite area, which is at most that of half a period, divided by an infinite duration, as $t \to \infty$; $F(0) = 0$ for $\sin\omega_0 t$ and $\cos\omega_0 t$, since the impulses occur at $\omega = \pm\omega_0$ and the FT is zero at $\omega = 0$.

$F(0)$ plays an important role in the FT, as illustrated by the following examples and by the differentiation-in-time and integration-in-time properties discussed later.

Primal Exercise 23.2

Determine $F(0)$ for (a) the function shown in Figure 23.5 and (b) $\delta(t)$.

Ans. (a) 5; (b) 1.

Example 23.3: Fourier Transform of Signum and Step Functions

It is required to derive the FTs of (a) $e^{-at}u(t)$, $a > 0$; (b) a signum function, sgn(t), defined as 1 for $t > 0$ and -1 for $t < 0$; and (c) a unit step function $u(t)$.

Solution:

(a) From Equation 23.3,

$$\mathfrak{F}\{e^{-at}u(t)\} = \int_0^{\infty} e^{-at}e^{-j\omega t}dt = \frac{1}{a+j\omega} \quad (23.18)$$

(b) $\text{sgn}(t) = u(t) - u(-t)$ (Figure 23.6). Since the Fourier integral of these step functions does not converge, we consider sgn(t) as $\lim_{\varepsilon\to 0}\left[e^{-\varepsilon t}u(t) - e^{\varepsilon t}u(-t)\right]$, $\varepsilon > 0$ (Figure 23.6). Hence,

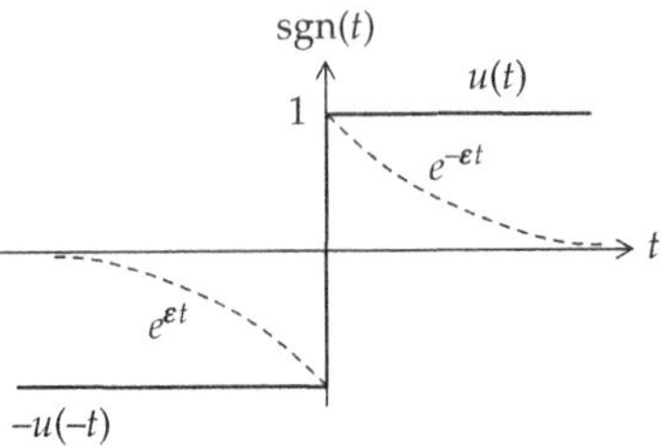

FIGURE 23.6
Figure for Example 23.3.

$$\mathfrak{F}\{\text{sgn}(t)\} = \lim_{\varepsilon\to 0}\left[\int_0^{\infty} e^{-\varepsilon t}e^{-j\omega t}dt - \int_{-\infty}^{0} e^{\varepsilon t}e^{-j\omega t}dt\right]$$

$$= \lim_{\varepsilon\to 0}\left[\frac{1}{\varepsilon + j\omega} - \frac{1}{\varepsilon - j\omega}\right] = \lim_{\varepsilon\to 0}\left[\frac{-2j\omega}{\varepsilon^2 + \omega^2}\right] = \frac{2}{j\omega} \quad (23.19)$$

Note that the signum function is an odd function of time. Hence, its FT should be imaginary and an odd function of ω, in accordance with Equation 23.19.

(c) $u(t) = \frac{1}{2}\text{sgn}(t) + \frac{1}{2}$ (Figure 23.7). Taking the FT of both sides and using Equations 23.9 and 23.19,

$$\mathfrak{F}\{u(t)\} = \pi\delta(\omega) + \frac{1}{j\omega} \quad (23.20)$$

It is seen that the FT of $u(t)$ has an impulse at $\omega = 0$ because of the dc component of 1/2. As explained previously, a dc signal has only a single frequency at $\omega = 0$; hence the impulse $\delta(\omega)$. The FTs of both $u(t)$ and sgn(t) have frequency components for $\omega > 0$, represented by the $1/j\omega$ term. These frequency components are due to the sudden jump of both functions at $t = 0$. sgn(t) does not have an impulse at $\omega = 0$ because it can be considered as the sum of $u(t)$ and $-u(-t)$. In effect, the impulses of these two functions cancel out at $\omega = 0$.

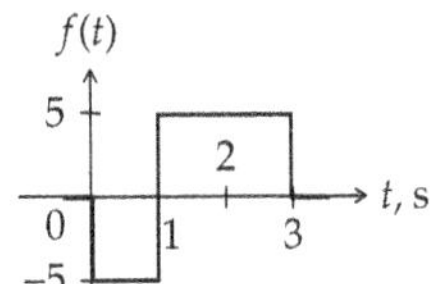

FIGURE 23.5
Figure for Primal Exercise 23.5.

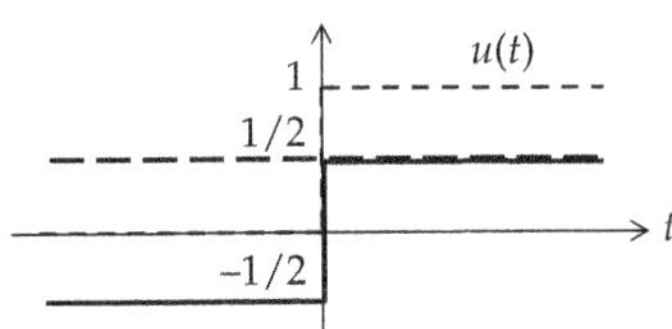

FIGURE 23.7
Figure for Example 23.3.

Exercise 23.3

The functions $e^{-at}\cos\omega_0 t u(t)$ and $e^{-at}\sin\omega_0 t u(t), a > 0$, have a finite $\int_{-\infty}^{\infty}|f(t)|dt$, so their FTs do not have impulses. Show that their FTs are given by

$$\mathfrak{F}\{e^{-at}\cos\omega_0 t u(t)\} = \frac{a + j\omega}{\omega_0^2 + (a + j\omega)^2}, \quad a > 0 \qquad (23.21)$$

$$\mathfrak{F}\{e^{-at}\sin\omega_0 t u(t)\} = \frac{\omega_0}{\omega_0^2 + (a + j\omega)^2}, \quad a > 0 \qquad (23.22)$$

Use two methods: (a) from the FT of $e^{-at}e^{j\omega_0 t}u(t)$ and (b) direct application of Equation 23.3.

Primal Exercise 23.4

Derive the FT of $f(t)$ in Figure 23.8

Ans. $2\pi\delta(\omega) + \frac{j}{\omega}\left(1 - e^{-j2\omega}\right)$.

Primal Exercise 23.5

Determine the FT of the gate function $f(t) = u(t) - u(t-1)$.

Ans. $\left(1 - e^{-j\omega}\right)\left(\pi\delta(\omega) + 1/j\omega\right)$.

Primal Exercise 23.6

Determine the FT of $(\text{sign}(t))^3 u(-t)$.

Ans. $-\pi\delta(\omega) + \frac{1}{j\omega}$.

23.2.3 Duality

Concept: *The symmetry between the expressions for the Fourier transform and its inverse underlies an important duality relationship:*

$$\text{If } \mathfrak{F}\{f(t)\} = F(\omega), \quad \text{then} \quad \mathfrak{F}\{F(t)\} = 2\pi f(-\omega) \qquad (23.23)$$

Proof: $f(t) = \mathfrak{F}^{-1}\{F(\omega)\} = \frac{1}{2\pi}\int_{-\infty}^{\infty} F(\omega)e^{j\omega t}d\omega$. Multiplying both sides by 2π and interchanging the variables

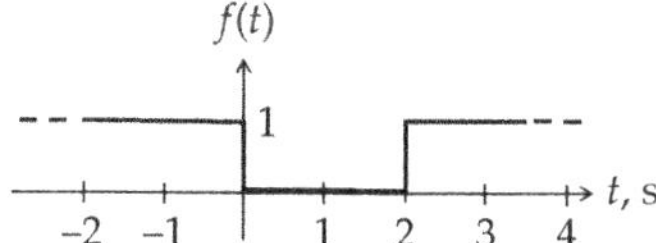

FIGURE 23.8
Figure for Primal Exercise 23.4.

t and ω, $2\pi f(\omega) = \int_{-\infty}^{\infty} F(t)e^{j\omega t}dt$. Replacing ω by $-\omega$ gives Equation 23.23.

$F(\omega)$ is the FT of $f(t)$, as usual. $F(t)$ is a time function that has the same waveform as $F(\omega)$, with ω replaced by t. $f(-\omega)$ is the FT of $F(t)$ and has the same waveform as $f(t)$ with t replaced by $-\omega$. If $f(t)$ is an even function, then $f(-\omega) = f(\omega)$, but if $f(t)$ is an odd function, then $f(-\omega) = -f(\omega)$.

To illustrate duality, consider $f(t) = \delta(t)$, the unit impulse at the origin (Figure 23.9a). Its FT is $F(\omega) = 1$ (Figure 23.9b). The time function that is of the same waveform as $F(\omega)$ is $F(t) = 1$ (Figure 23.9c). Its FT is $2\pi\delta(\omega)$, which is the same as $2\pi\delta(-\omega)$ (Figure 23.9d), since the unit impulse at the origin is an even function. The function $2\pi\delta(-\omega)$ is $2\pi f(-\omega)$, in accordance with Equation 23.23. A similar result is obtained if $f(t) = 1$, in which case $F(\omega) = 2\pi\delta(\omega)$. The function $F(t) = 2\pi\delta(t)$. Its FT is 2π, which is an even function and is $2\pi f(-\omega)$.

As an illustration of duality involving an odd function of time, consider $f(t) = \sin t$ (Figure 23.10a). Its FT is $F(\omega) = j\pi\delta(\omega + 1) - j\pi\delta(\omega - 1)$, illustrated in Figure 23.10b. The time function that is of the same waveform as

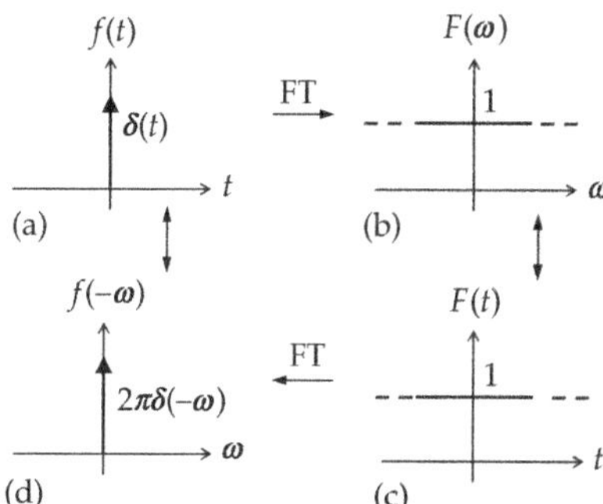

FIGURE 23.9
Duality between impulse and dc functions. An impulse function (a) and its FT (b), (c) the time function that is of the same waveform as the FT in (b), and (d) the FT of the function in (c).

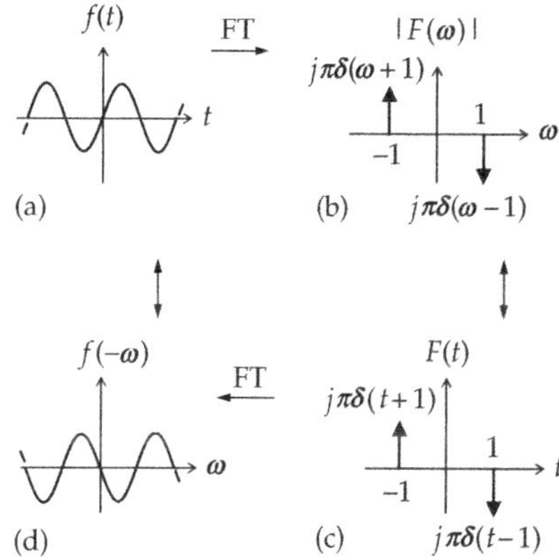

FIGURE 23.10
Duality between sinusoidal and impulse functions. A sine function (a) and its FT (b), (c) the time function that is of the same waveform as the FT in (b), and (d) the FT of the function in (c).

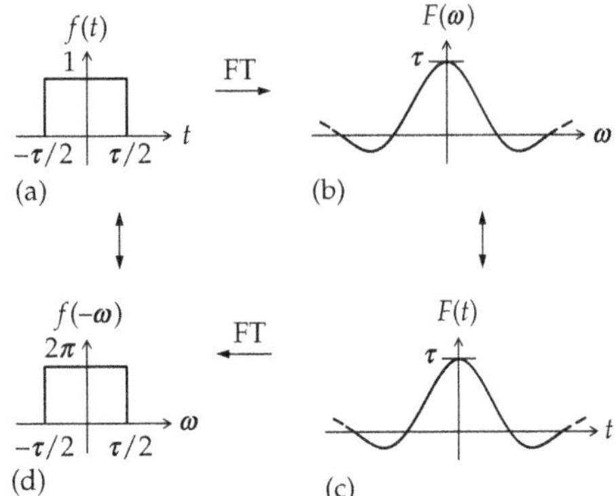

FIGURE 23.11
Duality between rectangular and sinc functions. A rectangular function (a) and its FT waveform (b), (c) the time function that is of the same waveform as the FT in (b), and (d) the FT of the function in (c).

$F(\omega)$ is $F(t) = j\pi\delta(t+1) - j\pi\delta(t-1)$, illustrated in Figure 23.10c, considering j to be just a constant. The FT of an impulse $A\delta(t-a)$ of strength A that is delayed by a is $Ae^{-j\omega a}$ (Equation 23.28). It follows that the FT of $F(t)$ is $j\pi e^{j\omega} - j\pi e^{-j\omega}$. Multiplying and dividing by $2j$, the FT becomes $2\pi\left(\dfrac{e^{-j\omega}-e^{j\omega}}{2j}\right) = -2\pi\sin\omega = 2\pi\sin(-\omega)$. This is $2\pi f(-\omega)$ (Figure 23.10d).

As another example of duality, consider $\text{rect}(t/\tau)$ discussed in Section 23.1 and shown in Figure 23.11a. Its FT is $\tau\text{sinc}(\omega\tau/2)$, as in Equation 23.2 and shown in Figure 23.11b. It follows from duality that

$$\mathfrak{F}\left\{\tau\,\text{sinc}\left(\frac{t\tau}{2}\right)\right\} = 2\pi\left[u\left(\omega+\frac{\tau}{2}\right) - u\left(\omega-\frac{\tau}{2}\right)\right] = 2\pi\,\text{rect}(\omega/\tau) \tag{23.24}$$

as shown in Figure 23.11c and d. Substituting $\beta = \tau/2$ and dividing both sides by 2β,

$$\mathfrak{F}\{\text{sinc}(\beta t)\} = \frac{\pi}{\beta}\left[u(\omega+\beta) - u(\omega-\beta)\right] = \frac{\pi}{\beta}\text{rect}\left(\frac{\omega}{2\beta}\right) \tag{23.25}$$

The sinc function of time is in fact used in magnetic resonance imaging to obtain a rectangular function of frequency.

23.3 Operational Properties of the Fourier Transform

Multiplication by a Constant. If $\mathfrak{F}\{f(t)\} = F(\omega)$, then

$$\mathfrak{F}\{Kf(t)\} = KF(\omega) \tag{23.26}$$

Proof: $\mathfrak{F}\{Kf(t)\} = \int_{-\infty}^{\infty} Kf(t)e^{-j\omega t}dt = K\int_{-\infty}^{\infty} f(t)e^{-j\omega t}ft = KF(\omega)$.

It follows that the FT of $K\delta(t)$ is K and the FT of a constant K is $2\pi K\delta(t)$.

Addition/Subtraction. If $\mathfrak{F}\{f_1(t)\} = F_1(\omega)$ and $\mathfrak{F}\{f_2(t)\} = F_2(\omega)$, then

$$\mathfrak{F}\{f_1(t) \pm f_2(t)\} = F_1(\omega) \pm F_2(\omega) \tag{23.27}$$

Proof: $\mathfrak{F}\{f_1(t) \pm f_2(t)\} = \int_{-\infty}^{\infty}\left[f_1(t) \pm f_2(t)\right]e^{-j\omega t}dt = \int_{-\infty}^{\infty} f_1(t)e^{-j\omega t}dt \pm \int_{-\infty}^{\infty} f_2(t)e^{-j\omega t}dt = F_1(\omega) \pm F_2(\omega)$. We have already used this property and the preceding one to derive the FTs of $\cos\omega t$ and $\sin\omega t$.

Time Scaling. If $\mathfrak{F}\{f(t)\} = F(\omega)$, then

$$\mathfrak{F}\{f(at)\} = \frac{1}{|a|}F\left(\frac{\omega}{a}\right) \tag{23.28}$$

where a is a constant.

Proof: $\mathfrak{F}\{f(at)\} = \int_{-\infty}^{\infty} f(at)e^{-j\omega t}dt$. Let $t' = at$, where $a > 0$. Then, $\int_{-\infty}^{\infty} f(at)e^{-j\omega t}dt = \frac{1}{a}\int_{-\infty}^{\infty} f(t')e^{-j\omega t'/a}dt' = \frac{1}{a}F\left(\frac{\omega}{a}\right)$. If $a < 0$, the integration limits become interchanged, so that $\mathfrak{F}\{f(at)\} = -\frac{1}{a}\int_{-\infty}^{\infty} f(t')e^{-j\omega t'/a}dt' = -\frac{1}{a}F\left(\frac{\omega}{a}\right)$. Combining these results gives Equation 23.28.

Concept: *If a function is compressed in the time domain, it expands in the frequency domain, and conversely.*

This follows from Equation 23.28. For example, the FT of a rectangular pulse of width τ, extending from $-\tau/2$ to $+\tau/2$, is $(A\tau)\text{sinc}(\omega\tau/2)$ (Equation 23.2), as illustrated in Figure 23.12a. If the pulse width is doubled to 2τ (Figure 23.12b), the width of the FT is halved, as indicated by

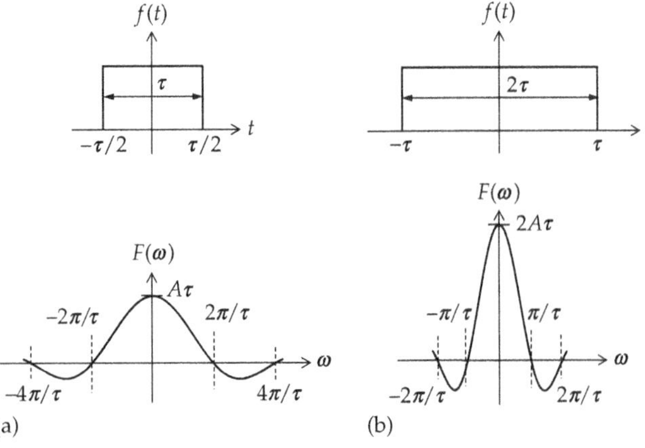

FIGURE 23.12
Relation between representation of a function in the time and frequency domains. (a) A rectangular function and its FT, and (b) if the duration of the time function is doubled, the FT is compressed, so that the zero crossings occur at half the frequencies in (a).

the successive zero crossings occurring at half the frequencies as in Figure 23.12a. Conversely, if the width of the rectangular pulse is halved, the width of the FT is doubled. As a practical application, if a voice recording is played back at double speed, for example, its duration is halved and its bandwidth is therefore doubled. This means that corresponding frequencies are doubled, so that the pitch of the voice is doubled.

Time Reversal. If $\mathfrak{F}\{f(t)\} = F(\omega)$, then, replacing t by $-t$ in Equation 23.3,

$$\mathfrak{F}\{f(-t)\} = \int_{\infty}^{-\infty} f(-t)e^{j\omega t}(-dt) = \int_{-\infty}^{\infty} f(-t)e^{j\omega t}(dt) = F(-\omega) \quad (23.29)$$

where the FT is obtained by replacing ω by $-\omega$. Applying Equation 23.15,

$$\mathfrak{F}\{f(-t)\} = F(-\omega) = A(\omega) - jB(\omega) = F^*(\omega) \quad (23.30)$$

In other words, a negation in time causes a negation in frequency. Equation 23.29 also follows from the time-scaling property when $a = -1$.

It follows from the time-reversal property that

$$\mathfrak{F}\{u(-t)\} = \pi\delta(\omega) - \frac{1}{j\omega} \quad (23.31)$$

so that $\mathfrak{F}\{\text{sgn}(t)\} = \mathfrak{F}\{u(t)\} - \mathfrak{F}\{u(-t)\} = 2/j\omega$, as in Equation 23.19.

Translation in Time. If $\mathfrak{F}\{f(t)\} = F(\omega)$, then

$$\mathfrak{F}\{f(t-a)\} = e^{-j\omega a}F(\omega) \quad (23.32)$$

In words, a delay of a in time reduces the phase angle of $F(\omega)$ by ωa without changing its magnitude, as is the case with the FSE (Equation 16.43).

Proof: $\mathfrak{F}\{f(t-a)\} = \int_{-\infty}^{\infty} f(t-a)e^{-j\omega t}dt$. Substituting $t' = t - a$, $\int_{-\infty}^{\infty} f(t-a)e^{-j\omega t}dt = \int_{-\infty}^{\infty} f(t')e^{-j\omega(t'+a)}dt' = e^{-j\omega a}$ $\int_{-\infty}^{\infty} f(t')e^{-j\omega t'}dt' = e^{-j\omega a}F(\omega)$.

An interesting application of the translation-in-time property is the *distortionless delay*. Consider a circuit having a magnitude frequency response K independent of frequency and a phase frequency response that is directly proportional to frequency that is $-a\omega$. If the FT of the input is denoted by $F(\omega) = |F(\omega)|\angle F(\omega)$, then the FT of the output is $K|F(\omega)|\angle F(\omega) - a\omega = e^{-j\omega a}F(\omega)$. It follows from Equation 23.32 that the output in the time domain is $Kf(t-a)$, that is, the output is scaled by K but delayed in time by a.

Example 23.4: Derivation of Fourier Transform of Biphasic Pulse from That of a Rectangular Pulse

It is required to obtain the FT of the function of Figure 23.3 from that of a symmetrical rectangular pulse.

Solution:

From Equation 23.2, the FT of a rectangular pulse of duration τ centered at $t = 0$ is $(2A/\omega)\sin(\omega\tau/2)$ (Equation 23.2). Replacing τ by $\tau/2$, the FT becomes $(2A/\omega)\sin(\omega\tau/4)$. If the pulse is delayed by $\tau/4$, its FT is $(2A/\omega)e^{-j\omega\tau/4}\sin(\omega\tau/4)$. If negated and advanced by $\tau/4$, its FT becomes $-(2A/\omega)e^{j\omega\tau/4}\sin(\omega\tau/4)$. The FT of the negative pulse can also be obtained from that of the positive pulse by negating the pulse and applying the time-reversal property. Adding the two FTs gives $(2A/\omega)\sin(\omega\tau/4)(e^{-j\omega\tau/4} - e^{j\omega\tau/4}) = -j(4A/\omega)\sin^2(\omega\tau/4) = -j(2A/\omega)(1-\cos(\omega\tau/2))$, as in Equation 23.17.

As an application of translation-in-time and duality, consider $\mathfrak{F}\{\delta(t-a)\} = e^{-ja\omega}$ (Equation 23.32). It follows from Equation 23.23 that $\mathfrak{F}\{e^{-jat}\} = 2\pi\delta(-\omega - a) = 2\pi\delta(\omega + a)$ because of the evenness of the impulse function. Setting $a = -\omega_0$ gives $\mathfrak{F}\{e^{j\omega_0 t}\} = 2\pi\delta(\omega - \omega_0)$, as in Equation 23.10 with $K = 1$.

The translation-in-time property is applied in Example 23.5 to derive the FT of a periodic signal.

Example 23.5: Fourier Transform of Periodic Function

It is required to obtain the FT of a periodic signal from the exponential form of its FSE.

Solution:

Consider the exponential form of the FSE of a periodic signal (Equation 16.30) $f(t) = \sum_{n=-\infty}^{\infty} C_n e^{jn\omega_0 t}$. Taking the FT of both sides and using Equation 23.10,

$$F(\omega) = \sum_{n=-\infty}^{\infty} 2\pi C_n \delta(\omega - n\omega_0) \quad (23.33)$$

As to be expected, the FT of the discrete frequencies of a periodic signal is an infinite series of impulses weighted by the Fourier coefficient C_n. As an example, consider the delayed square wave of Figure 16.14a. From Exercise 16.8, $C_n = 0$, for n even or zero, and $-j2A_m/\pi n$, for odd n. Its FT is therefore

$$F(\omega) = \sum_{\substack{n=-\infty \\ n\neq 0}}^{n=\infty} -(j4A_m/n)\delta(\omega - n\omega_0), \quad n \text{ odd}.$$

Figure 23.13 shows a plot of $|F(\omega)|$. The phase angle is $-\pi/2$ for $n > 0$ and $\pi/2$ for $n < 0$.

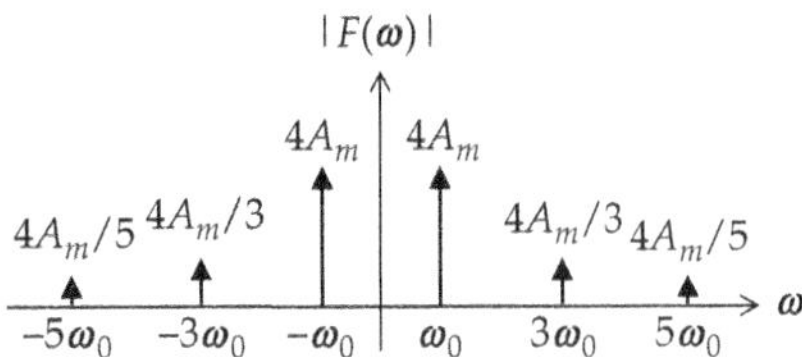

FIGURE 23.13
Figure for Example 23.5.

Note that for a cosine function, $\cos\omega t = \left(e^{j\omega t} + e^{-j\omega t}\right)/2$, which makes $C_n = 1/2$, $n = \pm 1$, and Equation 23.33 reduces to Equation 23.11. For a sine function, $\sin\omega t = \left(e^{j\omega t} + e^{-j\omega t}\right)/2j$, which makes $C_n = -j/2$ for $n = 1$ and $-j/2$ for $n = -1$. Equation 23.33 reduces to Equation 23.12. Note that $F(0) = 0$ in Figure 23.13.

Primal Exercise 23.7

Determine the FT of the periodic square waveform of Figure 23.14.

Ans. $5\pi\delta(t) + \frac{10}{j}\sum_{n=-\infty}^{\infty}\delta(\omega - n\pi)$, n is odd.

Differentiation in Time. If $\mathfrak{F}\{f(t)\} = F(\omega)$, then

$$\mathfrak{F}\left\{\frac{df(t)}{dt}\right\} = j\omega F(\omega) \quad (23.34)$$

Proof: $f(t) = \mathfrak{F}^{-1}\{F(\omega)\} = \frac{1}{2\pi}\int_{-\infty}^{\infty} F(\omega)e^{j\omega t}d\omega$. Differentiating both sides with respect to t: $\frac{df}{dt} = \frac{j\omega}{2\pi}\int_{-\infty}^{\infty} F(\omega)e^{j\omega t}d\omega$. In other words, $\frac{df}{dt} = j\omega\mathfrak{F}^{-1}\{F(\omega)\}$. Taking the FT of both sides gives Equation 23.34. Equation 23.34 is in accordance with differentiation in phasor analysis being equivalent to multiplication by $j\omega$.

From repeated application of Equation 23.34,

$$\mathfrak{F}\left\{\frac{df^n(t)}{dt}\right\} = (j\omega)^n F(\omega) \quad (23.35)$$

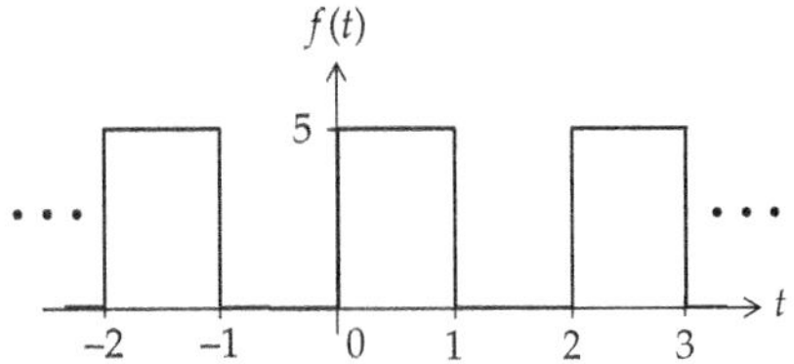

FIGURE 23.14
Figure for Primal Exercise 23.7

If $f(t)$ is a constant K, $df(t)/dt = 0$ for all values of K. $\mathfrak{F}\{K\} = 2\pi K\delta(\omega)$ and $\mathfrak{F}\{dK/dt\} = (j\omega)2\pi K\delta(\omega) = 0$. This is because $\delta(\omega) = 0$ for $\omega \neq 0$, and when $\omega = 0$, $0 \times \delta(\omega) = 0$. It follows that the differentiation-in-time property gives the same value of $\mathfrak{F}\{df(t)/dt\}$ when a constant K is added to $f(t)$. Thus, in the case of (1/2)sgn(t), the derivative is $\delta(t)$ so that the differentiation-in-time property gives $\mathfrak{F}\{\delta(t)\} = j\omega \times \frac{1}{2} \times \frac{2}{j\omega} = 1$. Adding 1/2 to (1/2)sgn(t) gives $u(t)$, whose derivative is still $\delta(t)$. $\mathfrak{F}\{\delta(t)\} = j\omega\mathfrak{F}\{u(t)\} = j\omega\pi\delta(\omega) + 1 = 0 + 1 = 1$.

Example 23.6: Application of Differentiation-In-Time Property

Given $f(t)$ of Figure 23.15, it is required to obtain the FT of its derivative $f^{(1)}(t)$ using the differentiation-in-time property.

Solution:

$f(t) = \frac{2A}{\tau}t$, $-\frac{\tau}{2} < t < \frac{\tau}{2}$. Hence, $F(\omega) = \frac{2A}{\tau}\int_{-\tau/2}^{\tau/2} te^{-j\omega t}dt = \frac{2A}{\tau}\left\{\left[\frac{te^{-j\omega t}}{-j\omega}\right]_{-\tau/2}^{\tau/2} + \frac{1}{\omega^2}\int_{-\tau/2}^{\tau/2} e^{-j\omega t}dt\right\} = -\frac{2A}{j\omega}\cos(\omega\tau/2) - \frac{j4A}{\tau\omega^2}\sin(\omega\tau/2) = \frac{2A}{j\omega^2}\left[-\omega\cos(\omega\tau/2) + \frac{2}{\tau}\sin(\omega\tau/2)\right]$. Since $f(t)$ is real and odd, $F(\omega)$ is imaginary and odd. Moreover, substituting $\omega = 0$ gives an indeterminate value for $F(0)$. To obtain $F(0)$, we apply L'Hopital's rule and differentiate the numerator and denominator with respect to ω, which gives $\frac{2A}{j2\omega}\left[\frac{\omega\tau}{2}\sin(\omega\tau/2) - \cos(\omega\tau/2) + \cos(\omega\tau/2)\right] = \frac{A\tau}{j2}\sin(\omega\tau/2)$. Substituting $\omega = 0$ gives $F(0) = 0$, as expected.

It follows that $\mathfrak{F}\{f^{(1)}(t)\} = j\omega F(\omega) = -2A\cos(\omega\tau/2) + \frac{4A}{\omega\tau}\sin(\omega\tau/2)$. It can be readily verified that this is the FT of $\mathfrak{F}\{f^{(1)}(t)\}$ shown in Figure 23.16. Using the time-shift property, the FT of the impulses at $t = -\tau/2$ and

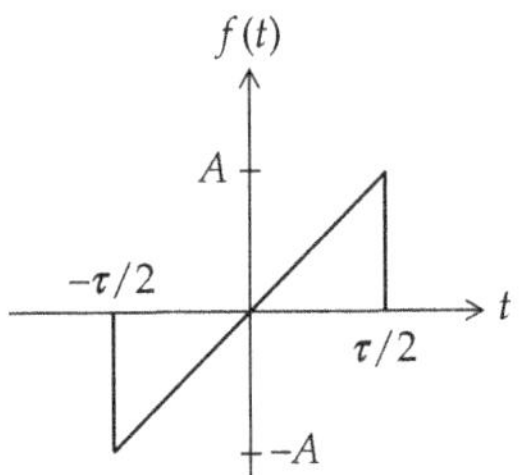

FIGURE 23.15
Figure for Example 23.6.

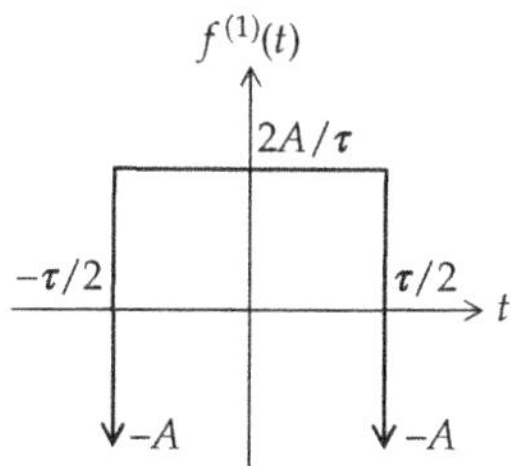

FIGURE 23.16
Figure for Example 23.6

$t = +\tau/2$ are $-Ae^{j\omega\tau/2}$ and $-Ae^{-j\omega\tau/2}$, respectively. From Equation 23.2, the FT of the pulse is $\frac{4A}{\omega\tau}\sin(\omega\tau/2)$. It follows that $f^{(1)}(t) = -Ae^{j\omega\tau/2} - Ae^{-j\omega\tau/2} + \frac{4A}{\omega\tau}\sin(\omega\tau/2) = -2A\cos(\omega\tau/2) + \frac{4A}{\omega\tau}\sin(\omega\tau/2)$, as derived earlier.

Exercise 23.8

Verify $F(\omega)$ of Figure 23.16 using Equation 23.3.

Primal Exercise 23.9

Consider $e^{-at}u(t)$, $a > 0$, whose FT is $1/(a + j\omega)$ (Equation 23.18). Since $\frac{d(e^{-at}u(t))}{dt} = -ae^{-at}u(t) + \delta(t)$, take the FT of both sides and apply the differentiation-in-time property to verify that $\mathcal{F}\left\{\frac{d(e^{-at}u(t))}{dt}\right\} = \frac{j\omega}{a + j\omega}$.

Integration in Time. If $\mathcal{F}\{f(t)\} = F(\omega)$, then

$$\mathcal{F}\left\{\int_{-\infty}^{t} f(t)dt\right\} = \frac{F(\omega)}{j\omega} + \pi F(0)\delta(\omega) \qquad (23.36)$$

where $F(0) = \int_{-\infty}^{\infty} f(t)dt$ (Equation 23.17).

***Proof*:** We will not give here a formal proof of Equation 23.36 but will justify it instead. Taking the derivative of both sides of Equation 23.36, the LHS is $\mathcal{F}\left\{\frac{d}{dt}\left[\int_{-\infty}^{t} f(t)dt\right]\right\} = \mathcal{F}\{f(t)\} = F(\omega)$. According to the differentiation-in-time property, multiplying the RHS of Equation 23.36 by $j\omega$ should give $F(\omega)$. This is clearly the case, for multiplying the RHS by $j\omega$ gives $F(\omega) + j\omega\pi F(0)\delta(\omega) = F(\omega) + 0$. Intuitively, if $f(t)$ includes a constant, the FT of its integral would include a term in $\delta(\omega)$. The form of $\pi F(0)\delta(\omega)$ in Equation 23.36 can be justified by considering $f(t) = \delta(t)$, so that $F(\omega) = 1$ and $F(0) = 1$. Applying Equation 23.36 gives $\mathcal{F}\{u(t)\} = 1/j\omega + \pi\delta(\omega)$, which is the correct expression.

Primal Exercise 23.10

Verify the differentiation-in-time and integration-in-time properties for $\cos\omega t$ and $\sin\omega t$. Note that multiplication or division by $j\omega$ changes an odd $F(\omega)$ to an even $F(\omega)$, and conversely.

Example 23.7: Application of Integration-in-Time Property

Given the function $g(t)$ of Figure 23.16, it is required to obtain the FT of its integral, the function of Figure 23.15, by applying the integration-in-time property.

Solution:

$g(t) = -A\delta(t + \tau/2) - A\delta(t - \tau/2) + 2A/\tau, -\tau/2^- < t < \tau/2^+$, and $g(t) = 0$ elsewhere. Hence, $G(\omega) = -Ae^{+j\omega\tau/2} - Ae^{-j\omega\tau/2} + \frac{4A}{\omega\tau}\sin(\omega\tau/2)$, using Equations 23.2 and 23.32, or $G(\omega) = -2A\cos(\omega\tau/2) + \frac{4A}{\omega\tau}\sin(\omega\tau/2)$. Moreover, $\int_{-\infty}^{\infty} g(t)dt = -A - A + 2A = 0$, which means that $G(0) = 0$. We can check this by expressing $G(\omega)$ as $\frac{2A}{\omega}\left[-\omega\cos(\omega\tau/2) + \frac{2}{\tau}\sin(\omega\tau/2)\right]$. Applying L'Hopital's rule by differentiating the numerator and denominator with respect to ω gives $2A\left[\frac{\omega\tau}{2}\sin(\omega\tau/2) - \cos(\omega\tau/2) + \cos(\omega\tau/2)\right] = A\omega\tau\sin(\omega\tau/2)$. Substituting $\omega = 0$ gives $G(0) = 0$. Using Equation 23.35, $\mathcal{F}\{G^{(-1)}(t)\} = \frac{G(\omega)}{j\omega} = j\frac{2A}{\tau\omega^2}\left[\omega\tau\cos(\omega\tau/2) - 2\sin(\omega\tau/2)\right]$, as in Example 23.6.

Primal Exercise 23.11

Derive the FT of the function of Figure 23.17 from that of Example 23.1 using the integration-in-time property.

Ans. $F(\omega) = \left(\frac{4A}{\omega^2\tau}\right)\left(1 - \cos\frac{\omega\tau}{2}\right) = \frac{A\tau}{2}\left(\text{sinc}\frac{\omega\tau}{4}\right)^2 \qquad (23.37)$

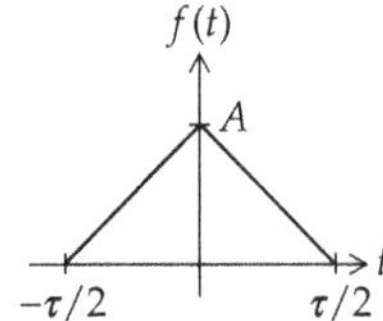

FIGURE 23.17
Figure for Primal Exercise 23.11.

Primal Exercise 23.12

Consider $e^{-at}u(t)$, $a > 0$, whose FT is $\frac{1}{a+j\omega}$. Since $\int_{-\infty}^{t} e^{-at}u(t) = \frac{1}{a}\left(1-e^{-at}\right)u(t)$, take the FT of both sides and apply the integration-in-time property to verify that $\mathfrak{F}\left\{\int_{-\infty}^{t} e^{-at}u(t)\right\} = \frac{1}{j\omega(a+j\omega)} + \frac{\pi\delta(\omega)}{a}$.

Multiplication by *t*. If $\mathfrak{F}\{f(t)\} = F(\omega)$, then

$$\mathfrak{F}\{tf(t)\} = j\frac{d}{d\omega}F(\omega) \tag{23.38}$$

Thus, multiplication by t corresponds to multiplying the derivative in frequency of $F(\omega)$ by j.

***Proof*:** $\frac{d}{d\omega}F(\omega) = \frac{d}{d\omega}\left[\int_{-\infty}^{\infty} f(t)e^{-j\omega t}dt\right] = \int_{-\infty}^{\infty} -jtf(t)e^{-j\omega_0 t}e^{-j\omega t}dt = -j\mathfrak{F}\{tf(t)\}$. Equation 23.38 then follows. Repeated application of Equation 23.38 gives

$$\mathfrak{F}\{t^n f(t)\} = j^n\frac{d^n}{d\omega^n}F(\omega) \tag{23.39}$$

As an application of Equation 23.38, consider $\mathfrak{F}\{e^{-at}u(t)\} = \frac{1}{a+j\omega}$ (Equation 23.18). It follows that

$$\mathfrak{F}\{te^{-at}u(t)\} = -\frac{j\times j}{(a+j\omega)^2} = \frac{1}{(a+j\omega)^2} \tag{23.40}$$

$$\mathfrak{F}\{t^2e^{-at}u(t)\} = -\frac{j\times 2j}{(a+j\omega)^3} = \frac{2}{(a+j\omega)^3} \tag{23.41}$$

and

$$\mathfrak{F}\{t^ne^{-at}u(t)\} = \frac{n!}{(a+j\omega)^{n+1}} \tag{23.42}$$

Translation in Frequency. If $\mathfrak{F}\{f(t)\} = F(\omega)$, then

$$\mathfrak{F}\{f(t)e^{j\omega_0 t}\} = F(\omega-\omega_0) \tag{23.43}$$

Hence, a translation in the frequency domain is equivalent to a phase shift in the time domain.

***Proof*:** $\mathfrak{F}\{f(t)e^{j\omega_0 t}\} = \int_{-\infty}^{\infty} f(t)e^{j\omega_0 t}e^{-j\omega t}dt = \int_{-\infty}^{\infty} f(t)e^{-j(\omega-\omega_0)t}dt = F(\omega-\omega_0)$.

Example 23.8: Fourier Transforms for sine and cosine Functions That Are Multiplied by Unit Step Functions

It is required to derive the FTs of (a) $\{(\cos\omega_0 t)u(t)\}$ and (b) $\{(\sin\omega_0 t)u(t)\}$.

Solution:

(a) We express $\cos\omega_0 t$ as $\frac{1}{2}\left(e^{j\omega_0 t}+e^{-j\omega_0 t}\right)$. Hence, $\mathfrak{F}\{(\cos\omega_0 t)u(t)\} = \mathfrak{F}\{(1/2)e^{j\omega_0 t}u(t)+(1/2)e^{-j\omega_0 t}u(t)\}$. Applying the translation-in-frequency property (Equation 23.43) to Equation 23.20 gives

$$\mathfrak{F}\{(\cos\omega_0 t)u(t)\} = \frac{\pi}{2}\left[\delta(\omega-\omega_0)+\delta(\omega+\omega_0)\right] + \frac{j\omega}{\omega_0^2-\omega^2} \tag{23.44}$$

(b) We express $\sin\omega_0 t$ as $\frac{1}{2j}\left(e^{j\omega_0 t}-e^{-j\omega_0 t}\right)$. Hence, $\mathfrak{F}\{(\sin\omega_0 t)\,u(t)\} = \mathfrak{F}\left\{\frac{1}{2j}e^{j\omega_0 t}u(t) - \frac{1}{2j}e^{-j\omega_0 t}u(t)\right\}$. Applying the translation-in-frequency property (Equation 23.43) to Equation 23.20 gives

$$\mathfrak{F}\{(\sin\omega_0 t)u(t)\} = \frac{\pi}{2j}\left[\delta(\omega-\omega_0)+\delta(\omega+\omega_0)\right] + \frac{\omega_0}{\omega_0^2-\omega^2} \tag{23.45}$$

Convolution in Time.

$$\mathfrak{F}\{f(t)*g(t)\} = F(\omega)G(\omega) \tag{23.46}$$

In words, the FT of the convolution of two time functions equals the product of their FTs.

***Proof*:** From Equation 20.1,

$$y(t) = f(t)*g(t) = \int_{-\infty}^{\infty} f(\lambda)g(t-\lambda)d\lambda \tag{23.47}$$

The FT of $y(t)$ is

$$Y(\omega) = \int_{-\infty}^{\infty}\left[\int_{-\infty}^{\infty} f(\lambda)g(t-\lambda)d\lambda\right]e^{-j\omega t}dt \tag{23.48}$$

Changing the order of integration and taking $f(\lambda)$ outside the integral with respect to t,

$$Y(\omega)=\int_{-\infty}^{\infty}f(\lambda)\left[\int_{-\infty}^{\infty}g(t-\lambda)dt\right]e^{-j\omega t}d\lambda \tag{23.49}$$

The inner integral is the FT of the function $g(t)$ translated in time by λ. From Equation 23.32, this FT is $e^{-j\omega\lambda}G(j\omega)$. Taking $G(\omega)$ outside the integral with respect to λ,

$$Y(\omega)=G(\omega)\int_{-\infty}^{\infty}f(\lambda)e^{-j\omega\lambda}d\lambda=G(\omega)F(\omega) \tag{23.50}$$

Equation 23.46 follows, bearing in mind that the ordering of the two functions on either side of the equation is immaterial because convolution, like multiplication, is commutative.

The convolution-in-time property can be applied to derive the integration-in-time property. From Equation 20.33,

$$f(t)*u(t)=\int_{0}^{t}f(\lambda)d\lambda \tag{23.51}$$

Taking the FT of both sides and applying Equation 23.46,

$$\mathfrak{F}\left\{\int_{0}^{t}f(\lambda)d\lambda\right\}=F(\omega)\mathfrak{F}\{u(t)\}$$

$$=F(\omega)\left(\pi\delta(\omega)+\frac{1}{j\omega}\right)=\frac{F(\omega)}{j\omega}+\pi F(0)\delta(\omega) \tag{23.52}$$

As an example of the convolution-in-time property, consider $f(t) = A\text{rect}(t/\tau)$, convolved with itself, as in Example 20.6. The convolution integral (Figure 20.35) is reproduced in Figure 23.18. This is the same triangular waveform considered in Exercise 23.11. Its FT can be obtained from Equation 23.37 by replacing $\tau/2$ by τ and A by $A^2\tau$. This gives $A^2\tau^2\text{sinc}^2(\omega\tau/2)$. The FT of $f(t) = A\text{rect}(t/\tau)$ is, from Equation 23.2, $A\tau\text{sinc}(\omega\tau/2)$. When the FTs of each $f(t)$ are multiplied together, in accordance with the convolution-in-time property, the product is $A^2\tau^2\text{sinc}^2(\omega\tau/2)$, as it should be.

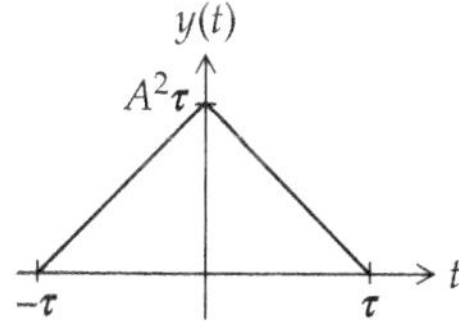

FIGURE 23.18
Convolution integral of the function $A\text{rect}(t/\tau)$ when convolved with itself.

Primal Exercise 23.13

Determine $y(t)=\text{sinc}\left(\frac{\omega_1}{2}t\right)*\cos\omega_2 t$ if (a) $\omega_2 > \omega_1/2$ and (b) $\omega_2 < \omega_1/2$.

Ans. (a) 0; (b) $(1/\omega_1)\cos\omega_2 t$.

Convolution in Frequency.

$$\mathfrak{F}\{f(t)g(t)\}=\frac{1}{2\pi}F(\omega)*G(\omega) \tag{23.53}$$

In words, the FT of the product of two time functions equals the convolution of their FTs divided by 2π.

Proof: The proof exploits the symmetry between the FT and the IFT. We first express the convolution of the FTs of two signals in the frequency domain as

$$Y(\omega)=F(\omega)*G(\omega)=\int_{-\infty}^{\infty}F(\lambda)G(\omega-\lambda)d\lambda \tag{23.54}$$

The IFT of $Y(\omega)$ is

$$y(t)=\frac{1}{2\pi}\int_{-\infty}^{\infty}\left[\int_{-\infty}^{\infty}F(\lambda)G(\omega-\lambda)d\lambda\right]e^{j\omega t}d\omega \tag{23.55}$$

Changing the order of integration, moving $e^{j\omega t}$ inside the square brackets, and taking $F(\lambda)$ outside the integral with respect to ω,

$$y(t)=\frac{1}{2\pi}\int_{-\infty}^{\infty}F(\lambda)\left[\int_{-\infty}^{\infty}G(\omega-\lambda)e^{j\omega t}d\omega\right]d\lambda \tag{23.56}$$

The inner integral is 2π times the IFT of the function $G(\omega)$ translated in frequency by λ. From Equation 23.43, this integral is $2\pi e^{j\lambda t}g(t)$. Taking $g(t)$ outside the integral with respect to λ,

$$y(t)=g(t)\int_{-\infty}^{\infty}F(\lambda)e^{jt\lambda}d\lambda=2\pi g(t)f(t) \tag{23.57}$$

Taking the FT of both sides and dividing by 2π gives Equation 23.51.

TABLE 23.1

Operational Properties of the Fourier Transform $\mathfrak{F}\{f(t)\} = F(\omega)$

$\mathfrak{F}\{Kf(t)\}$	$KF(j\omega)$
$\mathfrak{F}\{f_1(t) \pm f_2(t)\}$	$F_1(j\omega) \pm F_2(j\omega)$
$\mathfrak{F}\{f(at)\}$	$\frac{1}{\lvert a\rvert}F\left(\frac{j\omega}{a}\right)$
$\mathfrak{F}\{f(-t)\}$	$F(-j\omega) = F^*(j\omega)$
$\mathfrak{F}\{f(t-a)\}$	$e^{-j\omega a}F(j\omega)$
$\mathfrak{F}\left\{\frac{df(t)}{dt}\right\}$	$j\omega F(j\omega)$
$\mathfrak{F}\left\{\frac{df^n(t)}{dt}\right\}$	$(j\omega)^n F(j\omega)$
$\mathfrak{F}\left\{\int_{-\infty}^{t} f(t)dt\right\}$	$\frac{F(j\omega)}{j\omega} + \pi F(0)\delta(\omega)$
$\mathfrak{F}\{tf(t)\}$	$j\frac{d}{d\omega}F(j\omega)$
$\mathfrak{F}\{t^n f(t)\}$	$j^n \frac{d^n}{d\omega^n}F(j\omega)$
$\mathfrak{F}\{f(t)e^{j\omega_0 t}\}$	$F(j(\omega-\omega_0))$
$\mathfrak{F}\{F(t)\}$	$2\pi f(-\omega)$
$\mathfrak{F}\{f(t) * g(t)\}$	$F(j\omega)G(j\omega)$
$\mathfrak{F}\{f(t)g(t)\}$	$\frac{1}{2\pi}F(j\omega) * G(j\omega)$

As a simple application of the convolution-in-frequency property, consider a function $f(t)$ and $g(t) = K$, a constant. The FT of the product is $KF(\omega)$ (Equation 23.26). According to Equation 23.53,

$$KF(\omega) = \frac{1}{2\pi}\left(F(\omega) * \mathfrak{F}\{K\}\right)$$

$$= \frac{1}{2\pi}\left(F(\omega) * 2\pi K\delta(\omega)\right) = KF(\omega) * \delta(\omega) \quad (23.58)$$

The convolution of a function with an impulse at the origin is the function itself (Equation 20.32), so the RHS of Equation 23.58 reduces to $KF(\omega)$.

Table 23.1 summarizes the basic properties of the FT, and Table 23.2 lists some useful FT pairs.

TABLE 23.2

Fourier Transform Pairs

$f(t)$	$F(j\omega)$	Reference Equation[a]
$\delta(t)$	1	(23.5)
1	$2\pi\delta(\omega)$	(23.9)
$u(t)$	$\pi\delta(\omega) + \frac{1}{j\omega}$	(23.20)
$u(t+\tau/2) - u(t-\tau/2) = \mathrm{rect}(t/\tau)$	$\tau\,\mathrm{sinc}(\omega\tau/2)$	(23.2)
$\frac{1}{2\pi}\omega_0\mathrm{sinc}(\omega_0 t/2)$	$u(\omega+\omega_0/2) - u(\omega-\omega_0/2) = \mathrm{rect}(\omega/\omega_0)$	(23.25)
$\mathrm{sgn}(t)$	$= \frac{2}{j\omega}$	(23.19)
$e^{-at}u(t), a>0$	$\frac{1}{a+j\omega}$	(23.18)
$t^n e^{-at}u(t)$	$\frac{n!}{(a+j\omega)^{n+1}}$	(23.42)
$e^{-a\lvert t\rvert}, a>0$	$\frac{2a}{a^2+\omega^2}$	(23.14)
$e^{j\omega_0 t}$	$2\pi\delta(\omega-\omega_0)$	(23.10)
$\cos\omega_0 t$	$\pi[\delta(\omega+\omega_0)+\delta(\omega-\omega_0)]$	(23.11)
$\sin\omega_0 t$	$j\pi[\delta(\omega+\omega_0)-\delta(\omega-\omega_0)]$	(23.12)
$(\cos\omega_0 t)u(t)$	$\left\{\frac{\pi}{2}[\delta(\omega+\omega_0)+\delta(\omega-\omega_0)] + \frac{j\omega}{\omega_0^2-\omega^2}\right\}$	(23.44)
$(\sin\omega_0 t)u(t)$	$\left\{\frac{\pi}{2j}[\delta(\omega+\omega_0)+\delta(\omega-\omega_0)] + \frac{\omega_0}{\omega_0^2-\omega^2}\right\}$	(23.45)
$e^{-at}(\cos\omega_0 t)u(t), a>0$	$\frac{a+j\omega}{\omega_0^2+(a+j\omega)^2}$	(23.21)
$e^{-at}\sin\omega_0 t u(t), a>0$	$\frac{\omega_0}{\omega_0^2+(a+j\omega)^2}$	(23.22)

23.4 Circuit Applications of the Fourier Transform

Generally speaking, the LT transform is more useful than the FT transform in circuit applications because (1) it can easily account for initial conditions; (2) it provides a powerful tool for analyzing switched circuits, particularly those involving impulsive readjustment at the instant of switching; and (3) some functions have an LT but not an FT. On the other hand, the FT is useful in some circuit analysis problems in that it can readily handle functions defined over all time, $-\infty < t < \infty$.

The general procedure of applying the FT in circuit analysis is a generalization of the phasor method. The circuit is represented in the frequency domain, as in phasor analysis, where R, L, and C are represented in terms of their impedances as R, $j\omega L$, and $1/j\omega C$, respectively. Phasor voltages and currents, including excitations, are replaced by their corresponding FTs. The usual circuit techniques are applied to derive the FT of any desired response. The IFT then gives the desired response in the time domain.

The transfer function $H(\omega)$ is defined in terms of the FT as:

$$H(\omega)=\frac{Y(\omega)}{X(\omega)} \tag{23.59}$$

where $X(\omega)$ and $Y(\omega)$ are the FTs of the excitation and response, respectively. Equation 23.59 is the same as Equation 22.42 for the LT, with s replaced by ω. If $x(t) = \delta(t)$, $X(\omega) = 1$ and the FT of the response $y(t)$ is $H(\omega)$. Thus, the FT of the response $h(t)$ to a unit impulse at $t = 0$ is the transfer function $H(\omega)$, and the IFT of $H(\omega)$ is $h(t)$.

Example 23.9: Responses of *RC* Circuit

It is required to obtain the responses v_R and v_C in Figure 23.19 when v_{SRC} is (a) $\delta(t)$ and (b) $u(t)$.

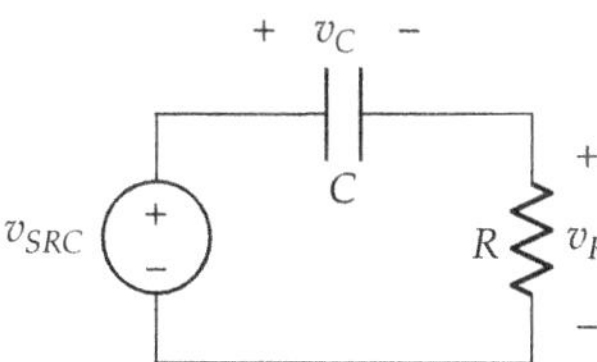

FIGURE 23.19
Figure for Example 23.9.

Solution:

The two transfer functions are independent of the type of excitation. Thus, $H_R(\omega)=\frac{V_R(\omega)}{V_{src}(\omega)}=\frac{R}{R+1/j\omega C}=\frac{j\omega}{(1/\tau)+j\omega}$; $H_C(\omega)=\frac{V_C(\omega)}{V_{src}(\omega)}=\frac{1/j\omega C}{R+1/j\omega C}=\frac{1}{\tau}\frac{1}{(1/\tau)+j\omega}$.

(a) $V_{src}(\omega)=1$. Hence, $V_C(\omega)=\frac{1}{\tau}\frac{1}{(1/\tau)+j\omega}$; from Table 23.2, $v_C(t)=\frac{1}{\tau}e^{-t/\tau}u(t)$. $V_R(\omega)=\frac{j\omega}{(1/\tau)+j\omega}$; dividing the numerator by the denominator, $V_R(\omega)=1-\frac{1/\tau}{(1/\tau)+j\omega}$, so that $v_R(t)=\delta(t)-\frac{1}{\tau}e^{-t/\tau}u(t)$.

These are evidently the correct responses, as follows from the natural response of the circuit. Note that an initial voltage equal to the strength of the impulse divided by τ can be easily accounted for in this simple case by applying an impulse at $t = 0$ and taking the responses at $t \geq 0^+$, after the impulse is over.

(b) $V_{src}(\omega)=\frac{1}{j\omega}+\pi\delta(\omega)$. Hence, $V_R(\omega)=\frac{1}{(1/\tau)+j\omega}+\frac{j\omega\pi\delta(\omega)}{(1/\tau)+j\omega}$. The IFT of the first term is $v_R(t)=e^{-t/\tau}u(t)$. The inverse of the second term is $\frac{1}{2\pi}\int_{-\infty}^{\infty}\frac{j\omega\pi\delta(\omega)}{(1/\tau)+j\omega}e^{j\omega t}d\omega=\frac{1}{2\pi}\int_{0^-}^{0^+}0\times\delta(\omega)d\omega=0\times 1=0.$

As for $V_C(\omega)$, we have $V_C(\omega)=\frac{1}{\tau}\frac{1}{(1/\tau)+j\omega}\times\left[\frac{1}{j\omega}+\pi\delta(\omega)\right]=\frac{1}{j\omega\tau}\frac{1}{(1/\tau)+j\omega}+\frac{\pi\delta(\omega)}{\tau}\frac{1}{(1/\tau)+j\omega}$.

To find the IFT of the first term, we express it in the form of partial fractions (Section 21.3). Although unnecessary in this case, it is usually more convenient when finding the partial fraction expansion (PFE) to substitute $s = j\omega$, determine the PFE, and substitute back for s. The first term in terms of partial fractions involving s becomes $\frac{1}{s\tau}\frac{1}{(1/\tau)+s}=\frac{K_1}{s}+\frac{K_2}{(1/\tau)+s}$, where K_1 and K_2 are constants to be determined. To determine K_1 we multiply both sides by s and substitute $s = 0$. This gives $K_1 = 1$. To determine K_2 we multiply both sides by $(1/\tau) + s$ and substitute $s = -1/\tau$, which gives $K_2 = -1$. Replacing s by $j\omega$, $\frac{1}{j\omega\tau}\frac{1}{(1/\tau)+j\omega}=\frac{1}{j\omega}-\frac{1}{(1/\tau)+j\omega}$.

As for the second term, it evaluates to zero for all ω except $\omega = 0$, for which it becomes $\pi\delta(\omega)$. Hence, $V_C(\omega)=\frac{1}{j\omega}+\pi\delta(\omega)-\frac{1}{(1/\tau)+j\omega}$. The IFT is $v_C(t)=\left(1-e^{-t/\tau}\right)u(t)$. Both $v_R(t)$ and $v_C(t)$ are in agreement with the step response of the circuit.

Evidently, there is no real advantage in obtaining the steady-state response to $\cos\omega_0 t$ using the FT, because the magnitude and phase of the output with respect to the input are given by $H(\omega)$.

Problem-Solving Tip

- When deriving the IFT by PFE, it is convenient and less error prone to replace $j\omega$ by s.

Primal Exercise 23.14

Determine v_R in Figure 23.19 in response to (a) v_{SRC} = sgn(t), both directly and by expressing sgn(t) as

$-1 + 2u(t)$, and (b) $\cos\omega_0 t$ using the FT, and verify using phasor analysis.

Ans. (a) $2e^{-t/\tau}$ V; (b) $\dfrac{-\tau\omega_0}{\sqrt{1+(\tau\omega_0)^2}}\sin(\omega_0 t-\beta)$ V, where $\tan\beta = \tau\omega_0$.

Example 23.10: Responses of *RL* Circuit

Consider a simple series *RL* circuit having $R = 1\ \Omega$ and $L = 1$ H. It is required to determine *i* for all *t* given that $v_{SRC}(t) = 10e^{-3|t|}$ V (Figure 23.20). Note that because $v_{SRC}(t)$ extends to $-\infty$, this problem cannot be analyzed by the single-sided LT.

Solution:

The transfer function of current in terms of applied voltage is $H(\omega) = \dfrac{I(\omega)}{V_{src}(\omega)} = \dfrac{1}{1+j\omega}$. From Exercise 23.1, $V_{src}(\omega) = \dfrac{60}{9+\omega^2}$. Hence, $I(\omega) = \dfrac{60}{(1+j\omega)(9+\omega^2)}$. To facilitate inverting $I(\omega)$, we substitute $s = j\omega$, as in Example 23.9, and express the rational function in *s* in terms of partial fractions of factors of the denominator. This gives $\dfrac{60}{(1+s)(9-s^2)} = \dfrac{K_1}{1+s} + \dfrac{K_2}{3+s} + \dfrac{K_3}{3-s}$. To find K_1, we multiply both sides by $(1 + s)$ and substitute $s = -1$, which gives $K_1 = 7.5$. In a similar manner, we find $K_2 = -5$ and $K_3 = 2.5$. Replacing *s* by $j\omega$, it follows that $I(\omega) = \dfrac{7.5}{1+j\omega} - \dfrac{5}{3+j\omega} + \dfrac{2.5}{3-j\omega}$. We now have to invert each of these terms. From Table 23.2, the inverse of the first two terms is $(7.5e^{-t} - 5e^{-3t})u(t)$. To invert the third term, we note that according to the time-reversal property (Equation 23.29), $\mathfrak{F}^{-1}\{F(-\omega)\} = f(-t)$. Hence, the IFT of the last term is $(2.5e^{3t})u(-t)$. It follows that $i(t) = (2.5e^{3t})u(-t) + (7.5e^{-t} - 5e^{-3t})u(t)$, which means that $i(t) = 2.5e^{3t}$, for $t \le 0$, and $i(t) = 7.5e^{-t} - 5e^{-3t}$, for $t \ge 0$. *i*(*t*) is plotted in Figure 23.21.

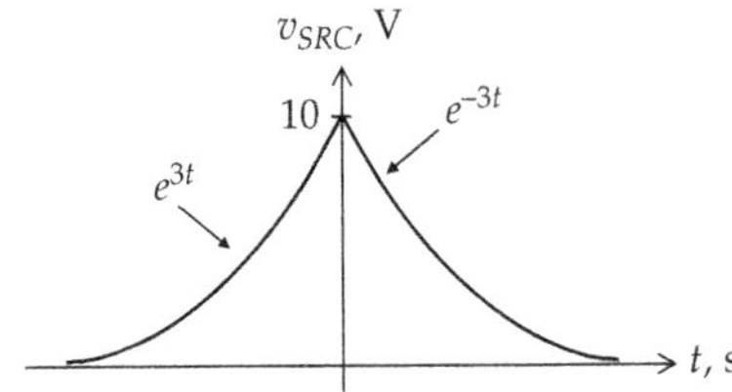

FIGURE 23.20
Figure for Example 23.10.

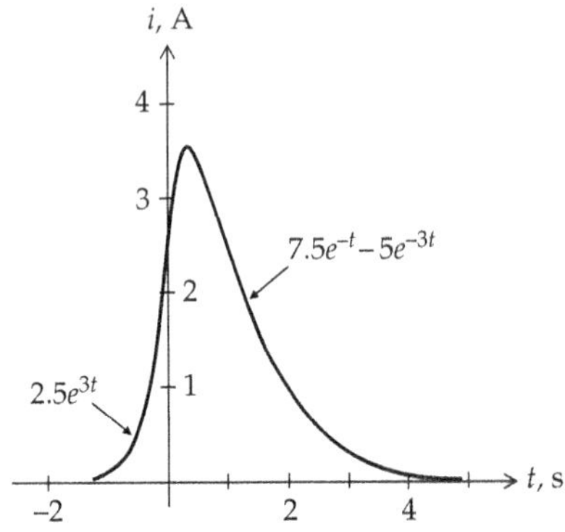

FIGURE 23.21
Figure for Example 23.10.

Exercise 23.15

Verify the expression for *i*(*t*) in Example 23.10 using the convolution integral.

23.5 Parseval's Theorem

The power dissipated in a resistor *R* by a voltage *v*(*t*) or a current *i*(*t*) is, respectively, $v^2(t)/R$ and $Ri^2(t)$. When $R = 1\ \Omega$, this power can be conveniently expressed as $f^2(t)$, where *f*(*t*) could be voltage or current. The energy dissipated in the 1 Ω resistor over all time is then $w(t) = \int_{-\infty}^{+\infty} f^2(t)dt$. Parseval's theorem states that

$$\int_{-\infty}^{\infty} f^2(t)dt = \frac{1}{2\pi}\int_{-\infty}^{\infty} |F(\omega)|^2 d\omega \tag{23.60}$$

That is, the energy dissipated in the 1 Ω resistor can be calculated either in the time domain or in the frequency domain using $|F(\omega)|$.

Proof:

$$\int_{-\infty}^{\infty} f^2(t)dt = \int_{-\infty}^{\infty} f(t)f(t)dt$$

$$= \int_{-\infty}^{\infty} f(t)\left[\frac{1}{2\pi}\int_{-\infty}^{\infty} F(\omega)e^{j\omega t}d\omega\right]dt \tag{23.61}$$

Moving *f*(*t*) inside the integral with respect to ω,

$$\int_{-\infty}^{\infty} f^2(t)dt = \frac{1}{2\pi}\int_{-\infty}^{\infty}\left[\int_{-\infty}^{\infty} F(\omega)f(t)e^{j\omega t}d\omega\right]dt \tag{23.62}$$

Interchanging the order of integration and moving $F(\omega)$ out of the integral with respect to t,

$$\int_{-\infty}^{\infty} f^2(t)dt = \frac{1}{2\pi}\int_{-\infty}^{\infty} F(\omega)\left[\int_{-\infty}^{\infty} f(t)e^{j\omega t}dt\right]d\omega \quad (23.63)$$

The inner integral is $F(-\omega) = F^*(j\omega)$. Substituting in Equation 23.63 gives Equation 23.60.

The following should be noted concerning Parseval's theorem:

1. Since $|F(\omega)|$ is even (Section 23.2), Equation 23.60 could be equally expressed as

$$\int_{-\infty}^{\infty} f^2(t)dt = \frac{1}{\pi}\int_{0}^{\infty} |F(\omega)|^2 d\omega \quad (23.64)$$

2. The plot of $|F(\omega)|^2$ vs. ω is the **energy spectrum** of the signal. The energy in any frequency band from ω_1 to ω_2 is, from Equation 23.64,

$$W_{12} = \frac{1}{\pi}\int_{\omega_1}^{\omega_2} |F(\omega)|^2 d\omega \quad (23.65)$$

Note that Equation 23.65 is used for W_{12}, and not Equation 23.60, because it accounts for positive and negative frequencies. To use Equation 23.60, we would have to integrate from $-\omega_2$ to $-\omega_1$ and from ω_1 to ω_2. These integrals are equal because $|F(\omega)|$ is even, which leads to Equation 23.65.

3. According to Equation 23.65, $\frac{|F(j\omega)|^2}{\pi}d\omega$ is the energy in an infinitesimal band of frequencies $d\omega$, so that $|F(\omega)|^2$ is π times the energy per radian of bandwidth or twice the energy per hertz.
4. Some types of time functions, such as random noise, are more conveniently represented in the frequency domain than in the time domain. Parseval's theorem allows calculation of the power associated with any band of noise frequencies and hence assesses the relative contribution of such a band to the total noise energy.
5. Because it establishes a direct relation between energy in the time domain and energy in the frequency domain, Parseval's theorem implies conservation of power and energy in the frequency domain.
6. Signals having a finite $\int_{-\infty}^{\infty} f^2(t)dt$ are referred to as **energy signals**. In the case of periodic signals, this integral tends to infinity.

Example 23.11: Parseval's Theorem

A voltage $4e^{-2t}u(t)$ V is applied to a 20 Ω resistor. It is required to determine (a) the total energy dissipated in both the time and frequency domains, (b) the energy associated with the frequency band $0 < f < 10$ Hz, and (c) the time interval over which an equal energy is dissipated.

Solution:

(a) In the time domain, $W_{1\Omega} = \int_0^{\infty} 16e^{-4t}dt = 4$ J. The energy dissipated in the 20 Ω resistor by the applied voltage is $4/20 = 0.2$ J. In the frequency domain, $F(\omega) = \frac{4}{2+j\omega}$, $|F(\omega)|^2 = \frac{16}{4+\omega^2}$, and $W_{1\Omega} = \frac{1}{\pi}\int_0^{\infty}\frac{16}{4+\omega^2}d\omega = \frac{16}{\pi}\left[\frac{1}{2}\tan^{-1}\frac{\omega}{2}\right]_0^{\infty} = \frac{16}{\pi}\times\frac{1}{2}\times\frac{\pi}{2} = 4$ J, as determined earlier.

(b) The energy content of the given frequency range is $\frac{1}{\pi}\int_0^{20\pi}\frac{16}{4+\omega^2}d\omega = \frac{16}{\pi}\left[\frac{1}{2}\tan^{-1}\frac{\omega}{2}\right] = \frac{8}{\pi}\left(\tan^{-1}10\pi\right) =$ 3.92 J. The relative energy content over the given frequency range is $= \frac{3.92}{4}\times 100 \cong 98\%$.

(c) The energy dissipated in a 1 Ω resistor from 0 to t is $\int_0^t 16e^{-4t}dt = 4\left(1-e^{-4t}\right)$. Equating this to 3.92 J gives $t = 0.98$ s.

Primal Exercise 23.16

Given that the FT of current in a 5 Ω resistor is $10e^{j\omega}$ As. Determine the total energy dissipated in the resistor.

Ans. $250/\pi$ J.

Primal Exercise 23.17

$v(t) = 10\text{sinc}(10\pi t)$ V is applied to a 6 Ω resistor. Determine the energy dissipated in the resistor in the frequency band from 2 Hz to infinity.

Ans. 1 J.

Example 23.12: Energies in a Lowpass Filter

The voltage $4e^{-2t}u(t)$ V of Example 23.11 is applied to a simple lowpass RC filter having $R = 10$ Ω and $C = 0.1$ F. It is required to determine the energy dissipated in the resistor and that supplied by the source as $t \to \infty$.

Solution:

The problem can be conveniently solved in the frequency domain. $V_i(\omega)=\dfrac{4}{2+j\omega}$, and the transfer function of the voltage across R is $H_R(\omega)=\dfrac{R}{R+1/j\omega C}=\dfrac{j\omega}{1+j\omega}; V_R(\omega)=\dfrac{j4\omega}{(2+j\omega)(1+j\omega)}$ and $|V_R(\omega)|^2=\dfrac{16\omega^2}{(4+\omega^2)(1+\omega^2)}=\dfrac{16}{3}\left[-\dfrac{1}{1+\omega^2}+\dfrac{4}{4+\omega^2}\right]$. It follows that $W_R=\dfrac{16}{3\pi}\int_0^{\infty}-\dfrac{1}{1+\omega^2}d\omega+\dfrac{16}{3\pi}\int_0^{\infty}\dfrac{4}{4+\omega^2}d\omega=$ $\dfrac{16}{3\pi}\left[-\tan^{-1}\omega+2\tan^{-1}\left(\dfrac{\omega}{2}\right)\right]=\dfrac{16}{3\pi}\left[-\dfrac{\pi}{2}+\pi\right]=\dfrac{8}{3}$ J. This is the energy dissipated in a 1 Ω resistor. The energy dissipated in the 10 Ω resistor by the applied voltage is 8/30 = 0.267 J.

Simulation: The schematic is entered as in Figure 23.22. A VEXP source is used, the source parameters being set as indicated for the $4e^{-2t}u(t)$V input (Appendix C). The voltage across R is the difference between the voltages at its two terminals, obtained using a subtracting DIFF part from the ABM library. The voltage is squared by applying it to both inputs of a multiplier MULT part from the ABM library. The multiplier output is integrated by the INTEG part from the ABM library, shown on the RHS of the figure. Setting the gain of the integrator to 0.1, with zero initial value, gives $W_R=\dfrac{1}{R}\int_0^t V_R^2 dt$, which is the energy dissipated in the resistor up to time t. Another multiplier multiplies the voltage across R, which is R times the current in the circuit, by the source voltage to give the instantaneous power delivered by the source. Integrating this by the integrator of gain 0.1 on the LHS of the figure gives $W_{SRC}=\int_0^t (V_{SRC})\left(\dfrac{V_R}{R}\right)dt$, which is the power delivered by the source up to time t. Figure 23.23 shows the plot of W_{SRC}, W_R, and their difference W_C, the energy in the capacitor at any time t. At large values of time, the source voltage approaches zero, the capacitor is discharged, and the energy supplied by the source is equal to that dissipated in the resistor. This energy is 0.27 J, in agreement with that calculated earlier.

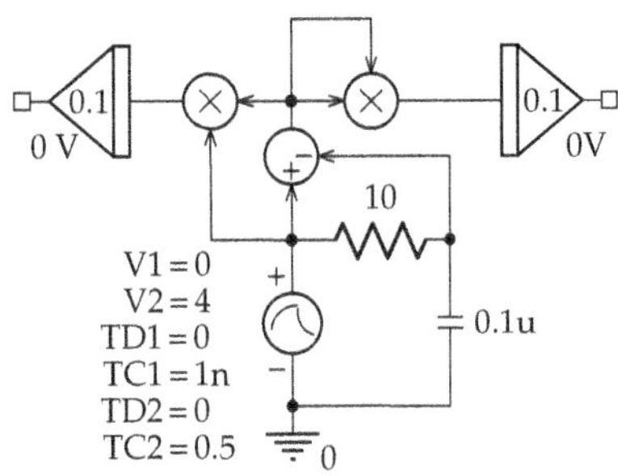

FIGURE 23.22
Figure for Example 23.12.

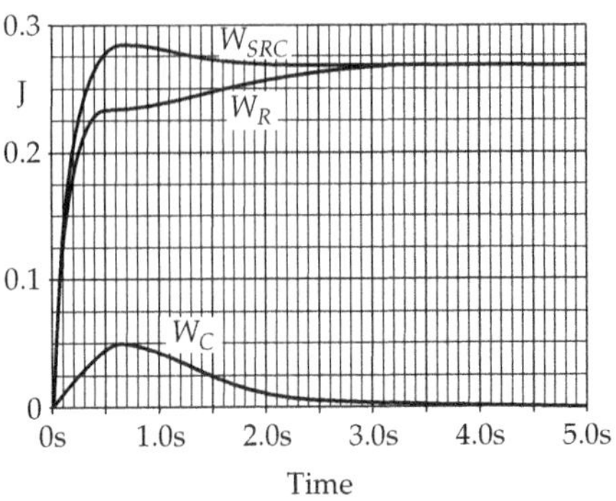

FIGURE 23.23
Figure for Example 23.12.

Primal Exercise 23.18

The voltage $4e^{-2t}u(t)$V is applied to the same filter of Example 23.12 used as a highpass filter. Using the results of Example 23.12, determine the energy dissipated in the resistor and that supplied by the source.

Ans. The same as in Example 23.11.

Learning Checklist: What Should Be Learned from This Chapter

- The FT is defined as $\mathfrak{F}\{f(t)\}=F(\omega)=\int_{-\infty}^{\infty} f(t)e^{-j\omega t}dt$, and the IFT as

$$\mathfrak{F}^{-1}\{F(\omega)\}=f(t)=\frac{1}{2\pi}\int_{-\infty}^{\infty}F(\omega)e^{j\omega t}d\omega.$$

- In practice, $f(t)$ is a real function of time, but $F(\omega)$ is in general a complex function of ω, whose real part is given by $A(\omega)=\int_{-\infty}^{\infty} f(t)\cos\omega t\,dt$ and whose imaginary part is given by $B(\omega)=-\int_{-\infty}^{\infty} f(t)\sin\omega t\,dt$.
- $F(\omega)$, $A(\omega)$, and $B(\omega)$ have the following properties:
  1. $A(\omega)$ is even, $B(\omega)$ is odd.
  2. $|F(\omega)|=\sqrt{A^2(\omega)+B^2(\omega)}$ is even, $\angle F(\omega)$ is odd.
  3. $F(-\omega)=F^*(\omega)$.

4. If $f(t)$ is even, $F(\omega)$ is real and even, with $A(\omega)=2\int_0^{\infty} f(t)\cos\omega t dt$.
5. If $f(t)$ is odd, $F(\omega)$ is imaginary and odd, with $B(\omega)=-2\int_0^{\infty} f(t)\sin\omega t dt$.

- $F(0)=\int_{-\infty}^{\infty} f(t)dt$.
- A duality relation exists between the FT and the IFT: if $\mathfrak{F}\{f(t)\}=F(\omega)$, then $\mathfrak{F}\{F(t)\}=2\pi f(-\omega)$.
- The operational properties of the FT are summarized in Table 23.1.
- If a function is compressed in the time domain, it expands in the frequency domain, and conversely.
- The FT of the convolution of two time functions equals the product of their FTs, and the FT of the product of two time functions equals the convolution of their FTs divided by 2π.
- The general procedure of applying the FT in circuit analysis is a generalization of the phasor method. The circuit is represented in the frequency domain, as in the phasor analysis, where R, L, and C are represented in terms of their impedances as R, $j\omega L$, and $1/j\omega C$, respectively. Phasor voltages and currents, including excitations, are replaced by the corresponding FTs. The usual circuit techniques are applied to derive the FT of any desired response. The IFT then gives the desired response in the time domain.
- The transfer function $H(\omega)$ is defined as in the general case of the LT as $H(\omega)=Y(\omega)/X(\omega)$.
- The FT of the response $h(t)$ to a unit impulse at $t=0$ is the transfer function $H(\omega)$, and the IFT of $H(\omega)$ is $h(t)$.
- The energy spectrum of a signal is a plot of $|F(\omega)|^2$ vs. ω. According to Parseval's theorem, the energy dissipated in a 1 Ω resistor over all time is $\int_{-\infty}^{\infty} f^2(t)dt=\frac{1}{\pi}\int_0^{\infty}|F(\omega)|^2 d\omega$.
- Parseval's theorem allows calculation of the power associated with any band of frequencies of a signal.

Problem-Solving Tips

1. The following checks on the FT are useful:
   - If $f(t)$ is even, $F(\omega)$ is real and even.
   - If $f(t)$ is odd, $F(\omega)$ is imaginary and odd.
   - If $\int_{-\infty}^{+\infty}|f(t)|dt$ is finite, if $f(t)$ is of finite duration, and if the dc component of $f(t)$ is zero, then $F(0)=0$. On the other hand, if $\int_{-\infty}^{+\infty}|f(t)|dt$ is infinite, and even if the dc component of $f(t)$ is zero, it does not follow that $F(0)=0$.
2. When deriving the IFT by PFE, it is convenient and less error prone to replace $j\omega$ by s.

Problems

Verify solutions using PSpice.

Fourier Transform and Its Properties

P23.1 Determine the FT of the following functions:
(a) $\delta(t)+\frac{3}{2}\delta(t-1)+\frac{1}{2}\delta(t-2)$
(b) $2u(t)+\sin(2000t)+4$
(c) $u(t)-u(t-4)$
(d) $t^2e^{-3t}u(t)$

Ans. (a) $1+\frac{3}{2}e^{-j\omega}+\frac{1}{2}e^{-j2\omega}$; (b) $10\pi\delta(\omega)+j\left(-\frac{2}{\omega}+\pi\delta(\omega+2000)-\pi\delta(\omega-2000)\right)$; (c) $\frac{1}{j\omega}\left(1-e^{-j4\omega}\right)$; (d) $\frac{2}{(3+j\omega)^3}$.

P23.2 Determine the FT of the following functions:
(a) $t\text{sgn}(t)$
(b) $|t|$ (*Hint*: $|t|=\int \text{sgn}(t)dt$)

Ans. (a) $-\frac{2}{\omega^2}$; (b) $-\frac{2}{\omega^2}$.

P23.3 Determine the FT of the following functions:
(a) $\left(\text{sgn}(t)e^{-|t|}\right)^2$
(b) $\frac{d}{dt}e^{-a|t|}$, by direct evaluation and by using Equation 23.14.
(c) $\frac{2\cos(at)}{t^2+1}$ (*Hint*: use Equation 23.14 and duality).

Ans. (a) $\frac{4}{4+\omega^2}$; (b) $\frac{j2a\omega}{a^2+\omega^2}$; (c) $\pi\left(e^{-|\omega-a|}+e^{-|\omega+a|}\right)$.

P23.4 Determine the FT of the following functions, defined for $-\pi\le t\le\pi$ and are zero outside this range:
(a) $\sin t$
(b) $\cos t$
(c) $1+\cos t$

Ans. (a) $\frac{j2}{\omega^2-1}\sin\pi\omega$; (b) $-\frac{2\omega\sin\pi\omega}{\omega^2-1}$; (c) $-\frac{2}{\omega(\omega^2-1)}\sin\pi\omega$.

P23.5 Determine the IFT of the following functions:

(a) $\dfrac{4}{\omega^2+1}$

(b) $\dfrac{1}{\left(1+\omega^2\right)^2}$

Ans. (a) $2e^{-|t|}$; (b) $\frac{1}{4}(t+1)e^{-t}u(t)+\frac{1}{4}(t-1)e^{t}u(t)$.

P23.6 Determine the IFT of the following functions:

(a) $2u(\omega+2)-2u(\omega-2)$

(b) $\dfrac{5}{j\omega(j\omega+5)}$

(c) $\dfrac{j\omega-2}{\omega^2-5j\omega-6}$

Ans. (a) $\frac{4}{\pi}\text{sinc}(2t)$; (b) $\frac{1}{2}\text{sgn}(t)-e^{-5t}u(t)$;
(c) $\left(4e^{-2t}-5e^{-3t}\right)u(t)$.

P23.7 Determine $F(\omega)$ of $f(t)$ in Figure P23.7.

Ans. $-\dfrac{j8}{\omega}$.

P23.8 Determine $F(\omega)$ of $f(t)$ in Figure P23.8, and verify the interpretation of $F(0)$.

Ans. $\dfrac{2A}{\omega}\sin\left(\dfrac{\tau\omega}{2}\right)-\dfrac{8A}{\tau\omega^2}\sin^2\left(\dfrac{\tau\omega}{4}\right)$.

P23.9 Determine $F(\omega)$ of $f^{(1)}(t)$, where $f(t)$ is that in Figure P23.8, and verify the result by applying the differentiation-in-time property to the result of Problem P23.8. Compare to Equation 23.13.

P23.10 Assume that the function shown in Figure 23.4 is in the time domain. Determine $F(\omega)$ and verify by applying duality to the result of Example 23.2.

P23.11 Determine $F(\omega)$ of $f(t)$ in Figure P23.11 from that of its second derivative.

Ans. $\dfrac{j30}{\omega^2}\sin 2\omega-\dfrac{j20}{\omega^2}\sin 3\omega$.

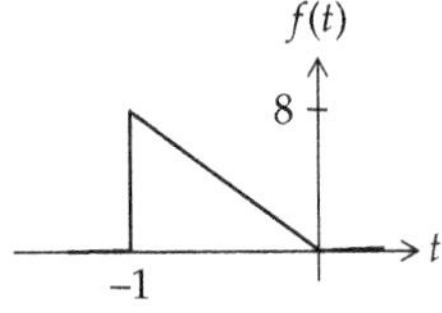

FIGURE P23.7

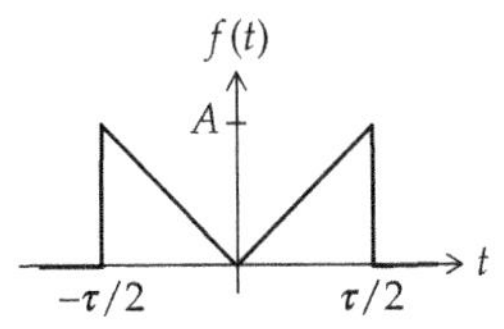

FIGURE P23.8

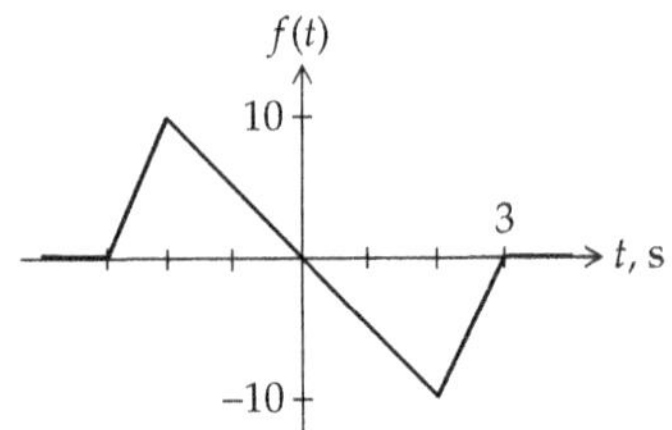

FIGURE P23.11

P23.12 Assume that the function shown in Figure 23.3 is in the frequency domain. Determine $f(t)$ and verify by applying duality to Equation 23.13.

P23.13 Assume that the function shown in Figure 23.15 is in the frequency domain. Determine $f(t)$ and verify by applying duality to the result of Example 23.6.

P23.14 Assume that the function shown in Figure 23.17 is in the frequency domain. Determine $f(t)$ and verify by applying duality to the result of Exercise 23.11.

P23.15 Assume that the function shown in Figure P23.8 is in the frequency domain. Determine $f(t)$ and verify by applying duality to the result of Problem P23.8.

P23.16 Determine $F(\omega)$ of $f(t) = (\cos 2t)u(-t) + 2(\cos 2t)u(t)$ (Figure P23.16).

Ans. $\dfrac{3\pi}{2}\left[\delta(\omega+2)+\delta(\omega-2)\right]+\dfrac{j\omega}{4-\omega^2}$.

P23.17 Determine the FT of a single period of $\sin t$ centered at the origin (Figure P23.17).

Ans. $\dfrac{j2A}{\omega^2-1}\sin\pi\omega$.

P23.18 Determine $F(\omega)$ of Figure P23.17 by considering it as the product of a sinusoidal function and a rectangular pulse of unit amplitude extending from $-\pi$ to $+\pi$.

P23.19 Determine $F(\omega)$ of $f(t)$ in Figure P23.19 in terms of step functions, impulse functions, and rectangular functions, and express it in the simplest possible form.

Ans. $8\text{sinc}(4\omega) - 12\text{sinc}(3\omega) + 4\text{sinc}(\omega)$.

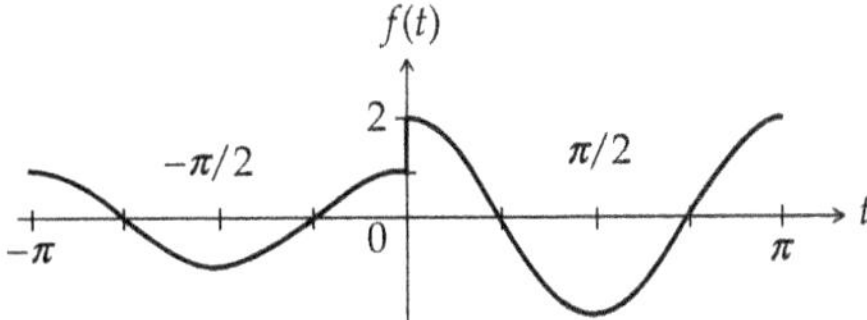

FIGURE P23.16

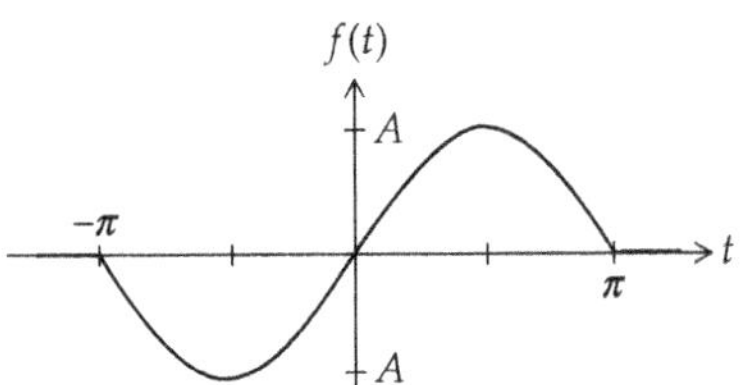

FIGURE P23.17

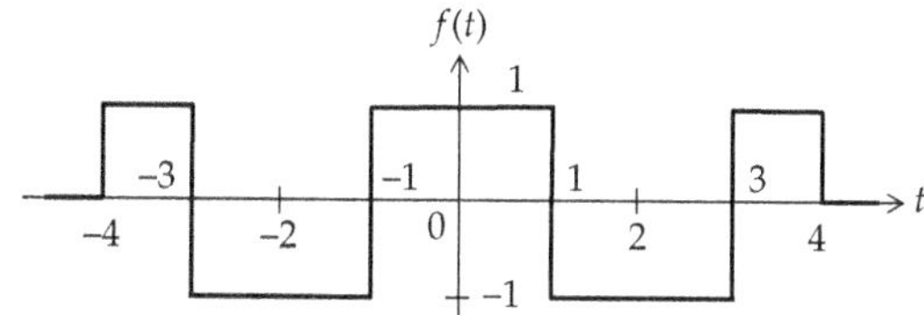

FIGURE P23.19

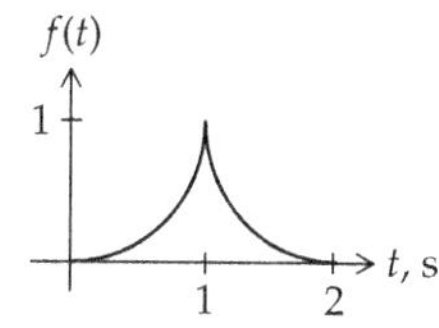

FIGURE P23.32

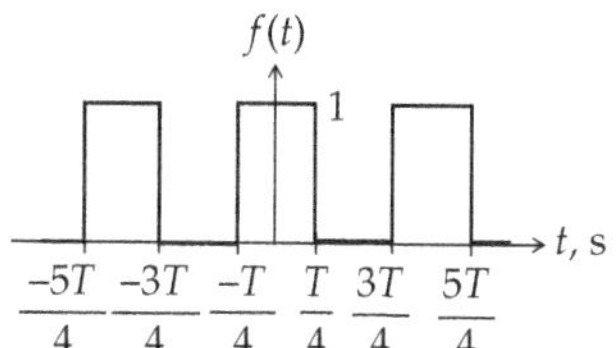

FIGURE P23.33

P23.20 Given that $f(t)$ is the periodic train of unit impulses: ..., $\delta(-2T)$, $-\delta(-T)$, $\delta(0)$, $-\delta(T)$, $\delta(2T)$, ..., with $T = 1$ ms, and that $g(t) = \text{sinc}(\pi t)$, where t is in ms. Determine the FT of the product $f(t) \times g(t)$.

Ans. 1.

P23.21 If $\mathfrak{F}\left\{\frac{1}{\sqrt{t}}\right\} = \sqrt{\frac{2\pi}{\omega}}$, determine the IFT of $\sqrt{\frac{\pi}{\omega - 1}}$.

Ans. $e^{jt}/\sqrt{2t}$.

P23.22 Determine the IFT of $\frac{2\pi\delta(\omega)}{2 + j\omega} + \frac{2}{j\omega(2 + j\omega)}$.

Ans. $(1 - e^{2t})u(t)$.

P23.23 Determine $\delta(\omega - a) * \delta(\omega - b)$ in both the time and frequency domains.

Ans. $\delta(\omega - (a + b))$, $\frac{1}{4\pi^2}e^{j(a+b)t}$.

P23.24 Determine the IFT of the convolution of $\frac{1 + j\omega}{4 + (1 + j\omega)^2}$ and $\frac{2}{4 + (1 + j\omega)^2}$.

Ans. $\pi\left(e^{-2t}\sin 4t\right)u(t)$.

P23.25 Determine the convolution of $F(\omega) = \delta(\omega) + 1/j\pi\omega$ and $G(\omega) = \frac{\pi}{2 + j\omega}$.

Ans. $\frac{2\pi}{2 + j\omega}$.

P23.26 Determine the FT of $f(t) = \sum_{n=0}^{\infty} r^n \delta(t - n)$, $|r| < 1$.

Ans. $\frac{1}{1 - re^{-j\omega}}$.

P23.27 Determine $F(\omega)$ of the periodic sawtooth of Figure 16.5.

Ans. $F(\omega) = \sum_{n=-\infty}^{\infty} \frac{jA}{n}\delta(\omega - n\omega_0), n \neq 0$, and $F(\omega) = 2\pi \times \frac{A}{2}\delta(\omega) = \pi A\delta(\omega)$ for $n = 0$.

P23.28 Determine $F(\omega)$ of the periodic sawtooth of Figure 16.9b.

Ans. $F(\omega) = -\sum_{n=-\infty}^{\infty} \frac{jA}{n}\delta(\omega - n\omega_0), n \neq 0$, and $F(\omega) = 2\pi \times \frac{A}{2}\delta(\omega) = \pi A\delta(\omega)$ for $n = 0$.

P23.29 Determine $F(\omega)$ of the periodic triangular waveform of Figure 16.24 and verify it by applying the integration-in-time property to the square wave of Primal Exercise 23.7.

Ans. $F(\omega) = \sum_{n=-\infty}^{\infty} -\frac{8A_m}{\pi n^2}\delta(\omega - n\omega_0)$, n odd.

P23.30 Determine $F(\omega)$ of the half-wave rectified waveform of Figure 16.27a, considering it as the product of a cosine function and a rectangular waveform, as in Example 16.6.

Ans. $F_{hw}(\omega) = \frac{A\pi}{2}\sum_{-\infty}^{\infty} \text{sinc}(n\pi/2)\left[\delta(\omega - \omega_0 - n\omega_0) + \delta(\omega + \omega_0 - n\omega_0)\right]$.

P23.31 Determine $F(\omega)$ of the full-wave rectified waveform of Figure 16.27b considering it as the product of a cosine function and a square wave of zero average centered at the origin.

Ans. $F_{fw}(\omega) = 4A\delta(\omega) + 2A\sum_{-\infty}^{\infty} \frac{1}{n}\left[\delta(\omega - \omega_0 - n\omega_0) + \delta(\omega + \omega_0 - n\omega_0)\right]$.

P23.32 Determine $F(\omega)$ of $f(t)$ of Figure P23.32, where $f(t) = t^2$, $0 \le t \le 1$, $f(t) = (2 - t)^2$, $1 \le t \le 2$, and $f(t) = 0$, elsewhere, using the second derivative.

Ans. $F(\omega) = \frac{4e^{-j\omega}}{\omega^2}(1 - \text{sinc}\omega)$

P23.33 Given $f(t)$ as in Figure P23.33, derive the convolution function $y(t) = f(t) * \cos\left(\frac{2\pi}{T}t\right)$.

Ans. $\frac{3T}{\pi}\cos\frac{2\pi}{T}t$.

Circuit Applications

P23.34 The impulse response of a certain circuit is $\delta(t) + 0.4\delta(t - 1)$. Determine the FT of the response of this circuit to an input $\cos\pi t$.

Ans. $F(\omega) = 0.6\pi(\delta(\omega + \pi) + \delta(\omega - \pi))$.

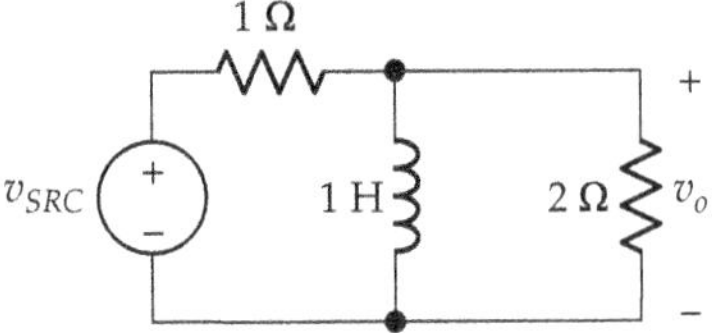

FIGURE P23.35

P23.35 Determine $v_O(t)$ in Figure P23.35 if $v_{SRC}(t) = (1/2)\text{sgn}(t)$ V. Deduce $v_O(t)$ for (a) $v_{SRC}(t) = \delta(t)$ V and (b) $v_{SRC}(t) = u(t)$ V. Verify the responses in (a) and (b) using the FT, and interpret the results for all three inputs.

Ans. $\frac{2}{3}e^{-2t/3}u(t)$; (a) $\frac{2}{3}\delta(t) - \frac{4}{9}e^{-2t/3}u(t)$; (b) $\frac{2}{3}e^{-2t/3}u(t)$.

P23.36 Determine $i_O(t)$ and $v_O(t)$ in Figure P23.36 if $v_{SRC}(t) = 5\text{sgn}(t)$ V and interpret the result.

Ans. $v_O(t) = \frac{25}{6}\text{sgn}(t) - \frac{25}{3}e^{-6t/25}u(t)$, $2e^{-6t/25}u(t)$ mA, t is in ms.

P23.37 Determine $i_O(t)$ and $v_O(t)$ in Figure P23.37 if $v_{SRC}(t) = 0.5\text{sgn}(t)$ V.

Ans. $i_O(t) = \frac{2}{3}\left(e^{-0.5t} - e^{-2t}\right)u(t)$ A

$v_O(t) = 0.5\text{sgn}(t) - \frac{4}{3}e^{-0.5t}u(t) + \frac{1}{3}e^{-2t}u(t)$ V.

P23.38 Determine $i_O(t)$ and $v_O(t)$ in Figure P23.37 if $v_{SRC}(t) = 10e^{-|t|}$ V.

Ans. $i_O(t) = -\frac{8}{9}e^{-2t}u(t) + \frac{16}{9}e^{-t}u(t) - \frac{8}{9}e^{-0.5t}u(t) + \frac{1}{9}e^{-|t|}$ A

$v_O(t) = \frac{4}{9}e^{-2t}u(t) - \frac{20}{9}e^{-t}u(t) + \frac{16}{9}e^{-0.5t}u(t) + \frac{1}{9}e^{-|t|}$ V.

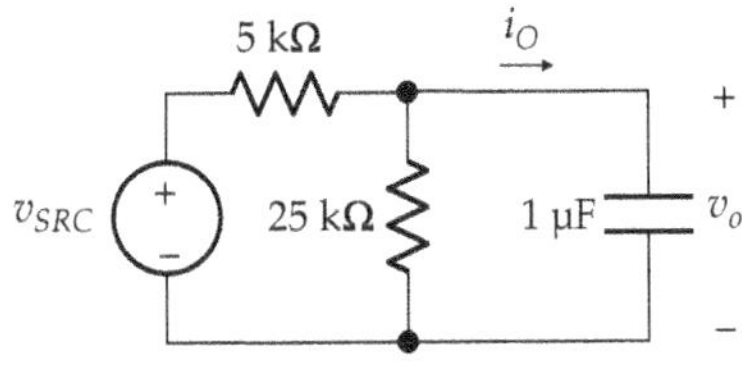

FIGURE P23.36

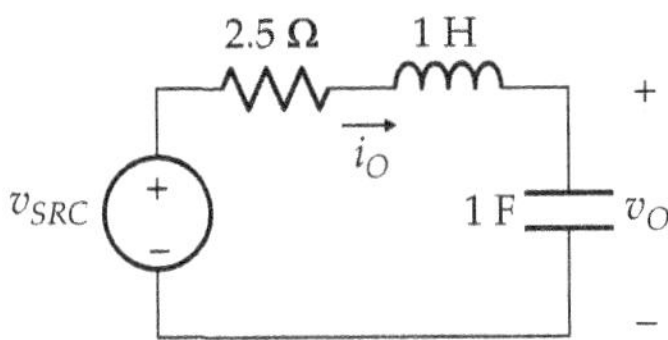

FIGURE P23.37

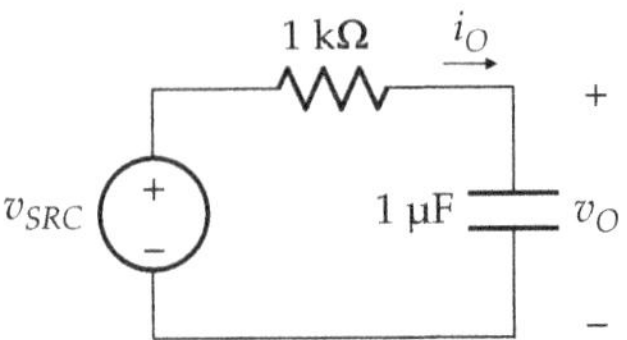

FIGURE P23.39

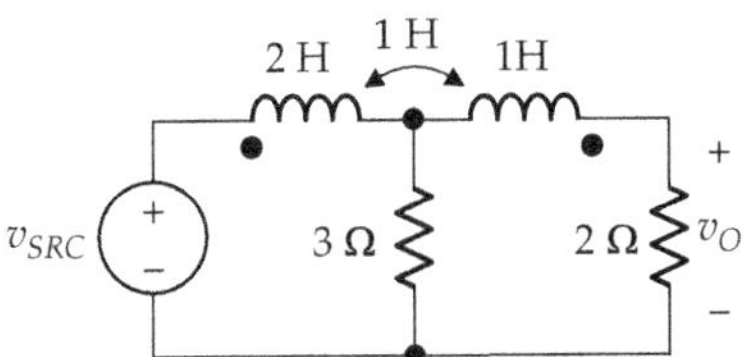

FIGURE P23.40

P23.39 Determine $v_O(t)$ and $i_O(t)$ in Figure P23.39 if $v_{SRC}(t)$ is the waveform of Figure 23.3 with $A = 10$ V and $\tau = 2$ ms.

Ans. $v_O(t) = 5\left(-\text{sgn}(t+1) + 2e^{-(t+1)}u(t+1) + 2\text{sgn}(t) - 4e^{-t}u(t) - \text{sgn}(t-1) + 2e^{-(t-1)}u(t-1)\right)$ V.

P23.40 Determine $v_O(t)$ in Figure P23.40 if $v_{SRC}(t) = e^{-2t}u(t)$ V.

Ans. $\left(\frac{4}{5}e^{-t} - \frac{1}{2}e^{-2t} - \frac{3}{10}e^{-6t}\right)u(t)$ V.

Parseval's Theorem

Refer to Table of Integrals in Appendix B when necessary.

P23.41 If $F(j\omega) = 1/(2 + j\omega)$, determine $W_{1\Omega}$ in both the frequency and time domains.

Ans. 1/4 J.

P23.42 If $f(t) = e^{-2t}(\sin 2t)u(t)$, determine $W_{1\Omega}$ in both the frequency and time domains.

Ans. 1/16 J.

P23.43 If $f(t) = te^{-2t}u(t)$, determine $W_{1\Omega}$ in both the frequency and time domains.

Ans. 1/32 J.

P23.44 If the voltage across a 50 Ω resistor is $f(t) = te^{-2t}u(t)$ V, determine the percentage of the total load energy that is associated with the frequency band 0–1 rad/s.

Ans. 81.83%.

P23.45 If the FT of the current in a 4 Ω resistor is $1/(1+\omega^2)$, determine the total energy dissipated in the resistor.

Ans. 1 J.

P23.46 If $i(t)$ = 2sinc($4t$), determine (a) the frequency band starting at $\omega = 0$ that contains half the energy of $i(t)$ and (b) the energy dissipated in a 2 Ω resistor due to $i(t)$ applied over all time.

Ans. (a) 0–2 rad/s; (b) 2π J.

P23.47 If $v_{SRC}(t) = 15e^{-|t|}$ V in Figure P23.47, determine the percentage of the 1 Ω energy of v_O in the frequency range $0 \le \omega \le 2$ rad/s.

Ans. 98.4%.

P23.48 If $v_{SRC}(t) = 20e^{-2t}u(t)$ V in Figure P23.48, what percentage of the 1 Ω energy of v_O is in the frequency range $0 \le \omega \le 2$ rad/s if (a) $R = 2$ Ω and (b) $R = 4$ Ω.

Ans. (a) 38.5%; (b) 44.0%.

P23.49 Repeat Problem P23.48 if $v_{SRC}(t) = 20e^{-|2|t}$ V.

Ans. (a) 72.5%; (b) 77.2%.

P23.50 A current $2\text{sgn}(t)e^{-2|t|}$ A is applied to a 10 Ω resistor. Determine the energy of the signal that is in the frequency band 0–10 Hz.

Ans. 19.2 J.

P23.51 Use Parseval's theorem to show that $\int_{-\infty}^{\infty} \delta^2(t)dt$ is infinite and therefore undefined.

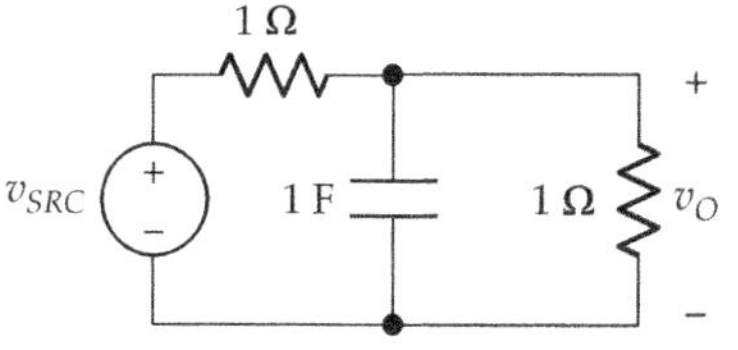

FIGURE P23.47

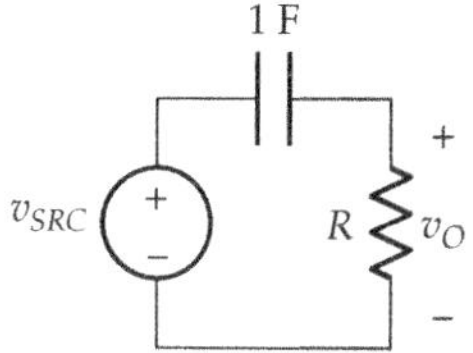

FIGURE P23.48

24

Two-Port Circuits

Objective and Overview

Any circuit that is excited by a single, independent source applied to one port and having a load connected to another port can be considered as a two-port circuit. Active, electronic devices and their associated circuits are the prime examples of two-port circuits. A two-port circuit is characterized, in general, by sets of four parameters defined in terms of open-circuit or short-circuit terminations at each port. The analysis based on this characterization provides a useful and powerful alternative to conventional circuit analysis discussed in previous chapters.

To utilize the full power of two-port analysis, matrix methods are needed. We will introduce some of these methods in the following sections, mainly as a convenient, compact notation that follows certain rules that are particularly simple for 2 × 2 matrices.

The chapter begins with presenting the six sets of two-port circuit equations, any one of which can be used to relate terminal voltages and currents of the circuit. Each of these sets of equations involves a corresponding set of parameters that characterize the circuit. The different sets of parameters are interpreted in terms of terminal conditions of the circuit, and the relations between these different sets are derived. Two special types of two-port circuits, namely, reciprocal and symmetric circuits, are discussed, followed by presenting some useful equivalent circuits based on two-port circuit equations.

The rest of the chapter is concerned with two-port circuit analysis, starting with the derivation of two-port circuit parameters for composite circuits that result from combining two-port circuits in one of five possible ways: cascading, parallel, series, series–parallel, and parallel–series connections. When single circuit elements are considered as two-port circuits, this approach leads to some powerful methods of analyzing many types of circuits. The chapter ends by analyzing a terminated two-port circuit in order to derive relations that are of particular interest in the analysis of electronic amplifier circuits.

24.1 Circuit Description

As its name implies, a two-port circuit has a pair of input terminals or input port and a pair of output terminals or output port (Figure 24.1), the assigned positive directions of input and output voltages and currents being as indicated. Voltages and currents in two-port circuits are considered, in general, to be Laplace transforms, unless indicated otherwise. These voltages and currents could equally well be dc quantities or phasors in the case of a sinusoidal steady state. By convention, the circuit may contain passive, linear circuit elements and dependent sources, but no independent sources.

In general, two of the four terminal variables, V_1, V_2, I_1, and I_2, can be specified independently, in which case the other two variables are determined by the parameters of the given circuit. This can be justified by considering a load connected to the output port. Evidently, V_2 and I_2 will be related by the load and will not be independent. Similarly, with the load connected, there will be an input impedance at the input port that relates V_1 and I_1, which means that these two variables are not independent. Hence, only two of the four terminal variables are, in general, independent. The two-port circuit may therefore be described in terms of two simultaneous equations. However, since four variables are involved, there are six ways of choosing two of these variables as independent, which results in six sets of simultaneous equations, listed in Table 24.1. In each of these equations, the two independent variables on the LHS are related to two dependent variables on the RHS by four coefficients or **parameters** that characterize the equation. Thus, one speaks of the z-parameter equation, the y-parameter equation, etc. The two equations in

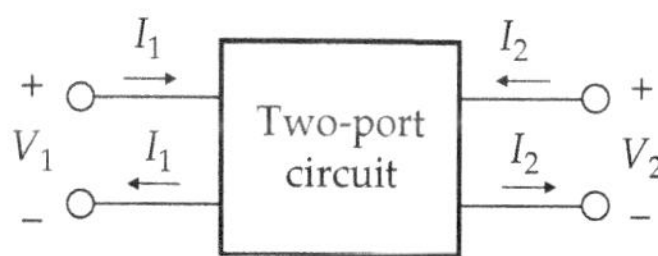

FIGURE 24.1
Terminal voltages and currents of two-port circuit.

TABLE 24.1

Two-Port Circuit Equations

$V_1 = z_{11}I_1 + z_{12}I_2$ $V_2 = z_{21}I_1 + z_{22}I_2$	$I_1 = y_{11}V_1 + y_{12}V_2$ $I_2 = y_{21}V_1 + y_{22}V_2$
$V_1 = a_{11}V_2 - a_{12}I_2$ $I_1 = a_{21}V_2 - a_{22}I_2$	$V_2 = b_{11}V_1 - b_{12}I_1$ $I_2 = b_{21}V_1 - b_{22}I_1$
$V_1 = h_{11}I_1 + h_{12}V_2$ $I_2 = h_{21}I_1 + h_{22}V_2$	$I_1 = g_{11}V_1 + g_{12}I_2$ $V_2 = g_{21}V_1 + g_{22}I_2$

the same row in Table 24.1 are inversely related, in the sense that the independent variables in one set are the dependent variables in the other. As discussed in the following text, this means that the matrices of parameters in the two cases are inversely related.

It is seen from Table 24.1 that a two-port circuit is specified, in general, by four nonzero parameters in one of the six equations. The z and y parameters, being impedances and admittances, respectively, are **immittance parameters,** where immittance refers to either an impedance or admittance. The z- and y-parameter equations express terminal voltages (or currents) in terms of terminal currents (or voltages). The a and b parameters are **transmission parameters** because the corresponding equations express voltage and current at one port in terms of voltage and current at the other port. The a and b parameters are also referred to as *ABCD* or *abcd* parameters, respectively. The h and g parameters are **hybrid parameters** because the corresponding equations express an input voltage (or current) and an output current (or voltage) in terms of an input current (or voltage) and an output voltage (or current).

Exercise 24.1

Of the six sets of equations in Table 24.1, which sets are dual relations?

Ans. z-parameter and y-parameter equations are dual relations, as are the h-parameter and g-parameter equations, provided the corresponding coefficients have equal numerical values.

24.2 Parameter Interpretation and Relations

24.2.1 Interpretation of Parameters

The parameters of two-port circuits can be interpreted in terms of voltage and current ratios under specified open-circuit and short-circuit terminations, which provide a convenient means of evaluation or measurement of these parameters. If excitation is applied to port 1, with port 2 open-circuited, then $I_2 = 0$, and the z-parameter equations reduce to

$$V_1 = z_{11}I_1 \quad \text{and} \quad V_2 = z_{21}I_1 \tag{24.1}$$

It follows that z_{11} and z_{21} can be defined as

$z_{11} = \left.\dfrac{V_1}{I_1}\right|_{I_2=0}$ Ω: Impedance looking into port 1 with port 2 open-circuited

$z_{21} = \left.\dfrac{V_2}{I_1}\right|_{I_2=0}$ Ω: Ratio of voltage at port 2 to current at port 1 with port 2 open-circuited

Similarly, if excitation is applied to port 2, with port 1 open-circuited, then $I_1 = 0$, and the z-parameter equations reduce to

$$V_1 = z_{12}I_2 \quad \text{and} \quad V_2 = z_{22}I_2 \tag{24.2}$$

It follows that z_{12} and z_{22} can be defined as follows:

$z_{12} = \left.\dfrac{V_1}{I_2}\right|_{I_1=0}$ Ω: Ratio of voltage at port 1 to current at port 2 with port 1 open-circuited.

$z_{22} = \left.\dfrac{V_2}{I_2}\right|_{I_1=0}$ Ω: Impedance looking into port 2 with port 1 open-circuited.

Whereas z_{11} and z_{22} are input impedances at ports 1 and 2, respectively, z_{12} and z_{21} are *transfer* impedances, being the ratio of voltage at one port to current at the other port. The other parameters may be similarly interpreted, as summarized in Table 24.2. Circuit parameters are, in general, frequency dependent and complex.

Since the same circuit variables are involved in the six equations, the parameters in these equations must be related. To find the relation between any two sets of parameters, we simply express one set of two simultaneous equations in the same form as the other set and compare coefficients. For example, to express the

TABLE 24.2

Interpretation of Circuit Parameters

$z_{11} = \left.\dfrac{V_1}{I_1}\right\|_{I_2=0}$ Ω	$z_{12} = \left.\dfrac{V_1}{I_2}\right\|_{I_1=0}$ Ω	$y_{11} = \left.\dfrac{I_1}{V_1}\right\|_{V_2=0}$ S	$y_{12} = \left.\dfrac{I_1}{V_2}\right\|_{V_1=0}$ S
$z_{21} = \left.\dfrac{V_2}{I_1}\right\|_{I_2=0}$ Ω	$z_{22} = \left.\dfrac{V_2}{I_2}\right\|_{I_1=0}$ Ω	$y_{21} = \left.\dfrac{I_2}{V_1}\right\|_{V_2=0}$ S	$y_{22} = \left.\dfrac{I_2}{V_2}\right\|_{V_1=0}$ S
$a_{11} = \left.\dfrac{V_1}{V_2}\right\|_{I_2=0}$	$a_{12} = -\left.\dfrac{V_1}{I_2}\right\|_{V_2=0}$ Ω	$b_{11} = \left.\dfrac{V_2}{V_1}\right\|_{I_1=0}$	$b_{12} = -\left.\dfrac{V_2}{I_1}\right\|_{V_1=0}$ Ω
$a_{21} = \left.\dfrac{I_1}{V_2}\right\|_{I_2=0}$ S	$a_{22} = -\left.\dfrac{V_1}{I_2}\right\|_{V_2=0}$	$b_{21} = \left.\dfrac{I_2}{V_1}\right\|_{I_1=0}$ S	$b_{22} = -\left.\dfrac{I_2}{I_1}\right\|_{V_1=0}$
$h_{11} = \left.\dfrac{V_1}{I_1}\right\|_{V_2=0}$ Ω	$h_{12} = \left.\dfrac{V_1}{I_2}\right\|_{I_1=0}$	$g_{11} = \left.\dfrac{I_1}{V_1}\right\|_{I_2=0}$ S	$g_{12} = \left.\dfrac{I_1}{I_2}\right\|_{V_1=0}$
$h_{21} = \left.\dfrac{I_2}{I_1}\right\|_{V_2=0}$	$h_{22} = \left.\dfrac{I_2}{V_2}\right\|_{I_1=0}$ S	$g_{21} = \left.\dfrac{V_2}{V_1}\right\|_{I_2=0}$	$g_{22} = \left.\dfrac{V_2}{I_2}\right\|_{V_1=0}$ Ω

z parameters in terms of the a parameters, we eliminate V_2 between the two equations of the a parameters to obtain

$$V_1 = \frac{a_{11}}{a_{21}} I_1 + \frac{\Delta a}{a_{21}} I_2 \quad (24.3)$$

where $\Delta a = a_{11}a_{22} - a_{12}a_{21}$. The second a-parameter equation can be rearranged as

$$V_2 = \frac{1}{a_{21}} I_1 + \frac{a_{22}}{a_{21}} I_2 \quad (24.4)$$

Comparing Equations 24.3 and 24.4 with the corresponding z-parameter equations gives $z_{11} = \frac{a_{11}}{a_{21}}$, $z_{12} = \frac{\Delta a}{a_{21}}$, $z_{21} = \frac{1}{a_{21}}$, and $z_{22} = \frac{a_{22}}{a_{21}}$. Table 24.3 shows the relations between the different sets of parameters.

Note that in certain cases, some parameters in a given set of the six sets of parameters may be zero, and some parameters in other sets of parameters may be infinite, in which case, the corresponding set of two-port circuit equations does not exist. For example, the ideal transformer may be considered as a two-port circuit in which V_2 and V_1 are related by a turns ratio independent of the currents and I_1 and I_2 are related by a turns ratio, independent of the voltages. For the dot markings indicated in Figure 24.2, the transformer equations are $V_1 = nV_2$ and $I_2 = -nI_1$. The two-port circuit equations are as follows:

a-parameter equations: $V_1 = nV_2$, $I_1 = -(1/n)I_2$ with $a_{11} = n$, $a_{22} = 1/n$, $a_{12} = 0 = a_{21}$

TABLE 24.3

Relations between Sets of Parameters

$z_{11} = \frac{y_{22}}{\Delta y} = \frac{a_{11}}{a_{21}} = \frac{b_{22}}{b_{21}} = \frac{\Delta h}{h_{22}} = \frac{1}{g_{11}}$	$y_{11} = \frac{z_{22}}{\Delta z} = \frac{a_{22}}{a_{12}} = \frac{b_{11}}{b_{12}} = \frac{1}{h_{11}} = \frac{\Delta g}{g_{22}}$
$z_{12} = -\frac{y_{12}}{\Delta y} = \frac{\Delta a}{a_{21}} = \frac{1}{b_{21}} = \frac{h_{12}}{h_{22}} = -\frac{g_{12}}{g_{11}}$	$y_{12} = -\frac{z_{12}}{\Delta z} = -\frac{\Delta a}{a_{21}} = -\frac{1}{b_{12}} = -\frac{h_{12}}{h_{11}} = \frac{g_{12}}{g_{22}}$
$z_{21} = -\frac{y_{21}}{\Delta y} = \frac{1}{a_{21}} = \frac{\Delta b}{b_{21}} = -\frac{h_{21}}{h_{22}} = \frac{g_{21}}{g_{11}}$	$y_{21} = -\frac{z_{21}}{\Delta z} = -\frac{1}{a_{12}} = -\frac{\Delta b}{b_{12}} = \frac{h_{21}}{h_{11}} = -\frac{g_{21}}{g_{22}}$
$z_{22} = \frac{y_{11}}{\Delta y} = \frac{a_{22}}{a_{21}} = \frac{b_{11}}{b_{21}} = \frac{1}{h_{22}} = \frac{\Delta g}{g_{11}}$	$y_{22} = \frac{z_{11}}{\Delta z} = \frac{a_{11}}{a_{12}} = \frac{b_{22}}{b_{12}} = \frac{\Delta h}{h_{11}} = \frac{1}{g_{22}}$
$a_{11} = \frac{z_{11}}{z_{21}} = -\frac{y_{22}}{y_{21}} = \frac{b_{22}}{\Delta b} = -\frac{\Delta h}{h_{21}} = \frac{1}{g_{21}}$	$b_{11} = \frac{z_{22}}{z_{12}} = -\frac{y_{11}}{y_{12}} = \frac{a_{22}}{\Delta a} = \frac{1}{h_{12}} = -\frac{\Delta g}{g_{12}}$
$a_{12} = \frac{\Delta z}{z_{21}} = -\frac{1}{y_{21}} = \frac{b_{12}}{\Delta b} = -\frac{h_{11}}{h_{21}} = \frac{g_{22}}{g_{21}}$	$b_{12} = \frac{\Delta z}{z_{12}} = -\frac{1}{y_{12}} = \frac{a_{12}}{\Delta a} = -\frac{h_{11}}{h_{12}} = -\frac{g_{22}}{g_{12}}$
$a_{21} = \frac{1}{z_{21}} = -\frac{\Delta y}{y_{21}} = \frac{b_{21}}{\Delta b} = -\frac{h_{22}}{h_{21}} = \frac{g_{11}}{g_{21}}$	$b_{21} = \frac{1}{z_{12}} = -\frac{\Delta y}{y_{12}} = \frac{a_{21}}{\Delta a} = \frac{h_{22}}{h_{12}} = -\frac{g_{11}}{g_{12}}$
$a_{22} = \frac{z_{22}}{z_{21}} = -\frac{y_{11}}{y_{21}} = \frac{b_{11}}{\Delta b} = -\frac{1}{h_{21}} = \frac{\Delta g}{g_{21}}$	$b_{22} = \frac{z_{11}}{z_{12}} = -\frac{y_{22}}{y_{12}} = \frac{a_{11}}{\Delta a} = \frac{\Delta h}{h_{12}} = -\frac{1}{g_{12}}$
$h_{11} = \frac{\Delta z}{z_{22}} = \frac{1}{y_{11}} = \frac{a_{12}}{a_{22}} = \frac{b_{12}}{b_{11}} = \frac{g_{22}}{\Delta g}$	$g_{11} = \frac{1}{z_{11}} = \frac{\Delta y}{y_{22}} = \frac{a_{21}}{a_{11}} = \frac{b_{21}}{b_{22}} = \frac{h_{22}}{\Delta h}$
$h_{12} = \frac{z_{12}}{z_{22}} = -\frac{y_{12}}{y_{11}} = \frac{\Delta a}{a_{22}} = \frac{1}{b_{11}} = -\frac{g_{12}}{\Delta g}$	$g_{12} = -\frac{z_{12}}{z_{11}} = \frac{y_{12}}{y_{22}} = -\frac{\Delta a}{a_{11}} = -\frac{1}{b_{11}} = -\frac{h_{12}}{\Delta h}$
$h_{21} = -\frac{z_{21}}{z_{22}} = \frac{y_{21}}{y_{11}} = -\frac{1}{a_{22}} = -\frac{\Delta b}{b_{11}} = -\frac{g_{21}}{\Delta g}$	$g_{21} = -\frac{z_{21}}{z_{11}} = -\frac{y_{12}}{y_{22}} = \frac{1}{a_{11}} = \frac{\Delta b}{b_{22}} = -\frac{h_{21}}{\Delta h}$
$h_{22} = \frac{1}{z_{22}} = \frac{\Delta y}{y_{11}} = \frac{a_{21}}{a_{22}} = \frac{b_{21}}{b_{11}} = \frac{g_{11}}{\Delta g}$	$g_{22} = \frac{\Delta z}{z_{11}} = \frac{1}{y_{22}} = \frac{a_{12}}{a_{11}} = \frac{b_{12}}{b_{22}} = \frac{h_{11}}{\Delta h}$
$\Delta z = z_{11}z_{22} - z_{12}z_{21}$	$\Delta y = y_{11}y_{22} - y_{12}y_{21}$
$\Delta a = a_{11}a_{22} - a_{12}a_{21}$	$\Delta b = b_{11}b_{22} - b_{12}b_{21}$
$\Delta h = h_{11}h_{22} - h_{12}h_{21}$	$\Delta g = g_{11}g_{22} - g_{12}g_{21}$

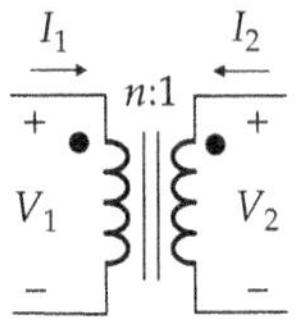

FIGURE 24.2
Ideal transformer.

b-parameter equations: $V_2 = (1/n)V_1$, $I_2 = -nI_1$ with $b_{11} = 1/n$, $b_{22} = n$, $b_{12} = 0 = b_{21}$

h-parameter equations: $V_1 = nV_2$, $I_2 = -nI_1$ with $h_{11} = 0 = h_{22}$, $h_{12} = n$, $h_{21} = -n$

g-parameter equations: $I_1 = -(1/n)I_2$, $V_2 = (1/n)V_1$ with $g_{11} = 0 = g_{22}$, $g_{12} = -(1/n)$, $g_{21} = (1/n)$

The *z*-parameter and *y*-parameter equations do not exist, as it is not possible to express the transformer voltages in terms of the transformer currents or conversely based on the transformer equations alone. Referring to Table 24.3, if $a_{12} = 0 = a_{21}$, then all the *z* parameters and *y* parameters are infinite.

Primal Exercise 24.2

Determine z_{12} and z_{21} in Figure 24.3.

Ans. $z_{12} = z_{21} = -j5\ \Omega$.

Primal Exercise 24.3

A two-port circuit is described by the equations $V_1 = 45I_1 + 2V_2$ and $I_2 = 10I_1 + V_2$. Determine the input resistance looking into the input port, with the output port open-circuited.

Ans. 25 Ω.

24.2.2 Inverse Relations

Any set of equations in Table 24.1, say, the *z*-parameter equations, may be expressed in matrix form as follows:

$$\begin{bmatrix} \mathbf{V}_1 \\ \mathbf{V}_2 \end{bmatrix} = \begin{bmatrix} z_{11} & z_{12} \\ z_{21} & z_{22} \end{bmatrix} \begin{bmatrix} \mathbf{I}_1 \\ \mathbf{I}_2 \end{bmatrix} \tag{24.5}$$

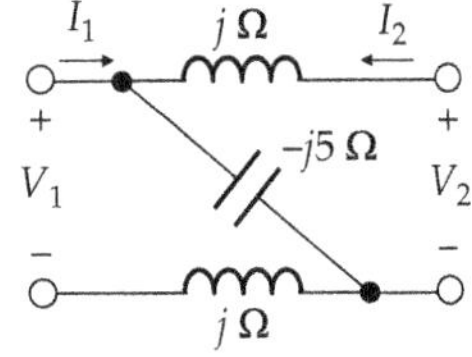

FIGURE 24.3
Figure for Primal Exercise 24.2.

Applying the rules of matrix multiplication to Equation 24.5 gives the *z*-parameter equations. Equation 24.5 is inverted by multiplying both sides by the inverse of the *z*-parameter matrix and rearranging the equation:

$$\begin{bmatrix} \mathbf{I}_1 \\ \mathbf{I}_2 \end{bmatrix} = \begin{bmatrix} z_{11} & z_{12} \\ z_{21} & z_{22} \end{bmatrix}^{-1} \begin{bmatrix} \mathbf{V}_1 \\ \mathbf{V}_2 \end{bmatrix} \tag{24.6}$$

Comparing this with the *y*-parameter equations, it is seen that

$$\begin{bmatrix} y_{11} & y_{12} \\ y_{21} & y_{22} \end{bmatrix} = \begin{bmatrix} z_{11} & z_{12} \\ z_{21} & z_{22} \end{bmatrix}^{-1} \tag{24.7}$$

Similar inverse relations hold between the parameter matrices of any two sets of equations in the same row of Table 24.1. Thus,

$$\begin{bmatrix} z_{11} & z_{12} \\ z_{21} & z_{22} \end{bmatrix} = \begin{bmatrix} y_{11} & y_{12} \\ y_{21} & y_{22} \end{bmatrix}^{-1}, \quad \begin{bmatrix} y_{11} & y_{12} \\ y_{21} & y_{22} \end{bmatrix} = \begin{bmatrix} z_{11} & z_{12} \\ z_{21} & z_{22} \end{bmatrix}^{-1} \tag{24.8}$$

$$\begin{bmatrix} a_{11} & -a_{12} \\ a_{21} & -a_{22} \end{bmatrix} = \begin{bmatrix} b_{11} & -b_{12} \\ b_{21} & -b_{22} \end{bmatrix}^{-1}, \quad \begin{bmatrix} b_{11} & -b_{12} \\ b_{21} & -b_{22} \end{bmatrix} = \begin{bmatrix} a_{11} & -a_{12} \\ a_{21} & -a_{22} \end{bmatrix}^{-1} \tag{24.9}$$

$$\begin{bmatrix} h_{11} & h_{12} \\ h_{21} & h_{22} \end{bmatrix} = \begin{bmatrix} g_{11} & g_{12} \\ g_{21} & g_{22} \end{bmatrix}^{-1}, \quad \begin{bmatrix} g_{11} & g_{12} \\ g_{21} & g_{22} \end{bmatrix} = \begin{bmatrix} h_{11} & h_{12} \\ h_{21} & h_{22} \end{bmatrix}^{-1} \tag{24.10}$$

Note the negative signs in the matrices of *a* and *b* parameters, which arise because of the negative signs in the equations of these parameters.

The inverse of a 2 × 2 parameter matrix is expressed as

$$\begin{bmatrix} m_{11} & m_{12} \\ m_{21} & m_{22} \end{bmatrix}^{-1} = \frac{1}{D} \begin{bmatrix} m_{22} & -m_{12} \\ -m_{21} & m_{11} \end{bmatrix} \tag{24.11}$$

where *D* is the determinant of the matrix and is equal to Δz, Δy, $-\Delta a$, $-\Delta b$, Δh, and Δg, as defined in Table 24.3. The relations in Table 24.3 are seen to conform to Equation 24.11.

Exercise 24.4

Derive the expressions for the z parameters in terms of the y, a, b, h, and g parameters in Table 24.3.

Example 24.1: Determination of a Parameters

It is required to determine the a parameters of the circuit of Figure 24.4 at $\omega = 1$ rad/s.

Solution:

With $I_2 = 0$, $V_2 = \frac{1/j5}{2+1/j5} \times 10I_1$, which gives $a_{21} = \left.\frac{I_1}{V_2}\right|_{I_2=0} = 0.1 + j$ S. On the input side, $V_1 = (20 + j10)I_1 + 10I_1$. Hence, $\left.\frac{V_1}{I_1}\right|_{I_2=0} = 30 + j10\ \Omega$. But $\left.\frac{V_1}{V_2}\right|_{I_2=0} = \left.\frac{V_1}{I_1}\right|_{I_2=0} \times \left.\frac{I_1}{V_2}\right|_{I_2=0}$. It follows that $a_{11} = \left.\frac{V_1}{V_2}\right|_{I_2=0} = (30 + j10)(0.1 + j) = -7 + 31j$.

With $V_2 = 0, I_2 = -\frac{10}{2}I_1$. Hence, $a_{22} = -\left.\frac{I_1}{I_1}\right|_{V_2=0} = 0.2$. On the input side, we have $-\left.\frac{V_1}{I_1}\right|_{V_2=0} = 30 + j10$ as before. Since $\left.\frac{V_1}{I_2}\right|_{V_2=0} = \left.\frac{V_1}{I_1}\right|_{V_2=0} \times \left.\frac{I_1}{I_2}\right|_{V_2=0}$, it follows that $a_{12} = -\left.\frac{V_1}{I_1}\right|_{V_2=0} = -(30 + j10)(-0.2) = 6 + j2\ \Omega$.

Example 24.2: Determination of h Parameters

The following dc measurements were made on a two-port resistive circuit:

Port 2 open-circuited: $V_1 = 10$ mV, $I_1 = 50\ \mu$A, $V_2 = 20$ V
Port 2 short-circuited: $V_1 = 40$ mV, $I_1 = 100\ \mu$A, $I_2 = -1$ mA
It is required to find the h parameters of the circuit.

Solution:

With port 2 short-circuited, $h_{11} = \left.\frac{V_1}{I_1}\right|_{V_2=0} = \frac{40 \text{ mV}}{100\ \mu\text{A}} \equiv 400\ \Omega$ and $h_{21} = \left.\frac{I_2}{I_1}\right|_{V_2=0} = -10$; $h_{12} = \left.\frac{V_1}{V_2}\right|_{I_1=0}$ and $h_{22} = \left.\frac{I_2}{V_2}\right|_{I_1=0}$ cannot be obtained from the given measurements because these do not include the case of port 1 open-circuited. However, we can obtain the a parameters from the given measurements. Thus,

$$a_{11} = \left.\frac{V_1}{V_2}\right|_{I_2=0} = \frac{10\times10^{-3}}{20} = 5\times10^{-4}$$

$$a_{12} = \left.\frac{V_1}{I_2}\right|_{V_2=0} = -\frac{40}{-1} = 40\ \Omega$$

$$a_{21} = \left.\frac{I_1}{V_2}\right|_{I_2=0} = \frac{50\times10^{-6}}{20} = 2.5\times10^{-6}\text{ S}$$

$$a_{22} = -\left.\frac{I_1}{I_2}\right|_{V_2=0} = -\frac{100\times10^{-6}}{-1\times10^{-3}} = 0.1$$

$$\Delta a = a_{11}a_{22} - a_{12}a_{21} = 5\times10^{-4}\times0.1 - 40\times2.5\times10^{-6} = -5\times10^{-5}$$

As a check, we find $h_{11} = \frac{a_{12}}{a_{22}} = \frac{40}{0.1}\ \Omega$ and $h_{21} = -\frac{1}{a_{22}} = -\frac{1}{0.1} = -10$, as derived earlier. It follows that $h_{12} = \frac{\Delta a}{a_{22}} = \frac{-5\times10^{-5}}{0.1} = -5\times10^{-4}$ and $h_{22} = \frac{a_{21}}{a_{22}} = \frac{2.5\times10^{-6}}{0.1} \equiv 25\ \mu$S.

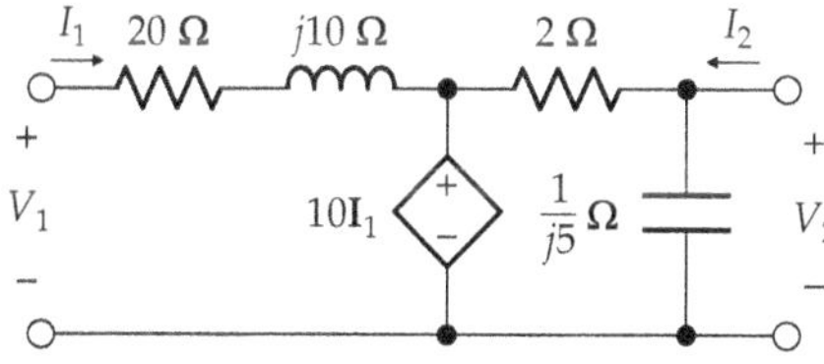

FIGURE 24.4
Figure for Example 24.1.

24.2.3 Reciprocal Circuits

Consider the T-circuit of Figure 24.5a. The mesh-current equations are

$$(Z_1 + Z_3)I_1 - Z_3I_2' = V_{SRC1} \quad (24.12)$$

$$-Z_3I_1 + (Z_2 + Z_3)I_2' = -V_{SRC2} \quad (24.13)$$

It is noteworthy, as discussed in Section 6.3, that in the absence of dependent sources, the coefficient of I_2'

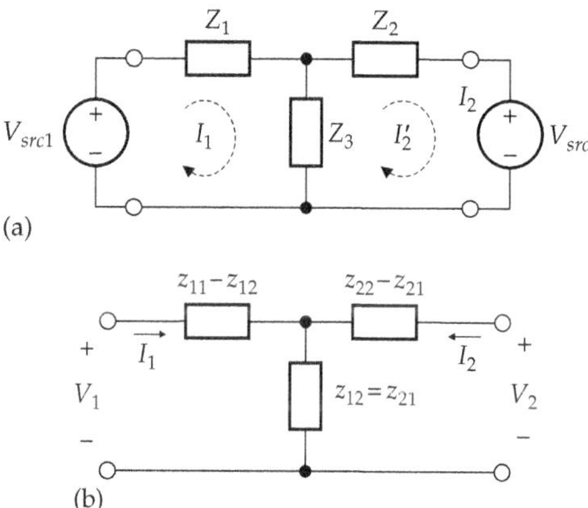

FIGURE 24.5
T-circuit (a) and its z parameters (b).

in Equation 24.12 and of I_1 in Equation 24.13 is simply the mutual impedance. Moreover, this mutual impedance, Z_3, is the same in the two mesh-current equations, because fundamentally the passive circuit elements of resistance, capacitance, and inductance are bilateral, that is, their values are independent of the direction of the current through them.

Next, let us substitute $I_2 = -I_2'$, identify V_{SRC1} with an input voltage V_1, in accordance with the substitution theorem, identify V_{SRC2} with an input voltage V_2, change the sign in Equation 24.13, and rearrange both equations to give

$$V_1 = (Z_1 + Z_3)I_1 + Z_3 I_2 \tag{24.14}$$

$$V_2 = Z_3 I_1 + (Z_2 + Z_3)I_2 \tag{24.15}$$

Equations 24.14 and 24.15 are the z-parameter equations of a two-port circuit having $z_{12} = z_{21} = Z_3$, $z_{11} = Z_1 + Z_3$ or $Z_1 = z_{11} - z_{12}$ and $z_{22} = Z_2 + Z_3$ or $Z_2 = z_{22} - z_{21}$, as shown in Figure 24.5b. We have thus shown that the z-parameter equations are of the form of mesh-current equations, where I_1 is a current flowing clockwise in an input mesh and I_2 is a current flowing counterclockwise in an output mesh. z_{11} and z_{22} are self-impedances of the meshes, whereas z_{12} and z_{21} are mutual impedances that are equal in the absence of dependent sources. Such a two-port circuit having $z_{12} = z_{21}$ is said to be **reciprocal**.

Similarly, the y-parameter equations can be shown to be of the form of node-voltage equations. Consider, for example, the π-circuit of Figure 24.6a. The node-voltage equations are

$$(Y_1 + Y_2)V_1 - Y_2 V_2 = I_{SRC1} \quad \text{and} \quad -Y_2 V_1 + (Y_2 + Y_3)V_2 = I_{SRC2} \tag{24.16}$$

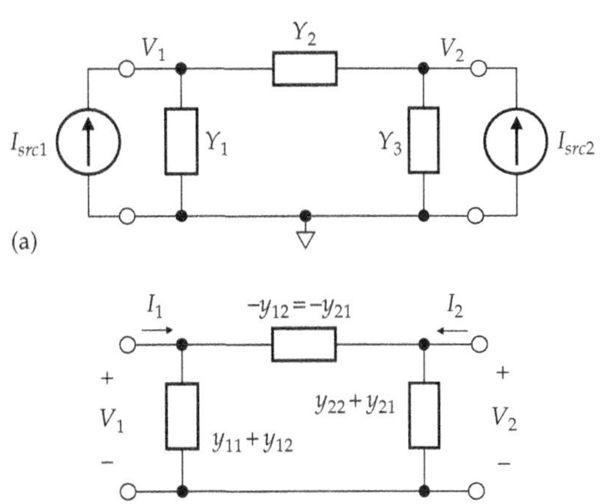

FIGURE 24.6
π-circuit (a) and its y-parameters (b).

Next, we may identify I_{SRC1} with an input current I_1, in accordance with the substitution theorem, identify I_{SRC2} with an input current I_2, and rearrange both equations to give

$$I_1 = (Y_1 + Y_2)V_1 - Y_2 V_2 \tag{24.17}$$

$$I_2 = -Y_2 I_V + (Y_2 + Y_3)V_2 \tag{24.18}$$

Equations 24.17 and 24.18 are the y-parameter equations of a two-port circuit having $y_{12} = y_{21} = -Y_2$, $y_{11} = Y_1 + Y_2$, or $Y_1 = y_{11} + y_{12}$ and $y_{22} = Y_2 + Y_3$ or $Y_3 = y_{22} + y_{21}$, as shown in Figure 24.6b. It is seen that the y-parameter equations are of the form of node-voltage equations, where y_{11} and y_{22} are self-admittances of the nodes, whereas y_{12} and y_{21} are mutual admittances that are equal in the absence of dependent sources. This again defines a reciprocal circuit in terms of the y parameters as a circuit having $y_{12} = y_{21}$.

Because of the relations between the various sets of two-port parameters, as shown in Table 24.3, the relations corresponding to $z_{12} = z_{21}$ or $y_{12} = y_{21}$ can be readily derived for the other sets of parameters. Thus, if $z_{12} = z_{21}$, it follows from Table 24.3 that $\Delta a/a_{21} = 1/a_{21}$, $1/b_{21} = \Delta b/b_{21}$, $h_{12}/h_{22} = -h_{21}/h_{22}$, and $-g_{12}/g_{11} = g_{21}/g_{11}$. These relations give $\Delta a = 1 = \Delta b$, $h_{12} = -h_{21}$, and $g_{12} = -g_{21}$. Parameter relations in reciprocal circuits are summarized in Table 24.4. It follows from these relations that in a reciprocal circuit, only three of the four circuit parameters are independent.

It should be noted that having no dependent sources in a two-port circuit is a *sufficient* condition for reciprocity but not a *necessary* one. In other words, a circuit that does not have dependent sources is reciprocal. But a circuit can have dependent sources and still be reciprocal, depending on the values of circuit parameters, as illustrated by Example 24.3. Hence, we define a reciprocal circuit as follows:

Definition: *A reciprocal circuit is one whose two-port parameters satisfy the reciprocity relations (Table 24.4).*

24.2.4 Symmetric Circuits

Definition: *A reciprocal circuit is symmetric if terminal voltages and currents remain the same when the two ports are interchanged.*

TABLE 24.4
Parameter Relations in Reciprocal Circuits

$z_{12} = z_{21}$	$y_{12} = y_{21}$
$\Delta a = a_{11}a_{22} - a_{12}a_{21} = 1$	$\Delta b = b_{11}b_{22} - b_{12}b_{21} = 1$
$h_{12} = -h_{21}$	$g_{12} = -g_{21}$

TABLE 24.5

Parameter Relations in Symmetric Circuits

$z_{11} = z_{22}$	$y_{11} = y_{22}$
$z_{12} = z_{21}$	$y_{12} = y_{21}$
$a_{11} = a_{22}$	$b_{11} = b_{22}$
$\Delta a = a_{11}a_{22} - a_{12}a_{21} = 1$	$\Delta b = b_{11}b_{22} - b_{12}b_{21} = 1$
$\Delta h = h_{11}h_{22} - h_{12}h_{21} = 1$	$\Delta g = g_{11}g_{22} - g_{12}g_{21} = 1$
$h_{12} = -h_{21}$	$g_{12} = -g_{21}$

To determine the relations between parameters in such a circuit, consider, to begin with, the z-parameter equations. If V_1 is interchanged with V_2 in these equations and I_1 is interchanged with I_2, the equations become

$$V_2 = z_{11}I_2 + z_{12}I_1 \quad \text{and} \quad V_1 = z_{21}I_2 + z_{22}I_1 \quad (24.19)$$

With $z_{12} = z_{21}$, since the circuit is reciprocal, Equations 24.19 are identical to the original equations if $z_{11} = z_{22}$. The same considerations apply to the y parameters so that the circuit is symmetric if $y_{11} = y_{22}$. In a symmetric circuit, therefore, only two of the four circuit parameters are independent.

The relations between the other two-port parameters in a symmetric circuit may be derived by equating $z_{11} = z_{22}$ or $y_{11} = y_{22}$ in Table 24.3. These relations are summarized in Table 24.5.

Exercise 24.5

(a) Verify the relations for the a, b, h, and g parameters in Table 24.5. (b) Show that the a and b parameters of a symmetric circuit are the same.

Example 24.3: Symmetric Two-Port Circuit

It is required to determine (a) R and α in Figure 24.7 so that the two-port circuit is symmetric and (b) the a parameters by simulation, with $R = 3\ \Omega$ and $\alpha = 0.5$.

Solution:

The procedure is to write Kirchhoff's voltage law (KVL) equations for the three meshes and then eliminate I_3 so as to have two equations of the form of the z-parameter equations.

It follows from the figure that

$$\text{Mesh 1: } V_1 + \alpha V_2 = RI_1 - RI_3 \quad (24.20)$$

$$\text{Mesh 2: } 0 = -(R+1)I_1 + I_2 + (R+2)I_3 \quad (24.21)$$

$$\text{Mesh 3: } V_2 = -I_1 + I_2 + I_3 \quad (24.22)$$

Eliminating I_3 between the last two equations gives

$$V_2 = -\frac{1}{R+2}I_1 + \frac{R+1}{R+2}I_2 \quad (24.23)$$

Eliminating I_3 by adding the first two equations, multiplying the third equation by 2, and subtracting,

$$V_1 + (\alpha - 2)V_2 = I_1 - I_2 \quad (24.24)$$

Substituting for V_2 from Equation 24.23 in Equation 24.24 and rearranging,

$$V_1 = \frac{\alpha + R}{R+2}I_1 - \frac{\alpha R + \alpha - R}{R+2}I_2 \quad (24.25)$$

Equations 24.23 and 24.25 are of the form of the z-parameter equations, from which it follows that $z_{11} = \frac{a+R}{R+2}$, $z_{12} = -\frac{\alpha R + \alpha - R}{R+2}$, $z_{21} = -\frac{1}{R+2}$, $z_{22} = \frac{R+1}{R+2}$

For a circuit to be symmetric, $z_{11} = z_{22}$ and $z_{12} = z_{21}$. Equating z_{11} and z_{22} gives $\alpha = 1$. With this value of α, any value of R makes $z_{12} = z_{21}$. If $R = 3\ \Omega$ and $\alpha = 0.5$, then $z_{11} = 0.7\ \Omega$, $z_{12} = 0.2\ \Omega$, $z_{21} = -0.2\ \Omega$, and $z_{22} = 0.8\ \Omega$. The circuit is neither symmetric nor reciprocal.

Simulation: The simulation is with the output port open-circuited and then short-circuited. The schematic for the former case is entered as in Figure 24.8. Although a 1 V source may be applied, it is convenient to use a 7 V in this case so as to obtain round figures for the voltages and currents. When the simulation is run, PSpice gives $V_2 = -2$ V and $I_1 = 10$ A. It follows that $a_{11} = -\frac{7}{2} = -3.5$ and $a_{21} = -\frac{10}{2} = -5$ S.

Figure 24.9 shows the schematic with the output port short-circuited by IPRINT2. It is convenient in this case to use a source voltage of 21 V so as to obtain round

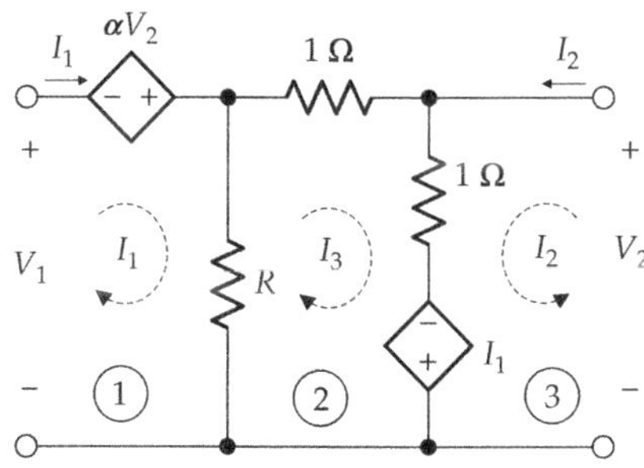

FIGURE 24.7
Figure for Example 24.3.

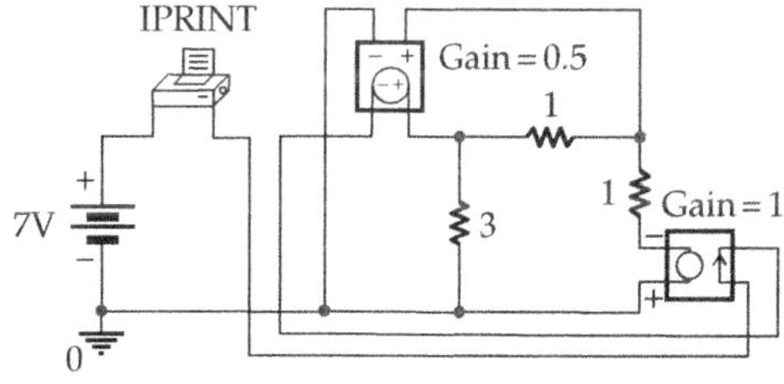

FIGURE 24.8
Figure for Example 24.3.

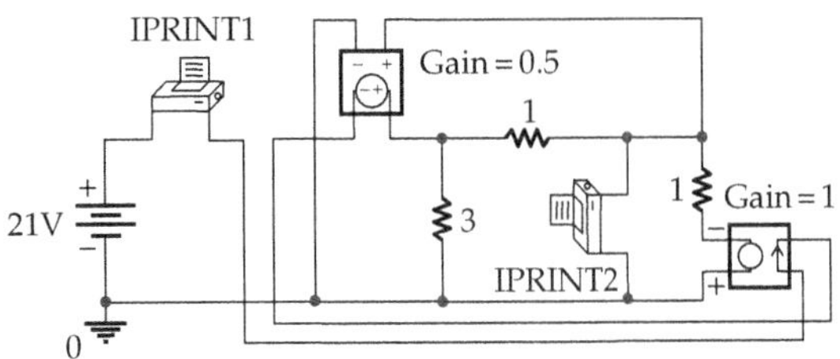

FIGURE 24.9
Figure for Example 24.3.

figures for the voltages and currents. When the simulation is run, PSpice gives $I_1 = 28$ A and $I_2 = 7$ A. It follows that $a_{12} = -\frac{21}{7} = -3\,\Omega$ and $a_{22} = -\frac{28}{7} = -4$. It is seen that $z_{11} = \frac{-3.5}{-5} = 0.7\,\Omega$, $z_{12} = \frac{(-3.5)(-4)-(-3)(-5)}{-5} = 0.2\,\Omega$, $z_{21} = \frac{1}{-5} = -0.2\,\Omega$ and $z_{22} = \frac{-4}{-5} = 0.8\,\Omega$, in agreement with the previous results.

Primal Exercise 24.6

Determine R_x in Figure 24.10 so that the two-port circuit is symmetric.

Ans. 15 Ω.

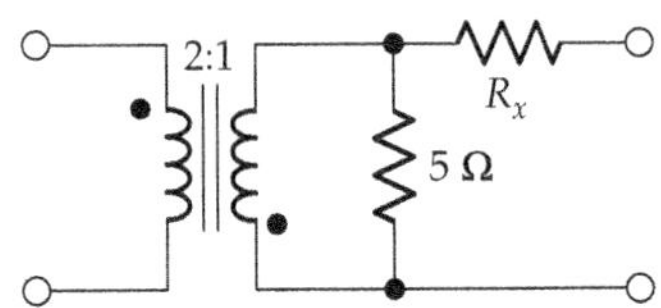

FIGURE 24.10
Figure for Primal Exercise 24.6.

Exercise 24.7

Verify that the two-port circuit of Figure 24.7 is symmetric if $\alpha = 1$ and $R = 0$ or $R = \infty$. Express Equations 24.20 through 24.22 in standard three-mesh form for $\alpha = 1$ and arbitrary R, and note that the matrix of coefficients is not symmetrical about the diagonal, although $z_{12} = z_{21}$.

24.3 Equivalent Circuits

In the *z*-parameter, *y*-parameter, *h*-parameter, and *g*-parameter equations, the first equation expresses an input variable in terms of the other input variable and an output variable, whereas the second equation expresses an output variable in terms of the other output variable and an input variable. These two-port equations may therefore be represented by an equivalent circuit with appropriate dependent sources on the input and output sides, as illustrated in Figure 24.11a through d.

The two-port parameters can be readily interpreted in terms of these circuits. For example, in the case of the *z*-parameter circuit, if port 2 is open-circuited, the dependent voltage source $z_{12}I_2 = 0$, so that the input impedance is z_{11}, and V_2 equals the source voltage $z_{21}I_1$. In general, the '11' subscript parameter describes an input impedance or admittance when the dependent source on the input side is set to zero, which imposes an open circuit or a short circuit at the output port. Similarly, the '22' subscript parameter describes an output impedance or admittance when the dependent source on the output side is set to zero, which imposes an open circuit or a short circuit at the input port.

The effect of the input on the output side, that is, the *forward transmission*, is expressed by the dependent

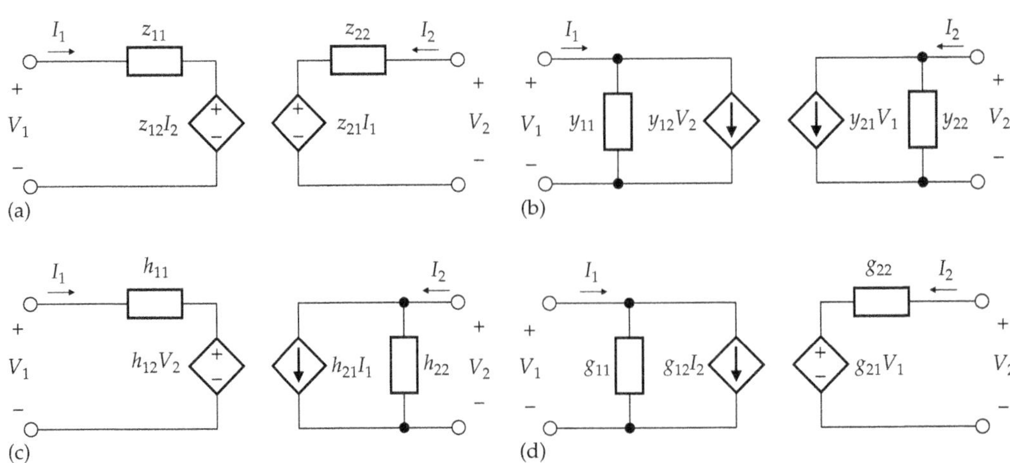

FIGURE 24.11
Equivalent circuits for *z*-parameter equations (a), *y*-parameter equations (b), *h*-parameter equations (c), and *g*-parameter equations (d).

source on the output side and its associated '21' subscript parameter. On the other hand, the effect of the output on the input side, that is, the *reverse transmission*, is expressed by the dependent source on the input side and its associated '12' subscript parameter.

The equivalent circuits of Figure 24.11 are of interest in that they represent the four possible types of electronic amplifiers, as represented by the '21' dependent sources. Thus, the circuit of Figure 24.11d is that of a *voltage amplifier* of input voltage V_1 and open-circuit output voltage $g_{21}V_1$, the voltage gain being g_{21}. The circuit of Figure 24.11c is that of a *current amplifier* of input current I_1 and short-circuit output current $h_{21}I_1$, the current gain being h_{21}. The circuit of Figure 24.11b is that of a *transconductance amplifier* of input voltage V_1 and short-circuit output current $y_{21}V_1$. Finally, the circuit of Figure 24.11a is that of a *transresistance amplifier* of input current I_1 and open-circuit output voltage $z_{21}I_1$.

As discussed in the preceding section, a reciprocal, two-port circuit having a common terminal between input and output may be represented by a T-equivalent circuit in terms of the z parameters (Figure 24.5b) or by a π-equivalent circuit in terms of the y-parameters (Figure 24.6b). If the circuit is nonreciprocal, dependent sources can be added to these T- and π-circuits to obtain the corresponding equivalent circuits, as illustrated by Example 24.4.

Example 24.4: T- and π-Equivalent Circuits of z- and y-Parameter Equations

It is required to derive the z-parameter and y-parameter equations, respectively, of the circuits shown in Figure 24.12a and b.

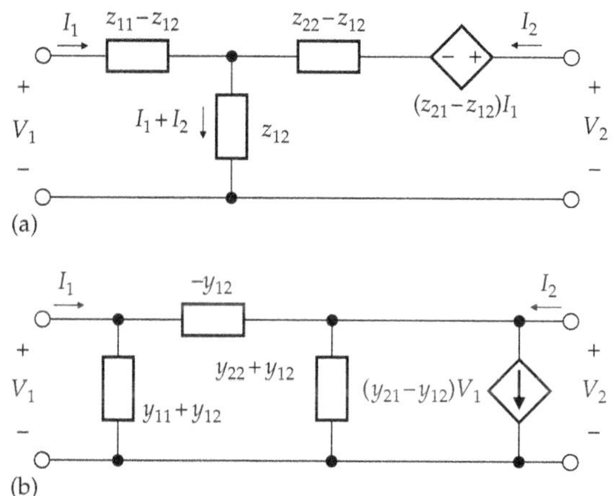

FIGURE 24.12
Figure for Example 24.4.

Solution:

Considering the circuit of Figure 24.12a, the KVL equations are

$$\text{Mesh 1: } \mathbf{V}_1 = (z_{11} - z_{12})\mathbf{I}_1 + z_{12}(\mathbf{I}_1 + \mathbf{I}_2) = z_{11}\mathbf{I}_1 + z_{12}\mathbf{I}_2 \quad (24.26)$$

$$\text{Mesh 2: } \mathbf{V}_2 = (z_{22} - z_{12})\mathbf{I}_2 + z_{12}(\mathbf{I}_1 + \mathbf{I}_2) + (z_{21} - z_{12})\mathbf{I}_1 = z_{21}\mathbf{I}_1 + z_{22}\mathbf{I}_2 \quad (24.27)$$

Equations 24.26 and 24.27 are the z-parameter equations for a two-port circuit in which z_{12} and z_{21} need not be equal.

Considering the circuit of Figure 24.12b, the node-voltage equations are

$$\text{Node 1: } \mathbf{I}_1 = (y_{11} + y_{12})\mathbf{V}_1 - y_{12}(\mathbf{V}_1 - \mathbf{V}_2) = y_{11}\mathbf{V}_1 + y_{12}\mathbf{V}_2 \quad (24.28)$$

$$\text{Node 2: } \mathbf{I}_2 = (y_{22} + y_{12} - y_{12})\mathbf{V}_2 + y_{12}\mathbf{V}_1 + (y_{21} - y_{12})\mathbf{V}_1 = y_{21}\mathbf{V}_1 + y_{22}\mathbf{V}_2 \quad (24.29)$$

Equations 24.28 and 24.29 are the y-parameter equations for a two-port circuit in which y_{12} and y_{21} need not be equal.

Exercise 24.8

Modify the circuits of Figure 24.28 so that the dependent sources appear at the input instead of the output.

24.4 Composite Two-Port Circuits

Concept: *An important feature of two-port circuit analysis is the derivation of the parameters of a composite two-port circuit that is a combination of individual two-port circuits.*

The individual two-port circuits may be combined in one of the five types of connection: cascade, parallel, series, series–parallel, or parallel–series.

24.4.1 Cascade Connection

When two or more circuits are cascaded, the output of any circuit other than the last is applied as the input of the following circuit, as illustrated in Figure 24.13 for two cascaded two-port circuits N′ and N″. A composite circuit is formed whose input variables are V_1 and I_1 and whose output variables are V_2 and I_2. It is desired to express the parameters of the composite circuit in terms of the parameters of the individual circuits.

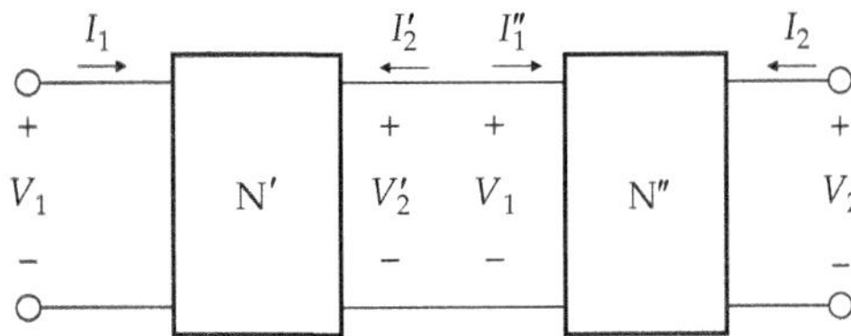

FIGURE 24.13
Cascade connection of two-port circuits.

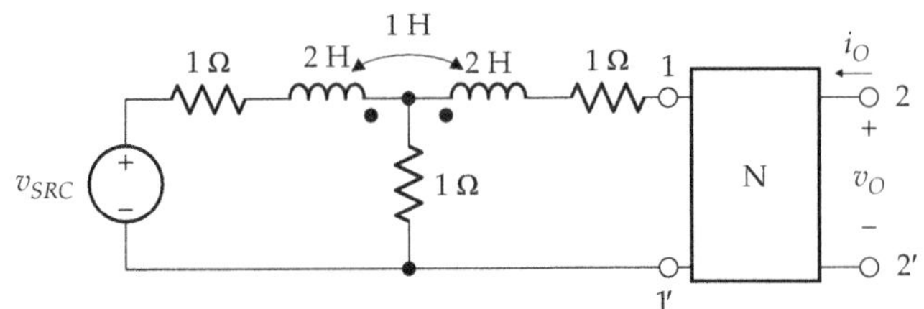

FIGURE 24.14
Figure for Example 24.5.

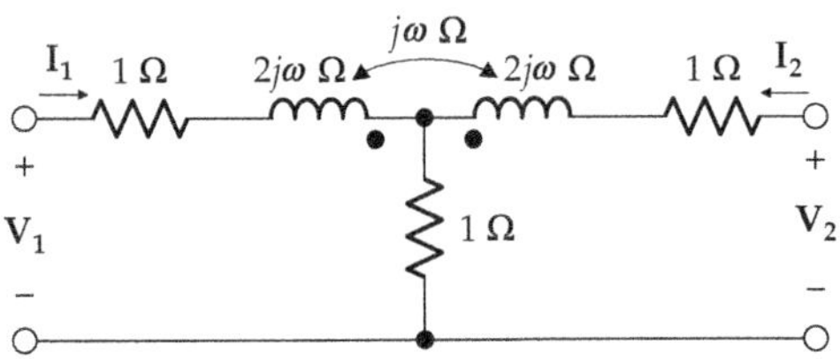

FIGURE 24.15
Figure for Example 24.5.

In analyzing the cascade connection, it is convenient to work with the a-parameter equations because they have the input variables on one side of the equation and the output variables on the other. One can therefore write

$$\begin{vmatrix} V_1 \\ I_1 \end{vmatrix} = \begin{bmatrix} a'_{11} & a'_{12} \\ a'_{21} & a'_{22} \end{bmatrix} \begin{vmatrix} V'_2 \\ -I'_2 \end{vmatrix} \qquad \begin{vmatrix} V''_1 \\ I''_1 \end{vmatrix} = \begin{bmatrix} a''_{11} & a''_{12} \\ a''_{21} & a''_{22} \end{bmatrix} \begin{vmatrix} V_2 \\ -I_2 \end{vmatrix} \tag{24.30}$$

Because of the cascade connection, $\begin{vmatrix} V'_2 \\ -I'_2 \end{vmatrix} = \begin{vmatrix} V''_1 \\ I''_1 \end{vmatrix}$. Substituting from the second equation in the first,

$$\begin{vmatrix} V_1 \\ I_1 \end{vmatrix} = \begin{bmatrix} a'_{11} & a'_{12} \\ a'_{21} & a'_{22} \end{bmatrix} \begin{bmatrix} a''_{11} & a''_{12} \\ a''_{21} & a''_{22} \end{bmatrix} \begin{vmatrix} V_2 \\ -I_2 \end{vmatrix} = \begin{bmatrix} a_{11} & a_{12} \\ a_{21} & a_{22} \end{bmatrix} \begin{vmatrix} V_2 \\ -I_2 \end{vmatrix} \tag{24.31}$$

or

$$\begin{bmatrix} a_{11} & a_{12} \\ a_{21} & a_{22} \end{bmatrix} = \begin{bmatrix} a'_{11} & a'_{12} \\ a'_{21} & a'_{22} \end{bmatrix} \begin{bmatrix} a''_{11} & a''_{12} \\ a''_{21} & a''_{22} \end{bmatrix} \tag{24.32}$$

Matrix multiplication gives

$$\begin{aligned} a_{11} &= a'_{11}a''_{11} + a'_{12}a''_{21} & a_{12} &= a'_{11}a''_{12} + a'_{12}a''_{22} \\ a_{21} &= a'_{21}a''_{11} + a'_{22}a''_{21} & a_{22} &= a'_{21}a''_{12} + a'_{22}a''_{22} \end{aligned} \tag{24.33}$$

This result can be generalized to the cascading of m two-port circuits:

$$[a] = [a_1][a_2]\ldots[a_m] \tag{24.34}$$

where

$[a]$ is the matrix of a parameters of the composite circuit
$[a_k]$ is the matrix of a parameters of the kth circuit, $k = 1, 2, \ldots, m$

Example 24.5: Cascaded Two-Port Circuits

In Figure 24.14, $v_{SRC}(t) = 10\cos t$ V. It is required to determine v_O with $i_O = 0$, given that circuit 'N' is symmetric and that it gave the following measurements, with port 2 open-circuited, $\mathbf{V}_2/\mathbf{V}_1 = 0.2$ and $\mathbf{V}_2/\mathbf{I}_1 = 5\ \Omega$.

Solution:

The procedure is to determine $a_{11} = \left.\frac{\mathbf{V}_{src}}{\mathbf{V}_o}\right|_{I_O=0}$ for the composite circuit from the a parameters of the two circuits.

The first circuit, redrawn in the frequency domain Figure 24.15, is symmetric also. With $\mathbf{I}_2 = 0$, $\mathbf{V}_1 = (2+2j\omega)\mathbf{I}_1$ and $\mathbf{V}_2 = (1 + j\omega)\mathbf{I}_1$. It follows that $a'_{21} = \left.\frac{\mathbf{I}_1}{\mathbf{V}_2}\right|_{\mathbf{I}_2=0} = \frac{1}{1+j\omega}$S and that $a'_{11} = \left.\frac{\mathbf{V}_1}{\mathbf{V}_2}\right|_{\mathbf{I}_2=0} = \frac{2(1+j\omega)}{1+j\omega} = 2$. From symmetry, $a'_{22} = a'_{11} = 2$ and $\Delta a' = a'_{11}a'_{22} - a'_{12}a'_{21} = 4 - \frac{a'_{12}}{s+1} = 1$, which gives $a'_{12} = 3(1+j\omega)\Omega$.

For circuit N, $a''_{11} = \left.\frac{\mathbf{V}_1}{\mathbf{V}_2}\right|_{\mathbf{I}_2=0} = \frac{1}{0.2} = 5$ and $z''_{11} = \left.\frac{\mathbf{V}_1}{\mathbf{V}_2}\right|_{\mathbf{I}_2=0} = 5\ \Omega$. From Table 24.2, $z''_{11} = \frac{a''_{11}}{a''_{21}}$. Hence, $a''_{21} = \frac{a''_{11}}{z''_{11}} = \frac{5}{5} = 1$ S. Since the circuit is symmetric, $a''_{22} = a''_{11} = 5$ and $\Delta a'' = a''_{11}a''_{22} - a''_{12}a''_{21} = 25 - a''_{12} \times 1 = 1$, which gives $a''_{12} = 24\ \Omega$. Note that the loading effect of circuit 'N' due to its finite input impedance is automatically taken care of.

From Equation 24.33, $a_{11} = a'_{11}a''_{11} + a'_{12}a''_{21} = 2 \times 5 + 3(1+j\omega) \times 1 = 13 + 3j\omega$. $\left.\frac{\mathbf{V}_o}{\mathbf{V}_{src}}\right|_{I_O=0} = \frac{1}{a_{11}} = \frac{1}{13+j3} = \frac{1}{\sqrt{178}\angle 13^\circ}$.

It follows that $v_O(t) = 0.75\cos(t - 13^\circ)$ V.

Problem-Solving Tip

- The open-circuit transfer function is readily determined by deriving a_{11}.

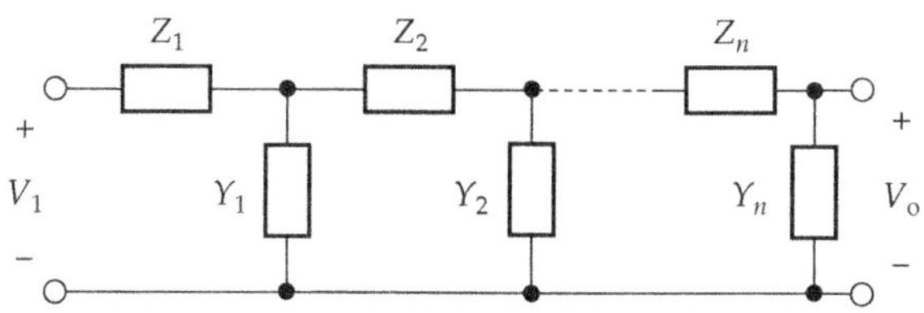

FIGURE 24.16
Ladder circuit as cascaded two-port circuits.

A useful special application of Equation 24.34 is in deriving the transfer function of ladder circuits of the general form shown in Figure 24.16. The idea is to consider each series impedance and each shunt admittance as a two-port circuit, derive the a-parameter matrix of each element, and then derive the a-parameter matrix of the ladder as the product of the matrices of the individual elements. The transfer function of the ladder network is then the $1/a_{11}$ parameter of the composite circuit. Example 24.6 illustrates the procedure.

Example 24.6: Transfer Function of Ladder Circuit

It is required to derive the transfer function of a ladder circuit as a cascade of two-port circuits.

Solution:

Consider a series impedance element Z in Figure 24.17a. The a parameters follow readily. With port 2 open-circuited, $a_{11} = \left.\frac{\mathbf{V}_1}{\mathbf{V}_2}\right|_{\mathbf{I}_2=0} = 1$ and $a_{21} = \left.\frac{\mathbf{I}_1}{\mathbf{V}_2}\right|_{\mathbf{I}_2=0} = 0$. With port 2 short-circuited, $a_{12} = -\left.\frac{\mathbf{V}_1}{\mathbf{I}_2}\right|_{\mathbf{V}_2=0} = Z$, and $a_{22} = -\left.\frac{\mathbf{I}_1}{\mathbf{I}_2}\right|_{\mathbf{V}_2=0} = 1$. The a-parameter matrix is therefore

$$[a] = \begin{bmatrix} 1 & Z \\ 0 & 1 \end{bmatrix} \tag{24.35}$$

For a shunt admittance element Y (Figure 24.17b), when port 2 is open-circuited, $a_{11} = \left.\frac{\mathbf{V}_1}{\mathbf{V}_2}\right|_{\mathbf{I}_2=0} = 1$, and

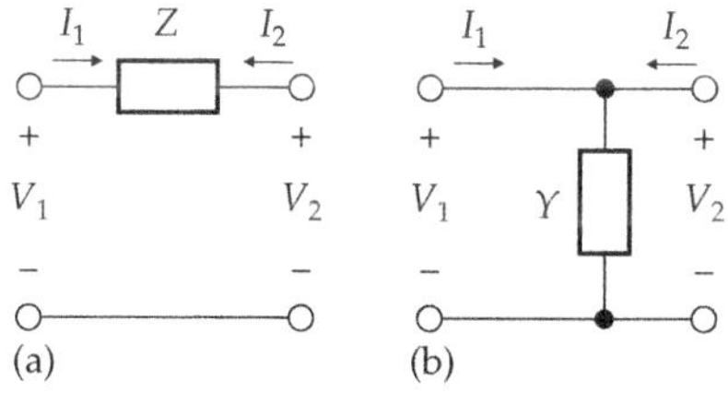

FIGURE 24.17
Figure for Example 24.6.

$a_{21} = \left.\frac{\mathbf{I}_1}{\mathbf{V}_2}\right|_{\mathbf{I}_2=0} = Y$. When port 2 is short-circuited, $\mathbf{V}_1 = \mathbf{V}_2 = 0$ and $\mathbf{I}_1 = -\mathbf{I}_2$. It follows that $a_{12} = 0$ and $a_{22} = 1$. The a-parameter matrix is therefore

$$[a] = \begin{bmatrix} 1 & 0 \\ Y & 1 \end{bmatrix} \tag{24.36}$$

Both two-port circuits of Figure 24.17 are symmetric so that $a_{11} = a_{22}$ and $\Delta a = 1$.

When elements Z and Y are cascaded, with Y following Z, the a-parameter matrix becomes

$$[a] = \begin{bmatrix} 1 & Z \\ 0 & 1 \end{bmatrix}\begin{bmatrix} 1 & 0 \\ Y & 1 \end{bmatrix} = \begin{bmatrix} 1+ZY & Z \\ Y & 1 \end{bmatrix} \tag{24.37}$$

according to the rules of matrix multiplication. Note that the resulting circuit is reciprocal but not symmetric.

As an example, consider the ladder circuit of Figure 24.18. The a-parameter matrix of the cascaded combination is

$$[a] = \begin{bmatrix} 1 & 1 \\ 0 & 1 \end{bmatrix}\begin{bmatrix} 1 & 0 \\ 2j\omega & 1 \end{bmatrix}\begin{bmatrix} 1 & j\omega \\ 0 & 1 \end{bmatrix}\begin{bmatrix} 1 & 0 \\ 2 & 1 \end{bmatrix} \tag{24.38}$$

To evaluate this product, we can multiply together the first two matrices, the last two matrices, and finally the two product matrices. Alternatively, MATLAB may be used. Thus,

$$\begin{bmatrix} 1 & 1 \\ 0 & 1 \end{bmatrix}\begin{bmatrix} 1 & 0 \\ 2\omega & 1 \end{bmatrix} = \begin{bmatrix} 2j\omega+1 & 1 \\ 2j\omega & 1 \end{bmatrix} \tag{24.39}$$

$$\begin{bmatrix} 1 & j\omega \\ 0 & 1 \end{bmatrix}\begin{bmatrix} 1 & 0 \\ 2 & 1 \end{bmatrix} = \begin{bmatrix} 2j\omega+1 & j\omega \\ 2 & 1 \end{bmatrix} \tag{24.40}$$

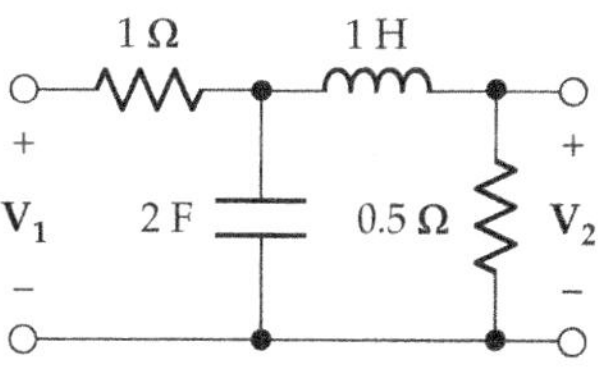

FIGURE 24.18
Figure for Example 24.6.

$$\begin{bmatrix} 2j\omega+1 & 1 \\ 2j\omega & 1 \end{bmatrix}\begin{bmatrix} 2j\omega+1 & j\omega \\ 2 & 1 \end{bmatrix}$$

$$= \begin{bmatrix} \left(3-4\omega^2\right)+4j\omega & \left(1-2\omega^2\right)+j\omega \\ \left(2-4\omega^2\right)+2j\omega & 1-2\omega^2 \end{bmatrix} \quad (24.41)$$

It follows that $\dfrac{\mathbf{V}_2}{\mathbf{V}_1} = \dfrac{1}{a_{11}} = \dfrac{1}{3-4\omega^2+4j\omega}$

Problem-Solving Tip

- Matrix multiplication is greatly facilitated using MATLAB.

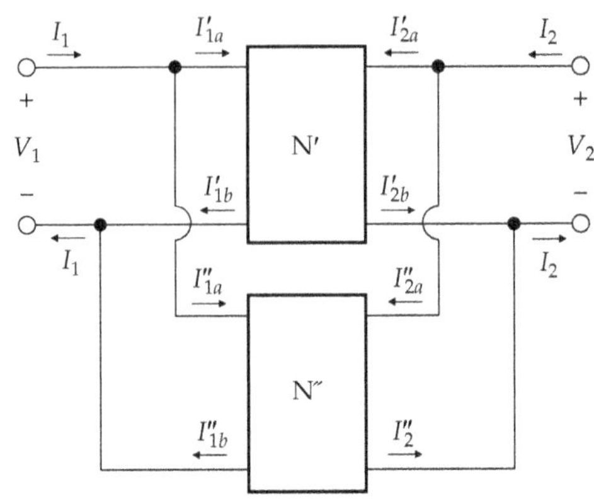

FIGURE 24.20
Two-port circuits connected in parallel.

Primal Exercise 24.9

Determine V_O in Figure 24.19a and b by considering each circuit to be a cascade of two-port circuits.

Ans. (a) 2 V; (b) 4 V.

24.4.2 Parallel Connection

When two-port circuits are connected in parallel or in series at their input or output ports, the two-port circuit equations no longer apply, in general, to the individual circuits. Thus, in the case of the two paralleled two-port circuits N′ and N″ in Figure 24.20, the currents entering and leaving the input and output ports are not, in general, equal. That is, $I'_{1a} \neq I'_{1b}, I'_{2a} \neq I'_{2b}, I''_{1a} \neq I''_{1b}$, and $I''_{2a} \neq I''_{2b}$. This invalidates the two-port circuit equations, which are based on the assumption of equality of the currents entering and leaving the input and output ports (Figure 24.1). This situation is illustrated in Figure 24.21 by two paralleled π-circuits shown in thick lines. The circuit can be readily analyzed in the conventional manner by noting that corresponding resistances in the two π-circuits are paralleled so that the composite circuit reduces to that shown in Figure 24.22, where $2||1 = 2/3\ \Omega$, $2||3 = 6/5\ \Omega$ and $1||4 = 4/5\ \Omega$. The current through the $2/3\ \Omega$ across the 10 V source is 15 A, and the current through the three

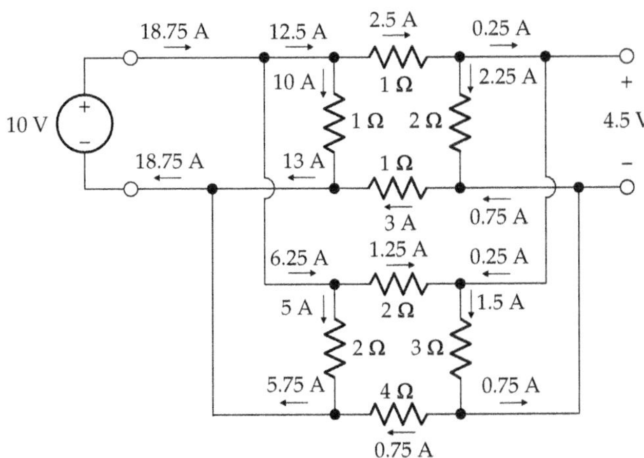

FIGURE 24.21
Illustration of inequality of terminal currents of two-port circuits.

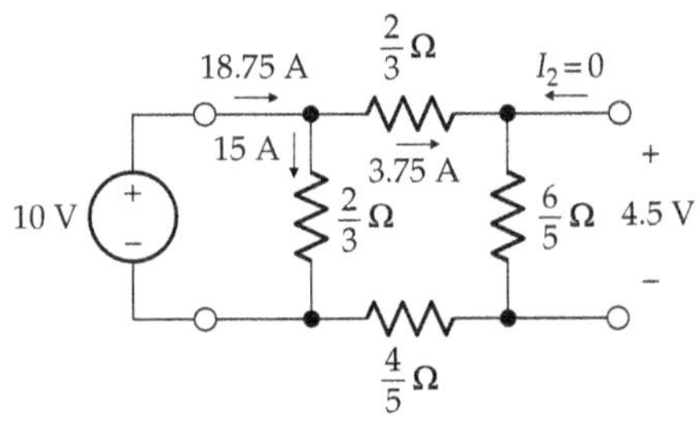

FIGURE 24.22
Equivalent circuit of Figure 24.21.

series-connected resistors is $10/(2/3 + 6/5 + 4/5) = 3.75$ A. The currents through the individual resistors follow from current division, as shown in Figure 24.21. It is seen that 12.5 A enter the input port of N′ but 13 A leave this port, and 0.25 A leave the output port of this circuit, but 0.75 A enter this port. Similarly, 6.25 A enter the input port of the lower π-circuit, but 5.75 A

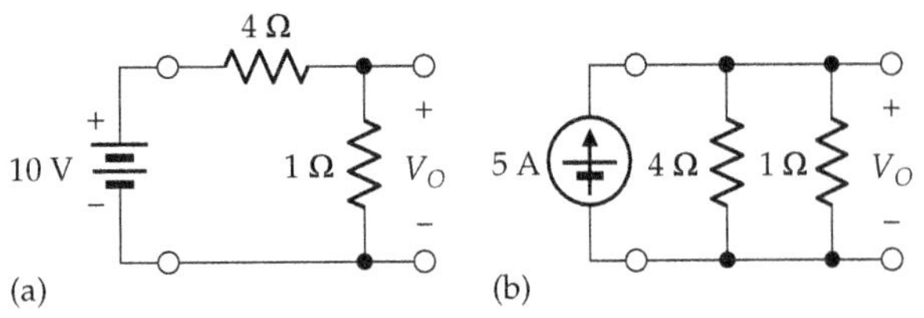

FIGURE 24.19
Figure for Primal Exercise 24.9.

leave this port, and 0.25 A enter the output port of this circuit, but 0.75 A leave the port.

The port currents for each individual circuit can be made equal by connecting a 1:1 ideal transformer at either port of either of the two circuits. The ideal transformer forces equality between the currents entering and leaving the port at which the transformer is connected. The equality of these currents at the other port of the same circuit then follows from Kirchhoff's current law (KCL). In Figure 24.23, for example, a 1:1 ideal transformer is connected at the input port of N′. From KCL, the currents entering and leaving the input or output terminals of the transformer are equal. With the same current entering and leaving the source, it follows from KCL at the source terminals that the same current enters and leaves the input port of N″. If either N′ or N″ is surrounded by a closed surface, it also follows from KCL that the same current enters and leaves the output port of N′ or N″. The conditions for applying the two-port equations are therefore satisfied.

By definition of a parallel connection of two-port circuits, (1) the input voltage is the same and the output voltage is the same, for all the paralleled circuits, and (2) the input and output currents of the composite circuit are the sums of the input and output currents of the individual circuits, respectively. The most appropriate two-port equations for describing the composite circuit are therefore the y-parameter equations, since the dependent variables in these equations are the same input and output voltages for all the paralleled circuits. In the case of the two-port circuits of Figure 24.23, the y-parameter equations are

$$\begin{vmatrix} I_1' \\ I_2' \end{vmatrix} = \begin{vmatrix} y_{11}' & y_{12}' \\ y_{21}' & y_{22}' \end{vmatrix} \begin{vmatrix} V_1' \\ V_2' \end{vmatrix} \qquad \begin{vmatrix} I_1'' \\ I_2'' \end{vmatrix} = \begin{vmatrix} y_{11}'' & y_{12}'' \\ y_{21}'' & y_{22}'' \end{vmatrix} \begin{vmatrix} V_1'' \\ V_2'' \end{vmatrix} \qquad (24.42)$$

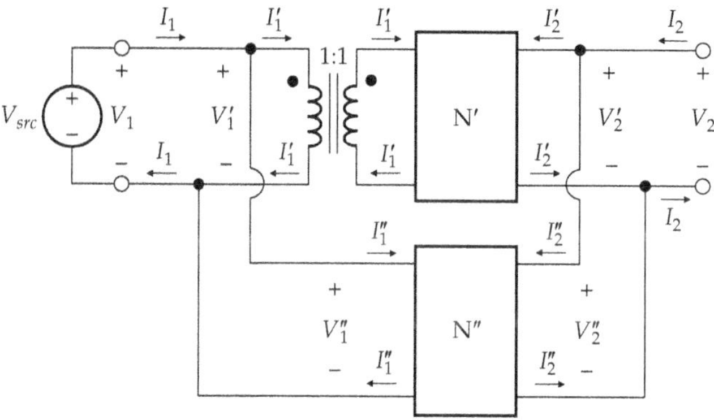

FIGURE 24.23
Ideal transformer added to a parallel connection of two-port circuits.

with $V_1 = V_1' = V_1''$, $V_2 = V_2' = V_2''$, $I_1 = I_1' + I_1''$, and $I_1 = I_2' + I_2''$. Substituting these relations in Equation 24.42, it follows that

$$\begin{vmatrix} I_1 \\ I_2 \end{vmatrix} = \begin{vmatrix} y_{11}' + y_{11}'' & y_{12}' + y_{12}'' \\ y_{21}' + y_{21}'' & y_{22}' + y_{22}'' \end{vmatrix} \begin{vmatrix} V_1 \\ V_2 \end{vmatrix} \qquad (24.43)$$

It is seen that the corresponding y parameters of the individual circuits simply add to give the y parameters of the composite circuit. In general,

$$[y] = [y_1] + [y_2] + \cdots + [y_m] \qquad (24.44)$$

where

$[y]$ is the matrix of y parameters of the composite circuit
$[y_k]$ is the matrix of y parameters of the kth circuit, $k = 1, 2, \ldots, m$

With the 1:1 ideal transformer connected, the composite circuit can be analyzed using Equation 24.44. The 1:1 ideal transformer does not change any of the two-port parameters of the circuit to which it is connected. It follows from applying the definition of the y parameters that these parameters for the two π-circuits of Figure 24.21 are (Exercise 24.10) as follows:

$$y_{11}' = 3/2\text{ S}, \quad y_{12}' = -1/2\text{ S}, \quad y_{21}' = -1/2\text{ S}, \quad y_{22}' = 1\text{ S}$$
$$y_{11}'' = 2/3\text{ S}, \quad y_{12}'' = -1/6\text{ S}, \quad y_{21}'' = -1/6\text{ S}, \quad y_{22}'' = 1/2\text{ S}$$

Hence, the y parameters of the composite circuit are

$$y_{11} = y_{11}' + y_{11}'' = 13/6\text{ S} \qquad y_{12} = y_{12}' + y_{12}'' = -2/3\text{ S}$$
$$y_{21} = y_{21}' + y_{21}'' = -2/3\text{ S} \qquad y_{22} = y_{22}' + y_{22}'' = 3/2\text{ S}$$

Note that in all cases, $y_{12} = y_{21}$ because of reciprocity. Once the y parameters of the composite circuit are known, terminal voltages and currents of the composite circuit can be derived for any given conditions. For example, if 10 V are applied at the input port (Figure 24.21), the open-circuit output voltage is $V_2 = V_1/a_{11} = V_1(-y_{21}/y_{22}) = 10 \times 4/9 = 40/9$ V.

The use of the ideal transformer precludes, of course, dc conditions. However, the two-port equations remain valid in this case if the paralleled two-port circuits are three-terminal, each having a common terminal between input and output, with the common terminals of all the paralleled circuits connected together. This is illustrated in Figure 24.24, which shows two π-circuits, as in Figure 24.21, but with the bottom resistor of each circuit replaced by a short circuit so as to make the circuits three-terminal. Applying the same procedure used in connection with Figure 24.21, it is seen that any arbitrarily assumed value of I_1', including 13 A, satisfies KCL at any of the six junctions representing the two common terminals that are connected together, that is, I_1' is indeterminate. The reason for this is that the six junctions are in reality a single node. The currents between the

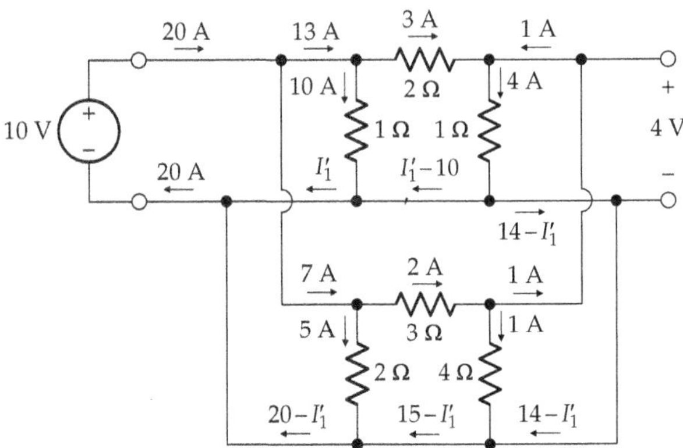

FIGURE 24.24
Three-terminal two-port circuits connected in parallel.

junctions are fictitious currents in assumed wiring connections that are short circuits. These currents do not flow through any circuit elements and therefore do not introduce any voltage drops in the circuit. It follows from applying the definition of the y parameters that these parameters for the two π-circuits of Figure 24.24 are (Exercise 24.10) as follows:

$$y'_{11} = 3/2\text{ S},\quad y'_{12} = -1/2\text{ S},\quad y'_{21} = -1/2\text{ S},\quad y'_{22} = 3/2\text{ S}$$
$$y''_{11} = 5/6\text{ S},\quad y''_{12} = -1/3\text{ S},\quad y''_{21} = -1/3\text{ S},\quad y''_{22} = 7/12\text{ S}$$

Hence, the y parameters of the composite circuit are

$$y_{11} = y'_{11} + y''_{11} = 7/3\text{ S} \qquad y_{12} = y'_{12} + y''_{12} = -5/6\text{ S}$$
$$y_{21} = y'_{21} + y''_{21} = -5/6\text{ S} \qquad y_{22} = y'_{22} + y''_{22} = 25/12\text{ S}$$

With 10 V applied at the input port (Figure 24.24), the open-circuit output voltage is $V_2 = V_1/a_{11} = V_1(-y_{21}/y_{22}) = 10 \times 2/5 = 4$ V.

Primal Exercise 24.10

Derive the y parameters for the individual two-port circuits of Figures 24.21 and 24.24.

Exercise 24.11

Verify by PSpice simulation that if a transformer is connected to any of the ports of Figure 24.21 and 10 V ac is applied at the input port, the output voltage is $40/9 = 4.44$ V.

Exercise 24.12

Argue that in order that the two-port circuit equations remain valid when n two-port circuits are paralleled, a 1:1 ideal transformer should be connected at the input or output of (n – 1) of the circuits.

Example 24.7: Two-Port Circuits Connected in Parallel

An application of the paralleling of two three-terminal circuits is the notch filter consisting of two paralleled T-circuits, as shown in Figure 24.25. These circuits are also part of the universal filter discussed in Section 15.6. It is required to derive the transfer function of the filter.

Solution:

Consider one of the T-circuits, with its output short-circuited (Figure 24.26). It follows from KVL that

$$V_1' = RI_1' - RI_2' \tag{24.45}$$

and, from current division, that

$$I_2' = -\frac{1/2sC}{R + 1/2sC} I_1' \tag{24.46}$$

Eliminating I_2' between these two equations and simplifying gives

$$y_{11}' = \left.\frac{I_1'}{V_1'}\right|_{V_2'=0} = \frac{1 + 2sCR}{2R(1 + sCR)} \tag{24.47}$$

Eliminating I_1' between these Equations 24.45 and 24.46 and simplifying gives

$$y_{21}' = \left.\frac{\mathbf{I}_2'}{\mathbf{V}_1'}\right|_{V_2'=0} = -\frac{1}{2R(1 + sCR)} \tag{24.48}$$

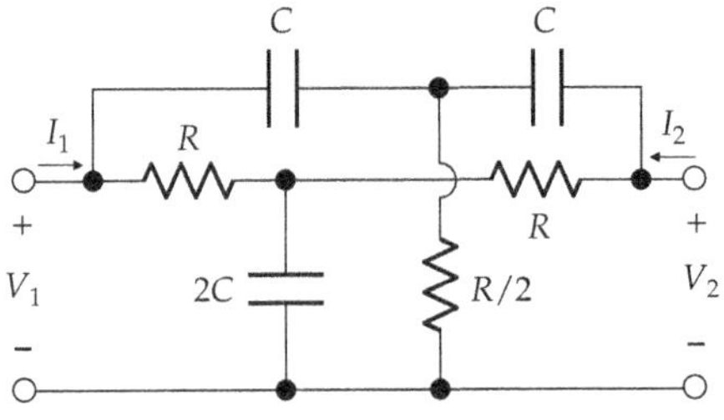

FIGURE 24.25
Figure for Example 24.7.

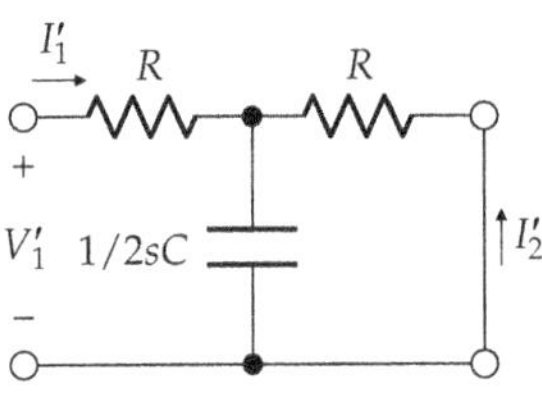

FIGURE 24.26
Figure for Example 24.7.

Since the circuit is symmetric, $y'_{11} = y'_{22}$ and $y'_{12} = y'_{21}$. The other T-circuit of Figure 24.25 may be derived from Equations 24.47 and 24.48 by replacing R by $1/sC$ and sC by $1/R$, which gives

$$y''_{11} = \frac{sC(2+sCR)}{2(1+sCR)} \text{ and } y''_{21} = -\frac{s^2C^2R^2}{2R(1+sCR)} \quad (24.49)$$

It follows that the y parameters of the notch filter (Figure 24.25) are

$$y_{11} = y_{22} = y'_{11} + y''_{22} = \frac{s^2C^2R^2 + 4sCR + 1}{2R(1+sCR)} = \frac{C}{2}\frac{s^2 + 4\omega_0 s + \omega_0^2}{s+\omega_0} \quad (24.50)$$

$$y_{12} = y_{21} = y'_{21} + y''_{21} = -\frac{s^2C^2R^2 + 1}{2R(1+sCR)} = -\frac{C}{2}\frac{s^2 + \omega_0^2}{s+\omega_0} \quad (24.51)$$

where $\omega_0 = 1/CR$.

The transfer function is $H(s) = \left.\frac{V_2}{V_1}\right|_{I_2=0} = a_{11} = -\frac{y_{21}}{y_{22}}$ (Tables 24.2 and 24.3) as follows:

$$H(s) = \frac{s^2 + \omega_n^2}{s^2 + 4\omega_n s + \omega_n^2} \quad (24.52)$$

From Table 14.3, this is the transfer function of a bandstop filter having $Q = 0.25$.

A useful application of paralleling two-port circuits is the derivation of the transfer function of a three-terminal circuit that is bridged by an admittance between the input and output terminals (Figure 24.27). The admittance Y is considered a symmetrical two-port circuit, in the same manner as the series impedance in Figure 24.17a. If $V_2 = 0$, then $I_1 = -I_2 = YV_1$ (Figure 24.28). It follows that the y-parameter matrix of the admittance Y is

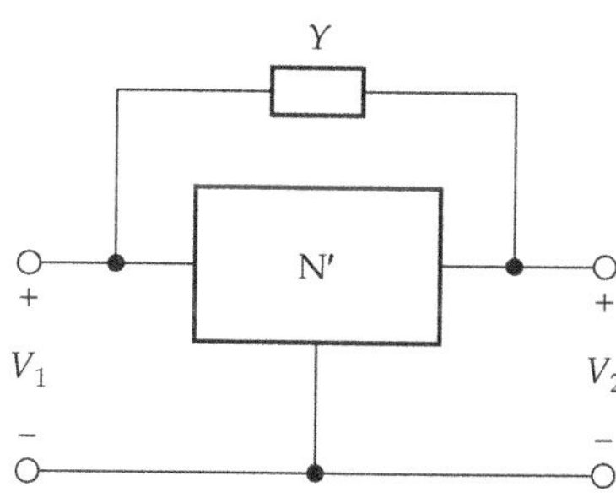

FIGURE 24.27
Bridged two-port circuit.

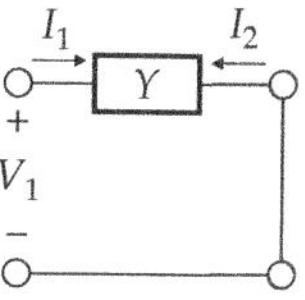

FIGURE 24.28
Bridging admittance as a two-port circuit.

$$\begin{bmatrix} Y & -Y \\ -Y & Y \end{bmatrix} \quad (24.53)$$

The y-parameter matrix of the composite circuit is then the sum of the matrix of Equation 24.53 and the y-matrix of the two-port circuit N′. Once the y parameters of the composite circuit are determined, any of the terminal conditions for the circuit can be derived using Table 24.3.

Primal Exercise 24.13

Derive the y-parameter matrix of Equation 24.53 from the a-parameter matrix of the series element of Figure 24.17a, using the conversion of a to y parameters of Table 24.3.

Example 24.8: Bridged Two-Port Circuits

It is required to determine the transfer function V_2/V_1 of the circuit in Figure 24.29.

Solution:

Consider the T-circuit consisting of the two resistors and inductor (Figure 24.30). It is seen that $V'_1 = I'_1 - I'_2$ and $I'_2 = -\frac{s}{s+1}I'_1$. It follows that $y'_{11} = \left.\frac{I'_1}{V'_1}\right|_{V'_2=0} = \frac{s+1}{2s+1}$ S and $y'_{21} = \left.\frac{I'_2}{V'_1}\right|_{V'_2=0} = -\frac{s}{2s+1}$ S. Since the circuit is symmetric,

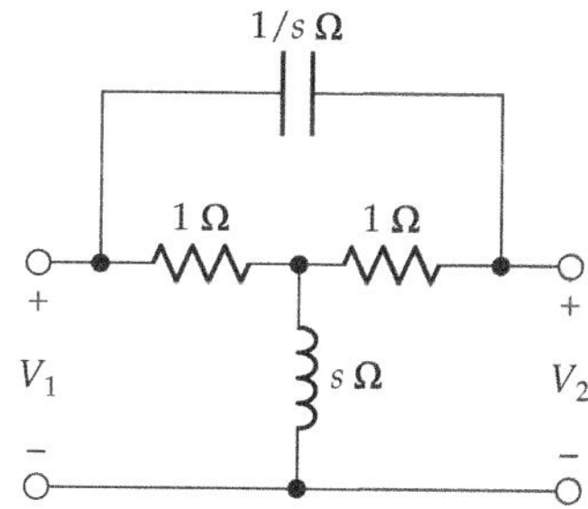

FIGURE 24.29
Figure for Example 24.8.

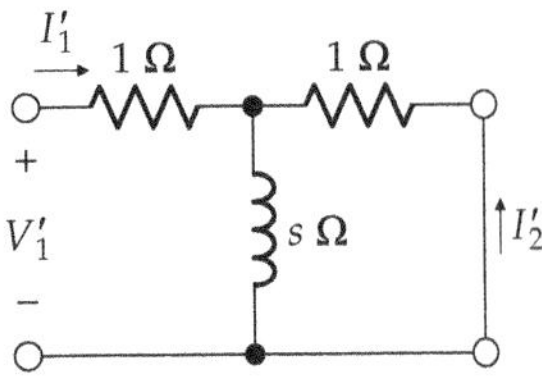

FIGURE 24.30
Figure for Example 24.8.

$y'_{11} = y'_{22}$ and $y'_{12} = y'_{21}$. The y-parameter matrix of the capacitor is, from Equation 24.53, $\begin{bmatrix} s & -s \\ -s & s \end{bmatrix}$. It follows that

$$y_{11} = y_{22} = \frac{s+1}{2s+1} + s = \frac{2s^2+2s+1}{2s+1}$$

$$\text{and} \quad y_{12} = y_{21} = -\frac{s}{2s+1} - s = -\frac{2s(s+1)}{2s+1}. \tag{24.54}$$

The transfer function is

$$\frac{V_2}{V_1} = -\frac{y_{21}}{y_{22}} = \frac{2s(s+1)}{2s^2+2s+1} \tag{24.55}$$

Primal Exercise 24.14

Assume that in Figure 24.27, circuit N′ has the following z parameters, $z_{11} = 8\ \Omega$, $z_{12} = 3\ \Omega$, $z_{21} = 5\ \Omega$, and $z_{22} = 2\ \Omega$. Determine the y parameters of N′ when bridged by a 1 H inductor.

Ans. $y_{11} = 2 + 1/j\omega$ S, $y_{12} = -(3 + 1/j\omega)$ S, $y_{21} = -(5 + 1/j\omega)$ S, and $y_{22} = 8 + 1/j\omega$ S.

24.4.3 Series Connection

In a series connection (Figure 24.31), (1) the input current is the same, and the output current is the same, for all the series-connected circuits, and (2) the input and output voltages of the composite circuit are the sums of the input and output voltages of the individual circuits. A 1:1 ideal transformer is connected at the input port of N′ so as to equalize the currents entering and leaving each port, which validates the use of the two port equations, as explained in connection with Figure 24.23.

The z-parameter equations are the most appropriate in this case, since the dependent variables are the input and output currents, which are equal for the series-connected circuits. The z-parameter equations are

$$\begin{vmatrix} V'_1 \\ V'_2 \end{vmatrix} = \begin{vmatrix} z'_{11} & z'_{12} \\ z'_{21} & z'_{22} \end{vmatrix} \begin{vmatrix} I'_1 \\ I'_2 \end{vmatrix} \qquad \begin{vmatrix} V''_1 \\ V''_2 \end{vmatrix} = \begin{vmatrix} z''_{11} & z''_{12} \\ z''_{21} & z''_{22} \end{vmatrix} \begin{vmatrix} I''_1 \\ I''_2 \end{vmatrix} \tag{24.56}$$

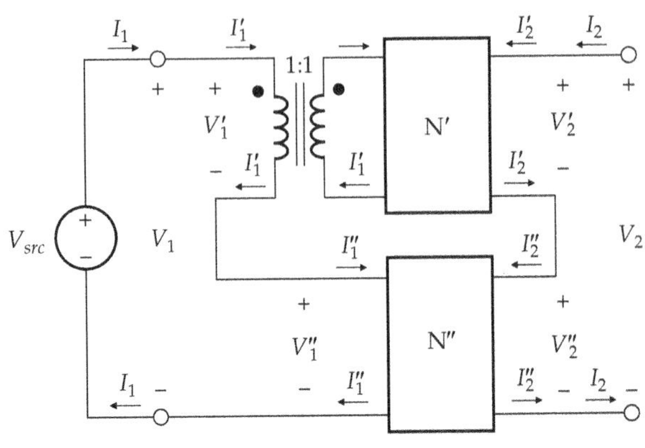

FIGURE 24.31
Ideal transformer added to a series connection of two-port circuits.

For the series connection, $V_1 = V'_1 + V''_1$, $V_2 = V'_2 + V''_2$, $I_1 = I'_1 = I''_1$, and $I_2 = I'_2 = I''_2$. Substituting in Equation 24.56, or using the z-parameter equations,

$$\begin{vmatrix} V_1 \\ V_2 \end{vmatrix} = \begin{vmatrix} z'_{11} + z''_{11} & z'_{12} + z''_{12} \\ z'_{21} + z''_{21} & z'_{22} + z''_{22} \end{vmatrix} \begin{vmatrix} I_1 \\ I_2 \end{vmatrix} \tag{24.57}$$

The corresponding z parameters of the individual circuits simply add to give the z parameters of the composite circuit. In general,

$$[z] = [z_1] + [z_2] + \cdots + [z_m] \tag{24.58}$$

where

$[z]$ is the matrix of z parameters of the composite circuit

$[z_k]$ is the matrix of z parameters of the kth circuit, $k = 1, 2, \ldots, m$

The 1:1 ideal transformer need not be used with three-terminal two-port circuits that have their common terminals connected together. Consider the two T-circuits of Figure 24.32. It follows from applying the definition of the z parameters that these parameters for the two T-circuits of Figure 24.32 are (Exercise 24.15) as follows:

$$z'_{11} = 2\ \Omega, \quad z'_{12} = 1\ \Omega, \quad z'_{21} = 1\ \Omega, \quad z'_{22} = 3\ \Omega$$
$$z''_{11} = 3\ \Omega, \quad z''_{12} = 1\ \Omega, \quad z''_{21} = 1\ \Omega, \quad z''_{22} = 2\ \Omega$$

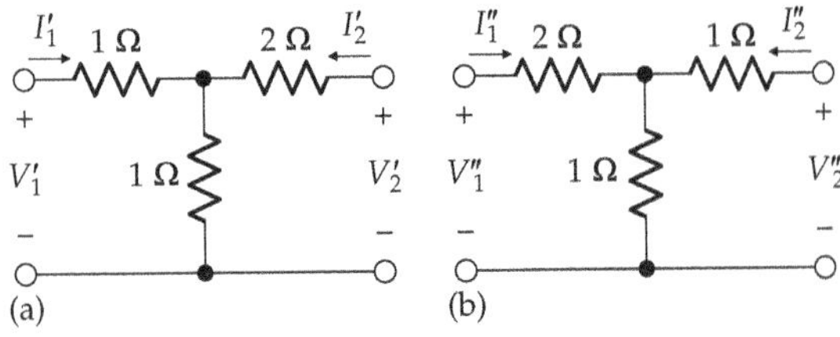

FIGURE 24.32
Two two-port, three-terminal circuits (a) and (b).

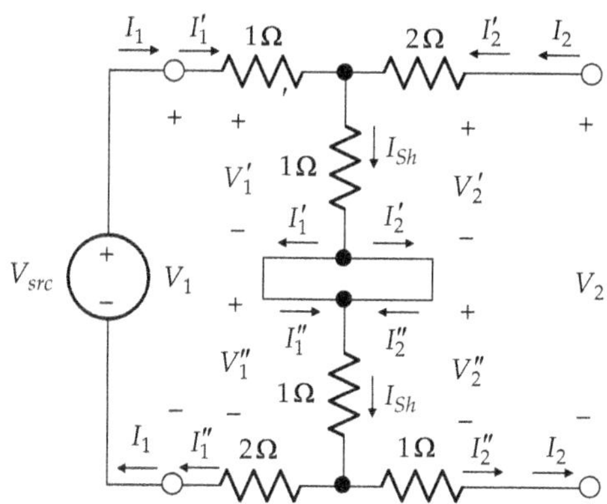

FIGURE 24.33
Series connection of the two-port circuits of Figure 24.32.

To connect these circuits in series with their common terminals connected together, it is convenient to connect the circuit of Figure 24.32b "upside down," as in Figure 24.33. This reverses the polarities of the current and voltage at *both* ports, as shown, which changes neither the two-port equations nor the z parameters. Hence, the z parameters of the composite circuit are

$$z_{11} = z_{11}' + z_{11}'' = 5\ \Omega; \quad z_{12} = z_{12}' + z_{12}'' = 2\ \Omega$$
$$z_{21} = z_{21}' + z_{21}'' = 2\ \Omega; \quad z_{22} = z_{22}' + z_{22}'' = 5\ \Omega$$

Note that according to the values of its z parameters, the composite circuit is symmetrical, although it is not geometrically symmetrical about a vertical midline. Note also that the currents in the 'wiring rectangle' in the center of Figure 24.33 are indeterminate. If this rectangle is enclosed by a closed figure, then I_{Sh} is the same in the two 1 Ω resistors connected to this rectangle. KCL at the upper junction is $I_1' + I_2' = I_{Sh}$. Substituting $I_1' = I_1''$ and $I_2' = I_2''$ gives $I_1'' + I_2'' = I_{Sh}$, which is KCL at the lower junction. Hence, given I_{Sh}, the currents in the wiring rectangle cannot be uniquely determined, since KCL at the two nodes does not give two independent equations. As pointed out earlier, this indeterminacy is because the currents in the rectangle are fictitious currents in assumed wiring connections that are short circuits, as explained in connection with Figure 24.24.

Suppose that the output terminals of the composite circuit of Figure 24.33 are short-circuited, and it is desired to determine the short circuit current from the z parameters when $V_{SRC} = 21$ V. With $V_2 = 0$, $y_{21} = I_2/V_1$ (Table 24.2), with $y_{21} = -z_{21}/\Delta z$, where $\Delta z = z_{11}z_{22} - z_{12}z_{21}$. Substituting gives $I_2 = -V_1(z_{21}/\Delta z) = -(21) \times 2/(25-4) = -2$ A so that the short-circuit current is 2 A.

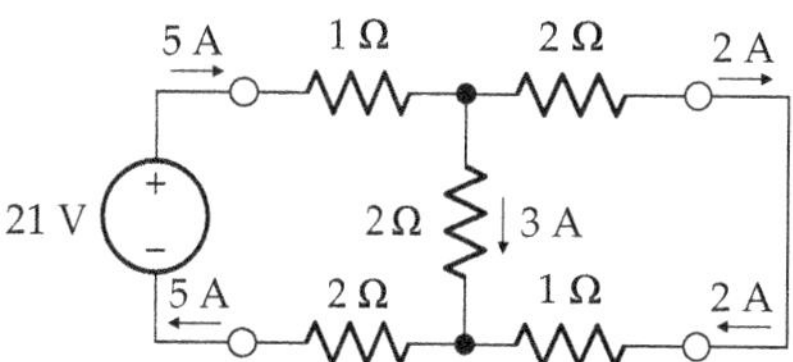

FIGURE 24.34
Equivalent circuit of Figure 24.33.

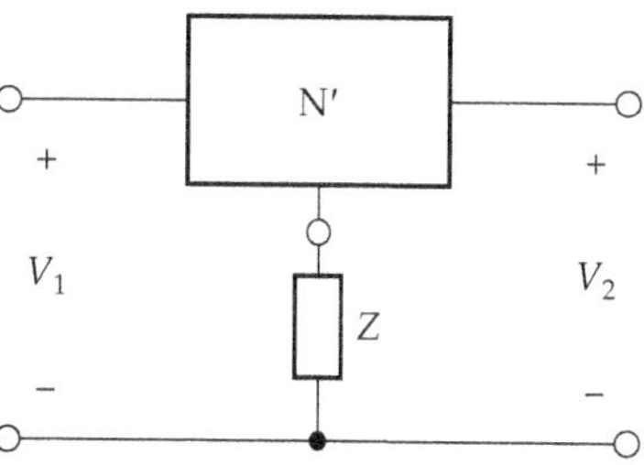

FIGURE 24.35
Impedance connected in series with a two-port circuit.

This result can be readily checked by conventional analysis of the composite circuit. The two subcircuits can be combined as in Figure 24.34. The resistance seen by the source is $1 + (2||3) + 2 = 21/5\ \Omega$. The source current is 5 A and divides into 2 and 3 A as shown. The short-circuit current is seen to be 2 A, as obtained earlier.

A useful application of the series connection of two-port circuits is the derivation of the transfer function of a three-terminal circuit having a coupling impedance as shown in Figure 24.35. The impedance Z may itself be considered a symmetrical three-terminal circuit, as the shunt admittance in Figure 24.17b. If $I_2 = 0$, then $V_1 = V_2 = ZI_1$. The z-parameter matrix is

$$\begin{bmatrix} Z & Z \\ Z & Z \end{bmatrix} \tag{24.59}$$

The z-parameter matrix of the composite circuit is then the sum of the matrix of Equation 24.42 and the z-parameter matrix of the two-port circuit N′. Once the z parameters of the composite circuit are determined, any of the terminal conditions for the circuit can be derived using Table 24.3.

Primal Exercise 24.15

Derive the z parameters for the two-port circuits of Figure 24.32a and b.

Primal Exercise 24.16

Derive the z-parameter matrix of Equation 24.59 from the a-parameter matrix of the shunt element of Figure 24.17b, using the conversion of a to z parameters of Table 24.3.

Example 24.9: Series Connection of Two-Port Circuits

It is required to derive the z parameters of the circuit of Figure 24.36 using Equation 24.58.

Solution:

For the linear transformer having the dot markings shown, the two-port equations are, from KVL (Figure 24.37),

$$sL_1I_1 + sMI_2 = V_1 \qquad sMI_1 + sL_2I_2 = V_2 \tag{24.60}$$

It follows that $z'_{11} = sL_1$, $z'_{12} = z'_{21} = sM$ and $z'_{22} = sL_2$. The z parameters of the composite circuit are then $z_{11} = sL_1 + 1/sC$, $z_{12} = z_{21} = sM + 1/sC$, and $z_{22} = sL_2 + 1/sC$.

24.4.4 Series–Parallel Connection

In a series–parallel connection, the input ports are connected in series and the output ports are connected in parallel, as illustrated in Figure 24.38, where a 1:1 ideal transformer is connected at the input port of one circuit to ensure equality of the current entering or leaving any of the ports, as explained previously. The h-parameter equations are the most appropriate in this case, since the dependent variables are the input current and the output voltage, which are equal for the series–parallel–connected circuits. The h-parameter equations are

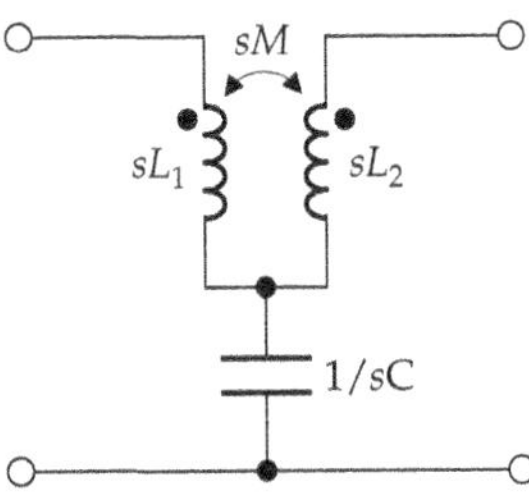

FIGURE 24.36
Figure for Example 24.9.

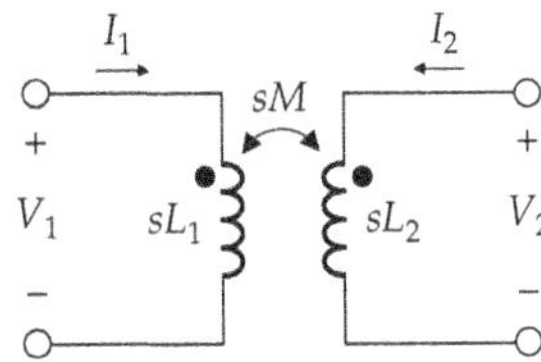

FIGURE 24.37
Figure for Example 24.9.

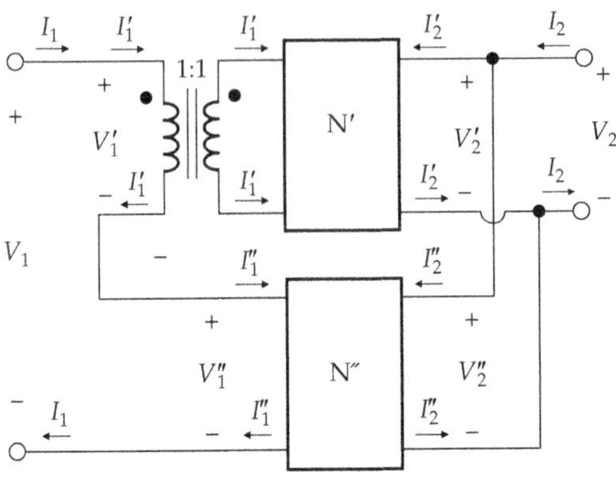

FIGURE 24.38
Ideal transformer added to a series–parallel connection of two-port circuits.

$$\begin{vmatrix} V'_1 \\ I'_2 \end{vmatrix} = \begin{vmatrix} h'_{11} & h'_{12} \\ h'_{21} & h'_{22} \end{vmatrix}\begin{vmatrix} I'_1 \\ V'_2 \end{vmatrix} \quad \begin{vmatrix} V''_1 \\ I''_2 \end{vmatrix} = \begin{vmatrix} h''_{11} & h''_{12} \\ h''_{21} & h''_{22} \end{vmatrix}\begin{vmatrix} I''_1 \\ V''_2 \end{vmatrix} \tag{24.61}$$

where $V_1 = V'_1 + V''_1$, $I_1 = I'_1 = I''_1$, $V_2 = V'_2 = V''_2$, and $I_2 = I'_2 + I''_2$. Substituting in Equation 24.59, or using the h-parameter equations,

$$\begin{vmatrix} V_1 \\ I_2 \end{vmatrix} = \begin{vmatrix} h'_{11} + h''_{11} & h'_{12} + h''_{12} \\ h'_{21} + h''_{21} & h'_{22} + h''_{22} \end{vmatrix}\begin{vmatrix} I_1 \\ V_2 \end{vmatrix} \tag{24.62}$$

The corresponding h parameters of the individual circuits simply add to give the h parameters of the composite circuit. In general,

$$[h] = [h_1] + [h_2] + \cdots + [h_m] \tag{24.63}$$

where
$[h]$ is the matrix of h parameters of the composite circuit
$[h_k]$ is the matrix of h parameters of the kth circuit, $k = 1, 2, \ldots, m$

Suppose next that the three-terminal circuits of Figure 24.32 are to be connected in series–parallel, with their common terminals joined together. This is shown in Figure 24.39, where the circuit of Figure 24.32b is drawn "upside down," as in Figure 24.33. As noted in connection with this figure, having the circuit upside down does not change the h-parameter equations in terms of the current and voltage at the input or output port nor does it change the h parameters. It follows from applying the definition of the h parameters that these parameters for the two T-circuits of Figure 24.32 are (Exercise 24.17) as follows:

$$h'_{11} = 5/3\,\Omega,\quad h'_{12} = 1/3,\quad h'_{21} = -1/3,\quad h'_{22} = 1/3\text{ S}$$
$$h''_{11} = 5/2\,\Omega,\quad h''_{12} = 1/2,\quad h''_{21} = -1/2,\quad h''_{22} = 1/2\text{ S}$$

The h-parameter equations for the individual circuits are still given by Equation 24.61, but the terminal conditions are $V_1 = V_1' + V_1''$, $I_1 = I_1' = I_1''$, $V_2 = V_2' = -V_2''$, and $I_2 = I_2' - I_2''$. Substituting in Equation 24.61, or using the h-parameter equations,

$$\begin{vmatrix} V_1 \\ I_2 \end{vmatrix} = \begin{vmatrix} h'_{11} + h''_{11} & h'_{12} - h''_{12} \\ h'_{21} - h''_{21} & h'_{22} + h''_{22} \end{vmatrix} \begin{vmatrix} I_1 \\ V_2 \end{vmatrix} \tag{24.64}$$

It follows that

$$h_{11} = h'_{11} + h''_{11} = 25/6\,\Omega \qquad h_{12} = h'_{12} - h''_{12} = -1/6$$
$$h_{21} = h'_{21} - h''_{21} = 1/6 \qquad h_{22} = h'_{22} + h''_{22} = 5/6\text{ S}$$

Note that the currents in the "wiring rectangle" in the center of Figure 24.39 are indeterminate. If this rectangle is enclosed by a closed figure, then $I'_{Sh} = I''_{Sh} + I_2$. KCL at the upper node is $I_1' + I_2' = I'_{Sh}$. Substituting $I_1' = I_1''$, $I_2' = I_2 + I_2''$, and $I'_{Sh} = I''_{Sh} + I_2$ gives $I_1'' + I_2'' = I''_{Sh}$, which is KCL at the lower junction. Hence, KCL at the two junctions does not give two independent equations. As in the preceding cases, the currents in the wiring rectangle are not independent and for the same reasons.

Suppose that it is desired to determine the open-circuit voltage of the composite circuit from the h parameters, with $V_{SRC} = 21$ V. If $I_2 = 0$, $V_2 = V_1/a_{11}$ (Table 24.2), or $V_2 = V_1(-h_{21}/\Delta h)$, where $\Delta h = h_{11}h_{22} - h_{12}h_{21} = 7/2$. Substituting, gives $V_2 = 21/-21 = -1$ V.

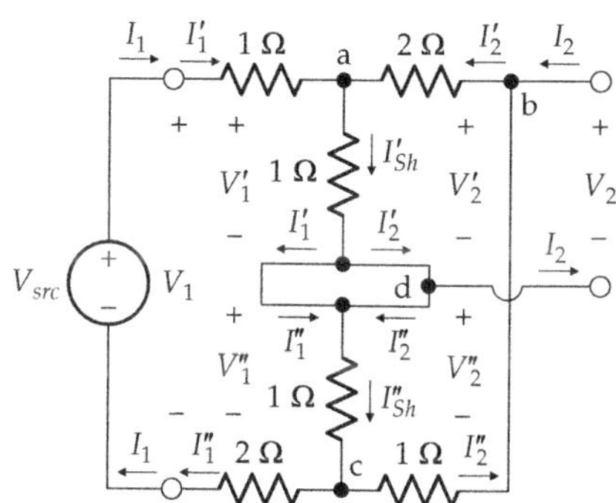

FIGURE 24.39
Series–parallel connection of the two-port circuits of Figure 24.32.

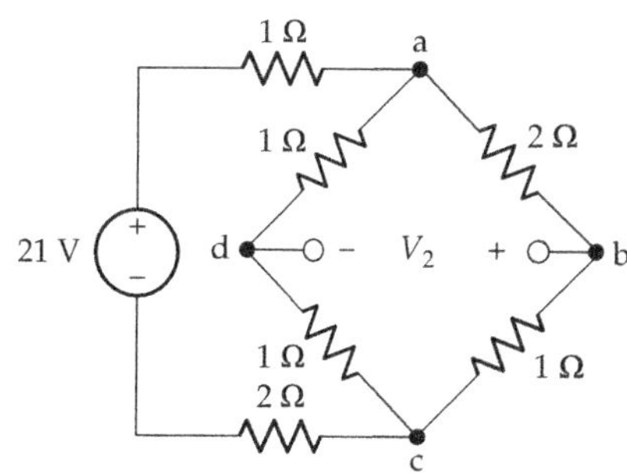

FIGURE 24.40
Circuit of Figure 24.39 redrawn as a bridge circuit.

This result can be checked by conventional analysis of the composite circuit. Figure 24.39 can be redrawn as a bridge circuit (Figure 24.40). $R_{ac} = 3||2 = 1.2\,\Omega$, $V_{ac} = 21(1.2/(3 + 1.2)) = 6$ V; $V_2 = 6(1/3 - 1/2) = -1$ V, as obtained earlier.

Primal Exercise 24.17

Derive the h parameters of the two-port circuits of Figures 24.39 and 24.40.

24.4.5 Parallel–Series Connection

In a parallel–series connection the input ports are connected in parallel and the output ports are connected in series, as illustrated in Figure 24.41, where a 1:1 ideal transformer is connected at the input port of one circuit to ensure equality of the current entering or leaving any of the ports, as explained previously. The g-parameter equations are the most appropriate in this case, since the dependent variables are the input voltage and the output current, which are equal

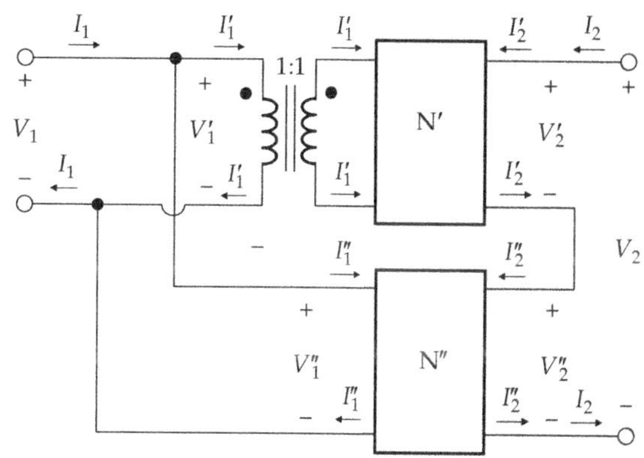

FIGURE 24.41
Ideal transformer added to a parallel–series connection of two-port circuits.

for the parallel- and series-connected circuits. The g-parameter equations are

$$\begin{vmatrix} I_1' \\ V_2' \end{vmatrix} = \begin{vmatrix} g_{11}' & g_{12}' \\ g_{21}' & g_{22}' \end{vmatrix} \begin{vmatrix} V_1' \\ I_2' \end{vmatrix} \quad \begin{vmatrix} I_1'' \\ V_2'' \end{vmatrix} = \begin{vmatrix} g_{11}'' & g_{12}'' \\ g_{21}'' & g_{22}'' \end{vmatrix} \begin{vmatrix} V_1'' \\ I_2'' \end{vmatrix} \quad (24.65)$$

where $V_1 = V_1' = V_1'', I_1 = I_1' + I_1'', V_2 = V_2' + V_2''$, and $I_2 = I_2' = I_2''$. Substituting in Equation 24.65,

$$\begin{vmatrix} I_1 \\ V_2 \end{vmatrix} = \begin{vmatrix} g_{11}' + g_{11}'' & g_{12}' + g_{12}'' \\ g_{21}' + g_{21}'' & g_{22}' + g_{22}'' \end{vmatrix} \begin{vmatrix} V_1 \\ I_2 \end{vmatrix} \quad (24.66)$$

The corresponding g parameters of the individual circuits simply add to give the g parameters of the composite circuit. In general,

$$[g] = [g_1] + [g_2] + \cdots + [g_m] \quad (24.67)$$

where
- $[g]$ is the matrix of g parameters of the composite circuit
- $[g_k]$ is the matrix of g parameters of the kth circuit, $k = 1, 2, \ldots, m$

Suppose next that the three-terminal circuits of Figure 24.32 are to be connected in parallel–series, with their common terminals joined together. This is shown in Figure 24.42, where the circuit of Figure 24.32b is drawn "upside down," as in Figure 24.39. As noted in connection with this figure, having the circuit upside down does not change the g-parameter equations in terms of the current and voltage at the input or output port nor does it change the g parameters. It follows from applying the definition of the g parameters that these parameters for the two T-circuits of Figure 24.42 are (Exercise 24.18) as follows:

$$g_{11}' = 1/2\text{ S}, \quad g_{12}' = -1/2, \quad g_{21}' = 1/2, \quad g_{22}' = 5/2\ \Omega$$
$$g_{11}'' = 1/3\text{ S}, \quad g_{12}'' = -1/3, \quad g_{21}'' = 1/3, \quad g_{22}'' = 5/3\ \Omega$$

The g-parameter equations for the individual circuits are still given by Equation 24.65, but the terminal conditions are $V_1 = V_1' = -V_1''$, $I_1 = I_1' - I_1''$, $V_2 = V_2' + V_2''$, and $I_2 = I_2' = I_2''$. Substituting in Equation 24.65,

$$\begin{vmatrix} I_1 \\ V_2 \end{vmatrix} = \begin{vmatrix} g_{11}' + g_{11}'' & g_{12}' - g_{12}'' \\ g_{21}' - g_{21}'' & g_{22}' + g_{22}'' \end{vmatrix} \begin{vmatrix} V_1 \\ I_2 \end{vmatrix} \quad (24.68)$$

It follows that

$$g_{11} = g_{11}' + g_{11}'' = 5/6\text{ S}, \quad g_{12} = g_{12}' - g_{12}'' = -1/6$$
$$g_{21} = g_{21}' - g_{21}'' = 1/6 \quad g_{22} = g_{22}' + g_{22}'' = 25/6\ \Omega$$

Note that the currents in the "wiring rectangle" in the center of Figure 24.42 are indeterminate. If this rectangle is enclosed by a closed figure, then $I_{Sh}' = I_{Sh}'' + I_1$. KCL at the upper node is $I_1' + I_2' = I_{Sh}'$. Substituting $I_1' = I_1 + I_1''$, $I_2' = I_2''$, and $I_{Sh}' = I_{Sh}'' + I_1$ gives $I_1'' + I_2'' = I_{Sh}''$, which is KCL at the lower junction. Hence, KCL at the two junctions does not give two independent equations. As in the preceding cases, the currents in the wiring rectangle are not independent and for the same reasons.

Suppose that it is desired to determine the short-circuit current of the composite circuit from the g parameters, with $V_{SRC} = 25$ V. If $V_2 = 0$, $I_2 = V_1 y_{21}$ (Table 24.2), or $I_2 = V_1(-g_{21}/g_{22})$. Substituting gives $I_2 = 25(-25) = -1$ A so that the short-circuit current is 1 A.

This result can be checked by conventional analysis of the composite circuit. Figure 24.42 can be redrawn as a bridge circuit (Figure 24.43). I_2 is determined using Thevenin's equivalent circuit (TEC). On open-circuit, $V_{Th} = V_{dc} - V_{bc} = 25(1/3 - 1/2) = -25/6$ V. With the 25 V

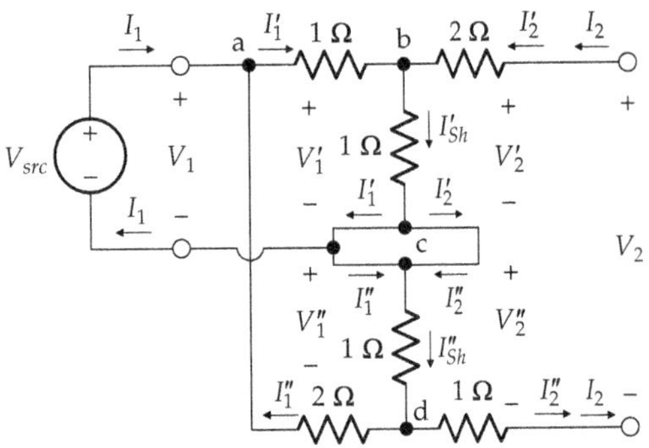

FIGURE 24.42
Parallel–series connection of the two-port circuits of Figure 24.32.

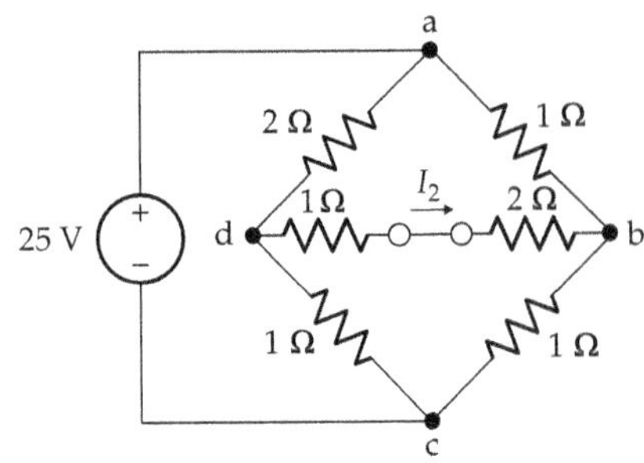

FIGURE 24.43
Circuit of Figure 24.39 redrawn as a bridge circuit.

replaced by a short circuit, $R_{Th} = 2 + 1\|1 + 1\|2 + 1 = 25/6$. It follows that $I_2 = (-25/6)/(25/6) = -1$ A, so that the short-circuit current is 1 A as obtained previously.

Primal Exercise 24.18

Derive the g parameters of the two-port circuits of Figures 24.43 and 24.44.

24.5 Analysis of Terminated Two-Port Circuits

A terminated two-port circuit has a source of impedance Z_{src} connected to one port, say port 1, and an impedance Z_L connected to the other port, as illustrated in Figure 24.44. It is required to analyze this circuit and derive the following expressions that are generally of interest in describing amplifier circuits:

1. The input impedance $Z_{in} = V_1/I_1$, with Z_L connected to port 2
2. The current gain I_2/I_1
3. TEC looking into port 2, where Z_{Th} is the output impedance looking into this port
4. The ratio I_2/V_{src}
5. The voltage gain V_2/V_1
6. The voltage gain V_2/V_{src}

It is straightforward enough to replace the circuit 'N' in Figure 24.44 by the equivalent z-parameter circuit of Figure 24.11a and to derive the expressions mentioned earlier from conventional mesh-current analysis (Exercise 24.19). What we will do instead in this section is to derive these expressions using two-port relations.

One approach is to consider 'N' and Z_L as two cascaded two-port circuits constituting a composite two-port circuit N′ (Figure 24.45). 'N' and N′ have the same V_1, I_1, and V_2 but different port 2 currents. When $I_2' = 0$, the same conditions prevail at port 2 of N′ as at the output in Figure 24.44.

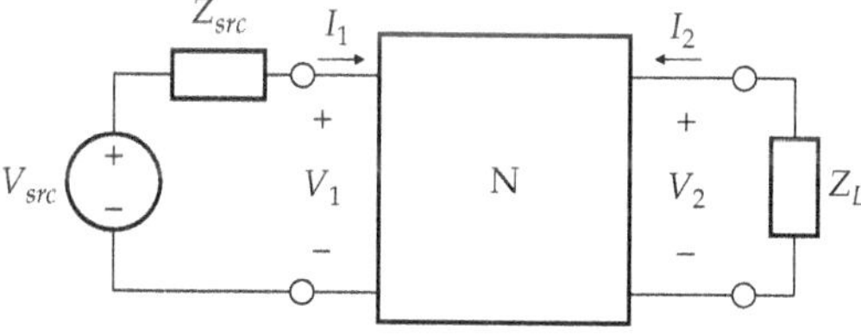

FIGURE 24.44
Terminated two-port circuit.

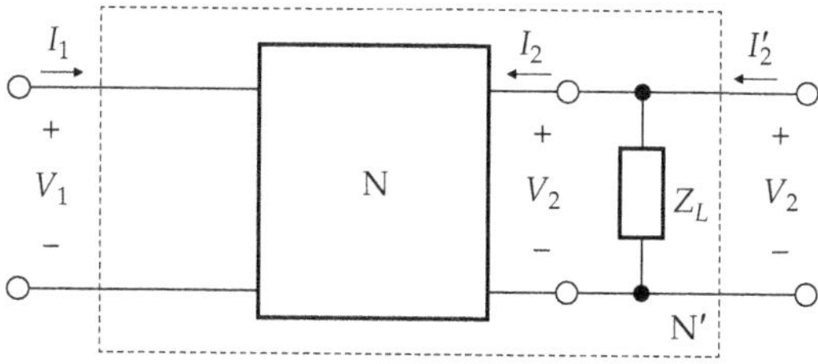

FIGURE 24.45
Terminated two-port circuit as a two-port circuit cascaded with load impedance.

The a-parameter matrix of Z_L is, from Equation 24.36, $\begin{bmatrix} 1 & 0 \\ 1/Z_L & 1 \end{bmatrix}$. It follows from Equation 24.32 that

$$\begin{bmatrix} a_{11}' & a_{12}' \\ a_{21}' & a_{22}' \end{bmatrix} = \begin{bmatrix} a_{11} & a_{12} \\ a_{21} & a_{22} \end{bmatrix} \begin{bmatrix} 1 & 0 \\ 1/Z_L & 1 \end{bmatrix} = \begin{bmatrix} a_{11} + a_{12}/Z_L & a_{12} \\ a_{21} + a_{22}/Z_L & a_{22} \end{bmatrix} \quad (24.69)$$

To find Z_{in}, we note that $Z_{in} = \left.\frac{V_1}{I_1}\right|_{I_2'=0} = z_{11}' = \frac{a_{11}'}{a_{21}'}$ (Table 24.3). Substituting from Equation 24.69 gives

$$Z_{in} = \frac{a_{11} + a_{12}/Z_L}{a_{21} + a_{22}/Z_L} = \frac{a_{11}Z_L + a_{12}}{a_{21}Z_L + a_{22}} \quad (24.70)$$

To find V_2/V_1, we note that $a_{11}' = \left.\frac{V_1}{V_2}\right|_{I_2'=0}$. Hence,

$$\frac{V_2}{V_1} = \frac{1}{a_{11}'} = \frac{1}{a_{11} + a_{12}/Z_L} = \frac{Z_L}{a_{11}Z_L + a_{12}} \quad (24.71)$$

To determine I_2/I_1, we note that $a_{21}' = \left.\frac{I_1}{V_2}\right|_{I_2'=0}$. But with $I_2' = 0$, $V_2 = -I_2Z_L$. Substituting, we obtain

$$\frac{I_2}{I_1} = -\frac{1}{a_{21}'Z_L} = -\frac{1}{a_{21}Z_L + a_{22}} \quad (24.72)$$

To derive TEC looking into port 2 in Figure 24.44, we transform V_{src} in series with Z_{src} to its equivalent current source and consider Z_{src} cascaded with 'N' to be a new two-port circuit N″ (Figure 24.46). 'N' and N″ have the same V_1, V_2, and I_2. However, $I_1'' = V_{src}/Z_{src}$.

The a-parameter matrix of N″ is given by

$$\begin{bmatrix} a_{11}'' & a_{12}'' \\ a_{21}'' & a_{22}'' \end{bmatrix} = \begin{bmatrix} 1 & 0 \\ 1/Z_L & 1 \end{bmatrix} \begin{bmatrix} a_{11} & a_{12} \\ a_{21} & a_{22} \end{bmatrix} = \begin{bmatrix} a_{11} & a_{12} \\ a_{11}/Z_{src} + a_{21} & a_{12}/Z_{src} + a_{22} \end{bmatrix} \quad (24.73)$$

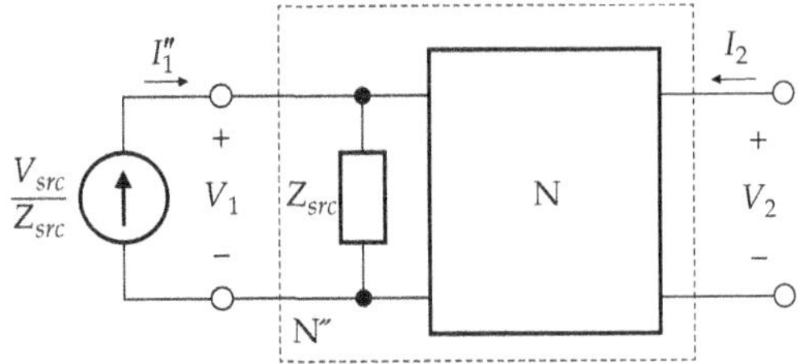

FIGURE 24.46
Terminated two-port circuit as a two-port circuit cascaded with source impedance.

Now $a_{21}'' = \frac{I_1''}{V_2}\Big|_{I_2=0}$, where $I_1'' = V_{src}/Z_{src}$ and $V_2 = V_{Th}$ when $I_2 = 0$. Substituting,

$$V_{Th} = \frac{1}{a_{21}''Z_{src}}V_{src} = \frac{1}{a_{11} + a_{21}Z_{src}}V_{src} \tag{24.74}$$

To find Z_{Th}, we note that $Z_{Th} = z_{22}'' = \frac{V_2}{I_2}\Big|_{I_1''=0} = \frac{a_{22}''}{a_{21}''}$ (Table 24.3). Hence,

$$Z_{Th} = \frac{a_{22}''}{a_{21}''} = \frac{a_{12}/Z_{src} + a_{22}}{a_{11}/Z_{src} + a_{21}} = \frac{a_{12} + a_{22}Z_{src}}{a_{11} + a_{21}Z_{src}} \tag{24.75}$$

Finally, to derive I_2/V_{src} and V_2/V_{src}, we make use of the fact that the source sees an input impedance Z_{in} at the input of the two-port circuit 'N' in Figure 24.44. This gives $\frac{I_1}{V_{src}} = \frac{1}{Z_{src} + Z_{in}}$. Substituting for I_2/I_1 from Equation 24.72 and for Z_{in} from Equation 24.70,

$$\frac{I_2}{V_{src}} = -\frac{1}{(Z_g + Z_{in})(a_{21}Z_L + a_{22})}$$

$$= -\frac{1}{Z_{src}(a_{21}Z_L + a_{22}) + a_{11}Z_L + a_{12}} \tag{24.76}$$

From Figure 24.46, $\frac{V_1}{V_{src}} = \frac{1}{1 + Z_{src}/Z_{in}}$. Substituting for V_2/V_1 from Equation 24.71 and for Z_{in} from Equation 24.70,

$$\frac{V_2}{V_{src}} = \frac{1}{(1 + Z_{src}/Z_{in})}\frac{Z_L}{(a_{11}Z_L + a_{12})}$$

$$= \frac{Z_L}{Z_{src}(a_{21}Z_L + a_{22}) + a_{11}Z_L + a_{12}} \tag{24.77}$$

Alternatively, since $V_2 = -Z_LI_2$, Equation 24.77 follows directly from Equation 24.76.

To express items 1–6 earlier in terms of parameters other than a parameters, we substitute in the expressions derived earlier for the a parameters in terms of any other set of parameters. The resulting expressions are listed in Table 24.6.

Exercise 24.19

Derive the expressions for items 1–6 listed at the beginning of the section directly in terms of the z parameters by using the equivalent circuit of Figure 24.11a for N in Figure 24.44 and solving the two mesh-current equations. Verify your results against Table 24.6.

Example 24.10: Analysis of Terminated Two-Port Circuit

In the circuit of Figure 24.47, 'N' has the following h parameters: $h_{11} = \frac{-\omega_0^2 + 2j\omega}{1 + j\omega}$, $h_{12} = \frac{1}{1 + j\omega}$, $h_{21} = -\frac{1}{1 + j\omega}$, $h_{22} = \frac{1}{1 + j\omega}$. It is required to find $\mathbf{V_O}$ if $v_{SRC} = 10\cos t$ V.

Solution:

From Table 24.6, $\frac{V_2}{V_{src}} = \frac{-h_{21}Z_L}{(h_{11} + Z_{src})(1 + h_{22}Z_L) - h_{12}h_{21}Z_L} = \frac{1 + j\omega}{3 + 7j\omega - 5\omega^2 - j\omega^3} = \frac{1 + j\omega}{(1 + j\omega)(3 - \omega^2 + j4\omega)}\frac{1}{3 - \omega^2 + j4\omega}$.

Substituting $\omega = 1$ rad/s, $\frac{\mathbf{V_O}}{\mathbf{V_{SRC}}} = \frac{1}{2 + j4} = \frac{1}{2\sqrt{5}}\angle -\tan^{-1}2$.

Hence, $v_O = \frac{1}{2\sqrt{5}}\cos(t - 63.4^\circ)$ V.

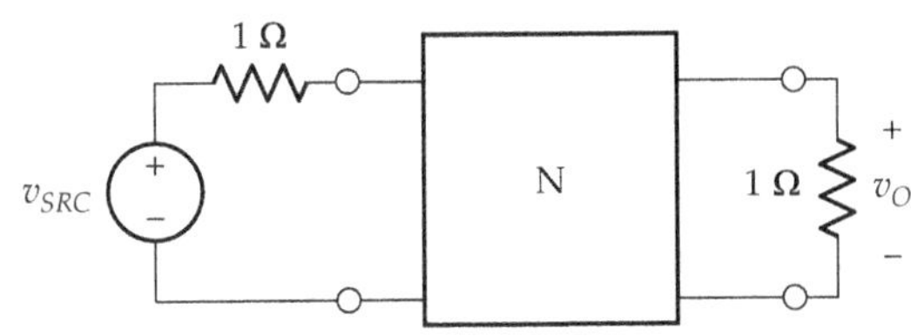

FIGURE 24.47
Figure for Example 24.10.

TABLE 24.6

Circuit Relations of Terminated Two-Port Circuits

Quantity	Expression	
	***z* Parameters**	***y* Parameters**
	$Z_{in} = z_{11} - \frac{z_{12}z_{21}}{z_{22}+Z_L}$	$Y_{in} = y_{11} - \frac{y_{12}y_{21}Z_L}{1+y_{22}Z_L}$
V_{Th}	$\frac{z_{21}}{z_{11}+Z_{src}}\mathbf{V_{src}}$	$-\frac{y_{21}}{y_{22}+\Delta y Z_{src}}\mathbf{V_{src}}$
Z_{Th}	$\frac{\Delta z + z_{22}Z_{src}}{z_{11}+Z_{src}} = z_{22} - \frac{z_{12}z_{21}}{z_{11}+Z_{src}}$	$\frac{1+y_{11}Z_{src}}{y_{22}+\Delta y Z_{src}}; Y_{Th} = y_{22} - \frac{y_{12}y_{21}Z_{src}}{1+y_{11}Z_{src}}$
$\frac{I_2}{V_{src}}$	$-\frac{z_{21}}{(z_{11}+Z_{src})(z_{22}+Z_L)-z_{12}z_{21}}$	$-\frac{y_{21}}{1+y_{11}+Z_{src}+y_{22}Z_L+\Delta y Z_{src}Z_L}$
$\frac{I_2}{I_1}$	$\frac{-z_{21}}{z_{22}+Z_L}$	$\frac{y_{21}}{y_{11}+\Delta y Z_L}$
$\frac{V_2}{V_1}$	$\frac{z_{21}Z_L}{z_{11}Z_L+\Delta z}$	$-\frac{y_{21}Z_L}{1+y_{22}Z_L}$
$\frac{V_2}{V_{src}}$	$\frac{z_{21}Z_L}{(z_{11}+Z_{src})(z_{22}+Z_L)-(z_{12}z_{21})}$	$\frac{y_{21}Z_L}{y_{12}y_{21}Z_{src}Z_L-(1+y_{11}Z_{src})(1+y_{22}Z_L)}$
	***a* Parameters**	***b* Parameters**
	$Z_{in} = \frac{a_{11}Z_L+a_{12}}{a_{21}Z_L+a_{22}}$	$Z_{in} = \frac{b_{22}Z_L+b_{12}}{b_{21}Z_L+b_{11}}$
V_{Th}	$\frac{1}{a_{11}+a_{21}Z_{src}}\mathbf{V_{src}}$	$\frac{\Delta b}{b_{22}+b_{21}Z_{src}}\mathbf{V_{src}}$
Z_{Th}	$\frac{a_{12}+a_{22}Z_{src}}{a_{11}+a_{21}Z_{src}}$	$\frac{b_{12}+b_{11}Z_{src}}{b_{22}+b_{21}Z_{src}}$
$\frac{I_2}{V_{src}}$	$-\frac{1}{Z_{src}(a_{21}Z_L+a_{22})+a_{11}Z_L+a_{12}}$	$-\frac{\Delta b}{Z_{src}(b_{21}Z_L+b_{11})+b_{22}Z_L+b_{12}}$
$\frac{I_2}{I_1}$	$-\frac{1}{a_{21}Z_L+a_{22}}$	$-\frac{\Delta b}{b_{21}Z_L+b_{11}}$
$\frac{V_2}{V_1}$	$\frac{Z_L}{a_{11}Z_L+a_{12}}$	$\frac{\Delta b Z_L}{a_{22}Z_L+b_{12}}$
$\frac{V_2}{V_{src}}$	$\frac{Z_L}{Z_{src}(a_{21}Z_L+a_{22})+a_{11}Z_L+a_{12}}$	$\frac{\Delta b Z_L}{Z_{src}(b_{21}Z_L+b_{11})+b_{22}Z_L+b_{12}}$
	***h* Parameters**	***g* Parameters**
	$Z_{in} = h_{11} - \frac{h_{12}h_{21}Z_L}{1+h_{22}Z_L}$	$Y_{in} = g_{11} - \frac{g_{12}g_{21}}{g_{22}+Z_L}$
V_{Th}	$-\frac{h_{21}}{h_{22}Z_{src}+\Delta h}\mathbf{V_{SRC}}$	$\frac{g_{21}}{1+g_{11}Z_{src}}\mathbf{V_{SRC}}$
Z_{Th}	$\frac{Z_{src}+h_{11}}{h_{22}Z_{src}+\Delta h}; Y_{Th} = h_{22} - \frac{h_{12}h_{21}}{h_{11}+Z_{src}}$	$\frac{g_{22}+\Delta g Z_{src}}{1+g_{11}Z_{src}} = g_{22} - \frac{g_{12}g_{21}Z_{src}}{1+g_{11}Z_{src}}$
$\frac{I_2}{V_{src}}$	$\frac{h_{21}}{(h_{11}+Z_{src})(1+h_{22}Z_L)-h_{12}h_{21}Z_L}$	$-\frac{g_{21}}{(1+g_{11}Z_{src})(g_{22}+Z_L)-g_{12}g_{21}Z_{src}}$
$\frac{I_2}{V_{src}}$	$\frac{h_{21}}{1+h_{22}Z_L}$	$-\frac{g_{21}}{g_{11}Z_L+\Delta g}$
$\frac{V_2}{V_1}$	$-\frac{h_{21}Z_L}{h_{11}+\Delta h Z_L}$	$\frac{g_{21}Z_L}{g_{22}+Z_L}$
$\frac{V_2}{V_{src}}$	$-\frac{h_{21}Z_L}{(h_{11}+Z_{src})(1+h_{22}Z_L)-h_{12}h_{21}Z_L}$	$\frac{g_{21}Z_L}{(1+g_{11}+Z_{src})(g_{22}+Z_L)-g_{12}g_{21}Z_{src}}$

Learning Checklist: What Should Be Learned from This Chapter

- A two-port circuit may be specified in terms of one of six sets of two simultaneous equations, each of which involves four parameters. In general, these parameters are nonzero and independent.
- The matrices of the z and y parameters are inversely related, as are the matrices of the a and b parameters and the matrices of the h and g parameters.
- The parameters of the two-port circuit equations can be interpreted in terms of voltage and current ratios under specified open-circuit and short-circuit terminations, and any one set of parameters can be expressed in terms of the other sets of parameters.
- In a reciprocal two-port circuit, $z_{12} = z_{21}$ and $y_{12} = y_{21}$, with corresponding relations for the other sets of parameters.
    1. A reciprocal circuit is specified in general by three independent, nonzero parameters.
    2. A circuit that does not have dependent sources is reciprocal. But a circuit can have dependent sources and still be reciprocal.
- A reciprocal circuit is symmetric if terminal voltages and currents remain the same when the two ports are interchanged.
    1. In a symmetric circuit and in addition to the reciprocity relations, $z_{12} = z_{22}$ and $y_{11} = y_{22}$, with corresponding relations for the other sets of parameters.
    2. A symmetric circuit is specified in general by two independent, nonzero parameters.
- The z-parameter, y-parameter, h-parameter, and g-parameter equations may be represented in terms of equivalent circuits in which forward and reverse transmissions are described by dependent sources on the output and input sides.
- Two-port circuits may be connected in cascade, in parallel, in series, in series–parallel, and in parallel–series. The matrix of parameters of the two-port composite circuit is a simple combination of the matrices of parameters of the individual circuits.
- Except in the cascade connection, the two-port circuit equations are no longer valid, in general, in the case of the four other composite connections, because of inequality of currents entering and leaving input or output ports. Equality of these currents can be forced by connecting a 1:1 ideal transformer at the input or output port of either of the two circuits.
- The two-port circuits remain valid in the case of the connections other than the cascade, if the two-port circuits are three-terminal, with their common terminals connected together.
- A two-port circuit that is terminated by source and load impedances can be analyzed by considering the two-port circuit proper to be cascaded with the load impedance, when looking into port 1, and to be cascaded with the source impedance when looking into port 2.

Problem-Solving Tips

1. The open-circuit transfer function is readily determined by deriving a_{11}.
2. Matrix multiplication is greatly facilitated using MATLAB.

Problems

Verify solutions by PSpice simulation.

Two-Port Circuit Parameters and Equations

P24.1 Determine the z parameters of the two-port circuit of Figure P24.1.

Ans. $z_{11} = z_{22} = 6\ \Omega$, $z_{12} = z_{21} = 2\ \Omega$.

P24.2 Determine h_{22} for the circuit in Figure P24.2.

Ans. 30 mS.

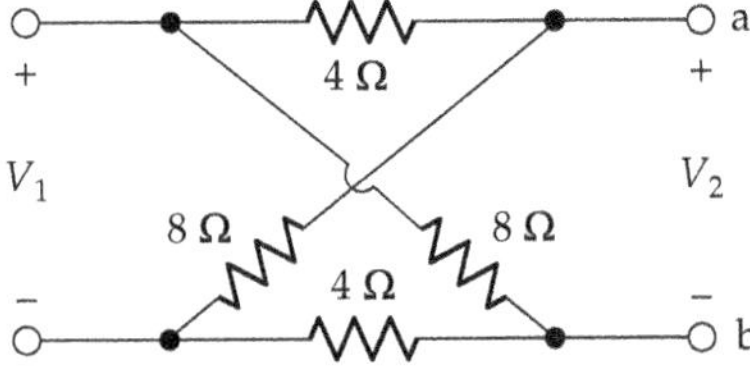

FIGURE P24.1

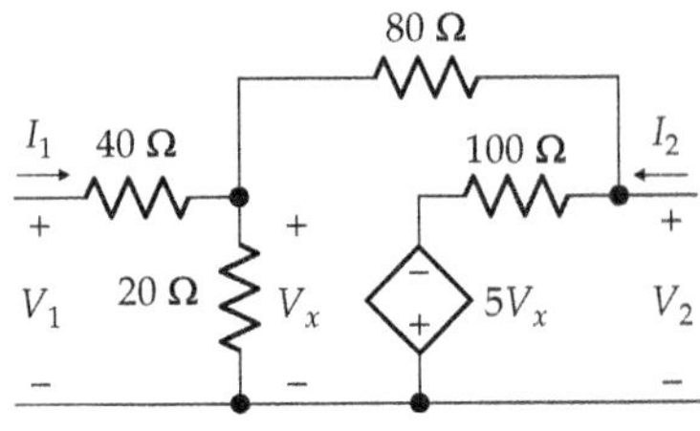

FIGURE P24.2

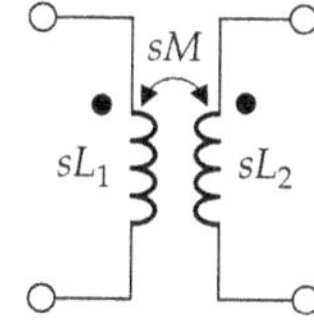

FIGURE P24.3

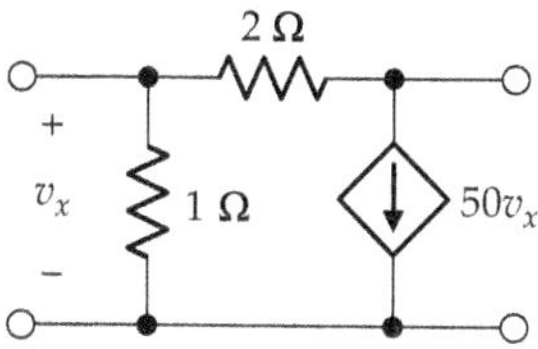

FIGURE P24.4

P24.3 Derive the z parameters of the linear transformer of Figure P24.3. Which set of two-port equations does not exist when the coupling is perfect?

Ans. $z_{11} = sL_1$, $z_{12} = z_{21} = sM$, $z_{22} = sL_2$, y-parameter equations.

P24.4 (a) Determine the z parameters and y parameters of the circuit in Figure P24.4 from the definition of these parameters. (b) Compare the y-parameter equations with the node-voltage equations. (c) Verify that the z and y matrices are the inverse of one another.

Ans. (a) $z_{11} = 1/51\ \Omega$, $z_{12} = 1/51\ \Omega$, $z_{21} = -99/51\ \Omega$, $z_{22} = 1/17\ \Omega$, $y_{11} = 1.5$ S, $y_{12} = -0.5$ S, $y_{21} = 49.5$ S, and $y_{22} = 0.5$ S.

P24.5 (a) Determine the h parameters and g parameters of the circuit in Figure P24.4 from the definition of these parameters. (b) Verify that the h- and g-parameter matrices are the inverse of one another.

Ans. (a) $h_{11} = 2/3\ \Omega$, $h_{12} = 1/3$, $h_{21} = 33$, $h_{22} = 17$ S, $g_{11} = 51$ S, $g_{12} = -1$, $g_{21} = -99$, $g_{22} = 2\ \Omega$.

P24.6 (a) Determine the a parameters and b parameters of the circuit in Figure P24.4 from the definition of these parameters. (b) Verify that the a- and b-parameter matrices are the inverse of one another.

Ans. (a) $a_{11} = -1/99$, $a_{12} = -2/99\ \Omega$, $a_{21} = -17/33$ S, $a_{22} = -1/33$, $b_{11} = 3$, $b_{12} = 2\ \Omega$, $b_{21} = 52$ S, $b_{22} = 1$.

P24.7 (a) Determine the z parameters and y parameters of the circuit shown in Figure P24.7 from the direct definition of these parameters. (b) Verify that the z and y matrices are the inverse of one another.

Ans. (a) $z_{11} = 1\ \Omega$, $z_{12} = 1\ \Omega$, $z_{21} = -0.5\ \Omega$, $z_{22} = 0.5\ \Omega$, $y_{11} = 0.5$ S, $y_{12} = -1$ S, $y_{21} = 0.5$ S, $y_{22} = 1$ S.

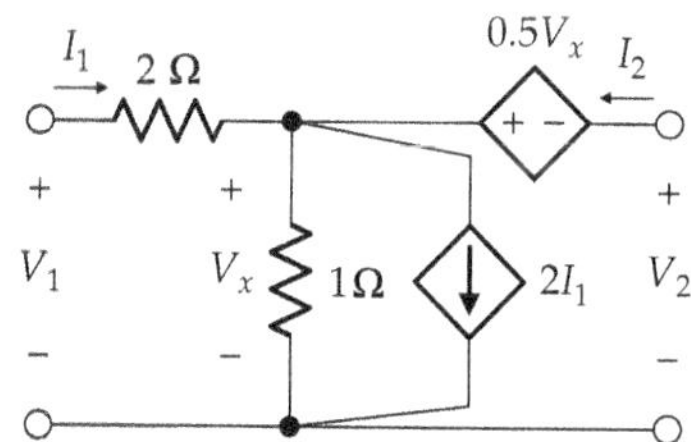

FIGURE P24.7

P24.8 (a) Determine the h parameters and g parameters of the circuit in Figure P24.7 from the definition of these parameters. (b) Verify that the h- and g-parameter matrices are the inverse of one another.

Ans. (a) $h_{11} = 2\ \Omega$, $h_{12} = 2$, $h_{21} = 1$, $h_{22} = 2$ S, $g_{11} = 1$ S, $g_{12} = -1$, $g_{21} = -0.5$, $g_{22} = 1\ \Omega$.

P24.9 (a) Determine the a parameters and b parameters of the circuit in Figure P24.7 from the definition of these parameters. (b) Verify that the a- and b-parameter matrices are the inverse of one another.

Ans. (a) $a_{11} = -2$, $a_{12} = -2\ \Omega$, $a_{21} = -2$ S, $a_{22} = -1$, $b_{11} = 0.5$, $b_{12} = 1\ \Omega$, $b_{21} = 1$ S, $b_{22} = 1$.

P24.10 Determine the h parameters of circuit in Figure P24.10, assuming $\omega = 1$ rad/s.

Ans. $h_{11} = 2(3-j)/5\ \Omega$, $h_{12} = (2+j)/5$, $h_{21} = -j$, $h_{22} = (1+j)$ S.

P24.11 (a) Determine the z parameters and y parameters of the circuit shown in Figure P24.11 from the direct definition of these parameters, assuming $\omega = 1$ rad/s. (b) Verify that the z- and y-parameter matrices are the inverse of one another.

Ans. (a) $z_{11} = j\ \Omega$, $z_{12} = j2\ \Omega$, $z_{21} = j2\ \Omega$, $z_{22} = 0$; (b) $y_{11} = 0$, $y_{12} = -j/2$ S, $y_{21} = -j/2$ S, $y_{22} = -j/4$ S.

P24.12 (a) Determine the h parameters and g parameters of the circuit in Figure P24.11 from the definition of these parameters. (b) Verify that the h- and g-parameter matrices are the inverse of one another.

Ans. (a) $h_{11} = \infty$, $h_{12} = \infty$, $h_{21} = \infty$, $h_{22} = \infty$, $g_{11} = j$ S, $g_{12} = 2$, $g_{21} = -2$, $g_{22} = j4\ \Omega$.

P24.13 (a) Determine the a parameters and b parameters of the circuit in Figure P24.11 from the definition of these parameters. (b) Verify that the a- and b-parameter matrices are the inverse of one another.

Ans. (a) $a_{11} = -1/2$, $a_{12} = -j2\ \Omega$, $a_{21} = -j/2$ S, $a_{22} = 0$; (b) $b_{11} = 0$, $b_{12} = -j2\ \Omega$, $b_{21} = -j/2$ S, $b_{22} = -1/2$.

P24.14 (a) Determine the z parameters and y parameters of the circuit shown in Figure P24.14 from the direct definition of these parameters, assuming $\omega = 1$ rad/s.

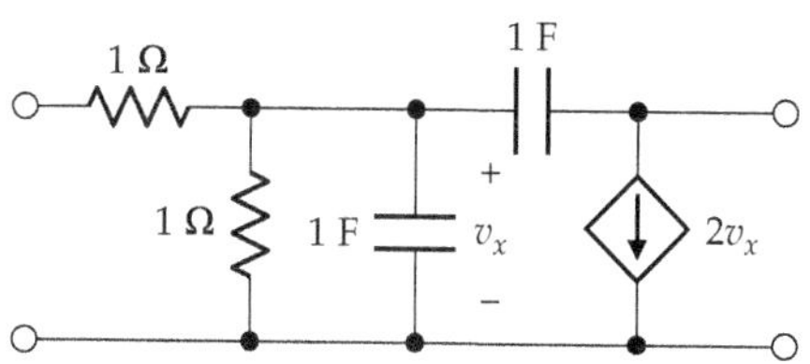

FIGURE P24.10

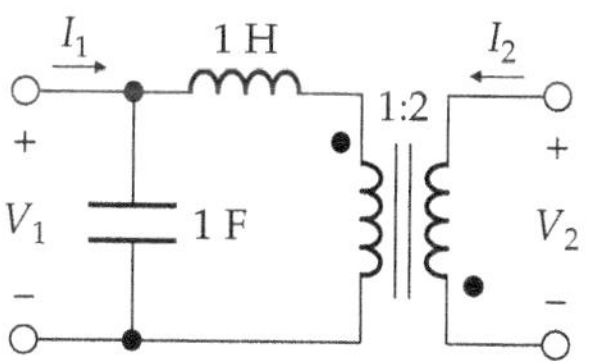

FIGURE P24.11

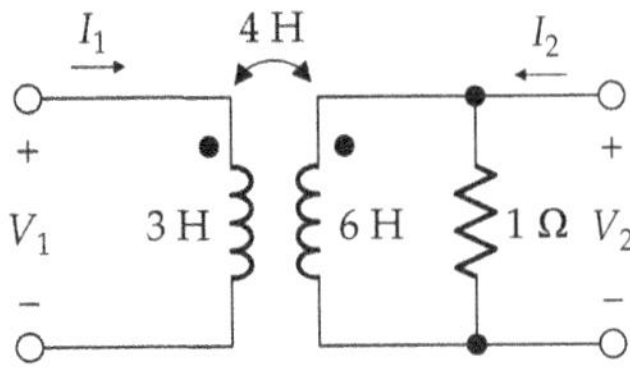

FIGURE P24.14

(b) Verify that the z- and y-parameter matrices are the inverse of one another.

Ans. (a) $z_{11} = (-2 + j3)/(1 + j6)\ \Omega$, $z_{12} = z_{21} = j2/(1 + j6)\ \Omega$, $z_{22} = j6/(1 + j6)\ \Omega$; (b) $y_{11} = -j3$ S, $y_{12} = y_{21} = j2$ S, and $y_{22} = (1 - j3/2)$ S.

P24.15 (a) Determine the h parameters and g parameters of the circuit in Figure P24.14 from the definition of these parameters. (b) Verify that the h- and g-parameter matrices are the inverse of one another.

Ans. (a) $h_{11} = j\omega/3\ \Omega$, $h_{12} = 2/3$, $h_{21} = -2/3$, $h_{22} = (1 - j/6\omega)$ S, $g_{11} = (1 + j6\omega)/(-2\omega^2 + j3\omega)$ S, $g_{12} = -4/(3 + j2\omega)$, $g_{21} = 4/(3 + j2\omega)$, $g_{22} = j2\omega/(3 + j2\omega)\ \Omega$.

P24.16 (a) Determine the a parameters and b parameters of the circuit in Figure P24.14 from the definition of these parameters. (b) Verify that the a- and b-parameter matrices are the inverse of one another.

Ans. (a) $a_{11} = (3 + j2)/4$, $a_{12} = j/2\ \Omega$, $a_{21} = (1 + j6)/j4$ S, $a_{22} = 3/2$. $b_{11} = 3/2$, $b_{12} = j/2\ \Omega$, $b_{21} = (1 + j6)/j4$ S, $b_{22} = (3 + j2)/4$.

P24.17 Determine the z parameters of the circuit in Figure P24.17. Verify by considering the circuit as a series connection of two three-terminal circuits across the dotted line.

Ans. $z_{11} = 2(s + 1/s)\ \Omega = z_{22}$; $z_{21} = (s + 1/s) = z_{12}\ \Omega$.

P24.18 Determine the y parameters of the circuit in Figure P24.18. Verify by considering the circuit as a series connection of two three-terminal circuits.

Ans. $y_{11} = 1\text{ S} = y_{22}$, $y_{12} = 1/(1 + s)\text{ S} = y_{21}$.

P24.19 Determine the z parameters of the circuit in Figure P24.19. Verify by considering the circuit as a cascade connection of two three-terminal circuits at the dotted line.

Ans. $z_{11} = \dfrac{s^2+2s+2}{s^3+2s^2+3s+1}\ \Omega$, $z_{12} = \dfrac{1}{s^3+2s^2+3s+1} = z_{21}\ \Omega$,

$$z_{22} = \frac{s^2+s+1}{s^3+2s^2+3s+1}\ \Omega.$$

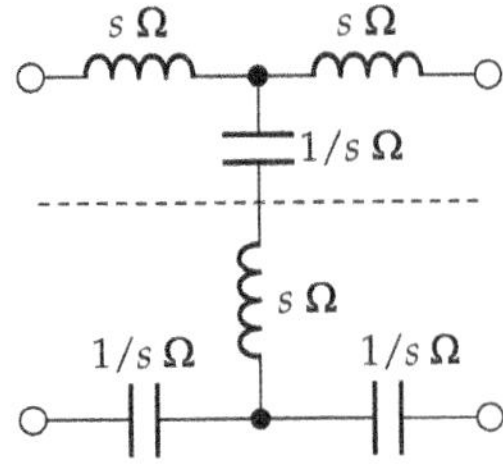

FIGURE P24.17

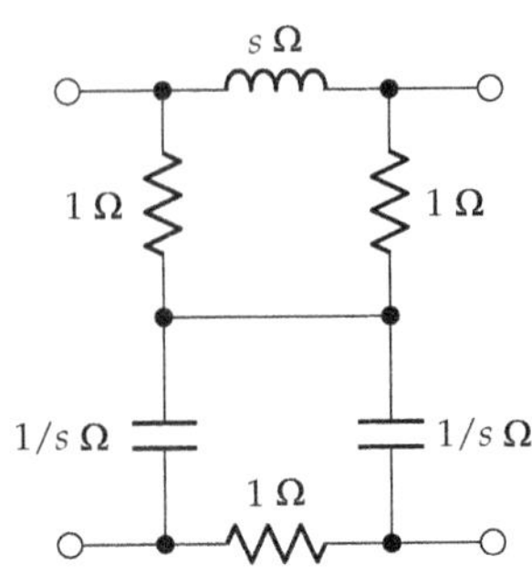

FIGURE P24.18

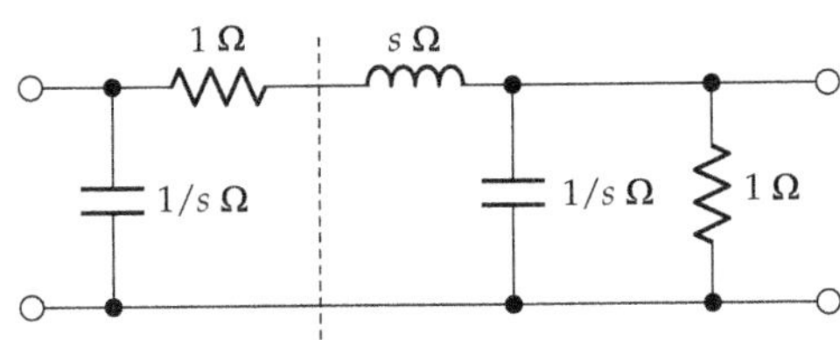

FIGURE P24.19

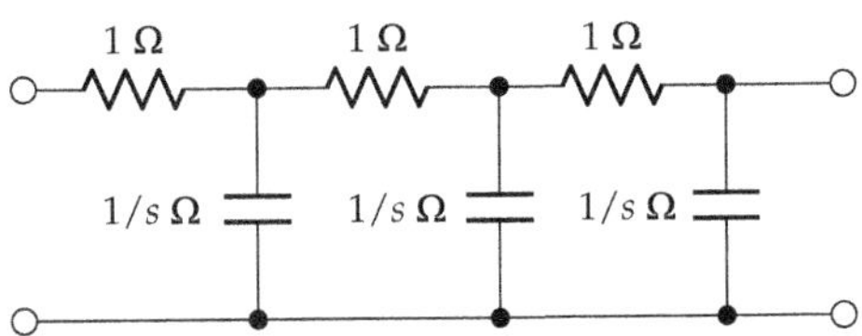

FIGURE P24.20

P24.20 Determine the z parameters of the circuit in Figure P24.20. Verify by considering it as a cascade of three-terminal circuits.

Ans. $z_{11} = \dfrac{s^3+5s^2+6s+1}{s\left(s^2+4s+3\right)}\ \Omega$, $z_{12} = \dfrac{1}{s\left(s^2+4s+3\right)} = z_{21}\ \Omega$,

$$z_{22} = \frac{s+2}{s(s+3)}\ \Omega.$$

P24.21 Determine the y parameters of the circuit in Figure P24.21. Verify by considering the circuit as a parallel connection of two three-terminal circuits.

Ans. $y_{11} = \left(\dfrac{s}{s+4} + \dfrac{s^2+1}{s^2+s+1}\right)$ S, $y_{12} = -\left(\dfrac{2s}{s+4} + \dfrac{1}{s^2+s+1}\right)$

$= y_{21}$ S, $y_{22} = \dfrac{4s}{s+4} + \dfrac{s+1}{s^2+s+1}$ S

P24.22 Determine the z parameters of the circuit in Figure P24.22.

Ans. $z_{11} = 2 + j10\ \Omega$, $z_{12} = 1 - j\ \Omega = z_{21}$, $z_{22} = 2(1 + j)\ \Omega$.

P24.23 $V_2 = 5$ V when the output terminals in Figure P24.23 are open-circuited, and $I_o = 1$ A when these terminals are short-circuited. In both cases, $I_i = 1$A. Determine y_{22}.

Ans. $y_{22} = 0.2$ S.

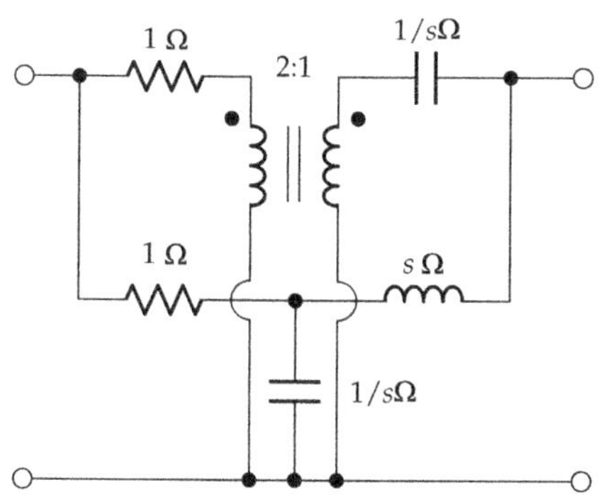

FIGURE P24.21

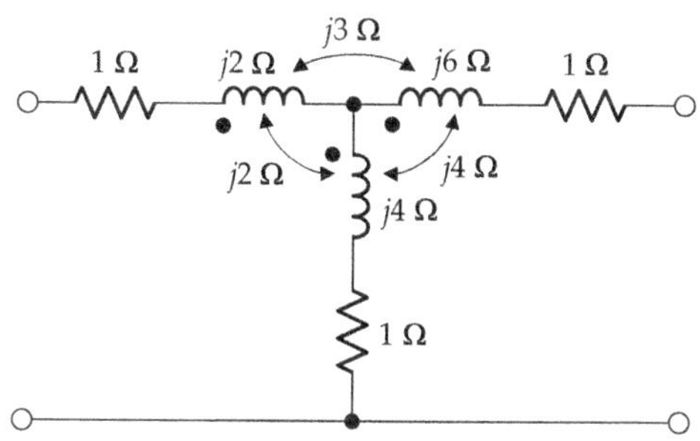

FIGURE P24.22

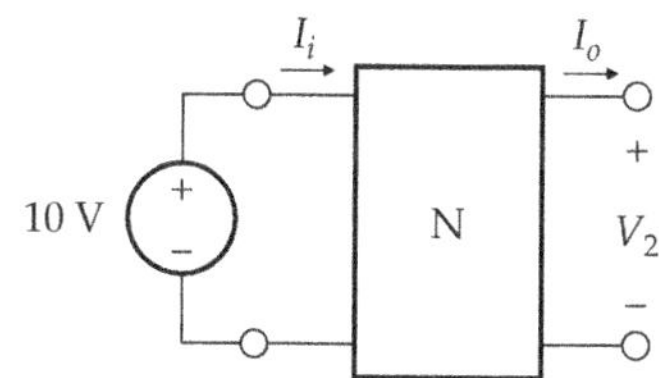

FIGURE P24.23

P24.24 The two two-port circuits shown in Figure P24.24 are identical and have $z_{11} = z_{22} = 2\ \Omega$, and $z_{12} = z_{21} = 1\ \Omega$. If $V_o = \rho I_{src} + \alpha V_{src}$, determine ρ and α.

Ans. $\rho = 1/4\ \Omega$, $\rho = 1/4$.

P24.25 The b parameters of the two-port circuit N in Figure P24.25 are $b_{11} = 1$, $b_{12} = 4s\ \Omega$, $b_{21} = 1/s$ S, and $b_{22} = 6$. Determine $v_O(t)$.

Ans. $v_O(t) = \sin t$ V

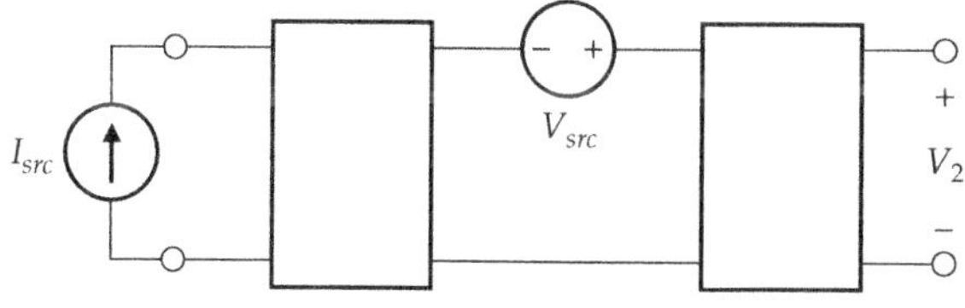

FIGURE P24.24

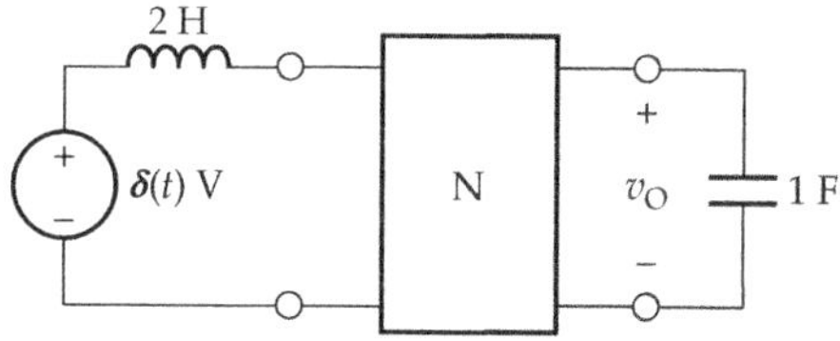

FIGURE P24.25

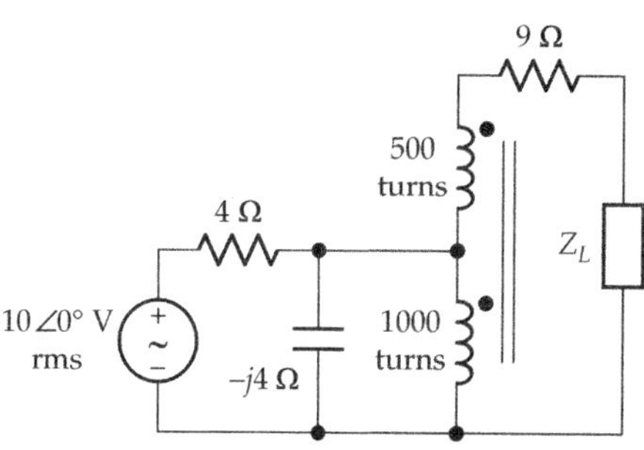

FIGURE P24.29

P24.26 A two-port circuit is described by its g parameters as $g_{11} = (1 - j)$ S, $g_{12} = (-1 - j)$, $g_{21} = (1 + j)$, and $g_{22} = (1 + j)\ \Omega$. Determine the complex power delivered by the source, assuming the output is short-circuited and $\mathbf{V_{SRC}} = 2\angle 45°$ V rms is applied to the input terminals.

Ans. 8 W.

P24.27 Determine the open-circuit voltage at the output port in Problem P24.26.

Ans. $2\sqrt{2}\angle 90°$ V.

P24.28 (a) Determine the short-circuit current, from terminal 2 to terminal 2′, in Problem P24.26 from the definitions of the two-port parameters; (b) verify the value of this current from the output impedance and the open-circuit voltage of P24.27.

Ans. (a) $2\angle 45°$ A; (b) output impedance = $(1 + j)\ \Omega$.

P24.29 (a) Reflect the circuit on the primary (source) side to the secondary (load) side in Figure P24.29 and determine the z parameters of the two-port circuit between the source and the load; (b) determine Z_L for maximum power transfer to Z_L and calculate this power.

Ans. (a) $z_{12} = z_{21} = -j9\ \Omega$, and $z_{11} = z_{22} = 9(1 - j)\ \Omega$; (b) $Z_{Lm} = 13.5 + j4.5\ \Omega$, 2.08 W.

Reciprocal and Symmetric Circuits

P24.30 Determine z_{11}/z_{22} for the circuit in Figure P24.30.

Ans. 1.

P24.31 Determine K so that the circuit in Figure P24.31 is symmetric.

Ans. $K = 0$.

P24.32 Determine K so that the circuit in Figure P24.32 is reciprocal.

Ans. $K = 1$.

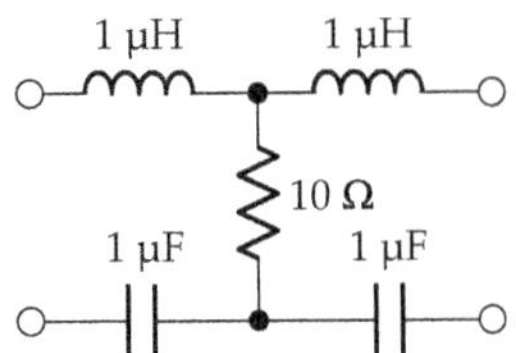

FIGURE P24.30

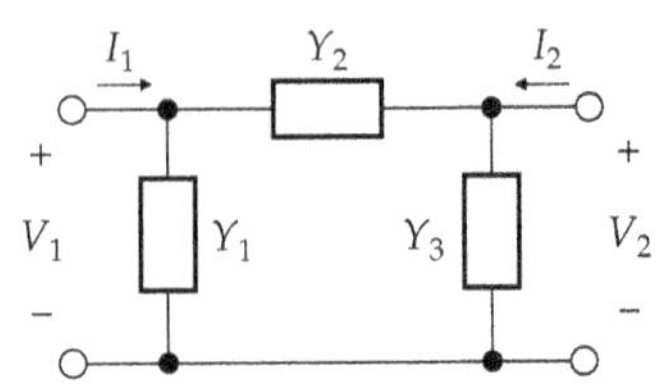

FIGURE P24.34

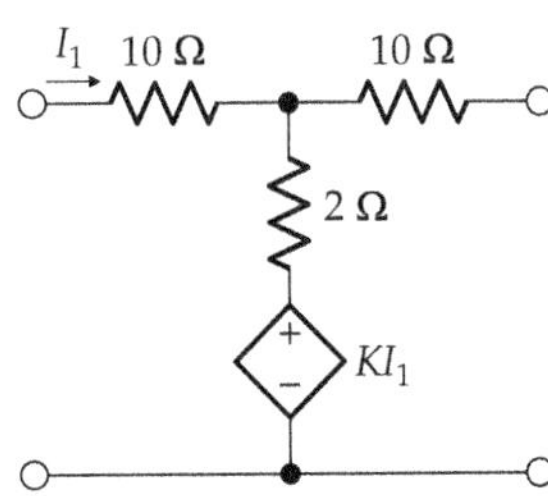

FIGURE P24.31

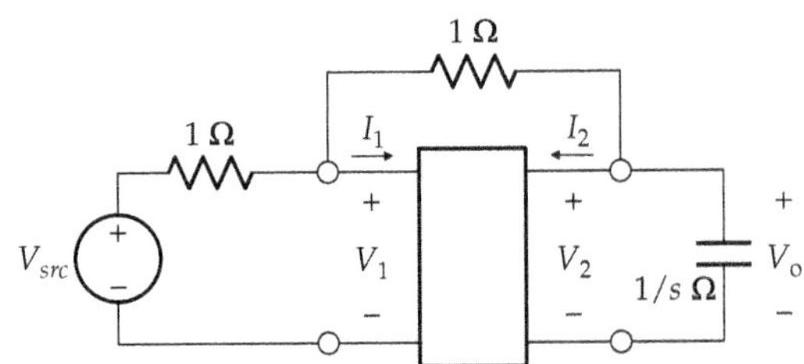

FIGURE P24.35

P24.33 A symmetric two-port circuit has an open-circuit input impedance of $(1 - j)$ Ω, and an open-circuit transfer impedance, that is V_2/I_1, of 1 Ω. Determine the load current if a load impedance of j Ω is connected to one port and a $10\angle 0°$ V source is connected to the other port.

Ans. $j10$ A.

P24.34 Given that the two-port circuit of Figure P24.34 is symmetric, with $y_{11} = 2/s$ S and $y_{12} = 1/s$ S, determine the input admittance $\mathbf{I}_1/\mathbf{V}_1$ when a load of $3/s$ S is connected at the output.

Ans. 9/5.

P24.35 The two-port circuit in Figure P24.35 is described in the s-domain by the equations $I_1 = 2sV_1 - sV_2$ and $I_2 = -sV_1 + 2sV_2$. (a) Show that the two-port circuit is symmetric and (b) determine the transfer function $H(s) = V_o/V_{src}$.

Ans. $H(s) = \frac{V_o}{V_{src}} = \frac{1}{5s+1}$.

P24.36 Determine the z parameters of the lattice circuit in Figure P24.36, assuming $\omega = 100$ rad/s.

Ans. $z_{11} = z_{22} = \frac{5}{2}(1+j)$ Ω and $z_{12} = z_{21} = \frac{5}{2}(-1+j)$ Ω.

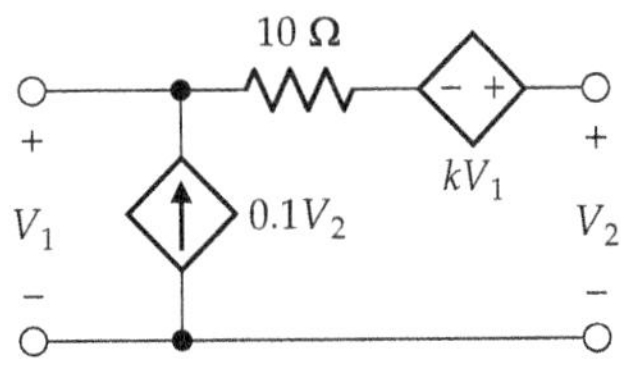

FIGURE P24.32

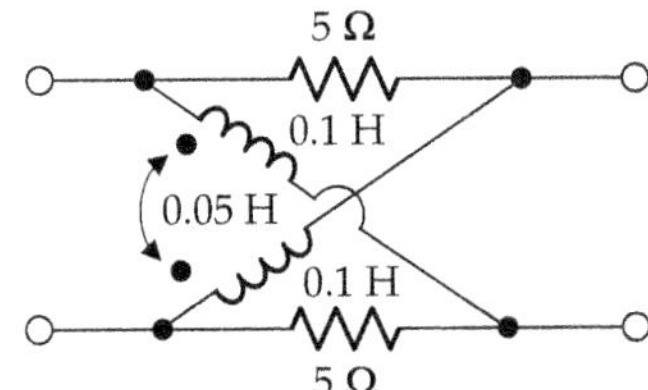

FIGURE P24.36

P24.37 (a) Redraw the circuit of Figure P24.36 as a bridge. Note that the symmetry of the circuit is more clearly shown in the lattice configuration than in the bridge configuration; (b) derive the a and b parameters of the circuit; (c) derive the a parameters of two identical, cascaded circuits; (d) if $1\angle 0°$ V is applied at the input of the composite circuit, determine the input current and the output voltage with the output open-circuited.

Ans. (b) The a-parameter matrix and the b-parameter matrix is $\begin{bmatrix} -j & 5(1-j) \\ -(1+j)/5 & -j \end{bmatrix}$; (c) the a-parameter matrix of the composite circuit is $\begin{bmatrix} -3 & -10(1+j) \\ 2(-1+j)/5 & -3 \end{bmatrix}$; (d) $-\frac{1}{3}\angle 0°$ V; $\frac{2\sqrt{2}}{15}\angle -45°$ A.

P24.38 A symmetric circuit N having $b_{11} = 1/2$ and $b_{12} = 2$ Ω is cascaded with the circuit of Figure P24.36. Determine $\mathbf{V}_2/\mathbf{V}_1$, where $\mathbf{V}_1$ is the input to circuit N and $\mathbf{V}_2$ is the open-circuit output of the circuit of Figure P24.36.

Ans. $\frac{\mathbf{V}_2}{\mathbf{V}_1} = \frac{10}{97}(-4+9j)$.

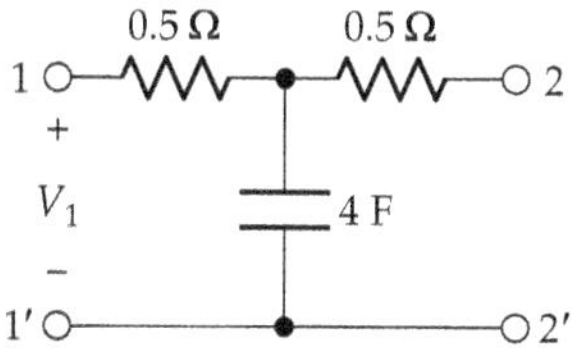

FIGURE P24.39

P24.39 Consider the filter circuit shown in Figure P24.39 as three elements in cascade, (a) determine the a parameters of each of these elements and the a parameters of the given filter and (b) derive the transfer function I_{sc}/V_1, where I_{sc} is the short-circuit current that flows from terminal 2 to terminal 2′.

Ans. (a) $[a_1]=[a_3]=\begin{bmatrix}1 & 0.5\\ 0 & 1\end{bmatrix}$, $[a_2]=\begin{bmatrix}1 & 0\\ 4s & 1\end{bmatrix}$,

$[a]=\begin{bmatrix}2s+1 & s+1\\ 4s & 2s+1\end{bmatrix}$; (b) $\dfrac{1}{s+1}$.

P24.40 Determine the z parameters of the circuit in Figure P24.40, assuming $\omega = 10^3$ rad/s.

Ans. $z_{11} = z_{22} = \dfrac{11+j3}{13}$ kΩ; $z_{12} = z_{22} = \dfrac{(2-j3)}{13}$ kΩ.

P24.41 A symmetric circuit 'N' having z_{11} = (1 + j) kΩ and z_{12} = (1 − j) kΩ is paralleled with the circuit of Figure P24.40. Determine Z_{in} with the output terminals open circuited, $Z_{out} = Z_{Th}$, and $\mathbf{V}_2/\mathbf{V}_1$.

Ans. $Z_{in} = \dfrac{46}{65} + \dfrac{21}{130} j$ kΩ, $Z_{out} = \dfrac{90}{137} + \dfrac{42}{137} j$ kΩ,

$\dfrac{\mathbf{V}_2}{\mathbf{V}_1} = 0.46\angle -41.8°$.

P24.42 A two-port resistive, symmetric circuit connected between a source and an 8 Ω load gave the measurements indicated in Figure P24.42. Determine the a parameters of the circuit.

Ans. $a_{11} = a_{22} = 2$, $a_{12} = 3$ Ω, and $a_{21} = 1$ S.

P24.43 Derive the delta–star transformation by determining the z parameters of a two-port circuit consisting of three impedances connected as a π-circuit, as in Figure P24.43. Deduce the values of the T-equivalent circuit from Figure 24.12a, assuming a reciprocal circuit.

P24.44 Derive the star–delta transformation by determining the y parameters of a two-port circuit consisting of three impedances connected as a T-circuit, as in Figure P24.44. Deduce the values of the delta equivalent circuit from Figure 24.12b, assuming a reciprocal circuit.

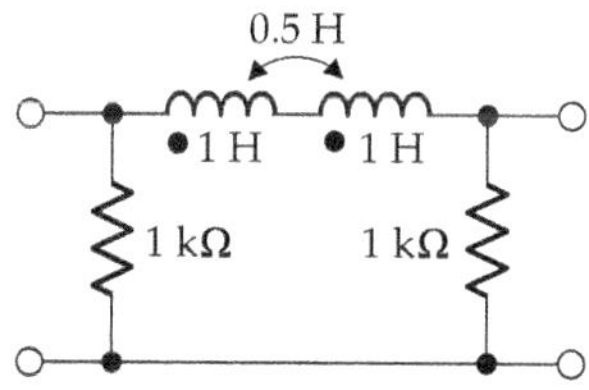

FIGURE P24.40

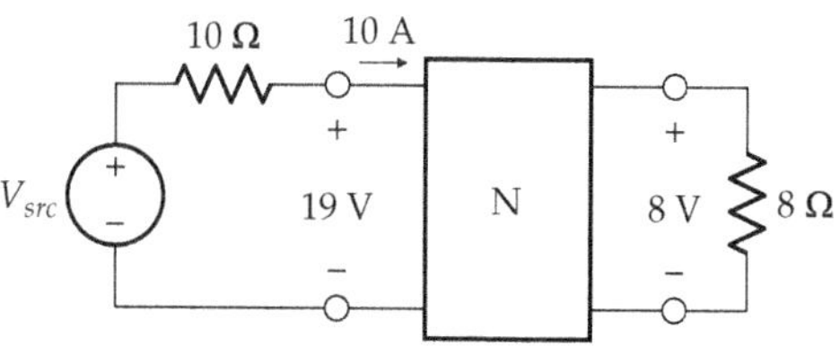

FIGURE P24.42

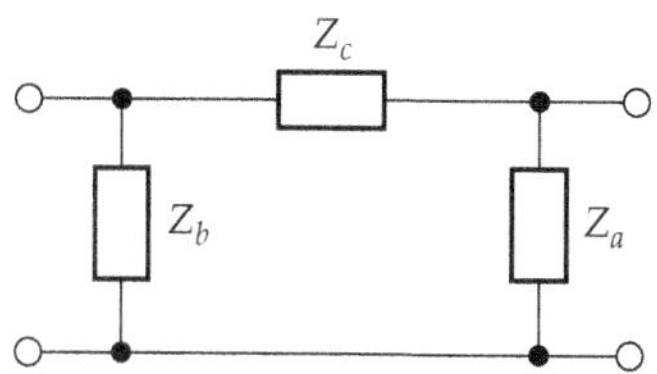

FIGURE P24.43

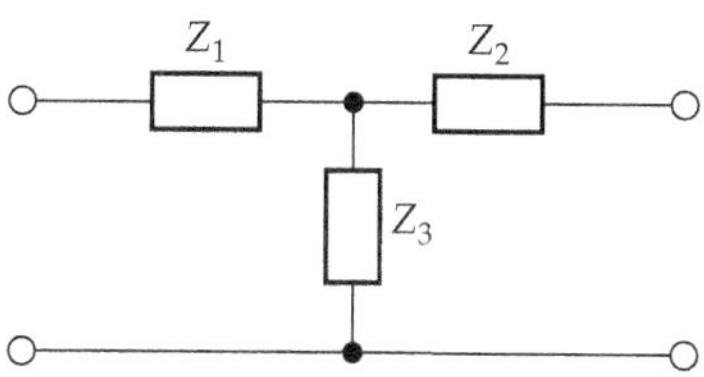

FIGURE P24.44

Two-Port Circuit Analysis

P24.45 Derive TEC seen by the load in Figure P24.45 and determine the steady-state load voltage, assuming $v_{SRC}(t) = 10\cos 1000t$ V.

Ans. $v_{Th} = 9.97\cos(1000t + 4.29°)$ V, $Z_{Th} = 605.0\angle 83.8°$ Ω, $v_2(t) = 1.6\cos(1000t - 70.4°)$ V.

P24.46 A two-port circuit has the following g parameters: $g_{11} = 1 - j$ S, $g_{12} = -2 + j2$, $g_{21} = 2 - j2$, and $g_{22} = 4 + j4$ Ω. It is connected at port 2 to a resistive load of 50 Ω and at port 1 to a source having an open-circuit voltage $v_{SRC}(t) = 20\cos t$ V and zero source resistance. Determine the real power delivered by the source to the two-port circuit and to the load.

Ans. 197.8 W, 27.3 W.

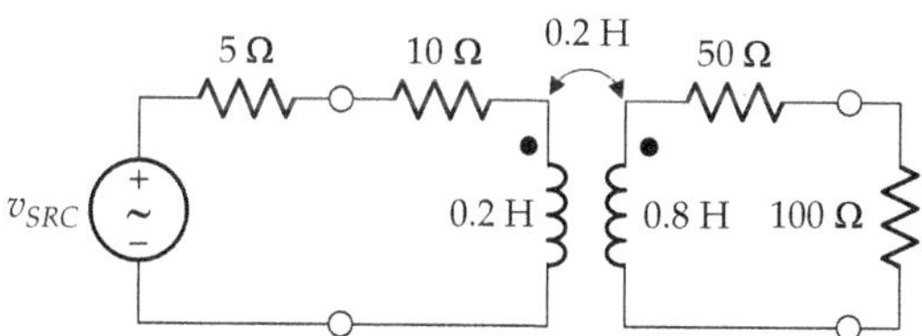

FIGURE P24.45

P24.47 Determine the load impedance in the preceding problem for maximum power transfer to the load and calculate this power.

Ans. $4 - j4\ \Omega$, 100 W.

P24.48 An amplifier has the following y parameters: y_{11} = 1 mS, y_{12} = −2 μS, y_{21} = 100 mS, and y_{22} = 50 μS. The amplifier is connected to a source of 10 mV rms open-circuit voltage and 10 kΩ source impedance. Determine the load resistance for maximum power transfer to the load and calculate this power.

Ans. 8.9 μW.

P24.49 A two-stage amplifier is shown in Figure P24.49. The y parameters of the two stages are as follows:

Stage 1: y_{11} = 1 mS, y_{12} = −2 μS, y_{21} = 500 mS, y_{22} = 0.1 μS.
Stage 2: y_{11} = 5 mS, y_{12} = −1 μS, y_{21} = 100 mS, y_{22} = 0.4 μS.

Determine the input impedances Z_{in1} and Z_{in2} and the overall voltage gain $\mathbf{V_L}/\mathbf{V_{SRC}}$.

Ans. Z_{in1} = 0.841 kΩ, Z_{in2} = 0.196 kΩ, $\mathbf{V_L}/\mathbf{V_{SRC}}$ = 8446.

P24.50 Determine R_L in Figure P24.49 for maximum power transfer.

Ans. 3 kΩ.

P24.51 If the two-port circuit N in Figure P24.51 is described by its y parameters, show that $\frac{\mathbf{V_2}}{\mathbf{V_1}} = -\frac{y_{21}}{y_{22} + sC}$. If the circuit is described by its h parameters, show that $\frac{\mathbf{V_2}}{\mathbf{I_1}} = -\frac{h_{21}}{h_{22} + sC}$. Obtain the input impedance by dividing the latter expression by the former and express the result in terms of (a) the y parameters and (b) the h parameters. Verify by comparing with Table 24.6.

P24.52 If the two-port circuit N in Figure P24.52 is described by its g parameters, show that $\frac{\mathbf{V_2}}{\mathbf{V_1}} = \frac{sL}{g_{22} + sL} g_{21}$. If the circuit is described by its z parameters, show that $\frac{\mathbf{V_2}}{\mathbf{I_1}} = \frac{sL}{z_{22} + sL} z_{21}$. Obtain the input impedance by dividing the latter expression by the former and express the result in terms of (a) the g parameters and (b) the z parameters. Verify by comparing with Table 24.6.

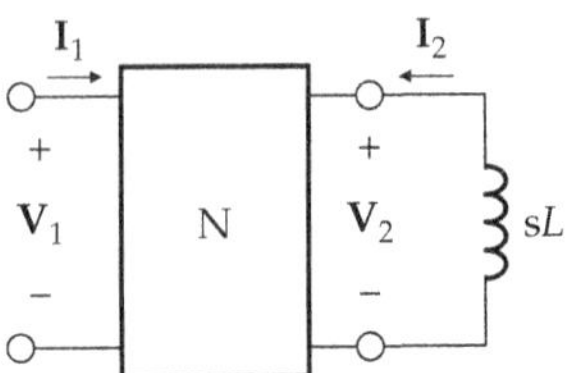

FIGURE P24.52

P24.53 Determine Z_{in}, Z_{Th}, and the steady-state value of v_2 in Figure P24.53, assuming $v_{SRC}(t)$ = 20cos1000t V.

Ans. Z_{in} = 14.38 − j0.0156 Ω, Z_{out} = 260 Ω, $v_2(t)$ = 12.9cos(1000t + 0.24°) V.

P24.54 Derive TEC seen by the load at terminals 2 and 3 in Figure P24.54 using the b parameters of the three-terminal circuit between the voltage source and the load, and determine v_2, assuming $v_{SRC}(t)$ = 20cos1000t V.

Ans. $\mathbf{V_{Th}}$ = 0, Z_{Th} = 20 − j10 Ω, $\mathbf{V_2}$ = 0.

P24.55 Derive TEC seen by the load at terminals 2 and 3 in Figure P24.55 using the g parameters of the three-terminal circuit between the current source and the load, assuming v_{SRC} = 10cos1000t V. Determine v_2.

Ans. $\mathbf{V_{Th}}$ = −10.181 + j0.533 V, Z_{Th} = 1.043 − j0.0546 Ω, $v_2(t)$ = 9.23cos(1000t + 177.3°) V.

P24.56 Determine $\mathbf{V_O}/\mathbf{V_{SRC}}$ in the circuit of Figure P24.56, where $s = j\omega$ and ω = 1 rad/s.

Ans. $\mathbf{V_O}/\mathbf{V_{SRC}}$ = 0.0068 − j0.0653.

P24.57 Determine the 3 dB frequency of the response $\mathbf{V_O}/\mathbf{V_{SRC}}$ in the circuit of Figure P24.57.

Ans. 911 krad/s

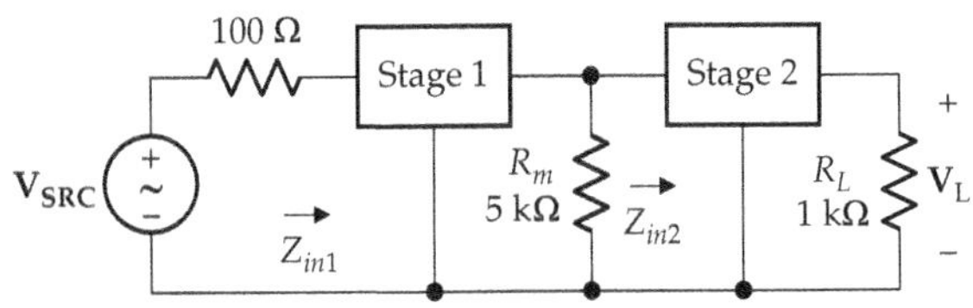

FIGURE P24.49

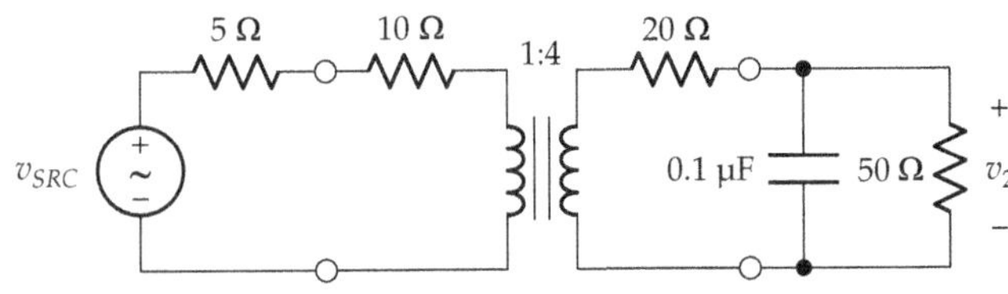

FIGURE P24.53

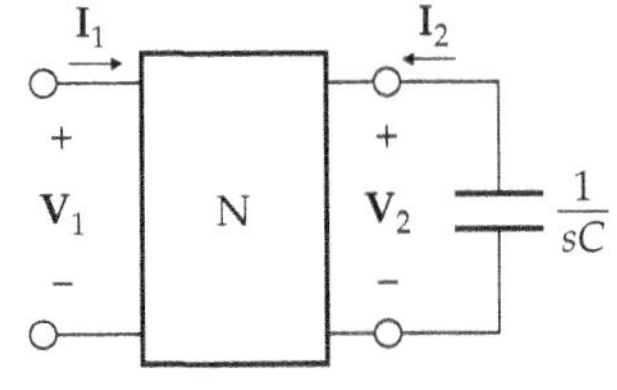

FIGURE P24.51

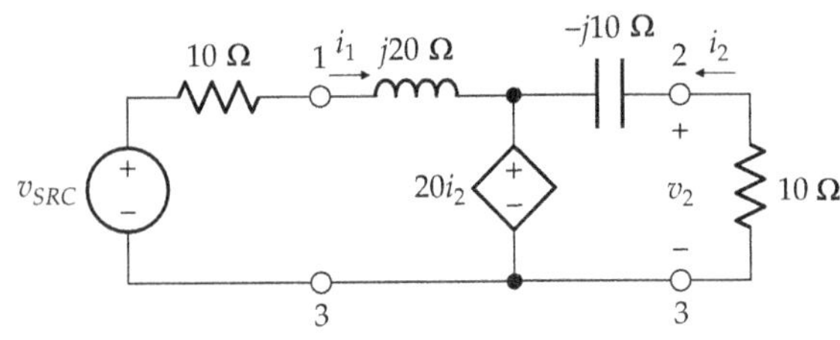

FIGURE P24.54

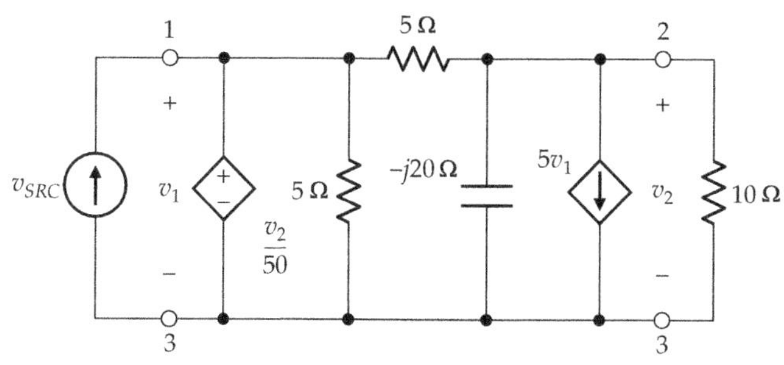

FIGURE P24.55

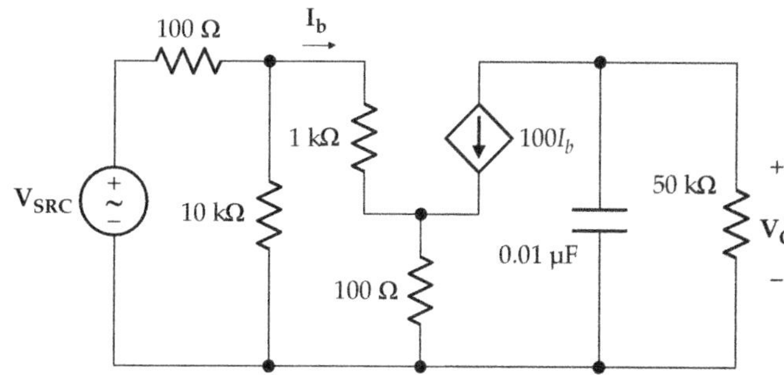

FIGURE P24.58

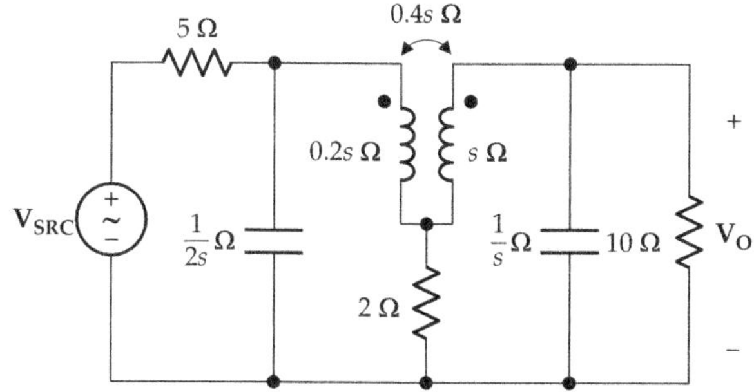

FIGURE P24.56

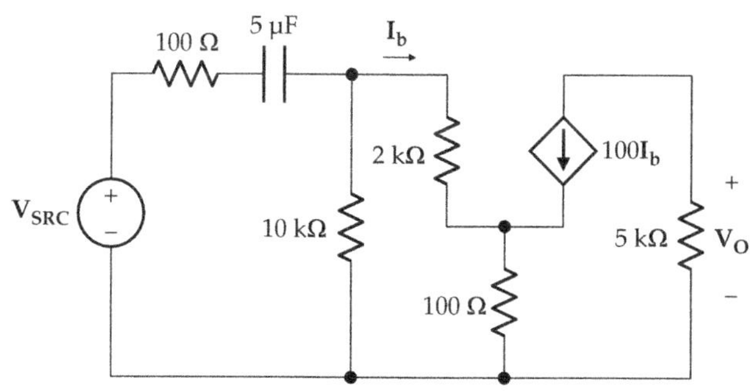

FIGURE P24.57

P24.58 Determine the 3 dB frequency of the response $\mathbf{V_O}/\mathbf{V_{SRC}}$ in the circuit of Figure P24.58.

Ans. 2 krad/s

P24.59 Two three-terminal circuits are connected in series–parallel to a load of 5 Ω and to a voltage source having a source resistance 2 Ω and an open-circuit voltage $\mathbf{V_{SRC}}$. The h parameters of the two circuits are

Circuit 1: $h_{11} = 2\ \Omega,\ h_{12} = -(j+2),\ h_{21} = -2,\ h_{22} = j\ \text{S}$
Circuit 2: $h_{11} = 4\ \Omega,\ h_{12} = -j,\ h_{21} = (j+1),\ h_{22} = j2\ \text{S}$

Determine the input impedance of the combination, the output impedance, and the voltage gain $\mathbf{V_O}/\mathbf{V_{SRC}}$, where $\mathbf{V_O}$ is the load voltage.

Ans. $Z_{in} = 5.204 + j1.947\ \Omega$, $Z_{out} = -0.3529 - j0.5882\ \Omega$, $\mathbf{V_O}/\mathbf{V_{SRC}} = -0.480 - j0.445$.

P24.60 Two three-terminal circuits connected in parallel–series as in Figure 24.42, to a load of 5 Ω and to a current source having a source resistance 2 Ω and a short-circuit current $\mathbf{I_{SRC}}$. The g parameters of the two circuits are

Circuit 1: $g_{11} = 1\ \text{S},\ g_{12} = j,\ g_{21} = j2,\ g_{22} = 2\ \Omega$
Circuit 2: $g_{11} = j\ \text{S},\ g_{12} = -(j+2),\ g_{21} = j,\ g_{22} = j2\ \Omega$

Determine the input impedance of the combination, the output impedance, and the gain $\mathbf{V_O}/\mathbf{V_{SRC}}$, where $\mathbf{V_O}$ is the load voltage.

Ans. $Z_{in} = 0.390 - j0.459\ \Omega$, $Z_{out} = 2.615 + j2.923\ \Omega$, $\mathbf{V_O}/\mathbf{V_{SRC}} = 0.131 + j0.146$.

P24.61 Perform the analysis of Section 24.5 using h parameters. Consider Z_{src} to add to h_{11} and Y_L to add to h_{22}.

P24.62 Perform the analysis of Section 24.5 using y parameters. Consider Y_L to add to h_{22} and transform the voltage source of impedance Z_{src} to an equivalent current source.

P24.63 Perform the analysis of Section 24.5 using g parameters. Consider Z_L to add to g_{22} and transform the voltage source of impedance Z_{src} to an equivalent current source.

Probing Further

P24.64 Show that the input and output impedances of a terminated circuit (Figure 24.44) may be expressed as follows: $IM_{in} = \lambda_{11} - \frac{\lambda_{12}\lambda_{21}}{\lambda_{22} + IM_L}, IM_{out} = \lambda_{22} - \frac{\lambda_{12}\lambda_{21}}{\lambda_{11} + IM_{src}}$, where IM_{in}, IM_{out}, IM_L, and IM_{src} are, respectively, the input, output, load, and source immittances (i.e., impedances or admittances), and the λ's may be z, y, h, or g parameters. The units of λ_{11} and λ_{22} just after the equality sign determine whether the expression is an impedance or admittance, and the units of λ_{11} and λ_{22} in the denominators determine whether a load or source impedance or admittance is to be used.

P24.65 Consider the circuit shown in Figure P24.65. When the input impedance equals the source resistance and the output impedance equals the load resistance, the circuit is said to be *image-terminated*. Note that under these conditions, maximum power is transferred to the load. The expressions for Z_{in} and Z_{Th} (Table 24.6) show that in an image-terminated circuit, $\frac{R_S}{R_L} = \frac{z_{11}}{z_{22}}$.

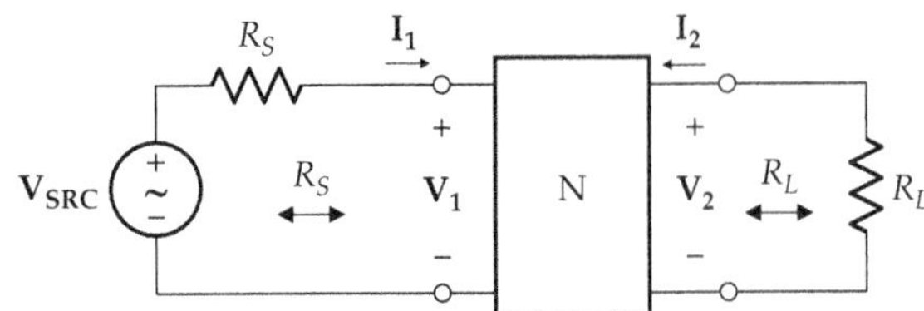

FIGURE P24.65

FIGURE P24.68

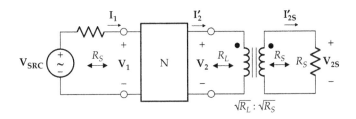

Substituting back in the expressions for Z_{in} and Z_{Th} shows that $R_S = \sqrt{z_{11}/y_{11}} = \sqrt{a_{11}a_{12}/a_{21}a_{22}}$ and that $R_L = \sqrt{z_{22}/y_{22}} = \sqrt{a_{12}a_{22}/a_{11}a_{21}}$.

P24.66 Consider an image-terminated circuit (Figure P24.65). The image transmission constant is defined as $\gamma = \alpha + j\beta = \frac{1}{2}\ln\frac{\mathbf{V}_1\mathbf{I}_1}{\mathbf{V}_2\mathbf{I}_2'}$, where $\mathbf{I}_2' = -\mathbf{I}_2$ is the current flowing out of port 2. Show that if the circuit is reciprocal, $\tanh\gamma = \frac{1}{\sqrt{z_{11}y_{11}}} = \frac{1}{\sqrt{z_{22}y_{22}}} = \sqrt{\frac{a_{12}a_{21}}{a_{11}a_{22}}}$, $\sinh\gamma = \sqrt{a_{12}a_{21}}$, and $\cosh\gamma = \sqrt{a_{11}a_{22}}$. Note that $\alpha = \frac{1}{2}\ln\frac{P_i}{P_L}$, where P_i is the power delivered to the two-port circuit and P_L is the power delivered to the load. Also $\beta = \frac{1}{2}\left[\angle(\mathbf{V}_1\mathbf{I}_1) - \angle(\mathbf{V}_2\mathbf{I}_2')\right]$.

P24.67 Show that if n image-terminated circuits are cascaded, the overall γ is the sum of the individual γ's, and that if the image-terminated circuit is reciprocal, γ is the same if the input and output are interchanged.

P24.68 Show that if an image-terminated circuit is symmetric, $R_S = \sqrt{\frac{a_{12}}{a_{21}}} = R_L$, $\gamma = \alpha + j\beta = \ln\frac{\mathbf{V}_1}{\mathbf{V}_{2S}} = \ln\frac{\mathbf{I}_1}{\mathbf{I}_{2S}'}$ so that $\alpha = \ln\left|\frac{\mathbf{V}_1}{\mathbf{V}_{2S}}\right| = \ln\left|\frac{\mathbf{I}_1}{\mathbf{I}_{2S}'}\right|$ and $\beta = \angle\mathbf{V}_1 - \angle\mathbf{V}_{2S}$, where the s subscript refers to the output of a symmetric circuit. A nonsymmetric circuit N can be made symmetric by inserting a transformer of turns ratio $\sqrt{R_L/R_S}$ between the output and load, as shown in Figure P24.68. Deduce that even in a nonsymmetric circuit N, $\beta = \angle\mathbf{V}_1 - \angle\mathbf{V}_2 = \angle\mathbf{I}_1 - \angle\mathbf{I}_2'$.

25

Balanced Three-Phase Systems

Objective and Overview

Modern electric power systems are gigantic networks that connect numerous power stations to a multitude of load centers dispersed over a wide geographical area that may span several neighboring countries. These networks incorporate sophisticated communications and control systems that are used to optimize the performance of the network. Three-phase systems are almost universally used for the generation, transmission, and distribution of electrical energy because of the many advantages they offer compared to single-phase systems. Practical three-phase systems are nominally balanced under normal operating conditions, that is, their voltages and currents possess a certain symmetry.

The chapter presents the fundamentals of balanced three-phase systems, starting with the two basic connections of these systems, namely, Y and Δ, and their characteristic features. This is followed by explaining the basic approach in analyzing balanced three-phase systems by transforming the Δ connection to Y and deriving an equivalent single-phase system. Because three-phase systems are primarily power systems, power relations in these systems are then discussed, including instantaneous and complex power, power measurement, and power factor correction. The chapter ends with highlighting the advantages of three-phase systems and considering the overall structure of power systems.

25.1 Three-Phase Variables

Consider the set y_a, y_b, and y_c representing either voltages or currents that vary sinusoidally with time as follows:

$$y_a(t) = Y_m\cos(\omega t + \theta) \tag{25.1}$$

$$y_b(t) = Y_m\cos(\omega t + \theta - 120°) \tag{25.2}$$

$$y_c(t) = Y_m\cos(\omega t + \theta + 120°) \tag{25.3}$$

y_a, y_b, and y_c are described as balanced three-phase variables, because of two distinguishing characteristics: they all have the same amplitude Y_m, and their phase angles differ by 120°, or 1/3 of the full angle of 360°. By extension, the set y_a, y_b, …, y_n represents balanced n-phase variables if all these variables have the same amplitude and their phase angles differ by 360°/n. If either condition is not satisfied, the variables are no longer balanced. Systems having $n > 2$ are described as **polyphase** systems.

Three-phase systems are by far the most important polyphase systems in practice. The basic three-phase system consists of a three-phase generator, represented by three sources, each having a source impedance, connected to a three-phase load. The generator and load may be connected in Y or Δ, which gives rise to four possibilities, illustrated in Figure 25.1a to d. The source voltages are three phase, in accordance with Equations 25.1 to 25.3, and the impedances of the three sources are assumed to be equal.

The common node of the Y connection is the **neutral**. Each connection between the generator and the load (aA, bB, or cC) in Figure 25.1a to d is a **line**. In power systems, the lines are often referred to as **feeders**. A line has a certain impedance, which in particular cases may be neglected. The line impedances are denoted by Z_{aA}, Z_{bB}, and Z_{cC} in Figure 25.1a to d. The current that flows in a given line is a **line current**. The voltage between any two lines, whether at the generator end or the load end, is a **line voltage**.

The generator or load branches that are connected in Y or Δ are the **phases**. In Figure 25.1a to d, each phase of the generator consists of an ideal voltage source in series with a source impedance. In the simplest cases, the source impedances can be neglected. The phases of the load are the impedances Z_A, Z_B, and Z_C. These may be single-phase loads such as lamps or heaters, connected in Y or Δ, or they may be an inherently three-phase load, such as a three-phase motor. A **phase voltage** and a **phase current** are associated with each phase.

A three-phase *system* is balanced if *all* of the following conditions are satisfied:

1. The generator voltages are balanced, in accordance with Equations 25.1 to 25.3:

$$v_{ga}(t) = V_{gm}\cos(\omega t + \theta_v) \tag{25.4}$$

$$v_{gb}(t) = V_{gm}\cos(\omega t + \theta_v - 120°) \tag{25.5}$$

$$v_{gc}(t) = V_{gm}\cos(\omega t + \theta_v + 120°) \tag{25.6}$$

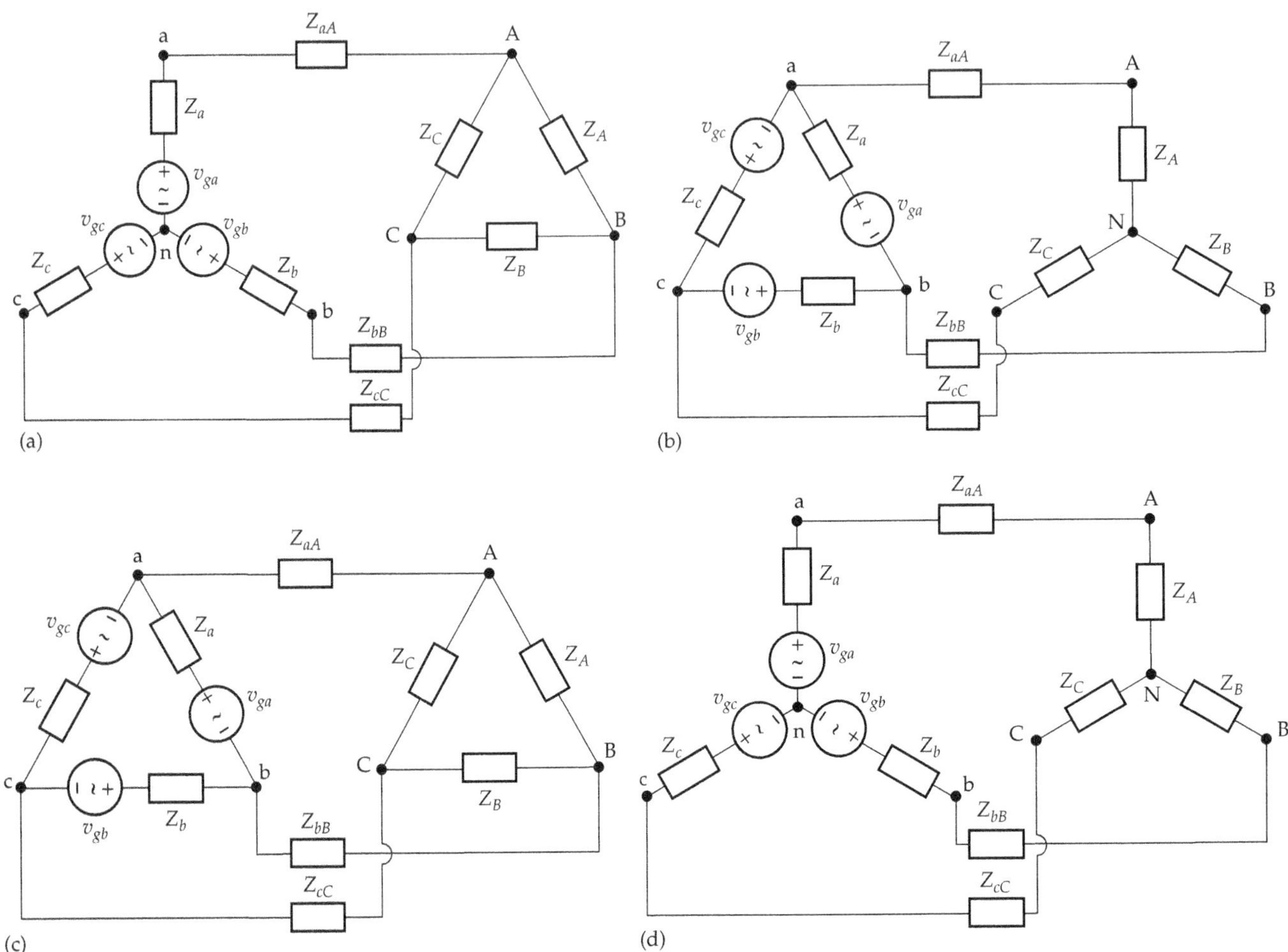

FIGURE 25.1
Δ and Y connections of three-phase systems. (a) Y-connected three-phase generator and Δ-connected load, (b) Δ-connected three-phase generator and Y-connected load, (c) Δ-connected three-phase generator and load, and (d) Y-connected three-phase generator and load.

2. Phase impedances are equal, both for the generator and the load:

$$Z_a = Z_b = Z_c \quad \text{and} \quad Z_A = Z_B = Z_C \tag{25.7}$$

3. Line impedances are equal:

$$Z_{aA} = Z_{bB} = Z_{cC} \tag{25.8}$$

Under these conditions, each of the sets of phase voltages, phase currents, line voltages, and line currents is a balanced set.

The three-phase systems illustrated in Figure 25.1a to d are **three-wire** systems because three lines connect the generator to the load. With both the generator and load connected in Y, the two neutral points may be connected together, resulting in a **four-wire** system (Figure 25.2).

25.1.1 Sum of Balanced Variables

An important characteristic of balanced voltages or currents is the following:

Concept: *The sum of a set of balanced voltages or currents is zero.*

Although this is true of any number of phases, we will illustrate it for the three-phase case. Thus, if y_a, y_b, and y_c are a balanced set given by Equations 25.1 to 25.3, their sum is zero:

$$y_a + y_b + y_c = 0 \tag{25.9}$$

Equation 25.9 can be proven in a number of ways. As phasors (Figure 25.3a), $\mathbf{Y_a} = Y_m\angle 0°$, $\mathbf{Y_b} = Y_m\angle -120°$, and $\mathbf{Y_c} = Y_m\angle 120°$ can be added by laying them end to end. Because their magnitudes are equal and their phase

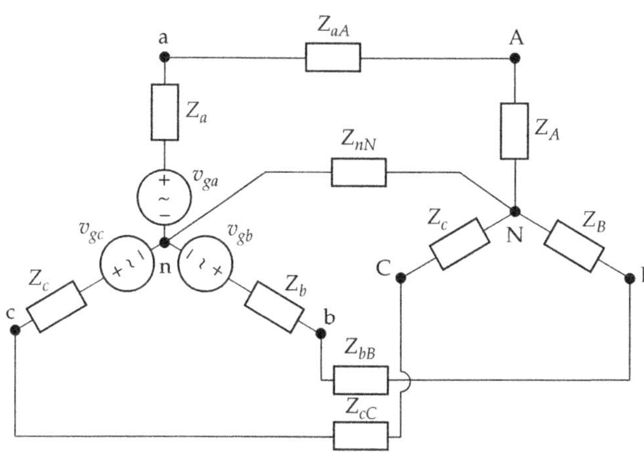

FIGURE 25.2
Four-wire three-phase system.

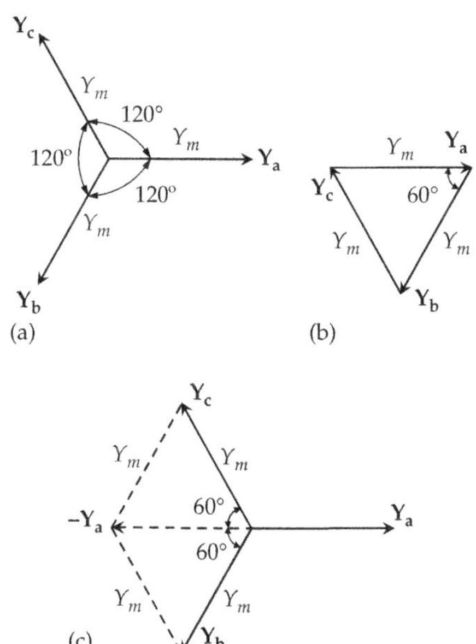

FIGURE 25.3
Sum of balanced three-phase variables. (a) Phasor diagram of balanced three-phase quantities, (b) phasor sum is zero, and (c) addition of two phasors is the negation of the third phasor.

angles differ by 120°, they form an equilateral triangle (Figure 25.3b). Since this is a closed figure,

$$\mathbf{Y_a} + \mathbf{Y_b} + \mathbf{Y_c} = 0 \tag{25.10}$$

Alternatively, we may add any two phasors and show that the sum is the negative of the third phasor. For example, if we add $\mathbf{Y_b}$ and $\mathbf{Y_c}$ in Figure 25.3c using the parallelogram construction, the resultant is $-\mathbf{Y_a}$.

In the time domain, adding y_b and y_c in Equations 25.2 and 25.3 gives $Y_m\cos(\omega t + \theta - 120°) + Y_m\cos(\omega t + \theta + 120°) = 2Y_m\cos(\omega t + \theta) \times \cos 120° = -Y_m\cos(\omega t + \theta) = -y_a$.

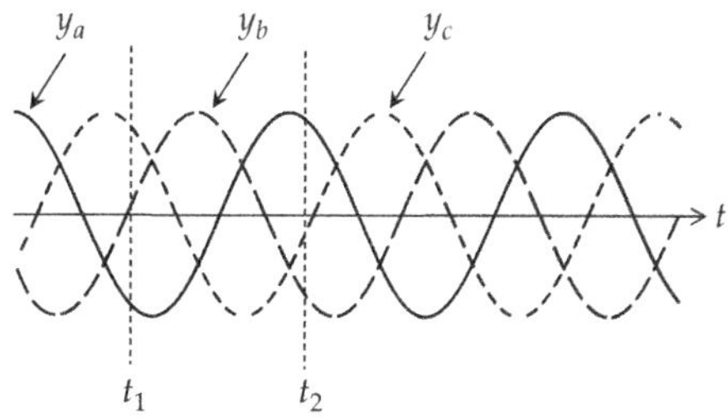

FIGURE 25.4
Time-variation of balanced three-phase variables.

Figure 25.4 illustrates the time variation of y_a, y_b, and y_c. At any instant of time, such as t_1 or t_2, the sum of y_a, y_b, and y_c is zero. At t_2, for example, if y_a represents current flowing toward a load, then the total current flowing away from the load, $(y_b + y_c)$, is equal and opposite. A useful way of looking at this is to consider that at any instant, such as t_2, two line conductors are acting as the return line of the third conductor or, alternatively, one conductor is acting as the return line for the other two.

25.1.2 Phase Sequence

In Equations 25.1 through 25.3, y_b lags y_a by 120° and y_c lags y_b by 120°, since $\cos(\omega t + \theta + 120°) = \cos(\omega t + \theta - 240°)$, or y_c leads y_a by 120°. The **phase sequence** is 'abc' and is a *positive* phase sequence. If the phase angles of y_b and y_c are interchanged,

$$y_b(t) = Y_m\cos(\omega t + \theta + 120°) \tag{25.11}$$

$$y_c(t) = Y_m\cos(\omega t + \theta - 120°) \tag{25.12}$$

then y_c lags y_a by 120° and y_b leads y_a by 120°. The phase sequence is now 'acb' and is a *negative* phase sequence.

The phasors in Figure 25.3a have a positive phase sequence. If we move around the origin in the *clockwise* direction, the order in which the phasors are encountered is the phase sequence, which is 'abc' for a positive sequence.

In a negative phase sequence, phasors $\mathbf{Y_b}$ and $\mathbf{Y_c}$ are interchanged, but the same interpretations apply. The phase sequence is now the order in which the phases are encountered in going clockwise around the origin, so that the phase sequence is 'acbacb', or 'cba', the reverse of 'abc'.

In practice, the phase sequence is reversed by interchanging any two line connections to a three-phase load. Assuming a positive phase sequence 'abc' in Figure 25.5a, the load, whose terminals are 'A', 'B', and 'C', sees the same phase sequence 'abc'. In Figure 25.5b, the same load connected to the same supply sees a

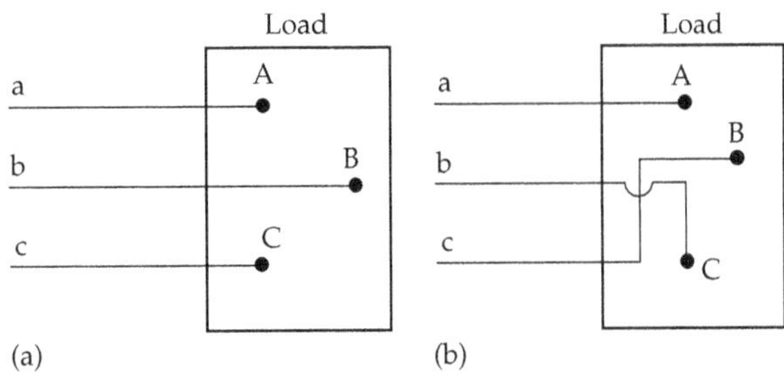

FIGURE 25.5
Reversal of phase sequence. The load sees a positive phase sequence in (a) and a negative phase sequence in (b).

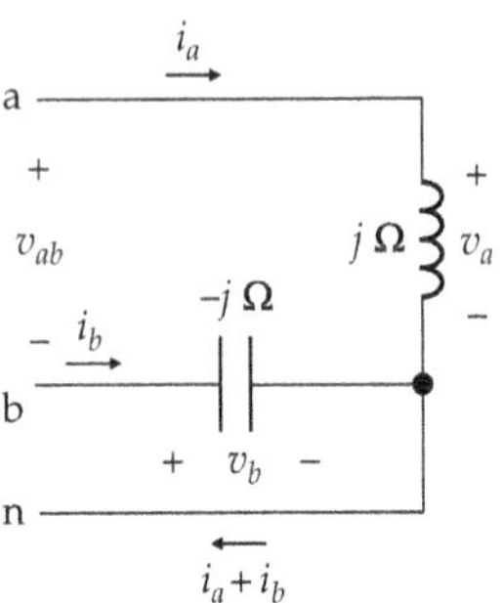

FIGURE 25.6
Figure for Exercise 25.2.

negative phase sequence 'acb'. If the load is a lighting or heating load, changing the phase sequence is generally of no consequence. However, if the load incorporates a three-phase motor, reversing the phase sequence reverses the direction of rotation of the motor. Although this provides a convenient way for reversing the direction of rotation of an ac motor, it may have disastrous mechanical consequences on loads already connected to the motor, such as compressors or machine tools. Moreover, it is common practice in medium and large three-phase power systems to have two or more three-phase generators connected in parallel, in which case they must have the same phase sequence.

Unless explicitly stated otherwise, a positive phase sequence is assumed in the rest of the chapter.

Primal Exercise 25.1

Given a set of balanced six-phase voltages of 10 V amplitude, in which one of these voltages, say that of phase 1, has a phase angle of 10°. Express the phase voltages in the time domain, with respect to phase 1, assuming the phase sequence is (a) positive and (b) negative.

Ans. (a) $v_1(t)=10\cos(\omega t+10°)$ V, $v_2(t)=10\cos(\omega t-50°)$ V, $v_3(t)=10\cos(\omega t-110°)$ V, $v_4(t)=10\cos(\omega t-170°)$ V, $v_5(t)=10\cos(\omega t+130°)$ V, and $v_6(t)=10\cos(\omega t+70°)$ V; (b) $v_1(t)=10\cos(\omega t+10°)$ V, $v_2(t)=10\cos(\omega t+70°)$ V, $v_3(t)=10\cos(\omega t+130°)$ V, $v_4(t)=10\cos(\omega t-170°)$ V, $v_5(t)=10\cos(\omega t-110°)$ V, and $v_6(t)=10\cos(\omega t-50°)$ V.

Primal Exercise 25.2

According to the definition of a polyphase system, a balanced two-phase system would be defined as $y_a(t)=Y_m\cos(\omega t+\theta)$ and $y_b(t)=Y_m\cos(\omega t+\theta-180°)$. However, a two-phase system is conventionally defined as $y_a(t)=Y_m\cos(\omega t+\theta)$ and $y_b(t)=Y_m\cos(\omega t+\theta\pm 90°)$. Assume that in the two-phase system shown in Figure 25.6, $v_a(t)=10\cos\omega t$ V and $v_b(t)=10\cos(\omega t-90°)$ V. Determine (a) $i_a(t)$; (b) $i_b(t)$; (c) $v_{ab}(t)$; and (d) $i_a(t)+i_b(t)$.

Ans. (a) $i_a(t)=10\cos(\omega t-90°)$ A; (b) $i_b(t)=10\cos\omega t$ A; (c) $v_{ab}(t)=10\sqrt{2}\cos(\omega t+45°)$ V; (d) $i_a(t)+i_b(t)=10\sqrt{2}\cos(\omega t-45°)$ A.

25.2 The Balanced Y Connection

25.2.1 Voltage Relations

Figure 25.7 shows a basic, four-wire Y–Y configuration for a balanced three-phase system, where $\mathbf{V_{ga}}=V_g\angle 0°$, $\mathbf{V_{gb}}=V_g\angle -120°$, and $\mathbf{V_{gc}}=V_g\angle 120°$ and the phase impedances Z_Y are all equal.

It follows from KVL that the phase load voltages are equal to the corresponding source voltages:

$$\mathbf{V_{ga}}=\mathbf{V_A},\quad \mathbf{V_{gb}}=\mathbf{V_B},\quad \mathbf{V_{gc}}=\mathbf{V_C} \tag{25.13}$$

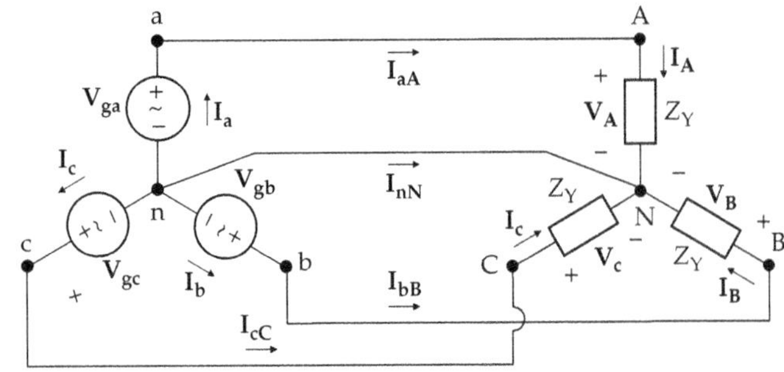

FIGURE 25.7
Balanced, four-wire Y–Y connection.

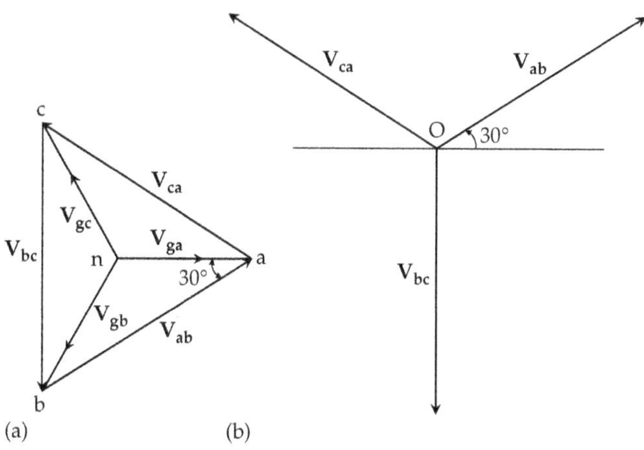

FIGURE 25.8
(a) Phase and (b) line voltages in a Y connection.

The line voltages at the load and generator are of course equal, since the line impedances are assumed to be zero. From KVL,

$$\mathbf{V}_{ab} = \mathbf{V}_{ga} - \mathbf{V}_{gb}, \quad \mathbf{V}_{bc} = \mathbf{V}_{gb} - \mathbf{V}_{gc}, \quad \mathbf{V}_{ca} = \mathbf{V}_{gc} - \mathbf{V}_{ga} \tag{25.14}$$

Figure 25.8a shows the phasors of the three phase voltages, of phase sequence abc, and the three line voltages derived from Equations 25.14. The line voltages are shown as position phasors drawn from the origin in Figure 25.8b.

Triangle 'anb' is isosceles having the two sides 'na' and 'nb' equal to V_g and acute angles of 30°. It follows that $\mathbf{V}_{ab} = \sqrt{3}V_g\angle 30°$; $\mathbf{V}_{bc}$ lags $\mathbf{V}_{ab}$ by 120°, so that its phase angle is −90°; $\mathbf{V}_{ca}$ lags $\mathbf{V}_{ab}$ by 120°, which makes its phase angle −210°, or +150° (Figure 25.8b). Thus,

$$\mathbf{V}_{ab} = \sqrt{3}V_g\angle 30°, \quad \mathbf{V}_{bc} = \sqrt{3}V_g\angle -90°, \quad \mathbf{V}_{ca} = \sqrt{3}V_g\angle 150° \tag{25.15}$$

Summary: *In a balanced Y connection, the line voltages are balanced three-phase voltages whose magnitude is $\sqrt{3}$ times that of the phase voltage and whose phase angle leads by 30° the phase voltage designated by the first subscript of the line voltage, where the assigned positive polarity of the line voltage is a voltage drop from the node designated by the first subscript to the node designated by the second subscript.*

Thus, $\mathbf{V}_{ab}$, for example, is $\sqrt{3}V_g$ and leads the phase voltage $\mathbf{V}_A$, or $\mathbf{V}_{ga}$, by 30°.

25.2.2 Current Relations

It follows from KCL in Figure 25.7 that the phase currents are the same as the line currents. Moreover, each of the phase currents is equal to the corresponding phase voltage of the generator or the load divided by $Z_Y = |Z_Y|\angle\theta$:

$$\mathbf{I}_a = \mathbf{I}_{aA} = \mathbf{I}_A = \frac{\mathbf{V}_A}{Z_Y} = \frac{V_g\angle 0°}{|Z_Y|\angle\theta} = \frac{V_g}{|Z_Y|}\angle -\theta \tag{25.16}$$

$$\mathbf{I}_b = \mathbf{I}_{bB} = \mathbf{I}_B = \frac{\mathbf{V}_B}{Z_Y} = \frac{V_g\angle -120°}{|Z_Y|\angle\theta} = \frac{V_g}{|Z_Y|}\angle(-120° - \theta) \tag{25.17}$$

$$\mathbf{I}_c = \mathbf{I}_{cC} = \mathbf{I}_C = \frac{\mathbf{V}_C}{Z_Y} = \frac{V_g\angle +120°}{|Z_Y|\angle\theta} = \frac{V_g}{|Z_Y|}\angle(120° - \theta) \tag{25.18}$$

It follows from Equations 25.16 to 25.18 that the line currents $\mathbf{I}_{aA}$, $\mathbf{I}_{bB}$, and $\mathbf{I}_{cC}$ form a balanced three-phase set, which means that their sum is zero:

$$\mathbf{I}_{aA} + \mathbf{I}_{bB} + \mathbf{I}_{cC} = 0 \tag{25.19}$$

However, if the generator is enclosed by a surface S, it follows from KCL, as discussed in connection with Figure 2.19, that $\mathbf{I}_{aA} + \mathbf{I}_{bB} + \mathbf{I}_{cC} + \mathbf{I}_{nN} = 0$. Substituting in Equation 25.19 gives $\mathbf{I}_{nN} = 0$. In other words, the current in the neutral conductor is zero, so that this conductor can be removed without affecting the voltages or currents in the rest of the system. Moreover, if the neutral node 'n' on the generator side is grounded, neutral node 'N' on the load side will also be at zero voltage.

Summary: *In a balanced Y–Y connection, the line currents are balanced three-phase currents that are equal to the phase currents. The neutral conductor does not carry any current. It may be removed, leaving the two neutral nodes at the same voltage.*

In practice, the neutral conductor carries some current due to any imbalance in the load, or due to a third-harmonic current (see Problem P25.14).

25.2.3 Power Relations

Let the rms value of the phase voltage at the load in Figure 25.7 be denoted by $V_{\phi rms}$ and the rms value of the phase current by $I_{\phi rms}$ in the direction of a voltage drop $V_{\phi rms}$. The real power per phase is $P_\phi = V_{\phi rms} \times I_{\phi rms}\cos\theta$. The total real power in the three phases of the load is $P_T = 3P_\phi$. If V_{lrms} and I_{lrms} are the rms values of the line voltage and the line current, respectively, then

$$P_T = 3V_{\phi rms} \times I_{\phi rms}\cos\theta = 3\frac{V_{lrms}}{\sqrt{3}}I_{lrms}\cos\theta = \sqrt{3}V_{lrms}I_{lrms}\cos\theta \tag{25.20}$$

where the substitutions $I_{lrms} = I_{\phi rms}$ and $V_{lrms} = \sqrt{3}V_{\phi rms}$ were made.

The reactive power per phase is $Q_\phi = V_{\phi rms} \times I_{\phi rms}\sin\theta$. The total reactive power in the three phases of the load is $Q_T = 3Q_\phi$, which may be written as

$$Q_T = 3V_{\phi rms} \times I_{\phi rms}\sin\theta = 3\frac{V_{lrms}}{\sqrt{3}}I_{lrms}\sin\theta = \sqrt{3}V_{lrms}I_{lrms}\sin\theta \quad (25.21)$$

The complex power is given by

$$\mathbf{S_T} = P_T + jQ_T = \sqrt{3}V_{lrms}I_{lrms}\angle\theta \quad (25.22)$$

The power factor angle of the balanced three-phase load is $\tan^{-1}Q_T/P_T = \theta$, the same as the phase angle of the load impedance. It follows that the power factor of the balanced three-phase load is the cosine of the phase angle of the load impedance.

Example 25.1: Phase Voltages and Currents in a Y-Connected Generator

A balanced three-phase load absorbs 72 kW at 0.8 p.f. lagging and is connected to a balanced, Y-connected, three-phase generator, the rms value of the line current being 100 A, the line impedances being negligible. Determine the phase voltages and phase currents of the generator.

Solution:

From Equation 25.20, $72{,}000 = \sqrt{3} \times V_{lrms} \times 100 \times 0.8$. This gives $V_{lrms} = 900/\sqrt{3}$ V. The magnitude of the generator phase voltage is $V_{\phi rms} = \frac{1}{\sqrt{3}}V_{lrms} = 300$ V.

Since no information is given regarding phase angles, we can consider the phase voltage $\mathbf{V_{ga}}$ to have a zero phase angle. Then,

$$\mathbf{V_{ga}} = 300\angle 0° \text{ V}, \quad \mathbf{V_{gb}} = 300\angle -120° \text{ V}, \quad \mathbf{V_{gc}} = 300\angle 120° \text{ V}$$

all being rms values.

From Equations 25.15, the line voltages are

$$\mathbf{V_{ab}} = \frac{900}{\sqrt{3}}\angle 30° \text{ V}, \quad \mathbf{V_{bc}} = \frac{900}{\sqrt{3}}\angle -90° \text{ V}, \quad \mathbf{V_{ca}} = \frac{900}{\sqrt{3}}\angle 150° \text{ V}$$

all being rms values.

The phase currents of the generator are the same as the corresponding line currents, whose magnitude is 100 A. To determine the phase angles, we note that the generator must deliver the same power as that absorbed by the load, that is, 72 kW at 0.8 p.f., or 24 kW per phase at 0.8 p.f. Consider phase 'a' of the generator; the power relation is 24,000 = 300 × 100 × cosθ, so cosθ = 0.8, or θ = 36.9°. It follows, since the power factor is lagging, that

$$\mathbf{I_{aA}} = 100\angle -\theta \text{ A}, \quad \mathbf{I_{bB}} = 100\angle(-120° - \theta) \text{ A},$$

$$\mathbf{I_{cC}} = 100\angle(120° - \theta) \text{ A}$$

Alternatively, we may assume that the load is Y connected, in which case the load phase voltages are the same as the generator phase voltages, the load phase currents are equal to the generator phase currents (Figure 25.7), and the load phase currents lag the corresponding phase voltages by θ. The relations derived earlier for the generator phase currents then follow.

Problem-Solving Tips

- When determining voltages or currents in a balanced, three-phase system, it is only necessary to evaluate the voltage or current for one phase or line. The remaining three-phase voltages and currents then follow from the fact that balanced three-phase voltages and currents have the same magnitude but with 120° phase difference between successive phases.
- In a Y connection, if the line voltages are known the phase voltages readily follow, since the phase voltage designated by the first subscript of the line voltage lags the line voltage by 30° and has $1/\sqrt{3}$ of its magnitude.

Primal Exercise 25.3

Assume that $\mathbf{V_{an}} = 240\angle 0°$ V for the generator in Example 25.1 and that the load is Y connected and of impedance 3 + j4 Ω per phase. Determine (a) the line voltages and (b) the line currents, assuming a positive phase sequence.

Ans. (a) $\mathbf{V_{ab}} = 416\angle 30°$ V, $\mathbf{V_{bc}} = 416\angle -90°$ V, $\mathbf{V_{ca}} = 416\angle 150°$ V; (b) $\mathbf{I_{aA}} = 48\angle -53.1°$ A, $\mathbf{I_{bB}} = 48\angle -173.1°$ A, $\mathbf{I_{cC}} = 48\angle 66.9°$ A.

Primal Exercise 25.4

Assume that $\mathbf{V_{an}} = 400\angle 0°$ V rms in Primal Exercise 25.3 and that the complex power absorbed by the load is 4 + j3 kVA. Determine (a) the power factor of the load, (b) $|\mathbf{I_{aA}}|$, and (c) the phase impedance of a Y-connected load.

Ans. (a) 0.8; (b) 4.17 A rms; (c) $96\angle 36.9°$ Ω.

Exercise 25.5

Determine the line voltages in Figure 25.7, assuming $\mathbf{V_{ga}} = V_g\angle 0°$ and a negative phase sequence.

Ans. $\mathbf{V_{ab}} = \sqrt{3}V_g\angle -30°$, $\mathbf{V_{ca}} = \sqrt{3}V_g\angle -150°$, $\mathbf{V_{bc}} = \sqrt{3}V_g\angle 90°$.

Exercise 25.6

Repeat Exercise 25.3 assuming a negative phase sequence.

Ans. (a) $\mathbf{V}_{ab} = 416\angle -30°$ V, $\mathbf{V}_{ca} = 416\angle -150°$ V, $\mathbf{V}_{bc} = 416\angle 90°$ V; (b) $\mathbf{I}_{aA} = 48\angle -53.1°$ A, $\mathbf{I}_{cC} = 48\angle -173.1°$ A, $\mathbf{I}_{bB} = 48\angle 66.9°$ A.

25.3 The Balanced Δ Connection

25.3.1 Voltage Relations

Consider a balanced Δ-connected load, where $\mathbf{I}_A = I_m\angle 0°$, $\mathbf{I}_B = I_m\angle -120°$, and $\mathbf{I}_C = I_m\angle 120°$ (Figure 25.9), and the phase angle of $\mathbf{I}_A$ is considered to be zero for convenience. The phase voltages of the load are the same as the corresponding line voltages at the generator and at the load:

$$\mathbf{V}_{AB} = \mathbf{V}_{ab}, \quad \mathbf{V}_{BC} = \mathbf{V}_{bc}, \quad \mathbf{V}_{CA} = \mathbf{V}_{ca} \tag{25.23}$$

25.3.2 Current Relations

Given the phase currents, the line currents may be obtained from KCL at nodes 'A', 'B', and 'C':

$$\mathbf{I}_{aA} = \mathbf{I}_A - \mathbf{I}_C, \quad \mathbf{I}_{bB} = \mathbf{I}_B - \mathbf{I}_A, \quad \mathbf{I}_{cC} = \mathbf{I}_C - \mathbf{I}_B \tag{25.24}$$

Figure 25.10a shows the phasors of the three-phase currents, of phase sequence 'abc', and the phasors of the three line currents satisfying Equations 25.24. Following an argument similar to that used in connection with Figure 25.8, it is seen that

$$\mathbf{I}_{aA} = \sqrt{3}I_m\angle -30°, \quad \mathbf{I}_{bB} = \sqrt{3}I_m\angle -150°, \quad \mathbf{I}_{cC} = \sqrt{3}I_m\angle 90° \tag{25.25}$$

Summary: *In a balanced Δ connection, the line voltages are equal to the phase voltages. The line currents are balanced three-phase currents that have $\sqrt{3}$ times the magnitude of the phase currents. The line current lags by 30° the phase current designated by either subscript of the line current, where the assigned positive direction of line current is from the node denoted by the first subscript toward the node designated by the second subscript.*

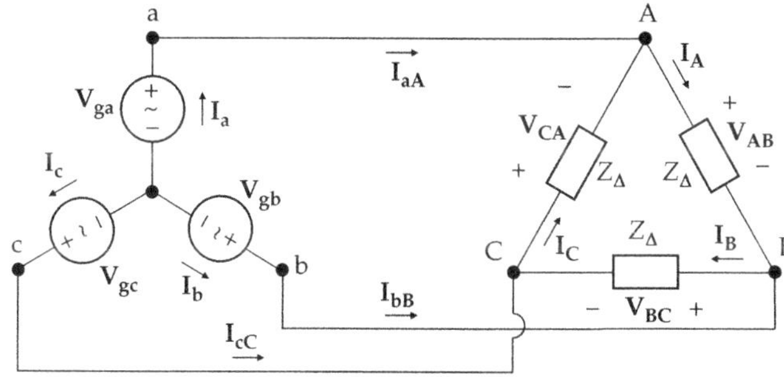

FIGURE 25.9
Balanced Y-Δ connection.

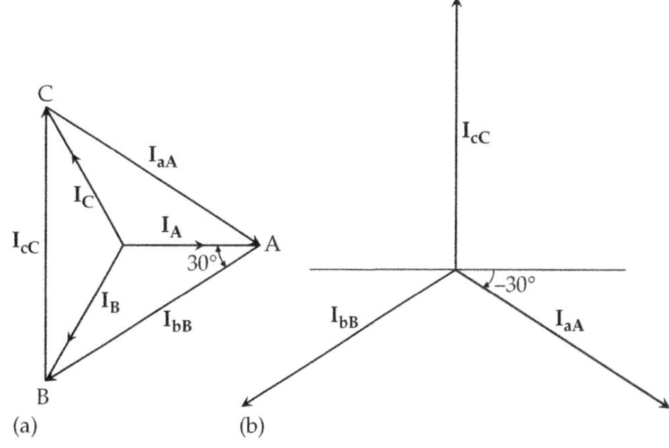

FIGURE 25.10
(a) Phase and (b) line currents in a Δ connection.

25.3.3 Power Relations

Consider the load in Figure 25.9. If $Z_\Delta = |Z_\Delta|\angle\theta$, the phase currents are related to the corresponding phase voltages as follows, assuming the voltage in phase 'A' to be zero:

$$\mathbf{I}_A = \frac{\mathbf{V}_{AB}}{Z_\Delta} = \frac{V_m\angle 0°}{|Z_\Delta|\angle\theta} = \frac{V_m}{|Z_\Delta|}\angle -\theta \tag{25.26}$$

$$\mathbf{I}_B = \frac{\mathbf{V}_{BC}}{Z_\Delta} = \frac{V_m\angle -120°}{|Z_\Delta|\angle\theta} = \frac{V_m}{|Z_\Delta|}\angle(-120° - \theta) \tag{25.27}$$

$$\mathbf{I}_C = \frac{\mathbf{V}_{CA}}{Z_\Delta} = \frac{V_m\angle +120°}{|Z_\Delta|\angle\theta} = \frac{V_m}{|Z_\Delta|}\angle(120° - \theta) \tag{25.28}$$

Let the rms value of the phase voltage be $V_{\phi rms}$ and the rms value of the phase current be $I_{\phi rms}$. The real power per phase is $P_\phi = V_{\phi rms} \times I_{\phi rms}\cos\theta$. The total real power in the three phases of the load is $P_T = 3P_\phi$, which may be written as

$$P_T = 3V_{\phi rms} \times I_{\phi rms}\cos\theta = 3V_{lrms}\frac{I_{lrms}}{\sqrt{3}}\cos\theta = \sqrt{3}V_{lrms}I_{lrms}\cos\theta \tag{25.29}$$

where $V_{lrms} = V_{\phi rms}$ and $I_{lrms} = \sqrt{3}I_{\phi rms}$. Comparing Equations 25.20 and 25.29, the expression for the total real power is the same in both cases. Following the same argument as in the case of the Y connection, the total reactive power and complex power are given by Equations 25.21 and 25.22, respectively. The power factor of the load is again the same as that of the phase impedance.

Summary: *For both balanced Y and Δ connections:*

$$\text{Real power} = \sqrt{3} \times (\text{rms line voltage}) \times (\text{rms line current}) \times (\text{p.f. of load phase})$$

$$\text{Reactive power} = \sqrt{3} \times (\text{rms line voltage}) \times (\text{rms line current}) \times (\text{reactive factor of load phase})$$

$$\text{Apparent power} = \sqrt{3} \times (\text{rms line voltage}) \times (\text{rms line current})$$

Example 25.2: Phase Currents in a Δ-Connected Load

Assume that the load in Example 25.1 is Δ connected (Figure 25.9). Determine the load phase currents.

Solution:

The phase currents have a magnitude of $100/\sqrt{3}$ A rms. According to Equations 25.25, the line currents lag the corresponding phase currents by 30°, or the phase currents *lead* the corresponding line currents by 30°. Hence, using the same phase angles of the line currents as in Example 25.1:

$$\mathbf{I}_{AB} = \frac{100}{\sqrt{3}}\angle(30° - \theta)\ \text{A},\quad \mathbf{I}_{BC} = \frac{100}{\sqrt{3}}\angle(-90° - \theta)\ \text{A},$$

$$\mathbf{I}_{CA} = \frac{100}{\sqrt{3}}\angle(150° - \theta)\ \text{A}$$

all being rms values.

Problem-Solving Tip

- In a Δ connection, if the line currents are known, the phase currents readily follow, since the phase current designated by the first subscript of the line current leads the line current by 30° and has $1/\sqrt{3}$ of its magnitude.

Primal Exercise 25.7

Assume that $\mathbf{I}_{aA} = 17.3\angle 0°$ A in Figure 25.9 and that the load is Δ connected, the impedance per phase being 3 + *j*4 Ω. Determine (a) the phase currents and (b) the line voltages, assuming a positive phase sequence.

Ans. (a) $\mathbf{I}_A = 10\angle 30°$ A, $\mathbf{I}_B = 10\angle -90°$ A, $\mathbf{I}_C = 10\angle 150°$ A; (b) $\mathbf{V}_{AB} = 50\angle 83.1°$ V, $\mathbf{V}_{BC} = 50\angle -36.9°$ V, $\mathbf{V}_{CA} = 50\angle -156.9°$ V.

Primal Exercise 25.8

Assume that in Figure 25.9, the load absorbs 9 kW at a lagging power factor of 0.8. If $\mathbf{V}_{AB} = 100\angle 60°$ V rms, determine (a) $\mathbf{I}_{aA}$ and (b) the complex power delivered by the source assuming the line impedance is 0.1 + *j*0.5 Ω.

Ans. (a) $65.0\angle -6.9°$ A rms; (b) 10.3 + *j*13.1 kVA.

Exercise 25.9

Determine the line currents in Figure 25.9, assuming a negative phase sequence.

Ans. $\mathbf{I}_{aA} = \sqrt{3}I_m\angle 30°$, $\mathbf{I}_{cC} = \sqrt{3}I_m\angle -90°$, $\mathbf{I}_{bB} = \sqrt{3}I_m\angle 150°$.

Exercise 25.10

Repeat Exercise 25.7 assuming a negative phase sequence.

Ans. (a) $\mathbf{I}_A = 10\angle -30°$ A, $\mathbf{I}_C = 10\angle -150°$ A, $\mathbf{I}_B = 10\angle 90°$ A.

25.4 Analysis of Balanced Three-Phase Systems

Three-phase systems, balanced or unbalanced, may be analyzed using the general methods discussed in previous chapters. The mesh-current method is advantageous since a three-wire Y–Y system has only two meshes, a Δ–Δ system four meshes. However, it is generally simpler to analyze balanced, three-phase systems by reducing them to a single-phase equivalent circuit based on a Y–Y representation.

25.4.1 Y–Y System

A balanced Y–Y system is shown in Figure 25.11a. Since nodes 'n' and 'N' are at the same voltage,

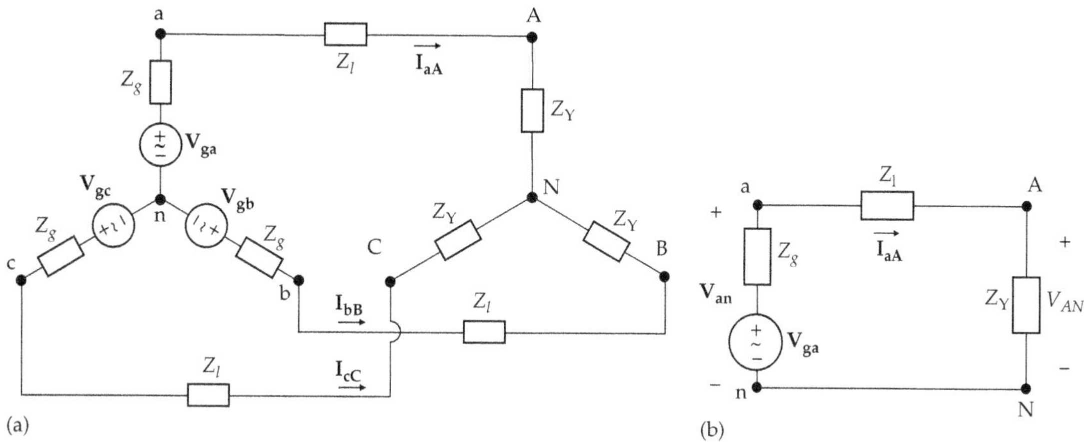

FIGURE 25.11
Single-phase equivalent circuit of Y–Y connection. (a) Y-connected three-phase generator and load and (b) single-phase equivalent circuit for phase 'a'.

as explained in connection with Figure 25.7, KVL around loop 'naAN' gives

$$\left(Z_g + Z_l + Z_Y\right)\mathbf{I}_{aA} = \mathbf{V}_{ga} \qquad (25.30)$$

Equation 25.30 also applies to the circuit shown in Figure 25.11b, which is the single-phase equivalent circuit for phase 'a'. Once $\mathbf{I}_{aA}$ is determined from Equation 25.30, the phase voltages $\mathbf{V}_{an}$ and $\mathbf{V}_{AN}$ are readily obtained as

$$\mathbf{V}_{an} = \mathbf{V}_{ga} - Z_g\mathbf{I}_{aA}, \quad \mathbf{V}_{AN} = Z_Y\mathbf{I}_{aA} \qquad (25.31)$$

Since all phase variables in a balanced three-phase system are balanced sets, then once $\mathbf{I}_{aA}$, $\mathbf{V}_{an}$, and $\mathbf{V}_{AN}$ are determined, the corresponding variables for the other two phases immediately follow; they have the same magnitude but are phase shifted by 120°. The line voltages are obtained as discussed in Section 25.2.

Example 25.3: Analysis of Balanced Y–Y System

It is required to analyze the balanced Y–Y system shown in Figure 25.11a given that $\mathbf{V}_{ga} = 120\angle 0°$ V, $\mathbf{V}_{gb} = 120\angle -120°$ V, $\mathbf{V}_{gc} = 120\angle 120°$ V, all rms values, $Z_g = 1\ \Omega$, $Z_l = j2\ \Omega$, and the load phase impedance consists of a resistance of 10 Ω in parallel with a capacitive reactance of $-j10\ \Omega$.

Solution:

The single-phase equivalent circuit is shown in Figure 25.12. The load impedance Z_Y is the parallel combination of 10 Ω and $-j10\ \Omega$:

$$Z_Y = \frac{10(-j10)}{10-j10} = \frac{-j100(10+j10)}{200} = 5 - j5 = 5\sqrt{2}\angle -45°\ \Omega.$$

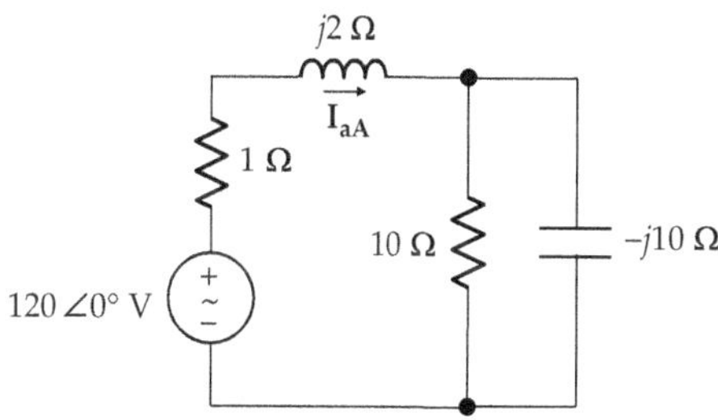

FIGURE 25.12
Figure for Example 25.3.

Hence, $\mathbf{I}_{aA} = \dfrac{120\angle 0°}{1+j2+5-j5} = \dfrac{120\angle 0°}{6-j3}$ A rms. It follows that

$$\mathbf{I}_{aA} = 16 + j8 = 17.9\angle 26.6° \text{ A rms}$$

$$\mathbf{I}_{bB} = 17.9\angle\left(26.6° - 120°\right) = 17.9\angle -93.4° \text{ A rms}$$

$$\mathbf{I}_{cC} = 17.9\angle\left(-93.4° - 120°\right) = 17.9\angle 146.6° \text{ A rms}$$

$$\mathbf{V}_{an} = \mathbf{V}_{ga} - Z_g\mathbf{I}_{aA} = 120\angle 0° - 1\times\left(16 + j8\right)$$

$$= 104.3\angle -4.40° \text{ V rms}$$

$$\mathbf{V}_{bn} = 104.3\angle\left(-4.40° - 120°\right) = 104.3\angle -124.4° \text{ V rms}$$

$$\mathbf{V}_{cn} = 104.3\angle\left(-124.4° - 120°\right) = 104.3\angle 115.6° \text{ V rms}$$

$$\mathbf{V}_{ab} = \sqrt{3}\mathbf{V}_{an}e^{j30°} = 180.7\angle 25.6° \text{ V rms}$$

$$\mathbf{V}_{bc} = 180.7\angle\left(25.6° - 120°\right) = 180.7\angle -94.4° \text{ V rms}$$

$$\mathbf{V}_{ca} = 180.7\angle\left(-94.4° - 120°\right) = 180.7\angle 145.6° \text{ V rms}$$

$$\mathbf{V}_{AN} = Z_Y\mathbf{I}_{aA} = 5\sqrt{2}\angle -45° \times 17.9\angle 26.6°$$

$$= 126.5\angle -18.4° \text{ V rms}$$

$$\mathbf{V}_{BN} = 126.5\angle(-18.4° - 120°) = 126.5\angle -138.4° \text{ V rms}$$

$$\mathbf{V}_{CN} = 126.5\angle(-138.4° - 120°) = 126.5\angle 101.6° \text{ V rms}$$

$$\mathbf{V}_{AB} = \sqrt{3}\mathbf{V}_{AN}e^{j30°} = 219.1\angle 11.6° \text{ V rms}$$

$$\mathbf{V}_{BC} = 219.1\angle(11.6° - 120°) = 219.1\angle -108.4° \text{ V rms}$$

$$\mathbf{V}_{CA} = 219.1\angle(-108.4° - 120°) = 219.1\angle 131.6° \text{ V rms}$$

Problem-Solving Tip

- The equivalent single-phase equivalent circuit provides a convenient means of analyzing Y–Y-connected, balanced, three-phase systems.

25.4.2 Δ–Δ System

When the generator, load, or both are Δ connected, the procedure is to transform the system to a Y–Y system, which is analyzed as discussed previously. Any desired phase current of the Δ-connected load can then be derived.

A balanced Δ–Δ system is shown in Figure 25.13a. The Δ-connected load may be readily transformed to an equivalent Y simply by substituting $Z_Y = Z_\Delta/3$. To transform the Δ-connected generator to its equivalent Y, we have to derive its TEC, which can be done from equivalence between every two of the three terminals 'abc' under two sets of conditions: (1) the same open-circuit line voltages (Figure 25.13b) and (2) the same equivalent impedance with the ideal voltage sources set to zero (Figure 25.13c). The latter condition is the usual Y equivalent of Δ-connected impedances, which is a set of impedances $Z_{gY} = Z_g/3$. To satisfy the first condition, we have to determine $\mathbf{I}_\Delta$, the current that circulates in the Δ under open-circuit conditions. From Figure 25.13b,

$$\mathbf{I}_\Delta = \frac{1}{3Z_g}\left(\mathbf{V}_{ga} + \mathbf{V}_{gb} + \mathbf{V}_{gc}\right) \tag{25.32}$$

The numerator is the sum of balanced three-phase voltages, which is zero. Hence, $\mathbf{I}_\Delta = 0$. It follows that the line voltages under open-circuit conditions are equal to the phase voltages:

$$\mathbf{V}_{aboc} = \mathbf{V}_{ga}, \quad \mathbf{V}_{bcoc} = \mathbf{V}_{gb}, \quad \mathbf{V}_{caoc} = \mathbf{V}_{gc} \tag{25.33}$$

From Equations 25.15, the corresponding Y generator phase voltages are $1/\sqrt{3}$ times the magnitude and *lag* the line voltages by 30°. Thus,

$$\mathbf{V}_{gaY} = \frac{\mathbf{V}_{ga}}{\sqrt{3}}e^{-j30°}, \quad \mathbf{V}_{gbY} = \frac{\mathbf{V}_{gb}}{\sqrt{3}}e^{-j30°}, \quad \mathbf{V}_{gcY} = \frac{\mathbf{V}_{gc}}{\sqrt{3}}e^{-j30°} \tag{25.34}$$

The equivalent Y connection will then have in each phase the corresponding voltage source in series with Z_{gY}, as illustrated in Figure 25.14. Clearly, this Y connection satisfies the required conditions: under open circuit, the line voltages $\mathbf{V}_{ab}$, $\mathbf{V}_{bc}$, and $\mathbf{V}_{ca}$ are the same as in Figure 25.13b; if the sources are set to zero in both cases, the resulting Y and Δ are equivalent, since $Z_{gY} = Z_g/3$. The system of Figure 25.14 can now be analyzed as was done for the system of Figure 25.11a. Once the line currents are found, the phase currents can be determined, as illustrated by the following example.

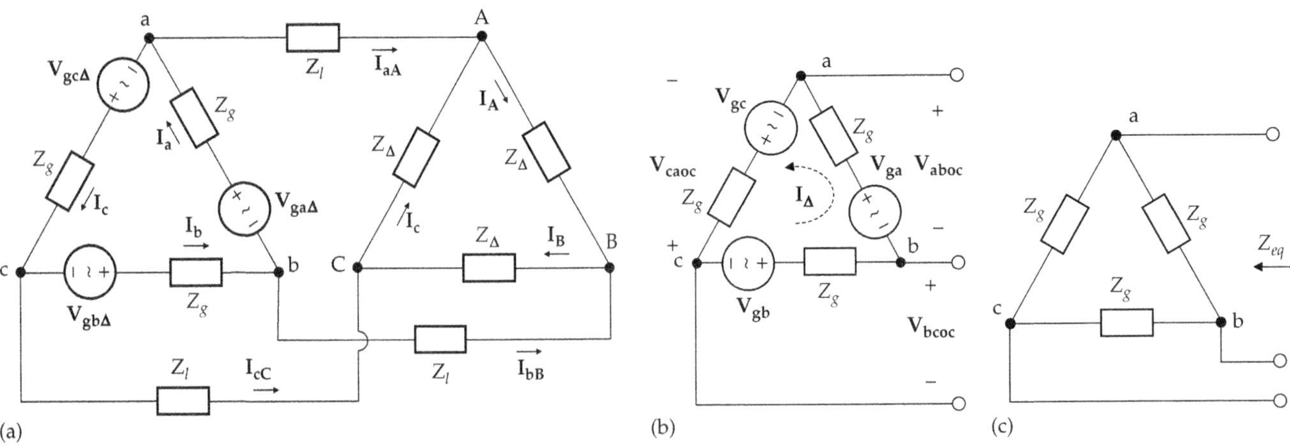

FIGURE 25.13
Thevenin's equivalent circuit of Δ–Δ connection. (a) Δ-connected three-phase generator and load, (b) the Δ-connected three-phase generator under open-circuit conditions, and (c) the Δ-connected three-phase generator with the ideal voltage source elements set to zero.

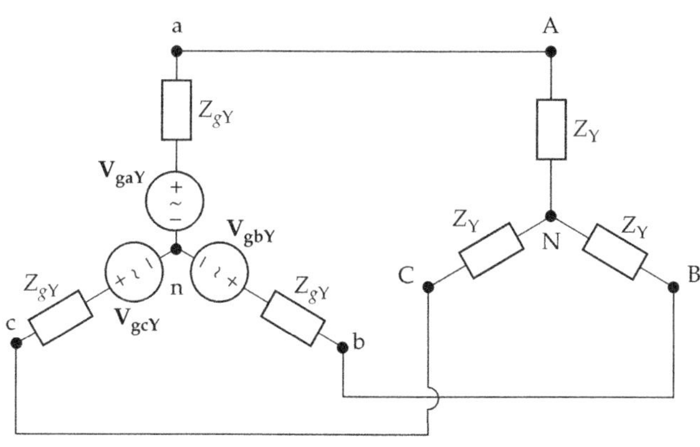

FIGURE 25.14
Y–Y connection equivalent to Δ–Δ connection.

Example 25.4: Analysis of Balanced Δ–Δ System

It is required to analyze the balanced Δ–Δ system shown in Figure 25.13a, where $\mathbf{V}_{ga\Delta} = 190\angle 120^\circ$ V, $\mathbf{V}_{gb\Delta} = 190\angle 0^\circ$ V, $\mathbf{V}_{gc\Delta} = 190\angle -120^\circ$ V, all rms values, $Z_{g\Delta} = 1.5 + j1.5\ \Omega$, $Z_l = 0.5 + j0.5$, and the load phase impedance Z_Δ consists of a 12 Ω resistance in parallel with a capacitive reactance of −12 Ω.

Solution:

Considering the load first, the phase impedance is $Z_\Delta = \dfrac{12(-j12)}{12-j12} = 6 - j6 = 6\sqrt{2}\angle -45^\circ\ \Omega$. The equivalent Y has a phase impedance of $Z_Y = 2 - j2 = 2\sqrt{2}\angle -45^\circ\ \ \Omega$.

As for the generator, the equivalent Y has a phase impedance $Z_{gY} = 0.5 + j0.5\ \Omega$. The ideal voltage sources are (Equation 25.34)

$$\mathbf{V}_{gaY} = \frac{190}{\sqrt{3}}\angle(120^\circ - 30^\circ) = 110\angle 90^\circ \text{ V rms}$$

$$\mathbf{V}_{gbY} = \frac{190}{\sqrt{3}}\angle(0^\circ - 30^\circ) = 110\angle -30^\circ \text{ V rms}$$

$$\mathbf{V}_{gcY} = \frac{190}{\sqrt{3}}\angle(-120^\circ - 30^\circ) = 110\angle -150^\circ \text{ V rms}$$

This resulting Y-Y system is shown in Figure 25.15a. The relationships between the voltage sources in the given Δ and the equivalent Y are shown in Figure 25.15b. Note the correspondence between the phasors in Figure 25.15b and the way the Δ and Y are drawn in Figures 25.13a and 25.15a.

The equivalent single-phase diagram for phase 'a' is shown in Figure 25.15c. The total impedance is $(0.5 + 0.5 + 2) + j(0.5 + 0.5 - 2) = (3 - j)\ \Omega$. The line currents are

$$\mathbf{I}_{aA} = \frac{110\angle 90^\circ}{3-j} = \frac{110\angle 90^\circ}{\sqrt{10}\angle -18.4^\circ} = 34.8\angle 108.4^\circ \text{ A rms}$$

$$\mathbf{I}_{bB} = 34.8\angle(108.4^\circ - 120^\circ) = 34.8\angle -11.6^\circ \text{ A rms}$$

$$\mathbf{I}_{cC} = 34.8\angle(-11.6^\circ - 120^\circ) = 34.8\angle -131.6^\circ \text{ A rms}$$

From Equations 25.25, the phase currents of the load are

$$\mathbf{I}_A = \frac{\mathbf{I}_{aA}}{\sqrt{3}}\angle(108.4^\circ + 30^\circ) = 20.1\angle 138.4^\circ \text{ A rms}$$

$$\mathbf{I}_B = \frac{\mathbf{I}_{bB}}{\sqrt{3}}\angle(-11.6^\circ + 30^\circ) = 20.1\angle 18.4^\circ \text{ A rms}$$

$$\mathbf{I}_C = \frac{\mathbf{I}_{cC}}{\sqrt{3}}\angle(-131.6^\circ + 30^\circ) = 20.1\angle -101.6^\circ \text{ A rms}$$

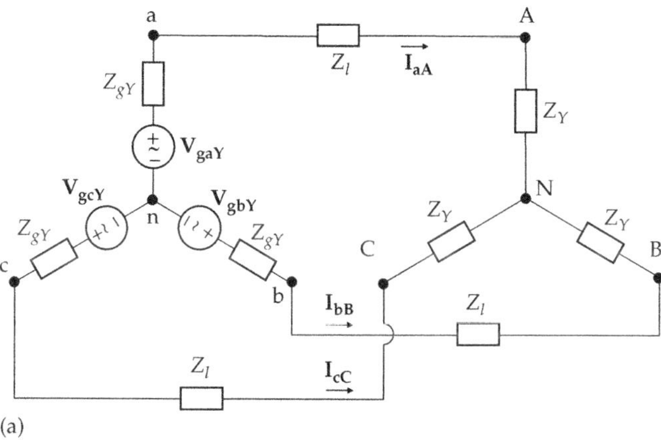

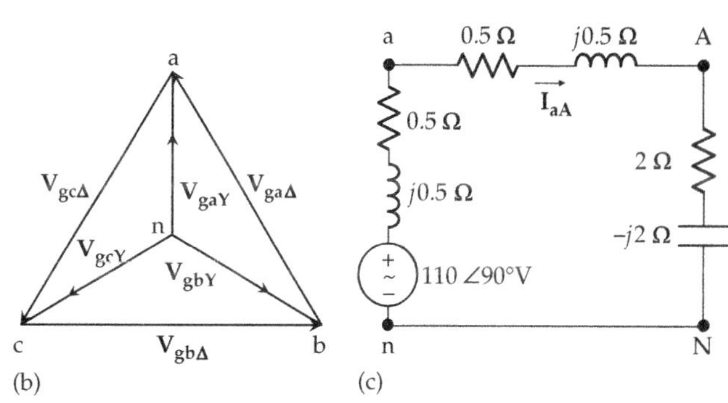

FIGURE 25.15
Figure for Example 25.4.

With the assignment of positive directions of phase currents in the generator and load as in Figure 25.13a, the corresponding phase currents are equivalent, since for the load $\mathbf{I}_{aA} = \mathbf{I}_A - \mathbf{I}_C$, for example, whereas for the generator $\mathbf{I}_{aA} = \mathbf{I}_a - \mathbf{I}_c$. It follows that $\mathbf{I}_a = \mathbf{I}_A$, $\mathbf{I}_b = \mathbf{I}_B$, and $\mathbf{I}_c = \mathbf{I}_C$.

The line voltages at the load are

$$\mathbf{V}_{AB} = Z_\Delta I_A = 6\sqrt{2}\angle -45° \times 20.1\angle 138.4° = 170.4\angle 93.4° \text{ V rms}$$

$$\mathbf{V}_{BC} = 170.5\angle(93.4° - 120°) = 170.4\angle -26.6° \text{ V rms}$$

$$\mathbf{V}_{CA} = 170.5\angle(-26.6° - 120°) = 170.4\angle -146.6° \text{ V rms}$$

From Figure 25.15c, the phase voltage is $\mathbf{V}_{an} = (2.5 - j1.5) \times \mathbf{I}_{aA} = 2.92\angle -31° \times 34.8\angle 108.4° = 101.4\angle 77.4°$ V rms, considering the voltage drop from node 'a' to node 'N' through node 'A'. It follows from Equations 25.15 that

$$\mathbf{V}_{ab} = \sqrt{3}\mathbf{V}_{an}e^{j30°} = 175.7\angle 107.4° \text{ V rms}$$

$$\mathbf{V}_{bc} = 176\angle(107.4° - 120°) = 175.7\angle -12.6° \text{ V rms}$$

$$\mathbf{V}_{ca} = 176\angle(-12.6° - 120°) = 175.7\angle -132.6° \text{ V rms}$$

Problem-Solving Tip

- The single-phase equivalent circuit provides a convenient means of analyzing Δ–Δ-connected, balanced, three-phase systems.

Primal Exercise 25.11

Consider the balanced Δ–Δ system of Figure 25.16. (a) Determine the load phase currents $\mathbf{I}_{AB}$, $\mathbf{I}_{BC}$, and $\mathbf{I}_{CA}$ from the corresponding line voltages and the load impedance per phase; (b) determine the line currents $\mathbf{I}_{aA}$, $\mathbf{I}_{bB}$, and $\mathbf{I}_{cC}$ from the phase currents and by using the single-phase equivalent circuit.

Ans. (a) $\mathbf{I}_{AB} = 100\angle -22.6°$ A, $\mathbf{I}_{BC} = 100\angle -142.6°$ A, $\mathbf{I}_{CA} = 100\angle 97.4°$ A; (b) $\mathbf{I}_{aA} = 100\sqrt{3}\angle -52.6°$ A, $\mathbf{I}_{bB} = 100\sqrt{3}\angle -172.6°$ A, $\mathbf{I}_{cC} = 100\sqrt{3}\angle 67.4°$ A.

Primal Exercise 25.12

Determine $\mathbf{I}_{aA}$ in Figure 25.16 if each line has an impedance of $(2 + j2)$ Ω.

Ans. $163.3\angle -53.96°$ A.

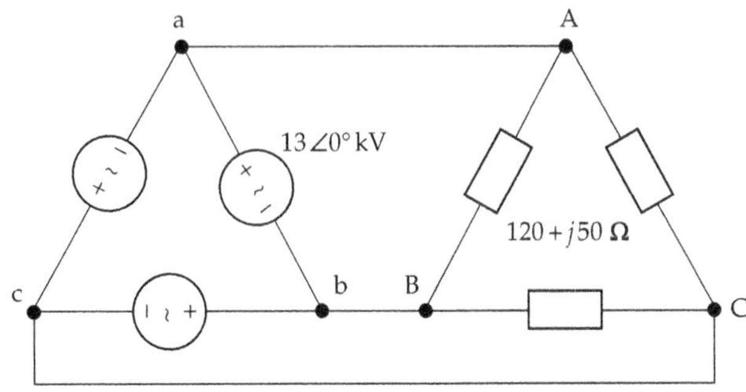

FIGURE 25.16
Figure for Primal Exercise 25.11.

Example 25.5: Analysis of Balanced Y–Δ System

It is required to determine (a) $\mathbf{I}_{aA}$, (b) $\mathbf{I}_{CA}$, and (c) $\mathbf{V}_{AB}$ in the balanced three-phase system of Figure 25.17, assuming the frequency is 50 Hz.

Solution:

$Z_Y = 10 + (j\omega \times 0.12)/3 = 10 + j100\pi \times 0.04 = 10 + j12.57$ Ω. The single-phase equivalent circuit is shown in Figure 25.18. It follows that

$$\text{(a) } \mathbf{I}_{aA} = \frac{100\angle 0°}{11 + j12.57} = \frac{100\angle 0°}{16.7\angle 48.8°} = 5.988\angle -48.8° \text{ A.}$$

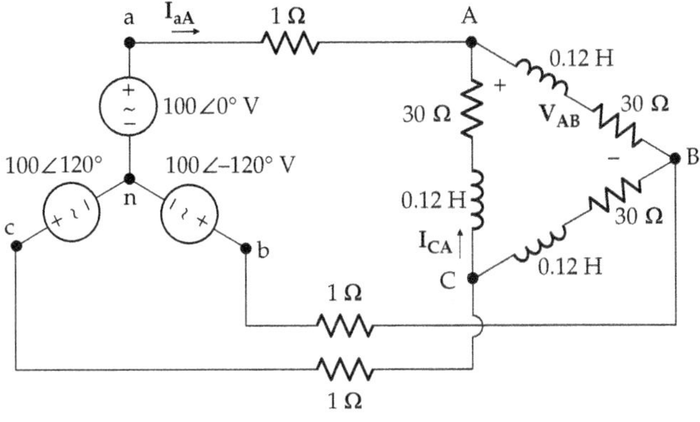

FIGURE 25.17
Figure for Example 25.5.

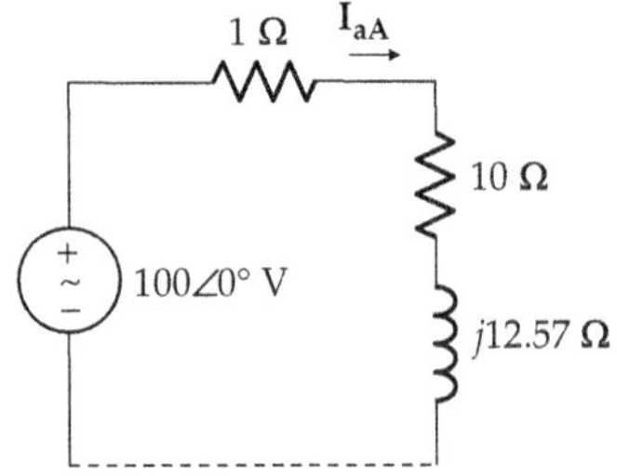

FIGURE 25.18
Figure for Example 25.5.

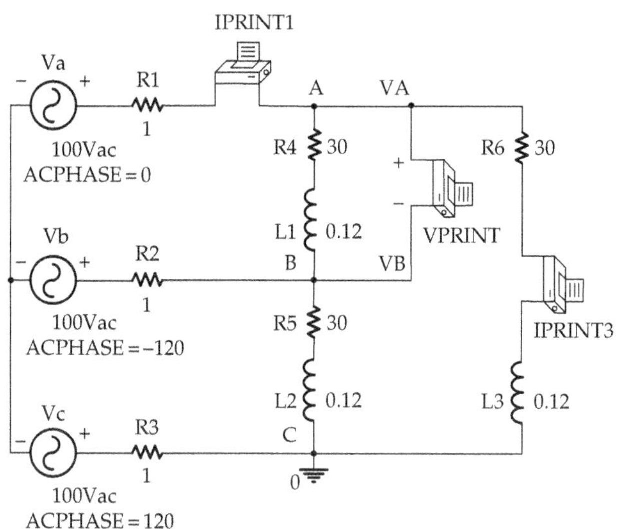

FIGURE 25.19
Figure for Example 25.5.

(b) $\mathbf{I_{AB}} = \frac{1}{\sqrt{3}}\mathbf{I_{aA}}\angle 30° = 3.457\angle -18.8°$ A; $\mathbf{I_{CA}} = \mathbf{I_{AB}}\angle 120°$
$= 3.457\angle 101.2°$ A.

(c) $Z_\Delta = 3(10 + j12.57)\ \Omega = 48.18\angle 51.49°\ \Omega$; $\mathbf{V_{AB}} = \mathbf{I_{AB}}Z_\Delta = 166.6\angle 32.69°$ V.

Simulation: The schematic is shown in Figure 25.19. Two IPRINT printers are used to indicate the line current $\mathbf{I_{aA}}$ and the phase current $\mathbf{I_{CA}}$. A VPRINT printer is used to indicate the voltage $\mathbf{V_{AB}}$. After the simulation is run, the following values are read from the output file: $\mathbf{I_{aA}} = 5.988\angle -48.8°$ A, $\mathbf{I_{CA}} = 3.457\angle 101.2°$ A, and $\mathbf{V_{AB}} = 166.6\angle 32.69°$ V.

Exercise 25.13

Determine the reading of the ideal voltmeter in Figure 25.20 if the line voltage is 220 V.

Ans. 190.5 V.

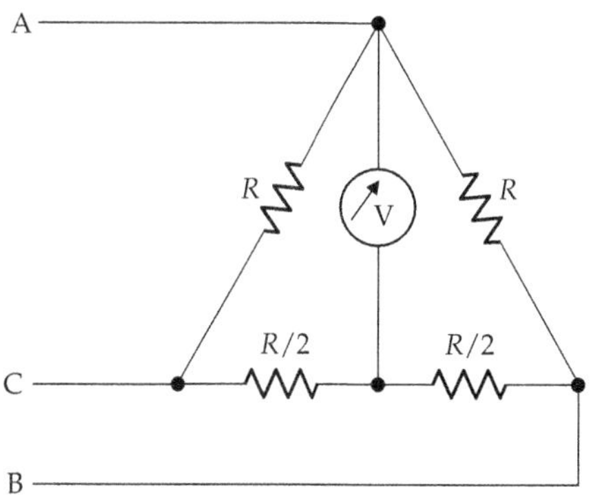

FIGURE 25.20
Figure for Exercise 25.13.

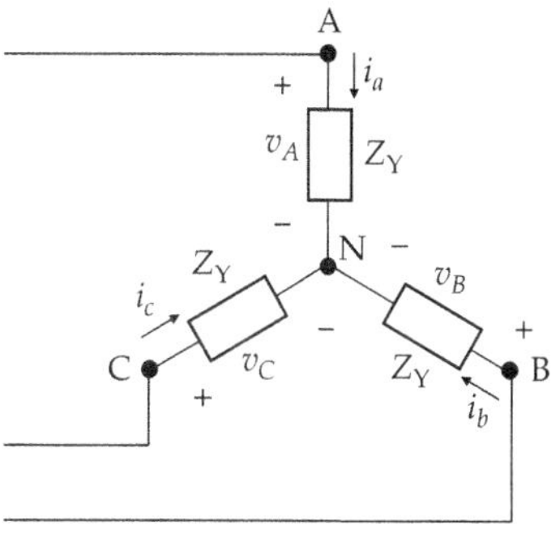

FIGURE 25.21
Instantaneous power.

25.5 Power in Balanced Three-Phase Systems

25.5.1 Instantaneous Power

Consider, for the sake of argument, a balanced, Y-connected load (Figure 25.21). If the instantaneous phase voltages are denoted by $v_A = \sqrt{2}V_{\phi rms}\cos\omega t$, $v_B = \sqrt{2}V_{\phi rms}\cos(\omega t - 120°)$, and $v_C = \sqrt{2}V_{\phi rms}\cos(\omega t + 120°)$, the instantaneous phase, or line, currents lag the corresponding phase voltages by the phase angle θ of Z_Y, assuming the load is inductive. The instantaneous power in each of the three phases is

$$\begin{aligned} p_A(t) &= 2V_{\phi rms}I_{\phi rms}\cos\omega t\cos(\omega t - \theta) \\ &= V_{\phi rms}I_{\phi rms}\left[\cos(2\omega t - \theta) + \cos\theta\right] \end{aligned} \tag{25.35}$$

$$\begin{aligned} p_B(t) &= 2V_{\phi rms}I_{\phi rms}\cos(\omega t - 120°)\cos(\omega t - 120° - \theta) \\ &= V_{\phi rms}I_{\phi rms}\left\{\cos\left[(2\omega t - \theta) + 120°\right] + \cos\theta\right\} \end{aligned} \tag{25.36}$$

$$\begin{aligned} p_C(t) &= 2V_{\phi rms}I_{\phi rms}\cos(\omega t + 120°)\cos(\omega t + 120° - \theta) \\ &= V_{\phi rms}I_{\phi rms}\left\{\cos\left[(2\omega t - \theta) - 120°\right] + \cos\theta\right\} \end{aligned} \tag{25.37}$$

In each phase, the instantaneous power has a steady component and an alternating component, just as in the single-phase case. The difference, however, is that the alternating components in the three phases form a balanced set. So when p_A, p_B, and p_C are added to obtain the total power p_T delivered through the lines, the alternating components sum to zero, whereas the steady components add to give

$$P_T = 3V_{\phi rms}I_{\phi rms}\cos\theta = \sqrt{3}V_{lrms}I_{lrms}\cos\theta \qquad (25.38)$$

where

$I_{lrms} = I_{\phi rms}$ is the rms value of the line current
V_{lrms} is the rms of the line voltage

The total instantaneous power is thus equal to the expression for the real power derived earlier for the Y and Δ connections.

Concept: *The total instantaneous power delivered to the balanced three-phase load is steady with respect to time; it is not pulsating as in the single-phase case.*

It is important to be clear about power relations in a three-phase system. Referring to Figure 25.21, the power relations per phase are exactly those for the single-phase case considered in Section 17.1. Thus, Equation 17.14, with $\theta_v = 0$ and $\theta_i = -\theta$, gives for the real power in terms of rms values per phase: $V_{\phi rms}I_{\phi rms}\cos\theta\,[1 + \cos 2(\omega t - \theta)]$. Equation 17.17 gives for the reactive power $-V_{\phi rms}I_{\phi rms}\sin\theta\sin 2(\omega t - \theta)$. Both the real power per phase and the reactive power per phase have time-varying alternating components. Adding these two components gives the instantaneous power per phase p_A (Equation 25.35), also having an alternating component that is present in phase 'A' *and* in the corresponding line. However, the alternating components of the three phases form a balanced set and cancel out for the three-phase load *as a whole*, leaving a steady component equal to the real power.

An important practical consequence of the steadiness of instantaneous power is that, because the electrical power input to a three-phase motor is steady, the output torque is ideally steady at constant speed. In contrast, the output torque of a single-phase motor inherently has a pulsating component that produces vibrations.

25.5.2 Complex Power

By analogy with the single-phase case, the alternating components of Equations 25.35 to 25.37 may be interpreted in terms of complex power in each phase. A power triangle may be constructed, as in the single-phase case, where the apparent power per phase $|S_\phi| = V_\phi I_\phi$ is the hypotenuse of a right triangle whose sides are the real power per phase, $P_\phi = V_\phi I_\phi\cos\theta$, and the reactive power per phase, $Q_\phi = V_\phi I_\phi\sin\theta$ (Figure 25.22). The load impedance angle θ is the angle between the hypotenuse and the side representing P_ϕ. Each of the total real, reactive, and complex power supplied to the load is three times the corresponding quantity in each phase and is the same as that derived in Section 25.3.

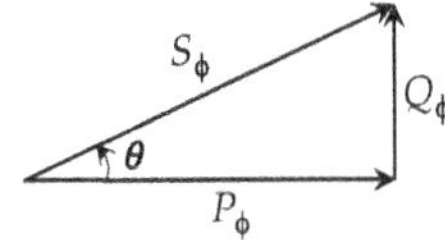

FIGURE 25.22
Complex power per phase.

As in the single-phase case, complex power can be very useful in solving three-phase problems, because of its conservation and the addition of real, reactive, and complex power branch by branch. This is illustrated by Example 25.6, which also demonstrates power factor correction in a three-phase system.

Example 25.6: Power Factor Correction Using Complex Power

A balanced three-phase system is shown in Figure 25.23a, where the load absorbs 50 kW at 0.8 p.f. lagging. A capacitor bank of three capacitors C is connected across the load terminals so that the p.f. at terminals 'abc' is unity. Given that the magnitude of the line current, with the capacitors connected, is 100 A rms, it is required to determine C assuming the frequency to be 50 Hz.

Solution:

From the complex power triangle for the load (Figure 25.23b), the reactive power of the load is $Q_L = 50 \times 0.6/0.8 = 37.5$ kVAR. The total reactive power in the line impedances is $Q_l = 3 \times (100)^2 \times 0.1 = 3$ kVAR. If the reactive power of the capacitors is Q_C, then $37.5 + 3 + Q_C = 0$, since the total reactive power at terminals 'abc' is zero. This gives $Q_C = -40.5$ kVAR.

In order to calculate C, we have to determine the line voltages at terminals 'ABC'. This can be done from the complex power. The total reactive power at ABC is $Q_{LC} = 37.5 - 40.5 = -3$ kVAR. Since the total

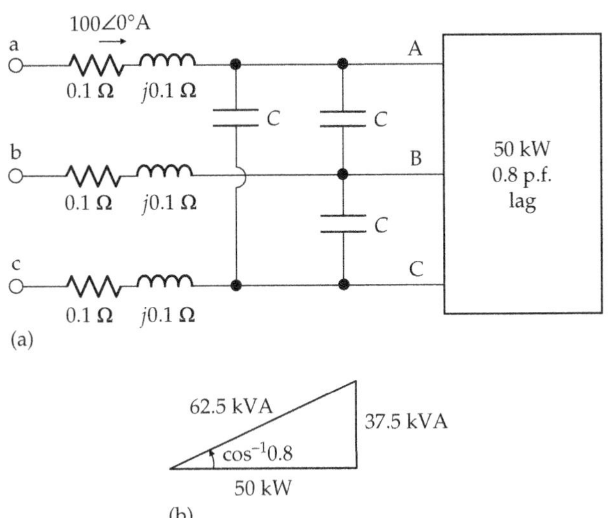

FIGURE 25.23
Figure for Example 25.6.

real power at terminals 'ABC' is 50 kW, the magnitude of the complex power at terminals 'ABC' is $S_{LC} = \sqrt{(50)^2 + (3)^2} = 50.1$ kVA. Let the line voltage at terminals 'ABC' be V_{LC} rms. Then, from Equation 25.22, $50.1 \times 10^3 = \sqrt{3} \times 100 \times V_{LC}$, which gives $V_{LC} = 289$ V rms.

The reactive power per phase of the capacitor bank is $40.5 \times 10^3/3 = 13.5 \times 10^3$ VAR. This is equal to $(V_{LC})^2 \times \omega C$, where $\omega = 2\pi \times 50 = 100\pi$ rad/s. It follows that $C = 0.514 \times 10^{-3} \equiv 514$ μF.

25.5.3 Two-Wattmeter Method of Power Measurement

In principle, the power in a three-phase system can be measured by using three wattmeters to measure the real power in each phase and adding the three readings. However, in a three-wire system only two wattmeters need to be used. This is because, as mentioned in connection with Figure 25.4, one of the lines may be considered as the return path for the currents in the two other lines. If, for the sake of argument, we consider lines 'aA' and 'cC' to be 'input' lines and line 'bB' to be the common return line, the two wattmeters may be connected as shown in Figure 25.24. The current coils are connected in accordance with the assigned positive directions of i_{aA} and i_{cC}, whereas the voltage coils are connected in accordance with the polarities of the line voltages v_{AB} and v_{CB}. The wattmeters are assumed ideal, so that the voltage drop across each current coil is zero and the current through each voltage coil is zero. The sum of the readings of the two wattmeters gives the total real power consumed by the load, irrespective of whether the load is balanced or unbalanced.

If the load is *balanced*, the readings of the two wattmeters can be readily related to the line voltages, the line currents, and the phase angle of the load. Assume, for the sake of argument, that the load is Y connected. The phasor diagram of Figure 25.8a relates the phase and line voltages. This is reproduced in Figure 25.25, with only the currents $\mathbf{I_{aA}}$ and $\mathbf{I_{cC}}$ and the voltages $\mathbf{V_{AB}}$ and $\mathbf{V_{CB}}$ indicated. The currents are shown lagging the corresponding phase voltages by θ.

The reading of each wattmeter equals the product of the rms magnitude of the current through the current coil, the rms magnitude of the voltage across the voltage coil, and the cosine of the phase angle between the current and the voltage. From the phasor diagram of Figure 25.25, the phase angle between $\mathbf{V_{AB}}$ and $\mathbf{I_{aA}}$ is $30° + \theta$, whereas the phase angle between $\mathbf{V_{CB}}$ and $\mathbf{I_{cC}}$ is $30° - \theta$. The readings of the two wattmeters are therefore

$$W_1 = V_{\text{lrms}} I_{\text{lrms}} \cos(30° + \theta) \text{ and } W_2 = V_{\text{lrms}} I_{\text{lrms}} \cos(30° - \theta) \quad (25.39)$$

The sum of the two readings is

$$W_1 + W_2 = \sqrt{3} V_{\text{lrms}} I_{\text{lrms}} \cos\theta \quad (25.40)$$

which is the same as the total real power P_T (Equation 25.38).

$\cos(30° + \theta)$ and $\cos(30° - \theta)$ are plotted in Figure 25.26 for $-90° \le \theta \le 90°$, where positive θ denotes a lagging p.f. whereas negative θ denotes a leading p.f. When $\theta = 0$, $W_1 = W_2$. If $|\theta| > 60°$, the reading of one of the wattmeters is negative, which means in practice that

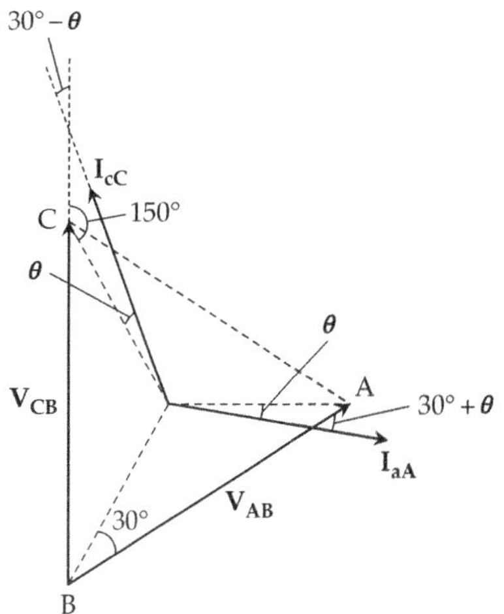

FIGURE 25.25
Phasor diagram for two-wattmeter method.

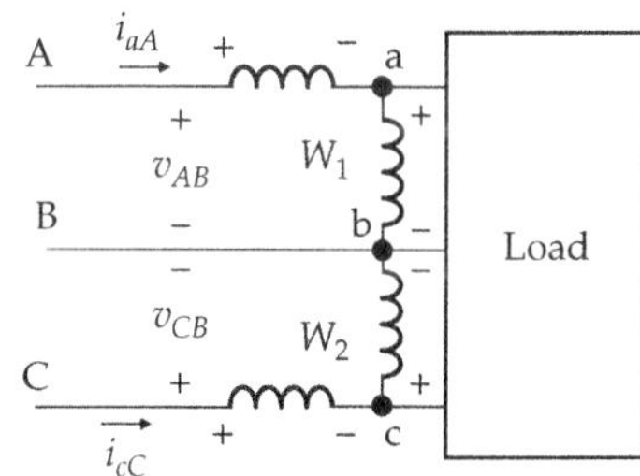

FIGURE 25.24
Two-wattmeter connection.

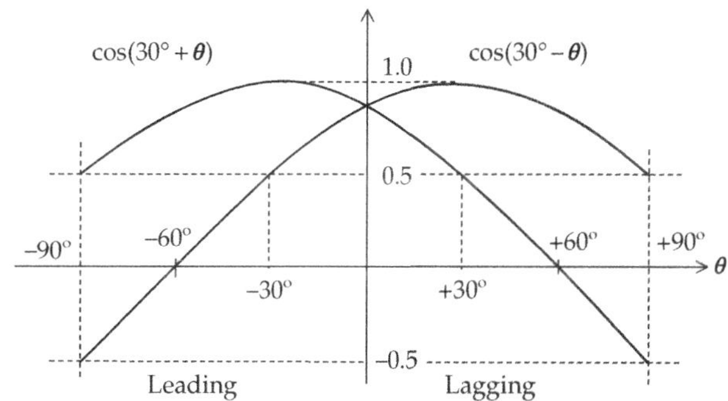

FIGURE 25.26
Variation of readings of the two wattmeters.

the connections of either the current coil or the voltage coil of that wattmeter are reversed and its reading subtracted from that of the other wattmeter. If $|\theta| = 90°$, the sum of the readings of the two wattmeters is zero.

The difference between the readings of the two wattmeters is

$$W_2 - W_1 = V_{lrms}I_{lrms}\sin\theta = \frac{Q_T}{\sqrt{3}} \quad (25.41)$$

Dividing Equation 25.41 by Equation 25.40 gives

$$\tan\theta = \sqrt{3}\frac{W_2 - W_1}{W_2 + W_1} \quad (25.42)$$

The power factor is determined from the identity $\sin^2\theta + \cos^2\theta = 1$. Dividing both sides by $\cos^2\theta$ and rearranging, the power factor is determined as

$$\text{p.f.} = \frac{1}{\sqrt{1+\tan^2\theta}} \quad (25.43)$$

Note that W_2, whose reading is proportional to $\cos(30° - \theta)$, is the wattmeter having its voltage coil connected to read a reverse line voltage, that is, $\mathbf{V}_{CB}$ instead of $\mathbf{V}_{BC}$ in Figure 25.24. This is true irrespective of which line is the common, or 'return', line.

Example 25.7: Two-Wattmeter Method of Measuring Power

In a balanced three-phase system, a load absorbs 30 kW at p.f. 0.75 lagging. If two wattmeters are connected to measure the real power in the load, as in Figure 25.24, determine the reading of each wattmeter.

Solution:

$\cos\theta = 0.75$, so $\theta = 41.4°$, and $\cos(30° + \theta) = 0.32$. The real power consumed by the load is 30 kW; hence, $30 = \sqrt{3}(V_{lrms}I_{lrms})\cos\theta$, where V_{lrms} and I_{lrms} are expressed in appropriate units to give the power in kW. From Equations 25.39, $W_1 = (V_{lrms}I_{lrms})\cos(30° + \theta)$.

Dividing these two equations gives $W_1 = \frac{30}{\sqrt{3}}\frac{\cos(30°+\theta)}{\cos\theta} = 7.5$ kW, so $W_2 = 22.5$ kW.

Problem-Solving Tip

- When determining the readings of wattmeters connected to three-phase systems, phasor diagrams of voltages and currents are very helpful.

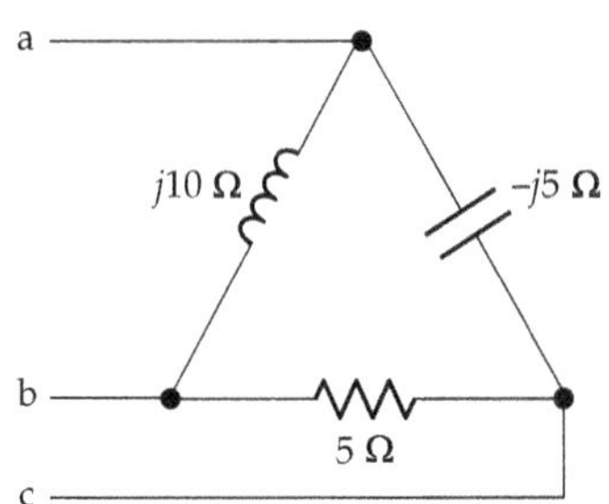

FIGURE 25.27
Figure for Primal Exercise 25.15.

Primal Exercise 25.14

The windings of a 10 kW three-phase motor are Δ connected to a balanced, three-phase supply of 380 V rms line voltage. If the line current is 19 A rms, determine the p.f. of the motor.

Ans. 0.8.

Primal Exercise 25.15

The load shown in Figure 25.27 is connected to a balanced, three-phase supply of 100 V rms line voltage. Determine the complex power absorbed by the load.

Ans. $2.24\angle-26.6°$ kVA.

Primal Exercise 25.16

The apparent power in a balanced Y-connected load is 30 kVA at a line current of 50 A rms, and the real power is 15 kW. Calculate (a) the phase voltage and (b) the impedance per phase.

Ans. (a) 200 V rms; (b) $2 + j3.46$ Ω.

Primal Exercise 25.17

A Δ-connected load is supplied from a balanced three-phase supply of 450 V rms line voltage. If the load absorbs 100 kVA at 0.65 p.f. lagging, what is the phase impedance?

Ans. $6.1\angle 49.5°$ Ω.

25.6 Advantages of Three-Phase Systems

An important advantage of three-phase systems is in power transmission. Two main "figures of merit" may be used in comparing a three-phase system with a

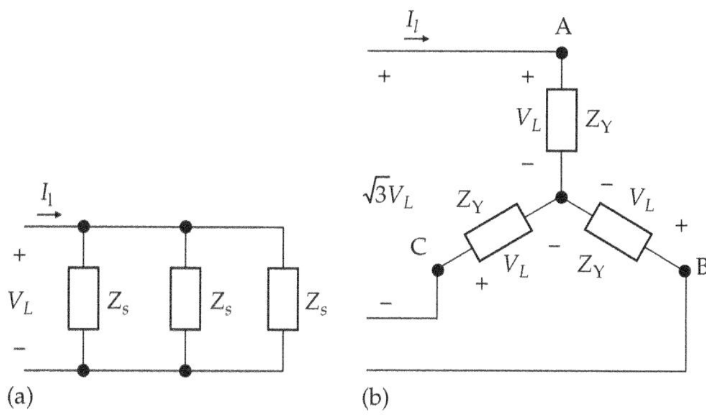

FIGURE 25.28
Comparison between (a) single-phase and (b) three-phase systems.

single-phase system that transmits the same power: the mass of line conductor used and the I^2R loss in these conductors. The mass of conductor represents an installation cost of the system; the lighter the line conductors, whether in the form of underground cables or overhead transmission lines, the lower is their cost, and, in the case of overhead transmission lines, the lighter and less expensive are the transmission towers required. The I^2R loss represents wasted power and is an operating cost of the system.

Consider a single-phase system supplying three loads Z_S (Figure 25.28a) and a three-phase system supplying three loads Z_Y connected in Y (Figure 25.28b). It is assumed that (1) the magnitude of the load voltage is V_L in both cases, which means that the magnitude of the line voltage in the Y connection is $\sqrt{3}V_L$; (2) $Z_S = 3Z_Y$, which implies that Z_S and Z_Y have the same phase angle and $|Z_S| = 3|Z_Y|$ (because $V_L = |Z_Y|I_l = |Z_S|I_l/3$, this means that the magnitude of the line current is the same in both cases); and (3) the line conductors in both systems are of the same length, type, and material, which in practice is either copper or aluminum. Because I_l is the same, this implies that the line conductors have the same current-carrying capacity and hence the same cross-sectional area, resistance, mass, and I^2R loss.

The power transmitted in the single phase is $P_S = V_L I_l \cos\theta$, and the power transmitted per line conductor is $P_{S/c} = P_S/2 = (V_L I_l \cos\theta)/2$. The power transmitted in the Y connection is $P_Y = 3V_L I_l \cos\theta$, and the power transmitted per line conductor is $P_{Y/c} = V_L I_l \cos\theta = 2P_{S/c}$. The Y connection transmits twice as much power per line conductor, or per unit mass of line conductor, as the single-phase connection. The reason for this is easy to see from Figure 25.28. *A separate return conductor is required in a single-phase system but not in a balanced three-phase system*, since in the latter case, each line conductor also acts as the return for the other two conductors. If each line conductor in the Y connection had its own return conductor, the power transmitted per line conductor will be $3V_L I_l \cos\theta/6$, which is the same as in single phase.

Since each line conductor in Figure 25.28 has the same I^2R loss, the power transmitted per kilowatt of I^2R loss varies between the two cases in the same manner as the power transmitted per line conductor. In other words, the three-phase system transmits twice as much power per kilowatt of I^2R loss as the single-phase system. Moreover, the total line voltage drop in the single-phase case has a magnitude of $2|Z_l|I_l$, where $|Z_l|$ is the magnitude of the line impedance. In the three-phase case, the line-to-line voltage drop is the line impedance multiplied by the phasor difference of any two line currents. The magnitude of this difference is $\sqrt{3}I_l$, so that the magnitude of the line-to-line voltage drop in the three-phase case is $\sqrt{3}|Z_l|I_l$, which is less than that in the single-phase case. The three-phase system of Figure 25.28b is therefore more advantageous than the single-phase system of Figure 25.28a for transmitting power in all of the respects considered.

Other advantages of three-phase systems may be listed:

1. As mentioned earlier, the total instantaneous power is steady with respect to time, which means that the output torque of a three-phase motor is steady and does not produce vibrations.
2. It can be shown that the magnetic field in a three-phase motor is rotating, so that a three-phase motor is inherently self-starting. On the other hand, the magnetic field in a single phase is pulsating rather than rotating, so that some additional means have to be provided for starting the motor.
3. Three-phase transformers, motors, and generators have smaller frame sizes than their single-phase counterpart of the same power rating because of the elimination of a separate return path for the magnetic flux.

Figure 25.29 illustrates a section through a three-phase transformer. The core has three limbs, one for each of the three-phase fluxes Φ_a, Φ_b, and Φ_c. Two sets of windings are shown around each limb, a primary winding, such as 'a_p', and a secondary winding, such as 'a_s'; similarly for phases 'b' and 'c'. Since the phase voltages form a balanced set, the three fluxes Φ_a, Φ_b, and Φ_c also form a balanced set. This means that their sum at any instant is zero, so no *separate return path is needed for each of these fluxes*. This is in contrast to three separate single-phase transformers that are connected in a three-phase system. In each transformer a return path must be provided for each flux. Consequently the three-phase transformer will have a smaller core than three separate single-phase transformers of the same voltage and total power rating.

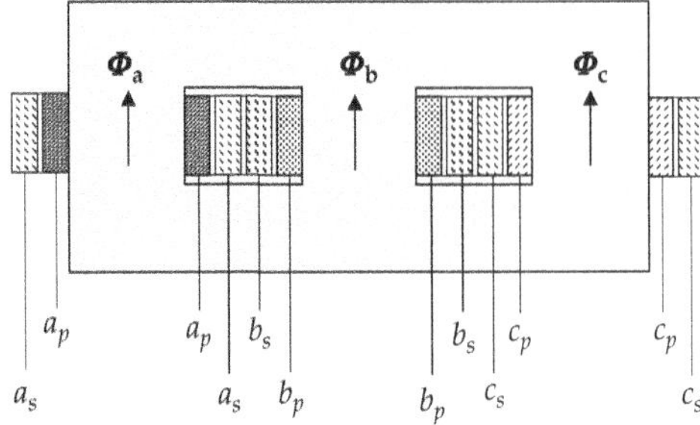

FIGURE 25.29
Cross section through the core and windings of a three-phase transformer.

25.7 Power Generation, Transmission, and Distribution

A power system may be divided into the following functional parts:

1. *Power generation*: Electric power is generated in power stations from hydroelectric power, or from burning fossil fuels, such as oil, natural gas, or coal, or from atomic energy. Electric power is being increasingly generated from solar energy and from wind power.
2. *Power transmission*: Since power stations are usually located a considerable distance from the load centers they serve, and in order to interconnect distant parts of the system, or different power systems, electric power is transmitted over long distances by means of overhead, high-voltage transmission lines.
3. *Medium-voltage distribution*: The high voltages are then stepped down to one or more medium voltages for distribution in the vicinity of load centers and within them.
4. *Low-voltage distribution*: The lowest medium voltage is eventually stepped down to a low voltage for utilization by commercial and domestic users.

In power generation, it is advantageous to have the highest practicable generator voltage, since this reduces the current for a given power rating, and hence the cross-sectional area of the conductors in the generator windings. However, too high a generator voltage requires heavy, and costly, insulation. As a compromise, three-phase voltages are usually generated at few tens of kilovolts. Power is generated at 50 Hz throughout most of the world but at 60 Hz in the United States.

For transmitting power over long distances, it is also advantageous to use the highest practicable voltages, because this reduces the current for a given transmitted power and hence the I^2R loss. Transmission voltages may be hundreds of thousands of volts or even a million volts. It is interesting to note that dc has some distinct advantages over ac for long-distance, high-voltage transmission (see Problem P25.61).

Medium-power distribution is normally at few tens of kilovolts or less, using underground cables, rather than the lower-cost overhead lines, in built-up areas. Low-voltage distribution is at standard voltages that differ between countries. In most countries, the low-voltage distribution transformer is a three-phase transformer having a Y-connected secondary, the phase voltage being 220–240 V rms, the line voltage being 380–415 V rms. The neutral point is grounded, and a ground wire connected to the neutral point is provided to consumers in some countries (Figure 25.30), in addition to the neutral line. Single-phase consumer loads are connected between line and neutral, and a three-phase supply is provided for relatively large loads.

In the United States, the low-voltage distribution transformer is usually a single-phase transformer that steps down the line voltage to a three-wire single-phase system (Figure 25.31). The midpoint of the secondary

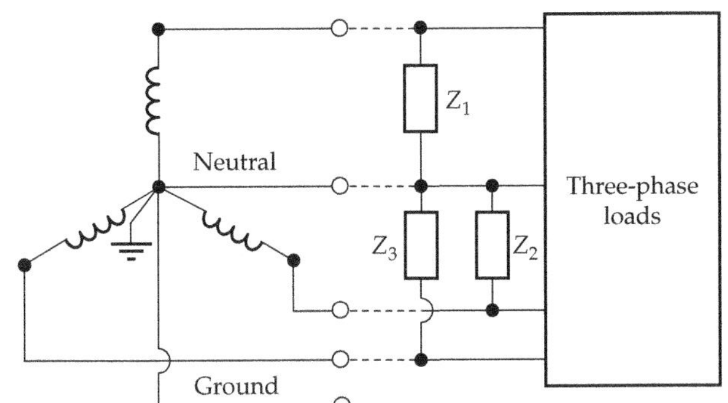

FIGURE 25.30
Three-phase, low-voltage distribution transformer.

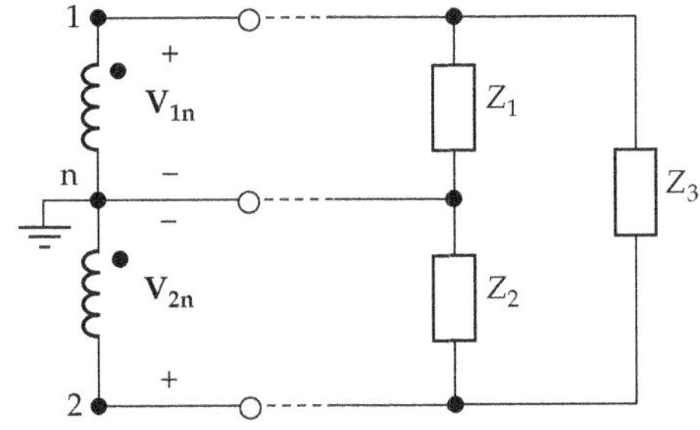

FIGURE 25.31
Single-phase, low-voltage distribution transformer.

winding is a grounded neutral point to which a neutral line is connected. The outer lines are at voltages of 110–120 V rms with respect to the neutral, and their phases are such that the voltage between lines is twice the line-to-neutral voltage, that is, 220–240 V. Higher voltage is used to supply relatively large loads, such as heaters and air-conditioning units, whereas small loads, such as lamps, are connected between one line and neutral.

Learning Checklist: What Should Be Learned from This Chapter

- The sum of a set of balanced variables is zero.
- In a balanced Y–Y connection, the neutral current is zero.
- In a balanced Y connection, the line currents are equal to the phase currents. The line voltages are balanced three-phase voltages that have $\sqrt{3}$ times the magnitude of the phase voltages and are of the same phase sequence. For a positive phase sequence, the line voltage leads by 30° the phase voltage designated by the first transcript of the line voltage, whereas for a negative phase sequence, the phase angle is 30° lag.
- In a balanced Δ connection, the line voltages are equal to the phase voltages. The line currents are balanced three-phase currents that have $\sqrt{3}$ times the magnitude of the phase currents and are of the same phase sequence. For a positive phase sequence, the line current lags by 30° the phase current designated by the subscripts of the line current. For a negative phase sequence, the phase angle is 30° lead.
- For both balanced Y and Δ connections:

$$\text{Real power} = \sqrt{3}\times(\text{rms line voltage})\times(\text{rms line current})\times(\text{p.f. of load phase}).$$

$$\text{Reactive power} = \sqrt{3}\times(\text{rms line voltage})\times(\text{rms line current})\times(\text{Reactive factor of load phase}).$$

$$\text{Apparent power} = \sqrt{3}\times(\text{rms line voltage})\times(\text{rms line current}).$$

- The total instantaneous power delivered to the balanced three-phase load is steady with respect to time; it is not pulsating as in the single-phase case.
- Three-phase real power can be measured using two wattmeters.

Problem-Solving Tips

1. When determining voltages or currents in a balanced, three-phase system, it is only necessary to evaluate the voltage or current for one phase or line. The remaining three-phase voltages and currents then follow from the fact that balanced three-phase voltages and currents have the same magnitude but with 120° phase difference between successive phases.
2. In a Y connection, if the line voltages are known, the phase voltages readily follow, since the phase voltage designated by the first subscript of the line voltage lags the line voltage by 30° and has $1/\sqrt{3}$ of its magnitude.
3. In a Δ connection, if the line currents are known, the phase currents readily follow, since the phase current designated by the first subscript of the line current leads the line current by 30° and has $1/\sqrt{3}$ of its magnitude.
4. The single-phase equivalent circuit provides a convenient means of analyzing balanced, three-phase systems, whether the generators and loads are Y or Δ connected.
5. When determining the readings of wattmeters connected to three-phase systems, phasor diagrams of voltages and currents are very helpful.

Problems

In the following problems, all voltages and currents are rms, and the phase sequence is positive, unless otherwise indicated.

Verify solutions by PSpice simulation.

Basic Y and Δ Connections

P25.1 The phase voltage of a Y-connected load supplied from a balanced three-phase system is $v_{AN}(t) = 220\cos(\omega t + 32°)$ V. Determine the line voltages, assuming a positive phase sequence.

Ans. $v_{ab}(t) = 381\cos(\omega t + 62°)$ V, $v_{bc}(t) = 381\cos(\omega t - 58°)$ V, $v_{ca}(t) = 381\cos(\omega t - 178°)$ V.

P25.2 If in the preceding problem, the phase impedance is $40\angle 20°$, determine the line currents.

Ans. $i_{aA} = 5.5\cos(\omega t + 12°)$ A, $i_{bB} = 26\sin(\omega t - 108°)$ A, $i_{cC} = 26\sin(\omega t + 132°)$ A.

P25.3 The phase current of a Δ-connected load supplied from a balanced three-phase system is $i_{AB}(t) = 15\sin(\omega t + 50°)$ A. Determine the line currents.

Ans. $i_{aA} = 26\sin(\omega t + 20°)$ A, $i_{bB} = 26\sin(\omega t - 100°)$ A, $i_{cC} = 26\sin(\omega t + 140°)$ A.

P25.4 If in the preceding problem, the phase impedance is $10\angle -20°$, determine the line voltages.

Ans. $v_{ab}(t) = 150\cos(\omega t + 30°)$ V, $v_{bc}(t) = 150\cos(\omega t - 90°)$ V, $v_{ca}(t) = 150\cos(\omega t + 150°)$ V.

P25.5 A balanced Y-connected load of $6\angle 20°$ Ω per phase is connected in parallel with a balanced Δ-connected load of $3\angle 40°$ Ω per phase. Determine the impedance per phase of the equivalent Δ-connected load.

Ans. $2.59\angle 37.2°$ Ω.

P25.6 A balanced three-phase system of line voltage 240 V supplies a parallel combination of a balanced Y-connected load and a balanced Δ-connected load, having phase impedances of $8 + j8$ Ω and $12 - j24$ Ω, respectively. Determine the line current.

Ans. $16.43\angle -11.6°$ A.

P25.7 A three-phase Δ-connected generator has an open-circuit line voltage of 380 V and a phase impedance of $0.1 + j0.5$ Ω. The generator supplies a balanced Y-connected resistive load of 50 Ω per phase. Determine the load phase voltage.

Ans. $379.7\angle -0.19°$V.

P25.8 Given a three-phase, four-wire system, the load in phase 'A' is a 360 W lamp, the current in phase 'B' is 4 A at a lagging p.f. of 0.966, and the impedance in phase 'C' is $60\angle -30°$ Ω. The supply is balanced and has a line voltage $\mathbf{V_{ab}} = 190\angle 0°$ V. Determine $\mathbf{I_{nN}}$.

Ans. $2.22\angle 29.4°$ A.

P25.9 A balanced Δ-connected load having $R_\Delta = 5$ Ω in parallel with $X_\Delta = -j5$ Ω is supplied from a balanced three-phase supply of negative phase sequence. If $\mathbf{V_{AB}} = 120\angle 0°$ V, determine $\mathbf{I_{aA}}$ and $\mathbf{I_{AB}}$.

Ans. $\mathbf{I_{AB}} = 33.9\angle 45°$ A, $\mathbf{I_{aA}} = 58.8\angle 75°$A.

P25.10 The impedances of a Δ-connected load are $Z_{AB} = 52\angle -30°$ Ω, $Z_{BC} = 52\angle 45°$ Ω, and $Z_{CA} = 104\angle 0°$ Ω. The load is supplied from a balanced three-phase supply of negative phase sequence. If $\mathbf{V_{AB}} = 208\angle 0°$ V, determine the magnitudes and phase angles of the three line currents.

Ans. $\mathbf{I_{aA}} = 5.82\angle 39.9°$ A, $\mathbf{I_{bB}} = 3.06\angle 142.5°$ A, $\mathbf{I_{cC}} = 5.95\angle -110°$ A.

P25.11 Given a three-phase generator having an open-circuit phase voltage of 400 V and a rated current of 50 A, determine the maximum magnitude of impedance per phase if the line voltage is not to drop by more than 2% when each phase is carrying its rated current, assuming that the generator phases are connected in (a) Y or (b) Δ. If the generator phases are connected in Δ, and there is a slight imbalance such that the open-circuit voltage of one of the phases is 395 V, what would be the magnitude of the circulating current in the windings with no load connected to the generator?

Ans. (a) or (b) 0.16 Ω; 10.42 A.

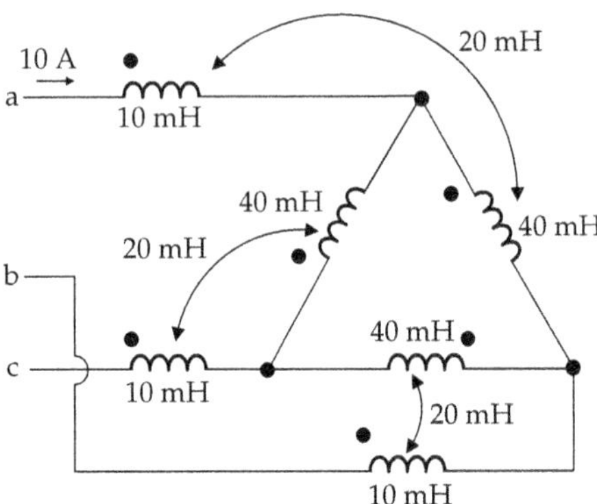

FIGURE P25.12

P25.12 Coils in the lines and phases of a balanced three-phase systems are magnetically coupled as shown in Figure P25.12. If the current in line 'a' is $10\angle 0°$ A, determine the current in the phase connected between lines ab.

Ans. $\dfrac{10}{\sqrt{3}}\angle 30°$ A.

P25.13 Consider a Y-connected load in a balanced three-phase system of positive phase sequence. Let the line voltages be denoted by $\mathbf{V_{ab}}$, $\mathbf{V_{bc}}$, and $\mathbf{V_{ca}}$ and the phase voltages be denoted by $\mathbf{V_{\phi a}}$, $\mathbf{V_{\phi b}}$, and $\mathbf{V_{\phi c}}$, the phase angle of $\mathbf{V_{\phi a}}$ being considered as 0°. Show that $\mathbf{V_{\phi a}} = (\mathbf{V_{ab}} - \mathbf{V_{ca}})/3$, $\mathbf{V_{\phi b}} = (\mathbf{V_{bc}} - \mathbf{V_{ab}})/3$, and $\mathbf{V_{\phi c}} = (\mathbf{V_{ca}} - \mathbf{V_{bc}})/3$.

P25.14 Consider the four-wire, Y-connected, three-phase system of Figure 25.7 with sinusoidal, balanced phase voltages. But suppose that the load includes ferromagnetic devices, such as fluorescent lamp ballasts, that introduce a third harmonic current magnetizing current, so that the current in phase 'A', for example, is $I_{m1}\cos\omega t + I_{m3}\cos 3\omega t$. Show that the third harmonic currents in the three phases are in phase, resulting in a current of $3I_{m3}\cos 3\omega t$ in the neutral conductor.

Analysis of Three-Phase Systems

P25.15 Determine the line currents and the line voltages at the load in Figure P25.15.

Ans. $\mathbf{I_{aA}} = 44\angle 53.13°$ A, $\mathbf{I_{bB}} = 44\angle -66.87°$ A, $\mathbf{I_{cC}} = 44\angle 173.13°$ A, $\mathbf{V_{AB}} = 426.0\angle 19.7°$ V, $\mathbf{V_{BC}} = 426.0\angle -100.3°$ V, $\mathbf{V_{CA}} = 426.0\angle 139.7°$ V.

P25.16 Determine the line currents and load phase currents in Figure P25.16 in two ways: (1) from the single-phase equivalent circuit and (2) from the line voltages, without transforming the Δ.

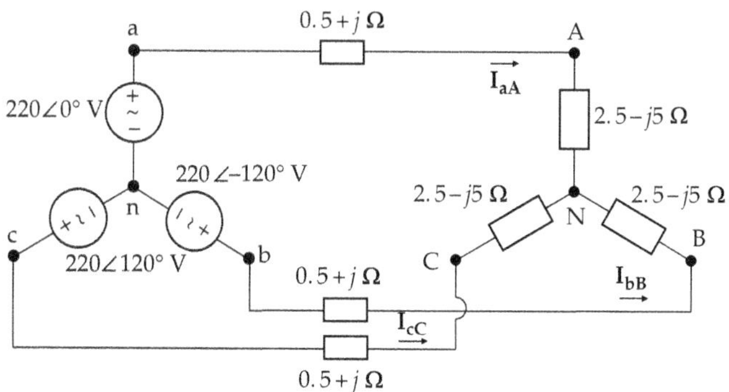

FIGURE P25.15

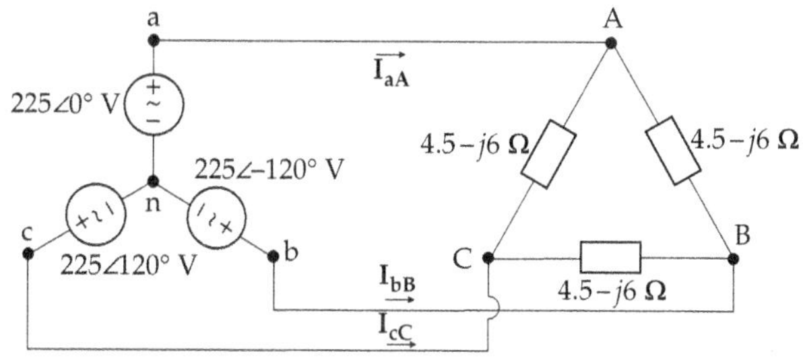

FIGURE P25.16

Ans. $\mathbf{I}_{aA} = 90\angle 53.13°$ A, $\mathbf{I}_{bB} = 90\angle -66.87°$ A, and $\mathbf{I}_{cC} = 90\angle -186.87°$ A; $\mathbf{I}_A = 30\sqrt{3}\angle 83.13°$ A, $\mathbf{I}_B = 30\sqrt{3}\angle -36.87°$ A, and $\mathbf{I}_C = 30\sqrt{3}\angle -156.87°$ A.

P25.17 Determine the line currents and the line voltages at the load in the preceding problem when a line impedance of $0.5 + j$ Ω is added in each line.

Ans. $\mathbf{I}_{aA} = 100.6\angle 26.57°$ A, $\mathbf{I}_{bB} = 100.6\angle -93.43°$, $\mathbf{I}_{cC} = 100.6\angle -213.43°$ A; $\mathbf{V}_{ab} = 435.7\angle 3.43°$ V, $\mathbf{V}_{bc} = 435.7\angle -116.6°$ V, and $\mathbf{V}_{ca} = 435.7\angle -236.6°$ V.

P25.18 Determine the line currents in Figure P25.18 in two ways: (1) from the phase currents and (2) from the single-phase equivalent circuit.

Ans. $\mathbf{I}_{aA} = 30\sqrt{3}\angle 83.13°$ A, $\mathbf{I}_{bB} = 30\sqrt{3}\angle 156.9°$ A, and $\mathbf{I}_{cC} = 30\sqrt{3}\angle -36.87°$ A.

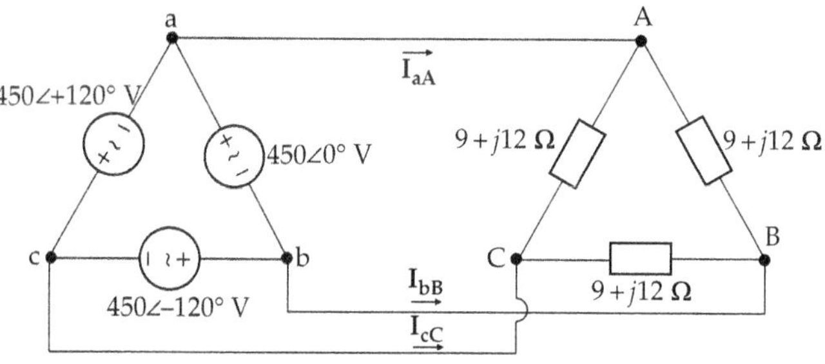

FIGURE P25.18

P25.19 A balanced Y-connected load consists in each phase of an inductive impedance of $50\angle 45°$ Ω in parallel with a 0.1 μF capacitor. The load is supplied from a balanced three-phase supply having a line voltage of 380 V, 50 Hz, through lines of $0.5 + j1.0$ Ω impedance. Determine the magnitude of the line current.

Ans. 4.3 A.

P25.20 A balanced Δ-connected load consists in each phase of an inductive impedance of $30 + j45$ Ω. The load is supplied from a balanced three-phase supply having a phase voltage of 220 V through lines of $0.1 + j0.2$ Ω impedance. Determine the magnitude of the line current.

Ans. 12.06 A.

P25.21 The load of Figure P25.21 is supplied from a balanced three-phase system of 450 V line voltage, 50 Hz. Determine the current in the neutral before and after phase 'B' is open circuited at x.

Ans. $204.0\angle -162.7°$ A, $176.8\angle 111.6°$ A.

P25.22 Three resistors of 6, 10, and 15 Ω are Y connected to a balanced three-wire, three-phase supply of 300 V line voltage. Determine the magnitude of the voltage across each resistor.

Ans. $|\mathbf{V}_{6\Omega}| = 6\mathbf{I}_1 = 130.8$ V, $|\mathbf{V}_{10\Omega}| = 10\mathbf{I}_2 = 187.3$ V, $|\mathbf{V}_{15\Omega}| = 15|\mathbf{I}_2 - \mathbf{I}_1| = 210$ V.

P25.23 The load of Figure P25.23 is connected to a balanced three-wire, three-phase system of 400 V line voltage. Determine the current in each phase of the load.

Ans. $-20\sqrt{3} + j20$ A, $-20 - j20\left(2+\sqrt{3}\right)$ A, $\mathbf{I}_{cC} = 20\left(\sqrt{3}+1\right)(1+j)$ A.

P25.24 Consider that the sources in Problem P25.23 are Y connected. Determine the current in the 10 Ω resistor by superposition.

Ans. $-20\sqrt{3} + j20$ A.

P25.25 A three-phase generator having an impedance of $0.9 + j0.9$ Ω per phase is Δ connected. The open-circuit terminal voltage of the generator is 13.2 kV. The generator supplies a Δ-connected load of $650 + j170$ Ω per phase through a transmission line of impedance $0.7 + j0.3$ Ω per phase. Determine the magnitude of the line voltage at the load end.

Ans. 13.13 kV.

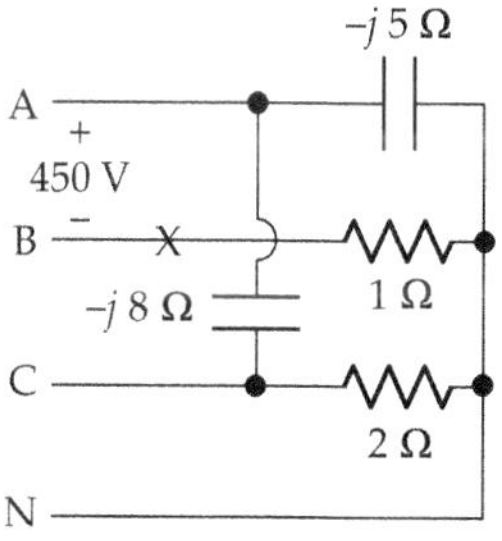

FIGURE P25.21

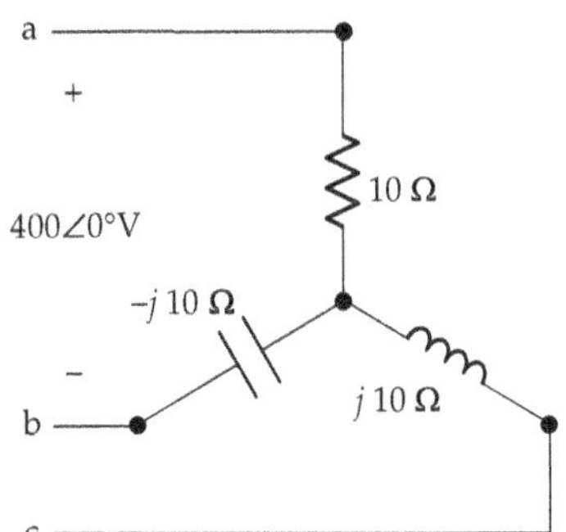

FIGURE P25.23

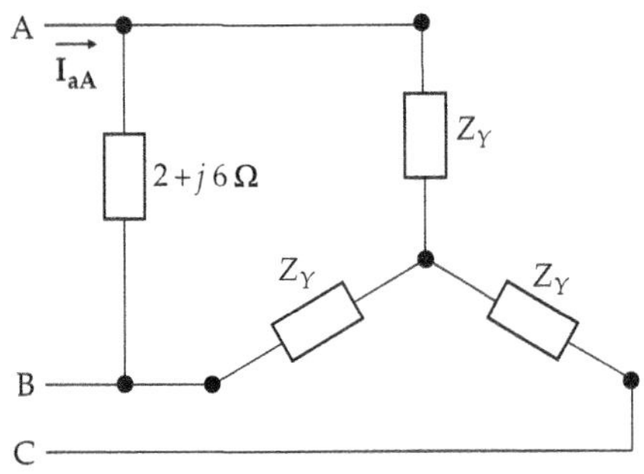

FIGURE P25.26

P25.26 The load of Figure P25.26 is connected to a balanced three-phase supply. If $Z_Y = 5 + j10\ \Omega$ and $\mathbf{V_{AB}} = 240\angle 0°$ V, determine $\mathbf{I_{aA}}$.

Ans. 26.17∠64.5° A.

P25.27 The sources in Figure P25.27 form a balanced set, with $\mathbf{V_{ga}} = 100\angle 90°$ V. Determine $\mathbf{I_{nN}}$.

Ans. −j6.1004 = 12.2∠−150° A.

P25.28 In Figure 25.20 (Exercise 25.13), what would be the reading of the voltmeter if it had a resistance of 5R?

Ans. 181.5∠−30° V.

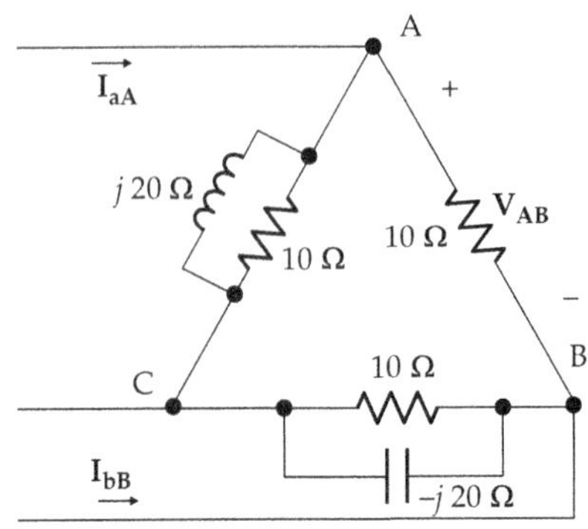

FIGURE P25.29

P25.29 Given that in Figure P25.29 $\mathbf{I_{aA}} = 10\angle 20°$ A and $\mathbf{I_{bB}} = 12\angle -120°$ A, determine $\mathbf{V_{AB}}$.

Ans. 75.2∠39.4° V.

P25.30 Given that in Figure P25.30 $\mathbf{V_{AB}} = 100\angle 20°$ V and $\mathbf{V_{BC}} = 120\angle -120°$ V, determine $\mathbf{I_C}$.

Ans. 1.36∠25.61° A.

P25.31 Determine the single-phase equivalent impedance of the circuit of Figure P25.12, assuming a frequency of 50 Hz.

Ans. $j13\pi/\sqrt{3}\ \Omega$.

P25.32 In the balanced three-phase system shown in Figure P25.32, the three voltage sources form a balanced set $\mathbf{V_{ga}} = 50\angle 0°$ V, $\mathbf{V_{gb}} = 50\angle -120°$ V, and $\mathbf{V_{gc}} = 50\angle 120°$ V, the current sources form a balanced set $\mathbf{I_A} = 10\angle 30°$ A, $\mathbf{I_B} = 10\angle -90°$ A, and $\mathbf{I_C} = 10\angle 150°$ A, and $R_\phi = 5\ \Omega$. Determine the single-phase equivalent circuit.

Ans. $\mathbf{V_{aeq}} = 25\left(2-\sqrt{3}\right) - j25$ V, with respect to n, in series with 5 Ω.

P25.33 In the balanced three-phase system shown in Figure P25.33, the three independent voltage sources form a balanced set $\mathbf{V_{ga}} = 50\angle 0°$ V, $\mathbf{V_{gb}} = 50\angle -120°$ V, and $\mathbf{V_{gc}} = 50\angle 120°$ V, the dependent sources form a balanced set with $K = 1 - j$, and $R_\phi = 5\ \Omega$. Determine the single-phase equivalent circuit.

Ans. $\mathbf{V_{ga}}$ in series with −j5 Ω.

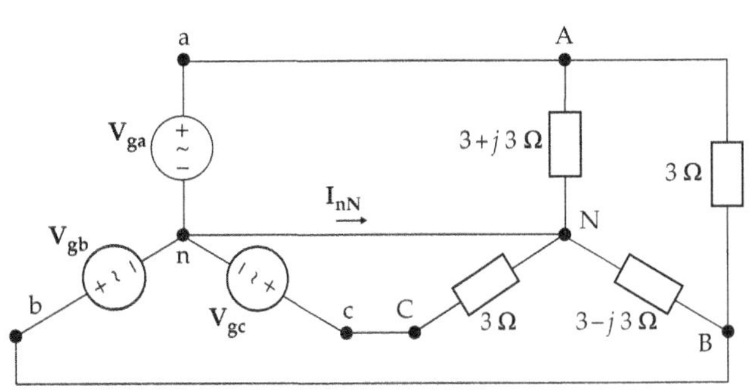

FIGURE P25.27

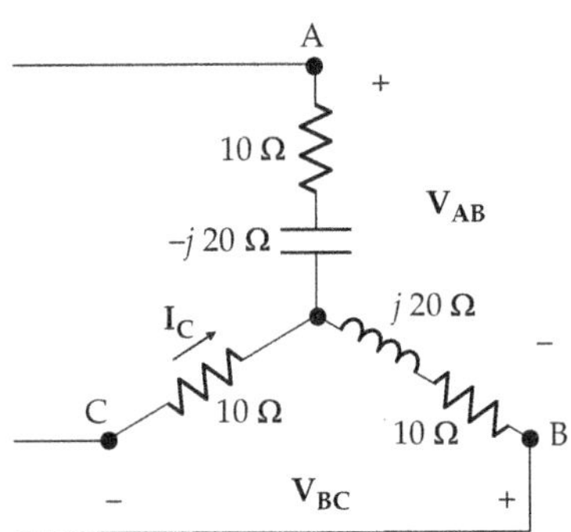

FIGURE P25.30

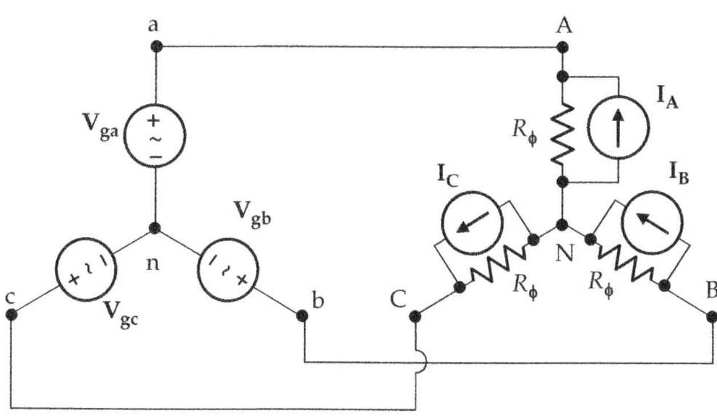

FIGURE P25.32

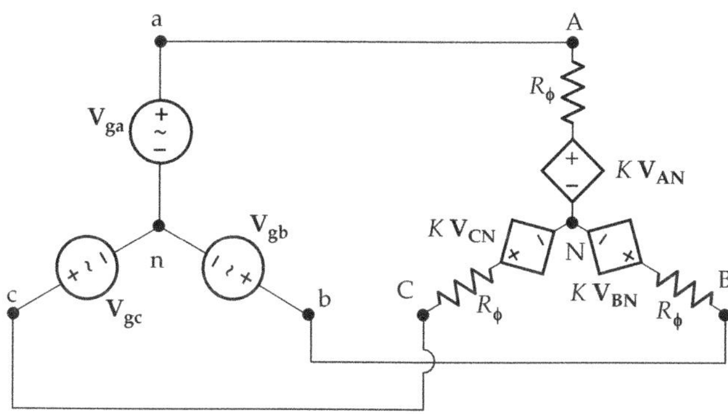

FIGURE P25.33

P25.34 Figure P25.34 shows a simple circuit for indicating the phase sequence using two incandescent lamps and a capacitor connected in Y. One lamp will glow brighter than the other. The phase sequence is then in the order of the line connections to the (bright lamp)–(dim lamp)–(capacitor), that is, 'abc' in Figure P25.34. Show this by assuming a 5 μF capacitor and two 60 W, 220 V lamps, having a constant resistance. Assume the circuit is connected to a three-phase system and analyze the circuit as a two-mesh circuit to show that the current in one lamp exceeds that in the other lamp, depending on the phase sequence.

P25.35 Verify the results of Example 25.1 using mesh-current analysis.

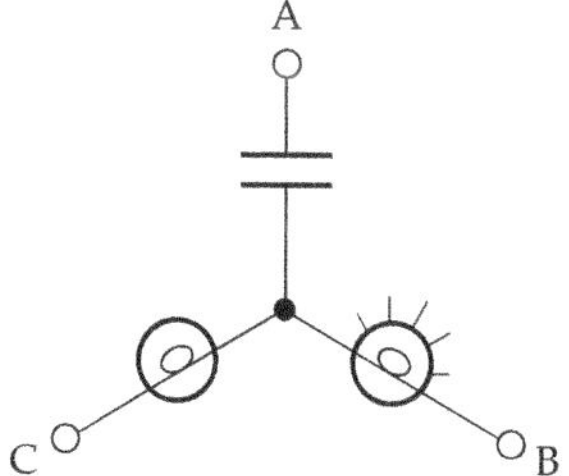

FIGURE P25.34

P25.36 Verify the results of Example 25.4 by solving the circuit of Figure 25.13a as a four-mesh circuit.

Power in Three-Phase Systems

P25.37 A balanced Δ-connected, series-connected inductive load draws a line current of 10 A rms and absorbs 3 kW of real power at a line voltage of 220 V rms. Determine the load impedance per phase.

Ans. 30 Ω + *j*23.49 Ω.

P25.38 A balanced three-phase system of 240 V rms line voltage, 50 Hz, supplies a 100 kW balanced load of 0.6 p.f. lagging. Determine the capacitance in each phase of a Y-connected capacitor bank that will give a power factor of 0.95 lagging.

Ans. 5.55 mF.

P25.39 In the circuit of Figure P25.39, the Y-connected load and the Δ-connected load are both balanced with $R_Y = 4\ \Omega$, $X_Y = j4\ \Omega$, $R_\Delta = 6\ \Omega$, and $Y_\Delta = -j8\ \Omega$. Determine the real and reactive powers absorbed if the line voltage is 190 V rms.

Ans. $\mathbf{S} = 11.0 - j4.15$ kVA.

P25.40 A three-phase AC generator is rated at 50 MVA, 11 kV line voltage. If the generator is operated at its rated voltage and current, determine the percentage increase in the real power delivered if the power factor is increased from 0.7 to 0.95 lagging.

Ans. 35.7%.

P25.41 Determine the real, reactive, and apparent powers absorbed by the load of Figure P25.41 when connected to a balanced three-phase supply of 220 V rms line voltage.

Ans. 18.392 kW, 1.936 kVAR, 18.49 kVA.

P25.42 A balanced three-phase supply of 380 V rms line voltage, 50 Hz, is connected to two paralleled, balanced three-phase inductive loads. One load absorbs 173 kW at 0.8 p.f., whereas the other load absorbs 110 kW at 0.7 p.f. A bank of equal Δ-connected capacitors is to be connected in parallel with the loads so as to bring the p.f. to unity. Determine the required capacitance per phase.

Ans. 1.78 mF.

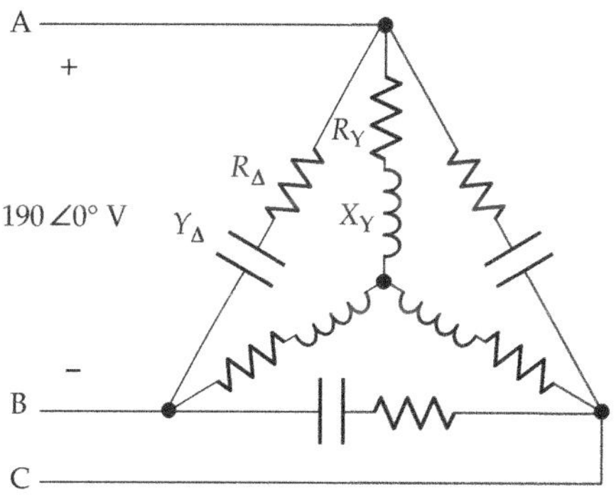

FIGURE P25.39

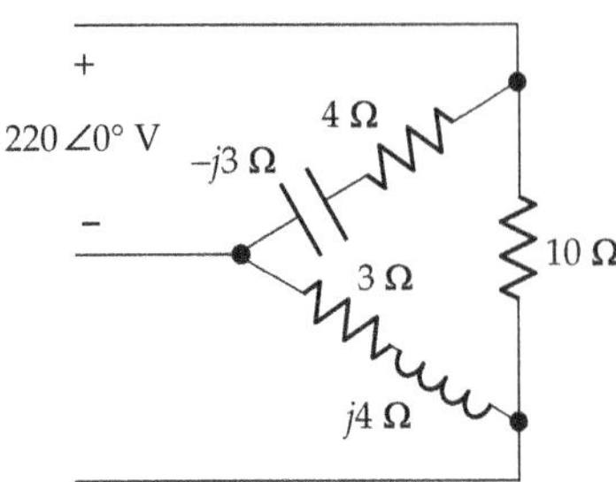

FIGURE P25.41

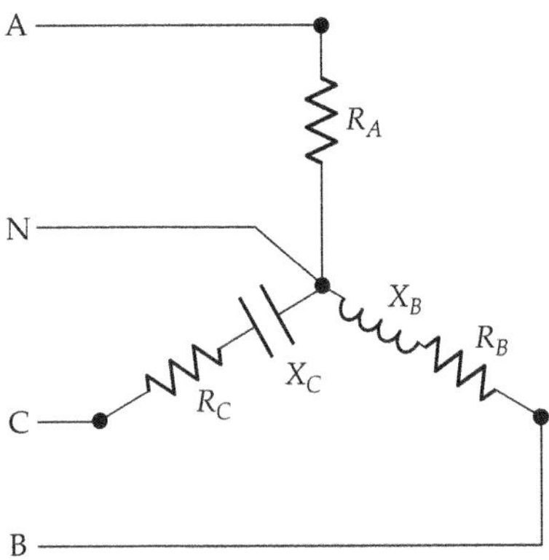

FIGURE P25.43

P25.43 The load in Figure P25.43 absorbs 5 kW when connected to a balanced, four-wire, three-phase supply of 380 V rms line voltage. The power factors of phases 'B' and 'C' are 0.5 lagging and 0.5 leading, respectively, and the magnitudes of the currents in phases 'A' and 'B' are 5 A and 10 A, respectively. Determine (a) the magnitude and phase angle of the current in phase 'C', assuming that the current in phase 'A' has a phase angle of zero, (b) R_C and X_C, and (c) the reactive power of phases 'B' and 'C'.

Ans. (a) 25.58∠180° A, (b) 4.29 and –7.43 Ω, (c) 1900 and –4860 VAR.

P25.44 Each phase of a Y-connected load consists of a resistance of 100 Ω in parallel with a capacitance of 31.8 μF. The load is connected to a balanced three-phase supply of 416 V rms line voltage, 50 Hz. Determine the real, reactive, and apparent powers absorbed by the load.

Ans. 1731 W, –1729 VAR, 2446 VA.

P25.45 A balanced bank of capacitors is connected in parallel with a balanced three-phase load in order to bring the p.f. at the supply end to unity. The apparent powers of the load and capacitor bank are 2 kVA and 500 VA, respectively. The load and capacitors are supplied from a balanced three-phase supply having a line impedance of 2 + *j*20 Ω per line, the total real power dissipated in the lines being 24 W. Determine the magnitude of the line voltage at the load end.

Ans. 577 V rms.

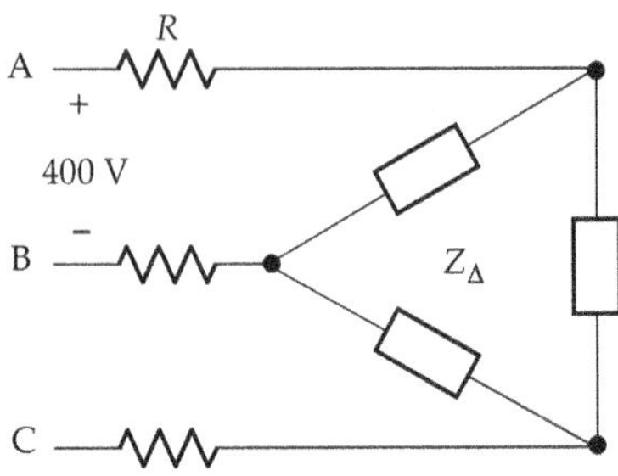

FIGURE P25.47

P25.46 Given a balanced three-phase system in which the load consumes 50 kW at 0.8 p.f. lagging, the line impedance being 0.5 + *j*0.5 Ω. Capacitors are Δ connected so that the power factor at the supply terminals is 0.95. If the magnitude of the line current is 120 A, 50 Hz, determine the magnitude of the line voltage at the supply terminals.

Ans. 363 V rms.

P25.47 In the balanced three-phase system shown in Figure P25.47, R = 0.5 Ω and Z_Δ consists of a 50 Ω resistance in parallel with an inductive reactance of 50 Ω. Determine the total real power consumed by the load. What is the p.f. seen at the supply end?

Ans. 9584 W; p.f. = 0.73.

P25.48 A balanced source supplies 100 kVA at a line voltage of 450 V rms and 0.9 p.f. leading to a balanced load, the line impedance being 0.1 + *j*0.5 Ω. Determine (a) the magnitude of the line current, (b) the magnitude of the line voltage at the load, and (c) the total apparent power at the load.

Ans. (a) 128.3 A, (b) 491 V rms, (c) 109.1 kVA.

P25.49 Verify that reversing the phase sequence interchanges the readings of the two wattmeters in the two-wattmeter method.

P25.50 Verify Equations 25.35 to 25.37 for a Δ connection.

P25.51 The power input to a three-phase synchronous motor is measured by the two-wattmeter method. When the p.f. of the motor is unity, each of the two wattmeters reads 50 kW. What would be the reading of each wattmeter if the power factor is changed to 0.866 leading, assuming the magnitudes of the line voltage and current stay the same?

Ans. 57.74 and 28.87 kW.

P25.52 A balanced load of 0.75 p.f. lagging is connected to a balanced three-phase, three-wire system. The sum of the readings of two wattmeters connected in the standard two-wattmeter connection is 26 kW. What is the reading of each wattmeter?

Ans. 19.62 and 6.38 kW.

P25.53 A balanced three-phase load is connected to a balanced three-phase supply of 220 V rms line voltage. A wattmeter reads 600 W when its current coil is connected in line 'a' and its voltage coil is connected between lines 'a' and 'b', with the positive terminal of the coil connected to line 'a'. When the voltage coil is connected between lines 'b' and 'c', with the positive terminal of the coil connected to phase 'b', and the current coil connected in line 'a' as before, the wattmeter again reads 600 W. What is the load p.f.?

Ans. 0.87.

P25.54 Two wattmeters W_1 and W_2 are connected to a balanced three-phase, inductive load of 0.8 p.f., as shown in Figure P25.54, the supply being also balanced. If W_1 reads 100 W, what is the reading of W_2?

Ans. 39.6 W.

P25.55 In Figure P25.55, the source voltages $\mathbf{V}_1$, $\mathbf{V}_2$, and $\mathbf{V}_3$ constitute a balanced set, with $\mathbf{V}_1 = 300\angle{-45°}$ V rms, $\mathbf{V}_{AB} = 381\angle 0°$ V rms, and $Z_\phi = 20\angle 30°$ Ω. The current coils C_1 and C_2 and voltage coils P_1 and P_2 of wattmeters W_1 and W_2 are connected as shown. Determine the readings of W_1 and W_2.

Ans. 8848 and 5572 W.

P25.56 Two wattmeters W_1 and W_2 are connected as shown in Figure P25.56 to measure the power in a balanced Δ-connected load, the line voltage being 380 V rms. If each wattmeter reads 1000 W, determine the load impedance per phase.

Ans. $-j125.1$ Ω.

P25.57 Consider a wattmeter connected as in Figure P25.57. Show that the reading of the wattmeter is $1/\sqrt{3}$ of the reactive power of the load.

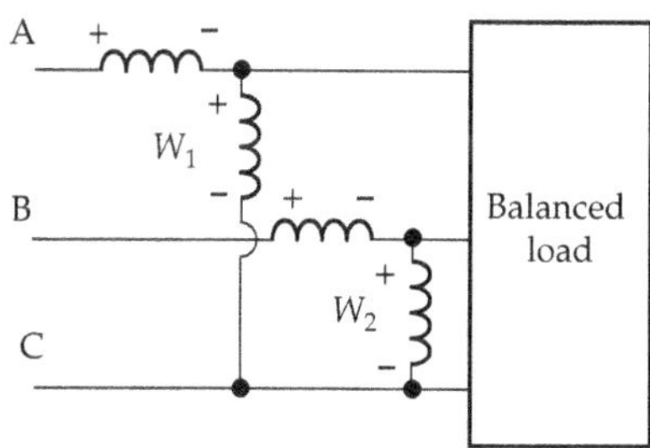

FIGURE P25.54

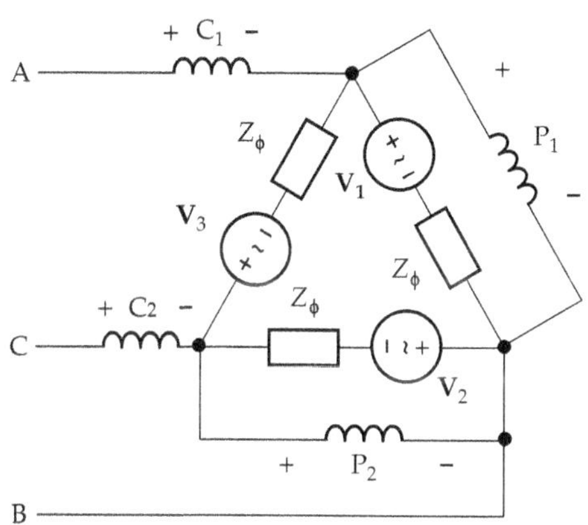

FIGURE P25.55

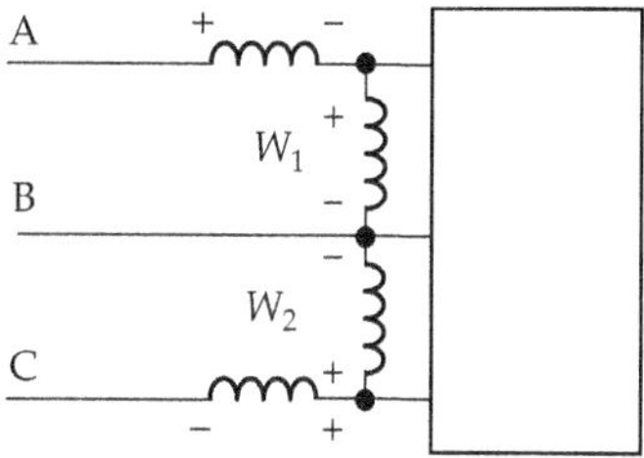

FIGURE P25.56

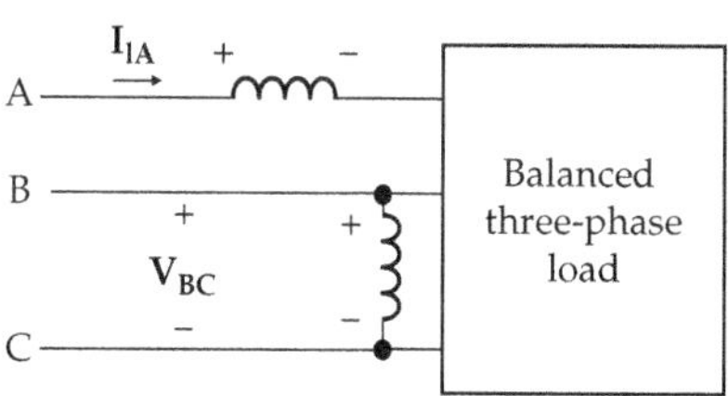

FIGURE P25.57

Probing Further

P25.58 Consider Figure P25.58, which shows a three-phase transformer in which the primary and secondary windings are Y connected. The neutral point of the secondary is grounded, as is commonly done in three-phase systems in order to define voltages in power systems with respect to the ground reference and to provide a path for current flow under fault conditions so as to operate protective devices. If the phase voltages are sinusoidal, the magnetic flux in the transformer core will be sinusoidal, but the magnetizing current in each phase of the primary will have a third harmonic component, as explained in Problem P25.14. However, since there is no connection to the neutral of the primary side, third harmonic currents, which are in phase, cannot flow. The phase currents must therefore be sinusoidal, which means that the phase voltages are not sinusoidal but contain a third harmonic component. The line-to-line voltage is the difference of two phase voltages, so that the third harmonics of the phase voltages cancel out, and the line voltages are

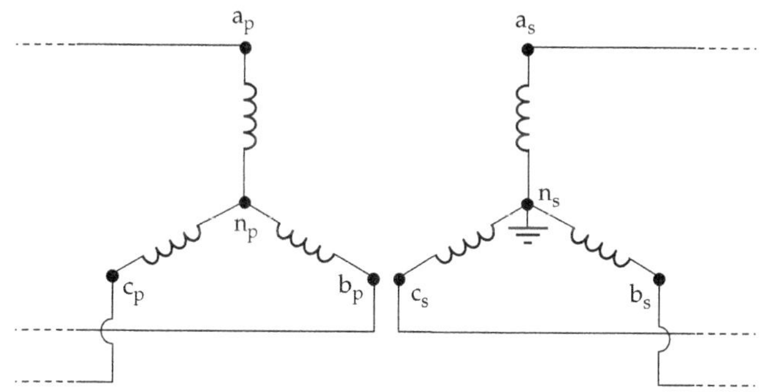

FIGURE P25.58

sinusoidal. With the line voltages sinusoidal and the three-phase voltages containing third harmonics that are in phase, argue that the primary neutral point, n_p, will oscillate at the third harmonic frequency. Such oscillations are undesirable. They could be avoided by grounding n_p, but third harmonic currents will flow through ground causing undesirable interference. The solution is to connect the primary or secondary windings in Δ, which allows the third harmonic currents to flow in the Δ-connected windings. The magnitude of this current is limited by the fact that the reactance of the transformer windings is multiplied by three at the third harmonic frequency. Hence, transformers in power systems invariably have either primary or secondary windings connected in Δ.

P25.59 Given that the primary or secondary windings of power transformers are connected in Δ, for reasons explained in the preceding problem, the other winding being connected in Y, argue that it is more advantageous to connect the primary windings in Δ when line voltages are to be stepped up, whereas it is more advantageous to connect the secondary windings in Δ when line voltages are to be stepped down.

P25.60 Y–Δ *starting of induction motors*: A popular type of ac motor is the three-phase, squirrel cage induction motor. A characteristic of this type of motor is that at starting, the motor appears like a transformer having a short-circuited secondary, which results in a starting current of low power factor and magnitude that is five to eight times the rated full-load current. This is objectionable as it causes an excessive, though momentary, voltage drop in the supply lines feeding the motor and other loads. A common method of avoiding this large starting current on medium-sized and small-sized motors is Y–Δ starting. The motor is designed to run with its stator windings connected in Δ. At starting, however, the windings are connected in Y. At a preset interval of a few tens of a second after starting, the windings are connected in Δ for normal running.

Assuming that the starting current is proportional to the phase voltage, show that the starting current with the windings connected in Y is 1/3 of the starting current with the windings connected in Δ.

P25.61 *dc vs. ac high-voltage transmission*: Consider a Y-connected power transmission system of phase voltage $\mathbf{V}_\phi$ and line current $\mathbf{I}_l$. Compare this to a dc power transmission system having the same line current I_l and line voltages of $+V_{DC}$ and $-V_{DC}$, with the midpoint grounded. To have the same insulation level in both systems, V_{DC} should equal the peak phase voltage, that is, $V_{DC} = \sqrt{2}V_\phi$. Show that under the best conditions, with unity p.f. in the three-phase system, the power transmitted in the ac case is $3/\left(2\sqrt{2}\right)$ that in the dc case, which amounts to only 1.06 times. However, the ac case uses 50% more conductor and has 50% more I^2R loss, compared to the dc case. In actual fact the conductors have to be of larger cross-sectional area in the ac case for the same current-carrying capacity, because under ac conditions, the current is not uniformly distributed across the cross section of the conductor due to the effect of the ac magnetic field inside the conductor, the so-called *skin effect*. Moreover, two line insulators per tower are required, compared to three for ac, and the dc towers are smaller since they have to support fewer conductors and insulators. An additional advantage of dc transmission is the absence of reactance, whereas the reactance of long ac transmission lines introduces problems of power system stability. The power systems connected by the dc line need not have exactly the same frequency. On the downside, since electric power is more conveniently generated and utilized as ac, ac must be converted to dc at high voltages for transmission, which is then converted back to ac for distribution. This introduces additional complexity and cost.

Appendix A: SI Units, Symbols, and Prefixes

A.1 The International System of Units

The International System of Units, referred to as the SI system (its acronym in French) is almost universally used for scientific and engineering purposes and is adopted throughout this book. It is a system of physical units in which the fundamental quantities are the seven listed in Table A.1 together with their corresponding units and symbols of these units.

In addition, circuit theory uses many units that are derived from these fundamental units. Table A.2 lists the more common of these derived units together with their relations to other quantities and their unit symbols.

In many practical problems, the SI units defined in the table are either too large or too small. Standard prefixes in powers of 10 are applied in order to bring the numeral preceding the power of 10 to a convenient value, generally between 1 and 10. Table A.3 lists the prefixes associated with the SI system of units. The prefixes centi, deci, deka, and hecto are not used with electrical quantities. The remaining prefixes progress in powers of 10 that are divisible by 3.

TABLE A.1

SI System of Units

Fundamental Quantity	Unit	Symbol
Length	meter	m
Time	second	s
Mass	kilogram	kg
Electric current	ampere	A
Temperature	degree Kelvin	°K
Luminous intensity	candela	cd
Amount of substance	mole	mol

TABLE A.2

SI Derived Quantities

			Relation to Other Quantities	
Quantity	**Unit**	**Symbol**	**Expression**	**Symbols**
Frequency	hertz	Hz	—	s^{-1}
Angular frequency	radians per second	rad/s	2π × (frequency)	s^{-1}
Energy or work	joule	J	Force × distance	N m
Power	watt	W	Energy/time	J/s or A V
Electric charge	coulomb	C	Current × time	A s
Electric potential difference (voltage)	volt	V	Power/current	V
Electric resistance	ohm	Ω	Voltage/current	V/A
Electric conductance	siemens (or mho)	S	Current/voltage	A/V
Electric capacitance	farad	F	Charge/voltage	A s/V
Magnetic flux	weber	Wb	Voltage × time	V s
Magnetic flux linkage	weber-turn	Wb-turns	(Magnetic flux) × (number of turns)	V s
Inductance	henry	H	Flux linkage/current	V s/A

TABLE A.3

Power of 10 Prefixes Used with the SI System

Prefix	Symbol	Power
atto	a	10^{-18}
femto	f	10^{-15}
pico	p	10^{-12}
nano	n	10^{-9}
micro	μ	10^{-6}
milli	m	10^{-3}
centi	c	10^{-2}
deci	d	10^{-1}
deka	da	10^{1}
hecto	h	10^{2}
kilo	k	10^{3}
mega	M	10^{6}
giga	G	10^{9}
tera	T	10^{12}
peta	P	10^{15}
exa	E	10^{18}

Appendix B: Useful Mathematical Relations

B.1 Trigonometric Relations

$$\sin(\alpha \pm \beta) = \sin\alpha\cos\beta \pm \cos\alpha\sin\beta$$

$$\cos(\alpha \pm \beta) = \cos\alpha\cos\beta \mp \sin\alpha\sin\beta$$

$$\tan(\alpha \pm \beta) = \frac{\tan\alpha \pm \tan\beta}{1 \mp \tan\alpha\tan\beta}$$

$$\sin 2\alpha = 2\sin\alpha\cos\alpha$$

$$\cos 2\alpha = 2\cos^2\alpha - 1 = 1 - 2\sin^2\alpha$$

$$\tan 2\alpha = \frac{2\tan\alpha}{1 - \tan^2\alpha}$$

$$\sin\alpha + \sin\beta = 2\sin\frac{\alpha+\beta}{2}\cos\frac{\alpha-\beta}{2}$$

$$\sin\alpha - \sin\beta = 2\cos\left(\frac{\alpha+\beta}{2}\right)\sin\left(\frac{\alpha-\beta}{2}\right)$$

$$\cos\alpha + \cos\beta = 2\cos\left(\frac{\alpha+\beta}{2}\right)\cos\left(\frac{\alpha-\beta}{2}\right)$$

$$\cos\alpha - \cos\beta = -2\sin\left(\frac{\alpha+\beta}{2}\right)\sin\left(\frac{\alpha-\beta}{2}\right)$$

$$2\sin\alpha\sin\beta = \cos(\alpha - \beta) - \cos(\alpha + \beta)$$

$$2\cos\alpha\cos\beta = \cos(\alpha - \beta) + \cos(\alpha + \beta)$$

$$2\sin\alpha\cos\beta = \sin(\alpha + \beta) + \sin(\alpha - \beta)$$

$$2\cos\alpha\sin\beta = \sin(\alpha + \beta) - \sin(\alpha - \beta)$$

$$\cos^2\alpha = \frac{1}{2} + \frac{1}{2}\cos 2\alpha$$

$$\sin^2\alpha = \frac{1}{2} - \frac{1}{2}\cos 2\alpha$$

$$\cos^3\alpha = \frac{1}{4}(3\cos\alpha + \cos 3\alpha)$$

$$\sin^3\alpha = \frac{1}{4}(3\sin\alpha - \sin 3\alpha)$$

$$\cos^4\alpha = \frac{1}{8}(3 + 4\cos 2\alpha + \cos 4\alpha)$$

$$\sin^4\alpha = \frac{1}{8}(3 - 4\cos 2\alpha + \cos 4\alpha)$$

Triangle rules: Let a, b, and c be the lengths of sides of a triangle and α, β, and γ be, respectively, the angles opposite these sides. Then,

$$\text{Sine rule:} \quad \frac{a}{\sin\alpha} = \frac{b}{\sin\beta} = \frac{c}{\sin\gamma}$$

$$\text{Cosine rule:} \quad a^2 = b^2 + c^2 - 2bc\cos\alpha$$
$$b^2 = a^2 + c^2 - 2ac\cos\beta$$
$$c^2 = a^2 + b^2 - 2ab\cos\lambda$$

Small angles: If α is a small angle expressed in radians,

$$\sin\alpha \cong \alpha, \quad \tan\alpha \cong \alpha, \quad \text{and} \quad \cos\alpha \cong 1 - \frac{\alpha^2}{2}$$

B.2 Useful Relations

Integration by parts: $\int u\left(\frac{dv}{dx}\right)dx = uv - \int v\left(\frac{du}{dx}\right)dx$

L'Hopital's rule: A function $f(x)/x$ that is indeterminate, that is, 0/0, when $x = 0$, can be evaluated by differentiating the numerator and denominator with respect to x any number of times until a finite answer is obtained.

Geometric series: $a + ar + ar^2 + \cdots + ar^{n-1} = \sum_{k=0}^{k=n-1} ar^k = a\frac{1-r^n}{1-r}$. The series converges as $n \to \infty$ if $|r| < 1$. In this case the sum is $a/(1-r)$. If $r = m$, $0 < m < 1$, the sum is $a/(1-m)$. If $r = -m$, $0 < m < 1$, the sum is $a/(1+m)$.

B.3 Table of Integrals

$$\int xe^{ax}dx = \frac{e^{ax}}{a^2}(ax - 1)$$

$$\int x^2 e^{ax} dx = \frac{e^{ax}}{a^3}\left(a^2x^2 - 2ax + 2\right)$$

$$\int_0^\infty x^{1/2} e^{-x} dx = \frac{\pi^{1/2}}{2}$$

$$\int x \sin ax dx = \frac{1}{a^2}\sin ax - \frac{x}{a}\cos ax$$

$$\int x \cos ax dx = \frac{1}{a^2}\cos ax + \frac{x}{a}\sin ax$$

$$\int e^{ax}\sin bx dx = \frac{e^{ax}}{a^2+b^2}\left(a\sin bx - b\cos bx\right)$$

$$\int e^{ax}\cos bx dx = \frac{e^{ax}}{a^2+b^2}\left(a\cos bx + b\sin bx\right)$$

$$\int \frac{1}{x^2+a^2} dx = \frac{1}{a}\tan^{-1}\frac{x}{a}$$

$$\int \frac{1}{ax^2 - x} dx = \ln\left(1 - \frac{1}{ax}\right)$$

$$\int \frac{1}{\left(x^2+a^2\right)^2} dx = \frac{1}{2a^2}\left(\frac{x}{x^2+a^2} + \frac{1}{a}\tan^{-1}\frac{x}{a}\right)$$

$$\int \sin ax \sin bx dx = \frac{\sin(a-b)x}{2(a-b)} - \frac{\sin(a+b)x}{2(a+b)}, \quad a^2 \neq b^2$$

$$\int \cos ax \cos bx dx = \frac{\sin(a-b)x}{2(a-b)} + \frac{\sin(a+b)x}{2(a+b)}, \quad a^2 \neq b^2$$

$$\int \sin ax \cos bx dx = -\frac{\cos(a-b)x}{2(a-b)} - \frac{\cos(a+b)x}{2(a+b)}, \quad a^2 \neq b^2$$

$$\int \sin^2 ax dx = \frac{x}{2} - \frac{\sin 2ax}{4a}$$

$$\int \cos^2 ax dx = \frac{x}{2} + \frac{\sin 2ax}{4a}$$

$$\int_0^\infty \frac{a}{a^2+x^2} dx = \begin{Bmatrix} \frac{\pi}{2}, & a>0 \\ 0, & a=0 \\ -\frac{\pi}{2}, & a<0 \end{Bmatrix}$$

$$\int_0^\infty \frac{\sin ax}{x} dx = \begin{Bmatrix} \frac{\pi}{2}, & a>0 \\ -\frac{\pi}{2}, & a<0 \end{Bmatrix}$$

$$\int x^2 \sin ax dx = \frac{2x}{a^2}\sin ax - \frac{a^2x^2-2}{a^3}\cos ax$$

$$\int x^2 \cos ax dx = \frac{2x}{a^2}\cos ax + \frac{a^2x^2-2}{a^3}\sin ax$$

$$\int e^{ax}\sin^2 bx dx = \frac{e^{ax}}{a^2+4b^2}\left[\left(a\sin bx - 2b\cos bx\right)\sin bx + \frac{2b^2}{a}\right]$$

$$\int e^{ax}\cos^2 bx dx = \frac{e^{ax}}{a^2+4b^2}\left[\left(a\cos bx + 2b\sin bx\right)\cos bx + \frac{2b^2}{a}\right]$$

$$\int_{-\infty}^{\infty} e^{-x^2/2\sigma^2} dx = \sigma\sqrt{2\pi}, \quad \sigma > 0$$

Appendix C: PSpice Simulation

C.1 General

SPICE—the acronym for Simulation Program with Integrated Circuit Emphasis—was developed by the Electronics Research Laboratory, University of California, Berkeley, CA, and made available to the public in 1975. Since then it underwent many enhancements and became available in a number of forms. OrCAD PSpice for Windows is widely used, particularly in education, and is supplied by Cadence Design Systems, Inc. The description in this appendix applies specifically to OrCAD 16.6 Demo.

OrCAD software can perform many types of analysis: dc, ac, transient, Fourier, transfer function, noise, distortion, and operating point analysis. It can be used for the simulation of linear and nonlinear circuits, transmission lines, semiconductor devices, digital circuits, and mixed analog and digital circuits, in addition to printed circuit board preparation.

There are three steps in using PSpice for circuit simulation: (1) describing the circuit to be analyzed by drawing it schematically using the Capture program, (2) performing the type of analysis desired, and (3) displaying, printing, or plotting the results.

Learning tutorials are included with the Capture program. After installing the software, invoke the Capture CIS program. CIS stands for Component Information System and includes all the features of Capture. When the OrCAD Capture window is displayed select Help/Learning OrCAD Capture CIS or Help/Learning PSpice and follow the instructions.

This appendix has been especially prepared to provide the information required for effective PSpice simulation of circuits in introductory courses on circuit analysis. Most of the difficulties commonly encountered are addressed. Additional details are given in the simulations of the solved examples in this book, as may be needed.

C.2 Starting a New Simulation

1. Run the Capture CIS program. An OrCAD Capture CIS—Lite (Start Page) is displayed with a Start Page tab.
2. Select Project/New. The New Project Dialog box is displayed. Enter a project name and select the 'Analog or Mixed A/D' method of creating a new project. Press the Browse button to specify a location of the directory where the simulation files will be saved. If you have not created such a directory, specify a path and press the Create Directory button. A Create Directory dialog box is opened. Enter a directory name of your choice.
3. After the OK buttons are pressed, a Create PSpice Project dialog box is displayed. Choose Create a blank project. Two tabs are created, one in the name of the project, and the other labeled Page 1. The Parts toolbar appears on the right-hand side of the screen.
4. Press the Page 1 tab to display the (SCHEMATIC1: PAGE1) window. You are now ready to place circuit components on the schematic page. This can be done in two ways: either from the Parts toolbar or by selecting items from the pull-down menu under Place in the menu bar.
5. To place a part using the toolbar, click on the Place Part button having the + sign at the top of the second column, which expands the toolbar window. Before parts can be placed, the proper libraries must be added. Click on the Add Library button, which appears as a dashed rectangle under Libraries. The Browse File dialog box appears. Select the PSpice directory. Add whatever libraries are needed for a particular simulation, such as SOURCE or ANALOG libraries.
6. To select a dc voltage source, for example, highlight the SOURCE library in the Place Part dialog box. A list of available sources appears under Part List. Scroll down the list and select VDC. A battery symbol appears in the lower right-hand corner of the dialog box. Press the Place Part button or double-click on the VDC symbol. The battery source symbol appears in the schematic window attached to an arrow cursor. Move the cursor to locate the source at an appropriate location in the window and press the left mouse button to anchor it in position. Another source can be placed by dragging the cursor and symbol to a new location and pressing the left mouse button to anchor it in position, and so on. To end the insertion process, either press the Esc key or the right mouse button and select End Mode from the drop-down menu.
7. To place resistors, for example, highlight the ANALOG library. Scroll down the list, highlight R,

then press the Place Part button or double-click on the R symbol. Place the resistor at an appropriate location in the window. After anchoring this resistance by pressing the left mouse button, move the arrow cursor to place another resistor, and so on. To rotate the resistor through 90° counterclockwise, select it by pressing the left mouse button on the resistor symbol and press the R key.

8. Any element or text can be moved by selecting it with the left mouse button, keeping this button pressed, and either dragging the selection to the desired location by moving the cursor or using the arrow keys. It is a good idea to save the schematic at this stage by selecting File/Save from the menu bar.
9. More than one component can be selected at the same time by pressing the Ctrl button while selecting with the left mouse button. Alternatively, the cursor is positioned outside one corner of the part of the circuit to be selected; the left mouse button is pressed and held while the cursor is moved across the parts to be selected until the rectangle attached to the cursor encloses, or its sides touch, these parts. When the left mouse button is released the desired parts of the circuit are selected.
10. To wire components, click on the Place wire button located as the second button in the first column of the toolbar. The cursor in the schematic window changes to a cross hair. Place the cursor at the terminal of a given component and press the left mouse button to anchor it. Move the cursor and a wire is traced. You can draw a straight connection or you can later change to a direction at right angles to the initial direction in order to draw a right-angled connection. If you have to change direction again, you must anchor the cursor before doing so by pressing the left mouse button. Keep moving the cursor till you reach the terminal of the component to which the connection is to be made. Press the left mouse button to finish the connection. The connection remains selected, as indicated by the handles, or filled squares, at both ends of the connection. You can deselect the connection by pressing the Esc key. Press the right mouse button and select End Wire to finish inserting wires or press the Esc key.

The following may be noted concerning wiring:

- When a wire crosses another wire, no connection is made between the two wires. If you want to make a connection, pause at the intersection, release the left mouse button, and continue. A junction is made, as indicated by a small, filled circle. Alternatively, a junction can be placed at the intersection by using the 'Place junction' part from second column of the toolbar.
- In order to delete a wire connection, select it by placing the cursor on it and pressing the left mouse button, and then press the Delete key.
- If a selected wire is dragged, with the left mouse button pressed, all other wires connected to the given wire will remain connected and will move. To isolate the given wire, press and hold the Alt key while dragging the wire. The same procedure applies to moving a group of selected objects.
- To place a wire at an angle other than 0° or 90°, press and hold the Shift key while drawing the wire at any desired angle.
- A part can be placed in the middle of a wire segment without redrawing the wire by placing the part over the wire such that the two pins of the part connect with the wire segment. Then click over the wire segment that overlaps the part, so that the pins of the part (filled circles) change to wire handles (filled squares). Press the Delete key to delete the overlapping wire segment. The same procedure is applied in the case of the common mistake of accidentally drawing a wire through a circuit element, thereby short-circuiting it.
- It is recommended that parts are connected using wires; that is, avoid connecting parts by having the pin of one part overlap the pin of another part.

11. To use the Auto-connect-two-points feature in OrCAD 16.6 Demo, click on the 'Auto connect two points' button in the first (leftmost) column of the toolbar. The cursor in the schematic window changes to an x. Move the cursor to one of the terminals, press, and release the left mouse button. Move the cursor to the other terminal, with the wire attached to it, and then click and release the left mouse button. A connection is automatically made between the two terminals. To use auto connect to wire multiple points, click on the 'Auto connect multi points' button in the second (rightmost) column of the toolbar. Move the cursor to one of the terminals, press, and release the left mouse button. Repeat at every terminal to be connected. Finally, right-click anywhere on the schematic page and choose Connect.

12. To change the value of a component from the default value, double-click on the component value in the schematic. A Display Properties window appears. Change the value in the Value field to the desired value. Scale factors can be added, as described later. Press the OK button. Once you finished wiring, save the schematic.
13. Before a simulation can be made, a *ground connection must be made and a zero value assigned to it.* Click on the ground button in the second column of the toolbar to display the place Ground dialog box. If a ground symbol with a 0 is not displayed in the RHS window, add the SOURCE library, select it, and then select the 0 in the middle window on the LHS. When the ground symbol with a 0 is displayed, press the OK button. The dialog box disappears and a ground symbol with a 0 is displayed in the schematic window attached to the cursor. Drag the symbol to where you want to place the ground and anchor it by pressing the left mouse button. To stop inserting ground, press the Esc button.
14. To simulate the circuit, select PSpice/New Simulation Profile or click on the New Simulation Profile button in the PSpice toolbar. (The toolbars can be identified by selecting Tools/Customize, then the Toolbars tab). The New Simulation dialog box is displayed. Enter a simulation name and then press the Create button. The Simulation Settings dialog box is displayed having the assigned name. From the pull-down menu under Analysis type, select the appropriate type of simulation and fill in the information required. For simple simulation of dc circuits, select Bias Point under Analysis type and press the OK button.
15. To run the simulation, select PSpice/Run from the menu bar or click on the Run PSpice button (the filled arrow head) in the PSpice toolbar.
16. After the simulation is completed, a Simulation Results page entitled SCHEMATIC1-(simulation name)-PSpice A/D Demo is displayed. The small window in the lower left corner indicates whether or not the simulation has run successfully. Press the third button in the bar to the left of the upper main window. A text file entitled SCHEMATIC1-(simulation name) is displayed in the main window. This file contains a PSpice circuit description, the *netlist*, and the simulation results.
17. To display dc voltages, currents, and power dissipated, return to the schematic page and press the V, I, and W buttons, respectively.
18. Note that some properties of circuit elements, such as initial conditions in capacitors and inductors, or default values of op amps and switches can be changed using the Property Editor spreadsheet of the given element. To display this spreadsheet, double-click with the left mouse button on the element symbol in the circuit.

C.3 Opening a Saved Simulation

1. If the folder containing the simulation is in compressed format (as in zip format or .rar extension), decompress it first.
2. Select File/Open/Project, choose the directory in which the simulation is saved, and open this directory. A subdirectory and a .opj file is displayed.
3. Double-click on the .opj file to display the Project Manager block.
4. In the Project Manager block, click on Design Resources all the way to SCHEMATIC1. Open PAGE1 under SCHEMATIC1 to display Page 1 of the saved project.
5. To enlarge the circuit, select any part of the circuit and click on the 'Zoom in' button in the Capture toolbar.

C.4 Scale Factors

PSpice uses the exponential forms for numbers and symbols to scale circuit parameters and variables, as indicated in Table C.1.

TABLE C.1

PSpice Numbers and Scale Factors

Value	Exponential Form	PSpice Symbol
10^{-15}	1E−15	F or f (femto)
10^{-12}	1E−12	P or p (pico)
10^{-9}	1E−9	N or n (nano)
10^{-6}	1E−6	U or u (micro)
10^{-3}	1E−3	M or m (milli)
10^{3}	1E3	K or k (kilo)
10^{6}	1E6	MEG or meg (mega)
10^{9}	1E9	G or g (giga)
10^{12}	1E12	T or t (tera)

C.5 Simulation Restrictions

- A PSpice simulation cannot be run unless the ground symbol is connected to a node and a value of zero is assigned to ground. Otherwise, floating node errors are indicated.
- Every node must have a dc connection to ground. Otherwise, a floating node error is indicated. Capacitors and current sources do not provide a dc connection to ground; resistors, inductors, and voltage sources do. When a node necessarily does not have a dc connection to ground, a floating node error can be avoided by connecting a very large resistor, say, 1 giga-ohm that does not affect the simulation results, between this node and ground.
- PSpice does not allow a loop of zero dc resistance, such as a loop that consists exclusively of voltage sources, inductors, or voltage sources and inductors. The loop can be broken by inserting a very small series resistance, say, 1 nano-ohm that does not affect the simulation results.
- When entering the value of a component, do not enter any space between the number and the scale factor. Fractions must be entered in decimal form, for example, 1.66667 and not 5/3.
- All PSpice entries are case independent. Hence, a resistance value of, say, 15 MΩ is entered as 15meg, and not as 15M. PSpice will interpret a 15M resistance as 15 milliohms.
- When entering circuit values, the basic unit, such as V (for voltage), A (for current), and L (for inductance), may be included or omitted. In the case of capacitance, omission is mandatory. Thus, a capacitance of 1.2 farads is entered simply as 1.2 and not as 1.2F. PSpice will interpret a 1.2F capacitance as 1.2 femtofarads, that is, 1.2×10^{-15} farads. Resistance is also entered without units. A 50 micro-ohm resistance is entered as 50u.
- To avoid strange results when using capacitors or inductors in time-domain analysis, always specify the initial conditions, even if these are zero. The initial conditions are entered in the IC column of the Property Editor spreadsheet for the circuit element.

C.6 Sign Conventions

- Positive voltage drop across a capacitor is from pin 1 of the element to pin 2. These pins can be determined from the Property Editor spreadsheet by pressing the pins tab along the bottom of the spreadsheet. The node designations of the pin numbers are indicated. Initial voltage across a capacitor can be assigned accordingly. Alternatively, the library ANALOG_P contains R and C elements with pin numbers.
- Positive current through an inductor flows from the marked pin to the unmarked pin. Alternatively, the library ANALOG_P contains an L element with pin numbers 1 and 2. Initial current in an inductor can be assigned accordingly.
- In both voltage sources and current sources, the positive direction of current *inside* the source is from the positively marked terminal to the negatively marked terminal.
- To be consistent with the positive directions of voltage and current, positive power is power absorbed. The power delivered by a source is therefore indicated as negative power.

C.7 Rotating Parts

- When a component is placed, it takes a default position. For example, a resistor, capacitor, or inductor will take a default horizontal position with its marked or "1" terminal to the left.
- Components may be rotated by pressing the following keys:
  H Flip part horizontally.
  R Rotate part 90° counterclockwise.
  V Flip part vertically.

C.8 Identifying and Labeling Nodes

- PSpice assigns a 0 node number to ground and arbitrary node numbers to the remaining nodes in the circuit. These numbers consist of the letter N followed by five numerals. The node number can be read by moving the cursor to any wire connected to the node.
- It is sometimes required to label a node with an arbitrary name. This may be necessary to identify the node in an entry in the simulation profile, or to easily identify it in the output file, or to simplify connections in a circuit.
- A node is most conveniently labeled using a net alias. To do this, click on the 'Place net alias' button marked 'abc' in the second column of the Parts toolbar. A Place Net alias window is displayed. Enter a name of your choice under Alias. Press the OK button and a rectangle

appears that is attached to the pointer. Place the rectangle next to the wire connecting to the node to be labeled and press the left mouse button to anchor it. The alias name appears. Press Esc to stop placing aliases. The node is now labeled with the alias name. The alias can be deleted by selecting it and pressing the Delete key.

- Note that nodes having the same name are considered to be connected together, whether in the same page or in different pages.

C.9 Working with Pages

- To add a new schematic page to a project, access the Project Manager, block, and click on Design Resources all the way to SCHEMATIC1. Press the right mouse button on SCHEMATIC1 and select New Page from the menu that pops up. Enter a page name, if desired, and press the OK button.
- A circuit can be copy-pasted from one page to another. The whole circuit is selected as described in item 9 of Section C.2. A copy-paste is then performed from one page to another.
- When a circuit is copy-pasted from one page to another, part numbers of components are automatically changed to preserve the integrity of the circuit in each page. Net aliases, however, are not automatically changed, so that PSpice would consider nodes on different pages having the same net alias to be connected together. To prevent this, the net aliases must be changed manually in the new page.
- A page can be renamed or edited by double-clicking on the page name.
- Circuits on different pages can be connected by power connectors or off-page connectors.
    - To connect a power connector to a node, click on the 'Place power' button in the second column of the Parts toolbar. The Place Power window appears. Select the CAPSYM library. Scroll down the list and select one of the last four entries: VCC_ARROW, VCC_BAR, VCC_CIRCLE, and VCC_WAVE. Each of these refers to a different graphic symbol for the connector. Suppose you choose VCC_CIRCLE. Click OK and the connector with this name appears attached to the cursor. Move it to the node to be labeled, press the left mouse button and press Esc to stop placing connectors. The node is now labeled with the name VCC_CIRCLE. To change this name, double-click on it and enter the name of your choice in the Value field of the Display properties window. Note that although PSpice refers to these connectors as power connectors, they can be used with any node.
    - To connect an off-page connector to a node, click on the 'Place off-page connector' button from the first column of the toolbar. The Place Off-Page Connector window appears. Select the CAPSYM library. Two choices are available: OFFPAGELEFT-L and OFFPAGELEFT-R that differ in the direction of the double arrow that will be attached to the node. Either one may be used. Select one of the symbols, drag it to the node to be labeled, anchor it, and rename it, as for the power connector.
- To clear garbage from the drawing page, select View/Zoom and click on Redraw or press the F5 key. The screen will be cleared and redrawn.

C.10 Markers

- Voltage and current markers can be added to the schematic and used in conjunction with the plot window. The markers obviate the need for labeling nodes and invoking Trace/Add Trace in the Simulation Results page when only relatively simple plots are required.
- Four markers are available identified as Voltage Level, Voltage Differential, Current into Pin, and Power Dissipation. These markers can be applied by pressing the appropriate button in the bottom row of the menu.
- Choosing any of these markers results in a marker symbol attached to the cursor, which can be moved and located at appropriate points in the schematic. After locating a marker, press the Esc button to stop locating a marker of a given type.
- Power dissipation markers are attached to the body of a PSpice device, current markers are attached to pins, and voltage markers are attached to pins, wires, or buses. The first attachment of a Voltage Differential marker is labeled V^+, whereas the second is labeled V^-. The value displayed is that of $(V^+ - V^-)$.
- Power markers indicate the power absorbed by the marked circuit element. Under ac conditions, this is the average power for a resistor, the reactive power for a capacitor or inductor, and the complex power for a source.

- Adding a marker automatically adds a plot to the plot window in the SCHEMATIC1 page, the color of each plot being that of the corresponding marker.
- Markers can be rotated just like circuit components.

C.11 Simulation Profile

- Every simulation must have a simulation profile. When a new simulation is started, clicking on the New Simulation Profile button in the PSpice toolbar opens a New Simulation dialog box. A name is entered for the simulation profile in the 'Name' field to allow performing different simulations on the same circuit. The 'Inherit From' field allows using entries from another simulation profile, the default entry being 'none'.
- Clicking on the Create button in the New Simulation dialog box opens the Simulation Settings window having the name chosen for the simulation profile. Four choices of 'Analysis type' are available:
  - *Bias Point*: Suitable for dc circuit analysis. Under 'Options', 'General Settings' is selected by default. No selections or entries are made under 'Output File Options' for straightforward dc circuit analysis.
  - *DC Sweep*: Used for sweeping the value of a current source, a voltage source, or a global parameter, as described in the solved examples in this book for deriving Thevenin's equivalent circuit, Norton's equivalent circuit, or maximum power transfer, all under dc conditions.
  - *AC Sweep/Noise*: Used for steady-state sinusoidal conditions, frequency responses, or maximum power transfer under ac conditions, as described in the solved examples in this book.
  - *Time Domain (Transient)*: Used for transient analysis and Fourier analysis, as described in the solved examples in this book.
- It should be emphasized that since PSpice simulations are performed at a succession of discrete values of the variable, or points, along the horizontal axis, then to obtain a smooth curve, a sufficient number of points must be included in the simulation. This number is determined by the 'Increment' entry in DC Sweep, the 'Total Points' or 'Points/Decade' entries in AC Sweep/Noise, and the 'Maximum step size' entry in Time-Domain (Transient) analysis.
- Too small a value for the number of points results in a "jagged" curve that looks like a series of straight-line segments and limits the accuracy of the reading of the variable on the horizontal axis. For example, in a time-domain analysis having a simulation time of 1 s and a 'Maximum step size' of 1 ms, the total number of simulation points is $1/10^{-3}$ or 1000. Successive points are separated by 1 ms, so it's not possible to read values of time to an accuracy better than 1 ms.
- On the other hand, too large a value for the number of points unnecessarily prolongs simulation time and increases the size of the simulation file. Generally speaking, it is a good idea to aim for about 2,000 to 5,000 points along the horizontal axis.

C.12 Linear and Ideal Transformers

- To simulate a two-winding linear transformer enter part XFRM_LINEAR, which is the last entry in the ANALOG library. In the Property Editor Spreadsheet of the transformer, enter under COUPLING the appropriate value of the coupling coefficient k. Enter under L1_VALUE the inductance value of the coil on the left, and under L2_VALUE the inductance value of the coil on the right.
- The default dot markings on the linear transformer are at the upper terminal of each coil. If the dot markings are to be reversed, either the connections to one of the coils should be reversed or a negative value entered for the coupling coefficient.
- If the terminals of one coil are open-circuited, PSpice will give an error that less than two connections are made to that node. Under these conditions connect a very large resistance, say 1 GΩ, across the coil terminals, or connect a voltage printer to the node.
- To simulate a two-winding ideal transformer use XFRM_LINEAR, with $k = 1$ and very large values, of the order of megahenries, for the inductances of the two coils but with the ratios of the inductances equal to the square of the turns ratio.
- XFRM_LINEAR does not have a provision for initial current values. It is therefore suitable for steady-state sinusoidal simulations. For time domain analysis, the initial value must be enforced in the circuit by switches or current

sources. In these cases, and in cases of transformers having two or more windings, the part K_Linear in the ANALOG library should be used. Each winding is entered as an inductor having its marked terminal in the same circuit location as the dot marking of the corresponding winding. A K_Linear entry is used to couple every two inductors together in accordance with the coupling coefficient between the two windings involved, using the part numbers of the inductors in the circuit.

C.13 Ideal Op Amp

- An ideal op amp is available from the Analog library as part OPAMP. The default gain is 10^6 and the default power supplies are +15 V and −15 V, but these can be changed in the Property Editor spreadsheet of the op amp, as may be desired.

C.14 Global Parameter

- PSpice allows the variation of a parameter such as a resistance value R_o.
- To allow variation of R_o, it has to be declared a global parameter, as follows:
  - Double-click on the value of R_o. In the Display Properties window, enter under Value: {R_val}, where R_val is a name arbitrarily assigned to R_o, and the curly brackets designate a parameter entry.
  - From the SPECIAL library, enter the part PARAM. The word *PARAMETERS*: is displayed on the schematic.
  - Double-click on *PARAMETERS*: to display its Property Editor spreadsheet. Click on New Column tab and enter in the Add New Column window R_val under Name and an arbitrary number, say 5, under value. Then click OK. It is convenient to have the Name and Value displayed. They would appear as R_val = 5 under *PARAMETERS*:.
- Select in the Simulation Settings the type of analysis required, such as DC Sweep, AC Sweep, or Transient analysis. Under Sweep variable, choose Global parameter and enter R_val under variable name. Under Sweep type, choose Linear and enter 1, say, for Start value, a convenient value, say, 25 for End value and 0.01 for Increment.
- Run the simulation. A graph will be displayed having the horizontal axis labeled R_val with values from 1 to 25. Select Trace/Add trace then choose W(R5), assuming the Part reference for R_o is R5. A graph of power dissipated vs. R_val will be displayed.

C.15 Switches

- PSpice allows the use of (1) a normally open switch that is closed at a certain time, the switch part name being Sw_tClose (read as switch to close), and (2) a normally closed switch that is opened at a certain time, the switch part name being Sw_tOpen (read as switch to open). Both of these parts are in the EVAL library.
- Switch parameters have to be set in the Property Editor spreadsheet as follows:
  - TCLOSE—for a Sw_tClose, this is the time at which the switch closes. You may leave this parameter at the default value of zero, unless a closing time other than zero is required.
  - TOPEN—for a Sw_tOpen, this is the time at which the switch opens. You may leave this parameter at the default value of zero, unless an opening time other than zero is required.
  - TTRAN—the transition time for switch closure. Leave this parameter at the default value of 1 μs. Since this is usually much smaller than the time constant of the circuit, the closing of the switch is essentially instantaneous. In cases of very fast transients, it may be necessary to enter a smaller value.
  - RCLOSED—the resistance of the switch when closed. Leave this parameter at the default value of 0.01 Ω. Since this is usually much smaller than the resistance in series with the switch, the switch appears as a short circuit when closed. In circuits involving small or zero resistance, it may be necessary to enter a smaller resistance value.
  - ROPEN—the resistance of the switch when open. Leave this parameter at the default value of 1 MΩ. Since this resistance is usually much larger than the resistance in series with the switch, the switch appears essentially as an open circuit. In circuits involving very large resistances, it may be necessary to enter a larger resistance value.

C.16 Printers

- In the sinusoidal steady state, printers are used to determine voltage and current values. For current measurement, a two-terminal IPRINT printer is available from the SPECIAL library. The printer is connected so that current in the assigned positive direction enters the unmarked terminal of the printer and leaves at the terminal marked with a minus sign.
- For voltage measurement, two printers are available from the SPECIAL library: (1) VPRINT1, which has one terminal only, and therefore measures voltage with respect to ground, and (2) VPRINT2, which has two terminals and measures the voltage drop from the unmarked terminal to the terminal marked with a minus sign.
- In order to use any of the printers, appropriate entries must be made in the Property Editor spreadsheet of the printer. For ac measurements, a Y must be entered under ac in the Property Editor spreadsheet of each printer. A Y is also entered under REAL and IMAG to express complex values in rectangular form, or a Y is entered under MAG and PHASE to express complex values in polar form.
- The values measured by the printers are read from the output file under the respective printer.

C.17 Evaluate Measurements

- An alternative to using printers for reading values of currents and voltages is to use the 'Evaluate Measurement' feature of PSpice. In the SCHEMATIC1-(simulation name)-page, select 'Trace/Evaluate Measurement'. Two windows are displayed. Under 'Functions or Macros' in the right-hand window, choose 'Analog Operators and Functions'. To read magnitudes, select the voltage or current required from the left-hand window labeled 'Simulation Output Variables'. To read phase angles, select P(), and then select the variable required. For example, to read the voltage of the node labeled 'a', V(a) and P(V(a)) are selected.
- The values are displayed in a window under the graph. This window can be displayed or hidden by clicking on the 'Toggle Measurement Results Window' button in the Probe toolbar.
- Evaluate Measurement can be used for measuring filter parameters, such as Q and 3-dB bandwidth. However, the measurements may not be reliable in the case of low-Q filters or filters having capacitors as the only energy storage elements. In these cases, the cursor should be used, as described in Section C.25.

C.18 Dependent Sources

Dependent sources are listed in the ANALOG library under the following part names:

E: Voltage-controlled voltage source (VCVS)
F: Current-controlled current source (CCCS)
G: Voltage-controlled current source (VCCS)
H: Current-controlled voltage source (CCVS)

Since they are controlled sources, dependent sources automatically assume the same time variation as the controlling quantity, whether dc, ac, etc.

C.19 Sources for the Sinusoidal Steady State

- For the sinusoidal steady state, a voltage source VAC and a current source IAC are available from the SOURCE library. Each of these sources, when entered, has default values of 0Vdc and 1Vac for VAC or 0Adc and 1Aac for IAC. The sources allow simultaneous simulation of dc and sinusoidal steady state. For a purely sinusoidal state, the dc value should be left at zero.
- The magnitude that is entered for VAC or IAC is interpreted by PSpice as an rms value. Since the responses are directly proportional to source values, the entered source values can be considered as peak values, in which case all magnitudes of voltages and currents will be peak values. However, if power values are required, the power value calculated by PSpice is based on considering source values to be rms.
- A phase value can be assigned to a source by entering the phase angle in degrees in the ACPHASE row of the Property Editor spreadsheet. If only one source is present in the circuit, the phase angles of all responses are relative to that of the source, which has a default value of zero, so no phase angle need be assigned to the source. However, if more than one ac source is present, with the sources having different phase angles, then the appropriate phase angle should be entered for each source.

- When performing the simulation, choose AC Sweep/Noise for Analysis type in the Simulation Settings. Choose Linear under AC Sweep Type and enter the source frequency in Hz, not in rad/s, under both Start Frequency and End Frequency. Since the start and end frequencies are the same, enter 1 for 'Total points'.
- Note that in some problems the frequency may not be specified but the reactance values may be given. To perform the simulation, assume any convenient frequency and enter the values of inductors or capacitors accordingly. For example, if the reactance of an inductor is given as 10 Ω, and $f = 1$ Hz is assumed, then enter the value of the inductance as 1.59155, which is $10/(2\pi)$.

C.20 Time-Varying Sinusoidal Sources

- PSpice ISIN and VSIN implement an exponentially decaying sinusoidal current source and an exponentially decaying sinusoidal voltage, respectively. When placed on the Schematic page, some parameters are attached to the source, namely, IOFF or VOFF, IAMPL or VAMPL, and FREQ. These parameters are explained in the following text for ISIN, with reference to Figure C.1, the meaning being similar for VSIN. In addition, VSIN has an AC parameter, which stands for the rms value. This parameter can be ignored when VAMPL is used.
- IOFF is the offset of the sinusoid from the 0 level, as indicated in the figure. IOFF = 0 for a sinusoidal function of zero average.
- IAMPL is the amplitude of the sinusoid.
- FREQ is the frequency in Hz.
- In addition, the following parameters can be entered using the Property Editor spreadsheet for the source, if needed:
  - PHASE is the phase angle in degrees, assuming a sine function. Thus the default value is zero, and PHASE = 90 for a cosine function.
  - TD is time of the start of the sinusoid with respect to $t = 0$. TD = 0 for a $\sin\omega t$ function and is the default value.
  - DF is the damping factor and is the reciprocal of the time constant of the exponential decay of the amplitude of the sinusoid. DF is entered in units of seconds. For a sinusoidal function of constant amplitude, DF = 0 and is the default value.

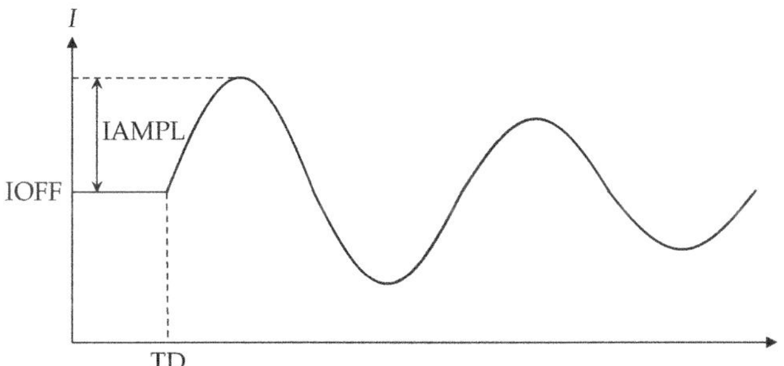

FIGURE C.1
ISIN source.

C.21 Pulse Sources

- The VPULSE and IPULSE sources in the SOURCE library allow the application of an excitation consisting of a single pulse, or any number of pulses, or a periodic pulse train. The pulse may be rectangular, triangular, or trapezoidal in shape, depending on the parameter values entered.
- The interpretation of the parameters of VPULSE is as follows (Figure C.2):

  V1: The lower level of the pulses.
  V2: The upper level of the pulses.
  TD: Time delay before the level changes from V1 to V2.
  TR: Time rise of the pulse; for a rectangular pulse, TR should be very small but should not be zero as this may produce strange spikes in the waveforms.
  TF: Time fall of the pulse; for a rectangular pulse, TF should be very small but should not be zero as this may produce strange spikes in the waveforms.

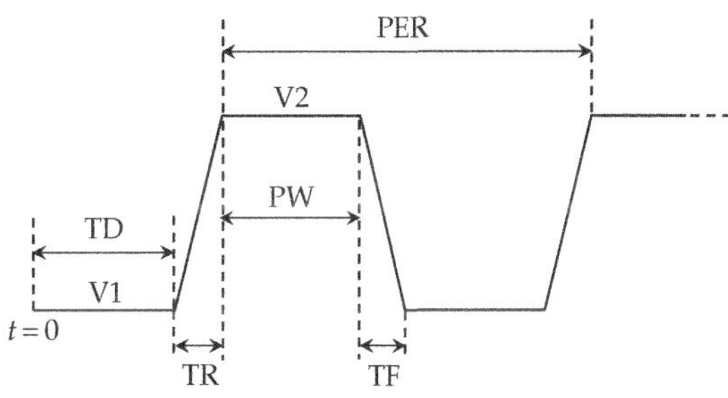

FIGURE C.2
VPULSE source.

PW: Duration of the V2 level; for a triangular pulse, PW should be very small but should not be zero.

PER: Time between repetition of pulses; to obtain a single pulse, PER is made larger than the duration of the simulation; similarly to obtain a specified number of pulses. For the response to a periodic pulse train, the simulation is started after a sufficient time is allowed for the circuit to reach a steady state.

C.22 Piecewise-Linear Sources

- The VPWL and IPWL, or piecewise-linear, sources in the SOURCE library allow the application of an excitation having a waveform consisting of straight-line segments between up to eight breakpoints. When the excitation level changes from one breakpoint to the next, the variation is linear.
- As an example, consider a waveform that suddenly changes from 0 to 2 V at $t = 0$, stays at 2 V until $t = 4$ s, and then decreases to 0 at $t = 6$ s. To generate this waveform, enter in the Property Editor spreadsheet of the source, the following breakpoints and the corresponding voltage levels:

T1 = 0	V1 = 0
T2 = 0	V2 = 2
T3 = 4	V3 = 2
T4 = 6	V4 = 0

  - The waveform has a step at $t = 0$. The voltage is maintained at 2 V till $t = 4$ s and then decreases linearly to zero at $t = 6$ s.
- The piecewise-linear waveform can be applied n times in succession, or continuously. In the former case, the sources VPWL_RE_N_TIMES and IPWL_RE_N_TIMES are used, whereas in the latter case the sources VPWL_RE_FOREVER and IPWL_RE_FOREVER are used. The use of VPWL_RE_FOREVER is described in a solved example on responses to periodic inputs.

C.23 Exponential Sources

- The exponential source in PSpice is IEXP or VEXP (Figure C.3).
- I1 is the level from which the rising exponential begins and toward which the decaying exponential falls.

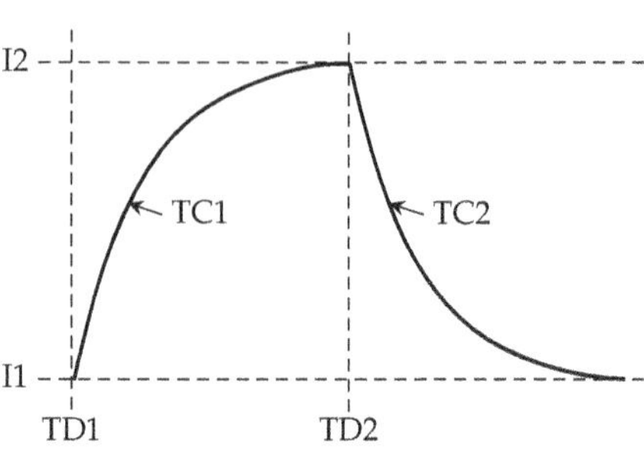

FIGURE C.3
IEXP source.

- I2 is the level which the rising exponential approaches.
- TD1 is the time at which the rising exponential begins.
- TD2 is the time at which the decaying exponential begins.
- TC1 and TC2 are the respective time constants.
- To simulate an exponential of the form $I = 10e^{-5t}$ A, choose I1 = 0, I2 = 10, TD1 = 0, TC1 = 0, TD2 = 0, and TC2 = 0.2.

C.24 Analog Behavioral Modules

- A number of useful functions are available from the Analog Behavioral Modules (ABM) PSpice library, which can be used to implement a variety of mathematical and other operations.
- Useful functions are the ABS function that gives the absolute value of a function, LIMIT that limits the maximum and minimum values of a function, SUM that adds two functions, MULT that multiplies two inputs, CONST that provides a constant value that can be used in addition or multiplication, and LAPLACE that implements an arbitrary transfer function.
- For example, to obtain a full-wave rectified waveform, enter VSIN or ISIN followed by the ABS block from the ABM PSpice library. The output of this block is the absolute value of the input. To obtain a half-wave rectified waveform, add a LIMIT block, or a GLIMIT to provide some gain.

C.25 Working with Plots

- The zoom feature can be used to enlarge a certain part of the plot for closer examination of details. Select View/Zoom. A menu appears that lists several ways of zooming in or out.

Suppose Area is chosen. The cursor changes to crosshairs. Position the cursor at a corner of the area to be enlarged. While pressing and holding the left mouse button, move the cursor so as to enclose the desired area within the rectangle attached to the cursor. When the mouse button is released the selected area appears enlarged to fit the screen. Further zooming of part of the selected area can be made. To return to a previous view, select View/Zoom/Previous. Zooming can also be done using the buttons on the Probe toolbar.

- Text can be placed on the screen by using the menus or the Parts toolbar. To use the menus, select Plot/Label/Text and enter in the window that appears the text it is desired to display. When you click OK, the text string will appear attached to the mouse. Drag the text to the desired location and press the left mouse button to anchor the text. Type CTRL-L to clear any undesired text fragments that may appear. The text can also be dragged to a new location. Alternatively, text can be placed on the screen by clicking on the 'abc' button in the Probe toolbar.
- The cursor can be displayed by choosing Trace/Cursor/Display or by clicking once on the 'Toggle cursor' button in the Probe toolbar. Once the cursor is displayed, cursor 1 can be moved by means of the left and right arrow keys. Pressing Shift while using these keys moves cursor 2. Clicking again on the 'Toggle cursor' button in the Probe toolbar hides the cursors.
- When the cursors are enabled, a number of buttons are highlighted in the Cursor toolbar that can be used for locating either cursor at various parts of the waveform. Clicking on the Cursor Max button, for example, moves the selected cursor to the maximum of the graph.
  - Moving the arrow cursor of the page to a particular point on the graph and clicking on the Cursor Point button moves the cursor to this point. Clicking on the Mark Label button displays the coordinates of the point.
  - The x- and y-values at a given cursor position are indicated in a cursor window in the bottom RHS of the page.
- When more than one trace is displayed, the following applies:
  - Each trace is identified by a small square of the same color as the corresponding trace. The variable whose variation is displayed by the trace appears next to this square. Clicking on the variable name highlights it. Pressing the Delete key deletes the trace.
  - Clicking on the small colored square moves cursor 1 from one trace to another, as indicated by a highlighted, dotted-line square surrounding the colored square.
  - The same numerical vertical scale applies to all traces. Thus, to display a current in mA simultaneously with a voltage in volts, where the numerical values of both traces are comparable, either the current must be multiplied by 10^3 or the voltage divided by 10^3.
- When using a single frequency, a small dash is displayed at the specified frequency. If it is attempted to read the value on the y-axis using the cursor, a "There is no valid trace to exam." message is displayed. The value can be read by selecting Trace/Evaluate Measurement (Section C.17).

C.26 Search Commands

The search command can be used with the cursor, after the cursor is displayed by selecting Trace/Cursor/Display or clicking on the 'Toggle cursor' button in the Probe toolbar. The search feature is invoked by selecting Trace/Cursor/Search Commands or clicking on the Cursor Search button in the Cursor toolbar. A window appears for entering the search command and for selecting whether the command is to apply to cursor 1 or 2. The syntax of the search command is as follows, without space separations:

$$\text{search}[\text{direction}][/\text{start\_point}/]$$
$$[\#\text{consecutive\_points}\#][(\text{range\_x}\ [,\text{range\_y}])]$$
$$[\text{for}][\text{repeat:}]\langle\text{condition}\rangle$$

Square brackets indicate optional arguments. Searches are case independent so that uppercase or lowercase characters can be used. Abbreviations can also be used for the various entries illustrated as follows:

- The command must start with the word 'search' or the letter 's'.
- *direction* could be forward (the default), entered with forward or f, or it could be backward, entered with backward or b.
- *condition* must be entered from one of the following options, each of which can be entered using the first two letters only, shown italicized:
  - *pe*ak, finds the data point with the [#consecutive_points#] on each side having lower y-values than the peak data point. For example, if the command is sf#2#pe, the peak searched for

is a data point with two neighboring data points of lower value.

- *max*, finds the greatest y-value for all points in the specified x-range. If there is more than one maximum having the same y-values, the nearest one is found. *max* is not affected by [direction], [#consecutive_points#], or [repeat:].
- *tr*ough, similar to *pe*ak, but in the opposite sense.
- *mi*n, similar to *ma*x, but in the opposite sense.
- *sl*ope[(posneg)], finds the maximum slope, which could be positive, and is specified as sl(p), or could be negative, specified as sl(n), or could be both, specified as sl(b).
- *po*int, finds the next data point in the given direction.
- *xv*alue<(value)>, finds the first point on the curve that has the specified x-value, where value can be a number (e.g., 10n or 1E5), a percentage (e.g., 50%), a marked point (e.g., x5), a value relative to max or min (e.g., max-2), a value relative to start value (e.g., .-3), where a dot denotes the start value, or a dB value relative to a max or min (e.g., max-3). *xv*alue is not affected by [direction], [#consecutive_points#], [(range_x [,range_y])], or [repeat:]. For example, sxv(5) searches for the point whose x-value is 5.
- *le*vel<(value[,posneg])> finds the next point at a certain value relative to the given level. [,posneg] has the same significance as in the *sl*ope[(posneg)] command and value has the same significance as in the *xv*alue<(value)> command. For example, sle(10) searches for the point whose y value is 10; sfle(max-3db) searches forward for the next point that is 3 dB below the maximum value of the trace.

- */start_point/* specifies the starting point to begin a search, the current point being the default. ^ and $ denote, respectively, the first and last point in the search range.
- *(range_x [, range_y])* specifies the range of values that the search is confined to, the search is inclusive of the range values specified. The range can be specified as floating-point values, as a percent of the full range, as marked points, or as an expression of marked points. The default range is all points available. The first comma separates the limits of the x-range. The second comma separates the x-range values from the y-range values, and the third comma separates the limits of the y-range. For example, (1n,200n) specifies an x-range of 1 to 200 nano and a full y-range. (1.5,20E−9,0,1m) specifies an x-range of 1.5–20 nano and a y-range from 0 to 1 milli. (,,1,3) specifies a full x-range and a y-range from 1 to 3.
- *repeat*: specifies which occurrence of <condition> to find. For example, sb2:le(3,p) searches backward for the second crossing of the positive 3 V level, assuming y-values are in volts.

As an example, the command

search forward#4#(1n,5n)for 5:level(3,positive)

or in abbreviated form sf#4#(1n,5n)5:le(3,p)

searches forward for the fifth occurrence of a 3 V, positive going, y level crossing that has at least four consecutive points at or beyond 3 volts, only searching within the x-value range of 1–5ns.

C.27 Frequency Responses

- AC sweep is used to plot the frequency response. The desired frequency range in Hz is entered in the simulation profile as a start frequency and an end frequency, usually extending over five decades. Use a logarithmic AC sweep type with at least 1000 points/decade to obtain a smooth curve.
- To obtain the transfer function, use a VAC or an IAC source, as appropriate, having a value of unity. Although a voltage marker can be used at the desired output, the resulting y-scale is linear and has to be changed to dB. It is usually more convenient to name the desired output, say vo, and select Trace/Add Trace in the schematic page. Select DB( ) from the list of Analog Operators and Functions in the RH window, and then select V(vo) from the list in the LH window so as to obtain the Trace Expression DB(V(vo)). Press the OK button and the plot appears in the plot window.
- Measurements can be made using cursor search commands, as explained in Section C.26. To obtain the maximum of a low-pass, high-pass, or bandpass response, press the 'Toggle cursor' button in the Probe toolbar, and then press the Cursor Max button in the Cursor toolbar. The cursor will move to the position of the maximum and the values can be read in the Probe

Cursor window. Press the OK button and the value appears in the Measurement Results window under the plot.

- The 3 dB frequency can be read using the command sle(max-3). In the case of a bandpass response, the command can be used with cursor 2 to obtain the lower 3 dB frequency and the command used twice with cursor 1 to obtain the upper 3 dB frequency. The frequencies and their difference, the bandwidth, can be read in the Probe Cursor window.
- Alternatively, the Evaluate Measurement feature can be used to obtain the bandwidth. Press the Evaluate Measurement button in the Probe toolbar. An Evaluate Measurement window is opened. To obtain the 3 dB bandwidth of a bandpass response, for example, select the second entry in the list in the RH window and select V(vo) from the list in the LH window so as to obtain the Trace Expression Bandwidth_Bandpass_3dB(V(vo)). Press the OK button to display the value in the Measurement Results window under the plot window.
- As noted previously, some results of Evaluate Measurement may not be reliable in the case of low-*Q* circuits or circuits having capacitors as the only energy storage elements.

C.28 Trace Expressions

- A plot can be added to the displayed plotted results by entering the appropriate expression in the Trace Expression field of the Add Traces window of the SCHEMATIC1-(simulation name) page.
- For example, to add an asymptote of slope −20 dB/decade to a low-pass response displayed in dB on a logarithmic frequency scale, enter the expression: $-20*\log10(\text{Frequency})-20*\log10(f_c)+A$, where f_c is the 3 dB cutoff frequency in Hz and A is the low-frequency gain in dB.
- In entering trace expressions, Analog Operators and Functions can be entered from the list in the RH window of the Add Traces window. It is useful to note that D( ) is the derivative and PWR(a,b) is equivalent to a^b. Thus, 10,000 can be entered as PWR(10,4).
- Frequency and Time appear in the list of Simulation Output Variables in the LH window of the Add Traces window. It is assumed in the trace expressions that the values of these variables are in Hz and seconds, respectively.

C.29 Fourier Analysis

- PSpice can be used to obtain a printout of the amplitude and phase of each frequency component of a periodic waveform and to display its amplitude spectrum.
- In performing Fourier analysis, when a periodic input is applied to a circuit at $t = 0$, sufficient time should be allowed for the output to reach a steady state. This time can be selected by first observing the output starting at $t = 0$, and determining the number of periods it takes the output to become very nearly periodic. PSpice is then instructed to collect data for one period after this initial time. For example, if the period is 1/120 s or 25/3 ms, data should be collected staring at, say, 300 ms in order to perform a Fourier analysis on the output voltage.
- It is convenient to assign names to the variables on which the Fourier analysis is to be performed. An input voltage, for example, can be labeled as VI, whereas an output voltage can be labeled as VO. Labeling can be done as described in Section C.8 on using a net alias.
- To perform Fourier analysis in the aforementioned example, select Time-Domain (Transient) analysis in the Simulation Settings. Enter 308.3333m for 'Run to time', 300m for 'Start saving data after', and 0.1m for 'Maximum step size'.
- Check the box for 'Skip the initial transient bias point calculation', then click the 'Output File Options' button to display the Transient Output File Options window.
- Check the Perform Fourier Analysis box. For Center Frequency, enter the frequency in Hz, which is 120 in the aforementioned example. Then enter the number of harmonics to be included in the Fourier analysis, say 5. Under 'Output Variables' enter V(VI) followed by a space then V(VO).
- After the simulation is run, the output file gives for VI and VO the dc value and the frequency, magnitude, and phase of each frequency component.
- To display the amplitude spectrum, select Trace/Fourier, and then Trace/Add Trace for V(VI) and V(VO). Change the scale of the axis by selecting Plot/Axis Settings. In the Axis Settings window, select User Defined under XAxis Data range and enter 0–600 Hz, for example. A plot of

triangles is displayed, the peaks of the triangles being at the frequencies of the dc component, fundamental, and harmonics.

- PSpice assumes a periodic function of the form $f(t) = c_0 + \sum_{n=1}^{\infty} c_n \sin(n\omega_0 t + \phi_n)$. Hence, a $\cos\omega_0 t$ term will be considered as $\sin(\omega_0 t + 90°)$ and assigned a phase angle of 90°.

C.30 Graph and Data Copying

- The whole graph window can be copied to the clipboard by selecting Window/Copy to Clipboard. A window is displayed in which various choices concerning colors of traces and background can be made. After making these choices, pressing the OK button copies the graph window to the clipboard. The graph window can then be pasted in another program such as Word or PowerPoint.
- Data can be copied from a Probe trace to Excel, where it can be processed or exported to other programs such as MATLAB. To copy data in this manner, simply click on the text label identifying the trace and located below the window displaying the trace. With the text highlighted and selected, copy to the clipboard using Edit/Copy, or CTRL C, and then paste in Excel. The data are copied in two columns: the first column contains the data from the independent, x-axis variable, whereas the second contains data from the y-axis variable. The labels of the two axes appear in the first row, but with leading spaces, which can be easily removed. The data from more than one trace can be copied at the same time by highlighting more than one text label, with the CTRL button pressed.

Appendix D: Complex Numbers and Algebra

D.1 Definitions and Notation

Imaginary numbers arise when taking the square roots of negative numbers. Thus, $\sqrt{9} = \pm 3$, since multiplying $+3$ or -3 by itself gives 9. But what about $\sqrt{-9}$? Whereas $\sqrt{9}$ is a real number (± 3), $\sqrt{-9}$ is said to be an **imaginary** number. It is evaluated by defining a quantity j as being equal to $\sqrt{-1}$. Then $j^2 = -1$, and $\sqrt{-9} = \sqrt{j^2 9}$. Now we have two positive quantities under the square root, so the square root becomes $\pm j3$. This is a valid answer, because $(j3)(j3) = j^2 9 = -9 = (-j3)(-j3)$.

Defined in this way, imaginary numbers are a perfectly valid set of numbers, just like real numbers, integers, or rational numbers. j, the basis of all imaginary numbers, is of course itself an imaginary number, and all imaginary numbers are multiplied by j. Imaginary numbers can be manipulated according to certain logical and consistent rules.

A **complex** number x is defined as the sum of a real number and an imaginary number:

$$x = a + jb \tag{D.1}$$

where a is referred to as the real part of x and b as the imaginary part. Complex numbers are commonly encountered in algebra and trigonometry. For example, the equation $x^2 + x + 1$ does not have real roots, but it does have complex roots, that is, roots that are complex numbers. The sine and cosine functions may be expressed in terms of complex quantities:

$$\cos x = \frac{e^{jx} + e^{-jx}}{2}, \quad \sin x = \frac{e^{jx} - e^{-jx}}{2j} \tag{D.2}$$

These relations can be readily verified using the infinite series representations of the exponential, sine, and cosine functions.

The **conjugate** of a complex number x, denoted as x^*, is the number that has the same real part but a negated imaginary part. Thus, $a + jb$ and $a - jb$ are conjugates.

Complex quantities play a central role in electric circuits. They are the basis for phasor notation. They are also encountered in Fourier series, Fourier and Laplace transforms, and in the extensive applications that derive from the theory of functions of complex variables.

D.2 Graphical Representation

A complex number may be represented in rectangular form as a point in the **complex plane** or **Argand diagram**. This is the familiar two-dimensional, Cartesian coordinate plane except that the vertical axis is denoted as the imaginary axis (Figure D.1). A complex number $a + jb$ is represented as a point whose horizontal coordinate is a and vertical coordinate is b. Figure D.1 illustrates several such examples where the number is real (such as 1, 4, −3), imaginary (such as $j2$, $-j4$), or complex (such as $4 + j3$, $4 - j3$, $-3 + j4$, $-5 - j4$). The numbers $4 + j3$ and $4 - j3$ are conjugates. Note that whereas in the real plane, a point is represented by a coordinate pair (x, y), the point is represented in the complex plane as a complex number, for example, $4 + j3$.

The rectangular form of the presentation is very convenient for addition or subtraction of complex numbers, since the real parts and the imaginary parts can be added, or subtracted, separately. It is not the most convenient, however, for operations such as multiplication or division. For such purposes, the polar form is preferred. This is similar to the polar coordinates used in the real plane, but with one important feature that will be discussed shortly.

In polar form, a complex number is represented as a line, of length equal to the magnitude of the complex number, drawn from the origin at a certain angle with respect to the horizontal axis. In the polar form, therefore,

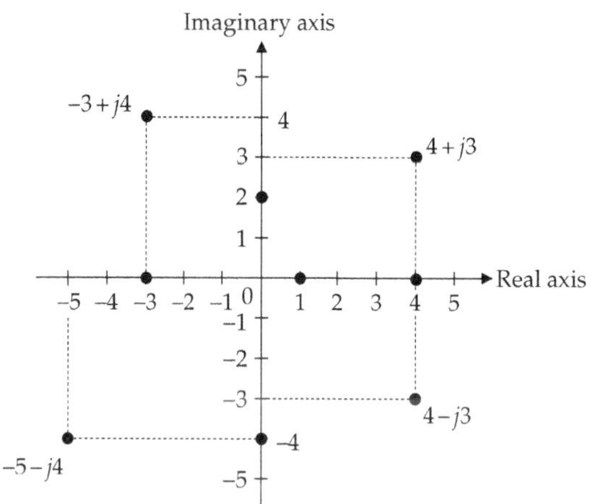

FIGURE D.1
Argand diagram.

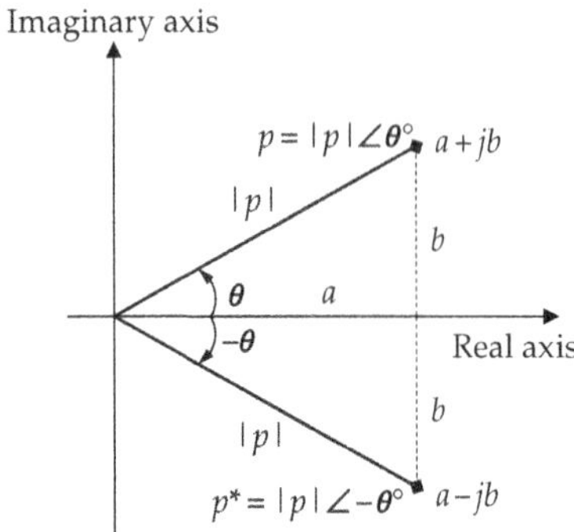

FIGURE D.2
Rectangular and polar coordinates.

two parameters are specified: a magnitude and a phase angle. A complex number p may be represented in polar form as $|p|\angle\theta°$, where $|p|$ is the magnitude of the complex number and θ is its phase angle (Figure D.2). The positive direction of angles is counterclockwise from the real axis, and the nagative direction is clockwise.

The conversion between rectangular and polar forms follows readily from Figure D.2. Given the rectangular form $p = a + jb$, it follows that

$$|p| = \sqrt{a^2 + b^2} \quad \text{and} \quad \theta = \tan^{-1}\frac{b}{a} \tag{D.3}$$

It can be readily verified that the complex numbers $4 + j3$, $4 - j3$, $-3 + j4$, $-5 - j4$ in Figure D.1 have the following representations in polar form:

$$4 + j3 \equiv 5\angle 36.9° \quad 4 - j3 \equiv 5\angle -36.9°$$

$$-3 + j4 \equiv 5\angle 126.9° \quad -5 - j4 \equiv 6.4\angle -141.3°$$

Given the polar form $p = |p|\angle\theta°$, the rectangular coordinates are

$$a = |p|\cos\theta \quad \text{and} \quad b = |p|\sin\theta \tag{D.4}$$

Figure D.2 also shows the conjugate of p in polar form. It has the same magnitude as p and a phase angle that is the negative of that of p.

The polar form of a complex number can also be represented in a very useful form that is not applicable to the polar form of real quantities. It follows from Equation D.4 that

$$p = a + jb = |p|(\cos\theta + j\sin\theta) \tag{D.5}$$

But according to Equation D.2: $e^{j\theta} = \cos\theta + j\sin\theta$, which is known as Euler's formula. Hence, a complex number X may be represented in polar form as

$$p = |p|e^{j\theta} \tag{D.6}$$

Note that $e^{j\theta}$ is a complex number in polar form whose magnitude is unity and whose phase angle is θ. Hence, a complex number whose phase angle is θ may be represented as in Equation D.6. In particular, $j = e^{j\frac{\pi}{2}}$ so the polar form of an imaginary number such as $j2$ is a line 2 units in length that lies along the imaginary axis.

D.3 Addition and Subtraction

Addition and subtraction of complex numbers are easy to perform in rectangular form. If $p = a + jb$ and $q = c + jd$, then

$$p \pm jq = (a \pm c) + j(b \pm d) \tag{D.7}$$

In other words, the real and imaginary parts are added, or subtracted, separately. Hence, if the numbers are given in polar form, it may be easier to convert them first to rectangular form, add or subtract them, and convert again to polar form, if desired. For example, if

$$p = 10\angle 35° \quad \text{and} \quad q = 5\angle -135°$$

then

$$p = 8.19 + j5.74 \quad \text{and} \quad q = -3.54 - j3.54$$

It follows that

$$p + q = 4.65 + j2.2 = 5.14\angle 25.3°$$
and
$$p - q = 11.73 + j9.28 = 14.96\angle 38.3°$$

If $p = a + jb$, then $p^* = a - jb$ and

$$p + p^* = 2a, \quad p - p^* = j2b \tag{D.8}$$

That is, the sum of a complex number and its conjugate is a real number equal to twice the real part of the number. The difference between a complex number and its conjugate is an imaginary number equal to twice the imaginary component of the minuend, that is, the number being subtracted from.

Addition and subtraction in polar form essentially follow the same procedure. To add p and q in Figure D.3, the parallelogram construction is followed. Suppose $p = |p|\angle\alpha$ and $q = |q|\angle\beta$, and it is desired to find the sum: $|p|\angle\alpha + |q|\angle\beta = |z|\angle\theta$. Each complex number is resolved into its real and imaginary parts: $p = |p|\cos\alpha + j|p|\sin\alpha$

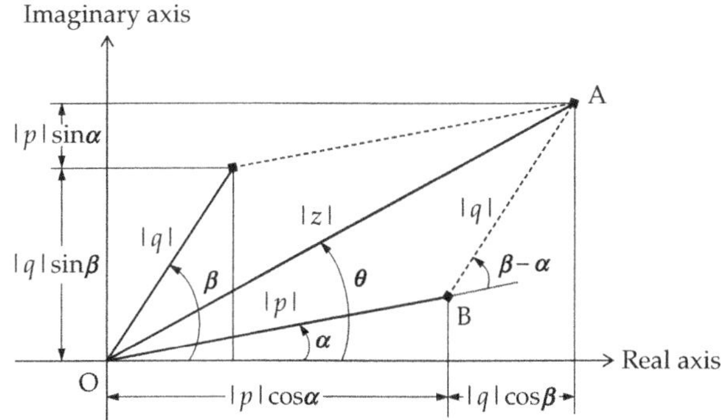

FIGURE D.3
Addition of complex quantities.

and $q = |q|\cos\beta + j|q|\sin\beta$. The real parts and imaginary parts of x and y are added separately to give the real and imaginary parts of z, respectively. Thus,

$$z = |p|\cos\alpha + |q|\cos\beta + j\left(|p|\sin\alpha + |q|\sin\beta\right)$$

The magnitude of z is given by the square root of the sum of the squares of the real and imaginary parts (Equation D.3):

$$\begin{aligned}|z| &= \sqrt{\left(|p|\cos\alpha + |q|\cos\beta\right)^2 + \left(|p|\sin\alpha + |q|\sin\beta\right)^2} \\ &= \sqrt{|p|^2 + |q|^2 + 2|p||q|\cos(\beta - \alpha)}\end{aligned} \tag{D.9}$$

which is a well-known cosine rule applied to the triangle OAB.

The phase angle θ of z is given by

$$\tan\theta = \frac{|p|\sin\alpha + |q|\sin\beta}{|p|\cos\alpha + |q|\cos\beta} \tag{D.10}$$

D.4 Multiplication and Division

Multiplication and division are most easily carried out on complex numbers in polar form. If $p = |p|\angle\alpha$ and $q = |q|\angle\beta$, the product $z = pq$ is given by

$$z = |p|e^{j\alpha} \times |q|e^{j\beta} = |p||q|e^{j(\alpha+\beta)} \tag{D.11}$$

In other words, when two complex numbers are multiplied together, the product has a magnitude that is equal to the products of the magnitudes of the two numbers and a phase angle equal to the sum of their phase angles. This is illustrated in Figure D.4a.

The quotient $z = \dfrac{p}{q}$ is given by

$$z = \frac{|p|e^{j\alpha}}{|q|e^{j\beta}} = \frac{|p|}{|q|}e^{j(\alpha-\beta)} \tag{D.12}$$

Thus, when two complex numbers are divided by one another, the quotient has a magnitude that is equal to the quotient of the magnitudes of the two numbers and a phase angle equal to the difference of their phase angles. This is illustrated in Figure D.4b.

For example, for the case $p = 10\angle 35°$ and $q = 5\angle -135°$,

$$pq = 10\angle 35° \times 5\angle -135° = 50\angle -100°$$

and

$$\frac{p}{q} = \frac{10\angle 35°}{5\angle -135°} = 2\angle 170°$$

An important special case is multiplication and division by j. Since $j = e^{j\frac{\pi}{2}}$, multiplication by j is equivalent to increasing the phase angle by 90°, that is, rotating the line representing the complex number by 90° in the counterclockwise direction. Similarly, division by j is equivalent to decreasing the phase angle by 90°, that is, rotating the line representing the complex number by 90° in the clockwise direction.

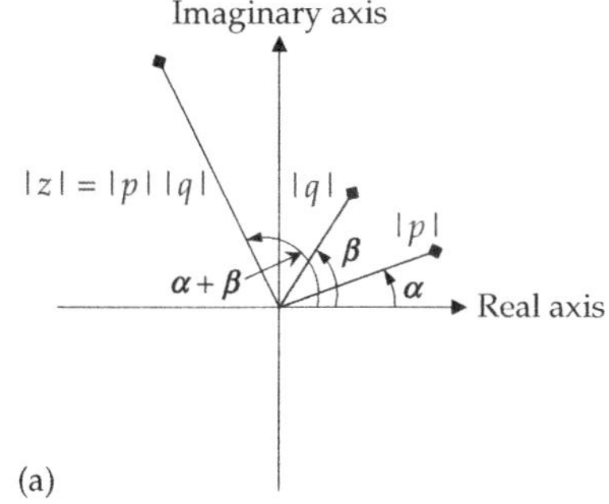

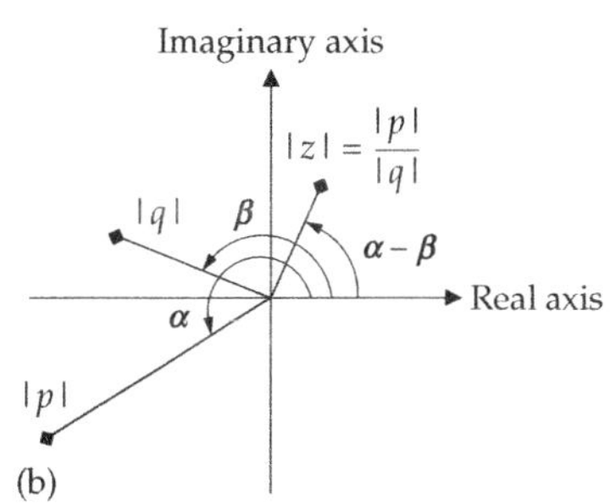

FIGURE D.4
Multiplication (a) and division (b) of complex quantities.

If $p = |p|e^{j\theta}$, then $p^* = |p|e^{-j\theta}$, so that $pp^* = |p|^2$. That is, multiplying a complex number by its conjugate gives a real number equal to the square of the magnitude of the complex number.

Multiplication and division may also be performed on complex numbers in rectangular form. If $p = a + jb$ and $q = c + jd$, then

$$pq = (a + jb)(c + jd) = ac - bd + j(ad + bc)$$

It follows that

$$|pq| = \sqrt{(ac - bd)^2 + (ad + bc)^2} = \sqrt{(a^2 + b^2)(c^2 + d^2)} = |p||q| \quad \text{(D.13)}$$

and that the phase angle θ of pq is given by

$$\tan\theta = \frac{ad + bc}{ac - bd} = \frac{\frac{d}{c} + \frac{b}{a}}{1 - \frac{d}{c}\frac{b}{a}} = \frac{\tan\alpha + \tan\beta}{1 - \tan\alpha\tan\beta} = \tan(\alpha + \beta) \quad \text{(D.14)}$$

The results of Equations D.13 and D.14 are of course in agreement with Equation D.11.

The quotient of p and q is given by $\frac{p}{q} = \frac{a + jb}{c + jd}$. In order to express this as the sum of real and imaginary quantities, the numerator and denominator are multiplied by the complex conjugate of the denominator, an operation referred to as **rationalization**. Thus,

$$\frac{p}{q} = \frac{a + jb}{c + jd} = \frac{a + jb}{c + jd}\frac{c - jd}{c - jd} = \frac{ac + bd + j(bc - ad)}{c^2 + d^2}$$

Hence,

$$\left|\frac{p}{q}\right| = \frac{\sqrt{(ac + bd)^2 + (bc - ad)^2}}{c^2 + d^2} = \frac{\sqrt{a^2 + b^2}}{\sqrt{c^2 + d^2}} = \frac{|p|}{|q|} \quad \text{(D.15)}$$

The phase angle θ of p/q is given by

$$\tan\theta = \frac{bc - ad}{ac + bd} = \frac{\frac{b}{a} - \frac{d}{c}}{1 - \frac{b}{a}\frac{d}{c}} = \frac{\tan\alpha - \tan\beta}{1 + \tan\alpha\tan\beta} = \tan(\alpha - \beta) \quad \text{(D.16)}$$

The results of Equations D.15 and D.16 are in agreement with Equation D.12.

D.5 Raising to a Power

A complex number in polar form can be raised to a power k, where k is a real number:

$$p^k = \left[|p|e^{j\theta}\right]^k = |p|^k e^{jk\theta} \quad \text{(D.17)}$$

In general, to raise a complex number to a real power k, its magnitude is raised to the power k and its phase angle is multiplied by k. For example, if $p = 4\angle 30°$, $p^2 = 16\angle 60°$, and $p^{-2} = \frac{1}{16}\angle -60°$.

However, if k is a positive integer less than unity, then this is tantamount to taking the root of a complex number, and care must be taken to include all the roots. Suppose, for example, that $p = 125\angle 75°$ and $k = 1/5$. Then the angle 75° will have to be divided by 5. However, in polar form the angle θ of a complex number is really $\theta + 2\pi n$, where n is any integer. When $\theta + 2\pi n$ is divided by 5, with $n = 0, 1, 2, 3, 4$, the result is an angle between 0° and 360°. This is as it should be, for taking the fifth root should give five, nonrepeating roots. The five values of n do indeed give these five roots. Each root will have a magnitude of $\sqrt[5]{125} = 5$. Thus, the five roots of $p = 125\angle 75°$ are

$$p_1 = 5\angle\left(\frac{75°}{5}\right) = 5\angle 15° = 5\cos 15° + j5\sin 15° = 4.83 + j1.34$$

$$p_2 = 5\angle\left(\frac{75° + 360°}{5}\right) = 5\angle 87° = 5\cos 87° + j5\sin 87°$$
$$= 0.26 + j4.99$$

$$p_3 = 5\angle\left(\frac{75° + 2 \times 360°}{5}\right) = 5\angle 159° = 5\cos 159° + j5\sin 159°$$
$$= -4.67 + j1.79$$

$$p_4 = 5\angle\left(\frac{75° + 3 \times 360°}{5}\right) = 5\angle 231° = 5\cos 231° + j5\sin 231°$$
$$= -3.15 - j3.89$$

$$p_5 = 5\angle\left(\frac{75° + 4 \times 360°}{5}\right) = 5\angle 303° = 5\cos 303° + j5\sin 303°$$
$$= 2.72 - j4.19$$

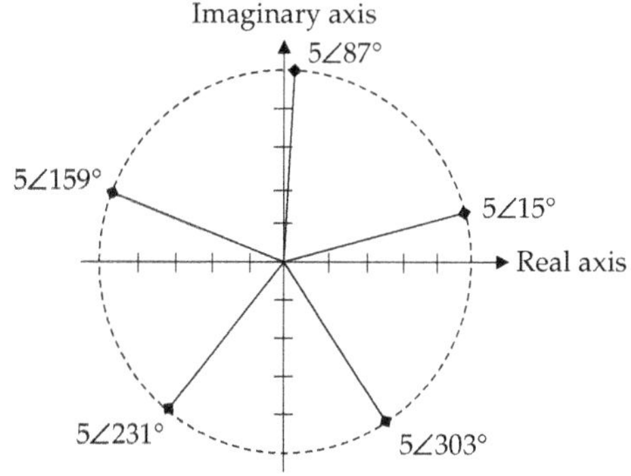

FIGURE D.5
Fifth root of 125∠75°.

The five roots are plotted in Figure D.5. Since they have the same magnitude, their ends all lie on the circle of diameter 5 units, centered at the origin. Raising any of these roots to the fifth power gives $p = 125\angle 75°$.

D.6 Some Useful Identities

$$\pm j^2 = \mp 1 \tag{D.18}$$

$$\frac{1}{\pm j} = \mp j \tag{D.19}$$

$$e^{\pm j\pi} = -1 \tag{D.20}$$

$$e^{\pm j\frac{\pi}{2}} = \pm j \tag{D.21}$$

Given that $p = a + jb = |p|\angle\alpha$ and $q = c + jd = |q|\angle\beta$,

$$pp^* = a^2 + b^2 = |p|^2 \tag{D.22}$$

$$p + p^* = j2a = j2|p|\cos\alpha \tag{D.23}$$

$$p - p^* = j2b = j2|p|\sin\alpha \tag{D.24}$$

$$\frac{p}{p^*} = 1\angle 2\alpha \tag{D.25}$$

$$pq = |p||q|\angle(\alpha + \beta) = (ac - bd) + j(bc + ad) \tag{D.26}$$

$$\frac{p}{q} = \frac{|p|}{|q|}\angle\alpha - \beta = \frac{(ac + bd) + j(bc - ad)}{c^2 + d^2} \tag{D.27}$$

Appendix E: Solution of Linear Simultaneous Equations

E.1 System of Linear Simultaneous Equations

A system of n linear simultaneous equations may be written in the standard form as follows:

$$
\begin{array}{ccccccccc}
a_{11}x_1 & + & a_{12}x_2 & + & a_{13}x_3 & +\cdots+ & a_{1n}x_n & = & B_1 \\
a_{21}x_1 & + & a_{22}x_2 & + & a_{23}x_3 & +\cdots+ & a_{2n}x_n & = & B_2 \\
a_{31}x_1 & + & a_{32}x_2 & + & a_{33}x_3 & +\cdots+ & a_{3n}x_n & = & B_3 \\
\vdots & & & & & & & & \\
a_{n1}x_1 & + & a_{n2}x_2 & + & a_{n3}x_3 & +\cdots+ & a_{nn}x_n & = & B_n
\end{array}
\qquad \text{(E.1)}
$$

where $x_1, x_2, \ldots, x_n$ are the unknowns or variables whose values are to be determined in terms of the B's and the a_{ij} coefficients. These coefficients, which in general could be complex numbers, may be ordered in an $n \times n$ array as follows:

$$
\Delta = \begin{vmatrix} a_{11} & a_{12} & a_{13} & \cdots & a_{1n} \\ a_{21} & a_{22} & a_{23} & \cdots & a_{2n} \\ \vdots & \vdots & \vdots & \vdots & \vdots \\ a_{n1} & a_{n2} & a_{n3} & \cdots & a_{nn} \end{vmatrix} \qquad \text{(E.2)}
$$

where the i and j subscripts in a_{ij} denote, respectively, the ith row and jth column in the array.

The array (E.2) between two parallel lines is known as the **determinant** of the set of simultaneous equations and is denoted by the symbol Δ. The determinant is a number whose value is evaluated according to certain rules.

E.2 Solution for Two Linear Simultaneous Equations

Consider the two simultaneous equations:

$$
\begin{aligned} a_{11}x_1 + a_{12}x_2 &= B_1 \\ a_{21}x_1 + a_{22}x_2 &= B_2 \end{aligned} \qquad \text{(E.3)}
$$

By simple elimination of variables these two equations may be solved to give

$$
x_1 = \frac{a_{22}B_1 - a_{12}B_2}{a_{11}a_{22} - a_{21}a_{12}} \quad x_2 = \frac{a_{11}B_2 - a_{21}B_1}{a_{11}a_{22} - a_{21}a_{12}} \qquad \text{(E.4)}
$$

The solutions for x_1 and x_2 may be written as the ratios of two determinants:

$$
x_1 = \frac{\begin{vmatrix} B_1 & a_{12} \\ B_2 & a_{22} \end{vmatrix}}{\begin{vmatrix} a_{11} & a_{12} \\ a_{21} & a_{22} \end{vmatrix}} \quad x_2 = \frac{\begin{vmatrix} a_{11} & B_1 \\ a_{21} & B_2 \end{vmatrix}}{\begin{vmatrix} a_{11} & a_{12} \\ a_{21} & a_{22} \end{vmatrix}} \qquad \text{(E.5)}
$$

The method of solving a system of linear simultaneous equations by means of determinants is known as **Cramer's rule**.

The solutions given by Equation E.4 are derived according to the following rules:

Step 1: The expression for each variable is the ratio of two determinants. The determinant in the denominator is always Δ, the determinant of the set of equations as defined by Equation (E.2). The determinant in the numerator of x_1 is obtained by replacing the coefficients of the first column in Δ, that is, the coefficients of x_1 in the equations, by the column representing the B's. Similarly, the determinant in the numerator of x_2 is obtained by replacing the coefficients of the second column in Δ, that is, the coefficients of x_2 in the equations, by the column representing the B's.

Step 2: Each determinant expands to the expression given by:

$$
\begin{vmatrix} B_1 & a_{12} \\ B_2 & a_{22} \end{vmatrix} = B_1a_{22} - B_2a_{12}
$$

$$
\begin{vmatrix} a_{11} & B_1 \\ a_{21} & B_2 \end{vmatrix} = a_{11}B_2 - a_{21}B_1
$$

$$
\Delta = \begin{vmatrix} a_{11} & a_{12} \\ a_{21} & a_{22} \end{vmatrix} = a_{11}a_{22} - a_{21}a_{12} \qquad \text{(E.6)}
$$

Each product term in the expansion of the determinant is obtained by multiplying each element in a given column or a given row by its **minor**, with the appropriate sign. The minor of a given element is what remains after deleting the row and the column in which the element occurs. For example, consider the determinant Δ. The minor of a_{11} is a_{22}, which is what remains after deleting the first column and the first row, in which a_{11} occurs. Similarly, the minor of a_{12} is a_{21}, the minor of a_{21} is a_{12}, and the minor of a_{22} is a_{11}. Considering the product terms $a_{11}a_{22}$ and $a_{12}a_{21}$ in the expression for Δ, it will be noted that they are obtained by going through any one column of the determinant, or any one row, and multiplying each element by its minor. Thus, if we go through the first column, multiplying a_{11} by its minor gives $a_{11}a_{22}$, and multiplying a_{21} by its minor gives $a_{12}a_{21}$. Alternatively, we may go through the elements of the first row. Again, multiplying a_{11} by its minor gives $a_{11}a_{22}$, and multiplying a_{12} by its minor gives $a_{12}a_{21}$. Similarly, we could have gone through the elements of the second column or the second row.

For a 2 × 2 determinant, the previous procedure reduces to the difference of the product of the elements in two diagonals: the first diagonal is that descending from the top leftmost element, whereas the second diagonal is that ascending from the bottom leftmost element.

It can be readily verified that the same procedure applies for evaluating the terms in the expressions for the first two determinants in Equation E.6. It remains to determine the signs of the terms in the expansions of the determinants.

The signs of the minors of each element are obtained from a checkerboard pattern of alternating plus and minus signs in the determinant Δ, starting with a plus sign in the upper left-hand position. Thus, for the 2 × 2 determinant under consideration, the signs are

$$\begin{vmatrix} + & - \\ - & + \end{vmatrix} \tag{E.7}$$

In other words, the minors of a_{11} and a_{22} have positive signs, whereas the minors of a_{12} and a_{21} have negative signs. It can be readily verified that the determinant expansions in Equation E.6 conform to this sign convention.

The signed minor of a given element is referred to as the **cofactor** of that element.

E.3 Solution for Three Linear Simultaneous Equations

The procedure outlined earlier can be extended to a system of linear simultaneous equation of any order. We will consider first the solution for three linear simultaneous equations, because a simplified procedure is applicable in this case. Let the three equations be expressed as

$$\begin{aligned} a_{11}x_1 + a_{12}x_2 + a_{13}x_3 &= B_1 \\ a_{21}x_1 + a_{22}x_2 + a_{23}x_3 &= B_2 \\ a_{31}x_1 + a_{32}x_2 + a_{33}x_3 &= B_3 \end{aligned} \tag{E.8}$$

Step 1 of this procedure gives the solution as

$$x_1 = \frac{\begin{vmatrix} B_1 & a_{12} & a_{13} \\ B_2 & a_{22} & a_{23} \\ B_3 & a_{32} & a_{33} \end{vmatrix}}{\Delta} \quad x_2 = \frac{\begin{vmatrix} a_{11} & B_1 & a_{13} \\ a_{21} & B_2 & a_{23} \\ a_{31} & B_3 & a_{33} \end{vmatrix}}{\Delta} \quad x_3 = \frac{\begin{vmatrix} a_{11} & a_{12} & B_1 \\ a_{21} & a_{22} & B_2 \\ a_{31} & a_{32} & B_3 \end{vmatrix}}{\Delta} \tag{E.9}$$

where

$$\Delta = \begin{vmatrix} a_{11} & a_{12} & a_{13} \\ a_{21} & a_{22} & a_{23} \\ a_{31} & a_{32} & a_{33} \end{vmatrix} \tag{E.10}$$

Each of the four determinants may be evaluated according to the procedure of Step 2 of the preceding section. Δ will be evaluated as an example. The signs of the cofactors of the elements are determined from the checkerboard pattern for a 3 × 3 determinant:

$$\begin{vmatrix} + & - & + \\ - & + & - \\ + & - & + \end{vmatrix} \tag{E.11}$$

Taking the cofactors of the first column,

$$\Delta = a_{11}\begin{vmatrix} a_{22} & a_{23} \\ a_{32} & a_{33} \end{vmatrix} - a_{21}\begin{vmatrix} a_{12} & a_{13} \\ a_{32} & a_{33} \end{vmatrix} + a_{31}\begin{vmatrix} a_{12} & a_{13} \\ a_{22} & a_{23} \end{vmatrix} \tag{E.12}$$

Note that because we have a 3 × 3 determinant, the cofactors of the elements are now 2 × 2 determinants which can be evaluated as the difference of the products of diagonal elements. This gives

$$\begin{aligned} \Delta = {} & a_{11}\left(a_{22}a_{33} - a_{32}a_{23}\right) - a_{21}\left(a_{12}a_{33} - a_{32}a_{13}\right) \\ & + a_{31}\left(a_{12}a_{23} - a_{22}a_{13}\right) \end{aligned} \tag{E.13}$$

The positive terms and negative terms may each be grouped together, so that

$$\begin{aligned} \Delta = {} & a_{11}a_{22}a_{23} + a_{21}a_{32}a_{13} + a_{31}a_{23}a_{12} - a_{31}a_{22}a_{13} - a_{32}a_{23}a_{11} \\ & - a_{33}a_{12}a_{21} \end{aligned} \tag{E.14}$$

This expression may be written by inspection as shown in the following diagram:

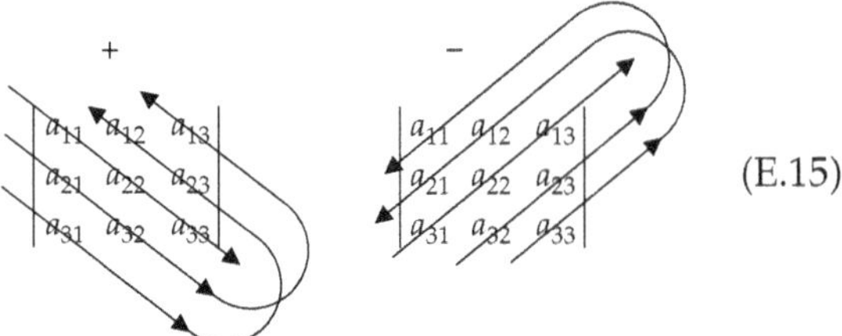

(E.15)

The positive terms are obtained by proceeding diagonally as shown in the diagram on the left. The first term is the product of the elements in the main diagonal pointing downwards from the upper left-hand corner. The second term is the product of the two elements in the smaller diagonal below the main diagonal and the element in the upper right-hand corner. The third term is the product of the element in the lower left-hand corner and the two elements in the smaller diagonal above the main diagonal. The negative terms are obtained in an analogous manner but proceeding upwards, as shown in the diagram on the right.

Example E.1

Solve the following simultaneous equations

$$\begin{aligned} 2x_1 + 4x_2 - 3x_3 &= 13 \\ x_1 - 3x_2 + 2x_3 &= -5 \\ 3x_1 + 2x_2 - 4x_3 &= 8 \end{aligned} \tag{E.16}$$

Solution:

Proceeding as indicated previously,

$$\Delta = \begin{vmatrix} 2 & 4 & -3 \\ 1 & -3 & 2 \\ 3 & 2 & -4 \end{vmatrix} = 2(-3)(-4) + 1\times 2(-3) + 3\times 2$$

$$\times 4 - 3(-3)(-3) - 2\times 2\times 2 - (-4)\times 4\times 1 = 23$$

$$x_1 = \frac{\begin{vmatrix} 13 & 4 & -3 \\ -5 & -3 & 2 \\ 8 & 2 & -4 \end{vmatrix}}{\Delta} \quad x_2 = \frac{\begin{vmatrix} 2 & 13 & -3 \\ 1 & -5 & 2 \\ 3 & 8 & -4 \end{vmatrix}}{\Delta} \quad x_3 = \frac{\begin{vmatrix} 2 & 4 & 13 \\ 1 & -3 & -5 \\ 3 & 2 & 8 \end{vmatrix}}{\Delta} \tag{E.17}$$

Evaluating these determinants according to the procedure of (E.15),

$$x_1 = \frac{13(-3)(-4) + (-5)2(-3) + 8\times 2\times 4 - 8(-3)(-3) - 2\times 2\times 13 - (-4)4(-5)}{23}$$

$$= \frac{46}{23} = 2$$

$$x_2 = \frac{2(-5)(-4) + 1\times 8(-3) + 3\times 2\times 13 - 3(-5)(-3) - 8\times 2\times 2 - (-4)13\times 1}{23}$$

$$= \frac{69}{23} = 3$$

$$x_3 = \frac{2(-3)8 + 1\times 2\times 13 + 3(-5)\times 4 - 3(-3)13 - 2(-5)\times 2 - 8\times 4\times 1}{23}$$

$$= \frac{23}{23} = 1$$

E.4 Evaluation of Determinants of Higher Order

The procedure indicated in (E.15) does not apply to determinants of order higher than 3. An alternative procedure, known as **pivotal condensation**, may be used to successively reduce the order of the determinant until it is a 3 × 3 or a 2 × 2 determinant and can be evaluated by the methods presented earlier.

The method of pivotal condensation may be illustrated by the 3 × 3 determinant Δ in Example E.1:

$$\Delta = \begin{vmatrix} 2 & 4 & -3 \\ 1 & -3 & 2 \\ 3 & 2 & -4 \end{vmatrix} = \frac{1}{2^1} \begin{vmatrix} \begin{vmatrix} 2 & 4 \\ 1 & -3 \end{vmatrix} & \begin{vmatrix} 2 & -3 \\ 1 & 2 \end{vmatrix} \\ \begin{vmatrix} 2 & 4 \\ 3 & 2 \end{vmatrix} & \begin{vmatrix} 2 & -3 \\ 3 & -4 \end{vmatrix} \end{vmatrix} = \frac{1}{2} \begin{vmatrix} -10 & 7 \\ -8 & 1 \end{vmatrix}$$

$$= \frac{46}{2} = 23 \tag{E.18}$$

The procedure is to form a determinant composed of a number of 2 × 2 determinants, where the first determinant in the first row is $\begin{vmatrix} a_{11} & a_{12} \\ a_{21} & a_{22} \end{vmatrix}$, the second determinant in the first row is $\begin{vmatrix} a_{11} & a_{13} \\ a_{21} & a_{23} \end{vmatrix}$, and so on. The first determinant in the second row is $\begin{vmatrix} a_{11} & a_{12} \\ a_{31} & a_{32} \end{vmatrix}$, the second determinant in the second row is $\begin{vmatrix} a_{11} & a_{13} \\ a_{31} & a_{33} \end{vmatrix}$, and so on. The reduced 2 × 2 determinant is divided by a_{11}^{n-2}, where n is the order of the original determinant to be condensed. The result is a determinant of order $n - 1$. This procedure is further illustrated by Example E.2.

Example E.2

Solve the following simultaneous equations

$$\begin{aligned} 2x_1 + x_2 + 3x_3 + 4x_4 &= 9 \\ x_1 + 2x_2 + x_3 + 3x_4 &= 5 \\ 5x_1 + x_2 + x_3 + 3x_4 &= 10 \\ 2x_1 + 3x_2 + 4x_3 + x_4 &= 1 \end{aligned} \tag{E.19}$$

Solution:

According to this procedure,

$$\Delta = \begin{vmatrix} 2 & 1 & 3 & 4 \\ 1 & 2 & 1 & 3 \\ 5 & 1 & 1 & 3 \\ 2 & 3 & 4 & 1 \end{vmatrix} = \frac{1}{2^2} \begin{vmatrix} \begin{vmatrix} 2 & 1 \\ 1 & 2 \end{vmatrix} & \begin{vmatrix} 2 & 3 \\ 1 & 1 \end{vmatrix} & \begin{vmatrix} 2 & 4 \\ 1 & 3 \end{vmatrix} \\ \begin{vmatrix} 2 & 1 \\ 5 & 1 \end{vmatrix} & \begin{vmatrix} 2 & 3 \\ 5 & 1 \end{vmatrix} & \begin{vmatrix} 2 & 4 \\ 5 & 3 \end{vmatrix} \\ \begin{vmatrix} 2 & 1 \\ 2 & 3 \end{vmatrix} & \begin{vmatrix} 2 & 3 \\ 2 & 4 \end{vmatrix} & \begin{vmatrix} 2 & 4 \\ 2 & 1 \end{vmatrix} \end{vmatrix} = \frac{1}{4} \begin{vmatrix} 3 & -1 & 2 \\ -3 & -13 & -14 \\ 4 & 2 & -6 \end{vmatrix}$$

$$= \frac{1}{4} \times \frac{1}{3} \begin{vmatrix} \begin{vmatrix} 3 & -1 \\ -3 & -13 \end{vmatrix} & \begin{vmatrix} 3 & 2 \\ -3 & -14 \end{vmatrix} \\ \begin{vmatrix} 3 & -1 \\ 4 & 2 \end{vmatrix} & \begin{vmatrix} 3 & 2 \\ 4 & -6 \end{vmatrix} \end{vmatrix} = \frac{1}{12} \begin{vmatrix} -42 & -36 \\ 10 & -26 \end{vmatrix} = 121 \tag{E.20}$$

The determinants for x_1 to x_4 will be evaluated by expansion in terms of cofactors of the first column:

$$x_1 = \frac{\begin{vmatrix} 9 & 1 & 3 & 4 \\ 5 & 2 & 1 & 3 \\ 10 & 1 & 1 & 3 \\ 1 & 3 & 4 & 1 \end{vmatrix}}{\Delta} = \frac{9\begin{vmatrix} 2 & 1 & 3 \\ 1 & 1 & 3 \\ 3 & 4 & 1 \end{vmatrix}}{\Delta} + \frac{-5\begin{vmatrix} 1 & 3 & 4 \\ 1 & 1 & 3 \\ 3 & 4 & 1 \end{vmatrix}}{\Delta} + \frac{10\begin{vmatrix} 1 & 3 & 4 \\ 2 & 1 & 3 \\ 3 & 4 & 1 \end{vmatrix}}{\Delta} + \frac{-1\begin{vmatrix} 1 & 3 & 4 \\ 2 & 1 & 3 \\ 1 & 1 & 3 \end{vmatrix}}{\Delta} = \frac{9(2+12+9-9-24-1)-5(1+16+27-12-12-3)}{121} + \frac{10(1+32+27-12-12-6)-(3+8+9-4-3-18)}{121} = \frac{121}{121} = 1.$$

$$x_2 = \frac{\begin{vmatrix} 2 & 9 & 3 & 4 \\ 1 & 5 & 1 & 3 \\ 5 & 10 & 1 & 3 \\ 2 & 1 & 4 & 1 \end{vmatrix}}{\Delta} = \frac{2\begin{vmatrix} 5 & 1 & 3 \\ 10 & 1 & 3 \\ 1 & 4 & 1 \end{vmatrix}}{\Delta} + \frac{-\begin{vmatrix} 9 & 3 & 4 \\ 10 & 1 & 3 \\ 1 & 4 & 1 \end{vmatrix}}{\Delta} + \frac{5\begin{vmatrix} 9 & 3 & 4 \\ 5 & 1 & 3 \\ 1 & 4 & 1 \end{vmatrix}}{\Delta} + \frac{-2\begin{vmatrix} 9 & 3 & 4 \\ 5 & 1 & 3 \\ 10 & 1 & 3 \end{vmatrix}}{\Delta} = \frac{2(5+120+3-3-60-10)-(9+160+9-4-108-30)}{121} + \frac{5(9+80+9-4-108-15)-2(27+20+90-40-27-45)}{121} = -\frac{121}{121} = -1.$$

$$x_3 = \frac{\begin{vmatrix} 2 & 1 & 9 & 4 \\ 1 & 2 & 5 & 3 \\ 5 & 1 & 10 & 3 \\ 2 & 3 & 1 & 1 \end{vmatrix}}{\Delta} = \frac{2\begin{vmatrix} 2 & 5 & 3 \\ 1 & 10 & 3 \\ 3 & 1 & 1 \end{vmatrix}}{\Delta} + \frac{-\begin{vmatrix} 1 & 9 & 4 \\ 1 & 10 & 3 \\ 3 & 1 & 1 \end{vmatrix}}{\Delta} + \frac{5\begin{vmatrix} 1 & 9 & 4 \\ 2 & 5 & 3 \\ 3 & 1 & 1 \end{vmatrix}}{\Delta} + \frac{-2\begin{vmatrix} 1 & 9 & 4 \\ 2 & 5 & 3 \\ 1 & 10 & 3 \end{vmatrix}}{\Delta} = \frac{2(20+3+45-90-6-5)-(10+4+81-120-3-9)}{121} + \frac{5(5+8+81-60-3-18)-2(15+80+27-20-30-54)}{121} = \frac{0}{121} = 0.$$

$$x_4 = \frac{\begin{vmatrix} 2 & 1 & 3 & 9 \\ 1 & 2 & 1 & 5 \\ 5 & 1 & 1 & 10 \\ 2 & 3 & 4 & 1 \end{vmatrix}}{\Delta} = \frac{2\begin{vmatrix} 2 & 1 & 5 \\ 1 & 1 & 10 \\ 3 & 4 & 1 \end{vmatrix}}{\Delta} + \frac{-\begin{vmatrix} 1 & 3 & 9 \\ 1 & 1 & 10 \\ 3 & 4 & 1 \end{vmatrix}}{\Delta} + \frac{5\begin{vmatrix} 1 & 3 & 9 \\ 2 & 1 & 5 \\ 3 & 4 & 1 \end{vmatrix}}{\Delta} + \frac{-2\begin{vmatrix} 1 & 3 & 9 \\ 2 & 1 & 5 \\ 1 & 1 & 10 \end{vmatrix}}{\Delta} = \frac{2(2+20+30-15-80-1)-(1+36+90-27-40-3)}{121} + \frac{5(1+72+45-27-20-6)-2(10+18+15-9-5-60)}{121} = \frac{242}{121} = 2.$$

E.5 Use of MATLAB

As the previous example illustrates, solving a system of more than three linear simultaneous equations is rather tedious. Fortunately, advanced calculators have the feature of solving such equations as does MATLAB.

To see how linear simultaneous equations may be solved using MATLAB, we note that Equations E.1 may be written in matrix notation as

$$\mathbf{AX} = \mathbf{B} \tag{E.21}$$

where

A is an $n \times n$ matrix of the a_{ij} coefficients
X is a column matrix of the variables $x_1, x_2, \ldots, x_n$
B is a column matrix of the B's

To solve for **X** we premultiply both sides of Equation E.21 by $\mathbf{A}^{-1}$, the inverse matrix of **A**, where $\mathbf{A}^{-1}\mathbf{A} = 1$. Equation E.21 becomes

$$\mathbf{X} = \mathbf{A}^{-1}\mathbf{B} \tag{E.22}$$

The procedure for solving the system of equations using MATLAB is as follows:

Step 1: Enter the numerical matrices **A** and **B**.

Step 2: Evaluate the product $\mathbf{A}^{-1}\mathbf{B}$; an alternative command is **A\B**.

Step 3: Read off the values of the variables as the elements of the resulting column matrix.

This procedure is illustrated in Example E.3 by solving the simultaneous equation of Example E.2 using MATLAB.

Example E.3

Using MATLAB, solve the following simultaneous equations:

$$\begin{aligned} 2x_1 + x_2 + 3x_3 + 4x_4 &= 9 \\ x_1 + 2x_2 + x_3 + 3x_4 &= 5 \\ 5x_1 + x_2 + x_3 + 3x_4 &= 10 \\ 2x_1 + 3x_2 + 4x_3 + x_4 &= 1 \end{aligned} \tag{E.23}$$

Solution:

At the MATLAB prompt, enter the matrix of a_{ij} coefficients as follows:

```
>>A = [2,1,3,4;1,2,1,3;5,1,1,3;2,3,4,1]
```

Note that a matrix is entered between square brackets. Elements in a row are separated by commas, whereas rows are separated by semicolons. After an entry is made, MATLAB displays the entry for checking purposes. At the next prompt enter the matrix **B**:

$$\gg B = [9;5;10;1]$$

Alternatively, **B** may be entered as

$$\gg B = [9,5,10,1]'$$

Here, the elements are separated by commas, so the matrix entered, without the apostrophe, is a row matrix. The apostrophe transposes the row matrix to a column matrix. At the next prompt enter

$$\gg A\backslash B$$

MATLAB displays the column matrix $[X]$:

```
 1.0000
-1.0000
-0.0000
 2.0000
```

where the elements of this matrix are x_1 to x_4, respectively.

MATLAB can also be used to handle complex quantities and to evaluate determinants. If matrix **A** has been entered as indicated earlier, entering det(A) causes MATLAB to display 121, the value of the determinant as determined earlier.

E.6 Properties of Determinants

The following properties of determinants are stated without proof:

Property 1. If all the rows of a determinant are changed into columns or if all the columns are changed into rows, the determinant does not change.

Example:

$$\begin{vmatrix} 2 & 4 & -3 \\ 1 & -3 & 2 \\ 3 & 2 & -4 \end{vmatrix} = \begin{vmatrix} 2 & 1 & 3 \\ 4 & -3 & 2 \\ -3 & 2 & -4 \end{vmatrix} = 23$$

Here, the first, second, and third rows of the first determinant have become the first, second, and third columns of the second determinant, respectively.

Property 2. If any two rows, or any two columns, of a determinant are interchanged, the sign of the determinant is changed, without affecting its value.

Example:

$$\begin{vmatrix} 2 & 4 & -3 \\ 1 & -3 & 2 \\ 3 & 2 & -4 \end{vmatrix} = 23; \quad \begin{vmatrix} 3 & 2 & -4 \\ 1 & -3 & 2 \\ 2 & 4 & -3 \end{vmatrix} = -23 = \begin{vmatrix} 2 & -3 & 4 \\ 1 & 2 & -3 \\ 3 & -4 & 2 \end{vmatrix}$$

where the first and third rows of the first determinant are interchanged in the second determinant, and the second and third columns of the first determinant are interchanged in the third determinant.

Property 3. If all the elements in a row (or in a column) are multiplied by a constant, the determinant is multiplied by the same constant.

$$\begin{vmatrix} 2\times 2 & 2\times 4 & 2(-3) \\ 1 & -3 & 2 \\ 3 & 2 & -4 \end{vmatrix} = 2\begin{vmatrix} 2 & 4 & -3 \\ 1 & -3 & 2 \\ 3 & 2 & -4 \end{vmatrix} = 46$$

Property 4. A determinant is unaltered if a multiple of the elements of a row are added to another row, or if a multiple of the elements of a column are added to another column.

Example:

$$\begin{vmatrix} 2 & 4 & -3 \\ (1+2\times 2) & (-3+2\times 4) & (2+2(-3)) \\ 3 & 2 & -4 \end{vmatrix} = \begin{vmatrix} 2 & 4 & -3 \\ 5 & 5 & -4 \\ 3 & 2 & -4 \end{vmatrix} = 23$$

Property 5. If the elements in one row are proportional to those of another row, or if the elements in one column are proportional to those of another column, the determinant is zero.

$$\begin{vmatrix} 2 & 4 & -3 \\ (-1)2 & (-1)4 & (-1)(-3) \\ 3 & 2 & -4 \end{vmatrix} = 0$$

As a special case, if any two rows, or any two columns, of a determinant are identical, the determinant is zero. The corresponding simultaneous equations are not independent.

Property 6. If all the elements of one row (or of one column) can be expressed as sums of two quantities, the determinant can be expressed as the sum of two determinants in which the other rows (or columns) are the same.

Example:

$$\begin{vmatrix} 2 & 4 & -3 \\ 1 & -3 & 2 \\ 3 & 2 & -4 \end{vmatrix} = 23 = \begin{vmatrix} 1+1 & 1+3 & -5+2 \\ 1 & -3 & 2 \\ 3 & 2 & -4 \end{vmatrix}$$

$$= \begin{vmatrix} 1 & 1 & -5 \\ 1 & -3 & 2 \\ 3 & 2 & -4 \end{vmatrix} + \begin{vmatrix} 1 & 3 & 2 \\ 1 & -3 & 2 \\ 3 & 2 & -4 \end{vmatrix} = -37 + 60 = 23$$

Property 7. If all the elements of one row (or of one column) are multiplied by the cofactors of the corresponding elements of another row (or another column) and the results are added, their sum is zero.

Example:

$$\begin{vmatrix} 2 & 4 & -3 \\ 1 & -3 & 2 \\ 3 & 2 & -4 \end{vmatrix} = 2\times(-1)\begin{vmatrix} 4 & -3 \\ 2 & -4 \end{vmatrix} + 4\begin{vmatrix} 2 & -3 \\ 3 & -4 \end{vmatrix} - 3(-1)\begin{vmatrix} 2 & 4 \\ 3 & 2 \end{vmatrix}$$

$$= 20 + 4 - 24 = 0$$

The elements of the first row have been multiplied by the cofactors of the corresponding elements of the second row. Note the signs of the cofactors in accordance with the checkerboard pattern of (E.11).

Index

G

H

I

J

K

L

Printed and bound by CPI Group (UK) Ltd, Croydon, CR0 4YY

01/11/2024

01782604-0019